Instructors: Assign and assess!

You can now easily assign powerful online tools—and have your assignments automatically graded!

Assign *Get Ready for Microbiology* quizzes so that your students come to your course prepared.

Assign quizzes and activities on 3-D movie-quality animations.

MicroFlix

Assign pre-lecture reading quizzes to make sure your students read the textbook before coming to class. Assign post-lecture quizzes to check students' understanding.

Assign pre-lab quizzes so that students are better prepared for lab. Assign post-lab quizzes to check that your lab objectives were met.

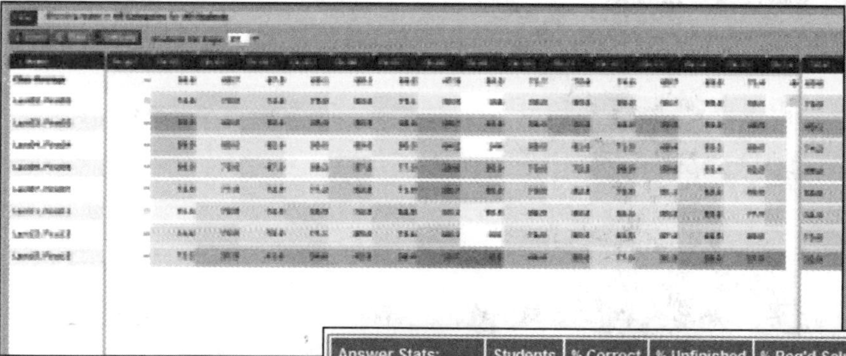

MasteringMicrobiology grades assignments automatically. Shades of red highlight students who are struggling or assignments that are difficult for the class as a whole. Easily export grades to a course management system or spreadsheet, or import grades from your exams, labs, or clicker questions.

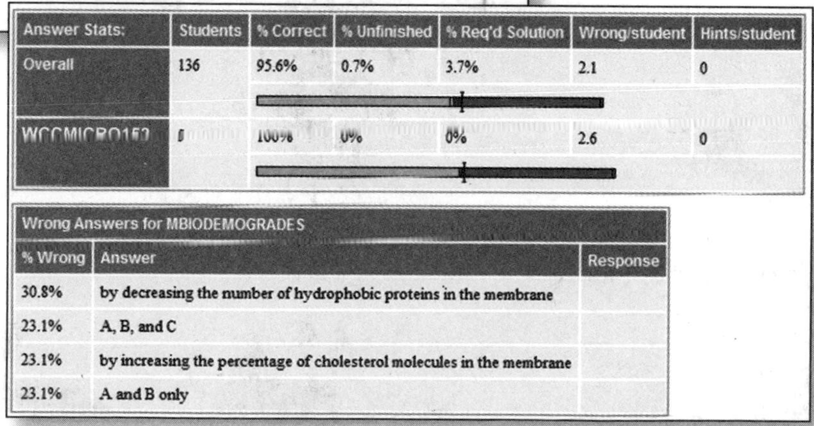

Answer Stats:	Students	% Correct	% Unfinished	% Req'd Solution	Wrong/student	Hints/student
Overall	136	95.6%	0.7%	3.7%	2.1	0
WCCMICRO152	5	100%	0%	0%	2.6	0

Wrong Answers for MBIODEMOGRADES		
% Wrong	Answer	Response
30.8%	by decreasing the number of hydrophobic proteins in the membrane	
23.1%	A, B, and C	
23.1%	by increasing the percentage of cholesterol molecules in the membrane	
23.1%	A and B only	

Gain unique insight into students' misunderstandings with at-a-glance statistics and wrong answer summaries and then make just-in-time teaching adjustments.

Customize your eText to fit the way you study

MasteringMICROBIOLOGY™

MasteringMicrobiology (www.masteringmicrobiology.com) also includes an eText. Now you can access your textbook whenever and wherever you're online.

eText pages look exactly like the printed text and offer powerful functionality. You can:

- Create notes
- Highlight text in different colors
- Create bookmarks
- Zoom in and out
- View in single-page or two-page view
- Click hyperlinked words and phrases to view definitions
- Link directly to relevant animations and videos as you read the text

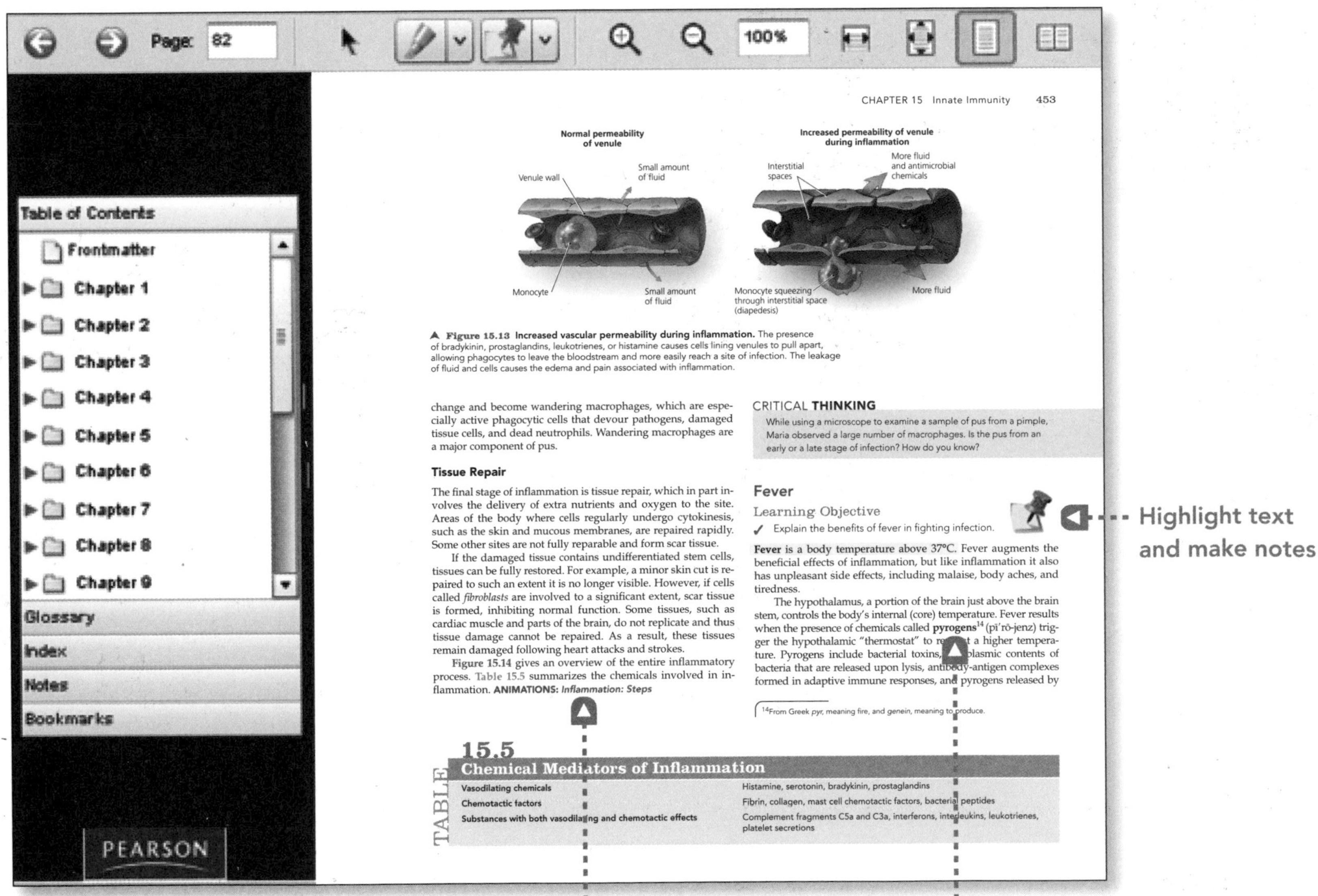

THIRD EDITION

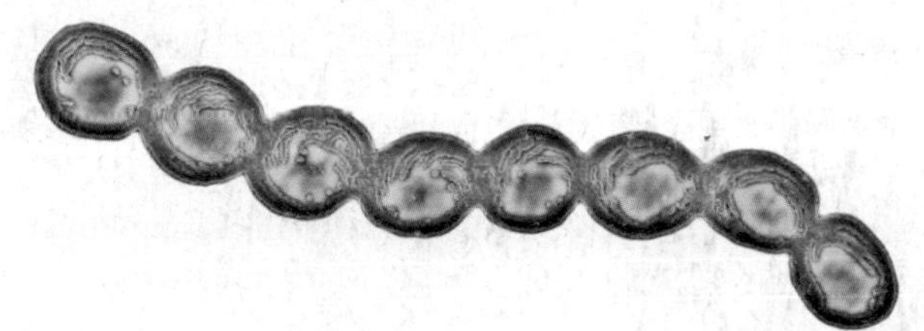

MICROBIOLOGY

WITH DISEASES BY TAXONOMY

ROBERT W. BAUMAN, Ph.D.

Amarillo College

Contributions by:

Elizabeth Machunis-Masuoka, Ph.D.

University of Virginia

Clinical Consultants:

Cecily D. Cosby, Ph.D., FNP-C, PA-C

Jean E. Montgomery, MSN, RN

Benjamin Cummings

Boston Columbus Indianapolis New York San Francisco Upper Saddle River
Amsterdam Cape Town Dubai London Madrid Milan Munich Paris Montréal Toronto
Delhi Mexico City São Paulo Sydney Hong Kong Seoul Singapore Taipei Tokyo

Executive Editor: *Leslie Berriman*
Project Editor: *Robin Pille*
Editorial Development Manager: *Barbara Yien*
Assistant Editor: *Kelly Reed*
Art Development Manager: *Laura Southworth*
Art Development Editor: *Elisheva Marcus*
Managing Editor: *Deborah Cogan*
Production Manager: *Michele Mangelli*
Production Supervisor: *David Novak*
Director, Media Development: *Lauren Fogel*
Media Producer: *Lucinda Bingham*
Copyeditor: *Anita Wagner*
Proofreader: *Betsy Dietrich*
Interior and Cover Designer: *Riezebos Holzbaur Design Group*
Art Coordinator: *Linda Jupiter*
Illustrators: *Precision Graphics*
Senior Art and Photo Manager: *Travis Amos*
Photo Researcher: *Maureen Spuhler*
Compositor: *Progressive Information Technologies*
Senior Manufacturing Buyer: *Stacey Weinberger*
Senior Marketing Manager: *Neena Bali*

Cover Photo Credit: Alfred Pasieka/Photo Researchers, Inc.

Credits and acknowledgments borrowed from other sources and reproduced, with permission, in this textbook appear on the appropriate page within the text or on p. CR-1.

Library of Congress Cataloging-in-Publication Data
Bauman, Robert W.
Microbiology : with diseases by taxonomy / Robert W. Bauman ; contributions by Elizabeth Machunis-Masuoka ; clinical consultants, Cecily D. Cosby, Jean E. Montgomery. – 3rd ed.
p. ; cm.
Includes index.
ISBN-13: 978-0-321-64043-7 (student ed.)
ISBN-10: 0-321-64043-8 (student ed.)
1. Microbiology. 2. Medical microbiology. I. Machunis-Masuoka, Elizabeth. II. Title.
[DNLM: 1. Microbiological Phenomena. 2. Microbiological Techniques. QW 4 B347m 2011]
QR41.2.B382 2011
616.9'041--dc22

2009038226

Benjamin Cummings
is an imprint of

www.pearsonhighered.com

ISBN 10: 0-321-64043-8 (Student edition)
ISBN 13: 978-0-321-64043-7 (Student edition)
ISBN 10: 0-321-67825-7 (Professional copy)
ISBN 13: 978-0-321-67825-6 (Professional copy)

1 2 3 4 5 6 7 8 9 10—WCV—12 11 10 09

To Michelle:
my best friend, my closest confidant, my cheerleader,
my partner, my love. I treasure my life with you.
—Robert

About the Author

ROBERT W. BAUMAN is a professor of biology and past chairman of the Department of Biological Sciences at Amarillo College in Amarillo, Texas. He teaches microbiology, human anatomy and physiology, and botany. In 2004, the students of Amarillo College selected Dr. Bauman as the recipient of the John F. Mead Faculty Excellence Award. He received an M.A. degree in botany from the University of Texas at Austin and a Ph.D. in biology from Stanford University. His research interests have included the morphology and ecology of freshwater algae, the cell biology of marine algae (particularly the deposition of cell walls and intercellular communication), and environmentally triggered chromogenesis in butterflies. He is a member of the American Society of Microbiology (ASM), Texas Community College Teacher's Association (TCCTA), American Association for the Advancement of Science (AAAS), Human Anatomy and Physiology Society (HAPS), and The Lepidopterist's Society. When he is not writing books, he enjoys spending time with his family: gardening, hiking, camping, rock climbing, backpacking, cycling, snow shoeing, skiing, and reading by a crackling fire.

About the Clinical Consultants

CECILY D. COSBY is nationally certified as both a family nurse practitioner and physician assistant. She is a professor of nursing, currently teaching at Samuel Merritt College in Oakland, California, and has been in clinical practice since 1980, most recently at the University of California, San Francisco, in a preoperative practice. She received her Ph.D. and MS from the University of California, San Francisco; her BSN from California State University, Long Beach; and her PA certificate from the Stanford Primary Care program. She was awarded the Paul C. Samson Clinical Nursing Professional Chair for 2007–2010.

JEAN E. MONTGOMERY is a registered nurse formerly teaching in the associate degree nursing program at Austin Community College in Texas. She received her MSN from the University of Texas Health Science Center at San Antonio, Texas.

Preface

The threat of swine flu and other emerging diseases; the progress of cutting-edge research into microbial genetics; the challenge of increasingly drug-resistant pathogens; the continual discovery of microorganisms previously unknown—these are just a few examples of why the study of microbiology has never been more exciting, or more important. Welcome!

I have taught microbiology to undergraduates for over 20 years and witnessed firsthand how students struggle with the same topics and concepts year after year. Students invariably come to class with different levels of preparation—while some have strong science backgrounds, others lack a foundation in chemistry and/or biology, making it a challenge to decide how to gear the course. In creating this textbook, my goal was to write a text that explains complex topics—especially metabolism, genetics, and immunology—in a way that beginning students can understand, while at the same time presenting a thorough and accurate overview of microbiology. I also wished to highlight the many positive effects of microorganisms on our lives, along with the medically important microorganisms that cause disease.

NEW TO THIS EDITION

In approaching the third edition, my goal was to build upon the strengths and success of the text by updating it with the latest scientific and educational research and data available and by incorporating the many terrific suggestions I have received from colleagues and students alike. The feedback from instructors who adopted the previous editions has been immensely gratifying and is much appreciated. The result is, once again, a collaborative effort of educators, students, editors, and top scientific illustrators: a textbook that, I hope, continues to improve upon conventional explanations and illustrations in substantive and effective ways. In this new edition:

- New **Emerging Diseases boxes** reflect the third edition's emphasis on this topic. Focused on a patient's experience with the disease, these boxes describe emerging and reemerging diseases such as *Hantavirus* pulmonary syndrome, babesiosis, and MRSA. (See p. xxxi for a full list.)
- New **Beneficial Microbes boxes** emphasize the practical and benevolent nature and uses of microbes and help students overcome the common misconception that most microbes are damaging and cause disease. (See p. xxx for a full list.)
- More than **150 new light and electron micrographs** enhance student understanding of the text and boxed features.
- The end-of-chapter review includes new **Concept Mapping sections,** guiding students to create their own concept maps from a list of key terms, and new **illustration-based questions** that help test students' visual understanding.
- **Chapter 3 (Cell Structure and Function) has been reorganized** to match the latest taxonomic research. The discussion deemphasizes the term "prokaryote" and emphasizes the three domains of living organisms. The newly separate section on the Archaea can be covered or easily skipped over, depending on instructor preference.
- **Chapter 22 (Pathogenic Fungi) has been revised, clarified, and updated** with the latest scientific advancements in this fast-changing field.
- **MasteringMicrobiology (www.masteringmicrobiology.com)** provides unprecedented, cutting-edge assessment resources for instructors as well as self-study tools for several text features, including Emerging Diseases boxes and Concept Mapping exercises.

The following section provides a detailed outline of this edition's chapter-by-chapter revisions followed by a visual walkthrough of its main themes and features.

Chapter-by-Chapter Revisions

Every chapter in this edition has been thoroughly revised, and data in the text, tables, and figures have been updated. The main changes for each chapter are summarized below.

THROUGHOUT THE PATHOGEN CHAPTERS (19–25)

- Updated disease diagnoses, treatments, and incidence and prevalence data
- Updated immunization recommendations and suggested treatments for all diseases
- Added "Clinical Case Study" boxes as noted below
- Added answers to "Clinical Case Study" boxes to Instructor's Manual

CHAPTER 1 A BRIEF HISTORY OF MICROBIOLOGY

- Revised section concerning pressing questions in microbiology based on American Association of Microbiologists' Council Policy committee to assess which topics will come to be of high interest over the next several years (*ASM News* 71:262–3)
- New "Beneficial Microbes" box on yeast in the service of humanity
- New "Emerging Diseases" box on variant Creutzfeldt-Jakob disease (prion disease)

CHAPTER 2 THE CHEMISTRY OF MICROBIOLOGY

- New chapter opener concerning the possibility of life on Enceladus, one of Saturn's moons
- Added coverage of recently discovered 21st and 22nd amino acids
- Added "Clinical Case Study" box appropriate for students' knowledge early in the course, on *V. vulnificus* food poisoning
- New "Beneficial Microbes" box on using bacteria to preserve ancient buildings

CHAPTER 3 CELL STRUCTURE AND FUNCTION

- Created new sections on domain Archaea to examine these microbes in more detail and independently from domain Bacteria, emphasizing that "prokaryote" is not a taxonomic grouping
- Expanded coverage of bacterial shapes and arrangements, but kept at a level appropriate for this early chapter
- Moved coverage of endospores from Chapter 11, at instructors' requests
- Reorganized discussion of eukaryotic flagella and cilia to emphasize that these structures are internal to the cytoplasmic membrane
- Revised comparison of eukaryotic and prokaryotic cytoplasmic membrane. For example:
  - Eukaryotic membranes are not uniform and fluid, but rather contain regions with unique lipid and protein molecules called membrane rafts. AIDS, Ebola, flu, and measles viruses use lipid rafts for replication and propagation
  - Eukaryotic membrane proteins are often glycosylated to mediate intercellular interactions
- Incorporated new discoveries concerning cell structure and function. For example:
  - Some prokaryotic cytoskeletons are contractile, allowing nonflagellated cells that possess them to be motile
  - Research has clarified the differences and similarities between fimbriae and pili
  - Conductive fimbriae allow electrical signaling among bacteria
  - Some archaea have hami—fimbriae-like cell extensions shaped like Ninja grappling hooks on barbed wire
- Expanded and updated coverage of bacterial flagella assembly, magnetosomes, the outer membrane of Gram-negative bacteria, and peptidoglycan structure
- Updated information on cell walls of algae, bacteria, and archaea
- Five new figures
- Revised and enhanced artwork in sixteen figures
- New critical thinking question
- New "Beneficial Microbes" box on bacterially produced plastics

CHAPTER 4 MICROSCOPY, STAINING, AND CLASSIFICATION

- Nine new photos
- Revised and enhanced eleven figures
- New "Beneficial Microbes" box on using fluorescent viruses to identify bacteria
- New "Emerging Diseases" box on necrotizing fasciitis

CHAPTER 5 MICROBIAL METABOLISM

- New art to illustrate relationships of catabolism, anabolism, ATP-ADP energy cycle, use of nutrients, precursor metabolites, and macromolecules
- Rearranged alternatives to Embden-Meyerhof glycolysis for greater clarity and better pedagogy
- Clarified definitions of aerobic respiration vs. anaerobic respiration vs. fermentation in text, figures, and critical thinking questions
- Upgraded nineteen figures for greater clarity and better pedagogy
- Simplified longer figure legends, at request of reviewers
- Added critical thinking question regarding photosynthesis
- New "Beneficial Microbes" box on gold-mining bacteria

CHAPTER 6 MICROBIAL NUTRITION AND GROWTH

- New chapter opener on photosynthetic bacteria that live miles below the surface—where sunlight does not penetrate
- Added material concerning definition, development, and prevalence of biofilms and quorum sensing
- Presented new hyperthermophile record holder, *Geogemma,* formerly called strain 121
- Increased coverage of archaea, serial dilutions, viable plate counting, the contrast between lithotrophy and organotrophy, nonculturable microbes, and methods to obtain pure cultures
- Fourteen new photos
- Upgraded ten figures for greater clarity, ease of reading, and better pedagogy
- New "Beneficial Microbes" box on using bacteria to clean nuclear waste
- New "Clinical Case Study" box on biofilms and pediatric dental caries

CHAPTER 7 MICROBIAL GENETICS

- Updated sections on bacterial chromosome number and bacterial plasmids
- Added actions of topoisomerase and gyrase
- Included brief discussion of structure and action of spliceosomes
- Covered regulation of genetic expression—antisense RNA, interference RNA (RNAi), and riboswitches—and CAP/cAMP-mediated, positive regulation of the *lac* operon
- Introduced quorum sensing as it relates to genetic control in infection
- Updated text to reflect new research indicating that sigma factor of RNA polymerase does not necessarily disengage during transcription
- Included newly discovered codons and tRNAs for 21st and 22nd amino acids
- Increased coverage of the overwhelming nature of horizontal (lateral) gene transfer among prokaryotes, including transformation, transduction, conjugation, Hfr conjugation, and intercellular transposition
- Included new research showing that DNA moves through hollow pili even over great distances
- Modified artwork to reflect changes in our understanding of molecular biology. For example, where possible, enzyme shapes are based on actual 3-D profiles as revealed by X-ray crystallography (e.g., Figures 7.5, 7.8, and 7.20)
- Upgraded nineteen figures for greater clarity, accuracy, ease of reading, and better pedagogy
- Two new figures
- New critical thinking question on nucleotide analogs in cancer research
- New "Beneficial Microbes" box on the hot springs bacterium that helped make DNA synthesis in the laboratory possible
- New "Emerging Diseases" box on tick-borne encephalitis

CHAPTER 8 RECOMBINANT DNA TECHNOLOGY

- Added coverage of use of recombinant DNA technology to produce antisense nucleic acid molecules
- Added coverage of DNA microarrays and fluorescent *in situ* hybridization (FISH)
- Added coverage of new recombinant agricultural crops, including blight-resistant potatoes
- Added section discussing use of recombinant DNA techniques to address environmental problems
- Upgraded nine figures for greater clarity, accuracy, ease of reading, and better pedagogy
- Five new figures

CHAPTER 9 CONTROLLING MICROBIAL GROWTH IN THE ENVIRONMENT

- Added discussion of newly developed and approved enzymatic disinfectant against prions
- Increased discussion of use of mercury (thimerosal) as a preservative in vaccines
- Increased discussion of antimicrobial uses of bromine and fluorine
- Added discussion of the standardized test used in Europe to determine efficacy of disinfectants
- Added descriptions of four biosafety levels (BSLs) established by the CDC
- One new photo for BSL-4
- Revised two figures for better pedagogy
- Added new critical thinking questions, including one based on the 2008 *Salmonella* outbreak associated with tomatoes and peppers
- New "Emerging Diseases" box on *Acanthamoeba* keratitis
- New "Highlight" box on the use of brass water pitchers to disinfect water in rural India
- New "Beneficial Microbes" box on the newly authorized decontamination of lunchmeats using bacteriophage against *Listeria*

CHAPTER 10 CONTROLLING MICROBIAL GROWTH IN THE BODY: ANTIMICROBIAL DRUGS

- Clarified etymology and use of the terms antimicrobial, antibiotic, and semisynthetic
- Added discussion of use of RNA interference (RNAi) and antisense nucleic acids as antimicrobial therapy
- Updated and revised tables of antimicrobials to include all antimicrobials mentioned in pathogen chapters
- Added coverage of antifungal candins (caspofungin), antibacterial oxazolidinones, new anti-HIV drug tenofovir, and new antibacterial drug mupirocin
- Emphasized contrast between minimum inhibitory concentration and minimum bactericidal concentration tests
- Discussed newly discovered resistance mechanism by which bacterium produces DNA look-alike that binds to DNA gyrase, protecting the latter from fluoroquinolone action
- Added section on research to discover novel antimicrobials
- Nineteen new and/or revised figures, affording greater clarity, accuracy, and ease of reading, as well as better pedagogy
- Two new "Clinical Case Study" boxes on antibiotics use
- New "Beneficial Microbes" box on probiotics
- New "Emerging Diseases" box on community-associated MRSA

CHAPTER 11 CHARACTERIZING AND CLASSIFYING PROKARYOTES

- Introduction of:
  - New archaeal phylum—Korarchaeota
  - New information on alternative methods of cell division among prokaryotes
  - Discovery of organelles bounded by lipid membranes, including a presumptive nucleus in some prokaryotes
  - Bacterial viviparity (production of live offspring within a mother)
  - New hyperthermophile record holder—*Geogemma* ("strain 121")
- New photos of bacteria, archaea, and actinomycete spores
- Six revised figures for better pedagogy
- New "Beneficial Microbes" box on Botox
- New "Beneficial Microbes" box on bacterial superglue
- New "Emerging Diseases" box on whooping cough

CHAPTER 12 CHARACTERIZING AND CLASSIFYING EUKARYOTES

- Updated algal, fungal, protozoan, water mold, and slime mold taxonomy
- New fungal life cycle artwork to more clearly delineate nuclear states
- Clarified cyst-trophozoite conversion for intestinal protozoa
- Clarified artwork of *Paramecium* life cycle
- Ten new photos
- Upgraded seven figures for greater clarity, accuracy, ease of reading, and better pedagogy

- Moved major discussion of vectors from Chapter 23
- New "Beneficial Microbes" box on truffles
- New "Emerging Diseases" box on aspergillosis

CHAPTER 13 CHARACTERIZING AND CLASSIFYING VIRUSES, VIROIDS, AND PRIONS

- Updated viral nomenclature to correspond to changes approved by the International Committee on Taxonomy of Viruses; for example, herpes simplex viruses are officially known as *human herpesvirus 1* and *2*
- Added a discussion of the newly approved enzymatic disinfectant to treat prion-contaminated medical equipment
- Expanded coverage of prions, including the spread of prion disease through consumption of skeletal muscles and blood transfusions
- Expanded coverage of RNA animal viruses, including a new figure
- New TEM photos of viruses
- Upgraded four figures for better pedagogy
- New "Beneficial Microbes" box on using bacteriophages against bacterial pathogens
- New "Beneficial Microbes" box on viruses that keep algal blooms in check, moderate global warming, and increase oxygen production by cyanobacteria
- New "Clinical Case Study" box on variant Creutzfeldtz-Jakob disease
- New "Emerging Diseases" box on chikungunya

CHAPTER 14 INFECTION, INFECTIOUS DISEASES, AND EPIDEMIOLOGY

- Reordered text in response to users' desire for more sequential coverage of disease transmission and process
- Clarified that symbiotic relationships form a continuum and that relationships can evolve
- Redrew incidence and prevalence figure to reflect actual AIDS data in U.S.
- Updated epidemiology charts, tables, and graphs
- Updated list of nationally notifiable infectious diseases
- Expanded coverage of roles of public health agencies
- Added figure on transmission of pathogens
- Revised seven figures for better pedagogy
- New "Beneficial Microbes" box on symbiosis and a natural insecticide
- Two new "Clinical Case Study" boxes, one on human carriers and one on tuberculosis
- New "Emerging Diseases" box on *Hantavirus* pulmonary syndrome

CHAPTER 15 INNATE IMMUNITY

- Included coverage of antimicrobial peptides and bradykinins (act in inflammation)
- Greatly enhanced coverage of Toll-like receptors (TLRs)
- Expanded coverage of pathogen-associated molecular patterns (PAMPs)
- Discussed neutrophil extracellular traps (NETs)
- Presented NOD receptor proteins
- Added latest discoveries in iron usage among pathogenic bacteria and sequestration of iron in the body as a defense
- Enhanced coverage of steps involved in phagocytosis
- Clarified artwork and discussion of pathways of complement activation, including the lectin pathway
- Clarified that cells and fluid move from venules (not arterioles or capillaries) during dilation and inflammation
- Modified seven figures for enhanced clarity and better pedagogy
- New "Beneficial Microbes" box on the role of microbes in decomposing human skin

CHAPTER 16 ADAPTIVE IMMUNITY

- Rewrote the entire chapter to make it more pedagogically efficient and an even better presentation of adaptive immunity
- Expanded coverage of T-independent humoral immunity, including coverage of its importance in pediatric medicine and immunity
- Updated coverage of adaptive immunity to reflect new discoveries and enhance accuracy and clarity:
  - Added direct killing of bacteria by antibodies via hydrogen peroxide and ozone production
  - Added coverage of immunological synapses
  - Added coverage of membrane rafts and their roles in immunological synapses
  - Added coverage of membrane rafts in IgE-triggered histamine release from mast cells
  - Updated information on T cell cancer therapy
- Eight new pieces of art, seventeen revised pieces of art, two new photos
- Modified and recolored retained artwork for greater clarity and pedagogical efficiency
- New "Highlight" box on BCR diversity
- New "Emerging Diseases" box on microsporidiosis

CHAPTER 17 IMMUNIZATION AND IMMUNE TESTING

- Updated coverage of types of vaccines, including newly approved combination vaccines
- Updated coverage of passive immunotherapy
- Added information regarding vaccines against agents of hepatitis A, human papillomavirus (cervical cancer), anthrax, cholera, plague, rotavirus, Japanese encephalitis, typhoid fever
- Updated to newly revised CDC 2009 vaccination schedule for children, adolescents, and adults
- Updated table of vaccine-preventable diseases in the U.S.
- Added methods of vaccine administration
- Added coverage of rapid immunochromatographic dipstick assays for diagnosis of infections with *E. coli* O157:H7, group A *Streptococcus*, respiratory syncytial virus, and influenzaviruses
- Expanded and clarified definition of herd immunity, contact immunity, immunization, vaccination, vaccine, titer, direct immune testing, and indirect immune testing
- Reorganized topics for better pedagogy
- Seventeen revised figures, two new figures, and two new photographs for better pedagogy
- New "Beneficial Microbes" box on the question of reinstating smallpox vaccinations

CHAPTER 18 IMMUNE DISORDERS

- Updated, simplified, and corrected material on Graves' disease, tissue transplants, and multiple sclerosis
- Twelve revised figures, five recolored figures, and three new photographs for better pedagogy
- New "Clinical Case Study" box on hypersensitivity reaction

CHAPTER 19 PATHOGENIC GRAM-POSITIVE BACTERIA

- Expanded coverage of methicillin-resistant and vancomycin-resistant *Staphylococcus aureus* (MRSA, VRSA), necrotizing fasciitis, and multi-drug-resistant tuberculosis (MDR-TB)
- Added coverage of extensively drug-resistant TB (XDR-TB)
- Expanded and updated coverage of action of anthrax toxins
- Moved coverage of low G + C, Gram-positive mycoplasmas (though pink-Gram-staining) from Chapter 21
- Fourteen new photos
- Six revisions to figures for consistency, accuracy, and better pedagogy
- New "Beneficial Microbes" box on probiotics
- New "Emerging Diseases" box on Buruli ulcer

CHAPTER 20 PATHOGENIC GRAM-NEGATIVE COCCI AND BACILLI

- Added discussion of CDC target incidence for gonorrhea to limit spread of the disease
- Added discussion of blebbing as it relates to meningococcal disease
- Added discussion of siderophores as virulence factor of enterobacteria
- Ten new figures
- Thirteen figures revised for better pedagogy
- Added real clinical case study of *E. coli* bacteremia
- New end-of-chapter question on transovarian transmission of pathogens
- New "Emerging Diseases" box on melioidosis
- New "Beneficial Microbes" box on bacterial-triggered angiogenesis
- New "Beneficial Microbes" box on using bacterial predators against coliforms and *Pseudomonas*

CHAPTER 21 RICKETTSIAS, CHLAMYDIAS, SPIROCHETES, AND VIBRIOS

- Updated to show changes in taxonomy: *Ehrlichia equi* is now known to be *Anaplasma phagocytophilum*, and its disease is now called anaplasmosis rather than human granulocytic ehrlichiosis; *Chlamydia pneumoniae* and *C. psittaci* are now *Chlamydophila* spp.
- Seven new figures
- Eleven figures revised for greater visual contrast, pedagogy, accuracy, currency, and general interest
- New "Emerging Diseases" box on *Vibrio vulnificus* infection

CHAPTER 22 PATHOGENIC FUNGI

- Major rewrite and reorganization of the chapter:
  - Expanded coverage of fungal taxonomy to briefly consider anamorphs and teleomorphs
  - New summary chart for agents of chromoblastomycosis, phaeohyphomycoses, mycetoma, and sporotrichosis
  - Enhanced discussion of the antifungal action of griseofulvin
  - Added discussion of relative lack of drug resistance by fungal pathogens
  - Updated diagnoses and treatment of fungal diseases and development of vaccines against fungi
  - Increased coverage of emerging fungal pathogens, especially as they relate to AIDS
- Twelve new figures
- Five figures revised for enhanced pedagogy
- New "Emerging Diseases" box on pulmonary blastomycosis

CHAPTER 23 PARASITIC PROTOZOA, HELMINTHS, AND ARTHROPOD VECTORS

- Expanded coverage of *Plasmodium* and malaria
- Expanded life cycle of *Toxoplasma gondii* to include latest research on human infections and effect on rodent-cat interactions
- Nine new figures
- Eleven revised, updated, enhanced, and pedagogically more effective figures
- New end-of-chapter critical thinking question and new figure legend questions on tapeworms and on risk factors for toxoplasmosis
- New "Clinical Case Study" box on leishmaniasis
- New "Emerging Diseases" box on babesiosis

CHAPTER 24 PATHOGENIC DNA VIRUSES

- Updated varicella-zoster vaccine recommendations
- Updated coverage of molluscum contagiosum
- Expanded coverage of all seven human herpesviruses
- Updated and expanded coverage of papillomaviruses and their treatment and prevention of their diseases
- Expanded coverage of the use of reverse transcription in the replication of hepatitis B virus
- Seven new figures
- Five revised, updated, or enhanced figures
- New "Highlight" box on a possible viral cause of obesity
- New "Beneficial Microbes" box on using a bacteriophage to combat cocaine addiction
- New "Emerging Diseases" box on monkeypox

CHAPTER 25 PATHOGENIC RNA VIRUSES

- Updated rotavirus vaccine efficacy
- Updated Nipah virus taxonomy
- Fourteen revised, updated, or enhanced figures
- Fourteen new figures
- Additional end-of-chapter questions
- New "Emerging Diseases" box on H1N1 influenza

CHAPTER 26 APPLIED AND ENVIRONMENTAL MICROBIOLOGY

- Added anammox reactions to nitrogen cycle
- Added discussion of problem of false-positive results from fecal coliform tests for sewage pollution
- Increased discussion of use of bacteria to directly generate electricity
- Added coverage of archaea in environment
- Clarified use of the term fermentation in biochemistry, food production, and industry
- Expanded coverage of pharmaceutical products produced by recombinant DNA technology
- Seven new photos
- Ten revised figures
- New "Emerging Diseases" box on *Norovirus* gastroenteritis

Reviewers for the Third Edition

First, I wish to thank the hundreds of instructors and students who participated in reviews, class tests, and focus groups for earlier editions of the textbook. Your comments have informed this book from beginning to end, and I am deeply grateful. For the Third Edition, I extend my deepest appreciation to the following reviewers:

Book Reviewers

Sandra Barnes
Housatonic Community Technical College

Susan Baxley
Troy University

Rita Connolly
Camden County Community College

Josephine Coursey
Tyler Junior College

Larry T. Crump
Joliet Junior College

Diane Dorsett
Georgia Gwinnett College

Ana Dowey
Palomar College

Wendy Owen Dusek
Wisconsin Indianhead Technical College

Lisa Ferrara
West Virginia University Institute of Technology

Joseph J. Gauthier
University of Alabama

Brinda Govindan
San Francisco State University

Michael T. Griffin
Angelo State University

John Lammert
Gustavus Adolphus College

Luis A. Materon
University of Texas, Pan American

Robin G. Maxwell
University of North Carolina, Greensboro

Virginia Meyer
Sacramento City College

Lori McGowan
Harrisburg Area Community College

Elizabeth McPherson
University of Tennessee, Knoxville

Bethanye Bagby Morgan
Tarrant County College

Gina Marie Morris
Frank Phillips College

Richard Myers
Missouri State University

Ruth Negley
Harrisburg Area Community College, Gettysburg Campus

Karl J. Roberts
Prince George's Community College

Michael Ruhl
Vernon College

Sarmad Saman
Massachusetts Bay Community College

Steve Schmidt
University of Colorado, Boulder

Timothy Secott
Minnesota State University, Mankato

Pramila Sen
Houston Community College

Alison Shakarian
Salve Regina University

Brian Shmaefsky
Lone Star College, Kingwood

Benjamin Simon
The Evergreen State College

Jane Slone
Cedar Valley College, Dallas County Community College District

Stephanie Rena Songer
North Georgia College & State University

Christopher Thompson
Loyola College

MasteringMicrobiology Faculty Advisory Board

Kris Dougherty
Valencia Community College

Judy Kandel
California State University—Fullerton

Suzanne Long
Monroe Community College

Mary Puglia
Central Arizona University

Jackie Reynolds
Richland College

Christopher Thompson
Loyola College

MicroFlix Reviewers

Lee Couch
University of New Mexico

Michael Griffin
Angelo State University

Suzanne Long
Monroe Community College

Mary Puglia
Central Arizona University

Jackie Reynolds
Richland College

Kathryn Sutton
Clarke College

Christopher Thompson
Loyola College

Acknowledgments

As was the case with the previous editions, this book has truly been a team effort. I am deeply grateful to Leslie Berriman of Benjamin Cummings and to the team she gathered to produce the third edition. Leslie, dedicated project editor Robin Pille, and Barbara Yien, project editor of the first two editions, helped develop the vision for the third edition, coming up with ideas for making it more effective and compelling. As project editor, Robin also had the task of coordinating everything and keeping me on track—thank you, Robin, for going beyond your job description. My gratefulness to Barbara Yien for years of support and pounds of chocolate truffles is constant. I am grateful to Frank Ruggirello for his support of my work and this book. I am also indebted to Daryl Fox, whose early support for this book never wavered.

The incomparable Anita Wagner edited the manuscript thoroughly and meticulously, suggesting important changes for clarity, accuracy, and consistency. Elisheva Marcus did a superb job as art development editor, helping to conceptualize new illustrations and suggesting ways to improve the art overall—thank you, Ellie. My friend Ken Probst is responsible for originally creating this book's amazingly beautiful biological illustrations. My thanks to Precision Graphics for rendering the art in this edition. David Novak and Michele Mangelli of Mangelli Productions expertly guided the project through production. David, you are amazing. Michelle Jones led the skilled team at Progressive Information Technologies in moving the book smoothly through composition. Maureen "Mo" Spuhler did an incredible job researching the photos for this text. Rich Robison and Brent Selinger supplied many of the text's wonderful and unique micrographs. The photos and illustrations also benefited from the wisdom and experience of Travis Amos and Laura Southworth, and the super-organized Linda Jupiter kept us on all track—thank you all. Yvo Riezebos created the stunning cover and wonderful interior design.

Thanks to Nichol Dolby of Amarillo College; Suzanne Long of Monroe Community College; Randall Harris of William Carey University; Mindy Miller-Kittrell of University of Tennessee, Knoxville; Jason Andrus of Meredith College; Tiffany Glaven of University of California, Davis; Kathryn Sutton of Clarke College; and Judy Meier Penn of Shoreline Community College for their work on the media and print supplements for this edition. Special thanks are due to Lucinda Bingham for her expert management of the extraordinary array of media resources for students and instructors, especially the exciting new MasteringMicrobiology website. I am grateful also for Kelly Reed's outstanding work as the editor of the supplements for this edition, in particular her editorial development of the concept mapping exercises. Thanks also to Nan Kemp and Jacob Price for their administrative, editorial, and research assistance. Betsy Dietrich proofread and checked pages—without her help the book would be less useful. I am grateful to Neena Bali in Marketing and the Benjamin Cummings sales representatives for continuing to do a terrific job of keeping in touch with the professors and students who provided so many wonderful suggestions for this textbook. You sales representatives inspire and humble me, and your role on the team deserves more praise than I can express here.

I am especially grateful to Phil Mixter of Washington State University, Mary Jane Niles of the University of San Francisco, Bronwen Steele of Estrella Mountain Community College, and Jane Reece for their expertise and advice on the cell and immunology chapters.

On the home front, I am grateful for Dr. Nichol Dolby and Dr. Michael Kopenits at Amarillo College and for Kendall Blythe; Jennie Bauman; Elizabeth McBride; Andy Roller; Larry Latham; Vance Esler, MD; and Mike Isley—all of whom were always supportive and helpful. My "secretarial staff," Michelle and Jeremy Bauman, were always there to photocopy, type, file, surf the web, run to the FedEx box, and provide emotional support. My life is not my own—I owe everything to others.

Robert W. Bauman
Amarillo, Texas

Cutting-Edge Science

NEW Beneficial Microbes boxes emphasize the practical and benevolent nature and uses of microbes and help students overcome the common misconception that most microbes are damaging and cause disease. A full list of these Beneficial Microbes boxes can be found on page xxx.

BENEFICIAL MICROBES

GOOD VIRUSES? WHO KNEW?

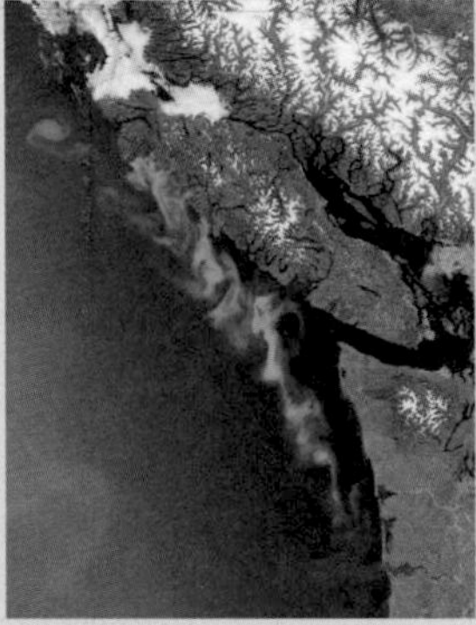

▲ *An algal bloom off the coast of Seattle, Washington.*

Viruses, though normally pathogenic to their host cells, do have positive influences, including what appear to be extensive roles in the environment. Recent discoveries by the United Kingdom's Marine and Freshwater Microbial Biodiversity program demonstrate important ways viruses impact our world.

First: Scientists found that a previously unknown virus attacks a tiny marine alga that form algal blooms consisting of hundreds of thousands to millions of algal cells per milliliter of water (see photo). Algal blooms like these can deplete the water of oxygen at night, potentially harming fish and other marine life. The newly discovered virus stops blooms by killing the algae, a result which is good for animal life.

Second: When the algae die by this means, they release an airborne sulfate compound that acts to seed clouds. The resulting increased cloudiness noticeably shades the ocean, measurably lowering water temperature. Thus, a marine virus helps to reduce global warming!

Third: The researchers discovered a bacteriophage of oceanic cyanobacteria that transfers genes for photosynthetic machinery into its hosts' cells, so that the cells' photosynthetic rate increases. There are up to 10 million of these viruses in a single milliliter of seawater, so researchers estimate that much of the oxygen we breathe may be attributable to the action of

BENEFICIAL MICROBES

WHAT HAPPENS TO ALL THAT SKIN?

▲ *Dust mite.* SEM

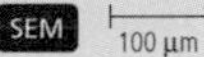

Your body sheds tens of thousands of skin flakes every time you walk or move, and you shed at only a slightly lower rate when you stand still. That comes to about 10 billion skin cells per day, or 250 grams (about half a pound) of skin every year! What happens to all that skin?

Much of household dust is skin that you and your housemates have shed as you go about your lives. The skin flakes fall to the rug and upholstery, where they become food for microscopic mites that live sedentary and harmless lives waiting patiently for meals to rain down on them from above. They dwell not only in the rug, but also in your mattress and pillow, and even in the hair follicles of your eyebrows, benefiting you by catching skin cells cascading down your forehead before they can irritate your eyes.

By the way, house dust also contains mite feces and mite skeletons, which can trigger allergies. So after reading this chapter, you just might want to clean your carpet.

BOTULISM AND BOTOX

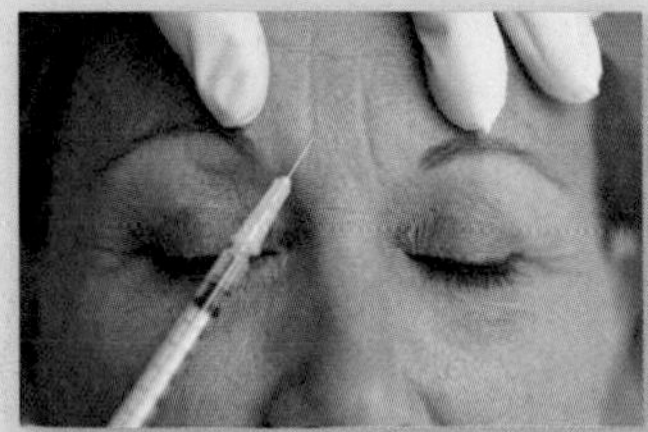

page 323

GOLD-MINING MICROBES

page 126

PLASTICS MADE PERFECT?

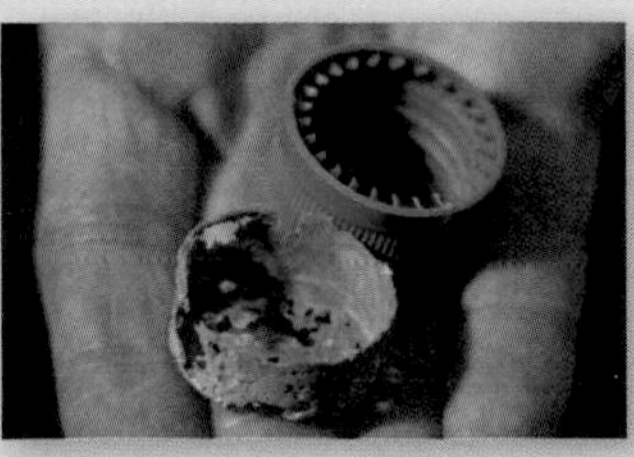

page 74

FUNGI FOR $3000 A POUND

page 361

NEW Emerging Diseases boxes are written in an engaging narrative voice that focuses on a patient's experience with a key emerging or re-emerging disease such as *Hantavirus* pulmonary syndrome, babesiosis, and MRSA. (See page xxxi for a full list.) Up-to-the-minute information about these emerging diseases can be found on the website for the book.

EMERGING DISEASES

CHIKUNGUNYA

An old man arrived at the doctor's office in Ravenna, Italy, with a combination of symptoms the physician had never heard of: a widespread, severe rash; difficulty in breathing; high fever; nausea; and extreme joint pain. Chikungunya (chik-en-gun' ya) had arrived in Europe.

Though scientists had known of chikungunya virus, which is related to

reported 47,000 cases of chikungunya in a single week! The same year it reemerged in India for the first time in four decades with more than 1.5 million reported cases. Why?

Aedes albopictus (Asian tiger mosquito), which carries the virus, has moved into temperate climates, including Europe and the United States, as the climate has warmed. With the mosquito

viral proliferation—the insects have

as far north as Italy.

? His crippling pain lasted for months,

he knows about mosquito-borne

it his family and friends use mosquito

in the rest of Europe and in the United

With the coming of *Ae. albopictus*, is

behind?

.com.

EMERGING DISEASES

VARIANT CREUTZFELDT-JAKOB DISEASE

Ellen screamed obscenities as she staggered from the room and collapsed in the hallway, unable to stand and jerking uncontrollably. Her parents were shocked that their kind, considerate, and lovable daughter had changed so drastically during the past year. Sadly, she couldn't even remember her siblings' names!

Ellen had joined the nearly 200 Europeans, and one Canadian, afflicted with variant Creutzfeldt-Jakob disease (vCJD) (what the media call "mad cow disease" because most humans with the condition acquired the pathogen from eating infected beef). Since vCJD affects the brain by slowly eroding nervous tissue and leaving the brain full of sponge-like holes, the signs and symptoms of vCJD are neurological. Ellen's disease started with insomnia, depression, and confusion, but eventually it led to uncontrollable emotional and verbal outbursts, inability to coordinate movements, coma, and death. Typically the disease lasts about a year, and there is no treatment.

Variant Creutzfeldt-Jakob disease is an *emerging disease;* that is, a disease arising in the past two decades, either because it is new to a population, or because it is newly recognized. Some investigators also include diseases that have been nearly eradicated but are now reemerging. Variant CJD resembles the rare genetic disorder Creutzfeldt-Jakob disease (named for its discoverers), which is caused by a mutation and occurs in the elderly. The difference is that the variant form of CJD results from an acquired infection and often strikes and kills college-aged people, like Ellen in our story. For more about vCJD, see p. 396.

Track vCJD online by going to the Study Area at www.masteringmicrobiology.com.

HANTAVIRUS PULMONARY SYNDROME

page 429

PERTUSSIS

page 331

ASPERGILLOSIS

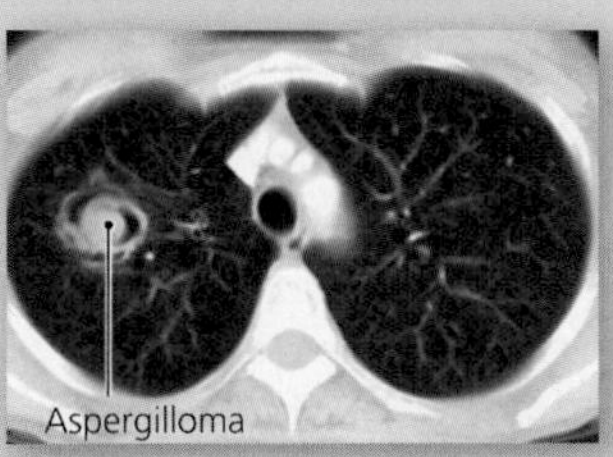

page 361

TICK-BORNE ENCEPHALITIS

page 204

Cutting-Edge Practice

Clinical Case Study boxes appear throughout the book and are particularly helpful to pre-nursing and allied health students. They ask students to apply material they have learned in the text to clinical scenarios and often feature a micrograph or clinical photo to interpret.

CLINICAL CASE STUDY

Cavities Gone Wild

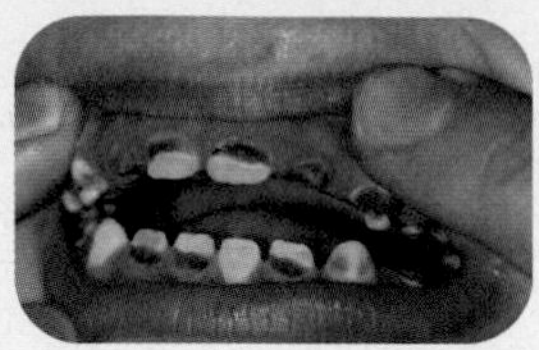

Five-year-old Daniel appears to be shy. He always looks at the floor, has no friends, never plays with the other children, will rarely speak to adults, and when he does speak, it difficult to understand broken enunciation. His skinny frame and the dark circles under his eyes make him appear malnourished. Daniel cries frequently and misses ma days of school.

A speech specialist at school finds that only two of Danie teeth are healthy; all the others have rotted away to the gum line. The little guy is in constant pain and it hurts to chew. A doctor later determines that bacteria from the cavities in his mouth have entered his bloodstream and infected his heart, causing an irregular heartbeat and poor blood circulation.

1. How does knowledge of biofilms help explain the bulk Daniel's problems?
2. What can Daniel, his parents, and health care professionals do to cure his diseases?
3. What nutrient should Daniel's parents eliminate from h diet to help prevent a repeat of his condition?
4. A recent study has shown that brushing does more tha clean plaque from the teeth; it also disrupts associatio between oral bacteria. How does this simple act help prevent the formation of biofilms?

MICROBE AT A GLANCE

Plasmodium falciparum

Taxonomy: Nomenclature is in flux; one possibility, used in this book: domain Eukarya, kingdom Alveolata, phylum Apicomplexa, class Sporozoea, order Eucoccidiida, family Plasmodiidae

Morphology: Sporozoite (slender, elongate single cell); merozoite (oval single cell); trophozoite (ring of cytoplasm with a dotlike nucleus; "ring form"); ookinete (motile wormlike form); oocyte (spherical)

Virulence factors: Lives intracellularly

Diseases caused: Malignant tertian malaria—a form of malaria called blackwater fever

Treatment for disease: Chloroquine, mefloquine, atovaquone, proguanil, artemisinin

Prevention of disease: Mosquito control, avoid mosquito bites

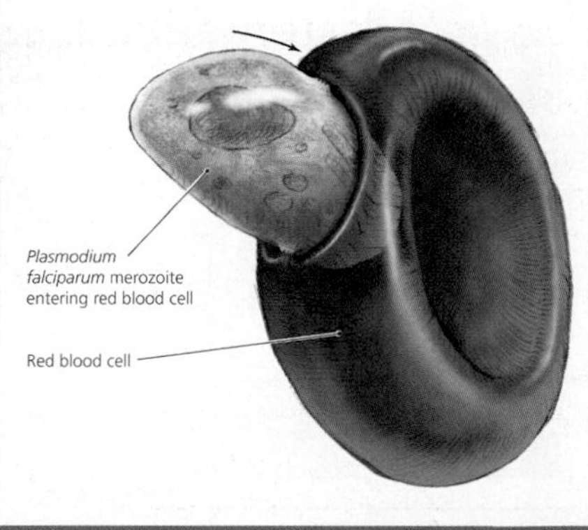

Download the Microbe at a Glance flashcards from the Study Area at www.masteringmicrobiology.com.

Microbe at a Glance boxes showcase representative microbes in each of the disease chapters (Chs 19–25). These boxed "snapshots" also appear as flashcards on the book website, giving students extra practice.

Critical Thinking questions, located throughout the text and at the end of each chapter, encourage students to apply what they have just read to an additional scenario or case study.

CRITICAL **THINKING**

Given that *Haemophilus ducreyi* is a chemoheterotrophic pathogen that requires heme as a growth factor, deduce how this bacterium phosphorylates most of its ADP to form ATP. Defend your answer.

CRITICAL **THINKING**

Solutions hypertonic to bacteria and fungi are used for food preservation. For instance, jams and jellies are hypertonic with sugar, and pickles are hypertonic with salt. How do hypertonic solutions kill …ould otherwise spoil these foods?

CRITICAL **THINKING**

After a man infected with the bacterium *Escherichia coli* was treated with the correct antibiotic for this pathogen, the bacterium was no longer found in the man's blood, but his symptoms of fever and inflammation worsened. What caused the man's response to the treatment? Why was his condition worsened by the treatment?

in the Book

NEW Concept Mapping sections

appear in the end-of-chapter material, guiding students to create their own concept maps from a list of key terms focused around an important chapter topic. Students can also complete interactive concept maps in the Study Area of MasteringMicrobiology (www.masteringmicrobiology.com) – see page xvi.

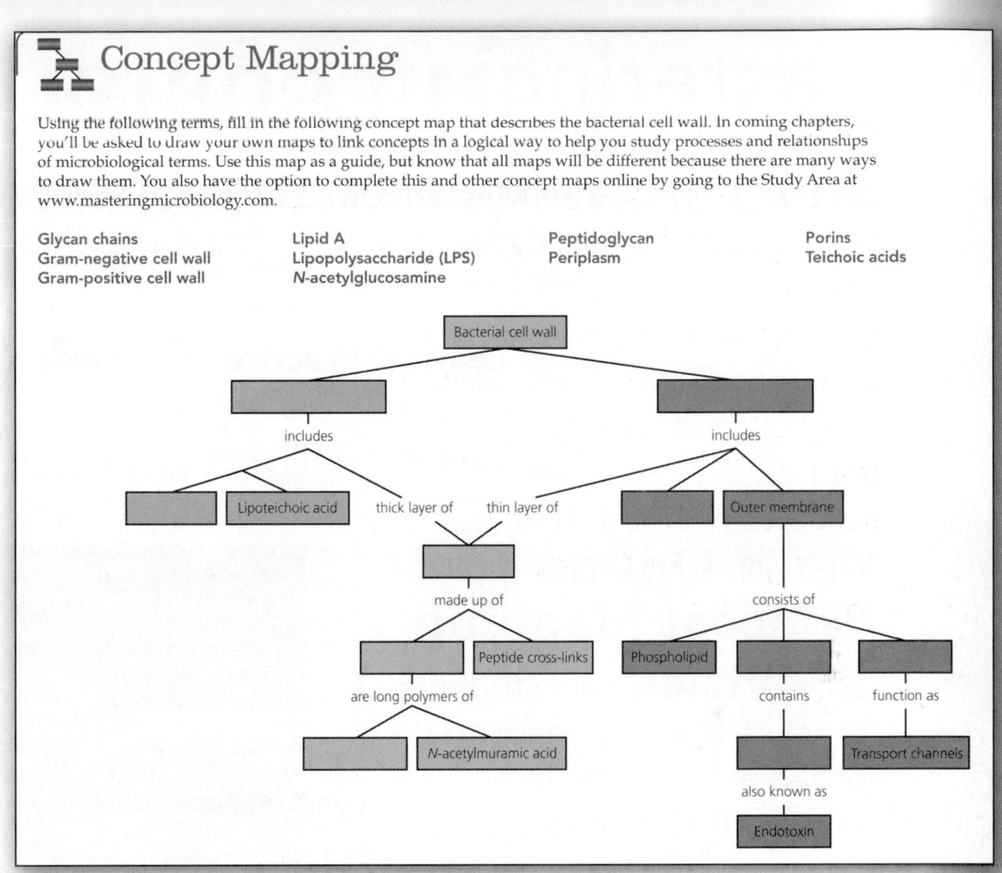

Concept Mapping
Using the following terms, fill in the following concept map that describes the bacterial cell wall. In coming chapters, you'll be asked to draw your own maps to link concepts in a logical way to help you study processes and relationships of microbiological terms. Use this map as a guide, but know that all maps will be different because there are many ways to draw them. You also have the option to complete this and other concept maps online by going to the Study Area at www.masteringmicrobiology.com.
Glycan chains
Gram-negative cell wall
Gram-positive cell wall
Lipid A
Lipopolysaccharide (LPS)
N-acetylglucosamine
Peptidoglycan
Periplasm
Porins
Teichoic acids
Bacterial cell wall
includes
includes
Lipoteichoic acid
thick layer of
thin layer of
Outer membrane
made up of
consists of
Peptide cross-links
Phospholipid
are long polymers of
contains
function as
N-acetylmuramic acid
Transport channels
also known as
Endotoxin

Labeling

Label the steps of phagocytosis.

1 ______

2 ______

3 ______

4 ______

5 ______

6 ______

NEW Art-based questions

check students' visual understanding by asking them to label and interpret figures from the chapter.

Cutting-Edge Practice

MasteringMICROBIOLOGY™

The MasteringMicrobiology Study Area helps students get ready for tests with its simple, three-step approach:

1. **Take a Pre-Test** and obtain a personalized Study Plan.
2. **Learn & Practice** with animations, activities, and MP3 Tutor Sessions.
3. **Test Yourself** with quizzes and a chapter post-test.

NEW Online Concept Mapping activities help students practice building maps to organize concepts in a meaningful, visual way.

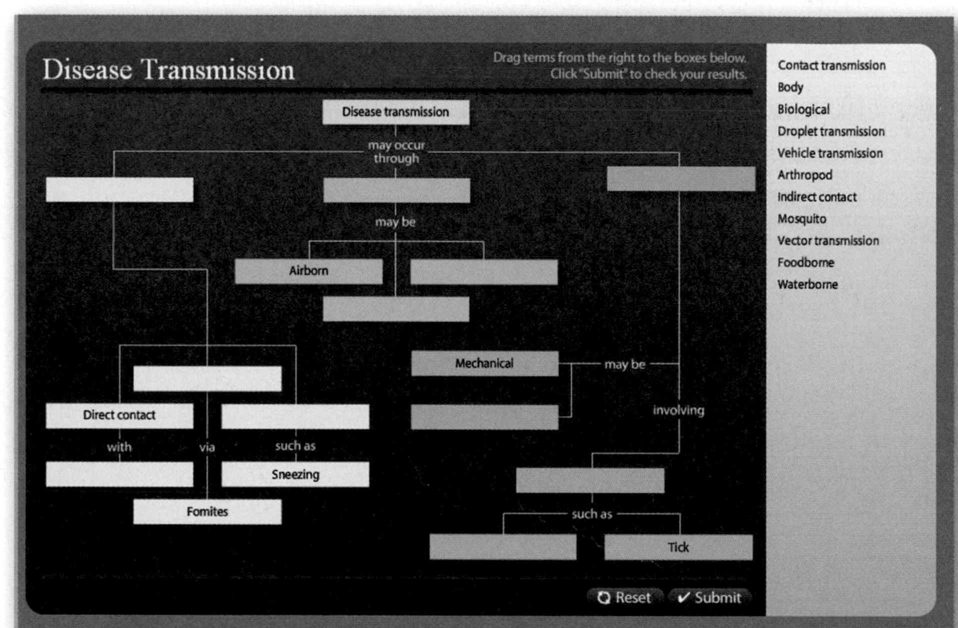

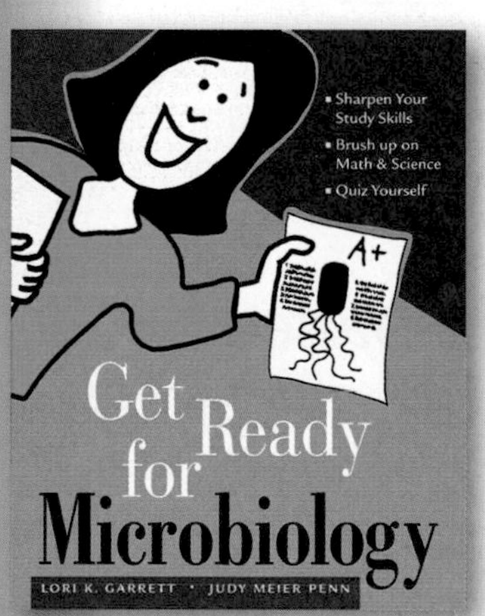

NEW Get Ready for Microbiology includes an eText and diagnostic tests. See page xviii for more information.

MP3 Tutor Sessions

Students can download MP3s for specific chapters of the textbook and study on the go. They can listen to mini-lectures about the tough topics and take audio quizzes to check their understanding.

Online

INSTRUCTORS:
See inside front cover for more about MasteringMicrobiology assignments and assessment!

NEW MicroFlix™ are 3-D, movie-quality animations with self-paced tutorials and gradable quizzes that help students master the three toughest topics in microbiology: metabolism, DNA replication, and immunology. Students can view the animations, complete the tutorial, print a study sheet, and take the quiz. Students also have access to BioFlix animations to help them review relevant concepts from general biology.

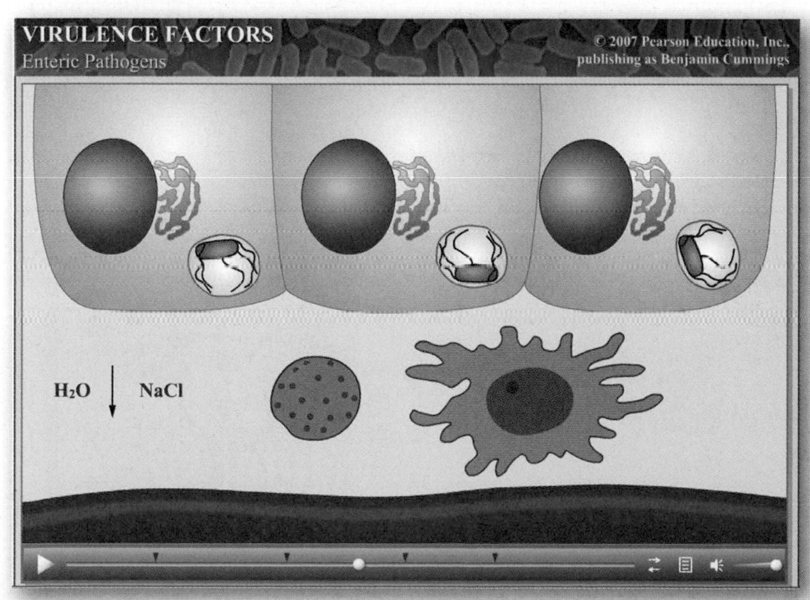

115 multi-step Microbiology Animations

explain and visually demonstrate core concepts, providing an additional chance for students to learn. They are accompanied by gradable quizzes. References to the Microbiology Animations appear throughout the chapters of the book.

Are your students ready?

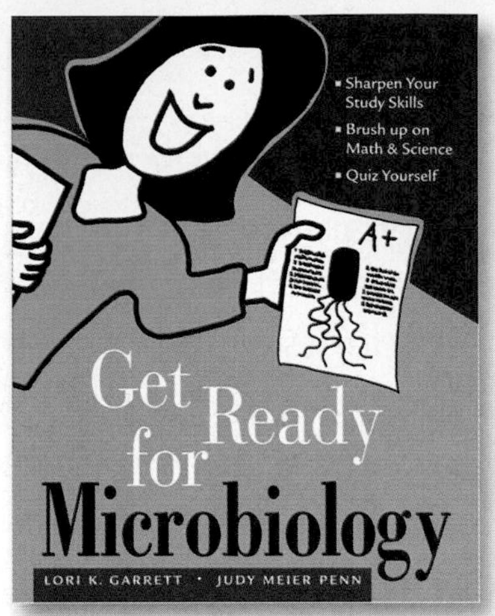

Get Ready for Microbiology quickly prepares students for the microbiology course, helping them brush up on the skills they need to succeed.

CONTENTS

YOUR STARTING POINT pre-tests students' grasp of chapter content before they start the chapter.

Your Starting Point

Answer the following questions to assess your chemistry knowledge.

1. The most basic unit of a chemical substance is the ____________
2. What are the three states of matter? ____________
3. An atom is made of what three subatomic particles? ____________

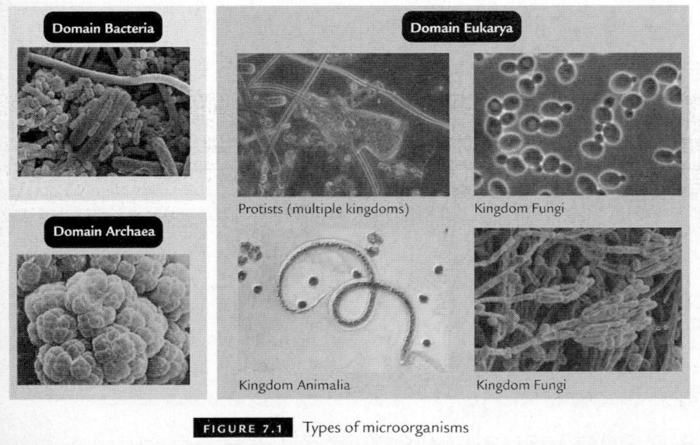

FIGURE 7.1 Types of microorganisms

Chapters include a variety of photographs and illustrations.

Engaging features like **TIME TO TRY** provide a simple experiment or quick question that gives students a chance to practice what they just learned.

TIME TO TRY

Examine the three drawings below. Circle the ones that you think are cells.

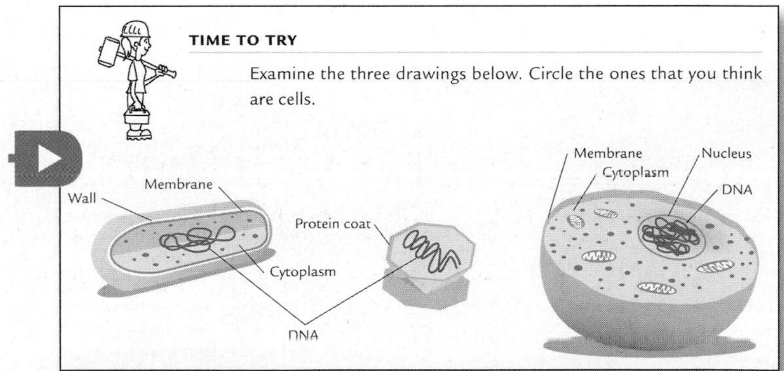

NEW MasteringMicrobiology offers assignable Get Ready quizzes. The Study Area offers additional practice material for students.

Options for the Lab

Choose a manual for your lab that features:

Stunning & easy-to-navigate design • Integrated color photographs • Step-by-step, color illustrations

Concise & Investigative

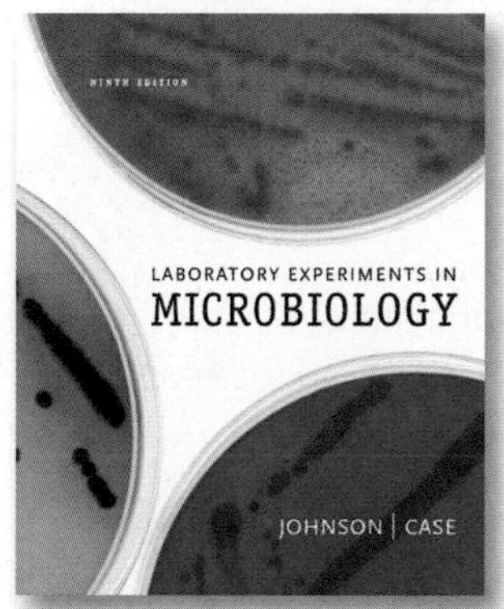

Laboratory Experiments in Microbiology,
Ninth Edition
by Ted R. Johnson
and Christine L. Case
978-0321-56028-5
/ 0-321-56028-0

This manual offers 57 exercises that encourage students to evaluate their results and draw conclusions. Questions placed within the procedure require students to pause and recall what they have learned. Critical thinking questions in the Laboratory Reports further promote analytical reasoning.

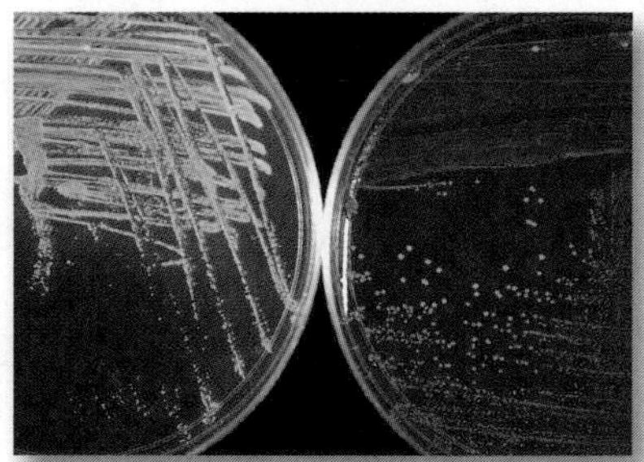

Rich, vibrant photographs and illustrations appear within the relevant exercise, allowing students to better interpret their results.

Versatile & Comprehensive

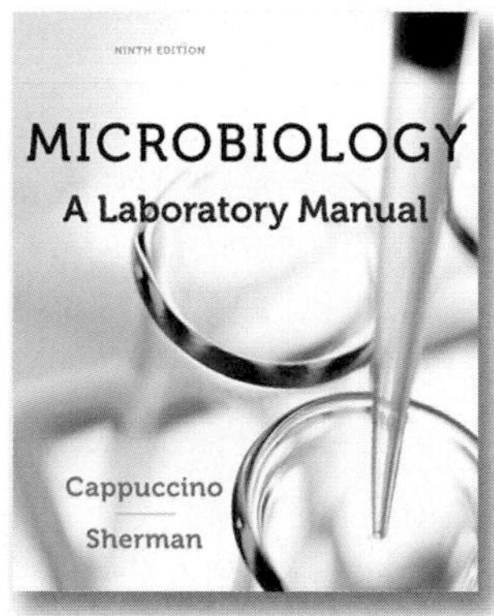

Microbiology:
A Laboratory Manual,
Ninth Edition
by James G. Cappuccino
and Natalie Sherman
978-0321-65133-4
/ 0-321-65133-2

This manual offers 75 straightforward, clearly explained experiments with minimal equipment requirements. Instructors can pick and choose from a variety of diverse experiments, including labs in the areas of biotechnology, genetics, immunology, and medical microbiology.

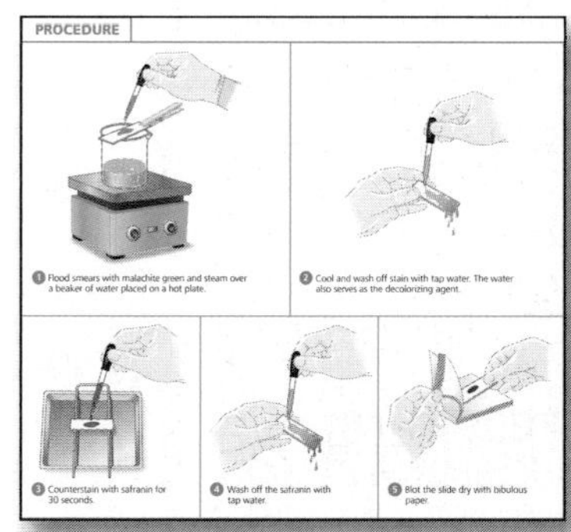

Clear, realistically-colored procedural figures guide students through each procedure, providing visual instructions along with narrative ones.

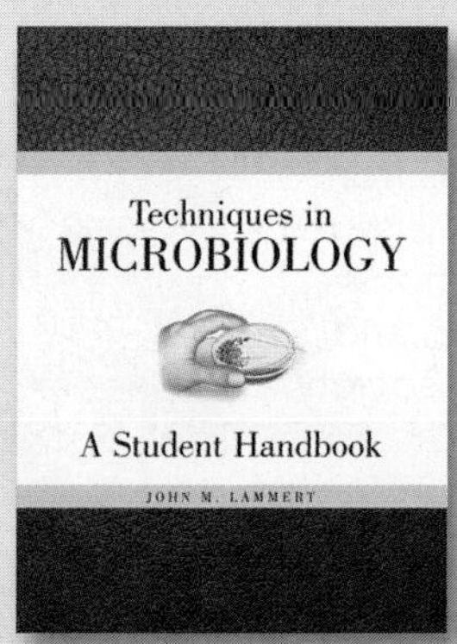

An excellent companion for your laboratory manual

Techniques in Microbiology: A Student Handbook
by John M. Lammert
978-0-13-224011-6 / 0-13-22401-4

An ideal complement to a complete lab manual, this vivid, full-color handbook guides students in manipulations and preparations that are fundamental to the microbiology laboratory.

The best support for instructors and students

Instructor Supplements

MasteringMicrobiology™

(www.masteringmicrobiology.com) This website helps instructors maximize class time with customizable, easy-to-assign, and automatically graded assessments that motivate students to learn outside of class. For more information, see the inside front cover of this book.

INSTRUCTOR RESOURCE DVD/CD-ROM

This media tool includes:

- All figures from the book with and without labels in both JPEG and PowerPoint® formats
- All figures from the book with the Label Edit feature in PowerPoint format
- Select "process" figures from the book with the Step Edit feature in PowerPoint format
- All tables from the book
- MicroFlix™ and BioFlix™ Animations, Microbiology Animations, and Microbiology Videos
- PowerPoint lecture outlines, including figures and tables from the book and links to the Microbiology Animations
- PRS-enabled Active Lecture Clicker Questions
- PRS-enabled Quiz Show Clicker Questions
- PDF files of Transparency Acetate masters
- The Instructor's Manual as editable Microsoft® Word files
- The Test Bank as editable Microsoft® Word files
- The Test Bank in TestGen® format

978-0-321-67737-2 / 0-321-67737-4

INSTRUCTOR'S MANUAL AND TEST BANK

This printed guide includes Chapter Outlines, Chapter Summaries, and answers to both the Critical Thinking questions and the Clinical Case Studies. Test items are tagged with a corresponding section title and Bloom's Taxonomy ranking. This supplement is also available in Microsoft Word® format on the IRDVD/CD-ROM and in the Instructor Resources area of MasteringMicrobiology.
978-0-321-67736-5 / 0-321-67736-6

COURSECOMPASS™/ WEBCT / BLACKBOARD

Pre-loaded book-specific content and test item files accompanying the text are available in several course management formats.

Student Supplements

MasteringMicrobiology™

(www.masteringmicrobiology.com) A three-step learning process in the Study Area of MasteringMicrobiology takes students through these simple steps: Pre-Test, Learn & Practice, and Test Yourself. Students have access to a variety of self-study tools. See page xvi for more information.

GET READY FOR MICROBIOLOGY

by Lori K. Garrett and Judy Meier Penn
This new brief primer saves classroom time and frustration by helping students quickly prepare for their microbiology course. See page xviii for details.
978-0-321-68347-2 / 0-321-68347-1

STUDY GUIDE FOR MICROBIOLOGY WITH DISEASES BY TAXONOMY THIRD EDITION

by Mindy Miller-Kittrell and Elizabeth Machunis-Masuoka
Students can master key concepts and earn a better grade with the help of the clear writing and creative, thought-provoking exercises in this study guide. It includes concise explanations of key concepts, definitions of important terms, critical thinking problems, and a variety of self-test questions, with answers.
978-0-321-67738-9 / 0-321-67738-2

STUDY CARD

The Study Card for Microbiology is a six-panel laminated card with an overview of diseases by body system on one side and an overview of diseases by taxonomy on the other side.

TECHNIQUES IN MICROBIOLOGY: A STUDENT HANDBOOK

by John Lammert
This vivid, full-color handbook guides students in manipulations and preparations needed in the microbiology laboratory. The techniques are the ones that are used frequently for studying microbes in the laboratory and include those identified by the American Society for Microbiology (ASM) in its recommendations for the Microbiology Core Lab Curriculum.
978-0-13-224011-6 / 0-13-224011-4

SCIENTIFIC AMERICAN: CURRENT ISSUES IN MICROBIOLOGY

Accessible, dynamic, and relevant articles from *Scientific American* magazine present key issues in microbiology, and end-of-article questions help students check their comprehension and make connections between science and society.
Vol.1 978-0-8053-4623-7 / 0-8053-4623-6
Vol.2 978-0-3215-3816-1 / 0-3215-3816-1

Table of Contents

Feature Boxes

BENEFICIAL MICROBES

HIGHLIGHTS

EMERGING DISEASES

CLINICAL CASE STUDIES

MICROBE AT A GLANCE

1 A Brief History of Microbiology

Life as we know it would not exist without microorganisms. Plants depend on microorganisms to help them obtain the nitrogen they need for survival. Animals such as cows and sheep need microorganisms in order to digest the cellulose in their plant-based diets. Our ecosystems rely on microorganisms to enrich soil, degrade wastes, and support life. We use microorganisms to make wine and cheese and to develop vaccines and antibiotics. The human body is home to billions of microorganisms, many of which help keep us healthy. Microorganisms are not only an essential part of our lives but quite literally a part of us.

Of course, some microorganisms do cause disease, from the common cold to more serious diseases such as tuberculosis, malaria, and AIDS. The threats of bioterrorism and new or reemerging infectious diseases are real. This textbook explores all the roles—both harmful and beneficial—that microorganisms play in our lives, as well as their sophisticated structures and processes. We begin with a look at the history of microbiology, starting with the invention of crude microscopes that revealed, for the first time, the existence of this miraculous, miniature world.

Take the pre-test for this chapter online. Visit the Study Area at www.masteringmicrobiology.com.

Aquatic microorganisms, such as these, thrilled early microscopists with their beauty and antics.

Science is the study of nature that proceeds by posing questions about observations. Why are there seasons? What is the function of the nodules at the base of this plant? Why does this bread taste sour? What does plaque from between teeth look like when magnified? Why are so many crows dying this winter? What causes new diseases?

Many early written records show that people have always asked questions like these. For example, the Greek physician Hippocrates (ca. 460–ca. 377 B.C.) wondered whether there is a link between environment and disease, and the Greek historian Thucydides (ca. 460–ca. 404 B.C.) questioned why he and other survivors of the plague could have intimate contact with victims and not fall ill again. For many centuries the answers to these and other fundamental questions about the nature of life remained largely unanswered. But then, about 350 years ago, the invention of the microscope began to provide some clues.

In this chapter we'll see how one man's determination to answer a fundamental question about the nature of life—What does life really look like?—led to the birth of a new science called *microbiology*. We'll then see how the search for answers to other questions, such as those concerning spontaneous generation, the reason fermentation occurs, and the cause of disease, prompted advances in this new science. Finally, we'll look briefly at some of the key questions microbiologists are asking today.

▲ **Figure 1.1 Antoni van Leeuwenhoek.** Leeuwenhoek reported the existence of protozoa in 1674 and of bacteria in 1676. *Why did Leeuwenhoek discover protozoa before bacteria?*

Figure 1.1 *Protozoa are generally larger than bacteria.*

The Early Years of Microbiology

The early years of microbiology brought the first observations of microbial life and the initial efforts to organize them into logical classifications.

What Does Life Really Look Like?

Learning Objectives

- ✓ Describe the world-changing scientific contributions of Leeuwenhoek.
- ✓ Define microbes in the words of Leeuwenhoek and as we know them today.

A few people have changed the world of science forever. We've all heard of Galileo, Newton, and Einstein, but the list also includes Antoni van Leeuwenhoek (lā'vĕn-huk; 1632–1723), a Dutch tailor, merchant, and lens grinder, and the man who first discovered the bacterial world **(Figure 1.1)**.

Leeuwenhoek was born in Delft, the Netherlands, and lived most of his 90 years in the city of his birth. What set Leeuwenhoek apart from most other men of his generation was an insatiable curiosity coupled with an almost stubborn desire to do everything for himself. His journey to fame began simply enough, when as a tailor he needed to examine the quality of cloth. Rather than merely buying one of the magnifying lenses already available, he learned to make glass lenses of his own **(Figure 1.2)**. Soon he began asking the question "What does it really look like?" of everything in his world: the stinger of a bee, the brain of a fly, the leg of a louse, a drop of blood, flakes of his own skin. To find answers, he spent hours examining, reexamining, and recording every detail of each object he observed.

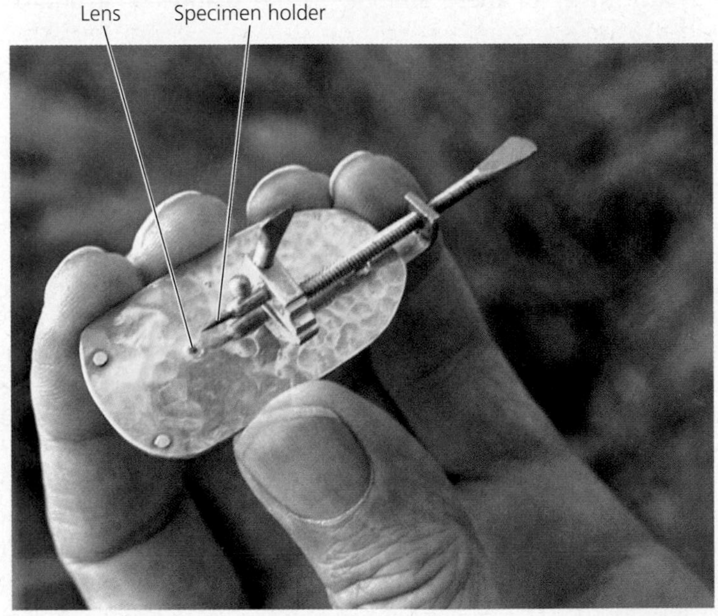

▲ **Figure 1.2 Reproduction of Leeuwenhoek's microscope.** This simple device is little more than a magnifying glass with screws for manipulating the specimen; yet with it, Leeuwenhoek changed the way we see our world. The lens, which is convex on both sides, is about the size of a pinhead. The object to be viewed was mounted either directly on the specimen holder or inside a small glass tube, which was then mounted on the specimen holder.

Making and looking through his simple microscopes, most really no more than magnifying glasses, became the overwhelming passion of his life. His enthusiasm and dedication are evident from the fact that he sometimes personally extracted the metal for his microscope from ore. Further, he often made a new microscope for each specimen, which remained mounted so that he could view it again and again. Then one day, he turned a lens onto a drop of water. We don't know what he expected to see, but certainly he saw more than he had anticipated. As he reported to the Royal Society of London[1] in 1674, he was surprised and delighted by

> some green streaks, spirally wound serpent-wise, and orderly arranged. . . . Among these there were, besides, very many little animalcules, some were round, while others a bit bigger consisted of an oval. On these last, I saw two little legs near the head, and two little fins at the hind most end of the body. . . . And the motion of most of these animalcules in the water was so swift, and so various, upwards, downwards, and round about, that 'twas wonderful to see.

Leeuwenhoek had discovered a previously unknown microbial world, which today we know to be populated with tiny animals, fungi, algae, and single-celled protozoa **(Figure 1.3)**. In a later report to the Royal Society, he noted that

> the number of these animals in the plaque of a man's teeth, are so many that I believe they exceed the number of men in a kingdom. . . . I found too many living animals therein, that I guess there might have been in a quantity of matter no bigger than the 1/100 part of a [grain of] sand.

From the figure accompanying this report and the precise description of the size of these organisms from between his teeth, we know that Leeuwenhoek was reporting the existence of bacteria. By the end of the 19th century, Leeuwenhoek's "beasties," as he sometimes dubbed them, were called **microorganisms,** and today we also know them as **microbes.** Both terms include all organisms that are too small to be seen without a microscope.

Because of the quality of his microscopes, his profound observational skills, his detailed reports over a 50-year period, and his report of the discovery of many types of microorganisms, Antoni van Leeuwenhoek was elected to the Royal Society in 1680. He and Isaac Newton were probably the most famous scientists of their time.

How Can Microbes Be Classified?

Learning Objectives

- ✓ List six groups of microorganisms.
- ✓ Explain why protozoa, algae, and nonmicrobial parasitic worms are studied in microbiology.
- ✓ Differentiate between prokaryotic and eukaryotic organisms.

Shortly after Leeuwenhoek made his discoveries, the Swedish botanist Carolus Linnaeus (1707–1778) developed a **taxonomic system**—that is, a system for naming plants and animals and grouping similar organisms together. For instance, Linnaeus and other scientists of the period grouped all organisms into either the animal kingdom or the plant kingdom. Today, biologists still use this basic system, but they have modified Linnaeus's scheme by adding categories that more realistically reflect the relationships among organisms. For example, scientists no longer classify yeasts, molds, and mushrooms as plants, but instead as fungi. We examine taxonomic schemes in more detail in Chapter 4.

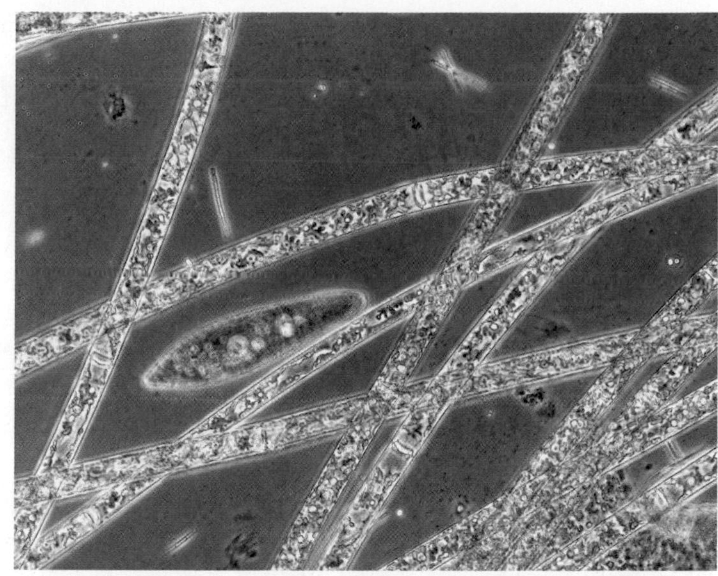

▲ **Figure 1.3 The microbial world.** Leeuwenhoek reported seeing a scene very much like this, full of numerous, fantastic, cavorting creatures.

The microorganisms that Leeuwenhoek described can be grouped into six basic categories: fungi, protozoa, algae, bacteria, archaea, and small multicellular animals. The only type of microbes not described by Leeuwenhoek are *viruses,*[2] which are too small to be seen without an electron microscope. We briefly consider organisms in the first five categories in the following sections.

Fungi

Fungi (fŭn′jī)[3] are organisms whose cells are **eukaryotic;**[4] that is, each of their cells contains a nucleus composed of genetic material surrounded by a distinct membrane. Fungi are different from plants because they obtain their food from other organisms (rather than making it for themselves). They differ from animals by having cell walls.

[1]The Royal Society of London for the Promotion of Natural Knowledge, granted a royal charter in 1662, is one of the older and more prestigious scientific groups in Europe.

[2]Technically, viruses are not "organisms" because they neither replicate themselves nor carry on the chemical reactions of living things. See Chapter 3 for a fuller discussion of this issue.

[3]Plural of the Latin *fungus,* meaning mushroom.

[4]From Greek *eu,* meaning true, and *karyon,* meaning kernel (which in this case refers to the nucleus of a cell).

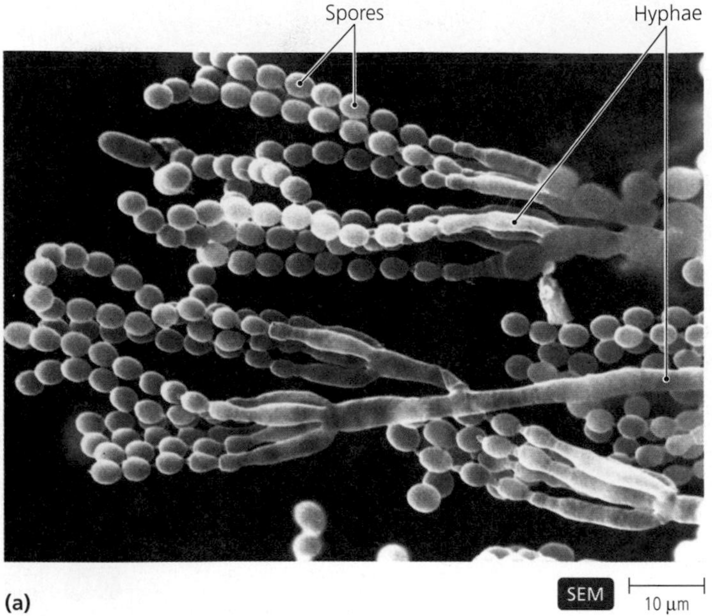

(a)

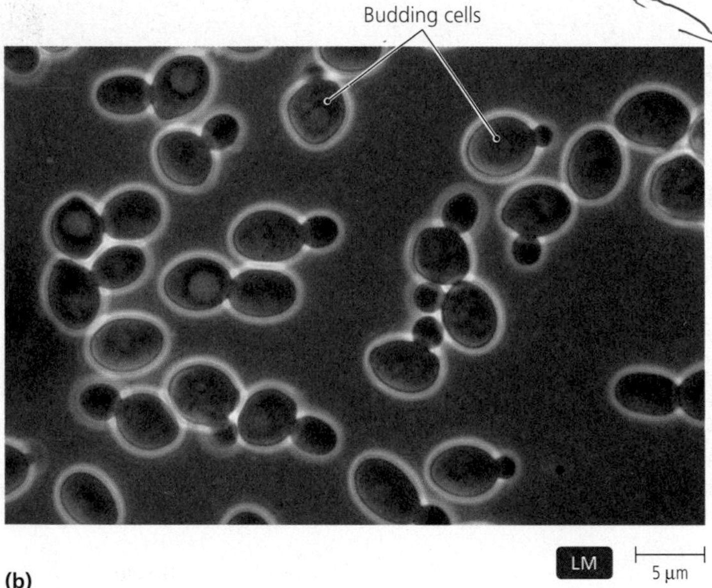

(b)

▲ **Figure 1.4 Fungi.** **(a)** The mold *Penicillium chrysogenum*, which produces penicillin, has long filamentous hyphae that intertwine to form its body. It reproduces by spores. **(b)** The yeast *Saccharomyces cerevisiae*. Yeasts are round to oval and typically reproduce by budding.

Microscopic fungi include some molds and yeasts. **Molds** are typically multicellular organisms that grow as long filaments, which intertwine to make up the body of the mold. Molds reproduce by sexual and asexual spores, which are cells that produce a new individual without fusing with another cell **(Figure 1.4a)**. The cottony growths on cheese, bread, and jams are examples of molds. *Penicillium chrysogenum* (pen-i-sil'ē-ŭm krī-so'jĕn-ŭm) is a mold that produces penicillin.

Yeasts are unicellular and typically oval to round. They reproduce asexually by *budding*, a process in which a daughter cell grows off the mother cell. Some yeasts also produce sexual spores. An example of a useful yeast is *Saccharomyces cerevisiae* (sak-ă-rō-mī'sēz se-ri-vis'ē-ī; **Figure 1.4b**), which causes bread to rise and produces alcohol from sugar (see **Beneficial Microbes: Bread, Wine, and Beer!** on p. 8). *Candida albicans* (kan'did-ă al'bi-kanz) is a yeast that causes most cases of yeast infections in women.

Fungi and their significance in the environment, in food production, and as agents of human disease are discussed in Chapters 12 and 22.

Protozoa

Protozoa are single-celled eukaryotes that are similar to animals in their nutritional needs and cellular structure. In fact, *protozoa* is Greek for "first animals," though scientists today classify them in their own groups rather than as animals. Most protozoa are capable of locomotion, and one way scientists categorize protozoa is according to their locomotive structures: *pseudopodia*,[5] *cilia*,[6] or *flagella*.[7] Pseudopodia are extensions of a cell that flow in the direction of travel **(Figure 1.5a)**. Cilia are numerous, short protrusions of a cell that beat rhythmically to propel the protozoan through its environment **(Figure 1.5b)**. Flagella are also extensions of a cell, but are fewer, longer, and more whiplike than cilia **(Figure 1.5c)**. Some protozoa, such as the malaria-causing *Plasmodium* (plaz-mō'dē-ŭm), are nonmotile in their mature forms.

Protozoa typically live freely in water, but some live inside animal hosts, where they can cause disease. Most protozoa reproduce asexually, though some are sexual as well. Chapters 12 and 23 further examine protozoa.

Algae

Algae[8] are unicellular or multicellular *photosynthetic* organisms **(Figure 1.6)**; that is, like plants they make their own food from carbon dioxide and water using energy from sunlight. They differ from plants in the relative simplicity of their reproductive structures. Algae are categorized on the basis of their pigmentation and the composition of their cell walls.

Large algae, commonly called seaweeds and kelps, are common in the world's oceans. Chemicals from their gelatinous cell walls are used as thickeners and emulsifiers in many food and cosmetic products, as well as in microbiological laboratory media.

Unicellular algae are common in freshwater ponds, streams, and lakes, and in the oceans as well. They are the

[5]From Greek *pseudes*, meaning false, and *podos*, meaning foot.
[6]Plural of the Latin *cilium*, meaning eyelid.
[7]Plural of the Latin *flagellum*, meaning whip.
[8]Plural of the Latin *alga*, meaning seaweed.

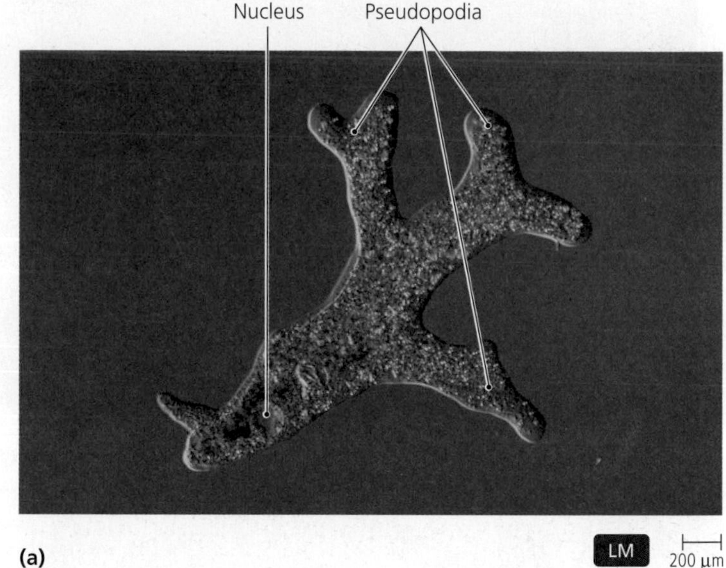

(a)

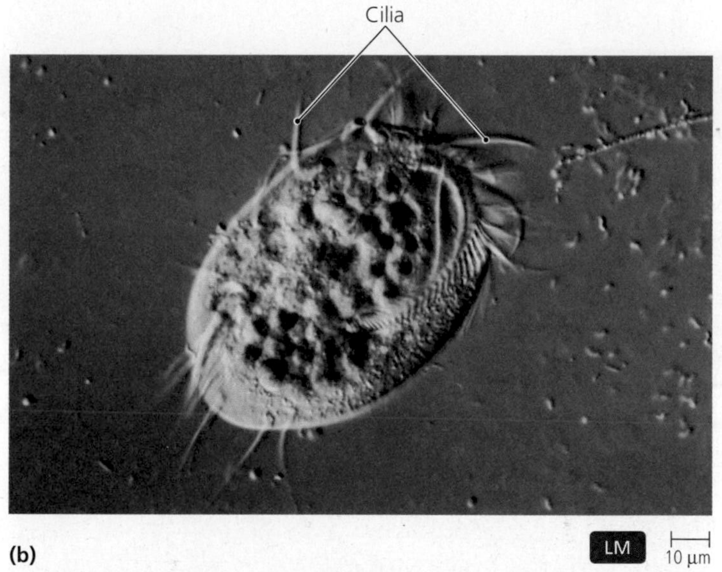

(b)

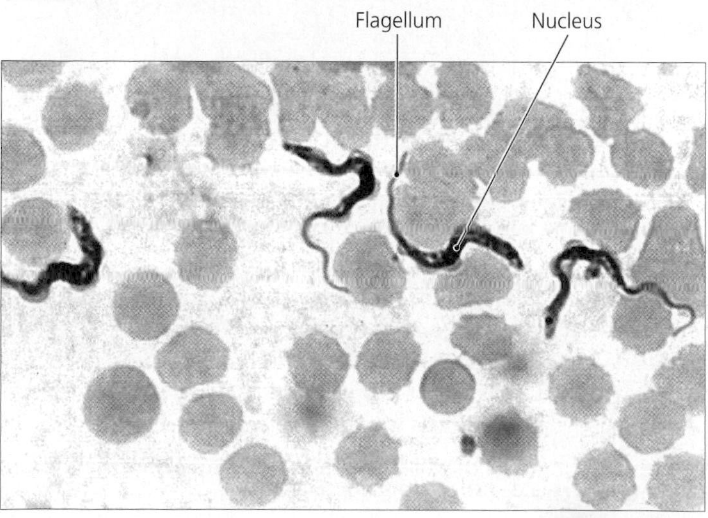

(c)

Figure 1.5 Locomotive structures of protozoa. **(a)** Pseudopodia are cellular extensions used for locomotion and feeding, as seen in *Amoeba proteus.* **(b)** Cilia are short, motile, hairlike extrusions, as seen in *Euplotes*, a protozoan. **(c)** Flagella are whiplike extensions that are less numerous and longer than cilia, as seen in *Trypanosoma*, a protozoan. *How do cilia and flagella differ?*

Figure 1.5 *Cilia are short, numerous, and often cover the cell, whereas flagella are long and relatively few in number.*

major food of small aquatic and marine animals and provide most of the world's oxygen as a by-product of photosynthesis. The glasslike cell walls of diatoms provide grit for many polishing compounds. Chapter 12 discusses other aspects of the biology of algae.

Bacteria and Archaea

Bacteria and **archaea** are **prokaryotic;**[9] that is, they lack nuclei. Bacterial cell walls are composed of a polysaccharide called *peptidoglycan,* though some bacteria lack cell walls. The cell walls of archaea lack peptidoglycan and instead are composed of other polymers. Members of both groups reproduce asexually. Chapters 3, 4, and 11 examine other differences between bacteria and archaea, and Chapters 19–21 discuss pathogenic (disease-causing) bacteria.

Most archaea and bacteria are much smaller than eukaryotic cells **(Figure 1.7)**. They live singly or in pairs, chains, or clusters in almost every habitat containing sufficient moisture. Archaea are often isolated from extreme environments, such as the highly saline Mono Lake in California, acidic hot springs in Yellowstone National Park, and oxygen-depleted mud at the bottom of swamps. No archaea are known to cause disease.

Though bacteria may have a poor reputation in our world, the great majority of bacteria do not cause disease in animals, humans, or crops. Indeed, bacteria are beneficial to us in many ways. For example, dead plants and animals are degraded by bacteria (and fungi) to release phosphorus, sulfur, nitrogen, and carbon back into the air, soil, and water to be used by new generations of organisms. Without microbial recyclers, the world would be buried under the petrified corpses of uncountable dead organisms.

CRITICAL **THINKING**

A few bacteria produce disease because they derive nutrition from human cells and produce toxic wastes. Algae do not cause disease. Why not?

Other Organisms of Importance to Microbiologists

Microbiologists also study parasitic worms, which range in size from microscopic forms **(Figure 1.8)** to adult tapeworms over 7 meters (approximately 23 feet) in length. Even though most of

[9] From Greek *pro*, meaning before, and *karyon*, meaning kernel.

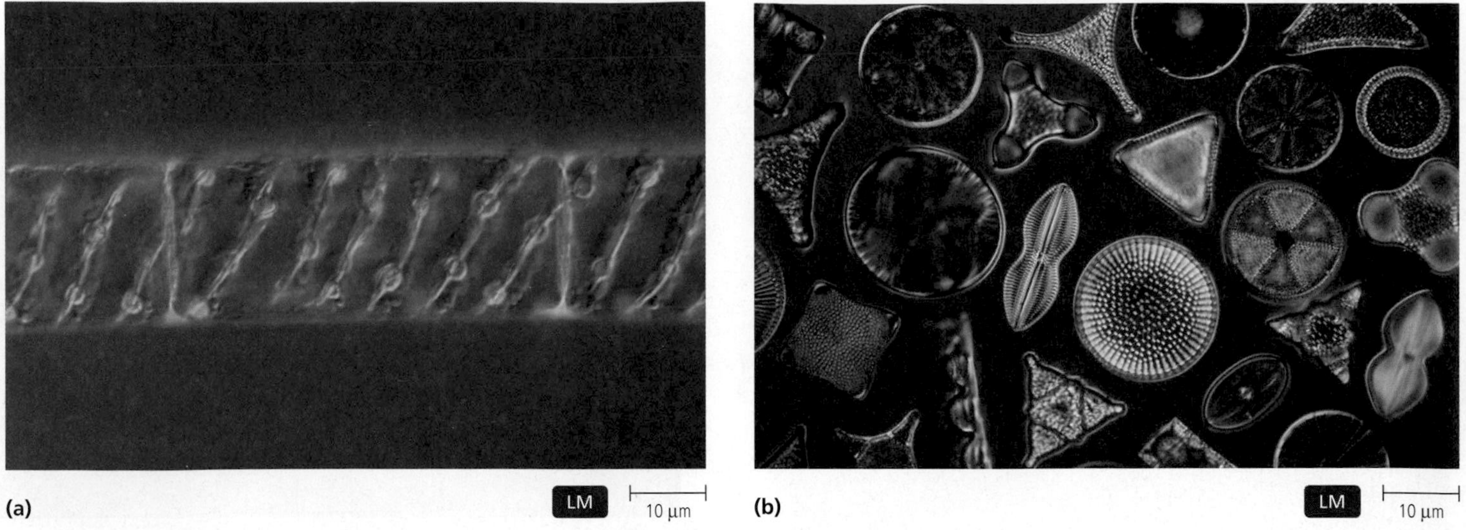

▲ **Figure 1.6 Algae.** **(a)** *Spirogyra*. These microscopic algae grow as chains of cells containing helical photosynthetic structures. **(b)** Diatoms. These beautiful algae have glasslike cell walls.

these worms are not microscopic as adults, many of them cause diseases that were studied by early microbiologists. Further, laboratory technicians diagnose infections of parasitic worms by finding microscopic eggs and immature stages in blood, fecal, urine, and lymph specimens. Chapter 23 discusses parasitic worms.

The only type of microbes that remained hidden from Leeuwenhoek and other early microbiologists were viruses, which are much smaller than the smallest prokaryote and are not visible by light microscopy **(Figure 1.9)**. Viruses could not be seen until the electron microscope was invented in 1932. All viruses are acellular (not composed of cells) obligatory parasites composed of small amounts of genetic material (either DNA or RNA) surrounded by a protein coat. Chapter 13 examines the general characteristics of viruses, and Chapters 24 and 25 discuss specific viral pathogens.

Leeuwenhoek first reported the existence of most types of microorganisms in the late 1600s, but microbiology did not develop significantly as a field of study for almost two centuries. There were a number of reasons for this delay. First, Leeuwenhoek

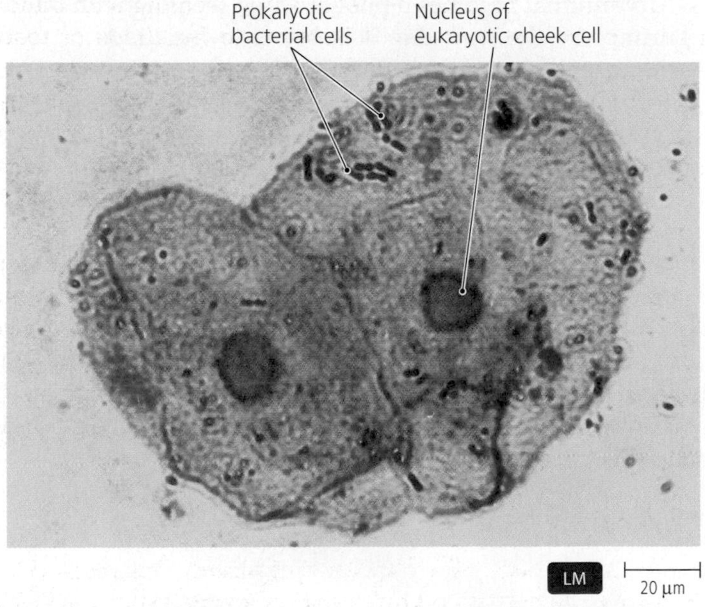

▲ **Figure 1.7 Cells of the bacterium *Streptococcus* (dark blue) and a human cheek cell.** Notice the size difference.

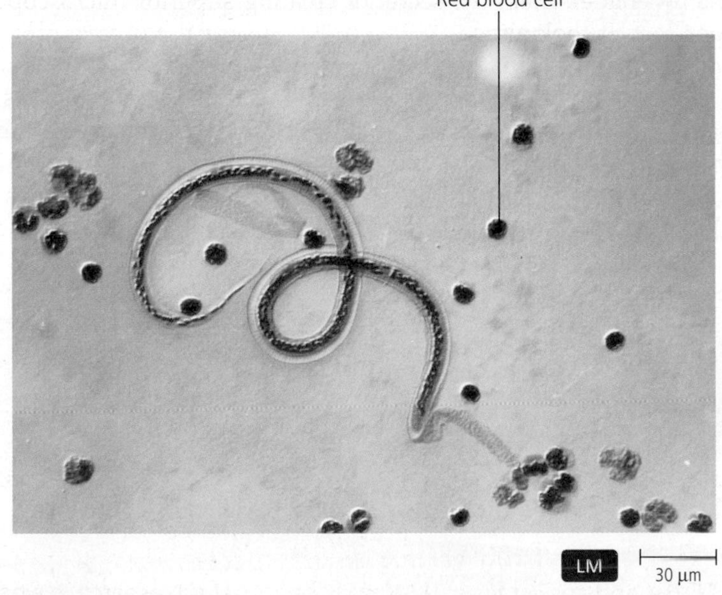

▲ **Figure 1.8 An immature stage of a parasitic worm in blood.**

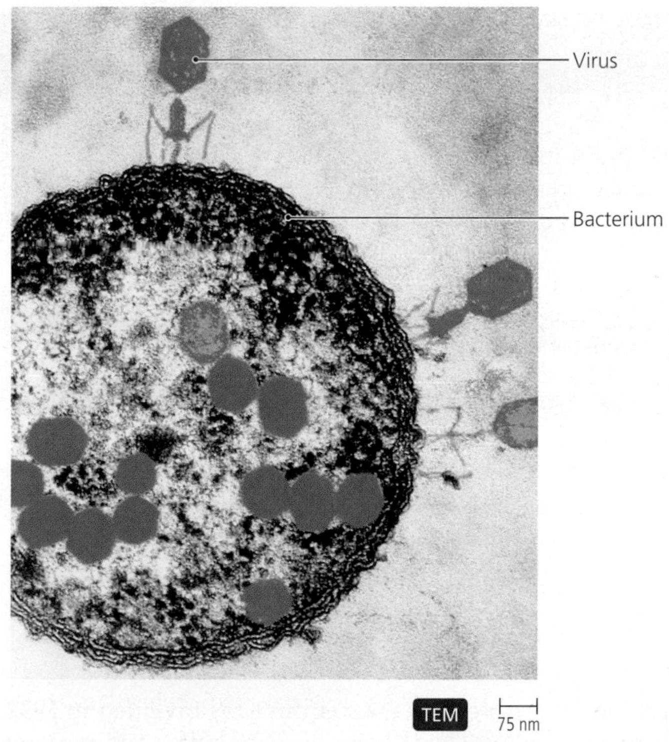

▲ **Figure 1.9 A colorized electron microscope image of viruses infecting a bacterium.** Viruses, which are acellular obligatory parasites, are too small to be seen with a light microscope. Notice how small the viruses are compared to the bacterium.

was a suspicious and secretive man. Though he built over 400 microscopes, he never trained an apprentice, and he never sold or gave away a microscope. In fact, he never let *anyone*—not his family or such distinguished visitors as the czar of Russia—so much as peek through his very best instruments. When Leeuwenhoek died, the secret of creating superior microscopes was lost. It took almost 100 years for scientists to make microscopes of equivalent quality.

Another reason that microbiology was slow to develop as a science is that scientists in the 1700s considered microbes to be curiosities of nature and insignificant to human affairs. But in the late 1800s, scientists began to adopt a new philosophy, one that demanded experimental proof rather than mere acceptance of traditional knowledge. This fresh philosophical foundation, accompanied by improved microscopes, new laboratory techniques, and a drive to answer a series of pivotal questions, propelled microbiology to the forefront as a scientific discipline.

The Golden Age of Microbiology

Learning Objective

✓ List and answer four questions that propelled research in what is called the "Golden Age of Microbiology."

For about 50 years, during what is sometimes called the "Golden Age of Microbiology," scientists and the blossoming field of microbiology were driven by the search for answers to the following four questions:

- Is spontaneous generation of microbial life possible?
- What causes fermentation?
- What causes disease?
- How can we prevent infection and disease?

Competition among scientists who were striving to be the first to answer these questions drove exploration and discovery in microbiology during the late 1800s and early 1900s. These scientists' discoveries and the fields of study they initiated continue to shape the course of microbiological research today.

In the next sections we consider these questions, and how the great scientists accumulated the experimental evidence that answered them.

Does Microbial Life Spontaneously Generate?

Learning Objectives

✓ Identify the scientists who argued in favor of spontaneous generation.

✓ Compare and contrast the investigations of Redi, Needham, Spallanzani, and Pasteur concerning spontaneous generation.

✓ List four steps in the scientific method of investigation.

A dry lakebed has lain under the relentless North African desert sun for eight long months. The cracks in the baked, parched mud are wider than a man's hand. There is no sign of life anywhere in the scorched terrain. With the abruptness characteristic of desert storms, rain falls in a torrent, and a raging flood of roiling water and mud crashes down the dry streambed and fills the lake. Within hours, what had been a lifeless, dry mudflat becomes a pool of water teeming with billions of shrimp; by the next day it is home to hundreds of toads. Where did these animals come from?

Many philosophers and scientists of past ages thought that living things arose via three processes: through asexual reproduction, through sexual reproduction, or from nonliving matter. The appearance of shrimp and toads in the mud of what so recently was a dry lakebed was seen as an example of the third process, which came to be known as *abiogenesis*[10] or **spontaneous generation.** The theory of spontaneous generation as promulgated by Aristotle (384–322 B.C.) was widely accepted for over 2000 years because it seemed to explain a variety of commonly observed phenomena, such as the appearance of maggots on spoiling meat. However, the validity of the theory came under challenge in the 17th century.

Redi's Experiments

In the late 1600s, the Italian physician Francesco Redi (1626–1697) demonstrated by a series of experiments that when

[10]From Greek *a*, meaning not, *bios*, meaning life, and *genein*, meaning to produce.

BENEFICIAL MICROBES

BREAD, WINE, AND BEER!

Microorganisms play important roles in humans' lives; for example, pathogens have undeniably altered the course of history. However, what may be the most important microbiological event—one that has had a greater impact upon culture and society than that of any disease or epidemic—was the domestication of the yeast used by bakers and brewers. Its name, *Saccharomyces cerevisiae*, means "sugar fungus [that makes] beer."

The earliest record of the use of yeast comes from Persia (modern Iran) where archeologists have found the remains of grapes and wine preservatives in pottery vessels more than 7000 years old. Brewing beer likely started even earlier, its beginnings undocumented. The earliest examples of leavened bread are from Egypt and show that bread-making was routine about 6000 years ago. Before that time, bread was unleavened and flat.

It is likely that making wine and brewing beer occurred earlier than the use of leavened bread because *Saccharomyces* is naturally found on grapes, which can begin to ferment while still on the vine. Historians hypothesize that early bakers may have exposed bread dough to circulating air, hoping that the invisible and inexplicable "fermentation principle" would inoculate the bread. Another hypothesis is that bakers learned to add small amounts of beer or wine to the bread, intentionally inoculating the dough with yeast. Of course, all those years before Leeuwenhoek and Pasteur, no one knew that the fermenting ingredient of wine was a living organism.

Besides its role in baking and in making alcoholic beverages, *S. cerevisiae* is an important tool for the study of cells. Scientists use yeast to delve into the mysteries of cellular function, organization, and genetics, making *Saccharomyces* the most intensely studied eukaryote. In fact, molecular biologists published the complete sequence of the genes of *S. cerevisiae* in 1996—a first for any eukaryotic cell.

Today, scientists are working toward using *S. cerevisiae* in novel ways. For example, some nutritionists and gastroenterologists are examining the use of *Saccharomyces* as a *probiotic;* that is, a microorganism intentionally taken to ward off disease and promote good health. Research suggests that the yeast helps in the treatment of diarrhea and colitis and may help prevent these and other gastrointestinal diseases.

decaying meat was kept isolated from flies, maggots never developed, whereas meat exposed to flies was soon infested **(Figure 1.10)**. As a result of experiments such as these, scientists began to doubt Aristotle's theory and adopt the view that animals only come from other animals.

Needham's Experiments

The debate over spontaneous generation was rekindled when Leeuwenhoek discovered microbes and showed that they appeared after a few days in freshly collected rainwater. Though scientists agreed that larger animals could not arise spontaneously, they disagreed about Leeuwenhoek's "wee animalcules"; surely they did not have parents, did they? They must arise spontaneously.

The proponents of spontaneous generation pointed to the careful demonstrations of British investigator John T. Needham (1713–1781). He boiled beef gravy and infusions[11] of plant material in vials, which he then tightly sealed with corks. Some days later, Needham observed that the vials were cloudy, and

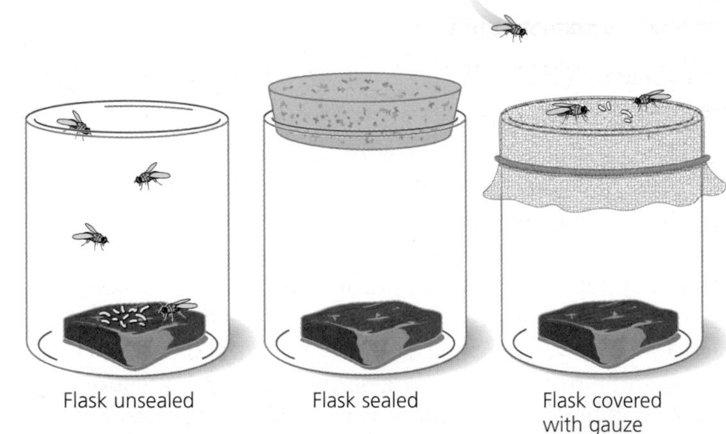

▲ **Figure 1.10 Redi's experiments.** When the flask remained unsealed, maggots covered the meat within a few days. When the flask was sealed, flies were kept away and no maggots appeared on the meat. When the flask opening was covered with gauze, flies were kept away and no maggots appeared on the meat, although a few maggots appeared on top of the gauze.

[11]Infusions are broths made by steeping plant or animal material in water.

examination revealed an abundance of "microscopical [*sic*] animals of most dimensions." As he explained it, there must be a "life force" that causes inanimate matter to spontaneously come to life, since he had heated the vials sufficiently to kill everything. Needham's experiments so impressed the Royal Society that they elected him a member.

Spallanzani's Experiments

Then, in 1799, the Italian scientist Lazzaro Spallanzani (1729–1799) reported results that contradicted Needham's findings. Spallanzani boiled infusions for almost an hour and sealed the vials by melting their slender necks closed. His infusions remained clear, unless he broke the seal and exposed the infusion to air, after which they became cloudy with microorganisms. He concluded three things:

- Needham had either failed to heat his vials sufficiently to kill all microbes, or he had not sealed them tightly enough.
- Microorganisms exist in the air and can contaminate experiments.
- Spontaneous generation of microorganisms does not occur; all living things arise from other living things.

Although Spallanzani's experiments would appear to have settled the controversy once and for all, it proved difficult to dethrone a theory that had held sway for 2000 years, especially when so notable a man as Aristotle had propounded it. One of the criticisms of Spallanzani's work was that his sealed vials did not allow enough air for organisms to thrive; another objection was that his prolonged heating destroyed the "life force." The debate continued until the French chemist Louis Pasteur **(Figure 1.11)** conducted experiments that finally laid the theory of spontaneous generation to rest.

Pasteur's Experiments

Louis Pasteur (1822–1895) was an indefatigable worker who pushed himself as hard as he pushed others. As he wrote his sisters, "To *will* is a great thing dear sisters, for Action and Work usually follow Will, and almost always Work is accompanied by Success. These three things, Work, Will, Success, fill human existence. Will opens the door to success both brilliant and happy; Work passes these doors, and at the end of the journey Success comes to crown one's efforts." When his wife complained about his long hours in the laboratory, he replied, "I will lead you to fame."

▲ **Figure 1.11 Louis Pasteur.** Often called the Father of Microbiology, he disproved spontaneous generation. In this depiction, Pasteur examines some bacterial cultures.

Pasteur's determination and hard work are apparent in his investigations of spontaneous generation. Like Spallanzani, he boiled infusions long enough to kill everything. But instead of sealing the flasks, he bent their necks into an S-shape, which allowed air to enter while preventing the introduction of dust and microbes into the broth **(Figure 1.12)**.

Crowded for space and lacking funds, he improvised an incubator in the opening under a staircase. Day after day he

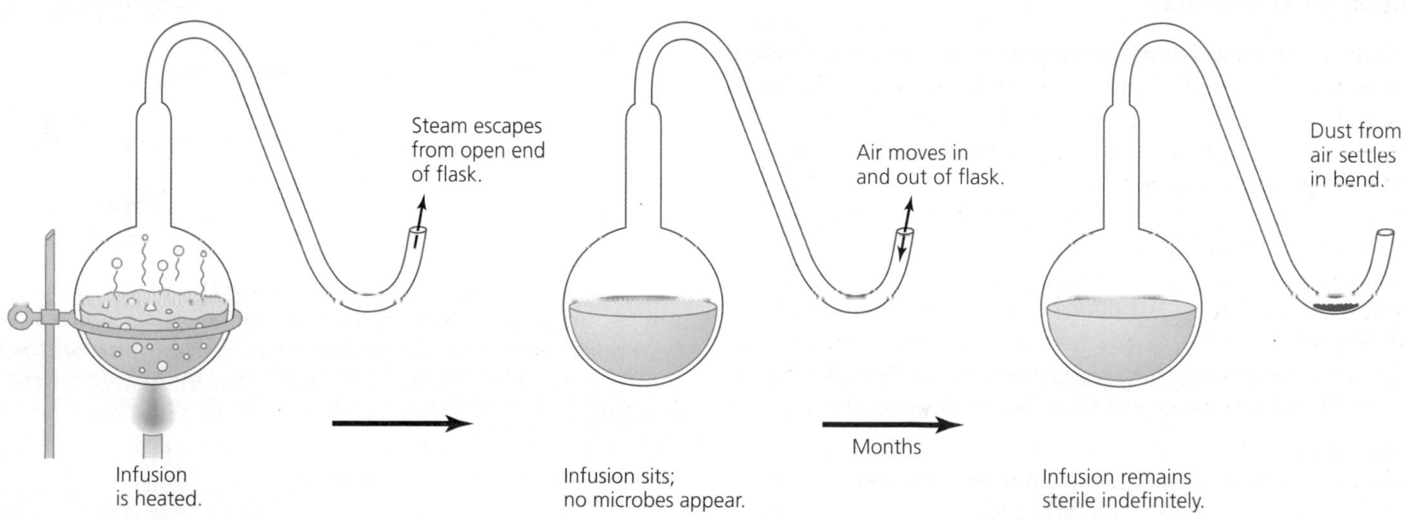

Steam escapes from open end of flask.
Air moves in and out of flask.
Dust from air settles in bend.
Months
Infusion is heated.
Infusion sits; no microbes appear.
Infusion remains sterile indefinitely.

▲ **Figure 1.12 Pasteur's experiments with "swan-necked flasks."** As long as the flask remained upright, no microbial growth appeared in the infusion.

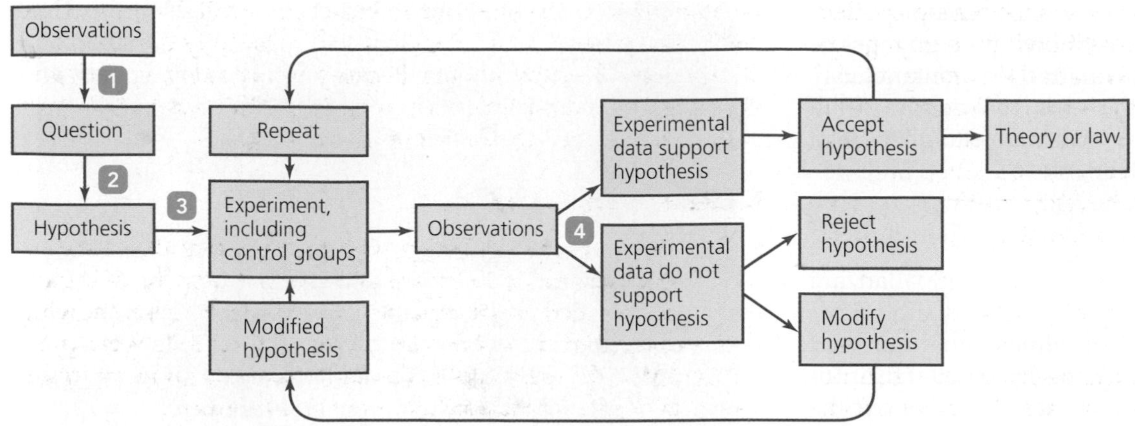

Figure 1.13 The scientific method, which forms a framework for scientific research.

crawled on hands and knees into this incommodious space and examined his flasks for the cloudiness that would indicate the presence of living organisms. In 1861, he reported that his "swan-necked flasks" remained free of microbes even 18 months later. Since they contained all the nutrients (including air) known to be required by living things, he concluded, "Never will spontaneous generation recover from the mortal blow of this simple experiment."

Pasteur followed this experiment with demonstrations that microbes in the air were the "parents" of Needham's microorganisms. He broke the necks off some flasks, exposing the liquid in them directly to the air, and he carefully tilted others so that the liquid touched the dust that had accumulated in their necks. The next day, all of these flasks were cloudy with microbes. He concluded that the microbes in the liquid were the progeny of microbes that had been on the dust particles in the air.

The Scientific Method

The debate over spontaneous generation led in part to the development of a generalized **scientific method** by which questions are answered through observations of the outcomes of carefully controlled experiments, instead of by conjecture or according to the opinions of any authority figure. The scientific method, which provides a framework for conducting an investigation rather than a rigid set of specific "rules," consists of four basic steps (Figure 1.13):

1. A group of observations leads a scientist to ask a question about some phenomenon.
2. The scientist generates a hypothesis—that is, a potential answer to the question.
3. The scientist designs and conducts an experiment to test the hypothesis.
4. Based on the observed results of the experiment, the scientist either accepts, rejects, or modifies the hypothesis.

As shown in Figure 1.13, the scientist then returns to earlier steps in the method, either modifying hypotheses and then testing them, or repeatedly testing accepted hypotheses, until the evidence for a hypothesis is convincing. Accepted hypotheses that explain many observations and are repeatedly verified by numerous scientists over many years are called *theories* or *laws*.

Note that in order for experiments (and their results) to be accepted as valid by the scientific community, they must include appropriate *control groups*—groups that are treated exactly the same as the other groups in the experiment, except for the one variable that the experiment is designed to test. In Pasteur's experiments on spontaneous generation, for example, his "control flasks" contained a sterile infusion composed of all the nutrients living things need, as well as air made available through the flasks' "swan necks." His "experimental flasks" for testing his hypothesis—that microbes would reach (and subsequently grow in) the infusion through contact with dust particles—were exposed to exactly the same conditions, *plus* contact with the dust in the bend in the neck. Because exposure to the dust was the *only* difference between the control and experimental groups, Pasteur was able to conclude that the microbes growing in the infusion arrived on the dust particles.

What Causes Fermentation?

Learning Objectives

- ✓ Discuss the significance of Pasteur's fermentation experiments to our world today.
- ✓ Explain why Pasteur may be considered the Father of Microbiology.
- ✓ Identify the scientist whose experiments led to the field of biochemistry and the study of metabolism.

The controversy over spontaneous generation was largely a philosophical exercise among men who conducted research to gain basic scientific knowledge. They had no practical goals in mind—except, perhaps, personal aggrandizement in the form of financial support, honor, and prestige. However, the second question that moved microbial studies forward in the 1800s had tremendous practical applications.

Our story resumes in 19th-century France where spoiled, acidic wine was threatening the livelihood of many grape growers. The initial question was, "Why is the wine spoiled?" but this led to a more fundamental question, "What causes the fermentation of grape juice into wine?" These questions were so important to vintners that they funded research concerning fermentation, hoping scientists could develop methods to promote the production of alcohol and prevent spoilage by acid during fermentation.

Pasteur's Experiments

Scientists of the 1800s used the word *fermentation* to mean not only the formation of alcohol from sugar, but also other chemical reactions such as the formation of lactic acid, the putrefaction of meat, and the decomposition of waste. Many scientists asserted that air caused fermentation reactions; others insisted that living organisms were responsible.

The debate over the cause of fermentation reactions was linked to the debate over spontaneous generation. Some scientists proposed that the yeasts observed in fermenting juices were nonliving globules of chemicals and gases. Others thought that yeasts were alive and were spontaneously generated during fermentation. Still others asserted that yeasts not only were living organisms, but also caused fermentation.

Pasteur conducted a series of careful observations and experiments that answered the question "What causes fermentation?" First, he observed yeast cells growing and budding in grape juice and conducted experiments showing that they arise only from other yeast cells. Then, by sealing some sterile flasks containing grape juice and yeast, and by leaving others open to the air, he demonstrated that yeast could grow with or without oxygen; that is, he discovered that yeasts are *facultative anaerobes*[12]—organisms that can live with or without oxygen. Finally, by introducing bacteria and yeast cells into different flasks of sterile grape juice, he proved that bacteria ferment grape juice to produce acids, and that yeast cells ferment grape juice to produce alcohol **(Figure 1.14)**.

Pasteur's discovery that *anaerobic* bacteria fermented grape juice into acids suggested a method for preventing the spoilage of wine. His name became a household word when he developed *pasteurization,* a process of heating the grape juice just enough to kill most contaminating bacteria without changing the juice's basic qualities, so that it could then be inoculated with yeast to ensure that alcohol fermentation occurred. Pasteur thus began the field of **industrial microbiology** (or **biotechnology**) in which microbes are intentionally used to manufacture products (**Table 1.1** on p. 13; see also Chapter 26). Today pasteurization is used routinely on milk to eliminate pathogens that cause diseases such as bovine tuberculosis and brucellosis; it is also used to eliminate pathogens in juices and other beverages.

These are just a few of the many experiments Pasteur conducted with microbes. While a few of Pasteur's successes can be attributed to the superior microscopes available in the late 1800s, his genius is clearly evident in his carefully designed and straightforward experiments. Because of his many, varied, and significant accomplishments in working with microbes, Pasteur may be considered the Father of Microbiology.

[12]From Greek *an*, meaning not, *aer*, meaning air (i.e., oxygen), and *bios*, meaning life.

Buchner's Experiments

Studies on fermentation began with the idea that fermentation reactions were strictly chemical and did not involve living organisms. This idea was supplanted by Pasteur's work showing that fermentation proceeded only when living cells were present, and that different types of microorganisms growing under varied conditions produced different end products.

In 1897, the German scientist Eduard Buchner (1860–1917) resurrected the chemical explanation by showing that fermentation does not require living cells. Buchner's experiments demonstrated the presence of *enzymes,* which are cell-produced proteins that promote chemical reactions. Buchner's work began the field of **biochemistry** and the study of **metabolism,** a term that refers to the sum of all chemical reactions within an organism.

CRITICAL **THINKING**

How might the debate over spontaneous generation have been different if Buchner had conducted his experiments in 1857 instead of 1897?

What Causes Disease?

Learning Objectives

- ✓ List at least seven contributions made by Koch to the field of microbiology.
- ✓ List the four steps that must be taken to prove the cause of an infectious disease.
- ✓ Describe the contribution of Gram to the field of microbiology.

You are a physician in London, and it is August 1854. It is past midnight, and you have been visiting patients since before dawn. As you enter the room of your next patient you observe with frustration and despair that this case is like hundreds of others you and your colleagues have attended in the neighborhood over the past month.

A five-year-old boy with a vacant stare lies in bed listlessly. As you watch, he is suddenly gripped by severe abdominal cramps, and his gastrointestinal tract empties in an explosion of watery diarrhea. The voided fluid is clear, colorless, odorless, and streaked with thin flecks of white mucus, reminiscent of water poured off a pot of cooking rice. His anxious mother changes his bedclothes as his father gives him a sip of water, but it is of little use. With a heavy heart you confirm the parents' fear—their child has cholera, and there is nothing you can do. He will likely die before morning. As you despondently turn to go, the question that has haunted you for two months is foremost in your mind: What causes such a disease?

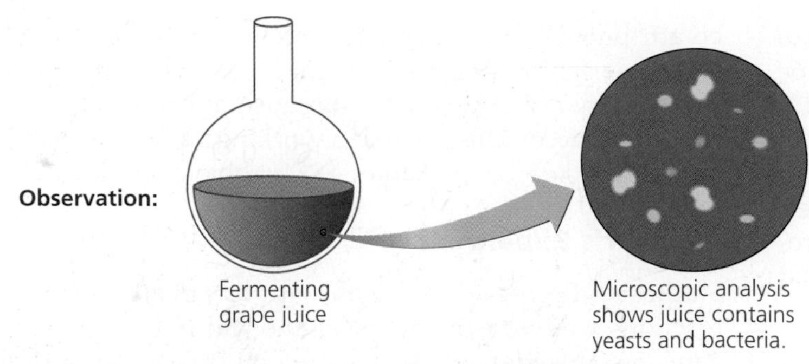

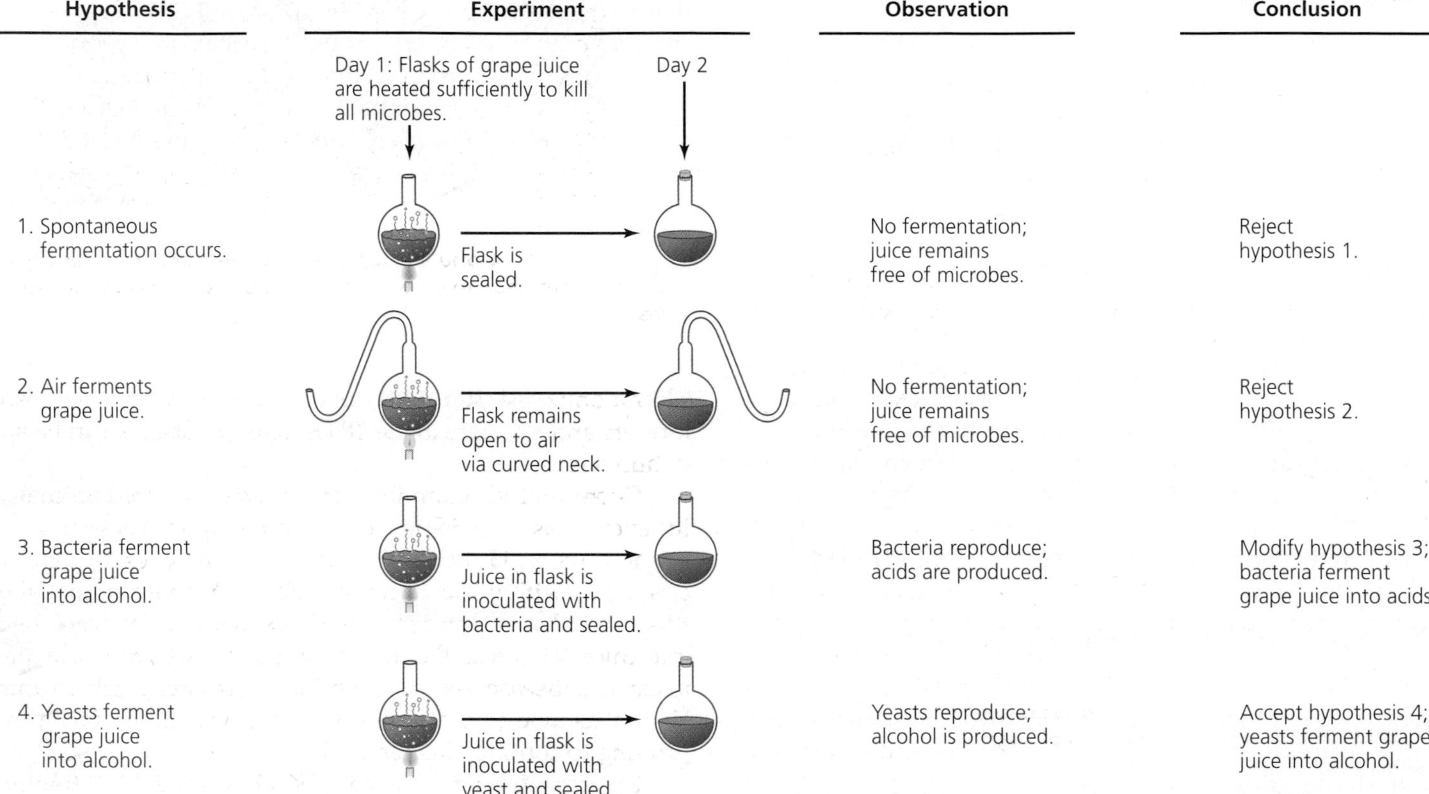

Hypothesis
Experiment
Observation
Conclusion
Day 1: Flasks of grape juice are heated sufficiently to kill all microbes.
Day 2
1. Spontaneous fermentation occurs.
Flask is sealed.
No fermentation; juice remains free of microbes.
Reject hypothesis 1.
2. Air ferments grape juice.
Flask remains open to air via curved neck.
No fermentation; juice remains free of microbes.
Reject hypothesis 2.
3. Bacteria ferment grape juice into alcohol.
Juice in flask is inoculated with bacteria and sealed.
Bacteria reproduce; acids are produced.
Modify hypothesis 3; bacteria ferment grape juice into acids.
4. Yeasts ferment grape juice into alcohol.
Juice in flask is inoculated with yeast and sealed.
Yeasts reproduce; alcohol is produced.
Accept hypothesis 4; yeasts ferment grape juice into alcohol.

▲ **Figure 1.14 How Pasteur applied the scientific method in investigating the nature of fermentation.** After observing that fermenting grape juice contained both yeasts and bacteria, Pasteur hypothesized that these organisms cause fermentation. Upon eliminating the possibility that fermentation could occur spontaneously or be caused by air (hypotheses 1 and 2), he concluded that fermentation requires the presence of living cells. The results of additional experiments (those testing hypotheses 3 and 4) indicated that bacteria ferment grape juice to produce acids, and that yeasts ferment grape juice to produce alcohol. *Which of Pasteur's flasks was the control?*

Figure 1.14 *The sealed flask that remained free of microorganisms served as the control.*

The third question that propelled the advance of microbiology concerned disease, defined generally as any abnormal condition in the body. Prior to the 1800s, disease was attributed to various factors, including evil spirits, astrological signs, imbalances in body fluids, and foul vapors. Although the Italian philosopher Girolamo Fracastoro (1478–1553) conjectured as early as 1546 that "germs[13] of contagion" cause disease, the idea that germs might be invisible living organisms awaited Leeuwenhoek's investigations 130 years later.

Pasteur's discovery that bacteria are responsible for spoiling wine led naturally to his hypothesis in 1857 that microorganisms are also responsible for diseases. This idea came to be known as the **germ theory of disease.** Since a particular disease is typically accompanied by the same symptoms in all affected individuals, early investigators suspected that diseases such as cholera,

[13]From Latin *germen*, meaning sprout.

TABLE 1.1 Some Industrial Uses of Microbes

Product or Process	Contribution of Microorganism
Foods and Beverages	
Cheese	Flavoring and ripening produced by bacteria and fungi; flavors dependent on the source of milk and the type of microorganism
Alcoholic beverages	Alcohol produced by bacteria or yeast by fermentation of sugars in fruit juice or grain
Soy sauce	Produced by fungal fermentation of soybeans
Vinegar	Produced by bacterial fermentation of sugar
Yogurt	Produced by bacteria growing in skim milk
Sour cream	Produced by bacteria growing in cream
Artificial sweetener	Amino acids synthesized by bacteria from sugar
Bread	Rising of dough produced by action of yeast; sourdough results from bacteria-produced acids
Other Products	
Antibiotics	Produced by bacteria and fungi
Human growth hormone, human insulin	Produced by genetically engineered bacteria
Laundry enzymes	Isolated from bacteria
Vitamins	Isolated from bacteria
Diatomaceous earth (used in polishes and buffing compounds)	Composed of cell walls of microscopic algae
Pest control chemicals	Insect pests killed or inhibited by bacterial pathogens
Drain opener	Protein-digesting and fat-digesting enzymes produced by bacteria

tuberculosis, and anthrax are each caused by a specific germ, called a **pathogen.**[14] Today we know that some diseases are genetic and that allergic reactions and environmental toxins cause others, so the germ theory applies only to *infectious*[15] *diseases.*

Just as Pasteur was the chief investigator in disproving spontaneous generation and determining the cause of fermentation, so investigations in **etiology**[16] (the study of causation of disease) were dominated by Robert Koch (1843–1910) **(Figure 1.15)**.

Figure 1.15 Robert Koch. Koch was instrumental in modifying the scientific method to prove that a given pathogen caused a specific disease.

Koch's Experiments

Koch was a country doctor in Germany when he began a race with Pasteur to discover the cause of anthrax, which is a potentially fatal disease, primarily of animals, in which toxins produce ulceration of the skin. Anthrax caused untold financial losses to farmers and ranchers in the 1800s, and the disease can be spread to humans.

Koch carefully examined the blood of infected animals, and in every case he identified a rod-shaped bacterium[17] that formed chains. He observed the formation of resting stages (endospores) within the bacterial cells and showed that the endospores always produced anthrax when they were injected into mice. This was the first time that a bacterium was proven to cause a disease. As a result of his successful work on anthrax, Koch was able to move to Berlin and was given facilities and funding to continue his research.

Heartened by his success, Koch turned his attention to other diseases. He had been fortunate when he chose anthrax for his initial investigations, because anthrax bacteria are quite large and easily identified with the microscopes of that time. However, most bacteria are very small, and different types exhibit few or no visible differences. Koch puzzled how to distinguish among these bacteria.

He solved the problem by taking specimens (for instance, blood, pus, or sputum) from disease victims and then smearing the specimens onto a solid surface such as a slice of potato or a gelatin medium. He then waited for bacteria and fungi present in the specimen to multiply and form distinct colonies **(Figure 1.16)**. Koch hypothesized that each colony consisted of the progeny of a single cell. He then inoculated samples from each colony into

[14]From Greek *pathos*, meaning disease, and *genein*, meaning to produce.
[15]From Latin *inficere*, meaning to taint (i.e., with a pathogen).
[16]From Greek *aitia*, meaning cause, and *logos*, meaning word or study.
[17]Now known as *Bacillus anthracis*—Latin for "the rod of anthrax."

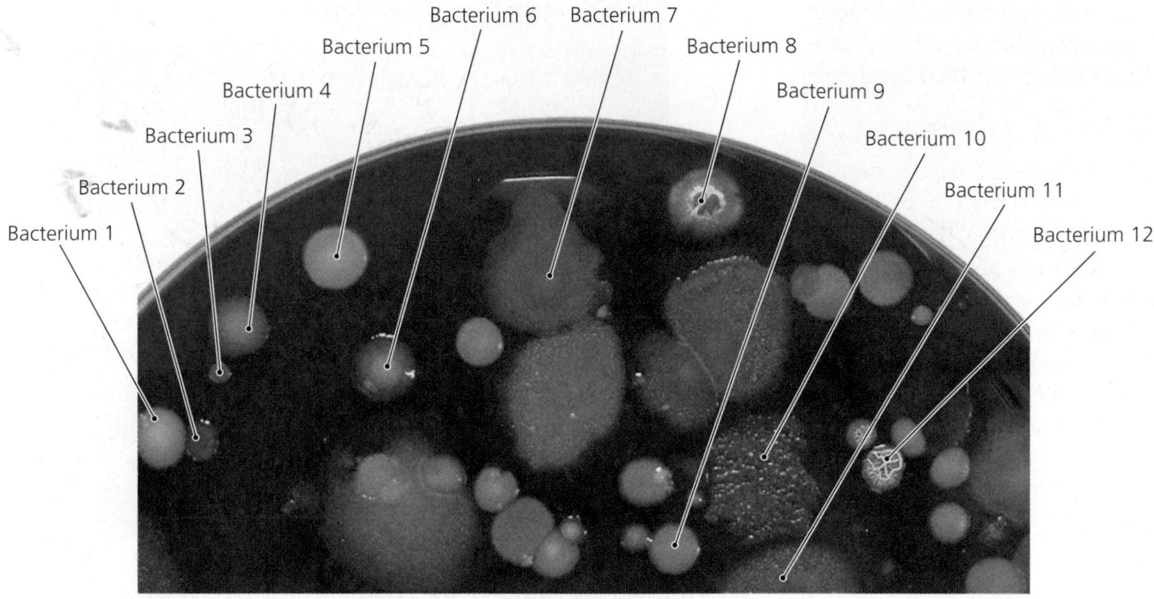

Figure 1.16
Bacterial colonies on a solid surface (agar). Differences in colony size, shape, and color indicate the presence of different species. Such differences allowed Koch to isolate specific types of microbes that could be tested for their ability to cause disease.

laboratory animals to see which caused disease. Koch's method of isolation is a standard technique in microbiological and medical labs to this day, though a gel called *agar,* derived from red seaweed, is used instead of gelatin or potato.

Koch and his colleagues are also responsible for many other advances in laboratory microbiology, including the following:

- Simple staining techniques for bacterial cells and flagella
- The first photomicrograph of bacteria
- The first photograph of bacteria in diseased tissue
- Techniques for estimating the number of bacteria in a solution based on the number of colonies that form after inoculation onto a solid surface
- The use of steam to sterilize growth media
- The use of Petri[18] dishes to hold solid growth media
- Laboratory techniques such as transferring bacteria between media using a platinum wire that had been heat-sterilized in a flame
- Elucidation of bacteria as distinct species

CRITICAL **THINKING**

French microbiologists, led by Pasteur, tried to isolate a single bacterium by diluting liquid media until only a single type of bacterium could be microscopically observed in a sample of the diluted medium. What advantages does Koch's method have over the French method?

Koch's Postulates

After discovering the anthrax bacterium, Koch continued to search for disease agents. In two pivotal scientific publications in 1882 and 1884, he announced that the cause of tuberculosis was a rod-shaped bacterium, *Mycobacterium tuberculosis* (mī′kō-bak-tēr′ē-ŭm too-ber-kyū-lō′sis). In 1905 he received the Nobel Prize in Physiology or Medicine for this work.

In his publications on tuberculosis, Koch elucidated a series of steps that must be taken to prove the cause of any infectious disease. These steps, now known as **Koch's postulates,** are one of his more important contributions to microbiology. His postulates, which we discuss in more detail in Chapter 14, are the following:

1. The suspected causative agent must be found in every case of the disease and be absent from healthy hosts.
2. The agent must be isolated and grown outside the host.
3. When the agent is introduced to a healthy, susceptible host, the host must get the disease.
4. The same agent must be found in the diseased experimental host.

We use the term *suspected causative agent* because it is merely "suspected" until the postulates have been fulfilled, and "agent" can refer to any fungus, protozoan, bacterium, virus, or other pathogen. There are practical and ethical limits in the application of Koch's postulates, but in almost every case they must be satisfied before the cause of an infectious disease is proven.

CRITICAL **THINKING**

Why aren't Koch's postulates always useful in proving the cause of a given disease? Consider a variety of diseases, such as cholera, pneumonia, Alzheimer's, AIDS, Down syndrome, and lung cancer.

During microbiology's "golden years," other scientists used Koch's postulates, as well as laboratory techniques introduced

[18]Named for Richard Petri, Koch's assistant, who invented them in 1887.

by Koch and Pasteur, to discover the causes of most protozoan and bacterial diseases, as well as some viral diseases. For example, Charles Laveran (1845–1922) showed that a protozoan is the cause of malaria, and Edwin Klebs (1834–1913) described the bacterium that causes diphtheria. Dmitri Ivanowski (1864–1920) and Martinus Beijerinck (1851–1931) discovered that a certain disease in tobacco plants is caused by a pathogen that passes through filters with such extremely small pores that bacteria cannot pass through. Beijerinck, recognizing that the pathogen was not bacterial, called it a *filterable virus.* Now such pathogens are simply called *viruses.* As previously noted, viruses couldn't be seen until electron microscopes were invented in 1932. The American physician Walter Reed (1851–1902) proved in 1900 that viruses can cause such diseases as yellow fever in humans. Chapter 13 deals with *virology,* and Chapters 24 and 25 deal with viral diseases.

A partial list of scientists and the pathogens they discovered is provided in Table 1.2 on p. 16.

Gram's Stain

The first of Koch's postulates demands that the suspected agent be found in every case of a given disease, which presupposes that minute microbes can be seen and identified. However, because most microbes are colorless and difficult to see, scientists began to use dyes to stain them and make them more visible under the microscope.

Though Koch reported a simple staining technique in 1877, the Danish scientist Hans Christian Gram (1853–1938) developed a more important staining technique in 1884. His procedure, which involves the application of a series of dyes, leaves some microbes purple and others pink. We now label the first group of cells as *Gram-positive* and the second as *Gram-negative,* and we use the Gram procedure to separate bacteria into these two large groups **(Figure 1.17)**.

The **Gram stain** is still the most widely used staining technique. It is one of the first steps carried out when bacteria are being identified, and is one of the procedures you will learn in microbiology lab. Chapter 4 discusses the full procedure.

How Can We Prevent Infection and Disease?

Learning Objectives

- ✓ Identify four health care practitioners who did pioneering research in the areas of public health microbiology and epidemiology.
- ✓ Name two scientists whose work with vaccines began the field of immunology.
- ✓ Describe the quest for a "magic bullet."

The last great question that drove microbiological research during the "Golden Age" was how to prevent infectious diseases. Though some methods of preventing or limiting disease were discovered even before it was understood that microorganisms caused contagious diseases, great advances occurred only after

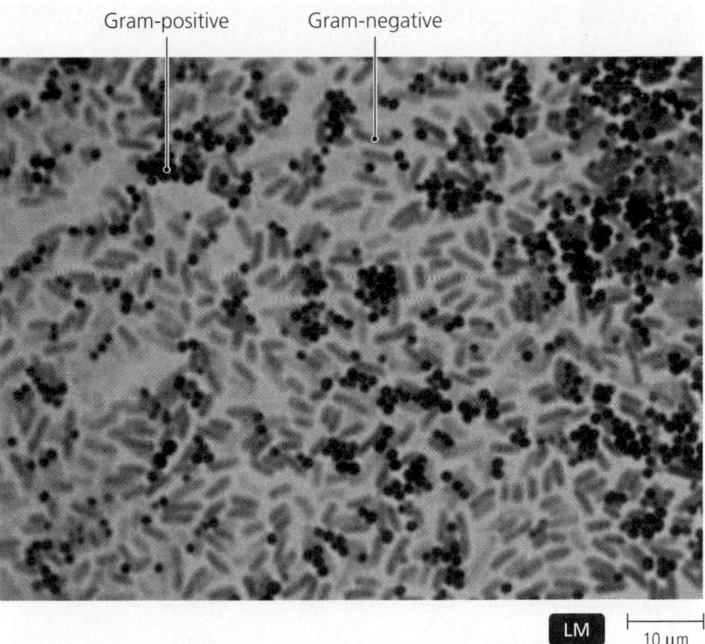

▲ **Figure 1.17 Results of Gram staining.** Gram-positive cells (in this case *Staphylococcus aureus*) are purple; Gram-negative cells (in this case *Escherichia coli*) are pink.

Pasteur and Koch showed that life comes from life and that microorganisms can cause diseases.

In the mid-1800s, modern principles of hygiene, such as those involving sewage and water treatment, personal cleanliness, and pest control, were not widely practiced. Typically, medical personnel and health care facilities lacked adequate cleanliness. *Nosocomial*[19] *infections*—infections acquired in a health care setting—were rampant. For example, surgical patients frequently succumbed to gangrene acquired while under their doctor's care, and many women who gave birth in hospitals died from puerperal[20] fever. Four health care practitioners who were especially instrumental in changing the way health care is delivered were Semmelweis, Lister, Nightingale, and Snow.

Semmelweis and Handwashing

Ignaz Semmelweis (1818–1865) was a physician on the obstetric ward of a teaching hospital in Vienna. In about 1848 he observed that women giving birth in the wing where medical students were trained died from puerperal fever at a rate 20 times higher than the mortality rates of either women attended by midwives in an adjoining wing or women who gave birth at home.

Though Pasteur had not yet elaborated his germ theory of disease, Semmelweis hypothesized that medical students carried "cadaver particles" from their autopsy studies into the delivery rooms, and that these "particles" resulted in puerperal

[19]From Greek *nosos,* meaning disease, and *komein,* meaning to care for (relating to a hospital).
[20]From Latin *puerperus,* meaning childbirth.

TABLE 1.2

Other Notable Scientists of the "Golden Age of Microbiology" and the Agents of Disease They Discovered

Scientist	Year	Disease	Agent
Albert Neisser	1879	Gonorrhea	*Neisseria gonorrhoeae* (bacterium)
Charles Laveran	1880	Malaria	*Plasmodium* species (protozoa)
Carl Eberth	1880	Typhoid fever	*Salmonella enterica* serotype Typhi (bacterium)
Edwin Klebs	1883	Diphtheria	*Corynebacterium diphtheriae* (bacterium)
Theodore Escherich	1884	Traveler's diarrhea Bladder infection	*Escherichia coli* (bacterium)
Albert Fraenkel	1884	Pneumonia	*Streptococcus pneumoniae* (bacterium)
David Bruce	1887	Undulant fever (brucellosis)	*Brucella melitensis* (bacterium)
Anton Weichselbaum	1887	Meningococcal meningitis	*Neisseria meningitidis* (bacterium)
A. A. Gartner	1888	Salmonellosis (form of food poisoning)	*Salmonella* species (bacterium)
Shibasaburo Kitasato	1889	Tetanus	*Clostridium tetani* (bacterium)
Dmitri Ivanowski and Martinus Beijerinck	1892 1898	Tobacco mosaic disease	*Tobamovirus tobacco mosaic virus*
William Welch and George Nuttall	1892	Gas gangrene	*Clostridium perfringens* (bacterium)
Alexandre Yersin and Shibasaburo Kitasato	1894	Bubonic plague	*Yersinia pestis* (bacterium)
Kiyoshi Shiga	1898	Shigellosis (a type of severe diarrhea)	*Shigella dysenteriae* (bacterium)
Walter Reed	1900	Yellow fever	*Flavivirus yellow fever virus*
Robert Forde and Joseph Dutton	1902	African sleeping sickness	*Trypanosoma brucei gambiense* (protozoan)

fever. Semmelweis gained support for his hypothesis when a doctor who sliced his finger during an autopsy died after showing symptoms similar to those of puerperal fever. Today we know that the primary cause of puerperal fever is a bacterium in the genus *Streptococcus* (strep-tō-kok'ŭs; see Figure 1.7), which is usually harmless on the skin or in the mouth but causes severe complications when it enters the blood.

Semmelweis began requiring medical students to wash their hands with chlorinated lime water, a substance long used to eliminate the smell of cadavers. Mortality in the subsequent year dropped from 18.3% to 1.3%. Despite his success, Semmelweis was ridiculed by the director of the hospital and eventually forced to leave. He returned to his native Hungary, where his insistence on handwashing met with general approval when it continued to produce higher patient survival rates.

Though his impressive record made it easier for later doctors to institute changes, Semmelweis was unsuccessful in gaining support for his method from most European doctors. He became severely depressed and was committed to a mental hospital, where he died from an infection of *Streptococcus*, the very organism he had fought for so long.

Lister's Antiseptic Technique

Shortly after Semmelweis was rejected in Vienna, the English physician Joseph Lister (1827–1912) modified and advanced the idea of *antisepsis*[21] in health care settings. As a surgeon, Lister was aware of the dreadful consequences that resulted from the infection of wounds. Therefore, he began spraying wounds, surgical incisions, and dressings with carbolic acid (phenol), a chemical that had previously proven effective in reducing odor and decay in sewage. Like Semmelweis, he initially met with some resistance, but when he showed that it reduced deaths among his patients by two-thirds, his method was accepted into common practice. In this manner, Lister vindicated Semmelweis, became the founder of antiseptic surgery, and opened new fields of research into antisepsis and disinfection.

Nightingale and Nursing

Florence Nightingale (1820–1910) **(Figure 1.18)** was a dedicated English nurse who succeeded in introducing cleanliness and other antiseptic techniques into nursing practice. She was instrumental in setting standards of hygiene that saved innumerable lives during the Crimean War of 1854–56. One of her first requisitions in the military hospital was for 200 scrubbing brushes, which she and her assistants used diligently in the squalid wards. She next arranged for each patient's filthy clothes and dressings to be replaced or cleaned at a different

[21]From Greek *anti*, meaning against, and *sepein*, meaning putrefaction.

location, thus removing many sources of infection. She thoroughly documented statistical comparisons to show that poor food and unsanitary conditions in the hospitals were responsible for the deaths of many soldiers.

After the war, Nightingale returned to England, where she actively exerted political pressure to reform hospitals and implement public health policies. Perhaps her greatest achievements were in nursing education. For example, she founded the Nightingale School for Nurses—the first of its kind in the world.

Snow and Epidemiology

Another English physician, John Snow (1813–1858), also played a key role in setting standards for good public hygiene to prevent the spread of infectious diseases. Snow had been studying the propagation of cholera and suspected that the disease was spread by a contaminating agent in water. In 1854, he mapped the occurrence of cholera cases during an epidemic in London and showed that they centered around a public water supply on Broad Street.

Though Snow did not know the cause of cholera, his careful documentation of the epidemic highlighted the critical need for adequate sewage treatment and a pure water supply. His study was the foundation for two branches of microbiology—**infection control** and **epidemiology,**[22] which is the study of the occurrence, distribution, and spread of disease in humans.

Jenner's Vaccine

In 1796, the English physician Edward Jenner (1749–1823) tested the hypothesis that a mild disease called cowpox provided protection against potentially fatal smallpox. After he intentionally inoculated a boy with pus collected from a milkmaid's cowpox lesion, the boy developed cowpox, which, of course, he survived. When Jenner then infected the boy with smallpox pus, he found that the boy had become immune[23] to smallpox. (Note that experiments that intentionally expose human subjects to deadly pathogens are unethical.) In 1798 Jenner reported similar results from additional experiments, demonstrating the validity of the procedure he named *vaccination* after *Vaccinia virus,*[24] the virus that causes cowpox. Jenner invented vaccination (the term *immunization* is often used synonymously today), established a safe treatment for preventing smallpox, and began the field of **immunology**—the study of the body's specific defenses against pathogens. Chapters 16–18 discuss immunology.

Pasteur later capitalized on Jenner's work by producing weakened strains of various pathogens for use in preventing the serious diseases they cause. In honor of Jenner's work with cowpox, Pasteur used the term *vaccine* to refer to all weakened, protective strains of pathogens. He subsequently developed successful vaccines against fowl cholera, anthrax, and rabies.

▲ **Figure 1.18 Florence Nightingale.** The founder of modern nursing, she was influential in introducing antiseptic technique into nursing practice.

Ehrlich's "Magic Bullets"

Gram's discovery that stained bacteria could be differentiated into two types suggested to the German microbiologist Paul Ehrlich (1854–1915) that chemicals could be used to kill microorganisms differentially. To investigate this idea, Ehrlich undertook an exhaustive survey of chemicals to find a "magic bullet" that would destroy pathogens while remaining nontoxic to humans. By 1908, he had discovered chemicals active against the protozoan parasites that cause sleeping sicknesses and against the causative agent of syphilis. His discoveries began the branch of medical microbiology known as **chemotherapy.**

In summary, the Golden Age of Microbiology was a time when researchers proved that living things come from other living things, that microorganisms can cause fermentation and disease, and that certain procedures and chemicals can limit, prevent, and cure infectious diseases. These discoveries were made by scientists who applied the scientific method to biological investigation, and they led to an explosion of knowledge in a number of scientific disciplines (Figure 1.19).

The Modern Age of Microbiology

Learning Objective

✓ List four major questions that drive microbiological investigations today.

[22]From Greek *epi*, meaning upon, *demos* meaning people, and *logos* meaning word or study.
[23]From Latin *immunis*, meaning free.
[24]From Latin *vacca*, meaning cow.

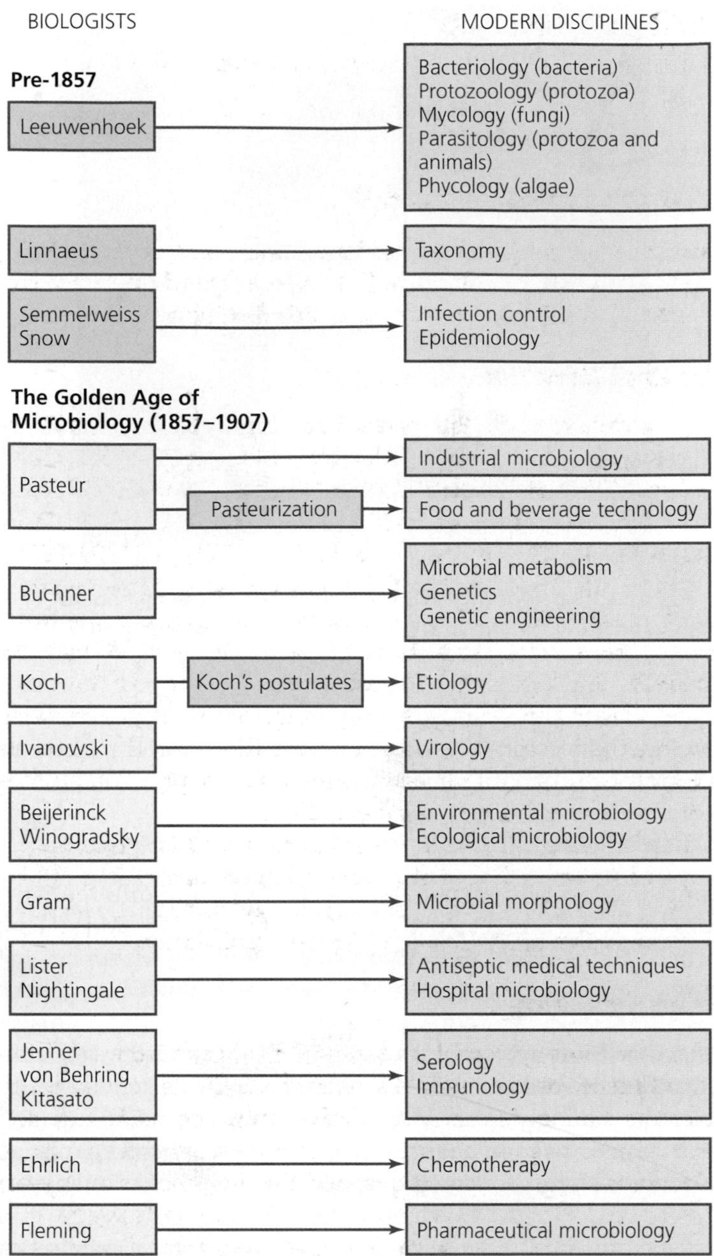

▲ **Figure 1.19 Some of the many scientific disciplines and applications that arose from the pioneering work of scientists just before and around the time of the Golden Age of Microbiology.**

The vast increase in the number of microbiological investigations and in scientific knowledge during the 1800s opened new fields of science, including disciplines called environmental science, immunology, epidemiology, chemotherapy, and genetic engineering (Table 1.3). Microorganisms played a significant role in the development of these disciplines because microorganisms are relatively easy to grow, take up little space, and are available by the trillions. Much of what has been learned about microbes also applies to other organisms, including humans. In

TABLE 1.3 Fields of Microbiology

Disciplines	Subject(s) of Study
Basic Research	
Microbe-Centered	
Bacteriology	Bacteria and archaea
Phycology	Algae
Mycology	Fungi
Protozoology	Protozoa
Parasitology	Parasitic protozoa and parasitic animals
Virology	Viruses
Process-Centered	
Microbial metabolism	Biochemistry: chemical reactions within cells
Microbial genetics	Functions of DNA and RNA
Environmental microbiology	Relationships between microbes, and among microbes, other organisms, and their environment
Applied Microbiology	
Medical Microbiology	
Serology	Antibodies in blood serum, particularly as an indicator of infection
Immunology	Body's defenses against specific diseases
Epidemiology	Frequency, distribution, and spread of disease
Etiology	Causes of disease
Infection control	Hygiene in health care settings and control of nosocomial infections
Chemotherapy	Development and use of drugs to treat infectious diseases
Applied Environmental Microbiology	
Bioremediation	Use of microbes to remove pollutants
Public health microbiology	Sewage treatment, water purification, and control of insects that spread disease
Agricultural microbiology	Use of microbes to control insect pests
Industrial Microbiology (Biotechnology)	
Food and beverage technology	Reduction or elimination of harmful microbes in food and drink
Pharmaceutical microbiology	Manufacture of vaccines and antibiotics
Recombinant DNA technology	Alteration of microbial genes to synthesize useful products

the rest of this text we examine advances made in these branches of microbiology, though it would require thousands of books this size to deal with all that is known.

Once the developing science of microbiology had successfully answered questions about spontaneous generation, fermentation, and disease, additional questions arose in each branch of the new science. Since the early 20th century, microbiologists have worked to answer these new questions. In this section we briefly consider some of the 20th century's overarching questions in both basic and applied research. The chapter concludes with a look at some of the questions that might propel microbiological research for the next 50 years.

What Are the Basic Chemical Reactions of Life?

Biochemistry is the study of metabolism—that is, the chemical reactions that occur in living organisms. Biochemistry began with Pasteur's work on fermentation by yeast and bacteria, and with Buchner's discovery of enzymes in yeast extract, but by the early 1900s many scientists thought that the metabolic reactions of microbes had little to do with the metabolism of plants and animals.

In contrast, microbiologists Albert Kluyver (1888–1956) and his student C. B. van Niel (1897–1985) proposed that basic biochemical reactions are shared by all living things, that these reactions are relatively few in number, and that their primary feature is the transfer of electrons and hydrogen ions. In adopting this view, scientists could use microbes as model systems to answer questions about metabolism in all organisms. Research during the 20th century validated this approach to understanding basic metabolic processes, but scientists have also documented an amazing metabolic diversity. Chapter 5 discusses basic metabolic processes, and Chapter 6 considers metabolic diversity.

Basic biochemical research has many practical applications, including:

- The design of herbicides and pesticides that are specific in their action and have no long-term adverse effects on the environment.
- The diagnosis of illnesses and the monitoring of a patient's responses to treatment. For example, physicians routinely monitor liver disease by measuring blood levels of certain enzymes and products of liver metabolism.
- The treatment of metabolic diseases. One example is treating phenylketonuria, a disease resulting from the inability to properly metabolize the amino acid phenylalanine, by eliminating foods containing phenylalanine from the diet.
- The design of drugs to treat leukemia, gout, bacterial infections, malaria, herpes, AIDS, asthma, and heart attacks.

CRITICAL THINKING

Albert Kluyver said, "From elephant to . . . bacterium—it is all the same!" What did he mean?

How Do Genes Work?

Genetics, the scientific study of inheritance, started in the mid-1800s as an offshoot of botany, but scientists studying microbes made most of the great advances in this discipline.

Microbial Genetics

While working with the bacterium *Streptococcus pneumoniae* (strep-tō-kok'ŭs nū-mō'nē-ī), Oswald Avery (1877–1955), Colin MacLeod (1909–1972), and Maclyn McCarty (1911–2005) determined that genes are contained in molecules of DNA. In 1958, George Beadle (1903–1989) and Edward Tatum (1909–1975), working with the bread mold *Neurospora crassa* (noo-ros'pōr-ă kras'ă), established that a gene's activity is related to the function of the specific protein coded by that gene. Other researchers, also working with microbes, determined the exact way in which genetic information is translated into a protein, the rates and mechanisms of genetic mutation, and the methods by which cells control genetic expression. Chapter 7 examines all of these aspects of microbial genetics.

Over the past 40 years, advances in microbial genetics developed into several new disciplines that are among the faster-growing areas of scientific research today, including *molecular biology, recombinant DNA technology,* and *gene therapy.*

Molecular Biology

Molecular biology combines aspects of biochemistry, cell biology, and genetics to explain cell function at the molecular level. Molecular biologists are particularly concerned with *genome*[25] *sequencing.* Using techniques perfected on microorganisms, molecular biologists have sequenced the genomes of many organisms, including humans and many of their pathogens. It is hoped that a fuller understanding of the genomes of organisms will result in practical ways to limit disease, repair genetic defects, and enhance agricultural yield.

The American Nobel laureate Linus Pauling (1901–1994) proposed in 1965 that gene sequences could provide a means of understanding evolutionary relationships and processes, establishing taxonomic categories that more closely reflect these relationships, and identifying the existence of microbes that have never been cultured in a laboratory. Two examples illustrate such uses of gene sequencing data:

- In the 1970s, Carl Woese (1928–) discovered that significant differences in nucleic acid sequences among organisms clearly reveal that cells belong to one of *three* major

[25] A genome is the total genetic information of an organism.

groups—bacteria, archaea, or eukaryotes—and not merely two groups (prokaryotes and eukaryotes) as previously thought.

- Scientists showed in 1990 that cat scratch disease is caused by a bacterium that could not be cultured. The bacterium was discovered by recognition of the sequence of a portion of its ribonucleic acid that differs from all other known ribonucleic acid sequences.

Recombinant DNA Technology

Molecular biology is applied in **recombinant DNA technology,**[26] commonly called *genetic engineering,* which was first developed using microbial models. Geneticists manipulate genes in microbes, plants, and animals for practical applications. For instance, once scientists have inserted the gene for human blood-clotting factor into the bacterium *Escherichia coli* (esh-ĕ-rik'ē-ă kō'lē), the bacterium produces the factor in a pure form. This technology is a boon to hemophiliacs, who previously depended on clotting factor isolated from donated blood, which was possibly contaminated by life-threatening viral pathogens.

Gene Therapy

An exciting new area of study is the use of recombinant DNA technology for **gene therapy,** a process that involves inserting a missing gene or repairing a defective one in human cells. In such procedures, researchers insert a desired gene into host cells, where it is incorporated into a chromosome and begins to function normally. Chapter 8 examines recombinant DNA technology and gene therapy in more detail.

What Roles Do Microorganisms Play in the Environment?

Learning Objectives

✓ Identify the field of microbiology that studies the role of microorganisms in the environment.

✓ Name the fastest-growing scientific disciplines in microbiology today.

Ever since Koch and Pasteur, most research in microbiology has focused on pure cultures of individual species; however, microorganisms are not alone in the "real world." Instead, they live in natural microbial communities in the soil, water, the human body, and other habitats, and these communities play critical roles in such processes as the production of vitamins and *bioremediation*—the use of living bacteria, fungi, and algae to detoxify polluted environments.

Microbial communities also play an essential role in the decay of dead organisms and the recycling of chemicals such as carbon, nitrogen, and sulfur. Martinus Beijerinck discovered bacteria capable of converting nitrogen gas (N_2) from the air into nitrate (NO_3), the form of nitrogen used by plants, and the Russian microbiologist Sergei Winogradsky (1856–1953) elucidated the role of microorganisms in the recycling of sulfur. Together these two microbiologists developed laboratory techniques for isolating and growing environmentally important microbes. Chapter 26 surveys their discoveries and other important aspects of **environmental microbiology.**

Another role of microbes in the environment is the causation of disease. Although most microorganisms are not pathogenic, in this book (particularly in Chapters 19–25), we focus on pathogenic microbes because of the threat they pose to human health. We examine their characteristics and the diseases they cause, as well as the steps we can take to limit their abundance and control their spread in the environment, such as sewage treatment, water purification, disinfection, pasteurization, and sterilization.

CRITICAL **THINKING**

The ability of farmers around the world to produce crops such as corn, wheat, and rice is often limited by the lack of nitrogen-based fertilizer. How might scientists use Beijerinck's discovery to increase world supplies of grain?

How Do We Defend Against Disease?

Why do some people get sick during the flu season while their close friends and family remain well? The germ theory of disease showed not only that microorganisms can cause diseases, but also that the body can defend itself—otherwise, everyone would be sick most of the time.

The work of Jenner and Pasteur on vaccines showed that the body can protect itself from repeated diseases by the same organism. The German bacteriologist Emil von Behring (1854–1917) and the Japanese microbiologist Shibasaburo Kitasato (1852–1931), working in Koch's laboratory, reported the existence in the blood of chemicals and cells that fight infection. Their studies developed into the fields of *serology,* the study of blood serum[27]—specifically, the chemicals in the liquid portion of blood that fight disease—and *immunology,* the study of the body's defense against specific pathogens. Chapters 15–18 cover these aspects of microbiology, which are of utmost importance to physicians, nurses, and other health care practitioners.

Ehrlich introduced the idea of a "magic bullet" that would kill pathogens, but it wasn't until Alexander Fleming (1881–1955) discovered penicillin (**Figure 1.20**) in 1929 and Gerhard Domagk (1895–1964) discovered sulfa drugs in 1935 that medical personnel finally had drugs effective against a wide range of bacteria. We study chemotherapy, and some physical and chemical agents used to control microorganisms in the environment, in Chapters 9 and 10.

[26]Recombinant DNA is DNA composed of genes from more than one organism.
[27]Latin, meaning whey. Serum is the liquid that remains after blood coagulates.

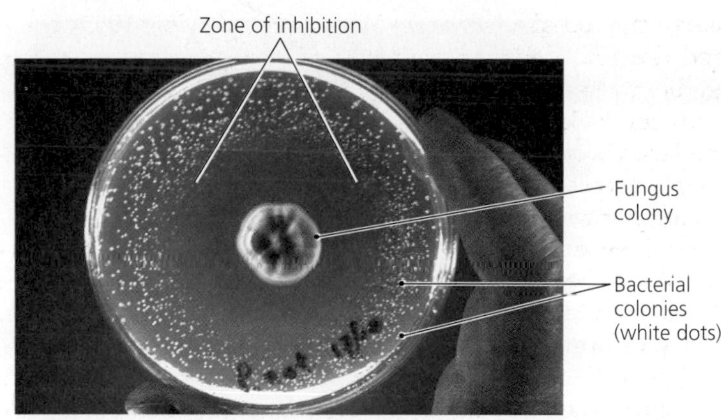

▲ **Figure 1.20 The effects of penicillin on a bacterial "lawn" in a Petri dish.** The clear area (zone of inhibition) surrounding the fungus colony, which is producing the antibiotic, is where the penicillin prevented bacterial growth.

What Will the Future Hold?

The science of microbiology is built on asking and answering questions. What began with the curiosity of a dedicated lens grinder in the Netherlands has come far in the past 350 years and has expanded into disciplines as diverse as immunology, recombinant DNA technology, and bioremediation. However, the adage remains true: *The more questions we answer, the more questions we have.*

What will microbiologists discover next? Among the questions for the next 50 years are the following:

- What is it about the physiology of life forms known only by their nucleic acid sequences that prevents those life forms from being grown in the laboratory?
- Can bacteria and archaea be used in ultraminiature technologies such as living computer circuit boards?
- How can an understanding of microbial communities help us understand the positive aspects of microbial action in preventing and curing diseases, recycling nutrients, degrading pollutants, and moderating climate changes?
- What genetic sequences make some microbes pathogenic, and what can we do at a genetic level to defend against these pathogens?
- What features of living organisms might allow them to live elsewhere in the universe?
- How can we reduce the threat from microbes resistant to antimicrobial drugs as well as conquer emerging and reemerging infectious diseases? (See **Highlight: "The New Normal": The Challenge of Emerging and Reemerging Diseases.**)

HIGHLIGHT

"THE NEW NORMAL": THE CHALLENGE OF EMERGING AND REEMERGING DISEASES

Severe acute respiratory syndrome (SARS). Monkeypox. West Nile encephalitis. These and diseases like them are emerging diseases—ones that appear in a population for the first time. Among them are H1N1 influenza ("swine flu"); Nipah encephalitis, a highly fatal disease carried by pigs; and mosquito-borne chikungunya, which causes severe joint pain and sometimes death. Indeed, unfamiliar diseases have become "the new normal" for health care workers, according to the Centers for Disease Control and Prevention.

Meanwhile, diseases once thought to be near eradication, such as polio, mumps, and tuberculosis, have reemerged in troubling outbreaks. Other near-vanquished pathogens such as smallpox or anthrax may become potential weapons in bioterrorist attacks.

How do emerging and reemerging diseases arise? Some are introduced to humans as we move into remote jungles and contact infected animals; some are carried by insects whose range is spreading as climate changes; some take advantage of the AIDS crisis, infecting immunocompromised patients; some previously harmless microbes acquire new genes that allow them to be infective and cause disease; some emerging pathogens spread with the speed of jet planes carrying infected people around the globe; and still others arise when previously treatable microbes develop resistance to our antibiotics.

However they arise, scientists are monitoring emerging and reemerging diseases that may develop into the next generation of high-profile infectious diseases. Throughout this textbook, you will encounter many boxed discussions of such emerging and reemerging diseases.

▲ *Workers dumping poultry suspected of harboring avian influenza virus.*

EMERGING DISEASES

VARIANT CREUTZFELDT-JAKOB DISEASE

Ellen screamed obscenities as she staggered from the room and collapsed in the hallway, unable to stand and jerking uncontrollably. Her parents were shocked that their kind, considerate, and lovable daughter had changed so drastically during the past year. Sadly, she couldn't even remember her siblings' names!

Ellen had joined the nearly 200 Europeans, and one Canadian, afflicted with variant Creutzfeldt-Jakob disease (vCJD) (what the media call "mad cow disease" because most humans with the condition acquired the pathogen from eating infected beef). Since vCJD affects the brain by slowly eroding nervous tissue and leaving the brain full of sponge-like holes, the signs and symptoms of vCJD are neurological. Ellen's disease started with insomnia, depression, and confusion, but eventually it led to uncontrollable emotional and verbal outbursts, inability to coordinate movements, coma, and death. Typically the disease lasts about a year, and there is no treatment.

Variant Creutzfeldt-Jakob disease is an *emerging disease;* that is, a disease arising in the past two decades, either because it is new to a population, or because it is newly recognized. Some investigators also include diseases that have been nearly eradicated but are now reemerging. Variant CJD resembles the rare genetic disorder Creutzfeldt-Jakob disease (named for its discoverers), which is caused by a mutation and occurs in the elderly. The difference is that the variant form of CJD results from an acquired infection and often strikes and kills college-aged people, like Ellen in our story. For more about vCJD, see p. 396.

Track vCJD online by going to the Study Area at www.masteringmicrobiology.com.

Chapter Summary

The Early Years of Microbiology (pp. 2–7)

1. Leeuwenhoek's observations of **microbes** introduced most types of **microorganisms** to the world and earned him the title "Father of Bacteriology and Protozoology." His discoveries were named and classified by Linnaeus in his **taxonomic system.**
2. Relatively large microscopic **eukaryotic fungi** include **molds** and **yeasts.**
3. Animal-like **protozoa** are single-celled eukaryotes. Some cause disease.
4. Plantlike eukaryotic **algae** are important providers of oxygen, serve as food for many marine animals, and make chemicals used in microbiological growth media.
5. Small **prokaryotes—bacteria** and **archaea—**live in a variety of communities and in most habitats. Even though some cause disease, most are beneficial.
6. Parasitic worms, the largest organisms studied by microbiologists, are often visible without a microscope, although their immature stages are microscopic.
7. Viruses, the smallest microbes, are so small they can be seen only by using an electron microscope.

The Golden Age of Microbiology (pp. 7–17)

1. The study of the Golden Age of Microbiology includes a look at the men who proposed or refuted the theory of **spontaneous generation:** Aristotle, Redi, Needham, Spallanzani, and Pasteur (the Father of Microbiology). The **scientific method** that emerged then remains the accepted sequence of study today.
2. The study of fermentation by Pasteur and Buchner led to the fields of **industrial microbiology** (**biotechnology**) and **biochemistry,** and to the study of **metabolism.**
3. Koch, Pasteur, and others proved that **pathogens** cause infectious diseases, an idea that is known as the **germ theory of disease. Etiology** is the study of the causation of diseases.
4. Koch initiated careful microbiological laboratory techniques in his search for disease agents. **Koch's postulates,** the logical steps he followed to prove the cause of an infectious disease, remain an important part of microbiology today.
5. The procedure for the **Gram stain** was developed in the 1880s and is still used to differentiate bacteria into two categories: Gram-positive and Gram-negative.
6. The investigations of Semmelweis, Lister, Nightingale, and Snow are the foundations upon which **infection control** and **epidemiology** are built.

7. Jenner's use of a cowpox-based vaccine for preventing smallpox began the field of **immunology.** Pasteur significantly advanced the field.
8. Ehrlich's search for "magic bullets"—chemicals that differentially kill microorganisms—laid the foundations for the field of **chemotherapy.**

The Modern Age of Microbiology (pp. 17–22)

1. Microbiology in the modern age has focused on answering questions regarding **biochemistry,** which is the study of metabolism; microbial genetics, which is the study of inheritance in microorganisms; and **molecular biology,** which involves investigations of cell function at the molecular level.
2. Scientists have applied knowledge from basic research to answer questions in **recombinant DNA technology** and **gene therapy.**
3. The study of microorganisms in their natural environment is **environmental microbiology.**
4. The discovery of chemicals in the blood that are active against specific pathogens advanced immunology and began the field of serology.
5. Advancements in chemotherapy were made in the 1900s with the discovery of numerous substances, such as penicillin and sulfa drugs, that inhibit pathogens.

Questions for Review

Answers to the Questions for Review (except Short Answer questions) begin on page A-1.

Multiple Choice

1. Which of the following microorganisms are not eukaryotic?
   a. bacteria
   b. yeasts
   c. molds
   d. protozoa
2. Which microorganisms are used to make microbiological growth media?
   a. bacteria
   b. fungi
   c. algae
   d. protozoa
3. In which habitat would you most likely find archaea?
   a. acidic hot springs
   b. swamp mud
   c. Great Salt Lake
   d. all of the above
4. Of the following scientists, who first promulgated the theory of abiogenesis?
   a. Aristotle
   b. Pasteur
   c. Needham
   d. Spallanzani
5. Which of the following scientists hypothesized that a bacterial colony arises from a single bacterial cell?
   a. Antoni van Leeuwenhoek
   b. Louis Pasteur
   c. Robert Koch
   d. Richard Petri
6. Which scientist first hypothesized that medical personnel can infect patients with pathogens?
   a. Edward Jenner
   b. Joseph Lister
   c. John Snow
   d. Ignaz Semmelweis
7. Leeuwenhoek described microorganisms as
   a. animalcules.
   b. prokaryotes.
   c. eukaryotes.
   d. protozoa.
8. Which of the following favored the theory of spontaneous generation?
   a. Spallanzani
   b. Needham
   c. Pasteur
   d. Koch
9. A scientist who studies the role of microorganisms in the environment is
   a. a genetic technologist.
   b. an earth microbiologist.
   c. an epidemiologist.
   d. an environmental microbiologist.
10. The laboratory of Robert Koch contributed which of the following to the field of microbiology?
    a. simple staining technique
    b. use of Petri dishes
    c. first photomicrograph of bacteria
    d. all of the above

Fill in the Blanks

Fill in the blanks with the name(s) of the scientist(s) whose investigations led to the following fields of study in microbiology.

1. Environmental microbiology __________ and __________
2. Biochemistry __________ and __________
3. Chemotherapy __________
4. Immunology __________
5. Public health microbiology __________
6. Etiology __________
7. Epidemiology __________
8. Biotechnology __________
9. Food microbiology __________

Labeling

On the photos below, label *cilium, flagellum, nucleus,* and *pseudopod.*

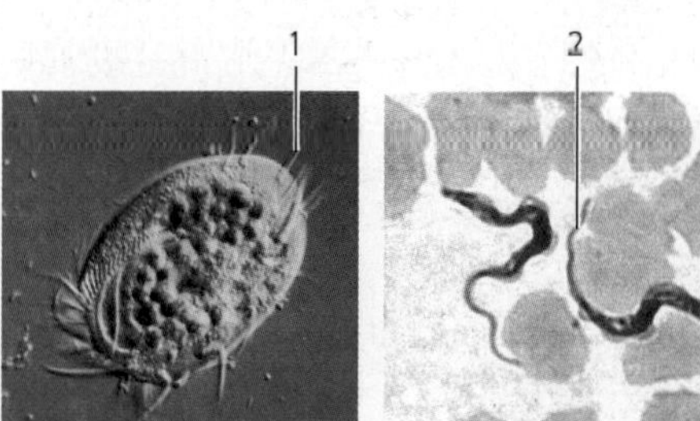

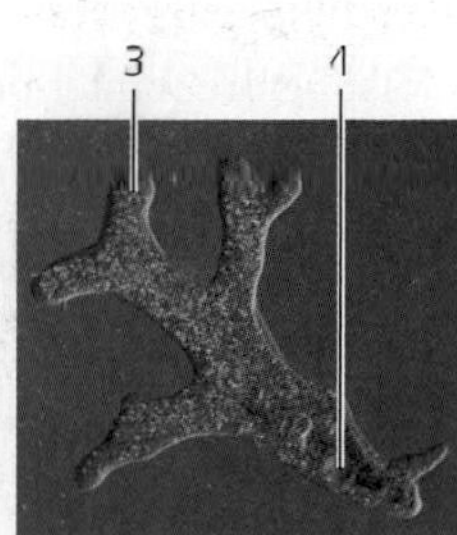

Matching

Match each of the following descriptions with the person it best describes. An answer may be used more than once.

1. ____ Developed smallpox immunization
2. ____ First photomicrograph of bacteria
3. ____ Germ theory of disease
4. ____ Germs cause disease
5. ____ Sought a "magic bullet" to destroy pathogens
6. ____ Early epidemiologist
7. ____ Father of Microbiology
8. ____ Classification system
9. ____ Discoverer of bacteria
10. ____ Discoverer of protozoa
11. ____ Founder of antiseptic surgery
12. ____ Developed the most widely used bacterial staining technique

A. John Snow
B. Paul Ehrlich
C. Louis Pasteur
D. Antoni van Leeuwenhoek
E. Carolus Linnaeus
F. John Needham
G. Eduard Buchner
H. Robert Koch
I. Joseph Lister
J. Edward Jenner
K. Girolamo Fracastoro
L. Hans Christian Gram

Short Answer

1. Why was the theory of spontaneous generation a hindrance to the development of the field of microbiology?
2. Discuss the significant difference between the flasks used by Pasteur and Spallanzani. How did Pasteur's investigation settle the dispute about spontaneous generation?
3. List six types of microorganisms.
4. Defend this statement: "The investigations of Antoni van Leeuwenhoek changed the world forever."
5. Why would a *macro*scopic tapeworm be studied in *micro*biology?
6. Describe what has been called the "Golden Age of Microbiology" with reference to four major questions that propelled scientists during that period.
7. List four major questions that drive microbiological investigations today.
8. Refer to the four steps in the scientific method in describing Pasteur's fermentation experiments.
9. List Koch's postulates, and explain why they are significant.
10. Why can Pasteur be honored with the title "Father of Microbiology"?

Critical Thinking

1. If Robert Koch had become interested in a viral disease such as influenza instead of anthrax (caused by a bacterium), how might his list of lifetime accomplishments be different? Why?
2. In 1911, the Polish scientist Casimir Funk proposed that a limited diet of polished white rice (rice without the husks) caused beriberi, a disease of the central nervous system. Even though history has proven him correct—beriberi is caused by a thiamine deficiency, which in his day resulted from unsophisticated milling techniques that removed the thiamine-rich husks—Funk was criticized by his contemporaries, who told him to find the microbe that caused beriberi. Explain how the prevailing scientific philosophy of the day shaped Funk's detractors' point of view.
3. *Haemophilus influenzae* does not cause flu, but it received its name because it was once thought to be the cause. Explain how a proper application of Koch's postulates would have prevented this error in nomenclature.
4. Just before winter break in early December, your roommate stocks the refrigerator with a gallon of milk, but both of you leave before opening it. When you return in January, the milk has soured. Your roommate is annoyed because the milk was pasteurized and thus should not have spoiled. Explain why your roommate's position is unreasonable.
5. Design an experiment to prove that microbes don't spontaneously generate in milk.
6. The British General Board of Health concluded in 1855 that the Broad Street cholera epidemic (see p. 17) resulted from fermentation of "nocturnal clouds of vapor" from the polluted Thames River. How could an epidemiologist prove or disprove this claim?
7. Compare and contrast the investigations of Redi, Needham, Spallanzani, and Pasteur in relation to the idea of spontaneous generation.
8. If you were a career counselor directing a student in the field of applied microbiology, describe three possible disciplines you could suggest.

Access more review material online in the Study Area at **www.masteringmicrobiology.com**. There, you'll find

- **MP3 Tutor Sessions**
- **Flashcards**
- **Quizzes**

and more to help you succeed.

2 The Chemistry of Microbiology

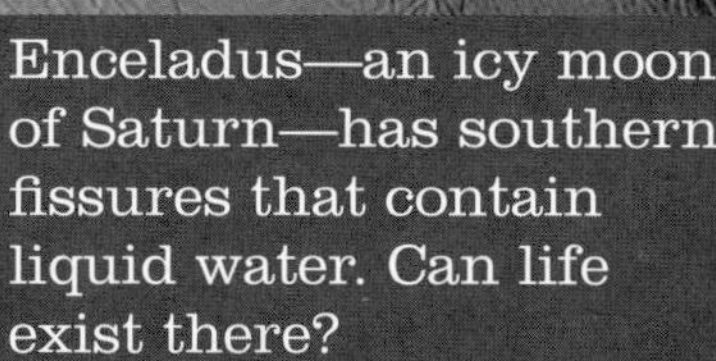

Enceladus—an icy moon of Saturn—has southern fissures that contain liquid water. Can life exist there?

Is there microbial life elsewhere in the solar system? Using telescopic observations and space probes of some of our nearest neighbors, we have identified chemicals necessary to life. For example, Venus and Mars have water and carbon dioxide; however, Enceladus, a small moon circling Saturn, is the strongest candidate for life beyond Earth.

Though the surface of Enceladus is covered with ice, it has warm liquid water below its cold exterior. Carbon (essential for all life forms) is available in CO_2 and in organic molecules such as methane and propane. Nitrogen is dissolved in the water. Thus, Enceladus has carbon, hydrogen, nitrogen, and oxygen, and water to act as an intermediary in chemical reactions. The south pole of Enceladus has fissures that spew water and dissolved chemicals hundreds of kilometers into space. These fissures are similar to oceanic hydrothermal vents that support life in the depths of Earth's oceans.

What environmental conditions are needed to support microbial life? That question applies on Earth too. So do these: How can we develop treatments against harmful bacteria and viruses? How do we differentiate one microorganism from another? The answers to these questions, and to many other important questions in microbiology, have their basis in chemistry.

Take the pre-test for this chapter online. Visit the Study Area at www.masteringmicrobiology.com.

Learning some basic concepts of chemistry will enable you to understand more fully the variety of interactions between microorganisms and their environment—which includes you. If you plan a career in health care, you will find microbial chemistry involved in the diagnosis of disease, the response of the immune system, the growth and identification of pathogenic microorganisms in the laboratory, and the function and selection of antimicrobial drugs. Even preserving your own health is aided by an understanding of the fundamentals of chemistry.

In this chapter we study atoms, which are the basic units of chemistry, and we consider how atoms react with one another to form chemical bonds and molecules. Then we examine the three major categories of chemical reactions. The chapter concludes with a look at the molecules of greatest importance to life: water, acids, bases, lipids, carbohydrates, proteins, nucleic acids, and ATP.

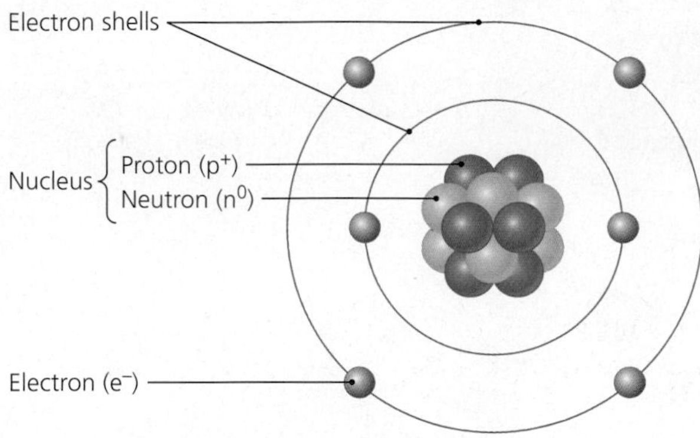

▲ **Figure 2.1 An example of a Bohr model of atomic structure.** This drawing is not to scale; for the electrons to be shown in scale with the greatly magnified nucleus, the electrons would have to occupy orbits located many miles from the nucleus. Put another way, the volume of an entire atom is about 100 trillion times the volume of its nucleus.

Atoms

Learning Objective

✓ Define *matter*, *atom*, and *element*, and explain how these terms relate to one another.

Matter is defined as anything that takes up space and has mass.[1] The smallest chemical units of matter are **atoms.** Atoms are extremely small, and only the very largest of them can be seen using the most powerful microscopes. Therefore, scientists have developed various models to conceptualize and illustrate the structure of atoms.

Atomic Structure

Learning Objective

✓ Draw and label an atom, showing the parts of the nucleus and orbiting electrons.

In 1913, the Danish physicist Niels H. D. Bohr (1885–1962) proposed a simple model in which negatively charged subatomic particles called **electrons** orbit a centrally located nucleus like planets in a miniature solar system **(Figure 2.1)**. A nucleus is composed of uncharged **neutrons** and positively charged **protons.** (The only exception to this description is the nucleus of a normal hydrogen atom, which is composed of only a single proton and no neutrons.) Protons and neutrons are extremely small. If a meterstick were stretched between the sun and the Earth, a neutron or proton would measure only about the width of a human hair! The number of electrons in an atom typically equals the number of protons, so overall atoms are electrically neutral.

An **element** is matter that is composed of a single type of atom. For example, gold is an element because it consists of only gold atoms. In contrast, the ink in your pen is not an element because it is composed of many different kinds of atoms.

Elements differ from one another in their **atomic number,** which is the number of protons in their nuclei. For example, the atomic numbers of hydrogen, carbon, and oxygen are 1, 6, and 8, respectively, because all hydrogen nuclei contain a single proton, all carbon nuclei have six protons, and all oxygen nuclei have eight protons.

The **atomic mass** of an atom (sometimes called its *atomic weight*) is the sum of the masses of its protons, neutrons, and electrons. Protons and neutrons each have a mass of approximately 1 *atomic mass unit,*[2] which is also called a *dalton.*[3] An electron is much less massive, with a mass of about 0.00054 dalton. Electrons are often ignored in discussions of atomic mass because their contribution to the overall mass is negligible. Therefore, the sum of the number of protons and neutrons approximates the atomic mass of an atom.

There are 93 naturally occurring elements known;[4] however, organisms typically utilize only about 20 elements, each of which has its own symbol that is derived from its English or Latin name (Table 2.1).

Isotopes

Learning Objective

✓ List at least four ways that radioactive isotopes are useful.

Every atom of an element has the same number of protons, but atoms of a given element can differ in the number of neutrons in their nuclei. Atoms that differ in this way are called **isotopes.** For example, there are three naturally occurring isotopes of carbon, each having six protons and six electrons **(Figure 2.2)**. Over 95% of carbon atoms also have six neutrons. Because these atoms have six protons and six neutrons, the atomic mass of

[1] *Mass* and *weight* are sometimes confused. Mass is the quantity of material in something, whereas weight is the effect of gravity on mass. Even though an astronaut is weightless in space, his mass is the same in space as on Earth.

[2] An atomic mass unit (dalton) is 1/597,728,630,000,000,000,000,000, or 1.673×10^{-24} grams

[3] Named for John Dalton, the British chemist who helped develop atomic theory around 1800.

[4] For many years, scientists thought that there were only 92 naturally occurring elements, but natural plutonium was discovered in Africa in 1997.

TABLE 2.1 Common Elements of Life

Element	Symbol	Atomic Number	Atomic Mass[a] (daltons)	Biological Significance
Hydrogen	H	1	1	Component of organic molecules and water; H^+ released by acids
Boron	B	5	11	Essential for plant growth
Carbon	C	6	12	Backbone of organic molecules
Nitrogen	N	7	14	Component of amino acids, proteins, and nucleic acids
Oxygen	O	8	16	Component of many organic molecules and water; OH^- released by bases; necessary for aerobic metabolism
Sodium (Natrium)	Na	11	23	Principal cation outside cells
Magnesium	Mg	12	24	Component of many energy-transferring enzymes
Silicon	Si	14	28	Component of cell wall of diatoms
Phosphorus	P	15	31	Component of nucleic acids and ATP
Sulfur	S	16	32	Component of proteins
Chlorine	Cl	17	35	Principal anion outside cells
Potassium (Kalium)	K	19	39	Principal cation inside cells; essential for nerve impulses
Calcium	Ca	20	40	Utilized in many intercellular signaling processes; essential for muscular contraction
Manganese	Mn	25	54	Component of some enzymes
Iron (Ferrum)	Fe	26	56	Component of energy-transferring proteins; transports oxygen in the blood of many animals
Cobalt	Co	27	59	Component of vitamin B_{12}
Copper (Cuprum)	Cu	29	64	Component of some enzymes
Zinc	Zn	30	65	Component of some enzymes
Molybdenum	Mo	42	96	Component of some enzymes
Iodine	I	53	127	Component of many brown and red algae

[a]Rounded to nearest whole number.

this isotope is about 12 daltons, and it is known as carbon-12, symbolized as ^{12}C. Atoms of carbon-13 (^{13}C) have seven neutrons per nucleus, and ^{14}C atoms each have eight neutrons.

Unlike the first two isotopes, the nucleus of ^{14}C is unstable due to the ratio of its protons and neutrons. Unstable atomic nuclei release energy and subatomic particles such as neutrons, protons, and electrons in a process called *radioactive decay*. Atoms that undergo radioactive decay are *radioactive isotopes*. Radioactive decay and radioactive isotopes play important roles in microbiological research, medical diagnosis, the treatment of disease, and the complete destruction of contaminating microbes (sterilization) of medical equipment and chemicals.

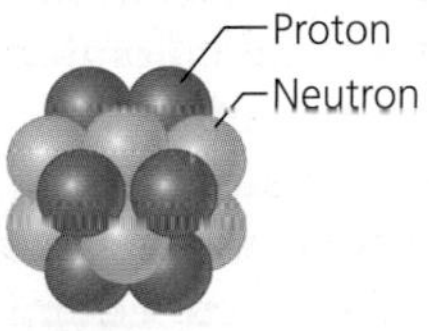

(a) Carbon-12
6 Protons
6 Neutrons

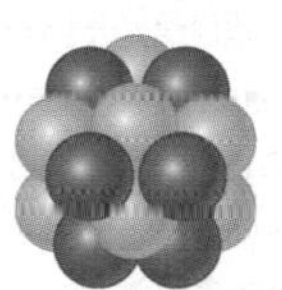

(b) Carbon-13
6 Protons
7 Neutrons

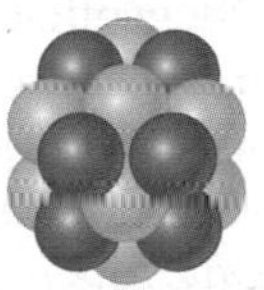

(c) Carbon-14
6 Protons
8 Neutrons

▲ **Figure 2.2 Nuclei of the three naturally occurring isotopes of carbon.** Each isotope also has six electrons, which are not shown. *What are the atomic number and atomic mass of each of these isotopes?*

Figure 2.2 *The atomic number of all three is 6; their atomic masses are 12, 13, and 14, respectively.*

Electron Configurations

Although the nuclei of atoms determine their identities, it is electrons that determine an atom's *chemical behavior*. Nuclei of different atoms almost never come close enough together to interact.[5] Typically, only the electrons of atoms interact. Thus, because all of the isotopes of carbon (for example) have the same number of electrons, all these isotopes behave the same way in chemical reactions, even though their nuclei are different.

Scientists know that electrons do not really orbit the nucleus in a two-dimensional circle, as indicated by a Bohr diagram; instead they speed around the nucleus 100 quadrillion

[5]Except during nuclear reactions, such as occur in nuclear power plants.

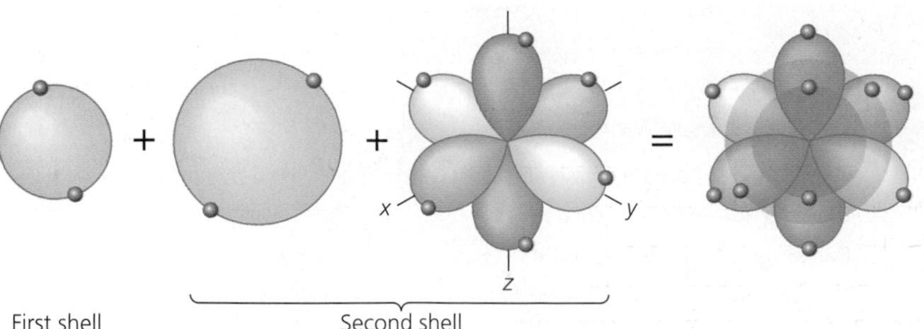

(a) Electron shells of neon: three-dimensional view

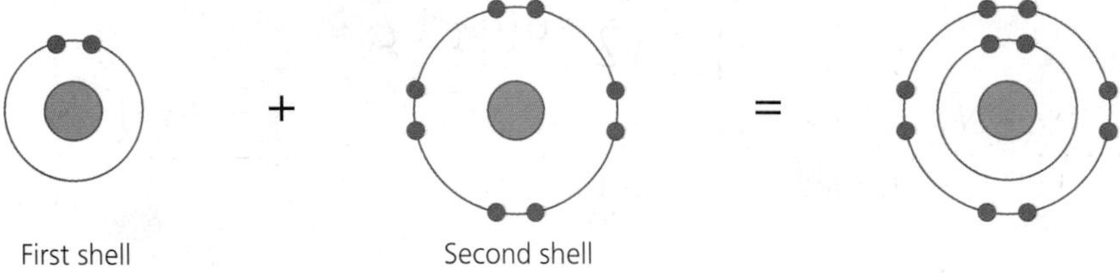

(b) Electron shells of neon: two-dimensional view

Figure 2.3 Electron configurations. **(a)** Three-dimensional model of the electron shells of neon. In this model, the first shell is a small sphere, whereas the second shell consists of a larger sphere plus three pairs of ellipses that extend from the nucleus at right angles. Larger shells (not shown) are even more complex. **(b)** Two-dimensional model (Bohr diagram) of the electron shells of neon.

times per second in unpredictable three-dimensional *electron shells* or *clouds* that assume unique shapes dependent on the energy of the electrons **(Figure 2.3a)**. More accurately put, an electron shell depicts the *probable* locations of electrons at a given time; nevertheless, it is simpler and more convenient to draw electron shells as circles **(Figure 2.3b)**.

Each electron shell can hold only a certain maximum number of electrons. For example, the first shell (the one nearest the nucleus) can accommodate a maximum of two electrons, and the second shell can hold no more than eight electrons. Atoms of hydrogen and helium have one and two electrons, respectively; thus, these two elements have only a single electron shell. A lithium atom, which has three electrons, has two shells.

Atoms with more than ten electrons require more shells. The third shell holds up to eight electrons when it is the outermost shell, though its capacity increases to eighteen when the fourth shell contains two electrons. Heavier atoms have even more shells, but these atoms do not play significant roles in the processes of life.

Electrons in the outermost shell of atoms are called *valence electrons*. **Figure 2.4** depicts the electron configurations of atoms of some elements important to microbial life. Notice that except for helium, atoms of all elements in a given column have the same number of valence electrons. Helium is placed in the far right-hand column with the other inert gases, because its outer shell is full, though it has two rather than eight valence electrons. Valence electrons are critical for interactions between atoms. Next we consider these interactions, which are called chemical bonds.

Chemical Bonds

Learning Objectives

- ✓ Describe the configuration of electrons in a stable atom.
- ✓ Contrast molecules and compounds.

Outer electron shells are stable when they contain eight electrons (except for the first electron shell, which is stable with only two electrons, since that is its maximum number). When atoms' outer shells are not filled with eight electrons, they either have room for more electrons or have "extra" electrons, depending on whether it is easier for them to gain electrons or lose electrons. For example, an oxygen atom, with six electrons in its outer shell, has two "unfilled spaces" (see Figure 2.4), because it requires less energy for the oxygen atom to gain two electrons than to lose six electrons. A calcium atom, by contrast, has two "extra" electrons in its outer (fourth) shell, because it requires less energy to lose these two electrons than to gain six new ones. When a calcium atom loses two electrons, its third shell, which is then its outer shell, is full and stable with eight electrons.

As previously noted, an atom's outermost electrons are called valence electrons, and thus the outermost shell of an atom is the *valence shell*. An atom's **valence,**[6] defined as its combining capacity, is considered to be positive if its valence shell has extra electrons to give up, and to be negative if its valence shell has spaces to fill. Thus a calcium atom, with two electrons in its valence shell, has a valence of +2, whereas an oxygen atom, with two spaces to fill in its valence shell, has a valence of −2.

Atoms combine with one another by either sharing or transferring valence electrons in such a way as to fill their valence shells. Such interactions between atoms are called **chemical bonds.** Two or more atoms held together by chemical bonds form a **molecule.** If a molecule contains atoms of more than one element, it is a **compound.** Two hydrogen atoms bonded together form a hydrogen molecule, which is not a compound because only one element is involved. However, two hydrogen atoms bonded to an oxygen atom form a molecule of water (H_2O), which is a compound.

[6]From Latin *valentia*, meaning strength.

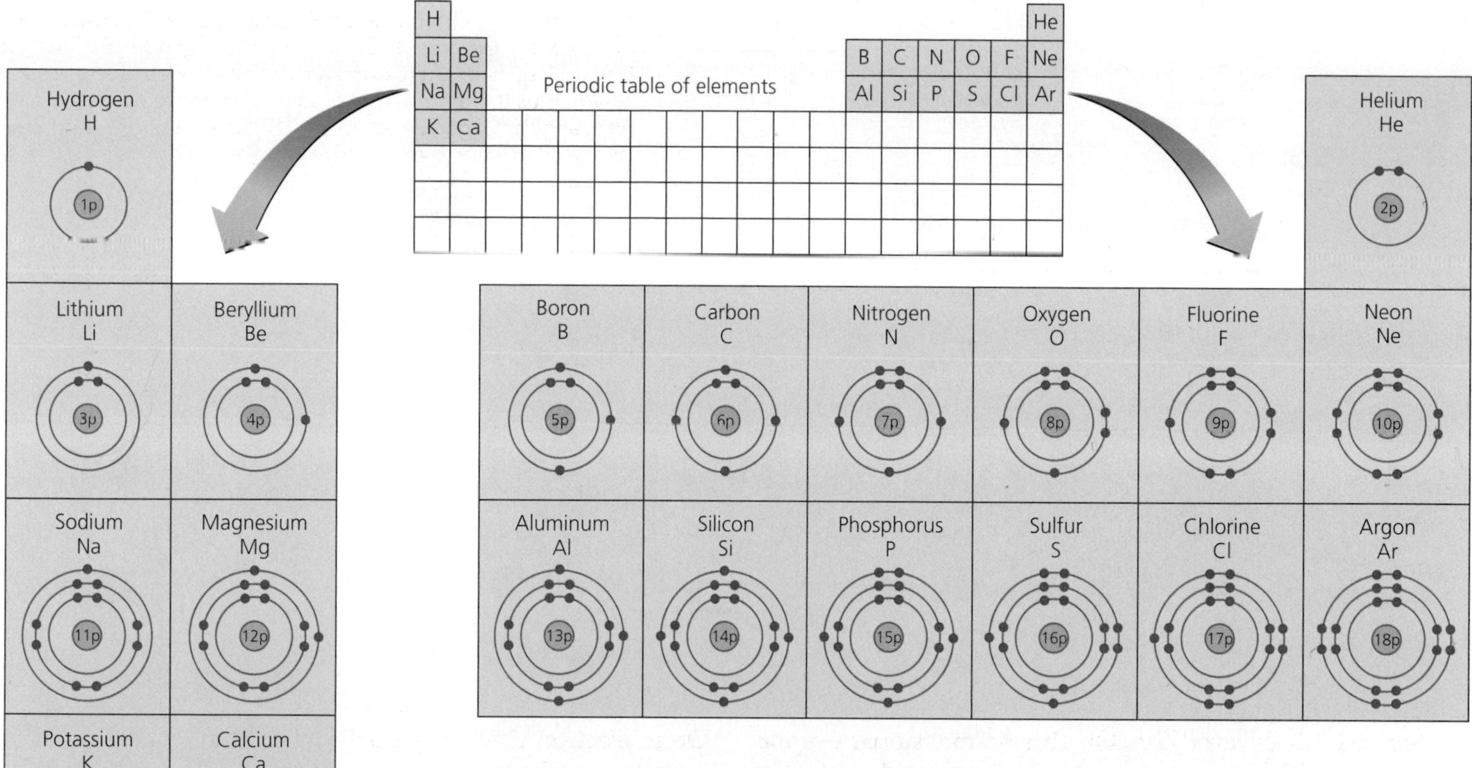

▲ Figure 2.4 Bohr diagrams of the first 20 elements, and their places within the chart known as the periodic table of the elements. Note that the number of valence electrons increases from left to right in each row, and that every element in a column has the same number of valence electrons (with the exception of helium). Heavier atoms have been omitted because most heavy elements are less important to living organisms. Neutrons are not shown because they have little effect on chemistry.

In this section, we discuss the three principal types of chemical bonds: *nonpolar covalent bonds, polar covalent bonds,* and *ionic bonds.* We also consider *hydrogen bonds,* which are weak forces that act with polar covalent bonds to give certain large chemicals their characteristic three-dimensional shapes.

CRITICAL **THINKING**

Neon (atomic mass 10) and argon (atomic mass 18) are *inert* elements, which means that they very rarely form chemical bonds. Give the electron configuration of their atoms and explain why these elements are inert.

Nonpolar Covalent Bonds

Learning Objective

✓ Contrast nonpolar covalent, polar covalent, and ionic bonds.

A **covalent**[7] **bond** is the sharing of a pair of electrons by two atoms. Consider, for example, what happens when two hydrogen atoms approach one another. Each hydrogen atom consists of a single proton orbited by a single electron. Since the valence shell of each atom requires two electrons to be filled, each atom shares its single electron with the other, forming a hydrogen molecule in which both atoms have full shells **(Figure 2.5a)**. Similarly, two oxygen atoms can share electrons, but they must share *two* pairs of electrons for their valence shells to be full **(Figure 2.5b)**. Since two pairs of electrons are involved, oxygen atoms form two covalent bonds, or a *double covalent bond,* with one another.

The attraction of an atom for electrons is called its **electronegativity.** The more electronegative an atom, the greater the pull its nucleus exerts on electrons. Note in **Figure 2.6**, which displays the electronegativities of atoms of several elements, that electronegativities tend to increase from left to right in the chart. The reason is that elements toward the right of the chart have more protons and thus exert a greater pull on electrons. Electronegativities of elements decrease from top to bottom in the chart because of the increasing distance between the nucleus and the valence shell as elements get larger.

Atoms with equal or nearly equal electronegativities, such as two hydrogen atoms or a hydrogen and a carbon, share electrons equally or nearly equally. In chemistry and physics, "poles" are opposed forces, such as north and south magnetic poles or positive and negative terminals of a battery. In the case of atoms with similar electronegativities, the shared electrons tend to spend an equal amount of time around each nucleus of the pair, and no poles exist; therefore, the bond between them is a **nonpolar covalent bond.** All the covalent bonds illustrated in Figure 2.5 are nonpolar.

[7]From Latin *co*, meaning with or together, and *valentia*, meaning strength.

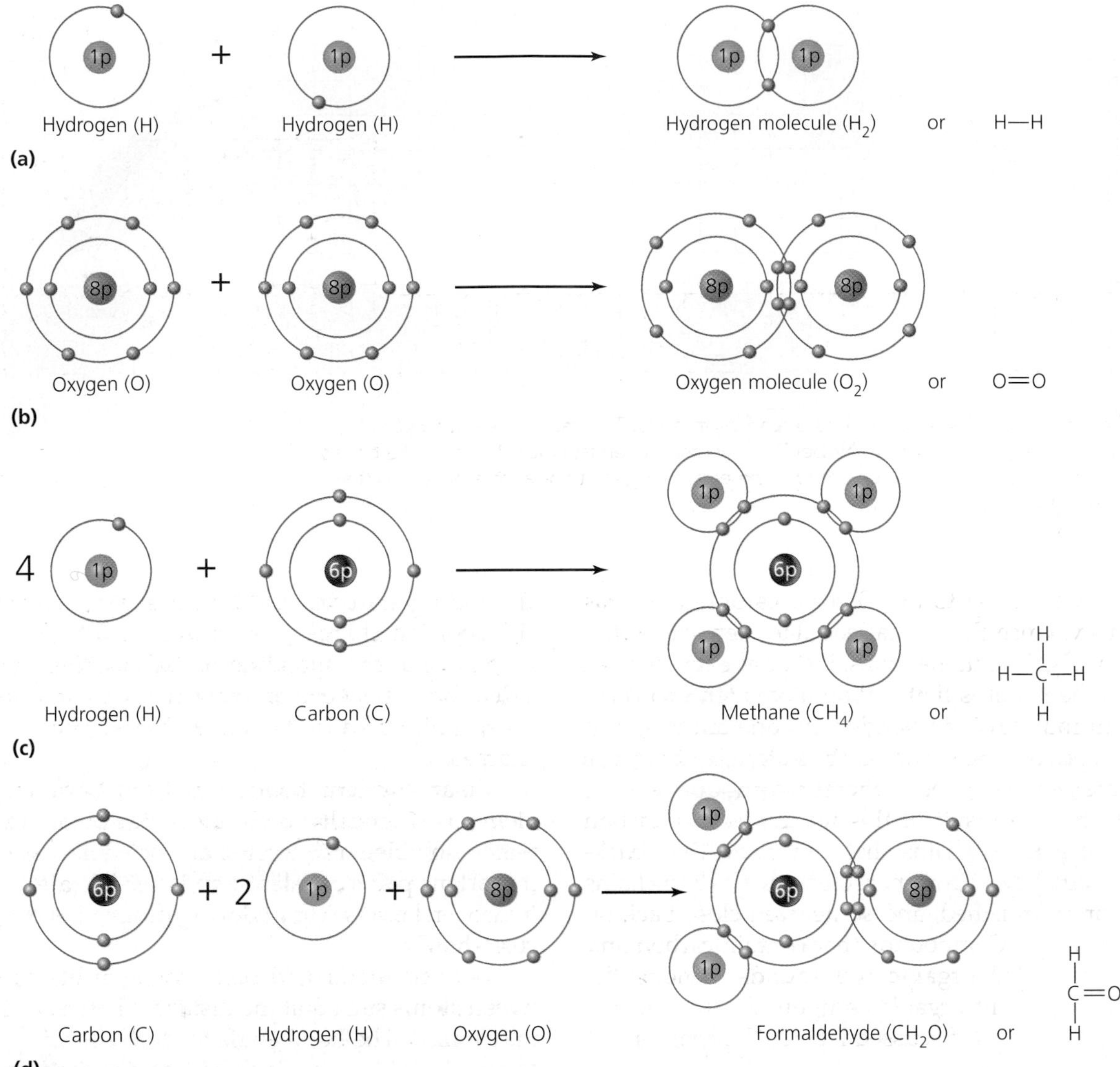

▲ **Figure 2.5 Four molecules formed by covalent bonds. (a)** Hydrogen. Each hydrogen atom needs another electron to have a full valence shell. The two atoms share their electrons, forming a covalent bond. **(b)** Oxygen. Oxygen atoms have six electrons in their valence shells; thus, they need two electrons each. When they share with each other, two covalent bonds are formed. Note that the valence electrons of oxygen atoms are in the second shell. **(c)** A methane molecule, which has four single covalent bonds. **(d)** Formaldehyde. The carbon atom forms a double bond with the oxygen atom and single bonds with two hydrogen atoms. *Which of these molecules are also compounds? Why?*

Figure 2.5 *Methane and formaldehyde molecules are also compounds because they are composed of more than one element.*

A hydrogen molecule can be symbolized a number of ways:

H—H H:H H_2

In the first symbol, the dash represents the chemical bond between the atoms. In the second symbol, the dots represent the electron pair of the covalent bond. These two symbols are known as *structural formulas.* In the third symbol, known as a *molecular formula,* the subscript "2" indicates the number of hydrogen atoms that are bonded, not the number of shared electrons. Each of these symbols indicates the same thing—two hydrogen atoms are sharing a pair of electrons.

Many atoms need more than one electron to fill their valence shell. For instance, a carbon atom has four valence electrons and needs to gain four more if it is to have eight in its valence shell. **Figure 2.5c** illustrates a carbon atom sharing with four hydrogen atoms. As before, a line in the structural formula represents a covalent bond formed from the sharing of two electrons. Two covalent bonds are formed between an oxygen atom and a carbon atom in formaldehyde **(Figure 2.5d)**. This fact is represented by a double line, which indicates that the carbon atom shares four electrons with the oxygen atom.

I	II											III	IV	V	VI	VII	Inert gases
H 2.1																	He 0.0
Li 1.0	Be 1.5											B 2.0	C 2.5	N 3.0	O 3.5	F 4.0	Ne 0.0
Na 0.9	Mg 1.2											Al 1.5	Si 1.8	P 2.1	S 2.5	Cl 3.0	Ar 0.0
K 0.8	Ca 1.0	Sc 1.3	Ti 1.5	V 1.6	Cr 1.6	Mn 1.5	Fe 1.8	Co 1.8	Ni 1.8	Cu 1.9	Zn 1.6	Ga 1.6	Ge 1.8	As 2.0	Se 2.4	Br 2.8	Kr 0.0

▲ **Figure 2.6 Electronegativity values of selected elements.** The values are expressed according to the Pauling scale, named for the Nobel Prize–winning chemist Linus Pauling, who based the scale on bond energies. Pauling chose to compare the electronegativity of each element to that of fluorine, to which he assigned a value of 4.0.

Carbon atoms are critical to life. Since a carbon atom has four electrons in its valence shell, it has equal tendency to either lose four electrons or gain four electrons. Either event produces a full outer shell. The result is that carbon atoms tend to share electrons and form four covalent bonds with one another, and with many other types of atoms. Each carbon atom in effect acts as a four-way intersection where different components of a molecule can attach. One result of this feature is that carbon atoms can form very large chains that constitute the "backbone" of many biologically important molecules. Carbon chains can be branched or unbranched, and some even close back on themselves to form rings. Compounds that contain carbon and hydrogen atoms are called **organic compounds.** Among the many biologically important organic compounds are proteins and carbohydrates, which are discussed later in the chapter.

CRITICAL **THINKING**

An article in the local newspaper about gangrene states that the tissue-destroying toxin, lecithinase, is "an organic compound. " But many people consider "organic" chemicals to mean something is good. Explain the apparent contradiction.

Polar Covalent Bonds

Learning Objective

✓ Explain the relationship between electronegativity and the polarity of a covalent bond.

If two covalently bound atoms have significantly different electronegativities, their electrons will not be shared equally. Instead, the electron pair will spend more time orbiting the nucleus of the atom with greater electronegativity. This type of bond, in which there is unequal sharing of electrons, is a **polar covalent bond.** An example of a molecule with polar covalent bonds is water **(Figure 2.7a)**.

Because oxygen is more electronegative than hydrogen, the electrons spend more time near the oxygen nucleus than near the hydrogen nuclei, and thus the oxygen atom acquires a partial negative charge (symbolized as δ^-). The hydrogen nuclei each have a corresponding partial positive charge (δ^+). The covalent bond between an oxygen atom and a hydrogen atom is called polar because the atoms have opposite partial electrical charges.

Polar covalent bonds can form between many different elements. Generally, molecules with polar covalent bonds are water soluble, and nonpolar molecules are not. The most important polar covalent bonds for life are those that involve hydrogen because they allow hydrogen bonding, which we discuss shortly.

Both nonpolar and polar covalent bonds form angles between atoms such that the distances between electron orbits are maximized. The bond angle for water is shown in **Figure 2.7b**. However, it is more convenient to simply draw molecules as if all the atoms were in one plane, as in **Figure 2.7c**.

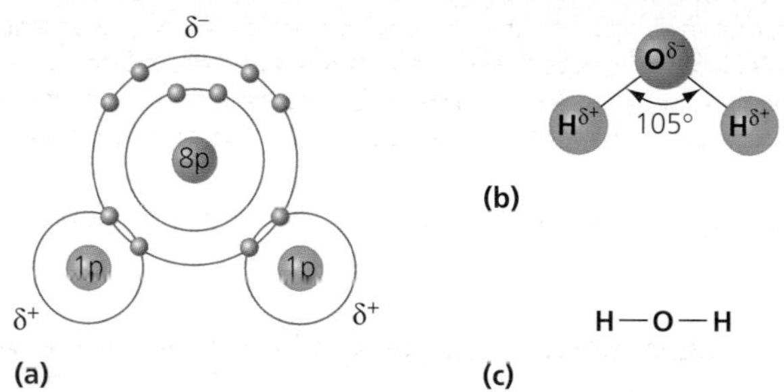

▲ **Figure 2.7 Polar covalent bonding in a water molecule.** **(a)** A Bohr model of a water molecule, which has two polar covalent bonds. When the electronegativities of two atoms are significantly different, the shared electrons of covalent bonds spend more time around the more electronegative atom, giving it a partial negative charge (δ^-). **(b)** The bond angle in a water molecule. Atoms maximize the distances between electron orbitals in polar and nonpolar covalent bonds. **(c)** A common, convenient way of representing water as a polar molecule without showing bond angles.

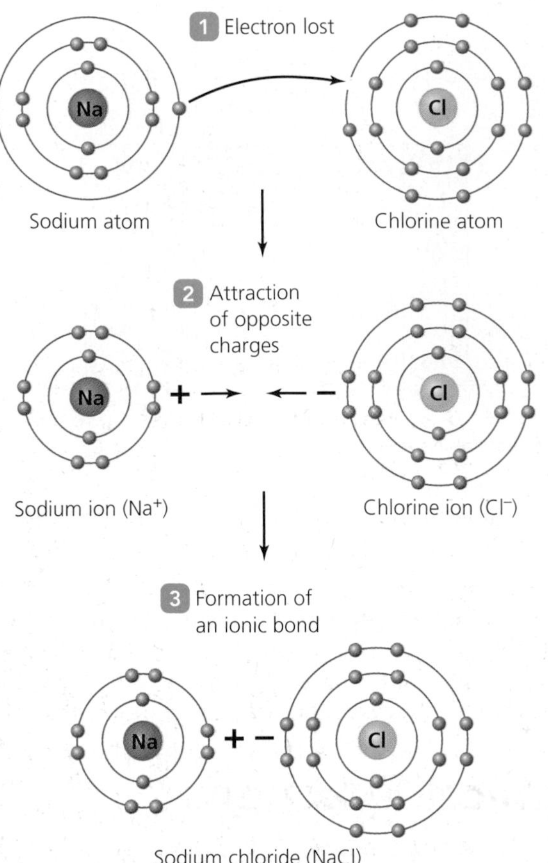

▲ **Figure 2.8 The interaction of sodium and chlorine to form an ionic bond.**

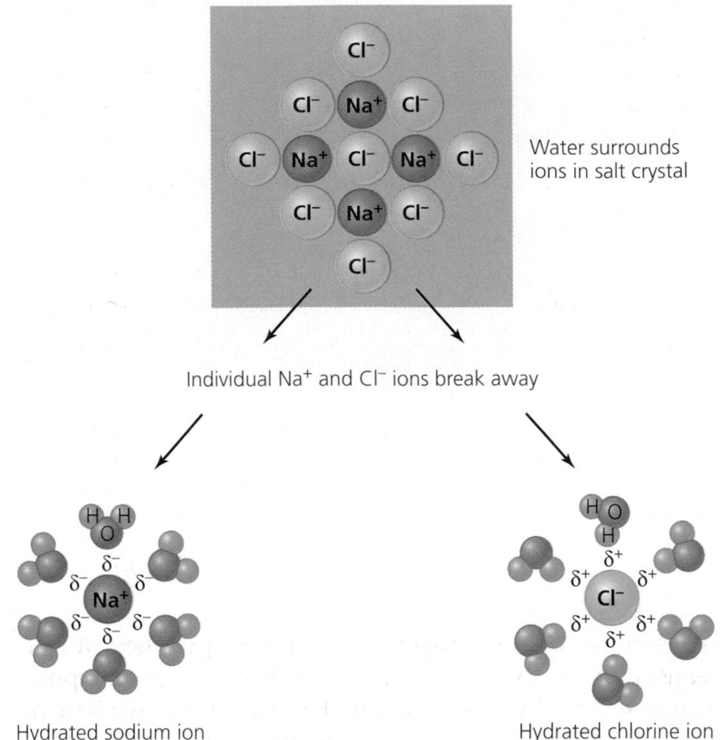

▲ **Figure 2.9 Dissociation of NaCl in water.** When water surrounds the ions in a NaCl crystal, the partial charges on water molecules are attracted to charged ions, and the water molecules hydrate the ions by surrounding them. The partial negative charges on oxygen atoms are attracted to cations (in this case, the sodium ions), and the partial positive charges on hydrogen atoms are attracted to anions (the chlorine ions). Because the ions no longer attract one another, the salt crystal dissolves. Hydrated ions are called electrolytes.

CRITICAL **THINKING**

The deadly poison hydrogen cyanide has the chemical formula H—C≡N. Describe the bonds between carbon and hydrogen, and between carbon and nitrogen, in terms of the number of electrons involved.

Triple covalent bonds are stronger and more difficult to break than single covalent bonds. Explain why by referring to the stability of a valence shell that contains eight electrons.

Ionic Bonds

Learning Objective

✓ Define *ionization* using the terms *cation* and *anion*.

Consider what happens when two atoms with vastly different electronegativities—for example, sodium, with one electron in its valence shell and an electronegativity of 0.9, and chlorine, with seven electrons in its valence shell and an electronegativity of 3.0—come together (Figure 2.8). Chlorine has such a higher electronegativity that it very strongly attracts sodium's valence electron, and the result is that the sodium loses that electron to chlorine (1).

Now that the chlorine atom has one more electron than it has protons, it has a full negative charge, and the sodium atom, which has lost an electron, now has a full positive charge (2).

An atom or group of atoms that has either a full negative charge or a full positive charge is called an *ion*. Positively charged ions are called **cations,** whereas negatively charged ions are called **anions.**

Because of their opposite charges, cations and anions attract each other and form what is termed an **ionic bond** (3). They form crystalline compounds composed of metallic and nonmetallic ions known as **salts,** such as sodium chloride (NaCl), also known as table salt, and potassium chloride (KCl, sodium-free table salt). Ionic bonds differ from covalent bonds in that ions do not share electrons. Instead, the bond is formed from the attraction of opposite electrical charges.

The polar bonds of water molecules interfere with the ionic bonds of salts, causing *dissociation* (also called *ionization*) (Figure 2.9). This occurs as the partial negative charge on the oxygen atom of water attracts cations, and the partial positive charge on hydrogen atoms attracts anions. Obviously, the presence of polar bonds interferes with the attraction between the cation and anion.

When cations and anions disassociate from one another and become surrounded by water molecules (are hydrated), they are called **electrolytes** because they can conduct electricity through the solution. Electrolytes are critical for life because they stabilize a variety of compounds, act as electron carriers, and allow electrical gradients to exist within cells. We examine these functions of electrolytes in later chapters.

In nature, chemical bonds range from nonpolar bonds to polar bonds to ionic bonds. The important thing to remember is that electrons are shared between atoms in covalent bonds and transferred from one atom to another in ionic bonds.

CRITICAL THINKING

According to the chart in Figure 2.6, what type of bond (nonpolar covalent, polar covalent, or ionic) would you expect between chlorine and potassium? Between carbon and nitrogen? Between phosphorus and oxygen? Explain your reasoning in each case.

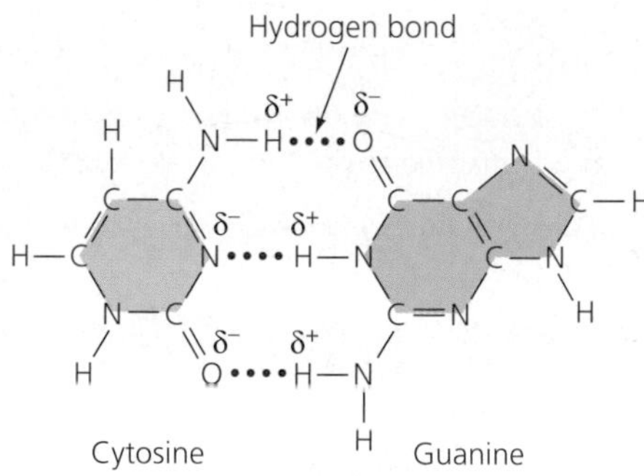

▲ **Figure 2.10 Hydrogen bonds.** Hydrogen bonds can hold together portions of the same molecule, or hold two different molecules together. In this case, three hydrogen bonds are holding molecules of cytosine and guanine together.

Hydrogen Bonds

Learning Objective

✓ Describe hydrogen bonds, and discuss their importance in living organisms.

As we have seen, hydrogen atoms bind to oxygen atoms by means of polar covalent bonds, which results in partial positive charges on the hydrogen atoms. Hydrogen atoms form polar covalent bonds with atoms of other elements as well.

The electrical attraction between a partially charged hydrogen atom and a full or partial negative charge on either a different region of the same molecule or another molecule is called a **hydrogen bond** (Figure 2.10). Hydrogen bonds can be likened to weak ionic bonds in that they arise from the attraction of positive and negative charges. Notice also that although they are a consequence of polar covalent bonds between hydrogen atoms and other, more electronegative atoms, hydrogen bonds themselves are not covalent bonds—they do not involve the sharing of electrons.

As we have seen, covalent bonds are essential for life because they strongly link atoms together to form molecules. Hydrogen bonds, though weaker than covalent bonds, are also essential. The cumulative effect of numerous hydrogen bonds is to stabilize the three-dimensional shapes of large molecules. For example, the familiar double-helix shape of DNA is due in part to the stabilizing effects of thousands of hydrogen bonds holding the molecule together. Exact shape is critical for the functioning of enzymes, antibodies, intercellular chemical messengers, and the recognition of target cells by pathogens. Further, because hydrogen bonds are weak, they can be overcome when necessary. For example, the two complementary halves of a DNA molecule are held together primarily by hydrogen bonds, and can be separated for DNA replication and other processes (see Figure 7.5).

Table 2.2 summarizes the characteristics of chemical bonds.

Chemical Reactions

Learning Objective

✓ Describe three general types of chemical reactions found in living things.

You are already familiar with many consequences of chemical reactions: you add yeast to bread dough and it rises; enzymes in your laundry detergent remove grass stains; and gasoline burned in your car releases energy to speed you on your way. What exactly is happening in these reactions? What is the precise definition of a chemical reaction?

We have discussed how bonds are formed via the sharing of electrons or the attraction of positive and negative charges. Scientists define **chemical reactions** as the making or breaking of such chemical bonds. All chemical reactions begin with **reactants**—the atoms, ions, or molecules that exist at the beginning of a reaction. Similarly, all chemical reactions result in

TABLE 2.2 Characteristics of Chemical Bonds

Type of Bond	Description	Relative Strength
Nonpolar covalent bond	Pair of electrons is nearly equally shared between two atoms	Strong
Polar covalent bond	Electrons spend more time around the more electronegative of two atoms	Strong
Ionic bond	Electrons are stripped from a cation by an anion	Weaker than covalent in aqueous environments
Hydrogen bond	Partial positive charges on hydrogen atoms are attracted to full and partial negative charges on other molecules or other regions of the same molecule	Weaker than ionic

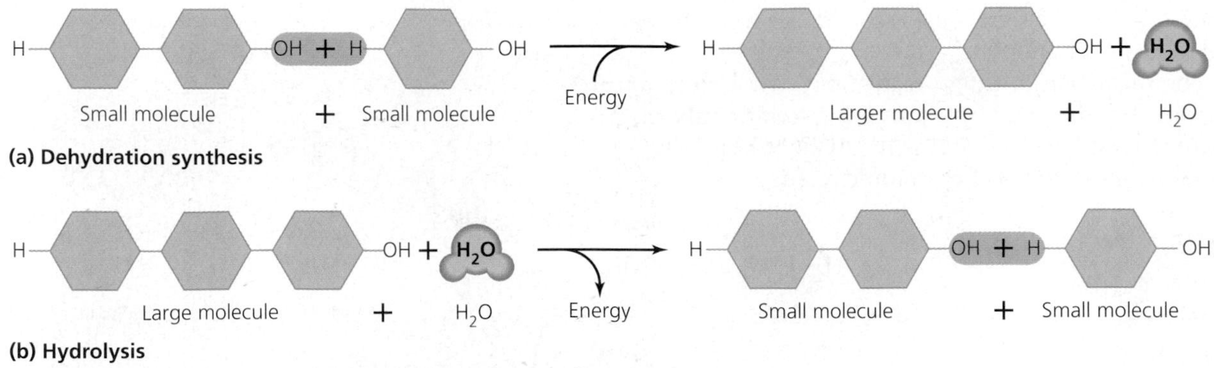

Small molecule + Small molecule → (Energy) Larger molecule + H_2O

(a) Dehydration synthesis

Large molecule + H_2O → (Energy) Small molecule + Small molecule

(b) Hydrolysis

▲ **Figure 2.11 Two types of chemical reactions in living things. (a)** Dehydration synthesis. In this energy-requiring reaction, a hydroxyl ion (OH^-) removed from one reactant and a hydrogen ion (H^+) removed from another reactant combine to form hydrogen hydroxide (HOH), which is water. **(b)** Hydrolysis, an energy-yielding reaction that is the reverse of a dehydration synthesis reaction. *What are the scientific words meaning "energy-requiring" and "energy-yielding"?*

Figure 2.11 Endothermic means "energy-requiring," and exothermic means "energy-releasing."

products—the atoms, ions, or molecules left after the reaction is complete. *Biochemistry* involves the chemical reactions of living things.

Reactants and products may have very different physical and chemical characteristics. For example, hydrogen and oxygen are gases and have very different properties from water, which is composed of hydrogen and oxygen atoms. However, the numbers and types of atoms never change in a chemical reaction; atoms are neither destroyed nor created, only rearranged.

Now let's turn our attention to three general categories of biochemical reactions (reactions that occur in organisms): *synthesis, decomposition,* and *exchange reactions.*

Synthesis Reactions

Learning Objectives

✓ Give an example of a synthesis reaction that involves the formation of a water molecule.

✓ Contrast endothermic and exothermic chemical reactions.

Synthesis reactions involve the formation of larger, more complex molecules. Synthesis reactions can be expressed symbolically as:

$$\text{Reactant} + \text{Reactant} \rightarrow \text{Product(s)}$$

The arrow indicates the direction of the reaction and the formation of new chemical bonds. For example, algae make their own glucose (sugar) using the following reaction:

$$6\,H_2O + 6\,CO_2 \rightarrow C_6H_{12}O_6 + 6\,O_2$$

The reaction is read, "Six molecules of water plus six molecules of carbon dioxide yield one molecule of glucose and six molecules of oxygen." Notice that the total number and kind of atoms are the same on both sides of the reaction.

Often a synthesis reaction in biochemistry is a **dehydration synthesis,** in which two smaller molecules are joined together by a covalent bond, and a water molecule is also formed (**Figure 2.11a**). The word *dehydration* in the name of this type of reaction refers to the fact that one of the products is a water molecule formed when a hydrogen ion (H^+) from one reactant combines with a hydroxyl ion (OH^-) from another reactant.

Synthesis reactions require energy to break bonds in the reactants and to form new bonds to make products. Reactions that require energy are said to be **endothermic**[8] **reactions** because they trap energy within new molecular bonds. As we will see in Chapter 6, an energy supply for fueling synthesis reactions is one common requirement of all living things.

Taken together, all of the synthesis reactions in an organism are called **anabolism.**

Decomposition Reactions

Learning Objective

✓ Give an example of a decomposition reaction that involves breaking the bonds of a water molecule.

Decomposition reactions are the reverse of synthesis reactions in that they break bonds within larger molecules to form smaller atoms, ions, and molecules. These reactions release energy and are therefore **exothermic.**[9] In general, decomposition reactions can be represented by the following formula:

$$\text{Reactant} \rightarrow \text{Product} + \text{Product}$$

An example of a biologically important decomposition reaction is the aerobic decomposition of glucose to form carbon dioxide and water:

$$C_6H_{12}O_6 + 6\,O_2 \rightarrow 6\,H_2O + 6\,CO_2$$

Note that this reaction is exactly the reverse of the synthesis reaction in algae that we examined previously. Synthesis and decomposition reactions are often reversible in living things.

[8]From Greek *endon*, meaning within, and *thermos*, meaning heat (energy).

[9]From Greek *exo*, meaning outside, and *thermos*, meaning heat (energy).

▲ **Figure 2.12 The cohesiveness of liquid water. (a)** Water molecules are cohesive because hydrogen bonds cause them to stick to one another. **(b)** One result of cohesiveness in water is surface tension, which can be strong enough to support the weight of insects known as water striders.

A common type of decomposition reaction in biochemistry is **hydrolysis,**[10] the reverse of dehydration synthesis (**Figure 2.11b**). In hydrolytic reactions, a covalent bond in a large molecule is broken, and the ionic components of water (H^+ and OH^-) are added to the products.

Collectively, all of the decomposition reactions in an organism are called **catabolism.**

Exchange Reactions

Learning Objective

✓ Compare exchange reactions to synthesis and decomposition reactions.

Exchange reactions (also called *transfer reactions*) have features similar to both synthesis and decomposition reactions. For instance, they involve breaking and forming covalent bonds, and they involve both endothermic and exothermic steps. As the name suggests, atoms are moved from one molecule to another. In general, these reactions can be represented as either

$$A + BC \rightarrow AB + C$$

or

$$AB + CD \rightarrow AD + BC$$

An important exchange reaction within organisms is the phosphorylation of glucose:

$$\underset{\text{Glucose}}{C_6H_{12}O_6} + \underset{\text{Adenosine triphosphate}}{A{-}Ⓟ{-}Ⓟ{-}Ⓟ} \rightarrow \underset{\text{Glucose phosphate}}{C_6H_{12}O_6{-}Ⓟ} + \underset{\text{Adenosine diphosphate}}{A{-}Ⓟ{-}Ⓟ}$$

The sum of all of the chemical reactions in an organism, including catabolic, anabolic, and exchange reactions, is called **metabolism.** We examine metabolism in more detail in Chapter 5.

[10]From Greek *hydor,* meaning water, and *lysis,* meaning loosing.

Water, Acids, Bases, and Salts

As previously noted, living things depend on organic compounds, those that contain carbon and hydrogen atoms. Living things also require a variety of **inorganic chemicals,** which typically lack carbon. Such inorganic substances include water, oxygen molecules, metal ions, and many acids, bases, and salts. In this section we examine the characteristics of some of these inorganic substances.

Water

Learning Objective

✓ Describe five qualities of water that make it vital to life.

Water is the most abundant substance in organisms, constituting 50–99% of their mass. Most of the special characteristics that make water vital result from the fact that a water molecule has two polar covalent bonds, which allows hydrogen bonding between water molecules and their neighbors. Among the special properties of water are the following:

- Water molecules are cohesive; that is, they tend to stick to one another through hydrogen bonding (**Figure 2.12**). This property generates many special characteristics of water, including *surface tension,* which allows water to form a thin layer on the surface of cells. This aqueous layer is necessary for the transport of dissolved materials into and out of a cell.
- Water is an excellent *solvent;* that is, it dissolves salts and other electrically charged molecules because it is attracted to both positive and negative charges (see Figure 2.9).
- Water remains a liquid across a wider range of temperatures than other molecules of its size. This is critical because living things require water in liquid form.
- Water can absorb significant amounts of heat energy without itself changing temperature. Further, when heated

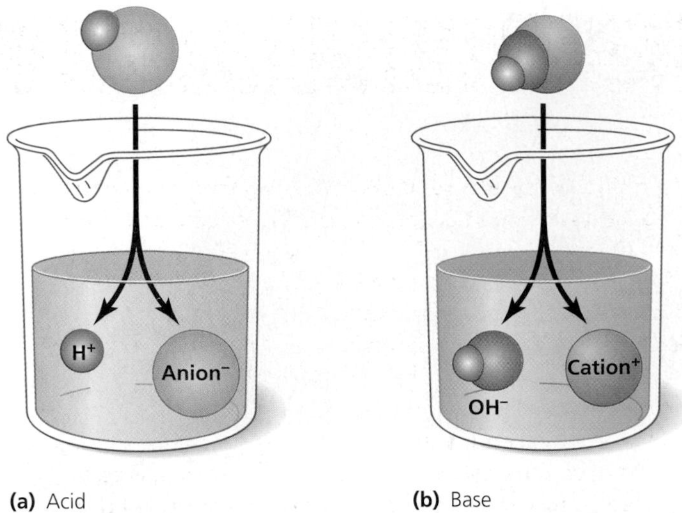

▲ **Figure 2.13 Acids and bases. (a)** Acids dissociate in water into hydrogen ions and anions. **(b)** Many bases dissociate into hydroxyl ions and cations.

water molecules eventually evaporate, they take much of this absorbed energy with them. These properties moderate temperature fluctuations that would otherwise damage organisms.

- Water molecules participate in many chemical reactions within cells, both as reactants in hydrolysis and as products of dehydration synthesis.

CRITICAL **THINKING**

How can hydrogen bonding between water molecules help explain water's ability to absorb large amounts of energy before evaporating?

Acids and Bases

Learning Objective

✓ Contrast acids, bases, and salts, and explain the role of buffers.

As we have seen, the polar bonds of water molecules dissociate salts into their component cations and anions. A similar process occurs with substances known as acids and bases.

An **acid** is a substance that dissociates into one or more hydrogen ions (H^+) and one or more anions **(Figure 2.13a)**. Acids can be inorganic molecules such as hydrochloric acid (HCl) and sulfuric acid (H_2SO_4), or organic molecules such as amino acids and nucleic acids. Familiar organic acids are found in lemon juice, black coffee, and tea. Of course, the anions of organic acids contain carbon, while those of inorganic acids do not.

A **base** is a molecule that binds with H^+ when dissolved in water. Some bases dissociate into cations and *hydroxyl ions* (OH^-) **(Figure 2.13b)**, which then combine with hydrogen ions to form water molecules:

$$H^+ + OH^- \rightarrow H_2O$$

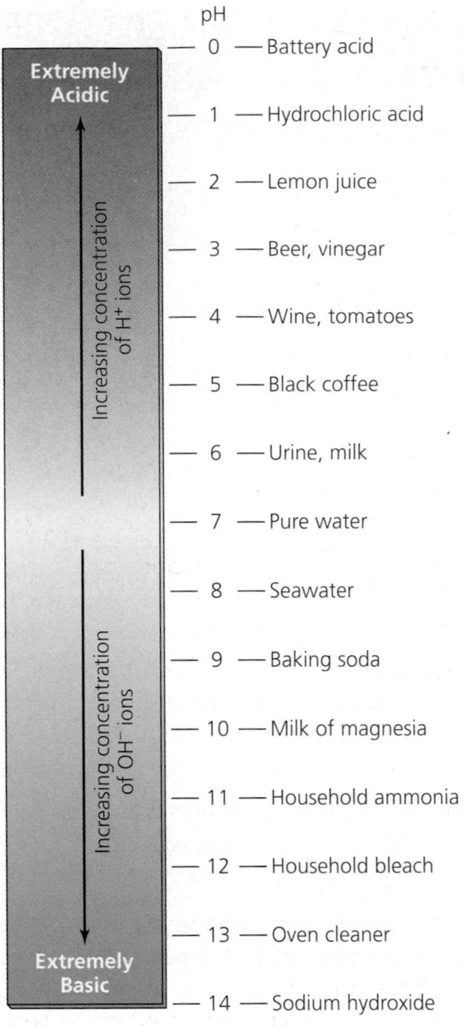

▲ **Figure 2.14 The pH scale.** Values below 7 are acidic; values above 7 are basic.

Other bases, such as household ammonia (NH_3), directly accept hydrogen ions and become compound ions such as NH_4^+ (ammonium). Another common household base is baking soda (sodium bicarbonate, $NaHCO_3$).

Metabolism requires a relatively constant balance of acids and bases because hydrogen ions and hydroxyl ions are involved in many chemical reactions. Further, many complex molecules such as proteins lose their functional shapes when acidity changes. If the concentration of either hydrogen ions or hydroxyl ions deviates too far from normal, metabolism ceases.

The concentration of hydrogen ions in a solution is expressed using a logarithmic **pH scale (Figure 2.14)**. The term *pH* comes from *potential hydrogen,* which is the negative of the logarithm of the concentration of hydrogen ions. In this logarithmic scale, it is important to notice that acidity increases as pH values decrease, and that each decrease by a whole number in pH indicates a 10-fold increase in acidity (hydrogen ion concentration). For example, a glass of grapefruit juice, which has a pH of 3.0, contains 10 times as many hydrogen ions as the same volume of tomato juice, which has a pH of 4.0. Similarly, tomato juice is 1000 times more acidic than pure water, which has a pH

BENEFICIAL MICROBES

ARCHITECTURE-PRESERVING BACTERIA

▲ *La Alhambra.*

La Alhambra, a Moorish palace constructed of limestone and marble beginning in the 9th century, was built to last. But not even stone lasts forever. Wind and rain wear away the surface. Acid rain reacts with the calcite crystals in limestone and marble. As years pass, stone slowly crumbles.

Those who would preserve La Alhambra and other historic structures face a dilemma. The microscopic pores that riddle limestone and marble make these materials particularly susceptible to weathering and decay. Sealing the stone's pores can reduce weathering but can also lock in moisture that speeds the stone's decay.

With the help of *Myxococcus xanthus*, a bacterium commonly found in soil, a team of researchers led by mineralogist Carlos Rodríguez-Navarro of the University of Granada may have found a way to protect the stone of structures like La Alhambra.

In many natural environments, bacteria instigate the formation of calcite crystals like the ones in limestone. In tests conducted using samples of the limestone commonly used in historic Spanish buildings, *Myxococcus xanthus* formed calcite crystals that lined the stone's pores, rather than plugging them. The crystals formed by the bacteria are even more durable than the original stone, offering the potential for long-term protection.

of 7.0 (neutral). Water is neutral because it dissociates into one hydrogen cation and one hydroxyl anion:

$$H_2O \rightarrow H^+ + OH^-$$

Alkaline (basic) substances have pH values greater than 7.0. They reduce the number of free hydrogen ions by combining with them. For bases that produce hydroxyl ions, the concentration of hydroxyl ions is inversely related to the concentration of hydrogen ions.

Organisms can tolerate only a certain, relatively narrow pH range. Fluctuations outside an organism's preferred range inhibit its metabolism and may even be fatal. Most organisms contain natural **buffers**—substances, such as proteins, that prevent drastic changes in internal pH. In a laboratory culture, the metabolic activity of microorganisms can change the pH of microbial growth solutions as nutrients are taken up and wastes are released; therefore, pH buffers are often added to them. One common buffer used in microbiological media is KH_2PO_4, which exists as either a weak acid or a weak base, depending on the pH of its environment. Under acidic conditions, KH_2PO_4 is a base that combines with H^+, neutralizing the acidic environment; in alkaline conditions, however, KH_2PO_4 acts as an acid, releasing hydrogen ions.

Microorganisms differ in their ability to tolerate various ranges of pH. Many grow best when the pH is between 6.5 and 8.5. Photosynthetic bacteria known as *cyanobacteria* grow well in more basic solutions. Fungi generally tolerate acidic environments better than most prokaryotes, though acid-loving prokaryotes, called *acidophiles,* require acidic conditions. One pathogenic bacterium that tolerates acidic conditions in the human body, allowing it to grow where other bacteria cannot, is *Propionibacterium acnes* (prō-pē-on-i-bak-tēr'ē-ŭm ak'nēz), a factor in acne in the skin, which normally has a pH of about 4.0. Another is *Helicobacter pylori* (hel'ĭ-kō-bak'ter pī'lō-rē),[11] a curved bacterium that has been shown to cause ulcers in the stomach, where pH can fall as low as 1.5 when acid is being actively secreted. **Clinical Case Study: Raw Oysters and Antacids: A Deadly Mix?** on p. 38 focuses on how the use of antacids may increase the survival rates of certain disease-causing bacteria in the stomach.

Microorganisms can change the pH of their environment by utilizing acids and bases and by producing acidic or basic wastes. For example, fermentative microorganisms form organic acids from the decomposition of sugar, and the bacterium *Thiobacillus* (thī-ō-bă-sil'ŭs) can reduce the pH of its environment to 0.0. Acid produced by this bacterium in mine water dissolves enough uranium and copper from low-grade ore to make some mines profitable.

Scientists measure pH with a pH meter or with test papers impregnated with chemicals (such as litmus or phenol red) that change color in response to pH. In a microbiological laboratory, changes in color of such pH indicators incorporated into microbial growth media are commonly used to distinguish among bacterial genera.

Salts

As we have seen, a salt is a compound that dissociates in water into cations and anions other than H^+ and OH^-. Acids and hydroxyl-yielding bases neutralize each other during exchange reactions that produce water and salt. For instance, milk of magnesia (magnesium hydroxide) is an antacid used to neutralize excess stomach acid. The chemical reaction is

[11]The name *pylori* refers to the pylorus, a region of the stomach.

CLINICAL CASE STUDY

Raw Oysters and Antacids: A Deadly Mix?

The highly acidic environment of the stomach kills most bacteria before they cause disease. One bacterium that can slightly tolerate conditions as it passes through the stomach is *Vibrio vulnificus*—a bacterium commonly ingested by eating raw tainted oysters. The bacterium cannot be seen, tasted, or smelled in food or water.

V. vulnificus is an emerging pathogen and a growing cause of food poisoning in the United States: it triggers vomiting, diarrhea, and abdominal pain. The pathogen can also infect the bloodstream, causing life-threatening illness characterized by fever, chills, skin lesions, and deadly loss of blood pressure. About 50% of patients with bloodstream infections die. *V. vulnificus* especially affects the immunocompromised and people with long-term liver disease.

Researchers have discovered that taking antacids may make people more susceptible to becoming ill from *V. vulnificus*. They discovered that antacids in a simulated gastric environment significantly increased the survival rate of *V. vulnificus*.

1. Why are patients who take antacids at greater risk for infections with *Vibrio vulnificus*?
2. Will antacids raise or lower the pH of the stomach?
3. Other than refraining from antacids, what can people do to reduce their risk of infection?

Reference: Adapted from *MMWR* 45:621–624. 1996.

$$\underset{\text{Magnesium hydroxide}}{Mg(OH)_2} + \underset{\text{Hydrochloric acid}}{2\,HCl} \rightarrow \underset{\text{Magnesium chloride (salt)}}{MgCl_2} + \underset{\text{Water}}{2\,H_2O}$$

Cations and anions of salts are electrolytes. A cell uses electrolytes to create electrical differences between its inside and outside, to transfer electrons from one location to another, and as important components of many enzymes. Certain organisms also use salts such as calcium carbonate ($CaCO_3$) to provide structure and support for their cells.

CRITICAL THINKING

How can a single molecule of magnesium hydroxide neutralize two molecules of hydrochloric acid?

Organic Macromolecules

Inorganic molecules play important roles in an organism's metabolism; however, water excluded, they compose only about 1.5% of its mass. Inorganic molecules are typically too small and too simple to constitute an organism's basic structures, or to perform the complicated chemical reactions required of life. These functions are fulfilled by organic molecules, which are generally larger and much more complex.

Functional Groups

Learning Objective

✓ Define *functional group* as it relates to organic chemistry.

As we have seen, organic molecules contain carbon and hydrogen atoms, and each carbon atom can form four covalent bonds with other atoms (see Figure 2.5c and d). Carbon atoms that are linked together in branched chains, unbranched chains, and rings provide the basic frameworks of organic molecules.

Atoms of other elements are bound to these carbon frameworks to form an unlimited number of compounds. Besides carbon and hydrogen, the most common elements in organic compounds are oxygen, nitrogen, phosphorus, and sulfur. Other elements, such as iron, copper, molybdenum, manganese, zinc, and iodine, are important in some proteins.

Atoms often appear in certain common arrangements called **functional groups.** For example, $-NH_2$, the amino functional group, is found in all amino acids, and —OH, the hydroxyl functional group,[12] is common to all alcohols. When a class of organic molecules is discussed, the letter **R** (for *residue*) designates atoms in the compound that vary from one molecule to another. The symbol R—OH, therefore, represents the general formula for an alcohol. Table 2.3 describes some common functional groups of organic molecules.

There is a great variety of organic compounds, but certain basic types are used by all organisms. These molecules—known as *macromolecules* because they are very large—are lipids, carbohydrates, proteins, and nucleic acids.

Lipids

Learning Objectives

✓ Describe the structure of a triglyceride molecule, and compare it to that of a phospholipid.

✓ Distinguish among saturated, unsaturated, and polyunsaturated fatty acids.

Lipids are a diverse group of organic macromolecules not composed of regular subunits. They have one common trait—they are **hydrophobic;**[13] that is, they are insoluble in water. Lipids have little or no affinity for water because they are composed almost entirely of carbon and hydrogen atoms linked by nonpolar covalent bonds. Because these bonds are nonpolar, they

[12]Note that the hydroxyl functional group is not the same thing as the hydroxyl *ion* because the former is covalently bonded to a carbon atom.

[13]From Greek *hydor*, meaning water, and *phobos*, meaning fear.

TABLE 2.3 **Functional Groups of Organic Molecules, and Some Classes of Compounds in Which They Are Found**

Structure	Name	Class of Compounds
—OH	Hydroxyl	Alcohol Monosaccharide Amino acid
R—CH_2—O—CH_2—R′	Ether	Disaccharide Polysaccharide
R—C(=O)—R′	Internal carbonyl—a carbon atom (in R group) on each side	Ketone Carbohydrate
R—C(=O)—H	Terminal carbonyl—a carbon atom (in R group) on only one side	Aldehyde
R—C(=O)—O—H	Carboxyl	Amino acid Nucleic acid
R—C(H)—NH_2	Amino	Amino acid Protein
R—C(=O)—O—R′	Ester	Fat Wax
R—CH_2—SH	Sulfhydryl	Amino acid Protein
R—CH_2—O—P(OH)(OH)=O	Organic phosphate	Phospholipid Nucleotide ATP

have no attraction to the polar bonds of water molecules. To look at it another way, the polar water molecules are attracted to each other and exclude the nonpolar lipid molecules. There are four major groups of lipids in cells: fats, phospholipids, waxes, and steroids.

Fats

Organisms make fats via dehydration synthesis reactions that form *esters* between three chainlike fatty acids and an alcohol named glycerol (**Figure 2.15a**). Fats are also called *triglycerides* because they contain three fatty acid molecules linked to a molecule of glycerol.

The three fatty acids in a fat molecule may be identical or different from one another, but each usually has 12 to 20 carbon atoms. An important difference among fatty acids is the presence and location of double bonds between the carbon atoms. When the carbon atoms are linked solely by single bonds, every carbon atom, with the exception of the terminal ones, is covalently linked to two hydrogen atoms. Such a fatty acid is **saturated** with hydrogen (**Figure 2.15b**). In contrast, **unsaturated fatty acids** contain at least one double bond between adjacent carbon atoms, and therefore contain at least one carbon atom bound to only a single hydrogen atom. If several double bonds exist in even one fatty acid of a molecule of fat, then it is a **polyunsaturated** fat.

Saturated fats (composed of saturated fatty acids), like those found in animals, are usually solid at room temperature because their fatty acids can be packed closely together. Unsaturated fatty acids, by contrast, are bent at every double bond, and so cannot be packed tightly; they remain liquid at room temperature. Most fats in plants are unsaturated or polyunsaturated. **Table 2.4** on p. 43 compares the structures and melting points of four common fatty acids.

Fats contain an abundance of energy stored in their carbon-carbon covalent bonds. Indeed, a major role of fats in organisms is to store energy. As we see in Chapter 5, fats can be catabolized to provide energy for movement, synthesis, and transport.

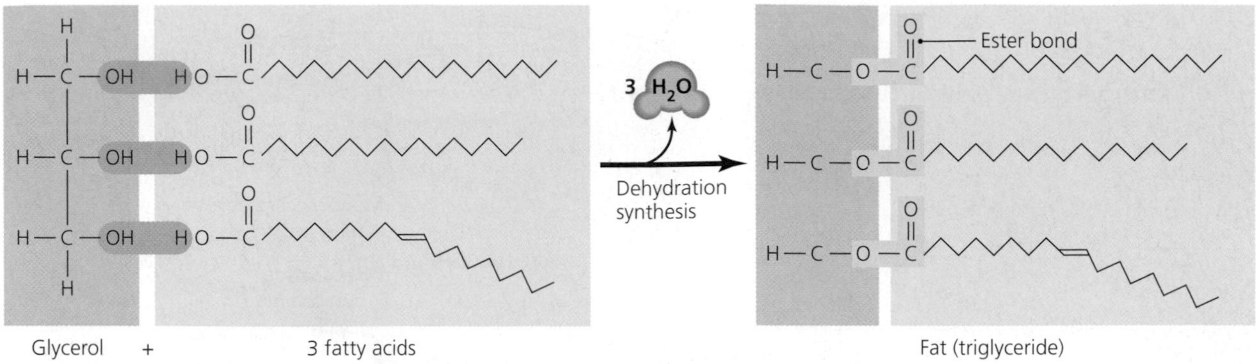

Glycerol + 3 fatty acids

$3\ H_2O$

Dehydration synthesis

Ester bond

Fat (triglyceride)

(a)

Saturated fatty acid

Monounsaturated fatty acid

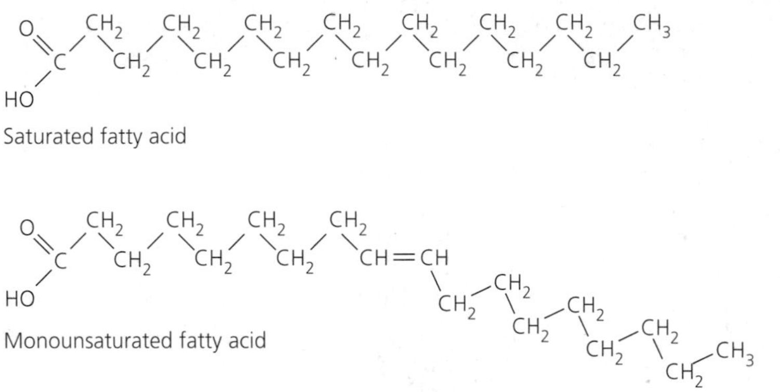

(b)

▲ **Figure 2.15 Fats (triglycerides). (a)** Fats are made in dehydration synthesis reactions that form ester bonds between a glycerol molecule and three fatty acids. **(b)** Saturated fatty acids have only single bonds between their carbon atoms, whereas unsaturated fatty acids have double bonds between carbon atoms. Scientists often use abbreviated diagrams of fatty acids in which each angle represents a carbon atom, and hydrogen atoms are omitted, as seen in part (a) of this figure. *According to Table 2.4 on p. 43, which fatty acids are shown in Figure 2.15a?*

Figure 2.15 *Stearic acid, palmitic acid, and oleic acid.*

Phospholipids

Phospholipids are similar to fats, but they contain only two fatty acid chains instead of three. In phospholipids, the third carbon atom of glycerol is linked to a phosphate (PO_4) functional group instead of to a fatty acid **(Figure 2.16a)**. Like fats, different phospholipids contain different fatty acids. Small organic groups linked to the phosphate group provide additional variety among phospholipid molecules.

The fatty acid "tail" portion of a phospholipid molecule is nonpolar and thus hydrophobic, whereas the phospholipid "head" is polar and is thus **hydrophilic**[14] **(Figure 2.16b)**. As a result, phospholipids placed in a watery environment will always self-assemble into forms that keep the fatty acid tails away from water. One way they do this is to form a spherical phospholipid bilayer, which resembles a two-ply ball **(Figure 2.16c)**.

The fatty acid tails, which are hydrophobic, congregate in the water-free interior of bilayers. The polar phosphate heads orient toward the water because they are hydrophilic. Phospholipid bilayers make up the outer membranes of all cells, as well as the internal membranes of plant, fungal, and animal cells.

Waxes

Waxes contain one long-chain fatty acid linked covalently to a long-chain alcohol by an ester bond. Waxes do not have a hydrophilic head; thus, they are completely water insoluble. Certain microorganisms, such as *Mycobacterium tuberculosis* (mī'kō-bak-tēr'ē-ŭm too-ber-kyū-lō'sis), are surrounded by a waxy wall, making them resistant to drying. Some marine microbes use waxes instead of fats as energy storage molecules.

Steroids

A final group of lipids are **steroids**. Steroids consist of four rings (each containing five or six carbon atoms) that are fused to one another and attached to various side chains and functional groups **(Figure 2.17a)**. Steroids play many roles in human metabolism. Some act as hormones; another steroid, *cholesterol,* is familiar to you as an undesirable component of food. However, cholesterol is also an essential part of the phospholipid bilayer membrane surrounding an animal cell. Cells of fungi, plants, and one group of bacteria (mycoplasmas) have similar sterol molecules in their membranes. Sterols, which are steroids with an —OH functional group, interfere with the tight packing of the fatty acid chains of phospholipids **(Figure 2.17b)**. This keeps the membranes fluid and flexible at low temperatures. Without steroids such as cholesterol, the membranes of cells would become stiff and inflexible in the cold.

CRITICAL **THINKING**

We have seen that it is important that biological membranes remain flexible. Most bacteria lack sterols in their membranes and instead incorporate unsaturated phospholipids in the membranes to resist tight packing and solidification. Reexamine Table 2.4 on p. 43. Which fatty acid might best protect the membranes of an ice-dwelling bacterium?

[14]From Greek *philos,* meaning love.

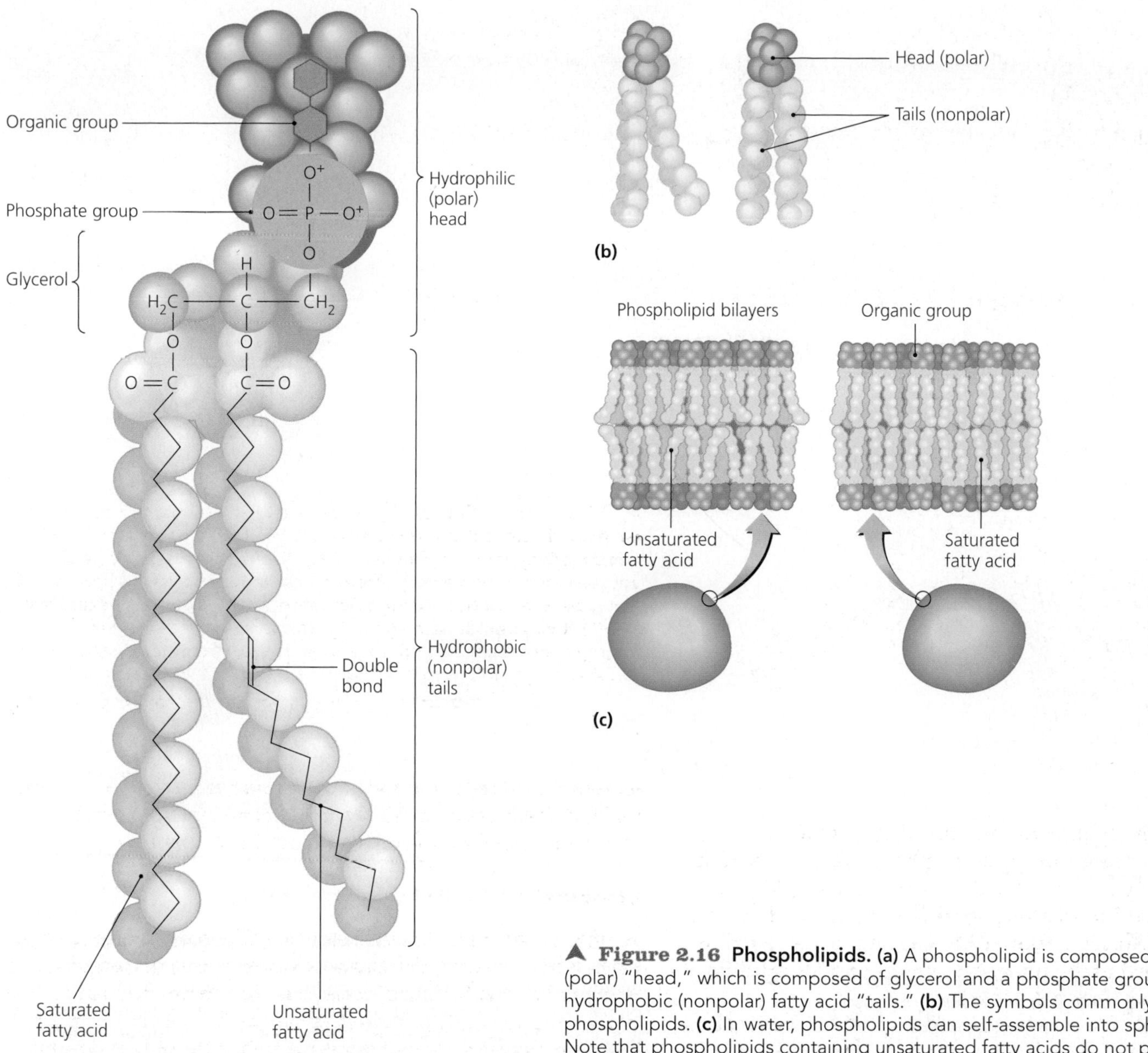

▲ **Figure 2.16 Phospholipids.** **(a)** A phospholipid is composed of a hydrophilic (polar) "head," which is composed of glycerol and a phosphate group, and two hydrophobic (nonpolar) fatty acid "tails." **(b)** The symbols commonly used to represent phospholipids. **(c)** In water, phospholipids can self-assemble into spherical bilayers. Note that phospholipids containing unsaturated fatty acids do not pack together as tightly as those containing saturated fatty acids.

Carbohydrates, proteins, and nucleic acid macromolecules are composed of simpler subunits known as **monomers,**[15] which are basic building blocks. The monomers of these macromolecules are joined together to form chains of monomers called **polymers.**[16] Some macromolecular polymers are composed of hundreds of thousands of monomers.

Carbohydrates

Learning Objective

✓ Discuss the roles of carbohydrates in living systems.

Carbohydrates are organic molecules composed solely of atoms of carbon, hydrogen, and oxygen. Most carbohydrate compounds contain an equal number of oxygen and carbon atoms, and twice as many hydrogen atoms as carbon atoms, so the general formula for a carbohydrate is $(CH_2O)_n$, where n indicates the number of CH_2O units.

Carbohydrates play many important roles in organisms. Large carbohydrates such as starch and glycogen are used for the long-term storage of chemical energy, while a smaller carbohydrate molecule—glucose—serves as a ready energy source in most cells. Carbohydrates also form part of the backbones of DNA and RNA, and other carbohydrates are converted routinely into amino acids. Additionally, polymers of carbohydrate form the cell walls of most fungi, plants, algae, and prokaryotes, and are involved in intercellular interactions between animal cells. For example, specific carbohydrate-protein combinations found on the surfaces of white blood cells determine which cells interact in immune responses against pathogens.

[15]From Greek *mono*, meaning one, and *meris*, meaning part.
[16]From Greek *poly*, meaning many, and *meris*, meaning part.

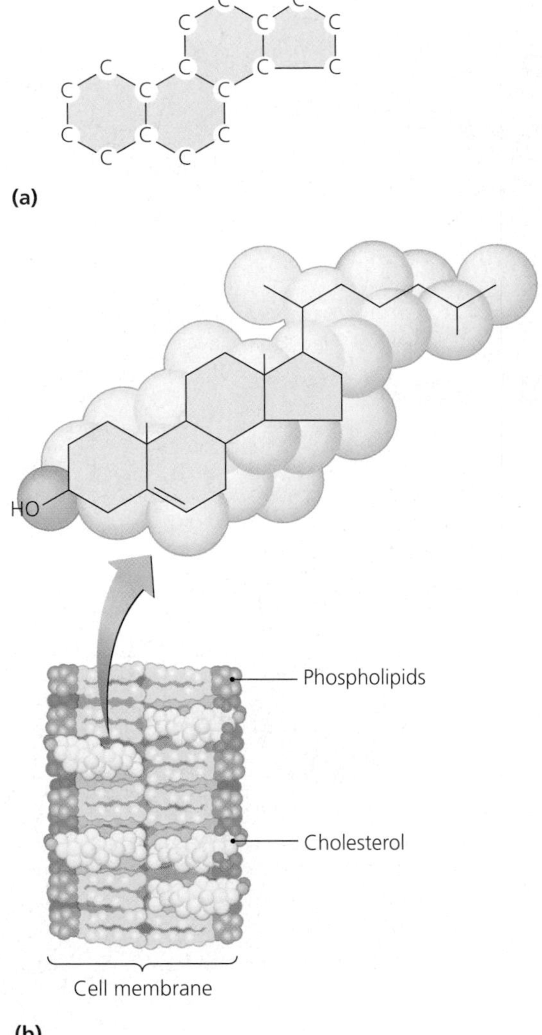

▲ **Figure 2.17 Steroids. (a)** Steroids are lipids characterized by four "fused" carbon rings. **(b)** The steroid cholesterol functions in animal and protozoan cell membranes to prevent packing of phospholipids, which keeps the membranes fluid at low temperatures.

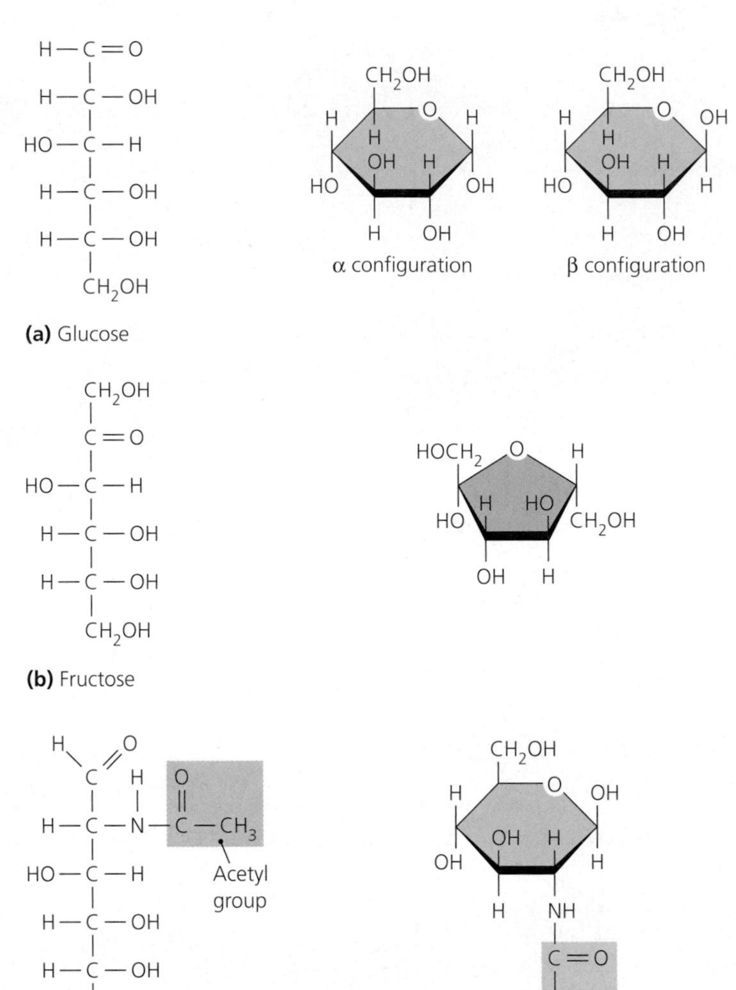

▲ **Figure 2.18 Monosaccharides (simple sugars).** Although simple sugars may exist as either linear molecules (at left) or rings (at right), energy dynamics generally favor ring forms. **(a)** Glucose, a hexose, is the primary energy source for cellular metabolism and an important monomer in many larger carbohydrates. Alpha and beta ring configurations exist. **(b)** Fructose, or fruit sugar, a hexose. **(c)** *N*-acetylglucosamine (NAG), a monomer in bacterial cell walls.

Monosaccharides

The simplest carbohydrates are **monosaccharides**[17]—simple sugars **(Figure 2.18)**. The general names for the classes of monosaccharides are formed from a prefix indicating the number of carbon atoms, and the suffix *-ose*. For example, pentoses are sugars with five carbon atoms, and *hexoses* are sugars with six carbon atoms. Pentoses and hexoses are particularly important in cellular metabolism. For example, deoxyribose, which is the sugar component of DNA, is a pentose. Glucose is a hexose and the primary energy molecule of cells, and fructose is a hexose found in fruit.

Monosaccharides may exist as linear molecules, but due to energy dynamics they usually take cyclic (ring) forms. In some cases, more than one cyclic structure may exist. For example, glucose can assume an alpha (α) configuration or a beta (β) configuration (see Figure 2.18a). As we will see, these configurations play important roles in the formation of different polymers.

Disaccharides

When two monosaccharide molecules are linked together via dehydration synthesis, the result is a **disaccharide.** For example, the linkage of two hexoses, glucose and fructose, forms sucrose (table sugar) and a molecule of water **(Figure 2.19a)**. Other disaccharides include maltose (malt sugar) and lactose (milk sugar). Disaccharides can be broken down via hydrolysis into their constituent monosaccharides **(Figure 2.19b)**.

[17] From Greek *sakcharon*, meaning sugar.

TABLE 2.4 Common Fatty Acids in Fats and Cell Membranes

Numbers of Carbon Atoms: Double Bonds	Type of Fatty Acid	Structure and Formula	Common Name	Melting Point
16:0	Saturated	$CH_3(CH_2)_{14}COOH$	Palmitic acid	63°C
18:0	Saturated	$CH_3(CH_2)_{16}COOH$	Stearic acid	70°C
18:1	Monounsaturated	$CH_3(CH_2)_7CH{=}CH(CH_2)_7COOH$	Oleic acid	16°C
18:2	Polyunsaturated	$CH_3(CH_2)_4(CH{=}CHCH_2)_2(CH_2)_6COOH$	Linoleic acid	−5°C

Hydrocarbon chain

Double bond

Carboxyl group

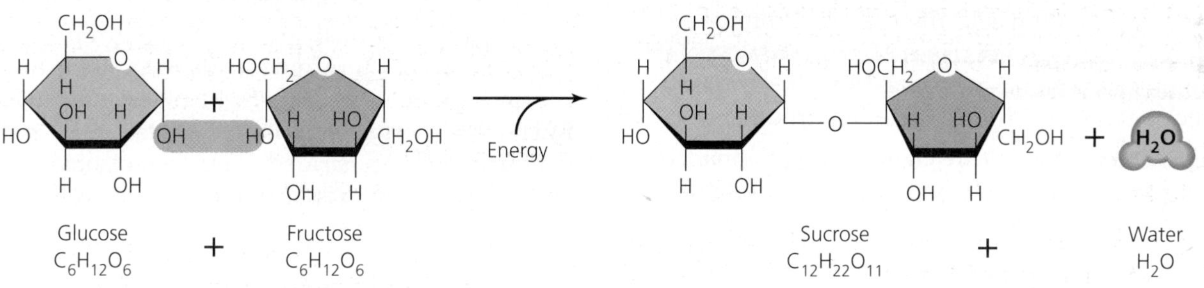

(a) Dehydration synthesis of sucrose

Sucrose $C_{12}H_{22}O_{11}$ + Water H_2O → (Energy) Glucose $C_6H_{12}O_6$ + Fructose $C_6H_{12}O_6$

(b) Hydrolysis of sucrose

▲ **Figure 2.19 Disaccharides. (a)** Formation of the disaccharide sucrose via dehydration synthesis. **(b)** Breakdown of sucrose via hydrolysis.

Polysaccharides

Polysaccharides are polymers composed of tens, hundreds, or thousands of monosaccharides that have been covalently linked in dehydration synthesis reactions. Even polysaccharides that contain only glucose monomers can be quite diverse because they can differ according to their monosaccharide monomer configurations (either alpha or beta) and their shapes (either branched or unbranched). Cellulose, the main constituent of the cell walls of plants and some green algae, is a long unbranched molecule that contains only β-monomers of glucose linked between carbons 1 and 4 of alternating monomers; such bonds are termed β-1,4 bonds (**Figure 2.20a**). Amylose, a starch storage compound in plants, has only α-1,4 bonds and is unbranched (**Figure 2.20b**); amylopectin, another plant starch, contains mostly α-1,4 monomers, but also has a few α-1,6 bonds and is branched (**Figure 2.20c**). Glycogen, a storage molecule formed in the liver and muscle cells of animals, is a highly branched molecule with both α-1,4 and α-1,6 bonds (**Figure 2.20d**).

The cell walls of bacteria are composed of *peptidoglycan*, which is made of polysaccharides and amino acids (see Figure 3.14). Polysaccharides may also be linked to lipids to form glycolipids, which can form cell markers such as those involved in the ABO blood typing system in humans.

Figure 2.20 Polysaccharides. All four polysaccharides shown here are composed solely of glucose but differ in the configuration of the glucose monomers and the amount of branching. **(a)** Cellulose, the major structural material in plants, is unbranched and contains only β-1,4 bonds. **(b)** Amylose is an unbranched plant starch with only α-1,4 bonds. **(c)** Amylopectin is a branched plant starch with mostly α-1,4 bonds but also a few α-1,6 branches. **(d)** Glycogen, a highly branched storage molecule in animals, is composed of glucose monomers linked by α-1,4 or α-1,6 bonds.

Proteins

Learning Objectives

- ✓ Describe five general functions of proteins in organisms.
- ✓ Sketch and label four levels of protein structure.

The most complex organic compounds are **proteins,** which are composed mostly of carbon, hydrogen, oxygen, nitrogen, and sulfur. Proteins perform many functions in cells, including:

- *Structure.* Proteins are structural components in cell walls, in membranes, and within cells themselves. Proteins are also the primary structural material of hair, nails, the outer cells of skin, muscle, and flagella and cilia (the last two act to move microorganisms through their environment).
- *Enzymatic catalysis.* Catalysts are chemicals that enhance the speed or likelihood of a chemical reaction. Protein catalysts in cells are called *enzymes.*
- *Regulation.* Some proteins regulate cell function by stimulating or hindering either the action of other proteins or the expression of genes. Hormones are examples of regulatory proteins.
- *Transportation.* As we will see in Chapter 3, certain proteins act as channels and "pumps" that move substances into or out of cells.
- *Defense and offense.* Antibodies and *complement* are examples of proteins that defend your body against microorganisms, and some bacteria produce proteins called *bacteriocins* that kill other bacteria.

A protein's function is dependent on its shape, which is determined by the molecular structures of its constituent parts.

Amino Acids

Proteins are polymers composed of monomers called **amino acids.** Amino acids contain a basic amino group ($—NH_2$), a hydrogen atom, and an acidic carboxyl group (—COOH). All attach to the same carbon atom, which is known as the α-carbon (**Figure 2.21**). A fourth bond attaches the α-carbon to a side group (—R) that varies among different amino acids. The side group may be a single hydrogen atom, various chains, or various complex ring structures. There are hundreds of amino acids, but most organisms use only 21 amino acids in the synthesis of proteins.[18] The different side groups affect the way amino acids interact with one another within a given protein, as well as how a protein interacts with other molecules. A change in an amino acid's side group may seriously interfere with a protein's normal function.

Because amino acids contain both an acidic carboxyl group and a basic amino group, they have both positive and negative charges and are easily soluble in water. Aqueous solutions of organic molecules such as amino acids and simple sugars bend light rays passing through the solution. Molecules known as D *forms*[19] bend light rays clockwise; other molecules bend light rays counterclockwise and are known as L *forms.*[20]

Many organic molecules exist as both D and L forms that are *stereoisomers* of one another; that is, they have the same atoms and functional groups but are mirror images of each other (**Figure 2.22**). Amino acids in proteins are almost always L forms—except for glycine, which does not have a stereoisomer. Interestingly, organisms almost always use D sugars in metabolism and polysaccharides. Rare stereoisomers—D amino acids and L sugars—do exist in some bacterial cell walls and in some antibiotics.

CRITICAL **THINKING**

Why isn't there a stereoisomer of glycine?

Peptide Bonds

Cells link amino acids together in chains that somewhat resemble beads on a necklace. By a dehydration synthesis reaction, a covalent bond is formed between the carbon of the carboxyl group of one amino acid and the nitrogen of the amino group

[18]The genes of a few prokaryotes code for 22 amino acids.
[19]From Latin *dexter*, meaning on the right.
[20]From Latin *laevus*, meaning on the left.

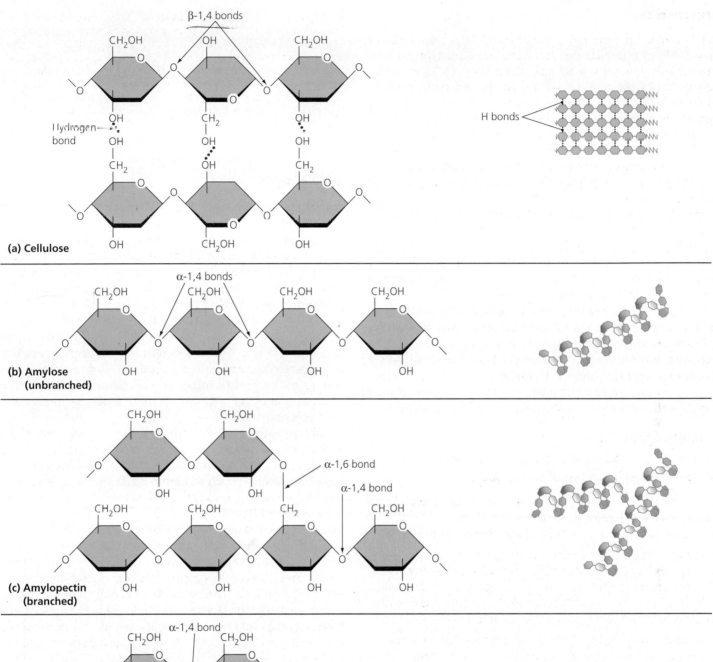

β-1,4 bonds
CH_2OH OH CH_2OH
Hydrogen bond
H bonds

(a) Cellulose

α-1,4 bonds

(b) Amylose (unbranched)

α-1,6 bond
α-1,4 bond

(c) Amylopectin (branched)

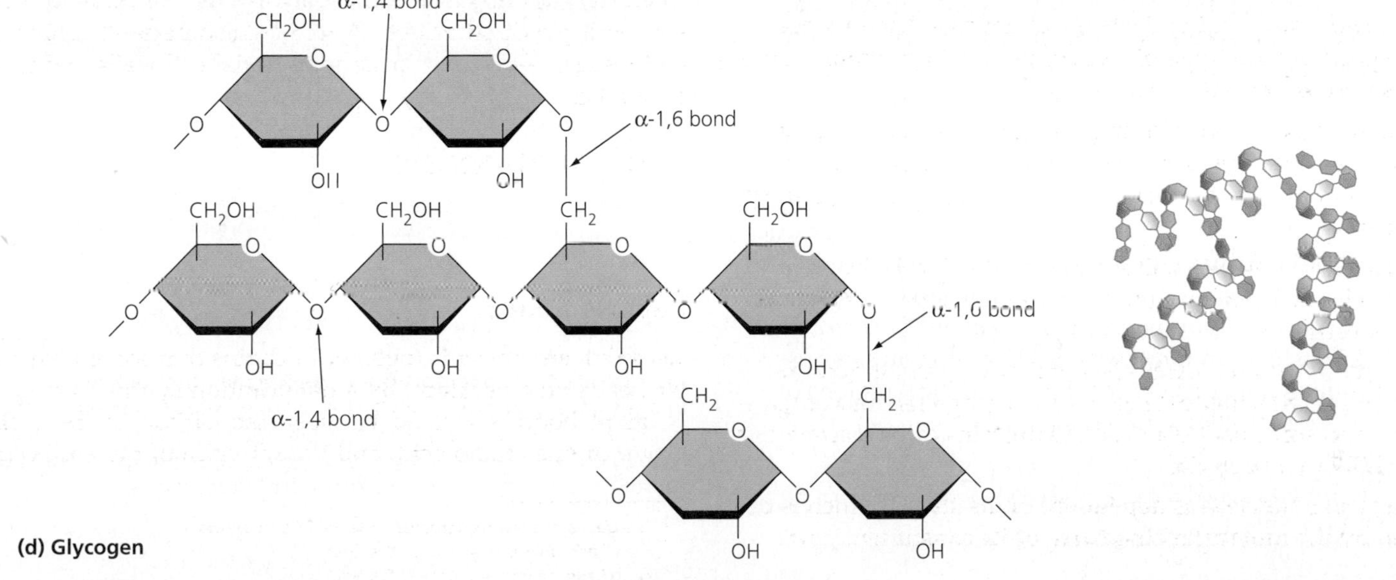

(d) Glycogen

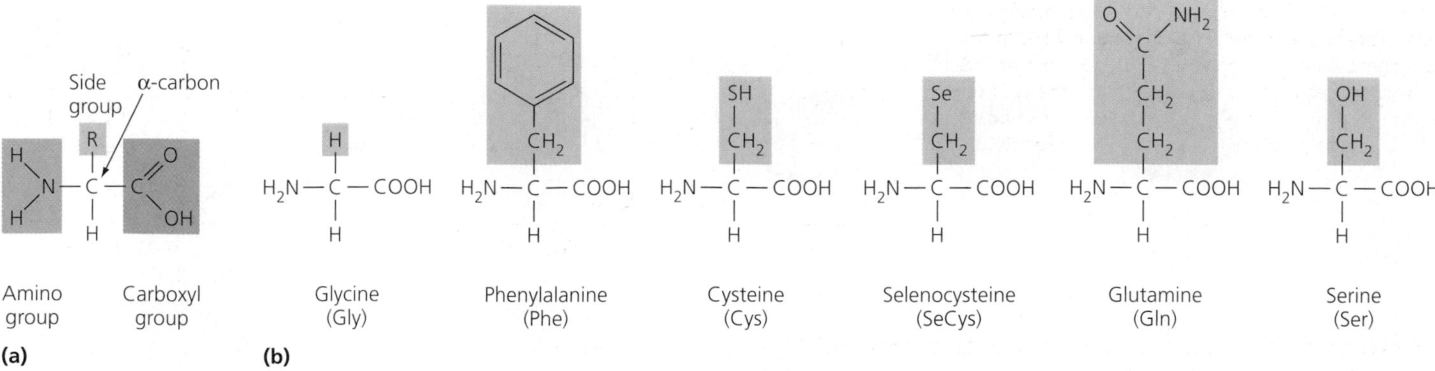

▲ **Figure 2.21 Amino acids.** **(a)** The basic structure of amino acids. The central α-carbon is attached to an amino group, a hydrogen atom, a carboxyl group, and a side group (—R group) that varies among amino acids. **(b)** Some selected amino acids, with their side groups highlighted. Note that each amino acid has a distinctive three-letter abbreviation.

of the next amino acid in the chain **(Figure 2.23)**. As we study in detail in Chapter 7, cells follow the organism's genetic instructions to link amino acids together in precise sequences.

Scientists refer to covalent bonds between amino acids by a special name: **peptide[21] bonds.** A molecule composed of two amino acids linked together by a single peptide bond is called a dipeptide; longer chains of amino acids are called *polypeptides.*

Protein Structure

Proteins are unbranched polypeptides composed of hundreds to thousands of amino acids linked together in specific patterns as determined by genes. The structure of a protein molecule is directly related to its function; therefore, an understanding of protein structure is critical to understanding certain specific chemical reactions, the action of antibiotics, and specific defense against pathogens. Every protein has at least three levels of structure, and some proteins have four levels.

- *Primary structure.* The primary structure of a protein is its sequence of amino acids **(Figure 2.24a)**. Cells use many different types of amino acids in proteins, though not every protein contains all types. The primary structures of proteins vary widely in length and amino acid sequence.

  A change in a single amino acid can drastically affect a protein's overall structure and function, though this is not always the case. For instance, the replacement of the amino acid valine by alanine in position 136 of the primary structure of a particular sheep brain protein, called cellular prion (prē′on) protein, may result in a disease called *scrapie.* The altered protein spread into cows, causing *mad cow disease,* and from cows into humans, causing *variant Creutzfeldt-Jakob* (kroytsfelt-yah-kŭp) *disease.*[22] However, numerous substitutions can be made in other, noncritical regions of cellular prion protein, with no ill effects.

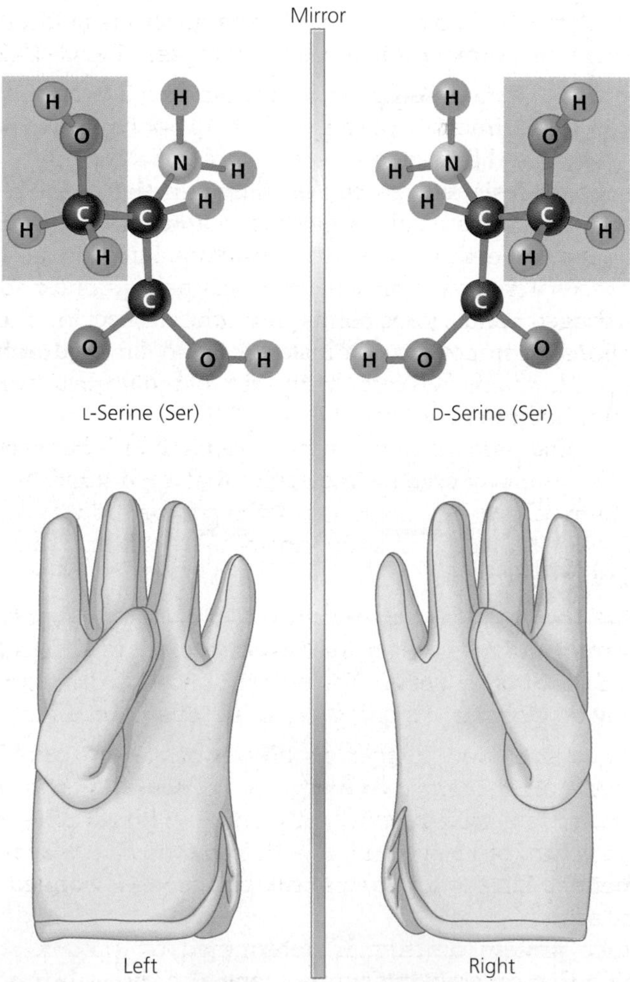

▲ **Figure 2.22 Stereoisomers, molecules that are mirror images of one another.** When dissolved in water, D isomers bend light clockwise, and L forms bend light counterclockwise. Just as a right-handed glove does not fit a left hand, so a D stereoisomer cannot be substituted for an L stereoisomer in metabolic reactions.

[21]From *peptone,* the name given to short chains of amino acids resulting from the partial digestion of protein.
[22]Named for the two German neurobiologists who first described the disease.

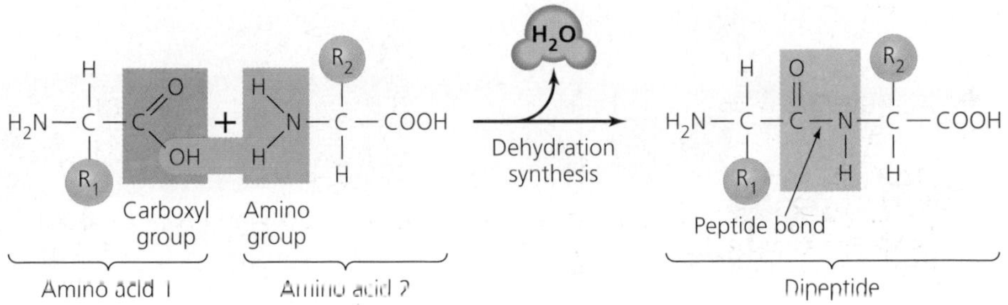

Figure 2.23 The linkage of amino acids by peptide bonds via a dehydration reaction. In this reaction, removal of a hydroxyl group from amino acid 1 and a hydrogen atom from amino acid 2 results in the production of a dipeptide—which is two amino acids linked by a single peptide bond—and a molecule of water.

- *Secondary structure.* Ionic bonds, hydrogen bonds, and hydrophobic and hydrophilic characteristics cause many polypeptide chains to fold into either coils called *α-helices,* or accordion-like structures called *β-pleated sheets* (**Figure 2.24b**). Proteins are typically composed of both α-helices and β-pleated sheets linked by short sequences of amino acids that do not show such secondary structure. Because of its primary structure, the protein that causes variant Creutzfeldt-Jakob disease has β-pleated sheets in locations where the normal protein has α-helices (see Figure 13.22).
- *Tertiary structure.* Polypeptides further fold into complex three-dimensional shapes that are not repetitive like α-helices and β-pleated sheets (**Figure 2.24c**), but are uniquely designed to accomplish the function of the protein. Scientists are only beginning to understand the interactions that determine tertiary structure, but it is clear that covalent bonds between —R groups of amino acids, hydrogen bonds, ionic bonds, and other molecular interactions are important. For instance, nonpolar (hydrophobic) side chains fold into the interior of molecules, away from the presence of water.

  Some proteins form strong covalent bonds between sulfur atoms of cysteine molecules that are brought into proximity by the folding of the polypeptide. These *disulfide bridges* are critical in maintaining tertiary structure of many proteins.
- *Quaternary structure.* Some proteins are composed of two or more polypeptide chains linked together by disulfide bridges or other bonds. The overall shape of such a protein may be globular (**Figure 2.24d**) or fibrous (threadlike).

Organisms may further modify proteins by combining them with other organic or inorganic molecules. For instance, *glycoproteins* are proteins covalently bound with carbohydrates, *lipoproteins* are proteins bonded with lipids, *metalloproteins* contain metallic ions, and *nucleoproteins* are proteins bonded with nucleic acids.

Since protein function is determined by protein shape, anything that severely interrupts shape also disrupts function. As we have seen, shape and function can be altered by amino acid substitution. Additionally, physical and chemical factors such as heat, changes in pH, and salt concentration can interfere with hydrogen and ionic bonding, which in turn disrupts the three-dimensional structure of proteins. This process is called **denaturation.** Denaturation can be temporary (if the denatured protein is able to return to its original shape again) or permanent.

Nucleic Acids

Learning Objectives

- ✓ Describe the basic structure of a nucleotide.
- ✓ Compare and contrast DNA and RNA.
- ✓ Contrast the structures of ATP, ADP, and AMP.

The nucleic acids **deoxyribonucleic acid (DNA)** and **ribonucleic acid (RNA)** are vital as the genetic material of cells and viruses. Moreover, RNA, acting as an enzyme, binds amino acids together to form polypeptides. DNA and RNA are both unbranched macromolecular polymers that differ primarily in the structures of their monomers, which we discuss next.

Nucleotides

Each monomer of nucleic acids is a **nucleotide** and consists of three parts (**Figure 2.25a**): (1) phosphate (PO_4^{3-}); (2) a pentose sugar, either deoxyribose or ribose (**Figure 2.25b**); and (3) one of five cyclic (ring-shaped) nitrogenous bases: **adenine (A), guanine (G), cytosine (C), thymine (T),** or **uracil (U)** (**Figure 2.25c**). Adenine and guanine are double-ringed molecules of a class called *purines,* whereas cytosine, thymine, and uracil have single rings and are *pyrimidines*. DNA contains A, G, C, and T bases, whereas RNA contains A, G, C, and U bases. As their names suggest, DNA nucleotides contain deoxyribose, and RNA nucleotides contain ribose. A nucleoside is a base attached to a sugar lacking phosphate.

Each nucleotide is also named for the base it contains. Thus a nucleotide made with ribose and uracil is a uracil RNA nucleotide (or uracil *ribonucleotide*). Likewise, a nucleotide composed of adenine and deoxyribose is an adenine DNA nucleotide (or adenine *deoxyribonucleotide*).

CRITICAL **THINKING**

A textbook states that only five nucleotide bases are found in cells, but a laboratory worker reports that she has isolated eight different nucleotides. Explain why both are correct.

Nucleic Acid Structure

Nucleic acids, like polysaccharides and proteins, are polymers. They are composed of nucleotides linked by covalent bonds between the phosphate of one nucleotide and the sugar of the next. Polymerization results in a linear spine composed of alternating sugars and phosphates, with bases extending from it

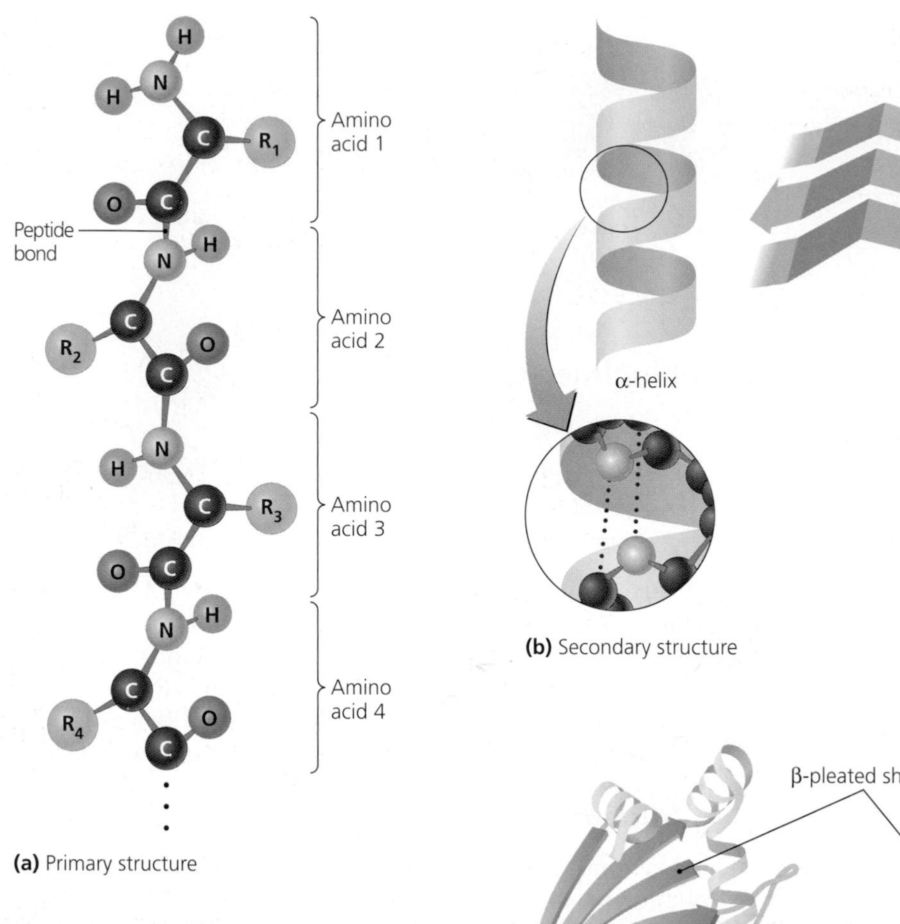

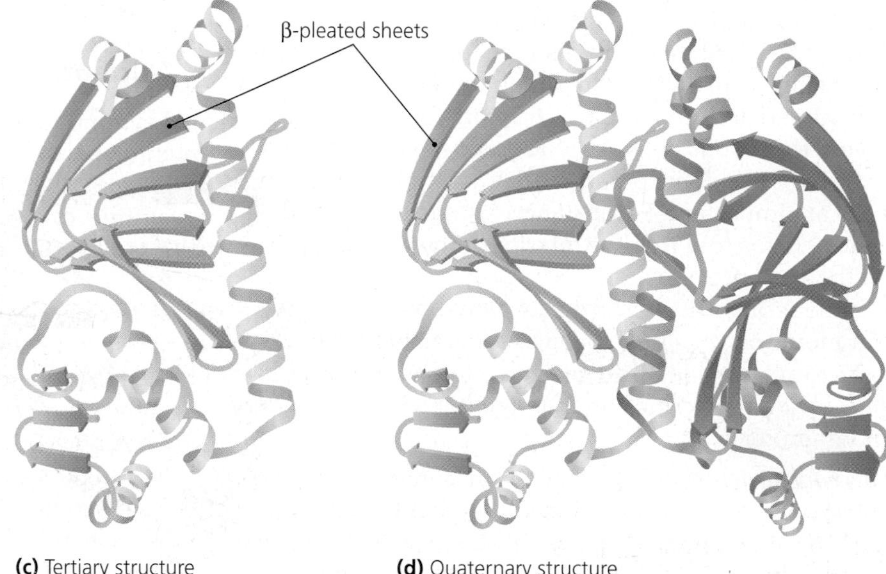

▲ **Figure 2.24 Levels of protein structure.** **(a)** A protein's primary structure is the sequence of amino acids in a polypeptide. **(b)** Secondary structure arises as a result of interactions, such as hydrogen bonding, between regions of the polypeptide. Secondary structure takes two basic shapes: α-helices and β-pleated sheets. **(c)** The more complex tertiary structure is a three-dimensional shape defined by further hydrogen bonding as well as covalent bonding between sulfur atoms of adjoining cysteine. **(d)** Those proteins that are composed of more than one polypeptide chain have a quaternary structure.

rather like the teeth of a comb **(Figure 2.26a)**. The two ends of a chain of nucleotides are different. At one end, called the 5′ end[23] (five prime end), carbon 5′ of the sugar is attached to a phosphate group. At the other end (3′ end), carbon 3′ of the sugar is not attached to a phosphate group.

The atoms of the bases in nucleotides are arranged in such a manner that hydrogen bonds readily form between specific bases of two adjacent nucleic acid chains. Three hydrogen bonds form between an adjacent pair composed of cytosine (C) and guanine (G), whereas two hydrogen bonds form between an adjacent pair composed of adenine (A) and thymine (T) in DNA **(Figure 2.26b)**, or between an adjacent pair composed of adenine (A) and uracil (U) in RNA. Hydrogen bonds do not readily form between other combinations of nucleotide bases; for example, adenine does not readily pair with cytosine, guanine, or another adenine nucleotide.

In cells and most viruses that use DNA as a genome, DNA molecules are double stranded. The two strands of DNA are complementary to one another; that is, the specificity of nucleotide base pairing ensures that opposite strands are composed of complementary nucleotides. For instance, if one strand has the sequence AATGCT, then its complement has TTACGA.

The two strands are also *antiparallel*; that is, they run in opposite directions. One strand runs from the 3′ end to the 5′ end,

[23]Carbon atoms in organic molecules are commonly identified by numbers. In a nucleotide, carbon atoms 1, 2, 3, etc. belong to the base, and carbon atoms 1′, 2′, 3′, etc. belong to the sugar.

Figure 2.25 Nucleotides. (a) The basic structure of nucleotides, each of which is composed of a phosphate, a pentose sugar, and a nitrogenous base. **(b)** The pentose sugars deoxyribose, which is found in deoxyribonucleic acid (DNA), and ribose, which is found in ribonucleic acid (RNA). **(c)** The nitrogenous bases, which are either the double-ringed purines adenosine or guanine, or the single-ringed pyrimidines thymine, cytosine, or uracil.

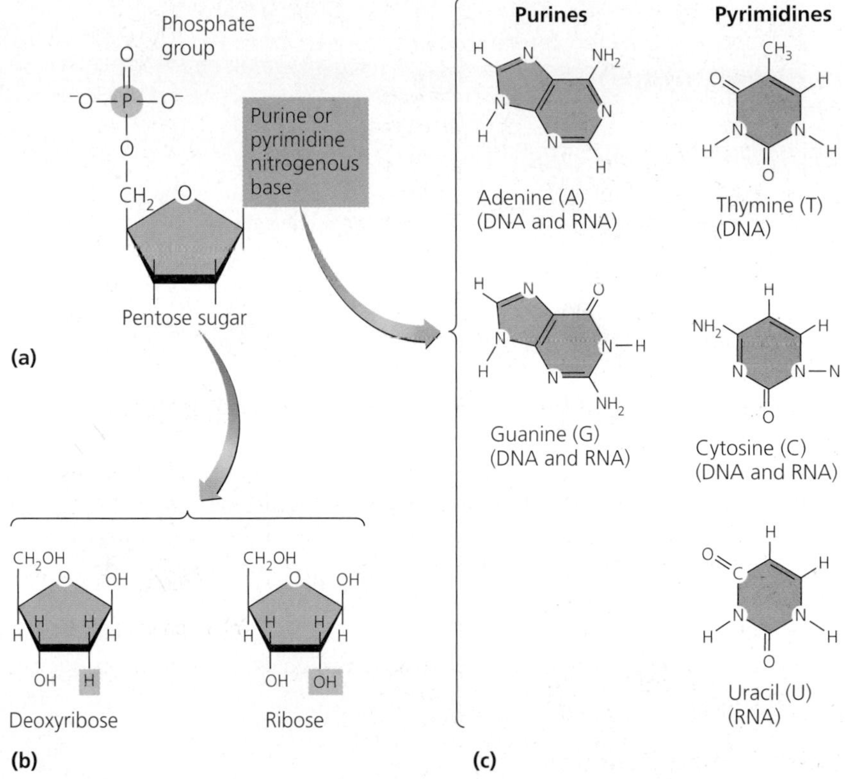

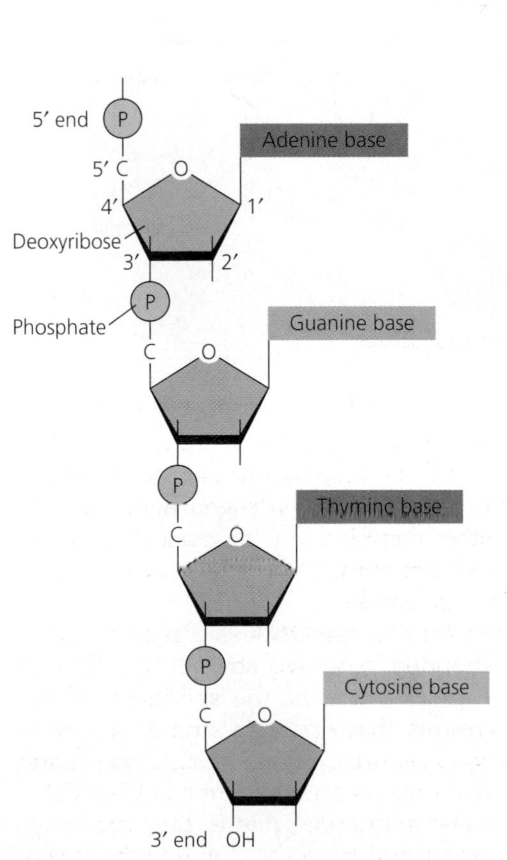

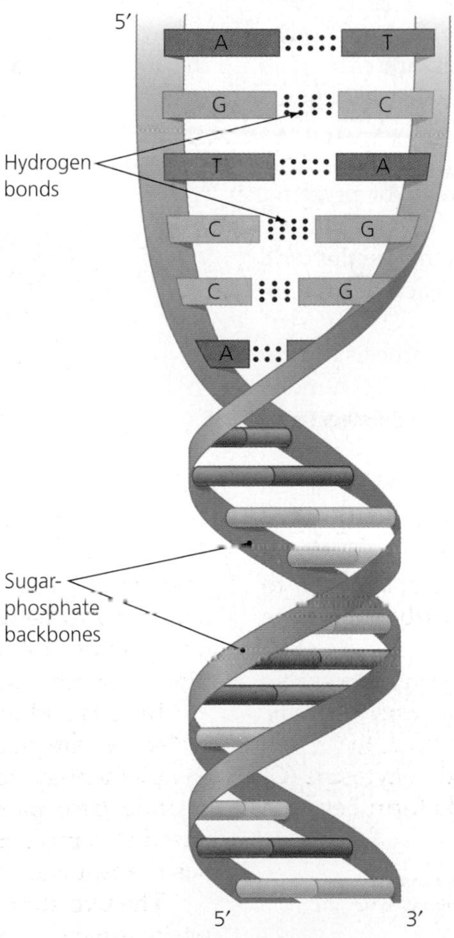

Figure 2.26 General nucleic acid structure. (a) Nucleotides are polymerized to form chains in which the nitrogenous bases extend from a sugar-phosphate backbone like the teeth of a comb. **(b)** Specific pairs of nitrogenous bases form hydrogen bonds between adjacent nucleotide chains to form the familiar DNA double helix. *How can you determine that the molecule in (a) is DNA and not RNA?*

Figure 2.26 *It is DNA because its nucleotides have deoxyribose sugar and because some of them have thymine bases (not uracil, as in RNA).*

TABLE 2.5 Comparison of Nucleic Acids

Characteristic	DNA	RNA
Sugar	Deoxyribose	Ribose
Purine nucleotides	A and G	A and G
Pyrimidine nucleotides	T and C	U and C
Number of strands	Double stranded in cells and in most DNA viruses; single stranded in parvoviruses	Single stranded in cells and in most RNA viruses; double stranded in reoviruses
Function	Genetic material of all cells and DNA viruses	Protein synthesis in all cells; genetic material of RNA viruses

while its complement runs in the opposite direction from its 5′ end to its 3′ end. Though hydrogen bonds are relatively weak bonds, thousands of them exist at normal temperatures, forming a stable, double-stranded DNA molecule, which looks much like a ladder: the two deoxyribose-phosphate chains are the side rails, and base pairs form the rungs. Hydrogen bonding also twists the phosphate-deoxyribose backbones into a helix (see Figure 2.26b). Thus, typical DNA is a double helix. Parvoviruses use single-stranded DNA, which is an exception to this rule.

Nucleic Acid Function

DNA is the genetic material of all organisms and of many viruses; it carries instructions for the synthesis of RNA molecules and proteins. By controlling the synthesis of enzymes and regulatory proteins, DNA controls the synthesis of all other molecules in an organism. Genetic instructions are carried in the sequence of nucleotides that make up the nucleic acid. Even though only four kinds of bases are found in DNA (A, T, G, and C), they can be sequenced in distinctive patterns that create genetic diversity and code for an infinite number of proteins, just as an alphabet of only four letters could spell a very large number of words. Cells replicate their DNA molecules and pass copies to their descendants, ensuring that each has the instructions necessary for life.

Ribonucleic acids play several roles in the synthesis of proteins, including catalyzing the synthesis of proteins. RNA molecules also function as structural components of ribosomes, and in place of DNA as the genome of RNA viruses.

Table 2.5 compares and contrasts RNA and DNA. We examine the synthesis and function of DNA and RNA in detail in Chapter 7.

ATP

Phosphate in nucleotides and other molecules is a highly reactive functional group and can form covalent bonds with other phosphate groups to make diphosphate and triphosphate molecules. Such molecules made from ribose nucleotides are important in many metabolic reactions. The names of these molecules indicate the nucleotide base and the number of phosphate groups they contain. Thus, cells make adenosine monophosphate (AMP) from the nitrogenous base adenine, ribose sugar, and one phosphate group; adenosine diphosphate (ADP), which has two phosphate groups; and **adenosine triphosphate** (ă-den′ō-sēn trī-fos′fāt) or **ATP**, which has three phosphate groups (**Figure 2.27**).

ATP is the principal, short-term, recyclable energy supply for cells. When the phosphate bonds of ATP are broken, a significant amount of energy is released; in fact, more energy is released from phosphate bonds than is released from most other covalent bonds. For this reason, the phosphate-phosphate bonds of ATP are known as *high-energy bonds*, and to show these specialized bonds, ATP is often symbolized as

$$A-Ⓟ\sim Ⓟ\sim Ⓟ$$

Energy is released when ATP is converted to ADP, and when phosphate is removed from ADP to form AMP, though the latter reaction is not as common in cells. Energy released

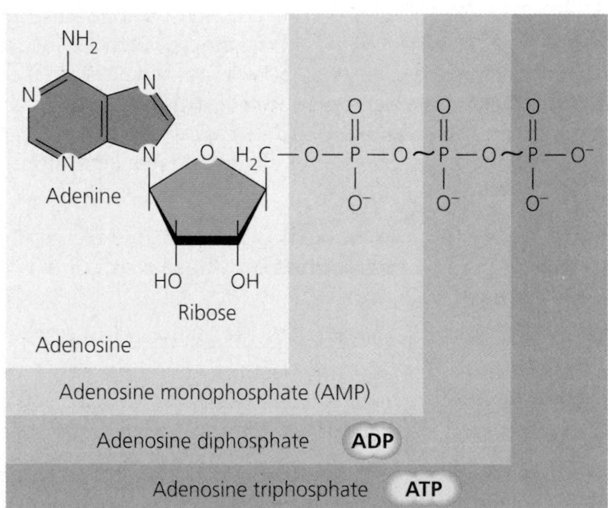

▲ **Figure 2.27 ATP.** Adenosine triphosphate (ATP), the main short-term, recyclable energy supply for cells. Energy is stored in high-energy bonds between the phosphate groups. *What is the relationship between AMP and adenine ribonucleotide?*

Figure 2.27 AMP and adenine ribonucleotide are two names for the same thing.

from the phosphate bonds of ATP is used for important life-sustaining activities, such as synthesis reactions, locomotion, and transportation of substances into and out of cells.

Cells also use ATP as a structural molecule in the formation of *coenzymes.* As we will see in Chapter 5, coenzymes such as *flavin adenine dinucleotide, nicotinamide nucleotide,* and *coenzyme A* function in many metabolic reactions.

A cell's supply of ATP is limited; therefore, an important part of cellular metabolism is replenishing ATP stores. We discuss the important ATP-generating reactions in Chapter 5.

Chapter Summary

Atoms (pp. 26–28)

1. **Matter** is anything that takes up space and has mass. Its smallest chemical units, **atoms,** contain negatively charged **electrons** orbiting a nucleus composed of uncharged **neutrons** and positively charged **protons.**
2. An **element** is matter composed of a single type of atom.
3. The number of protons in the nucleus of an atom is its **atomic number.** The sum of the masses of its protons, neutrons, and electrons is an atom's **atomic mass,** which is estimated by adding the number of neutrons and protons (because electrons have little mass).
4. **Isotopes** are atoms of an element that differ only in the numbers of neutrons they contain.

Chemical Bonds (pp. 28–33)

1. The region of space occupied by electrons is an electron shell. The number of electrons in the outermost or **valence shell** of an atom determines the atom's reactivity. Most valence shells hold a maximum of eight electrons. Sharing or transferring valence electrons to fill a valence shell results in **chemical bonds.**
2. A chemical bond results when two atoms share a pair of electrons. The **electronegativities** of each of the atoms, which is the strength of their attraction for electrons, determines whether the bond between them will be a **nonpolar covalent bond** (equal sharing of electrons), a **polar covalent bond** (unequal sharing of electrons), or an **ionic bond** (giving up of electrons from one atom to another).
3. If a **molecule** contains atoms of more than one element, it is a **compound. Organic compounds** are those that contain carbon and hydrogen atoms.
4. An **anion** is an atom with an extra electron and thus a negative charge. A **cation** has lost an electron and thus has a positive charge. Ionic bonding between the two types of ions makes **salt.** When salts dissolve in water, their ions are called **electrolytes.**
5. **Hydrogen bonds** are relatively weak but important chemical bonds. They hold molecules in specific shapes and confer unique properties to water molecules.

Chemical Reactions (pp. 33–35)

1. **Chemical reactions** result from the making or breaking of chemical bonds in a process in which **reactants** are changed into **products.** Biochemistry involves chemical reactions of life.
2. **Synthesis reactions** form larger, more complex molecules. In **dehydration synthesis,** a molecule of water is removed from the reactants as the larger molecule is formed. **Endothermic reactions** require energy. **Anabolism** is the sum of all synthesis reactions in an organism.
3. **Decomposition reactions** break larger molecules into smaller molecules and are **exothermic** because they release energy. **Hydrolysis** is a decomposition reaction that uses water as one of the reactants. The sum of all decomposition reactions in an organism is called **catabolism.**
4. **Exchange reactions** involve exchanging atoms between reactants.
5. **Metabolism** is the sum of all anabolic, catabolic, and exchange chemical reactions in an organism.

Water, Acids, Bases, and Salts (pp. 35–38)

1. **Inorganic** molecules typically lack carbon.
2. Water is a vital inorganic compound because of its properties as a solvent, its liquidity, its great capacity to absorb heat, and its participation in chemical reactions.
3. **Acids** release hydrogen ions. **Bases** release hydroxyl anions. The relative strength of each is assessed on a logarithmic **pH scale,** which measures the hydrogen ion concentration in a substance.
4. **Buffers** are substances that prevent drastic changes in pH.

Organic Macromolecules (pp. 38–51)

1. Certain groups of atoms in common arrangements, called **functional groups,** are found in organic macromolecules. **Monomers** are simple subunits that can be covalently linked to form chainlike **polymers.**
2. **Lipids,** which include fats, phospholipids, waxes, and steroids, are **hydrophobic** (insoluble in water) macromolecules.
3. **Fat** molecules are formed from a glycerol and three chainlike fatty acids. **Saturated fatty acids** contain more hydrogen in their structural formulas than **unsaturated fatty acids,** which contain double bonds between some carbon atoms. If several double bonds exist in the fatty acids of a molecule of fat, it is a **polyunsaturated** fat.
4. **Phospholipids** contain two fatty acid chains and a phosphate functional group. The phospholipid head is **hydrophilic,** whereas the fatty acid portion of the molecule is hydrophobic.

5. **Waxes** contain a long-chain fatty acid covalently linked to a long-chain alcohol. Waxes, which are water insoluble, are components of cell walls and are sometimes used as energy storage molecules.
6. **Steroid** lipids such as cholesterol help maintain the structural integrity of membranes as temperature fluctuates.
7. **Carbohydrates** such as **monosaccharides, disaccharides,** and **polysaccharides** serve as energy sources, structural molecules, and recognition sites during intercellular interactions.
8. **Proteins** are structural components of cells, enzymatic catalysts, regulators of various activities, molecules involved in the transportation of substances, and defensive molecules. They are composed of **amino acids** linked by **peptide bonds,** and they possess primary, secondary, tertiary, and (sometimes) quaternary structures that affect their function. **Denaturation** of a protein disrupts its structure and subsequently its function.
9. **Deoxyribonucleic acid (DNA)** and **ribonucleic acid (RNA)** are unbranched macromolecular polymers of **nucleotides,** each composed either of deoxyribose or ribose sugar, ionized phosphate, and a nitrogenous base. Five different bases exist: **adenine, guanine, cytosine, thymine,** and **uracil.** DNA contains A, G, C, and T nucleotides. RNA uses U nucleotides instead of T nucleotides.
10. The structure of nucleic acids allows for genetic diversity, correct copying of genes for their passage on to the next generation, and the accurate synthesis of proteins.
11. **Adenosine triphosphate (ATP),** which is related to adenine nucleotide, is the most important short-term energy storage molecule in cells. It is also incorporated into the structure of many coenzymes.

Questions for Review

Answers to the Questions for Review (except Short Answer questions) begin on page A-1.

Multiple Choice

1. Which of the following structures have no electrical charge?
   a. protons
   b. neutrons
   c. electrons
   d. ions
2. The atomic mass of an atom most closely approximates the sum of the masses of all its
   a. protons.
   b. electrons.
   c. isotopes.
   d. protons and neutrons.
3. One isotope of iodine differs from another in
   a. the number of protons.
   b. the number of electrons.
   c. the number of neutrons.
   d. atomic number.
4. Which of the following is *not* an organic compound?
   a. monosaccharide
   b. formaldehyde
   c. water
   d. steroid
5. Which of the following terms most correctly describes the bonds in a molecule of water?
   a. nonpolar covalent bond
   b. polar covalent bond
   c. ionic bond
   d. hydrogen bond
6. In water, cations and anions of salts disassociate from one another and become surrounded by water molecules. In this state, the ions are also called
   a. electrically negative.
   b. ionically bonded.
   c. electrolytes.
   d. hydrogen bonds.
7. Which of the following can be most accurately described as a decomposition reaction?
   a. $C_6H_{12}O_6 + 6\ O_2 \rightarrow 6\ H_2O + 6\ CO_2$
   b. glucose + ATP $\rightarrow$ glucose phosphate + ADP
   c. $6\ H_2O + 6\ CO_2 \rightarrow C_6H_{12}O_6 + 6\ O_2$
   d. $A + BC \rightarrow AB + C$
8. Which of the following statements about a carbonated cola beverage with a pH of 2.9 is true?
   a. It has a relatively high concentration of hydrogen ions.
   b. It has a relatively low concentration of hydrogen ions.
   c. It has equal amounts of hydroxyl and hydrogen ions.
   d. Cola is a buffered solution.
9. Proteins are polymers of
   a. amino acids.
   b. fatty acids.
   c. nucleic acids.
   d. monosaccharides.
10. Which of the following are hydrophobic organic molecules?
    a. proteins
    b. carbohydrates
    c. lipids
    d. nucleic acids

Fill in the Blanks

1. The outermost electron shell of an atom is known as the ________________ shell.
2. The type of chemical bond between atoms with nearly equal electronegativities is called a(n) ________________ bond.
3. The principal short-term energy storage molecule in cells is ________________.
4. The most common long-term energy storage molecule is ________________.
5. Groups of atoms such as NH_2 or OH that appear in certain common arrangements are called ________________.
6. The reverse of dehydration synthesis is ________________.
7. Reactions that release energy are called ________________ reactions.
8. All chemical reactions begin with reactants and result in new molecules called ________________.
9. The ________________ scale is a measure of the concentration of hydrogen ions in a solution.
10. A nucleic acid containing the base uracil would also contain ________________ sugar.

Labeling

Label a portion of the molecule below where the primary structure is visible; label two types of secondary structure; circle the tertiary structure.

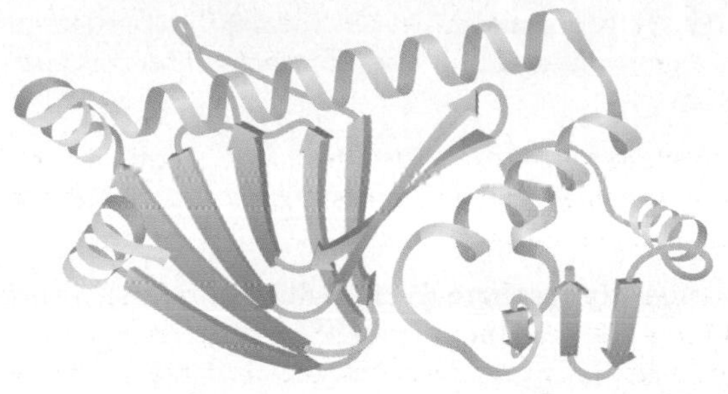

Short Answer

1. List three main types of chemical bonds, and give an example of each.
2. Name five properties of water that are vital to life.
3. Describe the difference(s) among saturated fatty acids, unsaturated fatty acids, and polyunsaturated fatty acids.
4. What is the difference between atomic oxygen and molecular oxygen?
5. Explain how the polarity of water molecules makes water an excellent solvent.

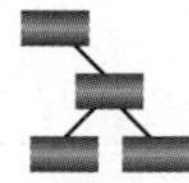

Concept Mapping

Using the following terms, draw a concept map that describes nucleic acids. You may use some terms more than once. For a sample concept map, see page 93. Or, complete this concept map online by going to the Study Area at www.masteringmicrobiology.com.

Adenine (2)
Cytosine (2)
Deoxyribonucleotides
Deoxyribose
DNA
Double-stranded
Guanine (2)
mRNA
Nitrogenous bases (2)
Phosphate
Ribonucleotides
Ribose
RNA
rRNA
Single-stranded
Thymine
tRNA
Uracil

Critical Thinking

1. Anthrax is caused by a bacterium, *Bacillus anthracis*, that avoids defenses against disease by synthesizing an outer glycoprotein covering made from D-glutamic acid. This covering is not digestible by white blood cells that normally engulf bacteria. Why is the covering indigestible?
2. Dehydrogenation is a chemical reaction in which a saturated fat is converted to an unsaturated fat. Explain why the name for this reaction is an appropriate one.
3. Two freshmen disagree about an aspect of chemistry. The nursing major insists that H^+ is the symbol for a hydrogen ion. The physics major insists that H^+ is the symbol for a proton. How can you help them resolve their disagreement?
4. When an egg white is heated, it changes from liquid to solid. When gelatin is cooled, it changes from liquid to solid. Both gelatin and egg white are proteins. From what you have learned about proteins, why can the gelatin be changed back to liquid but the cooked egg cannot?
5. When amino acids are synthesized in a test tube, D and L forms occur in equal amounts. However, cells use only L forms in their proteins. Occasionally, meteorites are found to contain amino acids. Based on these facts, how could NASA scientists determine whether the amino acids recovered from space are evidence of Earth-like extraterrestrial life or of nonmetabolic processes?
6. The poison glands of many bees and wasps contain acidic compounds. What common household chemical could be used to neutralize this poison?
7. Examine the molecules depicted in Figure 2.5. Which are polar? Which are nonpolar? Predict which are water soluble. Which are hydrophobic?

Access more review material online in the Study Area at **www.masteringmicrobiology.com**. There, you'll find

- **MP3 Tutor Sessions**
- **Concept Mapping Activities**
- **Flashcards**
- **Quizzes**

and more to help you succeed.

Cell Structure and Function

Can a microbe be a magnet? The answer is yes, if it is a magnetobacterium.

Magnetobacteria are microorganisms with an unusual feature: cellular structures called *magnetosomes.* Magnetosomes are stored deposits (also called inclusions) of the mineral magnetite. These deposits align magnetobacteria with the lines of the Earth's magnetic field, much like a compass. In the Southern Hemisphere, magnetobacteria exist as south-seeking varieties; in the Northern Hemisphere, they exist as north-seeking varieties.

How do these bacteria benefit from magnetosomes? Magnetobacteria prefer environments with little or no oxygen, such as those that exist below the surfaces of land and sea. The magnetosomes point toward the underground magnetic poles, helping magnetobacteria move toward regions with optimal oxygen content.

Take the pre-test for this chapter online. Visit the Study Area at www.masteringmicrobiology.com.

Aquaspirillum magnetotacticum, a spiral-shaped magnetobacterium, contains a clearly visible magnetosome composed of tiny square inclusions of magnetite.

All living things—including our bodies and the bacterial, protozoan, and fungal pathogens that attack us—are composed of living cells. If we want to understand disease and its treatment, therefore, we must first understand the life of cells. How pathogens attack our cells, how our bodies defend themselves, how current medical treatments assist our bodies in recovering—all of these activities have their basis in the biology of our, and our pathogens', cells.

In this chapter, we will examine cells and the structures within cells. We will discuss similarities and differences between the three major kinds of cells—bacterial, archaeal, and eukaryotic. The differences are particularly important because they allow researchers to develop treatments that inhibit or kill pathogens without adversely affecting a patient's own cells. We will also learn about cellular structures that allow pathogens to evade the body's defenses and cause disease.

Processes of Life

Learning Objective

✓ Describe four major processes of living cells.

As we discussed in Chapter 1, microbiology is the study of particularly small living things. What we have not yet discussed is the question of how we define life. Scientists once thought that living things were composed of special organic chemicals, such as glucose and amino acids, that carried a "life force" found only in living organisms. These organic chemicals were thought to be formed only by living things and to be very different from the inorganic chemicals of nonliving things.

The idea that organic chemicals could come only from living organisms had to be abandoned in 1828, when Friedrich Wöhler (1800–1882) synthesized urea, an organic molecule, using only inorganic reactants in his laboratory. Today we know that all living things contain both organic and inorganic chemicals, and that many organic chemicals can be made from inorganic chemicals by laboratory processes. If organic chemicals can be made even in the absence of life, what is the difference between a living thing and a nonliving thing? What is life?

At first this may seem a simple question. After all, you can usually tell when something is alive. However, defining "life" itself is difficult, so biologists generally avoid setting a definition, preferring instead to describe characteristics common to all living things. Biologists agree that all living things share at least four processes of life: growth, reproduction, responsiveness, and metabolism.

- **Growth.** Living things can grow; that is, they can increase in size.
- **Reproduction.** Organisms normally have the ability to reproduce themselves. Reproduction means that they increase in number, producing more organisms organized like themselves. Reproduction may be accomplished asexually (alone) or sexually with gametes (sex cells). Note that reproduction is an increase in number, whereas growth is an increase in size. Growth and reproduction often occur simultaneously. We consider several methods of reproduction when we examine microorganisms in detail in Chapters 11–13.
- **Responsiveness.** All living things respond to their environment. They have the ability to change internal and/or external properties in reaction to changing conditions around or within them. Many organisms also have the ability to move toward or away from environmental stimuli—a response called *taxis*.
- **Metabolism.** Metabolism can be defined as the ability of organisms to take in nutrients from outside themselves and use the nutrients in a series of controlled chemical reactions to provide the energy and structures needed to grow, reproduce, and be responsive. Metabolism is a unique process of living things; nonliving things cannot metabolize. Cells store metabolic energy in the chemical bonds of *adenosine triphosphate* (ă-den′ō-sēn trī-fos′fāt), or *ATP*. Major processes of microbial metabolism, including the generation of ATP, are discussed in Chapters 5–7.

Table 3.1 shows how these characteristics, along with cell structure, relate to various kinds of microbes.

Organisms may not exhibit these processes at all times. For instance, in some organisms, reproduction may be postponed or curtailed by age or disease, or in humans at least, by choice.

TABLE 3.1 Characteristics of Life and Their Distribution in Microbes

Characteristic	Bacteria, Archaea, Eukaryotes	Viruses
Growth: increase in size	Occurs in all	Does not occur
Reproduction: increase in number	Occurs in all	Host cell replicates the virus
Responsiveness: ability to react to environmental stimuli	Occurs in all	Reaction to host cells seen in some viruses
Metabolism: controlled chemical reactions of organisms	Occurs in all	Uses host cell's metabolism
Cellular structure: membrane-bound structure capable of all of the above functions	Present in all	Lacks cytoplasmic membrane or cellular structure

HIGHLIGHT

IT'S ALIVE!? . . . MAYBE.

An ongoing discussion in microbiology concerns the nature of viruses. Are they alive?

Viruses are noncellular; that is, they lack cell membranes, cell walls, and most other cellular components. However, they do contain genetic material in the form of either DNA or RNA. (Few viruses have both DNA and RNA.) While many viruses use DNA molecules for their genes, others such as HIV and poliovirus use RNA for genes instead of DNA. No cells use RNA molecules for their genes.

However, viruses have some characteristics of living cells. For instance, some demonstrate responsiveness to their environment, as when they inject their genetic material into susceptible host cells. Nevertheless, viruses lack most characteristics of life: they are unable to grow, reproduce, or metabolize outside of a host cell, although once they enter a cell they take control of the cell's metabolism and cause it to make more viruses. This takeover typically leads to the death of the cell and results in disease in an organism.

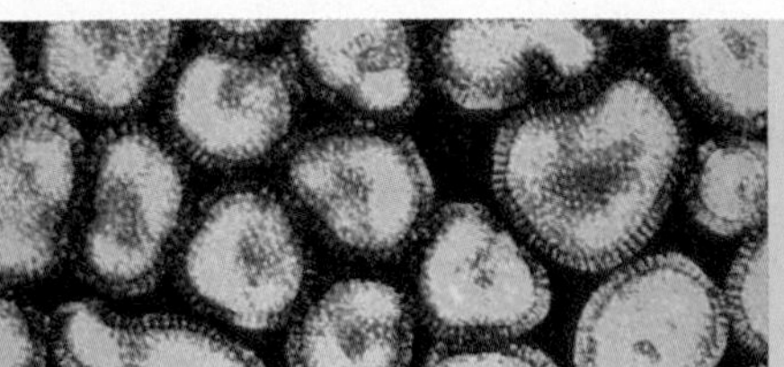

▲ ***Influenzaviruses.*** TEM 100 nm

Likewise, the rate of metabolism may be reduced, as occurs in a seed, a hibernating animal, or a bacterial endospore,[1] and growth often stops when an animal reaches a certain size. However, microorganisms typically grow, reproduce, respond, and metabolize as long as conditions are suitable. Chapter 6 discusses the proper conditions for the metabolism and growth of various types of microorganisms.

A group of pathogens essential to the debate regarding "What constitutes life?" are the viruses. **Highlight: It's Alive!? . . . Maybe** discusses whether viruses should be considered living things.

Prokaryotic and Eukaryotic Cells: An Overview

Learning Objective

✓ Compare and contrast prokaryotic and eukaryotic cells.

In the 1800s, two German biologists, Theodor Schwann (1810–1882) and Matthias Schleiden (1804–1881) developed the theory that all living things are composed of cells. *Cells* are living entities, surrounded by a membrane, that are capable of growing, reproducing, responding, and metabolizing. The smallest living things are single-celled microorganisms.

There are many different kinds of cells (**Figure 3.1**). Some cells are free-living, independent organisms; others live together in colonies or form the bodies of multicellular organisms. Cells also exist in various sizes, from the smallest bacteria to bird eggs, which are the largest of cells. All cells may be described as either *prokaryotes* (prō-kar'ē-ōts) or *eukaryotes* (yū-kar'ē-ōts).

Scientists categorize organisms based on shared characteristics into groups called *taxa.* "Prokaryotic" is a characteristic of organisms in two taxa—*Domain Archaea* and *Domain Bacteria*—but "prokaryote" itself is not a taxon. The distinctive feature of **prokaryotes** is that they can make proteins simultaneously to reading the genetic code because the typical prokaryote does not have a membrane surrounding its genetic material (DNA). In other words, a typical prokaryote does not have a nucleus (**Figure 3.2**). (Researchers have discovered a few prokaryotes with internal membranes that look like nuclei, but further investigation is needed to determine what these structures are.) The word *prokaryote* comes from Greek words meaning "before nucleus." Moreover, electron microscopy has revealed that prokaryotes typically lack various types of internal structures bound with phospholipid membranes that are present in eukaryotic cells.

Bacteria and archaea differ fundamentally in such ways as the type of lipids in their cytoplasmic membranes, and in the chemistry of their cell walls. In many ways, archaea are more like eukaryotes than they are like bacteria. Chapter 11 discusses archaea and bacteria in more detail.

Eukaryotes have a membrane surrounding their DNA, forming a nucleus (**Figure 3.3**), which sets eukaryotes in *Domain Eukarya*. Indeed, the term *eukaryote* comes from Greek words meaning "true nucleus." Besides the nuclear membrane, eukaryotes have numerous other internal membranes that compartmentalize cellular functions. These compartments are membrane-bound **organelles**—specialized structures that act like tiny organs to carry on the various functions of the cell. Organelles and their functions are discussed later in this chapter. The cells of algae, protozoa, fungi, animals, and plants are eukaryotic. Eukaryotes are usually larger and more complex than prokaryotes, which are typically 1.0 μm in diameter or smaller, as compared to 10–100 μm for eukaryotic cells (**Figure 3.4**). (See **Highlight: Giant Bacteria** on p. 62 for some interesting exceptions concerning the size of prokaryotes.)

Although there are many kinds of cells, they all share the characteristic processes of life as previously described, as well as certain physical features. In this chapter, we will distinguish between bacterial, archaeal, and eukaryotic "versions" of physical features common to cells, including (1) external structures, (2) the cell wall, (3) the cytoplasmic membrane, and (4) the cytoplasm. We will also discuss features unique to each type. Further details of prokaryotic and eukaryotic organisms, their classification, and their ability to cause disease are discussed in Chapters 11, 12, and 19–23.

Next, we explore characteristics of bacterial cells beginning with external features and working into the cell.

[1]Endospores are resting stages, produced by some bacteria, that are tolerant of environmental extremes.

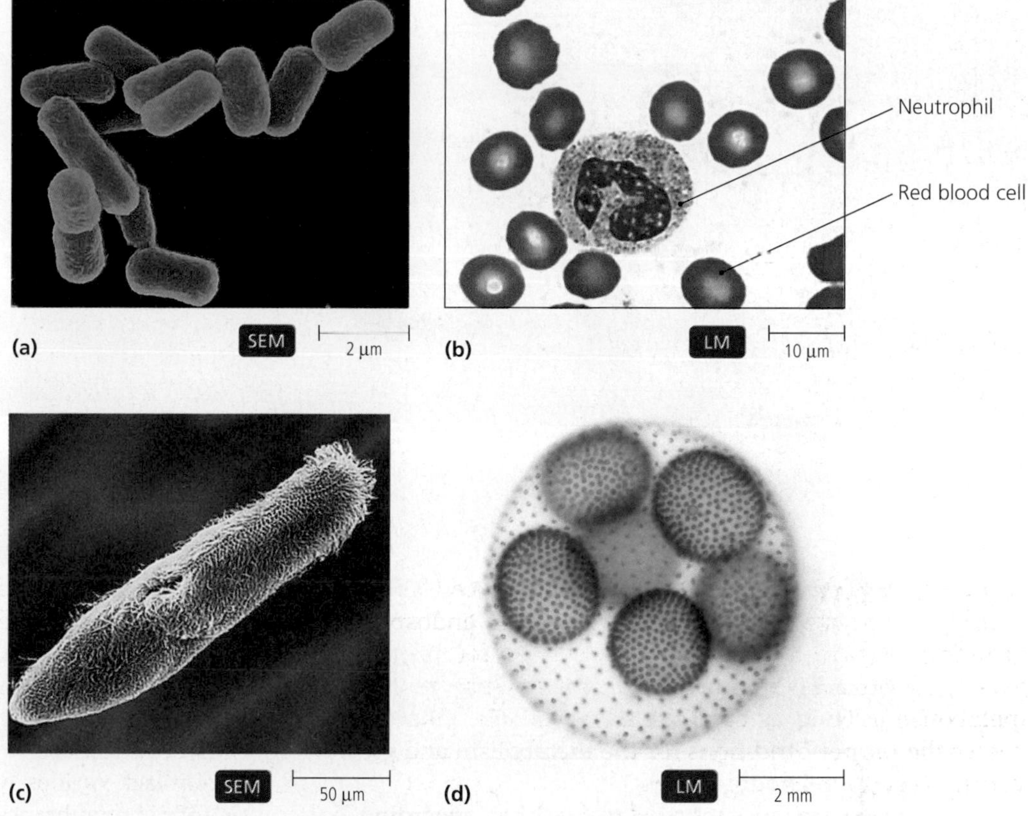

▶ **Figure 3.1 Examples of types of cells.** **(a)** *Escherichia coli* bacterial cells. **(b)** A neutrophil, a defensive white blood cell, amidst human red blood cells. **(c)** *Paramecium*, a single-celled eukaryote. **(d)** *Volvox*, a colonial, photosynthetic eukaryote composed of numerous small cells and daughter colonies. Note the differences in magnification.

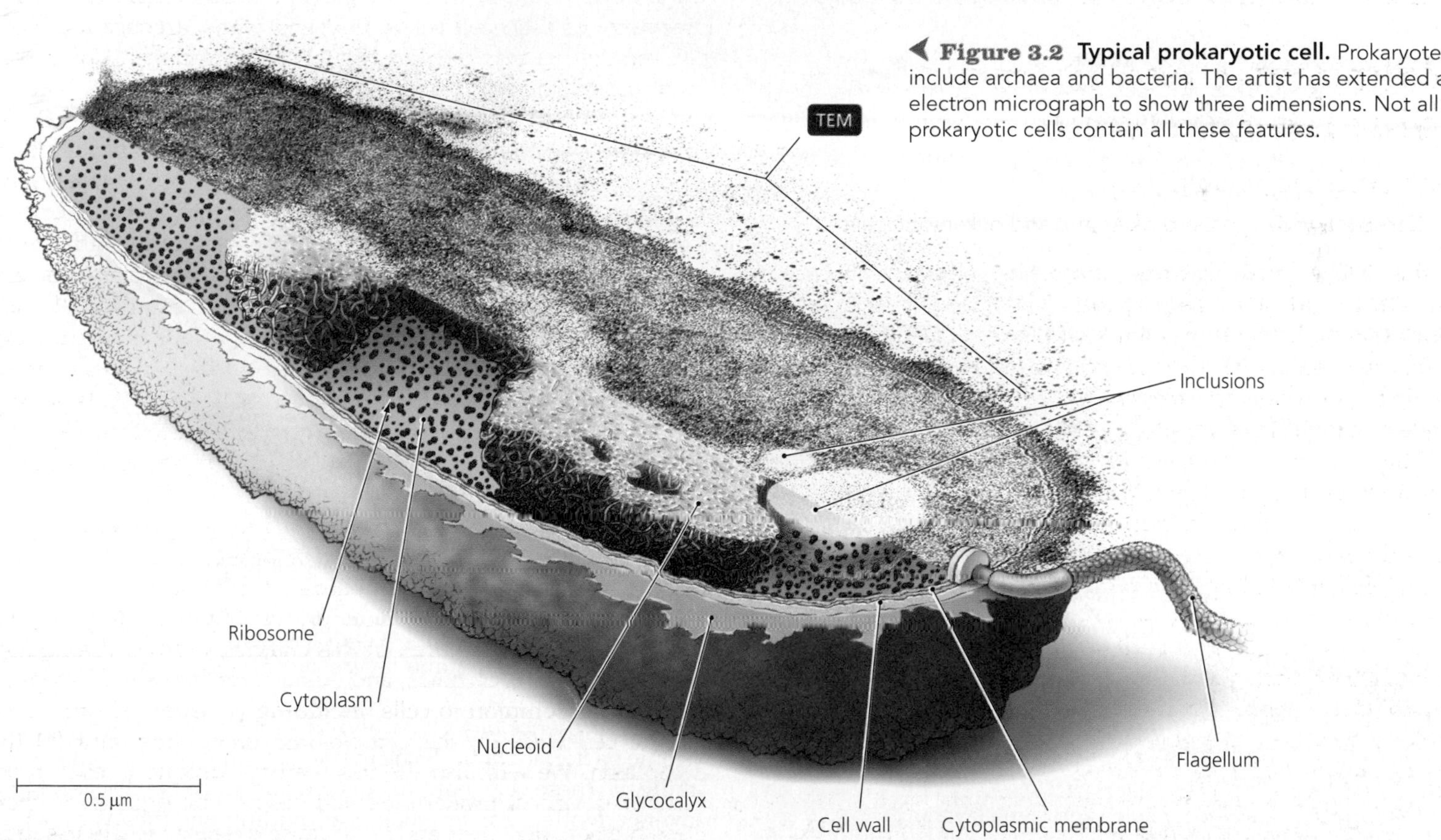

◀ **Figure 3.2 Typical prokaryotic cell.** Prokaryotes include archaea and bacteria. The artist has extended an electron micrograph to show three dimensions. Not all prokaryotic cells contain all these features.

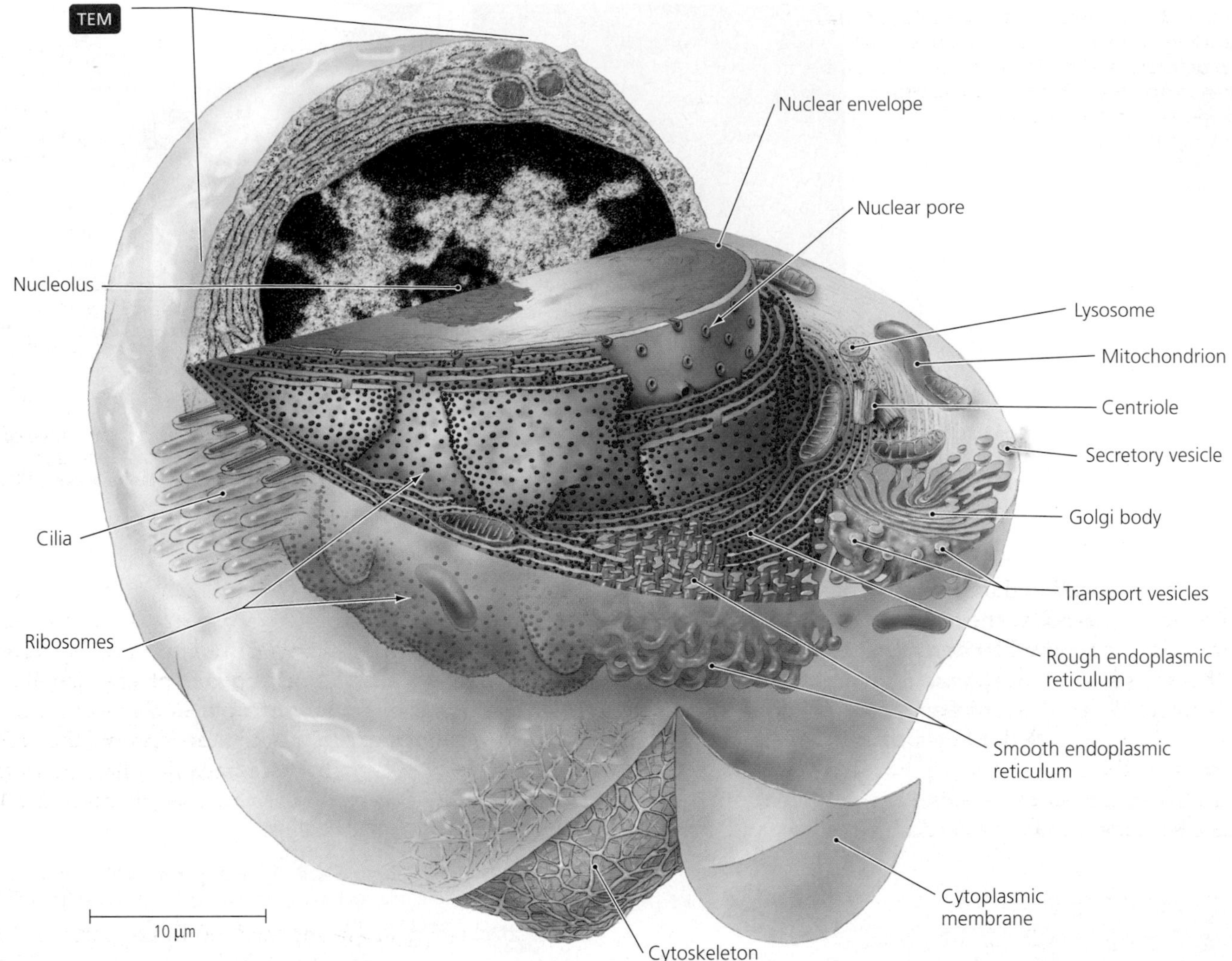

▲ **Figure 3.3 Typical eukaryotic cell.** Not all eukaryotic cells have all these features. The artist has extended the electron micrograph to show three dimensions. Note the difference in magnification between this cell and the prokaryotic cell in the previous figure. *Besides size, what major difference between prokaryotes and eukaryotes was visible to early microscopists?*

Figure 3.3 *Eukaryotic cells contain nuclei, which are visible with light microscopes, whereas prokaryotes lack nuclei.*

External Structures of Bacterial Cells

Many cells have special external features that enable them to respond to other cells and their environment. In bacteria, these features include glycocalyces, flagella, fimbriae, and pili.

Glycocalyces

Learning Objective

✓ Describe the composition, function, and relevance to human health of glycocalyces.

✓ Distinguish between capsules and slime layers.

Some cells have a gelatinous, sticky substance that surrounds the outside of the cell. This substance is known as a **glycocalyx** (plural: *glycocalyces*), which literally means "sugar cup." The glycocalyx may be composed of polysaccharides, polypeptides, or both. These chemicals are produced inside the cell and extruded onto the cell's surface.

When the glycocalyx of a bacterium is composed of organized repeating units of organic chemicals firmly attached to the cell surface, the glycocalyx is called a **capsule** (**Figure 3.5a**). A loose, water-soluble glycocalyx is called a **slime layer** (**Figure 3.5b**). Capsules and slime layers protect cells from desiccation (drying).

The presence of a glycocalyx is a feature of numerous pathogenic bacteria. Their glycocalyces play an important role in the ability of these cells both to survive and to cause disease. Slime layers are often *viscous* (sticky), providing one means by which bacteria attach to surfaces. For example, they enable oral bacteria to colonize the teeth, where they produce acid and cause

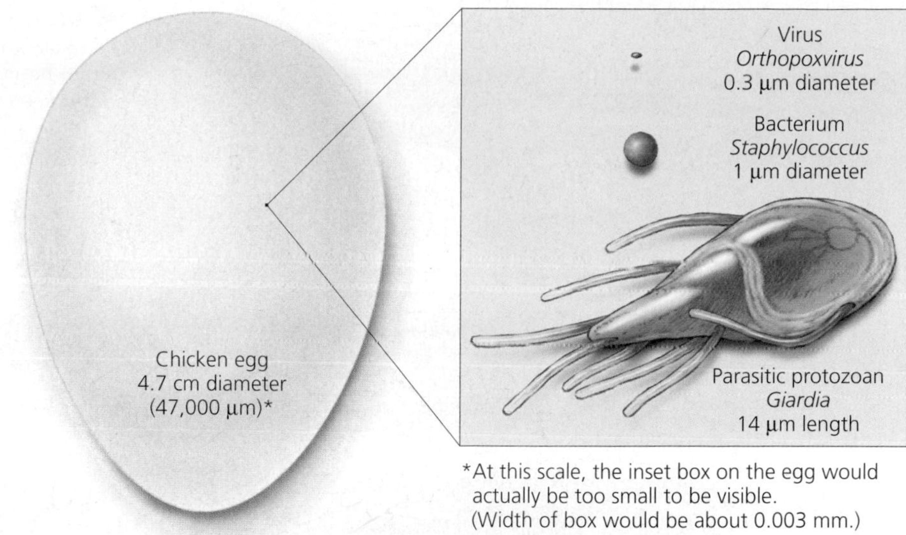

Figure 3.4 Approximate size of various types of cells. Birds' eggs are the largest cells. Note that *Staphylococcus*, a bacterium, is smaller than *Giardia*, a unicellular eukaryote. A chickenpox virus is shown only for comparison; viruses are not cellular.

decay. Since the chemicals in many capsules are similar to those normally found in the body, they may prevent bacteria from being recognized or devoured by defensive cells of the host. For example, the capsules of *Streptococcus pneumoniae* (strep-tō-kok'ŭs nū-mō'nē-ī) and *Klebsiella pneumoniae* (kleb-sē-el'ă nū-mō'nē-ī) enable these prokaryotes to avoid destruction by defensive cells in the respiratory tract and cause pneumonia. Unencapsulated strains of these same bacterial species do not cause disease because the body's defensive cells destroy them.

Flagella

Learning Objectives

✓ Discuss the structure and function of bacterial flagella.

✓ List and describe four bacterial flagellar arrangements.

A cell's motility may enable it to flee from a harmful environment or move toward a favorable environment such as one where food or light is available. The most notable structures responsible for such bacterial movement are flagella. **Flagella** (singular: *flagellum*) are long structures that extend beyond the surface of a cell and its glycocalyx and propel the cell through its environment. Not all bacteria have flagella, but for those that do, the flagella are very similar in composition, structure, and development. **ANIMATIONS:** ***Motility***

Structure

Bacterial flagella are composed of three parts: a long, thin *filament*, a *hook*, and a *basal body* **(Figure 3.6)**. The hollow filament is a long hollow shaft, about 20 nm in diameter, that

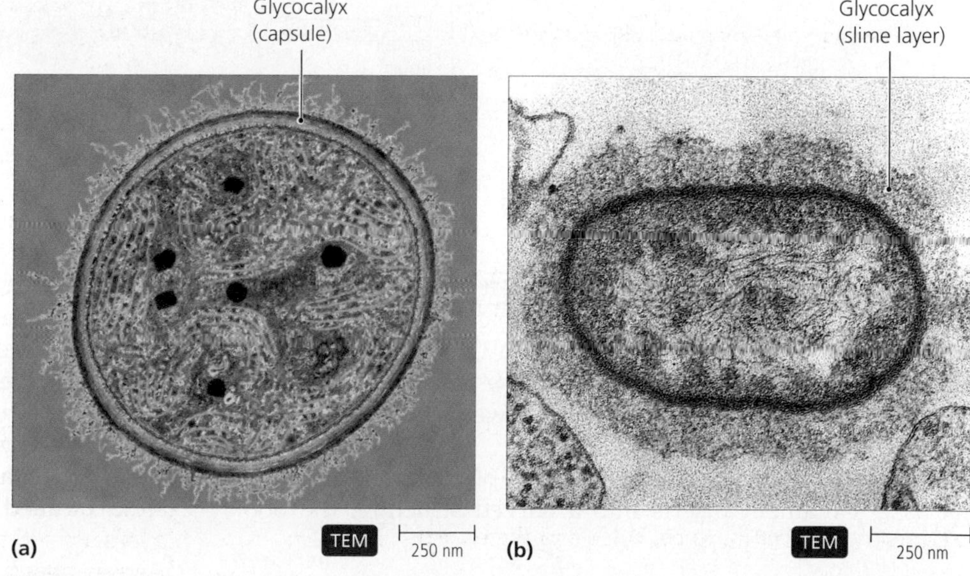

Figure 3.5 Glycocalyces. (a) Micrograph of *Streptococcus pneumoniae*, the common cause of pneumonia, showing a prominent capsule. **(b)** *Bacteroides*, a common fecal bacterium, has a slime layer surrounding the cell. *What advantage does a glycocalyx provide a cell?*

Figure 3.5 *A glycocalyx provides protection from drying and from being devoured; it may also help attach cells to one another and to surfaces in the environment.*

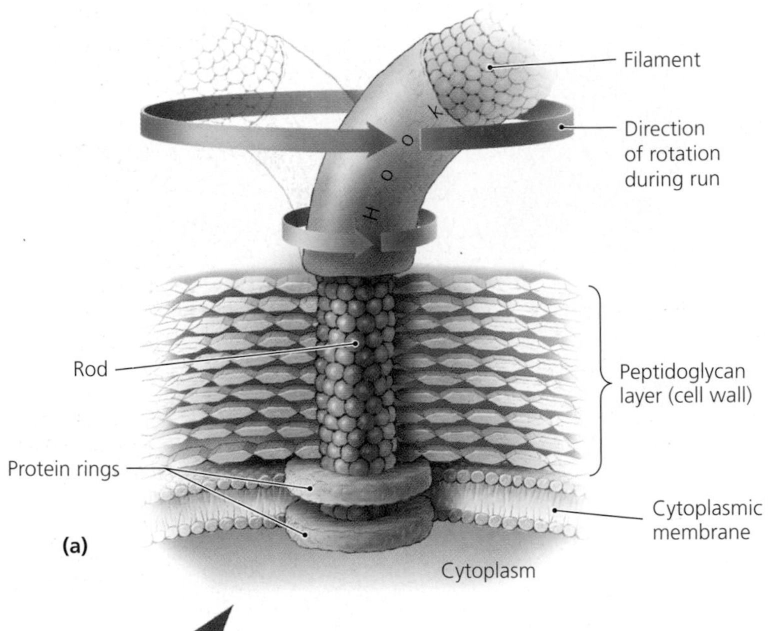

◀ **Figure 3.6 Proximal structure of bacterial flagella.** **(a)** Detail of flagellar structure of a Gram-positive cell. **(b)** Detail of the flagellum of a Gram-negative bacterium. *How do flagella of Gram-positive bacteria differ from those of Gram-negative bacteria?*

Figure 3.6 *Flagella of Gram-positive cells have a single pair of rings in the basal body, which function to attach the flagellum to the cytoplasmic membrane. The flagella of Gram-negative cells have two pairs of rings; one pair anchors the flagellum to the cytoplasmic membrane, the other pair to the cell wall.*

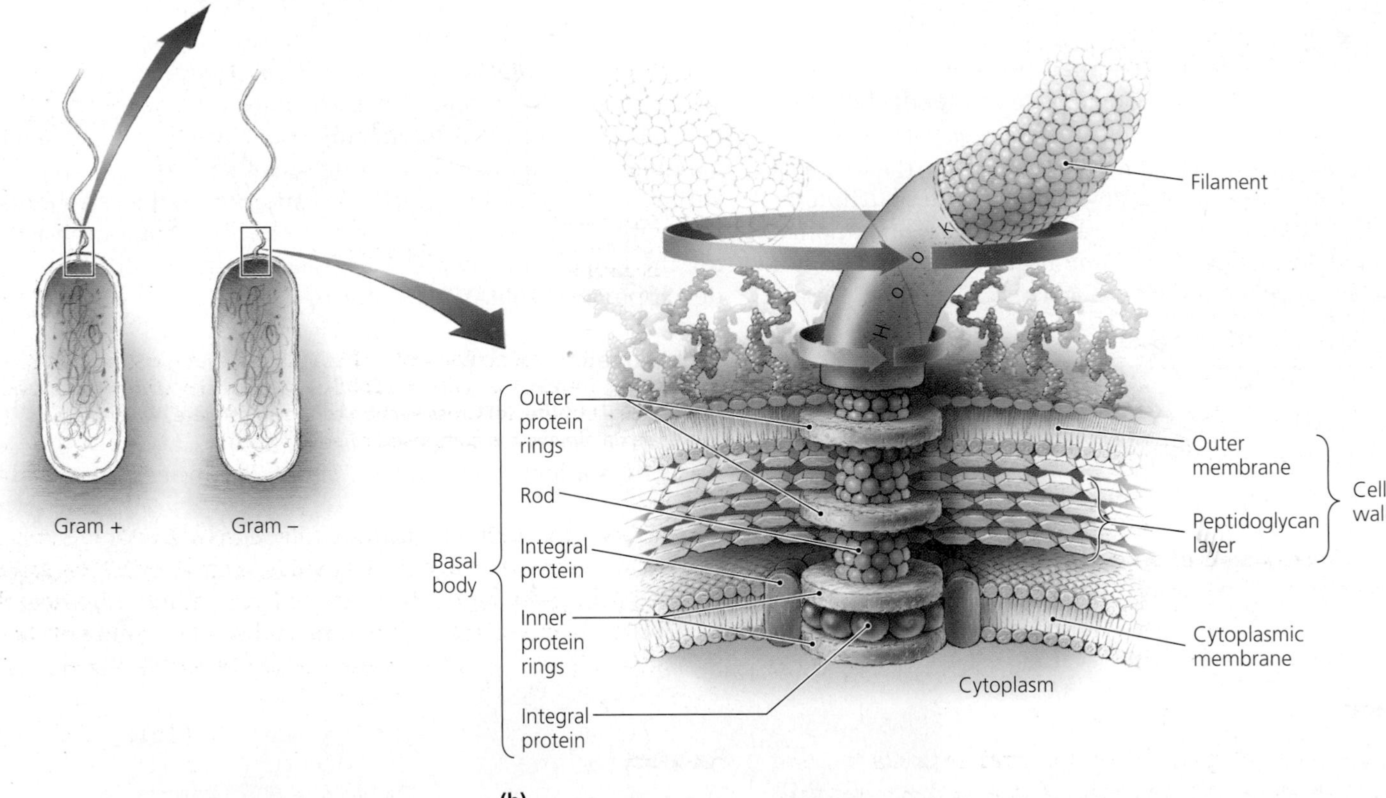

extends out into the cell's environment. It is composed of many identical globular molecules of a protein called *flagellin.* The cell extrudes molecules of flagellin through the hollow core of the flagellum to be deposited in a clockwise helix at the lengthening tip. Bacterial flagella sense external wetness, inhibiting their own growth in dry habitats.

No membrane covers the filament of bacterial flagella. At its base a filament inserts into a curved structure, the hook, which is composed of a different protein. The basal body, which is composed of still different proteins, anchors the filament and hook to the cell wall and cytoplasmic membrane by means of a rod and a series of either two or four rings of integral proteins. Together the hook, rod, and rings allow the filament to rotate 360°. Differences in the proteins associated with bacterial flagella vary enough to allow classification of species into groups (strains) called *serovars.* **ANIMATIONS:** ***Flagella: Structure***

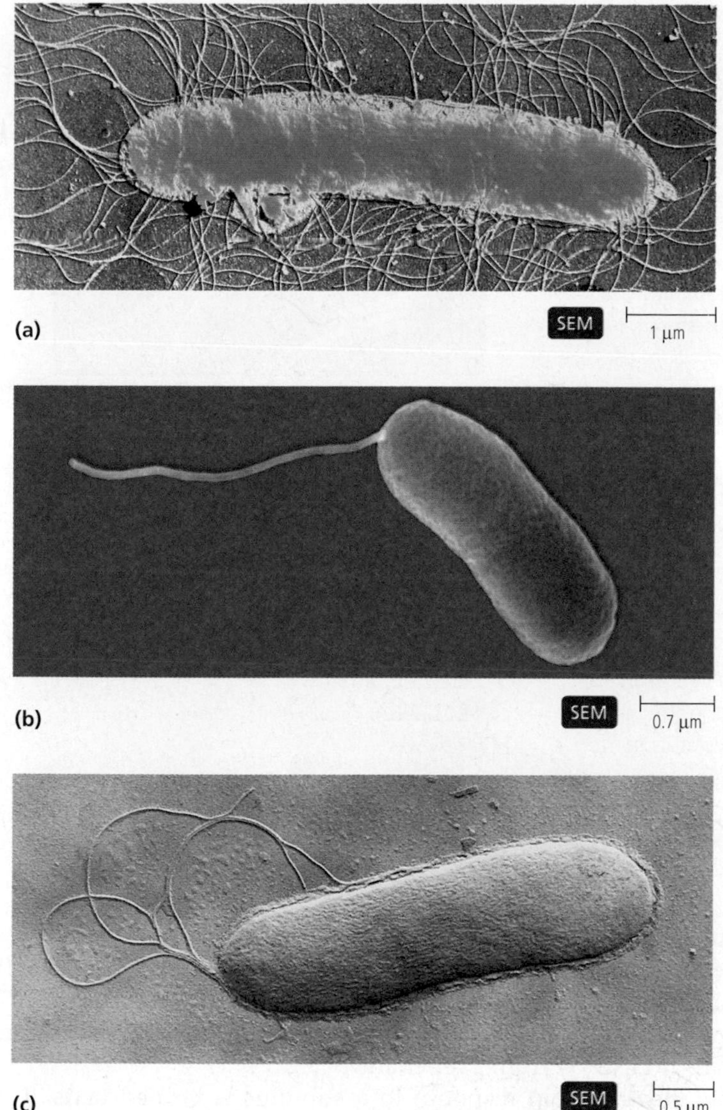

▲ **Figure 3.7 Micrographs of basic arrangements of bacterial flagella.** **(a)** Peritrichous. **(b)** Single polar flagellum. **(c)** Tuft of polar flagella.

Arrangement

Bacteria may have one of several flagellar arrangements **(Figure 3.7)**. Flagella that cover the surface of the cell are termed **peritrichous;**[2] in contrast, **polar** flagella are only at the ends. Some cells have tufts of polar flagella.

Some spiral-shaped bacteria, called *spirochetes* (spī′rō-kēts),[3] have flagella at both ends that spiral tightly around the cell instead of protruding into the surrounding medium. These flagella, called **endoflagella,** form an **axial filament** that wraps around the cell between its cytoplasmic membrane and an outer membrane **(Figure 3.8)**. Rotation of endoflagella evidently causes the axial filament to rotate around the cell, causing the spirochete to

[2]From Greek *peri*, meaning around, and *trichos*, meaning a hair.
[3]From Greek *speira*, meaning coil, and *chaeta*, meaning hair.

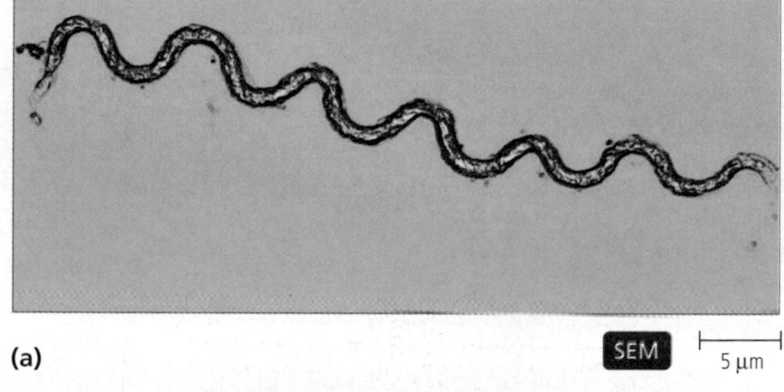

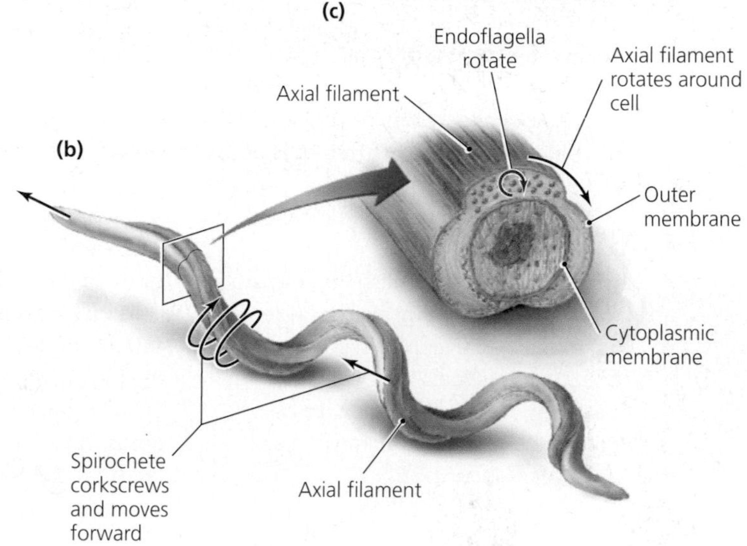

▲ **Figure 3.8 Axial filament.** **(a)** Scanning electron micrograph of spirochetes, *Treponema pallidum*. **(b)** Diagram of axial filament wrapped around a spirochete. **(c)** Cross section of the spirochete, which reveals that the axial filament is composed of endoflagella.

"corkscrew" through its medium. *Treponema pallidum* (trep-ō-nē′mă pal′li-dŭm), the agent of syphilis, and *Borrelia burgdorferi* (bō-rē′lē-ă burg-dōr′fer-ē), the cause of Lyme disease, are notable spirochetes. Some scientists think the corkscrew motility of these pathogens allows them to invade human tissues. **ANIMATIONS:** ***Flagella: Arrangement; Spirochetes***

Function

Although the precise mechanism by which bacterial flagella move is not completely understood, we do know that they rotate 360° like boat propellers rather than whipping from side to side. The flow of hydrogen ions (H^+) or of sodium ions (Na^+) through the cytoplasmic membrane near the basal body powers the rotation, propelling the bacterium through the environment at about 60 cell lengths per second—equivalent to a car traveling at 670 miles per hour! Flagella rotate at more than 100,000 rpm and can change direction from counterclockwise to clockwise.

Bacteria move with a series of "runs" punctuated by "tumbles." Counterclockwise flagellar rotation produces runs, which are movements of a cell in a single direction for some time. If more than one flagellum is present, the flagella align and rotate

HIGHLIGHT

GIANT BACTERIA

Most prokaryotes are small compared to eukaryotes. The dimensions of a typical bacterial cell—for example, a cell of *Escherichia coli*—are 1.0 μm × 2.0 μm. Typical eukaryotic cells are often tens of micrometers in diameter.

For many years biologists thought that the small size of prokaryotes was a necessary result of the nature of their cells. Without internal compartments, prokaryotes that were as large as eukaryotes could not isolate and control the metabolic reactions of life; nor could they efficiently distribute nutrients and eliminate wastes.

However, scientists have located a new unicellular organism from the intestines of a surgeonfish. Its cells are 0.6 mm (600 μm) long, a size visible to the unaided eye. Microscopic examination reveals the presence of a fine covering of what appear to be cilia, which are a feature of eukaryotic cells only. The new organism is named *Epulopiscium fishelsoni*,[a] but a surprise came when these cells were examined with an electron microscope. They contained no nuclei or other eukaryotic organelles, and the "cilia" more closely resembled short bacterial flagella. *Epulopiscium*, as it turned out, is indeed a giant prokaryote. (Compare the size of *Epulopiscium* with that of *Paramecium*, a eukaryote, in the box photo.)

It is not, however, the largest prokaryote. In 1997, a German microbiologist discovered an even larger bacterium, named *Thiomargarita namibiensis*,[b] which lives in chains of 2–50 cells in ocean sediments off the coast of Namibia, Africa. The spherical cells of *Thiomargarita* can grow to 750 μm in diameter, or about the size of the period at the end of this sentence. Such large cells have volumes over 66 times those of the largest *Epulopiscium*.

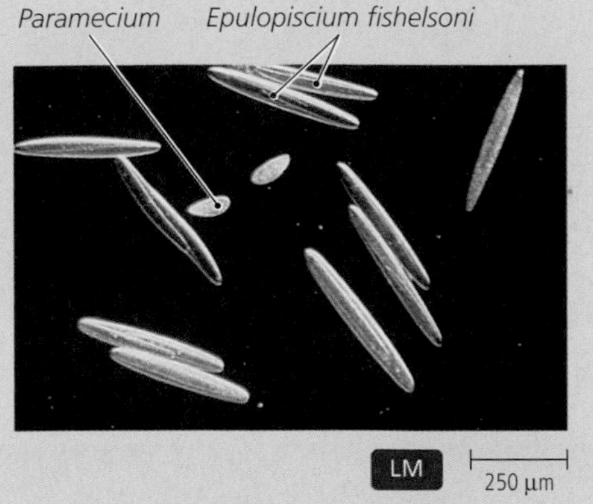

If a single cell of *Thiomargarita* were the size of an African elephant, then *Epulopiscium* would be the size of a cow, *Paramecium* would be the size of an adult Labrador retriever, and *Escherichia* would be the size of an ant.

[a] *Epulopiscium* in Latin means "guest at a banquet of fish."

[b] This name means "sulfur pearl of Namibia," because granules of sulfur stored in the cytoplasm cause the cells to glisten white, like a string of pearls.

together as a bundle. Tumbles are abrupt, random changes in direction resulting from clockwise flagellar rotation where each flagellum rotates independently. Both runs and tumbles occur in response to stimuli.

Receptors for light or chemicals on the surface of the cell send signals to the flagella, which then adjust their speed and direction of rotation. A bacterium can position itself in a more favorable environment by varying the number and duration of runs and tumbles. The presence of favorable stimuli increases the number of runs and decreases the number of tumbles; as a result, the cell tends to move toward an attractant **(Figure 3.9)**. Unfavorable stimuli increase the number of tumbles, which increases the likelihood that it will move randomly in another direction, away from a repellant.

Movement in response to a stimulus is termed **taxis.** The stimulus may be either light **(phototaxis)** or a chemical **(chemotaxis).** Movement toward a favorable stimulus is *positive taxis,* whereas movement away from an unfavorable stimulus is

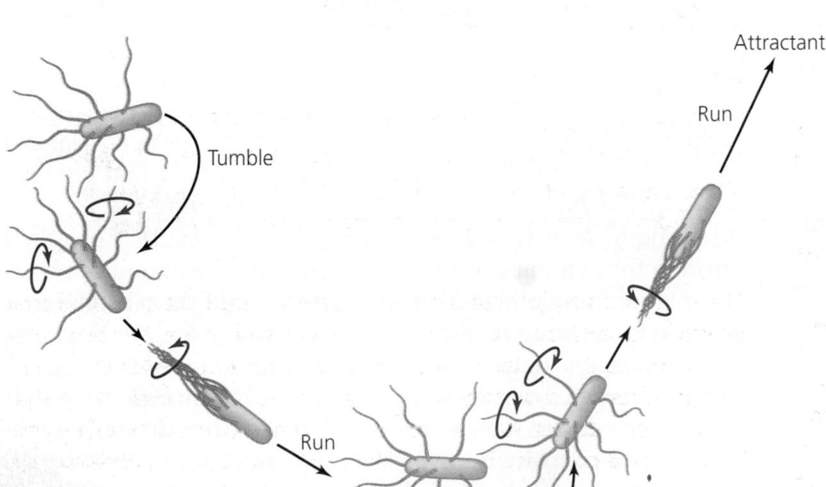

Figure 3.9 Motion of a peritrichous bacterium. In peritrichous bacteria, runs occur when all of the flagella rotate counterclockwise and become bundled. Tumbles occur when the flagella rotate clockwise, become unbundled, and the cell spins randomly. In positive chemotaxis (shown), runs last longer than tumbles, resulting in motion toward the chemical attractant. *What triggers a bacterial flagellum to rotate counterclockwise, producing a run?*

Figure 3.9 *Favorable environmental conditions induce runs.*

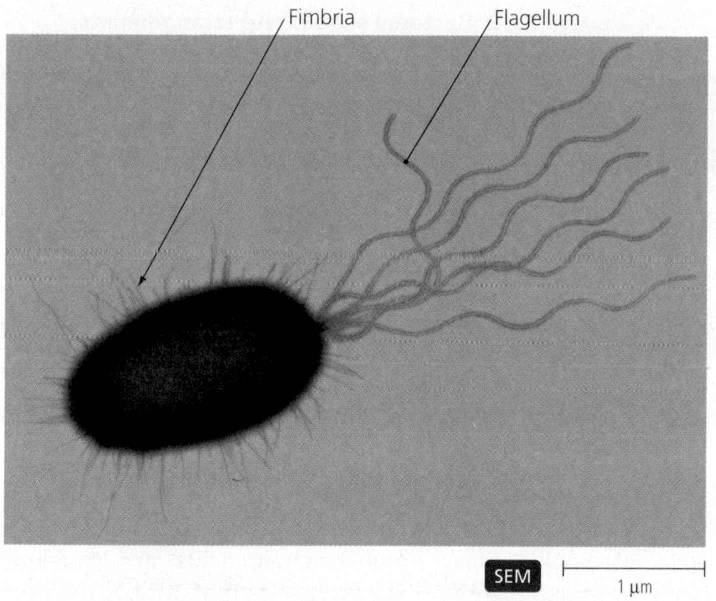

▲ **Figure 3.10 Fimbriae.** *Pseudomonas aeruginosa* has fimbriae and flagella.

negative taxis. For example, movement toward a nutrient would be *positive chemotaxis.* **ANIMATIONS:** ***Flagella: Movement***

Fimbriae and Pili

Learning Objective

✓ Compare and contrast the structures and functions of fimbriae, pili, and flagella.

Some bacteria have rodlike proteinaceous extensions called fimbriae and pili. Bacteria use **fimbriae** (fim′brē-ē)—sticky, bristlelike projections—to adhere to one another and to substances in the environment. There may be hundreds of fimbriae per cell, and they are usually shorter than flagella **(Figure 3.10)**. An example of a bacterium with fimbriae is *Neisseria gonorrhoeae* (nī-se′rē-ă go-nor-rē′ī), which causes gonorrhea. Such pathogens must be able to adhere to their hosts if they are to survive and cause disease. This bacterium is able to colonize the mucous membrane of the reproductive tract by attaching with fimbriae. *Neisseria* cells that lack fimbriae are nonpathogenic.

Bacteria may use fimbriae to move across a substrate or toward another bacterium via a process similar to pulling an object with a rope. The bacterium extrudes a fimbria, which attaches to the substrate or to another bacterium; then the bacterium retracts the fimbria, pulling itself toward the attachment point.

Fimbriae also serve an important function in **biofilms,** slimy masses of microbes adhering to a substrate by means of fimbriae and glycocalyces. Some fimbriae act as electrical wires, conducting electrical signals among cells in a biofilm. It has been estimated that at least 99% of bacteria in nature exist in biofilms. Researchers are interested in biofilms because of the

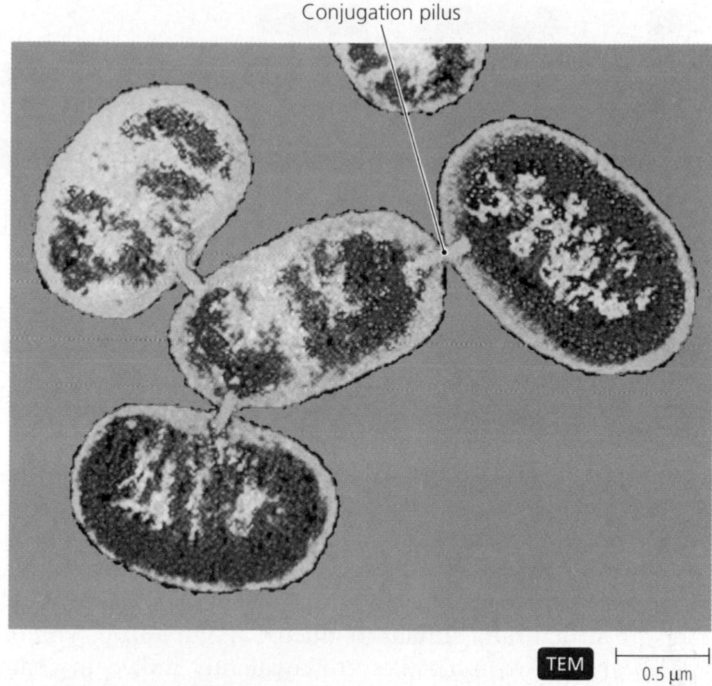

▲ **Figure 3.11 Pili.** Several *Escherichia coli* cells are connected by conjugation pili. *How are pili different from bacterial flagella?*

Figure 3.11 *Bacterial flagella are flexible structures that rotate to propel the cell; pili are hollow tubes that mediate transfer of DNA from one cell to another.*

roles they play in human diseases and in industry. (See **Highlight: Biofilms: Slime Matters** on p. 66.)

Pili (pī′lī; singular: *pilus*), which are also called *conjugation pili,* are tubules composed of a protein called *pilin.* Pili are longer than fimbriae, but usually shorter than flagella. Typically only one to a few pili are present per cell in bacteria that have them. Conjugation pili mediate the transfer of DNA from one cell to the other via a process termed *conjugation* **(Figure 3.11)**, which is discussed in detail in Chapter 7.

Bacterial Cell Walls

Learning Objectives

✓ Describe common shapes and arrangements of bacterial cells.
✓ Describe the sugar and peptide portions of peptidoglycan.
✓ Compare and contrast the cell walls of Gram-positive and Gram-negative bacteria in terms of structure and Gram staining.

The cells of most prokaryotes are surrounded by a **cell wall** that provides structure and shape to the cell and protects it from osmotic forces. In addition, a cell wall assists some cells in attaching to other cells or in resisting antimicrobial drugs. Note that animal cells do not have walls, a difference that plays a key role in treatment of many bacterial diseases with certain types of antibiotics. For example, penicillin attacks the cell wall of bacteria but is harmless to human cells, which lack walls.

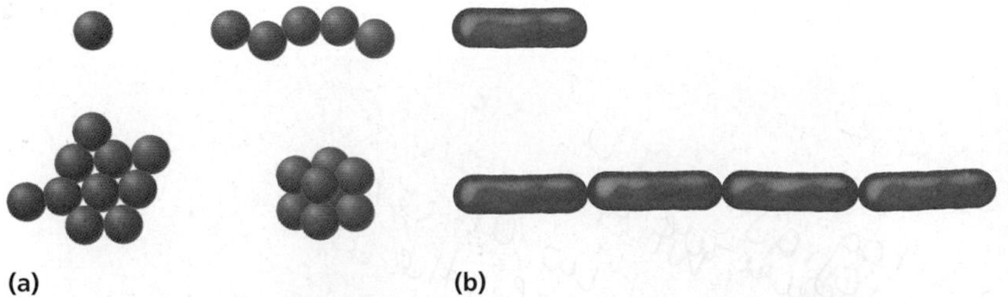

Figure 3.12 Bacterial shapes and arrangements. **(a)** Spherical cocci may be in arrangements such as single, chains (streptococci), clusters (staphylococci), and cuboidal packets. **(b)** Rod-shaped bacilli may also be single or in arrangements such as chains.

Cell walls give bacterial cells characteristic shapes. Spherical cells, called cocci (kok'sī), may appear in various arrangements, including singly or in chains (streptococci), clusters (staphylococci), or cuboidal packets (sarcinae, sar'si-nī) **(Figure 3.12)** depending on the planes of cell division. Rod-shaped cells, called bacilli (bă-sil'ī), typically appear singly or in chains.

Bacterial cell walls are composed of **peptidoglycan,** a complex polysaccharide. Peptidoglycan in turn is composed of two types of regularly alternating sugar molecules, called *N-acetylglucosamine (NAG)* and *N-acetylmuramic acid (NAM),* which are structurally similar to glucose **(Figure 3.13)**. Millions of NAG and NAM molecules are covalently linked in chains in which NAG alternates with NAM. These chains are the "glycan" portions of peptidoglycan.

Chains of NAG and NAM are attached to other chains by crossbridges of four amino acids (tetrapeptides). **Figure 3.14** illustrates one possible configuration. These peptide crossbridges are the "peptido" portion of peptidoglycan. Depending on the bacterium, tetrapeptide bridges are either covalently bonded to one another or are held together by *short connecting* chains of other amino acids as shown in Figure 3.14. Peptidoglycan covers the entire surface of a cell, which must insert millions of new subunits if it is to grow and divide.

Scientists describe two basic types of bacterial cell walls as *Gram-positive* cell walls or *Gram-negative* cell walls. They distinguish Gram-positive and Gram-negative cells by the use of the Gram staining procedure (described in Chapter 4), which was invented long before the structure and chemical nature of bacterial cell walls were known.

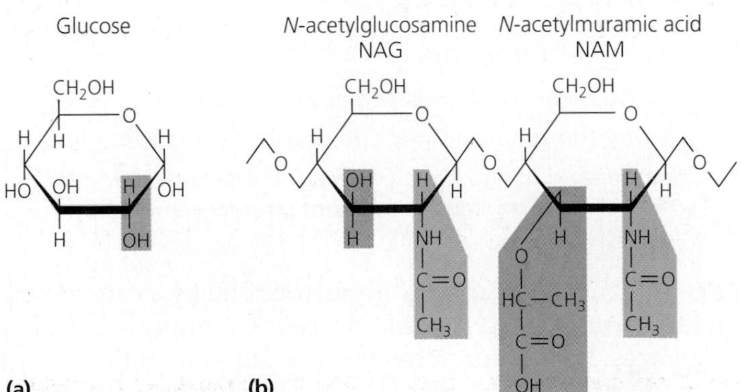

Figure 3.13 Comparison of the structures of glucose, NAG, and NAM. **(a)** Glucose. **(b)** *N*-acetylglucosamine (NAG) and *N*-acetylmuramic acid (NAM) molecules linked as in peptidoglycan. Blue shading indicates the differences between glucose and the other two sugars. Orange boxes highlight the difference between NAG and NAM.

Gram-Positive Bacterial Cell Walls

Learning Objective

✓ Compare and contrast the cell walls of acid-fast bacteria with typical Gram-positive cell walls.

Gram-positive bacterial cell walls have a relatively thick layer of peptidoglycan that also contains unique chemicals called *teichoic* (tī-kō'ik)[4] *acids*. Some teichoic acids are covalently linked to lipids, forming *lipoteichoic acids* that anchor the peptidoglycan to the cytoplasmic membrane **(Figure 3.15a)**. Teichoic acids have negative electrical charges, which help give the surface of a Gram-positive bacterium a negative charge and may play a role in the passage of ions through the wall. The thick cell wall of a Gram-positive bacterium retains the crystal violet dye used in the Gram staining procedure, so the stained cells appear purple under magnification.

Some additional chemicals are associated with the walls of some Gram-positive bacteria. For example, species of *Mycobacterium* (mī'kō-bak-tēr'ē-ŭm), which include the causative

[4]From Greek *teichos*, meaning wall.

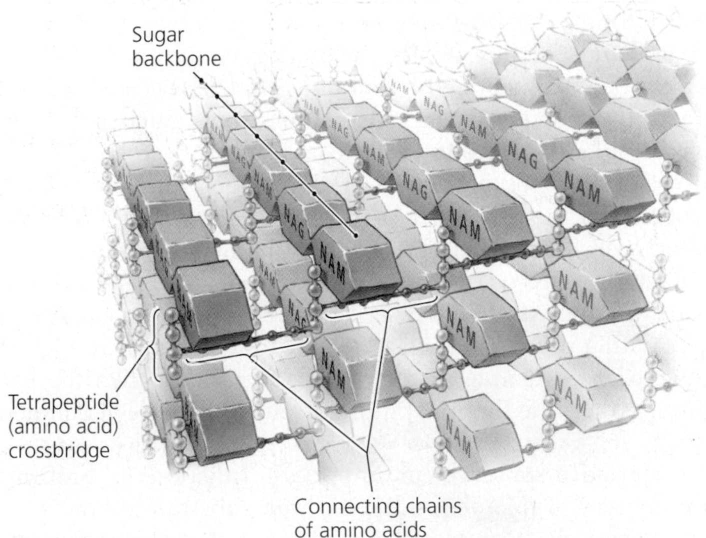

Figure 3.14 One possible structure of peptidoglycan. Peptidoglycan is composed of chains of NAG and NAM linked by tetrapeptide crossbridges and, in some cases, connecting chains of amino acids to form a tough yet flexible structure. The amino acids of the crossbridges differ among bacterial species.

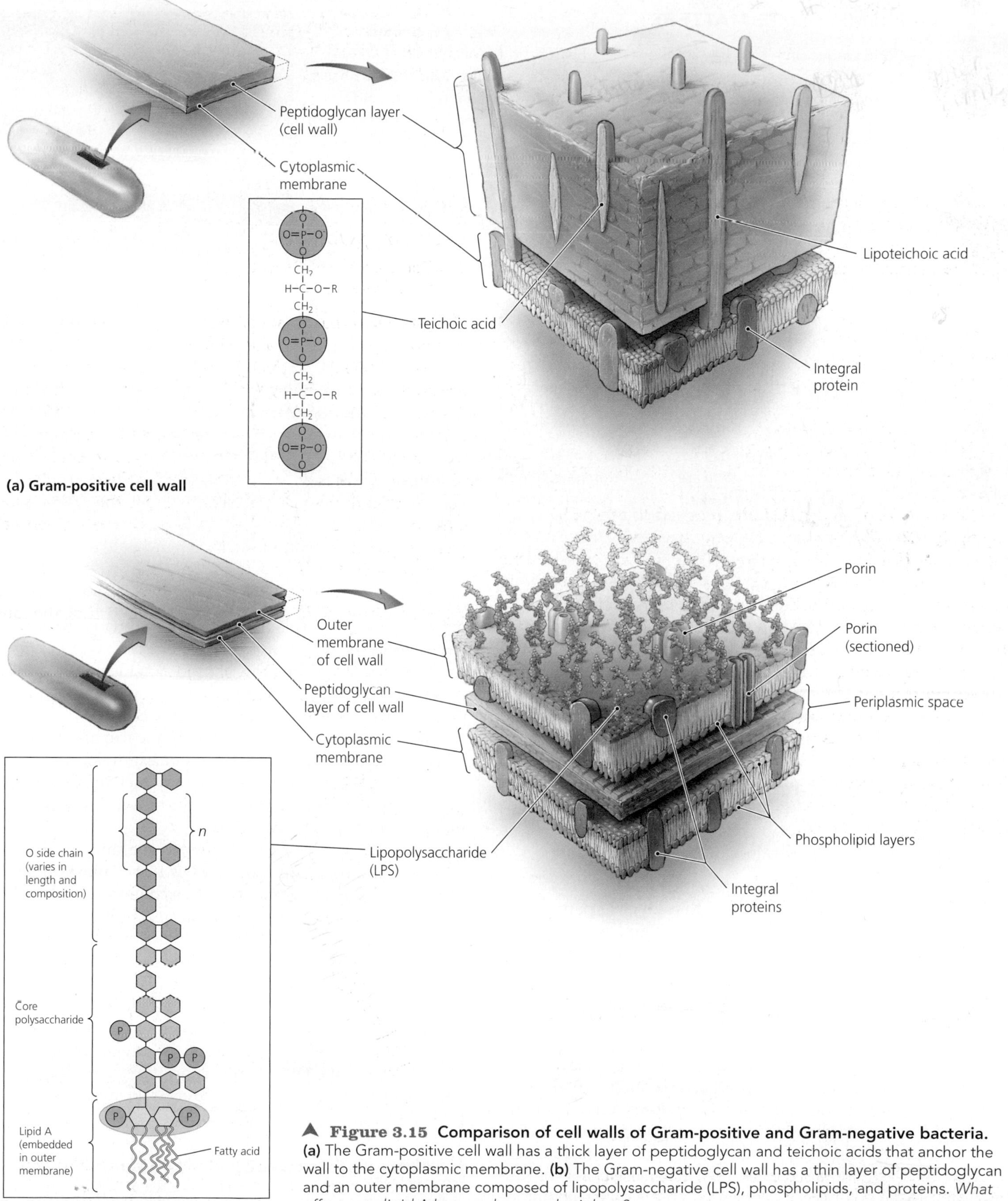

▲ **Figure 3.15 Comparison of cell walls of Gram-positive and Gram-negative bacteria.** **(a)** The Gram-positive cell wall has a thick layer of peptidoglycan and teichoic acids that anchor the wall to the cytoplasmic membrane. **(b)** The Gram-negative cell wall has a thin layer of peptidoglycan and an outer membrane composed of lipopolysaccharide (LPS), phospholipids, and proteins. *What effects can lipid A have on human physiology?*

Figure 3.15 *Lipid A can cause shock, blood clotting, and fever in humans.*

HIGHLIGHT

BIOFILMS: SLIME MATTERS

They form plaque on teeth; they are the slime on rocks in rivers and streams; and they can cause disease, clog drains, or aid in hazardous-waste cleanup. They are biofilms—organized, layered systems of bacteria and other microbes attached to a surface. Understanding biofilms holds the key to many important clinical and industrial applications.

Bacteria in biofilms behave in significantly different ways from individual, free-floating bacteria. For example, as a free-floating cell, the soil bacterium *Pseudomonas putida* propels itself through water with its flagella; however, once it becomes part of a biofilm, it turns off the genes for flagellar proteins and starts synthesizing pili instead. In addition, the genes for antibiotic resistance in *P. putida* are more active within cells in a biofilm than within a free-floating cell.

Bacteria in a biofilm also communicate via chemical and electrical signals that help them organize and form three-dimensional structures. The architecture of a biofilm provides protection that free-floating bacteria lack. For example, the lower concentrations of oxygen found in the interior of biofilms thwart the effectiveness of some antibiotics. Moreover, the presence of multiple bacterial species in most biofilms increases the likelihood that some bacteria within the biofilm community will remain resistant to a given antibiotic.

The more we understand biofilms, the more readily we can reduce their harmful effects or put them to good use. Biofilms account for two-thirds of bacterial infections in humans, including gum disease and the serious lung infections suffered by cystic fibrosis patients. Biofilms are also the culprits in many industrial problems, including corroded pipes and clogged water filters, which cause millions of dollars of damage each year. Fortunately, not all biofilms are detrimental; some show potential as aids in preventing and controlling certain kinds of industrial pollution.

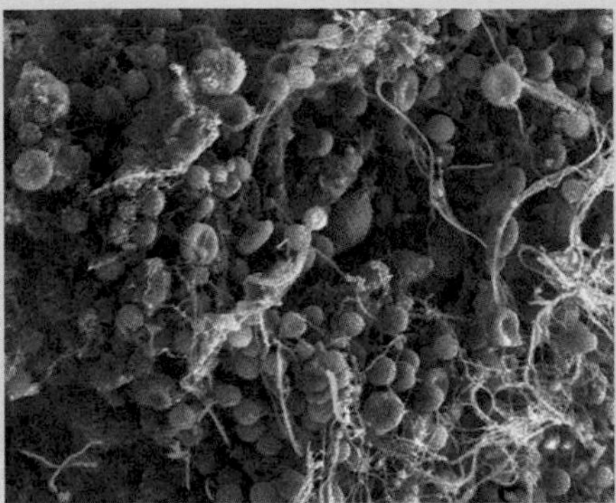

▲ **Biofilm on medical tubing.** SEM 4 μm

agents of tuberculosis and leprosy, have walls with up to 60% mycolic acid, a waxy lipid. Mycolic acid helps these cells survive desiccation and makes them difficult to stain with regular water-based dyes. Researchers have developed a special staining procedure called the *acid-fast stain* to stain these Gram-positive cells that contain large amounts of waxy lipids. Such cells are called *acid-fast bacteria* (see Chapter 4).

Gram-Negative Bacterial Cell Walls

Learning Objective

✓ Describe the clinical implications of the structure of the Gram-negative cell wall.

Gram-negative cell walls have only a thin layer of peptidoglycan **(Figure 3.15b)**, but outside this layer is an asymmetric bilayer membrane. The inner leaflet of the outer membrane is composed of phospholipids and proteins, while the outer leaflet is made of **lipopolysaccharide (LPS).** Integral proteins called *porins* form channels through both leaflets of the outer membrane, allowing for the movement of glucose and other monosaccharides across the membrane. The outer membrane is protective, allowing Gram-negative bacteria such as *Escherichia coli* (esh-ĕ-rik′ē-ă kō′lē) to better survive in harsh environments.

LPS is a union of lipid with sugar. The lipid portion of LPS is known as **lipid A.** The erroneous idea that lipid A is *inside* Gram-negative cells led to the use of the term *endotoxin*[5] for this chemical. A dead cell releases lipid A when the outer membrane disintegrates, and lipid A may trigger fever, vasodilation, inflammation, shock, and blood clotting in humans. Because killing large numbers of Gram-negative bacteria with antimicrobial drugs releases large amounts of lipid A, which might threaten the patient more than the live bacteria, any internal infection by Gram-negative bacteria is cause for concern.

The Gram-negative outer membrane can also be an impediment to the treatment of disease. For example, the outer membrane may prevent the movement of penicillin to the underlying peptidoglycan, thus rendering the drug ineffectual against many Gram-negative pathogens.

Between the cytoplasmic membrane and the outer membrane of Gram-negative bacteria is a **periplasmic space** (see Figure 3.15b). The periplasmic space contains the peptidoglycan and *periplasm,* the name given to the gel between the membranes of these Gram-negative cells. Periplasm contains water, nutrients, and substances secreted by the cell, such as digestive enzymes and proteins involved in specific transport. The enzymes function to catabolize large nutrient molecules into smaller molecules that can be absorbed or transported into the cell.

Because the cell walls of Gram-positive and Gram-negative bacteria differ, the Gram stain is an important diagnostic tool. After the Gram staining procedure, Gram-negative cells appear pink, and Gram-positive cells appear purple.

Bacteria Without Cell Walls

A few bacteria, such as *Mycoplasma pneumoniae* (mī′kō-plaz-mă nū-mō′nē-ī), lack cell walls entirely. In the past, these bacteria were often mistaken for viruses because of their small size and

[5]From Greek *endo*, meaning inside, and *toxikon*, meaning poison.

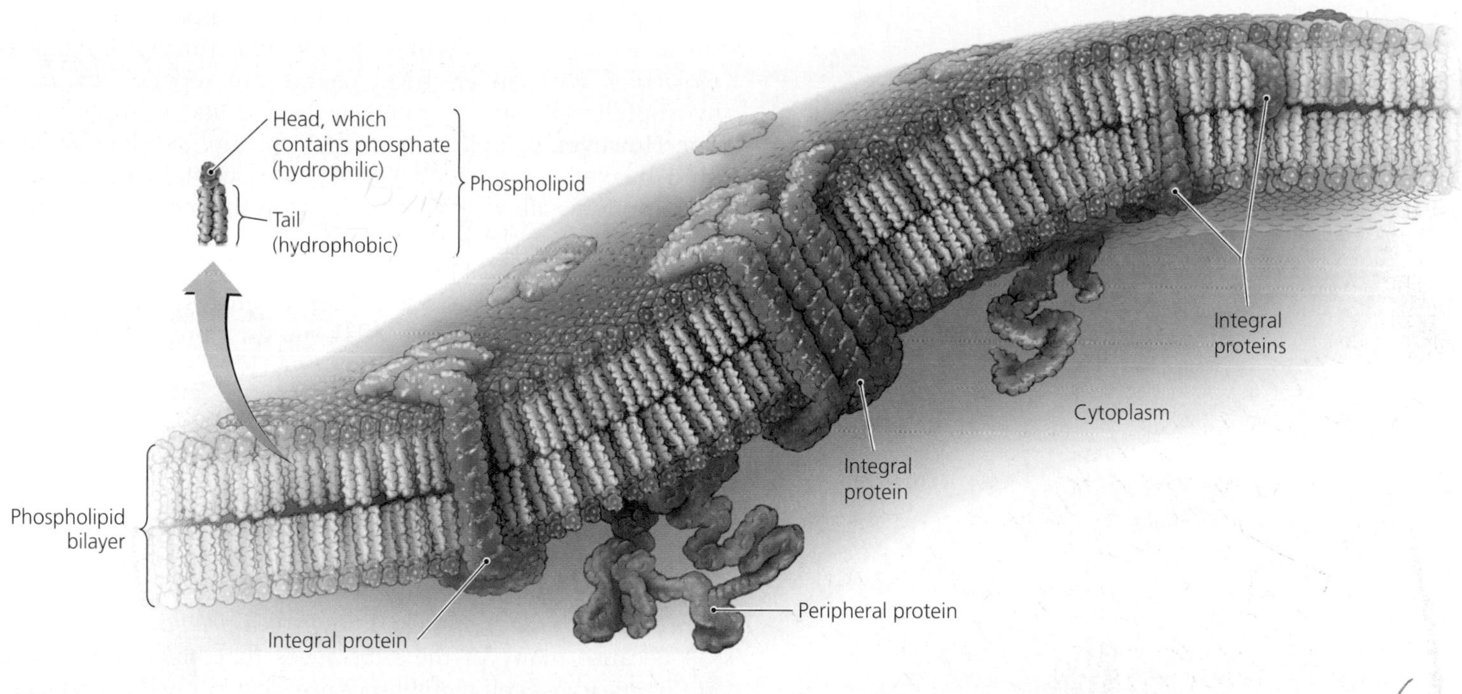

Figure 3.16
The structure of a prokaryotic cytoplasmic membrane: a phospholipid bilayer.

lack of walls. However, they do have other features of prokaryotic cells, such as prokaryotic ribosomes (discussed later in the chapter).

CRITICAL THINKING

After a man infected with the bacterium *Escherichia coli* was treated with the correct antibiotic for this pathogen, the bacterium was no longer found in the man's blood, but his symptoms of fever and inflammation worsened. What caused the man's response to the treatment? Why was his condition worsened by the treatment?

Bacterial Cytoplasmic Membranes

Beneath the glycocalyx and the cell wall is a **cytoplasmic membrane.** The cytoplasmic membrane may also be referred to as *the cell membrane* or a *plasma membrane.*

Structure

Learning Objective

- ✓ Diagram a phospholipid bilayer, and explain its significance in reference to a cytoplasmic membrane.
- ✓ Explain the fluid mosaic model of membrane structure.

Cytoplasmic membranes are about 8 nm thick and composed of phospholipids (see Figure 2.16) and associated proteins. Some bacterial membranes also contain sterol-like molecules, called *hopanoids,* that help stabilize the membrane.

The structure of a cytoplasmic membrane is referred to as a **phospholipid bilayer (Figure 3.16).** A phospholipid molecule is bipolar; that is, the two ends of the molecule are different. The phosphate-containing heads of each phospholipid molecule are *hydrophilic,*[6] that is, they are attracted to water at the two surfaces of the membrane. The hydrocarbon tails of each phospholipid molecule are *hydrophobic*[7] and huddle together with other tails in the interior of the membrane, away from water. Phospholipids placed in a watery environment naturally form a bilayer because of their bipolar nature.

About half of a bacterial cytoplasmic membrane is composed of *integral proteins* inserted amidst the phospholipids. Some integral proteins penetrate the entire bilayer; others are found in only half the bilayer. In contrast, *peripheral proteins* are loosely attached to the membrane on one side or the other. Proteins of cell membranes may act as recognition proteins, enzymes, receptors, carriers, or channels.

The **fluid mosaic model** describes our current understanding of membrane structure. The term *mosaic* indicates that the membrane proteins are arranged in a way that resembles the tiles in a mosaic, and *fluid* indicates that the proteins and lipids are free to flow laterally within a membrane. **ANIMATIONS:** ***Membrane Structure***

[6]From Greek *hydro,* meaning water, and *philos,* meaning love.
[7]From Greek *hydro,* meaning water, and *phobos,* meaning fear.

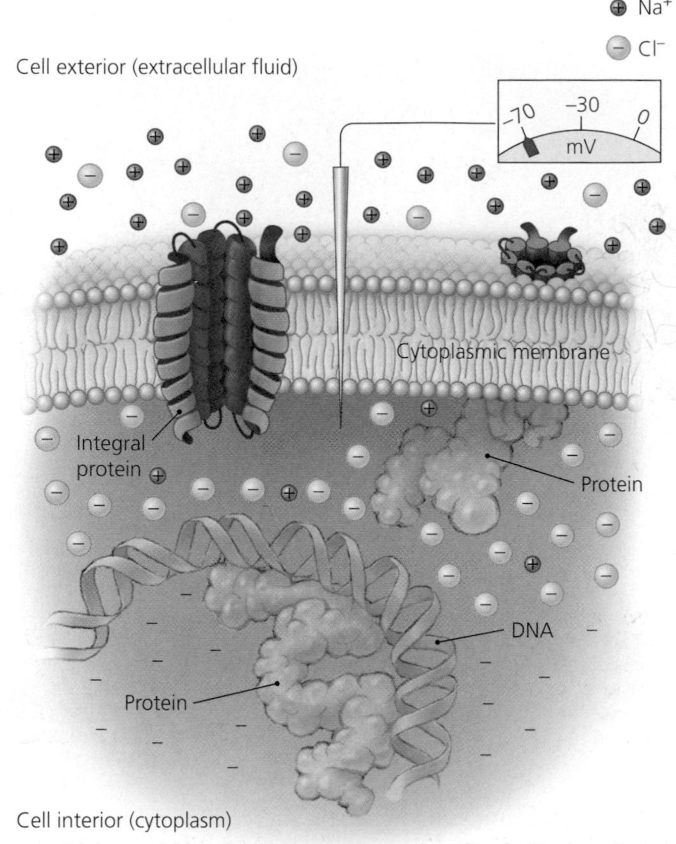

▲ **Figure 3.17 Electrical potential of a cytoplasmic membrane.** The electrical potential exists across a membrane because there are more negative charges inside the cell than outside it.

Function

Learning Objectives

✓ Describe the functions of a cytoplasmic membrane as they relate to permeability.

✓ Compare and contrast the passive and active processes by which materials cross a cytoplasmic membrane.

✓ Define *osmosis*, and distinguish among isotonic, hypertonic, and hypotonic solutions.

A cytoplasmic membrane does more than separate the contents of the cell from the outside environment. The cytoplasmic membrane controls the passage of substances into and out of the cell. Nutrients are brought into the cell and wastes are removed. The membrane also functions in production of molecules for energy storage, and to harvest light energy in photosynthetic bacteria. Energy storage and photosynthesis are discussed in Chapter 5.

In its function of controlling the contents of the cell, the cytoplasmic membrane is **selectively permeable;** that is, it allows some substances to cross it while preventing the crossing of others. How does a membrane exert control over the contents of the cell, and the substances that move across it? **ANIMATIONS:** ***Membrane Permeability***

A phospholipid bilayer is naturally impermeable to most substances. Large molecules cannot cross through it; ions and molecules with an electrical charge are repelled by it; and hydrophilic substances cannot easily cross its hydrophobic interior. However, cytoplasmic membranes, unlike plain phospholipid bilayers in a scientist's test tube, contain proteins, and these proteins allow substances to cross the membrane by functioning as pores, channels, or carriers.

Movement across the cytoplasmic membrane occurs either by passive or active processes. Passive processes do not require the expenditure of a cell's ATP energy store, whereas active processes require the expenditure of ATP, either directly or indirectly. Active and passive processes will be discussed shortly, but first you must understand another feature of selectively permeable cytoplasmic membranes: their ability to maintain a *concentration gradient.*

Membranes enable a cell to concentrate chemicals on one side of the membrane or the other. The difference in concentration of a chemical on the two sides of a membrane is its **concentration gradient** (also known as a *chemical gradient*).

Because many of the substances that have concentration gradients across cell membranes are electrically charged chemicals, a corresponding **electrical gradient,** or voltage, also exists across the membrane **(Figure 3.17)**. For example, a greater concentration of negatively charged proteins exists inside the membrane, and positively charged sodium ions are more concentrated outside the membrane. One result of the segregation of electrical charges by a membrane is that the interior of a cell is usually electrically negative compared to the exterior. This tends to repel negatively charged chemicals and attract positively charged substances into the cell. **ANIMATIONS:** ***Passive Transport: Principles of Diffusion***

Passive Processes

In passive processes, the electrochemical gradient provides a source of energy; the cell does not expend its ATP energy reserve. Passive processes include diffusion, facilitated diffusion, and osmosis.

Diffusion **Diffusion** is the net movement of a chemical down its concentration gradient—that is, from an area of higher concentration to an area of lower concentration. It requires no energy output by the cell, a common feature of all passive processes. In fact, diffusion occurs even in the absence of cells or their membranes. In the case of diffusion into or out of cells, only chemicals that are small or lipid soluble can diffuse through the lipid portion of the membrane **(Figure 3.18a)**. For example, oxygen, carbon dioxide, alcohol, and fatty acids can freely diffuse through the cytoplasmic membrane, but molecules such as glucose and proteins cannot.

Facilitated Diffusion The phospholipid bilayer blocks the movement of large or electrically charged molecules, so they do not cross the membrane unless there is a pathway for diffusion. As we have seen, cytoplasmic membranes contain integral proteins. Some of these proteins act as channels or carriers to allow certain molecules to diffuse into or out of the cell. This process is called **facilitated diffusion** because the proteins facilitate the

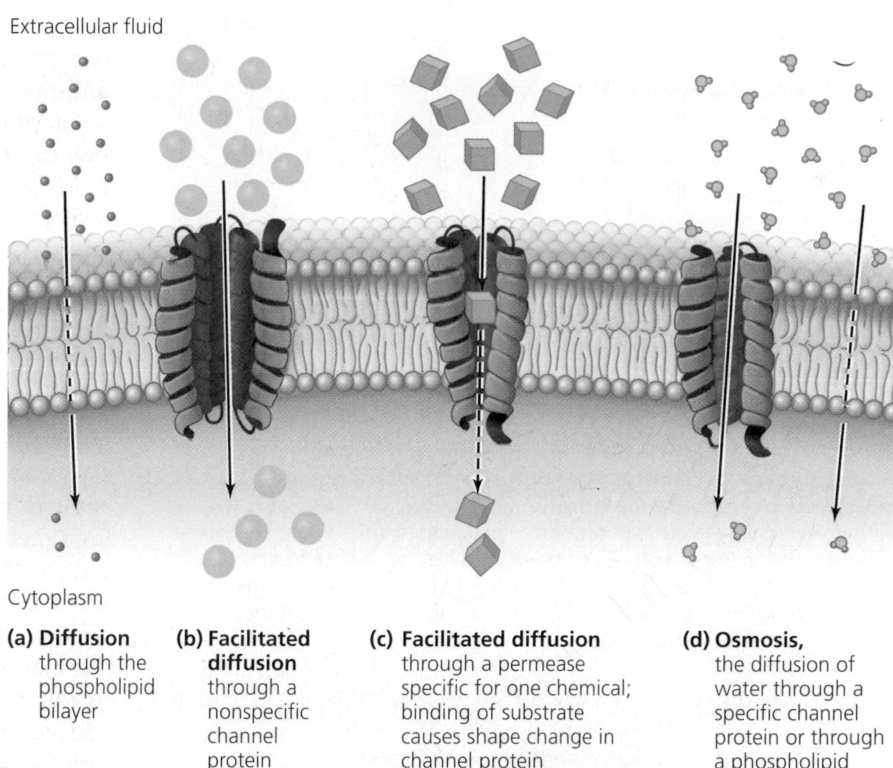
Extracellular fluid
Cytoplasm
(a) Diffusion through the phospholipid bilayer
(b) Facilitated diffusion through a nonspecific channel protein
(c) Facilitated diffusion through a permease specific for one chemical; binding of substrate causes shape change in channel protein
(d) Osmosis, the diffusion of water through a specific channel protein or through a phospholipid bilayer

▶ **Figure 3.18 Passive processes of movement across a cytoplasmic membrane.** Passive processes always involve movement down an electrochemical gradient. **(a)** Diffusion. **(b)** Facilitated diffusion through a nonspecific channel protein. **(c)** Facilitated diffusion through a specific channel protein. **(d)** Osmosis through a nonspecific channel protein or through a phospholipid bilayer.

process by providing a pathway for diffusion. The cell expends no energy in facilitated diffusion; the electrochemical gradient provides all of the energy necessary.

Some channel proteins allow the passage of a range of chemicals that have the right size or electrical charge **(Figure 3.18b)**. Other channel proteins, known as *permeases,* are more specific, carrying only certain substrates **(Figure 3.18c)**. A permease has a binding site that is selective for one substance.

CRITICAL THINKING

A scientist who is studying passive movement of chemicals across the cytoplasmic membrane of *Salmonella enterica* serotype Typhi measures the rate at which two chemicals diffuse into a cell as a function of external concentration. The results are shown in the following figure. Chemical A diffuses into the cell more rapidly than does B at lower external concentrations, but the rate levels off as the external concentration increases. The rate of diffusion of chemical B continues to increase as the external concentration increases.

1. How can you explain the differences in the diffusion rates of chemicals A and B?
2. Why does the diffusion rate of chemical A taper off?
3. How could the cell increase the diffusion rate of chemical A?
4. How could the cell increase the diffusion rate of chemical B?

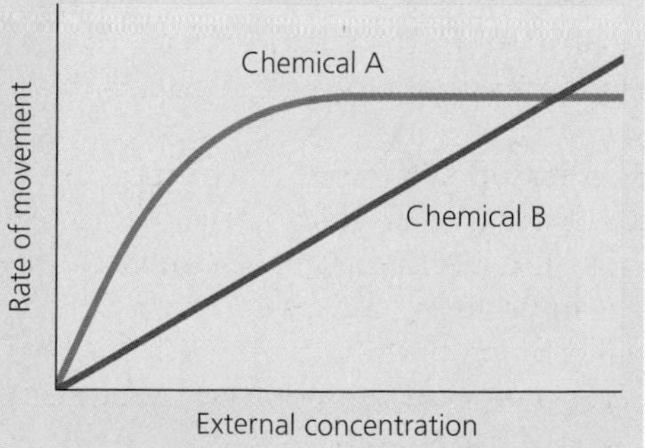
Chemical A
Chemical B
Rate of movement
External concentration

Osmosis When discussing simple and facilitated diffusion, we considered a solution in terms of the solutes (dissolved materials) it contains, because it is those solutes that move in and out of the cell. In contrast, with osmosis it is useful to consider the concentration of the solvent, which in organisms is always water. **Osmosis** is the special name given to the diffusion of water across a selectively permeable membrane—that is, across a membrane that is permeable to water molecules, but not to all solutes that are present, such as proteins, amino acids, salts, or glucose **(Figure 3.18d)**. Because these solutes cannot freely penetrate the membrane, they cannot diffuse no matter how unequal their concentrations on either side of the membrane may be. Instead, what diffuses is the water, which crosses from the side of the membrane that contains a higher concentration of water (lower concentration of solute) to the side that contains a lower concentration of water (higher concentration of solute). Osmosis continues until equilibrium is reached, or until the pressure of water is equal to the force of osmosis **(Figure 3.19)**.

We commonly classify solutions according to their concentrations of solutes. When solutions on either side of a selectively permeable membrane have the same concentration of solutes, the two solutions are said to be **isotonic**.[8] In an isotonic situation, neither side of a selectively permeable membrane will experience a net loss or gain of water **(Figure 3.20a)**.

When the concentrations of solutions are unequal, the solution with the higher concentration of solutes is said to be **hypertonic**[9] to the other. The solution with a lower concentration

[8]From Greek *isos,* meaning equal, and *tonos,* meaning tone.
[9]From Greek *hyper,* meaning more or over.

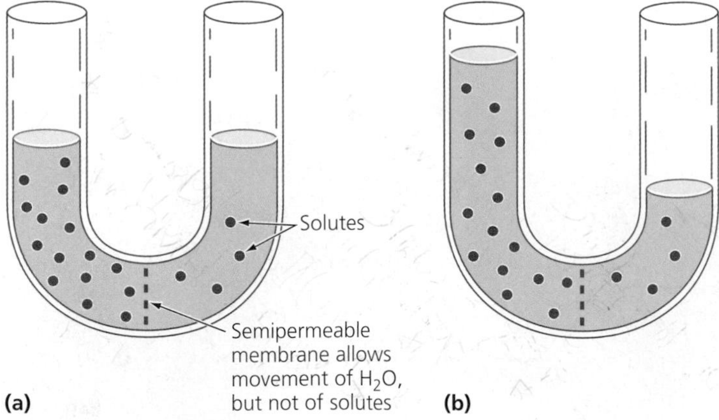

▲ **Figure 3.19 Osmosis, the diffusion of water across a selectively permeable membrane.** **(a)** A membrane separates two solutions of different concentrations in a U-shaped tube. The membrane is permeable to water, but not to the solute. **(b)** After time has passed, water has moved down its concentration gradient until water pressure prevented the osmosis of any additional water. *Which side of the tube more closely represents a living cell?*

Figure 3.19 *The left-hand side represents the cell, because cells are typically hypertonic to their environment.*

of solutes is **hypotonic**[10] in comparison. Note that the terms *hypertonic* and *hypotonic* refer to the concentration of solute, even though osmosis refers to the movement of the *solvent*, which, in cells, is water. The terms *isotonic, hypertonic,* and *hypotonic* are relative. For example, a glass of tap water is isotonic to another glass of the same water, but it is hypertonic compared to distilled water, and hypotonic when compared to seawater. In biology, the three terms are traditionally used relative to the interior of cells.

[10]From Greek *hypo*, meaning less or under.

Obviously, a higher concentration of solutes necessarily means a lower concentration of water; that is, a hypertonic solution has a lower concentration of water than does a hypotonic solution. Like other chemicals, water moves down its concentration gradient from a hypotonic solution into a hypertonic solution. A cell placed in a hypertonic solution will therefore lose water and shrivel **(Figure 3.20b)**.

On the other hand, water will diffuse into a cell placed in a hypotonic solution because the cell has a higher solutes-to-water concentration. As water moves into the cell, water pressure against its cytoplasmic membrane increases, and the cell expands **(Figure 3.20c)**. One function of a cell wall, such as the peptidoglycan of bacteria, is to resist further osmosis and prevent cells from bursting.

It is useful to compare solutions to the concentration of solutes in a patient's blood cells. Isotonic saline solutions administered to a patient have the same percent dissolved solute (in this case, salt) as do the patient's blood cells. Thus, the patient's intracellular and extracellular environments remain in equilibrium when an isotonic saline solution is administered. However, if the patient is infused with a hypertonic solution, water will move out of the patient's cells, and the cells will shrivel, a condition called *crenation*. Conversely, if a patient is infused with a hypotonic solution, water will move into the patient's cells, which will swell and possibly burst. **ANIMATIONS:** ***Passive Transport: Special Types of Diffusion***

CRITICAL **THINKING**

Solutions hypertonic to bacteria and fungi are used for food preservation. For instance, jams and jellies are hypertonic with sugar, and pickles are hypertonic with salt. How do hypertonic solutions kill bacteria and fungi that would otherwise spoil these foods?

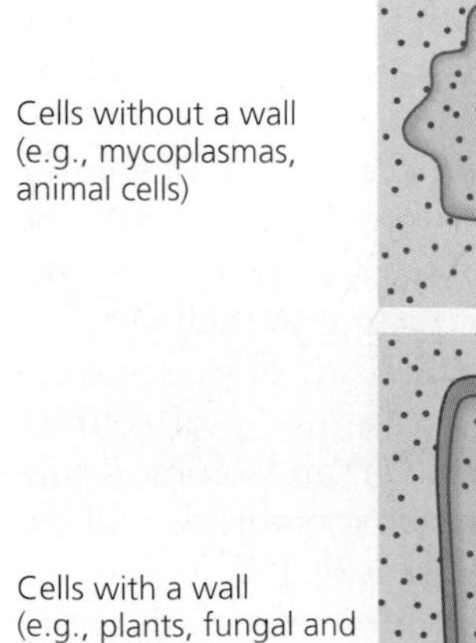

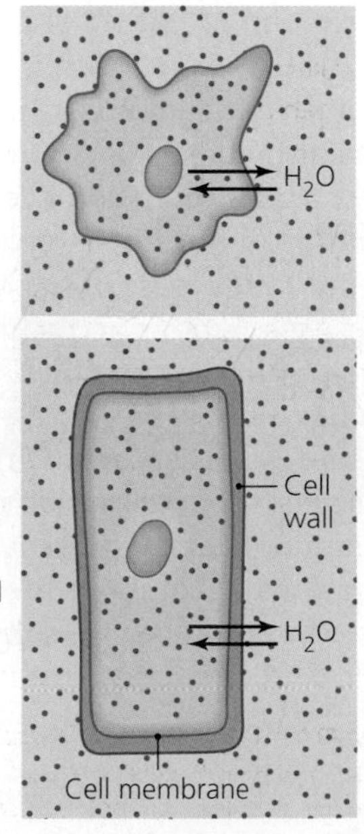

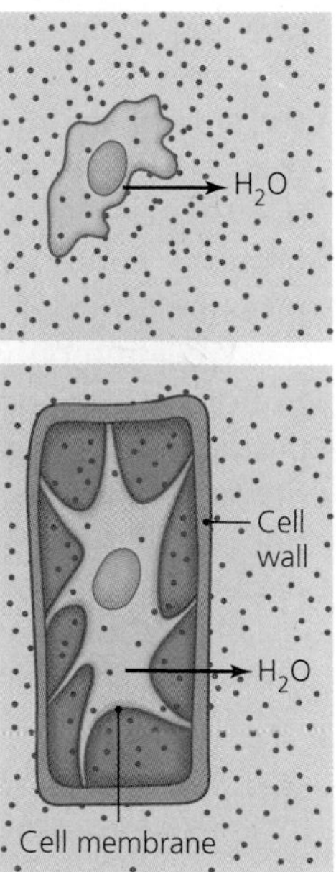

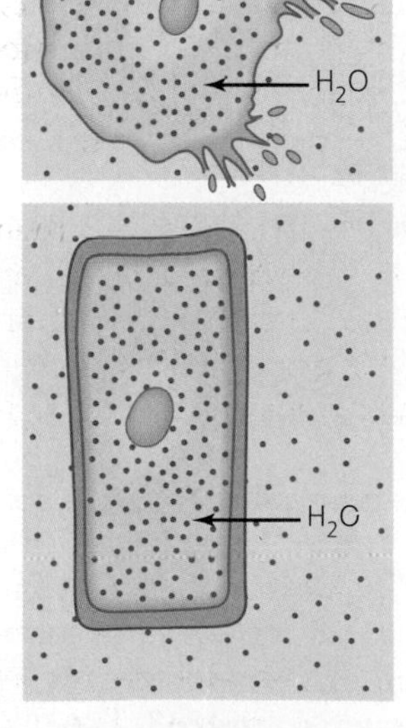

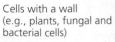

◀ **Figure 3.20 Effects of isotonic, hypertonic, and hypotonic solutions on cells.** **(a)** Cells in isotonic solutions experience no net movement of water. **(b)** Cells in hypertonic solutions shrink due to the net movement of water out of the cell. **(c)** Cells in hypotonic solutions undergo a net gain of water. Animal cells burst because they lack a cell wall; in cells with a cell wall, the pressure of water pushing against the interior of the wall eventually stops the movement of water into the cell.

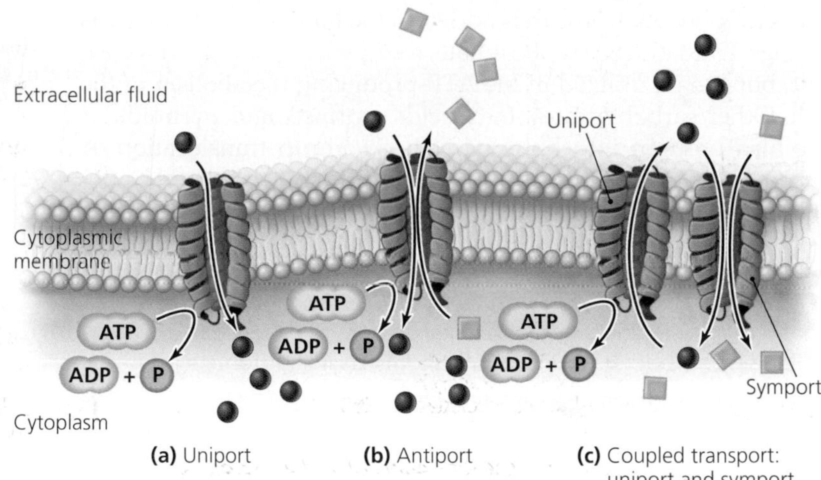

▶ **Figure 3.21 Mechanisms of active transport.** **(a)** Via a uniport. **(b)** Via an antiport. **(c)** Via a uniport coupled with a symport. In this example, the membrane uses ATP energy to pump one substance out through a uniport. As this substance flows back into the cell it brings another substance with it through the symport. *What is the usual source of energy for active transport?*

Figure 3.21 *ATP is the usual source of energy for active transport processes.*

Active Processes

As stated previously, active processes require the cell to expend energy stored in ATP molecules to move materials across the cytoplasmic membrane against their electrochemical gradient. This is analogous to moving water uphill. As we will see, ATP may be utilized directly during transport, or indirectly at some other site and at some other time. Active processes in bacteria include *active transport* by means of carrier proteins and a special process termed *group translocation*. **ANIMATIONS:** ***Active Transport: Overview***

Active Transport Like facilitated diffusion, **active transport** utilizes transmembrane permease proteins; however, the functioning of active transport proteins requires the cell to expend ATP to transport molecules across the membrane. Some such proteins are referred to as *gated channels* or *ports* because they are controlled. When the cell is in need of a substance, the protein becomes functional (the gate "opens"). At other times the gate is "closed."

If only one substance is transported at a time, the permease is called a *uniport* **(Figure 3.21a)**. In contrast, *antiports* simultaneously transport two chemicals, but in opposite directions; that is, one substance is transported into the cell at the same time that a second substance is transferred out of the cell **(Figure 3.21b)**. In other types of active transport, two substances move together in the same direction across the membrane by means of a single carrier protein. Such proteins are known as *symports* **(Figure 3.21c)**.

In all cases, active transport moves substances against their electrochemical gradient. Typically, the protein acts as an ATPase—an enzyme that breaks down ATP into ADP and inorganic phosphate during transport, releasing energy that is used to move the chemical against its electrochemical gradient across the membrane.

With symports and antiports, one chemical's electrochemical gradient may provide the energy needed to transport the second chemical, a mechanism called *coupled transport* (Figure 3.21c). For example, H^+ moving into a cell down its electrochemical gradient by facilitated diffusion provides energy to carry glucose into the cell, against the glucose gradient. The two processes are linked by a symport. However, ATP may still be utilized for transport because the H^+ gradient can be previously established by the active pumping of H^+ to the outside of the cell by an ATP-dependent H^+ uniport. The use of ATP is thus separated in time and space from the active transport of glucose, but ATP was still expended. **ANIMATIONS:** ***Active Transport: Types***

Group Translocation **Group translocation** is an active process that occurs only in some bacteria. In group translocation, the substance being actively transported across the membrane is chemically changed during transport **(Figure 3.22)**. The membrane is impermeable to the altered substance, trapping it inside the cell. Group translocation is very efficient at bringing substances into a cell. It can operate efficiently even if the external concentration of the chemical being transported is as low as 1 part per million (ppm).

One well-studied example of group translocation is the accumulation of glucose inside a bacterial cell. As glucose is transported across the bacterial cell membrane, it is phosphorylated;

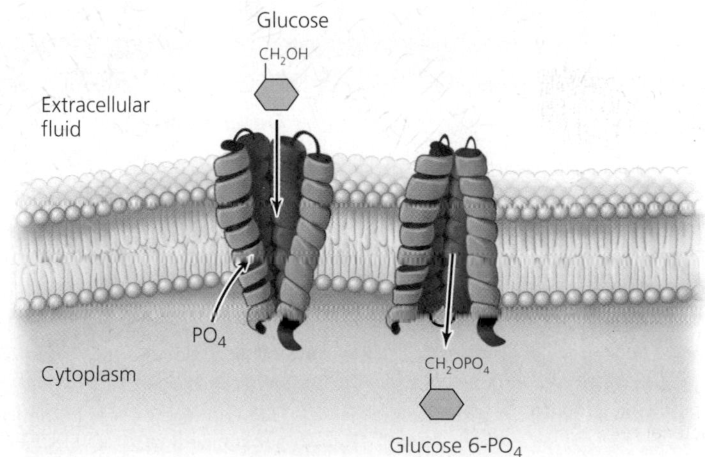

▲ **Figure 3.22 Group translocation.** This process involves a chemical change in a substance as it is being transported. This figure depicts glucose being transported into a bacterial cell via group translocation.

that is, a phosphate group is added to the glucose. The glucose is changed into glucose 6-phosphate, a sugar that cannot cross back out, but can be utilized in the ATP-producing metabolism of the cell. Other carbohydrates, fatty acids, purines, and pyrimidines are also brought into bacterial cells by group translocation. A summary of bacterial transport processes is shown in Table 3.2.

Cytoplasm of Bacteria

Learning Objectives

- ✓ Describe bacterial cytoplasm and its basic contents.
- ✓ Define *inclusion* and give two examples.
- ✓ Describe the formation and function of endospores.

Cytoplasm is the general term used to describe the gelatinous material inside a cell. Cytoplasm is semitransparent, fluid, elastic, and aqueous. It is composed of cytosol, inclusions, ribosomes, and in many cells, a cytoskeleton. Some bacterial cells produce internal, resistant, dormant forms called endospores.

Cytosol

The liquid portion of the cytoplasm is called **cytosol.** It is mostly water, but it also contains dissolved and suspended substances, including ions, carbohydrates, proteins (mostly enzymes), lipids, and wastes. The cytosol of prokaryotes also contains the cell's DNA in a region called the **nucleoid.** Recall that a distinctive feature of prokaryotes is lack of a phospholipid membrane surrounding this DNA.

Most bacteria have a single, circular DNA molecule organized as a chromosome. Some bacteria, such as *Vibrio cholerae* (vib'rē-ō kol'er-ī), the bacterium that causes cholera, are unusual in that they have two chromosomes.

The cytosol is the site of some chemical reactions. For example, enzymes within the cytosol function to produce amino acids and degrade sugar.

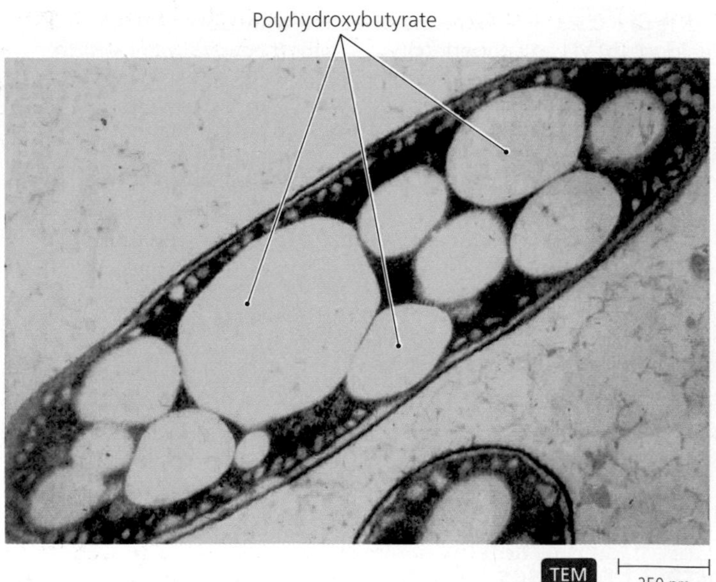

▲ **Figure 3.23** Granules of PHB in the bacterium *Azotobacter chroococcum.*

Inclusions

Deposits, called **inclusions,** are often found within bacterial cytosol. Rarely, a cell surrounds its inclusions with a polypeptide membrane. Inclusions may include reserve deposits of lipids, starch, or compounds containing nitrogen, phosphate, or sulfur. Such chemicals may be taken in and stored in the cytosol when nutrients are in abundance and then utilized when nutrients are scarce. The presence of specific inclusions is diagnostic for several pathogenic bacteria.

Many bacteria store carbon and energy in molecules of glycogen, which is a polymer of glucose molecules, or as a lipid polymer called *polyhydroxybutyrate (PHB)* **(Figure 3.23)**. Long

TABLE 3.2 Transport Processes Across Bacterial Cytoplasmic Membranes

	Description	Examples of Transported Substances
Passive transport processes	Processes require no use of energy by the cell; the electrochemical gradient provides energy.	
Diffusion	Molecules move down their electrochemical gradient through the phospholipid bilayer of the membrane.	Oxygen, carbon dioxide, lipid-soluble chemicals
Facilitated diffusion	Molecules move down their electrochemical gradient through channels or carrier proteins.	Glucose, fructose, urea, some vitamins
Osmosis	Water molecules move down their concentration gradient across a selectively permeable membrane.	Water
Active transport processes	Cell expends energy in the form of ATP to move a substance against its electrochemical gradient.	
Active transport	ATP-dependent carrier proteins bring substances into cell.	Na^{+}, K^{+}, Ca^{2+}, H^{+}, Cl^{-}
Group translocation	The substance is chemically altered during transport; found only in some bacteria.	Glucose, mannose, fructose

chains of PHB accumulate as inclusion granules in the cytoplasm. Slight chemical modification of PHB produces a plastic that can be used for packaging and other applications (see **Beneficial Microbes: Plastics Made Perfect?** on p. 74). PHB plastics are biodegradable, breaking down in a landfill in a few weeks rather than persisting for years as petroleum-based plastics do.

Many aquatic cyanobacteria (blue-green photosynthetic bacteria) contain inclusions called *gas vesicles* that store gases in protein sacs. The gases buoy the cells to the surface and into the light needed for photosynthesis. Other interesting inclusions are the small crystals of magnetite stored by *magnetobacteria,* featured at the beginning of this chapter. Invaginations of the cytoplasmic membrane surround the magnetite to form membrane-bound sacs.

Endospores

Some bacteria, notably *Bacillus* (ba-sil'ŭs) and *Clostridium* (klos-trid'ē-ŭm), are characterized by the ability to produce unique structures called **endospores,** which are important for several reasons, including their durability and potential pathogenicity. Though some people refer to endospores simply as "spores," endospores should not be confused with the reproductive spores of actinobacteria, algae, and fungi. A single bacterial cell, called a *vegetative* cell to distinguish it from an endospore, transforms into only one endospore, which then germinates to grow into only one vegetative cell; therefore, endospores are not reproductive structures. Instead, endospores constitute a defensive strategy against hostile or unfavorable conditions.

A vegetative cell normally transforms itself into an endospore only when one or more nutrients (such as carbon or nitrogen) are in limited supply. The process of endospore formation, called *sporulation,* requires 8 to 10 hours and proceeds in seven steps **(Figure 3.24)**. During the process, two membranes, a thick layer of peptidoglycan, and a spore coat form around a copy of the cell's DNA and a small portion of cytoplasm. The cell deposits large quantities of dipicolinic acid, calcium, and DNA-binding proteins within the endospore while removing most of the water. Depending on the species, a cell forms an endospore either *centrally, subterminally* (near one end), or *terminally* (at one end). Sometimes an endospore is so large it swells the vegetative cell.

Endospores are extremely resistant to drying, heat, radiation, and lethal chemicals. For example, they remain alive in boiling water for several hours; are unharmed by alcohol, peroxide, bleach, and other toxic chemicals; and can tolerate over 400 rad of radiation, which is more than five times the dose that is lethal to most humans. Endospores are stable resting stages that barely metabolize—they are essentially in a state of suspended animation—and they germinate only when conditions improve. Scientists do not know how endospores are able to resist harsh conditions, but it appears that the double membrane, spore coats, dipicolinic acid, calcium, and DNA-binding proteins serve to stabilize DNA and enzymes, protecting them from adverse conditions.

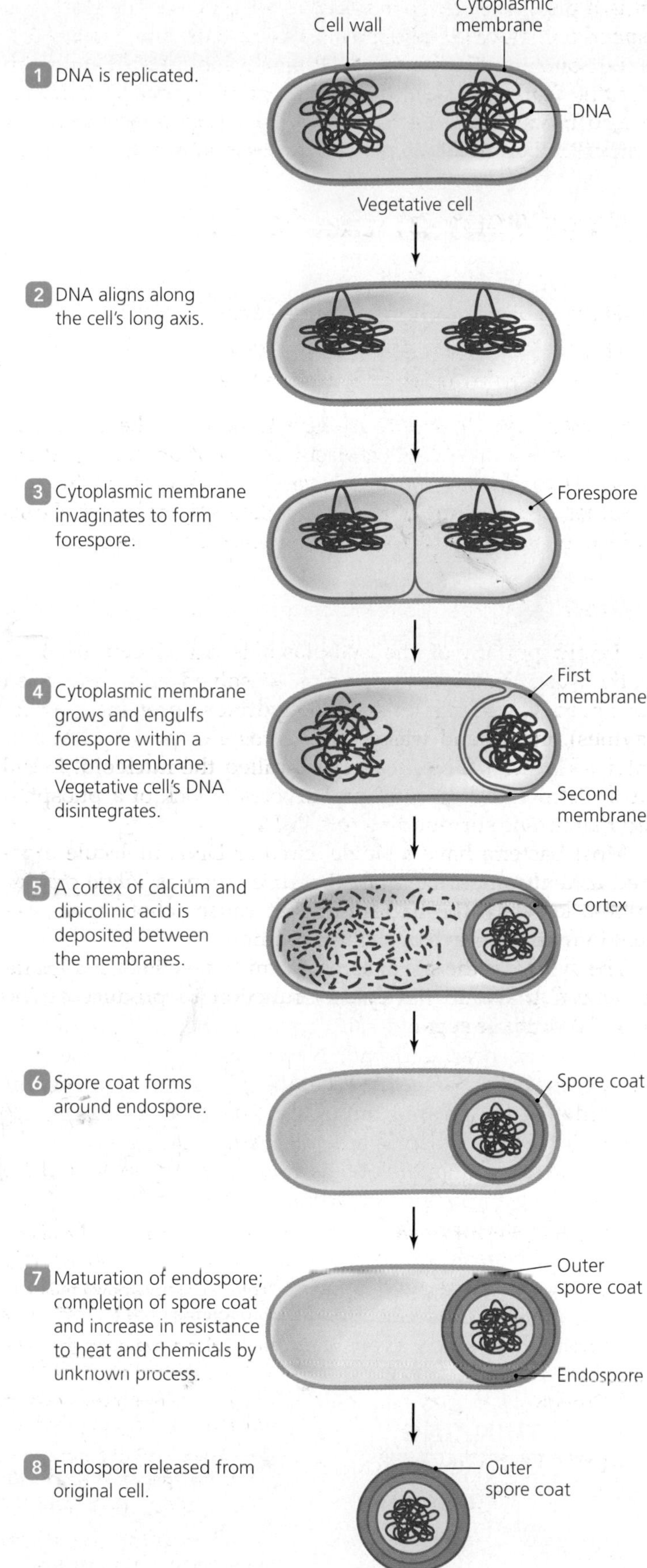
Cell wall
Cytoplasmic membrane
1 DNA is replicated.
DNA
Vegetative cell
2 DNA aligns along the cell's long axis.
3 Cytoplasmic membrane invaginates to form forespore.
Forespore
4 Cytoplasmic membrane grows and engulfs forespore within a second membrane. Vegetative cell's DNA disintegrates.
First membrane
Second membrane
5 A cortex of calcium and dipicolinic acid is deposited between the membranes.
Cortex
6 Spore coat forms around endospore.
Spore coat
7 Maturation of endospore; completion of spore coat and increase in resistance to heat and chemicals by unknown process.
Outer spore coat
Endospore
8 Endospore released from original cell.
Outer spore coat

▲ **Figure 3.24 The formation of an endospore.** The steps depicted occur over a period of 8–10 hours.

BENEFICIAL MICROBES

PLASTICS MADE PERFECT?

▲ *PHB bottle caps. One partially biodegraded in 60 days.*

Petroleum-based plastics play a considerable role in modern life, appearing in packaging, bottles, appliances, furniture, automobiles, disposable diapers, and many other synthetic goods. Despite the good that plastic brings to our lives, there are problems with this artificial polymer.

Manufacturers make plastic from oil, exacerbating the dependence of the U.S. on foreign supplies of crude oil. Consumers discard plastic, filling landfills with more than 15 million tons of plastic every year in the United States. Because plastic is artificial, microorganisms do not break it down effectively, and discarded plastic will remain in landfills for decades or centuries. What is needed is a functional "green" plastic—a plastic that is strong and light and that can be shaped and colored as needed, yet a plastic that is naturally biodegradable.

Enter the bacteria. Many bacterial cells, particularly Gram-negative bacteria, use polyhydroxybutyrate (PHB) as a storage molecule and energy source much as humans use fat. PHB and similar storage molecules turn out to be rather versatile plastics that are produced when bacteria metabolizing certain types of sugar are simultaneously deprived of an essential element such as nitrogen, phosphorus, or potassium. The bacteria, faced with such a nutritionally stressed environment, convert the sugar to PHB, which they store as intracellular inclusions. Scientists harvest these biologically created molecules by breaking the cells open and treating the cytoplasm with chemicals to isolate the plastics and remove dangerous endotoxin.

Purified PHB possesses many of the properties of petrochemically derived plastic—its melting point, crystal structure, molecular weight, and strength are very similar. Further, PHB has a singular, overwhelming advantage compared to artificial plastic: PHB is naturally and completely biodegradable—bacteria catabolize PHB into carbon dioxide and water. The positive effect this would have on our overtaxed landfills would be tremendous, and replacing just half of the oil-based plastic used in the United States with PHB could reduce oil imports by more than 250 million barrels per year, improving our foreign trade balance and reducing our dependence on overseas oil suppliers.

The ability to survive harsh conditions makes endospores the most resistant and enduring cells. In one case, scientists were able to revive endospores of *Clostridium* that had been sealed in a test tube for 34 years. This record pales, however, beside other researchers' claim to have revived *Bacillus* endospores from inside 250-million-year-old salt crystals retrieved from an underground site near Carlsbad, New Mexico. Some scientists question this claim, suggesting that the bacteria might be recent contaminants that entered through invisible cracks in the salt crystals. In any case, there is little doubt that endospores can remain viable for a minimum of tens, if not thousands, of years.

Endospore formation is a serious concern to food processors, health care professionals, and governments because endospores are resistant to treatments that inhibit other microbes, and because endospore-forming bacteria produce deadly toxins that cause such fatal diseases as anthrax, tetanus, and gangrene. Chapter 9 considers techniques for controlling endospore formers.

CRITICAL **THINKING**

Following the bioterrorist anthrax attacks in the fall of 2001, a news commentator suggested that people steam their mail for 30 seconds before opening it. Would the technique protect people from anthrax infections? Why or why not?

Nonmembranous Organelles

Learning Objective

✓ Describe the structure and function of ribosomes and the cytoskeleton.

As previously noted, prokaryotes do not usually have membranes surrounding their organelles. However, two types of *nonmembranous organelles* are found in direct contact with the cytosol in bacterial cytoplasm. Some investigators do not consider them to be true organelles because they lack a membrane, but other scientists consider them organelles. Nonmembranous organelles in bacteria include ribosomes and the cytoskeleton.

Ribosomes

Ribosomes are the sites of protein synthesis in cells. Bacterial cells have thousands of ribosomes in their cytoplasm, which gives cytoplasm a grainy appearance (see Figure 3.2). The approximate size of ribosomes—and indeed other cellular structures—is expressed in Svedbergs (S),[11] and is determined by their sedimentation rate—the rate at which they move to the bottom of a test tube during centrifugation. As you might expect, large, compact, heavy particles sediment faster than small,

[11]Svedberg units are named for Theodor Svedberg, a Nobel Prize winner and the inventor of the ultracentrifuge.

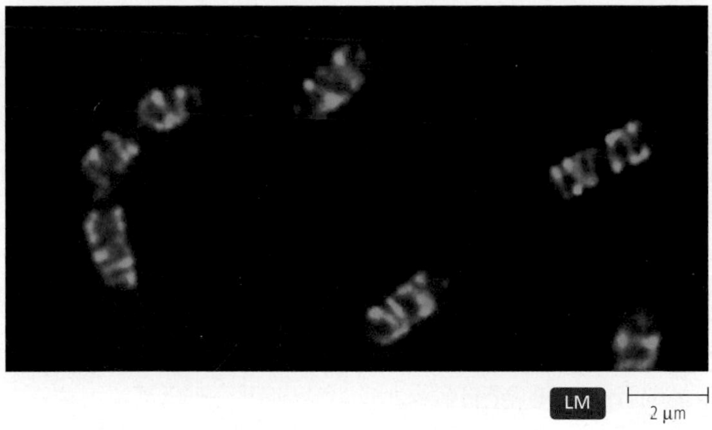

▲ **Figure 3.25 A simple helical cytoskeleton.** The rod-shaped bacterium *Bacillus subtilis* has a cytoskeleton composed of only a single protein, which has been stained with a fluorescent dye.

loosely packed, or light ones, and are assigned a higher number. Prokaryotic ribosomes are 70S; in contrast, the larger ribosomes of eukaryotes are 80S.

All ribosomes are composed of two subunits, each of which is composed of polypeptides and molecules of RNA called **ribosomal RNA (rRNA).** The subunits of prokaryotic 70S ribosomes are a smaller 30S subunit and a larger 50S subunit; the 30S subunit contains polypeptides and a single rRNA molecule, whereas the 50S subunit has polypeptides and two rRNA molecules. Because sedimentation rates are dependent not only on mass and size but also on shape, the sedimentation rates of subunits do not add up to the sedimentation rate of a whole ribosome.

Many antibacterial drugs act on bacterial 70S ribosomes or their subunits without deleterious effects on the larger 80S ribosomes of eukaryotic cells (see Chapter 10). This is why such drugs can stop protein synthesis in bacteria without affecting protein synthesis in a patient.

Cytoskeleton

Most cells contain an internal network of fibers called a **cytoskeleton** that plays a role in forming a cell's basic shape. Additionally, the contractile cytoskeleton of the nonflagellated bacterium *Spiroplasma* (spē-ro-plaz'mă) allows the cell to swim through its environment. Bacteria were long thought to lack cytoskeletons, but research has revealed that bacteria have simple ones **(Figure 3.25)**.

We have considered bacterial cells. Next we turn our attention to archaea—the other prokaryotic cells—and compare them to bacterial cells.

External Structures of Archaea

Archaeal cells have external structures similar to those seen in bacteria. These include glycocalyces, flagella, and fimbriae. Some archaea have another kind of proteinaceous appendage called a *hamus*. We consider each of these in order beginning with the outermost structures—glycocalyces.

Glycocalyces

Learning Objective

✓ Compare the structure and chemistry of archaeal and bacterial glycocalyces.

Like those of bacteria, archaeal glycocalyces are gelatinous, sticky, extracellular structures composed of polysaccharides, polypeptides, or both. Scientists have not studied archaeal glycocalyces as much as those of bacteria, but archaeal glycocalyces function at a minimum in the formation of biofilms—adhering cells to one another, to other types of cells, and to inanimate surfaces in the environment. Organized glycocalyces (capsules) of bacteria and bacterial biofilms are often associated with disease, but researchers have not demonstrated such a link between archaeal capsules or biofilms and disease. Though some research has demonstrated the presence of archaea in some biofilms associated with oral gum disease, no archaeon has been shown conclusively to be pathogenic.

Flagella

Learning Objectives

✓ Describe the structure and formation of archaeal flagella.

✓ Compare and contrast archaeal flagella with bacterial flagella.

Archaea use flagella to move through their environments, though at a slower speed than bacteria. An archaeal flagellum is superficially similar to a bacterial flagellum, consisting of a basal body, hook, and filament, each composed of protein. The flagellum extends outside the cell and is not covered by a membrane. The basal body anchors the flagellum in the cell wall and cytoplasmic membrane. As with bacterial flagella, archaeal flagella rotate like propellers.

However, scientists have discovered many differences between archaeal and bacterial flagella:

- Archaeal flagella are 10–14 nm in diameter, which is about half the thickness of bacterial flagella, and lack a central channel.
- Therefore, archaeal flagella grow with the addition of subunits at the base of the filament.
- The proteins making up archaeal flagella share common amino acid sequences across archaeal species. These are very different from the amino acid sequences common to bacterial flagella.
- Sugar molecules are attached to the filaments of many archaeal flagella, a condition which is rare in bacteria.
- Archaeal flagella are powered with energy stored in molecules of ATP, whereas the flow of hydrogen ions across the membrane powers bacterial flagella.

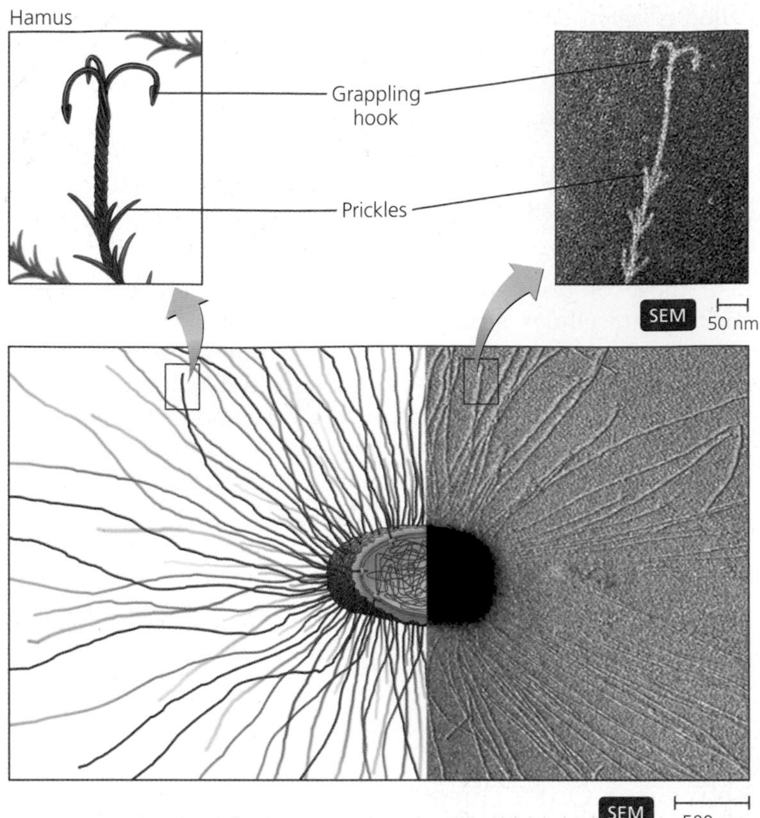

▲ **Figure 3.26 Archaeal hami.** Archaea use hami, which are shaped like grappling hooks on barbed wire, to attach themselves to structures in the environment.

- Archaeal flagella rotate together as a bundle both when they rotate clockwise and counterclockwise. In contrast, bacterial flagella operate independently when rotating clockwise.

These differences indicate that archaeal flagella arose independently of bacterial flagella; they are *analagous* structures—having similar structure without having a common ancestor.

Fimbriae and Hami

Learning Objectives

✓ Compare the structure and function of archaeal and bacterial fimbriae.

✓ Describe the structure and function of hami.

Many archaea have fimbriae—nonmotile, rodlike, sticky projections. As with bacteria, archaeal fimbriae are composed of protein and anchor the cells to one another and to environmental surfaces.

Some archaea make unique proteinaceous, fimbriae-like structures called **hami**[12] (singular: *hamus*). More than 100 hami may radiate from the surface of a single archaeon **(Figure 3.26)**. Each hamus is a helical filament with tiny prickles sticking out at regular intervals much like barbed wire. The end of the hamus is frayed into three distinct arms, each of which has a thickened end and bends back toward the cell to make the entire structure look like a grappling hook. Indeed, hami function to securely attach archaea to biological and inanimate surfaces.

[12]From Latin *hamus*, meaning prickle, claw, hook, or barb.

Archaeal Cell Walls and Cytoplasmic Membranes

Learning Objectives

✓ Contrast types of archaeal cell walls with each other and with bacterial cell walls.

✓ Contrast the archaeal cytoplasmic membrane with that of bacteria.

Most archaea, like most bacteria, have cell walls. All archaea have cytoplasmic membranes. However, there are distinct differences between archaeal and bacterial walls and membranes, further emphasizing the uniqueness of archaea.

Archaeal cell walls are composed of specialized proteins or polysaccharides. In some species, the outermost protein molecules form an array that coats the cell like chain mail. All archaeal walls lack peptidoglycan, which is common to all bacterial cell walls.

Gram-negative archaeal cells have an outer layer of protein rather than an outer lipid bilayer as seen in Gram-negative bacteria. Gram-negative archaea still appear pink when Gram stained. Gram-positive archaea have a thick cell wall and Gram stain purple, like Gram-positive bacteria.

Archaeal cells are typically spherical or rod shaped, though irregularly shaped, needle-like, rectangular, and flattened square archaea exist **(Figure 3.27)**.

Archaeal cytoplasmic membranes are composed of lipids that lack phosphate groups and have branched hydrocarbons linked to glycerol by ether linkages rather than the ester linkages seen in bacterial membranes (see Table 2.3 on p. 39). Ether linkages are stronger in many ways than ester linkages, allowing archaea to live in extreme environments such as near boiling water and in hypersaline lakes. Some archaea—particularly those that thrive in very hot water—have a single layer of lipid composed of two glycerol groups covalently linked with branched hydrocarbon chains.

The archaeal cytoplasmic membrane maintains electrical and chemical gradients in the cell. It also functions to control the import and export of substances from the cell using membrane proteins as ports and pumps, just as proteins are used in bacterial cytoplasmic membranes.

Cytoplasm of Archaea

Learning Objective

✓ Compare and contrast the cytoplasm of archaea with that of bacteria.

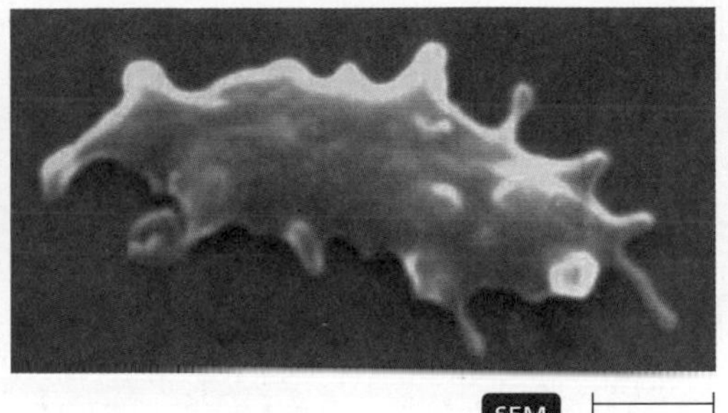

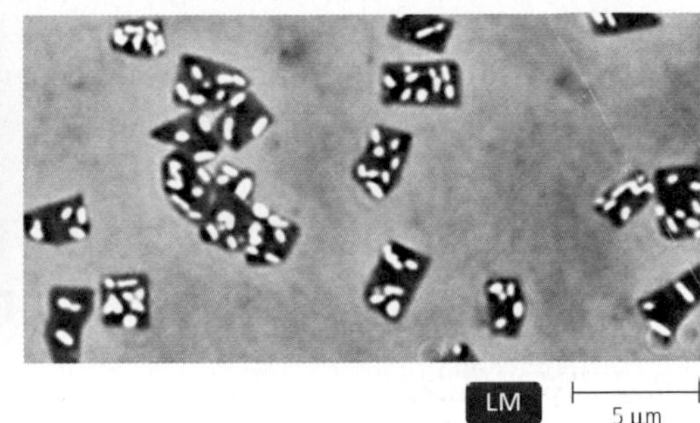

▲ **Figure 3.27 Representative shapes of archaea. (a)** Cocci, *Pyrococcus furiosus*, attached to rod-shaped *Methanopyrus kandleri*. **(b)** Irregularly shaped archaeon, *Thermoplasma acidophilum*. **(c)** Square archaeon, *Haloquadra walsbyi*.

Cytoplasm is the gel-like substance found in all cells, including archaea. Like bacteria, archaeal cells have 70S ribosomes, a fibrous cytoskeleton, and circular DNA suspended in a liquid cytosol. Also, like bacteria, they do not have membranous organelles.

However, archaeal cytoplasm differs from that of bacteria in several ways. For example, the ribosomes of archaea have different proteins than do the ribosomes of bacteria; indeed, archaeal ribosomal proteins are more like those of eukaryotes. Scientists further distinguish archaea from bacteria in that archaea use different metabolic enzymes to make RNA and use a genetic code more similar to the code used by eukaryotes. Chapter 7 discusses these genetic differences in more detail. Table 3.3 contrasts features of archaea and bacteria.

To this point, we have discussed basic features of bacterial and archaeal prokaryotic cells. Chapter 11 discusses the classification of prokaryotic organisms in more detail. Next we turn our attention to eukaryotic cells.

External Structure of Eukaryotic Cells

Some eukaryotic cells have glycocalyces, which are similar to those of prokaryotes.

Glycocalyces

Learning Objective

✓ Describe the composition, function, and importance of eukaryotic glycocalyces.

Animal and most protozoan cells lack cell walls, but they have sticky carbohydrate glycocalyces that are anchored to their cytoplasmic membranes via covalent bonds to membrane proteins and lipids. The functions of eukaryotic glycocalyces, which are never as structurally organized as prokaryotic capsules, include helping to anchor animal cells to each other, strengthening the cell surface, providing some protection

TABLE 3.3 **Some Structural Characteristics of Prokaryotes**

Feature	Archaea	Bacteria
Glycocalyx	Polypeptide or polysaccharide	Polypeptide or polysaccharide
Flagella	Present in some, 10–14 nm in diameter, grow at base, rotate both counterclockwise and clockwise as a bundle	Present in some, about 20 nm in diameter, grow at the tip, rotate independently clockwise to create tumbles
Fimbriae	Proteinaceous, used for attachment and in formation of biofilms	Proteinaceous, used for attachment, gliding motility, and in formation of biofilms
Pili	None discovered	Present in some, proteinaceous, used in bacterial exchange of DNA
Hami	Present in some, used for attachment	Absent
Cell walls	Present in most, composed of polysaccharides (not peptidoglycan) or proteins	Present in most, composed of peptidoglycan—a polysaccharide
Cytoplasmic membrane	Present in all, membrane lipids made with ether linkages, some have single lipid layer	Present in all, phospholipids made with ester linkages in bilayer
Cytoplasm	Cytosol contains circular DNA molecule and 70S ribosomes, ribosomal proteins similar to eukaryotic ribosomal proteins	Cytosol contains at least a circular DNA molecule and 70S ribosomes with bacterial proteins

against dehydration, and functioning in cell-to-cell recognition and communication. Glycocalyces are absent in eukaryotes that have cell walls such as plants and fungi.

Eukaryotic Cell Walls and Cytoplasmic Membranes

Learning Objectives

- ✓ Compare and contrast prokaryotic and eukaryotic cell walls and cytoplasmic membranes.
- ✓ Contrast exocytosis and endocytosis.
- ✓ Describe the role of pseudopodia in eukaryotic cells.

The eukaryotic cells of fungi, algae, plants, and some protozoa have cell walls. Recall that glycocalyces are absent from eukaryotes with cell walls; instead, the cell wall takes on one of the functions of a glycocalyx by providing protection from the environment. The wall also provides shape and support against osmotic pressure. Most eukaryotic cell walls are composed of various polysaccharides, but not the peptidoglycan seen in the walls of bacteria.

The walls of plant cells are composed of *cellulose,* a polysaccharide that is familiar to you as paper and dietary fiber. Fungi also have walls of polysaccharides, including cellulose, *chitin,* and/or *glucomannan.* The walls of algae **(Figure 3.28)** are composed of a variety of polysaccharides or other chemicals, depending on the type of alga. These chemicals include cellulose, proteins, *agar, carrageenan, silicates, algin, calcium carbonate,* or a combination of these substances. Chapter 12 discusses fungi and algae in more detail.

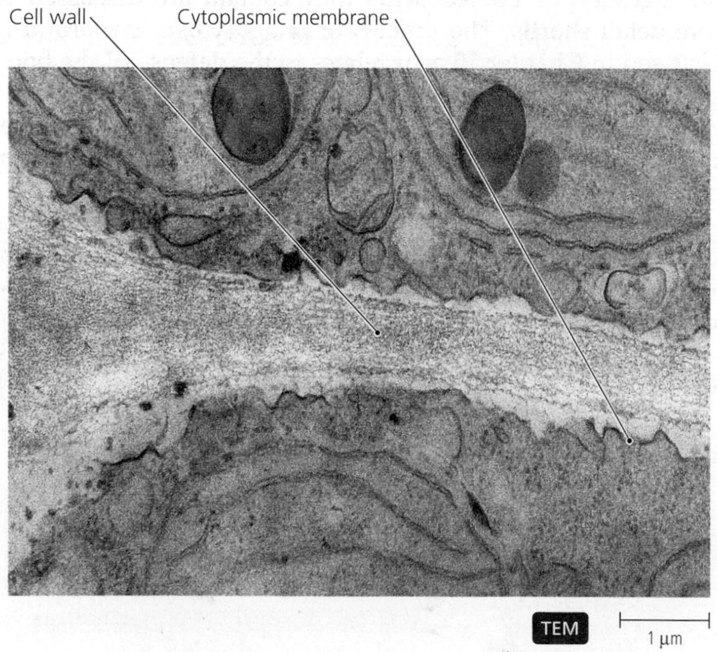

▲ **Figure 3.28 A eukaryotic cell wall.** The cell wall of the red alga *Gelidium* is composed of layers of the polysaccharide called agar. *What is the function of a cell wall?*

Figure 3.28 The cell wall provides support, protection, and resistance to osmotic forces.

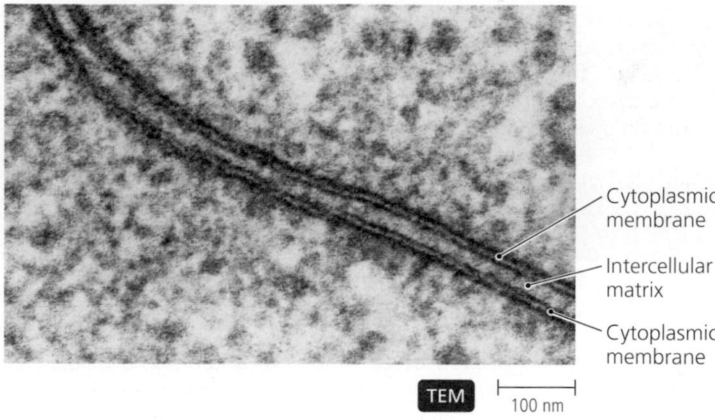

▲ **Figure 3.29 Eukaryotic cytoplasmic membrane.** Note that this micrograph depicts the cytoplasmic membranes of two adjoining cells.

All eukaryotic cells have cytoplasmic membranes **(Figure 3.29)**. (Botanists call the cytoplasmic membrane of an algal or plant cell a *plasmalemma.*) A eukaryotic cytoplasmic membrane, like those of bacteria, is a fluid mosaic of phospholipids and proteins, which act as recognition molecules, enzymes, receptors, carriers, or channels. Channel proteins for facilitated diffusion are more common in eukaryotes than in prokaryotes. Additionally, within multicellular organisms some membrane proteins serve to anchor cells to each other.

Eukaryotic cytoplasmic membranes may differ from prokaryotic membranes in several ways. Eukaryotic membranes contain steroid lipids *(sterols),* such as cholesterol in animal cells, that help maintain membrane fluidity. Paradoxically, at high temperatures sterols stabilize a phospholipid bilayer by making it less fluid, but at low temperatures sterols have the opposite effect—they prevent phospholipid packing, making the membrane more fluid.

Eukaryotic cytoplasmic membranes may contain small, distinctive assemblages of lipids and proteins that remain together as a functional group and do not flow independently amidst other membrane components. Such distinct regions are called **membrane rafts.** Eukaryotic cells appear to use membrane rafts to compartmentalize cellular processes including signaling the inside of the cell, protein sorting, and some kinds of cell movement. Some viruses, including those of AIDS, Ebola, measles, and flu, use membrane rafts to enter human cells or during viral replication and propagation. Researchers hope that blocking molecules in membrane rafts will provide a way to limit the spread of these viruses.

Eukaryotic cells frequently attach chains of sugar molecules to the outer surfaces of lipids and proteins in their cytoplasmic membranes; prokaryotes rarely do this. Sugar molecules may act in intercellular signaling, cellular attachment, and in other roles.

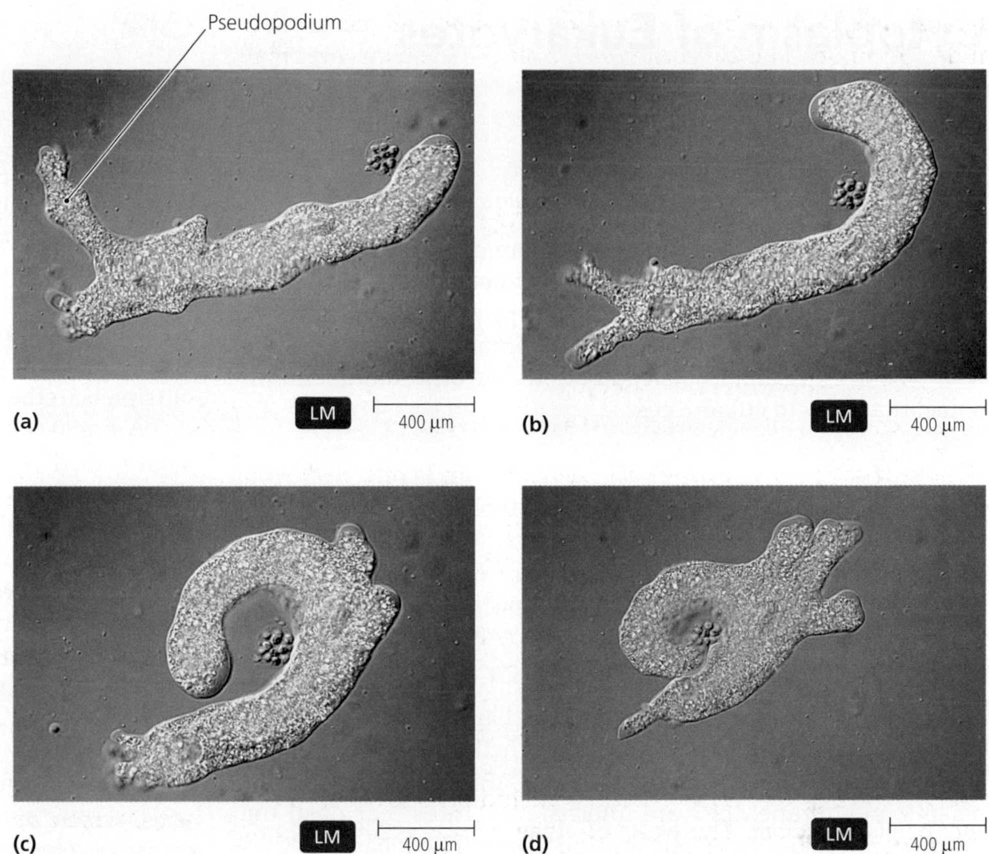

▶ **Figure 3.30 Endocytosis.** Pseudopodia extend to surround solid and/or liquid nutrients, which become incorporated into a food vesicle inside the cytoplasm. *What is the difference between phagocytosis and pinocytosis?*

Figure 3.30 *Phagocytosis is endocytosis of a solid; pinocytosis is endocytosis of a liquid.*

Like its prokaryotic counterpart, a eukaryotic cytoplasmic membrane controls the movement of materials into and out of a cell. Eukaryotic cytoplasmic membranes use both passive processes (diffusion, facilitated diffusion, and osmosis; see Figure 3.18) and active transport (see Figure 3.21). Eukaryotic membranes do not perform group translocation, which occurs only in some prokaryotes, but many perform another type of active transport—**endocytosis** (Figure 3.30), which involves physical manipulation of the cytoplasmic membrane around the cytoskeleton. Endocytosis occurs when the membrane distends to form **pseudopodia** (false feet) that surround a substance, bringing it into the cell. Endocytosis is termed **phagocytosis** if a solid is brought into the cell, and **pinocytosis** if only liquid is brought into the cell. Nutrients brought into a cell by endocytosis are then enclosed in a *food vesicle*. Vesicles and digestion of the nutrients they contain are discussed in more detail shortly. The process of phagocytosis is more fully discussed in Chapter 15 as it relates to the defense of the body against disease.

Some eukaryotes also use pseudopodia as a means of locomotion. The cell extends a pseudopod and then the cytoplasm streams into it, a process called *amoeboid action.*

Exocytosis, another solely eukaryotic process, is the reverse of endocytosis in that it enables substances to be exported from the cell. Not all eukaryotic cells can perform endocytosis or exocytosis.

Table 3.4 lists some of the features of endocytosis and exocytosis.

TABLE 3.4 Active Transport Processes Found Only in Eukaryotes: Endocytosis and Exocytosis

	Description	Examples of Transported Substances
Endocytosis: phagocytosis and pinocytosis	Substances are surrounded by pseudopodia and brought into the cell. Phagocytosis involves solid substances; pinocytosis involves liquids.	Bacteria, viruses, aged and dead cells; liquid nutrients in extracellular solutions
Exocytosis	Vesicles containing substances are fused with cytoplasmic membrane, dumping their contents to the outside.	Wastes, secretions

Cytoplasm of Eukaryotes

Learning Objectives

✓ Compare and contrast the cytoplasm of prokaryotes and eukaryotes.

✓ Identify nonmembranous and membranous organelles.

The cytoplasm of eukaryotic cells is more complex than that of either bacteria or archaea. The most distinctive difference is the presence of numerous membranous organelles in eukaryotes. However, before we discuss these membranous organelles, we will consider organelles of locomotion and other nonmembranous organelles in eukaryotes.

Flagella

Learning Objective

✓ Compare and contrast the structure and function of prokaryotic and eukaryotic flagella.

Structure and Arrangement

Some eukaryotic cells are flagellated. Flagella of eukaryotes **(Figure 3.31a)** differ structurally and functionally from flagella of prokaryotes. First, eukaryotic flagella are within the cytoplasmic membrane; they are internal structures that push the cytoplasmic membrane out around them. Their basal bodies are in the cytoplasm. Second, the shaft of a eukaryotic flagellum is composed of molecules of a globular protein called *tubulin* arranged in chains to form hollow *microtubules* **(Figure 3.31b)**. Nine pairs of microtubules surround two microtubules in the center. This "9 + 2" arrangement of microtubules is common to all flagellated eukaryotic cells, whether they are found in protozoa, algae, animals, or plants. The filaments of eukaryotic flagella are anchored in the cytoplasm by a basal body, but no hook connects the two parts, as in prokaryotes. The basal body has *triplets* of microtubules instead of pairs, and there are no microtubules in the center, so scientists say it has a "9 + 0" arrangement of microtubules. Eukaryotic flagella may be single or multiple and are generally found at one pole of the cell.

Function

The flagella of eukaryotes also move differently from those of prokaryotes. Rather than rotating like prokaryotic flagella, those of eukaryotes undulate rhythmically **(Figure 3.32a)**. Some eukaryotic flagella push the cell through the medium (as occurs in animal sperm), whereas others pull the cell through the medium (as occurs in many protozoa). Positive and negative phototaxis and chemotaxis are seen in eukaryotic cells, but such cells do not move in runs and tumbles.

Cilia

Learning Objectives

✓ Describe the structure and function of cilia.

✓ Compare and contrast eukaryotic cilia and flagella.

Other eukaryotic cells move by means of internal hairlike structures called **cilia,** which extend the surface of the cell and are shorter and more numerous than flagella **(Figure 3.31c)**. No prokaryotic cells have cilia. Like flagella, cilia are composed primarily of tubulin microtubules, which are arranged in a "9 + 2" arrangement of pairs in their shafts and a "9 + 0" arrangement of triplets in their basal bodies (Figure 3.31b).

A single cell may have hundreds or even thousands of cilia. Cilia beat rhythmically, much like a swimmer doing a butterfly stroke **(Figure 3.32b)**. Coordinated beating of cilia propels single-celled eukaryotes through their environment. Cilia are also used within multicellular eukaryotes to move substances in the local environment past the surface of the cell. For example, such movement of cilia helps cleanse the human respiratory tract of dust and microorganisms.

Other Nonmembranous Organelles

Learning Objectives

✓ Describe the structure and function of ribosomes, cytoskeletons, and centrioles.

✓ Compare and contrast the ribosomes of prokaryotes and eukaryotes.

✓ List and describe the three filaments of a eukaryotic cytoskeleton.

Here we discuss three nonmembranous organelles found in eukaryotes: ribosomes and cytoskeleton (both of which are also present in prokaryotes), and centrioles (which are present only in certain kinds of eukaryotic cells).

Ribosomes

The cytosol of eukaryotes, like that of prokaryotes, is a semitransparent fluid composed primarily of water containing dissolved and suspended proteins, ions, carbohydrates, lipids, and wastes. Within the cytosol of eukaryotic cells are ribosomes that are larger than prokaryotic ribosomes; instead of 70S ribosomes, eukaryotic ribosomes are 80S and are composed of 60S and 40S subunits. In addition to the 80S ribosomes found within the cytosol, many eukaryotic ribosomes are attached to the membranes of the endoplasmic reticulum (discussed shortly).

Cytoskeleton

Eukaryotic cells contain an extensive cytoskeleton composed of an internal network of fibers and tubules. The eukaryotic

Figure 3.31 Eukaryotic flagella and cilia. **(a)** Micrograph of *Euglena*, which possesses a single flagellum. **(b)** Details of the arrangement of microtubules of eukaryotic flagella and cilia. Both flagella and cilia have the same internal structure. **(c)** Light micrograph of a protozoan *Blepharisma*, which has numerous cilia. *How do eukaryotic flagella differ from cilia?*

Figure 3.31 *Flagella are longer and less numerous than cilia*

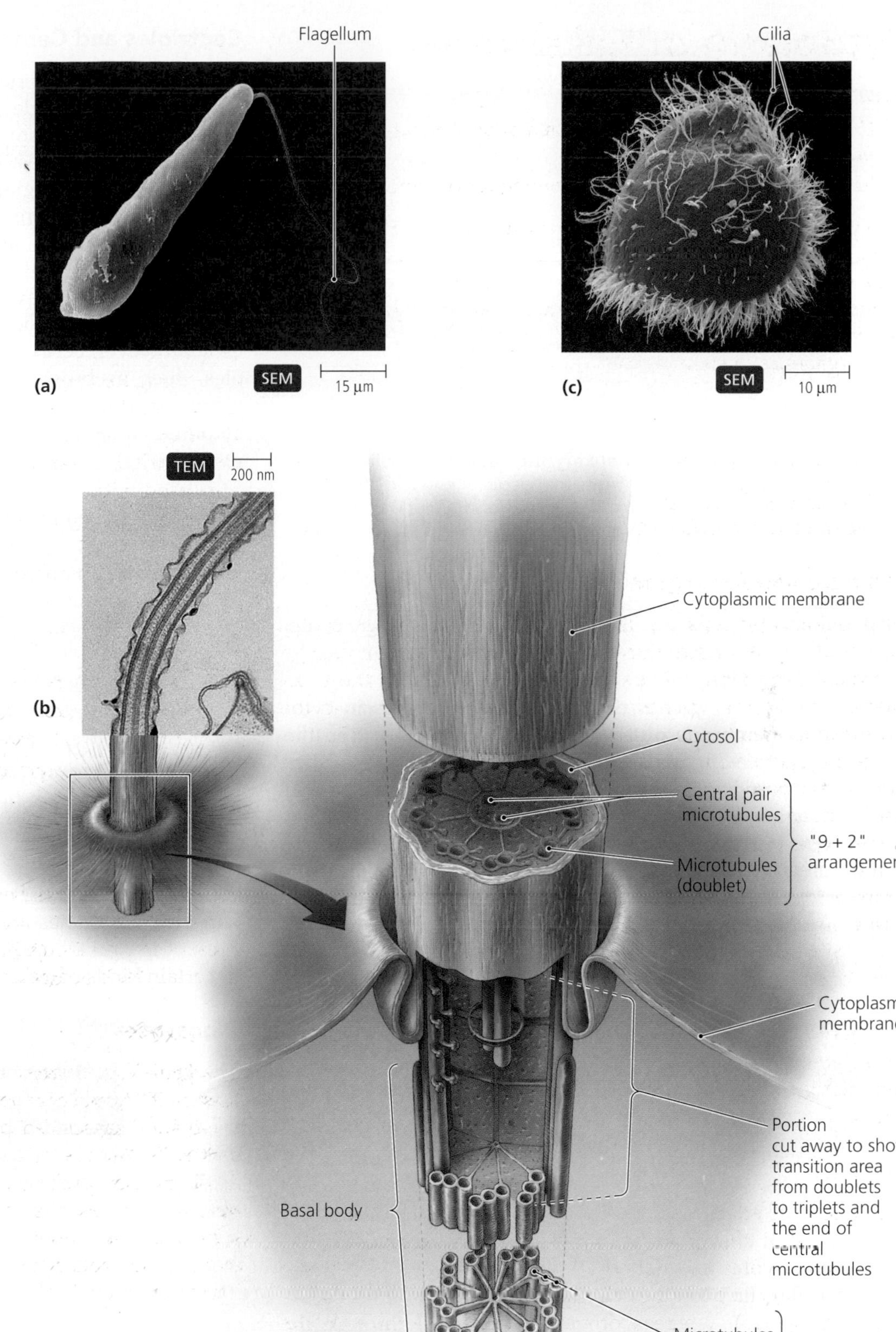

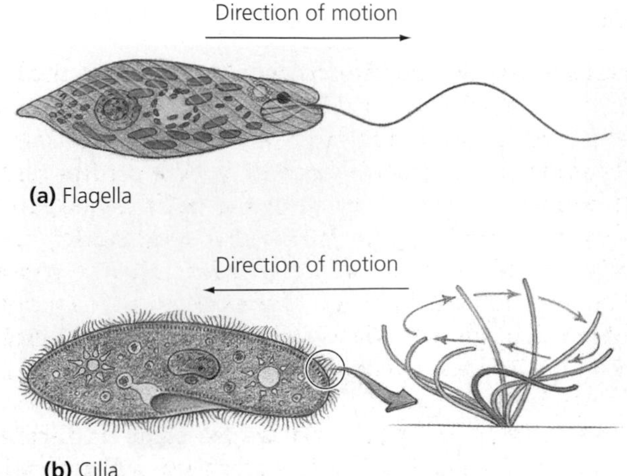

▲ Figure 3.32 Movement of eukaryotic flagella and cilia. **(a)** Eukaryotic flagella undulate in waves that begin at one end and traverse the length of the flagellum. **(b)** Cilia move with a power stroke followed by a return stroke. In the power stroke a cilium is stiff; it relaxes during the return stroke. *How is the movement of eukaryotic flagella different from that of prokaryotic flagella?*

Figure 3.32 Eukaryotic flagella undulate in a wave that moves down the flagellum; the flagella of prokaryotes rotate about the basal body.

cytoskeleton acts to anchor organelles and functions in cytoplasmic streaming and in movement of organelles within the cytosol. Cytoskeletons in some cells enable the cell to contract, move the cytoplasmic membrane during endocytosis and amoeboid action, and produce the basic shapes of the cells.

The eukaryotic cytoskeleton is made up of *tubulin microtubules* (also found in flagella and cilia), thinner *microfilaments* composed of *actin,* and *intermediate filaments* composed of various proteins **(Figure 3.33)**.

Centrioles and Centrosome

Animal cells and some fungal cells contain two **centrioles,** which lie at right angles to each other near the nucleus, in a region of the cytoplasm called the **centrosome (Figure 3.34)**. Plants, algae, and most fungi (and prokaryotes) lack centrioles but usually have a region of cytoplasm corresponding to a centrosome. Centrioles are composed of nine *triplets* of tubulin microtubules arranged in a way that resembles the "9 + 0" arrangement seen at the base of eukaryotic flagella and cilia.

Centrosomes play a role in *mitosis* (nuclear division), *cytokinesis* (cell division), and the formation of flagella and cilia. However, because many eukaryotic cells that lack centrioles, such as brown algal sperm and numerous one-celled algae, are still able to form flagella and undergo mitosis and cytokinesis, the function of centrioles is the subject of ongoing research.

Membranous Organelles

Learning Objectives

✓ Discuss the function of each of the following membranous organelles: nucleus, ER, Golgi body, lysosome, peroxisome, vesicle, vacuole, mitochondrion, and chloroplast.

✓ Label the structures associated with each of the membranous organelles.

Eukaryotic cells contain a variety of organelles that are surrounded by phospholipid bilayer membranes similar to the cytoplasmic membrane. These membranous organelles include the nucleus, endoplasmic reticulum, Golgi body, lysosomes, peroxisomes, vacuoles, vesicles, mitochondria, and chloroplasts. Prokaryotic cells lack these structures.

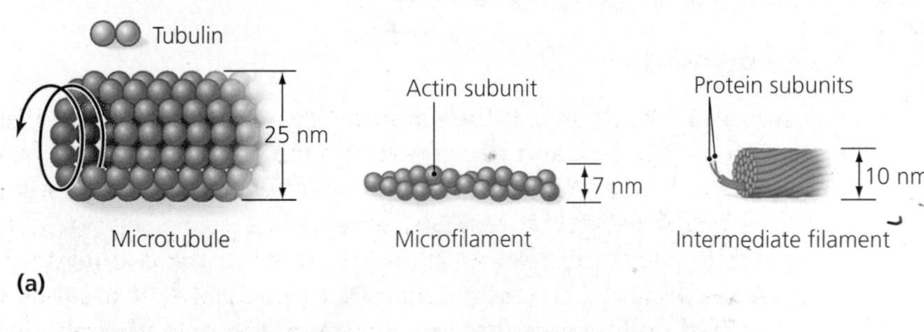

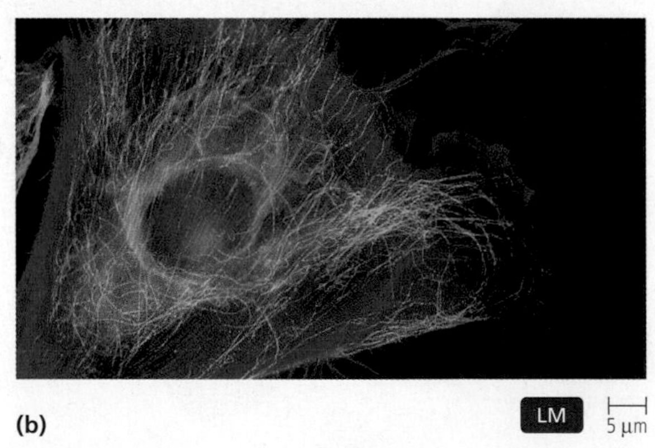

▲ Figure 3.33 Eukaryotic cytoskeleton. The cytoskeleton of eukaryotic cells serves to anchor organelles, provides a "track" for the movement of organelles throughout the cell, and provides shape to animal cells. Eukaryotic cytoskeletons are composed of microtubules, microfilaments, and intermediate filaments. **(a)** Artist's rendition of cytoskeleton filaments. **(b)** Various elements of the cytoskeleton shown here have been stained with different fluorescent dyes.

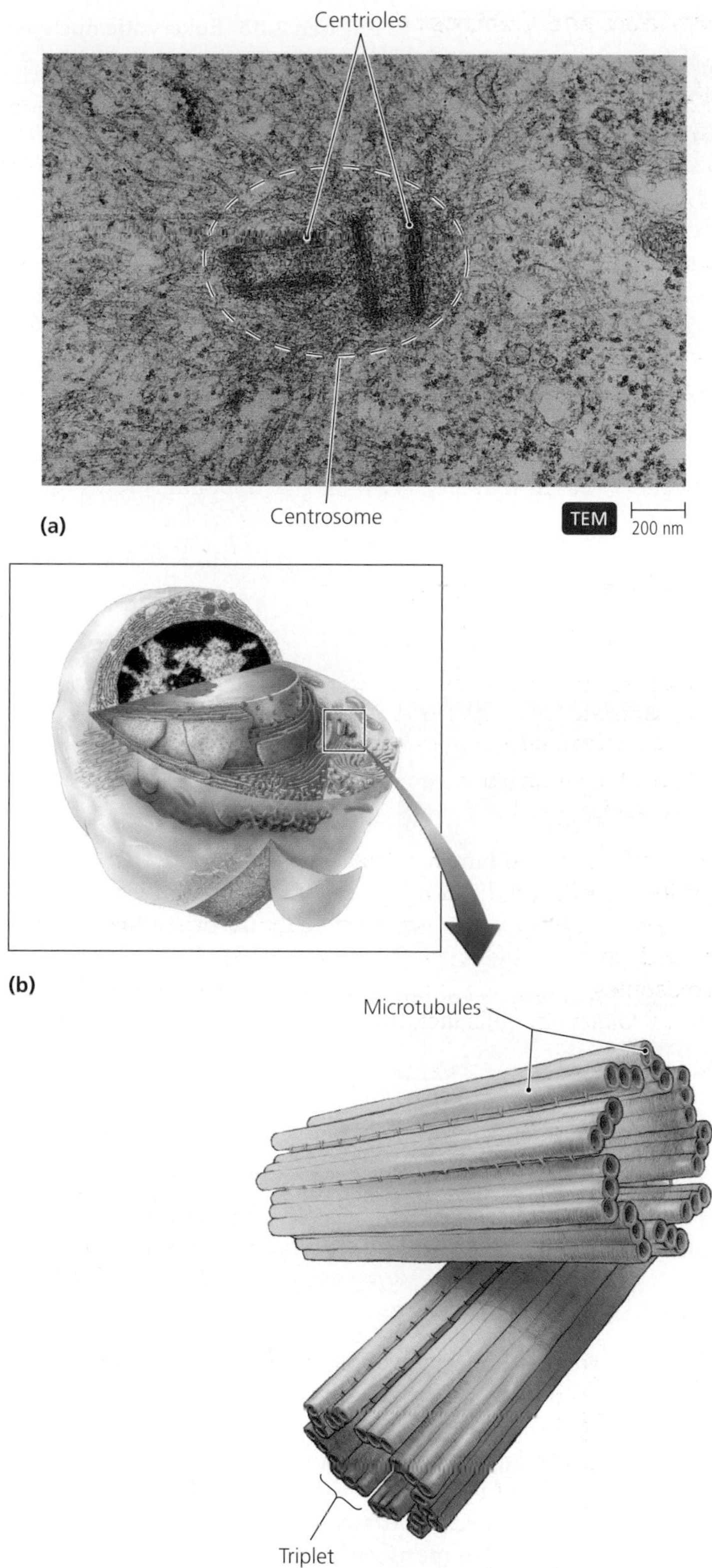

▲ **Figure 3.34 Centrosome.** A centrosome is a region of cytoplasm that in animal cells contains two centrioles at right angles to one another; each centriole has nine triplets of microtubules. **(a)** Transmission electron micrograph of centrosome and centrioles. **(b)** Artist's rendition of a centrosome. *How do centrioles compare with the basal body and shafts of eukaryotic flagella and cilia (see Figure 3.31b)?*

Figure 3.34 *Centrioles have the same "9 + 0" arrangement of microtubules that is found in the basal bodies of eukaryotic cilia and flagella.*

Nucleus

The **nucleus** is usually spherical to ovoid and is often the largest organelle in a cell[13] **(Figure 3.35)**. Some cells have a single nucleus; others are multinucleate, while still others lose their nuclei. The nucleus is often referred to as "the control center of the cell" because it contains most of the cell's genetic instructions in the form of DNA. Cells that lose their nuclei, such as mammalian red blood cells, can survive for only a few months.

Just as the semiliquid portion of the cell is called cytoplasm, the semiliquid matrix of the nucleus is called **nucleoplasm.** Within the nucleoplasm may be one or more **nucleoli** (noo-klē'ō-lī; singular: *nucleolus*), which are specialized regions where RNA is synthesized. The nucleoplasm also contains **chromatin,** which is a threadlike mass of DNA associated with special proteins called *histones* that play a role in packaging nuclear DNA. During mitosis (nuclear division) chromatin becomes visible as *chromosomes.* Chapter 12 discusses mitosis in more detail.

Surrounding the nucleus is a double membrane called the **nuclear envelope,** which is composed of two phospholipid bilayers, for a total of four phospholipid layers. The nuclear envelope contains **nuclear pores** (see Figure 3.35) that function to control the import and export of substances through the envelope.

Endoplasmic Reticulum

Continuous with the outer membrane of the nuclear envelope is a netlike arrangement of flattened hollow tubules called **endoplasmic reticulum (ER) (Figure 3.36)**. The ER traverses the cytoplasm of eukaryotic cells. Endoplasmic reticulum functions as a transport system and is found in two forms: **smooth endoplasmic reticulum (SER)** and **rough endoplasmic reticulum (RER).** SER plays a role in lipid synthesis as well as transport. Rough endoplasmic reticulum is rough because ribosomes adhere to its outer surface. Proteins produced by ribosomes on the RER are inserted into the lumen (central canal) of the RER and transported throughout the cell.

Golgi Body

A **Golgi**[14] **body** is like the shipping department of a cell: it receives, processes, and packages large molecules for export from the cell **(Figure 3.37)**. The Golgi body packages secretions in sacs called **secretory vesicles,** which then fuse with the cytoplasmic membrane before dumping their contents outside the cell via exocytosis. Golgi bodies are composed of a series of flattened hollow sacs that are circumscribed by a phospholipid bilayer. Not all eukaryotic cells contain Golgi bodies.

Lysosomes, Peroxisomes, Vacuoles, and Vesicles

Lysosomes, peroxisomes, vacuoles, and vesicles are membranous sacs that function to store and transfer chemicals within

[13]Historically, the nucleus was not considered an organelle because it is large and not considered part of the cytoplasm.

[14]Camillo Golgi was an Italian histologist who first described the organelle in 1898. This organelle is also known as a Golgi complex or Golgi apparatus and in plants and algae as a dictyosome.

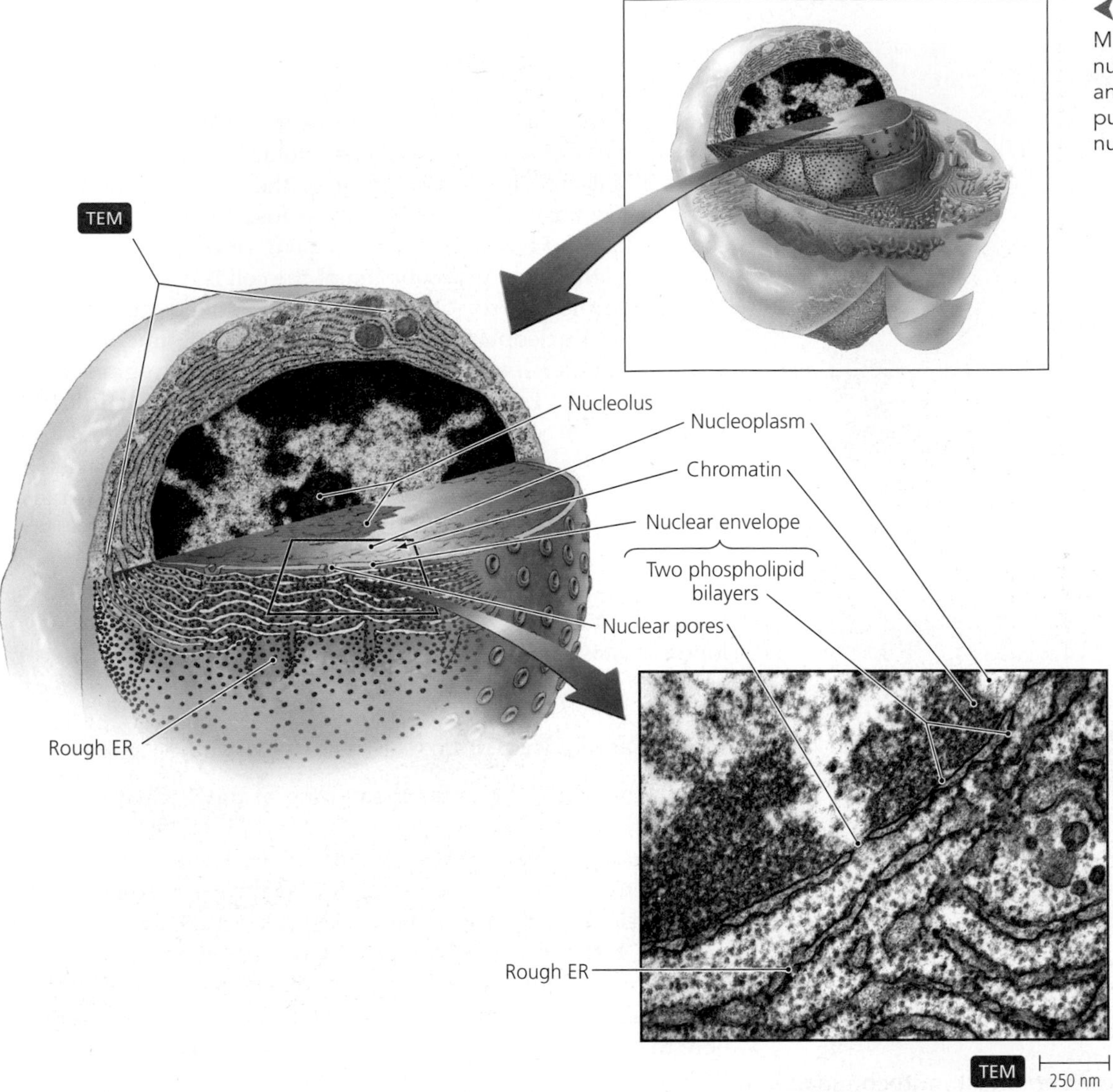

Figure 3.35 Eukaryotic nucleus. Micrograph and artist's conception of a nucleus showing chromatin, a nucleolus, and the nuclear envelope. Nuclear pores punctuate the two membranes of the nuclear envelope.

eukaryotic cells. Both **vesicle** and **vacuole** are general terms for such sacs. Large vacuoles are found in plant and algal cells that store starch, lipids, and other substances in the center of the cell. Often a central vacuole is so large that the rest of the cytoplasm is pressed against the cell wall in a thin layer (**Figure 3.38**).

Lysosomes, which are found in animal cells, contain catabolic enzymes that damage the cell if they are released from their packaging into the cytosol. The enzymes are used during the self-destruction of old, damaged, and diseased cells, and to digest nutrients that have been phagocytized. For example, the digestive enzymes in lysosomes are utilized by white blood cells to destroy phagocytized pathogens (**Figure 3.39**).

Peroxisomes are vesicles derived from ER. They contain *oxidase* and *catalase,* which are enzymes that degrade poisonous metabolic wastes (such as free radicals and hydrogen peroxide) resulting from some oxygen-dependent reactions. Peroxisomes are found in all types of eukaryotic cells but are especially prominent in the kidney and liver cells of mammals.

Mitochondria

Mitochondria are spherical to elongated structures found in most eukaryotic cells (**Figure 3.40**). Like nuclei, they have two membranes, each composed of a phospholipid bilayer. The inner membrane is folded into numerous *cristae* that increase the inner membrane's surface area. Mitochondria are often called the "powerhouses of the cell" because their cristae produce most of the ATP in many eukaryotic cells. The chemical reactions that produce ATP are discussed in Chapter 5.

The interior matrix of a mitochondrion contains ("prokaryotic") 70S ribosomes and a circular molecule of DNA. This DNA contains genes for some RNA molecules and for a few mitochondrial polypeptides that are manufactured

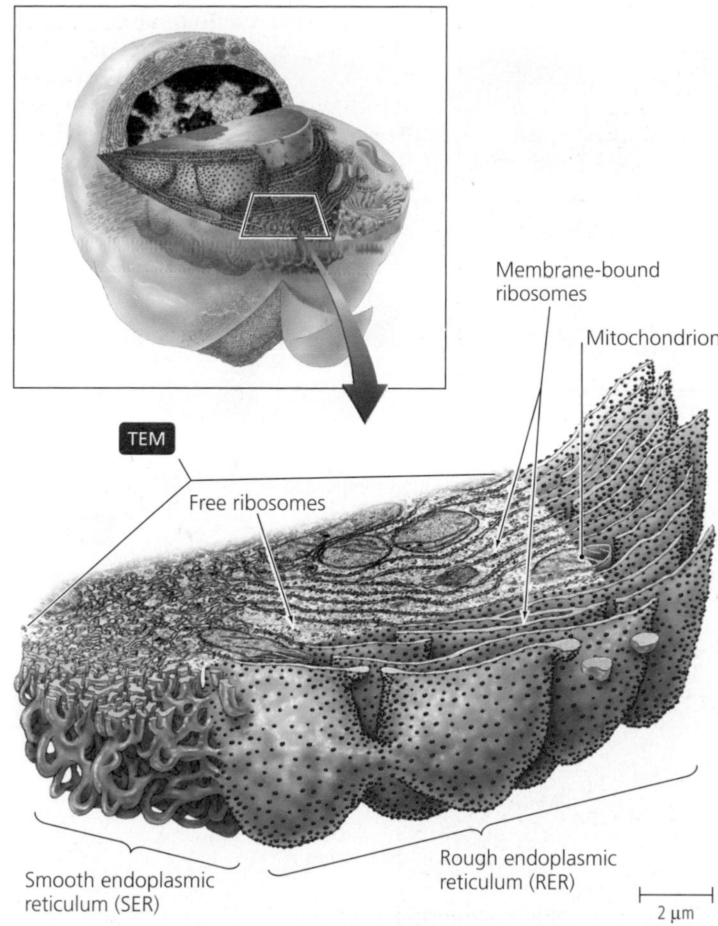

▲ **Figure 3.36 Endoplasmic reticulum.** ER functions in transport throughout the cell. Ribosomes are on the surface of rough ER; smooth ER lacks ribosomes.

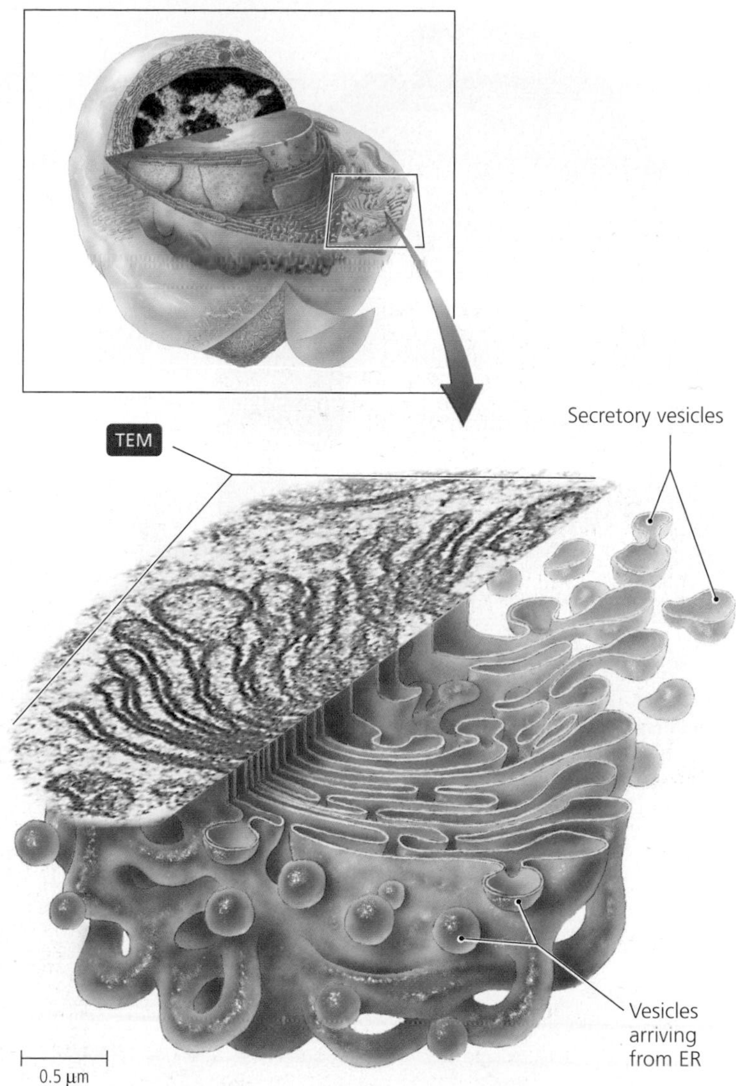

▲ **Figure 3.37 Golgi body.** A Golgi body is composed of flattened sacs. Proteins synthesized by ribosomes on RER are transported via vesicles to a Golgi body. The Golgi body then modifies the proteins and sends them via secretory vesicles to the cytoplasmic membrane, where they can be secreted from the cell by exocytosis.

by mitochondrial ribosomes; however, most mitochondrial proteins are coded by nuclear DNA and synthesized by cytoplasmic ribosomes.

Chloroplasts

Chloroplasts are light-harvesting structures found in photosynthetic eukaryotes (**Figure 3.41**). Like mitochondria and the nucleus, chloroplasts have two phospholipid bilayer membranes and DNA. Further, like mitochondria, chloroplasts can synthesize a few polypeptides with their own 70S ribosomes. The pigments of chloroplasts gather light energy to produce ATP and form sugar from carbon dioxide. Numerous membranous sacs called *thylakoids* form an extensive surface area for the biochemical and photochemical reactions of chloroplasts. The fluid between the thylakoids and the inner membrane is called the *stroma*. The space enclosed by the thylakoids is called the thylakoid space.

Photosynthetic prokaryotes lack chloroplasts and instead have infoldings of their cytoplasmic membranes called *photosynthetic lamellae*. Chapter 5 discusses the details of photosynthesis.

The functions of the nonmembranous and membranous organelles, and their distribution among prokaryotic and eukaryotic cells, are summarized in Table 3.5 on p. 87.

Endosymbiotic Theory

Learning Objectives

- ✓ Describe the endosymbiotic theory of the origin of mitochondria, chloroplasts, and eukaryotic cells.
- ✓ List evidence for the endosymbiotic theory.

Mitochondria and chloroplasts are semiautonomous; that is, they divide independently of the cell but remain dependent on the cell for most of their proteins. As we have seen, both

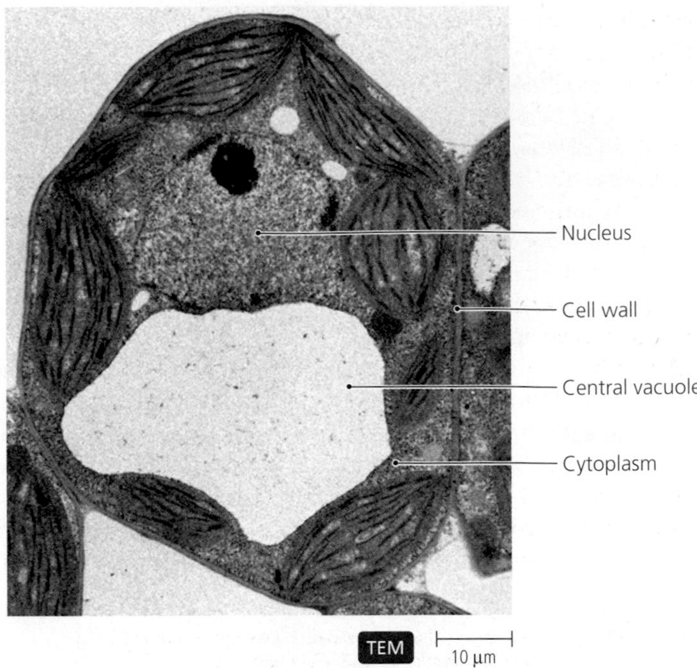

▲ **Figure 3.38 Vacuole.** The large central vacuole of a plant cell, which constitutes a storehouse for the cell, presses the cytoplasm against the cell wall.

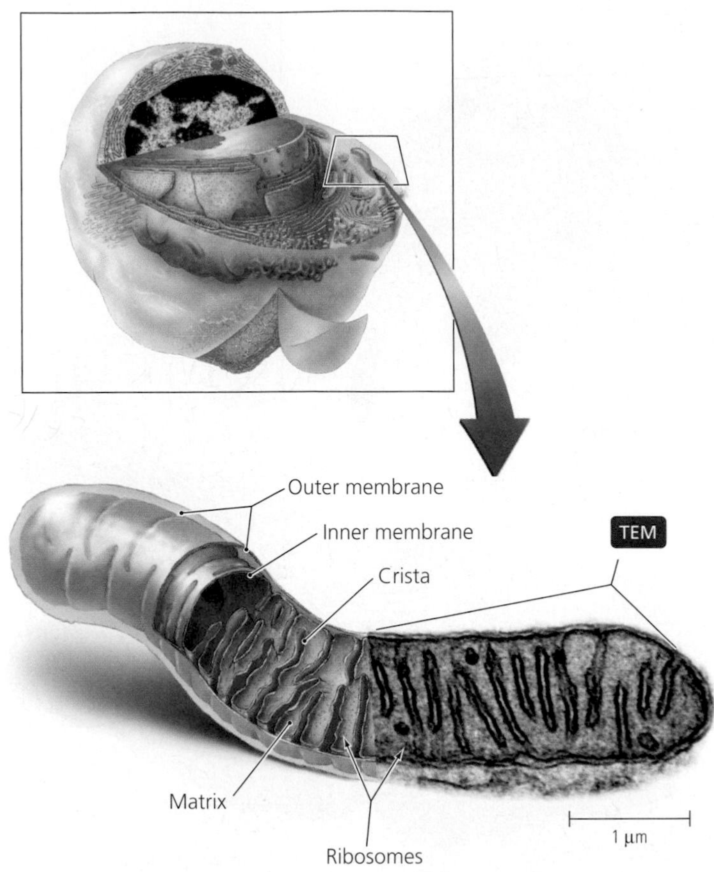

▲ **Figure 3.40 Mitochondrion.** Note the double membrane. The inner membrane is folded into cristae that increase its surface area. *What is the importance of the increased surface area of the inner membrane that results from having cristae?*

Figure 3.40 *The chemicals involved in aerobic ATP production are located on the inner membranes of mitochondria. Increased surface area provides more space for more chemicals.*

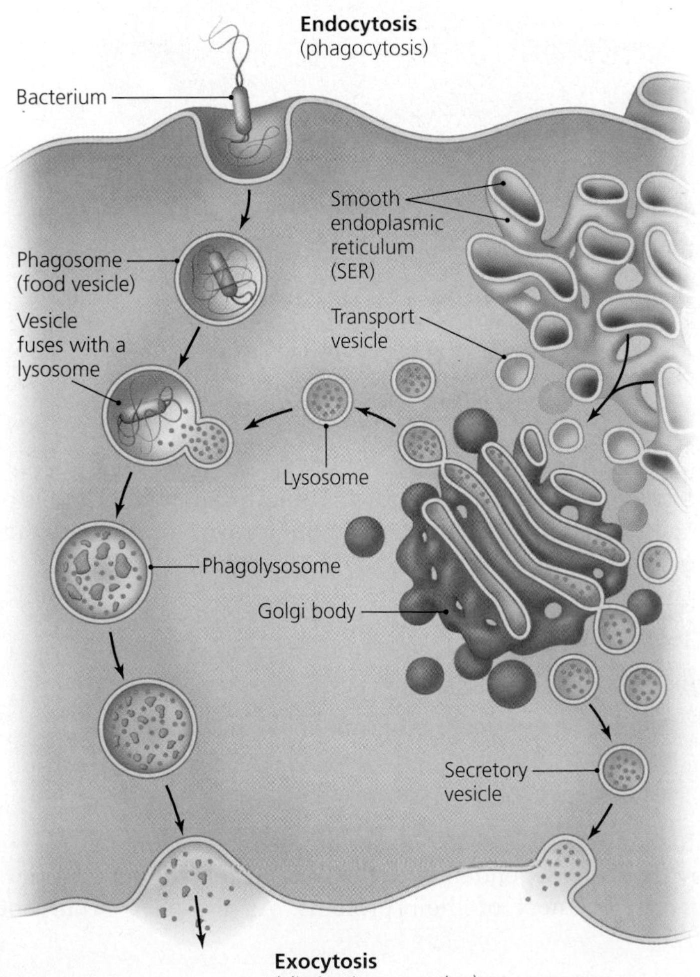

◀ **Figure 3.39 The roles of vesicles in the destruction of a phagocytized pathogen within a white blood cell.** Before endocytosis, vesicles from the SER deliver digestive enzymes to the Golgi body, which then packages them into lysosomes. During endocytosis, phagocytized particles (in this case, a bacterium) are enclosed within a vesicle called a phagosome (food vesicle), which then fuses with a lysosome to form a phagolysosome. Once digestion within the phagolysosome is complete, the resulting wastes can be expelled from the cell via exocytosis.

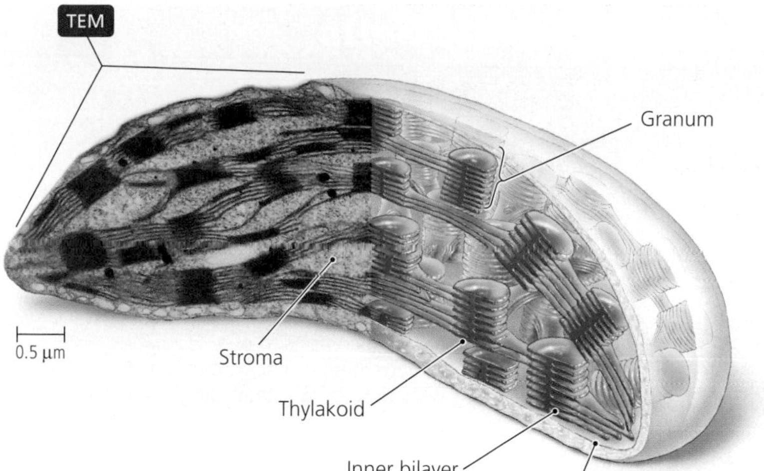

▲ **Figure 3.41 Chloroplast.** Chloroplasts have an ornate internal structure designed to harvest light energy for photosynthesis.

mitochondria and chloroplasts contain a small amount of DNA and 70S ribosomes, and each can produce a few polypeptides with its own ribosomes. The presence of circular DNA, 70S ribosomes, and two bilipid membranes in these semiautonomous organelles led scientist Lynn Margulis (1938–) to popularize the **endosymbiotic**[15] **theory** for the formation of eukaryotic cells. This theory suggests that eukaryotes formed from the union of small aerobic[16] prokaryotes with larger anaerobic prokaryotes. The smaller prokaryotes were not destroyed by the larger cells, but instead became internal parasites that remained surrounded by a vesicular membrane of the host.

According to the theory, the parasites eventually lost the ability to exist independently, but they retained a portion of their DNA, some ribosomes, and their cytoplasmic membranes. During the same time, the larger cell became dependent on the parasites for aerobic ATP production. According to the theory, the aerobic prokaryotes eventually evolved into mitochondria, and their cytoplasmic membranes became cristae. Margulis

[15]From Greek *endo*, meaning inside, and *symbiosis*, meaning to live with.
[16]*Aerobic* means requiring oxygen; *anaerobic* is the opposite.

TABLE 3.5 Nonmembranous and Membranous Organelles of Cells

	General Function	Prokaryotes	Eukaryotes
Nonmembranous organelles			
Ribosomes	Protein synthesis	Present in all	Present in all
Cytoskeleton	Shape in prokaryotes; support, cytoplasmic streaming, and endocytosis in eukaryotes	Present in some	Present in all
Centrosome	Appears to play a role in mitosis, cytokinesis, and flagella and cilia formation in animal cells	Absent in all	Present in animals
Membranous organelles	Sequester chemical reactions within the cell		
Nucleus	"Control center" of the cell	Absent in all	Present in all
Endoplasmic reticulum	Transport within the cell, lipid synthesis	Absent in all	Present in all
Golgi bodies	Exocytosis, secretion	Absent in all	Present in some
Lysosomes	Breakdown of nutrients, self-destruction of damaged or aged cells	Absent in all	Present in some
Peroxisomes	Neutralization of toxins	Absent in all	Present in some
Vacuoles	Storage	Absent in all	Present in some
Vesicles	Storage, digestion, transport	Absent in all	Present in all
Mitochondria	Aerobic ATP production	Absent in all	Present in most
Chloroplasts	Photosynthesis	Absent in all, though infoldings of cytoplasmic membrane called photosynthetic lamellae have same function in photosynthetic prokaryotes	Present in plants, algae, and some protozoa

TABLE 3.6 Comparison of Archaeal, Bacterial, and Eukaryotic Cells

Characteristic	Archaea	Bacteria	Eukaryotes
Nucleus	Absent in all	Absent	Present in all
Free organelles bound with phospholipid membranes	Absent in all	Present in few	Present; include ER, Golgi bodies, lysosomes, mitochondria, and chloroplasts
Glycocalyx	Present	Present as organized capsule or unorganized slime layer	Present, surrounding some animal cells
Motility	Present in some	Present in some	Some have complex undulating flagella and cilia composed of a "9 + 2" arrangement of microtubules; others move with amoeboid action using pseudopodia
Flagella	Some have flagella, each composed of basal body, hook, and filament; flagella rotate	Some have flagella, each composed of basal body, hook, and filament; flagella rotate	Present in some
Cilia	Absent in all	Absent in all	Present in some
Fimbriae or pili	Present in some	Present in some	Absent in all
Hami	Present in some	Absent in all	Absent in all
Cell wall	Present in most, lacking peptidoglycan	Present in most; composed of peptidoglycan	Present in plants, algae, and fungi
Cytoplasmic membrane	Present in all	Present in all	Present in all
Cytosol	Present in all	Present in all	Present in all
Inclusions	Present in most	Present in most	Present in some
Endospores	Absent in all	Present in some	Absent in all
Ribosomes	Small (70S)	Small (70S)	Large (80S) in cytosol and on ER, smaller (70S) in mitochondria and chloroplasts
Chromosomes	Commonly single and circular	Commonly single and circular	Linear and more than one chromosome per cell

described a similar scenario for the origin of chloroplasts from phagocytized photosynthetic prokaryotes. The theory provides an explanation for the presence of 70S ribosomes and circular DNA within mitochondria and chloroplasts, and it accounts for the presence of their two membranes.

The endosymbiotic theory is widely but not universally accepted, however, partly because it does not explain all of the facts. For example, the theory provides no explanation for the two membranes of the nuclear envelope; nor does it explain why only a few polypeptides of mitochondria and chloroplasts are made in the organelles while the bulk of their proteins come from nuclear DNA and cytoplasmic ribosomes.

Table 3.6 summarizes features of prokaryotic and eukaryotic cells.

CRITICAL THINKING

Eukaryotic cells are almost always larger than prokaryotic cells. What structures might allow for their larger size?

Chapter Summary

Processes of Life (pp. 55–56)

1. All living things have some common features, including **growth,** an increase in size; **reproduction,** an increase in number; **responsiveness,** reactions to environmental stimuli; **metabolism,** controlled chemical reactions in an organism; and cellular structure.
2. Viruses are acellular and do not grow, self-reproduce, or metabolize.

Prokaryotic and Eukaryotic Cells: An Overview (pp. 56–58)

1. All cells can be described as either **prokaryotic** or **eukaryotic.** These descriptive terms help scientists categorize organisms in groups called taxa. Generally, prokaryotic cells make proteins simultaneously to reading the genetic code, and lack a nucleus and organelles surrounded by phospholipid membranes. Domain Bacteria and Domain Archaea are prokaryotic taxa.
2. Eukaryotes (Domain Eukarya) have internal, membrane-bound **organelles,** including nuclei. Animals, plants, algae, fungi, and protozoa are eukaryotic.
3. Cells share common structural features. These include external structures, cell walls, cytoplasmic membranes, and cytoplasm.

External Structures of Bacterial Cells (pp. 58–63)

1. The external structures of bacterial cells include glycocalyces, flagella, fimbriae, and pili.
2. **Glycocalyces** are sticky external sheaths of cells. They may be loosely attached **slime layers** or firmly attached **capsules.** Glycocalyces prevent cells from drying out. Capsules protect cells from phagocytosis by other cells, and slime layers enable cells to stick to each other and to surfaces in their environment.
3. A prokaryotic **flagellum** is a long, whiplike protrusion of some cells composed of a basal body, hook, and filament. Flagella allow cells to move toward favorable conditions such as nutrients or light, or move away from unfavorable stimuli such as poisons.
ANIMATIONS: *Motility; Flagella: Structure*
4. Bacterial flagella may be **polar** (single or tufts), or cover the cell (**peritrichous**). **Endoflagella,** which are special flagella of a spirochete, form an **axial filament,** located in the periplasmic space.
ANIMATIONS: *Arrangement, Spirochetes*
5. **Taxis** is movement that may be either a positive response or a negative response to light **(phototaxis)** or chemicals **(chemotaxis).**
ANIMATIONS: *Flagella: Movement*
6. **Fimbriae** are extensions of some bacterial cells that function along with glycocalyces to adhere cells to one another and to environmental surfaces. A mass of such bacteria on a surface is termed a **biofilm.** Cells may also use fimbriae to pull themselves across a surface.
7. **Pili,** also known as conjugation pili, are hollow, nonmotile tubes of protein that allow bacteria to pull themselves forward and mediate the movement of DNA from one cell to another. Not all bacteria have fimbriae or pili.

Bacterial Cell Walls (pp. 63–67)

1. Most prokaryotic cells have **cell walls** that provide shape and support against osmotic pressure. Cell walls are composed primarily of polysaccharide chains.
2. Cell walls of bacteria are composed of a large, interconnected molecule of **peptidoglycan.** Peptidoglycan is composed of alternating sugar molecules called *N*-acetylglucosamine (NAG) and *N*-acetylmuramic acid (NAM).
3. A **Gram-positive** bacterial cell has a thick layer of peptidoglycan.
4. A **Gram-negative** bacterial cell has a thin layer of peptidoglycan and an external wall membrane with a **periplasmic space** between. This wall membrane contains **lipopolysaccharide (LPS),** which contains **lipid A.** During an infection with Gram-negative bacteria, lipid A can accumulate in the blood, causing shock, fever, and blood clotting.
5. Acid-fast bacteria have waxy lipids in their cell walls.

Bacterial Cytoplasmic Membranes (pp. 67–72)

1. A **cytoplasmic membrane** is typically composed of phospholipid molecules arranged in a double layer configuration called a **phospholipid bilayer.** Proteins associated with the membrane vary in location and function and are able to flow laterally within the membrane. The **fluid mosaic model** is descriptive of the current understanding of membrane structure.
ANIMATIONS: *Membrane Structure*
2. The **selectively permeable** cytoplasmic membrane prevents the passage of some substances while allowing other substances to pass through protein pores or channels, sometimes requiring carrier molecules.
ANIMATIONS: *Membrane Permeability*
3. The relative concentrations inside and outside the cell of a chemical create a **concentration gradient.** Differences of electrical charges on the two sides of a membrane create an **electrical gradient** across the membrane. The gradients have a predictable effect on the passage of substances through the membrane.
ANIMATIONS: *Passive Transport: Principles of Diffusion*
4. Passive processes that move chemicals across the cytoplasmic membrane require no energy expenditure by the cell. Molecular size and concentration gradients determine the rate of simple **diffusion. Facilitated diffusion** depends on the electrochemical gradient and carriers within the membrane that allow certain substances to pass through the membrane. **Osmosis** specifically refers to the diffusion of water molecules across a selectively permeable membrane.
5. The concentrations of solutions can be compared. **Hypertonic** solutions have a higher concentration of solutes than **hypotonic** solutions, which have a lower concentration of solutes. Two **isotonic solutions** have the same concentrations of solutes. In biology, comparisons are usually made with the cytoplasm of cells.
ANIMATIONS: *Passive Transport: Special Types of Diffusion*
6. Active transport processes require cell energy from ATP. **Active transport** moves a substance against its electrochemical gradient via carrier proteins. These carriers may move two substances in the same direction at once (symports) or move substances in opposite directions (antiports). **Group translocation** occurs in prokaryotes, during which the substance being transported is chemically altered in transit.
ANIMATIONS: *Active Transport: Overview, Types*

Cytoplasm of Bacteria (pp. 72–75)

1. **Cytoplasm** is composed of the liquid **cytosol** inside a cell plus nonmembranous organelles and inclusions. **Inclusions** in the cytosol are deposits of various substances.
2. Both prokaryotic and eukaryotic cells contain nonmembranous organelles.

3. The **nucleoid** is the nuclear region in prokaryotic cytosol. It has no membrane and usually contains a single circular molecule of DNA.
4. Inclusions include reserve deposits of lipids, starch, or compounds containing nitrogen, phosphate, or sulfur. Inclusions called gas vesicles store gases.
5. Some bacteria produce dormant resistant **endospores** within vegetative cells.
6. **Ribosomes,** composed of protein and **ribosomal RNA (rRNA),** are nonmembranous organelles, found in both prokaryotes and eukaryotes, that function to make proteins. The 70S ribosomes of prokaryotes are smaller than the 80S ribosomes of eukaryotes.
7. The **cytoskeleton** is a network of fibrils that appears to help maintain the basic shape of prokaryotes.

External Structures of Archaea (pp. 75–76)

1. Archaea form polysaccharide and polypeptide glycocalyces that function in attachment and biofilm formation but are evidently not associated with diseases.
2. Archaeal flagella differ from bacterial flagella. For example, archaeal flagella are thinner than bacterial flagella.
3. Archaeal flagella rotate together as a bundle in both directions and are powered by molecules of ATP.
4. Archaea may have fimbriae and grappling-hook-like **hami** that serve to anchor the cells to environmental surfaces.

Archaeal Cell Walls and Cytoplasmic Membranes (p. 76)

1. Archaeal cell walls are composed of protein or polysaccharides but not peptidoglycan.
2. Phospholipids in archaeal cytoplasmic membranes are built with ether linkages rather than ester linkages, which are seen in bacterial membranes.

Cytoplasm of Archaea (pp. 76–77)

1. Gel-like archaeal cytoplasm is similar to the cytoplasm of bacteria, having DNA, ribosomes, and a fibrous cytoskeleton, all suspended in the liquid cytosol.
2. 70S ribosomes of archaea have proteins more similar to those of eukaryotic ribosomes than to bacterial ribosomes.

External Structure of Eukaryotic Cells (pp. 77–78)

1. Eukaryotic animal and some protozoan cells lack cell walls but have glycocalyces that prevent desiccation, provide support, and enable cells to stick together.
2. Wall-less eukaryotic cells have glycocalyces, which are not found with eukaryotic cells that have walls.

Eukaryotic Cell Walls and Cytoplasmic Membranes (pp. 78–79)

1. Fungal, plant, algal, and some protozoan cells have cell walls composed of polysaccharides or other chemicals. Cell walls provide support, shape, and protection from osmotic forces.
2. Fungal cell walls are composed of chitin or other polysaccharides. Plant cell walls are composed of cellulose. Algal cell walls contain agar, carrageenan, algin, cellulose, or other chemicals.
3. Eukaryotic cytoplasmic membranes contain sterols such as cholesterol, which act to strengthen and solidify the membranes when temperatures rise and provide fluidity when temperatures fall.
4. **Membrane rafts** are distinct assemblages of certain lipids and proteins that remain together in the cytoplasmic membrane. Some viruses use lipid rafts during their infections of cells.
5. Some eukaryotic cells transport substances into the cytoplasm via **endocytosis,** which is an active process requiring the expenditure of energy from ATP. In endocytosis, **pseudopodia**—movable extensions of the cytoplasm and membrane of the cell—surround a substance and move it into the cell. When solids are brought into the cell, endocytosis is called **phagocytosis;** the incorporation of liquids by endocytosis is called **pinocytosis.**
6. **Exocytosis** is the active export of substances out of a cell.

Cytoplasm of Eukaryotes (pp. 80–88)

1. Eukaryotic cytoplasm is characterized by membranous organelles, particularly a nucleus. It also contains nonmembranous organelles and cytosol.
2. Some eukaryotic cells have long, whiplike flagella that differ from the flagella of prokaryotes. They have no hook, and the basal bodies and shafts are arrangements of microtubules. Further, eukaryotic flagella are internal to the cytoplasmic membrane.
3. Some eukaryotic cells have **cilia,** which have the same structure as eukaryotic flagella but are much shorter and more numerous. Cilia are internal to the cytoplasmic membrane.
4. The 80S ribosomes of eukaryotic cells are composed of 60S and 40S subunits. They are found free in the cytosol and attached to endoplasmic reticulum. The ribosomes within mitochondria and chloroplasts are 70S.
5. The eukaryotic cytoskeleton is composed of microtubules, intermediate filaments, and microfilaments. It provides an infrastructure and aids in movement of cytoplasm and organelles.
6. **Centrioles,** which are nonmembranous organelles in animal and some fungal cells only, are found in a region of the cytoplasm called the **centrosome** and are composed of triplets of microtubules in a "9 + 0" arrangement. Centrosomes function in the formation of flagella and cilia and in cell division.
7. The **nucleus,** a membranous structure in eukaryotic cells, contains **nucleoplasm** in which are found one or more **nucleoli** and **chromatin.** Chromatin consists of the multiple strands of DNA and associated histone proteins that become obvious as chromosomes during mitosis. **Nuclear pores** penetrate the four phospholipid layers of the **nuclear envelope** (membrane).
8. The **endoplasmic reticulum** functions as a transport system. It can be **rough ER (RER),** which has ribosomes on its surface, or **smooth ER (SER),** which lacks ribosomes.
9. A **Golgi body** is a series of flattened hollow sacs surrounded by phospholipid bilayers. It packages large molecules destined for export from the cell in **secretory vesicles,** which release these molecules from the cell via exocytosis.

10. **Vesicles** and **vacuoles** are general terms for membranous sacs that store or carry substances. More specifically, **lysosomes** of animal cells contain digestive enzymes, and **peroxisomes** contain enzymes that neutralize poisonous free radicals and hydrogen peroxide.

11. Four phospholipid layers surround **mitochondria,** site of production of ATP in a eukaryotic cell. The inner bilayer is folded into cristae, which greatly increase the surface area available for chemicals that generate ATP.

12. Photosynthetic eukaryotes possess **chloroplasts,** which are organelles containing membranous thylakoids that provide increased surface area for photosynthetic reactions.

13. The **endosymbiotic theory** has been suggested to explain why mitochondria and chloroplasts have 70S ribosomes, circular DNA, and two membranes. The theory states that the ancestors of these organelles were prokaryotic cells that were internalized by other prokaryotes and then lost the ability to exist outside their host—thus forming early eukaryotes.

Questions for Review

Answers to the Questions for Review (except Short Answer Questions) begin on page A-1.

Multiple Choice

1. A cell may allow a large or charged chemical to move across the cytoplasmic membrane, down the chemical's electrical and chemical gradients, in a process called
   a. active transport.
   b. facilitated diffusion.
   c. endocytosis.
   d. pinocytosis.

2. Which of the following statements concerning growth and reproduction is *not* true?
   a. Growth and reproduction may occur simultaneously in living organisms.
   b. A living organism must reproduce to be considered alive.
   c. Living things may stop growing and reproducing, yet still be alive.
   d. Normally, living organisms have the ability to grow and reproduce themselves.

3. A "9 + 2" arrangement of microtubules is seen in
   a. archaeal flagella.
   b. bacterial flagella.
   c. eukaryotic flagella.
   d. all prokaryotic flagella.

4. Which of the following is most associated with diffusion?
   a. symports
   b. antiports
   c. carrier proteins
   d. endocytosis

5. Which of the following is *not* associated with prokaryotic organisms?
   a. nucleoid
   b. glycocalyx
   c. cilia
   d. circular DNA

6. Which of the following is true of Svedbergs?
   a. They are not exact but are useful for comparisons.
   b. They are abbreviated "sv."
   c. They are prokaryotic in nature but exhibit some eukaryotic characteristics.
   d. They are an expression of sedimentation rate during high-speed centrifugation.

7. Which of the following statements is true?
   a. The cell walls of bacteria are composed of peptidoglycan.
   b. Peptidoglycan is a fatty acid.
   c. Gram-positive bacterial walls have a relatively thin layer of peptidoglycan anchored to the cytoplasmic membrane by teichoic acids.
   d. Peptidoglycan is found mainly in the cell walls of fungi, algae, and plants.

8. Which of the following is *not* a function of a glycocalyx?
   a. It forms pseudopodia for faster mobility of an organism.
   b. It can protect a bacterial cell from drying out.
   c. It hides a bacterial cell from other cells.
   d. It allows a bacterium to stick to a host.

9. Bacterial flagella
   a. are anchored to the cell by a basal body.
   b. are composed of hami.
   c. are surrounded by an extension of the cytoplasmic membrane.
   d. are composed of tubulin in hollow microtubules in a "9 + 2" arrangement.

10. Which cellular structure is important in classifying a bacterial species as Gram-positive or Gram-negative?
    a. flagella
    b. cell wall
    c. cilia
    d. glycocalyx

11. A Gram-negative cell is moving uric acid across the cytoplasmic membrane against its chemical gradient. Which of the following statements is true?
    a. The exterior of the cell is probably electrically negative compared to the interior of the cell.
    b. The acid probably moves by a passive means such as facilitated diffusion.
    c. The acid moves by an active process such as active transport.
    d. The movement of the acid requires phagocytosis.

12. Gram-positive bacteria
    a. have a thick cell wall, which retains crystal violet dye.
    b. contain teichoic acids in their cell walls.
    c. appear purple after Gram staining.
    d. all of the above

13. Endospores
    a. are reproductive structures of some bacteria.
    b. occur in some archaea.
    c. can cause shock, fever, and inflammation.
    d. are dormant, resistant cells.

14. Inclusions have been found to contain
    a. DNA.
    b. sulfur globules.
    c. dipicolinic acid.
    d. tubulin.

15. Dipicolinic acid is an important component of
    a. Gram-positive archaeal walls.
    b. cytoplasmic membranes in eukaryotes.
    c. endospores.
    d. Golgi bodies.

Matching

1. Match the structures on the left with the descriptions on the right. A letter may be used more than once or not at all, and more than one letter may be correct for each blank.

__D__ Glycocalyx
__B__ Flagella
__F__ Axial filaments
__E__ Cilia
__I__ Fimbriae
__H__ Pili
__A__ Hami

A. Bristlelike projections found in quantities of 100 or more
B. Long whip
C. Responsible for conjugation
D. "Sugar cup" composed of polysaccharides and/or polypeptides
E. Short, numerous, nonmotile projections
F. Responsible for motility of spirochetes
G. Extensions not used for cell motility
H. Made of tubulin in eukaryotes
I. Made of flagellin in bacteria

2. Match the term on the left with its description on the right. Only one description is intended for each term.

__C__ Ribosome
____ Cytoskeleton
____ Centriole
____ Nucleus
____ Mitochondrion
____ Chloroplast
____ ER
____ Golgi body
__B__ Peroxisome

A. Site of protein synthesis
B. Contains enzymes to neutralize hydrogen peroxide
C. Functions as the transport system within a eukaryotic cell
D. Allows contraction of the cell
E. Site of most DNA in eukaryotes
F. Contains microtubules in "9 + 0" arrangement
G. Light-harvesting organelle
H. Packages large molecules for export from the cell
I. Its internal membranes are sites for ATP production

Short Answer

1. Describe (or draw) an example of diffusion down a concentration gradient.
2. Sketch, name, and describe three flagellar arrangements in bacteria.
3. Define *cytosol.*
4. The term *fluid mosaic* has been used in describing the cytoplasmic membrane. How does each word of that phrase accurately describe our current understanding of a cell membrane?
5. A local newspaper writer has contacted you, an educated microbiology student from a respected college. He wants to obtain scientific information for an article he is writing about "life" and poses the following query: "What is the difference between a living thing and a nonliving thing?" Knowing that he will edit your material to fit the article, give an intelligent, scientific response.
6. What is the difference between growth and reproduction?
7. Compare bacterial cells and algal cells, giving at least four similarities and four differences.
8. Contrast a cell of *Streptococcus pyogenes* (a bacterium) with the unicellular protozoan *Entamoeba histolytica,* listing at least eight differences.
9. Differentiate among pili, fimbriae, and cilia using sketches and descriptive labels.
10. Can nonliving things metabolize? Explain your answer.
11. How do archaeal flagella differ from bacterial flagella and eukaryotic flagella?
12. Contrast bacterial and eukaryotic cells by filling in the following table.

Characteristic	Bacteria	Eukaryotes
Size		
Presence of nucleus		
Presence of membrane-bound organelles		
Structure of flagella		
Chemicals in cell walls		
Type of ribosomes		
Structure of chromosomes		

13. What is the function of glycocalyces and fimbriae in forming a biofilm?
14. What factors may prevent a molecule from moving across a cell membrane?
15. Compare and contrast three types of passive transport across a cell membrane.
16. Contrast the following active processes for transporting materials into or out of a cell: active transport, group translocation, endocytosis, exocytosis.
17. Contrast symports and antiports.
18. Describe the endosymbiotic theory. What evidence supports the theory? Which features of eukaryotic cells are not explained by the theory?

19. Label the structures of the following prokaryotic and eukaryotic cells. With a single word or short phrase, explain the function of each structure.

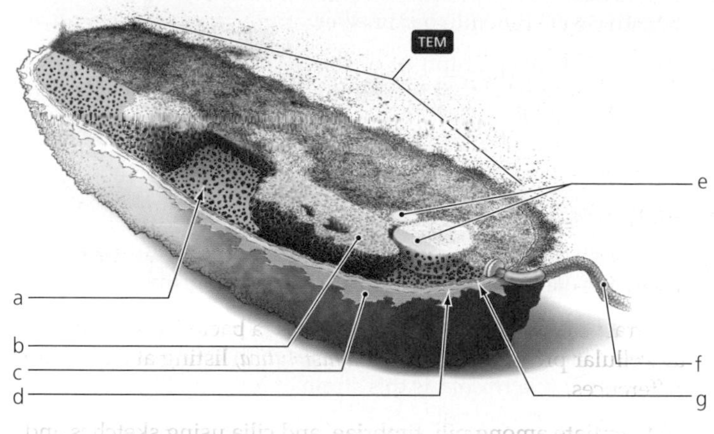

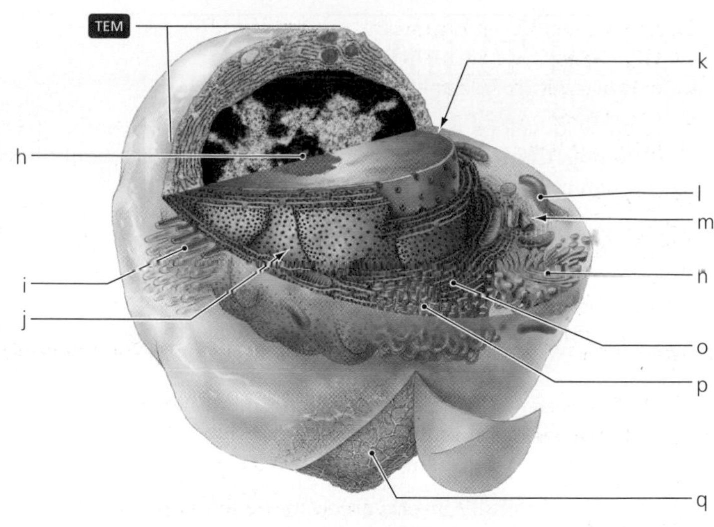

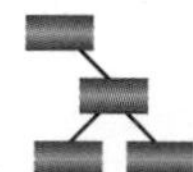

Concept Mapping

Using the following terms, fill in the following concept map that describes the bacterial cell wall. In coming chapters, you'll be asked to draw your own maps to link concepts in a logical way to help you study processes and relationships of microbiological terms. Use this map as a guide, but know that all maps will be different because there are many ways to draw them. You also have the option to complete this and other concept maps online by going to the Study Area at www.masteringmicrobiology.com.

Glycan chains
Gram-negative cell wall
Gram-positive cell wall
Lipid A
Lipopolysaccharide (LPS)
***N*-acetylglucosamine**
Peptidoglycan
Periplasm
Porins
Teichoic acids

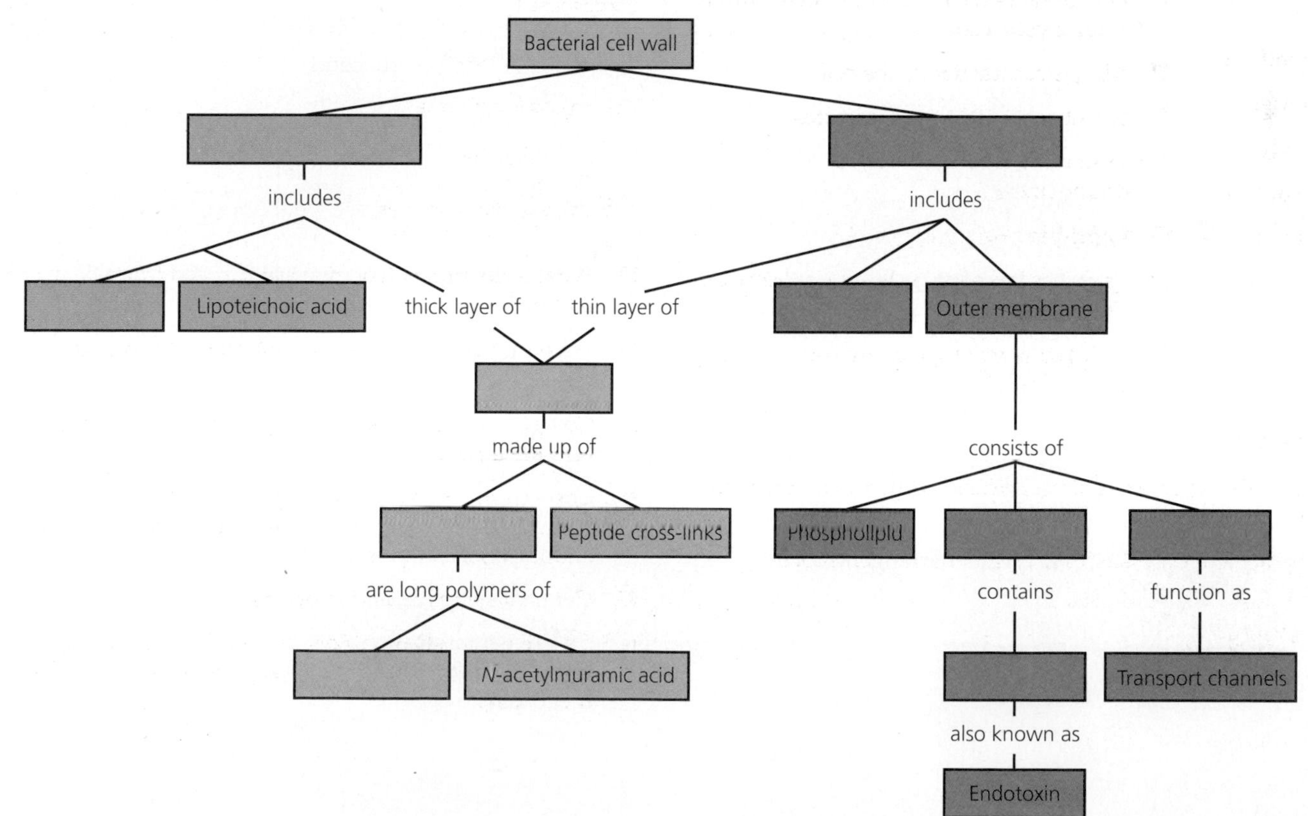

Critical Thinking

1. A scientist develops a chemical that prevents Golgi bodies from functioning. Contrast the specific effects the chemical would have on human cells versus bacterial cells.
2. Methylene blue binds to DNA. What structures in a yeast cell would be stained by this dye?
3. A new chemotherapeutic drug kills bacteria, but not humans. Discuss the possible ways the drug may act selectively on bacterial cells.
4. Some bacterial toxins cause cells lining the digestive tract to secrete ions, making the contents of the tract hypertonic. What effect does this have on a patient's water balance?
5. A researcher carefully inserts an electrode into a plant cell. He determines that the electrical charge across the cytoplasmic membrane is –70 millivolts. Then he slips the electrode deeper into the cell across another membrane and measures an electrical charge of –90 millivolts compared to the outside. What relatively large organelle is surrounded by the second membrane? Explain your answer.
6. Does Figure 3.2 on p. 57 best represent a Gram-positive or a Gram-negative cell? Defend your answer.
7. An electron micrograph of a newly discovered cell shows long projections with a basal body in the cell wall. What kind of projections are these? Is the cell prokaryotic or eukaryotic? How might this cell behave in its environment because of the presence of this structure?
8. An entry in a recent scientific journal reports positive phototaxis in a newly described species. What condition could you create in the lab to encourage the growth of these organisms?
9. A medical microbiological lab report indicates that a sample contained a biofilm, and that one species in the biofilm was identified as *Neisseria gonorrhoeae*. Is this strain of *Neisseria* likely to be pathogenic? Why or why not?
10. A researcher treats a cell to block the function of SER only. Describe the initial effects this would have on the cell.

Access more review material online in the Study Area at **www.masteringmicrobiology.com.** There, you'll find

- **Animations**
- **BioFlix**
- **Videos**
- **MP3 Tutor Sessions**
- **Concept Mapping Activities**
- **Flashcards**
- **Quizzes**

and more to help you succeed.

4 Microscopy, Staining, and Classification

When we want to weigh ourselves, we just step on a bathroom scale. But what can we use to measure the mass of something as delicate as the cell wall of a microorganism? The answer is an interference microscope, which uses a split beam of light to form vertical light and dark bands across a specimen (see the photo). Where light rays have traveled through a specimen and slowed, the pattern of light and dark bands is shifted, and the amount of shift is directly proportional to the change in speed. That change in speed can then be correlated with the specimen's density; the denser the specimen, the slower light travels through it. In this way, we can "weigh" a cell wall—or a nucleus or a chloroplast, or even a single chromosome.

This chapter explores the illuminating world of microscopes and the amazing lessons about microbial life that we continue to learn from them.

Take the pre-test for this chapter online. Visit the Study Area at www.masteringmicrobiology.com.

A piece of cell wall from the marine red alga *Griffithsia pacifica*, as seen through an interference microscope. The mass of the cell wall is 0.65 picogram per square micrometer. (A picogram equals 10^{-12} gram.)

Either the well was very deep, or she fell very slowly, for she had plenty of time as she went down to look about her, and to wonder what was going to happen next. . . . "Curiouser and curiouser!" cried Alice. . . .

—***Alice's Adventures in Wonderland*** **by Lewis Carroll**

Like Alice falling into Wonderland or traveling through a looking glass, scientists have entered the marvelous microbial world through advances in microscopy. With the invention of new laboratory techniques and the construction of new instruments, biologists are still discovering "curiouser and curiouser" wonders about the microbial world. In this chapter we will discuss some of the techniques microbiologists use to enter that world. We begin with a discussion of metric units as they relate to measuring the size of microbes. We then examine the instruments and staining techniques used in microbiology. Finally, we consider the classification schemes used to categorize the inhabitants of the microbial wonderland. **ANIMATIONS:** ***Microscopy and Staining: Overview***

Units of Measurement

Learning Objectives

- ✓ Identify the two primary metric units used to measure the diameters of microbes.
- ✓ List the metric units of length in order, from meter to nanometer.

Microorganisms are small. This may seem an obvious statement, but it is one that should not be taken for granted. Exactly how small are they? How can we measure the width and length of microbes?

Typically, a unit of measurement is smaller than the object being measured. For example, we measure a person's height in feet or inches, not in miles. Likewise, the diameter of a dime is measured in fractions of an inch, not in feet. So, measuring the size of a microbe requires units that are smaller than even the smallest interval on a ruler marked with English units (typically 1/16 inch). Even smaller units, such as 1/64 inch or 1/128 inch, become quite cumbersome and very difficult to use when we are dealing with microorganisms.

So that they can work with units that are simpler and in standard use the world over, scientists use metric units of measurement. Unlike the English system, the metric system is a decimal system, so each unit is one-tenth the size of the next largest unit. Even extremely small metric units are much easier to use than the fractions involved in the English system.

The unit of length in the metric system is the *meter (m)*, which is slightly longer than a yard. One-tenth of a meter is a *decimeter (dm)*, and one-hundredth of a meter is a *centimeter (cm)*, which is equivalent to about a third of an inch. One-tenth of a centimeter is a *millimeter (mm)*, which is the thickness of a dime. A millimeter is still too large to measure the size of most microorganisms, but in the metric system we continue to divide by multiples of 10 until we have a unit appropriate for use. Thus, one-thousandth of a millimeter is a *micrometer (μm)*, which is small enough to be useful in measuring the size of cells. One-thousandth of a micrometer is a *nanometer (nm)*, a unit used to measure the smallest cellular organelles and viruses. A nanometer is one-billionth of a meter.

Table 4.1 presents these metric units and some English equivalents. Refer to Figure 3.4 for a visual size comparison of a typical eukaryotic cell, prokaryotic cell, and virus particle.

TABLE 4.1 **Metric Units of Length**

Metric Unit (abbreviation)	Meaning of Prefix	Metric Equivalent	U.S. Equivalent	Representative Microbiological Application of the Unit
meter (m)	—[a]	1 m	39.37 in (about a yard)	Length of pork tapeworm, *Taenia solium* (e.g., 1.8 m–8.0 m)
decimeter (dm)	1/10	$0.1 \text{ m} = 10^{-1} \text{ m}$	3.94 in	—[b]
centimeter (cm)	1/100	$0.01 \text{ m} = 10^{-2} \text{ m}$	0.39 in; 1 in – 2.54 cm	Diameter of a mushroom cap (e.g., 12 cm)
millimeter (mm)	1/1000	$0.001 \text{ m} = 10^{-3} \text{ m}$	—	Diameter of a bacterial colony (e.g., 2.3 mm); length of a tick (e.g., 5.7 mm)
micrometer (μm)	1/1,000,000	$0.000001 \text{ m} = 10^{-6} \text{ m}$	—	Diameter of white blood cells (e.g., 5 μm-25 μm)
nanometer (nm)	1/1,000,000,000	$0.000000001 \text{ m} = 10^{-9} \text{ m}$	—	Diameter of a poliovirus (e.g., 25 nm)

[a]The meter is the standard metric unit of length.
[b]Decimeters are rarely used.

Microscopy

Learning Objective

✓ Define *microscopy.*

Microscopy[1] refers to the use of light or electrons to magnify objects. The science of microbiology began when Leeuwenhoek used primitive microscopes to observe and report the existence of microorganisms. Since that time, scientists and engineers have developed a variety of light and electron microscopes.

General Principles of Microscopy

Learning Objectives

✓ Explain the relevance of electromagnetic radiation to microscopy.

✓ Define *empty magnification.*

✓ List and explain two factors that determine resolving power.

✓ Discuss the relationship between contrast and staining in microscopy.

General principles involved in both light and electron microscopy include the wavelength of radiation, the magnification of an image, the resolving power of the instrument, and contrast in the specimen.

Wavelength of Radiation

Visible light is one part of a spectrum of electromagnetic radiation that includes X rays, microwaves, and radio waves **(Figure 4.1)**. Note that beams of radiation may be referred to as either rays or waves. These various forms of radiation differ in **wavelength**—the distance between two corresponding parts of a wave. The human eye discriminates among different wavelengths of visible light and sends patterns of nerve impulses to the brain, which interprets the impulses as different colors. For example, we see wavelengths of 400 nm as violet and wavelengths of 650 nm as red. White light, composed of many colors (wavelengths), has an average wavelength of 550 nm.

In Chapter 2 we learned that electrons are negatively charged particles that orbit the nuclei of atoms. Besides being particulate, moving electrons act as waves with wavelengths dependent upon the voltage of an electron beam. For example, the wavelength of electrons at 10,000 volts (V) is 0.01 nm, whereas the wavelength of electrons at 1,000,000 V is 0.001 nm. As we will see, using radiation of smaller wavelengths results in enhanced microscopy.

Magnification

Magnification is the apparent increase in size of an object. It is indicated by a number and "×", which is read "times." For example, the cells in Figure 4.13c are magnified 16,000×, or 16,000 times. Magnification results when a beam of radiation *refracts* (bends) as it passes through a lens. Curved glass lenses refract

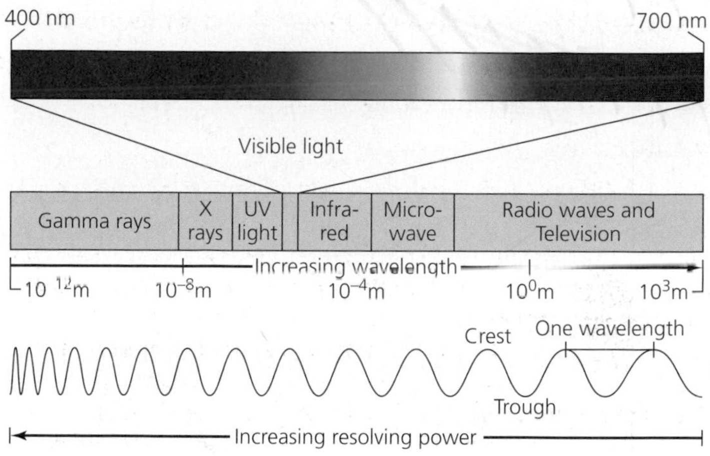

▲ **Figure 4.1 The electromagnetic spectrum.** Visible light comprises a narrow band of wavelengths of radiation. Visible and ultraviolet (UV) light are used in microscopy.

light, and magnetic fields (magnetic lenses) refract electron beams. Let's consider the magnifying power of a glass lens that is convex on both sides.

A lens refracts light because the lens is *optically dense* compared to the surrounding medium (such as air); that is, light travels more slowly through the lens than through air. This can be compared to a car moving at an angle from a paved road onto a dirt shoulder. As the right front tire leaves the pavement, it has less traction and slows down. Since the other wheels continue at their original speed, the car will veer toward the dirt, and the line of travel will bend to the right. Likewise, the leading edge of a light beam slows as it enters glass, and the beam bends **(Figure 4.2a)**. Light also bends as it leaves the glass and reenters the air.

Because of its curvature, a lens refracts light rays that pass through its periphery more than light rays that pass through its center so that the lens focuses light rays on a *focal point.* Importantly for the purpose of microscopy, light rays spread apart as they travel past the focal point and produce an enlarged, inverted image **(Figure 4.2b)**. The degree to which the image is enlarged depends on the thickness of the lens, its curvature, and the speed of light through its substance.

Microscopists could combine lenses to obtain an image magnified millions of times, but the image would be faint and blurry. Such magnification is said to be *empty magnification.* The properties that determine the clarity of an image, which in turn determines the useful magnification of a microscope, are *resolution* and *contrast.*

Resolution

Resolution, which is also called *resolving power,* is the ability to distinguish between objects that are close together. An optometrist's eye chart is a test of resolution at a distance of 20 feet (6.1 m). Leeuwenhoek's microscopes had a resolving power of about 1 μm; that is, he could distinguish between objects if they were more than about 1 μm apart, and objects closer together than 1 μm appeared as a single object. The better the resolution,

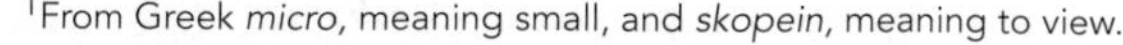

[1]From Greek *micro,* meaning small, and *skopein,* meaning to view.

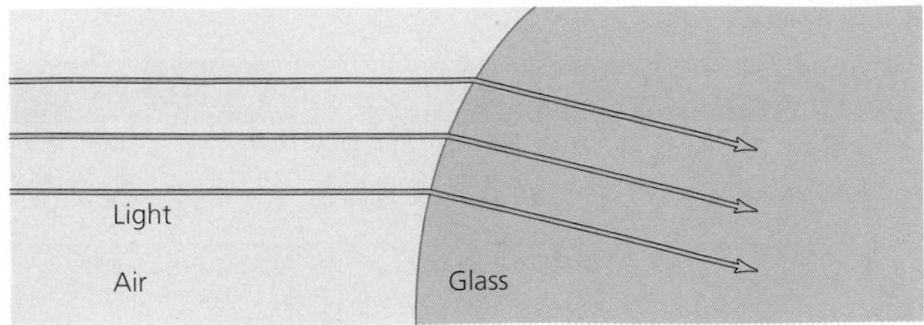

(a)

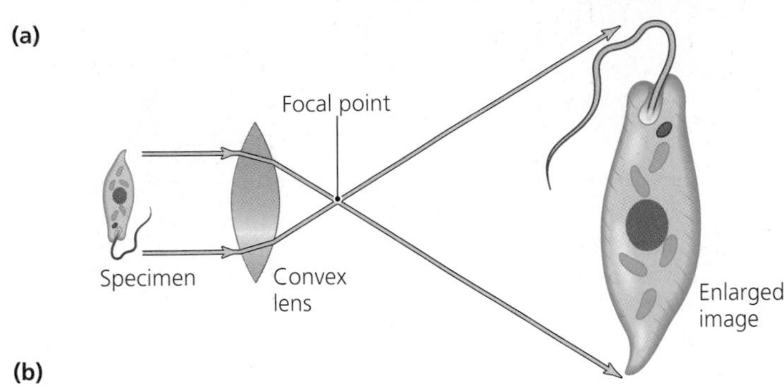

(b)

▲ **Figure 4.2 Light refraction and image magnification by a convex glass lens. (a)** Light passing through a lens refracts (bends) because light rays slow down as they enter the glass, and light at the leading edge of a beam that strikes the glass at an angle slows first. **(b)** A convex lens focuses light on a focal point. The image is enlarged and inverted as light rays pass the focal point and spread apart.

the better the ability to distinguish two objects that are close to one another. Modern microscopes have fivefold better resolution than Leeuwenhoek's; they can distinguish between objects as close together as 0.2 μm. **Figure 4.3** illustrates the size of various objects that can be resolved by the unaided human eye and by various types of microscopes.

Why do modern microscopes have better resolution than Leeuwenhoek's microscopes? A principle of microscopy is that resolution distance is dependent on (1) the wavelength of the electromagnetic radiation and (2) the **numerical aperture** of the lens, which refers to the ability of a lens to gather light.

Resolution distance is calculated using the following formula:

$$\text{resolution distance} = \frac{0.61 \times \text{wavelength}}{\text{numerical aperture}}$$

The resolution of today's microscopes is greater than that of Leeuwenhoek's microscopes because modern microscopes use shorter-wavelength radiation, such as blue light or electron beams, and because they have lenses with larger numerical apertures.

Contrast

Contrast refers to differences in intensity between two objects, or between an object and its background. Contrast is important in determining resolution. For example, although you can easily distinguish between two golf balls lying side by side on a putting green 15 m away, at that distance it is much more difficult to distinguish between them if they are lying on a white towel.

Most microorganisms are colorless and have very little contrast whether one uses light or electrons. One way to increase the contrast between microorganisms and their background is to stain them. Stains and staining techniques are covered later in the chapter. As we will see, the use of light that is in *phase*—that is, in which all of the waves' crests and troughs are aligned—can also enhance contrast.

Light Microscopy

Learning Objectives

- ✓ Contrast simple and compound microscopes.
- ✓ Compare and contrast bright-field microscopy, dark-field microscopy, and phase microscopy.
- ✓ Compare and contrast fluorescent and confocal microscopes.

Several classes of microscopes use various types of light to examine microscopic specimens. The most common microscopes are *bright-field microscopes,* in which the background (or *field*) is illuminated. In *dark-field microscopes,* the specimen is made to appear light against a dark background. *Phase microscopes* use the alignment or misalignment of light waves to achieve the desired contrast between a living specimen and its background. *Fluorescent microscopes* use invisible ultraviolet light to cause specimens to radiate visible light, a phenomenon called *fluorescence.* Microscopes that use lasers to illuminate fluorescent chemicals in a thin plane of a specimen are called *confocal microscopes.* Next we examine each of these kinds of light microscope in turn.

Bright-Field Microscopes

There are two basic types of bright-field microscopes: *simple microscopes* and *compound microscopes.*

Simple Microscopes As we learned in Chapter 1, Leeuwenhoek first reported his observations of microorganisms using a simple microscope in 1674. A **simple microscope,** which contains a single magnifying lens, is more similar to a magnifying glass than to a modern microscope (see Figure 1.2). Though Leeuwenhoek did not invent the microscope, he was the finest lens grinder of his day and produced microscopes of exceptional quality. They were capable of approximately 300× magnification and achieved excellent clarity, far surpassing other microscopes of his time.

Compound Microscopes Simple microscopes have been replaced in modern laboratories by compound microscopes. A **compound microscope** uses a series of lenses for magnification **(Figure 4.4a)**. Many scientists, including Galileo Galilei (1564–1642), made compound microscopes as early as 1590, but it was not until about 1830 that scientists developed compound microscopes that exceeded the clarity and magnification of Leeuwenhoek's simple microscope.

In a basic compound microscope, magnification is achieved as light rays pass through a specimen and into an **objective lens,** which is the lens immediately above the object being magnified

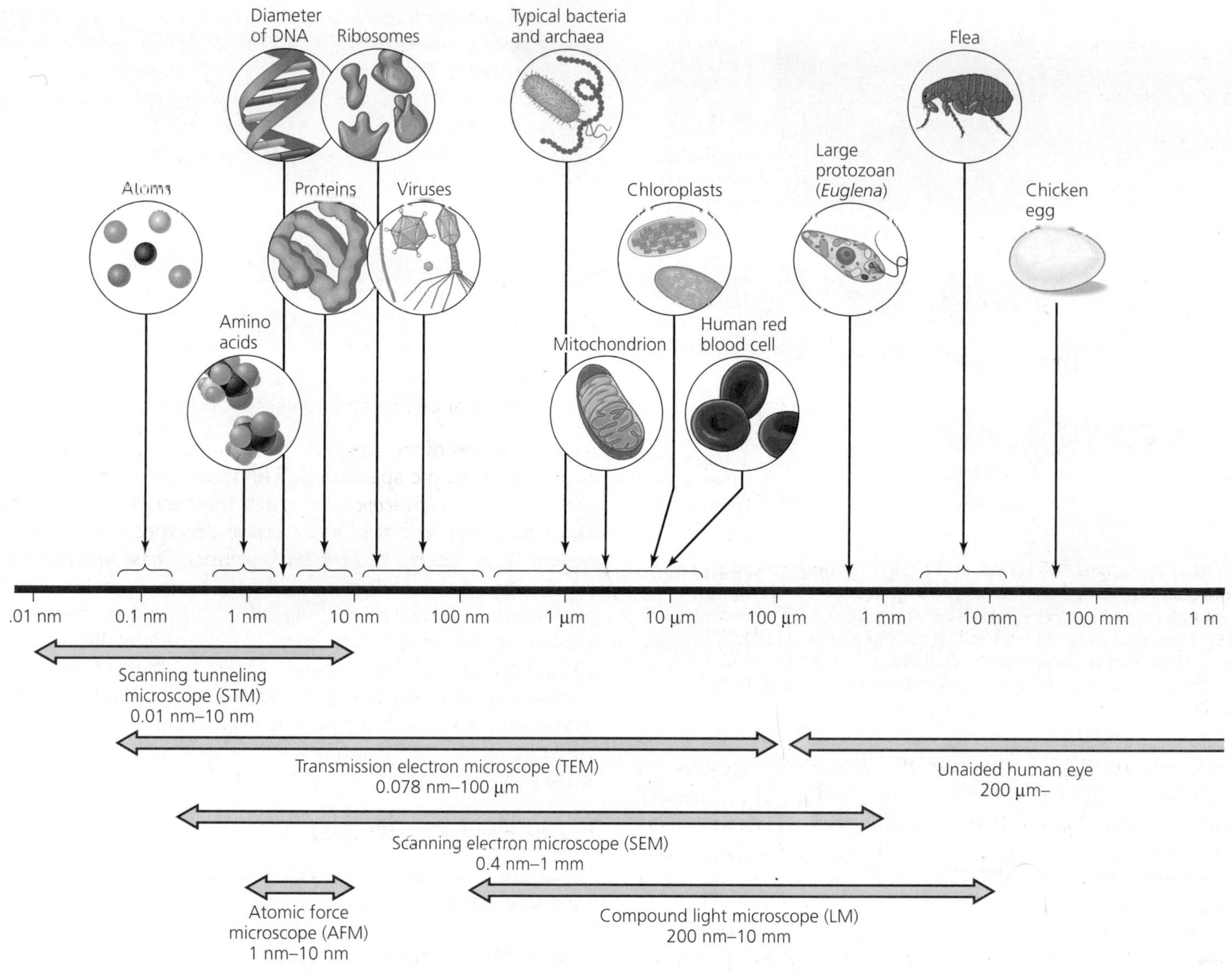

▲ **Figure 4.3 The limits of resolution (and some representative objects within those ranges) of the human eye and of various types of microscopes.**

(Figure 4.4b). An objective lens is really a series of lenses that not only create a magnified image, but are also engineered to reduce aberrations in the shape and color of the image. Most light microscopes used in biology have three or four objective lenses mounted on a **revolving nosepiece.** The objective lenses on a typical microscope are: *scanning objective lens* (4×), *low-power objective lens* (10×), *high-power lens* or *high dry objective lens* (40×), and *oil immersion objective lens* (100×).

Not only does an oil immersion lens increase magnification, it also increases resolution. As we have seen, light refracts as it travels from air into glass and also from glass into air; therefore, some of the light passing out of a glass slide is bent so much that it bypasses the lens **(Figure 4.5a)**. Placing *immersion oil* between the slide and an *oil immersion objective lens* enables the lens to capture this light because light travels through immersion oil at the same speed as through glass. Since light is traveling at a uniform speed through the slide, the immersion oil, and the glass lens, it does not refract **(Figure 4.5b)**. Immersion oil increases the numerical aperture, which increases resolution, because more light rays are gathered into the lens to produce the image. Obviously, the space between the slide and the lens can be filled with oil only if the distance between the lens and the specimen, called the *working distance*, is small.

An objective lens bends the light rays, which then pass up through one or two **ocular lenses,** which are the lenses closest to the eyes. Microscopes with a single ocular lens are *monocular*, and those with two are *binocular*. Ocular lenses magnify the image created by the objective lens, typically another 10 times (10×).

The **total magnification** of a compound microscope is determined by multiplying the magnification of the objective lens by the magnification of the ocular lens. Thus, total magnification using a 10× ocular lens and a 10× low-power objective lens is 100×. Using the same ocular and a 100× oil immersion objective produces 1000× magnification. Some light microscopes,

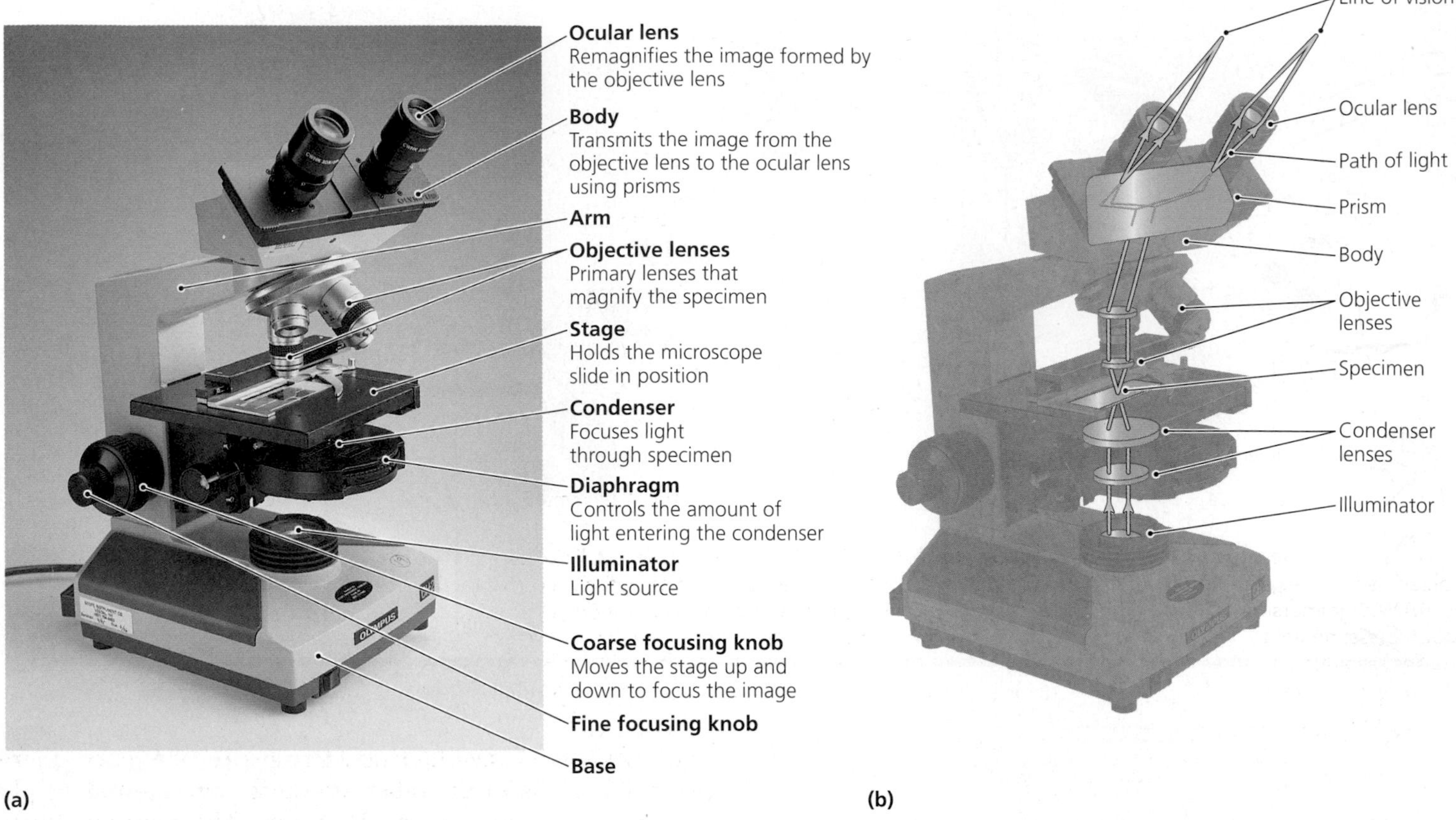

▲ **Figure 4.4 A bright-field, compound light microscope. (a)** The parts of a compound microscope, which uses a series of lenses to produce an image at up to 2000× magnification. **(b)** The path of light in a compound microscope; light travels from bottom to top. *Why can't light microscopes produce clear images that are magnified 10,000×?*

***Figure 4.4** Although it is possible for a light microscope to produce an image that is magnified 10,000×, magnification above 2000× is empty magnification because pairs of objects in the specimen are too close together to resolve with even the shortest-wavelength (blue) light.*

using higher-magnification oil immersion objective lenses and ocular lenses, can achieve 2000× magnification, but this is the limit of useful magnification for light microscopes because their resolution is restricted by the wavelength of visible light.

Modern compound microscopes also have a **condenser lens** (or lenses), which directs light through the specimen, as well as one or more mirrors or prisms that deflect the path of the light rays from an objective lens to the ocular lens (see Figure 4.4b). Some microscopes have mirrors or prisms that direct light to a camera through a special tube. Photographs of such a microscopic image are called light **micrographs.**

Dark-Field Microscopes

Pale objects are best observed with **dark-field microscopes.** These microscopes utilize a *dark-field stop* in the condenser that prevents light from directly entering the objective lens **(Figure 4.6)**. Instead, light rays are reflected inside the condenser so that they pass into the slide at such an oblique angle that they miss the objective lens. Only those light rays that are scattered by the specimen enter the objective lens and are seen, so the specimen appears light against a dark background. This increases contrast and enables observation of more details than are visible in bright-field microscopy. Dark-field microscopes are especially useful for examining small or colorless cells.

Phase Microscopes

Scientists use **phase microscopes** to examine living microorganisms or specimens that would be damaged or altered by attaching them to slides or staining them. Basically, phase microscopes treat one set of light rays differently from another set of light rays.

Light rays are said to be *in phase* when their crests and troughs are aligned, and *out of phase* when their crests and troughs are not aligned **(Figure 4.7a)**. Light rays that are in phase reinforce one another, producing a brighter image, and light rays that are out of phase interfere with one another and produce a darker image. Light rays passing through a specimen naturally slow down and are shifted about 1/4 wavelength out of phase. A special filter called a *phase plate,* which is mounted in a phase objective lens, retards these rays another 1/4 wavelength, so that they are 1/2 wavelength out of phase with their neighbors. When the phase microscope lens brings

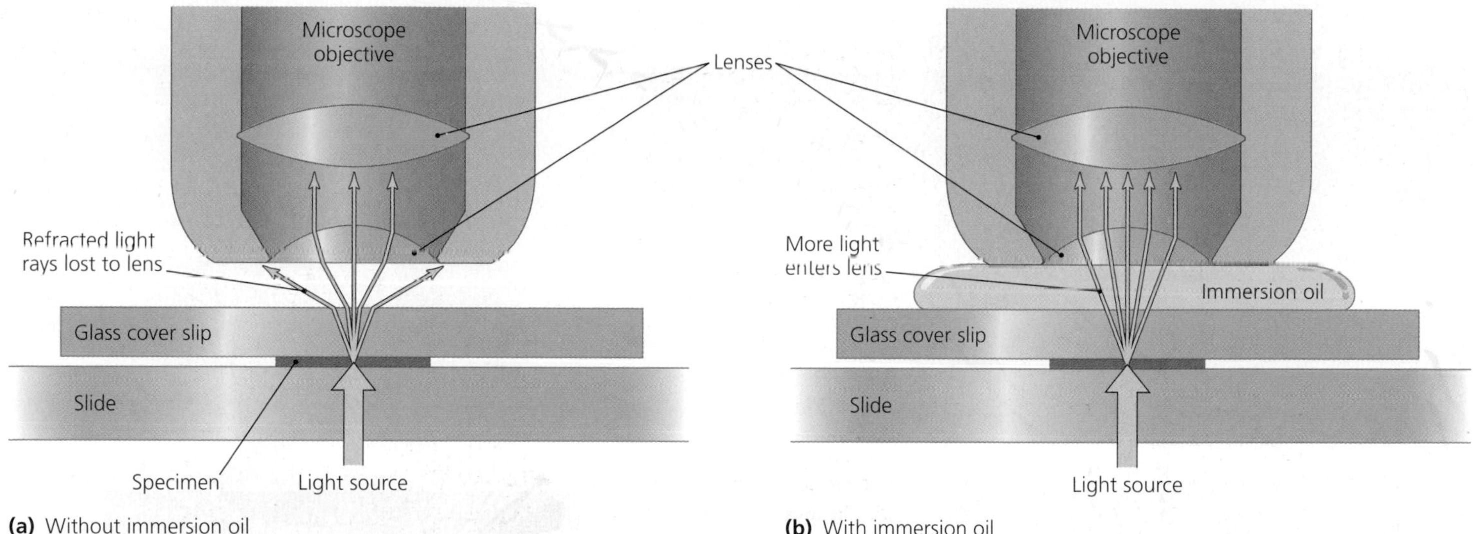

▲ **Figure 4.5 The effect of immersion oil on resolution.** **(a)** Without immersion oil, light is refracted as it moves from the cover glass into the air. Part of the scattered light misses the objective lens. **(b)** With immersion oil. Because light travels through the oil at the same speed as it does through glass, no light is refracted as it leaves the specimen, and more light enters the lens, which increases resolution.

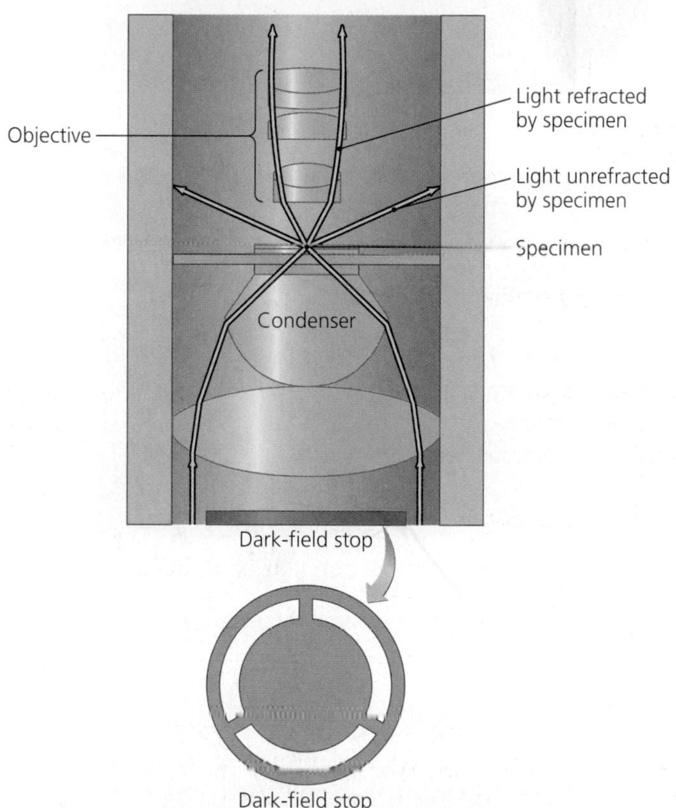

▲ **Figure 4.6 The light path in a dark-field microscope.** A dark-field stop prevents light from entering the specimen directly; only light rays that are scattered by the specimen reach the objective lens and can be seen. The resulting high-contrast image—a brightly lit specimen against a dark background—enhances resolution.

the two sets of rays together, troughs of one wave interfere with the crests of the other—because they are out of phase **(Figure 4.7b)**—and contrast is created. There are two types of phase microscopes: phase-contrast and differential interference contrast microscopes.

Phase-Contrast Microscopes The simplest phase microscopes, **phase-contrast microscopes,** produce sharply defined images in which fine structures can be seen in living cells. These microscopes are particularly useful for observing cilia and flagella.

Differential Interference Contrast Microscopes These microscopes (also called Nomarski[2] microscopes) create phase interference patterns. They also use prisms that split light beams into their component wavelengths (colors). This significantly increases contrast and gives the image a dramatic three-dimensional or shadowed appearance, almost as if light were striking the specimen from one side. This technique also produces unnatural colors, which enhance contrast.

Figure 4.8 illustrates the differences that can be observed in a single specimen when viewed using four different types of light microscopy.

Fluorescent Microscopes

Molecules that absorb energy from invisible radiation (such as ultraviolet light) and then radiate the energy back as a longer, visible wavelength are said to be *fluorescent.* **Fluorescent microscopes** use an ultraviolet (UV) light source to fluoresce objects.

[2]After the French physicist Georges Nomarski, who invented the differential interference contrast microscope.

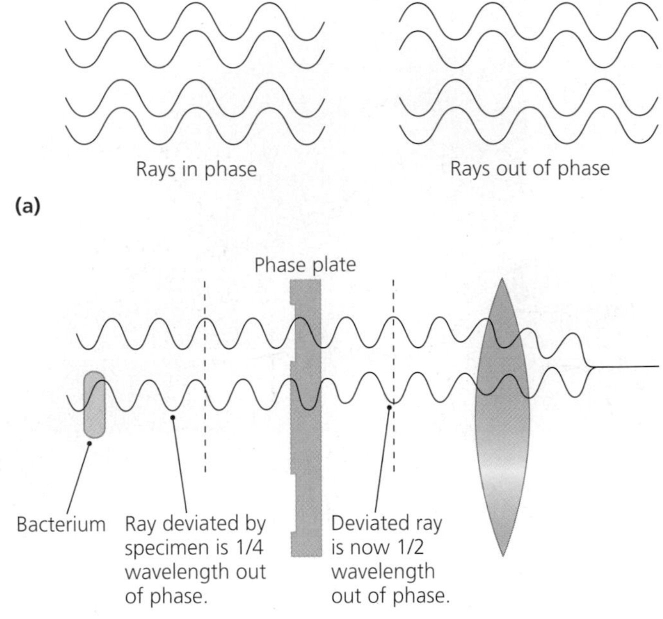

▲ **Figure 4.7 Principles of phase microscopy.** **(a)** Light rays that are in phase are aligned; their crests and troughs reinforce one another to produce a brighter image. Rays that are out of phase interfere with one another and produce a darker image. **(b)** Together, the regions of a specimen and a phase plate built into a phase objective lens slow some of the light rays such that they are 1/2 wavelength out of phase. These rays interfere with light rays that bypass the specimen, which produces contrast between the field and various regions of the specimen (see Figure 4.8c and d).

UV light increases resolution because it has a shorter wavelength than visible light, and contrast is improved because fluorescing structures are visible against a black background.

Some cells—for example, the pathogen *Pseudomonas aeruginosa* (soo-dō-mō′nas ā-roo-ji-nō′să)—and some cellular molecules (such as chlorophyll in photosynthetic organisms) are naturally fluorescent. Other cells and cellular structures can be stained with fluorescent dyes. When these dyes are bombarded with ultraviolet light, they emit visible light and show up as bright orange, green, yellow, or other colors against a black background (see Figure 3.33b).

Some fluorescent dyes are specific for certain cells. For example, the dye fluorescein isothiocyanate attaches to cells of *Bacillus anthracis* (ba-sil′ŭs an-thrā′sis), the causative agent of anthrax, and appears apple green when viewed in a fluorescent microscope. Another fluorescent dye, auramine O, stains *Mycobacterium tuberculosis* (mī′kō-bak-tēr′ē-ŭm too-ber-kyū-lō′sis; **Figure 4.9**).

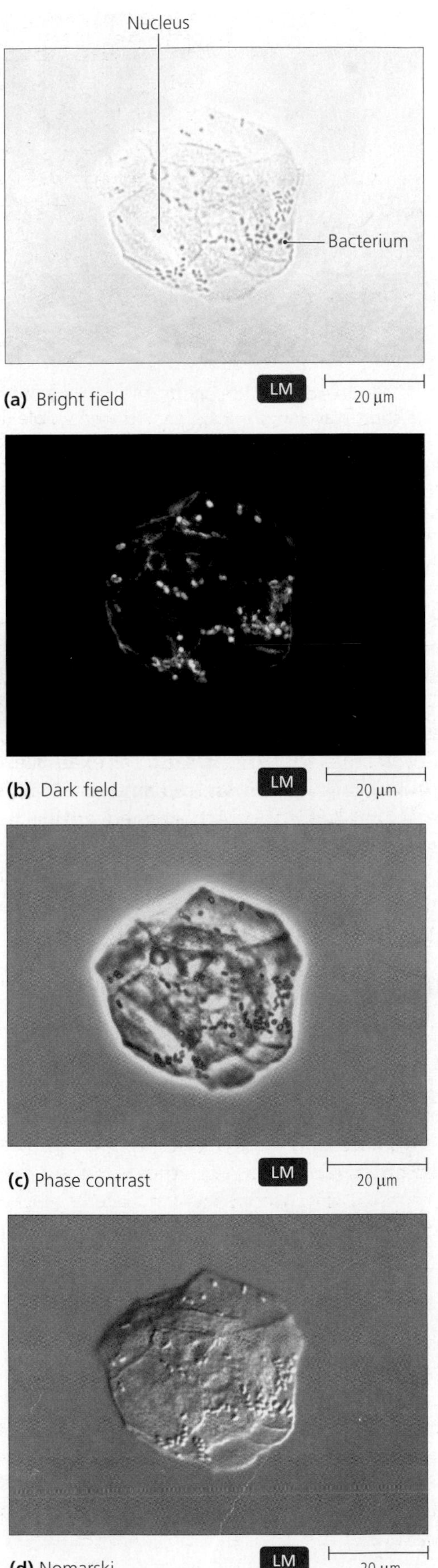

➤ **Figure 4.8 Four kinds of light microscopy.** All four photos show the same human cheek cell and bacteria. **(a)** Bright-field microscopy reveals some internal structures. **(b)** Dark-field microscopy increases contrast between some internal structures and between the edges of the cell and the surrounding medium. **(c)** Phase-contrast microscopy provides greater resolution of internal structures. **(d)** Differential interference contrast (Nomarski) microscopy produces a three-dimensional effect.

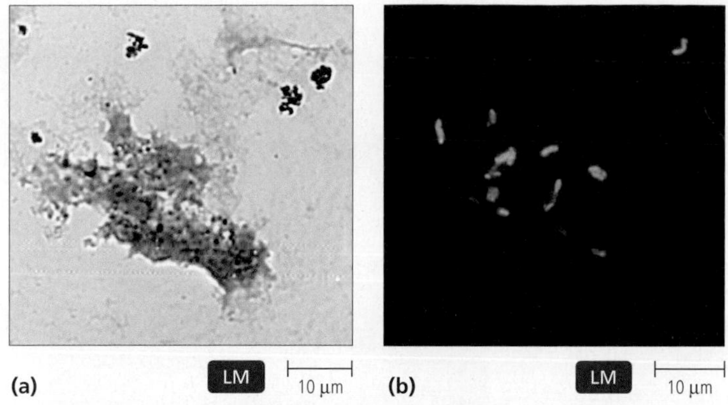

▲ **Figure 4.9 Fluorescent microscopy.** Fluorescent chemicals absorb invisible short-wavelength radiation and emit visible (longer wavelength) radiation. **(a)** When viewed under normal illumination, *Mycobacterium tuberculosis* cells stained with the fluorescent dye auramine O are invisible amid the mucus and debris in a sputum smear. **(b)** When the same smear is viewed under UV light, the bacteria fluoresce and are clearly visible.

Fluorescent microscopy is also used in a process called *immunofluorescence*. First, fluorescent dyes are covalently linked to Y-shaped immune system proteins called *antibodies* **(Figure 4.10a)**. Given the opportunity, these dye-tagged antibodies will bind specifically to complementary-shaped *antigens,* which are portions of molecules that are present, for example, on the surface of microbial cells. When viewed under UV light, a microbial specimen that has bound dye-tagged antibodies becomes visible **(Figure 4.10b)**. In addition to identifying pathogens, including those that cause syphilis, rabies, and Lyme disease, scientists use immunofluorescence to locate and make visible a variety of proteins of interest.

Confocal Microscopes

Confocal[3] **microscopes** also use fluorescent dyes or fluorescent antibodies, but these microscopes use ultraviolet lasers to illuminate the fluorescent chemicals in only a single plane that is no thicker than 1.0 μm; the rest of the specimen remains dark and out of focus. Visible light emitted by the dyes passes through a pinhole aperture, which helps eliminate blurring that can occur with other types of microscopes and increases resolution by up to 40%. Each image from a confocal microscope is thus an "optical slice" through the specimen, as if it had been thinly cut. Once individual images are digitized, a computer is used to construct a three-dimensional representation, which can be rotated and viewed from any direction. Confocal microscopes have been particularly useful for examining the relationships among various organisms within complex microbial communities called biofilms (see **Highlight: Studying Biofilms in Plastic "Rocks"** on p. 104). Regular light microscopy cannot produce clear images of structures within a living biofilm, and removing surface layers from a biofilm would change the dynamics of a biofilm community. **ANIMATIONS:** ***Light Microscopy***

[3]From *coinciding focal* points of (laser) light.

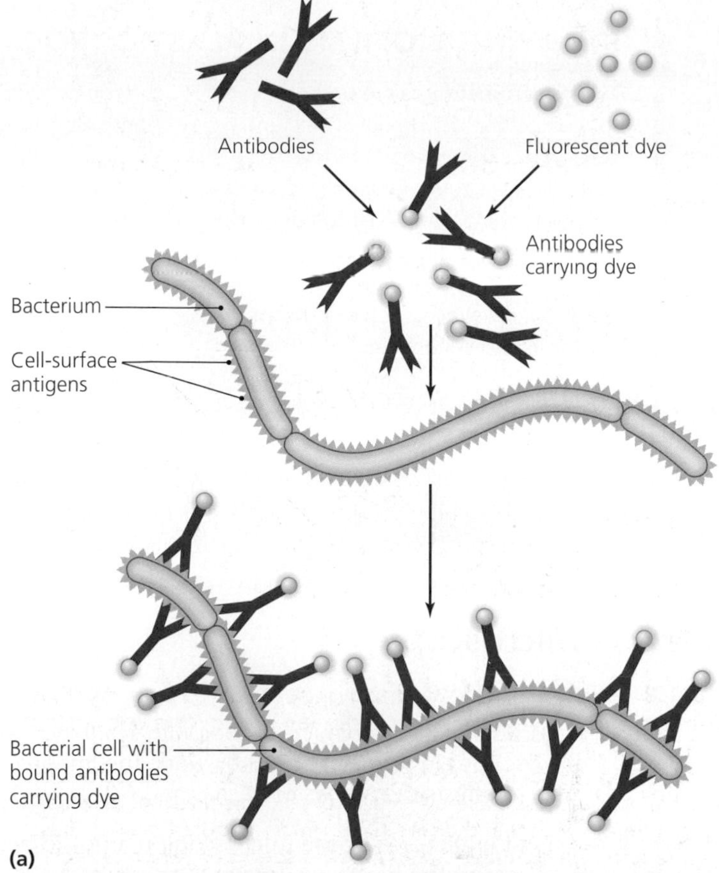

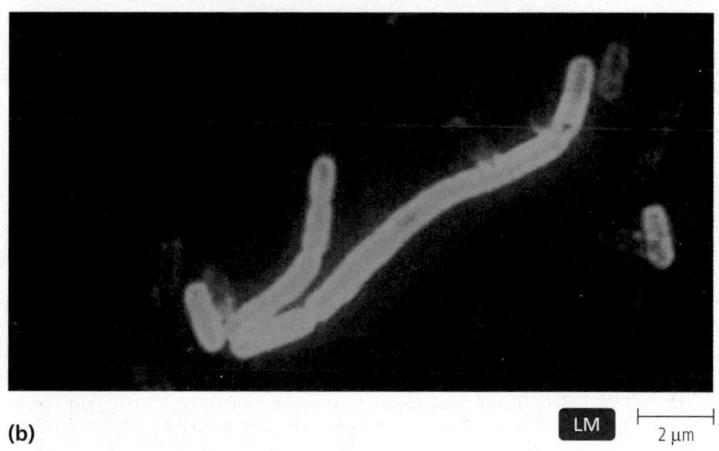

▲ **Figure 4.10 Immunofluorescence. (a)** After a fluorescent dye is covalently linked to an antibody, the dye-antibody combination binds to the antibody's target, making the target visible under fluorescent microscopy. **(b)** Immunofluorescent staining of *Yersinia pestis*, the causative agent of plague. The bacteria are brightly colored against a dark background.

HIGHLIGHT

STUDYING BIOFILMS IN PLASTIC "ROCKS"

Marine stromatolites are unique rock structures made up of calcium carbonate. The insides of these rock structures are teeming with bacterial communities organized into complex biofilms. These biofilms contain many different microorganisms, including cyanobacteria, aerobic heterotrophic bacteria, and sulfate-reducing bacteria.

To study how the tiny creatures in marine stromatolites interact with each other and their environment, researchers at the University of Southern California pulled samples of the biofilms out of the rock structures and fixed them in nontoxic resin. In some cases, specific sections of the microbial cells were stained using fluorescent probes, and then confocal microscopy was used to study cross sections of the communities. By embedding the bacteria in a resin while maintaining the biofilm structure, the researchers were able to watch the bacterial network in an almost natural state. This technique is versatile and can be used to study other microbial systems to gain further understanding of biofilm mechanics and composition.

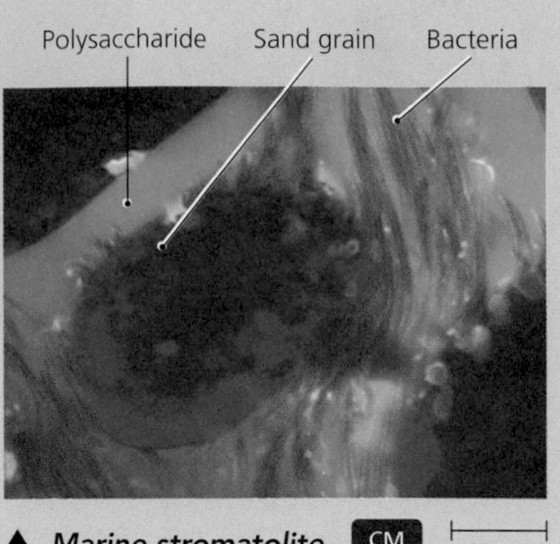

▲ *Marine stromatolite.* CM 25 μm

Electron Microscopy

Learning Objective

✓ Contrast transmission electron microscopes with scanning electron microscopes in terms of how they work, the images they produce, and the advantages of each.

Even with the most expensive phase microscope using the best oil immersion lens with the highest numerical aperture, resolution is still limited by the wavelength of visible light. Because the shortest visible radiation (violet) has a wavelength of about 400 nm, structures closer together than about 200 nm cannot be distinguished using even the best light microscope. By contrast, electrons traveling as waves have wavelengths between 0.01 nm and 0.001 nm, which is one ten-thousandth to one hundred-thousandth the wavelength of visible light. The resolving power of electron microscopes is therefore much greater than that of light microscopes, and with greater resolving power comes the possibility of greater magnification.

Generally, electron microscopes magnify objects 10,000× to 100,000×, though millions of times magnification with good resolution is possible. Electron microscopes provide detailed views of the smallest bacteria, viruses, internal cellular structures, and even molecules and large atoms. Cellular structures that can be seen only by using electron microscopy are referred to as a cell's *ultrastructure.* Ultrastructural details cannot be made visible by light microscopy because they are too small to be resolved.

There are two general types of electron microscopes: *transmission electron microscopes* and *scanning electron microscopes.*

Transmission Electron Microscopes

A **transmission electron microscope (TEM)** generates a beam of electrons that ultimately produces an image on a fluorescent screen **(Figure 4.11a)**. The path of electrons is similar to the path of light in a light microscope. From their source, the electrons pass through the specimen, through magnetic fields (instead of glass lenses) that manipulate and focus the beam, and then onto a fluorescent screen that absorbs electrons, thereby changing some of their energy into visible light **(Figure 4.11b)**. Dense areas of the specimen block electrons, resulting in a dark area on the screen. In regions where the specimen is less dense, the screen fluoresces more brightly. As with light microscopy, contrast and resolution can be enhanced through the use of electron-dense stains, which are discussed later. The brightness of each region of the screen corresponds to the number of electrons striking it. Therefore, the image on the screen is composed of light and dark areas much like a photographic negative.

The screen can be folded out of the way to enable the electrons to strike a photographic film, located in the base of the microscope. Prints made from the film are called *transmission electron micrographs* or *TEM images* **(Figure 4.11c)**. Such images can be colorized to emphasize certain features.

Matter, including air, absorbs electrons, so the column of a transmission electron microscope must be a vacuum, and the specimen must be very thin. Before thicker specimens such as whole cells can be examined, they must be dehydrated, embedded in plastic, and cut to a thickness of about 100 nm with a diamond or glass knife mounted in a slicing machine called an *ultramicrotome.* Such thin sections are placed on a small copper grid and inserted into the microscope through an air lock.

Because the vacuum and slicing of the specimens are required, transmission electron microscopes cannot be used to study living organisms. Dehydration and sectioning can also introduce shrinkage, distortion, and other *artifacts,* which are structures that appear in a TEM image but are not present in natural specimens.

Scanning Electron Microscopes

A **scanning electron microscope (SEM)** also uses magnetic fields within a vacuum tube to manipulate a beam of electrons,

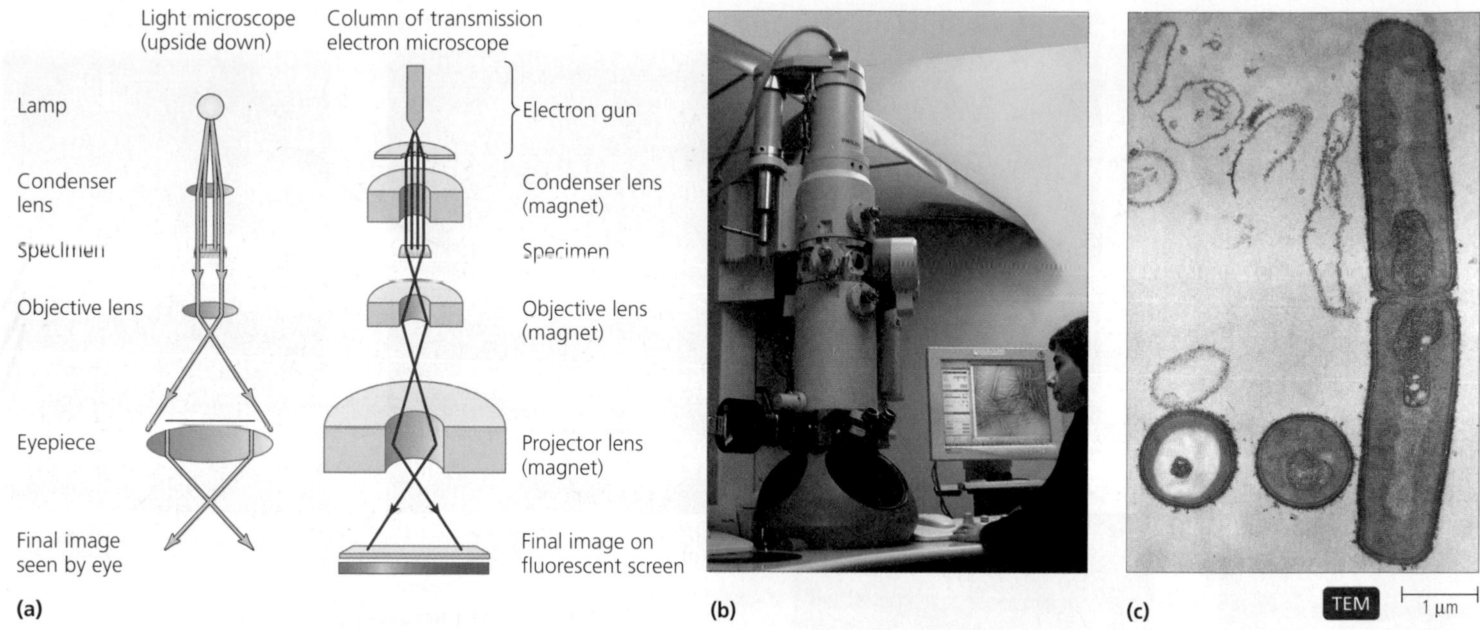

▲ **Figure 4.11 A transmission electron microscope (TEM).** **(a)** The path of electrons through a TEM, as compared to the path of light through a light microscope (at left, drawn upside down to facilitate the comparison). **(b)** TEMs are much larger than light microscopes. **(c)** Transmission electron image of a bacterium, *Bacillus subtilis.* A transmission electron micrograph reveals much internal detail not visible by light microscopy. *Why must air be evacuated from the column of an electron microscope?*

Figure 4.11 *Air would absorb electrons, so there would be no radiation to produce an image.*

called primary electrons **(Figure 4.12)**. However, rather than passing electrons through a specimen, the SEM rapidly focuses them back and forth across the specimen's surface, which has previously been coated with a metal such as platinum or gold. The primary electrons knock electrons off the surface of the coated specimen, and these scattered secondary electrons pass through a detector and a photomultiplier, producing an amplified signal that is displayed on a monitor. Typically, scanning microscopes are used to magnify up to 10,000× with a resolution of about 20 nm.

One advantage of scanning microscopy over transmission microscopy is that whole specimens can be observed, because sectioning is not required. Scanning electron micrographs can be beautifully realistic and three-dimensional **(Figure 4.13)**. Two disadvantages of a scanning electron microscope are that it magnifies only the external surface of a specimen and, like TEM, it requires a vacuum and thus can examine only dead organisms. **ANIMATIONS:** ***Electron Microscopy***

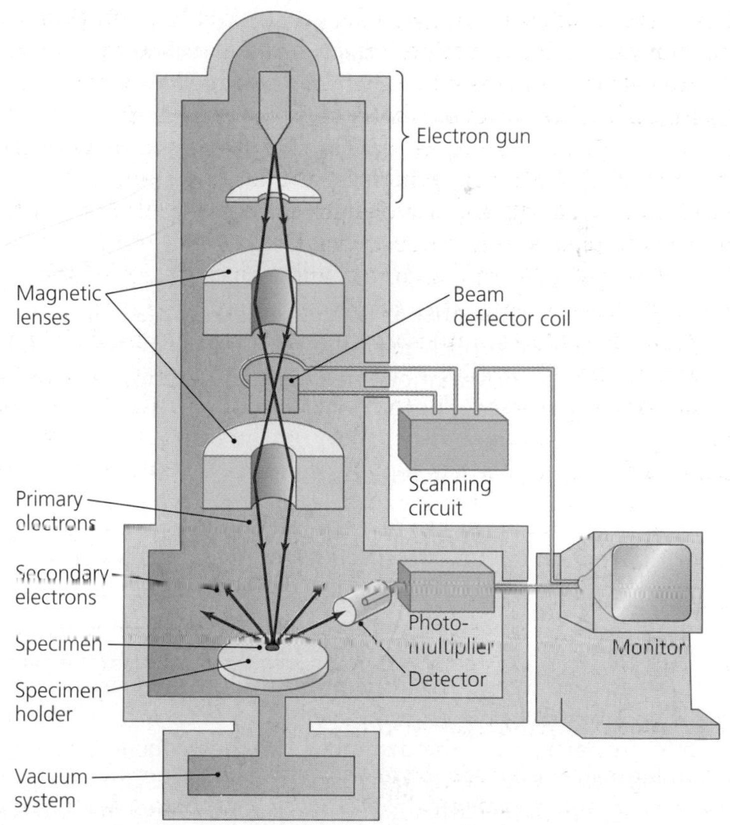

▶ **Figure 4.12 Scanning electron microscope (SEM).** The SEM uses magnetic lenses to focus a beam of primary electrons, which are scanned across the metal-coated surface of a specimen. Secondary electrons, knocked off the surface of the specimen by the primary electrons, are collected by a detector, and their signal is amplified and displayed on a monitor.

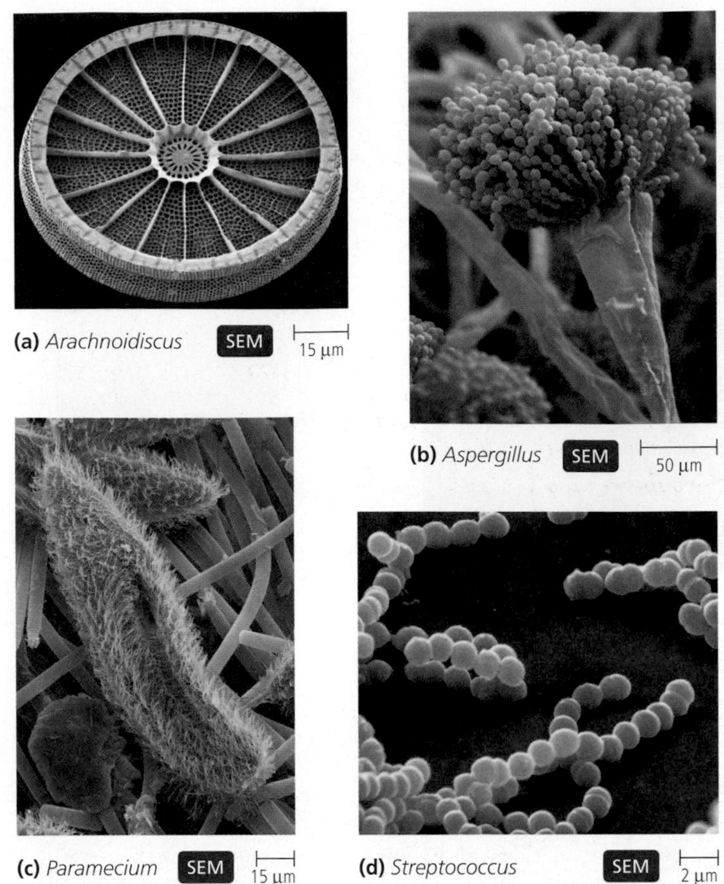

▲ **Figure 4.13 SEM images.** **(a)** *Arachnoidiscus*, a marine diatom (alga). **(b)** *Aspergillus*, a fungus. **(c)** *Paramecium*, a unicellular "animal" on top of rod-shaped bacteria. **(d)** *Streptococcus*, a bacterium.

Probe Microscopy

Learning Objective

✓ Describe two variations of probe microscopes.

A relatively recent advance in microscopy utilizes minuscule, pointed, electronic probes to magnify more than 100,000,000 times. There are two variations of probe microscopes: *scanning tunneling microscopes* and *atomic force microscopes.*

Scanning Tunneling Microscopes

A **scanning tunneling microscope (STM)** passes a metallic probe, sharpened to end in a single atom, back and forth across and slightly above the surface of a specimen. Rather than scattering a beam of electrons into a detector, as in scanning electron microscopy, a scanning tunneling microscope measures the flow of electrons to and from the probe and the specimen's surface. The amount of electron flow, called a *tunneling current,* is directly proportional to the distance from the probe to the specimen's surface. A scanning tunneling microscope can measure distances as small as 0.01 nm and reveal details on the surface of a specimen at the atomic level **(Figure 4.14a)**. A requirement for scanning tunneling microscopy is that the specimen be electrically conductive.

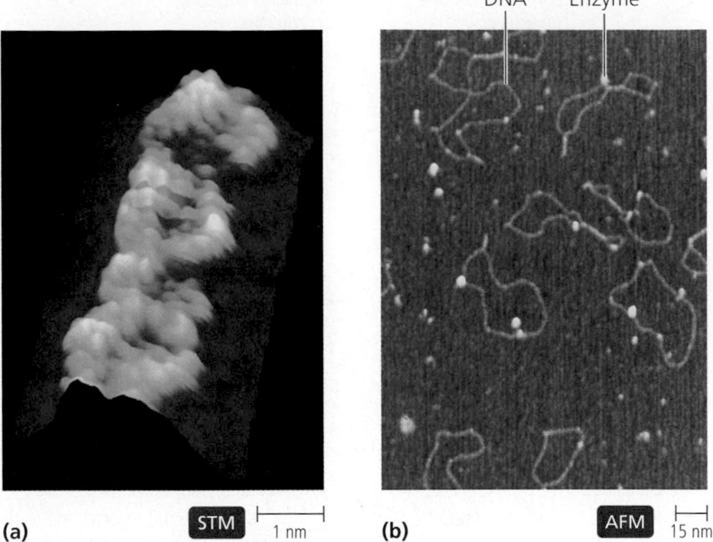

▲ **Figure 4.14 Probe microscopy.** **(a)** Scanning tunneling microscopes reveal surface detail; in this case, three turns of a DNA double helix. **(b)** Plasmid DNA being digested by an enzyme, viewed through atomic force microscopy.

Atomic Force Microscopes

An **atomic force microscope (AFM)** also uses a pointed probe, but it traverses the tip of the probe lightly on the surface of the specimen, rather than at a distance. This might be likened to the way a person reads Braille. Deflection of a laser beam aimed at the probe's tip measures vertical movements, which when translated by a computer reveals the atomic topography.

Unlike tunneling microscopes, atomic force microscopes can magnify specimens that do not conduct electrons. They can also magnify living specimens, because neither an electron beam nor a vacuum is required **(Figure 4.14b)**. Researchers have used these microscopes to magnify the surfaces of bacteria, viruses, proteins, and amino acids. Recent studies using atomic force microscopes have examined single living bacteria in three dimensions while they are dividing.

Table 4.2 summarizes the features of the various types of microscopes.

Staining

Earlier we discussed the difficulty of resolving two distant white golf balls viewed against a white background. If the balls were painted black, they could be distinguished more readily from the background, and from one another. This illustrates why staining increases contrast and resolution.

Most microorganisms are colorless and difficult to view with bright-field microscopes. Microscopists use stains to make microorganisms and their parts more visible because stains increase contrast between structures, and between a specimen and its background. Electron microscopy also requires the treatment of specimens with stains or coatings to enhance contrast.

TABLE 4.2 Comparison of Types of Microscopes

Type of Microscope	Typical Image	Description of Image	Special Features	Typical Uses
Light microscopes		Useful magnification 1× to 2000×; resolution to 200 nm	Use visible light; shorter, blue wavelengths provide better resolution	
Bright field		Colored or clear specimen against bright background	Simple to use; relatively inexpensive; stained specimens often required	Observation of killed stained specimens and naturally colored live ones; also used to count microorganisms
Dark field		Bright specimen against dark background	Use a special filter in the condenser that prevents light from directly passing through a specimen; only light scattered by the specimen is visible	Observation of living, colorless, unstained organisms
Phase-contrast		Specimen has light and dark areas	Use a special condenser that splits a polarized light beam into two beams, one of which passes through the specimen, and one of which bypasses the specimen; the beams are then rejoined before entering the oculars; contrast in the image results from the interactions of the two beams	Observation of internal structures of living microbes
Differential interference contrast (Nomarski)		Image appears three-dimensional	Use two separate beams instead of a split beam; false color and a three-dimensional effect result from interactions of light beams and lenses; no staining required	Observation of internal structures of living microbes
Fluorescent		Brightly colored fluorescent structures against dark background	An ultraviolet light source causes fluorescent natural chemicals or dyes to emit visible light	Localization of specific chemicals or structures; used as an accurate and quick diagnostic tool for detection of pathogens
Confocal		Single plane of structures or cells that have been specifically stained with fluorescent dyes	Use a laser to fluoresce only one plane of the specimen at a time	Detailed observation of structures of cells within communities
Electron microscopes		Typical magnification 1000× to 100,000×; resolution to 0.001 nm	Use electrons traveling as waves with short wavelengths; require specimens to be in a vacuum, so cannot be used to examine living microbes	
Transmission		Monotone, two-dimensional, highly magnified images; may be color-enhanced	Produce two-dimensional image of ultrastructure of cells	Observation of internal ultrastructural detail of cells and observation of viruses and small bacteria
Scanning		Monotone, three-dimensional, surface images; may be color-enhanced	Produce three-dimensional view of the surface of microbes and cellular structures	Observation of the surface details of structures
Probe microscopes		Magnification greater than 100,000,000× with resolving power greater than that of electron microscopes	Use microscopic probes that move over the surface of a specimen	
Scanning tunneling		Individual molecules and atoms visible	Measure the flow of electrical current between the tip of a probe and the specimen to produce an image of the surface at atomic level	Observation of the surface of objects; provide extremely fine detail, high magnification, and great resolution
Atomic force		Individual molecules and atoms visible	Measure the deflection of a laser beam aimed at the tip of a probe that travels across the surface of the specimen	Observation of living specimens at the molecular and atomic levels

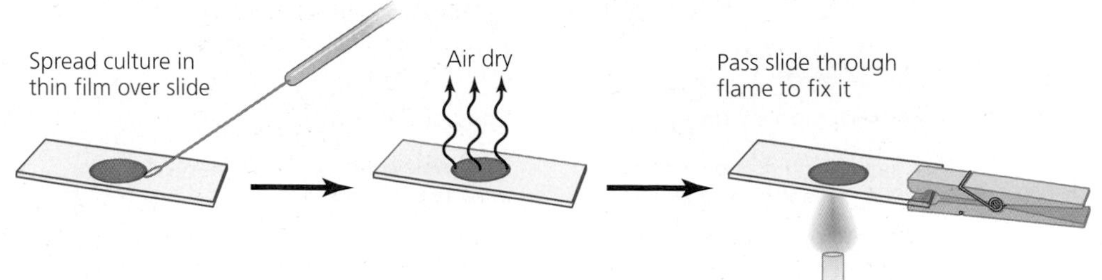
Spread culture in
thin film over slide
Air dry
Pass slide through
flame to fix it

▲ **Figure 4.15 Preparing a specimen for staining.** Microorganisms are spread in liquid across the surface of a slide using a circular motion. After drying in the air, the smear is passed through the flame of a Bunsen burner to fix the cells to the glass. Alternatively, chemical fixation can be used. *Why must a smear be fixed to the slide?*

Figure 4.15 *Fixation causes the specimen to adhere to the glass so that it does not easily wash off during staining.*

In this section we examine how scientists prepare specimens for staining and how stains work, and we consider five kinds of stains used for light microscopy. We conclude with a look at staining for electron microscopy.

Preparing Specimens for Staining

Learning Objective

✓ Explain the purposes of a smear, heat fixation, and chemical fixation in the preparation of a specimen for microscopic viewing.

Many investigations of microorganisms, especially those seeking to identify pathogens, begin with light microscopic observation of stained specimens. **Staining** simply means coloring specimens with stains, which are also called dyes.

Before microbiologists stain microorganisms, they must place them on and then firmly attach them to a microscope slide. Typically, this involves making a *smear* and *fixing* it to the slide **(Figure 4.15)**. If the organisms are growing in a liquid, a small drop is spread across the surface of the slide. If the organisms are growing on a solid surface, such as an agar plate, then they are mixed into a small drop of water on the slide. Either way, the thin film of organisms on the slide is called a **smear.**

The smear is air dried and then attached or fixed to the surface of the slide. In **heat fixation,** developed more than a hundred years ago by Robert Koch, the slide is gently heated by passing the slide, smear up, through the flame of a Bunsen burner. Alternatively, **chemical fixation** involves applying methyl alcohol (methanol) or formalin (a solution of formaldehyde in water) to the smear for one minute. Desiccation (drying) and fixation kill the microorganisms, attach them firmly to the slide, and generally preserve their shape and size. It is important to smear and fix specimens properly so that they are not lost during staining.

Specimens prepared for electron microscopy are also dried, because water vapor from a wet specimen would stop the electron beam. As we have seen, transmission electron microscopy requires that the desiccated sample also be sliced very thin, generally before staining. Specimens for scanning electron microscopy are coated, not stained.

Principles of Staining

Learning Objective

✓ Describe the uses of acidic and basic dyes, mentioning ionic bonding and pH.

Dyes used as microbiological stains for light microscopy are usually salts. As we saw in Chapter 2, a salt is composed of a positively charged *cation* and a negatively charged *anion.* At least one of the two ions in the molecular makeup of dyes is colored; this colored portion of a dye is known as the *chromophore.* Chromophores typically bind to chemicals via covalent, ionic, or hydrogen bonds. For example, methylene blue chloride is composed of a cationic chromophore, methylene blue, and a chloride anion. Since methylene blue is positively charged, it ionically bonds to negatively charged molecules in cells, including DNA and many proteins. In contrast, anionic dyes, for example, eosin, bind to positively charged molecules such as some amino acids.

Anionic chromophores are also called **acidic dyes** because they stain alkaline structures and work best in acidic (low pH) environments. Positively charged, cationic chromophores are called **basic dyes** because they combine with and stain acidic structures; further, they work best under basic (higher pH) conditions. In microbiology, basic dyes are used more commonly than acidic dyes because most cells are negatively charged. Acidic dyes are used in negative staining, which is discussed shortly.

Some stains do not form bonds with cellular chemicals, but rather function because of their solubility characteristics. For example, Sudan black selectively stains membranes because it is lipid soluble and accumulates in phospholipid bilayers.

Simple Stains

Learning Objective

✓ Describe the simple, Gram, acid-fast, and endospore staining procedures.

Simple stains are composed of a single basic dye such as crystal violet, safranin, or methylene blue. They are "simple" because they involve no more than soaking the smear in the dye for 30–60 seconds and then rinsing off the slide with water. (A properly

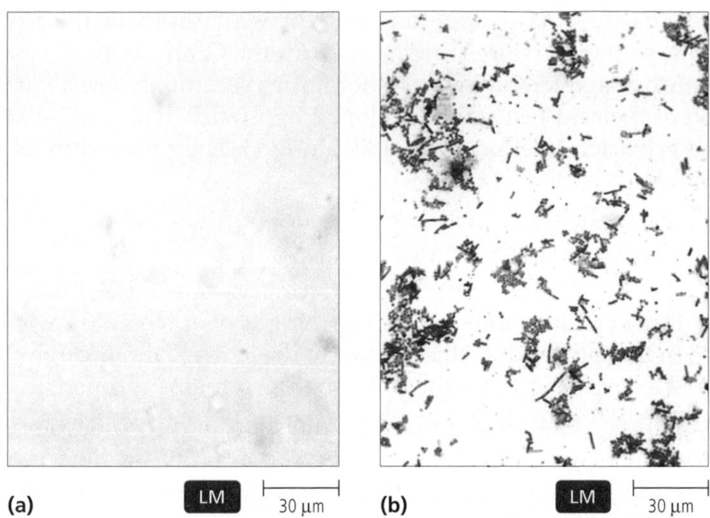

▲ **Figure 4.16 Simple stains.** Simply and quickly performed, simple stains increase contrast and allow determination of size, shape, and arrangement of cells. **(a)** Unstained *Escherichia coli and Staphylococcus aureus.* **(b)** Same mixture stained with crystal violet. Note that all cells, no matter their type, stain almost the same color with a simple stain because only one dye is used.

fixed specimen will remain attached to the slide despite this treatment.) After carefully blotting the slide dry, the microbiologist observes the smear under the microscope. Simple stains are used to determine size, shape, and arrangement of cells **(Figure 4.16)**.

Differential Stains

Most stains used in microbiology are **differential stains,** which use more than one dye so that different cells, chemicals, or structures can be distinguished when microscopically examined. Common differential stains are the *Gram stain,* the *acid-fast stain,* and the *endospore stain.*

Gram Stain

In 1884, the Danish scientist Hans Christian Gram developed the most frequently used differential stain, which now bears his name. The **Gram stain** differentiates between two large groups of microorganisms: purple-staining Gram-positive cells and pink-staining Gram-negative cells. Recall from Chapter 3 that these cells differ significantly in the chemical and physical structures of their cell walls (see Figure 3.15 on p. 65). Typically, a Gram stain is the first step a medical laboratory technologist performs to identify bacterial pathogens.

Let's examine the Gram staining procedure as it was originally developed and as it is typically performed today, over a hundred years later. For the purposes of our discussion, we'll assume that a smear has been made on a slide and heat fixed, and that the smear contains both Gram-positive and Gram-negative colorless bacteria. The classical Gram staining procedure has the following four steps **(Figure 4.17)**:

1 Flood the smear with the basic dye crystal violet for 1 min, and then rinse with water. Crystal violet, which is called the **primary stain,** colors all cells.

1 Slide is flooded with crystal violet for 1 min, then rinsed with water.

Result: All cells are stained purple.

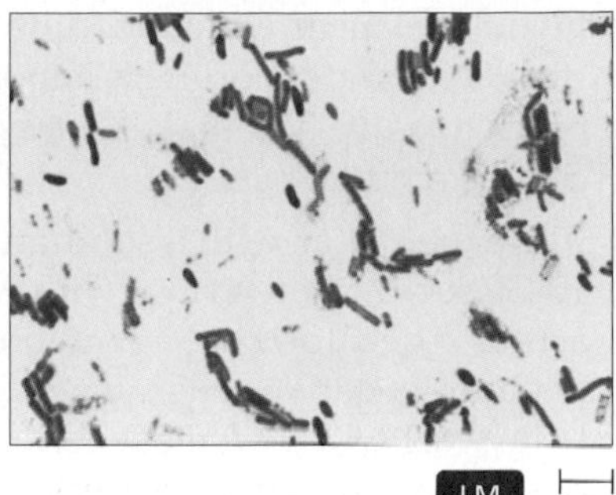

2 Slide is flooded with iodine for 1 min, then rinsed with water.

Result: Iodine acts as a mordant; all cells remain purple.

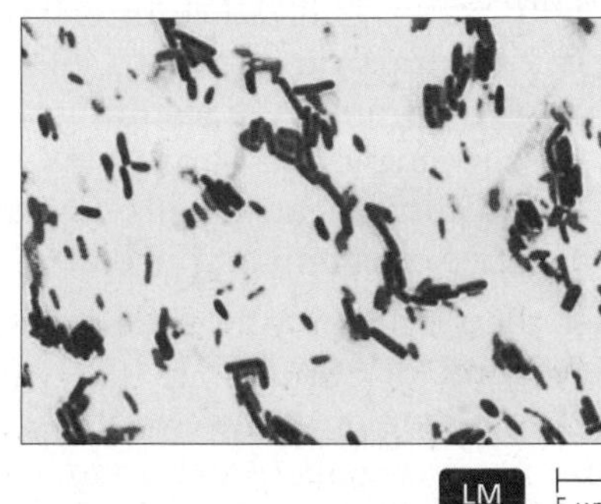

3 Slide is flooded with solution of ethanol and acetone for 10–30 sec, then rinsed with water.

Result: Smear is decolorized; Gram-positive cells remain purple, but Gram-negative cells are now colorless.

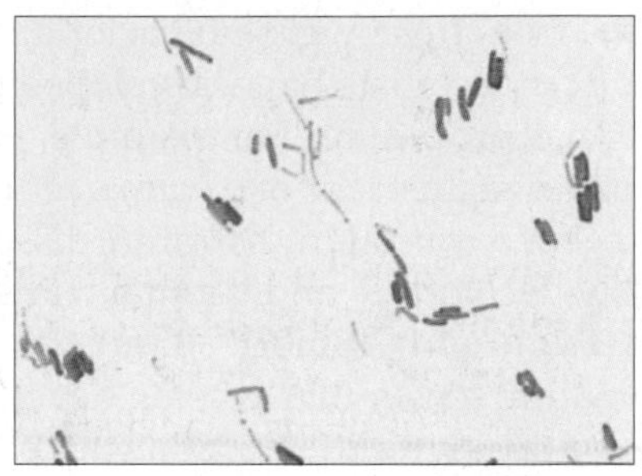

4 Slide is flooded with safranin for 1 min, then rinsed with water and blotted dry.

Result: Gram-positive cells remain purple, Gram-negative cells are pink.

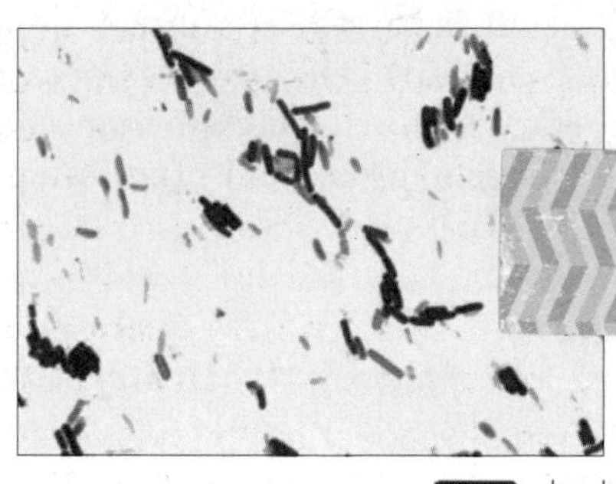

▲ **Figure 4.17 The Gram staining procedure.** A specimen is smeared and fixed to a slide. The classical procedure consists of four steps. Gram-positive cells (in this case, *Bacillus cereus*) remain purple throughout the procedure; Gram-negative cells (here, *Escherichia coli*) end up pink.

2 Flood the smear with an iodine solution for 1 min, and then rinse with water. Iodine is a **mordant,** a substance that binds to a dye and makes it less soluble. After this step, all cells remain purple.

3 Rinse the smear with a solution of ethanol and acetone for 10–30 sec, and then rinse with water. This solution, which acts as a **decolorizing agent,** breaks down the thin cell wall of Gram-negative cells, allowing the stain and mordant to be washed away; these cells are now colorless. Gram-positive cells, with their thicker cell walls, remain purple.

4 Flood the smear with safranin for 1 min, and then rinse with water. This red **counterstain** provides a contrasting color to the primary stain. Although all types of cells may absorb safranin, the resulting pink color is masked by the darker purple dye already in Gram-positive cells. After this step, Gram-negative cells now appear pink, whereas Gram-positive cells remain purple.

After the final step, the slide is blotted dry in preparation for microscopy.

The Gram procedure works best with young cells. Older Gram-positive cells bleach more easily than younger cells and can therefore stain pink, which makes them appear to be Gram-negative cells. Therefore, smears for Gram staining should come from freshly grown bacteria.

Microscopists have developed minor variations on Gram's original procedure. For example, 95% ethanol may be used to decolorize instead of Gram's ethanol-acetone mixture. In a three-step variation, safranin dissolved in ethanol simultaneously decolorizes and counterstains.

Highlight: "Smart" Bandages of the Future focuses on a miniature sensor that is being developed to differentiate between Gram-positive and Gram-negative bacteria.

Acid-Fast Stain

The **acid-fast** stain is another important differential stain because it stains cells of the genera *Mycobacterium* and *Nocardia* (nō-kar'dē-ă), which cause many human diseases, including tuberculosis, leprosy, and other lung and skin infections. Cells of these bacteria have large amounts of waxy lipid in their cell walls, so they do not readily stain with the Gram stain.

Modern microbiological laboratories commonly use a variation of the acid-fast stain developed by Franz Ziehl (1857–1926) and Friedrich Neelsen (1854–1894) in 1883. Their procedure is as follows:

1. Cover the smear with a small piece of tissue paper to retain the dye during the procedure.
2. Flood the slide with the red primary stain, carbolfuchsin, for several minutes while warming it over steaming water. In this procedure, heat is used to drive the stain through the waxy wall and into the cell, where it remains trapped.
3. Remove the tissue paper, cool the slide, and then decolorize the smear by rinsing it with a solution of hydrochloric acid ($pH < 1.0$) and alcohol. The bleaching action of acid-alcohol removes color from both non-acid-fast cells and the background. Acid-fast cells retain their red color because the acid cannot penetrate the waxy wall. The name of the procedure is derived from this step; that is, the cells are colorfast in acid.
4. Counterstain with methylene blue, which stains only bleached, non-acid-fast cells.

The Ziehl-Neelsen acid-fast staining procedure results in pink acid-fast cells, which can be differentiated from blue non-acid-fast cells, including human cells and tissue **(Figure 4.18)**. The presence of *acid-fast rods (AFRs)* in sputum is indicative of mycobacterial infection.

CRITICAL THINKING

In what ways are the Gram stain and the acid-fast staining procedures similar? In the acid-fast procedure, what takes the place of Gram's iodine mordant?

Endospore Stain

Some bacteria—notably those of the genera *Bacillus* and *Clostridium* (klos-trid'ē-ŭm), which contain species that cause such diseases as anthrax, gangrene, and tetanus—produce

HIGHLIGHT

"SMART" BANDAGES OF THE FUTURE

Imagine a bandage that not only covers an infected cut on your finger, but also identifies whether the infectious agent is Gram-positive or Gram-negative. The use of such "smart" bandages may not be far off. Researchers at the University of Rochester have developed a miniature sensor that can differentiate between Gram-positive and Gram-negative bacteria even more reliably than the traditional Gram staining procedure. The sensor (see photo) is a porous silicon chip that can detect lipid A, which is present only on the outer membrane of Gram-negative bacteria (see Figure 3.15). When lipid A molecules bind to the organic receptor molecules that coat the sensor, they cause a color change on the sensor that can be used to differentiate between the two main groups of bacteria. Although the sensor is not yet practical for household use, a version of it could

someday be incorporated into bandages to rapidly identify infectious agents, or be used in meat packaging plants to detect contaminants.

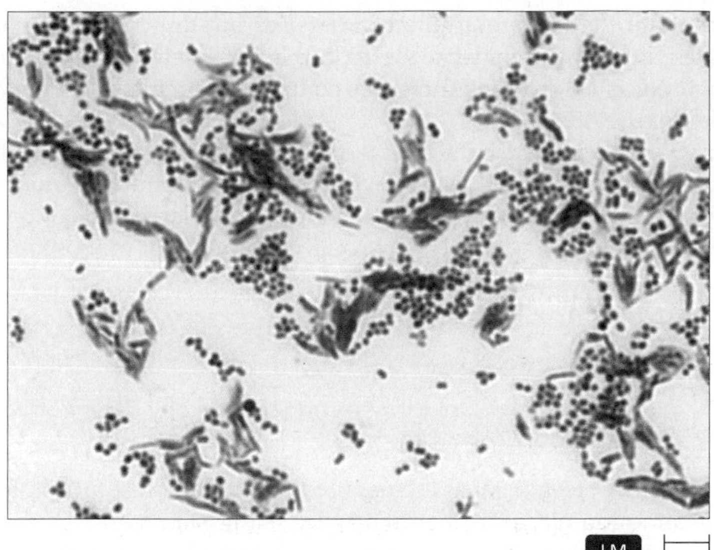

▲ **Figure 4.18 Ziehl-Neelsen acid-fast stain.** Acid-fast cells such as these rod-shaped *Mycobacterium bovis* cells stain pink or red. Non-acid-fast cells—in this case, *Staphylococcus*—stain blue. *Why isn't the Gram stain utilized to stain Mycobacterium?*

Figure 4.18 *Cell walls of Mycobacterium are composed of waxy materials that repel the water-based dyes of the Gram stain.*

endospores. These dormant, highly resistant cells form inside the cytoplasm of the bacteria and can survive environmental extremes such as desiccation, heat, and harmful chemicals. Endospores cannot be stained by normal staining procedures because their walls are practically impermeable to all chemicals. The **Schaeffer-Fulton endospore stain** uses heat to drive the primary stain, *malachite green*, into the endospore. After cooling, the slide is decolorized with water and counterstained with safranin. This staining procedure results in green-stained endospores and red-colored vegetative cells (**Figure 4.19**).

Special Stains

Special stains are simple stains designed to reveal special microbial structures. There are three types of special stains: *negative stains, flagellar stains,* and *fluorescent stains* (which we already discussed in the section on fluorescent microscopy).

Negative (Capsule) Stain

Most dyes used to stain bacterial cells, such as crystal violet, methylene blue, malachite green, and safranin, are basic dyes. These dyes stain cells by attaching to negatively charged molecules within them.

Acidic dyes, by contrast, are repulsed by the negative charges on the surface of cells and therefore do not stain them. Such stains are called **negative stains** because they stain the background and leave cells colorless. Eosin and nigrosin are examples of acidic dyes used for negative staining. A counterstain may be added to color the cells.

Negative stains are used primarily to reveal the presence of negatively charged bacterial capsules. Therefore, they are also

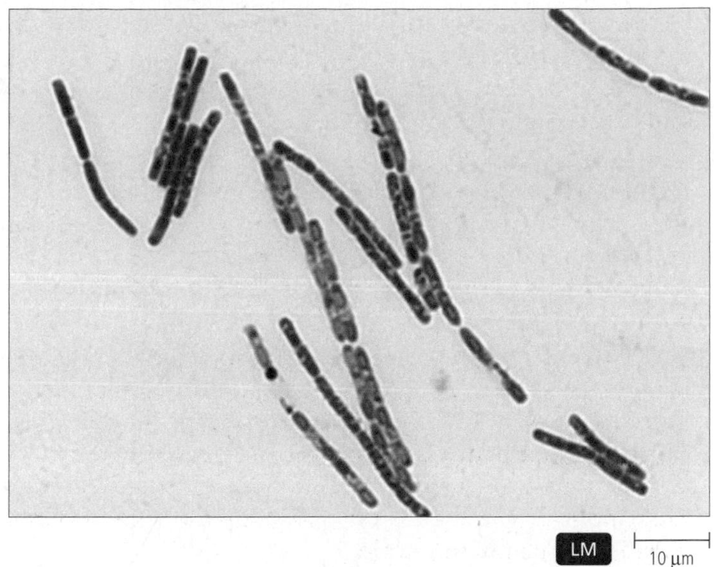

▲ **Figure 4.19 Schaeffer-Fulton endospore stain of *Bacillus anthracis*.** The nearly impermeable spore wall retains the green dye during decolorization. Vegetative cells, which lack spores, pick up the counterstain and appear red. *Why don't the spores stain red as well?*

Figure 4.19 *Heat from steam is used to drive the green primary stain into the endospores. Counterstaining is performed at room temperature, and the thick, impermeable walls of the endospores resist the counterstain.*

called **capsule stains.** Encapsulated cells appear to have a halo surrounding them (**Figure 4.20**).

Flagellar Stain

Bacterial flagella are extremely thin and thus normally invisible with light microscopy, but their presence, number, and arrangement are important in the identification of some species, including some pathogens. Flagellar stains, such as pararosaniline and carbolfuchsin, and mordants, such as tannic acid and

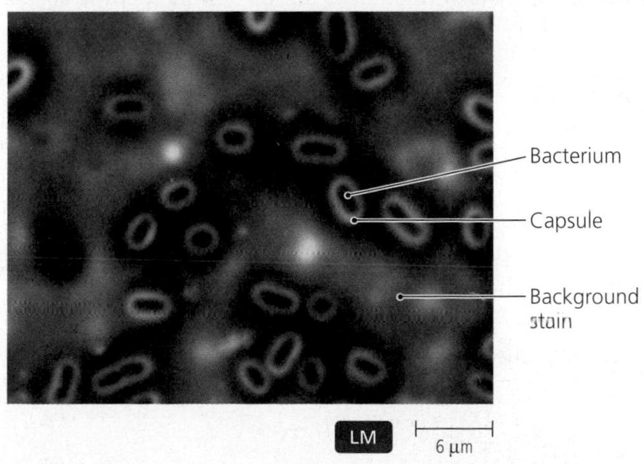

▲ **Figure 4.20 Negative (capsule) stain of *Klebsiella pneumoniae*.** Notice that the acidic dye stains the background and does not penetrate the capsule. The bacterial cells have been counterstained with a different (basic) dye.

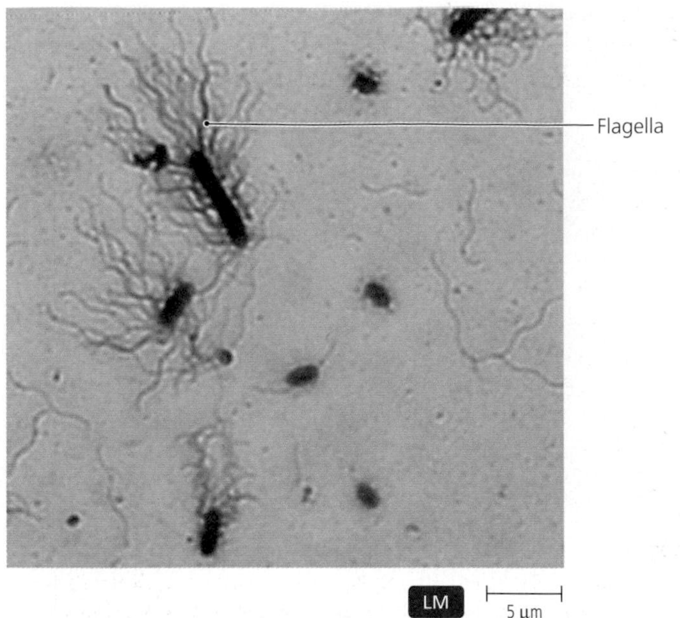

▲ **Figure 4.21 Flagellar stain of *Proteus vulgaris*.** Various bacteria have different numbers and arrangements of flagella, features that are important in identifying some species. *How can the flagellar arrangement shown here be described?*

Figure 4.21 Peritrichous.

potassium alum, are applied in a series of steps. These molecules bind to the flagella, increase their diameter, and change their color, all of which increases contrast and makes them visible **(Figure 4.21)**.

Stains used for light microscopy are summarized in **Table 4.3**. **Beneficial Microbes: Glowing Viruses** illustrates a unique kind of stain that uses viruses to stain particular strains of bacteria, which are then viewed through a fluorescent microscope.

Staining for Electron Microscopy

Learning Objective

✓ Explain how stains used for electron microscopy differ from those used for light microscopy.

Laboratory technicians increase contrast and resolution for transmission electron microscopy by using stains, just as they do for light microscopy. However, stains used for transmission electron microscopy are not colored dyes, but instead chemicals containing atoms of heavy metals such as lead, osmium, tungsten, and uranium, which absorb electrons. Electron-dense stains may bind to molecules within specimens, or they may stain the background. The latter type of negative staining is used to provide contrast for extremely small specimens such as viruses and molecules.

Stains for electron microscopy can be general, in that they stain most objects to some degree, or they may be highly specific. For example, osmium tetraoxide (OsO_4) has an affinity for lipids

TABLE 4.3 Some Stains Used for Light Microscopy

Type of Stain	Examples	Results	Typical Images	Representative Uses
Simple stains (use a single dye)	Crystal violet Methylene blue	Uniform purple stain Uniform blue stain		Reveals size, morphology, and arrangement of cells
Differential stains (use two or more dyes to differentiate between cells or structures)	Gram stain	Gram-positive cells are purple; Gram-negative cells are pink		Differentiates between Gram-positive and Gram-negative bacteria, which is typically the first step in their identification
	Ziehl-Neelsen acid-fast stain	Pink to red acid-fast cells and blue non-acid-fast cells		Distinguishes the genera *Mycobacterium* and *Nocardia* from other bacteria
	Schaeffer-Fulton endospore stain	Green endospores and pink to red vegetative cells		Highlights the presence of endospores produced by species in the genera *Bacillus* and *Clostridium*
Special stains	Negative stain for capsules	Background is dark, cells unstained or stained with simple stain		Reveals bacterial capsules
	Flagellar stain	Bacterial flagella become visible		Allows determination of number and location of bacterial flagella

BENEFICIAL MICROBES

GLOWING VIRUSES

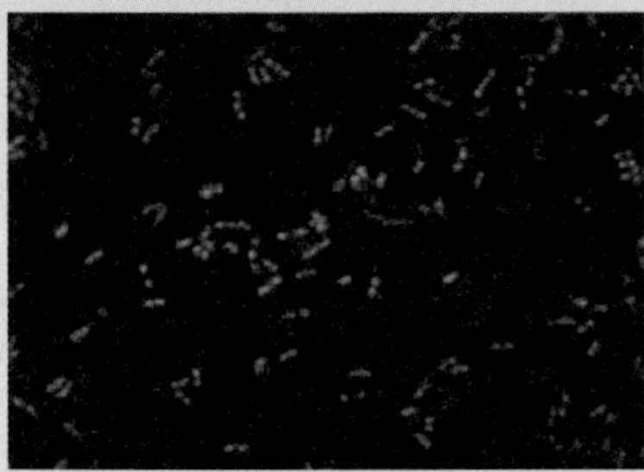

▲ ***Fluorescent phages light up bacteria.***

A bacteriophage is a virus that injects its DNA into a bacterium. Commonly called a phage, it adheres only to a select bacterial strain for which each phage type has a specific adhesion factor. Many phages are so specialized for their particular bacterial strain that scientists have used phages to identify and classify bacteria. Such identification is called phage typing.

Scientists at San Diego State University have taken phage specificity a step further. They successfully linked a fluorescent dye to the DNA of phages of the bacterium *Salmonella* and used the phages to detect and identify *Salmonella* species. Such fluorescent phages rapidly and accurately detect specific strains of *Salmonella* in mixed bacterial cultures.

Fluorescent phages have advantages over fluorescent antibodies: unlike antibodies, phages are not metabolized by bacteria. Phages are also more stable over time and are not as sensitive to vagaries in temperature, pH, and ionic strength. Further, fluorescent phages have a long shelf life; they protect the fluorescent dye inside their phage coat until the dyed DNA is injected.

There are numerous uses for test kits using fluorescent phages. Environmental scientists could use them to detect bacterial contamination of streams and lakes, food processors could identify potentially fatal *Escherichia coli* strain O157:H7 in meat and vegetables, or homeland security agents could positively establish the presence or absence of bacteria used for biological warfare. Antibody-based kits frequently failed to accurately detect *Bacillus anthracis* used in the 2001 terrorist attacks. Fluorescent phage field kits should be much more robust and precise.

and is thus used to enhance the contrast of membranes. Electron-dense stains can also be linked to antibodies to provide an even greater degree of staining specificity because antibodies bind only to their specific target molecules. **ANIMATIONS:** ***Staining***

Classification and Identification of Microorganisms

Learning Objective

✓ Discuss the purposes of classification and identification of organisms.

Biologists classify organisms for several reasons: to bring a sense of order and organization to the variety and diversity of living things, to enhance communication, to make predictions about the structure and function of similar organisms, and to uncover and understand potential evolutionary connections. They sort organisms on the basis of mutual similarities into nonoverlapping groups called **taxa.**[4] As we discussed in Chapter 1, **taxonomy**[5] is the science of classifying and naming organisms. Taxonomy consists of *classification,* which is assigning organisms to taxa based upon similarities; *nomenclature,* which is concerned with the rules of naming organisms; and *identification,* which is the practical science of determining that an isolated individual or population belongs to a particular taxon. In this book we concentrate on classification and identification.

Since all members of any given taxon share certain common features, taxonomy enables scientists to both organize large amounts of information about organisms and make predictions based on knowledge of similar organisms. For example, if one member of a taxon is important in recycling nitrogen in the environment, it is likely that others in the group will play a similar ecological role. Similarly, a clinician might suggest a treatment against one pathogen based on what has been effective against another pathogen in the same taxon.

Identification of organisms is an essential part of taxonomy because it enables scientists to communicate effectively and be confident that they are discussing the same organism. Further, identification is often essential for treatment of groups of diseases such as meningitis and pneumonia, which can be caused by pathogens as different as fungi, bacteria, and viruses.

In this section we examine the historical basis of taxonomy, consider modern advances in this field, and briefly consider various taxonomic methods. This chapter also presents a general overview of the taxonomy of prokaryotes (which are considered in greater detail in Chapter 11); of animals, protozoa, fungi, and algae (Chapter 12); and of viruses, viroids, and prions (Chapter 13).

Linnaeus, Whittaker, and Taxonomic Categories

Learning Objectives

✓ Discuss the difficulties in defining species of microorganisms.

✓ List the hierarchy of taxa from general to specific.

✓ Define binomial nomenclature.

✓ Describe a few modifications of the Linnaean system of taxonomy.

[4]From Greek *taxis,* meaning order.
[5]From *taxis* and Greek *nomos,* meaning rule.

Our current system of taxonomy began in 1758 with the publication of the tenth edition of *Systema Naturae* by the Swedish botanist Carolus Linnaeus. Until his time, the names of organisms were often strings of descriptive terms that varied from country to country and from one scientist to another. Linnaeus provided a system that standardized the naming and classification of organisms based on characteristics they have in common. He grouped similar organisms that can successfully interbreed into categories called **species.**

The definition of species as "a group of organisms that interbreed to produce viable offspring" works relatively well for more complex, sexually reproducing organisms, but it is not satisfactory for asexual organisms, including most microorganisms. As a result, for asexual organisms many scientists define a species as a collection of *strains*—populations of cells that arose from a single cell—that share many stable properties and differ from other strains. Not surprisingly, this definition sometimes results in disagreements and inconsistencies in the classification of microbial life.

In Linnaeus's system, which forms the basis of modern taxonomy, similar species are grouped into **genera,**[6] and similar genera into still larger taxonomic categories. That is, genera sharing common features are grouped together to form **families;** similar families are grouped into **orders;** orders are grouped into **classes;** classes into **phyla,**[7] and phyla into **kingdoms** (Figure 4.22).

All these categories, including species and genera, are taxa, which are hierarchical; that is, each successive taxon has a broader description than the preceding one, and each taxon includes all the taxa beneath it. The rules of nomenclature require that all taxa have Latin or Latinized names, in part because the language of science during Linnaeus's time was Latin, and in part because using Latin ensures that no country or ethnic group has priority in the language of taxonomy. The name *Chondrus crispus* (kon'drŭs krisp'ŭs) describes the exact same algal species all over the world, despite the fact that in England its common name is Irish moss, in Ireland it is carragheen, in North America it is curly moss, and it isn't really a moss at all!

When new microscopic, genetic, or biochemical techniques identify new or more detailed characteristics of organisms, a taxon may be split into two or more taxa. Alternatively, several taxa may be lumped together into a single taxon. For example, the genus "*Diplococcus*" has been united (synonymized) with the genus *Streptococcus* (strep-tō-kok'ŭs), and the name *Diplococcus pneumoniae* (nū-mō'nē-ī) has been changed to reflect this synonymy—to *Streptococcus pneumoniae.*

Linnaeus assigned each species a descriptive name consisting of its genus name and a **specific epithet.** The genus name is always a noun, and it is written first and capitalized. The specific epithet always contains only lowercase letters and is usually an adjective. Both names, together called a *binomial,* are either printed in italics or underlined. Because the Linnaean system assigns two names to every organism, it is said to use **binomial**[8] **nomenclature.**

Consider the following examples of binomials. *Enterococcus faecalis* (en'ter-ō-kok'ŭs fē-kă'lis)[9] is a fecal bacterium. Whereas *Enterococcus faecium* (fē-sē'ŭm) is in the same genus, it is a different species because of certain differing characteristics, so it is given a different specific epithet. Humans are classified as *Homo sapiens* (hō'mō sā'pē-enz); the genus name means "man" and the specific epithet means "wise." The genus *Homo* contains no other living species, but scientists have assigned some fossil remains to *Homo neanderthalensis* (nē-an'der-thol-en'sis).

Note that even though binomials are often descriptive of an organism, sometimes they can be misleading. For instance, *Haemophilus influenzae* (hē-mof'i-lŭs in-flu-en'zī) does not cause influenza. In some cases, binomials honor people. Examples include *Pasteurella haemolytica* (pas-ter-el'ă hē-mō-lit'i-kă), a bacterium named after the microbiologist Louis Pasteur; *Escherichia coli* (esh-ĕ-rik'ē-ă kō-lē), a bacterium named after the physician Theodor Escherich (1857–1911); and *Isabella abbottae* (iz-ă-bel'ă ab'ot-tī), a marine alga named after the phycologist and taxonomist Isabella Abbott (1919–).

CRITICAL **THINKING**

Examine the binomials of the species discussed in this section. What do the genus names and specific epithets indicate about the organisms?

Most scientists still use the Linnaean system today, though significant modifications have been adopted. For example, scientists sometimes use additional categories, such as tribes, sections, subfamilies, and subspecies.[10] Further, Linnaeus divided all organisms into only two kingdoms (Plantae and Animalia), and he did not know of the existence of viruses and smaller pathogens. As scientists learned more about organisms, they adopted taxonomic schemes to reflect their advances in knowledge. For example, in 1969, Robert Whittaker (1920–1980) proposed a widely accepted taxonomic approach based on five kingdoms: Animalia, Plantae, Fungi, Protista, and Prokaryotae (**Figure 4.23**). Though widely accepted, this scheme groups together in a single kingdom (Protista) such obviously disparate organisms as massive brown seaweeds (kelps) and unicellular protozoa. Further, because it does not address the taxonomy of viruses nor all the differences between microorganisms, scientists have proposed other taxonomic schemes that have from 4 to more than 50 kingdoms.

Sometimes students are upset that taxonomists do not agree about all the taxonomic categories or the species they contain, but it must be realized that classification of organisms reflects the state of our current knowledge and theories. Taxonomists change their schemes to accommodate new information, and not every expert agrees with every proposed modification.

[6]Plural of Latin *genus*, meaning race or birth.
[7]Plural of *phylum*, from Greek *phyllon*, meaning tribe. The term *phylum* is used for animals and bacteria; the corresponding taxon in mycology and botany is called a *division*.
[8]From Greek *bi*, meaning two, and *nomos*, meaning rule.
[9]From Greek *enteron*, meaning intestine, and *kokkos*, meaning berry, and Latin *faeces*.
[10]Subspecies, which are also called *varieties* or *strains*, differ only slightly from each other. Not all taxonomists agree on how much difference between two microbes constitutes a strain versus a species.

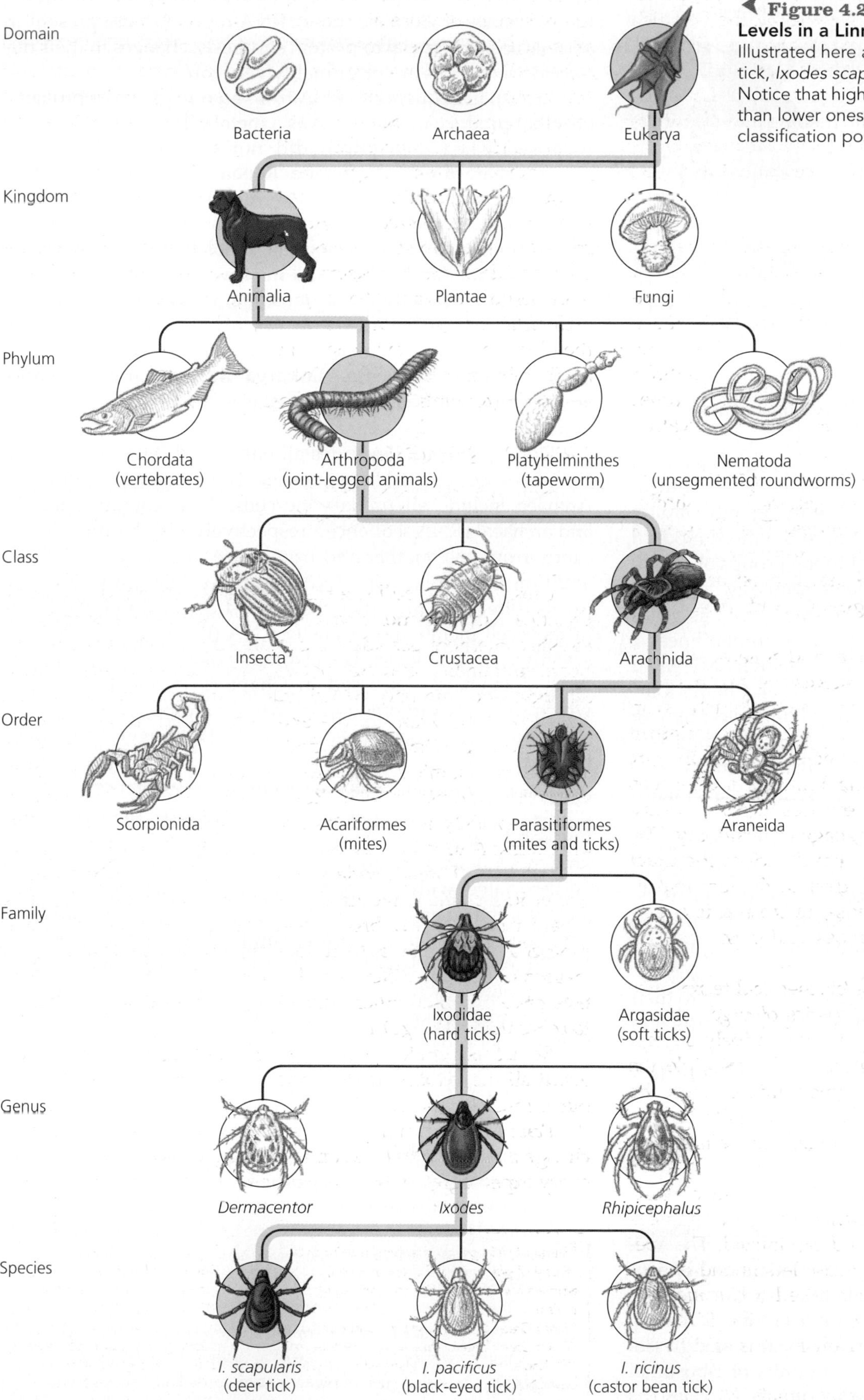

Figure 4.22
Levels in a Linnaean taxonomic scheme.
Illustrated here are the taxonomic categories for the deer tick, *Ixodes scapularis*, the primary vector of Lyme disease. Notice that higher taxonomic categories are more inclusive than lower ones. To simplify the diagram, only selected classification possibilities are shown.

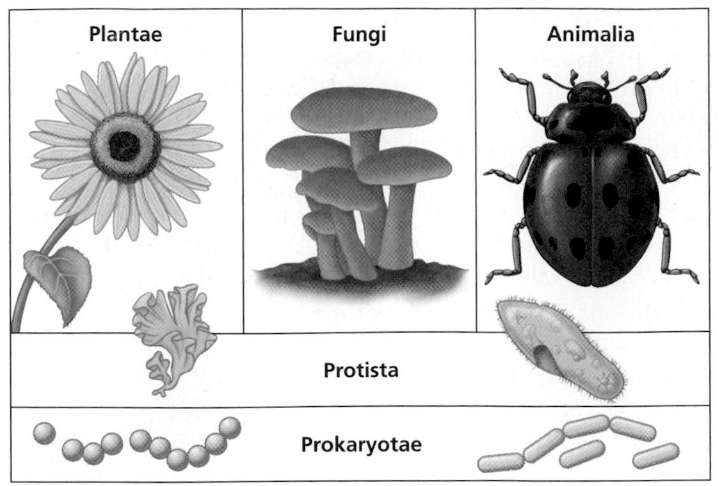

▲ **Figure 4.23** **Whittaker's five-kingdom taxonomic scheme.**

Another significant development in taxonomy is a shift in its basic goal—from Linnaeus's goal of classifying and naming organisms as a means of cataloging them, to the more modern goal of understanding the relationships among groups of organisms. Linnaeus based his taxonomic scheme primarily on organisms' structural similarities, and whereas such terms as *genus* and *family* may suggest the existence of some common lineage, he and his contemporaries thought of species as divinely created and generally immutable entities. However, when Charles Darwin (1809–1882) propounded his theory of the evolution of species by natural selection (a century after Linnaeus published his pivotal work on the taxonomy of plants), taxonomists came to consider that common ancestry largely explains the similarities among organisms in the various taxa. Today, most taxonomists agree that a major goal of modern taxonomy is to reflect a *phylogenetic*[11] *hierarchy;* that is, that the ways in which organisms are grouped should reflect their evolution from common ancestors. The spatial arrangement of kingdoms seen in Figure 4.23 (and the placement of smaller taxa within them, as well) reflects this phylogenetic approach to taxonomy.

Taxonomists' efforts to classify organisms according to their ancestry has resulted in reduced emphasis on comparisons of physical and chemical traits, and in greater emphasis on comparisons of their genetic material. Such work has led to a proposal to add a new, most inclusive taxon: the *domain.*

Domains

Learning Objective

✓ List and describe the three domains proposed by Carl Woese.

Carl Woese (1928–) labored for years to understand the taxonomic relationships among cells. Morphology (shape) and biochemical tests did not provide enough information to classify organisms fully, so for over a decade Woese painstakingly sequenced the nucleotides of the smaller subunits of ribosomal RNA (rRNA) in an effort to unravel the relationships among these organisms. Because these rRNA molecules are present in all cells and are crucial to protein synthesis, changes in their nucleotide sequences are presumably very rare.

In 1976 he sequenced rRNA from an odd group of prokaryotes that produce methane gas as a metabolic waste. Woese was surprised when their rRNA did not contain nucleotide sequences characteristic of bacteria. Repeated testing showed that methanogens, as they are called, were not like other prokaryotic or eukaryotic organisms. They were something new to science, a third branch of life, ushering microbiologists into a new and curious wonderland. Woese and his coworkers discovered that there are three basic types of ribosome, which led them to propose a new classification scheme in which a new taxon, called a **domain,** contains the Linnaean taxon of kingdom. The three domains identified by Woese—**Eukarya, Bacteria,** and **Archaea**—are based on three basic types of cells as determined by ribosomal nucleotide sequences.

Domain Eukarya includes all eukaryotic cells, all of which contain eukaryotic rRNA sequences. The domains Bacteria and Archaea include all prokaryotic cells. They contain bacterial and archaeal rRNA sequences, respectively, which differ significantly from one another and from those in eukaryotic cells. In addition to differences in rRNA sequences, cells of the three domains differ in many other characteristics, including the lipids in their cell membranes, transfer RNA (tRNA) molecules, and sensitivity to antibiotics. The taxonomy of organisms within the three domains is discussed in Chapters 11 and Chapter 12, which cover prokaryotes and eukaryotes, respectively.

Ribosomal nucleotide sequences further suggest that there may be at least 50 kingdoms of Bacteria and three kingdoms of Archaea. Further, scientists examining substances such as human saliva, water, soil, and rock regularly discover novel nucleotide sequences. When these sequences are compared to known sequences stored in a computer database, they cannot be associated with any previously identified organism. This suggests that many curious new forms of microbial life have never been grown in a laboratory and still await discovery.

Ribosomal nucleotide sequences have also given microbiologists a new way to define prokaryotic species. Some scientists propose that prokaryotes whose rRNA sequence differs from that of other prokaryotes by more than 3% be classified as a distinct species. While this definition has the advantage of being precise, not all taxonomists agree with it.

CRITICAL **THINKING**

Microbiologists at Stanford University have announced the discovery of 31 new species of bacteria that thrive between the teeth and gums of humans. The bacteria could be not be grown in the researchers' laboratories, nor were any of them ever observed via any kind of microscopy.

If they couldn't culture them or see them, how could the researchers know they had discovered 31 new species? If they couldn't examine the cells for the presence of a nucleus, how did they determine that the organisms were prokaryotes and not eukaryotes?

[11]From Greek *phyllon*, meaning tribe, and Latin *genus*, meaning birth (i.e., origin of a group).

Taxonomic and Identifying Characteristics

Learning Objective

✓ Describe five procedures used by taxonomists to identify and classify microorganisms.

Other criteria and laboratory techniques used for classifying and identifying microorganisms are quite numerous and include macroscopic and microscopic examination of physical characteristics, differential staining characteristics, growth characteristics, microorganisms' interactions with antibodies, microorganisms' susceptibilities to viruses, nucleic acid analysis, biochemical tests, and organisms' environmental requirements, including the temperature and pH ranges of their various types of habitats. Clearly, then, microbial taxonomy is too broad a subject to cover in one chapter, and thus the details of the criteria for the classification of major groups are provided in subsequent chapters.

It is important to note that even though scientists may use a given technique to either classify or identify microorganisms, the criteria used for identifying a particular organism are not always the same as those that were used to classify it. For example, even though laboratory technologists distinguish the genus *Escherichia* from other bacterial genera by its inability to utilize citric acid (citrate) as a sole carbon source, this identifying characteristic was not vital in the classification of *Escherichia.*

Bergey's Manual of Determinative Bacteriology, first published in 1923 and now in its 9th edition (1994), contains information used for the identification of prokaryotes. *Bergey's Manual of Systematic Bacteriology* (2nd edition, 2001, 2005, 2009) is a similar reference work that is used for classification. Its scheme is based on ribosomal RNA sequences, which taxonomists use to describe possible relationships among organisms. Each of these volumes is known as "Bergey's Manual."

Linnaeus did not know of the existence of viruses and thus did not include them in his original taxonomic hierarchy; nor are viruses assigned to any of Whittaker's five kingdoms or Woese's three domains because viruses are acellular and generally lack rRNA. Virologists do classify viruses into families and genera, but higher taxa are poorly defined for viruses. Chapter 13 further discusses viral taxonomy.

With this background, let's turn now to some brief discussions of five types of information microbiologists commonly use to distinguish among microorganisms: physical characteristics, biochemical tests, serological tests, phage typing, and analysis of nucleic acids.

Physical Characteristics

Many physical characteristics are used to identify microorganisms. Scientists can usually identify protozoa, fungi, algae, and parasitic worms based solely upon their *morphology* (shape). As we see in Chapter 6, lab technologists can also use the physical appearance of a bacterial colony[12] to help identify microorganisms. As we have discussed, stains are used to view the size and shape of individual bacterial cells and to show the presence or absence of identifying features such as endospores and flagella.

Linnaeus categorized prokaryotic cells into two genera based upon two prevalent shapes. He classified spherical prokaryotes in the genus "*Coccus*,"[13] and he placed rod-shaped cells in the genus *Bacillus.*[14] However, subsequent studies have revealed vast differences between many of the thousands of spherical and rod-shaped prokaryotes, and thus physical characteristics alone are not sufficient for the classification of prokaryotes. Instead, taxonomists rely primarily on genetic differences as revealed by metabolic dissimilarities, and more and more frequently on rRNA sequences.

CRITICAL THINKING

Why is the genus name "*Coccus*" placed within quotation marks, but not the genus name *Bacillus*?

Biochemical Tests

Microbiologists distinguish many prokaryotes that are similar in microscopic appearances and staining characteristics on the basis of differences in their ability to utilize or produce certain chemicals. Biochemical tests include procedures that determine an organism's ability to ferment various carbohydrates; utilize various substrates such as specific amino acids, starch, citrate, and gelatin; or produce waste products such as hydrogen sulfide (H_2S) gas (**Figure 4.24**). Differences in fatty acid composition of bacteria are also used to distinguish between bacteria. Obviously, biochemical tests can be used to identify only those microbes that can be grown under laboratory conditions.

Laboratory technicians utilize biochemical tests to identify pathogens, allowing physicians to prescribe appropriate treatments. Many tests require that the microorganisms be *cultured* (grown) for 12 to 24 hours, though this time can be greatly reduced by the use of rapid identification tools. Such tools exist for many groups of medically important pathogens, such as Gram-negative bacteria in the family Enterobacteriaceae, Gram-positive bacteria, yeasts, and filamentous fungi. Automated systems for identifying pathogens, such as the one shown in **Figure 4.25a**, read the results of a whole battery of biochemical tests performed in a plastic plate containing numerous small wells (**Figure 4.25b**). A color change in a well indicates the presence of a particular metabolic reaction, and the machine reads the pattern of colors in the plate to ascertain the identity of the pathogen.

Serological Tests

In the narrowest sense, serology is the study of serum, the liquid portion of blood after the clotting factors have been removed and an important site of antibodies. In its most practical application, serology is the study of antigen-antibody reactions in laboratory settings. As we have seen, antibodies are immune system proteins that bind very specifically to target antigens.

[12]A group of bacteria that has arisen from a single cell grown on a solid laboratory medium.

[13]From Greek *kokkos*, meaning berry. This genus name has been supplanted by many genera, including *Staphylococcus*, *Micrococcus*, and *Streptococcus*.

[14]From Latin *bacillum*, meaning small rod.

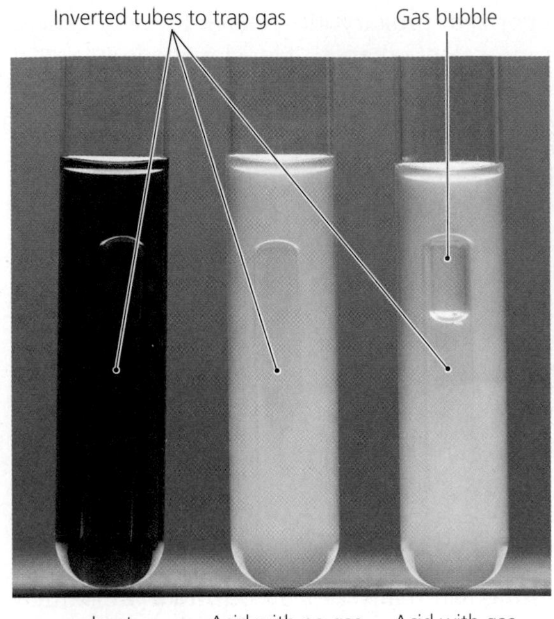

◀ **Figure 4.24 Two biochemical tests for identifying bacteria. (a)** A carbohydrate utilization test. At right is a tube in which the bacteria have metabolized a particular carbohydrate to produce acid (which changes the color of a pH indicator, phenol red, to yellow) and gas, as indicated by the bubble. At center is a tube with another bacterium that metabolized the carbohydrate to produce acid but no gas. At left is a tube inoculated with bacteria that are "inert" with respect to this test. **(b)** A hydrogen sulfide (H_2S) test. Bacteria that produce H_2S are identified by the black precipitate formed by the reaction of the H_2S with iron present in the medium.

Antibodies and antigens are discussed in detail in Chapter 16; in this section we briefly consider the use of serological testing to identify microorganisms.

Many microorganisms are *antigenic;* that is, within a host organism they trigger an immune response that results in the production of antibodies. Suppose, for example, that a scientist injects a sample of *Borrelia burgdorferi* (bō-rē′lē-ă burg-dōr′fer-ē), the bacterium that causes Lyme disease, into a rabbit. The bacterium has many surface proteins and carbohydrates that are antigenic because they are foreign to the rabbit. The rabbit responds to these foreign antigens by producing antibodies against them. These antibodies can be isolated from the rabbit's serum and concentrated into a solution known as an **antiserum.** Antisera bind to the antigens that triggered their production.

In a procedure called an **agglutination test,** antiserum is mixed with a sample that potentially contains its target cells. If the antigenic cells are present, antibodies in the antiserum will clump (*agglutinate*) the antigen (**Figure 4.26**). Other antigens, and therefore other organisms, remain unaffected because antibodies are highly specific for their targets.

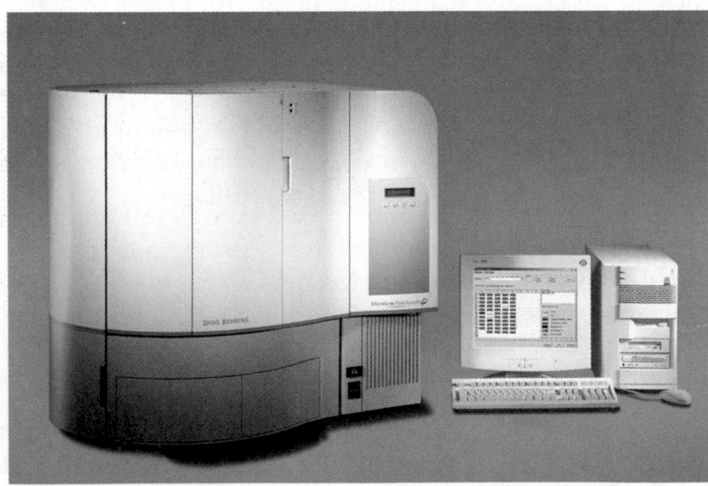

(a) MicroScan instrument

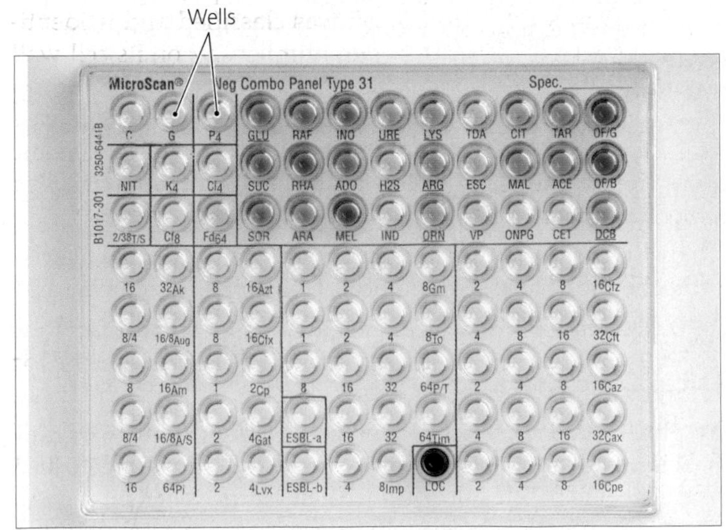

(b) MicroScan panel

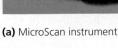

▲ **Figure 4.25 One tool for the rapid identification of bacteria, the automated MicroScan system. (a)** The MicroScan instrument. **(b)** A MicroScan panel, a plate containing numerous wells, each the site of a particular biochemical test. The instrument ascertains the identity of the organism by reading the pattern of colors in the wells after the biochemical tests have been performed.

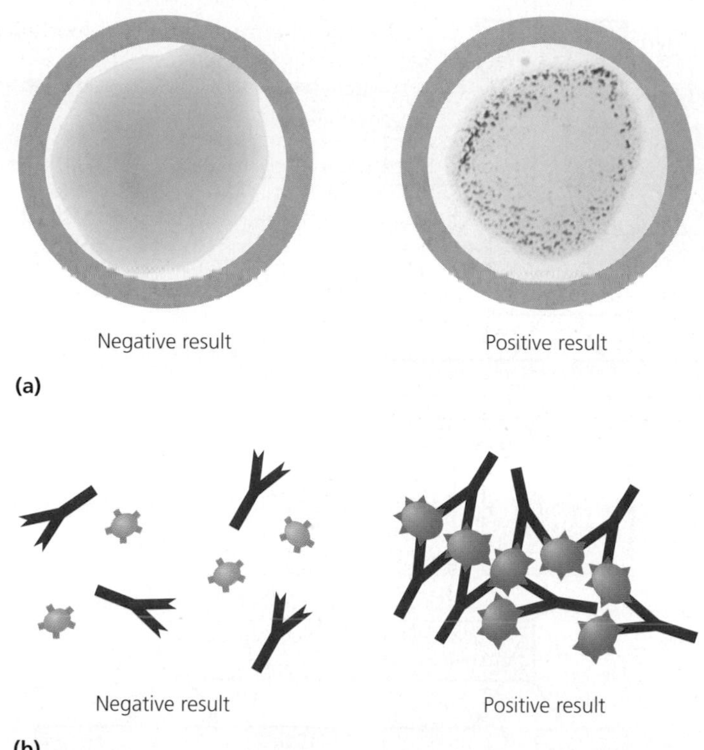

▲ **Figure 4.26 An agglutination test, one type of serological test.** **(a)** In a positive agglutination test, visible clumps are formed by the binding of antibodies to their target antigens present on cells. **(b)** The processes involved in agglutination tests. In a negative result, antibody binding cannot occur because its specific target is not present; in a positive result, specific binding does occur. Note that agglutination occurs because each antibody molecule can bind simultaneously to two antigen molecules.

Antisera can be used to distinguish among species, and even among strains of the same species. For example, a particularly pathogenic strain of *Escherichia coli* was classified and is identified by the presence of both antigen number 157 on its cell wall (designated O157) and antigen number 7 on its flagella (designated H7). This pathogen, known as E. coli O157:H7, has caused several deaths in the United States over the past few years. Chapter 17 examines other serological tests, such as *enzyme-linked immunosorbent assay (ELISA)* and *western blotting*.

Phage Typing

Bacteriophages (or simply **phages**) are viruses that infect and usually destroy bacterial cells. Just as antibodies are specific for their target antigens, phages are specific for the hosts they can infect. **Phage typing,** like serological testing, works because of such specificity. One bacterial strain may be susceptible to a particular phage while a related strain is not.

In phage typing, a technician spreads a solution containing the bacterium to be identified across a solid surface of growth medium and then adds small drops of solutions containing different types of bacteriophage. Wherever a specific phage is able to infect and kill bacteria, the resulting lack of bacterial growth produces within the bacterial lawn a clear area called a **plaque** (**Figure 4.27**). A microbiologist can identify an unknown bacterium by comparing the phages that form plaques with known phage-bacteria interactions.

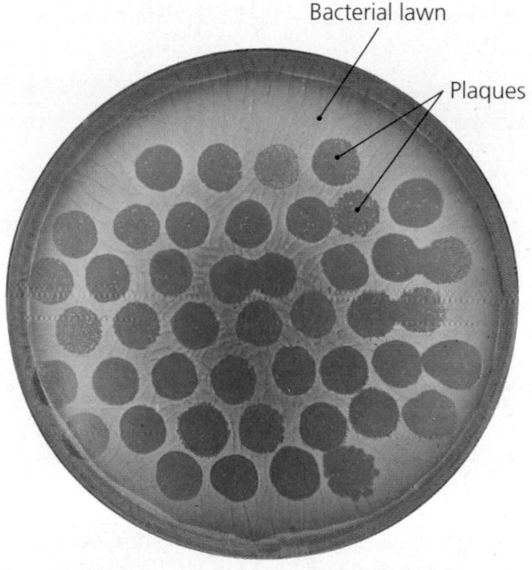

▲ **Figure 4.27 Phage typing.** Drops containing type A bacteriophages were added to this plate after its entire surface was inoculated with an unknown strain of *Salmonella*. After 12 hours of bacterial growth, clear zones, called plaques, developed where the phages killed bacteria. Given the great specificity of phage A for infecting and killing its host, the strain of bacterium can be identified as *Salmonella enterica* serotype Typhi.

Analysis of Nucleic Acids

As we have discussed, the sequence of nucleotides in nucleic acid molecules provides a powerful tool for the classification and identification of microbes. In many cases, nucleic acid analysis has confirmed classical taxonomic hierarchies. In other cases, as in Woese's discovery of domains, curious new organisms and relationships that were not obvious from classical methodologies have come to light. Techniques of nucleotide sequencing and comparison, such as *polymerase chain reaction (PCR)*, are best understood after we have discussed microbial genetics; therefore, these techniques are considered in Chapter 8.

Determination of the proportion of a cell's DNA that is guanine and cytosine, a quantity referred to as the cell's *G + C ratio,* has also become a part of prokaryotic taxonomy. Scientists express the ratio as follows:

$$\frac{G + C}{A + T + G + C} \times 100\%$$

G + C ratios vary from 20% to 80% among prokaryotes. Often (but not always) organisms that share characteristics have similar G + C ratios. Organisms that were once thought to be closely related but have widely different G + C ratios are invariably not as closely related as thought.

1a. Gram-positive cells.................................. Gram-positive bacteria
1b. Gram-negative cells................................ 2

2a. Rod-shaped cells..................................... 3
2b. Non-rod-shaped cells.............................. Cocci and pleomorphic bacteria

3a. Can tolerate oxygen................................ 4
3b. Cannot tolerate oxygen............................ Obligate anaerobes

4a. Ferments lactose..................................... 5
4b. Cannot ferment lactose........................... Non-lactose fermenters

5a. Can use citric acid as a sole carbon source... 6
5b. Cannot use citric acid alone.................... 8

6a. Produces hydrogen sulfide gas............... *Salmonella*
6b. Does not produce hydrogen sulfide gas.. 7

7a. Produces acetoin..................................... *Enterobacter*
7b. Does not produce acetoin........................ *Citrobacter*

8a. Produces gas from glucose...................... *Escherichia*
8b. Does not produce gas from glucose....... *Shigella*

(a)

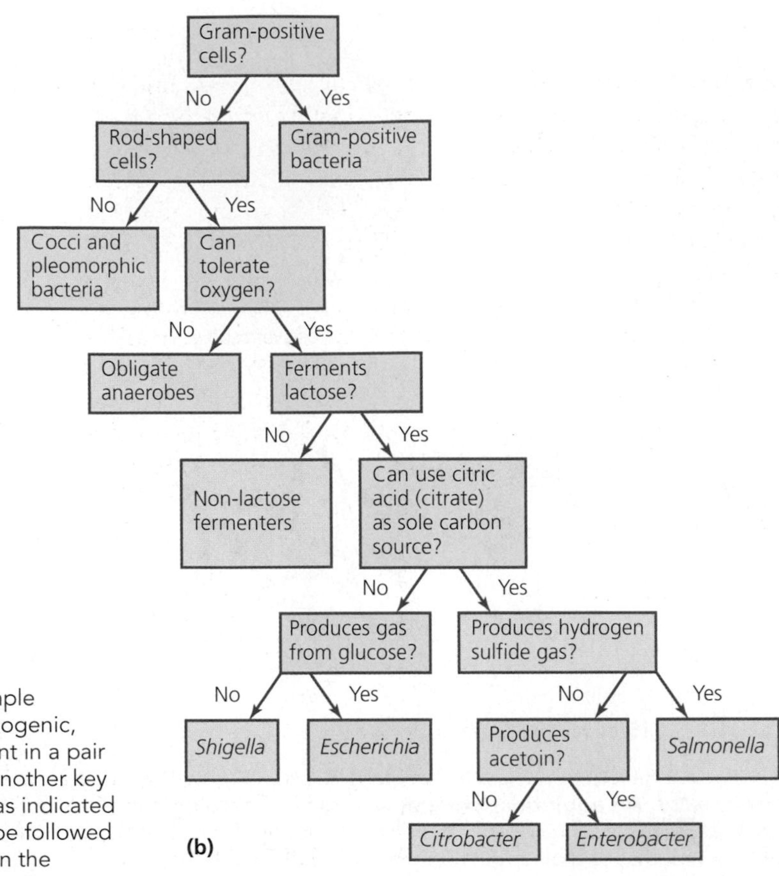

▲ **Figure 4.28 Use of a dichotomous taxonomic key.** The example presented here involves identification of the genera of potentially pathogenic, intestinal bacteria. **(a)** A sample key. To use it, choose the one statement in a pair that applies to the organism to be identified, and then either refer to another key (as indicated by words) or go to the appropriate place within this key (as indicated by a number). **(b)** A flowchart that shows the various paths that might be followed in using the key presented in part a. Highlighted is the path taken when the bacterium in question is *Escherichia*.

EMERGING DISEASES

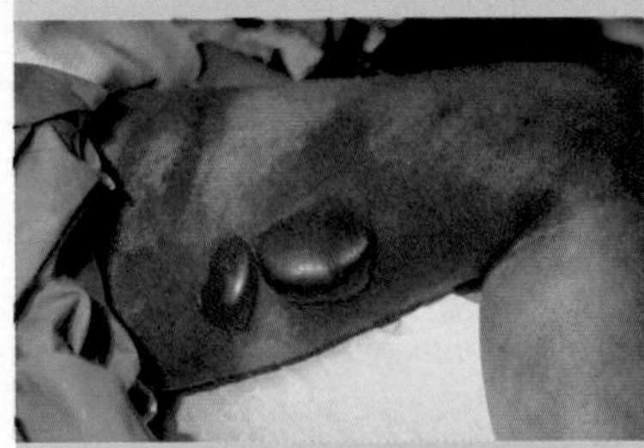

NECROTIZING FASCIITIS

Fever, chills, nausea, weakness, and general yuckiness. Carlos thought he was getting the flu. Further, he had pulled a cactus thorn from his leg the day before, and the tiny wound had swollen to a centimeter in diameter. It was red, extremely hot, and much more painful than such a puncture had a right to be. Everything was against him. He couldn't afford to miss days at work, but he had no choice.

He shivered in bed with fever for the next two days and suffered more pain than he had ever experienced, certainly more than the time he broke his leg. Even more than passing a kidney stone. The red, purple, and black inflammation on his leg had grown to the size of a baseball. It was hard to the touch and excruciatingly painful. He decided it was time to call his brother to take him to the doctor. That decision saved his life.

Carlos's blood pressure dropped severely, and he was unconscious by the time they arrived. The physician immediately admitted Carlos to the hospital, where the medical team raced to treat necrotizing fasciitis, commonly called "flesh-eating" disease. This reemerging disease is caused by Group A *Streptococcus*, a serotype of Gram-positive bacteria also known as *S. pyogenes*. Group A strep invades through a break in the skin and travels along the fascia—the protective covering of muscles—producing toxins that kill human tissues.

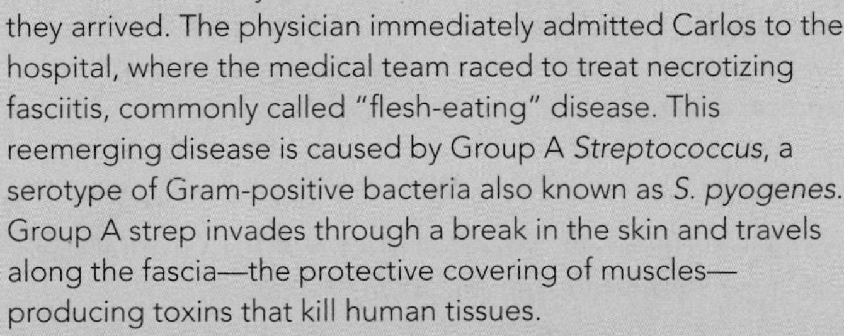

By cutting away all the infected tissue; using high-pressure, pure oxygen to inhibit bacterial growth; and applying antibiotics to kill the bacterium, the doctors stabilized Carlos. After months of skin grafts and rehabilitation, he returned to work, grateful to be alive. For more about necrotizing fasciitis, see p. 540.

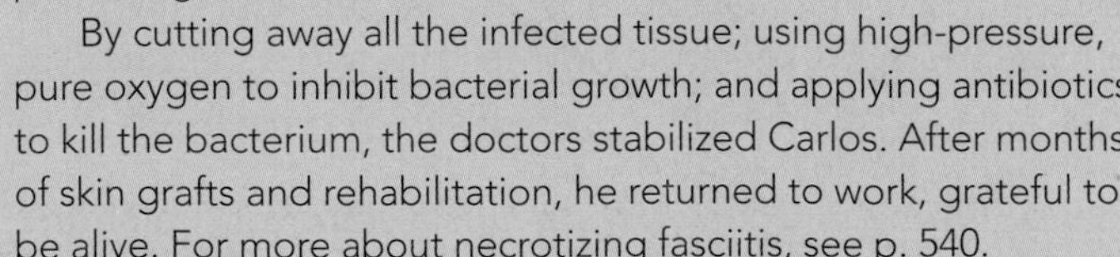

Track necrotizing fasciitis online by going to the Study Area at www.masteringmicrobiology.com.

Taxonomic Keys

As we have seen, taxonomists, medical clinicians, and researchers can use a wide variety of information, including morphology, chemical characteristics, and results from biochemical, serological, and phage typing tests, in their efforts to identify microorganisms, including pathogens. But how can all these characteristics and results be organized so that they can be used efficiently to identify an unknown organism? All this information is often arranged in **dichotomous keys,** which contain a series of paired statements worded so that only one of two choices applies to any particular organism **(Figure 4.28)**. Based on which of the two statements applies, the key either directs the user to another pair of statements, or it provides the name of the organism in question. Note that more than one key can be created to enable the identification of a given set of organisms, but all such keys involve mutually exclusive, "either/or" choices that send the user along a path that leads to the identity of the unknown organism. **ANIMATIONS:** ***Dichotomous Keys: Overview, Sample with Flowchart, Practice***

CRITICAL THINKING

A clinician obtains a specimen of urine from a patient suspected to have a bladder infection. From the specimen she cultures a Gram-negative, rod-shaped bacterium that ferments lactose in the presence of oxygen, utilizes citrate, and produces acetoin but not hydrogen sulfide. Using the key presented in Figure 4.28, identify the genus of the infective bacterium.

Chapter Summary

ANIMATIONS: ***Microscopy and Staining: Overview***

Units of Measurement (p. 96)

1. The metric system is a decimal system in which each unit is one-tenth the size of the next largest unit.
2. The basic unit of length in the metric system is the meter.

Microscopy (pp. 97–106)

1. **Microscopy** refers to the passage of light or electrons of various **wavelengths** through lenses to **magnify** objects and provide **resolution** and contrast so that those objects can be viewed and studied.
2. Immersion oil is used in light microscopy to fill the space between the specimen and a lens to reduce light refraction and thus increase the **numerical aperture** and resolution.
3. Staining techniques and polarized light may be used to enhance **contrast** between an object and its background.
4. **Simple microscopes** contain a single magnifying lens, whereas **compound microscopes** use a series of lenses for magnification.
5. The lens closest to the object being magnified is the **objective lens,** several of which are mounted on a **revolving nosepiece.** The lenses closest to the eyes are **ocular lenses. Condenser lenses** lie beneath the stage and direct light through the slide.
6. The magnifications of the objective lens and the ocular lens are multiplied together to give **total magnification.**
7. A photograph of a microscopic image is a **micrograph.**
8. **Dark-field microscopes** provide a dark background for small or colorless specimens.
9. **Phase microscopes,** such as **phase-contrast** and **differential interference contrast** (Nomarski) microscopes, cause light rays that pass through a specimen to be out of phase with light rays that pass through the field, producing contrast.
10. **Fluorescent microscopes** use ultraviolet light and fluorescent dyes to fluoresce specimens and enhance contrast.
11. A **confocal microscope** uses fluorescent dyes in conjunction with computers to provide three-dimensional images of a specimen. **ANIMATIONS:** ***Light Microscopy***
12. A **transmission electron microscope (TEM)** provides an image produced by the transmission of electrons through a thinly sliced, dehydrated specimen.
13. A **scanning electron microscope (SEM)** provides a three-dimensional image by scattering electrons from the metal-coated surface of a specimen.
14. Minuscule electronic probes are used in **scanning tunneling microscopes** and in **atomic force microscopes** to reveal details at the atomic level. **ANIMATIONS:** ***Electron Microscopy***

Staining (pp. 106–113)

1. Preparing to **stain** organisms with dyes for light microscopy involves making a smear or thin film of the specimens on a slide and then either passing the slide through a flame **(heat fixation)** or applying a chemical **(chemical fixation)** to attach the specimens to the slide. **Acidic dyes** or **basic dyes** are used to stain different portions of an organism to aid viewing and identification.
2. **Simple stains** involve the simple process of soaking the smear with one dye and then rinsing with water. **Differential stains** such as the **Gram stain, acid-fast stain,** and **endospore stain** use more than one dye to differentiate different cells, chemicals, or structures.
3. The Gram stain procedure includes use of a **primary stain,** a **mordant,** a **decolorizing agent,** and a **counterstain** that results in either purple (Gram-positive) or pink (Gram-negative) organisms, depending on the chemical structures of their cell walls.
4. The acid-fast stain is used to differentiate cells with waxy cell walls. **Endospores** are stained by the **Schaeffer-Fulton endospore** stain procedure.
5. Dyes that stain the background and leave the cells colorless are called **negative** (or **capsule**) **stains.** **ANIMATIONS:** ***Staining***

Classification and Identification of Microorganisms (pp. 113–121)

1. **Taxa** are nonoverlapping groups of organisms that are studied and named in **taxonomy.** Carolus Linnaeus invented a system of taxonomy, grouping similar interbreeding organisms into **species,** species into **genera,** genera into **families,** families into **orders,** orders into **classes,** classes into **phyla,** and phyla into **kingdoms.**
2. Linnaeus gave each species a descriptive name consisting of a genus name and **specific epithet.** This practice of naming organisms with two names is called **binomial nomenclature.**
3. Carl Woese proposed the existence of three taxonomic **domains** based on three cell types revealed by rRNA sequencing: **Eukarya, Bacteria,** and **Archaea.**
4. Taxonomists rely primarily on genetic differences revealed by morphological and metabolic dissimilarities to classify organisms. Species or strains within species may be distinguished by using **antisera, agglutination tests,** nucleic acid analysis, or **phage typing** with **bacteriophages,** in which unknown bacteria are identified by observing **plaques** (regions of a bacterial lawn where the phage has killed bacterial cells).
5. Microbiologists use **dichotomous keys,** which involve stepwise choices between paired characteristics, to assist them in identifying microbes.

   ANIMATIONS: *Dichotomous Keys: Overview, Sample with Flowchart, Practice*

Questions for Review

Answers to the Questions for Review below (except Short Answer Questions) begin on page A-1.

Multiple Choice

1. Which of the following is smallest?
   a. decimeter
   b. millimeter
   c. nanometer
   d. micrometer
2. A nanometer is ________________ than a micrometer.
   a. 10 times larger
   b. 10 times smaller
   c. 1000 times larger
   d. 1000 times smaller
3. Resolution is best described as
   a. the ability to view something that is small.
   b. the ability to magnify a specimen.
   c. the ability to distinguish between two adjacent objects.
   d. the difference between two waves of electromagnetic radiation.
4. Curved glass lenses ________________ light.
   a. refract
   b. bend
   c. magnify
   d. both a and b
5. Which of the following factors is important in making an image appear larger?
   a. the thickness of the lens
   b. the curvature of the lens
   c. the speed of the light passing through the lens
   d. all of the above
6. Which of the following is different between light microscopy and transmission electron microscopy?
   a. magnification
   b. resolution
   c. wavelengths
   d. all of the above
7. Which of the following types of microscopes produces a three-dimensional image with a shadowed appearance?
   a. simple microscope
   b. differential interference contrast microscope
   c. fluorescent microscope
   d. transmission electron microscope
8. Which of the following microscopes combines the greatest magnification with the best resolution?
   a. confocal microscope
   b. phase-contrast microscope
   c. dark-field microscope
   d. bright-field microscope
9. Negative stains such as eosin are also called
   a. capsule stains.
   b. endospore stains.
   c. simple stains.
   d. acid-fast stains.
10. In the binomial system of nomenclature, which term is always written in lowercase letters?
   a. kingdom
   b. domain
   c. genus
   d. specific epithet

Fill in the Blanks

1. If an objective magnifies 40×, and each binocular lens magnifies 15×, the total magnification of the object being viewed is ________________.
2. The type of fixation developed by Koch for bacteria is ________________.
3. Immersion oil ________________ (increases/decreases) the numerical aperture, which ________________ (increases/decreases) resolution because ________________ (more/fewer) light rays are involved.
4. ________________ refers to differences in intensity between two objects.
5. Cationic chromophores such as methylene blue ionically bond to ________________ (positively/negatively) charged chemicals such as DNA and proteins.

Labeling

Label each photograph below with the type of microscope used to acquire the image.

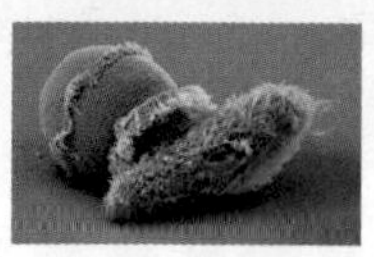

1.____________

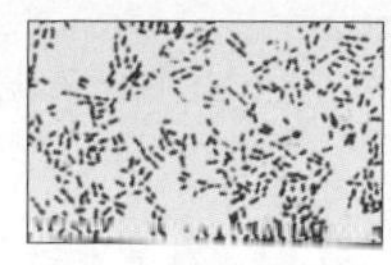

2.____________

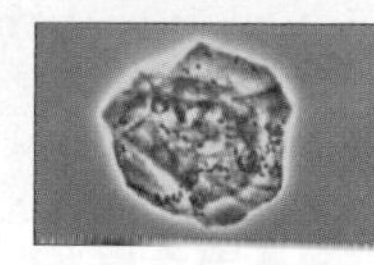

3.____________

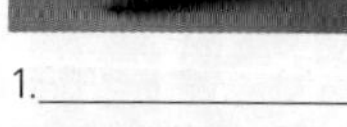

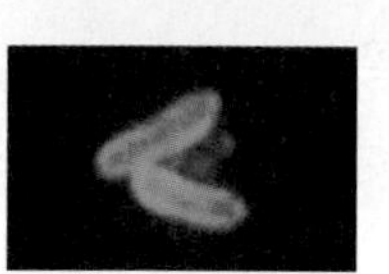

4.____________

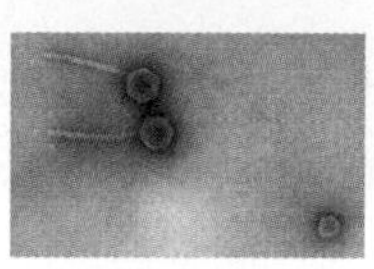

5.____________

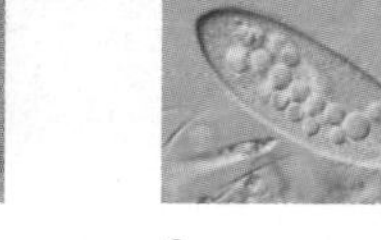

6.____________

Short Answer

1. Explain how the principle "electrons travel as waves" applies to microscopy.
2. Critique the following definition of magnification given by a student on a microbiology test: "Magnification makes things bigger."
3. Why can electron microscopes magnify only dead organisms?
4. Put the following substances in the order they are used in a Gram stain: counterstain, decolorizing agent, mordant, primary stain.
5. Why is Latin used in taxonomic nomenclature?
6. Give three characteristics of a "specific epithet."
7. How does the study of the nucleotide sequences of ribosomal RNA fit into a discussion of taxonomy?
8. An atomic force microscope can magnify a living cell, whereas electron microscopes and scanning tunneling microscopes cannot. What requirement of electron and scanning tunneling microscopes precludes the imaging of living specimens?

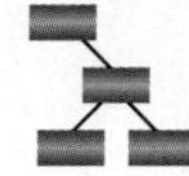

Concept Mapping

Using the following terms, draw a concept map that describes Gram stain. For a sample concept map, see p. 93. Or, complete this concept map online by going to the Study Area at www.masteringmicrobiology.com.

Counterstain
Crystal violet
Decolorizer
Ethanol and acetone
Gram negative bacteria
Gram positive bacteria
Iodine
Mordant
Outer membrane
Pink
Primary stain
Purple
Safranin
Thick peptidoglycan layer
Thin peptidoglycan layer

Critical Thinking

1. Miki came home from microbiology lab with very green fingers and a bad grade. When asked about this, she replied that she was doing a Gram stain but it never worked the way the book said it should. Boone overheard the conversation and said that she must have used the wrong chemicals. What dye was she probably using, and what structure does that chemical normally stain?
2. Why is the definition of *species* as "successfully interbreeding organisms" not satisfactory for most microorganisms?
3. With the exception of the discovery of new organisms, is it logical to assume that taxonomy as we currently know it will stay the same? Why or why not?
4. A novice microbiology student incorrectly explains that immersion oil increases the magnification of his microscope. What is the function of immersion oil?
5. A light microscope has 10× oculars and 0.3-μm resolution. Using the oil immersion lens (100×), will you be able to resolve two objects 400 nm apart? Will you be able to resolve two objects 40 nm apart?

Access more review material online in the Study Area at **www.masteringmicrobiology.com**. There, you'll find

- **Animations**
- **MP3 Tutor Sessions**
- **Concept Mapping Activities**
- **Flashcards**
- **Quizzes**

and more to help you succeed.

5 Microbial Metabolism

The next time you bite into a delicious piece of chocolate, consider this: microbial metabolism played a key role in how that chocolate became delicious.

Chocolate comes from cacao seeds, found inside the pods of *Theobroma cacao* trees. After the pods are split open, the seeds and the surrounding pulp are scooped out and placed in heaps on top of plantain or banana leaves. The heaps are then covered and left to ferment for two to seven days. Fermentation occurs as microorganisms—including yeast and several kinds of bacteria—grow on the fleshy, sugary pulp. Bathed in fermenting pulp, the cacao seeds (which start off tasting bitter) begin to develop the flavors and colors that we associate with chocolate. After fermentation, the seeds are dried, roasted, and then processed further by chocolate manufacturers before becoming one of our favorite desserts.

In this chapter we will learn about many metabolic processes of microorganisms, including fermentation.

Take the pre-test for this chapter online. Visit the Study Area at www.masteringmicrobiology.com.

This chapter has *MicroFlix*. Go to **www.masteringmicrobiology.com** to view movie-quality animations for metabolism.

Microorganisms will ferment these sun-dried cacao seeds to begin the transformation of cacao into chocolate.

How do pathogens acquire energy and nutrients at the expense of a patient's health? How does grape juice turn into wine, and how does yeast cause bread to rise? How do disinfectants, antiseptics, and antimicrobial drugs work? When laboratory personnel perform biochemical tests to identify unknown microorganisms and help diagnose disease, what exactly are they doing?

The answers to all of these questions require an understanding of microbial **metabolism,**[1] the collection of controlled biochemical reactions that takes place within the microbe. While it is true that metabolism in its entirety is complex, consisting of thousands of chemical reactions and control mechanisms, the reactions are nevertheless elegantly logical and can be understood in a simplified form. In this chapter we will concern ourselves only with central metabolic pathways and energy metabolism.

Your study of metabolism will be manageable if you keep in mind that the ultimate function of an organism's metabolism is to reproduce the organism, and that metabolic processes are guided by the following eight elementary statements:

- Every cell acquires *nutrients,* which are the chemicals necessary for metabolism.
- Metabolism requires energy from light or from the *catabolism* (kă-tab′ō-lizm; breakdown) of acquired nutrients.
- Energy is stored in the chemical bonds of *adenosine triphosphate (ATP).*
- Using *enzymes,* cells catabolize nutrient molecules to form elementary building blocks called *precursor metabolites.*
- Using these precursor metabolites, energy from ATP, and other enzymes, cells construct larger building blocks in *anabolic* (an-ă-bol′ik; biosynthetic) reactions.
- Cells use enzymes and additional energy from ATP to anabolically link building blocks together to form macromolecules in *polymerization* reactions.
- Cells grow by assembling macromolecules into cellular structures such as ribosomes, membranes, and cell walls.
- Cells typically reproduce once they have doubled in size.

We will discuss each aspect of metabolism in the chapters that most directly apply. For instance, we discussed the first step of metabolism—the active and passive transport of nutrients into cells—in Chapter 3. In this chapter we examine the importance of enzymes in catabolic and anabolic reactions, study the three ways that ATP molecules are synthesized, and show that catabolic and anabolic reactions are linked. We also examine the catabolism of nutrient molecules; the anabolic reactions involved in the synthesis of carbohydrates, lipids, amino acids, and nucleotides; and a few ways that cells control their metabolic activities. Genetic control of metabolism and the polymerization of DNA, RNA, and proteins are discussed in Chapter 7, and the specifics of cell division are covered in Chapters 11 and 12.

[1]From Greek *metabole,* meaning change.

Basic Chemical Reactions Underlying Metabolism

In the following sections we will examine the basic concepts of catabolism, anabolism, and a special class of reactions called *oxidation-reduction reactions.* The latter involve the transfer of electrons between molecules. Then we will turn our attention briefly to the synthesis of ATP and energy storage before we discuss the organic catalysts called *enzymes,* which make metabolism possible. **ANIMATIONS:** ***Metabolism: Overview***

Catabolism and Anabolism

Learning Objective

✓ Distinguish among metabolism, anabolism, and catabolism.

Metabolism, which is all of the chemical reactions in an organism, can be divided into two major classes of reactions: **catabolism** and **anabolism** **(Figure 5.1)**. A series of such reactions is called a *pathway.* Cells have *catabolic pathways,* which break larger molecules into smaller products, and *anabolic pathways,* which synthesize large molecules from the smaller products of catabolism. Even though catabolic and anabolic pathways are intimately linked in cells, it is often useful to study the two types of pathways as if they were separate.

When catabolic pathways break down large molecules, they release energy; that is, catabolic pathways are *exergonic* (ek-ser-gon′ik). Cells store some of this released energy in the bonds of ATP, though much of the energy is lost as heat. Another result of the breakdown of large molecules by catabolic pathways is the production of numerous smaller molecules, some of which are **precursor metabolites** of anabolism. Some organisms, such as *Escherichia coli* (esh-ĕ-rik′ē-ă kō′lē), can synthesize everything in their cells just from precursor metabolites; other organisms must acquire some anabolic building blocks from outside their cells as nutrients. Catabolic *pathways,* but not necessarily *individual* catabolic *reactions,* produce ATP, or metabolites, or both. An example of a catabolic pathway is the breakdown of lipids into glycerol and fatty acids.

Anabolic pathways are functionally the opposite of catabolic pathways in that they synthesize macromolecules and cellular structures. Because building anything requires energy, anabolic pathways are *endergonic* (en-der-gon′ik); that is, they require more energy than they release. The energy required for anabolic pathways usually comes from ATP molecules produced during catabolism. An example of an anabolic pathway is the synthesis of lipids for cell membranes from glycerol and fatty acids.

To summarize, a cell's metabolism involves both catabolic pathways that break down macromolecules to supply molecular building blocks and energy in the form of ATP, and anabolic pathways that use the building blocks and ATP to synthesize macromolecules needed for growth and reproduction.

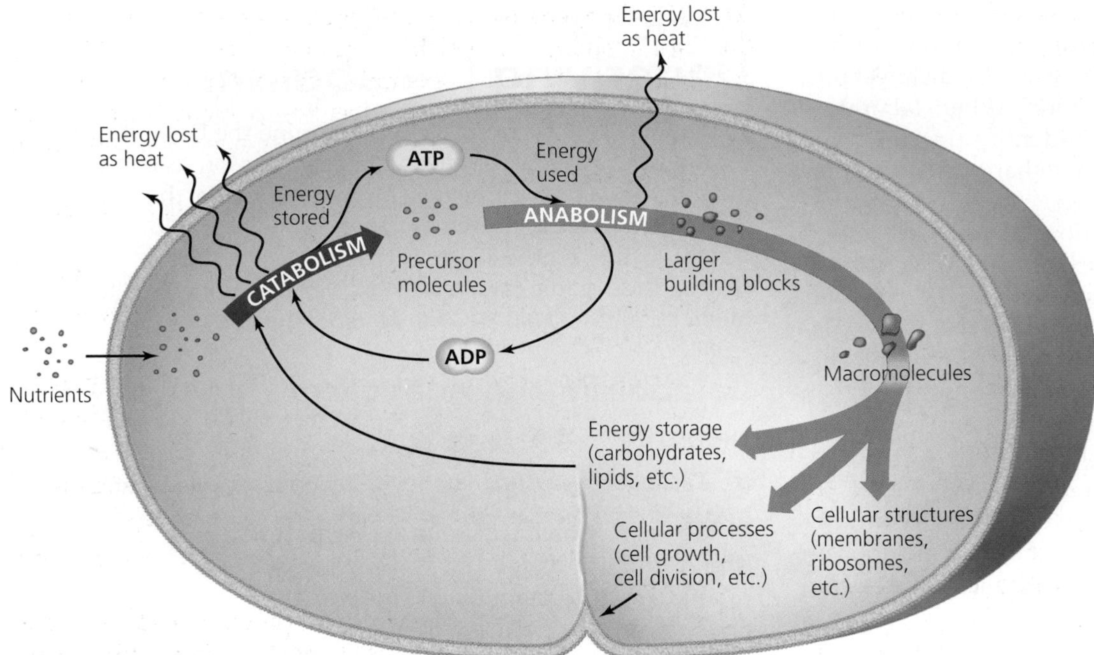

Figure 5.1 Metabolism is composed of catabolic and anabolic reactions. Some energy released in catabolism is stored in ATP molecules, but most is lost as heat. Anabolic reactions require energy, typically provided by ATP. There is some heat loss in anabolism as well. The products of catabolism provide many of the building blocks for anabolic reactions. These reactions produce macromolecules and cellular structures, leading to cell growth and division.

Oxidation and Reduction Reactions

Learning Objective

✓ Contrast reduction and oxidation reactions.

Many metabolic reactions involve the transfer of electrons from an *electron donor* (a molecule that donates an electron) to an *electron acceptor* (a molecule that accepts an electron). Such electron transfers are called **oxidation-reduction reactions,** or **redox reactions** (Figure 5.2). An electron acceptor is said to be *reduced.* This may seem backward, but electron acceptors are reduced because their gain in electrons reduces their overall electrical charge (that is, they are more negatively charged). Molecules that lose electrons are said to be *oxidized* because frequently their electrons are donated to oxygen atoms. An acronym to help you remember these concepts is OIL RIG: oxidation involves loss; reduction involves gain.

Reduction and oxidation reactions always happen simultaneously because every electron donated by one chemical is accepted by another chemical. A chemical may be reduced by gaining either a simple electron or an electron that is part of a hydrogen atom—which, as we saw in Chapter 2, is composed of one proton and one electron. **Beneficial Microbes: Gold-Mining Microbes** describes an interesting example of how some prokaryotes are able to reduce gold dissolved in solution.

In contrast, a molecule may be oxidized in one of three ways: by losing a simple electron, by losing a hydrogen atom, or by gaining an oxygen atom. Biological oxidations often involve the loss of

BENEFICIAL MICROBES

GOLD-MINING MICROBES

▲ *Solid gold is gold in its reduced form.*

Gold, as found in nature, exists in two forms: gold-ore deposits, which are gold in its reduced form, usually found near the Earth's crust, and gold dissolved in solution, as found in thermal springs and in seawater. Dissolved gold, which is gold in its oxidized forms, is largely useless to humans; it cannot be converted inexpensively into solid gold. Even though gold in either form is toxic when ingested by most living things, scientists have discovered that certain bacteria, such as *Ralstonia metallidurans,* can metabolize oxidized gold. When placed in a solution containing oxidized gold, these microorganisms reduce the gold and encase themselves in solid gold, which is their metabolic waste.

Entrepreneurial minds may wonder whether *Ralstonia* could be potentially profitable. While it is true that a great deal of dissolved gold is found in thermal springs and oceans, the gold is very dilute—only minute amounts are present in very large volumes of water. Moreover, were someone to perfect a way of using microorganisms to convert dissolved gold to great quantities of solid gold, they would be wise to keep it to themselves: so much solid gold could become available that its market value would plunge dramatically.

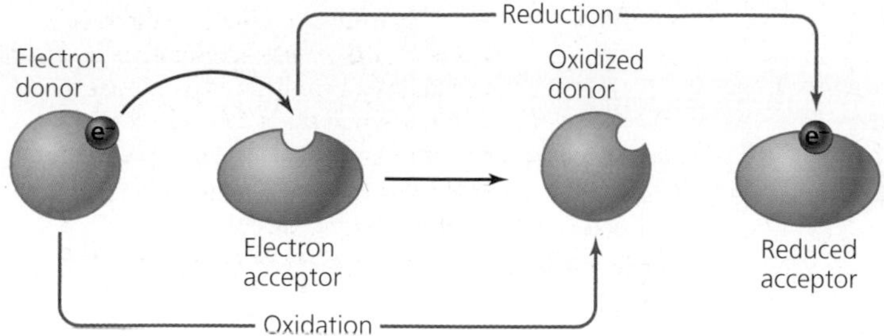

▲ **Figure 5.2 Oxidation-reduction, or redox, reactions.** When electrons are transferred from donor molecules to acceptor molecules, donors become oxidized and acceptors become reduced. *Why are acceptor molecules said to be reduced when they are gaining electrons?*

Figure 5.2 *"Reduction" refers to the overall electrical charge on a molecule. Because electrons have a negative charge, the gain of an electron reduces the molecule's overall charge.*

hydrogen atoms; such reactions are also called *dehydrogenation* (dē-hī'drō-jen-ā'shŭn) *reactions.*

Electrons rarely exist freely in cytoplasm; instead, they orbit atomic nuclei. Therefore, cells use electron carrier molecules to carry electrons (often in hydrogen atoms) from one location in a cell to another. Three important electron carrier molecules, which are derived from vitamins, are **nicotinamide adenine dinucleotide (NAD^+), nicotinamide adenine dinucleotide phosphate ($NADP^+$),** and **flavin adenine dinucleotide (FAD).** Cells use each of these molecules in specific metabolic pathways to carry pairs of electrons. One of the electrons carried by either NAD^+ or $NADP^+$ is part of a hydrogen atom, forming NADH or NADPH. FAD carries two electrons as hydrogen atoms ($FADH_2$). Many metabolic pathways, including those that synthesize ATP, require such electron carrier molecules. **ANIMATIONS:** ***Oxidation-Reduction Reactions***

CRITICAL **THINKING**

Arsenic is a poison that exists in two states in the environment—arsenite ($H_2AsO_3^-$) and arsenate ($H_2AsO_4^-$). Arsenite dissolves in water, making the water dangerous to drink. Arsenate is less soluble and binds to minerals, making this form of arsenic less toxic in the environment. Some strains of the bacterium *Thermus* oxidize arsenic in an aerobic environment but reduce arsenic under anaerobic conditions. In case of arsenic contamination of water, how could scientists use *Thermus* to remediate the problem?

ATP Production and Energy Storage

Learning Objective

✓ Compare and contrast the three types of ATP phosphorylation.

Nutrients contain energy, but that energy is spread throughout their chemical bonds and generally is not concentrated enough for use in anabolic reactions. During catabolism, organisms release energy from nutrients that can then be concentrated and stored in high-energy phosphate bonds of molecules such as ATP. This happens by a general process called *phosphorylation* (fos'fōr-i-lā'shŭn), in which inorganic phosphate (PO_4^{3-}) is added to a substrate. For example, cells phosphorylate adenosine diphosphate (ADP), which has two phosphate groups, to form adenosine triphosphate (ATP), which has three phosphate groups (see Figure 2.27).

As we will examine in the following sections, cells phosphorylate ADP to form ATP in three specific ways:

- *Substrate-level phosphorylation* (see p. 135), which involves the transfer of phosphate to ADP from another phosphorylated organic compound
- *Oxidative phosphorylation* (see p. 143), in which energy from redox reactions of respiration (described shortly) is used to attach inorganic phosphate to ADP
- *Photophosphorylation* (see p. 151), in which light energy is used to phosphorylate ADP with inorganic phosphate

We will investigate each of these in more detail as we proceed through the chapter.

After ADP is phosphorylated to produce ATP, anabolic pathways use some energy of ATP by breaking a phosphate bond (which re-forms ADP). Thus the cyclical interconversion of ADP and ATP functions somewhat like rechargeable batteries: ATP molecules store energy from light (in photosynthetic organisms) and from catabolic reactions and then release stored energy to drive cellular processes (including anabolic reactions, active transport, and movement). ADP molecules can be "recharged" to ATP again and again.

The Roles of Enzymes in Metabolism

Learning Objectives

✓ Make a table listing the six basic types of enzymes, their activities, and an example of each.

✓ Describe the components of a holoenzyme and contrast protein and RNA enzymes.

✓ Define *activation energy, enzyme, apoenzyme, cofactor, coenzyme, active site,* and *substrate,* and describe their roles in enzyme activity.

✓ Describe how temperature, pH, substrate concentration, and competitive and noncompetitive inhibition affect enzyme activity.

As we saw in Chapter 2, reactions occur when chemical bonds are broken or formed between atoms. In catabolic reactions, a bond must be destabilized before it will break, whereas in anabolic reactions reactants collide with sufficient energy for bonds to form between them. In anabolism, increasing either the concentrations of reactants or ambient temperatures increases the number of collisions and produces more chemical reactions; however, in living organisms, neither reactant concentration nor temperature is usually high enough to ensure that bonds will form. Therefore, the chemical reactions of life depend upon *catalysts,* which are chemicals that increase the likelihood of a reaction but are not permanently changed in the process. Organic catalysts are known as **enzymes.** **ANIMATIONS:** ***Enzymes: Overview***

TABLE 5.1 Enzyme Classification Based on Reaction Types

Class	Type of Reaction Catalyzed	Example
Hydrolase	Hydrolysis (catabolic)	Lipase—breaks down lipid molecules
Isomerase	Rearrangement of atoms within a molecule (neither catabolic nor anabolic)	Phosphoglucoisomerase—converts glucose 6-phosphate into fructose 6-phosphate during glycolysis
Ligase or polymerase	Joining two or more chemicals together (anabolic)	Acetyl-CoA synthetase—combines acetate and coenzyme A to form acetyl-CoA for the Krebs cycle
Lyase	Splitting a chemical into smaller parts without using water (catabolic)	Fructose-1,6-bisphosphate aldolase—splits fructose 1,6-bisphosphate into G3P and DHAP
Oxidoreductase	Transfer of electrons or hydrogen atoms from one molecule to another	Lactic acid dehydrogenase—oxidizes lactic acid to form pyruvic acid during fermentation
Transferase	Moving a functional group from one molecule to another (may be anabolic)	Hexokinase—transfers phosphate from ATP to glucose in the first step of glycolysis

Naming and Classifying Enzymes

The names of enzymes usually end with the suffix "-ase," and the name of each enzyme often incorporates the name of that enzyme's **substrate,** which is the molecule the enzyme acts upon. Based on their mode of action, enzymes can be grouped into six basic categories:

- *Hydrolases* catabolize molecules by adding water in a decomposition process known as *hydrolysis*. Hydrolases are used primarily in the depolymerization of macromolecules.
- *Isomerases*[2] rearrange the atoms within a molecule but do not add or remove anything (so they are neither catabolic nor anabolic).
- *Ligases,* or *polymerases,* join two molecules together (and are thus anabolic). They often use energy supplied by ATP.
- *Lyases* split large molecules (and are thus catabolic) without using water in the process.
- *Oxidoreductases* remove electrons from (oxidize) or add electrons to (reduce) various substrates. They are used in both catabolic and anabolic pathways.
- *Transferases* transfer functional groups, such as an amino group (NH_2), a phosphate group, or a two-carbon (acetyl) group, between molecules. Transferases can be anabolic.

Table 5.1 summarizes these types of enzymes and gives examples of each.

The Makeup of Enzymes

Many protein enzymes are complete in themselves, but others are composed of both protein and nonprotein portions. The proteins, called **apoenzymes** (ap'ō-en-zīms), are inactive if they are not bound to one or more of the nonprotein substances called **cofactors.** Cofactors are either inorganic ions (such as iron, magnesium, zinc, or copper ions) or certain organic molecules called coenzymes. All **coenzymes** (ko-en'zīms) are either vitamins or contain vitamins, which are organic molecules that are required for metabolism but cannot be synthesized by certain organisms (especially mammals). Some apoenzymes bind with inorganic cofactors, some bind with coenzymes, and some bind with both. The binding of an apoenzyme and its cofactor(s) forms an active enzyme, called a **holoenzyme** (hol-ō-en'zīm; **Figure 5.3**).

Table 5.2 lists several examples of inorganic cofactors and organic cofactors (coenzymes). Note that three important coenzymes are the electron carriers NAD^+, $NADP^+$, and FAD, which, as we have seen, carry electrons in hydrogen atoms from place to place within cells. We will examine more closely the roles of these coenzymes in the generation of ATP later in the chapter.

Not all enzymes are proteinaceous; some are RNA molecules called **ribozymes.** In eukaryotes, ribozymes process other RNA molecules by removing sections of RNA and splicing the remaining pieces together. Recently, researchers have discovered that the functional core of a ribosome is a ribozyme; therefore, given that ribosomes make all proteins, ribosomal enzymes make protein enzymes.

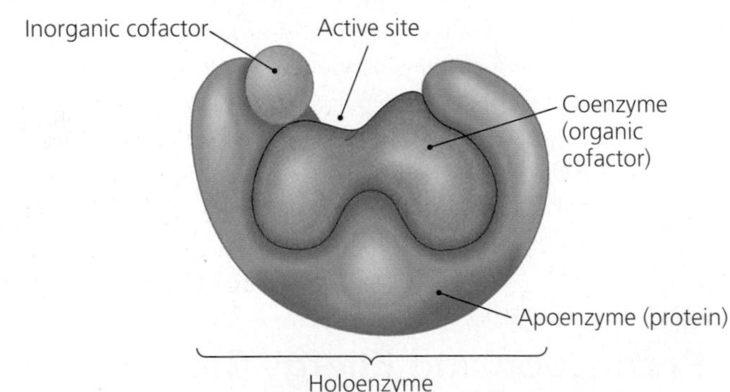

▲ **Figure 5.3 Makeup of a protein enzyme.** The combination of a proteinaceous apoenzyme with one or more cofactors forms a holoenzyme, which is the active form of an enzyme. A cofactor is either an inorganic ion or a coenzyme, which is an organic cofactor derived from a vitamin. The apoenzyme is inactive unless it is bound to its cofactors. *Name four metal ions that can act as cofactors.*

Figure 5.3 Iron, magnesium, zinc, and copper ions can act as cofactors.

[2] An isomer is a compound with the same molecular formula as another molecule, but with a different arrangement of atoms.

TABLE 5.2 Representative Cofactors of Enzymes

Cofactors	Examples of Use in Enzymatic Activity	Substance Transferred in Enzymatic Activity	Vitamin Source (of Coenzyme)
Inorganic (Metal Ion)			
Magnesium (Mg^{2+})	Forms bond with ADP during phosphorylation	Phosphate	None
Organic (Coenzymes)			
Nicotinamide adenine dinucleotide (NAD^+)	Carrier of reducing power	Two electrons and a hydrogen ion	Niacin
Nicotinamide adenine dinucleotide phosphate ($NADP^+$)	Carrier of reducing power	Two electrons and a hydrogen ion	Niacin
Flavin adenine dinucleotide (FAD)	Carrier of reducing power	Two hydrogen atoms	Riboflavin
Tetrahydrofolate	Used in synthesis of nucleotides and some amino acids	One-carbon molecule	Folic acid
Coenzyme A	Formation of acetyl-CoA in Krebs cycle and beta-oxidation	Two-carbon molecule	Pantothenic acid
Pyridoxal phosphate	Transaminations in the synthesis of amino acids	Amine group	Pyridoxine
Thiamine pyrophosphate	Decarboxylation of pyruvic acid	Aldehyde group (CHO)	Thiamine

Enzyme Activity

Within cells, enzymes catalyze reactions by lowering the **activation energy,** which is the amount of energy needed to trigger a chemical reaction (**Figure 5.4**). Whereas heat can provide energy to trigger reactions, the temperatures needed to reach activation energy for most metabolic reactions are often too high to allow cells to survive, so enzymes are needed if metabolism is to occur. This is true regardless of whether the enzyme is a protein or RNA, or whether the chemical reaction is anabolic or catabolic.

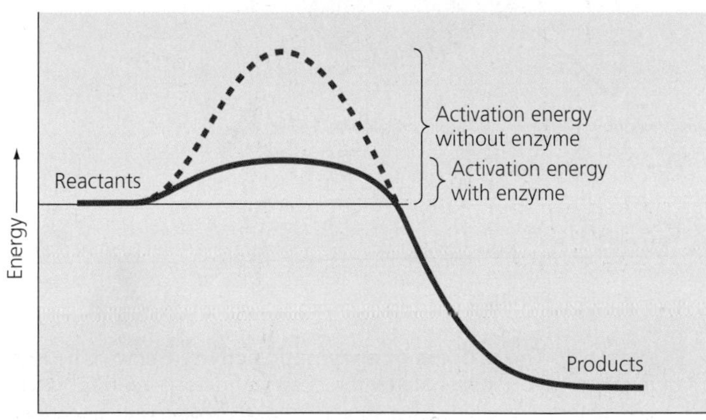

▲ **Figure 5.4 The effect of enzymes on chemical reactions.** Enzymes catalyze reactions by lowering the activation energy—that is, the energy needed to trigger the reaction.

The activity of enzymes depends on the closeness of fit between the functional sites of an enzyme and its substrate. The shape of an enzyme's functional site, called its **active site,** is complementary to the shape of the substrate. Generally, the shapes and locations of only a few amino acids or nucleotides determines the shape of an enzyme's active site. A change in a single component—for instance, through mutation—can render an enzyme less effective or even completely nonfunctional.

Enzyme-substrate specificity, which is critical to enzyme activity, has been likened to the fit between a lock and key. This analogy is not completely apt because enzymes change shape slightly when they bind to their substrate, almost as if a lock could grasp its key once it had been inserted. This latter description of enzyme-substrate specificity is called the **induced-fit model (Figure 5.5)**.

In some cases, several different enzymes possess active sites that are complementary to various portions of a single substrate molecule. For example, an important precursor metabolite called phosphoenolpyruvic acid (PEP) is the substrate for at least five enzymes; depending on the enzyme involved, various products are produced from PEP. In one catabolic pathway PEP is converted to pyruvic acid, whereas in a particular anabolic pathway PEP is converted to the amino acid phenylalanine.

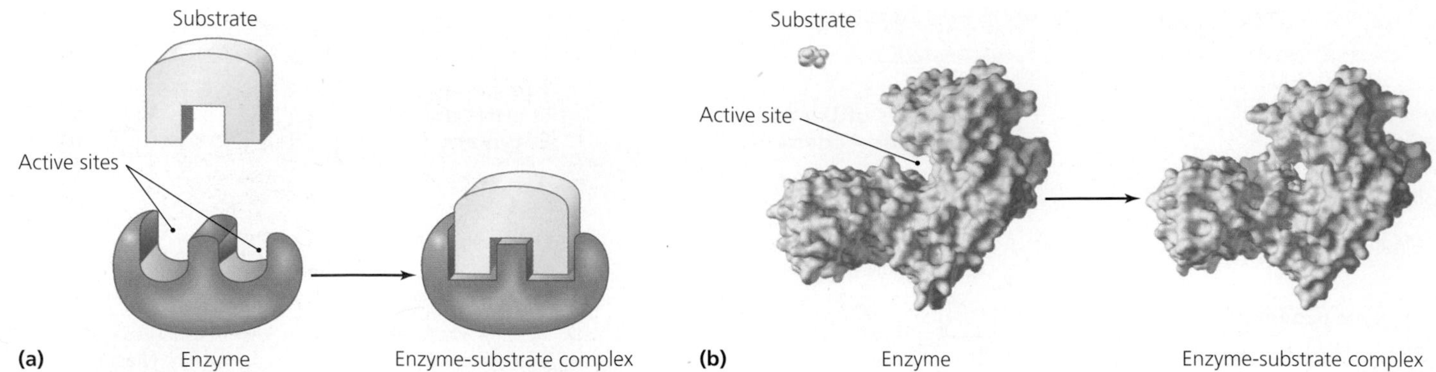

▲ **Figure 5.5 Enzymes fitted to substrates.** **(a)** The induced-fit model of enzyme-substrate interaction. An enzyme's active site is generally complementary to the shape of its substrate, but a perfect fit between them does not occur until the substrate and enzyme bind to form a complex. **(b)** Space-filling models of an enzyme that changes shape to bind its substrate more tightly.

Although the exact ways that enzymes lower activation energy are not known, it appears that several mechanisms are involved. Some enzymes appear to bring reactants into sufficiently close proximity to enable a bond to form, whereas other enzymes change the shape of a reactant, inducing a bond to be broken. In any case, enzymes increase the likelihood that bonds will form or break.

The activity of enzymes is believed to follow the process illustrated in **Figure 5.6**, which depicts the catabolic lysis of a molecule called fructose 1,6-bisphosphate:

1. An enzyme associates with a specific substrate molecule having a shape that is complementary to that enzyme's active site.
2. The enzyme and its substrate bind to form a temporary intermediate compound called an enzyme-substrate complex. The binding of the substrate induces the enzyme to fit the shape of the substrate even more closely.
3. Bonds within the substrate are broken, forming two (and in some other reactions, more than two) products. (This is a catabolic reaction; in anabolic reactions, reactants are linked together to form products.)
4. The enzyme dissociates from the newly formed molecules, which diffuse away from the site of the reaction, and the enzyme resumes its original configuration and is ready to associate with another substrate molecule.

Many factors influence the rate of enzymatic reactions, including temperature, pH, enzyme and substrate concentrations, and the presence of inhibitors. **ANIMATIONS:** ***Enzymes: Steps in a Reaction***

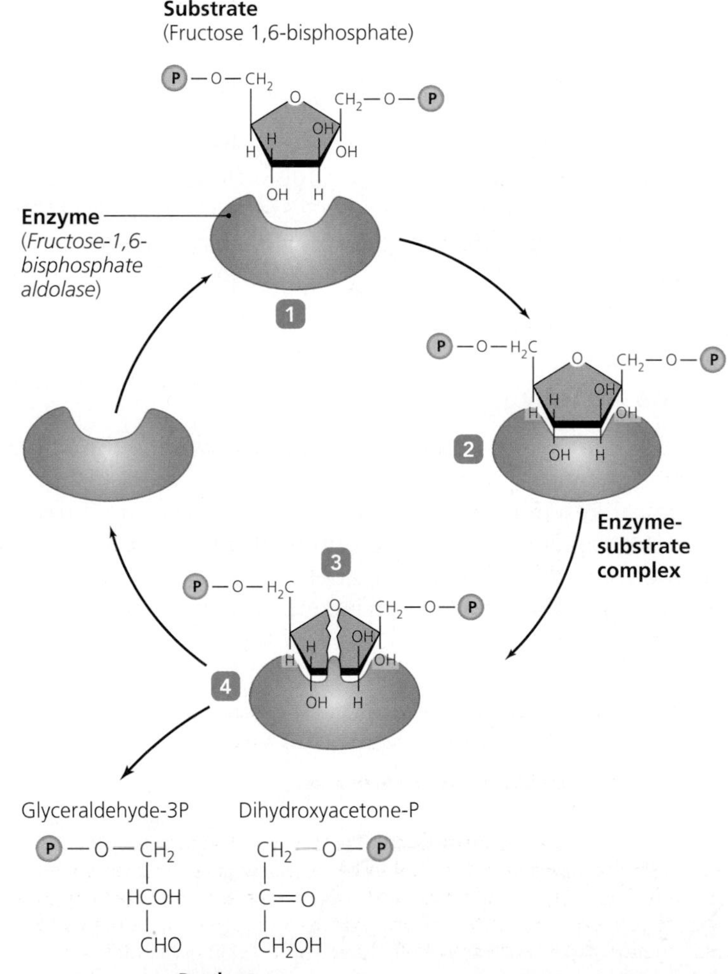

▲ **Figure 5.6 The process of enzymatic activity.** Shown here is the lysis of fructose 1,6-bisphosphate by the enzyme fructose-1,6-bisphosphate aldolase (a catabolic reaction). After the enzyme associates with the substrate (1), the two molecules bind to form an enzyme-substrate complex. (2), As a result of binding, the enzyme's active site is induced to fit the substrate even more closely. Next, bonds within the substrate are broken (3), after which the enzyme dissociates from the two new products (4). The enzyme resumes its initial configuration and is then ready to associate with another substrate molecule. This entire process occurs 14 times per second at 37°C.

Temperature As mentioned, higher temperatures tend to increase the rate of most chemical reactions because molecules are moving faster and collide more frequently, which encourages bonds to form or break. However, this is not entirely true of enzymatic reactions, because the active sites of enzymes change shape as temperature changes. If the temperature rises too high or falls too low, an enzyme is often no longer able to achieve a fit with its substrate.

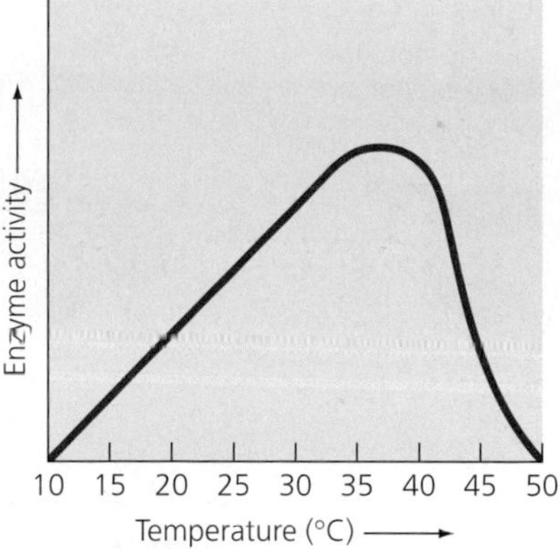

(a) Temperature

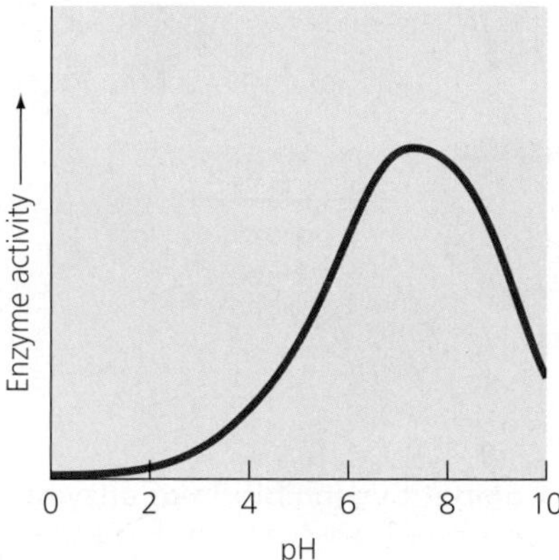

(b) pH

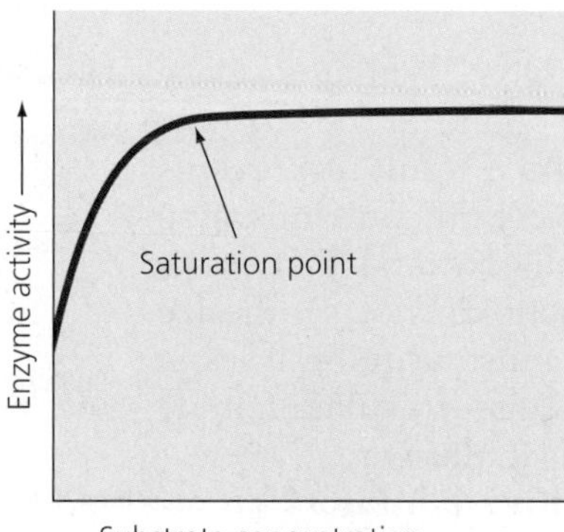

(c) Substrate concentration

▲ **Figure 5.7 Representative effects of temperature, pH, and substrate concentration on enzyme activity.** Effects on each enzyme will vary. **(a)** Rising temperature enhances enzymatic activity to a point, but above some optimal temperature an enzyme denatures and loses function. **(b)** Enzymes typically have some optimal pH, at which point enzymatic activity reaches a maximum. **(c)** At lower substrate concentrations, enzyme activity increases as the substrate concentration increases and as more and more active sites are utilized. At the substrate concentration at which all active sites are utilized, termed the saturation point, enzymatic activity reaches a maximum, and any additional increase in substrate concentration has no effect on enzyme activity. *What is the optimal pH of the enzyme shown in part (b)?*

Figure 5.7 *The enzyme's optimal pH is approximately 7.2.*

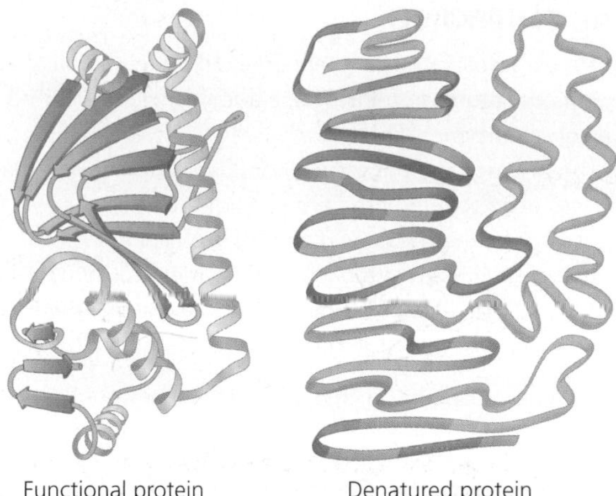

▲ **Figure 5.8 Denaturation of protein enzymes.** Breakage of noncovalent bonds (such as hydrogen bonds) causes the protein to lose its secondary and tertiary structure and become denatured; as a result, the enzyme is no longer functional.

Each enzyme has an optimal temperature for its activity **(Figure 5.7a)**. The optimum temperature for the enzymes in the human body is 37°C, which is normal body temperature. Part of the reason certain pathogens can cause disease in humans is that the optimal temperature for the enzymes in those microorganisms is also 37°C. The enzymes of some other microorganisms, however, function best at much higher temperatures; this is the case for *hyperthermophiles,* organisms that grow best at temperatures above 80°C.

If temperature rises beyond a certain critical point, the noncovalent bonds within an enzyme (such as the hydrogen bonds between amino acids) will break, and the enzyme will **denature** **(Figure 5.8)**. Denatured enzymes lose their specific three-dimensional structure, so they are no longer functional. Denaturation is said to be *permanent* when an enzyme cannot regain its original three-dimensional structure once conditions return to normal, much like the irreversible solidification of the protein albumin when egg whites are cooked and then cooled. In other cases denaturation is *reversible*—the denatured enzyme's noncovalent bonds reform upon the return of normal conditions.

CRITICAL **THINKING**

Explain why hyperthermophiles do not cause disease in humans.

pH Extremes of pH also denature enzymes when ions released from acids and bases interfere with hydrogen bonding and distort and disrupt an enzyme's secondary and tertiary structures. Therefore, each enzyme has an optimal pH **(Figure 5.7b)**.

Changing the pH provides a way to control the growth of unwanted microorganisms by denaturing their proteins. For example, vinegar (acetic acid, pH 3.0) acts as a preservative in dill pickles, and ammonia (pH 11.5) can be used as a disinfectant.

CRITICAL THINKING

In addition to extremes in temperature and pH, other chemical and physical agents denature proteins. These agents include ionizing radiation, alcohol, enzymes, and heavy-metal ions. For example, the first antimicrobial drug, arsphenamine, contained the heavy metal arsenic and was used to inhibit the enzymes of the bacterium *Treponema pallidum*, the causative agent of syphilis.

Given that both human and bacterial enzymes are denatured by heavy metals, how was arsphenamine used to treat syphilis without poisoning the patient? Why is syphilis no longer treated with arsenic-containing compounds?

Enzyme and Substrate Concentration Another factor that determines the rate of enzymatic activity within cells is the concentration of substrate present **(Figure 5.7c)**. As substrate concentration increases, enzymatic activity increases as more and more enzyme active sites bind more and more substrate molecules. Eventually, when all enzyme active sites have bound substrate, the enzymes have reached their saturation point, and the addition of more substrate will not increase the rate of enzymatic activity.

Obviously, the rate of enzymatic activity is also affected by the concentration of enzyme within cells. In fact, one way that organisms regulate their metabolism is by controlling the quantity and timing of enzyme synthesis. In other words, many enzymes are produced in the amounts and at the times they are needed to maintain metabolic activity. Chapter 7 discusses the role of genetic mechanisms in the regulation of enzyme synthesis. Additionally, eukaryotic cells control some enzymatic activities by compartmentalizing enzymes inside membranes so that certain metabolic reactions proceed physically separated from the rest of the cell. For example, white blood cells catabolize phagocytized pathogens using enzymes packaged within lysosomes.

Inhibitors Enzymatic activity can be influenced by a variety of inhibitory substances that block an enzyme's active site. Enzymatic inhibitors, which may be either competitive or noncompetitive, do not denature enzymes.

Competitive inhibitors are shaped such that they fit into an enzyme's active site and thus prevent the normal substrate from binding **(Figure 5.9a)**. However, such inhibitors do not undergo a chemical reaction to form products. Competitive inhibitors can bind permanently or reversibly to an active site. Permanent binding results in permanent loss of enzymatic activity; reversible competition can be overcome by an increase in the concentration of substrate molecules, which increases the likelihood that active sites will be filled with substrate instead of inhibitor **(Figure 5.9b)**. **ANIMATIONS:** ***Enzymes: Competitive Inhibition***

An example of competitive inhibition is the action of sulfanilamide, which has a shape similar to that of para-aminobenzoic acid (PABA).

Sulfanilamide has great affinity for the active site of an enzyme required in the conversion of PABA into folic acid, which is essential for DNA synthesis. Once sulfanilamide is bound to the enzyme, it stays bound. As a result, it prevents synthesis of folic acid. Sulfanilamide effectively inhibits bacteria that make folic acid. Humans do not synthesize folic acid—we must acquire it as a vitamin in our diets—so sulfanilamide does not affect us in this way.

SO_2NH_2 $COOH$

NH_2 NH_2

Sulfanilamide PABA

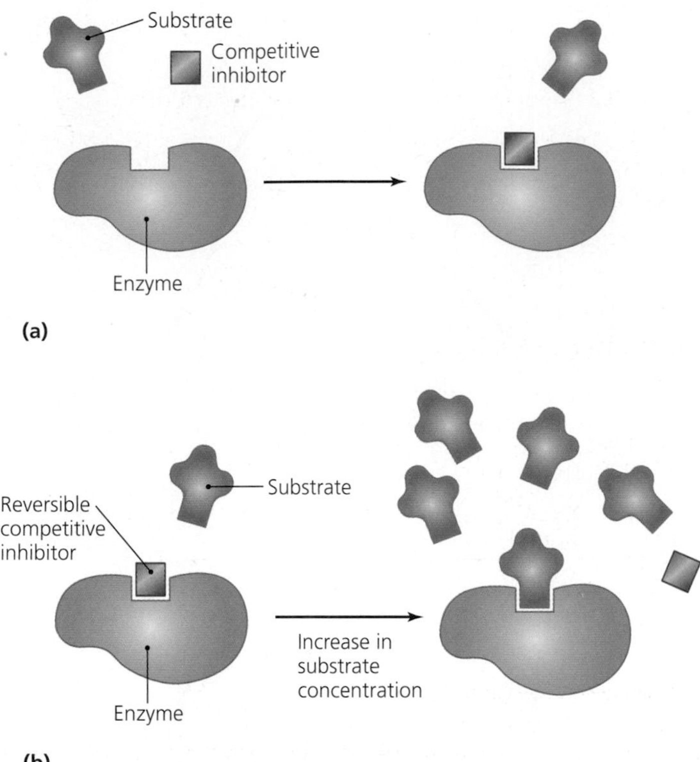

▲ **Figure 5.9 Competitive inhibition of enzyme activity.** **(a)** Inhibitory molecules, which are similar in shape to substrate molecules, compete for and block active sites. **(b)** Reversible inhibition can be overcome by an increase in substrate concentration.

Noncompetitive inhibitors do not bind to the active site but instead prevent enzymatic activity by binding to an *allosteric* (al-ō-stār′ik) *site* located elsewhere on the enzyme. Binding at an allosteric site alters the shape of the active site so that substrate cannot be bound. Allosteric control of enzyme activity can take two forms: inhibitory and excitatory. *Allosteric (noncompetitive) inhibition* halts enzymatic activity in the manner just described **(Figure 5.10a)**. In *excitatory allosteric control,* the binding of certain activator molecules (such as a heavy-metal ion cofactor) to an allosteric site causes a change in shape of the active site, which activates an otherwise inactive enzyme **(Figure 5.10b)**. Some enzymes have several allosteric sites, both inhibitory and excitatory, which allows their function to be closely regulated. **ANIMATIONS:** ***Enzyme-Substrate Interaction: Noncompetitive Inhibition***

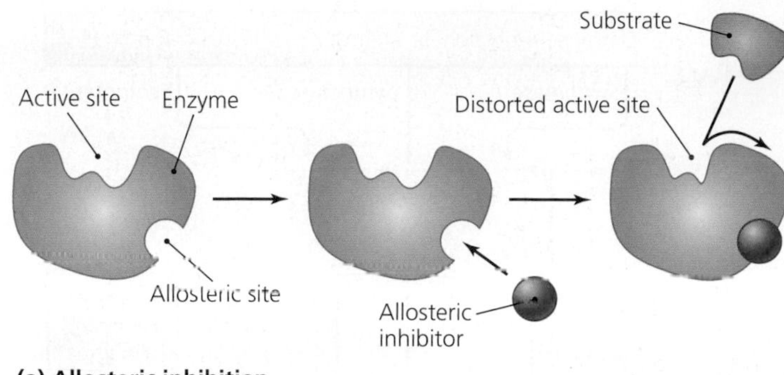

(a) Allosteric inhibition

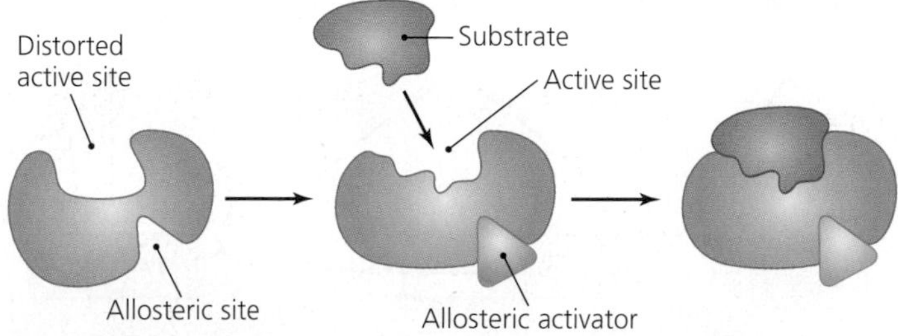

(b) Allosteric activation

▲ **Figure 5.10 Allosteric control of enzyme activity. (a)** Allosteric (noncompetitive) inhibition results from a change in the shape of the active site when an inhibitor binds to an allosteric site. **(b)** Allosteric activation results when the binding of an activator molecule to an allosteric site causes a change in the active site that makes it capable of binding substrate.

Cells often control the action of enzymes through **feedback inhibition** (also called *negative feedback* or *end-product inhibition*). Allosteric feedback inhibition functions in much the way a thermostat controls a heater. As the room gets warmer, a sensor inside the thermostat changes shape and sends an electrical signal that turns off the heater. Similarly, in metabolic feedback inhibition, the end-product of a series of reactions is an allosteric inhibitor of an enzyme in an earlier part of the pathway **(Figure 5.11a)**. Because the product of each reaction in the pathway is the substrate for the next reaction, inhibition of the first enzyme in the series inhibits the entire pathway, thereby saving the cell energy. For example, in *Escherichia coli,* the presence of the amino acid isoleucine allosterically inhibits the first enzyme in the anabolic pathway that produces isoleucine. In this manner, the bacterium prevents the synthesis of isoleucine when the amino acid is already available. When isoleucine is depleted, the enzyme is no longer inhibited, and isoleucine production resumes.

Feedback inhibition can occur in even more complex ways. For instance, even though the first step in the synthesis of the amino acids tyrosine, phenylalanine, and tryptophan is the same—the linkage of phosphoenolpyruvic acid (PEP) and erythrose 4-phosphate to form 3-deoxy-arabino-heptulosonic acid 7-phosphate (DAHAP)—three different synthetase enzymes are involved **(Figure 5.11b)**. In other words, enzymatic reactions convert DAHAP into the three amino acids via three different pathways. As shown, each of the end-product amino acids inhibits only one of the three synthetase enzymes. Therefore, an excess of all three amino acids is needed to completely hinder synthesis of DAHAP and thereby stop the production of these amino acids.

To this point we have viewed the concept of metabolism as a collection of chemical reactions (pathways) that can be categorized as either catabolic (breaking down) or anabolic (building up). Because enzymes are required to lower the activation energy of these reactions, we examined these catalysts in some detail.

Energy is also critical to metabolism, so we examined redox reactions as a means of transferring energy within cells. We saw, for example, that redox reactions and carrier molecules are used to transfer energy from catabolic pathways to ATP, a molecule that stores energy in cells.

We will now consider how cells acquire and utilize metabolites, which are used to synthesize the macromolecules necessary for growth and, eventually, reproduction—the ultimate goal of metabolism. We will also consider in more detail the phosphorylation of ADP to make ATP.

Carbohydrate Catabolism

Learning Objective

✓ In general terms, describe the three stages of aerobic glucose catabolism (glycolysis, the Krebs cycle, and the electron transport chain), including their substrates, products, and net energy production.

Many organisms oxidize carbohydrates as their primary energy source for anabolic reactions. They use glucose most commonly, though other sugars, amino acids, and fats are also utilized, often by first converting them into glucose. Glucose is catabolized via one of two processes: either via *cellular respiration*[3]—a process that results in the complete breakdown of glucose to carbon dioxide and water—or via *fermentation,* which results in organic waste products.

As shown in **Figure 5.12**, both cellular respiration and fermentation begin with *glycolysis* (glī-kol′i-sis), a process that catabolizes a single molecule of glucose to two molecules of pyruvic acid (also called pyruvate) and results in a small amount of ATP production. Respiration then continues via the *Krebs cycle* and the *electron transport chain,* which results in a significant amount of ATP production. Fermentation involves the conversion of pyruvic acid into other organic compounds. Because it lacks the Krebs cycle and electron transport chain, fermentation results in the production of much less ATP than does respiration.

[3]Cellular respiration is often referred to simply as *respiration*, which should not be confused with breathing, also called respiration.

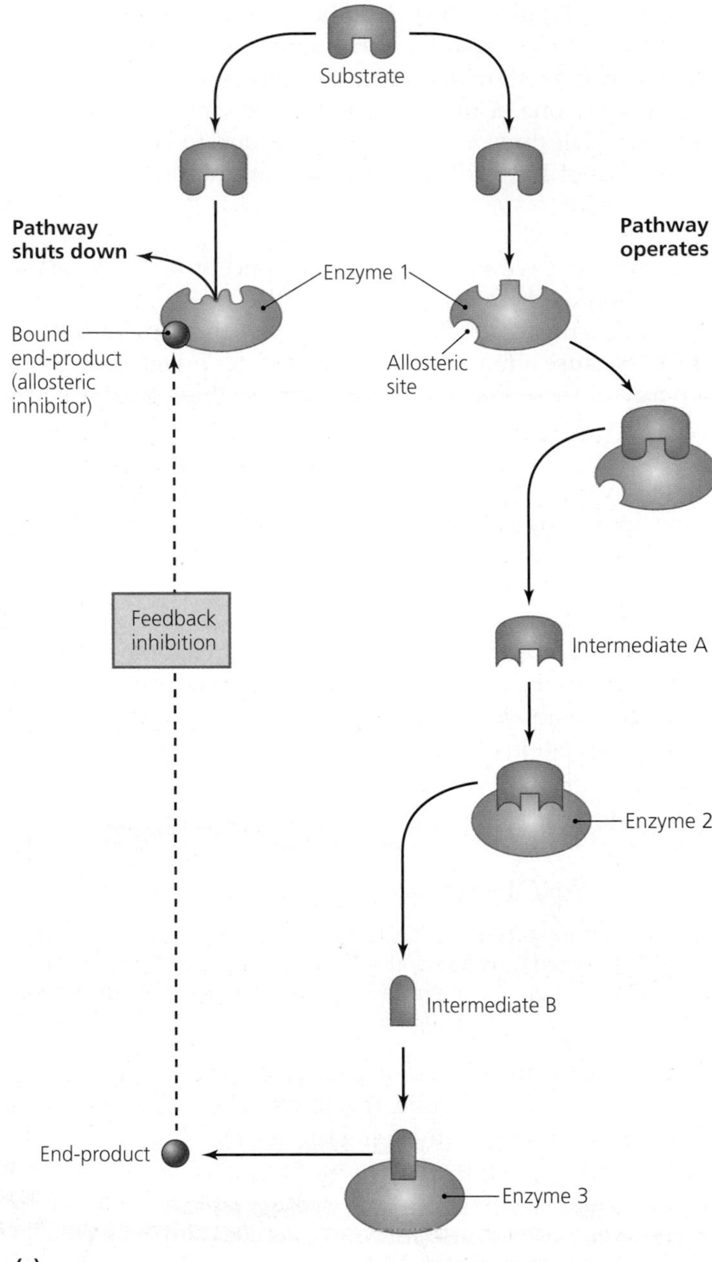

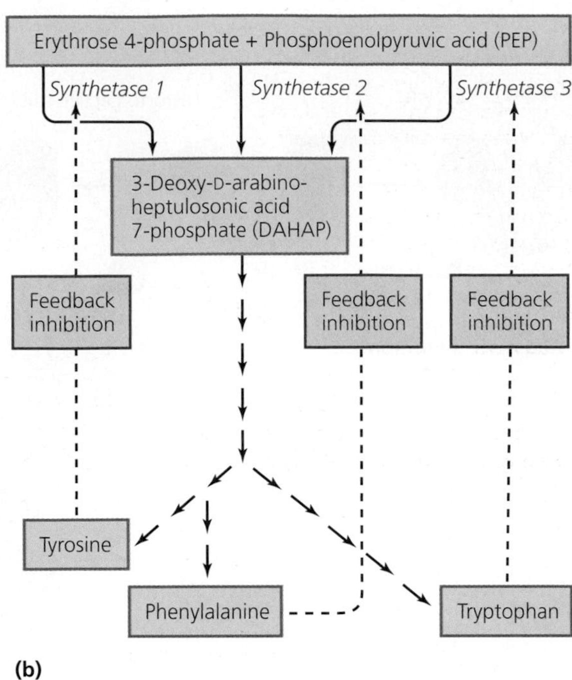

Figure 5.11 Feedback inhibition. (a) The end-product of a metabolic pathway allosterically inhibits the initial step, shutting down the pathway. **(b)** In this example, each of the three end-products (tyrosine, phenylalanine, and trypotophan) inhibits a different synthetase enzyme. An excess of all three end-products is needed to completely hinder the synthesis of 3-deoxy-arabino-heptulosonic acid 7-phosphate. Each arrow indicates an individual metabolic step.

The following is a simplified discussion of glucose catabolism. To help understand the basic reactions in each of the pathways of glucose catabolism, pay special attention to three things: the number of carbon atoms in each of the intermediate products, the relative numbers of ATP molecules produced in each pathway, and the changes in the coenzymes NAD^+ and FAD as they are reduced and then oxidized back to their original forms.

Glycolysis

Glycolysis,[4] also called the *Embden-Meyerhof pathway* after the scientists who discovered it, is the first step in the catabolism of glucose via both respiration and fermentation. Glycolysis occurs in most cells. In general, as its name implies, glycolysis involves the splitting of a six-carbon glucose molecule into two three-carbon sugar molecules. When these three-carbon molecules are oxidized to pyruvic acid, some of the energy released is stored in molecules of ATP. **ANIMATIONS:** ***Glycolysis: Overview***

Glycolysis, which occurs in the cytosol, can be divided into three stages involving a total of 10 steps **(Figure 5.13)**, each of which is catalyzed by its own enzyme. The three stages of glycolysis are:

1. *Energy-investment stage* (steps 1–3). As with money, one must invest before a profit can be made. In this case, the energy in two molecules of ATP is invested to phosphorylate a six-carbon glucose molecule and rearrange its atoms to form fructose 1,6-bisphosphate.
2. *Lysis stage* (steps 4 and 5). Fructose 1,6-bisphosphate is cleaved into glyceraldehyde 3-phosphate (G3P)[5] and dihydroxyacetone phosphate (DHAP). Each of these compounds contains three carbon atoms and is freely convertible into the other.
3. *Energy-conserving stage* (steps 6–10). G3P is oxidized to pyruvic acid, yielding two ATP molecules. DHAP is converted to G3P and also oxidized to pyruvic acid, yielding another two ATP molecules, for a total of four ATP molecules.

[4]From Greek *glukus*, meaning sweet, and *lusis*, meaning to loosen.
[5]G3P is also known as phosphoglyceraldehyde or PGAL.

Figure 5.12 Summary of glucose catabolism. Glucose catabolism begins with glycolysis, which forms pyruvic acid and two molecules of both ATP and NADH. Two pathways branch from pyruvic acid: respiration and fermentation. In aerobic respiration (shown here), the Krebs cycle and the electron transport chain completely oxidize pyruvic acid to CO_2 and H_2O, in the process synthesizing many molecules of ATP. Fermentation results in the incomplete oxidation of pyruvic acid to form organic fermentation products.

To see a 3-D animation on metabolism, go to the Study Area at **www.masteringmicrobiology.com** and watch the *MicroFlix*.

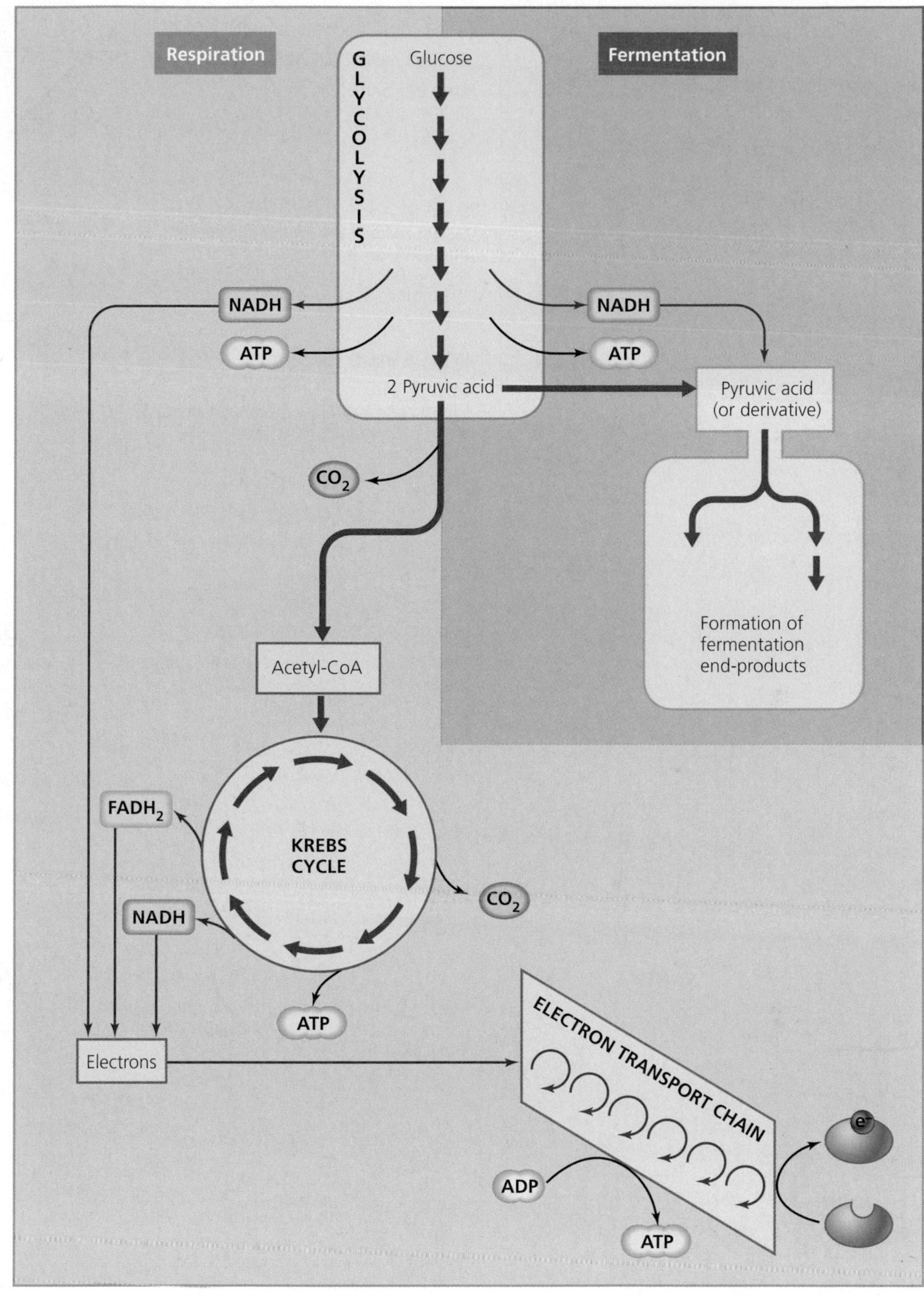

Details of the substrates and enzymes involved are provided in Appendix A on pp. A-4 to A-5. **ANIMATIONS:** ***Glycolysis: Steps***

Our study of glycolysis provides our first opportunity to study *substrate-level phosphorylation* (see steps 1, 3, 7, and 10). Let's examine this important process more closely by considering the 10th and final step of glycolysis.

Each of the two phosphoenolpyruvic acid (PEP) molecules produced in step 9 of glycolysis is a three-carbon compound containing a high-energy phosphate bond. In the presence of a specific holoenzyme (which requires a Mg^{2+} cofactor), the high-energy phosphate in PEP (one substrate) is transferred to an ADP molecule (a second substrate) to form ATP (step 10 and **Figure 5.14**); the direct transfer of the phosphate between the two substrates is the reason the process is called **substrate-level phosphorylation.** A variety of substrate-level phosphorylations occur in metabolism. As you might expect, each type has its own enzyme that recognizes both its substrate molecule and ADP.

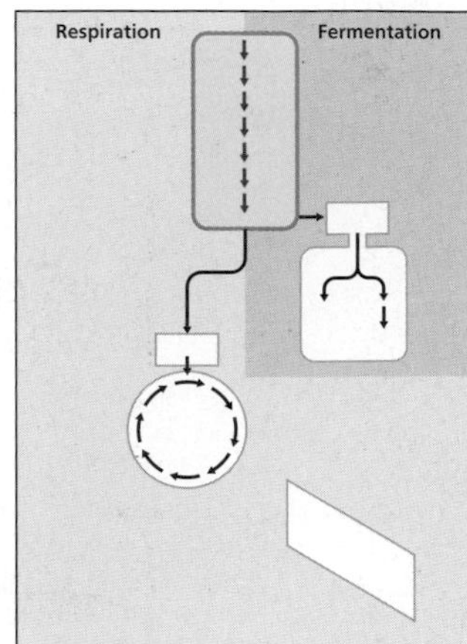

ENERGY-INVESTMENT STAGE

Step 1. Glucose is phosphorylated by ATP to form glucose 6-phosphate.

Steps 2 and 3. The atoms of glucose 6-phosphate are rearranged to form fructose 6-phosphate. Fructose 6-phosphate is phosphorylated by ATP to form fructose 1,6-bisphosphate.

LYSIS STAGE

Step 4. Fructose 1,6-bisphosphate is cleaved to form glyceraldehyde 3-phosphate (G3P) and dihydroxyacetone phosphate (DHAP).

Step 5. DHAP is rearranged to form another G3P.

ENERGY-CONSERVING STAGE

Step 6. Inorganic phosphates are added to the two G3P, and two NAD^+ are reduced.

Step 7. Two ADP are phosphorylated by substrate-level phosphorylation to form two ATP.

Steps 8 and 9. The remaining phosphates are moved to the middle carbons. A water molecule is removed from each substrate.

Step 10. Two ADP are phosphorylated by substrate-level phosphorylation to form two ATP. Two pyruvic acid are formed.

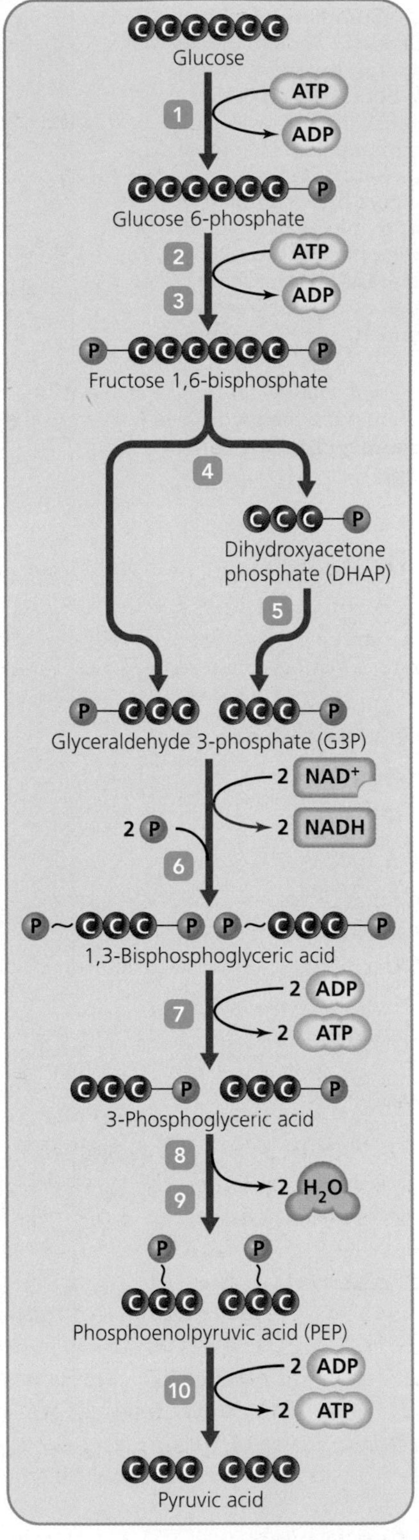

▲ **Figure 5.13 Glycolysis.** Glucose is cleaved and ultimately transformed into two molecules of pyruvic acid in this process (also known as the Embden-Meyerhof pathway). Four ATPs are formed and two ATPs are used, so a net gain of two ATPs results. Two molecules of NAD^+ are reduced to NADH.

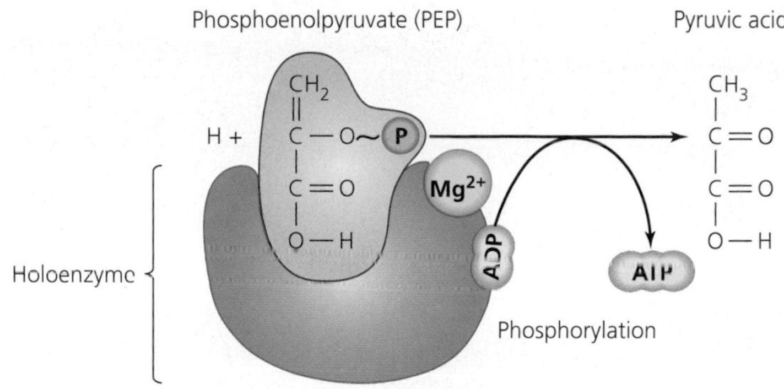

▲ **Figure 5.14 Substrate-level phosphorylation.** High-energy phosphate bonds are transferred from one substrate to another in this process. *What role does Mg^{2+} play in this reaction?*

Figure 5.14 *Mg^{2+} is a cofactor of the enzyme.*

In glycolysis, two ATP molecules are invested by substrate-level phosphorylation to prime glucose for lysis, and four molecules of ATP are produced by substrate-level phosphorylation. Therefore, a net gain of two ATP molecules occurs for each molecule of glucose that is oxidized to pyruvic acid. Glycolysis also yields two molecules of NADH.

Cellular Respiration

Learning Objectives

- ✓ Discuss the roles of acetyl-CoA, the Krebs cycle, and electron transport in carbohydrate catabolism.
- ✓ Contrast electron transport in aerobic and anaerobic respiration.
- ✓ Identify four classes of carriers in electron transport chains.
- ✓ Describe the role of chemiosmosis in oxidative phosphorylation of ATP.

After glucose has been oxidized via glycolysis or one of the alternate pathways considered shortly, a cell uses the resultant pyruvic acid molecules to complete either cellular respiration or fermentation (which we will discuss in a later section). Our topic here—**cellular respiration**—is a metabolic process that involves the complete oxidation of substrate molecules and then production of ATP by a series of redox reactions. The three stages of cellular respiration are: (1) synthesis of acetyl-CoA, (2) the Krebs cycle, and (3) a final series of redox reactions, called an electron transport chain, that passes electrons to a chemical not derived from the cell's metabolism.

Synthesis of Acetyl-CoA

Before pyruvic acid (generated by glycolysis or an alternate pathway) can enter the Krebs cycle for respiration, it must first be converted to *acetyl-coenzyme A* or **acetyl-CoA** (as'e-til kō-ā'; see Figure 5.12). Enzymes remove one carbon from pyruvic acid as CO_2 and join the remaining two-carbon acetate to *coenzyme A*

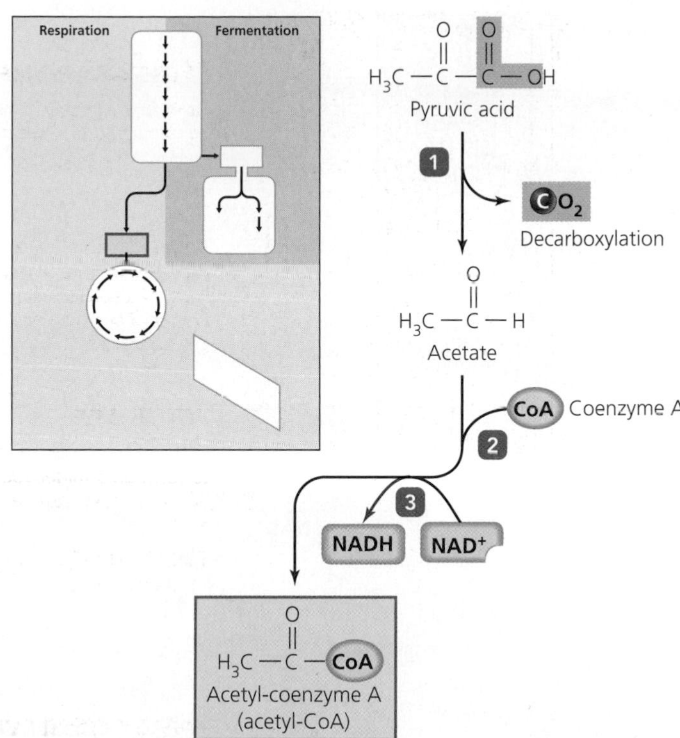

▲ **Figure 5.15 Formation of acetyl-CoA.** The responsible enzyme acts in a stepwise manner to **1** remove CO_2 from pyruvic acid, **2** attach the remaining two-carbon acetate to coenzyme A, and **3** reduce a molecule of NAD^+ to NADH.

with a high-energy bond **(Figure 5.15)**. The removal of CO_2, called *decarboxylation,* requires a coenzyme derived from the vitamin thiamine. One molecule of NADH is also produced during this reaction.

Recall that two molecules of pyruvic acid were derived from each molecule of glucose. Therefore, at this stage, two molecules of acetyl-CoA, two molecules of CO_2, and two molecules of NADH are produced.

The Krebs Cycle

At this point in the catabolism of a molecule of glucose, a great amount of energy remains in the bonds of acetyl-CoA. The **Krebs cycle**[6] is a series of eight enzymatically catalyzed reactions that transfer much of this stored energy to the coenzymes NAD^+ and FAD. The two carbon atoms in acetate are oxidized, and the coenzymes are reduced. The Krebs cycle, which occurs in the cytosol of prokaryotes and in the matrix of mitochondria in eukaryotes, is diagrammed in **Figure 5.16** and presented in more detail in Appendix A on p. A-8. It is also known as the *tricarboxylic acid (TCA) cycle,* because many of its compounds have three carboxyl (—COOH) groups, and as the *citric acid cycle,* for the first compound formed in the cycle.

[6]Named for biochemist Hans Krebs, who elucidated its reactions in the 1940s.

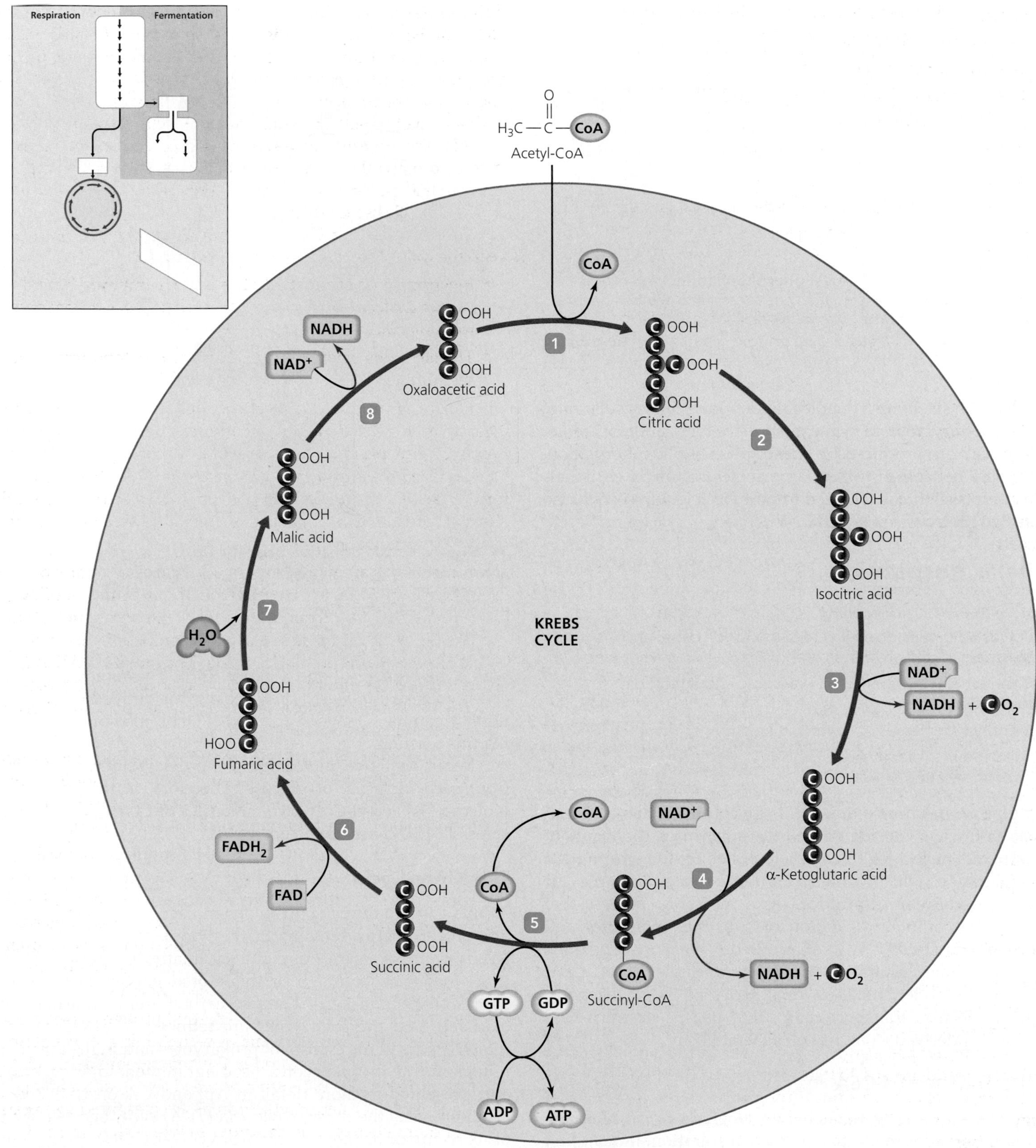

Figure 5.16 The Krebs cycle. 1 Acetyl-CoA enters the Krebs cycle by joining with oxaloacetic acid to form citric acid and coenzyme A. 2–4 Two oxidations and decarboxylations and the addition of coenzyme A yield succinyl-CoA. 5 Substrate-level phosphorylation produces ATP and regenerates coenzyme A. 6–8 Further oxidations and rearrangements regenerate oxaloacetic acid, and the cyle can begin anew.

There are six types of reactions in the Krebs cycle:

- Anabolism of citric acid (step 1)
- Isomerization reaction (step 2)
- Redox reactions (steps 3, 4, 6, and 8)
- Decarboxylations (steps 3 and 4)
- Substrate-level phosphorylation (step 5)
- Hydration reaction (step 7)

In the first step of the Krebs cycle, the splitting of the high-energy bond between acetate and coenzyme A releases enough energy to enable the binding of the freed two-carbon acetate to a four-carbon compound called oxaloacetic acid, forming the six-carbon compound citric acid.

As you study Figure 5.16, notice that after isomerization (step 2), the decarboxylations of the Krebs cycle release two molecules of CO_2 for each acetyl-CoA that enters (steps 3 and 4). Thus, for every two carbon atoms that enter the cycle, two are lost to the environment. At this juncture in the respiration of a molecule of glucose, all six carbon atoms have been lost to the environment: two as CO_2 molecules produced in decarboxylation of two molecules of pyruvic acid to form two acetyl-CoA molecules, and four in CO_2 molecules produced in decarboxylations in the *two* turns through the Krebs cycle. (One molecule of acetyl-CoA enters the cycle at a time.)

A small amount of ATP is also produced in the Krebs cycle. For every two molecules of acetyl-CoA that pass through the Krebs cycle, two molecules of ATP are generated by substrate-level phosphorylation (step 5). A molecule of guanosine triphosphate (GTP), which is similar to ATP, can serve as an intermediary in this process.

Redox reactions reduce FAD to $FADH_2$ (step 6) and NAD^+ to NADH (steps 3, 4, and 8), so that for every two molecules of acetyl-CoA that move through the cycle, six molecules of NADH and two of $FADH_2$ are formed. In the Krebs cycle, little energy is captured directly in high-energy phosphate bonds, but much energy is transferred via electrons to NADH and $FADH_2$. These coenzymes are the most important molecules of respiration because they carry a large amount of energy that is subsequently used to phosphorylate ADP to ATP. **ANIMATIONS:** ***Krebs Cycle: Overview, Steps***

Electron Transport

Some scientists estimate that each day an average human synthesizes his or her own weight in ATP molecules and uses them for metabolism, responsiveness, growth, and cell reproduction. ATP turnover in prokaryotes is relatively as copious. The most significant production of ATP does not occur through glycolysis or the Krebs cycle, but rather through the stepwise release of energy from a series of redox reactions known as an **electron transport chain** (**Figure 5.17**). **ANIMATIONS:** ***Electron Transport Chain: Overview***

An electron transport chain consists of a series of membrane-bound carrier molecules that pass electrons from one to another and ultimately to a *final electron acceptor.* Typically, as we have seen, electrons come from the catabolism of an organic molecule such as glucose; however, microorganisms called *lithotrophs* (lith'ō-trōfs) acquire electrons from inorganic sources such as H_2, NO^{2-}, or Fe^{2+}. Chapter 6 discusses lithotrophs further. In any case, carrier molecules pass electrons down the chain to the final acceptor like firefighters of old passed buckets of water from one to another until the last one threw the water on a fire. As with a bucket brigade, the final step of electron transport is irreversible. Energy from the electrons is used to actively transport (pump) protons (H^+) across the membrane, establishing a *proton gradient* that generates ATP via a process called *chemiosmosis,* which we will discuss shortly.

To avoid getting lost in the details of electron transport, keep the following critical concepts in mind:

- Electrons pass sequentially from one membrane-bound carrier molecule to another and eventually to a final acceptor molecule.
- The electrons' energy is used to pump protons across the membrane.

Electron transport chains are located in the inner mitochondrial membranes (cristae) of eukaryotes and in the cytoplasmic membrane of prokaryotes (**Figure 5.18**). Though NADH and $FADH_2$ donate electrons as hydrogen atoms (electrons and protons), many carrier molecules pass only the electrons down the chain. There are four categories of carrier molecules in electron transport chains. They are:

- *Flavoproteins* are integral membrane proteins, many of which contain flavin, a coenzyme derived from riboflavin (vitamin B_2). One form of flavin is flavin mononucleotide (FMN), which is the initial carrier molecule of electron transport chains of mitochondria. The familiar FAD is a coenzyme for other flavoproteins. Like all carrier molecules in the electron transport chain, flavoproteins alternate between the reduced and oxidized states.
- *Ubiquinones* (yū-bik'wi-nōns) are lipid-soluble, nonprotein carriers that are so named because they are ubiquitous in all cells. Ubiquinones are derived from vitamin K. In mitochondria, the ubiquinone is called *coenzyme Q.*
- *Metal-containing proteins* are a mixed group of integral proteins with a wide-ranging number of iron, sulfur, and copper atoms that can alternate between the reduced and oxidized states. Iron-sulfur proteins occur in various places in electron transport chains of various organisms. Copper proteins are found only in electron transport chains involved in photosynthesis (discussed shortly).
- *Cytochromes* (sī'tō-krōms) are integral proteins associated with *heme,* which is the same iron-containing, nonprotein, pigmented molecule found in the hemoglobin of blood. Iron can alternate between a reduced (Fe^{2+}) state and an oxidized (Fe^{3+}) state. Cytochromes are identified by letters and numbers based on the order in which they were identified, so their sequence in electron transport chains does not always seem logical. **ANIMATIONS:** ***Electron Transport Chain: The Process***

The carrier molecules in electron transport chains are diverse—bacteria typically have different carrier molecules arranged in different sequences than do archaea or the

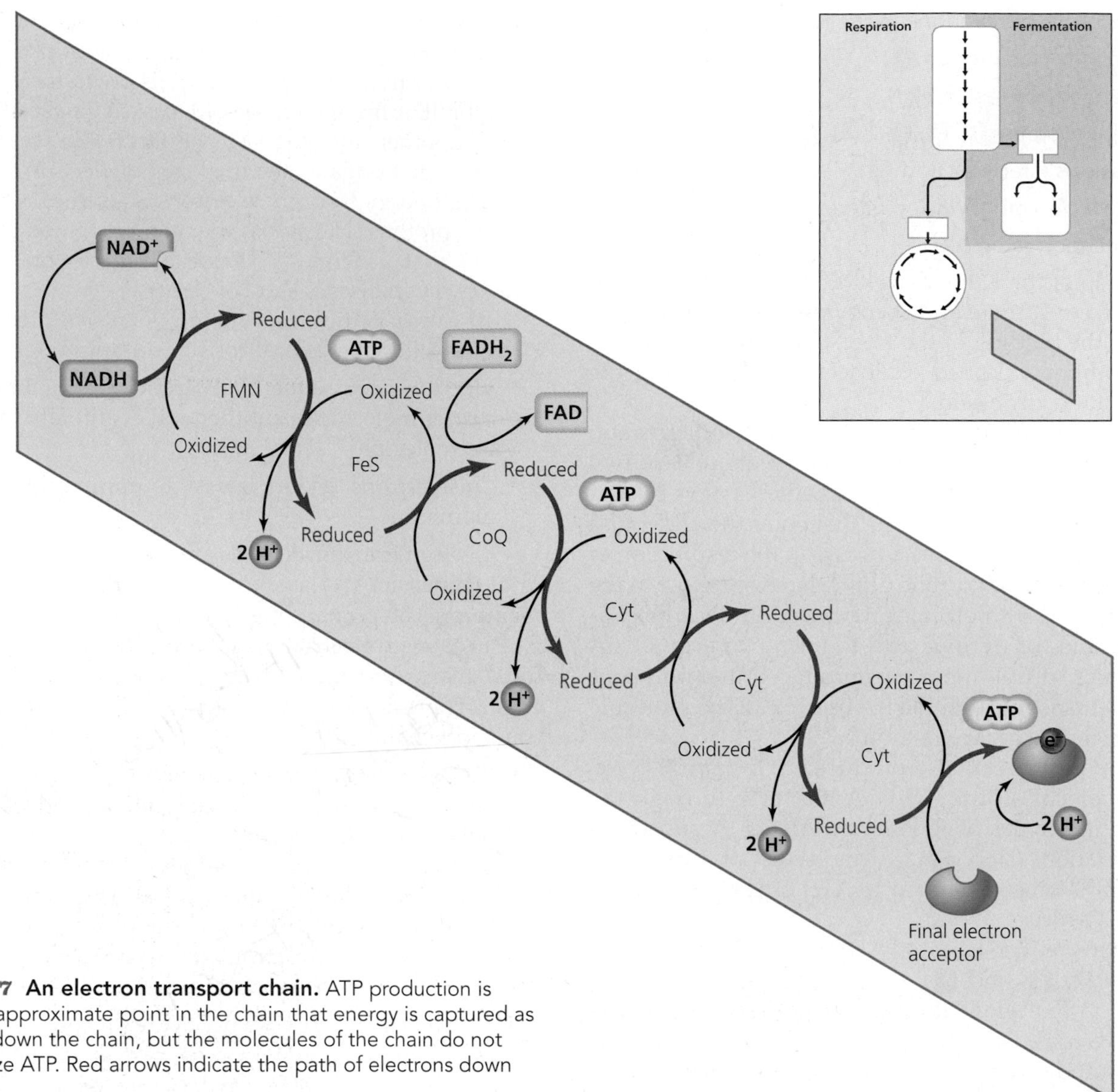

▲ **Figure 5.17 An electron transport chain.** ATP production is indicated at the approximate point in the chain that energy is captured as electrons move down the chain, but the molecules of the chain do not actually synthesize ATP. Red arrows indicate the path of electrons down the chain.

mitochondria of eukaryotes. Even among bacteria, the makeup of carrier molecules can vary. For example, the pathogens *Neisseria* (nī-se'rē-ă) and *Pseudomonas* (soo-dō-mō'nas) contain two cytochromes, *a* and a_3—together called *cytochrome oxidase*—that oxidize cytochrome *c;* such bacteria are said to be *oxidase positive.* In contrast, other bacterial pathogens, such as *Escherichia, Salmonella* (sal'mŏ-nel'ă), and *Proteus* (prō' tē-ŭs), lack cytochrome oxidase and are thus considered to be *oxidase negative.* Some bacteria including *E. coli* can even vary their carrier molecules under different environmental conditions.

Electrons carried by NADH enter the transport chain at a flavoprotein, and those carried by $FADH_2$ are introduced via a ubiquinone. This explains why more molecules of ATP are generated from NADH than from $FADH_2$. Researchers do not agree on which carrier molecules are the actual proton pumps, nor on the number of protons that are pumped. Figure 5.18 shows one possibility.

In some organisms, the final electron acceptors are oxygen atoms, which, with the addition of hydrogen ions, generate H_2O; these organisms conduct **aerobic**[7] **respiration** and are called *aerobes.* **Highlight: Glowing Bacteria** on p. 142 describes an unusual aerobic bacterial electron transport system that produces light instead of ATP.

Other organisms, called *anaerobes,*[8] use other inorganic molecules (or rarely an organic molecule) instead of oxygen as the final electron acceptor and perform **anaerobic respiration.** The anaerobic bacterium *Desulfovibrio* (dē'sul-fō-vib-rē-ō), for example, reduces sulfate (SO_4^{2-}) to hydrogen sulfide gas (H_2S), whereas other anaerobes in the genera *Bacillus* (ba-sil'ŭs) and *Pseudomonas* utilize nitrate (NO_3^-) to produce nitrite ions (NO^{2-}), nitrous oxide (N_2O), or nitrogen gas (N_2). Some prokaryotes—particularly archaea called methanogens—reduce carbonate

[7]From Greek *aer,* meaning air (that is, oxygen), and *bios,* meaning life.
[8]The Greek prefix *an* means not.

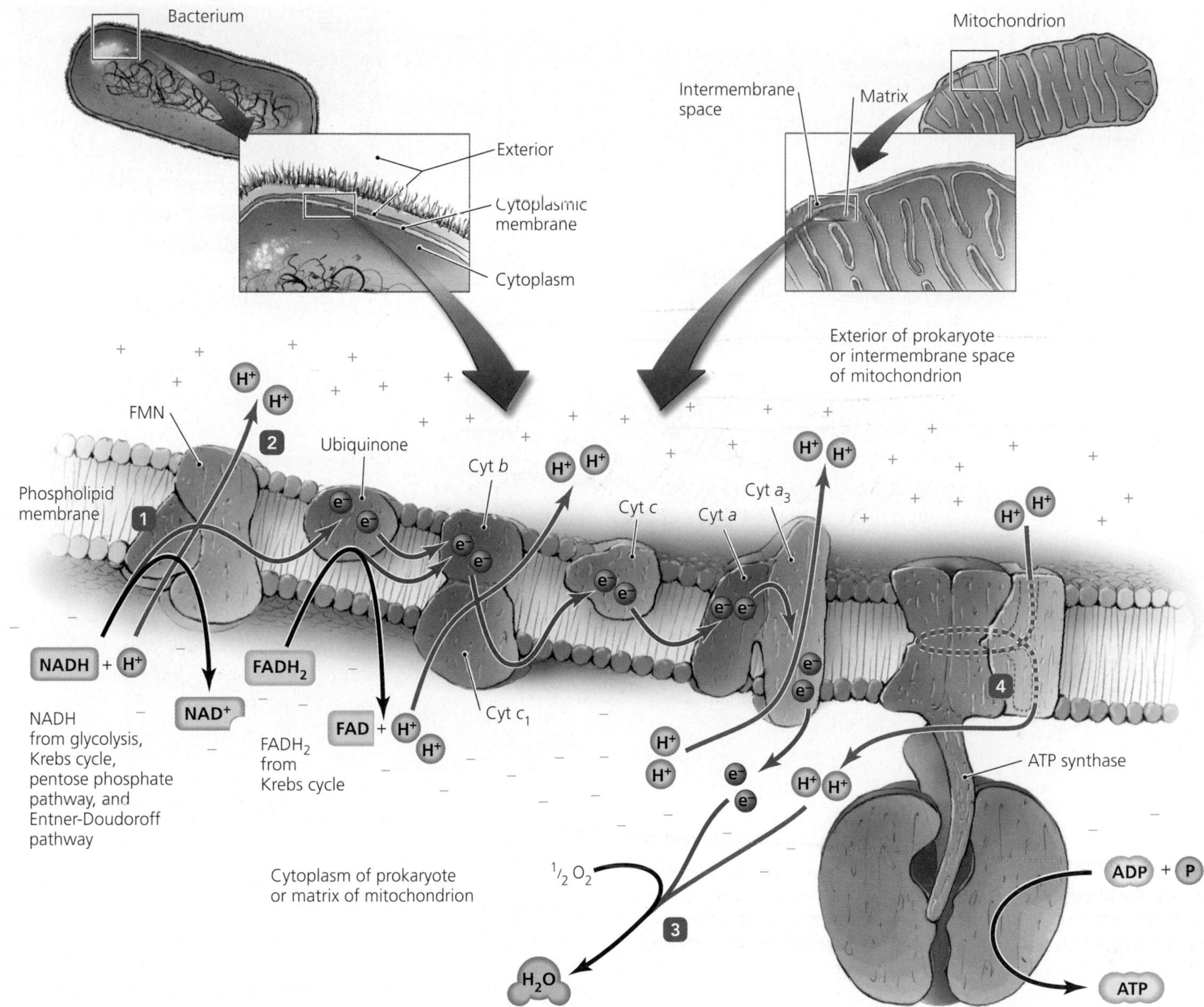

▲ **Figure 5.18 One possible arrangement of an electron transport chain.** Electron transport chains are located in the cytoplasmic membranes of prokaryotes and in the inner membranes of mitochondria in eukaryotes. The exact types and sequences of carrier molecules in electron transport chains vary among organisms. As electrons move down the chain (red arrows) (1), their energy is used to pump protons (H^+) across the membrane (2). Eventually the electrons pass to a final acceptor, in this case oxygen (3). The protons then flow through ATP synthase (ATPase), which synthesizes ATP (4). Approximately one molecule of ATP is generated for every two protons that cross the membrane. *Why is it essential that electron carriers of an electron transport chain be membrane bound?*

Figure 5.18 *The carrier molecules of electron transport chains are membrane bound for at least two reasons: (1) Carrier molecules need to be in fairly close proximity so they can pass electrons between each other. (2) The goal of the chain is the pumping of H^+ across a membrane to create a proton gradient; without a membrane, there could be no gradient.*

HIGHLIGHT

GLOWING BACTERIA

Bacteria in the genus *Photobacterium* possess an interesting electron tansport chain that generates light instead of ATP. These organisms can switch the flow of electrons from a standard electron transport chain (composed of cytochromes, proton pumps, and O_2 as the final electron acceptor) to an alternate chain. Whereas the standard chain establishes a proton gradient that is used to synthesize ATP, the alternate chain uncouples electron transport from ATP synthesis: instead of transferring electrons to proton pumps, the alternate chain shunts the electrons to the coenzyme flavin mononucleotide (FMN). Then, in the presence of an enzyme called luciferase and a long-chain hydrocarbon, the alternate chain emits light as it transfers electrons to O_2 (see the diagram). The exact mechanism of bioluminescence is not known, but both FMN and the hydrocarbon are oxidized as oxygen is reduced.

Interestingly, free-living *Photobacterium* are usually not bioluminescent. It is primarily when these bacteria colonize the tissues of marine animals such as squid and fish that they use their light-generating pathway. The animals gain from the association because the light produced serves as an attractant for mates and a warning against predators. One species, the "flashlight fish" (*Kryptophanaron alfredi*), has a special organ near its mouth that is specially adapted for the growth of luminescent bacteria. Enough light is generated from millions of bacteria that the fish can navigate over coral reefs at night and attract prey to their light. The light organ even has a membrane that descends like an eyelid to control the amount of light emitted.

It is not clear what the bacteria gain from this association. Presumably, the protection and nutrients the bacteria gain from the fish make up for the enormous metabolic cost the bacteria incur in the form of lost ATP synthesis.

▲ *Kryptophanaron alfredi.*

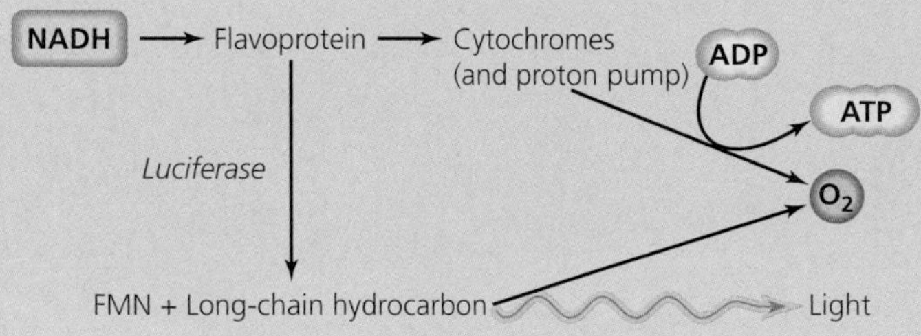

(CO_3^{2-}) to methane gas (CH_4). Laboratory technologists test for products of anaerobic respiration, such as nitrite, to aid in identification of some species of bacteria. As discussed more fully in Chapter 26, anaerobic respiration is also critical for the recycling of nitrogen and sulfur in nature. **ANIMATIONS:** ***Electron Transport Chain: Factors Affecting ATP Yield***

CRITICAL THINKING

Figure 5.18 illustrates events in aerobic respiration where oxygen acts as the final electron acceptor to yield water. How would the figure be changed to reflect anaerobic respiration?

In summary, a number of redox reactions in glycolysis or alternate pathways and in the Krebs cycle strip electrons, which carry energy, from glucose molecules and transfer them to molecules of NADH and $FADH_2$. In turn, NADH and $FADH_2$ pass the electrons to an electron transport chain. As the electrons move down the electron transport chain, proton pumps use the electrons' energy to actively transport protons (H^+) across the membrane, creating a proton concentration gradient.

Recall, however, that the significance of electron transport is not merely that it pumps protons across a membrane, but that it ultimately results in the synthesis of ATP. We turn now to the process by which cells synthesize ATP using the proton gradient.

Chemiosmosis

Chemiosmosis is a general term for the use of ion gradients to generate ATP; that is, ATP is synthesized utilizing energy released by the flow of ions down their electrochemical gradient across a membrane. The term should not be confused with osmosis of water. To understand chemiosmosis, we need to review several concepts from Chapter 3 concerning diffusion and phospholipid membranes.

Recall that chemicals diffuse from areas of high concentration to areas of low concentration, and toward an electrical charge opposite their own. We call the composite of differences in concentration and charge an *electrochemical gradient.* Chemicals diffuse down their electrochemical gradients. Recall as well that membranes of cells and organelles are impermeable to most chemicals unless a specific protein channel allows their passage across the membrane. A membrane maintains an electrochemical gradient by keeping one or more chemicals in a higher concentration on one side. The blockage of diffusion creates potential energy, like water behind a dam.

Chemiosmosis uses the potential energy of an electrochemical gradient to phosphorylate ADP into ATP. Even though chemiosmosis is a general principle with relevance to both *oxidative phosphorylation* and *photophosphorylation,* here we consider it as it relates to oxidative phosphorylation.

As we have seen, cells use the energy released in the redox reactions of electron transport chains to actively transport

TABLE 5.3
Summary of Ideal Prokaryotic Aerobic Respiration of One Molecule of Glucose

Pathway	ATP Produced	ATP Used	NADH Produced	$FADH_2$ Produced
Glycolysis	4	2	2	0
Synthesis of acetyl-CoA and Krebs cycle	2	0	8	2
Electron transport chain	34	0	0	0
Total	40	2		
Net total	**38**			

protons (H^+) across a membrane. Theoretically, an electron transport chain pumps three pairs of protons for each pair of electrons contributed by NADH, and pumps two pairs of protons for each electron pair delivered by $FADH_2$. This difference results from the fact that $FADH_2$ delivers electrons farther down the chain than does NADH; therefore, energy carried by $FADH_2$ is used to transport one-third fewer protons (see Figure 5.17). Because lipid bilayers are impermeable to protons, the transport of protons to one side of the membrane creates an electrochemical gradient known as a **proton gradient,** which has potential energy known as a *proton motive force.*

Protons, propelled by the proton motive force, flow down their electrochemical gradient through protein channels, called **ATP synthases (ATPases),** that phosphorylate molecules of ADP to ATP (see Figure 5.18 4). Such phosphorylation is called **oxidative phosphorylation** because the proton gradient is created by the oxidation of components of an electron transport chain.

In the past, scientists attempted to calculate the exact number of ATP molecules synthesized per pair of electrons that travel down an electron transport chain. However, it is now apparent that phosphorylation and oxidation are not directly coupled. In other words, chemiosmosis does not require exact constant relationships among the number of molecules of NADH and $FADH_2$ reduced, the number of electrons that move down an electron transport chain, and the number of molecules of ATP that are synthesized. Additionally, cells use proton gradients for other cellular processes, including active transport and bacterial flagellar motion, so not every transported electron results in ATP production.

Nevertheless, about 34 molecules of ADP per molecule of glucose are oxidatively phosphorylated to ATP via chemiosmosis: three from each of the 10 molecules of NADH generated from glycolysis, the synthesis of acetyl-CoA, and the Krebs cycle, and two from each of the two molecules of $FADH_2$ generated in the Krebs cycle. Given that glycolysis produces a net two molecules of ATP by substrate-level phosphorylation, and that the Krebs cycle produces two more, the complete aerobic oxidation of one molecule of glucose by a prokaryote can theoretically yield a net total of 38 molecules of ATP (Table 5.3). The theoretical net maximum for eukaryotic cells is generally given as 36 molecules of ATP because the energy from two ATP molecules is required to transport NADH generated by glycolysis in the cytoplasm into the mitochondria.

CRITICAL **THINKING**

Suppose you could insert a tiny pH probe into the space between mitochondrial membranes. Would the pH be above or below 7.0? Why?

Alternatives to Glycolysis

Learning Objective

✓ Compare the pentose phosphate pathway and the Entner-Doudoroff pathway with glycolysis in terms of energy production and products.

The initial part of the catabolism of glucose can also proceed via two alternate pathways: the pentose phosphate pathway and the Entner-Doudoroff pathway. Though they yield fewer molecules of ATP than glycolysis, these alternate pathways reduce coenzymes and yield substrate metabolites that are needed in anabolic pathways. Next we briefly examine each of these alternate pathways.

Pentose Phosphate Pathway

The **pentose phosphate pathway** (sometimes called the *phosphogluconate pathway*) is named for the phosphorylated pentose (five-carbon) sugars—ribulose, xylulose, and ribose—that are formed from glucose 6-phosphate by enzymes in the pathway **(Figure 5.19).** The pentose phosphate pathway is primarily used for the production of precursor metabolites used in anabolic reactions, including the synthesis of nucleotides for nucleic acids, of certain amino acids, and of glucose by *photosynthesis* (described in a later section). The pathway reduces two molecules of $NADP^+$ to NADPH and nets a single molecule of ATP from each molecule of glucose. NADPH is a necessary coenzyme for anabolic enzymes that synthesize DNA nucleotides, steroids, and fatty acids.

Page A-6 in Appendix A shows the details of the substrates and enzymes of the pentose phosphate pathway.

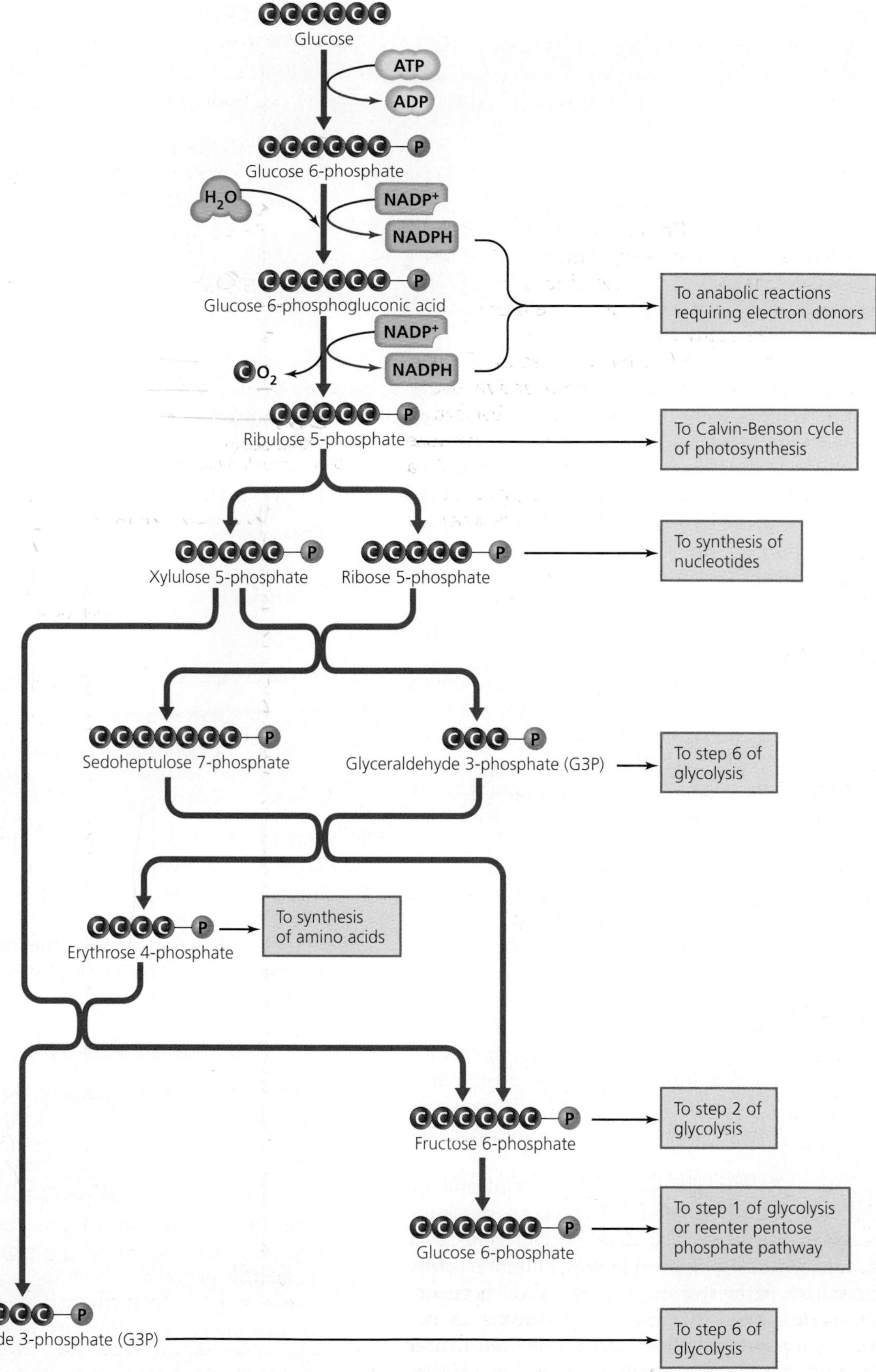

▲ **Figure 5.19 The pentose phosphate pathway.** Energy captured during the pentose phosphate pathway is less than that from glycolysis, but the pathway produces ribose 5-phosphate and erythrose 4-phosphate, two metabolites necessary for synthesis of nucleotides and certain amino acids, respectively. Ribulose 5-phosphate, necessary for glucose synthesis in photosynthetic organisms, and NADPH, an electron carrier for anabolic reactions, are also produced.

CRITICAL THINKING

Examine the biosynthetic pathway for the production of the amino acids tryptophan, tyrosine, and phenylalanine in Figure 5.11b. Where do the initial reactants (erythrose 4-phosphate and PEP) originate?

Entner-Doudoroff Pathway

Many bacteria use glycolysis and the pentose phosphate pathway, but a few substitute the **Entner-Doudoroff pathway** **(Figure 5.20)** for glycolysis. This pathway, named for its discoverers, is a series of reactions that catabolize glucose to pyruvic acid using different enzymes from those used in either glycolysis or the pentose phosphate pathway.

Among organisms, only a very few bacteria use the Entner-Doudoroff pathway. These include the Gram-negative bacterium *Pseudomonas aeruginosa* and the Gram-positive bacterium *Enterococcus faecalis* (en-ter-ō-kok'ŭs fē-kă'lis). Like the pentose phosphate pathway, the Entner-Doudoroff pathway nets only a single molecule of ATP for each molecule of glucose, but it does yield precursor metabolites and NADPH. The latter is unavailable from glycolysis.

Page A-7 in Appendix A shows the Entner-Doudoroff pathway in more detail.

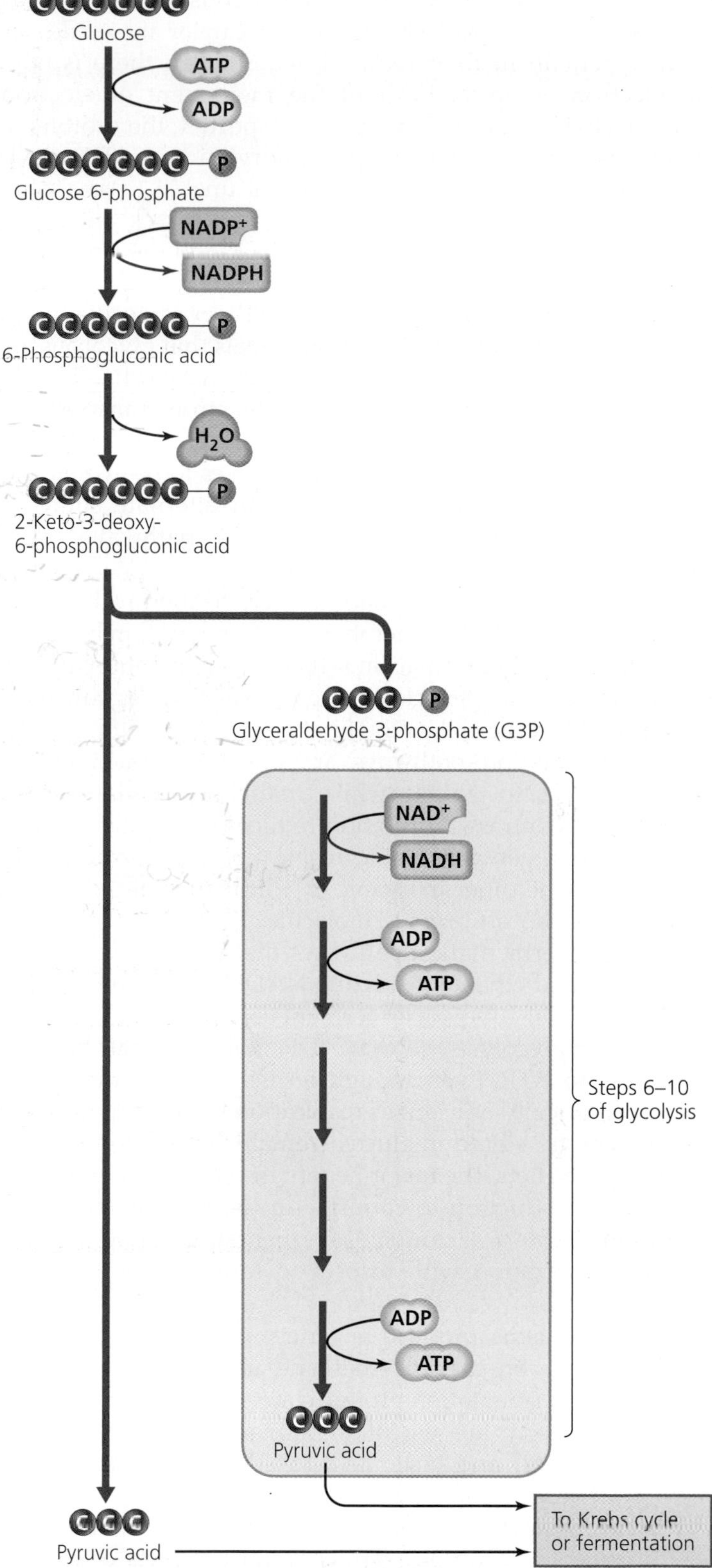

▲ **Figure 5.20** **Entner-Doudoroff pathway.** A few bacteria use this alternate pathway for the oxidation of glucose to pyruvic acid. *What potential pathogens use this pathway?*

CRITICAL THINKING

Even though *Pseudomonas aeruginosa* and *Enterococcus faecalis* usually grow harmlessly, they can cause disease. Because these bacteria use the Entner-Doudoroff pathway instead of glycolysis to catabolize glucose, investigators can use clinical tests that provide evidence of the Entner-Doudoroff pathway to identify the presence of these potential pathogens.

Suppose you were able to identify the presence of any specific organic compound. Name a substrate molecule you would find in *Pseudomonas* and *Enterococcus* cells, but not in human cells.

Fermentation

Learning Objectives

- ✓ Describe fermentation, and contrast it with respiration.
- ✓ Identify three useful end-products of fermentation, and explain how fermentation reactions are used in the identification of bacteria.
- ✓ Discuss the use of biochemical tests for metabolic enzymes and products in the identification of bacteria.

Sometimes cells cannot completely oxidize glucose by cellular respiration. For instance, they may lack sufficient final electron acceptors, as is the case, for example, for an aerobic bacterium in the anaerobic environment of the colon. Electrons cannot flow down an electron transport chain unless oxidized carrier molecules are available to receive them. Our bucket brigade analogy can help clarify this point.

Suppose the last person in the brigade did not throw the water, but instead held onto two full buckets. What would happen? The entire brigade would soon consist of firefighters holding full buckets of water. The analogous situation occurs in an electron transport chain: All the carrier molecules are forced to remain in their reduced states when there is not a final electron acceptor. Without the movement of electrons down the chain, protons cannot be transported, the proton motive force is lost, and oxidative phosphorylation of ADP to ATP ceases. Without sufficient ATP, a cell is unable to anabolize, grow, or divide.

ATP could be synthesized in glycolysis and the Krebs cycle by substrate-level phosphorylation. After all, together these pathways produce four molecules of ATP per molecule of glucose. However, careful consideration reveals that glycolysis and the Krebs cycle require a continual supply of oxidized NAD^+ molecules (see Figures 5.13 and 5.16). Electron transport produces the required NAD^+ in respiration, but without a final electron acceptor, this source of NAD^+ ceases to be available. A cell in such a predicament must use an alternate source of NAD^+ provided by alternative metabolic pathways, called *fermentation pathways*.

In the vernacular, fermentation refers to the production of alcohol from sugar, but in microbiology, fermentation has an expanded meaning. **Fermentation** is the partial oxidation of sugar (or other metabolites) to release energy using an organic molecule from within the cell as the final electron acceptor. In other words, fermentation pathways are metabolic reactions that oxidize NADH to NAD^+ while reducing cellular organic molecules. In contrast, respiration reduces externally acquired substances—oxygen in aerobic respiration, and in anaerobic respiration, some other inorganic chemical such as sulfate and nitrate or (rarely) an organic molecule. **Figure 5.21** illustrates two common fermentation pathways that reduce pyruvic acid to lactic acid and ethanol, oxidizing NADH in the process.

The essential function of fermentation is the regeneration of NAD^+ for glycolysis, so that ADP molecules can be phosphorylated to ATP. Even though fermentation pathways are not as energetically efficient as respiration because much of the potential energy stored in glucose remains in the bonds of fermentation products, the major benefit of fermentation is that it allows ATP production to continue in the absence of cellular respiration. **Table 5.4** compares fermentation to aerobic and anaerobic respiration with respect to four crucial aspects of these processes.

Microorganisms produce a variety of fermentation products depending on the enzymes and substrates available to each. Though fermentation products are wastes to the cells that make them, many are useful to humans, including ethanol (drinking alcohol) and lactic acid (used in the production of cheese, sauerkraut, and pickles) **(Figure 5.22)**.

Other fermentation products are harmful to human health and industry. For example, fermentation products of the bacterium *Clostridium perfringens* (klos-trid′ē-um per-frin′jens) are involved in the necrosis (death) of muscle tissue associated with gangrene. Also, as you may recall from Chapter 1, Pasteur discovered that bacterial contaminants in grape juice fermented the sugar into unwanted products such as acetic acid and lactic acid, which spoiled the wine.

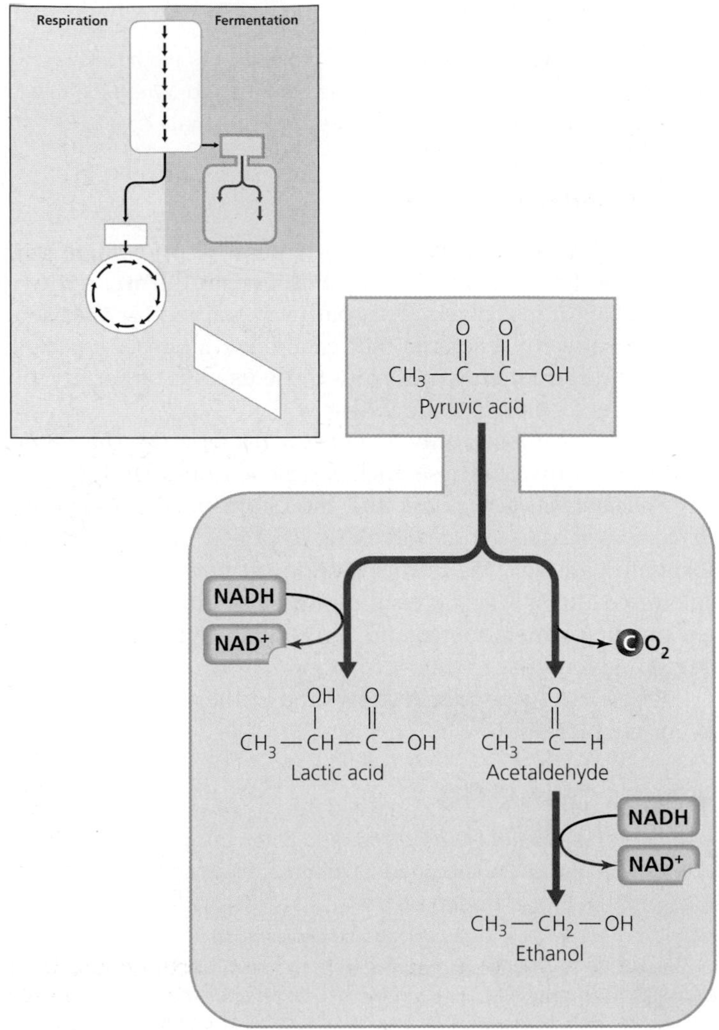

▲ **Figure 5.21 Fermentation.** In the simplest fermentation reaction, NADH reduces pyruvic acid to form lactic acid. Another simple fermentation pathway involves a decarboxylation reaction and reduction to form ethanol.

Laboratory personnel routinely use the detection of fermentation products in the identification of microbes. For example, *Proteus* ferments glucose but not lactose, whereas *Escherichia* and *Enterobacter* ferment both. Further, glucose fermentation by *Escherichia* produces mixed acids (acetic, lactic, succinic, and formic), whereas *Enterobacter* produces 2,3-butanediol. Common fermentation tests contain a carbohydrate and a pH indicator, which is a molecule that changes color as the pH changes. An organism that utilizes the carbohydrate causes a change in pH, causing the pH indicator to change color. **ANIMATIONS: *Fermentation***

In this section on carbohydrate catabolism, we have spent some time examining glycolysis, alternatives to glycolysis, the Krebs cycle, and electron transport because these pathways are central to metabolism. They generate all of the precursor metabolites and most of the ATP needed for anabolism. We have seen that

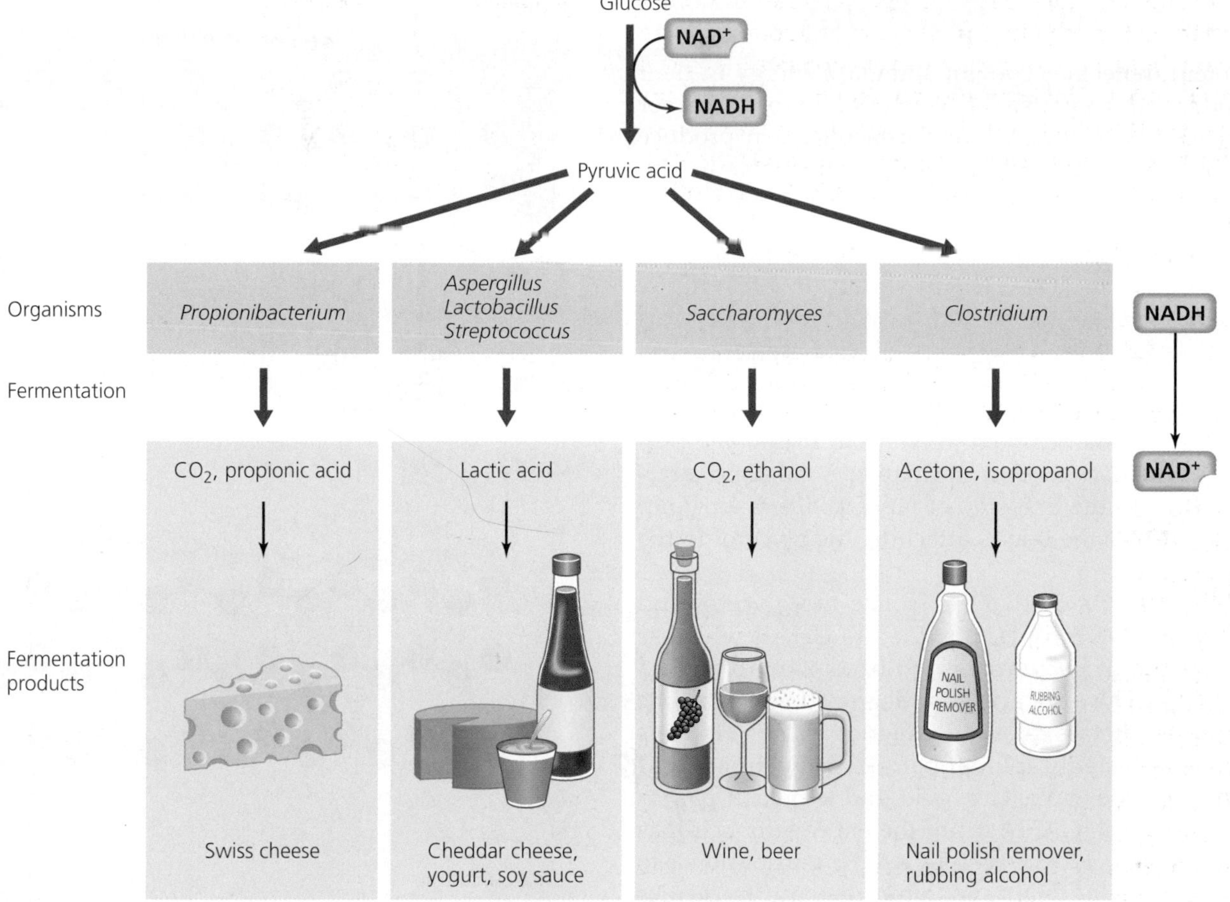

▲ **Figure 5.22 Representative fermentation products and the organisms that produce them.** All of the organisms are bacteria except *Saccharomyces* and *Aspergillus*, which are yeasts.
Why does Swiss cheese have holes, but cheddar does not?

Figure 5.22 *Swiss cheese is a product of Propionibacterium, which ferments pyruvic acid to produce CO_2; bubbles of this gas cause the holes in the cheese. Cheddar is a product of Lactobacillus, which does not produce gas.*

some ATP is generated in respiration by substrate-level phosphorylation (in both glycolysis and the Krebs cycle), but that most ATP is generated by oxidative phosphorylation via chemiosmosis utilizing the reducing power of NADH and $FADH_2$. We also saw that some microorganisms use fermentation to provide an alternate source of NAD^+.

Thus far we have concentrated on the catabolism of glucose as a representative carbohydrate, but microorganisms can also use other molecules as energy sources. In the next section we will examine catabolic pathways that utilize lipids and proteins.

TABLE 5.4 Comparison of Aerobic Respiration, Anaerobic Respiration, and Fermentation

	Aerobic Respiration	Anaerobic Respiration	Fermentation
Oxygen required	Yes	No	No
Type of phosphorylation	Substrate-level and oxidative	Substrate-level and oxidative	Substrate-level
Final electron (hydrogen) acceptor	Oxygen	NO_3^-, SO_4^{2-}, CO_3^{2-}, or externally acquired organic molecules	Cellular organic molecules
Potential molecules of ATP produced per molecule of glucose	38 in prokaryotes, 36 in eukaryotes	2–36	2

Other Catabolic Pathways

Lipid and protein molecules contain abundant energy in their chemical bonds and can also be converted into precursor metabolites. These molecules are first catabolized to produce their constituent monomers, which serve as substrates in glycolysis and the Krebs cycle.

Lipid Catabolism

Learning Objective

✓ Explain how lipids are catabolized for energy and metabolite production.

The most common lipids involved in ATP and metabolite production are fats, which, as we saw in Chapter 2, consist of glycerol and fatty acids. In the first step of fat catabolism, enzymes called *lipases* hydrolyze the bonds attaching the glycerol to the fatty acid chains **(Figure 5.23a)**.

Subsequent reactions further catabolize the glycerol and fatty acid molecules. Glycerol is converted to DHAP, which as one of the substrates of glycolysis is oxidized to pyruvic acid (see Figure 5.13 4). The fatty acids are degraded in a catabolic process known as **beta-oxidation**[9] **(Figure 5.23b)**. In this process, enzymes repeatedly split off pairs of the hydrogenated carbon atoms that make up a fatty acid and join each pair to coenzyme A to form acetyl-CoA, until the entire fatty acid has been converted to molecules of acetyl-CoA. Beta-oxidation also generates NADH and $FADH_2$, and more of these molecules are generated when the acetyl-CoA is utilized in the Krebs cycle to generate ATP (see Figure 5.16). As is the case for the Krebs cycle, the enzymes involved in beta-oxidation are located in the cytosol of prokaryotes and in the mitochondria of eukaryotes.

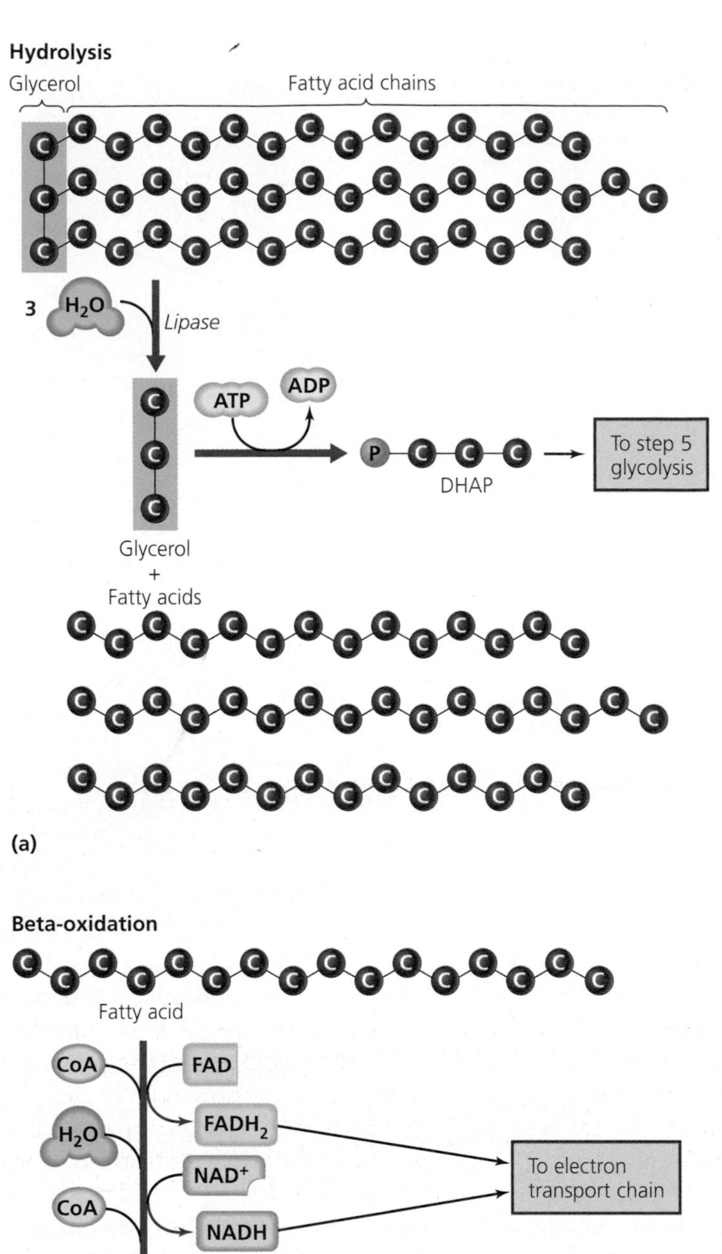

▲ **Figure 5.23 Catabolism of a fat molecule. (a)** Lipase breaks fats into glycerol and three fatty acids by hydrolysis. Glycerol is converted to DHAP, which can be catabolized via glycolysis and the Krebs cycle. **(b)** Fatty acids are catabolized via beta-oxidation reactions that produce molecules of acetyl-CoA and reduced coenzymes (NADH and $FADH_2$).

CRITICAL **THINKING**

How and where do cells use the molecules of NADH and $FADH_2$ produced during beta-oxidation?

Protein Catabolism

Learning Objective

✓ Explain how proteins are catabolized for energy and metabolite production.

Some microorganisms, notably food-spoilage bacteria, pathogenic bacteria, and fungi, normally catabolize proteins as an important source of energy and metabolites. Most cells catabolize proteins and their constituent amino acids only when carbon sources such as glucose and fat are not available.

Generally, proteins are too large to cross cytoplasmic membranes, so prokaryotes typically conduct the first step in the process of protein catabolism outside the cell by secreting

[9] "Beta" is part of the name of this process because enzymes break the bond at the second carbon atom from the end of a fatty acid, and beta is the second letter in the Greek alphabet.

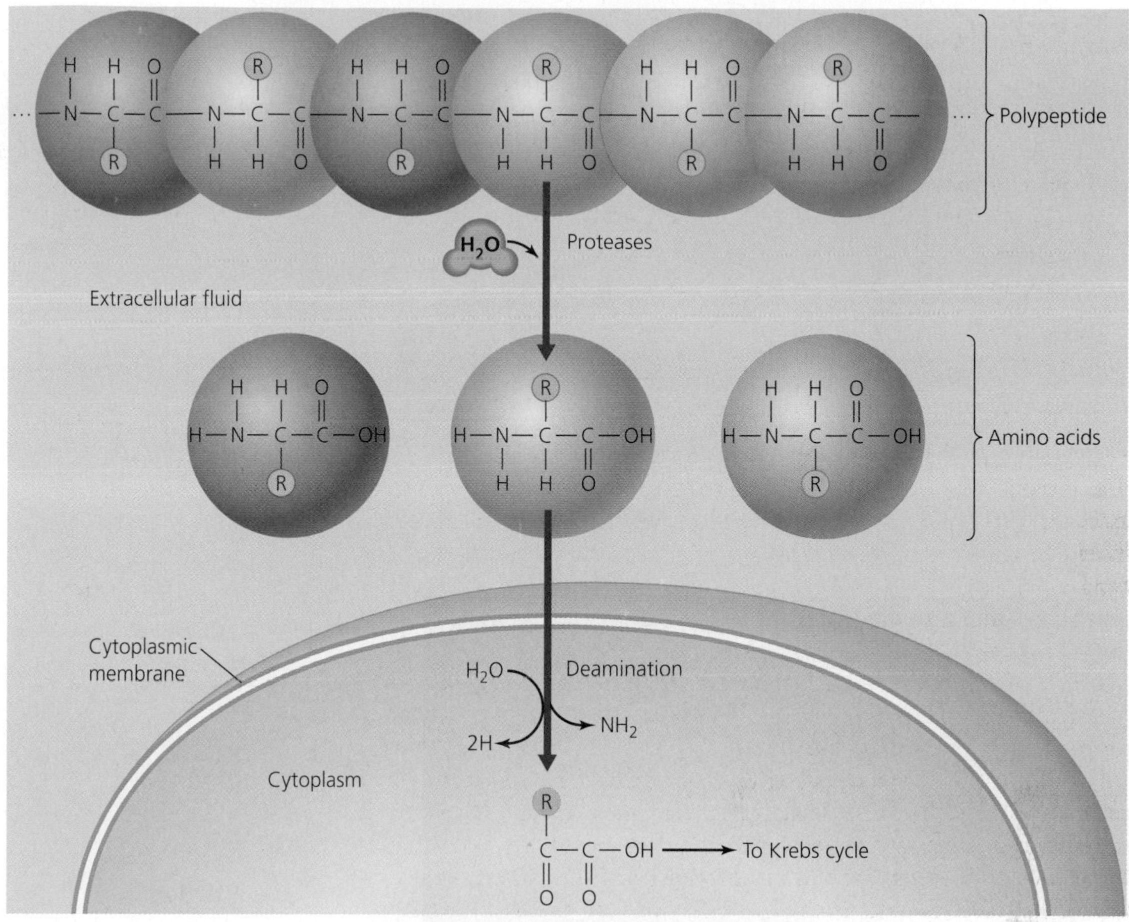

➤ **Figure 5.24 Protein catabolism.** Secreted proteases hydrolyze proteins, releasing amino acids, which are deaminated after uptake to produce molecules used as substrates in the Krebs cycle. R indicates the side group, which varies among amino acids.

proteases (prō′ tē-ās-ez)—enzymes that split proteins into their constituent amino acids **(Figure 5.24)**. Once released by the action of proteases, amino acids are transported into the cell, where special enzymes split off amino groups in a reaction called **deamination.** The resulting altered molecules enter the Krebs cycle, and the amino groups are either recycled to synthesize other amino acids or excreted as nitrogenous wastes such as ammonia (NH_3), ammonium ion (NH_4^+), or trimethylamine oxide (which has the chemical formula $(CH_3)_3NO$ and is abbreviated TMAO). As **Highlight: What's That Fishy Smell?** describes, TMAO plays an interesting role in the production of the odor we describe as "fishy."

Thus far, we have examined the catabolism of carbohydrates, lipids, and proteins. Now we turn our attention to the synthesis of these molecules, beginning with the anabolic reactions of photosynthesis.

HIGHLIGHT

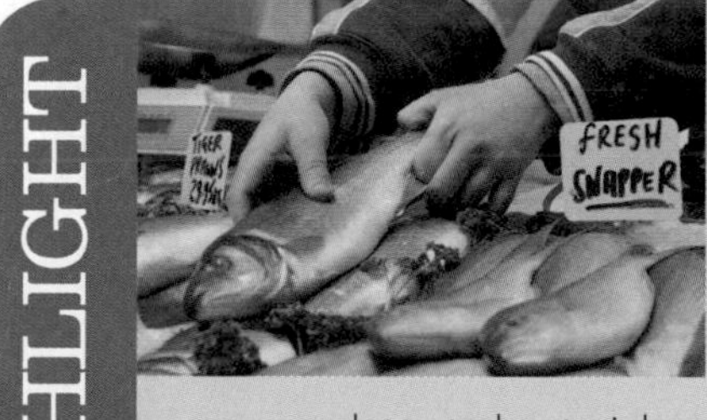

WHAT'S THAT FISHY SMELL?

Ever wonder why fish at the supermarket sometimes have that "fishy" smell? The reason relates to bacterial metabolism.

Trimethylamine oxide (TMAO) is an odorless, nitrogenous waste product of fish metabolism used to dispose of excess nitrogen (as amine groups) produced in the catabolism of amino acids. TMAO does not affect the smell, appearance, or taste of fresh fish; however, some bacteria use TMAO as a final electron acceptor in anaerobic respiration, reducing TMAO to trimethyl amine (TMA), a compound with a very definite—and, for most people, unpleasant—"fishy" odor. Thus, the "fishy" odor is really the odor of contaminating bacteria. A human nose is able to detect even a few molecules of TMA. Because bacterial degradation of fish begins as soon as a fish is dead, a fish that smells fresh probably is fresh, or was frozen while still fresh.

Photosynthesis

Learning Objective

✓ Define *photosynthesis.*

Many organisms use only organic molecules as a source of energy and metabolites, but where do they acquire organic molecules? Ultimately every food chain begins with anabolic pathways in organisms that synthesize their own organic

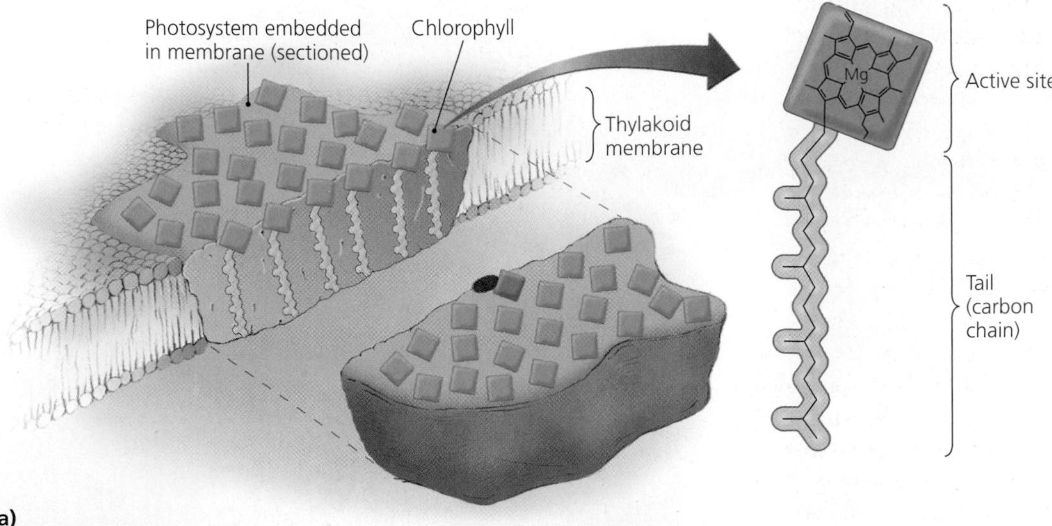

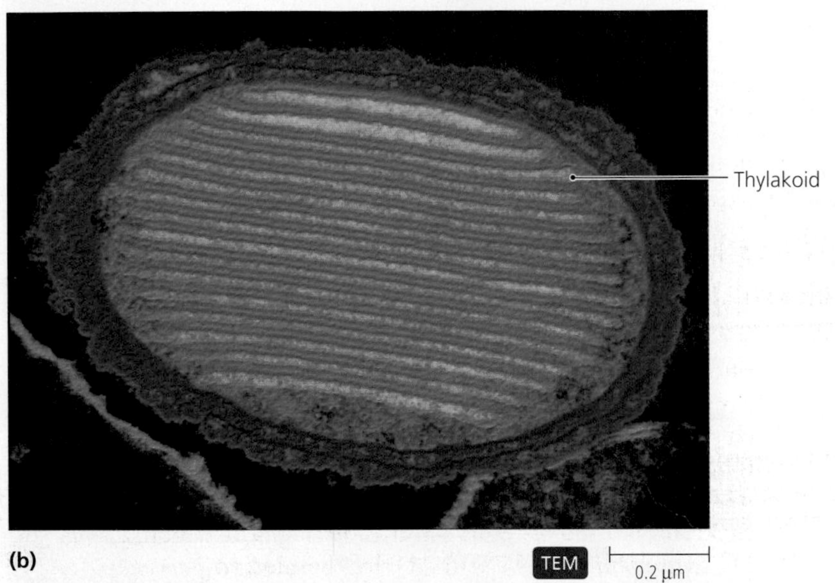

Figure 5.25 Photosynthetic structures in a prokaryote. **(a)** Chlorophyll molecules are grouped together in thylakoid membranes to form photosystems. Chlorophyll contains a light-absorbing active site that is connected to a long hydrocarbon tail. The chlorophyll shown here is bacteriochlorophyll *a*; other chlorophyll types differ in the side chains protruding from the central ring. **(b)** Prokaryotic thylakoids are infoldings of the cytoplasmic membrane, here in a cyanobacterium.

molecules from inorganic carbon dioxide. Most of these organisms capture light energy from the sun and use it to drive the synthesis of carbohydrates from CO_2 and H_2O by a process called **photosynthesis.** Cyanobacteria, purple sulfur bacteria, green sulfur bacteria, green nonsulfur bacteria, purple nonsulfur bacteria, algae, green plants, and a few protozoa are photosynthetic. **ANIMATIONS:** ***Photosynthesis: Overview***

CRITICAL **THINKING**

Photosynthetic organisms are rarely pathogenic. Why?

Chemicals and Structures

Learning Objective

✓ Compare and contrast the basic chemicals and structures involved in photosynthesis in prokaryotes and eukaryotes.

Photosynthetic organisms capture light energy with pigment molecules, the most important of which are **chlorophylls.** Chlorophyll molecules are composed of a hydrocarbon tail attached to a light-absorbing *active site* centered around a magnesium ion (Mg^{2+}) **(Figure 5.25a)**. The active sites are structurally similar to the cytochrome molecules found in electron transport chains, except chlorophylls use Mg^{2+} rather than Fe^{2+}. Chlorophylls, typically designated with letters—for example, chlorophyll *a*, chlorophyll *b*, and bacteriochlorophyll *a*—vary slightly in the lengths and structures of their hydrocarbon tails and in the atoms that extend from their active sites. Green plants, algae, photosynthetic protozoa, and cyanobacteria principally use chlorophyll *a*, whereas green and purple bacteria use bacteriochlorophylls.

The slight structural differences among chlorophylls cause them to absorb light of different wavelengths. For example, chlorophyll *a* from algae best absorbs light with wavelengths of about 425 nm and 660 nm (violet and red), whereas bacteriochlorophyll *a* from purple bacteria best absorbs light with wavelengths of about 350 nm and 880 nm (ultraviolet and infrared). Because they best use light with differing wavelengths, algae and purple bacteria successfully occupy different ecological niches.

Cells arrange numerous molecules of chlorophyll and other pigments within a protein matrix to form light-harvesting matrices called **photosystems** that are embedded in cellular membranes called **thylakoids.** Thylakoids of photosynthetic prokaryotes are invaginations of their cytoplasmic membranes **(Figure 5.25b)**. The thylakoids of eukaryotes appear to be formed from infoldings of the inner membranes of chloroplasts, though thylakoid membranes and the inner membranes are not connected in mature chloroplasts (see Figure 3.41). Thylakoids of chloroplasts are arranged in stacks called *grana.* An outer chloroplast membrane surrounds the grana, and the space between the outer membrane and the thylakoid membrane is known as the *stroma.* The thylakoids enclose a narrow, convoluted cavity called the *thylakoid space*.

There are two types of photosystems, named photosystem I (PS I) and photosystem II (PS II), in the order of their discovery. Photosystems absorb light energy and use redox reactions to store this energy in molecules of ATP and NADPH. Because they depend upon light energy, these reactions of photosynthesis are classified as **light-dependent reactions.** Photosynthesis also involves **light-independent reactions** that actually synthesize glucose from carbon dioxide and water. Historically, these reactions were called *light* and *dark reactions,* but this older terminology implies that light-independent reactions occur only in the dark, which is not the case. We first consider the light-dependent reactions of the two photosystems.

Light-Dependent Reactions

Learning Objectives

- ✓ Describe the components and function of the two photosystems, PS I and PS II.
- ✓ Contrast cyclic and noncyclic photophosphorylation.

The pigments of photosystem I absorb light energy and transfer it to a neighboring molecule within the photosystem until the energy eventually arrives at a special chlorophyll molecule called the **reaction center chlorophyll (Figure 5.26)**. Light energy from hundreds of such transfers excites electrons in the reaction center chlorophyll, which passes its excited electrons to the reaction center's electron acceptor, which is the initial carrier of an electron transport chain. As electrons move down the chain, their energy is used to pump protons across the membrane, creating a proton motive force. In prokaryotes, protons are pumped out of the cell; in eukaryotes they are pumped from the stroma into the interior of the thylakoids.

The proton motive force is used in chemiosmosis, in which, you should recall, protons flow down their electrochemical gradient through ATPases, which generate ATP. In photosynthesis, this process is called *photophosphorylation,* which can be either cyclic or noncyclic.

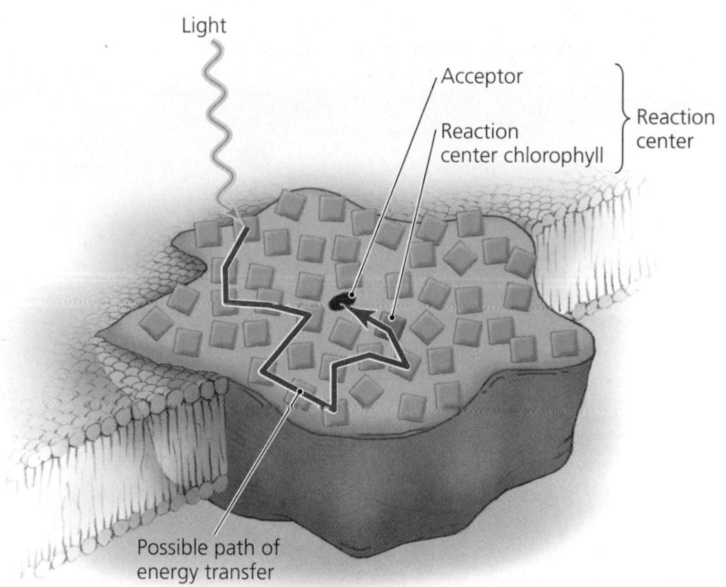

▲ **Figure 5.26 Reaction center of a photosystem.** Light energy absorbed by pigments anywhere in the photosystem is transferred to the reaction center chlorophyll, where it is used to excite electrons for delivery to the reaction center's electron acceptor.

Cyclic Photophosphorylation

Electrons moving from one carrier molecule to another in a thylakoid must eventually pass to a final electron acceptor. In **cyclic photophosphorylation,** which occurs in all photosynthetic organisms, the final electron acceptor is the original reaction center chlorophyll that donated the electrons **(Figure 5.27a)**. In other words, when light energy excites electrons in PS I, they pass down an electron transport chain and return to PS I. The energy from the electrons is used to establish a proton gradient that drives the phosphorylation of ADP to ATP by chemiosmosis. **ANIMATIONS:** ***Photosynthesis: Light Reaction: Cyclic Photophosphorylation***

Noncyclic Photophosphorylation

Some photosynthetic bacteria and all plants, algae, and photosynthetic protozoa utilize **noncyclic photophosphorylation** as well. (Exceptions are green and purple sulfur bacteria.) Noncyclic photophosphorylation, which requires both PS I and PS II, not only generates molecules of ATP but also reduces molecules of coenzyme $NADP^+$ to NADPH **(Figure 5.27b)**.

When light energy excites electrons of PS II, they are passed to PS I through an electron transport chain. (Note that photosystem II occurs first in the pathway, and photosystem I second, a result of the fact that the photosystems were named for the order in which they were discovered, not the order in which they operate.) PS I further energizes the electrons with additional light energy and transfers them through an electron transport chain to $NADP^+$, which is thereby reduced to NADPH. The hydrogen ions added to NADPH come from the stroma or cytosol. NADPH subsequently participates in the synthesis of glucose in the light-independent reactions, which we will examine in the next section.

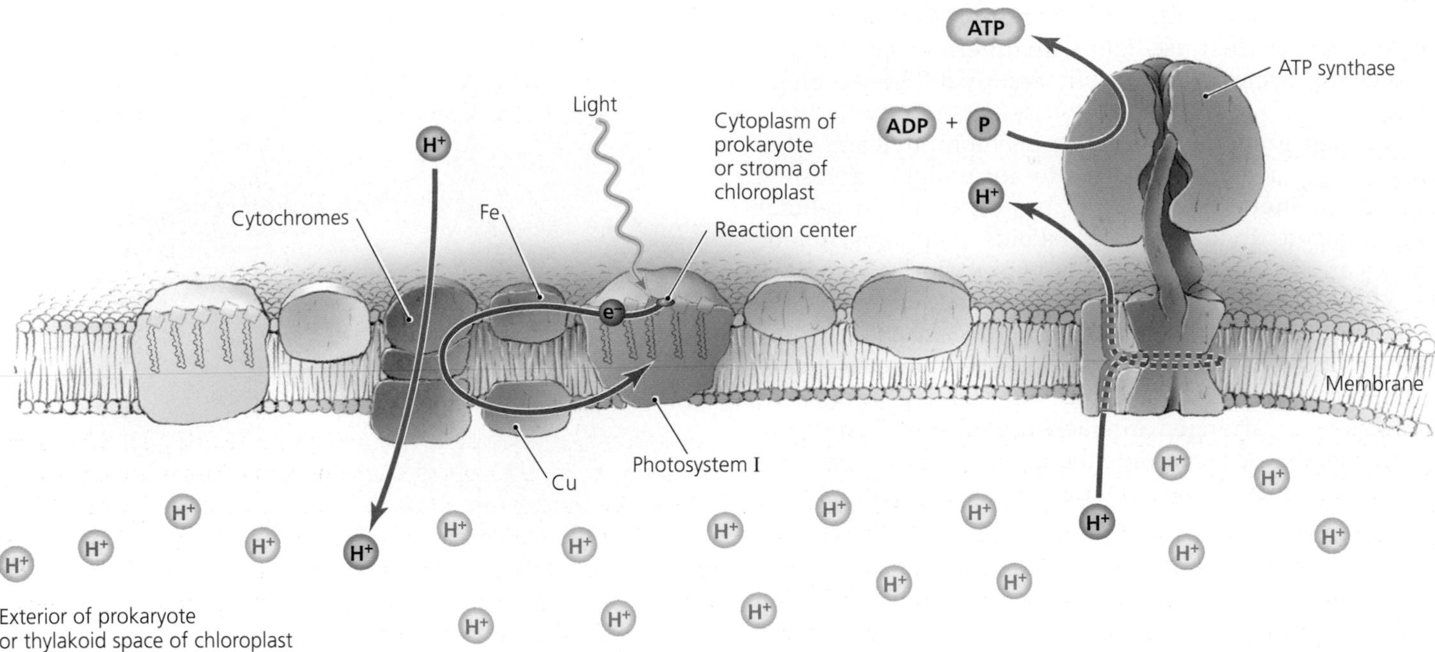

(a) Cyclic photophosphorylation

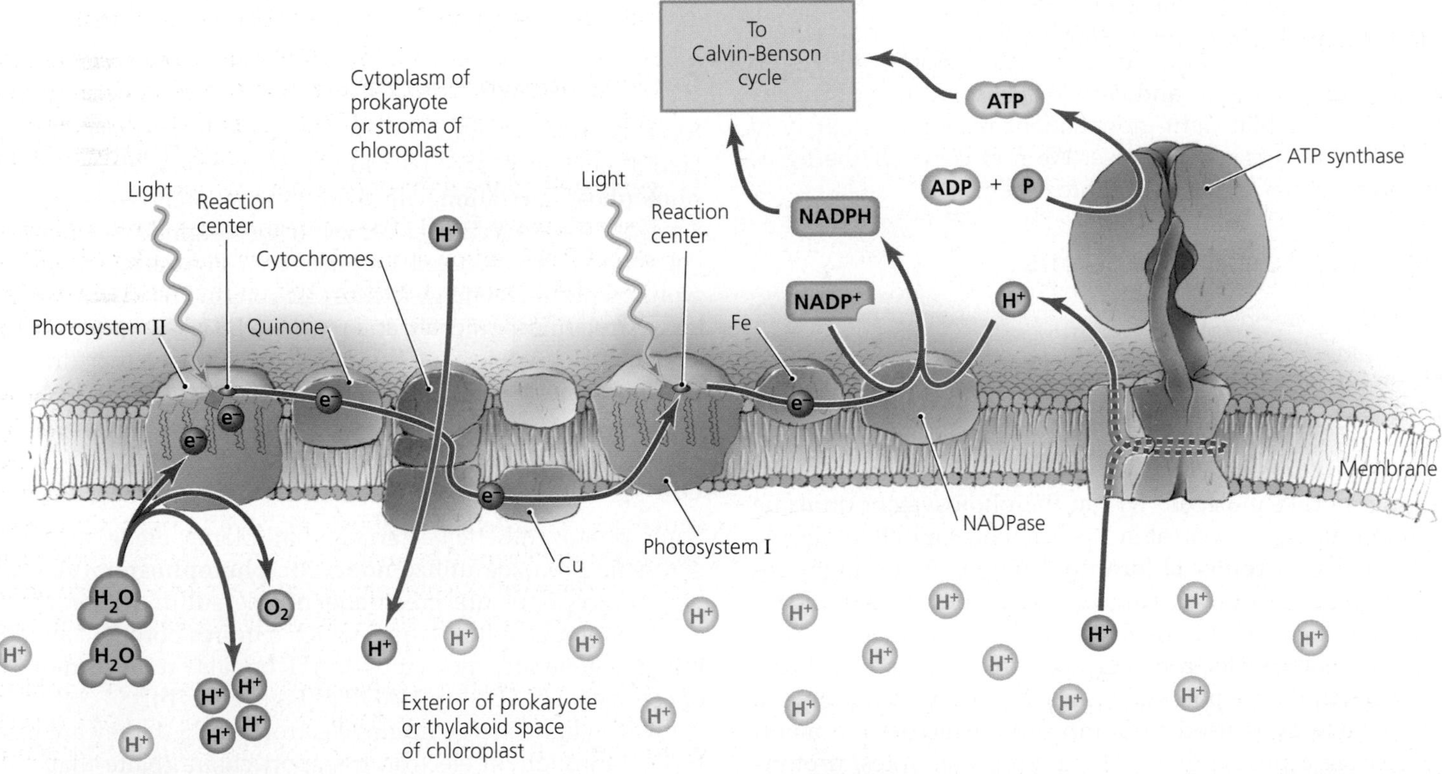

(b) Noncyclic photophosphorylation

▲ **Figure 5.27 The light-dependent reactions of photosynthesis: Cyclic and noncyclic photophosphorylation.** **(a)** In cyclic photophosphorylation, electrons excited by light striking photosystem I travel down an electron transport chain (red arrows), and then return to the reaction center. **(b)** In noncyclic photophosphorylation, light striking photosystem II excites electrons that are passed to photosystem I, simultaneously establishing a proton gradient. Light energy collected by photosystem I further excites the electrons, which are used to reduce $NADP^+$ to NADPH via an electron transport chain. In oxygenic organisms, new electrons are provided by the lysis of H_2O (bottom left). In both types of photophosphorylation, the proton gradient drives the phosphorylation of ADP by ATP synthase (chemiosmosis).

TABLE 5.5 A Comparison of the Three Types of Phosphorylation

	Source of Phosphate	Source of Energy	Location in Eukaryotic Cell	Location in Prokaryotic Cell
Substrate-level phosphorylation	Organic molecule	High-energy phosphate bond of donor	Cytosol and mitochondrial matrix	Cytosol
Oxidative phosphorylation	Inorganic phosphate (PO_4^{3-})	Proton motive force	Inner membrane of mitochondrion	Cytoplasmic membrane
Photophosphorylation	Inorganic phosphate (PO_4^{3-})	Proton motive force	Thylakoid of chloroplast	Thylakoid of cytoplasmic membrane

In noncyclic photophosphorylation, a cell must constantly replenish electrons to the reaction center of photosystem II. *Oxygenic* (oxygen-producing) organisms, such as algae, green plants, and cyanobacteria, derive electrons from the dissociation of H_2O (see Figure 5.27b). In these organisms, two molecules of water give up their electrons, producing molecular oxygen (O_2) as a waste product during photosynthesis. *Anoxygenic* photosynthetic bacteria get electrons from inorganic compounds such as H_2S, which results in a nonoxygen waste such as sulfur.

Table 5.5 compares photophosphorylation to substrate-level and oxidative phosphorylation.

CRITICAL **THINKING**

We have seen that of the two ways ATP is generated via chemiosmosis—photophosphorylation and oxidative phosphorylation—the former can be cyclical, but the latter is never cyclical. Why can't oxidative phosphorylation be cyclical; that is, why aren't electrons passed back to the molecules that donated them?

To this point, we have examined the use of photosynthetic pigments and thylakoid structure to harvest light energy to produce both ATP and reducing power in the form of NADPH. Next we examine the light-independent reactions of photosynthesis. **ANIMATIONS:** ***Photosynthesis: Light Reaction: Noncyclic Photophosphorylation***

Light-Independent Reactions

Learning Objectives

- ✓ Contrast the light-dependent and light-independent reactions of photosynthesis.
- ✓ Describe the reactants and products of the Calvin-Benson cycle.

Light-independent reactions of photosynthesis do not require light directly; instead, they use ATP and NADPH generated by the light-dependent reactions. The key reaction of the light-independent pathway of photosynthesis is **carbon fixation** by the **Calvin-Benson cycle,**[10] which involves the attachment of molecules of CO_2 to molecules of a five-carbon organic compound called ribulose 1,5-bisphosphate (RuBP). RuBP is derived initially from phosphorylation of a precursor metabolite produced by the pentose phosphate pathway (see Figure 5.19).

The Calvin-Benson cycle is reminiscent of the Krebs cycle in that the substrates of the cycle are regenerated. It is helpful to notice the number of carbon atoms during each part of the three steps of the Calvin-Benson cycle **(Figure 5.28)**:

1. *Fixation of CO_2.* An enzyme attaches three molecules of carbon dioxide (3 carbon atoms) to three molecules of RuBP (15 carbon atoms), which are then split to form six molecules of 3-phosphoglyceric acid (18 carbon atoms).
2. *Reduction.* Molecules of NADPH reduce the six molecules of 3-phosphoglyceric acid to form six molecules of glyceraldehyde 3-phosphate (G3P) (18 carbon atoms). These reactions require six molecules each of ATP and NADPH generated by the light-dependent reactions.
3. *Regeneration of RuBP.* The cell regenerates three molecules of RuBP (15 carbon atoms) from five molecules of G3P (15 carbon atoms). It uses the remaining molecule of glyceraldehyde 3-phosphate to synthesize glucose by reversing the reactions of glycolysis.

In summary, ATP and NADPH from the light-dependent reactions drive the synthesis of glucose from CO_2 in the light-independent reactions of the Calvin-Benson cycle. For every three molecules of CO_2 that enter the Calvin-Benson cycle, a molecule of glyceraldehyde 3-phosphate (G3P) leaves. Glycolysis is subsequently reversed to anabolically combine two molecules of G3P to synthesize glucose 6-phosphate. A more detailed account of the Calvin-Benson cycle is given in Appendix A on p. A-9. **ANIMATIONS:** ***Photosynthesis: Light-Independent Reaction***

The processes of oxygenic photosynthesis and aerobic respiration complement one another to complete both a carbon cycle and an oxygen cycle. During the synthesis of glucose in oxygenic photosynthesis, water and carbon dioxide are used, and oxygen is released as a waste product; in aerobic respiration, oxygen serves as the final electron acceptor in the oxidation of glucose to carbon dioxide and water.

In this section we examined an essential anabolic pathway for life on Earth—photosynthesis, which produces glucose. Next we will consider anabolic pathways involved in the synthesis of other organic molecules.

[10]Named for the men who elucidated its pathways.

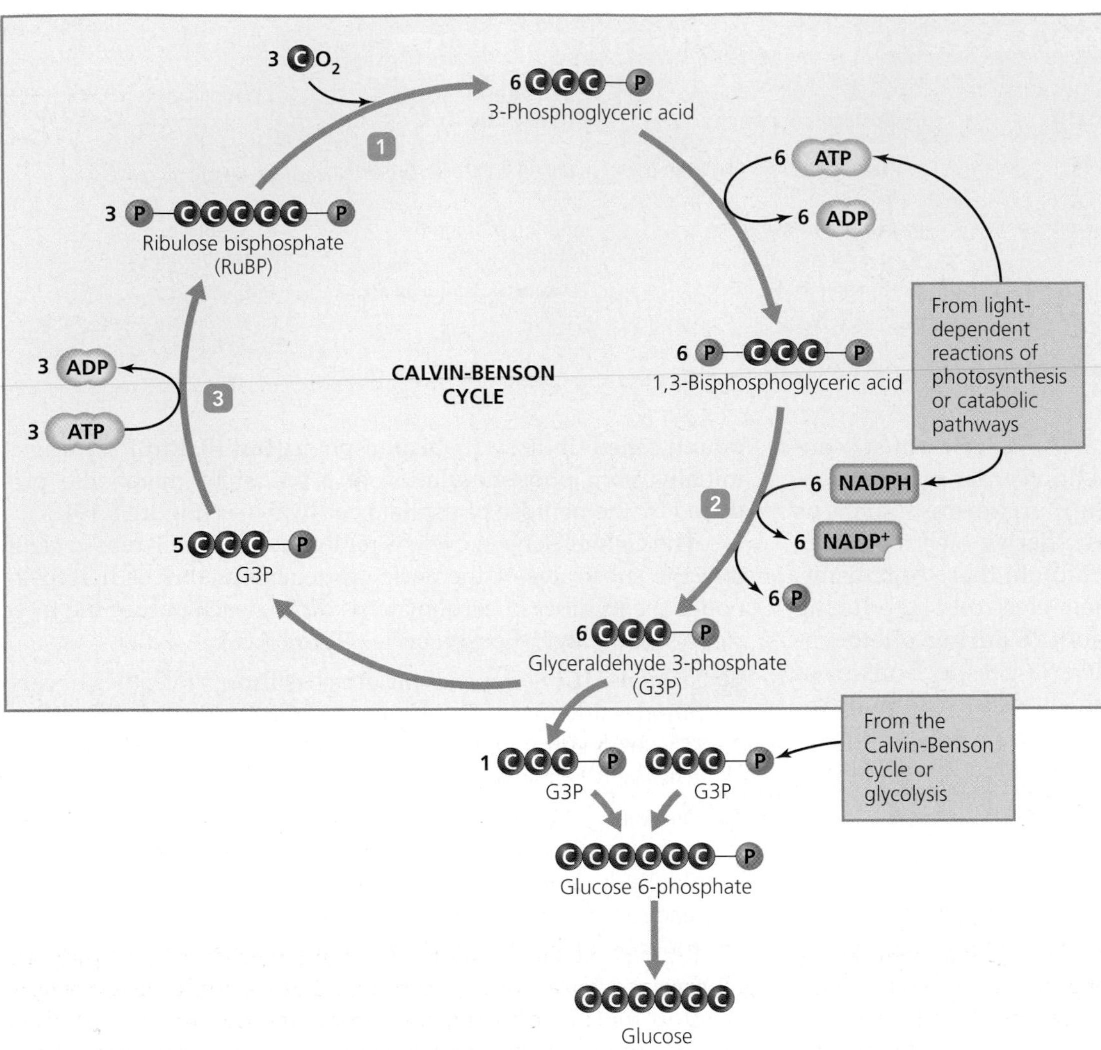

Figure 5.28 Simplified diagram of the Calvin-Benson cycle. 1 Fixation of CO_2. Three molecules of RuBP combine with three molecules of CO_2. 2 Reduction. The resulting molecules are reduced to form six molecules of G3P. 3 Generation of RuBP. Five molecules of G3P are converted to three molecules of RuBP, which completes the cycle. Two turns of the cycle also yield two molecules of G3P, which are polymerized to synthesize glucose 6-phosphate.

Other Anabolic Pathways

Learning Objective

✓ Define *amphibolic reaction.*

Anabolic reactions are synthesis reactions. As such, they require energy and a source of metabolites. Energy for anabolism is provided by ATP generated in the catabolic reactions of aerobic respiration, anaerobic respiration, and fermentation, and by the initial redox reactions of photosynthesis. Glycolysis, the Krebs cycle, and the pentose phosphate pathway provide 12 basic precursor metabolites from which all macromolecules and cellular structures can be made (Table 5.6). Some microorganisms, such as *E. coli,* can synthesize all 12 precursors, while other organisms, such as humans, must acquire some precursors in their diets.

Many anabolic pathways are the reversal of the catabolic pathways we have discussed; therefore, much of the material that follows has *in a sense* been discussed in the sections on catabolism. Reactions that can proceed in either direction—toward catabolism or toward anabolism—are said to be **amphibolic.** The following sections discuss the synthesis of carbohydrates, lipids, amino acids, and nucleotides. Chapter 7 covers anabolic reactions that are closely linked to genetics—the polymerizations of amino acids into proteins and of nucleotides into RNA and DNA.

Carbohydrate Biosynthesis

Learning Objective

✓ Describe the biosynthesis of carbohydrates.

As we have seen, anabolism begins in photosynthetic organisms with carbon fixation by the enzymes of the Calvin-Benson cycle to form molecules of G3P. Enzymes use G3P as the starting point for synthesizing sugars, complex polysaccharides such as starch, cellulose for cell walls in algae, and peptidoglycan for cell walls of bacteria. Animals and protozoa synthesize the storage molecule glycogen.

Some cells are able to synthesize sugars from noncarbohydrate precursors such as amino acids, glycerol, and fatty acids by

TABLE 5.6
The Twelve Precursor Metabolites

	Pathway That Generates the Metabolite	Examples of Macromolecule Synthesized from Metabolite[a]	Examples of Functional Use
Glucose 6-phosphate	Glycolysis	Lipopolysaccharide	Outer membrane of cell wall
Fructose 6-phosphate	Glycolysis	Peptidoglycan	Cell wall
Glyceraldehyde 3-phosphate (G3P)	Glycolysis	Glycerol portion of lipids	Fats—energy storage
Phosphoglyceric acid	Glycolysis	Amino acids: cysteine, selenocysteine, glycine, and serine	Enzymes
Phosphoenolpyruvic acid (PEP)	Glycolysis	Amino acids: phenylalanine, tryptophan, and tyrosine	Enzymes
Pyruvic acid	Glycolysis	Amino acids: alanine, leucine, and valine	Enzymes
Ribose 5-phosphate	Pentose phosphate pathway	DNA, RNA, amino acid, and histidine	Genome, enzymes
Erythrose 4-phosphate	Pentose phosphate pathway	Amino acids: phenylalanine, tryptophan, and tyrosine	Enzymes
Acetyl-CoA	Krebs cycle	Fatty acid portion of lipids	Cytoplasmic membrane
α-Ketoglutaric acid	Krebs cycle	Amino acids: arginine, glutamic acid, glutamine, and proline	Enzymes
Succinyl-CoA	Krebs cycle	Heme	Cytochrome electron carrier
Oxaloacetate	Krebs cycle	Amino acids: aspartic acid, asparagine, isoleucine, lysine, methionine, and threonine	Enzymes

[a]Examples given apply to the bacterium *E. coli*.

pathways collectively called *gluconeogenesis*[11] (glū′kō-nē-ō-jen′ĕ-sis; **Figure 5.29**). Most of the reactions of gluconeogenesis are amphibolic, using enzymes of glycolysis in reverse, but four of the reactions require unique enzymes. Gluconeogenesis is highly endergonic and can proceed only if there is an adequate supply of energy.

CRITICAL **THINKING**

A scientist moves a green plant grown in sunlight to a room with 24 hours of artificial green light. Will this increase or decrease the plant's rate of photosynthesis? Why?

Lipid Biosynthesis

Learning Objective

✓ Describe the biosynthesis of lipids.

As we saw in Chapter 2, lipids are a diverse group of organic molecules that function as energy-storage compounds and as components of membranes. *Carotenoids,* which are reddish pigments found in many bacterial and plant photosystems, are also lipids.

[11]From Greek *glukus*, meaning sweet, *neo*, meaning new, and *genesis*, meaning generate.

Because of their variety, it is not surprising that lipids are synthesized by a variety of routes. For example, fats are synthesized in anabolic reactions that are the reverse of their catabolism—cells polymerize glycerol and three fatty acids **(Figure 5.30)**. Glycerol is derived from G3P generated by the Calvin-Benson cycle and glycolysis; the fatty acids are produced by the linkage of two-carbon acetyl-CoA molecules to one another by a sequence of endergonic reactions that effectively reverse the catabolic reactions of beta-oxidation. Other lipids, such as steroids, are synthesized in complex pathways involving polymerizations and isomerizations of sugar and amino acid metabolites.

Mycobacterium tuberculosis (mī′kō-bak-tēr′ē-ŭm too-ber-kyū-lō′sis), the pathogen that causes tuberculosis, makes copious amounts of a waxy lipid called *mycolic acid,* which is incorporated into its cell wall. The arduous, energy-intensive process of making long lipid chains explains why this organism grows slowly, and why tuberculosis requires a long course of treatment with antimicrobial drugs.

Amino Acid Biosynthesis

Learning Objective

✓ Describe the biosynthesis of amino acids.

Cells synthesize amino acids from precursor metabolites derived from glycolysis, the Krebs cycle, and the pentose phosphate pathway, and from other amino acids. Some organisms (such as

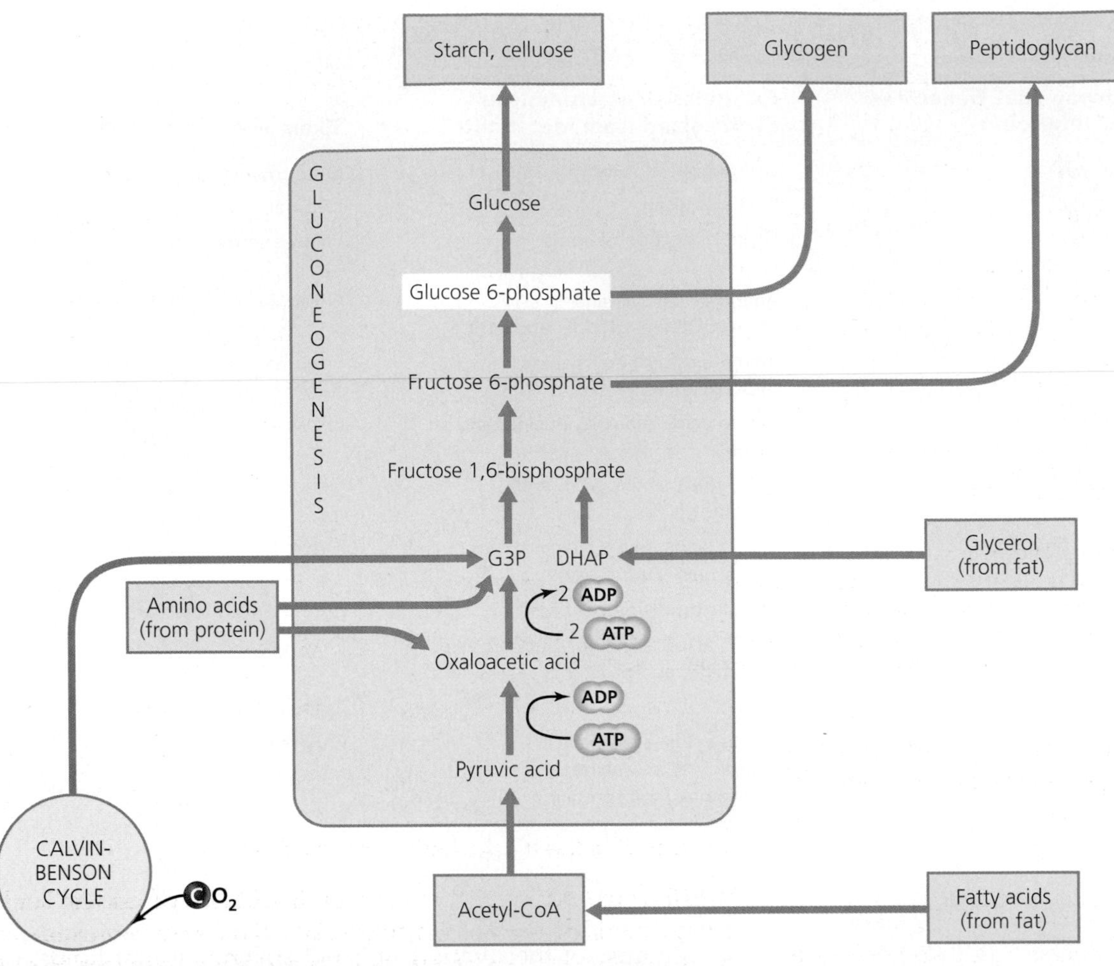

Figure 5.29 The role of gluconeogenesis in the biosynthesis of complex carbohydrates. Complex carbohydrates are synthesized from simple sugar molecules such as glucose, glucose 6-phosphate, and fructose 6-phosphate. Starch and cellulose are found in algae; glycogen is found in animals and protozoa; and peptidoglycan is found in bacteria.

E. coli, and most plants and algae) synthesize all their amino acids from precursor metabolites. Other organisms, including humans, cannot synthesize certain amino acids, called *essential amino acids;* they must be acquired in the diet. One extreme example is *Lactobacillus* (lak′-tō-bă-sil′ŭs), a bacterium that ferments milk and produces some cheeses. This microorganism cannot synthesize any amino acids; it acquires all of them by catabolizing proteins in its environment.

Precursor metabolites are converted to amino acids by the addition of an amine group. This process is called **amination** when the amine group comes from ammonia (NH_3); an example is the formation of aspartic acid from NH_3 and the Krebs cycle intermediate oxaloacetic acid **(Figure 5.31a)**. Amination reactions are the reverse of the catabolic deamination reactions we discussed previously. More commonly, however, a cell moves the amine group from one amino acid and adds it to a metabolite, producing a different amino acid. This process is called **transamination** because the amine group is transferred from one amino acid to another **(Figure 5.31b)**. All transamination enzymes use a coenzyme, *pyridoxal phosphate,* which is derived from vitamin B_6.

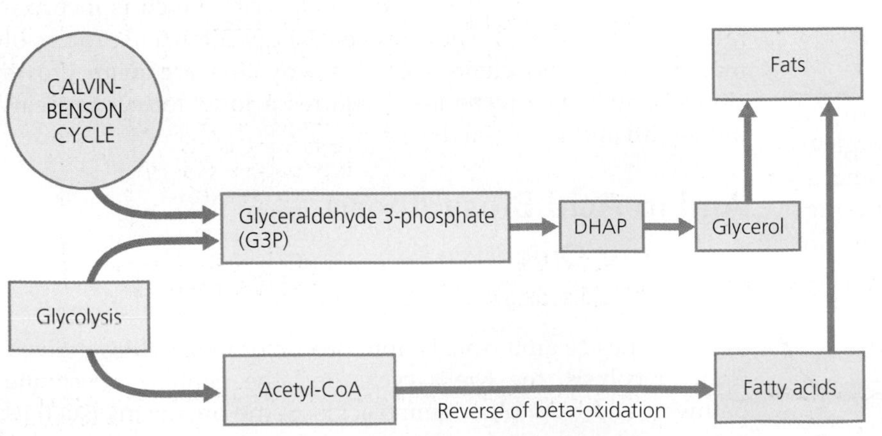

Figure 5.30 Biosynthesis of fat, a lipid. A fat molecule is synthesized from glycerol and three molecules of fatty acid, the precursors of which are produced in glycolysis.

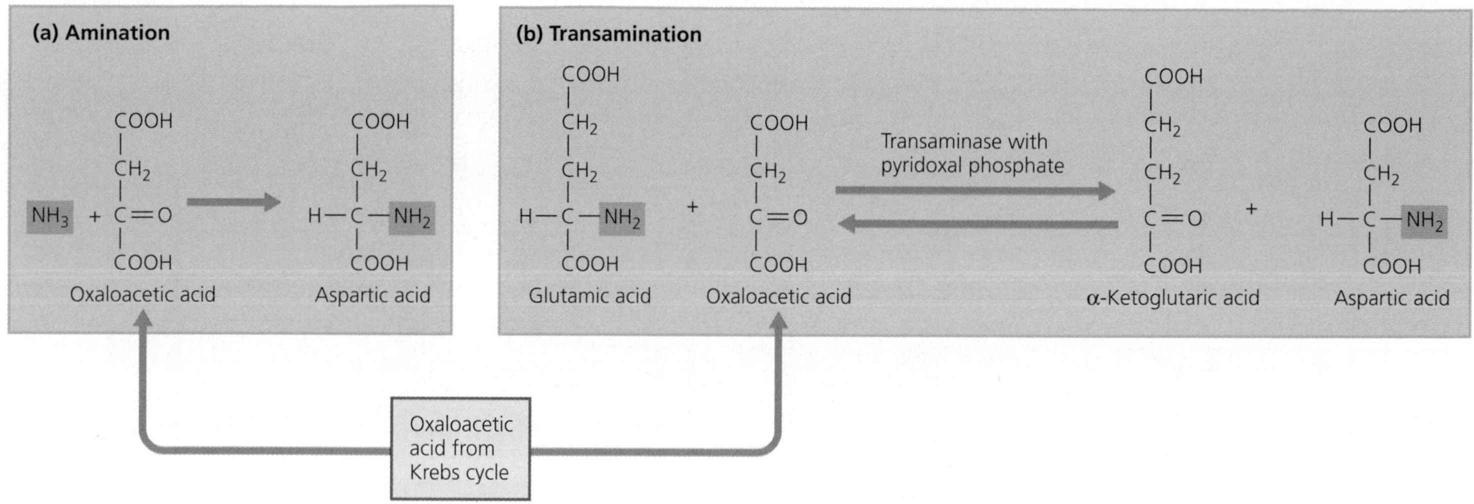

▲ **Figure 5.31 Examples of the synthesis of amino acids via amination and transamination.** **(a)** In amination, an amine group from ammonia is added to a precursor metabolite. In this example, oxaloacetic acid is converted into the amino acid aspartic acid. **(b)** In transamination, the amine group is derived from an existing amino acid. In the example shown here, the amino acid glutamic acid donates its amine group to oxaloacetic acid, which is converted into aspartic acid. Note that transamination is a reversible reaction.

Ribozymes of ribosomes polymerize amino acids into proteins. Chapter 7 examines this energy-demanding process, because it is intimately linked with genetics.

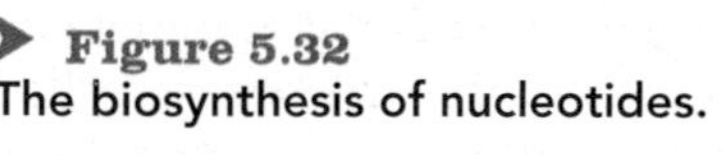

CRITICAL **THINKING**

What class of enzyme is involved in amination reactions? What class of enzyme catalyzes transaminations?

Nucleotide Biosynthesis

Learning Objective

✓ Describe the biosynthesis of nucleotides.

Recall from Chapter 2 that the building blocks of nucleic acids are nucleotides, each of which consists of a five-carbon sugar, a phosphate group, and a purine or pyrimidine base (see Figure 2.25a). Nucleotides are produced from precursor metabolites of glycolysis and the Krebs cycle **(Figure 5.32)**:

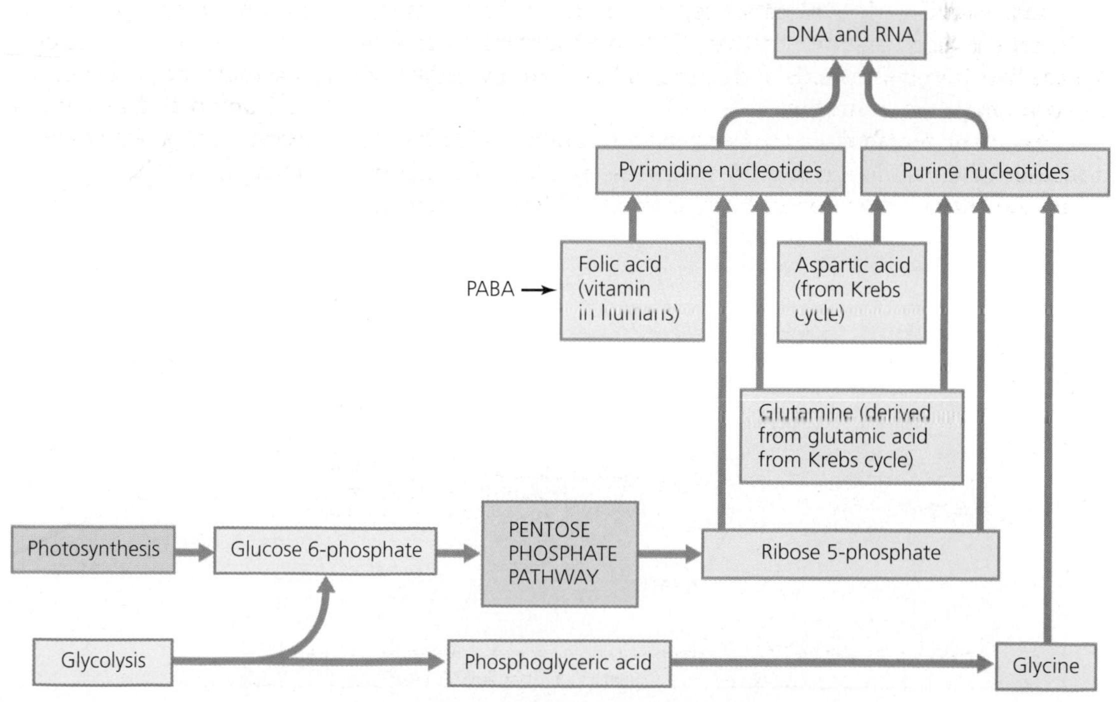

➤ **Figure 5.32**
The biosynthesis of nucleotides.

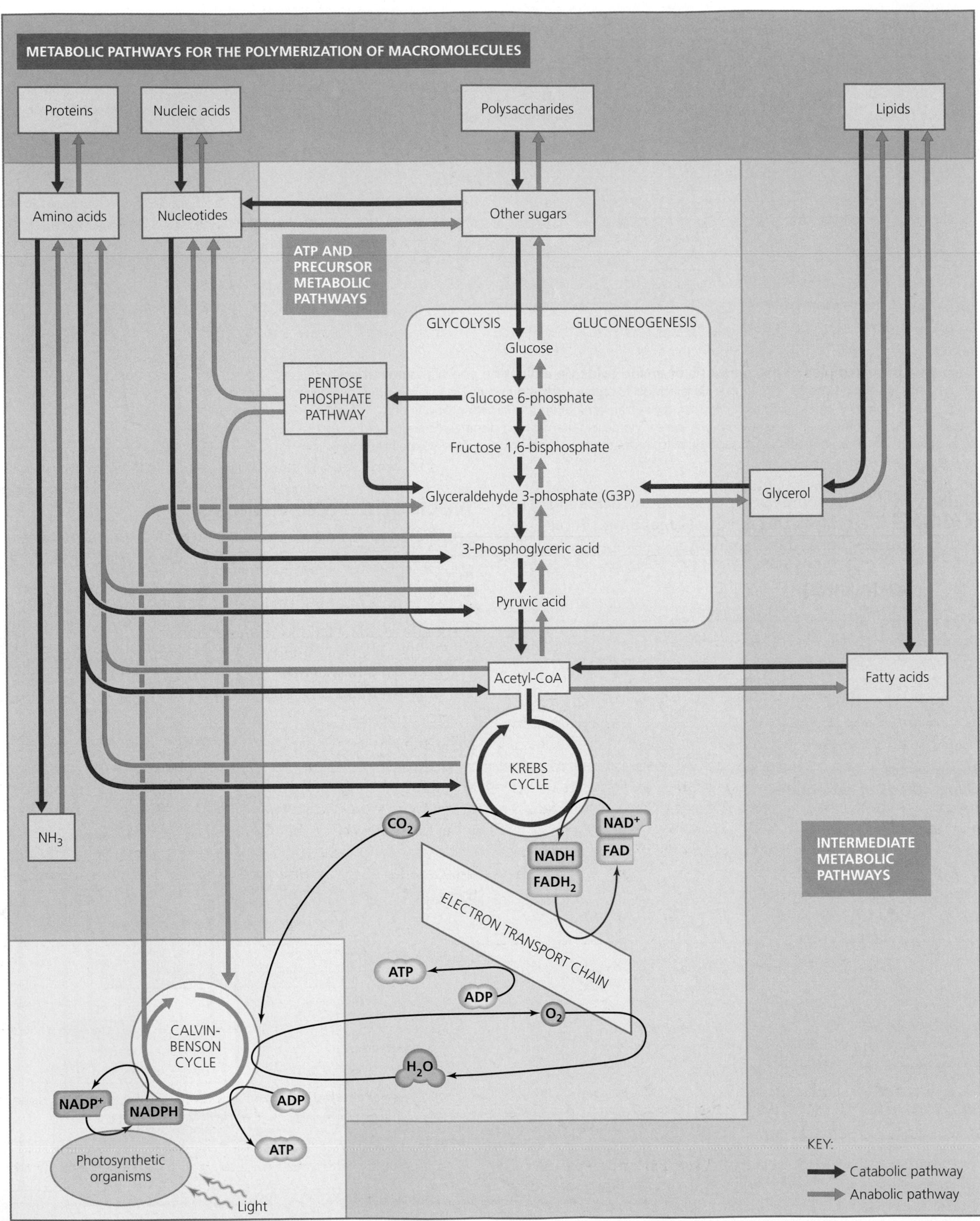
METABOLIC PATHWAYS FOR THE POLYMERIZATION OF MACROMOLECULES
Proteins
Nucleic acids
Polysaccharides
Lipids
Amino acids
Nucleotides
Other sugars
ATP AND PRECURSOR METABOLIC PATHWAYS
GLYCOLYSIS
GLUCONEOGENESIS
Glucose
PENTOSE PHOSPHATE PATHWAY
Glucose 6-phosphate
Fructose 1,6-bisphosphate
Glyceraldehyde 3-phosphate (G3P)
Glycerol
3-Phosphoglyceric acid
Pyruvic acid
Acetyl-CoA
Fatty acids
KREBS CYCLE
CO_2
NAD^+
FAD
NADH
$FADH_2$
NH_3
INTERMEDIATE METABOLIC PATHWAYS
ELECTRON TRANSPORT CHAIN
ATP
ADP
O_2
CALVIN-BENSON CYCLE
H_2O
$NADP^+$
NADPH
ADP
ATP
Photosynthetic organisms
Light
KEY:
Catabolic pathway
Anabolic pathway

◀ **Figure 5.33 Integration of cellular metabolism (shown in an aerobic organism).** Cells possess three major categories of metabolic pathways: pathways for the polymerization of macromolecules (proteins, nucleic acids, polysaccharides, and lipids), intermediate pathways, and ATP and precursor pathways (glycolysis, Krebs cycle, the pentose phosphate pathway, and the Entner-Doudoroff pathway [not shown]). Cells of photosynthetic organisms also have the Calvin-Benson cycle.

- The five-carbon sugars—ribose in RNA and deoxyribose in DNA—are derived from ribose 5-phosphate from the pentose phosphate pathway.
- The phosphate group is derived ultimately from ATP.
- Purines and pyrimidines are synthesized in a series of ATP-requiring reactions from the amino acids glutamine and aspartic acid derived from Krebs cycle intermediates, ribose 5-phosphate, and *folic acid.* (The latter is synthesized by many bacteria and protozoa but is a vitamin for humans.)

Chapter 7 examines the anabolic reactions by which polymerases polymerize nucleotides to form DNA and RNA.

Integration and Regulation of Metabolic Functions

Learning Objectives

✓ Describe interrelationships between catabolism and anabolism in terms of ATP and substrates.

✓ Discuss regulation of metabolic activity.

As we have seen, catabolic and anabolic reactions interact with one another in several ways. First, ATP molecules produced by catabolism are used to drive anabolic reactions. Second, catabolic pathways produce precursor metabolites to use as substrates for anabolic reactions. Additionally, most metabolic pathways are amphibolic; they function as part of either catabolism or anabolism as needed.

Cells regulate metabolism in a variety of ways to maximize efficiency in growth and reproductive rate. Among the mechanisms involved are the following:

- Cells synthesize or degrade channel and transport proteins to increase or decrease the concentration of chemicals in the cytosol or organelles.
- Cells often synthesize the enzymes needed to catabolize a particular substrate only when that substrate is available. For instance, the enzymes of beta-oxidation are not produced when there are no fatty acids to catabolize.
- If two energy sources are available, cells catabolize the more energy efficient of the two. For example, a bacterium growing in the presence of both glucose and lactose will produce enzymes only for the transport and catabolism of glucose. Once the supply of glucose is depleted, lactose-utilizing proteins are produced.
- Cells synthesize the metabolites they need, but they typically cease synthesis if a metabolite is available as a nutrient. For instance, bacteria grown with an excess of aspartic acid will cease the amination of oxaloacetic acid (see Figure 5.31).
- Eukaryotic cells keep metabolic processes from interfering with each other by isolating particular enzymes within membrane-bounded organelles. For example, proteases sequestered within lysosomes digest phagocytized proteins without destroying vital proteins in the cytosol.
- Cells use inhibitory and excitatory allosteric sites on enzymes to control the activity of enzymes (see Figure 5.10).
- Feedback inhibition slows or stops anabolic pathways when the product is in abundance (see Figure 5.11).
- Cells regulate catabolic and anabolic pathways that use the same substrate molecules by requiring different coenzymes for each. For instance, NADH is used almost exclusively with catabolic enzymes, whereas NADPH is typically used for anabolism.

Note that these regulatory mechanisms are generally of two types: *control of gene expression,* in which cells control the amount and timing of protein (enzyme) production, and *control of metabolic expression,* in which cells control the activity of proteins (enzymes) once they have been produced.

Figure 5.33 schematically diagrams some of the numerous interrelationships among the metabolic pathways discussed in this chapter. **ANIMATIONS:** ***Metabolism: The Big Picture***

Chapter Summary

This chapter has *MicroFlix*. Go to the Study Area at **www.masteringmicrobiology.com** to view movie-quality animations for metabolism.

Basic Chemical Reactions Underlying Metabolism (pp. 125–133)

1. **Metabolism** is the sum of biochemical reactions within the cells of an organism, including **catabolism,** which breaks down molecules and releases energy, and **anabolism,** which synthesizes molecules and uses energy.
   ANIMATIONS: ***Metabolism: Overview***
2. **Precursor metabolites,** often produced in catabolic reactions, are used to synthesize all other organic compounds.
3. **Reduction** reactions are those in which electrons are added. The molecule that donates an electron is **oxidized.** If the electron is

part of a hydrogen atom, an oxidation reaction is also called dehydrogenation. Oxidation and reduction reactions always occur in pairs called **oxidation-reduction (redox) reactions.**
ANIMATIONS: *Oxidation-Reduction Reactions*

4. Three important electron carrier molecules are **nicotinamide adenine dinucleotide (NAD^+), nicotinamide adenine dinucleotide phosphate ($NADP^+$),** and **flavin adenine dinucleotide (FAD).**

5. Phosphorylation is the addition of phosphate to a molecule. Three types of phosphorylation form ATP: **Substrate-level phosphorylation** involves the transfer of phosphate from a phosphorylated organic compound to ADP. In **oxidative phosphorylation,** energy from redox reactions of respiration is used to attach inorganic phosphate (PO_4^{3-}) to ADP. **Photophosphorylation** is the phosphorylation of ADP with inorganic phosphate using energy from light.

6. Catalysts increase the rates of chemical reactions and are not permanently changed in the process. **Enzymes,** which are organic catalysts, are often named for their **substrates**—the molecules upon which they act. Enzymes can be classified as hydrolases, isomerases, ligases (polymerases), lyases, oxidoreductases, or transferases, reflecting their mode of action.
ANIMATIONS: *Enzymes: Overview*

7. **Apoenzymes** are the portions of enzymes that may require one or more **cofactors** such as inorganic ions or organic cofactors (also called **coenzymes**). The combination of both apoenzyme and its cofactors is a **holoenzyme.** RNA molecules functioning as enzymes are called **ribozymes.**

8. **Activation energy** is the amount of energy required to initiate a chemical reaction.

9. Substrates fit into the specifically shaped **active sites** of the enzymes that catalyze their reactions.
ANIMATIONS: *Enzymes: Steps in a Reaction*

10. Enzymes may be **denatured** by physical and chemical factors such as heat and pH. Denaturation may be reversible or permanent.

11. Enzyme activity proceeds at a rate proportional to the concentration of substrate molecules until all the active sites are filled.

12. **Competitive inhibitors** block active sites and thereby block enzyme activity. **Noncompetitive inhibitors** attach to an allosteric site on an enzyme, altering the active site so that it is no longer functional.
ANIMATIONS: *Enzymes: Competitive Inhibition, Noncompetitive Inhibition*

13. **Feedback inhibition** (negative feedback) occurs when the final product of a series of reactions is an allosteric inhibitor of some previous step in the series. Thus, accumulation of the end-product "feeds back" into the series a signal that stops the process.

Carbohydrate Catabolism (pp. 133–147)

1. **Glycolysis** involves the splitting of a glucose molecule in a three-stage, 10-step process that ultimately results in two molecules of pyruvic acid and a net gain of two ATP and two NADH molecules.
ANIMATIONS: *Glycolysis: Overview, Steps*

2. The **pentose phosphate** and **Entner-Doudoroff pathways** are alternative means for the catabolism of glucose that yield fewer ATP molecules than does glycolysis. However, they produce precursor metabolites not produced in glycolysis.

3. **Cellular respiration** is a metabolic process that involves the complete oxidation of substrate molecules and the production of ATP following a series of redox reactions.

4. Two carbons from pyruvic acid join coenzyme A to form acetyl-coenzyme A (**acetyl-CoA**), which then enters the **Krebs cycle,** a series of eight enzymatic steps that transfer electrons from acetyl-CoA to coenzymes NAD^+ and FAD.
ANIMATIONS: *Krebs Cycle: Overview, Steps*

5. An **electron transport chain** is a series of redox reactions that pass electrons from one membrane-bound carrier to another, and then to a final electron acceptor. The energy from these electrons is used to pump protons across the membrane.
ANIMATIONS: *Electron Transport Chain: Overview*

6. The four classes of carrier molecules in electron transport systems are flavoproteins, ubiquinones, metal-containing proteins, and cytochromes.
ANIMATIONS: *Electron Transport Chain: The Process*

7. Aerobes use oxygen atoms as final electron acceptors in their electron transport chains in a process known as **aerobic respiration,** whereas anaerobes use other inorganic molecules (such as NO_3^-, SO_4^{2-}, and CO_3^{2-}, or rarely an externally acquired organic molecule) as the final electron acceptor in **anaerobic respiration.**
ANIMATIONS: *Electron Transport Chain: Factors Affecting ATP Yield*

8. In **chemiosmosis,** ions flow down their electrochemical gradient across a membrane through **ATP synthase (ATPase)** to synthesize ATP.

9. A **proton gradient** is an electrochemical gradient of hydrogen ions across a membrane. It has potential energy known as a proton motive force.

10. Oxidative phosphorylation and photophosphorylation use chemiosmosis.

11. **Fermentation** is the partial oxidation of sugar to release energy using a cellular organic molecule rather than an electron transport chain as the final electron acceptor. End-products of fermentation, which are often useful to humans and aid in identification of microbes in the laboratory, include acids, alcohols, and gases.
ANIMATIONS: *Fermentation*

Other Catabolic Pathways (pp. 148–149)

1. Lipids and proteins can be catabolized into smaller molecules, which can be used as substrates for glycolysis and the Krebs cycle.

2. **Beta-oxidation** is a catabolic process in which enzymes split pairs of hydrogenated carbon atoms from a fatty acid and join them to coenzyme A to form acetyl-CoA.

3. **Proteases** secreted by microorganisms digest proteins outside the microbes' cell walls. The resulting amino acids are moved into the cell and used in anabolism, or **deaminated** and catabolized for energy.

Photosynthesis (pp. 149–154)

1. **Photosynthesis** is a process in which light energy is captured by pigment molecules called **chlorophylls** (bacteriochlorophylls in

some bacteria) and transferred to ATP and metabolites. **Photosystems** are networks of light-absorbing chlorophyll molecules and other pigments held within a protein matrix on membranes called **thylakoids.**
ANIMATIONS: *Photosynthesis: Overview*

2. The redox reactions of photosynthesis are classified as **light-dependent reactions** and **light-independent reactions.**
3. A **reaction center chlorophyll** is a special chlorophyll molecule in a photosystem in which electrons are excited by light energy and passed to an acceptor molecule of an electron transport chain.
4. In **cyclic photophosphorylation,** the electrons return to the original reaction center after passing down the electron transport chain.
ANIMATIONS: *Photosynthesis: Light Reaction: Cyclic Photophosphorylation*
5. In **noncyclic photophosphorylation,** photosystem II works with photosystem I, and the electrons are used to reduce $NADP^+$ to NADPH. In oxygenic photosynthesis, cyanobacteria, algae, and green plants replenish electrons to the reaction center by dissociation of H_2O molecules, resulting in the release of O_2 molecules. Anoxygenic bacteria derive electrons from inorganic compounds such as H_2S, producing waste such as sulfur.
ANIMATIONS: *Photosynthesis: Light Reaction: Noncyclic Photophosphorylation*
6. In the light-independent pathway of photosynthesis, **carbon fixation** occurs in the **Calvin-Benson cycle,** in which CO_2 is reduced to produce glucose.
ANIMATIONS: *Photosynthesis: Light-Independent Reaction*

Other Anabolic Pathways (pp. 154–159)

1. **Amphibolic reactions** are metabolic reactions that are reversible—they can operate catabolically or anabolically.
2. Some cells are able to synthesize glucose from amino acids, glycerol, and fatty acids via a process called gluconeogenesis.
3. **Amination** reactions involve adding an amine group from ammonia to a metabolite to make an amino acid. **Transamination** occurs when an amine group is transferred from one amino acid to another.
4. Nucleotides are synthesized from precursor metabolites produced by glycolysis, the Krebs cycle, and the pentose phosphate pathway.

Integration and Regulation of Metabolic Functions (p. 159)

1. Cells regulate metabolism by control of gene expression or metabolic expression. They control the latter in a variety of ways, including synthesizing or degrading channel proteins and enzymes, sequestering reactions in membrane-bounded organelles (seen only in eukaryotes), and feedback inhibition.
ANIMATIONS: *Metabolism: The Big Picture*

Questions for Review

Answers to the Questions for Review (except Short Answer questions) begin on page A-1.

Multiple Choice

For each of the phrases in questions 1–7, indicate the type of metabolism referred to, using the following choices:

a. anabolism only
b. both anabolism and catabolism (amphibolic)
c. catabolism only

1. Breaks a large molecule into smaller ones
2. Includes dehydration synthesis reactions
3. Is exergonic
4. Is endergonic
5. Involves the production of cell membrane constituents
6. Includes hydrolytic reactions
7. Includes metabolism
8. Redox reactions
   a. transfer energy.
   b. transfer electrons.
   c. involve oxidation and reduction.
   d. are involved in all of the above.
9. A reduced molecule
   a. has gained electrons.
   b. has become more positive in charge.
   c. has lost electrons.
   d. is an electron donor.
10. Activation energy
    a. is the amount of energy required during an activity such as flagellar motion.
    b. requires the addition of nutrients in the presence of water.
    c. is lowered by the action of organic catalysts.
    d. results from the movement of molecules.
11. Coenzymes
    a. are types of apoenzymes.
    b. are proteins.
    c. are inorganic cofactors.
    d. are organic cofactors.
12. Which of the following statements best describes ribozymes?
    a. Ribozymes are proteins that aid in the production of ribosomes.
    b. Ribozymes are nucleic acids that produce ribose sugars.
    c. Ribozymes store enzymes in ribosomes.
    d. Ribozymes process RNA molecules in eukaryotes.

13. Which of the following does *not* affect the function of enzymes?
    a. ubiquinone
    b. substrate concentration
    c. temperature
    d. competitive inhibitors

14. Most oxidation reactions in bacteria involve the
    a. removal of hydrogen ions and electrons.
    b. removal of oxygen.
    c. addition of hydrogen ions and electrons.
    d. addition of hydrogen ions.

15. Under ideal conditions, the fermentation of one glucose molecule by a bacterium allows a net gain of how many ATP molecules?
    a. 2 b. 4 c. 38 d. 0

16. Under ideal conditions, the complete aerobic oxidation of one molecule of glucose by a bacterium allows a net gain of how many ATP molecules?
    a. 2 b. 4 c. 38 d. 0

17. Which of the following statements is *not* true of the Entner-Doudoroff pathway?
    a. It is a series of reactions that synthesizes glucose.
    b. Its products are sometimes used to determine the presence of *Pseudomonas*.
    c. It is a pathway of chemical reactions that catabolizes glucose.
    d. It is an alternative pathway to glycolysis.

18. Reactions involved in the light-independent reactions of photosynthesis constitute the
    a. Krebs cycle.
    b. Entner-Doudoroff pathway.
    c. Calvin-Benson cycle.
    d. pentose phosphate pathway.

19. The glycolysis pathway is basically
    a. catabolic. c. anabolic.
    b. amphibolic. d. cyclical.

20. A major difference between anaerobic respiration and anaerobic fermentation is
    a. in the use of oxygen.
    b. that the former requires breathing.
    c. the latter uses organic molecules within the cell as final electron acceptors.
    d. fermentation only produces alcohol.

Matching

1. ____ Occurs when energy from a compound containing phosphate reacts with ADP to form ATP
2. ____ Involves formation of ATP via reduction of coenzymes in the electron transport chain
3. ____ Begins with glycolysis
4. ____ Occurs when all active sites on substrate molecules are filled

A. Saturation
B. Oxidative phosphorylation
C. Substrate-level phosphorylation
D. Photophosphorylation
E. Carbohydrate catabolism

Fill in the Blanks

1. The final electron acceptor in cyclic photophosphorylation is ________________.
2. Two ATP molecules are used to initiate glycolysis. Enzymes generate molecules of ATP for each molecule of glucose that undergoes glycolysis. Thus a net gain of ________________ molecules of ATP is produced in glycolysis.
3. The initial catabolism of glucose occurs by glycolysis and/or the ________________ and ________________ pathways.
4. ________________ is a cyclic series of eight reactions involved in the catabolism of acetyl-CoA that yields eight molecules of NADH and two molecules of $FADH_2$.
5. The final electron acceptor in aerobic respiration is ________________ .
6. Three common inorganic electron acceptors in anaerobic respiration are ________________, ________________, and ________________.
7. Chemolithotrophs *acquire* electrons from (organic/inorganic) ________________ compounds.
8. Complete the following chart:

Category of Enzymes	Description
	Catabolizes substrate by adding water
Isomerase	
Ligase/polymerase	
	Moves functional groups such as an acetyl group
	Adds or removes electrons
Lyase	

9. The use of a proton motive force to generate ATP is ________________.
10. The main coenzymes that carry electrons in catabolic pathways are ________________ and ________________.

Short Answer

1. Why are enzymes necessary for anabolic reactions to occur in living organisms?
2. How do organisms control the rate of metabolic activities in their cells?
3. How does a noncompetitive inhibitor at a single allosteric site affect a whole pathway of enzymatic reactions?
4. Explain the mechanism of negative feedback with respect to enzyme action.
5. Facultative anaerobes can live under either aerobic or anaerobic conditions. What metabolic pathways allow these organisms to continue to harvest energy from sugar molecules in the absence of oxygen?

6. How does oxidation of a molecule occur without oxygen?
7. List at least four groups of microorganisms that are photosynthetic.
8. Why do we breathe oxygen and give off carbon dioxide?
9. Why do cyanobacteria and algae take in carbon dioxide and give off oxygen?
10. What happens to the carbon atoms in sugar catabolized by *Escherichia coli*?
11. How do yeast cells make alcohol and cause bread to rise?
12. Where specifically does the most significant production of ATP occur in prokaryotic and eukaryotic cells?
13. Why are vitamins essential metabolic factors for microbial metabolism?
14. A laboratory scientist notices that a certain bacterium does not utilize lactose when glucose is available in its environment. Describe a cellular regulatory mechanism that would explain this observation.
15. How does amination differ from transamination?
16. Label the diagram on the right to indicate acetyl-CoA, electron transport chain, $FADH_2$, fermentation, glycolysis, Krebs cycle, NADH, and respiration. Indicate the net number of molecules of ATP that could be synthesized at each stage during bacterial respiration of one molecule of glucose.

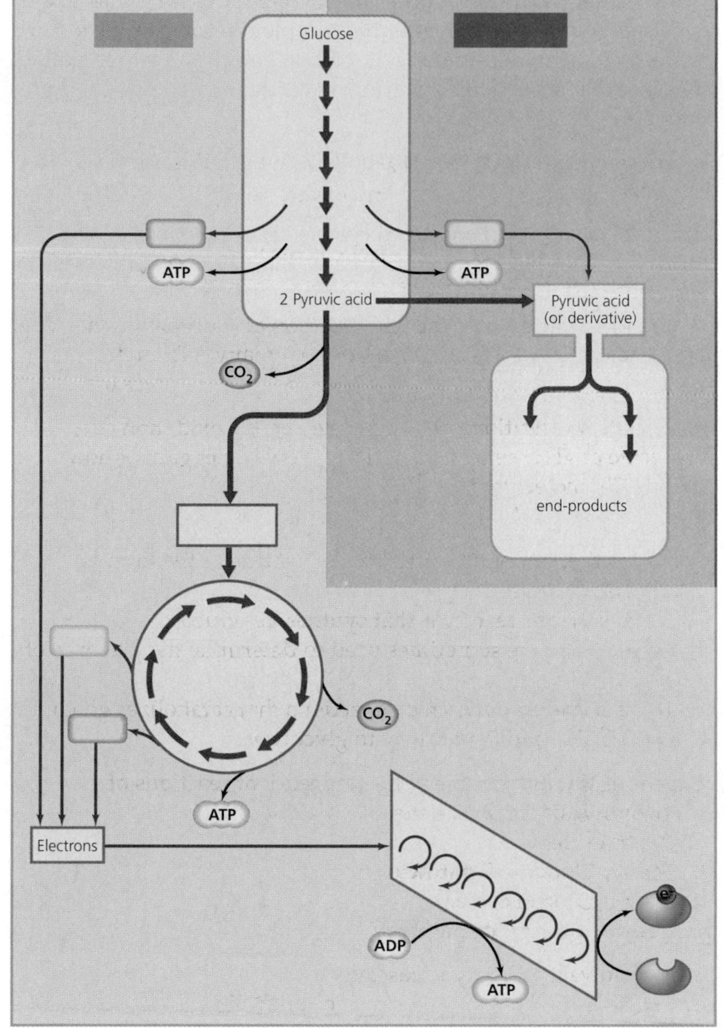

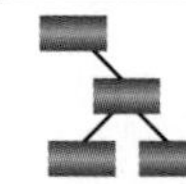

Concept Mapping

Using the following terms, draw a concept map that describes aerobic respiration. For a sample concept map, see p. 93. Or, complete this concept map online by going to the Study Area at www.masteringmicrobiology.com.

2 ATP **(2)**	Electron transport chain	Oxidative phosphorylation	Substrate-level phosphorylation **(2)**
34 ATP	Glycolysis		Synthesis of acetyl-CoA
Chemiosmosis	Krebs cycle		

Critical Thinking

1. Why might an organism that uses glycolysis and the Krebs cycle also need the pentose phosphate pathway?
2. Describe how bacterial fermentation causes milk to sour.
3. *Giardia intestinalis* and *Entamoeba histolytica* are protozoa that live in the colons of mammals and can cause life-threatening diarrhea. Interestingly, these microbes lack mitochondria. What kind of pathway must they have for carbohydrate catabolism?
4. Two cultures of a facultative anaerobe are grown in the same type of medium, but one is exposed to air and the other is maintained under anaerobic conditions. Which of the two cultures will contain more cells at the end of a week? Why?

5. What is the maximum number of molecules of ATP that can be generated by a bacterium after the complete aerobic oxidation of a fat molecule containing three 12-carbon chains? (Assume that all the available energy released during catabolism goes to ATP production.)

6. In terms of its effects on metabolism, why is a fever over 40°C often life threatening?

7. Cyanide is a potent poison because it irreversibly blocks cytochrome a_3. What effect would its action have on the rest of the electron transport chain? What would be the redox state (reduced or oxidized) of ubiquinone in the presence of cyanide?

8. How are photophosphorylation and oxidative phosphorylation similar? How are they different?

9. Members of the pathogenic bacterial genus *Haemophilus* require NAD^+ and heme from their environment. For what purpose does *Haemophilus* use these growth factors?

10. Compare and contrast aerobic respiration, anaerobic respiration, and fermentation.

11. Scientists estimate that up to one-third of Earth's biomass is composed of methanogenic prokaryotes in ocean sediments (*Science* 295:2067–2070). Describe the metabolism of these organisms.

12. A young student was troubled by the idea that a bacterium is able to control its diverse and complex metabolic activities, even though it lacks a brain. How would you explain its metabolic control?

13. If a bacterium uses beta-oxidation to catabolize a molecule of the fatty acid arachidic acid, which contains 20 carbon atoms, how many acetyl-CoA molecules will be generated?

14. Some desert rodents rarely have water to drink. How do they get enough water for their cells without drinking it?

15. Why do fatty acids typically contain an even number of carbon atoms?

16. We have examined the total ATP, NADH, and $FADH_2$ production in the Krebs cycle for each molecule of glucose coming through Embden-Meyerhof glycolysis. How many of each of these molecules would be produced if the Entner-Doudoroff pathway were used instead of Embden-Meyerhof glycolysis?

Access more review material online in the Study Area at **www.masteringmicrobiology.com.** There, you'll find

- **Animations**
- **Microflix**
- **MP3 Tutor Sessions**
- **Concept Mapping Activities**
- **Flashcards**
- **Quizzes**

and more to help you succeed.

6 Microbial Nutrition and Growth

Green sulfur bacteria are obligate anaerobes that must derive the energy they need for metabolism from light. So what are green sulfur bacteria doing on the bottom of the Pacific Ocean one and a half miles from the surface? How is it possible that they can live at this inky depth where sunlight never penetrates?

Oceanographers have discovered that water around deep-sea hydrothermal vents emits an eerie glow in the visible spectrum of light. No one knows how this strange light forms—perhaps from chemical reactions in the vent seepage, perhaps from imploding gas bubbles released by the superheated water from the vent. Though the light is too dim to be seen by human eyes, green sulfur bacteria have pigments suitable for absorbing the specific wavelengths of vent light. Presumably the bacteria absorb enough energy to support their metabolism.

Under what conditions do microorganisms grow and survive? In this chapter, we will study the requirements for microbial nutrition and growth.

Take the pre-test for this chapter online. Visit the Study Area at www.masteringmicrobiology.com.

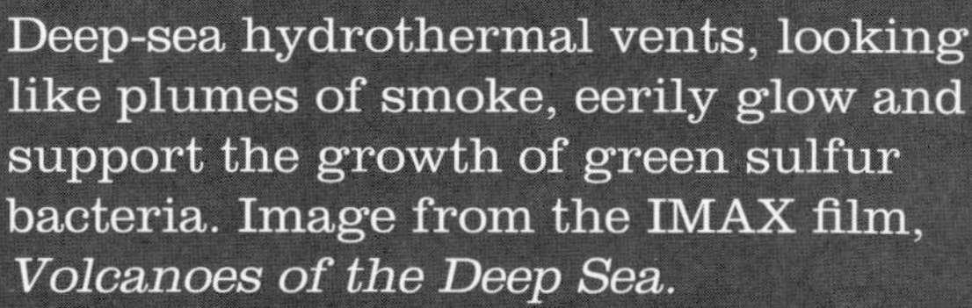

Deep-sea hydrothermal vents, looking like plumes of smoke, eerily glow and support the growth of green sulfur bacteria. Image from the IMAX film, *Volcanoes of the Deep Sea*.

We saw in Chapter 5 that metabolism—the set of controlled chemical reactions within cells—is a major characteristic of all living things. Recall as well that the ultimate outcome of metabolic activity is reproduction, an increase in the number of individual cells or organisms. When speaking of the reproductive activities of microbes in general, and of bacteria in particular, microbiologists typically use the term *growth,* referring to an increase in the size of a *population* of microbes rather than to an increase in size of an individual. The result of such microbial growth is either a discrete *colony,* which is an aggregation of cells arising from a single parent cell, or a *biofilm,* which is a collection of microbes living on a surface in a complex community. Put another way, the *reproduction* of individual microorganisms results in the *growth* of a colony or biofilm. Further, the common expression "The microorganisms *grow* in salt-containing media" is widely understood to mean that they metabolize and reproduce rather than that they increase in size.

In this chapter we consider the characteristics of microbial growth from two different but related perspectives: We examine the requirements of microbes living in their natural settings, including their chemical, physical, and energy requirements, and we explore how microbiologists try to create similar conditions to grow microorganisms in the laboratory so that they can be transported, identified, and studied. We conclude by examining laboratory analysis of bacterial population dynamics and some techniques for measuring bacterial population growth.

Growth Requirements

As we discussed in Chapter 5, organisms use a variety of chemicals—called **nutrients**—to meet their energy needs and to build organic molecules and cellular structures. The most common of these nutrients are compounds containing necessary elements such as carbon, oxygen, nitrogen, and hydrogen. Like all organisms, microbes obtain nutrients from a variety of sources in their environment, and they must bring nutrients into their cells by the passive and active transport processes we discussed in Chapter 3. When they acquire their nutrients by living in or on another organism, they may cause disease as they interfere with their hosts' metabolism and nutrition.

Nutrients: Chemical and Energy Requirements

Learning Objectives

- ✓ Describe the roles of carbon, hydrogen, oxygen, nitrogen, trace elements, and vitamins in microbial growth and reproduction.
- ✓ Compare the four basic categories of organisms based on their carbon and energy sources.
- ✓ Distinguish among anaerobes, aerobes, aerotolerant anaerobes, facultative anaerobes, and microaerophiles.
- ✓ Explain how oxygen can be fatal to organisms by discussing singlet oxygen, superoxide radical, peroxide anion, and hydroxyl radical, and describe how organisms protect themselves from toxic forms of oxygen.
- ✓ Define *nitrogen fixation* and explain its importance.

We begin our examination of microbial growth requirements by considering three things all cells need for metabolism: a carbon source, a source of energy, and a source of electrons or hydrogen atoms.

Sources of Carbon, Energy, and Electrons

Organisms can be categorized into two broad groups based on their source of carbon. Organisms that utilize an inorganic source of carbon (that is, carbon dioxide) as their sole source of carbon are called *autotrophs*[1] (aw'tō-trōfs), so named because they "feed themselves." More precisely, autotrophs make organic compounds from CO_2 and thus need not acquire carbon from organic compounds from other organisms. In contrast, organisms called *heterotrophs*[2] (het'er-ō-trōfs) catabolize reduced organic molecules (such as proteins, carbohydrates, amino acids, and fatty acids) they acquire from other organisms.

Organisms can also be categorized according to whether they use chemicals or light as a source of energy for such cellular processes as anabolism, intracellular transport, and motility. Organisms that acquire energy from redox reactions involving inorganic and organic chemicals are called *chemotrophs* (kēm'ō-trōfs). Recall from Chapter 5 that these reactions are either aerobic respiration, anaerobic respiration, or fermentation, depending on the final electron acceptor. Organisms that use light as their energy source are called *phototrophs*[3] (fō'tō-trōfs).

Thus we see that organisms can be categorized based on their carbon and energy sources into one of four basic groups: **photoautotrophs**, **chemoautotrophs**, **photoheterotrophs**, and **chemoheterotrophs** (**Figure 6.1**). Plants, some protozoa, and algae are photoautotrophs, and animals, fungi, and other protozoa are chemoheterotrophs. Bacteria and archaea exhibit greater metabolic diversity than any other group, with members in all four groups.

Additionally, the cells of all organisms require electrons or hydrogen atoms for redox reactions. Hydrogen is the most common chemical element in cells, and it is so common in organic molecules and water that it is never a *limiting nutrient;* that is, metabolism is never interrupted by a lack of hydrogen. Hydrogen is essential for hydrogen bonding and in electron transfer. Heterotrophs acquire electrons (typically as part of hydrogen atoms) from the same organic molecules that provide them carbon and are called **organotrophs** (ōr'gān-ō-trōfs); alternatively, autotrophic organisms acquire electrons or hydrogen atoms from inorganic molecules (such as H_2, NO^{2-}, H_2S, and Fe^{2+}) and are called **lithotrophs**[4] (lith'ō-trōfs).

[1]From Greek *auto*, meaning self, and *trophe*, meaning nutrition.
[2]From Greek *hetero*, meaning other, and *trophe*, meaning nutrition.
[3]From Greek *photos*, meaning light, and *trophe*, meaning nutrition.
[4]From Greek *lithos*, meaning rock, and *trophe*, meaning nutrition.

Figure 6.1 Four basic groups of organisms based on their carbon and energy sources. Additionally, organotrophs utilize electrons from organic molecules, and lithotrophs utilize electrons from inorganic molecules.

Carbon Source	Energy Source: Light *(photo-)*	Energy Source: Chemical compounds *(chemo-)*
Carbon dioxide *(auto-)*	***Photoautotrophs*** • Plants, algae, and cyanobacteria use H_2O to reduce CO_2, producing O_2 as a by-product • Green sulfur bacteria and purple sulfur bacteria do not use H_2O nor produce O_2	***Chemoautotrophs*** • Hydrogen, sulfur, and nitrifying bacteria, some archaea
Organic compounds *(hetero-)*	***Photoheterotrophs*** • Green nonsulfur bacteria and purple nonsulfur bacteria, some archaea	***Chemoheterotrophs*** • Aerobic respiration: most animals, fungi, and protozoa, and many bacteria • Anaerobic respiration: some animals, protozoa, bacteria, and archaea • Fermentation: some bacteria, yeasts, and archaea

CRITICAL **THINKING**

The filamentous bacterium *Beggiatoa* gets its carbon from carbon dioxide and its electrons and energy from hydrogen sulfide. What is its nutritional classification?

Not all organisms are easy to classify. For instance, the single-celled eukaryote *Euglena granulata* typically uses light energy and gets its carbon from carbon dioxide. However, when it is cultured on a suitable medium in the dark, this microbe utilizes energy and carbon solely from organic compounds. What is the nutritional classification of *Euglena*?

Oxygen Requirements

As we learned in Chapter 5, oxygen is essential for **obligate aerobes** because it serves as the final electron acceptor of electron transport chains, which produce most of the ATP in these organisms. By contrast, oxygen is a deadly poison for **obligate anaerobes**. How can oxygen be essential for one group of organisms and yet be a fatal toxin for others?

The key to understanding this apparent incongruity is understanding that neither atmospheric oxygen (O_2) nor covalently bound oxygen in compounds such as carbohydrates and water is poisonous. Rather, the toxic forms of oxygen are those that are highly reactive. They are toxic for the same reason that oxygen is the final electron acceptor for aerobes: They are excellent oxidizing agents, so they steal electrons from other compounds, which in turn steal electrons from still other compounds. The resulting chain of vigorous oxidations causes irreparable damage to cells by oxidizing important compounds, including proteins and lipids.

There are four toxic forms of oxygen:

- **Singlet oxygen ($^1O_2$).** Singlet oxygen is molecular oxygen with electrons that have been boosted to a higher energy state, typically during aerobic metabolism. Singlet oxygen is a very reactive oxidizing agent. Phagocytic cells, such as certain human white blood cells, use it to oxidize pathogens. Because singlet oxygen is also photochemically produced by the reaction of oxygen and light, phototrophic microorganisms often contain pigments called **carotenoids** (ka-rot′e-noyds) that prevent toxicity by removing the excess energy of singlet oxygen.
- **Superoxide radical (O_2^-).** A few superoxide radicals form during the incomplete reduction of O_2 during electron transport in aerobes and during metabolism by anaerobes in the presence of oxygen. Superoxide radicals are so reactive and toxic that aerobic organisms must produce enzymes called *superoxide dismutases* (dis′-myu-tās-es) to detoxify them. These enzymes, which have active sites that contain metal ions—Zn^{2+}, Mn^{2+}, Fe^{2+}, Ni^{2+}, or Cu^{2+}, depending on the organism—combine two superoxide radicals and two protons to form hydrogen peroxide (H_2O_2) and molecular oxygen (O_2):

$$2\,O_2^- + 2\,H^+ \rightarrow H_2O_2 + O_2$$

One reason that anaerobes are susceptible to oxygen is that they lack superoxide dismutase; they die as a result of the oxidizing reactions of superoxide radicals formed in the presence of oxygen.

- **Peroxide anion (O_2^{2-}).** Hydrogen peroxide formed during reactions catalyzed by superoxide dismutase (and during other metabolic reactions) contains peroxide anion, another highly reactive oxidant. It is peroxide anion that makes hydrogen peroxide an antimicrobial agent. Aerobes contain either catalase or peroxidase, enzymes that detoxify peroxide anion.

Catalase converts hydrogen peroxide to water and molecular oxygen:

$$2\,H_2O_2 \rightarrow 2\,H_2O + O_2$$

A simple test for catalase involves adding a sample from a bacterial colony to a drop of hydrogen peroxide. The

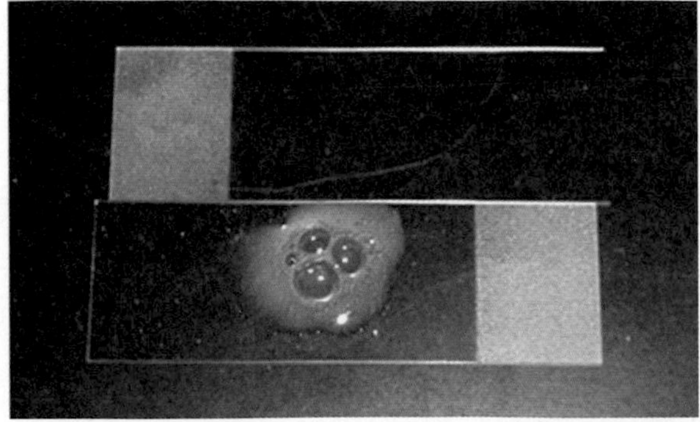

▲ **Figure 6.2 Catalase test.** The enzyme catalase converts hydrogen peroxide into water and oxygen, the latter of which can be seen as visible bubbles. *Enterococcus faecalis* (above) is catalase negative, whereas *Staphylococcus epidermidis* (below) is catalase positive.

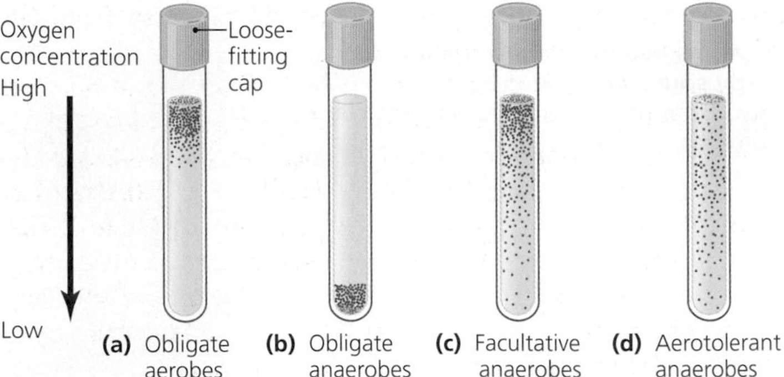

▲ **Figure 6.3 Using a liquid thioglycollate growth medium to identify the oxygen requirements of organisms.** The surface is exposed to atmospheric oxygen and is aerobic. Oxygen concentration decreases with depth; the bottom of the tube is anaerobic. **(a)** Obligate aerobes cannot survive below the depth to which oxygen penetrates the medium. **(b)** Obligate anaerobes cannot tolerate any oxygen. **(c)** Facultative anaerobes can grow with or without oxygen, but their ability to use aerobic respiration pathways enhances their growth near the surface. **(d)** Aerotolerant aerobes can grow equally well with or without oxygen; their growth is relatively evenly distributed throughout the medium. *Where in such a test tube would the growth zone be for a microaerophilic aerobe?*

Figure 6.3 *Microaerophiles would be found slightly below the surface, but neither directly at the surface nor in the depths of the tube.*

production of bubbles of oxygen indicates the presence of catalase **(Figure 6.2)**.

Peroxidase breaks down hydrogen peroxide without forming oxygen, using a reducing agent such as the coenzyme NADH:

$$H_2O_2 + NADH + H^+ \rightarrow 2\,H_2O + NAD^+$$

Obligate anaerobes either lack both catalase and peroxidase or have only a small amount of them, so they are susceptible to the toxic action of hydrogen peroxide.

- **Hydroxyl radical (OH·).** Hydroxyl radicals result from ionizing radiation and from the incomplete reduction of hydrogen peroxide:

$$H_2O_2 + e^- + H^+ \rightarrow H_2O + OH\cdot$$

Hydroxyl radicals are the most reactive of the four toxic forms of oxygen, but because hydrogen peroxide does not accumulate in aerobic cells (due to the action of catalase and peroxidase), the threat of hydroxyl radical is virtually eliminated in aerobic cells.

Besides the enzymes superoxide dismutase, catalase, and peroxidase, aerobes use other antioxidants, such as vitamins C and E, to protect themselves against toxic oxygen products. These antioxidants provide electrons that reduce toxic forms of oxygen.

Not all organisms are either strict **aerobes** or **anaerobes;** many organisms can live in various oxygen concentrations between these two extremes. For example, some aerobic organisms can maintain life via fermentation or anaerobic respiration, though their metabolic efficiency is often reduced in the absence of oxygen. Such organisms are called **facultative anaerobes.** *Escherichia coli* (esh-ĕ-rik'ē-ā kō'lē) is an example of a facultatively anaerobic bacterium.

Aerotolerant anaerobes do not use aerobic metabolism, but they tolerate oxygen by having some of the enzymes that detoxify oxygen's poisonous forms. The lactobacilli that transform cucumbers into pickles, and milk into cheese, are aerotolerant. These organisms can be kept in a laboratory without the special conditions required by obligate anaerobes.

Microaerophiles, such as the ulcer-causing pathogen *Helicobacter pylori*[5] (hel'ĭ-kō-bak'ter pī'lō-rē), require oxygen levels of 2% to 10%. This concentration of oxygen is found in the stomach. Microaerophiles are damaged by the 21% concentration of oxygen in the atmosphere, presumably because they have limited ability to detoxify hydrogen peroxide and superoxide radicals.

Microbial groups contain members with each of the five types of oxygen requirement. Algae, most fungi and protozoa, and many prokaryotes are obligate aerobes. A few yeasts and numerous prokaryotes are facultative anaerobes. Many prokaryotes and a few protozoa are aerotolerant, microaerophilic, or obligate anaerobes. The oxygen requirement of an organism can be identified by growing it in a medium that contains an oxygen gradient from top to bottom **(Figure 6.3)**.

Nitrogen Requirements

Another essential element is nitrogen, which is contained in many organic compounds, including the amine group of amino acids and as part of nucleotide bases. Nitrogen makes up about 14% of the dry weight of microbial cells.

Nitrogen is often a growth-limiting nutrient for many organisms; that is, their anabolism ceases because they do not have sufficient nitrogen to build proteins and nucleotides. Organisms acquire nitrogen from organic and inorganic nutrients. For example, most photosynthetic organisms can reduce nitrate (NO_3^-) to ammonium (NH_4^+), which can then be used for

[5]From semihelical shape of the cell, and Greek *pyle,* meaning gate, in reference to *pylorus,* the distal portion of the stomach, which is the gate to the small intestine.

biosynthesis. In addition, all cells recycle nitrogen from their amino acids and nucleotides.

Though nitrogen constitutes about 79% of the atmosphere, relatively few organisms can utilize nitrogen gas. A few bacteria, notably many cyanobacteria and *Rhizobium* (rī-zō′ bē-ŭm), reduce nitrogen gas (N_2) to ammonia (NH_3) via a process called **nitrogen fixation.** Nitrogen fixation is essential for life on Earth because nitrogen-fixers provide nitrogen in a usable form to other organisms. Chapters 11 and 26 discuss nitrogen-fixing prokaryotes, as well as nitrifying prokaryotes—those that oxidize nitrogenous compounds to acquire electrons for electron transport.

Other Chemical Requirements

Together, carbon, hydrogen, oxygen, and nitrogen make up more than 95% of the dry weight of cells; phosphorus, sulfur, calcium, manganese, magnesium, copper, iron, and a few other elements constitute the rest. Phosphorus is a component of phospholipid membranes, DNA, RNA, ATP, and some proteins. Sulfur is a component of sulfur-containing amino acids, which bind to one another via disulfide bonds that are critical to the tertiary structure of proteins, and in vitamins such as thiamine (B_1) and biotin.

Other elements are called **trace elements** because they are required in very small ("trace") amounts. For example, a few atoms of selenium dissolved out of the walls of glass test tubes provide the total requirement for the growth of green algae in a laboratory. Other trace elements are usually found in sufficient quantities dissolved in water. For this reason, tap water can sometimes be used instead of distilled or deionized water to grow microorganisms in the laboratory.

Some microorganisms—for example, algae and photosynthetic bacteria—are lithotrophic photoautotrophs; that is, they can synthesize all of their metabolic and structural needs from inorganic nutrients. They have every enzyme and cofactor they need to produce all their cellular components. Most organisms, however, require small amounts of certain organic chemicals that they cannot synthesize, in addition to those that provide carbon and energy. These necessary organic chemicals are called **growth factors** (Table 6.1). For example, vitamins are growth factors for some microorganisms. Recall that vitamins constitute all or part of many coenzymes. (Note that vitamins are not growth factors for microorganisms that can manufacture them, such as *E. coli*.) Growth factors for various microbes

TABLE 6.1 Some Growth Factors of Microorganisms and Their Functions

Growth Factor	Function
Amino acids	Components of proteins
Cholesterol	Used by mycoplasmas (bacteria) for cell membranes
Heme	Functional portion of cytochromes in electron transport system
NADH	Electron carrier
Niacin (nicotinic acid, vitamin B_3)	Precursor of NAD^+ and $NADP^+$
Pantothenic acid (vitamin B_5)	Component of coenzyme A
Para-aminobenzoic acid (PABA)	Precursor of folic acid, which is involved in metabolism of one-carbon compounds and nucleic acid synthesis
Purines, pyrimidines	Components of nucleic acids
Pyridoxine (vitamin B_6)	Utilized in transamination syntheses of amino acids
Riboflavin (vitamin B_2)	Precursor of FAD
Thiamine (vitamin B_1)	Utilized in some decarboxylation reactions

HIGHLIGHT

HYDROGEN-LOVING MICROBES IN YELLOWSTONE'S HOT SPRINGS

If you have ever visited the geothermal springs at Yellowstone National Park, you may recall a "rotten-egg" odor caused by sulfur in the environment. Until recently, it was believed that microorganisms living in these springs used sulfur as their primary source of energy. Researchers at the University of Colorado at Boulder, however, have discovered that most of these microorganisms actually seem to live off hydrogen.

The researchers learned that the gene sequences of bacteria collected from the springs closely matched the gene sequences of other bacteria known to metabolize hydrogen. This was a surprise, because many people assumed that the bacteria were sulfur dependent. But the results also made sense, given the high-temperature environment of the springs. Sulfur-metabolizing microbes require oxygen, which is poorly soluble at high temperatures. Since water temperatures in Yellowstone's springs often surpass 70°C, it is understandable that microbes living in this environment would rely on hydrogen instead of sulfur.

▲ ***Thermophilic bacteria, which can be distinguished by their orange-colored carotenoids, surround the Grand Prismatic Spring in Yellowstone National Park.***

include some amino acids, purines, pyrimidines, cholesterol, NADH, and heme.

CRITICAL THINKING

Given that *Haemophilus ducreyi* is a chemoheterotrophic pathogen that requires heme as a growth factor, deduce how this bacterium phosphorylates most of its ADP to form ATP. Defend your answer.

Physical Requirements

Learning Objective

✓ Explain how extremes of temperature, pH, and osmotic and hydrostatic pressure limit microbial growth.

In addition to chemical nutrients, organisms have physical requirements for growth, including specific conditions of temperature, pH, osmolarity, and pressure.

Temperature

Temperature plays an important role in microbial life through its effects on the three-dimensional configurations of biological molecules. Recall that to function properly, proteins require a specific three-dimensional shape that is determined in part by temperature-sensitive hydrogen bonds, which are more likely to form at lower temperatures, and more likely to break at higher temperatures. When hydrogen bonds break, proteins denature and lose function. Additionally, lipids, such as those that are components of the membranes of cells and organelles, are temperature sensitive. If the temperature is too low, membranes become rigid and fragile; if the temperature is too high, the lipids become too fluid, and the membrane cannot contain the cell or organelle.

Thus, because temperature plays an important role in the three-dimensional structure of many types of biological molecules, different temperatures have different effects on the survival and growth of microbes (**Figure 6.4**). The lowest temperature at which an organism is able to conduct metabolism is called the *minimum growth temperature*. Note, however, that many microbes, particularly bacteria, survive but do not thrive at temperatures far below this temperature, despite the fact that cell membranes are less fluid and transport processes are too slow to support metabolic activity. The highest temperature at which an organism continues to metabolize is called the *maximum growth temperature;* when the temperature exceeds this value, the organism's proteins are permanently denatured, and it dies. The temperature at which an organism's metabolic activities produce the highest growth rate is the **optimum growth temperature.** Each organism thus survives over a *temperature range,* within which its growth and metabolism are supported.

CRITICAL THINKING

Examine the graph in Figure 6.4. Note that the growth rate increases slowly until the optimum is reached, and then it declines steeply at higher temperatures. In other words, organisms tolerate a wider range of temperatures below their optimal temperature than they do above the optimum. Explain this observation.

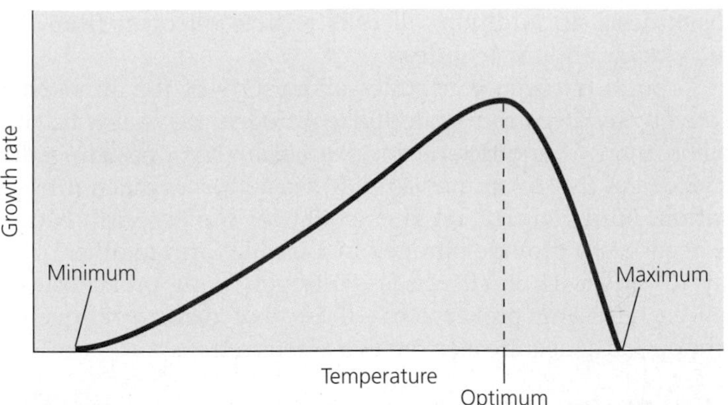

(a)

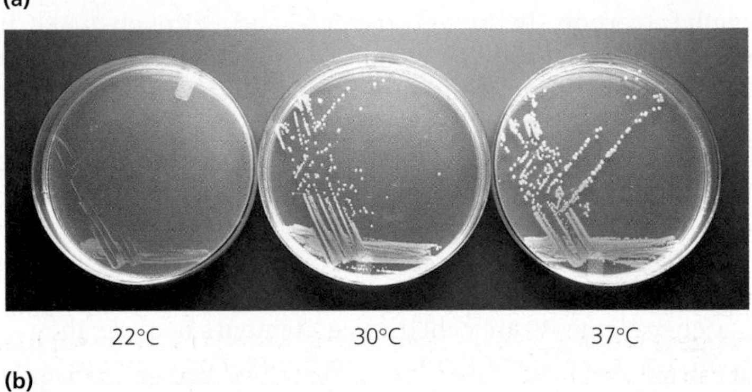

(b)

▲ **Figure 6.4 The effects of temperature on microbial growth.** **(a)** Minimum, optimum, and maximum growth temperatures, as determined from a graph showing growth rate plotted against temperature. **(b)** Growth of *Escherichia coli* on nutrient agar after 18 hours of incubation at three different temperatures. *If microorganisms can survive at temperatures lower than their minimum growth temperature, then why is it called "minimum"?*

Figure 6.4 *The minimum growth temperature is defined as the lowest temperature that supports metabolism. Many organisms can survive at low temperatures but do not actively metabolize, grow, or reproduce.*

Based on their preferred temperature ranges—the temperatures within which their metabolic activity and growth are best supported—microbes can be categorized into four overlapping groups (**Figure 6.5**). **Psychrophiles**[6] (sī'krō-fīls) grow best at temperatures below about 15°C and can even continue to grow at temperatures below 0°C. They die at temperatures much above 20°C. In nature, psychrophilic algae, fungi, archaea, and bacteria live in snowfields, ice, and cold water (**Figure 6.6**). They do not cause disease in humans because they cannot survive at body temperature; some do cause food spoilage in refrigerators. Psychrophiles present unique challenges to laboratory investigations, because they must be kept at cold temperatures. For example, microscope stages must be refrigerated, and the air temperatures needed to maintain living psychrophiles are uncomfortably cold for lab personnel.

Mesophiles[7] (mez'ō-fīls) are organisms that grow best in temperatures ranging from 20°C to about 40°C (see Figure 6.5),

[6]From Greek *psuchros*, meaning cold, and *philos*, meaning love.
[7]From Greek *mesos*, meaning middle, and *philos*, meaning love.

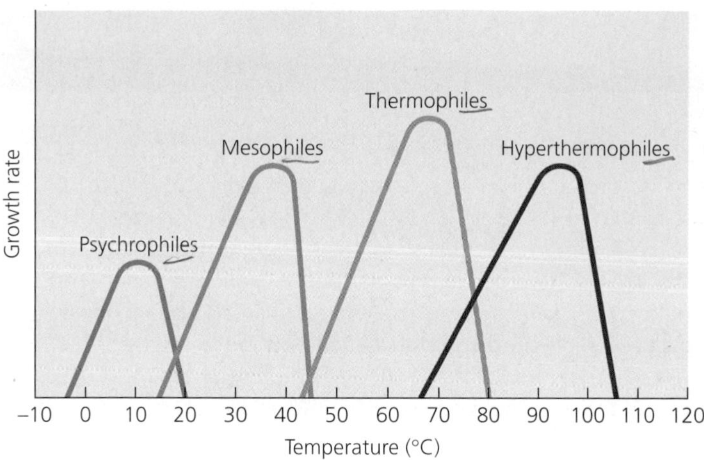

▲ **Figure 6.5 Four categories of microbes based on temperature ranges for growth.** *Categorize the bacterium* Vibrio marinus*, which has an optimum growth temperature near 10°C.*

Figure 6.5 *Vibrio marinus is a psychrophile.*

though they can survive at higher and lower temperatures. Because normal body temperature is approximately 37°C, human pathogens are mesophiles. *Thermoduric*[8] *organisms* are mesophiles that can survive brief periods at higher temperatures. Inadequate heating during pasteurization and canning can result in food spoilage by thermoduric mesophiles.

Thermophiles[9] (ther′mō-fīls) grow at temperatures above 45°C in habitats such as compost piles and hot springs. Some members of the Archaea, called **hyperthermophiles**, grow in water above 80°C; others live at temperatures above 100°C.[10] The current record holder is an archaeon, *Geogemma barossii* (jē′ō-jem-a ba-rōs′ē-ē). *Geogemma* grows and reproduces near submarine hot springs at temperatures between 85°C and 121°C and can survive for at least two hours at 130°C! Thermophiles and hyperthermophiles stabilize their proteins with extra hydrogen and covalent bonds between amino acids. Heat-stable enzymes are useful in industrial, engineering, and research applications. Heat-loving organisms do not cause disease because they "freeze" at body temperature. **Beneficial Microbes: A Nuclear-Waste-Eating Microbe?** on p. 172 highlights an unusual thermophile that can also withstand radiation.

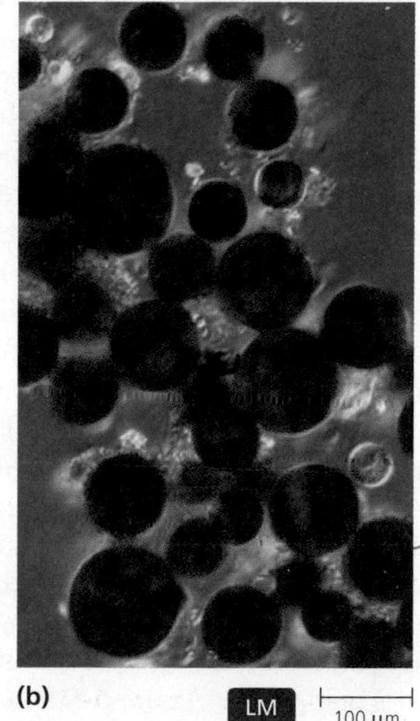

(a) (b) LM 100 μm

▲ **Figure 6.6 An example of a psychrophile. (a)** The alga *Chlamydomonas nivalis* colors this summertime snowbank on Couverville Island, Antarctica. **(b)** Microscopic view of the red-pigmented spores of *C. nivalis.*

CRITICAL **THINKING**

Over 100 years ago, doctors infected syphilis victims with malaria parasites to induce a high fever. Surprisingly, such treatment often cured the syphilis infection. Explain how this could occur.

pH Organisms are sensitive to changes in acidity because hydrogen ions and hydroxyl ions interfere with hydrogen bonding within proteins and nucleic acids; as a result, organisms have ranges of acidity that they prefer and can tolerate. Recall from Chapter 2 that pH is a measure of the concentration of hydrogen ions in a solution; that is, it is a measure of the acidity or alkalinity of a substance. A pH below 7.0 is acidic; the lower the pH value, the more acidic a substance is. Alkaline (basic) pH values are higher than 7.0.

Most bacteria and protozoa, including most pathogens, grow best in a narrow range around a neutral pH—that is, between pH 6.5 and pH 7.5, which is also the pH range of most tissues and organs in the human body; such microbes are thus called **neutrophiles** (nū′trō-fīls). By contrast, other bacteria and many fungi are **acidophiles** (ā-sid′ō-phīls), organisms that grow best in acidic habitats. One example of acidophilic microbes are the chemoautotrophic prokaryotes that live in mines and in water that runs through mine tailings (waste rock), habitats that have pHs as low as 0.0. These prokaryotes oxidize sulfur to sulfuric acid, further lowering the pH of their environment. Whereas *obligate acidophiles* require an acidic environment and die if the pH approaches 7.0, *acid-tolerant microbes* merely survive in acid without preferring it.

Many organisms produce acidic waste products that accumulate in their environment until eventually they inhibit further growth. For example, many cheeses are acidic because of lactic acid produced by fermenting bacteria and fungi. The low pH of these cheeses then acts as a preservative by preventing any further microbial growth. Other acidic foods, such as sauerkraut and dill pickles, are also kept from spoiling because most organisms cannot tolerate their low pH.

The normal acidity of certain regions of the body inhibits microbial growth and retards many kinds of infection. At one site, the vaginas of adult women, acidity results from the fermentation

[8]From Greek *therme*, meaning hot. The organisms are so named because of their ability to endure or tolerate heat.
[9]From Greek *therme*, meaning hot, and *philos*, meaning love.
[10]Water can remain a liquid above 100°C if it has a high salt content or is under pressure, such as occurs in geysers or deep ocean troughs.

BENEFICIAL MICROBES

A NUCLEAR-WASTE-EATING MICROBE?

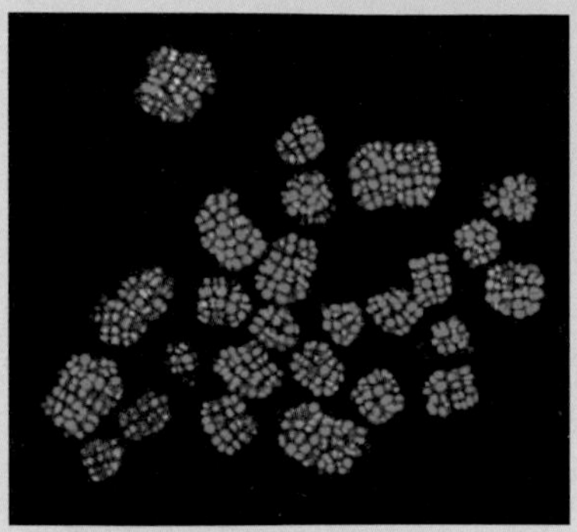

▲ *Kineococcus radiotolerans.* SEM 10 μm

Gamma rays emitted by radioactive decay are usually deadly. However, the bacterium *Kineococcus radiotolerans* can survive not only exposure to gamma rays but also toxic chemicals and desiccation. Currently, scientists at the U.S. Department of Energy (DOE) are studying this bacterium in the hope that it will prove useful in cleaning up nuclear wastes.

K. radiotolerans is what scientists call an *extremophile*, a microbe that can survive in extremely hostile environments. *K. radiotolerans* is notable among extremophiles because it thrives while bombarded with radiation thousands of times what it would take to kill a person. The microbe can also break down herbicides, chlorinated compounds, and other toxic substances. DOE researchers would like to shape *K. radiotolerans* into a biological tool that can clean up environments contaminated with radioactive wastes. Using microbes to break down toxic chemicals in the environment, a process known as bioremediation, is often cheaper and quicker than conventional methods. *K. radiotolerans* could potentially slash the cost of nuclear cleanup.

of carbohydrates by normal resident bacteria. If the growth of these normal residents is disrupted—for instance, by antibiotic therapy—the resulting higher pH may allow yeasts to grow and lead to a yeast infection. Another site, the stomach, is inhospitable to most microbes because of the normal production of stomach acid. However, the acid-tolerant bacterium *Helicobacter pylori* neutralizes stomach acid by secreting bicarbonate and urease, an enzyme that converts urea to ammonia, which is alkaline. The growth of *Helicobacter* is the cause of most gastric ulcers.

Alkaline conditions also inhibit the growth of most microbes, but **alkalinophiles** live in alkaline soils and water up to pH 11.5. For example, *Vibrio cholerae* (vib'rē-ō kol'er-ī), the causative agent of cholera, grows best outside of the body in water at pH 9.0.

Physical Effects of Water

Microorganisms require water; they must be in a moist environment if they are to be metabolically active. Water is needed to dissolve enzymes and nutrients; also, it is an important reactant in many metabolic reactions. Even though most cells die in the absence of water, some microorganisms—for example, the bacterium *Mycobacterium tuberculosis* (mī'kō-bak-tēr'ē-ŭm too-ber-kyū-lō'sis)—have cell walls that retain water, allowing them to survive for months under dry conditions. Additionally, the spores and cysts of some other single-celled microbes cease most metabolic activity in a dry environment for years; these cells are in essence in a state of suspended animation because they neither grow nor reproduce in their dry condition.

We now consider the physical effects of water on microbes by examining two topics: osmotic pressure and hydrostatic pressure.

Osmotic Pressure As we saw in Chapter 3, *osmosis* is the diffusion of water across a semipermeable membrane and is driven by unequal solute concentrations on the two sides of such a membrane. The *osmotic pressure* of a solution is the pressure exerted on a semipermeable membrane by a solution containing solutes (dissolved material) that cannot freely cross the membrane. Osmotic pressure is related to the concentration of dissolved molecules and ions in a solution. Solutions with greater concentrations of such solutes are *hypertonic* relative to those with a lower solute concentration, which are *hypotonic*.

Osmotic pressure can have dire effects on cells. For example, a cell placed in freshwater (a hypotonic solution relative to the cell's cytoplasm) gains water from its environment and swells to the limit of its cell wall. Cells that lack a cell wall—animal cells and some bacterial, fungal, and protozoan cells—will swell until they burst in hypotonic solutions. By contrast, a cell placed in seawater, which is a solution containing about 3.5% solutes and thus hypertonic to most cells, loses water into the surrounding saltwater. Such a cell can die from **crenation,** or shriveling of its cytoplasm. Osmotic pressure accounts for the preserving action of salt in jerky and salted fish, and of sugar in jellies, preserves, and honey. In those foods, the salt and sugar are solutes that draw water out of any microbial cells that are present, preventing growth and reproduction.

Osmotic pressure restricts organisms to certain environments. Some microbes, called **obligate halophiles,**[11] are adapted to growth under high osmotic pressure such as exists in the Great Salt Lake and smaller salt ponds. They may grow in up to 30% salt and will burst if placed in freshwater. Other microbes are *facultative halophiles;* that is, although they do not require high salt concentrations, they can tolerate them. One potential bacterial pathogen, *Staphylococcus aureus* (staf'i-lō-kok'ŭs o'rē-ŭs), can tolerate up to 20% salt, which allows it to colonize the surface of the skin—an environment that is too salty for

[11] From Greek *halos*, meaning salt, and *philos*, meaning love.

most microbes. *S. aureus* causes a number of different skin and mucous membrane diseases ranging from pimples, sties, and boils to life-threatening scalded skin and toxic shock syndromes. These diseases are covered more fully in Chapter 19.

Hydrostatic Pressure Water exerts pressure in proportion to its depth. For every additional 10 m of depth, water pressure increases 1 atmosphere (atm). Therefore, the pressure at 100 m below the surface is 10 atm—ten times greater than at the surface. Obviously, the pressure in deep ocean basins and trenches, which are thousands of meters below the surface, is tremendous. Organisms that live under such extreme pressure are called **barophiles**[12] (bar'ō-fīls). Their membranes and enzymes do not merely tolerate pressure but depend on pressure to maintain their three-dimensional, functional shapes. Thus barophiles brought to the surface quickly die because their proteins denature. Obviously, barophiles cannot cause diseases in humans, plants, or animals that do not live at great depths.

Associations and Biofilms

Learning Objective

✓ Describe how quorum sensing can lead to formation of a biofilm.

Organisms in a laboratory environment are living very differently than organisms in nature, which live in association with other individuals of their own and different species. The relationships between organisms can be viewed as falling along a continuum stretching from causing harm to providing benefits.

Relationships in which one organism harms or even kills another organism are considered *antagonistic relationships.* As discussed in Chapter 13, viruses are especially clear examples of antagonistic microbes; they require a cell in which to replicate themselves and almost always kill their cellular hosts. Beneficial relationships take at least two forms: synergistic relationships and symbiotic relationships. In *synergistic relationships,* the individual members of an association cooperate such that each receives benefits that exceed those that would result if each lived by itself, even though each member could live separately. In *symbiotic relationships,* organisms live in such close nutritional or physical contact that they become interdependent, such that the members rarely (if ever) live outside the relationship. Symbiotic relationships are discussed in greater detail in Chapter 14, particularly as they relate to the production of disease.

Biofilms are examples of complex relationships among numerous microorganisms, often different species, attached to surfaces such as teeth (dental plaque), rocks in streams, shower curtains ("soap scum" is really a biofilm), implanted medical devices (e.g., catheters), and mucous membranes of the digestive system. Many scientists consider that biofilms are the primary residence of microorganisms in nature. For example, one study showed that more than 10 billion bacteria per square centimeter form slippery biofilms on rocks in stream beds. The Centers for Disease Control and Prevention (CDC) estimates that biofilms cause up to 70% of bacterial diseases in industrialized countries, including prostatitis, kidney infections, tooth and gum decay, cystic fibrosis, and infections associated with implantation of medical devices.

Biofilms develop an extracellular *matrix,* composed of DNA, proteins, and primarily the tangled fibers of polysaccharides of the cells' glycocalyces. A matrix adheres cells to one another, sticks a biofilm to its substrate, forms microenvironments within a biofilm, sequesters nutrients, and may protect individuals in a biofilm from environmental stresses, including ultraviolet radiation, antimicrobial drugs, and changes in pH, temperature, and humidity. A matrix not only attaches a biofilm and positions the cells, it may also allow members of the biofilm to concentrate and conserve digestive enzymes, directing them against the underlying structure rather than having them diffuse away in the surrounding medium. Further, the synergistic relationships allowed in a biofilm organize the biofilm community, allowing individual members to display metabolic and structural traits different from those expressed by the same cells living individually. Members assume different roles in different areas of a biofilm, much like cells and tissues of multicellular organisms have different functions in different parts of the body.

Biofilms often form as a result of a process called **quorum sensing**, in which microorganisms respond to the density of nearby microorganisms. Cells secrete molecules into their environment that act as signals; many cells also possess receptors for these signal molecules. When the density of microorganisms increases, the concentration of signal molecules also increases, such that more and more receptors bind signal molecules. Once the binding exceeds a certain threshold amount, the expression of previously suppressed genes is triggered, and the result is that the microorganisms have new characteristics, such as the production of enzymes, changes in cell shape, formation of mating types, and the ability to form biofilms. Some researchers estimate quorum sensing may regulate up to 10% of the genes in a cell.

Given that many microorganisms become more harmful when they are part of a biofilm, scientists are seeking ways to prevent biofilms from forming in the first place. Researchers have learned that drugs that block cell receptors, thereby disrupting communication among bacterial cells, will inhibit biofilm formation and thus prevent disease. Such receptor-blocking drugs have successfully blocked biofilm formation and prevented disease in mice and are being considered for use in humans.

Another possible approach to preventing biofilm formation involves artificially amplifying quorum sensing while bacteria are still relatively few in number. Some pathogens hide within capsules and in blood clots, and only after they have multiplied significantly and formed a biofilm do they emerge and become "visible" to the immune system. With amplified quorum sensing, bacteria might produce certain proteins earlier in an infection cycle than normal, thus "revealing" themselves sooner to the immune system, which may then eliminate them before they can form biofilms and cause disease.

Dental plaque is a common biofilm that can lead to dental caries (cavities, **Figure 6.7**). Plaque formation usually begins with

[12]From Greek *baros*, meaning weight, and *philos*, meaning love.

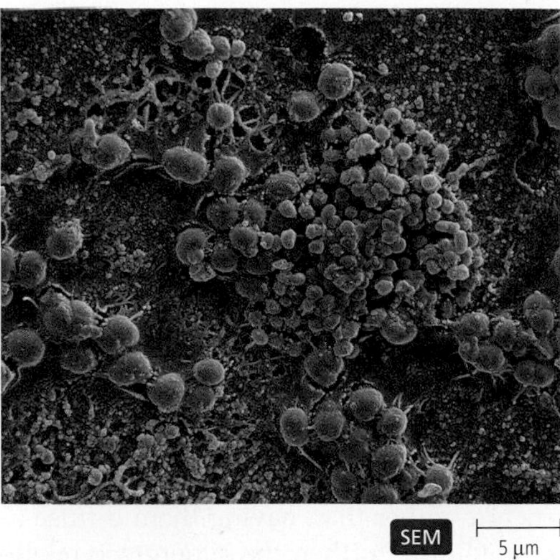

▲ **Figure 6.7** **Plaque (biofilm) on a human tooth.**

CLINICAL CASE STUDY

Cavities Gone Wild

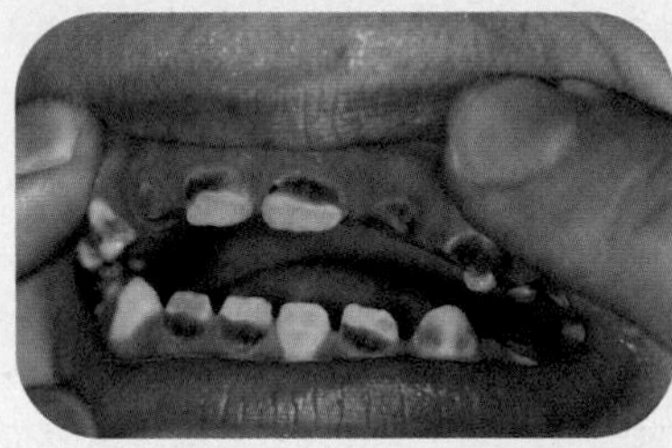

Five-year-old Daniel appears to be shy. He always looks at the floor, has no friends, never plays with the other children, will rarely speak to adults, and when he does speak, it is difficult to understand his broken enunciation. His skinny frame and the dark circles under his eyes make him appear malnourished. Daniel cries frequently and misses many days of school.

A speech specialist at school finds that only two of Daniel's teeth are healthy; all the others have rotted away to the gum line. The little guy is in constant pain and it hurts to chew. A doctor later determines that bacteria from the cavities in his mouth have entered his bloodstream and infected his heart, causing an irregular heartbeat and poor blood circulation.

1. How does knowledge of biofilms help explain the bulk of Daniel's problems?
2. What can Daniel, his parents, and health care professionals do to cure his diseases?
3. What nutrient should Daniel's parents eliminate from his diet to help prevent a repeat of his condition?
4. A recent study has shown that brushing does more than clean plaque from the teeth; it also disrupts associations between oral bacteria. How does this simple act help prevent the formation of biofilms?

colonization of the teeth by *Streptococcus mutans* (strep-tō-kok'ŭs mū'tanz). This bacterium breaks down carbohydrates, particularly the disaccharide sucrose (table sugar), to provide itself with nutrition and a glycocalyx. One of its enzymes catabolizes sucrose into its component monosaccharides—glucose and fructose—which the cell uses as energy sources. A second enzyme releases fructose as an energy source but polymerizes glucose into long, insoluble polysaccharide strands called glucan molecules, which form a sticky glycocalyx matrix around the bacterium. Glucan adheres *S. mutans* to the tooth, provides a home for other species of oral bacteria, and traps food particles. A biofilm has formed.

Bacteria in the biofilm digest nutrients and release acid, which is held against the teeth by the biofilm's matrix. The acid gradually eats away the minerals that compose the tooth, resulting in dental caries and eventually total loss of the teeth. **Clinical Case Study: Cavities Gone Wild** deals with an especially severe case of a biofilm running amok in a small boy's mouth.

Culturing Microorganisms

As we saw in Chapter 1, the second of Koch's postulates for demonstrating that a certain agent causes a specific disease requires that microorganisms be isolated and cultivated. Medical laboratory personnel must also grow pathogens as a step in the diagnosis of many diseases. To cultivate or *culture* microorganisms, a sample called an **inoculum** (plural: *inocula*) is introduced into a collection of nutrients called a **medium.** Microorganisms that grow from an inoculum are also called a *culture;* thus, **culture** can refer to the act of cultivating microorganisms or to the microorganisms that are cultivated.

Cultures can be grown in liquid media called **broths,** or on the surface of solid media. Cultures that are visible on the surface of solid media are called **colonies.** Bacterial and fungal colonies often have distinctive characteristics—including color, size, shape, elevation, texture, and appearance of the colony's margin (edge)—that taken together aid in identifying the microbial species that formed the colony **(Figure 6.8)**.

Microbiologists obtain inocula from a variety of sources. *Environmental specimens* are taken from such sources as ponds, streams, soil, and air. *Clinical specimens* are taken from patients and handled in ways that facilitate the examination of or testing for the presence of microorganisms. Another source of inocula is a culture originally grown from an environmental or clinical specimen and maintained in storage in a laboratory. Next we briefly examine clinical sampling.

Clinical Sampling

Learning Objective

✓ Describe methods for collecting clinical specimens from the skin and from the respiratory, reproductive, and urinary tracts.

Diagnosis and treatment of disease often depend on the isolation and correct identification of pathogens. Health care professionals must properly obtain samples from their patients and must then transport them quickly and correctly to a microbiology laboratory for culture and identification. They must take care to prevent the contamination of samples with microorganisms from the

Shape	Circular, Rhizoid, Irregular, Filamentous, Spindle
Margin	Entire, Undulate, Lobate, Curled, Rhizoid, Filamentous
Elevation	Flat, Raised, Convex, Pulvinate, Umbonate
Size	Punctiform, Small, Moderate, Large
Texture	Smooth or rough
Appearance	Glistening (shiny) or dull
Pigmentation	Nonpigmented (e.g., cream, tan, white) Pigmented (e.g., purple, red, yellow)
Optical property	Opaque, translucent, transparent

(a)

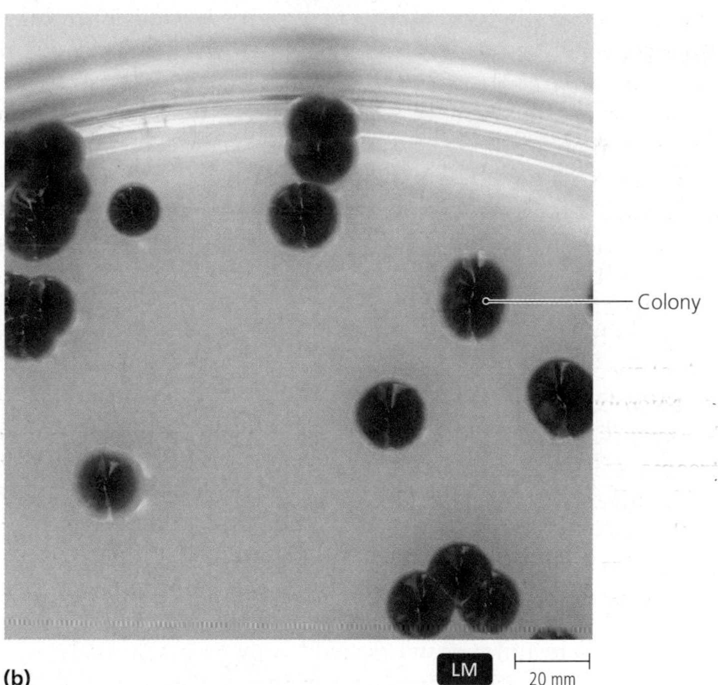

(b)

▲ **Figure 6.8 Characteristics of bacterial colonies.** **(a)** Shape, margin, elevation (side view), size, texture, appearance, pigmentation (color), and optical properties are described by a variety of terms. **(b)** *Serratia marcescens* growing on an agar surface. These colonies are circular, entire, convex, large, smooth, shiny, red, and opaque.

TABLE 6.2 Clinical Specimens and the Methods Used to Collect Them

Type or Location of Specimen	Collection Method
Skin, accessible membrane (including eye, outer ear, nose, throat, vagina, cervix, urethra) or open wounds	Sterile swab brushed across the surface; care should be taken not to contact neighboring tissues
Blood	Needle aspiration from vein; anticoagulants are included in the specimen transfer tube
Cerebrospinal fluid	Needle aspiration from subarachnoid space of spinal column
Stomach	Intubation, which involves inserting a tube into the stomach, often via a nostril
Urine	In aseptic collection, a catheter is inserted into the bladder through the urethra; in the "clean catch" method, initial urination washes the urethra, and the specimen is midstream urine
Lungs	Collection of sputum either dislodged by coughing or acquired via a catheter
Diseased tissue	Surgical removal (biopsy)

environment or other regions of the patient's body, and they must prevent infecting themselves with pathogens while sampling. In this regard, the Centers for Disease Control and Prevention (CDC) has established a set of guidelines, called *standard precautions,* to protect health care professionals from contamination by pathogens.

In clinical microbiology, a **clinical specimen** is a sample of human material, such as feces, saliva, cerebrospinal fluid, or blood, that is examined or tested for the presence of microorganisms. As summarized in Table 6.2, health care professionals collect clinical specimens using a variety of techniques and equipment. Specimens must be properly labeled and promptly transported to a microbiological laboratory to avoid death of the pathogens and to minimize the growth of normal organisms. Clinical specimens are often transported in special *transport media* that are chemically formulated to maintain the relative abundance of different microbial species or to maintain an anaerobic environment.

Obtaining Pure Cultures

Learning Objective

✓ Describe the two most common methods by which microorganisms can be isolated for culture.

Clinical specimens are collected in order to identify a suspected pathogen, but they also contain *normal microbiota,* which are microorganisms associated with a certain area of the body without

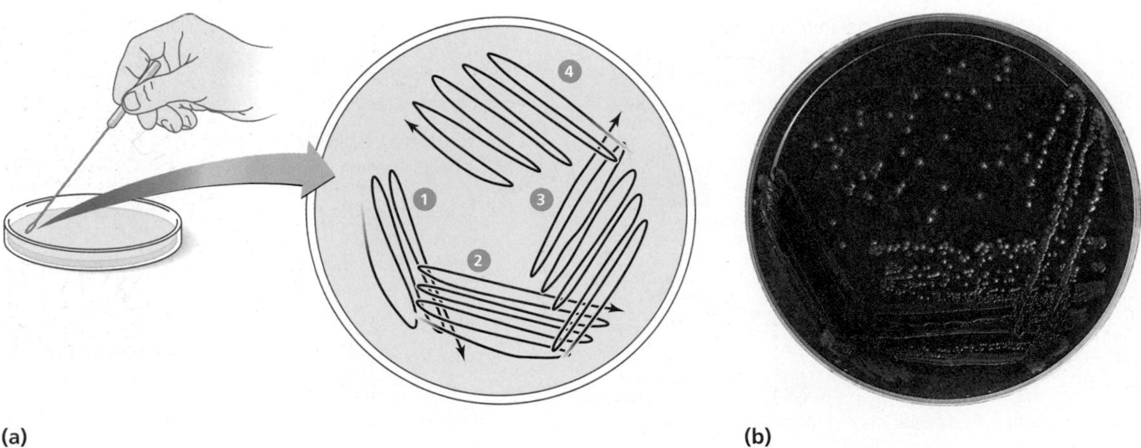

▲ **Figure 6.9 The streak-plate method of isolation.** **(a)** An inoculum is spread across the surface of an agar plate in a sequential pattern of streaks (as indicated by the numbers and arrows). The loop is sterilized between streaks. In streaks 2, 3, and 4, bacteria are picked up from a previous streak, diluting the number of cells each time. **(b)** A streak plate showing colonies of *Escherichia coli* on blood agar.

causing diseases (see Table 14.2 on p. 405). As a result, the suspected pathogen in a specimen must be isolated from the normal microbiota in culture. Scientists use several techniques to isolate organisms in **pure cultures,** that is, cultures composed of cells arising from a single progenitor. The word *axenic*[13] (ā-zen′ik) is also used to refer to a pure culture. The progenitor from which a particular pure culture is derived may be either a single cell or a group of related cells; therefore, the progenitor is termed a **colony-forming unit (CFU).**

In all microbiological procedures, care must be taken to reduce the chance of contamination as occurs when instruments or air currents carry foreign microbes into culture vessels. All media, vessels, and instruments must be **sterile**—that is, free of any microbial contaminants. Sterilization and *aseptic techniques,* which are designed to limit contamination, are discussed in Chapter 9. Now we examine two common isolation techniques: streak plates and pour plates. We consider another method, *serial dilution,* in a later section.

Streak Plates

The most commonly used isolation technique in microbiological laboratories is the **streak-plate** method. In this technique, a sterile inoculating loop (or sometimes a needle) is used to spread an inoculum across the surface of a solid medium in *Petri dishes,* which are clear, flat culture dishes with loose-fitting lids. The loop is used to lightly streak a set pattern that gradually dilutes the sample to a point that CFUs are isolated from one another **(Figure 6.9a)**. After an appropriate period of time called **incubation,** colonies develop from each isolate **(Figure 6.9b)**. The various types of organisms present are distinguished from one another by differences in colonial characteristics (see Figure 6.10b). Samples from each variety can then be inoculated in new media to establish axenic cultures.

[13]From Greek *a,* meaning no, and *xenos,* meaning stranger.

CRITICAL **THINKING**

Using the terms in Figure 6.8a, describe the shape, margin, pigmentation, and optical properties of two bacterial colonies seen in Figure 6.10b.

Pour Plates

In the **pour-plate** technique, CFUs are separated from one another using a series of dilutions. There are various ways to perform pour-plate isolations. In one method, an initial 1-milliliter sample is mixed into 9.0 ml of medium in a test tube. After mixing, a new sample from this medium is then used to inoculate a second tube of liquid medium. The process is repeated to establish a series of dilutions **(Figure 6.10a)**. Samples from the more diluted media are mixed in Petri dishes with sterile, warm medium containing *agar*—a gelling agent derived from the cell walls of red algae. After the agar cools and solidifies, and the Petri plates are incubated, isolated colonies—colonies that are separate and distinct from all others—form in the dishes from CFUs that have been separated via the dilution series **(Figure 6.10b)**. One difference between this method and the streak-plate technique is that colonies form both at and below the surface of the medium. As before, pure cultures can be established from distinct colonies.

Isolation techniques work well only if a relatively large number of CFUs of the organism of interest are present in the initial sample, and if the medium supports the growth of that microbe. As discussed later, special media and enriching techniques can be used to increase the likelihood of success.

Other Isolation Techniques

Streak plates and pour plates are used primarily to establish pure cultures of bacteria, but they can also be used for some fungi, particularly yeasts. Protozoa and motile unicellular algae are not usually cultured on solid media because they do not remain in one location to form colonies. Instead, they are isolated

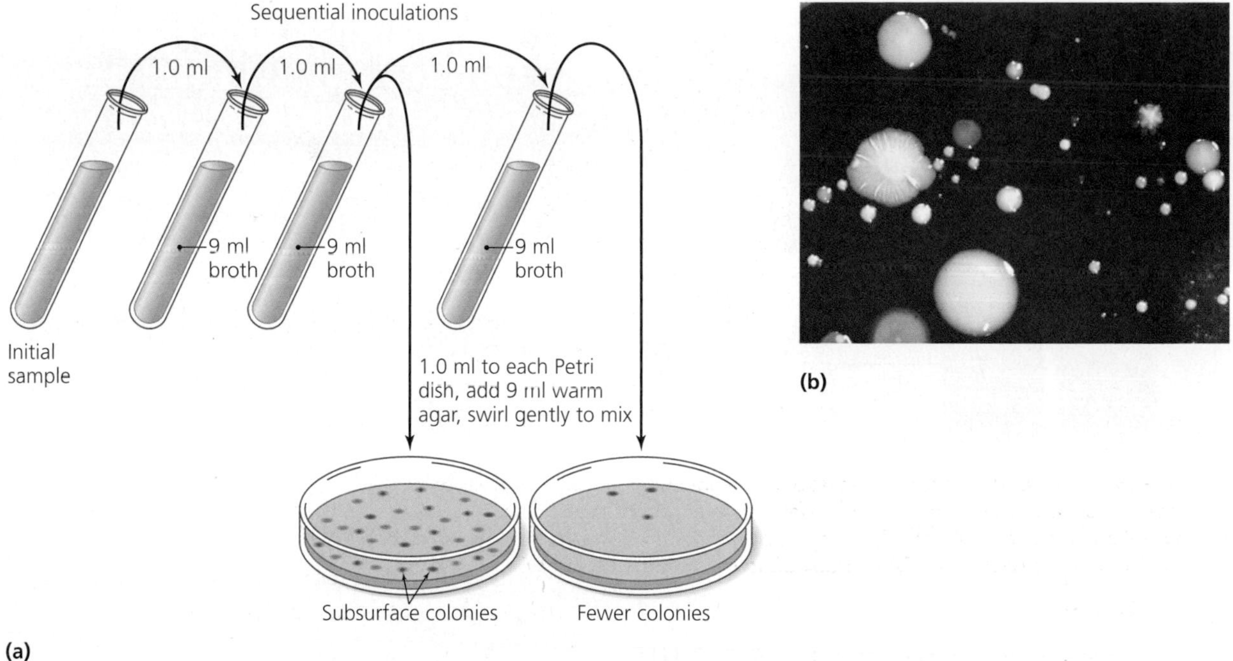

▲ **Figure 6.10 The pour-plate method of isolation.** **(a)** After an initial sample is diluted through a series of transfers, the final dilutions are mixed with warm agar in Petri plates. Individual CFUs form colonies in and on the agar. **(b)** A portion of a plate showing the results of isolation.

through a series of dilutions, but remain in broth culture media. In cases of fairly large microorganisms such as the protists *Euglena* (yū-glēn'ă) and *Amoeba* (am-ē'bă), hollow tubes with small diameters called *micropipettes* can be used to pick up a single cell, which is then used to establish a culture.

Culture Media

Learning Objective

✓ Describe six types of general culture media available for bacterial culture.

Culturing microorganisms can be an exacting science. Although some microbes, such as *E. coli,* are not particular about their nutritional needs and can be grown in a variety of media, others, such as *Neisseria gonorrhoeae* (nī-sē' rē-ă go-nor-rē' ī) and *Haemophilus influenzae* (hē-mof' i-lŭs in-flu-en' zī), require specific nutrients, including specific growth factors. The majority of prokaryotes have never been successfully grown in any culture medium, in part because scientists have concentrated their efforts on culturing commercially important species and pathogens. However, some pathogens, such as the syphilis bacterium *Treponema pallidum* (trep-ō-nē' mă pal' li-dŭm), have never been cultured in any laboratory medium, despite over a century of effort.

A variety of media are available for microbiological cultures, and more are developed each year to support the needs of food, water, industrial, and clinical microbiologists. Most media are available from commercial sources and come in powdered forms that require only the addition of water to make broths. A common medium, for example, is *nutrient broth,* which contains powdered beef extract and peptones (short chains of amino acids produced by enzymatic digestion of protein) dissolved in water. For some purposes broths are adequate, but if solid media are needed, dissolving about 1.5% agar into hot broth, pouring the liquid mixture into an appropriate vessel, and allowing it to cool provides a solid surface to support colonial growth. Media made solid by the addition of agar to a broth have the word *agar* in their names; thus *nutrient agar* is nutrient broth to which 1.5% agar has been added.

Agar, a complex polysaccharide derived from the cell walls of certain red algae, is a useful compound in microbiology for several reasons:

- Most microbes cannot digest agar; therefore, agar media remain solid even when bacteria and fungi are growing on them.
- Powdered agar dissolves in water at 100°C, a temperature at which most nutrients remain undamaged.
- Agar solidifies at temperatures below 40°C, so temperature-sensitive, sterile nutrients such as vitamins and blood can be added without detriment to cooling agar before it solidifies. Further, cooling liquid agar can be poured over most bacterial cells without harming them. The latter technique plays a role in the pour-plate isolation technique.
- Solid agar does not melt below 100°C; thus, it can be used to culture many thermophiles.

As we have seen, still-warm liquid agar media can be poured into Petri dishes, which once the agar solidifies are called **Petri plates.** When warm agar media are poured into test tubes that are then placed at an angle and left to cool until the

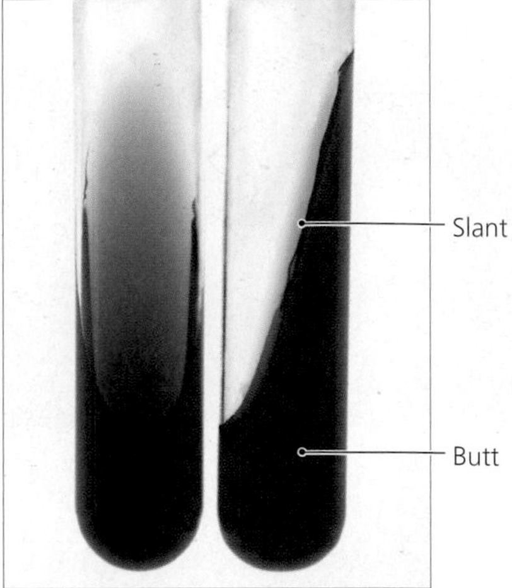

Figure 6.11 Slant tube containing solid media. In this case, citrate agar is the medium.

agar solidifies, the result is **slant tubes**, or **slants** (Figure 6.11). The slanted surface provides a larger surface area for aerobic microbial growth while the butt of the tube remains almost anaerobic. If the tubes are kept vertical until the agar solidifies, they are called *deeps.*

Next we examine six types of general culture media: defined media, complex media, selective media, differential media, anaerobic media, and transport media. It is important to note that these types of media are not mutually exclusive categories; that is, in some cases a given medium can belong to more than one category.

Defined Media

If microorganisms are to grow and multiply in culture, the medium must provide essential nutrients (including an appropriate energy source for chemotrophs), water, an appropriate oxygen level, and the required physical conditions (such as the correct pH and suitable osmotic pressure and temperature). A **defined medium** (also called a **synthetic medium**) is one in which the exact chemical composition is known. Table 6.3 gives a recipe for one example. Relatively simple defined media containing inorganic salts and a source of CO_2 (such as sodium bicarbonate) are available for autotrophs, particularly cyanobacteria and algae. Chemoheterotrophs require organic molecules such as glucose, amino acids, and vitamins, which supply carbon and energy or are vital growth factors.

Organisms that require a relatively large number of growth factors are termed *fastidious.* Such organisms may be used as living assays for the presence of growth factors. For example, a scientist needing to know if a sample contains vitamin B_{12} could inoculate the sample with *Euglena granulata* (gran-yū-lă'tă), an organism that requires the vitamin. If the microbe grows, the vitamin is present in the sample. The amount of growth provides an estimate of the amount of vitamin present, scant growth indicating a small amount.

TABLE 6.3 Ingredients of a Representative Defined (Synthetic) Medium for Culturing *E. coli*

Ingredient	Amount
Glucose	1.0 g
Na_2HPO_4	16.4 g
KH_2PO_4	1.5 g
$(NH_4)_3PO_4$	2.0 g
$MgSO_4 \cdot 7H_2O$	0.2 g
$CaCl_2$	0.01 g
$FeSO_4 \cdot 7H_2O$	0.005 g
Distilled or deionized water	Enough to bring volume to 1 L

CRITICAL **THINKING**

Why have scientists been unable to axenically culture *Treponema pallidum* in a laboratory medium?

Complex Media

For most clinical cultures, defined media are unnecessarily troublesome to prepare. Most chemoheterotrophs, including pathogens, are routinely grown on **complex media** that contain nutrients released by the partial digestion of yeast, beef, soy, or proteins such as casein from milk. The exact chemical composition of a complex medium is unknown because partial digestion releases many different chemicals in a variety of concentrations.

Complex media have advantages over defined media. Because a complex medium contains a variety of nutrients, including growth factors, it can support a wider variety of different microorganisms. Complex media are also used to culture organisms whose exact nutritional needs are unknown. Nutrient broth, Trypticase soy agar, and MacConkey agar are some common complex media. Blood is often added to complex media to provide additional growth factors, such as NADH and heme. Such a fortified medium is said to be *enriched* and can support the growth of many fastidious microorganisms.

Selective Media

Selective media typically contain substances that either favor the growth of particular microorganisms or inhibit the growth of unwanted ones. Eosin, methylene blue, and crystal violet dyes as well as bile salts are included in media to inhibit the growth of Gram-positive bacteria without adversely affecting Gram-negatives. A high concentration of NaCl (table salt) in a medium selects for halophiles and for salt-tolerant bacteria such as the pathogen *S. aureus*. Sabouraud dextrose agar has a slightly low pH, which by inhibiting the growth of bacteria is selective for fungi (Figure 6.12).

A medium can also become a selective medium when a single crucial nutrient is left out of it. For example, leaving glucose out of Trypticase soy agar makes the resulting medium selective for organisms that can meet all their carbon requirements by catabolizing amino acids.

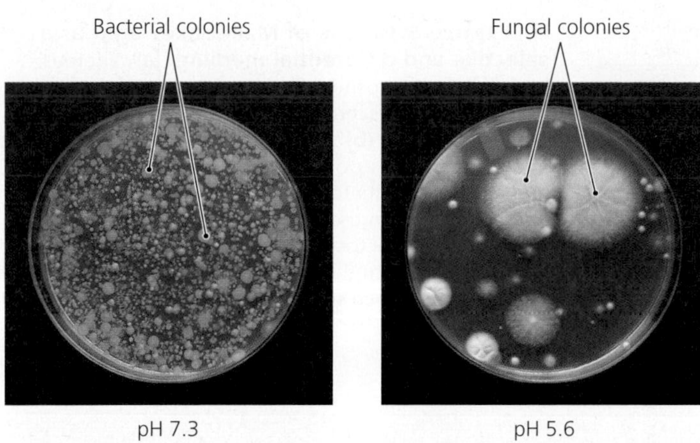

▲ **Figure 6.12 An example of the use of a selective medium.** After the medium is inoculated with a diluted soil sample, acidic pH in Sabouraud dextrose agar (right) makes the medium selective for fungi by inhibiting the growth of bacteria. At left for comparison is a nutrient agar plate inoculated with an identical sample.

Differential Media

Differential media are formulated such that either the presence of visible changes in the medium or differences in the appearance of colonies help microbiologists differentiate among different kinds of bacteria growing on the medium. Such media take advantage of the fact that different bacteria utilize the ingredients of any given medium in different ways. One example of the use of a differential medium involves the differences in organisms' utilization of red blood cells in blood agar (**Figure 6.13**). *Streptococcus pneumoniae* (nū-mō′nē-ī) partially digests (lyses) red blood cells, producing around its colonies a greenish-brown discoloration denoted *alpha-hemolysis.* By contrast, *Streptococcus pyogenes* (pī-oj′en-ēz) completely digests red blood cells, producing around its colonies clear zones termed *beta-hemolysis.* *Enterococcus faecalis* (en′ter-ō-kok′ŭs fē-kă′lis) does not digest red blood cells, so the agar appears unchanged, a reaction called *gamma-hemolysis* even though no lysis occurs. In some differential media, such as carbohydrate utilization broth tubes, a pH-sensitive dye changes color when bacteria metabolizing sugars produce acid waste products (**Figure 6.14**). Some common differential complex media are described in Table 6.4 on p. 180.

Many media are both selective and differential; that is, they enhance the growth of certain species that can then be distinguished from other species by variations in their effect on the medium or by the color of colonies they produce. For example, bile salts and crystal violet in MacConkey agar both inhibit the growth of Gram-positive bacteria and differentiate between lactose-fermenting and non-lactose-fermenting Gram-negative bacteria (**Figure 6.15**).

CRITICAL **THINKING**

Examine the ingredients of MacConkey agar as listed in Table 6.4 on p. 180. Does this medium select for Gram-positive or Gram-negative bacteria? Explain your reasoning.

The sole carbon source in citrate medium is citric acid (citrate). Why might a laboratory microbiologist use this medium?

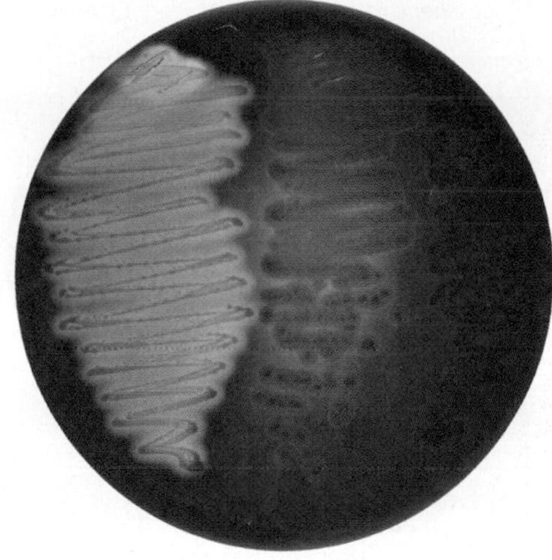

▲ **Figure 6.13 The use of blood agar as a differential medium.** *Streptococcus pyogenes* (left) completely uses red blood cells, producing a clear zone termed beta-hemolysis. *Streptococcus pneumoniae* (middle) partially uses red blood cells, producing a discoloration termed alpha-hemolysis. *Enterococcus faecalis* (right) does not use red blood cells; the lack of any change in the medium around colonies is termed gamma-hemolysis even though no red cells are hemolyzed.

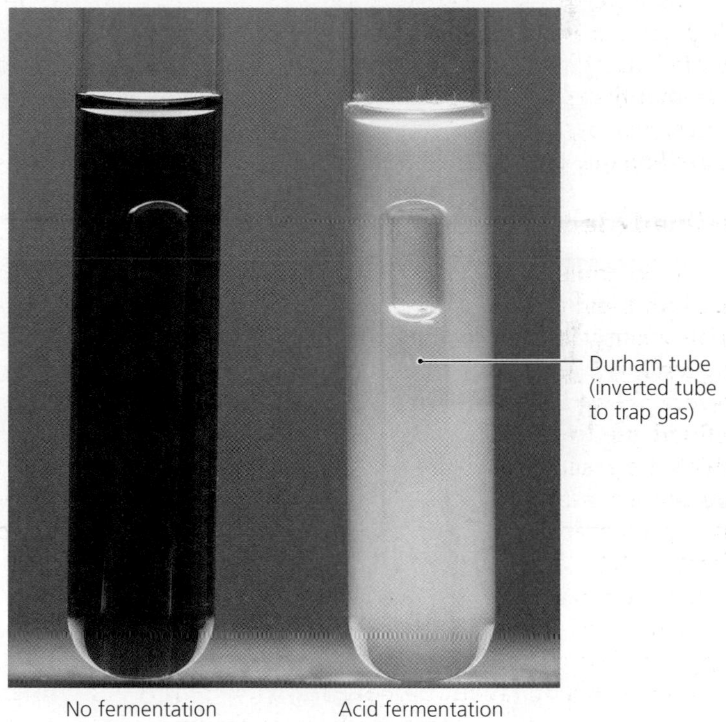

▲ **Figure 6.14 The use of carbohydrate utilization tubes as differential media.** Each tube contains a single kind of simple carbohydrate (a sugar) as a carbon source, and the dye phenol red as a pH indicator. *Alcaligenes faecalis* in the tube on the left did not ferment this carbohydrate; because no acid was produced, the medium did not turn yellow. *Escherichia coli* in the tube on the right fermented the sugar, producing acid and lowering the pH enough to cause the phenol red to turn yellow. This bacterium also produced gas, which is visible as a small bubble in the Durham tube.

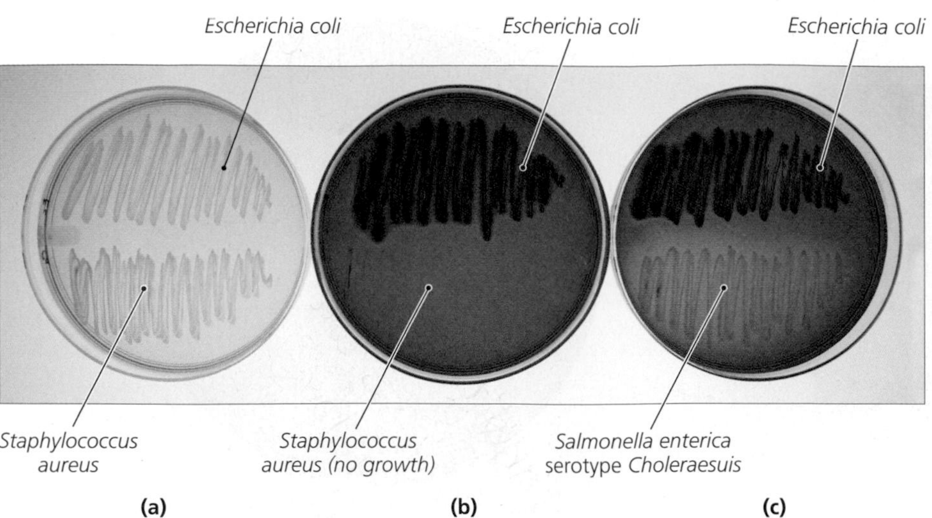

Figure 6.15 Use of MacConkey agar as a selective and differential medium. **(a)** Whereas both the Gram-positive *Staphylococcus aureus* and the Gram-negative *Escherichia coli* grow on nutrient agar, MacConkey agar **(b)** selects for Gram-negative bacteria and inhibits Gram-positive bacteria. **(c)** MacConkey agar also differentiates between Gram-negative bacteria based on their ability to ferment lactose. The colonies of the lactose-fermenting *E. coli* are easily distinguished from those of the non-lactose-fermenting *Salmonella enterica* serotype Choleraesuis.

CRITICAL **THINKING**

Using as many of the following terms as apply—selective, differential, broth, solid, defined, complex—categorize each of the media listed in Table 6.3 on p. 178 and Table 6.4 below.

TABLE 6.4 Representative Differential Complex Media

Medium and Ingredients	Use and Interpretation of Results
MacConkey Medium	
Peptone (20.0 g) Agar (12.0 g) Lactose (10.0 g) Bile salts (5.0 g) NaCl (5.0 g) Neutral red (0.075 g) Crystal violet (0.001 g) Water to bring volume to 1 L	For the culture and differentiation of enteric bacteria based on the ability to ferment lactose Lactose fermenters produce red to pink colonies; non-lactose-fermenters form colorless or transparent colonies
Blood Agar	
Agar (15.0 g) Pancreatic digest of casein (15.0 g) Papaic digest of soybean meal (5.0 g) NaCl (5.0 g) Sterile blood (50.0 ml) Water to bring volume to 950.0 ml (Blood is added to medium after autoclaving and cooling.)	For culture of fastidious microorganisms and differentiation of hemolytic microorganisms Partial digestion of blood: alpha-hemolysis; complete digestion of blood: beta-hemolysis; no digestion of blood: gamma-hemolysis

Anaerobic Media

Obligate anaerobes require special culture conditions in that their cells must be protected from free oxygen. Anaerobes can be introduced with a straight inoculating wire into the anoxic (oxygen-free) depths of solid media to form a *stab culture*, but special media called **reducing media** provide better anaerobic culturing conditions. These media contain compounds, such as sodium thioglycollate, that chemically combine with free oxygen and remove it from the medium. Heat is used to drive absorbed oxygen from thioglycollate immediately before such a medium is inoculated.

The use of Petri plates presents special problems for the culture of anaerobes because each dish has a loose-fitting lid that allows the entry of air. For the culture of anaerobes, inoculated Petri plates are placed in sealable containers containing reducing chemicals (**Figure 6.16**). Of course, the airtight lids of anaerobic culture vessels must be sealed so that oxygen cannot enter. Only anaerobes that can tolerate exposure to oxygen can be cultured by this method because inoculation and transfer occur outside of the anaerobic environment. Laboratories that routinely study strict anaerobes have large anaerobic glove boxes, which are transparent, airtight chambers with special airtight rubber gloves, chemicals that remove oxygen, and air locks. These chambers allow scientists to manipulate equipment and anaerobic cultures in an oxygen-free environment.

Transport Media

Hospital personnel use special **transport media** to carry clinical specimens of feces, urine, saliva, sputum, blood, and other bodily fluids in such a way as to ensure that people are not infected and that the specimens are not contaminated. Speed in transporting clinical specimens to the laboratory is extremely important because pathogens often do not survive outside the body as long as normal microbiota. Stool and other specimens are transported in buffered media designed to maintain the ratios among different microorganisms. Anaerobic clinical specimens can be transported for less than an hour in syringes from which the

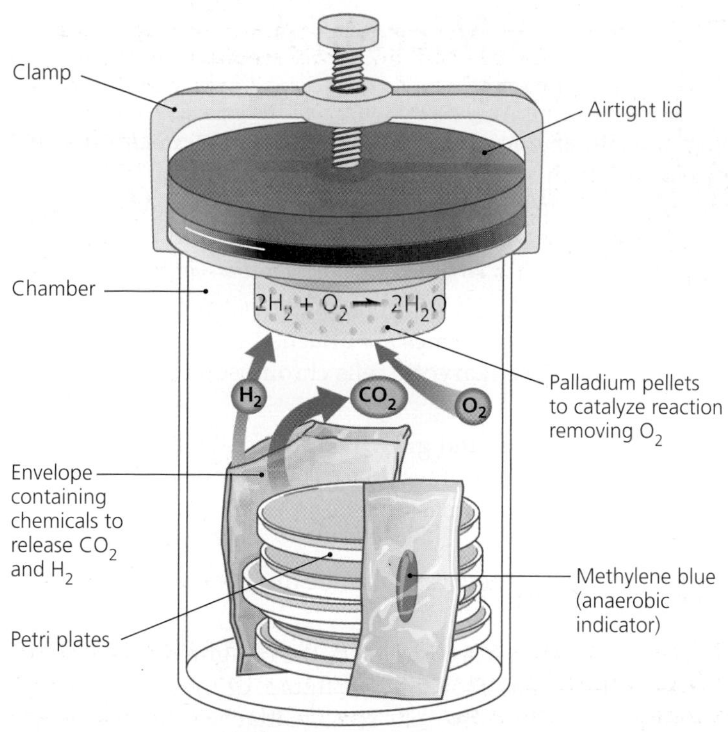

▲ **Figure 6.16 An anaerobic culture system.** The system utilizes chemicals to create an anaerobic environment inside a sealable, airtight jar. Methylene blue, which turns colorless in the absence of oxygen, indicates when the environment within the jar is anaerobic.

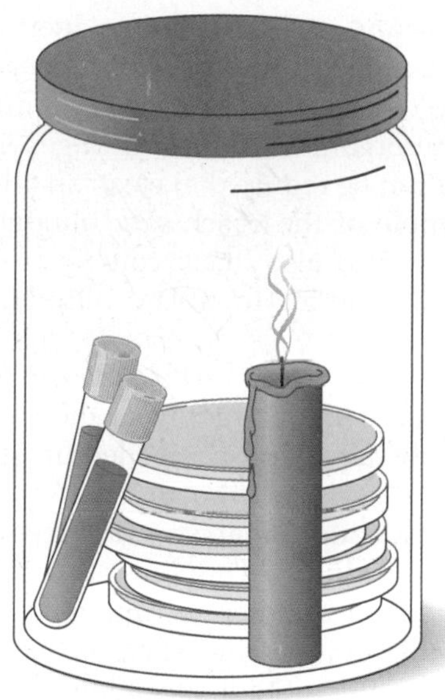

▲ **Figure 6.17 A candle jar.** The burning candle consumes oxygen and releases carbon dioxide until it extinguishes itself; the resulting environment is suitable for growing microaerophiles and capnophiles.

needles have been removed, but longer transport times require that the specimens be injected into anaerobic transport media.

Special Culture Techniques

Learning Objective

✓ Discuss the use of special culture methods, including animal and cell culture, low-oxygen culture, and enrichment culture.

Not all organisms can be grown under the culture conditions we have discussed. Scientists have developed other techniques to culture many of these organisms.

Animal and Cell Culture

Microbiologists have developed animal and cell culture techniques for growing microbes for which artificial media are inadequate. The causative agents of leprosy and syphilis, for example, must be grown in animals because all attempts to grow them using standard culture techniques have been unsuccessful. *Mycobacterium leprae* (mī'kō-bak-tēr'ē-ŭm lep'rī) is cultured in armadillos, whose internal conditions (including a relatively low body temperature) provide the conditions this microbe prefers. Rabbits meet the culture needs for *Treponema pallidum,* the bacterium that causes syphilis. Because viruses and small bacteria called rickettsias and chlamydias are obligate intracellular parasites—that is, they grow and reproduce only within living cells—bird eggs and cultures of living cells are used to culture these organisms.

Low-Oxygen Culture

As we have discussed, many types of organisms prefer oxygen conditions that are intermediate between strictly aerobic and anaerobic environments. *Carbon dioxide incubators,* machines that electronically monitor and control CO_2 levels, provide atmospheres that mimic the environments of the intestinal tract, the respiratory tract, and other body tissues and thus are useful for culturing these kinds of organisms. Smaller and much less expensive alternatives to CO_2 incubators are *candle jars* (**Figure 6.17**). In these simple but effective devices, culture plates are sealed in a jar along with a lit candle; the flame consumes much of the O_2, replacing it with CO_2. The candle eventually extinguishes itself, creating an environment that is ideal for aerotolerant anaerobes, microaerophiles, and **capnophiles,** which are organisms such as *Neisseria gonorrhoeae* that grow best with a relatively high concentration of carbon dioxide (3–10%) in addition to low oxygen levels. Remaining oxygen in the jar prevents the growth of strict anaerobes. The use of packets of chemicals that remove oxygen from the jar has largely replaced candles in modern microbiology labs.

Enrichment Culture

Bacteria that are present in small numbers may be overlooked on a streak plate or overwhelmed by faster-growing, more abundant strains. This is especially true of organisms in soil and fecal samples that contain a wide variety of microbial species. To isolate potentially important microbes that might otherwise be overlooked, microbiologists enhance the growth of less abundant organisms by a variety of techniques.

In the late 1800s, the Dutch microbiologist Martinus Beijerinck (1851–1931) introduced the most common of these methods, called simply **enrichment culture.** Enrichment cultures use a selective medium and are designed to increase very small numbers of a chosen microbe to observable levels. For example, suppose a microbiologist specializing in environmental cleanup wanted to isolate an organism capable of digesting crude oil to have on hand should it be required to clean an oil-soaked beach. Even though a sample of the beach sand might contain a few such organisms, it would also likely contain many millions of unwanted common bacteria. To isolate oil-utilizing microbes, the scientist would inoculate a sample of the sand into a tube of selective medium containing oil as the sole carbon source, and then incubate it. Then a small amount of the culture would be transferred into a new tube of the same medium, to be incubated again. After a series of such enrichment transfers, any remaining bacteria will be oil-utilizing organisms. Different species could be isolated by either streak-plate or pour-plate methods.

Cold enrichment is another technique used to enrich a culture with cold-tolerant species, such as *Vibrio cholerae,* the bacteria that cause cholera. Stool specimens or water samples suspected of containing the bacterium are incubated in a refrigerator instead of at 37°C. Cold enrichment works because *Vibrio* cells are much less sensitive to cold than are more common fecal bacteria such as *E. coli;* therefore, *Vibrio* continues to grow in the cold while the other species are inhibited. The result of cold enrichment is a culture with a greater percentage of *Vibrio* cells than the original sample; the *Vibrio* cells can then be isolated by other methods.

CRITICAL **THINKING**

Beijerinck used the concept of enrichment culture to isolate aerobic and anaerobic nitrogen-fixing bacteria, sulfate-reducing bacteria, and sulfur-oxidizing bacteria. What kind of selective media could he have used for isolating each of these four types of microbes?

Preserving Cultures

Learning Objective

✓ Contrast refrigeration, deep freezing, and lyophilization as methods for preserving cultures of microbes.

To store living cells, a scientist slows the cells' metabolism to prevent the excessive accumulation of waste products and the exhaustion of all nutrients in a medium. **Refrigeration** is often the best technique for storing bacterial cultures for short periods of time.

Deep-freezing and lyophilization are used for long-term storage of bacterial cultures. **Deep-freezing** involves freezing the cells at temperatures from –50°C to –95°C. Deep-frozen cultures can be restored years later by thawing them and placing a sample in an appropriate medium.

Lyophilization (lī-of′i-li-zā′shŭn; freeze-drying) involves removing water from a frozen culture using an intense vacuum. Under these conditions, ice sublimates (directly becomes a gas) and is removed from cells without permanently damaging cellular structures and chemicals. Lyophilized cultures can last for decades and are revived by adding lyophilized cells to liquid culture media.

Growth of Microbial Populations

Most unicellular microorganisms reproduce by *binary fission,* a process in which a cell grows to twice its normal size and divides in half to produce two daughter cells of equal size. Binary fission generally involves four steps, as illustrated in **Figure 6.18** for a prokaryotic cell. **ANIMATIONS:** ***Bacterial Growth: Overview***

1. The cell replicates its chromosome (DNA molecule). The duplicated chromosomes are attached to the cytoplasmic membrane. (In eukaryotic cells chromosomes are attached to microtubules.)
2. The cell elongates and growth between attachment sites pushes the chromosomes apart. (Eukaryotic cells segregate their chromosomes by *mitosis,* a process described in Chapter 12.)
3. The cell forms a new cytoplasmic membrane and wall (septum) across the midline.
4. When the septum is completed, the daughter cells may remain attached as shown in the figure, or they may separate completely. When the cells remain attached, further binary fission in parallel planes produces a chain. When further divisions are in different planes, the cells become a cluster (as shown in the figure).
5. The process repeats. **ANIMATIONS:** ***Binary Fission***

Other reproductive strategies of prokaryotes and eukaryotes are discussed in Chapters 11 and 12. Here we consider only the growth of populations by binary fission, using bacterial cultures as examples. We begin with a brief discussion of the mathematics of population growth.

Mathematical Considerations in Population Growth

Learning Objective

✓ Describe logarithmic growth.

With binary fission, any given cell divides to form two cells; then each of these new cells divides in two, to make four, and then four become eight, and so on. This type of growth, called **logarithmic growth** or **exponential growth**, produces very different results from simple addition, known as arithmetic growth. We can compare these two types of growth by considering what would happen over time to two identical hypothetical populations, as shown in **Figure 6.19**. In this case we assume that a population of species A increases by adding one new cell every 20 minutes, whereas the cells of species B divide by binary fission every 20 minutes. After 20 minutes, each population, which started with a single cell, would have two cells; after 40 minutes, species A would have three cells, while species B would have four cells. At this point there is little difference in the growth of the two populations, but after 2 hours, the arithmetically growing species A would have only seven cells, whereas the logarithmically growing species B would have increased to 64 cells. Clearly, logarithmic growth can increase a

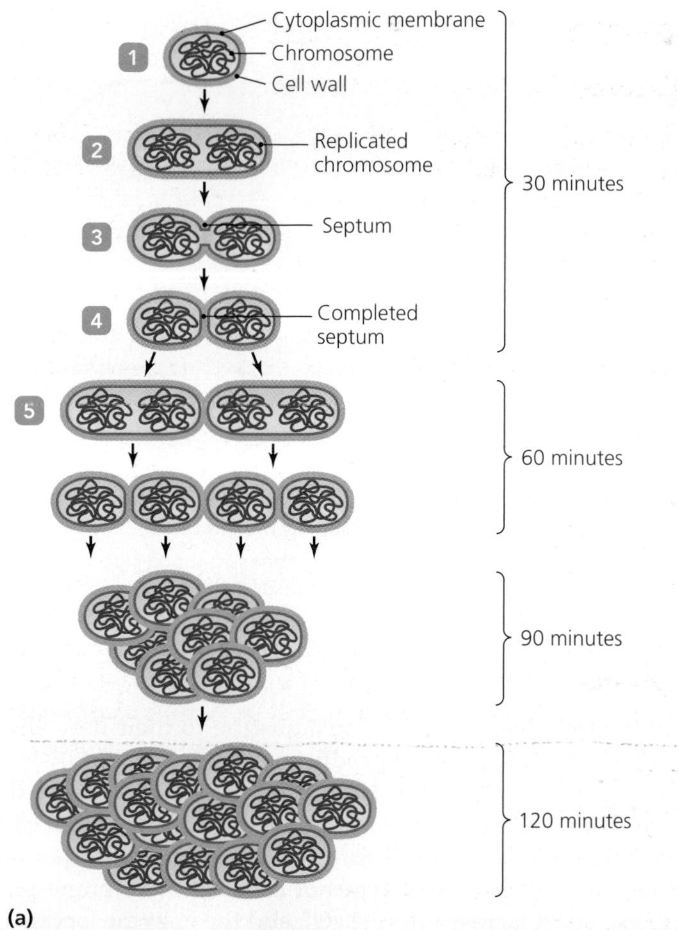

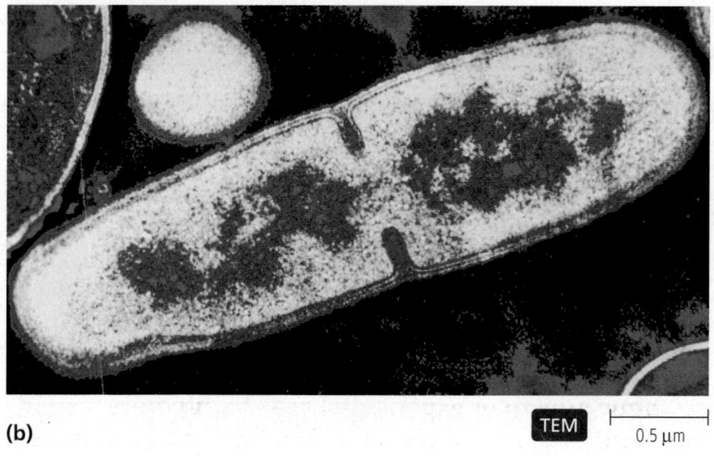

▲ **Figure 6.18 Binary fission.** **(a)** The events in binary fission. All the cells may divide in parallel planes and remain attached to form a chain, or they may divide in different planes to form a cluster (as shown here). **(b)** Transmission electron micrograph of *Bacillus licheniformis* undergoing binary fission.

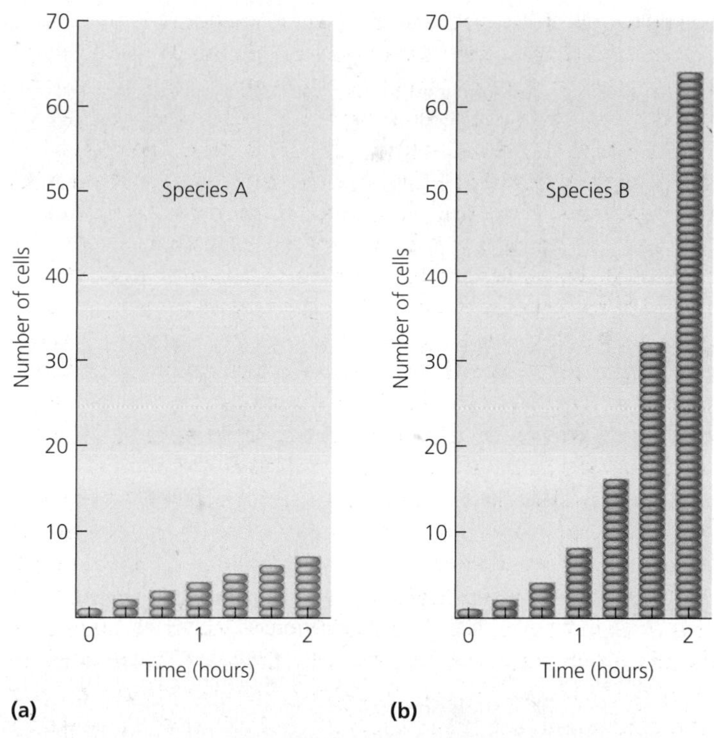

▲ **Figure 6.19 A comparison of arithmetic and logarithmic growth.** Given two hypothetical initial populations consisting of a single cell each, after 2 hours the arithmetically growing species A will have seven cells, whereas the logarithmically growing species B will have 64 cells.

population's size dramatically—after only 7 hours, species B will have over 2 million cells!

The number of cells arising from a single cell reproducing by binary fission is calculated as 2^n, where n is the number of generations; in other words, multiply 2 times itself n number of times. To calculate the total number of cells in a population, we multiply the original number of cells by 2^n. If, for example, species B had begun with three cells instead of one, then after 2 hours it would have 192 cells ($3 \times 2^6 = 3 \times 64 = 192$).

A visible culture of bacteria may consist of trillions of cells, so microbiologists use scientific notation to deal with the huge numbers involved. One advantage of scientific notation is that large numbers are expressed as powers of 10, making them easier to read and write. For example, consider our culture of species B. After 10 hours (30 generations) it would have 1,073,741,824 (2^{30}) cells. This large number can be rounded off and expressed more succinctly in scientific notation as 1.07×10^9. After 30 more generations, scientific notation would be the only practical way to express the huge number of cells in the culture, which would be 1.15×10^{18} (2^{60}). Such a number, if written out (the digits 115 followed by 16 zeros), would be impractically large. Appendix B presents scientific notation in more detail.

Generation Time

Learning Objective

✓ Explain what is meant by the generation time of bacteria.

The time required for a bacterial cell to grow and divide is its **generation time.** Viewed another way, generation time is also the time required for a population of cells to double in number. Generation times vary among populations and are dependent on chemical and physical conditions. Under optimal conditions, some bacteria (such as *E. coli* and *S. aureus*) have a generation

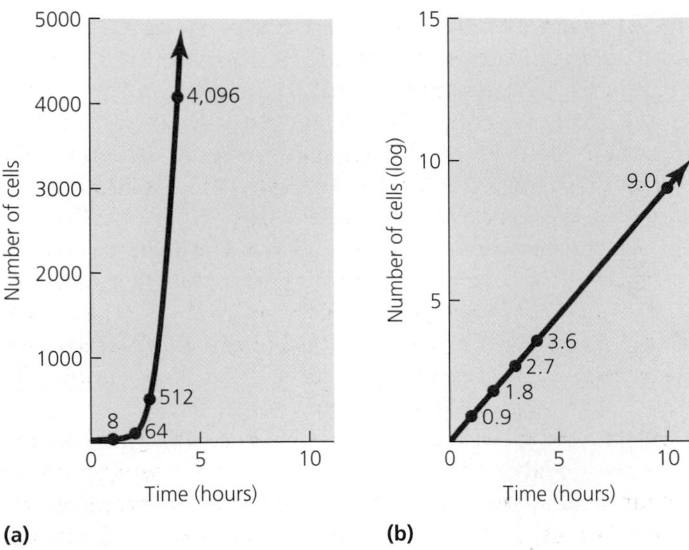

▲ **Figure 6.20 Two growth curves of logarithmic growth.** The generation time for this *E. coli* population is 20 minutes. **(a)** An arithmetic graph. Using an arithmetic scale for the *y*-axis makes it difficult to ascertain actual numbers of cells near the beginning, and impossible to plot points after only a short time. **(b)** A semilogarithmic graph. Using a logarithmic scale for the *y*-axis solves both of these problems. Note that a plot of logarithmic population growth using a logarithmic scale produces a straight line.

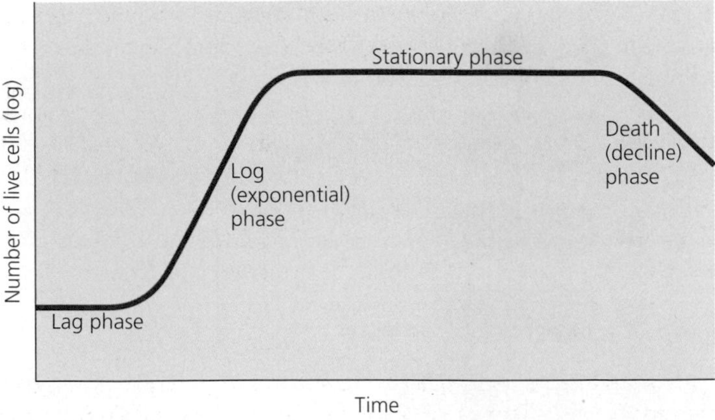

▲ **Figure 6.21 A typical microbial growth curve.** The curve shows the four phases of population growth. *Why do cells trail behind their optimum reproductive potential during the lag phase?*

Figure 6.21 *During lag phase, cells are synthesizing the metabolic machinery and chemicals required for optimal reproduction.*

time of 20 minutes or less. For this reason, food contaminated with only a few of these organisms can cause food poisoning if not properly refrigerated and cooked. Most bacteria have a generation time of 1–3 hours, though some slow-growing species such as *Mycobacterium leprae* require more than 10 days before they double. Appendix B presents the math required to calculate generation time for a population.

Phases of Microbial Growth

Learning Objectives

✓ Draw and label a bacterial growth curve.

✓ Describe what occurs at each phase of a population's growth.

A graph that plots the numbers of organisms in a growing population over time is known as a **growth curve.** When drawn using an arithmetic scale on the *y*-axis, a plot of exponential growth presents two problems **(Figure 6.20a)**: it is difficult or impossible to distinguish numbers in early generations from the baseline, and as the population grows it becomes impossible to accommodate the graph on a single page.

The solution to these problems is to replace the arithmetic scale on the *y*-axis with a logarithmic (log) scale **(Figure 6.20b)**. Such a log scale, in which each division is 10 times larger than the preceding one, can accommodate small numbers at the lower end of the graph, and very large numbers at the upper end. This kind of graph is *semilogarithmic,* because only one axis uses a log scale.

When bacteria are inoculated into a liquid medium, there are four distinct phases to a population's growth curve: the lag, log, stationary, and death phases **(Figure 6.21)**. **ANIMATIONS:** ***Bacterial Growth Curve***

Lag Phase

During the **lag phase** the cells are adjusting to their new environment; most cells do not reproduce immediately, but instead actively synthesize enzymes to utilize novel nutrients in the medium. For example, bacteria inoculated from a medium containing glucose as a carbon source into a medium containing lactose must synthesize two types of proteins: membrane proteins to transport lactose into the cell, and the enzyme lactase to catabolize lactose. The lag phase can last less than an hour or for days depending on the species and the chemical and physical conditions of the medium.

Log Phase

Eventually, the bacteria synthesize the necessary chemicals for conducting metabolism in their new environment, and they then enter a phase of rapid chromosome replication, growth, and reproduction. This is the **log phase,** so named because the population increases logarithmically, and the reproductive rate reaches a constant as DNA and protein syntheses are maximized.

Researchers are interested in the log phase for many reasons. Populations in log phase are more susceptible to antimicrobial drugs that interfere with metabolism, such as erythromycin, and to drugs that interfere with the formation of cell structures, such as the inhibition of cell wall synthesis by penicillin. Populations in log phase are preferred for Gram staining because most cells' walls are intact—an important characteristic for correct staining. Further, because the metabolic rate of individual cells is at a maximum during log phase, this phase is sometimes preferred for industrial and laboratory purposes.

Stationary Phase

If bacterial growth continued at the exponential rate of the log phase, bacteria would soon overwhelm the Earth. This does not occur, because as nutrients are depleted and wastes accumulate, the rate of reproduction decreases. Eventually, the number of dying cells equals the number of cells being produced, and the size of the population becomes stationary—hence the name **stationary phase.** During this phase the metabolic rate of surviving cells declines.

The onset of the stationary phase can be postponed indefinitely (and thus exponential growth can be maintained indefinitely) by a special apparatus called a *chemostat,* which establishes a continuous culture by continually removing wastes (along with old medium and some cells) and adding fresh medium. Chemostats are used in industrial fermentation processes.

Death Phase

If nutrients are not added and wastes are not removed, a population reaches a point at which cells die at a faster rate than they are produced. Such a culture has entered the **death phase** (or *decline phase*). Bear in mind that during the death phase, some cells remain alive and continue metabolizing and reproducing, but the number of dying cells exceeds the number of new cells produced, so that eventually the population decreases to a fraction of its previous abundance. In some cases, all the cells die, while in others a few survivors may remain indefinitely. The latter case is especially true for cultures of bacteria that can develop resting structures called *endospores* (see Chapter 3).

Measuring Microbial Reproduction

Learning Objective

✓ Contrast direct and indirect methods of measuring bacterial reproduction.

We have discussed the concepts of population growth and have seen that large numbers result from logarithmic growth, but we have not discussed practical methods of determining the size of a microbial population. Because of each cell's small size and incredible rate of reproduction, it is not possible to actually count every one in a population. For one thing, they grow so rapidly that their number changes during the count. Therefore, laboratory personnel must estimate the number of cells in a population by counting the number in a small, representative sample and then multiplying to estimate the number in the whole specimen. For example, if there are 25 cells in a microliter (μl) sample of urine, then there are approximately 25 million cells in a liter of urine.

Estimating the number of microorganisms in a sample is useful for determining such things as the severity of urinary tract infections, the effectiveness of pasteurization and other methods of food preservation, the degree of fecal contamination of water supplies, and the effectiveness of particular disinfectants and antibiotics.

Microbiologists use either direct or indirect methods to estimate the number of cells. We begin with direct methods of measuring bacterial reproduction.

Direct Methods Requiring Incubation

Among the many direct techniques are techniques requiring incubation—viable plate counts following dilution, membrane filtration, and the most probable number method. Direct techniques not requiring incubation include microscopic counts and electronic counting. We will consider each in turn.

Serial Dilution and Viable Plate Counts What if the number of cells in even a very small sample is still too great to count? If, for example, a 1-ml sample of milk containing 20,000 bacterial cells per ml were plated on a Petri plate, there would be too many colonies to count. In such cases, microbiologists make a **serial dilution,** which is the stepwise dilution of a liquid culture. Typically, the dilution factor at each step is constant. The scientists plate a set amount of each dilution onto an agar surface and count the number of colonies resulting on a plate from each dilution. They count the colonies on plates with 25–250 colonies and multiply the number by the reciprocal of the dilution to estimate the number of bacteria per ml of the original culture. This method is called a **viable plate count** (Figure 6.22).

When a plate has fewer than 25 colonies, it is not used to estimate the number of bacteria in the original sample because the chance of underestimating the population increases when the number of colonies is small. Recall that the number of colonies on a plate indicates the number of *colony-forming units* that were inoculated onto the plate. This number differs from the actual number of cells when the colony-forming units are composed of more than one cell. In such cases, a viable plate count underestimates the number of cells present in the sample.

The accuracy of a viable plate count is also dependent on the homogeneity of the dilutions, the ability of the bacteria to grow on the medium used, the number of cell deaths, and the growth phase of the sample population. Thoroughly mixing each dilution, inoculating multiple plates per dilution, and using log-phase cultures minimize errors.

Membrane Filtration Viable plate counts allow scientists to estimate the number of microorganisms when the population is very large, but if the population density is very small—as is the case, for example, for fecal bacteria in a stream or lake—microbes are more accurately counted by **membrane filtration** (Figure 6.23). In this method, a large sample (perhaps as large as several liters) is poured (or drawn under a vacuum) through a membrane filter with pores small enough to trap the cells. The membrane is then transferred onto a solid medium, and the colonies present after incubation are counted. In this case, the number of colonies is equal to the number of CFUs in the original large sample.

CRITICAL **THINKING**

Viable plate counts are used to estimate population size when the density of microorganisms is high, whereas membrane filtration is used when the density is low. Why is a viable plate count appropriate when the density is high, but not when the density is low?

Most Probable Number The **most probable number (MPN)** method is a statistical estimation technique based on the fact that the more bacteria are in a sample, the more dilutions are required to reduce their number to zero.

Let's consider an example of the use of the MPN method to estimate the number of fecal bacteria contaminating a stream. A researcher inoculates a set of test tubes of a broth medium with a sample of stream water. Even though the more tubes that are used, the more accurate is the MPN method, accuracy

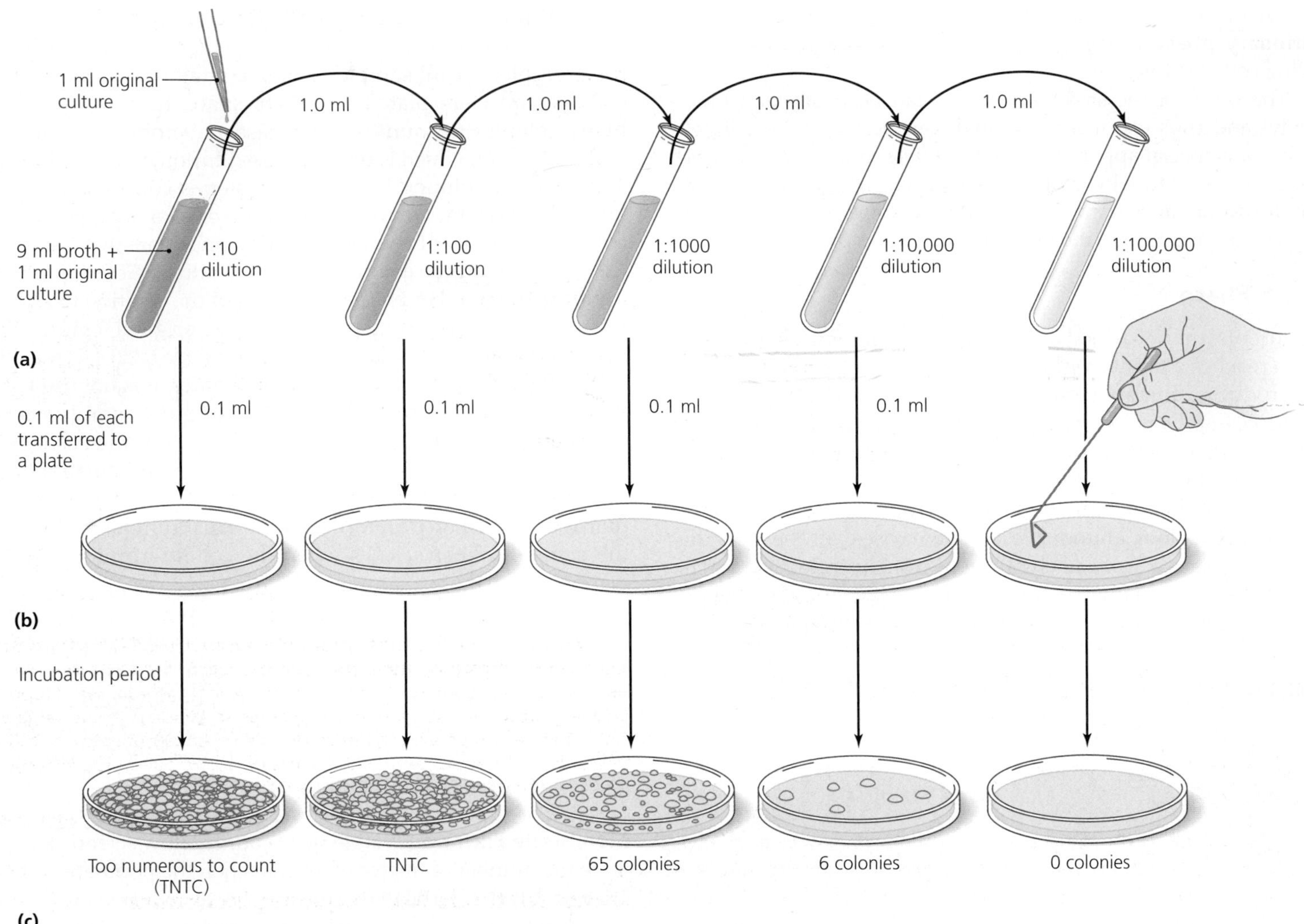

▲ **Figure 6.22 A serial dilution and viable plate count for estimating microbial population size.** **(a)** Serial dilutions. A series of 10-fold dilutions is made. **(b)** Plating. A 0.1-ml sample from each dilution is poured onto a plate and spread with a sterile rod. Alternatively, 0.1 ml of each dilution can be mixed with melted agar medium and poured into plates. **(c)** Counting. Plates are examined after incubation. Some plates may contain so many colonies that they are too numerous to count (TNTC). The number of colonies is multiplied by 10 (because 0.1 ml was plated instead of 1 ml) and then by the reciprocal of the dilution to estimate the concentration of bacteria in original culture—in this case, 65 colonies × 10 × 1000 = 650,000 bacteria/ml.

must be balanced against the time and cost involved in inoculating and incubating numerous tubes. Typically, a set of five tubes is inoculated.

The researcher also inoculates a set of five tubes with a 1:10 dilution, and another set of five tubes with a 1:100 dilution of stream water. Thus, there are 15 test tubes—the first set of five tubes inoculated with undiluted sample, the second set with a 1:10 dilution, and the third set with a 1:100 dilution (Figure 6.24).

After incubation for 48 hours, the researcher counts the number of test tubes in each set that show growth as determined by some method, such as cloudiness of the broth, gas production, or pH changes. This generates three numbers, in this case—4, 2, 1—that are compared to the numbers in an MPN table (Table 6.5) on p. 188. How statisticians develop MPN tables is beyond the scope of our discussion, but they are constructed in such a way that they accurately estimate the number of cells in a culture 95% of the time. Thus, when growth occurs in four of the undiluted broth tubes, two of the 1:10 tubes, and only one of the 1:100 tubes (4, 2, 1), the MPN table estimates that there were 26 bacteria/100 ml of stream water.

The most probable number method is useful for counting microorganisms that do not grow on solid media, when bacterial counts are required routinely, and when samples of wastewater, drinking water, and food samples contain too few organisms to use a viable plate count. The MPN method is also used to count algal cells because algae seldom form distinct colonies on solid media.

We have considered direct methods requiring incubation. Now we consider methods without incubation.

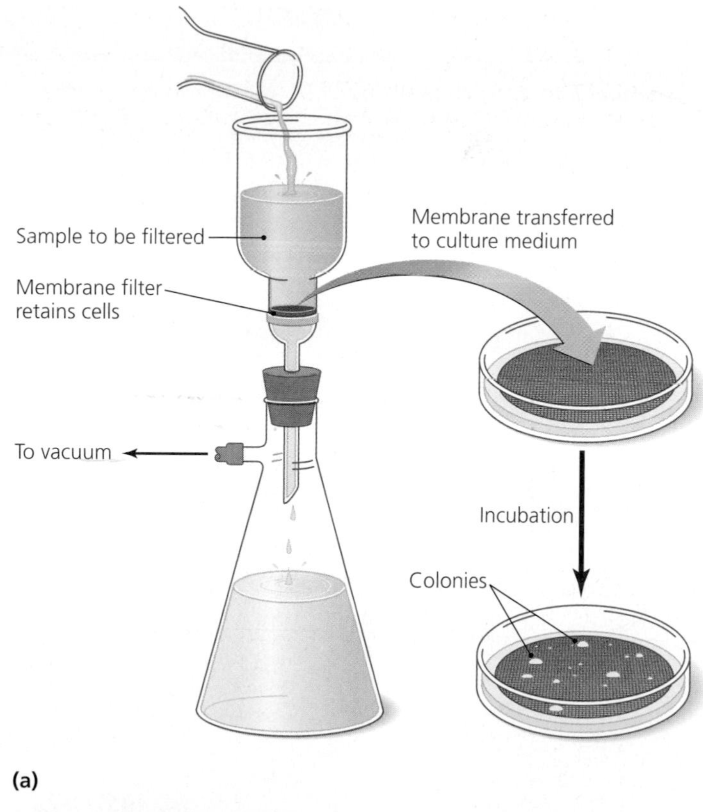

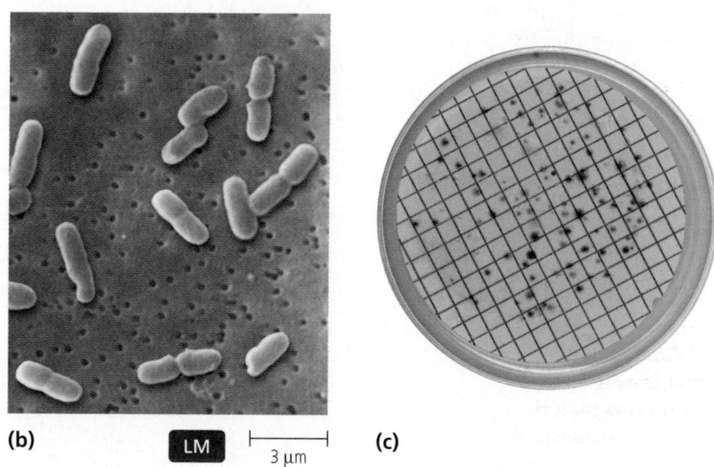

▲ **Figure 6.23 The use of membrane filtration to estimate microbial population size.** **(a)** After all the bacteria in a given volume of sample are trapped on a membrane filter, the filter is transferred onto an appropriate medium and incubated. The microbial population is estimated by multiplying the number of colonies counted by the volume of sample filtered. **(b)** Bacteria trapped on the surface of a membrane filter. **(c)** Colonies growing on a solid medium after being transferred from a membrane filter. Scientists use a superimposed grid to help them count the colonies. *If the colonies in (c) resulted from filtering 2.5 liters of stream water, what is the minimum number of bacteria per liter in the stream?*

Figure 6.23 $\frac{25 \text{ colonies}}{2.5 \text{ L}} = 10 \text{ colonies/L}$

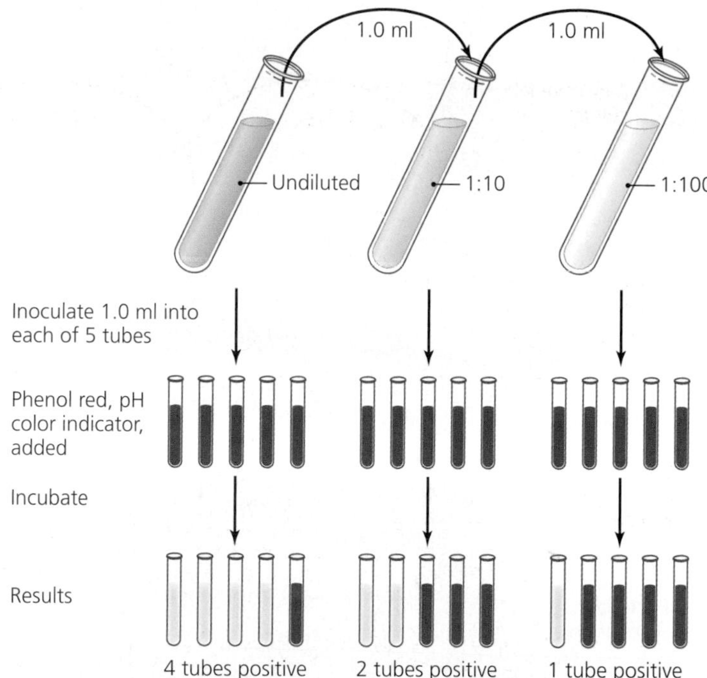

▲ **Figure 6.24 The most probable number (MPN) method for estimating microbial numbers.** Typically, sets of five test tubes are used for each of three dilutions. After incubation, the number of tubes showing growth in each set is used to enter an MPN table (see Table 6.5 on p. 188), which provides an estimate of the number of cells per 100 ml of liquid. *If the results were 5, 3, 1, what would be the most probable number of microorganisms in the original broth?*

Figure 6.24 *The MPN is 110/100 ml.*

Direct Methods Not Requiring Incubation

It is possible to directly count cells without having to incubate cultures. Here we consider two such direct methods.

Microscopic Counts Microbiologists can also count microorganisms directly through a microscope rather than inoculating them onto the surface of a solid medium. In this method, particularly suitable for stained prokaryotes and relatively large eukaryotes, a sample is placed on a *cell counter* (also called a *Petroff-Hausser counting chamber*), which is a glass slide composed of an etched grid positioned beneath a glass cover slip (**Figure 6.25**). Because the cover slip is 0.02 mm above the grid, the volume of bacterial suspension over a 1 mm^2 portion of the grid is $1 \text{ mm} \times 1 \text{ mm} \times 0.02 \text{ mm} = 0.02 \text{ mm}^3$. Each 1 mm^2 grid contains 25 large squares, so a microbiologist can count the number of bacteria in several of the large squares and then calculate the mean number of bacteria per square. The number of bacteria per milliliter (cm^3) can be calculated as follows:

$$\text{mean no. of bacteria per square} \times 25 \text{ squares} = \text{no. of bacteria per } 0.02 \text{ mm}^3$$

$$\text{no. of bacteria per } 0.02 \text{ mm}^3 \times 50 = \text{no. of bacteria per mm}^3$$

$$\text{no. of bacteria per mm}^3 \times 1000 = \text{no. of bacteria per cm}^3 \text{ (ml)}$$

This means that one needs only to multiply the mean number of bacteria per square by 1,250,000 (25 × 50 × 1000) to calculate the number of bacteria per ml of bacterial suspension.

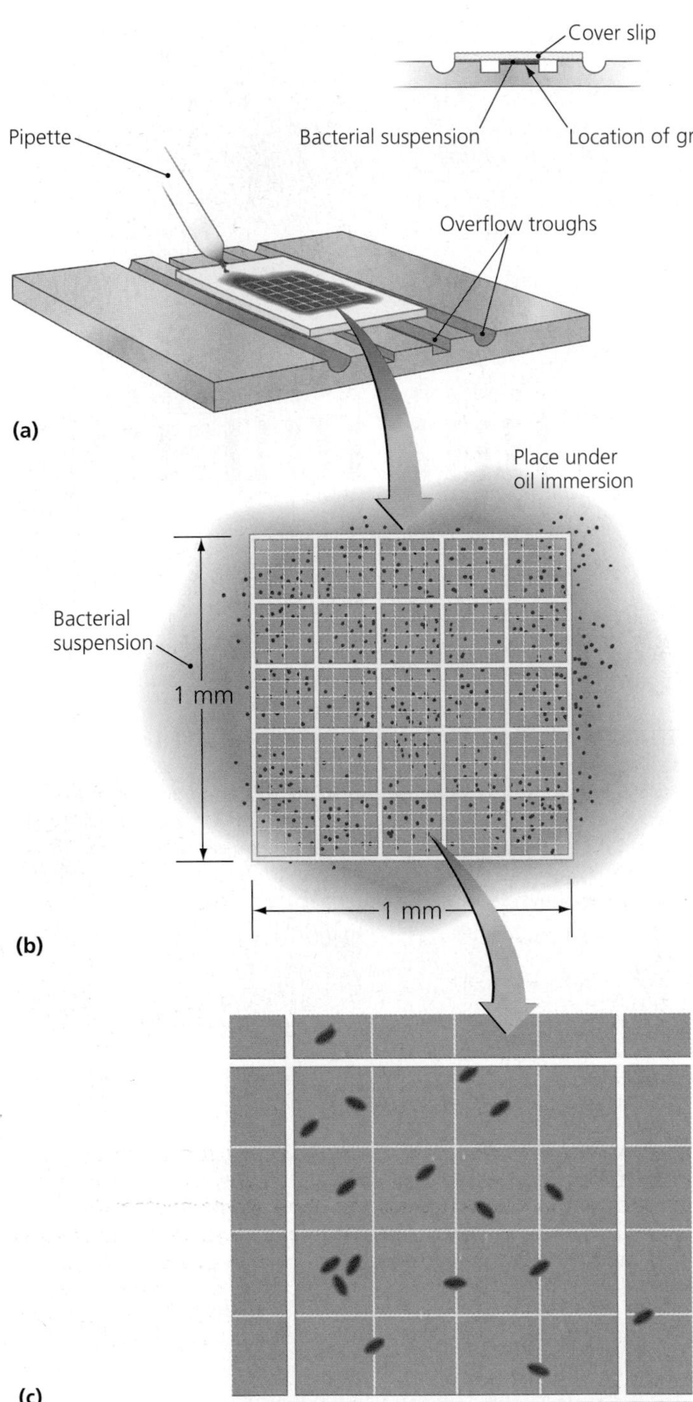

▲ Figure 6.25
The use of a cell counter for estimating microbial numbers.
(a) The counter is a glass slide with an etched grid that is exactly 0.02 mm lower than the bottom of the cover slip. A bacterial suspension placed next to the cover slip through a pipette moves under the cover slip and over the grid by capillary action. **(b)** View of a 1 mm^2 portion of the grid through the microscope. Each square millimeter of the grid has 25 large squares, each of which is divided into 16 small squares. **(c)** Enlarged view of one large square containing 15 cells. The number of bacteria in several large squares is counted and averaged. The calculations involved in estimating the number of bacteria per milliliter (cm^3) of suspension are described in the text.

TABLE 6.5 Most Probable Number Table

Number Out of 5 Tubes Giving Positive Results in Three Dilutions			Most Probable Number of Bacteria per 100 ml
0	0	0	1.8
0	0	1	2
0	1	0	2
0	2	0	4
1	0	0	2
1	0	1	4
1	1	0	4
1	1	1	6
1	2	0	6
2	0	0	4
2	0	1	7
2	1	0	7
2	1	1	9
2	2	0	9
2	3	0	12
3	0	0	8
3	0	1	11
3	1	0	11
3	1	1	14
3	2	0	14
3	2	1	17
4	0	0	13
4	0	1	17
4	1	0	17
4	1	1	21
4	1	2	26
4	2	0	22
4	2	1	26
4	3	0	27
4	3	1	33
4	4	0	34
5	0	0	23
5	0	1	30
5	0	2	40
5	1	0	30
5	1	1	50
5	1	2	60
5	2	0	50
5	2	1	70
5	2	2	90
5	3	0	80
5	3	1	110
5	3	2	140
5	3	3	170
5	4	0	130
5	4	1	170
5	4	2	220
5	4	3	280
5	4	4	350
5	5	0	240
5	5	1	300
5	5	2	500
5	5	3	900
5	5	4	1,600
5	5	5	≥1,600

Direct microscopic counts are advantageous when there are more than 10,000,000 cells per ml or when a speedy estimate of population size is required. However, direct counts can be problematic because it is often difficult to differentiate between living and dead cells, and it is difficult to count motile microorganisms.

Electronic Counters A *Coulter counter* is a device that directly counts cells as they interrupt an electrical current flowing across a narrow tube held in front of an electronic detector. This device is useful for counting the larger cells of yeast, unicellular algae, and protozoa; it is less useful for bacterial counts because of debris in the media and the presence of filaments and clumps of cells.

Flow cytometry is a variation of counting with a Coulter counter. A cytometer uses a light-sensitive detector to record changes in light transmission through the tube as cells pass. Scientists use this technique to distinguish among cells that have been differentially stained with fluorescent dyes or tagged with fluorescent antibodies. They can count bacteria in a solution and even count host cells that contain fluorescently stained intracellular parasites.

Indirect Methods

It is not always necessary to count microorganisms to estimate population size or density. Industrial and research microbiologists use indirect methods that measure such variables as metabolic activity, dry weight, and turbidity instead of counting microorganisms, colonies, or MPN tubes. Scientists can also estimate population size and diversity by analyzing the unique sequences of DNA present in a sample.

Metabolic Activity Under standard temperature conditions, the rate at which a population of cells utilizes nutrients and produces wastes is dependent on their number. Once they establish the metabolic rate of a microorganism, scientists can indirectly estimate the number of cells in a culture by measuring changes in such things as nutrient utilization, waste production, or pH.

Dry Weight The abundance of some microorganisms, particularly filamentous microorganisms, is difficult to measure by direct methods. Instead, these organisms are filtered from their culture medium, dried, and weighed. The *dry weight method* is suitable for broth cultures, but growth cannot be followed over time because the organisms are killed during the process.

Turbidity As bacteria reproduce in a broth culture, the broth often becomes *turbid* (cloudy) **(Figure 6.26a)**. Generally, the greater the bacterial population, the more turbid a broth will be. An indirect method for estimating the growth of a microbial population involves measuring changes in turbidity using a device called a *spectrophotometer* **(Figure 6.26b)**.

A spectrophotometer measures the amount of light transmitted through a culture under standardized conditions **(Figure 6.26c)**. The greater the concentration of bacteria within a broth, the more light will be absorbed and scattered, and the less light will pass through and strike a light-sensitive detector. Generally, transmission is inversely proportional to the population size; that is, the larger the population grows, the less light will reach the detector.

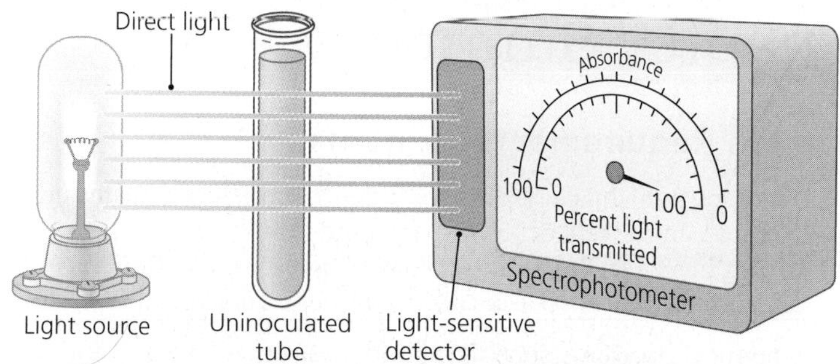

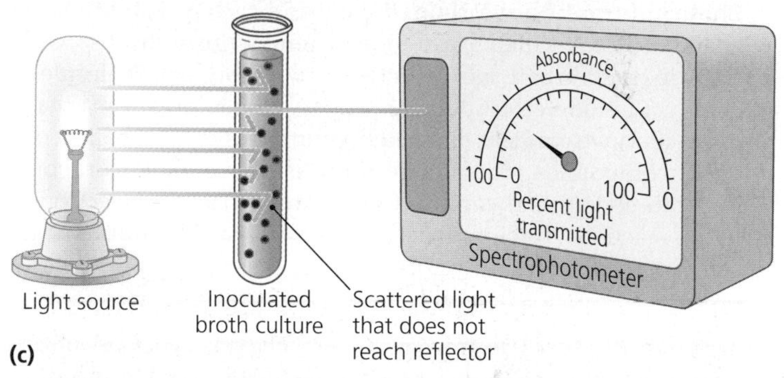

▲ **Figure 6.26 Turbidity and the use of spectrophotometry in indirectly measuring population size.** **(a)** Turbidity (right), an increased optical density or cloudiness of a solution. **(b)** A spectrophotometer. **(c)** The principle of spectrophotometry. After passing a light beam through an uninoculated sample of the culture medium, the scale is set at 100% transmission. In an inoculated sample, the microbial cells absorb and scatter light, reducing the amount reaching the detector. The percentage of light transmitted is inversely proportional to population density.

Scales on the gauge of a spectrophotometer report *percentage of transmission* and *absorbance*. These are two ways of looking at the same things, for example, 25% transmission is the same thing as 75% absorbance. Direct counts must be calibrated with transmission and absorbance readings to provide estimates of population size. Once these values are determined, spectrophotometry provides estimates of population size more quickly than any direct method.

The benefits of measuring turbidity to estimate population growth include ease of use and speed. However, the technique is useful only if the concentration of cells exceeds 1 million per milliliter; densities below this value generally do not produce turbidity. Further, the technique is accurate only if the cells are suspended uniformly in the medium. If they form either a *pellicle* (a film of cells at the surface) or a *sediment* (an accumulation of cells at the bottom), their number will be underestimated. Further, spectrophotometry does not distinguish between living and dead cells.

Genetic Methods The majority of bacteria and archaea have not or cannot be grown in the laboratory, and representatives of most species are too few in number to study by direct observation. How do scientists estimate the number of such unculturable microbes?

Scientists can isolate unique DNA sequences representing unculturable prokaryotic species using genetic techniques such as *polymerase chain reaction (PCR)* and *hybridization* of DNA that codes for ribosomal RNA. For example, one study estimated that more than 100 billion bacteria and archaea, representing more than 10 million different species, are in a single gram of garden soil. Chapter 8 discusses genetic methods in more detail.

Chapter Summary

Growth Requirements (pp. 166–174)

1. A colony, which is a visible population of microorganisms arising from a single cell or colony-forming unit living in one place, grows in size as the number of cells increases. Most microbes live as biofilms, that is, in association with one another on surfaces.
2. Chemical **nutrients** such as carbon, hydrogen, oxygen, and nitrogen are required for the growth of microbial populations.
3. **Photoautotrophs** use carbon dioxide as a carbon source and light energy to make their own food; **chemoautotrophs** use carbon dioxide as a carbon source but catabolize organic molecules for energy. **Photoheterotrophs** are photosynthetic organisms that acquire energy from light and acquire nutrients via catabolism of organic compounds; **chemoheterotrophs** use organic compounds for both energy and carbon. **Organotrophs** acquire electrons for redox reactions from organic sources, whereas lithotrophs acquire electrons from inorganic sources.
4. **Obligate aerobes** require oxygen molecules as the final electron acceptor of their electron transport chains, whereas **obligate anaerobes** cannot tolerate oxygen and must use an electron acceptor other than oxygen.
5. The four toxic forms of oxygen are **singlet oxygen ($^1O_2$)**, which is neutralized by pigments called **carotenoids; superoxide radicals (O_2^-)**, which are detoxified by superoxide dismutase; **peroxide anion (O_2^{2-})**, which is detoxified by catalase or peroxidase; and **hydroxyl radicals (OH·)**, the most reactive of the toxic forms of oxygen.
6. Microbes are described in terms of their oxygen requirements and limitations as strict **aerobes,** which require oxygen; as strict **anaerobes,** which cannot tolerate oxygen; as facultative anaerobes, which can live with or without oxygen; as **aerotolerant anaerobes,** which prefer anaerobic conditions but can tolerate exposure to low levels of oxygen; or as **microaerophiles,** which require low levels of oxygen.
7. Nitrogen, acquired from organic or inorganic sources, is an essential element for microorganisms. Some bacteria can reduce nitrogen gas into a more usable form via a process called **nitrogen fixation.**
8. In addition to the main elements found in microbes, very small amounts of **trace elements** are required. Vitamins are among the **growth factors,** which are organic chemicals required in small amounts for metabolism.
9. Though microbes survive within the limits imposed by a minimum growth temperature and a maximum growth temperature, an organism's metabolic activities produce the highest growth rate at the **optimum growth temperature.**
10. Microbes are described in terms of their temperature requirements as (from coldest to warmest) **psychrophiles, mesophiles, thermophiles,** or **hyperthermophiles.**
11. **Neutrophiles** grow best at neutral pH, **acidophiles** grow best in acidic surroundings, and **alkalinophiles** live in alkaline habitats.
12. Osmotic pressure can cause cells to die from either swelling and bursting or from **crenation** (shriveling). The cell walls of some microorganisms protect them from osmotic shock. **Obligate halophiles** require high osmotic pressure, whereas facultative halophiles do not require but can tolerate such conditions.
13. **Barophiles,** organisms that normally live under the extreme hydrostatic pressure at great depth below the surface of a body of water, often cannot live at the pressure found at the surface.
14. **Quorum sensing** is the process by which bacteria respond to changes in microbial density by utilizing signal and receptor molecules. **Biofilms,** which are communities of cells attached to surfaces, use quorum sensing.

Culturing Microorganisms (pp. 174–182)

1. Microbiologists culture microorganisms by transferring an **inoculum** from a clinical or environmental **specimen** into a **medium** such as **broth** or solid media. The microorganisms grow into a **culture.** On solid surfaces, cultures are seen as **colonies.**
2. A **clinical specimen** is a sample of human material. Standard precautions are the guidelines to protect health-care professionals from infection.
3. **Pure cultures** (axenic cultures) contain cells of only one species and are derived from a **colony-forming unit (CFU)** composed of a single cell or group of related cells. To obtain pure cultures, **sterile** equipment and use of aseptic techniques are critical.
4. The **streak-plate** method allows CFUs to be isolated by streaking. The **pour-plate** technique isolates CFUs via a series of dilutions.
5. Petri dishes that are filled with solid media are called **Petri plates. Slant tubes (slants)** are test tubes containing agar media that solidified while the tube was resting at an angle.

6. A **defined medium** (also known as **synthetic medium**) provides exact known amounts of nutrients for the growth of a particular microbe. **Complex media** contain a variety of growth factors. **Selective media** either inhibit the growth of unwanted microorganisms or favor the growth of particular microbes. Microbiologists use **differential media** to distinguish among groups of bacteria. **Reducing media** provide conditions conducive to culturing anaerobes. **Transport media** are designed to move specimens safely from one location to another while maintaining the relative abundance of organisms and preventing contamination of the specimen or environment.
7. Special culture techniques include the use of animal and cell cultures, low-oxygen cultures, **enrichment cultures,** and **cold enrichment cultures.** A **capnophile** grows best with high CO_2 levels in addition to low oxygen levels.
8. Cultures can be preserved in the short term by **refrigeration** and in the long term by **deep-freezing** and **lyophilization.**

Growth of Microbial Populations (pp. 182–190)

1. Bacteria grow by **logarithmic,** or **exponential, growth.**
   ANIMATIONS: *Bacterial Growth: Overview, Binary Fission*
2. A population of microorganisms doubles during its **generation time**—the time also required for a single cell to grow and divide.
3. A graph that plots the number of organisms growing in a population over time is called a **growth curve.** When organisms are grown in a broth and the growth curve is plotted on a semilogarithmic scale, the population's growth curve has four phases. In the **lag phase,** the organisms are adjusting to their environment. In the **log phase,** the population is most actively growing. In the **stationary phase,** new organisms are being produced at the same rate at which they are dying. In the **death phase,** the organisms are dying more quickly than they can be replaced by new organisms.
   ANIMATIONS: *Bacterial Growth Curve*
4. Direct methods for estimating population size include methods requiring incubation—**viable plate counts, membrane filtration,** and the **most probable number (MPN)** method. Direct methods that do not require incubation are microscopic counts and electronic counters, including flow cytometry.
5. Indirect methods include measurements of metabolic activity, dry weight, and turbidity, and analysis of numbers and kinds of unique genetic sequences.

Questions for Review

Answers to the Questions for Review (except Short Answer questions) begin on page A-1.

Multiple Choice

1. Which of the following can grow in a Petri plate on a laboratory table?
   a. an anaerobe
   b. a colony on an agar surface
   c. viruses on an agar surface
   d. barophiles
2. Which of the following terms best describes an organism that cannot exist in the presence of oxygen?
   a. obligate aerobe
   b. facultative aerobe
   c. obligate anaerobe
   d. facultative anaerobe
3. Superoxide dismutase
   a. causes hydrogen peroxide to become toxic.
   b. detoxifies superoxide radicals.
   c. neutralizes singlet oxygen.
   d. is missing in aerobes.
4. The most reactive of the four toxic forms of oxygen is
   a. the hydroxyl radical.
   b. the peroxide anion.
   c. the superoxide radical.
   d. singlet oxygen.
5. Microaerophiles that grow best with a high concentration of carbon dioxide in addition to a low level of oxygen are called
   a. aerotolerant.
   b. capnophiles.
   c. facultative anaerobes.
   d. fastidious.
6. Which of the following is *not* a growth factor for various microbes?
   a. cholesterol
   b. water
   c. vitamins
   d. heme
7. Organisms that preferentially may thrive in icy waters are described as
   a. barophiles.
   b. thermophiles.
   c. mesophiles.
   d. psychrophiles.
8. Barophiles
   a. cannot cause diseases in humans.
   b. live at normal barometric pressure.
   c. die if put under high pressure.
   d. thrive in warm air.
9. This statement, "In the laboratory, a sterile inoculating loop is moved across the agar surface in a culture dish, thinning a sample and isolating individuals," describes which of the following?
   a. broth culture
   b. pour plate
   c. streak plate
   d. dilution plate
10. In a defined medium,
    a. the exact chemical composition of the medium is known.
    b. agar is available for microbial nutrition.
    c. blood may be included.
    d. organic chemicals are excluded.

11. Which of the following is most useful in representing population growth on a graph?
    a. logarithmic reproduction of the growth curve
    b. a semilogarithmic graph using a log scale on the *y*-axis
    c. an arithmetic graph of the lag phase followed by a logarithmic section for the log, stationary, and death phases
    d. none of the above would best represent a population growth curve

12. Which of the following methods is best for counting fecal bacteria from a stream to determine the safety of the water for drinking?
    a. dry weight
    b. turbidity
    c. viable plate counts
    d. membrane filtration

13. A Coulter counter is
    a. a statistical estimation using 15 dilution tubes and a table of numbers to estimate the number of bacteria per milliliter.
    b. an indirect method of counting microorganisms.
    c. a device that directly counts microbes as they pass through a tube in front of an electronic detector.
    d. a device that directly counts microbes that are differentially stained with fluorescent dyes.

14. Lyophilization can be described as
    a. freeze-drying.
    b. deep-freezing.
    c. refrigeration.
    d. pickling.

15. Quorum sensing is
    a. the ability to respond to changes in population density.
    b. a characteristic of most bacteria.
    c. dependent on direct contact among cells.
    d. associated with colonies on an agar plate.

Fill in the Blanks

1. All cells require sources of _______________, _______________, and _______________.
2. A toxic form of oxygen, _______________ oxygen, is molecular oxygen with electrons that have been boosted to a higher energy state.
3. All cells recycle the essential element _______________ from amino acids and nucleotides.
4. _______________ are small organic molecules that are required in minute amounts for metabolism.
5. The lowest temperature at which a microbe continues to metabolize is called its _______________.
6. Cells that shrink in hypertonic solutions such as saltwater are responding to _______________ pressure.
7. Obligate _______________ exist in salt ponds because of their ability to withstand high osmotic pressure.
8. _______________ pigments protect many phototrophic organisms from photochemically produced singlet oxygen.
9. Microbes that reduce N_2 to NH_3 engage in nitrogen _______________.
10. A student observes a researcher streaking a plate numerous times, flaming the loop between streaks. The researcher is likely using the _______________ method to isolate microorganisms.

Labeling

Label each of these thioglycollate tubes to indicate the oxygen requirements of the microbes growing in them.

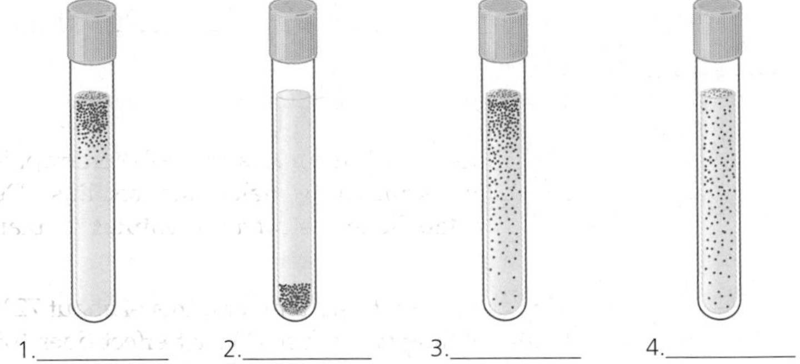

Short Answer

1. High temperature affects the shape of particular molecules. How does this affect the life of a microbe?
2. Support or refute the following statement: microbes cannot tolerate the low pH of the human stomach.
3. Explain quorum sensing and describe how it is related to biofilm formation.
4. Why must media, vessels, and instruments be sterilized before they are used for microbiological procedures?
5. Why is agar used in microbiology?
6. What is the difference between complex media and defined media?
7. Draw and label the four distinct phases of a bacterial growth curve. Describe what is happening within the culture as it passes through the phases.
8. If there are 47 cells in 1 μl of sewage, how many cells are there in a liter?
9. List three indirect methods of counting microbes.
10. List five direct methods of counting microbes.
11. Explain the differences among photoautotrophs, chemoautotrophs, photoheterotrophs, chemoheterotrophs, organotrophs, and lithotrophs.
12. Contrast the media described in Tables 6.3 and 6.4 (pp. 178 and 180). Why is *E. coli* medium described as defined, whereas MacConkey medium and blood agar are defined as complex?

Concept Mapping

Using the following terms, draw a concept map that describes culture media. For a sample concept map, see p. 93. Or, complete this concept map online by going to the Study Area at www.masteringmicrobiology.com.

Blood agar
Differential
Enriched
Fastidious microorganisms
Fermentation broths (Phenol red)
General purpose
MacConkey agar
Nutrient broth
Sabouraud agar
Selective
Selective and differential
Trypticase soy agar
Unwanted microorganisms
Visible differences between microorganisms

Critical Thinking

1. A microbiologist describes an organism as a chemoheterotrophic, aerotolerant, mesophilic, facultatively halophilic bacillus. Describe the organism's metabolic and structural features in plain English.
2. Pasteurization is a technique that uses temperatures of about 72°C to neutralize potential pathogens in foods. What effect does this temperature have on the enzymes and cellular metabolism of pathogens? Why does the heat of pasteurization kill some microorganisms yet fail to affect thermophiles?
3. If two cultures of a facultative anaerobe were grown under identical conditions except that one was exposed to oxygen and the other was completely deprived of oxygen, what differences would you expect to see between the dry weights of the cultures? Why?
4. Some organisms require riboflavin (vitamin B_2) to make FAD. For what purpose do they use FAD?
5. A scientist inoculates a bacterium into a complex nutrient slant tube. The bacterium forms only a few colonies on the slanted surface but grows prolifically in the depth of the agar. Describe the oxygen requirements of the bacterium.
6. A scientific article describes a bacterium as an obligate microaerophilic chemoorganoheterotroph. Describe the oxygen and nutritional characteristics of the bacterium in everyday language.
7. Microorganisms require phosphorus, sulfur, iron, and magnesium for metabolism. What specifically are these elements used for in microbial metabolism? (Review the information provided in Chapters 2 and 5.)

Access more review material online in the Study Area at **www.masteringmicrobiology.com.** There, you'll find

- **Animations**
- **MP3 Tutor Sessions**
- **Concept Mapping Activities**
- **Flashcards**
- **Quizzes**

and more to help you succeed.

7 Microbial Genetics

Do genes hold the secrets to life? If we can identify all the genes in an organism and determine the functions of each, can we explain all of biological behavior? What genes do organisms have in common, and what genes make them unique? What genes cause certain microorganisms to be harmful or even deadly, and how can we develop drugs or techniques to target a pathogen's genes?

These are the kinds of questions that drive the dynamic world of genetic research. We now have complete genome maps, or genetic blueprints, of hundreds of viruses, bacteria, and other organisms. We even have mapped the human genome, which our knowledge of microbial genetics is helping us to analyze. For example, by studying the genes of the *Escherichia coli* bacterium and then identifying which genes we share, we can determine the roles these same genes play in humans. Genetically speaking, you may have much more in common with microorganisms than you think!

Take the pre-test for this chapter online. Visit the Study Area at www.masteringmicrobiology.com.

This chapter has MicroFlix. Go to **www.masteringmicrobiology.com** to view movie-quality animations for DNA replication.

The familiar double helix of DNA, here illustrated by computer graphics, contains information needed to govern a cell's life.

Genetics is the study of inheritance and inheritable traits as expressed in an organism's genetic material. Geneticists study many aspects of inheritance, including the physical structure and function of genetic material, mutations, and the transfer of genetic material among organisms. In this chapter, we will examine these topics as they apply to microorganisms, the study of which has formed much of the basis of our understanding of human, animal, and plant genetics.

The Structure and Replication of Genomes

Learning Objective

✓ Compare and contrast the genomes of prokaryotes and eukaryotes.

The **genome** (je'nōm) of a cell or virus is its entire genetic complement, including both its **genes**—specific sequences of nucleotides that code for polypeptides or RNA molecules—and nucleotide sequences that connect genes to one another. The genomes of cells and DNA viruses are composed solely of molecules of deoxyribonucleic acid (DNA), whereas RNA viruses use ribonucleic acid instead. We will examine the genomes of viruses in more detail in Chapter 13. The remainder of this chapter focuses on bacterial genomes—their structure, replication, function, mutation, and repair, and how they compare and contrast with eukaryotic genomes and with the genomes of archaea. We begin by examining the structure of nucleic acids.

The Structure of Nucleic Acids

Learning Objective

✓ Describe the structure of DNA, and discuss how it facilitates the ability of DNA to act as genetic material.

As we studied in Chapter 2, nucleic acids are polymers of nucleotides, each of which contains a pentose sugar (deoxyribose in DNA, ribose in RNA), a phosphate, and one of five nitrogenous bases (guanine, cytosine, adenine, thymine, or uracil). These bases hydrogen-bond in specific ways called **base pairs (bp):** In DNA and in RNA, the complementary bases guanine and cytosine bond to one another with three hydrogen bonds **(Figure 7.1a)**. In DNA, the complementary bases adenine and thymine bond to one another with two hydrogen bonds **(Figure 7.1b)**, whereas in RNA, uracil (not thymine) bonds with adenine **(Figure 7.1c)**.

Deoxyribonucleotides are linked through their sugars and phosphates to form the two backbones of a helical, double-stranded DNA (dsDNA) molecule **(Figure 7.1d)**. The carbon atoms of deoxyribose are numbered 1′ (pronounced "one prime") through 5′. One end of a DNA strand is called the 5′ end because it terminates in a phosphate group attached to a 5′ carbon; the opposite (3′) end terminates with a hydroxyl group bound to a 3′ carbon of deoxyribose. The two strands are oriented in opposite directions to each other; one strand runs in a 3′ to 5′ direction, while the other runs in a 5′ to 3′ direction. Scientists say the two strands are *antiparallel.* The base pairs extend into the middle of the molecule in a way reminiscent of the steps of a spiral staircase.

The lengths of DNA molecules are not usually given in metric units; instead the length of a DNA molecule is expressed in base pairs (bp). For example, the genome of *Carsonella ruddii* (kar-son-el'ă rŭd'ē-ē) is 160,000 bp long, making it the smallest known cellular genome.

The structure of DNA helps explain its ability to act as genetic material. First, the linear sequence of nucleotides carries the instructions for the synthesis of polypeptides and RNA molecules—in much the way a sequence of letters carries information used to form words and sentences. Second, the complementary structure of the two strands allows a cell to make exact copies to pass to its progeny. We will examine the genetic code and DNA replication shortly.

CRITICAL **THINKING**

The chromosome of *Mycobacterium tuberculosis* is 4,411,529 bp long. A scientist who isolates and counts the number of nucleotides in its DNA molecule discovers that there are 2,893,963 molecules of guanine. How many molecules of the other three nucleotides are in the original DNA?

The amount of DNA in a genome can be extraordinary, as some examples will illustrate. The bacterium *Escherichia coli* (esh-ĕ-rik'ē-ă kō'lē) is approximately 2 μm long and 1 μm in diameter, but its genome consists primarily of a 4.6×10^6 bp DNA molecule that is about 1600 μm long—800 times longer than the cell. The human genome has about 6 billion base pairs in 46 nuclear DNA molecules and numerous copies of a unique mitochondrial DNA molecule, and the entire genome would be about 3 meters (3,000,000 μm) long if all 47 DNA molecules from a single cell were laid end to end. Most of a human cellular genome is packed into a nucleus that is typically only 5 μm in diameter. This is like packing 45 miles of thread into a golf ball! To understand how cells package such prodigious amounts of DNA into such small spaces, we must first understand that bacteria, archaea, and eukaryotes package DNA in different ways. We begin by examining the structure of prokaryotic genomes.

The Structure of Prokaryotic Genomes

The DNA of prokaryotic genomes is found in two structures: chromosomes and plasmids.

Prokaryotic Chromosomes

Prokaryotic cells, both bacterial and archaeal, package the main portion of their DNA, along with associated molecules of protein and RNA, as one or two distinct **chromosomes.**[1] Prokaryotic cells have a single copy of each chromosome and are called *haploid* cells.

[1]From Greek *chroma*, meaning color (because they typically stain darkly in eukaryotes, where they were first discovered), and *soma*, meaning body.

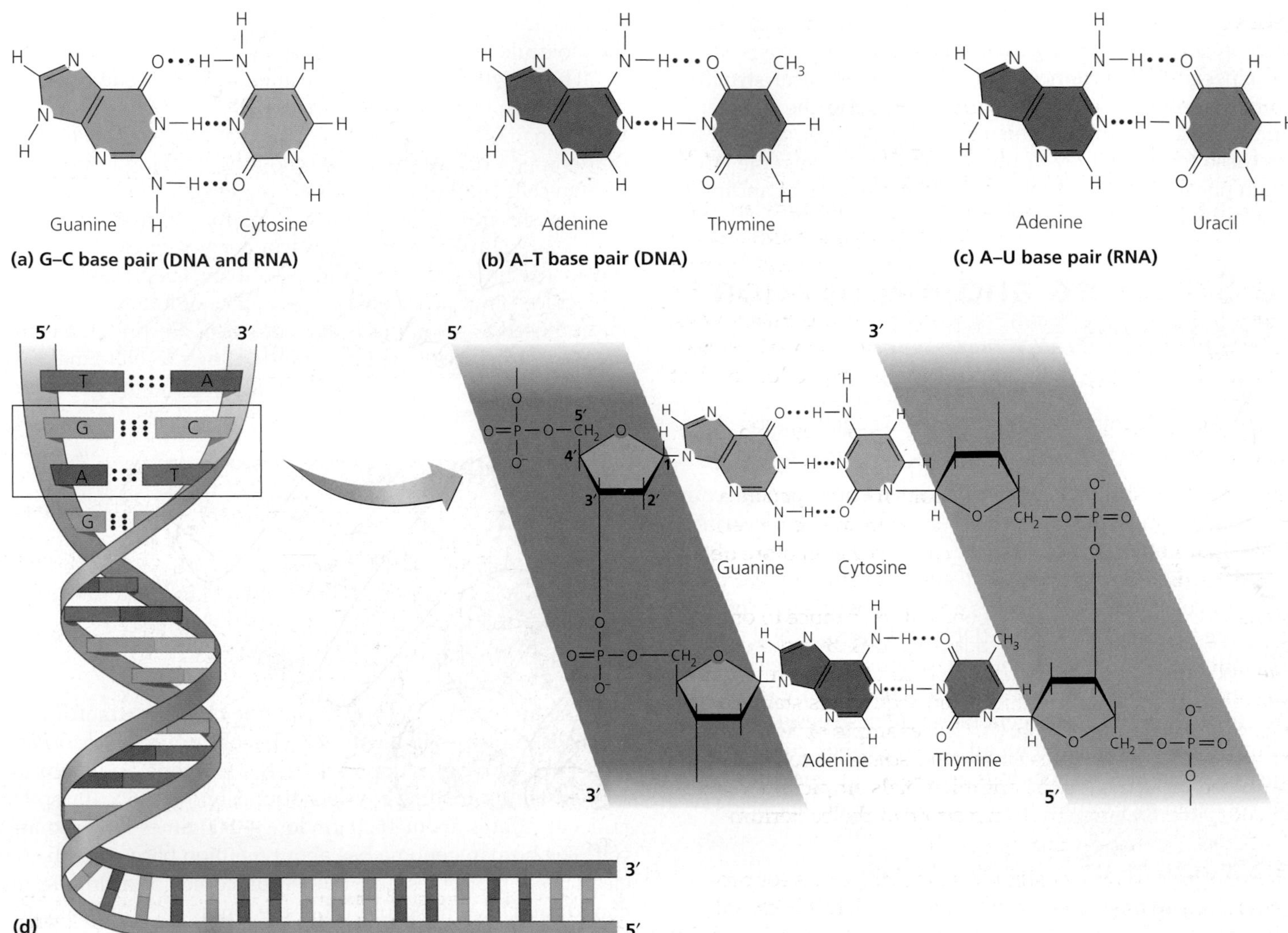

▲ **Figure 7.1 The structure of nucleic acids.** Nucleic acids are polymers of nucleotides consisting of a pentose sugar, a phosphate, and a nitrogenous base. **(a)** Base pairing between the complementary bases guanine (G) and cytosine (C) formed by three hydrogen bonds, found in both DNA and RNA. **(b)** Base pairing between the complementary bases adenine (A) and thymine (T) formed by two hydrogen bonds, found in DNA only. **(c)** Base pairing between adenine and uracil (U), found in RNA only. Notice the structural similarities between thymine and uracil. **(d)** Double-stranded DNA, which consists of antiparallel strands of nucleotides held to one another by the hydrogen bonding between complementary bases. *What structures do DNA nucleotides and RNA nucleotides have in common?*

Figure 7.1 Both DNA and RNA nucleotides are each composed of a pentose sugar, a phosphate, and a nitrogenous base.

A typical prokaryotic chromosome **(Figure 7.2a)** consists of a circular molecule of DNA localized in a region of the cytoplasm called the **nucleoid.** With few exceptions, no membrane surrounds a nucleoid, though the chromosome is packed in such a way that a distinct boundary is visible between the nucleoid and the rest of the cytoplasm. Chromosomal DNA is folded into loops that are 50,000–100,000 bp long **(Figure 7.2b)** held in place by molecules of protein and RNA. Archaeal DNA is wrapped around globular proteins called **histones.** The enzyme *gyrase* further folds and supercoils the entire prokaryotic chromosome like a skein of yarn into a compact mass.

For many years scientists thought that each prokaryote had only a single circular chromosome, but we now know that there are exceptions. For example, *Epulopiscium* (ep′yoo-lō-pis′-sē-ŭm), the giant bacterium introduced on p. 62, may have tens of thousands of identical chromosomes. Some bacterial species contain two different chromosomes, and at least one member of such a pair may be linear. *Agrobacterium tumefaciens* (ag′rō-bak-tēr′ē-um tū′me-fāsh-enz), a bacterium used to transfer genes into plants, is an example of a prokaryote with two chromosomes, one circular and one linear.

Plasmids

Learning Objective

✓ Describe the structure and function of plasmids.

In addition to chromosomes, many prokaryotic cells contain one or more **plasmids,** which are small molecules of DNA that replicate independently of the chromosome. Plasmids are usually circular and 1–5% of the size of a prokaryotic chromosome (see Figure 7.2b), ranging in size from a few thousand bp to a few million bp. Each plasmid carries information required for its own replication, and often for one or more cellular traits. Typically, genes carried on plasmids are not essential for normal metabolism, for growth, or for cellular reproduction but can confer advantages to the cells that carry them.

Researchers have identified many types of plasmids, (sometimes also called *factors*), including the following:

- *Fertility (F) plasmids* carry instructions for *conjugation,* a process involved in transferring genes from one bacterial cell to another. We will consider conjugation in more detail near the end of this chapter.
- *Resistance (R) plasmids* carry genes for resistance to one or more antimicrobial drugs or heavy metals. By processes we will discuss shortly, certain cells can transfer resistance plasmids to other cells, which then acquire resistance to the same antimicrobial chemicals. One example of the effects of an R plasmid involves strains of *Escherichia coli* that have acquired resistance to the antimicrobials ampicillin, tetracycline, and kanamycin from a strain of the bacterium *Pseudomonas* (soo-dō-mō′nas).
- *Bacteriocin* (bak-tēr′ē-ō-sin) *plasmids* carry genes for proteinaceous toxins called *bacteriocins,* which kill bacterial cells of the same or similar species that lack the plasmid. In this way a bacterium containing this plasmid can kill its competitors.
- *Virulence plasmids* carry instructions for structures, enzymes, or toxins that enable a bacterium to become pathogenic. For example, *E. coli,* a normal resident of the human gastrointestinal tract, causes diarrhea only when it carries plasmids that code for certain toxins.

Now that we have examined the structure of prokaryotic genomes, we turn to the structure of eukaryotic genomes.

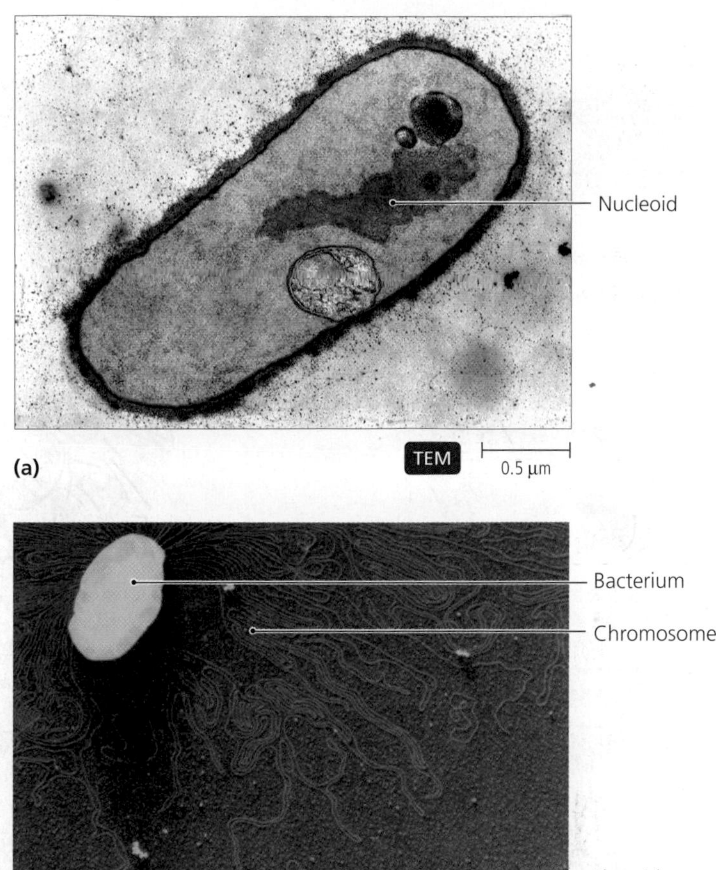

▲ **Figure 7.2 Bacterial genome.** **(a)** Bacterial chromosomes are packaged in a region of the cytosol called the nucleoid, which is not surrounded by a membrane. **(b)** The packing of a circular bacterial chromosome into loops, as seen after the cell was gently broken open to release the chromosome. Extrachromosomal DNA in the form of plasmids is also visible.

The Structure of Eukaryotic Genomes

Eukaryotic genomes consist of both nuclear and extranuclear DNA.

Nuclear Chromosomes

Learning Objective

✓ Compare and contrast prokaryotic and eukaryotic chromosomes.

Typically, eukaryotic cells have more than one nuclear chromosome in their genomes, though one species of Australian ant has a single chromosome per nucleus, and some eukaryotic cells such as mammalian red blood cells lose their chromosomes as they mature. Eukaryotic cells are often *diploid;* that is, they have two copies of each chromosome.

Eukaryotic chromosomes differ from their typical prokaryotic counterparts in that they are sequestered within a nucleus and are linear (rather than circular). As we saw in Chapter 3, the nucleus is an organelle surrounded by two membranes, which together are called the *nuclear envelope*. Given that a typical eukaryotic cell must package substantially more DNA than its prokaryotic counterpart, it is not surprising that nuclear chromosomes are more elaborate than those of prokaryotes.

Eukaryotic chromosomes are composed of DNA and globular eukaryotic histones, which are similar to archaeal histones. DNA, which has an overall negative electrical charge, wraps around the positively charged histones to form 10-nm-diameter beads called **nucleosomes** **(Figure 7.3a)**. Nucleosomes clump with other proteins to form **chromatin fibers** that are about 30 nm in diameter **(Figure 7.3b)**. Except during *mitosis* (nuclear

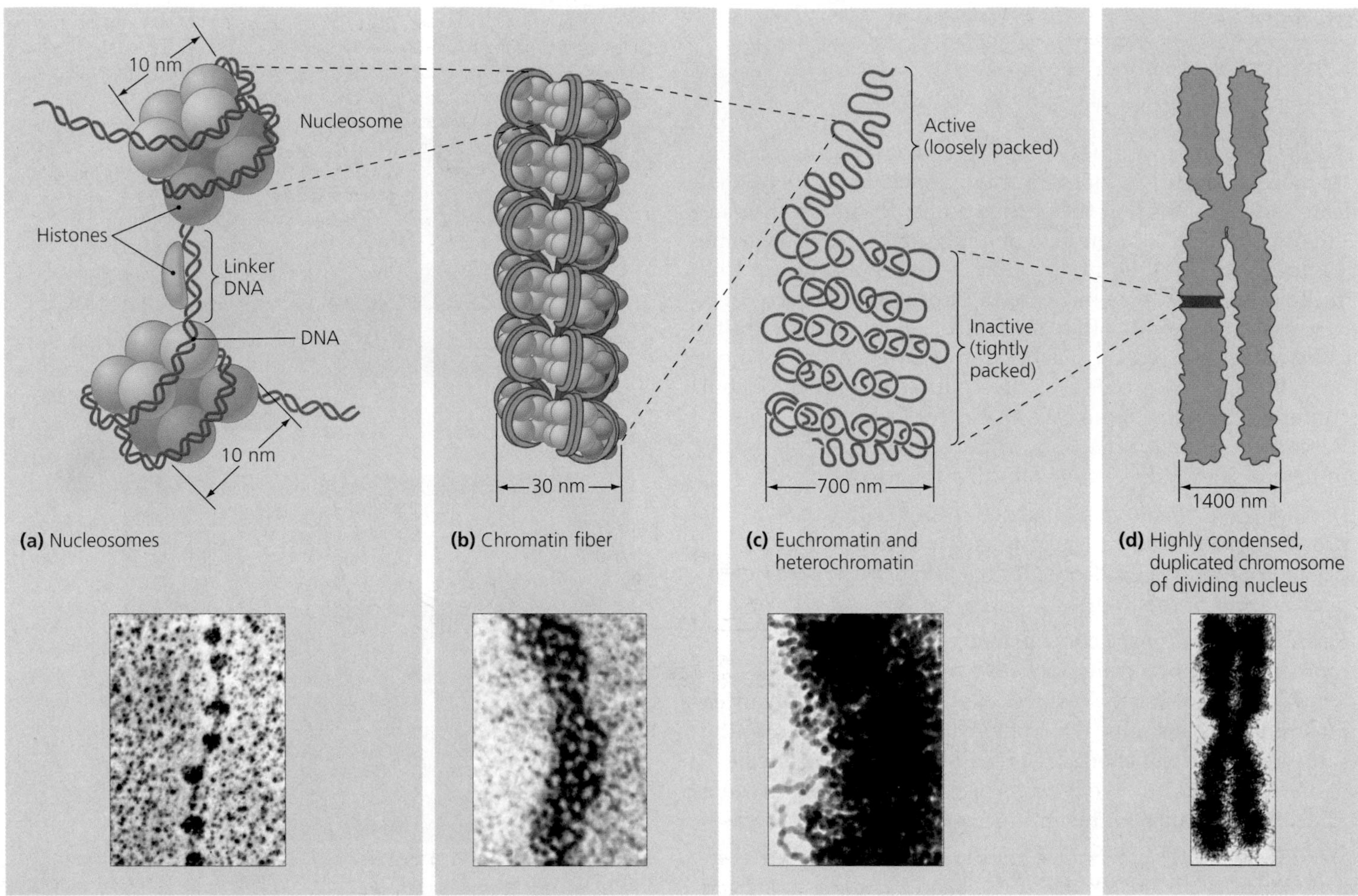

▲ **Figure 7.3 Eukaryotic nuclear chromosomal packaging.** **(a)** Histones stabilize and package DNA to form nucleosomes connected by linker DNA. **(b)** Nucleosomes clump to form chromatin fibers. **(c)** Chromatin fibers fold and are organized into active euchromatin and inactive heterochromatin. **(d)** During nuclear division (mitosis), duplicated chromatin fully condenses into a mitotic chromosome that is visible by light microscopy. If the nucleosomes were actually the size shown in the artist's illustration (a), the chromosome in (d) would be 20 m (about 65 feet) long.

division), chromatin fibers are dispersed throughout the nucleus and are too thin to be resolved without the extremely high magnification of electron microscopes. In regions of the chromosome where genes are active, the chromatin fibers are loosely packed to form *euchromatin* (yū-krō′ mă-tin) or *open chromatin;* inactive DNA is more tightly packed and is called *heterochromatin* (het′-er-ō-krō′ mă-tin) or *closed chromatin* **(Figure 7.3c)**.

Prior to mitosis, a cell replicates its chromosomes and then condenses them into pairs of chromosomes visible by light microscopy **(Figure 7.3d)**. One molecule of each pair is destined for each daughter nucleus. Chapter 12 discusses mitosis in more detail. The net result is that each DNA molecule is packaged as a mitotic chromosome that is 50,000× shorter than its extended length.

Extranuclear DNA of Eukaryotes

Not all of the DNA of a eukaryotic genome is contained in its nuclear chromosomes; most eukaryotic cells also have mitochondria, and plant, algal, and some protozoan cells have chloroplasts that also contain DNA. DNA molecules of mitochondria and chloroplasts are circular and resemble the circular chromosomes of prokaryotes. Genes located on these "prokaryotic" chromosomes code for about 5% of the RNA and polypeptides required for the organelle's replication and function; nuclear DNA codes for the remaining 95% of polypeptides and RNA molecules. Recall from Chapter 2 that some proteins have a quaternary structure formed from the association of individual polypeptides. Interestingly, polypeptides coded by mitochondrial or chloroplast chromosomes do not alone constitute any functional proteins. Rather, they become functional only when associated with polypeptides coded by nuclear chromosomes.

In addition to the extranuclear DNA in their mitochondria, some fungi and protozoa carry plasmids. For instance, most

Guanosine triphosphate deoxyribonucleotide (dGTP)

Guanine nucleotide (dGMP)

Guanine base

Deoxyribose

(a)

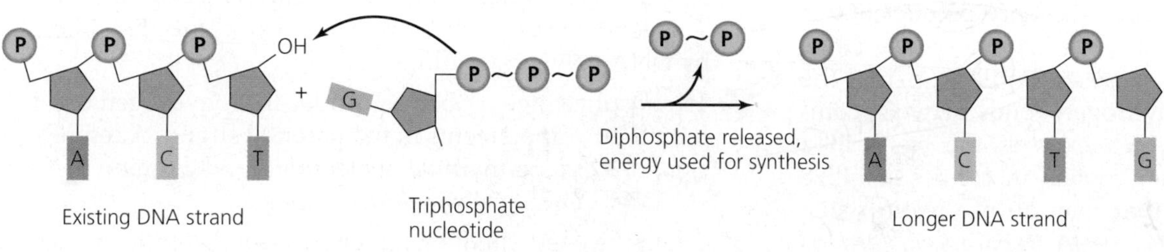

(b)

Figure 7.4 The dual role of triphosphate deoxyribonucleotides as building blocks and energy sources in DNA synthesis. **(a)** Guanosine triphosphate deoxyribonucleotide (dGTP), like all the triphosphate monomers of DNA, is a nucleotide to which two additional phosphate groups are attached. **(b)** The energy required for DNA polymerization (the addition of nucleotide building blocks to a DNA strand) is carried by each triphosphate nucleotide in the high-energy bonds between phosphate groups. *What is the difference between dGTP and guanosine triphosphate ribonucleotide (rGTP)?*

Figure 7.4 *This molecule (dGTP) contains deoxyribose; rGTP contains ribose.*

strains of the yeast *Saccharomyces cerevisiae* (sak-ă-rō-mī′sēz se-ri-vis′ē-ī) contain about 70 copies of a plasmid known as a *2-μm circle*. Each 2-μm circle is about 6300 bp long and has four protein-encoding genes that are involved solely in replicating the plasmid and confer no other traits to the cell.

In summary, the haploid genome of a prokaryotic cell consists of both chromosomal DNA, which is usually in a single circular chromosome, and all extrachromosomal DNA in the form of plasmids that are present. In contrast, a eukaryotic genome consists of nuclear chromosomal DNA in one or more linear chromosomes, plus all the extranuclear DNA in mitochondria, chloroplasts, and any plasmids that are present. The genomes of prokaryotes and eukaryotes are compared and contrasted in Table 7.1 on p. 200.

CRITICAL **THINKING**

In Chapter 3 we learned that the endosymbiotic theory proposes that mitochondria and chloroplasts evolved from prokaryotes living within other prokaryotes. What aspects of the eukaryotic genome support this theory? What aspects do not support the theory?

DNA Replication

Learning Objectives

- ✓ Describe the replication of DNA as a semiconservative process.
- ✓ Compare and contrast the synthesis of leading and lagging strands in DNA replication

DNA replication is an anabolic polymerization process that allows a cell to pass copies of its genome to its descendants. Though bacterial, archaeal, and eukaryotic cells package DNA differently, all three types employ similar mechanisms for DNA replication. **ANIMATIONS:** ***DNA Replication: Overview***

As discussed in Chapter 2, all polymerization processes require monomers (building blocks) and energy. *Triphosphate deoxyribonucleotides*—DNA nucleotides with three phosphate groups linked together by two high-energy bonds—serve both functions in DNA replication. In other words, the building blocks of DNA carry within themselves the energy required for DNA synthesis **(Figure 7.4)**. The structure of guanosine triphosphate deoxyribonucleotide (dGTP), shown in Figure 7.4a, differs from that of cytidine triphosphate (dCTP), thymidine triphosphate (dTTP), and adenosine triphosphate (dATP) only in the kind of base present. dATP has a structure similar to that of the energy-storage molecule ATP, except that ATP is a ribonucleotide rather than a deoxyribonucleotide (see Figure 2.27).

The key to DNA replication is the complementary structure of the two strands: Adenine and guanine in one strand bond with thymine and cytosine, respectively, in the other. DNA replication is a simple concept—a cell separates the two original strands and uses each as a template for the synthesis of a new complementary strand. Biologists say that DNA replication is *semiconservative* because each daughter DNA molecule is composed of one original strand and one new strand. The following sections focus on bacterial DNA replication and then consider small differences in the process in eukaryotes. Archaeal processes are not as well characterized.

TABLE 7.1 **Characteristics of Microbial Genomes**

	Bacteria	Archaea	Eukarya
Number of chromosomes	Single (haploid) copies of one or rarely two	One (haploid)	With one exception, two or more, typically diploid
Plasmids present?	In some cells; frequently more than one per cell	In some cells	In some fungi and protozoa
Type of nucleic acid	Circular or linear dsDNA	Circular dsDNA	Linear dsDNA in nucleus; circular dsDNA in mitochondria, chloroplasts, and plasmids
Location of DNA	In nucleoid of cytoplasm and in plasmids	In nucleoid of cytoplasm and in plasmids	In nucleus and in mitochondria, chloroplasts, and plasmids in cytosol
Histones present?	No, though chromosome is associated with a small amount of nonhistone protein	Yes	Yes

Initial Processes in DNA Replication

DNA replication begins at a specific sequence of nucleotides called an *origin* (not shown). First, proteins expose the DNA helix. Next, an enzyme called DNA *helicase* locally "unzips" the DNA molecule by breaking the hydrogen bonds between complementary nucleotide bases, which exposes the bases in a *replication fork* **(Figure 7.5a)**. Other protein molecules stabilize the separated single strands so that they do not rejoin while replication proceeds. **ANIMATIONS:** ***DNA Replication: Forming the Replication Fork***

After helicase untwists and separates the strands, a molecule of an enzyme called *DNA polymerase* (po-lim′er-ās) binds to each strand. Scientists have identified five kinds of prokaryotic DNA polymerase. These five enzymes vary in their specific functions, but all of them share one important feature—they catalyze synthesis of DNA by the addition of new nucleotides only to a hydroxyl group at the 3′ end of a nucleic acid. **Beneficial Microbes: Hot to Replicate** on p. 203 illustrates a type of DNA polymerase that has become a mainstay of genomic investigations. All DNA polymerases replicate DNA in only one direction—5′ to 3′—like a jeweler stringing pearls to make a necklace, adding them one at a time, always moving from one end of the string to the other. DNA polymerase III is the usual enzyme of DNA replication in bacteria.

Because the two original (template) strands are antiparallel, cells synthesize new strands in two different ways. One new strand, called the **leading strand,** is synthesized continuously—5′ to 3′—as a single long chain of nucleotides. The other new strand, called the **lagging strand,** is also synthesizd 5′ to 3′ but in short segments that are later joined. We will consider synthesis of the leading strand before examining replication of the lagging strand, even though the two processes occur simultaneously. **ANIMATIONS:** ***DNA Replication: Replication Proteins***

Synthesis of the Leading Strand

A cell synthesizes a leading strand toward the replication fork in the following series of five steps, the first three of which are shown in **Figure 7.5b**:

1. An enzyme called *primase* synthesizes a short RNA molecule that is complementary to the template DNA strand. This *RNA primer* provides the 3′ hydroxyl group required by DNA polymerase III.
2. Triphosphate deoxyribonucleotides form hydrogen bonds with their complements in the parental strand. Adenine nucleotides bind to thymine nucleotides, and guanine nucleotides bind to cytosine nucleotides.
3. Using the energy in the high-energy bonds of the triphosphate deoxyribonucleotides, DNA polymerase III covalently joins them one at a time to the leading strand. DNA polymerase III can add about 500–1000 nucleotides per second to a new strand.
4. DNA polymerase III also performs a proofreading function (not shown). About one out of every 100,000 nucleotides is mismatched with its template; for instance, a guanine might become incorrectly paired with a thymine. DNA polymerase III recognizes most such errors and removes the incorrect nucleotides before proceeding with synthesis. This role, known as the *proofreading exonuclease* function, acts like the backspace key on a keyboard, removing the most recent error. Because of this proofreading exonuclease function, and other repair strategies beyond the scope of this discussion, only about one error remains for every 10 billion (10^{10}) base pairs replicated.
5. Another DNA polymerase—DNA polymerase I—replaces the RNA primer with DNA (not shown). Note that researchers named DNA polymerase enzymes in the order of their discovery, not the order of their actions.

Synthesis of the Lagging Strand

Because DNA polymerase III adds nucleotides only to the 3′ end of the new strand, the enzyme moves away from the replication fork as it synthesizes a lagging strand. As a result, the lagging strand is synthesized discontinuously and always lags behind the process occurring in the leading strand. The steps in the synthesis of a lagging strand are as follows **(Figure 7.5c)**:

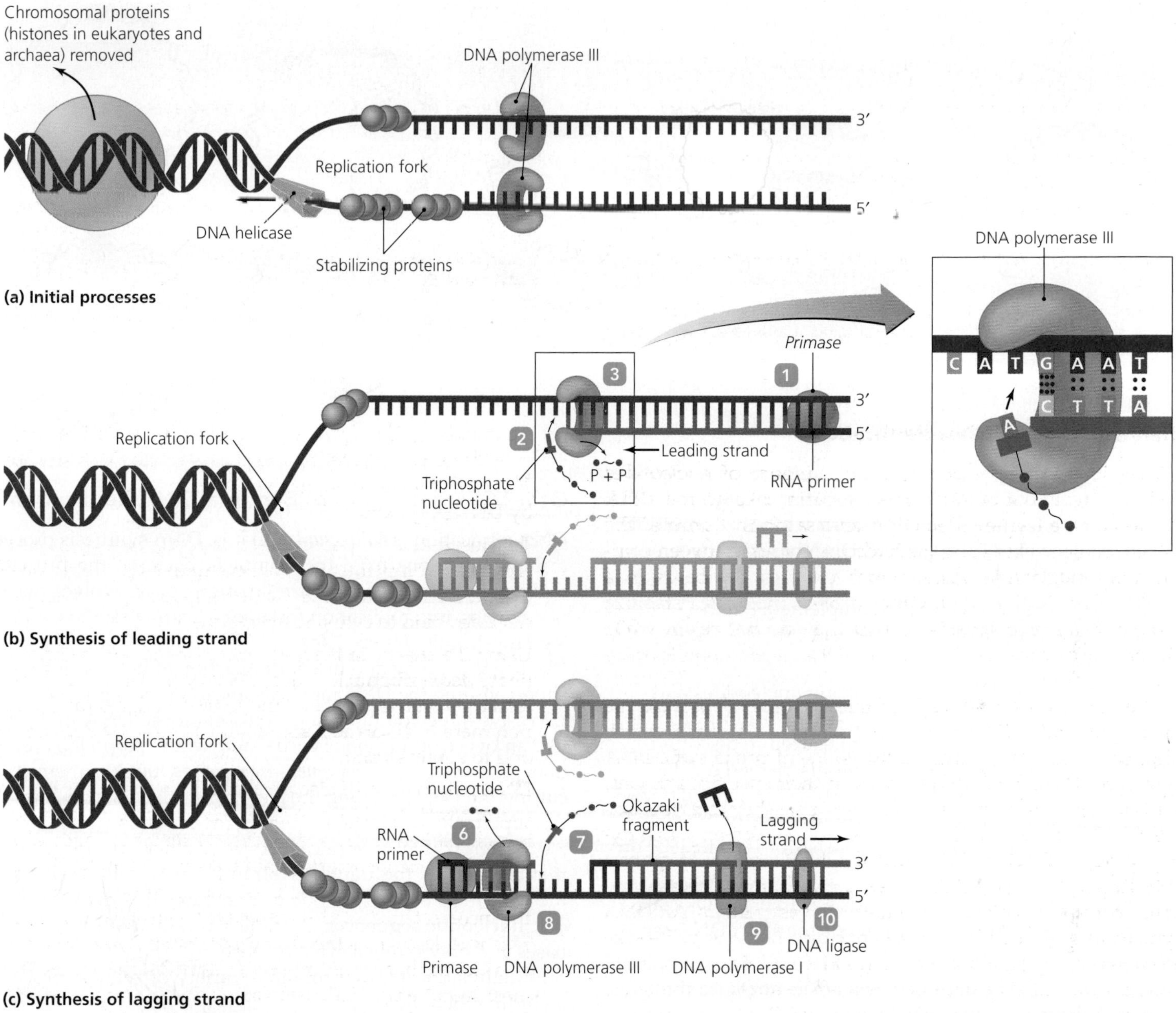

Chromosomal proteins (histones in eukaryotes and archaea) removed

DNA polymerase III

3′

Replication fork

5′

DNA helicase

Stabilizing proteins

(a) Initial processes

DNA polymerase III

Primase

C A T G A A T

C T T A

A

3

1

3′

Replication fork

2

5′

Leading strand

Triphosphate nucleotide

P + P

RNA primer

(b) Synthesis of leading strand

Replication fork

Triphosphate nucleotide

Okazaki fragment

Lagging strand

RNA primer

6

7

3′

5′

8

10

9

Primase

DNA polymerase III

DNA polymerase I

DNA ligase

(c) Synthesis of lagging strand

▲ **Figure 7.5 DNA replication. (a)** Initial processes. The cell removes proteins (histones in eukaryotes and archaea) from the DNA molecule. Helicase unzips the double helix—breaking hydrogen bonds between complementary base pairs—to form a replication fork. **(b)** Continuous synthesis of the leading strand. DNA synthesis always moves in the 5′ to 3′ direction, so the leading strand is synthesized toward the replication fork. The numbers refer to the steps in the process, which are described in the text. (Steps 4 and 5, the proofreading function of DNA polymerase and the replacement of the RNA primer with DNA, are not shown.) **(c)** Discontinuous synthesis of the lagging strand, which proceeds moving away from the replication fork. Actual Okazaki fragments are about 1000 nucleotides long. *Why is DNA replication termed "semiconservative"?*

Figure 7.5 *"Semiconservative" refers to the fact that each of the daughter molecules retains one parental strand and has one new strand; in other words, each is half new and half old.*

To see a 3-D animation on DNA replication, go to the Study Area at **www.masteringmicrobiology.com** and watch the *MicroFlix*.

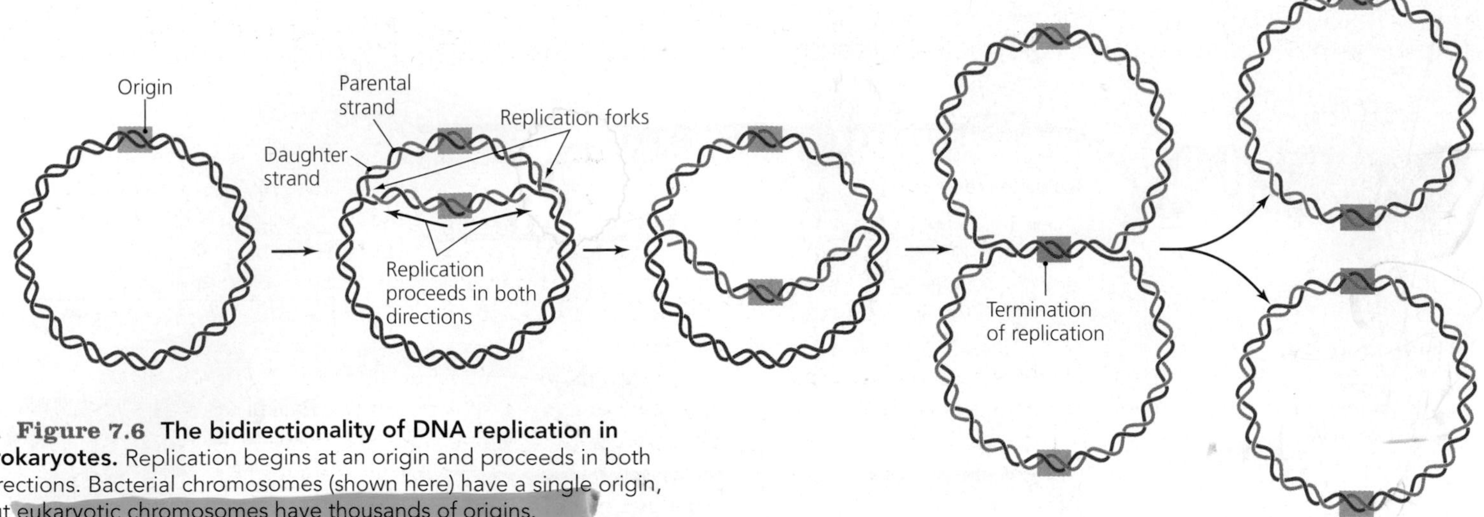

▲ **Figure 7.6 The bidirectionality of DNA replication in prokaryotes.** Replication begins at an origin and proceeds in both directions. Bacterial chromosomes (shown here) have a single origin, but eukaryotic chromosomes have thousands of origins.

6 Primase synthesizes RNA primers, but in contrast to its action on the leading strand, primase synthesizes multiple primers—one every 1000 to 2000 DNA bases of the template strand.

7 Nucleotides pair up with their complements in the template—adenine with thymine, and cytosine with guanine.

8 DNA polymerase III joins neighboring nucleotides and proofreads. In contrast to synthesis of the leading strand, however, the lagging strand is synthesized in discontinuous segments called *Okazaki fragments,* named for the Japanese scientist Reiji Okazaki (1930–1975), who first identified them. Each Okazaki fragment uses one of the new RNA primers, so each fragment consists of 1000 to 2000 nucleotides.

9 DNA polymerase I replaces the RNA primers of Okazaki fragments with DNA and proofreads the short DNA segment it has synthesized.

10 *DNA ligase* seals the gaps between adjacent Okazaki fragments to form a continuous DNA strand.

In summary, synthesis of the leading strand proceeds continuously toward the replication fork from a single RNA primer at the origin, following helicase and the replication fork down the DNA. The lagging strand is synthesized away from the replication fork, discontinuously as a series of Okazaki fragments, each of which begins with its own RNA primer. All the primers are eventually replaced with DNA nucleotides, and ligase joins the Okazaki fragments.

As noted earlier, DNA replication is semiconservative; each daughter molecule is composed of one parental strand and one daughter strand. The replication process produces double-stranded daughter molecules with a nucleotide sequence identical to that in the original double helix, ensuring that the integrity of an organism's genome is maintained each time it is copied. **ANIMATIONS:** ***DNA Replication: Synthesis***

Other Characteristics of Bacterial DNA Replication

DNA replication is *bidirectional;* that is, DNA synthesis proceeds in both directions from the origin. In bacteria, the process of replication proceeds from a single origin, so it involves two sets of enzymes, two replication forks, two leading strands, and two lagging strands (**Figure 7.6**).

The unzipping and unwinding action of helicase introduces supercoils into the DNA molecule ahead of the replication forks. Excessive supercoiling creates tension on the DNA molecule—like an overwound phone cord—and would stop DNA replication. The enzyme *topoisomerase* removes these supercoils by cutting the DNA, rotating the cut ends in the direction opposite the supercoiling, and then rejoining the cut ends.

Bacterial DNA replication is further complicated by **methylation** of the daughter strands, in which a cell adds a methyl group (—CH_3) to one or two bases that are part of specific nucleotide sequences. Bacteria typically methylate adenine bases and only rarely a cytosine base.

Methylation plays a role in a variety of cellular processes, including the following:

- *Control of genetic expression.* In some cases, genes that are methylated are "turned off" and are not transcribed, whereas in other cases methylated genes are "turned on" and are transcribed.
- *Initiation of DNA replication.* In many bacteria, methylated nucleotide sequences play a role in initiating DNA replication.
- *Protection against viral infection.* Methylation at specific sites in a nucleotide sequence enables cells to distinguish their DNA from viral DNA, which lacks methylation. The cells can then selectively degrade viral DNA.
- *Repair of DNA.* The role of methylation in some DNA repair mechanisms is discussed on pp. 221–222.

BENEFICIAL MICROBES

HOT TO REPLICATE

Cells replicate their DNA in order to pass copies to each of their offspring; scientists replicate the DNA of cells for a variety of tasks, including the study of gene action and regulation, elucidation of relationships among various kinds of cells, detection of hereditary diseases, determination of "genetic fingerprints" in such things as paternity tests, detection of pathogens, and diagnosis of infectious diseases. All such studies use millions or billions of identical copies of DNA produced using a process called polymerase chain reaction (PCR). PCR enzymatically replicates DNA without using living cells.

The concept of PCR is relatively simple. As its inventor wrote in *Scientific American,* "Beginning with a single molecule of the genetic material DNA, the PCR can generate 100 billion similar molecules in an afternoon. The reaction is easy to execute. It requires no more than a test tube, a few simple reagents, and a source of heat. " In the latter, however, lies a problem.

The temperature required to perform PCR is about 94°C. This temperature, which is almost that of boiling water, is the temperature required to break the hydrogen bonds of DNA and unzip the double helix, but this temperature also permanently denatures most DNA polymerase enzymes.

Enter *Thermus aquaticus,* a bacterium that thrives in hot springs such as those of Yellowstone National Park. Since this bacterium loves hot water, it is not surprising that its enzymes are heat-stable, and its DNA polymerase—called Taq polymerase or Taq—was the first polymerase used for PCR replication of DNA. Though *Science* magazine declared Taq "Molecule of the Year" in 1989, scientists now have many other heat-stable polymerases from bacterial and archaeal hyperthermophiles available for PCR.

Replication of Eukaryotic DNA

Eukaryotes replicate DNA in much the same way as do bacteria; helicases and topoisomerases unwind DNA, protein molecules stabilize single-stranded DNA, and molecules of DNA polymerase synthesize leading and lagging strands simultaneously. However, eukaryotic replication differs from prokaryotic replication in some significant ways:

- Eukaryotic cells use four different DNA polymerases to replicate DNA. DNA polymerase α initiates replication, including synthesis of a primer—the function performed by primase in bacteria. DNA polymerase δ elongates the leading strand, and DNA polymerase ε appears to be responsible for replicating the lagging strand. DNA polymerase γ replicates mitochondrial DNA.[2]
- The large size of eukaryotic chromosomes necessitates thousands of origins per molecule, each generating two replication forks; otherwise, the replication of eukaryotic genomes would take days instead of hours.
- Eukaryotic Okazaki fragments are shorter than those of bacteria—100 to 400 nucleotides long.
- Plant and animal cells methylate cytosine bases exclusively.

CRITICAL **THINKING**

Hydrogen bonds between complementary nucleotides are crucial to the structure of dsDNA because they hold the two strands together. Why couldn't the two strands be effectively linked by covalent bonds?

We have examined the physical structure of cellular genes—the specific sequences of DNA nucleotides—and the way cells replicate their genes. Now we will consider how genes function and how cells control genetic expression.

Gene Function

The first topic we must consider if we are to understand gene function is the relationship between an organism's genotype and its phenotype.

The Relationship Between Genotype and Phenotype

Learning Objective

✓ Explain how the genotype of an organism determines its phenotype.

The **genotype**[3] (jen′ō-tīp) of an organism is the actual set of genes in its genome. A genotype differs from a genome in that a genome also includes nucleotides that are not part of genes, such as the nucleotide sequences that link genes together. At the molecular level, the genotype consists of all the series of DNA nucleotides that carry instructions for an organism's life.

[2]Greek letters α, δ, ε and γ (alpha, delta, epsilon, and gamma) are equivalent to the numbers 1, 4, 5, and 3, corresponding to the order in which the polymerases were elucidated, not the order in which they act.

[3]From *Greek genos,* meaning race, and *typos,* meaning type.

EMERGING DISEASES

▲ *Ixodes ricinus* tick.

TICK-BORNE ENCEPHALITIS

Analiesa's head hurt; rather, her head felt as if it were being pounded from within by gorillas with sledgehammers. This, in addition to last week's fever, nausea, and vomiting, indicated she was more than worn down by her climb two weeks ago to Krimml waterfalls—Austria's highest. There must be something else causing her muscle aches, back pain, and inability to move her shoulders properly.

Indeed, Analiesa is a victim of an emerging disease in Europe and Asia—tick-borne encephalitis (TBE). Several viruses of rodents, transmitted to humans by *Ixodes* ticks, cause TBE. Genetic analysis reveals that the viruses are related to one another and are in the family *Flaviviridae*. Doctors diagnose about 10,000 cases of TBE worldwide each year, but TBE is not a reportable disease, so many more cases likely occur. As global warming allows ticks to survive in higher latitudes and at higher elevations, and as more and more people trek into wilderness areas, scientists expect that tick-borne encephalitis will become more common.

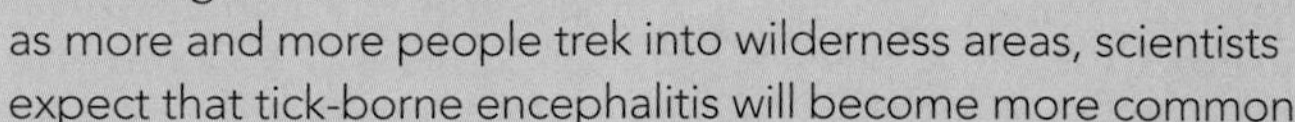

Analiesa's physician explained that there is no specific drug therapy for TBE other than painkiller for the muscle and headaches and corticosteroids for the inflammation. Her body's immune response would have to clear the virus from her system, which would take several weeks. Fortunately, it is unlikely that the virus would cause long-term complications, and TBE is not considered a fatal disease.

And should Analiesa forgo treks in the wilderness? No, but she should take advantage of the TBE vaccine, use chemical tick repellent, and avoid consumption of raw milk from goats or cows—infected animals pass the virus in their milk.

Track tick-borne encephalitis online by going to the Study Area at www.masteringmicrobiology.com.

Phenotype[4] (fe'nō-tīp) refers to the physical features and functional traits of an organism, including characteristics such as structures, morphology, and metabolism. For example, the shape of a cell, the presence and location of flagella, the enzymes and cytochromes of electron transport chains, and membrane receptors that trigger chemotaxis are all phenotypic traits.

Genotype determines phenotype by specifying what kinds of RNA and which structural, enzymatic, and regulatory protein molecules are produced. Though genes do not code *directly* for such molecules as phospholipids or for behaviors such as chemotaxis, ultimately phenotypic traits result from the actions of RNA and protein molecules that are themselves coded by DNA.

Not all genes are active at all times; that is, the information of a genotype is not always expressed as a phenotype. For example, *E. coli* activates genes for lactose catabolism only when it detects lactose in its environment.

The Transfer of Genetic Information

Learning Objective

✓ State the central dogma of genetics, and explain the roles of DNA and RNA in polypeptide synthesis.

Cells must continually synthesize proteins required for growth, reproduction, metabolism, and regulation. This synthesis requires that they accurately transfer the genetic information contained in DNA nucleotide sequences to the amino acid sequences of polypeptides. However, cells do not transfer the information coded in DNA directly but first make an RNA copy of the gene. In this copying process, called **transcription,**[5] the information is copied as RNA nucleotide sequences; RNA molecules in ribosomes then synthesize polypeptides in a process called **translation.**[6] These processes make up the **central dogma** of genetics: DNA is transcribed to RNA, which is translated to form polypeptides **(Figure 7.7)**.

An analogy serves to illustrate the central dogma. Suppose you were trying to understand the following message (which is a portion of the oath of Hippocrates, written in the Greek alphabet):

ΔΙΑΙΤΗΜΑΣΙΤΕΧΡΗΣΟΜΑΙΕΠΩΦΕΛΕΙΝ
ΚΑΜΝΟΝΤΩΝΚΑΤΑΔΥΝΑΜΙΝΚΑΙΚΡΙΣΙΝΕΜΗΝ
ΕΠΙΔΗΛΗΣΕΙΔΕΚΑΙΑΔΙΚΙΗ

If the Greek alphabet is foreign to you, you might have the Greek characters *transcribed* into the familiar English alphabet as a first step in understanding the message:

Diaiteimasi te chreisomai ep ophelein kamnonton kata dunamin kai krisin emein epi deileisei de kai adikiei eirzein

[4]From *Greek phainein*, meaning to show.
[5]From *Latin trans*, meaning across, and *scribere*, meaning to write—that is, to transfer in writing.
[6]From Latin *translatus*, meaning transferred.

Then you could begin the process of having the Greek words, now expressed in English letters, *translated* into English words:

> I will prescribe treatment to the best of my ability and judgment to help the sick and never for a harmful or illicit purpose

To a ribosome, DNA is like a foreign language written in a foreign alphabet. Thus, a cell must use processes analogous to those just described: it must first *transcribe* the "foreign alphabet" of DNA nucleotides (genes) into the more "familiar alphabet" of RNA nucleotides; then it must *translate* the message formed by these "letters" into the "words" (amino acids) that make up the "message" (a polypeptide). In this way a genotype can be expressed as a phenotype. There are a few exceptions to the central dogma. For example, some RNA viruses transcribe DNA from an RNA template—a process that is the reverse of cellular transcription.

In the following sections, we will examine the processes of transcription and translation. **ANIMATIONS:** ***Transcription: Overview; Translation: Overview***

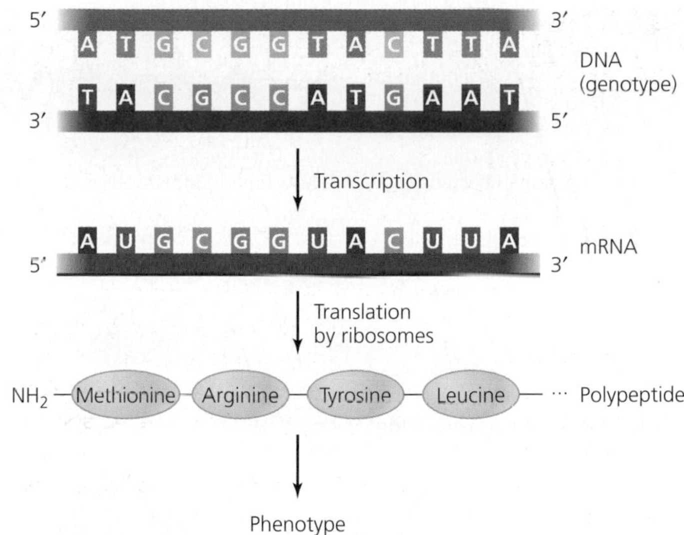

▲ **Figure 7.7 The central dogma of genetics.** A cell transcribes RNA from a DNA gene and then translates polypeptides using the code carried by the RNA molecules. Polypeptides determine phenotype by acting as structural, enzymatic, and regulatory proteins.

The Events in Transcription

Learning Objective

✓ Describe three steps in RNA transcription, mentioning the following: DNA, RNA polymerase, promoter, 5′ to 3′ direction, and terminator.

Cells transcribe four main types of RNA from DNA:

- **RNA primer** molecules for DNA polymerase to use during DNA replication
- **messenger RNA (mRNA)** molecules, which carry genetic information from chromosomes to ribosomes
- **ribosomal RNA (rRNA)** molecules, which combine with ribosomal polypeptides to form ribosomes—the organelles that synthesize polypeptides
- **transfer RNA (tRNA)** molecules, which deliver the correct sequence of amino acids to ribosomes based on the sequence of nucleotides in mRNA

We have already considered the role of RNA primer in DNA replication and will more closely examine the functions of the other types of RNA shortly. Next we examine transcription in bacteria; archaeal processes are not as well known.

Transcription occurs in the nucleoid region of the cytoplasm in bacteria. The steps of RNA transcription are: *initiation of transcription, elongation of the RNA transcript,* and *termination of transcription.* **Figure 7.8** depicts the events in transcription. **ANIMATIONS:** ***Transcription: The Process***

Initiation of Transcription

RNA polymerases—the enzymes that synthesize RNA—bind to specific nucleotide sequences called **promoters,** each of which is located near the beginning of a gene and initiates transcription (1 in Figure 7.8a). In bacteria, a polypeptide subunit of RNA polymerase called the *sigma factor* is necessary for recognition of a promoter. Once it adheres to a promoter sequence, RNA polymerase unzips and unwinds the DNA molecule in the promoter region and then travels along the DNA, unzipping the double helix as it moves (2). Primase transcribes RNA primer, and RNA polymerase transcribes mRNA, rRNA, and tRNA.

A cell uses different sigma factors and different promoter sequences to provide some control over the relative amount of transcription. RNA polymerases using different sigma factors do not adhere equally strongly to all promoters; there is about a 100-fold difference between the strongest attraction and weakest one. The greater the attraction between a particular sigma factor and a promoter, the more likely it is that transcription will proceed. Ultimately, variations in sigma factors and promoters affect the amounts and kinds of polypeptides produced.

Elongation of the RNA Transcript

RNA transcription does not actually begin in the promoter region, but at a spot 10 nucleotides away. There, triphosphate ribonucleotides (rATP, rUTP, rGTP, and rCTP) align opposite their complements in the open DNA. RNA polymerase links together two adjacent ribonucleotide molecules using energy from the phosphate bonds of the first ribonucleotide (3 in Figure 7.8b). The enzyme then moves down the DNA strand, elongating RNA by repeating the process.

Many molecules of RNA polymerase may concurrently transcribe the same gene **(Figure 7.9)**. In this way, a prokaryotic cell simultaneously produces numerous identical copies of RNA from a single gene—much as many identical prints can be made from a single photographic negative.

Like DNA polymerase, RNA polymerase links nucleotides only to the 3′ end of the growing molecule; however, RNA polymerase differs from DNA polymerase in the following ways:

1 RNA polymerase attaches nonspecifically to DNA and travels down its length until it recognizes a promoter sequence. Sigma factor enhances promoter recognition in bacteria.

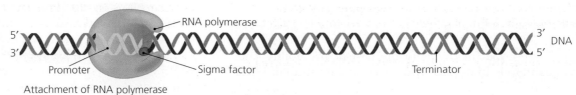

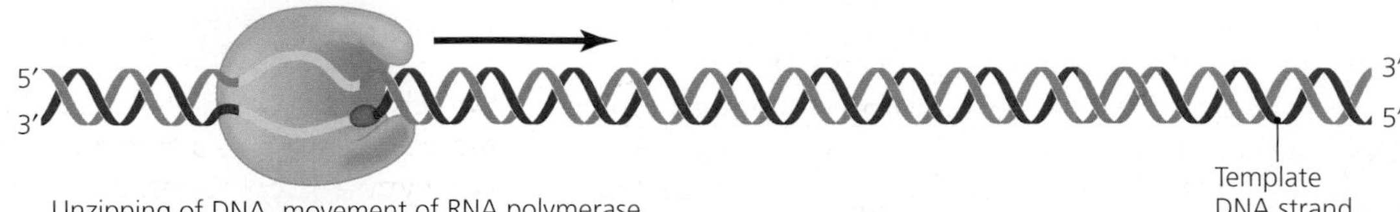

Attachment of RNA polymerase

2 Upon recognition of the promoter, RNA polymerase unzips the DNA molecule beginning at the promoter.

Unzipping of DNA, movement of RNA polymerase

(a) Initiation of transcription

3 Triphosphate ribonucleotides align with their DNA complements and RNA polymerase links them together, synthesizing RNA. No primer is needed. The triphosphate ribonucleotides also provide the energy required for RNA synthesis.

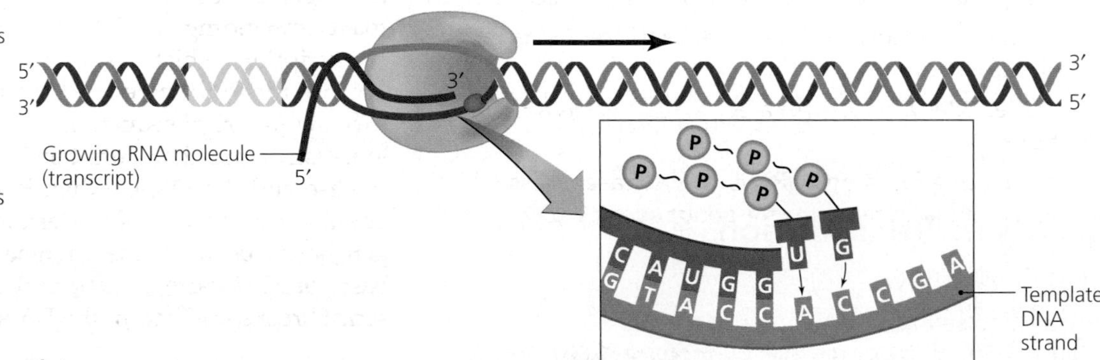

(b) Elongation of the RNA transcript

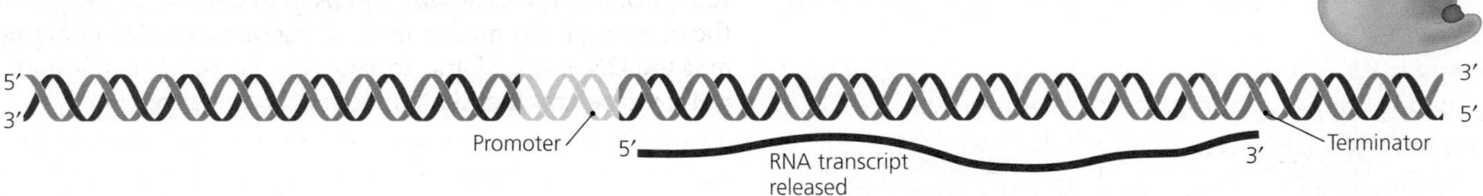

4 Self-termination: transcription of DNA terminator sequences causes the RNA to fold, loosening the grip of polymerase on DNA.

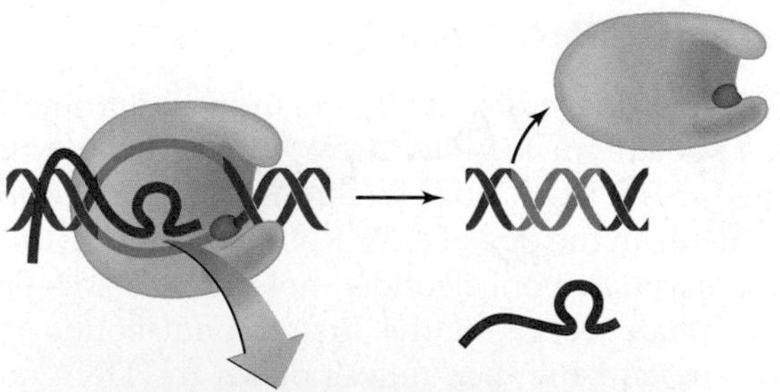

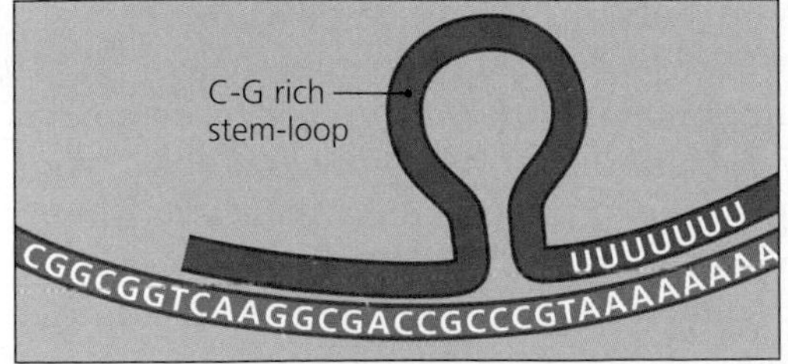

5 Enzyme-dependent termination: Rho pushes between polymerase and DNA, releasing polymerase, RNA transcript and Rho.

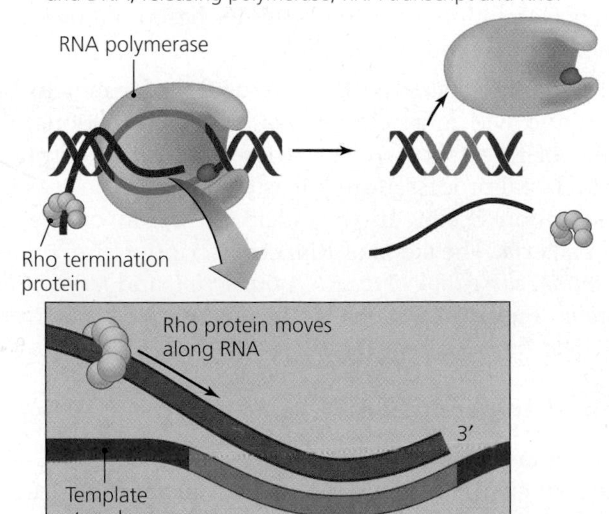

(c) Termination of transcription

Figure 7.8 The events in the transcription of RNA in prokaryotes. The helical dsDNA molecule is depicted as straight, parallel strands for clarity. **(a)** Initiation of transcription. **(b)** Elongation of the RNA transcript. **(c)** Termination of transcription, which is effected by the release of RNA polymerase. *What is the difference between a promoter sequence and an origin?*

Figure 7.8 *A promoter is a DNA sequence that initiates transcription; an origin is a point where DNA replication begins.*

- RNA polymerase unwinds and opens DNA by itself; helicase is not required.
- RNA polymerase does not need a primer.
- RNA polymerase transcribes only one of the DNA strands.
- RNA polymerase is slower than DNA polymerase III, proceeding at a rate of about 50 nucleotides per second.
- RNA polymerase incorporates ribonucleotides instead of deoxyribonucleotides.
- Uracil nucleotides are incorporated instead of thymine nucleotides.
- The proofreading function of RNA polymerase is less efficient, leaving a base-pair error about every 10,000 nucleotides.

CRITICAL **THINKING**

On average, RNA polymerase makes one error for every 10,000 nucleotides it incorporates in RNA. By contrast, only one base-pair error remains for every 10 billion base pairs during DNA replication. Explain why the accuracy of RNA transcription is not as critical as the accuracy of DNA replication.

Termination of Transcription

Transcription terminates when RNA polymerase and the transcribed RNA are released from DNA (see Figure 7.8c). RNA polymerase is tightly associated with the DNA molecule and cannot be removed easily; therefore, the termination of transcription is complicated. Scientists have elucidated two types of termination processes in bacteria—those that are self-terminating and those that depend on the action of an additional termination enzyme. These processes of transcription termination should not be confused with termination of translation examined in a later section.

Self-Termination Self-termination occurs when RNA polymerase transcribes a **terminator** sequence of DNA composed of two symmetrical series: one that is very rich in guanine and cytosine bases, followed by a region rich in adenine bases (see 4 in Figure 7.8c). RNA polymerase slows down during transcription of the GC-rich portion of the terminator because the three hydrogen bonds between each guanine and cytosine base pair make unwinding the DNA helix more difficult. This pause in transcription, which lasts about 60 seconds, provides enough time for the RNA molecule to form hydrogen bonds between its own symmetrical sequences, forming a stem and loop (hairpin loop) structure that puts tension on the union of RNA polymerase and the DNA. When RNA polymerase transcribes the adenine-rich portion of the terminator, the relatively few hydrogen bonds between the adenine bases of DNA and the uracil bases of RNA cannot withstand the tension, and the RNA transcript breaks away from the DNA, releasing RNA polymerase.

Rho-Dependent Termination The second type of termination depends on a termination protein, called Rho, that binds to a specific RNA sequence near the end of an RNA transcript. The Rho protein moves toward RNA polymerase at the 3′ end of the growing RNA molecule, pushing between RNA polymerase and the DNA strand and forcing them apart; this releases RNA polymerase and the RNA transcript (see Figure 7.8c 5).

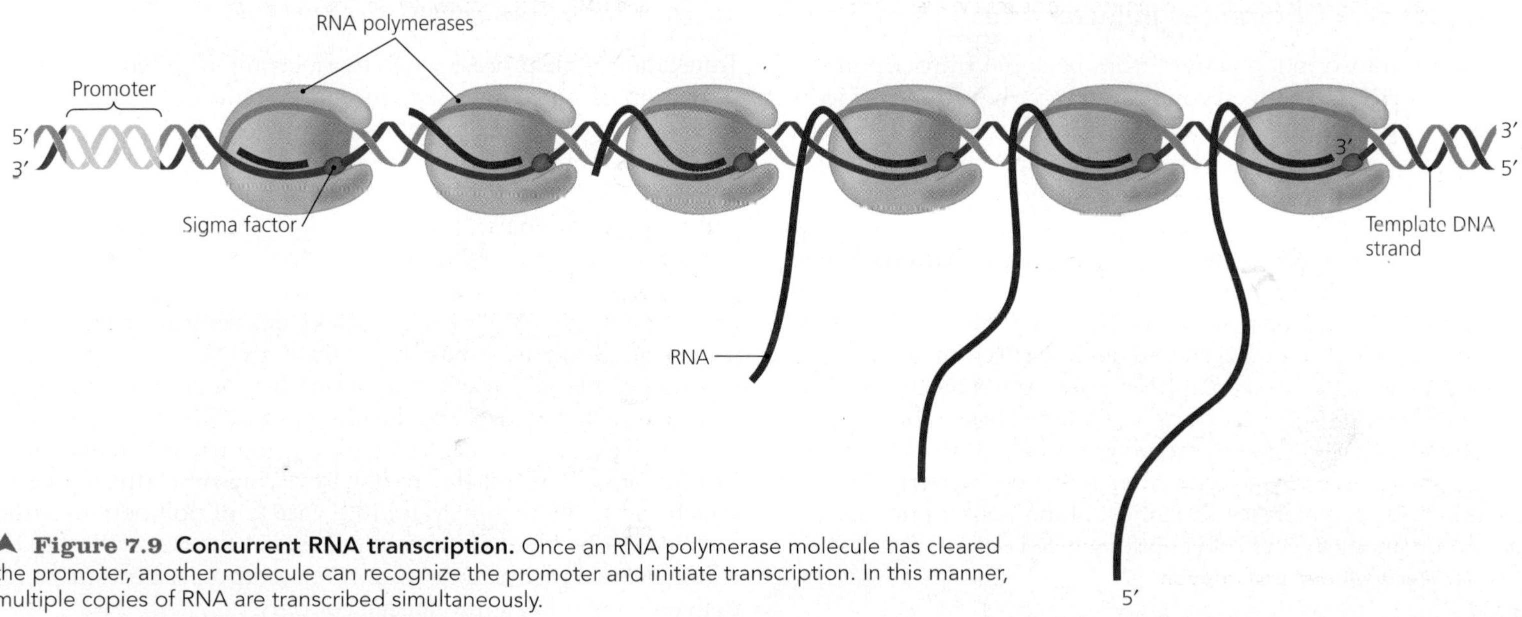

Figure 7.9 Concurrent RNA transcription. Once an RNA polymerase molecule has cleared the promoter, another molecule can recognize the promoter and initiate transcription. In this manner, multiple copies of RNA are transcribed simultaneously.

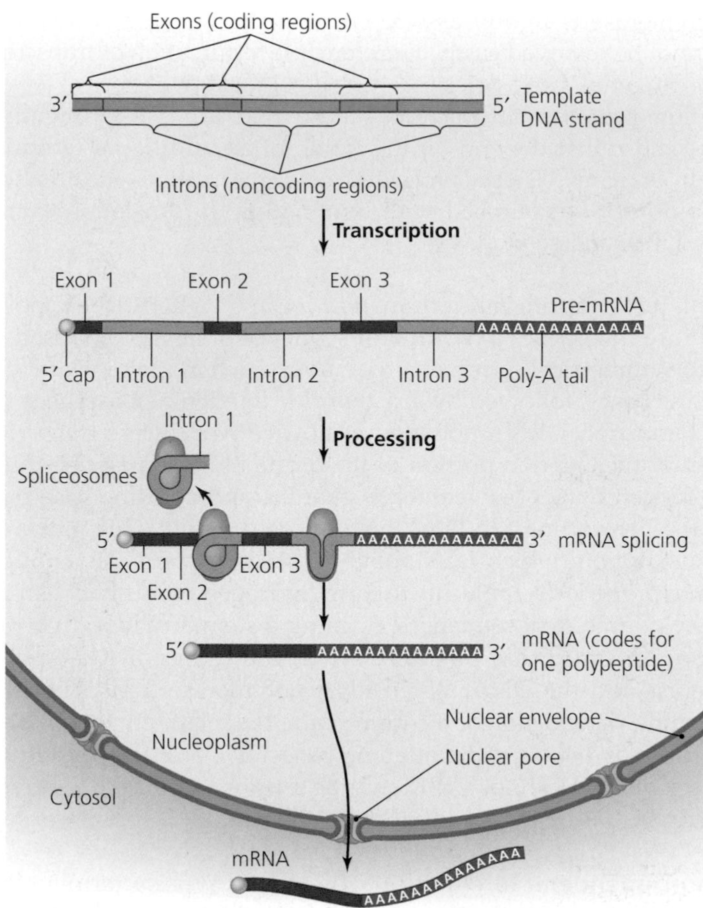

▲ **Figure 7.10 Processing eukaryotic mRNA.** Within the nucleus, transcription produces pre-mRNA, which contains coding exons and noncoding introns. Enzymes cap the 5′ end with a modified guanine nucleotide and add hundreds of adenine nucleotides to the 3′ end, a process known as polyadenylation. Ribozymes further process pre-mRNA by removing introns and splicing together exons to form a molecule that codes for a single polypeptide. Eukaryotic mRNA then moves from the nucleus to the cytoplasm.

Transcriptional Differences in Eukaryotes

Eukaryotic transcription differs from bacterial transcription in several ways. First, a eukaryotic cell transcribes RNA inside its nucleus, primarily in the region of the nucleus called the nucleolus, as well as inside any mitochondria and chloroplasts that are present. In contrast, transcription in prokaryotes occurs in the cytosol.

Another difference between eukaryotes and bacteria is that eukaryotes have three types of nuclear RNA polymerase—one for transcribing mRNA, one for transcribing the major rRNA gene, and one for transcribing tRNA and smaller rRNA molecules. Mitochondria use a fourth type of RNA polymerase. Further, several separate protein *transcription factors* at a time assist in binding eukaryotic RNA polymerase to promoter sequences, in contrast to a single sigma factor in bacteria. After initiating transcription, eukaryotic RNA polymerases shed most of the transcription factors and recruit another set of polypeptides called *elongation factors.*

Finally, eukaryotic cells must process mRNA before beginning polypeptide translation **(Figure 7.10)**. RNA processing involves three events:

1. *Capping.* The cell adds a modified guanine nucleotide to the "front" end (5′ end) of the mRNA when the RNA molecule is about 30 nucleotides long.
2. *Polyadenylation.* When RNA polymerase reaches the end of a gene, termination proteins cleave the RNA molecule and add 100 to 250 adenine nucleotides, depending on the organism, to the 3′ end. Polyadenylation occurs without a DNA template.
3. *Splicing.* Newly capped and polyadenylated mRNA molecules are called *pre-messenger RNA* because they contain **introns,** which are noncoding sequences that may be thousands of nucleotides long. (Few prokaryotic mRNA molecules contain introns.) A cell removes introns to make functional mRNA containing only coding regions called **exons,** each of which is about 150 nucleotides long. The "in" in *intron* refers to *intervening* sequences (that is, they lie between coding regions), whereas the "ex" in *exon* refers to the fact that these regions are expressed. Five small RNA molecules associate with about 300 polypeptides to form a *spliceosome* that acts as a ribozyme (ribosomal enzyme) to splice pre-mRNA into mRNA—it removes introns and splices the exons to produce a functional mRNA molecule that exits the nucleus.

Now that we have discussed how cells use DNA as the genetic material, maintain the integrity of their genomes through semiconservative replication, and transcribe RNA from DNA genes, we turn to the process of translation and the role of each type of RNA.

Translation

Learning Objectives

✓ Describe the genetic code in general, and identify the relationship between codons and amino acids.

✓ Describe the translation of polypeptides, identifying the roles of the three types of RNA.

Translation is the process whereby ribosomes use the genetic information of nucleotide sequences to synthesize polypeptides composed of specific amino acid sequences. As we studied in Chapter 2, some proteins are simple polypeptides, whereas other proteins are composed of several polypeptides bound together in a quaternary structure.

Ribosomes can be thought of as "polypeptide factories," so consider the following analogy between translation and a hypothetical automobile factory. Trucks deliver auto parts to the factory at the correct times and in the correct order to manufacture one of a large variety of automobile models, depending on instructions from corporate headquarters delivered by special courier. Similarly, molecules of tRNA (the trucks) deliver preformed amino acids (the parts) to a ribosome (the factory), which can manufacture an infinite variety of polypeptides (the car models) by assembling amino acids in the correct order according to the instructions from DNA (corporate headquarters) delivered via mRNA (the special courier).

First nucleotide base (5′ position)	Second nucleotide base: U	Second nucleotide base: C	Second nucleotide base: A	Second nucleotide base: G	Third nucleotide base (3′ position)
U	UUU Phenylalanine (Phe)	UCU Serine (Ser)	UAU Tyrosine (Tyr)	UGU Cysteine (Cys)	U
U	UUC Phenylalanine (Phe)	UCC Serine (Ser)	UAC Tyrosine (Tyr)	UGC Cysteine (Cys)	C
U	UUA Leucine (Leu)	UCA Serine (Ser)	UAA STOP	UGA STOP Selenocysteine (SeCys)	A
U	UUG Leucine (Leu)	UCG Serine (Ser)	UAG STOP*	UGG Tryptophan (Trp)	G
C	CUU Leucine (Leu)	CCU Proline (Pro)	CAU Histidine (His)	CGU Arginine (Arg)	U
C	CUC Leucine (Leu)	CCC Proline (Pro)	CAC Histidine (His)	CGC Arginine (Arg)	C
C	CUA Leucine (Leu)	CCA Proline (Pro)	CAA Glutamine (Gln)	CGA Arginine (Arg)	A
C	CUG Leucine (Leu)	CCG Proline (Pro)	CAG Glutamine (Gln)	CGG Arginine (Arg)	G
A	AUU Isoleucine (Ile)	ACU Threonine (Thr)	AAU Asparagine (Asn)	AGU Serine (Ser)	U
A	AUC Isoleucine (Ile)	ACC Threonine (Thr)	AAC Asparagine (Asn)	AGC Serine (Ser)	C
A	AUA Isoleucine (Ile)	ACA Threonine (Thr)	AAA Lysine (Lys)	AGA Arginine (Arg)	A
A	AUG START Methionine (Met)	ACG Threonine (Thr)	AAG Lysine (Lys)	AGG Arginine (Arg)	G
G	GUU Valine (Val)	GCU Alanine (Ala)	GAU Aspartic acid (Asp)	GGU Glycine (Gly)	U
G	GUC Valine (Val)	GCC Alanine (Ala)	GAC Aspartic acid (Asp)	GGC Glycine (Gly)	C
G	GUA Valine (Val)	GCA Alanine (Ala)	GAA Glutamic acid (Glu)	GGA Glycine (Gly)	A
G	GUG Valine (Val)	GCG Alanine (Ala)	GAG Glutamic acid (Glu)	GGG Glycine (Gly)	G

*also codes for a 22nd amino acid, pyrrolysine, in some prokaryotes.

▲ **Figure 7.11 The genetic code.** The table shows the set of mRNA codons and the amino acids for which they code. AUG is not only the start codon but also specifies methionine (Met) in eukaryotes and *N*-formylmethionine (fMet) in prokaryotes, mitochondria, and chloroplasts. Two codons (UAA and UAG) are stop codons that do not typically specify amino acids. UGA functions as a stop codon and also specifies selenocysteine.

How do ribosomes interpret the nucleotide sequence of mRNA to determine the correct order in which to assemble amino acids? To answer this question, we will consider the genetic code, examine in more detail the RNA molecules that participate in translation, and describe the specific steps of translation.

The Genetic Code

When geneticists in the early 20th century began to consider that DNA might be the genetic molecule, they were confronted with a problem: How can four kinds of DNA nucleotide bases—adenine, thymine, guanine, and cytosine (abbreviated A, T, G, and C)—specify the 21 different amino acids commonly found in proteins? If each nucleotide coded for a single amino acid, only 4 amino acids could be specified. Even if pairs of nucleotides served as the code—for instance, if AA, AT, and TA each specified a different amino acid—only 16 amino acids (that is, 4^2) could be accommodated. Eventually scientists showed that genes are composed of sequences of three nucleotides that specify amino acids. For example, the DNA nucleotide sequence AAA specifies the amino acid phenylalanine, and GTA codes for histidine. There are 64 possible arrangements of the four nucleotides in triplets (4^3)—more than enough to specify 21 amino acids.

These examples are DNA triplets, but ribosomes do not directly access genetic information on a DNA molecule. Instead, molecules of mRNA carry the code to the ribosomes; therefore, scientists define the genetic code **(Figure 7.11)** as triplets of mRNA nucleotides called **codons** (kō′donz) that code for specific amino acids. UUU is a codon for phenylalanine, and CAU is a codon for histidine.

In most cases, 61 codons specify amino acids and 3 codons—UAA, UAG, and UGA—instruct ribosomes to stop translating; though, under some conditions, UGA codes for the 21st amino acid, selenocysteine. Codon AUG also has a dual function, acting as both a start signal and the codon for the amino acid methionine. In bacteria, mitochondria, and chloroplasts, AUG initially codes for *N*-formylmethionine (fMet), a modified amino acid.

$$\mathrm{O{=}C(H){-}N(H){-}C(H)(COOH){-}CH_2{-}CH_2{-}S{-}CH_3}$$

N-formylmethionine (fMet)

$$\mathrm{H{-}N(H){-}C(H)(COOH){-}CH_2{-}CH_2{-}S{-}CH_3}$$

Methionine (Met)

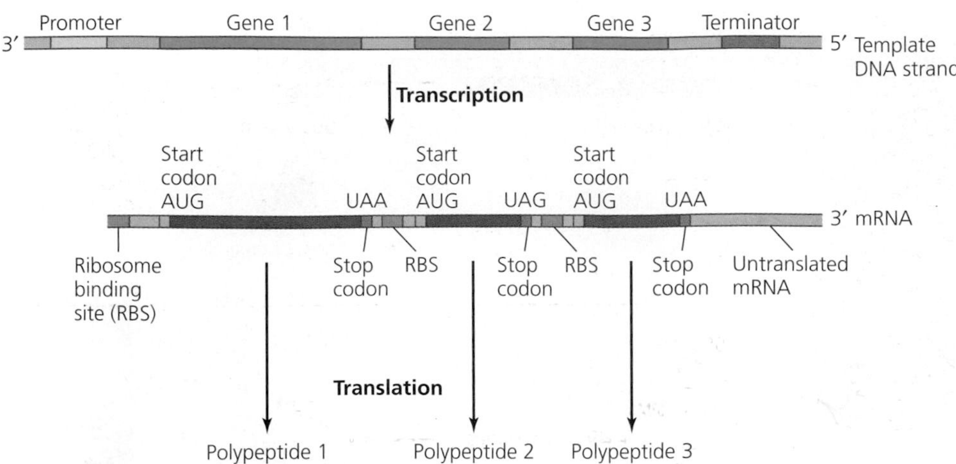

Figure 7.12 A single prokaryotic mRNA can code for several polypeptides. Prokaryotes typically code for several related polypeptides via a single mRNA molecule. The mRNA molecule shown here has transcripts of three genes encoding three polypeptides. The transcript of each gene begins with a start codon and ends with a stop codon.

As you examine the genetic code, notice that it is redundant; that is, more than one codon is associated with every amino acid except methionine and tryptophan. With most redundant codons, the first two nucleotides determine the amino acid, and the third nucleotide is inconsequential. For example, the codons GUU, GUC, GUA, and GUG all specify the amino acid valine.

Interestingly, the genetic code is nearly universal; that is, with few exceptions, ribosomes in archaeal, bacterial, plant, fungal, protozoan, and animal cells use the same genetic code. Some exceptions are listed in Table 7.2. **ANIMATIONS:** ***Translation: The Genetic Code***

CRITICAL THINKING

A scientist isolates a molecule of mRNA with the following base sequence: AUGUACGACAUAUGCAUA. What is the sequence of amino acids in the polypeptide synthesized by a bacterial ribosome from this message? What would be different if the message were translated in a mitochondrion instead?

TABLE 7.2 Some Exceptions to the Genetic Code

Codon	Usual Use	Alternative Use
AUA	Codes for isoleucine	Codes for methionine in mitochondria
UAG	STOP	Codes for glutamine in some protozoa and algae and for pyrrolysine, a 22nd amino acid found in some prokaryotes
CGG	Codes for arginine	Codes for tryptophan in plant mitochondria
UGA	STOP, selenocysteine	Codes for tryptophan in mitochondria and mycoplasmas (type of bacteria)

Participants in Translation

As we have discussed, transcription produces messenger RNA, transfer RNA, and ribosomal RNA—each of which is involved in translation. We now discuss each kind of RNA in turn.

Messenger RNA Messenger RNA carries genetic information from a chromosome to ribosomes as triplets of RNA nucleotides (codons) that encode amino acid sequences. In prokaryotes a basic mRNA molecule contains sequences of nucleotides that are recognized by ribosomes: an AUG start codon, sequential codons for other amino acids in the polypeptide, and at least one of the three stop codons. A single molecule of prokaryotic mRNA often contains start codons and instructions for more than one polypeptide arranged in series (**Figure 7.12**). Because both transcription and the subsequent events of translation occur in the cytosol of prokaryotes, prokaryotic ribosomes can begin translation before transcription is finished.

Eukaryotic mRNA differs from prokaryotic mRNA in several ways:

- As we have seen, eukaryotic cells extensively process pre-mRNA to make mRNA (see Figure 7.10).
- A molecule of eukaryotic mRNA contains instructions for only one polypeptide.
- Eukaryotic mRNA is not translated until it is fully transcribed, processed, and has left the nucleus. In other words, transcription and translation of a molecule of eukaryotic mRNA do not occur simultaneously because eukaryotic ribosomes are located in the cytoplasm, while transcription occurs in the nucleus.

Transfer RNA A transfer RNA (tRNA) molecule is a sequence of about 75 ribonucleotides that curves back on itself to form three main hairpin loops held in place by hydrogen bonding between complementary nucleotides. Although transfer RNA molecules can be modeled simplistically by a cloverleaf structure (**Figure 7.13a**), their three-dimensional shape is more complex (**Figure 7.13b**). For simplicity, tRNA will be represented in subsequent figures by an icon shaped like the 3-D icon in Figure 7.13b.

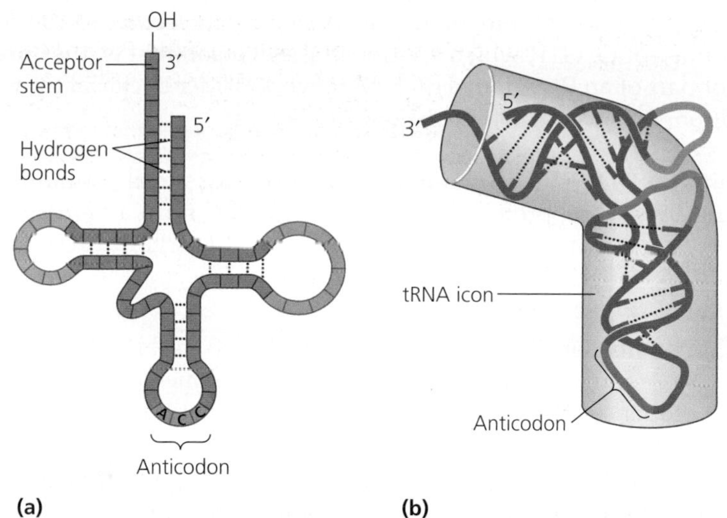

▲ **Figure 7.13 Transfer RNA.** **(a)** A two-dimensional "cloverleaf" representation of tRNA showing three hairpin loops held in place by intramolecular hydrogen bonding. **(b)** A three-dimensional drawing of the same tRNA. A specific amino acid attaches to the 3′ acceptor stem; the anticodon is a nucleotide triplet that is complementary to the mRNA codon for that amino acid. *Which amino acid would be attached to the acceptor stem of this tRNA?*

Figure 7.13 *UGG is the codon for tryptophan.*

A molecule of tRNA transfers the correct amino acid to a ribosome during polypeptide synthesis. To this end, tRNA has an **anticodon** (an-tē-kō′ don) triplet in its bottom loop and an *acceptor stem* for a specific amino acid at its 3′ end. Specific enzymes in the cytoplasm *charge* each tRNA molecule; that is, they attach the appropriate amino acid to the acceptor stem.

Anticodons are complementary to mRNA codons, and each acceptor stem is designed to carry one particular amino acid, which varies with the tRNA. In other words, each transfer RNA carries a specific amino acid and recognizes mRNA codons only for that amino acid. A tRNA molecule is designated by a superscript abbreviation of its amino acid. For example, $tRNA^{Phe}$ carries phenylalanine, and $tRNA^{Ser}$ transfers serine.

The fact that 62 codons specify the 21 amino acids used by cells does not mean that there must be 62 anticodons on 62 different types of tRNA, because many tRNA molecules recognize more than one codon. *E. coli,* for example, has only about 40 different tRNAs. The variability in codon recognition by tRNA is due to "wobble" of the anticodon's third nucleotide. Wobble, which is a change of angle from the normal axis of the molecule, allows the third nucleotide to hydrogen bond to a nucleotide other than its usual complement. For example, a guanine nucleotide in the third position normally bonds to cytosine, but it can wobble and also pair with uracil; therefore, whether a codon has cytosine or uracil in the third position makes no difference, because the same tRNA recognizes either nucleotide in the third position. For example, the codons UUU and UUC both specify the amino acid phenylalanine because the anticodon AAG recognizes both of them. Similarly, UCU and UCC code for serine, and UAU and UAC code for tyrosine. This redundancy in the genetic code helps protect cells against the effects of errors in replication and transcription.

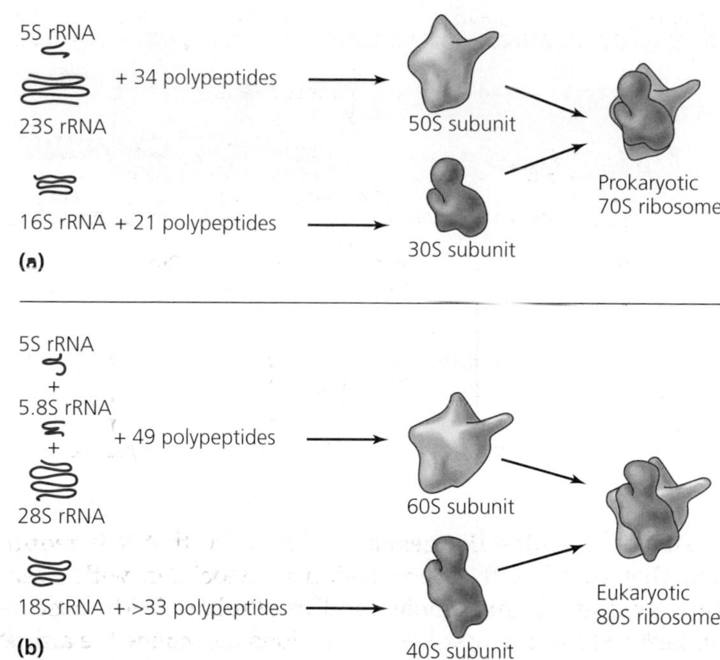

▲ **Figure 7.14 Ribosomal structures.** **(a)** The 70S prokaryotic ribosome, which is composed of polypeptides and three rRNA molecules arranged in 50S and 30S subunits. **(b)** The 80S eukaryotic ribosome, which is composed of molecules of rRNA and polypeptides, arranged in 60S and 40S subunits.

CRITICAL **THINKING**

We have seen that wobble makes the genetic code redundant in the third position for C and U. After reexamining the genetic code in Figure 7.11, state what other nucleotides in the third position appear to accommodate anticodon wobbling.

Ribosomes and Ribosomal RNA Prokaryotic ribosomes, which are also called 70S ribosomes based on their sedimentation rate in an ultracentrifuge, are extremely complex associations of ribosomal RNAs and polypeptides. Each ribosome is composed of two subunits: 50S and 30S **(Figure 7.14a)**. The 50S subunit is in turn composed of two rRNA molecules (23S and 5S) and about 34 different polypeptides, whereas the 30S subunit consists of one molecule of 16S rRNA and 21 ribosomal polypeptides. The ribosomes of mitochondria and chloroplasts are also 70S ribosomes composed of similar subunits and polypeptides.

In contrast, both the cytosol and the rough endoplasmic reticulum (RER) of eukaryotic cells have 80S ribosomes composed of 60S and 40S subunits **(Figure 7.14b)**. These subunits contain larger molecules of rRNA and more polypeptides than the corresponding prokaryotic subunits, though researchers do not agree on the exact number of polypeptides. The term *eukaryotic ribosome* is understood to mean only the 80S ribosomes of the cytosol and RER. Since the ribosomes of mitochondria and chloroplasts are 70S, they are called *prokaryotic ribosomes* even though they are in eukaryotic cells.

The structural differences between prokaryotic and eukaryotic ribosomes play a crucial role in the efficacy and safety of antimicrobial drugs. Because erythromycin, for example, binds

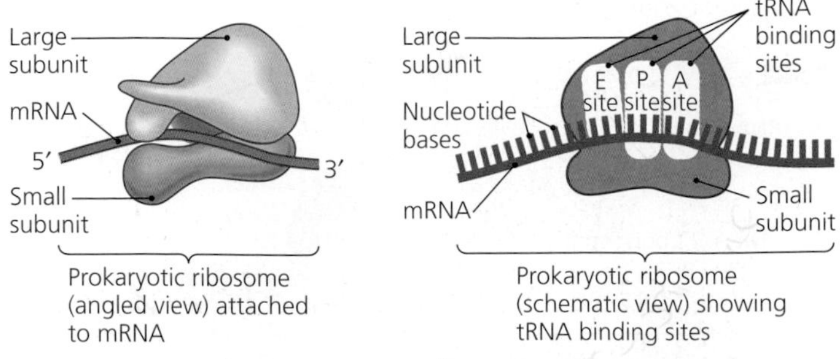

▲ **Figure 7.15 Assembled ribosome and its tRNA-binding sites.** The A site accepts the tRNA carrying the next amino acid to be added to the growing polypeptide, whereas the P site holds the tRNA carrying the polypeptide. Empty tRNA molecules exit from the E site.

only to the 23S rRNA found in prokaryotic ribosomes, it has no effect on eukaryotic 80S ribosomes and thus little deleterious effect on a patient. Chapter 10 discusses antimicrobial drugs in more detail.

The smaller subunit of a ribosome is shaped to accommodate three codons at one time—that is, nine nucleotide bases of a molecule of mRNA. Each ribosome also has three tRNA-binding sites that are named for their function **(Figure 7.15)**:

- The **A site** accommodates a tRNA delivering an *amino acid.*
- The **P site** holds a tRNA and the growing *polypeptide.*
- Discharged tRNAs *exit* from the **E site.**

We will now examine translation, the process whereby ribosomes actually synthesize polypeptides using amino acids delivered by tRNAs and following the genetic instructions of mRNA.

Stages of Translation

Molecular biologists divide translation into three stages: *initiation, elongation,* and *termination.* All three stages require additional protein factors that assist the ribosomes. Initiation and elongation also require energy provided by molecules of the ribonucleotide GTP, which are free in the cytosol (that is, they are not part of an RNA molecule). Here we consider bacterial translation. **ANIMATIONS:** ***Translation: The Process***

Initiation During initiation, the two ribosomal subunits, mRNA, several protein factors, and $tRNA^{fMet}$ form an *initiation complex*. Initiation may occur while the cell is still transcribing mRNA from DNA. The events of initiation in a bacterium are as follows **(Figure 7.16)**:

1. The smaller ribosomal subunit attaches to mRNA at a ribosome-binding site (also known as a Shine-Dalgarno sequence after its discoverers), so as to position a start codon (AUG) at its P site.
2. $tRNA^{fMet}$ (whose anticodon—UAC— is complementary to the start codon) attaches at the ribosome's P site.
3. The larger ribosomal subunit then attaches to form a complete initiation complex.

Elongation Elongation of a polypeptide is a cyclical process that involves the sequential addition of amino acids to a polypeptide chain growing at the P site. **Figure 7.17** illustrates several cycles of the process. The steps of each cycle occur as follows:

4. The transfer RNA whose anticodon matches the next codon—in this case, phenylalanine (Phe)—delivers its amino acid to the A site. Proteins called *elongation factors* (not shown) escort the tRNA along with a molecule of GTP. Energy from GTP (not shown) is used to stabilize each tRNA as it is added to the A site.
5. A ribozyme in the larger ribosomal subunit forms a peptide bond between the terminal amino acid of the growing polypeptide chain (in this case, *N*-formylmethionine) and the newly introduced amino acid. The polypeptide is now attached to the tRNA occupying the A site.

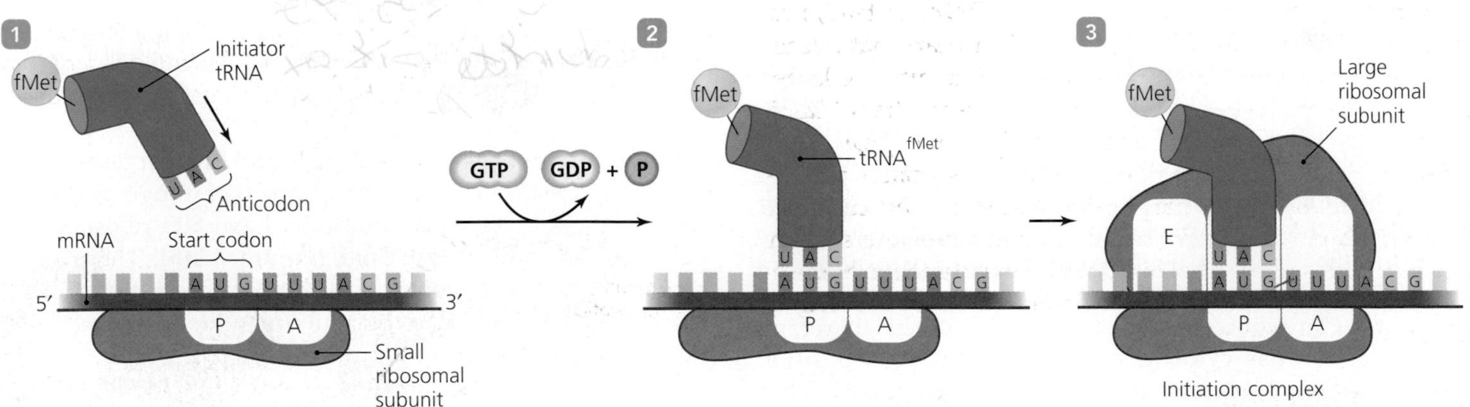

▲ **Figure 7.16 The initiation of translation in prokaryotes.**
1 The smaller ribosomal subunit attaches to mRNA at a ribosome binding site near a start codon (AUG). 2 The anticodon of $tRNA^{fMet}$ aligns with the start codon on the mRNA; energy from GTP (not shown) is used to bind the tRNA in place. 3 The larger ribosomal subunit attaches to form an initiation complex—a complete ribosome attached to mRNA.

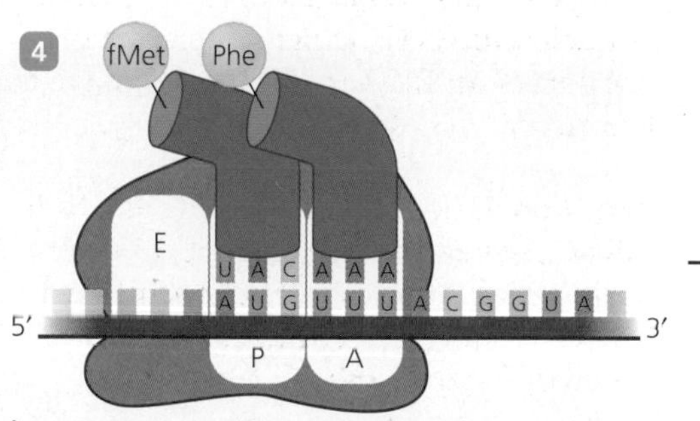

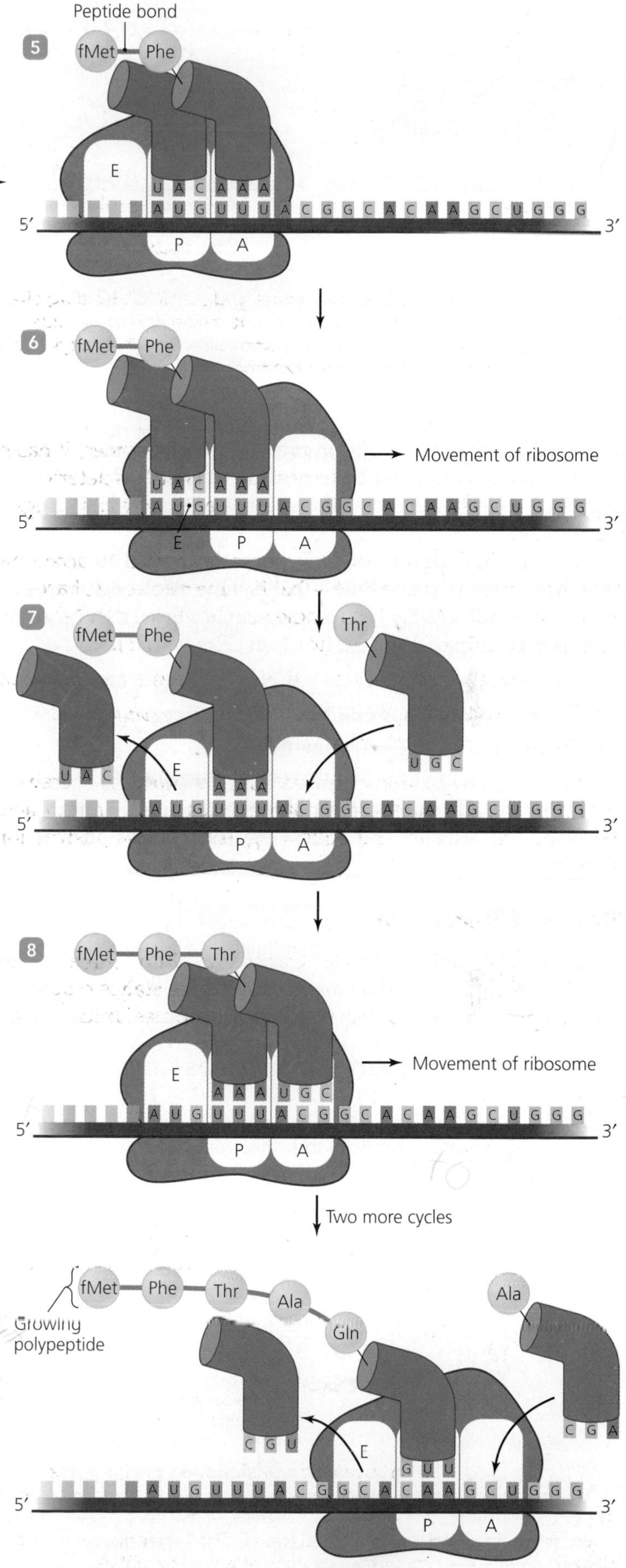

Figure 7.17 The elongation stage of translation. Transfer RNAs sequentially deliver amino acids as directed by the codons of the mRNA. Ribosomal RNA in the large ribosomal subunit catalyzes a peptide bond between the amino acid at the A site and the growing polypeptide at the P site. The steps in the process are described in the text. *How would eukaryotic translation differ?*

Figure 7.17 *The initial amino acid in eukaryotic polypeptide is methionine.*

6 Using energy supplied by more GTP, the ribosome moves one codon down the mRNA. This transfers each tRNA to the adjacent binding site; that is, the first tRNA moves from the P site to the E site, and the second tRNA (with the attached polypeptide) moves to the vacated P site.

7 The ribosome releases the "empty" tRNA from the E site. In the cytosol, the appropriate enzyme recharges it with another molecule of its specific amino acid.

8 The cycle repeats, each time adding another amino acid, at a rate of about 15 amino acids per second (in this case, threonine, then alanine, and then glutamine).

As elongation proceeds, ribosomal movement exposes the start codon, allowing another ribosome to attach behind the first one. In this way, one ribosome after another attaches at the start codon and begins to translate identical polypeptide molecules from the same message. Such a group of ribosomes, called a *polyribosome,* resembles beads on a string **(Figure 7.18)**.

Termination Termination does not involve tRNA; instead, proteins called *release factors* halt elongation. It appears that release factors somehow recognize stop codons and modify the larger ribosomal subunit in such a way as to activate another of its ribozymes, which severs the polypeptide from the final tRNA (resident at the P site). The ribosome then dissociates into its subunits. Termination of translation should not be confused with termination of transcription covered in a previous section. The polypeptides released at termination may function alone as proteins, or they may function with other polypeptides in quaternary protein structures.

CRITICAL **THINKING**

If a scientist synthesizes a DNA molecule with the nucleotide base sequence TACGGGGGAGGGGGAGGGGGA and then uses it for transcription and translation, what would be the amino acid sequence of the product?

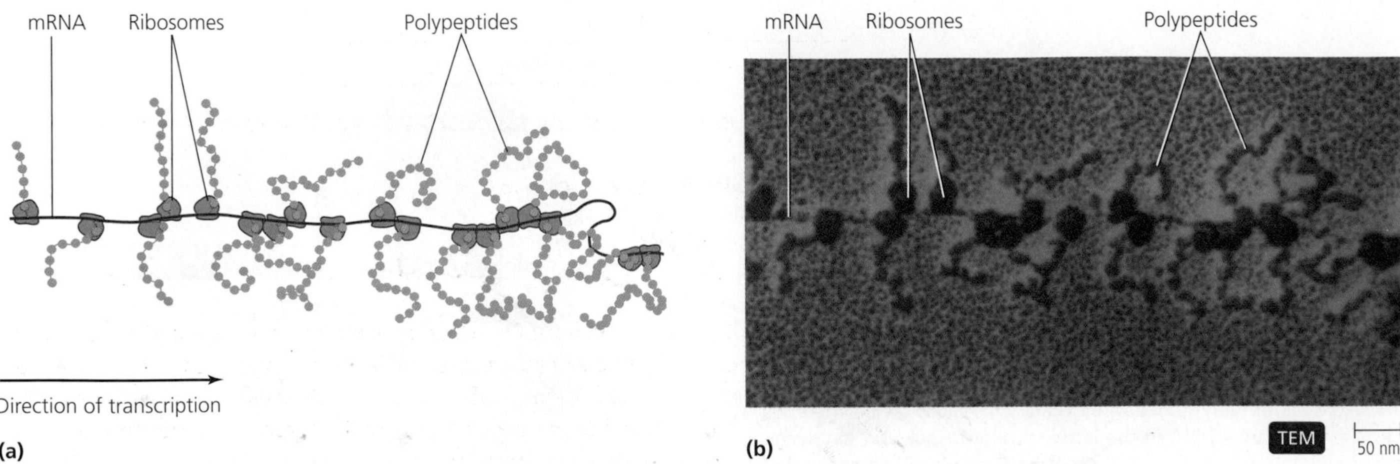

Figure 7.18 In prokaryotes, one mRNA, many ribosomes and polypeptides. **(a)** As a ribosome moves down the mRNA, the start codon (AUG) becomes available to another ribosome. In this manner, numerous identical polypeptides are translated simultaneously from a single mRNA molecule. **(b)** A polyribosome in a prokaryotic cell.

Translational Differences in Eukaryotes

Eukaryotic translation is similar to that of bacteria with some notable differences, including:

- Initiation of translation in eukaryotes occurs when the small ribosomal subunit binds to the 5′ guanine cap rather than a specific nucleotide sequence.
- The first amino acid in eukaryotic polypeptides is methionine rather than formylmethionine.

Archaeal translation is more similar to that of eukaryotes than to that of bacteria.

The processes we have examined thus far—how a cell replicates DNA, transcribes RNA, and translates RNA into polypeptides—are summarized in Table 7.3. Next we examine the way cells control the process of transcription.

Regulation of Genetic Expression

Learning Objectives

✓ Explain the operon model of transcriptional control in prokaryotes.

✓ Contrast the regulation of an inducible operon with that of a repressible operon, and give an example of each.

✓ Describe the use of microRNA, short interference RNA, and riboswitches in genetic control.

Most of a bacterium's genes are expressed at all times; that is, they are constantly transcribed and translated and play a persistent role in the phenotype. Such genes code for RNAs and polypeptides that are needed in large amounts by the cell—for example, integral proteins of the cytoplasmic membrane, structural proteins of ribosomes, and enzymes of glycolysis.

Other genes are regulated so that the polypeptides they encode are synthesized only in response to a change in the environment. Protein synthesis requires a large amount of energy, which can be conserved if a cell forgoes production of unneeded polypeptides. For example, *Pseudomonas aeruginosa* (ā-roo-ji-nō′sa), a potential pathogen of cystic fibrosis patients, can synthesize harmful proteins. Early in an infection, *Pseudomonas* does not produce the proteins because the body would respond defensively and eliminate the bacterium. Instead, the pathogen uses *quorum sensing*—a process whereby cells measure their density in an environment—so that it synthesizes the proteins only after there are numerous bacterial cells, overwhelming the body's defenses.

Cells regulate protein synthesis in many ways. They may initiate or stop transcription of mRNA or may stop translation directly.

Much of our knowledge of regulation has come from the study of microorganisms such as *E. coli*. We will examine two types of regulation of transcription in this bacterium—*induction*

TABLE 7.3 Comparison of Genetic Processes

Process	Purpose	Beginning Point	Ending Point (termination)
Replication	To duplicate the cell's genome	Origin	Origin or the end of a linear DNA molecule
Transcription	To synthesize RNA	Promoter	Terminator
Translation	To synthesize polypeptides	AUG start codon	UAA, UAG, or UGA stop codons

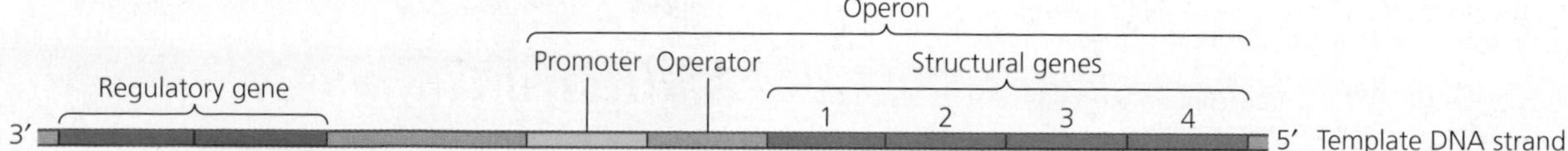

▲ **Figure 7.19 An operon.** Each operon consists of genes, their promoter, and a contiguous operator. The genes code for enzymes and structures such as channel and carrier proteins. A regulatory gene codes for a protein that controls the operon.

and *repression*. But before we examine these processes, we must first consider *operons*—special arrangements of prokaryotic genes that play roles in gene regulation. (Eukaryotes do not have operons.)

The Nature of Prokaryotic Operons

As originally described, an **operon** consists of a promoter, a series of genes that code for enzymes and structures such as channel proteins needed in response to changed environmental conditions, and an adjacent regulatory element called an **operator** **(Figure 7.19)**, which acts like a traffic light to control movement of RNA polymerase. A *repressor protein* binds to the operator, repressing transcription. In the absence of functional repressor, transcription is induced. A *regulatory gene* located outside the operon codes for the repressor. **Inducible operons** are not usually transcribed and must be activated by *inducers*. **Repressible operons** operate in reverse fashion—they are transcribed continually until deactivated by repressors. To clarify these concepts, let's examine an inducible operon and a repressible operon found in *E. coli*. **ANIMATIONS:** ***Operons: Overview***

The Lactose Operon, an Inducible Operon

The lactose (lac) *operon of E. coli* is an inducible operon and the first operon whose structure and action were elucidated. It includes a promoter, an operator, and three genes that encode proteins involved in the transport and catabolism of lactose **(Figure 7.20a)**.

Repression and Induction The *lac* operon is controlled by a regulatory gene that is constantly transcribed and translated to produce a repressor protein that attaches to DNA at the *lac*

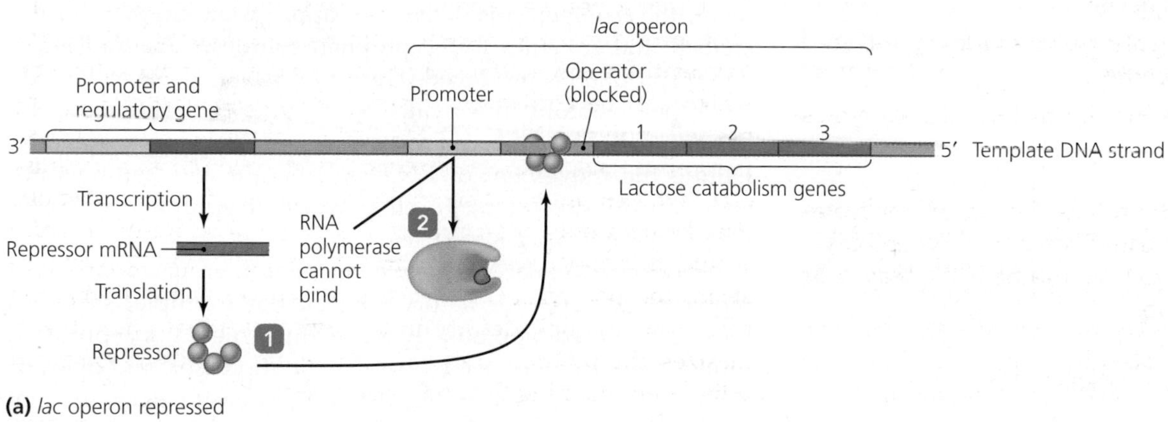

(a) *lac* operon repressed

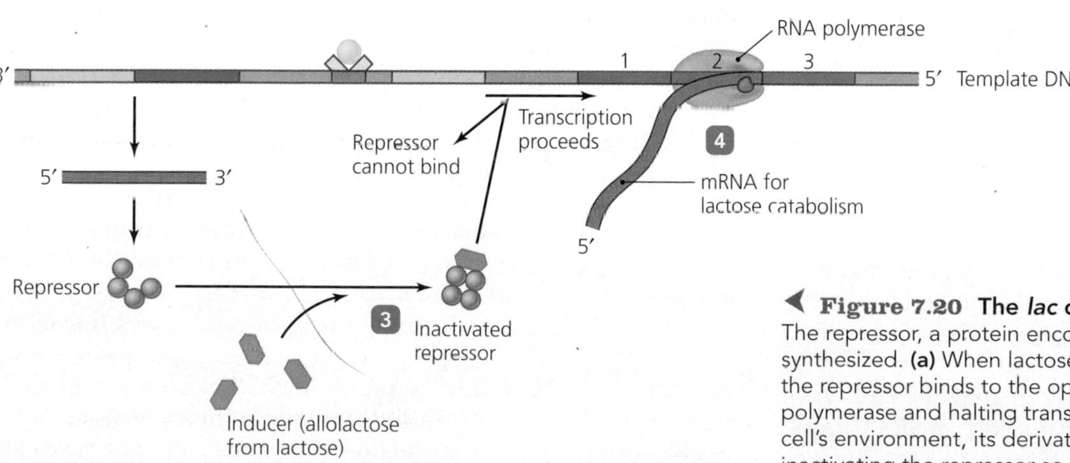

(b) *lac* operon induced

◄ **Figure 7.20 The *lac* operon, an inducible operon.** The repressor, a protein encoded by a regulatory gene, is constantly synthesized. **(a)** When lactose is absent from the cell's environment, the repressor binds to the operator, blocking the movement of RNA polymerase and halting transcription. **(b)** When lactose is present in the cell's environment, its derivative, allolactose, acts as an inducer by inactivating the repressor so that the repressor cannot bind to the operator, allowing transcription to proceed.

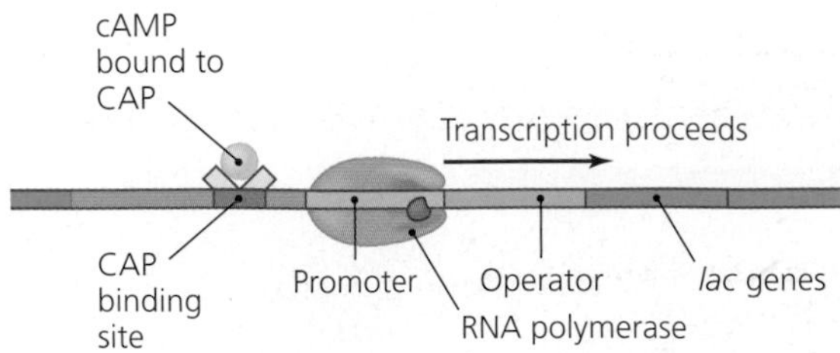

▲ **Figure 7.21 CAP-cAMP enhances *lac* transcription.** Cyclic adenosine monophosphate (cAMP), accumulating in *E. coli* when glucose is absent, binds to catabolic activator protein (CAP). CAP-cAMP binds to a CAP-binding site of DNA, allowing RNA polymerase to effectively bind to the *lac* promoter and begin transcription.

operator (1). This repressor prevents RNA polymerase from binding to the promoter, stopping synthesis of mRNA (2). Thus, the *lac* operon is usually inactive.

Under certain conditions, discussed shortly, *E. coli* takes in lactose whenever it becomes available and converts it to allolactose—an inducer that inactivates the repressor by changing its quaternary structure so that it can no longer attach to DNA (**Figure 7.20b** 3). This allows transcription of the three structural genes to proceed—the operon has been induced and can become active (4). Ribosomes then translate the newly synthesized mRNA to produce enzymes that catabolize lactose. Once the lactose supply has been depleted, there is no more inducer, and the repressor once again becomes active, suppressing transcription of the *lac* operon. In this manner, *E. coli* cells conserve energy by synthesizing enzymes for the catabolism of lactose only when lactose is available to them.

Positive Regulation By CAP Another condition must be met before *E. coli* transcribes its *lac* operon—glucose must be absent. *Lac* genes are not transcribed when glucose is available, because glucose is more efficiently catabolized than is lactose. However, glucose does not directly inhibit transcription. Instead, the small molecule *cyclic adenosine monophosphate (cAMP)* is involved. When glucose is present, the cell does not synthesize cAMP, and cAMP levels will be low; however, cAMP accumulates in *E. coli* when glucose is absent.

Cyclic AMP binds to an allosteric site of a regulatory protein, *catabolic activator protein (CAP)*, which is then able to bind to a CAP-binding site on DNA near the *lac* operon's promoter. This action is necessary for RNA polymerase to bind effectively. Once bound, RNA polymerase transcribes the *lac* genes (**Figure 7.21**). Thus, CAP-cAMP positively enhances *lac* transcription, but only when the operon is also induced by allolactose.

Inducible operons are often involved in controlling catabolic pathways whose polypeptides are not needed unless a particular nutrient is available. Such operons can also be associated with production of harmful proteins (virulence proteins) by pathogens. A different situation occurs with anabolic pathways such as those that synthesize amino acids. **ANIMATIONS:** ***Operons: Induction***

TABLE 7.4 The Roles of Operons in the Regulation of Transcription

Type of Regulation	Type of Metabolic Pathway Regulated	Regulating Condition
Inducible operons	Catabolic pathways; production of virulence proteins	Presence of substrate of pathway
Repressible operons	Anabolic pathways	Presence of product of pathway

The Tryptophan Operon, a Repressible Operon

E. coli can synthesize all the amino acids it needs for polypeptide synthesis; however, it can save energy by using amino acids available in its environment. In such cases, E. coli represses the genes for a given amino acid's synthetic pathway.

The *tryptophan operon,* which consists of a promoter, an operator, and five genes that code for the enzymes involved in the synthesis of tryptophan, is an example of such a repressible operon. Just as with the *lac* operon, a regulatory gene codes for a repressor molecule that is constantly synthesized. In contrast to inducible operons, however, the repressor of repressible operons is normally *inactive*. Thus, in the case of the repressible *trp* operon, whenever tryptophan is not present in the environment, the *trp* operon is active: the appropriate mRNA is transcribed, the enzymes for tryptophan synthesis are translated, and tryptophan is produced **(Figure 7.22a)**.

When tryptophan is available, it activates the repressor by binding to it. The activated repressor then binds to the operator, halting the movement of RNA polymerase and halting transcription **(Figure 7.22b)**. In other words, tryptophan acts as a *corepressor* of its own synthesis.

Biochemical analyses have revealed numerous variations of repressor-operator regulatory control. For example, scientists have discovered operons that use dual regulatory proteins, multiple operons controlled by a single repressor, and operons with multiple operators. Furthermore, some operons are not merely on-off systems, but can be fine-tuned so that transcription rates vary with the concentration of corepressors.

Table 7.4 summarizes how the basic characteristics of inducible and repressible operons relate to the regulation of transcription. **ANIMATIONS:** ***Operons: Repression***

Control of Translation by RNA Molecules

Cells can also use molecules of RNA to control translation of proteins. Regulatory RNA molecules include *microRNA, short interference RNA,* and *riboswitches*.

Eukaryotic cells and some viruses of eukaryotes transcribe single-stranded RNA molecules about 22 nucleotides long called **micro RNAs (miRNAs).** A micro RNA molecule folds back on itself in a hairpin loop giving it the appearance of a double-stranded RNA molecule. Ribosomes do not translate miRNA molecules; rather, a micro RNA joins with regulatory proteins to form an *RNA silencing complex (RISC).*

Figure 7.22 The *trp* operon, a repressible operon. **(a)** When tryptophan is absent from the cell's environment, the repressor is inactive, so the structural genes are transcribed and translated, and the five enzymes needed in the synthesis of tryptophan are produced. **(b)** When tryptophan is present in the cell's environment, it acts as a corepressor, activating the repressor and inhibiting its own synthesis.

RISC binds to messenger RNA but only to messenger RNA that is complementary to the micro RNA within the RISC. Once bound, RISC performs one of two functions: In some cases, RISC cleaves the messenger RNA molecule, rendering it useless. In other cases, RISC remains bound to messenger RNA, blocking the RNA molecule from ribosomes. In either case, ribosomes do not translate polypeptide from messenger RNA molecules bound by RISC; thus, miRNAs as a part of RISC regulate gene expression by blocking translation. Eukaryotic cells use such miRNAs to regulate a number of processes, including embryogenesis, cell division, apoptosis (programmed cell death), blood cell formation, and development of cancer.

Another method of regulation involving RNA molecules uses **small interfering RNA (siRNA).** siRNAs are about the same length as miRNAs, but differ from miRNAs in that siRNAs are double-stranded RNA. Further, siRNAs may be complementary to transfer RNA or to the coding strand of DNA as well as to messenger RNA. Most known siRNAs are non-natural, being created by scientists to artificially regulate gene expression in laboratory studies. siRNAs function in a manner similar to micro RNA molecules of cells—siRNA becomes incorporated into RISC. RISC then binds to a complementary nucleic acid sequence, rendering the target inactive.

HIGHLIGHT

FLIPPING THE SWITCH: RNA INTERFERENCE

One of the most effective ways to find out what a gene does is to disable it and see what happens. Researchers know how to cut out portions of DNA to turn off genes, a process called "knocking out." They also have a method for silencing specific genes in cells via RNA. By stopping expression at the RNA level, researchers can see what happens when the gene is turned on but its resultant protein is not produced. The approach is called RNA interference (RNAi).

RNAi uses miRNAs or siRNAs that pinpoint a messenger RNA molecule, attach to it, and target it for destruction. In this way, researchers use RNAi to selectively terminate the generation of a protein and study the cellular response. For example, scientists have used RNAi to induce protection against hepatitis virus in laboratory rodents—perhaps RNAi may provide defense for people in the future.

A **riboswitch** is another RNA molecule that helps regulate translation. Riboswitches change shape in response to environmental conditions such as changes in temperature or shifts in the concentration of specific nutrients, including vitamins, nucleotide bases, or amino acids. Some mRNA molecules themselves act as riboswitches. When conditions warrant, riboswitch mRNA folds to either favor or block translation, depending on the need by the cell for the polypeptide it encodes. For example, messenger RNA for the virulence regulator of the plague bacterium *Yersinia pestis* (yer-sin'ē-a pes'tis) folds in such a way as to prevent translation when the temperature is below human body temperature (37°C). When the bacterium enters a human, the mRNA refolds into a shape that allows translation, and virulence regulator is synthesized; plague ensues.

Mutations of Genes

Learning Objective

✓ Define *mutation.*

The phenotype of a cell is dependent upon both the integrity and accurate control of its genes; however, the nucleotide sequences of genes are not always accurately maintained. A **mutation** is a permanent change in the nucleotide base sequence of a genome, particularly its genes. Mutations of genes are almost always deleterious, though a few make no difference to the organism. Even more rarely a mutation leads to a novel property that improves the ability of an organism and its descendants to survive and reproduce. This is evolution in action. Mutations in unicellular organisms are passed on to the organism's progeny, but mutations in multicellular organisms typically are passed to offspring only if a mutation occurs in gametes or gamete-producing cells.

Types of Mutations

Learning Objective

✓ Define *point mutation* and describe three types.

Mutations range from large changes in an organism's genome, such as the loss or gain of an entire chromosome, to the most common type of mutation—**point mutations**—in which just one nucleotide base pair is affected. Point mutations include base pair **insertions, deletions,** and **substitutions.** The following analogy illustrates some types of mutations. Suppose that the DNA code was represented by the letters THECATATEELK. Grouping the letters into triplets (like codons) yields THE CAT ATE ELK. The substitution of a single letter could either change the meaning of the sentence, as in THE RAT ATE ELK, or result in a meaningless phrase, such as THE CAT RTE ELK. Insertion or deletion of a letter produces more serious changes, such as TRH ECA TAT EEL K or TEC ATA TEE LK. Insertions and deletions are also called **frameshift mutations** because nucleotide triplets following the mutation are displaced, creating new sequences of codons that result in vastly altered polypeptide sequences. Frameshift mutations affect proteins much more seriously than mere substitutions because a frame shift affects all codons subsequent to the mutation.

Mutations can also involve inversion (THE ACT ATE KLE), duplication (THE CAT CAT ATE ELK ELK), or transposition (THE ELK ATE CAT). Such mutations and even larger deletions and insertions are *gross mutations.* **ANIMATIONS:** ***Mutations: Types***

Effects of Mutations

Learning Objective

✓ List three effects of mutations.

Some base-pair substitutions produce **silent mutations** because the substitution does not change the amino acid sequence due to the redundancy of the genetic code (**Figure 7.23a**). For example, when the DNA triplet AAA is changed to AAG, the mRNA codon will be changed from UUU to UUC; however, because both codons specify phenylalanine, there is no change in the phenotype—the mutation is silent because it affects the genotype only.

Of greater concern are substitutions that change a codon for one amino acid into a codon for a different amino acid. A change that specifies a different amino acid is called a **missense mutation** (**Figure 7.23b**); what gets transcribed and translated makes sense, but not the right sense. The effect of missense mutations depends on where in the protein the changed amino acid occurs. When the different amino acid is in a critical region of a protein, the protein becomes nonfunctional; however, when the different amino acid is in a less important region, the mutation may have no adverse effect.

A third type of mutation occurs when a base-pair substitution changes an amino acid codon into a stop codon. This is called a **nonsense mutation** (**Figure 7.23c**). Nearly all nonsense mutations result in nonfunctional proteins.

Frameshift mutations (that is, insertions or deletions) typically result in drastic missense and nonsense mutations (**Figure 7.23d, e**), except when the insertion or deletion is very close to the end of a gene.

Table 7.5 on p. 220 summarizes the types and effects of point mutations.

CRITICAL **THINKING**

What DNA nucleotide triplet codes for codon UGU? Identify a base-pair substitution that would produce a silent mutation at this codon. Identify a base-pair substitution that would result in a missense mutation at this codon. Identify a base-pair substitution that would produce a nonsense mutation at this codon.

Mutagens

Learning Objectives

✓ Discuss how different types of radiation cause mutations in a genome.

✓ Describe three kinds of chemical mutagens and their effects.

Mutations occur naturally during the life of an organism. Such *spontaneous mutations* result from errors in replication and repair as well as from *recombination* in which relatively long

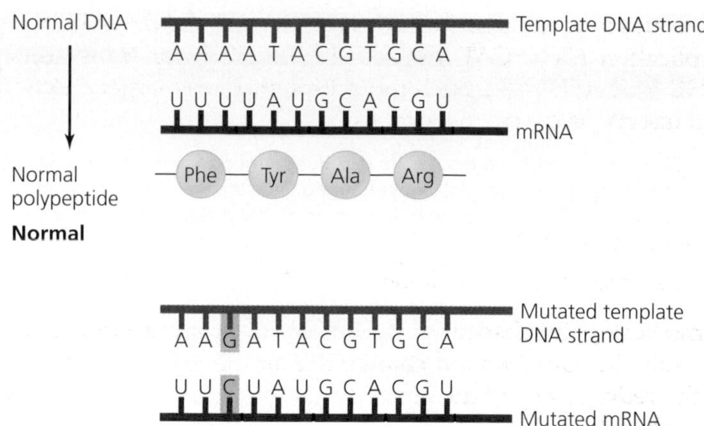

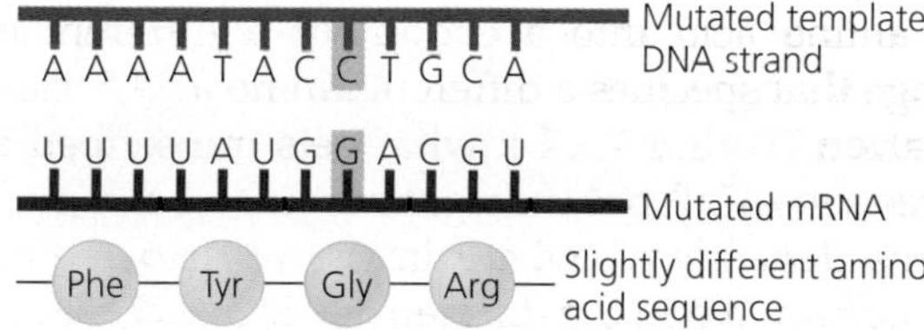

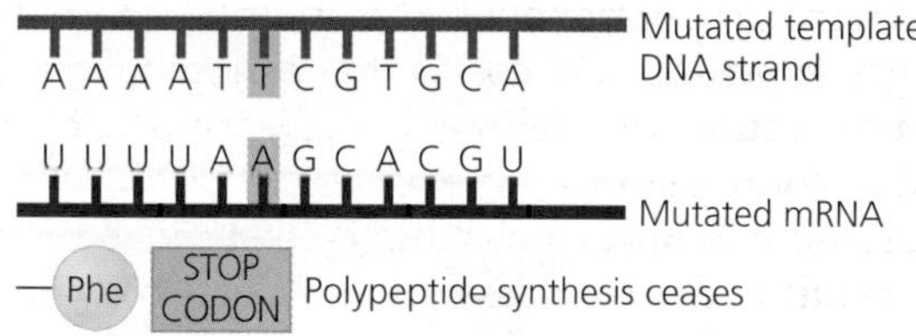

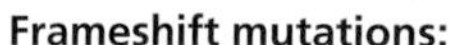

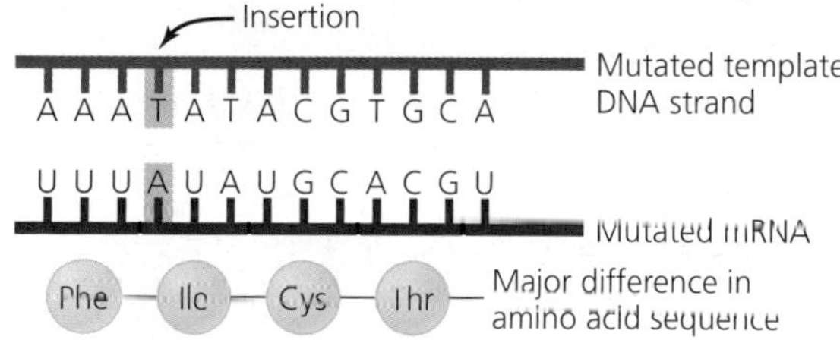

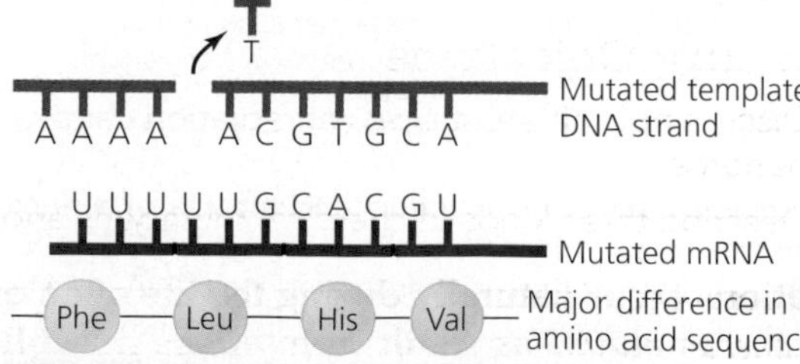

Figure 7.23
The effects of the various types of point mutations. Base-pair substitutions can result in silent mutations **(a)**, missense mutations **(b)**, or nonsense mutations **(c)**. Frameshift insertions **(d)** and frameshift deletions **(e)** usually result in severe missense or nonsense mutations because all codons downstream from the mutation are altered.

stretches of DNA move among chromosomes, plasmids, and viruses, introducing frameshift mutations. For example, we have seen that mismatched base pairing during DNA replication results in one error in every 10 billion (10^{10}) base pairs. Since an average gene has 10^3 base pairs, about one of every 10^7 (10 million) genes contains an error. Further, though cells have repair mechanisms to reduce the effect of mutations, the repair process itself can introduce additional errors. Physical or chemical agents called **mutagens** (myū′tă-jenz), which include radiation and several types of DNA-altering chemicals, induce mutations. **ANIMATIONS:** ***Mutagens***

Radiation

In the 1920s, Hermann Muller (1890–1967) discovered that X rays increased phenotypic variability in fruit flies by causing mutations. *Gamma rays* also damage DNA. As discussed in Chapter 9, X rays and gamma rays are *ionizing radiation;* that is, they energize electrons in atoms, causing some of the electrons to escape from their atoms. These free electrons strike other atoms, producing ions that can react with the structure of DNA, creating mutations. More seriously, electrons and ions can break the covalent bonds between the sugars and phosphates of a DNA backbone, causing physical breaks in chromosomes and complete loss of cellular control.

Nonionizing radiation in the form of *ultraviolet (UV) light* is also mutagenic because it causes adjacent pyrimidine bases to covalently bond to one another, forming **pyrimidine dimers** (**Figure 7.24**). The presence of dimers prevents hydrogen bonding with nucleotides in the complementary strand, distorts the sugar-phosphate backbone, and prevents proper replication and transcription. Cells have several methods of repairing dimers (discussed shortly).

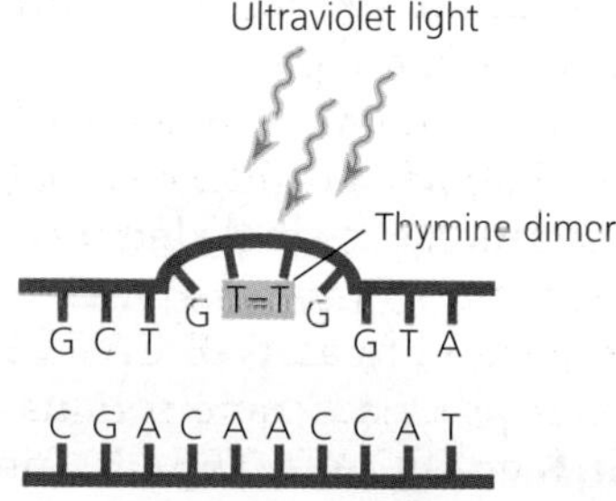

Figure 7.24 A pyrimidine (in this case thymine) dimer. Ultraviolet light causes adjacent pyrimidine bases (in this case thymine bases) to covalently bond to each other, preventing hydrogen bonding with bases in the complementary strand. The resulting distortion of the sugar-phosphate backbone prevents proper replication and transcription.

TABLE 7.5 The Types of Point Mutations and Their Effects

Type of Point Mutation	Description	Effects
Substitution	Mismatching of nucleotides or replacement of one base pair by another	*Silent mutation* if change results in redundant codon, as amino acid sequence in polypeptide is not changed. *Missense mutation* if change results in codon for a different amino acid; effect depends on location of different amino acid in polypeptide. *Nonsense mutation* if codon for an amino acid is changed to a stop codon.
Frameshift (insertion)	Addition of one or a few nucleotide pairs creates new sequence of codons	Missense and nonsense mutations
Frameshift (deletion)	Removal of one or a few nucleotide pairs creates new sequence of codons	Missense and nonsense mutations

Chemical Mutagens

Here we consider three of the many basic types of mutagenic chemicals.

Nucleotide Analogs **Nucleotide analogs** are compounds that are structurally similar to normal nucleotides **(Figure 7.25a)**. When nucleotide analogs are available to replicating cells, they may be incorporated into DNA in place of normal nucleotides, where their structural differences either inhibit nucleic acid polymerases or result in mismatched base pairing. When, for example, thymine is replaced by 5′-bromouracil, a wrong complement can form—5′-bromouracil pairs with guanine rather than adenine, resulting in a point mutation **(Figure 7.25b)**. Figure 10.7 illustrates other nucleotide analogs.

Nucleotide analogs make potent antiviral and anticancer drugs because viruses and cancer cells typically replicate faster than normal cells. Chapter 10 considers the use of nucleotide analogs as antimicrobial agents.

Nucleotide-Altering Chemicals Some chemical mutagens alter the structure of nucleotides. For example, a group of nucleotide-altering chemicals, called *aflatoxins,* are produced by *Aspergillus* (as-per-jil′ŭs) molds growing on grains and nuts. Aflatoxins catabolized in the liver can convert guanine nucleotides into thymine nucleotides, so that a GC base pair is converted to a TA base pair, resulting in missense mutations and possibly cancer. The Food and Drug Administration (FDA) prohibits excessive amounts of aflatoxins in human and animal food. Another example is *nitrous acid* (HNO_2), which removes the amine group of adenine, converting adenine into a guanine analog. When a cell replicates DNA containing this analog, an AT base pair is changed to a GC base pair in one daughter molecule—a base-pair-substitution mutation.

Frameshift Mutagens Still other mutagenic chemical agents insert or delete nucleotide base pairs, resulting in frameshift mutations. Examples of frameshift mutagens are *benzopyrene,* which is found in smoke; *ethidium bromide,* which is used to stain DNA; and *acridine,* one of a class of dyes commonly used as mutagens in genetic research. These chemicals are exactly the right size to slip between adjoining nucleotides in DNA, producing a bulge in the molecule **(Figure 7.26)**. When DNA

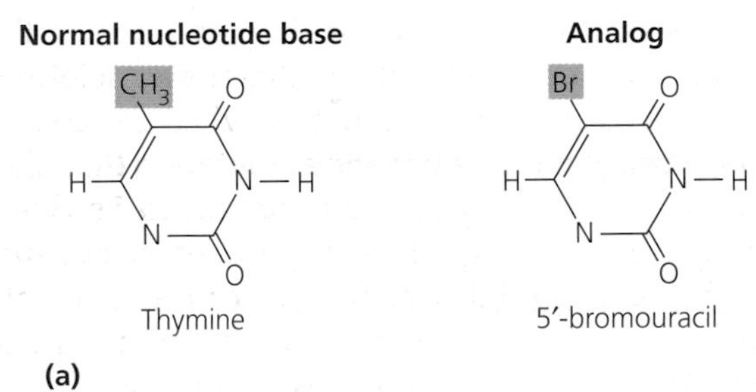

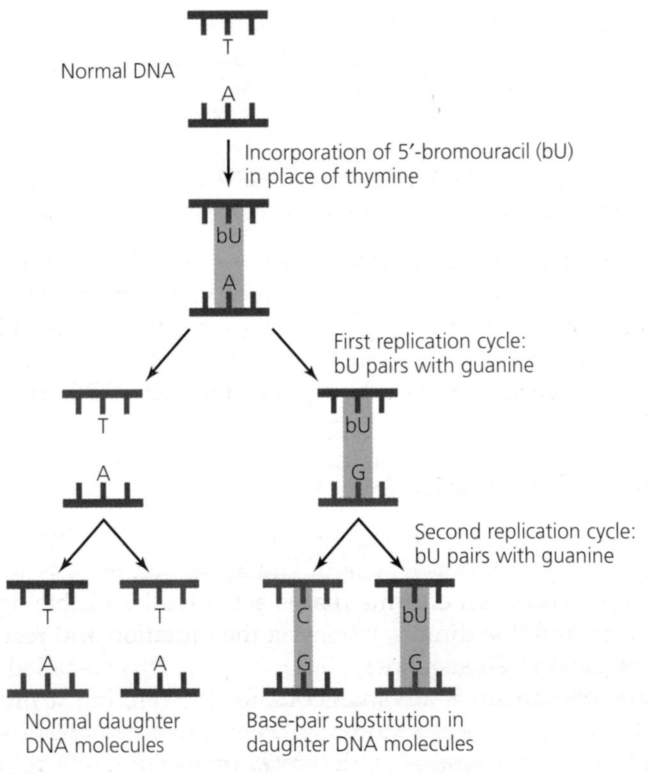

▲ **Figure 7.25 The structure and effects of a nucleotide analog.** **(a)** The structures of thymine and its nucleotide analog, 5′-bromouracil (bU). **(b)** When 5′-bromouracil is incorporated into DNA, the result is a point mutation. A single replication cycle results in one normal DNA molecule and one mutated molecule in which bU can pair with guanine. After a second replication cycle, a complete base-pair substitution has occurred in one of the four DNA molecules: CG has been substituted for TA.

polymerase copies the misshapen strands, one or more base pairs may be inserted or deleted in the daughter strand.

Frequency of Mutation

Learning Objective

✓ Discuss the relative frequency of deleterious and useful mutations.

Mutations are rare events. If they were not, organisms could not live or effectively reproduce themselves. As we have seen, about one of every 10 million (10^7) genes contains an error. Mutagens typically increase the mutation rate by a factor of 10–1000 times; that is, mutagens induce an error in one of every 10^6 to 10^4 genes.

Many mutations are deleterious because they code for nonfunctional proteins or stop transcription entirely. Cells without functional proteins cannot metabolize; therefore, deleterious mutations are removed from the population when the cells die. Rarely, however, a cell acquires a beneficial mutation that allows it to survive, reproduce, and pass the mutation to its descendants. This change in gene frequency is the basis of evolution. For example, a bacterium might randomly acquire a mutation that confers resistance to an antibiotic. In an environment containing the antibiotic, cells without resistance die, but the mutated cell survives and reproduces. As long as the antibiotic is present, cells with such a mutation have an advantage over cells without the mutation.

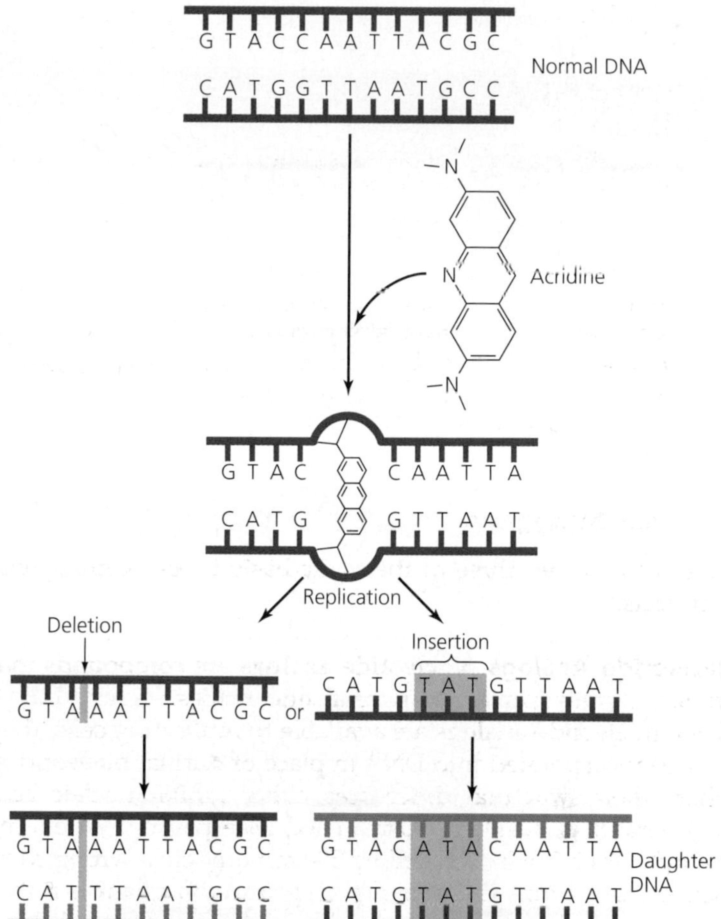

▲ **Figure 7.26 The action of a frameshift mutagen.** When DNA polymerase III passes the bulge caused by the insertion of acridine between the two DNA strands, it incorrectly synthesizes a daughter strand that contains either a deletion or an insertion mutation.

DNA Repair

Learning Objective

✓ Describe light and dark repair of pyrimidine dimers, base-excision repair, mismatch repair, and the SOS response.

We have seen that mutations rarely convey an advantage; most mutations are deleterious. To respond to the dangers mutations pose, cells have numerous methods for repairing damaged DNA, including light and dark repair of pyrimidine dimers, base-excision repair, mismatch repair, and an SOS response. **ANIMATIONS:** ***Mutations: Repair***

Repair of Pyrimidine Dimers

Pyrimidine dimers resulting from exposure to UV light constitute the most common type of mutation. Many cells contain DNA *photolyase,* an enzyme that is activated by visible light to break pyrimidine dimers, reversing the mutation and restoring the original DNA sequence **(Figure 7.27a)**. This so-called **light repair** mechanism is advantageous for the cell, but it presents a difficulty to scientists studying UV-induced mutations—they must keep such strains in the dark, or the mutants revert to their normal form.

So-called **dark repair** involves a different repair enzyme—one that doesn't require light. Dark repair enzymes cut the damaged section of DNA from the molecule, creating a gap that is repaired by DNA polymerase I and DNA ligase **(Figure 7.27b)**. Though called dark repair, this mechanism operates either in light or in the dark.

Base-Excision Repair

Rarely, DNA polymerase III incorporates an incorrect nucleotide during DNA replication. If the proofreading function of the polymerase does not repair the error, cells may use another enzyme system in a process called **base-excision repair.** This enzyme system excises the erroneous base, and then DNA polymerase I fills in the gap **(Figure 7.27c)**.

Mismatch Repair

A similar repair mechanism is called **mismatch repair.** Mismatch repair enzymes scan newly synthesized DNA looking for mismatched bases, which they remove and replace **(Figure 7.27d)**. How does the mismatch repair system determine which strand to repair? If it chose randomly, 50% of the time it would choose the wrong strand and introduce mutations. Mismatch repair enzymes, however, do not choose randomly. They distinguish between a new DNA strand and an old strand because old strands are methylated. Recognition of an error as far as 1000 base pairs away from an unmethylated portion of DNA

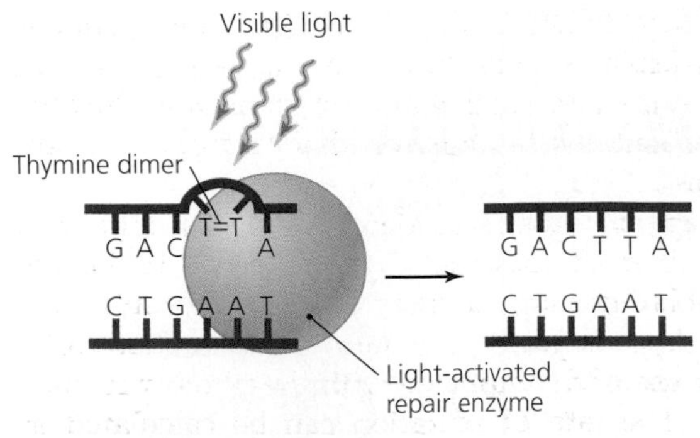

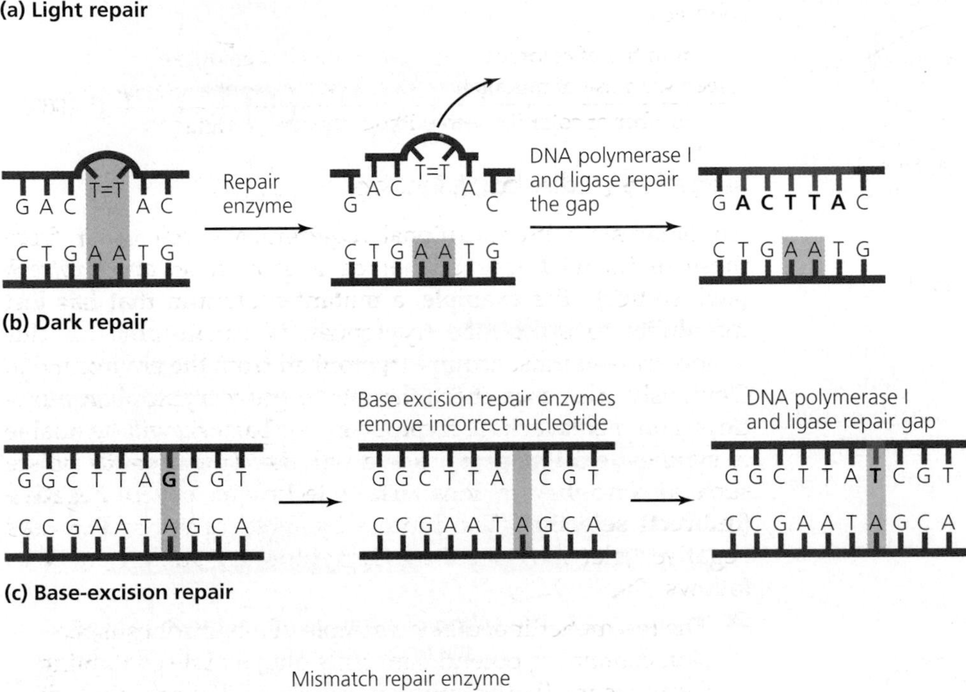

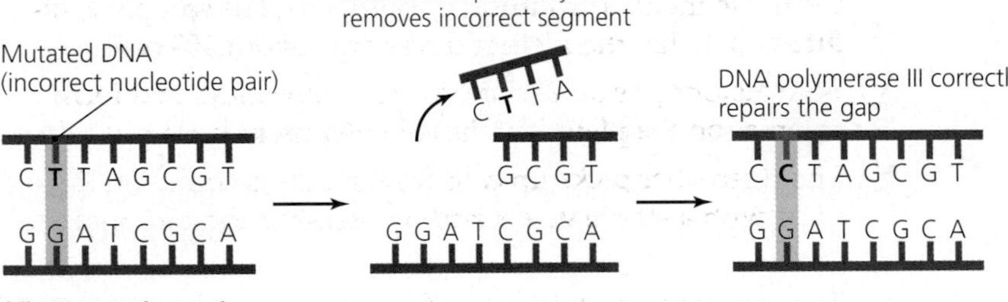

Figure 7.27 DNA repair mechanisms. **(a)** Light repair of pyrimidine dimers. A light-activated enzyme breaks pyrimidine-to-pyrimidine bonds. **(b)** Dark repair of dimers. After the repair enzyme removes the entire damaged section from one strand of DNA, DNA polymerase I and DNA ligase repair the breach. This mechanism is called dark repair because it does not require light; in fact, it operates in either light or darkness. **(c)** Base-excision repair, in which enzymes remove a segment with an incorrect base and DNA polymerase I fills the gap. **(d)** Mismatch repair, which involves total excision of an incorrect nucleotide.

triggers the mismatch repair enzymes. Once a new DNA strand is methylated, mismatch repair enzymes cannot correct any errors that remain.

SOS Response

Sometimes damage to DNA is so extreme that regular repair mechanisms cannot cope with the damage. In such cases, bacteria resort to what geneticists call an **SOS response** involving a variety of processes, such as the production of novel DNA polymerases (IV and V) capable of copying less-than-perfect DNA. These polymerases replicate DNA with little regard to the base sequence of the template strand. Of course, this introduces many new and potentially fatal mutations, but presumably SOS repair allows a few offspring of these bacteria to survive.

Identifying Mutants, Mutagens, and Carcinogens

Learning Objectives

- ✓ Contrast the positive and negative selection techniques for isolating mutants.
- ✓ Describe the Ames test, and discuss its use in discovering carcinogens.

If a cell does not successfully repair a mutation, it and its descendants are called **mutants.** In contrast, cells normally found in nature (in the wild) are called **wild-type** cells. Scientists distinguish mutants from wild-type cells by observing or testing for altered phenotypes. Since mutations are rare, and nonfatal

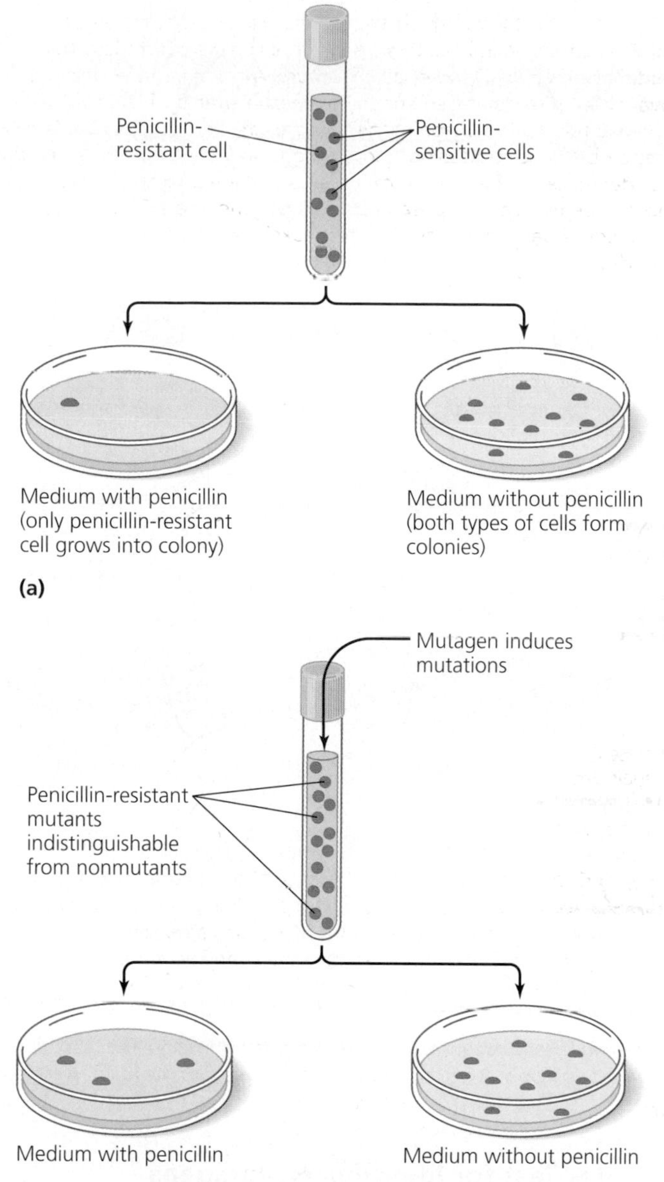

▲ **Figure 7.28 Positive selection of mutants.** Only mutants that are resistant to penicillin can survive on the plate containing the antibiotic. **(a)** Normally a population includes very few mutants. **(b)** Introduction of a mutagen increases the number of mutants and thus the number of colonies that grow in the presence of penicillin. *What is the rate of mutation induced by the mutagen in (b)?*

Figure 7.28 $\frac{3-1}{1} \times 100\% = 200\%$

mutations are even rarer, mutants can easily be "lost in the crowd." Therefore, researchers have developed methods to recognize mutants amidst their wild-type neighbors.

Positive Selection

Positive selection involves selecting a mutant by eliminating wild-type phenotypes. Assume, for example, that researchers want to isolate penicillin-resistant bacterial mutants from a liquid culture. To do so they spread the liquid medium, which contains mostly penicillin-sensitive cells but also the few penicillin-resistant mutants, onto medium that includes penicillin. Only the penicillin-resistant mutants multiply on this medium and produce visible colonies **(Figure 7.28a)**.

When a mutagenic agent is added to a liquid culture, it increases the number of mutants **(Figure 7.28b)**. While the researchers are isolating mutants, they can also determine the rate of mutation by comparing the number of mutant colonies formed after use of the mutagen with the number formed before treatment. The rate of mutation can be calculated as follows.

$$\frac{\text{number of colonies seen with use of mutagen} - \text{number of colonies seen without use of mutagen}}{\text{number of colonies seen without the use of mutagen}} \times 100\%$$

Negative (Indirect) Selection

An organism with nutritional requirements that differ from those of its wild-type phenotype is known as an *auxotroph* (awk'sō-trōf).[7] For example, a mutant bacterium that has lost the ability to synthesize tryptophan is auxotrophic for this amino acid—it must acquire tryptophan from the environment. Obviously, if a researcher attempts to grow tryptophan auxotrophs on media lacking tryptophan, the bacteria will be unable to synthesize all its proteins and will die. Therefore, to isolate such auxotrophs we must use a technique called **negative (indirect) selection.** The process by which a researcher uses negative selection to culture a tryptophan auxotroph is as follows **(Figure 7.29)**.

1. The researcher inoculates a sample of a bacterial suspension containing potential mutants onto a plate containing complete media (including tryptophan). The sample is diluted such that the plate receives only about 100 cells.
2. Both auxotrophs and wild-type cells reproduce and form colonies on the plate, but the colonies are indistinguishable.
3. The researcher picks up cells from all the colonies on the plate with a sterile velvet pad by pressing the pad onto the plate.
4. The researcher inoculates two new plates—one containing tryptophan, the other lacking tryptophan—by pressing the pad onto each of them. This technique is called *replica plating*.
5. After the plates have incubated for several hours, the researcher compares the two replica plates. Tryptophan auxotrophs growing on medium containing tryptophan are revealed by the absence of a corresponding colony on the plate lacking tryptophan.
6. The researcher takes cells of the auxotroph colony from the replica plate and inoculates them into a complete medium. The auxotroph is now isolated.

[7]From Greek *auxein*, meaning to increase, and *trophe*, meaning nutrition.

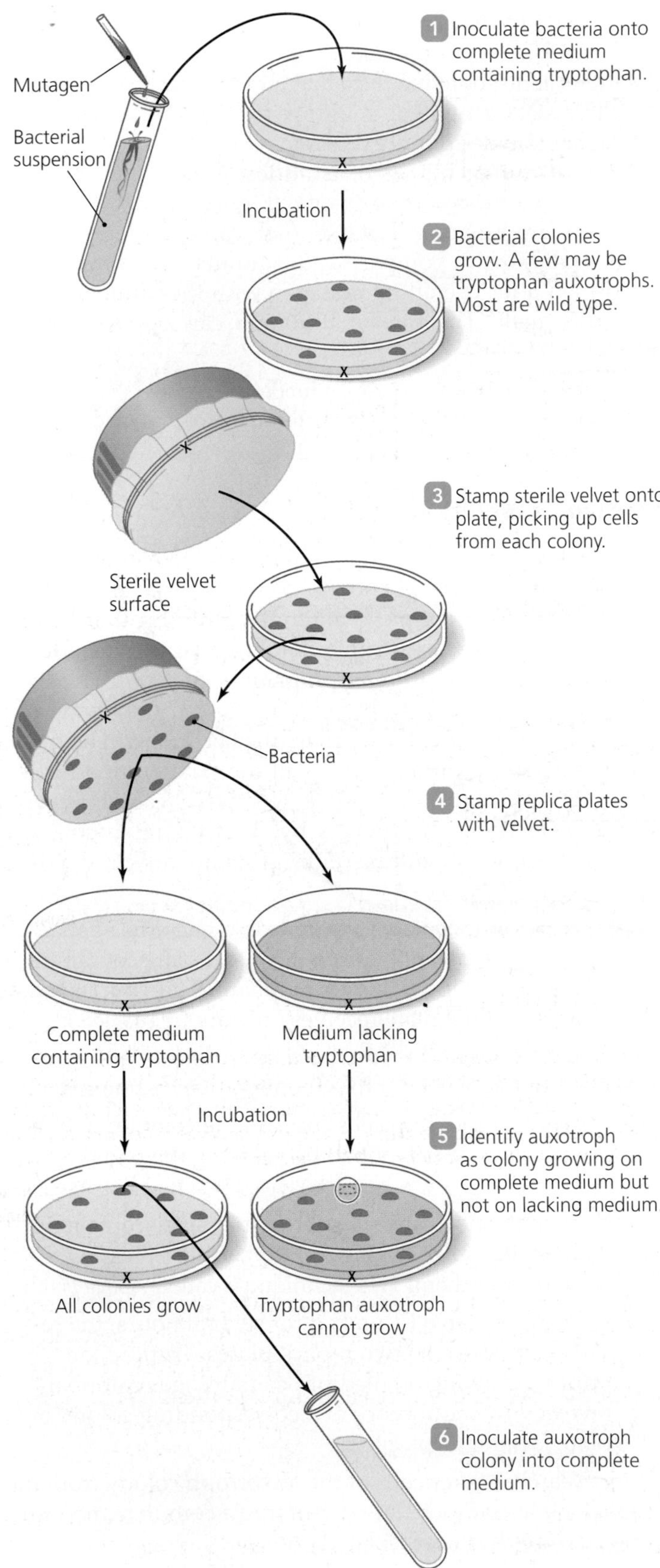
Mutagen
Bacterial suspension
1 Inoculate bacteria onto complete medium containing tryptophan.
Incubation
2 Bacterial colonies grow. A few may be tryptophan auxotrophs. Most are wild type.
3 Stamp sterile velvet onto plate, picking up cells from each colony.
Sterile velvet surface
Bacteria
4 Stamp replica plates with velvet.
Complete medium containing tryptophan
Medium lacking tryptophan
Incubation
5 Identify auxotroph as colony growing on complete medium but not on lacking medium.
All colonies grow
Tryptophan auxotroph cannot grow
6 Inoculate auxotroph colony into complete medium.

▲ **Figure 7.29 The use of negative (indirect) selection to isolate a tryptophan auxotroph.** The plates are marked (in this case, with an X) so that their orientation can be maintained throughout the procedure. Researchers may have to inoculate hundreds of such plates to identify a single mutant.

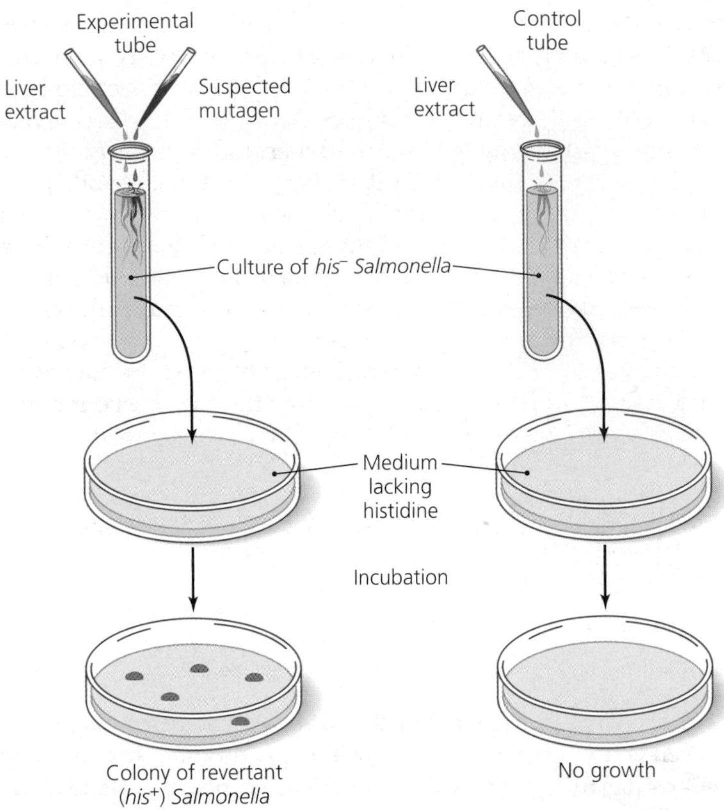
Experimental tube
Control tube
Liver extract
Suspected mutagen
Liver extract
Culture of *his⁻ Salmonella*
Medium lacking histidine
Incubation
Colony of revertant (*his⁺*) *Salmonella*
No growth

▲ **Figure 7.30 The Ames test.** A mixture containing *his*$^-$ *Salmonella* mutants, rat liver extract, and the suspected mutagen is inoculated onto a plate lacking histidine. Colonies will form only if a mutagen reverses the *his*$^-$ mutation, producing revertant *his*$^+$ organisms with the ability to synthesize histidine. A control tube that lacks the suspected mutagen demonstrates that reversion did not occur in the absence of the mutagen. *What is the purpose of liver extract in an Ames test?*

Figure 7.30 *Liver extract simulates conditions in the body by providing enzymes that may degrade harmless substances into mutagens.*

The Ames Test for Identifying Mutagens

Numerous chemicals in food, the workplace, and the environment in general have been suspected of being **carcinogenic** (kar′si-nō-jen′ik) mutagens; that is, of causing mutations that result in cancer. Because animal tests to prove that they are indeed carcinogenic are expensive and time consuming, researchers have used a fast and inexpensive method for screening mutagens called an **Ames test,** which is named for its inventor, Bruce Ames (1928–).

An Ames test uses mutant *Salmonella* (sal′mŏ-nel′ă) bacteria possessing a point mutation that prevents the synthesis of the amino acid histidine; in other words, they are histidine auxotrophs, indicated by the abbreviation *his*$^-$. To perform the test, an investigator mixes *his*$^-$ mutants with liver extract and the substance suspected to be a mutagen **(Figure 7.30)**. The presence of liver extract simulates the conditions in the body under which liver enzymes can turn harmless chemicals into mutagens. The researcher then spreads the treated bacteria on a solid medium lacking histidine. If the suspected substance does in fact cause mutations, some of the mutations will likely reverse the effect of the original mutation, producing revertant cells (designated *his*$^+$) that have regained the ability to synthesize histidine and thus can survive on a medium lacking histidine. Thus, the presence of colonies during an Ames test reveals that the suspected substance is mutagenic in *Salmonella*.

Given that DNA in all cells is very similar, the ability of a chemical to cause mutations in *Salmonella* indicates that it is likely to cause mutations in humans as well. Some mutations cause cancers, so a mutagenic chemical may also be carcinogenic. To prove that the substance can in fact cause cancer, scientists must test it in laboratory animals, a process that usually is warranted only if a chemical is mutagenic in *Salmonella*. Ames testing reduces the cost and time that would be required to assay every chemical for carcinogenicity in animals.

Genetic Recombination and Transfer

Learning Objective

✓ Define *genetic recombination*.

Genetic recombination refers to the exchange of nucleotide sequences between two DNA molecules and often involves segments that are composed of identical or nearly identical nucleotide sequences called *homologous sequences*. Scientists have discovered a number of molecular mechanisms for genetic recombination, one of which is illustrated simply in **Figure 7.31**. In this type of recombination, enzymes nick one strand of DNA at the homologous sequence, and another enzyme inserts the nicked strand into the second DNA molecule. Ligase then reconnects the strands in new combinations, and the molecules resolve themselves into novel molecules. Such DNA molecules that contain new arrangements of nucleotide sequences (and the cells that contain them) are called **recombinants**. Chapter 12 discusses crossing over and the formation of gametes in more detail.

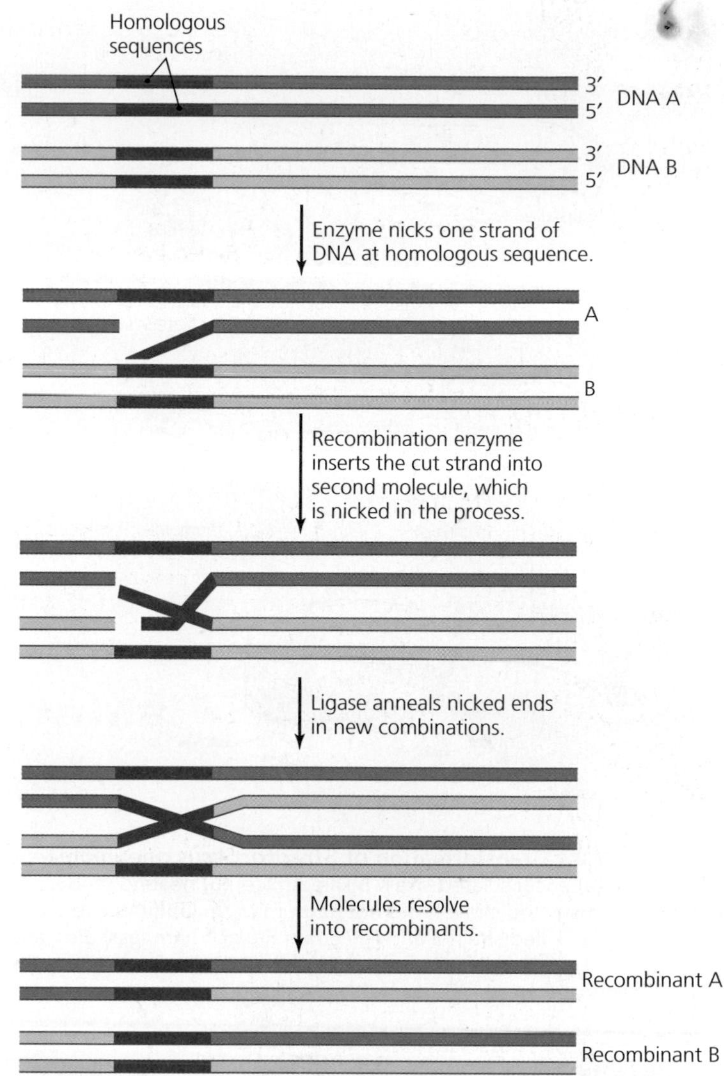

▲ **Figure 7.31 Genetic recombination.** Simplified depiction of one type of recombination between two DNA molecules. After an enzyme nicks one strand (here, strand A), a recombination enzyme rearranges the strands, and ligase seals the gaps to form recombinant molecules. *What is the function of ligase during DNA replication?*

Figure 7.31 Ligase functions to anneal Okazaki fragments during replication of the lagging strand.

Horizontal Gene Transfer Among Prokaryotes

Learning Objectives

✓ Contrast vertical gene transfer with horizontal gene transfer.

✓ Explain the roles of an F factor, F$^+$ cells, and Hfr cells in bacterial conjugation.

✓ Describe the structures and actions of simple and complex transposons.

✓ Compare and contrast crossing over, transformation, transduction, and conjugation.

As we have discussed, both prokaryotes and eukaryotes replicate their genomes and supply copies to their descendants. This is known as *vertical gene transfer*—the passing of genes to the next generation. In addition, many prokaryotes acquire genes from other microbes of the same generation—a process termed **horizontal (lateral) gene transfer.** In horizontal gene transfer, a **donor cell** contributes part of its genome to a **recipient cell,** which may be of a different species from the donor. Typically, the recipient cell inserts part of the donor's DNA into its own chromosome, becoming a recombinant cell. Cellular enzymes then usually degrade remaining unincorporated DNA. Horizontal gene transfer is a rare event, typically occurring in less than 1% of a population of prokaryotes. **ANIMATIONS:** ***Horizontal Gene Transfer: Overview***

Here we consider the three types of horizontal gene transfer: *transformation, transduction,* and *bacterial conjugation*.

Transformation

In **transformation,** a recipient cell takes up DNA from the environment, such as DNA that might be released by dead organisms. Frederick Griffith (1879–1941) discovered this process in 1928

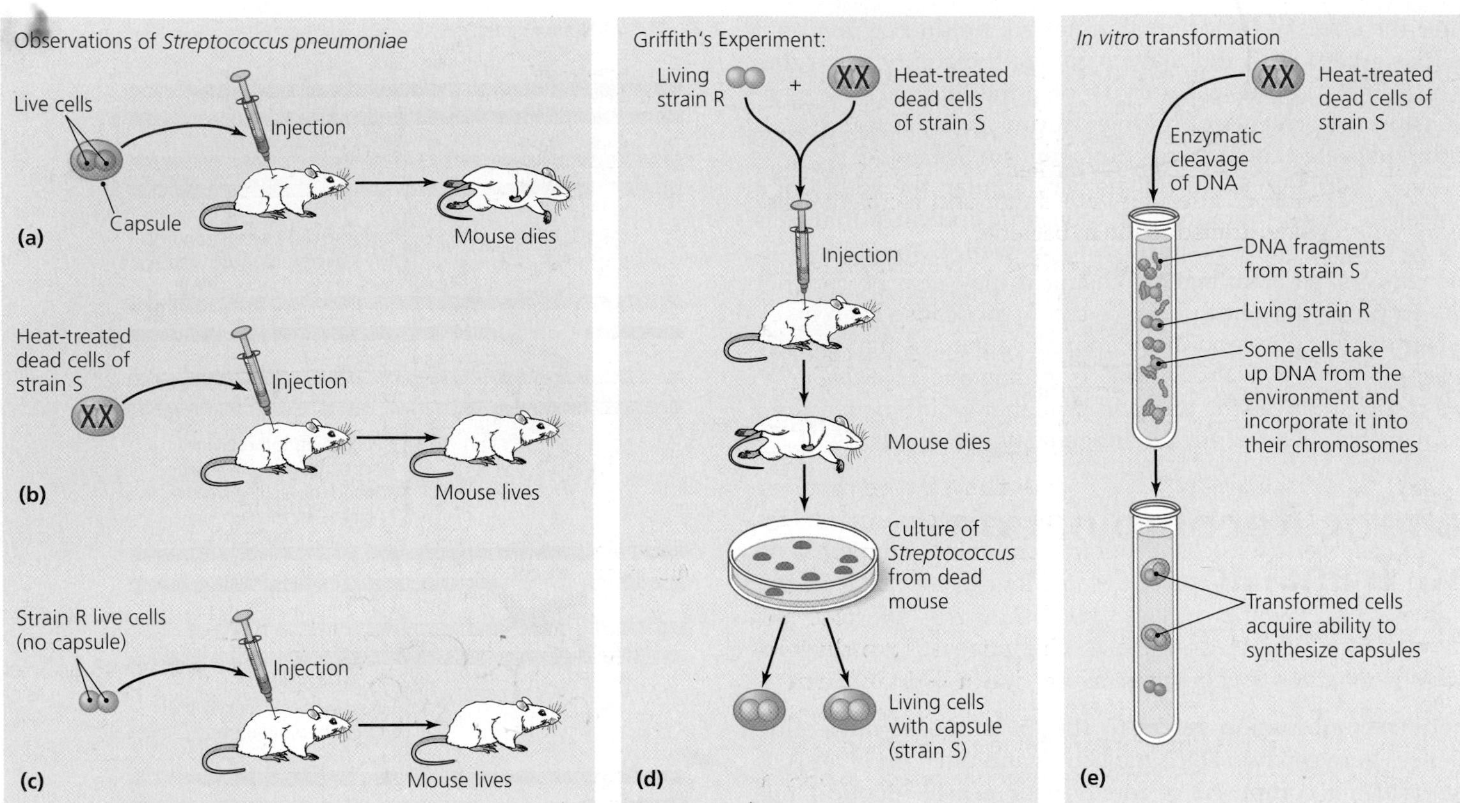

▲ **Figure 7.32 Transformation of *Streptococcus pneumoniae*.** Griffith's observations revealed that **(a)** encapsulated strain S killed mice, **(b)** heating renders strain S harmless to mice, and **(c)** unencapsulated strain R did not harm mice. In Griffith's experiment **(d),** a mouse injected concurrently with killed strain S and live strain R (each harmless) died and was found to contain numerous living, encapsulated bacteria. **(e)** A demonstration that transformation of R cells to S cells also occurs *in vitro*.

while studying pneumonia caused by *Streptococcus pneumoniae* (strep-tō-kok'ŭs nū-mō'nē-ī). Griffith worked with two strains of *Streptococcus*. Cells of the first strain have protective capsules that enable them to escape a body's defensive white blood cells; thus, these encapsulated cells cause deadly pneumonia when injected into mice **(Figure 7.32a)**. They are called *strain S* because they form *smooth* colonies on an agar surface. The application of heat kills such encapsulated cells and renders them harmless when injected into mice **(Figure 7.32b)**. In contrast, cells of the second strain (called *strain R* because they form *rough* colonies) are mutants that cannot make capsules. The unencapsulated cells of strain R do not cause disease **(Figure 7.32c)** because a mouse's defensive white blood cells quickly devour them.

Griffith discovered that when he injected both heat-killed strain S and living strain R into a mouse, the mouse died, even though neither of the injected strains was harmful when administered alone **(Figure 7.32d)**. Further, and most significantly, Griffith isolated numerous living, *encapsulated* cells from the dead mouse. He realized that harmless, unencapsulated strain R bacteria had been transformed into deadly, encapsulated strain S bacteria. Subsequent investigations showed that transformation also occurs *in vitro*[8] **(Figure 7.32e)**.

[8]Latin, meaning within glassware.

The fact that the living encapsulated cells retrieved at the end of this experiment outnumbered the dead encapsulated cells injected at the beginning indicated that strain R cells were not merely appropriating capsules released from dead strain S cells. Instead, strain R cells had acquired the capability of producing their own capsules by assimilating the capsule-coding genes of strain S cells. In 1944, Oswald Avery (1877–1955), Colin MacLeod (1909–1972), and Maclyn McCarty (1911–2005) extracted various chemicals from S cells and determined that the transforming agent was DNA. This discovery was one of the conclusive pieces of evidence that DNA is the genetic material of cells.

Cells that have the ability to take up DNA from their environment are said to be **competent.** Competence results from alterations in the cell wall and cytoplasmic membrane that allow DNA to enter the cell. Natural competency occurs in only a few types of bacteria, including pathogens in the genera *Streptococcus, Haemophilus* (hē-mof'i-lŭs), *Neisseria* (nī-se'rē-ă), *Bacillus* (ba-sil'ŭs), *Staphylococcus* (staf'i-lō-kok'ŭs), and *Pseudomonas*. Scientists can also generate competency artificially in *Escherichia* and other bacteria by manipulating the temperature and salt content of the medium. Because competent cells take up DNA from any donor genome, competency and transformation are important tools in recombinant DNA technology, commonly called *genetic engineering*. **ANIMATIONS:** ***Transformation***

Transduction

A second method of horizontal gene transfer, called **transduction,** involves the transfer of DNA from one cell to another via a replicating virus. Transduction can occur either between prokaryotic cells or between eukaryotic cells; it is limited only by the availability of a virus capable of infecting both donor and recipient cells. Here we will consider transduction in bacteria.

A virus that infects bacteria is called a **bacteriophage** or simply a **phage** (fāj).[9] The process by which a phage participates in transduction is depicted in **Figure 7.33**. To replicate, a bacteriophage attaches to a bacterial host cell and injects its genome into the cell (1). Phage enzymes, translated by bacterial ribosomes, degrade the cell's DNA (2). The phage genome now controls the cell's functions and directs it to synthesize new phage DNA and phage proteins. Normally, phage proteins assemble around phage DNA to form new phage particles, but some phages mistakenly incorporate remaining fragments of bacterial DNA that are about the same length as phage DNA. This forms **transducing phages** (3). Eventually the host cell lyses, releasing daughter and transducing phages. Transduction occurs when a transducing phage injects donor DNA into a new host cell (the recipient) (4). The recipient host cell incorporates the donated DNA into its chromosome by recombination (5).

In *generalized transduction,* the transducing phage carries a random DNA segment from a donor host cell's chromosome or plasmids to a recipient host cell. The transduction is not limited to a particular DNA sequence. In *specialized transduction,* only certain host sequences are transferred (along with phage DNA). In nature, specialized transduction is important in transferring genes encoding for certain bacterial toxins—including those responsible for diphtheria, scarlet fever, and the bloody, life-threatening diarrhea caused by *E. coli* O157:H7—into cells that would otherwise be harmless. Chapter 8 discusses the use of specialized transduction to intentionally insert genes into cells. **ANIMATIONS:** ***Transduction: Generalized Transduction, Specialized Transduction***

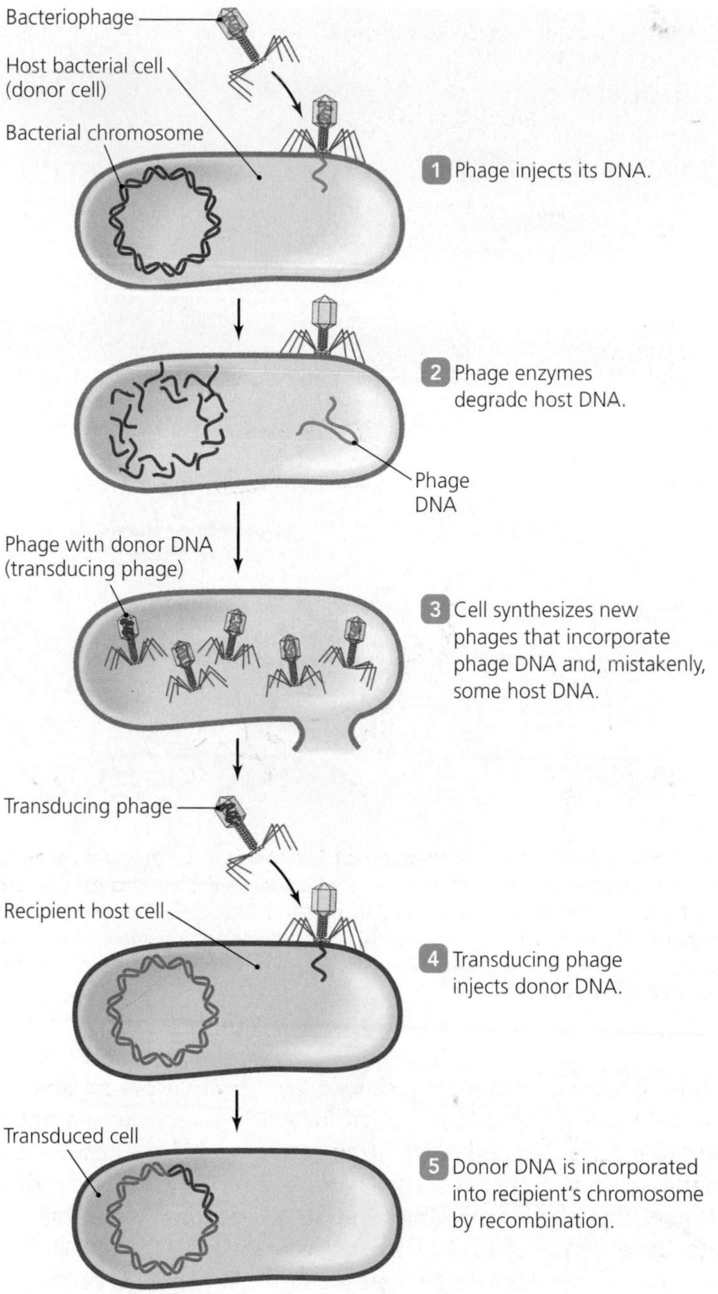

▲ **Figure 7.33 Transduction.** After a virus called a bacteriophage (phage) attaches to a host bacterial cell, it injects its genome into the cell and directs the cell to synthesize new phages. During assembly of new phages, some host DNA may be incorporated, forming transducing phages, which subsequently carry donor DNA to a recipient host cell.

Bacterial Conjugation

A third method of genetic transfer in bacteria is **conjugation.**[10] Unlike the typical donor cells in transformation and transduction, a donor cell in conjugation remains alive. Further, conjugation requires physical contact between donor and recipient cells. Scientists discovered conjugation between cells of *E. coli,* and it is best understood in this species. Thus, the remainder of our discussion will focus on conjugation in this bacterium. **ANIMATIONS:** ***Conjugation: Overview***

Conjugation is mediated by **conjugation pili** (pīlī; singular: *pilus*), also called *sex pili,*[11] which are thin, proteinaceous tubes extending from the surface of a cell. The gene coding for conjugation pili is located on a plasmid called an **F (fertility) plasmid.** (Recall that a plasmid is a small, circular, extrachromosomal molecule of DNA; see Figure 7.2b.) Cells that contain an F plasmid are called F^+ cells, and they serve as donors during conjugation. Recipient cells are F^-; that is, they lack an F plasmid and therefore have no conjugation pili. **ANIMATIONS:** ***Conjugation: F Factor***

The process of bacterial conjugation is illustrated in **Figure 7.34**. First, a sex pilus connects a donor cell (F^+) to a recipient cell (F^-) (1). The pilus may draw the cells together, though DNA transfer may occur when the cells are still more than 10 μm apart (2). A single strand of the F plasmid DNA transfers to the

[9]From Greek *phagein*, meaning to eat.
[10]From Latin *conjugatus*, meaning yoked together.
[11]Bacterial conjugation may resemble intercourse, but because no gametes are involved, it is not true sex.

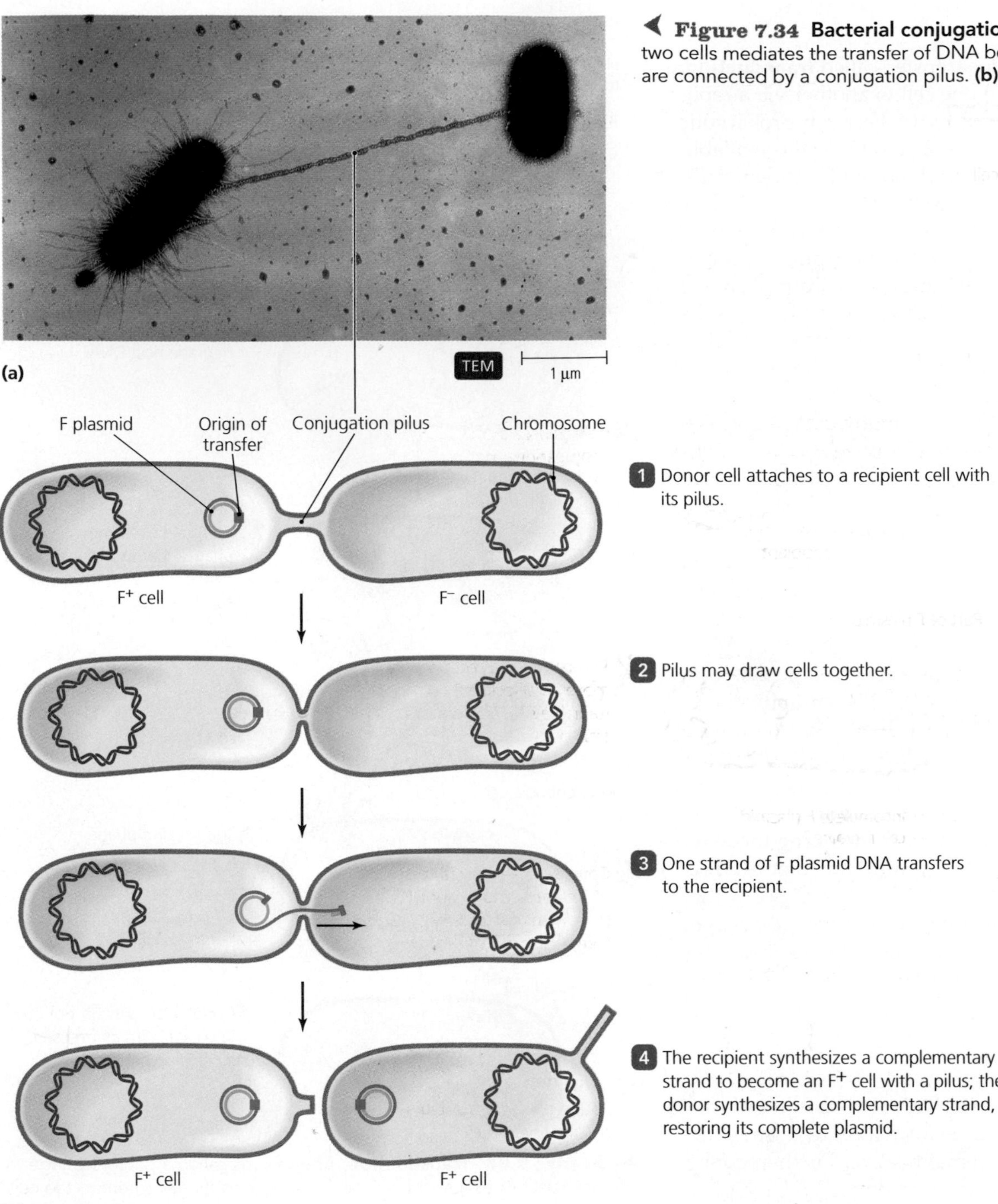

TEM
1 µm
(a)
F plasmid
Origin of transfer
Conjugation pilus
Chromosome
F⁺ cell
F⁻ cell
1 Donor cell attaches to a recipient cell with its pilus.
2 Pilus may draw cells together.
3 One strand of F plasmid DNA transfers to the recipient.
4 The recipient synthesizes a complementary strand to become an F⁺ cell with a pilus; the donor synthesizes a complementary strand, restoring its complete plasmid.
F⁺ cell
F⁺ cell
(b)

◀ **Figure 7.34 Bacterial conjugation.** A conjugation pilus connecting two cells mediates the transfer of DNA between the cells. **(a)** Two *E. coli* cells are connected by a conjugation pilus. **(b)** Artist's rendition of the process.

recipient beginning with a section called the *origin of transfer* (3). The F^- recipient then synthesizes a complementary strand of F plasmid DNA, becoming an F^+ cell (4). The donor cell also synthesizes a complementary plasmid DNA strand.

In some bacterial cells, an F plasmid does not remain independent in the cytosol but instead integrates at a specific DNA sequence in the cellular chromosome. Such cells, which are called **Hfr (high frequency of recombination) cells,** can conjugate with an F^- cell **(Figure 7.35)**. After the F plasmid has integrated (1) and the Hfr and F^- cells join via a sex pilus (2), DNA transfer begins at the origin of transfer of the F Plasmid, carrying with it a copy of the donor's chromosome (3). In most cases, movement of the cells breaks the intercellular connection before an entire donor chromosome is transferred (4). Because the recipient receives only a portion of the F plasmid, it remains an F^- cell; however, it also acquires some chromosomal genes from the donor. Recombination can integrate the donor DNA into the recipient's chromosome (5). The recipient is now a recombinant cell that contains its own genes as well as some donor genes. **ANIMATIONS:** ***Conjugation: Hfr Conjugation***

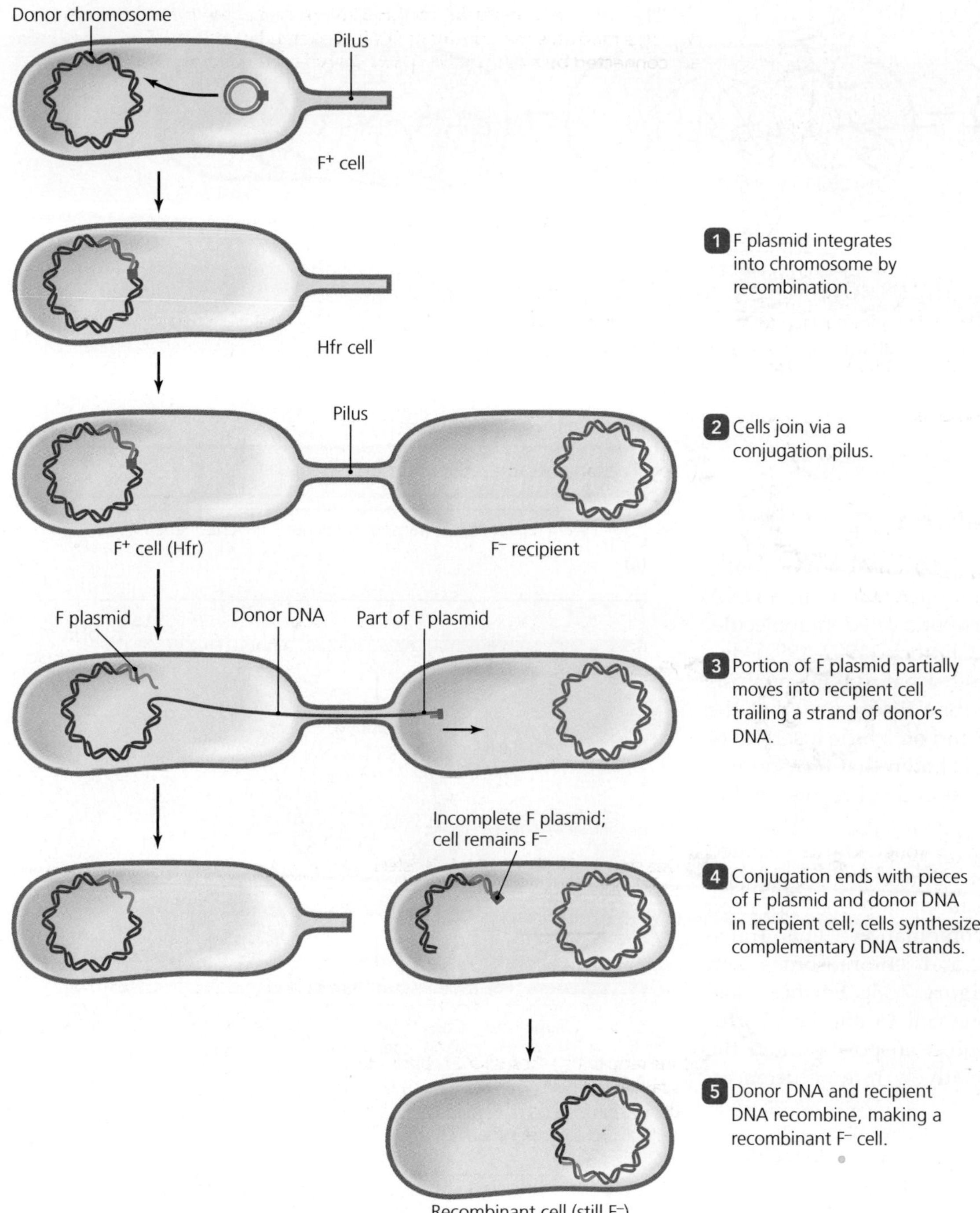
Donor chromosome
Pilus
F+ cell
1 F plasmid integrates into chromosome by recombination.
Hfr cell
Pilus
2 Cells join via a conjugation pilus.
F+ cell (Hfr)
F− recipient
F plasmid
Donor DNA
Part of F plasmid
3 Portion of F plasmid partially moves into recipient cell trailing a strand of donor's DNA.
Incomplete F plasmid; cell remains F−
4 Conjugation ends with pieces of F plasmid and donor DNA in recipient cell; cells synthesize complementary DNA strands.
5 Donor DNA and recipient DNA recombine, making a recombinant F− cell.
Recombinant cell (still F−)

▲ **Figure 7.35 Conjugation involving an Hfr cell.** An Hfr cell is formed when an F^+ cell integrates its F plasmid into its chromosome. Hfr cells donate a partial copy of their DNA and a portion of the F plasmid to a recipient, which is rendered a recombinant cell but remains F^-.

Because an F plasmid integrates into chromosomes at only a few locations, the order in which genes are transferred is consistent. Scientists produced the first gene maps of bacterial chromosomes by noting the time, in minutes, required for a particular gene to transfer from donor cells to recipient cells. **ANIMATIONS:** ***Conjugation: Chromosome Mapping***

Conjugation occurs in several species of bacteria, which can be quite promiscuous; that is, conjugation can occur among bacteria of widely varying kinds and even between a bacterium and a yeast cell or between a bacterium and a plant cell. For example, the crown gall bacterium, *Agrobacterium,* transfers some of its genes into the chromosome of its host plant.

The natural transfer of genes by conjugation among diverse organisms heightens some scientists' concerns about the spread of resistance (R) plasmids among pathogens. These scientists note that antibiotic resistance developed by one pathogen can spread to other pathogens.

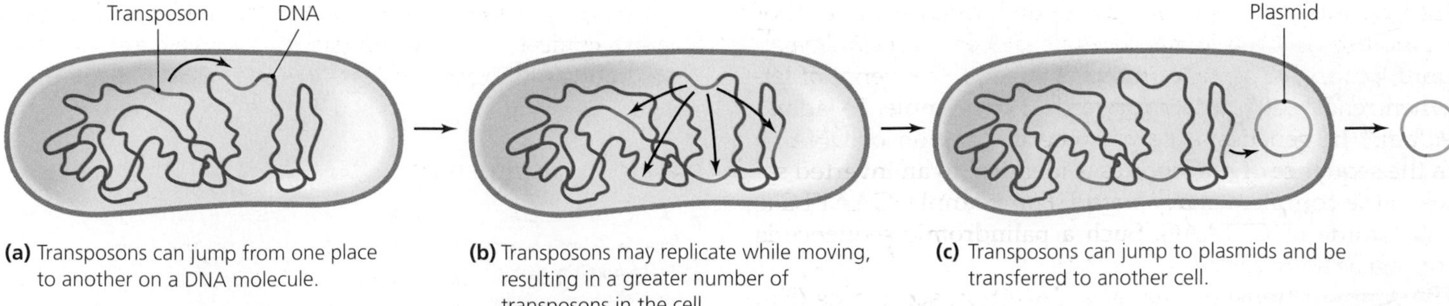

Transposon
DNA
Plasmid
(a) Transposons can jump from one place to another on a DNA molecule.
(b) Transposons may replicate while moving, resulting in a greater number of transposons in the cell.
(c) Transposons can jump to plasmids and be transferred to another cell.

▲ **Figure 7.36 Transposition.** Transposons move from place to place within, among, and between chromosomes and plasmids. Plasmids can carry transposons to and from cells.

Table 7.6 summarizes the mechanisms of horizontal gene transfer in bacteria.

Transposons and Transposition

Transposons[12] are segments of DNA, 700–40,000 bp in length, that transpose (move) themselves from one location in a DNA molecule to another location in the same or a different molecule.

American geneticist Barbara McClintock (1902–1992) discovered these "jumping genes" through a painstaking analysis of the colors of the kernels of corn. She discovered that the genes for kernel color were turned on and off by the insertion of transposons. Subsequent research has shown that transposons are found in many, if not all, prokaryotes and eukaryotes and in many viruses.

The result of the action of a transposon is termed **transposition;** in effect it is a kind of frameshift insertion (see Figure 7.23d). This "illegitimate" recombination does not need a region of homology, unlike other recombination events. Transposition occurs between plasmids and chromosomes and within and among chromosomes (**Figure 7.36**). Further, plasmids can carry transposons from one cell to another. Fortunately, while transposons are common, transposition and the frameshift mutations it causes are relatively rare occurrences. **ANIMATIONS:** ***Transpons: Overview***

[12]From *transposable* elements.

TABLE 7.6 Natural Mechanisms of Horizontal Genetic Transfer in Bacteria

Mechanism	Requirements
Transformation	Free DNA in the environment and a competent recipient
Transduction	Bacteriophage
Conjugation	Cell-to-cell contact and F plasmid (either in cytosol or incorporated into chromosome of donor)

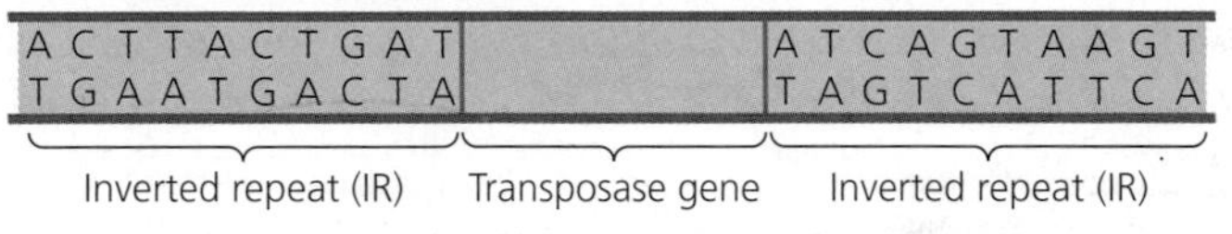

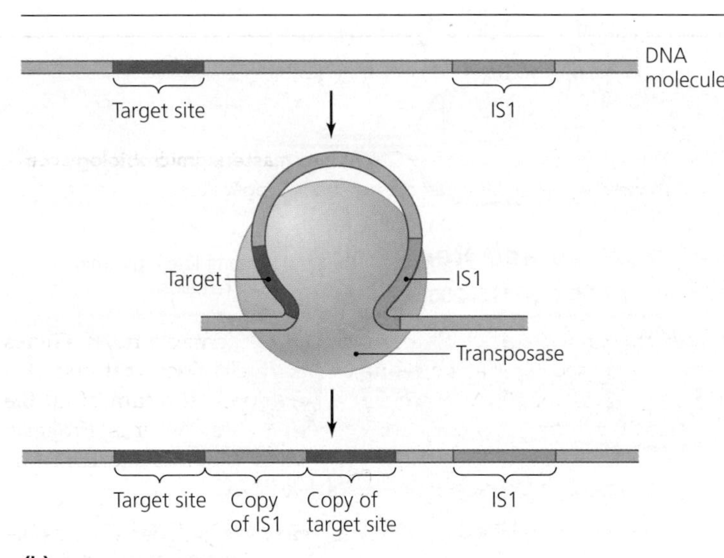

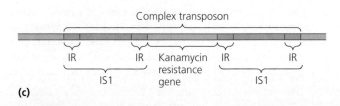

▲ **Figure 7.37 Transposons. (a)** A simple transposon, or insertion sequence. Shown here is Insertion Sequence 1 (IS1), which consists of a gene for the enzyme transposase bounded by identical (though inverted) repeats of nucleotides. **(b)** Transposase recognizes its target site (elsewhere in the same DNA molecule, as shown here, or in a different DNA molecule; then it moves the transposon (or, as in the case of IS1, a copy of the transposon) to its new site. The target site is duplicated in the process. **(c)** A complex transposon, which contains genes not related to transposition. Shown here is Tn5, which consists of a gene for kanamycin resistance between two IS1 transposons.

Transposons vary in their nucleotide sequences, but all of them contain palindromic sequences at each end. A *palindrome*[13] is a word, phrase, or sentence that has the same sequence of letters when read backward or forward—for example, "Madam, I'm Adam." In genetics, a palindrome is a region of DNA in which the sequence of nucleotides is identical to an inverted sequence in the complementary strand. For example, GAATTC is the palindrome of CTTAAG. Such a palindromic sequence is also known as an **inverted repeat (IR).**

The simplest transposons, called **insertion sequences (IS),** consist of no more than two inverted repeats and a gene that encodes the enzyme *transposase* (**Figure 7.37a**). Transposase recognizes its own inverted repeat in a target site, cuts the DNA at that site, and inserts the transposon (or a copy of it) into the DNA molecule at that site (**Figure 7.37b**). Such transposition also produces a duplicate copy of the target site.

Complex transposons contain one or more genes not connected with transposition, such as genes for antibiotic resistance (**Figure 7.37c**). *R factors* often contain transposons. R factors are of great clinical concern because they spread antibiotic resistance among pathogens. **ANIMATIONS:** ***Transposons: Insertion Sequences, Complex Transposons***

[13]From Greek *palin*, meaning again, and *dramein*, meaning to run.

CRITICAL **THINKING**

Suppose you are a scientist who wants to insert into your dog a gene that encodes a protein that protects dogs from heartworms. A dog's cells are not competent, so they cannot take up the gene from the environment; but you have a plasmid, a competent bacterium, and a related (though incompetent) F^+ bacterium that lives as an intracellular parasite in dogs. Describe a possible scenario by which you could use natural processes to genetically alter your dog to be heartworm resistant.

Chapter 8 discusses how scientists have adapted the natural processes of transformation, transduction, conjugation, and transposition to manipulate the genes of organisms.

Chapter Summary

This chapter has ***MicroFlix***. Go to **www.masteringmicrobiology.com** to view movie-quality animations for DNA replication.

The Structure and Replication of Genomes (pp. 195–203)

1. **Genetics** is the study of inheritance and inheritable traits. **Genes** are composed of specific sequences of nucleotides that code for polypeptides or RNA molecules. A **genome** is the sum of all the genes and linking nucleotide sequences in a cell or virus. Prokaryotic and eukaryotic cells use DNA as their genetic material; some viruses use DNA, and other viruses use RNA.
2. The two strands of DNA are held together by hydrogen bonds between complementary **base pairs (bp)** of nucleotides. Adenine bonds with thymine, and guanine bonds with cytosine.
3. Bacterial and archaeal genomes consist of one (or rarely, two or more) **chromosomes,** which are typically circular molecules of DNA associated with protein and RNA molecules, localized in a region of the cytosol called the **nucleoid**. Archaeal DNA organizes around globular proteins called **histones**. Prokaryotic cells may also contain one or more extrachromosomal DNA molecules called **plasmids,** which contain genes that regulate nonessential life functions, such as bacterial conjugation and resistance to antibiotics.
4. In addition to DNA, eukaryotic chromosomes contain eukaryotic histones and are arranged as **nucleosomes** (beads of DNA) that clump to form **chromatin fibers**. Eukaryotic cells may also contain extranuclear DNA in mitochondria, chloroplasts, and plasmids.
5. DNA replication is semiconservative; that is, each newly synthesized strand of DNA remains associated with one of the parental strands. After helicase unwinds and unzips the original molecule, synthesis of each of the two daughter strands—called the **leading strand** and the **lagging strand**—occurs from 5′ to 3′. Synthesis is mediated by enzymes that prime, join, and proofread the pairing of new nucleotides.
   ANIMATIONS: ***DNA Replication: Overview, Forming the Replication Fork, Replication Proteins, Synthesis***
6. After DNA replication, **methylation** occurs. Methylation plays several roles, including the control of gene expression, the initiation of DNA replication, recognition of a cell's own DNA, and repair.

Gene Function (pp. 203–218)

1. The **genotype** of an organism is the actual set of genes in its genome, whereas its **phenotype** refers to the physical and functional traits expressed by those genes.
2. RNA has several forms. These include **RNA primer; messenger RNA (mRNA),** which carries genetic information from DNA to a ribosome; **transfer RNA (tRNA),** which carries amino acids to the ribosome; and **ribosomal RNA (rRNA),** which, together with polypeptides, makes up the structure of ribosomes.
3. The **central dogma** of genetics states that genetic information is transferred from DNA to RNA to polypeptides, which function alone or in conjunction as proteins.
4. The transfer of genetic information begins with **transcription** of the genetic code from DNA to RNA, in which **RNA polymerase** links RNA nucleotides that are complementary to genetic sequences in DNA. Transcription begins at a region of DNA called a **promoter** (recognized by RNA polymerase) and ends with a sequence called a **terminator**. In bacteria, Rho protein may assist in termination, or termination may depend solely on the nucleotide sequence of the transcribed RNA.
   ANIMATIONS: ***Transcription: Overview, The Process***

5. Eukaryotic mRNA is synthesized as pre-messenger RNA. Before translation can occur, a spliceosome removes noncoding **introns** from pre-mRNA and splices together the **exons,** which are the coding sections.

6. In **translation,** the sequence of genetic information carried by mRNA is used by ribosomes to construct polypeptides with specific amino acid sequences.
ANIMATIONS: *Translation: Overview, Genetic Code, The Process*

7. The genetic code consists of triplets of mRNA nucleotides, called **codons**. These bind with complementary **anticodons** on transfer RNAs (tRNAs), which are molecules that carry specific amino acids. Ribosomal RNA (rRNA) catalyzes the bonding of one amino acid to another to form a polypeptide. A sequence of nucleotides thus codes for a sequence of amino acids.

8. A ribosome contains three tRNA binding sites: an **A site** (associated with incoming amino acids), a **P site** (associated with elongation of the polypeptide), and an **E site** from which tRNA exits the ribosome.

9. An **operon** is a series of prokaryotic genes, a promoter, and in some cases an **operator** sequence, all controlled by one regulatory gene. The operon model explains gene regulation in prokaryotes.
ANIMATIONS: *Operons: Overview*

10. **Inducible operons** are normally "turned off" and are activated by inducers that block repressors from binding to the operator, whereas **repressible operons** are normally "on" and are deactivated by repressors that bind to the operator.
ANIMATIONS: *Operons: Induction, Repression*

11. Cells and some viruses use short microRNAs (miRNAs) in conjunction with proteins to form RNA silencing complexes (RISC) that attach to mRNA sequences to inhibit polypeptide translation.

12. Genetic expression can also be controlled with short interference RNA (siRNA) associated with protein in RISC or with riboswitches.

Mutations of Genes (pp. 218–225)

1. A **mutation** is a change in the nucleotide base sequence of a genome. **Point mutations** involve a change in a nucleotide base pair and include **substitutions** and two types of **frameshift mutations: insertions** and **deletions.** Mutations can be categorized by their effects as **silent, missense,** or **nonsense mutations**.
ANIMATIONS: *Mutations: Types*

2. Physical or chemical agents called **mutagens** can increase the normal rate of mutation. Physical mutagens include ionizing radiation, such as X rays and gamma rays, and nonionizing ultraviolet light. Ultraviolet light causes adjacent pyrimidine bases to bond to one another to form **pyrimidine dimers**. Mutagenic chemicals include **nucleotide analogs,** chemicals that are structurally similar to nucleotides and can result in mismatched base pairing, and chemicals that insert or delete nucleotide base pairs, producing frameshift mutations.
ANIMATIONS: *Mutagens*

3. Cells repair damaged DNA via **light repair** and **dark repair** of pyrimidine dimers, **base-excision repair,** and **mismatch repair**. When damage is so extensive that these mechanisms are overwhelmed, bacterial cells may resort to an **SOS response**.
ANIMATIONS: *Mutations: Repair*

4. Researchers have developed methods to distinguish **mutants,** which carry mutations, from normal **wild-type** cells. These methods include **positive selection, negative (indirect) selection,** and the **Ames test,** which is used to identify mutagens, which may be potential **carcinogens**.

Genetic Recombination and Transfer (pp. 225–231)

1. Organisms acquire new genes through **genetic recombination,** which is the exchange of segments of DNA. Crossing over occurs during gamete formation, part of sexual reproduction in eukaryotes.

2. Vertical gene transfer is the transmission of genes from parents to offspring. In **horizontal (lateral) gene transfer,** DNA from a **donor cell** is transmitted to a **recipient cell.** A **recombinant cell** results from genetic recombination between donated and recipient DNA. Transformation, transduction, and bacterial conjugation are types of horizontal gene transfer.
ANIMATIONS: *Horizontal Gene Transfer: Overview*

3. In **transformation,** a **competent** recipient prokaryote takes up DNA from its environment. Competency is found naturally or can be created artificially in some cells.
ANIMATIONS: *Transformation*

4. In **transduction,** a virus such as a **bacteriophage,** or **phage,** carries DNA from a donor cell to a recipient cell. Donor DNA is accidentally incorporated in such **transducing phages**.
ANIMATIONS: *Transduction: Generalized Transduction, Specialized Transduction*

5. In **conjugation,** an F^+ bacterium—that is, one containing an **F (fertility) plasmid (factor)**—forms **a conjugation pilus** that attaches to an F^- recipient bacterium. Plasmid genes are transferred to the recipient, which becomes F^+ as a result.
ANIMATIONS: *Conjugation: Overview, F Factor, Hfr Conjugation, Chromosome Mapping*

6. **Hfr (high frequency of recombination) cells** result when an F plasmid integrates into a prokaryotic chromosome. Hfr cells form conjugation pili and transfer cellular genes more frequently than normal F^+ cells do.

7. **Transposons** are DNA segments that code for the enzyme transposase and have palindromic sequences known as **inverted repeats (IR)** at each end. Transposons move among locations in chromosomes in eukaryotes and prokaryotes—a process called **transposition**. The simplest transposons, known as **insertion sequences (IS),** consist only of inverted repeats and transposase. **Complex transposons** contain other genes as well.
ANIMATIONS: *Transposons: Overview, Insertion Sequences, Complex Transposons*

Questions for Review

Answers to the Questions for Review (except Short Answer questions) begin on page A-1.

Multiple Choice

1. Which of the following is most likely the number of base pairs in a bacterial chromosome?
   a. 4,000,000
   b. 4000
   c. 400
   d. 40

2. Which of the following is a true statement concerning prokaryotic chromosomes?
   a. They typically have two or three origins of replication.
   b. They contain single-stranded DNA.
   c. They are located in the cytosol.
   d. They are associated in linear pairs.

3. A plasmid is
   a. a molecule of RNA found in bacterial cells.
   b. distinguished from a chromosome by being circular.
   c. a structure in bacterial cells formed from plasma membrane.
   d. extrachromosomal DNA.

4. Which of the following forms ionic bonds with eukaryotic DNA and stabilizes it?
   a. chromatin
   b. bacteriocin
   c. histone
   d. nucleosome

5. Nucleotides used in the replication of DNA
   a. carry energy.
   b. are found in four forms, each with a deoxyribose sugar, a phosphate, and a base.
   c. are present in the cytosol of cells as triphosphate nucleotides.
   d. all of the above

6. Which of the following molecules functions as a "proofreader" for a newly replicated strand of DNA?
   a. DNA polymerase III
   b. primase
   c. helicase
   d. ligase

7. The addition of $-CH_3$ to a cytosine nucleotide after DNA replication is called
   a. methylation.
   b. restriction.
   c. transcription.
   d. transversion.

8. In translation, the binding site through which tRNA molecules leave is called the
   a. A site.
   b. X site.
   c. P site.
   d. E site.

9. The Ames test
   a. uses auxotrophs and liver extract to reveal potential mutagens.
   b. is time intensive and costly.
   c. involves the isolation of a mutant by eliminating wild-type phenotypes with specific media.
   d. proves that suspected chemicals are carcinogenic.

10. Which of the following methods of DNA repair involves enzymes that recognize and correct nucleotide errors in unmethylated strands of DNA?
    a. light repair of thymine dimers
    b. dark repair of pyrimidine dimers
    c. mismatch repair
    d. SOS response

11. Which of the following is *not* a mechanism of natural genetic transfer and recombination?
    a. transduction
    b. transformation
    c. transcription
    d. conjugation

12. Cells that have the ability to take up DNA from their environment are said to be
    a. Hfr cells.
    b. transposing.
    c. genomic.
    d. competent.

13. Which of the following statements is true?
    a. Conjugation requires a sex pilus extending from the surface of a cell.
    b. Conjugation involves a C factor.
    c. Conjugation is an artificial genetic engineering technique.
    d. Conjugation involves DNA that has been released into the environment from dead organisms.

14. Which of the following are called "jumping genes"?
    a. Hfr cells
    b. transducing phages
    c. palindromic sequences
    d. transposons

15. Although cells P and Q are totally unrelated, cell Q receives DNA from cell P and incorporates this new DNA into its chromosome. This process is
    a. crossing over of P and Q.
    b. vertical gene transfer.
    c. horizontal gene transfer.
    d. transposition.

16. Which of the following is part of each molecule of mRNA?
    a. palindrome
    b. codon
    c. anticodon
    d. base pair

17. A nucleotide is composed of
    a. a five-carbon sugar.
    b. phosphate.
    c. a nitrogenous base.
    d. all of the above

18. In DNA, adenine forms _______________ hydrogen bonds with _______________.
    a. three/uracil
    b. two/uracil
    c. two/thymine
    d. three/thymine

19. A sequence of nucleotides formed during replication of the lagging DNA strand is
    a. a palindrome.
    b. an Okazaki fragment.
    c. a template strand.
    d. an operon.

20. Which of the following is *not* part of an operon?
    a. operator
    b. promoter
    c. origin
    d. gene

21. Repressible operons are important in regulating prokaryotic
    a. DNA replication.
    b. RNA transcription.
    c. rRNA processing.
    d. sugar catabolism.

22. Transcription produces
    a. DNA molecules.
    b. RNA molecules.
    c. polypeptides.
    d. palindromes.

23. Ligase plays a major role in
    a. lagging strand replication.
    b. mRNA processing in eukaryotes.
    c. polypeptide synthesis by ribosomes.
    d. RNA transcription.

24. Before mutations can affect a population permanently, they must be
    a. lasting.
    b. inheritable.
    c. beneficial.
    d. all of the above

25. The *trp* operon is repressible. This means it is usually _______________ and is directly controlled by _______________.
    a. active/an inducer
    b. active/a repressor
    c. inactive/an inducer
    d. inactive/a repressor

Fill in the Blanks

1. The three steps in RNA transcription are _______________, _______________, and _______________.
2. A triplet of mRNA nucleotides that specifies a particular amino acid is called a _______________.
3. Three effects of point mutations are _______________, _______________, and _______________.
4. Insertions and deletions in the genetic code are also called _______________ mutations.
5. An operon consists of _______________, _______________, and _______________ and is associated with a regulatory gene.
6. In general, _______________ operons are inactive until the substrate of their genes' polypeptides is present.
7. A daughter DNA molecule is composed of one original strand and one new strand because DNA replication is _______________.
8. A gene for antibiotic resistance can move horizontally among bacterial cells by _______________, _______________, and _______________.
9. _______________ are nucleotide sequences containing palindromes and genes for proteins that cut DNA strands.
10. _______________ _______________ is a recombination event that occurs during gamete formation in eukaryotes.
11. _______________ RNA carries amino acids.
12. _______________ RNA and _______________ RNA are antisense; that is, they are complementary to another nucleic acid molecule.

Labeling

On the figure below, label DNA polymerase I, DNA polymerase III, helicase, lagging strand, leading strand, ligase, nucleotide (triphosphate), Okazaki fragment, primase, replication fork, RNA primer, and stabilizing proteins.

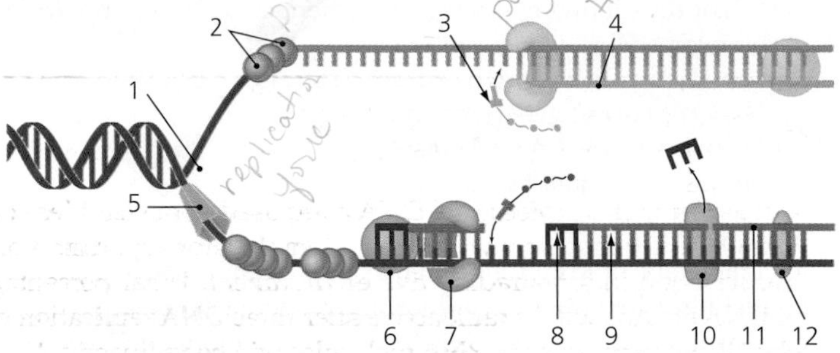

Short Answer

1. How does the genotype of a bacterium determine its phenotype? Use the terms *gene, mRNA, ribosome,* and *polypeptide* in your answer.
2. List several ways in which eukaryotic messenger RNA differs from prokaryotic mRNA.
3. Compare and contrast introns and exons.
4. Polypeptide synthesis requires large amounts of energy. How do cells regulate synthesis to conserve energy? Describe one specific example.
5. Describe the operon model of gene regulation.
6. Compare and contrast the structure and components of DNA and RNA in prokaryotes.
7. Besides the fact that it synthesizes RNA, how does RNA polymerase differ in function from DNA polymerase?
8. Describe the formation and function of mRNA, rRNA, and tRNA in prokaryotes and eukaryotes.
9. Describe how DNA is packaged in both prokaryotes and eukaryotes.
10. Explain the central dogma of genetics.
11. Compare and contrast the processes of transformation, transduction, and conjugation.
12. Fill in the following table:

Process	Purpose	Beginning Point	Ending Point
Replication			Origin or end of molecule
Transcription		Promoter	
Translation	Synthesis of polypeptides		

Concept Mapping

Using the following terms, draw a concept map that describes point mutations. For a sample concept map, see p. 93. Or, complete this concept map online by going to the Study Area at www.masteringmicrobiology.com.

Change in DNA sequence
Deleted
Deletion mutation
Effect on amino acid sequence
Frameshift
Incorrect amino acid substituted
Inserted
Insertion mutation
Missense mutation
No change in amino acid sequence
Nonsense mutation
One or few nucleotide base pairs
Premature termination of polypeptide
Silent mutation
Substituted
Substitution mutation

Critical Thinking

1. A scientist uses a molecule of DNA composed of nucleotides containing radioactive deoxyribose as a template for replication and transcription in a nonradioactive environment. What percentage of DNA strands will be radioactive after three DNA replication cycles? What percentage of RNA molecules will be radioactive?

2. If molecules of mRNA have the following nucleotide base sequences, what will be the sequence of amino acids in polypeptides synthesized by eukaryotic ribosomes?
   a. AUGGGGAUACGCUACCCC
   b. CCGUACAUGCUAAUCCCU
   c. CCGAUGUAACCUCGAUCC
   d. AUGCGGUCAGCCCCGUGA

3. The drugs ddC and AZT are used to treat AIDS.

   ddC (2′, 3′-dideoxycytidine) AZT (3′-azido-2′, 3′-dideoxythymidine)

   Based on their chemical structures, what is their mode of action?

4. Explain why an insertion of three nucleotides is less likely to result in a deleterious effect than an insertion of a single nucleotide.

5. Suppose that *E. coli* sustains a mutation in its gene for the *lac* operon repressor such that the repressor is ineffective. What effect would this have on the bacterium's ability to catabolize lactose? Would the mutant strain have an advantage over wild-type cells? Explain your answer.

6. A student claims that nucleotide analogs can be carcinogenic. Another student in the study group insists that nucleotide analogs are used to treat cancer. Explain why both students are correct.

7. Why is DNA polymerase so named?

8. *Corynebacterium diphtheriae,* the causative agent of diphtheria, secretes a toxin that enzymatically inactivates all molecules of elongation factor in a eukaryotic cell. What immediate and long-term effects does this have on cellular metabolism?

9. How could scientists use siRNA to turn off a cancer-inducing gene?

10. How can knowledge of nucleotide analogs be useful to a cancer researcher?

Access more review material online in the Study Area at **www.masteringmicrobiology.com.** There, you'll find

- **Animations**
- **MicroFlix**
- **MP3 Tutor Sessions**
- **Concept Mapping Activities**
- **Flashcards**
- **Quizzes**

and more to help you succeed.

8 Recombinant DNA Technology

Recombinant DNA technology affects our lives in many ways. Genetically altered corn, soybeans, or canola are ingredients in over 60% of processed foods in the United States. Scientists now have the ability to take a gene from a bacterium and insert it into a corn plant, enabling the plant to produce a toxin that kills insects but is harmless to corn or humans. Geneticists have inserted daffodil and bacterial genes into rice to boost the grain's nutritional content, and they can insert growth-hormone genes into salmon to make the fish grow faster. The possibilities are limitless, intriguing, and frequently controversial, as questions about safety and environmental impact arise. Could crops genetically altered to resist pests spread resistance genes to weeds? Could genetically modified foods introduce new allergens? What are the effects of these technologies on wildlife and human populations?

Despite the concerns, recombinant DNA technology holds much promise, not only for the food industry and agriculture, but for medicine as well. This chapter explores the fascinating field of recombinant DNA technology and the roles of microorganisms within it.

Take the pre-test for this chapter online. Visit the Study Area at www.masteringmicrobiology.com.

▲ Vitamin A deficiency is responsible for millions of deaths worldwide. Genetically modified rice plants can provide vitamin A in rice grains, saving millions of lives. However, such genetic modification is not without controversy.

This chapter examines how genetic researchers have adapted the natural enzymes and processes of DNA recombination, replication, transcription, transformation, transduction, and conjugation to manipulate genes for industrial, medical, and agricultural purposes. Together these techniques are termed *recombinant DNA technology,* commonly called "genetic engineering." We end the chapter with a discussion of the ethics and safety of these techniques.

The Role of Recombinant DNA Technology in Biotechnology

Learning Objectives

✓ Define *biotechnology* and *recombinant DNA technology.*

✓ List several examples of useful products made possible by biotechnology.

✓ Identify the three main goals of recombinant DNA technology.

Biotechnology—the use of microorganisms to make practical products—is not a new field. For thousands of years, humans have used microbes to make products such as bread, cheese, soy sauce, and alcohol. During the 20th century, scientists industrialized the natural metabolic reactions of bacteria to make large quantities of acetone, butanol, and antibiotics. More recently, scientists have adapted microorganisms for use in the manufacture of paper, textiles, and vitamins; to assist in cleaning up industrial wastes, oil spills, and radioactive isotopes; and to aid in mining copper, gold, uranium, and other metals. Chapter 26 discusses such applications of biotechnology.

Until recently, microbiologists were limited to working with naturally occurring organisms and their mutants for achieving such industrial and medical purposes. Since the 1990s, however, scientists have become increasingly adept at intentionally modifying the genomes of organisms, by natural processes, for a variety of practical purposes. This is **recombinant DNA technology,** and it has expanded the possibilities of biotechnology in ways that seemed like science fiction only a few years ago. Today, scientists isolate specific genes from almost any so-called donor organism, such as a human, a plant, or a bacterium, and insert it into the genome of almost any kind of recipient organism.

Scientists who manipulate genomes have three main goals:

- *To eliminate undesirable phenotypic traits in humans, animals, plants, and microbes.* For example, scientists have inserted genes from microbes into plants to make them resistant to pests or freezing, and since 1999 they have cured some children born with a fatal and previously untreatable genetic disorder called severe combined immunodeficiency disease (SCID).
- *To combine beneficial traits of two or more organisms to create valuable new organisms,* such as laboratory animals that mimic human susceptibility to HIV.
- *To create organisms that synthesize products that humans need,* such as vaccines, antibiotics, hormones, and enzymes. For instance, gene therapists have successfully inserted the human gene for insulin into bacteria so that the bacteria synthesize human insulin, which is cheaper and safer than insulin derived from animals.

Recombinant DNA technology is not a single procedure or technique, but rather a collection of tools and techniques scientists use to manipulate the genomes of organisms. In general, they isolate a gene from a cell, manipulate it *in vitro,*[1] and insert it into another organism. **Figure 8.1** illustrates the basic processes involved in recombinant DNA technology.

The Tools of Recombinant DNA Technology

Scientists use a variety of physical agents, naturally occurring enzymes, and synthetic molecules to manipulate genes and genomes. These tools of recombinant DNA technology include *mutagens, reverse transcriptase, synthetic nucleic acids, restriction enzymes,* and *vectors.* Scientists use these molecular tools to create *gene libraries,* which are a time-saving tool for genetic researchers.

Mutagens

Learning Objective

✓ Describe how gene researchers use mutagens.

As we saw in Chapter 7, **mutagens** are physical and chemical agents that produce mutations. Scientists deliberately use mutagens to create changes in microbes' genomes so that the microbes' phenotypes are changed. They then select for and culture cells with characteristics considered beneficial for a given biotechnological application. For example, scientists exposed the fungus *Penicillium* (pen-i-sil'ē-ŭm) to mutagenic agents and then selected strains that produce greater amounts of penicillin. In this manner, they developed a strain of *Penicillium* that secretes over 25 times as much penicillin as the strain originally isolated by Alexander Fleming. Today, with recombinant DNA techniques (discussed shortly), researchers can isolate mutated genes rather than dealing with entire organisms.

The Use of Reverse Transcriptase to Synthesize cDNA

Learning Objective

✓ Explain the function and use of reverse transcriptase in synthesizing cDNA.

As we saw in Chapter 7, transcription involves the transmission of genetic information from molecules of DNA to molecules of RNA. The discovery of retroviruses, which have genomes consisting of RNA instead of DNA, led to the discovery of an unusual enzyme—**reverse transcriptase.** Reverse transcriptase

[1]Latin, meaning within glassware.

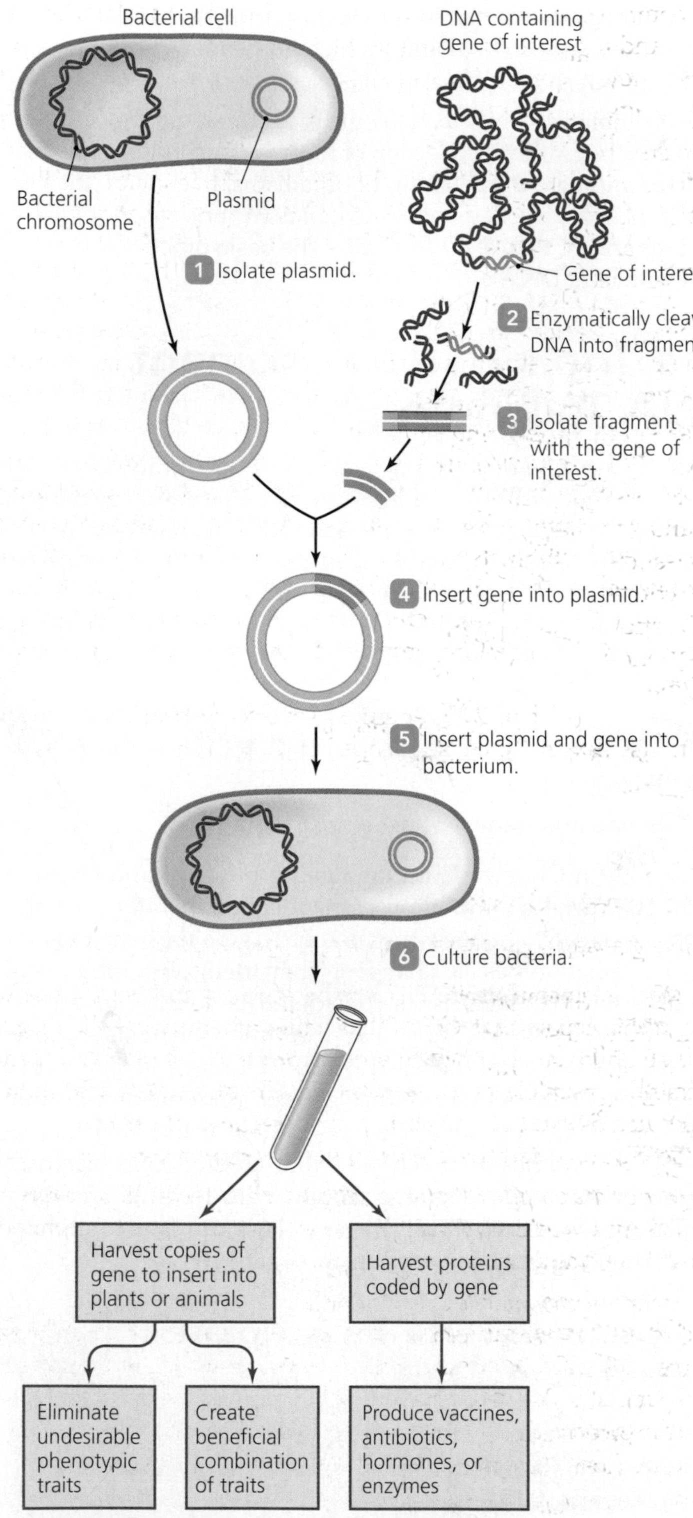

▲ **Figure 8.1 Overview of recombinant DNA technology.**

creates a flow of genetic information in the opposite direction from the flow in conventional transcription: it uses an RNA template to transcribe a molecule of DNA, which is called **complementary DNA (cDNA)** because it is complementary to an RNA template.

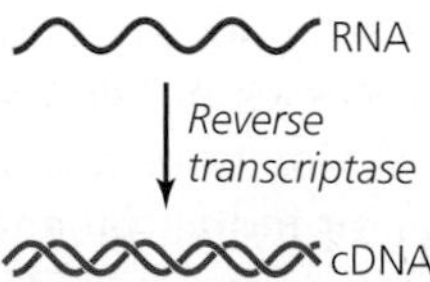

Because hundreds to millions of copies of mRNA exist for every active gene, it is frequently easier to produce a desired gene by first isolating the mRNA molecules that code for a particular polypeptide and then use reverse transcription to synthesize a cDNA gene from the mRNA template. Further, eukaryotic DNA is not normally expressible by prokaryotic cells, which cannot remove the introns (noncoding sequences) present in eukaryotic pre-mRNA. However, since eukaryotic mRNA has already been processed to remove introns, cDNA produced from it lacks noncoding sequences. Therefore, scientists can successfully insert cDNA into prokaryotic cells, making it possible for the prokaryotes to produce eukaryotic proteins such as human growth factor, insulin, or blood-clotting factors.

Synthetic Nucleic Acids

Learning Objectives

✓ Explain how gene researchers synthesize nucleic acids.

✓ Describe three uses of synthetic nucleic acids.

Not only do the enzymes of DNA replication and RNA transcription function *in vivo*;[2] they also function *in vitro*; making it possible for scientists to produce molecules of DNA and RNA in cell-free solutions for genetic research. In fact, scientists have so mechanized the processes of nucleic acid replication and transcription that they can produce molecules of DNA and RNA with any nucleotide sequence; all they must do is enter the desired sequence into a synthesis machine's four-letter keyboard. A computer controls the actual synthesis, using a supply of nucleotides and other required reagents. Nucleic acid synthesis machines synthesize molecules over 100 nucleotides long in a few hours, and scientists can join two or more of these molecules end to end with ligase to create even longer synthetic molecules.

Researchers have used synthetic nucleic acids in many ways, including:

- *Elucidating the genetic code.* Using synthetic molecules of varying nucleotide sequences and observing the amino acids in the resulting polypeptides, scientists elucidated the genetic code. For example, synthetic DNA consisting only of adenine nucleotides yields a polypeptide consisting solely of the amino acid phenylalanine. Therefore, the mRNA codon UUU (transcribed from the DNA triplet AAA) must code for phenylalanine.
- *Creating genes for specific proteins.* Once they know the genetic code and the amino acid sequence of a protein, scientists can create a gene for that protein. In this manner, scientists synthesized a gene for human insulin. Of course, such a synthetic gene likely consists of a nucleotide

[2]Latin, meaning in life (that is, within a cell).

sequence different from that of its cellular counterpart because of the redundancy in the genetic code (see Figure 7.11).

- *Synthesizing DNA and RNA probes to locate specific sequences of nucleotides.* **Probes** are nucleic acid molecules with a specific nucleotide sequence that have been labeled with radioactive or fluorescent chemicals so that their locations can be detected. The use of probes to locate specific sequences of nucleotides is based on the fact that any given nucleotide sequence will preferentially bond to its complementary sequence. Thus a probe constructed with the nucleotide sequence ATGCT will bond to a DNA strand with the sequence TACGA, and the probe's label allows researchers to then detect the complementary site. Probes are essential tools for locating specific nucleic acid sequences such as genes for particular polypeptides.
- *Synthesizing antisense nucleic acid molecules.* Antisense nucleic acid molecules have nucleotide sequences that bind to and interfere with genes and mRNA molecules. Scientists are researching the use of antisense molecules to control genetic diseases.

CRITICAL **THINKING**

Even though some students correctly synthesize a fluorescent cDNA probe complementary to mRNA for a particular yeast protein, they find that the probe does not attach to any portion of the yeast's genome. Explain why the students' probe does not work.

Restriction Enzymes

Learning Objectives

✓ Explain the source and names of restriction enzymes.

✓ Describe the importance and action of restriction enzymes.

An important development in recombinant DNA technology was the discovery of **restriction enzymes** in bacterial cells. Such enzymes cut DNA molecules and are restricted in their action—they cut DNA only at locations called *restriction sites.* Restriction sites are specific nucleotide sequences, which are usually *palindromes*[3]—they have the same sequence when read forward or backward. In nature, bacterial cells use restriction enzymes to protect themselves from phages by cutting phage DNA into nonfunctional pieces. (Recall from Chapter 7 that bacterial cells protect their own DNA by methylation of some of their nucleotides, hiding the DNA from the restriction enzymes.)

Researchers name restriction enzymes with three letters (denoting the genus and specific epithet of the source bacterium) and Roman numerals (to indicate the order in which enzymes from the same bacterium were discovered). In some cases, a fourth letter denotes the strain of the bacterium. Thus *Escherichia coli* (esh-ĕ-rik'ē-ă kō'lē) strain R produces the restriction enzymes *Eco*RI and *Eco*RII. *Hin*dIII is the third restriction enzyme isolated from *Haemophilus influenzae* (hē-mof'i-lŭs in-flu-en'zī) strain Rd.

[3]From Greek *palin,* meaning again, and *dramein,* meaning to run.

Scientists have discovered several hundred restriction enzymes, and categorize them in two groups based on the types of cuts they make. The first type, as exemplified by *Eco*RI, makes staggered cuts of the two strands of DNA, producing fragments that terminate in mortise-like *sticky ends.* Each sticky end is composed of up to four nucleotides that form hydrogen bonds with its complementary sticky end **(Figure 8.2a)**. Scientists can use these bits of single-stranded DNA to combine pieces of DNA from different organisms into a single recombinant DNA molecule (the enzyme ligase unites the sugar-phosphate backbones of the pieces) **(Figure 8.2b)**. Other restriction enzymes, such as *Hin*dII and *Sma*I (from *Serratia marcescens,* ser-rat'ē-a mar-ses'enz), cut both strands of DNA at the same point, resulting in *blunt ends* **(Figure 8.2c)**. It is more difficult to make recombinant DNA from blunt-ended fragments because they are not sticky, but they have a potential advantage—blunt ends are nonspecific. This enables any two blunt-ended fragments, even those produced by different restriction enzymes, to be combined easily **(Figure 8.2d)**. In contrast, sticky-ended fragments bind only to complementary, sticky-ended fragments produced by the same restriction enzyme.

Table 8.1 on p. 240 identifies several restriction enzymes and their target DNA sequences. **ANIMATIONS:** ***Recombinant DNA Technology***

Vectors

Learning Objective

✓ Define *vector* as the term applies to genetic manipulation.

One goal of recombinant DNA technology is to insert a useful gene into a cell so that the cell has a new phenotype—for example, the ability to synthesize a novel protein. To deliver a gene into a cell, researchers use **vectors,** which are nucleic acid molecules such as viral genomes, transposons, and plasmids.

Genetic vectors share several useful properties:

- *Vectors are small enough to manipulate in a laboratory.* Large DNA molecules the size of entire chromosomes are generally too fragile to serve as vectors.
- *Vectors survive inside cells.* Plasmids, which are circular DNA, make good vectors because they are more stable than are linear fragments of DNA, which are typically degraded by cellular enzymes. However, some linear vectors such as transposons and certain viruses insert themselves rapidly into a host's chromosome before they can be degraded.
- *Vectors contain a recognizable genetic marker* so that researchers can identify the cells that have received the vector and thereby the specific gene of interest. Genetic markers can either be phenotypic markers, such as those that confer antibiotic resistance or code for enzymes that metabolize a unique nutrient, or radioactive or fluorescent labels.
- *Vectors can ensure genetic expression by* providing required genetic elements such as promoters.

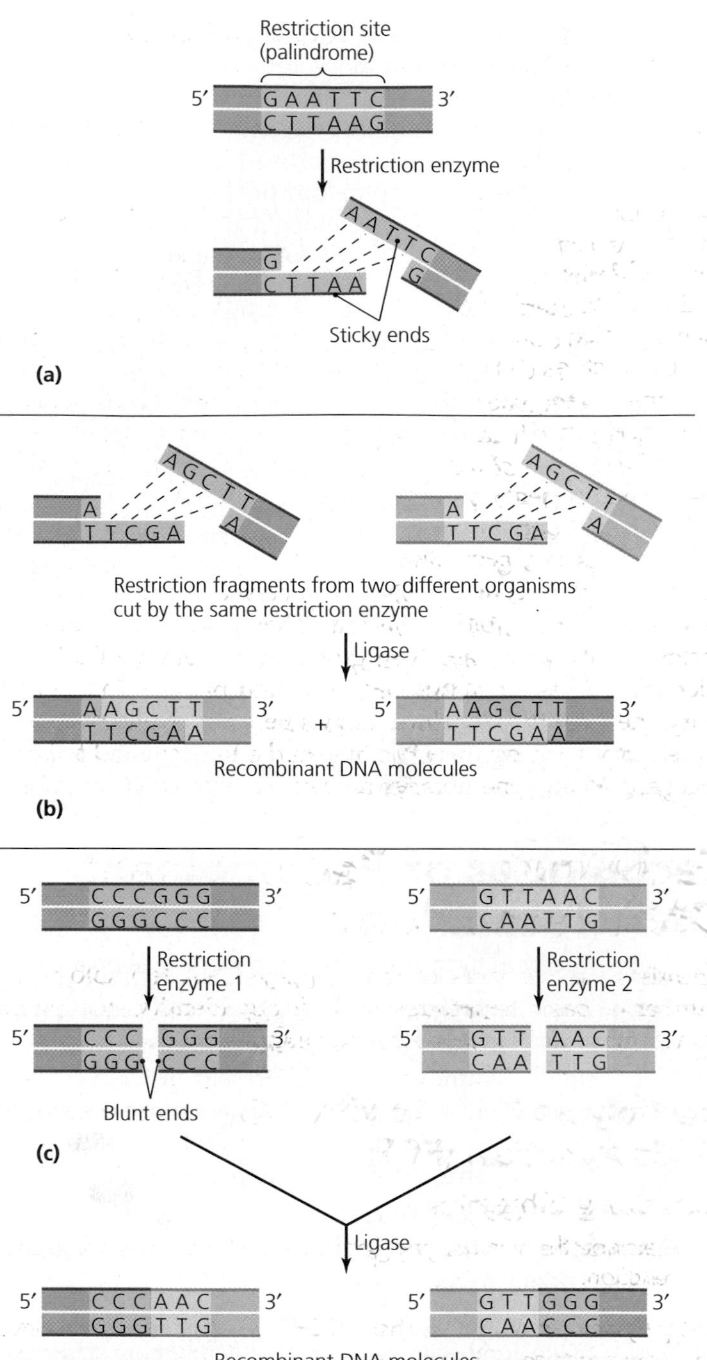

▲ **Figure 8.2 Actions of restriction enzymes.** Restriction enzymes recognize and cut both strands of a DNA molecule at a specific (usually palindromic) restriction site. **(a)** Certain restriction enzymes produce staggered cuts with complementary "sticky ends." **(b)** When two complementary sticky-ended fragments come from different organisms, their bonding (catalyzed by ligase) produces recombinant DNA. **(c)** Other restriction enzymes produce blunt-ended fragments. **(d)** A lack of specificity enables blunt-ended fragments produced by different restriction enzymes to be combined easily into recombinant DNA. *Which restriction enzymes act at the restriction sites shown?*

Figure 8.2 (a) EcoRI, (b) HindIII, and (c) SmaI and HpaI.

TABLE 8.1 Properties of Some Restriction Enzymes

Enzyme	Bacterial Source	Restriction Site[a]
BamHI	*Bacillus amyloliquefaciens* H	G↓GATCC CCTAG↑G
EcoRI	*Escherichia coli* RY13	G↓AATTC CTTAA↑G
EcoRII	*E. coli* R245	CC↓GG GG↑CC
HindII	*Haemophilus influenzae* Rd	GTPy↓PuAC CAPu↑PyTG
HindIII	*H. influenzae* Rd	A↓AGCTT TTCGA↑A
Hinfl	*H. influenzae* Rf	G↓ANTC CTNA↑G
HpaI	*H. parainfluenzae*	GTT↓AAC CAA↑TTG
MspI	*Moraxella* sp.	CC↓GG GG↑CC
SmaI	*Serratia marcescens*	CCC↓GGG GGG↑CCC

[a] Arrows indicate sites of cleavage; Py = pyrimidine (either T or C); Pu = purine (either A or G); N = any nucleotide (A, T, G, or C).

An example of the process used to produce a vector containing a specific gene is depicted in **Figure 8.3**. After a given restriction enzyme cuts both the DNA molecule containing the gene of interest (in this example, the human growth hormone gene) and the vector DNA (here a plasmid containing a gene for antibiotic resistance as a marker) into fragments with sticky ends (1), ligase anneals the fragments to produce a recombinant plasmid (2). After the recombinant plasmid has been inserted into a bacterial cell (3), the bacteria are grown on a medium containing the antibotic (4); only those cells that contain the recombinant plasmid (and thus the human growth hormone gene as well) can grow on the medium.

Generally, viruses and transposons are able to carry larger genes than can plasmids. Researchers are developing vectors from adenoviruses, poxviruses, and a genetically modified form of the human immunodeficiency virus (HIV). HIV in particular might make an excellent vector because HIV inserts itself directly into human chromosomes; however, scientists must ensure that viral vectors do not insert DNA into the middle of a necessary gene, mutating and possibly killing their target cells.

Gene Libraries

Learning Objective

✓ Explain the significance of gene libraries.

Suppose you were a scientist investigating the effects of the genes for 24 different kinds of interleukins (proteins that mediate certain aspects of immunity). Having to isolate the specific

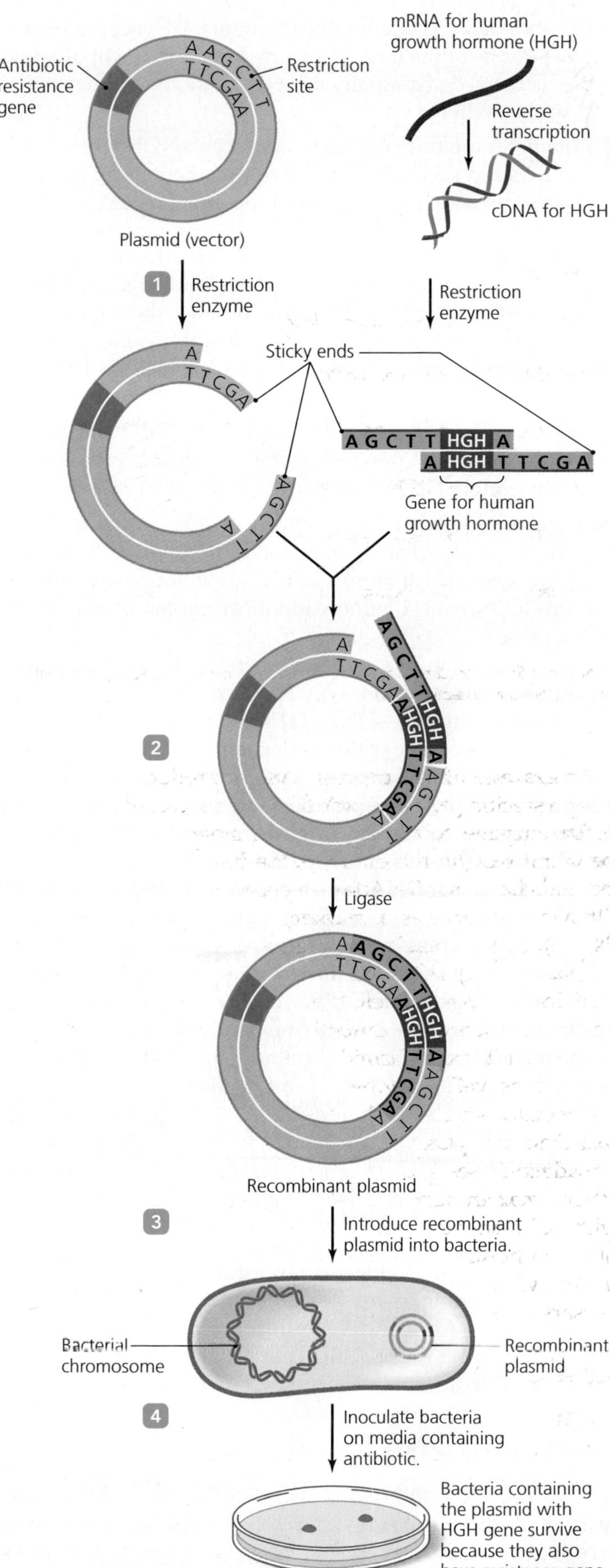

◀ **Figure 8.3 An example of the process for producing a recombinant vector.** In this case, the vector is a plasmid. *Which restriction enzyme was used in this example?*

Figure 8.3 *HindIII.*

genes for each type of interleukin would require much time, labor, and expense. Your task would be made much easier if you could obtain the genes you need from a **gene library,** a collection of bacterial or phage clones—identical descendants—each of which contains a portion of the genetic material of interest. In effect, each clone is like one book in a library in that it contains one fragment (typically a single gene) of an organism's entire genome. Alternatively, a gene library may contain clones with all the genes of a single chromosome or of the set of cDNA that is complementary to an organism's mRNA.

As depicted in **Figure 8.4**, genetic researchers can create each of the clones in a gene library by using restriction enzymes to generate fragments of the DNA of interest and then using ligase to synthesize recombinant vectors. They insert the vectors into bacterial cells, which are then grown on culture media. Once a scientist isolates a recombinant clone and places it in a gene library, the gene that the clone carries becomes available to other investigators, saving them the time and effort required to isolate that gene. Many gene libraries are now commercially available.

Techniques of Recombinant DNA Technology

Scientists use the tools of recombinant DNA technology in a number of basic techniques to multiply, identify, manipulate, isolate, map, and sequence the nucleotides of genes.

Multiplying DNA *in vitro:* The Polymerase Chain Reaction (PCR)

Learning Objective

✓ Describe the purpose and application of the polymerase chain reaction.

The **polymerase chain reaction (PCR)** is a technique by which scientists produce a large number of identical molecules of DNA *in vitro.* Using PCR, researchers start with a single molecule of DNA and generate billions of exact replicas within hours. Such rapid amplification of DNA is critical in a variety of situations. For example, epidemiologists used PCR to amplify the genome of a previously unknown pathogen that killed people in Hong Kong in 2003 with severe acute respiratory syndrome (SARS). The large number of identical DNA molecules produced by PCR allowed scientists to determine the nucleotide sequence, which was found to be similar to that of coronaviruses, until then thought to cause only mild colds. **ANIMATIONS:** ***Polymerase Chain Reaction (PCR): Overview, Components***

PCR is a repetitive process that alternately separates and replicates the two strands of DNA. Each cycle of PCR consists of the following three steps **(Figure 8.5a)**:

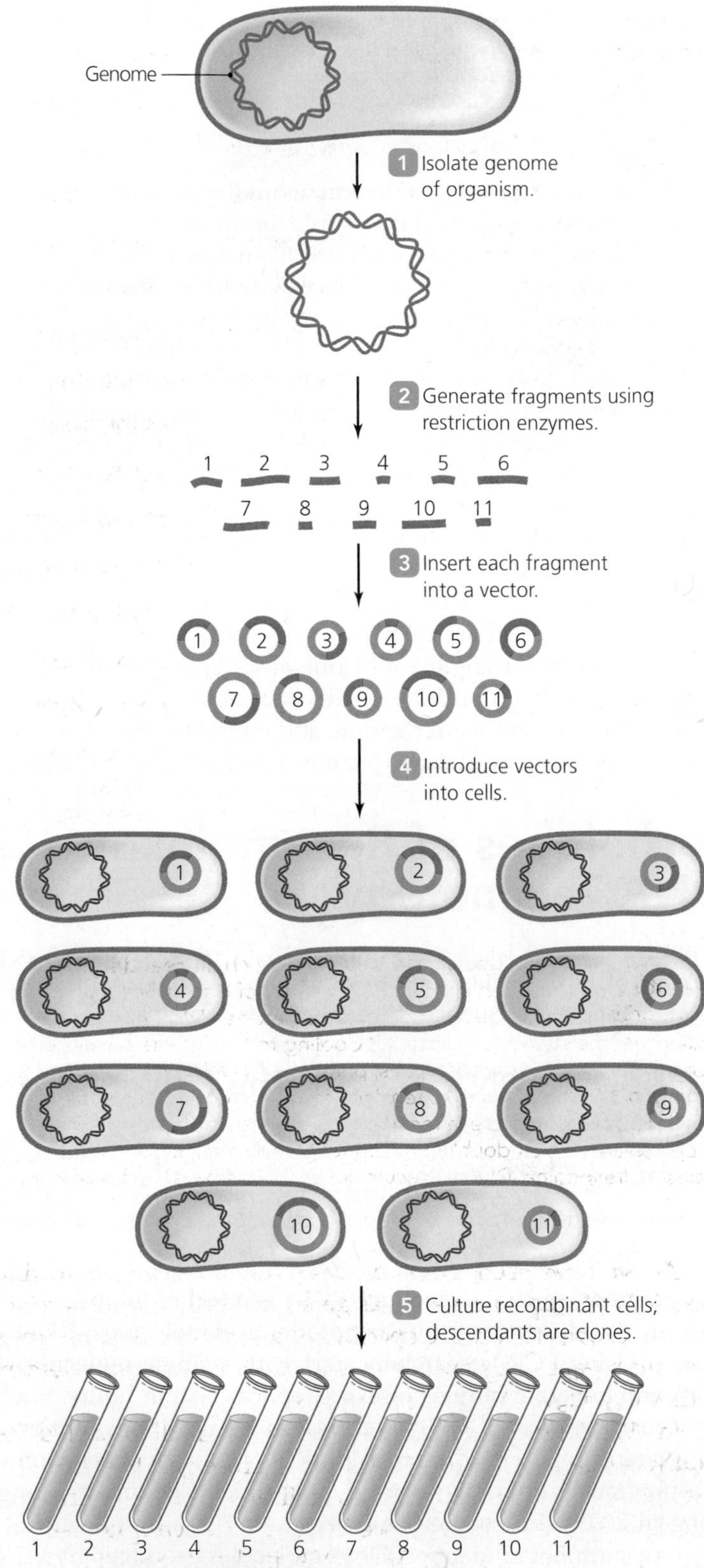
Genome
1 Isolate genome of organism.
2 Generate fragments using restriction enzymes.
1 2 3 4 5 6
7 8 9 10 11
3 Insert each fragment into a vector.
1 2 3 4 5 6
7 8 9 10 11
4 Introduce vectors into cells.
1 2 3
4 5 6
7 8 9
10 11
5 Culture recombinant cells; descendants are clones.
1 2 3 4 5 6 7 8 9 10 11

▲ **Figure 8.4 Production of a gene library.** A gene library is the population of all cells or phages that together contain all of the genetic material of interest. In this figure, each clone of cells carries a portion of a bacterium's genome.

1 *Denaturation.* Exposure to heat (about 94°C) separates the two strands of the target DNA by breaking the hydrogen bonds between base pairs but otherwise leaves the two strands unaltered.

2 *Priming.* A mixture containing an excess of DNA primers (synthesized such that they are complementary to nucleotide sequences near the ends of the target DNA), DNA polymerase, and an abundance of the four deoxyribonucleotide triphosphates (A, T, G, and C) is added to the target DNA. This mixture is then cooled to about 65°C, enabling double-stranded DNA to re-form. Because there is an excess of primers, single strands are more likely to bind to a primer than to one another. The primers provide DNA polymerase with the 3′ hydroxyl group it requires for DNA synthesis.

3 *Extension.* Raising the temperature to about 72°C increases the rate at which DNA polymerase replicates each strand to produce more DNA. **ANIMATIONS:** ***PCR: The Process***

These steps are repeated over and over (**4**), so the number of DNA molecules increases exponentially **(Figure 8.5b)**. After only 30 cycles—which requires only a few hours to complete—PCR produces over 1 billion identical copies of the original DNA molecule.

The process can be automated using a *thermocycler,* a device that automatically performs PCR by continuously cycling all the necessary reagents—DNA, DNA polymerase, primers, and triphosphate deoxynucleotides—through the three temperature regimes. A thermocycler uses DNA polymerase derived from hyperthermophilic archaea or bacteria such as *Thermus aquaticus* (ther′mŭs a-kwa′ti-kŭs). This enzyme, called *Taq DNA polymerase* or simply *Taq,* is not denatured at 94°C, so the machine need not be replenished with DNA polymerase after each cycle.

Selecting a Clone of Recombinant Cells

Learning Objective

✓ Explain how researchers use DNA probes to identify recombinant cells.

Before recombinant DNA technology can have practical application, a scientist must be able to select and isolate recombinant cells that contain particular genes of interest. For example, once researchers have created a gene library, they must find the clone containing the DNA of interest. To do so, scientists use probes—which, you may recall, bind specifically and exclusively to their complementary nucleotide sequences and have either radioactive or fluorescent markers. Researchers then isolate and culture cells that have the radioactive or fluorescent marker, which also aids in identifying the specific location of the genes of interest, as performed in a technique called *gel electrophoresis.*

Separating DNA Molecules: Gel Electrophoresis and the Southern Blot

Learning Objective

✓ Describe the process and use of gel electrophoresis, particularly as it is used in a Southern blot.

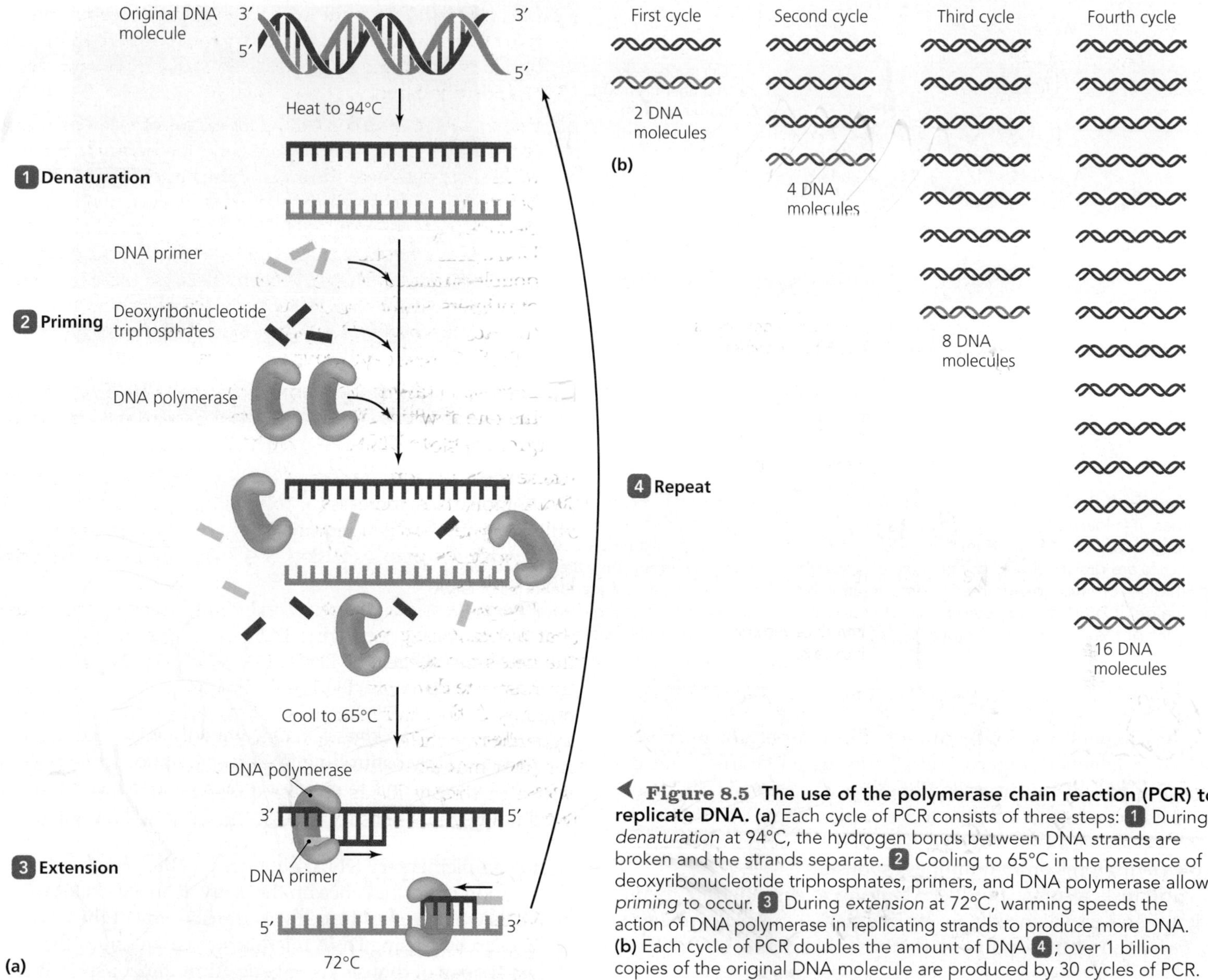

Figure 8.5 The use of the polymerase chain reaction (PCR) to replicate DNA. **(a)** Each cycle of PCR consists of three steps: 1 During *denaturation* at 94°C, the hydrogen bonds between DNA strands are broken and the strands separate. 2 Cooling to 65°C in the presence of deoxyribonucleotide triphosphates, primers, and DNA polymerase allows *priming* to occur. 3 During *extension* at 72°C, warming speeds the action of DNA polymerase in replicating strands to produce more DNA. **(b)** Each cycle of PCR doubles the amount of DNA 4; over 1 billion copies of the original DNA molecule are produced by 30 cycles of PCR.

Electrophoresis (ē-lek-trō-fōr-ē′sis) is a technique that involves separating molecules based on their electrical charge, size, and shape. In recombinant DNA technology, scientists use **gel electrophoresis** to isolate fragments of DNA molecules that can then be inserted into vectors, multiplied by PCR, or preserved in a gene library.

In gel electrophoresis, DNA molecules, which have an overall negative charge, are drawn through a semisolid gel by an electric current toward the positive electrode within an electrophoresis chamber **(Figure 8.6)**. The gel is typically composed of a purified sugar component of agar, called *agarose,* which acts as a molecular sieve that retards the movement of DNA fragments down the chamber and separates the fragments by size. Smaller DNA fragments move faster and farther than larger ones. Scientists can determine the size of a fragment by comparing the distance it travels to the distances traveled by standard DNA fragments of known sizes.

As we have seen, DNA probes allow a researcher to find specific DNA sequences such as genes in a cell. Scientists could also use probes to localize specific sequences in electrophoresis gels, but because gels are flimsy and easily broken, and deform as they dry, it is difficult to probe gels.

In 1975, Ed Southern (1938–) devised a method, called the **Southern blot,** to transfer DNA from agarose gels to nitrocellulose membranes, which are less delicate. The Southern blot technique begins with the procedures of gel electrophoresis just described (**Figure 8.7** 1). The DNA is denatured into single strands with NaOH. Once the DNA fragments have been separated by size, the liquid in the electrophoresis gel is blotted out (2). DNA is transferred and bonded with heat to a nitrocellulose membrane (3). Radioactive probes complementary to DNA sequences of interest are added (4). The probes expose photographic film, revealing the DNA of interest (5). A *northern blot* is a similar technique used to detect specific RNA molecules.

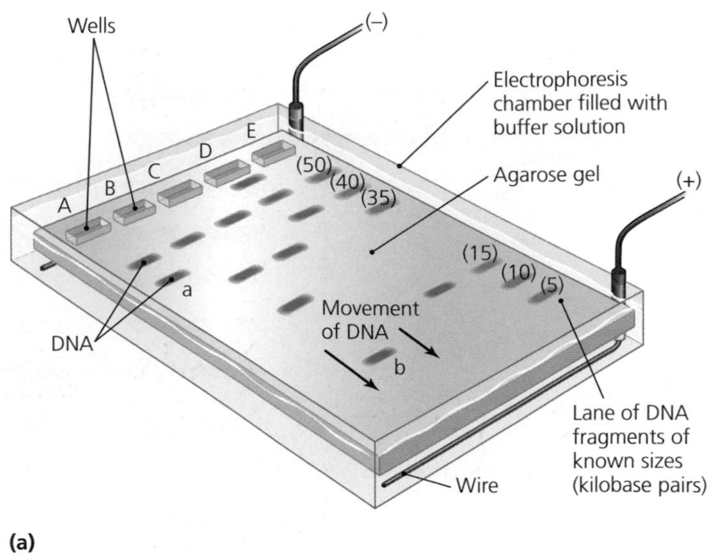

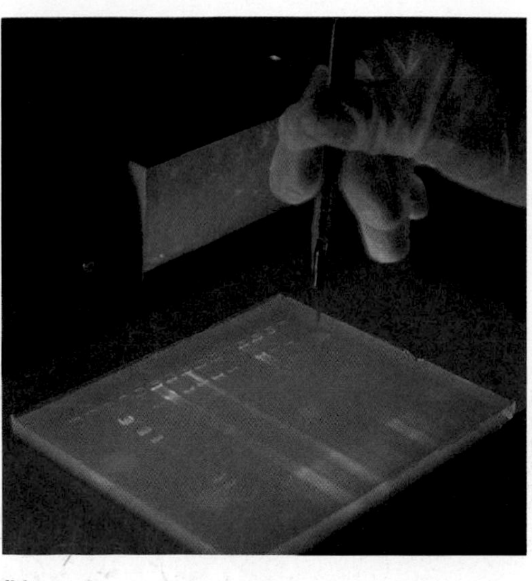

▲ **Figure 8.6 Gel electrophoresis.** **(a)** After DNA is cleaved into fragments by restriction enzymes, it is loaded into wells, which are small holes cut into the agarose gel. DNA fragments of known sizes are often loaded into one well (in this case, E) to serve as standards. After the DNA fragments are drawn toward the positive electrode by an electric current, they are stained with a dye. **(b)** Ethidium bromide dye fluoresces under ultraviolet illumination to reveal the locations of DNA within a gel. *Compare the positions of the fragments in lanes A and B of the diagram to the positions of the fragments of known sizes. What sizes are the fragments labeled a and b?*

Figure 8.6 *a, 40 kilobase pairs. b, 10 kilobase pairs.*

Researchers use Southern blots for a variety of purposes, including genetic "fingerprinting" (discussed shortly) and diagnosis of infectious diseases. For example, scientists can detect the presence of genetic sequences unique to hepatitis B virus in a blood sample of an infected patient even before the patient shows symptoms or an immune response.

Scientists also use Southern blotting to demonstrate the incidence and prevalence in an environmental sample of archaea, bacteria, and viruses, particularly those that cannot be cultured.

DNA Microarrays

Learning Objective

✓ Describe the manufacture and use of DNA microarrays.

Another tool of biotechnology is a **DNA microarray.** An array consists of molecules of single-stranded DNA, either genetic DNA or cDNA, immobilized on glass slides, silicon chips, or nylon membranes. Robots, similar to those that construct computer chips, deposit PCR-derived copies of hundreds of thousands of different DNA sequences in precise locations on the array **(Figure 8.8)**. An array may consist of DNA from a single species (for example, DNA microarrays containing sequences from all the genes of *E. coli* are available commercially), or a DNA array may contain sequences from numerous species. In any case, single strands of fluorescently labeled DNA in a sample washed over an array adhere only to locations on the array where there are complementary DNA sequences.

Scientists use DNA microarrays in a number of ways, including:

- *Monitoring gene expression.* One way organisms control metabolism is by controlling RNA transcription. Scientists use DNA microarrays to monitor which genes a cell is transcribing at a particular time by making fluorescently labeled cDNA from mRNA in the cell. These DNA strands bind to complementary DNA sequences on the array, and the location of fluorescence on the array at specific sites reveals which genes the cell was transcribing at the time. Researchers using DNA microarrays can monitor the expression of thousands of genes simultaneously and can compare and contrast genetic expression under different conditions. In the latter type of experiment, a different color of fluorescent dye is used to label DNA from microbes grown in each condition (see Figure 8.8).
- *Diagnosis of infection.* DNA microarrays made with DNA sequences of numerous pathogens reveal the presence of those pathogens in medical samples.
- *Identification of organisms in an environmental sample.* Microbial ecologists monitor the presence or absence of microbes in an environment by using microarrays of DNA from the organisms.

Inserting DNA into Cells

Learning Objective

✓ List and explain three artificial techniques for introducing DNA into cells.

A goal of recombinant DNA technology is the insertion of a gene into a cell. In addition to using vectors and the natural

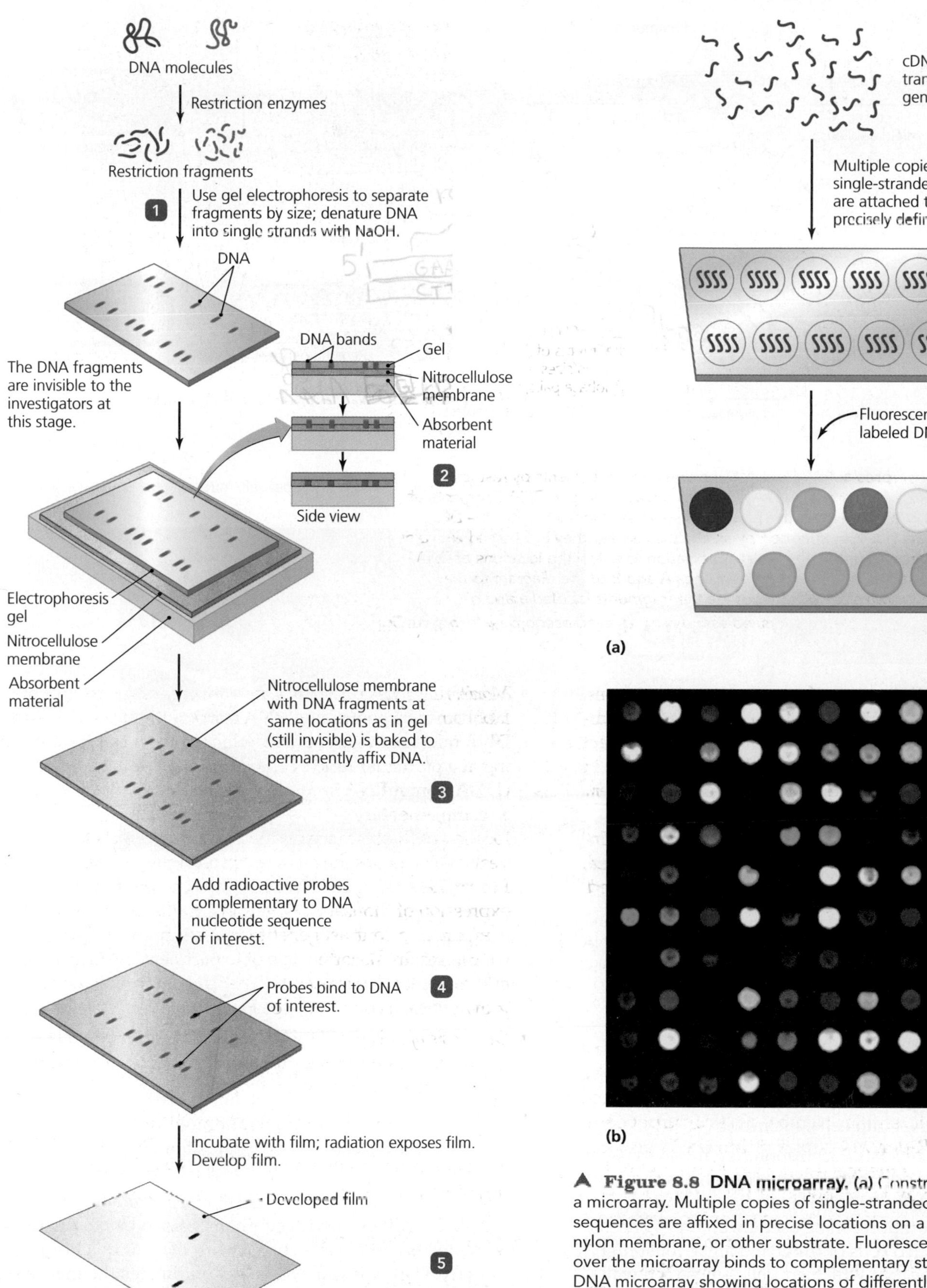
DNA molecules
Restriction enzymes
Restriction fragments
1
Use gel electrophoresis to separate fragments by size; denature DNA into single strands with NaOH.
DNA
The DNA fragments are invisible to the investigators at this stage.
DNA bands
Gel
Nitrocellulose membrane
Absorbent material
2
Side view
Electrophoresis gel
Nitrocellulose membrane
Absorbent material
Nitrocellulose membrane with DNA fragments at same locations as in gel (still invisible) is baked to permanently affix DNA.
3
Add radioactive probes complementary to DNA nucleotide sequence of interest.
Probes bind to DNA of interest.
4
Incubate with film; radiation exposes film. Develop film.
Developed film
5

▲ **Figure 8.7 The Southern blot technique.** This method enables scientists to locate DNA sequences of interest.

cDNA sequences from reverse transcription, a microbe, or a gene library
Multiple copies of each single-stranded sequence are attached to substrate in precisely defined locations.
DNA microarray
Fluorescently labeled DNA
(a)
(b)

▲ **Figure 8.8 DNA microarray.** **(a)** Construction and use of a microarray. Multiple copies of single-stranded DNA with known sequences are affixed in precise locations on a glass slide, silicon chip, nylon membrane, or other substrate. Fluorescently labeled DNA washed over the microarray binds to complementary strands. **(b)** Photograph of a DNA microarray showing locations of differently labeled cDNA molecules.

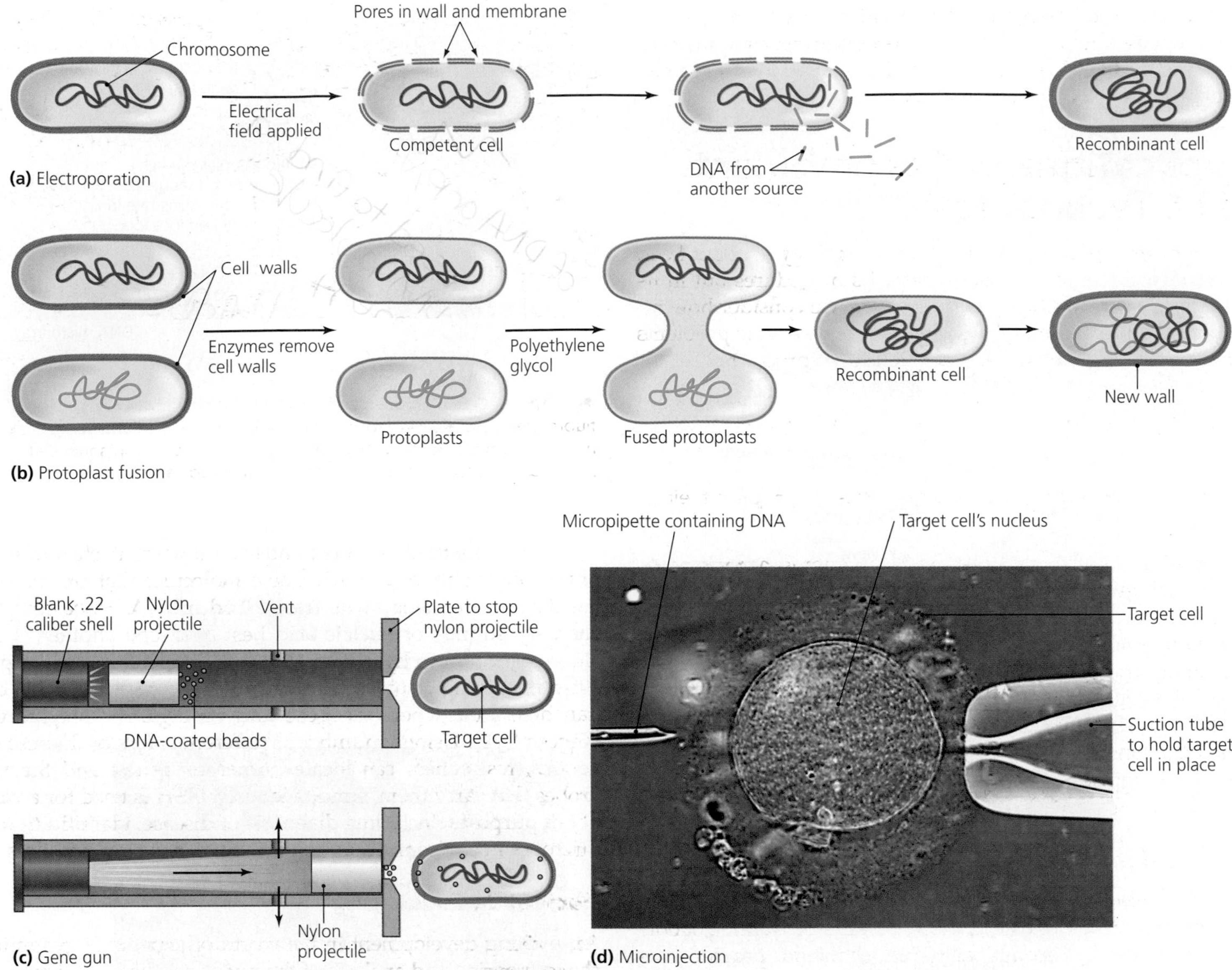

▲ **Figure 8.9 Artificial methods of inserting DNA into cells.** **(a)** Electroporation, in which an electrical current applied to a cell makes it competent to take up DNA. **(b)** Protoplast fusion, in which enzymes digest cell walls to create protoplasts that fuse at a high rate when treated with polyethylene glycol. **(c)** A gene gun, which fires DNA-coated beads into a cell. **(d)** Microinjection, in which a solution of DNA is introduced into a cell through a micropipette.

methods of transformation of competent cells, transduction, and conjugation, scientists have developed several artificial methods to introduce DNA into cells, including:

- *Electroporation* **(Figure 8.9a)**. Electroporation involves using an electrical current to puncture microscopic holes through a cell's membrane so that DNA can enter the cell from the environment. Electroporation can be used on all types of cells, though the thick-walled cells of fungi and algae must first be converted to *protoplasts,* which are cells whose cell walls have been enzymatically removed. Cells treated by electroporation repair their membranes and cell walls after a time.
- *Protoplast fusion* **(Figure 8.9b)**. When protoplasts encounter one another, their cytoplasmic membranes may fuse to form a single cell that contains the genomes of both "parent" cells. Exposure to polyethylene glycol increases the rate of fusion. The DNA from the two fused cells recombines to form a recombinant molecule. Scientists often use protoplast fusion for the genetic modification of plants.
- *Injection.* Two types of injection are used with larger eukaryotic cells. Researchers use a *gene gun* powered by a blank .22-caliber cartridge or compressed gas to fire tiny tungsten or gold beads coated with DNA into a target cell **(Figure 8.9c)**. The cell eventually eliminates the inert metal beads. In *microinjection,* a geneticist inserts DNA into a target cell with a glass micropipette having a tip diameter smaller than that of the cell or nucleus **(Figure 8.9d)**. Unlike electroporation and protoplast fusion, injection can be used on intact tissues such as in plant seeds.

In every case, foreign DNA that enters a cell remains in a cell's progeny only if the DNA is self-replicating, as in the case of plasmid and viral vectors, or if the DNA integrates into a cellular chromosome by recombination.

Applications of Recombinant DNA Technology

The importance of recombinant DNA technology does not lie in the novelty, cleverness, or elegance of its procedures but in its wide range of applications. In this section we consider how recombinant DNA technology is used to solve various problems and create research, medical, and agricultural products.

Genetic Mapping

Learning Objective

✓ Describe genetic mapping and genomics, and explain their usefulness.

One application of these tools and techniques is **genetic mapping,** which involves locating genes on a nucleic acid molecule. Genetic maps provide scientists with useful facts, including information concerning an organism's metabolism and growth characteristics, as well as its potential relatedness to other microbes. For example, scientists have discovered a virus with a genetic map similar to those of certain hepatitis viruses. They named the new discovery *hepatitis G virus* because it presumably causes hepatitis, though it has not been demonstrated that the virus actually causes the disease.

Locating Genes

Until about 1970, scientists identified the specific location of genes on chromosomes by cumbersome, time-consuming, labor-intensive methods. Recombinant DNA techniques provide simpler and universal methods for genetic mapping.

One technique for locating genes, called *restriction fragmentation,* was one of the earliest applications of restriction enzymes. In this technique, which is used for mapping the relative locations of genes in plasmids and viruses, researchers compare DNA fragments resulting from cleavages by several restriction enzymes to determine each fragment's location relative to the others. If the researchers know the locations of specific genes on specific fragments, then elucidation of the correct arrangement of the fragments will reveal the relative locations of the genes on the entire DNA molecule.

Using this method, scientists first completed the entire gene map of a cellular microbe—the bacterium *H. influenzae*—in 1995. Since then, geneticists have elucidated complete gene maps of numerous viruses and prokaryotic and eukaryotic organisms.

Often a scientist wants to know where in the environment, clinical sample, or biofilm a particular microbial species is located. When researchers know of a particular gene exclusive to that organism, they can locate the gene, and thereby the microbe, using *fluorescent* in situ *hybridization (FISH).*

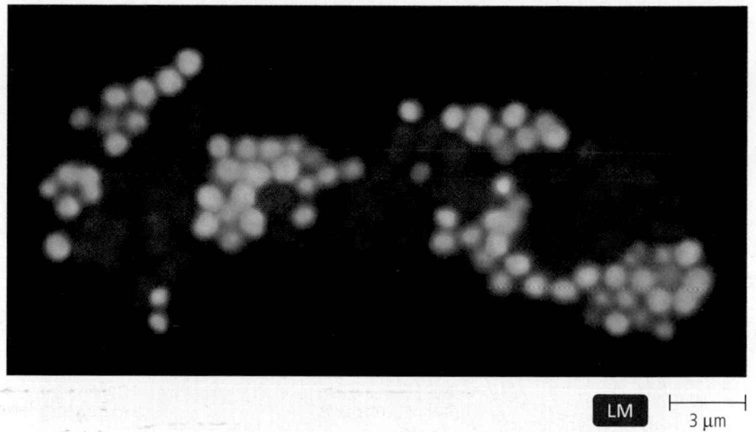

▲ **Figure 8.10 Fluorescent *in situ* hybridization (FISH).** Green-fluorescing cells are *Staphylococcus aureus,* while red-fluorescing cells are non-aureus staphylococci. The absence of fluorescence in some cells indicates another species is present in this blood sample.

In this method, scientists attach fluorescent chemicals to short, single strands of nucleic acid molecules that are complementary to the gene or its transcribed mRNA. Since complementary strands of nucleic acid best bind one another, these fluorescent probes hybridize with their complementary target. Scientists using fluorescent microscopes to view such probes can determine where the gene and its organism are located (**Figure 8.10**). Using a number of different colors of fluorescent probes, researchers can locate numerous genes, and the microbes that carry them, simultaneously. FISH is used for a variety of purposes, including diagnosis of disease, identification of microbes in environmental samples, and analysis of biofilms.

Nucleotide Sequencing

An exciting development in the world of genetics is **genomics,** the sequencing and analysis of the nucleotide bases of genomes. At first, scientists sequenced DNA molecules by selectively cleaving DNA at A, T, G, or C bases, separating the fragments by gel electrophoresis, and mapping the order in which the fragments occur in a complete DNA molecule. Such time-consuming, labor-intensive, and cumbersome sequencing was limited to short DNA molecules such as those of plasmids.

Today, scientists use a faster technique that utilizes cDNA synthesized with nucleotides that have been tagged with four different fluorescent dyes—a different color for each nucleotide base; then an automated DNA sequencer determines the sequence of base colors emitted by the dyes (**Figure 8.11**). Such machines, often running 24 hours a day for months, have sequenced the entire genomes of numerous viruses, bacteria, and eukaryotic organisms. Scientists reached a milestone in 2001 by sequencing the 3 billion nucleotide base pairs that constitute the human genome.

Elucidation of the gene sequences of pathogens, particularly those affecting hundreds of millions of people and those with potential bioterrorist uses, is a current priority of researchers. Scientists hope to use the information to develop novel drugs and more effective therapies and vaccines.

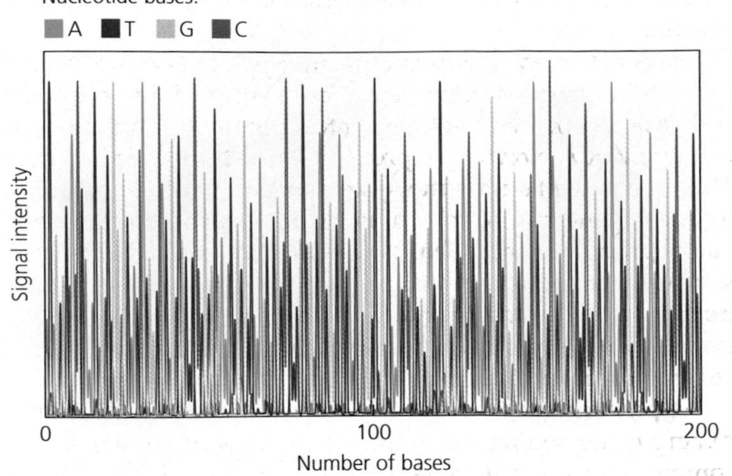

▲ **Figure 8.11 Automated DNA sequencing.** Each colored line corresponds to a different nucleotide base; each peak indicates the location of a particular base. *What is the nucleotide in position 100?*

Figure 8.11 Thymine.

Another use for genomics is to relate DNA sequence data to protein function. For instance, scientists are investigating the genes and proteins of *Deinococcus radiodurans* (dī-nō-kok'ŭs rā-dē-ō-dur'anz), a microorganism that is remarkably resistant to damage of its DNA by radiation. Such studies may lead to methods of reversing genetic damage in cancer patients undergoing radiation therapy. Researchers are also investigating the genetic basis of the enzymes of psychrophiles, which are microorganisms that thrive at temperatures below 20°C. Such enzymes have potential applications in food processing and in the manufacture of drugs.

Table 8.2 summarizes the tools and techniques of recombinant DNA technology.

Environmental Studies

It is estimated that more than 99% of microorganisms have never been grown in a laboratory; indeed, scientists know them only by unique DNA patterns in electrophoresis gels and Southern blot membranes. For example, based on such unique DNA sequences, sometimes called signatures or DNA fingerprints, scientists have isolated over 500 species of bacteria from human

TABLE 8.2 Tools and Techniques of Recombinant DNA Technology

Tool or Technique	Description	Potential Application
Mutagen	Chemical or physical agent that creates mutations	Creating novel genotypes and phenotypes
Reverse transcriptase	Enzyme from RNA retrovirus that synthesizes cDNA from an RNA template	Synthesizing a gene using an mRNA template
Synthetic nucleic acid	DNA molecule prepared *in vitro*	Creating DNA probes to localize genes within a genome
Restriction enzyme	Bacterial enzyme that cleaves DNA at specific sites	Creating recombinant DNA by joining fragments
Vector	Transposon, plasmid, or virus that carries DNA into cells	Altering the genome of a cell
Gene library	Collection of cells or viruses, each of which carries a portion of a given organism's genome	Providing a ready source of genetic material
Polymerase chain reaction (PCR)	Produces multiple copies of a DNA molecule	Multiplying DNA for various applications
Gel electrophoresis	Uses electrical charge to separate molecules according to their size	Separating DNA fragments by size
Electroporation	Uses electrical current to make cells competent	Inserting a novel gene into a cell
Protoplast fusion	Fuses two cells to create recombinants	Inserting a novel gene into a cell
Gene gun	Blasts genes into target cells	Inserting a novel gene into a cell
Microinjection	Uses micropipette to inject genes into cells	Inserting a novel gene into a cell
Southern blot	Localizes specific DNA sequences on a stable membrane	Identifying a strain of pathogen
Nucleic acid probes	RNA or DNA molecules labeled with radioactive or fluorescent tags	Localizing specific genes in a Southern blot
Genetic mapping	Uses restriction enzymes to locate relative positions of restriction sites	Locating genes in an organism's genome
DNA sequencing	Determines the sequence of nucleotide bases in DNA	Comparing genomes of organisms
DNA microarray	Reveals presence of specific DNA or RNA molecules in a sample	Diagnosis of infection

mouths; however, they have been able to identify only about 150 of these. An understanding of the biology of the other 350 species may lead to a better understanding of tooth and gum decay, diagnosis of disease, and advances in oral health care.

Another application of genetics to environmental studies may have ramifications for global warming. Rice agriculture is possibly the largest human contributor of the so-called greenhouse gas methane to the atmosphere. Rice paddies, concentrated in Asia, contribute 50 to 100 million metric tons of methane to the environment every year, though neither rice nor humans directly cause this deluge of methane. Scientists, having analyzed the DNA signatures of microbes in the soil, have determined that mud-dwelling, methane-producing archaea feed on carbohydrates released by the rice plants' roots. These archaea have not been isolated or grown in a laboratory. They are known only by their DNA signatures. Thus, the tools and techniques of recombinant DNA technology have revealed the source of a problem. Discovering that these organisms exist is certainly the first step in developing methods to reduce their impact on the environment.

Pharmaceutical and Therapeutic Applications

Learning Objectives

- ✓ Describe six potential medical applications of recombinant DNA technology.
- ✓ Describe the steps and uses of genetic fingerprinting.
- ✓ Define *gene therapy*.

Researchers now supplement traditional biotechnology with recombinant DNA technology to produce a variety of pharmaceutical and therapeutic substances, and to perform a host of medically important tasks. Here we explore the use of recombinant DNA technology to synthesize selected proteins, produce vaccines, screen for genetic diseases, match DNA specimens to the organisms from which they came, treat genetic illnesses, and aid in organ transplantation.

Protein Synthesis

Scientists have inserted synthetic genes for insulin, for interferon (a natural antiviral chemical), and for other proteins into bacteria and yeast cells so that the microbes synthesize these proteins in vast quantities. In the past, such proteins were isolated from donated blood or from animals—labor-intensive processes that carry the risk of inducing allergies or of transferring pathogens such as hepatitis B and HIV. "Genetically engineered" proteins are safer and less expensive than their naturally occurring counterparts.

Vaccines

Vaccines contain *antigens*—foreign substances such as weakened bacteria, viruses, and toxins that stimulate the body's immune system to respond to and subsequently remember these foreign materials. In effect, a vaccine primes the immune system to respond quickly and effectively when confronted with pathogens and their toxins. However, the use of some vaccines entails a risk—they may cause the disease they are designed to prevent.

Scientists now use recombinant DNA technology to produce safer vaccines. Once they have inserted the gene that codes for a pathogen's antigens into a vector, they can inject the recombinant vector or the proteins it produces into a patient. Thus the patient's immune system is exposed to a subunit of the pathogen—one of the pathogen's antigens—but not to the pathogen itself. Such **subunit vaccines** are especially useful in safely protecting against pathogens that either cannot be cultured or cause incurable fatal diseases. Hepatitis B vaccine is an example of a successful subunit vaccine. Scientists are also pursuing subunit vaccines against HIV.

A promising future approach to vaccination involves introducing genes coding for antigenic proteins of pathogens into common fruits or vegetables such as bananas or beans. The immune systems of people or animals eating such altered produce would be exposed to the pathogen's antigens and theoretically would develop immunological memory against the pathogen. Such a vaccine would have the advantages of being painless and easy to administer, and vaccination would not require a visit to a health care provider. **Highlight: Vaccines on the Menu** on p. 250 focuses on such a vaccine developed to protect cattle against a disease commonly known as "shipping fever."

Another type of vaccination involves producing a recombinant plasmid carrying a gene from a pathogen and injecting the plasmid into a human, whose body then synthesizes polypeptides characteristic of the pathogen. The polypeptides stimulate immunological memory within the human body, readying it to mount a vigorous immune response and prevent infection should it subsequently be exposed to the real pathogen. Clinical trials of such a vaccine against malaria have shown some promise.

Genetic Screening

Genetic mutations cause some diseases, such as inherited forms of breast cancer and Huntington's disease. Laboratory technicians use DNA microarrays to screen patients, prospective parents, and fetuses for such mutant genes. This procedure, called **genetic screening,** can also identify viral DNA sequences in a patient's blood or other tissues. For instance, genetic screening can identify HIV in a patient's cells even before the patient shows any other sign of infection.

DNA Fingerprinting

Medical laboratory technicians and forensic investigators use gel electrophoresis and Southern blotting for so-called **genetic fingerprinting,** or **DNA fingerprinting**—identifying individuals or organisms by their unique DNA sequences.

DNA fingerprinting involves procuring a sample of DNA, making multiple copies of it via PCR, cutting the copies with restriction enzymes, and separating the fragments by gel electrophoresis to produce a unique pattern. The process is analogous to standard fingerprinting in that the pattern resulting from a particular DNA sample is unique, and it must be compared to patterns produced from other DNA molecules **(Figure 8.12)**, much like a standard fingerprint must be compared to known

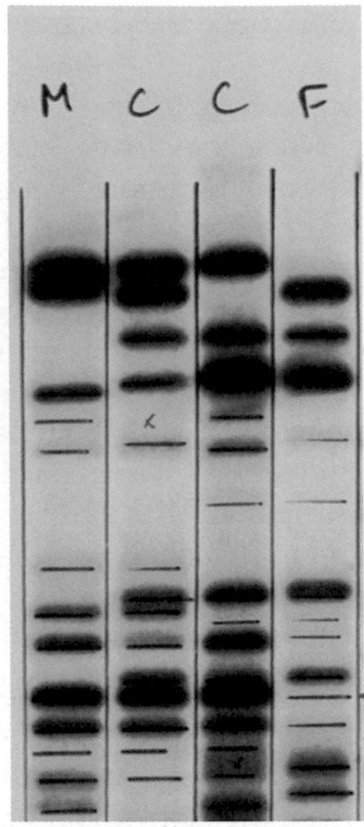

▲ **Figure 8.12 DNA fingerprinting.** Shown here is a partial X ray of bands of DNA from four family members: a mother (M), father (F), and two children (C's). Both children share some bands with each parent, proving they are indeed related. A similar process can be used to compare DNA bands from microbial specimens in order to identify a particular specimen.

fingerprints. For example, the patterns from DNA collected at a crime scene either match or do not match a suspect's or victim's DNA; or the pattern from an environmental sample matches or does not match patterns from known organisms. Genetic fingerprinting is used to determine paternity; to connect blood, semen, or even single skin cells to a particular crime suspect; to identify badly damaged human remains; and to identify pathogens.

Gene Therapy

An exciting use of recombinant DNA technology is **gene therapy,** in which missing or defective genes are replaced with normal copies. Scientists remove a few genetically defective cells—for example, cells that produce a defective protein—from a patient, insert normal genes, and replace the cells into the patient, curing the disease. Alternatively, plasmid or viral vectors could deliver genes directly to target cells within a patient.

Unfortunately, gene therapy has proven difficult in practice because of unexpected results. Specifically, some patients' immune systems react uncontrollably to the presence of vectors, resulting in the death of these patients. Nevertheless, doctors have successfully treated patients for severe combined immunodeficiency disease and a form of blindness. Other diseases that may respond well to gene therapy are cystic fibrosis, sickle-cell anemia, and some types of hemophilia and diabetes.

Medical Diagnosis

Clinical microbiologists use PCR, fluorescent genetic probes, and DNA microarrays in diagnostic applications. They examine specimens from patients for the presence of gene sequences unique to certain pathogens, such as particular hepatitis viruses, cytomegalovirus, human immunodeficiency virus, or the bacterial pathogens of gonorrhea, tuberculosis, and trachoma.

Xenotransplants

Xenotransplants[4] are animal cells, tissues, or organs introduced into the human body. For years physicians have performed xenotransplants, for instance, using valves from pig hearts to repair

[4]From Greek *xenos*, meaning stranger.

HIGHLIGHT

VACCINES ON THE MENU

Wouldn't it be great if instead of receiving painful needle-administered shots, we could be immunized simply by eating an antigen-laced chocolate bar? We aren't quite there yet, but scientists are making progress in developing genetically modified oral vaccines.

In one case, researchers developed a plant-based oral vaccine to protect cattle against a disease called pneumonic pasteurellosis, commonly known as "shipping fever" because cattle experience this stress-induced disease while being shipped from cow/calf operations to feedlots. During this respiratory illness, caused by the bacterium *Mannheimia haemolytica,* diseased cattle experience decreased appetite, fever, and nasal discharge. An injectable vaccine against the disease is available, but it is costly and labor intensive, and the injection is so stressful for cattle that it may actually contribute to the problem.

The oral vaccine is a white clover genetically modified so that the plant expresses one of *M. haemolytica*'s proteins. The idea is that when cattle are fed the genetically altered clover, the bacterial protein will trigger a protective immune response. To date, the vaccine has worked successfully in laboratory animals.

▲ *White clover.*

severely damaged human hearts. However, recombinant DNA technology may expand the possibilities. It is theoretically feasible to insert functional human genes into animals to direct them to produce organs and tissues for transplantation into humans. For example, scientists could induce pigs to produce humanlike cytoplasmic membrane proteins so that entire organs from pigs would not be rejected as foreign tissue by a transplant recipient.

Agricultural Applications

Learning Objective

✓ Identify five agricultural applications of recombinant DNA technology.

Recombinant DNA technology has been applied to the realm of agriculture to produce **transgenic** organisms—recombinant plants and animals that have been altered for specific purposes by the addition of genes from other organisms. The purposes for which transgenic organisms have been produced are many and varied and include herbicide resistance, tolerance to salty soils, resistance to freezing and pests, and improvements in nutritional value and yield. More than 10 million farmers in 22 countries are growing transgenic crops.

Herbicide Resistance

The biodegradable herbicide *glyphosate* (Roundup) normally kills all plants—weeds and crops alike—by blocking an enzyme that is essential for plants to synthesize several amino acids. After scientists discovered and isolated an *Agrobacterium* (ag'rō-bak-tēr'ē-um) gene that conveys resistance to glyphosate, they produced transgenic crop plants containing the gene. As a result of this application of recombinant DNA technology, farmers can now apply glyphosate to a field of transgenic plants to kill weeds without damaging the crop. An added benefit is that farmers do not need to till the soil to suppress weeds during the growing season, reducing soil erosion by 80%. Most of the soybeans, corn, and cotton grown in the United States are genetically modified in this manner to be "Roundup ready." Glyphosate-resistant rice, wheat, sugarbeets, and alfalfa strains are also available, as are soybeans and corn resistant to other herbicides.

Salt Tolerance

Years of irrigation have resulted in excessive salt buildup in farmland throughout the world, rendering the land useless for farming. Though salt-tolerant plants can grow under these conditions, they are not edible.

Scientists have now successfully removed the gene for salt tolerance and inserted it into tomato and canola plants to create food crops that can grow in soil so salty it would poison normal crops. Not only do such transgenic plants survive and produce fruit; they also remove salt from the soil, restoring the soil and making it suitable to grow unmodified crops as well. Researchers are now attempting to insert the gene for salt tolerance into canola, rice, cotton, tomatoes, wheat, and corn.

Freeze Resistance

Ice crystals form more readily when bacterial proteins (from natural bacteria present in a field) are available as crystallization nuclei. Scientists have modified strains of the bacterium

HIGHLIGHT

NATURE'S OWN DECAF MACHINE

Coffee beans need to be industrially processed to remove the caffeine they contain. But someday, with the help of *Escherichia coli*, coffee beans may be naturally decaffeinated while they are still on the plant.

Coffee plants naturally produce an enzyme that breaks caffeine down into a compound called theophylline, which is subsequently converted to innocuous CO_2 and NH_3. However, the enzyme works very slowly, and scientists don't know which gene controls its production. That's where the bacteria come in. To find the gene that codes for the caffeine-destroying enzyme, researchers "genetically engineered" *E. coli* to need theophylline to survive. Next, the researchers hope to use these bacteria to help identify which gene in the coffee plant codes for the caffeine-destroying enzyme.

To do this, they will introduce various coffee-plant genes into these modified bacteria. They then will culture the bacteria on caffeine. Only bacteria with the right gene will be able to make the caffeine-destroying enzyme and convert caffeine into the theophylline that they need to survive. Once they've identified the proper gene, researchers plan to boost its efficiency by creating hundreds of mutated versions of the gene, which they will insert into bacterial hosts. The cells that break caffeine down fastest will outperform the others and live longer. Theoretically, the winning genes could then be inserted into coffee plants that will in the future produce virtually caffeine-free beans.

▲ *Coffee plant.*

Pseudomonas (soo-dō-mō′nas) with a gene for a polypeptide that prevents ice crystals from forming. Crops sprayed with genetically modified bacteria can tolerate mild freezes, so the farmers no longer lose their crops to unseasonable cold snaps.

Pest Resistance

Strains of the bacterium *Bacillus thuringiensis* (ba-sil′ŭs thur-in-jē-en′sis) produce a protein that, when modified by enzymes in the intestinal tracts of insects, becomes **Bt toxins (Bt).** Bt binds to receptors lining the insect's digestive tract and causes the tissue to dissolve. Unlike some insecticides, Bts are naturally occurring, harmful only to insects, and biodegradable. Farmers, particularly organic growers, have used Bts for over 30 years to reduce insect damage to their crops.

Now, genes for Bt toxins have been inserted into a variety of crop plants, including potatoes, cotton, rice, and corn, so that they produce Bt for themselves. Insects feeding on such plants are killed, while humans and other animals that eat them are unharmed. By inserting genes for more than one type of Bt, scientists hope to forestall evolution of resistant insects.

The water mold *Phytophthora infestans* (fī-tof′tho-ră in-fes′-tanz) is the most devasting potato pathogen; it caused the great Irish potato famine of the 19th century, resulting in the deaths of at least a million people. The mold still causes many billions of dollars of crop damage each year. Scientists have now cloned genes from potato species resistant to *Phytophthora;* the cloned genes, multiplied by PCR, can be inserted into potato crops to reduce losses to farmers and increase available food for a growing world population.

Improvements in Nutritional Value and Yield

Genetic researchers have increased crop and animal yields in several ways. For example, MacGregor tomatoes remain firm after harvest because the gene for the enzyme that breaks down pectin has been suppressed. This allows farmers to let the tomatoes ripen on the vine before harvesting and increases the tomatoes' shelf life. Scientists suppressed the gene indirectly by inserting a promoter upstream from the gene's complement in the noncoding DNA strand. The promoter allows transcription of antisense RNA that binds to the gene's mRNA, which makes it impossible to translate into protein.

Another example of agricultural improvement involves bovine growth hormone (BGH), which when injected into cattle enables them to more rapidly gain weight, have meat with a reduced fat content, and produce 10% more milk. Though BGH can be derived from animal tissue, it is more economical to insert the BGH gene into bacteria so that they produce the hormone, which is then purified and injected into farm and ranch animals.

In yet another application, scientists have improved the nutritional value of rice by adding a gene for beta-carotene, which is a precursor to vitamin A. Vitamin A is required for human embryonic development and for vision in adults, and it is an important antioxidant that plays a role in ameliorating cancer and atherosclerosis (hardening of arteries).

Recombinant DNA technology has progressed to the point that scientists are now considering transplanting genes coding for entire metabolic pathways, rather than merely genes encoding single proteins. For instance, researchers are attempting to transfer into corn and rice all the genes bacteria use to convert atmospheric nitrogen into nitrogenous fertilizer, in effect allowing the recombinant plants to produce their own fertilizer.

Recombinant DNA tools and techniques allow scientists to examine, compare, and manipulate the genomes of microorganisms, plants, animals, and humans for a variety of purposes including gene mapping and forensic, medical, and agricultural applications. However, as with many scientific advances, concerns arise about the ethics and safety of genetic manipulations. The next section deals with these issues.

The Ethics and Safety of Recombinant DNA Technology

Learning Objective

✓ Discuss the pros and cons concerning the safety and ethics of recombinant DNA technology.

Recombinant DNA technology provides the opportunity to transfer genes among unrelated organisms, even among organisms in different kingdoms, but how safe and ethical is it? "Frankenfood" and "biological Russian roulette" are some of the terms opponents use to denigrate transgenic agricultural products and gene therapy. Some opponents question the ethics of raising genetically altered animals solely for creating products for human use. They contend that this exemplifies a supremacist view—the view that humans are of greater intrinsic value than animals. Other critics of transgenic crops and animals correctly state that the long-term effects of transgenic manipulations are unknown, and that unforeseen problems arise from every new technology and procedure. Recombinant DNA technology may burden society with complex and unforeseen regulatory, administrative, financial, legal, social, medical, and environmental problems.

Critics also argue that natural genetic transfer through sexual reproduction and processes such as transformation and transduction could deliver genes from transgenic plants and animals into other organisms. For example, if a herbicide-resistant plant cross-pollinates with a related weed species, we might be cursed with a weed that is more difficult to kill. Opponents further express concern that transgenic organisms could trigger allergies or cause harmless organisms to become pathogenic. Some opponents of recombinant DNA technology desire a ban on all genetically modified products.

The U.S. National Academy of Sciences, the U.S. National Research Council, and 81 research projects conducted between 1985 and 2001 by the European Union have not revealed any risks to human health or the environment from genetically modified agricultural products beyond the usual uncertainties inherent in conventional plant breeding. In fact, the European Union concluded in 2001 that "the use of more precise technology and the greater regulatory scrutiny probably make them [genetically modified foods] even safer than conventional plants and foods."

As the debate continues, governments continue to impose standards on laboratories involved in recombinant DNA technology. These are intended to prevent the accidental release of altered organisms or exposure of laboratory workers to potential dangers. Additionally, genetic researchers often design organisms to lack a vital gene so that they cannot survive for long outside of a laboratory.

Unfortunately, biologists can apply the procedures used to create beneficial crops and animals to create biological weapons that are more infective and more resistant to treatment than their natural counterparts are. Though international treaties prohibit the development of biological weapons, *B. anthracis* spores were used in bioterrorist attacks in the United States in 2001, though, thankfully, the strain utilized was not genetically altered to realize its deadliest potential.

Emergent recombinant DNA technologies raise numerous other ethical issues. Should people be routinely screened for diseases that are untreatable or fatal? Who should pay for these procedures: individuals, employers, prospective employers, insurance companies, HMOs, government agencies? What rights do individuals have to genetic privacy? If entities other than individuals pay the costs involved in genetic screening, should those entities have access to *all* the genetic information that results? Should businesses be allowed to have patents on and make profits from living organisms they have genetically altered? Should governments be allowed to require genetic screening and then force genetic manipulations on individuals to correct perceived genetic abnormalities that some claim are the bases of criminality, manic depression, risk-taking behavior, and alcoholism? Should HMOs, physicians, or the government demand genetic screening and then refuse to provide services related to the birth or care of supposedly "defective" children?

We as a society will have to confront these and other ethical considerations as the genomic revolution continues to affect people's lives in many unpredictable ways.

Chapter Summary

The Role of Recombinant DNA Technology in Biotechnology (p. 237)

1. **Biotechnology** is the use of microorganisms to make useful products. Historically these include bread, wine, beer, and cheese.
2. **Recombinant DNA technology** is a new type of biotechnology in which scientists change the genotypes and phenotypes of organisms to benefit humans.

The Tools of Recombinant DNA Technology (pp. 237–241)

1. The tools of recombinant DNA technology include mutagens, reverse transcriptase, synthetic nucleic acids, restriction enzymes, vectors, and gene libraries.
2. **Mutagens** are chemical and physical agents used to create changes in a microbe's genome to effect desired changes in the microbe's phenotype.
3. The enzyme **reverse transcriptase** transcribes DNA from an RNA template; genetic researchers use reverse transcriptase to make **complementary DNA (cDNA).**
4. Scientists used synthetic nucleic acids to elucidate the genetic code, and they now use them to create genes for specific proteins and to synthesize DNA and RNA **probes** labeled with radioactive or fluorescent markers.
5. **Restriction enzymes** cut DNA at specific (usually palindromic) nucleotide sequences and are used to produce recombinant DNA molecules.
   ANIMATIONS: *Recombinant DNA Technology*
6. In recombinant DNA technology, a **vector** is a small DNA molecule (such as a viral genome, transposon, or plasmid) that carries a particular gene and a recognizable genetic marker into a cell.
7. A **gene library** is a collection of bacterial or phage clones, each of which carries a fragment (typically a single gene) of an organism's genome.

Techniques of Recombinant DNA Technology (pp. 241–247)

1. The **polymerase chain reaction (PCR)** allows researchers to replicate molecules of DNA rapidly.
   ANIMATIONS: *Polymerase Chain Reaction (PCR): Overview, Components, The Process*
2. **Gel electrophoresis** is a technique for separating molecules (including fragments of nucleic acids) by size, shape, and electrical charge.
3. The **Southern blot** technique allows researchers to stabilize DNA sequences from an electrophoresis gel and then localize them using DNA dyes or probes.
4. **DNA microarrays,** containing nucleotide sequences of thousands of genes, are used to monitor gene activity and the presence of microbes in patients and the environment.
5. Geneticists artificially insert DNA into cells by electroporation, protoplast fusion, or injection.
6. Fluorescent *in situ* hybridization (FISH) uses fluorescent nucleic acid probes to localize specific genetic sequences.

Applications of Recombinant DNA Technology (pp. 247–252)

1. **Genomics** is the sequencing (**genetic mapping**), analysis, and comparison of genomes. Genetic sequencing has been speeded up by an automated machine that distinguishes among fluorescent dyes attached to each type of nucleotide base.
2. Unique DNA sequences reveal the presence of microbes that have never been cultured in a laboratory.
3. Scientists synthesize **subunit vaccines** by introducing genes for a pathogen's polypeptides into cells or viruses. When the cells, the viruses, or the polypeptides they produce are injected into a human, the body's immune system is exposed to and reacts against relatively harmless antigens instead of the potentially harmful pathogen.

4. **Genetic screening** can detect infections and inherited diseases before a patient shows any sign of disease.
5. **Genetic fingerprinting (DNA fingerprinting),** which identifies unique sequences of DNA, is used in paternity investigations, crime scene forensics, diagnostic microbiology, and epidemiology.
6. **Gene therapy** cures various diseases by replacing defective genes with normal genes.
7. In **xenotransplants** involving recombinant DNA technology, human genes would be inserted into animals to produce cells, tissues, or organs for introduction into the human body.
8. **Transgenic** plants and animals have been genetically altered by the inclusion of genes from other organisms.
9. Agricultural uses of recombinant DNA technology include advances in herbicide resistance, salt tolerance, freeze resistance, and pest resistance, as well as improvements in nutritional value and yield.

The Ethics and Safety of Recombinant DNA Technology (p. 252–253)

1. Among the ethical and safety issues surrounding recombinant DNA technology are concerns over the accidental release of altered organisms into the environment, the ethics of altering animals for human use, and the potential for creating genetically modified biological weapons.

Questions for Review

Answers to the Questions for Review (except Short Answer questions) begin on page A-1.

Multiple Choice

1. Which of the following statements is true concerning recombinant DNA technology?
   a. It will replace biotechnology in the future.
   b. It is a single technique for genetic manipulation.
   c. It is useful in manipulating genotypes but not phenotypes.
   d. It involves modification of an organism's genome.
2. A DNA gene synthesized from an RNA template is
   a. reverse transcriptase.
   b. complementary DNA.
   c. recombinant DNA.
   d. probe DNA.
3. After scientists exposed cultures of *Penicillium* to agents X, Y, and Z, they examined the type and amount of penicillin produced by the altered fungi to find the one that is most effective. Agents X, Y, and Z were probably
   a. recombinant cells.
   b. competent.
   c. mutagens.
   d. phages.
4. Which of the following is *false* concerning vectors in recombinant DNA technology?
   a. Vectors are small enough to manipulate outside a cell.
   b. Vectors contain a recognizable genetic marker.
   c. Vectors survive inside cells.
   d. Vectors must contain genes for self-replication.
5. Which recombinant DNA technique is used to replicate copies of a DNA molecule?
   a. PCR
   b. gel electrophoresis
   c. electroporation
   d. reverse transcription
6. Which of the following would be most useful in following gene expression in a yeast cell?
   a. Southern blot
   b. reverse transcription
   c. DNA microarray
   d. restriction enzymes
7. Which of the following techniques is used regularly in the study of genomics?
   a. Clones are selected using a vector with two genetic markers.
   b. Genes are inserted to produce an antigenic protein from a pathogen.
   c. Fluorescent nucleotide bases are sequenced.
   d. Defective organs are replaced with those made in animal hosts.
8. Restriction enzyme *Hha*I
   a. recombines DNA.
   b. cuts DNA at a specific nucleotide sequence.
   c. is likely derived from *Haemophilus influenzae.*
   d. all of the above.
9. Which application of recombinant DNA technology involves the production of a distinct pattern of DNA fragments on a gel?
   a. genetic fingerprinting
   b. gene therapy
   c. genetic screening
   d. protein synthesis
10. A DNA microarray consists of
    a. a series of clones containing the entire genome of a microbe.
    b. recombinant microbial cells.
    c. restriction enzyme fragments of DNA molecules.
    d. single-stranded DNA localized on a substrate.

Modified True/False

Indicate which of the following are true and which are false. Rewrite any false statements to make them true by changing the italicized words.

____ 1. Restriction enzymes *inhibit the movement of DNA.*

____ 2. Restriction enzymes act at *specific* nucleotide sequences within a double-stranded DNA molecule.

____ 3. *A thermocycler* separates molecules based on their size, shape, and electrical charge.

____ 4. Protoplast fusion is often used in the genetic modification of *plants.*

____ 5. Gel electrophoresis is used in *DNA microarrays.*

Labeling

Label the reagents and steps of PCR on the figure below. Indicate the temperature of the chemicals at each numbered step.

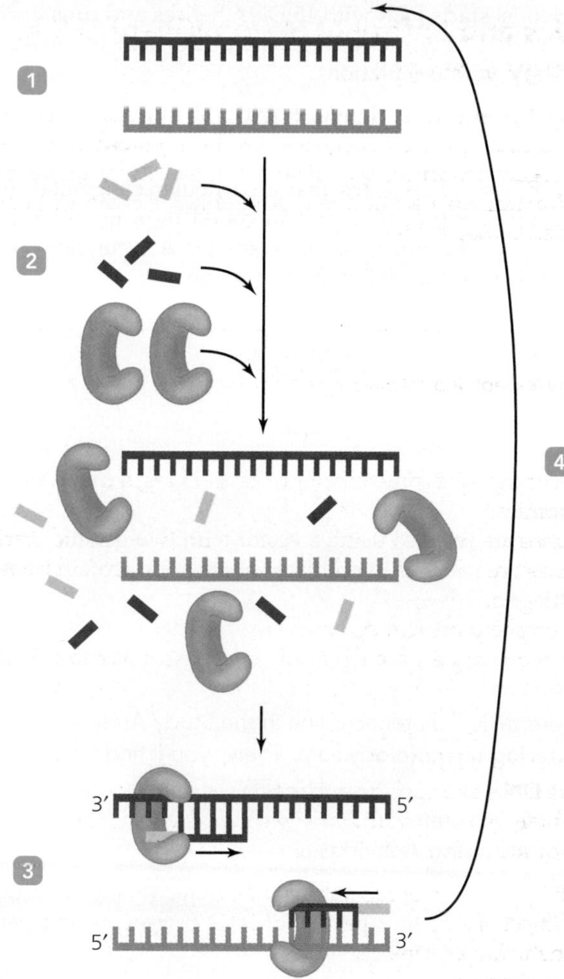

Short Answer

1. Describe three artificial methods of introducing DNA into cells.
2. Why is cloning a practical technique for medical researchers?
3. Describe three ways scientists use synthetic nucleic acids.
4. Describe a gene library and its usefulness.
5. List three potential problems of recombinant DNA technology.

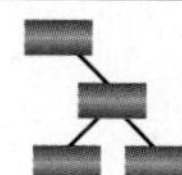

Concept Mapping

Using the following terms, draw a concept map that describes the polymerase chain reaction. For a sample concept map, see p. 93. Or, complete this concept map online by going to the Study Area at www.masteringmicrobiology.com.

Cooling to ~65°C
Denaturation
DNA strands
Extension
Primers
Priming
Raising temperature to ~72°C
Raising temperature to ~94°C
Repeated in multiple cycles
Specific sequences of DNA
***Taq* polymerase**
Target DNA (2)
Thermocycler

Critical Thinking

1. Examine the restriction sites listed in Table 8.1 Which restriction enzymes produce restriction fragments with sticky ends? Which produce fragments with blunt ends?
2. A cancer-inducing virus, HTLV-1, inserts itself into a human chromosome, where it remains. How can a laboratory technician prove that a patient is infected with HTLV-1 even when there is no sign of cancer?
3. A thermocycler uses DNA polymerase from hyperthermophilic prokaryotes, but it cannot use DNA polymerase derived from *E. coli*. Why not?
4. How is the result of a Southern blot similar to the result of surveillance using a DNA microarray?
5. *Hha*I recognizes and cuts this DNA sequence at the sites indicated:

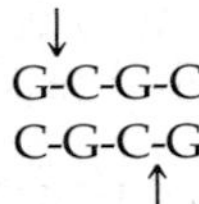

Describe the fragments resulting from the use of this enzyme.

6. PCR replication of DNA is similar to bacterial population growth. If a scientist starts PCR with 15 DNA helices and runs the reaction for 15 cycles, how many DNA molecules will be present at the end? Show your calculations.
7. If a gene contains the sequence TACAATCGCATTGAA, what antisense RNA could be used to stop translation directly?
8. Suppose researchers learn that a particular congenital disease is caused by synthesis of a protein coded by a mutated gene. Describe a way in which recombinant DNA technology might be used to prevent translation of the protein.

9 Controlling Microbial Growth in the Environment

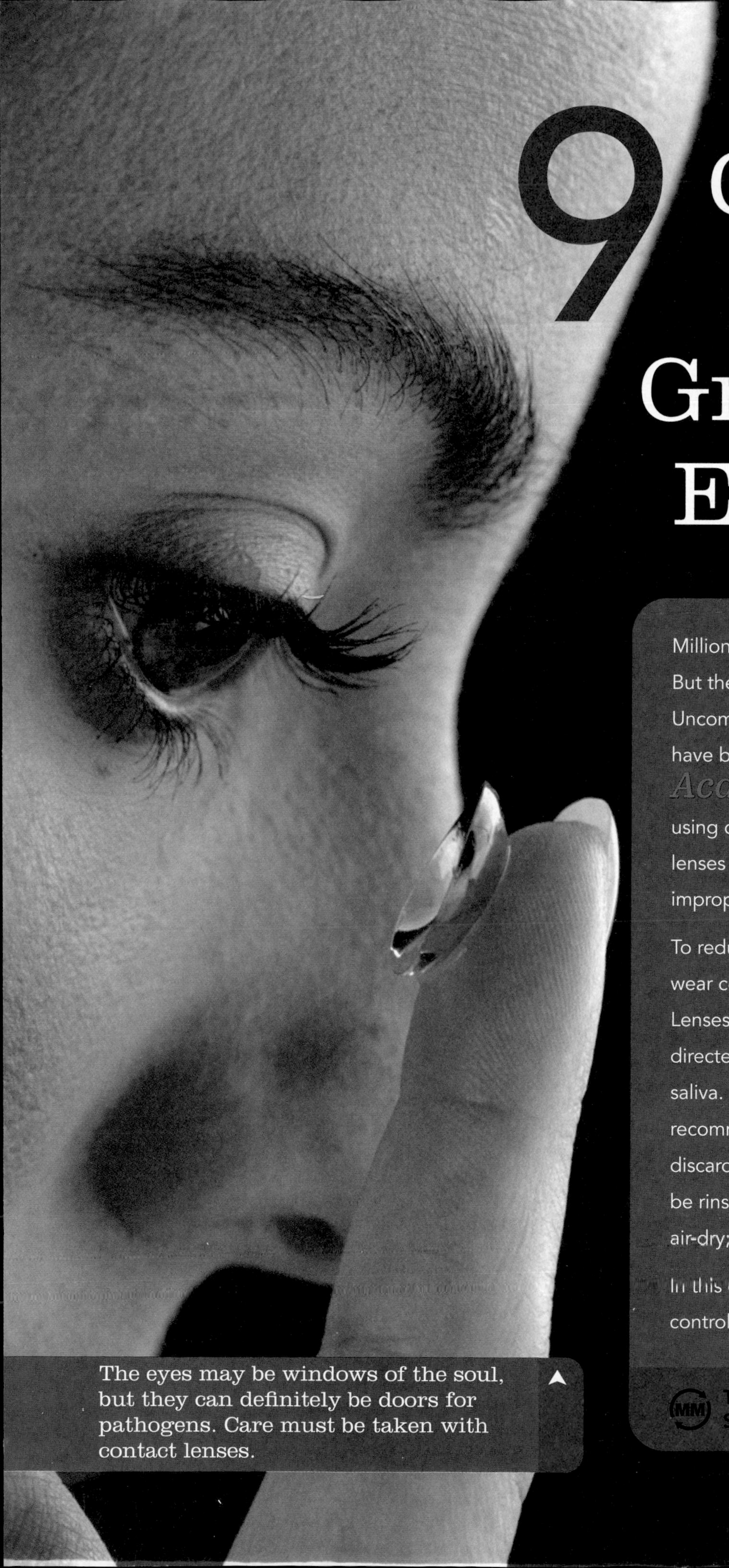

Millions of people wear contact lenses without complications. But the use of contact lenses is not completely without risk. Uncommon but serious infections, such as bacterial keratitis, have been linked to contact-lens use. Infection by *Acanthamoeba*, a protozoan, can result from using contaminated lens-care solutions or rinsing contact lenses in tap water. Bacteria can also proliferate on improperly cleaned lens-storage cases, forming slimy biofilms.

To reduce the risk of microbial infection, people who wear contact lenses must care for and use them properly. Lenses should be chemically disinfected as directed by an optometrist and never rinsed in tap water or saliva. Lenses should not be left in the eyes longer than recommended. Lens-cleaning solutions should be discarded after their expiration dates. Storage cases should be rinsed daily with contact-lens cleaning solution and left to air-dry; periodically, they should be replaced altogether.

In this chapter we will study a wide variety of methods used to control microorganisms in our environment.

Take the pre-test for this chapter online. Visit the Study Area at www.masteringmicrobiology.com.

▲ The eyes may be windows of the soul, but they can definitely be doors for pathogens. Care must be taken with contact lenses.

The control of microbes in health care facilities, in laboratories, and at home is a significant and practical aspect of microbiology. In this chapter we study the terminology and principles of microbial control, consider the factors affecting the efficacy of microbial control, and examine the control of microorganisms and viruses by various chemical and physical means. One important aspect of microbial control—the use of antimicrobial drugs to assist the body's defenses against pathogens—will be considered in Chapter 10.

Basic Principles of Microbial Control

Scientists, health care professionals, researchers, and government workers should use precise terminology in reference to microbial control in the environment. In the following sections we consider the terminology of microbial control, examine the concept of microbial death rates, and discuss the action of antimicrobial agents.

Terminology of Microbial Control

Learning Objectives

- ✓ Contrast sterilization, disinfection, and antisepsis, and describe their practical uses.
- ✓ Contrast the terms *degerming*, *sanitization*, and *pasteurization*.
- ✓ Compare the effects of *-static* versus *-cidal* control agents on microbial growth.

It is important for microbiologists, health care workers, and others to use correct terminology for describing microbial control. While many of these terms are familiar to the general public, they are often misused.

In its strictest sense, **sterilization** refers to the removal or destruction of *all* microbes, including viruses and bacterial endospores, in or on an object. (The term does not apply to *prions,* which are infectious proteins, because standard sterilizing techniques do not destroy them.)

In practical terms, sterilization indicates only the eradication of harmful microorganisms and viruses; some innocuous microbes may still be present and viable in an environment that is considered sterile. For instance, *commercial sterilization* of canned food does not kill all hyperthermophilic microbes; however, since they do not cause disease and they cannot grow and spoil food at ambient temperatures, they are of no practical concern. Likewise, some hyperthermophiles may survive sterilization by laboratory methods (discussed shortly), but they are of no practical concern to technicians because they cannot grow or reproduce under normal laboratory conditions.

The term **aseptic**[1] (ā-sep′tik) describes an environment or procedure that is free of contamination by *pathogens.* For example, vegetables and fruit juices are available in aseptic packaging, and surgeons and laboratory technicians use aseptic techniques to avoid contaminating a surgical field or laboratory equipment.

Disinfection[2] refers to the use of physical or chemical agents known as **disinfectants,** including ultraviolet light, heat, alcohol, and bleach, to inhibit or destroy microorganisms, especially pathogens. Unlike sterilization, disinfection does not guarantee that all pathogens are eliminated; indeed, disinfectants alone cannot inhibit endospores or some viruses. Further, the term *disinfection* is used only when discussing treatment of inanimate objects. When a chemical is used on skin or other tissue, the process is called **antisepsis**[3] (an-tē-sep′sis), and the chemical is called an **antiseptic.** Antiseptics and disinfectants often have the same components, but disinfectants are more concentrated or can be left on a surface for longer periods of time. Of course, some disinfectants, such as steam or concentrated bleach, are not suitable for use as antiseptics.

Degerming is the removal of microbes from a surface by scrubbing, such as when you wash your hands or a nurse prepares an area of skin for an injection. Though chemicals such as soap or alcohol are commonly used during degerming, the action of thoroughly scrubbing the surface may be more important than the chemical in removing microbes.

Sanitization[4] is the process of disinfecting places and utensils used by the public to reduce the number of pathogenic microbes to meet accepted public health standards. For example, steam, high-pressure hot water, and scrubbing are used to sanitize restaurant utensils and dishes, and chemicals are used to sanitize public toilets. Thus, the difference between *disinfecting* dishes at home and *sanitizing* dishes in a restaurant is the arena—private versus public—in which the activity takes place.

Pasteurization[5] is the use of heat to kill pathogens and reduce the number of spoilage microorganisms in food and beverages. Milk, fruit juices, wine, and beer are commonly pasteurized.

So far, we have seen that there are two major types of microbial control—sterilization, which is the elimination of all microbes, and antisepsis or disinfection, which are the destruction of vegetative (nonspore) cells and many viruses. Modifications of disinfection include degerming, sanitization, and pasteurization. Some scientists and clinicians apply these terms only to pathogenic microorganisms.

Additionally, scientists and health care professionals use the suffixes *-stasis/-static*[6] to indicate that a chemical or physical agent inhibits microbial metabolism and growth, but doesn't necessarily kill microbes. Thus, refrigeration is bacteriostatic for most bacterial species; it inhibits their growth, but they can resume metabolism when the optimal temperature is restored. By contrast, words ending in *-cide/-cidal*[7] refer to agents that destroy or permanently inactivate a particular type of microbe; *virucides* inactivate viruses, *bactericides* kill bacteria, and *fungicides* kill fungal hyphae, spores, and yeasts. *Germicides* are chemical agents that destroy pathogenic microorganisms in general.

[1]From Greek *a*, meaning not, and *sepsis*, meaning decay.
[2]From Latin *dis*, meaning reversal, and *inficere*, meaning to corrupt.
[3]From Greek *anti*, meaning against, and *sepsis*, meaning putrefaction.
[4]From Latin *sanitas*, meaning healthy.
[5]Named for Louis Pasteur, inventor of the process.
[6]Greek, meaning to stand—that is, to remain relatively unchanged.
[7]From Latin *cidium*, meaning a slaying.

Table 9.1 summarizes the terminology used to describe the control of microbial growth.

CRITICAL **THINKING**

A student inoculates *Escherichia coli* into two test tubes containing the same sterile liquid medium, except the first tube also contains a drop of a chemical with an antimicrobial effect. After 24 hours of incubation, the first tube remains clear while the second tube becomes cloudy with bacteria. Design an experiment to determine if this amount of the antimicrobial chemical is *bacteriostatic* or *bactericidal* against *E. coli*.

Microbial Death Rates

Learning Objective

✓ Define *microbial death rate*, and describe its significance in microbial control.

Scientists define **microbial death** as the permanent loss of reproductive ability under ideal environmental conditions. One technique for evaluating the efficacy of an antimicrobial agent is to calculate the **microbial death rate,** which is usually found to be constant over time for any particular microorganism under a particular set of conditions **(Figure 9.1)**. Suppose for example that a scientist treats a broth containing 1 billion (10^9) microbes with an agent that kills 90% of them in 1 minute. The most susceptible cells die first, leaving 100 million (10^8) hardier cells after the first minute. After another minute of treatment, another 90% die, leaving 10 million (10^7) cells that have even greater resistance to and require longer exposure to the agent before they die. Notice that in this case, each full minute decreases the number of living cells 10-fold. The broth will be sterile when all the cells are dead. When these results are plotted on

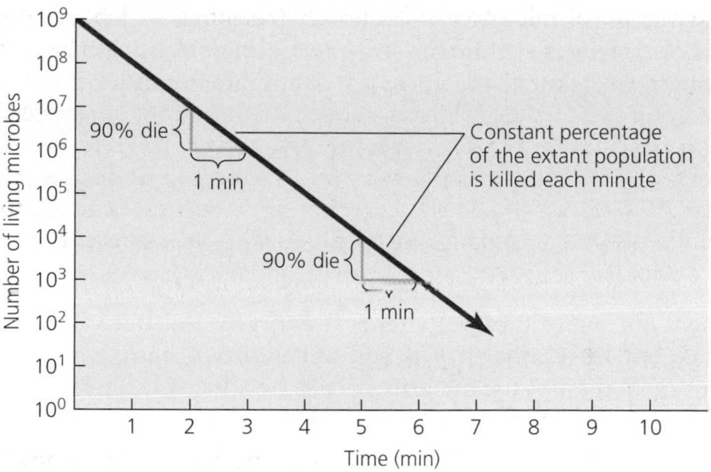

▲ **Figure 9.1 A plot of microbial death rate.** Microbicidal agents do not simultaneously kill all cells. Rather they kill a constant percentage of cells over time—in this case 90% per minute. On this semilogarithmic graph, a constant death rate is indicated by a straight line. *How many minutes are required for sterilization in this case?*

Figure 9.1 *For this microbe under these conditions, this microbicidal agent requires 9 minutes to achieve sterilization.*

TABLE 9.1 Terminology of Microbial Control

Term	Definition	Examples	Comments
Antisepsis	Reduction in the number of microorganisms and viruses, particularly potential pathogens, on living tissue	Iodine; alcohol	Antiseptics are frequently disinfectants whose strength has been reduced to make them safe for living tissues.
Aseptic	Refers to an environment or procedure free of pathogenic contaminants	Preparation of surgical field; handwashing; flame sterilization of laboratory equipment	Scientists, laboratory technicians, and health care workers routinely follow standardized aseptic techniques.
-cide **-cidal**	Suffixes indicating destruction of a type of microbe	Bactericide; fungicide; germicide; virucide	Germicides include ethylene oxide, propylene oxide, and aldehydes.
Degerming	Removal of microbes by mechanical means	Handwashing; alcohol swabbing at site of injection	Chemicals play a secondary role to the mechanical removal of microbes.
Disinfection	Destruction of most microorganisms and viruses on nonliving tissue	Phenolics; alcohols; aldehydes; soaps	The term is used primarily in relation to pathogens.
Pasteurization	Use of heat to destroy pathogens and reduce the number of spoilage microorganisms in foods and beverages	Pasteurized milk and fruit juices	Heat treatment is brief to reduce alteration of taste and nutrients; microbes still remain and eventually cause spoilage.
Sanitization	Removal of pathogens from objects to meet public health standards	Washing tableware in scalding water in restaurants	Standards of sanitization vary among governmental jurisdictions.
-stasis **-static**	Suffixes indicating inhibition, but not complete destructhon, of a type of microbe	Bacteriostatic; fungistatic; virustatic	Germistatic agents include some chemicals, refrigeration, and freezing.
Sterilization	Destruction of all microorganisms and viruses in or on an object	Preparation of microbiological culture media and canned food	Typically achieved by steam under pressure, incineration, or ethylene oxide gas.

a semilogarithmic graph—in which the y-axis is logarithmic, and the x-axis is arithmetic—the plot of microbial death rate is a straight line; that is, the microbial death rate is constant.

Action of Antimicrobial Agents

Learning Objective

✓ Describe how antimicrobial agents act against cell walls, cytoplasmic membranes, proteins, and nucleic acids.

There are many types of chemical and physical microbial controls, but their modes of action fall into two basic categories: those that disrupt the integrity of cells by adversely altering their cell walls or cytoplasmic membranes, and those that interrupt cellular metabolism and reproduction by interfering with the structures of proteins and nucleic acids.

Alteration of Cell Walls and Membranes

As we saw in Chapter 3, a cell wall maintains cellular integrity by counteracting the effects of osmosis when the cell is in a hypotonic solution. If the wall is disrupted by physical or chemical agents, it no longer prevents the cell from bursting as water moves into the cell by osmosis.

Beneath a cell wall, the cytoplasmic membrane essentially acts as a bag that contains the cytoplasm and controls the passage of chemicals into and out of the cell. Extensive damage to a membrane's proteins or phospholipids by any physical or chemical agent allows the cellular contents to leak out—which, if not immediately repaired, causes death.

In enveloped viruses, the envelope is a membrane composed of proteins and phospholipids that is responsible for the attachment of the virus to its target cell; thus damage to the envelope by physical or chemical agents fatally interrupts viral replication. The lack of an envelope in nonenveloped viruses accounts for their greater tolerance of harsh environmental conditions, including antimicrobial agents.

Damage to Proteins and Nucleic Acids

Proteins regulate cellular metabolism, function as enzymes in most metabolic reactions, and form structural components in membranes and cytoplasm. As we have seen, a protein's function depends on an exact three-dimensional shape, which is maintained by hydrogen and disulfide bonds between amino acids. When these bonds are broken by extreme heat or certain chemicals, the protein's shape changes (see Figure 5.8). Such *denatured* proteins cease to function, bringing about cellular death.

Chemicals, radiation, and heat can also alter and even destroy nucleic acids. Given that the genes of a cell or virus are composed of nucleic acids, disruption of these molecules can produce fatal mutations. Additionally, that portion of a ribosome that actually catalyzes the synthesis of proteins is a *ribozyme*—that is, an enzymatic RNA molecule—so physical or chemical agents that interfere with nucleic acids also stop protein synthesis.

CRITICAL **THINKING**

Would you expect Gram-negative bacteria or Gram-positive bacteria to be more susceptible to antimicrobial chemicals that act against cell walls? Explain your answer, which you should base solely upon the nature of the cells' walls (see Figure 3.15).

Scientists and health care workers have at their disposal many chemical and physical agents to control microbial growth and activity. In the next section we consider the factors and conditions that should be considered in choosing a particular control method, as well as some ways to evaluate a method's effectiveness.

The Selection of Microbial Control Methods

Ideally, agents used for the control of microbes should be inexpensive, fast-acting, and stable during storage. Further, a perfect agent would control the growth and reproduction of every type of microbe while being harmless to humans, animals, and objects. Unfortunately, such ideal products and procedures do not exist—every agent has limitations and disadvantages. In the next two subsections we consider the factors that affect the efficacy of antimicrobial methods and some ways to evaluate disinfectants.

Factors Affecting the Efficacy of Antimicrobial Methods

Learning Objectives

✓ List factors to consider in selecting a microbial control method.

✓ Identify the three most-resistant groups of microbes, and explain why they are resistant to many antimicrobial agents.

✓ Discuss environmental conditions that can influence the effectiveness of antimicrobial agents.

In each situation, microbiologists, laboratory personnel, and medical staff must consider at least three factors: the nature of the sites to be treated, the degree of susceptibility of the microbes involved, and the environmental conditions that pertain.

Site to Be Treated

In many cases, the choice of an antimicrobial method depends on the nature of the site to be treated. For example, harsh chemicals and extreme heat cannot be used on humans, animals, and fragile objects such as artificial heart valves and plastic utensils. Moreover, when performing medical procedures, medical personnel must choose a method and level of microbial control based on the site of the procedure, because the site greatly affects the potential for subsequent infection. For example, the use of medical instruments that penetrate the outer defenses of the body such as needles and scalpels carries a greater potential

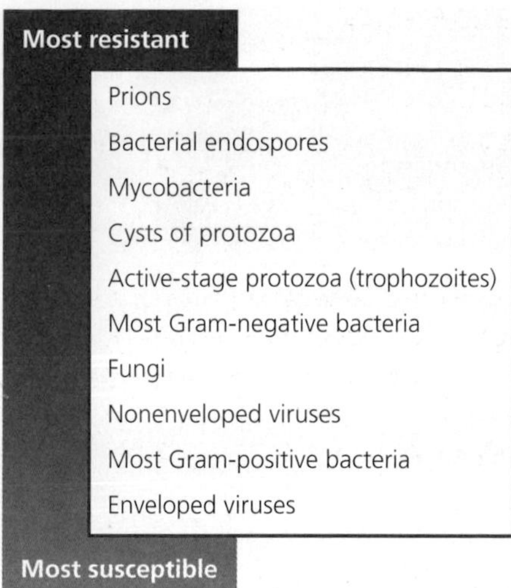

▲ **Figure 9.2 Relative susceptibilities of microbes to antimicrobial agents.** *Why are nonenveloped viruses generally more resistant than enveloped viruses?*

Figure 9.2 *A phospholipid envelope is typically more fragile than a protein capsid.*

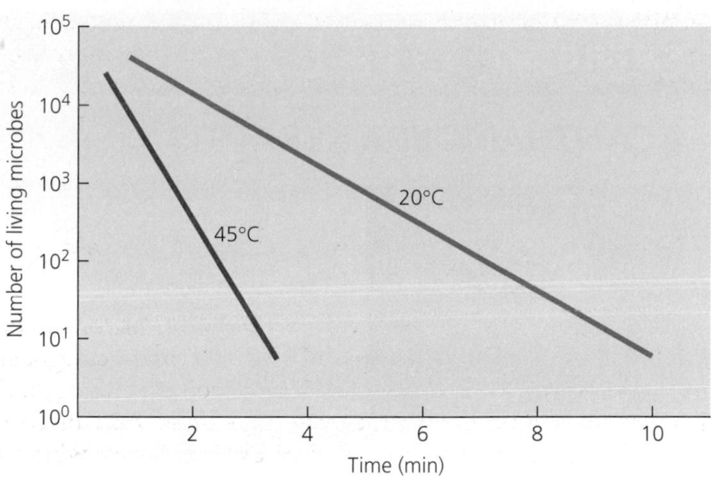

▲ **Figure 9.3 Effect of temperature on the efficacy of an antimicrobial chemical.** This semilogarithmic graph shows that the microbial death rate is higher at higher temperatures; to kill the same number of microbes, this disinfectant required only 4 minutes at 45°C, but 12 minutes at 20°C.

for infection, so they must be sterilized, whereas items that contact only the surface of a mucous membrane or the skin may be disinfected. In the latter case sterilization is required only if the patient is immunocompromised.

Relative Susceptibility of Microorganisms

Though microbial death rate is usually constant for a particular agent acting against a single microbe, death rates do vary—sometimes dramatically—among microorganisms and viruses. Microbes fall along a continuum from most susceptible to most resistant to antimicrobial agents. For example, *enveloped* viruses, such as HIV, are more susceptible to antimicrobial agents and heat than are *nonenveloped viruses,* such as poliovirus, because viral envelopes are more easily disrupted than the protein coats of nonenveloped viruses. The relative susceptibility of microbes to antimicrobial agents is illustrated in **Figure 9.2**.

Often, scientists and medical personnel select a method to kill the hardiest microorganisms present, assuming that such a treatment will kill more fragile microbes as well. The most resistant microbes include the following:

- Bacterial endospores. The endospores of *Bacillus* (ba-sil'ŭs) and *Clostridium* (klos-trid'ē-ŭm) are the most resistant forms of life. They can survive environmental extremes of temperature and acidity, and many chemical disinfectants. For example, endospores can survive more than 20 years in 70% alcohol, and scientists have recovered viable endospores that were embalmed with Egyptian mummies thousands of years ago.
- Species of *Mycobacterium.* The cell walls of members of this genus, such as *Mycobacterium tuberculosis* (mī'kō-bak-tēr'-ē-ŭm too-ber-kyū-lō'sis), contain large amounts of waxy lipids. The wax allows these bacteria to survive drying and protects them from most water-based chemicals; therefore, medical personnel must use strong disinfectants or heat to treat whatever comes into contact with tuberculosis patients, including utensils, equipment, and patients' rooms.
- Cysts of protozoa. A protozoan cyst's wall prevents entry of most disinfectants, protects against drying, and shields against radiation and heat.

Prions, infectious proteins that cause degenerative diseases of the brain, are more resistant than any living thing.

The effectiveness of germicides can be classified as high, intermediate, or low depending on their proficiency in inactivating or destroying microorganisms on medical instruments that cannot be sterilized with heat. *High-level germicides* kill all pathogens, including bacterial endospores. Health care professionals use them to sterilize invasive instruments such as catheters, implants, and parts of heart-lung machines. *Intermediate-level germicides* kill fungal spores, protozoan cysts, viruses, and pathogenic bacteria, but not bacterial endospores. They are used to disinfect instruments that come in contact with mucous membranes but are noninvasive, such as respiratory equipment and endoscopes. *Low-level germicides* eliminate vegetative bacteria, fungi, protozoa, and some viruses; they are used to disinfect items that only contact the skin of patients, such as furniture and electrodes.

Environmental Conditions

Temperature and pH affect microbial death rates and the efficacy of antimicrobial methods. Warm disinfectants, for example, generally work better than cool ones because chemicals react faster at higher temperatures (**Figure 9.3**). The antimicrobial effect of heat is enhanced by acidic conditions. Some chemical disinfectants such as household chlorine bleach are more effective at low pH.

Organic materials such as fat, feces, vomit, blood, and the intercellular secretions in biofilms interfere with the penetration

EMERGING DISEASES

ACANTHAMOEBA KERATITIS

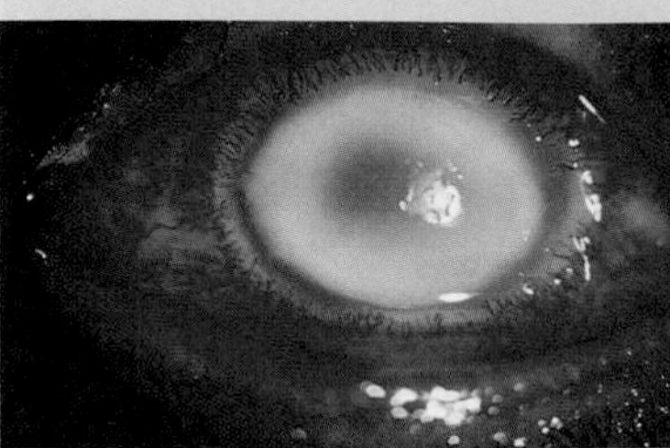

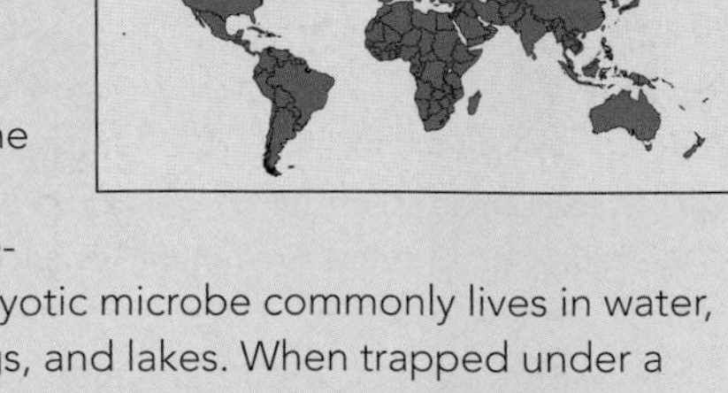

Greg liked the lake; in fact, his girlfriend suggested he was more fish than man. They spent all of their free time swimming, water skiing, diving, and sunbathing, until Greg met *Acanthamoeba*.

Greg noticed something was wrong when his right eye began to hurt, turned red, and he could not stand to be outside because the light was too bright. Two days later, the pain was excruciating, like nothing Greg had ever experienced. It felt as if someone was pounding pieces of broken glass into the front of his eye while quickly inserting a thousand tiny needles into the back of the eye. With his eye swelled shut and tears flowing down his face, Greg sought medical aid.

The doctor diagnosed *Acanthamoeba* keratitis, which is inflammation of the covering of the eye (the cornea) caused by a single-celled amoeba. This eukaryotic microbe commonly lives in water, including rivers, hot springs, and lakes. When trapped under a contact lens, the amoeba can penetrate the eye to cause keratitis. Very occasionally, *Acanthamoeba* may also enter the body through the nasal mucous membrane or through a cut in the skin. It has become an emerging menace in our modern society because it can live in hot tubs, pools, shower heads, and sink taps.

The physician prescribed a solution of antiseptic agent, which Greg had to drop into his eyes every 30 minutes, day and night, for three weeks. The treatment is painful and time consuming, but at least Greg retained his sight without having to receive a corneal transplant. He also learned to remove his contact lenses at the lake! For more about *Acanthamoeba* keratitis, see p. 654.

Track *Acanthamoeba* keratitis online by going to the Study Area at www.masteringmicrobiology.com.

of heat, chemicals, and some forms of radiation, and in some cases these materials inactivate chemical disinfectants. For this reason, it is important to clean objects before sterilization or disinfection so that antimicrobial agents can thoroughly contact all the object's surfaces.

Methods for Evaluating Disinfectants and Antiseptics

Learning Objective

✓ Compare and contrast four methods used to measure the effectiveness of disinfectants and antiseptics.

With few exceptions, higher concentrations and fresher solutions of a disinfectant are more effective than more dilute, older solutions. We have also seen that longer exposure times ensure the deaths of more microorganisms. However, anyone using disinfectants must consider whether higher concentrations and longer exposures may damage an object or injure a patient.

Scientists have developed several methods to measure the efficacy of antimicrobial agents. These include the phenol coefficient, the use-dilution test, the Kelsey-Sykes capacity test, and the in-use test.

Phenol Coefficient

Recall from Chapter 1 that Lister used phenol (also known as carbolic acid) as an antiseptic during surgery in the late 1800s. Since then, researchers have evaluated the efficacy of various disinfectants and antiseptics by calculating a ratio that compares a given agent's ability to control microbes to that of phenol under standardized conditions. This ratio is referred to as the **phenol coefficient.** A phenol coefficient greater than 1.0 indicates that an agent is more effective than phenol, and the larger the ratio, the greater the effectiveness. For example, *chloramine,* a mixture of chlorine and ammonia, has a phenol coefficient of 133.0 when used against the bacterium *Staphylococcus aureus* (staf′i-lō-kok′ŭs o′rē-ŭs), and a phenol coefficient of 100.0 when used against *Salmonella enterica* (sal′mŏ-nel′ă en-ter′i-kă). This indicates that chloramine is at least 133 times more effective than phenol against *Staphylococcus* but only 100 times more effective against *Salmonella.* Measurement of an agent's phenol coefficient has been replaced by newer methods because scientists have developed disinfectants and antiseptics much more effective than phenol.

CRITICAL **THINKING**

What is the phenol coefficient of phenol when used against *Staphylococcus?*

Use-Dilution Test

Another method for measuring the efficacy of disinfectants and antiseptics against specific microbes is the **use-dilution test.** In this test, a researcher dips several metal cylinders into broth cultures of bacteria and briefly dries them at 37°C. The bacteria used in the standard test are *Pseudomonas aeruginosa* (soo-dō-mō′nas ā-roo-ji-nō′să), *Salmonella enterica* serotype Choleraesuis (sal′mŏ-nel′ă en-ter′i-kă kol-er-a-su′is), and *S. aureus.* The researcher then immerses each contaminated cylinder into a different dilution of the disinfectants being evaluated. After 10 minutes, each cylinder is removed, rinsed with water to remove excess chemical, and placed into a fresh tube of sterile medium for 48 hours of incubation. The most effective agent is the one that entirely prevents microbial growth at the highest dilution.

The use-dilution test is the current standard test in the United States, though it was developed several decades ago, before the appearance of many of today's pathogens, including hepatitis C virus, HIV, and antibiotic-resistant bacteria and protozoa. Moreover, the disinfectants in use at the time were far less powerful than many used today. Some government agencies have expressed concern that the test is neither accurate, reliable, nor relevant; therefore, the American Official Analytical Chemists are developing a new standard procedure for use in the United States.

Kelsey-Sykes Capacity Test

The **Kelsey-Sykes capacity test** is the standard alternative assessment approved by the European Union to determine the capacity of a given chemical to inhibit bacterial growth. In this test, researchers add a suspension of a bacterium such as *P. aeruginosa* or *S. aureus* to a suitable concentration of the chemical being tested. Then at predetermined times, they move samples of the mixture into growth medium containing a disinfectant deactivator. After incubation for 48 hours, turbidity in the medium indicates that bacteria survived treatment. Lack of turbidity, indicating lack of bacterial reproduction, reveals the minimum time required for the disinfectant to be effective.

In-Use Test

Though phenol coefficient, use-dilution, and Kelsey-Sykes capacity tests can be beneficial for initial screening of disinfectants, they can also be misleading. These types of evaluation are measures of effectiveness under controlled conditions against one, or at most a few, species of microbes, but disinfectants are generally used in various environments against a diverse population of organisms that are often associated with one another in complex biofilms affording mutual protection.

A more realistic (though more time-consuming) method for determining the efficacy of a chemical is called an **in-use test.** In this procedure, swabs are taken from actual objects, such as operating room equipment, both before and after the application of a disinfectant or an antiseptic. The swabs are then inoculated into appropriate growth media, which after incubation are examined for microbial growth. The in-use test allows a more accurate determination of the proper strength and application procedure of a given disinfection agent for each specific situation.

Now that we have studied the terminology and general principles of microbial control, we turn our attention to the actual physical and chemical agents available to scientists, medical personnel, and the general public to control microbial growth.

Physical Methods of Microbial Control

Learning Objective

✓ Describe five types of physical methods of microbial control.

Physical methods of microbial control include exposure of the microbes to extremes of heat and cold, desiccation, filtration, osmotic pressure, and radiation.

Heat-Related Methods

Learning Objectives

✓ Discuss the advantages and disadvantages of using moist heat in an autoclave and dry heat in an oven for sterilization.

✓ Explain the use of *Bacillus stearothermophilus* endospores in sterilization techniques.

✓ Explain the importance of pasteurization, and describe three different pasteurization methods.

Heat is one of the older and more common means of microbial control. High temperatures denature proteins, interfere with the integrity of cytoplasmic membranes and cell walls, and disrupt the function and structure of nucleic acids. Heat can be used for sterilization, in which case all cells and viruses are deactivated, or for commercial preparation of canned goods. In so-called commercial sterilization, hyperthermophilic prokaryotes remain viable but are harmless because they cannot grow at the normal (room) temperatures in which canned foods are stored.

Though microorganisms vary in their susceptibility to heat, it can be an important agent of microbial control. As a result, scientists have developed concepts and terminology to convey these differences in susceptibility. **Thermal death point** is the lowest temperature that kills all cells in a broth in 10 minutes, while **thermal death time** is the time it takes to completely sterilize a particular volume of liquid at a set temperature.

As we have discussed, cell death occurs logarithmically. When measuring the effectiveness of heat sterilization, researchers calculate the **decimal reduction time (D),** which is the time required to destroy 90% of the microbes in a sample (**Figure 9.4**). This concept is especially useful to food processors because they must heat foods to eliminate all the endospores of anaerobic *Clostridium botulinum* (bo-tū-lī'num), which could germinate and produce botulism toxin inside sealed cans. The standard in food processing is to apply heat such that a population of 10^{12} *C. botulinum* endospores is reduced to 10^0 (that is, 1) endospore (a 12-fold reduction), which leaves only a very small chance that any particular can of food contains an endospore. Researchers have calculated that the D value for *C. botulinum* endospores at 121°C is 0.204 minute, so it takes 2.5 minutes (0.204×12) to reduce 10^{12} endospores to 1 endospore.

Moist Heat

Moist heat, which is commonly used to disinfect, sanitize, sterilize, and pasteurize, kills cells by denaturing proteins and destroying cytoplasmic membranes. Moist heat is more effective

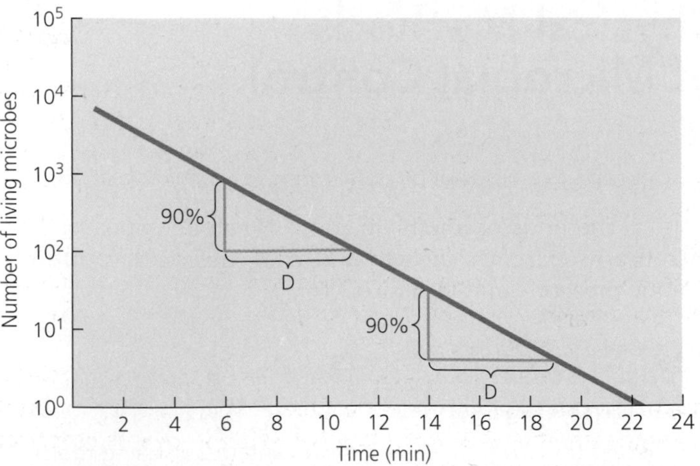

▲ **Figure 9.4 Decimal reduction time (D) as a measure of microbial death rate.** D is defined as the time it takes to kill 90% of a microbial population. Note that D is a constant that is independent of the initial density of the population. *What is the decimal reduction time of this heat treatment against this organism? What is the thermal death time?*

Figure 9.4 D = 5 minutes; thermal death time = 22.5 minutes.

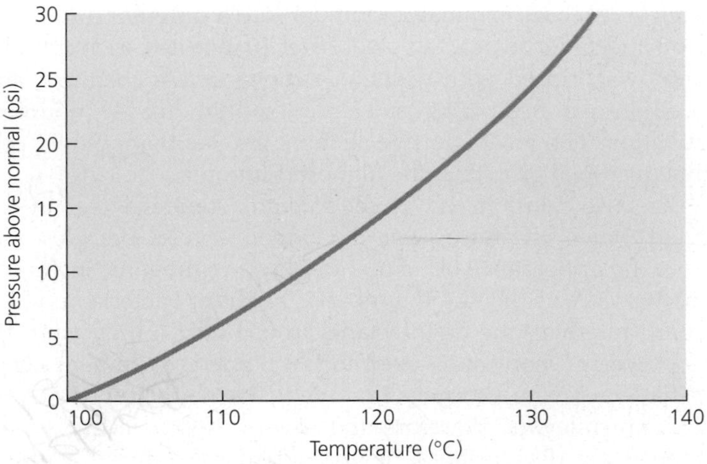

▲ **Figure 9.5 The relationship between temperature and pressure.** Note that higher temperatures—and in consequence, greater antimicrobial action—are associated with higher pressures. *Ultrahigh-temperature pasteurization of milk requires a temperature of 134°C; what pressure must be applied to the milk to achieve this temperature?*

Figure 9.5 29 psi.

in microbial control than dry heat because water is a better conductor of heat than air. This is easily seen with a kitchen example: you can safely stick your hand into an oven at 350°F for a few moments, but putting it into boiling water at the lower temperature of 212°F would burn you severely.

The first method we consider for controlling microbes using moist heat is boiling.

Boiling Boiling kills the vegetative cells of bacteria and fungi, the trophozoites of protozoa, and most viruses within 10 minutes at sea level. Contrary to popular belief, water at a rapid boil is no hotter than that at a slow boil; boiling water at normal atmospheric pressure cannot exceed boiling temperature (100°C at sea level) because escaping steam carries excess heat away. It is impossible to boil something more quickly simply by applying more heat; the added heat is carried away by the escaping steam. Boiling *time* is the critical factor. Further, it is important to realize that water boils at lower temperatures at higher elevations because atmospheric pressure is lower; thus a longer boiling time is required in Denver than in Los Angeles to get the same antimicrobial effect.

Bacterial endospores, protozoan cysts, and some viruses (such as hepatitis viruses) can survive boiling at sea level for many minutes or even hours. In fact, because bacterial endospores can withstand boiling for more than 20 hours, boiling is not recommended when true sterilization is required. Boiling is effective for sanitizing restaurant tableware or disinfecting baby bottles.

Autoclaving Practically speaking, true sterilization using heat requires higher temperatures than that of boiling water. To achieve the required temperature, pressure is applied to boiling water to prevent the escape of heat in steam. The reason that applying pressure succeeds in achieving sterilization is that the temperature at which water boils (and steam is formed) increases as pressure increases (**Figure 9.5**). Scientists and medical personnel routinely use a piece of equipment called an *autoclave* to sterilize chemicals and objects that can tolerate moist heat. Alternate techniques (discussed shortly) must be used for items such as some plastics and vitamins that are damaged by heat or water.

An **autoclave** consists of a pressure chamber, pipes to introduce and evacuate steam, valves to remove air and control pressure, and pressure and temperature gauges to monitor the procedure (**Figure 9.6**). As steam enters an autoclave chamber, it forces air out, raises the temperature of the contents, and increases the pressure, until a set temperature and pressure are reached.

Scientists have determined that a temperature of 121°C, which requires the addition of 15 pounds per square inch (psi)[8] of pressure above that of normal air pressure (see Figure 9.5), destroys all microbes in a small volume in about 10 minutes. Typically, an autoclave holds the pressure and temperature for 15 minutes to provide a margin of safety. The presence of large volumes of liquids or solids to be sterilized slows the process because they require more time for heat to penetrate. Thus, it requires more time to sterilize 1 liter of fluid in a flask than the same volume of fluid distributed into smaller tubes. Autoclaving requires extra time to sterilize solid substances such as meat because it takes longer for heat to penetrate to their centers.

Sterilization in an autoclave requires that steam be able to contact all liquids and surfaces that might be contaminated with microbes; therefore, solid objects must be wrapped in porous cloth or paper, not sealed in plastic or aluminum foil, which are impermeable to steam. Containers of liquids must be sealed

[8]The Standard International (SI) equivalent of 1 psi is 6.9×10^3 pascals.

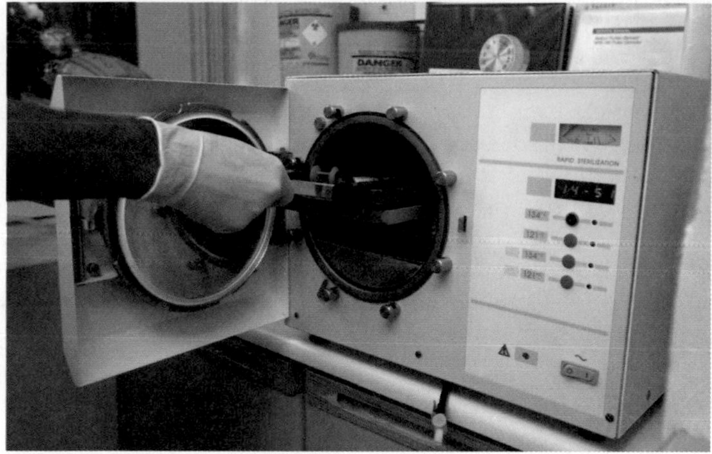

(a)

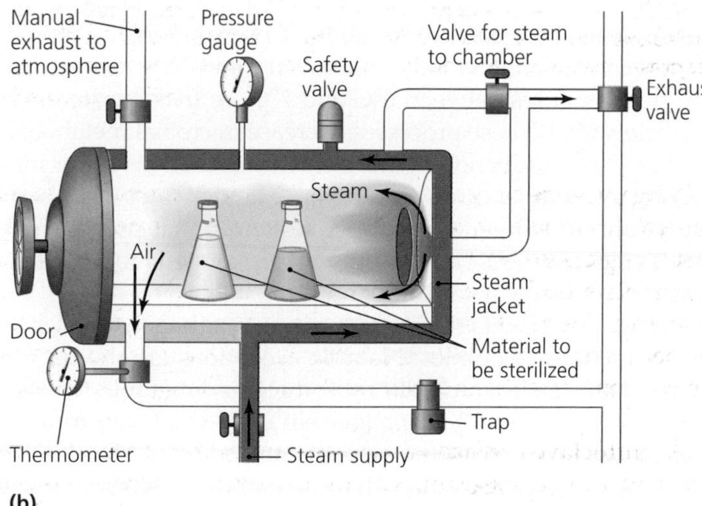

(b)

▲ **Figure 9.6 An autoclave.** **(a)** A photo of a laboratory autoclave. **(b)** A schematic of an autoclave, showing how it functions.

loosely enough to allow steam to circulate freely, and all air must be forced out by steam. Since steam is lighter than air, it cannot force air from the bottom of an empty vessel; therefore, empty containers must be tipped so that air can flow out of them.

Scientists use several means to ensure that an autoclave has sterilized its contents. A common one is a chemical that changes color when the proper combination of temperature and time have been reached. Often such a color indicator is impressed in a pattern on tape or paper so that the word *sterile* or a pattern or design appears. Another technique uses plastic beads that melt when proper conditions are met.

A biological indicator of sterility uses endospores of the bacterium *Bacillus stearothermophilus* (ba-sil′ŭs ste-rō-ther-ma′fil-ŭs) impregnated into tape. After autoclaving, the tape is aseptically inoculated into sterile broth. If no bacterial growth appears, the original material is considered sterile. In a variation on this technique, the endospores are on a strip in one compartment of a vial that also includes a growth medium containing a pH color indicator. After autoclaving, a barrier between the two compartments is broken, putting the endospores into contact with the medium **(Figure 9.7)**. In this case, the absence of a color change after incubation indicates sterility.

CRITICAL **THINKING**

Where should you place a sterilization indicator within an autoclave? Explain your reasoning.

Pasteurization In Chapter 1 we learned that Louis Pasteur developed a method of heating beer and wine just enough to destroy the microorganisms that cause spoilage without raising the temperature so much that the taste was ruined. Today, pasteurization is also used to kill pathogens in milk, ice cream, yogurt, and fruit juices. *Brucella melitensis* (broo-sel′lă me-li-ten′sis), *Mycobacterium bovis* (bō′vis), and *Escherichia coli* (esh-ĕ-rik′ē-ă kō′lē), the causative agents of undulant fever, bovine tuberculosis, and one kind of diarrhea respectively, are controlled in this manner.

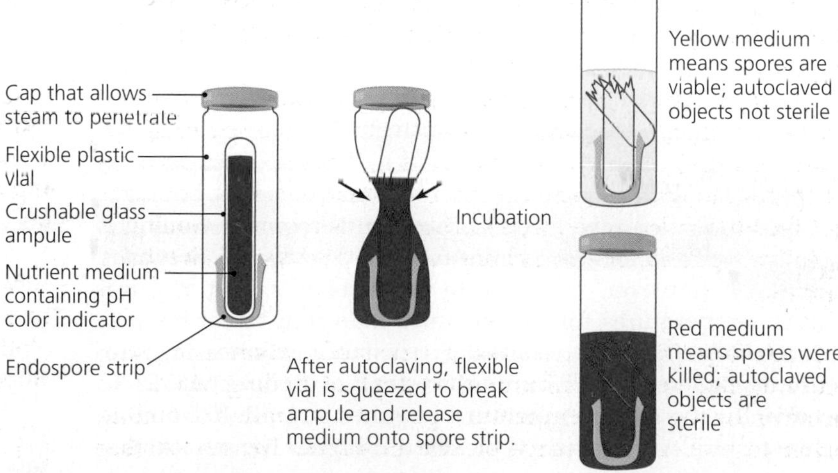

▶ **Figure 9.7 Sterility indicators.** A commercial endospore-test ampule, which is included among objects to be sterilized. After autoclaving is complete, the medium, which contains a pH color indicator, is released onto the endospore strip by breaking the ampule. If the endospores are still alive, their metabolic wastes lower the pH, changing the color of the medium.

TABLE 9.2 Moist Heat Treatments of Milk

Process	Treatment
Historical (batch) pasteurization	63°C for 30 minutes
Flash pasteurization	72°C for 15 seconds
Ultrahigh-temperature pasteurization	134°C for 1 second
Ultrahigh-temperature sterilization	140°C for 1–3 seconds

Pasteurization is not sterilization. *Thermoduric* and *thermophilic*—heat-tolerant and heat-loving—prokaryotes survive pasteurization, but they do not cause spoilage over the relatively short times during which properly refrigerated and pasteurized foods are stored before consumption. In addition, such prokaryotes are generally not pathogenic.

The combination of time and temperature required for effective pasteurization varies with the product. Because milk is the most familiar pasteurized product, we consider the pasteurization of milk in some detail. Historically, milk was pasteurized by the *batch method* for 30 minutes at 63°C, but most milk processors today use a high-temperature, short-time method known as *flash pasteurization*, in which milk flows through heated tubes that raise its temperature to 72°C for only 15 seconds. This treatment effectively destroys all pathogens. *Ultrahigh-temperature pasteurization* heats the milk to 134°C for only 1 second, but some consumers claim it adversely affects the taste.

Ultrahigh-Temperature Sterilization The dairy industry and other food processors can also use *ultrahigh-temperature sterilization*, which involves flash heating milk or other liquids to rid them of all living microbes. The process involves passing the liquid through superheated steam at 140°C for 1–3 seconds, and then cooling it rapidly. Treated liquids can be stored indefinitely at room temperature without microbial spoilage, though after months of storage chemical degradation results in flavor changes. Small packages of dairy creamer served in restaurants are often sterilized by the ultrahigh-temperature method. Table 9.2 summarizes the dairy industry's use of moist heat for controlling microbes in milk.

Dry Heat

For substances such as powders and oils that cannot be sterilized by boiling or with steam, or for materials that can be damaged by repeated exposure to steam (such as some metal objects), sterilization can be achieved by the use of dry heat, as occurs in an oven.

Hot air is an effective sterilizing agent because it denatures proteins and fosters the oxidation of metabolic and structural chemicals; however, in order to sterilize, dry heat requires higher temperatures for longer times than moist heat because dry heat penetrates more slowly. For instance, whereas an autoclave needs less than 15 minutes to sterilize an object at 121°C, an oven at the same temperature requires at least 16 hours to achieve sterility. Scientists typically use higher temperatures—171°C for 1 hour, or 160°C for 2 hours—to sterilize objects in an oven, but objects made of rubber, paper, and many types of plastic oxidize rapidly (combust) under these conditions.

Complete incineration is the ultimate means of sterilization. As part of standard aseptic technique in microbiological laboratories, inoculating loops are sterilized by heating them in the flame of a Bunsen burner or with an electric heating coil until they glow red (about 1500°C). Health care workers incinerate contaminated dressings, bags, and paper cups; field epidemiologists incinerate the carcasses of animals that have diseases such as anthrax or bovine spongiform encephalopathy (mad cow disease).

Refrigeration and Freezing

Learning Objective

✓ Describe the use and importance of refrigeration and freezing in limiting microbial growth.

In many situations, particularly in food preparation and storage, the most convenient method of microbial control is either refrigeration (temperatures between 0°C and 7°C) or freezing (temperatures below 0°C). These processes decrease microbial metabolism, growth, and reproduction because chemical reactions occur more slowly at low temperatures, and because liquid water is not available at subzero temperatures. Note, however, that psychrophilic (cold-loving) microbes can multiply in refrigerated food and spoil its taste and suitability for consumption.

Refrigeration halts the growth of most pathogens, which are predominantly mesophiles. Notable exceptions are the bacteria *Listeria* (lis-tēr′ē-ă), which can reproduce to dangerous levels in refrigerated food, and *Yersinia* (yer-sin′ē-ă), which can multiply in refrigerated blood products and be passed on to blood recipients. Chapters 19 and 20 discuss these pathogens in more detail.

Slow freezing, during which ice crystals have time to form and puncture cell membranes, is more effective than quick freezing in inhibiting microbial metabolism, though microorganisms also vary in their susceptibility to freezing. Whereas the cysts of tapeworms perish after several days in frozen meat, many vegetative bacterial cells, bacterial endospores, and viruses can survive subfreezing temperatures for years. In fact, scientists store many bacteria and viruses in low-temperature freezers at −30°C to −80°C and are able to reconstitute the microbes into viable populations by warming them in media containing proper nutrients. Therefore, we must take care in thawing and cooking frozen food, because it can still contain many pathogenic microbes.

Highlight: Microbes in Sushi? on p. 268 describes how freezing is but one method of microbial control used in the preparation of sushi.

Desiccation and Lyophilization

Learning Objective

✓ Compare and contrast desiccation and lyophilization.

Desiccation, or drying, has been used for thousands of years to preserve such foods as fruits, peas, beans, grain, nuts, and yeast (**Figure 9.8**). Desiccation inhibits microbial growth because metabolism requires liquid water. Drying inhibits the spread of

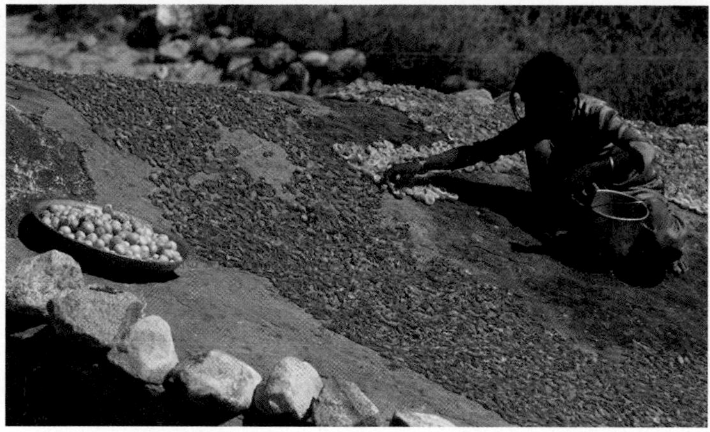

▲ **Figure 9.8 The use of desiccation as a means of preserving apricots in Pakistan.** In this time-honored practice, drying inhibits microbial growth in the food by removing the water that microbes need for metabolism.

most pathogens, including the bacteria that cause syphilis, gonorrhea, and the more common forms of bacterial pneumonia and diarrhea. However, most molds can grow on dried raisins and apricots, which have as little as 16% water content.

Scientists use **lyophilization** (lī-of'i-li-zā'shŭn), a technique combining freezing and drying, to preserve microbes and other cells for many years. In this process, scientists instantly freeze a culture in liquid nitrogen or frozen carbon dioxide (dry ice); then, they subject it to a vacuum that removes frozen water through a process called sublimation, in which the water is transformed directly from a solid to a gas. Lyophilization prevents the formation of large, damaging ice crystals. Although not all cells survive, enough are viable to enable the culture to be reconstituted many years later.

CRITICAL **THINKING**

Why is liquid water necessary for microbial metabolism?

Filtration

Learning Objective

✓ Describe the use of filters for disinfection and sterilization.

Filtration is the passage of a fluid (either a liquid or a gas) through a sieve designed to trap particles—in this case, cells or viruses—and separate them from the fluid. Researchers often use a vacuum to assist the movement of fluid through the filter **(Figure 9.9a)**. Filtration traps microbes larger than the pore size, allowing smaller microbes to pass through. In the late 1800s, filters were able to trap cells, but their pores were too large to trap the pathogens of such diseases as rabies and measles. These pathogens were thus named *filterable viruses,* which today has been shortened to *viruses.*[9] Now, filters with pores small enough to trap even viruses are available, so filtration can be used to sterilize such heat-sensitive materials as ophthalmic solutions, antibiotics, vaccines, liquid vitamins, enzymes, and culture media.

Over the years, filters have been constructed from porcelain, glass, cotton, asbestos, and diatomaceous earth, a substance composed of the innumerable glasslike cell walls of single-celled algae called diatoms. Scientists today typically use thin (only 0.1 mm thick), circular **membrane filters** manufactured of nitrocellulose or plastic and containing specific pore sizes ranging from 25 μm to less than 0.01 μm in diameter **(Figure 9.9b)**. The pores of the latter filters are small enough to trap small viruses and even some large protein molecules. Microbiologists also use filtration to estimate the number of microbes in a fluid by counting the number deposited on the filter after passing a

[9]Latin, meaning poisons.

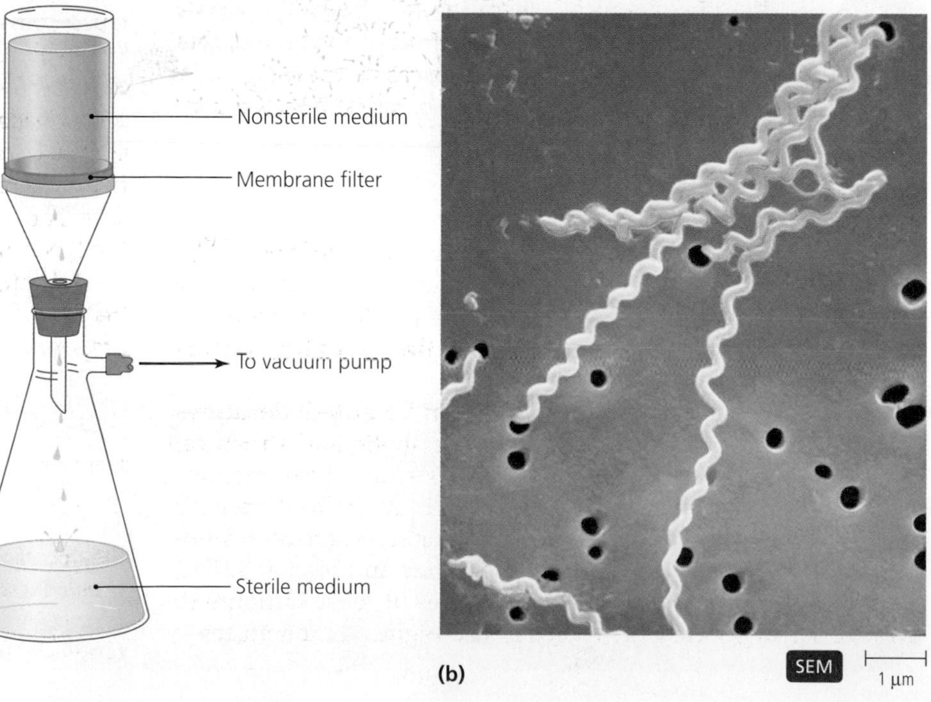

▶ **Figure 9.9 Filtration equipment used for microbial control. (a)** Assembly for sterilization by vacuum filtration. **(b)** Membrane filters composed of various substances and with pores of various sizes can be used to trap diverse microbes, here bacteria known as spirochetes.

TABLE 9.3 Membrane Filters

Pore Size (μm)	Smallest Microbes That Are Trapped
5	Multicellular algae, animals, and fungi
3	Yeasts and larger unicellular algae
1.2	Protozoa and small unicellular algae
0.45	Largest bacteria
0.22	Largest viruses and most bacteria
0.025	Larger viruses and pliable bacteria (mycoplasmas, rickettsias, chlamydias, and some spirochetes)
0.01	Smallest viruses

given volume through the filter (see Figure 6.23). Table 9.3 lists some pore sizes of membrane filters and the microbes they allow through.

Health care and laboratory workers routinely use filtration to prevent airborne contamination by microbes. The use of surgical masks prevents contamination of the environment by microbes exhaled by medical personnel, and cotton plugs in culture vessels prevent contamination of cultures by airborne microbes. Additionally, *high-efficiency particulate air (HEPA) filters* are crucial parts of biological safety cabinets **(Figure 9.10)**, and HEPA filters are mounted in the air ducts of some operating rooms, rooms occupied by patients with airborne diseases such as tuberculosis, and rooms of immunocompromised patients such as burn victims and AIDS patients.

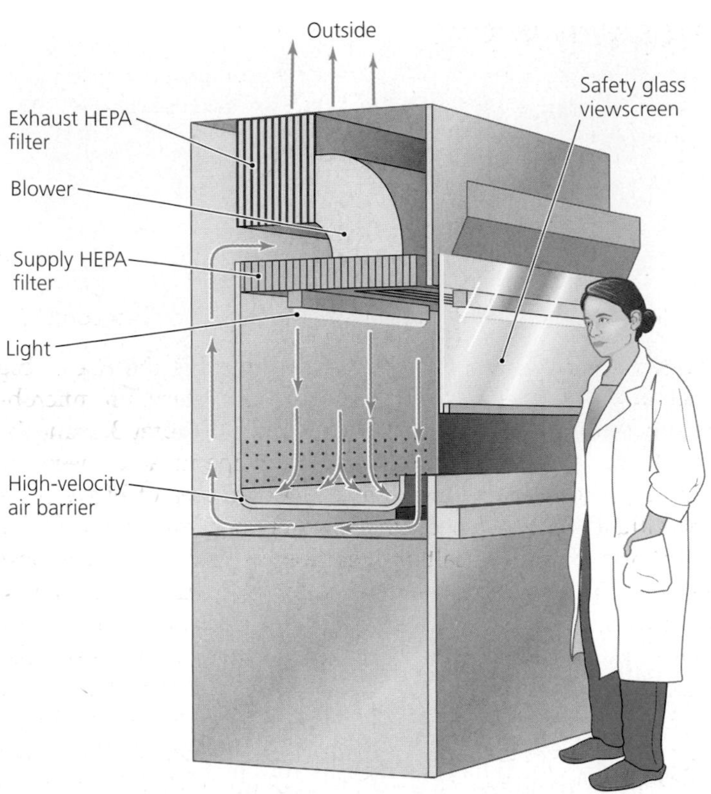

▲ **Figure 9.10 The roles of high-efficiency particulate air (HEPA) filters in biological safety cabinets.** HEPA filters protect workers from exposure to microbes (by maintaining a barrier of filtered air across the opening of the cabinet). Hospital also use HEPA filters in air ducts of operating rooms and of the rooms of highly contagious or immunocompromised patients.

HIGHLIGHT

MICROBES IN SUSHI?

Sushi—it either makes you squirm or salivate! Although technically the term refers to rice, it's generally understood to refer to bite-sized slices of raw fish served with rice. Long a staple of Japanese cuisine, sushi has become popular throughout the United States. But isn't eating raw fish dangerous? What methods of microbial control are applied in the preparation of sushi?

It's true that raw fish can contain harmful microorganisms. Parasitic roundworms called anisakids are commonly found in fish and can cause gastrointestinal symptoms in humans. Accordingly, before fish can be served raw, the Food and Drug Administration requires that it be frozen at 20°C for 7 days, or at −37°C for 15 hours. Unfortunately, freezing does not kill all bacterial or viral pathogens. Consumers should be aware that they always assume some risk of food poisoning caused by such bacteria as *S. aureus*, *E. coli*, and species of *Salmonella* and *Vibrio* whenever they eat any kind of raw food, including sushi, raw oysters, ceviche, or carpaccio.

Even though diners tend to fixate on the safety of raw fish, cooked rice left sitting at room temperature is also vulnerable to the growth of pathogens. To counter this potential problem, sushi rice is prepared with vinegar, which acidifies the rice. At pH values below 4.6, rice becomes too acidic to support the growth of most pathogens. Sushi bars can also reduce the risk posed by pathogens by keeping restaurant temperatures cool. Additionally, it is believed that wasabi, the fiery horseradish-like green paste commonly

eaten with sushi, contains antimicrobial properties, although its antimicrobial action is not well understood.

With these antimicrobial precautions in place, the vast majority of diners consume sushi safely meal after meal (although pregnant women and people with compromised immune systems should avoid all raw seafood). When properly prepared, sushi is beautiful, low in calories, a source of heart-healthy omega-3 fatty acids—and very delicious.

CRITICAL **THINKING**

A virologist needs to remove all bacteria from a solution containing viruses without removing the viruses. What size membrane filter should the scientist use?

Osmotic Pressure

Learning Objective

✓ Discuss the use of hypertonic solutions in microbial control.

Another ancient method of microbial control is the use of high concentrations of salt or sugar in foods to inhibit microbial growth by **osmotic pressure.** As we saw in Chapter 3, osmosis is the net movement of water across a semipermeable membrane (such as a cytoplasmic membrane) from an area of higher water concentration to an area of lower water concentration. Cells in a hypertonic solution of salt or sugar lose water, and the cell desiccates (see Figure 3.20b). The removal of water inhibits cellular metabolism because enzymes are fully functional only in aqueous environments. Thus, osmosis preserves honey, jerky, jams, jellies, salted fish, and some types of pickles from most microbial attacks.

Fungi have a greater ability than bacteria to tolerate hypertonic environments with little moisture, which explains why jelly in your refrigerator may grow a colony of *Penicillium* (pen-i-sil'ē-ŭm) mold but is not likely to grow the bacterium *Salmonella.*

Radiation

Learning Objective

✓ Differentiate between ionizing radiation and nonionizing radiation as they relate to microbial control.

Another physical method of microbial control is the use of **radiation.** There are two types of radiation: particulate radiation and electromagnetic radiation. Particulate radiation consists of high-speed subatomic particles, such as protons, that have been freed from their atoms. Electromagnetic radiation can be defined as energy without mass traveling in waves at the speed of light (3×10^5 km/sec). Electromagnetic energy is released from atoms that have undergone internal changes. The *wavelength* of electromagnetic radiation, defined as the distance between two crests of a wave, ranges from very short gamma rays, through X rays, ultraviolet light, and visible light, to long infrared rays, and finally to very long radio waves (see Figure 4.1). Though they are particles, electrons also have a wave nature with wavelengths that are even shorter than gamma rays.

The shorter the wavelength of an electromagnetic wave, the more energy it carries; therefore, shorter-wavelength radiation is more suitable for microbial control than longer-wavelength radiation, which carries less energy and is less penetrating. Scientists describe all types of radiation as either *ionizing* or *nonionizing* according to its effects on the chemicals within cells.

Ionizing Radiation

Electron beams, gamma rays, and X rays, all of which have wavelengths shorter than 1 nm and are **ionizing radiation** because when they strike molecules, they have sufficient energy

▲ **Figure 9.11 A demonstration of the increased shelf life of food achieved by ionizing radiation.** The circular radura symbol is used in the United States to label irradiated foods.

to eject electrons from atoms, creating ions. Such ions disrupt hydrogen bonding, oxidize double covalent bonds, and create highly reactive hydroxyl radicals (see Chapter 6). These ions in turn denature other molecules, particularly DNA, causing fatal mutations and cell death.

Electron beams are produced by *cathode ray machines.* Electron beams are highly energetic and therefore very effective in killing microbes in just a few seconds, but they cannot sterilize thick objects or objects coated with large amounts of organic matter. They are used to sterilize spices, meats, microbiological plastic ware, and dental and medical supplies such as gloves, syringes, and suturing material.

Gamma rays, which are emitted by some radioactive elements such as radioactive cobalt, penetrate much farther than electron beams but require hours to kill microbes. The FDA has approved the use of gamma irradiation for microbial control in meats, spices, and fresh fruits and vegetables **(Figure 9.11)**. Irradiation with gamma rays kills not only microbes but also the larvae and eggs of insects; it also kills the cells of fruits and vegetables, preventing both microbial spoilage and overripening.

Consumers have been reluctant to accept irradiated food. A number of reasons have been cited, including fear that radiation makes food radioactive, and claims that it changes the taste and nutritive value of foods or produces potentially carcinogenic (cancer-causing) chemicals. Supporters of irradiation reply that gamma radiation passes through food and cannot make it radioactive any more than a dental X ray produces radioactive teeth, and they cite numerous studies that conclude that irradiated foods are tasty, nutritious, and safe.

TABLE 9.4 Physical Methods of Microbial Control

Method	Conditions	Action	Representative Use(s)
Moist heat			
Boiling	10 min at 100°C	Denatures proteins and destroys membranes	Disinfection of baby bottles and sanitization of restaurant cookware and tableware
Autoclaving (pressure cooking)	15 min at 121°C	Denatures proteins and destroys membranes	Autoclave: sterilization of medical and laboratory supplies that can tolerate heat and moisture; pressure cooker: sterilization of canned food
Pasteurization	15 sec at 72°C	Denatures proteins and destroys membranes	Destruction of all pathogens and most spoilage microbes in dairy products, fruit juices, beer, and wine
Ultrahigh-temperature sterilization	1–3 sec at 140°C	Denatures proteins and destroys membranes	Sterilization of dairy products
Dry heat			
Hot air	2 h at 160°C or 1 h at 171°C	Denatures proteins, destroys membranes, oxidizes metabolic compounds	Sterilization of water-sensitive materials such as powders, oils, and metals
Incineration	1 sec at more than 1000°C	Oxidizes everything completely	Sterilization of inoculating loops, flammable contaminated medical waste, and diseased carcasses
Refrigeration	0–7°C	Inhibits metabolism	Preservation of food
Freezing		Inhibits metabolism	Long-term preservation of foods, drugs, and cultures
Desiccation (drying)	Varies with amount of water to be removed	Inhibits metabolism	Preservation of food
Lyophilization (freeze drying)	−196°C for a few minutes while drying	Inhibits metabolism	Long-term storage of bacterial cultures
Filtration	Filter retains microbes	Physically separates microbes from air and liquids	Sterilization of air and heat-sensitive ophthalmic and enzymatic solutions, vaccines, and antibiotics
Osmotic pressure	Exposure to hypertonic solutions	Inhibits metabolism	Preservation of food
Ionizing radiation (electron beams, gamma rays, X rays)	Seconds to hours of exposure (depending on wavelength of radiation)	Destroys DNA	Sterilization of medical and laboratory equipment and preservation of food
Nonionizing radiation (ultraviolet light)	Irradiation with 260-nm-wavelength radiation	Formation of thymine dimers inhibits DNA transcription and replication	Disinfection and sterilization of surfaces and of transparent fluids and gases

X rays travel the farthest through matter, but they have less energy than gamma rays and require a prohibitive amount of time to make them practical for microbial control.

Nonionizing Radiation

Electromagnetic radiation with a wavelength greater than 1 nm does not have enough energy to force electrons out of orbit, so it is **nonionizing radiation.** However, such radiation does contain enough energy to excite electrons and cause them to make new covalent bonds, which can affect the three-dimensional structure of proteins and nucleic acids.

Ultraviolet (UV) light, visible light, infrared radiation, and radio waves are nonionizing radiation. Of these, only UV light has sufficient energy to be a practical antimicrobial agent. Visible light and microwaves (extremely short wavelength radio waves) have little value in microbial control, though microwaves heat food, which can inhibit microbial growth and reproduction if the food gets hot enough.

UV light with a wavelength of 260 nm is specifically absorbed by adjacent pyrimidine nucleotide bases in DNA, causing them to form covalent bonds with each other rather than forming hydrogen bonds with bases in the complementary DNA strand (see Figure 7.24). Such *pyrimidine* dimers distort the shape of DNA, making it impossible for the cell to accurately transcribe or replicate its genetic material. If dimers remain uncorrected, an affected cell may die.

The effectiveness of UV irradiation is tempered by the fact that UV light does not penetrate well. UV light is therefore suitable primarily for disinfecting air, transparent fluids, and the surfaces of objects such as barber's shears and operating tables. Some cities use UV irradiation in sewage treatment. By passing wastewater past banks of UV lights, they reduce the number of bacteria without using chlorine, which might damage the environment.

Table 9.4 summarizes the physical methods of microbial control discussed in the previous pages.

CRITICAL **THINKING**

During the bioterrorist attack in which anthrax endospores were sent through the mail in the fall of 2001, one news commentator suggested that people should iron all their incoming mail with a regular household iron as a means of destroying endospores. Would you agree that this is a good way to disinfect mail? Explain your answer. Which disinfectant methods would be both more effective and practical?

Biosafety Levels

Learning Objective

✓ Describe four levels of biosafety and give examples of microbes handled at each level.

The Centers for Disease Control and Prevention (CDC) has established guidelines for four levels of safety in microbiological laboratories dealing with pathogens. Each level raises personnel and environmental safety by specifying increasingly strict laboratory techniques, use of safety equipment, and design of facilities.

Biosafety Level 1 (BSL-1) is suitable for handling microbes, such as *E. coli,* not known to cause disease in healthy humans. Precautions in BSL-1 are minimal and include handwashing with antibacterial soap and washing surfaces with disinfectants.

BSL-2 facilities are similar to those of BSL-1 but are designed for handling moderately hazardous agents such as hepatitis and influenza viruses and methicillin-resistant *Staphylococcus aureus* (MRSA). Access to BSL-2 labs is limited when work is being conducted, extreme precautions are taken with contaminated sharp objects, and procedures that might produce aerosols are conducted within safety cabinets (see Figure 9.10).

BSL-3 is stricter, requiring that all manipulations be done within HEPA safety cabinets and specifying special design features for the laboratory. These include entry through double sets of doors and ventilation such that air only moves into the room through an open door. Air leaving the room is HEPA-filtered before being discharged outside the room. BSL-3 is designed for experimentation on microbes such as tuberculosis and anthrax bacteria and viruses of yellow fever and Rocky Mountain spotted fever.

The most secure laboratories are BSL-4 facilities, designated for working with dangerous or exotic microbes that cause severe or fatal diseases in humans, such as Ebola, smallpox, and Lassa fever viruses. BSL-4 labs are either separate buildings or are completely isolated from all other areas of their buildings. Entrance and exit is strictly controlled through electronically sealed airlocks with multiple showers, a vacuum room, an ultraviolet light room, and other safety precautions designed to destroy all traces of the biohazard. All air and water entering and leaving the facility are filtered to prevent accidental release. Personnel wear "space suits" supplied with air hoses **(Figure 9.12)**. Suits and the laboratory itself are pressurized such that microbes are swept away from workers.

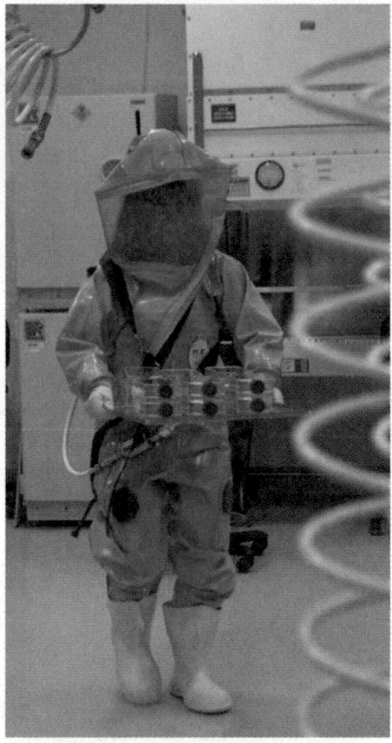

▲ **Figure 9.12** A BSL-4 worker carries Ebola virus cultures.

Chemical Methods of Microbial Control

Learning Objective

✓ Compare and contrast nine major types of antimicrobial chemicals, and discuss the positive and negative aspects of each.

Although physical agents are sometimes used for disinfection, antisepsis, and preservation, more often chemical agents are used for these purposes. As we have seen, chemical agents act to adversely affect microbes' cell walls, cytoplasmic membranes, proteins, or DNA. As with physical agents, the effect of a chemical agent varies with temperature, length of exposure, and the amount of contaminating organic matter in the environment. The effect also varies with pH, concentration, and freshness of the chemical. Chemical agents tend to destroy or inhibit the growth of enveloped viruses and the vegetative cells of bacteria, fungi, and protozoa more than fungal spores, protozoan cysts, or bacterial endospores. The latter are particularly resistant to chemical agents, as demonstrated by numerous failed attempts to decontaminate a United States Senate office building of anthrax endospores sent there by bioterrorists in 2001.

In the following sections we discuss nine major categories of antimicrobial chemicals used as antiseptics and disinfectants: *phenols, alcohols, halogens, oxidizing agents, surfactants, heavy metals, aldehydes, gaseous agents,* and *enzymes.* Some chemical agents combine one or more of these. Additionally, researchers and food processors sometimes use antimicrobials—substances normally used to treat diseases—as disinfectants.

▲ **Figure 9.13 Phenol and phenolics. (a)** Phenol, a naturally occurring molecule that is also called carbolic acid. **(b)** Phenolics, which are compounds synthesized from phenol, have greater antimicrobial efficacy with fewer side effects. Bisphenols are paired, covalently linked phenolics.

Phenol and Phenolics

Learning Objective

✓ Distinguish between phenol and the types of phenolics, and discuss their action as antimicrobial agents.

In 1867, Dr. Joseph Lister began using phenol **(Figure 9.13a)** to reduce infection during surgery. As stated previously, the efficacy of phenol remains one standard to which the actions of other antimicrobial agents can be compared.

Phenolics are compounds derived from phenol molecules that have been chemically modified by the addition of halogens or organic functional groups **(Figure 9.13b)**. For instance, chlorinated phenolics contain one or more atoms of chlorine and have enhanced antimicrobial action and a less annoying odor than phenol. Natural oils such as pine and clove oils are also phenolics and can be used as antiseptics.

Bisphenolics are composed of two covalently linked phenolics. Two examples of bisphenolics are *orthophenylphenol,* which is the active ingredient in the disinfectant Lysol, and *triclosan,* which is incorporated into numerous consumer products, including garbage bags, diapers, and cutting boards.

Phenol and phenolics are intermediate- to low-level disinfectants that denature proteins and disrupt cell membranes in a wide variety of pathogens. They are effective even in the presence of contaminating organic material such as vomit, pus, saliva, and feces, and they remain active on surfaces for a prolonged time. For these reasons, phenolics are commonly used in health care settings, laboratories, and households.

Negative aspects of phenolics include their disagreeable odor and possible side effects; for example, phenolics irritate the skin of some individuals. *Hexachlorophene* (see Figure 9.13b), which was once a popular household bisphenolic, was found to cause brain

BENEFICIAL MICROBES

HARD TO SWALLOW?

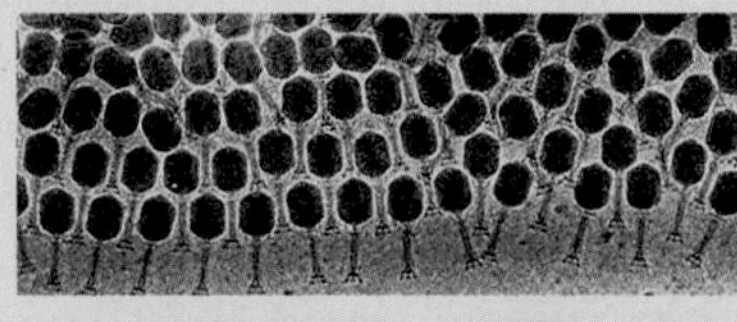

▲ *Bacteriophages.* SEM 200 nm

Controlling bacteria in the environment is becoming more difficult due to the development of resistance to common disinfectants and antiseptics, so scientists are turning to natural parasites of bacteria—bacteriophages, also simply called phages—to control bacterial contamination. *Bacteriophages,* which literally means "bacteria eaters," is the term for viruses that specifically attack particular strains of bacteria. Phages are like smart bombs; unlike disinfectants and most antimicrobial drugs, phages attack specific bacterial strains and leave neighboring bacteria unharmed. A phage injects its genetic material into a bacterial cell, causing the bacterial cells to produce hundreds of new phages before bursting out of the bacterium and killing it. Researchers are developing phage solutions to control bacteria in medical settings, in food, and in patients.

In 2006, the Food and Drug Administration (FDA) approved the nonmedical use of a phage that specifically kills *Listeria monocytogenes,* which is frequently a bacterial contaminant of cheese and lunchmeat. *Listeria* kills about 20% of infected people. The approved anti-*Listeria* phage is available in a solution that food processors, delicatessen owners, and consumers can spray on food to reduce the number of *Listeria* cells. Some people find the idea hard to swallow—deliberately contaminating food and equipment with viruses sounds like poor hygiene.

Proponents of using phages point out that an individual consumes millions of phages daily in water and food without ill effect. Indeed, phages in restaurants are more common than mustard and mayonnaise. Medical professionals in the former Soviet Union, particularly the country of Georgia, have used phages to successfully treat disease for over six decades without deleterious side effects.

damage in infants. Now it is available only by prescription and is used in nurseries only in response to severe staphylococcal contamination.

Alcohols

Learning Objective

✓ Discuss the action of alcohols as antimicrobial agents, and explain why solutions of 70% to 90% alcohol are more effective than pure alcohols.

Alcohols are bactericidal, fungicidal, and virucidal against enveloped viruses; however, they are not effective against fungal spores or bacterial endospores. Alcohols are considered intermediate-level disinfectants. Commonly used alcohols include rubbing alcohol (isopropanol) and drinking alcohol (ethanol):

$$CH_3—CH(OH)—CH_3 \qquad CH_3—CH_2OH$$

Isopropanol   Ethanol

Isopropanol is slightly superior to ethanol as a disinfectant and antiseptic. *Tinctures* (tingk′chūrs), which are solutions of other antimicrobial chemicals in alcohol, are often more effective than the same chemicals dissolved in water.

Alcohols denature proteins and disrupt cytoplasmic membranes. Surprisingly, pure alcohol is not an effective antimicrobial agent because the denaturation of proteins requires water; therefore, solutions of 70% to 90% alcohol are typically used to control microbes. Alcohols evaporate rapidly, which is advantageous in that they leave no residue, but disadvantageous in that they may not contact microbes long enough to be effective. Nevertheless, alcohol-based cleansers are more effective than soap in removing bacteria from hands. Swabbing the skin with alcohol prior to an injection removes more microbes by physical action (degerming) than by chemical action.

Halogens

Learning Objective

✓ Discuss the types and uses of halogen-containing antimicrobial agents.

Halogens are the four very reactive, nonmetallic chemical elements: iodine, chlorine, bromine, and fluorine. Halogens are intermediate-level antimicrobial chemicals that are effective against vegetative bacterial and fungal cells, fungal spores, some bacterial endospores and protozoan cysts, and many viruses. Halogens are used both alone and combined with other elements in organic and inorganic compounds. Scientists do not know exactly how halogens exert their antimicrobial effect, but it is thought that they damage enzymes via oxidation or by unfolding them, denaturing them.

Iodine is a well-known antiseptic. In the past, backpackers and campers disinfected water with iodine tablets, but experience has shown that protozoan cysts can survive iodine treatment unless the iodine concentration is so great that the water is undrinkable. Knowledgeable campers now filter stream and lake water or carry bottled water.

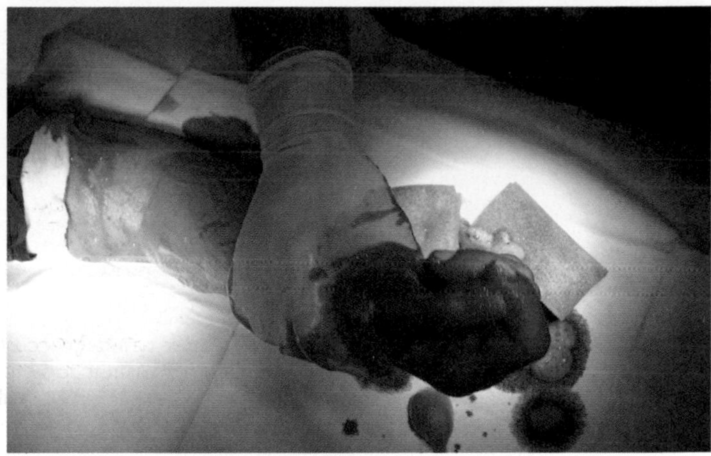

▲ **Figure 9.14 Degerming in preparation for surgery on a hand.** Betadine, an iodophor, is the antiseptic used here.

Medically, iodine is used either as a tincture or as an *iodophor,* which is an iodine-containing organic compound that slowly releases iodine. Iodophors have the advantage of being long lasting and nonirritating to the skin. Betadine is an example of an iodophor used in medical institutions to prepare skin for surgery **(Figure 9.14)** and injections and to treat burns.

Municipalities commonly use *chlorine* in its elemental form (Cl_2) to treat drinking water, swimming pools, and wastewater from sewage treatment plants. Compounds containing chlorine are also effective disinfectants. Examples include *sodium hypochlorite* (NaOCl), which is household chlorine bleach, and *calcium hypochlorite.* The dairy industry and restaurants use these compounds to disinfect utensils, and the medical field uses them to disinfect hemodialysis systems. Household bleach diluted by adding two drops to a liter of water can be used in an emergency to make water safer to drink, but it does not kill all protozoan cysts, bacterial endospores, or viruses. *Chlorine dioxide* (ClO_2) is a gas that can be used to disinfect large spaces; for example, it was used in the federal office buildings contaminated with anthrax spores following the 2001 bioterrorism attack. Chloramines—chemical combinations of chlorine and ammonia—are used in wound dressings, as skin antiseptics, and in some municipal water supplies. Chloramines are less effective antimicrobial agents than other forms of chlorine, but they release chlorine slowly and are thus longer lasting.

Bromine is an effective disinfectant in hot tubs because it evaporates more slowly than chlorine at high temperatures. Bromine is also used as an alternative to chlorine in the disinfection of swimming pools, cooling towers, and other water containers.

Fluorine in the form of fluoride is antibacterial in drinking water and toothpastes and can help reduce the incidence of dental caries (cavities). Fluorine works in part by disrupting metabolism in the biofilm of dental plaque.

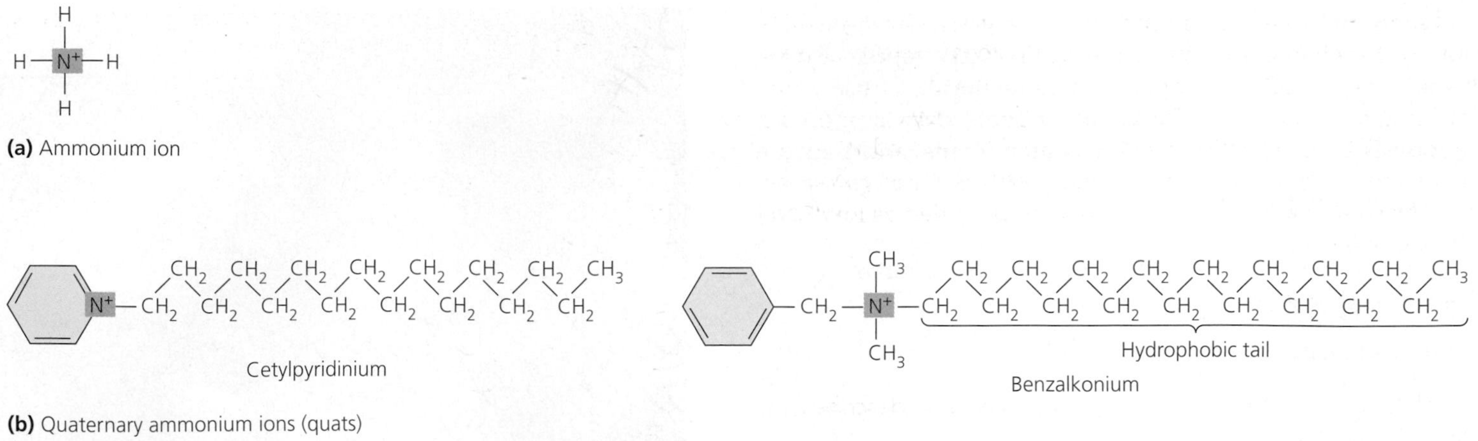

(a) Ammonium ion

Cetylpyridinium

Hydrophobic tail

Benzalkonium

(b) Quaternary ammonium ions (quats)

▲ **Figure 9.15 Quaternary ammonium compounds (quats).** Quats are surfactants in which the hydrogen atoms of an ammonium ion **(a)** are replaced by other functional groups **(b)**.

Oxidizing Agents

Learning Objective

✓ Describe the use and action of oxidizing agents in microbial control.

Peroxides, ozone, and *peracetic acid* kill microbes by oxidizing their enzymes, thereby preventing metabolism. **Oxidizing agents** are high-level disinfectants and antiseptics that work by releasing oxygen radicals, which are particularly effective against anaerobic microorganisms. Health care workers use oxidizing agents to kill anaerobes in deep puncture wounds.

Hydrogen peroxide is a common household chemical that can disinfect and even sterilize the surfaces of inanimate objects such as contact lenses, but it is often mistakenly used to treat open wounds. Hydrogen peroxide does not make a good antiseptic for open wounds because *catalase*—an enzyme released from damaged human cells—quickly neutralizes hydrogen peroxide by breaking it down into water and oxygen gas, which can be seen as escaping bubbles. Though aerobes and facultative anaerobes on inanimate surfaces also contain catalase, the volume of peroxide used as a disinfectant overwhelms the enzyme, making hydrogen peroxide a useful disinfectant. Food processors use hot hydrogen peroxide to sterilize packages such as juice boxes.

Ozone (O_3) is a reactive form of oxygen that is generated when molecular oxygen (O_2) is subjected to electrical discharge. Ozone gives air its "fresh smell" after a thunderstorm. Some Canadian and European municipalities treat their drinking water with ozone rather than chlorine. Ozone is a more effective antimicrobial agent than chlorine, but it is more expensive, and it is difficult to maintain an effective concentration of ozone in water.

Peracetic acid is an extremely effective sporicide that can be used to sterilize surfaces. Food processors and medical personnel use peracetic acid to sterilize equipment because it is not adversely affected by organic contaminants, and it leaves no toxic residue.

Surfactants

Learning Objective

✓ Define *surfactants* and describe their antimicrobial action.

Surfactants are "surface active" chemicals. One of the ways surfactants act is to reduce the surface tension of solvents such as water by decreasing the attraction among molecules. One result of this reduction in surface tension is that the solvent becomes more effective at dissolving solute molecules.

Two common surfactants involved in microbial control are soaps and detergents. One end of a soap molecule is hydrophobic because it is composed of fatty acids, and the other end is hydrophilic and negatively charged. When soap is used to wash skin, for instance, the hydrophobic ends of soap molecules are effective at breaking oily deposits into tiny droplets, and the hydrophilic ends attract water molecules; the result is that the tiny droplets of oily material—and any bacteria they harbor—are more easily dissolved in and washed away by water. Thus soaps by themselves are good degerming agents, though poor antimicrobial agents; when household soaps are antiseptic, it is largely because they contain antimicrobial chemicals.

Synthetic **detergents** are positively charged organic surfactants that are more soluble in water than soaps. The most popular detergents for microbial control are **quaternary ammonium compounds** or **quats,** which are composed of an ammonium cation (NH_4^+) in which the hydrogen atoms are replaced by other functional groups or hydrocarbon chains **(Figure 9.15)**. Quats are not only antimicrobial; they are colorless, tasteless, and harmless to humans (except at high concentrations), making them ideal for many industrial and medical applications. If your mouthwash foams, it probably contains a quaternary ammonium compound. Examples of quats are benzalkonium chloride (Zephiran) and cetylpyridinium chloride (used in Cepacol mouthwash).

Quats function by disrupting cellular membranes so that affected cells lose essential internal ions, such as potassium ions

(K^+). Quats are bactericidal (particularly against Gram-positive bacteria), fungicidal, and virucidal against enveloped viruses; but they are not effective against nonenveloped viruses, mycobacteria, or endospores. The action of quaternary ammonium compounds is retarded by organic contaminants, and they are deactivated by soaps. Some pathogens, such as *P. aeruginosa*, actually thrive in quats; therefore, quats are classified as low-level disinfectants.

Heavy Metals

Learning Objective

✓ Define *heavy metals*, give several examples, and describe their use in microbial control.

Heavy-metal ions such as ions of arsenic, zinc, mercury, silver, and copper are antimicrobial because they combine with sulfur atoms in molecules of cysteine, an amino acid. Such bonding denatures proteins, inhibiting or eliminating their function. Heavy-metal ions are low-level bacteriostatic and fungistatic agents, and with few exceptions their use has been superseded by more effective antimicrobial agents. **Figure 9.16** illustrates the effectiveness of heavy metals in inhibiting bacterial reproduction on a Petri plate.

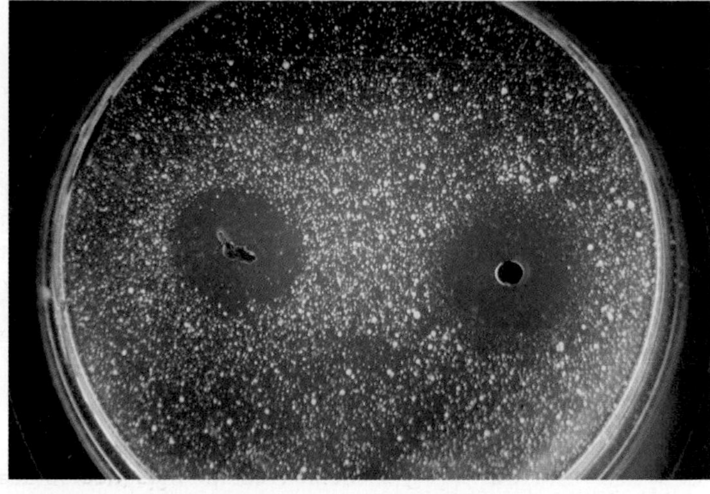

▲ **Figure 9.16 The effect of heavy-metal ions on bacterial growth.** The zones of inhibition formed because ions from gold (left) and silver alloy (right) inhibit bacterial reproduction through their effects on protein function.

At one time, many states required that the eyes of newborns be treated with a cream containing 1% *silver nitrate* ($AgNO_3$) to prevent blindness caused by *Neisseria gonorrhoeae* (nī-se′rē-ă go-nor-re′ī), which can enter babies' eyes while they pass through an infected birth canal. Today, silver nitrate has largely been displaced by other antimicrobial ointments that are less irritating and are also effective against other pathogens. Silver still plays an antimicrobial role in some surgical dressings, burn creams, and catheters.

For over 70 years, drug companies used *thimerosal,* a mercury-containing compound, to preserve vaccines. In 1999, the U.S. Public Health Service recommended that alternatives be used because mercury is a metabolic poison, though the small amount of mercury in vaccines is considered safe. Today only a few adult vaccines contain thimerosal. These include whole-cell pertussis and some vaccines against tetanus, flu, and meningococcal meningitis.

Copper, which interferes with chlorophyll, is used to control algal growth in reservoirs, fish tanks, swimming pools, and water storage tanks. In the absence of organic contaminants, copper is an effective algicide in concentrations as low as 1 ppm (part per million). In addition to copper, zinc and mercury are used to control mildew in paint. **Highlight: Brass to the Rescue** examines the use of copper to disinfect drinking water.

CRITICAL THINKING

What common household antiseptic contains a heavy metal as its active ingredient?

HIGHLIGHT

BRASS TO THE RESCUE

Indian tradition holds that water stored in brass pitchers prevents disease. British and Indian scientists have discovered recently that there is some truth in the tradition. The researchers found that river water samples collected in India had fecal bacterial counts as high as 1 million bacteria per milliliter. However, the scientists could detect no bacteria in the water after storage for two days in traditional brass pitchers. Bacterial levels in plastic or earthenware containers remained high over the same period. How can brass make water safer to drink?

Brass is an alloy of copper mixed with zinc. Brass vessels release a small amount of copper ions into stored water, and copper ions are antibacterial. Evidently there is enough antibacterial action from the copper ions to render the water safer, though it must be noted that the researchers did not examine the action of brass on protozoan or viral pathogens.

Aldehydes

Learning Objective

✓ Compare and contrast formaldehyde and glutaraldehyde as antimicrobial agents.

Aldehydes are compounds containing terminal —CHO groups (see Table 2.3 on p. 39). *Glutaraldehyde,* which is a liquid, and *formaldehyde,* which is a gas, are highly reactive chemicals with the following structural formulas:

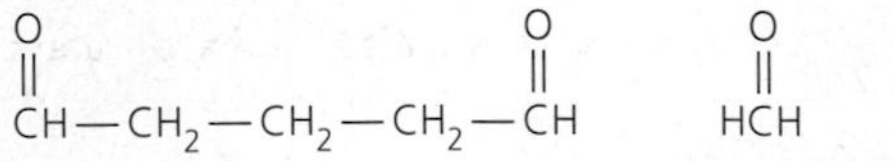

$$\overset{\displaystyle O}{\underset{}{\overset{\|}{CH}}}-CH_2-CH_2-CH_2-\overset{\displaystyle O}{\overset{\|}{CH}} \qquad \overset{\displaystyle O}{\overset{\|}{HCH}}$$

Glutaraldehyde Formaldehyde

Aldehydes function in microbial control by cross-linking amino, hydroxyl, sulfhydryl, and carboxyl organic functional groups, thereby denaturing proteins and inactivating nucleic acids.

Hospital personnel and scientists use 2% solutions of glutaraldehyde to kill bacteria, viruses, and fungi; a 10-minute treatment effectively disinfects most objects, including medical and dental equipment. When the time of exposure is increased to 10 hours, glutaraldehyde sterilizes. Although glutaraldehyde is less irritating and more effective than formaldehyde, it is more expensive as well.

Morticians and health care workers use formaldehyde dissolved in water to make a 37% solution called *formalin.* They use formalin for embalming and to disinfect isolation rooms, exhaust cabinets, surgical instruments, and reusable kidney dialysis machines. Formaldehyde must be handled with care because it irritates mucous membranes and is carcinogenic.

Gaseous Agents

Learning Objective

✓ Describe the advantages and disadvantages of gaseous agents of microbial control.

Many items, such as heart-lung machine components, sutures, plastic laboratory ware, mattresses, pillows, artificial heart valves, catheters, electronic equipment, and dried or powdered foods, cannot be sterilized easily with heat or water-soluble chemicals; nor is irradiation always practical for large or bulky items. However, they can be sterilized within a closed chamber containing highly reactive microbicidal and sporicidal gases such as *ethylene oxide, propylene oxide,* and *beta-propiolactone:*

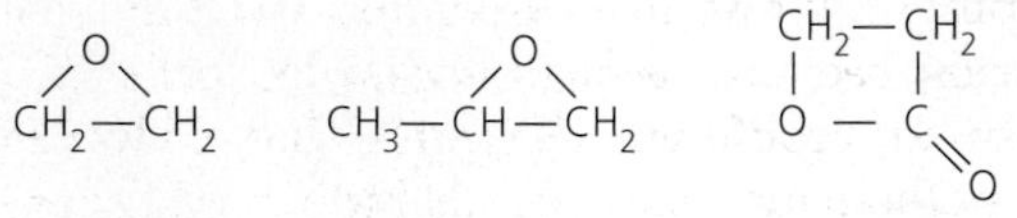

$$\overset{O}{\overbrace{CH_2-CH_2}} \qquad CH_3-\overset{O}{\overbrace{CH-CH_2}} \qquad \begin{matrix} CH_2 - CH_2 \\ | \quad\quad | \\ O \; - \; C \!=\! O \end{matrix}$$

Ethylene oxide Propylene oxide Beta-propiolactone

These gases rapidly penetrate paper and plastic wraps and diffuse into every crack. Over time (usually 4–18 hours), they denature proteins and DNA by cross-linking organic functional groups, thereby killing everything they contact without harming inanimate objects.

Ethylene oxide is frequently used as a gaseous sterilizing agent in hospitals and dental offices, and NASA uses the gas to sterilize spacecraft designed to land on other worlds, lest they accidentally export earthly microbes. Large hospitals often use ethylene oxide chambers, which are similar in appearance to autoclaves, to sterilize instruments and equipment sensitive to heat.

Despite their advantages, gaseous agents are far from perfect: They can be extremely hazardous to the people using them. They must be administered by combining them with 10% to 20% nitrogen gas or carbon dioxide, because they are often highly explosive. Moreover, they are extremely poisonous, so workers must extensively flush sterilized objects with air to remove every trace of the gas (which adds to the time required to use them). Finally, gaseous agents, especially beta-propiolactone, are potentially carcinogenic.

Enzymes

Learning Objective

✓ Describe the use of an enzyme to remove most bacteria from food and to remove prions from medical instruments.

Many organisms produce chemicals that inhibit or destroy a variety of fungi, bacteria, or viruses. Among these are **antimicrobial enzymes,** which are enzymes that act against microorganisms. For example, human tears contain the enzyme *lysozyme,* which is a protein that digests the peptidoglycan cell wall of bacteria, causing the cell to rupture due to osmotic pressure and thus protecting the eye from most bacterial infections.

Scientists, food processors, and medical personnel are researching ways to use natural and chemically modified antimicrobial enzymes to control microbes in the environment, inhibit microbial decay of foods and beverages, and reduce the number and kinds of microbes on medical equipment. For example, food processors use lysozyme to reduce the number of bacteria in cheese, and vintners use lysozyme instead of poisonous sulfur dioxide (SO_2) to remove bacteria that would spoil wine.

One exciting development is the use of an enzyme to eliminate the prion that causes variant Creutzfeldt-Jakob disease, also called mad cow disease. The brain, spinal cord, placenta, eye, liver, kidney, pituitary gland, spleen, lung, and lymph nodes, as well as cerebrospinal fluid, can harbor prions. Medical instruments contaminated by these highly infectious and deadly proteins may remain infectious even after normal autoclaving; boiling; exposure to formaldehyde, glutaraldehyde, or ethylene oxide; or 24 hours of dry heat at 160°C. Until recently, harsh methods such as autoclaving in sodium hydroxide for 30 minutes or complete incineration were required to eliminate prions. In 2006, the European Union approved the use of the enzyme Prionzyme to safely and completely remove prions on medical instruments. Prionzyme is the first certified, noncaustic chemical to target prions.

TABLE 9.5
Chemical Methods of Microbial Control

Method	Action(s)	Level of Activity	Some Uses
Phenol (carbolic acid)	Denatures proteins and disrupts cell membranes	Intermediate to low	Original surgical antiseptic; now replaced by less odorous and injurious phenolics
Phenolics (chemically altered phenol; bisphenols are composed of a pair of linked phenolics)	Denature proteins and disrupt cell membranes	Intermediate to low	Disinfectants and antiseptics
Alcohols	Denature proteins and disrupt cell membranes	Intermediate	Disinfectants, antiseptics, and as a solvent in tinctures
Halogens (iodine, chlorine, bromine, and fluorine)	Presumably denature proteins	Intermediate	Disinfectants, antiseptics, and water purification
Oxidizing agents (peroxides, ozone, and peracetic acid)	Denature proteins by oxidation	High	Disinfectants, antiseptics for deep wounds, water purification, and sterilization of food-processing and medical equipment
Surfactants (soaps and detergents)	Decrease surface tension of water and disrupt cell membranes	Low	Soaps: degerming; detergents: antiseptic
Heavy metals (arsenic, zinc, mercury, silver, copper, etc.)	Denature proteins	Low	Fungistats in paints; silver nitrate cream: surgical dressings, burn creams, and catheters; copper: algicide in water reservoirs, swimming pools, and aquariums
Aldehydes (glutaraldehyde and formaldehyde)	Denature proteins	High	Disinfectant and embalming fluid
Gaseous agents (ethylene oxide, propylene oxide, and beta-propiolactone)	Denature proteins	High	Sterilization of heat- and water-sensitive objects
Enzymes	Denature proteins	High against target substrate	Removal of prions on medical instruments
Antimicrobials	Act against cell walls, cell membranes, protein synthesis, and DNA transcription and replication	Intermediate to low	Disinfectants and treatment of infectious diseases

Antimicrobials

Learning Objective

✓ Describe the types of antimicrobials and their use in environmental control of microorganisms.

Antimicrobials include antibiotics, semisynthetics, and synthetics. *Antibiotics* are antimicrobial chemicals produced naturally by microorganisms. When scientists chemically modify an antibiotic, the agent is called a *semisynthetic*. Scientists have also developed wholly *synthetic* antimicrobial drugs. The main difference between these antimicrobials and the chemical agents we have discussed in this chapter is that antimicrobials are typically used for treatment of disease, and not for environmental control of microbes. Nevertheless, some antimicrobials *are* used for antimicrobial control outside the body. For example, the antimicrobials *nisin* and *natamycin* are used to reduce the growth of bacteria and fungi, respectively, in cheese. Chapter 10 discusses in more detail the nature and use of antimicrobials to treat infectious diseases.

Table 9.5 summarizes the chemical methods of microbial control discussed in this chapter.

Development of Resistant Microbes

Many scientists are concerned that Americans have become overly preoccupied with antisepsis and disinfection, as evidenced by the proliferation of products containing antiseptic and disinfecting chemicals. For example, one can now buy hand soap, shampoo, toothpaste, hand lotion, foot pads for shoes, deodorants, and bath sponges that contain antiseptics, as well as kitty litter, cutting boards, scrubbing pads, garbage bags, children's toys, and laundry detergents that contain disinfectants. There is little evidence that the extensive use of such products adds to human or animal health, but it does promote the development of strains of microbes resistant to antimicrobial chemicals because, while susceptible cells die, resistant cells remain to proliferate. Scientists have already isolated strains of pathogenic bacteria, including *M. tuberculosis, P. aeruginosa, E. coli,* and *S. aureus,* that are less susceptible to common disinfectants and antiseptics.

Highlight: Antibacterial Soap: Too Much of a Good Thing? on p. 278 discusses a controversy regarding the use of antibacterial soap.

HIGHLIGHT

ANTIBACTERIAL SOAP: TOO MUCH OF A GOOD THING?

Although soaps containing antimicrobial drugs are more effective than plain soaps in reducing the presence of microbes, there is concern that overuse of antimicrobials contributes to the evolution of resistant microorganisms: antibacterial soaps kill off weaker bacteria, leaving stronger, more resistant strains to multiply. The CDC has taken a cautious stance, acknowledging the benefits of antimicrobial soaps in certain circumstances while agreeing that we need further research to determine if such products may actually do more harm than good.

Who knew cleanliness could be so complicated? And in the meantime, what kind of soap should you use? Experts can't seem to agree on a single guideline, but the CDC recommends using mild, regular soap and washing in warm running water for at least 10–15 seconds in most cases, reserving antimicrobial soaps for limited applications: for use in handling food, for the care of newborns, and for the care of high-risk patients by health care workers.

Chapter Summary

Basic Principles of Microbial Control (pp. 258–260)

1. **Sterilization** is the eradication of microorganisms and viruses; the term is not usually applied to the destruction of prions.
2. An **aseptic** environment or procedure is free of contamination by pathogens.
3. **Antisepsis** is the inhibition/killing of microorganisms (particularly pathogens) on skin or tissue by the use of a chemical **antiseptic,** whereas **disinfection** refers to the use of agents to inhibit microbes on inanimate objects.
4. **Degerming** refers to the removal of microbes from a surface by scrubbing.
5. **Sanitization** is the reduction of a prescribed number of pathogens from surfaces and utensils in public settings.
6. **Pasteurization** is a process using heat to kill pathogens and control microbes that cause spoilage of food and beverages.
7. The suffixes *-stasis* and *-static* indicate that an antimicrobial agent inhibits microbes, whereas the suffixes *-cide* and *-cidal* indicate that the agent kills or permanently inactivates a particular type of microbe.
8. **Microbial death** is the permanent loss of reproductive capacity. **Microbial death rate** measures the efficacy of an antimicrobial agent.
9. Antimicrobial agents destroy microbes either by altering their cell walls and membranes or by interrupting their metabolism and reproduction via interference with proteins and nucleic acids.

The Selection of Microbial Control Methods (pp. 260–263)

1. Factors affecting the efficacy of antimicrobial methods include the site to be treated, the relative susceptibility of microorganisms, and environmental conditions.
2. Four methods for evaluating the effectiveness of a disinfectant or antiseptic are the **phenol coefficient;** the **use-dilution test;** the **Kelsey-Sykes capacity test;** and the **in-use test,** which provides a more accurate determination of efficacy under real-life conditions.

Physical Methods of Microbial Control (pp. 263–271)

1. **Thermal death point** is the lowest temperature that kills all cells in a broth in 10 minutes, whereas **thermal death time** is the time it takes to completely sterilize a particular volume of liquid at a set temperature. **Decimal reduction time (D)** is the time required to destroy 90% of the microbes in a sample.
2. An **autoclave** uses steam heat under pressure to sterilize chemicals and objects that can tolerate moist heat.
3. Pasteurization, a method of heating foods to kill pathogens and control spoilage organisms without altering the quality of the food, can be achieved by several methods: the historical (batch) method, flash pasteurization, and ultrahigh-temperature pasteurization. The methods differ in their combinations of temperature and time of exposure.
4. Under certain circumstances, microbes can be controlled using ultrahigh-temperature sterilization, dry-heat sterilization, incineration, refrigeration, or freezing.
5. Antimicrobial methods involving drying are **desiccation,** used to preserve food, and **lyophilization** (freeze drying), used for the long-term preservation of cells or microbes.
6. When used as a microbial control method, **filtration** is the passage of air or a liquid through a material that traps and removes microbes. Some **membrane filters** have pores small enough to trap the smallest viruses. HEPA (high-efficiency particulate air) filters remove microbes and particles from air.

7. The high **osmotic pressure** exerted by hypertonic solutions of salt or sugar can preserve foods such as jerky and jams by removing the water from microbes that they need to carry out their metabolic functions.

8. **Radiation** includes high-speed subatomic particles and even more energetic electromagnetic waves released from atoms. **Ionizing radiation** (wavelengths shorter than 1 nm) produces ions that denature important molecules and kill cells. **Nonionizing radiation** (wavelengths longer than 1 nm), is less effective in microbial control, although UV light causes pyrimidine dimers, which can kill affected cells.

9. The CDC designates biosafety level 1–4 standards for working with microbes in laboratories; BSL-4 is strictest.

Chemical Methods of Microbial Control (pp. 271–278)

1. **Phenolics,** which are chemically modified phenol molecules, are intermediate- to low-level disinfectants that denature proteins and disrupt cell membranes in a wide variety of pathogens.

2. **Alcohols** are intermediate-level disinfectants that denature proteins and disrupt cell membranes; they are used either as 70% to 90% aqueous solutions or in a tincture, which is a combination of an alcohol and another antimicrobial chemical.

3. **Halogens** (iodine, chlorine, bromine, and fluorine) are used as intermediate-level disinfectants and antiseptics to kill microbes in water or on medical instruments or skin. Although their exact mode of action is unknown, they are believed to denature proteins.

4. **Oxidizing agents** such as hydrogen peroxide, ozone, and peracetic acid are high-level disinfectants and antiseptics that release oxygen radicals, which are toxic to many microbes, especially anaerobes.

5. **Surfactants** include soaps, which act primarily to break up oils during degerming, and **detergents** such as **quaternary ammonium compounds (quats),** which are low-level disinfectants.

6. **Heavy-metal ions** such as arsenic, silver, mercury, copper, and zinc are low-level disinfectants that denature proteins. For most applications they have been superseded by less toxic alternatives.

7. **Aldehydes** are high-level disinfectants that cross-link organic functional groups in proteins and nucleic acids. A 2% solution of glutaraldehyde or a 37% aqueous solution of formaldehyde (called formalin) is used to disinfect or sterilize medical or dental equipment and in embalming fluid.

8. **Gaseous agents** of microbial control, which include ethylene oxide, propylene oxide, and beta-propiolactone, are high-level disinfecting agents used to sterilize heat-sensitive equipment and large objects. These gases are explosive and potentially carcinogenic.

9. Many organisms use **antimicrobial enzymes** to combat microbes. Humans use them commercially in food preservation and as a noncaustic, nondestructive way to eliminate prions on medical instruments.

10. **Antimicrobials,** which include antibiotics, semisynthetics, and synthetics, are compounds that are typically used to treat diseases but can also function as intermediate-level disinfectants.

Questions for Review

Answers to the Questions for Review (except Short Answer questions) begin on page A-1.

Multiple Choice

1. In practical terms, which of the following statements provides the definition of sterilization?
   a. Sterilization eliminates all organisms and their spores or endospores.
   b. Sterilization eliminates harmful microorganisms and viruses.
   c. Sterilization eliminates prions.
   d. Sterilization eliminates hyperthermophiles.

2. Which of the following substances or processes kills microorganisms on laboratory surfaces?
   a. antiseptics
   b. disinfectants
   c. degermers
   d. pasteurization

3. Which of the following terms best describes the disinfecting of cafeteria plates?
   a. pasteurization
   b. antisepsis
   c. sterilization
   d. sanitization

4. The microbial death rate is used to measure
   a. the efficiency of a detergent.
   b. the efficiency of an antiseptic.
   c. the efficiency of sanitization techniques.
   d. all of the above

5. Which of the following statements is true concerning the selection of an antimicrobial agent?
   a. An ideal antimicrobial agent is stable during storage.
   b. An ideal antimicrobial agent is fast acting.
   c. Ideal microbial agents do not exist.
   d. all of the above

6. The spores of which organism are used as a biological indicator of sterilization?
   a. *Bacillus stearothermophilus*
   b. *Salmonella enterica*
   c. *Mycobacterium tuberculosis*
   d. *Staphylococcus aureus*

7. A company that manufactures an antimicrobial cleaner for kitchen counters claims that its product is effective when used in a 50% water solution. By what means might scientists best verify this statement?
   a. disk-diffusion test
   b. phenol coefficient
   c. filter paper test
   d. in-use test

8. Which of the following items functions most like an autoclave?
   a. a boiling pan
   b. an incinerator
   c. a desiccator
   d. a pressure cooker

9. The preservation of beef jerky from microbial growth relies on which method of microbial control?
   a. filtration
   b. lyophilization
   c. desiccation
   d. radiation

10. Which of the following types of radiation is more widely used as an antimicrobial technique?
    a. electron beams
    b. visible light waves
    c. radio waves
    d. microwaves

11. Which of the following substances would most effectively inhibit anaerobes?
    a. phenol
    b. silver
    c. ethanol
    d. hydrogen peroxide

12. Which of the following adjectives best describes a surgical procedure that is free of microbial contaminants?
    a. disinfected
    b. sanitized
    c. degermed
    d. aseptic

13. Biosafety Level 3 includes
    a. double sets of entry doors
    b. pressurized suits
    c. showers in entryways
    d. all of the previous

14. A sample of *E. coli* has been subjected to heat for a specified time, and 90% of the cells have been destroyed. Which of the following terms best describes this event?
    a. thermal death point
    b. thermal death time
    c. decimal reduction time
    d. none of the above

15. Which of the following substances is least toxic to humans?
    a. carbolic acid
    b. glutaraldehyde
    c. hydrogen peroxide
    d. formalin

16. Which of the following chemicals is active against bacterial endospores?
    a. copper ions
    b. ethylene oxide
    c. ethanol
    d. triclosan

17. Which of the following disinfectants acts against cell membranes?
    a. phenol
    b. peracetic acid
    c. silver nitrate
    d. glutaraldehyde

18. Which of the following disinfectants contains alcohol?
    a. an iodophor
    b. a quat
    c. formalin
    d. a tincture of bromine

19. Which antimicrobial chemical has been used to sterilize spacecraft?
    a. phenol
    b. alcohol
    c. heavy metal
    d. ethylene oxide

20. Which class of surfactant is most soluble in water?
    a. quaternary ammonium compounds
    b. alcohols
    c. soaps
    d. peracetic acids

Short Answer

1. Describe three types of microbes that are extremely resistant to antimicrobial treatment, and explain why they are resistant.
2. Compare and contrast four tests that have been developed to measure the efficacy of disinfectants.
3. Why is it necessary to use strong disinfectants in areas exposed to tuberculosis patients?
4. Why do warm disinfectant chemicals generally work better than cool ones?
5. Why are Gram-negative bacteria more susceptible to heat than Gram-positive bacteria?
6. Describe five physical methods of microbial control.
7. What is the difference between thermal death point and thermal death time?
8. Defend the following statement: "Pasteurization is not sterilization."
9. Compare and contrast desiccation and lyophilization.
10. Compare and contrast the action of alcohols, halogens, and oxidizing agents in controlling microbial growth.
11. Hyperthermophilic prokaryotes may remain viable in canned goods after commercial sterilization. Why is this situation not dangerous to consumers?
12. Why are alcohols more effective in a 70% solution than in a 100% solution?
13. Contrast the structures and actions of soaps and quats.
14. What are some advantages and disadvantages of using ionizing radiation to sterilize food?
15. How can campers effectively treat stream water to remove pathogenic protozoa, bacteria, and viruses?

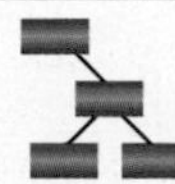

Concept Mapping

Using the following terms, draw a concept map that describes moist heat applications to control microorganisms. For a sample concept map, see p. 93. Or, complete this concept map online by going to the Study Area at www.masteringmicrobiology.com.

100°C, ≥10 min.
121°C, 15 psi, ≥15 min.
134°C, 1 sec.
140°C, 1–3 sec.
63°C, 30 min.
72°C, 15 sec.
Batch method (classic)
Boiling water
Equivalent treatments
Flash pasteurization
Fungi
Most viruses
Pasteurization
Protozoan trophozoites
Autoclave
Sterilization technique (2)
Thermoduric microorganisms (2)
Ultrahigh-temperature pasteurization
Ultrahigh-temperature sterilization
Vegetative bacterial cells

Critical Thinking

1. In 2004 a casino paid $28,000 for a grilled cheese sandwich that was purported to have an image of the Virgin Mary on it. The seller had stored the sandwich in a less-than-airtight box for ten years without decay or the growth of mold. What antimicrobial chemical and physical agents might account for the longevity of the sandwich?
2. Is desiccation the only antimicrobial effect operating when grapes are dried in the sun to make raisins? Explain.
3. How long would it take to reduce a population of 100 trillion (10^{14}) bacteria to 10 viable cells if the D value of the treatment is 3 minutes?
4. Some potentially pathogenic bacteria and fungi, including strains of *Enterococcus, Staphylococcus, Candida,* and *Aspergillus,* can survive for 1–3 months on a variety of materials found in hospitals, including the cotton blends of scrub suits, nurses' clothes, lab coats, and plastics from splash aprons and computer keyboards. What can hospital personnel do to reduce the spread of these pathogens?
5. Over 1400 people developed severe diarrhea from the Saintpaul strain of *Salmonella enterica* in the summer of 2008; 286 were hospitalized, and at least two died. CDC epidemiologists determined that infection resulted from consumption of raw tomatoes, jalapeño peppers, or serrano peppers. Based on this chapter, what types of treatment are available to produce growers and packers? What other precautions could consumers have taken?
6. An over-the-counter medicated foot powder contains camphor, eucalyptus oil, lemon oil, and zinc oxide. Only one of the ingredients is a proven antimicrobial. Which one? How does it act against fungi?
7. The 2004 tsunami in the Indian Ocean and hurricanes in 2005 and 2008 in the U.S. severely contaminated water wells and disrupted water supply lines. What immediate steps should the people have taken to lessen the spread of waterborne illnesses such as cholera?
8. In what ways might it be argued that the widespread commercial use of antiseptics and disinfectants has hurt rather than helped American health?
9. Explain why quaternary ammonium compounds are not very effective against mycobacteria such as *Mycobacterium tuberculosis.*
10. Calculate the decimal reduction time (D) for the two temperatures in the following graph.

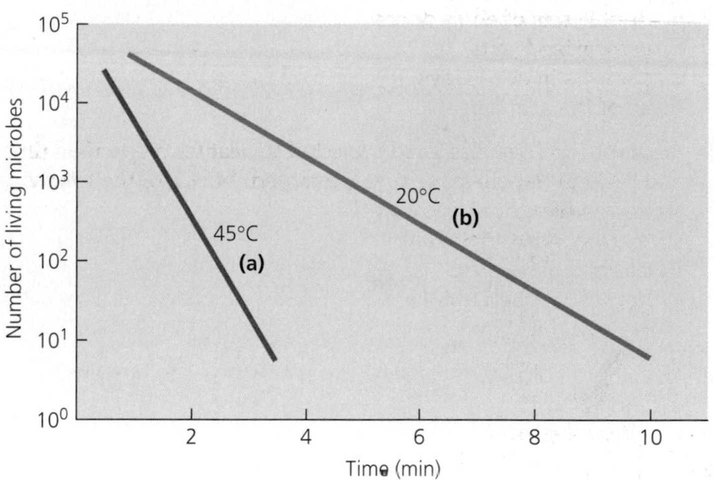

10 Controlling Microbial Growth in the Body: Antimicrobial Drugs

Meet *Staphylococcus aureus*, a common bacterium that is the number one cause of hospital-acquired infections in the United States. The particular strain of *S. aureus* shown here, however, is remarkable in some very important and alarming ways. It is not only resistant to methicillin, the antimicrobial that is traditionally used to treat staphylococcal infections; it is also resistant to vancomycin, a drug that was long considered the last line of defense against methicillin-resistant *S. aureus*. Although other drugs may be used to fight it, this "superbug" strain of *S. aureus* leaves patients few options for defense—and indeed it can cause death.

How do antimicrobial drugs work? Why do they work on some microorganisms and not on others? What can be done about the increasing problem of superbugs that resist a wide variety of existing drugs? This chapter focuses on the chemical control of pathogens in the body.

Take the pre-test for this chapter online. Visit the Study Area at www.masteringmicrobiology.com.

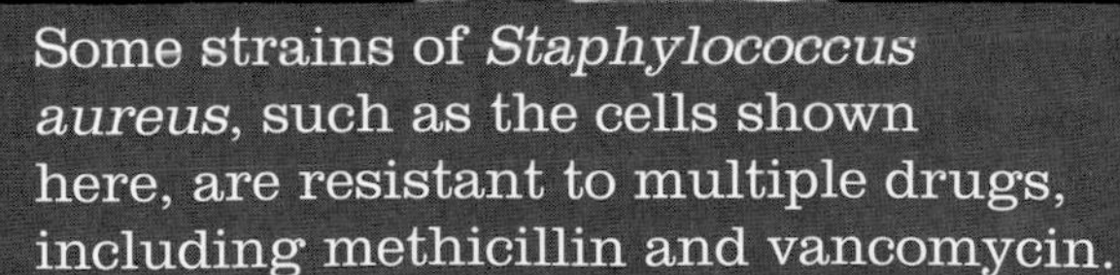

Some strains of *Staphylococcus aureus*, such as the cells shown here, are resistant to multiple drugs, including methicillin and vancomycin.

Chemicals that affect physiology in any manner, such as caffeine, alcohol, and tobacco, are called *drugs.* Drugs that act against diseases are called *chemotherapeutic agents.* Examples include insulin, anticancer drugs, and drugs for treating infections—called **antimicrobial agents (antimicrobials),** the subject of this chapter.

In the pages that follow we'll examine the mechanisms by which antimicrobial agents act, the factors that must be considered in the use of antimicrobials, and several issues surrounding resistance to antimicrobial agents among microorganisms. First, however, we begin with a brief history of antimicrobial chemotherapy.

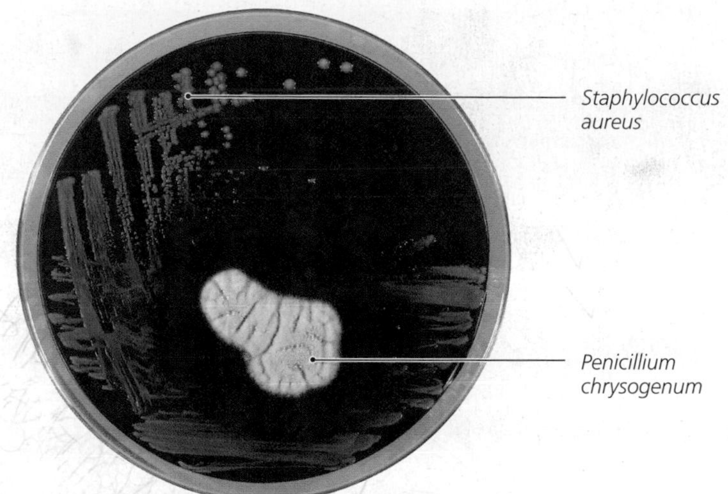

▲ **Figure 10.1 Antibiotic effect of the mold *Penicillium chrysogenum.*** Alexander Fleming observed that this mold secretes penicillin, which inhibits the growth of bacteria, as is apparent with *Staphylococcus aureus growing* on this blood agar plate.

The History of Antimicrobial Agents

Learning Objectives

- ✓ Describe the contributions of Paul Ehrlich, Alexander Fleming, and Gerhard Domagk in the development of antimicrobials.
- ✓ Explain how semisynthetic and synthetic antimicrobials differ from antibiotics.

The little girl lay struggling to breathe as her parents stood mutely by, willing the doctor to do something—anything—to relieve the symptoms that had so quickly consumed their four-year-old daughter's vitality. Sadly, there was little the doctor could do. The thick "pseudomembrane" of diphtheria, composed of bacteria, mucus, blood-clotting factors, and white blood cells, adhered tenaciously to her pharynx, tonsils, and vocal cords. He knew that trying to remove it could rip open the underlying mucous membrane, resulting in bleeding, possibly additional infections, and death. In 1902, there was little medical science could offer for the treatment of diphtheria; all physicians could do was wait and hope.

At the beginning of the 20th century, much of medicine involved diagnosing illness, describing its expected course, and telling family members either how long a patient might be sick or when they might expect her to die. Even though physicians and scientists had recently accepted the germ theory of disease and knew the causes of many diseases, very little could be done to inhibit pathogens, including *Corynebacterium diphtheriae* (kō-rī'nē-bak-tēr'ē-ŭm dif-thi'rē-ī), and alter the course of infections. In fact, one-third of children born in the early 1900s died from infectious diseases before the age of five.

It was at this time that Paul Ehrlich (1854–1915), a visionary German scientist, proposed the term *chemotherapy* to describe the use of chemicals that would selectively kill pathogens while having little or no effect on a patient. He wrote of "magic bullets" that would bind to receptors on germs to bring about their death while ignoring host cells, which lacked the receptor molecules.

Ehrlich's search for antimicrobial agents resulted in the discovery of one arsenic compound that killed trypanosome parasites and another that worked against the bacterial agent of syphilis. A few years later, in 1929, the British bacteriologist Alexander Fleming (1881–1955) reported the antibacterial action of penicillin released from *Penicillium* (pen-i-sil'ē-ŭm) mold **(Figure 10.1)**.

Though arsenic compounds and penicillin were discovered first, they were not the first antimicrobials in widespread use: Ehrlich's arsenic compounds are toxic to humans, and penicillin was not available in large enough quantities to be useful until the late 1940s. Instead, *sulfanilamide,* discovered in 1932 by the German chemist Gerhard Domagk (1895–1964), was the first practical antimicrobial agent efficacious in treating a wide array of bacterial infections.

Selman Waksman (1888–1973) discovered other microorganisms that are sources of useful antimicrobials, most notably

CLINICAL CASE STUDY

Antibiotic Overkill

A young woman was taking antibiotic pills for a urinary infection. Several days into her course of medication, she began to experience peculiar symptoms. At first they were hardly noticeable. Very quickly, however, they worsened and became embarrassing and unbearable.

She noticed a white coating on her tongue, bad breath, and an awful taste in her mouth. Despite persistent brushing and mouthwash applications, she was unable to completely remove the film. Furthermore, she had excessive vaginal discharges consisting of a cheeselike white substance. When she began to have vaginal itching, she finally decided it was time to seek help.

Reluctantly she revisited her personal physician and described the symptoms. Her doctor explained the symptoms and provided additional prescriptions to alleviate her distress.

1. What happened to the young woman in this situation?
2. How had her body's defenses been violated?
3. How can she avoid a repeat of this situation?

TABLE 10.1 Sources of Some Common Antibiotics and Semisynthetics

Microorganism	Antimicrobial
Fungi	
Penicillium chrysogenum	Penicillin
Penicillium griseofulvum	Griseofulvin
Cephalosporium spp.[a]	Cephalothin
Bacteria	
Amycolatopsis orientalis	Vancomycin
Amycolatopsis rifamycinica	Rifampin
Bacillus licheniformis	Bacitracin
Bacillus polymyxa	Polymyxin
Micromonospora purpurea	Gentamicin
Pseudomonas fluorescens	Mupirocin
Streptomyces griseus	Streptomycin
Streptomyces fradiae	Neomycin
Streptomyces aureofaciens	Tetracycline
Streptomyces venezuelae	Chloramphenicol
Streptomyces erythraeus	Erythromycin
Streptomyces nodosus	Amphotericin B
Streptomyces avermitilis	Ivermectin

[a]spp. is the abbreviation for multiple species of a genus.

species of soil-dwelling bacteria in the genus *Streptomyces* (strep-tō-mī'sēz). Waksman coined the term **antibiotics** to describe antimicrobial agents that are produced naturally by an organism **(Highlight: Microbe Altruism: Why Do They Do It?).** In common usage today "antibiotic" means "an antibacterial agent," including synthetic compounds and excluding agents with antiviral and antifungal activity.

Other scientists produced **semisynthetics**—chemically altered antibiotics—that are more effective, longer lasting, or easier to administer than naturally occurring antibiotics. Antimicrobials that are completely synthesized in a laboratory are called **synthetics.** Most antimicrobials are either natural or semisynthetic. Table 10.1 provides a partial list of common antibiotics and semisynthetics and their sources.

CRITICAL **THINKING**

Why aren't antibiotics effective against the common cold?

Mechanisms of Antimicrobial Action

Learning Objectives

- ✓ Explain the principle of selective toxicity.
- ✓ List five mechanisms by which antimicrobial drugs affect the growth of pathogens.

HIGHLIGHT

MICROBE ALTRUISM: WHY DO THEY DO IT?

We all know that antibiotics benefit humans, but what good are they to the microorganisms that secrete them? From the viewpoint of evolutionary theory, the answer is obvious: antibiotics are weapons that confer an advantage to the secreting organisms in their struggle for survival. In reality, however, the answer is not so simple.

Antibiotics are members of an extremely diverse group of metabolic products known as *secondary metabolites,* which typically are complex organic molecules that are not essential for normal cell growth and reproduction and are produced only after an organism has already established itself in its environment. The production of secondary metabolites results in a metabolic cost for the cell; that is, antibiotic production consumes energy and raw materials that could be used for growth and reproduction. Tetracycline, for example, is the end result of 72 separate enzymatic steps, and erythromycin requires 28 different chemical reactions—none of which contribute to the normal growth or reproduction of *Streptomyces.* Therefore, the question can be modified: Of what use are metabolically "expensive" antibiotics to organisms that are already secure in their environment?

Adding to the conundrum is the fact that antimicrobials against bacteria have never been discovered in natural soil at high enough concentrations to be inhibitory to neighboring cells. For example, it is almost impossible to detect antibiotics produced by *Streptomyces* except when the bacteria are grown in a laboratory, perhaps because numerous soil bacteria digest antibiotics. Minimal and inconsequential quantities of antibiotics hardly give an adaptive edge.

Some scientists suggest that antibiotics are evolutionary vestiges—leftovers of metabolic pathways that were once useful but no longer have a

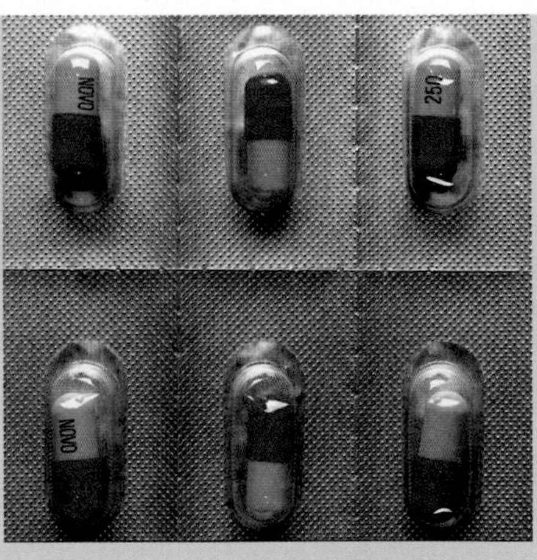

significant role. However, there should be tremendous selective pressure against the slightest continued manufacture of complex antibiotics if they truly have little purpose for the microorganism. Other research indicates that antibiotics are signals used for interbacterial communication, and their antimicrobial action is coincidental. Further research is required before we may satisfactorily answer the question, "Why do microbes make antibiotics?"

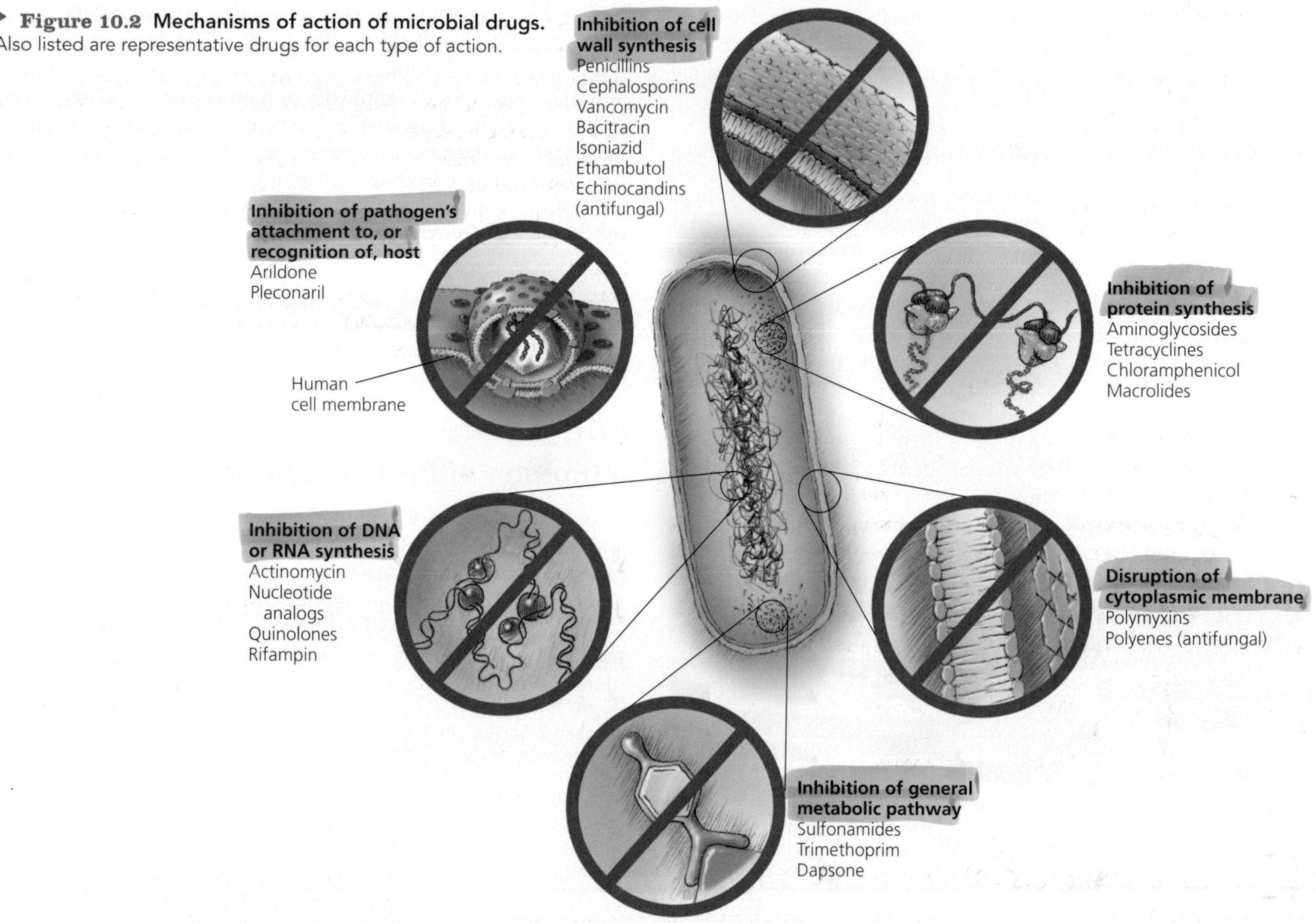

▶ **Figure 10.2 Mechanisms of action of microbial drugs.** Also listed are representative drugs for each type of action.

As Ehrlich foresaw, the key to successful chemotherapy against microbes is **selective toxicity;** that is, an effective antimicrobial agent must be more toxic to a pathogen than to the pathogen's host. Selective toxicity is possible because of differences in structure or metabolism between the pathogen and its host. Typically, the more differences, the easier it is to discover or create an effective antimicrobial agent.

Because there are many differences between the structure and metabolism of pathogenic bacteria and their eukaryotic hosts, antibacterial drugs constitute the greatest number and diversity of antimicrobial agents. Fewer antifungal, antiprotozoan, and anthelmintic drugs are available because fungi, protozoa, and helminths—like their animal and human hosts—are eukaryotic and thus share many common features. The number of effective antiviral drugs is also limited, despite major differences in structure, because viruses utilize their host cells' enzymes and ribosomes to metabolize and replicate. Therefore, drugs that are effective against viral replication are likely toxic to the host as well.

Although they can have a variety of effects on pathogens, antimicrobial drugs can be categorized into several general groups according to their mechanisms of action **(Figure 10.2)**:

- Drugs that inhibit cell wall synthesis. These drugs are selectively toxic to certain fungal or bacterial cells, which have cell walls, but not to animals, which lack cell walls.
- Drugs that inhibit protein synthesis (translation) by targeting the differences between prokaryotic and eukaryotic ribosomes.
- Drugs that disrupt unique components of the cytoplasmic membrane.
- Drugs that inhibit general metabolic pathways not used by humans.
- Drugs that inhibit nucleic acid synthesis.
- Drugs that block a pathogen's recognition of or attachment to its host.

In the following sections we examine these mechanisms in turn.

ANIMATIONS: ***Chemotherapeutic Agents: Modes of Action***

Inhibition of Cell Wall Synthesis

Learning Objective

✓ Describe the actions and give examples of drugs that affect the cell walls of bacteria and fungi.

A cell wall protects a cell from the effects of osmotic pressure. Both pathogenic bacteria and fungi have cell walls, which animals and humans lack. First, we examine drugs that act against bacterial cell walls.

Inhibition of Synthesis of Bacterial Walls

The major structural component of a bacterial cell wall is its peptidoglycan layer. As we discussed in Chapter 3, peptidoglycan is a huge macromolecule composed of polysaccharide chains of alternating *N*-acetylglucosamine (NAG) and *N*-acetylmuramic acid (NAM) molecules that are cross-linked by short peptide chains extending between NAM subunits (see Figure 3.14). To enlarge or divide, a cell must synthesize more peptidoglycan by adding new NAG and NAM subunits to existing NAG-NAM chains, and the new NAM subunits must then be bonded to neighboring NAM subunits (**Figure 10.3a** and **b**).

The most common antibacterial agents act by preventing the cross-linkage of NAM subunits. Most prominent among these drugs are **beta-lactams,** such as penicillins and cephalosporins, which are antimicrobials whose functional portions are called *beta-lactam (β-lactam) rings* (**Figure 10.3c**). Beta-lactams inhibit peptidoglycan formation by irreversibly binding to the enzymes that cross-link NAM subunits (**Figure 10.3d**). In the absence of correctly formed peptidoglycan, growing bacterial cells have weakened cell walls that are less resistant to the effects of osmotic pressure. The underlying cytoplasmic membrane bulges through the weakened portions of cell wall as water moves into the cell, and eventually the cell lyses (**Figure 10.3e**).

Chemists have made alterations to natural beta-lactams, such as penicillin G, to create semisynthetic derivatives such as methicillin and cephalothin (see Figure 10.3c), which are more stable in the acidic environment of the stomach, more readily absorbed in the intestinal tract, less susceptible to deactivation by bacterial enzymes, or more active against more types of bacteria. The simplest beta-lactams are *monobactams,* which are seldom used because they are effective only against aerobic Gram-negative bacteria.

Other antimicrobials such as **vancomycin** (van-kō-mī'sin), which is obtained from *Amycolatopsis orientalis* (am-ē-kō'la-top-sis o-rē-en-tal'is), and **cycloserine,** a semisynthetic, disrupt cell wall formation in a different manner. They directly interfere with particular alanine-alanine bridges that link the NAM subunits in many Gram-positive bacteria. Those bacteria that lack alanine-alanine crossbridges are naturally resistant to these drugs. Still another drug that prevents cell wall formation, **bacitracin** (bas-i-trā'sin), blocks the transport of NAG and NAM from the cytoplasm out to the wall. Like beta-lactams, vancomycin, cycloserine, and bacitracin result in cell lysis due to the effects of osmotic pressure.

Since all these drugs prevent bacteria from *increasing* the amount of cell wall material but have no effect on existing peptidoglycan, they are effective only on bacterial cells that are growing or reproducing; dormant cells are unaffected.

Bacteria of the genus *Mycobacterium* (mī'kō-bak-tēr'ē-ŭm), notably the agents of leprosy and tuberculosis, are characterized by unique, complex cell walls that have a layer of arabinogalactan-mycolic acid in addition to the usual peptidoglycan of prokaryotic cells. **Isoniazid** (ī-sō-nī'ă-zid), or **INH,**[1] and **ethambutol** (eth-am'boo-tol) disrupt the formation of this extra layer. Mycobacteria typically only reproduce every 12–24 hours, in part because of the complexity of their cell walls, so antimicrobial agents that act against mycobacteria must be administered for months or even years to be effective. It is often difficult to ensure that patients continue such a long regimen of treatment.

Inhibition of Synthesis of Fungal Walls

Fungal cell walls are composed of various polysaccharides containing a sugar, 1,3-D-glucan, that is not found in mammalian cells. A new class of antifungal drugs called **echinocandins** such as *caspofungin,* inhibit the enzyme that synthesizes glucan, and without glucan fungal cells cannot make cell walls, leading to osmotic rupture.

Inhibition of Protein Synthesis

As we discussed in Chapter 2, cells use proteins for structure and regulation, as enzymes in metabolism, and as channels and pumps to move materials across cell membranes. Thus, a consistent supply of proteins is vital for the active life of a cell. Given that all cells, including human cells, use ribosomes to translate proteins using information from messenger RNA templates, it is not immediately obvious that drugs could selectively target differences related to protein synthesis. Recall, however, that prokaryotic ribosomes differ from eukaryotic ribosomes in structure and size: prokaryotic ribosomes are 70S and composed of 30S and 50S subunits, whereas eukaryotic ribosomes are 80S with 60S and 40S subunits (see Figure 7.14).

Many antimicrobial agents take advantage of the differences between ribosomes to selectively target bacterial protein translation without significantly affecting eukaryotes. Note, however, that because some of these drugs affect eukaryotic mitochondria, which also contain 70S ribosomes like those of prokaryotes, such drugs may be harmful to animals and humans, especially the very active cells of the liver and bone marrow.

Understanding the actions of antimicrobials that inhibit protein synthesis requires an understanding of the process of translation, because various parts of ribosomes are the targets of antimicrobial drugs. Recall from the discussion of translation in Chapter 7 that both the 30S and 50S subunits of a prokaryotic ribosome play a role in the initiation of protein synthesis, in codon recognition, and in the docking of tRNA–amino acid complexes, and that the 50S subunit contains the enzymatic portion that actually forms peptide bonds.

Among the antimicrobials that target the 30S ribosomal subunit are aminoglycosides and tetracyclines. **Aminoglycosides** (am'i-nō-glī'kō-sīds), such as *streptomycin* and *gentamicin,* change the shape of the 30S subunit, making it impossible for the ribosome to read the codons of mRNA correctly (**Figure 10.4a**). Other aminoglycosides and **tetracyclines** (tet-ră-sī'klēns) block the tRNA docking site (A site), which then prevents the incorporation of additional amino acids into a growing polypeptide (**Figure 10.4b**).

[1]From *isonicotinic acid hydrazide,* the correct chemical name for isoniazid.

A bacterial cell wall is composed of a macromolecule of peptidoglycan composed of NAG-NAM chains that are cross-linked by peptide bridges between the NAM subunits.

New NAG and NAM subunits are inserted into the wall by enzymes, allowing the cell to grow. Normally, other enzymes link new NAM subunits to old NAM subunits with peptide cross-links.

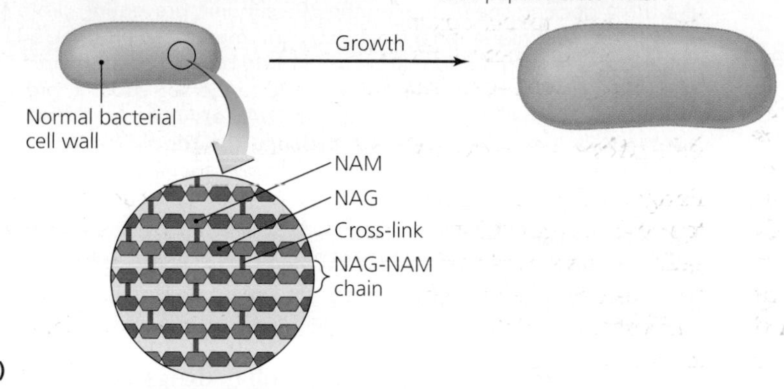

(a)

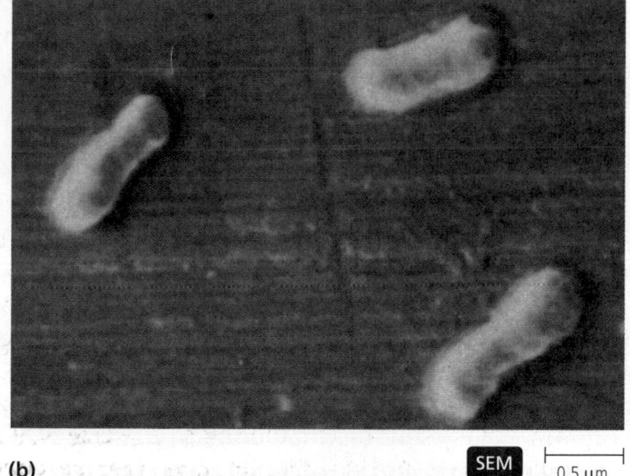

(b)

Penicillin G (natural)

Methicillin (semisynthetic)

Penicillins

β-lactam ring

Cephalothin (semisynthetic)

Cephalosporin

$(C_8H_{10}N_3S_1O_3)$—C—N—

Aztreonam (semisynthetic)

Monobactam

(c)

Penicillin interferes with the linking enzymes, and NAM subunits remain unattached to their neighbors. However, the cell continues to grow as it adds more NAG and NAM subunits.

The cell bursts from osmotic pressure because the integrity of peptidoglycan is not maintained.

Growth

New cross-links inhibited by penicillin

Penicillin prevents NAM-NAM linkages

Previously formed cross-link

(d)

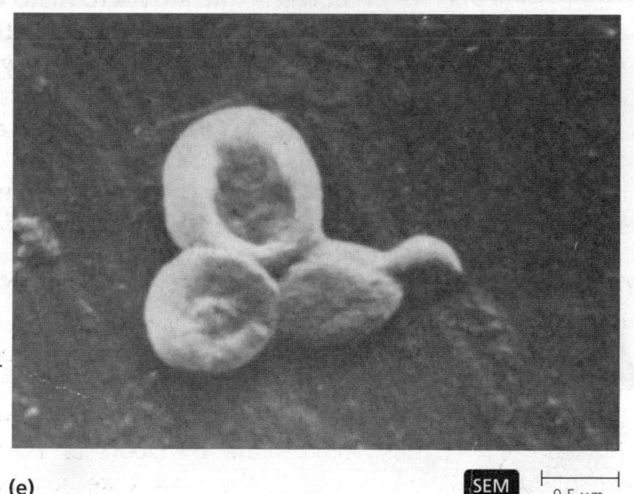

(e)

▲ **Figure 10.3 Bacterial cell wall synthesis, and the inhibitory effects of beta-lactams on it. (a)** A schematic depiction of a normal peptidoglycan cell wall showing NAG-NAM chains and cross-linked NAM subunits. **(b)** Bacterial cells with normal cell walls. **(c)** Structural formulas of some beta-lactam drugs. Their functional portion is the beta-lactam ring. **(d)** A schematic depiction of the effect of penicillin on peptidoglycan in preventing NAM-NAM cross-links. **(e)** Bacterial lysis due to the effects of penicillin. *After a beta-lactam weakens a peptidoglycan molecule by preventing NAM-NAM cross-linkages, what force actually kills an affected bacterial cell?*

Figure 10.3 *Cells lyse because of osmotic movement of water into the cell.*

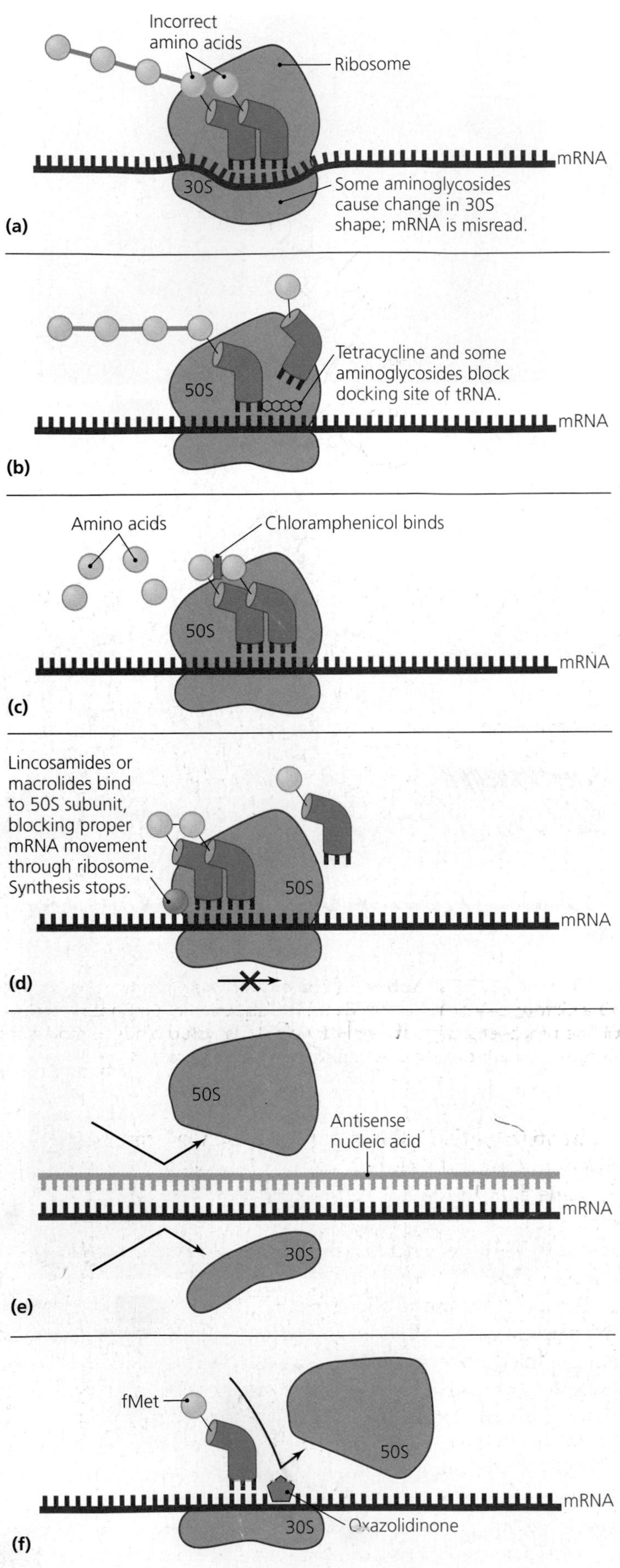
Incorrect amino acids
Ribosome
30S
mRNA
Some aminoglycosides cause change in 30S shape; mRNA is misread.
(a)
50S
Tetracycline and some aminoglycosides block docking site of tRNA.
mRNA
(b)
Amino acids
Chloramphenicol binds
50S
mRNA
(c)
Lincosamides or macrolides bind to 50S subunit, blocking proper mRNA movement through ribosome. Synthesis stops.
50S
mRNA
(d)
50S
Antisense nucleic acid
mRNA
30S
(e)
fMet
50S
mRNA
30S
Oxazolidinone
(f)

Figure 10.4 The mechanisms by which antimicrobials inhibit protein synthesis by targeting prokaryotic ribosomes. **(a)** Aminoglycosides change the shape of the 30S subunit, causing incorrect pairing of tRNA anticodons with mRNA codons. **(b)** Tetracyclines block the tRNA docking site (A site) on the 30S subunit, preventing protein elongation. **(c)** Chloramphenicol blocks enzymatic activity of the 50S subunit, preventing the formation of peptide bonds between amino acids. **(d)** Lincosamides or macrolides bind to the 50S subunit, preventing movement of the ribosome along the mRNA. **(e)** Antisense nucleic acids bind to mRNA, blocking ribosomal subunits. **(f)** Oxazolidinones inhibit initiation of translation. *Which tRNA anticodon should align with the codon CUG?*

Figure 10.4 Anticodon GAC is the correct complement for the codon CUG.

Other antimicrobials interfere with the function of the 50S subunit. **Chloramphenicol** and similar drugs block the enzymatic site of the 50S subunit **(Figure 10.4c)**, which prevents translation. **Lincosamides**, **streptogramins**, and **macrolides** (mak'rō-līds), including *erythromycin*, bind to a different portion of the 50S subunit, preventing movement of the ribosome from one codon to the next **(Figure 10.4d)**; as a result, translation is frozen and protein synthesis is halted.

Mupirocin is a unique drug that selectively binds to the bacterial tRNA that carries the amino acid isoleucine ($tRNA^{Ile}$). It does not bind to eukaryotic tRNA. Binding prevents the incorporation of isoleucine into polypeptides, effectively crippling the bacterium's protein production. Physicians prescribe mupirocin in topical creams to treat skin infections.

Other drugs that block protein synthesis are **antisense nucleic acids (Figure 10.4e)**. These RNA or single-stranded DNA molecules are designed to be complementary to specific mRNA molecules of pathogens. They block ribosomal subunits from attaching to that mRNA with no effect on human mRNA. *Fomiversen* is the first of this class of drugs to be approved. It inactivates cytomegalovirus and is used to treat eye infections.

Oxazolidinones are antimicrobial drugs that work to stop protein synthesis by blocking initiation of translation. Oxazolidinones are used as a last resort in treating infections of Gram-positive bacteria resistant to other antimicrobials, including vancomycin and methicillin-resistant *Staphylococcus aureus* (staf'i-lō-kok'ŭs o'rē-ŭs) mentioned in the chapter opener.

Disruption of Cytoplasmic Membranes

Some antibacterial drugs, such as the short polypeptide *gramicidin*, disrupt the cytoplasmic membrane of a targeted cell, often by forming a channel through the membrane, damaging its integrity. This is also the mechanism of action of a group of antifungal drugs called **polyenes** (pol-ē-ēns'). The polyenes *nystatin* and *amphotericin B* **(Figure 10.5a)** are fungicidal because they attach to *ergosterol*, a lipid constituent of fungal membranes **(Figure 10.5b)**, in the process disrupting the membrane and causing lysis of the cell. The cytoplasmic membranes of humans are somewhat susceptible to amphotericin B because they contain cholesterol, which is similar to ergosterol, though cholesterol does not bind amphotericin B as well as does ergosterol.

Azoles, such as *fluconazole,* and **allylamines,** such as *terbinafine,* are two other classes of antifungal drugs that disrupt cytoplasmic membranes. They act by inhibiting the synthesis of ergosterol; without ergosterol, the cell's membrane does not remain intact, and the fungal cell dies. Azoles and allylamines are generally harmless to humans, because human cells do not manufacture ergosterol.

Most bacterial membranes lack sterols, so these bacteria are naturally resistant to polyenes, azoles, and allylamines; however, there are other agents that disrupt bacterial membranes. An example of these antibacterial agents is *polymyxin,* produced by *Bacillus polymyxa* (ba-sil'ŭs po-lē-miks'a). Polymyxin is effective against Gram-negative bacteria, particularly *Pseudomonas* (soo-dō-mō'nas), but because it is toxic to human kidneys it is usually reserved for use against external pathogens that are resistant to other antibacterial drugs.

Pyrazinamide disrupts transport across the cytoplasmic membrane of *M. tuberculosis* (too-ber-kyū-lō'sis). The pathogen uniquely activates and accumulates the drug. Unlike many other antimicrobials, pyrazinamide is most effective against intracellular, nonreplicating bacterial cells.

Some antiparasitic drugs also act against cytoplasmic membranes. For example, *praziquantel* and *ivermectin* change the permeability of cell membranes of several types of parasitic worms.

Inhibition of Metabolic Pathways

As we discussed in Chapter 5, metabolism can be defined simply as the sum of all chemical reactions that take place within an organism. Whereas most living things share certain metabolic reactions—for example, glycolysis—other chemical reactions are unique to certain organisms. Whenever differences exist between the metabolic processes of a pathogen and its host, *antimetabolic agents* can be effective.

Various kinds of antimetabolic agents are available, including *atovaquone,* which interferes with electron transport in protozoa and fungi; heavy metals (such as arsenic, mercury, and antimony), which inactivate enzymes; agents that disrupt tubulin polymerization and glucose uptake by many protozoa and parasitic worms; drugs that block the activation of viruses; and metabolic antagonists such as sulfanilamide, the first commercially available antimicrobial agent.

Sulfanilamide and similar compounds, collectively called **sulfonamides,** act as antimetabolic drugs because they are **structural analogs** of—that is, are chemically very similar to—*para-aminobenzoic acid* (*PABA;* **Figure 10.6a**). PABA is crucial in the synthesis of nucleotides required for DNA and RNA synthesis. Many organisms, including some pathogens, enzymatically convert PABA into dihydrofolic acid, and then dihydrofolic acid into tetrahydrofolic acid (THF), a form of folic acid that is used as a coenzyme in the synthesis of purine and pyrimidine nucleotides (**Figure 10.6b**). As analogs of PABA, sulfonamides compete with PABA molecules for the active site of the enzyme involved in the production of dihydrofolic acid (**Figure 10.6c**). This competition leads to a decrease in the production of THF, and thus of DNA and RNA. The end result of sulfonamide competition with PABA is the cessation of cell metabolism, which leads to cell death.

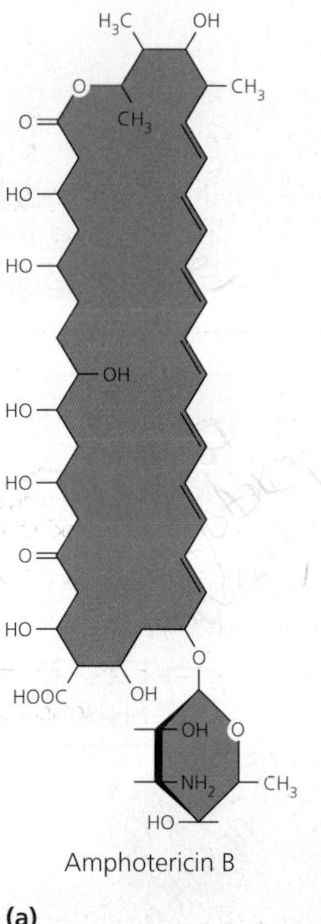

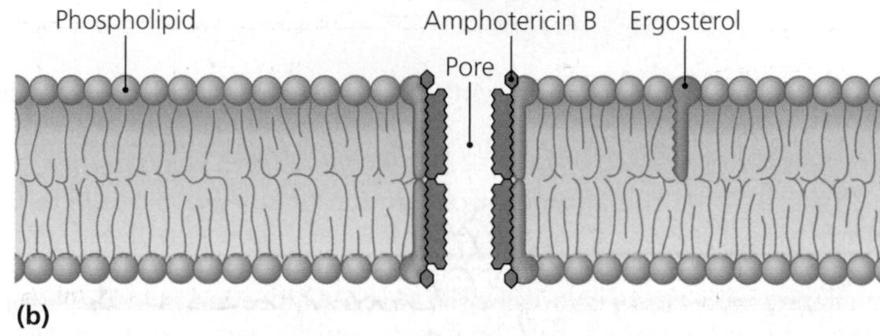

▲ **Figure 10.5 Disruption of the cytoplasmic membrane by the antifungal amphotericin B. (a)** The structure of amphotericin B. **(b)** The proposed action of amphotericin B. The drug binds to molecules of ergosterol, which then congregate, forming a pore.

Note that humans do not synthesize THF from PABA; instead, we take simple folic acids found in our diets and convert them into THF. As a result, human metabolism is unaffected by sulfonamides.

Another antimetabolic agent, *trimethoprim,* also interferes with nucleic acid synthesis. However, instead of binding to the enzyme that converts PABA to dihydrofolic acid, trimethoprim binds to the enzyme involved in the conversion of dihydrofolic acid to THF, the second step in this metabolic pathway.

Some antiviral agents target the unique aspects of the metabolism of viruses. After attachment to a host cell, viruses must penetrate the cell's membrane and be uncoated to release viral genetic instructions and assume control of the cell's metabolic machinery. Some viruses of eukaryotes are uncoated as a result

PABA Sulfanilamide Sulfamethoxazole Sulfisoxazole

(a) Para-aminobenzoic acid (PABA) and its structural analogs, the sulfonamides

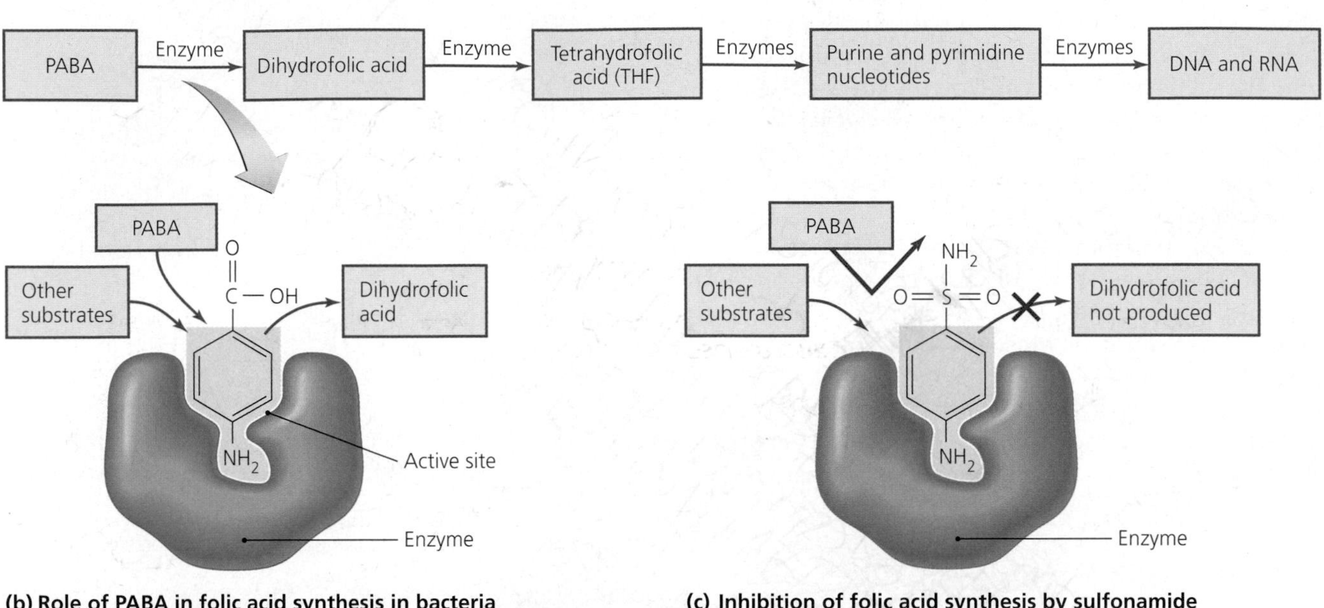

(b) Role of PABA in folic acid synthesis in bacteria and protozoa

(c) Inhibition of folic acid synthesis by sulfonamide

▲ **Figure 10.6 The antimetabolic action of sulfonamides in inhibiting nucleic acid synthesis.** **(a)** Para-aminobenzoic acid (PABA) and representative members of its structural analogs, the sulfonamides. The analogous portions of the compounds are shaded. **(b)** The metabolic pathway in bacteria and protozoa by which folic acid is synthesized from PABA. **(c)** The inhibition of folic acid synthesis by the presence of a sulfonamide, which deactivates the enzyme by binding irreversibly to the enzyme's active site.

of the acidic environment within phagolysosomes. *Amantadine, rimantadine,* and weak organic bases can neutralize the acid of phagolysosomes and thereby prevent viral uncoating; thus, these are antiviral drugs. Amantadine is used exclusively to prevent infections by influenza type A virus.

Protease inhibitors interfere with the action of protease—an enzyme that HIV needs near the end of its replication cycle. These drugs, when used as part of a "cocktail" of drugs including reverse transcriptase inhibitors (discussed shortly), have revolutionized treatment of AIDS patients in industrialized countries. Researchers have reduced the number of pills in the daily cocktail to just a few that contain all the drugs formerly found in 35 or more daily pills.

CRITICAL **THINKING**

It would be impractical and expensive for every American to take amantadine during the entire flu season to prevent influenza infections. For what group of people might amantadine prophylaxis be cost effective?

Inhibition of Nucleic Acid Synthesis

As we saw in Chapter 7, the nucleic acids DNA and RNA are built from purine and pyrimidine nucleotides and are critical to the survival of cells. Several drugs function by blocking either the replication of DNA or its transcription into RNA.

Because only slight differences exist between the DNA of prokaryotes and eukaryotes, drugs that affect DNA replication often act against both types of cells. For example, *actinomycin* binds to DNA and effectively blocks DNA synthesis and RNA transcription not only in bacterial pathogens but in their hosts as well. Generally, drugs of this kind are not used to treat infections, though they are used in research of DNA replication and may be used judiciously to slow replication of cancer cells.

Other compounds that can act as antimicrobials by interfering with the function of nucleic acids are called **nucleotide analogs** because of the compounds' structural similarities to the normal nucleotide building blocks of nucleic acids **(Figure 10.7)**. The structures of certain nucleotide analogs enable them to be incorporated into the DNA or RNA of pathogens, where they

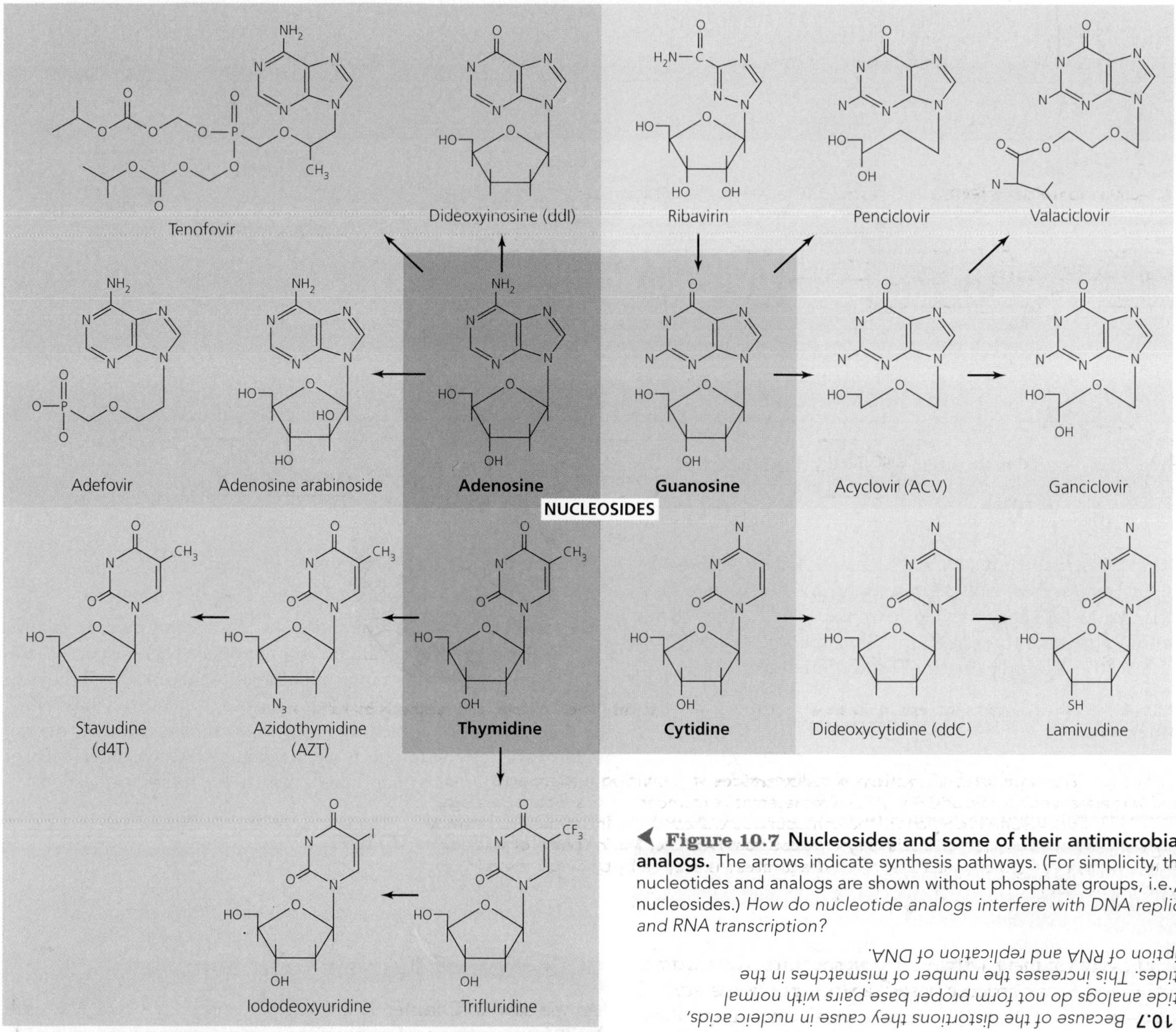

Figure 10.7 Nucleosides and some of their antimicrobial analogs. The arrows indicate synthesis pathways. (For simplicity, the nucleotides and analogs are shown without phosphate groups, i.e., as nucleosides.) *How do nucleotide analogs interfere with DNA replication and RNA transcription?*

Figure 10.7 *Because of the distortions they cause in nucleic acids, nucleotide analogs do not form proper base pairs with normal nucleotides. This increases the number of mismatches in the transcription of RNA and replication of DNA.*

distort the shapes of the nucleic acid molecules and prevent further replication, transcription, or translation. Nucleotide analogs are most often used against viruses because viral DNA polymerases are tens to hundreds of times more likely to incorporate nonfunctional nucleotides into nucleic acids than is human DNA polymerase. Additionally, complete viral nucleic acid synthesis is more rapid than cellular nucleic acid synthesis. These characteristics make viruses more susceptible to nucleotide analogs than their hosts are, though nucleotide analogs are also effective against rapidly dividing cancer cells.

The synthetic drugs called *quinolones* and *fluoroquinolones* are unusual because they are active against prokaryotic DNA specifically. These antibacterial agents inhibit *DNA gyrase,* an enzyme necessary for correct coiling and uncoiling of replicating bacterial DNA; they typically have little effect on eukaryotes or viruses.

Other antimicrobial agents function by binding to and inhibiting the action of RNA polymerases during the synthesis of RNA from a DNA template. Several drugs, including *rifampin* (rif'am-pin), bind more readily to prokaryotic RNA polymerase than to eukaryotic RNA polymerase; as a result, rifampin is more toxic to prokaryotes than to eukaryotes. Rifampin is used primarily against *M. tuberculosis* and other pathogens that metabolize slowly and thus are less susceptible to antimicrobials targeting active metabolic processes.

Clofazimine binds to the DNA of *Mycobacterium leprae* (lep'rī), the causative agent of leprosy, and prevents normal replication and transcription. It is also used to treat tuberculosis and other mycobacterial infections. *Pentamidine* and *propamidine isethionate* bind to protozoan DNA, inhibiting the pathogen's reproduction and development.

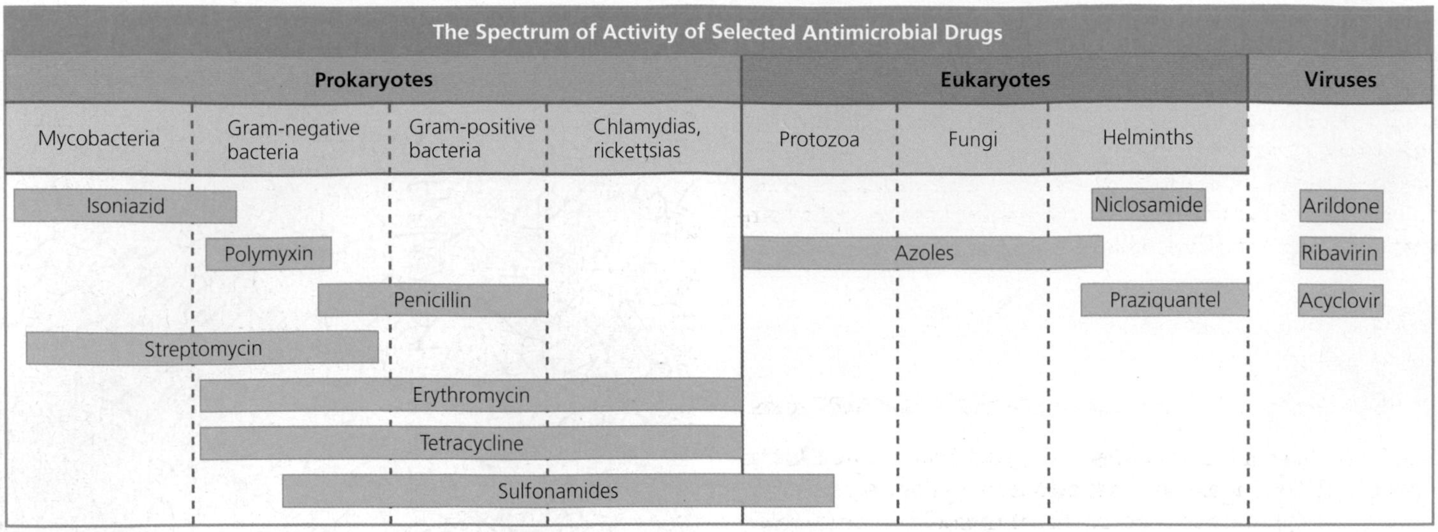

The Spectrum of Activity of Selected Antimicrobial Drugs

Prokaryotes				Eukaryotes			Viruses
Mycobacteria	Gram-negative bacteria	Gram-positive bacteria	Chlamydias, rickettsias	Protozoa	Fungi	Helminths	

Isoniazid
Polymyxin
Penicillin
Streptomycin
Erythromycin
Tetracycline
Sulfonamides
Azoles
Niclosamide
Praziquantel
Arildone
Ribavirin
Acyclovir

▲ **Figure 10.8 Spectrum of action for selected antimicrobial agents.** The more kinds of pathogens a drug affects, the broader its spectrum of action.

Reverse transcriptase inhibitors, which are part of AIDS cocktails, act against reverse transcriptase, which is an enzyme HIV uses early in its replication cycle to make DNA copies of its RNA genome (reverse transcription). Since people lack reverse transcriptase, the inhibitor does not harm patients.

Prevention of Virus Attachment

Many pathogens, particularly viruses, must attach to their host's cells via the chemical interaction between attachment proteins on the pathogen and complementary receptor proteins on a host cell. Attachment of viruses can be blocked by peptide and sugar analogs of either attachment or receptor proteins. When these sites are blocked by analogs, viruses can neither attach to nor enter their hosts' cells. The use of such substances, called *attachment antagonists,* is an exciting new area of antimicrobial drug development. *Arildone* and *pleconaril* are antagonists of the receptor of polioviruses and some cold viruses. They block attachment of these viruses and deter infections.

Clinical Considerations in Prescribing Antimicrobial Drugs

Even though antibiotics are produced commonly by many fungi and bacteria, most of these chemicals are not effective for treating diseases because they are toxic to humans and animals, are too expensive, are produced in minute quantities, or lack adequate potency. The ideal antimicrobial agent to treat an infection or disease would be one that is:

- Readily available
- Inexpensive
- Chemically stable (so that it can be transported easily and stored for long periods of time)
- Easily administered
- Nontoxic and nonallergenic
- Selectively toxic against a wide range of pathogens

No agent has all of these qualities, so doctors and medical laboratory technicians must evaluate antimicrobials with respect to several characteristics: the range of pathogens against which they are effective; their efficacy, including the dosages required to be effective; the routes by which they can be administered; their overall safety; and any side effects they produce. We consider each of these characteristics of antimicrobials in the following sections.

Spectrum of Action

Learning Objective

✓ Distinguish between narrow-spectrum drugs and broad-spectrum drugs in terms of their targets and side effects.

The number of different kinds of pathogens a drug acts against is known as its **spectrum of action (Figure 10.8)**; drugs that work against only a few kinds of pathogens are **narrow-spectrum drugs,** whereas those that are effective against many different kinds of pathogens are **broad-spectrum drugs.** For instance, because tetracycline acts against many different kinds of bacteria, including Gram-negative, Gram-positive, chlamydias, and rickettsias, it is considered a broad-spectrum antibiotic. In contrast, penicillin cannot easily penetrate the outer membrane of a Gram-negative bacterium to reach and prevent the formation of peptidoglycan, so its efficacy is largely limited to Gram-positive bacteria. Thus penicillin has a narrower spectrum of action than erythromycin, which acts against protein synthesis in numerous bacteria.

The use of broad-spectrum antimicrobials is not always as desirable as it might seem. Broad-spectrum antimicrobials can also open the door to serious secondary infections by transient pathogens or *superinfections* by members of the normal microbiota unaffected by the antimicrobial. This results because the killing of normal microbiota reduces *microbial antagonism,* the competition between normal microbes and pathogens for nutrients and space. Microbial antagonism reinforces the body's defense by limiting the ability of pathogens to colonize the skin and mucous membranes. Thus a woman using erythromycin to treat strep throat (a bacterial disease) could develop vaginitis resulting from the excessive growth of *Candida albicans* (kan'did-ă al'bi-kanz), a yeast that is unaffected by erythromycin and is freed from microbial antagonism when the antibiotic kills normal bacteria in the vagina.

Efficacy

Learning Objective

✓ Compare and contrast Kirby-Bauer, Etest, MIC, and MBC tests.

To effectively treat infectious diseases, physicians must know which antimicrobial agent is most effective against a particular pathogen. To ascertain the efficacy of antimicrobials, microbiologists conduct a variety of tests, including diffusion susceptibility tests, the minimum inhibitory concentration test, and the minimum bactericidal concentration test.

Diffusion Susceptibility Test

Diffusion susceptibility tests, also known as *Kirby-Bauer* tests, involve uniformly inoculating a Petri plate with a standardized amount of the pathogen in question. Then small disks of paper containing standard concentrations of the drugs to be tested are firmly arranged on the surface of the plate. The plate is incubated, and the bacteria grow and reproduce to form a "lawn" everywhere but the areas where effective antimicrobial drugs diffuse through the agar. After incubation, the plates are examined for the presence of a **zone of inhibition**—that is, a clear area where bacteria do not grow (Figure 10.9). A zone of inhibition is measured as the diameter (to the closest millimeter) of the clear region.

If all drugs were equal, then the larger the zone of inhibition, the more effective that drug is; however, the size of a zone depends on the rate of diffusion of the antimicrobial; for example, drugs with lower molecular weights generally diffuse more quickly than those with higher molecular weights. The size of a zone of inhibition must be compared to a standard table for that particular drug before accurate comparisons can be made. Diffusion susceptibility tests enable scientists to classify pathogens as *susceptible, intermediate,* or *resistant* to each drug.

CRITICAL **THINKING**

Sometimes it is not possible to conduct a susceptibility test, either because of a lack of time or an inability to access the bacteria (from an inner-ear infection, for instance). How could a physician select an appropriate therapeutic agent in such cases?

Minimum Inhibitory Concentration (MIC) Test

Once scientists identify an effective antimicrobial agent, they quantitatively express its potency as a **minimum inhibitory concentration (MIC).** As the name suggests, the MIC is the smallest amount of the drug that will *inhibit* growth and reproduction of the pathogen. The MIC can be determined via a **broth dilution test,** in which a standardized amount of bacteria is

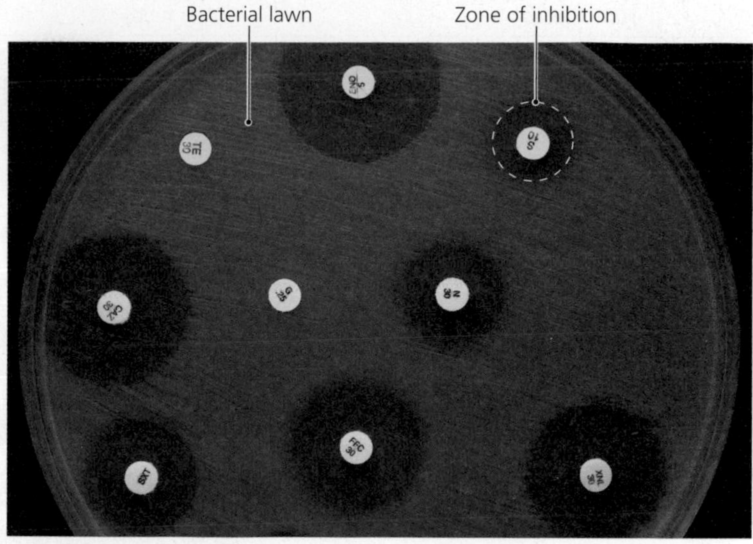

▲ **Figure 10.9 Zones of inhibition in a diffusion susceptibility (Kirby-Bauer) test.** In general, the larger the zone of inhibition around disks, which are impregnated with an antimicrobial agent, the more effective that antimicrobial is against the organism growing on the plate. The organism is classified as either susceptible, intermediate, or resistant to the antimicrobials tested based on the sizes of the zones of inhibition. *If all of these antimicrobial agents diffuse at the same rate and are equally safe and easily administered, which one would be the drug of choice for killing this pathogen?*

Figure 10.9 *The drug ENO, a fluoroquinolone found in the uppermost disk, is most effective.*

added to serial dilutions of antimicrobial agents in tubes or wells containing broth. After incubation, turbidity (cloudiness) indicates bacterial growth; lack of turbidity indicates that the bacteria were either inhibited or killed by the antimicrobial agent (Figure 10.10). Dilution tests can be conducted simultaneously in wells, and the entire process can be automated, with turbidity measured by special scanners connected to computers.

Another test that determines minimum inhibitory concentration combines aspects of an MIC test and a diffusion susceptibility test. This test, called an **Etest,**[2] involves placing a plastic strip containing a gradient of the antimicrobial agent being tested on a plate uniformly inoculated with the organism of interest (Figure 10.11). After incubation, an elliptical zone of inhibition indicates antimicrobial activity, and the minimum inhibitory concentration can be noted where the zone of inhibition intersects a scale printed on the strip.

Minimum Bactericidal Concentration (MBC) Test

Similar to the MIC test is a **minimum bactericidal concentration (MBC) test,** though an MBC test determines the amount of drug required to kill the microbe rather than just the amount to inhibit it, as the MIC does. In an MBC test, samples taken from clear MIC tubes (or alternatively, from zones of inhibition from a series of diffusion susceptibility tests) are transferred to plates containing a drug-free growth medium (Figure 10.12). The appearance of bacterial growth in these subcultures after appropriate incubation

[2]The name *Etest* has no specific origin.

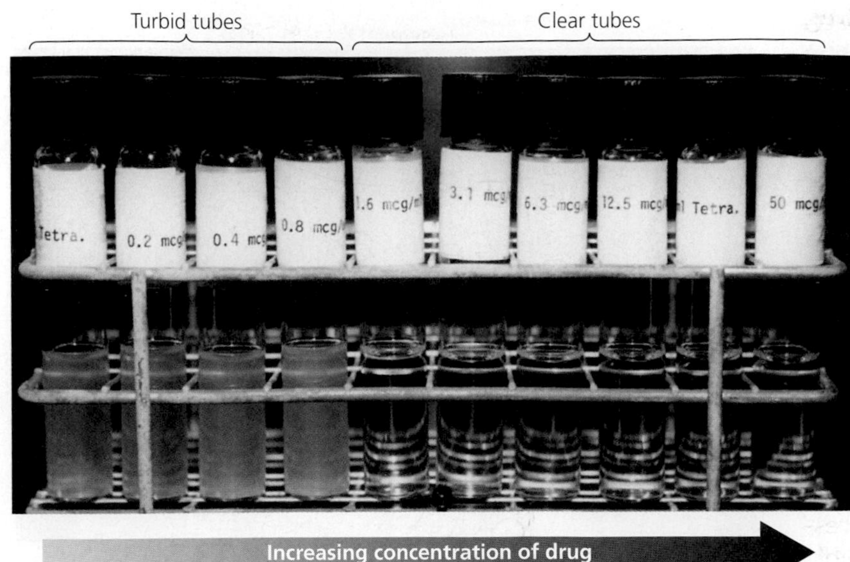

Figure 10.10 Minimum inhibitory concentration (MIC) test in test tubes. *What is the MIC for the drug acting against this bacterium?*

Figure 10.10 The MIC is 1.6 mcg (μg) per ml.

indicates that at least some bacterial cells survived that concentration of the antimicrobial drug and were able to grow and multiply once placed in a drug-free medium. Any drug concentration at which growth occurs in subculture is *bacteriostatic,* not *bactericidal,* for that bacterium. The lowest concentration of drug for which no growth occurs in the subcultures is the minimum bactericidal concentration (MBC).

Routes of Administration

Learning Objective

✓ Discuss the advantages and disadvantages of the different routes of administration of antimicrobial drugs.

An adequate amount of an antimicrobial agent must reach a site of infection if it is to be effective. For external infections such as athlete's foot, drugs can be applied directly. This is known as *topical* or *local* administration. For internal infections, drugs can be administered *orally, intramuscularly (IM),* or *intravenously (IV).* Each route has advantages and disadvantages.

Even though the oral route is simplest (it requires no needles and is self-administered), the drug concentrations achieved in the body are lower than occur via other routes of administration

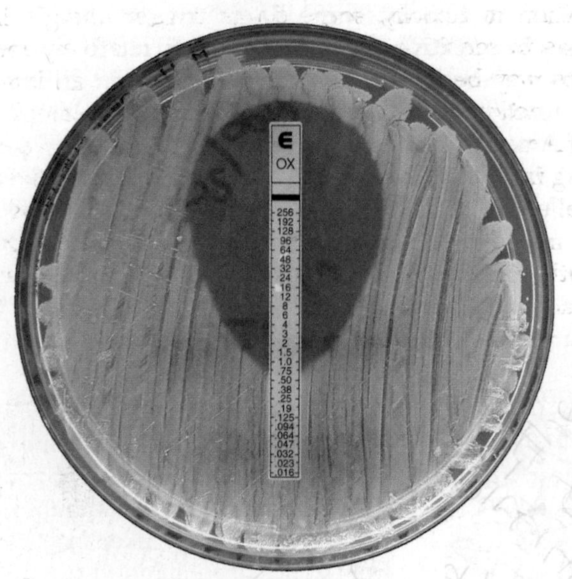

Figure 10.11 An Etest, which combines aspects of Kirby-Bauer and MIC tests. The plastic strip contains a gradient of the antimicrobial agent of interest. The MIC is estimated to be the concentration printed on the strip where the zone of inhibition intersects the strip. In this example, the MIC is 0.75 μg/ml.

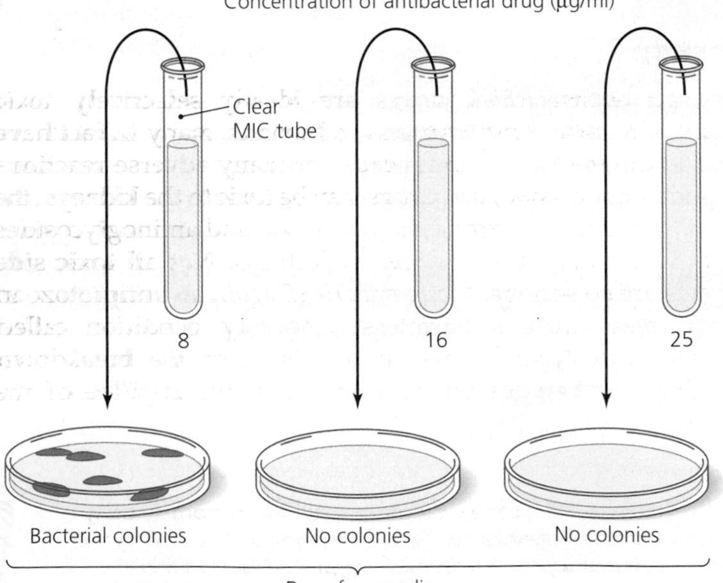

Figure 10.12 A minimum bactericidal concentration (MBC) test. In this test, plates containing a drug-free growth medium are inoculated with samples taken from zones of inhibition or from clear MIC tubes. After incubation, growth of bacterial colonies on a plate indicates that the concentration of antimicrobial drug (in this case, 8 μg/ml) is bacteriostatic. The lowest concentration for which no bacterial growth occurs on the plate is the minimum bactericidal concentration; in this case, the MBC is 16 μg/ml.

(**Figure 10.13**). Further, because patients administer the drug themselves, they do not always follow prescribed timetables.

IM administration via a hypodermic needle allows a drug to diffuse slowly into the many blood vessels within muscle tissue, but the concentration of the drug in the blood is never as high as that achieved by IV administration, which delivers the drug directly into the bloodstream through either a needle or a catheter (a plastic or rubber tube). Even though the amount of the drug in the blood is initially very high for the IV route, the concentration can rapidly diminish as the liver and kidneys remove the drug from the circulation, unless the drug is continuously administered.

In addition to considering the route of administration, physicians must consider how antimicrobial agents are distributed to infected tissues by the blood. For example, an agent removed rapidly from the blood by the kidneys might be the drug of choice for a bladder infection but would not be chosen to treat an infection of the heart. Finally, given that blood vessels in the brain, spinal cord, and eye are almost impermeable to many antimicrobial agents (because the tight structure of capillary walls in these structures creates the blood-brain barrier), infections there are often difficult to treat.

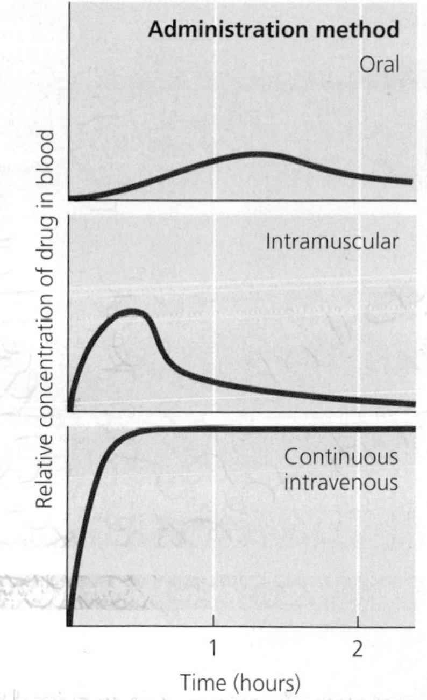

▲ **Figure 10.13 The effect of route of administration on blood levels of a chemotherapeutic agent.** Although intravenous (IV) and intramuscular (IM) administration achieve higher drug concentrations in the blood, oral administration has the advantage of simplicity.

Safety and Side Effects

Learning Objective

✓ Identify three main categories of side effects of antimicrobial therapy.

Another aspect of chemotherapy that physicians must consider is the possibility of adverse side effects. These fall into three main categories—toxicity, allergies, and disruption of normal microbiota.

Toxicity

Though antimicrobial drugs are ideally selectively toxic against microbes and harmless to humans, many in fact have toxic side effects. The exact cause of many adverse reactions is poorly understood, but drugs may be toxic to the kidneys, the liver, or nerves. For example, polymyxin and aminoglycosides can have fatally toxic effects on kidneys. Not all toxic side effects are so serious. *Metronidazole (Flagyl)*, an antiprotozoan drug, may cause a harmless temporary condition called "black hairy tongue," which results when the breakdown products of hemoglobin accumulate in the papillae of the tongue (**Figure 10.14a**).

Doctors must be especially careful when prescribing drugs for pregnant women, as many drugs that are safe for adults can have adverse affects when absorbed by a fetus. For instance, tetracyclines form complexes with calcium that can become incorporated into bones and developing teeth, causing malformation of the skull and stained, weakened tooth enamel (**Figure 10.14b**).

Allergies

In addition to toxicity, some drugs trigger allergic immune responses in sensitive patients. Although relatively rare, such reactions may be life threatening, especially in an immediate, violent reaction called *anaphylactic shock*. For example, about 0.1% of Americans have an anaphylactic reaction to penicillin, resulting in approximately 300 deaths per year. However, not every allergy to an antimicrobial agent is so serious. Recent studies indicate that patients with mild allergies to penicillin frequently lose their sensitivity to it over time. Thus, an initial mild reaction to penicillin need not preclude its use in treating future infections. Chapter 18 discusses allergies in more detail.

▶ **Figure 10.14 Some side effects resulting from toxicity of antimicrobial agents. (a)** "Black hairy tongue," caused by the antiprotozoan drug metronidazole (Flagyl). **(b)** Discoloration and damage to tooth enamel caused by tetracycline. *Who should avoid taking tetracycline?*

Figure 10.14 *Pregnant women and children should not use tetracycline.*

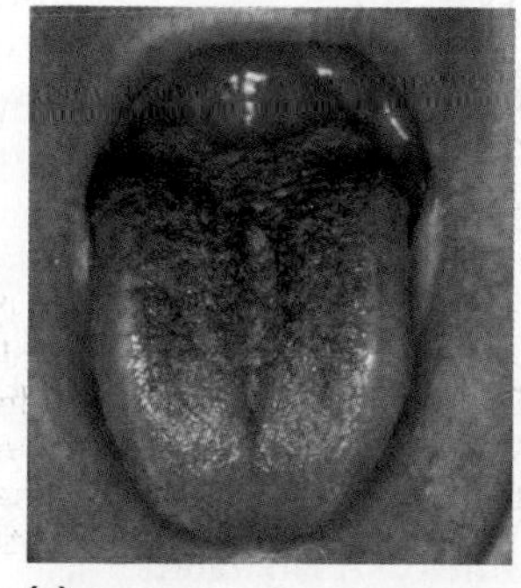

(a)

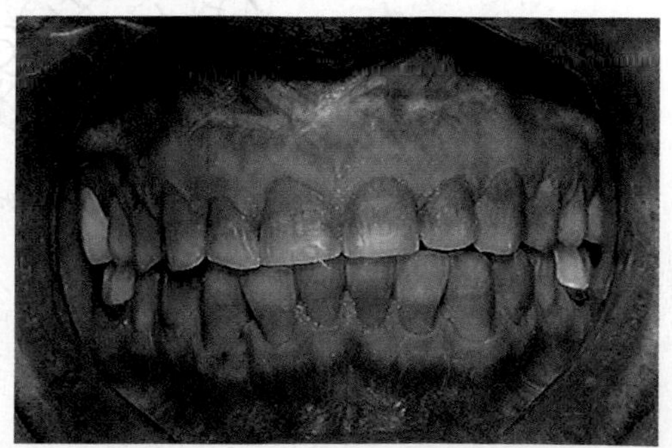

(b)

EMERGING DISEASES

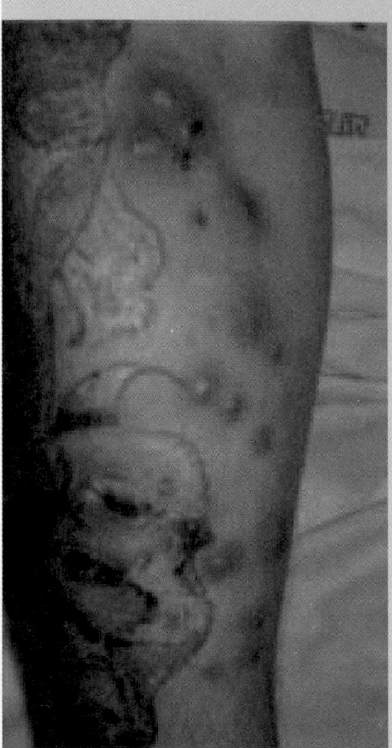

COMMUNITY-ASSOCIATED MRSA

Julie was proud and excited about her new tattoo; it was crunk! Jason, her boyfriend, had bought it as a gift for her 18th birthday, and she loved the way it covered her arm. A week later something else covered her arm . . . and his leg.

Julie and Jason had fallen victim to an emerging disease—community-associated, methicillin-resistant *Staphylococcus aureus* (CA-MRSA) infection. Julie's arm was covered with red, swollen, painful, pus-containing lesions, and she had a fever.

Jason thought he had a spider bite on Wednesday, but by Thursday morning a bright red line extended from his ankle to his groin. He could barely walk, and then his condition worsened. By the weekend, he was hospitalized, and his fever hit 107°F during his 10 days there. MRSA had entered his blood (bacteremia) and infected his bones (osteomyelitis). Jason's pain was so severe that he wondered only when he might die to be free of the agony.

For years health care workers have battled nosocomial MRSA, which commonly afflicts patients in hospitals. Now, researchers are concerned that victims who have never been in a hospital are succumbing; MRSA has escaped hospitals and now travels in the community, including among athletes, students in middle schools, and customers of unsafe tattooists. The bacterium can be spread between individuals who share fomites—towels, razors, clothing, or sheets.

Physicians drained Julie's lesions and prescribed oral antimicrobials, including trimethoprim, sulfamethoxazole, levofloxacin, and clindamycin. Jason received intravenous vancomycin. Both eventually recovered, but her tattoo is now a reminder of a terrible experience and not a happy birthday. For more about MRSA, see p. 538.

Track community-associated MRSA online by going to the Study Area at www.masteringmicrobiology.com.

Disruption of Normal Microbiota

As we have seen, drugs that disrupt normal microbiota and their microbial antagonism of opportunistic pathogens may result in

secondary infections. In instances when a member of the normal microbiota is not affected by a drug, it can overgrow, causing a superinfection. For example, long-term use of broad-spectrum antibacterials often results in explosions in the growth rate of *Candida albicans* in the vagina (vaginitis) or mouth (thrush) and the multiplication of *Clostridium difficile* (klos-trid'ē-ŭm di-fi'sil-ē) in the colon, causing a potentially fatal condition called *pseudomembranous colitis*. Such superinfections are of greatest concern for hospitalized patients, who are often not only debilitated but also more likely to be exposed to pathogens with resistance to antimicrobial drugs—the topic of the next section.

Beneficial Microbes: Probiotics: The New Sheriff in Town focuses on how normal microbiota can help keep pathogens in check.

Resistance to Antimicrobial Drugs

Among the major challenges facing microbiologists today are the problems presented by pathogens that are resistant to antimicrobial agents (see **Emerging Diseases: Community–Associated MRSA**). In the sections that follow we examine the development of resistant populations of pathogens, the mechanisms by which pathogens are resistant to antimicrobials, and some ways that resistance can be retarded.

The Development of Resistance in Populations

Learning Objectives

- ✓ Describe how populations of resistant microbes can arise.
- ✓ Describe the relationship between R-plasmids and resistant cells.

Not all pathogens are equally sensitive to a given therapeutic agent; a population may contain a few organisms that are naturally either partially or completely resistant. Among bacteria, individual cells can acquire such resistance in two ways: through new mutations of chromosomal genes or by acquiring resistance genes on extrachromosomal pieces of DNA called **R-plasmids** (or *R-factors*) via the processes of horizontal gene transfer—transformation, transduction, or conjugation. We focus here on resistance in populations of bacteria, but resistance is known to occur among protozoa and viruses.

The process by which a resistant strain of bacteria develops is depicted in **Figure 10.15**. In the absence of an antimicrobial drug, resistant cells are usually less efficient than their normal

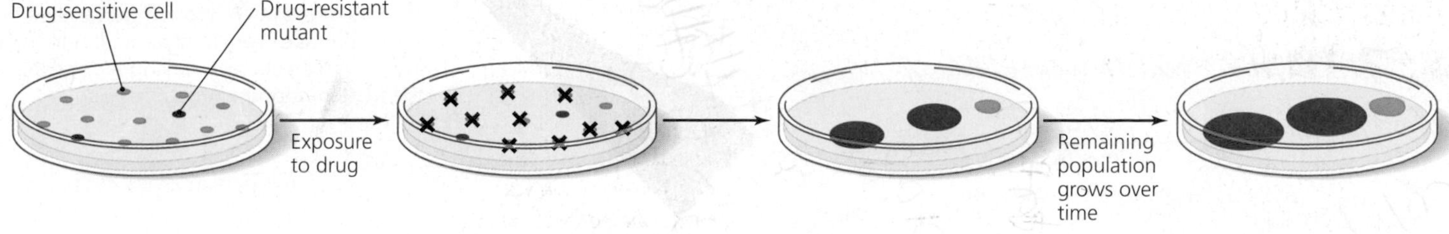

▲ **Figure 10.15 The development of a resistant strain of bacteria. (a)** A bacterial population contains both drug-sensitive and drug-resistant cells, although sensitive cells constitute the vast majority of the population. **(b)** Exposure to an antimicrobial drug inhibits the sensitive cells; so long as the drug is present, reduced competition from sensitive cells facilitates the multiplication of resistant cells. **(c)** Eventually resistant cells constitute the majority of the population. *Why do resistant strains of bacteria more often develop in hospitals and nursing homes than in college dormitories?*

Figure 10.15 *Resistant strains are more likely to develop in hospitals and other health care facilities because the extensive use of antimicrobial agents in those places inhibits the growth of sensitive strains and selects for the growth of resistant strains.*

neighbors because they must expend extra energy to maintain resistance genes and proteins. Under these circumstances, resistant cells remain the minority in a population because they reproduce more slowly. However, when an antimicrobial agent is present, the majority of cells (which are sensitive to the antimicrobial) are inhibited or die while the resistant cells continue to grow and multiply, often more rapidly because they then face less competition. The result is that resistant cells soon replace the sensitive cells as the majority in the population. The bacterium has evolved resistance. It should be noted that the presence of the antimicrobial agent does not *produce* resistance but instead selects for the replication of resistant cells that were already present in the population. **ANIMATIONS:** ***Antibiotic Resistance: Origins of Resistance***

CRITICAL **THINKING**

Enterococcus faecium is frequently resistant to vancomycin. Why might this be of concern in a hospital setting in terms of developing resistant strains of *other* genera of bacteria?

Mechanisms of Resistance

Learning Objective

✓ List six ways by which microorganisms can be resistant to antimicrobial drugs.

The problem of resistance to antimicrobial drugs is a major health threat to our world. An "alphabet soup" of resistant pathogens and diseases plague health care professionals: MRSA,[3] VRSA,[4] VISA,[5] VRE,[6] and MDR-TB[7] are just some of these.

How do microbes gain resistance to antimicrobial drugs? Consider the path a typical antimicrobial drug must take to affect a microbe: The drug must cross the cell's wall, then cross the cytoplasmic membrane to enter the cell; there the antimicrobial

[3]Methicillin-resistant *S. aureus*
[4]Vancomycin-resistant *S. aureus*
[5]*S. aureus* with intermediate level of resistance to vancomycin
[6]Vancomycin-resistant enterococci
[7]Multi-drug-resistant tuberculosis

BENEFICIAL MICROBES

PROBIOTICS: THE NEW SHERIFF IN TOWN

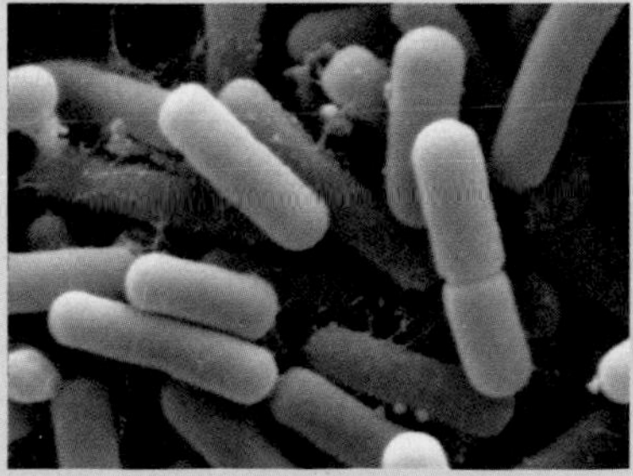

▲ ***Lactobacillus reuteri.***

As overuse of antimicrobials has allowed more bacteria that are drug resistant to thrive, scientists are investigating alternative methods of combating microbial infections. One growing field of interest is *probiotics*, the use of microorganisms for health benefits.

Probiotics include bacteria such as *Lactobacillus*, which may help reduce symptoms of diarrhea in children, relieve milk allergies, and alleviate certain respiratory infections. One of the central ideas behind probiotics is that "good" bacteria, such as *Lactobacillus* (which lives naturally and harmlessly in our intestines and other parts of the body), compete with harmful microorganisms for resources, keeping pathogenic microbes in check. Some evidence suggests that *Lactobacillus* also induces the expression of certain proteins that protect the intestines against harmful bacteria and viruses. Much more research remains to be done, but the preliminary findings are encouraging.

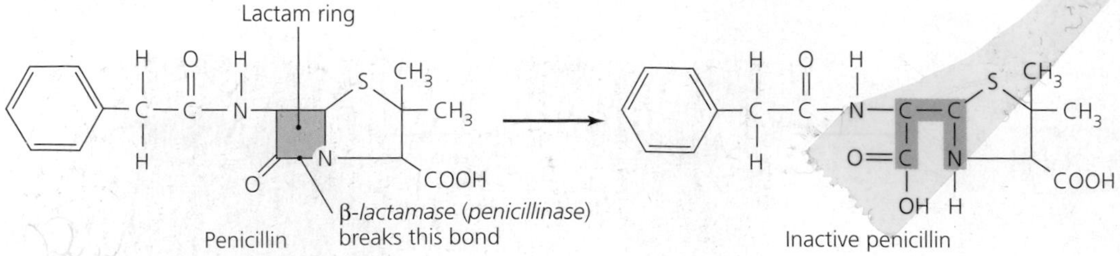

Figure 10.16 **How β-lactamase (penicillinase) renders penicillin inactive.** The enzyme acts by breaking a bond in the lactam ring, the functional portion of the drug.

binds to its target (receptor) molecule. Only then can it inhibit or kill the microbe. Microbes gain resistance by blocking some point in this pathway. There are at least six mechanisms of resistance:

- Resistant cells may produce an enzyme that destroys or deactivates the drug. This common mode of resistance is exemplified by **beta-(β) lactamases** (penicillinases), which are enzymes that break the beta-lactam rings of penicillin and similar molecules, rendering them inactive (**Figure 10.16**). Many MRSA strains have evolved resistance to penicillin-derived antimicrobials in this manner. Over 200 different lactamases have been identified. Frequently their genes are located on R-plasmids.
- Resistant microbes may slow or prevent the entry of the drug into the cell. This mechanism typically involves changes in the structure or electrical charge of the cytoplasmic membrane proteins that constitute channels or pores. Such proteins in the outer membranes of Gram-negative bacteria are called *porins* (see Figure 3.15b). Altered pore proteins result from mutations in chromosomal genes. Resistance against tetracycline and penicillin are known to occur via this mechanism.
- Resistant cells may alter the target of the drug so that the drug either cannot attach to it or binds it less effectively. This form of resistance is often seen against antimetabolites (such as sulfonamides) and against drugs that thwart protein translation (such as erythromycin).
- Resistant cells may alter their metabolic chemistry, or they may abandon the sensitive metabolic step altogether. For example, a cell may become resistant to a drug by producing more enzyme molecules for the affected metabolic pathway, effectively reducing the power of the drug. Alternatively, cells become resistant to sulfonamides by abandoning the synthesis of folic acid, absorbing it from the environment instead.
- Resistant cells may pump the antimicrobial out of the cell before the drug can act. So-called *resistance pumps,* which are typically powered by ATP, are often able to pump more than one type of antimicrobial from a cell. Some microbes become multi-drug resistant (perhaps to as many as 10 or more drugs) by utilizing resistance pumps.
- Some resistant strains of the bacterium *Mycobacterium tuberculosis* have a novel method of resistance against fluoroquinolone drugs that bind to DNA gyrase. These strains synthesize an unusual protein that forms a negatively charged, rodlike helix about the width of a DNA molecule. This protein, called *MfpA protein,* binds to DNA gyrase in place of DNA, depriving fluoroquinolone of its target site. This is the first method of antibiotic resistance that involves protecting the target of an antimicrobial drug rather than, say, changing the target or deactivating the drug. MfpA protein probably slows down cellular division of *M. tuberculosis,* but that is better for the bacterium than being killed. **ANIMATIONS:** ***Antibiotic Resistance: Forms of Resistance***

Multiple Resistance and Cross Resistance

Learning Objective

✓ Define *cross resistance,* and distinguish it from multiple resistance.

A given pathogen can acquire resistance to more than one drug at a time, especially when resistance is conferred by R-plasmids, which are exchanged readily among bacterial cells. Such multiresistant strains of bacteria frequently develop in hospitals and nursing homes, where the constant use of many kinds of antimicrobial agents eliminates sensitive cells and encourages the development of resistant strains.

Multiple-drug-resistant pathogens (commonly called *superbugs*) are resistant to three or more types of antimicrobial agents. Multiple-drug-resistant strains of *Staphylococcus, Streptococcus* (strep-tō-kok'ŭs), *Enterococcus* (en'ter-ō-kok'ŭs), *Pseudomonas, Mycobacterium tuberculosis,* and *Plasmodium* (plaz-mō'dē-ŭm) (the protozoan that causes malaria) pose unique problems for health care professionals, who must treat infected patients without effective antimicrobials while taking extra care to protect themselves and other patients from infection.

Resistance to one antimicrobial agent may confer resistance to similar drugs, a phenomenon called **cross resistance.** Cross resistance typically occurs when drugs are similar in structure. For example, resistance to one aminoglycoside drug, such as streptomycin, may confer resistance to similar aminoglycoside drugs.

Retarding Resistance

Learning Objective

✓ Describe four ways that development of resistance can be retarded.

The development of resistant populations of pathogens can be averted in at least four ways. First, sufficiently high concentrations of the drug can be maintained in a patient's body for a long enough time to kill all sensitive cells and inhibit others long enough for the body's defenses to defeat them. Discontinuing a drug before all of the pathogens have been neutralized promotes the development of resistant strains. For this reason, it is important that patients finish their entire antimicrobial prescription and resist the temptation to "save some for another day."

CLINICAL CASE STUDY

To Treat or Not to Treat?

A young Hispanic mother brought her frail infant to a South Texas emergency room. While she waited, her infant began to have seizures. The staff stabilized the baby, while the mother explained that the child had been ill for many days. Feverish at times, the infant had lost weight because she was too short of breath to nurse. Other family members were ill with bad coughs.

The baby girl was later diagnosed with a dangerous form of tuberculosis (TB) that affected her brain. She was hospitalized and placed in isolation. Health department officials investigated and discovered that the infant had been exposed to TB from two individuals: an infected uncle and her TB-infected father whom the mother had visited in a Mexican jail. Inmates from the jail carried a potent strain of TB that was resistant to several standard anti-tuberculosis drugs. It would take many weeks to determine which of the two strains was affecting her brain.

Doctors had to make a tough decision: Should they immediately begin treating the infant for the multi-drug-resistant strain (MDR-TB), which involves using five to seven different drugs with multiple and painful side effects for many months, or treat her for the more normal form of TB using a more typical and less stressful drug regimen for an equal length of time?

1. What issues must be considered in order to determine the drug therapy for the infant?
2. What treatment would you guess was used for the infant?

Reference: Adapted from *MMWR* 40:373–375. 1991.

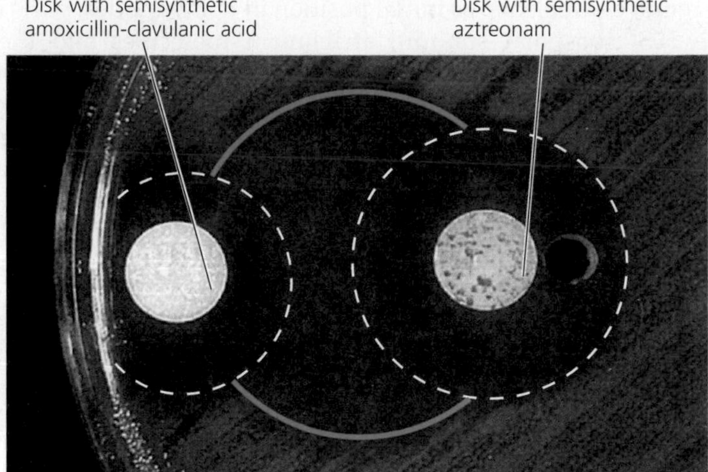

▲ **Figure 10.17 An example of synergism between two antimicrobial agents.** The portion of the zone of inhibition outlined in green represents the synergistic enhancement of antimicrobial activity beyond the activities of the individual drugs (outlined in white). *What does clavulanic acid do?*

Figure 10.17 *Clavulanic acid deactivates β-lactamase, allowing penicillins to work.*

A second way to avert resistance is to use antimicrobial agents in combination so that pathogens resistant to one drug will be killed by other drugs, and vice versa. Additionally, one drug sometimes enhances the effect of a second drug in a process called **synergism** (sin′er-jizm) **(Figure 10.17)**. In one example of a synergistic drug combination, the inhibition of cell wall formation by penicillin makes it easier for streptomycin molecules to enter bacteria and interfere with protein synthesis. Synergism can also result from combining an antimicrobial drug and a chemical, as occurs when *clavulanic acid* enhances the effect of penicillin by deactivating β-lactamase. (Not all drugs act synergistically; some combinations of drugs can be *antagonistic*—interfering with each other. For example, drugs that slow bacterial growth are antagonistic to the action of penicillin, which acts only against growing and dividing cells.)

A third way to reduce the development of resistance is to limit the use of antimicrobials to necessary cases. Unfortunately, many antimicrobial agents are used indiscriminately, both in developed countries and in less developed regions where many are available without a physician's prescription. In the United States, an estimated 50% of prescriptions for antibacterial agents to treat sore throats, and 30% of prescriptions for ear infections, are inappropriate because the diseases are viral, not bacterial. Likewise, because antibacterial drugs have no effect on cold and flu viruses, 100% of antibacterial prescriptions for treating these diseases are superfluous. As discussed previously, the use of antimicrobial agents encourages the reproduction of resistant bacteria by limiting the growth of sensitive cells; therefore, inappropriate use of such drugs increases the likelihood that resistant strains of bacteria will multiply.

Finally, scientists can combat resistant strains by developing new variations of existing drugs, in some cases by adding novel side chains to the original molecule. In this way, scientists develop semisynthetic *second-generation* drugs. If resistance develops to these drugs, *third-generation* drugs may be developed to replace them.

Alternatively, scientists are searching for new antibiotics, semisynthetics, and synthetics. Researchers are exploring diverse habitats such as peat bogs, ocean sediments, marine organisms, garden soil, and people's mouths for organisms that produce novel antibiotics. For example, a compound produced by parasitic hookworms that interferes with blood coagulation may mitigate the deadly effects of Ebola. Some researchers see potential in *bacteriocins*—antibacterial proteins coded by bacterial plasmids. Bacteria use bacteriocins to inhibit other bacterial strains; perhaps we can do the same. Tinkering with these and other antibiotics should yield promising new semisynthetics.

With the advent of genome sequencing and enhanced understanding of protein folding, some scientists predict that we are moving into a new golden era of antimicrobial drug discovery and development. They point out that researchers who know the exact shapes of microbial proteins should be able to design drugs complementary to those shapes—drugs that will inhibit microbial proteins without affecting humans. Currently scientists are examining ways to inhibit the following: secretion systems; peptide formylase, the critical enzyme used by bacter-

to remove fMet from its initial position in polypeptides; parts of the 50S ribosomal subunit; attachment molecules and their receptors; biofilm signaling molecules; bacterial cell division; and unique metabolic pathways of bacteria.

Despite researchers' best efforts, other health care professionals and scientists are concerned about how long drug developers can stay ahead of the development of resistance by pathogens.

Selected antimicrobial agents, their modes of action, clinical considerations, and other features are summarized in Tables 10.2–10.6. Particular use of antimicrobial drugs against specific pathogens is covered in the relevant chapters of this book.

TABLE 10.2 Antibacterial Drugs

Drug	Description and Mode of Action	Clinical Considerations	Method of Resistance
Antibacterial Drugs That Inhibit Cell Wall Synthesis			
Bacitracin	Isolated from *Bacillus licheniformis* growing on a patient named Tracy; appears to have three modes of action: • Interference with the movement of peptidoglycan precursors through the bacterial cell membrane to the cell wall • Inhibition of RNA transcription • Damage to the bacterial cytoplasmic membrane The latter two modes of action have not been proven definitely	**Spectrum of action:** Gram-positive (G+) bacteria **Route of administration:** Topical **Adverse effects:** Toxic to kidneys	Resistance most often involves changes in bacterial cell membranes that prevent bacitracin from entering the cell
Beta-lactams Representative natural penicillins: Penicillin G Penicillin V Representative semisynthetic penicillin: Ampicillin Methicillin Dicloxacillin Representative natural cephalosporin: Cephalothin Representative semisynthetic cephalosporins: Cefixime Ceftriaxone Cefuroxime Representative semisynthetic monobactam: Aztreonam	Large number of natural and semisynthetic derivatives from *Penicillium* (penicillins) and *Cephalosporium* (cephalosporins); bind to and deactivate the enzyme that cross-links the NAM subunits of peptidoglycan Monobactams have only a single ring instead of the two rings seen in other beta-lactams	**Spectrum of action:** Natural drugs have limited action against most Gram-negative (G−) bacteria because they do not readily cross the outer membrane; semisynthetics have broader spectra of action Monobactams have a limited spectrum of action, affecting only aerobic, G− bacteria **Route of administration:** Penicillin V, a few cephalosporins (e.g., cephalexin), and monobactams: oral; penicillin G, and many semisynthetics (e.g., methicillin, ampicillin, carbenicillin, cephalothin): IM or IV **Adverse effects:** Allergic reactions against beta-lactams in some adults; monobactams are least allergenic	Develops in three ways in G− bacteria: • Change their outer membrane structure to prevent entrance of the drug • Modify the enzyme so that the drug no longer binds • Synthesize beta-lactamases that cleave the functional lactam ring of the drug; genes for lactamases are often carried on R-plasmids
Cycloserine	Analog of alanine that interferes with the formation of alanine-alanine bridges between NAM subunits	**Spectrum of action:** Some G+ bacteria, mycobacteria **Route of administration:** Oral **Adverse effects:** Toxic to nervous system, producing depression, aggression, confusion, and headache	Some G+ bacteria enzymatically deactivate the drug
Ethambutol	Prevents the formation of mycolic acid; used in combination with other antimycobacterial drugs	**Spectrum of action:** Mycobacteria, including *M. tuberculosis* and *M. leprae* **Route of administration:** Oral **Adverse effects:** None	Resistance is due to random mutations of bacterial chromosomes that result in alteration of target site

TABLE 10.2 *(continued)*

Drug	Description and Mode of Action	Clinical Considerations	Method of Resistance
Antibacterial Drugs That Inhibit Cell Wall Synthesis			
Isoniazid (isonicotinic acid hydrazide, INH)	Blocks the gene for an enzyme that forms mycolic acid; analog of the vitamins nicotinamide and pyridoxine	**Spectrum of action:** Mycobacteria, including *M. tuberculosis* and *M. leprae* **Route of administration:** Oral **Adverse effects:** May be toxic to liver	Resistance is due to random mutations of bacterial chromosomes that result in alteration of target site or overproduction of target molecules
Vancomycin	Produced by *Amycolatopsis orientalis;* directly interferes with the formation of alanine-alanine bridges between NAM subunits	**Spectrum of action:** Effective against most G+ bacteria but generally reserved for use against strains resistant to other drugs such as methicillin-resistant *Staphylococcus aureus* (MRSA) **Route of administration:** IV **Adverse effects:** Damage to ears and kidneys, allergic reactions	G− bacteria are naturally resistant because the drug is too large to pass through the outer membrane; some G+ bacteria (e.g., *Lactobacillus*) are naturally resistant because they do not form alanine-alanine bonds between NAM subunits
Antibacterial Drugs That Inhibit Protein Synthesis			
Aminoglycosides Representatives: Gentamicin Kanamycin Neomycin Paromomycin Spiramycin Streptomycin	Compounds in which two or more amino sugars are linked with glycosidic bonds; were originally isolated from species of the bacterial genera *Streptomyces* and *Micromonospora.* Inhibit protein synthesis by irreversibly binding to the 30S subunit of prokaryotic ribosomes, this either causing the ribosome to mistranslate mRNA, producing aberrant proteins, or causing premature release of the ribosome from mRNA, which stops synthesis. Also bactericidal by destroying outer membranes of G− bacteria	**Spectrum of action:** Broad: effective against both G+ and G− bacteria **Route of administration:** IV; do not traverse blood-brain barrier **Adverse effects:** May be toxic to kidneys or auditory nerves, causing deafness	Uptake of these drugs is energy dependent, so anaerobic bacteria with less ATP available are less susceptible; aerobic bacteria alter membrane pores to prevent uptake, or synthesize enzymes that alter or degrade the drug once it enters; rarely, bacteria alter the binding site on the ribosome; some bacteria make biofilms when exposed to the drugs
Chloramphenicol	Rarely used drug that prevents prokaryotic ribosomes from moving along mRNA by binding to their 50S subunits	**Spectrum of action:** Broad, but rarely used except in treatment of typhoid fever **Route of administration:** Oral; traverses blood-brain barrier **Adverse effects:** In 1 of 24,000 patients, causes aplastic anemia, a potentially fatal condition in which blood cells fail to form; can also cause neurological damage	Develops via gene carried on an R-plasmid that codes for an enzyme that deactivates drug
Lincosamides Representative: Clindamycin	Binds to 50S ribosomal subunit and stops protein elongation	**Spectrum of action:** Effective against G+ and anaerobic G− bacteria **Route of administration:** Oral or IV; does not traverse blood-brain barrier **Adverse effects:** Gastrointestinal distress, including nausea, diarrhea, vomiting, and pain	Develops via changes in ribosomal structure that prevent drug from binding; resistance genes are same as those of aminoglycosides
Macrolides Representatives: Azithromycin Erythromycin Telithromycin	Group of antimicrobials typified by a macrocyclic lactone ring; the most prescribed is erythromycin, which is produced by *Streptomyces erythraeus;* act by binding to the 50S subunit of prokaryotic ribosomes and preventing the elongation of the nascent protein	**Spectrum of action:** Effective against G+ and a few G− bacteria **Route of administration:** Oral; do not traverse blood-brain barrier **Adverse effects:** Nausea, mild gastro-intestinal pain, vomiting; erythromycin increases risk of cardiac arrest	Develops via changes in ribosomal RNA that prevent drugs from binding, or via R-plasmid genes coding for the production of macrolide-digesting enzymes; resistance genes are same as those of lincosamides

continued

TABLE 10.2

Antibacterial Drugs *(continued)*

Drug	Description and Mode of Action	Clinical Considerations	Method of Resistance
Antibacterial Drugs That Inhibit Protein Synthesis			
Mupirocin	Produced by *Pseudomonas fluorescens*; binds to bacterial tRNA^{Ile}, which prevents delivery of isoleucine to ribosomes, blocking polypeptide synthesis	**Spectrum of action:** Primarily effective against G+ bacteria **Route of administration:** Topical cream **Adverse effects:** None reported	Resistance develops from mutations that change the shape of tRNA^{Ile}
Oxazolidinones Representative: Linezolid	Synthetic; inhibits initiation of polypeptide synthesis	**Spectrum of action:** G+ bacteria **Route of administration:** Oral, IV **Adverse effects:** Rash, diarrhea, loss of appetite, constipation, fever	Method of resistance not known
Streptogramin Representatives: Quinupristin Dalfopristin	Binds to 50S ribosomal subunit, stops protein synthesis; the synergistic drugs quinupristin and dalfopristin are taken together	**Spectrum of action:** Broad but reserved for use against multiple-drug-resistant strains **Route of administration:** IV **Adverse effects:** Muscle and joint pain	Not known
Tetracyclines Representatives: Doxycycline Tetracycline	Composed of four hexagonal rings with various side groups; prevent tRNA molecules, which carry amino acids, from binding to ribosomes at the 30S subunit's docking site	**Spectrum of action:** Most are broad: effective against many G+ and G− bacteria as well as against bacteria that lack cell walls, such as *Mycoplasma* **Route of administration:** Oral, cross poorly into brain **Adverse effects:** Nausea, diarrhea, sensitivity to light; forms complexes with calcium, which stains developing teeth and adversely affects the strength and shape of bones	Develops in three ways; bacteria may: • Alter gene for pores in outer membrane; new pore prevents drug from entering cell • Alter binding site on the ribosome to allow tRNA to bind even in presence of drug • Actively pump drug from cell
Antibacterial Drugs That Alter Cytoplasmic Membranes			
Gramicidin	Short polypeptide that forms pore across cytoplasmic membrane, allowing single-charged cations to cross freely	**Spectrum of action:** G+ bacteria **Route of administration:** Topical **Adverse effects:** Toxic (also forms pores in eukaryotic membranes)	Not known
Polymyxin	Produced by *Bacillus polymyxa;* destroys cytoplasmic membranes of susceptible cells	**Spectrum of action:** Effective against G+ bacteria, particularly *Pseudomonas,* and some amoebae **Route of administration:** Topical **Adverse effects:** Toxic to kidneys	Results from changes in cell membrane that prohibit entrance of the drug
Pyrazinamide	Disrupts membrane transport	**Spectrum of action:** *Mycobacterium tuberculosis* **Route of administration:** Oral **Adverse effects:** Malaise, nausea, diarrhea	Results from point mutations in bacterial gene for enzyme necessary to activate drug

TABLE 10.2 (continued)

Drug	Description and Mode of Action	Clinical Considerations	Method of Resistance
Antibacterial Drugs That Are Antimetabolites			
Dapsone	Interferes with synthesis of folic acid	**Spectrum of action:** *M. leprae, M. tuberculosis* **Route of administration:** Oral **Adverse effects:** Insomnia, headache, nausea, vomiting, increased heart rate	Not known
Sulfonamides Representatives: Sulfadiazine Sulfadoxine Sulfanilamide	Synthetic drugs; first produced as a dye; analogs of PABA that bind irreversibly to enzyme that produces dihydrofolic acid; synergistic with trimethoprim	**Spectrum of action:** Broad: effective against G+ and G− bacteria and some protozoa and fungi; however, resistance is widespread **Route of administration:** Oral **Adverse effects:** Rare: allergic reactions, anemia, jaundice, mental retardation of fetus if administered in last trimester of pregnancy	*Pseudomonas* is naturally resistant due to permeability barriers; cells that require folic acid as a vitamin are also naturally resistant; chromosomal mutations result in lowered affinity for the drugs
Trimethoprim	Blocks second metabolic step in the formation of folic acid from PABA; synergistic with sulfonamides	**Spectrum of action:** Broad: effective against G+ and G− bacteria and some protozoa and fungi; however, resistance is widespread **Route of administration:** Oral **Adverse effects:** Allergic reactions or liver damage in some patients	*Pseudomonas* is naturally resistant due to permeability barriers; cells that require folic acid as a vitamin are also naturally resistant; chromosomal mutations result in lowered affinity for the drug
Antibacterial Drugs That Inhibit Nucleic Acid Synthesis			
Clofazimine	Binds to DNA, preventing replication and transcription	**Spectrum of action:** Mycobacteria, especially *M. tuberculosis, M. leprae,* and *M. ulcerans* **Route of administration:** Oral **Adverse effects:** Diarrhea, discoloration of skin and eyes	Not known
Fluoroquinolones Representatives: Ciprofloxacin Moxifloxacin Ofloxacin	Synthetic agents that inhibit DNA gyrase, which is needed to correctly replicate bacterial DNA; penetrate cytoplasm of cells	**Spectrum of action:** Broad: G+ and G− bacteria are affected **Route of administration:** Oral **Adverse effects:** Tendonitis, tendon rupture	Results from chromosomal mutations that lower affinity for drug, reduce its uptake, or protect gyrase from drug
Nitroimidazoles Representative: Metronidazole	Anaerobic conditions reduce the molecule, which then damages DNA and prevents its correct replication	**Spectrum of action:** Obligate anaerobic bacteria	Not known
Rifamycin Representatives: Rifampin Rifaximin	Natural and semisynthetic derivatives from *Amycolatopsis rifamycinica* that bind to bacterial RNA polymerase, preventing transcription of RNA; used with other antimicrobial bacterial drugs	**Spectrum of action:** Bacteriostatic against aerobic G+ bacteria; bactericidal against mycobacteria **Route of administration:** Oral **Adverse effects:** None of major significance	Results from chromosomal mutation that alters binding site on enzyme; G− bacteria are naturally resistant due to poor uptake

TABLE 10.3 Antiviral Drugs

Drug	Description and Mode of Action	Clinical Considerations	Method of Resistance
Attachment Antagonists			
Arildone **Pleconaril**	Blocks attachment molecule on host cell or pathogen	**Spectrum of action:** Picornaviruses (e.g., poliovirus, some cold viruses) **Route of administration:** Oral **Adverse effects:** None	Not known
Neuraminidase inhibitors Representatives: Oseltamivir Zanamivir	Prevent influenzaviruses from attaching to or exiting from cells	**Spectrum of action:** Influenzavirus **Route of administration:** Oral (oseltamivir) or aerosol (zanamivir) **Adverse effects:** None	Not known
Antiviral Drugs That Inhibit Viral Uncoating			
Amantadine	Neutralizes acid environment within phagolysosomes that is necessary for viral uncoating	**Spectrum of action:** Influenza A virus **Route of administration:** Oral **Adverse effects:** Toxuc to central nervous system; results in nervousness, irritability, insomnia, and blurred vision	Mutation resulting in a single amino acid change in a membrane ion channel leads to viral resistance
Rimantadine	Neutralizes phagolysosomal acid, preventing viral uncoating	**Spectrum of action:** Influenza A virus **Route of administration:** Oral, adults only **Adverse effects:** Toxic to central nervous system; results in nervousness, irritability, insomnia, and blurred vision	Mutation resulting in a single amino acid change in a membrane ion channel leads to viral resistance
Antiviral Drugs That Inhibit Nucleic Acid Synthesis			
Acyclovir (ACV) Representative: Ganciclovir	Phosphorylation by virally coded kinase enzyme activates the drug; inhibits DNA and RNA synthesis	**Spectrum of action:** Viruses that code for kinase enzymes: herpes, Epstein-Barr, cytomegalovirus, varicella viruses **Route of administration:** Oral **Adverse effects:** None	Mutations in genes for kinase enzymes may render them ineffective at drug activation
Adenosine arabinoside	Phosphorylation by cell-coded kinase enzyme activates the drug: inhibits DNA synthesis; viral DNA polymerase more likely to incorporate the drugs than human DNA polymerase	**Spectrum of action:** Herpesvirus **Route of administration:** IV **Adverse effects:** Fatal to host cells that incorporate the drug into cellular DNA; anemia	Results from mutation of viral DNA polymerase
Nucleotide analogs Representatives (see also Figure 10.7): Adefovir Azidothymidine (AZT) Entecavir Lamivudine Tenofovir Valaciclovir	Phosphorylation by cell-coded kinase enzyme activates these drugs: inhibits DNA synthesis; viral reverse transcriptase more likely to incorporate these drugs; used in conjunction with protease inhibitor to treat HIV	**Spectrum of action:** HIV, hepatitis B virus **Route of administration:** Oral **Adverse effects:** Nausea, bone marrow toxicity	Results from mutation of viral reverse transcriptase
Ribavirin	Phosphorylation by virally coded kinase enzyme activates the drug; inhibits DNA and RNA synthesis; viral DNA polymerase more likely to incorporate the drugs	**Spectrum of action:** Respiratory syncytial, hepatitis C, influenza A, measles, some hemorrhagic fever viruses **Route of administration:** Oral, aerosol, IV **Adverse effects:** Perhaps harmful to developing fetus	Not known

TABLE 10.3 (continued)

Drug	Description and Mode of Action	Clinical Considerations	Method of Resistance
Antiviral Drugs That Inhibit Protein Synthesis			
Antisense nucleic acids Representative: Fomiversen	Complementary to mRNA; binding prevents protein synthesis by blocking ribosomes	**Spectrum of action:** Specific to species with complementary mRNA; fomiversen specific against cytomegalovirus **Route of administration:** Fomiversen injected weekly into eyes **Adverse effects:** Possible glaucoma	Not known
Antiviral Drugs That Inhibit Viral Proteins			
Protease inhibitors	Computer-assisted modeling of protease enzyme, which is unique to HIV, allowed the creation of drugs that block the active site; used in conjunction with drugs active against nucleic acid synthesis	**Spectrum of action:** HIV **Route of administration:** Oral **Adverse effects:** None	Result from mutation in protease gene

TABLE 10.4 Antimicrobials Against Eukaryotes: Antifungal Drugs

Drug	Description and Mode of Action	Clinical Considerations	Method of Resistance
Antifungal Drugs That Inhibit Cell Membranes			
Allylamines Representative: Terbinafine	Antifungal action due to inhibition of ergosterol synthesis	**Spectrum of action:** Fungi **Route of administration:** Oral, IV **Adverse effects:** Headache, nausea, vomiting, diarrhea, liver damage, rash	Not known
Azoles Representatives: Fluconazole Itraconazole Ketoconazole Voriconazole	Antifungal action due to inhibition of synthesis of ergosterol, an essential component of fungal cytoplasmic membranes	**Spectrum of action:** Fungi and protozoa **Route of administration:** Topical, IV **Adverse effects:** Possibly causes cancer in humans	Mutation in gene for target enzyme
Polyenes Representatives: Amphotericin B Nystatin	Associate with molecules of ergosterol, forming a pore through the fungal membrane, which leads to leakage of essential ions from the cell; amphotericin B is produced by *Streptomyces nodosus*	**Spectrum of action:** Fungi, some amoebae **Route of administration:** Amphotericin B: IV; nystatin: topical **Adverse effects:** Chills, vomiting, fever	Rare; decrease in amount or change in chemistry of ergosterol
Other Antifungal Drugs			
Echinocandins Representative: Caspofungin	Inhibits synthesis of glucan subunit of fungal cell walls	**Specimen of action:** *Candida, Aspergillus* **Route of administration:** IV **Adverse effects:** Rash, facial swelling, respiratory spasms, gastrointestinal distress	Result from mutation in glucan synthase gene
5-Fluorocytosine	Fungi, but not mammals, have an enzyme that converts this drug into 5-fluorouracil, an analog of uracil that inhibits RNA function	**Spectrum of action:** *Candida, Cryptococcus, Aspergillus* **Route of administration:** Oral **Adverse effects:** None	Develops from mutations in the genes for enzymes necessary for utilization of uracil
Griseofulvin	Isolated from *Penicillium griseofulvum;* deactivates tubulin, preventing cytokinesis and segregation of chromosomes during mitosis (see Chapter 12)	**Spectrum of action:** Molds of ringworm (tinea) **Route of administration:** Topical, oral **Adverse effects:** None	Not known

TABLE 10.5 Antimicrobials Against Eukaryotes: Anthelmintic Drugs

Drug	Description and Mode of Action	Clinical Considerations	Method of Resistance
Anthelmintic Drugs That Are Antimetabolites			
Benzimidazole derivatives Representatives: Albendazole Mebendazole Thiabendazole Triclabendazole	Inhibit microtubule formation and glucose uptake	**Spectrum of action:** Helminths, protozoa **Route of administration:** Oral **Adverse effects:** Possible diarrhea	Not known
Iodoquinol	Halogenated (iodine-containing), possibly works by sequestering iron ions required by protozoan	**Spectrum of action:** *Entamoeba* **Route of administration:** Oral **Adverse effects:** Neuropathy and blindness with prolonged use	Not known
Ivermectin **Metrifonate**	Produce flaccid paralysis by blocking neurotransmitters	**Spectrum of action:** Helminths **Route of administration:** Oral **Adverse effects:** Allergic reactions may result from antigens of dead helminths	Not known
Niclosamide	Inhibits oxidative phosphorylation of ATP by mitochondria	**Spectrum of action:** Cestodes **Route of administration:** Oral **Adverse effects:** Abdominal pain, nausea, diarrhea	Not known
Praziquantel	Changes membrane permeability to calcium ions, which are required for muscular contraction; induces complete muscular contraction in helminths	**Spectrum of action:** Cestodes, trematodes **Route of administration:** Oral **Adverse effects:** None	Not known
Pyrantel pamoate **Diethylcarbamazine**	Bind to neurotransmitter receptors, causing complete muscular contraction of helminths	**Spectrum of action:** Nematodes **Route of administration:** Oral **Adverse effects:** None	Not known
Anthelmintic Drugs That Inhibit Nucleic Acid Synthesis			
Niridazole	When partially catabolized by schistosome enzymes, binds to DNA, preventing replication	**Spectrum of action:** *Schistosoma*	Not known
Oltipraz	Possibly acts by reducing the supply of deoxyribonucleotides	**Spectrum of action:** *Schistosoma*	Not known
Oxamniquine	Schistosome enzyme activates drug, which then inhibits DNA synthesis	**Spectrum of action:** *Schistosoma*	Not known

TABLE 10.6 Antimicrobials Against Eukaryotes: Antiprotozoan Drugs

Drug	Description and Mode of Action	Clinical Considerations	Method of Resistance
Antiprotozoan Drugs That Are Antimetabolites			
Artemisinin	Derived from Chinese wormwood shrub; interferes with heme detoxification and with Ca^{2+} transport	**Spectrum of action:** *Plasmodium* **Route of administration:** Oral **Adverse effects:** Nausea, vomiting, itching, dizziness	Mutation in Ca^{2+} transporter gene
Atovaquone	Analog of coenzyme Q or cytochrome *b* of several protozoa and of *Pneumocystis;* interrupts electron transport	**Spectrum of action:** Protozoa, *Pneumocystis* **Route of administration:** Oral **Adverse effects:** Possible rash, diarrhea, headache	Cells modify the structure of their electron transport chain proteins

TABLE 10.6 (continued)

Drug	Description and Mode of Action	Clinical Considerations	Method of Resistance
Antiprotozoan Drugs That Are Antimetabolites			
Benzimidazole derivatives Representative: Mebendazole	Inhibit microtubule formation and glucose uptake	**Spectrum of action:** Helminths, protozoa **Route of administration:** Oral **Adverse effects:** Possible diarrhea	Not known
Furazolidone	Appear to block a number of metabolic pathways, including carbohydrate metabolism and initiation of translation	**Spectrum of action:** Protozoa, Gram+, Gram− bacteria **Route of administration:** Oral **Adverse effects:** Possible nausea and vomiting	Not known
Heavy metals (e.g., Hg, As, Cr, Sb) Representatives: Meglumine antimonate Melarsoprol Salvarsan (contains As) Sodium stibogluconate (contains Sb)	Deactivate enzymes by breaking hydrogen bonds necessary for effective tertiary structure; drugs containing arsenic were the first recognized selectively toxic chemotherapeutic agents	**Spectrum of action:** Metabolically active cells **Route of administration:** Topical, oral **Adverse effects:** Toxic to active cells, such as those of the brain, kidney, liver, and bone marrow	Not known
Iodoquinol	Mode of action unknown	**Spectrum of action:** Intestinal amoebae **Route of administration:** Oral **Adverse effects:** Fever, chills, rash	Not known
Nifurtimox	Interferes with electron transport	**Spectrum of action:** *Trypanosoma* **Route of administration:** Oral **Adverse effects:** Abdominal pain, nausea	Not known
Proguanil **Pyrimethamine**	Block second metabolic step in the formation of folic acid from PABA; synergistic with sulfonamides	**Spectrum of action:** Broad: effective against G+ and G− bacteria and some protozoa and fungi; however, resistance is widespread **Route of administration:** Oral **Adverse effects:** Allergic reactions in some patients	Cells that require folic acid as a vitamin are also naturally resistant; chromosomal mutations result in lowered affinity for the drugs
Sulfonamides Representatives: Sulfadiazine Sulfadoxine Sulfanilamide	Synthetic drugs; first produced as a dye; analogs of PABA that bind irreversibly to enzyme that produces dihydrofolic acid; synergistic with trimethoprim	**Spectrum of action:** Broad: effective against G+ and G− bacteria and some protozoa and fungi; however, resistance is widespread **Route of administration:** Oral **Adverse effects:** Rare: Allergic reactions, anemia, jaundice, mental retardation of fetus if administered in last trimester of pregnancy	Cells that require folic acid as a vitamin are also naturally resistant; chromosomal mutations result in lowered affinity for the drugs
Suramin	Inhibits specific enzymes in some protozoa	**Spectrum of action:** *Trypanosoma* **Routine of administration:** Oral **Adverse effects:** None	Not known
Trimethoprim	Block second metabolic step in the formation of folic acid from PABA; synergistic with sulfonamides	**Spectrum of action:** Broad: effective against some protozoa, fungi, and some G+ and G− bacteria; however, resistance is widespread **Route of administration:** Oral **Adverse effects:** Allergic reactions in some patients	Cells that require folic acid as a vitamin are also naturally resistant; chromosomal mutations result in lowered affinity for the drugs

continued

TABLE 10.6 **Antimicrobials Against Eukaryotes: Antiprotozoan Drugs** ***(continued)***

Drug	Description and Mode of Action	Clinical Considerations	Method of Resistance
Antiprotozoan Drugs That Inhibit DNA Synthesis			
Eflornithine	Inhibits synthesis of precursors of nucleic acids	**Spectrum of action:** Primarily *Trypanosoma brucei* **Route of administration:** Oral **Adverse effects:** Anemia, inhibition of blood clotting, nausea, vomiting	Not known
Nitroimidazoles Representatives: Benznidazole Metronidazole	Anaerobic conditions reduce the drug, which then damages DNA, preventing correct replication and transcription	**Spectrum of action:** Protozoa **Route of administration:** Oral **Adverse effects:** Metronidazole causes cancer in laboratory rodents	Not known
Pentamidine	Binds to nucleic acids, inhibiting replication, transcription, and translation	**Spectrum of action:** Protozoa and *Pneumocystis* (fungus) **Route of administration:** IM, IV **Adverse effects:** Rash, low blood pressure, irregular heartbeat, kidney and liver failure	Not known
Quinolones Representatives: Natural quinine Semisynthetic quinines: Chloroquine Mefloquine Primaquine	Natural and semisynthetic drugs derived from the bark of cinchona tree; inhibit metabolism of malaria parasites by one or more unknown methods	**Spectrum of action:** *Plasmodium* **Route of administration:** Oral **Adverse effects:** Allergic reactions, visual disturbances	Results from the presence of quinoline pumps that remove the drugs from parasite's cells

Chapter Summary

The History of Antimicrobial Agents (pp. 283–284)

1. Chemotherapeutic agents are chemicals used to treat diseases. Among them are **antimicrobial agents (antimicrobials),** which include **antibiotics** (biologically produced agents), **semisynthetics** (chemically modified antibiotics), and **synthetic** agents.

Mechanisms of Antimicrobial Action (pp. 284–292)

1. Successful chemotherapy against microbes is based on **selective toxicity;** that is, using antimicrobial agents that are more toxic to pathogens than to the patient.
2. Antimicrobial drugs affect pathogens by inhibiting cell wall synthesis, inhibiting the translation of proteins, disrupting cytoplasmic membranes, inhibiting general metabolic pathways, inhibiting the replication of DNA, blocking the attachment of viruses to their hosts, or by blocking a pathogen's recognition of its host. **ANIMATIONS:** ***Chemotherapeutic Agents: Modes of Action***
3. **Beta-lactams**—penicillins, cephalosporins, and monobactams—have a functional lactam ring. They prevent bacteria from cross-linking NAM subunits of peptidoglycan in the bacterial cell wall during growth. **Vancomycin** and **cycloserine** also disrupt cell wall formation in many Gram-positive bacteria. **Bacitracin** blocks NAG and NAM transport from the cytoplasm. **Isoniazid (INH)** and **ethambutol** block mycolic acid synthesis in the walls of mycobacteria. **Echinocandins** block synthesis of fungal cell walls.
4. Antimicrobial agents that inhibit protein synthesis include **aminoglycosides** and **tetracyclines,** which inhibit functions of the 30S ribosomal subunit, and **chloramphenicol, lincosamides, streptogramins,** and **macrolides,** which inhibit 50S subunits. Mupirocin stops polypeptide synthesis by binding to tRNA molecules that carry isoleucine. **Oxazolidinones** block initiation of translation. **Antisense nucleic acid** molecules also inhibit protein synthesis.
5. **Polyenes, azoles,** and **allylamines** disrupt the cytoplasmic membranes of fungi. Polymyxin acts against the membranes of Gram-negative bacteria.
6. **Sulfonamides** are **structural analogs** of para-aminobenzoic acid (PABA), a chemical needed by some microorganisms but not by humans. The substitution of sulfonamides in the metabolic pathway leading to nucleic acid synthesis kills those organisms. Trimethoprim also blocks this pathway.
7. Drugs that inhibit nucleic acid replication in pathogens include actinomycin, **nucleotide analogs,** fluoroquinolones, quinolones, and rifampin.

Clinical Considerations in Prescribing Antimicrobial Drugs (pp. 292–296)

1. Chemotherapeutic agents have a **spectrum of action** and may be classed as either **narrow-spectrum drugs or broad-spectrum drugs** depending on how many kinds of pathogens they affect.
2. **Diffusion susceptibility tests,** such as the Kirby-Bauer test, reveal which drug is most effective against a particular pathogen; in general, the larger the **zone of inhibition** around a drug-soaked disk on a Petri plate, the more effective the drug.
3. The **minimum inhibitory concentration (MIC),** usually determined by either a **broth dilution test** or an **Etest,** is the smallest amount of a drug that will inhibit a pathogen.
4. A **minimum bactericidal concentration (MBC) test** ascertains whether a drug is bacteriostatic and the lowest concentration of a drug that is bactericidal.
5. In choosing antimicrobials, physicians must consider how a drug is best administered—orally, intramuscularly, or intravenously—and possible side effects, including toxicity and allergic responses.

Resistance to Antimicrobial Drugs (pp. 296–308)

1. Some members of a pathogenic population may develop resistance to a drug due to extra DNA pieces called **R-plasmids** or to the mutation of genes. Microorganisms may resist a drug by producing enzymes such as **β-lactamase** that deactivate the drug, by inducing changes in the cell membrane that prevent entry of the drug, by altering the drug's target to prevent its binding, by altering the cell's metabolic pathways, by pumping the drug out of the cell, or by protecting the drug's target by binding another molecule to it.

   ANIMATIONS: *Antibiotic Resistance: Origins of Resistance, Forms of Resistance*
2. **Cross resistance** occurs when resistance to one chemotherapeutic agent confers resistance to similar drugs. **Multiple-drug-resistant pathogens** are resistant to three or more types of antimicrobial drugs.
3. **Synergism** describes the interplay between drugs that results in efficacy that exceeds the efficacy of either drug alone. Some drug combinations are antagonistic.

Questions for Review

Answers to the Questions for Review (except Short Answer questions) begin on page A-1.

Multiple Choice

1. Diffusion and dilution tests that expose pathogens to antimicrobials are designed to
   a. determine the spectrum of action of a drug.
   b. determine which drug is most effective against a particular pathogen.
   c. determine the amount of a drug to use against a particular pathogen.
   d. both b and c
2. In a Kirby-Bauer susceptibility test, the presence of a zone of inhibition around disks containing antimicrobial agents indicates
   a. that the microbe does not grow in the presence of the agents.
   b. that the microbe grows well in the presence of the agents.
   c. the smallest amount of the agent that will inhibit the growth of the microbe.
   d. the minimum amount of an agent that kills the microbe in question.
3. The key to successful chemotherapy is
   a. selective toxicity.
   b. a diffusion test.
   c. the minimum inhibitory concentration test.
   d. the spectrum of action.
4. Which of the following statements is relevant in explaining why sulfonamides are effective?
   a. Sulfonamides attach to sterol lipids in the pathogen, disrupt the membranes, and lyse the cells.
   b. Sulfonamides prevent the incorporation of amino acids into polypeptide chains.
   c. Humans and microbes use folic acid and PABA differently in their metabolism.
   d. Sulfonamides inhibit DNA replication in both pathogens and human cells.
5. Cross resistance is
   a. the deactivation of an antimicrobial agent by a bacterial enzyme.
   b. alteration of the resistant cells so that an antimicrobial agent cannot attach.
   c. the mutation of genes that affect the cell membrane channels so that antimicrobial agents cannot cross into the cell's interior.
   d. resistance to one antimicrobial agent because of its similarity to another antimicrobial agent.
6. Multiple-drug-resistant microbes
   a. are resistant to all antimicrobial agents.
   b. respond to new antimicrobials by developing resistance.
   c. frequently develop in hospitals.
   d. all of the above
7. Which of the following is most closely associated with a beta-lactam ring?
   a. penicillin
   b. vancomycin
   c. bacitracin
   d. isoniazid
8. Drugs that act against protein synthesis include
   a. beta-lactams.
   b. trimethoprim.
   c. polymyxin.
   d. aminoglycosides.
9. Which of the following statements is *false* concerning antiviral drugs?
   a. Macrolide drugs block attachment sites on the host cell wall and prevent viruses from entering.
   b. Drugs that neutralize the acidity of phagolysosomes prevent viral uncoating.
   c. Nucleotide analogs in antiviral drugs can be used to stop viral replication.
   d. Drugs containing protease inhibitors retard viral growth by blocking the production of essential viral proteins.

10. PABA is
    a. a substrate used in the production of penicillin.
    b. a type of β-lactamase.
    c. molecularly similar to cephalosporins.
    d. used to synthesize folic acid.

Labeling

Label each figure below to indicate the class of drug that is stopping polypeptide translation.

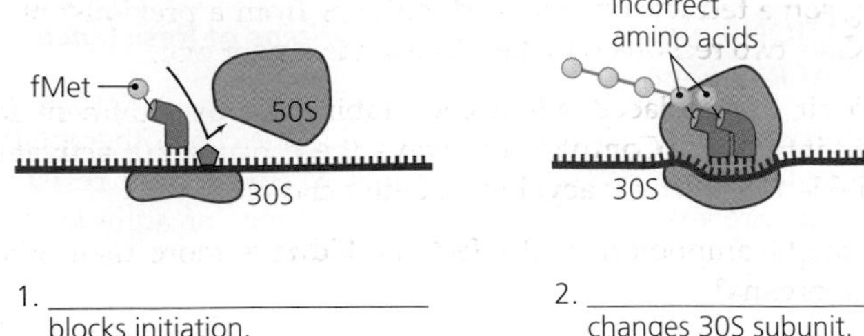

1. ______________________
blocks initiation.

2. ______________________
changes 30S subunit.

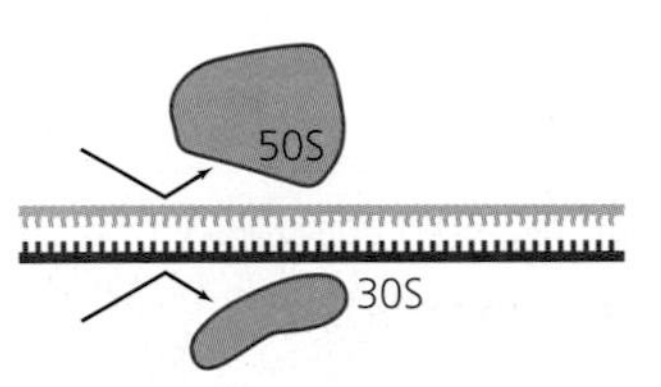

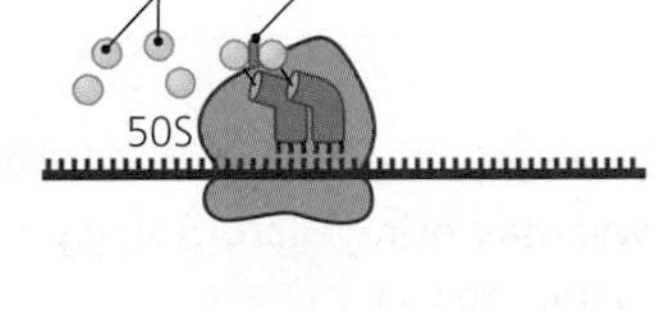

3. ______________________
blocks ribosome attachment.

4. ______________________
inhibits peptide bonding.

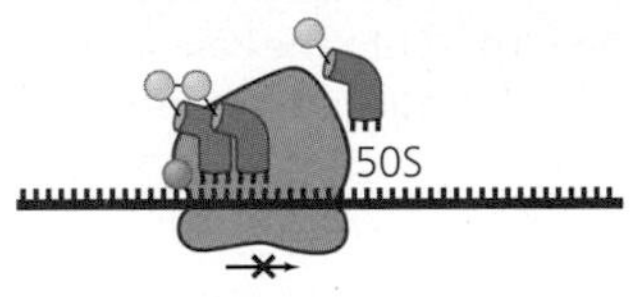

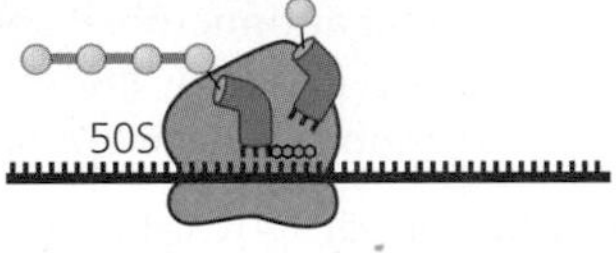

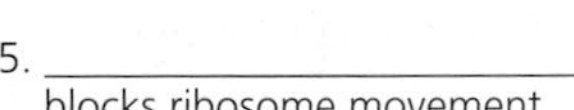

5. ______________________
blocks ribosome movement.

6. ______________________
blocks tRNA docking.

Short Answer

1. What characteristics would an ideal chemotherapeutic agent have? Which drug has these qualities?
2. Contrast narrow-spectrum and broad-spectrum drugs. Which are more effective?
3. Why is the fact that drug Z destroys the NAM portions of a cell's wall structure an important factor in considering the drug for chemotherapy?
4. Given that both human cells and pathogens synthesize proteins at ribosomal sites, how can antimicrobial agents that target this process be safe to use in humans?
5. Support or refute the following statement: antimicrobial agents produce resistant cells.
6. Given that resistant strains of pathogens are a concern to the general health of a population, what can be done to prevent their development?
7. Why are antiviral drugs difficult to develop?
8. A man has been given a broad-spectrum antibiotic for his stomach ulcer. What unintended consequences could arise from this therapy?
9. Compare and contrast the actions of polyenes, azoles, allylamines, and polymyxin.
10. What is the difference in drug action of synergists contrasted with that of antagonists?

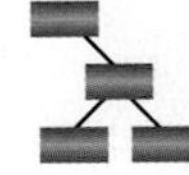

Concept Mapping

Using the following terms, draw a concept map that describes antimicrobial resistance. For a sample concept map, see p. 93. Or, complete this concept map online by going to the Study Area at www.masteringmicrobiology.com.

Altered pore proteins
Altered target
Antibiotics
Bacterial enzymes
Beta lactamase
Binding of antibiotic
Cell division
Conjugation
Entry of antibiotic into bacterial cell
Mutation
Penicillin
Pumping antimicrobial out of the cell
Resistance pumps
Transduction

Critical Thinking

1. AIDS is treated with a "cocktail" of several antiviral agents at once. Why is the cocktail more effective than a single agent? What is a physician trying to prevent by prescribing several drugs at once?
2. How does *Penicillium* escape the effects of the penicillin it secretes?
3. How might a colony of *Bacillus licheniformis* escape the effects of its own bacitracin?
4. Fewer than 1% of known antibiotics have any practical value in treatment of disease. Why is this so?
5. In the summer issue of *News of the Lepidopterists' Society* in 2000, a recommendation was made to moth and butterfly collectors to use antimicrobials to combat disease in the young of these insects. What are the possible ramifications for human health of such usage of antimicrobials?
6. Even though aminoglycosides such as gentamicin can cause deafness, there are still times when they are the best choice for treating some infections. What laboratory test would a clinical scientist use to show that gentamicin is the best choice to treat a particular *Pseudomonas* infection?
7. Your pregnant neighbor has a sore throat and tells you that she is taking some tetracycline she had left over from a previous infection. Give two reasons why her decision is a poor one.
8. Acyclovir has replaced adenosine arabinoside as treatment for herpes infections. Compare the ways these drugs are activated (see Table 10.3). Why is acyclovir a better choice?
9. Why might amphotericin B affect the kidneys more than other human organs?
10. Antiparasitic drugs in the benzimidazole family inhibit the polymerization of tubulin. What effect might these drugs have on mitosis and flagella?

Access more review material online in the Study Area at **www.masteringmicrobiology.com**. There, you'll find

- **Animations**
- **MP3 Tutor Sessions**
- **Concept Mapping Activities**
- **Flashcards**
- **Quizzes**

and more to help you succeed.

11 Characterizing and Classifying Prokaryotes

Microbiologist Tony Walsby startled the scientific world by announcing the discovery of something no one had seen before: a rectangular-shaped prokaryote. This novel archaeon was named *Haloarcula*, meaning "salt box," because of its unusual boxy shape and its habitat, the salt-encrusted Dead Sea. Until this extraordinary find, scientists had only seen spherical, rod-shaped, or spiral prokaryotes. *Haloarcula* was a surprise, one that forced scientists to reconsider some of their assumptions about prokaryotic life forms.

More recently, we have discovered species of archaea that may offer clues to whether or not there is life on other planets. And among the millions of unknown microorganisms that live on Earth, there may very well exist entirely new categories of life. The vast and diverse world of prokaryotes continues to surprise and amaze us.

Take the pre-test for this chapter online. Visit the Study Area at www.masteringmicrobiology.com.

The Dead Sea, the lowest point on the Earth's surface, is also one of the saltiest, but it isn't really dead—rectangular-shaped halophilic archaea live there.

Prokaryotes are by far the most numerous and diverse group of cellular microbes. Scientists estimate there are more than 5×10^{30} prokaryotes on Earth. They thrive in various habitats: from Antarctic glaciers to thermal hot springs, from the colons of animals to the cytoplasm of other prokaryotes, from distilled water to supersaturated brine, and from disinfectant solutions to basalt rocks thousands of meters below the Earth's surface. In part because of such great diversity, only a very few prokaryotes have enzymes, toxins, or cellular structures that enable them to colonize humans and cause disease. In this chapter we will begin by examining general prokaryotic characteristics and conclude with a survey of specific prokaryotic taxa. We will briefly mention human pathogens throughout the chapter but reserve detailed discussion of these important microbes for Chapters 19–21.

General Characteristics of Prokaryotic Organisms

In previous chapters we considered the general characteristics of prokaryotic cells, including their cellular structure, metabolism, growth, and genetics. Here we will consider prokaryotes not merely as cells, but as distinct organisms, focusing on their cellular shapes, their reproductive processes, their spatial arrangements, and the ability of some to survive unfavorable conditions by forming resistant structures within themselves.

Morphology of Prokaryotic Cells

Learning Objective

✓ Identify six basic shapes of prokaryotic cells.

Prokaryotic cells exist in a variety of shapes, or morphologies (**Figure 11.1**). The three basic shapes are **cocci** (kok′sī, roughly spherical), **bacilli** (bă-sil′ī, rod-shaped), and **spirals**. Cocci are not all perfectly spherical; for example, there are pointed, kidney-shaped, and oval cocci. Similarly, bacilli vary in shape; for example, some bacilli are pointed, spindle shaped, or threadlike (filamentous). Spiral-shaped prokaryotes are either **spirilla,** which are stiff, or **spirochetes** (spī′rō-kētz), which are flexible. Slightly curved rods are **vibrios,** and the term **coccobacillus** is used to describe cells that are intermediate in shape between cocci and bacilli; that is, when it is difficult to ascertain if a cell is an elongated coccus or a short bacillus. In addition to these basic shapes, there are star-shaped, triangular, and rectangular prokaryotes, as well as prokaryotes that are **pleomorphic**[1] (plē-ō-mōr′fik); that is, they vary in shape and size (see Figure 11.10).

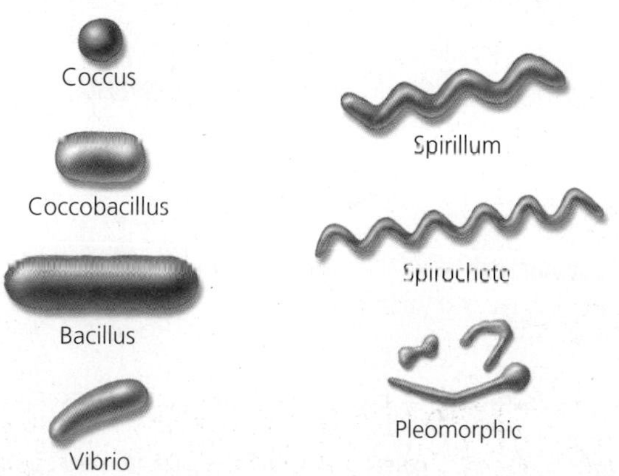
Coccus
Coccobacillus
Bacillus
Vibrio
Spirillum
Spirochete
Pleomorphic

▲ **Figure 11.1 Typical prokaryotic morphologies.** *What is one difference between a spirillum and a spirochete?*

Figure 11.1 *Generally, spirilla are stiff, whereas spirochetes are flexible.*

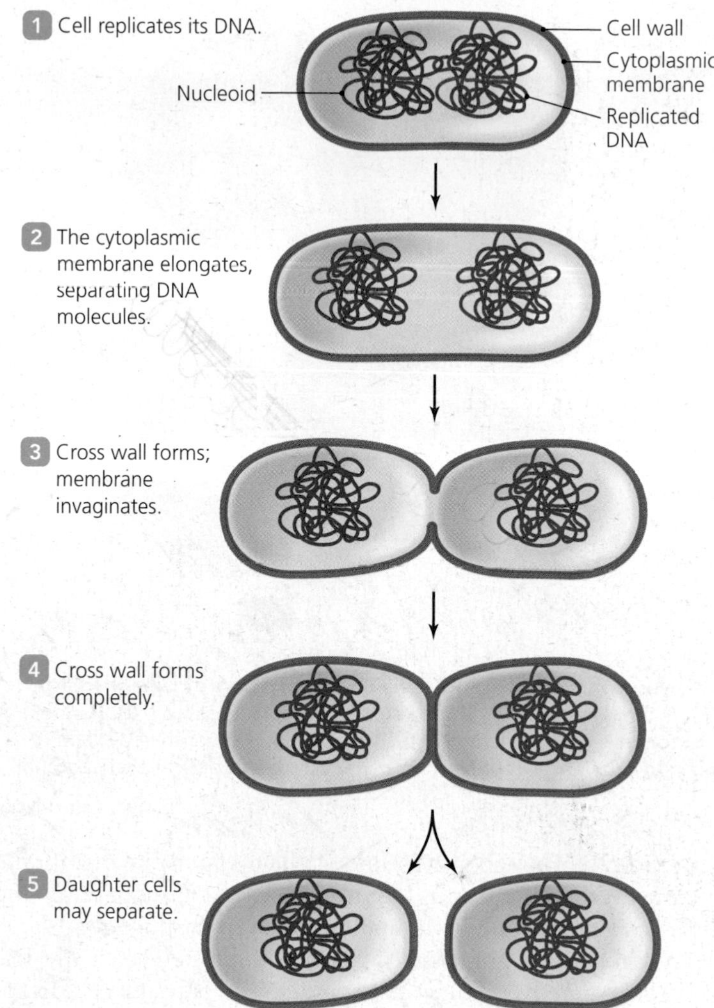
1 Cell replicates its DNA.
Nucleoid
Cell wall
Cytoplasmic membrane
Replicated DNA
2 The cytoplasmic membrane elongates, separating DNA molecules.
3 Cross wall forms; membrane invaginates.
4 Cross wall forms completely.
5 Daughter cells may separate.

▲ **Figure 11.2 Binary fission.** The cell replicates its DNA, elongates, forms a cross wall, and divides into two equal-sized daughter cells.

Reproduction of Prokaryotic Cells

Learning Objectives

✓ List three common types of reproduction in prokaryotes.

✓ Describe snapping division as a type of binary fission.

All prokaryotes reproduce asexually; none reproduce sexually. The most common method of asexual reproduction is **binary fission,** which proceeds as follows (**Figure 11.2**): 1 The cell replicates its DNA; each DNA molecule is attached to the cytoplasmic

[1]From Greek *pleon,* meaning more, and *morphe,* meaning form.

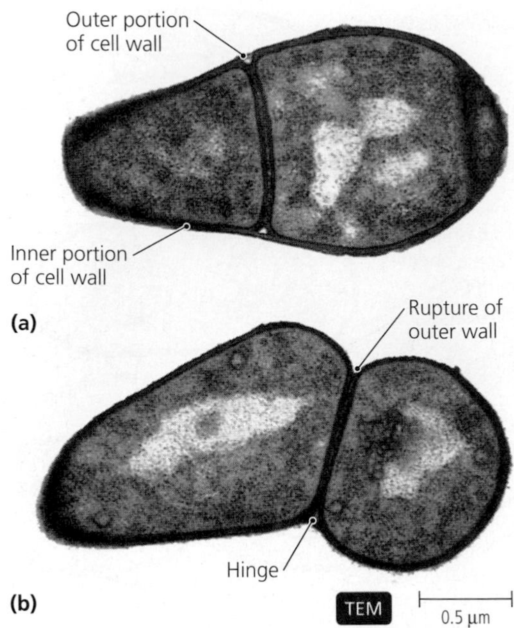

▲ **Figure 11.3 Snapping division, a variation of binary fission.** **(a)** Only the inner portion of the cell wall forms a cross wall. **(b)** As the daughter cells grow, tension snaps the outer portion of the cell wall, leaving the daughter cells connected by a hinge of old cell wall material.

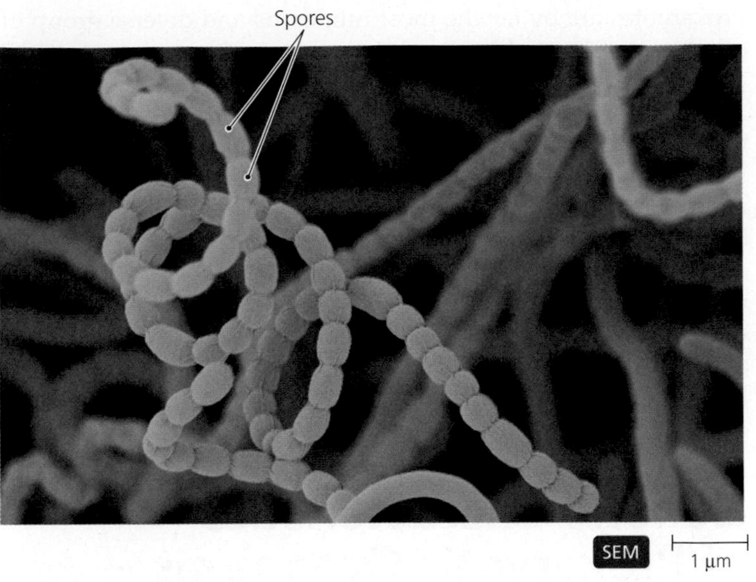

▲ **Figure 11.4 Spores of actinomycetes.** Filamentous vegetative cells produce chains of spores, shown here in *Streptomyces*.

membrane. **2** The cell grows; and as the cytoplasmic membrane elongates, it moves the daughter molecules of DNA apart. **3** The cell forms a cross wall, invaginating the cytoplasmic membrane. **4** The cross wall completely divides daughter cells. **5** The daughter cells may or may not separate. The parental cell disappears with the formation of progeny. **ANIMATIONS:** ***Bacterial Growth: Overview***

A variation of binary fission called **snapping division** occurs in some Gram-positive bacilli **(Figure 11.3)**. In snapping division, only the inner portion of the cell wall is deposited across the dividing cell. The thickening of this new transverse wall puts tension on the outer layer of the old cell wall, which still holds the two cells together. Eventually, as the tension increases, the outer wall breaks at its weakest point with a snapping movement that tears it most of the way around. The daughter cells then remain hanging together, held at an angle by a small remnant of the original outer wall that acts like a hinge.

A few prokaryotes have other methods of reproduction. The parental cell retains its identity during and after these methods. The *actinomycetes* (ak′ti-nō-mī-sētz) produce reproductive cells called **spores** at the ends of their filamentous cells **(Figure 11.4)**. Each spore can develop into a clone of the original organism. Some *cyanobacteria* reproduce by fragmentation into small motile filaments that glide away from the parental strand. Still other prokaryotes reproduce by **budding,** in which an outgrowth of the original cell (a bud) receives a copy of the genetic material and enlarges. Eventually the bud is cut off from the parental cell, typically while it is still quite small **(Figure 11.5)**.

Epulopiscium (ep′yoo-lō-pis′sē-ŭm), the giant bacterial symbiont of surgeonfish introduced in Chapter 3 (see p. 62), and many of its relatives have a truly unique method of reproduction among prokaryotes: They give "birth" to as many as 12 live offspring that emerge from the body of a dead mother cell. The production of live offspring within a mother is called *viviparity,* and this is the first documented case of viviparous behavior in the prokaryotic world. In these bacteria, formation of internal offspring proceeds in a manner similar to the early stages of endospore formation (see Figure 3.24).

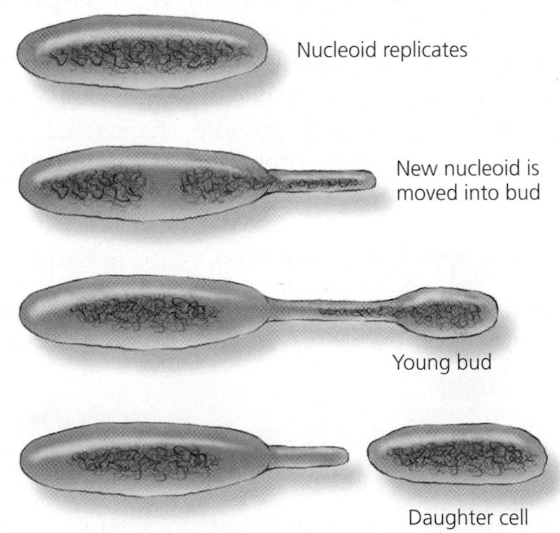

▲ **Figure 11.5 Budding.** *How does budding differ from binary fission?*

Figure 11.5 *In binary fission, the parent cell disappears with the formation of two equal-sized offspring; in contrast, a bud is often much smaller than its parent, and the parent remains to produce more buds.*

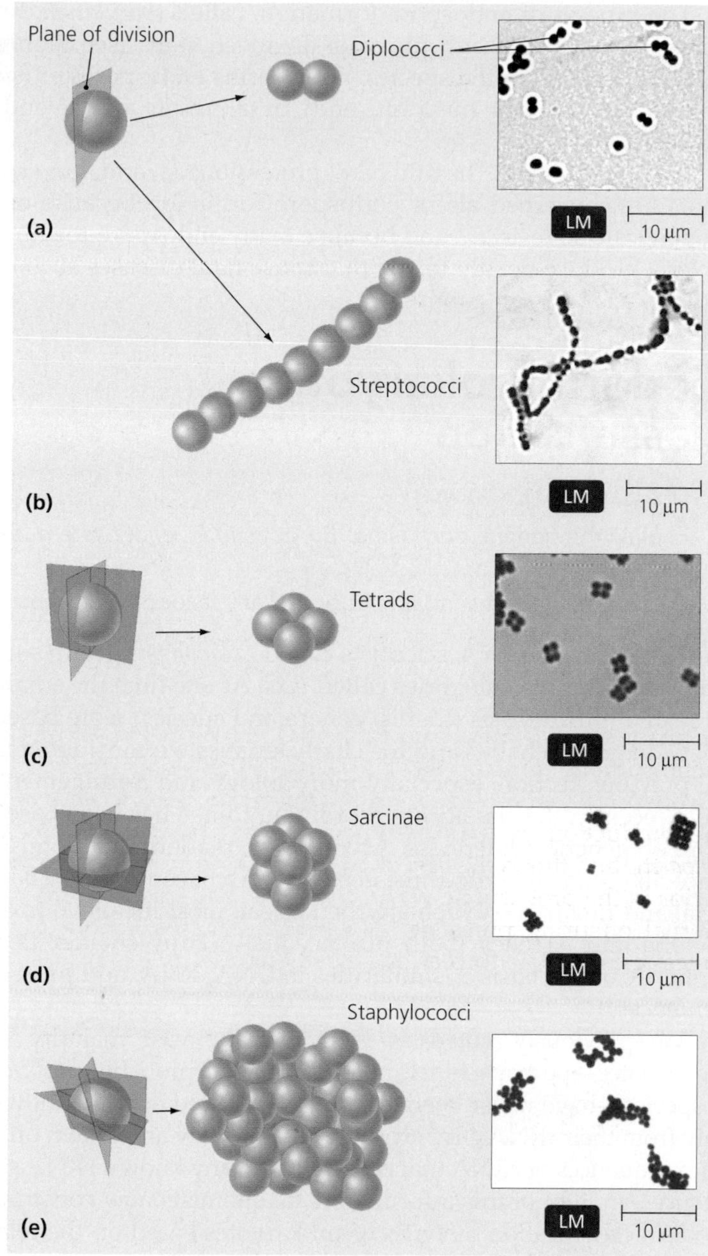

▲ **Figure 11.6 Arrangements of cocci.** **(a)** The diplococci of *Neisseria gonorrhoeae.* **(b)** The streptococci of *Streptococcus pyogenes.* **(c)** Tetrads, in this case of *Micrococcus luteus.* **(d)** The genus *Sarcina* is characterized by sarcinae. **(e)** The staphylococci of *Staphylococcus aureus.*

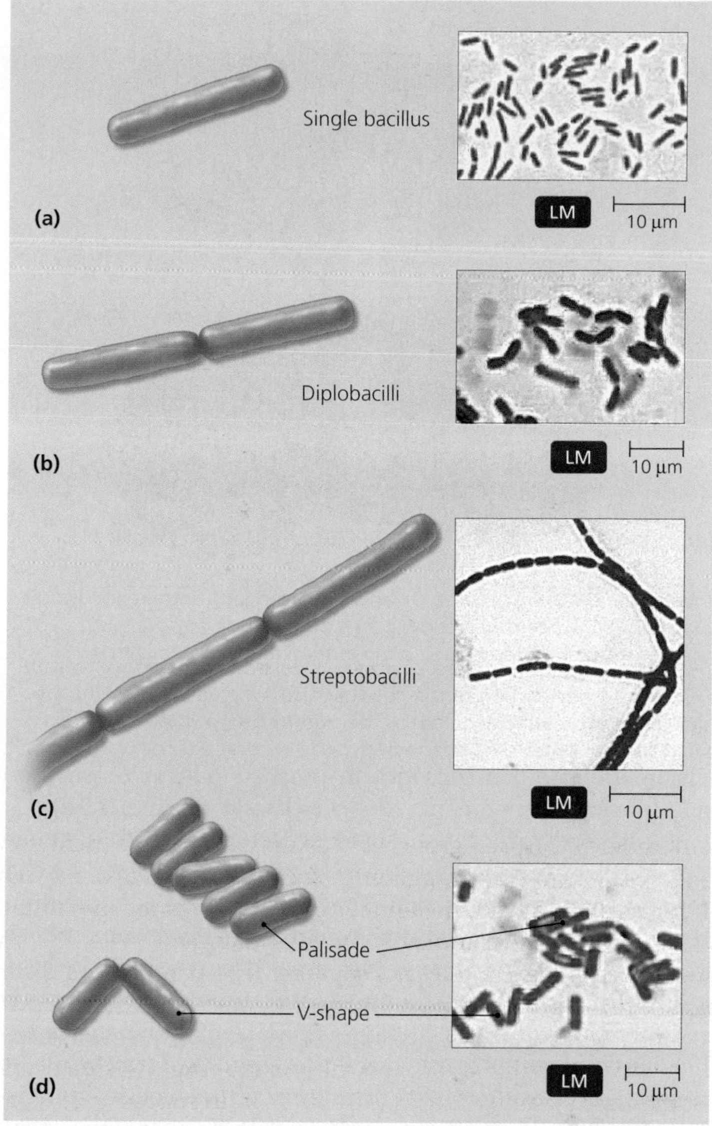

▲ **Figure 11.7 Arrangements of bacilli.** **(a)** A single bacillus of *Escherichia coli.* **(b)** Diplobacilli in a young culture of *Bacillus cereus.* **(c)** Streptobacilli in an older culture of *Bacillus cereus.* **(d)** V-shape and a palisade of *Corynebacterium diphtheriae.*

Arrangements of Prokaryotic Cells

Learning Objective

✓ Draw and label five arrangements of prokaryotes.

The arrangements of prokaryotic cells result from two aspects of division during binary fission: the planes in which cells divide and whether or not daughter cells separate completely or remain attached to each other. Cocci that remain attached in pairs are **diplococci (Figure 11.6a)**, and long chains of cocci are called **streptococci**[2] **(Figure 11.6b)**. Some cocci divide in two planes and remain attached to form **tetrads (Figure 11.6c)**; others divide in three planes to form cuboidal packets called **sarcinae**[3] (sar'si-nī) **(Figure 11.6d)**. Clusters called **staphylococci**[4] (staf'i-lo-kok-sī), which look like bunches of grapes, form when the planes of cell division are random **(Figure 11.6e)**.

Bacilli are less varied in their arrangements than cocci because bacilli divide transversely—that is, perpendicular to the long axis. Daughter bacilli may separate to become single cells or stay attached as either pairs or chains **(Figure 11.7a–c)**. Because the cells of *Corynebacterium diphtheriae* (kŏ-rī'nē-bak-tēr'ē-ŭm dif-thi'rē-ī), the causative agent of diphtheria, divide by snapping

[2]From Greek *streptos*, meaning twisted, because long chains tend to twist.
[3]Latin for bundles.
[4]From Greek *staphyle*, meaning bunch of grapes.

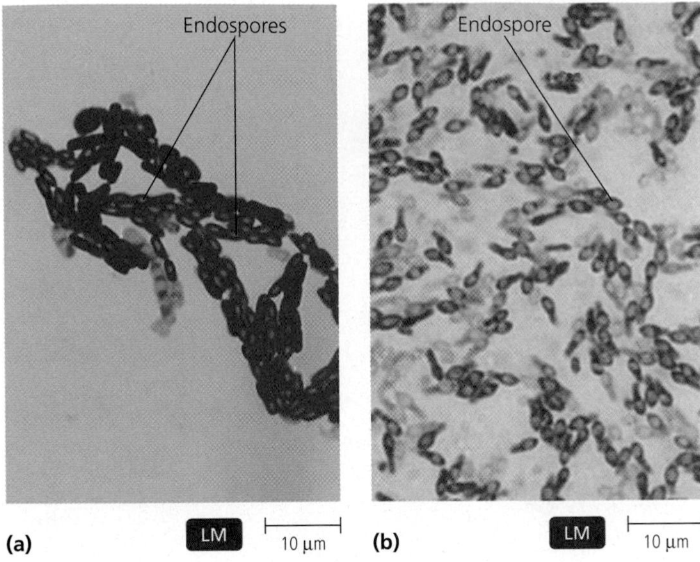

▲ **Figure 11.8 Locations of endospores.** **(a)** Central endospores of *Bacillus.* **(b)** Subterminal endospores of *Clostridium botulinum.* The enlarged endospores have swollen the vegetative cells that produced them.

division, the daughter cells remain attached to form V-shapes and a side-by-side arrangement called a **palisade**[5] **(Figure 11.7d)**.

The same word can be used to refer either to a general shape and/or arrangement or to a specific genus. Thus the characteristic shape of the genus *Bacillus* (ba-sil'ŭs) is a bacillus, a rod-shaped bacterium, and the characteristic arrangement of bacteria in the genus *Sarcina* (sar'si-nă) is cuboidal. In such potentially confusing cases the meaning can be distinguished because genus names are always capitalized and italicized. In other cases a genus name uses the singular form while the arrangement uses the plural form; thus streptococci—spherical cells arranged in a chain—are characteristic of the genus *Streptococcus* (strep-tō-kok'ŭs).

Endospores

Learning Objective

✓ Describe the formation and function of bacterial endospores.

The Gram-positive bacteria *Bacillus* and *Clostridium* (klos-trid'-ē-ŭm) produce **endospores,** which are important for several reasons, including their durability and potential pathogenicity. Endospores constitute a defensive strategy against hostile or unfavorable conditions. They are stable resting stages that barely metabolize and germinate only when conditions improve.

Though some people refer to endospores as "spores," endospores should not be confused with the reproductive spores of actinomycetes, algae, and fungi. A single bacterial cell, called a *vegetative* cell to distinguish it from an endospore, transforms into only one endospore, which then germinates to grow into a single vegetative cell; therefore, endospores are not reproductive structures.

The process of endospore formation, called *sporulation,* requires 8 to 10 hours and proceeds in seven steps (see Figure 3.24). Depending on the species, a cell forms endospores either *centrally, subterminally* (near one end), or *terminally* (at one end) **(Figure 11.8)**.

Food processors, health care professionals, and governments are concerned about endospore formation because endospores are so resistant and because many endospore-forming bacteria produce deadly toxins that cause fatal diseases such as anthrax, tetanus, and gangrene.

Modern Prokaryotic Classification

Learning Objectives

✓ Explain the general purpose of *Bergey's Manual of Systematic Bacteriology.*

✓ Discuss the veracity and limitations of any taxonomic scheme.

As we saw in Chapter 4, scientists called *taxonomists* group similar organisms into categories called *taxa.* At one time the smallest taxa of prokaryotes (that is, genera and species) were based solely on growth habits and the characteristics we considered in the previous section, especially morphology and arrangement. More recently, the classification of living things has been based more on genetic relatedness. Accordingly, modern taxonomists place all organisms into three *domains*—Archaea (ar'kē-ă), Bacteria, and Eukarya—which are the largest, most inclusive taxa. Bacteria and Archaea (both prokaryotic) occupy smaller taxa primarily on the basis of similarities in DNA, RNA, and protein sequences.

We previously noted as well that the vast majority of prokaryotes—perhaps as many as 99.5%, and probably millions of species—have never been isolated or cultured and are known only from their rRNA "fingerprints"; that is, they are known only from sequences of rRNA that do not match any known rRNA sequences. In light of this information, taxonomists now construct modern classification schemes of prokaryotes based on the relative similarities of rRNA sequences found in various prokaryotic groups **(Figure 11.9)**.

Perhaps the most authoritative reference in modern prokaryotic systematics is *Bergey's Manual of Systematic Bacteriology,* which classifies prokaryotes into 27 phyla—3 in Archaea and 24 in Bacteria. The five volumes of the second edition of *Bergey's Manual* discuss the great diversity of prokaryotes based in large part (but not exclusively) on their possible evolutionary relationships as reflected in their rRNA sequences.

Our examination of prokaryotic diversity in this text is for the most part organized to reflect the taxonomic scheme that appears in *Bergey's Manual,* but it is important to note that as authoritative as *Bergey's Manual* is, it is not an "official" list of prokaryotic taxa. The reason is that taxonomy is partly a matter of opinion and judgment, and not all taxonomists agree. Because there is room in taxonomy for honest differences of

[5]From Latin *palus,* meaning stake, referring to a fence made of adjoining stakes.

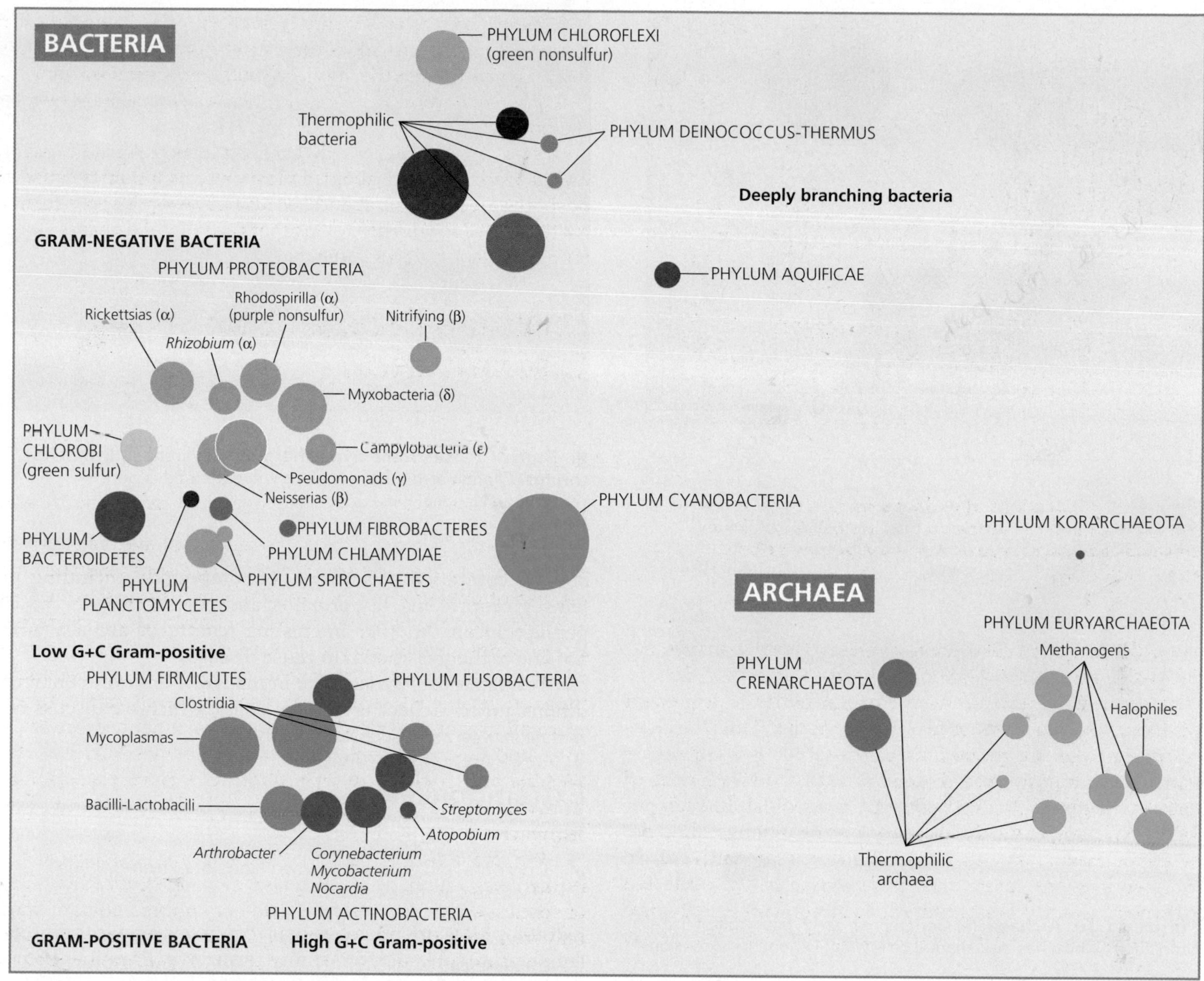

▲ **Figure 11.9 Prokaryotic taxonomy.** This scheme is based on relatedness according to rRNA sequences. The closer together the disks, the more similar are the rRNA sequences of the species within the group. The sizes of disks are proportional to the number of species known for that group. Note that archaea are distinctly separate from bacteria. The discussion in this chapter is largely based on the scheme depicted in this figure. Not all phyla are shown.
(In phylum Proteobacteria, classes are indicated by a Greek letter, α = alpha, β = beta, γ = gamma, δ = delta, and ε = epsilon.)
(Adapted from *Road Map to Bergey's*. 2002, Bergey's Manual Trust.)

opinion, and because legitimately differing views often change as more information is uncovered and examined, *Bergey's Manual* is merely a consensus of experts at a given time. More information about *Bergey's Manual* can be found on the textbook website at www.masteringmicrobiology.com.

In the following sections we examine representative prokaryotes in each major phylum. We begin our exploration of prokaryotic diversity with a survey of Archaea.

Survey of Archaea

Learning Objective

✓ Identify the common features of microbes in the domain Archaea.

Scientists originally identified archaea as a distinct type of prokaryotes on the basis of unique rRNA sequences. Archaea also share other common features that distinguish them from bacteria:

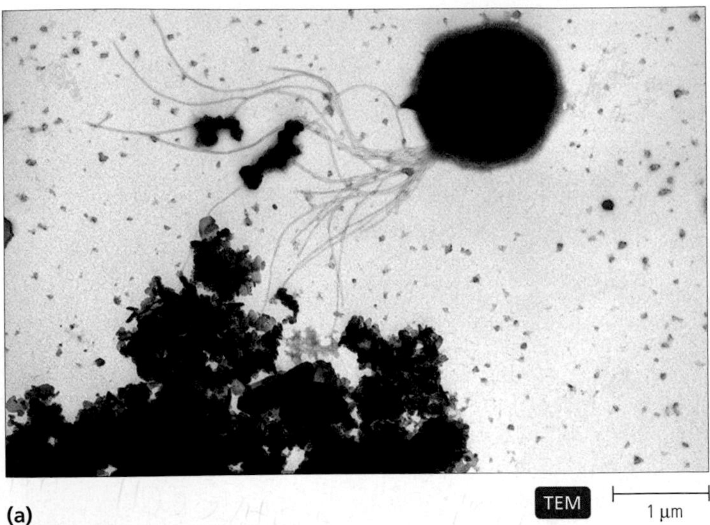

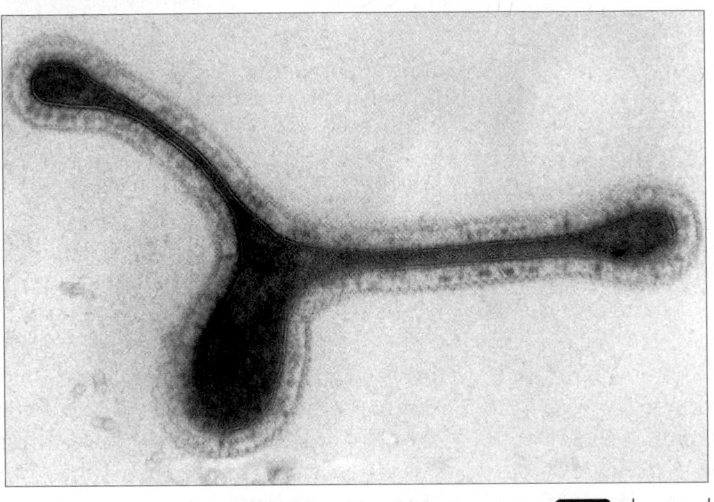

▲ **Figure 11.10 Archaea. (a)** *Geogemma,* which has a tuff of flagella. **(b)** *Pyrodictium,* which has disk-shaped cells with filamentous extensions.

- Archaea lack true peptidoglycan in their cell walls.
- Their cell membrane lipids have branched hydrocarbon chains.
- The initial amino acid in their polypeptide chains, coded by the AUG start codon, is methionine (as in eukaryotes and in contrast to the *N*-formylmethionine used by bacteria).

Archaea are currently classified in three phyla—Crenarchaeota (kren-ar′kē-ō-ta), Euryarchaeota (ŭ-rē-ar′kē-ō-ta), and Korarchaeota (kor-ar′kē-ō-ta)—based primarily upon rRNA sequences. In fact, the latter phylum is known only from environmental RNA samples; no members have been observed visually. Researchers have discovered RNA from another unculturable archaeon—*Nanoarchaeum* (nan-o-ar′kē-um)—that may represent a fourth archaeal phylum, though there is not consensus on this taxon.

Archaea reproduce by binary fission, budding, or fragmentation. Known archaeal cells are cocci, bacilli, spirals, or pleomorphic **(Figure 11.10)**. Archaeal cell walls vary among taxa

▲ **Figure 11.11 Some hyperthermophilic archaea live in hot springs.** Orange archaea thrive along the edge of this pool in Yellowstone National Park.

and are composed of a variety of compounds, including proteins, glycoproteins, lipoproteins, and polysaccharides; all lack peptidoglycan. Another interesting feature of archaea is that not one of them is known to cause diseases.

Though most Archaea live in moderate environmental conditions, noted archaea are *extremophiles,* which we discuss next, and *methanogens* (discussed shortly).

Extremophiles

Learning Objective

✓ Compare and contrast the two kinds of extremophiles discussed in this section.

Extremophiles are microbes that require extreme conditions of temperature, pH, and/or salinity to survive. Prominent among the extremophiles are *thermophiles* and *halophiles.*

Thermophiles

As discussed in Chapter 6, **thermophiles**[6] are prokaryotes whose DNA, RNA, cytoplasmic membranes, and proteins do not function properly at temperatures lower than 45°C. Prokaryotes that require temperatures over 80°C are called **hyperthermophiles.** Most thermophilic archaea are in the phylum Crenarchaeota, though some are also found in the phylum Euryarchaeota.

Two representative genera of thermophiles are *Geogemma* (jē′ō-jem-a) and *Pyrodictium* (pī-rō-dik′tē-um; see Figure 11.10). These microorganisms live in acidic hot springs such as those found in deep ocean rifts and similar terrestrial volcanic habitats **(Figure 11.11)**. *Geogemma* is the current record holder for surviving high temperatures—it can survive 2 hours at 130°C! The cells of *Pyrodictium,* which live in deep-sea hydrothermal

[6]From Greek *thermos,* meaning heat, and *philos,* meaning love.

vents, are irregular disks with elongated protein tubules that attach them to grains of sulfur, which they use as final electron acceptors in respiration.

Scientists use thermophiles and their enzymes in recombinant DNA technology applications because thermophiles' cellular structure and enzymes are stable and functional at temperatures that denature most proteins and nucleic acids and kill other cells. As discussed in Chapter 8, DNA polymerase from hyperthermophilic archaea makes possible the automated amplification of DNA in a thermocycler. Heat-stable enzymes are also ideal for many industrial applications, including their use as additives in laundry detergents.

Halophiles

Halophiles[7] are classified in the phylum Euryarchaeota. They inhabit extremely saline habitats such as the Dead Sea, the Great Salt Lake, and solar evaporation ponds used to concentrate salt for use in seasoning and for the production of fertilizer (**Figure 11.12**). Halophiles can also colonize and spoil such foods as salted fish, sausages, and pork.

The distinctive characteristic of halophiles is their absolute dependence on a concentration of NaCl greater than 9% (that is, a 1.5 molar solution) to maintain the integrity of their cell walls. Most halophiles grow and reproduce within an optimum range of 17–23% NaCl, and many species can survive in a saturated saline solution (35% NaCl). Many halophiles contain red to orange pigments that probably play a role in protecting them from intense visible and ultraviolet light.

The most studied halophile is *Halobacterium salinarium* (hā′lō-bak-tēr′ē-ŭm sal-ē-nar′ē-um), which is an archaeon despite its name. It is a photoheterotroph, using light energy to drive the synthesis of ATP, but deriving carbon from organic compounds. *Halobacterium* lacks photosynthetic pigments—chlorophylls and bacteriochlorophylls. Instead, it synthesizes purple proteins, called **bacteriorhodopsins** (bak-tēr′ē-ō-rō-dop′sinz), that absorb light energy to pump protons across the cytoplasmic membrane to establish a proton gradient. (Recall from Chapter 5 that cells use the energy of proton gradients to produce ATP via chemiosmosis.) *Halobacterium* also rotates its flagella with energy from the proton gradient so as to position itself at the proper water depth for maximum light absorption.

▲ **Figure 11.12 The habitat of halophiles: highly saline water.** These are solar evaporation ponds near San Francisco. Halophiles often contain red to orange pigments, possibly to protect them from intense solar energy.

Methanogens

Learning Objective

✓ List at least four significant roles played by methanogens in the environment.

Methanogens are obligate anaerobes in the phylum Euryarchaeota that convert CO_2, H_2, and organic acids into methane gas (CH_4). These microbes constitute the largest known group of archaea. Most species are mesophilic, though a few thermophilic methanogens are known. For example, *Methanopyrus*[8] (meth′a-nō-pī′rŭs) has an optimum growth temperature of 98°C and grows in 110°C seawater around submarine hydrothermal vents. Scientists have also discovered halophilic methanogens.

Methanogens play significant roles in the environment by converting organic wastes in pond, lake, and ocean sediments into methane. Other methanogens living in the colons of animals are one of the primary sources of environmental methane. Methanogens dwelling in the intestinal tract of a cow, for example, can produce 400 liters of methane a day. Methanogens have produced about 10 trillion tons of methane—twice the known amount of oil, natural gas, and coal combined—that lies buried in mud on the ocean floor. Sometimes the production of methane in swamps and bogs is so great that bubbles rise to the surface as "swamp gas." Methane is a so-called *greenhouse gas;* that is, methane in the atmosphere traps heat, which adds to global warming. It is about 25 times more potent as a greenhouse gas than carbon dioxide. If all the methane trapped in ocean sediments were released, it would wreak havoc with the world's climate.

Methanogens also have useful industrial applications. An important step in sewage treatment is the digestion of sludge by methanogens, and some sewage treatment plants burn methane to heat buildings and generate electricity.

Though we have concentrated our discussion on extremophiles and methanogens, many archaea live in more moderate habitats. For example, archaea make up about a third of the prokaryotic biomass in coastal Antarctic water, providing food for marine animals.

CRITICAL **THINKING**

A scientist who discovers a prokaryote living in a hot spring at 100°C suspects that it belongs to the archaea. Why does he think it might be archaeal? How could he prove that it is not bacterial?

[7]From Greek *halos*, meaning salt.
[8]From Greek *pyrus*, meaning fire.

Survey of Bacteria

As we noted previously, our survey of prokaryotes in this chapter reflects the classification scheme that is featured in the still-incomplete second edition of *Bergey's Manual.* Whereas the classification scheme for bacteria in the first edition of the *Manual* emphasized morphology, Gram reaction, and biochemical characteristics, the second edition largely bases its classification of bacteria on differences in 16S rRNA sequences. We begin our survey of bacteria by considering the deeply branching and phototrophic bacteria.

Deeply Branching and Phototrophic Bacteria

Learning Objectives

✓ Provide a rationale for the name "deeply branching bacteria."

✓ Explain the function of heterocysts in terms of both photosynthesis and nitrogen fixation.

Deeply Branching Bacteria

The **deeply branching bacteria** are so named because their rRNA sequences and growth characteristics lead scientists to conclude that these organisms are similar to the earliest bacteria; that is, they appear to have branched off the "tree of life" at an early stage. For example, the deeply branching bacteria are autotrophic, and early organisms must have been autotrophs because heterotrophs by definition must derive their carbon from autotrophs. Further, many of the deeply branching bacteria live in habitats similar to those some scientists think existed on the early Earth—hot, acidic, anaerobic, and exposed to intense ultraviolet radiation from the sun.

One representative of these microbes—the Gram-negative, microaerophilic *Aquifex* (ăk'wē-feks), a bacterium in the phylum Aquificae—is considered to represent the earliest branch of bacteria. It is chemoautotrophic, hyperthermophilic, and anaerobic, deriving energy and carbon from inorganic sources in very hot habitats containing little oxygen.

Another representative of deeply branching bacteria is *Deinococcus* (dī-nō-kok'ŭs), in phylum Deinococcus-Thermus which grows as tetrads of Gram-positive cocci. Interestingly, the cell wall of *Deinococcus* has an outer membrane similar to that of Gram-negative bacteria, but the cells stain purple like typical Gram-positive microbes. *Deinococcus* is extremely resistant to radiation because of the way it packages its DNA and the presence of radiation-absorbing pigments, unique lipids within its membranes, and high cytoplasmic levels of manganese, which protect its DNA repair proteins from radiation damage. Even when exposed to 5 million rad of radiation, which is enough energy to shatter its chromosome into hundreds of fragments, its enzymes can repair the damage. Not surprisingly, researchers have isolated *Deinococcus* from sites severely contaminated with radioactive wastes.

Phototrophic Bacteria

Phototrophic bacteria acquire the energy needed for anabolism by absorbing light with pigments located in thylakoids called *photosynthetic lamellae*. They lack the membrane-bound thylakoids seen in eukaryotic chloroplasts. Most phototrophic bacteria are also autotrophic—they produce organic compounds from carbon dioxide.

Phototrophs are a diverse group of microbes that are taxonomically confusing. Based on their pigments and their source of electrons for photosynthesis, phototrophic bacteria can be divided into the following five groups:

- Blue-green bacteria (cyanobacteria)
- Green sulfur bacteria
- Green nonsulfur bacteria
- Purple sulfur bacteria
- Purple nonsulfur bacteria

These five groups of organisms are classified into four phyla. We consider all of them in the following sections because of their common phototrophic metabolism.

Cyanobacteria **Cyanobacteria** are Gram-negative phototrophs that vary greatly in shape, size, and method of reproduction. They range in size from 1 μm to 10 μm in diameter and are either coccoid or disk shaped. Coccal forms can be single or arranged in pairs, tetrads, chains, or sheets **(Figure 11.13a, b)**; disc-shaped forms are often tightly appressed end to end to form filaments, which can be either straight, branched, or helical and are frequently contained in a gelatinous glycocalyx called a *sheath* **(Figure 11.13c)**. Some filamentous cyanobacteria are motile, moving along surfaces by *gliding*. Cyanobacteria generally reproduce by binary fission, with some species also reproducing by motile fragments or by thick-walled spores called akinetes (ā-kin-ēts'; see Figure 11.13a).

Like plants and algae, cyanobacteria utilize chlorophyll *a* and are oxygenic (generate oxygen) during photosynthesis:

$$12\,H_2O + 6\,CO_2 \xrightarrow{\text{light}} C_6H_{12}O_6 + 6\,H_2O + 6\,O_2 \qquad (1)$$

For this reason, cyanobacteria were formerly called *blue-green algae;* however, the name *cyanobacteria* properly emphasizes their true prokaryotic nature.

Photosynthesis by cyanobacteria is thought to have transformed the anaerobic atmosphere of the early Earth into our oxygen-containing one, and according to the endosymbiotic theory, chloroplasts developed from cyanobacteria. Indeed, chloroplasts and cyanobacteria have similar rRNA and structures such as 70S ribosomes and photosynthetic membranes.

Nitrogen is an essential element in proteins and nucleic acids. Though nitrogen constitutes about 79% of the atmosphere, relatively few organisms can utilize this gas. A few species of filamentous cyanobacteria and proteobacteria (discussed later in the chapter) reduce nitrogen gas (N_2) to ammonia (NH_3) via a process called **nitrogen fixation.** Nitrogen fixation is essential for life on Earth because nitrogen-fixers not only are able to enrich their own growth but also provide nitrogen in a usable form to other organisms.

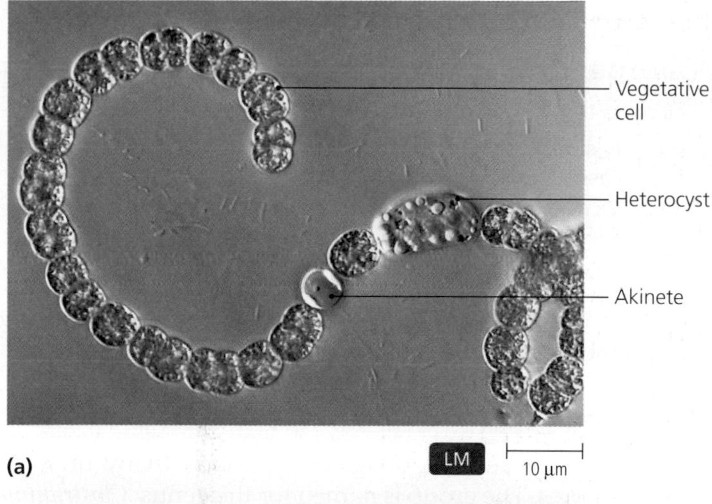

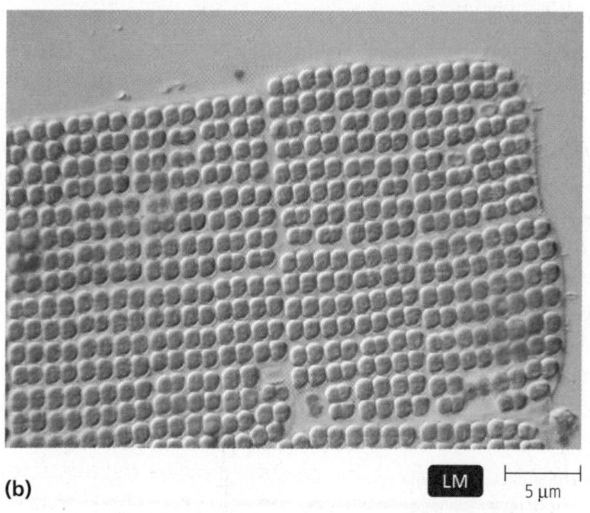

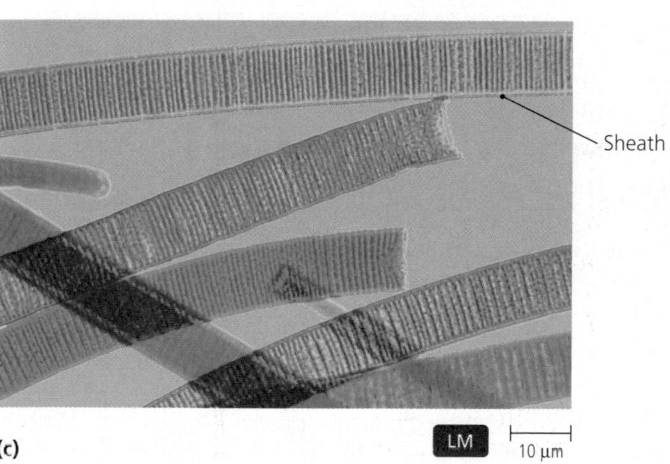

▲ **Figure 11.10 Examples of cyanobacteria with different growth habits.** **(a)** *Anabaena*, which grows as a filament of cocci with differentiated cells. Heterocysts fix nitrogen; akinetes are reproductive cells. **(b)** *Merismopedia*, which grows as a flat sheet of cocci surrounded by a gelatinous glycocalyx. **(c)** *Oscillatoria*, which forms a filament of tightly appressed disk-shaped cells.

Because the enzyme responsible for nitrogen fixation is inhibited by oxygen, nitrogen-fixing cyanobacteria are faced with a problem—how to segregate nitrogen fixation, which is inhibited by oxygen, from oxygenic photosynthesis, which produces oxygen. Nitrogen-fixing cyanobacteria solve this problem in one of two ways. Most cyanobacteria isolate the enzymes of nitrogen fixation in specialized, thick-walled, nonphotosynthetic cells called **heterocysts** (see Figure 11.13a). Heterocysts transport reduced nitrogen to neighboring cells in exchange for glucose. Other cyanobacteria photosynthesize during daylight hours and fix nitrogen at night, thereby separating nitrogen fixation from photosynthesis in time rather than in space.

Green and Purple Phototrophic Bacteria Green and purple bacteria differ from plants, algae, and cyanobacteria in two ways: They use *bacteriochlorophylls* for photosynthesis instead of chlorophyll *a*, and they are *anoxygenic;* that is, they do not generate oxygen during photosynthesis. Green and purple phototrophic bacteria commonly inhabit anaerobic muds rich in hydrogen sulfide at the bottoms of ponds and lakes. These microbes are not necessarily green and purple in color; rather, the terms refer to pigments in some of the better-known members of the groups.

As previously indicated, the green and purple phototrophic bacteria include both sulfur and nonsulfur forms. Whereas nonsulfur bacteria derive electrons for the reduction of CO_2 from organic compounds such as carbohydrates and organic acids, sulfur bacteria derive electrons from the oxidation of hydrogen sulfide to sulfur, as follows:

$$12\,H_2S + 6\,CO_2 \xrightarrow{\text{light}} C_6H_{12}O_6 + 6\,H_2O + 12\,S \qquad (2)$$

Green sulfur bacteria deposit the resultant sulfur outside their cells, whereas purple sulfur bacteria deposit sulfur within their cells **(Figure 11.14)**.

At the beginning of the 20th century, a prominent question in biology concerned the origin of the oxygen released by photosynthetic plants. It was initially thought that oxygen was derived from carbon dioxide, but a comparison of photosynthesis in cyanobacteria (Equation 1 on p. 320) with that in sulfur bacteria (Equation 2) provided evidence that free oxygen is derived from water.

Whereas green sulfur bacteria are placed in phylum Chlorobi, green nonsulfur bacteria are members of phylum Chloroflexi. The purple bacteria (both sulfur and nonsulfur) are placed in three classes of phylum Proteobacteria, which is composed of Gram-negative bacteria and is discussed shortly. **Table 11.1** on p. 322 summarizes the characteristics of phototrophic bacteria. **ANIMATIONS:** ***Photosynthesis: Comparing Prokaryotes and Eukaryotes***

Now we turn our attention to various groups of Gram-positive bacteria, and to a different characteristic of microbes that is used in the classification of Gram-positive bacteria—*G + C ratio*. This ratio is the percentage of all base pairs in a genome that are guanine-cytosine base pairs—and is a useful criterion in classifying microbes. Bacteria with G + C ratios below 50% are considered "low G + C bacteria"; the remainder are

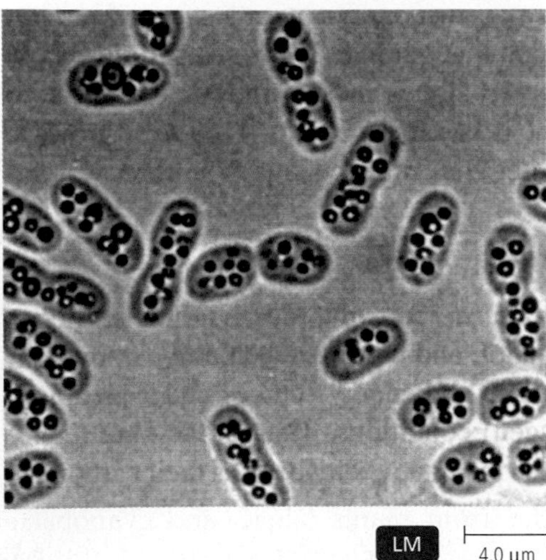

▲ **Figure 11.14 Deposits of sulfur within purple sulfur bacteria in the genus *Chromatium*.** These bacteria oxidize H_2S to produce the granules of elemental sulfur evident in this photomicrograph. *Where do green sulfur bacteria deposit sulfur grains?*

Figure 11.14 Green sulfur bacteria deposit sulfur grains externally.

considered "high G + C bacteria." Because taxonomists have discovered that Gram-positive bacteria with low G + C ratios have similar sequences in their 16S rRNA, and that those with high G + C ratios also have rRNA sequences in common, they have assigned low G + C bacteria and high G + C bacteria to different phyla. We discuss the low G + C bacteria first.

Low G + C Gram-Positive Bacteria

Learning Objectives

✓ Discuss the lack of cell walls in mycoplasmas.

✓ Identify significant beneficial or detrimental effects of the genera *Clostridium, Bacillus, Listeria, Lactobacillus, Streptococcus*, and *Staphylococcus*.

The low G + C Gram-positive bacteria are classified within phylum Firmicutes (fer-mik'ū-tēz), which includes three groups: clostridia, mycoplasmas, and other low G + C Gram-positive bacilli and cocci. Next we consider these three groups in turn.

Clostridia

Clostridia are rod-shaped, obligate anaerobes, many of which form endospores. The group is named for the genus *Clostridium*,[9] which is important both in medicine—in large part because its members produce potent toxins that cause a variety of diseases in humans—and in industry because their endospores enable them to survive harsh conditions, including many types of disinfection and antisepsis. Examples of clostridia include *C. tetani* (te'tan-ē, which causes tetanus), *C. perfringens* (per-frin'jens, gangrene), *C. botulinum* (bo-tū-lī'num, botulism), and *C. difficile* (di-fi'sil-ē, severe diarrhea). Chapter 19 examines these pathogens and the diseases they cause in greater detail; **Beneficial Microbes: Botulism and Botox** describes how a deadly toxin produced by *C. botulinum* has been put to use for cosmetic purposes.

Microbes related to *Clostridium* include sulfate-reducing microbes, which produce H_2S from elemental sulfur during

[9]From Greek *kloster*, meaning spindle.

TABLE 11.1 Characteristics of the Major Groups of Phototrophic Bacteria

	Phylum				
	Cyanobacteria	**Chlorobi**	**Chloroflexi**	**Proteobacteria**	**Proteobacteria**
Class	Cyanobacteria	Chlorobia	Chloroflexi	Gammaproteobacteria	Alphaproteobacteria and one genus in betaproteobacteria
Common name(s)	Blue-green bacteria ("blue-green algae")	Green sulfur bacteria	Green nonsulfur bacteria	Purple sulfur bacteria	Purple nonsulfur bacteria
Major photosynthetic pigments	Chlorophyll *a*	Bacteriochlorophyll *a* plus *c, d*, or *e*	Bacteriochlorophylls *a* and *c*	Bacteriochlorophyll *a* or *b*	Bacteriochlorophyll *a* or *b*
Types of photosynthesis	Oxygenic	Anoxygenic	Anoxygenic	Anoxygenic	Anoxygenic
Electron donor in photosynthesis	H_2O	H_2, H_2S, or S	Organic compounds	H_2, H_2S, or S	Organic compounds
Sulfur deposition	None	Outside of cell	None	Inside of cell	None
Nitrogen fixation	Some species	None	None	None	None
Motility	Nonmotile or gliding	Nonmotile	Gliding	Motile with polar or peritrichous flagella	Nonmotile or motile with polar flagella

anaerobic respiration, and *Veillonella* (vī-lō-nel'ă), a genus of anaerobic cocci that live as part of the biofilm (plaque) that forms on the teeth of warm-blooded animals. *Veillonella* is unusual because even though it has a typical Gram-positive cell wall structure, it has a negative Gram reaction—it stains pink.

Mycoplasmas

A second class of low G + C bacteria are the **mycoplasmas**[10] (mī'kō-plaz'mas). These facultative or obligate anaerobes lack cell walls, which means they stain pink when Gram stained. Indeed, until their nucleic acid sequences proved their similarity to Gram-positive organisms, mycoplasmas were classified as Gram-negative microbes, instead of in phylum Firmicutes with other low G + C Gram-positive bacteria.

Mycoplasmas are able to survive without cell walls in part because they colonize osmotically protected habitats such as animal and human bodies, and because they have tough cytoplasmic membranes, many of which contain lipids called *sterols* that give the membranes strength and rigidity. Because they lack cell walls, they are pleomorphic. They were named "mycoplasmas" because their filamentous forms resemble the filaments of fungi. Mycoplasmas have diameters ranging from 0.2 μm to 0.8 μm, making them the smallest free-living cells, and many mycoplasmas have a terminal structure that is used for attachment to eukaryotic cells and that gives the bacterium a pearlike shape. They require organic growth factors, such as cholesterol, fatty acids, vitamins, amino acids, and nucleotides, which they acquire from their host or which must be added to laboratory media. When growing on solid media, most species form a distinctive "fried egg" appearance because cells in the center of the colony grow into the agar while those around the perimeter only spread across the surface **(Figure 11.15)**.

In animals, mycoplasmas colonize mucous membranes of the respiratory and urinary tracts and are associated with pneumonia and urinary tract infections. Pathogenic mycoplasmas and the diseases they cause are discussed more fully in Chapter 19.

[10]From Greek *mycos*, meaning fungus, and *plassein*, meaning to mold.

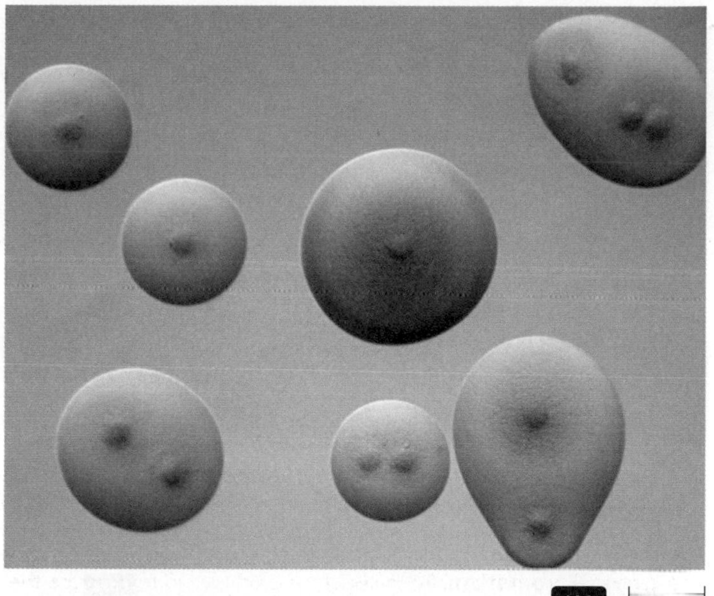

▲ **Figure 11.15 The distinctive "fried egg" appearance of *Mycoplasma* colonies.** This visual feature is unique to this group of bacteria, growing on an agar surface.

Other Low G + C Bacilli and Cocci

A third group of low G + C Gram-positive organisms is composed of bacilli and cocci that are significant in environmental, industrial, and health care settings. Among the genera in this group are *Bacillus, Listeria* (lis-tēr'ē-ă), *Lactobacillus* (lak'tō-bă-sil'ŭs), *Streptococcus, Enterococcus* (en'ter-ō-kok'ŭs), and *Staphylococcus* (staf'i-lō-kok'ŭs). Chapter 19 discusses the pathogens in these genera in greater detail.

Bacillus The genus *Bacillus* includes endospore-forming aerobes and facultative anaerobes that typically move by means of peritrichous flagella. Numerous species of *Bacillus* are common in soil.

Bacillus thuringiensis (thur-in-jē-en'sis) is beneficial to farmers and gardeners. During sporulation, this bacterium produces a

BENEFICIAL MICROBES

BOTULISM AND BOTOX

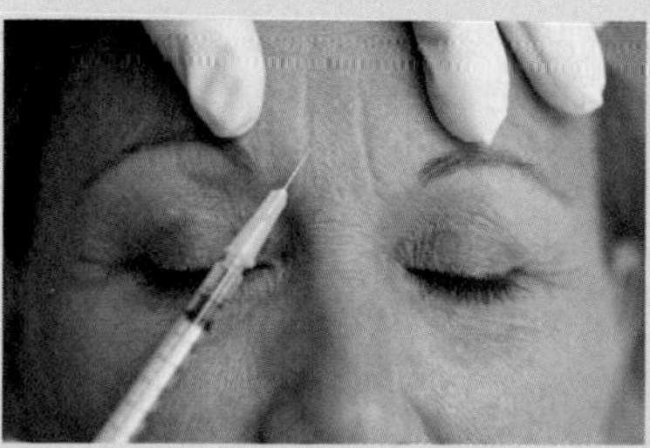

Clostridium botulinum produces botulinum toxins, some of the deadliest toxins known. When absorbed in the body, botulism toxins interfere with the release of acetylcholine, the neurotransmitter that signals muscles to contract. As a result, muscle cells cannot contract, and a slow but progressive paralysis spreads throughout the body. Death occurs when paralysis of respiratory muscles results in respiratory failure. This disease, called botulism, is discussed in more detail in Chapter 19.

Purified type A botulinum toxin is marketed as Botox, extremely small doses of which are injected into facial muscles that cause skin wrinkles. The toxin paralyzes or weakens the muscles, smoothing the skin. Such treatments last approximately 6 months and must be repeated in order to maintain the desired effects.

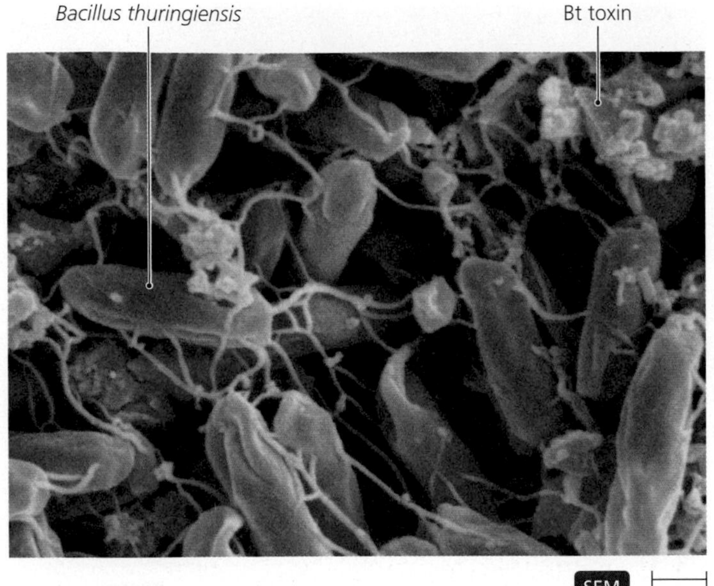

▲ **Figure 11.16 Crystals of Bt toxin, produced by the endospore-forming *Bacillus thuringiensis*.** The crystalline protein kills caterpillars that ingest it.

crystalline protein that is toxic to caterpillars that ingest it (**Figure 11.16**). Gardeners spray *Bt toxin,* as preparations of the bacterium and toxin are known, on plants to protect them from caterpillars. Scientists have achieved the same effect, without the need of spraying, by introducing the gene for Bt toxin into plants' chromosomes. Other beneficial species of *Bacillus* include *B. polymyxa* (po-lē-miks'ă) and *B. licheniformis* (lī-ken-i-for'mis), which synthesize the antibiotics polymyxin and bacitracin, respectively. Chapter 10 discusses the production and effects of antibiotics in more detail.

Bacillus anthracis (an-thrā'sis), which causes anthrax, gained notoriety in 2001 as an agent of bioterrorism. Its endospores are either inhaled or enter the body through breaks in the skin. When they germinate, the vegetative cells produce toxins that kill surrounding tissues. Untreated *cutaneous anthrax* is fatal in 20% of patients; untreated inhalational *anthrax* is generally 100% fatal without prompt aggressive treatment. As discussed in Chapter 9, refrigeration prevents the excessive growth of this and other contaminants.

Listeria Another pathogenic low G + C Gram-positive rod is *Listeria monocytogenes* (mo-nō-sī-tah'je-nēz), which can contaminate milk and meat products. This microbe, which does not produce endospores, is notable because it continues to reproduce under refrigeration, and it can survive inside phagocytic white blood cells. *Listeria* rarely causes disease in adults, but it can kill a fetus in an infected woman when it crosses the placental barrier. It also causes meningitis[11] and bacteremia[12] when it infects immunocompromised patients such as the aged and patients with AIDS, cancer, or diabetes.

Lactobacillus Organisms in the genus *Lactobacillus* are non-spore-forming rods normally found growing in the human mouth, stomach, intestinal tract, and vagina. These organisms rarely cause disease; instead, they protect the body by inhibiting the growth of pathogens—a situation called *microbial antagonism.* Lactobacilli are used in industry in the production of yogurt, buttermilk, pickles, and sauerkraut. The Beneficial Microbes box on p. 297 discusses the growing role of *Lactobacillus* in promoting human health.

Streptococcus and Enterococcus The genera *Streptococcus* and *Enterococcus* are diverse groups of Gram-positive cocci associated in pairs and chains (see Figure 11.6b). They cause numerous human diseases, including pharyngitis (strep throat), scarlet fever, impetigo, fetal meningitis, wound infections, pneumonia, and diseases of the inner ear, skin, blood, and kidneys. In recent years, health care providers have become concerned over strains of multi-drug-resistant streptococci. Of particular concern are so-called flesh-eating streptococci, which produce toxins that destroy muscle and fat tissue.

Staphylococcus Among the common inhabitants of humans is *Staphylococcus aureus*[13] (o'rē-ŭs), which is typically found growing harmlessly in clusters on the skin. A variety of toxins and enzymes allow some strains of *S. aureus* to invade the body and cause such diseases as bacteremia, pneumonia, wound infections, food poisoning, toxic shock syndrome, and diseases of the joints, bones, heart, and blood.

The characteristics of the low G + C Gram-positive bacteria are summarized in the first part of Table 11.2. These bacteria, which are classified into three classes within phylum Firmicutes, include the anaerobic endospore-forming rod *Clostridium,* the pleomorphic *Mycoplasma,* the aerobic and facultative aerobic endospore-forming rod *Bacillus,* the non-endospore-forming rods *Listeria* and *Lactobacillus,* and the cocci *Streptococcus, Enterococcus,* and *Staphylococcus.*

Next we consider Gram-positive bacteria that have high G + C ratios.

High G + C Gram-Positive Bacteria

Learning Objectives

✓ Explain the slow growth of *Mycobacterium.*

✓ Identify significant beneficial or detrimental properties of the genera *Corynebacterium, Mycobacterium, Actinomyces, Nocardia,* and *Streptomyces.*

Taxonomists classify Gram-positive bacteria with a G + C ratio greater than 50% in the phylum Actinobacteria, which includes species with rod-shaped cells (many of which are significant human pathogens) and filamentous bacteria, which resemble fungi in their growth habit and in the production of reproductive spores. Here we examine briefly some prominent high G + C Gram-positive bacteria; the numerous pathogens in this group are discussed more fully in Chapter 19.

[11]Inflammation of the membranes covering the brain and spinal cord; from Greek *meninx*, meaning membrane.
[12]The presence of bacteria, particularly those that produce disease symptoms, in the blood.
[13]From Latin *aurum*, meaning gold, because it produces yellow pigments.

TABLE 11.2 Characteristics of Selected Gram-Positive Bacteria

Phylum/Class	G + C Ratio	Representative Genera	Special Characteristics	Diseases
Firmicutes				
Clostridia	Low (less than 50%)	*Clostridium*	Obligate anaerobic rods; endospore formers	Tetanus Botulism Gangrene Severe diarrhea
		Epulopiscium	Giant rods	
		Veillonella	Part of oral biofilm on human teeth; stain like Gram-negative bacteria (pink)	Dental caries
Mollicutes	Low (less than 50%)	*Mycoplasma*	Lack cell walls; pleomorphic; smallest free-living cells; stain like Gram-negative bacteria (pink)	Pneumonia Urinary tract infections
Bacilli	Low (less than 50%)	*Bacillus*	Facultative anaerobic rods; endospore formers	Anthrax
		Listeria	Contaminates dairy products	Listeriosis
		Lactobacillus	Produce yogurt, buttermilk, pickles, sauerkraut	Rare blood infections
		Streptococcus	Cocci in chains	Strep throat, scarlet fever, and others
		Staphylococcus	Cocci in clusters	Bacteremia, food poisoning, and others
Actinobacteria				
Actinobacteria	High (greater than 50%)	*Corynebacterium*	Snapping division; metachromatic granules in cytoplasm	Diphtheria
		Mycobacterium	Waxy cell walls (mycolic acid)	Tuberculosis and meningitis
		Actinomyces	Filaments	Actinomycosis
		Nocardia	Filaments; degrade pollutants	Lesions
		Streptomyces	Produce antibiotics	Rare sinus infections

Corynebacterium

Members of the genus *Corynebacterium* are pleomorphic—though generally rod-shaped—aerobes and facultative anaerobes. They reproduce by snapping division, which often causes the cells to form V-shapes and palisades (see Figures 11.3 and 11.7d). Corynebacteria are also characterized by their stores of phosphate within inclusions called **metachromatic granules,** which stain differently from the rest of the cytoplasm when the cells are stained with methylene blue or toluidine blue. The best-known species is *C. diphtheriae,* which causes diphtheria.

Mycobacterium

The genus *Mycobacterium* (mī′kō-bak-tēr′ē-um) is composed of aerobic species that are slightly curved to straight rods that sometimes form filaments. Mycobacteria grow very slowly, often requiring a month or more to form a visible colony on an agar surface. Their slow growth is partly due to the time and energy required to enrich their cell walls with high concentrations of long carbon-chain waxes called **mycolic acids,** which make the cells resistant to desiccation and to staining with water-based dyes. As discussed in Chapter 4, microbiologists developed the *acid-fast stain* for mycobacteria because they are difficult to stain by standard staining techniques such as the Gram stain. Though some mycobacteria are free-living, the most prominent species are pathogens of animals and humans, including *Mycobacterium tuberculosis* (too-ber-kyū-lō′sis) and *Mycobacterium leprae* (lep′rī), which cause tuberculosis and leprosy, respectively. Mycobacteria should not be confused with the low G + C mycoplasmas discussed earlier.

Actinomycetes

Actinomycetes (ak′ti-nō-mī-sētz) are high G + C Gram-positive bacteria that form branching filaments resembling fungi (**Figure 11.17**). Of course, in contrast to fungi, the filaments of actinomycetes are composed of prokaryotic cells. As we have seen, some actinomycetes also resemble fungi in the production of chains of reproductive spores at the ends of their filaments (see Figure 11.4). These spores should not be confused with endospores, which are resting stages and not reproductive cells. Actinomycetes may cause disease, particularly in immunocompromised patients. Among the important actinomycete genera are *Actinomyces* (which gives this group its name), *Nocardia,* and *Streptomyces.*

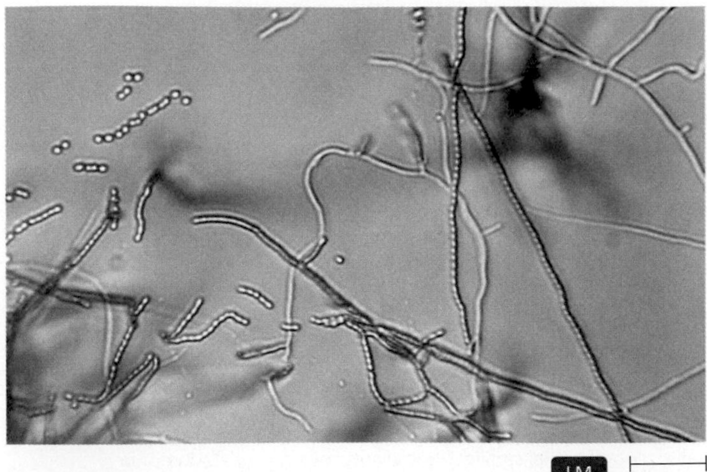

▲ **Figure 11.17 The branching filaments of actinomycetes.** This photograph shows filaments of a colony of *Streptomyces* sp. growing on agar. *How do the filaments of actinomycetes compare to the filaments of fungi?*

Figure 11.17 *Filaments of actinomycetes are thinner than those of fungi, and they are composed of prokaryotic cells.*

Actinomyces Species of *Actinomyces* (ak′ti-nō-mī′sēz) are facultative capneic[14] filaments that are normal inhabitants of the mucous membranes lining the oral cavity and throats of humans. *Actinomyces israelii* (is-rā′el-ē-ē) growing as an opportunistic pathogen in humans destroys tissue to form abscesses and can spread throughout the abdomen, consuming every vital organ.

Nocardia Species of *Nocardia* (nō-kar′dē-ă) are soil- and water-dwelling aerobes that typically form aerial and subterranean filaments that make them resemble fungi. *Nocardia* is notable because it can degrade many pollutants of landfills, lakes, and streams, including waxes, petroleum hydrocarbons, detergents, benzene, polychlorinated biphenyls (PCBs), pesticides, and rubber. Some species of *Nocardia* cause lesions in humans.

Streptomyces Bacteria in the genus *Streptomyces* (strep-tō-mī′-sēz) are important in several realms. Ecologically, they recycle nutrients in the soil by degrading a number of carbohydrates, including cellulose, lignin (the woody part of plants), chitin (the skeletal material of insects and crustaceans), latex, aromatic chemicals (organic compounds containing a benzene ring), and keratin (the protein that forms hair, nails, and horns). The metabolic by-products of *Streptomyces* give soil its musty smell. Medically, *Streptomyces* produce most of the important antibiotics, including chloramphenicol, erythromycin, and tetracycline, as discussed in Chapter 10. **Highlight: *Streptomyces* and the European Beewolf** discusses a species of *Streptomyces* that resides in the antennal glands of a wasp.

Table 11.2 on p. 325 includes a summary of the characteristics of the genera of high G + C Gram-positive bacteria (phylum Actinobacteria) discussed in this chapter.

HIGHLIGHT

STREPTOMYCES AND THE EUROPEAN BEEWOLF

The female European beewolf is a wasp that takes motherhood seriously. True to its name, the beewolf catches bees, paralyzes them with its sting, and then tucks them into holes in the ground, where the (still living) bees serve as food for the beewolf's young. To protect her young, the female beewolf also enlists the help of bacteria. She secretes a goolike substance from her antennal glands, which, as scientists have recently learned, contains a previously unidentified species of *Streptomyces* bacteria. It appears that the goo, with the help of these bacteria, somehow defends young beewolves from threats such as pathogens in their environment. Studies have shown that beewolf larvae deprived of contact with the goo were less likely to live to adulthood than larvae allowed contact with the goo—suggesting that the European beewolf may have bacteria to thank for keeping her young secure.

CRITICAL **THINKING**

In Chapter 4, we discussed the identification of microbes using dichotomous taxonomic keys (see p. 121). Design a key for all the genera of Gram-positive bacteria listed in Table 11.2 (p. 325).

To this point we have discussed archaea and deeply branching, phototrophic, and Gram-positive bacteria. Now we turn our attention to the Gram-negative bacteria that are grouped together within the phylum Proteobacteria.

Gram-Negative Proteobacteria

Phylum **Proteobacteria**[15] constitutes the largest and most diverse group of bacteria (see Figure 11.9). Though they have a variety of shapes, reproductive strategies, and nutritional types, they are all Gram-negative and share common 16S rRNA nucleotide sequences. The G + C ratio of Gram-negative species is not critical in delineating taxa of most Gram-negative organisms, so we will not consider this characteristic in our discussion of these bacteria.

There are five distinct classes of proteobacteria, designated by the first five letters of the Greek alphabet—alpha, beta, gamma, delta, and epsilon. These classes are distinguished by minor differences in their rRNA sequences. Here we will focus

[14]Meaning they grow best with a relatively high concentration of carbon dioxide.
[15]Named for the Greek god *Proteus*, who could assume many shapes.

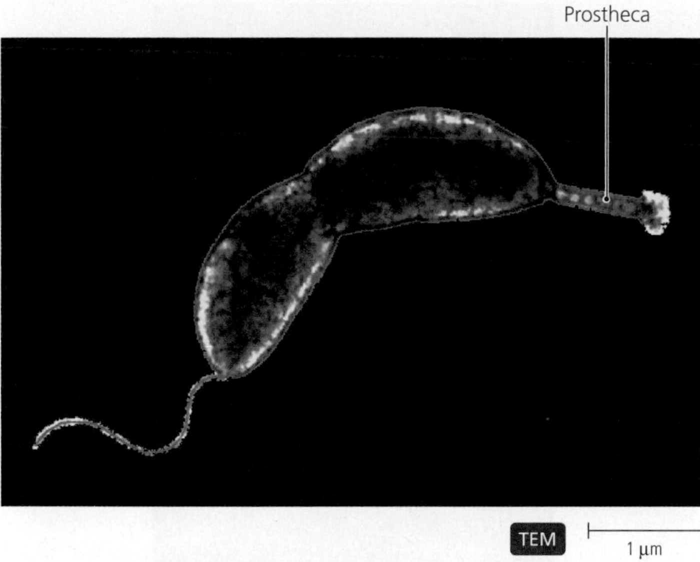

▲ **Figure 11.18 A prostheca.** This extension of an alphaproteobacterial cell increases surface area for absorbing nutrients and serves as an organ of attachment. This prosthecate bacterium, *Hyphomicrobium facilis*, also produces buds from its prostheca.

▲ **Figure 11.19 Nodules on pea roots.** *Rhizobium*, a nitrogen-fixing alphaproteobacterium, stimulates the growth of such nodules.

our attention on species with novel characteristics as well as species with practical importance.

Alphaproteobacteria

Learning Objective

✓ Describe the appearance and function of prosthecae in alphaproteobacteria.

Alphaproteobacteria are typically aerobes capable of growing at very low nutrient levels. Many have unusual methods of metabolism, as we will see shortly. They may be rods, curved rods, spirals, coccobacilli, or pleomorphic. Many species have unusual extensions called *prosthecae* (pros-thē'kē), which are composed of cytoplasm surrounded by the cytoplasmic membrane and cell wall **(Figure 11.18)**. They use prosthecae for attachment and to increase surface area for nutrient absorption. Some prosthecate species produce buds at the ends of the extensions.

Nitrogen Fixers Two genera of nitrogen-fixers in the class Alphaproteobacteria—*Azospirillum* (ā-zō-spī'ril-ŭm) and *Rhizobium* (ri-zō'be-ŭm)—are important in agriculture. They grow in association with the roots of plants, where they make atmospheric nitrogen (N_2) available to the plants as ammonia (NH_3), often called fixed nitrogen. *Azospirillum* associates with the outer surfaces of roots of tropical grasses, such as sugar cane. In addition to supplying nitrogen to the grass, this bacterium also releases chemicals that stimulate the plant to produce numerous root hairs, which increase a root's surface area and thus its uptake of nutrients.

Rhizobium grows within the roots of leguminous plants such as peas, beans, and clover, stimulating the formation of nodules on their roots **(Figure 11.19)**. *Rhizobium* cells within the nodules make ammonia available to the plant, encouraging growth. Scientists are actively seeking ways to successfully insert the genes of nitrogen fixation into plants such as corn, which require large amounts of nitrogen.

Nitrifying Bacteria As discussed in Chapter 6, chemoautotrophic organisms derive their carbon from carbon dioxide and their energy from inorganic chemicals. Those that derive electrons from the oxidation of nitrogenous compounds are called **nitrifying bacteria.** These microbes are important in the environment and in agriculture because they convert reduced nitrogen compounds, such as ammonia (NH_3), and ammonium (NH_4^+), into nitrate (NO_3^-)—a process called **nitrification.** Nitrate moves more easily through soil than reduced nitrogenous compounds and is thereby more available to plants. Nitrifying alphaproteobacteria oxidize ammonia or ammonium and are in the genus *Nitrobacter* (nī-trō-bak'ter).

CRITICAL THINKING

Contrast the processes of nitrogen fixation and nitrification.

Purple Nonsulfur Phototrophs With one exception (the betaproteobacterium *Rhodocyclus*), purple nonsulfur phototrophs are classified as alphaproteobacteria. Purple nonsulfur bacteria grow in the upper layer of mud at the bottoms of lakes and ponds. As we discussed earlier, they harvest light as an energy source by using bacteriochlorophylls, and they do not generate oxygen during photosynthesis. Morphologically, they may be rods, curved rods, or spirals; some species are prosthecate. Refer to Table 11.1 (p. 322) to review the characteristics of all phototrophic bacteria.

Pathogenic Alphaproteobacteria Notable pathogens among the alphaproteobacteria include *Rickettsia* (ri-ket'sē-ă), and *Brucella* (broo-sel'lă).

Rickettsia is a genus of small, Gram-negative, aerobic rods that live and reproduce inside mammalian cells. They cause a number

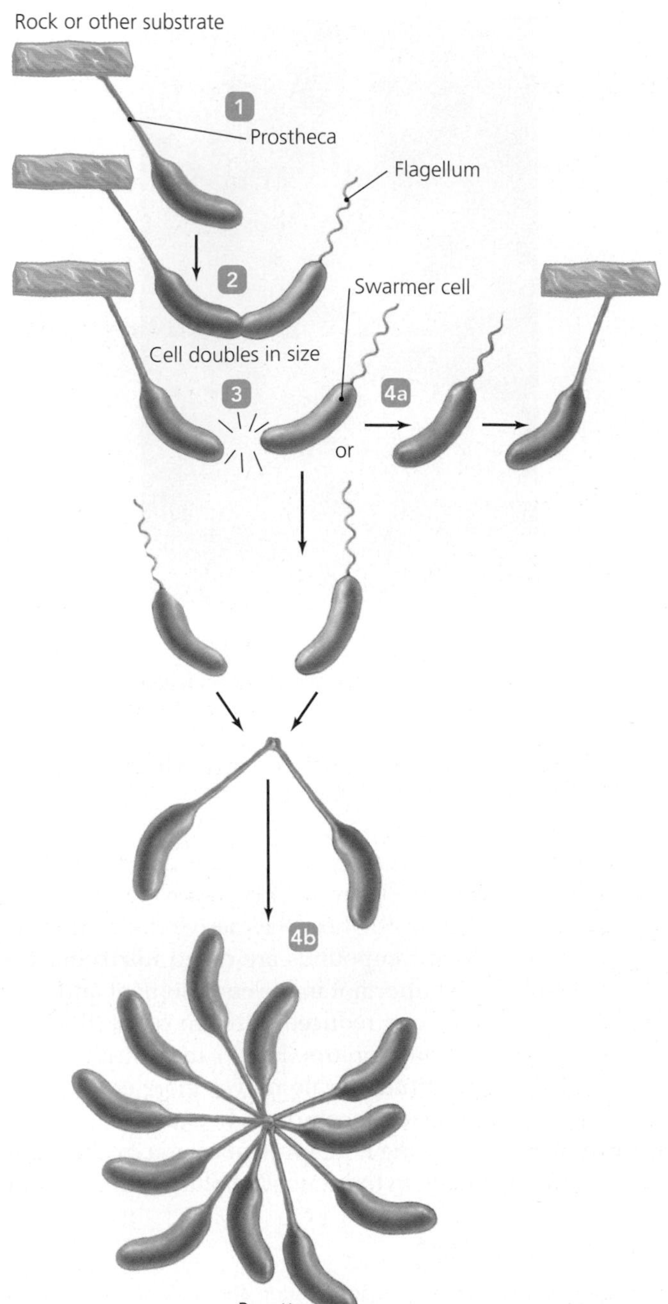

▲ **Figure 11.20 Growth and reproduction of *Caulobacter*.** 1 A cell attached to a substrate by its prostheca. 2 Growth and production of an apical flagellum. 3 Division by asymmetrical binary fission to produce a flagellated daughter cell called a swarmer cell. 4a Attachment of a swarmer cell to a substrate by a prostheca that replaces the flagellum. 4b Attachment of a swarmer cell to other cells to form a rosette.

of human diseases, including typhus and Rocky Mountain spotted fever. Rickettsias cannot use glucose as a nutrient; instead they oxidize amino acids and Krebs cycle intermediates such as glutamic acid and succinic acid. For this reason they are obliged to live within other cells where these nutrients are produced.

Brucella is a coccobacillus that causes *brucellosis,* a disease of mammals characterized by spontaneous abortions and sterility in animals. In contrast, infected humans suffer chills, sweating, fatigue, and fever. *Brucella* is notable because it survives phagocytosis by white blood cells—normally an important step in the body's defense against disease.

▲ **Figure 11.21 A plant gall.** Infecting cells of *Agrobacterium* have inserted into a plant chromosome a plasmid carrying a plant growth hormone gene. The hormone causes the proliferation of undifferentiated plant cells. These gall cells synthesize nutrients for the bacteria.

Other Alphaproteobacteria Other alphaproteobacteria are important in industry and the environment. For example, *Acetobacter* (a-sē′tō-bak-ter) and *Gluconobacter* (gloo-kon′ō-bak-ter) are used to synthesize acetic acid (vinegar). *Caulobacter* (kaw′lō-bak-ter) is a common prosthecate rod-shaped microbe that inhabits nutrient-poor seawater and freshwater; it can also be found in laboratory water baths.

Caulobacter has a unique reproductive strategy (**Figure 11.20**). A cell that has attached to a substrate with its prostheca (1) grows until it has doubled in size, at which time it produces a flagellum at its apex (2); it then divides by asymmetric binary fission (3). The flagellated daughter cell, which is called a swarmer cell, then swims away. The swarmer cell can either attach to a substrate with a new prostheca that replaces the flagellum (4a) or attach to other swarmer cells to form a rosette of cells (4b). The process of reproduction repeats about every 2 hours. **Beneficial Microbes: A Microtube of Super Glue** on p. 330 examines an amazing property of *Caulobacter* prosthecae.

Scientists are very interested in the usefulness of another alphaproteobacterium, *Agrobacterium* (ag′ro-bak-tēr′ē-ŭm), which infects plants to form tumors called *galls* (**Figure 11.21**). The bacterium inserts a plasmid called Ti, which carries a gene for a plant growth hormone, into a chromosome of the plant. The growth hormone causes the cells of the plant to proliferate into a gall and to produce nutrients for the bacterium. Scientists have discovered that they can insert almost any DNA sequence into the plasmid, making it an ideal vector for genetic manipulation of plants.

Characteristics of selected members of the alphaproteobacteria are listed in Table 11.3.

TABLE 11.3 Characteristics of Selected Gram-Negative Bacteria

Phylum/Class	Representative Members	Special Characteristics	Diseases
Proteobacteria			
Alphaproteobacteria	*Azospirillum*	Nitrogen fixer	
	Rhizobium	Nitrogen fixer	
	Nitrobacter	Nitrifying bacterium	
	Purple nonsulfur bacteria	Anoxygenic phototrophs	
	Rickettsia	Intracellular pathogen	Typhus and Rocky Mountain spotted fever
	Brucella	Coccobacillus	Brucellosis
	Acetobacter, Gluconobacter	Synthesize acetic acid	
	Caulobacter	Prosthecate bacterium	
	Agrobacterium	Causes galls in plants; vector for gene transfer in plants	
Betaproteobacteria	*Nitrosomonas*	Nitrifying bacterium	
	Neisseria	Diplococcus	Gonorrhea and meningitis
	Bordetella		Pertussis
	Burkholderia		Lung infection of cystic fibrosis patients
	Thiobacillus	Colorless sulfur bacterium	
	Zoogloea	Used in sewage treatment	
	Sphaerotilus	Blocks sewage treatment pipes	
Gammaproteobacteria	Purple sulfur bacteria		
	Legionella	Intracellular pathogen	Legionnaires' disease
	Coxiella	Intracellular pathogen	Q fever
	Methylococcus	Oxidizes methane	
	Glycolytic facultative anaerobes	Facultative anaerobes that catabolize carbohydrates via glycolysis and the pentose phosphate pathway	See Table 11.4 on p. 332
	Pseudomonas	Aerobe that catabolizes carbohydrates via Entner-Doudoroff and pentose phosphate pathways	Urinary tract infections, external otitis
	Azotobacter *Azomonas*	Nitrogen fixers not associated with plant roots	
Deltaproteobacteria	*Desulfovibrio*	Sulfate reducer	
	Bdellovibrio	Pathogen of Gram-negative bacteria	
	Myxobacteria	Reproduces by forming differentiated fruiting bodies	
Epsilonproteobacteria	*Campylobacter*	Curved rod	Gastroenteritis
	Helicobacter	Spiral	Gastric ulcers
Chlamydiae			
Chlamydiae	*Chlamydia*	Intracellular pathogen; lacks peptidoglycan	Neonatal blindness and lymphogranuloma venereum
Spirochaetes			
"Spirochaetes"[a]	*Treponema*	Motile by axial filaments	Syphilis
	Borrelia	Motile by axial filaments	Lyme disease
"Bacteroidetes"			
"Bacteroidetes"	*Bacteroides*	Anaerobe that lives in animal colons	Abdominal infections
"Sphingobacteria"	*Cytophaga*	Digests complex polysaccharides	

[a]The names of taxa in quotations are not officially recognized.

BENEFICIAL MICROBES

A MICROTUBE OF SUPERGLUE

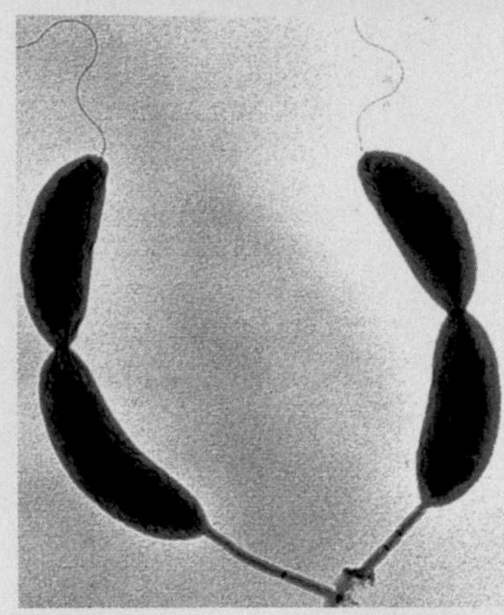

▲ *Caulobacter crescentus.* SEM 1 μm

A swarmer cell of the Gram-negative, alphaproteobacterium *Caulobacter crescentus* attaches itself to an environmental substrate by secreting an organic adhesive from its prostheca as if it were a tube of glue. This polysaccharide-based bonding agent is the strongest known glue of biological origin, beating out such contenders as barnacle glue, mussel glue, and the adhesion of gecko lizard bristles.

One way scientists gauge adhesive strength is to measure the force required to break apart two glued objects. Commercial superglues typically lose their grip when confronted with a shear force of 18 to 28 newtons (N) per square millimeter (1 newton is the amount of force needed to accelerate a 1-kg mass 1 meter per second per second). Dental cements bond with strengths up to 30 N/mm^2, but *Caulobacter* glue is more than twice as adhesive. It maintains its grip up to 68 N/mm^2! That is equivalent to being able to hang an adult female elephant on a wall with a spot of glue the size of an American quarter. And remember, this bacterium lives in water; its glue works even when submerged.

Scientists are researching the biophysical and chemical mechanisms that give this biological glue such incredible gripping power. One critical component of the glue is *N*-acetylglucosamine, one of the sugar subunits of peptidoglycan found in bacterial cell walls. Scientists are struggling to characterize the other molecules that make up the glue. The problem? They cannot pry the glue free to analyze it.

Some potential applications of such a bacterial superglue include use as a biodegradable suture in surgery, as a more durable dental adhesive, or as an adhesive to stick anti-biofilm disinfectants onto surfaces such as medical devices and ships' hulls.

Betaproteobacteria

Learning Objective

✓ Name three pathogenic and three useful betaproteobacteria.

Betaproteobacteria are another diverse group of Gram-negative bacteria that thrive in habitats with low levels of nutrients. They differ from alphaproteobacteria in their rRNA sequences, though metabolically the two groups overlap. One example of the metabolic overlap is *Nitrosomonas* (nī-trō-sō-mō′nas), an important nitrifying bacterium of soils; instead of oxidizing ammonia or ammonium to nitrate, however, it oxidizes nitrite to nitrate. In this section we will discuss a few interesting betaproteobacteria.

Pathogenic Betaproteobacteria Species of *Neisseria* (ni-se′rē-ă) are Gram-negative diplococci that inhabit the mucous membranes of mammals and cause such diseases as gonorrhea, meningitis, pelvic inflammatory disease, and inflammation of the cervix, pharynx, and external lining of the eye.

Other pathogenic betaproteobacteria include *Bordetella* (bōr-dĕ-tel′ă), which is the cause of pertussis (whooping cough) (see **Emerging Diseases: Pertussis**), and *Burkholderia* (burk-hol-der′ē-ă), which recycles numerous organic compounds in nature. *Burkholderia* commonly colonizes moist environmental surfaces (including laboratory and medical equipment) and the respiratory passages of patients with cystic fibrosis.

Chapter 20 discusses these pathogens in more detail.

Other Betaproteobacteria Members of the genus *Thiobacillus* (thī-ō-bă-sil′ŭs) are colorless sulfur bacteria that are important in recycling sulfur in the environment by oxidizing hydrogen sulfide (H_2S) or elemental sulfur (S^0) to sulfate (SO_4^{2-}). Miners use *Thiobacillus* to leach metals from low-grade ore, though the bacterium does cause extensive pollution when it releases metals and acid from mine wastes. Chapter 26 discusses the sulfur cycle.

Sewage treatment supervisors are interested in *Zoogloea* (zō′ō-glē-ă) and *Sphaerotilus* (sfēr-ō′til-us), two genera that form *flocs*—slimy, tangled masses of bacteria and organic matter in sewage. *Zoogloea* forms compact flocs that settle to the bottom of treatment tanks and assist in the purification process. *Sphaerotilus,* in contrast, forms loose flocs that do not settle and thus impede the proper flow of waste through a treatment plant.

The characteristics of the genera of betaproteobacteria discussed in this section are summarized in Table 11.3 on p. 329.

Gammaproteobacteria

Learning Objective

✓ Describe the gammaproteobacteria.

The **gammaproteobacteria** make up the largest and most diverse class of proteobacteria; almost every shape, arrangement of cells, metabolic type, and reproductive strategy is represented in this group. Ribosomal studies indicate that gammaproteobacteria can be divided into several subgroups:

- Purple sulfur bacteria
- Intracellular pathogens
- Methane oxidizers
- Facultative anaerobes that utilize Embden-Meyerhof glycolysis and the pentose phosphate pathway
- Pseudomonads, which are aerobes that catabolize carbohydrates by the Entner-Doudoroff and pentose phosphate pathways

Here we will examine some representatives of these groups.

Purple Sulfur Bacteria Whereas the purple *non*sulfur bacteria are distributed among the alpha- and betaproteobacteria, **purple sulfur bacteria** are all gammaproteobacteria (**Figure 11.22**). Purple sulfur bacteria are obligate anaerobes that oxidize hydrogen sulfide to sulfur, which they deposit as internal granules. They are found in sulfur-rich zones in lakes, bogs, and oceans. Some species form intimate relationships with marine worms, covering the body of a worm like strands of hair.

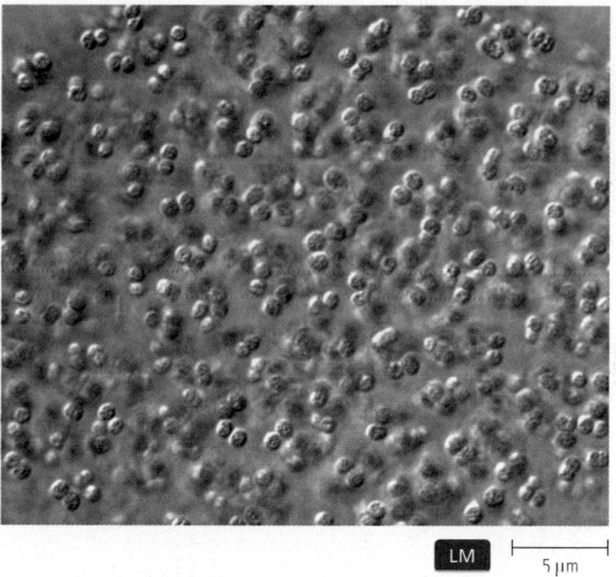

▲ **Figure 11.22** Purple sulfur bacteria.

Intracellular Pathogens Organisms in the genera *Legionella* (lē-jŭ-nel′lă) and *Coxiella* (kok-sē-el′ă) are pathogens of humans that avoid digestion by white blood cells, which are normally part of a body's defense; in fact, they thrive inside these defensive cells. *Legionella* derives energy from the metabolism of amino acids, which are more prevalent inside cells than outside. *Coxiella* grows best at low pH, such as is found in the phagolysomes of white blood cells. The bacteria in these genera cause Legionnaires' disease and Q fever, respectively.

Methane Oxidizers Methane oxidizers, such as *Methylococcus* (meth-i-lō-kok′ŭs), are Gram-negative bacteria that utilize methane as a carbon source and as an energy source. Like

EMERGING DISEASES

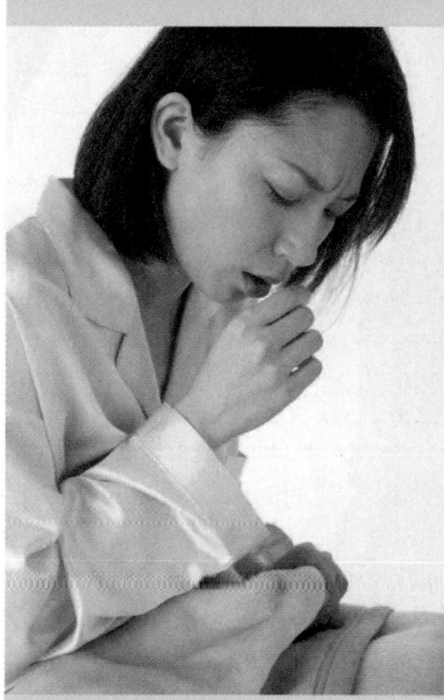

PERTUSSIS

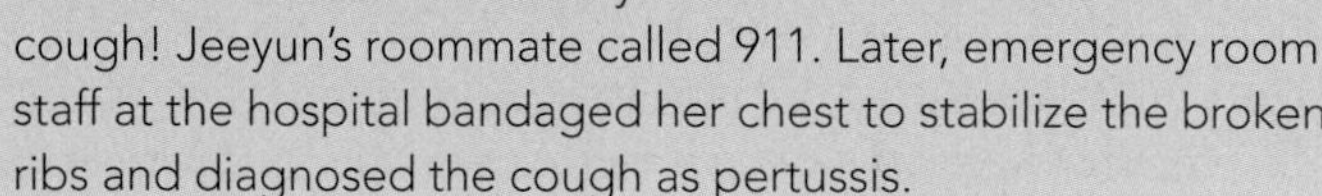

Jeeyun was coughing again. She had been coughing off and on for two weeks. If she had thought of it, she might have noticed that the coughing spells began soon after her flight from New York, where she had visited her grandmother. Within a week of her return to California, she had developed coldlike signs—runny nose, sneezing, and a slight fever—but she didn't remember these things. What she did know was that the coughing was worse; her chest hurt from the constant hacking. Then, the coughing broke two ribs.

The surprising pain caused Jeeyun to involuntarily urinate. When the coughing stopped, she vomited and fainted in a heap on the floor. This was no ordinary cough! Jeeyun's roommate called 911. Later, emergency room staff at the hospital bandaged her chest to stabilize the broken ribs and diagnosed the cough as pertussis.

Pertussis, commonly known as whooping cough, is usually considered a childhood disease, but it can strike adults. Older patients seldom develop the characteristic "whooping" sound associated with gasping inhalation, but in Jeeyun's case, the severity and length of the coughing led to a speedy diagnosis.

Immunization is the only way to control pertussis, but adults have been lax in vaccinating children and receiving boosters for themselves. As a result, whooping cough is reemerging as a major problem in the industrialized world. Infected people spread *Bordetella* in respiratory droplets to their neighbors, as happened to Jeeyun in an airplane's cabin on a long flight. For more about pertussis, see p. 588.

 Track pertussis online by going to the Study Area at www.masteringmicrobiology.com.

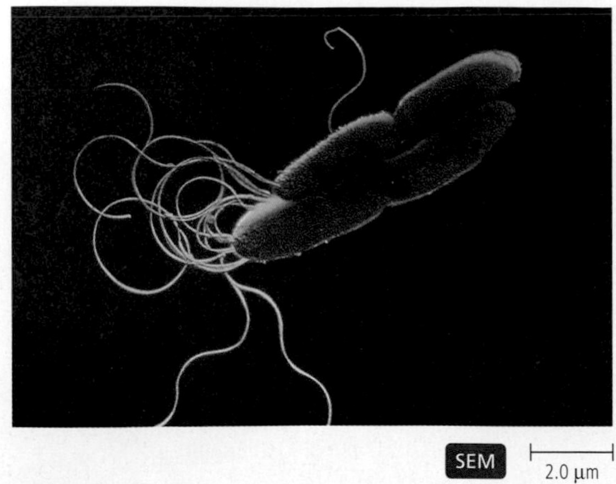

▲ **Figure 11.23** Two dividing *Pseudomonas* cells and their characteristic polar flagella.

archaeal methanogens, methane oxidizers inhabit anaerobic environments worldwide, growing just above the anaerobic layers that contain methanogens. Although methane is one of the so-called greenhouse gases that retains heat in the atmosphere, methane oxidizers digest most of the methane in their local environment before it can adversely affect the world's climate.

Glycolytic Facultative Anaerobes The largest group of gammaproteobacteria is composed of Gram-negative, facultatively anaerobic rods that catabolize carbohydrates by glycolysis and the pentose phosphate pathway. This group, which is divided into three families (Table 11.4), contains numerous human pathogens. Members of the family Enterobacteriaceae, including *Escherichia coli* (esh-ĕ-rik'ē-ă kō'lē), are frequently used for laboratory studies of metabolism, genetics, and recombinant DNA technology. Chapter 20 examines pathogenic gammaproteobacteria in detail.

Pseudomonads Bacteria called **pseudomonads** are Gram-negative, aerobic, flagellated, straight to slightly curved rods that catabolize carbohydrates by the Entner-Doudoroff and pentose phosphate pathways. These organisms are noted for their ability to break down numerous organic compounds. Many of them are important pathogens of humans and animals and are involved in the spoilage of refrigerated milk, eggs, and meat because they can grow and catabolize proteins and lipids at 4°C. The pseudomonad group is named for its most important genus: *Pseudomonas* (soo-dō-mō'nas; **Figure 11.23**), which causes diseases such as urinary tract infections and external otitis (swimmer's ear). Other pseudomonads such as *Azotobacter* (ā-zō-tō-bak'ter) and *Azomonas* (ā-zō-mō'nas) are soil-dwelling, nonpathogenic nitrogen fixers; however, in contrast to nitrogen-fixing alphaproteobacteria, these gammaproteobacteria do not associate with the roots of plants.

The characteristics of the members of the gammaproteobacteria discussed in this section are summarized in Table 11.3 on p. 329.

Deltaproteobacteria

Learning Objective

✓ List several members of deltaproteobacteria.

The **deltaproteobacteria** are not a large assemblage, but like other proteobacteria they include a wide variety of metabolic types. *Desulfovibrio* (dē'sul-fō-vib'rē-ō) is a sulfate-reducing microbe that is important in the sulfur cycle. It is also an important member of bacterial communities living in the sediments of polluted streams and sewage treatment lagoons, where its presence is often apparent by the odor of hydrogen sulfide that it releases during anaerobic respiration. Hydrogen sulfide reacts

TABLE 11.4 Representative Glycolytic Facultative Anaerobes of the Class Gammaproteobacteria

Family	Special Characteristics	Representative Genera	Typical Human Diseases
Enterobacteriaceae	Straight rods; oxidase negative; peritrichous flagella or nonmotile	*Escherichia*	Gastroenteritis
		Enterobacter	(Rarely pathogenic)
		Serratia	(Rarely pathogenic)
		Salmonella	Enteritis
		Proteus	Urinary tract infection
		Shigella	Shigellosis
		Yersinia	Plague
		Klebsiella	Pneumonia
Vibrionaceae	Vibrios; oxidase positive; polar flagella	*Vibrio*	Cholera
Pasteurellaceae	Cocci or straight rods; oxidase positive; nonmotile	*Haemophilus*	Meningitis in children

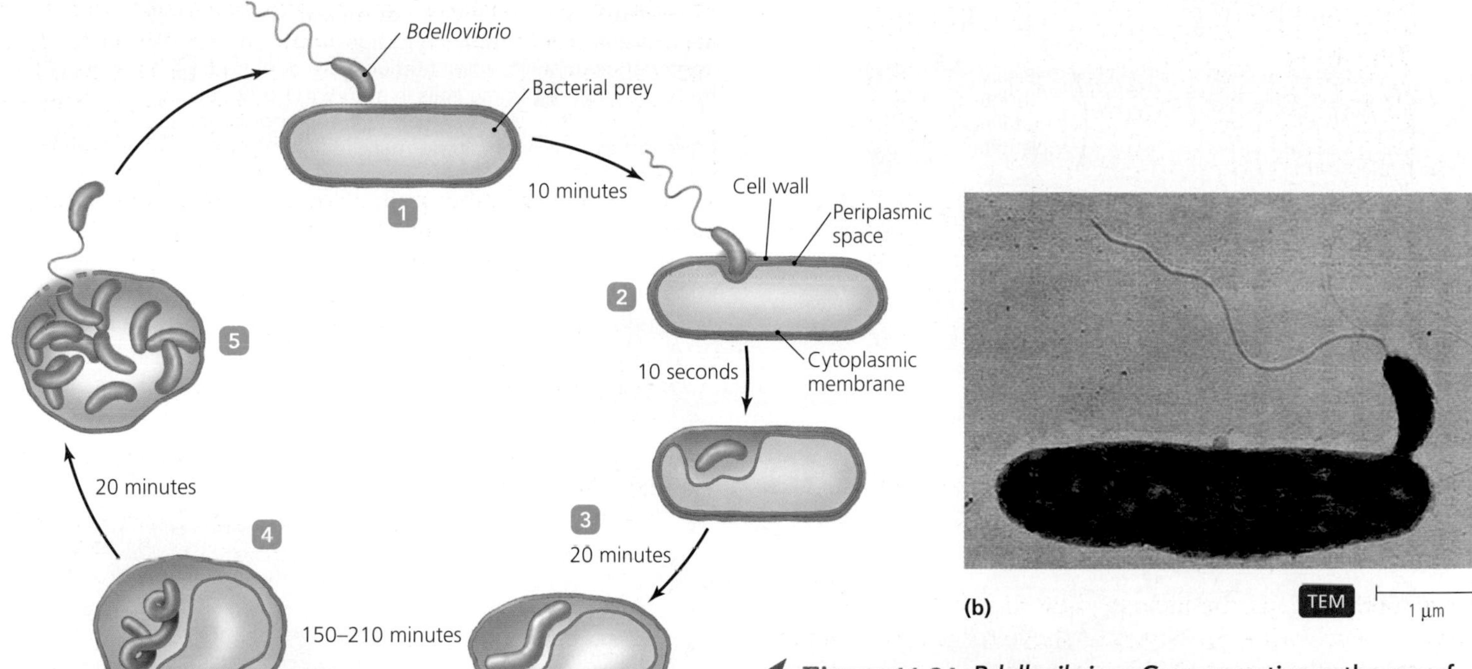

Figure 11.24 ***Bdellovibrio*, a Gram-negative pathogen of other Gram-negative bacteria.** **(a)** Life cycle, including the elapsed times between events. **(b)** *Bdellovibrio* invading the periplasmic space of its host.

with iron to form iron sulfide, so sulfate-reducing bacteria play a primary role in the corrosion of iron pipes in heating systems, sewer lines, and other structures.

Bdellovibrio (del-lō-vib′rē-ō) is not pathogenic to eukaryotes, but instead attacks other Gram-negative bacteria. It has a complex and unusual life cycle **(Figure 11.24a)**:

1 A free *Bdellovibrio* swims rapidly through the medium until it attaches via fimbriae to a Gram-negative bacterium.

2 It rapidly drills through the cell wall of its prey by secreting hydrolytic enzymes and rotating in excess of 100 revolutions per second **(Figure 11.24b)**.

3 Once inside, *Bdellovibrio* lives in the periplasmic space—the space between the cytoplasmic membrane and the cell wall. It kills its host by disrupting the host's cytoplasmic membrane and inhibiting DNA, RNA, and protein synthesis.

4 The invading bacterium uses the nutrients released from its dying prey and grows into a long filament.

5 Eventually, the filament divides into many smaller cells, each of which, when released from the dead cell, produces a flagellum, and swims off to repeat the process. Multlple fissions to produce many offspring is a rare form of reproduction.

Myxobacteria are Gram-negative, aerobic, soil-dwelling bacteria with a unique life cycle for prokaryotes in that individuals cooperate to produce differentiated reproductive structures. The life cycle of myxobacteria can be summarized as follows **(Figure 11.25a)**:

1 Vegetative myxobacteria glide on slime trails through their environment, digesting yeasts and other bacteria or scavenging nutrients released from dead cells. When nutrients and cells are plentiful, the myxobacteria divide by binary fission; when nutrients are depleted, however, they aggregate by gliding into a mound of cells.

2 Myxobacteria within the mound differentiate to form a macroscopic *fruiting body* ranging in height from 50 µm to 700 µm.

3 Some cells within the fruiting body develop into dormant *myxospores* that are enclosed within walled structures called *sporangia* (singular: *sporangium;* **Figure 11.25b)**.

4 The sporangia release the myxospores, which can resist desiccation and nutrient deprivation for a decade or more.

5 When nutrients are again plentiful, the myxospores germinate and become vegetative cells.

Myxobacteria live worldwide in soils that have decaying plant material or animal dung. Though certain species live in the arctic and others in the tropics, most myxobacteria live in temperate regions.

CRITICAL **THINKING**

What do the names *Desulfovibrio* and *Bdellovibrio* tell you about the morphology of these deltaproteobacteria?

Epsilonproteobacteria

Epsilonproteobacteria are Gram-negative rods, vibrios, or spirals. Important genera are *Campylobacter* (kam′pi-lō-bak′ter),

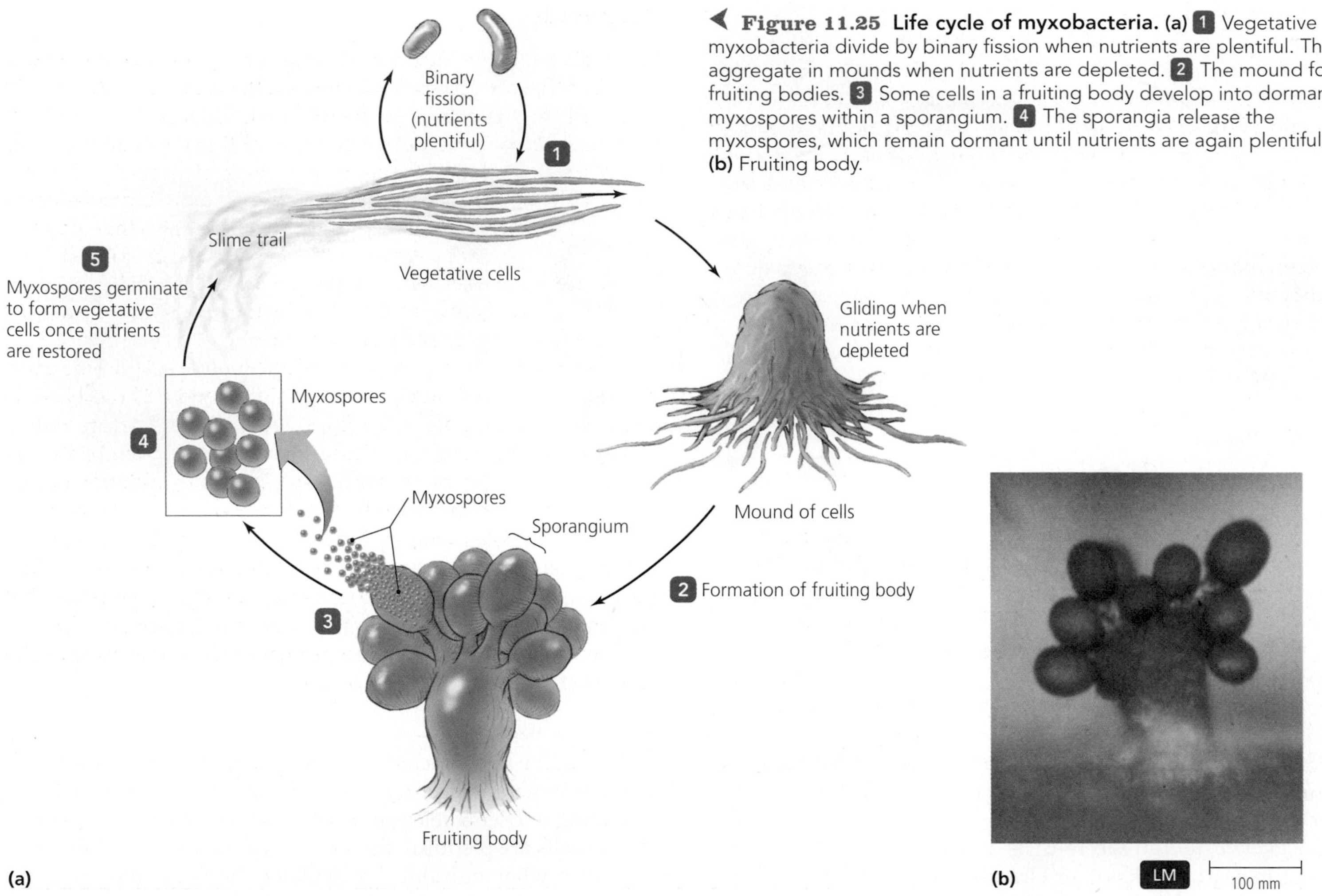

Figure 11.25 Life cycle of myxobacteria. (a) 1 Vegetative myxobacteria divide by binary fission when nutrients are plentiful. They aggregate in mounds when nutrients are depleted. 2 The mound forms fruiting bodies. 3 Some cells in a fruiting body develop into dormant myxospores within a sporangium. 4 The sporangia release the myxospores, which remain dormant until nutrients are again plentiful. **(b)** Fruiting body.

which causes blood poisoning and inflammation of the intestinal tract, and *Helicobacter* (hel'ĭ-kō-bak'ter), which causes ulcers. Chapter 21 examines these and other pathogens more fully.

The characteristics of the delta- and epsilonproteobacteria are summarized in Table 11.3 on p. 329.

CRITICAL **THINKING**

Design a dichotomous key for the proteobacteria discussed in this chapter.

Other Gram-Negative Bacteria

Learning Objectives

- ✓ Describe the unique features of chlamydias and spirochetes.
- ✓ Describe the ecological importance of bacteroids.

In the final section of this chapter we consider an assortment of Gram-negative bacteria that are classified in the second edition of *Bergey's Manual* into nine phyla that are grouped together for convenience rather than because of genetic relatedness. Species in six of the nine phyla are of relatively minor importance. Here we discuss representatives from the three phyla that are either of particular ecological concern or significantly affect human health: the chlamydias (phylum Chlamydiae), the spirochetes (phylum Spirochaetes), and the bacteroids (phylum Bacteroidetes) (see Figure 11.9).

Chlamydias

Microorganisms called **chlamydias** (kla-mid'ē-ăz) are small, Gram-negative cocci that grow and reproduce only within the cells of mammals, birds, and a few invertebrates. The smallest chlamydias—0.2 μm in diameter—are smaller than the largest viruses; however, in contrast to viruses, chlamydias have both DNA and RNA, cytoplasmic membranes, functioning ribosomes, reproduction by binary fission, and metabolic pathways. Like other Gram-negative prokaryotes, chlamydias have two membranes, but in contrast they lack peptidoglycan.

Because chlamydias and rickettsias share an obvious characteristic—both have a requirement for intracellular life—these two types of organisms were grouped together in a single taxon in the first edition of *Bergey's Manual.* Now, however, the rickettsias are classified with the alphaproteobacteria, and the chlamydias are in their own phylum. Chlamydias cause neonatal blindness, pneumonia, and a sexually transmitted disease called *lymphogranuloma venereum;* in fact, chlamydias are the most common sexually transmitted bacteria in the United States.

Chlamydias have a unique method of reproduction. After invading a host cell, a chlamydial cell forms an **initial body** (also called a *reticulate body*). The initial body grows and undergoes repeated binary fissions until the host cell is filled with reticulate bodies. These then change into **elementary bodies,** which contain electron-dense material and have rigid outer boundaries formed when they cross-link their two membranes with disulfide bonds. Each elementary body, which is relatively resistant to drying, is an infective stage. When the host cell dies, the elementary bodies are released to drift until they contact and attach to other host cells, triggering their own endocytosis by the new host cells. Once inside a host cell, elementary bodies transform back into initial bodies, and the cycle repeats. The chlamydial life cycle is illustrated in Figure 21.6.

Spirochetes

Spirochetes are unique helical bacteria that are motile by means of axial filaments—flagella-like structures that lie within the periplasmic space (see Figure 3.8c). When the axial filaments rotate, the entire cell corkscrews through the medium.

Spirochetes have a variety of types of metabolism and live in diverse habitats. They are frequently isolated from the human mouth, marine environments, moist soil, and the surfaces of protozoa that live in termites' guts. In the latter case, they may coat the protozoan so thickly that they look, and act, like cilia. The spirochetes *Treponema* (trep-ō-ne'mă) and *Borrelia* (bo-rē'lē-ā) cause syphilis and Lyme disease, respectively, in humans.

Bacteroids

Bacteroids are yet another diverse group of Gram-negative microbes that are grouped together on the basis of similarities in their rRNA nucleotide sequences. The group is named for *Bacteroides* (bak-ter-oy'dēz), a genus of obligately anaerobic rods that normally inhabit the digestive tracts of humans and animals. Bacteroids assist in digestion by catabolizing substances such as cellulose and other complex carbohydrates that are indigestible by mammals. About 30% of the bacteria isolated from human feces are *Bacteroides.* Some species of *Bacteroides* cause abdominal, pelvic, blood, and other infections; in fact, they are the most common anaerobic human pathogen.

Bacteroids in the genus *Cytophaga* (sī-tof'ă-gă) are aquatic, gliding, rod-shaped aerobes with pointed ends. These bacteria degrade complex polysaccharides such as agar, pectin, chitin, and even cellulose, so they cause damage to wooden boats and piers; they also play an important role in the degradation of raw sewage. Their ability to glide allows these bacteria to position themselves at sites with optimum nutrients, pH, temperature, and oxygen levels. Organisms in *Cytophaga* differ from other gliding bacteria, such as cyanobacteria and myxobacteria, in that they are nonphotosynthetic and do not form fruiting bodies.

The characteristics of these groups of Gram-negative bacteria are listed in Table 11.3 on p. 329.

CRITICAL **THINKING**

Design a dichotomous key for the genera of Gram-negative bacteria listed in Table 11.3 on p. 329.

Chapter Summary

General Characteristics of Prokaryotic Organisms (pp. 313–316)

1. Three basic shapes of prokaryotic cells are spherical **cocci,** rod-shaped **bacilli,** and **spirals.** Spirals may be stiff **(spirilla)** or flexible **(spirochetes).**
2. Other variations in shapes include **vibrios** (slightly curved rods), **coccobacilli** (intermediate to cocci and bacilli), and **pleomorphic** (variable shape and size).
3. Cocci may typically be found in groups, including long chains **(streptococci),** pairs **(diplococci),** foursomes **(tetrads),** cuboidal packets **(sarcinae),** and clusters **(staphylococci).**
4. Bacilli are found singly, in pairs, in chains, or in a **palisade** arrangement.
5. Prokaryotes reproduce asexually by **binary fission, snapping division** (a type of binary fission), **spore** formation, and **budding.** **ANIMATIONS:** ***Bacterial Growth: Overview***
6. Environmentally resistant **endospores** are produced within vegetative cells of the Gram-positive genera *Bacillus* and *Clostridium.* Depending on the species in which they are formed, the endospores may be terminal, subterminal, or centrally located.

Modern Prokaryotic Classification (pp. 316–317)

1. Living things are now classified into three domains—Archaea, Bacteria, and Eukarya—based largely on genetic relatedness.
2. The most authoritative reference in modern prokaryotic systematics is *Bergey's Manual of Systematic Bacteriology,* 2nd ed., which classifies prokaryotes into 3 phyla of Archaea and 24 phyla of Bacteria. The organization of this text's survey of prokaryotes largely follows Bergey's classification scheme.

Survey of Archaea (pp. 317–319)

1. The domain Archaea includes **extremophiles,** microbes that require extreme conditions of temperature, pH, and/or salinity to survive. One archaeal phylum is known only from rRNA sequences.
2. **Thermophiles** and **hyperthermophiles** (in the phyla Crenarchaeota and Euryarchaeota) live at temperatures above 45°C and 80°C, respectively, because their DNA, membranes, and proteins do not function properly at lower temperatures.
3. **Halophiles** (phylum Euryarchaeota) depend on high concentrations of salt to keep their cell walls intact. Halophiles such as *Halobacterium salinarium* synthesize purple proteins called **bacteriorhodopsins** that harvest light energy to synthesize ATP.

4. **Methanogens** (phylum Euryarchaeota) are obligate anaerobes that produce methane gas and are useful in sewage treatment.

Survey of Bacteria (pp. 320–335)

1. **Deeply branching bacteria** have rRNA sequences thought to be similar to those of earliest bacteria. They are autotrophic and live in hot, acidic, and anaerobic environments, often with intense exposure to sun.
2. Phototrophic bacteria trap light energy with photosynthetic lamellae. The five groups of phototrophic bacteria are cyanobacteria, green sulfur bacteria, green nonsulfur bacteria, purple sulfur bacteria, and purple nonsulfur bacteria.
3. Many **cyanobacteria** reduce atmospheric N_2 to NH_3 via a process called **nitrogen fixation.** Cyanobacteria must separate (in either time or space) the metabolic pathways of nitrogen fixation from those of oxygenic photosynthesis because nitrogen fixation is inhibited by the oxygen generated during photosynthesis. Many cyanobacteria fix nitrogen in thick-walled cells called **heterocysts.**
4. Green and purple bacteria use bacteriochlorophylls for anoxygenic photosynthesis. Nonsulfur forms derive electrons from organic compounds; sulfur forms derive electrons from H_2S.
   ANIMATIONS: *Photosynthesis: Comparing Prokaryotes and Eukaryotes*
5. The phylum Firmicutes contains bacteria with a G + C ratio (the proportion of all base pairs that are guanine-cytosine base pairs) of less than 50%. Firmicutes includes clostridia, mycoplasmas, and other low G + C cocci and bacilli.
6. Clostridia include the genus *Clostridium* (pathogenic bacteria that cause gangrene, tetanus, botulism, and diarrhea), *Epulopiscium* (which is large enough to be seen without a microscope), and *Veillonella* (often found in dental plaque).
7. **Mycoplasmas** are Gram-positive, pleomorphic, facultative anaerobes and obligate anaerobes that lack cell walls and therefore stain pink with Gram stain. They are frequently associated with pneumonia and urinary tract infections.
8. Low G + C Gram-positive bacilli and cocci important to human health and industry include *Bacillus* (which contains species that cause anthrax and food poisoning, and includes beneficial Bt-toxin bacteria), *Listeria* (which causes bacteremia and meningitis), *Lactobacillus* (used to produce yogurt and pickles), *Streptococcus* (which causes strep throat and other diseases), *Enterococcus* (which can cause endocarditis and other diseases), and *Staphylococcus* (which causes a number of human diseases).
9. High G + C bacteria (*Corynebacterium, Mycobacterium,* and actinomycetes) are classified in phylum Actinobacteria.
10. Bacteria in *Corynebacterium* store phosphates in **metachromatic granules;** *C. diphtheriae* causes diphtheria.
11. Members of the genus *Mycobacterium,* including species that cause tuberculosis and leprosy, grow slowly and have unique, resistant cell walls containing waxy **mycolic acids.**
12. **Actinomycetes** resemble fungi in that they produce spores and form filaments; this group includes *Actinomyces* (normally found in human mouths), *Nocardia* (useful in degradation of pollutants), and *Streptomyces* (produces important antibiotics).
13. Phylum **Proteobacteria** is a very large group of Gram-negative bacteria divided into five classes—the alpha-, beta-, gamma-, delta-, and epsilonproteobacteria.
14. The **alphaproteobacteria** include a variety of aerobes, many of which have unusual cellular extensions called prosthecae. *Azospirillum* and *Rhizobium* are nitrogen fixers that are important in agriculture.
15. Some members of the alphaproteobacteria are **nitrifying bacteria,** which oxidize NH_3 to NO_3 via a process called **nitrification.** Nitrifying alphaproteobacteria are in the genus *Nitrobacter.*
16. Most purple nonsulfur phototrophs are alphaproteobacteria.
17. Pathogenic alphaproteobacteria include *Rickettsia* (typhus and Rocky Mountain spotted fever) and *Brucella* (brucellosis).
18. There are many beneficial alphaproteobacteria, including *Acetobacter* and *Gluconobacter,* both of which are used to synthesize acetic acid. *Caulobacter* is of interest in reproductive studies, and *Agrobacterium* is used in genetic recombination in plants.
19. The **betaproteobacteria** include the nitrifying *Nitrosomonas* and pathogenic species such as *Neisseria* (gonorrhea), *Bordetella* (whooping cough), and *Burkholderia* (which colonizes the lungs of cystic fibrosis patients).
20. Other betaproteobacteria include *Thiobacillus* (ecologically important), *Zoogloea* (useful in sewage treatment), and *Sphaerotilus* (hampers sewage treatment).
21. The **gammaproteobacteria** constitute the largest class of proteobacteria; they include **purple sulfur bacteria,** intracellular pathogens, facultative anaerobes that utilize glycolysis and the pentose phosphate pathway, and pseudomonads.
22. Both *Legionella* and *Coxiella* are intracellular, pathogenic gammaproteobacteria.
23. Numerous human pathogens are facultatively anaerobic gammaproteobacteria that catabolize carbohydrates by glycolysis.
24. **Pseudomonads,** including pathogenic *Pseudomonas* and nitrogen-fixing *Azotobacter* and *Azomonas,* utilize the Entner-Doudoroff and pentose phosphate pathways for catabolism of glucose.
25. The **deltaproteobacteria** include *Desulfovibrio* (important in the sulfur cycle and in corrosion of pipes), *Bdellovibrio* (pathogenic to bacteria), and **myxobacteria.** The latter form stalked fruiting bodies containing resistant, dormant myxospores.
26. The **epsilonproteobacteria** include some important human pathogens, including *Campylobacter* and *Helicobacter.*
27. **Chlamydias** are Gram-negative cocci typified by the genus *Chlamydia;* they cause neonatal blindness, pneumonia, and a sexually transmitted disease. Within a host cell, chlamydias form **initial bodies,** which change into smaller **elementary bodies** released when the host cell dies.
28. **Spirochetes** are flexible, helical bacteria that live in diverse environments. *Treponema* (syphilis) and *Borrelia* (Lyme disease) are important spirochetes.
29. **Bacteroids** include *Bacteroides,* an obligate anaerobic rod that inhabits the digestive tract, and *Cytophaga,* an aerobic rod that degrades wood and raw sewage.

Questions for Review

Answers to the Question for Review (except Short Answers questions) begin on page A-1.

Modified True/False

For each of the following statements that is true, write "true" in the blank. For each statement that is false, write the word(s) that should be substituted for the italicized word(s) to make the statement correct.

1. ____ All prokaryotes reproduce *sexually*.
2. ____ A *bacillus* is a bacterium with a slightly curved rod shape.
3. ____ If you were to view *staphylococci*, you should expect to see clusters of cells.
4. ____ *Initial* bodies are stable resting stages that do not metabolize but will germinate when conditions improve.
5. ____ Archaea are classified into phyla based primarily on *tRNA sequences*.
6. ____ *Halophiles* inhabit extremely saline habitats such as the Great Salt Lake.
7. ____ Pigments located in thylakoids in phototrophic bacteria trap *light* energy for metabolic processes.
8. ____ Most *cyanobacteria* form heterocysts in which nitrogen fixation occurs.
9. ____ A giant bacterium that is large enough to be seen without a microscope is *Veillonella*.
10. ____ When environmental nutrients are depleted, *myxobacteria* aggregate in mounds to form fruiting bodies.

Matching

Match the bacterium on the left with the term with which it is most closely associated.

1. ____ *Bacillus anthracis*	A. wood damage
2. ____ *Veillonella*	B. dental biofilm (plaque)
3. ____ *Clostridium perfringens*	C. gangrene
4. ____ *Clostridium botulinum*	D. botox
5. ____ *Bacillus licheniformis*	E. anthrax
6. ____ *Streptococcus*	F. lymphogranuloma venereum
7. ____ *Streptomyces*	G. leprosy
8. ____ *Corynebacterium*	H. tetracycline
9. ____ *Gluconobacter*	I. vinegar
10. ____ *Bordetella*	J. yogurt
11. ____ *Zoogloea*	K. impetigo
12. ____ *Azotobacter*	L. bacitracin
13. ____ *Desulfovibrio*	M. iron pipe corrosion
14. ____ *Chlamydia*	N. pertussis
15. ____ *Cytophaga*	O. nitrogen fixation
	P. floc formation
	Q. diphtheria

Multiple Choice

1. The type of reproduction in prokaryotes that results in a palisade arrangement of cells is called
   a. pleomorphic division.
   b. endospore formation.
   c. snapping division.
   d. binary fission.
2. The thick-walled reproductive spores produced in the middle of cyanobacterial filaments are called
   a. akinetes.
   b. terminal endospores.
   c. metachromatic granules.
   d. heterocysts.
3. Which of the following terms best describes stiff, spiral-shaped prokaryotic cells?
   a. cocci
   b. bacilli
   c. spirilla
   d. spirochetes
4. Endospores
   a. can remain alive for decades.
   b. can remain alive in boiling water.
   c. live in a state of suspended animation.
   d. all of the above
5. *Halobacterium salinarium* is distinctive because
   a. it is absolutely dependent on high salt concentrations to maintain its cell wall.
   b. it is found in terrestrial volcanic habitats.
   c. it photosynthesizes without chlorophyll.
   d. it can survive 5 million rad of radiation.
6. Photosynthetic bacteria that also fix nitrogen are
   a. mycoplasmas.
   b. spirilla.
   c. bacteroids.
   d. cyanobacteria.
7. Which genus is the most common anaerobic human pathogen?
   a. *Bacteroides*
   b. *Spirochetes*
   c. *Chlamydia*
   d. *Methanopyrus*
8. Flexible spiral-shaped prokaryotes are
   a. spirilla.
   b. spirochetes.
   c. vibrios.
   d. rickettsias.
9. Bacteria that convert nitrogen gas into ammonia are
   a. nitrifying bacteria.
   b. nitrogenous.
   c. nitrogen fixers.
   d. nitrification bacteria.
10. The presence of mycolic acid in the cell wall characterizes
    a. *Corynebacterium*.
    b. *Listeria*.
    c. *Nocardia*.
    d. *Mycobacterium*.

Labeling

Label the shapes of these prokaryotic cells.

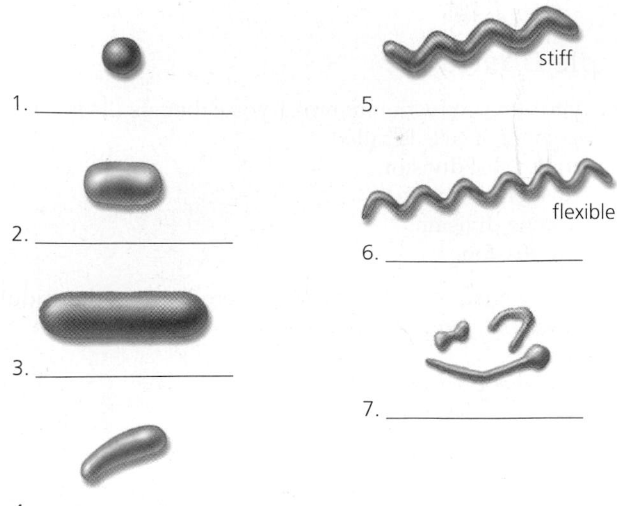

Short Answer

1. Whereas the first edition of *Bergey's Manual* relied on morphological and biochemical characteristics to classify microbes, the new edition focuses on ribosomal RNA sequences. List several other criteria for grouping and classifying bacteria.
2. What are extremophiles? Describe two kinds and give examples.
3. Name and describe three types of bacteria mentioned in this chapter that "glide."
4. Name three groups of low G + C Gram-positive bacteria.
5. Sketch and label six shapes of bacteria.
6. A student was memorizing the arrangements of bacteria and noticed that there are more arrangements for cocci than for bacilli. Why might this be so?
7. How is *Agrobacterium* used in recombinant DNA technology?
8. Name and describe five distinct classes of phylum Proteobacteria.
9. Explain why organisms formerly known as blue-green algae are now called cyanobacteria.
10. Contrast the processes of nitrification and nitrogen fixation.

Concept Mapping

Using the following terms, draw a concept map that describes the domain archaea. For a sample concept map, see p. 93. Or, complete this concept map online by going to the Study Area at www.masteringmicrobiology.com.

>45°C	Acidophiles	Extremophiles	Hydrothermal vents	Methane	Prokaryotes
>80°C	Animal colons	Great Salt Lake	Hyperthermophiles	Methanogens	Sewage treatment
17–25% salt	Disease	Halophiles	Low pH	Peptidoglycan	Thermophiles

Critical Thinking

1. A microbiology student described "deeply branching bacteria" as having a branched filamentous growth habit akin to *Streptomyces*. Do you agree with this description? Why or why not?
2. Iron oxide (rust) forms when iron is exposed to oxygen, particularly in the presence of water. Nevertheless, iron pipes typically corrode more quickly when they are buried in moist *anaerobic* soil than when they are buried in soil containing oxygen. Explain why this is the case.
3. Why is it that Gram-positive species don't have axial filaments?
4. Even though *Clostridium* is strictly an anaerobic bacterium, it can be isolated easily from the exposed surface of your skin. Explain how this can be.
5. Louis Pasteur said, "The role of the infinitely small in nature is infinitely large." Explain what he meant by using examples of the roles of microorganisms in health, industry, and the environment.
6. How are bacterial endospores different from the spores of actinomycetes?

Access more review material online in the Study Area at **www.masteringmicrobiology.com.** There, you'll find

- **Animations**
- **MP3 Tutor Sessions**
- **Concept Mapping Activities**
- **Flashcards**
- **Quizzes**

and more to help you succeed.

12 Characterizing and Classifying Eukaryotes

The mushrooms in this photo constitute only the tips of what may be the largest living organism on Earth—a fungus known as *Armillaria ostoyae* that lives 3 feet underground in Oregon's Malheur National Forest. Also known as a honey mushroom, it is 3.5 miles long, covers an area equivalent to 1665 football fields, and is estimated to be at least 2400 years old!

Amazingly, this "humongous fungus" began life as a microscopic spore. During its life span of over two millennia, it has used rootlike structures called rhizomorphic hyphae to draw water and nutrients from tree roots. Despite its gigantic proportions, its visible portions are ordinary mushrooms.

Fungi are eukaryotes—a vast category that also includes protozoa, algae, parasitic helminths, as well as all plants and animals. In this chapter we will take a closer look at the eukaryotic organisms of microbiological importance.

Take the pre-test for this chapter online. Visit the Study Area at www.masteringmicrobiology.com.

The world's largest organism is a fungus that lives underground. Its above-ground mushrooms are just its sporulating appendages.

Eukaryotic microbes include a fascinating and almost bewilderingly diverse assemblage. Eukaryotic microbes include unicellular and multicellular protozoa,[1] fungi,[2] algae,[3] water molds, and slime molds. Additionally, microbiologists study parasitic helminths[4] because they have microscopic stages, and they study arthropod vectors because they are intimately involved in the transmission of microbial pathogens. Eukaryotes include both human pathogens and organisms that are vital for human life. For example, one group of marine algae called diatoms and a set of protozoa called dinoflagellates (dī'nō-flaj'ĕ-lātz) provide the basis for the oceans' food chains and produce most of the world's oxygen. Eukaryotic fungi produce penicillin, and tiny baker's and brewer's yeasts are essential for making bread and alcoholic beverages.

Among the 20 most frequent microbial causes of death worldwide, six are eukaryotic, including the agents of malaria, African sleeping sickness, and amebic dysentery. *Pneumocystis* pneumonia, toxoplasmosis, and cryptosporidiosis—common afflictions of AIDS patients—are all caused by eukaryotic pathogens.

In previous chapters we discussed characteristics of *cells*—their metabolism, growth, and genetics. In this chapter we discuss eukaryotic *organisms* of interest to microbiologists—protozoa (single-celled "animals"), fungi, algae, water molds, and slime molds—and conclude with a brief discussion of the relationship of parasitic helminths and vectors to microbiology.

We begin by discussing general features of eukaryotic reproduction and classification; the following sections survey some representative members of microbiologically important eukaryotic groups, focusing on beneficial, environmentally significant, and unusual species. Chapters 22–23 discuss fungal and parasitic agents and vectors of human disease in more detail.

General Characteristics of Eukaryotic Organisms

Our discussion of the general characteristics of eukaryotes begins with a survey of the events in eukaryotic reproduction; then we consider some aspects of the complex matter of classifying the great variety of eukaryotic organisms.

Reproduction of Eukaryotes

Learning Objectives

- ✓ State four reasons why eukaryotic reproduction is more complex than prokaryotic reproduction.
- ✓ Describe the phases of mitosis, mentioning chromosomes, chromatids, centromeres, and spindle.
- ✓ Contrast meiosis with mitosis, mentioning homologous chromosomes, tetrads, and crossing over.
- ✓ Distinguish among nuclear division, cytokinesis, and schizogony.

As discussed in Chapter 3, a unique characteristic of living things is the ability to reproduce themselves. Prokaryotic reproduction typically involves replication of DNA and binary fission of the cytoplasm to produce two identical offspring. Reproduction of eukaryotes is more complicated and varied than reproduction in prokaryotes for a number of reasons:

- Most of the DNA in eukaryotes is packaged with histone proteins as *chromosomes* in the form of *chromatin* (krō'ma-tin) *fibers* located within nuclei. The remaining DNA in eukaryotic cells is found in mitochondria and chloroplasts, organelles that reproduce by binary fission in a manner similar to prokaryotic reproduction. In this chapter we will discuss only the nuclear portion of eukaryotic genomes.
- Eukaryotes have a variety of methods of asexual reproduction, including binary fission, budding, fragmentation, spore formation, and *schizogony* (ski-zog'ō-nē) (discussed later).
- Many eukaryotes reproduce sexually—that is, via a process that involves the formation of sexual cells called *gametes,* and the subsequent fusion of two gametes to form a cell called a *zygote.*
- Additionally, algae, fungi, and some protozoa reproduce both sexually and asexually. (Animals generally reproduce only one way or the other.)

Eukaryotic reproduction involves two types of division: nuclear division and cytoplasmic division (also called cytokinesis). After we discuss the various aspects of these two types of division, we will consider schizogony.

Nuclear Division

Typically, a eukaryotic nucleus has either one or two complete copies of the chromosomal portion of a cell's genome. A nucleus with a single copy of each chromosome is called a **haploid**[5] or $1n$ nucleus, and one with two sets of chromosomes is a **diploid**[6] or $2n$ nucleus. Generally, each organism has a consistent number of chromosomes. For example, each haploid cell of the brewer's yeast, *Saccharomyces cerevisiae*[7] (sak'ă-rō-mī'sēz se-ri-vis'ē-ī), has 16 chromosomes.

Whereas the cells of most fungi, many algae, and some protozoa are haploid, the cells of most plants and animals and the remaining fungi, algae, and protozoa are diploid. Typically, gametes are haploid, and a zygote (formed from the union of gametes) is diploid.

A cell divides its nucleus so as to pass a copy of its chromosomal DNA to each of its descendants, so that each new generation has the necessary genetic instructions to carry on life. There are two types of nuclear division—*mitosis* (mī-tō'sis) and *meiosis* (mī-ō'sis).

Mitosis Cells have two main stages in their life cycle: a stage called *interphase,*[8] during which cells grow and eventually

[1] From Greek *protos,* meaning first, and *zoion,* meaning animal.
[2] Plural of Latin *fungus,* meaning mushroom.
[3] Plural of Latin *alga,* meaning seaweed.
[4] From Greek *helmins,* meaning worm.
[5] From Greek *haploos,* meaning single.
[6] From Greek *diploos,* meaning double.
[7] From Greek *sakcharon,* meaning sugar, and *mykes,* meaning fungus, and Latin *cerevisiae,* meaning beer.
[8] Latin, meaning between phases.

replicate their DNA, and a stage during which the cell's nucleus divides. In the type of nuclear division called **mitosis,**[9] which begins after the cell has duplicated its DNA such that there are two exact DNA copies (see Figure 7.5), the cell partitions its replicated DNA equally between two nuclei. Thus mitosis maintains the ploidy of the parent nucleus; that is, a haploid nucleus that undergoes mitosis forms two haploid nuclei, and a diploid nucleus that undergoes mitosis produces two diploid nuclei.

Mitosis has four phases: prophase,[10] metaphase,[11] anaphase,[12] and telophase.[13] The events of mitosis proceed as follows **(Figure 12.1a)**:

1. **Prophase.** The cell condenses its DNA molecules into visible threads called *chromatids* (krō′mă-tidz). Two identical chromatids, sister DNA molecules, are joined together in a region called a *centromere* to form one chromosome. Also during prophase, a set of microtubules is constructed in the cytosol to form a *spindle.* In most cells, the nuclear envelope disintegrates during prophase so that mitosis occurs freely in the cytosol; however, many fungi and some unicellular microbes (for example, diatoms and dinoflagellates) maintain their nuclear envelopes so that mitosis occurs inside their nuclei.
2. **Metaphase.** The chromosomes line up on a plane in the middle of the cell and attach near their centromeres to microtubules of the spindle.
3. **Anaphase.** Sister chromatids separate and crawl along the microtubules toward opposite poles of the spindle. Each chromatid is now called a chromosome.
4. **Telophase.** The cell restores its chromosomes to their less compact, nonmitotic state, and nuclear envelopes form around the daughter nuclei. Though a cell may divide during telophase, as shown in Figure 12.1a, mitosis is nuclear division, not cell division.

Though certain specific events distinguish each of the four phases of mitosis, the phases are not discrete steps; that is, mitosis is a continuous process, and there are no clear boundaries between succeeding phases—one phase leads seamlessly to the next. For example, late anaphase and early telophase are indistinguishable.

Students sometimes confuse the terms *chromosome* and *chromatid,* in part because early microscopists used the word *chromosome* for two different things. During prophase and metaphase, a chromosome consists of two chromatids (DNA molecules) joined at a centromere. However, during anaphase and telophase, the chromatids separate, and each chromatid is then called a chromosome. In other words, a "chromosome" is a pair of chromatids during the first two phases, while "chromatid" and "chromosome" are synonymous terms during the latter two phases of mitosis.

Meiosis In contrast to mitosis, **meiosis**[14] is nuclear division that involves the partitioning of chromatids into four nuclei such that each nucleus receives only half the original amount of DNA. Thus, diploid nuclei use meiosis to produce haploid daughter nuclei. Meiosis is a necessary condition for sexual reproduction (in which nuclei from two different cells fuse to form a single nucleus), because if cells lacked meiosis, each nuclear fusion to form a zygote would cause the number of chromosomes to double, and their number would soon become unmanageable.

Meiosis occurs in two stages known as *meiosis I* and *meiosis II* **(Figure 12.1b)**. As in mitosis, each stage has four phases, named prophase, metaphase, anaphase, and telophase. The events in meiosis as they occur in a diploid nucleus proceed as follows:

1. Early prophase I (prophase of meiosis I). As with mitosis, DNA replication during interphase has resulted in pairs of identical chromatids, forming chromosomes. But now an additional pairing occurs: *homologous chromosomes*—that is, chromosomes carrying similar or identical genetic sequences—line up side by side. Because these are prophase chromosomes, each of them consists of two identical chromatids; therefore, four DNA molecules are involved in this pairing. An aligned pair of homologous chromosomes is known as a *tetrad.*
2. Late prophase I. Once tetrads have formed, the homologous chromosomes exchange sections of DNA in a random fashion via a process called *crossing over.* This results in recombinations of their DNA. It is because of meiotic crossing over that the offspring produced by sexual reproduction have different genetic makeups from their siblings. Prophase I can last for days or longer.
3. Metaphase I. Tetrads align on a plane in the center of the cell and attach to spindle microtubules. Metaphase I differs from metaphase of mitosis in that homologous chromosomes remain as tetrads.
4. Anaphase I. Chromosomes of the tetrads move apart from one another; however, in contrast to mitotic anaphase, sister chromatids remain attached to one another.
5. Telophase I. The first stage of meiosis is completed as the spindle disintegrates. Typically, the cell divides at this phase to form two cells. Nuclear envelopes may form. Each daughter nucleus is haploid, though each haploid chromosome consists of two chromatids.
6. Prophase II. Nuclear envelopes disintegrate, and new spindles form.
7. Metaphase II. The chromosomes align in the middle of each cell and attach to microtubules of the spindles.
8. Anaphase II. Sister chromatids separate as in mitosis.
9. Telophase II. Daughter nuclei form. The cells divide, yielding four haploid cells.

[9]From Greek *mitos*, meaning thread, after the threadlike appearance of chromosomes during nuclear division.
[10]From Greek *pro*, meaning before, and *phasis*, meaning appearance.
[11]From Greek *meta*, meaning in the middle.
[12]From Greek *ana*, meaning back.
[13]From Greek *telos*, meaning end.
[14]From Greek *meioun*, meaning to make smaller.

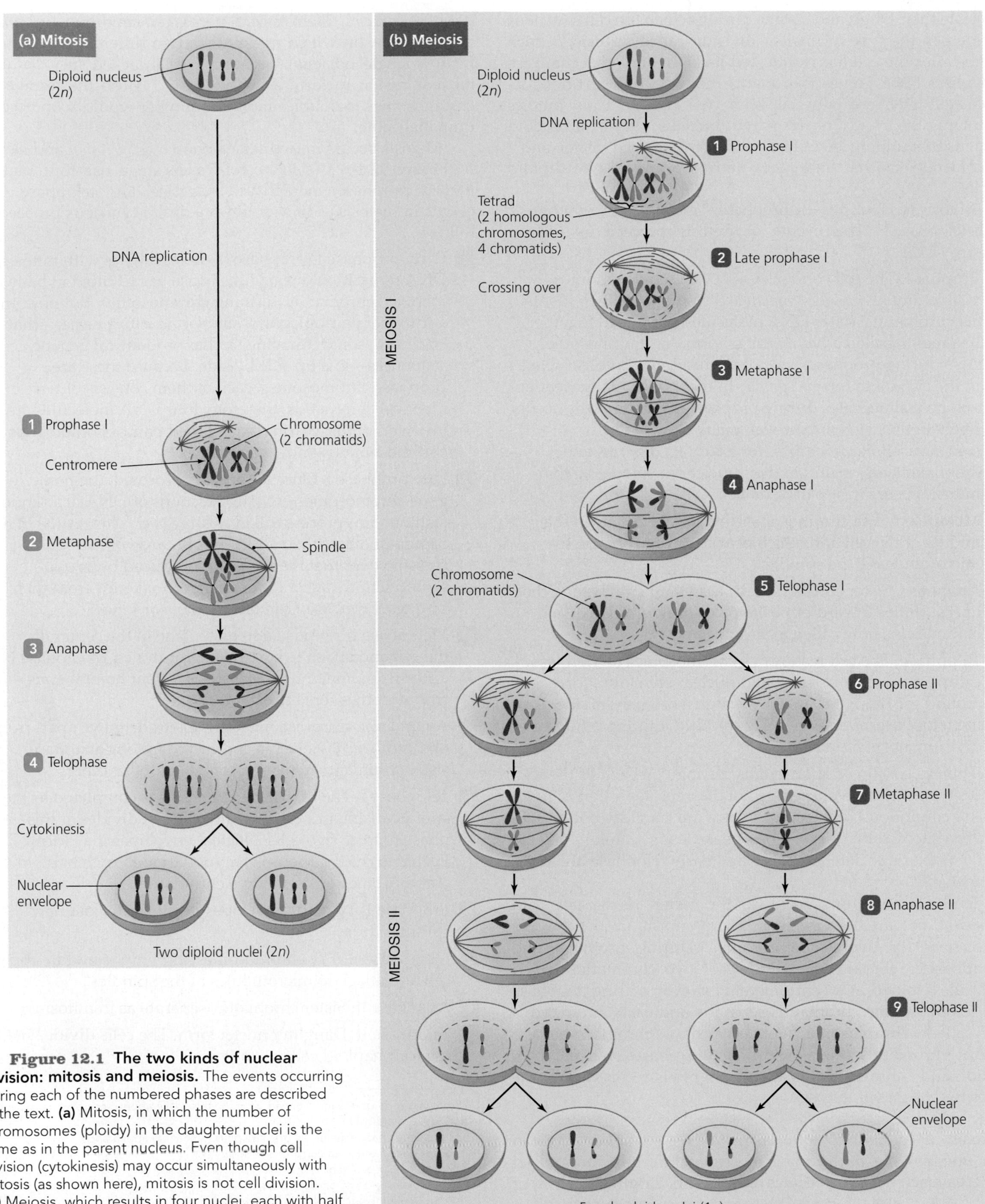

▲ **Figure 12.1 The two kinds of nuclear division: mitosis and meiosis.** The events occurring during each of the numbered phases are described in the text. **(a)** Mitosis, in which the number of chromosomes (ploidy) in the daughter nuclei is the same as in the parent nucleus. Even though cell division (cytokinesis) may occur simultaneously with mitosis (as shown here), mitosis is not cell division. **(b)** Meiosis, which results in four nuclei, each with half the number of chromosomes of the parent nucleus.

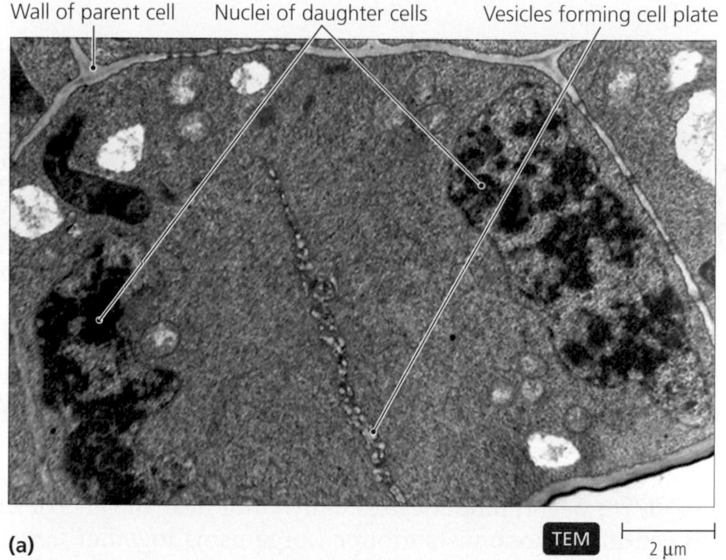

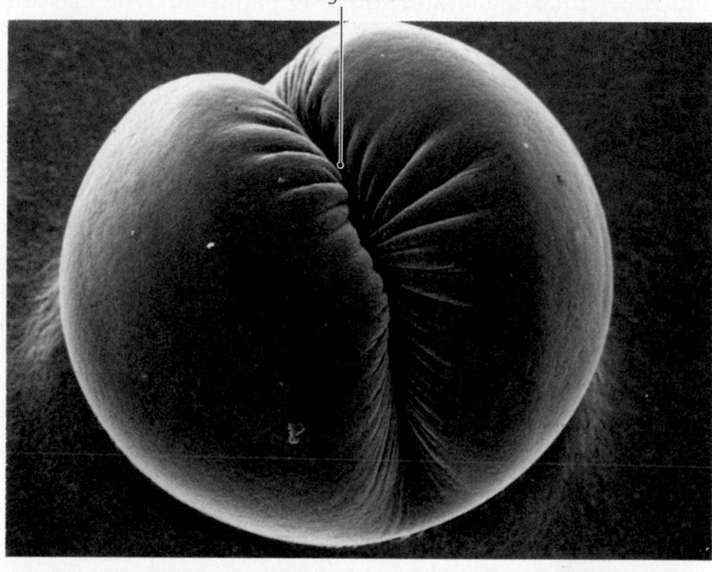

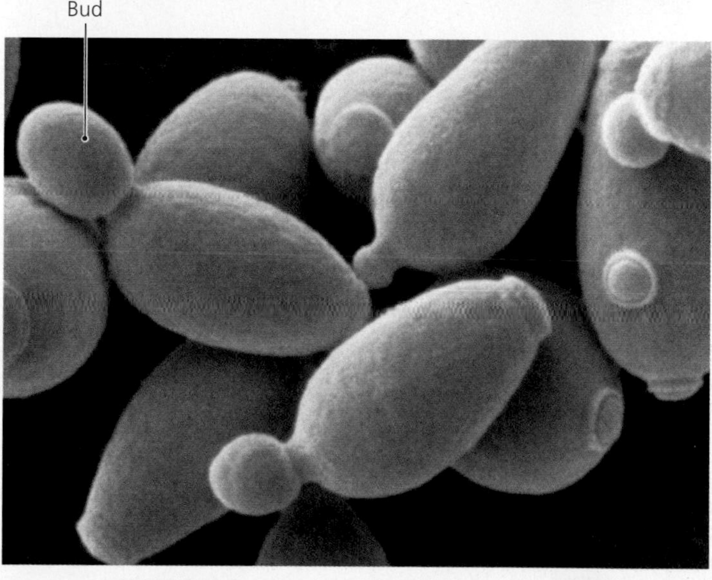

▶ **Figure 12.2 Different types of cytoplasmic division.** (a) Cytokinesis in a plant cell, in which vesicles form a cell plate. (b) Cytokinesis as it occurs in animals, protozoa, and some fungi. (c) Budding in yeast cells.

In summary, meiosis produces four haploid nuclei from a single diploid nucleus. Meiosis can be considered back-to-back mitoses without the DNA replication of interphase between them, though the four phases of meiosis I differ from those of mitosis. The phases of meiosis II are equivalent to those in mitosis. Additionally, crossing over during meiosis I produces genetic recombinations, which ensures that the chromosomes resulting from meiosis are different from the parental chromosomes. This provides genetic variety in the next generation. Table 12.1 on p. 345 compares and contrasts mitosis and meiosis.

Cytokinesis (Cytoplasmic Division)

Cytoplasmic division—also called **cytokinesis** (sī'tō-ki-nē'sis)—typically occurs simultaneously with telophase of mitosis, though in some algae and fungi it may be postponed or may not occur at all. In these cases, mitosis produces multinucleate cells called **coenocytes** (sē'nō-sītz).

In plant and algal cells, cytokinesis occurs as vesicles deposit wall material at the equatorial plane between nuclei to form a *cell plate,* which eventually becomes a transverse wall between daughter cells **(Figure 12.2a)**. Cytokinesis of protozoa and some fungal cells occurs when an equatorial ring of actin microfilaments contracts just below the cytoplasmic membrane, pinching the cell in two **(Figure 12.2b)**. Single-celled fungi called *yeasts* form a bud, which receives one of the daughter nuclei and pinches off from the parent cell **(Figure 12.2c)**.

Schizogony

Some protozoa, such as *Plasmodium* (plaz-mō'dē-ŭm)—the cause of malaria—reproduce asexually within red blood cells and liver cells via a special type of reproduction called **schizogony** (ski-zog'ō-nē; **Figure 12.3**). In schizogony, multiple mitoses form a multinucleate **schizont** (skiz'ont); only then does cytokinesis occur, simultaneously releasing numerous uninucleate daughter cells called *merozoites* (mer-ō-zō'ītz). The body of an infected host responds to the release of huge numbers of merozoites with the cyclic fever and chills characteristic of malaria. Chapter 23 discusses *Plasmodium* and malaria in more detail.

The Classification of Eukaryotic Organisms

Learning Objectives

✓ Briefly describe the major groups of eukaryotes as they were first classified in the late 18th century and as they were classified in the late 20th century.

✓ List some of the problems involved in the classification of protists in particular.

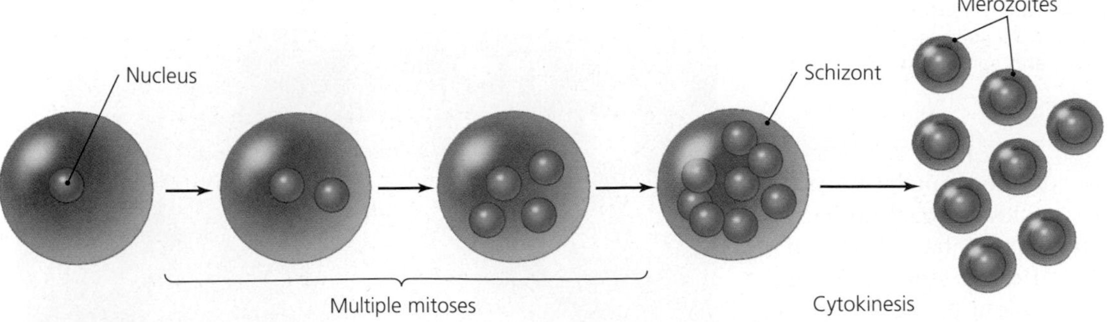

Figure 12.3 Schizogony. Sequential mitoses without intervening cytokineses produce a multinucleate schizont, which later undergoes cytokinesis to produce many daughter cells.

Historically, the classification of many of the eukaryotic microbes has been fraught with difficulty and characterized by change. Since the late 18th century, when Carolus Linnaeus (1707–1778) began modern taxonomy, until near the end of the 20th century, taxonomists grouped organisms together largely according to readily observable structural traits. Linnaeus classified unicellular algae and fungi as plants, and he classified protozoa—as the name suggests—as animals **(Figure 12.4a)**. By the late 20th century, some taxonomists, including Robert Whittaker (1920–1980), had placed fungi in their own kingdom and grouped protozoa and algae together within the kingdom Protista **(Figure 12.4b)**; other taxonomists, however, kept the green algae in the kingdom Plantae.

This scheme is troublesome, in part because the Protista included both large, photosynthetic, multicellular algae (such as kelps) and nonphotosynthetic unicellular protozoa. Adding to the confusion is the fact that taxonomists who classify plants and fungi use the term *divisions* to refer to the same taxonomic level that zoologists call *phyla*.

More recently, many taxonomists have abandoned classification schemes that are so strongly grounded in large-scale structural similarities in favor of schemes based on similarities in nucleotide sequences and cellular ultrastructure as revealed by electron microscopy. One of the most evident results of such taxonomic studies is that modern schemes no longer include a taxon "Protista"; instead, such eukaryotic microbes belong in several kingdoms.

Though no one classification scheme has garnered universal support, and more thorough understanding based on new information will almost certainly dictate changes, many taxonomists favor a scheme similar to the one shown in **Figure 12.4c**. In this scheme, on which the discussions of eukaryotic microbes in this chapter are largely based, the organisms we commonly refer to as protozoa are classified in seven kingdoms: Parabasala, Alveolata, Cercozoa, Radiolaria, Amoebozoa; Euglenozoa, and Diplomonadida; fungi are in the kingdom Fungi; algae are distributed among the kingdoms Stramenopila, Rhodophyta, and Plantae; water molds are in kingdom Stramenopila, and slime molds are in kingdom Amoebozoa. As we study the eukaryotic microbes discussed in this chapter, bear in mind that because the relationships among eukaryotic microbes are not fully understood, not all taxonomists would completely agree with this

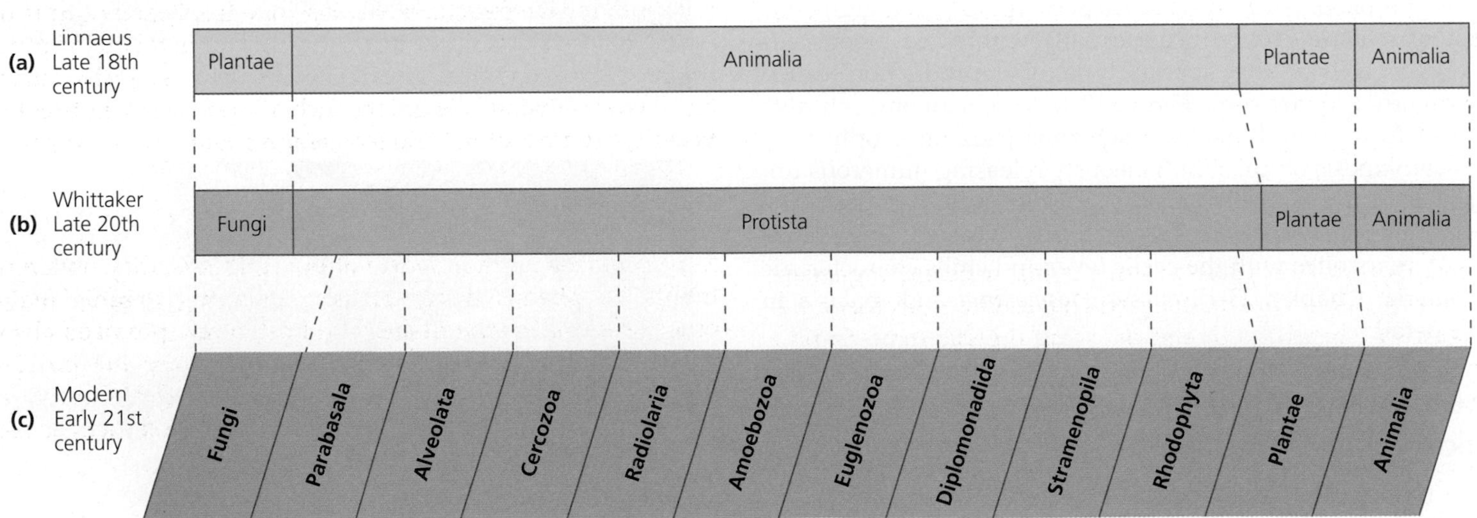

Figure 12.4 The changing classification of eukaryotes over the centuries. **(a)** In the late 18th century, Linnaeus classified all organisms as either plants or animals. **(b)** In the late 20th century, Whittaker placed fungi in their own group and recognized a new kingdom Protista. **(c)** Today, microbial eukaryotes are classified into numerous kingdoms based largely on their genetic relatedness. Though not all taxonomists would agree about every detail of this scheme, it forms the basis for the discussion of eukaryotic organisms in this chapter.

TABLE 12.1 Characteristics of the Two Types of Nuclear Division

	Mitosis	Meiosis
DNA replication	During interphase, before nuclear division	During interphase, before meiosis I begins
Phases	Prophase, metaphase, anaphase, telophase	Meiosis I—prophase I, metaphase I, anaphase I, telophase I Meiosis II—prophase II, metaphase II, anaphase II, telophase II
Formation of tetrads (alignment of homologous chromosomes)	Does not occur	Early in prophase I
Crossing over	Does not occur	Following formation of tetrads during prophase I
Number of accompanying cytoplasmic divisions that may occur	One	Two
Resulting nuclei	Two nuclei with same ploidy as the original	Four nuclei with half the ploidy of the original

scheme, and new information will shape future alterations in the taxonomy of eukaryotic microbes.

We begin our survey of eukaryotic microbes with the group of organisms commonly known as protozoa.

Protozoa

Learning Objective

✓ List three characteristics shared by all protozoa.

The microorganisms called **protozoa** (prō-tō-zō'ă) are a disparate group that are defined by three characteristics: They are eukaryotic, are unicellular, and lack a cell wall. Note that "protozoa" is not a currently accepted taxon. With the exception of one subgroup (called apicomplexans), protozoa are motile by means of cilia, flagella, and/or pseudopodia. By these criteria, protozoa include a diverse assemblage of microbes. The scientific study of protozoa is *protozoology,* and scientists who study these microbes are *protozoologists.*

In the following sections we discuss the distribution, morphology, nutrition, reproduction, and classification of various groups of protozoa.

Distribution of Protozoa

Protozoa require moist environments; most species live worldwide in ponds, streams, lakes, and oceans, where they are critical members of the *plankton*—free-living, drifting organisms that form the basis of aquatic food chains. Other protozoa live in moist soil, beach sand, and decaying organic matter, and a very few are pathogens—that is, disease-causing microbes—of animals and humans.

Morphology of Protozoa

Though protozoa have most of the features of eukaryotic cells discussed in Chapter 3 and illustrated in Figure 3.3, this group of eukaryotic microbes is characterized by great morphological diversity. Indeed, taxonomists once used the variety in locomotory structures as a basis for classification. Locomotory structures no longer figure prominently in the taxonomic classification of protozoa because the presence of a given structure may not indicate evolutionary relatedness.

Some ciliates have two nuclei: a larger *macronucleus,* which contains many copies of the genome (often more than $50n$) and controls metabolism, growth, and sexual reproduction, and a smaller *micronucleus,* which is involved in genetic recombination, sexual reproduction, and regeneration of macronuclei.

Protozoa also show variety in the number and kind of mitochondria they contain. Several groups completely lack mitochondria, while all the others have mitochondria with discoid or tubular cristae rather than the platelike cristae seen in animals, plants, fungi, and many algae. Additionally, some protozoa have *contractile vacuoles* that actively pump water from the cells, protecting them from osmotic lysis **(Figure 12.5)**.

All free-living aquatic and pathogenic protozoa exist as a motile feeding stage called a **trophozoite** (trof-ō-zō'īt), and many have a hardy resting stage called a **cyst,** which is characterized by a thick capsule and a low metabolic rate. Cysts of protozoa are not reproductive structures, because one trophozoite forms one cyst, which later becomes one trophozoite. Such cysts allow intestinal protozoa to pass from one host to another and to survive harsh environmental conditions such as desiccation, nutrient deficiency, extremes of pH and temperature, and lack of oxygen.

Nutrition of Protozoa

Most protozoa are chemoheterotrophic; that is, they obtain nutrients by phagocytizing bacteria, decaying organic matter, other protozoa, or the tissues of a host; a few protozoa absorb nutrients from the surrounding water. Because the protozoa called dinoflagellates and euglenids (discussed shortly) are photoautotrophic, botanists historically classified them as algal plants rather than as protozoa.

Reproduction of Protozoa

Most protozoa reproduce asexually only, by binary fission or schizogony; a few protozoa also have sexual reproduction in which two individuals exchange genetic material. Some

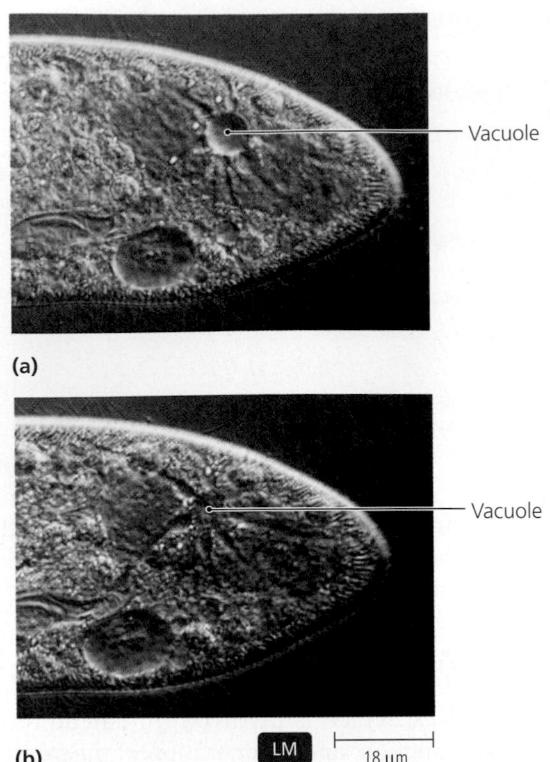

▲ **Figure 12.5 A contractile vacuole.** Many protozoa such as *Paramecium* have this prominent feature. **(a)** A vacuole being filled by water that entered the cell via osmosis. **(b)** A vacuole that is contracting to pump water out of the cell. *Is the environment in this case hypertonic, hypotonic, or isotonic to the cell? Explain.*

Figure 12.5 *The environment is hypotonic to the cell; water moves down its concentration gradient—into the cell—by osmosis.*

sexually reproducing protozoa become **gametocytes** (gametes) that fuse with one another to form a diploid **zygote.** Ciliates, such as *Paramecium* (par-ă-mē′sē-ŭm), reproduce sexually via a complex process called *conjugation* **(Figure 12.6)**, which involves the coupling of two compatible mating cells (1), meiosis of diploid micronuclei (2), loss of some haploid micronuclei (3), exchange of micronuclei between the coupled cells (4), uncoupling of the cells (5), fusion of haploid micronuclei to form a diploid micronucleus (6), three mitoses of the micronucleus to form eight micronuclei (7), disintegration of the macronucleus and the subsequent formation of a new macronucleus from four micronuclei (8), and three cytokineses to produce four daughter cells, each with one macronucleus and one micronucleus (9).

Classification of Protozoa

Learning Objectives

- ✓ Discuss the reasons for the many different taxonomic schemes for protozoa.
- ✓ Compare and contrast three types of alveolate.
- ✓ Compare and contrast three types of amoeba.
- ✓ Describe the life cycles of plasmodial and acellular slime molds.
- ✓ Identify several features of a typical euglenid.
- ✓ Describe characteristic features of diplomonads, parabasalids, cercozoa, radiolarians, and amoebozoa.

As we have seen, over two centuries ago Linnaeus classified protozoa as animals; later taxonomists grouped protozoa into kingdom Protista. Furthermore, some taxonomists divided the protozoa into four groups based on the organisms' mode of locomotion: Sarcodina (which are motile by means of pseudopodia), Mastigophora (flagella), Ciliophora (cilia), and Sporozoa (which are nonmotile). Other taxonomists lumped the first two groups together into a single group called Sarcomastigophora. Grouping of the protozoa according to locomotory features is still in common usage for many practical applications.

Taxonomists today recognize that these schemes do not reflect genetic relationships, either between protozoa and other organisms or among protozoa. Accordingly, taxonomists continue to revise and refine the classification of protozoa based on 18S rRNA nucleotide sequencing and features made visible by electron microscopy. One such genetic scheme classifies protozoa into the seven taxa Parabasala through Diplomonadida shown in Figure 12.4c, which different scientists consider kingdoms (as here), subkingdoms, or phyla.

In the following sections we will briefly discuss members of these seven taxa of protozoa, formed largely according to similarities in nucleotide sequences and ultrastructure. We begin with the parabasalids.

Parabasala

Parabasalids lack mitochondria, but each has a single nucleus and a *parabasal body,* which is a Golgi body–like structure. *Trichonympha* (trik-ō-nimf′ă), a parabasalid with numerous flagella **(Figure 12.7)**, inhabits the guts of termites, where it assists in the digestion of wood. Another well-known parabasalid is *Trichomonas* (trik-ō-mō′nas), which lives in the human vagina (see Figure 23.9). When the normally acidic pH of the vagina is raised, *Trichomonas* proliferates and causes severe inflammation that can lead to sterility. It is spread by sexual intercourse and is usually asymptomatic in males.

Alveolates

Alveolates (al-vē′ō-lātz) are protozoa with small membrane-bound cavities called alveoli[15] (al-vē′ō-lī) beneath their cell surfaces **(Figure 12.8)**. Scientists do not know the purpose of alveoli. Alveolates share at least one other characteristic—tubular mitochondrial cristae. This group is further divided into three subgroups: *ciliates, apicomplexans,* and *dinoflagellates.*

Ciliates As their name indicates, **ciliate** (sil′ē-āt) alveolates have cilia by which they either move themselves or move water past their cell surfaces. The structure and function of cilia were discussed on p. 80. Some ciliates are covered with cilia, whereas others have only a few isolated tufts. All ciliates are chemoheterotrophs and have two nuclei—one macronucleus and one micronucleus. Some taxonomists consider them the sole members of phylum Ciliophora.

Notable ciliates include *Vorticella* (vōr-ti-sel′ă), whose apical cilia create a whirlpool-like current to direct food into its

[15]Latin, meaning small hollows.

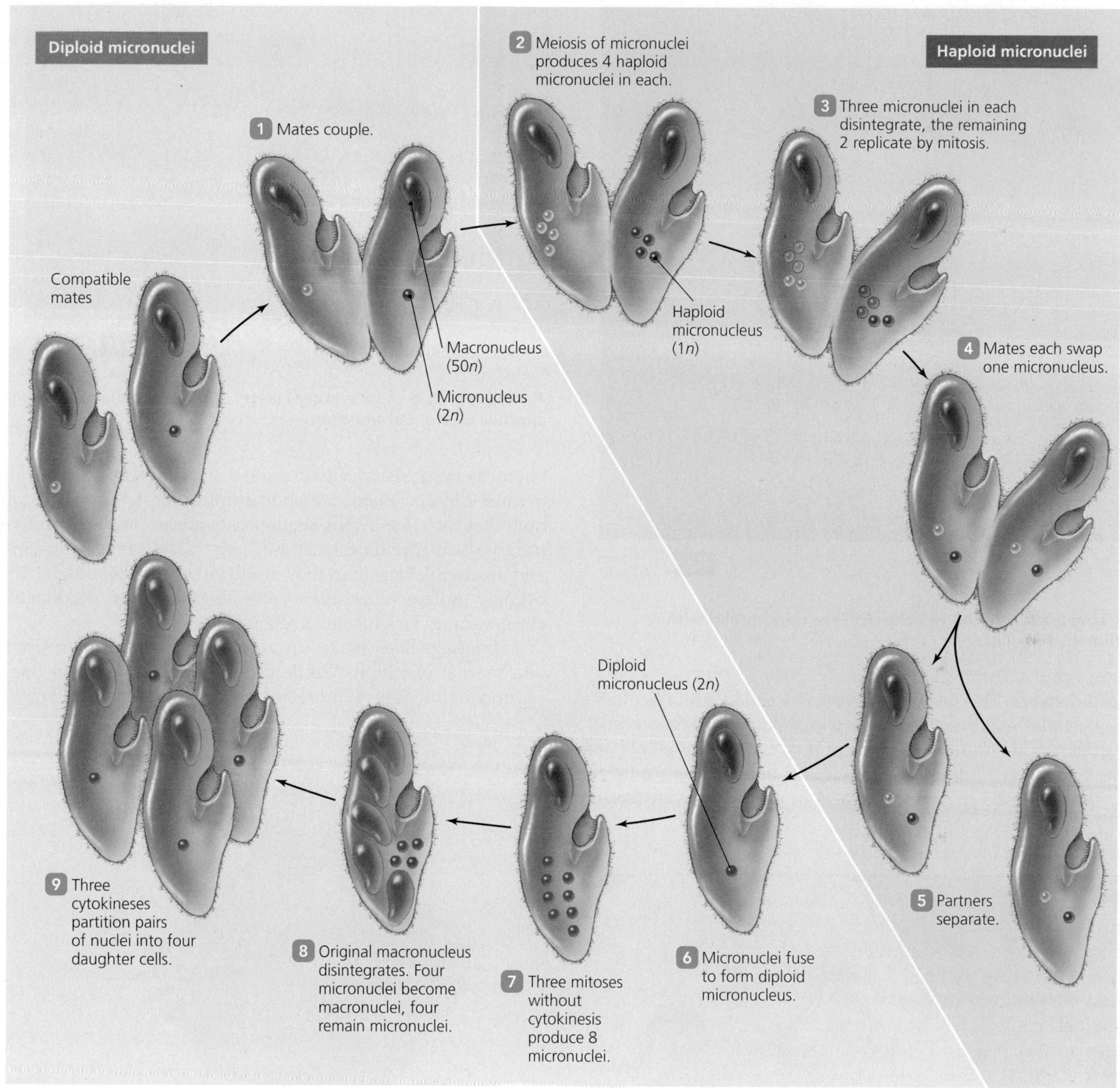
Diploid micronuclei
Haploid micronuclei
Compatible mates
1 Mates couple.
Macronucleus (50n)
Micronucleus (2n)
2 Meiosis of micronuclei produces 4 haploid micronuclei in each.
Haploid micronucleus (1n)
3 Three micronuclei in each disintegrate, the remaining 2 replicate by mitosis.
4 Mates each swap one micronucleus.
5 Partners separate.
Diploid micronucleus (2n)
6 Micronuclei fuse to form diploid micronucleus.
7 Three mitoses without cytokinesis produce 8 micronuclei.
8 Original macronucleus disintegrates. Four micronuclei become macronuclei, four remain micronuclei.
9 Three cytokineses partition pairs of nuclei into four daughter cells.

▲ **Figure 12.6 Sexual reproduction via conjugation in ciliates.** Shown here for *Paramecium*. Cells with haploid micronuclei are shown against a blue background; those with diploid micronuclei, against a green background.

"mouth"; *Balantidium* (bal-an-tid′ē-ŭm), which is the only ciliate pathogenic to humans; and the carnivorous *Didinium* (dī-di′-nē-ŭm), which phagocytizes other protozoa such as the well-known pond-water ciliate *Paramecium* (**Figure 12.9**). Chapter 23 discusses *Balantidium* in more detail.

Apicomplexans The alveolates called **apicomplexans** (ap-i-kom-plek′sănz) are all chemoheterotrophic pathogens of animals. The name of this group refers to the *complex* of special intracellular organelles, located at the *apices* of the infective stages of these microbes, that enables them to penetrate host cells. Examples of apicomplexans are *Plasmodium, Cryptosporidium* (krip-tō-spō-rid′-ē-ŭm), and *Toxoplasma* (tok-sō-plaz′mă), which cause malaria, cryptosporidiosis, and toxoplasmosis, respectively. Representative apicomplexans and the diseases they cause are discussed in more detail in Chapter 23.

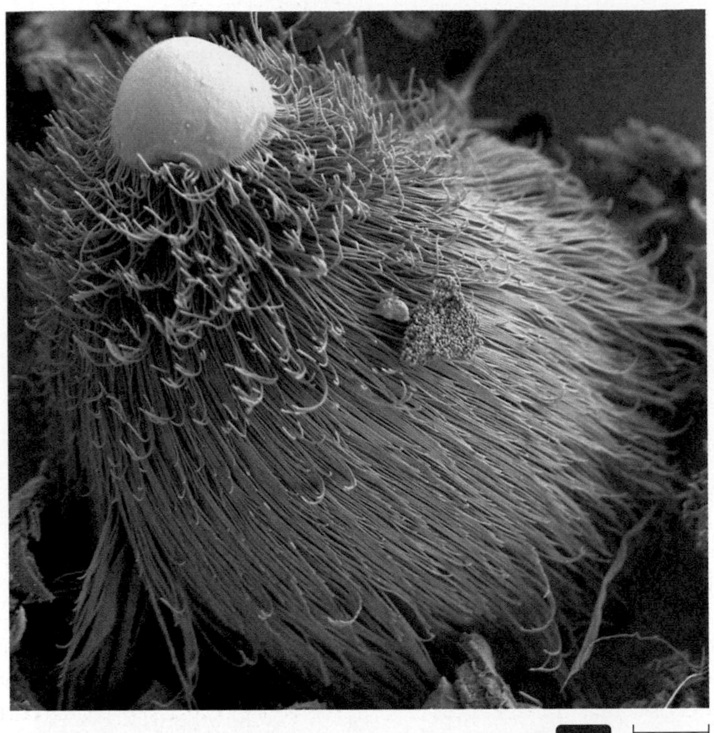

▲ **Figure 12.7** ***Trichonympha acuta*, a parabasalid with prodigious flagella.**

Dinoflagellates The group of alveolates called **dinoflagellates** are unicellular microbes that have photosynthetic pigments such as carotene and chlorophylls *a*, c_1, and c_2. Like many plants and algae, their food reserves are starch and oil, and their cells are often strengthened by internal plates of cellulose. Even though

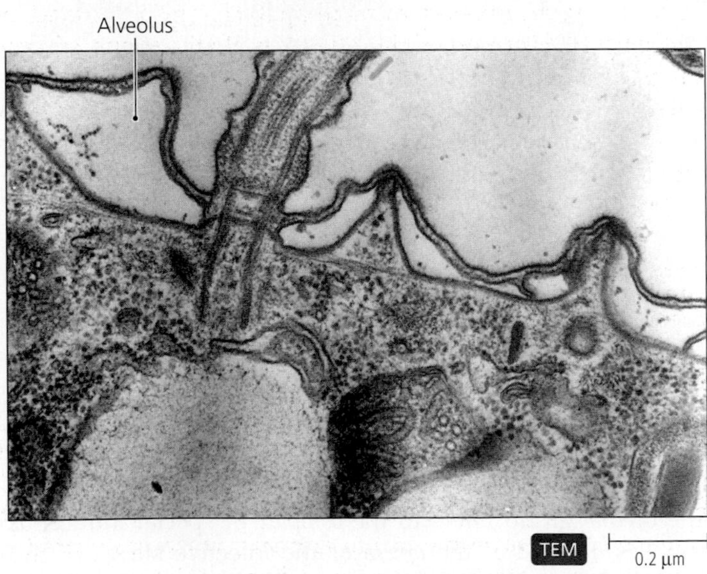

▲ **Figure 12.8 Alveoli membrane-bound structures found in some protozoa.** Even though the alveolus's function is not yet known, it is present in eukaryotic microbes with similar 18S rRNA sequences, indicating genetic relatedness and forming the basis of the group of eukaryotes called alveolates.

▲ **Figure 12.9 A predatory ciliate, *Didinium* (on left), devouring another ciliate, *Paramecium*.**

botanists have historically classified the dinoflagellates as algae because dinoflagellates are photoautotrophic, taxonomists today note that their 18S rRNA sequences and the presence of alveoli indicate that dinoflagellates are more closely related to ciliates and apicomplexans than they are to either plants or algae. Interestingly, unlike other eukaryotic chromosomes, dinoflagellate chromosomes lack histone proteins.

Dinoflagellates make up a large proportion of freshwater and marine plankton. Motile dinoflagellates have two flagella of unequal length **(Figure 12.10)**. The transverse flagellum

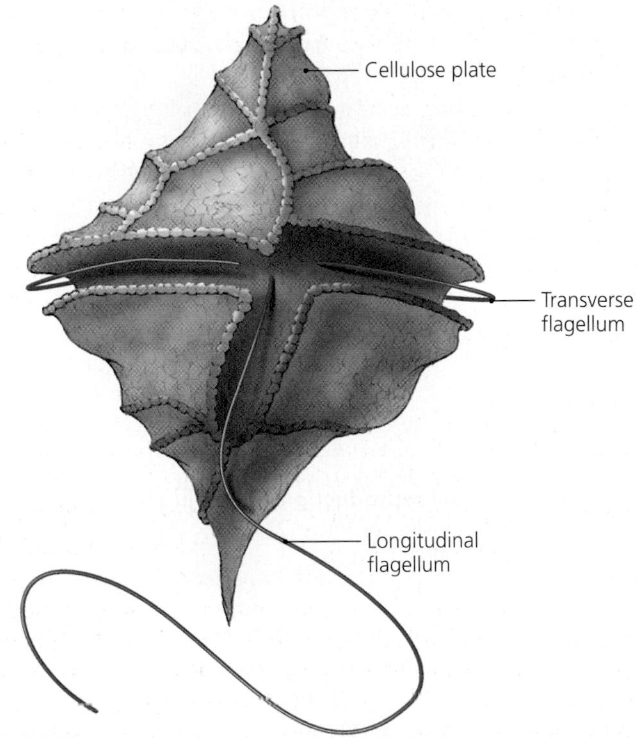

▲ **Figure 12.10 *Peridinium*, a motile armored dinoflagellate.** Dinoflagellates have two flagella. The transverse flagellum spins the cell; the longitudinal flagellum propels the cell forward.

▲ **Figure 12.11 Foraminifera have multichambered, snail-like shells of calcium carbonate.** Pseudopodia not visible. *Source:* R. Kessell and G. Shih/Visuals Unlimited.

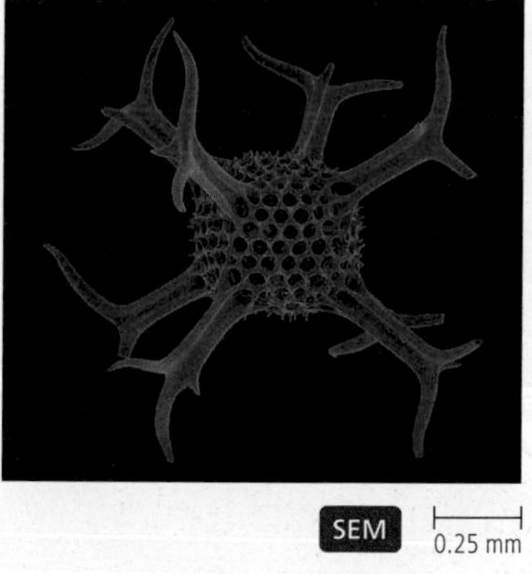

▲ **Figure 12.12 Radiolaria have ornate shells of silica.** The pseudopodia (not present in this dead specimen) extend through the holes.

wraps around the equator of the cell in a groove in the cell wall, and its beat causes the cell to spin; the second flagellum extends posteriorly and propels the cell forward.

Many dinoflagellates are bioluminescent—that is, able to produce light via metabolic reactions. When luminescent dinoflagellates are present in large numbers, the ocean water lights up with every crashing wave, passing ship, or jumping fish. Other dinoflagellates produce a red pigment, and their abundance in marine water is one cause of a phenomenon called a **red tide.**

Some dinoflagellates, such as *Gymnodinium* (jīm-nō-din'-ē-um) and *Gonyaulax* (gon-ē-aw'laks), produce *neurotoxins*—poisons that act against the nervous system. In 1991, over a billion fish died from dinoflagellate poisoning off the coast of North Carolina. Dinoflagellates can also poison humans who eat shellfish that have ingested planktonic dinoflagellates and concentrated their toxins.

The neurotoxin of another dinoflagellate, *Pfiesteria*[16] (fes-tēr'ē-ă), may be even more potent: It has been claimed the toxin poisons people who merely handle infected fish or breathe air laden with the microbes, resulting in memory loss, confusion, headache, respiratory difficulties, skin rash, muscle cramps, diarrhea, nausea, and vomiting. The Centers for Disease Control and Prevention (CDC) calls such poisoning *possible estuary-associated syndrome (PEAS).*

Cercozoa

Unicellular eukaryotes called **amoebae**[17] are protozoa that move and feed by means of pseudopodia (see Figure 3.30). Beyond this common feature and the fact that they all reproduce via binary fission (an unusual aspect of which is explored in **Highlight: Amoeba Midwives** on p. 351), amoebae exhibit little uniformity. Some taxonomists currently classify amoebae in three kingdoms: Cercozoa, Radiolaria, and Amoebozoa. We consider them in order.

Cercozoa is a group of amoebae with threadlike pseudopodia. A major taxon is composed of armored marine amoebae known as **foraminifera.** A foraminiferan has a porous shell composed of calcium carbonate arranged on an organic matrix in a snail-like manner **(Figure 12.11)**. Pseudopodia extend through holes in the shell. Commonly foraminifera live attached to sand grains on the ocean floor. Most foraminifera are microscopic, though scientists have discovered species several centimeters in diameter.

Over 90% of known foraminifera are fossil species, some of which form layers of limestone hundreds of meters thick. The great pyramids of Giza outside Cairo, Egypt, are built of foraminiferan limestone. Geologists correlate the ages of sedimentary rocks from different parts of the world by finding identical foraminiferan fossils embedded in them.

[16]Named for dinoflagellate biologist Lois Pfiester.
[17]From Greek *ameibein,* meaning to change.

Radiolaria

Amoebae of kingdom Radiolaria make up another taxon with threadlike pseudopodia, but they have ornate shells composed of silica (SiO_2, the mineral found in opal) **(Figure 12.12)** and live in marine water as part of the plankton. Radiolarians reinforce their pseudopodia with stiff internal bundles of microtubules so that the pseudopodia radiate from the central body like spokes of a spherical wheel. The dead bodies of radiolarians settle to the bottom of the ocean where they form ooze that is hundreds of meters thick in some locations.

Amoebozoa

Amoebozoa constitute a third kingdom of amoebae distinguished from the other two taxa of amoebae by having lobe-shaped pseudopodia and no shells (see Figure 3.30). Amoebozoa include the normally free-living amoebae *Naegleria* (nā-glē'-rē-ă) and *Acanthamoeba* (ă-kan-thă-mē'bă), which can each cause diseases of the eyes or brains of humans and animals that swim in water containing them. Other amoebozoa, such as *Entamoeba* (ent-ă-mē'bă), always live inside animals, where they produce potentially fatal amebic dysentery. Chapter 23 examines these pathogenic amoebae in more detail.

Taxonomists formerly considered another group of amoebozoa—**slime molds**—to be fungi, but the lobe-shaped pseudopodia by which they feed and move as well as their nucleotide sequences show that they are amoebozoa. Scientists

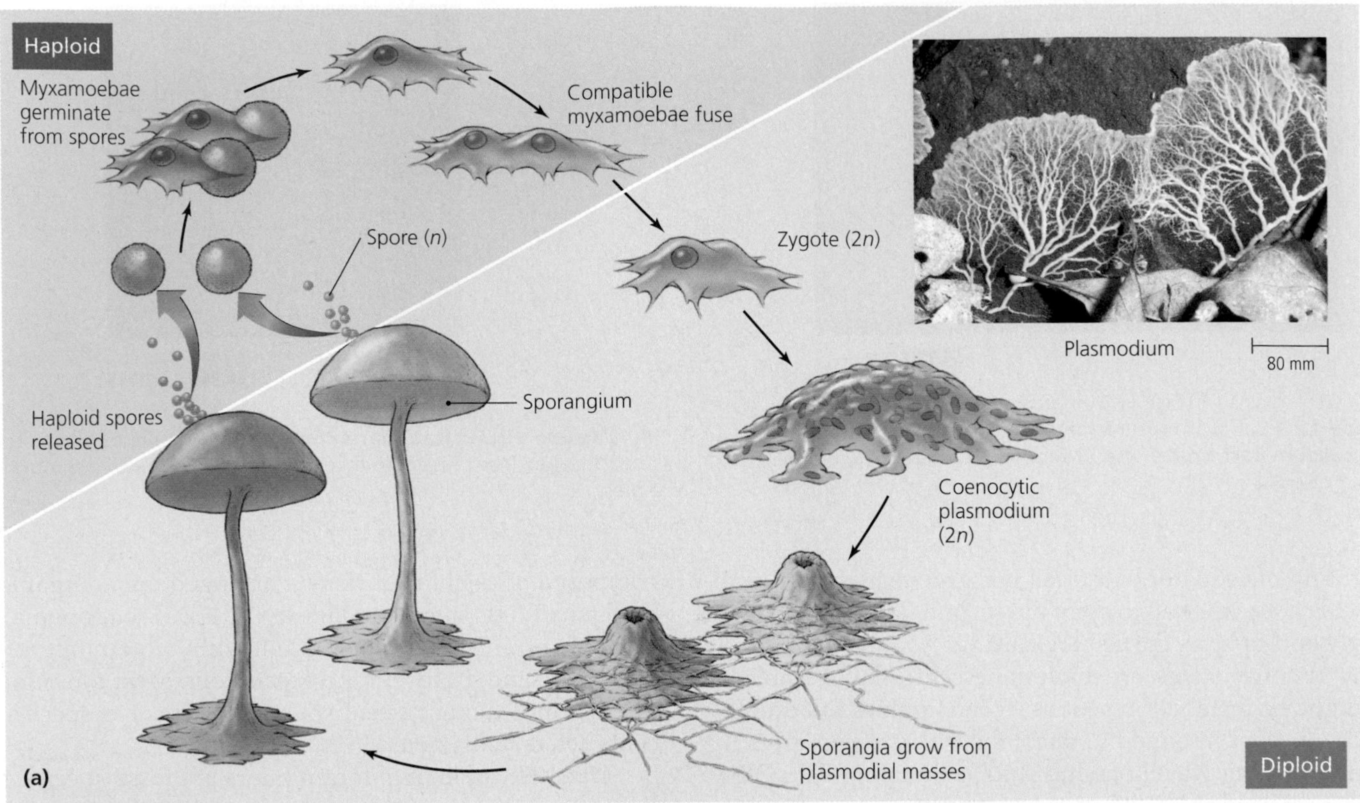

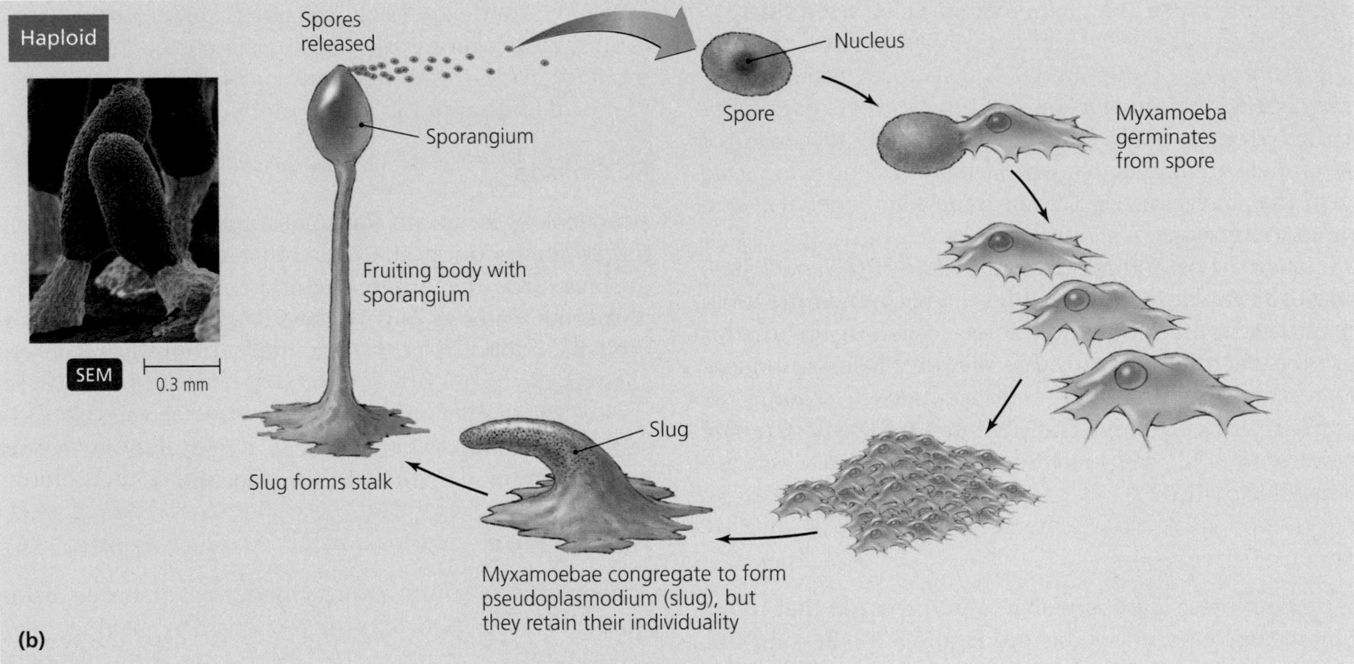

▲ **Figure 12.13 Life cycles of slime molds.** **(a)** A plasmodial (acellular) slime mold. Myxamoebae fuse to form a coenocytic plasmodium that feeds on organic debris and bacteria as the plasmodium creeps through its environment. Under adverse conditions, it produces sporangia that undergo meiosis to produce haploid spores. Compatible spores fuse to form a zygote that undergoes multiple mitoses, but not cytokinesis, to form a new plasmodium. **(b)** A cellular slime mold. Individual myxamoebae congregate during times of starvation to form a multicellular slug, which produces a stalked sporangium. Myxamoebae retain their individuality. The sporangium releases spores that germinate when conditions are more favorable. All phases are haploid.

HIGHLIGHT

AMOEBA MIDWIVES

Amoebae divide by binary fission, during which a cell splits into two daughter cells. Occasionally, however, an amoeba fails to divide completely, and the two daughter cells remain attached, as has been observed in studies of *Entamoeba invadens*, an amoeba found in the guts of snakes. When stalled in the division process, *E. invadens* signals for the assistance of neighboring amoebae. Like an overaggressive midwife helping with the birthing process, an assisting amoeba will advance toward a dividing amoeba and barge into it—completing the separation of the two daughter cells in a fascinating example of microbial cooperation.

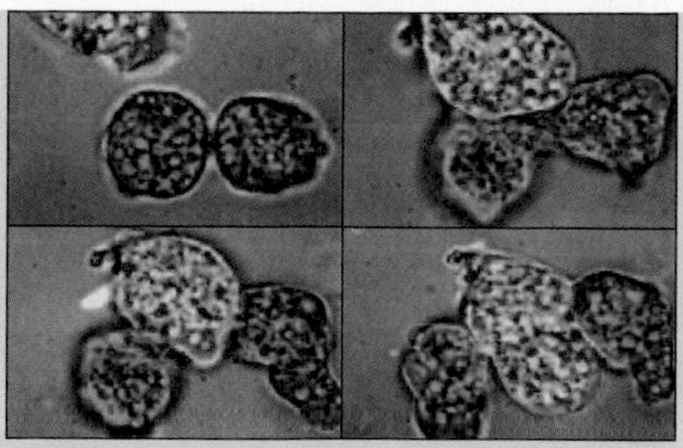

▲ *An "assisting" amoeba (yellow) barges into a dividing amoeba (blue).*

have identified two types of slime molds: *plasmodial molds* and *cellular slime molds.*

Slime molds differ from true fungi in two main ways:

- They lack cell walls, more closely resembling the amoebae in this regard.
- They are phagocytic rather than absorptive in their nutrition.

Species in the two groups of slime molds differ based on their morphology, reproduction, and 18S rRNA sequences. Slime molds are important to humans primarily as excellent laboratory systems for the study of developmental and molecular biology.

Plasmodial (Acellular) Slime Molds **Plasmodial slime molds,** also known as *acellular* slime molds (e.g., *Physarum,* fi-sar'um), exist as streaming, coenocytic, colorful filaments of cytoplasm that creep as amoebae through forest litter, feeding by phagocytizing organic debris and bacteria. The body, called a *plasmodium,* may contain millions of diploid nuclei and cover many square centimeters **(Figure 12.13a)**. Nutrients are distributed throughout the plasmodium by cytoplasmic streaming.

When food or water are in short supply, the plasmodium divides into individual masses of cytoplasm, each of which produces a stalked sporangium. Meiosis occurs within the sporangia to generate haploid spores. These spores germinate to produce *myxamoebae,* which look and act like other unicellular amoebae; in the presence of water, however, myxamoebae produce flagella and swim about (not shown). When the water disappears, they become amoeboid again.

Compatible myxamoebae of opposite mating types fuse to form a diploid zygote. The nucleus of the zygote undergoes numerous mitoses—without cytokineses—to form a new coenocytic plasmodium.

Cellular Slime Molds **Cellular slime molds,** such as *Dictyostelium* (dik-tē-ō-stē'lē-um), exist as individual haploid myxamoebae that phagocytize bacteria, yeasts, dung, and decaying vegetation. **Figure 12.13b** illustrates their life cycle, all of which is haploid—there is no diploid phase. Myxamoebae reproduce by mitosis and cytokinesis when food is abundant; however, in scarcity, some secrete cyclic adenosine monophosphate (cAMP), which acts as a chemotactic attractant for other myxamoebae. The myxamoebae congregate into a sluglike *pseudoplasmodium,* which can migrate for several days. Unlike the true plasmodium of acellular slime molds, the cells of a pseudoplasmodial slug retain their individuality and can be separated mechanically.

Some cells of a pseudoplasmodium form a stalked sporangium; the remaining cells climb the stalk and become spores. In contrast to the spores of plasmodial slime molds, the spores of cellular slime molds do not result from meiosis and are not enclosed in a common wall.

CRITICAL **THINKING**

Why are cellular slime molds called "cellular"?

Euglenozoa

Part of the reason that taxonomists established the kingdom Protista in the 1960s was to create a "dumping ground" for *euglenids,* eukaryotic microbes that share certain characteristics of both plants and animals. More recently, based on similar 18S rRNA sequences, the presence of a crystalline rod of unknown function in the flagella, and the presence of mitochondria with disk-shaped cristae, some taxonomists have created a new taxon: kingdom Euglenozoa. The euglenozoa include euglenids and some flagellated protozoa called *kinetoplastids.*

Euglenids The group of euglenozoa called **euglenids,** which are named for the genus *Euglena* (yū-glēn'ă; **Figure 12.14a**), are photoautotrophic, unicellular microbes with chloroplasts containing light-absorbing pigments—chlorophylls *a* and *b,* and carotene. For this reason, botanists historically classified euglenids in the kingdom Plantae. However, one reason for not including euglenids with plants is that euglenids store food as a unique polysaccharide called *paramylon* instead of as starch. Euglenids are similar to animals in that they lack cell walls, have flagella, are chemoheterotrophic phagocytes (in the dark), and move by using their flagella as well as by flowing, contracting,

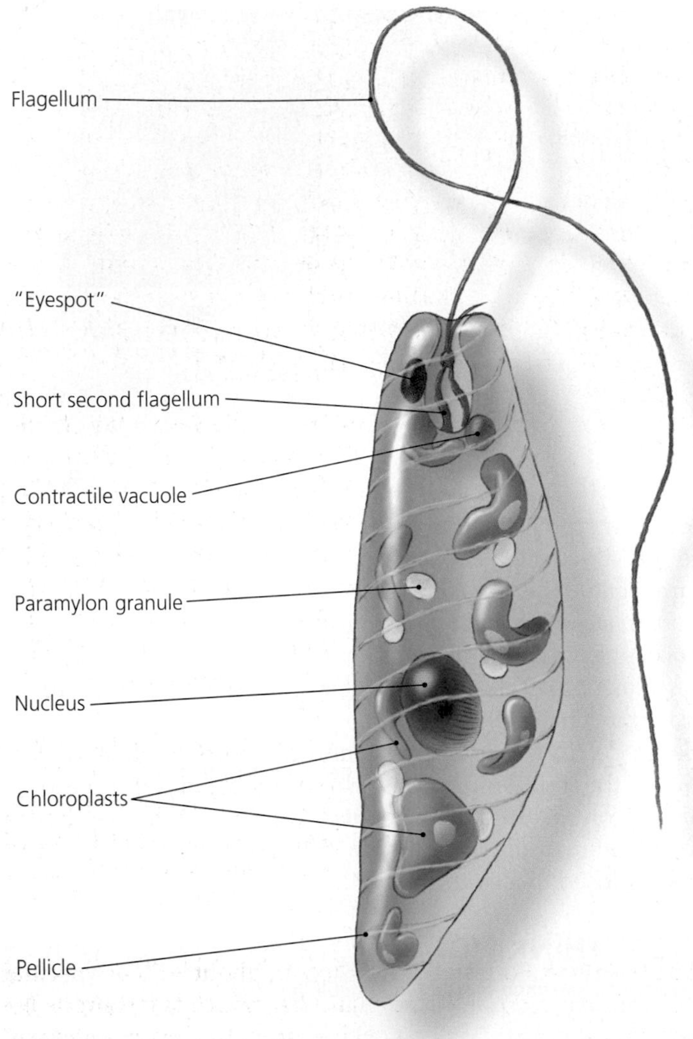

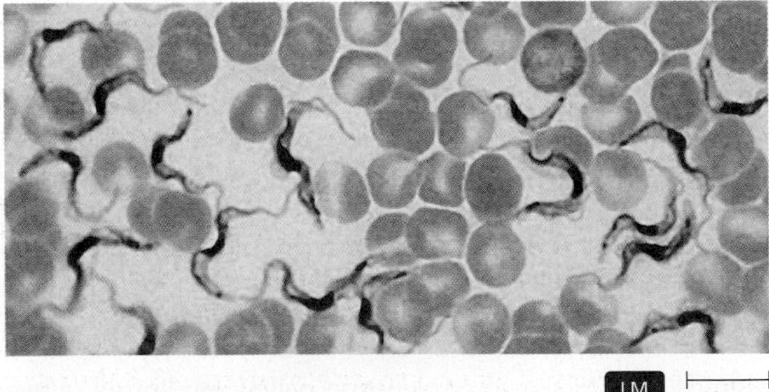

▲ **Figure 12.14 Two representatives of the kingdom Euglenozoa.** **(a)** The euglenid *Euglena*. Euglenids have characteristics that are similar to both plants and animals. **(b)** The kinetoplastid *Trypanosoma*. *What is the function of the "eyespot" in* Euglena?

Figure 12.14 *The "eyespot" functions in positive phototaxis by acting as a sun visor, shading the flagellar photoreceptor.*

and expanding their cytoplasm. Such a squirming movement, which is similar to amoeboid movement but does not involve pseudopodia, is called *euglenoid movement.*

A euglenid has a flexible, proteinaceous, helical *pellicle* that underlies its cytoplasmic membrane and helps maintain its shape. Typically each euglenid also has a red "eyespot," which plays a role in positive phototaxis by casting a shadow on a photoreceptor at the flagellar base, triggering movement in that direction. Euglenids reproduce by mitosis followed by longitudinal cytokinesis. They form cysts when exposed to harsh conditions.

Kinetoplastids Euglenozoa called **kinetoplastids** (**Figure 12.14b**) each have a single large mitochondrion that contains a unique region of mitochondrial DNA called a *kinetoplast.* Kinetoplastids live inside animals, and some are pathogenic. Among the latter are the genera *Trypanosoma* and *Leishmania,* certain species of which cause potentially fatal diseases of mammals, including humans (see Chapter 23).

Diplomonadida

Because members of the group Diplomonadida[18] lack mitochondria, Golgi bodies, and peroxisomes, biologists once thought these organisms were descended from ancient eukaryotes that had not yet phagocytized the prokaryotic ancestors of mitochondria. More recently, however, geneticists have discovered rudimentary *mitosomes* in the cytoplasm and mitochondrial genes in the nuclear chromosomes, a finding that suggests that diplomonads might be descended from typical eukaryotes that somehow lost their organelles.

Diplomonads have two equal-sized nuclei and multiple flagella. A prominent example is *Giardia* (jē-ar′dē-ă), a diarrhea-causing pathogen of animals and humans that is spread to new hosts when they ingest resistant *Giardia* cysts.

In summary, protozoa are a heterogeneous collection of single-celled, mostly chemoheterotrophic organisms that lack cell walls. Some taxonomists classify them in Parabasala, Alveolata, Cercozoa, Radiolaria, Amoebozoa, Euglenozoa, and Diplomonadida, though their relationships with one another and with other eukaryotic organisms are still unclear. Table 12.2 summarizes the incredible diversity of these microbes.

We next turn our attention to another group of chemoheterotrophs: the fungi, which chiefly differ from the protozoa in that they have cell walls.

CRITICAL **THINKING**

In Chapter 4 (see p. 121), we discussed identifying microbes using dichotomous taxonomic keys. Design a key for the kingdoms of protozoa discussed in this section.

[18]From Greek *diploos,* meaning double, and *monas,* meaning unit; refering to two nuclei.

TABLE 12.2 Characteristics of Protozoa

Category	Distinguishing Features	Representative Genera Mentioned in the Text
Parabasala	Parabasal body; single nucleus; lack mitochondria	*Trichomonas*
Alveolates	Alveoli (membrane-bound cavities underlying the cytoplasmic membrane); tubular cristae in mitochondria	
Ciliates	Cilia	*Balantidium, Paramecium, Didinium*
Apicomplexans	Apical complex of organelles	*Plasmodium, Cryptosporidium, Toxoplasma*
Dinoflagellates	Photosynthesis; two flagella; internal cellulose plates	*Gymnodinium, Gonyaulax, Pfiesteria*
Cercozoa	Threadlike pseudopodia	
Foraminifera	Shells of calcium carbonate	
Radiolarians	Threadlike pseudopodia, shells of silica	
Amoebozoa	Lobe-shaped pseudopodia; no shells	
Free-living and parasitic forms	Do not form aggregates	*Naegleria, Acanthamoeba, Entamoeba*
Plasmodial (acellular) slime molds	Multinucleate body called plasmodium	*Physarum*
Cellular slime molds	Cells aggregate to form pseudoplasmodium, but retain individual nature	*Dictyostelium*
Euglenozoa	Flagella with internal crystalline rod; disk-shaped mitochondrial cristae	
Euglenids	Photosynthesis; pellicle; "eyespot"	*Euglena*
Kinetoplastids	Single mitochondrion with DNA localized in kinetoplast	*Trypanosoma, Leishmania*
Diplomonadida	Two equal-sized nuclei; lack mitochondria, Golgi bodies, and peroxisomes	
Diplomonads	Multiple flagella	*Giardia*

Fungi

Learning Objective

✓ Cite at least three characteristics that distinguish fungi from other groups of eukaryotes.

Organisms in the kingdom **Fungi** (fŭn'jī), such as molds, mushrooms, and yeasts, are like most protozoa in that they are chemoheterotrophic; however, unlike protozoa they have cell walls, which typically are composed of a strong, flexible, nitrogenous polysaccharide called **chitin.** (The chitin in fungi is chemically identical to that in the exoskeletons of insects and other arthropods such as grasshoppers, lobsters, and crabs.) Fungi differ from plants in that they lack chlorophyll and do not perform photosynthesis; they differ from animals by having cell walls, although genetic sequencing of fungal and animal genomes has shown that fungi and animals are related. The study of fungi is *mycology,*[19] and scientists who study fungi are *mycologists.*

The Significance of Fungi

Learning Objective

✓ List five ways in which fungi are beneficial.

Fungi are extremely beneficial microorganisms. In nature, they decompose dead organisms (particularly plants) and recycle their nutrients. Additionally, the roots of about 90% of vascular plants form *mycorrhizae,*[20] which are beneficial associations between roots and fungi that assist the plants to absorb water and dissolved minerals.

Humans use fungi for food (mushrooms and truffles), in religious ceremonies (because of their hallucinogenic properties), and in the manufacture of foods and beverages, including bread, alcoholic beverages, citric acid (the basis of the soft drink industry), soy sauce, and some cheeses. Fungi also produce antibiotics such as penicillin and cephalosporin; the immunosuppressive drug *cyclosporine,* which makes organ transplants possible; and *mevinic acids,* which are cholesterol-reducing agents.

Fungi are also important research tools in the study of metabolism, growth, and development, and in genetics and biotechnology. For instance, based on their work with *Neurospora* (noo-ros'pōr-ă) in the 1950s, George Beadle (1903–1989) and Edward Tatum (1909–1975) developed their Nobel Prize–winning theory that one gene codes for one enzyme. Because of similar research, *Saccharomyces* (brewer's yeast) is the best understood eukaryote and the first eukaryote to have its entire genome sequenced. Chapter 26 highlights the uses of fungi in agriculture and industry.

[19]From Greek *mykes,* meaning mushroom, and *logos,* meaning discourse.
[20]From Greek *rhiza,* meaning root.

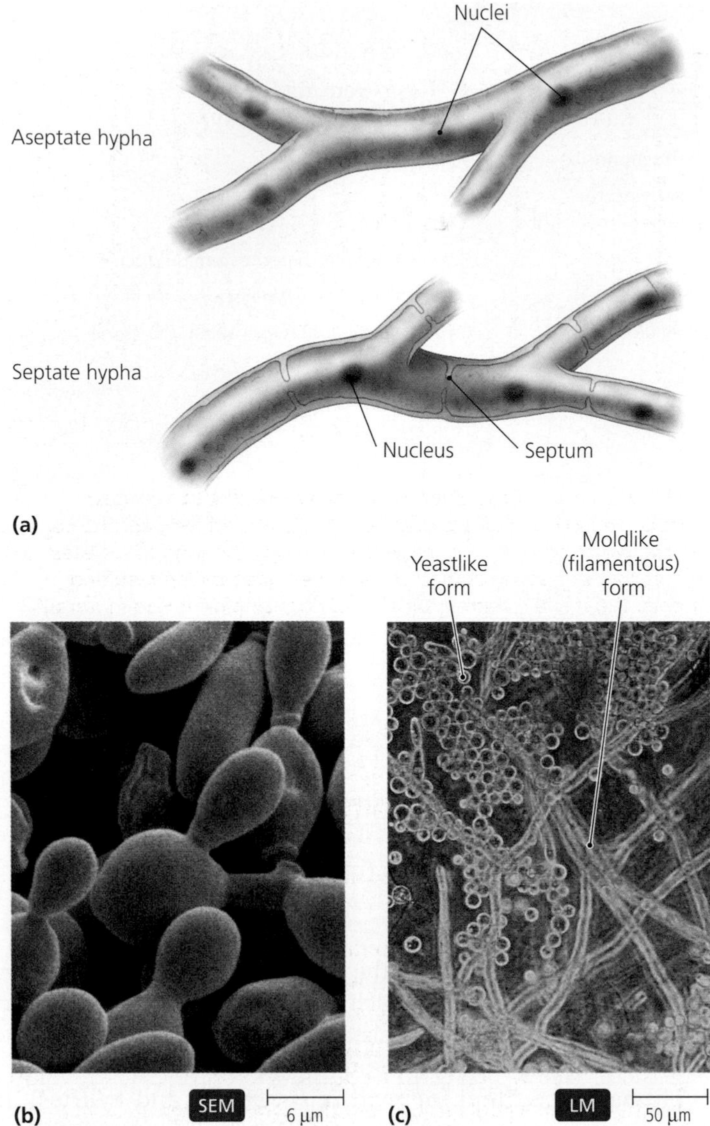

▲ **Figure 12.15 Fungal morphology.** **(a)** Artist's conception of septate and aseptate hyphae. **(b)** The thalli of *Saccharomyces* (baker's or brewer's yeast), which are unicellular and spherical to irregularly oval in shape. **(c)** The thalli of a dimorphic fungus, *Mucor rouxii*, showing both yeastlike and moldlike growth in response to environmental conditions.

Not all fungi are beneficial—about 30% of known fungal species produce **mycoses** (mī-kō′sēz), which are fungal diseases of plants, animals, and humans. For example, Dutch elm disease is a mycosis of elm trees, and athlete's foot is a fungal disease of humans. Because fungi tolerate concentrations of salt, acid, and sugar that inhibit bacteria, fungi are responsible for the spoilage of fruit, pickles, jams, and jellies exposed to air.

In the following sections we will consider the basic characteristics of fungal morphology, nutrition, and reproduction before turning to a brief survey of the major groups of fungi.

Morphology of Fungi

Learning Objective

✓ Distinguish among septate hyphae, aseptate hyphae, and mycelia.

The vegetative (nonreproductive) body of a fungus is called its **thallus**[21] (plural: *thalli*). The morphology of fungal thalli is variable. The thalli of *molds* are large and composed of long, branched, tubular filaments called **hyphae.**[22] Hyphae are either **septate** (divided into cells by cross walls called *septa*[23]) or **aseptate** (not divided by septa; **Figure 12.15a**). Aseptate hyphae are *coenocytic* (multinucleate). The thalli of *yeasts* are typically small, globular, and composed of a single cell **(Figure 12.15b)**.

In response to environmental conditions such as temperature or carbon dioxide concentration, some fungi produce both yeastlike thalli and moldlike thalli **(Figure 12.15c)**; fungi that produce two types of thalli are said to be **dimorphic** (which means "two-shaped"). Many medically important fungi are dimorphic, including *Histoplasma capsulatum* (his-tō-plaz′mă kap-soo-lā′tŭm), which causes a respiratory disease called histoplasmosis, and *Coccidioides immitis* (kok-sid-ē-oy′dēz im′-mi-tis), which causes a flulike disease called coccidioidomycosis (kok-sid-ē-oy′dō-mī-kō′sis). Generally, the yeast form of a dimorphic fungal pathogen causes disease, whereas the filamentous form does not.

The thallus of a mold is composed of hyphae intertwined to form a tangled mass called a **mycelium** (plural: *mycelia;* **Figure 12.16**). Mycelia are typically subterranean and thus usually escape our notice, though they can be very large. In fact, as mentioned in the chapter opener, the largest known organisms on Earth are fungi in the genus *Armillaria,* the mycelia of which can spread through thousands of acres of forest to a depth of several feet and weigh many hundreds of tons. (In contrast, blue whales, the largest living animals, weigh only about 150 tons.) *Fruiting bodies,* such as puffballs and mushrooms, are the reproductive structures of molds and are only small visible extensions of vast underground mycelia.

Nutrition of Fungi

Fungi acquire nutrients by absorption; that is, they secrete catabolic enzymes outside their thalli to break large organic molecules into smaller molecules, which they then transport into their thalli. Most fungi are **saprobes**[24] (sap′rōbz)—they absorb nutrients from the remnants of dead organisms—though some species trap and kill microscopic soil-dwelling nematodes (worms; **Figure 12.17**). Fungi that derive their nutrients from living plants and animals usually have modified hyphae called **haustoria**[25] (haw-stō′rē-ă), which penetrate the tissue of the host to withdraw nutrients. Absorptive nutrition is important in the role that fungi play as decomposers and recyclers of organic waste. Cytoplasmic streaming frequently transports nutrients and organelles, including nuclei, throughout a mycelium. Streaming between cells of septate mycelia occurs through pores in the septa.

[21]From Greek *thallos,* meaning young shoot.
[22]From Greek *hyphe,* meaning weaving or web.
[23]Latin, meaning partitions or fences.
[24]From Greek *sapros,* meaning rotten, and *bios,* meaning life.
[25]From Latin *haustor,* meaning someone who draws water from a well.

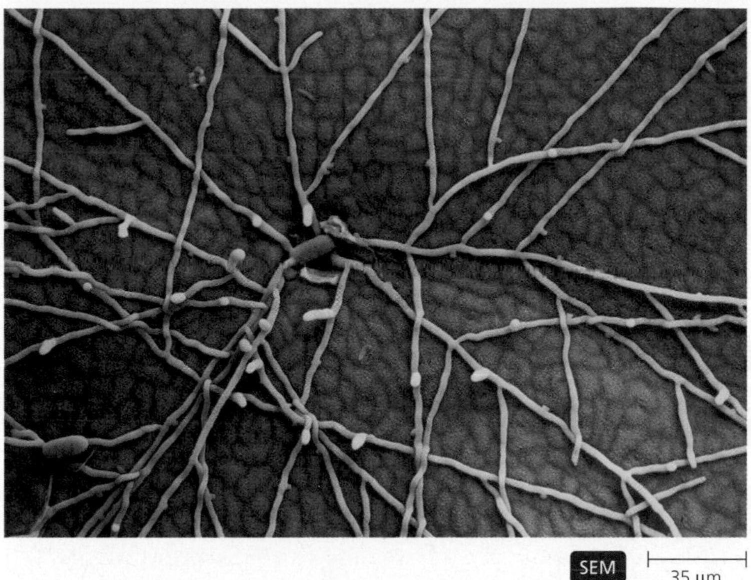

▲ **Figure 12.16** **A fungal mycelium growing on a leaf.**

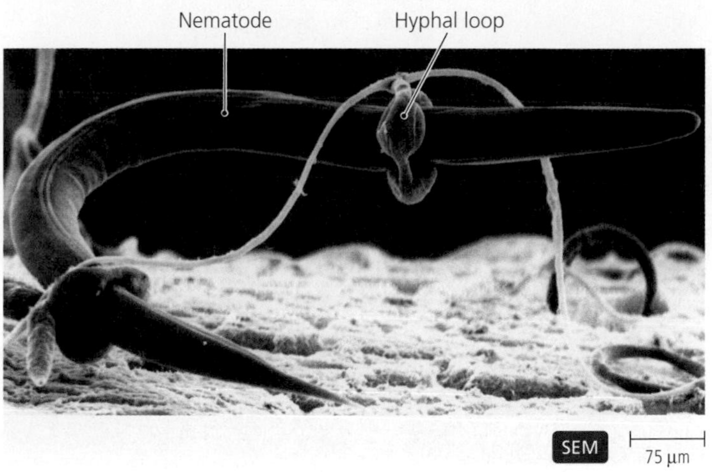

▲ **Figure 12.17** **Predation of a nematode by the fungus *Arthrobotrys.*** The fungus produces special looped hyphae that constrict when the worm contacts the inside of the loop. The fungus secretes enzymes that digest the nematode and then absorbs the resulting nutrients. *What is the more typical mode of nutrition found in fungi?*

Figure 12.17 *Most fungi are saprobic.*

Most fungi are aerobic, though many yeasts (for example, *Saccharomyces*) are facultative anaerobes that obtain energy from fermentation, such as occurs in the reactions that produce alcohol. Anaerobic fungi are found in the digestive systems of many herbivores, such as cattle and deer, where they assist in the catabolism of plant material.

Reproduction of Fungi

Learning Objectives

✓ Describe asexual reproduction in fungi.

✓ List three basic types of asexual spores found in molds.

Whereas all fungi have some means of asexual reproduction involving mitosis followed by cytokinesis, most fungi also reproduce sexually. In the next sections we briefly examine asexual and sexual reproduction in fungi.

Budding and Asexual Spore Formation

Yeasts typically bud in a manner similar to prokaryotic budding. Following mitosis, one daughter nucleus is sequestered in a small bleb (a blisterlike outgrowth) of cytoplasm that is isolated from the parent cell by the formation of a new wall (see Figure 12.15b). In some species, especially *Candida albicans* (kan′did-ă al′bi-kanz), which causes human oral thrush and vaginal yeast infections, a series of buds remain attached to one another and to the parent cell, forming a long filament called a *pseudohypha. Candida* invades human tissues by means of such pseudohyphae, which can penetrate intercellular cracks.

Filamentous fungi reproduce asexually by producing lightweight spores, which enable the fungi to disperse vast distances on the wind. Researchers have isolated fungal spores from wind currents many miles above the surface of the Earth. Scientists categorize the asexual spores of molds according to their mode of development:

- *Sporangiospores* form inside a sac called a *sporangium,*[26] which is often borne on a spore-bearing stalk, called a *sporangiophore,*[27] at either the tips or sides of hyphae **(Figure 12.18a)**.
- *Chlamydospores* form with a thickened cell wall inside hyphae **(Figure 12.18b)**.
- *Conidiospores* (also called *conidia*) are produced at the tips or sides of hyphae, but not within a sac. There are many types of conidia, including *arthroconidia,* which develop from hyphae that fragment into individual spores; *blastoconidia,* which form as buds, and others that develop in chains on stalks called *conidiophores* **(Figure 12.18c)**.

Medical lab technologists use the presence and type of asexual spores in clinical samples to identify many fungal pathogens.

Sexual Spore Formation

Scientists designate fungal mating types as "+" and "−" rather than as male and female, in part because their thalli are morphologically indistinguishable. The process of sexual reproduction in fungi has four basic steps **(Figure 12.19)**:

1. Haploid (n) cells from a + thallus and a − thallus fuse to form a *dikaryon,* a cell containing both + and − nuclei. The dikaryotic stage is neither diploid nor haploid, but instead is designated ($n + n$).
2. After a period of time that typically ranges from hours to years but can be centuries, a pair of nuclei within a dikaryon fuse to form one diploid ($2n$) nucleus.
3. Meiosis of the diploid nucleus restores the haploid state.
4. The haploid nuclei are partitioned into + and − spores, which reestablish + and − thalli.

[26]From Greek *spora,* meaning seed, and *angeion,* meaning vessel.
[27]From Greek *phoros,* meaning bearing.

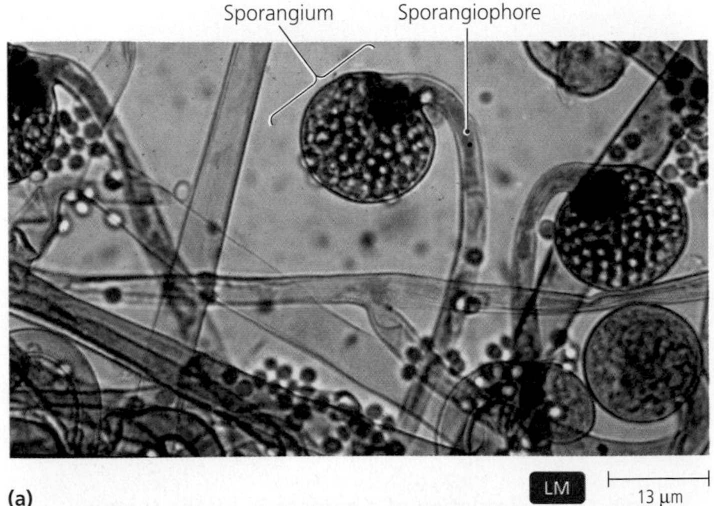

(a)

Chlamydospore

LM 10 μm

(b)

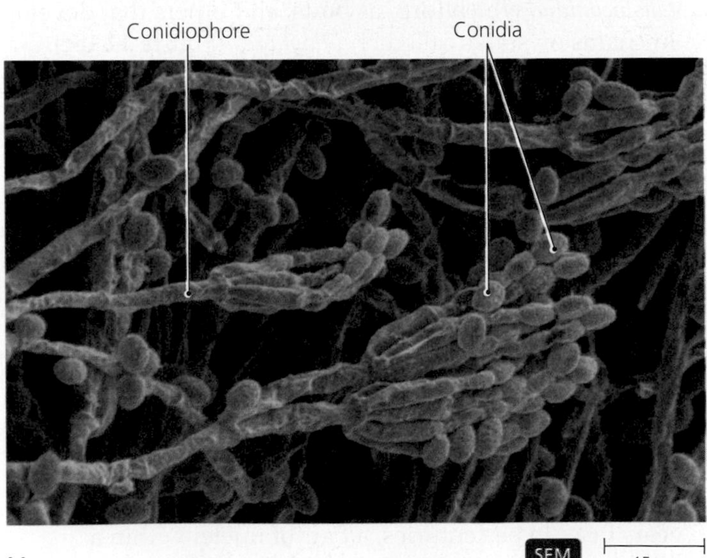

(c)

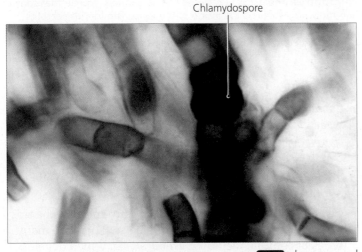

◀ **Figure 12.18 Representative asexual spores of molds.** **(a)** Sporangiospores, which develop within a sac called a sporangium that is borne on a sporangiophore, here of *Circinella*. **(b)** Chlamydospores are thick-walled spores that form inside hyphae, here of *Aureobasidium*. **(c)** Conidiospores (conidia) develop on conidiophores at the ends of hyphae, here of *Penicillium*. *How do conidia differ from sporangiospores?*

Figure 12.18 Conidia are never enclosed in a sac.

Fungi differ in the ways they form dikaryons and in the site at which meiosis occurs.

CRITICAL **THINKING**

Fungi tend to reproduce sexually when nutrients are limited or other conditions are unfavorable, but reproduce asexually when conditions are more ideal. Why is this a successful strategy?

Classification of Fungi

Learning Objectives

- ✓ Compare and contrast the three divisions of fungi with respect to the formation of sexual spores.
- ✓ Describe the deuteromycetes, and explain why this group no longer constitutes a formal taxon.
- ✓ List several beneficial roles or functions of lichens.

In the following sections we will consider the four major subgroups into which taxonomists traditionally divided the kingdom Fungi. Three of these subgroups, which are taxa called *divisions* that are equivalent to phyla in other kingdoms, are based on the type of sexual spore produced (divisions Zygomycota, Ascomycota, and Basidiomycota); the fourth (the deuteromycetes) was a repository of fungi for which no sexual stage is known. We begin by considering the Zygomycota.

Division Zygomycota

Fungi in the division **Zygomycota** are coenocytic molds called zygomycetes (zī′gō-mī-sēts). Of the approximately 1100 species known, most are saprobes; the rest are obligate parasites of insects and other fungi.

Figure 12.20 illustrates the life cycle of a typical zygomycete: the black bread mold *Rhizopus nigricans* (rī-zō′pŭs ni′gri-kans).[28] Zygomycetes reproduce asexually via sporangiospores 1 to 4, but the distinctive feature of most zygomycetes is the formation of sexual structures called **zygosporangia** (sometimes incorrectly termed zygospores). Zygosporangia of *R. nigricans* are black, rough-walled structures that develop from the fusion of sexually compatible hyphal tips 5 to 8. Like fungal spores, zygosporangia can withstand desiccation and other harsh environmental conditions.

Nuclei from one hypha (+) fuse with nuclei from the other hypha (−) to form many diploid nuclei within the zygosporangium. Each nucleus undergoes meiosis, but only one of the

[28]From Greek *rhiza*, meaning root, and *pous*, meaning foot, and Latin *niger*, meaning black.

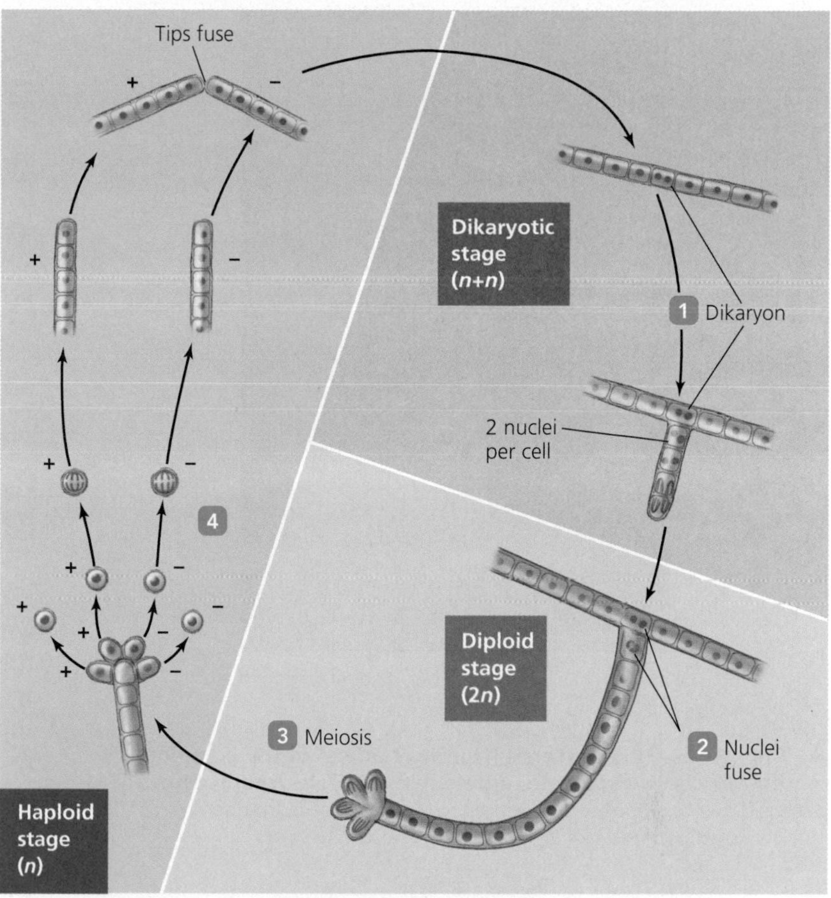

Tips fuse
+
−
Dikaryotic stage (n+n)
1 Dikaryon
2 nuclei per cell
Diploid stage (2n)
2 Nuclei fuse
3 Meiosis
4
Haploid stage (n)

▶ **Figure 12.19 The process of sexual reproduction in fungi.** The steps in the process are described in the text. Haploid cells appear against a blue background, diploid cells against a green background, and dikaryotic cells against an orange background.

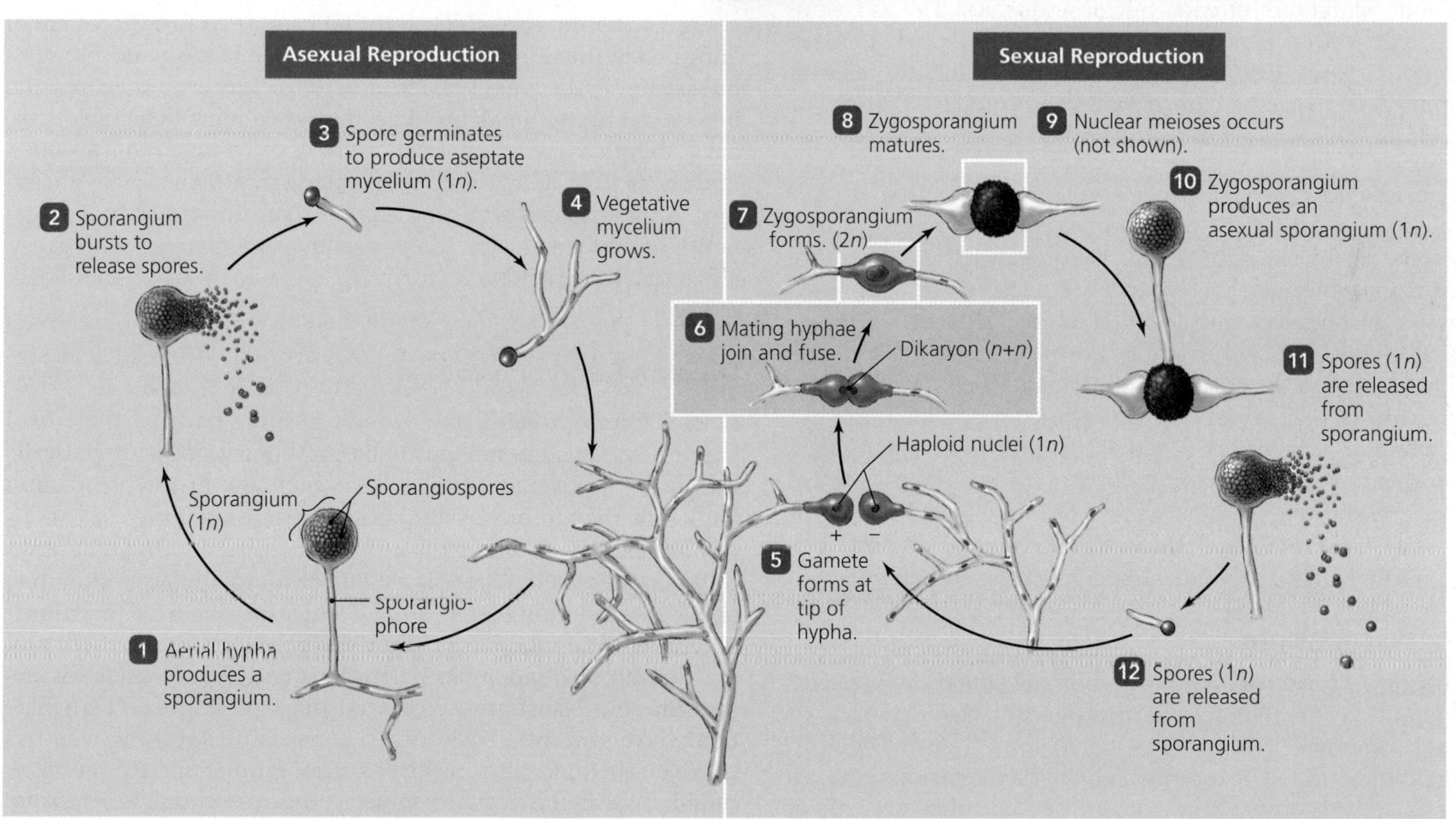

Asexual Reproduction
Sexual Reproduction
1 Aerial hypha produces a sporangium.
2 Sporangium bursts to release spores.
3 Spore germinates to produce aseptate mycelium (1n).
4 Vegetative mycelium grows.
Sporangium (1n)
Sporangiospores
Sporangio-phore
5 Gamete forms at tip of hypha.
6 Mating hyphae join and fuse.
Dikaryon (n+n)
Haploid nuclei (1n)
7 Zygosporangium forms. (2n)
8 Zygosporangium matures.
9 Nuclear meioses occurs (not shown).
10 Zygosporangium produces an asexual sporangium (1n).
11 Spores (1n) are released from sporangium.
12 Spores (1n) are released from sporangium.

▲ **Figure 12.20 Life cycle of the zygomycete *Rhizopus*.** During the asexual cycle, the fungus reproduces via sporangiospores that germinate and produce hyphae. In the sexual cycle, the tips of + and − hyphae fuse and form a diploid zygosporangium that matures, undergoes meiosis, germinates, and produces a sporangium containing haploid sporangiospores, which germinate to form new mycelia.

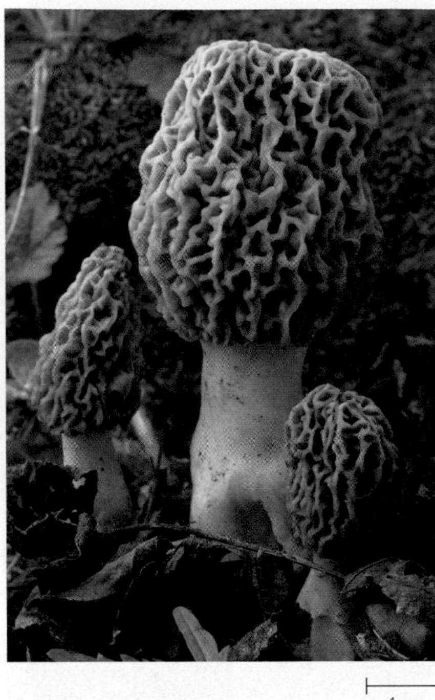

▲ **Figure 12.21 Ascocarps (fruiting bodies) of the common morel, *Morchella esculenta*, a delectable edible ascomycete.** The pits visible in this photograph are lined with asci, sacs that contain numerous ascospores.

four meiotic daughters of each nucleus survives. The zygosporangium then produces a haploid sporangium, which is filled with haploid spores (true zygospores). The sporangium releases these spores, each of which germinates to produce either a + or a − mycelium. This completes the life cycle.

Microsporidia Microsporidia are small organisms that are difficult to classify. Until 2003, taxonomists thought microsporidia were protozoa, but genetic analysis indicates they are more similar to zygomycetes.

They are obligatory intracellular parasites; that is, organisms that must live within their hosts' cells. Microsporidia spread from host to host as small, resistant spores. An example is *Nosema* (nō-sē'mă), which is parasitic on insects such as silkworms and honeybees. The Environmental Protection Agency has approved one species of *Nosema* as a biological control agent for grasshoppers. Seven genera of microsporidia, including *Nosema* and *Microsporidium* (mī-krō-spor-i'dē-ŭm), are known to cause diseases in immunocompromised patients.

Division Ascomycota

The division **Ascomycota** contains about 32,000 known species of molds and yeasts that are characterized by the formation of haploid **ascospores** within sacs called **asci.**[29] Asci occur in fruiting bodies called *ascocarps,* which have various shapes (**Figure 12.21**). Ascomycetes (as'kō-mī-sēts), as they are called, also reproduce asexually by conidiospores, as illustrated for a representative ascomycete *Penicillium* in **Figure 12.22** 1 to 4.

[29]From Greek *askus*, meaning wineskin.

Figure 12.22 also illustrates the sexual reproduction of an ascomycete, which proceeds as follows:

5 Multinucleate, hyphal tips of opposite mating types fuse to form a dikaryon.

6 The dikaryon reproduces to form hyphae whose cells are all dikaryotic.

7 In the dikaryon, nuclei of opposite mating types fuse to form a diploid nucleus.

8 Each diploid nucleus undergoes meiosis and cytokinesis to form four haploid cells within an ascus.

9 Each haploid cell may undergo mitosis and cytokinesis to form two haploid ascospores, resulting in eight ascospores, which line up inside the ascus.

10 The asci open to release their ascospores.

11 Each ascospore germinates to produce a + or − hypha.

Ascomycetes are familiar and economically important fungi. For example, most of the fungi that spoil food are ascomycetes. This group also includes plant pathogens such as the causative agents of Dutch elm disease and chestnut blight, which have almost eliminated their host trees in many parts of the United States. *Claviceps purpurea* (klav'i-seps poor-poo'rē-ă) growing on grain produces *lysergic acid,* which causes abortions in cattle and hallucinations in humans. *Aspergillus* can also infect humans (**Emerging Diseases: Aspergillosis** on p. 361).

On the other hand, many ascomycetes are beneficial. For example, *Penicillium* (pen-i-sil'ē-ŭm) mold is the source of penicillin; *Saccharomyces,* which ferments sugar to produce alcohol and carbon dioxide gas, is the basis of the baking and brewing industries; and *truffles* (varieties of *Tuber*) grow as mycorrhizae in association with oak and beech trees to form delectable culinary delights (see **Beneficial Microbes: Fungi for $3000 a Pound** on p. 361). As previously noted, another ascomycete, the pink bread mold *Neurospora,* has been an important tool in genetics and biochemistry. Many ascomycetes partner with green algae or cyanobacteria to form *lichens,* which are discussed in more detail shortly.

Division Basidiomycota

A walk through fields and woods in most parts of the world may reveal mushrooms, puffballs, stinkhorns, jelly fungi, bird's nest fungi, or bracket fungi, all of which are the visible fruiting bodies of the almost 22,000 known species of fungi in the division **Basidiomycota.** Poisonous mushrooms are sometimes called *toadstools,* but there is no sure way to always distinguish between an edible "mushroom" and a poisonous "toadstool" except by eating them—a truly risky practice!

Mushrooms and other fruiting bodies of *basidiomycetes* (ba-sid'ē-ō-mi-sēts) are called **basidiocarps (Figure 12.23)**. The entire structure of a basidiocarp consists of tightly woven hyphae that extend into multiple, often club-shaped projections called *basidia,* the ends of which produce sexual *basidiospores* (typically four on each basidium). **Figure 12.24** illustrates the life cycle of a poisonous mushroom (toadstool), *Amanita muscaria* (am-ă-nī'tă mus-ka'rē-ă).

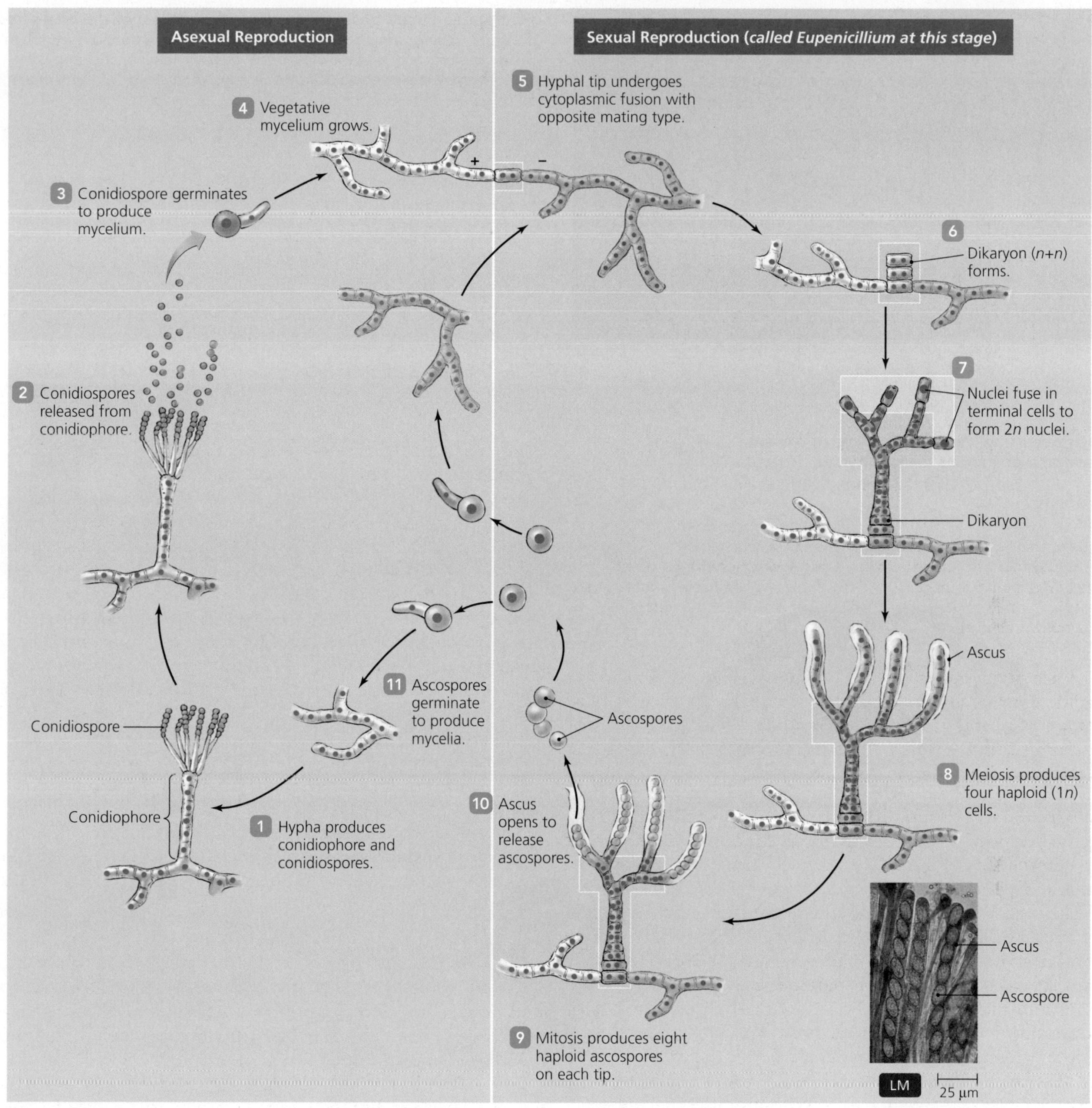

Asexual Reproduction
Sexual Reproduction (called Eupenicillium at this stage)
4 Vegetative mycelium grows.
5 Hyphal tip undergoes cytoplasmic fusion with opposite mating type.
+
−
3 Conidiospore germinates to produce mycelium.
6 Dikaryon (n+n) forms.
2 Conidiospores released from conidiophore.
7 Nuclei fuse in terminal cells to form 2n nuclei.
Dikaryon
Ascus
11 Ascospores germinate to produce mycelia.
Conidiospore
Ascospores
8 Meiosis produces four haploid (1n) cells.
10 Ascus opens to release ascospores.
Conidiophore
1 Hypha produces conidiophore and conidiospores.
Ascus
Ascospore
9 Mitosis produces eight haploid ascospores on each tip.
LM
25 μm

▲ **Figure 12.22** **Life cycle of an ascomycete.**

Besides the edible mushrooms—most notably, cultivated *Agaricus* (a-gär'i-kus)—basidiomycetes affect humans in several ways. Most basidiomycetes are important decomposers that digest chemicals such as cellulose and lignin in dead plants and return nutrients to the soil. Many mushrooms produce toxins or hallucinatory chemicals. An example of the latter is *Psilocybe cubensis* (sil-ō-sī'bē kū-ben'sis), which produces *psilocybin,* a hallucinogenic chemical. The basidiomycete yeast *Cryptococcus neoformans* (krip-tō-kok'us nē-ō-for'manz) is the leading cause of fungal meningitis. Other basidiomycetes are *rusts* and *smuts,* which cause millions of dollars in crop loss each year.

1 mm

Figure 12.23 Basidiocarps (fruiting bodies) of the bird's nest fungus, *Crucibulum*. Basidiospores, looking like flattened eggs, develop inside basidiocarps that look like birds' nests. The familiar shapes of mushrooms are also basidiocarps of extensive mycelia.

Figure 12.24 Sexual life cycle of *Amanita muscaria*, a poisonous basidiomycete. *How can a novice distinguish between edible and poisonous mushrooms?*

Figure 12.24 There is no sure way to distinguish between edible and poisonous mushrooms except by assuming the risk of eating them.

Cross section of gill showing hyphae.
SEM
35 μm
5 Pair of haploid nuclei fuse.
2n
6 Meiosis produces four haploid nuclei.
7 Four basidiospores (1n) develop.
Basidium
Gills
1 Basidiospore released.
2 Basidiospores germinate to produce mycelia.
+
–
Dikaryon (n+n)
3 Hyphae of opposite mating types fuse belowground to produce dikaryotic mycelium.
4 Dikaryotic mycelium growing in soil produces basidiocarp (mushroom).

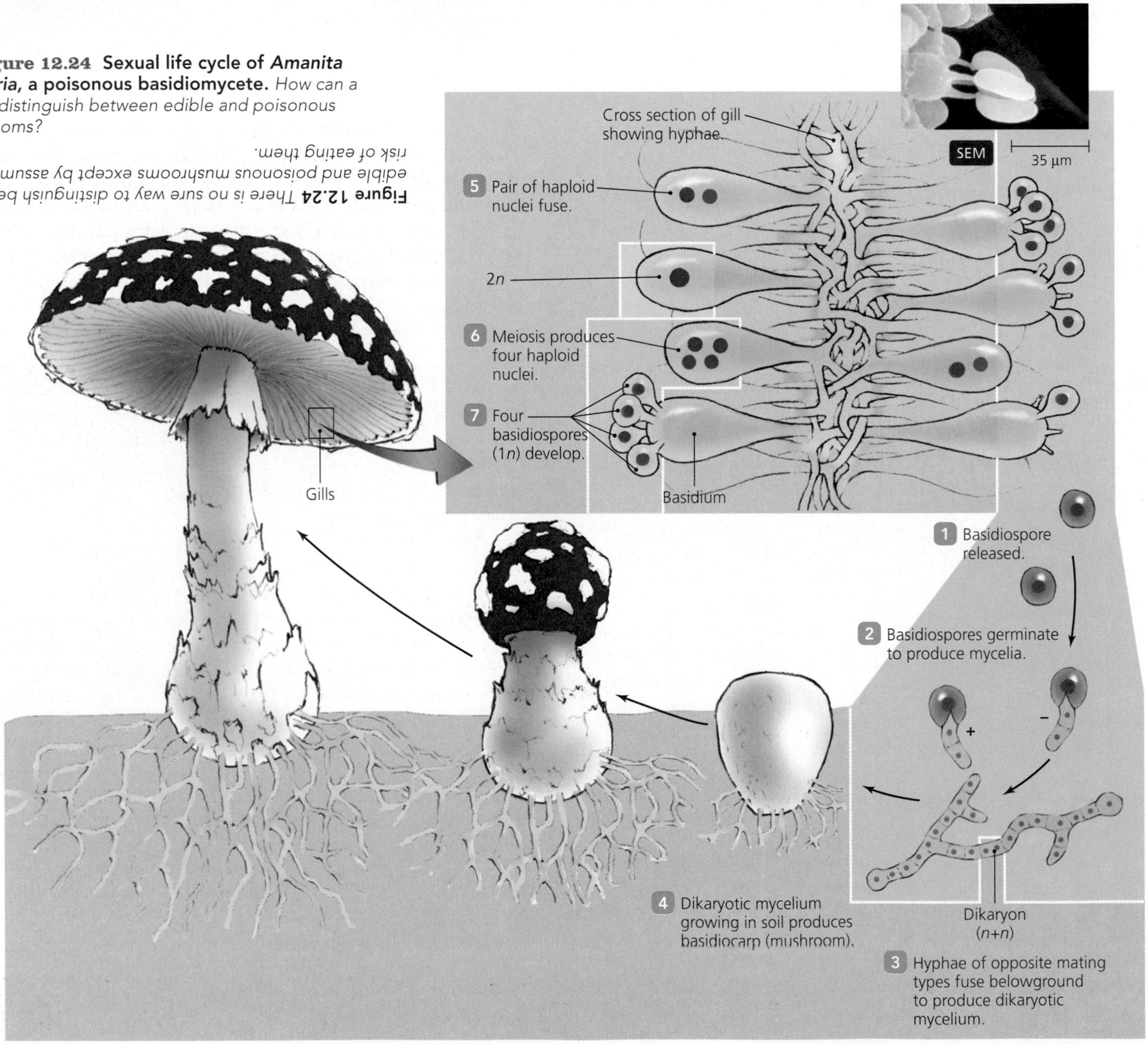

BENEFICIAL MICROBES

FUNGI FOR $3000 A POUND

▲ *Tuber melanosporum*

Truffles—rare, intensely flavored ascomycetes that grow underground—are one of the most luxurious and expensive foods on Earth, selling on average for more than $800 per pound. There are many different varieties of truffles. The most coveted include *Tuber melanosporum*, a black truffle that is also known as the "black diamond," and *Tuber magnatum*, a white truffle that can sell for $3000 per pound!

Because truffles are very difficult to find, truffle hunters often use pigs and dogs trained to sniff them out. (Dogs are preferred, because pigs are more likely to eat the truffles.) Dogs may be trained at the University of Truffle Hunting Dogs founded in 1880.

The underground habitat of truffles requires them to form symbiotic relationships with trees (for nutrients) and with animals such as squirrels and chipmunks (for spore dispersal; the animals dig up the truffles and thus help spread the spores to other locations).

Incidentally, chocolate truffles derive only their name from their prized fungal counterparts.

Deuteromycetes

As noted previously, the divisions Zygomycota, Ascomycota, and Basidiomycota are based on type of sexual spore produced. Because scientists have not observed sexual reproduction in all fungi, taxonomists in the middle of the 20th century created the division *Deuteromycota* to contain the fungi whose sexual stages are unknown—either because they do not produce sexual spores or because their sexual spores have not been observed. More recently, however, the analysis of rRNA sequences has revealed that most deuteromycetes in fact belong in the division Ascomycota, and thus modern taxonomists have abandoned Deuteromycota as a formal taxon. Nevertheless, many medical laboratory technologists, health care practitioners, and scientists continue to refer to "deuteromycetes" because it is a traditional name.

EMERGING DISEASES

ASPERGILLOSIS

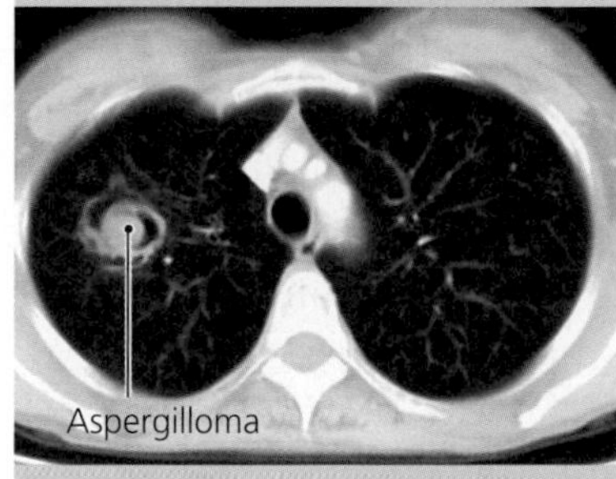

Matt was not in good shape, and it had nothing to do with his physique or time spent at the gym; it had to do with the ball of fungus in his right lung. That fungal sphere had started as a single spore of an ascomycete fungus called *Aspergillus*. The mold had formed a spherical mass of fungal hyphae that was invading the airways of his lung and slowly killing him. Such bronchopulmonary aspergillosis is a rare but increasingly frequent pathogen of the immunocompromised. Besides experiencing difficulty in breathing, fever, and chest pain, Matt most hated coughing up wads of bloody mucus—as if he were expelling the very fabric of his life. In fact, he literally was coughing up pieces of his lung.

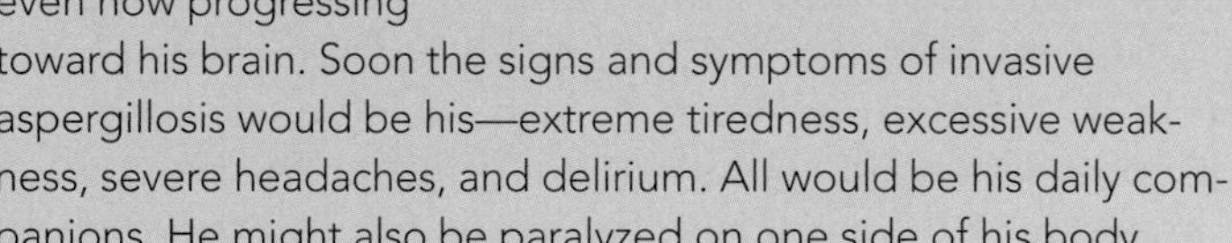

Unfortunately for Matt, the disease had not yet peaked. *Aspergillus* had invaded his blood and was even now progressing toward his brain. Soon the signs and symptoms of invasive aspergillosis would be his—extreme tiredness, excessive weakness, severe headaches, and delirium. All would be his daily companions. He might also be paralyzed on one side of his body.

Matt spent four weeks in hospital. The worst days were those when he was aware. The sheer terror of knowing his brain was being pierced by thin hyphal threads, digesting his personality away, was almost more than Matt could stand. He looked forward to the times he would lapse into unconsciousness, even though he knew that each period of wakefulness might be his last.

A medical miracle in the form of a new antifungal drug (voriconazole) brought Matt back to life. The invasive mold was defeated, and Matt returned home. Grateful. Aware that life is precious. And hopeful that no more spores floated his way. For more about aspergillosis, see p. 637.

Track aspergillosis online by going to the Study Area at www.masteringmicrobiology.com.

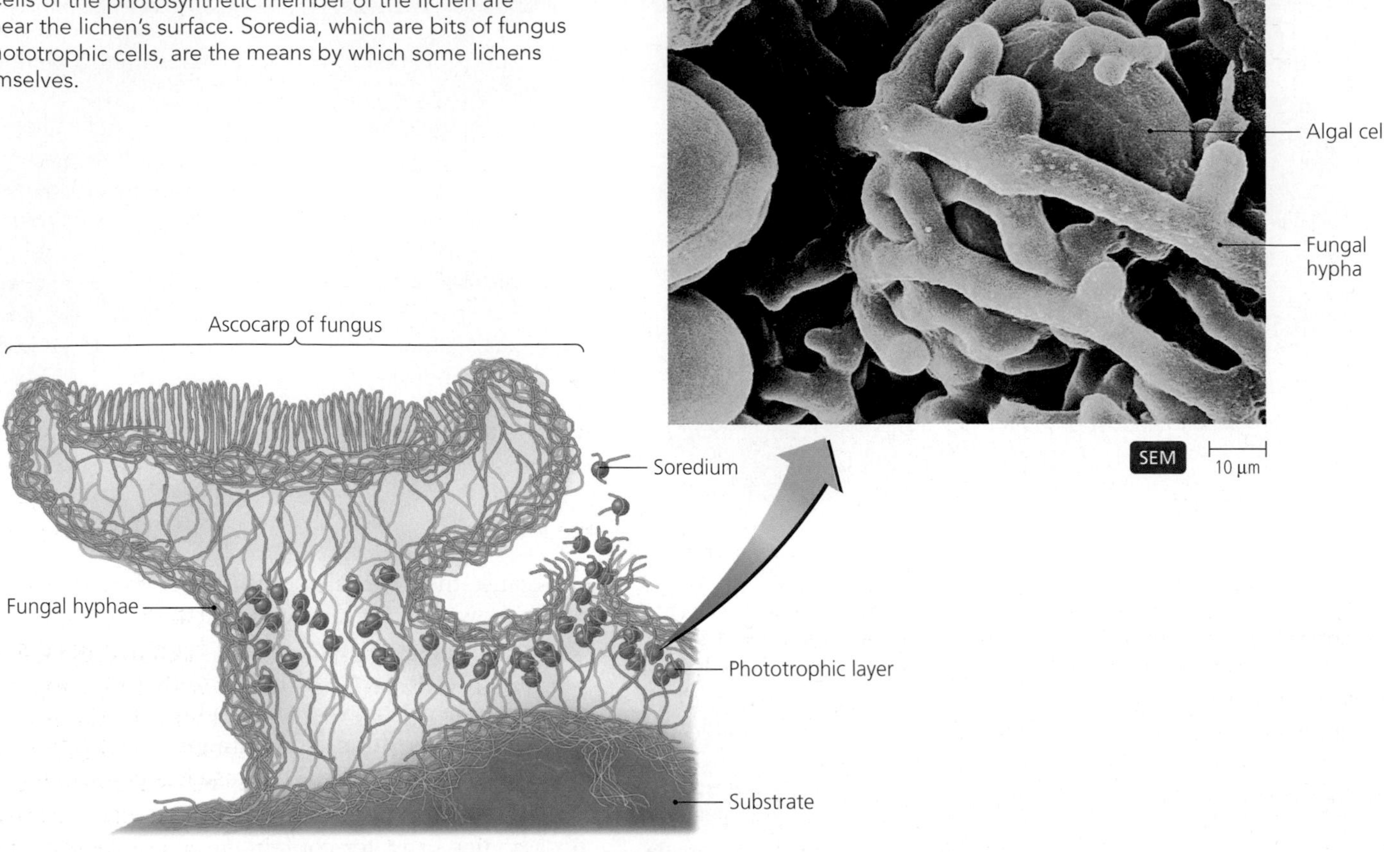

Figure 12.25 Makeup of a lichen. The hyphae of a fungus (most commonly an ascomycete) constitute the major portion of the thallus of the lichen. Cells of the photosynthetic member of the lichen are concentrated near the lichen's surface. Soredia, which are bits of fungus surrounding phototrophic cells, are the means by which some lichens propagate themselves.

Lichens

A discussion of fungi is incomplete without considering **lichens,** which are partnerships between fungi and photosynthetic microbes—commonly cyanobacteria or, less frequently, green algae. In a lichen, the hyphae of the fungus, which is usually an ascomycete, surround the photosynthetic cells **(Figure 12.25)** and provide them nutrients, water, and protection from desiccation and harsh light. In return, each alga or cyanobacterium provides the fungus with products of photosynthesis—carbohydrates and oxygen. In some lichens, the phototroph releases 60% of its carbohydrates to the fungus.

The partnership in a lichen is not always mutually beneficial; in some lichens, the fungus produces haustoria that penetrate and kill the photosynthetic member. Such lichens are maintained only because the phototroph's cells reproduce faster than the fungus can devour them.

The fungus of a lichen reproduces by spores, which must germinate and develop into hyphae that capture an appropriate cyanobacterium or alga. Alternatively, wind, rain, and small animals disperse bits of lichen called *soredia,* which contain both phototrophs and fungal hyphae, to new locations where they can establish a new lichen if there is suitable substrate.

Scientists have identified over 14,000 species of lichens, which are abundant throughout the world, particularly in pristine unpolluted habitats, growing on soil, rocks, leaves, tree bark, other lichens, and even the backs of tortoises. Indeed, lichens grow in almost every habitat—from high-elevation alpine tundra to submerged rocks on the oceans' shores, from frozen Antarctic soil to hot desert climes. The only unpolluted places where lichens do not consistently grow are in the dark depths of the oceans and the black world of caves—after all, lichens require light.

Lichens grow slowly but they can live for hundreds and possibly thousands of years. They occur in three basic shapes **(Figure 12.26)**:

- *Fruticose* lichens are either erect or hanging cylinders.
- *Crustose* lichens grow appressed to their substrates and may extend into the substrate for several millimeters.
- *Foliose* lichens are leaflike, with margins that grow free from the substrate.

Lichens create soil from weathered rocks, and lichens containing nitrogen-fixing cyanobacteria provide nitrogen to nutrient-poor environments. Many animals eat lichens; for example, reindeer and caribou subsist primarily on lichens throughout the winter. Birds use lichens for nesting materials, and some insects camouflage themselves with bits of living lichen. Humans also use lichens in the production of foods, dyes, clothing, perfumes, medicines, and the litmus of indicator paper. Because lichens will not grow well in polluted environments, ecologists use them as sensitive living assays for monitoring air pollution.

In the preceding sections, we have seen that fungi are chemoheterotrophic yeasts and molds that function primarily to

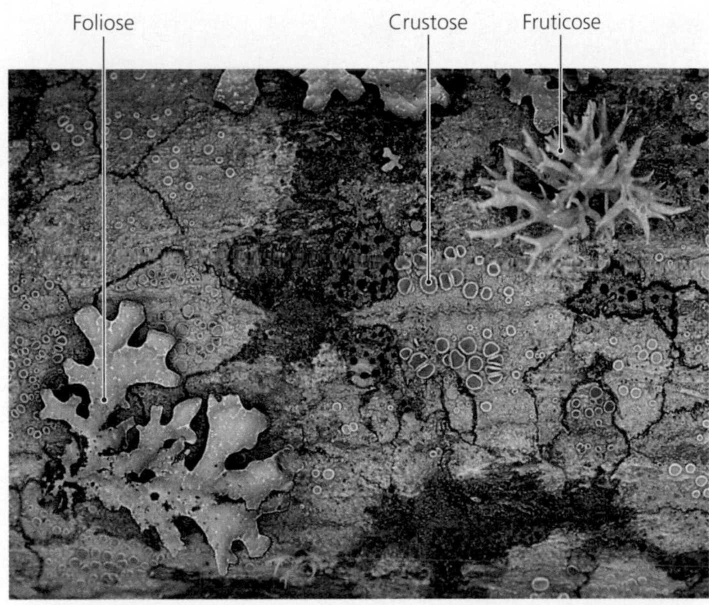

▲ **Figure 12.26 Gross morphology of lichens.** Crustose forms are flat and tightly joined to the substrate, here a branch; foliose forms are leaflike with free margins; and fruticose forms are either erect (as shown) or pendulant.

decompose and degrade dead organisms. Some fungi are pathogenic, and others associate with cyanobacteria or algae to form lichens. All fungi reproduce asexually either by budding or via asexual spores, and most fungi also produce sexual spores, by which taxonomists classify them. Table 12.3 summarizes the characteristics of fungi.

CRITICAL **THINKING**

Design a key for the genera of fungi discussed in this section.

TABLE 12.3 **Characteristics of Fungi**

Division and Type of Sexual Spore	Comments	Representative Genera
Zygomycota Zygospores	Coenocytic (aseptate)	*Rhizopus*
Ascomycota Ascospores	Septate; some associated with cyanobacteria or green algae to form lichens	*Claviceps, Neurospora, Penicillium, Saccharomyces, Tuber*
Basidiomycota Basidiospores	Septate	*Agaricus, Amanita, Cryptococcus*

Algae

Learning Objective

✓ Describe the distinguishing characteristic of algae.

The Romans used the word *alga* (al'ga) to refer to any simple aquatic plant, particularly one found in marine habitats. Their usage thus included the organisms we recognize as algae (al'jē), cyanobacteria, sea grasses, and other aquatic plants. Today, the word *algae* properly refers to simple, eukaryotic, phototrophic organisms that, like plants, carry out oxygenic photosynthesis using chlorophyll *a*. Algae differ from plants such as sea grass in having sexual reproductive structures in which every cell becomes a gamete. In plants, by contrast, a portion of the reproductive structure always remains vegetative.

Algae are not a unified group; rather, they differ widely in distribution, morphology, reproduction, and biochemical traits. Moreover, the word *algae* is not synonymous with any taxon; in the taxonomic scheme shown in Figure 12.4c, algae can be found in kingdoms Alveolata, Euglenozoa, Stramenopila, Rhodophyta, and Plantae. The study of algae is *phycology*,[30] and the scientists that study them are called *phycologists*.

Distribution of Algae

Even though some algae grow in such diverse habitats as in soil and ice, in intimate association with fungi as lichens, and on plants, most algae are aquatic, living in the *photic zone* (penetrated by sunlight) of fresh, brackish, and salt bodies of water. This watery environment provides some benefits and also presents some difficulties for photosynthetic organisms. Whereas most bodies of water contain sufficient dissolved chemicals to provide nutrients for algae, water also differentially absorbs longer wavelengths of light (including red light), so only shorter (blue) wavelengths penetrate more than a meter below the surface. This is problematic for algae because their primary photosynthetic pigment—chlorophyll *a*—captures red light. Thus to grow in deeper waters, algae must have *accessory photosynthetic pigments* that trap the energy of penetrating, short-wavelength light and pass that energy to chlorophyll *a*. Members of the group of algae known as red algae, for example, contain a red pigment that absorbs blue light, enabling red algae to inhabit even the deepest parts of the photic zone.

Morphology of Algae

Algae can be unicellular or colonial, or they can have simple multicellular bodies called thalli, which are commonly composed of branched filaments or sheets. The thalli of large marine algae, commonly called seaweeds, can be relatively complex, with branched *holdfasts* to anchor them to rocks, stemlike *stipes*, and leaflike *blades*. The thalli of many of the larger marine algae are buoyed in the water by gas-filled bulbs called *pneumocysts* (see Figure 12.30). Though the thalli of some marine algae can surpass land plants in length, they lack the well-developed transport systems common to vascular plants.

[30]From Greek *phykos*, meaning seaweed, and *logos*, meaning discourse.

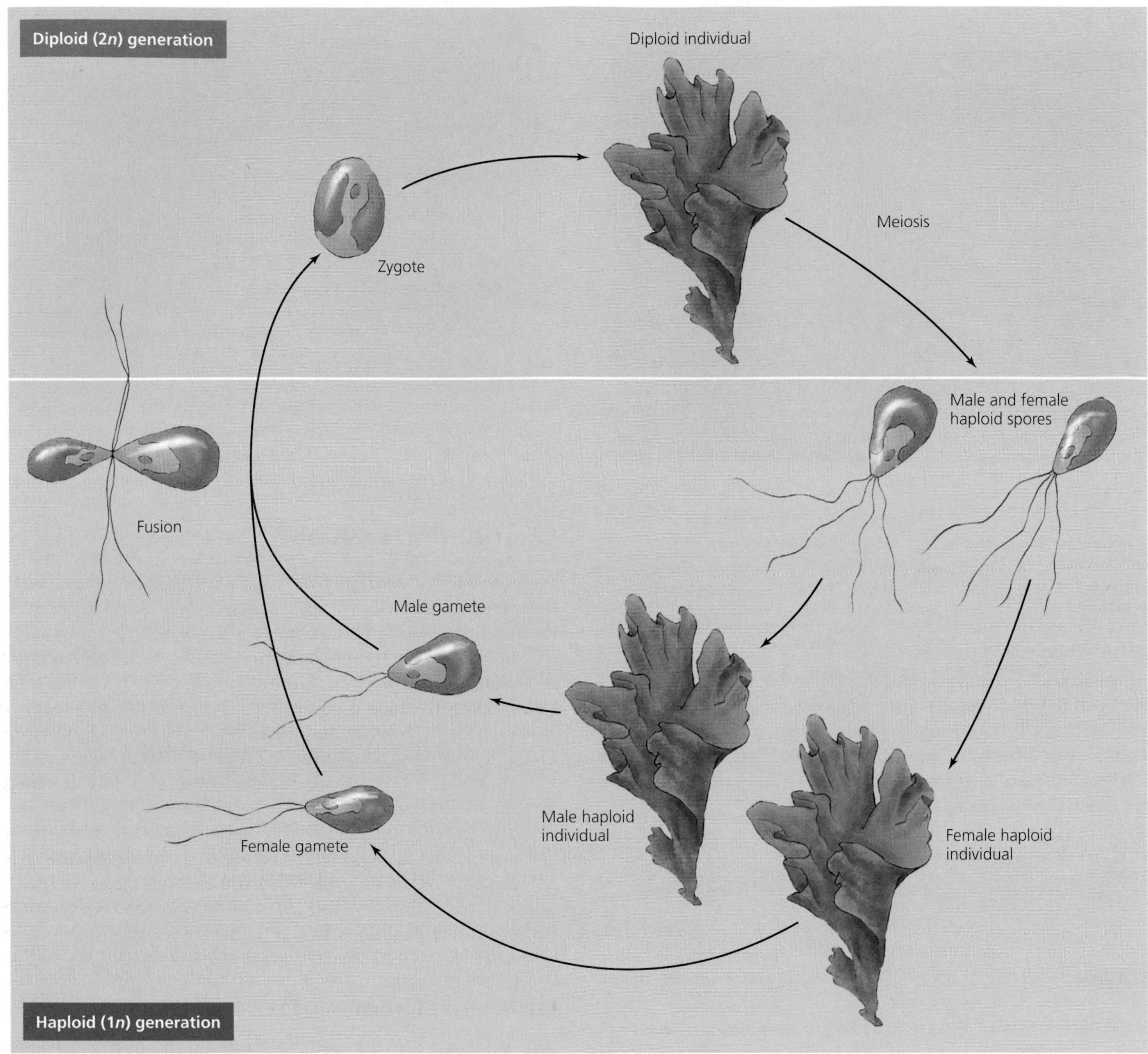

▲ **Figure 12.27 Alternation of generations in algae, as occurs in the green alga *Ulva*.** A diploid individual meiotically produces haploid spores that germinate and grow into male and female thalli. These haploid algae produce gametes that fuse to form a diploid zygote, which grows into a new diploid individual. Each generation can also reproduce asexually by fragmentation and by spores.

Reproduction of Algae

Learning Objective

✓ Describe the alternation of generations in algae.

In unicellular algae, asexual reproduction involves mitosis followed by cytokinesis. In unicellular algae that reproduce sexually, each algal cell acts as a gamete and fuses with another such gamete to form a zygote, which then undergoes meiosis to return to the haploid state.

Multicellular algae may reproduce asexually by fragmentation, in which each piece of a parent alga develops into a new individual, or by motile or nonmotile asexual spores. As noted previously, in multicellular algae that reproduce sexually, every cell in the reproductive structures of the alga becomes a gamete—a feature that distinguishes algae from all other photosynthetic eukaryotes.

Many multicellular algae reproduce sexually with an **alternation of generations** of haploid and diploid individuals **(Figure 12.27)**. In such life cycles, diploid individuals undergo meiosis to produce male and female haploid spores that develop into haploid male and female thalli, which may look identical to the diploid thallus. In some algae, each of these thalli produces gametes that fuse to form a zygote, which grows into a new diploid thallus. Both haploid and diploid thalli may reproduce asexually as well.

Classification of Algae

Learning Objectives

- ✓ List four groups of algae and describe the distinguishing characteristics of each.
- ✓ List several economic benefits derived from algae.

The classification of algae is not yet settled. Historically, taxonomists have used differences in photosynthetic pigments, storage products, and cell wall composition to classify algae into several groups that are named for the colors of their photosynthetic pigments: green algae, red algae, brown algae, golden algae, and yellow-green algae. The following sections present some of these groups. We begin with the green algae of the division Chlorophyta.

Division Chlorophyta (Green Algae)

Chlorophyta[31] are green algae that share numerous characteristics with plants—they have chlorophylls *a* and *b,* and use sugar and starch as food reserves. Many have cell walls composed of cellulose, while others have walls of glycoprotein or lack walls entirely. In addition, the 18S rRNA sequences of green algae and plants are comparable. Because of the similarities, green algae are often considered to be the progenitors of plants, and in some taxonomic schemes the Chlorophyta are placed in the kingdom Plantae.

Most green algae are unicellular or filamentous (see Figure 1.6a) and live in freshwater ponds, lakes, and pools, where they form green to yellow scum. Some multicellular forms grow in the marine intertidal zone—that is, in the region exposed to air during low tide.

Prototheca (prō-tō-thē'kă) is an unusual green alga in that it lacks pigments, making it colorless. This chemoheterotrophic alga causes a skin rash in sensitive individuals. *Codium* (kō'dē-ŭm) is a member of a group of marine green algae that do not form cross walls after mitosis; thus, its entire thallus is a single, large, multinucleate cell. Some Polynesians dry and grind *Codium* for use as seasoning pepper. The green alga *Trebouxia* (tre-book'sē-a) is the most common alga found in association with fungi in lichens.

CRITICAL THINKING

Since *Prototheca* is colorless, how do scientists know that it is really a green alga?

Kingdom Rhodophyta (Red Algae)

Algae of division **Rhodophyta,**[32] which had been placed historically in kingdom Plantae and then Protista, are now in their own kingdom—Rhodophyta. They are characterized by the red accessory pigment **phycoerythrin;** the storage molecule *glycogen* (also known as *floridean*[33] *starch*); cell walls of **agar** or **carrageenan** (kar-ă-gē'nan), sometimes supplemented with calcium carbonate; and nonmotile male gametes called *spermatia.* Phycoerythrin allows red algae to absorb short-wavelength blue light and photosynthesize at depths greater than 100 m. Because the relative proportions of phycoerythrin and chlorophyll *a* vary, red algae range in color from green to black in the intertidal zone to red in deeper water **(Figure 12.28)**. Most red algae are marine, though a few freshwater genera are known.

▲ **Figure 12.28** ***Antithamnion*, a red alga.**

The gel-like polysaccharides agar and carrageenan, once they have been isolated from red algae such as *Gelidium* (jel-li'dē-ŭm) and *Chondrus* (kon'drŭs), are used as thickening agents for the production of solid microbiological media and numerous consumer products, including ice cream, toothpaste, syrup, salad dressings, and snack foods.

Phaeophyta (Brown Algae)

The **Phaeophyta**[34] are in kingdom Stramenopila[35] based in large part on their gametes being motile by means of two flagella—one "hairy" and one whiplike **(Figure 12.29)**. They have chlorophylls *a* and *c,* carotene, and brown pigments called *xanthophylls* (zan'thō-fils). Depending on the relative amounts of these pigments, brown algae may appear dark brown, tan, yellow-brown, greenish brown, or green. Most brown algae are marine organisms, and some of the giant

[31]From Greek *chloros,* meaning green, and *phyton,* meaning plant.
[32]From Greek *rhodon,* meaning rose.
[33]Named for a taxon of red algae, Florideophycidae.
[34]From Greek *phaeo,* meaning brown.
[35]From Latin *stramen,* meaning straw, and *pilos,* meaning hair.

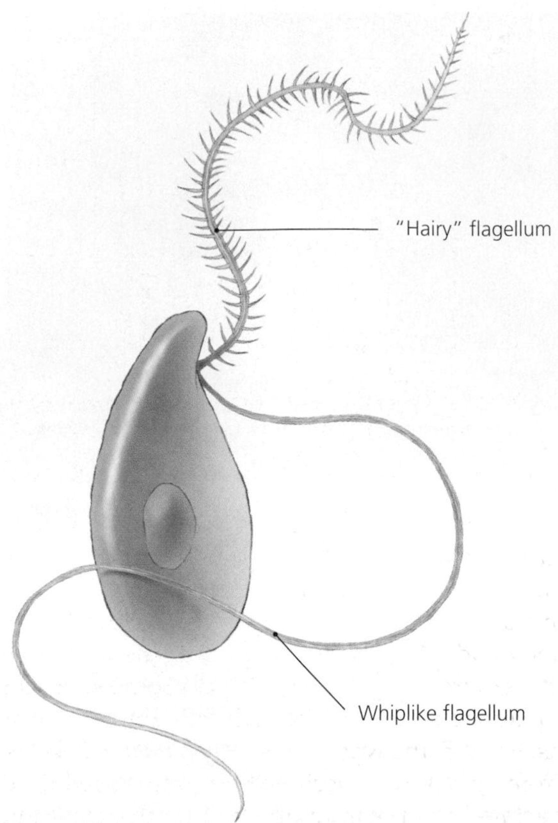

▲ **Figure 12.29 The two types of flagella of the sperm of the brown alga *Fucus*.**

kelps, such as *Macrocystis* **(Figure 12.30)**, rival the tallest trees in length, though not in girth.

Brown algae use the polysaccharide *laminarin* and oils as food reserves and have cell walls composed of cellulose and **alginic acid** (alginate). Alginic acid is used in numerous foods as a thickening agent and emulsifier.

▲ **Figure 12.30 The giant kelp *Macrocystis*, a brown alga.** A kelp's blades are kept afloat by pneumocysts.

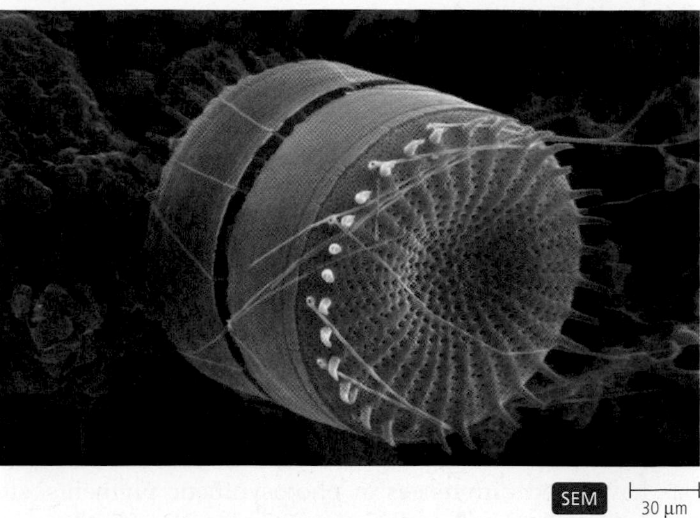

▲ **Figure 12.31 *Stephanodiscus*, a diatom.** Diatoms have frustules, composed of silica and cellulose, that fit together like a Petri dish.

Chrysophyta (Golden Algae, Yellow-Green Algae, and Diatoms)

Chrysophyta[36] is a group of algae that are diverse with respect to cell wall composition and pigments. They are unified in using the polysaccharide *chrysolaminarin* as a storage product. Some additionally store oils. Modern taxonomists group these algae with brown algae and water molds (discussed shortly) in the kingdom Stramenopila based on similarities in nucleotide sequences and flagellar structure. Whereas some chrysophytes lack cell walls, others have ornate external coverings such as scales or plates. One taxon of chrysophytes, Bacillariophyceae—the **diatoms** (dī′ă-tomz)—are unique in having silica cell walls composed of two halves called *frustules* that fit together like a Petri dish **(Figure 12.31)**.

Most chrysophytes are unicellular or colonial. All chrysophytes contain more orange-colored *carotene* pigment than they do chlorophylls *a* and *c*, which accounts for the common names of two major classes of chrysophytes—*golden algae* and *yellow-green algae.*

Diatoms are a major component of marine *phytoplankton:* free-floating photosynthetic microorganisms that form the basis of food chains in the oceans. Further, because of their enormous number, diatoms are the major source of the world's oxygen. The silica frustules of diatoms contain minute holes for the exchange of gases, nutrients, and wastes with the environment. Organic gardeners use *diatomaceous earth,* composed of innumerable frustules of dead diatoms, as a pesticide against harmful insects and worms. Diatomaceous earth is also used in polishing compounds, detergents, paint removers, and as a component of firebrick, soundproofing products, swimming pool filters, and reflective paints.

In summary, algae are unicellular or multicellular photoautotrophs characterized by sexual reproductive structures in which every cell becomes a gamete. The colors produced by the combi-

[36]From Greek *chrysos*, meaning gold.

nation of their primary and accessory photosynthetic pigments give them their common names and provide the basis of at least one classification scheme. Table 12.4 summarizes the characteristics of the major groups of algae; the dinoflagellates and euglenids are included in the table because botanists historically classified these phototrophic protozoa as algae.

CRITICAL **THINKING**

Design a dichotomous key for the genera of algae discussed in this section.

Water Molds

Learning Objective

✓ List four ways in which water molds differ from true fungi.

Scientists once classified the microbes commonly known as **water molds** as fungi because they resemble filamentous fungi in having finely branched filaments; however, water molds are not true molds; they are not fungi. Water molds differ from fungi in the following ways:

- They have tubular cristae in their mitochondria.
- They have cell walls of cellulose instead of chitin.
- Their spores have two flagella—one whiplike and one "hairy."
- They have true diploid bodies rather than haploid bodies.

▲ **Figure 12.32** **An example of the important role of water molds in recycling organic nutrients in aquatic habitats.**

Because water molds have "hairy" flagella and certain similarities in rRNA sequence to sequences of diatoms, other chrysophytes, and brown algae, taxonomists classify all these organisms in kingdom Stramenopila (see Figure 12.4c).

Water molds decompose dead animals and return nutrients to the environment (Figure 12.32). Some species are detrimental pathogens of crops such as grapes, tobacco, and soybeans. In 1845, the water mold *Phytophthora infestans* (fī-tof'-tho-ră in-fes'tanz) was accidentally introduced into Ireland and devastated the potato crop, causing the great famine that killed over 1 million people and forced a greater number to immigrate to the United States and Canada.

TABLE 12.4 **Characteristics of Various Algae**

Group (Common Name)	Kingdom	Pigments	Storage Product(s)	Cell Wall Component(s)	Habitat	Representative Genera
Chlorophyta (green algae)	Plantae	Chlorophylls *a* and *b*, carotene, xanthophylls	Sugar, starch	Cellulose or glycoprotein; absent in some	Fresh, brackish, and saltwater; terrestrial	*Spirogyra* *Prototheca* *Codium* *Trebouxia*
Rhodophyta (red algae)	Rhodophyta	Chlorophyll *a*, phycoerythrin, phycocyanin, xanthophylls	Glycogen (floridean starch)	Agar or carrageenan, some with calcium carbonate	Mostly saltwater	*Chondrus* *Gelidium* *Antithamnion*
Chrysophyta (golden algae, yellow-green algae, diatoms)	Stramenopila	Chlorophylls *a*, c_1 and c_2; carotene, xanthophylls	Chrysolaminarin, oils	Cellulose, silica, calcium carbonate	Fresh, brackish, and saltwater; terrestrial; ice	*Stephanodiscus*
Phaeophyta (brown algae)	Stramenopila	Chlorophylls *a* and *c*, carotene, xanthophylls	Laminarin, oils	Cellulose and alginic acid	Brackish and saltwater	*Macrocystis*
Pyrrhophyta (dinoflagellates)	Alveolata	Chlorophylls *a*, c_1, c_2; carotene	Starch, oils	Cellulose	Fresh, brackish, and saltwater	*Gymnodinium* *Gonyaulax* *Pfiesteria*
Euglenophyta (euglenids)	Euglenozoa	Chlorophylls *a* and *b*, carotene	Paramylon, oils, sugar	Absent	Fresh, brackish, and saltwater; terrestrial	*Euglena*

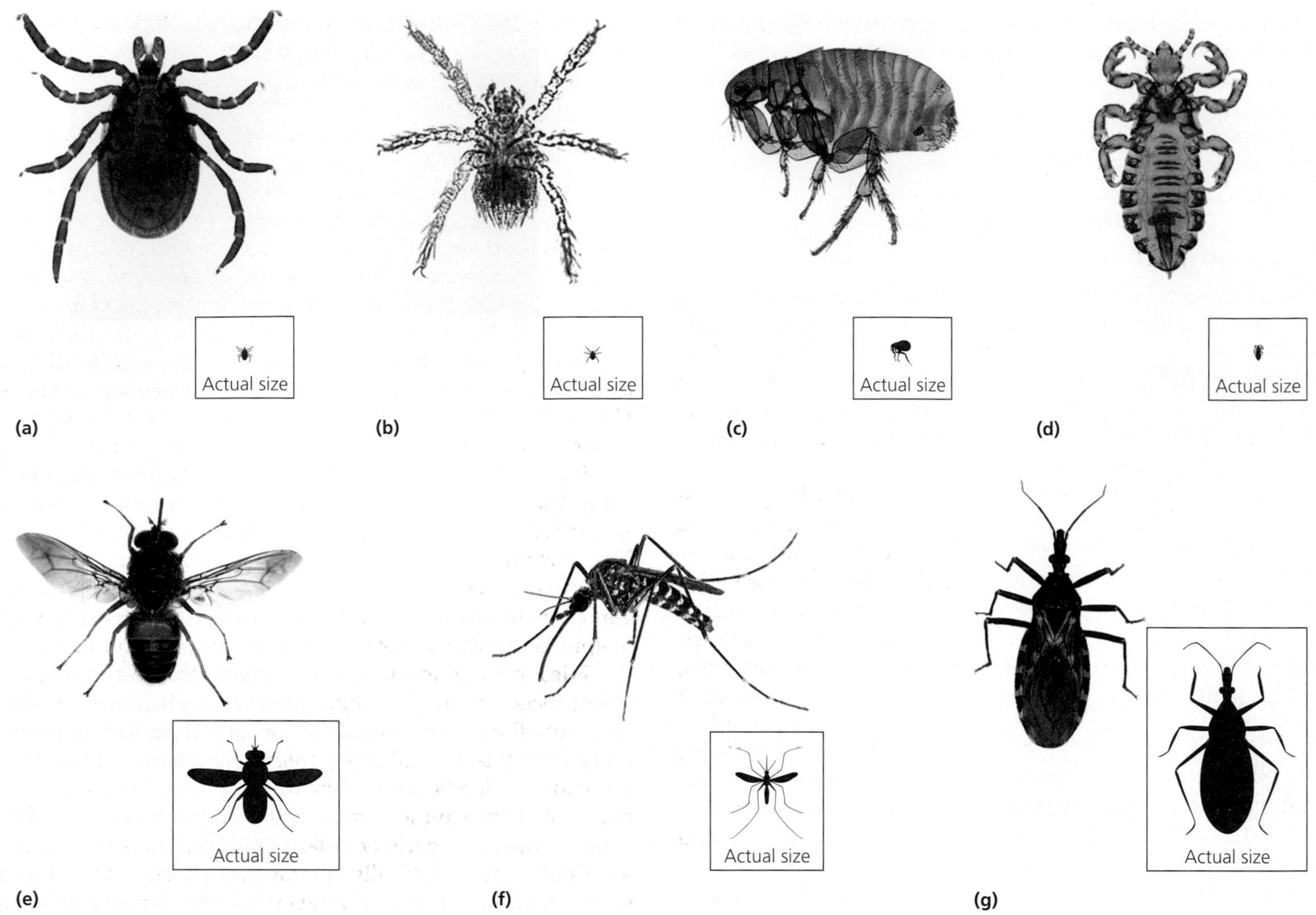

▲ **Figure 12.33 Representative arthropod vectors.** Arachnid vectors include **(a)** ticks and **(b)** mites; insect vectors include **(c)** fleas, **(d)** lice, **(e)** true flies, such as this tsetse fly, **(f)** mosquitoes (a type of fly), and **(g)** true bugs.

Other Eukaryotes of Microbiological Interest: Parasitic Helminths and Vectors

Learning Objectives

- ✓ Explain why microbiologists study large organisms such as parasitic worms.
- ✓ Discuss the inclusion of vectors in a study of microbiology.

Microbiologists are interested also in two other groups of eukaryotes, although they are not microorganisms. The first group are the parasitic *helminths,* commonly called parasitic worms. Microbiologists became interested in parasitic helminths because they observed the microscopic infective and diagnostic stages of the helminths—usually eggs or larvae (immature forms)—in samples of blood, feces, and urine. Thus microbiologists study parasitic helminths in part because they must distinguish the parasites' microscopic forms from other microbes. Chapter 14 discusses parasitism and other relationships that exist among microbes and other organisms.

Microbiologists are also interested in **arthropod vectors**[37]—animals that carry pathogens and have segmented bodies, hard external skeletons, and jointed legs. Some arthropods are *mechanical vectors,* meaning they merely carry pathogens; others are *biological vectors,* meaning they also serve as hosts for microbial pathogens. Given that arthropods are small organisms (so small that we generally don't notice them until they bite us), and given that they produce numerous offspring, controlling arthropod vectors to eliminate their role in the transmission of important human diseases is an almost insurmountable task.

Disease vectors belong to two classes of arthropods: *Arachnida* and *Insecta.* Ticks and mites are arachnoid vectors. (Though spiders are also arachnids, they do not transmit diseases.) Insects account for the greatest number of vectors, and within this group are fleas, lice, flies (such as tsetse flies and mosquitoes), and true bugs (such as kissing bugs). **Figure 12.33** shows representatives of major types of arthropod vectors. Most vectors are found on a host only when they are actively feeding.

[37]From Latin *vectus,* meaning carried.

Lice are the only arthropods that may spend their entire lives in association with a single individual host.

Arachnids

Learning Objectives

- ✓ Describe the distinctive features of arachnids.
- ✓ List five diseases vectored by ticks and two diseases vectored by mites.

All adult **arachnids** (ă-rak′nidz) have four pairs of legs. **Ticks** and **mites** (commonly known as chiggers) both go through a six-leg stage when they are juveniles, but they display the characteristic eight legs as adults. Ticks and mites resemble each other morphologically, having disk-shaped bodies. Ticks are roughly the size of a small rice grain, whereas mites are usually the size of sand grains.

Ticks are the most important arachnid vectors. They are distributed worldwide and serve as vectors for bacterial, viral, and protozoan diseases. Ticks are second only to mosquitoes in the number of diseases that they transmit. Hard ticks—those with a hard plate on their dorsal surfaces—are the most prominent tick vectors. They wait on stalks of grasses and brush for their hosts to come by. When a human, for example, walks past them, the ticks leap onto the person and begin searching for exposed skin. They use their mouthparts to cut holes in the skin and attach themselves with a gluelike compound to prevent being dislodged. As they feed on blood, their bodies swell to several times normal size. Some tick-borne diseases are Lyme disease, Rocky Mountain spotted fever, tularemia, relapsing fever, and tick-borne encephalitis.

Parasitic mites of humans live around the world, wherever humans and animals coexist. A few mite species transmit rickettsial diseases (rickettsial pox and scrub typhus) among animals and humans.

Insects

Learning Objectives

- ✓ Describe the general physical features of insects and the features specific to the various groups of insect vectors.
- ✓ List diseases transmitted by fleas, lice, true flies, mosquitoes, and kissing bugs.

As adults, all **insects** have three pairs of legs and three body regions—head, thorax (chest), and abdomen. Adult insects are, however, far from uniform in appearance. Some have two wings, others have four wings, and others are wingless; some have long legs, some short; and some have biting mouthparts, whereas other have sucking mouthparts. Many insects have larval stages that look very different from the adults, complicating their identification.

The fact that many insects can fly has epidemiological implications. Flying insects have broader home ranges than nonflying insects, and some migrate, making control difficult.

Fleas are small, vertically flattened, wingless arthropods that are found worldwide, though some species have geographically limited ranges. Most are found in association with wild rodents, bats, and birds and are not encountered by humans. A few species, however, feed on humans. Cat and dog fleas are usually just pests (to the animal and its owner), but they can also serve as the intermediate host for a dog tapeworm, *Dipylidium* (dip-ĭ-lid′ē-ŭm). The most significant microbial disease transmitted by fleas is plague, carried by rat fleas.

Lice (singular: *louse*) are parasites that can also transmit disease. They are horizontally flat, soft-bodied, wingless insects found worldwide among humans and their habitations. They live in clothing and bedding and move onto humans to feed. Lice are most common among the poor and those living in severely overcrowded communities. Lice are the vectors involved in epidemic outbreaks of typhus in developing countries.

Flies are among the more common insects, and many different species are found around the world. Flies differ greatly in size, but all have at least two wings and fairly well-developed body segments. Not all flies transmit disease, but those that do are usually bloodsuckers. Female sand flies (*Phlebotomus*, flebot′ō-mŭs) transmit leishmaniasis in North Africa, the Middle East, Europe, and parts of Asia. Tsetse flies (*Glossina*, glo-sī′nă) are limited geographically to tropical Africa, where they are found in brushy areas and transmit African sleeping sickness.

Mosquitoes are a type of fly, though they are morphologically distinct from other fly species. Female mosquitoes are thin and have wings, elongated bodies, long antennae, long legs, and a long proboscis for feeding on blood. Male mosquitoes do not feed on blood. Mosquitoes are found throughout the world, but particular species are geographically limited. Mosquitoes are the most important arthropod vectors of diseases, and they carry the pathogens that cause malaria, yellow fever, dengue fever, filariasis, viral encephalitis, and Rift Valley fever.

Kissing bugs are relatively large, winged, true bugs with cone-shaped heads and wide abdomens. They are called kissing bugs because of their tendency to take blood meals near the mouths of their hosts. Both sexes feed nocturnally while their victims sleep. Kissing bugs transmit disease in Central and South America. The most important disease they transmit is Chagas′ disease.

Chapter Summary

General Characteristics of Eukaryotic Organisms (pp. 340–345)

1. A typical eukaryotic nucleus may be **haploid** (having a single copy of each chromosome) or **diploid** (having two copies). It divides by **mitosis** in four phases—**prophase, metaphase, anaphase,** and **telophase**—resulting in two nuclei with the same ploidy as the original.
2. **Meiosis** is nuclear division that results in four nuclei, each with half the ploidy of the original.

3. A cell's cytoplasm divides by **cytokinesis** either during or after nuclear division.
4. **Coenocytes** are multinucleate cells resulting from repeated mitoses but postponed or no cytokinesis.
5. Some microbes undergo multiple mitoses by **schizogony** to form a multinucleate **schizont,** which then undergoes cytokinesis.
6. The classification of eukaryotic microbes is problematic and has changed frequently. Historical schemes based on similarities in morphology and chemistry have been replaced with schemes based on nucleotide sequences and ultrastructural features.

Protozoa (pp. 345–353)

1. **Protozoa** (studied by protozoologists) are eukaryotic, unicellular organisms that lack cell walls. Most of them are chemoheterotrophs.
2. A motile **trophozoite** is the feeding stage of a typical protozoan. A **cyst,** a resting stage that is resilient to environmental changes, is formed by some protozoa.
3. A few protozoa undergo sexual reproduction by forming **gametocytes** that fuse to form a **zygote.**
4. Protozoa may be classified into seven groups: parabasalids, alveolates, cercozoa, radiolaria, amoebozoa, euglenozoa, and diplomonads.
5. Parabasalids (e.g., *Trichomonas*) are characterized by a Golgi body–like structure called a parabasal body.
6. Alveolates, with cavities called alveoli beneath their cell surfaces, include **ciliate** alveolates (characterized by cilia), **apicomplexans** (all are pathogenic), and **dinoflagellates** (responsible for **red tides**).
7. Protozoa that move and feed with pseudopodia are **amoebae.** They make up three taxa. Cercozoa, such as **foraminifera,** have threadlike pseudopodia and calcium carbonate shells. Members of Radiolaria have threadlike pseudopodia but silica shells. Amoebozoa have lobe-shaped pseudopodia. The latter include free-living amoebae, parasitic amoebae, and slime molds.
8. **Slime molds** lack cell walls and are phagocytic in their nutrition. They are amoebozoa.
9. **Plasmodial** (acellular) **slime molds** are composed of multinucleate cytoplasm. **Cellular slime molds** are composed of myxamoebae that phagocytize bacteria and yeasts.
10. Unicellular flagellated **euglenids** are euglenozoa that store food as paramylon, lack cell walls, and have eyespots used in positive phototaxis. Because they exhibit characteristics of both animals and plants, they are a taxonomic problem.
11. A **kinetoplastid** is a euglenozoan with a single, large, apical mitochondrion that contains a kinetoplast, which is a region of DNA.
12. Members of the Diplomonadida lack mitochondria, Golgi bodies, and peroxisomes.

Fungi (pp. 353–363)

1. **Fungi** (studied by mycologists) are chemoheterotrophic eukaryotes with cell walls usually composed of **chitin.**
2. Most fungi are beneficial, but some produce **mycoses** (fungal diseases).
3. The nonreproductive body of a filamentous fungus (mold) or yeast (unicellular fungus) is a **thallus.** Mold thalli are composed of tubular filaments called **hyphae.**
4. Hyphae are described as either **septate** or **aseptate** depending on the presence of cross walls. A **mycelium** is a tangled mass of hyphae.
5. A **dimorphic** fungus has either type of thallus, depending on environmental conditions.
6. Most fungi are **saprobes**—they acquire nutrients by absorption from dead organisms; others get nutrients from living organisms using **haustoria** that penetrate host tissues.
7. Fungi reproduce asexually either by budding or via asexual spores, which are categorized according to their mode of development. Most fungi also reproduce sexually via sexual spores.
8. Most fungi in the division **Zygomycota** produce rough-walled **zygosporangia. Microsporidia** are intracellular parasites formerly classified as protozoa, but now classed with zygomycetes based on genetic analysis.
9. Fungi in the division **Ascomycota,** a group of economically important fungi, produce **ascospores** within sacs called **asci.**
10. Fungi of the division **Basidiomycota** have fruiting bodies called **basidiocarps** that include mushrooms, puffballs, and bracket fungi. Basidiocarps produce basidiospores at the ends of basidia.
11. Deuteromycota is an informal grouping of fungi having no known sexual stage.
12. **Lichens** are economically and environmentally important organisms composed of fungi living in partnership with photosynthetic microbes, either green algae or cyanobacteria.

Algae (pp. 363–367)

1. Algae (studied by phycologists) typically reproduce by an **alternation of generations** in which a haploid thallus alternates with a diploid thallus.
2. Large algae have multicellular thalli with stemlike stipes, leaflike blades, and holdfasts that attach them to substrates.
3. Division **Chlorophyta** contains green algae, which are metabolically similar to land plants.
4. **Rhodophyta,** red algae, contain the pigment **phycoerythrin,** the storage molecule floridean starch, and cell walls of **agar** or **carrageenan,** substances used as thickening agents.
5. **Phaeophyta,** brown algae, contain xanthophylls, laminarin, and oils. They have cell walls composed of cellulose and **alginic acid,** which is another thickening agent. A brown algal spore is motile by means of one "hairy" flagellum and one whiplike flagellum.
6. **Chrysophyta**—the golden algae, yellow-green algae, and **diatoms**—contain chrysolaminarin as a storage product. The silica cell walls of diatoms are arranged in nesting halves called frustules.

Water Molds (p. 367)

1. **Water molds** have tubular cristae in their mitochondria, cell walls of cellulose, spores having two different flagella, and diploid thalli. They are placed in the kingdom Stramenopila along with chrysophytes and brown algae.

Other Eukaryotes of Microbiological Interest: Parasitic Helminths and Vectors (pp. 368–369)

1. Parasitic helminths are significant to microbiologists in part because their infective stages are usually microscopic.
2. Also important to microbiologists are **arthropod vectors,** animals that carry and transmit pathogens. Mechanical vectors merely carry microbes; biological vectors also serve as microbial hosts.
3. **Ticks** and **mites** (chiggers) are **arachnids**—eight-legged arthropods.
4. **Insect** vectors include **fleas, lice,** bloodsucking **flies** including **mosquitoes,** and **kissing bugs.**

Questions for Review

Answers to the Questions for Review (except Short Answer questions) begin on page A-1.

Multiple Choice

1. Haploid nuclei
   a. contain one set of chromosomes.
   b. contain two sets of chromosomes.
   c. are found in zygotes.
   d. are found in the cytosol of eukaryotic organisms.
2. Which of the following sequences reflects the correct order of events in mitosis?
   a. telophase, anaphase, metaphase, prophase
   b. prophase, anaphase, metaphase, telophase
   c. telophase, prophase, metaphase, anaphase
   d. prophase, metaphase, anaphase, telophase
3. Which of the following statements accurately describes prophase?
   a. The cell appears to have a line of chromosomes across the midregion.
   b. The nuclear envelope becomes visible.
   c. The cell constructs microtubules to form a spindle.
   d. Chromatids separate and become known as chromosomes.
4. Multiple nuclear divisions without cytoplasmic divisions result in cells called
   a. mycoses.
   b. coenocytes.
   c. haustoria.
   d. pseudohypha.
5. Branched, tubular filaments with cross walls found in large fungi are
   a. septate hyphae.
   b. aseptate hyphae.
   c. aseptate haustoria.
   d. dimorphic mycelia.
6. The type of asexual fungal spore that forms within hyphae is called a
   a. sporangiospore.
   b. conidiospore
   c. blastospore.
   d. chlamydospore.
7. A phycologist studies which of the following?
   a. classification of eukaryotes
   b. alternation of generations in algae
   c. rusts, smuts, and yeasts
   d. parasitic worms
8. The stemlike portion of a seaweed is called its
   a. thallus.
   b. holdfast.
   c. stipe.
   d. blade.
9. Carrageenan is found in the cell walls of which group of algae?
   a. red algae
   b. green algae
   c. dinoflagellates
   d. yellow-green algae
10. Chrysolaminarin is a storage product found in which group of microbes?
    a. dinoflagellates
    b. euglenids
    c. golden algae
    d. brown algae
11. Which of the following features characterizes diatoms?
    a. laminarin and oils as food reserves
    b. protective plates of cellulose in their cells
    c. chlorophylls *a* and *c* and carotene
    d. paramylon as a food storage molecule
12. Water molds differ from true fungi in which of the following ways?
    a. They have tubular cristae in their mitochondria.
    b. Their cell walls are made of chitin instead of cellulose.
    c. Their spores are encased in silica shells.
    d. Their thalli are haploid rather than diploid.
13. The motile feeding stage of a protozoan is called
    a. an apicomplexan.
    b. a gametocyte.
    c. a cyst.
    d. a trophozoite.
14. Which of the following is common to mitosis and meiosis?
    a. spindle
    b. crossing over
    c. tetrad of chromatids
    d. cytokinesis
15. Which taxon is characterized by "hairy" flagella?
    a. Apicomplexa
    b. Euglenozoa
    c. Alveolata
    d. Stramenopila

Matching

1. ____	Mitosis	A. Cytoplasmic division
2. ____	Meiosis	B. Diploid nuclei producing haploid nuclei
3. ____	Homologous chromosomes	C. Results in genetic variation
4. ____	Crossing over	D. Carry similar genes
5. ____	Cytokinesis	E. Diploid nuclei producing diploid nuclei

1. ____	Chitin	A. Fungal cell wall component
2. ____	Basidiospore	B. Fungus + alga or bacterium
3. ____	Zygospore	C. Fungal body
4. ____	Thallus	D. Fungal spore formed in a sac
5. ____	Ascospore	E. Diploid fungal zygote with a thick wall
6. ____	Lichen	F. Fungal spore formed on club-shaped hypha

1. ____	Chlorophyta	A.	Foraminifera
2. ____	Rhodophyta	B.	Yellow-green algae
3. ____	Chrysophyta	C.	Green algae
4. ____	Phaeophyta	D.	Brown algae
5. ____	Cercozoa	E.	Red algae

Labeling

Label the photos below with the type of fungal spore and indicate whether the spore is asexual or sexual.

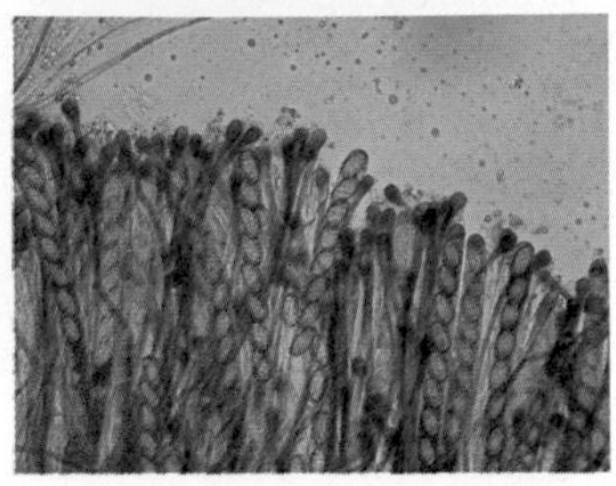

1. ______________________

2. ______________________

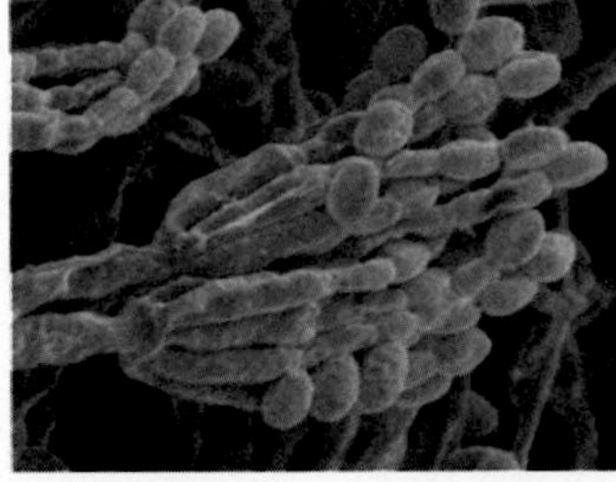

3. ______________________

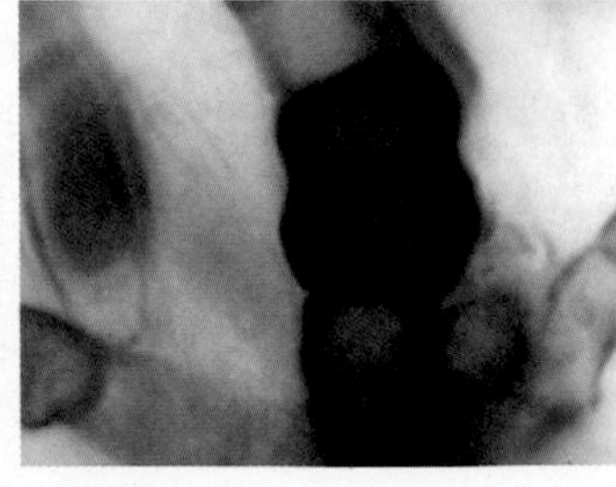

4. ______________________

Short Answer

1. Compare and contrast the following closely related terms:
   Chromatid and chromosome
   Mitosis and meiosis II
   Hypha and mycelium
   Algal thallus and fungal thallus
   Water mold and slime mold
2. How do fungi transport nutrients?
3. How are lichens useful in environmental protection studies?
4. What are the taxonomic challenges in classifying euglenids?
5. List several economic benefits of algae.
6. Why are relatively large animals such as parasitic worms studied in microbiology?
7. Why are microbiologists interested in macroscopic ticks, fleas, lice, and mosquitoes?
8. Name two ways that slime molds differ from true fungi.
9. What is the role of rRNA sequencing in the classification of eukaryotic microbes?
10. Describe the nuclear divisions that produce eight ascospores in an ascus.

Fill in the Blanks

1. The study of protozoa is called ______________.
2. The study of fungi is called ______________.
3. The study of algae is called ______________.
4. Fungal diseases are called ______________.
5. Amoebae with stiff pseudopodia and silica shells are ______________.

Concept Mapping

Using the following terms, draw a concept map that describes eukaryotic microorganisms. For a sample concept map, see p. 93. Or, complete this concept map online by going to the Study Area at www.masteringmicrobiology.com.

Algae	**Colonial**	**Mold**	**Protozoa**
Cell walls (3)	***Cryptosporidium***	**Multicellular (2)**	**Unicellular (3)**
Cellulose	**Fungi**	**Photosynthetic**	**Yeast**
Chitin	***Giardia***	***Plasmodium***	

Critical Thinking

1. How are cysts of protozoa similar to bacterial endospores? How are they different?
2. The host of a home improvement show suggests periodically emptying a package of yeast into a drain leading to a septic tank. Explain why this would be beneficial.
3. Why doesn't penicillin act against any of the pathogens discussed in this chapter?
4. How can one distinguish between a filamentous fungus and a colorless alga?

5. Why do scientists as a group spend more time and money studying protozoa than they do studying algae?
6. Why are there more antibacterial drugs than antifungal drugs?
7. Which type of metabolic pathways are present in protozoa that lack mitochondria (amoebae, diplomonads, and parabasalids)? Which metabolic pathways are absent?
8. Without reference to genetic sequences, mycologists are certain that none of the septate, filamentous deuteromycetes will be shown to make zygospores. How can they be certain when the sexual stages of deuteromycetes are still unknown?
9. Twenty years ago, *Pneumocystis jiroveci,* a pathogen of immunocompromised patients that causes pneumonia, was classified as a protozoan because it is a chemoheterotroph that lacks a cell wall. However, taxonomists today classify *Pneumocystis* as a fungus. Why do you think this pathogen has been reclassified as a fungus?
10. Explain why both dinoflagellates and euglenids were originally classified by zoologists as protozoa and by botanists as algae.

Access more review material online in the Study Area at **www.masteringmicrobiology.com**. There, you'll find

- **MP3 Tutor Sessions**
- **Concept Mapping Activities**
- **Flashcards**
- **Quizzes**

and more to help you succeed.

13 Characterizing and Classifying Viruses, Viroids, and Prions

In April 2007, the chief of a village in the Democratic Republic of the Congo (DRC) developed a severe fever, and then he bled to death over a few days. Over a hundred people who attended his funeral also contracted the same disease. Eight months later in a neighboring village, a mother delivered her baby but then both died of bloody diarrhea, and the nurse who cared for them was showing the same signs. Deadly Ebola hemorrhagic fever was on the prowl once again in the DRC.

Hemorrhagic fevers, smallpox, AIDS, SARS, influenza, common colds—many of the world's deadliest and most feared, as well as many common diseases are caused by viruses. However, not all viruses are harmful to humans. Some, such as bacteriophages, attack pathogens and have clinical use. This chapter is an introduction to viruses and other pathogenic particles called viroids and prions: how they infect cells, how they multiply, and how they differ from cellular pathogens.

Take the pre-test for this chapter online. Visit the Study Area at www.masteringmicrobiology.com.

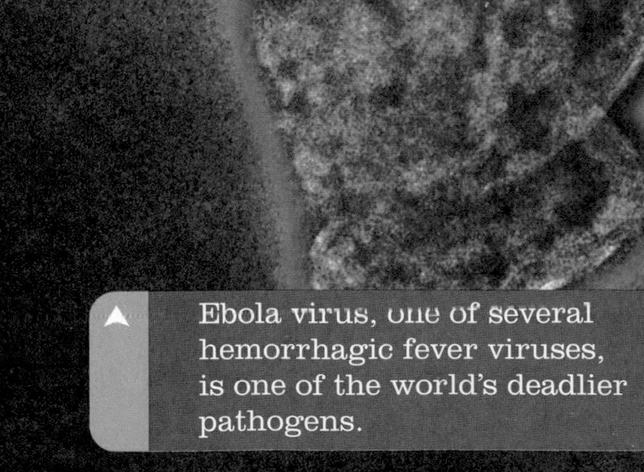

Ebola virus, one of several hemorrhagic fever viruses, is one of the world's deadlier pathogens.

Not all pathogens are cellular. Many infections of humans, animals, and plants, and even of bacteria, are caused by **acellular** (noncellular) agents, including viruses and other pathogenic particles called viroids and prions. Although these agents are like some eukaryotic and prokaryotic microbes in that they cause disease when they invade susceptible cells, they are simple compared to a cell—lacking cell membranes and composed of only a few organic molecules. In addition to lacking a cellular structure, they lack most of the characteristics of life described in Chapter 3: They cannot carry out any metabolic pathway, they can neither grow nor respond to the environment, and they cannot reproduce independently but instead must utilize the chemical and structural components of the cells they infect. They must recruit the cell's metabolic pathways in order to increase their numbers.

In this chapter we first examine a range of topics concerning viruses: what their characteristics are, how they are classified, how they replicate, what role they play in some kinds of cancers, how they are maintained in the laboratory, and whether or not viruses are alive. Then we consider the nature of viroids and prions.

Characteristics of Viruses

Viruses cause most of the diseases that still plague the industrialized world: the common cold, influenza, herpes, SARS, and AIDS, to name a few. Although we have immunizations against many viral diseases and are adept at treating the symptoms of others, the characteristics of viruses and the means by which they attack their hosts make cures for viral diseases elusive. Throughout this section, consider the clinical implications of viral characteristics.

We begin by looking at the characteristics viruses have in common. A **virus** is a minuscule, acellular, infectious agent having one or several pieces of nucleic acid—either DNA or RNA. The nucleic acid is the genetic material (genome) of the virus. Being acellular, viruses have no cytoplasmic membrane (though, as we will see, some viruses possess a membrane-like *envelope*). Viruses also lack cytosol and functional organelles. They are not capable of metabolic activity on their own; instead, once viruses have invaded a cell, they take control of the cell's metabolic machinery to produce more molecules of viral nucleic acid and viral proteins, which then assemble into new viruses via a process we will examine shortly.

Viruses have an extracellular and an intracellular state. Outside of a cell, in the extracellular state, a virus is called a **virion** (vir'ē-on). Basically, a virion consists of a protein coat, called a **capsid,** surrounding a nucleic acid core (**Figure 13.1a**). Together the nucleic acid and its capsid are also called a *nucleocapsid,* which in many cases can crystallize like crystalline chemicals (**Figure 13.1b**). Some virions have a phospholipid membrane called an **envelope** surrounding the nucleocapsid. The outermost layer of a virion (capsid or envelope) provides the virus both protection and recognition sites that bind to complementary chemicals on the surfaces of their specific host cells. Once a virus is inside, the intracellular state is initiated, and the capsid is removed. A virus without its capsid exists solely as nucleic acid, but is still referred to as a virus.

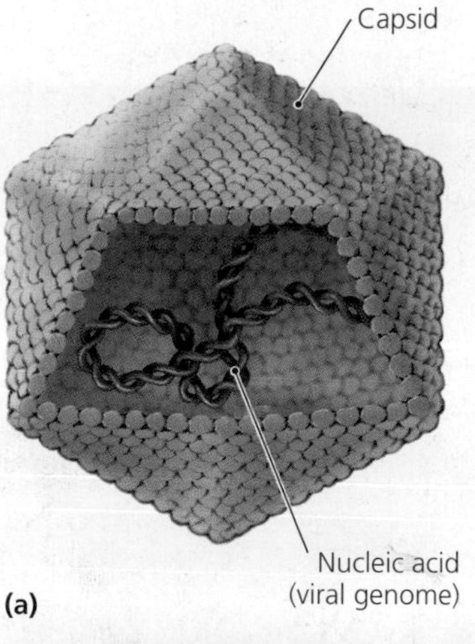

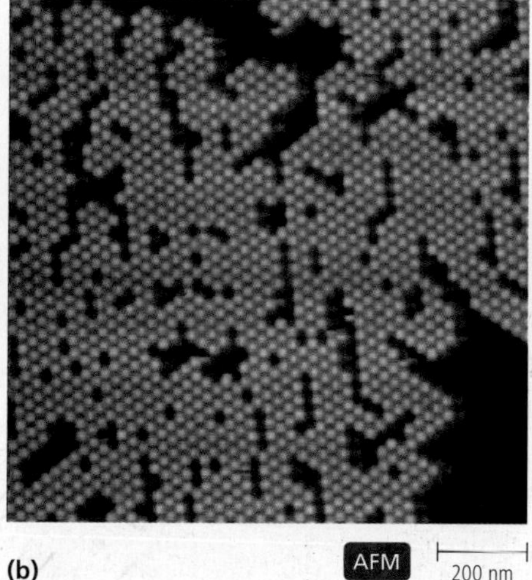

▲ **Figure 13.1 Virions, complete virus particles, include a nucleic acid, a capsid, and in some cases an envelope.** **(a)** A drawing of a nonenveloped polyhedral virus containing DNA. **(b)** An atomic force microscope image of crystallized tobacco mosaic virus. Like many chemicals and unlike cells, some viruses can form crystals.

Now that we have examined ways in which viruses are alike, we will consider the characteristics that are used to distinguish different viral groups. Viruses differ in the type of genetic material they contain, the kinds of cells they attack, their size, the nature of their capsid coat, their shapes, and the presence or absence of an envelope.

Genetic Material of Viruses

Learning Objective

✓ Discuss viral genomes in terms of dsDNA, ssDNA, ssRNA, dsRNA, and number of segments of nucleic acid.

Viruses show more variety in the nature of their genomes than do cells. Whereas the genome of every cell is double-stranded DNA, the genome of a virus may be either DNA or RNA. The primary

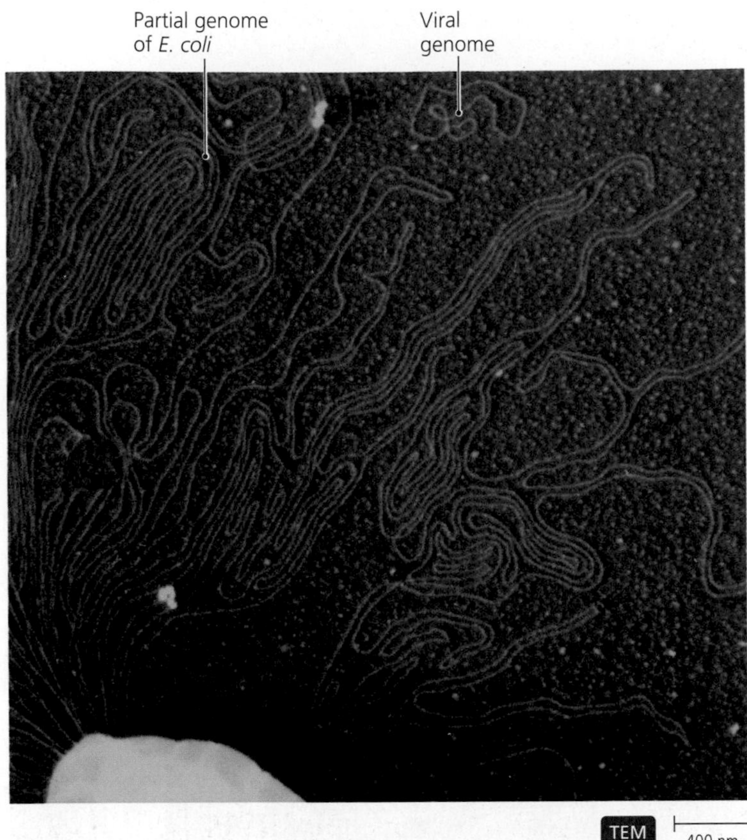

▲ **Figure 13.2 The relative sizes of genomes.** *Escherichia coli* was ruptured to release the DNA.

way in which scientists categorize and classify viruses is based on the type of genetic material that makes up the viral genome.

Some viral genomes, such as those of herpesvirus and chickenpox virus, are double-stranded DNA (dsDNA), like the genomes of cells. Other viruses use either single-stranded RNA (ssRNA), single-stranded DNA (ssDNA), or double-stranded RNA (dsRNA) as their genomes. These molecules never function as a genome in any cell—in fact, ssDNA and dsRNA are almost nonexistent in cells. Further, the genome of any particular virus may be either linear and composed of several molecules of nucleic acid, as in eukaryotic cells, or circular and singular, as in most prokaryotic cells. For example, the genome of an influenzavirus is composed of eight linear segments of single-stranded RNA, whereas the genome of poliovirus is one molecule of single-stranded RNA.

Viral genomes are usually much smaller than the genomes of cells. For example, the genome of the smallest chlamydial bacterium has almost 1000 genes, whereas the genome of virus MS2 has only three genes. **Figure 13.2** compares the genome of a virus with the genome of the bacterium *Escherichia coli* (esh-ĕ-rik′ē-ă kō′lē), which contains over 4000 genes.

CRITICAL **THINKING**

Some viral genomes, composed of single-stranded RNA, act as mRNA. What advantage might these viruses have over other kinds of viruses?

Hosts of Viruses

Learning Objectives

✓ Explain the mechanism by which viruses are specific for their host cells.

✓ Compare and contrast viruses of fungi, plants, animals, and bacteria.

Most viruses infect only a particular host's cells. This specificity is due to the precise affinity of viral surface proteins or glycoproteins for complementary proteins or glycoproteins on the surface of the host cell. Viruses may be so specific they infect not only a particular host, but a particular kind of cell in that host. For example, HIV (human immunodeficiency virus, the agent that causes AIDS) specifically attacks helper T lymphocytes (a type of white blood cell) in humans and has no effect on, say, human muscle or bone cells. By contrast, some viruses are *generalists;* they infect many kinds of cells in many different hosts. An example of a generalist virus is West Nile virus, which can infect most species of birds, several mammalian species, and some reptiles.

All types of organisms are susceptible to some sort of viral attack. There are viruses that infect archaeal, bacterial, plant, protozoan, fungal, and animal cells **(Figure 13.3)**. There is even a tiny virus that attacks a large virus. Most viral research and scientific study has focused on bacterial and animal viruses. A virus that infects bacteria is referred to as a **bacteriophage** (bak-tēr′ē-ō-fāj), or simply a **phage** (fāj). Scientists have determined that bacteriophages outnumber all bacteria, archaea, and eukaryotes put together. We will return our attention to bacteriophages and animal viruses later in this chapter.

Viruses of plants are less well known than bacterial and animal viruses, even though viruses were first identified and isolated from tobacco plants. Plant viruses infect many food crops, including corn, beans, sugar cane, tobacco, and potatoes, resulting in billions of dollars in losses each year. Viruses of plants are introduced into plant cells either through abrasions of the cell wall or by plant parasites such as nematodes and aphids. After entry, plant viruses follow the replication cycle discussed below for animal viruses.

Fungal viruses have been little studied. We do know that fungal viruses are different from animal and bacterial viruses in that fungal viruses exist only within cells; that is, they seemingly have no extracellular state. Presumably, fungal viruses cannot penetrate a thick fungal cell wall. However, because fusion of cells is typically a part of a fungal life cycle, viral infections can easily be propagated by the fusion of an infected fungal cell with an uninfected one.

Not all viruses are deleterious. The box, **Beneficial Microbes: Good Viruses? Who Knew?**, on p. 378 illustrates some useful aspects of viruses in the environment.

Sizes of Viruses

In the late 1800s, scientists hypothesized that the cause of many diseases, including polio and smallpox, was an agent smaller than a bacterium. They named these tiny agents "viruses," from

Figure 13.3 Some examples of plant, bacterial, and human hosts of viral infections. **(a)** Left, a tobacco leaf infected with tobacco mosaic virus, the first virus discovered. **(b)** A bacterial cell under attack by bacteriophages (pink). **(c)** A human white blood cell's cytoplasmic membrane, to which HIV (blue) is attached.

(a)

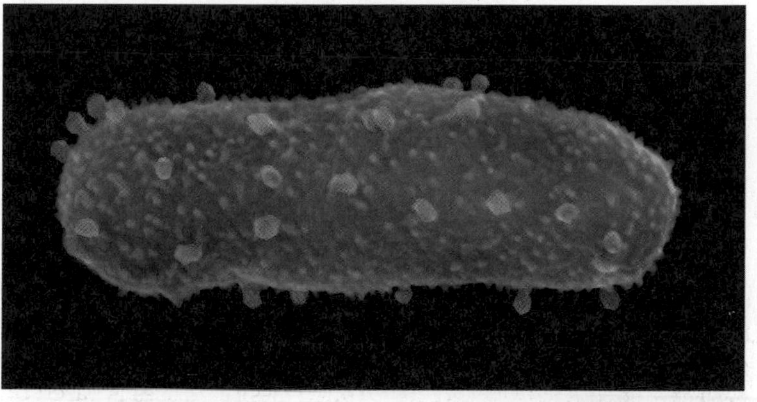

(b)

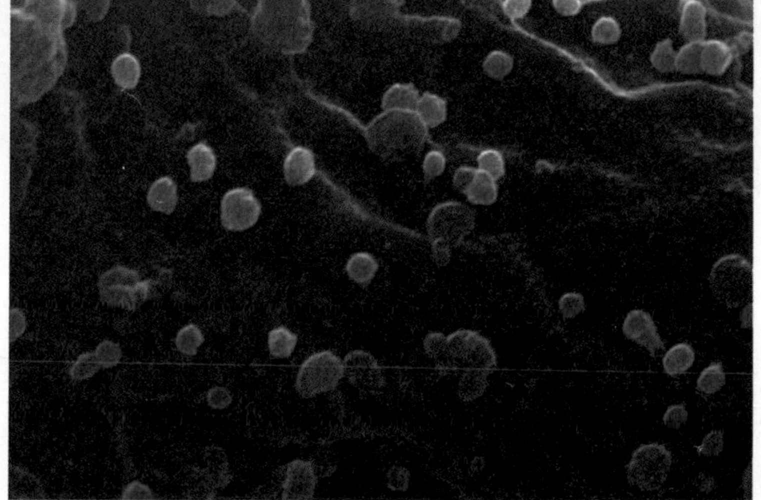

(c)

the Latin word for "poison." Viruses are so small that only a few can be seen by light microscopy. One hundred million polioviruses could fit side by side on the period at the end of this sentence. The smallest viruses have a diameter of 10 nm, whereas the largest are approximately 400 nm in diameter, which is about the size of the smallest bacterial cells. **Figure 13.4** compares the sizes of selected viruses to *E. coli* and a human red blood cell.

In 1892, Russian microbiologist Dmitri Ivanowski (1864–1920) first demonstrated that viruses are acellular with an experiment designed to elucidate the cause of tobacco mosaic disease. He filtered the sap of infected tobacco plants through a porcelain filter fine enough to trap even the smallest of bacterial cells. Viruses, however, were not trapped but instead passed through the filter with the liquid, which remained infectious to tobacco plants. This experiment proved the existence of an acellular disease-causing entity smaller than a bacterium. Tobacco mosaic virus (TMV) was isolated and characterized in 1935 by an American chemist, Wendell Stanley (1904–1971). The invention of electron microscopy allowed scientists to finally see TMV and other viruses.

Capsid Morphology

Learning Objective

✓ Discuss the structure and function of the viral capsid.

As we have seen, viruses have capsids—protein coats that provide both protection for viral nucleic acid and a means by which many viruses attach to their hosts' cells. The capsid of a virus is composed of proteinaceous subunits called **capsomeres** (or *capsomers*). Some capsomeres are composed of only a single type of protein, whereas others are composed of several different kinds of proteins. Recall that viral nucleic acid surrounded by its capsid is termed a *nucleocapsid*.

CRITICAL **THINKING**

In some viruses, the capsomeres act enzymatically as well as structurally. What advantage might this provide the virus?

Viral Shapes

The shapes of virions are also used to classify viruses. There are three basic types of viral shapes: helical, polyhedral, and complex (**Figure 13.5**). The capsid of a helical virus is composed of capsomeres that bond together in a spiral fashion to form a tube around the nucleic acid. The capsid of a polyhedral virus is roughly spherical, with a shape similar to a geodesic dome. The most common type of polyhedral capsid is an icosahedron, which has 20 sides.

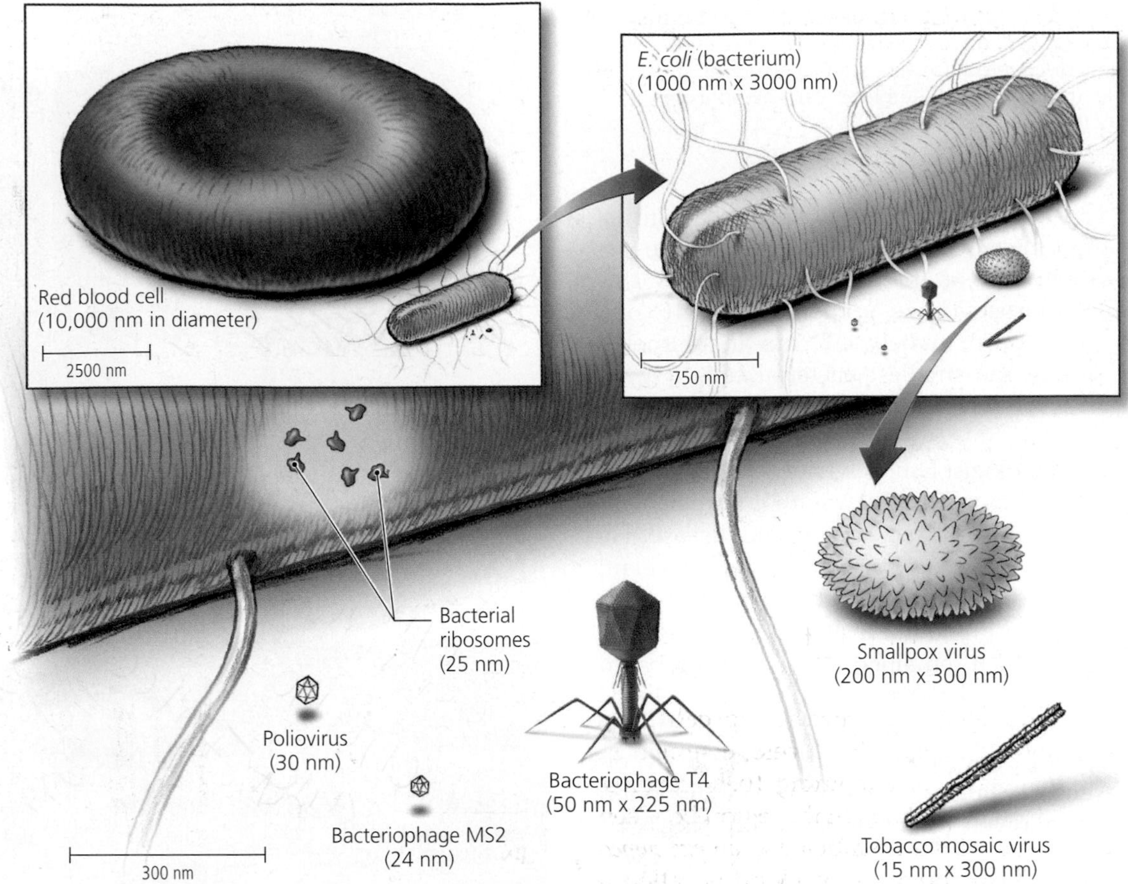

▲ **Figure 13.4 Sizes of selected virions.** Selected viruses are compared in size to a bacterium, *Escherichia coli*, and a human red blood cell. *How can viruses be so small and yet still be pathogenic?*

Figure 13.4 *Viruses utilize a host cell's enzymes, organelles, and membranes to complete their replication cycle.*

BENEFICIAL MICROBES

GOOD VIRUSES? WHO KNEW?

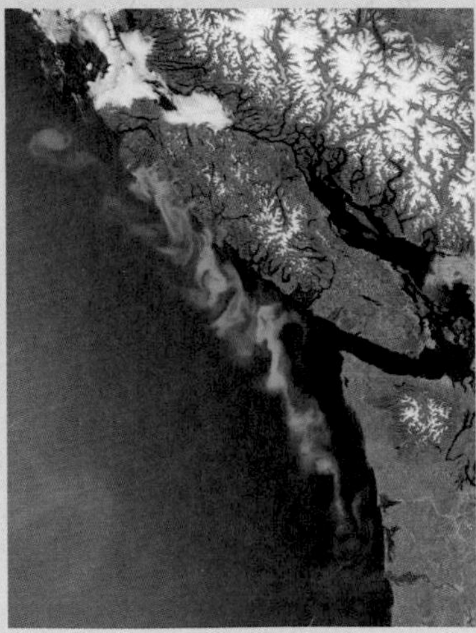

▲ ***An algal bloom off the coast of Seattle, Washington.***

Viruses, though normally pathogenic to their host cells, do have positive influences, including what appear to be extensive roles in the environment. Recent discoveries by the United Kingdom's Marine and Freshwater Microbial Biodiversity program demonstrate important ways viruses impact our world.

First: Scientists found that a previously unknown virus attacks a tiny marine alga that multiplies to form algal blooms consisting of hundreds of thousands to millions of algal cells per milliliter of water (see photo). Algal blooms like these can deplete the water of oxygen at night, potentially harming fish and other marine life. The newly discovered virus stops blooms by killing the algae, a result which is good for animal life.

Second: When the algae die by this means, they release an airborne sulfate compound that acts to seed clouds. The resulting increased cloudiness noticeably shades the ocean, measurably lowering water temperature. Thus, a marine virus helps to reduce global warming!

Third: The researchers discovered a bacteriophage of oceanic cyanobacteria that transfers genes for photosynthetic machinery into its hosts' cells, so that the cells' photosynthetic rate increases. There are up to 10 million of these viruses in a single milliliter of seawater, so researchers estimate that much of the oxygen we breathe may be attributable to the action of these viruses on the blue-green bacteria.

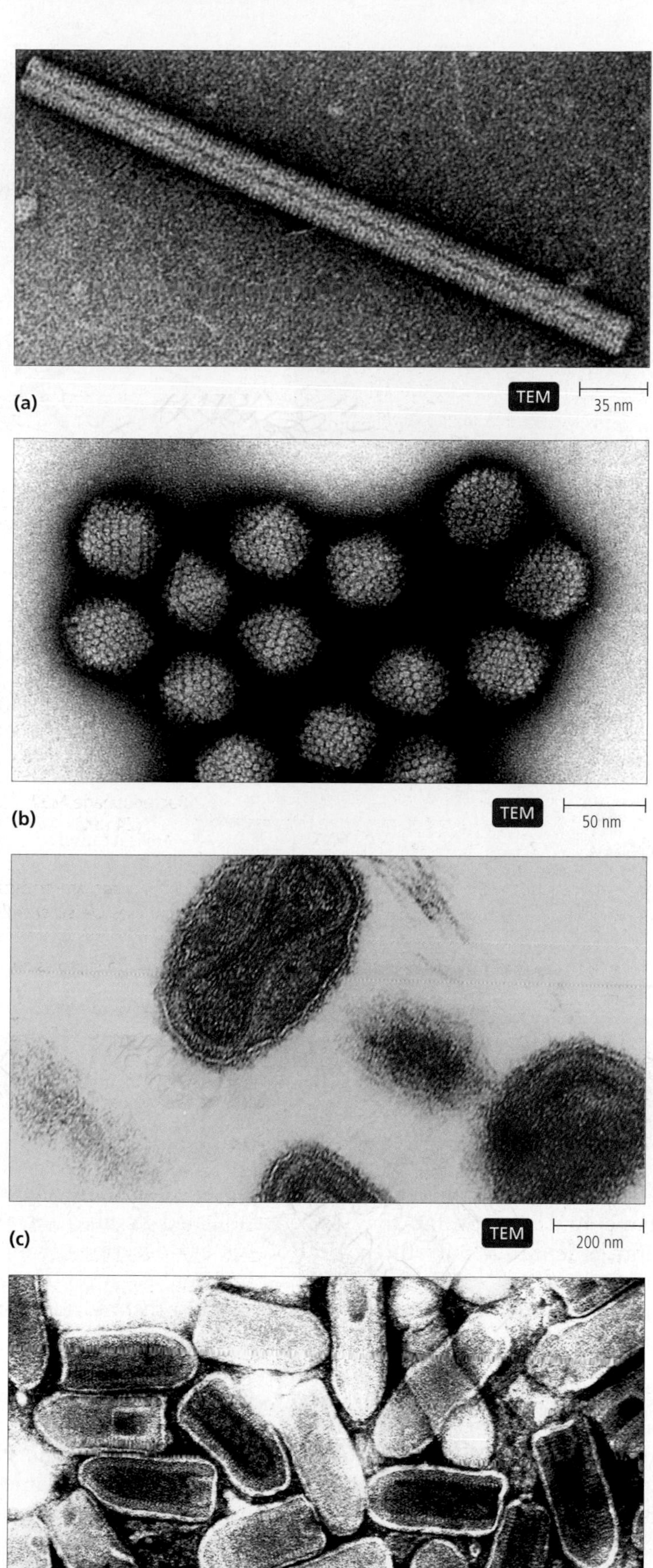

Figure 13.5 The shapes of virions. **(a)** A helical virus, tobacco mosaic virus. The tubular shape of the capsid results from the tight arrangement of several rows of helical capsomeres. **(b)** Polyhedral virions of a virus that causes common colds. **(c)** Complex virions of smallpox virus. **(d)** The complex shape of rabies virus, which results from the shapes of the capsid and bullet-shaped envelope.

Complex viruses have capsids of many different shapes that do not readily fit into either of the other two categories. An example of a complex virus is smallpox virus, which has several covering layers (including lipid) and no easily identifiable capsid. The complex shapes of many bacteriophages include icosahedral heads, which contain the genome, attached to helical tails with tail fibers. The complex capsids of such bacteriophages somewhat resemble NASA's lunar lander **(Figure 13.6)**.

The Viral Envelope

Learning Objective

✓ Discuss the origin, structure, and function of the viral envelope.

All viruses lack cell membranes (after all, they are not cells), but some, particularly animal viruses, have an envelope similar in composition to a cell membrane surrounding their capsids. Other viral proteins called *matrix proteins* fill the region between capsid and envelope. A virus with a membrane is an *enveloped virion* **(Figure 13.7)**; a virion without an envelope is called a *nonenveloped* or *naked virion*.

An enveloped virus acquires its envelope from its host cell during viral replication or release (discussed shortly). Indeed, the envelope of a virus is a portion of the membrane system of a host cell. Like a cytoplasmic membrane, a viral envelope is composed of a phospholipid bilayer and proteins. Some of the proteins are virally coded glycoproteins, which appear as spikes protruding outward from the envelope's surface (see Figure 13.7). Host DNA carries the genetic code required for the assembly of the phospholipids and some of the proteins in the envelope, while the viral genome specifies the other membrane proteins.

An envelope's proteins and glycoproteins often play a role in the recognition of host cells. A viral envelope does not perform other physiological roles of a cytoplasmic membrane such as endocytosis or active transport.

Table 13.1 on p. 381 summarizes the novel properties of viruses, and how those properties differ from the corresponding characteristics of cells. Next we turn our attention to the criteria by which virologists classify viruses.

Classification of Viruses

Learning Objective

✓ List the characteristics by which viruses are classified.

The International Committee on Taxonomy of Viruses (ICTV) was established in 1966 to provide a single taxonomic scheme for viral classification and identification. Virologists classify

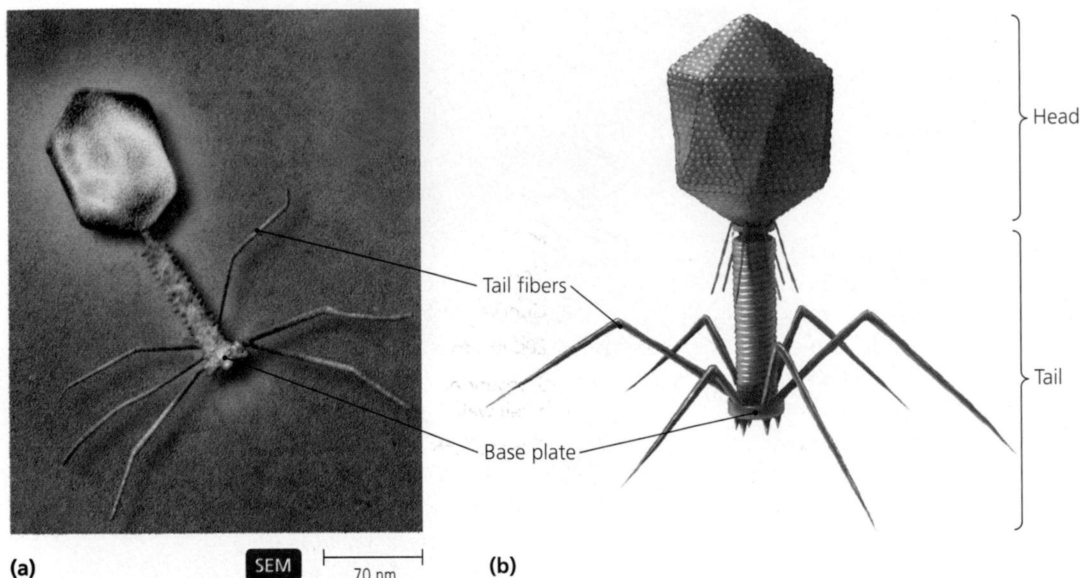

Figure 13.6 The complex shape of bacteriophage T4. It includes an icosahedral head and an ornate tail that enables viral attachment and penetration.

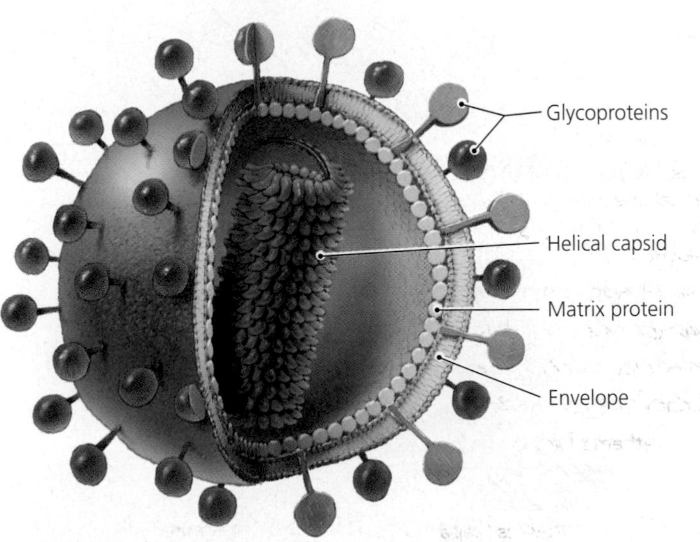

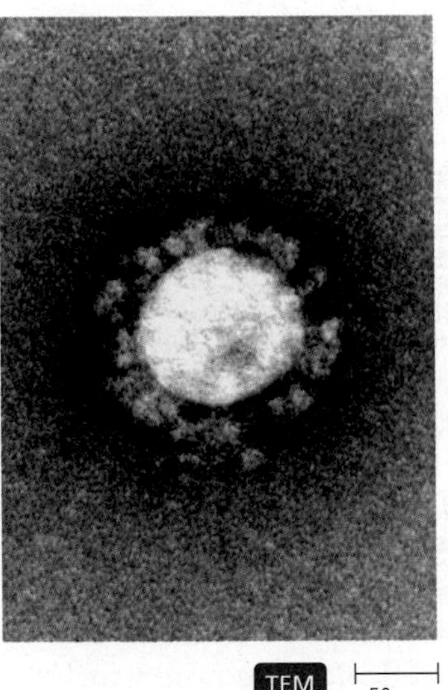

Figure 13.7 Enveloped virion. Artist's rendition and electron micrograph of severe acute respiratory syndrome (SARS) virus, an enveloped virus with a helical capsid.

viruses by their type of nucleic acid, presence of an envelope, shape, and size. So far, they have established families for all viral genera, but only three viral orders are described. Kingdoms, divisions, and classes have not been determined for viruses because the relationships among viruses are not well understood.

Family names are typically derived either from special characteristics of viruses within the family or from the name of an important member of the family. For example, family *Picornaviridae* contains very small[1] RNA viruses, and *Hepadnaviridae* contains a DNA virus that causes hepatitis B. Family *Herpesviridae* is named for herpes simplex, a virus that can cause genital herpes. Table 13.2 lists the major families of human viruses, grouped according to the type of nucleic acid each contains.

[1]Pico means one-trillionth, 10^{-12}.

Specific epithets for viruses are their common English designations written in italics. Accordingly, the nomenclature for two important viral pathogens, HIV and rabies virus, is as follows:

	HIV	Rabies virus
Order	Not yet established	Mononegavirales
Family	*Retroviridae*	*Rhabdoviridae*
Genus	*Lentivirus* (len′ti-vī-rŭs)	*Lyssavirus* (lis′ă-vī-rŭs)
Specific epithet	*human immunodeficiency virus*	*rabies virus*

CRITICAL **THINKING**

Why has it been difficult to develop a complete taxonomy for viruses?

TABLE 13.1

The Novel Properties of Viruses

Viruses	Cells
Inert macromolecules outside of a cell, but become active inside a cell	Metabolize on their own
Do not divide or grow	Divide and grow
Acellular	Cellular
Obligate intracellular parasites	Most are free-living
Contain either DNA or RNA, except *Cytomegalovirus* and *Minivirus* contain both	Contain both DNA and RNA
Genome can be dsDNA, ssDNA, dsRNA, or ssRNA	Genome is dsDNA
Ultramicroscopic in size, ranging from 10 nm to 400 nm	200 nm to 12 cm in diameter
Have a proteinaceous capsid around genome; some have an envelope around the capsid	Surrounded by a phospholipid membrane and often a cell wall
Replicate in an assembly-line manner using the enzymes and organelles of a host cell	Self-replicating by asexual and/or sexual means

TABLE 13.2

Families of Human Viruses

Family	Strand Type	Representative Genera (Diseases)
DNA Viruses		
Poxviridae	Double	*Orthopoxvirus* (smallpox)
Herpesviridae	Double	*Simplexvirus*, Herpes type 1 (fever blisters, respiratory infections), Herpes type 2 (genital infections); *Varicellovirus* (chickenpox); *Lymphocryptovirus*, Epstein-Barr virus (infectious mononucleosis, Burkitt's lymphoma); *Cytomegalovirus* (birth defects); *Roseolavirus* (roseola)
Papillomaviridae	Double	*Papillomavirus* (benign tumors, warts, cervical and penile cancers)
Polyomaviridae	Double	*Polyomavirus* (progressive multifocal leukoencephalopathy)
Adenoviridae	Double	*Mastadenovirus* (conjunctivitis, respiratory infections)
Hepadnaviridae	Partial single and partial double	*Orthohepadnavirus* (hepatitis B)
Parvoviridae	Single	*Erythrovirus* (erythema infectiosum)
RNA Viruses		
Picornaviridae	Single, +[a]	*Enterovirus* (polio); *Hepatovirus* (hepatitis A); *Rhinovirus* (common cold)
Caliciviridae	Single, +	*Norovirus* (gastroenteritis)
Astroviridae	Single, +	*Astrovirus* (gastroenteritis)
Hepeviridae	Single, +	*Hepevirus* (hepatitis E)
Togaviridae	Single, +	*Alphavirus* (encephalitis); *Rubivirus* (rubella)
Flaviviridae	Single, +	*Flavivirus* (yellow fever); Japanese encephalitis virus (encephalitis); *Hepacivirus* (hepatitis C)
Coronaviridae	Single, +	*Coronavirus* (common cold, severe acute respiratory syndrome)
Retroviridae	Single, +, segmented	Human T cell leukemia virus (leukemia); *Lentivirus* (AIDS)
Orthomyxoviridae	Single, −[b], segmented	*Influenzavirus* (flu)
Paramyxoviridae	Single, −	*Paramyxovirus* (common cold, respiratory infections); *Pneumovirus* (pneumonia, common cold); *Morbillivirus* (measles); *Rubulavirus* (mumps)
Rhabdoviridae	Single, −	*Lyssavirus* (rabies)
Bunyaviridae	Single, −, segmented	*Bunyavirus* (California encephalitis virus); *Hantavirus* (pneumonia)
Filoviridae	Single, −	*Filovirus* (Ebola hemorrhagic fever); Marburg virus (hemorrhagic fever)
Arenaviridae	Single, −, segmented	*Lassavirus* (hemorrhagic fever)
Reoviridae	Double, segmented	*Orbivirus* (encephalitis); *Rotavirus* (diarrhea); *Coltivirus* (Colorado tick fever)

[a]Positive-sense (+RNA) is equivalent to mRNA; i.e., it instructs ribosomes in protein translations.
[b]Negative-sense (−RNA) is complementary to mRNA; it cannot be directly translated.

BENEFICIAL MICROBES

PRESCRIPTION BACTERIOPHAGES?

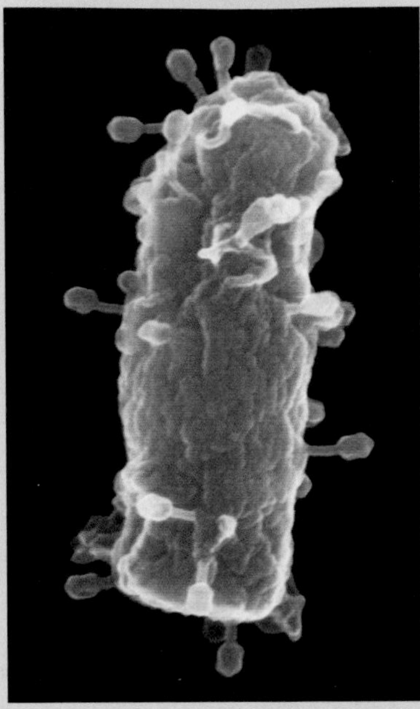

▲ *Escherichia coli* infected with bacteriophages.

In 1917, Canadian biologist Felix d'Herelle published a paper announcing the discovery of the *bacteriophage*, a virus that preyed on bacteria. In fact, half the bacteria on Earth succumb to phages every two days! D'Herelle felt that phages can be natural weapons against bacterial pathogens.

Phage therapy was used in the early 1900s to combat dysentery, typhus, and cholera, but was largely abandoned in the 1940s in the United States, eclipsed by the development of antibiotics such as penicillin. Phage therapy continued in the USSR and Eastern Europe, where research is still centered. Today, motivated by the growing problem of antibiotic-resistant bacteria, scientists in the U.S. and Western Europe have renewed interest in investigating phage therapy.

A phage reproduces by inserting genetic material into a bacterium, causing the bacterium to build copies of the virus that burst out of the cell to infect other bacteria. A single phage can become 10 trillion phages within two hours, killing 99.9% of its host bacteria. Each type of phage attacks a specific strain of bacteria. This means that phage treatment is effective only if the phages are carefully matched to the disease-causing bacterium. It also means that phage treatment, unlike the use of antibiotics, can be effective without killing the body's helpful bacteria.

Introducing an active microbe into a patient does present some dangers, however. Phages can kill bacteria, but they can also make bacteria more lethal. A strain of *Escherichia coli* that is responsible for a deadly form of food poisoning, for example, has been observed to gain the ability to produce a toxic chemical from a phage genome that integrates itself into the bacterium's DNA. If a phage being used in therapy picked up a toxin-coding gene, the attempted cure could become lethal.

Viral Replication

As previously noted, viruses cannot reproduce themselves because they lack the genes for all the enzymes necessary for replication, nor do they possess functional ribosomes for protein synthesis. Instead, viruses are dependent on their hosts' enzymes and organelles to produce new virions. Once a host cell falls under control of a viral genome, it is forced to replicate viral genetic material and translate viral proteins, including viral capsomeres and viral enzymes.

The replication cycle of a virus usually results in the death and lysis of the host cell. Because the cell undergoes lysis near the end of the cycle, this type of replication is termed **lytic replication.** In general, a lytic replication cycle consists of the following five stages:

- **Attachment** of the virion to the host cell
- **Entry** of the virion or its genome into the host cell
- **Synthesis** of new nucleic acids and viral proteins by the host cell's enzymes and ribosomes
- **Assembly** of new virions within the host cell
- **Release** of the new virions from the host cell

ANIMATIONS: *Viral Replication: Overview*

In the following sections we examine the events that occur in the replication of bacteriophages and animal viruses. We begin with lytic replication in bacteriophages, turn to a modification of replication (called lysogenic replication), and then consider the replication of animal viruses.

Lytic Replication of Bacteriophages

Learning Objective

✓ Sketch and describe the five stages of the lytic replication cycle as it typically occurs in bacteriophages.

Studies of phages revealed the basics of viral biology. Indeed, bacteriophages make excellent tools for the general study of viruses because they are easier and less expensive to culture than animal or human viruses. **Beneficial Microbes: Prescription Bacteriophages?** is an interesting side note on the potential use of bacteriophages as an alternative to antibiotics.

Here we examine the replication of a much-studied dsDNA phage of *E. coli* called *type 4 (T4)*. T4 virions are complex, having the polyhedral heads and helical tails seen in many bacteriophages. We begin with attachment, the first stage of replication **(Figure 13.8)**.

Attachment 1

Because phages, like all virions, are nonmotile, contact with a bacterium occurs by purely random collision, brought about as molecular bombardment and currents move virions through the environment. The structures responsible for the attachment of T4

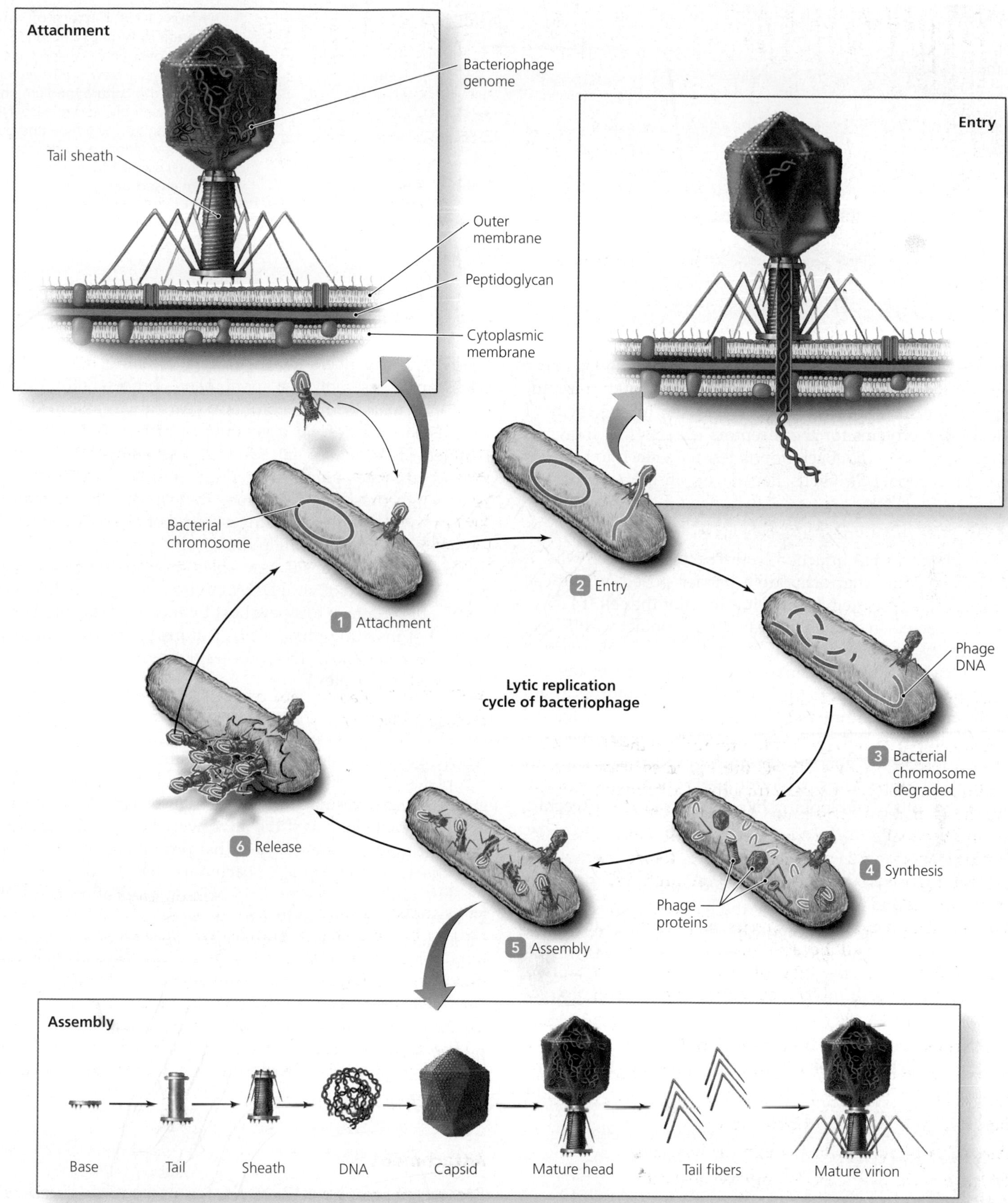

▲ **Figure 13.8 The lytic replication cycle in bacteriophages.** The phage shown in this illustration is T4, and the bacterium shown is *E. coli*. The circular bacterial chromosome is represented diagrammatically; in reality it would be much longer.

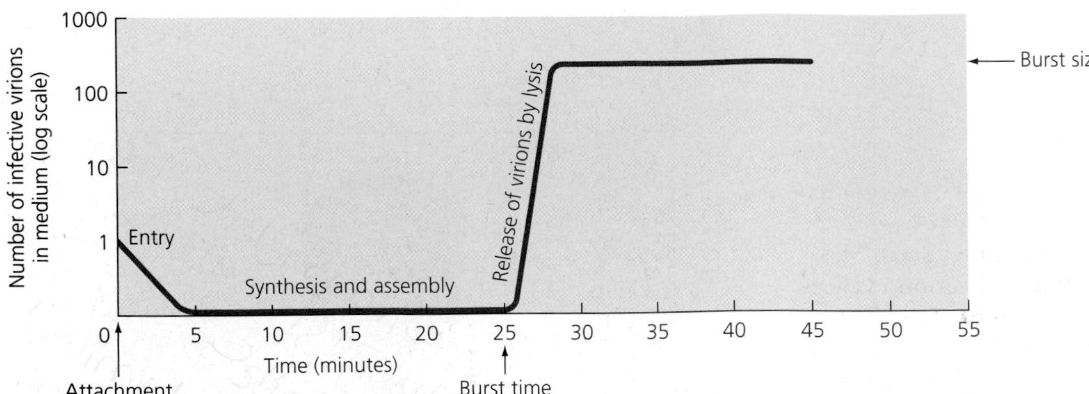

Figure 13.9 Pattern of virion abundance in lytic cycle. Virion abundance over time for single lytic replication cycle. New virions are not observed in the culture medium until synthesis, assembly, and release (lysis) are complete, at which time (the burst time) the new virions are released all at once. Burst size is the number of new virions released per lysed host cell.

to its host bacterium are its tail fibers. Attachment is dependent on the chemical attraction and precise fit between attachment proteins on the phage's tail fibers and complementary receptor proteins on the surface of the host's cell wall. The specificity of the attachment proteins for the receptors ensures that the virus will attach only to *E. coli*. Bacteriophages may attach to receptor proteins on bacterial cells' walls, flagella, or pili.

Entry 2

Now that phage T4 has attached to the bacterium's cell wall, it must still overcome the formidable barrier posed by the cell wall and cytoplasmic membrane if it is to enter the cell. T4 overcomes this obstacle in an elegant way. Upon contact with *E. coli,* T4 releases *lysozyme* (lī'sō-zīm), a protein enzyme carried within the capsid that weakens the peptidoglycan of the cell wall. The phage's tail sheath then contracts, which forces an internal hollow tube within the tail through the cell wall and membrane, much as a hypodermic needle penetrates the skin. The phage injects the genome through the tube and into the bacterium. The empty capsid, having performed its task, is left on the outside of the cell looking like an abandoned spacecraft.

Synthesis 3 – 4

After entry, viral enzymes (either carried within the capsid or coded by viral genes and made by the bacterium) degrade the bacterial DNA into its constituent nucleotides. As a result, the bacterium stops synthesizing its own molecules and begins synthesizing new viruses under control of the viral genome.

For dsDNA viruses like T4, protein synthesis is straightforward and similar to cellular transcription and translation, except that mRNA is transcribed from viral DNA instead of cellular DNA. Translation by the host cell's ribosomes results in viral proteins, including head capsomeres, components of the tail, viral DNA polymerase (which replicates viral DNA), and lysozyme (which weakens the bac terial cell wall from within, enabling the virions to leave the cell once they have been assembled).

Assembly 5

Scientists do not understand completely how phages are assembled inside a host cell, but it appears that as capsomeres accumulate within the cell, they spontaneously attach to one another to form new capsid heads. Likewise, tails assemble and attach to heads, and tail fibers attach to tails, forming mature virions. Such capsid assembly is a spontaneous process, requiring little or no enzymatic activity. For many years it was assumed that all capsids formed around a genome in just such a spontaneous manner. However, recent research has shown that for some viruses, enzymes pump the genome into the assembled capsid under high pressure—five times that used in a paintball gun. This process resembles stuffing a strand of cooked spaghetti into a matchbox through a single small hole.

Sometimes a capsid assembles around leftover pieces of host DNA instead of viral DNA. A virion formed in this manner is still able to attach to a new host by means of its tail fibers, but instead of inserting phage DNA it transfers DNA from the first host into a new host. This process, known as *transduction,* was described in Chapter 7.

Release 6

Newly assembled virions are released from the cell as lysozyme completes its work on the cell wall and the bacterium disintegrates. Areas of disintegrating bacterial cells in a lawn of bacteria in a Petri plate look as if the lawn were being eaten, and it was the appearance of these *plaques* that prompted early scientists to give the name *bacteriophage,* "bacterial eater," to these viruses.

For phage T4, the process of lytic replication takes about 25 minutes and can produce as many as 100–200 new virions for each bacterial cell lysed **(Figure 13.9)**. For any phage undergoing lytic replication, the period of time required to complete the entire process, from attachment to release, is called the *burst time,* and the number of new virions released from each lysed bacterial cell is called the *burst size.* **ANIMATIONS: *Viral Replication: Virulent Bacteriophages***

CRITICAL **THINKING**

If a colony of 1.5 billion *E. coli* cells were infected with a single phage T4, and each lytic replication cycle of the phage produced 200 new phages, how many replication cycles would it take for T4 phages to overwhelm the entire bacterial colony? (Assume for the sake of simplicity that every phage completes its replication cycle in a different cell and that the bacteria themselves do not reproduce.)

Lysogeny

Learning Objective

✓ Compare and contrast the lysogenic replication cycle of viruses with the lytic cycle.

Not all viruses follow the lytic pattern of phage T4 we just examined. Some bacteriophages have a modified replication cycle in which infected host cells grow and reproduce normally for many generations before they lyse. Such a replication cycle is called a **lysogenic replication cycle** or **lysogeny** (lī-soj′ĕ-nē), and the phages are called **temperate phages** or *lysogenic phages.*

Here we examine lysogenic replication as it occurs in a much-studied temperate phage, *lambda phage,* which is another parasite of *E. coli.* A lambda phage has a linear molecule of dsDNA in a complex capsid consisting of an icosahedral head attached to a tail that lacks tail fibers **(Figure 13.10)**.

Figure 13.11 illustrates lysogeny with lambda phage. First, the virion randomly contacts an *E. coli* cell and attaches via its tail (1). The viral DNA enters the cell, just as occurs with phage T4, but the host cell's DNA is not destroyed, and the phage's genome does not immediately assume control of the cell. Instead, the virus remains inactive. Such an inactive bacteriophage is called a **prophage** (prō′ fāj) (2). A prophage remains inactive

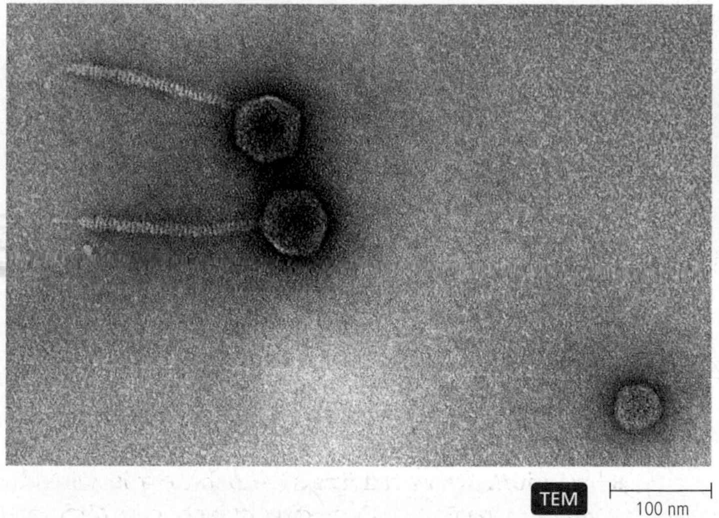

▲ **Figure 13.10 Bacteriophage lambda.** Note the absence of tail fibers. *Phage T4 attaches by means of molecules on its tail fibers. How does lambda phage, which lacks fibers, attach?*

Figure 13.10 Lambda has attachment molecules at the end of its tail rather than on its tail fibers.

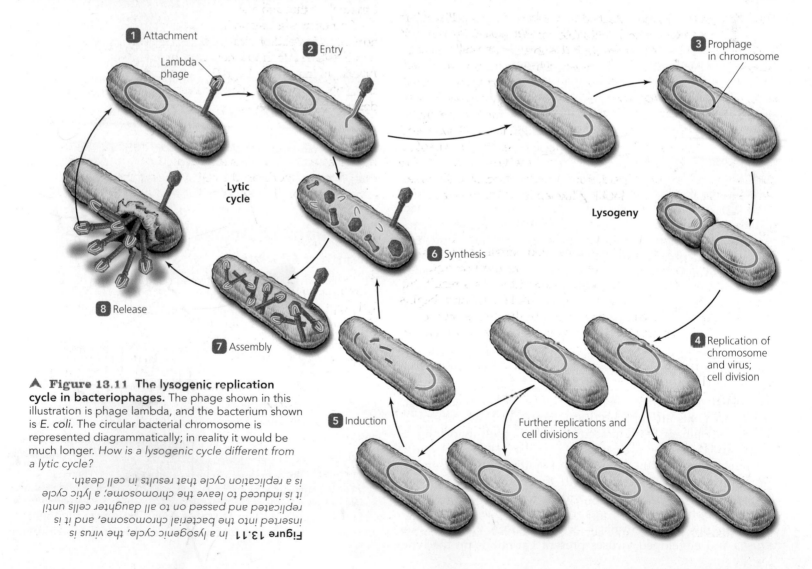

▲ **Figure 13.11 The lysogenic replication cycle in bacteriophages.** The phage shown in this illustration is phage lambda, and the bacterium shown is *E. coli.* The circular bacterial chromosome is represented diagrammatically; in reality it would be much longer. *How is a lysogenic cycle different from a lytic cycle?*

Figure 13.11 In a lysogenic cycle, the virus is inserted into the bacterial chromosome, and it is replicated and passed on to all daughter cells until it is induced to leave the chromosome; a lytic cycle is a replication cycle that results in cell death.

HIGHLIGHT

THE THREAT OF AVIAN INFLUENZA

In 1997, 18 people in Hong Kong contracted avian influenza, caused by the H5N1 strain of influenzavirus that spreads easily among chickens and other birds. At least half of these people caught the disease directly from birds, something that scientists had previously thought improbable. Human cases have had a 60% death rate. When people began dying of the illness, Hong Kong officials slaughtered all 1.5 million chickens within three days. That stopped a potential epidemic in Hong Kong, but it didn't stop the spread of the avian flu.

Avian flu virus very rarely spreads from one person to another. However, the possibility that this may change is of great concern to health officials worldwide. Avian flu viruses mutate quickly and can pick up genes from other flu viruses. If an avian flu virus picks up genes from a human flu virus, it could become a strain that spreads easily from person to person. In a worst-case scenario such a strain of avian/human flu, in this age of jet travel, could cause a pandemic killing 2–50 million people worldwide, according to the World Health Organization.

What should be done? In Asia, domesticated fowl have been slaughtered to prevent the virus from spreading. But while governments concentrated on culling domestic poultry, the virus spread to wild birds, such as geese, gray herons, and feral pigeons, and via wild birds it has spread throughout Asia, Europe, and central Africa.

Scientists have developed a vaccine that they believe could protect against the H5N1 strain of the virus. Unfortunately, a government could stockpile vaccine for one strain of flu only to have the virus mutate into a new form against which the vaccine is ineffective.

by coding for a protein that suppresses prophage genes. A side effect of this repressor protein is that it renders the bacterium resistant to additional infection by other viruses of the same type.

Another difference between a lysogenic cycle and a lytic cycle is that the prophage is inserted into the DNA of the bacterium, becoming a physical part of the bacterial chromosome (3). For DNA viruses like lambda phage, this is a simple process of fusing two pieces of DNA: one piece of DNA, the virus, is fused to another piece of DNA, the chromosome of the cell. Every time the cell replicates its infected chromosome, the prophage is also replicated (4). All daughter cells of a lysogenic cell are thus infected with the quiescent virus. A prophage and its descendants may remain a part of bacterial chromosomes for generations, or forever.

Lysogenic phages can change the phenotype of a bacterium, for example from a harmless form into a pathogen—a process called **lysogenic conversion.** Bacteriophage genes are responsible for toxins and other disease-evoking proteins found in the bacterial agents of diphtheria, cholera, rheumatic fever, and certain severe cases of diarrhea caused by *E. coli.*

At some later time a prophage might be excised from the chromosome by recombination or some other genetic event; it then reenters the lytic phase. The process whereby a prophage is excised from the host chromosome is called **induction** (5). Inductive agents are typically the same physical and chemical agents that damage DNA molecules, including ultraviolet light, X rays, and carcinogenic chemicals.

After induction, the lytic steps of synthesis (6), assembly (7), and release (8) resume from the point at which they stopped. The cell becomes filled with virions and breaks open.

Bacteriophages T4 and lambda demonstrate two replication strategies that are typical for many DNA viruses. RNA viruses and enveloped viruses present variations on the lytic and lysogenic cycles we have examined. We will next examine some of these variations as they occur with animal viruses. **ANIMATIONS:** ***Viral Replication: Temperate Bacteriophages***

CRITICAL **THINKING**

What differences would you expect in the replication cycles of RNA phages from those of DNA phages? (Hints: Think about the processes of transcription, translation, and replication of nucleic acids. Also, note that RNA is not normally inserted into a DNA molecule.)

Replication of Animal Viruses

Learning Objectives

- ✓ Explain the differences between bacteriophage replication and animal viral replication.
- ✓ Compare and contrast the replication and synthesis of DNA, −RNA, and +RNA viruses.
- ✓ Compare and contrast the release of viral particles by lysis and budding.
- ✓ Compare and contrast latency in animal viruses with phage lysogeny.

Animal viruses have the same five basic steps in their replication pathways as bacteriophages—that is, attachment, entry, synthesis, assembly, and release. However, there are significant differences in the replication of animal viruses that result in part from the presence of envelopes around some of the viruses, and in part from the eukaryotic nature of animal cells as well as their lack of a cell wall. **Highlight: The Threat of Avian Influenza** highlights an animal virus that is of great concern to health officials worldwide.

In this section we examine the replication processes that are shared by DNA and RNA animal viruses, compare these processes with those of bacteriophages, and discuss how the synthesis of DNA and RNA viruses differ.

Attachment of Animal Viruses

As with bacteriophages, attachment of an animal virus is dependent on the chemical attraction and exact fit between proteins or glycoproteins on the virion and complementary protein or glycoprotein receptors on the animal cell's cytoplasmic membrane. Unlike the bacteriophages we have examined, animal viruses lack both tails and tail fibers. Instead, animal viruses typically have glycoprotein spikes or other attachment molecules on their capsids or envelopes.

Entry and Uncoating of Animal Viruses

Animal viruses enter a host cell shortly after attachment. Even though entry of animal viruses is not as well understood as entry of bacteriophages, there appear to be at least three different mechanisms: direct penetration, membrane fusion, and endocytosis.

Some naked viruses enter their hosts' cells by *direct penetration*—a process in which the viral capsid attaches and sinks into the cytoplasmic membrane, creating a pore through which the genome alone enters the cell **(Figure 13.12a)**. Poliovirus infects host cells via direct penetration.

With other animal viruses, by contrast, the entire capsid and its contents (including the genome) enter the host cell by membrane fusion or endocytosis. With viruses using *membrane fusion*, including the measles and AIDS viruses, the viral envelope and the host cell membrane fuse, releasing the capsid into the cell's cytoplasm and leaving the envelope glycoproteins as part of the cell membrane **(Figure 13.12b)**.

Most enveloped viruses and some naked viruses enter host cells by triggering *endocytosis*. Attachment of the virus to receptor molecules on the cell's surface stimulates the cell to endocytize the entire virus **(Figure 13.12c)**. Adenoviruses (naked) and herpesviruses (enveloped) enter human host cells via endocytosis.

For those viruses that penetrate a host cell with their capsids intact, the capsids must be removed to release their genomes before the viruses can continue to replicate. The removal of a viral capsid within a host cell is called **uncoating**, a process that remains poorly understood. It apparently occurs via different means in different viruses; some viruses are uncoated within vesicles by cellular enzymes, whereas others are uncoated by enzymes within the cell's cytosol.

Synthesis of Animal Viruses

Synthesis of animal viruses also differs from synthesis of bacteriophages. Each type of animal virus requires a different strategy for synthesis that depends on the kind of nucleic acid involved—whether it is DNA or RNA, and whether it is double stranded or single stranded. DNA viruses typically enter the nucleus, whereas most RNA viruses are replicated in the cytoplasm. As we discuss the synthesis and assembly of each type of animal virus, consider the following two questions:

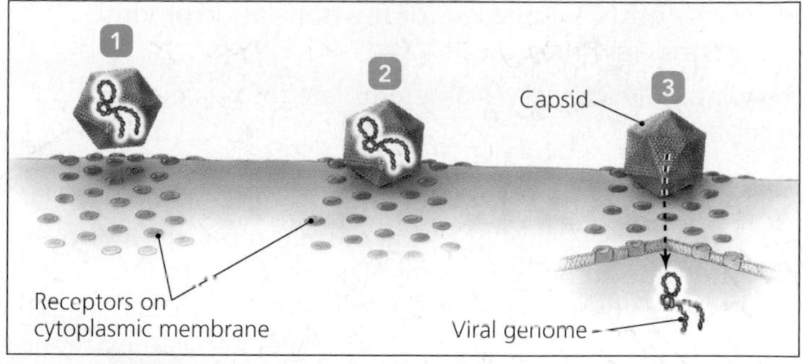

(a) Direct penetration

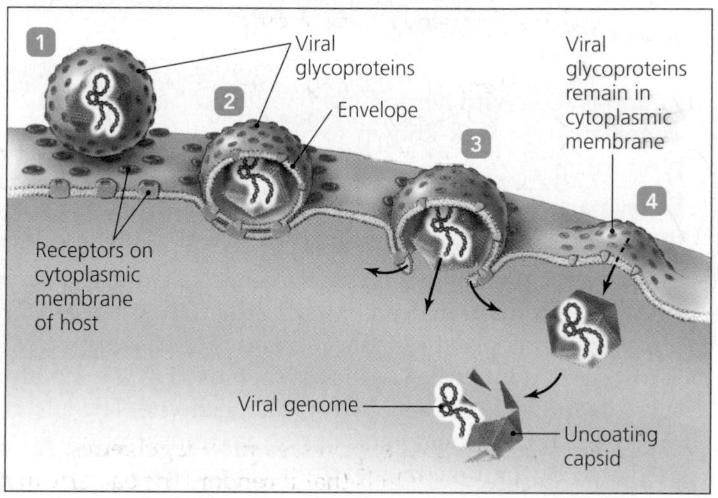

(b) Membrane fusion

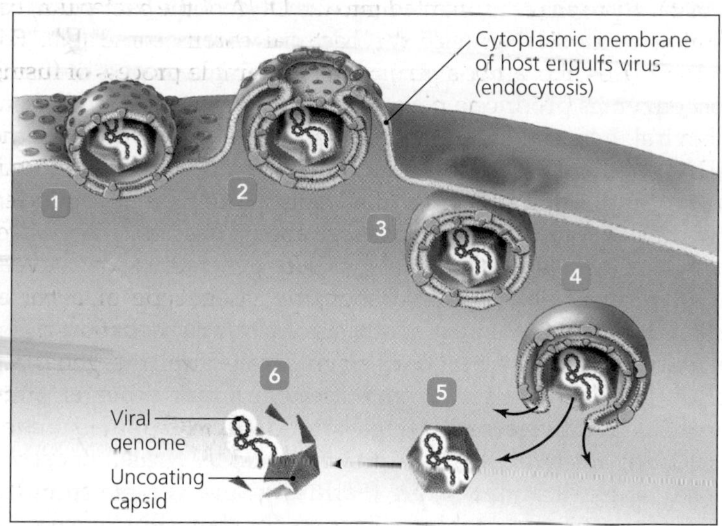

(c) Endocytosis

▲ **Figure 13.12 Three mechanisms of entry of animal viruses.** **(a)** Direct penetration, a process whereby naked virions inject their genomes into their animal cell hosts. **(b)** Membrane fusion, in which fusion of the viral envelope and cell membrane dumps the capsid into the cell. **(c)** Endocytosis, in which attachment of a naked or an enveloped virus stimulates the host cell to engulf the entire virus. *After penetration, many animal viruses must be uncoated, but bacteriophages need not be. Why is this so?*

Figure 13.12 *Generally bacteriophages inject their DNA during penetration, so the capsid does not enter the cell.*

- How is mRNA—needed for the translation of viral proteins—synthesized?
- What molecule serves as a template for nucleic acid replication?

dsDNA Viruses Synthesis of new double-stranded DNA (dsDNA) virions is similar to the normal replication of cellular DNA and translation of proteins. The genomes of most dsDNA viruses enter the nucleus of the cell, where cellular enzymes replicate the viral genome in the same manner as they replicate host dsDNA—using each strand of viral DNA as a template for its complement. After messenger RNA is transcribed from viral DNA in the nucleus and capsomere proteins are made in the cytoplasm by host ribosomes, capsomeres enter the nucleus, where new virions spontaneously assemble. This method of replication is seen with herpes and papilloma (wart) viruses.

There are two well-known exceptions to this regimen of dsDNA viruses:

- Every part of a poxvirus is synthesized and assembled in the cytoplasm of the host's cell; the nucleus is not involved.
- The genome of hepatitis B viruses is replicated using an RNA intermediary instead of replicating DNA from a DNA template. In other words, the genome of hepatitis B virus is transcribed into RNA, which is then used as a template to make multiple copies of viral DNA genome. The latter process, which is the reverse of normal transcription, is mediated by a viral enzyme, *reverse transcriptase.*

Chapter 24 discusses diseases of these two viruses.

ssDNA Viruses A human virus with a genome composed of single-stranded DNA (ssDNA) is a parvovirus (see Table 13.2 on p. 381). When a parvovirus enters the nucleus of a host cell, host enzymes produce a new strand of DNA comple mentary to the viral genome. This complementary strand binds to the ssDNA of the virus to form a dsDNA molecule. Transcription of mRNA, replication of new ssDNA, and viral assembly then follow the DNA virus pattern just described.

As previously noted, RNA is not used as genetic material in cells, so it follows that the synthesis of RNA viruses must differ significantly from typical cellular processes, and from the replication of DNA viruses as well. There are four types of RNA viruses: positive-sense, single-stranded RNA (designated +ssRNA); retroviruses (a kind of +ssRNA virus); negative-sense, single-stranded RNA (−ssRNA); and double-stranded RNA (dsRNA). The synthesis process for these RNA viruses is varied and rather complex. We start with the synthesis of +ssRNA viruses.

Positive-Sense ssRNA Viruses Single-stranded viral RNA that can act directly as mRNA is called **positive-strand RNA (+RNA).** Ribosomes translate polypeptides using the codons of such RNA. An example of a +ssRNA virus is poliovirus. In many +ssRNA viruses, a complementary **negative-strand RNA (−RNA)** is transcribed from the +ssRNA genome by viral RNA polymerase; −RNA then serves as the template for the transcription of multiple +ssRNA genomes. Such transcription of RNA from RNA is unique to viruses; no cell transcribes RNA from RNA.

Retroviruses Unlike other +ssRNA viruses, the +ssRNA viruses called **retroviruses** do not use their genome as mRNA. Instead, retroviruses use a DNA intermediary that is transcribed from +RNA by reverse transcriptase carried within the capsid. This DNA intermediary then serves as the template for the synthesis of additional +RNA molecules, which act both as mRNA for protein synthesis and as genomes for new virions. Human immunodeficiency virus (HIV) is a prominent retrovirus.

Negative-Sense ssRNA Viruses Other single-stranded RNA virions are −ssRNA viruses, which must overcome a unique problem. In order to synthesize a protein, a ribosome can use only mRNA (i.e., +RNA), because −RNA is not recognized by ribosomes. The virus overcomes this problem by carrying within its capsid an enzyme, *RNA-dependent RNA transcriptase,* which is released into the host cell's cytoplasm during uncoating and then transcribes +RNA molecules from the virus's −RNA genome. Translation of proteins can then occur as usual. The newly transcribed +RNA also serves as a template for transcription of additional copies of −RNA. Diseases caused by −ssRNA viruses include rabies and flu.

dsRNA Viruses Viruses that have double-stranded RNA use yet another method of synthesis. The positive strand of the molecule serves as mRNA for the translation of proteins, one of which is an RNA polymerase that transcribes dsRNA. Each strand of RNA acts as a template for transcription of its opposite, which is reminiscent of DNA replication in cells. Double-stranded RNA rotaviruses cause most cases of diarrhea in infants.

Figure 13.13 illustrates and **Table 13.3** summarizes the various strategies by which animal viruses are synthesized.

CRITICAL **THINKING**

Although many +ssRNA viruses use their genome directly as messenger RNA, +ssRNA retroviruses do not. Instead, their +RNA is transcribed into DNA by reverse transcriptase. What advantage do retroviruses gain by using reverse transcriptase?

Assembly and Release of Animal Viruses

As with bacteriophages, once the components of animal viruses are synthesized, they assemble into virions that are then released from the host cell. Most DNA viruses assemble in and are released from the nucleus into the cytosol, whereas most RNA viruses develop solely in the cytoplasm. The number of viruses produced and released depends on both the type of virus and the size and initial health of the host cell.

Replication of animal viruses takes more time than replication of bacteriophages. Herpesviruses, for example, require almost 24 hours to replicate, as compared to 25 minutes for hundreds of copies of bacteriophage T4.

Enveloped animal viruses are often released via a process called **budding (Figure 13.14)**. As virions are assembled, they are extruded through one of the cell's membranes—the nuclear,

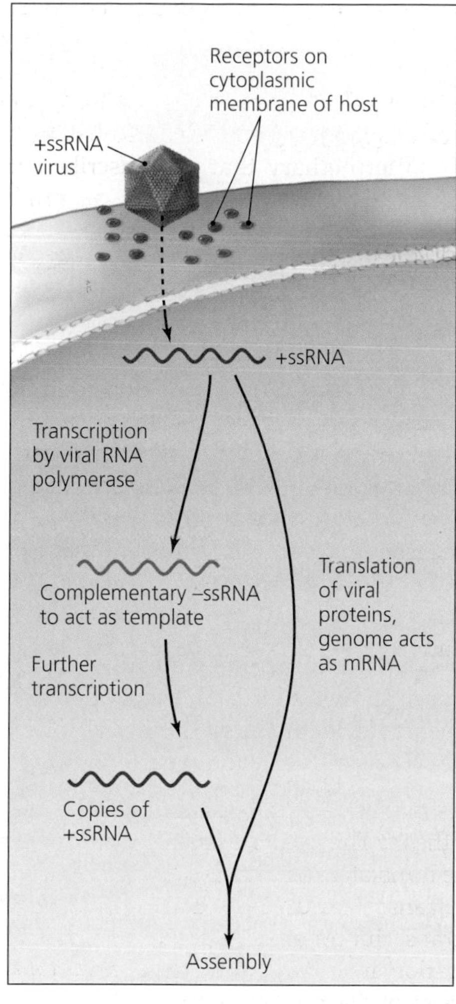

(a) Positive-sense ssRNA virus

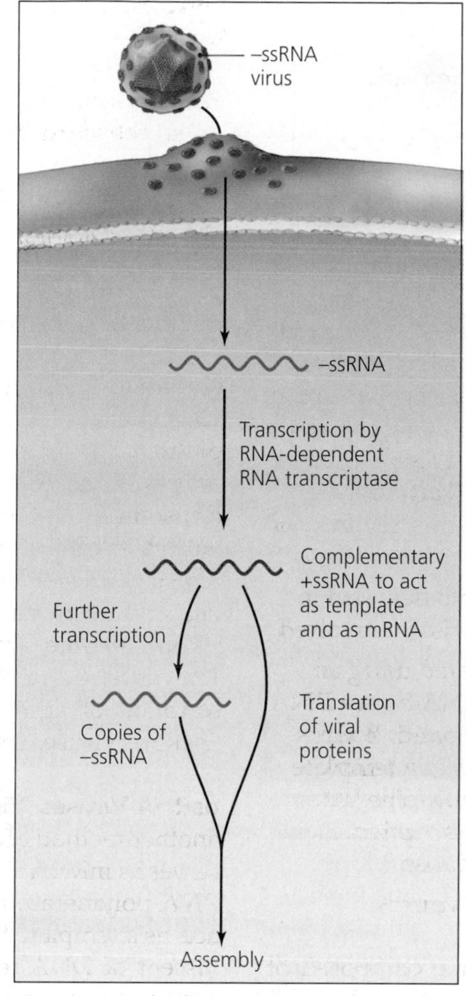

(b) Negative-sense ssRNA virus

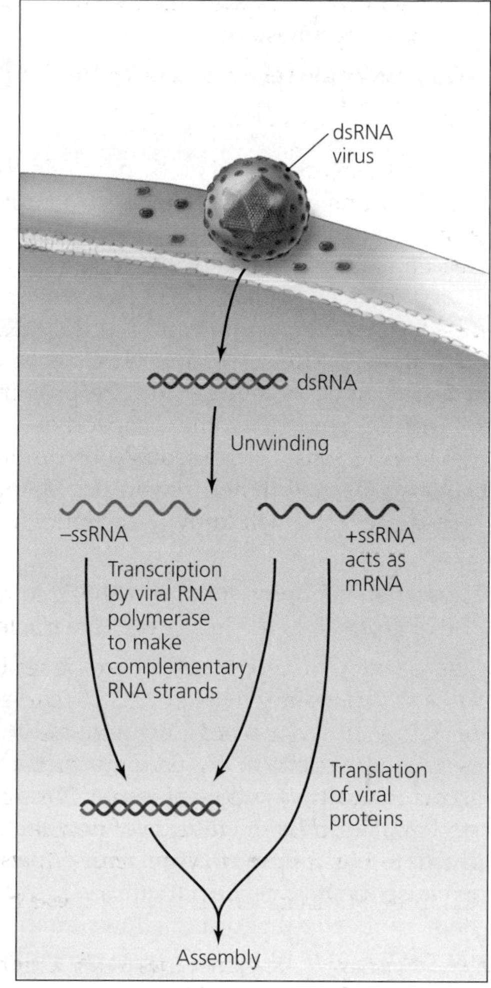

(c) Double-stranded RNA virus

▲ **Figure 13.13** **Synthesis of proteins and genomes in animal RNA viruses.** (a) Positive-sense ssRNA virus, in which +ssRNA acts as mRNA and −ssRNA is the genome template. (b) Negative-sense ssRNA virus: transcription forms +ssRNA to serve both as mRNA and template. (c) dsRNA virus genome unwinds so that the positive-sense strand serves as mRNA, and each strand serves as a template for its complement.

TABLE 13.3 Synthesis Strategies of Animal Viruses

Genome	How Is mRNA Synthesized?	What Molecule Is the Template for Genome Replication?
dsDNA	By RNA polymerase (in nucleus or cytoplasm of cell)	Each strand of DNA serves as template for its complement (except for hepatitis B, which synthesizes RNA to act as the template for new DNA)
ssDNA	By RNA polymerase (in nucleus of cell)	Complementary strand of DNA is synthesized to act as template
+ssRNA	Genome acts as mRNA	−RNA complementary to the genome is synthesized to act as template
+ssRNA (*Retroviridae*)	DNA is synthesized from RNA by reverse transcriptase; mRNA is transcribed from DNA by RNA polymerase	DNA
−ssRNA	By RNA-dependent RNA transcriptase	+RNA (mRNA) complementary to the genome
dsRNA	Positive strand of genome acts as mRNA	Each strand of genome acts as template for its complement

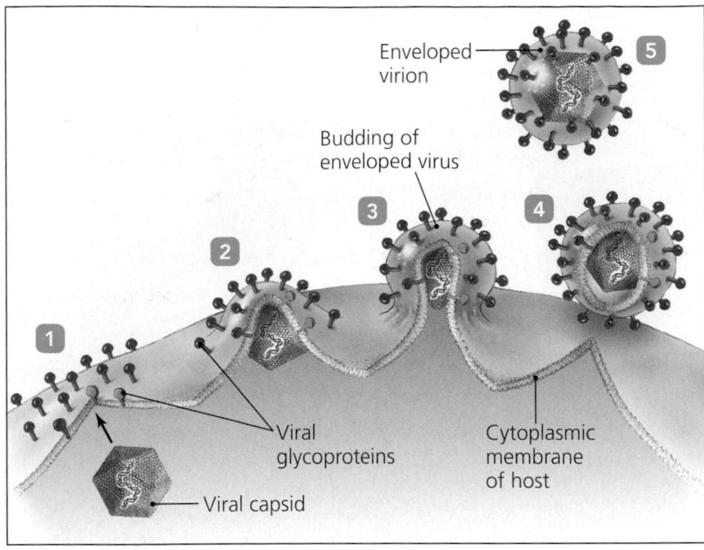

▲ **Figure 13.14 The process of budding in enveloped viruses.** *What term describes a nonenveloped virus?*

Figure 13.14 Naked.

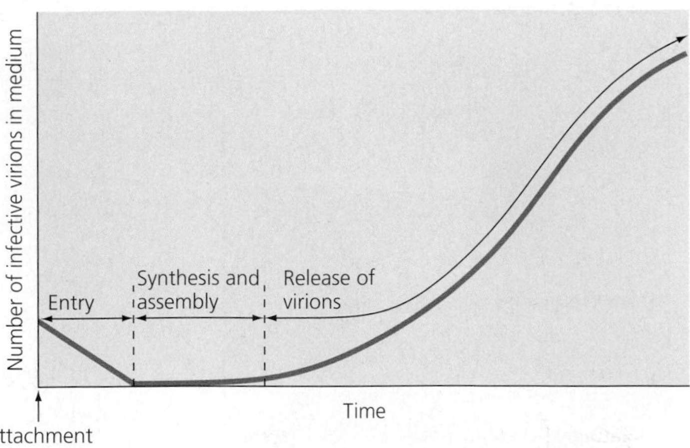

▲ **Figure 13.15 Pattern of virion abundance in persistent infections.** A generalized curve of virion abundance for persistent infections by budding enveloped viruses. Because the curve does not represent any actual infection, units for the graph's axes are omitted.

endoplasmic reticulum, or the cytoplasmic membrane. Each virion acquires a portion of membrane, which becomes the viral envelope. During synthesis, some viral glycoproteins are inserted into cellular membranes, and these proteins become the glycoprotein spikes on the surface of the viral envelope.

Because the host cell is not quickly lysed, as occurs in bacteriophage replication, budding allows an infected cell to remain alive for some time. Infections with enveloped viruses in which host cells shed viruses slowly and relatively steadily are called *persistent infections;* a curve showing virus abundance over time during a persistent infection lacks the burst of new virions seen in lytic replication cycles **(Figure 13.15)** (compare to Figure 13.9).

Naked animal viruses may be released in one of two ways: They may either be extruded from the cell by exocytosis, in a manner similar to budding, but without the acquisition of an envelope, or they may cause lysis and death of the cell, reminiscent of bacteriophage release. **ANIMATIONS:** ***Viral Replication: Animal Viruses***

EMERGING DISEASES

CHIKUNGUNYA

An old man arrived at the doctor's office in Ravenna, Italy, with a combination of symptoms the physician had never heard of: a widespread, severe rash; difficulty in breathing; high fever; nausea; and extreme joint pain. Chikungunya (chik-en-gun' ya) had arrived in Europe.

Though scientists had known of chikungunya virus, which is related to equine encephalitis viruses, for over 50 years, most considered the tropical disease benign—a limited, mild irritation, not a catastrophe. Therefore, few researchers studied chikungunya virus or its disease. Now, they know better.

Over the past decade, chikungunya virus has spread throughout the nations of the Indian Ocean. In 2006, officials on the French-owned island of La Réunion in the Indian Ocean reported 47,000 cases of chikungunya in a single week! The same year it reemerged in India for the first time in four decades with more than 1.5 million reported cases. Why?

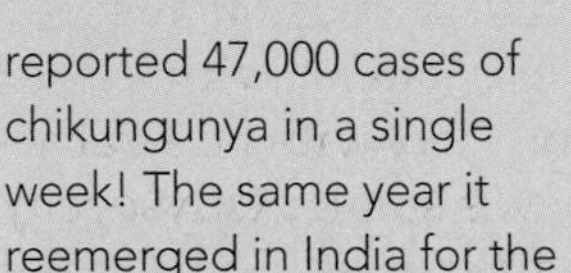

Aedes albopictus (Asian tiger mosquito), which carries the virus, has moved into temperate climates, including Europe and the United States, as the climate has warmed. With the mosquito comes the possibility of viral proliferation—the insects have spread the tropical disease as far north as Italy.

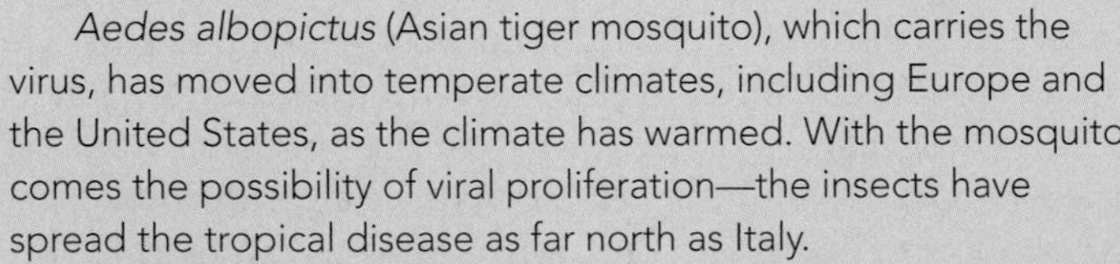

And our Italian patient? His crippling pain lasted for months, but he survived. Now that he knows about mosquito-borne chikungunya, he insists that his family and friends use mosquito repellant liberally. Officials in the rest of Europe and in the United States join in his concern: With the coming of *Ae. albopictus*, is incurable chikungunya far behind?

Track chikungunya online by going to the Study Area at www.masteringmicrobiology.com.

TABLE 13.4

A Comparison of Bacteriophage and Animal Virus Replication

	Bacteriophage	Animal Virus
Attachment	Proteins on tails attach to proteins on cell wall	Spikes, capsids, or envelope proteins attach to proteins or glycoproteins on cell membrane
Penetration	Genome is injected into cell or diffuses into cell	Capsid enters cell by direct penetration, fusion, or endocytosis
Uncoating	None	Removal of capsid by cell enzymes
Site of synthesis	In cytoplasm	RNA viruses in cytoplasm; most DNA viruses in nucleus
Site of assembly	In cytoplasm	RNA viruses in cytoplasm; most DNA viruses in nucleus
Mechanism of release	Lysis	Naked virions: exocytosis or lysis; enveloped virions: budding
Nature of chronic infection	Lysogeny, always incorporated into host chromosome, may leave host chromosome	Latency, with or without incorporation into host DNA; incorporation is permanent

CRITICAL **THINKING**

If an enveloped virus were somehow released from a cell without budding, it would not have an envelope. What effect would this have on the virulence of the virus? Why?

Because viral replication uses cellular structures and pathways involved in the growth and maintenance of healthy cells, any strategy for the treatment of viral diseases that involves disrupting viral replication may disrupt normal cellular processes as well. This is one reason it is difficult to treat viral diseases. The modes of action of some available antiviral drugs are discussed in Chapter 10; the body's naturally produced antiviral chemicals—interferons and antibodies—are discussed in Chapters 15 and 16.

Latency of Animal Viruses

Some animal viruses, including chickenpox and herpes viruses, may remain dormant in cells in a process known as **latency;** the viruses involved in latency are called **latent viruses** or **proviruses.** Latency may be prolonged for years with no viral activity, signs, or symptoms. Though latency is similar to lysogeny as seen with bacteriophages, there are differences. Some latent viruses do not become incor porated into the chromosomes of their host cells, whereas lysogenic phages always do.

On the other hand, some animal viruses (e.g., HIV) are more like lysogenic phages in that they do become integrated into a host chromosome as a provirus. However, when a provirus is incorporated into its host DNA, the condition is permanent; induction does not occur in eukaryotes. Thus an incorporated provirus becomes a permanent, physical part of the host's chromosome, and all descendants of the infected cell will carry the provirus.

Given that RNA cannot be incorporated directly into a chromosome molecule, how does the ssRNA of HIV become a provirus incorporated into the DNA of its host cell? HIV can become a permanent part of a host's chromosome because it, like all retroviruses, carries reverse transcriptase, which transcribes the genetic information of the +RNA molecule to a DNA molecule—which *can* become incorporated into the host cell's genome.

CRITICAL **THINKING**

A latent virus that is incorporated into a host cell's chromosome is never induced; that is, it never emerges from the host cell's chromosome to become a free virus. Given that it cannot emerge from the host cell's chromosome, can such a latent virus be considered "safe"? Why or why not?

Table 13.4 compares the features of the replication of bacteriophages and animal viruses. Next we turn our attention to the part viruses can play in cancer, beginning with a brief consideration of the terminology needed to understand the basic nature of cancer.

The Role of Viruses in Cancer

Learning Objectives

- ✓ Define the terms *neoplasia, tumor, benign, malignant, cancer,* and *metastasis.*
- ✓ Explain in simple terms how a cell may become cancerous, with special reference to the role of viruses.

Under normal conditions, the division of cells in a mature multicellular animal is under strict genetic control; that is, the animal's genes dictate that some types of cells can no longer divide at all, and that those that can divide are prevented from unlimited division. In this genetic control, either genes for cell division are "turned off," or genes that inhibit division are "turned on," or some combination of both these genetic events occur.

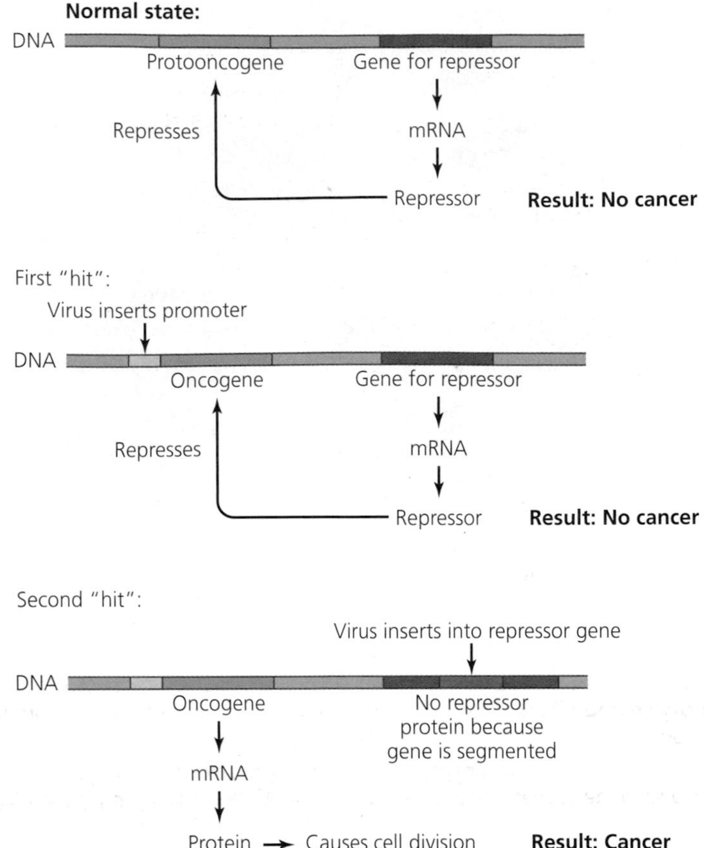

▲ **Figure 13.16 The oncogene theory of the induction of cancer in humans.** The theory suggests that more than one "hit" to the DNA (that is, any change or mutation), whether caused by a virus (as shown here) or various physical or chemical agents, is required to induce cancer.

However, if something upsets the genetic control, cells begin to divide uncontrollably. This phenomenon of uncontrolled cell division in a multicellular animal is called **neoplasia**[2] (nē-ō-plā'zē-ă). Cells undergoing neoplasia are said to be neoplastic, and a mass of neoplastic cells is a **tumor.**

Some tumors are **benign;** that is, they remain in one place and are not generally harmful, although occasionally such noninvasive tumors are painful and rob adjacent normal cells of space and nutrients. Other tumors are **malignant,** invading neighboring tissues and even traveling throughout the body to invade other organs and tissues to produce new tumors—a process called **metastasis** (mĕ-tas'tă-sis). Malignant tumors are also called **cancers.** Cancers rob normal cells of space and nutrients and cause pain; in some kinds of cancer, malignant cells derange the function of the affected tissues, until eventually the body can no longer withstand the loss of normal function and dies.

Several theories have been proposed to explain the role viruses play in the development of cancers. These theories revolve around the presence of *protooncogenes* (prō-tō-ong'kō-jēnz)—genes that play a role in cell division. So long as protooncogenes are repressed, no cancer results. However, activity of oncogenes (their name when they are active) or inactivation of oncogene repressors can cause cancer to develop. In most cases, several genetic changes must occur before cancer develops. Put another way, "multiple hits" to the genome must occur for cancer to result **(Figure 13.16)**.

A variety of environmental factors contribute to the inhibition of oncogene repressors and the activation of oncogenes. Ultraviolet light, radiation, certain chemicals called *carcinogens* (kar-si'nō-jenz), and viruses have all been implicated in the development of cancer.

Viruses cause 20–25% of human cancers in several ways. Some viruses carry copies of oncogenes as part of their genomes; other viruses promote oncogenes already present in the host; still other viruses interfere with normal tumor repression when they insert (as proviruses) into repressor genes.

That viruses cause some animal cancers is well established. In the first decade of the 1900s, virologist F. Peyton Rous (1879–1970) proved that viruses induce cancer in chickens. Though several DNA and RNA viruses are known to cause about 15% of human cancers, the link between viruses and most human cancers has been difficult to document. Among the virally induced cancers in humans are Burkitt's lymphoma, Hodgkin's disease, Kaposi's sarcoma, and cervical cancer. DNA viruses in the families *Adenoviridae, Herpesviridae, Hepadnaviridae, Papillomaviridae,* and *Polyomaviridae,* and two RNA viruses in the family *Retroviridae,* cause these and other human cancers. Chapters 24 and 25 discuss diseases caused by DNA viruses and RNA viruses.

CRITICAL **THINKING**

Why are DNA viruses more likely to cause neoplasias than are RNA viruses?

Culturing Viruses in the Laboratory

Learning Objectives

- ✓ Describe some ethical and practical difficulties to overcome in culturing viruses.
- ✓ Describe three types of media used for culturing viruses.

Scientists must culture viruses in order to conduct research and develop vaccines and treatments, but because viruses cannot metabolize or replicate by themselves, they cannot be grown in standard microbiological broths or on agar plates. Instead, they must be cultured inside suitable host cells, a requirement that complicates the detection, identification, and characterization of viruses. Virologists have developed three types of media for culturing viruses: media consisting of mature organisms (bacteria, plants, or animals), embryonated (fertilized) eggs, and cell cultures. We begin by considering the culture of viruses in organisms.

Culturing Viruses in Mature Organisms

Learning Objectives

- ✓ Explain the use of a plaque assay in culturing viruses in bacteria.
- ✓ List three problems with growing viruses in animals.

[2]From Greek *neo,* meaning new, and *plassein,* meaning to mold.

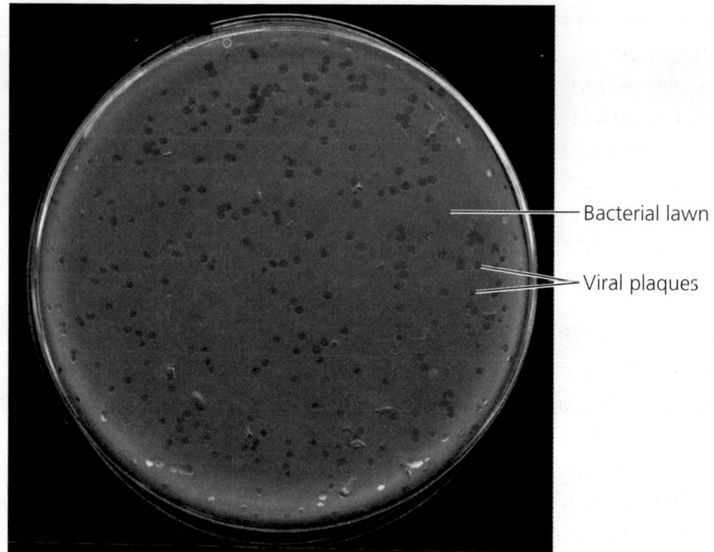

▲ **Figure 13.17 Viral plaques in a lawn of bacterial growth on the surface of an agar plate.** *What is the cause of viral plaques?*

Figure 13.17 *Each plaque is an area in a bacterial lawn where bacteria have succumbed to phage infections.*

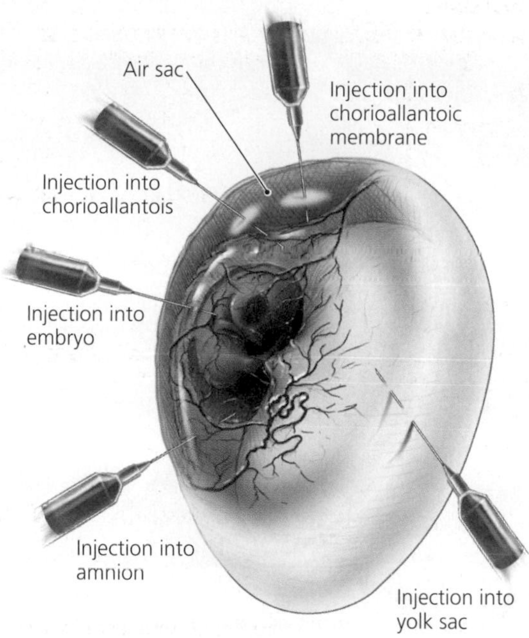

▲ **Figure 13.18 Inoculation sites for the culture of viruses in embryonated chicken eggs.** *Why are eggs often used to grow animal viruses?*

Figure 13.18 *Eggs are large, sterile, self-sufficient cells that contain a number of different sites suitable for viral replication.*

In the following sections we consider the use of bacterial cells as a virus culture medium before considering the issues involved in growing viruses in living animals.

Culturing Viruses in Bacteria

Most of our knowledge of viral replication has been derived from research on bacteriophages, which are relatively easy to culture because some bacteria are easily grown and maintained. Phages can be grown in bacteria maintained in either liquid cultures or on agar plates. In the latter case, bacteria and phages are mixed with warm (liquid) nutrient agar and poured in a thin layer across the surface of an agar plate. During incubation, bacteria infected by phages lyse and release new phages that infect nearby bacteria, while uninfected bacteria grow and reproduce normally. After incubation, the appearance of the plate includes a uniform bacterial lawn interrupted by clear zones called **plaques,** which are areas where phages have lysed the bacteria (**Figure 13.17**). Such plates enable the estimation of phage numbers via a technique called plaque assay, in which virologists assume that each plaque corresponds to a single phage in the original bacterium-virus mixture.

Culturing Viruses in Plants and Animals

Plant and animal viruses can be grown in laboratory plants and animals. Recall that the first discovery and isolation of a virus was the discovery of tobacco mosaic virus in tobacco plants. Rats, mice, guinea pigs, rabbits, pigs, and primates have been used to culture and study animal viruses.

However, maintaining laboratory animals can be difficult and expensive, and this practice raises ethical issues for some. Growing viruses that infect only humans raises additional ethical complications. Therefore, scientists have developed alternative ways of culturing animal and human viruses using fertilized chicken eggs or cell cultures.

Culturing Viruses in Embryonated Chicken Eggs

Chicken eggs are a useful culture medium for viruses because they are inexpensive, are among the largest of cells, are free of contaminating microbes, and contain a nourishing yolk (which makes them self-sufficient). Most suitable for culturing viruses are chicken eggs that have been fertilized and thus contain a developing embryo, the tissues of which (called membranes, which should not be confused with cellular membranes) provide ideal inoculation sites for growing viruses (**Figure 13.18**). Researchers inject samples of virus into embryonated eggs at the sites that are best suited for the particular virus's replication.

Vaccines against some viruses can also be prepared in egg cultures. You may have been asked if you are allergic to eggs before you received such a vaccine, because egg protein may remain as a contaminant in the vaccine.

Culturing Viruses in Cell (Tissue) Culture

Learning Objective

✓ Compare and contrast diploid cell culture and continuous cell culture.

Viruses can also be grown in **cell culture,** which consists of cells isolated from an organism and grown on the surface of a

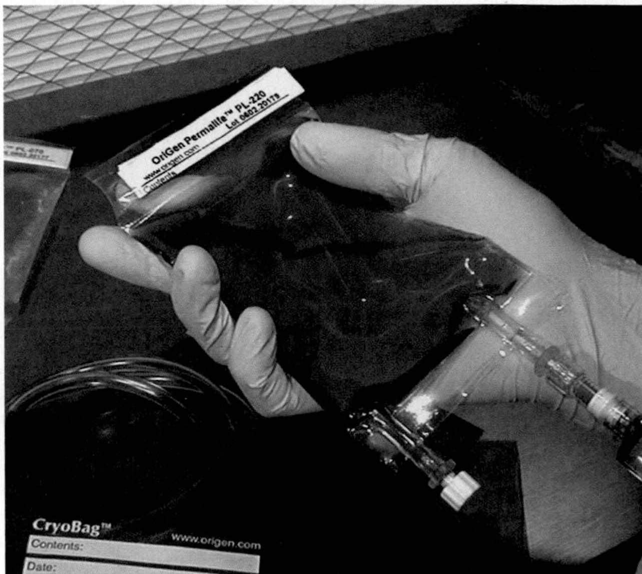

▲ **Figure 13.19 An example of cell culture.** The bag contains a colored nutrient medium for growing cells in which viruses can be cultured.

medium or in broth **(Figure 13.19)**. Such cultures became practical when antibiotics provided a way to limit the growth of contaminating bacteria. Cell culture can be less expensive than maintaining research animals, plants, or eggs, and it avoids some of the moral problems associated with experiments performed on animals and humans. Cell cultures are sometimes called *tissue cultures,* but the term *cell culture* is more accurate because only a single type of cell is used in the culture. (By definition, a tissue is composed of at least two kinds of cells.)

Cell cultures are of two types. The first type, **diploid cell cultures,** are created from embryonic animal, plant, or human cells that have been isolated and provided appropriate growth conditions. The cells in diploid cell culture generally last no more than about 100 generations (cell divisions) before they die.

The second type of culture, **continuous cell cultures,** are longer lasting because they are derived from tumor cells. Recall that a characteristic of neoplastic cells is that they divide relentlessly, providing a never-ending supply of new cells. One of the more famous continuous cell cultures is of HeLa cells, derived from a woman named *He*nrietta *La*cks who died of cervical cancer in 1951. Though she is dead, Mrs. Lacks's cells live on in laboratories throughout the world.

It is interesting that HeLa cells have lost some of their original characteristics. For example, they are no longer diploid because they have lost many chromosomes. HeLa cells provide a semistandard[3] human tissue culture medium for studies on cell metabolism, aging, and (of course) viral infection.

[3]HeLa cells are "semistandard" because different strains have lost different chromosomes, and mutations have occurred over the years. Thus, HeLa cells in one laboratory may be slightly different from HeLa cells in another laboratory.

CRITICAL **THINKING**

HIV replicates only in certain types of human cells, and one early problem in AIDS research was culturing those cells. How do you think scientists are now able to culture HIV?

Are Viruses Alive?

Learning Objective

✓ Discuss aspects of viral replication that are life-like and non-life-like.

Now that we have studied the characteristics and replication processes of viruses, let's return to a question first posed in Chapter 3: Are viruses alive?

To be able to wrestle with the answer, we must first recall the characteristics of life: growth, self-reproduction, responsiveness, and the ability to metabolize, all within structures called cells. According to these criteria, viruses seem to lack the qualities of living things, prompting some scientists to consider them nothing more than complex pathogenic chemicals. For other scientists, however, at least three observations—that viruses use sophisticated methods to invade cells, have the means of taking control of their host cells, and possess genomes containing instructions for replicating themselves—indicate that viruses are the ultimate parasites, because they use cells to make more viruses. According to this viewpoint, viruses are the least complex living entities.

In any case, viruses are right on the threshold of life—outside of cells they do not appear to be alive, but within cells they direct the synthesis and assembly required to make copies of themselves.

What do you think? Are viruses alive?

Other Parasitic Particles: Viroids and Prions

Viruses are not the only submicroscopic entities capable of causing disorders within cells. In this section we will consider the characteristics of two molecular particles that infect cells: viroids and prions.

Characteristics of Viroids

Learning Objectives

✓ Define and describe viroids.

✓ Compare and contrast viroids and viruses.

Viroids are extremely small, circular pieces of RNA that are infectious and pathogenic in plants **(Figure 13.20)**. Viroids are similar to RNA viruses except that they lack capsids. Even though they are circular, viroids may appear linear because of hydrogen bonding within the molecule. Several plant diseases, including some of coconut palm, chrysanthemum, potato, cucumber, and avocado, are caused by viroids, including the stunting shown in **Figure 13.21**.

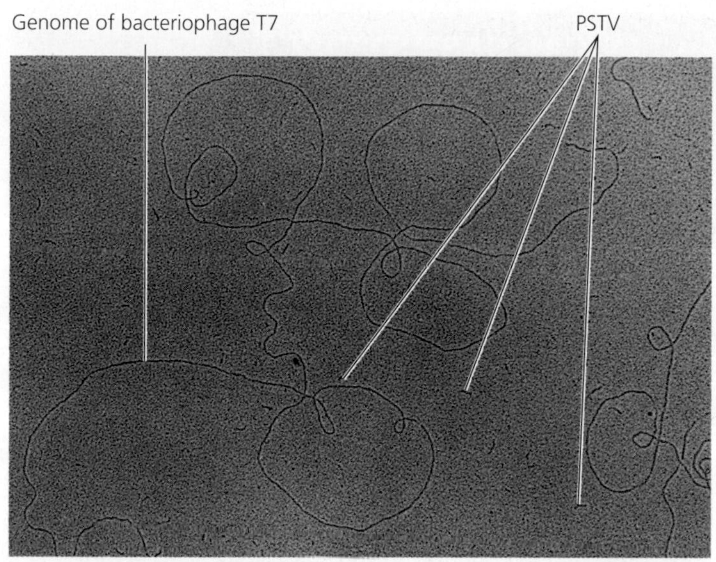

▲ **Figure 13.20 The RNA strand of the small potato spindle tuber viroid (PSTV).** Also shown for comparison is the longer DNA genome of bacteriophage T7. *Compare both to the size of a bacterial genome in Figure 13.2. How are viroids similar to and different from viruses?*

Figure 13.20 *Viroids are similar to certain viruses in that they are infectious and contain a single strand of RNA; they are different from viruses in that they lack a proteinaceous capsid.*

Viroidlike agents—infectious, pathogenic RNA particles that lack capsids but do not infect plants—affect some fungi. (They are not called viroids because they do not infect plants.) No animal diseases are known to be caused by viroidlike molecules, though the possibility exists that infectious RNA may be responsible for some diseases in humans.

Characteristics of Prions

Learning Objectives

- ✓ Define and describe prions, including their replication process.
- ✓ Compare and contrast prions and viruses.
- ✓ List four diseases caused by prions.

In 1982, Stanley Prusiner (1942–) described a proteinaceous infectious agent that was different from any other known infectious agent in that it lacked nucleic acid. Prusiner named such agents of disease **prions** (prē′onz), for *pro*teinaceous *in*fective particles. Before his discovery, the diseases now known to be caused by prions were thought to be caused by what were known as "slow viruses," which were so named because of the long delay between infection and the onset of signs and symptoms. Through experiments, Prusiner and his colleagues showed that prions are not viruses because they lack any nucleic acid. **ANIMATIONS:** ***Prions: Overview***

Some scientists resisted the concept of prions because particles that lack any nucleic acid violate the "universal" rule of protein synthesis—that proteins are translated from a molecule of mRNA. Since "infectious proteins" lack nucleic acid, how can they carry the information required to replicate themselves?

All mammals make a cytoplasmic membrane protein called *PrP*, which plays a role in the normal activity of the brain, though the exact function of PrP is unknown. The amino acid sequence in PrP is such that the protein can fold into two stable tertiary structures: the normal, functional structure of *cellular PrP* has several prominent α-helices, while a disease-causing form—*prion PrP*—is characterized by β-pleated sheets (**Figure 13.22**).

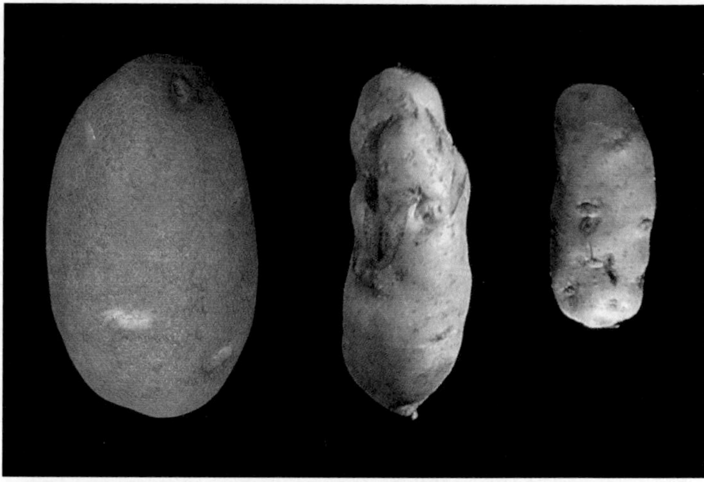

▲ **Figure 13.21 One effect of viroids on plants.** The potatoes at right are stunted as the result of infection with PSTV viroids.

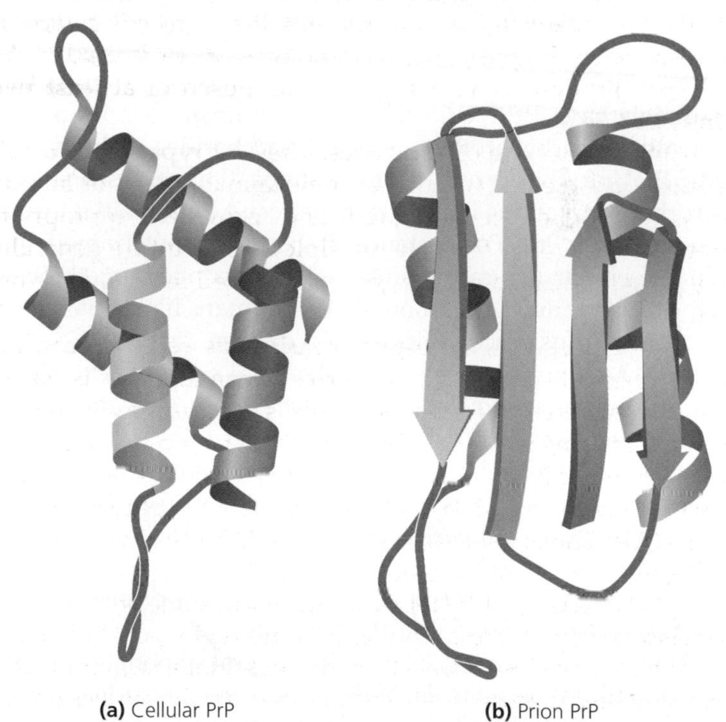

▲ **Figure 13.22 The two stable, three-dimensional forms of prion protein (PrP). (a)** Cellular prion protein found in normal functional cells has a preponderance of alpha-helices. **(b)** Prion PrP, which has the same amino acid sequence, is folded to produce a preponderance of beta-pleated sheets.

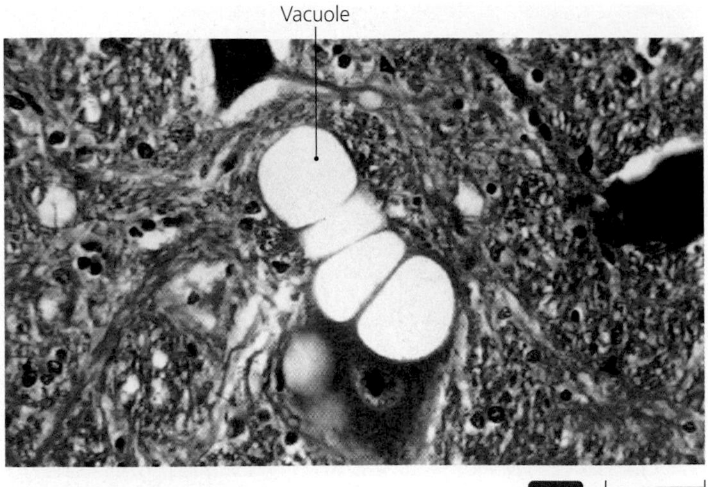

▲ **Figure 13.23 A brain showing the large vacuoles and spongy appearance typical in prion-induced diseases.** Shown here is the brain of a sheep with the prion disease called scrapie.

Prusiner determined that prion PrP acts enzymatically to convert cellular PrP into prion PrP by inducing a conformational change in the shape of cellular PrP. In other words, prion PrP refolds cellular PrP into a prion shape. As cellular ribosomes synthesize more cellular PrP, prions already present cause these new molecules to refold, becoming more prions. Thus, the rule of protein synthesis is maintained, because prions do not code for new prions but instead convert extant cellular PrP into prions. For his work, Prusiner was awarded a Nobel Prize in 1997. **ANIMATIONS:** ***Prions: Characteristics***

Why don't prions develop in all mammals, given that all mammals have PrP? Under normal circumstances, it appears that other nearby proteins and polysaccharides in lipid rafts force PrP into the correct (cellular) shape. Only excess PrP production or mutations in the PrP gene result in the initial formation of prion PrP. For instance, it has been determined that human cellular PrP can misfold only if it contains methionine as the 129th amino acid. About 40% of humans have this type of PrP and are thus susceptible to prion disease.

All known prion diseases involve fatal neurological degeneration, the deposition of fibrils in the brain, and the loss of brain matter such that eventually large vacuoles form. The latter process results in a characteristic spongy appearance **(Figure 13.23)**, which is why as a group, prion diseases are called *spongiform encephalopathies* (spŭn'ji-fōrm en-sef'ă-lop'ă-thēz). The way in which prion PrP causes these changes in the brain is not totally understood, but it appears prions interupt signaling events that cellular PrP controls.

Prions are associated with several spongiform encephalopathies of animals and of humans, including *bovine spongiform encephalitis* (*BSE*, so-called mad cow disease), *scrapie* in sheep, *kuru* (a disease that afflicted Papua New Guinea in the 1950s but has since been eliminated), *chronic wasting disease (CWD)* in deer and elk, and *variant Creutzfeldt-Jakob disease*[4] (*vCJD*) in humans. All these diseases are transmitted through either the ingestion of infected tissue, including brain, spinal cord, eyes, and, in some cases, skeletal muscle tissue; transplants of infected tissue, including nervous tissue and blood; or contact between infected tissue and mucous membranes or skin abrasions. **ANIMATIONS:** ***Prions: Diseases***

Normal cooking or sterilization procedures do not deactivate prions, though they are destroyed by incineration or autoclaving in 1 normal NaOH. The European Union recently approved the use of special enzymes developed using biotechnology to remove infectious prions from medical equipment.

There is no treatment for any prion disease, though scientists have determined that the antimalarial drug quinacrine and the antipsychotic drug chlorpromazine forestall spongiform encephalopathy in prion-infected mice. Human trials of these drugs are ongoing.

PrP proteins of different species are slightly different in primary structure, and at one time it was thought unlikely that prions of one species could cross over into another species to cause disease. However, an epidemic of BSE in Great Britain in the late 1980s and in France in the 2000s resulted in the spread of

CLINICAL CASE STUDY

Invasion from Within or Without?

A 32-year-old husband and father of two small children lived in the midwestern United States. An avid hunter and hearty meat eater since childhood, the man visited annually with family and friends in Colorado for elk hunting. His job required frequent international travel (including Europe) where he enjoyed exotic foods.

Early in the spring of 1988, his wife recalls that he began having problems. Frequently he forgot to pick up things from the store or even that his wife had called him. Later that summer, he was unable to complete his paperwork at his business and had great difficulty performing even basic math. In England on business, he had forgotten his home phone number in the U.S. and couldn't remember how to spell his last name to ask for directory assistance.

By September he could not perform his job and his wife insisted he seek medical care. All the standard blood tests came back normal. A psychologist diagnosed depression, but after a brain biopsy, the man was found to have spongiform changes in his brain. He was given 6–8 weeks to live as there is no treatment for this condition.

1. What is the likely diagnosis?
2. The man's wife was concerned for her children's welfare and asked, "Can we catch this disease from my husband?" How would you respond?
3. Where and how was the man probably infected?

[4]"Variant" because it is derived from BSE prions in cattle, as opposed to the regular form of CJD, which is a genetic disease.

TABLE 13.5

Comparison of Viruses, Viroids, and Prions to Bacterial Cells

	Bacteria	Viruses	Viroids	Prions
Width	200–2000 nm	10–400 nm	2 nm	5 nm
Length	200–550,000 nm	20–800 nm	40–130 nm	5 nm
Nucleic acid?	Both DNA and RNA	Either DNA or RNA, never both	RNA only	None
Protein?	Present	Present	Absent	Present (PrP)
Cellular?	Yes	No	No	No
Cytoplasmic membrane?	Present	Absent (though some viruses do have a membranous envelope)	Absent	Absent
Functional ribosomes?	Present	Absent	Absent	Absent
Growth?	Present	Absent	Absent	Absent
Self-replicating?	Yes	No	No	Yes; transform PrP protein already present in cell
Responsiveness?	Present	Some bacteriophages respond to a host cell by injecting their genomes	Absent	Absent
Metabolism?	Present	Absent	Absent	Absent

prions not only to other cattle, but also to humans who ate infected beef. To prevent further infection, most countries ban the use of animal-derived protein supplements in animal feed. Unfortunately, this step was too late to prevent the spread of bovine prions to more than 150 Europeans who have developed fatal vCJD. (See **Clinical Case Study: Invasion from Within or Without?**)

Prions composed of different proteins may lie behind other muscular and neuronal degenerative diseases, such as Alzheimer's disease, Parkinson's disease, and amyotrophic lateral sclerosis (ALS, or "Lou Gehrig's disease").

CRITICAL **THINKING**

Why did most scientists initially resist the idea of an infectious, replicating protein?

In this chapter we have seen that humans, animals, plants, fungi, bacteria, and archaea are susceptible to infection by acellular pathogens: viruses, viroids, and prions. Table 13.5 summarizes the differences and similarities between these pathogenic agents and bacterial pathogens.

Chapter Summary

1. Viruses, viroids, and prions are **acellular** disease-causing agents that lack cell structure and cannot metabolize, grow, self-reproduce, or respond to their environment.

Characteristics of Viruses (pp. 375–379)

1. A **virus** is a tiny infectious agent with nucleic acid surrounded by proteinaceous **capsomeres** that form a coat called a **capsid.** A virus exists in an extracellular state and an intracellular state. A **virion** is a complete viral particle, including a nucleic acid and a capsid, outside of a cell.
2. The genomes of viruses include either DNA or RNA. Viral genomes may be dsDNA, ssDNA, dsRNA, or ssRNA. They may exist as linear or circular and singular or multiple molecules of nucleic acid, depending on the type of virus.
3. **A bacteriophage** (or **phage**) is a virus that infects a bacterial cell.
4. Virions can have a membranous **envelope** or be naked—that is, without an envelope.

Classification of Viruses (pp. 379–381)

1. Viruses are classified based on type of nucleic acid, presence of an envelope, shape, and size.
2. The International Committee on Taxonomy of Viruses (ICTV) has recognized viral family and genus names. With the exception of three orders, higher taxa are not established.

Viral Replication (pp. 382–391)

1. Viruses depend on random contact with a specific host cell type for replication. Typically a virus in a cell proceeds with a lytic replication cycle with five stages: **attachment, entry, synthesis, assembly,** and **release.**
   ANIMATIONS: *Viral Replication: Overview*
2. Once attachment has been made between virion and host cell, the nucleic acid enters the cell. With phages, only the nucleic acid enters the host cell. With animal viruses, the entire virion often enters the cell, where the capsid is then removed in a process called **uncoating.**
   ANIMATIONS: *Viral Replication: Animal Viruses*

3. Within the host cell, the viral nucleic acid directs synthesis of more viruses using metabolic enzymes and ribosomes of the host cell.
4. Assembly of synthesized virions occurs in the host cell, typically as capsomeres surround replicated or transcribed nucleic acids to form new virions.
5. Virions are released from the host cell either by lysis of the host cell (seen with phages and animal viruses) or by the extrusion of enveloped virions through the host's cytoplasmic membrane (called **budding**), a process seen only with certain animal viruses. If budding continues over time, the infection is persistent. An envelope is derived from a cell membrane.
   ANIMATIONS: *Viral Replication: Virulent Bacteriophages*
6. **Temperate** (lysogenic) **phages** enter a bacterial cell and remain inactive, in a process called **lysogeny** or a **lysogenic replication cycle.** Such inactive phages are called **prophages** and are inserted into the chromosome of the cell and passed to its daughter cells. **Lysogenic conversion** results when phages carry genes that alter the phenotype of a bacterium. At some point in the generations that follow, a prophage may be excised from the chromosome in a process known as **induction.** At that point the prophage again becomes a lytic virus.
   ANIMATIONS: *Viral Replication: Temperate Bacteriophages*
7. In **latency,** a process similar to lysogeny, an animal virus remains inactive in a cell, possibly for years, as part of a chromosome or in the cytosol. A **latent** virus is also known as a **provirus.** A provirus that has become incorporated into a host's chromosome remains there.
8. With the exception of hepatitis B virus, dsDNA viruses act like cellular DNA in transcription and replication.
9. Some ssRNA viruses have **positive-strand RNA (+RNA),** which can be directly translated by ribosomes to synthesize protein. From the +RNA, complementary **negative-strand RNA (−RNA)** is transcribed to serve as a template for more +RNA.
10. **Retroviruses,** such as HIV, are +ssRNA viruses that carry reverse transcriptase, which transcribes DNA from RNA. This reverse process (DNA transcribed from RNA) is reflected in the name retrovirus.
11. −ssRNA viruses carry an RNA-dependent RNA transcriptase for transcribing mRNA from the −RNA genome so that protein can then be translated. Transcription of RNA from RNA is not found in cells.
12. In dsRNA viruses, one strand of RNA functions as a genome, and the other strand functions as a template for RNA replication.

The Role of Viruses in Cancer (pp. 391–392)

1. **Neoplasia** is uncontrolled cellular reproduction in a multicellular animal. A mass of neoplastic cells, called a **tumor,** may be relatively harmless **(benign)** or invasive **(malignant).** Malignant tumors are also called **cancer. Metastasis** describes the spreading of malignant tumors. Environmental factors or oncogenic viruses may cause neoplasia.

Culturing Viruses in the Laboratory (pp. 392–394)

1. In the laboratory, viruses must be cultured inside mature organisms, in embryonated chicken eggs, or in cell cultures because viruses cannot metabolize or replicate alone.
2. When a mixture of bacteria and phages is grown on an agar plate, bacteria infected with phages lyse, producing clear areas called **plaques** on the bacterial lawn. A technique called **plaque assay** enables the estimation of phage numbers.
3. Viruses can be grown in two types of **cell cultures.** Whereas **diploid cell cultures** last about 100 generations, **continuous cell cultures,** derived from cancer cells, last longer.

Are Viruses Alive? (p. 394)

1. Outside of cells, viruses do not appear to be alive, but within cells, they exhibit lifelike qualities such as the ability to replicate themselves.

Other Parasitic Particles: Viroids and Prions (pp. 394–397)

1. **Viroids** are small circular pieces of RNA with no capsid that infect and cause disease in plants. Similar pathogenic RNA molecules have been found in fungi.
2. **Prions** are infectious protein particles that lack nucleic acids and replicate by converting similar, normal proteins into new prions. Diseases caused by prions are spongiform encephalopathies, which involve fatal neurological degeneration.
   ANIMATIONS: *Prions: Overview, Characteristics, and Diseases*

Questions for Review

Answers to the Questions for Review (except Short Answer questions) begin on page A-1.

Multiple Choice

1. Which of the following is *not* an acellular agent?
   a. viroid
   b. virus
   c. rickettsia
   d. prion
2. Which of the following statements is true?
   a. Viruses move toward their host cells.
   b. Viruses are capable of metabolism.
   c. Viruses lack a cell membrane.
   d. Viruses grow in response to their environmental conditions.
3. A virus that is specific for a bacterial host is called a
   a. phage.
   b. prion.
   c. virion.
   d. viroid.
4. A naked virus
   a. has no membranous envelope.
   b. has injected its DNA or RNA into a host cell.
   c. is devoid of capsomeres.
   d. is one that is unattached to a host cell.

5. Which of the following statements is *false*?
   a. Viruses may have circular DNA.
   b. dsRNA is found in bacteria more often than in viruses.
   c. Viral DNA may be linear.
   d. Typically, viruses have DNA or RNA, but not both.
6. When a eukaryotic cell is infected with an enveloped virus and sheds viruses slowly over time, this infection is
   a. called a lytic infection.
   b. a prophage cycle.
   c. called a persistent infection.
   d. caused by a quiescent virus.
7. Another name for a complete virus is
   a. virion.
   b. viroid.
   c. prion.
   d. capsid.
8. Which of the following viruses can be latent?
   a. HIV
   b. chickenpox virus
   c. herpesviruses
   d. all of the above
9. Which of the following is *not* a criterion for specific family classification of viruses?
   a. the type of nucleic acid present
   b. envelope structure
   c. capsid type
   d. lipid composition
10. A clear zone of phage infection in a bacterial lawn is
   a. a prophage.
   b. a plaque.
   c. naked.
   d. a capsomere.

Matching

Match each numbered term with its description.

1. ____ uncoating — A. dormant virus in a eukaryotic cell
2. ____ prophage — B. a virus that infects a bacterium
3. ____ retrovirus — C. transcribes DNA from RNA
4. ____ bacteriophage — D. protein coat of virus
5. ____ capsid — E. a membrane on the outside of a virus
6. ____ envelope — F. complete viral particle
7. ____ virion — G. inactive virus within bacterial cell
8. ____ provirus — H. removal of capsomeres from a virion
9. ____ benign tumor — I. invasive neoplastic cells
10. ____ cancer — J. harmless neoplastic cells

Labeling

Label each step in the bacterioplage replication cycle below.

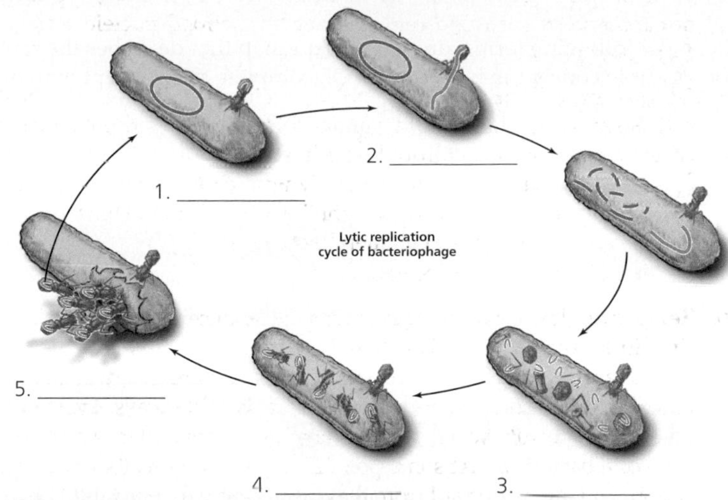

Short Answer

1. Compare and contrast a bacterium and a virus by writing either "Present" or "Absent" for each of the following structures.

Structure	Bacterium	Virus
Cell membrane		
Functional ribosome		
Cytoplasm		
Nucleic acid		
Nuclear membrane		

2. Describe the five phases of a generalized lytic replication cycle.
3. Why is it difficult to treat viral infections?
4. Describe four different ways that viral nucleic acid can enter a host cell.
5. Contrast lysis and budding as means of release of virions from a host cell.
6. What is the difference between a virion and a virus particle?
7. How is a provirus like a prophage? How is it different?
8. Describe lysogeny.
9. How are viruses specific for their host's cells?
10. Compare and contrast diploid cell culture and continuous cell culture.

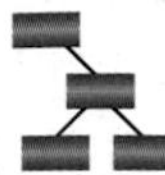

Concept Mapping

Using the following terms, draw a concept map that describes the replication of animal viruses. For a sample concept map, see p. 93. Or, complete this concept map online by going to the Study Area at www.masteringmicrobiology.com.

+ssRNA
−ssRNA
Assembly
Attachment
Budding
dsDNA
dsRNA
Endocytosis
Entry
Exocytosis
Fusion
Host cells
Lysis
New virions
Release
ssDNA
Synthesis
Uncoating
Viral genome
Viral nucleic acid
Viral proteins

Critical Thinking

1. Small viruses typically have a single-stranded genome, while larger viruses usually have a double-stranded genome. What reasonable explanation can you offer for this observation?
2. What are the advantages and disadvantages to bacteriophages of the lytic and lysogenic reproductive strategies?
3. How are computer viruses similar to biological viruses? Are computer viruses alive? Why or why not?
4. Compare and contrast lysogeny by a prophage and latency by a provirus.
5. An agricultural microbiologist wants to stop the spread of a viral infection of a crop. Is stopping viral attachment a viable option? Why or why not?

14 Infection, Infectious Diseases, and Epidemiology

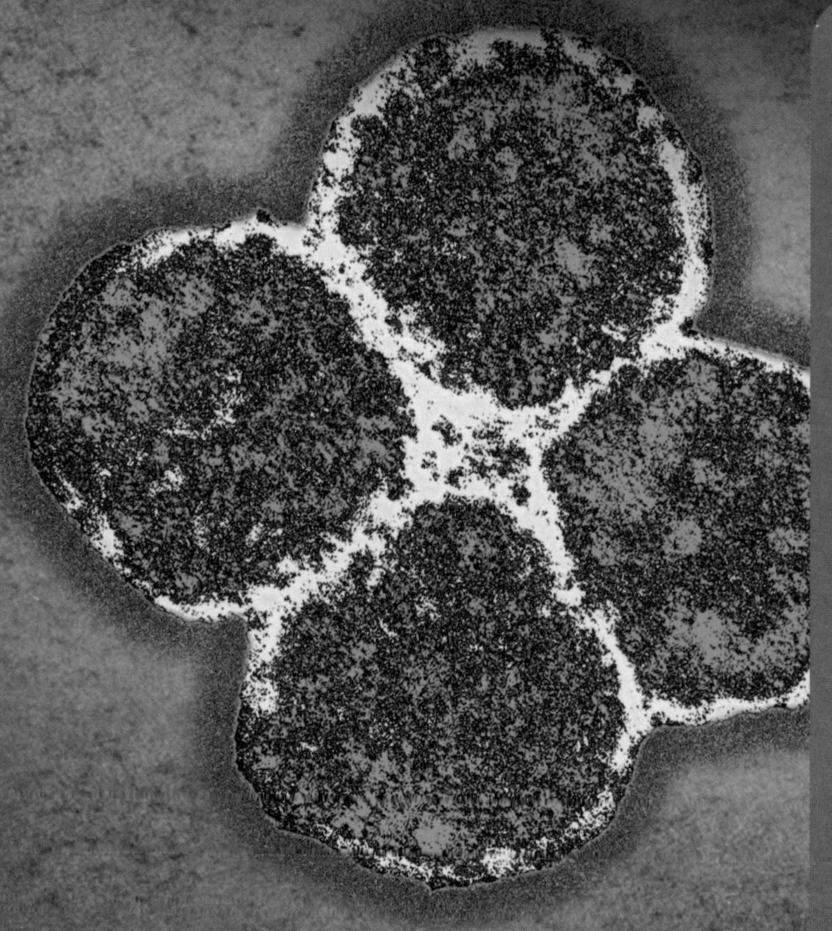

Severe acute respiratory syndrome (SARS) virus traveled halfway around the world in just weeks, providing health care workers and governments with unique opportunities to monitor and contain an epidemic.

A new mysterious illness began appearing in China's Guangdong province in the winter of 2002–2003. Characterized by fever, shortness of breath, and atypical pneumonia, the highly infectious illness spread rapidly, particularly among food handlers and hospital workers tending to the already stricken. Initially confined to China, the disease is believed to have begun its worldwide spread in February 2003, when a physician who had contracted the disease in Guangdong traveled to Hong Kong. During his stay in a Hong Kong hotel, the physician passed the disease on to at least 12 other hotel guests, including travelers who continued on to Canada, Singapore, and Vietnam, unknowingly advancing the disease's spread to these countries. By March 2003, the disease had a name—severe acute respiratory syndrome (SARS)—and the World Health Organization (WHO) issued a global alert, triggering quarantines, travel cancellations, and other measures intended to curb the spread of this new, sometimes-fatal disease. (See Chapter 25 for more information on SARS and the virus that causes it.)

How do microorganisms infect and cause disease in people? And how do experts investigate the transmission and spread of infectious diseases in populations? Read on.

Take the pre-test for this chapter online. Visit the Study Area at www.masteringmicrobiology.com.

In previous chapters we have discussed characteristics of microorganisms and their metabolism. We examined ways microbes can be controlled, viewed, identified, and used for our benefit. Now it is time to examine how microbes directly affect our bodies. In this discussion, continually bear in mind that most microorganisms are neither harmful nor expressly beneficial to humans—they live their lives, and we live ours. Relatively few microbes either directly benefit us or harm us.

In this chapter we first examine the general types of relationships that microbes can have with the bodies of their hosts. Then we discuss sources of infectious diseases of humans; how microbes enter and attach to their hosts; the nature of infectious diseases; and how microbes leave hosts to become available to enter new hosts. Next we explore the ways that infectious diseases are spread among hosts. Finally, we consider *epidemiology*, the study of the occurrence and spread of diseases within groups of humans, and the methods by which we can limit the spread of pathogens within society.

▲ **Figure 14.1 Mutualism.** Wood-eating termites in the genus *Reticulitermes* cannot digest the cellulose in wood, but the protozoan *Trichonympha*, which lives in their intestines, can digest cellulose with the help of bacteria. The three organisms maintain a mutualistic relationship that is crucial for the life of the termite. *What benefits accrue to each of the symbionts in this mutalistic relationship?*

Figure 14.1 *The termite gets digested cellulose from the protozoa, which get a constant supply of wood pulp to digest, so the bacteria get a food source as well.*

Symbiotic Relationships Between Microbes and Their Hosts

Symbiosis (sim-bī-o′sis) means "to live together." Each of us has symbiotic relationships with countless microorganisms, although often we are completely unaware of them. The 10 trillion or so cells in your body provide homes for more than 100 trillion bacteria and fungi and even more viruses. We begin this section by considering the types of relationships between microbes and their hosts.

Types of Symbiosis

Learning Objectives

✓ Distinguish among the types of symbiosis, listing them in order from most beneficial to most harmful for the host.

✓ Describe the relationships among the terms *parasite*, *host*, and *pathogen*.

Biologists see symbiotic relationships as a continuum from cases that are beneficial to both members of a pair to situations in which one member lives at a damaging expense to the other. In the following sections, we examine three conditions along the continuum: mutualism, commensalism, and parasitism.

Mutualism

In **mutualism** (mū′tū-ăl-izm), both members benefit from their interaction. For example, bacteria in your colon receive a warm, moist, nutrient-rich environment in which to thrive, while you absorb vitamin precursors and other nutrients released from the bacteria. In this example, the relationship is beneficial to microbe and human alike but is not required by either. Many of the bacteria could live elsewhere, and you could get vitamins from your diet.

Some mutualistic relationships provide such important benefits that one or both of the parties cannot live without the other. For instance, termites, which cannot digest the cellulose in wood by themselves, would die without a mutualistic relationship with colonies of wood-digesting protozoa and bacteria living in their intestines (**Figure 14.1**). The termites provide wood pulp and a home to the protozoa, while the protozoa break down the wood within the termite intestines using enzymes from the bacteria, which live on the protozoa. The microbes share the nutrients released with the termites. **Beneficial Microbes: A Bioterrorist Worm** describes another mutualistic relationship.

Commensalism

In **commensalism** (kŏ-men′săl-izm), a second type of symbiosis, one member of the relationship benefits without significantly affecting the other. For example, *Staphylococcus epidermidis* (staf′i-lō-kok′us ep-i-der-mid′is) growing on the skin typically causes no measurable harm to a person. An absolute example of commensalism is difficult to prove because the host may experience unobserved benefits. In this example, *Staphylococcus* may inhibit pathogenic microbes from colonizing the skin.

Parasitism

Of concern to health care professionals is a third type of symbiosis called **parasitism**[1] (par′ă-si-tizm). A **parasite** derives benefit from its **host** while harming it, though some hosts sustain only slight damage. In the most severe cases, a parasite kills its host, in the process destroying its own home, which is not

[1]From Greek *parasitos*, meaning one who eats at the table of another.

TABLE 14.1 The Three Types of Symbiotic Relationships

	Organism 1	Organism 2	Example
Mutualism	Benefits	Benefits	Bacteria in human colon
Commensalism	Benefits	Neither benefits nor is harmed	*Staphylococcus* on skin
Parasitism	Benefits	Is harmed	Tuberculosis bacteria in human lung

beneficial for the parasite. Therefore, parasites that allow their hosts to survive are more likely to spread. Similarly, hosts that tolerate a parasite are more likely to reproduce. The result over time will be *coevolution* toward commensalism or mutualism. Any parasite that causes disease is called a **pathogen** (path'ō-jen).

A variety of protozoa, fungi, and bacteria are microscopic parasites of humans. Larger parasites of humans include parasitic worms and biting arthropods,[2] including mites (chiggers), ticks, mosquitoes, fleas, and bloodsucking flies.

You should realize that microbes that are parasitic may become mutualistic or vice versa; that is, the relationships between and among organisms can change over time. Table 14.1 summarizes three types of symbiosis.

CRITICAL **THINKING**

Corals are colonial marine animals that feed by filtering small microbes from seawater in tropical oceans worldwide. Biologists have discovered that the cells of most corals are hosts to microscopic algae called zooxanthellae. Design an experiment to ascertain whether corals and zooxanthellae coexist in a mutual, commensal, or parasitic relationship.

Normal Microbiota in Hosts

Learning Objective

✓ Describe the normal microbiota, including resident and transient members.

Even though many parts of your body are *axenic*[3] (ā-zen'ik) environments—that is, sites that are free of any microbes—other parts of your body shelter millions of mutualistic and commensal symbionts. Each square centimeter of your skin, for example, contains about 3 million bacteria, and your large intestine contains 400 to 1000 kinds of bacteria outnumbering your own cells many times. The microbes that colonize the surfaces of the body without normally causing disease constitute the body's **normal microbiota (Figure 14.2)**, also sometimes called the *normal flora*[4]

[2]From Greek *arthros*, meaning jointed, and *pod*, meaning foot.
[3]From Greek *a*, meaning no, and *xenos*, meaning foreigner.
[4]This usage of the term *flora*, which literally means plants, derives from the fact that bacteria and fungi were considered plants in early taxonomic schemes. *Microbiota* is preferable because it properly applies to all microbes (whether eukaryotic or prokaryotic) and to viruses.

BENEFICIAL MICROBES

A BIOTERRORIST WORM

Killed insect *Steinernema*

LM 4 mm

▲ **The worm *Steinernema* (white) kills by infecting insects with lethal *Xenorhabdus* bacteria.**

Bioterrorism has been defined as the deliberate release of viruses, bacteria, or other germs to cause illness or death. A nematode worm, *Steinernema*, is by that definition a bioterrorist, releasing its mutualistic bacterial symbiont, *Xenorhabdus*.

Nematodes are microscopic, unsegmented, round worms that live in soil worldwide. *Steinernema* preys on insects, including ants, termites, and the immature stages of various beetles, weevils, worms, fleas, ticks, and gnats. The bioterrorist worms crawl into an insect's mouth or anus and then cross the intestinal walls to enter the insect's blood. The nematodes then release their symbiotic *Xenorhabdus*. This bacterium inactivates the insect's defensive systems; generates antibacterial compounds, which eliminate other bacteria; produces insecticidal toxins to kill the insect; and secretes digestive enzymes. The bacterial enzymes turn the insect's body into a slimy porridge of nutrients within 48 hours. Meanwhile, *Steinernema* nematodes mature, mate, and reproduce within the insect's liquefying body. The nematodes' offspring feed on the gooey fluid until they are old enough to emerge from the insect's skeleton, but not before taking up a supply of bacteria for their own future bioterroist raids on new insect hosts.

This is a true mutualistic symbiosis: The nematode depends upon the bacterium to kill and digest the insect host, while the bacterium depends on the nematode to deliver it to new hosts. Both benefit. Farmers too can benefit. They can use *Steinernema* and its bacterium to control insect pests without damaging their crops, animals, or people.

▲ **Figure 14.2 An example of normal microbiota.** Bacteria (pink) colonize the human nasal cavity.

or the *indigenous microbiota*. The normal microbiota are of two main types: resident microbiota and transient microbiota.

Resident Microbiota

Resident microbiota remain a part of the normal microbiota of a person throughout life. These organisms are found on the skin and on the mucous membranes of the digestive tract, upper respiratory tract, distal[5] portion of the urethra,[6] and vagina. Table 14.2 illustrates these environments and lists some of the resident bacteria, fungi, and protozoa that live there. Most of the resident microbiota are commensal; that is, they feed on excreted cellular wastes and dead cells without causing harm.

Transient Microbiota

Transient microbiota remain in the body for only a few hours, days, or months before disappearing. They are found in the same locations as the resident members of the normal microbiota but cannot persist because of competition from other microorganisms, elimination by the body's defense cells, or chemical and physical changes in the body that dislodge them.

Acquisition of Normal Microbiota

You developed in your mother's womb without a normal microbiota because you had surrounded yourself with an amniotic membrane and fluid, which generally kept microorganisms at bay, and because your mother's uterus was essentially an axenic environment.

[5]*Distal* refers to the end of some structure, as opposed to *proximal*, which refers to the beginning of a structure.
[6]The tube that empties the bladder.

Your normal microbiota began to develop when your surrounding amniotic membrane ruptured and microorganisms came in contact with you during birth. Microbes entered your mouth and nose as you passed through the birth canal, and your first breath was loaded with microorganisms that quickly established themselves in your upper respiratory tract. Your first meals provided the progenitors of resident microbiota for your colon, while *Staphylococcus* and other microbes transferred from the skin of both the medical staff and your parents began to colonize your skin. Although you continue to add to your transient microbiota, most of the resident microbiota was initially established during your first months of life.

How Normal Microbiota Become Opportunistic Pathogens

Learning Objective

✓ Describe three conditions that create opportunities for normal microbiota to cause disease.

Under ordinary circumstances, normal microbiota do not cause disease. However, these same microbes may become harmful if an opportunity to do so arises. In this case, normal microbiota (or other normally harmless microbes from the environment) become **opportunistic pathogens** or *opportunists.* Conditions that create opportunities for pathogens include the following:

- *Introduction of a member of the normal microbiota into an unusual site in the body*. As we have seen, normal microbiota are present only in certain body sites, and each site has only certain species of microbiota. If a member of the normal microbiota in one site is introduced into a site it normally does not inhabit, the organism may become an opportunistic pathogen. In the colon, for example, *Escherichia coli* (esh-ĕ-rik'ē-ă kō'lē) is mutualistic, but should it enter the urethra it becomes an opportunist that can produce disease.
- *Immune suppression.* Anything that suppresses the body's immune system—including disease, malnutrition, emotional or physical stress, extremes of age (either very young or very old), the use of radiation or chemotherapy to combat cancer, or the use of immunosuppressive drugs in transplant patients—can enable opportunistic pathogens. AIDS patients often die from opportunistic infections that are typically controlled by a healthy immune system, because HIV infection suppresses immune system function.
- *Changes in the normal microbiota.* Normal microbiota use nutrients, take up space, and release toxic waste products, all of which make it less likely that arriving pathogens can compete well enough to become established and produce disease. This situation is known as **microbial antagonism** or **microbial competition.** However, changes in the relative abundance of normal microbiota, for whatever reason, may allow a member of the normal microbiota to become an opportunistic pathogen and thrive. For example, when a woman must undergo long-term antimicrobial treatment for a bacterial blood infection, the antimicrobial may also

TABLE 14.2 Some Resident Microbiota[a]

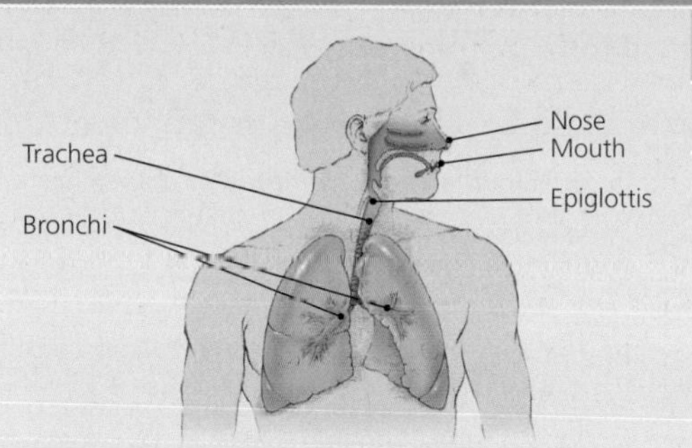

Upper Respiratory Tract

Genera	Notes
Fusobacterium, Haemophilus, Lactobacillus, Moraxella, Staphylococcus, Streptococcus, Veillonella, Candida (fungus)	The nose is cooler than the rest of the respiratory system and has some unique microbiota. The trachea and bronchi have a sparse microbiota compared to the nose and mouth. The alveoli of the lungs, which are too small to depict in this illustration, have no natural microbiota—they are axenic.

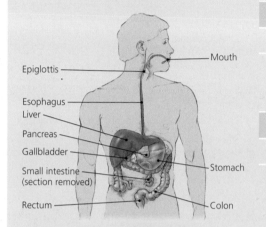

Upper Digestive Tract

Genera	Notes
Actinomyces, Bacteroides, Corynebacterium, Haemophilus, Lactobacillus, Neisseria, Staphylococcus, Treponema, Entamoeba (protozoan), *Trichomonas* (protozoan)	Microbes colonize surfaces of teeth, gingiva, lining of cheeks, and pharynx, and are found in saliva in large numbers. Dozens of species have never been identified.

Lower Digestive Tract

Genera	Notes
Bacteroides, Bifidobacterium, Clostridium, Enterococcus, Escherichia, Fusobacterium, Lactobacillus, Proteus, Shigella, Candida (fungus), *Entamoeba* (protozoan), *Trichomonas* (protozoan)	The bacteria are mostly strict anaerobes, though some facultative anaerobes are also resident.

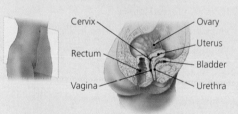

Female Urinary and Reproductive Systems

Genera	Notes
Bacteroides, Clostridium, Lactobacillus, Staphylococcus, Streptococcus, Candida (fungus), *Trichomonas* (protozoan)	Microbiota change as acidity in the vagina changes during menstrual cycle. The flow of urine prevents extensive colonization of the urinary bladder or urethra.

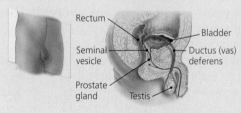

Male Urinary and Reproductive Systems

Genera	Notes
Bacteroides, Fusobacterium, Lactobacillus, Mycobacterium, Peptostreptococcus, Staphylococcus, Streptococcus	The flow of urine prevents extensive colonization of the urinary bladder or urethra.

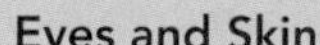

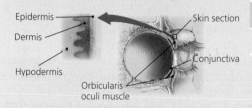

Eyes and Skin

Genera	Notes
Skin: *Corynebacterium, Micrococcus, Propionibacterium, Staphylococcus, Candida* (fungus), *Malassezia* (fungus)	Microbiota live on the outer, dead layers of the skin and in hair follicles and pores of glands. The deeper layers (dermis and hypodermis) are axenic.
Conjunctiva: *Staphylococcus*	Tears wash most microbiota from the eyes, so there are few compared to the skin.

[a]Genera are bacteria unless noted.

kill normal bacterial microbiota in the vagina. In the absence of competition from bacteria, *Candida albicans* (kan′did-ă al′bi-kanz), a yeast and also a member of the normal vaginal microbiota, grows prolifically, producing an opportunistic vaginal yeast infection. Other conditions that can disrupt the normal microbiota are hormonal changes, stress, changes in diet, and exposure to overwhelming numbers of pathogens.

Now that we have considered the relationships of the normal microbiota to their hosts, we turn our attention to the

TABLE 14.3 Some Common Zoonoses

Disease	Causative Agent	Animal Reservoir	Mode of Transmission
Helminthic			
Tapeworm infestation	*Dipylidium caninum*	Dogs	Ingestion of larvae transmitted in dog saliva
Fasciola infestation	*Fasciola hepatica*	Sheep, cattle	In gestion of contaminated vegetation
Protozoan			
Malaria	*Plasmodium* spp.	Monkeys	Bite of *Anopheles* mosquito
Toxoplasmosis	*Toxoplasma gondii*	Cats and other animals	Ingestion of contaminated meat, inhalation of pathogen, direct contact with infected tissues
Fungal			
Ringworm	*Trichophyton* sp. *Microsporum* sp. *Epidermophyton* sp.	Domestic animals	Direct contact
Bacterial			
Anthrax	*Bacillus anthracis*	Domestic livestock	Direct contact with infected animals, inhalation
Bubonic plague	*Yersinia pestis*	Rodents	Flea bites
Lyme disease	*Borrelia burgdorferi*	Deer	Tick bites
Salmonellosis	*Salmonella* spp.	Birds, rodents, reptiles	Ingestion of fecally contaminated water or food
Typhus	*Rickettsia prowazekii*	Rodents	Louse bites
Viral			
Rabies	*Lyssavirus* sp.	Bats, skunks, foxes, dogs	Bite of infected animal
Hantavirus pulmonary syndrome	*Hantavirus* sp.	Deer mice	Inhalation of viruses in dried feces and urine
Yellow fever	*Flavivirus* sp.	Monkeys	Bite of *Aedes* mosquito

movement of microbes into new hosts by considering aspects of human diseases.

Reservoirs of Infectious Diseases of Humans

Learning Objective

✓ Describe three types of reservoirs of infection in humans.

Most pathogens of humans cannot survive for long in the relatively harsh conditions they encounter outside their hosts. Thus if these pathogens are to enter new hosts, they must survive in some site from which they can infect new hosts. Sites where pathogens are maintained as a source of infection are called **reservoirs of infection.** In this section we discuss three types of reservoirs: animal reservoirs, human carriers, and nonliving reservoirs.

Animal Reservoirs

Many pathogens that normally infect either domesticated or sylvatic[7] (wild) animals can also affect humans. The more similar an animal's physiology is to human physiology, the more likely its pathogens are to affect human health. Diseases that spread naturally from their usual animal hosts to humans are called **zoonoses** (zō-ō-nō′sēz). Over 150 zoonoses have been identified throughout the world. Well-known examples include yellow fever, anthrax, bubonic plague, and rabies.

Humans may acquire zoonoses from animal reservoirs via a number of routes, including various types of direct contact with animals and their wastes, by eating animals, or via bloodsucking arthropods. Human infections with zoonoses are difficult to eradicate because extensive animal reservoirs are often involved. The larger the animal reservoir (that is, the greater the number and types of infected animals) and the greater the contact between humans and the animals, the more difficult and costly it is to control the spread of the disease to humans. This is especially true when the animal reservoir consists of both sylvatic and domesticated animals. In the case of rabies, for example, the disease typically spreads from a sylvatic reservoir (often bats, foxes, and skunks) to domestic pets from which humans may be infected. The wild animals constitute a reservoir for the rabies virus, but transmission to humans can be limited by vaccinating domestic pets.

Table 14.3 provides a brief view of some common zoonoses. Humans are usually dead-end hosts for zoonotic pathogens—that is, humans do not act as significant reservoirs

[7]From Latin *sylva*, meaning woodland.

CLINICAL CASE STUDY

A Deadly Carrier

In 1937, a man employed to lay water pipes in Croyden, England, was found to be the source of a severe epidemic of typhoid fever. The man, an asymptomatic carrier of *Salmonella enterica* serotype Typhi, the bacterium that causes typhoid, habitually urinated at his job site. In the process, he contaminated the town's water supply with bacteria from his bladder. Over 300 cases of typhoid fever developed, and 43 people died before the man was identified as the carrier.

1. How do you think health officials were able to identify the source of this typhoid epidemic?
2. Given that antibiotics were not generally available in 1937, how could health officials end the epidemic short of removing the man from the job site?

for the reinfection of animal hosts—largely because the circumstances under which zoonoses are transmitted favor movement from animals to humans but not in the opposite direction. For example, animals do not often eat humans these days, and animals less frequently have contact with human wastes than humans have contact with animal wastes. Zoonotic diseases transmitted via the bites of bloodsucking arthropods are the most likely type to be transmitted back to animal hosts.

Human Carriers

Experience tells you that humans with active diseases are important reservoirs of infection for other humans. What may not be so obvious is that people with no obvious symptoms before or after an obvious disease may also be infective in some cases. Further, some infected people remain both asymptomatic and infective for years. This is true of tuberculosis, syphilis, and AIDS, for example. Whereas some of these **carriers** incubate the pathogen in their body and eventually develop the disease, others remain a continued source of infection without ever becoming sick. Presumably many such healthy carriers have defensive systems that protect them from illness. An example of such a carrier is given in **Clinical Case Study: A Deadly Carrier.**

Nonliving Reservoirs

Soil, water, and food can be **nonliving reservoirs** of infection. Soil, especially if fecally contaminated, can harbor *Clostridium* (klos-trid'ē-ŭm) bacteria, which cause botulism, tetanus, and other diseases. Water can be contaminated with feces and urine containing parasitic worm eggs, pathogenic protozoa, bacteria, and viruses. Meats and vegetables can also harbor pathogens. Milk can contain many pathogens, which is why it is routinely pasteurized in the United States.

The Movement of Microbes into Hosts: Infection

In this section we examine events that occur when hosts are exposed to microbes from a reservoir, the sites at which microbes can gain entry into hosts, and the ways entering microbes become established in new hosts.

Exposure to Microbes: Contamination and Infection

Learning Objective

✓ Describe the relationship between contamination and infection.

In the context of the interaction between microbes and their hosts, **contamination** refers to the mere presence of microbes in or on the body. Some microbial contaminants reach the body in food, drink, or the air, whereas others are introduced via wounds, biting arthropods, or sexual intercourse. Several outcomes of contamination by microbes are possible. Some microbial contaminants remain where they first contacted the body (such as the skin or mucous membranes) without causing harm and subsequently become part of the resident microbiota; other microbial contaminants remain on the body for only a short time, as part of the transient microbiota. Still others overcome the body's external defenses, multiply, and become established in the body; such a successful invasion of the body by a pathogen is called an **infection.** An infection may or may not result in disease; that is, it may not adversely affect the body.

Next we consider in greater detail the various sites through which pathogens gain entry into the body.

Portals of Entry

Learning Objective

✓ Identify and describe the portals through which pathogens invade the body.

The sites through which most pathogens enter the body can be likened to the great gates or portals of a castle, because those sites constitute the routes by which microbes gain entry. Pathogens thus enter the body at several sites, called **portals of entry (Figure 14.3)**, which are of three major types: the skin, the mucous membranes, and the placenta. A fourth entry point, the so-called *parenteral* (pă-ren'ter-ăl) *route,* is not a portal but a way of circumventing the usual portals. Next we consider each type in turn.

Skin

Because the outer layer of skin is composed of relatively thick layers of tightly packed, dead, dry cells **(Figure 14.4)**, it forms a formidable barrier to most pathogens so long as it remains intact. Still, some pathogens can enter the body through natural

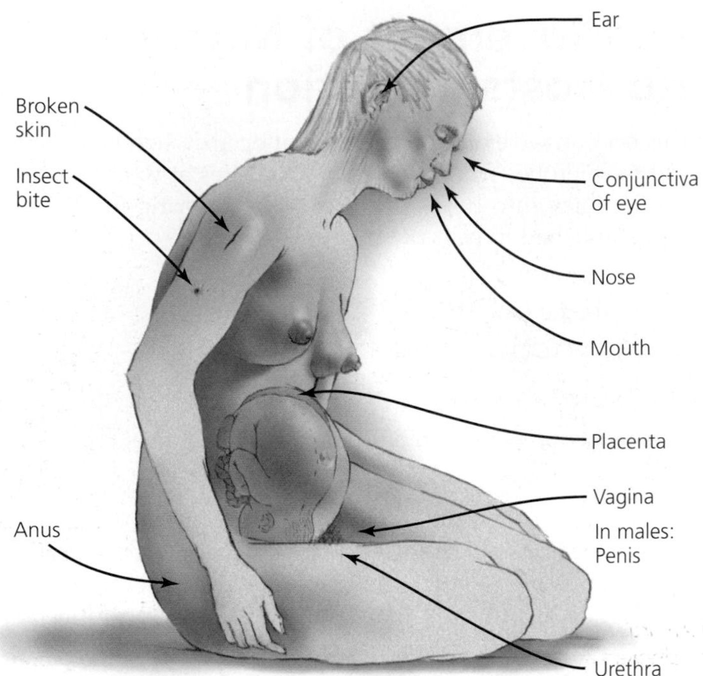

▲ **Figure 14.3 Routes of entry for invading pathogens.** The portals of entry include the skin, placenta, conjunctiva, and mucous membranes of the respiratory, gastrointestinal, urinary, and reproductive tracts. The parenteral route involves a puncture through the skin.

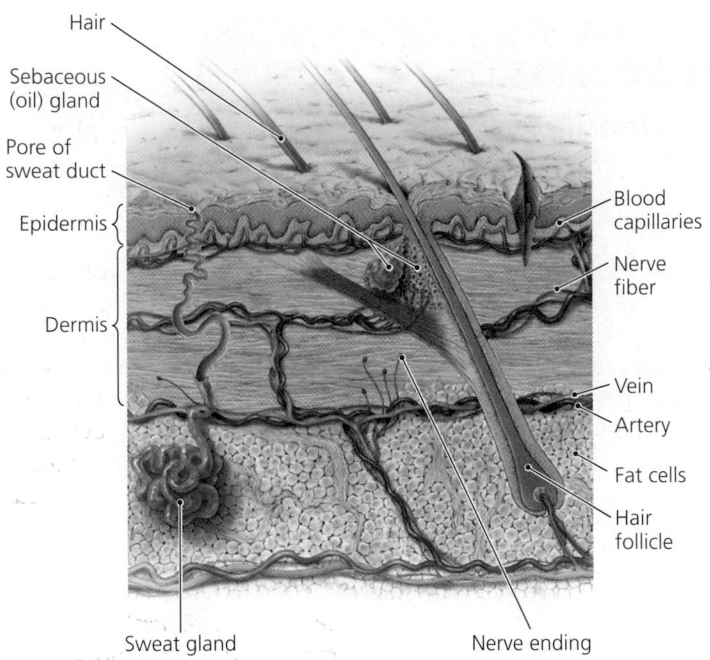

▲ **Figure 14.4 A cross section of skin.** The layers of cells constitute a barrier to most microbes so long as the skin remains intact. Some pathogens can enter the body through hair follicles, the ducts of sweat glands, and parenterally.

openings in the skin, such as hair follicles and sweat glands. Abrasions, cuts, bites, scrapes, stab wounds, and surgeries open the skin to infection by con taminants. Additionally, the larvae of some parasitic worms are capable of burrowing through the skin to reach underlying tissues, and some fungi can digest the dead outer layers of skin, thereby gaining access to deeper, moister areas within the body.

Mucous Membranes

The major portals of entry for pathogens are the *mucous membranes,* which line all the body cavities that are open to the outside world. They include the linings of the respiratory, gastrointestinal, urinary, and reproductive tracts, as well as the *conjunctiva* (kon-jŭnk-tī'vă), the thin membrane covering the surface of the eyeball and the underside of each eyelid. Like the skin, mucous membranes are composed of tightly packed cells; but unlike the skin, mucous membranes are relatively thin, moist, and warm, and their cells are living. Therefore, pathogens find mucous membranes more hospitable and easier portals of entry.

The respiratory tract is the most frequently used portal of entry. Pathogens enter the mouth and nose in the air, on dust particles, and in droplets of moisture. For example, the bacteria that cause whooping cough, diphtheria, pneumonia, strep throat, and meningitis, as well as some fungi, viruses, and protozoa, enter through the respiratory tract.

Surprisingly, many viruses enter the respiratory tract via the eyes. They are introduced onto the conjunctiva by contaminated fingers and are washed into the nasal cavity with tears. Cold and influenza viruses typically enter the body in this manner.

Some parasitic protozoa, helminths, bacteria, and viruses infect the body through the gastrointestinal mucous membranes. These parasites are able to survive the acidic pH of the stomach and the digestive juices of the intestinal tract. Noncellular pathogens called *prions* enter the body through oral mucous membranes.

Placenta

A developing embryo forms an organ, called the *placenta,* through which it obtains nutrients from the mother. The placenta is in such intimate contact with the wall of the mother's uterus that nutrients and wastes diffuse between the blood vessels of the developing child and of the mother, but because the two blood supplies do not actually contact each other, the placenta typically forms an effective barrier to most pathogens. However, in about 2% of pregnancies, pathogens cross the placenta and infect the embryo or fetus, sometimes causing spontaneous abortion, birth defects, or premature birth. Some pathogens that can cross the placenta are listed in Table 14.4.

The Parenteral Route

The **parenteral route** is not a portal of entry, but instead a means by which the portals of entry can be circumvented. To enter the body by the parenteral route, pathogens must be deposited directly into tissues beneath the skin or mucous membranes, such as occurs in punctures by a nail, thorn, or hypodermic needle. Some experts include breaks in the skin by cuts, bites, stab wounds, deep abrasions, or surgery in the parenteral route.

The Role of Adhesion in Infection

Learning Objectives

- ✓ List the types of adhesion factors and the roles they play in infection.
- ✓ Explain how a biofilm may facilitate contamination and infection.

After entering the body, symbionts must adhere to cells if they are to be successful in establishing colonies. The process by which microorganisms attach themselves to cells is called **adhesion** (ad-hē'zhŭn) (or *attachment*). To accomplish adhesion, pathogens use **adhesion factors,** which are either specialized structures or attachment proteins. Examples of such specialized structures are adhesion disks in some protozoa and suckers and hooks in some helminths (see Chapter 23). In contrast, viruses and many bacteria have surface lipoprotein and glycoprotein molecules called *ligands* that enable them to bind to complementary receptors on host cells **(Figure 14.5)**. Ligands are also called *adhesins* on bacteria and *attachment proteins* on viruses. Adhesins are found on fimbriae, flagella, and glycocalyces of many pathogenic bacteria. Receptor molecules on host cells are typically glycoproteins containing sugar molecules such as mannose and galactose. If ligands or their receptors can be changed or blocked, infection can often be prevented.

The specific interaction of adhesins and receptors with chemicals on host cells often determines the specificity of pathogens for particular hosts. For example, *Neisseria gonorrhoeae* (nī-se'rē-ă go-nor-rē'ī) has adhesins on its fimbriae that adhere to cells lining the urethra and vagina of humans. Thus, this pathogen cannot affect other hosts.

Some bacteria, including *Bordetella* (bōr-dě-tel'ă), the cause of whooping cough, have more than one type of adhesin. Other pathogens, such as *Plasmodium* (plaz-mō'dē-um; the cause of malaria), change their adhesins over time, which helps the pathogen evade the body's immune system and allows the pathogen to attack more than one kind of cell. Bacterial cells and viruses that have lost the ability to make ligands—whether as the result of some genetic change (mutation) or exposure to certain physical or chemical agents (as occurs in the production of some vaccines)—become harmless, or **avirulent** (ā-vir'ū-lent).

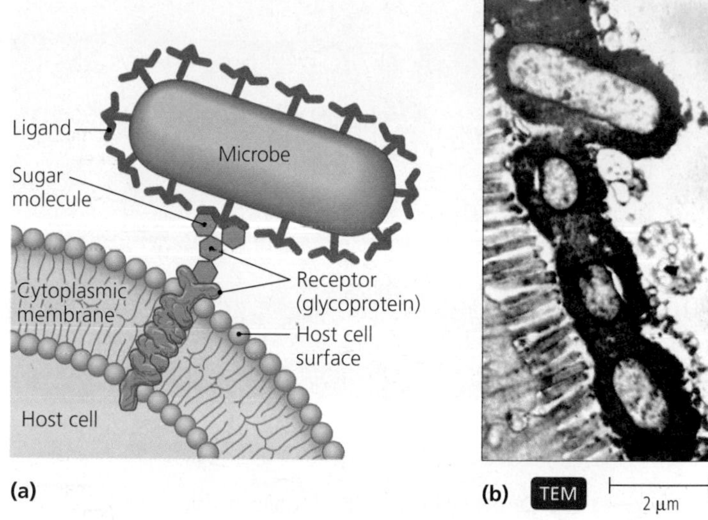

▲ **Figure 14.5 The adhesion of pathogens to host cells.** **(a)** An artist's rendition of the attachment of a microbial ligand to a complementary surface receptor on a host cell. **(b)** A photomicrograph of cells of a pathogenic strain of *E. coli* attached to the mucous membrane of the intestine. Although the thick glycocalyces (black), of the bacteria are visible, the bacteria's ligands are too small to be seen at this magnification.

Some bacterial pathogens do not attach to host cells directly, but instead interact with each other to form a sticky web of bacteria and polysaccharides called a **biofilm,** which adheres to a surface within a host. A prominent example of a biofilm is dental plaque **(Figure 14.6)**, which contains the bacteria that cause dental caries (tooth decay).

CRITICAL **THINKING**

If a mutation occurred in *Escherichia coli* that deleted the gene for an adhesin, what effect might it have on the ability of *E. coli* to cause urinary tract infections?

TABLE 14.4 Some Pathogens That Cross the Placenta

	Pathogen	Condition in the Adult	Effect on Embryo or Fetus
Protozoan	*Toxoplasma gondii*	Toxoplasmosis	Abortion, epilepsy, encephalitis, microcephaly, mental retardation, blindness, anemia, jaundice, rash, pneumonia, diarrhea, hypothermia, deafness
Bacteria	*Treponema pallidum*	Syphilis	Abortion, multiorgan birth defects, syphilis
	Listeria monocytogenes	Listeriosis	Granulomatosis infantiseptica (nodular inflammatory lesions and infant blood poisoning), death
DNA viruses	*Cytomegalovirus*	Usually asymptomatic	Deafness, microcephaly, mental retardation
	Parvovirus B19	Erythema infectiosum	Abortion
RNA viruses	*Lentivirus* (HIV)	AIDS	Immunosuppression (AIDS)
	Rubivirus	German measles	Severe birth defects or death

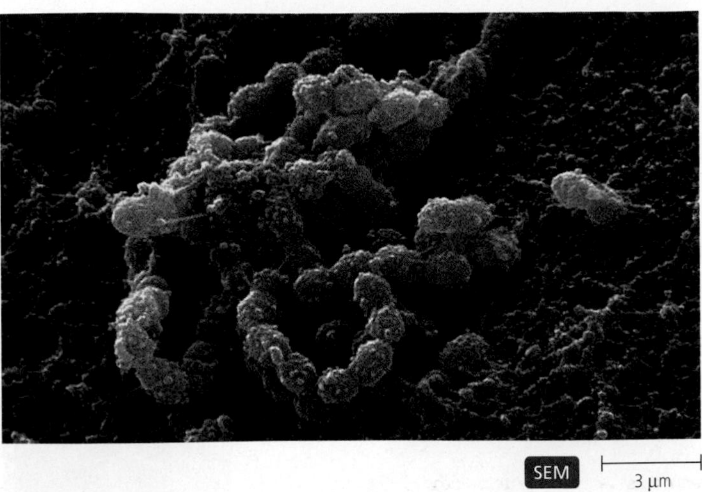

▲ **Figure 14.6 Dental plaque.** This biofilm consists of bacteria that adhere to one another and to teeth by means of sticky capsules, slime layers, and fimbriae.

The Nature of Infectious Disease

Learning Objective

✓ Compare and contrast the terms *infection, disease, morbidity, pathogenicity,* and *virulence.*

Most infections succumb to the body's defenses (discussed in detail in Chapters 15 and 16), but some infectious agents not only evade the defenses and multiply, but also affect body function. By definition, all parasites injure their hosts. When the injury is significant enough to interfere with the normal functioning of the body, the result is termed **disease.** Thus, disease, also known as **morbidity,** is any change from a state of health.

Infection and disease are not the same thing. Infection is the invasion of a pathogen; disease results only if the pathogen multiplies sufficiently to adversely affect the body. Some illustrations will help to clarify this point. While caring for a patient, a nurse may become contaminated with the bacterium *Staphylococcus aureus* (o'rē-ŭs). If the pathogen is able to gain access to his body, perhaps through a break in his skin, he will become infected. Then, if the multiplication of *Staphylococcus* results in the production of a boil, the nurse will have a disease. Another example: A drug addict who shares needles may become contaminated and subsequently infected with HIV without experiencing visible change in body function for years. Such a person is infected but not yet diseased.

Manifestations of Disease: Symptoms, Signs, and Syndromes

Learning Objective

✓ Contrast symptoms, signs, and syndromes.

Diseases may become manifest in different ways. **Symptoms** are *subjective* characteristics of a disease that can be felt by the patient alone, whereas **signs** are *objective* manifestations of disease that can be observed or measured by others. Symptoms include pain, headache, dizziness, and fatigue; signs include swelling, rash, redness, and fever. Sometimes pairs of symptoms and signs reflect the same underlying cause: Nausea is a symptom, and vomiting a sign; chills are a symptom, whereas shivering is a sign. Note, however, that even though signs and symptoms may have the same underlying cause, they need not occur together. Thus, for example, a viral infection of the brain may result in both a headache (a symptom) and the presence of viruses in the cerebrospinal fluid (a sign), but viruses in the cerebrospinal fluid are not invariably accompanied by headache. Symptoms and signs are used in conjunction with laboratory tests to make diagnoses. Some typical symptoms and signs are listed in Table 14.5.

TABLE 14.5 Typical Manifestations of Disease

Symptoms (sensed by the patient)	Signs (detected or measured by an observer)
Pain, nausea, headache, chills, sore throat, fatigue or lethargy (sluggishness, tiredness), malaise (discomfort), itching, abdominal cramps	Swelling, rash or redness, vomiting, diarrhea, fever, pus formation, anemia, leukocytosis/leukopenia (increase/decrease in the number of circulating white blood cells), bubo (swollen lymph node), tachycardia/bradycardia (increase/decrease in heart rate)

A **syndrome** is a group of symptoms and signs that collectively characterizes a particular disease or abnormal condition. For example, acquired immunodeficiency syndrome (AIDS) is characterized by malaise, loss of certain white blood cells, diarrhea, weight loss, pneumonia, toxoplasmosis, and tuberculosis.

Some infections go unnoticed because they have no symptoms. Such cases are **asymptomatic** or **subclinical** infections. Note that even though asymptomatic infections by definition lack symptoms, in some cases certain signs may still be detected if the proper tests are performed. For example, leukocytosis (an excess of white blood cells) may be detected in a blood sample from an individual who feels completely healthy.

Table 14.6 lists some prefixes and suffixes used to describe various aspects of diseases and syndromes.

Causation of Disease: Etiology

Learning Objectives

✓ Define *etiology.*

✓ List Koch's postulates, explain their function, and describe their limitations.

Even though our focus in this chapter is infectious disease, it's obvious that not all diseases result from infections. Some diseases are *hereditary,* which means they are genetically transmitted from

TABLE 14.6

Terminology of Disease

Prefix/Suffix	Meaning	Example
carcino-	cancer	carcinogenic: giving rise to cancer
col-, colo-	colon	colitis: inflammation of the colon
dermato-	skin	dermatitis: inflammation of the skin
-emia	pertaining to the blood	viremia: viruses in the blood
endo-	inside	endocarditis: inflammation of lining of heart
-gen, gen-	give rise to	pathogen: giving rise to disease
hepat-	liver	hepatitis: inflammation of the liver
idio-	unknown	idiopathic: pertaining to a disease of unknown cause
-itis	inflammation of a structure	meningitis: inflammation of the meninges (covering of the brain); endocarditis
-oma	tumor or swelling	papilloma: wart
-osis	condition of	toxoplasmosis: being infected with *Toxoplasma*
-patho, patho-	abnormal	pathology: study of disease
septi-	literally, *rotting;* refers to presence of pathogens	septicemia: pathogens in the blood
terato-	defects	teratogenic: causing birth defects
tox-	poison	toxin: harmful compound

parents to offspring. Other diseases, called *congenital diseases,* are diseases that are present at birth, regardless of the cause (whether hereditary, environmental, or infectious). Still other diseases are classified as *degenerative, nutritional, endocrine, mental, immunological,* or *neoplastic.*[8] The various categories of diseases are described in Table 14.7 on p. 413. Note that some diseases can fall into more than one category. For example, liver cancer, which is usually associated with infection by hepatitis B and D viruses, can be classified as both neoplastic and infectious.

The study of the cause of a disease is called **etiology** (ē-tē-ol′ō-jē). Because our focus here and in subsequent chapters is on infectious diseases, we next examine how microbiologists investigate the causation of infectious diseases, beginning with a reexamination of the work of Robert Koch, described in Chapter 1.

Using Koch's Postulates

In our modern world, we take for granted the idea that specific microbes cause specific diseases, but this has not always been the case. In the past, disease was thought to result from a variety of causes, including bad air, imbalances in body fluids, or astrological forces. In the 19th century, Louis Pasteur, Robert Koch, and other microbiologists proposed the **germ theory of disease,** which states that disease is caused by infections of pathogenic microorganisms (at the time, called *germs*).

But which pathogen causes a specific disease? How can we distinguish between the pathogen, which causes a disease, and all the other biological agents (fungi, bacteria, protozoa, and viruses) that are part of the normal microbiota and are in effect "innocent bystanders"?

Koch developed a series of essential conditions, or *postulates,* that scientists must demonstrate or satisfy to prove that a particular microbe is pathogenic and causes a particular disease. Using his postulates, Koch proved that *Bacillus anthracis* (ba-sil′ŭs an-thrā′sis) causes anthrax and that *Mycobacterium tuberculosis* (mī′kō-bak-tēr′ē-ŭm too-ber-kyū-lō′sis) causes tuberculosis.

To prove that a given infectious agent causes a given disease, a scientist must satisfy all of **Koch's postulates (Figure 14.7):**

1. The suspected agent (bacterium, virus, etc.) must be present in every case of the disease.
2. That agent must be isolated and grown in pure culture.
3. The cultured agent must cause the disease when it is inoculated into a healthy, susceptible experimental host.
4. The same agent must be reisolated from the diseased experimental host.

It is critical that all the postulates be satisfied in order. The mere presence of an agent does not prove that it causes a disease. Although Koch's postulates have been used to prove the cause of many infectious diseases in humans, animals, and plants, inadequate attention to the postulates has resulted in incorrect conclusions concerning some disease causation. For instance, in the early 1900s *Haemophilus influenzae* (he-mof′i-lŭs in-flu-en′zī) was found in the respiratory systems of flu victims and identified as the causative agent of influenza based on its presence. Later, flu victims that lacked *H. influenzae* in their lungs were discovered. This discovery violated the first postulate, so *H. influenzae* cannot be the cause of flu. Today we know that an RNA virus causes flu, and that *H. influenzae* was part of

[8]Neoplasms are tumors, which may either remain in one place (benign tumors) or spread (cancers).

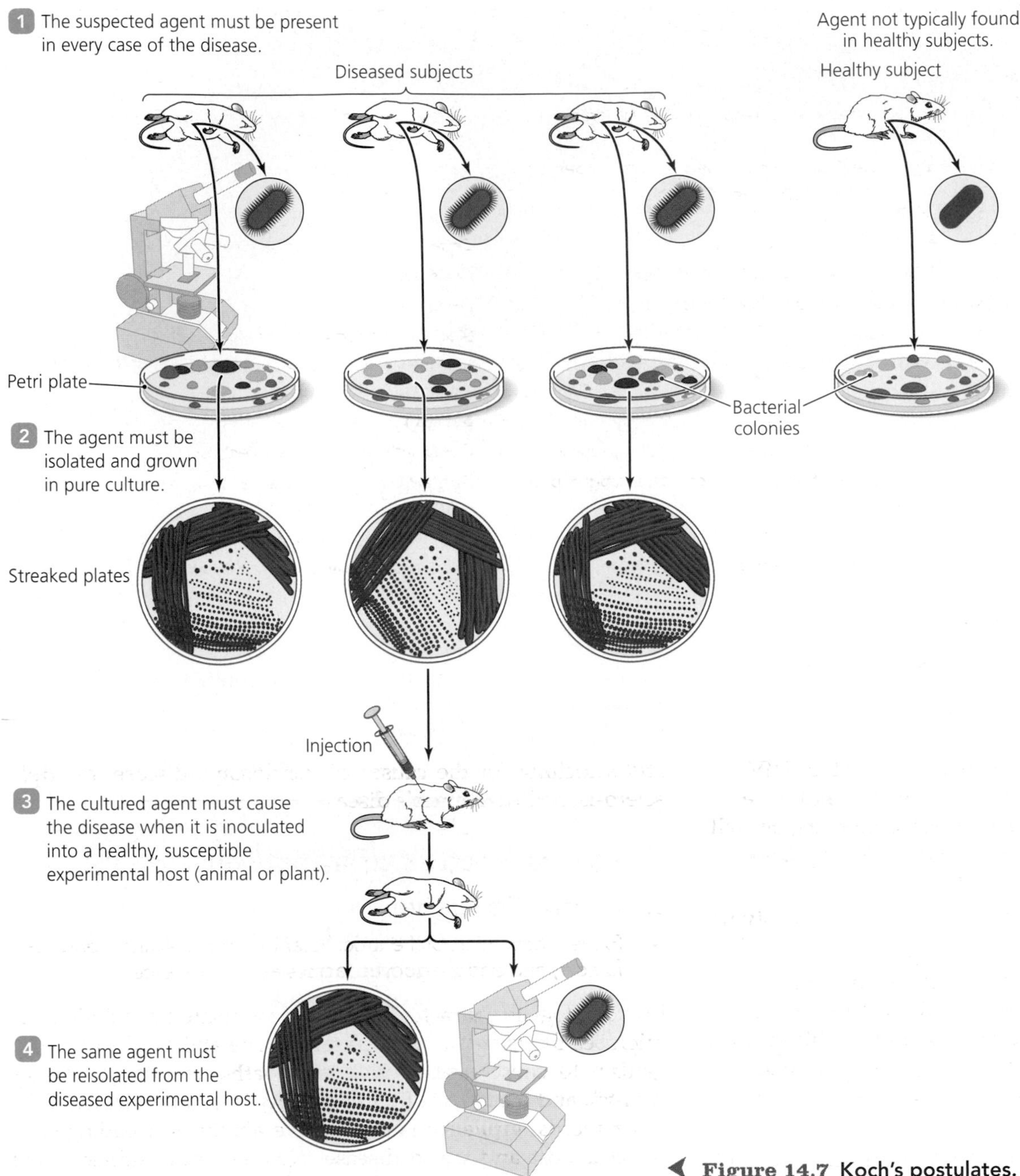
1 The suspected agent must be present in every case of the disease.
Diseased subjects
Agent not typically found in healthy subjects.
Healthy subject
Petri plate
Bacterial colonies
2 The agent must be isolated and grown in pure culture.
Streaked plates
Injection
3 The cultured agent must cause the disease when it is inoculated into a healthy, susceptible experimental host (animal or plant).
4 The same agent must be reisolated from the diseased experimental host.

Figure 14.7 Koch's postulates.

the normal microbiota of those early flu patients. (Later studies, correctly using Koch's postulates, showed that *H. influenzae* can cause an often-fatal meningitis in children.)

Exceptions to Koch's Postulates

Clearly, Koch's postulates are a cornerstone of the etiology of infectious diseases, but using them is not always feasible, for the following reasons:

- Some pathogens cannot be cultured in the laboratory. For example, pathogenic strains of *Mycobacterium leprae* (lep'rī), which causes leprosy, have never been grown on laboratory media.
- Some diseases are caused by a combination of pathogens, or by a combination of a pathogen and physical, environmental, or genetic cofactors. In such cases, the pathogen alone is avirulent, but when accompanied by another pathogen or the appropriate cofactor, disease results. For example, liver cancer can result when liver cells are infected by both the hepatitis B and hepatitis D viruses, but seldom when only one of the viruses infects the cells.
- Ethical considerations prevent applying Koch's postulates to diseases and pathogens that occur in humans only. In such cases the third postulate, which involves inoculation of a healthy susceptible host, cannot be satisfied within ethical boundaries. For this reason, scientists have never

TABLE 14.7 Categories of Diseases[a]

	Description	Examples
Hereditary	Caused by errors in the genetic code received from parents	Sickle-cell anemia, diabetes mellitus, Down syndrome
Congenital	Anatomical and physiological (structural and functional) defects present at birth; caused by drugs (legal and illegal), X-ray exposure, or infections	Fetal alcohol syndrome, deafness from rubella infection
Degenerative	Result from aging	Renal failure, age-related farsightedness
Nutritional	Result from lack of some essential nutrients in diet	Kwashiorkor, rickets
Endocrine (hormonal)	Due to excesses or deficiencies of hormones	Dwarfism
Mental	Emotional or psychosomatic	Skin rash, gastrointestinal distress
Immunological	Hyperactive or hypoactive immunity	Allergies, autoimmune diseases, agammaglobulinemia
Neoplastic (tumor)	Abnormal cell growth	Benign tumors, cancers
Infectious	Caused by an infectious agent	Colds, influenza, herpes infections
Iatrogenic[b]	Caused by medical treatment or procedures; are a subgroup of hospital-acquired diseases	Surgical error, yeast vaginitis resulting from antimicrobial therapy
Idiopathic[c]	Unknown cause	Alzheimer's disease, multiple sclerosis
Nosocomial[d]	Disease acquired in health care setting	*Pseudomonas* infection in burn patient

[a]Some diseases may fall in more than one category.
[b]From Greek *iatros*, meaning physician.
[c]From Greek *idiotes*, meaning ignorant person, and *pathos*, meaning disease.
[d]From Greek *nosokomeion*, meaning hospital.

attempted to apply Koch's postulates to prove that HIV causes AIDS; however, observations of fetuses that were naturally exposed to HIV by their infected mothers, as well as accidentally infected health care and laboratory workers, have in effect satisfied the third postulate.

Additionally, some circumstances may make satisfying Koch's postulates difficult:

- It is not possible to establish a single cause for such infectious diseases as pneumonia, meningitis, and hepatitis, because the names of these diseases refer to conditions that can be caused by more than one pathogen. For these diseases, laboratory technicians must identify the etiologic agent involved in any given case.
- Some pathogens have been ignored. For example, gastric ulcers were long thought to be caused by excessive production of stomach acid in response to stress, but the majority of such ulcers are now known to be caused by a long-overlooked bacterium, *Helicobacter pylori* (hel'ĭ-kō-bak'ter pī'lō-rē).

If Koch's postulates cannot be applied to a disease condition for whatever reason, how can we positively know the causative agent of a disease? *Epidemiological* studies, discussed later in this chapter, can give statistical support to causation theories, but not absolute proof. For example, some researchers have proposed that the bacterium *Chlamydophila pneumoniae* (kla-mē-dof'ĭ-lă nū-mō'nē-ī) causes many cases of arteriosclerosis on the basis of the bacterium's presence in patients with the disease. Debate among scientists about such cases fosters a continued drive for knowledge and discovery. For example, we are still searching for the causes of Parkinson's disease, multiple sclerosis, and Alzheimer's disease.

Virulence Factors of Infectious Agents

Learning Objective

✓ Explain how microbial extracellular enzymes, toxins, adhesion factors, and antiphagocytic factors affect virulence.

Microbiologists characterize the disease-related capabilities of microbes by using two related terms. The ability of a microorganism to cause disease is termed **pathogenicity** (path'ō-jĕ-nis'i-tē), and the *degree* of pathogenicity is termed **virulence.** In other words, virulence is the relative ability of a pathogen to infect a host and cause disease. Neither term addresses the severity of a disease; the pathogen causing rabbit fever is highly virulent but the disease is relatively mild. Organisms can be placed along a virulence continuum **(Figure 14.8)**; highly virulent organisms almost always cause disease, while less virulent organisms (including opportunistic pathogens) cause disease only in weakened hosts or when present in overwhelming numbers.

Pathogens have a variety of traits that interact with a host and enable the pathogen to enter a host, adhere to host cells, gain access to nutrients, and escape detection or removal by the immune system. These traits are collectively called **virulence factors.** Virulent pathogens have one or more virulence factors that nonvirulent microbes lack. We discussed two virulence factors—adhesion factors and biofilm formation—previously;

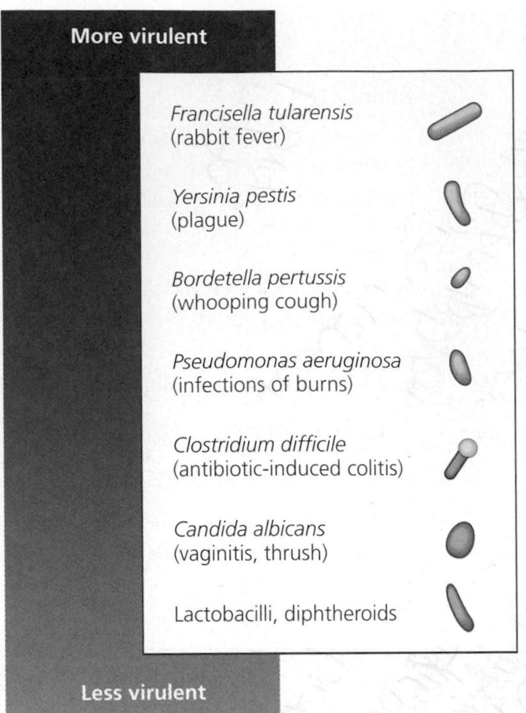

▲ **Figure 14.8 Relative virulence of some microbial pathogens.** Virulence involves ease of infection and the ability of a pathogen to cause disease; the term does not indicate the seriousness of a disease.

now we examine three other virulence factors: extracellular enzymes, toxins, and antiphagocytic factors. Other virulence factors can be examined online at www.masteringmicrobiology.com. **ANIMATIONS:** ***Virulence Factors: Inactivating Host Defenses***

Extracellular Enzymes

Many pathogens secrete enzymes that enable them to dissolve structural chemicals in the body and thereby maintain an infection, invade further, and avoid body defenses. **Figure 14.9a** illustrates the action of some extracellular enzymes of bacteria:

- Hyaluronidase and collagenase degrade specific molecules to enable bacteria to invade deeper tissues. Hyaluronidase digests hyaluronic acid, the "glue" that holds animal cells together, and collagenase breaks down collagen, the body's chief structural protein.
- Coagulase causes blood proteins to clot, providing a "hiding place" for bacteria within a clot.
- Kinases such as staphylokinase and streptokinase digest blood clots, allowing subsequent invasion of damaged tissues.

Many bacteria with these enzymes are virulent; mutant strains of the same species that have defective genes for these extracellular enzymes are usually avirulent.

Pathogenic eukaryotes also secrete enzymes that contribute to virulence. For example, fungi that cause "ringworm" produce *keratinase,* which enzymatically digests keratin—the main component of skin, hair, and nails. *Entamoeba histolytica* (ent-ă-mē′bă his-tō-li′ti-kă) secretes *mucinase* to digest the mucus lining the intestinal tract, allowing the amoeba entry to the underlying cells where it causes amebic dysentery. **ANIMATIONS:** ***Virulence Factors: Penetrating Host Tissues***

Toxins

Toxins are chemicals that either harm tissues or trigger host immune responses that cause damage. The distinction between extracellular enzymes and toxins is not always clear, because many enzymes are toxic, and many toxins have enzymatic action. In a condition called **toxemia** (tok-sē′mē-ā), toxins enter the bloodstream and are carried to other parts of the body, including sites that may be far removed from the site of infection. There are two types of toxin: exotoxins and endotoxin.

Exotoxins Many microorganisms secrete **exotoxins** that are central to their pathogenicity in that they destroy host cells or interfere with host metabolism. Exotoxins are of three principal types:

- *Cytotoxins,* which kill host cells in general or affect their function **(Figure 14.9b)**
- *Neurotoxins,* which specifically interfere with nerve cell function
- *Enterotoxins,* which affect cells lining the gastrointestinal tract

Examples of pathogenic bacteria that secrete exotoxins are the clostridia that cause gangrene, botulism, and tetanus; pathogenic strains of *S. aureus* that cause food poisoning and other ailments; and diarrhea-causing *E. coli, Salmonella enterica* (sal′mŏ-nel′ă en-ter′i-kă), and *Shigella* (shē-gel′lă) species. Some fungi and marine dinoflagellates (protozoa) also secrete exotoxins. Specific exotoxins are discussed in the chapters that examine specific diseases. **ANIMATIONS:** ***Virulence Factors: Exotoxins***

The body protects itself with **antitoxins,** which are protective molecules called *antibodies* that bind to specific toxins and neutralize them. Health care workers stimulate the production of antitoxins by administering immunizations composed of *toxoids,* which are toxins that have been treated with heat, formaldehyde, chlorine, or other chemicals to make them nontoxic but still capable of stimulating the production of antibodies. Chapters 16 and 17 further discuss antibodies, toxoids, and immunizations.

Endotoxin Recall from Chapter 3 that Gram-negative bacteria have an outer (cell wall) membrane composed of lipopoly saccharide, phospholipids, and proteins (see Figure 3.15). **Endotoxin,** also called **lipid A,** is the lipid portion of the membrane's lipopolysaccharide.

Endotoxin can be released when Gram-negative bacteria divide, die naturally, or are digested by phagocytic cells such as macrophages (see Figure 14.9b). Many types of lipid A stimulate the body to release chemicals that cause fever, inflammation, diarrhea, hemorrhaging, shock, and blood coagulation.

Hyaluronidase and collagenase

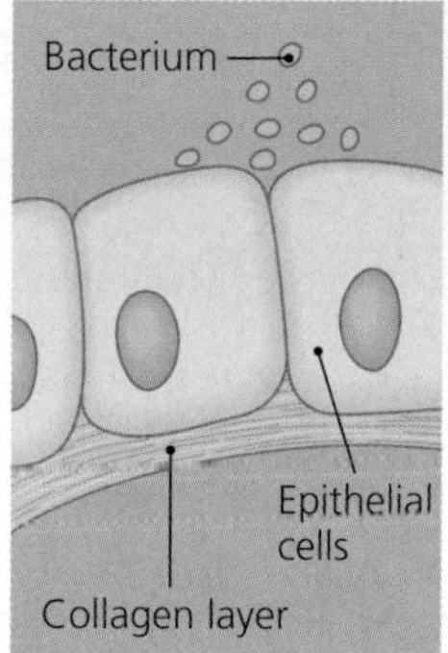

Invasive bacteria reach epithelial surface.

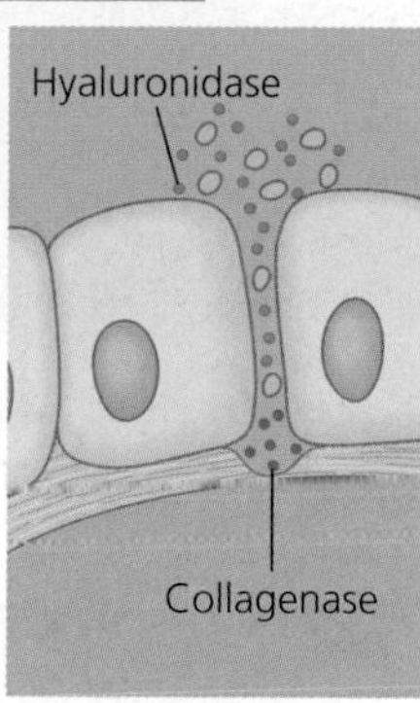

Bacteria produce hyaluronidase and collagenase.

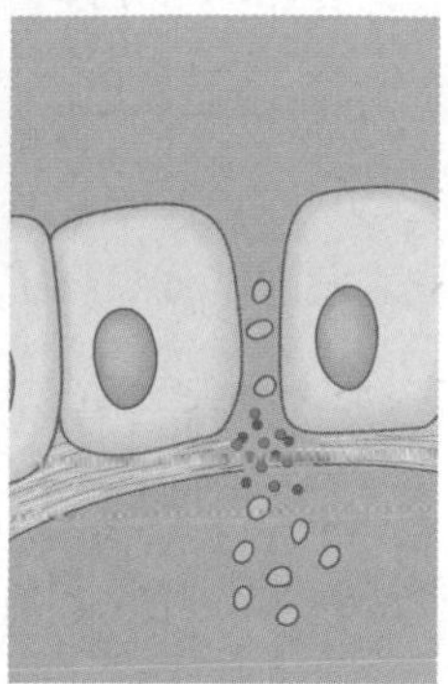
Bacteria invade deeper tissues.

Coagulase and kinase

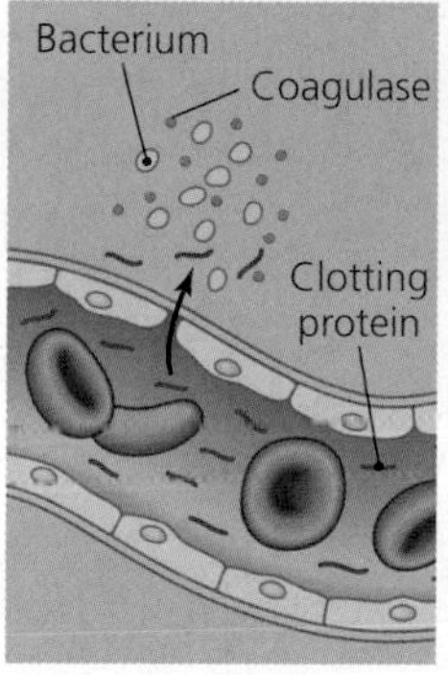

Bacteria produce coagulase.

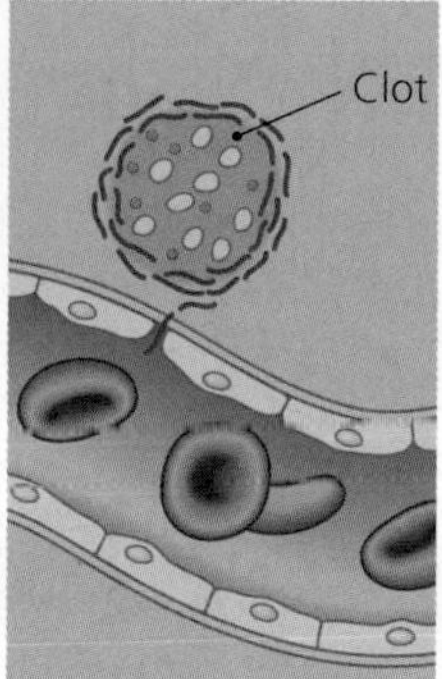

Clot forms.

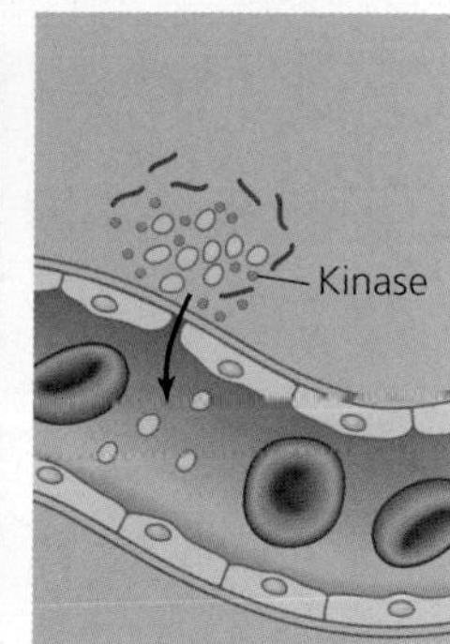

Bacteria later produce kinase, dissolving clot and releasing bacteria.

(a) Extracellular enzymes

Exotoxin

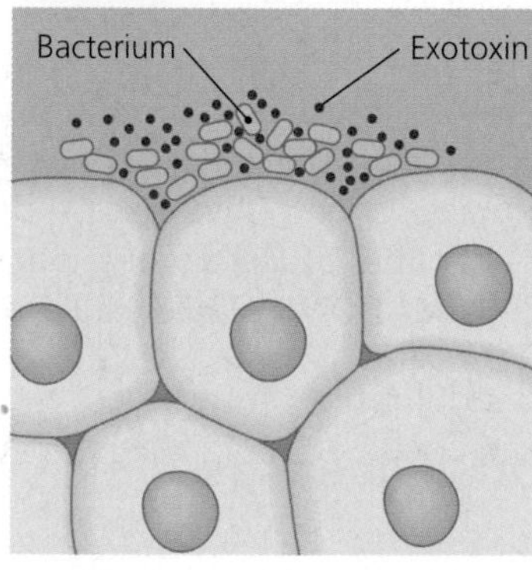

Bacteria secrete exotoxins, in this case a cytotoxin.

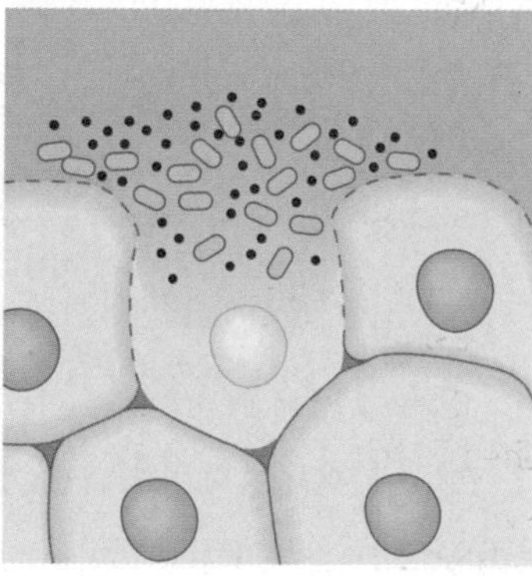
Cytotoxin kills host's cells.

(b) Toxins

Endotoxin

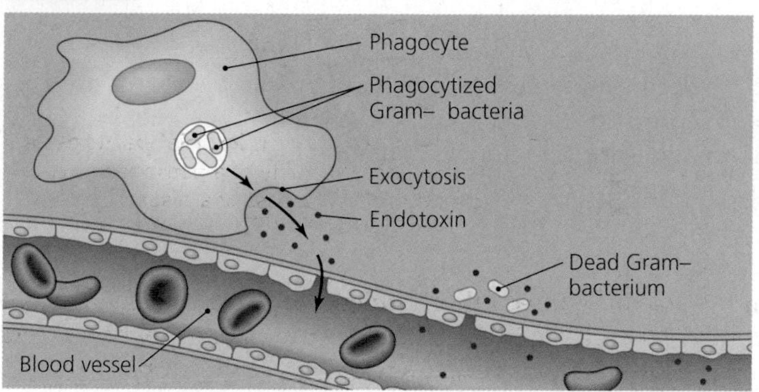

Dead Gram-negative bacteria release endotoxin (lipid A) which induces effects such as fever, inflammation, diarrhea, shock, and blood coagulation.

Phagocytosis blocked by capsule

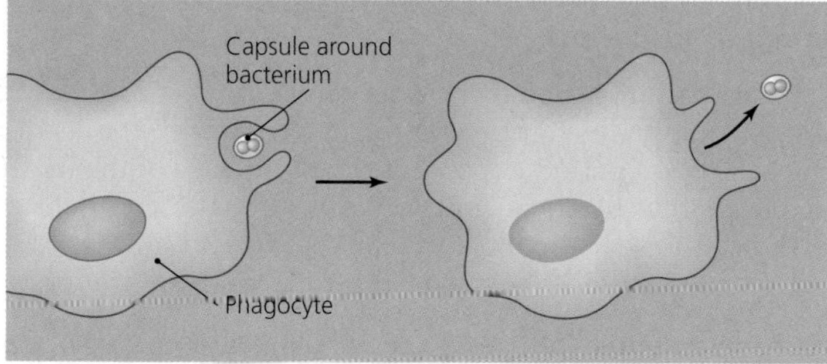

Incomplete phagocytosis

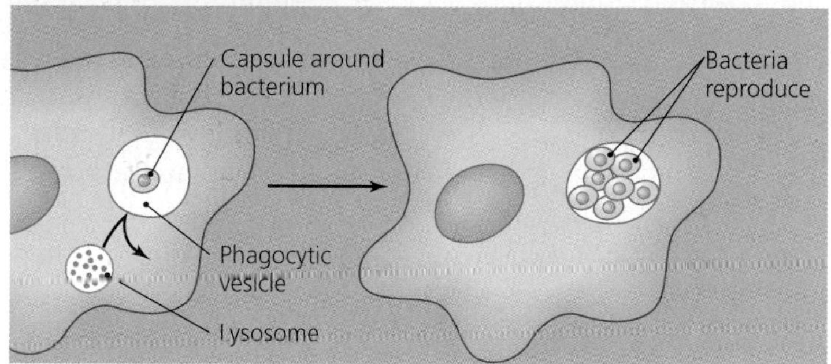

(c) Antiphagocytic factors

▲ **Figure 14.9 Some virulence factors.** **(a)** Extracellular enzymes. Hyaluronidase and collagenase digest structural materials in the body. Coagulase in effect "camouflages" bacteria inside a blood clot, whereas kinases digest clots to release bacteria. **(b)** Toxins. Exotoxins (including cytotoxin, shown here) are released from living pathogens and harm neighboring cells. Endotoxin is released from many dead Gram-negative bacteria and can trigger widespread disruption of normal body functions. **(c)** Antiphagocytic factors. Capsules, one antiphagocytic factor, can prevent phagocytosis or stop digestion by a phagocyte. *How can bacteria prevent a phagocytic cell from digesting them once they have been engulfed by pseudopodia?*

Figure 14.9 *Some bacteria secrete a chemical that prevents the fusion of lysosomes with the phagosome containing the bacteria.*

TABLE 14.8 A Comparison of Bacterial Exotoxins and Endotoxin

	Exotoxins	Endotoxin
Source	Mainly Gram-positive and Gram-negative bacteria	Gram-negative bacteria
Relation to bacteria	Metabolic product secreted from living cell	Portion of outer (cell wall) membrane released upon cell death
Chemical nature	Protein or short peptide	Lipid portion of lipopolysaccharide (lipid A) of outer (cell wall) membrane
Toxicity	High	Low, but may be fatal in high doses
Heat stability	Typically unstable at temperatures above 60°C	Stable for up to 1 hour at autoclave temperature (121°C)
Effect on host	Variable depending on source; may be cytotoxin, neurotoxin, enterotoxin	Fever, lethargy, malaise, shock, blood coagulation
Fever producing?	No	Yes
Antigenicity[a]	Strong: stimulates antitoxin (antibody) production	Weak
Toxoid formation for immunization?	By treatment with heat or fomaldehyde	Not feasible
Representative diseases	Botulism, tetanus, gas gangrene, diphtheria, cholera, plague, staphylococcal food poisoning	Typhoid fever, tularemia, endotoxic shock, urinary tract infections, meningococcal meningitis

[a]Refers to the ability of a chemical to trigger a specific immune response, particularly the formation of antibodies.

Although infections by bacteria that produce exotoxins are generally more serious than infections with other bacteria, most Gram-negative pathogens can be potentially life threatening because the release of endotoxin from dead bacteria can produce serious, systemic effects in the host. **ANIMATIONS:** ***Virulence Factors: Endotoxins***

Table 14.8 summarizes differences between exotoxins and endotoxins.

Antiphagocytic Factors

Typically, the longer a pathogen remains in a host, the greater the damage and the more severe the disease. To limit the extent and duration of infections, the body's phagocytic cells, such as the white blood cells called macrophages, engulf and remove invading pathogens. Here we consider some virulence factors related to the evasion of phagocytosis, beginning with bacterial capsules. **ANIMATIONS:** ***Phagocytosis: Microbes That Evade It***

Capsules The capsules of many pathogenic bacteria (see Figure 3.5a) are effective virulence factors because many capsules are composed of chemicals normally found in the body (including polysaccharides); as a result, they do not stimulate a host's immune response. For example, hyaluronic acid capsules in effect deceive phagocytic cells into treating them and the enclosed bacteria as if they were a normal part of the body. Additionally, capsules are often slippery, making it difficult for phagocytes to surround and phagocytize them—their pseudopodia cannot grip the capsule, much as wet hands have difficulty holding a wet bar of soap **(Figure 14.9c)**.

Antiphagocytic Chemicals Some bacteria, including the cause of gonorrhea, produce chemicals that prevent the fusion of lysosomes with phagocytic vesicles, which allows the bacteria to survive inside of phagocytes (see Figure 14.9c). *Streptococcus pyogenes* (strep-tō-kok′ŭs pī-oj′en-ēz) produces a protein on its cell wall and fimbriae, called *M protein,* that resists phagocytosis and thus increases virulence. Other bacteria produce *leukocidins,* which are chemicals capable of destroying phagocytic white blood cells outright. **ANIMATIONS:** ***Virulence Factors: Hiding from Host Defenses***

The Stages of Infectious Diseases

Learning Objective

✓ List and describe the five typical stages of infectious diseases.

Following exposure and infection, a sequence of events called the **disease process** can occur. Many infectious diseases have five stages following infection: an incubation period, a prodromal period, illness, decline, and convalescence **(Figure 14.10)**.

Incubation Period

The **incubation period** is the time between infection and occurrence of the first symptoms or signs of disease. The length of the incubation period depends on the virulence of the infective agent, the infective dose (initial number of pathogens), the state and health of the patient's immune system, the nature of the pathogen and its reproduction time, and the site of infection. Some diseases have typical incubation periods, whereas for others the incubation period varies considerably. Table 14.9 lists incubation periods for selected diseases.

Prodromal Period

The **prodromal**[9] (prō-drō′măl) period is a short time of generalized, mild symptoms (such as malaise and muscle aches) that

[9]From Greek *prodromos*, meaning forerunner.

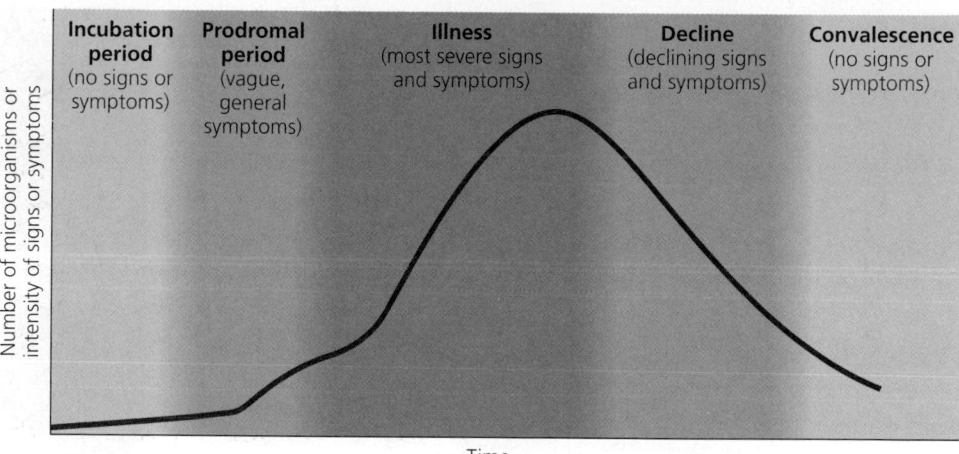

Figure 14.10
The stages of infectious diseases.
Not every stage occurs in every disease.

precedes illness. Not all infectious diseases have a prodromal stage.

Illness

Illness is the most severe stage of an infectious disease. Signs and symptoms are most evident during this time. Typically the patient's immune system has not yet fully responded to the pathogens, and their presence is harming the body. This stage is usually when a physician first sees the patient.

Decline

During the period of **decline,** the body gradually returns to normal as the patient's immune response and/or medical treatment vanquish the pathogens. Fever and other signs and symptoms subside. Normally the immune response and its products (such as antibodies in the blood) peak during this stage. If the patient doesn't recover, then the disease is fatal.

TABLE 14.9 Incubation Periods of Selected Infectious Diseases

Disease	Incubation Period
Staphylococcus foodborne infection	<1 day
Influenza	About 1 day
Cholera	2 to 3 days
Genital herpes	About 5 days
Tetanus	5 to 15 days
Syphilis	10 to 21 days
Hepatitis B	70 to 100 days
AIDS	1 to >8 years
Leprosy	10 to >30 years

Convalescence

During **convalescence** (kon-vă-les'ens), the patient recovers from the illness; tissues are repaired and returned to normal. The length of a convalescent period depends on the amount of damage, the nature of the pathogen, the site of infection, and the overall health of the patient. Thus, whereas recovery from staphylococcal food poisoning may take less than a day, recovery from Lyme disease may take years.

A patient is likely to be infectious during every stage of disease. Even though most of us realize we are infective during the symptomatic periods, many people are unaware that infections can be spread during incubation and convalescence as well. For example, a patient who no longer has any obvious herpes sores is always capable of transmitting herpesviruses. Good aseptic technique can limit the spread of many pathogens from recovering patients.

The Movement of Pathogens Out of Hosts: Portals of Exit

Just as infections occur through portals of entry, so pathogens must leave infected patients through **portals of exit** in order to infect others **(Figure 14.11)**. Many portals of exit are essentially identical to portals of entry. However, pathogens often exit hosts in materials that the body secretes or excretes. Thus, pathogens may leave hosts in secretions (earwax, tears, nasal secretions, saliva, sputum, and respiratory droplets), in blood (via arthropod bites, hypodermic needles, or wounds), in vaginal secretions or semen, in milk produced by the mammary glands, and in excreted bodily wastes (feces and urine). As we will see, health care personnel must consider the portals of entry and exit in their efforts to understand and control the spread of diseases within populations.

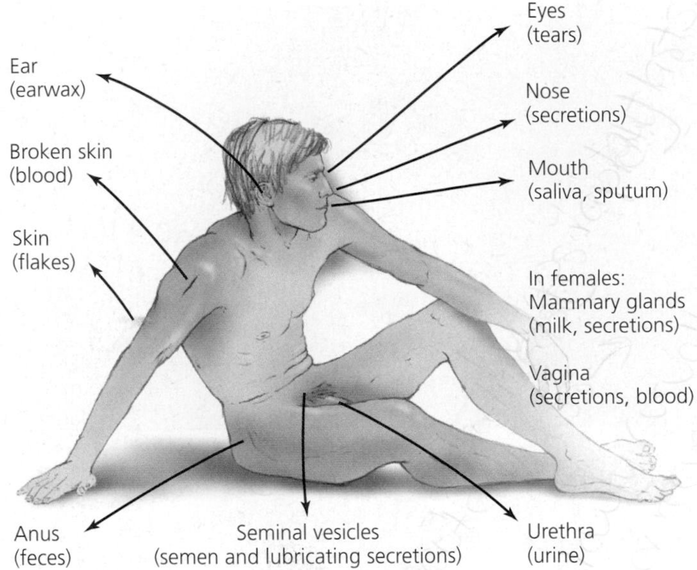

▲ **Figure 14.11 Portals of exit.** Many portals of exit are the same sites as portals of entry. Pathogens often leave the body via bodily secretions and excretions produced at those sites.

▲ **Figure 14.12 Droplet transmission.** In this case, droplets are propelled, primarily from the mouth, during a sneeze. By convention, such transmission is considered contact transmission only if droplets transmit pathogens to a new host within 1 meter of their source. *By what portal of entry does airborne transmission most likely occur?*

Figure 14.12 The most likely portal of entry of airborne pathogens is the respiratory mucous membrane.

Modes of Infectious Disease Transmission

Learning Objectives

- ✓ Contrast contact, vehicle, and vector transmission of pathogens.
- ✓ Contrast *droplet transmission* and *airborne transmission.*
- ✓ Contrast *mechanical* and *biological* vectors.

By definition, an infectious disease agent must be transmitted from either a reservoir or a portal of exit to another host's portal of entry. Transmission can occur by numerous modes that are somewhat arbitrarily categorized into three groups: contact transmission, vehicle transmission, and vector transmission. **ANIMATIONS:** ***Epidemiology: Transmission of Disease***

Contact Transmission

Contact transmission is the spread of pathogens from one host to another by direct contact, indirect contact, or respiratory droplets.

Direct contact transmission, including *person-to-person spread,* typically involves body contact between hosts. Touching, kissing, and sexual intercourse are involved in the transmission of such diseases as warts, herpes, and gonorrhea. Touching, biting, or scratching can transmit zoonoses such as rabies, ringworm, and tularemia from an animal reservoir to a human. The transfer of pathogens from an infected mother to a developing baby across the placenta is another form of direct contact transmission. Direct transmission within a single individual can also occur if the person transfers pathogens from a portal of exit directly to a portal of entry—as occurs, for example, when people with poor personal hygiene unthinkingly place fingers contaminated with fecal pathogens into their mouths.

Indirect contact transmission occurs when pathogens are spread from one host to another by **fomites** (fōm'i-tēz; singular: *fomes,* fō'mēz), which are inanimate objects that are inadvertently used to transfer pathogens to new hosts. Fomites include needles, toothbrushes, paper tissues, toys, money, diapers, drinking glasses, bedsheets, medical equipment, and other objects that can harbor or transmit pathogens. Contaminated needles are a major source of infection of the hepatitis B and AIDS viruses.

Droplet transmission is a third type of contact transmission. Pathogens can be transmitted within *droplet nuclei* (droplets of mucus) that exit the body during exhaling, coughing, and sneezing **(Figure 14.12)**. Pathogens such as cold and flu viruses may be spread in this manner. If pathogens travel more than 1 meter in respiratory droplets, the mode is considered to be *airborne transmission* (discussed shortly), rather than contact transmission.

Vehicle Transmission

Vehicle transmission is the spread of pathogens via air, drinking water, and food, as well as bodily fluids being handled outside the body.

Airborne transmission involves the spread of pathogens farther than 1 meter to the respiratory mucous membranes of a new host via an **aerosol** (ār'ō-sol)—a cloud of small droplets and solid particles suspended in the air. Aerosols may contain pathogens either on dust or inside droplets. (Recall that transmission via droplet nuclei that travel less than 1 meter is considered to be a form of direct contact transmission.) Aerosols can come from

CLINICAL CASE STUDY

TB in the Nursery

In the early fall, a neonatal nurse from a large metropolitan hospital became ill with a cough and fever. Her physician believed she had seasonal allergies and so treated her with cough suppressant, antihistamines, and aerosol steroids. She returned to work in the hospital's nursery.

Three weeks later her condition worsened; her symptoms were complicated by shortness of breath and bloody sputum. Upon further questioning, her physician noted that she was working in the United States on a work visa and was a native of the Philippines. She had a positive skin test for tuberculosis (TB) but had always believed this was her body's natural reaction to the TB vaccine she had received as a child. Her chest X rays in the past had always been clear of infection.

This time, however, her sputum smear tested positive for acid-fast bacilli. She was diagnosed with active tuberculosis and began a standard drug regimen for TB. She was restricted from work and placed in respiratory isolation for six weeks, but during the three weeks that she had continued to work in the nursery, she exposed over 900 obstetric patients including 620 newborns to TB, an airborne infectious disease.

1. How can private physicians quickly assess their clients for the possibility of an infectious disease?
2. What policies should be in place at hospitals to protect patients from exposure to infectious staff members?
3. How could doctors' offices improve public knowledge and protect the public from infectious diseases like tuberculosis?

Reference: Adapted from *MMWR* 54:1280–1283. 2005.

sneezing and coughing, or they can be generated by such things as air conditioning systems, sweeping, mopping, changing clothes or bed linens, or even from flaming inoculating loops in microbiology labs. Dust particles can carry *Staphylococcus, Streptococcus,* and *Hantavirus* (han'tă-vī-rŭs), whereas measles virus and tuberculosis bacilli can be transmitted in dried, airborne droplets. Fungal spores of *Histoplasma* (his tō plaz'mă) and *Coccidioides* (kok-sid-ē-oy'dēz) are typically inhaled.

Waterborne transmission is important in the spread of many gastrointestinal diseases, including giardiasis, amebic dysentery, and cholera. Note that water can act as a reservoir as well as a vehicle of infection. **Fecal-oral infection** is a major source of disease in the world. Some waterborne pathogens, such as *Schistosoma* (skis-tō-sō'mă) worms (see Chapter 23) and enteroviruses (see Chapter 25), are shed in the feces, enter through the gastrointestinal mucous membrane or skin, and subsequently can cause disease elsewhere in the body.

Foodborne transmission involves pathogens in and on foods that are inadequately processed, undercooked, or poorly refrigerated (**Figure 14.13**). Foods may be contaminated with normal microbiota (e.g., *E. coli* and *S. aureus*), with zoonotic pathogens such as *Mycobacterium bovis* (bō'vis) and *Toxoplasma* (tok-sō-plaz'mă), and with parasitic worms that alternate between human and animal hosts. Contamination of food with feces and pathogens such as hepatitis A virus is another kind of fecal-oral transmission. Because milk is particularly rich in nutrients that microorganisms use (protein, lipids, vitamins, and sugars), it would be associated with the transmission of many diseases from infected animals and milk handlers if it were not properly pasteurized.

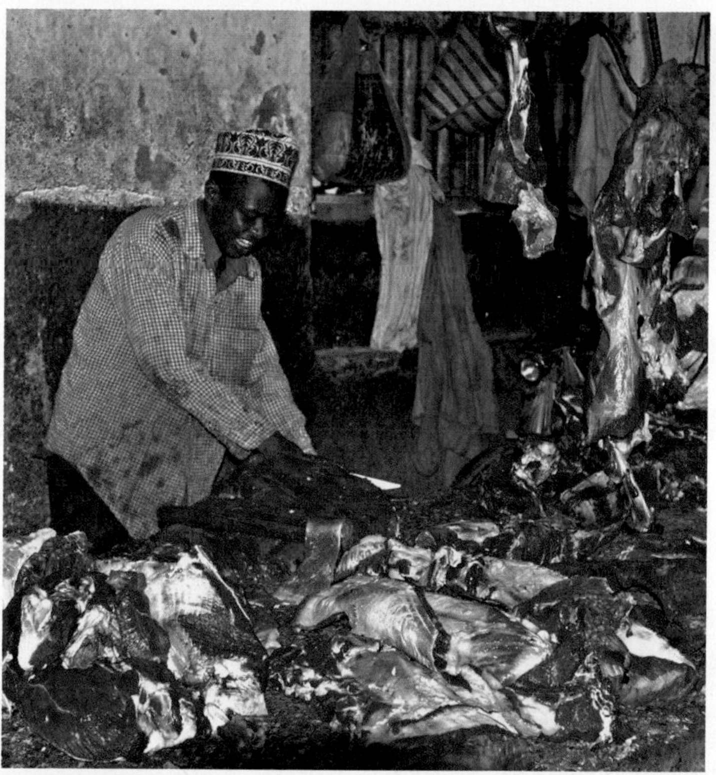

▲ **Figure 14.13 Poorly refrigerated foods can harbor pathogens and transmit diseases.**

Because blood, urine, saliva, and other bodily fluids can contain pathogens, everyone—but especially health care workers—must take precautions when handling these fluids in order to prevent **bodily fluid transmission.** Special care must be taken to prevent such fluids—all of which should be considered potentially contaminated with pathogens—from contacting the conjunctiva or any breaks in the skin or mucous membranes. Examples of diseases that can be transmitted via bodily fluids are AIDS, hepatitis, and herpes, which as we have seen can also be transmitted via direct contact. The **Clinical Case Study** on p. 420 examines the transmission of West Nile virus via blood and infected organs.

Vector Transmission

Vectors are animals that transmit diseases from one host to another. Vectors can be either biological or mechanical.

CLINICAL CASE STUDY

Unusual Transmission of West Nile Virus

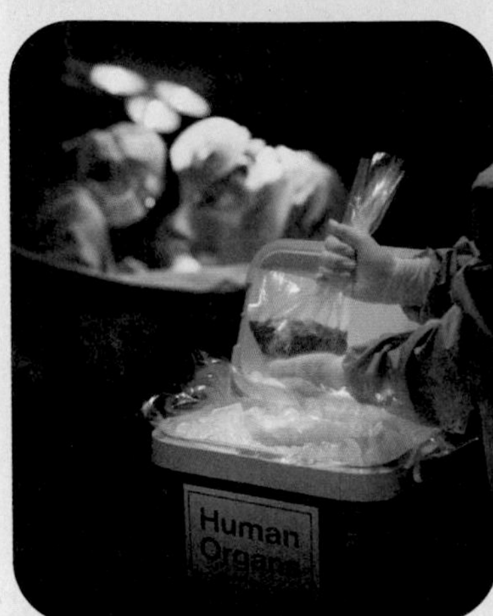

Although West Nile virus is primarily transmitted through mosquito bites, it can also be transmitted in at least two other ways. On August 1, 2002, four organs harvested from a single donor were transplanted into four patients. Within 17 days of transplantation, all four recipients became sick with West Nile virus–associated illnesses. Three of the patients developed meningoencephalitis, an inflammation of the brain and of the membranes surrounding the brain and spinal cord; the fourth patient developed West Nile virus fever. The organ donor's plasma tested positive for West Nile virus. It was determined that the donor herself was infected by a transfusion of WNV-positive blood a day before the organs were harvested.

In September 2005, another organ donor was likely infected by a mosquito rather than through blood transfusion. Lung and liver transplant recipients had severe West Nile virus encephalitis and still remain comatose. One kidney transplant recipient had a positive plasma test for West Nile virus but remains without symptoms. The other kidney recipient had no evidence of West Nile virus infection.

1. In addition to patients who had received organs from the infected donor, what other patients might be at risk of West Nile virus infection?
2. As a scientist, what questions would you seek to answer about the infected donor?
3. What precautions would you advise instituting in order to prevent transmission of West Nile virus infection via blood transfusion and organ transplantation?

References: Adapted from: *MMWR* 51:833–836. 2002. and *MMWR* 54:1021–1023. 2005.

Biological vectors not only transmit pathogens, they also serve as hosts for the multiplication of a pathogen during some stage of the pathogen's life cycle. The biological vectors of diseases affecting humans are typically biting arthropods, including mosquitoes, ticks, lice, fleas, bloodsucking flies, bloodsucking bugs, and mites (see Figure 12.33). After pathogens replicate within a biological vector, often in its gut or salivary gland, the pathogens enter a new host through a bite. The bite site becomes contaminated with the vector's feces, or the vector's bite directly introduces pathogens into the new host.

Mechanical vectors are not required as hosts by the pathogens they transmit; such vectors only passively carry pathogens to new hosts on their feet or other body parts. Mechanical vectors such as houseflies and cockroaches may introduce pathogens such as *Salmonella* and *Shigella* into drinking water and food or onto the skin.

Table 14.10 lists some arthropod vectors and the diseases they transmit. Table 14.11 on p. 424 summarizes the modes of disease transmission.

Classification of Infectious Diseases

Learning Objectives

✓ Describe the basis for each of the various classification schemes of infectious diseases.

✓ Distinguish among acute, subacute, chronic, and latent diseases.

✓ Distinguish among communicable, contagious, and noncommunicable infectious diseases.

Infectious diseases can be classified in a number of ways. No one way is "*the* correct way"; each has its own advantages. One scheme groups diseases based upon the body systems affected. A difficulty with this method of classification is that many diseases involve more than one organ system. For example, AIDS begins as a sexually transmitted infection (reproductive system) or a parenteral infection of the blood (cardiovascular system). It then becomes an infection of the lymphatic system as viruses invade lymphocytes. Finally, the syndrome involves diseases and degeneration of the respiratory, nervous, digestive, cardiovascular, and lymphatic systems.

Another classification system deals with diseases according to taxonomic categories. Chapters 19 to 25 examine diseases based on this approach.

Every disease (not just infectious diseases) can also be classified according to its longevity and severity. If a disease develops rapidly but lasts a relatively short time, it is called an **acute disease.** An example is the common cold. In contrast, **chronic diseases** develop slowly (usually with less severe symptoms) and are continual or recurrent. Infectious mononucleosis, hepatitis C, tuberculosis, and leprosy are chronic diseases. **Subacute diseases** have durations and severities that lie somewhere between acute and chronic. Subacute bacterial endocarditis, a disease of heart valves, is one example. **Latent**

TABLE 14.10 Selected Arthropod Vectors

	Disease	Causative Agent (bacteria unless otherwise indicated)
Biological Vectors		
Mosquitoes		
Anopheles, Aedes	Malaria	*Plasmodium* spp. (protozoan)
	Yellow fever	*Flavivirus* sp. (virus)
	Elephantiasis	*Wuchereria bancrofti* (helminth)
	Dengue	*Flavivirus* spp. (virus)
	Viral encephalitis	*Alphavirus* spp. (virus)
Ticks		
Ixodes	Lyme disease	*Borrelia burgdorferi*
Dermacentor	Rocky Mountain spotted fever	*Rickettsia rickettsii*
Flea		
Xenopsylla	Bubonic plague	*Yersinia pestis*
	Endemic typhus	*Rickettsia prowazekii*
Louse		
Pediculus	Epidemic typhus	*Rickettsia typhi*
Bloodsucking flies		
Glossina	African sleeping sickness	*Trypanosoma brucei*
Simulium	River blindness	*Onchocerca volvulus* (helminth)
Bloodsucking bug		
Triatoma	Chagas' disease	*Trypanosoma cruzi* (protozoan)
Mite (chigger)		
Leptotrombidium	Scrub typhus	*Orientia tsutsugamushi*
Mechanical Vectors		
Housefly		
Musca	Foodborne infections	*Shigella* spp., *Salmonella* spp., *Escherichia coli*
Cockroaches		
Blatella, Periplaneta	Foodborne infections	*Shigella* spp., *Salmonella* spp., *Escherichia coli*

diseases are those in which a pathogen remains inactive for a long period of time before becoming active. Herpes is an example of a latent disease.

When an infectious disease comes from another infected host, either directly or indirectly, it is a **communicable** disease. Influenza, herpes, and tuberculosis are examples of communicable diseases. If a communicable disease is easily transmitted between hosts, as is the case for chickenpox or measles, it is also called a **contagious** disease. **Noncommunicable** diseases arise outside of hosts or from normal microbiota. In other words, they are not spread from one host to another, and diseased patients are not a source of contamination for others. Tooth decay, acne, and tetanus are examples of noncommunicable diseases.

Table 14.12 on p. 424 defines these and other terms used to classify infectious diseases.

Yet another way in which all infectious diseases may be classified is by the effects they have on populations, rather than on individuals. Is a certain disease consistently found in a given group of people or geographic area? Under what circumstances is it more prevalent than normal in a given geographic area? How prevalent is "normal"? How is the disease transmitted throughout a population? We next examine issues related to diseases at the population level.

Epidemiology of Infectious Diseases

Learning Objective

✓ Define *epidemiology*.

Our discussion so far has centered on the negative impact of microorganisms on *individuals*. Now we turn our attention to the effects of pathogens on *populations*. **Epidemiology**[10] (ep-i-dē-mē-ol'ō-jē) is the study of where and when diseases occur, and how they are transmitted within populations. During the 20th century, epidemiologists expanded the scope of their work beyond infectious diseases to also consider injuries and deaths related to automobile and fireworks accidents, cigarette smoking, lead poisoning, and other causes; however, we will limit our discussion primarily to the epidemiology of infectious diseases. **ANIMATIONS:** ***Epidemiology: Overview***

Frequency of Disease

Learning Objectives

✓ Contrast between incidence and prevalence.

✓ Differentiate among the terms *endemic, sporadic, epidemic,* and *pandemic.*

Epidemiologists keep track of the occurrence of diseases by using two measures: incidence and prevalence. **Incidence** is the number of *new* cases of a disease in a given area or population during a given period of time; **prevalence** is the *total number* of cases, both new and already existing, in a given area or population during a given period of time. In other words, prevalence is a cumulative number. Thus, for example, the reported number of new cases—the incidence—of AIDS in the United States in 2006 was 36,828. However, the prevalence of AIDS in 2006 was 436,690 because almost 400,000 patients who got the disease prior to 2006 still survived with the disease. **Figure 14.14** illustrates this relationship between incidence and prevalence.

Epidemiologists report their data in many ways including maps, graphs, charts, and tables **(Figure 14.15)**. Why do they report their data in so many different ways? Using a variety of formats enables epidemiologists to observe patterns that may give clues about the causes of or ways to prevent diseases. For example, shigellosis (an infection of *Shigella* characterized by

[10]From Greek *epidemios*, meaning among the people, and *logos*, meaning study of.

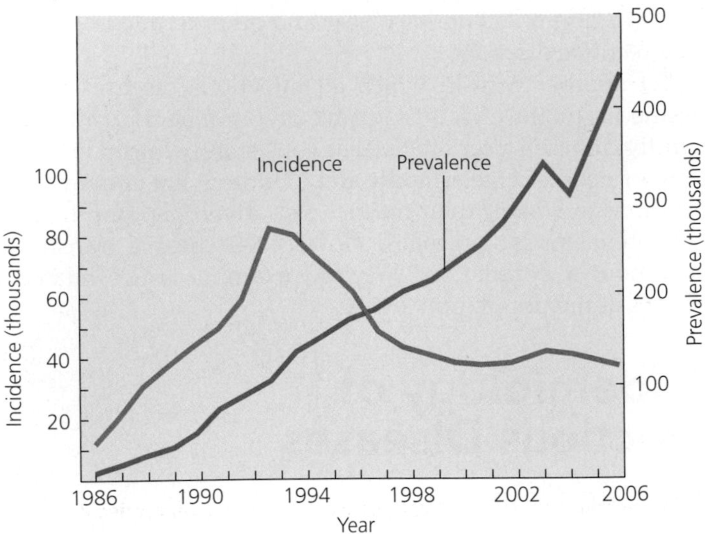

▲ **Figure 14.14 Curves representing the incidence and the estimated prevalence of AIDS among U.S. adults.** Note that the scales for the two curves are different. *Why can the incidence of a disease never exceed the prevalence of that disease?*

Figure 14.14 Because prevalence includes all cases, both old and new, prevalence must always be larger than incidence.

diarrhea, abdominal pain, and fever) had an overall prevalence of 85 cases per million in the U.S. population in 2000, but when the data are reported by age group, it becomes apparent that the highest prevalence (350 cases per million) occurs in young children (see Figure 14.15c).

The occurrence of a disease can also be considered in terms of a combination of frequency and geographic distribution **(Figure 14.16)**. A disease that normally occurs continually (at moderately regular intervals) at a relatively stable incidence within a given population or geographical area is said to be **endemic**[11] to that population or region. A disease is considered **sporadic** when only a few scattered cases occur within an area or population. Whenever a disease occurs at a greater frequency than is usual for an area or population, the disease is said to be **epidemic** within that area or population. **ANIMATIONS:** ***Epidemiology: Occurrence of Disease***

The commonly held belief that a disease must infect thousands or millions to be considered an epidemic is mistaken. The time period and the number of cases necessary for an outbreak of disease to be classified as an epidemic are not specified; the important fact is that there are more cases than historical statistics indicate are expected. For example, less than 200 cases of *Hantavirus* pulmonary syndrome occurred in the Four Corners area[12] of the United States in 1993, but because they were the first cases of this disease ever reported for the region, the outbreak was considered an epidemic. At the same place, thousands of cases of flu occurred, but there was no flu epidemic because the number of cases observed did not exceed the number expected.

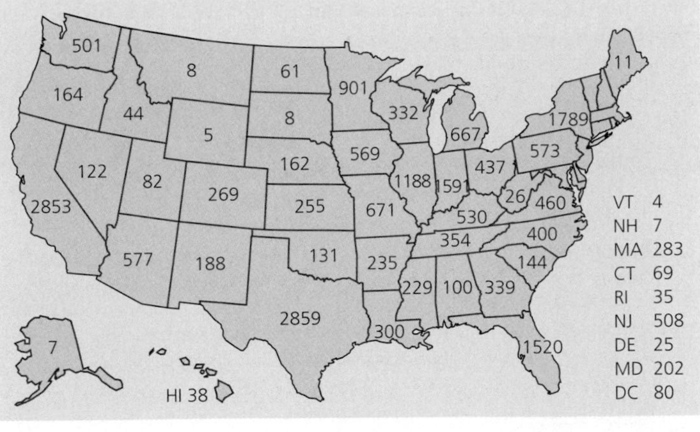

(a)

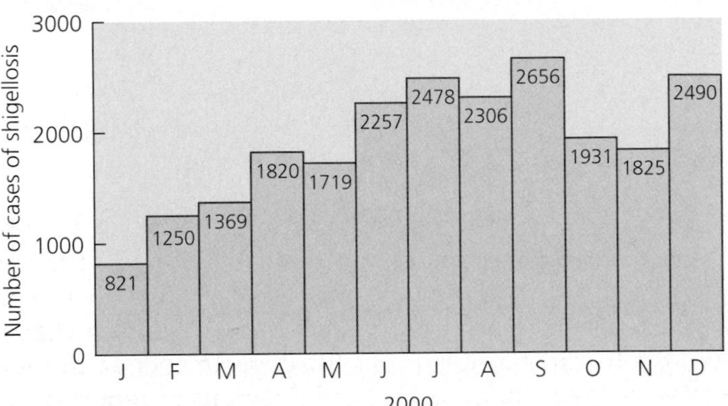

(b)

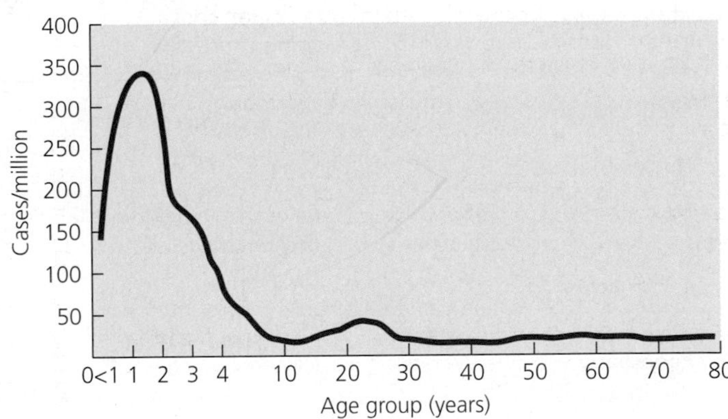

(c)

▲ **Figure 14.15 Epidemiologists report data in a number of ways.** Here incidence of shigellosis in the U.S. in 2000 is presented **(a)** on a map by state, **(b)** by month, and **(c)** as cases per million by age group. The latter graph reveals that shigellosis in young children is much more prevalent (as high as 350 cases per million) than in the U.S. population as a whole, a fact not apparent unless the data are sorted by age.

Figure 14.17 illustrates how epidemics are defined according to the number of expected cases, and not according to the absolute number of cases.

If an epidemic occurs simultaneously on more than one continent, it is referred to as a **pandemic** (see Figure 14.16d). H1N1 flu, so-called swine flu, became pandemic worldwide in 2009.

[11]From Greek *endemos*, meaning native.
[12]Four Corners is the geographic region surrounding the point at which Arizona, Utah, New Mexico, and Colorado meet.

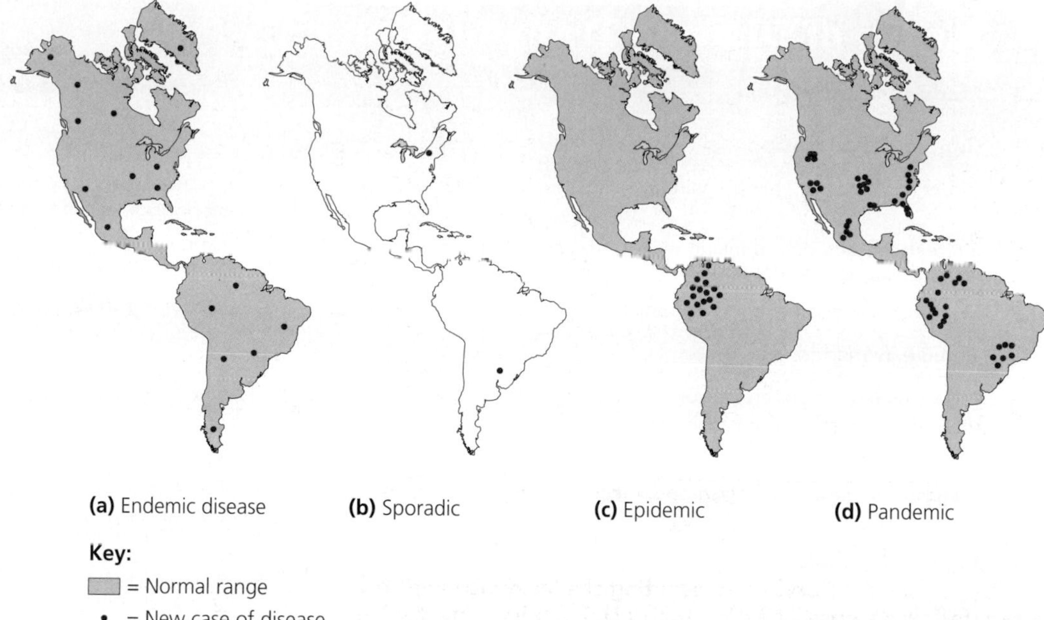

Figure 14.16 Illustrations of the different terms for the occurrence of disease. **(a)** An endemic disease, which is normally present in a region. **(b)** A sporadic disease, which occurs irregularly and infrequently. **(c)** An epidemic disease, which is present in greater frequency than is usual. **(d)** A pandemic disease, which is an epidemic disease occurring on more than one continent at a given time. *Regardless of where you live, name a disease that is endemic, one that is sporadic, and one that is epidemic in your state.*

Figure 14.16 *Some possible answers: Flu is endemic in every state; tuberculosis is sporadic in most states; AIDS is epidemic in every state.*

Obviously, for disease prevalence to be classified as either endemic, sporadic, epidemic, or pandemic, good records must be kept for each region and population. From such records, incidence and prevalence can be calculated, and then changes in these data can be noted. Health departments at the local and state levels require doctors and hospitals to report certain infectious diseases. Some are also nationally notifiable; that is, their occurrence must be reported to the Centers for Disease Control and Prevention (CDC) in Atlanta, Georgia, which is the headquarters and clearinghouse for national epidemiological research. Nationally notifiable diseases are listed in Table 14.13 on p. 427. Each week the CDC reports the number of cases of

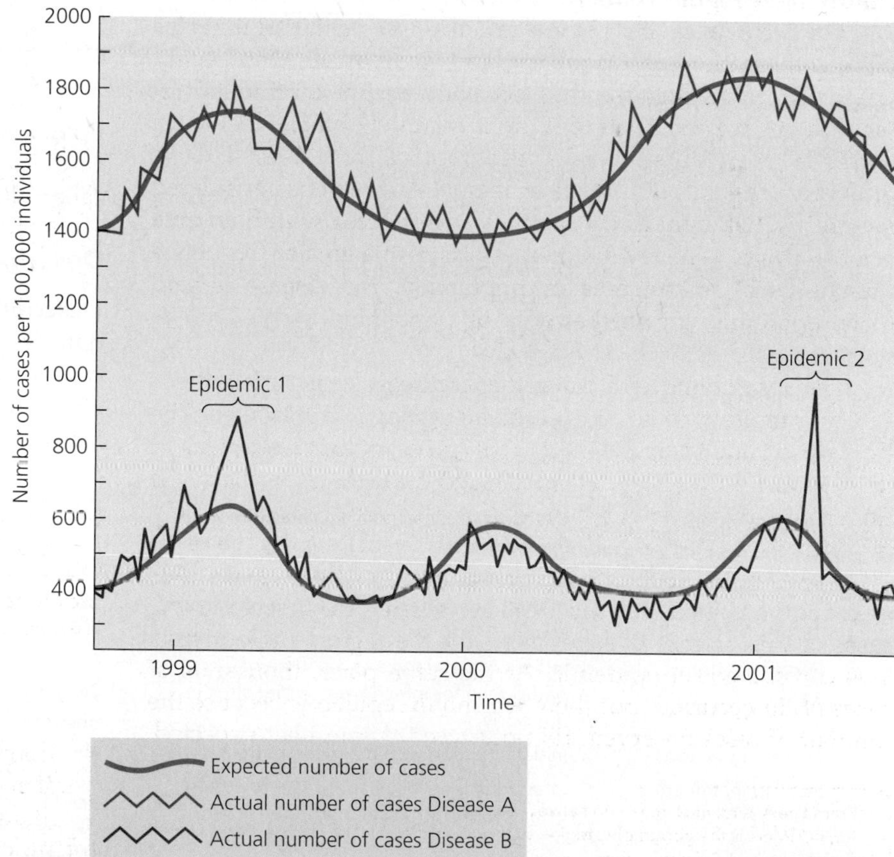

Figure 14.17 Epidemics may have fewer cases than nonepidemics. These two graphs demonstrate the independence between the absolute number of cases and a disease's designation as an epidemic. Even though the number of cases of disease A always exceeds the number of cases of disease B, only disease B is ever considered epidemic—at those times when the number of cases observed exceeds the number of cases expected. *How can scientists know the normal prevalence of a given disease?*

Figure 14.17 *Scientists record every case of the disease so that they will have a "baseline" prevalence, which then becomes the expected prevalence for each disease.*

TABLE 14.11 Modes of Disease Transmission

Mode of Transmission	Examples of Diseases Spread
Contact Transmission	
Direct contact: e.g., handshaking, kissing, sexual intercourse, bites	Cutaneous anthrax, genital warts, gonorrhea, herpes, rabies, staphylococcal infections, syphilis
Indirect contact: e.g., drinking glasses, toothbrushes, toys, punctures	Common cold, enterovirus infections, influenza, measles, Q fever, pneumonia, tetanus
Droplet transmission: e.g., droplets from sneezing (within 1 meter)	Whooping cough, streptococcal pharyngitis (strep throat)
Vehicle Transmission	
Airborne: e.g., dust particles or droplets carried more than 1 meter	Chickenpox, coccidioidomycosis, histoplasmosis, influenza, measles, pulmonary anthrax, tuberculosis
Waterborne: e.g., streams, swimming pools	*Campylobacter* infections, cholera, *Giardia* diarrhea
Foodborne: e.g., poultry, seafood, meat	Food poisoning (botulism, staphylococcal); hepatitis A, listeriosis, tapeworms, toxoplasmosis, typhoid fever
Vector Transmission	
Mechanical: e.g., on bodies of flies, roaches	*E. coli* diarrhea, salmonellosis, trachoma
Biological: e.g., lice, mites, mosquitoes, ticks	Chagas' disease, Lyme disease, malaria, plague, Rocky Mountain spotted fever, typhus fever, yellow fever

TABLE 14.12 Terms Used to Classify Infectious Diseases

Term	Definition
Acute disease	Disease in which symptoms develop rapidly and that runs its course quickly
Chronic disease	Disease with usually mild symptoms that develop slowly and last a long time
Subacute disease	Disease with time course and symptoms between acute and chronic
Asymptomatic disease	Disease without symptoms
Latent disease	Disease that appears a long time after infection
Communicable disease	Disease transmitted from one host to another
Contagious disease	Communicable disease that is easily spread
Noncommunicable disease	Disease arising from outside of hosts or from opportunistic pathogen
Local infection	Infection confined to a small region of the body
Systemic infection	Widespread infection in many systems of the body; often travels in the blood or lymph
Focal infection	Infection that serves as a source of pathogens for infections at other sites in the body
Primary infection	Initial infection within a given patient
Secondary infection	Infections that follow a primary infection; often by opportunistic pathogens

most of the nationally notifiable diseases in the *Morbidity and Mortality Weekly Report (MMWR)* **(Figure 14.18)**.

Epidemiological Studies

Learning Objective

✓ Explain three approaches epidemiologists use to study diseases in populations.

Epidemiologists conduct research to study the dynamics of diseases in populations by taking three different approaches, called descriptive, analytical, and experimental epidemiology.

Descriptive Epidemiology

Descriptive epidemiology involves the careful tabulation of data concerning a disease. Relevant information includes the location and time of cases of the disease, as well as information about the patients, such as ages, gender, occupations, health histories, and socioeconomic groups. Because the time course and chains of transmission of a disease are an important part of descriptive epidemiology, epidemiologists strive to identify the **index case** (the first case) of the disease in a given area or population. Sometimes it is difficult or impossible to identify an index case because the patient has recovered, moved, or died.

▶ **Figure 14.18 A page from the *MMWR*.** The CDC's *Morbidity and Mortality Weekly Report (MMWR)* provides epidemiological data state by state for the current week, for the current year to date, and for the previous year to date. The *MMWR* also publishes reports on epidemiological case studies. This page shows the incidence of three diseases in one week in 2008–2009.

1426 MMWR January 9, 2009

TABLE II, (*Continued*) Provisional cases of selected notifiable diseases, United States, weeks ending January 3, 2009, and December 29, 2007 (53rd week)*

	Pertussis					Rabies, animal					Rocky Mountain spotted fever				
Reporting area	Current week	Previous 52 weeks Med	Previous 52 weeks Max	Cum 2008	Cum 2007	Current week	Previous 52 weeks Med	Previous 52 weeks Max	Cum 2008	Cum 2007	Current week	Previous 52 weeks Med	Previous 52 weeks Max	Cum 2008	Cum 2007
United States	164	174	315	10,007	10,454	17	101	164	4,911	5,975	14	31	145	2,276	2,221
New England	4	11	35	620	1,552	6	7	20	372	522	—	0	2	4	10
Connecticut	—	0	4	34	89	3	4	17	203	219	—	0	0	—	—
Maine†	—	0	5	47	83	1	1	5	64	86	N	0	0	N	N
Massachusetts	—	8	32	420	1,178	N	0	0	N	N	—	0	1	1	9
New Hampshire	4	1	4	48	80	—	0	3	35	53	—	0	1	1	1
Rhode Island	—	0	7	59	59	N	0	0	N	N	—	0	2	2	—
Vermont†	—	0	2	12	63	2	1	6	70	164	—	0	0	—	—
Mid. Atlantic	—	19	41	1,051	1,314	6	28	63	1,536	997	—	1	5	80	85
New Jersey	—	1	6	71	229	—	0	0	—	—	—	0	2	12	32
New York (Upstate)	—	7	24	426	549	6	9	20	500	514	—	0	2	17	7
New York City	—	0	5	46	150	—	0	2	19	44	—	0	2	24	28
Pennsylvania	—	9	34	508	386	—	18	48	1,017	439	—	0	2	27	18
E.N. Central	84	30	189	1,919	1,495	—	3	28	247	414	—	1	15	150	60
Illinois	—	6	39	517	199	—	1	21	103	113	—	1	10	104	39
Indiana	27	1	15	139	68	—	0	2	10	13	—	0	3	8	6
Michigan	—	6	14	294	292	—	0	8	73	202	—	0	1	3	4
Ohio	57	10	176	846	609	—	1	7	61	86	—	0	4	34	10
Wisconsin	—	2	7	123	327	N	0	0	N	N	—	0	1	1	1
W.N. Central	18	17	120	1,378	909	—	3	13	206	276	1	4	32	456	369
Iowa	—	2	20	209	150	—	0	5	29	31	—	0	2	7	17
Kansas	1	1	13	78	104	—	0	0	—	110	—	0	0	—	12
Minnesota	—	2	26	224	393	—	0	10	65	40	—	0	0	—	6
Missouri	15	6	50	535	118	—	1	8	65	38	1	4	31	426	315
Nebraska†	2	2	35	281	70	—	0	0	—	—	—	0	4	20	14
North Dakota	—	0	1	1	14	—	0	7	34	30	—	0	0	—	—
South Dakota	—	0	7	50	60	—	0	2	23	27	—	0	1	3	5
S. Atlantic	16	17	44	937	978	4	37	101	2,024	2,184	13	12	71	928	1,020
Delaware	—	0	3	18	11	—	0	0	—	—	—	0	5	33	17
District of Columbia	—	0	1	7	9	—	0	0	—	—	—	0	2	8	3
Florida	—	5	20	306	211	—	0	77	139	128	—	0	3	20	19
Georgia	—	1	6	91	37	—	5	42	339	300	—	1	8	73	60
Maryland†	1	2	8	130	118	—	8	17	420	431	—	1	7	72	63
North Carolina	15	0	15	94	330	4	9	16	454	472	12	2	55	511	665
South Carolina†	—	2	10	129	102	—	0	0	—	46	1	1	9	55	64
Virginia†	—	3	10	152	128	—	11	24	591	730	—	2	15	149	123
West Virginia	—	0	2	10	32	—	1	9	81	77	—	0	1	7	6
E.S. Central	3	7	28	395	463	—	3	7	165	156	—	3	23	324	276
Alabama†	—	1	5	59	91	—	0	0	—	—	—	1	8	90	96
Kentucky	1	2	11	136	33	—	0	4	45	21	—	0	1	1	5
Mississippi	1	2	5	92	255	—	0	1	2	3	—	0	3	12	20
Tennessee†	1	1	14	108	84	—	2	6	118	132	—	2	19	221	155
W.S. Central	10	27	106	1,631	1,303	—	1	11	92	1,086	—	1	41	286	361
Arkansas	9	1	19	93	173	—	0	6	48	33	—	0	14	68	122
Louisiana	—	1	7	77	21	—	0	0	—	6	—	0	1	5	4
Oklahoma	1	0	21	56	58	—	0	10	42	78	—	0	26	170	186
Texas†	—	23	98	1,405	1,051	—	0	1	2	969	—	1	6	43	49
Mountain	8	15	34	814	1,137	—	1	8	77	97	—	1	3	43	37
Arizona	—	4	10	204	210	N	0	0	N	N	—	0	2	17	10
Colorado	6	3	6	160	307	—	0	0	—	—	—	0	1	1	3
Idaho†	2	0	5	38	45	—	0	0	—	12	—	0	1	1	4
Montana†	—	1	11	83	53	—	0	2	9	21	—	0	1	3	1
Nevada†	—	0	7	19	37	—	0	4	5	13	—	0	2	2	—
New Mexico†	—	1	8	68	74	—	0	3	25	15	—	0	1	2	6
Utah	—	4	13	226	397	—	0	6	14	16	—	0	1	7	—
Wyoming	—	0	2	16	24	—	0	3	24	20	—	0	2	10	13
Pacific	21	22	83	1,262	1,303	1	3	13	192	243	—	0	1	5	3
Alaska	1	3	21	258	89	—	0	4	15	45	N	0	0	N	N
California	1	7	23	392	590	1	3	12	100	186	—	0	1	2	1
Hawaii	—	0	2	17	19	—	0	0	—	—	N	0	0	N	N
Oregon†	—	3	10	176	123	—	0	4	14	12	—	0	1	3	2
Washington	19	5	63	419	482	—	0	0	—	—	N	0	0	N	N
American Samoa	—	0	0	—	—	N	0	0	N	N	N	0	0	N	N
C.N.M.I.	—	—	—	—	—	—	—	—	—	—	—	—	—	—	—
Guam	—	0	0	—	—	—	0	0	—	—	N	0	0	N	N
Puerto Rico	—	0	0	—	—	—	1	5	59	48	N	0	0	N	N
U.S. Virgin Islands	—	0	0	—	—	N	0	0	N	N	N	0	0	N	N

C.N.M.I.: Commonwealth of Northern Mariana Islands.
U: Unavailable. —: No reported cases. N: Not notifiable. Cum: Cumulative year-to-date counts. Med: Median. Max: Maximum.
* Incidence data for reporting year 2008 is provisional.
† Contains data reported through the National Electronic Disease Surveillance System (NEDSS).

▲ **Figure 14.19 A map showing cholera deaths in a section of London, 1854.** From the map he compiled, Dr. John Snow showed that cholera cases centered around the Broad Street pump. Snow's work was a landmark in epidemiological research.

The earliest descriptive epidemiological study was by John Snow (1813–1858), who studied a cholera outbreak in London in 1854. By carefully mapping the locations of the cholera cases in a particular part of the city, Snow found that the cases were clustered around the Broad Street water pump **(Figure 14.19)**. This distribution of cases, plus the voluminous watery diarrhea of cholera patients, suggested that the disease was spread via contamination of drinking water by sewage.

Analytical Epidemiology

Analytical epidemiology investigates a disease in detail, including analysis of data acquired in descriptive epidemiological studies, to determine the probable cause, mode of transmission, and possible means of prevention of the disease. Analytical epidemiology may be used in situations where it is not ethical to apply Koch's postulates. Thus, even though Koch's third postulate has never been fulfilled in the case of AIDS (because it is unethical to intentionally inoculate a human with HIV), analytical epidemiological studies indicate that HIV causes AIDS and that it is transmitted primarily sexually.

Often analytical studies are *retrospective;* that is, they attempt to identify causation and mode of transmission after an outbreak has occurred. Epidemiologists compare a group of people who had the disease with a group who did not. The groups are carefully matched by factors such as gender, environment, and diet, and then compared to determine which pathogens and factors may play a role in morbidity.

Experimental Epidemiology

Experimental epidemiology involves testing a hypothesis concerning the cause of a disease. The application of Koch's postulates is an example of experimental epidemiology. Experimental epidemiology also involves studies to test a hypothesis resulting from an analytical study such as the efficacy of a preventive measure or certain treatment. For example, analytical epidemiological studies suggested that the bacterium *Chlamydophila pneumoniae* may contribute to arteriosclerosis, resulting in heart attacks. Some scientists have hypothesized that antibacterial drugs could prevent heart attacks by killing the causative agent. To test this hypothesis the scientists administered either the antimicrobial drug azithromycin or a medicinally inactive placebo to 7700 patients, all of whom had a history of heart disease and were infected with *C. pneumoniae.* The researchers observed no significant difference in the number of heart attacks suffered by patients in the two groups over a 2-year period. Thus an experimental epidemiological study disproved a hypothesis suggested by an analytical epidemi ological analysis.

The **Clinical Case Study: *Legionella* in the Produce Aisle** illustrates the work of epidemiologists.

Hospital Epidemiology: Nosocomial Infections

Learning Objectives

- ✓ Explain how nosocomial infections differ from other infections.
- ✓ Describe the factors that influence the development of nosocomial infections.
- ✓ Describe three types of nosocomial infections and how they may be prevented.

Of special concern to epidemiologists and health care workers are nosocomial (nos-ō-kō′mē-ăl) infections and nosocomial diseases. **Nosocomial infections** are those acquired by patients or health care workers while they are in health care facilities, including hospitals, dental offices, nursing homes, and doctors' waiting rooms. The CDC estimates that about 10% of American patients acquire a nosocomial infection each year. **Nosocomial diseases**—those acquired in a health care setting—increase the duration and cost of medical care and result in some 90,000 deaths annually in the U.S. **ANIMATIONS:** ***Nosocomial Infections: Overview***

TABLE 14.13

Nationally Notifiable Infectious Diseases[a]

Acquired immunodeficiency syndrome (AIDS)	*Hantavirus* pulmonary syndrome	Poliomyelitis, paralytic	*Streptococcus pneumoniae,* drug resistant
Anthrax	Hemolytic uremic syndrome, postdiarrheal	Poliovirus infection	*Streptococcus pneumoniae,* non-drug-resistant and invasive in children
Arboviral diseases	Hepatitis A	Psittacosis	Syphilis, congenital
Botulism	Hepatitis B	Q fever	Tetanus
Brucellosis	Hepatitis C	Rabies, animal and human	Toxic-shock syndrome, non-streptococcal
Chancroid	HIV infection	Rocky Mountain spotted fever	Trichinellosis
Chlamydia trachomatis, genital infections	Influenza-associated pediatric mortality	Rubella	Tuberculosis
Cholera	Legionellosis	Rubella, congenital syndrome	Tularemia
Coccidioidomycosis	Listeriosis	Salmonellosis	Typhoid fever
Cryptosporidiosis	Lyme disease	Severe acute respiratory syndrome (SARS)	Vancomycin-resistant *Staphylococcus aureus*
Cyclosporiasis	Malaria	Shiga-toxin-producing *Escherichia coli*	Varicella
Diphtheria	Measles	Shigellosis	Vibriosis
Ehrlichiosis/anaplasmosis	Meningococcal disease	Smallpox	Yellow fever
Giardiasis	Mumps	Streptococcal disease, invasive, Group A	
Gonorrhea	Novel influenza A infections	Streptococcal toxic shock syndrome	
Haemophilus influenzae, invasive disease	Pertussis		
Hansen disease (leprosy)	Plague		

[a]Diseases for which hospitals, physicians, and other health care workers are required to report cases to state health departments, who then forward the data to the CDC.

CLINICAL CASE STUDY 

Legionella in the Produce Aisle

The Lousiana state health department has received reports of 33 cases of Legionnaires' disease in the town of Bogalusa (population 16,000). Legionnaires' disease, or legionellosis, is a potentially fatal respiratory disease caused by the growth of a bacterium, *Legionella pneumophila,* in the lungs of patients. The bacterium enters humans via the respiratory portal in aerosols produced by cooling towers, air conditioners, whirlpool baths, showers, humidifiers, and respiratory therapy equipment.

Epidemiologists begin trying to ascertain the source of Bogalusa's Legionnaires' disease outbreak by interviewing the victims and their relatives to develop complete histories and to identify areas of commonality among the victims that were lacking among nonvictims. Victims include a range of ages, occupations, hobbies, religions, and types and locations of dwellings. Although no significant differences are identified among the lifestyles, ages, or smoking habits of victims and nonvictims, one curious fact is discovered: All the victims did their grocery shopping at the same store. However, healthy individuals also shopped at that store.

The air conditioning system of the grocery store proves to be free of *Legionella,* but the vegetable misting machine does not. The strain of *Legionella* isolated from the misting machine is identical to the strain recovered from the victims' lungs.

1. Would this outbreak be classified as endemic, epidemic, or pandemic?
2. Is this a descriptive, analytical, or experimental epidemiological study?
3. Knowing the epidemiology and causative agent of Legionnaires' disease, what questions would you ask of the victims or of their surviving relatives?
4. What, as an epidemiologist, would you examine at the store?
5. How did the victims become contaminated? Why didn't everyone who bought vegetables at the store get legionellosis? What could the owners of the store do to limit or prevent future infections?

Reference: Adapted from *MMWR* 39: 108–109. 1990.

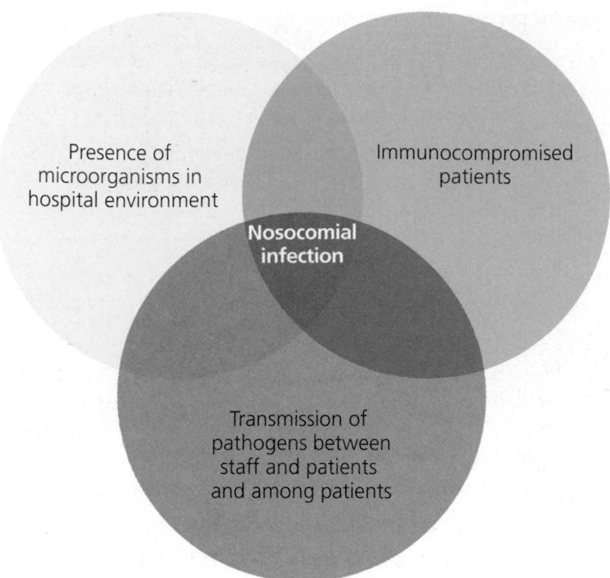

▲ **Figure 14.20 The interplay of factors that result in nosocomial infections.** Although nosocomial infections can result from any one of the factors shown, most nosocomial infections are the product of the interaction of all three factors.

Types of Nosocomial Infections

When most people think of nosocomial infections, what likely comes to mind are **exogenous** (eks-oj′ĕ-nŭs) infections, which are caused by pathogens acquired from the health care environment. After all, hospitals are filled with sick people shedding pathogens from every type of portal of exit. However, we have seen that members of the normal microbiota can become opportunistic pathogens as a result of hospitalization or medical treatments such as chemotherapy. Such opportunists cause **endogenous** (en-doj′ĕ-nŭs) nosocomial infections; that is, they arise from normal microbiota within the patient that become pathogenic because of factors within the health care setting.

Iatrogenic (ī-at-rō-jen′ik) **infections** (literally meaning "doctor induced" infections) are a subset of nosocomial infections that ironically are the direct result of modern medical procedures such as the use of catheters, invasive diagnostic procedures, and surgery.

Superinfections may result from the use of antimicrobial drugs that, by inhibiting some resident microbiota, allow others to thrive in the absence of competition. For instance, long-term antimicrobial therapy to inhibit a bacterial infection may allow *Clostridium difficile* (dif-fi′sil-ē), a transient microbe of the colon, to grow excessively and cause a painful condition called pseudomembranous colitis. Such superinfections are not limited to health care settings.

Nosocomial infections most often occur in the urinary, respiratory, cardiovascular, and integumentary (skin) systems, though surgical wounds can become infected and result in nosocomial infections in any part of the body.

Factors Influencing Nosocomial Infections

Nosocomial infections arise from the interaction of several factors in the health care environment, which include the following **(Figure 14.20)**:

- Exposure to numerous pathogens present in the health care setting, including many that are resistant to antimicrobial agents
- The weakened immune systems of patients who are ill, making them more susceptible to opportunistic pathogens
- Transmission of pathogens among patients and health care workers—from staff and visitors, to patients, and even from one patient to another via activities of staff members (including invasive procedures and other iatrogenic factors)

Control of Nosocomial Infections

Aggressive control measures can noticeably reduce the incidence of nosocomial infections. These include disinfection; medical asepsis, including good housekeeping, handwashing, bathing, sanitary handling of food, proper hygiene, and precautionary measures to avoid the spread of pathogens among patients; surgical asepsis and sterile procedures, including thorough cleansing of the surgical field, use of sterile instruments, and use of sterile gloves, gowns, caps, and masks; isolation of particularly contagious or susceptible patients; and establishment of a nosocomial infection control committee charged with surveillance of nosocomial diseases and review of control measures. **ANIMATIONS:** ***Nosocomial Infections: Prevention***

Numerous studies have shown that the single most effective way to reduce nosocomial infections is effective handwashing by all medical and support staff. In one study, deaths from nosocomial infections were reduced by over 50% when hospital personnel followed strict guidelines about washing their hands frequently.

Epidemiology and Public Health

Learning Objective

✓ List three ways public health agencies work to limit the spread of diseases.

As you have likely realized by now, epidemiologists gather information concerning the spread of disease within populations so that they can take steps to reduce the number of cases and improve the health of individuals within a community. In the following sections we will examine how the various public health agencies share epidemiological data, facilitate the interruption of disease transmission, and educate the public about public health issues. Public health agencies also implement immunization programs; immunization is discussed in Chapter 17.

EMERGING DISEASES

▲ The deer mouse, *Peromyscus maniculatus.*

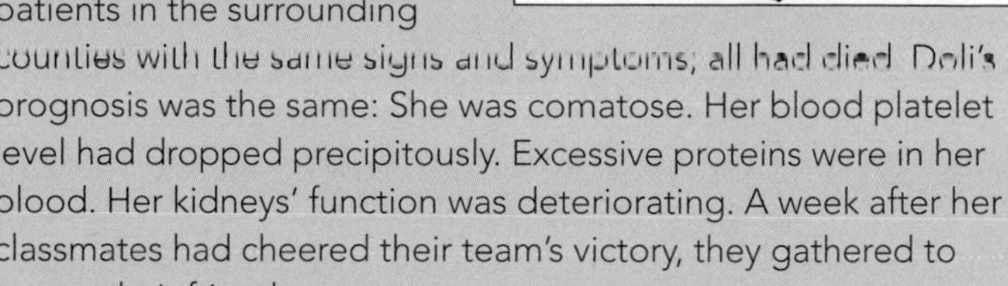

HANTAVIRUS PULMONARY SYNDROME

Doli was excited. Her Navajo basketball team had won the divisional championship for 1993, and she was high point (the high scorer) for the championship game. The thrill of the moment made her forget the argument with her parents earlier in the week about sweeping out the storage shed. They were champions, and life was good.

The next morning, Doli woke with deep muscle pain that she attributed to the exertions of the game. By noon she had a headache and was nauseous and experiencing the chill associated with a fever. "The flu," her mother concluded, and sent Doli to bed with acetaminophen and plenty of fluids. Over the next few days, Doli's lungs began to congest; she struggled to breathe, and her heart raced. Doli was drowning in her own bodily fluid.

Desperately worried, Doli's parents drove her 60 miles to the Navajo medical center, where bewildered doctors provided respiratory and cardiac care while searching for answers. They discovered three patients in the surrounding counties with the same signs and symptoms; all had died. Doli's prognosis was the same: She was comatose. Her blood platelet level had dropped precipitously. Excessive proteins were in her blood. Her kidneys' function was deteriorating. A week after her classmates had cheered their team's victory, they gathered to mourn their friend.

As heartbreaking as the deaths in this epidemic were, the cases advanced our medical understanding and put to use the power of modern epidemiology and genetic analysis. Within eight days of the initial case, epidemiologists had isolated a suspect virus, sequenced its genes, identified it as a previously unknown species of *Hantavirus*, and shown that the virus was the cause of the condition, which is now known as *Hantavirus* pulmonary syndrome (HPS). Scientists also showed that HPS is acquired when victims inhale the virus in aerosolized deer-mouse urine or feces stirred up by sweeping, enabling other residents of the region to take preventive measures. For more about *Hantavirus* pulmonary syndrome, see p. 737.

Track *Hantavirus* pulmonary syndrome online by going to the Study Area at www.masteringmicrobiology.com.

The Sharing of Data Among Public Health Organizations

Numerous agencies at the local, state, national, and global levels work together with the entire spectrum of health care personnel to promote public health. By submitting reports on incidence and prevalence of disease to public officials, physicians can subsequently learn of current disease trends. Additionally, public health agencies often provide physicians with laboratory and diagnostic assistance.

City and county health departments report data on disease incidence to state agencies. Because state laws govern disease reporting, state agencies play a vital role in epidemiological studies. States accumulate data similar to that in the *MMWR* and assist local health departments and medical practitioners with diagnostic testing for diseases such as rabies and Lyme disease.

Data collected by the states are forwarded to the CDC, which is but one branch of the United States Public Health Service, our national public health agency. In addition to epidemiological studies, the CDC and other branches of the Public Health Service conduct research in disease etiology and prevention, make recommendations concerning immunization schedules, and work with public health organizations of other countries.

The World Health Organization (WHO) coordinates efforts to improve public health throughout the world, particularly in poorer countries, and the WHO has undertaken ambitious projects to eradicate such diseases as polio, measles, and mumps. Some other current campaigns involve AIDS education, malaria control, and childhood immunization programs in poor countries.

The Role of Public Health Agencies in Interrupting Disease Transmission

As we have seen, pathogens can be transmitted in air, food, and water, as well as by vectors and via fomites. Public health agencies work to limit disease transmission by a number of methods:

- Enforce standards of cleanliness in water and food supplies
- Work to reduce the number of disease vectors and reservoirs
- Establish and enforce immunization schedules (see Chapter 17)
- Locate and prophylactically treat individuals exposed to contagious pathogens
- Establish isolation and quarantine measures to control the spread of pathogens

A water supply that is **potable** (fit to drink) is vital to good health. Organisms that cause dysentery, cholera, and typhoid fever are just some of the pathogens that can be spread in water contaminated by sewage. Filtration and chlorination processes are used to reduce the number of pathogens in water supplies. Local, state, and national agencies work to ensure that water supplies remain clean and healthful by monitoring both sewage treatment facilities and the water supply.

Food can harbor infective stages of parasitic worms, protozoa, bacteria, and viruses. National and state health officials ensure the safety of the food supply by enforcing standards in the use of canning, pasteurization, irradiation, and chemical preservatives, and by insisting that food preparers and handlers wash their hands and use sanitized utensils. The Department of Agriculture also provides for the inspection of meats for the presence of pathogens such as *E. coli* and tapeworms.

Milk is an especially rich food, and the same nutrients that nourish us also facilitate the growth of many microorganisms. In the past, contaminated milk has been responsible for epidemics of tuberculosis, brucellosis, typhoid fever, scarlet fever, and diphtheria. Today, public health agencies require the pasteurization of milk, and as a result disease transmission via milk has been practically eliminated in the United States.

Individuals should assume responsibility for their own health by washing their hands before and during food preparation, using disinfectants on kitchen surfaces, and using proper refrigeration and freezing procedures. It is also important to thoroughly cook all meats.

Public health officials also work to control vectors, especially mosquitoes and rodents, by eliminating breeding grounds such as stagnant pools of water and garbage dumps. Insecticides have been used with some success to control insects.

Public Health Education

Diseases that are transmitted sexually or through the air are particularly difficult for public health officials to control. In these cases, individuals must take responsibility for their own health, and health departments can only educate the public to make healthy choices.

Colds and flu remain the most common diseases in the United States because of the ubiquity of the viral pathogens and their mode of transmission in aerosols and via fomites. Health departments encourage afflicted people to remain at home, use disposable tissues to reduce the spread of viruses, and avoid crowds of coughing, sneezing people. As we have discussed, handwashing is also important in preventing the introduction of cold and flu viruses onto the conjunctiva.

We as a society are faced with several epidemics of sexually transmitted diseases. Syphilis, gonorrhea, genital warts, and sexually transmitted AIDS are completely preventable if the chain of transmission is interrupted by abstinence or mutually faithful monogamy. Their incidence can be reduced, but not eliminated, by the use of condoms. Based on the premise that "an ounce of prevention is worth a pound of cure," public health agencies expend considerable effort in public campaigns to educate people to make good choices—those that can result in healthier individuals and improved health for the public at large.

Chapter Summary

Symbiotic Relationships Between Microbes and Their Hosts (pp. 402–406)

1. Microbes live with their hosts in **symbiotic** relationships, including **mutualism,** in which both members benefit; **parasitism,** in which a parasite benefits while the **host** is harmed; and more rarely **commensalism,** in which one member benefits while the other is relatively unaffected. Any parasite that causes disease is called a **pathogen.**
2. Organisms called **normal microbiota** live in and on the body. Some of these microbes are resident, whereas others are transient.
3. **Opportunistic pathogens** cause disease when the immune system is suppressed, when normal **microbial antagonism (competition)** is affected by certain changes in the body, and when a member of the normal microbiota is introduced into an area of the body unusual for that microbe.

Reservoirs of Infectious Diseases of Humans (pp. 406–407)

1. Living and nonliving continuous sources of infectious disease are called **reservoirs of infection.** Animal reservoirs harbor agents of **zoonoses,** which are diseases of animals that may be spread to humans via direct contact with the animal or its waste products, or via an arthropod. Humans may be asymptomatic **carriers.**
2. **Nonliving reservoirs** of infection include soil, water, and inanimate objects.

The Movement of Microbes into Hosts: Infection (pp. 407–409)

1. Microbial **contamination** refers to the mere presence of microbes in or on the body or object. Microbial contaminants include harmless resident and transient members of the microbiota, as well as pathogens, which after a successful invasion cause an **infection.**
2. **Portals of entry** of pathogens into the body include skin, mucous membranes, and the placenta. These portals may be bypassed via the **parenteral route,** by which microbes are directly deposited into deeper tissues.
3. Pathogens attach to cells—a process called **adhesion**—via a variety of structures or attachment proteins called **adhesion factors.** Some bacteria and viruses lose the ability to make adhesion factors called adhesins and thereby become **avirulent.**

4. Some bacteria interact to produce a sticky web of cells and polysaccharides called a **biofilm** that adheres to a surface.

The Nature of Infectious Disease (pp. 410–417)

1. **Disease,** also known as **morbidity,** is a condition sufficiently adverse to interfere with normal functioning of the body.
2. **Symptoms** are subjectively felt by a patient, whereas an outside observer can observe **signs.** A **syndrome** is a group of symptoms and signs that collectively characterizes a particular abnormal condition.
3. **Asymptomatic** or **subclinical** infections may go unnoticed because of the absence of symptoms, even though clinical tests might reveal signs of disease.
4. **Etiology** is the study of the cause of a disease.
5. Nineteenth-century microbiologists proposed the **germ theory of disease,** and Robert Koch developed a series of essential conditions called **Koch's postulates** to prove the cause of infectious diseases. Certain circumstances can make the use of these postulates difficult or even impossible.
6. **Pathogenicity** is a microorganism's ability to cause disease; **virulence** is a measure of pathogenicity. **Virulence factors** such as adhesion factors, extracellular enzymes, toxins, and antiphagocytic factors affect the relative ability of a pathogen to infect and cause disease.
   ANIMATIONS: *Virulence Factors: Hiding from Host Defenses, Inactivating Host Defenses, Penetrating Host Tissues; Phagocytosis: Microbes That Evade It*
7. **Toxemia** is the presence in the blood of poisons called **toxins. Exotoxins** are secreted by pathogens into their environment. **Endotoxin,** also known as **lipid A,** is released from the cell wall of dead and dying Gram-negative bacteria and can have fatal effects.
   ANIMATIONS: *Virulence Factors: Exotoxins, Endotoxins*
8. **Antitoxins** are antibodies the host forms against toxins.
9. The **disease process**—the stages of infectious diseases—typically consists of the **incubation period, prodromal period, illness, decline,** and **convalescence.**

The Movement of Pathogens Out of Hosts: Portals of Exit (p. 417)

1. **Portals of exit,** such as the nose, mouth, and urethra, allow pathogens to leave the body and are of interest in studying the spread of disease.

Modes of Infectious Disease Transmission (pp. 418–420)

ANIMATIONS: *Epidemiology: Transmission of Disease*

1. **Direct contact transmission** of infectious diseases involves person-to-person spread by body contact. When pathogens are transmitted via inanimate objects (called **fomites**), it is called **indirect contact transmission.**
2. **Droplet transmission** (a third type of contact transmission) occurs when pathogens travel less than 1 meter in droplets of mucus to a new host as a result of speaking, coughing, or sneezing.
3. **Vehicle transmission** involves **airborne, waterborne,** and **foodborne** transmission. **Aerosols** are clouds of water droplets, which travel more than 1 meter in airborne transmission. **Fecal-oral infection** can result from sewage-contaminated drinking water or from ingesting fecal contaminants. **Bodily fluid transmission** is the spread of pathogens via blood, urine, saliva, or other fluids.
4. **Vectors** transmit pathogens between hosts. **Biological** vectors are animals, usually biting arthropods, that serve as both host and vector of pathogens. **Mechanical vectors** are not hosts to the pathogens they carry.

Classification of Infectious Diseases (pp. 420–421)

1. There are various ways in which infectious disease may be grouped and studied. When grouped by time course and severity, disease may be described as **acute, subacute, chronic,** or **latent.**
2. When an infectious disease comes either directly or indirectly from another host, it is considered a **communicable** disease. If a communicable disease is easily transmitted from a reservoir or patient, it is called a **contagious** disease. **Noncommunicable** diseases arise either from outside of hosts or from normal microbiota.

Epidemiology of Infectious Diseases (pp. 421–430)

1. **Epidemiology** is the study of where and when diseases occur, and how they are transmitted within populations.
   ANIMATIONS: *Epidemiology: Overview*
2. Epidemiologists track the **incidence** (number of new cases) and **prevalence** (total number of cases) of a disease, and classify disease outbreaks as **endemic** (usually present), **sporadic** (occasional), **epidemic** (more cases than usual), or **pandemic** (epidemic on more than one continent).
   ANIMATIONS: *Epidemiology: Occurrence of Disease*
3. **Descriptive epidemiology** is the careful recording of data concerning a disease; it often includes detection of the **index case**—the first case of the disease in a given area or population. **Analytical epidemiology** seeks to determine the probable cause of a disease. **Experimental epidemiology** involves testing a hypothesis resulting from analytical studies.
4. **Nosocomial infections** and **nosocomial diseases** are acquired by patients or workers in health care facilities. They may be **exogenous** (acquired from the health care environment), **endogenous** (derived from normal microbiota that become opportunistic while in the hospital setting), or **iatrogenic** (induced by treatment or medical procedures).
   ANIMATIONS: *Nosocomial Infections: Overview*
5. Health care workers can help protect their patients and themselves from exposure to pathogens by handwashing and other aseptic and disinfecting techniques.
   ANIMATIONS: *Nosocomial Infections: Prevention*
6. Public health organizations such as the World Health Organization (WHO) use epidemiological data to promulgate rules and standards for clean, **potable** water and safe food, to prevent disease by controlling vectors and animal reservoirs, and to educate people to make healthy choices concerning the prevention of disease.

Questions for Review

Answers to the Questions for Review (except Short Answer questions) begin on page A-1.

Multiple Choice

1. In which type of symbiosis do both members benefit from their interaction?
   a. mutualism
   b. parasitism
   c. commensalism
   d. pathogenesis
2. An axenic environment is one that
   a. exists in the human mouth.
   b. contains only one species.
   c. exists in the human colon.
   d. both a and c
3. Which of the following is *false* concerning microbial contaminants?
   a. Contaminants may become opportunistic pathogens.
   b. Most microbial contaminants will eventually cause harm.
   c. Contaminants may be a part of the transient microbiota.
   d. Contaminants may be introduced by a mosquito bite.
4. The most frequent portal of entry for pathogens is
   a. the respiratory tract.
   b. the skin.
   c. the conjunctiva.
   d. a cut or wound.
5. The process by which microorganisms attach themselves to cells is
   a. infection.
   b. contamination.
   c. disease.
   d. adhesion.
6. Which of the following is the correct sequence of events in infectious diseases?
   a. incubation, prodromal period, illness, decline, convalescence
   b. incubation, decline, prodromal period, illness, convalescence
   c. prodromal period, incubation, illness, decline, convalescence
   d. convalescence, prodromal period, incubation, illness, decline
7. Which of the following are most likely to cause disease?
   a. opportunistic pathogens in a weakened host
   b. pathogens lacking the enzyme kinase
   c. pathogens lacking the enzyme collagenase
   d. highly virulent organisms
8. The nature of bacterial capsules
   a. causes widespread blood clotting.
   b. allows phagocytes to readily engulf these bacteria.
   c. affects the virulence of these bacteria.
   d. has no effect on the virulence of bacteria.
9. When pathogenic bacterial cells lose the ability to make adhesins, they typically
   a. become avirulent.
   b. produce endotoxin.
   c. absorb endotoxin.
   d. increase in virulence.
10. A disease in which a pathogen remains inactive for a long period of time before becoming active is termed a(n)
    a. subacute disease.
    b. acute disease.
    c. chronic disease.
    d. latent disease.
11. Which of the following statements is the best definition of a pandemic disease?
    a. It normally occurs in a given geographic area.
    b. It is a disease that occurs more frequently than usual for a geographical area or group of people.
    c. It occurs infrequently at no predictable time scattered over a large area or population.
    d. It is an epidemic that occurs on more than one continent at the same time.
12. Which of the following types of epidemiologists is most like a detective?
    a. a descriptive epidemiologist
    b. an analytical epidemiologist
    c. an experimental epidemiologist
    d. a reservoir epidemiologist
13. Consider the following case. An animal was infected with a virus. A mosquito bit the animal, was contaminated with the virus, and proceeded to bite and infect a person. Which was the vector?
    a. animal
    b. virus
    c. mosquito
    d. person
14. A patient contracted athlete's foot after long-term use of a medication. His physician explained that the malady was directly related to the medication. Such infections are termed
    a. nosocomial infections.
    b. exogenous infections.
    c. iatrogenic infections.
    d. endogenous infections.
15. Which of the following phrases describes a contagious disease?
    a. a disease arising from fomites
    b. a disease that is easily passed from host to host in aerosols
    c. a disease that arises from opportunistic, normal microbiota
    d. both a and b

Fill in the Blanks

1. A microbe that causes disease is called a ________.
2. Infections that may go unnoticed due to the absence of symptoms are called ________ infections.
3. The study of the cause of a disease is ________.
4. The study of where and when diseases occur and how they are transmitted within populations is ________.
5. Diseases that are naturally spread from their usual animal hosts to humans are called ________.
6. Nonliving reservoirs of disease, such as a toothbrush, drinking glass, and needle, are called ________.
7. ________ infections are those acquired by patients or staff while in health care facilities.
8. The total number of cases of a disease in a given area is its ________.
9. An animal that carries a pathogen and also serves as host for the pathogen is a ________ vector.
10. Endotoxin is the part of the cell wall known as ________ in a Gram-negative bacterium.

Labeling

Each map below shows the locations (dots) of cases of a disease that normally occurs in the Western Hemisphere. Label each map with a correct epidemiological description of the disease's occurrence.

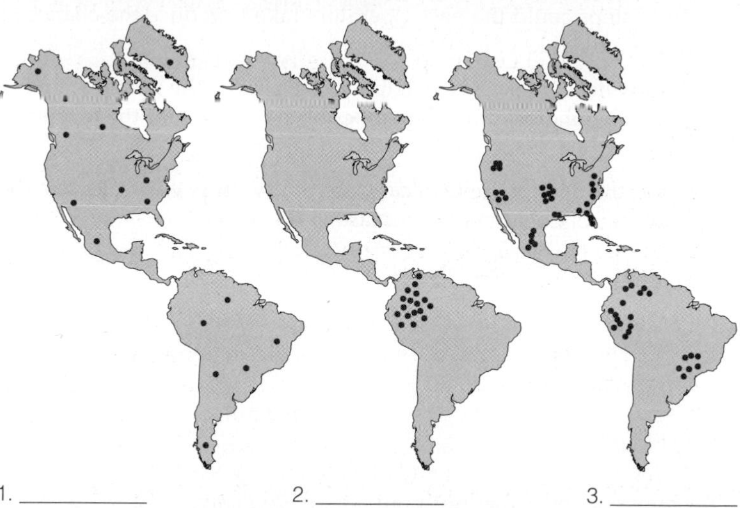

Short Answer

1. List three types of symbiotic relationships, and give an example of each.
2. List three conditions that create opportunities for pathogens to become harmful in a human.
3. List three portals through which pathogens enter the body.
4. List Koch's four postulates, and describe situations in which not all may be applicable.
5. List in the correct sequence the five stages of infectious diseases.
6. Describe three modes of disease transmission.
7. Describe the parenteral route of infection.
8. In general, contrast transient microbiota with resident microbiota.
9. Contrast the terms *infection* and *morbidity.*
10. Contrast iatrogenic and nosocomial diseases.

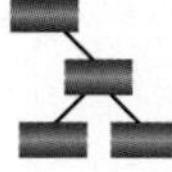

Concept Mapping

Using the following terms, draw a concept map that describes disease transmission. For a sample concept map, see p. 93. Or, complete this concept map online by going to the Study Area at www.masteringmicrobiology.com.

Airborne
Arthropods
Biological
Body
Contact transmission
Direct contact
Droplet transmission
Fomites
Foodborne
Indirect contact
Mechanical
Mosquito
Sneezing
Tick
Vector transmission
Vehicle transmission
Waterborne

Critical Thinking

1. Explain why Ellen H., a menopausal woman, may have developed gingivitis from normal microbiota.
2. Will P. died of *E. coli* infection after an intestinal puncture. Explain why this microbe, which normally lives in the colon, could kill this patient.
3. Examine the graph of the red epidemic and the blue epidemic. Neither red disease nor blue disease is treatable. A person with either disease is ill for only 1 day and recovers fully. Both epidemics began at the same time. Which epidemic affected more people during the first 3 days? What could explain the short time course for the red epidemic? Why was the blue epidemic longer lasting?

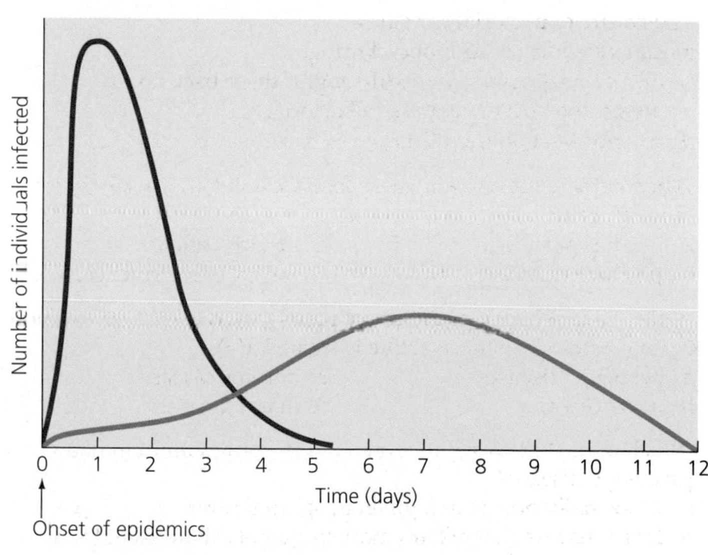

4. A 27-year-old female came to her doctor's office with a widespread rash, fever, malaise, and severe muscle pain. The symptoms had begun with a mild headache 3 days previously. She reported being bitten by a tick 1 week prior to that.

   The doctor correctly diagnosed Rocky Mountain spotted fever (RMSF) and prescribed tetracycline. The rash and other signs and symptoms disappeared in a couple of days, but she continued on her antibiotic therapy for 2 weeks.

   Draw a graph showing the course of disease and the relative numbers of pathogens over time. Label the stages of the disease.

5. Over 30 children younger than 3 years of age developed gastroenteritis after visiting a local water park. These cases represented 44% of the park visitors in this age group on the day in question. No older individuals were affected. The causative agent was determined to be a member of the bacterial genus *Shigella*. The disease resulted from oral transmission to the children.

   Based only on the information given, can you classify this outbreak as an epidemic? Why or why not? If you were an epidemiologist, how would you go about determining which pools in the water park were contaminated? What factors might account for the fact that no older children or adults developed disease? What steps could the park operators take to reduce the chance of future outbreaks of gastroenteritis?

6. Review lichen biology in Chapter 12. Describe the relationship between the photosynthetic member of the lichen and the fungus in terms of the type of symbiosis.

7. Using the data in the Clinical Case Study on p. 427, calculate the incidence of Legionnaires' disease in Bogalusa, Louisiana.

Access more review material online in the Study Area at **www.masteringmicrobiology.com.** There, you'll find

- **Animations**
- **MP3 Tutor Sessions**
- **Concept Mapping Activities**
- **Flashcards**
- **Quizzes**

and more to help you succeed.

15 Innate Immunity

Each day the equivalent of a small room full of air enters your respiratory tract through your nose. With that air come dust, smoke, bacteria, viruses, fungi, pollen, soot, fuzz, sand, and more. Acting as a first line of defense, your respiratory mucous membrane uses nose hairs, ciliated epithelium, and mucus to cleanse the inhaled air of pathogens and certain harmful pollutants. Each day you swallow and subsequently digest about a liter of mucus, along with the trapped pathogens and pollutants it contains. Still more nasal mucus clumps around microbes and pollutants to form masses of mucus that may dry out or remain slimy, depending on how rapidly you're breathing and the humidity of the air. Yellowish or greenish mucus contains a large population of trapped bacteria, their waste products, and dead or dying defensive white blood cells.

Mucus is just one example of the body's general defenses against pathogens. The skin and certain protective cells, chemicals, and processes within the body are other ways the body limits early stages of infection by microbes. In this chapter we will focus on each of these aspects of the body's defenses.

Take the pre-test for this chapter online. Visit the Study Area at www.masteringmicrobiology.com.

Cilia, such as these lining the upper respiratory tract, move mucus and contaminants out of the body.

As we saw in Chapter 14, a pathogen can cause a disease only if it can (1) gain access, either by penetrating the surface of the skin or by entering through some other portal of entry; (2) attach itself to host cells; and (3) evade the body's defense mechanisms long enough to produce harmful changes. In this chapter we will examine the structures, processes, and chemicals that respond in a general way to protect the body from all types of pathogens.

An Overview of the Body's Defenses

Learning Objectives

✓ List and briefly describe the three lines of defense in the human body.

✓ Explain the phrases *species resistance* and *innate immunity*.

Because the cells and certain basic physiological processes of humans are incompatible with those of most plant and animal pathogens, humans have what is termed **species resistance** to these pathogens. In many cases the chemical receptors these pathogens require for attachment to a host cell do not exist in the human body; in other cases the pH or temperature of the human body are incompatible with the conditions under which these pathogens can survive. Thus, for example, all humans have species resistance to both tobacco mosaic virus and to the virus that causes feline immunodeficiency syndrome in members of the cat family.

Nevertheless, we are confronted every day with pathogens that can cause disease in humans. Bacteria, viruses, fungi, protozoa, and parasitic worms come in contact with your body in the air you breathe, in the water you drink, in the food you eat, and during the contacts you have with other people. Your body must defend itself from these potential pathogens, and in some cases from members of your normal microbiota, which may become opportunistic pathogens.

It is convenient to cluster the structures, cells, and chemicals that act against pathogens into three main lines of defense, each of which overlaps and reinforces the other two. The first line of defense is chiefly composed of external physical barriers to pathogens, especially the skin and mucous membranes. The second line of defense is internal and is composed of protective cells, bloodborne chemicals, and processes that inactivate or kill invaders. Together, the first two lines of defense are called **innate immunity** because they are present at birth prior to contact with infectious agents or their products. Innate immunity is rapid and works against a wide variety of pathogens, including parasitic worms, protozoa, fungi, bacteria, and viruses.

By contrast, the third line of defense, **adaptive immunity,** responds against unique species or strains of pathogens and alters the body's defenses such that they act more effectively upon subsequent infection with the specific strain. Chapter 16 examines adaptive immune responses and the white blood cells called *lymphocytes* that moderate them. Here we turn our attention to the two lines of innate immunity. **ANIMATIONS:** ***Host Defenses: Overview***

▲ **Figure 15.1 A scanning electron micrograph of a section of skin.** Epidermal cells are dead, dry, and slough off, providing an effective barrier to most microorganisms.

The Body's First Line of Defense

The body's initial line of defense is made up of structures, chemicals, and processes that work together to prevent pathogens from entering the body in the first place. Here we discuss the main components of the first line of defense: the skin and the mucous membranes of the respiratory, digestive, urinary, and reproductive systems. These structures provide a formidable barrier to the entrance of microorganisms. As we discussed in Chapter 14, if these barriers are pierced, broken, or otherwise damaged, they become portals of entry for pathogens. In this section we examine aspects of the first line of defense, including the role of the normal microbiota.

The Role of Skin in Innate Immunity

Learning Objective

✓ Identify the physical and chemical aspects of skin that enable it to prevent the entrance of pathogens.

The skin—the organ of the body with the greatest surface area—is composed of two major layers: an outer **epidermis,** and a deeper **dermis,** which contains hair follicles, glands, and nerve endings (see Figure 14.4). Both the physical structure and the chemical components of skin enable it to act as an effective defense.

The epidermis is composed of multiple layers of tightly packed cells. It constitutes a physical barrier to most bacteria, fungi, and viruses. Very few pathogens can penetrate the layers of epidermal cells unless the skin has been burned, broken, or cut.

The deepest cells of the epidermis continually divide, pushing their daughter cells toward the surface. As the daughter cells are pushed toward the surface, they flatten and die and are eventually shed in flakes **(Figure 15.1)**. Microorganisms that attach to the skin's surface are sloughed off with the flakes of

BENEFICIAL MICROBES

WHAT HAPPENS TO ALL THAT SKIN?

▲ *Dust mite.*

Your body sheds tens of thousands of skin flakes every time you walk or move, and you shed at only a slightly lower rate when you stand still. That comes to about 10 billion skin cells per day, or 250 grams (about half a pound) of skin every year! What happens to all that skin?

Much of household dust is skin that you and your housemates have shed as you go about your lives. The skin flakes fall to the rug and upholstery, where they become food for microscopic mites that live sedentary and harmless lives waiting patiently for meals to rain down on them from above. They dwell not only in the rug, but also in your mattress and pillow, and even in the hair follicles of your eyebrows, benefiting you by catching skin cells cascading down your forehead before they can irritate your eyes.

By the way, house dust also contains mite feces and mite skeletons, which can trigger allergies. So after reading this chapter, you just might want to clean your carpet.

dead cells. **Beneficial Microbes: What Happens to All That Skin?** describes the fate of lost epidermal cells.

The epidermis also contains phagocytic cells called **dendritic**[1] **cells.** The slender, fingerlike processes of dendritic cells extend among the surrounding cells, forming an almost continuous network to intercept invaders. Dendritic cells both phagocytize pathogens nonspecifically and play a role in adaptive immunity, as we will see in Chapter 16.

The combination of the barrier function of the epidermis, its continual replacement, and the presence of phagocytic dendritic cells provides significant nonspecific defense against colonization and infection by pathogens.

The dermis also defends nonspecifically. It contains tough fibers of a protein called collagen. These give the skin strength and pliability to prevent jabs and scrapes from penetrating the dermis and introducing microorganisms. Blood vessels in the dermis deliver defensive cells and chemicals, which will be discussed shortly.

In addition to its physical structure, the skin has a number of chemical substances that nonspecifically defend against pathogens. Dermal cells secrete antimicrobial peptides and sweat glands secrete perspiration, which contains salt, antimicrobial peptides, and lysozyme. Salt draws water osmotically from invading cells, which inhibits their growth and kills them.

Antimicrobial peptides (sometimes called *defensins*) are positively charged chains of 20–50 amino acids that act against microorganisms. Sweat glands secrete a class of antimicrobial peptides called *dermicidins.* Dermicidins are broad-spectrum antimicrobials that are active against many Gram-negative and Gram-positive bacteria and fungi. As expected of a peptide active on the surface of the skin, dermicidins are insensitive to low pH and salt. The exact mechanism of dermicidin action is not known.

Lysozyme (lī'sō-zīm) is an enzyme that destroys the cell walls of bacteria by cleaving the bonds between the sugar subunits of the walls. Bacteria without cell walls are more susceptible to osmotic shock and digestion by other enzymes within phagocytes (discussed later in the chapter as part of the second line of defense).

The skin also contains sebaceous (oil) glands that secrete **sebum** (sē'bŭm), an oily substance that not only helps keep the skin pliable and less sensitive to breaking or tearing but also contains fatty acids that lower the pH of the skin's surface to about pH 5, which is inhibitory to many bacteria.

Although salt, defensins, lysozyme, and acidity make the surface of the skin an inhospitable environment for most microorganisms, some bacteria, such as *Staphylococcus epidermidis* (staf'i-lō-kok'ŭs ep-i-der-mid'is), find the skin a suitable environment for growth and reproduction. Bacteria are particularly abundant in crevices around hairs and in the ducts of glands; usually they are nonpathogenic.

In summary, the skin is a complex barrier that limits access by microbes.

CRITICAL THINKING

Some strains of *Staphylococcus aureus* produce exfoliative toxin, a chemical that causes portions of the entire outer layer of the skin to be sloughed off in a disease called scalded skin syndrome. Given that cells of the outer layer are going to fall off anyway, why is this disease dangerous?

The Role of Mucous Membranes in Innate Immunity

Learning Objectives

- ✓ Identify the locations of the body's mucous membranes.
- ✓ Explain how mucous membranes protect the body both physically and chemically.

Mucus-secreting (mucous) membranes, a second part of the first line of defense, cover all body cavities that are open to the outside environment. Thus mucous membranes line the

[1]Form Greek *dendron,* meaning tree, referring to their branched appearance.

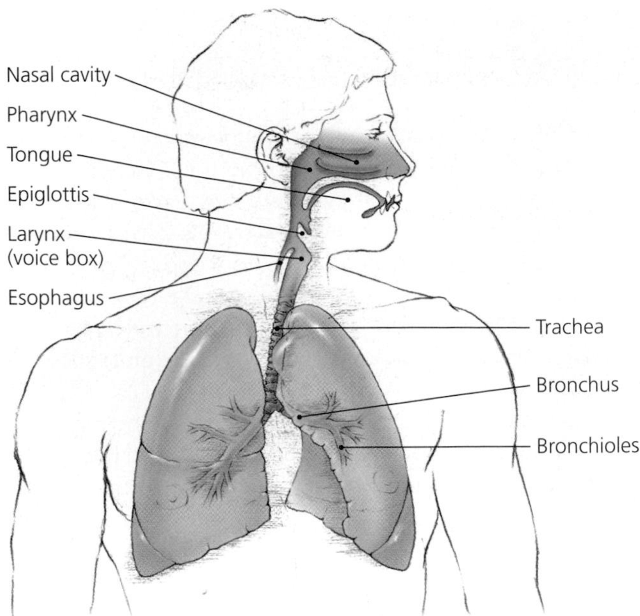

▲ **Figure 15.2 The structure of the respiratory system, which is lined with a mucous membrane.** The epithelium of the trachea contains mucus-secreting goblet cells and ciliated cells whose cilia propel the mucus (and the microbes trapped within it) up to the larynx for removal. *What is the function of stem cells within the respiratory epithelium?*

Figure 15.2 *Stem cells in the respiratory epithelium undergo cytokinesis to form both ciliated cells and goblet cells to replace those lost during normal shedding.*

TABLE 15.1 The First Line of Defense: A Comparison of the Skin and Mucous Membranes

	Skin	Mucous Membrane
Number of cell layers	Many	One to a few
Cells tightly packed?	Yes	Yes
Cells dead or alive?	Outer layers: dead; inner layers: alive	Alive
Mucus present?	No	Yes
Relative water content	Dry	Moist
Defensins present?	Yes	With some
Lysozyme present?	Yes	With some
Sebum present?	Yes	No
Cilia present?	No	Trachea, uterine tubes
Constant shedding and replacement of cells?	Yes	Yes

lumens[2] of the respiratory, urinary, digestive, and reproductive tracts. Like the skin, mucous membranes act nonspecifically to limit infection both physically and chemically.

Mucous membranes are moist and have two distinct layers: the *epithelium,* in which cells from a covering that is superficial (closest to the surface, in this case the lumen), and a deeper connective tissue layer that provides mechanical and nutritive support for the epithelium. Epithelial cells of mucous membranes are packed closely together, like those of the epidermis, but they form only a thin layer. Indeed, in some mucous membranes, the epithelium is only a single cell thick. Unlike surface epidermal cells, surface cells of mucous membranes are alive and play roles in the diffusion of nutrients and oxygen (in the digestive, respiratory, and female reproductive systems) and in the elimination of wastes (in the urinary, respiratory, and female reproductive systems).

The thin epithelium on the surface of a mucous membrane provides a less efficient barrier to the entrance of pathogens than the multiple layers of dead cells found at the skin's surface. So how are microorganisms kept from invading through these thin mucous membranes? In some cases they are not, which is why some mucous membranes, especially those of the respiratory and reproductive systems, are common portals of entry for pathogens. Nevertheless, the epithelial cells of mucous membranes are tightly packed to prevent the entry of many pathogens, and the cells are continually shed and then replaced via the cytokinesis of **stem cells,** which are generative cells capable of dividing to form daughter cells of various types. One effect of such shedding is that it carries attached microorganisms away.

Dendritic cells reside below the mucous epithelium to phagocytize invaders. These cells are also able to extend pseu dopodia between epithelial cells to "sample" the contents of the lumen, which helps prepare adaptive immune responses against particular pathogens that might breach the mucosal barrier—a subject covered more fully in Chapter 16.

In addition, the epithelia of some mucous membranes have still other means of removing pathogens. In the mucous membrane of the trachea, for example, the stem cells produce both *goblet cells,* which secrete an extremely sticky mucus that traps bacteria and other pathogens, and *ciliated columnar cells,* whose cilia propel the mucus and its trapped particles and pathogens up from the lungs **(Figure 15.2)**. The effect of the action of the cilia is often likened to that of an escalator. Mucus carried into the throat is coughed up and either swallowed or expelled. Because the poisons and tars in tobacco smoke damage cilia, the lungs of smokers are not properly cleared of mucus, so smokers may develop severe coughs as their respiratory tracts attempt to expel excess mucus from the lungs. Smokers also typically succumb to more respiratory pathogens, because they are unable to effectively clear pathogens from their lungs.

In addition to these physical actions, mucous membranes produce chemicals that defend against pathogens. Nasal mucus contains lysozyme, which chemically destroys bacterial cell walls. Mucus also contains antimicrobial peptides (defensins). Table 15.1 compares the physical and chemical actions of the skin and mucous membranes in the body's first line of defense.

[2]A lumen is a cavity or channel within any tubular structure or organ.

The Role of the Lacrimal Apparatus in Innate Immunity

Learning Objective

✓ Describe the lacrimal apparatus and the role of tears in combating infection.

The lacrimal apparatus is a group of structures that produce and drain away tears **(Figure 15.3)**. Lacrimal glands, located above and to the sides of the eyes, secrete tears into lacrimal gland ducts and onto the surface of the eyes. The tears either evaporate or drain into small lacrimal canals, which carry them into nasolacrimal ducts that empty into the nose. There, the tears join the nasal mucus and flow into the pharynx, where they are swallowed. The blinking action of eyelids spreads the tears and washes the surfaces of the eyes. Normally, evaporation and flow into the nose balance the flow of tears onto the eye. However, if the eyes are irritated, increased tear production floods the eyes, carrying the irritant away. In addition to their washing action, tears contain lysozyme, which destroys bacteria.

The Role of Normal Microbiota in Innate Immunity

Learning Objective

✓ Define *normal microbiota* and explain how they help provide protection against disease.

As we saw in Chapter 14, the skin and mucous membranes of the body are normally home to a variety of protozoa, fungi, bacteria, and viruses. This **normal microbiota** plays a role in protecting the body by competing with potential pathogens in a variety of ways, a situation called **microbial antagonism.**

A variety of activities of the normal microbiota make it less likely that a pathogen can compete with them and produce disease. Microbiota consume nutrients, making them unavailable to pathogens. Additionally, normal microbiota can change the pH, creating an environment that is favorable for themselves but unfavorable to other microorganisms.

Further, the presence of microbiota stimulates the body's second line of defense (discussed shortly). Researchers have observed that animals raised in an *axenic*[3] (ā-zēn′ik) environment—that is, one free of all other organisms or viruses—are slower to defend themselves when exposed to a pathogen. Recent studies have shown that members of the normal microbiota in the intestines boost the production of anti microbial substances by the body.

Finally, the resident microbiota of the intestines improve overall health by providing several vitamins, including biotin and pantothenic acid (vitamin B_5), which are important in glucose metabolism; folic acid, which is essential for the production of the purine and pyrimidine bases of nucleic acids; and the precursor of vitamin K, which has an important role in blood clotting.

[3]From Greek *a*, meaning no, and *xenos*, meaning foreigner.

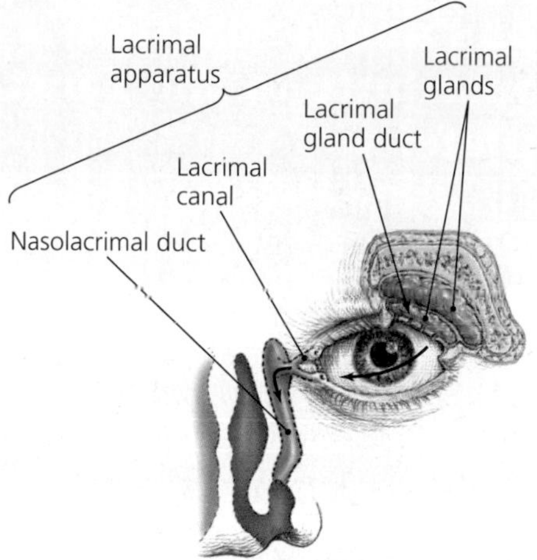

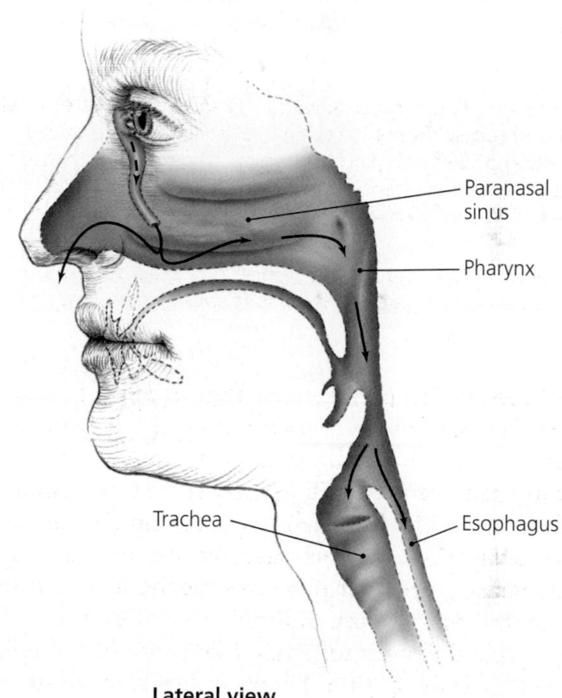

▲ **Figure 15.3 The lacrimal apparatus.** These structures function in the body's first line of defense by bathing the eye with tears. Arrows indicate the route tears take across the eye and into the throat. *Name an antimicrobial protein found in tears.*

Figure 15.3 *Tears contain lysozyme, an antimicrobial protein that acts against the peptidoglycan of bacterial cell walls.*

Other First-Line Defenses

Learning Objective

✓ Describe antimicrobial peptides as part of the body's defenses.

Besides the physical barrier of the skin and mucous membranes, there are other hindrances to microbial invasion. Among these are additional antimicrobial peptides and other processes and chemicals.

TABLE 15.2 Secretions and Activities That Contribute to the First Line of Defense

Secretion/Activity	Function
Digestive System	
Saliva	Washes microbes from teeth, gums, tongue, and palate; contains lysozyme, an antibacterial enzyme
Stomach acid	Digests and/or inhibits microorganisms
Gastroferritin	Sequesters iron being absorbed, making it unavailable for microbial use
Bile	Inhibitory to most microorganisms
Intestinal secretions	Digests and/or inhibits microorganisms
Peristalsis	Moves gastrointestinal contents through GI tract, constantly eliminating potential pathogens
Defecation	Eliminates microorganisms
Vomiting	Eliminates microorganisms
Urinary System	
Urine	Contains lysozyme; urine's acidity inhibits microorganisms; may wash microbes from ureters and urethra during urination
Reproductive System	
Vaginal secretions	Acidity inhibits microorganisms; contains iron-binding proteins that sequester iron, making it unavailable for microbial use
Menstrual flow	Cleanses uterus and vagina
Prostate secretion	Contains iron-binding proteins that sequester iron, making it unavailable for microbial use
Cardiovascular System	
Blood flow	Removes microorganisms from wounds
Coagulation	Prevents entrance of many pathogens
Transferrin	Binds iron for transport, making it unavailable for microbial use

Antimicrobial Peptides

As we saw in our examination of skin and mucous membranes, antimicrobial peptides (sometimes called defensins) act against microorganisms. Scientists have discovered hundreds of these antimicrobial peptides in organisms as diverse as silkworms, frogs, and humans. Besides being secreted onto the surface of the skin, antimicrobial peptides are found in mucous membranes and in neutrophils. These peptides act against a variety of potential pathogens, being triggered by sugar and protein molecules on the external surfaces of microbes. Some antimicrobial peptides act only against Gram-positive bacteria or Gram-negative bacteria, others act against both, and still others act against protozoa, enveloped viruses, or fungi.

Researchers have elucidated several ways in which antimicrobial peptides work. Some punch holes in the cytoplasmic membranes of the pathogens, while others interrupt internal signaling or enzymatic action. Some antimicrobial peptides are chemotactic factors that recruit leukocytes to the site.

Other Processes and Chemicals

Many other body organs contribute to the first line of defense by secreting chemicals with antimicrobial properties that are secondary to their prime function. For example, stomach acid is primarily present to aid digestion of proteins, but it also prevents the growth of many potential pathogens. Likewise, saliva contains lysozyme as well as a digestive enzyme; further, saliva physically washes microbes from the teeth. The contributions of these and other processes and chemicals to the first line of defense are listed in Table 15.2.

The Body's Second Line of Defense

Learning Objective

✓ Compare and contrast the body's first and second lines of defense against disease.

When pathogens succeed in penetrating the skin or mucous membranes, the body's second line of innate defense comes into play. Like the first line of defense, the second line operates against a wide variety of pathogens, from parasitic worms to viruses. But unlike the first line of defense, the second line includes no barriers; instead, it is composed of cells (especially phagocytes), antimicrobial chemicals (peptides, complement, interferons), and processes (inflammation, fever). Some cells and chemicals from the first line of defense play additional roles in the second line of defense. We will consider each component of the second line of defense in some detail shortly, but because many of them are either contained in or originate in the blood, we first consider the components of blood.

Defense Components of Blood

Learning Objectives

✓ Discuss the components of blood and their functions in the body's defense.

✓ Explain how macrophages are named.

Blood is a complex liquid tissue composed of cells and portions of cells within a fluid called *plasma*. We begin our discussion of the defense functions of blood by briefly considering plasma.

Plasma

Plasma is mostly water containing electrolytes (ions), dissolved gases, nutrients, and—most relevant to the body's defenses—a variety of proteins. Some plasma proteins are involved in inflammation (discussed later) and in blood clotting, a defense

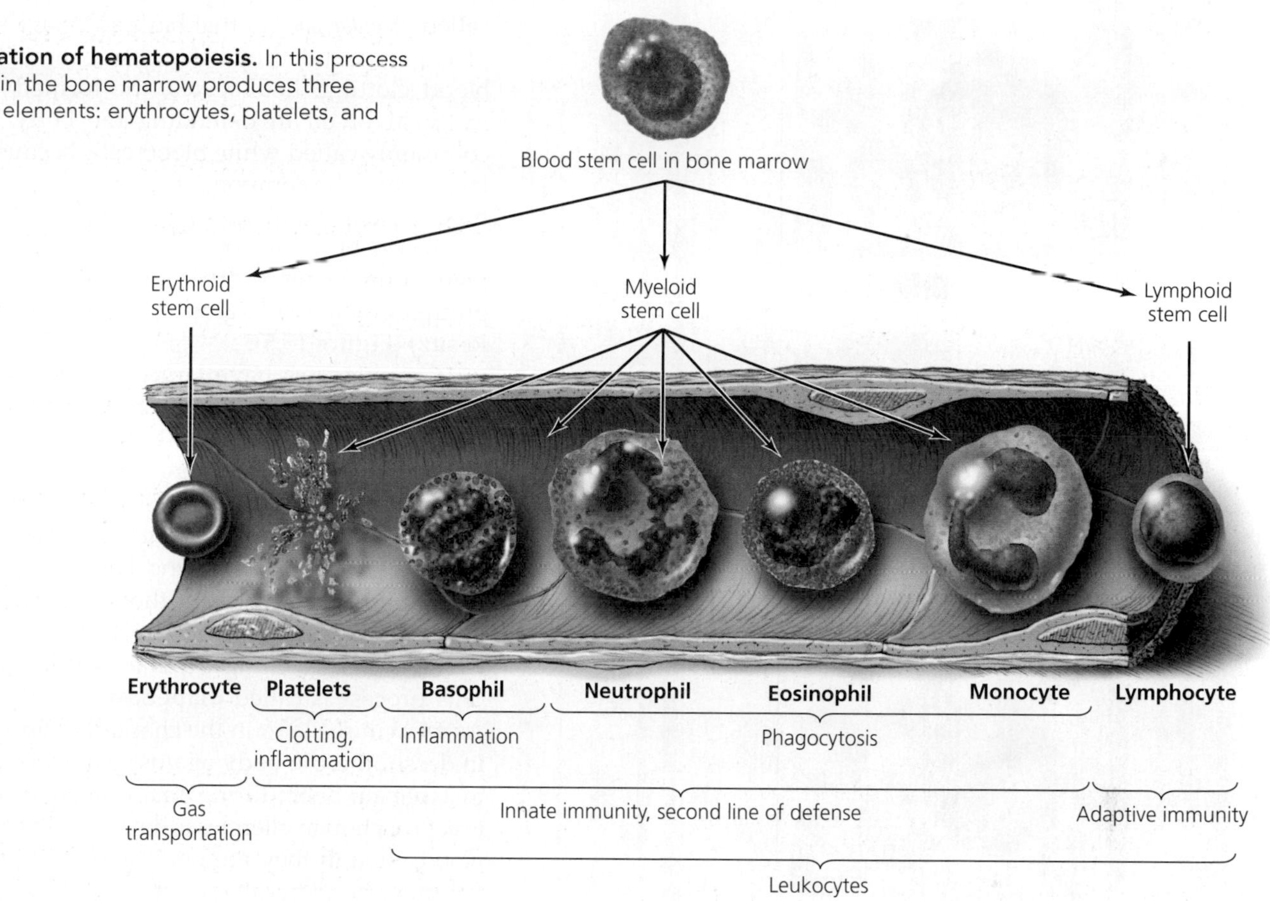

➤ **Figure 15.4**
A schematic representation of hematopoiesis. In this process the division of stem cells in the bone marrow produces three types of formed (cellular) elements: erythrocytes, platelets, and leukocytes.

mechanism that reduces both blood loss and the risk of infection. When clotting factors have been removed from the plasma, as for instance when blood clots, the remaining liquid is called *serum.*

Most cells require iron for metabolism: it is a component of cytochromes of electron transport chains, functions as an enzyme cofactor, and is an essential part of hemoglobin—the oxygen-carrying protein of erythrocytes. Because iron is relatively insoluble, in humans it is transported in plasma to cells by a transport protein called *transferrin.* When transferrin-iron complexes reach cells with receptors for transferrin, the binding of the protein to the receptor stimulates the cell to take up the iron via endocytosis. Excess iron is stored in the liver bound to another protein called *ferritin.*

Though the main function of iron-binding proteins is transporting and storing iron, they play a secondary, defensive role—sequestering iron so that it is unavailable to microorganisms.

Some bacteria, such as *Staphylococcus aureus* (o'rē-ŭs), respond to a shortage of iron by secreting their own iron-binding proteins called *siderophores.* Because siderophores have a greater affinity for iron than does transferrin, bacteria that produce siderophores can in effect steal iron from the body. In response, the body produces *lactoferrin,* which retakes the iron from the bacteria by its even greater affinity. Thus, the body and the pathogens engage in a kind of chemical "tug-of-war" for the possession of iron.

Some pathogens bypass this contest altogether. For example, *Neisseria meningitidis* (nī-se'rē-ă me-nin-ji'ti-dis), a pathogen that causes often fatal meningitis, produces receptors for transferrin and plucks iron from the bloodstream as it flows by. *S. aureus* and related pathogens can secrete the protein *hemolysin,* which punches holes in the cytoplasmic membranes of red blood cells, releasing hemoglobin. Other bacterial proteins then bind hemoglobin to the bacterial membrane and strip it of its iron.

Another group of plasma proteins, called *complement proteins,* is an important part of the second line of defense and is discussed shortly. Still other plasma proteins, called *antibodies* or *immunoglobulins,* are a part of adaptive immunity, the body's third line of defense, which we will explore in Chapter 16.

Defensive Blood Cells: Leukocytes

Cells and cell fragments suspended in the plasma are called **formed elements.** In a process called *hematopoiesis,*[4] blood stem cells located principally in the bone marrow within the hollow cavities of the large bones produce three types of formed elements: **erythrocytes**[5] (ĕ-rith'rō-sītz), **platelets**[6] (plāt'letz), and **leukocytes**[7] (loo'kō-sīts) **(Figure 15.4)**. Erythrocytes, the most numerous of the formed elements, carry oxygen and carbon dioxide in the blood. Platelets, which are pieces of large cells

[4]From Greek *haima,* meaning blood, and *poiein,* meaning to make.
[5]From Greek *erythro,* meaning red, and *cytos,* meaning cell.
[6]French for small plates. Platelets are also called thrombocytes, from Greek *thrombos,* meaning lump, and *cytos,* meaning cell, though they are technically not cells, but instead pieces of cells.
[7]From Greek *leuko,* meaning white, and *cytos,* meaning cell.

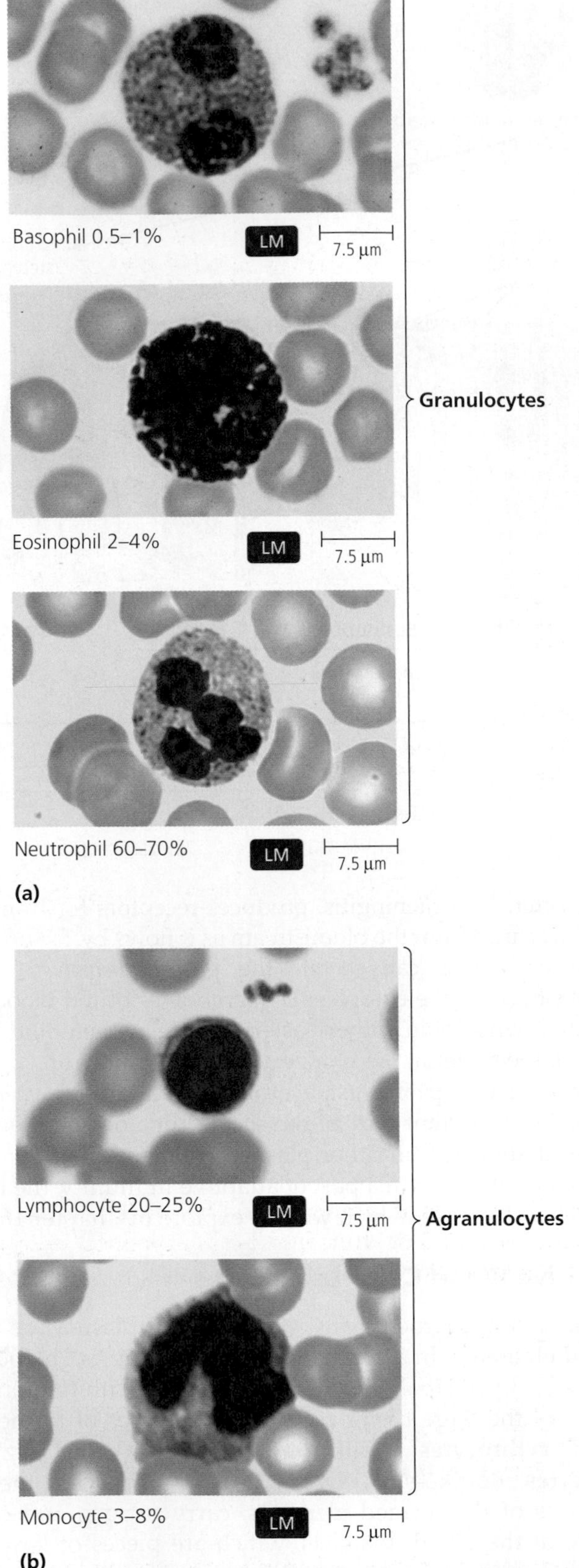

▲ **Figure 15.5 Leukocytes as seen in stained blood smears.** **(a)** Granulocytes: basophil, eosinophil, and neutrophil. **(b)** Agranulocytes: lymphocyte and monocyte. The numbers are the normal percentages of each cell type among all leukocytes.

called *megakaryocytes* that have split into small portions of cytoplasm surrounded by cytoplasmic membranes, are involved in blood clotting. Leukocytes, the formed elements that are directly involved in defending the body against invaders, are commonly called white blood cells because they form a whitish layer when the components of blood are separated within a test tube.

Based on their appearance in stained blood smears when viewed under the microscope, leukocytes are divided into two groups: *granulocytes* (gran′ū-lō-sītz) and *agranulocytes* (ā-gran′ū-lō-sītz) **(Figure 15.5)**.

Granulocytes have large granules in their cytoplasm that stain different colors depending on the type of granulocyte and the dyes used: **basophils** (bā′sō-fils) stain blue with the basic dye methylene blue; **eosinophils** (ē-ō-sin′ō-fils) stain red to orange with the acidic dye eosin; and **neutrophils** (noo′trō-fils), also known as *polymorphonuclear leukocytes* (PMNs), stain lilac with a mixture of acidic and basic dyes. Both neutrophils and eosinophils phagocytize pathogens, and both can exit the blood to attack invading microbes in the tissues by squeezing between the cells lining capillaries (the smallest blood vessels). This process is called **diapedesis**[8] (dī′ă-pĕ-dē′sis) or *emigration.* As we will see later in the chapter, eosinophils are also involved in defending the body against parasitic worms and are present in large number during many allergic reactions, though their exact function in allergies is disputed. Basophils can also leave the blood, though they are not phagocytic; instead, they release inflammatory chemicals, an aspect of the second line of defense that will be discussed shortly.

The cytoplasm of agranulocytes appears uniform when viewed via light microscopy, though granules do become visible with an electron microscope. Agranulocytes are of two types: **lymphocytes** (lim′fō-sītz), which are the smallest leukocytes and have nuclei that nearly fill the cells, and **monocytes** (mon′ō-sītz), which are large agranulocytes with slightly lobed nuclei. Although most lymphocytes are involved in adaptive immunity (see Chapter 16), *natural killer (NK) lymphocytes* function in innate defense and thus are discussed later in this chapter. Monocytes leave the blood and mature into **macrophages** (mak′rō-fāj-ĕz), which are phagocytic cells of the second line of defense. Their initial function is to devour foreign objects, including bacteria, fungi, spores, and dust, as well as dead body cells.

Macrophages are named for their location in the body. *Wandering macrophages* leave the blood via diapedesis and perform their scavenger function while traveling throughout the body, including extracellular spaces. Other macrophages are fixed and do not wander. These include *alveolar* (al-vē′ō-lăr) *macrophages*[9] of the lungs and *microglia* (mī-krog′lē-ă) of the central nervous system. Fixed macrophages generally phagocytize within specific organs, such as the heart chambers, blood vessels, and lymphatic[10] vessels.

[8]From Greek *dia*, meaning through, and *pedan*, meaning to leap.
[9]Alveoli are small pockets at the end of respiratory passages where oxygen and carbon dioxide exchange occurs between the lungs and the blood.
[10]The lymphatic system is discussed in Chapter 16.

CLINICAL CASE STUDY

Evaluating an Abnormal CBC

CBC Profile

Name: Brown, Roger   Age/Sex: 61/M   Attend Dr: Kevin, Larry
Acct#: 04797747   Status: ADM IN
Reg: 11/27/09

SPEC #: 0303:AS:H00102T   COLL: 12 / 03 / 09-0620   STATUS: COMP   REQ #: 01797367
RECD: 12 / 03 / 09-0647   SUBM DR: Kevin, Larry

ENTERED: 12 / 03 / 09-0002   OTHER DR: NONE, PER PT
ORDERED: CBC W/ MAN DIFF

Test	Result	Flag	Reference	Site
CBC				
WBC (white blood cells)	0.8	*L*	4.8–10.8 K/mm3	ML
RBC (red blood cells)	3.09	*L*	4.20–5.40 M/mm3	ML
HGB (hemoglobin)	9.6	*L*	12.0–16.0 g/dL	ML
HCT (hematocrit)	28.2	*L*	37.0–47.0 %	ML
MCV	91.3		81.0–99.0 fL	ML
MCH	31.1	*H*	27.0–31.0 pg	ML
MCHC	34.1		32.0–36.0 g/dl	ML
RDW	17.1	*H*	11.5–14.5 %	ML
PLT (platelets)	21	*L*	150–450 K/mm3	ML
MPV	8.7		7.4–10.4 fL	ML
DIFF				
CELLS COUNTED	100		#CELLS	ML
SEGS	39		%	ML
BAND	4		%	ML
LYMPH (lymphocytes)	41		%	ML
MONO (monocytes)	15		%	ML
EOS (eosinophils)	1		%	ML
NEUT# (# neutrophils)	0.3	*L*	1.9–8.0 K/mm3	ML
LYMPH#	0.3	*L*	0.9–5.2 K/mm3	ML
MONO#	0.1		0.1–1.2 K/mm3	ML
EOS#	0.0		0–0.8 K/mm3	ML
PLATELET EST	DECREASED	*		ML
RBC MORPHOLOGY	ABNORMAL	*		ML
CELL MORPHOLOGY				
ANISOCYTOSIS	MODERATE	*		ML
POIKILOCYTOSIS	MODERATE			ML
TEAR DROP CELLS	PRESENT	*		ML

ML - MAIN LABORATORY

Roger Brown, an African American cancer patient, received a chemotherapeutic agent as a treatment for his disease. The drug used to destroy the cancer also produced an undesirable condition known as bone marrow depression. The complete blood count (CBC) profile shown here indicates that this patient is in trouble. Review the lab values and answer the following questions.

1. Note that the platelet count is very low. How does this affect the patient? Discuss measures to protect him.
2. Note that the white blood cell count is abnormally low. With the second line of defense impaired, how should the first line of defense be protected?

A special group of phagocytes are not white blood cells. These are the dendritic cells, mentioned previously, which are multibranched cells plentiful throughout the body, particularly in the skin and mucous membranes. Dendritic cells await microbial invaders, phagocytize them, and inform cells of adaptive immunity that there is a microbial invasion (see Chapter 16).

Lab Analysis of Leukocytes Analysis of blood for diagnostic purposes, including white blood cell counts, is one task of medical lab technologists. The proportions of leukocytes, as determined in a **differential white blood cell count,** can serve as a sign of disease. For example, an increase in the percentage of eosinophils can indicate allergies or infection with parasitic worms; bacterial diseases typically result in an increase in the number of leukocytes and an increase in the percentage of neutrophils, while viral infections are associated with an increase in the relative number of lymphocytes. The ranges for the normal values for each kind of white blood cell, expressed as a percentage of the total leukocyte population, are shown in Figure 15.5.

CRITICAL **THINKING**

A medical laboratory technologist argues that granulocytes are a natural group, whereas agranulocytes are an artificial grouping. Based on Figure 15.4, do you agree or disagree with the lab tech? What evidence can you cite to justify your conclusion?

Now that we have some background concerning the defensive properties of plasma components and leukocytes, we turn our attention to the details of the body's second line of defense: phagocytosis, nonphagocytic killing by leukocytes, nonspecific chemical defenses, inflammation, and fever.

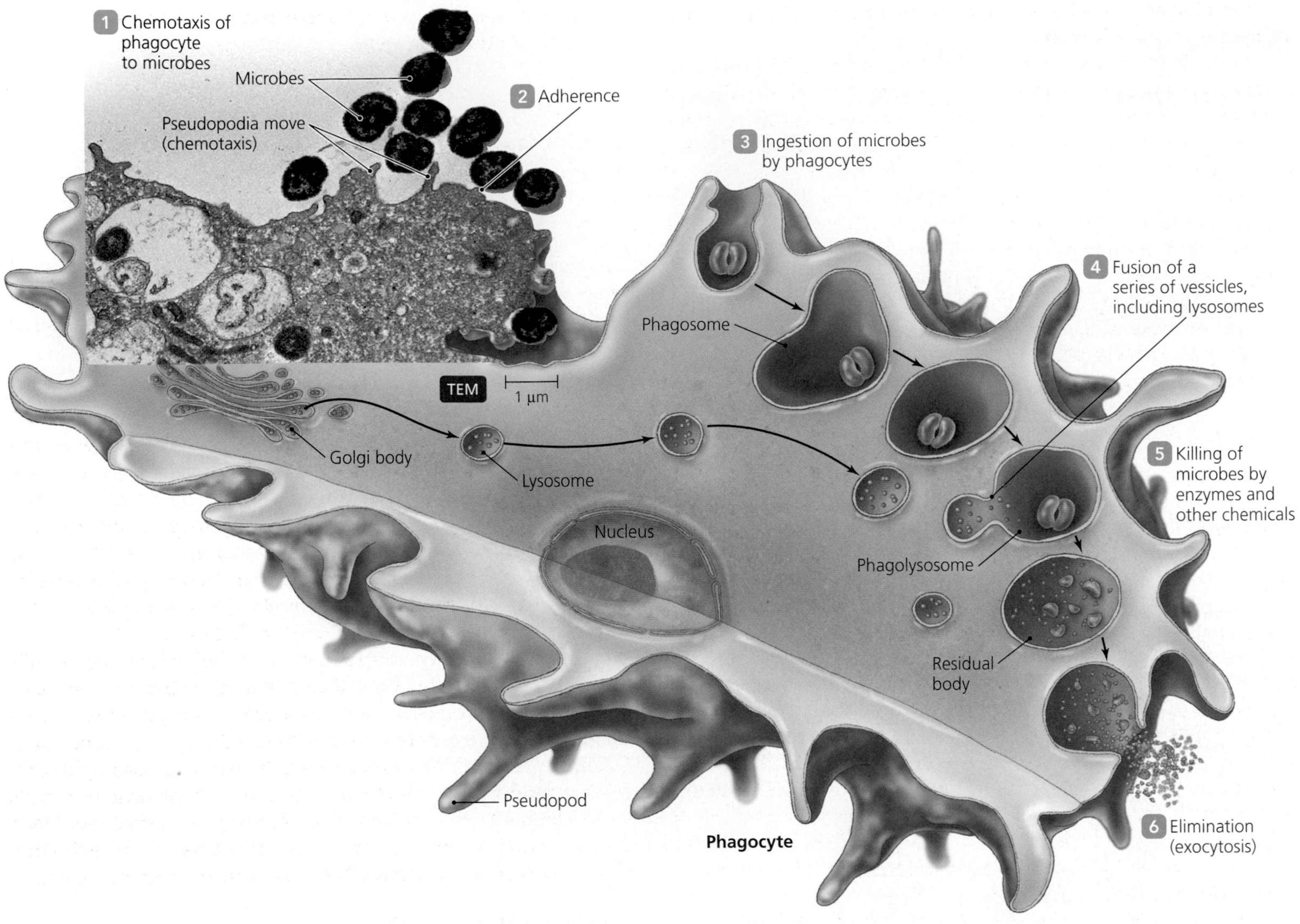

▲ **Figure 15.6 The events in phagocytosis.**
Here a neutrophil phagocytizes *Neisseria gonorrhoeae.*

Phagocytosis

Learning Objective

✓ Name and describe the six stages of phagocytosis.

In Chapter 3 we examined phagocytosis as a way by which a cell actively transports substances across the cytoplasmic membrane and into the cytoplasm. Here we consider how those cells of the body that are capable of phagocytosis—collectively known as **phagocytes** (fag′ō-sītz)—play a role against pathogens that get past the body's first line of defense. **ANIMATIONS:** ***Phagocytosis: Overview***

Phagocytosis is a complex process that is still not completely understood. For the purposes of our discussion, we will divide the continuous process of phagocytosis into six steps: chemotaxis, adherence, ingestion, maturation, killing, and elimination (**Figure 15.6**).

Chemotaxis

Recall from Chapter 3 that *chemotaxis* is movement of a cell either toward a chemical stimulus (positive chemotaxis) or away from a chemical stimulus (negative chemotaxis). In the case of phagocytes, positive chemotaxis involves the use of *pseudopodia* (sū-dō-pō′dē-ă) to crawl toward microorganisms at the site of an infection (1). Chemicals that attract phagocytic leukocytes include microbial components and secretions, components of damaged tissues and white blood cells, and **chemotactic** (kem-ō-tak′tik) **factors.** Chemotactic factors include defensins, peptides derived from complement (discussed later in this chapter), and chemicals called **chemokines** (kē′mō-kīnz), which are released by leukocytes already at a site of infection.

Adherence

After arriving at the site of an infection, phagocytes attach to microorganisms through the binding of complementary chemicals such as glycoproteins found on the membranes of cells (2). This process is called **adherence.**

Some bacteria have virulence factors, such as M protein of *Streptococcus pyogenes* (strep-tō-kok'ŭs pī-oj'en-ēz), or slippery capsules that hinder adherence of phagocytes and thereby increase the virulence of the bacteria. Such bacteria are more readily phagocytized if they are pushed up against a surface such as connective tissue, the wall of a blood vessel, or a blood clot.

All pathogens are more readily phagocytized if they are first covered with antimicrobial proteins such as complement proteins (discussed later) or the specific antimicrobial proteins called antibodies (discussed in Chapter 16). This coating process is called **opsonization**[11] (op'sŭ-nī-zā'shun), and the proteins are called **opsonins.** Generally, opsonins increase the number and kinds of binding sites on a microbe's surface.

Ingestion

After phagocytes adhere to pathogens, they extend *pseudopodia* to surround the microbe (3). The encompassed microbe is internalized as the pseudopodia fuse to form a food vesicle called a **phagosome.**

Phagosome Maturation and Microbial Killing

A series of membranous organelles within the phagocyte fuse with newly formed phagosomes to form digestive vesicles. One organelle, the lysosome, adds digestive chemicals to the maturing phagosome, which is now called a **phagolysosome** (fag-ŏ-lī'sō-sōm) (4). Phagolysosomes contain antimicrobial substances, such as highly reactive, toxic forms of oxygen, in an environment with a pH of about 5.5 due to the active pumping of H^+ from the cytosol. These factors, along with 30 or so different enzymes such as lipases, proteases, nucleases, and a variety of others, destroy the engulfed microbes (5).

Most pathogens are dead within 30 minutes, though some bacteria contain virulence factors (such as M protein or waxy cell walls) that resist a lysosome's action. In the end, a phagolysosome is known as a *residual body.*

Elimination

Digestion is not always complete, and phagocytes eliminate remnants of microorganisms via *exocytosis,* a process that is essentially the reverse of ingestion (6). Some microbial components are specially processed and remain attached to the cytoplasmic membrane of some phagocytes, particularly dendritic cells, a phenomenon that plays a role in the adaptive immune response (discussed in Chapter 16). **ANIMATIONS:** ***Phagocytosis: Mechanism***

How is it that phagocytes destroy invading pathogens, and leave the body's own healthy cells unharmed? At least two mechanisms are responsible for this:

- Some phagocytes have cytoplasmic membrane receptors for various microbial surface components lacking on the body's cells, such as cell wall components or flagellar proteins.
- Opsonins such as complement and antibody provide a signal to the phagocyte.

[11]From Greek *opsonein,* meaning to supply food, and *izein,* meaning to cause; thus, loosely, "to prepare for dinner."

Nonphagocytic Killing

Learning Objective

✓ Describe the role of eosinophils, NK cells, and neutrophils in nonphagocytic killing of microorganisms and parasitic helminths.

Phagocytosis involves killing a pathogen once it has been ingested—that is, once it is inside the phagocyte. In contrast, eosinophils, natural killer cells, and neutrophils can accomplish *killing* without phagocytosis.

Killing by Eosinophils

As discussed earlier, eosinophils can phagocytize; however, this is not their usual mode of attack. Instead, eosinophils secrete antimicrobial chemicals. They attack parasitic helminths (worms) by attaching to the worm's surface, where they secrete extracellular protein toxins onto the surface of the parasite. These weaken the helminth and may even kill it. **Eosinophilia** (ē-ō-sin'ō-fil-e-ă), an abnormally high number of eosinophils in the blood, is often indicative of helminth infestation.

Besides their attacks against parasitic helminths, eosinophils have recently been discovered to use a never-before-seen tactic against bacteria: Lipopolysaccharide from Gram-negative bacterial cell walls triggers eosinophils to rapidly eject mitochondrial DNA, which combines with previously extruded eosinophil proteins to form a physical barrier. This extracellular structure binds to and then kills the bacteria. This is the first evidence that DNA can have antimicrobial activity, and scientists are investigating exactly how mitochondrial DNA acts as an antibacterial agent.

Killing by Natural Killer Lymphocytes

Natural killer lymphocytes (or **NK cells**) are another type of defensive leukocyte of innate immunity that works by secreting toxins onto the surfaces of virally infected cells and neoplasms (tumors). NK cells identify and spare normal body cells because the latter express membrane proteins similar to those on the NK cells. The ability to distinguish one's own healthy cells from diseased cells and pathogens is discussed more fully in Chapter 16.

Killing by Neutrophils

Neutrophils do not always devour pathogens; they can destroy nearby microbial cells without phagocytosis. They can do this in at least two ways. Enzymes in a neutrophil's cytoplasmic membrane add electrons to oxygen, creating highly reactive superoxide radical O_2^- and hydrogen peroxide (H_2O_2). Another enzyme converts these into hypochlorite, the active antimicrobial ingredient in household bleach. These chemicals can kill nearby invaders. Yet another enzyme in the membrane makes nitric oxide, which is a powerful inducer of inflammation.

Scientists have recently discovered another way that neutrophils disable microorganisms in their vicinity. They generate

TABLE 15.3 **Toll-Like Receptors and Their Natural Microbial Binding Partners**

TLR/Location	PAMP (Microbial Molecule)
TLR1 Cytoplasmic membrane	Bacterial lipopeptides and certain proteins in multicellular parasites
TLR2 Cytoplasmic membrane	Bacterial lipopeptides, lipoteichoic acid (found in Gram-positive cell wall), and cell wall of yeast
TLR3 Phagosome membrane	Double-stranded RNA (found only in viruses)
TLR4 Cytoplasmic membrane	Lipid A (found in Gram-negative cell wall)
TLR5 Cytoplasmic membrane	Flagellin (bacterial flagella)
TLR6 Cytoplasmic membrane	Bacterial lipopeptides, lipoteichoic acid (found in Gram-positive cell wall), and cell wall of yeast
TLR7 Phagosome membrane	Single-stranded viral RNA
TLR8 Phagosome membrane	Single-stranded viral RNA
TLR9 Phagosome membrane	Unmethylated cytosine-guanine pairs of viral and bacterial DNA
TLR10 Unknown	Unknown

webs of extracellular fibers nicknamed *NETs* for *neutrophil extracellular traps*. Neutrophils synthesize NETs via a unique form of cellular suicide involving the disintegration of their nuclei. As the nuclear envelope breaks down, DNA and histones are released into the cytosol, and the mixing of nuclear components with cytoplasmic granule membranes and proteins forms NET fibers. Reactive oxygen species—superoxide and peroxide—then kill the neutrophil. The NETs are released from the dying cell as its cytoplasmic membrane ruptures. NETs trap both Gram-positive and Gram-negative bacteria, immobilizing them and sequestering them along with antimicrobial peptides, which kill the bacteria. Thus, even in their dying moments, neutrophils fulfill their role as defensive cells.

Nonspecific Chemical Defenses Against Pathogens

Learning Objectives

- ✓ Define *Toll-like receptors* and describe their action in relation to pathogen-associated molecular patterns.
- ✓ Describe the location and functions of NOD proteins.
- ✓ Explain the roles of interferons in innate immunity.
- ✓ Describe the complement system, including its three activation pathways.

Chemical defenses augment phagocytosis in the second line of defense. The chemicals assist phagocytic cells either by enhancing other features of innate immunity or by directly attacking pathogens. Defensive chemicals include lysozyme and defensins (examined previously) as well as Toll-like receptors, NOD proteins, interferons, and complement.

Toll-Like Receptors (TLRs)

Toll-like receptors (TLRs)[12] are integral membrane proteins produced by phagocytic cells. TLRs act as an early warning system, triggering your body's responses to a number of molecules that are shared by various bacterial or viral pathogens and are absent in humans. These microbial molecules include peptidoglycan, lipopolysaccharide, flagellin, unmethylated pairs of cytosine and guanine nucleotides from bacteria and viruses, double-stranded RNA, and single-stranded viral RNA. Such microbial components are collectively referred to as **pathogen-associated molecular patterns (PAMPs).**

Ten TLRs are known for humans. TLRs 1, 2, 4, 5, and 6 are found spanning cytoplasmic membranes, while TLRs 3, 7, 8, and 9 span phagosome membranes. Some TLRs act alone; others act in pairs to recognize a particular PAMP. For example, TLR3 binds to double-stranded RNA from viruses such as West Nile virus, and TLR2 and TLR6 in conjunction bind to lipoteichoic acid—a component of Gram-positive cell walls. Table 15.3 summarizes the PAMPs and the membrane locations of the ten known TLRs of humans.

Binding of a PAMP to a Toll-like receptor initiates a number of defensive responses, including apoptosis (cell suicide) of an infected cell, secretion of inflammatory mediators or interferons (both discussed shortly), or production of chemical stimulants of adaptive immune responses (discussed in Chapter 16). If TLRs fail, much of immune response collapses, leaving the body open to attack by a myriad of pathogens.

Scientists are actively seeking ways to stimulate TLRs, so as to enhance the body's immune response to pathogens and immunizations. In contrast, methods to inhibit TLRs may provide us with ways to counter inflammatory disorders and some hyperimmune responses.

NOD Proteins

NOD[13] **proteins** are another set of receptors for PAMPs, but NOD proteins are located in a host cell's cytosol rather than in its membranes. Scientists have mostly studied NOD proteins that bind to components of bacterial cell walls. NOD proteins trigger inflammation, apoptosis, and other innate immune responses against bacterial pathogens, though researchers are still elucidating their exact method of action. Mutations in NOD genes are associated with several inflammatory bowel diseases, including Crohn's disease.

[12]German, meaning fantastic. Originally referring to a gene of fruit flies, mutations of which cause the flies to be bizarre looking. Toll-like proteins are similar to fruit fly Toll in their amino acid sequence, though not in their function.
[13]Nucleotide-oligomerization domains, referring to their ability to bind a region of a finite number (oligomer) of DNA nucleotides.

Interferons

So far in this chapter we have focused primarily on how the body defends itself against bacteria and eukaryotes. Now we consider how chemicals in the second line of defense act against viral pathogens.

As mentioned in Chapter 13, viruses use a host's metabolic machinery to produce new viruses. For this reason, it is often difficult to interfere with virus replication without also producing deleterious effects on the host. **Interferons** (in-ter-fēr'onz) are protein molecules released by host cells to nonspecifically inhibit the spread of viral infections. Their lack of specificity means that interferons produced against one viral invader protect somewhat against infection by other types of viruses as well. However, interferons also cause malaise, muscle aches, chills, headache, and fever, which are typically associated with viral infections.

Different cell types produce one of two basic types of interferon when stimulated by viral nucleic acid binding to TLR3, TLR7, or TLR8. Interferons within any given type share certain physical and chemical features, though they are specific to the species that produces them. In general, type I interferons—also known as alpha and beta interferons—are present early in viral infections, whereas type II (gamma) interferon appears somewhat later in the course of infection. Because their actions are identical, we examine alpha and beta interferons together before discussing gamma interferon.

Type I (Alpha and Beta) Interferons Within hours after infection, virally infected monocytes, macrophages, and some lymphocytes secrete small amounts of **alpha interferon (IFN-α);** similarly, fibroblasts, which are undifferentiated cells in such connective tissues as cartilage, tendon, and bone, secrete small amounts of **beta interferon (IFN-β)** when infected by viruses. The structures of alpha and beta interferons are similar, and their actions are identical.

Interferons do not protect the cells that secrete them—these cells are already infected with viruses. Instead, interferons activate natural killer lymphocytes and trigger protective steps in neighboring uninfected cells. Alpha and beta interferons bind to interferon receptors on the cytoplasmic membranes of neighboring cells. Such binding triggers the production of **antiviral proteins (AVPs),** which remain inactive within these cells until AVPs bind to viral nucleic acids, particularly double-stranded RNA, a molecule that is common among viruses but generally absent in eukaryotic cells (**Figure 15.7**).

At least two types of AVPs are produced: *oligoadenylate synthetase,* the action of which results in the degradation of mRNA, and *protein kinase,* which inhibits protein synthesis by ribosomes. Between them, these AVP enzymes essentially destroy the protein production system of the cell, preventing viruses from being replicated. Of course, cellular metabolism is also affected negatively. The antiviral state lasts three to four days, which may be long enough for a cell to rid itself of viruses but still a short enough period for the cell to survive without protein production.

Type II (Gamma) Interferon Gamma interferon (IFN-γ) is produced by activated T lymphocytes and NK lymphocytes. Because T lymphocytes are usually activated as part of an adaptive immune response (see Chapter 16) days after an infection has occurred, gamma interferon appears later than either alpha or beta interferon. Its action in stimulating the activity of macrophages gives IFN-γ its other name: *macrophage activation factor.* Gamma interferon plays a small role in protecting the body against viral infections; mostly IFN-γ regulates the immune system, as in its activation of phagocytic activity.

Table 15.4 on p. 448 summarizes various properties of interferons in humans.

As scientists learned more about the effects of interferons, many thought these proteins might be a "magic bullet" against viral infections, but viruses can interfere with the effects of interferon, as discussed in **Highlight: How Do Viruses Thwart Interferon?** on p. 449. Many variations of interferons in all three classes have been produced in laboratories using recombinant DNA technology, in the hopes that antiviral therapy can be improved.

Complement

The **complement system**—or **complement** for short—is a set of serum proteins designated numerically according to the order of their discovery. These proteins initially act as opsonins and chemotactic factors, and indirectly trigger inflammation and fever. The end result of full complement activation is lysis of foreign cells. **ANIMATIONS:** ***Complement: Overview***

Complement is activated in three ways (**Figure 15.8**):

- In the *classical pathway,* antibodies activate complement.
- In the *alternative pathway,* pathogens or pathogenic products (such as bacterial endotoxins and glycoproteins) activate complement.
- In the *lectin pathway,* microbial polysaccharides bind to activating molecules.

As Figure 15.8 shows, the three pathways merge. Complement proteins react with one another in an amplifying sequence of chemical reactions, in which the product of each reaction becomes an enzyme that catalyzes the next reaction many times over. Such reactions are called *cascades* because they progress in a way that can be likened to a rock avalanche in which one rock dislodges several other rocks, each of which dislodges many others until a whole cascade of rocks is tumbling down the mountain. The products of each step in the complement cascade initiate other reactions, often with wide-ranging effects in the body.

The Classical Pathway Complement got its name from events in the originally discovered "classical" pathway. In this pathway the various proteins act to "complement," or act in conjunction with, the action of antibodies, which we now understand are part of adaptive immunity. As you study the

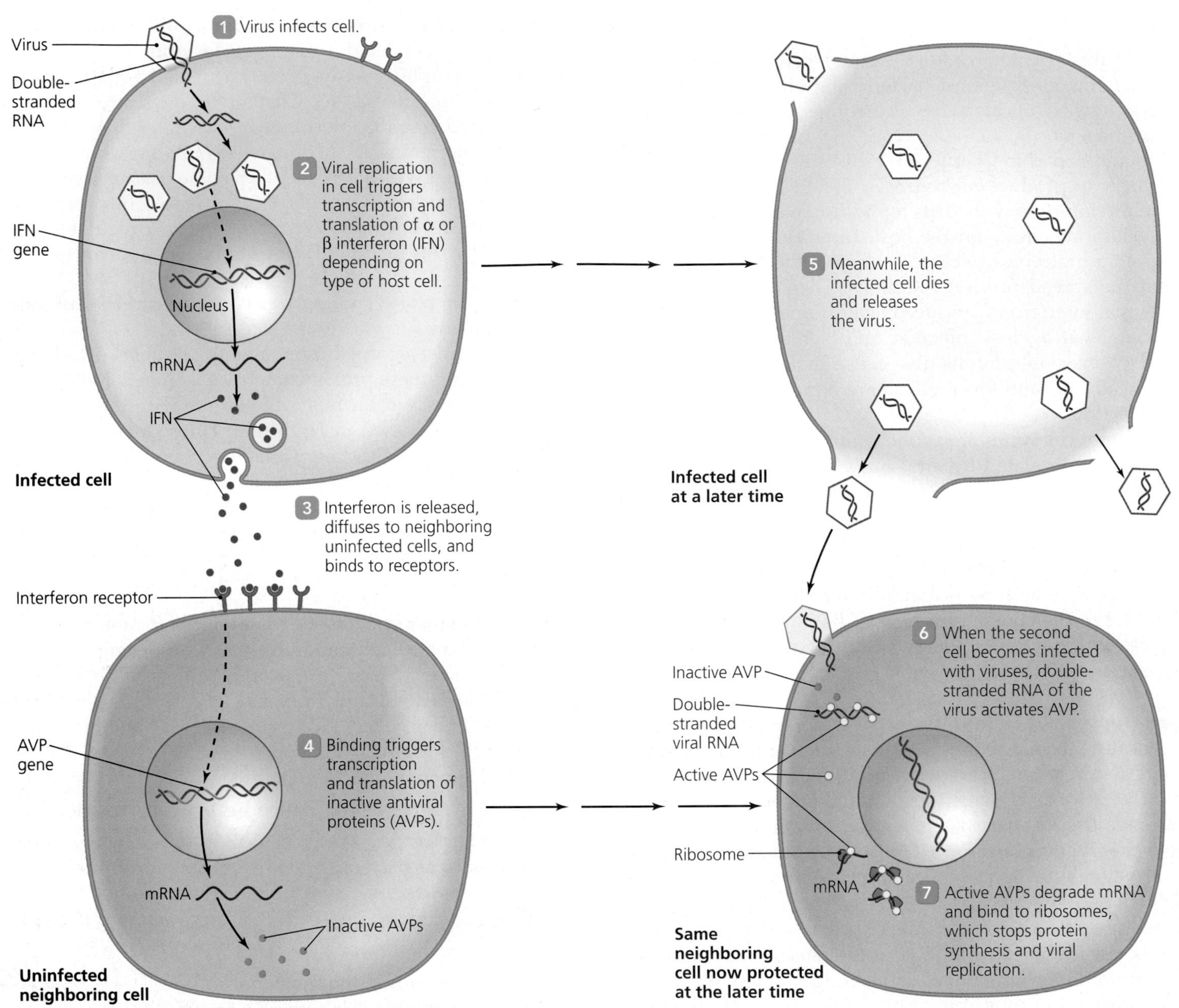
Virus
Double-stranded RNA
1 Virus infects cell.
IFN gene
2 Viral replication in cell triggers transcription and translation of α or β interferon (IFN) depending on type of host cell.
Nucleus
mRNA
IFN
Infected cell
3 Interferon is released, diffuses to neighboring uninfected cells, and binds to receptors.
Interferon receptor
AVP gene
4 Binding triggers transcription and translation of inactive antiviral proteins (AVPs).
mRNA
Inactive AVPs
Uninfected neighboring cell
5 Meanwhile, the infected cell dies and releases the virus.
Infected cell at a later time
Inactive AVP
Double-stranded viral RNA
Active AVPs
Ribosome
mRNA
6 When the second cell becomes infected with viruses, double-stranded RNA of the virus activates AVP.
7 Active AVPs degrade mRNA and bind to ribosomes, which stops protein synthesis and viral replication.
Same neighboring cell now protected at the later time

▲ **Figure 15.7** The actions of alpha and beta interferons.

TABLE 15.4 The Characteristics of Human Interferons

	Type I		Type II
Property	**Alpha Interferon (IFN-α)**	**Beta Interferon (IFN-β)**	**Gamma Interferon (IFN-γ)**
Principal source	Epithelium, leukocytes	Fibroblasts	Activated T lymphocytes and NK lymphocytes
Inducing agent	Viruses	Viruses	Adaptive immune responses
Action	Stimulates production of antiviral proteins	Stimulates production of antiviral proteins	Stimulates phagocytic activity of macrophages and neutrophils
Other names	Leukocyte-IFN	Fibroblast-IFN	Immune-IFN, macrophage activation factor

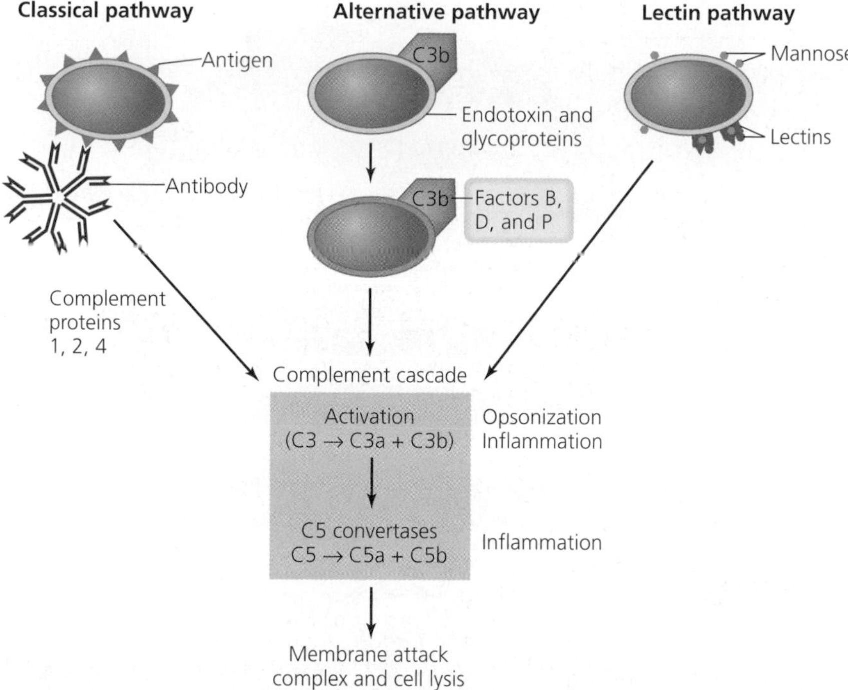

Figure 15.8
Pathways by which complement is activated. In the classical pathway, the binding of antibodies to antigens activates complement. In the alternative pathway the binding of factors B, D, and P to endotoxin or glycoproteins in the cell walls of bacteria or fungi activates complement. The lectin pathway activates when lectins bind to microbial carbohydrates. *How did complement get that name?*

Figure 15.8 *Complement proteins add to—or complement—the action of antibodies.*

depiction of the classical complement cascade in **Figure 15.9**, keep the following concepts in mind:

- Complement enzymes in early events cleave other complement molecules to form *fragments*, which are designated with lowercase letters. For example, inactive complement protein 3 (C3) is cleaved into active fragments C3a and C3b.
- Most fragments have specific and important roles in achieving the functions of the complement system. Some combine together to form new enzymes; some act to increase vascular permeability, which increases diapedesis; others enhance inflammation; still others are involved as chemotactic factors or in opsonization.

HIGHLIGHT

HOW DO VIRUSES THWART INTERFERON?

Hepatitis C is a viral disease that chronically infects nearly 200 million people worldwide. In an effort to shed light on how viruses thwart the human body's defense mechanisms, researchers at the University of Texas have studied the tactics that the hepatitis C virus uses when infecting a host.

In the human body, interferon is part of the defense against viral infections. When a virus invades a cell, the cell's defense systems produce interferon, which inhibits the spread of the viral infection to neighboring cells. One way that viruses attempt to overcome the body's defenses is by preventing the production of interferon.

Researchers have found that the human body's defense against hepatitis C begins with a gene called *RIG-I*. This gene codes for a protein with a novel shape. The new shape signals other proteins to activate interferon regulatory factor 3 (IRF-3), which turns on the genes that produce interferon.

Hepatitis C virus interrupts this process by producing a protease. Protease chops up the proteins that are needed to carry RIG-I's signal to IRF-3. Since IRF-3 doesn't get the message, no interferon is produced.

Drugs known as protease inhibitors disable the viral protease. Widely used against the human immunodeficiency virus (HIV), protease inhibitors have shown promise in treating patients with hepatitis C. If researchers can identify the exact proteins that are attacked by the hepatitis C protease, they may be able to develop protease inhibitors that are more effective against hepatitis C.

The University of Texas discoveries have far-reaching implications. Tactics used by the hepatitis C virus are likely to be similar to those of the West Nile, influenza, and common cold viruses. Understanding of these mechanisms may lead to the development of treatments against these viruses as well.

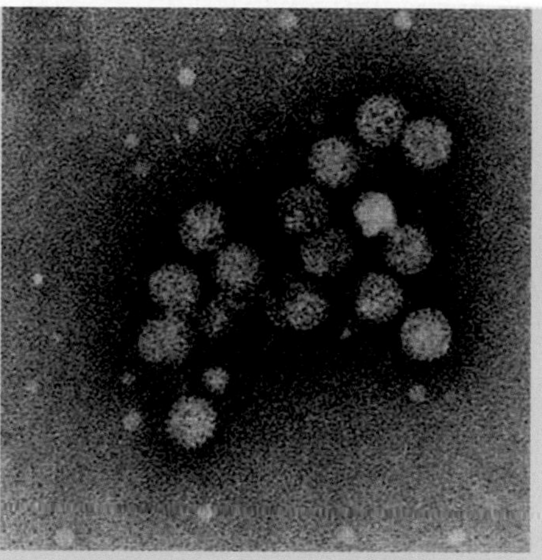

▲ ***Hepatitis C viruses.*** TEM 150 nm

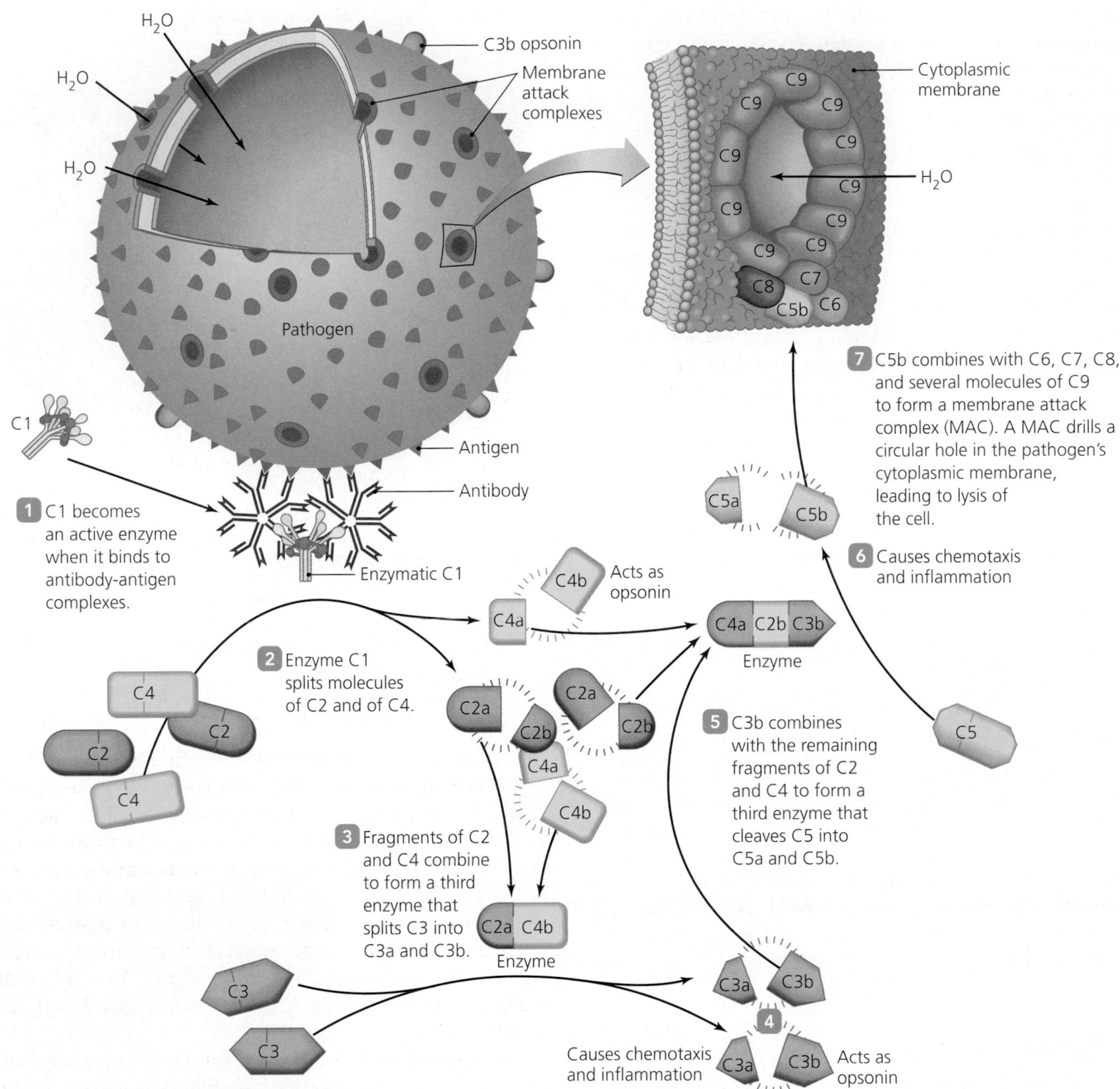

▲ **Figure 15.9 The complement cascade.** The major functions of complement are opsonization and mediation of chemotaxis, and inflammation. A membrane attack complex is a potent antimicrobial weapon that can form against a wide variety of bacterial and eukaryotic pathogens. *What proteins would be involved in activating a complement cascade if this were the alternative pathway?*

Figure 15.9 *Whereas the classical pathway of complement activation involves proteins C1, C2, and C4, the alternative pathway involves factors B, D, and P (properdin).*

- One end product of a full cascade is a **membrane attack complex (MAC),** which forms a circular hole in a pathogen's membrane. The production of numerous MACs **(Figure 15.10)** leads to lysis in a wide variety of bacterial and eukaryotic pathogens. Gram-negative bacteria, such as the bacterium causing gonorrhea, are particularly sensitive to the production of MACs via the complement cascade because their outer membranes are exposed and susceptible. In contrast, a Gram-positive bacterium, which has a thick layer of peptidoglycan overlying its cytoplasmic membrane, is typically resistant to the MAC-induced lytic properties of complement, though it is susceptible to the other effects of the complement cascade.

In addition to its enzymatic role, fragment C3b acts as an opsonin. Fragment C4b also acts as an opsonin. Fragments C3a and C5a function as chemotactic factors, attracting phagocytes to the site of infection, and are also inflammatory agents that trigger increased vascular permeability and dilation. The inflammatory roles of these fragments are discussed in more detail shortly.

The Alternative Pathway The alternative pathway was so named because scientists discovered it second. As previously mentioned, antibodies bound to antigens are necessary for the classical activation of complement, whereas activation of the alternative pathway occurs independently of antibodies. The alternative pathway begins with the cleavage of C3 into C3a and C3b. This naturally occurs at a slow rate in the plasma but proceeds no further because C3b is cleaved into smaller fragments almost immediately. However, when C3b binds to microbial surfaces, it stabilizes long enough for a protein called factor B to adhere. Another plasma protein, factor D, then cleaves factor B, creating an enzyme composed of C3b and Bb. This enzyme, which is stabilized by a third protein—factor P (properdin)—cleaves more molecules of C3 into C3a and C3b, continuing the complement cascade and the formation of MACs.

The alternative pathway is useful in the early stages of an infection, before the adaptive immune response has created the antibodies needed to activate the classical pathway.

The Lectin Pathway Researchers have discovered a third pathway for complement activation that acts through the use of *lectins*. Lectins are chemicals that bind to specific sugar subunits of polysaccharide molecules; in this case, to mannose sugar in mannan polysaccharide on the surfaces of fungi, bacteria, or viruses. Mannose is rare in mammals. Lectins bound to mannose act to trigger a complement cascade by cleaving C2 and C4. The cascade then proceeds like the classical pathway (see steps 3 to 7 in Figure 15.9). **ANIMATIONS:** ***Complement: Activation, Results***

CRITICAL **THINKING**

A patient has a genetic disorder that makes it impossible for her to synthesize complement protein 8 (C8). Is her complement system nonfunctional? What major effects of complement could still be produced?

Inactivation of Complement We have seen that the complement system is nonspecific, and that MACs can form on any cell's exposed membrane. How do the body's own cells withstand the action of complement? Membrane-bound proteins on the body's cells bind with and break down activated complement proteins, which interrupts the complement cascade before damage can occur.

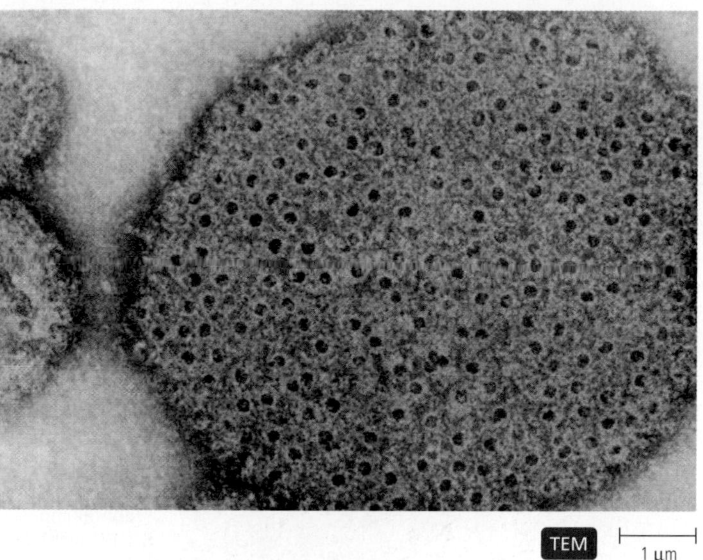

▲ **Figure 15.10 Membrane attack complexes.** Transmission electron micrograph of a cell damaged by numerous punctures produced by membrane attack complexes.

Inflammation

Learning Objective

✓ Discuss the process and benefits of inflammation.

Inflammation is a general, nonspecific response to tissue damage resulting from a variety of causes, including heat, chemicals, ultraviolet light (sunburn), abrasions, cuts, and pathogens. **Acute inflammation** develops quickly, is short lived, is typically beneficial, and results in the elimination or resolution of whatever condition precipitated it. Long-lasting **chronic inflammation** causes damage (even death) to tissues, resulting in disease. Both acute and chronic inflammation exhibit similar signs and symptoms, including redness in light-colored skin (rubor), localized heat (calor), edema (swelling), and pain (dolor). **ANIMATIONS:** ***Inflammation: Overview***

It may not be obvious from this list of signs and symptoms that acute inflammation is beneficial; however, acute inflammation is an important part of the second line of defense because it results in (1) dilation and increased permeability of blood vessels, (2) migration of phagocytes, and (3) tissue repair. Although the chemical details of inflammation are beyond the scope of our study, we now consider these three aspects of acute inflammation.

Dilation and Increased Permeability of Blood Vessels

Part of the body's initial response to an injury or invasion of pathogens is localized dilation (increase in diameter) of blood vessels in the affected region. The process of blood clotting triggers the conversion of a soluble plasma protein into a nine-amino-acid peptide chain called **bradykinin** (brad-e-ki′nin), which is a potent mediator of inflammation. Further, patrolling macrophages, using Toll-like receptors and NOD proteins to

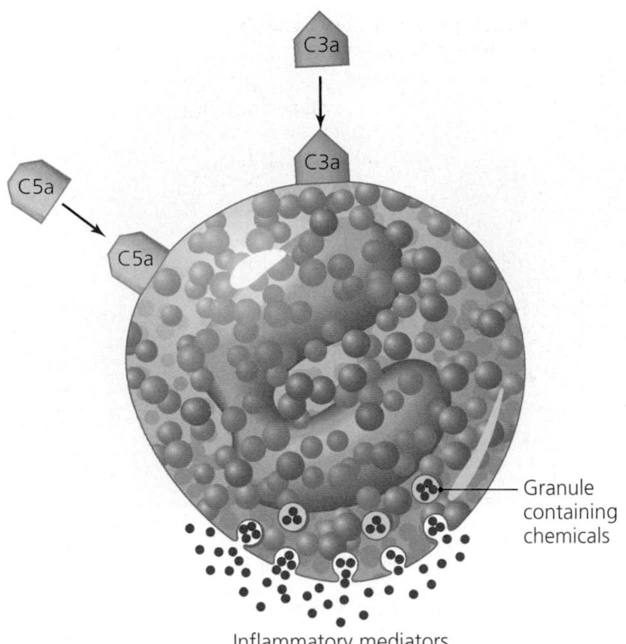

▲ **Figure 15.11 The stimulation of inflammation by complement.** The complement fragments C3a and C5a bind to platelets, basophils, and mast cells, causing them to release histamine, which in turn stimulates the dilation of arterioles.

identify invaders, release other inflammatory chemicals, including **prostaglandins** (pros-tă-glan′dinz) and **leukotrienes** (loo-kō-tri′ēnz). Basophils, platelets, and specialized cells located in connective tissue—called **mast cells**—also release inflammatory mediators, such as **histamine** (his′tă-mēn), when they are exposed to complement fragments C3a and C5a **(Figure 15.11)**. Recall that these complement peptides were cleaved from larger polypeptides during the complement cascade.

Bradykinin and histamine cause vasodilation of the body's smallest arteries (arterioles) **(Figure 15.12)**. Vasodilation results in more blood being delivered to the site of infection, which in turn delivers more phagocytes, oxygen, and nutrients to the site. Inflammatory mediators cause cells that line blood vessels to make adhesion molecules, which are receptors for leukocytes. Bradykinin, prostaglandins, leukotrienes, and (to some extent) histamine also make small veins more permeable—that is, they cause cells lining the vessels to contract and pull apart, leaving gaps in the walls through which phagocytes can move into the damaged tissue and fight invaders **(Figure 15.13)**. Increased permeability also allows delivery of more bloodborne antimicrobial chemicals to the site.

Dilation of blood vessels in response to inflammatory mediators results in the redness and localized heat associated with inflammation. At the same time, prostaglandins and leukotrienes cause fluid to leak from the more permeable blood vessels and accumulate in the surrounding tissue, resulting in edema, which is responsible for much of the pain of inflammation as pressure is exerted on nerve endings.

Vasodilation and increased permeability also deliver fibrinogen, the blood's clotting protein. Clots forming at the site of injury or infection wall off the area and help prevent pathogens and their toxins from spreading. One result is the formation of *pus,* a fluid containing dead tissue cells, leukocytes, and pathogens in the walled-off area. Pus may push up toward the surface and erupt, or it may remain isolated in the body, where it is slowly absorbed over a period of days. Such an isolated site of infection is called an **abscess.** Pimples, boils, and pustules are examples of abscesses.

The signs and symptoms of inflammation can be treated with antihistamines, which block histamine receptors on blood vessel walls, or with antiprostaglandins. One of the ways aspirin and ibuprofen reduce pain is by acting as antiprostaglandins.

Migration of Phagocytes

Increased blood flow due to vasodilation delivers monocytes and neutrophils to a site of infection. As they arrive, these leukocytes roll along the inside walls of blood vessels until they adhere to the receptors lining the vessels, in a process called **margination.** They then squeeze between the cells of the vessel's wall (diapedesis) and enter the site of infection, usually within an hour of tissue damage. The phagocytes then destroy pathogens via phagocytosis.

As mentioned previously, phagocytes are attracted to the site of infection by chemotactic factors, including C3a, C5a, leukotrienes, and microbial components and toxins. The first phagocytes to arrive are often neutrophils, which are then followed by monocytes. Once monocytes leave the blood, they

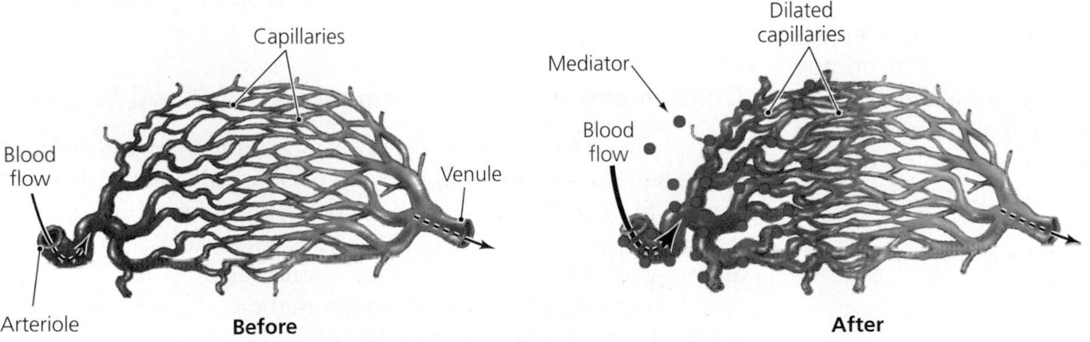

◀ **Figure 15.12 The dilating effect of inflammatory mediators on small blood vessels.** The release of mediators from damaged tissue causes nearby arterioles to dilate, which causes capillaries to expand and enables more blood to be delivered to the affected site. The increased blood flow causes the reddening and heat associated with inflammation.

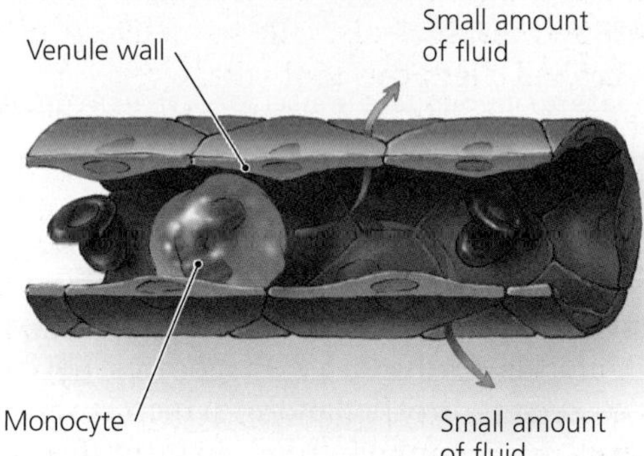

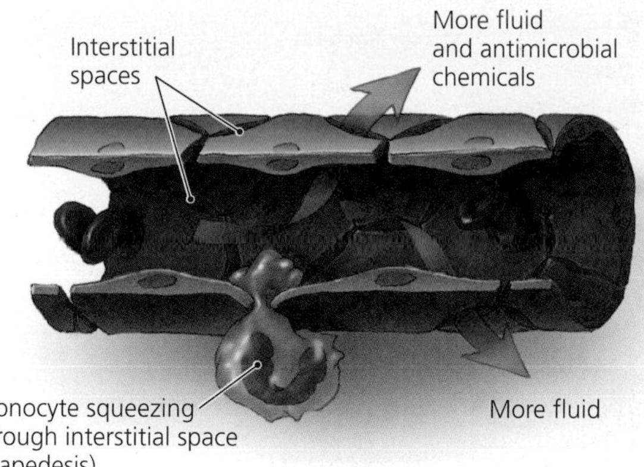

▲ **Figure 15.13 Increased vascular permeability during inflammation.** The presence of bradykinin, prostaglandins, leukotrienes, or histamine causes cells lining venules to pull apart, allowing phagocytes to leave the bloodstream and more easily reach a site of infection. The leakage of fluid and cells causes the edema and pain associated with inflammation.

change and become wandering macrophages, which are especially active phagocytic cells that devour pathogens, damaged tissue cells, and dead neutrophils. Wandering macrophages are a major component of pus.

Tissue Repair

The final stage of inflammation is tissue repair, which in part involves the delivery of extra nutrients and oxygen to the site. Areas of the body where cells regularly undergo cytokinesis, such as the skin and mucous membranes, are repaired rapidly. Some other sites are not fully reparable and form scar tissue.

If the damaged tissue contains undifferentiated stem cells, tissues can be fully restored. For example, a minor skin cut is repaired to such an extent it is no longer visible. However, if cells called *fibroblasts* are involved to a significant extent, scar tissue is formed, inhibiting normal function. Some tissues, such as cardiac muscle and parts of the brain, do not replicate and thus tissue damage cannot be repaired. As a result, these tissues remain damaged following heart attacks and strokes.

Figure 15.14 gives an overview of the entire inflammatory process. Table 15.5 summarizes the chemicals involved in inflammation. **ANIMATIONS:** ***Inflammation: Steps***

CRITICAL **THINKING**

While using a microscope to examine a sample of pus from a pimple, Maria observed a large number of macrophages. Is the pus from an early or a late stage of infection? How do you know?

Fever

Learning Objective

✓ Explain the benefits of fever in fighting infection.

Fever is a body temperature above 37°C. Fever augments the beneficial effects of inflammation, but like inflammation it also has unpleasant side effects, including malaise, body aches, and tiredness.

The hypothalamus, a portion of the brain just above the brain stem, controls the body's internal (core) temperature. Fever results when the presence of chemicals called **pyrogens**[14] (pī′rō-jenz) trigger the hypothalamic "thermostat" to reset at a higher temperature. Pyrogens include bacterial toxins, cytoplasmic contents of bacteria that are released upon lysis, antibody-antigen complexes formed in adaptive immune responses, and pyrogens released by

[14]From Greek *pyr*, meaning fire, and *genein*, meaning to produce.

TABLE 15.5 Chemical Mediators of Inflammation

Vasodilating chemicals	Histamine, serotonin, bradykinin, prostaglandins
Chemotactic factors	Fibrin, collagen, mast cell chemotactic factors, bacterial peptides
Substances with both vasodilating and chemotactic effects	Complement fragments C5a and C3a, interferons, interleukins, leukotrienes, platelet secretions

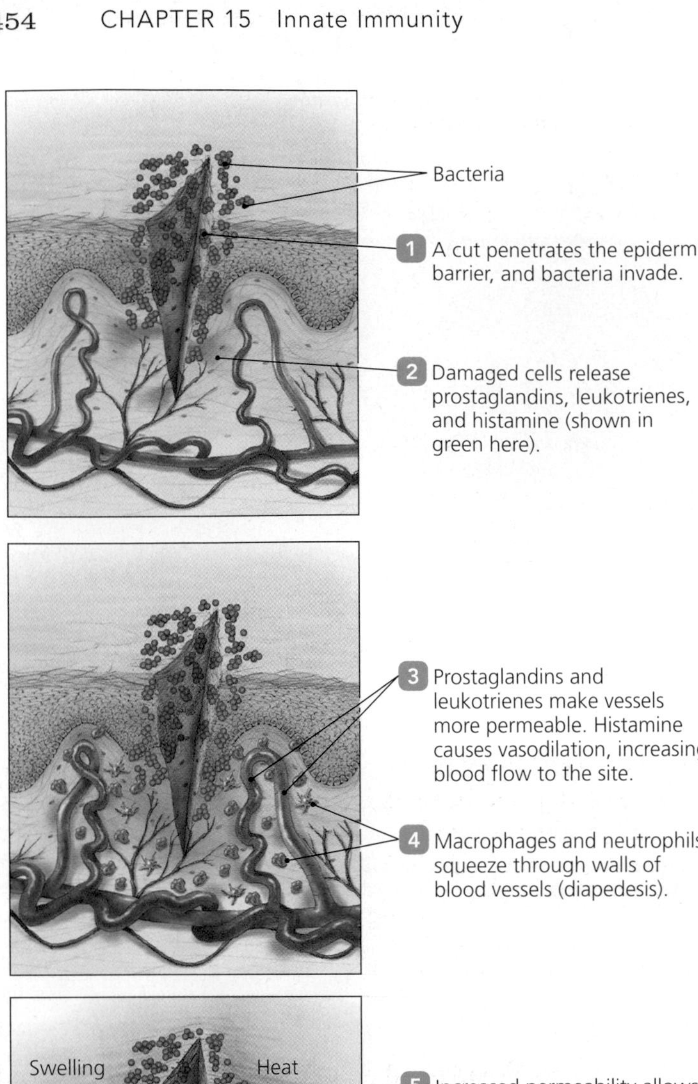
Bacteria
1 A cut penetrates the epidermis barrier, and bacteria invade.
2 Damaged cells release prostaglandins, leukotrienes, and histamine (shown in green here).
3 Prostaglandins and leukotrienes make vessels more permeable. Histamine causes vasodilation, increasing blood flow to the site.
4 Macrophages and neutrophils squeeze through walls of blood vessels (diapedesis).

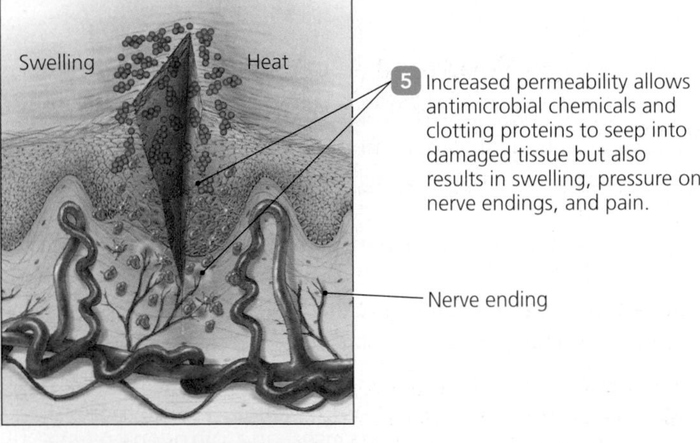
Swelling
Heat
5 Increased permeability allows antimicrobial chemicals and clotting proteins to seep into damaged tissue but also results in swelling, pressure on nerve endings, and pain.
Nerve ending

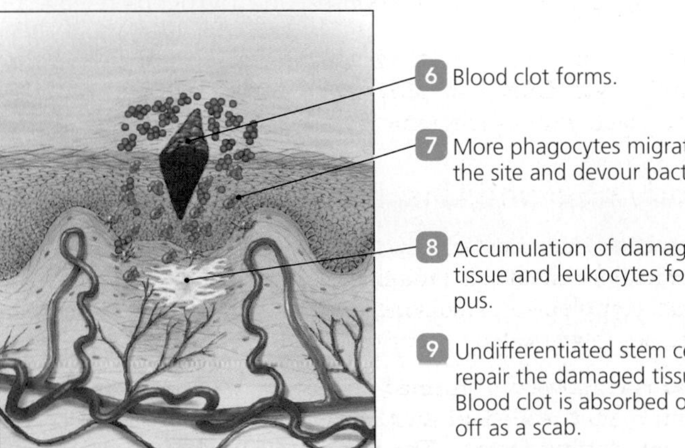
6 Blood clot forms.
7 More phagocytes migrate to the site and devour bacteria.
8 Accumulation of damaged tissue and leukocytes forms pus.
9 Undifferentiated stem cells repair the damaged tissue. Blood clot is absorbed or falls off as a scab.

◀ **Figure 15.14 An overview of the events in inflammation.** The process, which is characterized by redness, swelling, heat, and pain, ends with tissue repair. *In general, what types of cells are involved in tissue repair?*

Figure 15.14 *Tissue repair is effected by cells that are capable of cytokinesis and differentiation. If fibroblasts are among them, scar tissue is laid down.*

phagocytes that have phagocytized bacteria. Although the exact mechanism of fever production is not known, the following discussion and **Figure 15.15** present one possible explanation.

Chemicals produced by phagocytes (1) cause the hypothalamus to secrete prostaglandin, which resets the hypothalamic thermostat by an unknown mechanism (2). The hypothalamus then communicates the new temperature setting to other parts of the brain, which initiate nerve impulses that produce rapid and repetitive muscle contractions (shivering), an increase in metabolic rate, and constriction of blood vessels of the skin (3). These processes combine to raise the body's core temperature until it equals the prescribed temperature setting (4). Because blood vessels in the skin constrict as fever progresses, one effect of inflammation (vasodilation) is undone. The constricted vessels carry less blood to the skin, causing it to appear paler and feel cold to the touch, even though the body's core temperature is higher. This symptom is the *chill* associated with fever.

Fever continues as long as pyrogens are present. As an infection comes under control and fewer active phagocytes are involved, the level of pyrogens decreases, the thermostat is reset to 37°C, and the body begins to cool by perspiring, lowering the metabolic rate, and dilating blood vessels in the skin. These

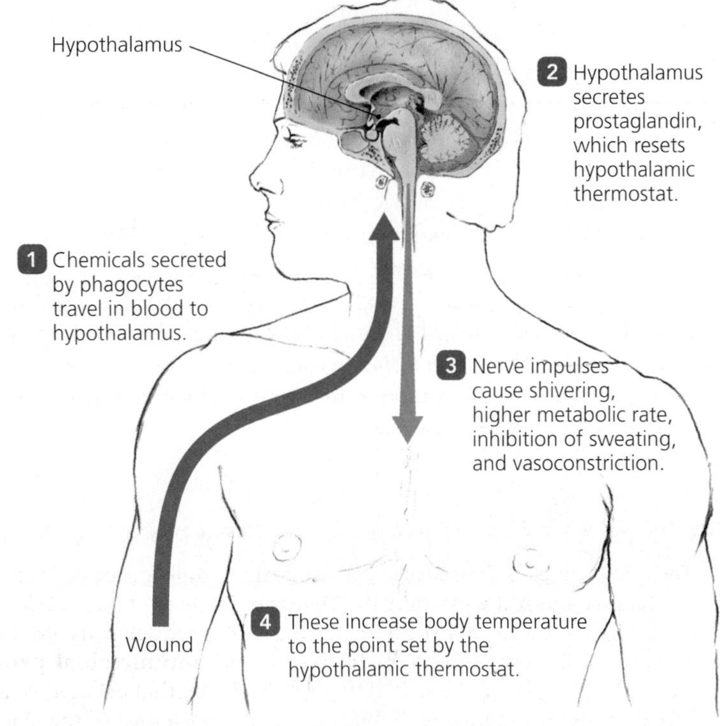
Hypothalamus
2 Hypothalamus secretes prostaglandin, which resets hypothalamic thermostat.
1 Chemicals secreted by phagocytes travel in blood to hypothalamus.
3 Nerve impulses cause shivering, higher metabolic rate, inhibition of sweating, and vasoconstriction.
Wound
4 These increase body temperature to the point set by the hypothalamic thermostat.

▲ **Figure 15.15 One theoretical explanation for the production of fever in response to infection.**

TABLE 15.6 A Summary of Some Nonspecific Components of the First and Second Lines of Defense (Innate Immunity)

First Line	Second Line						
Barriers and Associated Chemicals	Phagocytes	Extracellular Killing	Complement	Interferons	Antimicrobial Peptides	Inflammation	Fever
Skin and mucous membranes prevent the entrance of pathogens; chemicals (e.g., sweat, acid, lysozyme, mucus) enhance the protection	Macrophages, neutrophils, and eosinophils ingest and destroy pathogens	Eosinophils and NK lymphocytes kill pathogens without phagocytizing them	Components attract phagocytes, stimulate inflammation, and attack a pathogen's cytoplasmic membrane	Increase resistance of cells to viral infection, slow the spread of disease	Interfere with membranes, internal signaling, and metabolism; act against pathogens	Increases blood flow, capillary permeability, and migration of leukocytes into infected area; walls off infected region, increases local temperature	Mobilizes defenses, accelerates repairs, inhibits pathogens

processes, collectively called the *crisis* of a fever, are a sign that the infection has been overcome and that body temperature is returning to normal.

The increased temperature of fever enhances the effects of interferons, inhibits the growth of some microorganisms, and is thought to enhance the performance of phagocytes, the activity of cells of specific immunity, and the process of tissue repair. However, if fever is too high, critical proteins are denatured; additionally, nerve impulses are inhibited, resulting in hallucinations, coma, and even death.

Because of the potential benefits of fever, many doctors recommend that patients refrain from taking fever-reducing drugs unless the fever is prolonged or extremely high. Other physicians believe that the benefits of fever are too slight to justify enduring the adverse symptoms.

CRITICAL **THINKING**

How do drugs such as aspirin and ibuprofen act to reduce fever? Should you take fever-reducing drugs or let a fever run its course?

Table 15.6 summarizes the barriers, cells, chemicals, and processes involved in the body's first two, nonspecific lines of defense.

Chapter Summary

An Overview of the Body's Defenses (p. 436)

1. Humans have **species resistance** to certain pathogens, as well as three overlapping lines of defense. The first two lines of defense compose **innate immunity,** which is generally nonspecific and protects the body against a wide variety of potential pathogens. A third line of defense is **adaptive immunity,** which is a specific response to a particular pathogen.
   ANIMATIONS: *Host Defenses: Overview*

The Body's First Line of Defense (pp. 436–440)

1. The first line of defense includes the skin, composed of an outer **epidermis** and a deeper **dermis. Dendritic cells** of the epidermis devour pathogens. Sweat glands of the skin produce salty sweat containing the enzyme called **lysozyme** and **antimicrobial peptides** (defensins), which are small peptide chains that act against a broad range of pathogens. **Sebum** is an oily substance of the skin that lowers pH, which deters the growth of many pathogens.
2. The mucous membranes, another part of the body's first line of defense, are composed of tightly packed cells that are replaced frequently by **stem cell** division and often coated with sticky mucus secreted by goblet cells.
3. **Microbial antagonism,** the competition between **normal microbiota** and potential pathogens, also contributes to the body's first line of defense.
4. Tears contain antibacterial lysozyme and also flush invaders from the eyes. Saliva similarly protects the teeth. The low pH of the stomach inhibits most microbes that are swallowed.

The Body's Second Line of Defense (pp. 440–455)

1. The second line of defense includes cells (especially **phagocytes**), antimicrobial chemicals (Toll-like receptors, NOD proteins, interferons, complement, lysozyme, and antimicrobial peptides), and processes (phagocytosis, information, and fever).
2. Blood is composed of **formed elements** (cells and parts of cells) within a fluid called **plasma.** Serum is that portion of plasma without clotting factors. The formed elements are **erythrocytes** (red blood cells), **leukocytes** (white blood cells), and **platelets.**

3. Based on their appearance in stained blood smears, leukocytes are grouped as either granulocytes (**basophils, eosinophils,** and **neutrophils**) or agranulocytes (**lymphocytes, monocytes).** When monocytes leave the blood they become macrophages.

4. Basophils function to release histamine during inflammation, whereas eosinophils and neutrophils phagocytize pathogens. They exit capillaries via **diapedesis** (emigration).

5. **Macrophages,** neutrophils, and dendritic cells are phagocytic cells of the second line of defense. Many are named for their location in the body, for example, alveolar macrophages (the lungs) and microglia (the nervous system).

6. A **differential white blood cell count** is a lab technique that indicates the relative numbers of leukocyte types; it can be helpful in diagnosing disease.

7. **Chemotactic factors,** such as chemicals called **chemokines,** attract phagocytic leukocytes to the site of damage or invasion. Phagocytes attach to pathogens via a process called **adherence.**
ANIMATIONS: *Phagocytosis: Overview*

8. **Opsonization,** the coating of pathogens by proteins called **opsonins,** makes those pathogens more vulnerable to phagocytes. A phagocyte's pseudopodia then surround the microbe to form a sac called a **phagosome,** which fuses with a lysosome to form a **phagolysosome,** in which the pathogen is killed.
ANIMATIONS: *Phagocytosis: Mechanism*

9. Leukocytes can distinguish between the body's normal cells and foreign cells because leukocytes have receptor molecules for foreign cells' components or because the foreign cells are opsonized by complement or antibodies.

10. Eosinophils and **natural killer (NK) lymphocytes** attack nonphagocytically, especially in the case of helminth infections and cancerous cells. **Eosinophilia**—an abnormally high number of eosinophils in the blood—typically indicates such a helminth infection.

11. **Interferons (IFNs)** are protein molecules that inhibit the spread of viral infections. **Alpha interferons** and **beta interferons,** which are released within hours of infection, trigger **antiviral proteins** to prevent viral reproduction in neighboring cells. **Gamma interferons,** produced days after initial infection, activate macrophages and neutrophils.

12. Microbial molecules called **pathogen-associated molecular patterns (PAMPs)** bind to **Toll-like receptors (TLRs)** on host cells' membranes or to **NOD proteins** inside cells, triggering innate immune responses.

13. The **complement system** is a set of proteins that act as chemotactic attractants, trigger inflammation and fever, and ultimately can effect the destruction of foreign cells via the formation of **membrane attack complexes (MACs),** which result in multiple, fatal holes in pathogens' membranes. Complement is activated by a classical pathway involving antibodies, by an alternative pathway triggered by bacterial chemicals, or by a lectin pathway triggered by mannose found on microbial surfaces.
ANIMATIONS: *Complement: Overview, Activation, Results*

14. **Acute inflammation** develops quickly and damages pathogens, whereas **chronic inflammation** develops slowly and can cause tissue damage that can lead to disease. Signs and symptoms of inflammation include redness, heat, swelling, and pain.
ANIMATIONS: *Inflammation: Overview, Steps*

15. The process of blood clotting triggers formation of **bradykinin**—a potent mediator of inflammation.

16. Macrophages with Toll-like receptors or NOD proteins release **prostaglandins** and **leukotrienes,** which increase permeability of blood vessels. **Mast cells,** basophils, and platelets release **histamine** when exposed to peptides from the complement system. Blood clots may isolate an infected area to form an **abscess** such as a pimple or boil.

17. When leukocytes rolling along blood vessel walls reach a site of infection, they stick to the wall in a process called **margination** and then undergo diapedesis to arrive at the site of tissue damage. The increased blood flow of inflammation also brings extra nutrients and oxygen to the infection site to aid in repair.

18. **Fever** results when chemicals called **pyrogens,** including substances released by bacteria and phagocytes, affect the hypothalamus in a way that causes it to reset body temperature to a higher level. The exact process of fever and its control are not fully understood.

Questions for Review

Answers to the Questions for Review (except Short Answer questions) begin on page A-1

Multiple Choice

1. Phagocytes of the epidermis are called
   a. microglia.
   b. goblet cells.
   c. alveolar macrophages.
   d. dendritic cells.

2. Mucus-secreting membranes are found in the
   a. urinary system.
   b. digestive cavity.
   c. respiratory passages.
   d. all of the above

3. The complement system involves
   a. the production of antigens and antibodies.
   b. serum proteins involved in nonspecific defense.
   c. a set of genes that distinguish foreign cells from body cells.
   d. the elimination of undigested remnants of microorganisms.

4. The alternative complement activation pathway involves
   a. factors B, D, and P.
   b. the cleavage of C5 to form C9.
   c. binding to mannose sugar.
   d. recognition of antigens bound to specific antibodies.

5. Complement must be inactivated because if it were not,
   a. viruses could continue to multiply inside host cells using the host's own metabolic machinery.
   b. necessary interferons would not be produced.
   c. protein synthesis would be inhibited, thus halting important cell processes.
   d. it would make holes in the body's own cells.

6. The type of interferon present late in an infection is
   a. alpha interferon.
   b. beta interferon.
   c. gamma interferon.
   d. delta interferon.
7. Interferons
   a. do not protect the cell that secretes them.
   b. stimulate the activity of macrophages.
   c. cause muscle aches, chills, and fever.
   d. all of the above
8. Which of the following is not targeted by a Toll-like receptor?
   a. lipid A
   b. eukaryotic flagellar protein
   c. single-stranded RNA
   d. lipoteichoic acid
9. Toll-like receptors (TLRs) act to
   a. bind microbial proteins and polysaccharides.
   b. induce phagocytosis.
   c. cause phagocytic chemotaxis.
   d. destroy microbial cells.
10. Which of the following binds iron?
   a. lactoferrin
   b. siderophores
   c. transferrin
   d. all of the above

Modified True/False

Indicate which statements are true. Correct all false statements by changing the italicized words.

1. ____ The surface cells of the epidermis of the skin are *alive.*
2. ____ The surface cells of mucous membranes are *alive.*
3. ____ Wandering macrophages experience *diapedesis.*
4. ____ *Monocytes* are immature macrophages.
5. ____ *Lymphocytes* are large agranulocytes.
6. ____ *Phagocytes* exhibit chemotaxis toward a pathogen.
7. ____ In phagocytosis, adherence involves the binding between complementary chemicals on a phagocyte and on the membrane of a *body cell.*
8. ____ *Opsonization* occurs when a phagocyte's pseudopodia surround a microbe and fuse to form a sac.
9. ____ Lysosomes fuse with phagosomes to form *peroxisomes.*
10. ____ A membrane attack complex drills circular holes in a *macrophage.*
11. ____ Rubor, calor, swelling, and dolor are associated with *fever.*
12. ____ Acute and chronic inflammation exhibit *similar* signs and symptoms.
13. ____ The *hypothalamus* of the brain controls body temperature.
14. ____ Defensins are *phagocytic cells* of the second line of defense.
15. ____ NETs are webs produced by *neutrophils* to trap microbes.

Matching

In the blank beside each cell, chemical, or process in the left column, write the letter of the line of defense that first applies. Each letter may be used several times.

1. ____ Inflammation	A. First line of defense
2. ____ Monocytes	B. Second line of defense
3. ____ Lactoferrin	C. Third line of defense
4. ____ Fever	
5. ____ Dendritic cells	
6. ____ Alpha interferon	
7. ____ Mucous membrane of the digestive tract	
8. ____ Neutrophils	
9. ____ Epidermis	
10. ____ Lysozyme	
11. ____ Goblet cells	
12. ____ Phagocytes	
13. ____ Sebum	
14. ____ T lymphocytes	
15. ____ Antimicrobial peptides	

Write the letter of the description that applies to each of the following terms.

1. ____ Goblet cell	A. Leukocyte that primarily attacks parasitic worms
2. ____ Lysozyme	B. Phagocytic cell in lungs
3. ____ Stem cell	C. Secretes sebum
4. ____ Dendritic cell	D. Devours pathogens in epidermis
5. ____ Cell from sebaceous gland	E. Breaks bonds in bacterial cell wall
6. ____ Bone marrow stem cell	F. Phagocytic cell in central nervous system
7. ____ Eosinophil	G. Generative cell with many types of offspring
8. ____ Alveolar macrophage	H. Develops into formed elements of blood
9. ____ Microglia	I. Intercellular scavenger
10. ____ Wandering macrophage	J. Secretes mucus

Labeling

Label the steps of phagocytosis.

1 ____________
2 ____________
3 ____________
4 ____________
5 ____________
6 ____________

Short Answer

1. In order for a pathogen to cause disease, what three things must happen?
2. How does a phagocyte "know" it is in contact with a pathogen instead of another body cell?
3. Give three characteristics of the epidermis that make it an intolerable environment for most microorganisms.
4. What is the role of Toll-like receptors in innate immune responses?
5. Describe the classical complement cascade pathway from C1 to the MAC.
6. How do NOD proteins differ from Toll-like receptors?

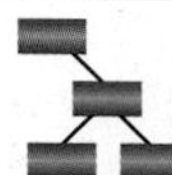

Concept Mapping

Using the following terms, draw a concept map that describes phagocytosis. For a sample concept map, see p. 93. Or, complete this concept map online by going to the Study Area at www.masteringmicrobiology.com.

Adherence
Chemotactic factors
Chemotaxis
Dendritic cells
Elimination
Eosinophils
Exocytosis
Ingestion
Killing
Lysosome
Macrophages
Neutrophils
Opsonins
Phagolysosome
Phagosome

Critical Thinking

1. John received a chemical burn on his arm and was instructed by his physician to take an over-the-counter, anti-inflammatory medication for the painful, red, swollen lesions. When Charles suffered pain, redness, and swelling from an infected cut on his foot, he decided to take the same anti-inflammatory drug because his symptoms matched John's symptoms. How is Charles's inflammation like that of John? How is it different? Is it appropriate for Charles to medicate his cut with the same medicine John used?
2. What might happen to someone whose body did not produce C3? C5?
3. My-lim, age 65, has had diabetes for 40 years, with resulting damage to the small blood vessels in her feet and toes. Her circulation is impaired. How might this condition affect her vulnerability to infection?
4. A patient's chart shows that eosinophils make up 8% of his white blood cells. What does this lead you to suspect? Would your suspicions change if you learned that the patient had spent the previous three years as an anthropologist living among an African tribe? What is the normal percentage of eosinophils?
5. There are two kinds of agranulocytes in the blood—monocytes and lymphocytes. Janice noted that monocytes are phagocytic and that lymphocytes are not. She wondered why two agranulocytes would be so different. What facts of hematopoiesis can help her answer her question?
6. A patient has a genetic disorder that prevents him from synthesizing C8 and C9. What effect does this have on his ability to resist bloodborne Gram-negative and Gram-positive bacteria? What would happen if C3 and C5 fragments were also inactivated?
7. Sweat glands in the armpits secrete perspiration with a pH close to neutral (7.0). How does this fact help explain body odor in this area as compared to other parts of the skin?
8. Scientists can raise "germ-free" animals in axenic environments. Would such animals be as healthy as their worldly counterparts?
9. Compare and contrast the protective structures and chemicals of the skin and mucous membranes.
10. Scientists are interested in developing antimicrobial drugs that act like the body's normal antimicrobial peptides. What advantage might such a drug have over antibiotics?

Access more review material online in the Study Area at **www.masteringmicrobiology.com.** There, you'll find

- **Animations**
- **MP3 Tutor Sessions**
- **Concept Mapping Activities**
- **Flashcards**
- **Quizzes**

and more to help you succeed.

16 Adaptive Immunity

Imagine that your friend has been bitten on the hand by a dog. The skin is broken and the wound is deep. Your friend washes the wound with soap and water but does not use any of the more powerful antimicrobial agents we discussed in Chapter 9. Within 24 hours, his hand and arm are swollen, red, and painful and he is feeling seriously ill. His body's inflammatory response is in full force as manifested by the swelling and pain, but it isn't strong enough to overcome the virulent invading microorganisms.

Fortunately, infections like this trigger the body's *adaptive immunity* in addition to the innate defense responses discussed in Chapter 15. In this chapter we will explore what happens when facets of adaptive immunity—especially lymphocytes—enter the battle against microbial invaders.

Take the pre-test for this chapter online. Visit the Study Area at www.masteringmicrobiology.com.

This chapter has MicroFlix. Go to **www.masteringmicrobiology.com** for 3-D movie-quality animations on immunology.

White blood cells called lymphocytes roll along blood vessel walls. Lymphocytes are the main cells of adaptive immune defenses.

In Chapter 15 we discussed innate immunity, which includes two lines of rapid defense against microbial pathogens. The first line includes intact skin and mucous membranes, whereas the second line includes phagocytosis, nonspecific chemical defenses, inflammation, and fever. As in the chapter opening story, innate defenses do not always offer enough protection in defending the body. Although the mechanisms of innate immunity are readily available and fast-acting, they do not adapt to enhance the effectiveness of response to the great variety of pathogens confronting us—a fever is a fever, whether it is triggered by a mild cold virus or the deadly Ebola virus. The body augments the mechanisms of innate immunity with a third line of defense that destroys invaders while becoming more effective in the process. This response is called *adaptive immunity.*

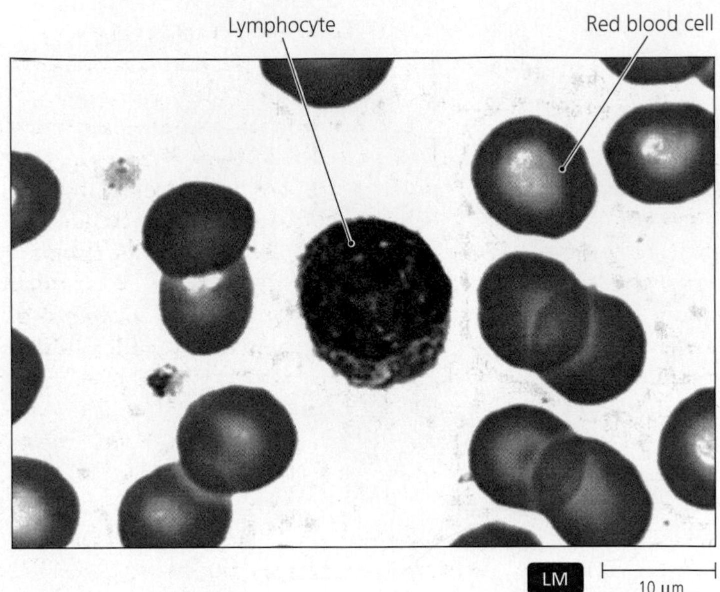

▲ **Figure 16.1 Lymphocytes play a central role in adaptive immunity.** A resting lymphocyte is the smallest leukocyte—slightly larger than a red blood cell.

Overview of Adaptive Immunity

Learning Objectives

- ✓ Describe five distinctive attributes of adaptive immunity.
- ✓ List the two basic types of white blood cells involved in adaptive immunity.
- ✓ List two basic divisions of adaptive immunity and describe their targets.

Adaptive immunity is the body's ability to recognize and then mount a defense against distinct invaders and their products, whether they are protozoa, fungi, bacteria, viruses, or toxins. *Immunologists*—scientists who study the cells and chemicals involved in immunity—are continually refining and revising our knowledge of adaptive immunity. In this chapter, we will examine some of what they have discovered.

Adaptive immunity has five distinctive attributes:

- *Specificity.* Any particular adaptive immune response acts only against one particular molecular shape and not against others. Adaptive immune responses are precisely tailored reactions against specific attackers, whereas innate immunity involves more generalized responses to pathogen-associated molecular patterns (PAMPs)—molecular shapes common to many microbes.
- *Inducibility.* Cells of adaptive immunity activate only in response to specific pathogens.
- *Clonality.* Once induced, cells of adaptive immunity proliferate to form many generations of nearly identical cells, which are collectively called *clones.*
- *Unresponsiveness to self.* As a rule, adaptive immunity does not act against normal body cells; in other words, adaptive immune responses are self-tolerant. Several mechanisms help ensure that immune responses do not attack the body itself.
- *Memory.* An adaptive immune response has "memory" about specific pathogens; that is, it adapts to respond faster and more effectively in subsequent encounters with a particular type of pathogen or toxin.

These aspects of adaptive immunity involve the activities of **lymphocytes** (lim′fō-sītz), which are a type of leukocyte (white blood cell) that acts against specific pathogens. Lymphocytes in their resting state are the smallest white blood cells, and each is characterized by a large, round, central nucleus surrounded by a thin rim of cytoplasm **(Figure 16.1)**. Initially, lymphocytes of humans form in the *red bone marrow,* located in the ends of long bones in juveniles and in the centers of adult flat bones such as the ribs and hip bones. These sites contain *blood stem cells,* which are cells that give rise to all types of blood cells including lymphocytes (see Figure 15.4).

Although lymphocytes appear identical in the microscope, scientists make distinctions between two main types—*B lymphocytes* and *T lymphocytes*—according to integral surface proteins that are part of each lymphocyte's cytoplasmic membrane. These proteins allow lymphocytes to recognize specific pathogens and toxins by their molecular shapes, and the proteins play roles in intercellular communication among immune cells.

Lymphocytes must undergo a maturation process. **B lymphocytes,** which are also called **B cells,** arise and mature in the red bone marrow of adults. (The designation "B" does not stand for bone marrow, but instead for the *bursa of Fabricius,* a unique lymphoid organ located near the end of the intestinal tract in birds, which is where B lymphocytes were originally identified. Mammals do not have this bursa, but the name "B cell" has been retained.)

T lymphocytes, also known as **T cells,** begin in bone marrow as well but do not mature there. Instead, T cells travel to and mature in the *thymus,* located in the chest near the heart. T lymphocytes are so called because of the role the thymus plays in their maturation.

Adaptive immunity consists of two basic types of responses. Historically, bodily fluids such as blood were called *humors* (the Latin word for moisture), and thus immunologists call immune activity centered in such fluids **humoral immune responses.** Descendants of activated B cells are the main defensive cells of humoral immunity. Once induced, they secrete soluble proteins called *antibodies* that act against *extracellular pathogens* such as bacteria and fungi in the body's fluids. In contrast, descendants of T cells regulate adaptive immune responses or attack *intracellular pathogens,* such as viruses replicating inside a cell. Such T lymphocytes mount **cell-mediated immune responses,** which do not involve antibodies. Long-lived descendants of B and T lymphocytes maintain the ability to mount adaptive immunological response against specific pathogens—an ability sometimes called immunological memory. **ANIMATIONS:** *Humoral Immunity: Overview* and *Cell-Mediated Immunity: Overview*

Both humoral immune responses against extracellular pathogens and cell-mediated immune responses against intracellular pathogens are powerful defensive reactions that have the potential to severely and fatally attack the body's own cells. The body must regulate adaptive immune responses to prevent damage; for example, an immune response requires multiple chemical signals before proceeding, thus reducing the possibility of randomly triggering an immune response against uninfected healthy tissue. Autoimmune disorders, hypersensitivities, or immunodeficiency diseases result when regulation is insufficient or overexcited. Chapter 18 deals with such disorders.

Elements of Adaptive Immunity

Just as the program at a dramatic presentation might present a synopsis of the performance and introduce the actors and their roles, the following sections present elements of adaptive immunity by describing the "stage" and introducing the "cast of characters" involved in adaptive immunity. We will examine the *lymphatic system*—the organs, tissues, and cells of adaptive immunity; then we consider the molecules called *antigens* that trigger adaptive immune responses; next, we take a look at *antibodies;* and finally we examine special *chemical signals* and *mediators* involved in coordinating and controlling a specific immune response.

First, we turn our attention to the "stage"—the lymphatic system, which plays an important role in the production, maturation, and housing of the cells that function in adaptive immunity.

The Tissues and Organs of the Lymphatic System

Learning Objectives

- ✓ Contrast the flow of lymph with the flow of blood.
- ✓ Describe the primary and secondary organs of the lymphatic system.
- ✓ Describe the importance of red bone marrow, the thymus, lymph nodes, and other lymphoid tissues.

The **lymphatic system** is composed of the *lymphatic vessels,* which conduct the flow of a liquid called *lymph;* and lymphatic cells, tissues, and organs, which are directly involved in adaptive immunity **(Figure 16.2a)**. Taken together, the components of the lymphatic system constitute a surveillance system that screens the tissues of the body—particularly possible points of entry such as the throat and intestinal tract—for foreign molecules. We begin by examining lymphatic vessels.

The Lymphatic Vessels and the Flow of Lymph

Lymphatic vessels form a one-way system that conducts **lymph** (pronounced limf) from local tissues and returns it to the circulatory system. Lymph is a colorless, watery liquid similar in composition to blood plasma; indeed, lymph arises from fluid that has leaked out of blood vessels into the surrounding intercellular spaces. Unlike blood which supplies nutrients, lymph contains wastes such as degraded proteins and toxins. Lymph is first collected by remarkably permeable *lymphatic capillaries* **(Figure 16.2b)**, which are located in most parts of the body (exceptions include the bone marrow, brain, and spinal cord). From the lymphatic capillaries, lymph passes into increasingly larger lymphatic vessels until it finally flows via two large *lymphatic ducts* into blood vessels near the heart. One-way valves ensure that lymph flows only toward the heart as skeletal muscular activity squeezes the lymphatic vessels. Unlike the cardiovascular system, the lymphatic system has no unique pump and is not circular; that is, lymph flows in only one direction—toward the heart. Located at various points within the system of lymphatic vessels are about 1000 *lymph nodes,* which house leukocytes that recognize and attack foreigners present in the lymph, allowing for immune system surveillance and interactions.

Lymphoid Organs

Once lymphocytes have arisen and matured in the *primary lymphoid organs* of the red bone marrow and thymus, they migrate to *secondary lymphoid organs* and *tissues,* including lymph nodes, spleen, and other less-organized accumulations of lymphoid tissue, where they in effect lie in wait for foreign microbes. The hundreds of **lymph nodes** are located throughout the body but concentrated in the cervical (neck), inguinal (groin), axillary (armpit), and abdominal regions (see Figure 16.2a). Each lymph node receives lymph from numerous *afferent* (inbound) *lymphatic vessels* and drains lymph into just one or two *efferent* (outbound) *lymphatic vessels* **(Figure 16.2c)**. Essentially, lymph nodes are sites to facilitate interactions among immune cells and between immune cells and material in the lymph arriving from throughout the body.

A node has a medullary (central) maze of passages, which filter the lymph passing through and house numerous lymphocytes that survey the lymph for foreign molecules and mount specific immune responses against them. The cortex (outer) portion of a lymph node consists of a tough capsule surrounding *germinal centers,* which is where clones of B cells replicate.

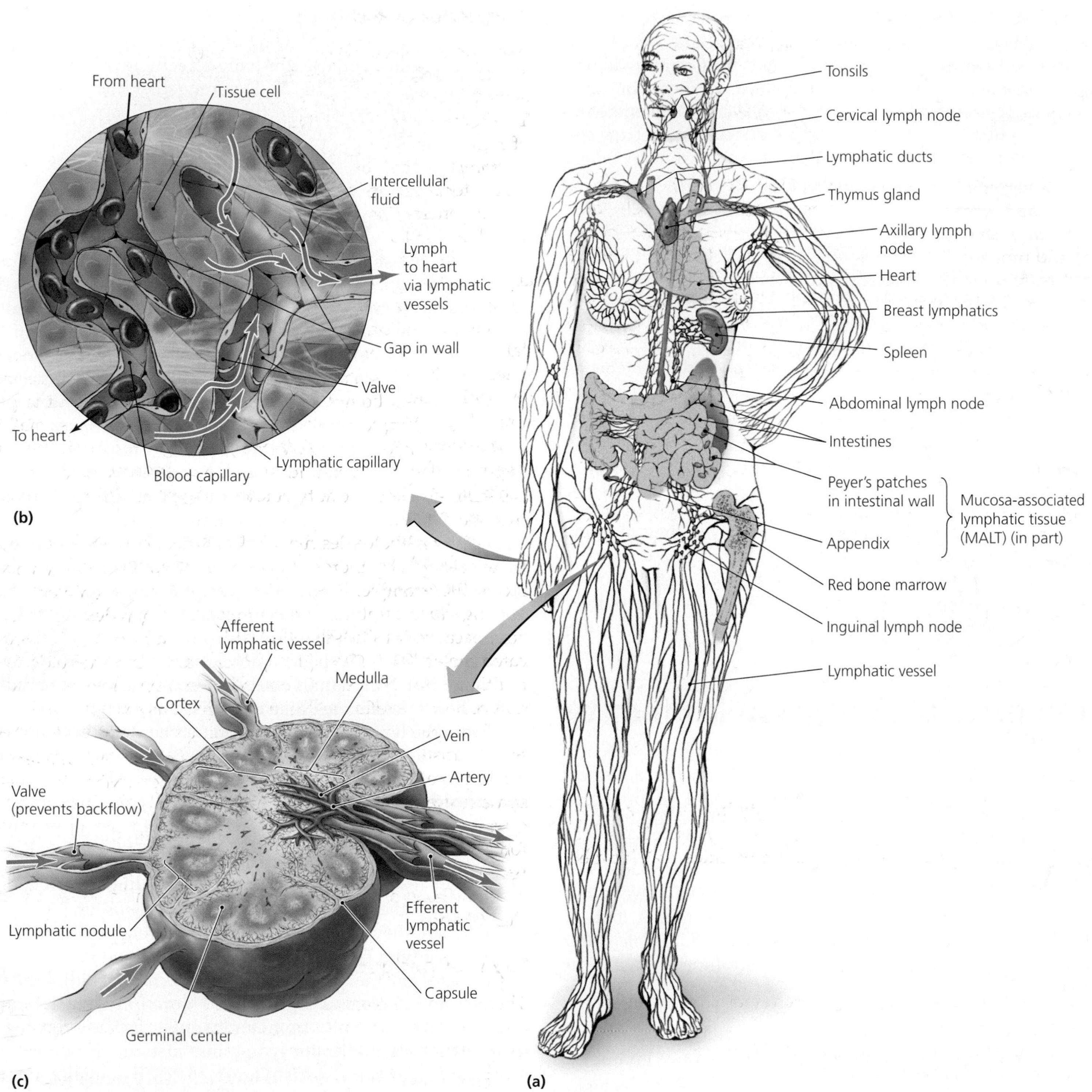

▲ **Figure 16.2 The lymphatic system.** **(a)** The system consists of primary lymphoid organs—bone marrow (not shown) and thymus gland—and secondary lymphoid organs, including lymphatic vessels, lymph nodes, tonsils, and other lymphatic tissue. **(b)** Lymphatic capillaries collect lymph from intercellular spaces. **(c)** Afferent lymphatic vessels carry lymph into lymph nodes; efferent vessels carry it away. The cortex contains germinal centers, where B lymphocytes proliferate; the lymphocytes in the medulla encounter foreign molecules. *Why do lymph nodes enlarge during an infection?*

Figure 16.2 *During an infection, lymphocytes multiply profusely in lymph nodes. This proliferation and swelling cause lymph nodes to enlarge.*

CRITICAL **THINKING**

The cross-sectional area of the afferent lymphatic vessels arriving at a lymph node is greater than the cross-sectional area of the efferent lymphatics exiting the lymph node. The result is that lymph moves slowly through a lymph node. What advantage does this provide the body?

The lymphatic system contains other secondary lymphoid tissues and organs, including the spleen, the tonsils, and *mucosa-associated lymphatic tissue.* The spleen is similar in structure and function to lymph nodes, except that it filters blood instead of lymph. The spleen removes bacteria, viruses, toxins, and other foreign matter from the blood. It also cleanses the blood of old and damaged blood cells, stores blood platelets (which are required for the proper clotting of blood), and stores blood components such as iron.

The tonsils and MALT lack the tough outer capsules of lymph nodes and the spleen, but they function in somewhat the same way by physically trapping foreign particles and microbes. MALT includes the appendix; lymphoid tissue of the respiratory tract, vagina, urinary bladder, and mammary glands; and discrete bits of lymphoid tissue called *Peyer's patches* in the wall of the small intestine. MALT contains most of the body's lymphocytes.

CRITICAL **THINKING**

As part of the treatment for some cancers, physicians kill the cancer patients' dividing cells, including the stem cells that produce leukocytes, and then give the patients a bone marrow transplant from a healthy donor. Which cell is the most important cell in such transplanted marrow?

We are considering adaptive immunity in terms of a "play" taking place on the stage of the lymphatic system. Now, let's meet the "villain actors"—the foreign molecules that lymphocytes recognize.

Antigens

Learning Objectives

- ✓ Identify the characteristics of antigens that stimulate effective immune responses.
- ✓ Distinguish among exogenous antigens, endogenous antigens, and autoantigens.

Adaptive immune responses are not directed against whole bacteria, fungi, protozoa, or viruses, but instead against portions of cells, viruses, and even parts of single molecules that the body recognizes as foreign and worthy of attack. Immunologists call these biochemical shapes **antigens**[1] (an'ti-jenz). They bind to lymphocytes and can trigger adaptive immune responses. Antigens from pathogens and toxins are the "villains" of our story.

Properties of Antigens

Not every molecule is an effective antigen. Among the properties that make certain molecules more effective at provoking adaptive immunity are a molecule's *shape, size,* and *complexity.* The body recognizes antigens by the three-dimensional shapes of regions called **epitopes,** which are also known as *antigenic determinants,* because they are the actual part of an antigen that determines an immune response **(Figure 16.3a)**.

In general, larger molecules with molecular masses (often called molecular weights) between 5000 and 100,000 daltons are better antigens than smaller ones. The most effective antigens are large foreign macromolecules such as proteins and glycoproteins, but carbohydrates and lipids can be antigenic. Small molecules, especially those with a molecular mass under 5000 daltons, make poor antigens by themselves because they evade detection. Such small molecules, called *haptens,* can become antigenic when bound to larger, carrier molecules (often proteins). For example, the fungal product penicillin is too small by itself (molecular mass: 302 daltons) to trigger a specific immune response. However, bound to a carrier protein in the blood, penicillin can become antigenic and trigger an allergic response in some patients.

Complex molecules make better antigens than simple ones because they have more epitopes, like a gemstone with its many facets. For example, starch, which is a very large polymer of repeating glucose subunits, is not a good antigen despite its large size, because it lacks structural complexity. In contrast, complicated molecules, such as glycoproteins and phospholipids, have multiple distinctive shapes and novel combinations of subunits that cells of the immune system recognize as foreign.

Examples of antigens include components of bacterial cell walls, capsules, pili, and flagella, as well as the external and internal proteins of viruses, fungi, and protozoa. Many toxins and some nucleic acid molecules are also antigenic. Invading microorganisms are not the only source of antigens; for example, food may contain antigens called *allergens* that provoke allergic reactions, and inhaled dust contains mite feces, pollen grains, dander (flakes of skin), and other antigenic and allergenic particles. Chapter 18 covers allergies in more detail.

Types of Antigens

Though immunologists categorize antigens in various ways, one especially important way is to group antigens according to their relationship to the body:

- **Exogenous**[2] **antigens (Figure 16.3b)**. Exogenous (eks-oj'en-us) antigens include toxins and other secretions and components of microbial cell walls, membranes, flagella, and pili.
- **Endogenous**[3] **antigens (Figure 16.3c)**. Protozoa, fungi, bacteria, and viruses that reproduce inside a body's cells produce endogenous antigens. The immune system cannot

[1]From *anti*body *gen*erator; antibodies are proteins secreted during a humoral immune response that bind to specific regions of antigens.
[2]From Greek *exo*, meaning without, and *genein*, meaning to produce.
[3]From Greek *endon*, meaning within.

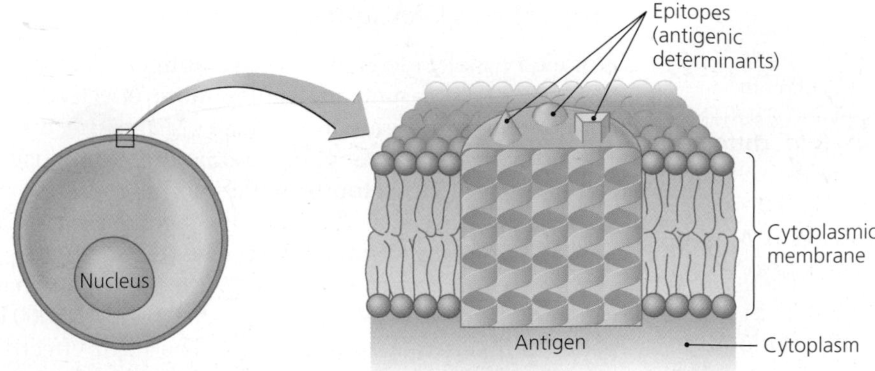

(a) Epitopes (antigenic determinants)

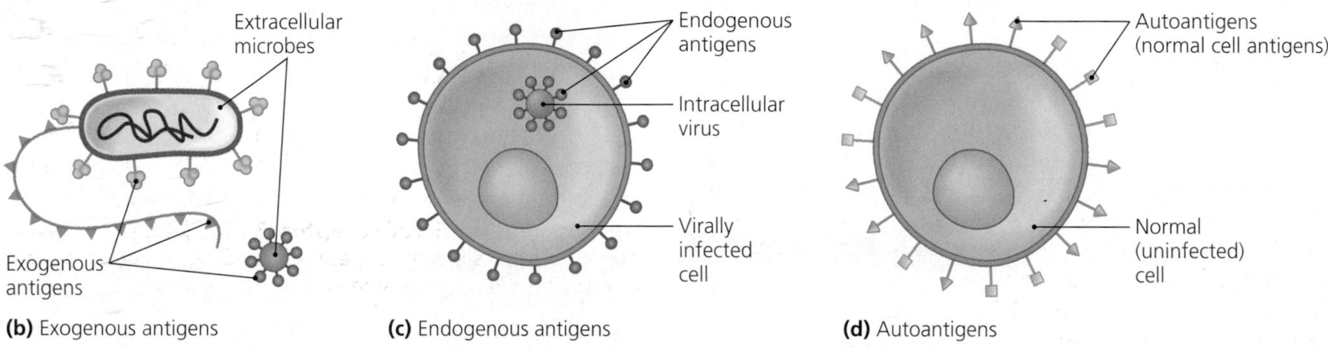

(b) Exogenous antigens **(c)** Endogenous antigens **(d)** Autoantigens

Figure 16.3 Antigens, molecules that provoke a specific immune response. **(a)** Epitopes, or antigenic determinants, are three-dimensional regions of antigens whose shapes are recognized by cells of the immune system. **(b–d)** Categories of antigens based on their relationship to the body. Exogenous antigens originate from microbes located outside the body's cells; endogenous antigens are produced by intracellular microbes and are typically incorporated into a host cell's cytoplasmic membrane; autoantigens are components of normal body cells.

assess the health of the body's cells; it responds to endogenous antigens only if the body's cells incorporate such antigens into their cytoplasmic membranes, leading to their external display.

- **Autoantigens**[4] **(Figure 16.3d).** Antigenic molecules derived from normal cellular processes are autoantigens (or *self-antigens*). As we will discuss more fully in a later section, immune cells that treat autoantigens as if they were foreign are normally eliminated during the development of the immune system. This phenomenon, called *self-tolerance,* prevents the body from mounting an immune response against itself.

So far, we have examined two aspects of immune responses in terms of an analogy to a stage play—the lymphoid organs and tissues provide the stage, and antigens of pathogens are the villains that induce an immune response with their epitopes (antigenic determinants). Next, we examine the activities of lymphocytes in more detail. Recall that B and T lymphocytes act against antigens; they are the "heroes" of the action. We begin by considering B lymphocytes.

[4]From Greek *autos,* meaning self.

B Lymphocytes (B cells) and Antibodies

Learning Objectives

- ✓ Describe the characteristic of B lymphocytes that furnishes them specificity.
- ✓ Describe the basic structure of an immunoglobulin molecule.
- ✓ Contrast the structure and function of the five classes of immunoglobulins.

B lymphocytes are found in the spleen, MALT, and primarily in the germinal centers of lymph nodes. A small percentage of B cells circulate in the blood. The major function of B cells is the secretion of soluble antibodies, which we examine in more detail shortly. As we have established, B cells function in humoral immune responses, each of which is against only a particular epitope. How is such specificity developed and maintained? Specificity comes from membrane proteins called *B cell receptors.*

Specificity of the B Cell Receptor (BCR)

The surface of each B lymphocyte is covered with about 500,000 identical copies of a protein called the **B cell receptor (BCR).** A BCR is a type of *immunoglobulin* (im′yū-nō-glob′yū-lin) *(Ig).*

A simple immunoglobulin contains four polypeptide chains—two identical longer chains called *heavy chains,* and two identical shorter *light chains* **(Figure 16.4)**. The terms *heavy* and *light* refer to their relative molecular masses. Disulfide bonds, which are covalent bonds between sulfur atoms in two different amino acids, link the light chains to the heavy chains in such a way that a simple immunoglobulin looks like the letter Y. A BCR has two *arms* and a *transmembrane portion.* Each arm's end is a *variable region,* because terminal ends of each heavy and each light chain vary in amino acid sequence among B cells. The transmembrane portion anchors the BCR in the cytoplasmic membrane and consists of the stem of the Y (composed of the tails of the two heavy chains) and two other polypeptides.

Each B cell randomly generates a single BCR gene during its development in the bone marrow. Scientists estimate that there are fewer than 25,000 genes in a human cell, yet there are billions of different BCR proteins in an individual. Obviously, a person cannot have a separate gene for each BCR; instead, a developing B cell randomly recombines segments from three immunoglobulin regions of DNA and combines these segments to develop novel BCR genes. B cells may also randomly change BCR genes to develop even more diversity. Each newly formed BCR gene codes for a specific and unique BCR. **Highlight: BCR Diversity: The Star of the Show** on p. 468 expands on the genetic explanation for the extensive BCR diversity.

All the BCRs of any particular cell are identical because the variable regions of every BCR on a single cell are identical—the two light chain variable regions are identical and the two heavy chain variable regions are identical. Together the two variable regions form **antigen-binding sites** (see Figure 16.4). Antigen-binding sites are complementary in shape to the three-dimensional shape of an epitope and bind precisely to it. Exact binding between antigen-binding site and epitope accounts for the specificity of a humoral immune response.

Though all of the BCRs on a single B cell are the same, the BCRs of one cell differ from the BCRs of all other B cells, much as each snowflake is distinct from all others. Scientists estimate that each person forms no less than 10^9 and likely as many as 10^{13} B lymphocytes—each with its own BCR. Since an antigen, for example a bacterial protein, typically has numerous epitopes of various shapes, many different BCRs will recognize any particular antigen's epitopes, but each BCR recognizes only one epitope. BCR genes are randomly generated in sufficient numbers that the entire repertoire of BCRs (each carried by a particular B lymphocyte) is capable of recognizing the entire repertoire of thousands of millions of different epitopes. In other words, at least one BCR is fortuitously complementary to any given specific epitope the body may or may not encounter.

An analogy will serve to clarify this point. Imagine a locksmith who has a copy of every possible key to fit every possible lock. If a customer arrives at the shop with a lock needing a key, the locksmith can provide it (though it may take a while to find the correct key). Similarly, you have a BCR complementary to every possible epitope in the environment, though you will not be infected with every epitope. For example, you have lymphocytes with BCRs complementary to epitopes of stingray venom, though it is unlikely you will ever be stabbed by a stingray.

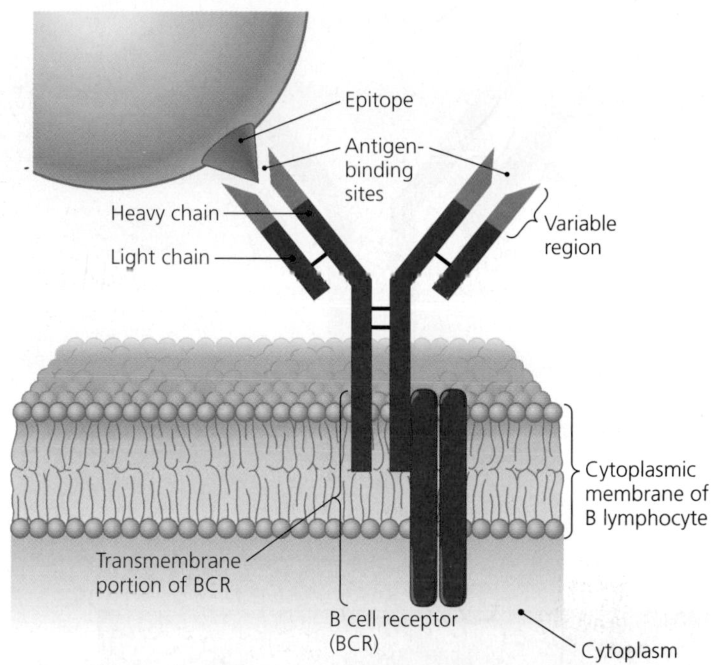

▲ **Figure 16.4 B cell receptor (BCR).** The B cell receptor is composed of a symmetrical, epitope-binding, Y-shaped protein in association with two transmembrane polypeptides.

When an antigenic epitope stimulates a specific B cell via the B cell's unique BCR, the B cell responds by undergoing cell division, giving rise to nearly identical offspring that secrete immunoglobulins into the blood or lymph. The immunoglobulins act against the epitope shape that stimulated the B cell. Activated, immunoglobulin-secreting B lymphocytes are called **plasma cells.** They have extensive rough endoplasmic reticulum and many Golgi bodies involved in the synthesis, packaging, and secreting of the immunoglobulins. Next we consider the structure and functions of the secreted immunoglobulins, which are called antibodies.

Specificity and Antibody Structure

Antibodies are immunoglobulins and similar to BCRs in shape, though antibodies are secreted and lack the bulk of the transmembrane portions of BCRs **(Figure 16.5)**. Thus, a basic antibody molecule is Y-shaped with two identical heavy chains and two identical light chains and is nearly identical to the secreting cell's BCR. The antigen-binding sites of antibodies from a given plasma cell are identical to one another and to the antigen-binding sites of that cell's BCR. Antibodies carry the same specificity for an epitope as the BCR of the activated B cell.

Because the arms of an antibody molecule contain antigen-binding sites, they are also known as the F_{ab} *regions* (*fragment, antigen-binding*). The angle between the F_{ab} regions and the stem can change because the point at which they join is hinge-like. An antibody stem, which is formed of the lower portions of the two heavy chains, is also called the F_c *region* (because it forms a *fragment* that is *crystallizable*).

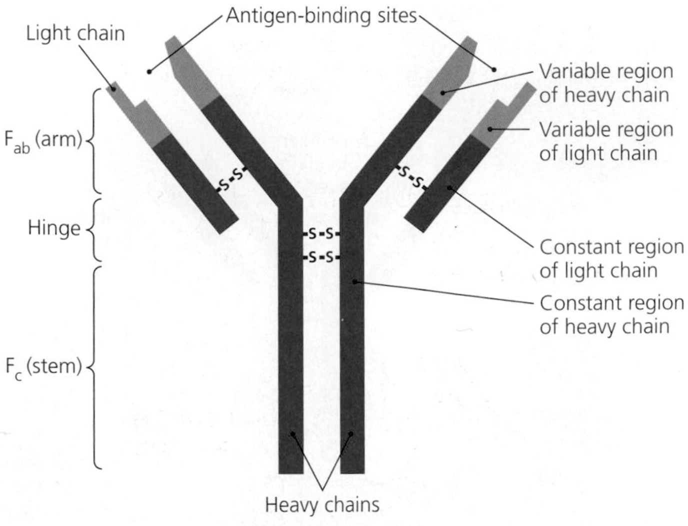

(a)

F_{ab} (arm)

F_c (stem)

(b)

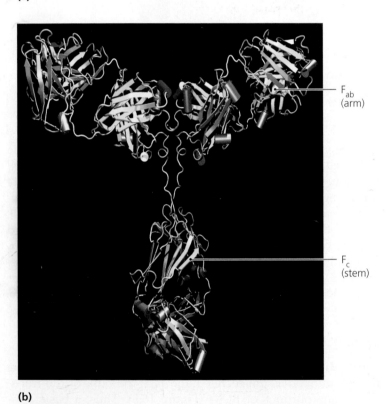

▲ **Figure 16.5 Basic antibody structure.** **(a)** Artist's rendition. Each antibody molecule, which is shaped like the letter Y, consists of two identical heavy chains and two identical light chains held together by disulfide bonds. Five different kinds of heavy chains form an antibody's stem (F_c region). The arms (F_{ab} regions) terminate in variable regions to form two antigen-binding sites. The hinge region is flexible, allowing the arms to bend in almost any direction. **(b)** Three-dimensional shape of an antibody based on X-ray crystallography. *Why are the two antigen-binding sites on a typical antibody molecule identical?*

Figure 16.5 The amino acid sequences of the two light chains of an antibody molecule are identical, as are the sequences of the two heavy chains, so the binding sites—each composed of the ends of one light chain and one heavy chain—must be identical.

There are five basic types of F_c region, designated by the Greek letters *mu, gamma, alpha, epsilon,* and *delta.* A plasma cell attaches the gene for its heavy chain variable region to one of the five F_c region genes to form one of five classes of antibodies known as IgM, IgG, IgA, IgE, and IgD.

Antibody Function

As we have seen, antigen-binding sites of antibodies are complementary to epitopes; in fact, the shapes of the two can match so very closely that most water molecules are excluded from the area of contact, producing a strong, noncovalent, hydrophobic interaction. Additionally, hydrogen bonds and other molecular attractions mediate antibody binding to epitope. The binding of antibody to epitope is the central functional feature of humoral adaptive immune responses. Once bound, antibodies function in several ways. These include activation of complement and inflammation, neutralization, opsonization, direct killing, agglutination, and antibody-dependent cytoxicity.

Activation of Complement and Inflammation F_c regions of two or more antibodies bind to complement protein 1 (C1), activating it to become enzymatic. This begins the classical complement pathway, which releases inflammatory mediators. Some types of antibodies leave the blood during inflammation. Chapter 15 more fully examines the defense reactions of complement activation and inflammation (see Figures 15.9 and 15.12).

Neutralization Antibodies can neutralize a toxin by binding to a critical portion of the toxin so that it can no longer function against the body. Similarly, antibodies can block adhesion molecules on the surface of a bacterium or virus, neutralizing the pathogen's virulence because it cannot adhere to its target cell **(Figure 16.6a)**.

Opsonization Antibodies act as **opsonins**[5]—molecules that stimulate phagocytosis. Neutrophils and macrophages have receptors for the F_c regions of IgG molecules; therefore, these leukocytes bind to the stems of antibodies. Once antibodies are so bound, the leukocytes phagocytize them, along with the antigens they carry, at a faster rate compared to antigens lacking bound antibody. Changing the surface of an antigen so as to enhance phagocytosis is called **opsonization** (op′sŏ-nī-zā′shŭn) **(Figure 16.6b)**.

Killing by Oxidation Recently, scientists have shown that some antibodies have catalytic properties that allow them to kill bacteria directly **(Figure 16.6c)**. Specifically, antibodies catalyze the production of hydrogen peroxide, ozone, and other potent oxidants that kill bacteria.

Agglutination Because each basic antibody has two antigen-binding sites, each can attach to two epitopes at once. Numerous antibodies can aggregate antigens together—a state called **agglutination** (ă-glū-ti-nā′shŭn) **(Figure 16.6d)**. Agglutination

[5]From Greek *opsonein*, meaning to supply food, and *izein*, meaning to cause; thus, loosely, "to prepare for dinner."

of soluble molecules typically causes them to become insoluble and precipitate. Agglutination may hinder the activity of pathogenic organisms and increases the chance that they will be phagocytized or filtered out of the blood by the spleen. Chapter 17 examines some uses scientists make of the agglutinating nature of antibodies.

Antibody-Dependent Cellular Cytoxicity (ADCC) Antibodies often coat a target cell by binding to epitopes all over the target's surface. The antibodies' F_c regions can then bind to receptors on special lymphocytes called *natural killer lymphocytes (NK cells)*, which are neither B nor T cells. NK lymphocytes lyse target cells with proteins called **perforin** (per′fōr-in) and **granzyme** (gran′zīm). Perforin molecules form into a tubular structure in the target cell's membrane, forming a channel through which granzyme enters the cell and triggers **apoptosis**[6] (programmed cell suicide) **(Figure 16.6e)**. ADCC is similar to opsonization in that antibodies cover the target cell; however, with ADCC the target dies by apoptosis, whereas with opsonization the target is phagocytized. **ANIMATIONS:** ***Humoral Immunity: Antibody Function***

Classes of Antibodies

Threats confronting the immune system can be extremely variable, so it is not surprising that there are several classes of antibody. The class involved in any given humoral immune response depends on the type of invading foreign antigens, the portal of entry involved, and the antibody function required. Here, we consider the structure and functions of the five classes of antibodies.

Every B cell begins by attaching its variable region gene to the gene for the mu F_c region and thus begins by making class M—**immunoglobulin M (IgM).** Some simple IgM molecules are put into the cytoplasmic membrane to act as a BCR, but most IgM is secreted during the initial stages of an immune response. A secreted IgM molecule is more than five times larger than the basic Y-shape because secreted IgM is a pentamer, consisting of five basic units linked together in a circular fashion via disulfide bonds and a short polypeptide *joining (J) chain* (see Table 16.1 on p. 470). Each IgM subunit has a conventional immunoglobulin structure, consisting of two light chains and two mu heavy chains. IgM is most efficient at complement activation, which also triggers inflammation, and can be involved in agglutination and neutralization.

In a process called **class switching,** a plasma cell then combines its variable region gene to the gene for a different F_c region and begins secreting a new class of antibodies. The most common switch is to the gene for heavy chain gamma; that is, the plasma cell switches to synthesizing immunoglobulin G.

Immunoglobulin G (IgG) is the most common and longest-lasting class of antibody in the blood, accounting for about 80% of serum antibodies, possibly because IgG has many functions. Each molecule of IgG has the basic Y-shaped antibody structure.

[6]Greek, meaning falling off.

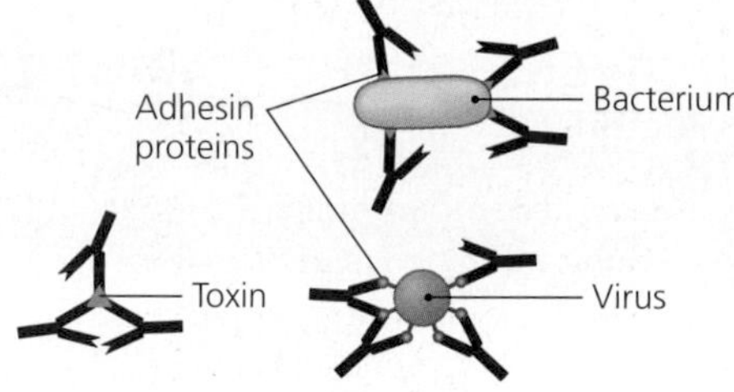

(a) Neutralization

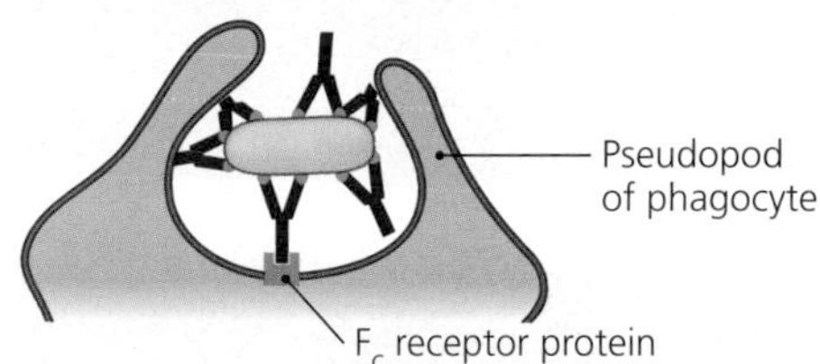

(b) Opsonization

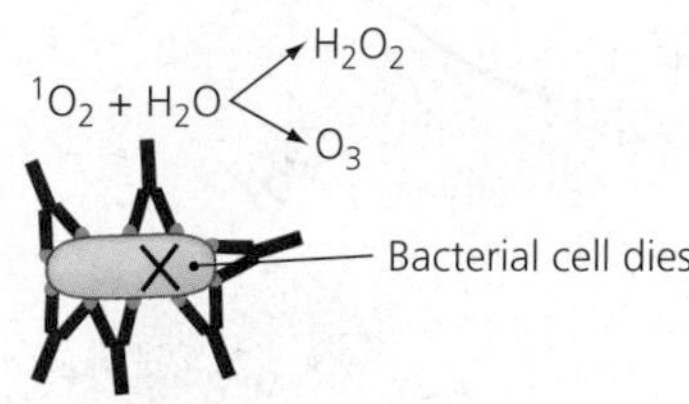

(c) Oxidation

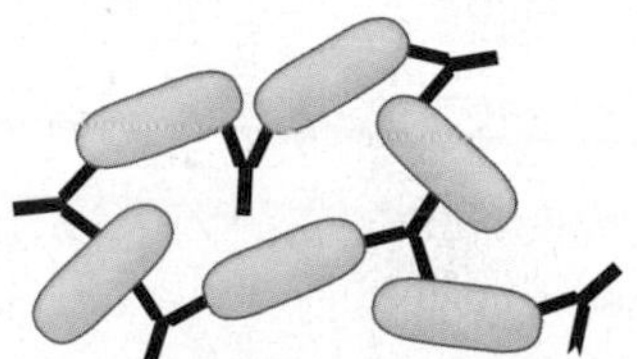

(d) Agglutination

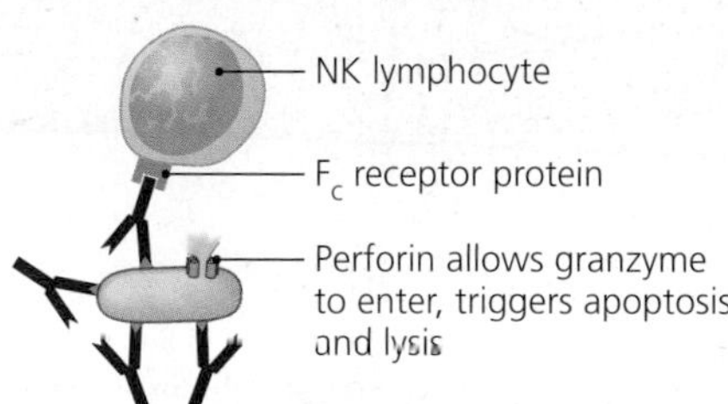

(e) Antibody-dependent cellular cytotoxicity (ADCC)

▲ **Figure 16.6 Four functions of antibodies.** Drawings are not to scale. **(a)** Neutralization of toxins and microbes. **(b)** Opsonization. **(c)** Oxidation by toxic forms of oxygen such as $^1O_2$ (singlet oxygen), H_2O_2 (hydrogen peroxide), and O_3 (ozone). **(d)** Agglutination. **(e)** Antibody-dependent cellular cytotoxicity. Two other functions of antibodies—activation of complement and participation in inflammation—are illustrated in Chapter 15.

HIGHLIGHT

BCR DIVERSITY: THE STAR OF THE SHOW

How does your body generate billions of unique B cell receptor proteins (BCRs) so as to recognize the thousand of millions of foreign epitopes on pathogens that attack you? There isn't enough DNA in a cell to have individual genes for so many receptors. It would require several thousand times more DNA than you have. The answer to the problem of BCR diversity lies in the ingenious way in which B cells use their relatively small number of diverse BCR genes.

Genes for BCRs occur in discrete stretches of DNA called *loci* (singular: *locus*). Genes for the constant region are located downstream from loci for the variable regions. We will consider the variable region genes first, because it is in the variable region that BCR diversity is greatest.

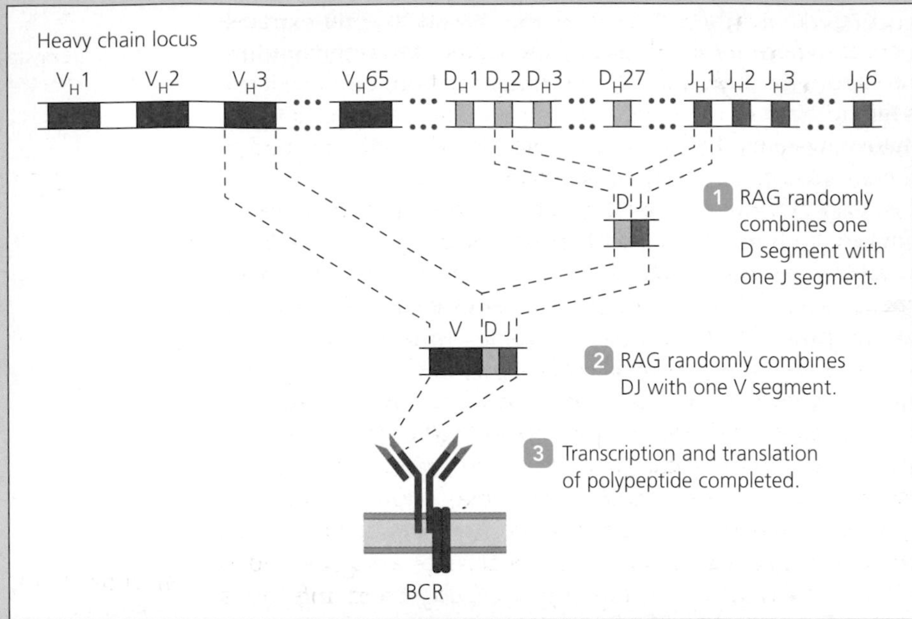
Heavy chain locus
V_H1 V_H2 V_H3 V_H65 D_H1 D_H2 D_H3 D_H27 J_H1 J_H2 J_H3 J_H6
D J
1 RAG randomly combines one D segment with one J segment.
V D J
2 RAG randomly combines DJ with one V segment.
3 Transcription and translation of polypeptide completed.
BCR

BCR variable region genes occur in three loci on each of two chromosomes—one locus for the heavy chain variable region and two loci (called *kappa* and *lambda*) for the light chain variable region. Each cell is diploid—having two of each type of chromosome—so there are six loci all together. However, a developing B cell uses only one chromosome's loci for each of its heavy and light chains. Once a locus on a particular chromosome is used, the other chromosome's corresponding locus is inhibited.

Each locus is divided further into distinct genetic segments coding for portions of its respective chain. The heavy chain variable region locus has three segments called *variable (V_H), diversity (D_H),* and *junction (J_H) segments.* In each of your developing B cells there are 65 variable segment genes, 27 diversity segment genes, and 6 junction segment genes. Each light chain variable region locus has an additional *variable segment (V_L)* and a *junction segment (J_L).* For the kappa locus, there are 40 variable segment genes and 5 junction genes. The lambda locus is less variable, having only 30 V genes and a single J gene.

A great variety of BCRs form during B cell maturation as the B cell uses an enzyme called RAG—the *recombination activating gene* protein—to randomly combine one of each kind of the various segments to form its BCR gene. An analogy will help your understanding of this concept. Imagine that you have a wardrobe consisting of 65 different pairs of shoes, 27 different shirts, and 6 different pairs of pants that can be combined to create outfits. You can choose any one pair of shoes, any one shirt, and any one pair of pants each day, so with these 98 pieces of clothing, you will potentially have 10,530 different outfits ($65 \times 27 \times 6$). Your roommate might have the same numbers of items, but in different colors and styles; so between you, you will have twice as many potential outfits—21,060! Similarly, each developing B cell has 10,530 possible combinations of V_H, D_H, and J_H segments on each of its two chromosomes for a total of 21,060 possible heavy chain variable region gene combinations.

The developing B cell also uses RAG to recombine the segments of the light chain variable loci. Since there are 200 (40×5) possible kappa genes and 30 (30×1) possible lambda genes on each chromosome, there are a total of 460 possible light chain genes on the two chromosomes. Therefore, each B cell can make one of a possible 9,687,600 ($21{,}060 \times 460$) different BCRs using only 348 genetic segments in the heavy and light chain loci on its two chromosomes.

Still, these nearly 10 million different BCRs do not account for all the variability seen among BCRs. Each B cell creates additional diversity: RAG randomly removes portions of D and J segments before joining the two together. Another enzyme randomly adds nucleotides to each heavy chain VDJ combination, and B cells in the germinal centers of lymph nodes undergo random point mutations in their V regions. The result of RAG recombinations, random deletions, random insertions, and point mutations is tremendous potential variability. Scientists estimate that you may have about 10^{23} B cell receptors possibilities. That's one hundred billion trillion—a number ten times greater than all the stars in the universe! Your B cells are indeed stars of the show.

IgG molecules play a major role in antibody-mediated defense mechanisms, including complement activation, opsonization, neutralization, and antibody-dependent cellular cytotoxicity (ADCC). IgG molecules can leave blood vessels to enter extracellular spaces more easily than can the other immunoglobulins. This is especially important during inflammation, because it enables IgG to bind to invading pathogens before they get into the circulatory systems. IgG molecules are also the only antibodies that cross a placenta to protect a developing child.

Immunoglobulin A (IgA), which has alpha heavy chains, is the immunoglobulin most closely associated with various body secretions. Some of the body's IgA is a monomer with the basic Y-shape that circulates in the blood, constituting about 12% of total serum antibody. However, plasma cells in the tear ducts, mammary glands, and mucous membranes synthesize **secretory IgA,** which is composed of two monomeric IgA molecules linked via a J chain and another short polypeptide (called a *secretory component*). Plasma cells add secretory component during the transport of secretory IgA across mucous membranes. Secretory component protects secretory IgA from digestion by intestinal enzymes.

Secretory IgA agglutinates and neutralizes antigens and is of critical importance in protecting the body from infections arising in the gastrointestinal, respiratory, urinary, and reproductive tracts. IgA provides nursing newborns some protection against foreign antigens because mammary glands secrete IgA into milk. Thus, nursing babies receive antibodies directed against antigens that have infected their mothers and are likely to infect them as well.

Immunoglobulin E (IgE) is a typical Y-shaped immunoglobulin with two epsilon heavy chains. Because it is found in extremely low concentrations in serum (less than 1% of total antibody), it is not critical for most antibody functions. Instead, IgE antibodies act as signal molecules—they attach to receptors on eosinophil cytoplasmic membranes to trigger the release of cell-damaging molecules onto the surface of parasites, particularly parasitic worms. IgE antibodies also trigger mast cells and basophils to release inflammatory chemicals such as histamine. In developed countries, IgE is more likely associated with allergies than with parasitic worms.

Immunoglobulin D (IgD) is characterized by delta heavy chains. IgD molecules are not secreted but are membrane-bound antigen receptors on B cells that are often seen during the initial phases of a humoral immune response. In this regard, IgD antibodies are like BCRs. Not all mammals have IgD, and animals that lack IgD show no observable ill effects; therefore, scientists do not know the exact function or importance of this class of antibody.

Table 16.1 on p. 470 compares the different classes of immunoglobulins and adds some details of antibody structure.

CRITICAL **THINKING**

Two students are studying for an exam on the body's defensive systems. One of them insists that complement is part of the nonspecific second line of defense, but the partner insists that complement is part of a humoral immune response in the third line of defense. How would you explain to them that they are both correct?

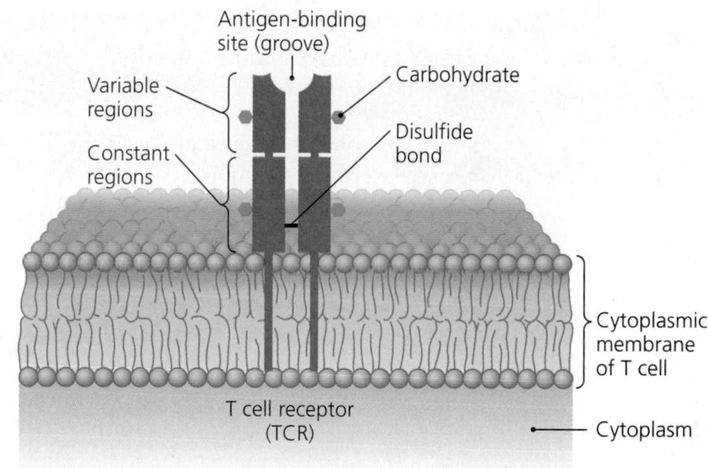

▲ **Figure 16.7 A T cell receptor (TCR).** A TCR is an asymmetrical surface molecule composed of two polypeptides containing a single antigen-binding site between them.

In most cases, B cells do not act alone to mount a humoral immune response. Instead, they respond to antigens only with the assistance of certain T lymphocytes. We will continue discussion of the details of humoral immune responses after we consider the various T cells.

T Lymphocytes (T Cells)

Learning Objectives

- ✓ Describe the importance of the thymus to the development of T lymphocytes.
- ✓ Describe the basic characteristics common to T lymphocytes.
- ✓ Compare and contrast three types of T cells.

A human adult's red bone marrow produces T lymphocytes. Sticky, adhesive molecules coupled with the action of chemotactic molecules attract these lymphocytes to the thymus from blood vessels. T cells mature in the thymus under the influence of molecular signals from the thymus. Following maturation, T cells circulate in the lymph and blood and migrate to the lymph nodes, spleen, and Peyer's patches. They account for about 70–85% of all lymphocytes in the blood.

T lymphocytes are like B cells in their specificity. Each T cell has about half a million copies of a **T cell receptor (TCR)** on its cytoplasmic membrane. Each T cell randomly chooses and combines segments of DNA in TCR gene regions to create a novel gene that codes for that cell's unique and specific TCR. The random production of diverse TCRs accomplished by recombining TCR genes is similar to the way BCRs are produced.

Specificity of the T Cell Receptor (TCR)

In contrast to a BCR, a TCR is composed of only two glycopolypeptide chains; however, like BCRs, the terminal ends of TCRs are composed of variable regions that grant each cell's TCR specific binding properties. The groove between the polypeptides acts as the antigen-binding site (**Figure 16.7**). Similarly to binding by a BCR, a TCR recognizes and binds to a

TABLE 16.1 Characteristics of the Five Classes of Antibodies

	IgG	IgA	IgM	IgE	IgD
	Antigen-binding site; Disulfide bond; γ; Carbohydrate	α; Monomer; Secretory component; J Chain; α; Secretory (dimer)	μ; J Chain	ε	δ
Structure, number of binding sites	Monomer, 2	Monomer, 2 Dimer, 4	Pentamer, 10	Monomer, 2	Monomer, 2
Type of heavy chain	Gamma (γ)	Alpha (α)	Mu (μ)	Epsilon (ε)	Delta (δ)
Functions	Complement activation, neutralization, opsonization, production of hydrogen peroxide, agglutination, and antibody-dependent cellular toxicity (ADCC); crosses placenta to protect fetus	Neutralization and agglutination; dimer is secretory antibody	Monomer can act as BCR; pentamer acts in complement activation, neutralization, agglutination	Triggers release of antiparasitic molecules from eosinophils and of histamines from basophils and mast cells (allergic reactions)	Unknown, but perhaps acts as BCR
Locations	Serum, mast cell surfaces	Monomer: serum Dimer: mucous membrane secretions (e.g. tears, saliva, mucus); milk	Serum, B cell surface	Serum, mast cell surfaces	B cell surface
Approximate half-life (time it takes for concentration to reduce by half) in blood	20 days	6 days	10 days	2 days	3 days
Percentage in serum	80	10–15	5–10	<1	<0.05
Size (mass in kilodaltons)	150	Monomer: 160 Dimer: 385	970	188	184

complementary shape, but TCRs do not recognize epitopes directly. A TCR binds only to an epitope associated with a particular protein called *MHC protein,* which will be discussed shortly. There are at least 10^9 different TCRs—each specific TCR type on a different T cell—which is enough for every epitope–MHC protein.

T lymphocytes act primarily against body cells that harbor intracellular pathogens such as viruses, though they also can act against body cells that produce abnormal cell-surface proteins (such as cancer cells). Because T cells act directly against antigens (they do not secrete immunoglobulins), their immune activities are called cell-mediated immune responses.

Types of T Lymphocytes

Immunologists recognize types of T cells based on surface glycoproteins and characteristic functions. These are *cytotoxic T cells, helper T cells,* and *regulatory T cells.*

Cytotoxic T Lymphocyte Every **cytotoxic T cell (Tc or CD8 cell)** is distinguished by copies of its own unique TCR as well as the presence of CD8 cell-surface glycoprotein. CD (for *cluster of differentiation*) glycoproteins are named with internationally accepted designations consisting of a number following the prefix. These numbers reflect the order in which the glycoproteins were discovered, not the order in which they are produced or function. As the name *cytotoxic T cell* implies, these lymphocytes directly kill other cells—those infected with viruses and other intracellular pathogens, as well as abnormal cells, such as cancer cells.

Helper T Lymphocyte Immunologists distinguish **helper T cells (Th or CD4 cells)** by the presence of the CD4 glycoproteins. These cells are called "helpers" because their function is to assist in regulating the activity of B cells and cytotoxic T cells during immune responses by providing necessary signals and growth factors. There are two main subpopulations of helper T cells: *type 1 helper T cells (Th1 cells),* which assist cytotoxic T cells and innate macrophages, and *type 2 helper T cells (Th2 cells),* which function in conjunction with B cells. Immunologists distinguish between Th1 and Th2 cells on the basis of their secretions and by characteristic cell-surface proteins.

Helper T cells secrete various soluble protein messengers called *cytokines* that regulate the entire immune system, both adaptive and innate portions. We will consider the types and effects of cytokines shortly. **Highlight: The Loss of Helper T Cells in AIDS Patients** describes some of the effects of the destruction of helper T cells in individuals infected with the human immunodeficiency virus (HIV). **ANIMATIONS:** ***Cell-Mediated Immunity: Helper T Cells***

Regulatory T Lymphocyte **Regulatory T cells (Tr cells),** previously known as *suppressor T cells,* repress adaptive immune responses and prevent autoimmune diseases. Tr cells express CD4 and CD25 glycoproteins. Scientists have not fully characterized the manner in which Tr cells work, but it is known that they are activated by contact with other immune cells and they secrete some immunologically active chemicals called *cytokines,* which we examine in a subsequent section.

Table 16.2 on p. 472 compares and contrasts the features of various types of lymphocytes. We have considered the primary lymphocytes involved in immune responses. Now, we turn our attention to the way in which the body eliminates T cells and B cells that recognize normal body antigens by means of their BCRs and TCRs, respectively.

Clonal Deletion

Learning Objectives

✓ Describe apoptosis and explain its role in lymphocyte editing by clonal deletion.

✓ Compare and contrast clonal deletion of T cells and clonal deletion of B cells.

Given that both T and B lymphocytes randomly generate the shapes of their receptors (TCRs and BCRs respectively), every population of maturing lymphocytes includes numerous cells with receptors complementary to normal body components—the autoantigens mentioned earlier. It is vitally important that specific immune responses not be directed against autoantigens; the immune system must be tolerant of "self." When self-tolerance is impaired, the result is an *autoimmune disease,* which we examine in Chapter 18.

The body eliminates self-reactive lymphocytes via **clonal deletion,** so named because elimination of a cell deletes its potential offspring (clones). In this process, lymphocytes are exposed to autoantigens, and those lymphocytes that react to autoantigens undergo apoptosis (programmed cell suicide) and

HIGHLIGHT

THE LOSS OF HELPER T CELLS IN AIDS PATIENTS

Neither B cells nor cytotoxic T cells respond effectively to most antigens without the participation of helper T cells, because signals are passed more effectively among leukocytes when the CD4 molecules of helper T cells bind to certain leukocytes.

We also saw in Chapter 13 that in order for viruses to enter cells and replicate, they must first attach to a specific protein on their target cells' surface. In the case of human immunodeficiency virus (HIV), the causative agent of AIDS, the surface protein to which it attaches is CD4, and its target cells are helper T cells. Once HIV enters a helper T cell, it, like other viruses, takes over the cell's protein-synthesizing machinery and directs the cell to produce more viruses.

Because they are essential to mounting an effective immune response, the body normally has a surplus of helper T cells—usually about three times as many as it needs. As a result, individuals infected with HIV can typically lose up to about two-thirds of their helper T cells before signs of immune deficiency appear. Cell-mediated immunity is usually affected first; even though B cells require the assistance of helper T cells to function optimally, they can still be stimulated directly by large quantities of antigen. Normal human blood typically contains about three CD4 helper T cells for every two CD8 cytotoxic T cells—a CD4 to CD8 ratio of 3:2. During the course of HIV infection, however, CD4 helper T cells are lost, reducing the CD4 to CD8 ratio such that individuals with full-blown AIDS may have a ratio lower than 1:7.

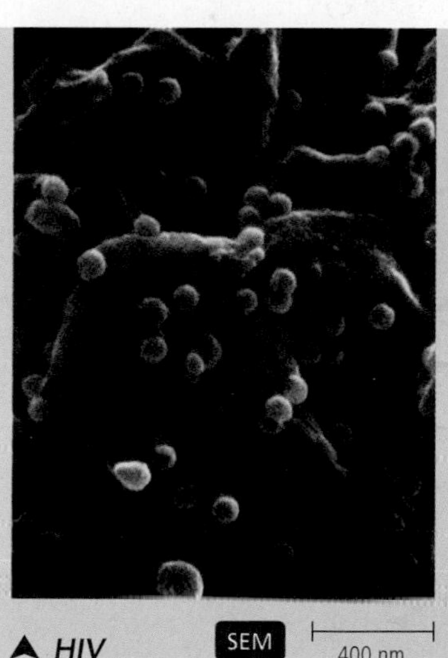

▲ *HIV budding from Th cells.*

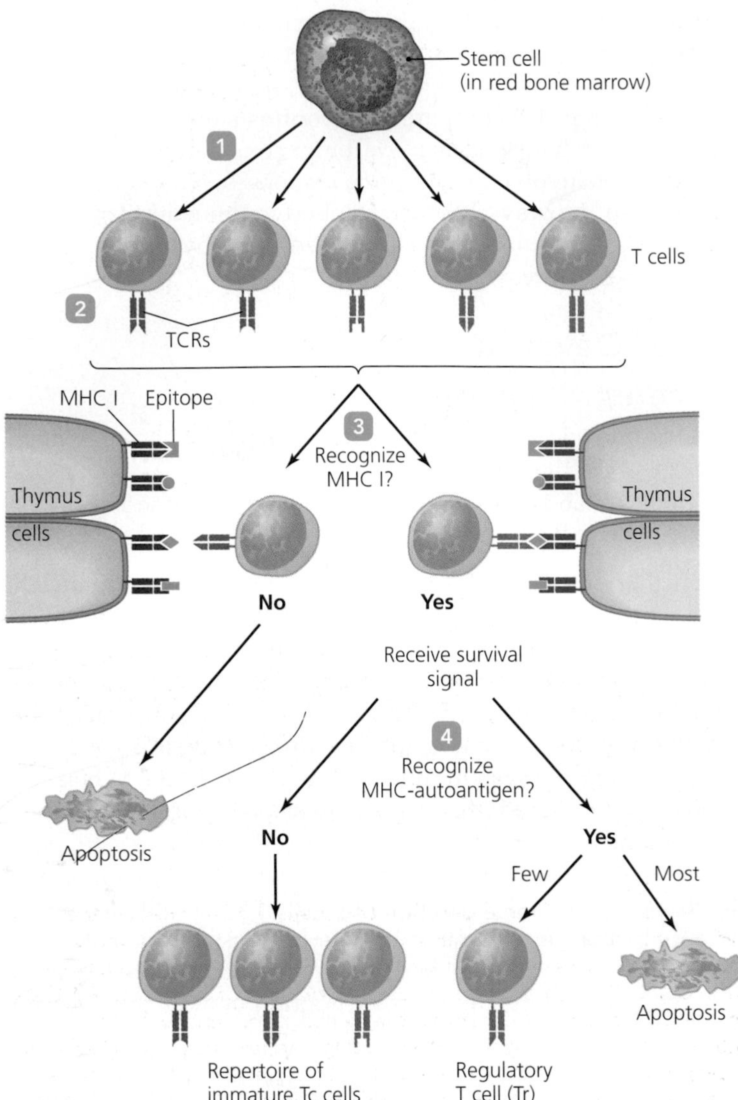

▲ **Figure 16.8 Clonal deletion of T cells.** 1 Stem cells in the red bone marrow generate a host of lymphocytes that move to the thymus. 2 In the thymus, each lymphocyte randomly generates a TCR with a particular shape. Note that each cell's TCR binding sites differ from those of other cells. 3 T cells pass through a series of "decision questions" in the thymus: Are their TCRs complementary to the body's MHC I protein? If "no," they undergo apoptosis—*clonal deletion*. If "yes," they survive. 4 Do the surviving cells recognize MHC I protein bound to any autoantigen? If "no," they survive and become the repertoire of immature T cells. If "yes," then most undergo apoptosis (more clonal deletion); a few survive as regulatory T cells (Tr).

are thereby deleted from the repertoire of lymphocytes. Apoptosis is the critical feature of clonal deletion and the development of self-tolerance. The result of clonal deletion is that surviving lymphocytes respond only to foreign antigens.

In humans, clonal deletion occurs in the thymus for T lymphocytes and in the bone marrow for B lymphocytes. Lymphocyte clonal deletion is slightly different in T cells than it is in B cells. We will examine each in turn.

Clonal Deletion of T Cells

As we have seen, T cells recognize epitopes only when the epitopes are bound to MHC protein. Young T lymphocytes spend about a week in the thymus being exposed to all of the body's natural epitopes though a unique feature of thymus cells. As a group, these cells express all of the body's normal proteins, including proteins that have no function in the thymus. For example, some thymus cells synthesize lysozyme, hemoglobin, and muscle cell proteins, though these proteins are not expressed externally to the cells. Rather, thymus cells process these autoantigens so as to express their epitopes in association with an MHC protein. Since the cells collectively synthesize polypeptides from the body's proteins, together they process and present all the body's autoantigens to young T cells.

Immature T cells undergo one of four fates **(Figure 16.8)**:

- Those T cells that do not recognize the body's MHC protein undergo apoptosis, in other words, clonal deletion. Since they do not recognize the body's own MHC protein, they will be of no use identifying foreign epitopes carried by MHC protein. T cells that do recognize the body's MHC protein receive a signal to survive.
- Those that subsequently recognize autoantigen in conjunction with MHC protein mostly die by apoptosis, further clonal deletion.
- A few of these "self-recognizing" T cells remain alive to become regulatory T cells.
- The remaining T cells are those that will recognize the body's own MHC protein in conjunction with foreign epitopes and not with autoantigens. These T lymphocytes become the repertoire of protective T cells, which leave the thymus to circulate in the blood and lymph.

TABLE 16.2 Characteristics of Selected Lymphocytes

Lymphocyte	Site of Maturation	Representative Cell-Surface Glycoproteins	Selected Secretions
B cell	Red bone marrow	CD40 and distinctive BCR	Antibodies
Helper T cell type 1 (Th1)	Thymus	CD4, CCR5, and distinctive TCR	Interleukin 2, IFN-γ
Helper T cell type 2 (Th2)	Thymus	CD4, CCR3, CCR4, and distinctive TCR	Interleukin 4
Cytotoxic T cell (Tc)	Thymus	CD8, CD95L, and distinctive TCR	Perforin, granzyme
Regulatory T cell (Tr)	Thymus	CD4, CD25, and distinctive TCR	Cytokines such as interleukin 10

Clonal Deletion of B Cells

Clonal deletion of B cells occurs in the bone marrow in a similar fashion **(Figure 16.9)**, though self-reactive B cells may become inactive or change their BCR rather than undergo apoptosis. In any case, self-reactive B cells are removed from the active B cell repertoire, so that humoral immunity does not act against autoantigens. Tolerant B cells leave the bone marrow and travel to the spleen where they undergo further maturation before circulating in the blood and lymph.

Surviving B and T lymphocytes move into the blood and lymph, where they form the lymphocyte repertoire that scans for antigens. They communicate among one another and with other body cells via chemical signals called *cytokines.* In terms of our analogy of a stage play, there is dialogue among the cast of characters, but in our story the dialogue consists of chemical signals.

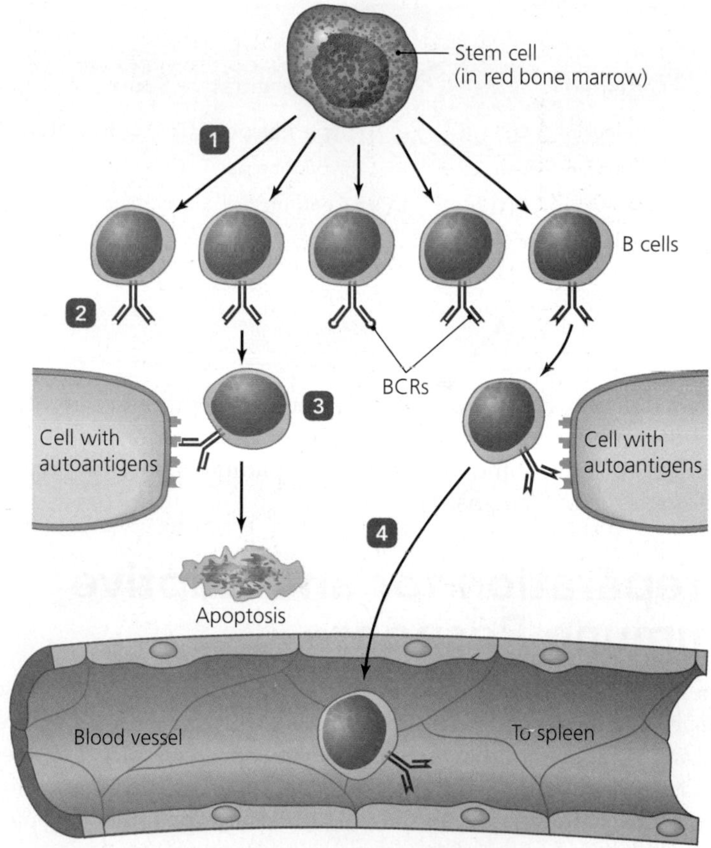

▲ **Figure 16.9 Clonal deletion of B cells.** **1** Stem cells in the red bone marrow generate a host of B lymphocytes. **2** Each newly formed B cell randomly generates a BCR with a particular shape. Note that each cell's BCR binding sites differ from those of other cells. **3** Cells whose BCR is complementary to some autoantigen bind with that autoantigen, stimulating the cell to undergo apoptosis. Thus, an entire set of potential daughter B cells (a clone) that are reactive with the body's own cells are eliminated—*clonal deletion.* **4** B cells with a BCR that is not complementary to any autoantigen are released from the bone marrow and into the blood. *Of the B cells shown, which is likely to undergo apoptosis?*

Figure 16.9 *The second and third B lymphocytes from the left will undergo apoptosis; their active sites are complementary to the second and fourth autoantigens from the top, respectively.*

Immune Response Cytokines

Learning Objective

✓ Describe five types of cytokines.

Cytokines (sī′tō-kīnz) are soluble regulatory proteins that act as intercellular messages when released by certain body cells including those of the kidney, skin, and immunity. Here we are concerned with cytokines that signal among various immune leukocytes. For example, cytotoxic T cells (Tc) do not respond to antigens unless they are first signaled by cytokines.

Immune system cytokines are secreted by various leukocytes and affect diverse cells. Many cytokines are redundant; that is, they have almost identical effects. Such complexity has given rise to the concept of a *cytokine network*—a complex web of signals among all the cell types of the immune system. The nomenclature of cytokines is not based on a systematic relationship among them; instead, scientists named cytokines after their cells of origin, their function, and/or the order in which they were discovered. Cytokines of the immune system include the following substances:

- **Interleukins**[7] (in-ter-lū′kinz) **(ILs).** As their name suggests, ILs signal among leukocytes, though cells other than leukocytes may also use interleukins. Immunologists named interleukins sequentially as they were discovered. Currently, scientists have identified about 35 interleukins.
- **Interferons** (in-ter-fēr′onz) **(IFNs).** These antiviral proteins, discussed in Chapter 15, may also act as cytokines. The most important interferon with such a dual function is gamma interferon (IFN-γ), which is a potent phagocytic activator secreted by Th1 cells.
- **Growth factors.** These proteins stimulate leukocyte stem cells to divide, ensuring that the body is supplied with sufficient white blood cells of all types. The body can control the progression of an adaptive immune response by limiting the production of growth factors.
- **Tumor necrosis**[8] **factor (TNF).** Macrophages and T cells secrete TNF to kill tumor cells and to regulate immune responses and inflammation.
- **Chemokines** (kē′mō-kīnz). Chemokines are chemotactic cytokines; that is, they signal leukocytes to move—for example, to rush to a site of inflammation or infection, or to move within tissues.

Table 16.3 on p. 474 summarizes some properties of selected cytokines.

We have been considering adaptive immunity as a stage play. To this point, we have examined the "stage" and the "cast of characters" involved in adaptive immunity. The latter are: antigens with their epitopes, B cells with their BCRs, plasma cells and their antibodies, and T cells with their TCRs. The dialogue consists of cytokines. Actors who would disrupt the play—those T cells and B cells that recognize autoantigens—have been eliminated by

[7]From Latin *inter,* meaning between, and Greek *leukos,* meaning white.
[8]From Latin *necare,* meaning to kill.

TABLE 16.3 Selected Immune Response Cytokines

Cytokine	Representative Source	Representative Target	Representative Action
Interleukin 2 (IL-2)	Type 1 helper T (Th1) cell, cytotoxic T (Tc) cell	Tc cell	Cloning of Tc cell
Interleukin 4 (IL-4)	Type 2 helper T (Th2) cell	B cell	B cell differentiates into plasma cell
Interleukin 12 (IL-12)	Dendritic cell	Helper T (Th) cell	Th cell differentiates into Th1 cell
Gamma interferon (IFN-γ)	Th1 cell	Macrophage	Increases phagocytosis
Tumor necrosis factor (TNF)	Macrophages, T cells	Body tissues	Triggers inflammation or apoptosis

clonal deletion. Before we finish examining the processes involved in humoral immune and cell-mediated immune responses, we need to consider some other initial preparations the body makes before the "play" begins.

Preparation for an Adaptive Immune Response

The body equips itself for specific immune responses by making *major histocompatibility complex proteins* (MHC proteins) and processing antigens so that T lymphocytes can recognize epitopes. We will examine each of these preparations of adaptive immunity, beginning with the roles of the major histocompatibility complex.

The Roles of the Major Histocompatibility Complex

Learning Objective

✓ Describe the two classes of major histocompatibility complex (MHC) proteins with regard to their location and function.

When scientists first began grafting tissue from one animal into another so as to determine a method for treating burn victims, they discovered that if the animals were not closely related, the recipients swiftly rejected the grafts. When they analyzed the reason for such rapid rejection, they found that a graft recipient mounted a very strong immune response against a specific type of antigen found on the cells of unrelated grafts. When the antigen on a graft's cells was sufficiently dissimilar from antigens on the host's cells, as seen with unrelated animals, the grafts were rejected. This is how scientists came to understand how the body is able to distinguish "self" from "nonself."

Immunologists named these types of antigens *major histocompatibility*[9] *antigens* to indicate their importance in determining the compatibility of tissues in successful grafting. Further research revealed that major histocompatibility antigens are glycoproteins found in the membranes of most cells of vertebrate animals. Major histocompatibility antigens are coded by a cluster of genes called the **major histocompatibility complex (MHC).** In humans, an MHC is located on each copy of chromosome 6.

Because organ grafting is a modern surgical procedure with no counterpart in nature, scientists reasoned that MHC proteins must have some other "real" function. Indeed, immunologists have determined that MHC proteins in cytoplasmic membranes function to hold and position epitopes for presentation to T cells. Each MHC molecule has an *antigen-binding groove* that lies between two polypeptides. Inherited variations in the amino acid sequences of the polypeptides modify the shapes of MHC binding sites and determine which epitopes can be bound and presented. It is important to recall that TCRs recognize only epitopes that are bound to MHC molecules.

MHC proteins are of two classes **(Figure 16.10)**. Class I MHC molecules are found on the cytoplasmic membranes of all nucleated cells. Thus red blood cells, which do not have nuclei, do not express MHC proteins, whereas cells such as those of skin and muscles do express MHC class I molecules. Special cells called **antigen-presenting cells (APCs)** also have class II MHC proteins. Professional antigen-presenting cells—those that regularly present antigen—are B cells, macrophages, and most importantly **dendritic cells,** which are so named because they have many long, thin cytoplasmic processes called *dendrites*[10] **(Figure 16.11)**. Phagocytic dendritic cells are found under the surface of the skin and mucous membranes. Some dendritic cells extend dentrites to a mucous surface to sample antigens, much like a submarine extends a periscope to get a view of surface activity. After acquiring antigens, dendritic cells migrate to lymph nodes to interact with B and T lymphocytes. Certain other phagocytes, such as *microglia* in the brain and *Kupffer cells* in the liver, may also present antigen under certain conditions. These cells are termed *nonprofessional antigen-presenting cells.*

The cytoplasmic membrane of a professional APC has about 100,000 MHC II molecules, each of which varies in the epitope it can bind. Their diversity is dependent on an individual's genotype. If an antigen fragment cannot be bound to an MHC molecule, it typically does not trigger an immune response. Thus, MHC molecules determine which antigen fragments might trigger immune responses.

Antigen Processing

Learning Objectives

✓ Explain the roles of antigen-presenting cells (e.g., dendritic cells and macrophages) and MHC molecules in antigen processing and presentation.

✓ Contrast endogenous antigen processing with exogenous antigen processing.

[9]From Greek *histos*, meaning tissue, and Latin *compatibilis*, meaning agreeable.
[10]From Greek *dendron*, meaning tree, referring to long cellular extensions that look like branches of a tree.

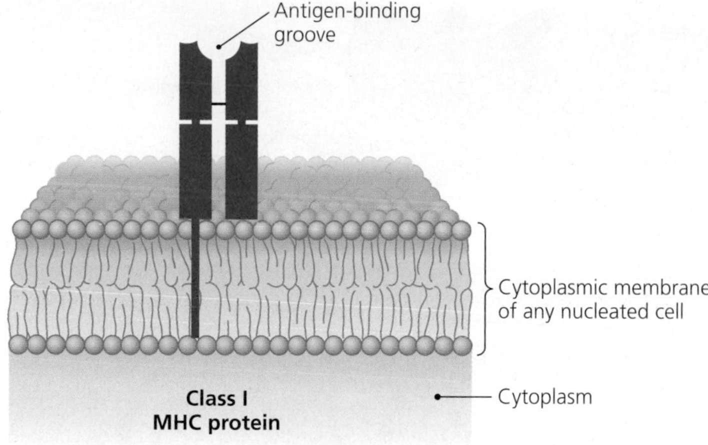

(a)

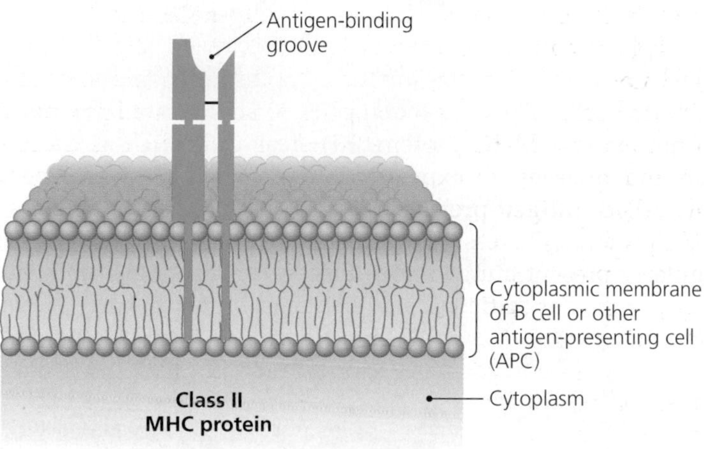

(b)

▲ **Figure 16.10 The two classes of major histocompatibility complex (MHC) proteins.** Each is composed of two polypeptides that form an antigen-binding groove. **(a)** MHC class I glycoproteins, which are found on all cells except red blood cells. **(b)** MHC class II glycoproteins, which are expressed only by B cells and special antigen-presenting cells (APCs).

Before MHC proteins can display epitopes, antigens must be processed. Antigen processing occurs via somewhat different processes according to whether the antigen is endogenous or exogenous. Recall that endogenous antigens come from the cells' cytoplasm or from pathogens living within the cells, whereas exogenous antigens have extracellular sources, such as lymph. **ANIMATIONS:** ***Antigen Processing and Presentation: Overview***

Processing Endogenous Antigens

A few molecules of new polypeptides produced within nucleated cells—including some polypeptides produced by intracellular bacteria or polypeptides coded by viruses—are catabolized into smaller pieces containing about 8–12 amino acids. These pieces will include epitopes from larger polypeptides. Peptides containing epitopes bind onto complementary antigen-binding

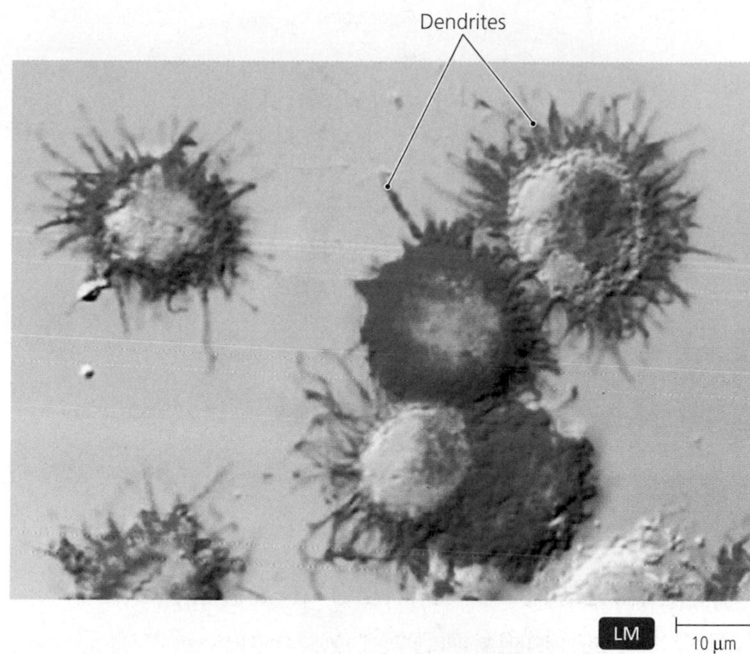

▲ **Figure 16.11 Dendritic cells.** These important antigen-presenting cells in the body are found in the skin and mucous membranes. They have numerous thin cytoplasmic processes called dendrites.

grooves of MHC class I molecules during biosynthesis in the membrane of the endoplasmic reticulum. This membrane, loaded with MHC class I proteins and epitopes, is then packaged into a vesicle by a Golgi body and inserted into the cytoplasmic membrane. The result is that the cell displays the MHC class I protein–epitope complex on the cell's surface (**Figure 16.12**). All nucleated cells express MHC class I molecules and epitopes in this manner.

Processing Exogenous Antigens

For exogenous antigens, a professional antigen-presenting cell (APC), which is usually a dendritic cell, internalizes the invading pathogen and enzymatically catabolizes the pathogen's molecules, producing peptide epitopes, which are contained in a phagolysosome. A vesicle already containing MHC class II molecules in its membrane fuses with the phagolysosome, and each peptide then attempts to bind to an antigen-binding groove of a complementary MHC class II molecule. The fused vesicle inserts successful MHC class II protein–epitope complexes into the cytoplasmic membrane such that the epitopes are presented on the outside (**Figure 16.13**). Empty MHC molecules are not stable on a cell's surface and are degraded. **ANIMATIONS:** ***Antigen Processing and Presentation: Steps, MHC***

So far, we have set the stage for the immune system "play" by examining the organs, cells, secretions, and signaling molecules of the immune system, as well as the preparatory steps of antigen processing, antigen presentation, and clonal deletion. We have seen that adaptive immunity is specific, due to the precise and accurate fit of lymphocyte receptors with their complements; that adaptive immunity involves clones of B and T cells,

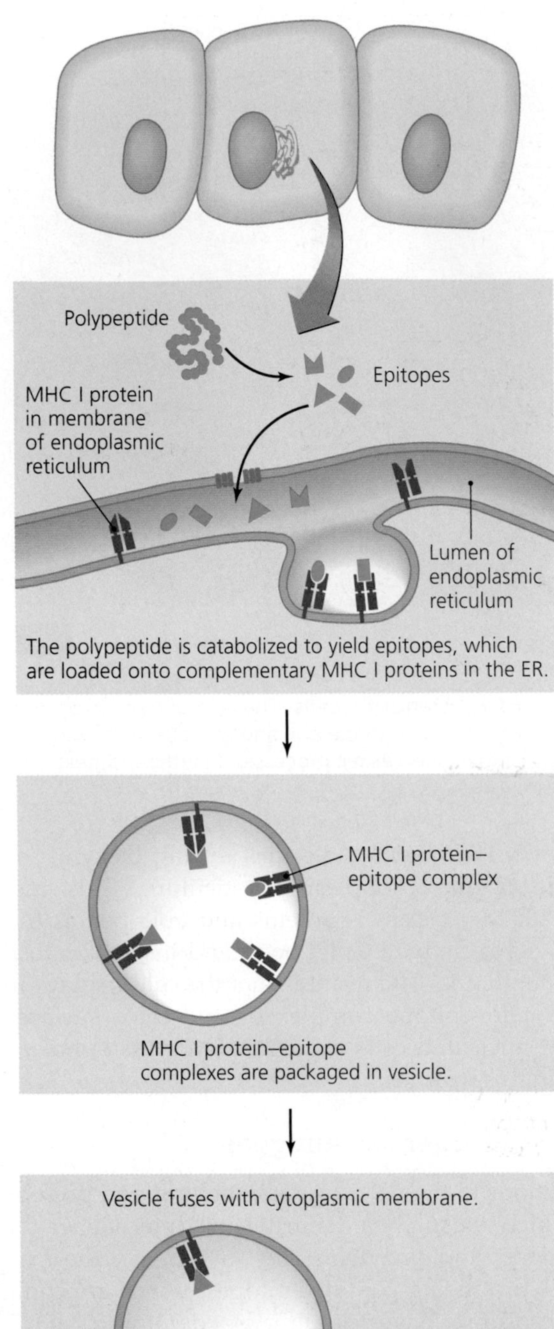

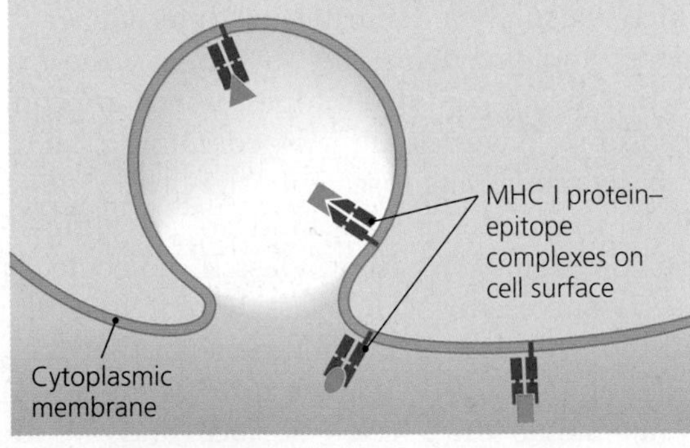

MHC I protein–epitope complexes displayed on cytoplasmic membranes of all nucleated cells

▲ **Figure 16.12 The processing of T-dependent endogenous antigens.** Epitopes from all polypeptides synthesized within a nucleated cell load onto complementary MHC I proteins, which are exported to the cytoplasmic membrane.

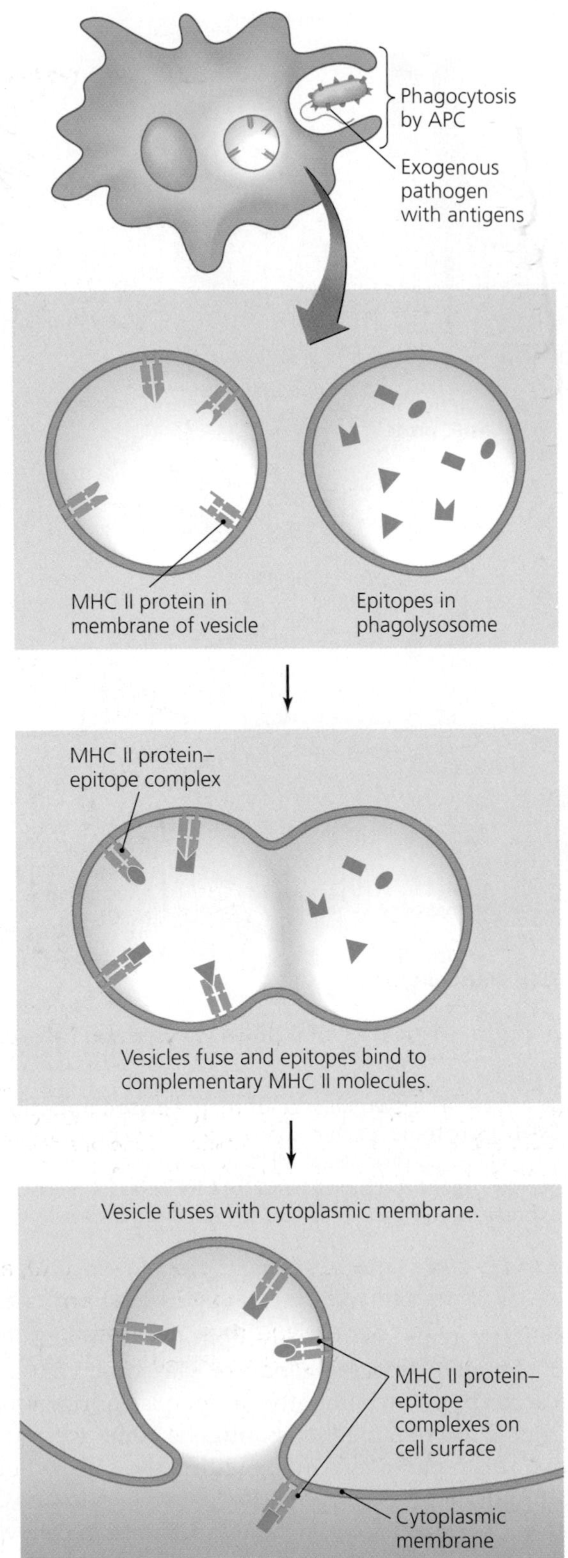

MHC II protein–epitope complexes displayed on cytoplasmic membranes of antigen-presenting cell

▲ **Figure 16.13 The processing of T-dependent exogenous antigens.** Antigens arising outside the body's cells are phagocytized by an APC, which then loads epitopes into complementary antigen-binding grooves of MHC II molecules. The MHC II protein–epitope complexes are then displayed on the outside of the APC's cytoplasmic membrane.

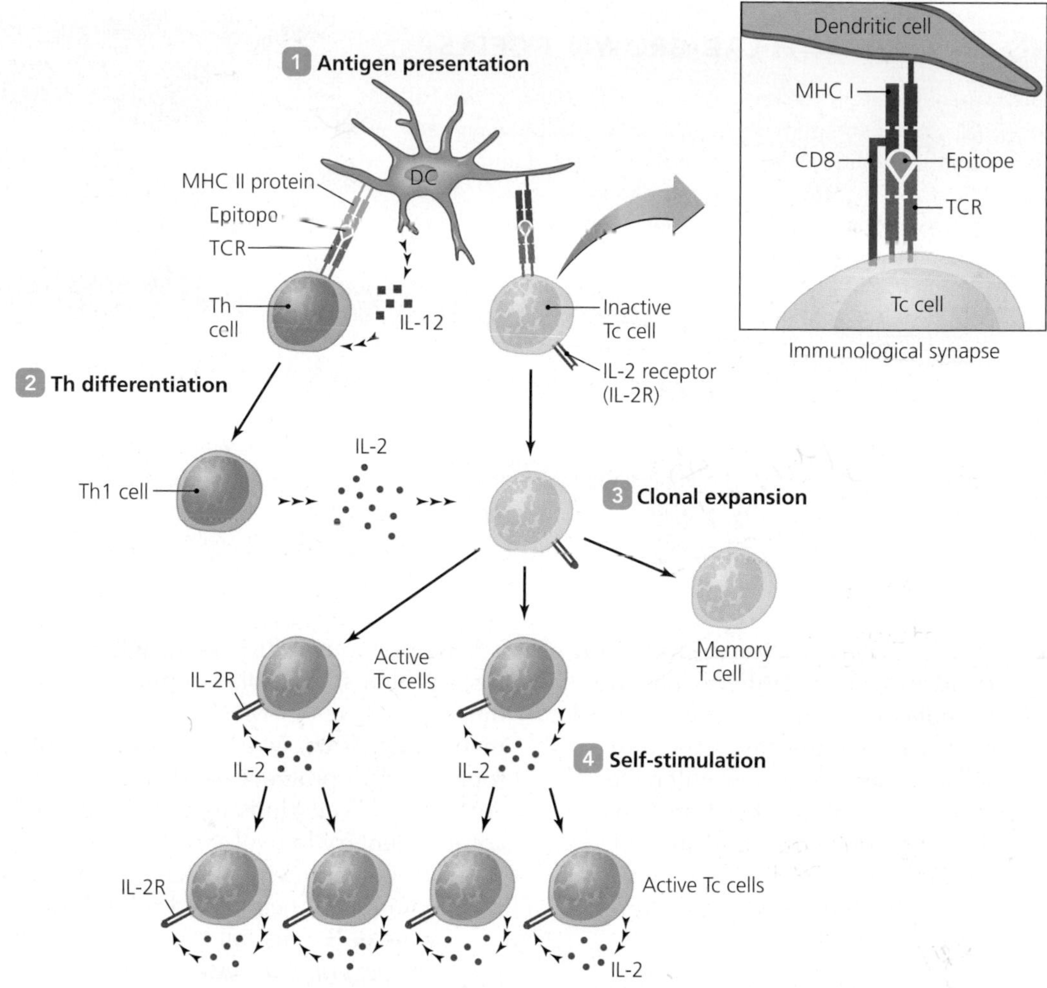

▲ **Figure 16.14 Activation of a clone of cytotoxic T (Tc) cells.** 1 Antigen-presenting cells, here a dendritic cell, present epitopes in conjunction with MHC I protein to inactive T cells. 2 Infected APCs secrete IL-12, which causes helper T cells to differentiate into Th1 cells. 3 Signaling from the APC and IL-2 from the Th1 cell activate Tc cells that recognize the MHC I protein–epitope complex. IL-2 triggers Tc cells to divide, forming a clone of active Tc cells as well as memory T cells. 4 Active Tc cells secrete IL-2, becoming self-stimulatory.

which are unresponsive to self (due to clonal deletion); and that immune responses are inducible against foreign antigens. Now, we can examine cell-mediated and humoral immune responses in more detail. We begin with cell-mediated immunity.

Cell-Mediated Immune Responses

Learning Objectives

- ✓ Describe a cell-mediated immune response.
- ✓ Compare and contrast the two pathways of cytotoxic T cell action.

The body uses cell-mediated immune responses to fight intracellular pathogens and abnormal body cells. Recall that inducibility and specificity are two hallmark characteristics of adaptive immunity. The body induces cell-mediated immune responses only against specific endogenous antigens. Given that many common intracellular invaders are viruses, our examination of cell-mediated immunity will focus on these pathogens. However, cell-mediated immune responses are also mounted against cancer cells, intracellular parasitic protozoa, and intracellular bacteria such as *Mycobacterium tuberculosis* (mī′kō-bak-tēr′-ē-ŭm too-ber-kyū-lō′sis), which causes tuberculosis. **Highlight: Attacking Cancer with Lab-Grown T Cells** on p. 478 describes an experimental use of T cells to treat one form of cancer.

Activation of T Cell Clones and Their Functions

The body does not initiate adaptive immune responses at the site of an infection but rather in lymphoid organs, usually lymph nodes, where antigen-presenting cells interact with lymphocytes. The initial event in cell-mediated immunity is the activation of a specific clone of cytotoxic T cells as depicted in **Figure 16.14**:

HIGHLIGHT

ATTACKING CANCER WITH LAB-GROWN T CELLS

The body naturally manufactures T cells that attack cancer cells, but it sometimes does not make enough of them to shrink tumors and effectively halt the tumor's progress. Recent studies have focused on patients with malignant melanoma, an aggressive form of skin cancer. Researchers took samples of the patients' own cancer-fighting T cells and cloned them in a laboratory until they numbered in the billions. These billions of identical T cells were then injected back into the individuals that first produced them. The results were encouraging. Many of the patients became "virtually cancer free," while tumors were substantially reduced in other patients.

Such T cell therapy remains experimental and is still some time away from becoming a generally accepted cancer treatment. Scientists do not yet understand why the therapy works in some patients and not others. To date, the therapy has primarily been tested against malignant melanoma, but planning to test it against other types of cancer is under way, and some scientists believe that a similar therapy may be effective against viral diseases such as hepatitis.

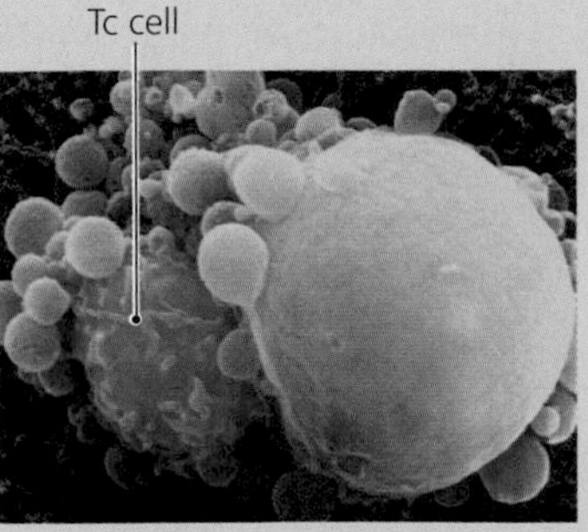

▲ *T cell (green) attacking a cancer cell.*

1 **Antigen presentation.** A virus-infected antigen-presenting cell (APC), typically a dendritic cell, migrates to a nearby lymph node where it presents virus epitopes in conjunction with MHC I protein. Because of the vast repertoire of randomly generated TCRs, at least one Tc cell will have a TCR complementary to the presented MHC I protein–epitope complex. This Tc cell binds to the dendritic cell to form a cell-cell contact site called an *immunological synapse.* CD8 glycoprotein of the Tc cell, which specifically binds to MHC I protein, stabilizes the synapse.

2 **Helper T cell differentiation.** When the interaction between Tc and APC is not particularly strong, a nearby CD4-bearing helper T (Th) lymphocyte helps. The Th cell binds to the APC via the TCR of the helper cell (which is complementary to an MHC II protein–epitope complex presented on the APC) and induces the APC to more vigorously signal the Tc cell. Viruses and some intracellular bacteria induce dendritic cells to secrete interleukin 12 (IL-12), which stimulates the helper T cells to become a clone of type 1 helper T (Th1) cells. Th1 cells in turn secrete IL-2. Lacking the assistance of Th cells, an immunological synapse between the APC and the Tc cell fails to progress. This limits improper immune responses.

3 **Clonal expansion.** The dendritic cell imparts a second required signal in the immunological synapse. This signal, in conjunction with any IL-2 from a Th1 cell that may be present, activates the cytotoxic T (Tc) cell to secrete its own IL-2. Interleukin 2 triggers cell division and differentiation in Tc cells. Activated Tc cells reproduce to form memory T cells (discussed shortly) and more Tc progeny—a process known as **clonal expansion.**

4 **Self-stimulation.** Daughter Tc cells activate and produce both IL-2 receptors (IL-2R) and more IL-2, thereby becoming self-stimulating—they no longer require either an APC or a helper T cell. They leave the lymph node and are now ready to attack virally infected cells.

As previously discussed, when any nucleated cell synthesizes proteins, it displays epitopes from them in the antigen-binding grooves of MHC class I molecules on its cytoplasmic membrane. Thus, when viruses are replicated inside cells, epitopes of viral proteins are displayed on the host cell's surface. An active Tc cell binds to an infected cell via its TCR, which is complementary to the MHC I protein–epitope complex, and via its CD8 glycoprotein, which is complementary to the MHC class I protein of the infected cell **(Figure 16.15a)**.

Cytotoxic T cells kill their targets through one of two pathways: the *perforin-granzyme pathway,* which involves the synthesis of special killing proteins, or the *CD95 pathway,* which is mediated through a glycoprotein found on the body's cells. **ANIMATIONS:** ***Cell-Mediated Immunity: Cytotoxic T Cells***

The Perforin-Granzyme Cytotoxic Pathway

The cytoplasm of cytotoxic T cells has vesicles containing two key protein cytotoxins—*perforin* and *granzyme,* which are also used by NK cells in conjunction with antibody-dependent cellular cytotoxicity (discussed previously). When a cytotoxic T cell first attaches to its target, vesicles containing the cytotoxins release their contents. Perforin molecules aggregate into a channel through which granzyme enters, activating apoptosis in the target cell **(Figure 16.15b)**. Having forced its target to commit suicide, the cytotoxic T cell disengages and moves on to another infected cell.

CRITICAL **THINKING**

Why did scientists give the name "perforin" to a molecule secreted by Tc cells?

The CD95 Cytotoxic Pathway

The **CD95 pathway** of cell-mediated cytotoxicity involves an integral glycoprotein called *CD95* that is present in the cytoplasmic membranes of many body cells. Activated Tc cells insert *CD95L*—

Figure 16.15 A cell-mediated immune response. **(a)** The binding of a virus-infected cell by an active cytotoxic T (Tc) cell. **(b)** The perforin-granzyme cytotoxic pathway. After perforins and granzymes have been released from Tc cell vesicles, granzymes enter the infected cell through the perforin complex pore and activate the enzymes of apoptosis. **(c)** The CD95 cytotoxic pathway. Binding of CD95L on the Tc cell activates the enzymatic portion of the infected cell's CD95 such that apoptosis is induced.

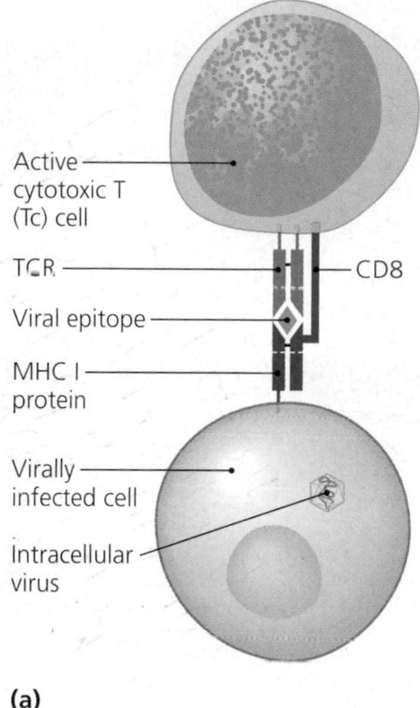

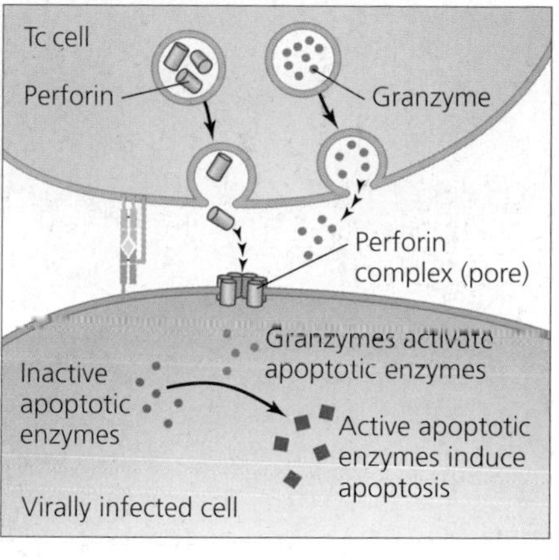

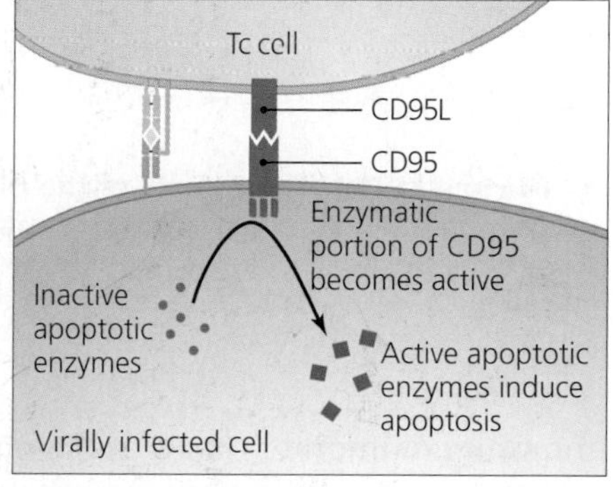

the receptor for CD95—into their cytoplasmic membranes. When an activated Tc cell comes into contact with its target, its CD95L binds to CD95 on the target, which then activates enzymes that trigger apoptosis, killing the target cells **(Figure 16.15c)**.

Memory T Cells

Learning Objective

✓ Describe the establishment of memory T cells.

Some activated T cells become **memory T cells,** which may persist for months or years in lymphoid tissues. If a memory T cell subsequently contacts an epitope–MHC I protein complex matching its TCR, it responds immediately (without a need for interaction with APCs) and produces cytotoxic T cell clones that recognize the offending epitope. These cells need fewer regulatory signals and become functional immediately. Further, since the number of memory T cells is greater than the number of T cells that recognized the antigen during the initial exposure, a subsequent cell-mediated immune response to a previously encountered antigen is much more effective than a primary response. An enhanced cell-mediated immune response upon subsequent exposure to the same antigen is called a **memory response.**[11]

[11]Sometimes also called anamnestic responses, from Greek *ana*, meaning again, and *mimneskein*, meaning to call to mind.

T Cell Regulation

Learning Objective

✓ Explain the process and significance of the regulation of cell-mediated immunity.

The body carefully regulates cell-mediated immune responses so that T cells do not respond to autoantigens. As we have seen, T cells require several signals from an antigen-presenting cell to activate. If the T cells do not receive these signals in a specific sequence—like the sequence of numbers in a combination padlock—they will not respond. Thus, when a T cell and an antigen-presenting cell interact in an immunological synapse, the two cell types have a chemical dialogue that stimulates the T cell to fully respond to the antigen. If a T cell does not receive the signals required for its activation, it will "shut down" as a precaution against autoimmune responses.

Regulatory T (Tr) cells also moderate cytotoxic T cells by mechanisms that are beyond the scope of our discussion. Suffice it to say that Tr cells provide one more level of control over potentially dangerous cell-mediated immune responses.

Humoral Immune Responses

As we have discussed, the body induces humoral immune responses against the antigens of exogenous pathogens. Recall that inducibility is one of the main characteristics of adaptive

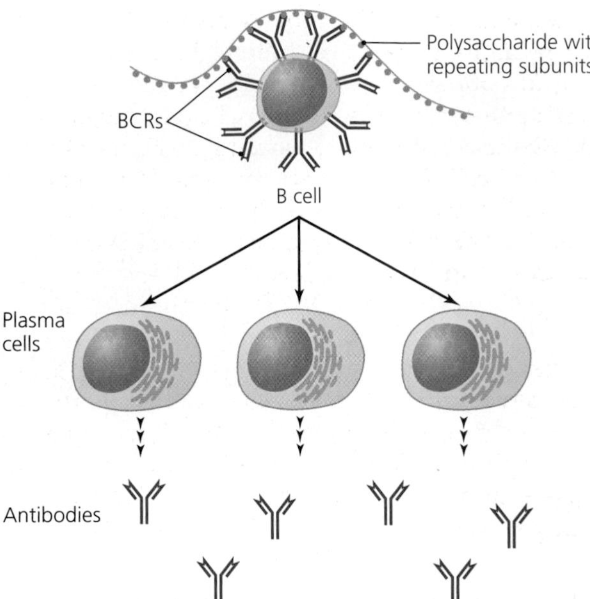

▲ **Figure 16.16 The effects of the binding of a T-independent antigen by a B cell.** When a molecule with multiple repeating epitopes (such as the polysaccharide shown here) cross-links the BCRs on a B cell, the cell is activated: it proliferates, and its daughter cells become plasma cells that secrete antibodies.

immunity: humoral immunity activates only in response to specific pathogens. The following sections examine the activity of B lymphocytes in humoral immune responses.

Inducement of T-Independent Humoral Immunity

Learning Objectives

- ✓ Contrast T-dependent and T-independent antigens in terms of size and repetition of subunits.
- ✓ Describe the inducement and action of a T-independent humoral response.

A few large antigens have many identical, repeating epitopes. These antigens can induce a humoral immune response without the assistance of a helper T cell (Th cell); therefore, these antigens are called *T-independent antigens,* and they trigger response of **T-independent humoral immunity** (Figure 16.16).

The repeating subunits of T-independent antigens allow extensive cross-linking between numerous BCRs on a B cell, stimulating the B cell to proliferate. Simultaneous interaction between Toll-like receptors on the B cell, innate mediators, and/or bacterial chemicals may facilitate activation of the B cell. Clones of the activated B cell become plasma cells, which have an extensive cytoplasm rich in rough endoplasmic reticulum and Golgi bodies and which secrete antibodies (Figure 16.17). Though these events occur without the direct involvement of Th cells, cytokines such as tumor necrosis factor (TNF) are required. T-independent humoral immune responses are fast because they do not depend on interactions with Th cells. Their speed is similar to that of innate immunity, but T-independent humoral immunity is relatively weak, disappears quickly, and induces little immunological memory.

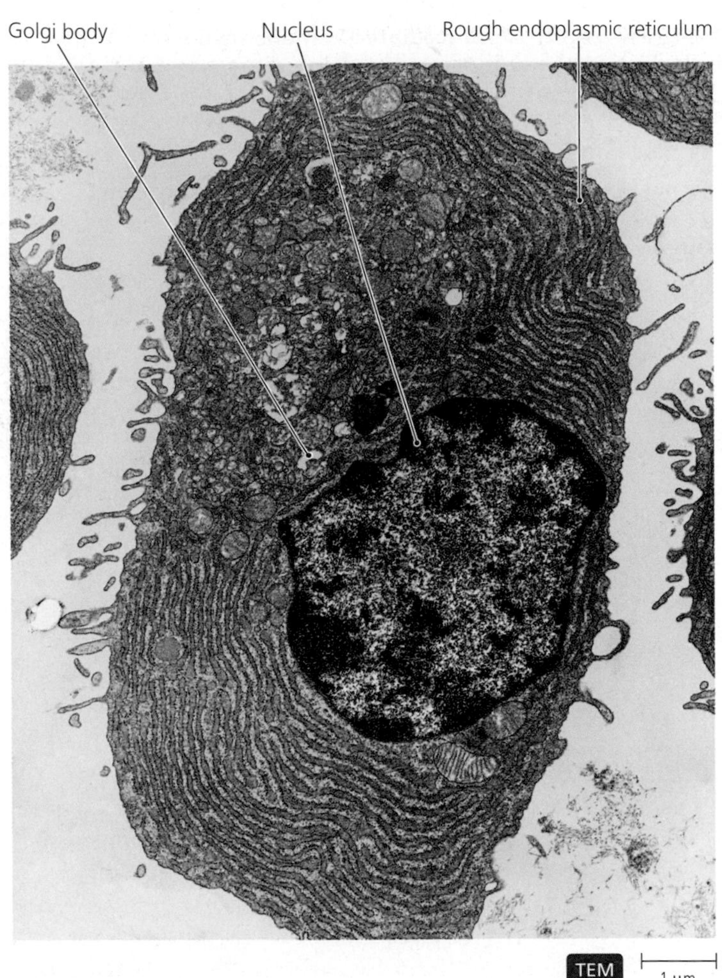

▲ **Figure 16.17 A plasma cell.** Plasma cells are almost twice as large as inactive B cells and are filled with endoplasmic reticulum and Golgi bodies for the synthesis and secretion of antibodies.

T-independent responses are stunted in children, possibly because the repertoire and abundance of B cells is not fully developed in children; therefore, pathogens displaying T-independent antigens can cause childhood diseases, which are rare in adults. An example of such a T-independent antigen and its disease is the capsule of *Haemophilus influenzae* (hē-mof′i-lŭs in-flu-en′zī) type B that causes most cases of meningitis in unvaccinated children. The capsules of other bacteria, lipopolysaccharide of Gram-negative cell walls, bacterial flagella, and the capsids (outer covering) of some viruses constitute other T-independent antigens.

Most humoral immune responses are of the T-dependent type. We consider them next.

Inducement of T-Dependent Humoral Immunity with Clonal Selection

Learning Objectives

- ✓ Describe the formation and functions of plasma cells and memory B cells.
- ✓ Describe the steps and effect of clonal selection.

T-dependent antigens lack the numerous, repetitive, and identical epitopes and the large size of T-independent antigens, and immunity against them requires the assistance of type 2 helper T (Th2) cells. **T-dependent humoral immunity** begins with the action of an antigen-presenting cell, which is usually a dendritic cell. After endocytosis and processing of the antigen, the dendritic cell presents epitopes in conjunction with MHC II proteins. This APC will induce the specific helper T (CD4) lymphocyte with a TCR complementary to the presented MHC II–epitope complex. The activated Th2 cell must in turn induce the equally rare B cell that recognizes the same antigen. Lymph nodes facilitate and cytokines mediate interactions among the antigen-presenting cells and lymphocytes, increasing the chance that the appropriate cells find each other.

Thus, a T-dependent humoral immune response involves a series of interactions among antigen-presenting cell, helper T cells, and B cells, all of which are mediated and enhanced by cytokines. Now we will examine each step in more detail (**Figure 16.18**):

1 **Antigen presentation for Th activation and cloning.** A dendritic cell, after acquiring antigens in the skin or mucous membrane, moves via the lymph to a local lymph node. The trip takes about a day. As helper T (Th, CD4) cells pass through the lymph node, they survey all the resident APCs for complementary epitopes in conjunction with MHC II proteins. Antigen presentation depends on chance encounters between Th cells and the dendritic cells, but immunologists estimate that every lymphocyte browses the dendritic cells in every lymph node every day; therefore, complementary cells eventually find each other. Once they have established an immunological synapse, CD4 molecules in membrane rafts of the Th cell's cytoplasmic membrane recognize and bind to MHC II, stabilizing the synapse.

   As we saw in cell-mediated immune responses, helper T cells need further stimulation before they activate. The requirement for a second signal helps prevent accidental inducement of an immune response. As before, the APC imparts the second signal by displaying an integral membrane protein in the immunological synapse. This induces the Th cell to proliferate, producing a clone.

2 **Differentiation of helper T cells into Th2 cells.** In humoral immune responses, the cytokine interleukin 4 (IL-4) acts as a signal to the Th cells to become type 2 helper T cells (Th2 cells). Immunologists do not know the source of IL-4, but it may be secreted initially by innate cells such as mast cells or secreted later in a response by the Th cells themselves.

3 **Clonal selection of B cell.** A B cell attaches to and phagocytizes a complementary antigen, which is a BCR-mediated event. It then displays the pathogen's epitopes in association with MHC II proteins. Antigen binding primes these B cells to participate in humoral immune responses. This is called **clonal selection.**

4 **Activation of B cell.** The repertoire of B cells and newly formed Th2 cells survey one another. A Th2 cell binds to the B cell with an MHC II protein–epitope complex that is complementary to the TCR of the Th2 cell. CD4 glycoprotein again stabilizes the immunological synapse.

   Th2 cells secrete more chemical signal molecules, which induce the selected B cell to move to the cortex of the lymph node. A Th2 cell in contact with an MHC II protein–epitope on a B cell is stimulated, expresses new gene products, and inserts a protein called CD40L into its cytoplasmic membrane. CD40L binds to CD40, which is found on B cells. This provides a second signal in the immunological synapse, triggering B cell activation. The B cell moves to the cortex of the lymph node.

   The activated B cell proliferates rapidly to produce a population of cells (clone) that make up a germinal center in the lymph node. The clone differentiates into two types of cells—*memory B cells* (discussed shortly) and antibody-secreting *plasma cells.* **ANIMATIONS:** ***Humoral Immunity: Clonal Selection and Expansion***

Most members of a clone become plasma cells. The initial plasma cell descendants of any single activated B cell secrete antibodies with binding sites identical to one another and complementary to the specific antigen recognized by their parent cell. However, as the plasma cell clones replicate, the cells slightly modify their antigen-binding-site genes such that they secrete antibodies with slightly different variable regions. Plasma cells that secrete antibodies with a higher affinity for the epitope have a selective survival advantage over plasma cells secreting antibodies with a less good fit; that is, active B cells with BCRs that bind the epitope more closely survive at a higher rate. Thus, as the humoral immune response progresses, there are more and more plasma cells, secreting antibodies whose fitness gets progressively better.

Each plasma cell produces about 2000 molecules of antibody per second. They begin by secreting IgM and then, through class switching, they secrete IgG. Some cells later switch a second time and begin secreting IgA or IgE. As discussed previously, antibodies activate complement, trigger inflammation, agglutinate and neutralize antigen, act as opsonins, directly kill pathogens, and induce antibody-dependent cytoxicity.

Individual plasma cells are short lived, at least in part because of their high metabolic rate; they die within a few days of activation, although their antibodies can remain in body fluids for several weeks. Providentially, their descendants persist for years to maintain long-term adaptive responses.

CRITICAL **THINKING**

In general, what sorts of pathogens would successfully attack a patient with an inability to synthesize B lymphocytes?

Memory B Cells and the Establishment of Immunological Memory

Learning Objective

✓ Contrast primary and secondary immune responses.

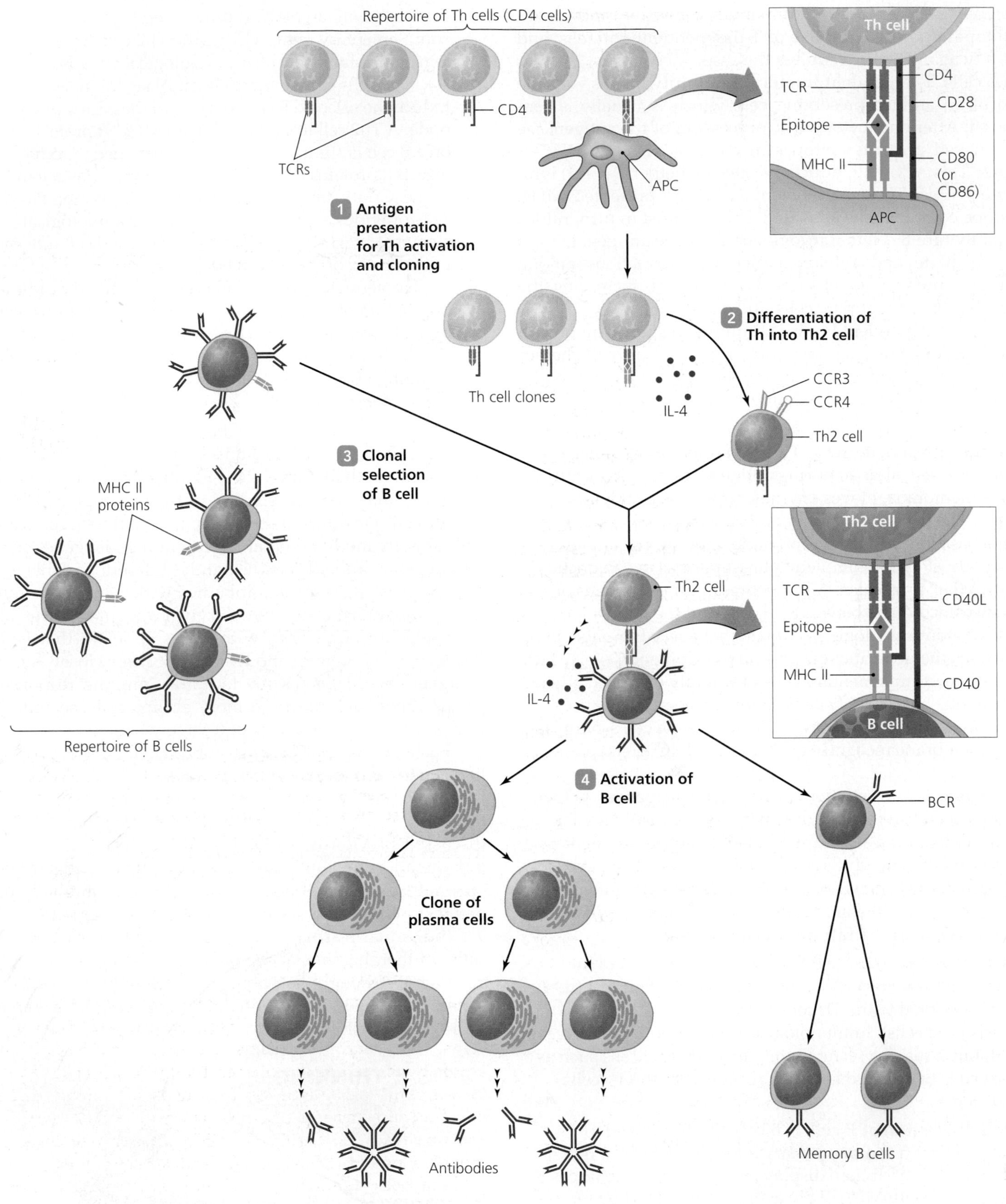
Repertoire of Th cells (CD4 cells)
CD4
TCRs
APC
1 Antigen presentation for Th activation and cloning
Th cell
TCR
CD4
CD28
Epitope
MHC II
CD80 (or CD86)
APC
Th cell clones
IL-4
2 Differentiation of Th into Th2 cell
CCR3
CCR4
Th2 cell
3 Clonal selection of B cell
MHC II proteins
Repertoire of B cells
Th2 cell
IL-4
Th2 cell
TCR
CD40L
Epitope
MHC II
CD40
B cell
4 Activation of B cell
BCR
Clone of plasma cells
Antibodies
Memory B cells

◀ **Figure 16.18 A T-dependent humoral immune response.** 1 Antigen presentation, in which an APC, typically a dendritic cell, presents antigen to a complementary Th cell. 2 Differentiation of the Th cell into a Th2 cell. 3 Clonal selection, in which the Th2 cell binds to the B cell clone that recognizes the antigen. 4 Activation of the B cell in response to secretion of IL-4 by the Th2 cell, which causes the B cell to differentiate into antibody-secreting plasma cells and long-lived memory cells.

To see a 3-D animation on immunology, go to the Study Area at **www.masteringmicrobiology.com** and watch the *MicroFlix*.

A small percentage of the cells produced during B cell proliferation do not secrete antibodies but survive as **memory B cells**—that is, long-lived cells with BCRs complementary to the specific epitope that triggered their production. In contrast to plasma cells, memory cells retain their BCRs and persist in lymphoid tissues, surviving for more than 20 years, ready to initiate antibody production if the same epitope is encountered again. Let's examine how memory cells provide the basis for immunization to prevent disease, using tetanus immunization as an example.

Because the body produces an enormous variety of B cells (and therefore BCRs), a few Th cells and B cells bind to and respond to epitopes of *tetanus toxoid* (deactivated tetanus toxin), which is used for tetanus immunization. In a **primary response (Figure 16.19a)**, relatively small amounts of antibodies are produced, and it may take days before sufficient antibodies are made to completely eliminate the toxoid from the body. Though some antibody molecules may persist for three weeks, a primary immune response basically ends when the plasma cells have lived out their normal life spans. **ANIMATIONS:** ***Humoral Immunity: Primary Immune Response***

Memory B cells, surviving in lymphoid tissue, constitute a reserve of antigen-sensitive cells that become active when there is another exposure to the antigen, in this case, toxin from infecting tetanus bacteria. Exposure may be many years later. Thus, tetanus toxin produced during the course of a bacterial infection will restimulate a population of memory cells, which proliferate and differentiate rapidly into plasma cells. The newly differentiated plasma cells produce large amounts of antibody within a few days **(Figure 16.19b)**, and the tetanus toxin is neutralized before it can cause disease. Since many memory cells recognize and respond to the antigen, such a **secondary immune response** is much faster and more effective than the primary response. **ANIMATIONS:** ***Humoral Immunity: Secondary Immune Response***

As you might expect, a third exposure (whether to tetanus toxin or to toxoid in an immunization booster) results in an even more effective response. Enhanced immune responses triggered by subsequent exposure to antigens are memory responses, which are the basis of *immunization*. Chapter 17 discusses immunization in more detail.

CRITICAL **THINKING**

Plasma cells are vital for protection against infection, but memory B cells are not. Why not?

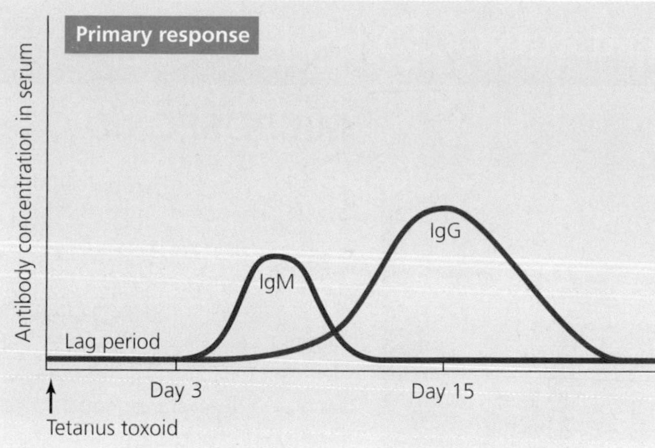

(a)

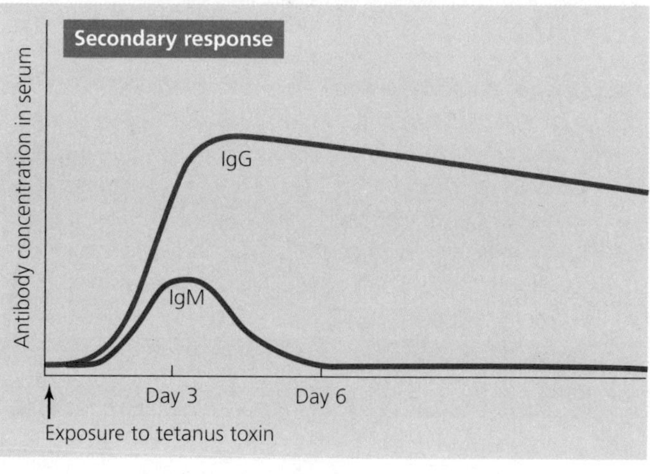

(b)

▲ **Figure 16.19 The production of primary and secondary humoral immune responses.** This example depicts some events following the administration of a tetanus toxoid in immunization. **(a)** Primary response. After the tetanus toxoid is introduced into the body, the body slowly removes the toxoid while producing memory B cells. **(b)** Secondary response. Upon exposure to active tetanus toxin during the course of an infection, memory B cells immediately differentiate into plasma cells and proliferate, producing a response that is faster and results in greater antibody production than occurs in the primary response.

In summary, the body's response to infectious agents seldom relies on one mechanism alone, because this course of action would be far too risky. Therefore, the body typically uses several different mechanisms to combat infections. In the example from the chapter opener, the body's initial response to intruders was inflammation (a nonspecific, innate response), but a specific immune response against the invading microorganisms was also necessary. APCs phagocytize some of the invaders, process their epitopes, and induce clones of lymphocytes in both humoral and cell-mediated immune responses. Key to enduring protection are the facts that adaptive immunity is unresponsive to self and involves immunological memory brought about by long-lived memory B and T cells.

Cell-mediated adaptive immune responses involve the activity of cytotoxic T lymphocytes in killing cells infected with

EMERGING DISEASES

MICROSPORIDIOSIS

Darius is sick, which is not surprising for an HIV-infected man. But he is sick in several new ways. Sick of having to stay within 20 feet of a toilet. Sick of the cramping, the gas, the pain, and the nausea. Sick with irregular, but persistent, watery diarrhea. He is losing weight because food is passing through him undigested. Most days over the past seven months have been disgusting despite his use of over-the-counter remedies, which provide a few days of intermittent relief. His belief that these normal days signaled the end of the ordeal have kept him from the doctor. But now his eyes have begun to hurt, and his vision is blurry. Whatever it is, it's attacking him at both ends. Time to get stronger drugs from his doctor.

Microscopic examination of Darius's stool sample reveals that he is being assaulted by *Encephalitozoon intestinalis*, a member of a group of emerging pathogens called microsporidia. The single-celled pathogens are also seen on smears from Darius's nose and eyes. Microsporidia were long thought to be simple single-celled animals, but genetic analysis and comparison with other organisms reveal that they are closer to zygomycete yeasts.

Microsporidia appear to infect humans who engage in unprotected sexual activity, eat contaminated food, or drink or swim in contaminated water. People with active T cells rarely have symptoms; but people with suppressed immunity become easy targets for the fungus.

Microsporidia attack by uncoiling a flexible, hollow filament that stabs into a host cell and serves as a conduit for the microsporidium's cytoplasm to invade. In this way, the pathogens become intracellular parasites. They can destroy the intestinal lining, causing diarrhea, and spread to infect the eyes, muscles, or lungs.

Fortunately for Darius, an antimicrobial, albendazole, kills the parasite, and the effects of the infection are reversed. Unfortunately for Darius, the loss of helper T cells in AIDS means that another emerging, reemerging, or opportunistic infection is sure to follow.

Track microsporidiosis online by going to the Study Area at www.masteringmicrobiology.com.

intracellular bacteria and viruses. Humoral adaptive immunity involves the secretion of specific antibodies that have a variety of functions. T-independent humoral immune responses are rare, but can occur when an adult is challenged with T-independent antigens such as bacterial flagella or capsules. In contrast, T-dependent humoral immune responses are more common. A T-dependent humoral immune response occurs when an APC binds to a specific Th cell and signals the Th cell to proliferate.

The relative importance of each of these pathways depends on the type of pathogen involved and on the mechanisms by which they cause disease. In any case, adaptive immune responses are specific, inducible, involve clones, are unresponsive to self, and give the body long-term memory against their antigenic triggers. **ANIMATIONS:** ***Host Defenses: The Big Picture***

CRITICAL **THINKING**

What sorts of pathogens could successfully attack a patient with an inability to produce T lymphocytes?

Types of Acquired Immunity

Learning Objective

✓ Contrast active versus passive acquired immunity, and naturally acquired versus artificially acquired immunity.

As we have seen, adaptive immunity is acquired during an individual's life. Immunologists categorize immunity as either naturally or artificially acquired. Naturally acquired immunity occurs when the body mounts an immune response against antigens, such as influenzaviruses or food antigens, encountered during the course of daily life. Artificial immunity is the body's response to antigens introduced in vaccines, as occurs with immunization against tetanus and flu. Immunologists further distinguish acquired immune responses as either *active* or *passive;* that is, the immune system either responds actively to antigens via humoral or cell-mediated responses, or the body passively receives antibodies from another individual. Next we consider each of four types of acquired immunity.

Naturally Acquired Active Immunity

Naturally acquired active immunity occurs when the body responds to exposure to pathogens and environmental antigens by mounting specific immune responses. The body is naturally and actively engaged in its own protection. As we have seen, once an immune response occurs, immunological memory persists—on subsequent exposure to the same antigen, the immune response will be rapid and powerful and often provides the body complete protection.

TABLE 16.4 A Comparison of the Types of Acquired Immunity

	Active	Passive
Naturally acquired	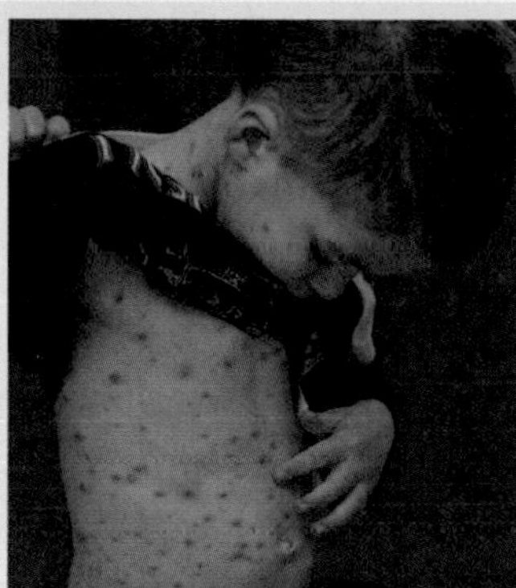 The body responds to antigens that enter naturally, such as during infections.	 Antibodies are transferred from mother to offspring, either across the placenta (IgG) or in breast milk (secretory IgA).
Artificially acquired	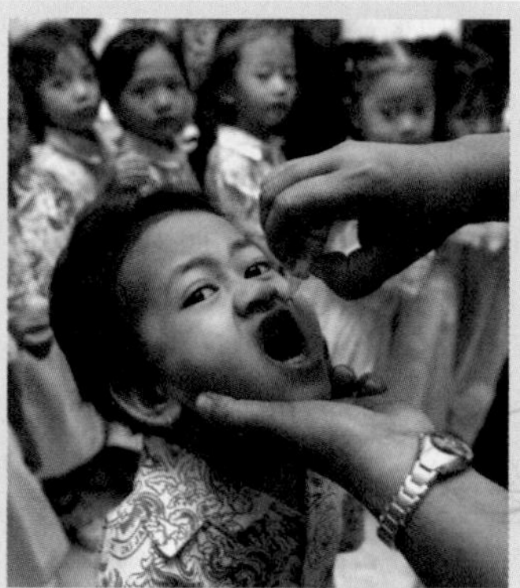 Health care workers introduce antigens in vaccines; the body responds with humoral or cell-mediated immune responses, including the production of memory cells.	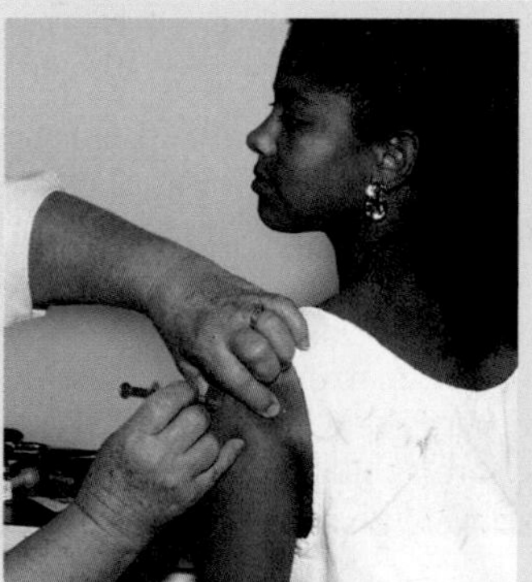 Health care workers give patients antisera or antitoxins, which are preformed antibodies obtained from immune individuals or animals.

Naturally Acquired Passive Immunity

Although newborns possess the cells and tissues needed to mount an immune response, they respond slowly to antigens. If required to protect themselves solely via naturally acquired active immunity, they might die of infectious disease before their immune systems were mature enough to respond adequately. However, they are not on their own; in the womb, IgG molecules cross the placenta from the mother's bloodstream to provide protection, and after birth, children receive secretory IgA in breast milk. Via these two processes, a mother provides her baby with antibodies that protect it during its early months. Because the baby is not actively producing its own antibodies, this type of protection is known as **naturally acquired passive immunity.**

Artificially Acquired Active Immunity

Physicians induce immunity in their patients by introducing antigens in the form of vaccines. The patients' own immune systems then mount active responses against the foreign antigens, just as if the antigens were part of a naturally acquired pathogen. Such **artificially acquired active immunity** is the basis of immunization, which is discussed in detail in Chapter 17.

Artificially Acquired Passive Immunotherapy

Active immunity usually requires days to weeks to develop fully, and in some cases such a delay can prove detrimental or even fatal. For instance, an active immune response may be too slow to protect against infection with rabies or exposure to rattlesnate venom. Therefore, medical personnel routinely harvest antibodies specific for toxins and pathogens that are so deadly or so fast-acting that an individual's active immune response is inadequate. They acquire these antibodies from the blood of immune humans or animals, typically a horse. Physicians then inject such *antisera* or *antitoxins* into infected patients to confer **artificially acquired passive immunotherapy.** Chapter 17 discusses this type of treatment in greater detail.

Active immune responses, whether naturally or artificially induced, are advantageous because they result in immunological memory and protection against future infections. However, they are slow-acting. Passive processes, in which individuals are provided fully formed antibodies, has the advantage of speed but does not confer immunological memory because B and T lymphocytes are not activated. Table 16.4 summarizes the four types of acquired immunity.

Chapter Summary

 This chapter has MicroFlix. Go to www.masteringmicrobiology.com for 3-D movie-quality animations on immunology.

Overview of Adaptive Immunity (pp. 460–461)

1. **Adaptive immunity** is the ability of a vertebrate to recognize and defend against distinct species or strains of invaders. Adaptive immunity is characterized by specificity, inducibility, clonality, unresponsiveness to self, and memory.
2. **B lymphocytes (B cells)** attack extracellular pathogens in **humoral immune responses,** involving soluble proteins called antibodies. **T lymphocytes (T cells)** carry out **cell-mediated immune responses** against intracellular pathogens.
   ANIMATIONS: *Humoral Immunity: Overview; Cell-Mediated Immunity: Overview; Host Defenses: The Big Picture*

Elements of Adaptive Immunity (pp. 461–474)

1. The **lymphatic system** is composed of **lymphatic vessels,** which conduct the flow of **lymph,** and lymphoid tissues and organs that are directly involved in specific immunity. The latter include **lymph nodes,** the thymus, the spleen, the tonsils, and mucosa-associated lymphoid tissue (MALT). Lymphocytes originate and mature in the red bone marrow and express characteristic membrane proteins. They migrate to and persist in various lymphoid organs, where they are available to encounter foreign invaders in the blood and lymph.
2. **Antigens** are substances that trigger specific immune responses. Effective antigen molecules are large, usually complex, stable, degradable, and foreign to their host. An **epitope** (or antigenic determinant) is the three-dimensional shape of a region of an antigen that is recognized by the immune system.
3. **Exogenous** antigens are found on microorganisms that multiply outside the cells of the body; **endogenous** antigens are produced by pathogens multiplying inside the body's cells.
4. Ideally, the body does not attack antigens on the surface of its normal cells, called **autoantigens;** this phenomenon is called self-tolerance.
5. B lymphocytes (B cells), which mature in the red bone marrow, make immunoglobulins (Ig) of two types—**B cell receptors (BCRs)** and **antibodies.** Immunoglobulins are complementary to epitopes and consist of two light chains and two heavy chains joined via disulfide bonds to form Y-shaped molecules. BCRs are inserted into the cytoplasmic membranes of B cells via a transmembrane polypeptide, and antibodies are secreted.
6. Together the variable regions of a heavy and a light chain form an **antigen-binding site,** and the upper portions of antibody molecules are called F_{ab} regions. Each B cell randomly chooses (once in its life) genes for its F_{ab}, region; therefore, the F_{ab} regions are called variable regions, because they differ from cell to cell. Each basic antibody molecule has two antigen-binding sites and can potentially bind two epitopes.
7. Antibodies function in complement activation, inflammation, **neutralization** (blocking the action of a toxin or attachment of a pathogen), **opsonization** (enhanced phagocytosis), direct killing by oxidation, **agglutination,** and **antibody-dependent cellular cytotoxicity (ADCC).**
   ANIMATIONS: *Humoral Immunity: Antibody Function*
8. Antibodies are of five basic classes based upon their stems (F_c regions), which differ in their type of heavy chain.
9. **IgM,** a pentamer with 10 antigen-binding sites, is the predominant class of antibody produced first during a primary humoral response. **IgG** is the predominant antibody found in the bloodstream and is largely responsible for defense against invading bacteria. IgG can cross a placenta to protect the fetus. Two molecules of **IgA** are attached via J chains and a polypeptide secretory component to produce **secretory IgA,** which is found in milk, tears, and mucous membrane secretions. **IgE** triggers inflammation and allergic reactions. It also functions during helminth infections. Immunoglobulin D **(IgD)** is found in cytoplasmic membranes of some animals.
10. Through a process called **class switching,** antibody-producing cells change the class of antibody they secrete, beginning with IgM and then producing IgG and then possibly IgA or IgE.
11. T cells have **T cell receptors (TCRs)** for antigens, mature under influence of signals from the thymus, and attack cells that harbor endogenous pathogens during cell-mediated immune responses.
12. In cell-mediated immunity, **cytotoxic T cells (Tc** or **CD8 cells)** act against infected or abnormal body cells, including virus-infected cells, bacteria-infected cells, some fungal- or protozoan-infected cells, some cancer cells, and foreign cells that enter the body as a result of organ transplantation.
13. Two types of **helper T (Th) cells**—Th1 and Th2—are characterized by **CD4.** They direct cell-mediated and humoral immune responses respectively.
    ANIMATIONS: *Cell-Mediated Immunity: Helper T Cells*
14. T cells that do not recognize MHC I protein and most T cells that recognize MHC I protein in conjunction with autoantigens are removed by apoptosis. This is **clonal deletion.** A few self-recognizing T cells are retained and become **regulatory T cells (Tr cells**). T cells that recognize MHC I protein but not autoantigens become the repertoire of immature T cells.
15. B cells with B cell receptors that respond to autoantigens are selectively killed via apoptosis—further clonal deletion. Only B cells that respond to foreign antigens survive to defend the body.
16. **Cytokines** are soluble regulatory proteins that act as intercellular signals to direct activities in immune responses. Cytokines include **interleukin (ILs), interferons (IFNs), growth factors, tumor necrosis factors (TNFs),** and **chemokines.**

Preparation for an Adaptive Immune Response (pp. 474–477)

1. Nucleated cells display epitopes of their own proteins and epitopes from intracellular pathogens such as viruses on **major histocompatibility complex (MHC)** class I proteins.

2. The initial step in mounting an immune response is that antigens are captured, ingested, and degraded into epitopes by **antigen-presenting cells (APCs)** such as B cells, macrophages, and **dendritic cells.** Epitopes are inserted into major histocompatibility complex (MHC) class II proteins.
ANIMATIONS: *Antigen Processing and Presentation: Overview, Steps, MHC*

Cell-Mediated Immune Responses (pp. 477–479)

1. Once activated by dendritic cells, cytotoxic T cells (Tc cells) recognize abnormal molecules presented by MHC I protein on the surface of infected, cancerous, or foreign cells. Sometimes cytotoxic T cells require cytokines from Th1 cells.
2. Activated Tc cells reproduce to form memory T cells and more Tc progeny in a process called **clonal expansion.**
3. Cytotoxic T cells destroy their target cells via two pathways: the perforin-granzyme pathway, which kills the affected cells by secreting **perforins** and **granzymes,** or the **CD95 pathway,** in which CD95L binds to CD95 on the target cell, triggering target cell apoptosis. Cytotoxic T cells may also form **memory T cells,** which function in **memory responses.**
ANIMATIONS: *Cell-Mediated Immunity: Cytotoxic T Cells*

Humoral Immune Responses (pp. 479–484)

1. T-independent antigens, such as bacterial capsules, trigger **T-independent humoral immune responses,** which are more common in adults than in children.
2. In **T-dependent humoral immunity,** an APC's MHC II protein–epitope complex activates a helper T cell (Th cell) bearing a complementary TCR. CD4 stabilizes the connection between the cells, which is an example of an immunological synapse. Interleukin 4 (IL-4) then induces the Th cell to become a type 2 helper T cell (Th2).
3. In **clonal selection,** an immunological synapse forms between the Th2 cell and a B cell bearing a complementary MHC II protein–epitope complex. The Th2 cell secretes IL-4, which induces the B cell to divide. Its offspring, collectively called a clone, become **plasma cells** or **memory B cells.**
ANIMATIONS: *Humoral Immunity: Clonal Selection and Expansion*
4. Plasma cells live for only a short time but secrete large amounts of antibodies, beginning with IgM and class switching as they get older. Memory B cells migrate to lymphoid tissues to await a subsequent encounter with the same antigen.
5. The **primary response** to an antigen is slow to develop and of limited effectiveness. When that antigen is encountered a second time, the activation of memory cells ensures that the immune response is rapid and strong. This is a **secondary immune response.** Such enhanced humoral immune responses are memory responses.
ANIMATIONS: *Humoral Immunity: Primary Immune Response, Secondary Immune Response*

Types of Acquired Immunity (pp. 484–485)

1. When the body mounts a specific immune response against an infectious agent, the result is called **naturally acquired active immunity.**
2. The passing of maternal IgG to the fetus and the transmission of secretory IgA in milk to a baby are examples of **naturally acquired passive immunity.**
3. **Artificially acquired active immunity** is achieved by deliberately injecting someone with antigens in vaccines to provoke an active response, as in the process of immunization.
4. **Artificially acquired passive immunotherapy** involves the administration of preformed antibodies in antitoxins or antisera to a patient.

Questions for Review

Answers to the Questions for Review (except Short Answer questions) begin on page A-1.

Multiple Choice

1. Antibodies function to
   a. directly destroy foreign organ grafts.
   b. mark invading organisms for destruction.
   c. kill intracellular viruses.
   d. promote cytokine synthesis.
   e. stimulate T cell growth.
2. MHC class II molecules bind to ________________ and trigger ________________.
   a. endogenous antigens, cytotoxic T cells
   b. exogenous antigens, cytotoxic T cells
   c. antibodies, B cells
   d. endogenous antigens, helper T cells
   e. exogenous antigens, helper T cells
3. Rejection of a foreign skin graft is an example of
   a. destruction of virus-infected cells.
   b. tolerance.
   c. antibody-mediated immunity.
   d. a secondary immune response.
   e. a cell-mediated immune response.
4. An autoantigen is
   a. an antigen from normal microbiota.
   b. a normal body component.
   c. an artificial antigen.
   d. any carbohydrate antigen.
   e. a nucleic acid.

5. Among the key molecules that mediate cell-mediated cytotoxicity are
   a. perforin.
   b. immunoglobulins.
   c. complement.
   d. cytokines.
   e. interferons.
6. Which of the following lymphocytes predominates in blood?
   a. T cells
   b. B cells
   c. plasma cells
   d. memory cells
   e. all are about equally prevalent
7. The major class of immunoglobulin found on the surfaces of the walls of the intestines and airways is secretory
   a. IgG.
   b. IgM.
   c. IgA.
   d. IgE.
   e. IgD.
8. Which cells express MHC class I molecules?
   a. red blood cells
   b. antigen-presenting cells only
   c. neutrophils only
   d. all nucleated cells
   e. dendritic cells only
9. In which of the following sites in the body can B cells be found?
   a. lymph nodes
   b. spleen
   c. red bone marrow
   d. intestinal wall
   e. all of the above
10. Tc cells recognize epitopes only when the latter are held by
   a. MHC proteins.
   b. B cells.
   c. interleukin 2.
   d. granzyme.

Modified True/False

Mark each statement as either true or false. Rewrite false statements to make them true by changing the italicized words.

1. ____ MHC class II molecules are found on *T cells*.
2. ____ *Apoptosis* is the term used to describe cellular suicide.
3. ____ Lymphocytes with CD8 glycoprotein are *helper* T cells.
4. ____ *Cytotoxic T cells* secrete immunoglobulin.
5. ____ Secretion of antibodies by activated B cells is a form of *cell-mediated* immunity.

Matching

1. Match each cell in the left column with its associated protein from the right column.

____ Plasma cell	A. MHC II molecule
____ Cytotoxic T cell	B. Interleukin 4
____ Th2 cell	C. Perforin and granzyme
____ Dendritic cell	D. Immunoglobulin

2. Match each type of immunity in the left column with its associated example from the right column.

____ Artificially acquired passive immunotherapy	A. Production of IgE in response to pollen
____ Naturally acquired active immunity	B. Acquisition of maternal antibodies in breast milk
____ Naturally acquired passive immunity	C. Administration of tetanus toxoid
____ Artificially acquired active immunity	D. Administration of antitoxin

Labeling

Label the parts of the immunoglobulin below.

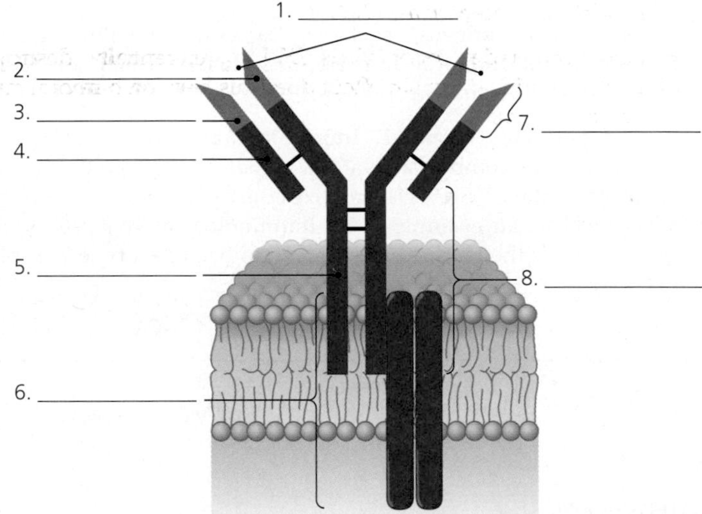

Short Answer

1. When is antigen processing an essential prerequisite for an immune response?
2. Why does the body have both humoral and cell-mediated immune responses?

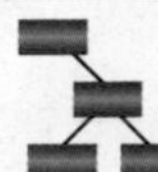 Concept Mapping

Using the following terms, draw a concept map that describes antibodies. For a sample concept map, see p. 93.
Or, complete this concept map online by going to the Study Area at www.masteringmicrobiology.com.

Agglutination
Antigens
Antigen-stimulated B cells
Death by Oxidation
IgA
IgD
IgE
IgG
IgM
Immunoglobulins
Inflammation
Neutralization
Opsonization
Phagocytosis (2)
Plasma cells
Target bacteria
Toxins
Viruses

Critical Thinking

1. Why is it advantageous for the lymphatic system to lack a pump?
2. Contrast innate defenses with adaptive immunity.
3. What is the benefit to the body of requiring the immune system to process antigen?
4. Scientists can develop genetically deficient strains of mice. Describe the immunological impairments that would result in mice deficient in each of the following: class I MHC, class II MHC, TCR, BCR, IL-2 receptor, and IFN-γ.
5. Human immunodeficiency virus (HIV) preferentially destroys CD4 cells. Specifically, what effect does this have on humoral and cell-mediated immunity?
6. What would happen to a person who failed to make MHC molecules?
7. Why does the body make five different classes of immunoglobulins?
8. Some materials such as metal bone pins and plastic heart valves can be implanted into the body without fear of rejection by the patient's immune system. Why is this? What are the ideal properties of any material that is to be implanted?
9. What nonmembranous organelle is prevalent in plasma cells? What membranous organelle is prevalent?

Access more review material online in the Study Area at **www.masteringmicrobiology.com.** There, you'll find

- **Animations**
- **Microflix**
- **MP3 Tutor Sessions**
- **Concept Mapping Activities**
- **Flashcards**
- **Quizzes**

and more to help you succeed.

17 Immunization and Immune Testing

A 13-month-old girl is brought into the hospital, coughing and crying. Her mother tells the pediatrician that she had noticed a mucous discharge pooling in the corners of her child's eyes the previous evening. The pediatrician checks the girl's medical record and notes that she has not yet received her first measles/mumps/rubella (MMR) vaccination. The doctor then swabs the child's throat, arranges for a blood sample to be taken, and tells the mother that they are going to have the lab perform a "special immune test" to confirm the diagnosis: measles. Because the pediatrician has seen several children with measles in the past few weeks and the child's symptoms support the diagnosis, a nurse administers an intramuscular injection of immunoglobulin against measles. The nurse explains that this injection will protect the child against the most severe form of the illness, which can be fatal. The physician also schedules the daughter for a routine measles/mumps/rubella immunization. Mother and daughter return home to rest and recuperate.

How do immunoglobulins and vaccines work? How do immunological tests aid in the diagnosis of specific diseases? This chapter is an introduction to the important topics of immunization and immune testing.

Take the pre-test for this chapter online. Visit the Study Area at www.masteringmicrobiology.com.

Immunizations have done more to decrease human disease than any other medical advancement. Here, vaccine vials await shipment.

In this chapter we will discuss three applications of immunology: active immunization (vaccination); passive immunotherapy using immunoglobulins (antibodies); and immune testing. Vaccination has proven the most efficient and cost-effective method of controlling infectious diseases. Without the use of effective vaccines, millions more people worldwide would suffer each year from potentially fatal infectious diseases, including measles, mumps, and polio. The administration of immunoglobulins has further reduced morbidity and mortality from certain infectious diseases, such as hepatitis A and yellow fever, in unvaccinated individuals. Medical personnel also make practical use of the immune response as a diagnostic procedure. For example, the detection of antibodies to HIV in a person's blood indicates that the individual has been exposed to that virus and may develop AIDS. The remarkable specificity of antibodies also enables the detection of drugs in urine, recognition of pregnancy at early stages, and the identification or characterization of other biological material. The many tests developed for these purposes are the focus of the discipline of *serology* (sĕ-rol'ō-jē) and are discussed in the second half of this chapter.

Immunization

As we saw in Chapter 16, an individual may be made immune to an infectious disease by two artificial methods: *active immunization,* which involves administering antigens to a patient so that the patient actively mounts an adaptive immune response, and *passive immunotherapy,* in which a patient acquires temporary immunity through the transfer of antibodies formed by other individuals or animals.

In the following sections we will review the history of immunization before examining immunization and immunotherapy in more detail.

Brief History of Immunization

Learning Objective

✓ Discuss the history of vaccination from the 12th century through the present.

As early as the 12th century, the Chinese noticed that children who recovered from smallpox never contracted the disease a second time. They therefore adopted a policy of deliberately infecting young children with particles of ground smallpox scabs from children who had survived mild cases. By doing so, they succeeded in significantly reducing the population's overall morbidity and mortality from the disease. News of this procedure, called *variolation* (var'ē-ō-lā'shŭn), spread westward through central Asia, and the technique was widely adopted.

Lady Mary Montagu (1689–1762), the wife of the English ambassador to the Ottoman Empire, learned of the procedure, had it performed on her own children, and told others about it upon her return to England in 1721. As a result, variolation came into use in England and in the American colonies. Although effective and usually successful, variolation caused death from smallpox in 1–2% of recipients and in people exposed to recipients, so in time the procedure was outlawed.

Thus, when the English physician Edward Jenner demonstrated in 1796 that protection against smallpox could be conferred by inoculation with crusts from a person infected with cowpox—a related but very mild disease—the new technique was adopted. Because cowpox was also called *vaccinia*[1] (vak-sin'ē-ă), Jenner called the new technique **vaccination** (vak'si-nā'shŭn), and the protective inoculum a **vaccine** (vak-sēn'). Today we use the term **immunization** to refer to the administration of any antigenic inoculum, which are all called vaccines. For many years thereafter, vaccination against smallpox was widely practiced, even though no one understood how it worked or whether similar techniques could protect against other diseases.

In 1879, Louis Pasteur conducted experiments on the bacterium *Pasteurella multocida* (pas-ter-el'ă mul-tŏ'si-da) and demonstrated that he could make an effective vaccine against this organism (which causes a disease in birds called fowl cholera). Once the basic principle of vaccine manufacture was understood, vaccines against anthrax and rabies rapidly followed. After it was discovered that these vaccines provide protection through the actions of antibodies, the technique of transferring protective antibodies to susceptible individuals—that is, *passive immunotherapy* (im'ū-nō-thār'ă-pē)—was developed soon thereafter.

By the late 1900s, immunologists and health care providers had formulated vaccines that significantly reduced the number of cases of many infectious diseases **(Figure 17.1)**. We also have successful vaccines against some types of cancer. Health care providers, governments, and international organizations working together have rid the world of naturally occurring smallpox, and we hope for the worldwide eradication of polio, measles, mumps, and rubella. **Highlight: Why Isn't There a Cold Vaccine?** on p. 492 discusses why a vaccine for the common cold does not yet exist.

Even though immunologists have produced vaccines that protect people against many deadly diseases, a variety of political, social, economic, and scientific problems prevent vaccines from reaching all those who need them. In developing nations worldwide, over 3 million children still die each year from vaccine-preventable infectious diseases, primarily because of political obstacles. Additionally, some pathogens, such as the protozoa of malaria and the virus of AIDS, still frustrate attempts to develop effective vaccines against them. Furthermore, the existence of vaccine-associated risks—both medical risks (the low but persistent incidence of vaccine-caused diseases) and financial risks (the high costs of developing and producing vaccines, and the risk of lawsuits by vaccine recipients who have adverse reactions)—has in recent years discouraged investment in new vaccines. Thus, although the history of immunization is marked by stunning advancements in public health, the future of immunization poses immense challenges.

Next we take a closer look at active immunization, commonly known as vaccination.

[1]From Latin *vacca*, meaning cow.

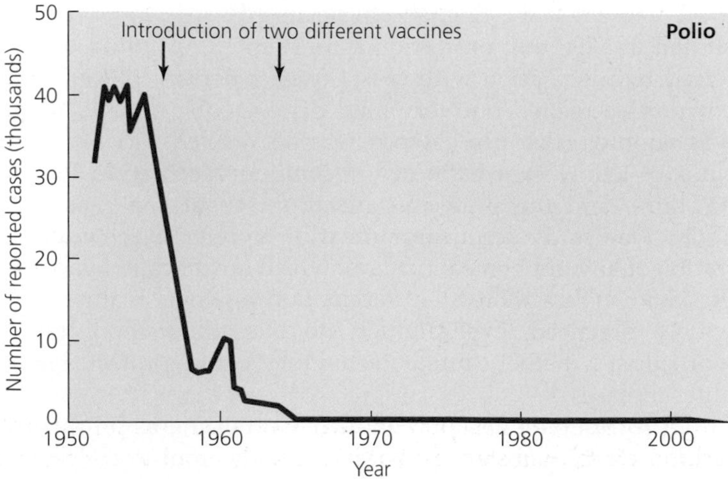

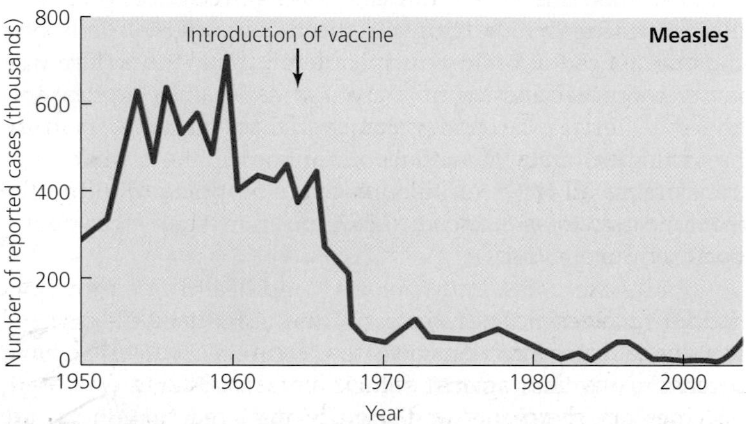

▲ **Figure 17.1 The effect of immunization in reducing the prevalence of two infectious diseases in the United States.** Polio is no longer endemic in the United States. Measles is nearly eradicated.

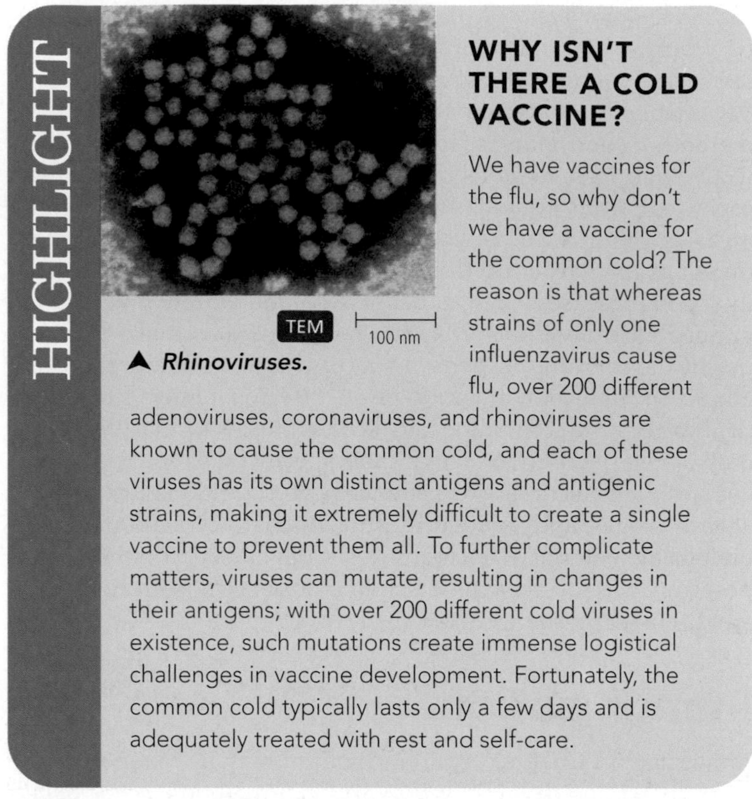

▲ *Rhinoviruses.*

WHY ISN'T THERE A COLD VACCINE?

We have vaccines for the flu, so why don't we have a vaccine for the common cold? The reason is that whereas strains of only one influenzavirus cause flu, over 200 different adenoviruses, coronaviruses, and rhinoviruses are known to cause the common cold, and each of these viruses has its own distinct antigens and antigenic strains, making it extremely difficult to create a single vaccine to prevent them all. To further complicate matters, viruses can mutate, resulting in changes in their antigens; with over 200 different cold viruses in existence, such mutations create immense logistical challenges in vaccine development. Fortunately, the common cold typically lasts only a few days and is adequately treated with rest and self-care.

Active Immunization

Learning Objectives

- ✓ Describe the advantages and disadvantages of five types of vaccines.
- ✓ Describe three methods by which recombinant genetic techniques can be used to develop improved vaccines.
- ✓ Delineate the risks and benefits of routine vaccination in healthy populations, mentioning contact immunity and herd immunity.

In the following subsections we examine types of vaccines, the roles of technology in producing modern vaccines, and issues concerning vaccine safety.

Vaccine Types

Scientists are constantly striving to develop vaccines of maximal efficacy and safety. In each case, a pathogen is altered or inactivated so that it is less likely to cause illness; however, not all types of vaccines are equally safe or effective. Effectiveness can be checked by measuring the antibody (IgG and IgM) level—called the **titer** (tī′ter)—in the blood. When the titer is low, antibody production can be bolstered by administration of more antigen—a *booster immunization.*

The general types of vaccines, each of which has its own combination of strengths and weaknesses, are attenuated (live) vaccines, killed (or inactivated) vaccines, toxoid vaccines, combination vaccines, and recombinant gene vaccines. Each of these is named for the type of antigen used in the inoculum.

Attenuated (Live) Vaccines Virulent microbes are normally not used in vaccines, because they cause disease. Instead, immunologists reduce virulence so that, although still active, the pathogens no longer cause disease. The process of reducing virulence is called **attenuation** (ă-ten-ū-ā′shŭn). The most common method for attenuating viruses involves raising them for numerous generations in tissue culture cells until the viruses lose the ability to produce disease. For example, rabies viruses, which preferentially attack nerve cells, are subjected to prolonged tissue culture until their virulence to nerve cells is lost; the resulting avirulent viruses can be used in vaccines. Bacteria may be made avirulent by culturing them under unusual conditions or by using genetic manipulation.

Attenuated vaccines—those containing attenuated microbes—are also called *modified live vaccines.* Because they contain active but avirulent organisms or viruses, these vaccines cause very mild infections but no serious disease under normal conditions. Attenuated viruses in such a vaccine infect host cells and replicate; the infected cells then process endogenous viral antigens. As a result, modified live viral vaccines trigger a cell-mediated immune response dominated by type 1 helper T cells (Th1) and cytotoxic T cells (Tc). Because modified live vaccines contain active microbes, a large number of antigen molecules are available to stimulate an immune response. Further, vaccinated individuals can infect those around them, providing **contact**

TABLE 17.1 Some Common Adjuvants

Adjuvant	Effects
Aluminum phosphate (alum)	Slows processing and degradation of antigen
Saponin (soaplike plant product)[a]	Stimulates T cell responses
Mineral oil[a]	Slows processing and degradation of antigen
Freund's complete adjuvant (mineral oil containing killed mycobacteria)[a]	Slows processing and degradation of antigen; stimulates T cell responses

[a]Considered too toxic for humans; used in vaccinating animals only.

immunity—that is, immunity beyond the individual receiving the vaccine.

Although usually very effective, attenuated vaccines can be hazardous because modified microbes may retain enough residual virulence to cause disease in immunosuppressed people. Pregnant women should not receive live vaccines because of the danger that the attenuated pathogen will cross the placenta and harm the fetus. Occasionally, modified viruses actually revert to wild type or mutate to a form that causes persistent infection or disease. For example, in 2000 a polio epidemic in the Dominican Republic and Haiti resulted from the reversion of an attenuated live virus in oral polio vaccine to a virulent poliovirus. For this reason, we no longer use live polio vaccine to immunize children in the United States.

Health care providers and government agencies must carefully balance the benefits of attenuated vaccines against their risks. As a result, they may change their recommended immunization schedules.

Inactivated (Killed) Vaccines For some diseases, live vaccines have been replaced by **inactivated vaccines,** which are of two types: *whole agent vaccines* are produced with deactivated but whole microbes, whereas *subunit vaccines* are produced with antigenic fragments of microbes. Because neither whole agent nor subunit vaccines can replicate, revert, mutate, or retain residual virulence, they are safer than live vaccines. However, because they cannot replicate, several "booster" doses must be administered to achieve full immunity, and immunized individuals do not stimulate contact immunity. Also, with whole agent vaccines, nonantigenic portions of the microbe occasionally stimulate a painful inflammatory response in some individuals. As a result, whole agent pertussis vaccine is now being replaced with a subunit vaccine called acellular pertussis vaccine.

When microbes are killed for use in vaccines, it is important that their antigens remain as similar to those of living organisms as possible. If chemicals are used for killing, they must not alter the antigens responsible for stimulating protective immunity. A commonly used inactivating agent is *formaldehyde,* which denatures proteins and nucleic acids.

Because the microbes of inactivated vaccines cannot reproduce, they do not present as many antigenic molecules to the body as do live vaccines; therefore, inactivated vaccines are antigenically weak. They are administered in high doses or in multiple doses, or incorporated with materials called **adjuvants** (ad'joo-vănts), substances that increase the effective antigenicity of the vaccine by stimulating Toll-like receptors and their actions. Unfortunately, high individual doses and multiple dosing increase the risk of producing allergies, and the use of adjuvants to increase antigenicity may stimulate local inflammation. Table 17.1 lists some common adjuvants and their effects in enhancing the efficacy of vaccines.

Because all types of killed vaccines are recognized by the immune system as exogenous antigens, they stimulate an antibody immune response.

Toxoid Vaccines For some bacterial diseases, notably tetanus and diphtheria, it is more efficacious to induce an immune response against toxins than against cellular antigens. **Toxoid** (tok'soyd) **vaccines** are chemically or thermally modified toxins that are used in vaccines to stimulate active immunity. As with killed vaccines, toxoids stimulate antibody-mediated immunity. Because toxoids have few antigenic determinants, effective immunization requires multiple childhood doses as well as reinoculations every 10 years for life. **ANIMATIONS:** ***Vaccines: Function, Types***

Combination Vaccines The Centers for Disease Control and Prevention (CDC) has approved several **combination vaccines** for routine use. These vaccines combine antigens from several toxoids and inactivated pathogens that are administered simultaneously. Examples include MMR—vaccine against measles, mumps, and rubella—and Pentacel, which is a vaccine against diphtheria, tetanus, pertussis (whooping cough), polio, and diseases of *Haemophilus influenzae* (hē-mof'i-lŭs in-flu-en'zī).

Vaccines Using Recombinant Gene Technology Although live, inactivated, and toxoid vaccines have been highly successful in controlling infectious diseases, researchers are always seeking ways to make vaccines more effective, cheaper, and safer and to make new vaccines against pathogens that have been difficult to protect against. For example, scientists have developed a recombinant DNA vaccine against a fungus, *Blastomyces* (blas-tō-mī'sēz)—the first vaccine against a fungal pathogen. Scientists can also use a variety of genetic recombinant techniques to make improved vaccines. For example, they can selectively delete virulence genes from a pathogen, producing an irreversibly attenuated microbe, one that cannot revert to a virulent pathogen **(Figure 17.2a)**.

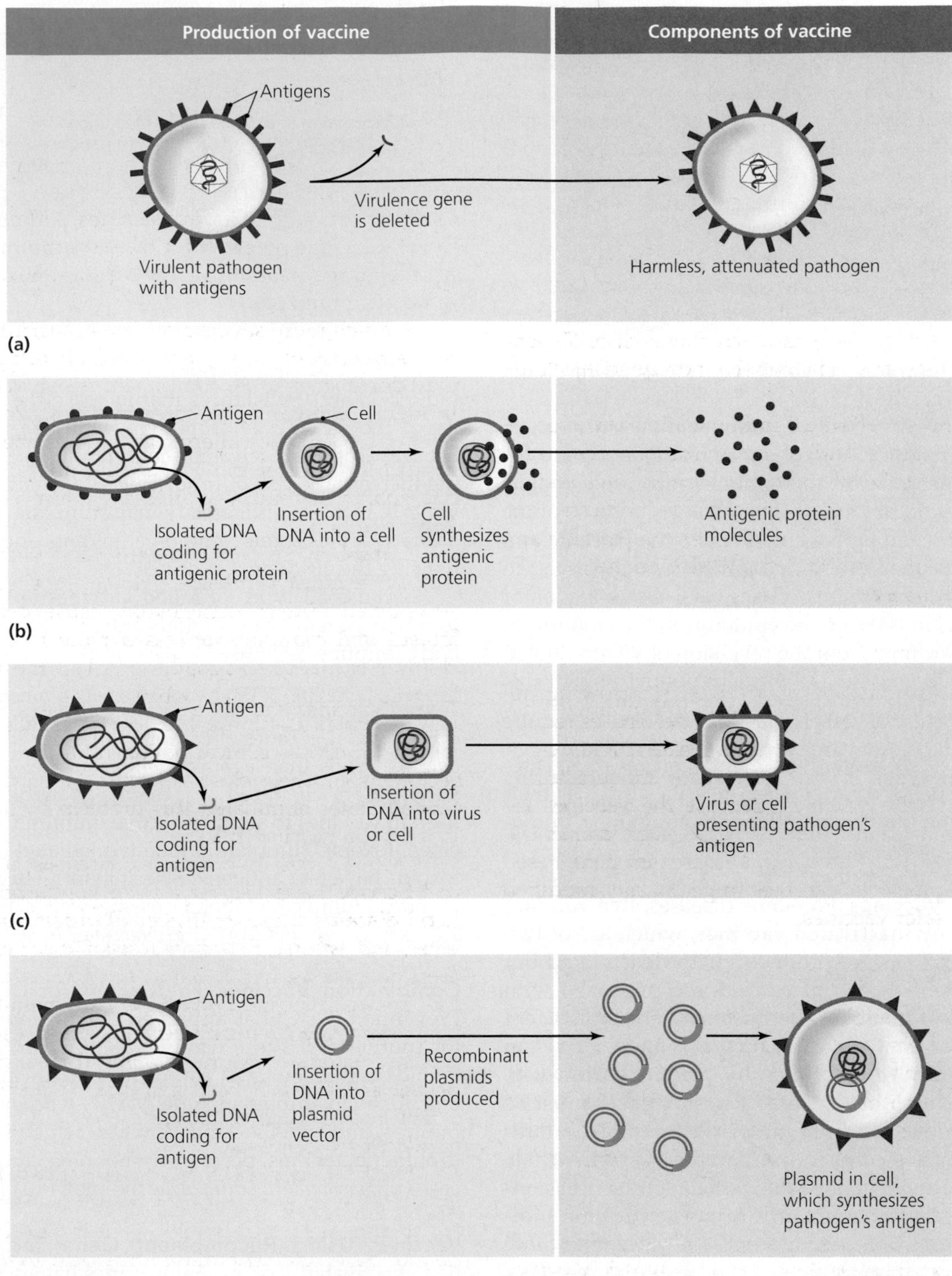

▲ **Figure 17.2 Some uses of recombinant DNA technology for making improved vaccines.** **(a)** Deletion of virulence gene(s) to create an attenuated pathogen for use in a vaccine. **(b)** Insertion of gene that codes for a selected antigenic protein into a cell, which then produces large quantities of the antigen for use in a vaccine. **(c)** Insertion of a gene that codes for a selected antigenic protein into a cell or virus, which displays the antigen. The entire recombinant is used in a vaccine. **(d)** Injection of DNA containing a selected gene (in this case, as part of a plasmid) into an individual. Once some of this DNA is incorporated into the genome of a patient's cells, those cells synthesize and process the antigen, which stimulates a cell-mediated immune response.

Scientists also use recombinant techniques to produce large quantities of very pure viral or bacterial antigens for use in vaccines. In this process, scientists isolate the gene that codes for an antigen and insert it into a bacterium, yeast, or other cell, which then expresses the antigen **(Figure 17.2b)**. Vaccine manufacturers produce hepatitis B vaccine in this manner using recombinant yeast cells.

Alternatively, a genetically altered microbial cell or virus may itself be used as a live recombinant vaccine **(Figure 17.2c)**. Experimental recombinant vaccines of this type have used adenoviruses, herpesviruses, poxviruses, and bacteria such as *Salmonella* (sal′mŏ-nel′ă). Vaccinia virus (a poxvirus) is used because it is easy to administer by dermal scratching or orally, and because its large genome makes inserting a new gene into it relatively easy.

Another innovative method of immunization involves injection not of antigens but instead of the DNA that codes for the antigen. For example, the DNA coding for a pathogen's antigen can be inserted into a plasmid vector, which is then injected into the body **(Figure 17.2d)**. The body's cells then transcribe and translate the gene to produce antigen, which triggers a cell-mediated immune response.

Vaccine Manufacture

Manufacturers mass-produce many vaccines by growing microbes in laboratory culture vessels, but because viruses require a host cell to reproduce, they are cultured inside chicken eggs. Availability of sterile eggs is thus critical for manufacturing viral vaccines such as flu vaccines. Because the vaccines are produced in eggs, physicians must withhold such immunizations from patients with egg allergies. Research on gene-based vaccines and development of vaccines in genetically modified plants may result in safer vaccines.

Recommended Immunizations

The CDC and medical associations publish recommended immunization schedules for children, adults, and special populations such as health care workers and HIV-positive individuals. The recommendations are frequently modified to reflect changes in the relationships between pathogens and the human population. **Figure 17.3** highlights the general 2009 immunization schedules recommended by the CDC. In **Table 17.2** on p. 497, facts concerning the types of vaccines available for immunizing against each of the diseases in the vaccination schedule as well as some other available vaccines are listed. Vaccines against anthrax, cholera, plague, tuberculosis, and other diseases are available, but the CDC does not recommend them for the general U.S. population.

It is important that patients follow the recommended immunization schedule not only to protect themselves but also to provide society with **herd immunity.** Herd immunity is the protection provided all individuals in a population due to the inability of a pathogen to effectively spread when a large proportion of individuals (typically more than 75%) are resistant. When immunization compliance in a population has fallen, local epidemics have resulted.

Vaccine Safety

Health care providers must carefully weigh the risks associated with vaccines against their benefits. A common vaccine-associated problem is mild toxicity. Some vaccines—especially whole agent vaccines that contain adjuvants—may cause pain at the injection site for several hours or days after injection. In rare cases, toxicity may result in general malaise and possibly a fever high enough to induce seizures. Although not usually life threatening, the potential for these symptoms may be sufficient to discourage people from being immunized or having their infants immunized.

A much more severe problem associated with immunization is the risk of *anaphylactic shock,* an allergic reaction, which may develop to some component of the vaccine, such as egg proteins, adjuvants, or preservatives. Because people are rarely aware of such allergies ahead of time, recipients should remain for several minutes in the physician's office, where epinephrine is readily available to counter any signs of an allergic reaction.

A third major problem associated with immunization is that of residual virulence, which we previously discussed. Attenuated viruses occasionally cause disease, not only in fetuses and immunosuppressed patients but also in healthy children and adults. A good example is the attenuated oral poliovirus vaccine (OPV), which was commonly used in the United States until the late 1990s. Though a very effective vaccine, it causes clinical poliomyelitis in 1 of every 2 million recipients or their close contacts. Medical personnel in the United States eliminated this problem by switching to inactivated polio vaccine (IPV).

Over the past two decades, lawsuits in the United States and Europe have alleged that certain vaccines against childhood diseases cause or trigger disorders such as autism, diabetes, and asthma. Extensive research has failed to substantiate these allegations, and to date no conclusive evidence for such a link has been found. Indeed, vaccine manufacturing methods have improved tremendously in recent years, ensuring that modern vaccines are much safer than those in use even a decade ago. The U.S. Food and Drug Administration (FDA) has established a Vaccine Adverse Event Reporting System for monitoring vaccine safety.

The CDC and FDA have determined that the problems associated with immunization are far less serious than the suffering and death that would result if we stopped immunizing people. **Beneficial Microbes: Smallpox: To Vaccinate or Not to Vaccinate?** on p. 499 discusses the issues surrounding the administration of smallpox vaccinations to the general public.

Passive Immunotherapy

Learning Objectives

- ✓ Identify two sources of antibodies for use in passive immunotherapy.
- ✓ Compare the relative advantages and disadvantages of active immunization and passive immunotherapy.

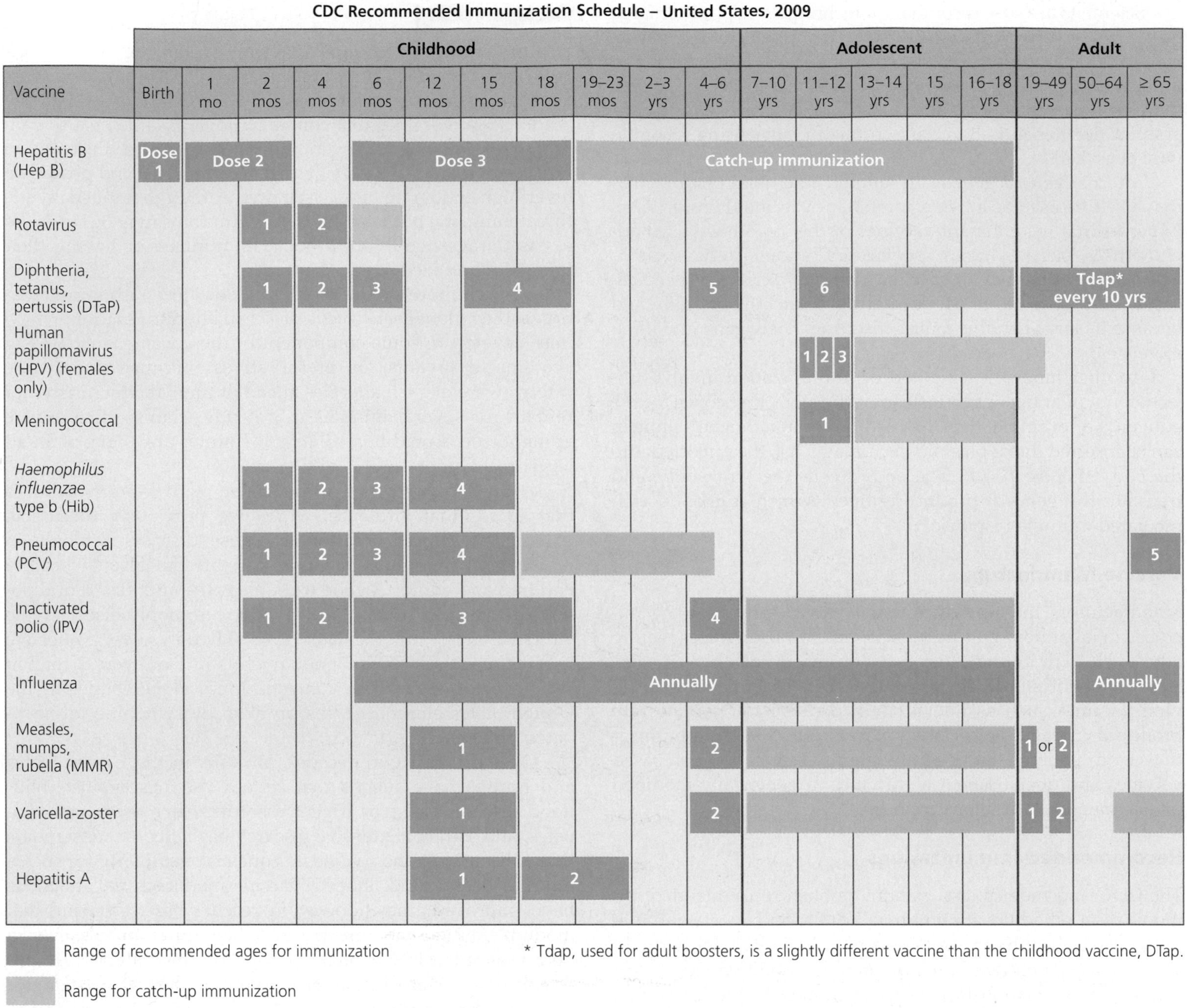

CDC Recommended Immunization Schedule – United States, 2009

<table>
<tr><th></th><th colspan="11">Childhood</th><th colspan="5">Adolescent</th><th colspan="3">Adult</th></tr>
<tr><th>Vaccine</th><th>Birth</th><th>1 mo</th><th>2 mos</th><th>4 mos</th><th>6 mos</th><th>12 mos</th><th>15 mos</th><th>18 mos</th><th>19–23 mos</th><th>2–3 yrs</th><th>4–6 yrs</th><th>7–10 yrs</th><th>11–12 yrs</th><th>13–14 yrs</th><th>15 yrs</th><th>16–18 yrs</th><th>19–49 yrs</th><th>50–64 yrs</th><th>≥ 65 yrs</th></tr>
<tr><td>Hepatitis B (Hep B)</td><td>Dose 1</td><td colspan="2">Dose 2</td><td></td><td colspan="4">Dose 3</td><td colspan="8">Catch-up immunization</td><td></td><td></td><td></td></tr>
<tr><td>Rotavirus</td><td></td><td></td><td>1</td><td>2</td><td></td><td></td><td></td><td></td><td></td><td></td><td></td><td></td><td></td><td></td><td></td><td></td><td></td><td></td><td></td></tr>
<tr><td>Diphtheria, tetanus, pertussis (DTaP)</td><td></td><td></td><td>1</td><td>2</td><td>3</td><td></td><td colspan="2">4</td><td></td><td></td><td>5</td><td></td><td>6</td><td colspan="3"></td><td colspan="3">Tdap* every 10 yrs</td></tr>
<tr><td>Human papillomavirus (HPV) (females only)</td><td></td><td></td><td></td><td></td><td></td><td></td><td></td><td></td><td></td><td></td><td></td><td></td><td>1 2 3</td><td colspan="4"></td><td></td><td></td></tr>
<tr><td>Meningococcal</td><td></td><td></td><td></td><td></td><td></td><td></td><td></td><td></td><td></td><td></td><td></td><td></td><td>1</td><td colspan="3"></td><td></td><td></td><td></td></tr>
<tr><td>Haemophilus influenzae type b (Hib)</td><td></td><td></td><td>1</td><td>2</td><td>3</td><td colspan="2">4</td><td></td><td></td><td></td><td></td><td></td><td></td><td></td><td></td><td></td><td></td><td></td><td></td></tr>
<tr><td>Pneumococcal (PCV)</td><td></td><td></td><td>1</td><td>2</td><td>3</td><td colspan="2">4</td><td colspan="4"></td><td></td><td></td><td></td><td></td><td></td><td></td><td></td><td>5</td></tr>
<tr><td>Inactivated polio (IPV)</td><td></td><td></td><td>1</td><td>2</td><td colspan="4">3</td><td></td><td></td><td>4</td><td colspan="5"></td><td></td><td></td><td></td></tr>
<tr><td>Influenza</td><td></td><td></td><td></td><td></td><td colspan="12">Annually</td><td></td><td colspan="2">Annually</td></tr>
<tr><td>Measles, mumps, rubella (MMR)</td><td></td><td></td><td></td><td></td><td></td><td colspan="2">1</td><td></td><td></td><td></td><td>2</td><td colspan="5"></td><td>1 or 2</td><td></td><td></td></tr>
<tr><td>Varicella-zoster</td><td></td><td></td><td></td><td></td><td></td><td colspan="2">1</td><td></td><td></td><td></td><td>2</td><td colspan="5"></td><td>1 2</td><td></td><td></td></tr>
<tr><td>Hepatitis A</td><td></td><td></td><td></td><td></td><td></td><td colspan="2">1</td><td colspan="2">2</td><td></td><td></td><td></td><td></td><td></td><td></td><td></td><td></td><td></td><td></td></tr>
</table>

Range of recommended ages for immunization

Range for catch-up immunization

* Tdap, used for adult boosters, is a slightly different vaccine than the childhood vaccine, DTap.

▲ **Figure 17.3**
The CDC's recommended immunization schedule for the general population.
Adult meningococcal vaccine recommended for college students who live in dormitories.
Three doses of human papillomavirus vaccine (HPV) recommended for women aged 19–26.

Passive immunotherapy (sometimes called passive immunization) involves the administration of preformed antibodies to a patient. Physicians use passive immunotherapy when protection against a recent infection or an ongoing disease is needed quickly. Rapid protection is achieved because passive immunotherapy does not require the body to mount a response; instead, preformed antibodies are immediately available to bind to antigen, enabling neutralization and opsonization to proceed without delay. For example, in a case of botulism poisoning [caused by the toxin of *Clostridium botulinum* (klos-trid'ē-ŭm bo-tū-lī'num)], passive immunotherapy with preformed antibodies against the toxin can prevent death.

Antibodies directed against toxins are also called *antitoxins* (an-tē-tok'sinz); *antivenin* used to treat snakebites is an antitoxin. In some cases, infections with certain viruses—hepatitis A and B, measles, rabies, Ebola, chickenpox, and shingles—are treated with antibodies directed against the causative viruses.

For much of the time since passive immunotherapy was first developed, immunologists harvested the desired antibodies from human or animal donors that have either experienced natural

TABLE 17.2

Principal Vaccines to Prevent Human Diseases

Vaccine	Disease Agent	Disease	Vaccine Type	Method of Administration
Recommended by CDC				
Hepatitis B	Hepatitis B virus	Hepatitis B	Inactive subunit from recombinant yeast	Intramuscular
Rotavirus	*Rotavirus*	Gastroenteritis	Attenuated, recombinant	Oral
Diphtheria/ tetanus/ acellular pertussis (DTaP)	Diphtheria toxin Tetanus toxin *Bordetella pertussis*	Diphtheria Tetanus Whooping cough	Toxoid Toxoid Inactivated subunit (inactivated whole also available)	Intramuscular
Human papillomavirus (HPV)	Human papillomaviruses	Genital warts, cervical cancer	Inactive recombinant	Intramuscular
Meningococcal	*Neisseria meningiditis*	Meningitis	Inactive	Subcutaneous or intramuscular
Haemophilus influenzae type b (Hib)	*Haemophilus influenzae*	Meningitis, pneumonia, epiglottitis	Inactivated subunit	Intramuscular
Pneumococcal (PCV)	*Streptococcus pneumoniae*	Pneumonia	Inactivated subunit	Intramuscular
Polio	Poliovirus	Poliomyelitis	Inactivated (attenuated also available)	Subcutaneous or intramuscular (attenuated: oral)
Influenza	Influenzaviruses	Flu	Inactivated subunit	Intramuscular or oral
Measles/ mumps/ rubella (MMR)	Measles virus Mumps virus Rubella virus	Measles Mumps Rubella (German measles)	Attenuated Attenuated Attenuated	Subcutaneous
Varicella-zoster	Chickenpox virus	Chickenpox, shingles	Attenuated	Subcutaneous
Hepatitis A	Hepatitis A virus	Hepatitis A	Inactivated whole	Intramuscular
Available But Not Recommended for General Population in the U.S.				
Anthrax	*Bacillus anthracis*	Anthrax	Inactivated whole	Subcutaneous
BCG (bacillus of Calmette and Guérin)	*Mycobacterium tuberculosis, M. leprae*	Tuberculosis, leprosy	Attenuated	Intradermal
Japanese encephalitis vaccine	Japanese encephalitis virus	Encephalitis	Inactive	Subcutaneous
Rabies	Rabies virus	Rabies	Inactivated whole	Intramuscular or intradermal
Typhoid fever vaccine	*Salmonella enterica*	Typhoid fever	Attenuated (inactive also available)	Oral (inactive: subcutaneous or intramuscular)
Vaccinia (cowpox)	Smallpox virus, monkeypox virus	Smallpox, monkeypox	Attenuated	Subcutaneous
Yellow fever	Yellow fever virus	Yellow fever	Attenuated	Subcutaneous

causes of a disease or were actively immunized against it. To do so, immunologists remove the cells and clotting factors from the blood to acquire *serum,* which contains a variety of antibodies. When used for passive immunotherapy, such serum is called **antiserum** (an-tē-sē′rŭm) or sometimes *immune serum.* Antisera are typically collected from human blood plasma donors, or from horses intentionally exposed to the disease agent of interest because their large blood volume contains more antibodies than can be obtained from smaller animals. Pooled antisera from a group of human donors can be administered intravenously (intravenous immunoglobulins, IVIg) to treat immunodeficiencies and some autoimmune and inflammatory diseases.

Antisera have the following limitations:

- Repeated injections of horse-derived antisera can trigger a serious allergic response called *serum sickness,* in which the recipient mounts an immune response against horse antigens found in the antisera.
- Viral pathogens may contaminate antisera.
- The antibodies of horse-derived antisera are degraded relatively quickly by the recipient. The half-life of such antibodies is about 3 weeks; that is, half the molecules are degraded every 3 weeks.

Scientists have overcome the limitations of antisera by developing **hybridomas** (hī-brid-ō′măz), which are tumor cells created

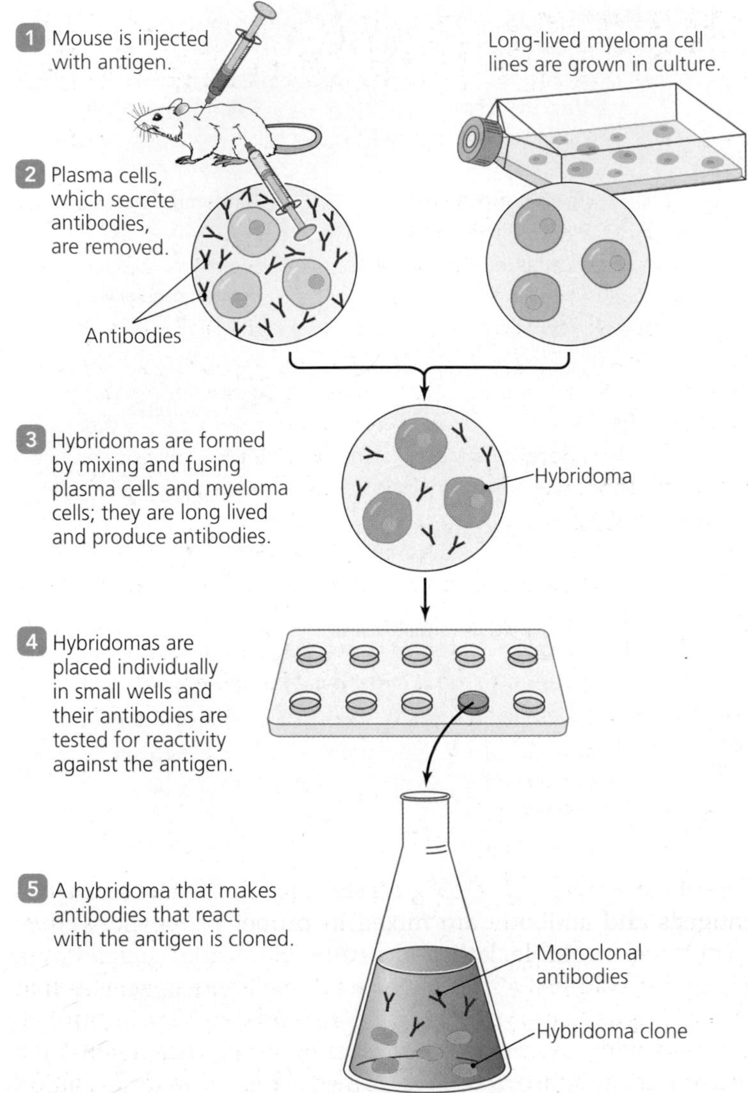

1 Mouse is injected with antigen.
Long-lived myeloma cell lines are grown in culture.
2 Plasma cells, which secrete antibodies, are removed.
Antibodies
3 Hybridomas are formed by mixing and fusing plasma cells and myeloma cells; they are long lived and produce antibodies.
Hybridoma
4 Hybridomas are placed individually in small wells and their antibodies are tested for reactivity against the antigen.
5 A hybridoma that makes antibodies that react with the antigen is cloned.
Monoclonal antibodies
Hybridoma clone

▲ **Figure 17.4 The production of hybridomas.** After a laboratory animal is injected with the antigen of interest 1, plasma cells are removed from the animal and isolated 2. When these plasma cells are fused with cultured cancer cells called myelomas, the results are hybridomas 3. Once the hybridomas are cultured individually and the hybridoma that produces antibodies against the antigen of interest is identified 4, it is cloned to produce a large number of hybridomas, all of which secrete identical antibodies called monoclonal antibodies 5.

by fusing antibody-secreting plasma cells with cancerous plasma cells called *myelomas* (mī-ĕ-lō′măz) **(Figure 17.4)**. Each hybridoma divides continuously (because of the cancerous plasma cell component) to produce clones of itself, and each clone secretes large amounts of a single antibody molecule. These identical antibodies are called **monoclonal** (mon-ō-klō′năl) **antibodies** because all of them are secreted by clones originating from a single plasma cell. Once scientists have identified the hybridoma that secretes antibodies complementary to the antigen of interest, they maintain it in tissue culture to produce the antibodies needed for passive immunotherapy. For example, physicians use such a monoclonal antibody to treat newborns infected with respiratory syncytial virus.

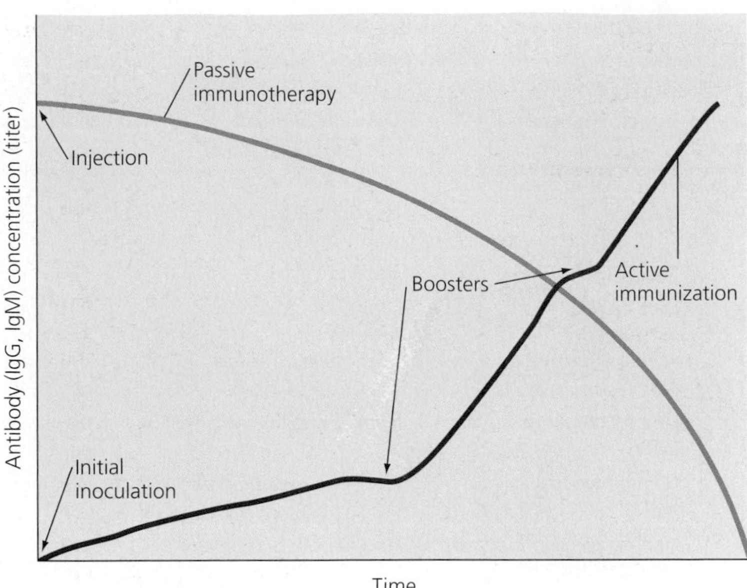

▲ **Figure 17.5 The characteristics of immunity produced by active immunization (red) and passive immunotherapy (green).** Passive immunotherapy provides strong and immediate protection, but it disappears relatively quickly. Active immunity takes some time and may require additional booster inoculations to reach protective levels, but it is long lasting and capable of restimulation.

Active immunization and passive immunotherapy are used in different circumstances because they provide protection with different characteristics **(Figure 17.5)**. As just noted, passive immunotherapy with preformed antibodies is used whenever immediate protection is required. However, because preformed antibodies are removed rapidly from the blood and no memory B cells are produced, protection is temporary, and the recipient becomes susceptible again. Active immunization provides long-term protection that is capable of restimulation. Thus, when initiated before any exposure to *C. tetani* has occurred, active immunization using a tetanus toxoid develops long-lasting protection that is readily available upon exposure to the toxin.

Antibody-Antigen Immune Testing

Learning Objectives

✓ Distinguish between direct and indirect immune testing using antibody-antigen interactions.

✓ Define *serology*.

✓ In general terms, compare and contract precipitation, agglutination, neutralization, complement fixation, and labeled antibody tests.

There are two basic categories of immune testing using antibody-antigen interactions. For *direct testing*, the investigator is looking for the presence of an antigen, often a pathogen, in a specimen, which is typically taken from a site of infection. *Indirect testing* involves checking the blood or serum for antibodies that have formed against a particular antigen. Thus, in each category the

BENEFICIAL MICROBES

SMALLPOX: TO VACCINATE OR NOT TO VACCINATE?

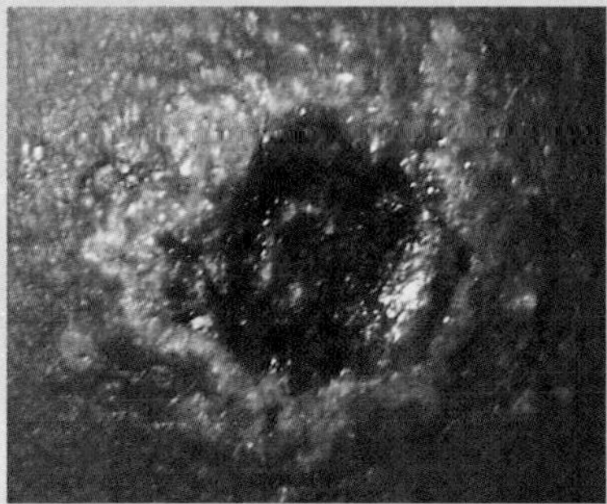

▲ *Vaccinia necrosum.*

Dr. Edward Jenner developed an early use for a beneficial microbe in medicine. Medical personnel in the United States followed Jenner's example by regularly administering cowpox virus—as the smallpox vaccine—to the general public until 1971, at which time the risk of contracting smallpox was deemed too low to justify required vaccinations. Indeed, in 1980 the World Health Assembly declared smallpox successfully eradicated from the natural world. However, recent concerns about the potential use of smallpox virus as an agent of bioterrorism has sparked debate about whether or not citizens should once again be vaccinated against it.

Although safe and effective for most healthy adults, for others the attenuated cowpox virus can result in serious side effects—even death. Individuals with compromised immune systems (such as AIDS patients or cancer patients undergoing chemotherapy) are considered to be at particularly high risk for developing adverse reactions. Pregnant women, infants, and individuals with a history of the skin condition eczema are also considered poor candidates for the vaccine. Though rare, adverse reactions to the vaccine may also develop in certain otherwise healthy individuals. The more serious side effects include vaccinia necrosum (characterized by progressive cell death in the area of vaccination) and encephalitis (inflammation of the brain). Approximately 1 in every million individuals receiving cowpox virus as a vaccine for the first time develops a fatal reaction to it.

Is the risk of a bioterrorist smallpox attack great enough to warrant the exposure to the known risks of administering smallpox vaccine to the general population? If you were a public health official, what would you decide?

investigator uses either antibody or antigen to test for the presence of its complement—antibody locates antigen in direct testing, whereas antigen locates antibody in indirect testing. The latter process can be used either to monitor the spread of an infection through a population or to establish a diagnosis in an individual. For example, the presence of antibodies against HIV in an adult's serum is strong evidence that this individual has been infected with HIV, and a patient's immune response to tuberculosis antigen indicates exposure to *Mycobacterium tuberculosis* (mī'kō-bak-tē'rē-ŭm too-ber-kyū-lō'sis) (see Figure 18.11).

The study and diagnostic use of antigen-antibody interactions in blood serum is called **serology** (sĕ-rol'ō-jē). Researchers have developed a wide variety of serologic tests to visualize antibody-antigen interactions, ranging from simple, automated processes to complex tests requiring skilled technicians. Physicians and medical laboratory technologists choose a particular test based on the suspected diagnosis, the cost to perform the test, and the speed with which a result can be obtained. In the following subsections we will consider several types of serologic tests: precipitation tests, agglutination tests, neutralization tests, and several tagged antibody tests. Some of these procedures are presented for historical reasons—more accurate and faster modern tests have replaced them.

Precipitation Tests

Learning Objectives

- ✓ Describe the general principles of precipitation testing.
- ✓ Describe the technique of immunodiffusion.
- ✓ Discuss the production of anti-antibodies for immune testing.

One of the simplest of serologic tests relies on the fact that when antigens and antibody are mixed in proper proportions, they form huge, insoluble, lattice-like complexes called precipitates. When, for example, a solution of a soluble antigen, such as that of the fungus *Coccidioides immitis* (kok-sid-ē-oy'dēz im'mi-tis), is mixed with an antiserum containing antibodies against the antigen, the mixture quickly becomes cloudy due to formation of a precipitate consisting of antigen-antibody complexes (*also called* **immune complexes**).

When a given amount of antibody is added to each of a series of test tubes containing increasing amounts of antigen, the amount of precipitate increases gradually until it reaches a maximum **(Figure 17.6a)**. In test tubes containing still more antigen molecules, the amount of precipitate declines; in fact, in test tubes containing antigen in great excess over antibody, no precipitate at all develops. Thus, a graph of the amount of precipitate versus the amount of antigen has a maximum in the middle and lower values on either side.

The reasons behind this pattern of precipitation reactions are simple. Complex antigens are generally multivalent—each possesses many epitopes—and antibodies have pairs of active sites and therefore can simultaneously cross-link the same epitope on two antigen molecules. When there is excess antibody, each antigen molecule is covered with many antibody molecules, preventing extensive cross-linkage and thus precipitation **(Figure 17.6b)**. Such soluble immune complexes can activate complement in the kidneys, leading to inflammatory damage to blood vessels in the kidneys.

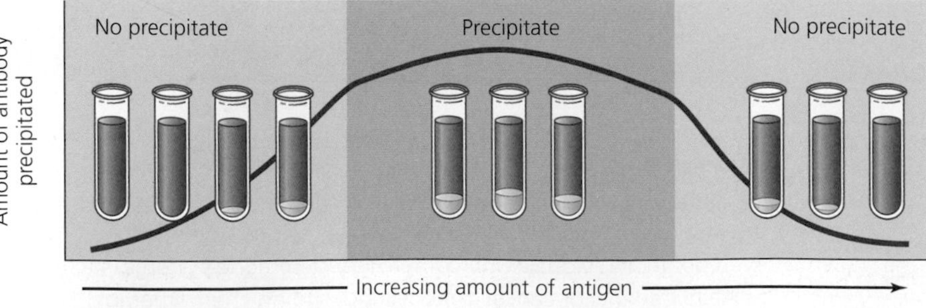

(a)

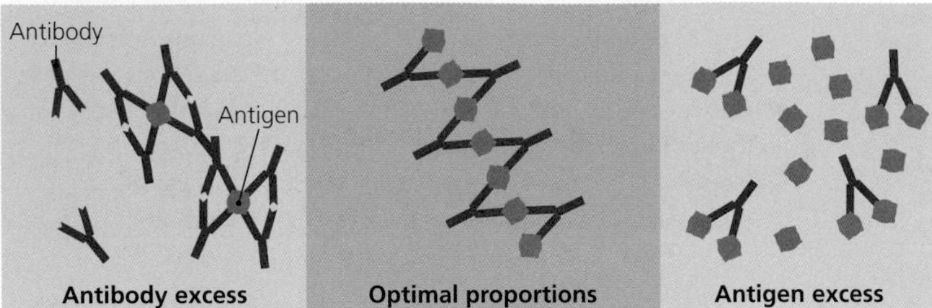

(b)

◀ **Figure 17.6 Characteristics of precipitation reactions.** **(a)** When increasing amounts of antigen are placed in tubes containing a constant amount of antibody, precipitate forms only in tubes having moderate amounts of antigen. The overlay graph shows the amount of precipitate formed versus the amount of antigen present. **(b)** In the presence of excess antigen or antibody, immune complexes are small and soluble; only when antigen and antibody are in optimal proportions do large complexes form and precipitate.

When the reactants are in optimal proportions, the ratio of antigen to antibody is such that cross-linking and lattice formation are extensive. As this lattice grows, it precipitates.

In mixtures in which antigen is in excess, each antigen molecule is bound to only two antibody molecules. There is no cross-linkage; because these complexes are small and soluble, no precipitation occurs. In the body, phagocytic cells do not easily remove these small immune complexes. As a result, immune complexes may be deposited in the joints and in the tiny blood vessels of the kidneys, where they trigger allergic reactions, as discussed in Chapter 18.

Because precipitation requires the mixing of antigen and antibody in optimal proportions, it is not possible to perform a precipitation test by combining just any two solutions containing these reagents. To ensure that the optimal concentrations of antibody and antigen come together, scientists can use a technique involving movement of the molecules through an agar gel: immunodiffusion.

Immunodiffusion

In the precipitation technique called **immunodiffusion** (im'ū-nō-di-fu'zhŭn), also known as *double immunodiffusion* or an *Ouchterlony* (ok'ter-lō-nē) *test,* a researcher cuts cylindrical holes called *wells* in an agar plate. One well is filled with a solution of antigen and the other with a solution of antibodies against the antigen. The antigen and antibody molecules diffuse in all directions out of the wells and into the surrounding agar, and where they meet in optimal proportions, a line of precipitation appears **(Figure 17.7a)**. If the solutions contain many different antigens and antibodies, each complementary pair of reactants reaches optimal proportions at different positions, and numerous lines of precipitation are produced—one for each interacting antigen-antibody pair **(Figure 17.7b)**. Such an immunodiffusion test has

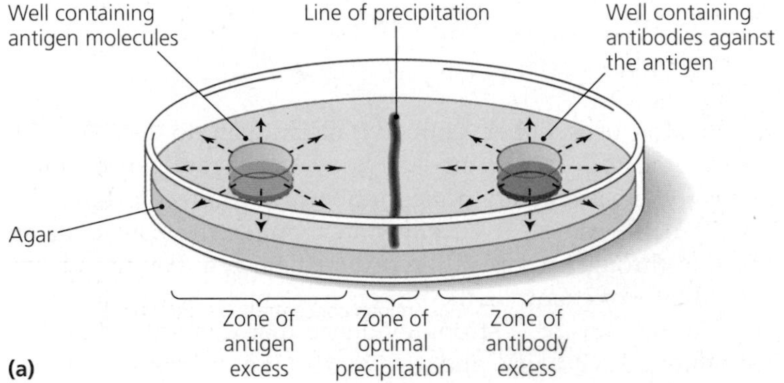

(a)

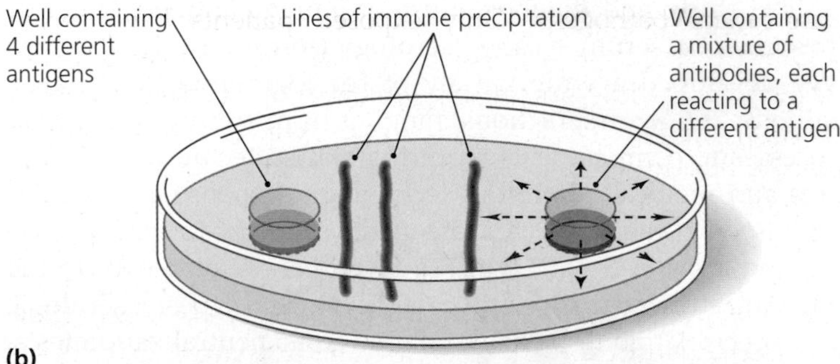

(b)

▲ **Figure 17.7 Immunodiffusion, a type of precipitation reaction.** **(a)** Antigen and antibody placed in wells diffuse out and through the agar; where they meet in optimal proportions, a line of precipitation forms. **(b)** When multiple antigens and antibodies are placed in the wells, multiple lines of precipitation mark the sites where different antigen-antibody combinations occurred in optimal proportions. *Why did only three lines of precipitation occur when the antigen well contained four antigens?*

Figure 17.7 None of the antibodies used in the test was complementary to the fourth antigen.

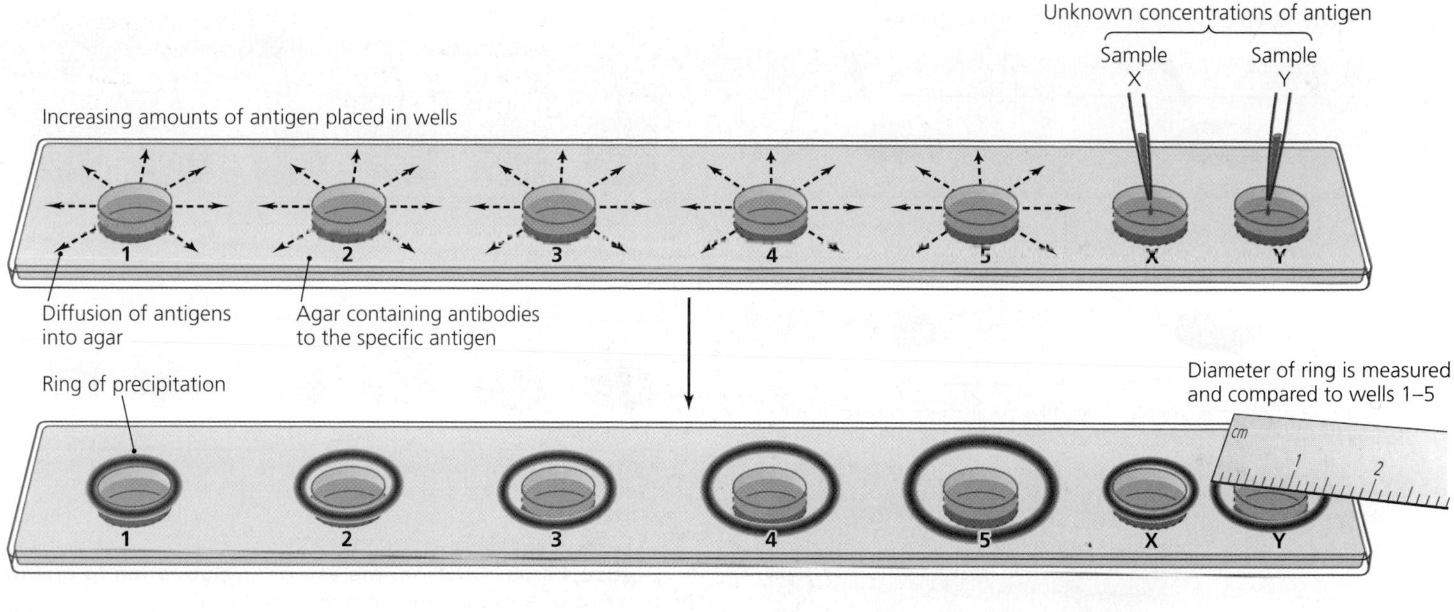

(a)

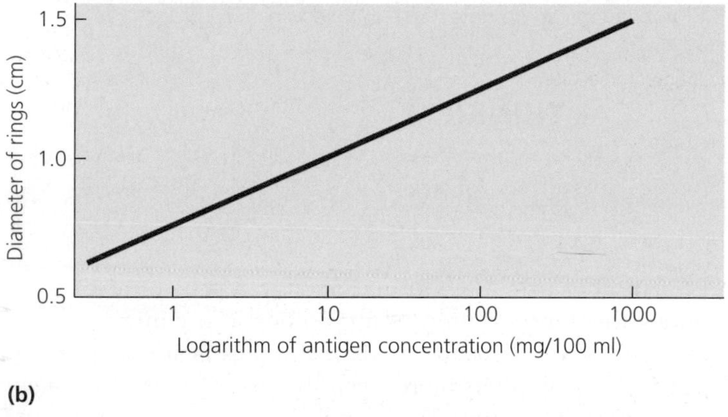

(b)

Figure 17.8
Radial immunodiffusion, a type of precipitation reaction.
(a) Antigen placed into wells in increasing, known concentrations diffuses into agar containing a known concentration of antibody. The diameter of each ring of precipitation formed is proportional to the concentration of antigen in the well. **(b)** Plotting ring diameter versus known antigen concentration produces a standard curve; subsequently, the concentration of antigen in a solution can be determined by comparing the ring diameter produced by radial diffusion to the standard curve. *What is the antigen concentration of sample Y?*

Figure 17.8 *Based on the graph in (b), the concentration in sample Y is about 900 mg/100 ml.*

been used to indicate exposure to complex mixtures of antigens from fungal pathogens. Only exposed patients have serum antibodies—and show precipitation—against the fungal antigens.

Another useful variation on precipitation is called *radial immunodiffusion* **(Figure 17.8a)**. In this technique, an antigen in solution diffuses from a well into agar that contains a known concentration of specific antibodies. The diameter of the ring of precipitate that forms around the antigen well is directly proportional to the concentration of antigen in the well. By using a series of wells containing increasing but known concentrations of antigen, researchers construct a standard curve that relates ring diameter to antigen concentration **(Figure 17.8b)**. Subsequently, the amount of antigen in a solution can be accurately assayed by comparing the ring diameter produced in other radial diffusion tests to the standard line. For example, investigators use radial immunodiffusion to measure the level of complement protein 3 (C3), which is often low in patients with lupus.

Scientists can also use radial immunodiffusion to measure the concentrations of specific antibodies or immunoglobulins in a person's serum. To do this, they take advantage of the fact that, because antibodies are complex proteins, they are antigenic when injected into an individual of another species. Thus, purified human antibodies injected into a rabbit stimulate the rabbit to produce rabbit antibodies against the human antibodies. Such antibodies directed against other antibodies are called *anti-antibodies*. So, in a radial immunodiffusion test to measure the concentration of a specific human antibody, the "antigen" in the test is the human immunoglobulin, and the antibody in the test is rabbit anti–human antibody.

Agglutination Tests

Learning Objectives

- ✓ Contrast agglutination and precipitation tests.
- ✓ Describe how agglutination is used in immunological testing, including titration.

Not all antigens are soluble proteins that can be precipitated by antibody. Because of their multiple antigen-binding sites, antibodies can also cross-link particles, such as whole bacteria or antigen-coated latex beads, causing **agglutination** (ă-gloo-ti-nā'shŭn) (clumping). The difference between agglutination and

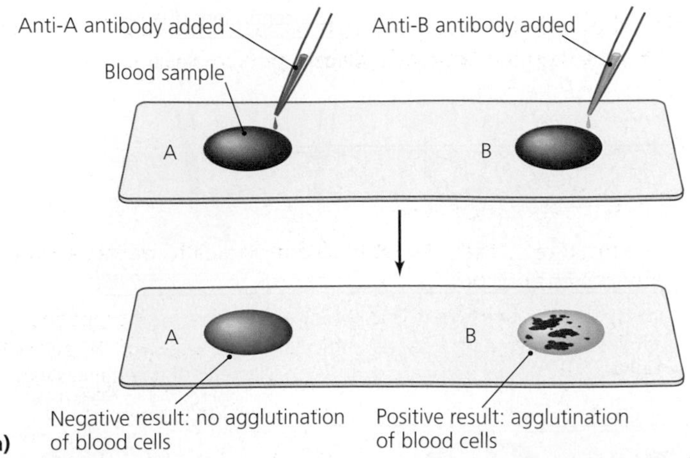

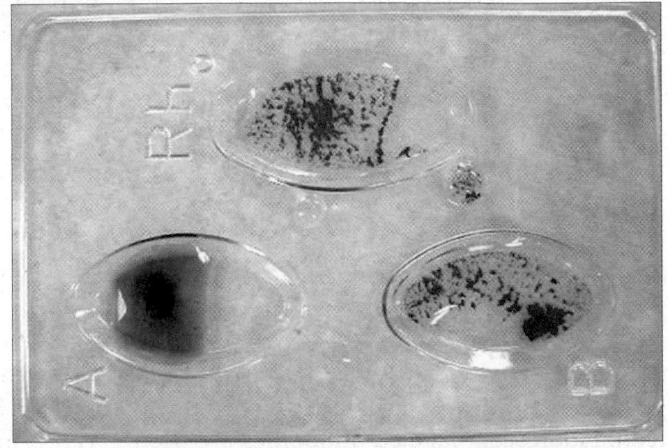

▲ **Figure 17.9 The use of hemagglutination to determine blood types in humans.** **(a)** Antibodies with active sites that bind to either of two red blood cell surface antigens (antigen A or antigen B) are added to portions of a given blood sample. Where the antibodies react with the surface antigens, the blood cells can be seen to agglutinate or clump together. **(b)** Photo of actual test. *What is the blood type of the person whose blood was used in this hemagglutination reaction?*

Figure 17.9 *The individual who donated the blood sample has type B+ blood.*

precipitation is that agglutination involves the clumping of insoluble particles, whereas precipitation involves the aggregation of soluble molecules. Agglutination reactions are easy to see and interpret with the unaided eye. IgM antibodies, which have 10 active sites, are more efficient than IgG antibodies (with only 2 active sites) at causing agglutination.

When the particles agglutinated are red blood cells, the reaction is called *hemagglutination* (hē-mă-gloo′ti-nā′shŭn). One use of hemagglutination is to determine blood type in humans. Blood is considered type A if the red blood cells possess surface antigens called A antigen, type B if they possess B antigens, type AB if they possess both antigens, and type O if they have neither antigen. In a hemagglutination reaction to determine blood type **(Figure 17.9)**, two portions of a given blood sample are placed on a slide. Anti-A antibodies are added to one portion, and anti-B antibodies to the other; the antibodies agglutinate those blood cells that possess complementary antigens.

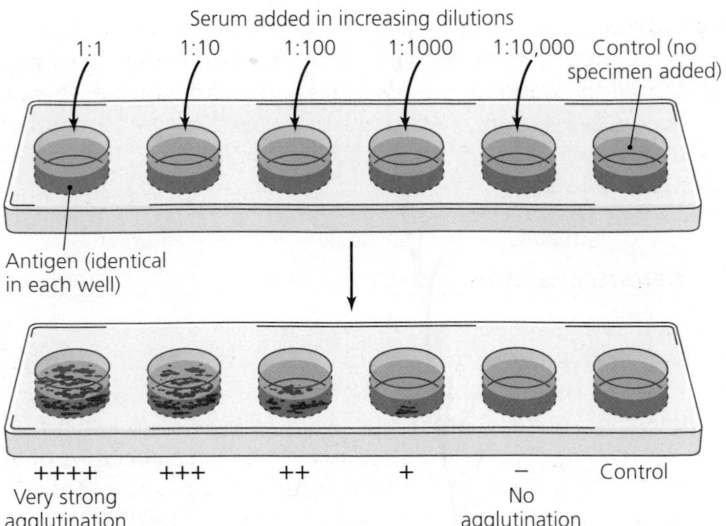

▲ **Figure 17.10 Titration, the use of agglutination to quantify the amount of antibody in a serum sample.** Serial dilutions of serum are added to wells containing a constant amount of antigen. At lower serum dilutions (higher concentrations of antibody), agglutination occurs; at higher serum dilutions, antibody concentration is too low to produce agglutination. The serum's titer is the highest dilution at which agglutination can be detected—in this case, 1:1000.

CRITICAL **THINKING**

Draw a picture showing, at *both the molecular and cellular levels*, IgM agglutinating red blood cells.

Another use of agglutination is in a type of test that determines the concentration of antibodies in a clinical sample. Although the simple *detection* of antibodies is sufficient for many purposes, it is often more desirable to measure the *amount* of antibodies in serum. By doing so, clinicians can determine whether a patient's antibody levels are rising, as occurs in response to the presence of active infectious disease, or falling, as occurs during the successful conclusion of a fight against an infection. One way of measuring antibody levels in blood sera is by **titration** (tī-trā′shŭn). In titration, the serum being tested undergoes a regular series of dilutions, and each dilution is then tested for agglutinating activity **(Figure 17.10)**. Eventually, the antibodies in the serum become so dilute that they can no longer cause agglutination. The highest dilution of serum giving a positive reaction is the titer. Thus, a serum that must be greatly diluted before agglutination ceases (for example, has a titer of 1000) contains more antibodies than a serum that no longer agglutinates after minimal dilution (has a titer of 10).

Neutralization Tests

Learning Objectives

- ✓ Explain the purpose of neutralization tests.
- ✓ Contrast a viral hemagglutination inhibition test with a hemagglutination test.

Neutralization tests work because antibodies can *neutralize* the biological activity of many pathogens and their toxins. For example, combining antibodies against tetanus toxin with a sample of toxin renders the sample harmless to mice because the antibodies have reacted with and neutralized the toxin. Next we briefly consider two neutralization tests that, although not simple to perform, effectively reveal the biological activity of antibodies.

Viral Neutralization

One neutralization test is **viral neutralization,** which is based on the fact that many viruses introduced into appropriate cell cultures will invade and kill the cells, a phenomenon called a *cytopathic effect* (seen in plaque formation, see Figure 13.17). However, if the viruses are first mixed with specific antibodies against them, their ability to kill culture cells is neutralized. In a viral neutralization test, the lack of cytopathic effects when a mixture containing serum and a known pathogenic virus is introduced into a cell culture indicates the presence of antibodies against that virus in the serum. For example, if a mixture containing an individual's serum and a sample of hantavirus produces no cytopathic effect in a culture of susceptible cells, then it can be concluded that the individual's serum contains antibodies to hantavirus, and these antibodies neutralized the virus. Viral neutralization tests are sufficiently sensitive and specific to ascertain whether an individual has been exposed to a particular virus or viral strain.

Viral Hemagglutination Inhibition Test

Because not all viruses are cytopathic—they do not kill their host cell—a neutralization test cannot be used to identify all viruses. However, many viruses (including influenzaviruses) have surface proteins that naturally clump red blood cells. (This natural process, called *viral hemagglutination,* must not be confused with the hemagglutination test we discussed previously—viral hemagglutination is not an antibody-antigen reaction.) Antibodies against influenzavirus inhibit viral hemagglutination; therefore, if serum from an individual stops viral hemagglutination, we know that the individual's serum contains antibodies to the influenzavirus. Such **viral hemagglutination inhibition tests** can be used to detect antibodies against influenza, measles, mumps, and other viruses that naturally agglutinate red blood cells.

The Complement Fixation Test

Learning Objective

✓ Briefly explain the phenomenon that is the basis for a complement fixation test.

As we discussed in Chapter 15, activation of the classical complement system by antibody leads to the generation of membrane attack complexes (MACs) that disrupt cytoplasmic membranes (see Figure 15.9). This phenomenon is the basis for the **complement fixation** (kom′plĕ-ment fik-sā′shŭn) **test,** which is a complex assay used to detect the presence of specific antibodies in an individual's serum. The test can detect the presence of small amounts of antibody—amounts too small to detect by agglutination—though complement fixation tests have been replaced by other serological methods such as ELISA (discussed shortly) or genetic analysis using polymerase chain reaction (PCR; see Figure 8.5).

Labeled Antibody Tests

Learning Objectives

✓ List three tests that use labeled antibodies to detect either antigen or antibodies.

✓ Compare and contrast the direct and indirect fluorescent antibody tests, and identify at least three uses for these tests.

✓ Compare and contrast the methods, purposes, and advantages of ELISA and the western blot tests.

A different form of serologic testing involves *labeled* (or *tagged*) *antibody tests,* so named because these tests use antibody molecules that are linked to some molecular "label" that enables them to be detected easily. Labeled antibody tests using radioactive or fluorescent labels can be used to detect either antigens or antibodies. In the following sections we will consider fluorescent antibody tests, ELISA, and the western blot test.

Fluorescent Antibody Tests

Fluorescent dyes are used as labels in several important serologic tests. One of these dyes is *fluorescein* (flōr-es′ē-in), which can be chemically linked to an antibody without affecting the antibody's ability to bind antigen. When exposed to ultraviolet light (as in a fluorescent microscope), fluorescein glows bright green. Fluorescein-labeled antibodies are used in direct and indirect fluorescent antibody tests.

Direct fluorescent antibody tests identify the presence of antigen in a tissue. The test is straightforward: A scientist floods a tissue sample suspected of containing the antigen with labeled antibody, waits a short time to allow the antibody to bind to the antigen, washes the preparation to remove any unbound antibody, and examines it with a fluorescent microscope. If the suspected antigen is present, labeled antibody will adhere to it, and the scientist will see fluorescence. This is not a quantitative test—the amount of fluorescence observed is not directly related to the amount of antigen present.

Scientists use direct fluorescent antibody tests to identify small numbers of bacteria in patient tissues. This technique has been used to detect *Mycobacterium tuberculosis* in sputum and rabies viruses infecting a brain. In one use, medical laboratory technologists employ a direct fluorescent antibody test to detect the presence of yeast in a sample **(Figure 17.11)**.

Indirect fluorescent antibody tests are used to detect the presence of specific antibodies in an individual's serum via a two-step process **(Figure 17.12a)**:

1. After an antigen of interest is fixed to a microscope slide, the individual's serum is added for long enough to allow serum antibodies, if present, to bind to the antigen. The serum is then washed off, leaving the antibodies bound to the antigen (but not yet visible).
2. Antibodies against human antibodies (anti–human antibody antibodies; in this example, anti-IgG) labeled with a

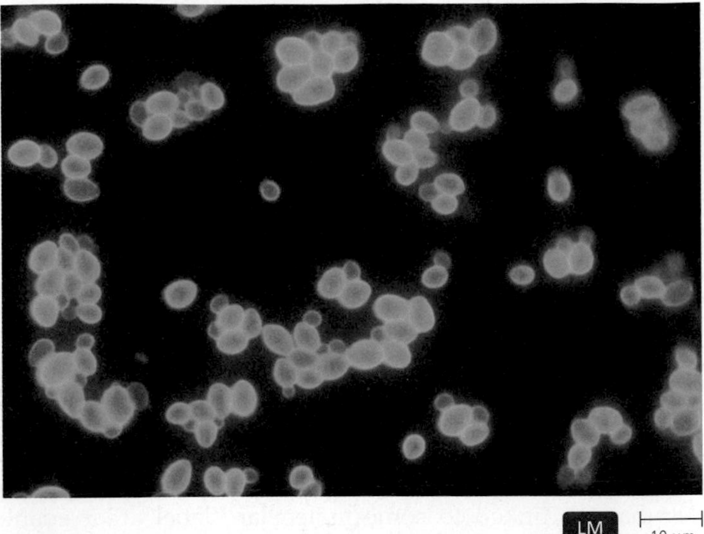

▲ **Figure 17.11 The direct fluorescent antibody test.** Fluorescence from labeled antibodies against the opportunistic yeast pathogen *Candida albicans* clearly reveals budding yeast cells.

fluorescent dye are added to the slide and bind to the antibodies already bound to the antigen. After washing to remove unbound anti-antibodies, the slide is examined with a fluorescent microscope.

The presence of fluorescence indicates the presence of the labeled anti-antibodies, which are bound to serum antibodies bound to the fixed antigen; thus fluoresence indicates that the individual has serum antibodies against the antigen of interest.

Indirect fluorescent antibody testing is used to detect antibodies against many viruses and some bacterial pathogens, including *Neisseria gonorrhoeae* (nī-se′rē-ă go-nor-rē′ī), the causative agent of gonorrhea **(Figure 17.12b)**.

Scientists routinely identify and separate B and T lymphocytes by using specific monoclonal antibodies produced against each cell type. The researchers can attach differently colored fluorescent dyes to the antibodies, allowing them to differentiate between types of lymphocytes. Such identification tests can quantify the numbers and ratios of lymphocyte subsets, information critical in diagnosing and monitoring disease progression and effectiveness of treatment in patients with AIDS and other immunodeficiency diseases.

1 Antigen is attached to slide and flooded with patient's serum.

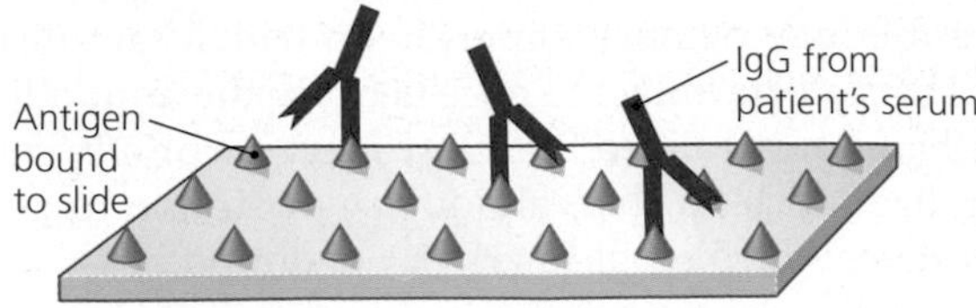

2 Fluorescent-labeled anti-Ig antiglobulin is added.

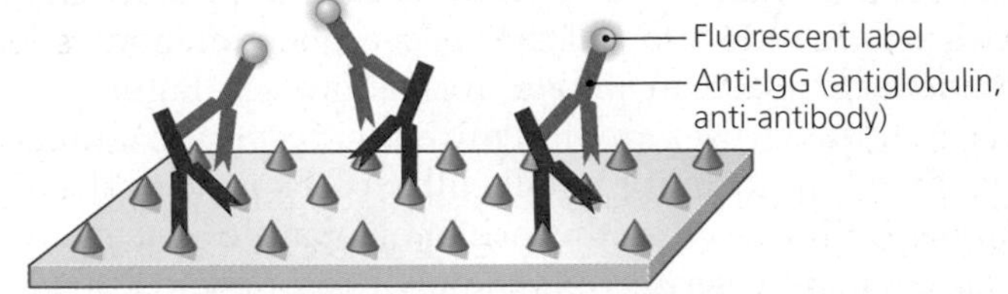

(a)

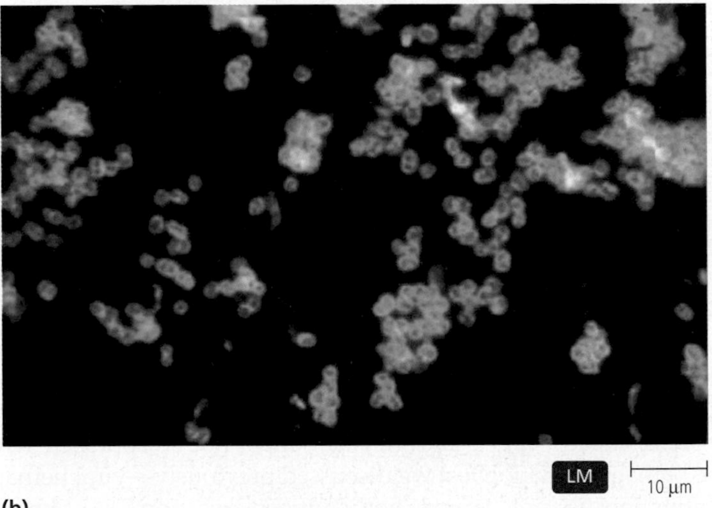

(b)

▲ **Figure 17.12 The indirect fluorescent antibody test.** This test detects the presence of a specific antibody in a patient's serum. **(a)** The test procedure. 1 Antigen is attached to the slide, which is then flooded with an individual's serum to allow specific antibodies in the serum to bind to the antigen. 2 After fluorescein-labeled anti-antibodies are added and then washed off, the slide is examined with a fluorescent microscope. **(b)** A positive indirect antibody test, in which fluorescence indicates the presence of antibodies against *Neisseria gonorrhoeae* (the agent of gonorrhea) in the individual's serum.

ELISAs

In another type of labeled antibody test, called an **enzyme-linked immunosorbent assay** (im′ū-nō-sōr′bent as′sā; ELISA), the label is not a dye, but instead an enzyme that reacts with its substrate to produce a colored product that indicates a positive test. One form of ELISA is used to detect the presence and quantify the abundance of antibodies in serum. This test, which often takes place in commercially produced plates, has five steps **(Figure 17.13)**:

1. Each of the wells in the plate is coated with antigen molecules in solution.
2. Excess antigen molecules are washed off, and another protein (such as gelatin) is added to the well to completely coat any of the surface not coated with antigen.
3. A sample of each of the sera being tested is added to a separate well. Whenever a serum sample contains antibodies against the antigen, they bind to the antigen affixed to the plate.
4. Anti-antibodies labeled with an enzyme are added to each well.

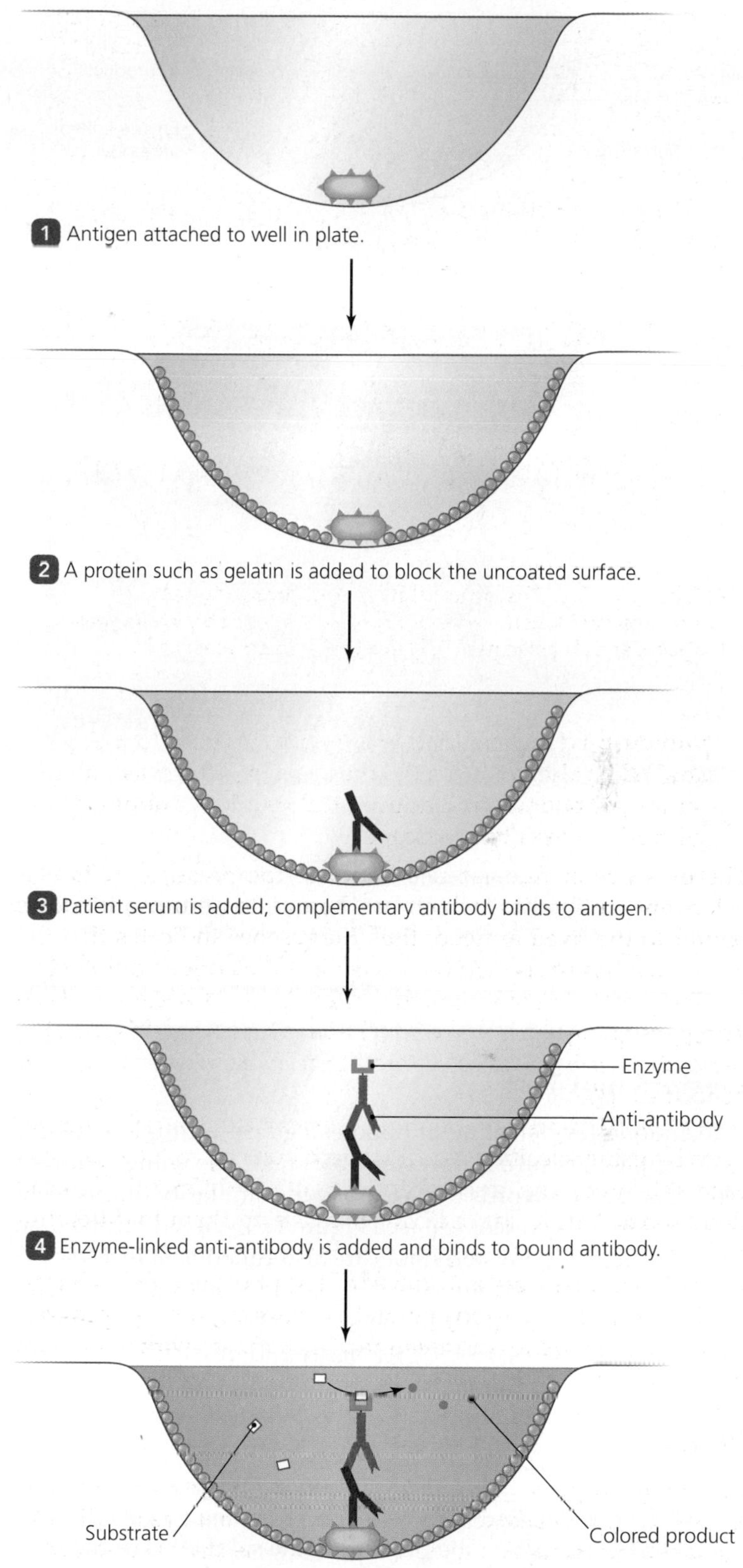
1 Antigen attached to well in plate.
2 A protein such as gelatin is added to block the uncoated surface.
3 Patient serum is added; complementary antibody binds to antigen.
Enzyme
Anti-antibody
4 Enzyme-linked anti-antibody is added and binds to bound antibody.
Substrate
Colored product
5 Enzyme's substrate is added, and reaction produces a visible color change.

Figure 17.13 The enzyme-linked immunosorbent assay (ELISA). Shown is one well in a plate. 1 Antigen added to the well attaches to it irreversibly. 2 After excess antigen is removed by washing, gelatin is added to cover any portion of the well not covered by antigen. 3 Test serum is added to the well; any specific antibodies in it bind to the antigen. 4 Enzyme-labeled anti-antibodies are added to the well and bind to any bound antibody. 5 The enzyme's substrate is added to the well, and the enzyme converts the substrate into a colored product; the amount of color, which can be measured via spectrophotometry, is directly proportional to the amount of antibody bound to the antigen.

5 The enzyme's substrate is added to each well. The enzyme and substrate are chosen because their reaction results in products that cause a visible color change.

A positive reaction in a well, indicated by the development of color, can occur only if the labeled anti-antibody has bound to antibodies attached to the antigen of interest. The intensity of the color, which can be estimated visually or measured accurately using a spectrophotometer, is proportional to the amount of antibody present in the serum.

ELISA has become a test of choice for many diagnostic procedures, such as determination of HIV infection, because of its many advantages:

- Like other labeled antibody tests, ELISA can detect either antibody or antigen.
- ELISAs are sensitive, able to detect very small amounts of antibody (or antigen).
- Unlike some diffusion and fluorescent tests, ELISA can quantify amounts of antigen or antibody.
- ELISAs are easy to perform.
- ELISAs are relatively inexpensive.
- ELISAs can simultaneously test many samples quickly.
- ELISAs lend themselves to efficient automation and can be read easily, either by direct observation or by machine.
- Plates coated with antigen and gelatin can be stored for testing whenever they are needed.

A modification of the ELISA technique, called an *antibody sandwich ELISA,* is commonly used to detect antigen (**Figure 17.14**). In testing for the presence of HIV in blood serum, for example, the plates are first coated with antibody against HIV (instead of antigen). Then the sera from individuals being tested for HIV are added to the wells, and any HIV in the sera will bind to the antibody attached to the well. Finally, each well is flooded with enzyme-labeled antibodies specific to the antigen. The name "antibody sandwich ELISA" refers to the fact that the antigen being tested for is "sandwiched" between two antibody molecules. Such tests can also be used to quantify the amount of antigen in a given sample.

CRITICAL **THINKING**

A diagnostician used an ELISA to show that a newborn had antibodies against HIV in her blood. However, six months later the same test was negative. How can this be?

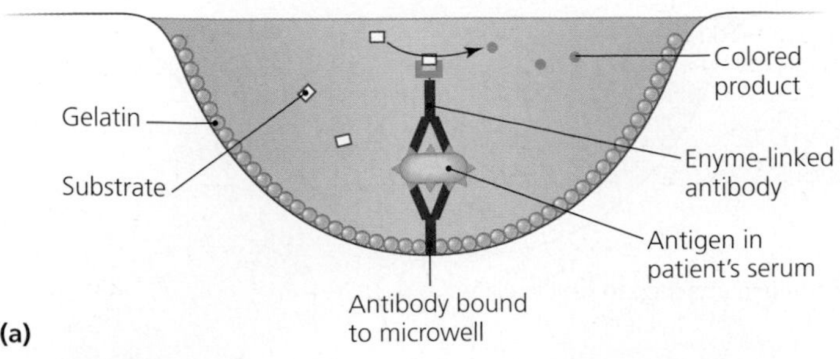

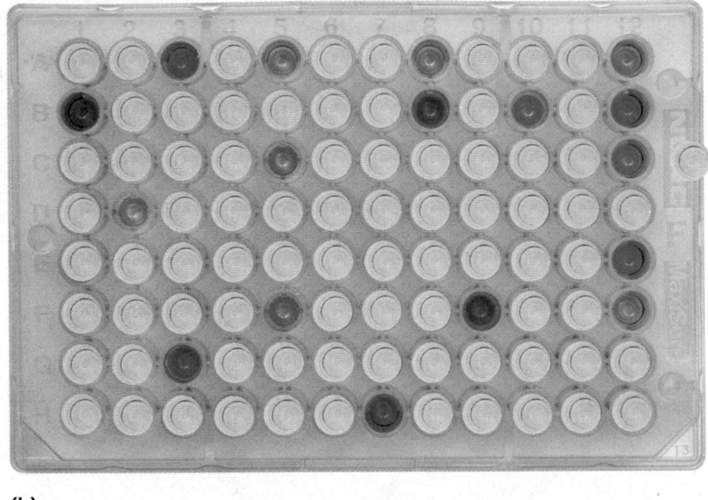

▲ **Figure 17.14 An antibody sandwich ELISA.** Because this variation of ELISA is used to test for the presence of antigen, antibody is attached to the well in the initial step. A second antibody sandwiches the antigen. **(a)** Artist's rendition. **(b)** Actual results. *How many wells are positive?*

Figure 17.14 *Seventeen wells are positive.*

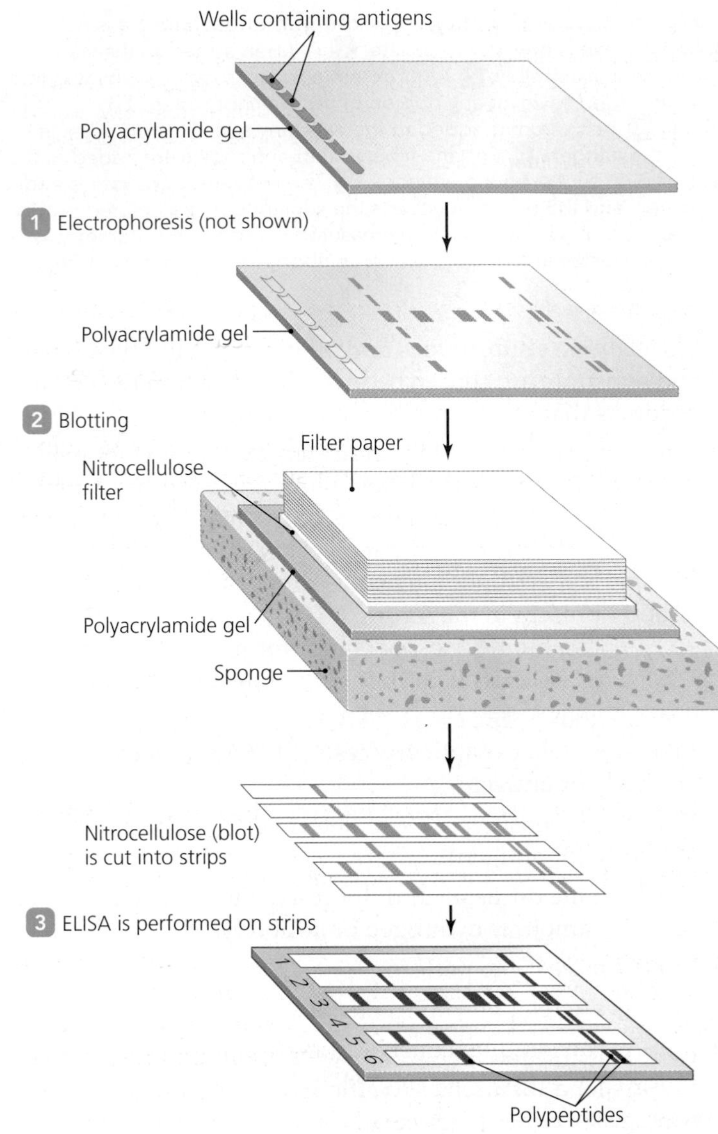

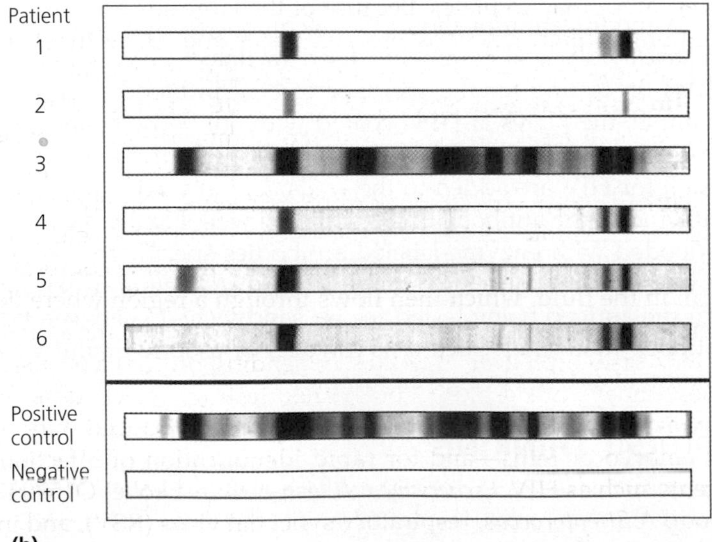

Western Blot Test

A technique for detecting antibodies against multiple antigens in a complex mixture is a **western blot test.** The name "western blot" is a play on words that refers to the similarity of this technique to a Southern blot test (see Figure 8.7), named for the man who developed it. Western blots are also called *immunoblots.*

Western blot tests are currently used to verify the presence of antibodies against HIV in the serum of individuals who are antibody-positive by ELISA. Compared to other tests, western blot tests can detect more types of antibodies and are less subject to misinterpretation. A western blot test has three steps **(Figure 17.15a)**:

1. Electrophoresis. Antigens in a solution (in this example, HIV proteins) are placed into wells and separated by gel electrophoresis. Each of the proteins in the solution is resolved into a single band, producing invisible protein bands.
2. Blotting. The protein bands are transferred to an overlying nitrocellulose membrane by absorbing the solution into absorbent paper—a process called blotting. The nitrocellulose membrane is then cut into strips.
3. ELISA. Each nitrocellulose strip is incubated with a test solution—in this example, those from each of six individuals

◀ **Figure 17.15 A western blot.** This technique demonstrates the presence of antibodies against multiple antigens in a complex mixture. **(a)** Steps in a western blot. **1** Antigens are separated by gel electrophoresis. **2** Separated proteins are transferred to a nitrocellulose membrane. **3** Test solutions, enzyme-labeled anti-antibody, and the enzyme's substrate are added; color changes are detected wherever antibody in the test solutions has bound to HIV proteins. **(b)** The results of a western blot. Patient 3 tested positive for antibodies.

who are being tested for antibodies against HIV. After the strips are washed, an enzyme-labeled anti-antibody solution is added for a time; then the strips are washed again and exposed to the enzyme's substrate.

Color develops wherever antibodies against the HIV proteins in the test solutions have bound to their substrates, as shown in the positive control. In this example, the individual tested in strip 3 is positive for antibodies against HIV, whereas the other five individuals are negative for antibodies against HIV. Colored bands common to all patients are normal serum proteins.

Recent Developments in Antibody-Antigen Immune Testing

Learning Objectives

- ✓ Discuss the benefits of using immunofiltration assays rather than an ELISA.
- ✓ Contrast immunofiltration and immunochromatographic assays.

Recent years have seen the development of simple immunoassays that give clinicians useful results within minutes. The most commonly used of these are *immunofiltration* and *immunochromatography assays*. These tests are not quantitative but rapidly give a positive or negative result, making them very useful in arriving at a diagnosis.

Immunofiltration (im′ū-nō-fil-trā′shŭn) **assays** are rapid ELISAs based on the use of antibodies bound to a membrane filter rather than to plates. Because of the large surface area of a membrane filter, reactions proceed faster and assay times are significantly reduced as compared to a traditional ELISA.

Immunochromatographic (im′ū-nō-krō′mat-ō-graf′ik) **assays** are still faster and easier to read immunoassays. In these systems, an antigen solution (such as diluted blood or sputum) flowing through a porous material encounters antibody labeled with either pink colloidal[2] gold or blue colloidal selenium. Where antigen and antibody bind, colored immune complexes form in the fluid, which then flows through a region where the complexes encounter antibody against them, resulting in a clearly visible pink or blue line, depending on the label used. These assays are used for pregnancy testing, which tests for *human chorionic growth hormone*—a hormone produced only by an embryo or fetus—and for rapid identification of infectious agents such as HIV, *Escherichia coli* (esh-ĕ-rik′ē-ă kō′lē) O157:H7, group A *Streptococcus*, respiratory syncytial virus (RSV), and influenzaviruses. In one adaptation, the antibodies are coated on membrane strips, which serve as dipsticks. At one end, anti-antibodies are fixed in a line so that they cannot move in the membrane. The lower portion of the membrane is coated with antibodies against the antigen in question. These antibodies are linked to a color indicator in the form of a colloidal metal and are free to move in the membrane by capillary action.

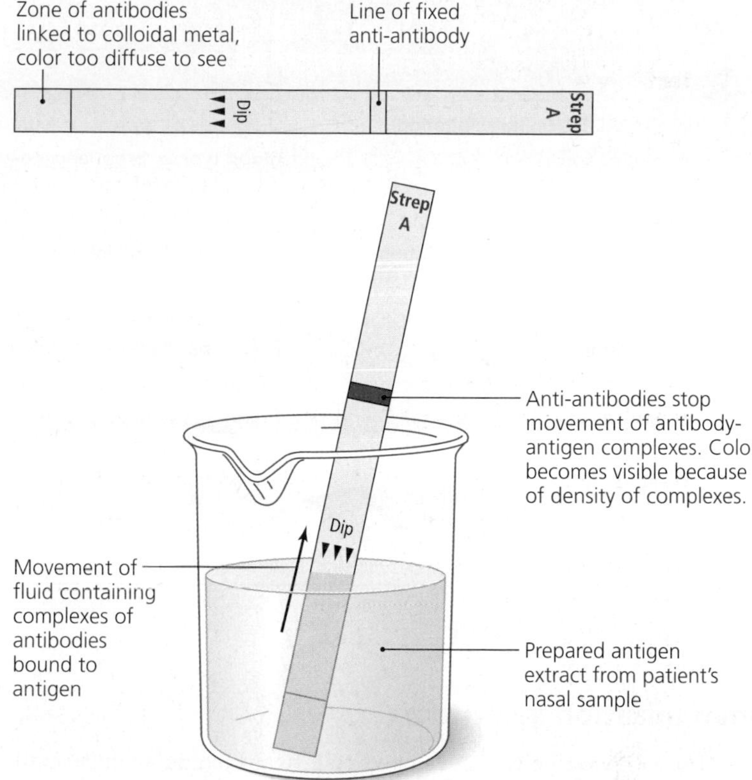

▲ **Figure 17.16 Immunochromatographic dipstick.** The dipstick is impregnated with colloidal metal particles linked to movable antibodies against particular antigens at one end, and anti-antibodies fixed in a line closer to the other end of the membrane.

Figure 17.16 illustrates the procedure used to test for the presence of group A *Streptococcus* in the nasal secretion of a patient. A technician prepares a nasal swab from the patient so as to release *Streptococcus* antigens if they are present. She then dips the membrane into the solution. The membrane's antibodies bind to streptococcal antigens, forming complexes. The complexes move up the membrane by capillary action until they reach the line of anti-antibodies, where they bind and must stop because the anti-antibodies are chemically bound to the strip. Previously the complexes were invisible because they were dilute; now they are concentrated at the line of anti-antibodies and become visible, indicating that this patient had group A *Streptococcus* in his nose. The procedure from antigen preparation to diagnosis takes less than 10 minutes.

Table 17.3 on p. 508 lists some antibody-antigen immune tests that can be used to diagnose selected bacterial and viral diseases.

[2] *Colloidal* refers to small particles suspended in a liquid or gas.

TABLE 17.3 Antibody-Antigen Immunological Tests and Some of Their Uses

Test	Use
Immunodiffusion (precipitation)	Diagnosis of syphilis, pneumococcal pneumonia
Agglutination	Blood typing; pregnancy testing; diagnosis of salmonellosis, brucellosis, gonorrhea, rickettsial infection, mycoplasma infection, yeast infection, typhoid fever, meningitis caused by *Haemophilus*
Viral neutralization	Diagnosis of infections by specific strains of viruses
Viral hemagglutination inhibition	Diagnosis of viral infections including influenza, measles, mumps, rubella, mononucleosis
Complement fixation	In the past, diagnosis of measles, influenza A, syphilis, rubella, rickettsial infections, scarlet fever, rheumatic fever, infections of respiratory syncytial virus and *Coxiella*
Direct fluorescent antibody	Diagnosis of rabies, infections of group A *Streptococcus*, identification of lymphocyte subsets
Indirect fluorescent antibody	Diagnosis of syphilis, mononucleosis
ELISA	Pregnancy testing; presence of drugs in urine; diagnosis of hepatitis A, hepatitis B, rubella; initial diagnosis of HIV infection
Western blot	Verification of infection with HIV; diagnosis of Lyme disease

Chapter Summary

Immunization (pp. 491–498)

1. The first **vaccine** was developed by Edward Jenner against smallpox. He called the technique **vaccination. Immunization** is a more general term referring to the use of vaccines against rabies, anthrax, measles, mumps, rubella, polio, and other diseases.
2. Individuals can be protected against many infections by either active immunization or passive immunotherapy.
   ANIMATIONS: *Vaccines: Function*
3. Active immunization involves giving antigen in the form of either **attenuated vaccines, inactivated (killed) vaccines, toxoid vaccines,** or recombinant gene vaccines. Antibody **titer** refers to the amount of antibody produced.
   ANIMATIONS: *Vaccines: Types*
4. Pathogens in attenuated vaccines are weakened so that they no longer cause disease, though they are still alive or active and can provide **contact immunity** in unimmunized individuals who associate with immunized people.
5. Inactivated vaccines are either whole agent or subunit vaccines and often contain **adjuvants,** which are chemicals added to increase their ability to stimulate active immunity.
6. Toxoid vaccines use modified toxins to stimulate antibody-mediated immunity.
7. A **combination vaccine** is composed of antigens from several pathogens so they can be administered to a patient at once.
8. Having a large proportion of immunized individuals (>75%) in a population interrupts disease transmission, providing protection to unimmunized individuals. Such protection is called **herd immunity.**
9. Passive immunotherapy (a type of passive immunization) involves administration of an **antiserum** containing preformed antibodies. Serum sickness results when the patient makes antibodies against the antiserum.
10. The fusion of myelomas (cancerous plasma cells) with plasma cells results in **hybridomas,** the source of **monoclonal antibodies,** which can be used in passive immunization.

Antibody-Antigen Immune Testing (pp. 498–508)

1. **Serology** is the study and use of immunological tests to diagnose and treat disease or identify antibodies or antigens. Direct testing involves using antibodies to find an antigen in a specimen. Indirect testing involves using an antigen to find antibodies in serum or blood.
2. The simplest of the serologic tests is a precipitation test, in which antigen and antibody meet in optimal proportions to form immune complexes, which are often insoluble. Often this test is performed in clear gels, where it is called **immunodiffusion.**
3. **Agglutination tests** involve the clumping of antigenic particles by antibodies. The amount of these antibodies, called the titer, is measured by diluting the serum in a process called **titration.**
4. Antibodies to viruses or toxins can be measured using a **neutralization test,** such as a **viral neutralization** test. Infection by viruses that naturally agglutinate red blood cells can be demonstrated using a **viral hemagglutination inhibition test.**
5. The **complement fixation test** is a complex assay used to determine the presence of specific antibodies.
6. Fluorescently labeled antibodies—those chemically linked to a fluorescent dye—can be used in a variety of **direct** and **indirect fluorescent antibody** tests. The presence of labeled antibodies is visible through a fluorescent microscope.
7. **Enzyme-linked immunosorbent assays (ELISAs)** are a family of simple tests that can be readily automated and read by machine. These tests are among the most common serologic tests used. A variation of the ELISA is the **western blot test,** which is used to detect antibodies against multiple antigens in a mixture.
8. **Immunofiltration assays** and **immunochromatographic assays** are modifications of ELISA tests that can give much more rapid diagnostic results.

Questions for Review

Answers to the Questions for Review (except Short Answer questions) begin on page A-1.

Multiple Choice

1. In order to obtain immediate immunity against tetanus, a patient should receive
   a. an attenuated vaccine of *Clostridium tetani.*
   b. a modified live vaccine of *C. tetani.*
   c. tetanus toxoid.
   d. immunoglobulin against tetanus toxin (antitoxin).
   e. a subunit vaccine against *C. tetani.*
2. Which of the following vaccine types is commonly given with an adjuvant?
   a. an attenuated vaccine
   b. a modified live vaccine
   c. a chemically killed vaccine
   d. an immunoglobulin
   e. an agglutinating antigen
3. Which of the following viruses was widely used in living vaccines?
   a. coronavirus
   b. vaccinia virus
   c. influenzavirus
   d. retrovirus
   e. myxovirus
4. When antigen and antibodies combine, maximal precipitation occurs when
   a. antigen is in excess.
   b. antibody is in excess.
   c. antigen and antibody are at equivalent concentrations.
   d. antigen is added to the antibody.
   e. antibody is added to the antigen.
5. An anti-antibody is used when
   a. an antigen is not precipitating.
   b. an antibody is not agglutinating.
   c. an antibody does not activate complement.
   d. an antigen is insoluble.
   e. the antigen is an antibody.
6. The many different proteins in serum can be analyzed by
   a. an anti-antibody test.
   b. a complement fixation test.
   c. a precipitation test.
   d. an agglutination test.
   e. an immunodiffusion test.
7. A direct fluorscent antibody test requires which of the following?
   a. heat-inactivated serum
   b. fluorescent serum
   c. immune complexes
   d. heated plasma
   e. antibodies against the antigen
8. An ELISA uses which of the following reagents?
   a. an enzyme-labeled anti-antibody
   b. a radioactive anti-antibody
   c. a source of complement
   d. an enzyme-labeled antigen
   e. an enzyme-labeled antibody
9. A direct fluorescent antibody test can be used to detect the presence of
   a. hemagglutination.
   b. specific antigens.
   c. antibodies.
   d. complement.
   e. precipitated antigen-antibody complexes.
10. Which of the following is a good test to detect rabies virus in the brain of a dog?
    a. agglutination
    b. hemagglutination inhibition
    c. virus neutralization
    d. western blot
    e. direct immunofluorescence
11. Attenuation is
    a. the process of reducing virulence.
    b. a necessary step in vaccine manufacture.
    c. a form of variolation.
    d. similar to an adjuvant.
12. An antiserum is
    a. an anti-antibody.
    b. an inactivated vaccine.
    c. formed of monoclonal antibodies.
    d. the liquid portion of blood used for immunization.
13. Monoclonal antibodies
    a. are produced by hybridomas.
    b. are secreted by clone cells.
    c. can be used for passive immunization.
    d. all of the above
14. The study of antibody-antigen interaction in the blood is
    a. attenuation.
    b. agglutination.
    c. precipitation.
    d. serology.
15. Anti–human antibody antibodies are
    a. found in immunocompromised individuals.
    b. used in direct fluorescent antibody tests.
    c. formed by animals reacting to human immunoglobulins.
    d. an alternative method in ELISA.

True/False

1. ____ Passive immunotherapy provides more prolonged immunity than active immunization.
2. ____ It is standard to attenuate killed virus vaccines.
3. ____ One single serologic test is inadequate for an accurate diagnosis of HIV infection.
4. ____ ELISA is very easily automated.
5. ____ ELISA has basically replaced the western blot test.

Matching

Match the characteristic in the first column with the therapy it most closely describes in the second column. Some choices may be used more than once.

1. ____ Induces rapid onset of immunity
2. ____ Induces mainly an antibody response
3. ____ Induces good cell-mediated immunity
4. ____ Increases antigenicity
5. ____ Uses antigen fragments
6. ____ Uses attenuated microbes

A. Attenuated viral vaccine
B. Adjuvant
C. Subunit vaccine
D. Immunoglobulin
E. Residual virulence

Labeling

Identify the chemicals represented by this artist's conception of an ELISA.

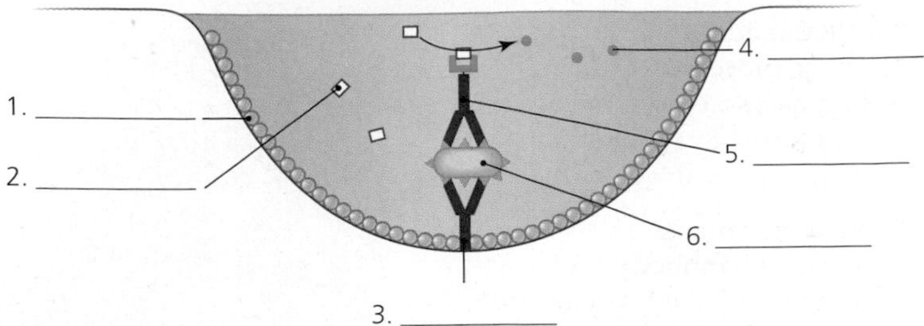

Short Answer

1. Compare and contrast the Chinese practice of variolation with Jenner's vaccination procedure.
2. What are the advantages and disadvantages of attenuated vaccines?
3. Compare the advantages and disadvantages of passive immunotherapy and active immunization.
4. How does precipitation differ from agglutination?
5. Explain how a pregnancy test works at the molecular level.
6. Compare and contrast herd immunity and contact immunity.

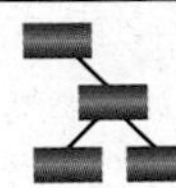

Concept Mapping

Using the following terms, draw a concept map that describes vaccines. For a sample concept map, see p. 93. Or, complete this concept map online by going to the Study Area at www.masteringmicrobiology.com.

Active immunity
Antigenic fragments
Attenuated whole agents
Hepatitis B vaccine
Immunizations
Inactivated whole agents
Inactivated polio vaccine
Killed vaccine
Live vaccines
Measles vaccine
Microorganisms
Modified toxins
Reduced virulence
Subunits
Tetanus toxoid
Toxoids
Varicella vaccine

Critical Thinking

1. Is it ethical to approve the use of a vaccine that causes significant illness in 1% of patients, if it protects immunized survivors against a serious disease?
2. Which is worse: to use a diagnostic test for HIV that may falsely indicate that a patient is not infected (false negative), or to use one that sometimes falsely indicates that a patient is infected (false positive)? Defend your choice.
3. Discuss the importance of costs and technical skill in selecting a practical serologic test. Under what circumstances does automation become important?
4. What bodily fluids, in addition to blood serum, might be usable for immune testing?
5. Why might a serologic test give a false positive result?
6. Some researchers want to distinguish between B cells and T cells in a mixture of lymphocytes. How could they do this without killing the cells?
7. Describe three ways by which genetic recombinant techniques could be used to develop safer, more effective vaccines.
8. How does a toxoid vaccine differ from an attenuated vaccine?
9. Explain why many health organizations promote breast feeding of newborns. What risks are involved in such nursing?
10. Contrast a hemagglutination test with a viral hemagglutination inhibition test.
11. Sixty years ago parents would have done almost anything to get a protective vaccine against polio for their children. Now parents fear the vaccine, not the disease. Why?

Access more review material online in the Study Area at **www.masteringmicrobiology.com.** There, you'll find

- **Animations**
- **MP3 Tutor Sessions**
- **Concept Mapping Activities**
- **Flashcards**
- **Quizzes**

and more to help you succeed.

18 Immune Disorders

It is summer, and you are hiking along a densely wooded trail wearing shorts and a T-shirt. You spot a beautiful wildflower just a few feet away and wander briefly off the trail to get a better look, wading through knee-high foliage. The next morning, an itchy red rash has developed on both of your legs. Within days, the rash has progressed to painful, oozing blisters. You realize, unhappily, that you have contacted poison ivy and will be treating the blisters until they heal—typically within 14 to 20 days.

A poison ivy rash is caused by an allergic response to urushiol, a chemical that coats poison ivy leaves. About 85% of the population experiences an allergic reaction when exposed to urushiol. The body's immune system essentially overreacts to the urushiol: it identifies the chemical as a foreign substance and initiates an immune response that results in allergic contact dermatitis (inflammation of the skin). In the absence of an immune response, poison ivy would be harmless.

In this chapter we examine a variety of immune system disorders, including allergies and other immune hypersensitivities, autoimmune diseases, and immune deficiencies.

Take the pre-test for this chapter online. Visit the Study Area at www.masteringmicrobiology.com.

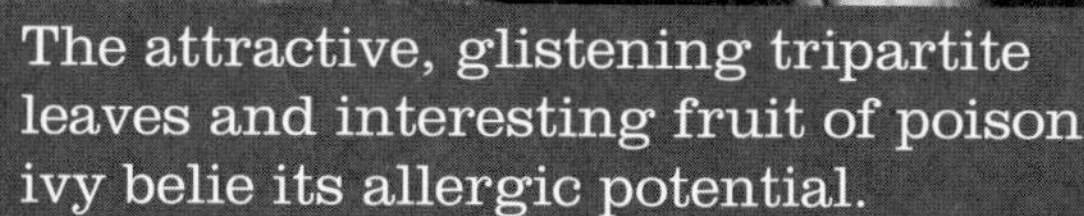

The attractive, glistening tripartite leaves and interesting fruit of poison ivy belie its allergic potential.

As we have seen in Chapter 16, the immune system is an absolutely essential component of the body's defenses. However, if the immune system functions abnormally—either by over reacting or underreacting—the malfunction may cause significant disease, even death. If the immune system functions excessively, the body develops any of a variety of *immune hypersensitivities*. An important example of a hypersensitivity is an allergy, which occurs when the immune system reacts to generally harmless substances such as pollen. Further, *autoimmune diseases* develop when the immune system attacks the body's own tissues. Finally, diseases that result from a failure of the immune system are called *immunodeficiency* (im'ū-nō-dē-fish'en-sē) *diseases*. In this chapter we discuss each of these three categories of derangements of the immune system.

Hypersensitivities

Learning Objective

✓ Compare and contrast the four types of hypersensitivity.

Hypersensitivity (hī'per-sen-si-tiv'i-tē) may be defined as any immune response against a foreign antigen that is exaggerated beyond the norm. For example, most people can wear wool, smell perfume, or dust furniture without experiencing itching, wheezing, runny nose, or watery eyes. When these symptoms do occur, the person is said to be experiencing a hypersensitivity response. In the following sections we examine each of the four main types of hypersensitivity response, designated as type I through type IV.

Type I (Immediate) Hypersensitivity

Learning Objectives

✓ Describe the two-part mechanism by which type I hypersensitivity occurs.

✓ Explain the roles of three inflammatory chemicals (mediators) released from mast cell granules.

✓ Describe three disease conditions resulting from type I hypersensitivity mechanisms.

Type I hypersensitivities are localized or systemic (whole body) reactions that result from the release of inflammatory molecules (such as histamine) in response to an antigen. These reactions are termed *immediate hypersensitivity* because they develop within seconds or minutes following exposure to the antigen. They are also commonly called **allergies,**[1] and the antigens that stimulate them are called **allergens** (al'er-jenz).

In the next two subsections we will examine the two-part mechanism of a type I hypersensitivity reaction: sensitization upon an initial exposure to an allergen, and the degranulation of sensitized cells.

[1]From Greek *allos*, meaning other, and *ergon*, meaning work.

Sensitization upon Initial Exposure to an Allergen

All of us are exposed to antigens in the environment, against which we typically mount immune responses that result in the production of IgG. But when cytokines (especially interleukin 4) from type 2 helper T (Th2) cells stimulate B cells in allergic individuals, the B cells become plasma cells that produce not IgG, but instead switch to IgE **(Figure 18.1a)**. Typically, IgE is directed against parasitic worms, which rarely infect most Americans, so IgE is found at low levels in the blood serum. However, in allergic individuals plasma cells produce IgE in response to allergens.

The precise reason that only some people produce high levels of IgE (and thereby suffer from allergies) is a matter of intense research. Some researchers think that environmental factors—including exposure during infancy and childhood to high levels of indoor antigens such as cigarette smoke, dust mites, and molds—sensitize people, making them hypersensitive (allergic). Conversely, other researchers believe that childhood exposure to such antigens is protective, that young children who experience multiple bacterial and viral infections as well as exposure to pet dander (flakes of animal skin) are less likely to develop allergies than those who have been relatively sheltered from infectious diseases and unexposed to pets. **Highlight: Can Pets Help Decrease Children's Allergy Risks?** on p. 514 examines the second hypothesis.

In any case, following initial exposure to allergens, the plasma cells of allergic individuals secrete IgE, which binds very strongly via its stem (F_c region) to three types of defense cells—*mast cells*, *basophils*, and *eosinophils*—sensitizing these cells to respond to subsequent exposures to the allergen.

Degranulation of Sensitized Cells

When the same allergen reenters the body, it binds to the active sites of IgE molecules on the surfaces of sensitized cells (described shortly). This binding triggers a cascade of internal biochemical reactions that causes the sensitized cells to release the inflammatory chemicals from their granules into the surrounding intercellular space—an event called *degranulation* (dē-gran-ū-lā'shŭn) **(Figure 18.1b)**. It is these inflammatory mediators that generate the characteristic symptoms of type I hypersensitivity reactions: respiratory distress, rhinitis (inflammation of the nasal mucous membranes commonly called "runny nose"), watery eyes, inflammation, and reddening of the skin.

The Roles of Degranulating Cells in an Allergic Reaction

Mast cells are specialized relatives of white blood cells, deriving from other stem cells in the bone marrow. They are distributed throughout the body in connective tissues other than blood. These large, round cells are most often found in sites close to body surfaces, including the skin and the walls of the

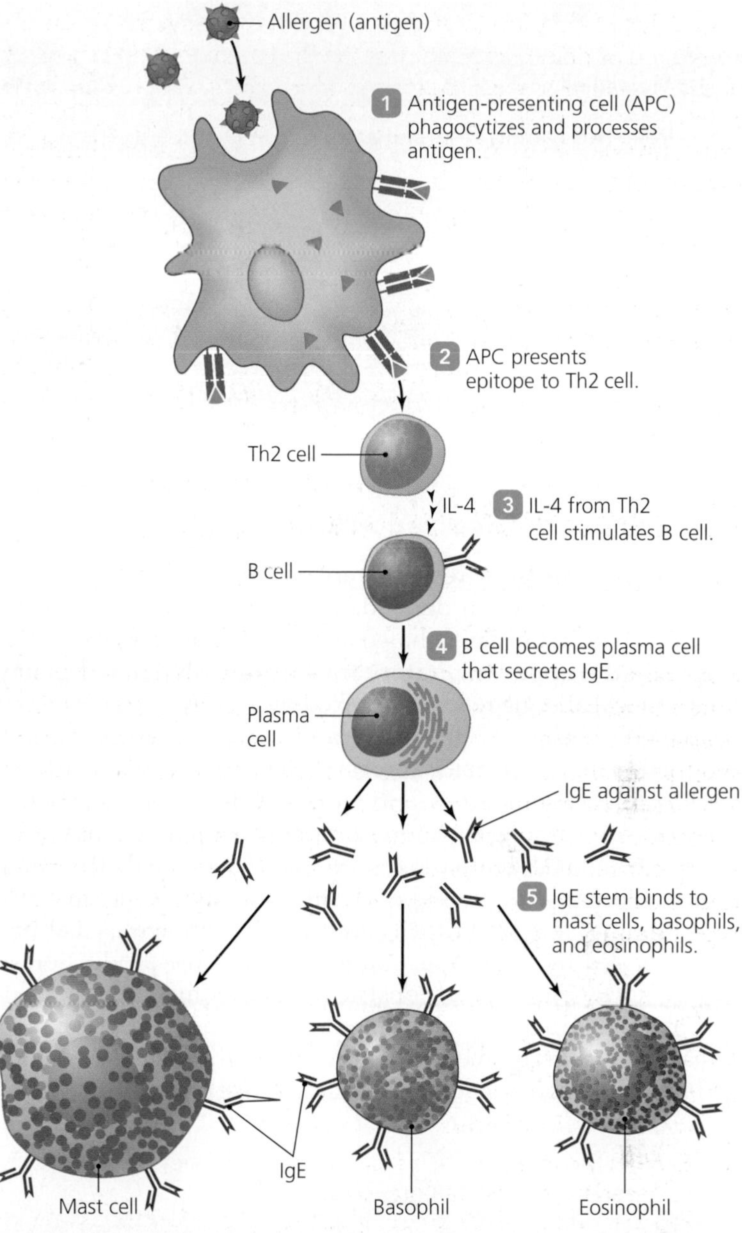

Allergen (antigen)
1 Antigen-presenting cell (APC) phagocytizes and processes antigen.
2 APC presents epitope to Th2 cell.
Th2 cell
IL-4
3 IL-4 from Th2 cell stimulates B cell.
B cell
4 B cell becomes plasma cell that secretes IgE.
Plasma cell
IgE against allergen
5 IgE stem binds to mast cells, basophils, and eosinophils.
IgE
Mast cell
Basophil
Eosinophil
(a) Sensitization

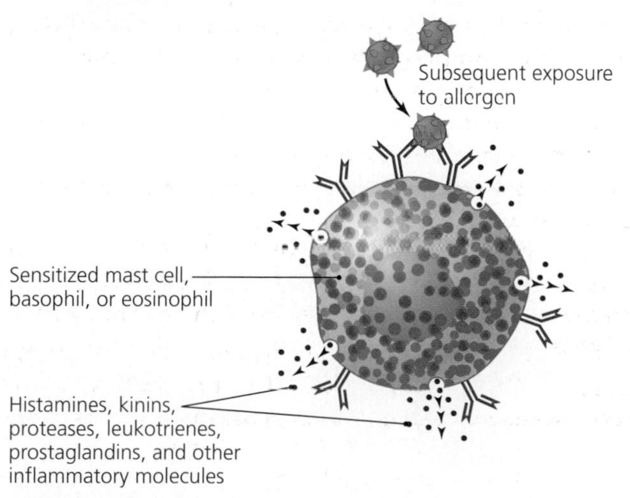

Subsequent exposure to allergen
Sensitized mast cell, basophil, or eosinophil
Histamines, kinins, proteases, leukotrienes, prostaglandins, and other inflammatory molecules
(b) Degranulation

Figure 18.1
The mechanisms of a type I hypersensitivity reaction.
(a) Sensitization. After normal processing of an antigen (allergen) by an antigen-presenting cell (1), Th2 cells are stimulated to secrete IL-4 (2). This cytokine stimulates a B cell (3), which becomes a plasma cell that secretes IgE (4), which then binds to and sensitizes mast cells, basophils, and eosinophils (5). **(b)** Degranulation. When the same allergen is subsequently encountered, its binding to the IgE molecules on the surfaces of sensitized cells triggers rapid degranulation and the release of inflammatory chemicals from the cells.

intestines and airways. Their characteristic feature is cytoplasm packed with large granules that are loaded with a mixture of potent inflammatory chemicals.

One significant chemical released from mast cell granules is **histamine** (his′tă-mēn), a small molecule related to the amino acid histidine. Histamine stimulates strong contractions in the smooth muscles of the bronchi, gastrointestinal tract, uterus, and bladder, and it also makes small blood vessels dilate (expand) and become leaky. As a result, tissues in which mast cells degranulate become red and swollen. Histamine also stimulates nerve endings, causing itching and pain. Finally, histamine is a potent stimulator of bronchial mucus secretion, tear formation, and salivation.

Other mediators of allergies released by degranulating mast cells include **kinins** (kī′ninz), which are powerful inflammatory chemicals, and **proteases** (prō′tē-ās-ez)—enzymes that destroy nearby cells, activating the complement system, which in turn results in the release of still more inflammatory chemicals. Proteases account for more than half the proteins in a mast cell granule. In addition, the binding of allergens to IgE on mast cells activates other enzymes that trigger the production of **leukotrienes** (loo-kō-trī′ēnz) and **prostaglandins** (pros-tă-glan′dinz), lipid molecules that are potent inflammatory agents. Table 18.1 on p. 514 summarizes the inflammatory molecules released from mast cells.

Basophils, the least numerous type of leukocyte in blood, contain cytoplasmic granules that stain intensely with basic dyes (see Figure 15.5a) and are filled with inflammatory chemicals similar to those found in mast cells. Sensitized basophils bind IgE and degranulate in the same way as mast cells when they encounter allergens.

The blood and tissues of allergic individuals also accumulate many **eosinophils,** leukocytes that contain numerous cytoplasmic granules that stain intensely with the red dye eosin and function primarily to destroy parasitic worms. The process during type I hypersensitivity reactions that results in the accumulation of eosinophils in the blood—a condition termed *eosinophilia*—begins with mast cell degranulation, which releases peptides that stimulate the release of eosinophils from the bone marrow. Once in the bloodstream, eosinophils are attracted to the site of mast cell degranulation, where they themselves degranulate. Eosinophil granules contain unique inflammatory mediators and produce large amounts of leukotrienes, which increase vascular permeability and stimulate smooth muscle contraction, thereby contributing greatly to the severity of a hypersensitivity response.

TABLE 18.1 Inflammatory Molecules Released from Mast Cells

Molecules	Role in Hypersensitivity Reactions
Released During Degranulation	
Histamine	Causes smooth muscle contraction, increased vascular permeability, and irritation
Kinins	Cause smooth muscle contraction, inflammation, and irritation
Proteases	Damage tissues and activate complement
Synthesized in Response to Inflammation	
Leukotrienes	Cause slow, prolonged smooth muscle contraction, inflammation, and increased vascular permeability
Prostaglandins	Some contract smooth muscle; others relax it

Clinical Signs of Localized Allergic Reactions

Type I hypersensitivity reactions are usually mild and localized. The site of the reaction depends on the portal of entry of the antigens. For example, inhaled allergens may provoke a response in the upper respiratory tract commonly known as **hay fever,** a local allergic reaction marked by a profuse watery nasal discharge, sneezing, itchy throat and eyes, and excessive tear production. Fungal (mold) spores; pollens from grasses, flowering plants, and some trees; and feces and dead bodies of house dust mites are among the more common allergens (**Figure 18.2**).

If inhaled allergen particles are sufficiently small, they may reach the lungs. A type I hypersensitivity in the lungs can cause an episode of severe difficulty in breathing known as **asthma,** characterized by wheezing, coughing, excessive production of a thick, sticky mucus, and constriction of the smooth muscles of the bronchi. Although a localized reaction, asthma can be life threatening: without medical intervention, the increased mucus and bronchial constriction can quickly cause suffocation.

Other allergens—including latex, wool, certain metals, and the venom or saliva of wasps, bees, fire ants, deer flies, fleas, and other stinging or biting insects—may cause a local dermatitis (inflammation of the skin). As a result of the release of histamine and other mediators, plus the ensuing leakage of serum from local blood vessels, the individual suffers raised, red areas called *hives* or **urticaria**[2] (er′ti-kar′i-ă; **Figure 18.3**). These lesions are very itchy because histamine irritates local nerve endings.

Clinical Signs of Systemic Allergic Reactions

Following a sensitized individual's contact with an allergen, many mast cells may degranulate simultaneously, releasing massive amounts of histamine and other inflammatory mediators into the bloodstream. The release of chemicals may exceed the body's ability to adjust, resulting in **acute anaphylaxis**[3] (an′ă-fī-lak′sis), or **anaphylactic shock.**

[2]From Latin *urtica*, meaning nettle.
[3]From Greek *ana*, meaning away from, and *phylaxis*, meaning protection.

HIGHLIGHT

CAN PETS HELP DECREASE CHILDREN'S ALLERGY RISKS?

A study headed by Dennis Ownby of the Medical College of Georgia suggests that children who grow up in households with two or more cats or dogs may be less likely to develop common allergies than children raised without pets. Researchers suspect that exposure to bacteria harbored by cats and dogs can suppress allergic reactions—not only to pets but also to other common allergens such as dust mites and grass. Ownby and his team studied 474 children from birth until six to seven years of age, at which point the children were tested for IgE against common allergens. They found that children exposed to two or more dogs or cats during their first year of life were, on average, 66–77% less likely to have antibodies to common allergens than were children exposed to only one or no pets during the same period. Exactly how the pets and/or their bacteria may suppress the allergic response remains unclear; perhaps stimulation of Th1 cells by pet allergens balances or counteracts the production of Th2 cells that would stimulate humoral immunity.

▲ **Figure 18.2 Some common allergens.** **(a)** Spores of the fungus, *Aspergillus* on stalks. **(b)** Pollen of *Ambrosia trifida* (ragweed), the most common cause of hay fever in the United States. **(c)** A house dust mite. Mites' fecal pellets and dead bodies may become airborne and trigger allergic responses.

A common cause of acute anaphylaxis is a bee sting in an allergic and sensitized individual. The first sting produces sensitization and the formation of IgE antibodies, and successive stings—which may occur years later—produce an anaphylactic reaction. Other allergens commonly implicated in acute anaphylaxis are certain foods (notoriously peanuts), vaccines, antibiotics such as penicillin, iodine dyes, local anesthetics, blood products, and certain narcotics such as morphine.

The clinical signs of acute anaphylaxis are those of rapid suffocation. Bronchial smooth muscle, which is highly sensitive to histamine, contracts violently. In addition, increased leakage of fluid from blood vessels causes swelling of the larynx and other tissues. The patient also experiences contraction of the smooth muscle of the intestines and bladder. Without prompt administration of *epinephrine* (ep'i-nef'rin) (discussed shortly), an individual in anaphylactic shock may suffocate, collapse, and die within minutes.

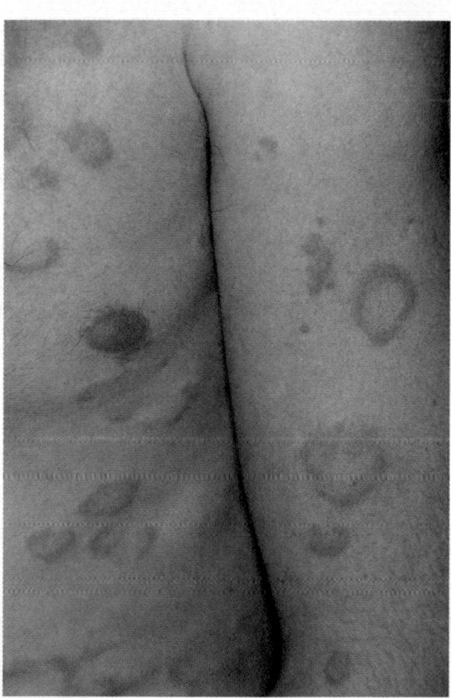

▲ **Figure 18.3 Urticaria.** These red, itchy patches on the skin are prompted by the release of histamine in response to an allergen. *What causes fluid to accumulate in the patches seen in urticaria?*

Figure 18.3 *The fluid accumulates because histamine from degranulated mast cells causes blood capillaries to become more permeable.*

Diagnosis of Type I Hypersensitivity

Clinicians diagnose type I hypersensitivity with a test variously called *ImmunoCAP Specific IgE blood test, CAP RAST,* or *Pharmacia CAP,* in which suspected allergens are mixed with samples of the patient's blood. The specifics of the test are beyond the scope of this chapter, but basically the test detects the amount of IgE directed against each allergen. High levels of a specific IgE indicate a hypersensitivity against that allergen.

Alternatively, physicians diagnose type I hypersensitivity by injecting a very small quantity of a dilute solution of the allergens being tested into the skin. In most cases the individual is screened for more than a dozen potential allergens simultaneously, using the forearms as injection sites. When the individual tested is sensitive to an allergen, local histamine release causes redness and swelling at the injection site within a few minutes **(Figure 18.4)**.

Prevention of Type I Hypersensitivity

Prevention of type I hypersensitivity begins with *identification* and *avoidance* of the allergens responsible. Filtration of air and avoidance of rural areas during pollen season can reduce upper respiratory allergies provoked by some pollens. Encasing

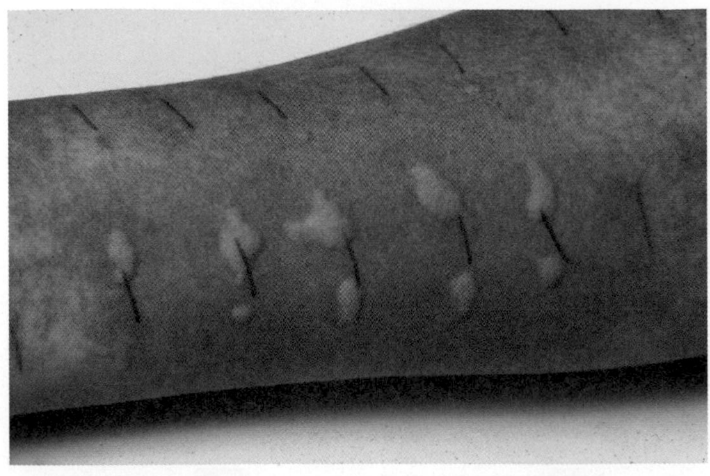

▲ **Figure 18.4 Skin tests for diagnosing type I hypersensitivity.** The presence of redness and swelling at an injection site indicates sensitivity to the allergen injected at that site.

bedding in mite-proof covers, frequent vacuuming, and avoidance of home furnishings that trap dust (such as wall-to-wall carpeting and heavy drapes) can reduce the severity of other household allergies. A dehumidifier can reduce mold in a home. The best way for individuals to avoid allergic reactions to pets is to find out whether they are allergic before they get a pet. If an allergy develops afterward, they may have to find a new home for the animal.

Food allergens can be identified and then avoided by using a medically supervised elimination diet in which foods are removed one at a time from the diet to see when signs of allergy cease. Peanuts and shellfish are among the foods more commonly implicated in anaphylactic shock; allergic individuals must scrupulously avoid consuming even minute amounts (see **Highlight: When Kissing Triggers Allergic Reactions**). For example, individuals sensitized against shellfish can suffer anaphylactic shock from consuming meat sliced on an unwashed cutting board previously used to prepare shellfish. Some vaccines contain small amounts of egg proteins and should not be given to individuals allergic to eggs.

Health care workers should assess a patient's medical history carefully before administering penicillin, iodine dyes, and other potential allergens. Therapies such as antidotes and respiratory stimulants should be immediately available if a patient's sensitivity status is unknown.

In addition to avoidance of allergens, type I hypersensitivity reactions can be prevented by *immunotherapy* (im′ū-nō-thār′-ă-pē), commonly called "allergy shots," which involves the administration of a series of injections of dilute allergen, usually once a week for many months. It is unclear just how allergy shots work, but they may alter the helper T cell balance from Th2 cells to Th1 cells, reducing the production of antibodies, or they may stimulate the production of IgG, which binds antigen before the antigen can react with IgE on mast cells, basophils, and eosinophils. Immunotherapy reduces the severity of allergy symptoms by roughly 50% in about two-thirds of patients with upper respiratory allergies; however, the series of injections must be repeated every two to three years. Immunotherapy is not effective in treating asthma.

Treatment of Type I Hypersensitivity

One way to treat type I hypersensitivity is to administer drugs that specifically counteract the inflammatory mediators released by degranulating cells. Thus, **antihistamines** are administered to counteract histamine. However, because histamine is but one of many mediators released in type I hypersensitivity, antihistamines do not completely eliminate all clinical signs. Indeed, because antihistamines are essentially useless in patients with asthma, asthmatics are typically prescribed an inhalant containing *glucocorticoid* (gloo-kō-kōr′ti-koyd) and a *bronchodilator* (brong-kō-dī-lā′ter), which counteract the effects of inflammatory mediators.

The hormone *epinephrine* quickly neutralizes many of the lethal mechanisms of anaphylaxis by relaxing smooth muscle tissue in the lungs, contracting smooth muscle of blood vessels, and reducing vascular permeability. Epinephrine is thus the drug of choice for the emergency treatment of both severe asthma and anaphylactic shock. Patients who suffer from severe type I hypersensitivities may carry a prescription epinephrine kit so that they can inject themselves before their allergic reactions become life threatening.

HIGHLIGHT

WHEN KISSING TRIGGERS ALLERGIC REACTIONS

Individuals with particularly severe food allergies not only need to watch what they eat—they also need to watch who kisses them. In a study from the University of California at Davis, researchers found that of 316 patients with severe allergies to peanuts, tree nuts, and/or seeds, 20 (about 6%) reported developing allergic reactions after being kissed. In nearly all of the cases, the kisser had eaten nuts to which the allergic individual was sensitive. In some of the cases, the allergic individuals were so sensitive that they developed reactions even though the kissers had brushed their teeth or had consumed the nuts several hours earlier. Food allergists believe that the duration and intensity of a kiss may also be a factor—that is, longer kisses involving exchange of saliva may be more likely to elicit allergic reactions than a quick buss on the cheek.

The moral of the story: if you have a severe food allergy, you might want to tell your romantic partner about it!

CLINICAL CASE STUDY

The First Time's Not the Problem

Steven, an eight-year-old boy, is brought to your office Monday morning by his father to have his upper arm checked for a possible infection. Dad is worried because the area of a bee sting on the boy's arm is getting more red, itchy, and tender. The father gave him some children's acetaminophen yesterday, which relieved the discomfort somewhat.

There is no history of medical problems or allergies, and the child takes no regular medication. The child is otherwise feeling well, and his father tells you he is playing and eating normally. There is no previous history of bee stings, and Steven proudly tells you he "hardly even cried" when he got stung.

There is a half-dollar-sized area on his left upper arm that is puffy and red, but there is no streaking or drainage, and the area does not appear to to be infected. Steven's temperature is normal and his lungs are clear.

1. What type of hypersensitivity reaction is Steven manifesting?
2. What other over-the-counter medication might relieve the itching?
3. What mechanism is causing the signs and symptoms you are seeing?
4. Since the area is not infected, what future health risk for his son should the father be made aware of?
5. What can be done to determine future risk from a bee sting?
6. How would you recognize a severe allergic reaction?
7. What precautions can the family take to protect the boy from future reactions?

Type II (Cytotoxic) Hypersensitivity

Learning Objectives

- ✓ Discuss the mechanisms underlying transfusion reactions.
- ✓ Construct a table comparing the key features of the four blood types in the ABO blood group system.
- ✓ Describe the mechanisms and treatment of hemolytic disease of the newborn.

The second major form of hypersensitivity results when cells are destroyed by an immune response—typically by the combined activities of complement and antibodies. This *cytotoxic hypersensitivity* is part of many autoimmune diseases, which we will discuss later in this chapter, but the most significant examples of type II hypersensitivity are the destruction of donor red blood cells following an incompatible blood transfusion and the destruction of fetal red blood cells in hemolytic disease of the newborn. We begin our discussion by focusing on these two manifestations of type II hypersensitivity.

The ABO System and Transfusion Reactions

Red blood cells have many different glycoprotein and glycolipid molecules on their surface. Some surface molecules of red blood cells, called **blood group antigens,** have various functions, including transportation of glucose and ions across the cytoplasmic membrane.

There are several sets of blood group antigens that vary in complexity. The ABO group system is most famous and consists of just two antigens arbitrarily given the names A antigen and B antigen. Each person's red blood cells have either A antigen, B antigen, both A and B antigens, or neither antigen. Individuals with neither antigen are said to have blood type O.

As you probably know, blood can be transfused from one person to another; however, if blood is transfused to an individual with a different blood type, then the donor's blood group antigens may stimulate the production of antibodies in the recipient. These bind to and eventually destroy the transfused cells. The result can be a potentially life-threatening *transfusion reaction.* Note that it is a blood recipient's own immune system that causes problems; the donated cells merely trigger the response.

Transfusion reactions, the most problematic of which involve the ABO group, develop as follows:

- If the recipient has preexisting antibodies to foreign blood group antigens, then the donated blood cells will be destroyed immediately—either the antibody-bound cells will be phagocytized by macrophages and neutrophils, or the antibodies will agglutinate the cells, and complement will rupture them, a process called *hemolysis* **(Figure 18.5)**. Hemolysis releases hemoglobin into the bloodstream, which may cause severe kidney damage. At the same time, the membranes of the ruptured blood cells trigger blood clotting within blood vessels, blocking them and causing circulatory failure, fever, difficulty in breathing, coughing, nausea, vomiting, and diarrhea. If the patient survives, recovery follows the elimination of all the foreign red blood cells.
- If the recipient has no preexisting antibodies to the foreign blood group antigens, then the transfused cells circulate and function normally, but only for a while—that is, until the recipient's immune system mounts a primary response against the foreign antigens and produces enough antibody to destroy the foreign cells. This happens gradually over a long enough time that the severe symptoms and signs mentioned above do not occur.

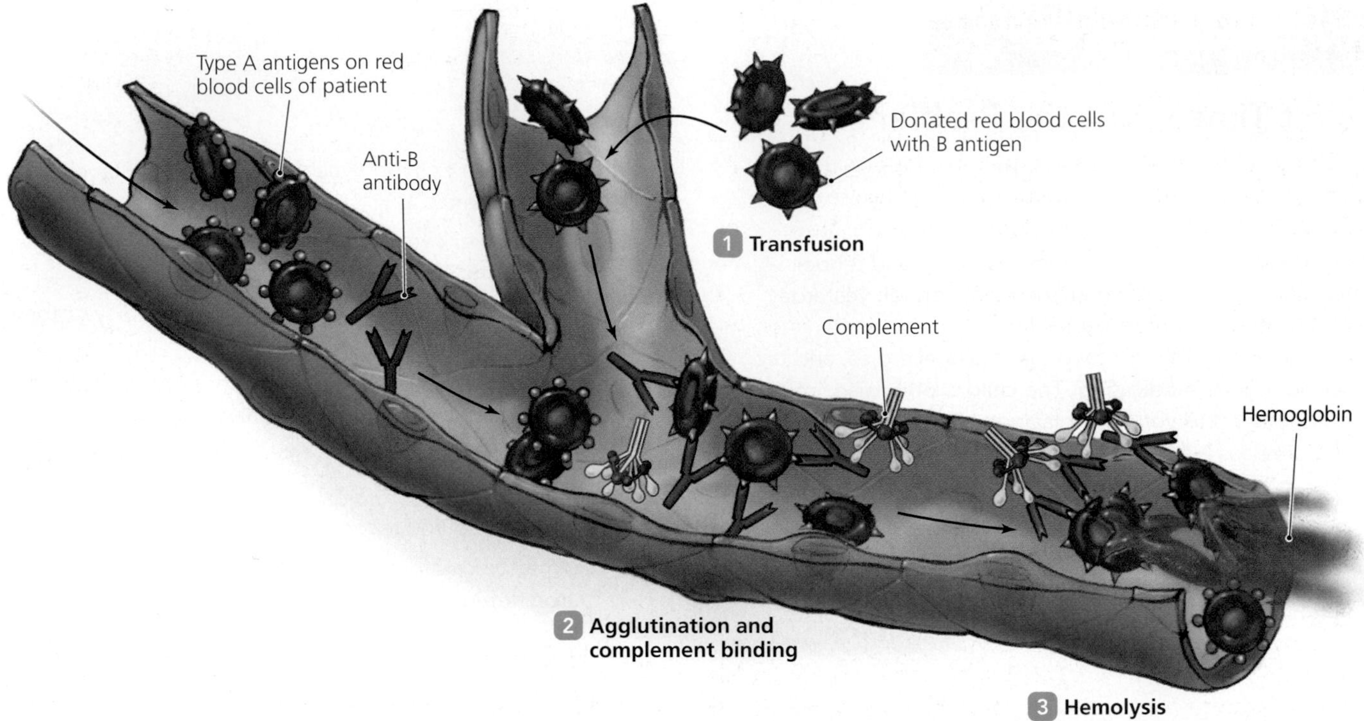

▲ **Figure 18.5 Events leading to hemolysis.** One of the negative consequences of a transfusion reaction (here illustrated by a transfusion of type B blood into a patient with type A blood): When antibodies against a foreign ABO antigen combine with the antigen on transfused red blood cells (1), the cells are agglutinated, and the complexes bind complement (2). The resulting hemolysis releases large amounts of hemoglobin into the bloodstream (3), and produces additional negative consequences throughout the body.

To prevent transfusion reactions, laboratory personnel must cross-match ABO blood types between donors and recipients. Before a recipient receives a blood transfusion, the red cells and serum of the donor can be mixed with the serum and red cells of the recipient. If any signs of agglutination are seen, then that blood is not used, and an alternative donor is sought. Table 18.2 presents the characteristics of the ABO blood group, and the compatible donor/recipient matches that can be made.

In most people, production of antibodies against foreign ABO antigens is not stimulated by exposure to foreign blood cells but instead by exposure to antigenically similar molecules found on a wide range of plants and bacteria. Individuals encounter and make antibodies against these antigens on a daily basis, but only upon receipt of a mismatched blood transfusion are they problematic.

CRITICAL **THINKING**

During the Iraq war in 2009, an Army corporal with type AB blood received a life-saving blood transfusion from his sergeant who had type O blood. Later the sergeant was involved in a traumatic accident and needed blood desperately. The corporal wanted to help but was told his blood was incompatible. Explain the immunological reasons the corporal could receive from but could not give to the sergeant.

TABLE 18.2 ABO Blood Group Characteristics and Donor/Recipient Matches

ABO Blood Group	ABO Antigen(s) Present	Antibodies Present	Can Donate To	Can Receive From
A	A	Anti-B	A or AB	A or O
B	B	Anti-A	B or AB	B or O
AB	A and B	None	AB	A, B, AB, or O (universal recipient)
O	None	Both anti-A and anti-B	A, B, AB, or O (universal donor)	O

The Rh System and Hemolytic Disease of the Newborn

Many decades ago, researchers discovered the existence of an antigen common to the red blood cells of humans and rhesus monkeys. They called this antigen, which transports anions and glucose across the cytoplasmic membrane, *rhesus* (rē'sŭs) antigen or **Rh antigen.** Laboratory analysis of human blood samples eventually showed that the Rh antigen is present on the red blood cells of about 85% of humans; that is, about 85% of the population is *Rh positive* (*Rh*+), and about 15% of the population is *Rh negative* (*Rh*−).

In contrast to the situation with ABO antigens, preexisting antibodies against Rh antigen do not occur. Although a transfusion reaction can occur in an Rh-negative patient who receives more than one transfusion of Rh-positive blood, such a reaction is usually minor because Rh antigen molecules are less abundant than A or B antigens. Instead, the primary problem posed by incompatible Rh antigen is the risk of **hemolytic** (hē-mō-lit'ik) **disease of the newborn** (**Figure 18.6**).

This hypersensitivity reaction develops when an Rh− mother is pregnant with an Rh+ baby (who inherited an Rh gene from its father). Normally, the placenta keeps fetal red blood cells separate from the mother's blood, so fetal cells do not enter the mother's bloodstream. However, in 20–50% of pregnancies—especially during the later weeks of pregnancy, during a clinical or spontaneous abortion, or during childbirth—fetal red blood cells escape into the mother's blood. The Rh− mother's immune system recognizes these Rh+ fetal cells as foreign and initiates a humoral immune response by developing antibodies against the Rh antigen. Initially, only IgM antibodies are produced, and because IgM is a very large molecule that cannot cross the placenta, no problems arise during this first pregnancy.

However, IgG antibodies can cross the placenta, so if during subsequent pregnancies the fetus is Rh+, anti-Rh IgG molecules produced by the mother cross the placenta and destroy the fetus's Rh+ red blood cells. Such destruction may be limited, or it may be severe enough—especially in a third or subsequent pregnancy—to cause grave problems. One classic feature of hemolytic disease of the newborn is severe jaundice from excessive *bilirubin* (bil-i-roo'bin), a yellowish blood pigment released during the degradation of hemoglobin from lysed red blood cells. The liver is normally responsible for removing bilirubin, but because of the immaturity of the fetal liver and overload from the hemolytic process, the bilirubin may instead be deposited in the brain, causing severe neurological damage or death during the latter weeks of pregnancy or shortly after the baby's birth.

In the past, when prevention of hemolytic disease of the newborn was impossible, this terrible disease occurred in about 1 of every 300 births. Today, however, physicians can routinely and drastically reduce the prevalence of the disease by administering anti-Rh immunoglobulin (IgG), called *RhoGAM,* to Rh-negative pregnant women at 28 weeks' gestation and also within 72 hours following abortion, miscarriage, or childbirth. Any fetal red cells that may have entered the mother's body are destroyed by RhoGAM before the fetal cells can trigger an immune response. As a result, sensitization of the mother does not occur, and subsequent pregnancies are safer.

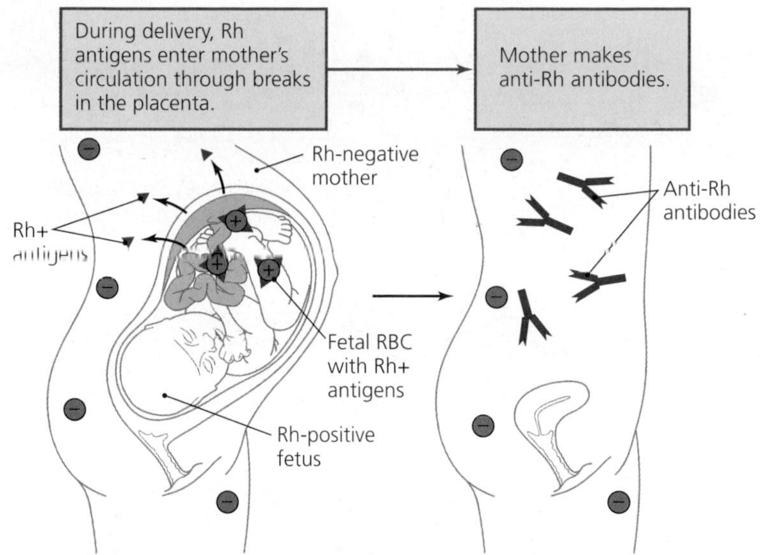

(a) First pregnancy

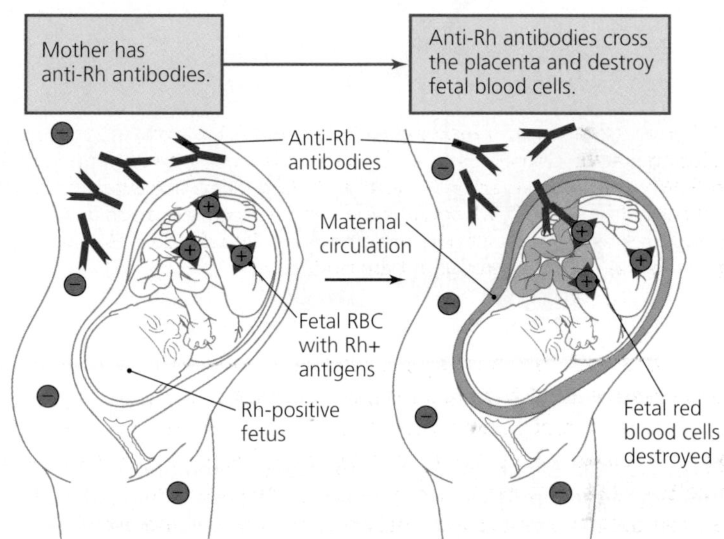

(b) Subsequent pregnancy

▲ **Figure 18.6 Events in the development of hemolytic disease of the newborn.** **(a)** During an initial pregnancy, Rh+ red blood cells from the fetus enter the Rh− mother's circulation, often during childbirth. As a result, the mother produces anti-Rh antibodies. **(b)** During a subsequent pregnancy with another Rh+ child, anti-Rh IgG molecules cross the placenta and trigger destruction of the fetus's red blood cells. *How is it possible that the child of an Rh− mother is Rh+?*

Figure 18.6 *The father of the child is Rh positive.*

Drug-Induced Cytotoxic Reactions

Another kind of type II hypersensitivity involves cytotoxic reactions to drugs. Although the molecules of such drugs as quinine, penicillin, or sulfanilamide are too small to trigger an

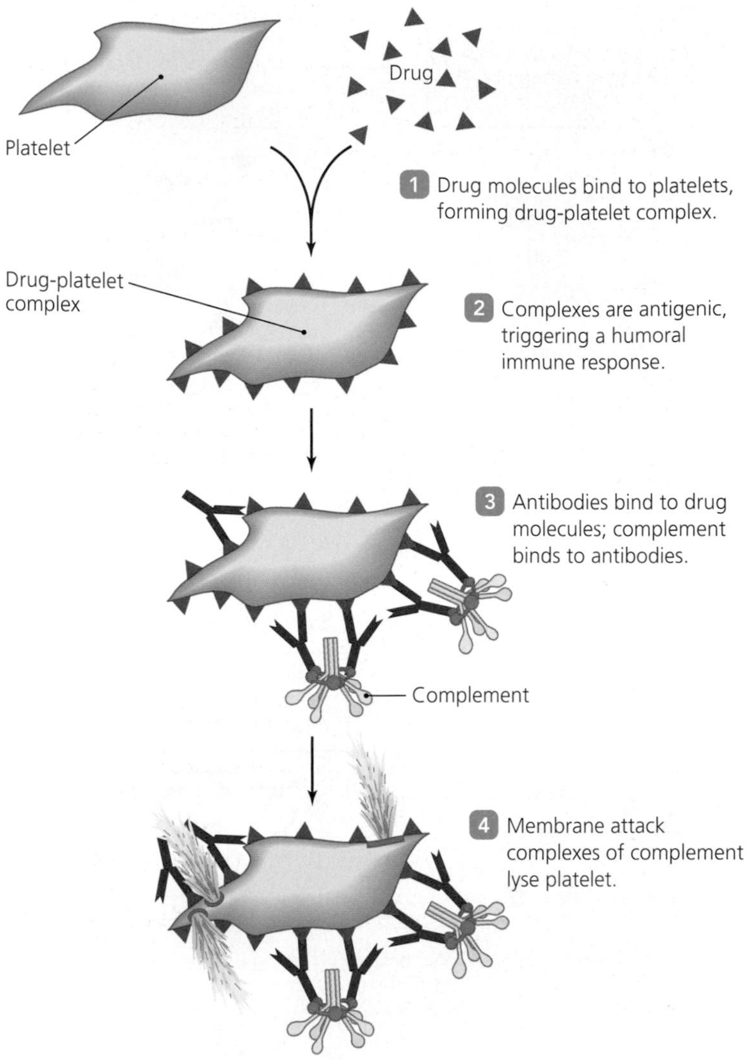
Platelet
Drug
1 Drug molecules bind to platelets, forming drug-platelet complex.
Drug-platelet complex
2 Complexes are antigenic, triggering a humoral immune response.
3 Antibodies bind to drug molecules; complement binds to antibodies.
Complement
4 Membrane attack complexes of complement lyse platelet.

▲ **Figure 18.7 Events in the development of immune thrombocytopenic purpura.** This disease results from a drug-induced type II (cytotoxic) hypersensitivity reaction. The destruction of platelets through the action of bound antibodies and complement produces the inhibition of blood clotting characteristic of this disease. *What similar disease results from type II (cytotoxic) destruction of red blood cells?*

Figure 18.7 *Hemolytic anemia is the type II (cytotoxic) destruction of red blood cells.*

immune response by themselves, they can be *haptens;* that is, when they bind to larger molecules, they become antigenic and stimulate production of antibodies (a humoral immune response).

When such antibodies and then complement bind to drug molecules already bound to blood platelets, the platelets are lysed, producing a disease called **immune thrombocytopenic[4] purpura** (throm′bō-sī-tō-pē′nik pūr′pū-ră) **(Figure 18.7)**. The destruction of the platelets inhibits the ability of the blood to clot correctly, which leads to the production of purple hemorrhages, called *purpura*[5] (pŭr′poo-ră), under the skin. Similar destruction of leukocytes is one form of *agranulocytosis,* and that of red blood cells is called *hemolytic anemia* (ă-nē′mē-ă).

[4] *Thrombocyte* is another name for platelet.
[5] From Latin, meaning purple.

Type III (Immune Complex–Mediated) Hypersensitivity

Learning Objectives

- ✓ Outline the basic mechanism of type III hypersensitivity.
- ✓ Describe hypersensitivity pneumonitis and glomerulonephritis.
- ✓ List the signs and symptoms of rheumatoid arthritis.
- ✓ Discuss the cause and signs of systemic lupus erythematosus.

The formation of complexes of antigen bound to antibody, called **immune complexes,** initiates several molecular processes, including complement activation. Normally, immune complexes are eliminated from the body via phagocytosis. However, in *type III (immune complex–mediated) hypersensitivity* reactions, the immune complexes escape phagocytosis, and so circulate in the bloodstream until they become trapped in organs, joints, and tissues (such as the walls of blood vessels). In these sites they trigger mast cells to degranulate, releasing inflammatory chemicals that damage the tissues. **Figure 18.8** illustrates the mechanism of type III reactions.

Type III hypersensitivities may be localized or may affect a number of body systems simultaneously. There is no cure for these diseases, though steroids that suppress the immune system may provide some relief. Two localized conditions resulting from immune complex–mediated hypersensitivity are hypersensitivity pneumonitis and glomerulonephritis; two systemic disorders are rheumatoid arthritis and systemic lupus erythematosus.

Hypersensitivity Pneumonitis

Type III hypersensitivities can affect the lungs, causing a form of pneumonia called **hypersensitivity pneumonitis** (noo-mō-nī′tus). Individuals become sensitized when minute mold spores or other antigens are inhaled deep into the lungs, stimulating the production of antibodies. A hypersensitivity reaction occurs when the subsequent inhalation of the same antigen stimulates the formation of immune complexes that then activate complement.

One form of hypersensitivity pneumonitis, called *farmer's lung,* occurs in farmers chronically exposed to spores from moldy hay. Many other syndromes in humans develop via a similar mechanism and are usually named after the source of the inhaled allergen. Thus, *pigeon breeder's lung* arises following exposure to the dust from pigeon feces, *mushroom grower's lung* is a response to the soil or fungal spores encountered when cultivating mushrooms, and *librarian's lung* results from inhalation of dust from old books.

Glomerulonephritis

Glomerulonephritis (glō-mār′ū-lō-nef-rī′tis) occurs when immune complexes circulating in the bloodstream are deposited in the walls of *glomeruli,* which are networks of minute blood

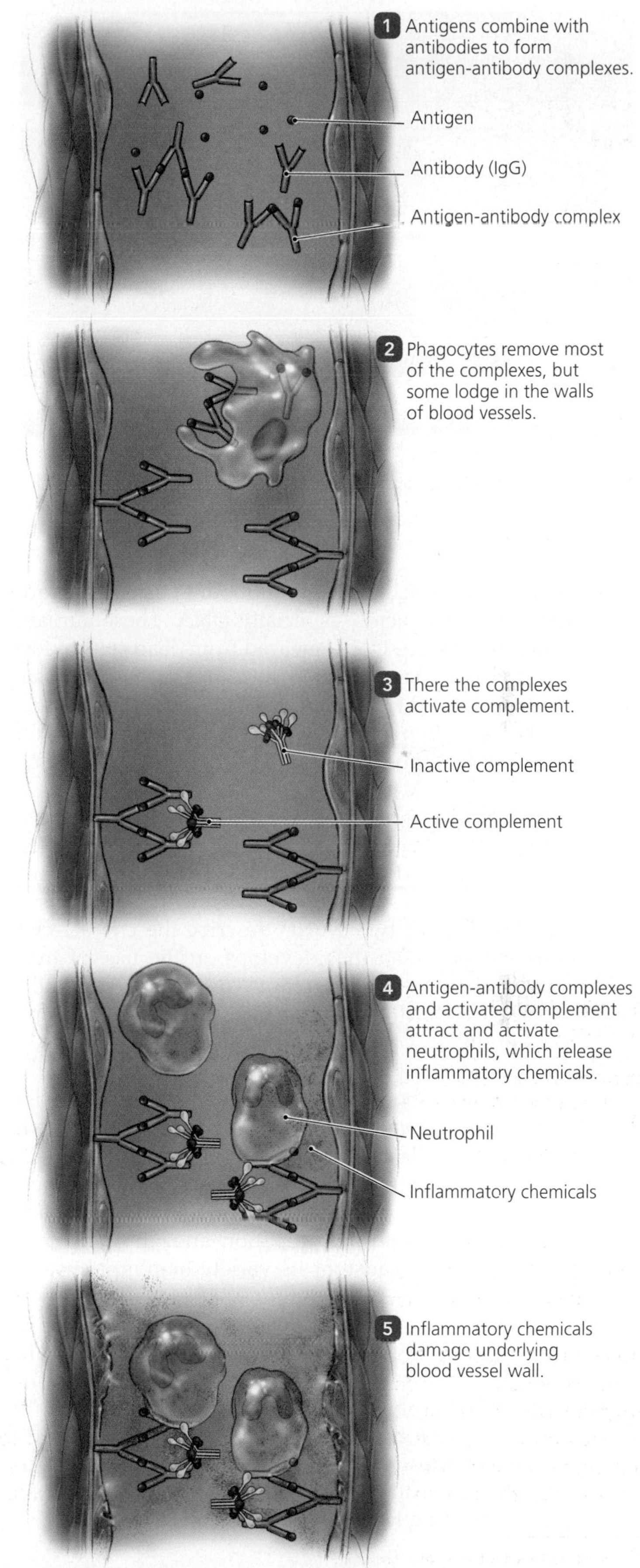
1 Antigens combine with antibodies to form antigen-antibody complexes.
Antigen
Antibody (IgG)
Antigen-antibody complex
2 Phagocytes remove most of the complexes, but some lodge in the walls of blood vessels.
3 There the complexes activate complement.
Inactive complement
Active complement
4 Antigen-antibody complexes and activated complement attract and activate neutrophils, which release inflammatory chemicals.
Neutrophil
Inflammatory chemicals
5 Inflammatory chemicals damage underlying blood vessel wall.

➤ **Figure 18.8 The mechanism of type III (immune complex–mediated) hypersensitivity.** After immune complexes form (1), the complexes that are not phagocytized lodge in certain tissues (in this case, the wall of a blood vessel) (2). Trapped complexes then activate complement (3), which attracts and activates neutrophils (4). The release of inflammatory chemicals from the activated neutrophils leads to the destruction of tissue (5).

vessels in the kidneys. Immune complexes damage the glomerular cells, leading to enhanced local production of cytokines that stimulate nearby cells to produce more of the intercellular proteins that underlie the cells, impeding blood filtration. Sometimes, the immune complexes are deposited in the center of the glomeruli, where they stimulate local cells to proliferate and compress nearby blood vessels, again interfering with kidney function. The net result is kidney failure; the glomeruli lose their ability to filter wastes from the blood, ultimately resulting in death.

Rheumatoid Arthritis

Rheumatoid arthritis (roo′mă-toyd ar-thrī′tis) commences when B cells secrete IgM that binds to certain IgG molecules. IgM-IgG complexes are deposited in the joints, where they activate complement and mast cells, which release inflammatory chemicals. The resulting inflammation causes the tissues to swell, thicken, and proliferate into the joint, resulting in severe pain. As the altered tissue extends into the joint, the inflammation further erodes and destroys joint cartilage and the neighboring bony structure, until the joints begin to break down and fuse; as a result, affected joints become distorted and lose their range of motion **(Figure 18.9)**. The course of rheumatoid arthritis is often intermittent; however, with each recurrence, the lesions and damage get progressively more severe.

The trigger of rheumatoid arthritis is not well understood. The fact that there are no animal models (because the disease appears to affect humans only) significantly hinders research on its cause. Many cases demonstrate that rheumatoid arthritis commonly follows an infectious disease in a genetically susceptible individual. Possession of certain MHC genes appears to increase susceptibility.

Physicians treat rheumatoid arthritis by administering anti-inflammatory drugs such as ibuprofen to prevent additional joint damage and immunosuppressive drugs to inhibit the humoral immune response.

Systemic Lupus Erythematosus

An example of type III hypersensitivity disease that affects multiple organs is **systemic lupus erythematosus** (loo′pŭs er-ĭ-thē′mă-tō-sus) **(SLE)**, often shortened to *lupus*. Patients with this generalized immunological disorder make antibodies against numerous self-antigens found in normal organs and tissues, giving rise to many different pathological lesions and clinical manifestations. One consistent feature of SLE is the

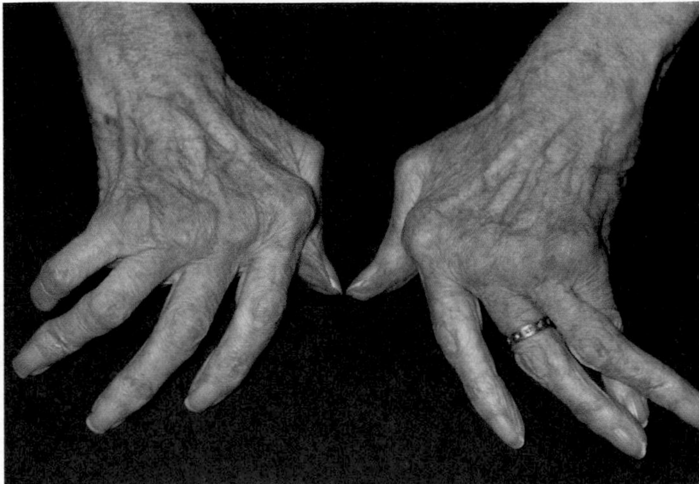

▲ **Figure 18.9 The crippling distortion of joints characteristic of rheumatoid arthritis.**

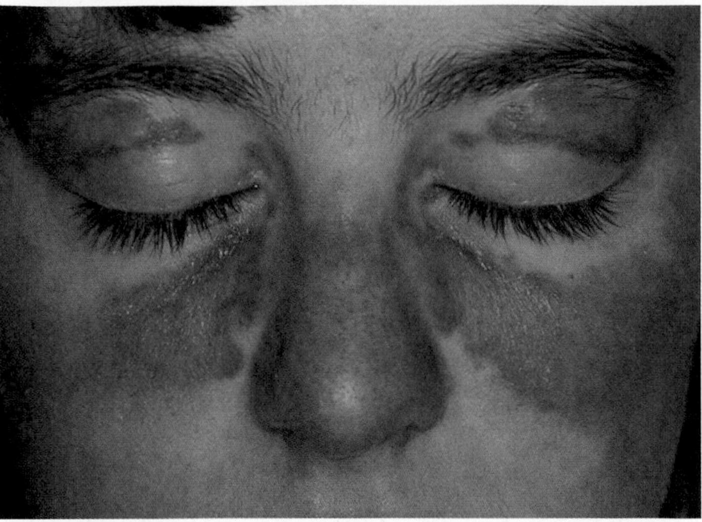

▲ **Figure 18.10 The characteristic facial rash of systemic lupus erythematosus.** Its shape corresponds to areas most exposed to sunlight, which worsens the condition.

development of such self-reactive antibodies, called autoantibodies, against nucleic acids, especially DNA. These autoantibodies combine with free DNA released from dead cells to form immune complexes that are deposited in glomeruli, causing glomerulonephritis and kidney failure. Thus, SLE is an immune complex–mediated hypersensitivity reaction. Immune complexes may also be deposited in joints, where they give rise to arthritis.

The disease's curious name—systemic lupus erythematosus—stems from two features of the disease: *Systemic* simply reflects the fact that it affects different organs throughout the body; *lupus* is Latin for wolf, and *erythematosus* refers to a redness of the skin, so these latter two words describe the characteristic red, butterfly-shaped rash that develops on the face of many patients, giving them what is sometimes described as a wolflike appearance **(Figure 18.10)**. This rash is caused by deposition of nucleic acid–antibody complexes in the skin and is worse in skin areas exposed to sunlight.

Although autoantibodies to nucleic acids are characteristic of SLE, many other autoantibodies are also produced. Autoantibodies to red blood cells cause hemolytic anemia; autoantibodies to platelets give rise to bleeding disorders; antilymphocyte antibodies alter immune reactivity; and autoantibodies against muscle cells cause muscle inflammation and, in some cases, damage to the heart. Because of its variety of symptoms, SLE can be misdiagnosed.

The trigger of lupus is unknown, though a lupus-like disease can be induced by some drugs. It is likely that SLE has many causes.

Physicians treat lupus with immunosuppressive drugs that reduce autoantibody formation and with glucocorticoids that reduce the inflammation associated with the deposition of immune complexes. Scientists are currently testing a battery of novel drugs for treating lupus.

Type IV (Delayed or Cell-Mediated) Hypersensitivity

Learning Objectives

- ✓ Outline the mechanism of type IV hypersensitivity.
- ✓ Describe the significance of the tuberculin test.
- ✓ Identify four types of grafts.
- ✓ Compare four types of drugs commonly used to prevent graft rejection.

When certain antigens contact the skin of sensitized individuals, they provoke inflammation that begins to develop at the site only after 12–24 hours. Such **delayed hypersensitivity reactions** result not from the action of antibodies but rather from interactions among antigen, antigen-presenting cells, and T cells; thus, a type IV reaction is also called *cell-mediated hypersensitivity.* The delay in this cell-mediated response reflects the time it takes for macrophages and T cells to migrate to and proliferate at the site of the antigen. We begin our discussion of type IV reactions by considering two common examples: the tuberculin response and allergic contact dermatitis. Then we discuss two type IV hypersensitivity reactions involving the interactions between the body and tissues grafted to (transplanted into) it—graft rejection and graft-versus-host disease—before considering donor-recipient matching and tissue typing.

The Tuberculin Response

The **tuberculin** (too-ber′kū-lin) **response** is an important example of a delayed hypersensitivity reaction in which the skin of an individual exposed to tuberculosis or tuberculosis vaccine reacts to a subcutaneous injection of *tuberculin*—a protein solution obtained from *Mycobacterium tuberculosis* (mī′kō-bak-tēr′ē-ŭm too-ber-kyū-lō′sis). Health care providers use the tuberculin test, also called a

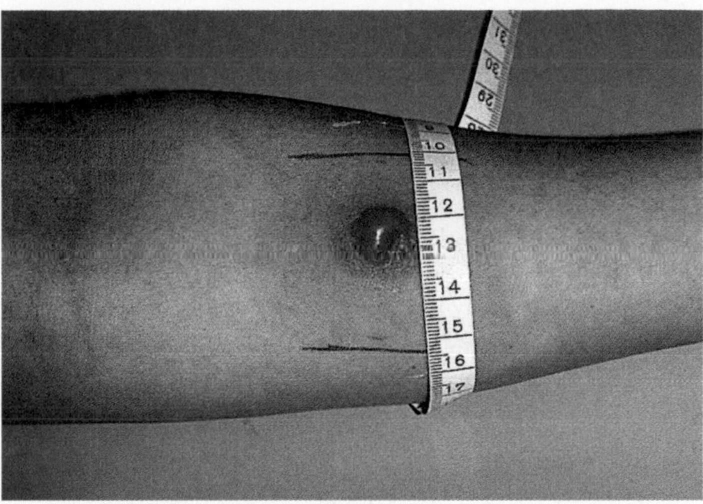

▲ **Figure 18.11 A positive tuberculin test.** The hard, red swelling 10 mm or greater in diameter that is characteristic of the tuberculin response indicates that the individual has been vaccinated against *Mycobacterium tuberculosis* or is now or has been previously infected with the bacterium.

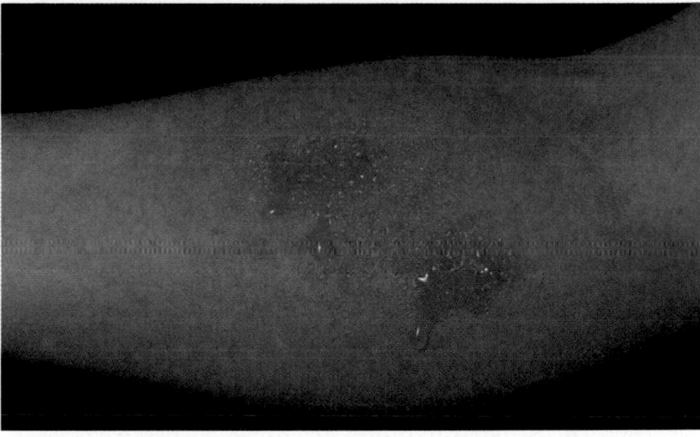

▲ **Figure 18.12 Allergic contact dermatitis, a type IV hypersensitivity response.** The response in this case is to poison ivy. Note the large, acellular, fluid-filled blisters that result from the destruction of skin cells.

Mantoux (mahn-too′) *test* after the French physician who perfected it, to diagnose contact with antigens of *M. tuberculosis*.

When tuberculin is injected into the skin of a healthy, never-infected or unvaccinated individual, no response occurs. In contrast, when tuberculin is injected into someone currently or previously infected with *M. tuberculosis* or an individual previously immunized with tuberculosis vaccine, a red, hard swelling (10 mm or greater in diameter) indicating a positive tuberculin test develops at the site (**Figure 18.11**). Such inflammation reaches its greatest intensity within 24–72 hours and may persist for several weeks before fading. Microscopic examination of the lesion reveals that it is infiltrated with lymphocytes and macrophages.

A tuberculin response is mediated by memory T cells. When an individual is first infected by or immunized against *M. tuberculosis,* the resulting cell-mediated immune response generates memory T cells that persist in the body. When a sensitized individual is subsequently injected with tuberculin, phagocytic cells migrate to the site and attract memory T cells, which secrete a mixture of cytokines that attract still more T cells and macrophages, giving rise to a slowly developing inflammation. The macrophages ingest and destroy the injected tuberculin, allowing the tissues eventually to return to normal.

Allergic Contact Dermatitis

Urushiol (ŭ-rū′shē-ŏl), the oil of poison ivy (*Toxicodendron radicans,* toks′si-kō-den′dron rā′dē-kanz) and related plants, is a highly reactive hapten that binds to almost any protein it contacts—including proteins in the skin of anyone who rubs against the plant. The body regards these chemically modified skin proteins as foreign, triggering a cell-mediated immune response and resulting in an intensely irritating skin rash called **allergic contact dermatitis** (der-mă-tī′tis). In severe cases, Tc cells destroy so many skin cells that acellular, fluid-filled blisters develop (**Figure 18.12**).

Other haptens that combine chemically with skin proteins can also induce allergic contact dermatitis. Examples include formaldehyde; some cosmetics, dyes, drugs, and metal ions; and chemicals used in the production of latex for hospital gloves and tubing.

Because T cells mediate allergic contact dermatitis, epinephrine and other drugs used for the treatment of immediate hypersensitivity reactions are ineffective. T cell activities and inflammation can, however, be suppressed by corticosteroid treatment. Good strategies for dealing with exposure to poison ivy include washing the area thoroughly and immediately with a strong soap and washing all exposed clothes as soon as possible.

Graft Rejection

A special case of type IV hypersensitivity is the rejection of **grafts** (or transplants), which are tissues or organs, such as livers, kidneys, or hearts, that have been transplanted, whether between sites within an individual or between a donor and an unrelated recipient. Even though advances in surgical technique enable surgeons to move grafts freely from site to site, grafts perceived as foreign by a recipient undergo rejection, a highly destructive phenomenon that can severely limit the success of organ and tissue transplantation. As discussed in Chapter 16, graft rejection is a normal immune response against foreign major histocompatibility complex (MHC) proteins on the surface of graft cells. The likelihood of graft rejection depends on the degree to which the graft is foreign to the recipient, which in turn is related to the type of graft.

Scientists name graft types according to the degree of relatedness between the donor and the recipient (**Figure 18.13**). A graft is called an **autograft** (aw′tō-graft) when tissues are moved to a different location within the same individual. Autografts do not trigger immune responses because they do not express foreign antigens. Examples of autografts include the grafting of

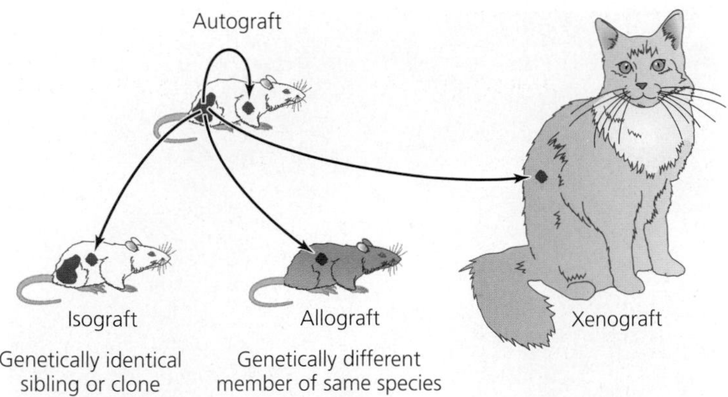

▲ **Figure 18.13 Types of grafts.** The names are based on the degree of relatedness between donor and recipient. Autografts are grafts moved from one location to another within a single individual. In isografts, the donor and recipient are either genetically identical siblings or clones. In allografts, the donor and recipient are genetically distinct individuals of the same species. In xenografts, the donor and recipient are of different species.

skin from one area of the body to another to cover a burn area, or the use of a leg vein to bypass blocked coronary arteries.

Isografts (ī′sō-graftz) are grafts transplanted between two genetically identical individuals; that is, between identical siblings or clones. Because these individuals have identical MHC proteins, the immune system of the recipient cannot differentiate between the grafted cells and its own normal body cells. As a result, isografts are not rejected.

Allografts (al′ō-graftz) are grafts transplanted between genetically distinct members of the same species. Most grafts performed in humans are allografts. Because the MHC proteins of the allograft are different from those of the recipient, allografts typically induce a strong type IV hypersensitivity, resulting in graft rejection. Rejection must be suppressed with immunosuppressive drugs if the graft is to survive.

Xenografts (zēn′ō-graftz) are grafts transplanted between individuals of different species. Thus, the transplant of a baboon's heart into a human is a xenograft. Because xenografts are usually very different from the tissues of the recipient, both biochemically and immunologically, they usually provoke a rapid, intense rejection that is very difficult to suppress. Therefore, xenografts from mature animals are not commonly used therapeutically.

Graft-Versus-Host Disease

Physicians often use bone marrow allografts as a component of the treatment for leukemias and lymphomas (cancers of leukocytes). In this procedure, physicians use total body irradiation combined with cytotoxic drugs to kill tumor cells in the patient's bone marrow. In the process, the patient's existing leukocytes are also destroyed, which completely eliminates the body's ability to mount any kind of immune response. Physicians then inject the patient with donated bone marrow which produces a new set of leukocytes. Ideally, within a few months, a fully functioning bone marrow is restored.

Unfortunately, when they are transplanted, donated bone marrow T cells may regard the patient's cells as foreign, mounting an immune response against them and giving rise to a condition called **graft-versus-host disease.** If the donor and recipient mainly differ in MHC class I molecules, the grafted T cells attack all of the recipient's tissues, producing especially destructive lesions in the skin and intestine. If the donor and recipient differ mainly in MHC class II molecules, then the grafted T cells attack the antigen-presenting cells of the host, leading to immunosuppression and leaving the recipient vulnerable to infections. The same immunosuppressive drugs used to prevent graft rejection (discussed shortly) can limit graft-versus-host disease.

Donor-Recipient Matching and Tissue Typing

Although it is usually not difficult to ensure that donor and recipient have identical blood groups, MHC compatibility is much harder to achieve. The reason is the very high degree of MHC variability, which ensures that unrelated individuals differ widely in their MHCs. In general, the more closely donor and recipient are related, the smaller their MHC difference. Given that in most cases an identical sibling is not available, it is usually preferable that grafts be donated by a parent or sibling possessing MHC antigens similar to those of the recipient.

In practice, of course, closely related donors are not always available. And even though a paired organ (such as a kidney) or a portion of an organ (such as liver tissue) may be available from a living donor, unpaired organs (such as the heart) almost always come from an unrelated cadaver. In such cases, attempts are made to match donor and recipient as closely as possible by means of *tissue typing*. Physicians examine the white cells of potential graft recipients to determine what MHC proteins they have. When a donor organ becomes available, it too is typed. Then the individual whose MHC proteins most closely match those of the donor is chosen to receive the graft. Though a perfect match is rarely achieved, the closer the match, the less intense the rejection process, and the greater the chance of successful grafting. A match of 50% or less of the MHC proteins is usually acceptable for most organs, but near absolute matches are required for successful bone marrow transplants.

Table 18.3 summarizes the major characteristics of the four types of hypersensitivity.

CRITICAL **THINKING**

A patient arrives at the doctor's office with a rash covering her legs. How could you determine whether the rash is a type I or a type IV hypersensitivity?

Next we turn our attention to the actions of various types of immunosuppressive drugs used in situations in which the immune system is overactive.

TABLE 18.3 The Characteristics of the Four Types of Hypersensitivity Reactions

Descriptive	Name	Cause	Time Course	Characteristic Cells Involved
Type I	Immediate hypersensitivity	Antibody (IgE) on sensitized cells' membranes binds antigen, causing degranulation	Seconds to minutes	Mast cells, basophils, and eosinophils
Type II	Cytotoxic hypersensitivity	Antibodies and complement lyse target cells	Minutes to hours	Red blood cells
Type III	Immune complex–mediated hypersensitivity	Nonphagocytized complexes of antibodies and antigens trigger mast cell degranulation	Several hours	Neutrophils
Type IV	Delayed hypersensitivity	T cells attack the body's cells	Several days	Activated T cells

The Actions of Immunosuppressive Drugs

The development of potent immunosuppressive drugs has played a large role in the dramatic success of modern transplantation procedures. These drugs can also be effective in combating certain autoimmune diseases (discussed shortly). In the following discussion we consider four classes of immunosuppressive drugs: glucocorticoids, cytotoxic drugs, cylosporine, and lymphocyte-depleting therapies.

Glucocorticoids, sometimes called *corticosteroids* and commonly referred to as steroids, have been used as immunosuppressive agents for many years. Glucocorticoids such as *prednisone* (pred′ni-sōn) and *methylprednisolone* suppress the response of T cells to antigen and inhibit such mechanisms as T cell cytotoxicity and cytokine production. These drugs have a much smaller effect on B cell function.

Cytotoxic (sī-tō-tok′sik) **drugs** inhibit mitosis and cytokinesis (cell division). Given that lymphocyte proliferation is a key feature of specific immunity, blocking cellular reproduction is a potent, although very nonspecific, method of immunosuppression. Among the cytotoxic drugs that have been used, the following are noteworthy:

- *Cyclophosphamide* (sī-klō-fos′fă-mīd) cross-links daughter DNA molecules in mitotic cells, preventing their separation and blocking mitosis. It impairs both B cell and T cell responses.
- *Azathioprine* (ā-ză-thī′ō-prēn) is a purine analog; it competes with purines during the synthesis of nucleic acids, thus blocking DNA replication and suppressing both primary and secondary antibody responses.

Three other cytotoxic drugs used for immunosuppression include *mycophenolate mofetil,* which inhibits purine synthesis, and *brequinar sodium* and *leflunomide,* which each inhibit pyrimidine synthesis and thereby inhibit cellular replication.

Drugs such as **cyclosporine** (sī-klō-spōr′ēn), a polypeptide derived from fungi, prevent production of interleukins and interferons by T cells, thereby blocking Th1 responses. Because cyclosporine acts only on activated T cells and has no effect on resting T cells, it is far less toxic than the nonspecific drugs previously described. When it is given to prevent allograft rejection, only activated T cells attacking the graft are suppressed. Because steroids have a similar effect, the combination of glucocorticoids and cyclosporine is especially potent and can enhance survival of allografts.

Scientists have developed techniques involving relatively specific *lymphocyte-depleting therapies* in an attempt to reduce the many adverse side effects associated with the use of less-specific immunosuppressive drugs. One technique involves administering an antiserum called *antilymphocyte globulin,* which is specific for lymphocytes. Another, more specific antilymphocyte technique uses monoclonal antibodies against CD3, which is found only on T cells. An even more specific monoclonal antibody is directed against the interleukin 2 receptor (IL-2R), which is expressed mainly on activated T cells. Such immunosuppressive therapies are effective in reversing graft rejection. Because they target a narrow range of cells, they produce fewer undesirable side effects than less-specific drugs.

Table 18.4 on p. 526 lists some drugs, and their actions, for each of the four classes of the immunosuppressive drugs.

Autoimmune Diseases

Just as today's military must control its arsenal of sophisticated weapons to avert losses of its own soldiers from "friendly fire," the immune system must be carefully regulated so that it does not damage the body's own tissues. However, an immune system does occasionally produce antibodies and cytotoxic T cells that target normal body cells—a phenomenon called *autoimmunity.* Although such responses are not always damaging, they can give rise to **autoimmune diseases,** some of which are life threatening.

Causes of Autoimmune Diseases

Learning Objective

✓ Briefly discuss seven hypotheses concerning the causes of autoimmunity and autoimmune diseases.

Most autoimmune diseases appear to develop spontaneously and at random, with no obvious predisposing factor. Nevertheless, scientists have noted some common features of autoimmune disease. For example, they occur more often in older individuals, and they are also much more common in women

TABLE 18.4 The Four Classes of Immunosuppressive Drugs

Class	Examples	Action
Glucocorticoids	Prednisone, methylprednisolone	Anti-inflammatory; kills T cells
Cytotoxic drugs	Cyclophosphamide, azathioprine, mycophenolate mofetil, brequinar sodium, leflunomide	Blocks cell division nonspecifically
Cyclosporine	Cyclosporine	Blocks T cell responses
Lymphocyte-depleting therapies	Antilymphocyte globulin, monoclonal antibodies	Kills T cells nonspecifically, kills activated T cells, inhibits IL–2 reception

than in men, although the reasons for this gender difference are unclear.

Hypotheses to explain the etiology of autoimmunity abound. They include the following:

- Estrogen may stimulate the destruction of tissues by cytotoxic T cells.
- During pregnancy, some maternal cells may cross the placenta and colonize the fetus. These cells are more likely to survive in a daughter than in a son and might trigger an autoimmune disease later in the daughter's life.
- Conversely, fetal cells may also cross the placenta and trigger autoimmunity in the mother.
- Environmental factors may contribute to the development of autoimmune disorders. Some autoimmune diseases—type I diabetes mellitus (dī-ă-bē′tēz mĕ-lī′tĕs) and rheumatoid arthritis, for example—develop in a few patients following their recovery from viral infections, though other individuals who develop these diseases have no history of such viral infections.
- Genetic factors may play a role in autoimmune diseases. MHC genes that in some way promote autoimmunity are found in individuals with autoimmune diseases, whereas MHC genes that somehow protect against autoimmunity may dominate in other individuals. MHC genes may also determine susceptibility to autoimmune disease by regulating the elimination of self-reactive T cells in the thymus.
- Some autoimmune diseases develop when T cells encounter self-antigens that are normally "hidden" in sites where T cells rarely go. For example, because sperm develop within the testes during puberty, long after the body has selected its T cell population, men may have T cells that recognize their own sperm as foreign. This is normally of little consequence because sperm are sequestered from the blood. But if the testes are injured, T cells may enter the site of damage and mount an autoimmune response against the sperm, resulting in infertility.
- Infections with a variety of microorganisms may trigger autoimmunity as a result of **molecular mimicry,** which occurs when an infectious agent has an epitope that is very similar or identical to a self-antigen. In responding to the invader, the body produces antibodies that are *autoantibodies* (antibodies against self-antigens), which damage body tissues. For example, children infected with some strains of *Streptococcus pyogenes* (strep-tō-kok′ŭs pī-oj′en-ēz) may produce antibodies to heart muscle and so develop heart disease. Other strains of streptococci trigger the production of antibodies that cross-react with glomerular basement membranes and so cause kidney disease. It is possible that some virally triggered autoimmune diseases may also result from molecular mimicry.
- Other autoimmune responses may result from failure of the normal control mechanisms of the immune system. Thus even though harmful, self-reactive T lymphocytes are normally destroyed via apoptosis triggered through the cell-surface receptor CD95 (see Chapter 16), defects in CD95 can cause autoimmunity by permitting abnormal T cells to survive and cause disease.

Examples of Autoimmune Diseases

Learning Objective

✓ Describe a serious autoimmune disease associated with each of the following: blood cells, endocrine glands, nervous tissue, and connective tissue.

Regardless of the specific mechanism that causes an autoimmune disease, immunologists categorize them into two major categories: systemic autoimmune diseases such as lupus (discussed previously) and single-organ autoimmune diseases, which affect a single organ or tissue. Among the many single-organ autoimmune diseases recognized, common examples affect blood cells, endocrine glands, nervous tissue, or connective tissue.

Autoimmunity Affecting Blood Cells

Individuals with **autoimmune hemolytic anemia** produce antibodies against their own red blood cells. These autoantibodies hasten the destruction of the red blood cells, and the patient becomes severely anemic. Different hemolytic anemia patients make antibodies of different classes. Some patients make IgM autoantibodies, which bind to red blood cells and activate the classical complement pathway; the red blood cells are lysed, and degration products from hemoglobin are released into the bloodstream. Other hemolytic anemia patients make IgG autoantibodies, which serve as opsonins that promote phagocytosis. In this case, red blood cells are removed by macrophages in the liver, spleen, and bone marrow. Even though these latter patients have no hemoglobin in their urine, they are still severely anemic.

The precise causes of all cases of autoimmune hemolytic anemia are unknown, but some cases follow infections with viruses or treatment with certain drugs, both of which alter the surface of red blood cells such that they are recognized as foreign and trigger an immune response.

Autoimmunity Affecting Endocrine Organs

Other common targets of autoimmune attack are the endocrine (hormone-producing) organs. For example, patients can develop autoantibodies or produce T cells against cells of the islets of Langerhans within the pancreas or against cells of the thyroid gland. In most cases, the ensuing autoimmune reaction results in damage to or destruction of the gland and in hormone deficiencies as endocrine cells are killed.

Immunological attack on the islets of Langerhans results in a loss of the ability to produce the hormone *insulin,* which leads to the development of **type I diabetes mellitus** (also known as *juvenile-onset diabetes*). As with other autoimmune diseases, the exact trigger of type I diabetes is unknown, but many patients endured a severe viral infection some months before the onset of diabetes. Additionally, some patients are known to have a genetic predisposition to developing type I diabetes that is associated with the possession of certain class I MHC molecules. Some physicians have been successful in delaying the onset of type I diabetes by treating at-risk patients with immunosuppressive drugs before damage to the islets of Langerhans becomes apparent.

An autoimmune response can lead to stimulation rather than to inhibition or destruction of glandular tissue. An example of this is **Graves' disease,** which involves the thyroid gland. This major endocrine gland located in the neck secretes iodine-containing hormones, which help regulate metabolic rate in the body. Like other autoimmune diseases, Graves' disease may be triggered by a viral infection in individuals with certain genetic backgrounds. Affected patients (usually women) make autoantibodies that bind to and stimulate receptors on the cytoplasmic membranes of thyroid cells, which elicits excessive production of thyroid hormone and growth of the thyroid gland. Such patients develop an enlarged thyroid gland—called a goiter—protruding eyes, rapid heartbeat, fatigue, and weight loss despite increased appetite. Physicians treat Graves' disease with antithyroid medicines, radioactive iodine, or, in nonresponsive patients, by surgically removing most of the thyroid tissue.

Autoimmunity Affecting Nervous Tissue

Of the group of autoimmune diseases affecting nervous tissue, the most frequent is **multiple sclerosis** (sklĕ-rō′sis) **(MS).** The exact cause of MS is unknown, but it appears that a cell-mediated immune response against a bacterium or virus generates cytotoxic T cells that mistakenly attack and destroy the myelin sheaths that normally insulate brain and spinal cord neurons and increase the speed of nerve impulses along the length of the neurons. Consequently, MS patients experience deficits in vision, speech, and neuromuscular function that may be quite mild and intermittent or may ultimately lead to death.

Autoimmunity Affecting Connective Tissue

Rheumatoid arthritis (discussed previously) is another crippling autoimmune disease resulting from a type III hypersensitivity. Autoantibodies are formed against connective tissue in joints.

CRITICAL **THINKING**

A 43-year-old woman has been diagnosed with rheumatoid arthritis. Unable to find relief from her symptoms, she seeks treatment from a doctor in another country who injects antibodies and complement into her afflicted joints. Do you expect the treatment to improve her condition? Explain your reasoning.

Immunodeficiency Diseases

Learning Objective

✓ Differentiate between primary and acquired immunodeficiencies, and cite one disease caused by each form of immunodeficiency.

You may have noticed that during periods of increased emotional or physical stress you are more likely to succumb to a cold or a bout of the flu. You are not imagining this phenomenon; stress has long been known to decrease the efficiency of the immune system. Similarly, chronic defects in the immune system typically first become apparent when affected individuals show increased susceptibility to infections of opportunistic pathogens (or even of organisms not normally considered pathogens). Such opportunistic infections are the hallmarks of *immunodeficiency diseases,* which are conditions resulting from defective immune mechanisms.

Researchers have characterized a large number of immunodeficiency diseases in humans, which are of two general types:

- **Primary immunodeficiency diseases,** which are detectable near birth and develop in infants and young children, result from some genetic or developmental defect.
- **Acquired (secondary) immunodeficiency diseases** develop in later life as a direct consequence of some other recognized cause, such as malnutrition, severe stress, or infectious disease.

Next we consider each general type of immunodeficiency disease in turn.

Primary Immunodeficiency Diseases

Many different inherited defects have been identified in all of the body's lines of defense, affecting first and second lines of defense as well as humoral and cell-mediated immune responses.

HIGHLIGHT

SCID: "BUBBLE BOY" DISEASE

As a result of the widely publicized case of David Vetter—a Houston, Texas, resident commonly known as the "bubble boy" because he lived from birth until age 12 in a sterile, plastic-enclosed environment —*severe combined immunodeficiency disease (SCID)* has been known as "bubble boy" disease since the 1970s. Like all SCID patients, David had no immune system—his body produced neither B cells nor T cells—so the slightest infection could be lethal. As a result, contact with him could occur only through a pair of antiseptic rubber gloves built into one of the walls of his plastic enclosure. At age 12, David underwent an experimental bone marrow transplant in the hope that marrow donated by his older sister would enable him to build up an immune system. Unfortunately, the donated cells turned out to be infected with Epstein-Barr virus (a common virus that causes mononucleosis), ultimately killing David.

Today, there is much more hope for patients with SCID. Virus-free bone marrow transplants are now possible. More recently, more than 20 children have undergone gene therapy, in which the gene for an enzyme (adenine deaminase) missing in SCID patients is inserted, via retroviral vectors, into clones of the patient's bone marrow cells. These genetically altered cells are then returned to the child, where they grow and synthesize sufficient enzyme to normalize the immune response. Though successful in several children studied in a French trial, the therapy was halted in 2002 after three children developed leukemia following treatment. Investigations into whether or not the illness was caused by the gene therapy are under way; however, scientists remain hopeful that gene therapy can achieve a cure for SCID—without side effects—in the near future.

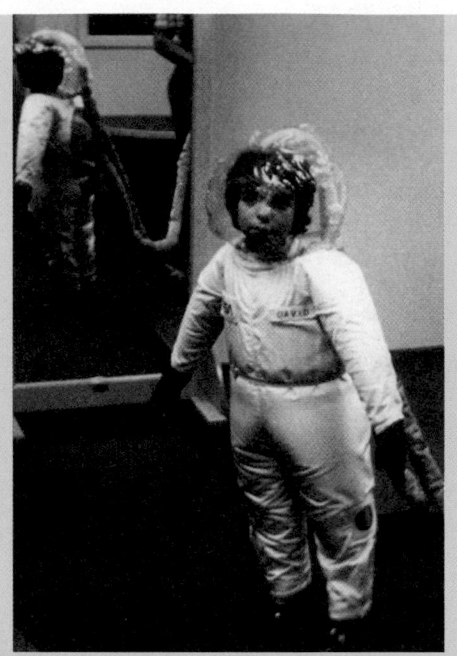

▲ *David Vetter.*

One of the more important inherited defects in the second line of defense is **chronic granulomatous disease,** in which children have recurrent infections characterized by the inability of their phagocytes to destroy bacteria. The reason is that these children have an inherited inability to make reactive forms of oxygen, which are necessary to destroy phagocytized bacteria.

Most primary immunodeficiencies are associated with defects in the components of the third line of defense—adaptive immunity. For example, some children fail to develop any lymphoid stem cells whatsoever, and as a result they produce neither B cells nor T cells and cannot mount immune responses. The resulting defects in the immune system cause **severe combined immunodeficiency disease (SCID),** which is discussed in **Highlight: SCID: "Bubble Boy" Disease.**

Other children suffer from T cell deficiencies alone. For example, **DiGeorge syndrome** results from a failure of the thymus to develop. Consequently, there are no T cells. The importance of T cells in protecting against viruses is emphasized by the observation that individuals with DiGeorge syndrome generally die of viral infections while remaining resistant to most bacteria. Physicians treat DiGeorge syndrome with a thymic stem cell transplant.

B cell deficiencies also occur in children. The most severe of the B cell deficiencies, called **Bruton-type agammaglobulinemia**[6] (ā-gam′ă-glob′ū-li-nē′mē-ă), is an inherited disease in which affected babies, usually boys, cannot make immunoglobulins. These children experience recurrent bacterial infections but are usually resistant to viral, fungal, and protozoan infections.

Inherited deficiencies of individual immunoglobulin classes are more common than a deficiency of all classes. Among these, IgA deficiency is the most common. Because affected children cannot produce secretory IgA, they experience recurrent infections in the respiratory and gastrointestinal tracts.

CRITICAL THINKING

Two boys have autoimmune diseases: One has Bruton-type agammaglobulinemia, and the other has DiGeorge syndrome. On a camping trip, each boy is stung by a bee and each falls into poison ivy. What hypersensitivity reactions might each boy experience as a result of his camping mishaps?

Table 18.5 summarizes the major primary immunodeficiency diseases.

Acquired Immunodeficiency Diseases

Unlike inherited primary immunodeficiency diseases, acquired immunodeficiency diseases affect older individuals who had a previously healthy immune system.

[6]Agammaglobulinemia means an absence of gamma globulin (IgG).

TABLE 18.5 Some Primary Immunodeficiency Diseases

Disease	Defect	Manifestation
Chronic granulomatous disease	Ineffective phagocytes	Uncontrolled infections
Severe combined immunodeficiency disease (SCID)	A lack of T cells and B cells	No resistance to any type of infection, leading to rapid death
Bruton-type agammaglobulinemia	A lack of B cells and thus a lack of immunoglobulins	Death from overwhelming bacterial infections
DiGeorge syndrome	A lack of T cells and thus no cell-mediated immunity	Death from overwhelming viral infections

Acquired immunodeficiencies result from a number of causes. In all humans the immune system (but especially T cell production) deteriorates with increasing age; as a result, older individuals normally have less effective immunity, especially cell-mediated immunity, than younger individuals, leading to an increased incidence of both viral diseases and certain types of cancer. Severe stress can also lead to immunodeficiencies by prompting the secretion of increased quantities of glucocorticoids, which are toxic to T cells and thus suppress cell-mediated immunity. This is why, for example, cold sores may "break out" on the faces of students during final exams: the causative herpes simplex viruses, which are controlled by a fully functioning cell-mediated immune response, escape immune control in stressed individuals. Malnutrition and certain environmental toxins can also cause acquired immunodeficiency diseases by inhibiting the normal production of B cells and T cells.

Of course, the most significant example of the result of an acquired immunodeficiency is **acquired immunodeficiency syndrome (AIDS).** From the time of its discovery in 1981 to its emergence as a worldwide pandemic, no affliction has affected modern life as much.

AIDS is not a single disease but a **syndrome;** that is, a group of signs, symptoms, and diseases associated with a common pathology. Currently, epidemiologists define this syndrome as the presence of several opportunistic or rare infections associated with a severe decrease in the number of CD4 cells and a positive test showing the presence of human immunodeficiency virus (HIV). The infections include diseases of the skin, such as shingles and herpes; diseases of the nervous system, including meningitis, toxoplasmosis, and *Cytomegalovirus* (sī-tō-meg′ă-lō-vī′rŭs) disease; diseases of the respiratory system, such as tuberculosis, *Pneumocystis* (nū-mō-sis′tis) pneumonia, histoplasmosis, and coccidioidomycosis; and diseases of the digestive system, including chronic diarrhea, thrush, and oral hairy leukoplakia. A rare cancer of blood vessels called Kaposi's sarcoma is also commonly seen in AIDS patients. AIDS often results in dementia during the final stages. Chapter 25 examines AIDS in detail.

Chapter Summary

1. Immunological responses may give rise to inflammatory reactions called **hypersensitivities.** An immunological attack on normal tissues gives rise to autoimmune disorders. A failure of the immune system to function normally may give rise to **immunodeficiency diseases.**

Hypersensitivities (pp. 512–525)

1. Type I hypersensitivity gives rise to **allergies.** Antigens that trigger this response are called **allergens.**
2. Allergies result when allergens bind to IgE molecules that are already bound to **mast cells, basophils,** and **eosinophils.** This causes the sensitized cells to degranulate and release **histamine, kinins, proteases, leukotrienes,** and **prostaglandins.**
3. Depending on the amount of these molecules and the site at which they are released, the result can produce various clinical syndromes, including **hay fever, asthma, urticaria** (hives), or various other allergies. When the inflammatory mediators exceed the body's coping mechanisms, **acute anaphylaxis (anaphylactic shock)** may occur. The specific treatment for anaphylaxis is epinephrine.
4. Type I hypersensitivity can be diagnosed by skin testing and can be partially prevented by avoidance of allergens and immunotherapy. Some type I hypersensitivities are treated with **antihistamines.**
5. Type II cytotoxic hypersensitivities, such as incompatible blood transfusions and hemolytic disease of the newborn, result when cells are destroyed by an immune response.
6. Red blood cells have **blood group antigens** on their surface. If incompatible blood is transfused into a recipient, a severe transfusion reaction can result.
7. The most important of the blood group antigens is the ABO group, which is largely responsible for transfusion reactions.
8. Approximately 85% of the human population carries **Rh antigen,** which is also found in rhesus monkeys. If an Rh-negative pregnant woman is carrying an Rh-positive fetus, the fetus may be at risk of **hemolytic disease of the newborn,** in which antibodies made by the mother against the Rh antigen may cross the placenta and destroy the fetus's red blood cells. RhoGAM administered to pregnant Rh– women may prevent this disease.

9. Drugs bound to blood platelets may subsequently bind antibodies and complement, causing **immune thrombocytopenic purpura,** in which the platelets lyse.
10. In type III hypersensitivity, excessive amounts of **immune complexes** are deposited in tissues, where they cause significant tissue damage. Immune complexes deposited in the lung cause a **hypersensitivity pneumonitis,** of which the most common example is farmer's lung. If large amounts of immune complexes form in the bloodstream, they may be filtered out by the glomeruli of the kidney, causing **glomerulonephritis,** which can result in kidney failure.
11. In **rheumatoid arthritis,** immune complexes result in the growth of inflammatory tissue within joints.
12. **Systemic lupus erythematosus** (SLE, lupus) is a systemic, autoimmune, type III hypersensitivity in which autoantibodies bind to many autoantigens, especially the patient's DNA.
13. Type IV hypersensitivity, also known as **delayed hypersensitivity,** is a T cell–mediated inflammatory reaction that takes 24–72 hours to reach maximal intensity.
14. A good example of a delayed hypersensitivity reaction is the **tuberculin response,** generated when tuberculin, a protein extract of *Mycobacterium tuberculosis,* is injected into the skin of an individual who has been infected with or vaccinated against *M. tuberculosis.*
15. Another example of a type IV hypersensitivity reaction is **allergic contact dermatitis,** which is T cell–mediated damage to chemically modified skin cells. The best-known example is a reaction to poison ivy.
16. An organ or tissue **graft** can be made between different sites within a single individual (an **autograft**), between genetically identical individuals (an **isograft**), between genetically dissimilar individuals (an **allograft**), or between individuals of different species (a **xenograft**). Most surgical organ grafting involves allografts, which, if not treated with immunosuppressive drugs, lead to **graft rejection.**
17. In **graft-versus-host disease,** an organ donor's cells attack the recipient's body.
18. Commonly used immunosuppressive drugs include **glucocorticoids, cytotoxic drugs, cyclosporine,** and lymphocyte-depleting therapies, which involve treatment with antibodies against T cells or their receptors.

Autoimmune Diseases (pp. 525–527)

1. **Autoimmune diseases** may result when an individual begins to make autoantibodies or cytotoxic T cells against normal body components.
2. There are many hypotheses concerning the cause of autoimmune disease. One involves **molecular mimicry,** in which microorganisms with epitopes similar to self-antigens trigger autoimmune tissue damage. Others implicate estrogen, pregnancy, and environmental and genetic factors.
3. One group of autoimmune diseases involve only a single organ or cell type. Examples of such diseases include **autoimmune hemolytic anemia, type I diabetes mellitus, Graves' disease, multiple sclerosis**, and rheumatoid arthritis.
4. A second group of autoimmune diseases, such as systemic lupus erythematosus, involves multiple organs or body systems.

Immunodeficiency Diseases (pp. 527–529)

1. Immunodeficiency diseases may be classified as **primary immunodeficiency diseases,** which result from mutations or developmental anomalies and occur in young children, and **acquired** or **secondary immunodeficiency diseases,** which result from other known causes such as viral infections.
2. Examples of primary immunodeficiency diseases include **chronic granulomatous disease,** in which a child's neutrophils are incapable of killing ingested bacteria. Inability to produce both T cells and B cells is called **severe combined immunodeficiency disease.** In **DiGeorge syndrome,** the thymus fails to develop. In **Bruton-type agammaglobulinemia,** B cells fail to function.
3. **Acquired immunodeficiency syndrome (AIDS),** a condition marked by the presence of antibodies against HIV in conjunction with certain opportunistic infections, is the most common acquired immunodeficiency disease.
4. A **syndrome** is a complex of signs, symptoms, and diseases with a common cause. AIDS is defined by a presence of a number of opportunistic infections in the presence of antibodies against HIV.

Questions for Review

Answers to the Questions for Review (except Short Answer questions) begin on page A-1.

Multiple Choice

1. The immunoglobulin class that mediates type I hypersensitivity is
   a. IgA.
   b. IgM.
   c. IgG.
   d. IgD.
   e. IgE.
2. The major inflammatory mediator released by degranulating mast cells in type I hypersensitivity is
   a. immunoglobulin.
   b. complement.
   c. histamine.
   d. interleukin.
   e. prostaglandin.

3. Hemolytic disease of the newborn is caused by antibodies against which major blood group antigen?
   a. MHC protein
   b. MN antigen
   c. ABO antigen
   d. rhesus antigen
   e. type II protein
4. Farmer's lung is a hypersensitivity pneumonitis resulting from
   a. a type I hypersensitivity reaction to grass pollen.
   b. a type II hypersensitivity to red cells in the lung.
   c. a type III hypersensitivity to mold spores.
   d. a type IV hypersensitivity to bacterial antigens.
   e. none of the above
5. A positive tuberculin skin test indicates that a patient not immunized against tuberculosis
   a. is free of tuberculosis.
   b. is shedding *Mycobacterium*.
   c. has been exposed to tuberculosis antigens.
   d. is susceptible to tuberculosis.
   e. is resistant to tuberculosis.
6. Which of the following is an autoimmune disease?
   a. a heart attack
   b. acute anaphylaxis
   c. farmer's lung
   d. graft-versus-host disease
   e. systemic lupus erythematosus
7. When a surgeon conducts a cardiac bypass operation by transplanting a piece of vein from a patient's leg to the same patient's heart, this is
   a. a rejected graft.
   b. an autograft.
   c. an allograft.
   d. a type IV hypersensitivity.
   e. a cardiograft.
8. A deficiency of both B cells and T cells is most likely
   a. a secondary immunodeficiency.
   b. a complex immunodeficiency.
   c. an acquired immunodeficiency.
   d. a primary immunodeficiency.
   e. an induced immunodeficiency.
9. Infection with HIV causes
   a. primary immunodeficiency disease.
   b. acquired hypersensitivity syndrome.
   c. acquired immunodeficiency syndrome.
   d. anaphylactic immunodeficiency diseases.
   e. combined immunodeficiency diseases.
10. What do medical personnel administer to counteract various type I hypersensitivities?
   a. antihistamine
   b. bronchodilator
   c. glucocorticoid
   d. epinephrine
   e. all of the above

Modified True/False

Indicate whether each statement is true or false. If the statement is false, change the italicized word or phrase to make the statement true.

1. ____ *Cyclosporine is* released by degranulating mast cells.
2. ____ *Type III* hypersensitivity reactions may lead to the development of glomerulonephritis.
3. ____ ABO blood group antigens are found on *nucleated cells*.
4. ____ The tuberculin reaction is a *type I* hypersensitivity.
5. ____ Graft-versus-host disease can follow a bone marrow *isograft*.

Matching

Match the immune system complication in the first column with the types of hypersensitivities in the second column. Choices may be used more than once or not at all.

1. ____ Acute anaphylaxis
2. ____ Allergic contact dermatitis
3. ____ Systemic lupus erythematosus
4. ____ Allograft rejection
5. ____ AIDS
6. ____ Graft-versus-host disease
7. ____ Milk allergy
8. ____ Farmer's lung
9. ____ Asthma
10. ____ Hay fever

A. Type I hypersensitivity
B. Type II hypersensitivity
C. Type III hypersensitivity
D. Type IV hypersensitivity
E. Not a hypersensitivity

Labeling

Label the four types of grafts on the figure below.

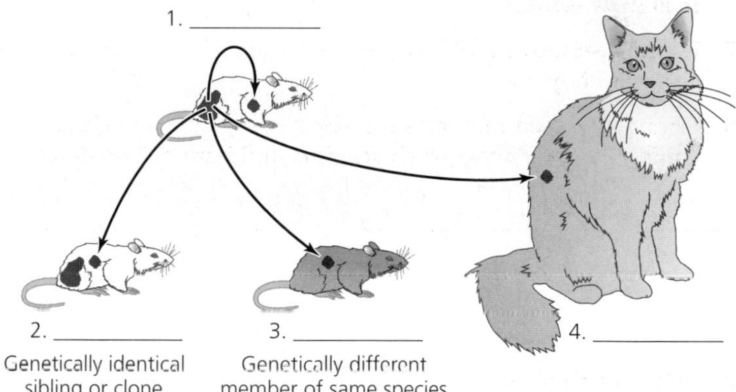

Short Answer

1. Why is AIDS more accurately termed a "syndrome" rather than a mere disease?
2. Why is a child born to an Rh+ mother not susceptible to Rh-related hemolytic disease of the newborn?

3. Why is a person who produces a large amount of IgE more likely to experience anaphylactic shock than a person who instead produces a large amount of IgG?
4. Contrast autografts, isografts, allografts, and xenografts.
5. Compare and contrast the functions of four classes of immunosuppressive drugs.

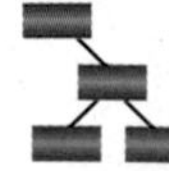

Concept Mapping

Using the following terms, draw a concept map that describes immediate hypersensitivity. For a sample concept map, see p. 93. Or, complete this concept map online by going to the Study Area at www.masteringmicrobiology.com.

Allergens
Allergy symptoms
Anaphylactic shock
Anaphylaxis
Antihistamine
Asthma
Basophils
Bee venom
Dust mites
Epinephrine
Hay fever
Histamine
IgE
Increased vascular permeability
Localized
Mast cells
Peanuts
Pollen
Smooth muscle contraction
Systemic
Type I hypersensitivity
Urticaria
Vasodilation

Critical Thinking

1. What possible advantages might an individual gain from making class E antibodies (IgE)?
2. Why can't physicians use skin tests similar to the tuberculin reaction to diagnose other bacterial diseases?
3. In both Graves' disease and juvenile onset diabetes mellitus, autoantibodies are directed against cytoplasmic membrane receptors. Speculate on the clinical consequences of an autoimmune response to estrogen receptors.
4. In general, people with B cell defects acquire numerous bacterial infections, whereas those with T cell defects get viral diseases. Explain why this is so.
5. What types of illnesses cause death in patients with combined immunodeficiencies or AIDS?
6. Because of the severe shortage of organ donors for transplants, many scientists are examining the possibility of using organs from nonhuman species such as pigs. What special clinical problems might be encountered when using these xenografts?
7. Why do the blisters of positive tuberculin reactions resemble the blisters of poison ivy?

Access more review material online in the Study Area at **www.masteringmicrobiology.com.** There, you'll find

- **MP3 Tutor Sessions**
- **Concept Mapping Activities**
- **Flashchards**
- **Quizzes**

and more to help you succeed.

19 Pathogenic Gram-Positive Bacteria

We are all familiar with acne, the common skin disorder characterized by pimples and other blemishes. But did you know that acne is often caused by a Gram-positive bacterium called *Propionibacterium acnes*? *P. acnes* grows in the sebaceous (oil) glands of the skin. Teenagers are especially prone to developing acne because the hormonal changes of adolescence can trigger the production of excess oil, stimulating bacterial growth. The bacteria secrete chemicals that attract the body's leukocytes, which phagocytize the bacteria and trigger inflammation. Pus-filled pimples result when lymphocytes and dead bacteria clog the skin's pores.

In this chapter we will learn more about *P. acnes* and about other medically important Gram-positive pathogens, including *Staphylococcus, Streptococcus,* and *Bacillus.* A general characteristic of Gram-positive pathogens is that the disease symptoms they cause often result from the toxins they secrete. In Chapter 20 we will turn to Gram-negative pathogens, which not only secrete toxins but also possess a toxic outer membrane that can produce disease symptoms in humans.

Take the pre-test for this chapter online. Visit the Study Area at www.masteringmicrobiology.com.

Gram-positive *Propionibacterium acnes* is a major cause of teenagers' acne.

In Chapter 11 we examined the general features of prokaryotes, including some pathogens. In this chapter we explore the major Gram-positive bacterial pathogens in more detail. Recall that Gram-positive bacteria stain purple when Gram stained and generally fall within the phylum Firmicutes. We will discuss the major pathogens of this group.

Taxonomists currently group Gram-positive bacteria into two major groups: low G + C Gram-positive bacteria and high G + C Gram-positive bacteria. As the names indicate, these classifications are based on the ratio of guanine-cytosine nucleotide base pairs to adenine-thymine nucleotide base pairs in these organisms' DNA. Low G + C Gram-positive bacteria include three genera of pathogenic spherical cells (cocci): *Staphylococcus, Streptococcus,* and *Enterococcus;* three genera of pathogenic rod-shaped cells (bacilli): *Bacillus, Clostridium,* and *Listeria;* and the *mycoplasmas,* a group of bacteria that lack cell walls. Mycoplasmas have historically been classified as Gram-negative bacteria because they stain pink when Gram stained. However, studies of their nucleotide sequences have revealed that they are genetically more similar to low G + C Gram-positive bacteria.

The high G + C Gram-positive pathogens include rod-shaped genera *Corynebacterium, Mycobacterium,* and *Propionibacterium,* and the filamentous, fungus-like *Nocardia* and *Actinomyces.*

We begin our discussion of Gram-positive bacteria with *Staphylococcus.*

Staphylococcus

Bacteria in the genus *Staphylococcus* (staf′i-lō-kok′ŭs) are living and reproducing on almost every square inch of your skin right now. They are normal members of every human's microbiota, which usually go unnoticed. However, they can be opportunistic pathogens, causing minor to life-threatening diseases.

Structure and Physiology

Learning Objective

✓ Contrast the virulence of *S. aureus* with that of *S. epidermidis* in humans.

Staphylococcus[1] is a genus of Gram-positive, facultatively anaerobic prokaryotes whose spherical cells are typically clustered in grapelike arrangements **(Figure 19.1)**. As we saw in Chapter 11, this arrangement results from two characteristics of cell division: cell divisions occur in successively different planes, and daughter cells remain attached to one another. Staphylococcal cells are 0.5–1.0 μm in diameter, nonmotile, and salt tolerant—they are capable of growing in media that are up to 10% NaCl, which explains how they tolerate the salt deposited on human skin by sweat glands. Additionally, they are tolerant of desiccation, radiation, and heat (up to 60°C for 30 min), allowing them to survive on environmental surfaces in addition to skin. Further, *Staphylococcus* synthesizes catalase—a characteristic that distinguishes this genus from other low G + C Gram-positive cocci.

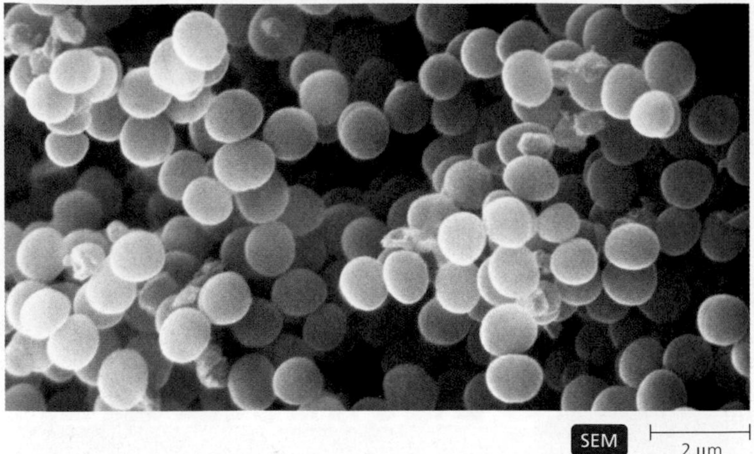

▲ **Figure 19.1** ***Staphylococcus.*** Gram-positive cells appear in grapelike clusters.

Two species are commonly associated with staphylococcal diseases in humans:

- *Staphylococcus aureus*[2] (o′rē-ŭs) is the more virulent, producing a variety of disease conditions and symptoms depending on the site of infection.
- *Staphylococcus epidermidis* (e-pi-der-mid′is), as its name suggests, is a part of the normal microbiota of human skin, but it is an opportunistic pathogen in immunocompromised patients or when introduced into the body via intravenous catheters or on prosthetic devices such as artificial heart valves.

Pathogenicity

Learning Objective

✓ Discuss the structural and enzymatic features and toxins of *Staphylococcus* that enable it to be pathogenic.

What are commonly called "staph" infections result when staphylococci breach the body's physical barriers (skin or mucous membranes); entry of only a few hundred bacteria can ultimately result in disease. The pathogenicity of *Staphylococcus* results from three features: structures that enable it to evade phagocytosis, the production of enzymes, and the production of toxins.

Structural Defenses Against Phagocytosis

The cells of *S. aureus* are uniformly coated with a protein, called *protein A,* that interferes with humoral immune responses by binding to the F_c regions (stems) of class G antibodies (IgG). Recall from Chapter 16 that antibodies are opsonins—they enhance phagocytosis—precisely because phagocytic cells have receptors for antibody stems; therefore, protein A, by binding stems, effectively inhibits opsonization. Protein A also inhibits

[1]From Greek *staphle,* meaning small bunch of grapes, and *kokkos,* meaning a berry.
[2]Latin for golden, for the golden color of its colonies growing on the surface of a solid medium.

the complement cascade, which is triggered by antibody molecules bound to antigen (see Figure 15.9).

The outer surfaces of most pathogenic strains of *S. aureus* also contain *bound coagulase,* an enzyme that converts the soluble blood protein fibrinogen into long, insoluble fibrin molecules, which are threads that form blood clots around the bacteria. Fibrin clots in effect hide the bacteria from phagocytic cells.

Both *S. aureus* and *S. epidermidis* also evade the body's defenses by synthesizing loosely organized polysaccharide slime layers (sometimes called capsules) that inhibit chemotaxis of and endocytosis by leukocytes, particularly neutrophils. The slime layer also facilitates attachment of *Staphylococcus* to artificial surfaces such as catheters, shunts, artificial heart valves, and synthetic joints.

In summary, protein A, bound coagulase, and a slime layer allow *S. aureus* to evade the body's defenses, whereas *S. epidermidis* relies almost exclusively on its slime layer.

Enzymes

Staphylococci produce a number of enzymes that contribute to their survival and pathogenicity:

- *Cell-free coagulase,* like bound coagulase, triggers blood clotting. Cell-free coagulase does not act on fibrin directly but instead combines with a blood protein before becoming enzymatic and converting fibrinogen to fibrin threads. Only *S. aureus* synthesizes coagulase; *S. epidermidis* and other species of *Staphylococcus* are coagulase negative.
- *Hyaluronidase* breaks down hyaluronic acid, which is a major component of the matrix between cells. Hyaluronidase, found in 90% of *S. aureus* strains, enables the bacteria to spread between cells throughout the body.
- *Staphylokinase* (produced by *S. aureus*) dissolves fibrin threads in blood clots, allowing *S. aureus* to free itself from clots. Thus, *Staphylococcus* can escape the immune system by enclosing itself in a fibrin clot (via coagulase), and then, when space and nutrients become limiting, it can digest its way out of the clot with staphylokinase and spread to new locations.
- *Lipases* digest lipids, allowing staphylococci to grow on the surface of the skin and in cutaneous oil glands. All staphylococci produce lipases.
- *β-lactamase (penicillinase),* now present in over 90% of *S. aureus* strains, breaks down penicillin. Though β-lactamase plays no role in inhibiting the natural defenses of the body, it does allow the bacteria to survive treatment with beta-lactam antimicrobial drugs such as penicillin and cephalosporin.

Toxins

Staphylococcus aureus and (less frequently) *S. epidermidis* also possess several toxins that contribute to their pathogenicity, including the following:

- *Cytolytic toxins.* So-called alpha, beta, gamma, and delta toxins are proteins, coded by chromosomal genes, that disrupt the cytoplasmic membranes of a variety of cells including leukocytes. Leukocidin is a fifth cytolytic toxin that lyses leukocytes specifically, providing *Staphylococcus* with some protection against phagocytosis.
- *Exfoliative toxins.* Each of two distinct proteins causes the dissolution of epidermal *desmosomes* (intercellular bridge proteins that hold adjoining cytoplasmic membranes together), causing the patient's skin cells to separate from each other and slough off the body.
- *Toxic shock syndrome (TSS) toxin.* This protein causes toxic shock syndrome (discussed shortly).
- *Enterotoxins.* These five proteins (designated A through E) stimulate the intestinal muscle contractions, nausea, and intense vomiting associated with staphylococcal food poisoning. Enterotoxins are heat stable, remaining active at 100°C for up to 30 minutes.

TABLE 19.1 A Comparison of the Virulence Factors of Staphylococcal Species

	S. aureus	*S. epidermidis*
Protein A	+	–
Coagulase	+	–
Catalase	+	+
Hyaluronidase	+	–
Staphylokinase	+	–
Lipase	+	+
β-Lactamase (penicillinase)	+	–

Table 19.1 compares and contrasts the virulence factors of *S. aureus* and *S. epidermidis.*

Epidemiology

Staphylococcus epidermidis is ubiquitous on human skin, whereas *S. aureus* is commonly found only on moist skin folds. Both species also grow in the upper respiratory, gastrointestinal, and urogenital tracts of humans. Both bacteria are transmitted through direct contact between individuals, as well as via fomites such as contaminated clothing, bed sheets, and medical instruments; as a result, proper handwashing and aseptic techniques are essential in preventing their transfer in health care settings.

Staphylococcal Diseases

Learning Objectives

- ✓ Describe the symptoms and prevention of staphylococcal food poisoning.
- ✓ List and describe six pyogenic lesions caused by *Staphylococcus aureus.*
- ✓ Discuss five systemic and potentially fatal diseases caused by *Staphylococcus.*

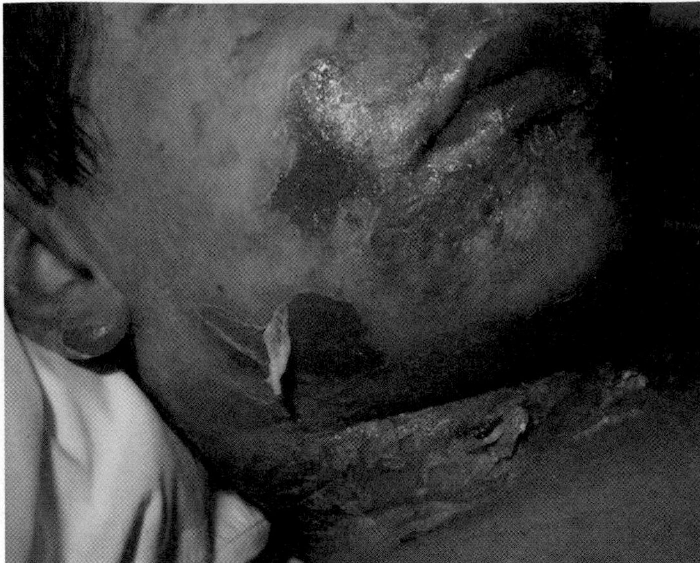

▲ **Figure 19.2 Staphylococcal scalded skin syndrome.** Exfoliative toxin, produced by some strains of *Staphylococcus aureus*, causes reddened patches of the epidermis to slough off.

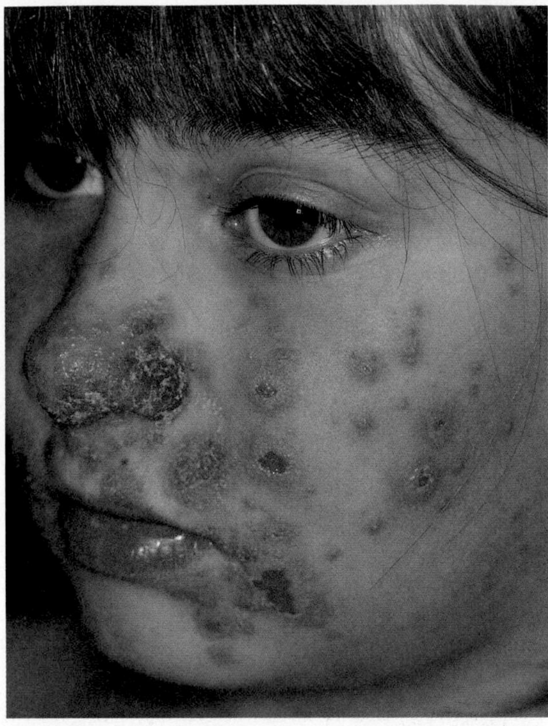

▲ **Figure 19.3 Impetigo.** Reddened patches of skin become pus-filled vesicles that eventually crust over.

Staphylococcus causes a variety of medical problems, depending on the site of infection, the immune state of its host, and the toxins and enzymes a particular species or strain secretes. Staphylococcal syndromes and diseases can be categorized as noninvasive, cutaneous, and systemic diseases.

Noninvasive Disease

Staphylococcus aureus is one of the more common causes of food poisoning (more specifically, this is food intoxication, because disease is caused by enterotoxin-contaminated food rather than by invasion of bacteria). Commonly affected foods include processed meats, custard pastries, potato salad, and ice cream that have been contaminated with bacteria from human skin. (In *S. aureus* food poisoning, unlike many other forms of food poisoning, animals are not involved.) The food must remain at room temperature or warmer for several hours for the bacteria to grow, reproduce, and secrete toxin. Warming or reheating inoculated food does not inactivate enterotoxins, which are heat stable, although heating does kill the bacteria. Food contaminated with staphylococci does not appear or taste unusual.

Symptoms, which include nausea, severe vomiting, diarrhea, headache, sweating, and abdominal pain, usually appear within four hours following ingestion. Consumed staphylococci do not continue to produce toxins, so the course of the disease is rapid, usually lasting 24 hours or less.

Cutaneous Diseases

Staphylococcus aureus causes localized *pyogenic*[3] (pī-ō-jen′ik) lesions. **Staphylococcal scalded skin syndrome** is a reddening of the skin that typically begins near the mouth, spreads over the entire body, and is followed by large blisters that contain clear fluid lacking bacteria or white blood cells—consistent with the fact that the disease is caused by a toxin released by bacteria growing on the skin. Within two days the affected outer layer of skin (epidermis) peels off in sheets, as if it had been dipped into boiling water (**Figure 19.2**). The seriousness of scalded skin syndrome results from secondary bacterial infections in denuded areas.

Small, flattened, red patches on the face and limbs, particularly of children whose immune systems are not fully developed, characterize **impetigo** (im-pe-tī′gō) (**Figure 19.3**). The patches develop into pus-filled vesicles that eventually crust over. The pus is filled with bacteria and white blood cells, which distinguishes impetigo from scalded skin syndrome. *S. aureus* acting alone causes about 80% of impetigo cases; about 20% of cases also involve streptococci.

Folliculitis (fo-lik′ū-lī′tis) is an infection of a hair follicle in which the base of the follicle becomes red, swollen, and pus filled. When this condition occurs at the base of an eyelid, it is called a **sty.** A **furuncle** (fū′rŭng-kl) or boil is a large, painful, raised nodular extension of folliculitis into surrounding tissue. When several furuncles coalesce, they form a **carbuncle** (kar′-bŭng-kl), which extends deeper into the tissues, triggering the fever and chills that are characteristic of innate immunity. As staphylococci spread into underlying tissues, multiple organs and systems can become involved.

Systemic Diseases

Staphylococcus aureus, and to a lesser extent *S. epidermidis,* cause a wide variety of potentially fatal systemic infections when they

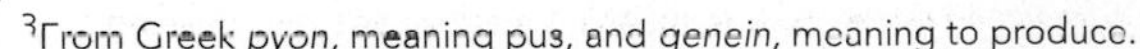

[3]From Greek *pyon*, meaning pus, and *genein*, meaning to produce.

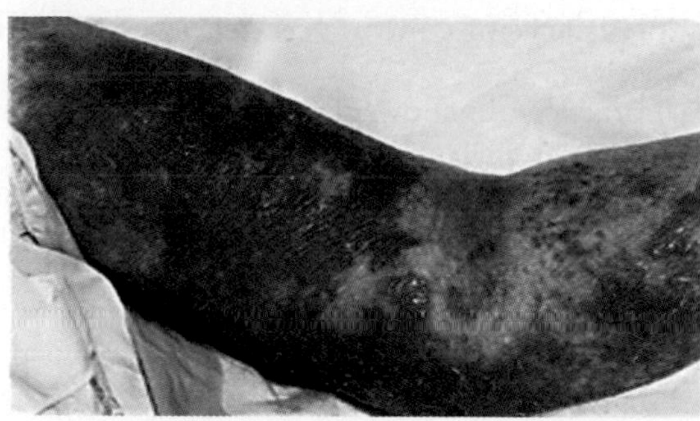

▲ **Figure 19.4 Staphylococcal toxic shock syndrome (STSS).** Fatal infections involve not only red rash (as shown here) but internal organs as well.

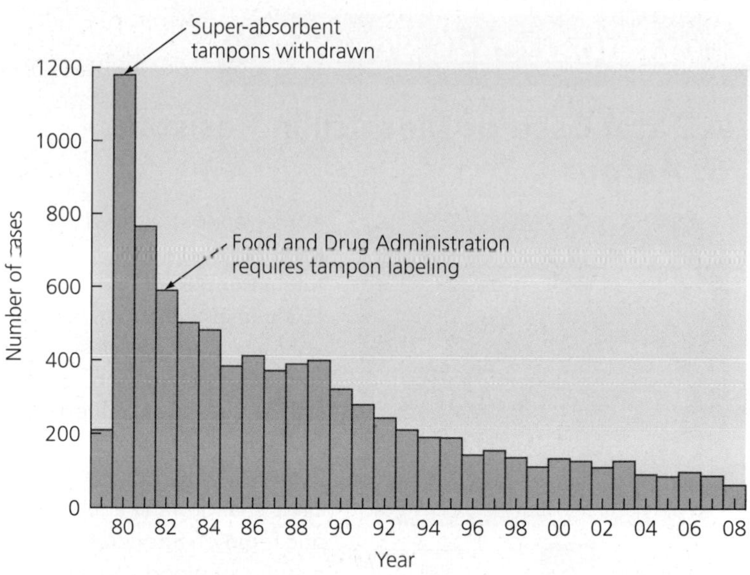

▲ **Figure 19.5 The incidence of staphylococcal toxic shock syndrome in the United States since 1979.** *Use of what product puts women at increased risk of STSS?*

Figure 19.5 *Use of tampons is a risk factor for STSS.*

are introduced into deeper tissues of the body, including the blood, heart, lungs, and bones.

Staphylococcal Toxic Shock Syndrome When strains of *Staphylococcus* that produce TSS toxin grow in a wound or in an abraded vagina, the toxin can be absorbed into the blood and cause **staphylococcal toxic shock syndrome (STSS),** characterized by fever, vomiting, red rash, extremely low blood pressure, and loss of sheets of skin **(Figure 19.4)**. STSS is fatal to 5% of patients when their blood pressure falls so low that the brain, heart, and other vital organs have an inadequate supply of oxygen—a condition known as *shock*.

Staphylococcal toxic shock syndrome occurs in both males and females, but in 1980 epidemiologists noted an epidemic of STSS among menstruating women. Researchers subsequently discovered that *S. aureus* grows exceedingly well in super-absorbent tampons, especially when the tampon remains in place for a prolonged period. As a result of the withdrawal of this type of tampon from the market, plus government-mandated reduction in the absorbency of all tampons and mandatory inclusion of educational information concerning the risks of STSS on every package of tampons, the number of cases of STSS declined rapidly **(Figure 19.5)**.

Bacteremia *S. aureus* is a common cause of **bacteremia** (bak-tēr-ē′mē-ă), the presence of bacteria in the blood. After staphylococci enter the blood from a site of infection, they travel to other organs of the body, which may become infected. Furuncles, vaginal infections, infected surgical wounds, and contaminated medical devices such as intravascular catheters have all been implicated in cases of bacteremia. Nosocomial (hospital-acquired) infections account for about half of all cases of staphylococcal bacteremia. Physicians fail to identify the initial site of infection in about a third of patients, but it is presumed to be an innocuous skin infection.

Endocarditis *S. aureus* may attack the lining of the heart (including its valves), producing a condition called **endocarditis**[4] (en′dō-kar-dī′tis). Typically, patients with endocarditis have nonspecific, flulike symptoms, but their condition quickly deteriorates as the amount of blood pumped from the heart drops precipitously. About 50% of patients with endocarditis do not survive.

Pneumonia and Empyema *Staphylococcus* in the blood can invade the lungs, causing **pneumonia** (noo-mō′nē-ă)—an inflammation of the lungs in which the alveoli (air sacs) and bronchioles (smallest airways) become filled with fluid. In 10% of patients with staphylococcal pneumonia this fluid is pus—a condition known as **empyema**[5] (em-pī-ē′mă).

Osteomyelitis When *Staphylococcus* invades a bone, either through a traumatic wound or via the blood during bacteremia, it causes **osteomyelitis**[6] (os′tē-ō-mī-e-lī′tis)—inflammation of the bone marrow and the surrounding bone. Osteomyelitis is characterized by pain in the infected bone accompanied by high fever. In children the disease typically occurs in the growing regions of long bones, which are areas with well-developed blood supplies. In adults, osteomyelitis is more commonly seen in vertebrae.

Diagnosis, Treatment, and Prevention

Learning Objectives

- ✓ Describe how staphylococcal species are distinguished from one another during diagnosis.
- ✓ Discuss briefly the history of staphylococcal resistance to antimicrobial drugs.

Physicians diagnose staphylococcal infection by detecting grape-like arrangements of Gram-positive bacteria isolated from pus, blood, or other fluids. If staphylococci isolated from an infection

[4]From Greek *endon*, meaning within; *kardia*, meaning heart; and *itis*, meaning inflammation.
[5]From Greek *en*, meaning in, and *pyon*, meaning pus.
[6]From Greek *osteon*, meaning bone; *myelos*, meaning marrow; and *itis*, meaning inflammation.

CLINICAL CASE STUDY

A Fatal Case of Methicillin-Resistant *S. Aureus*

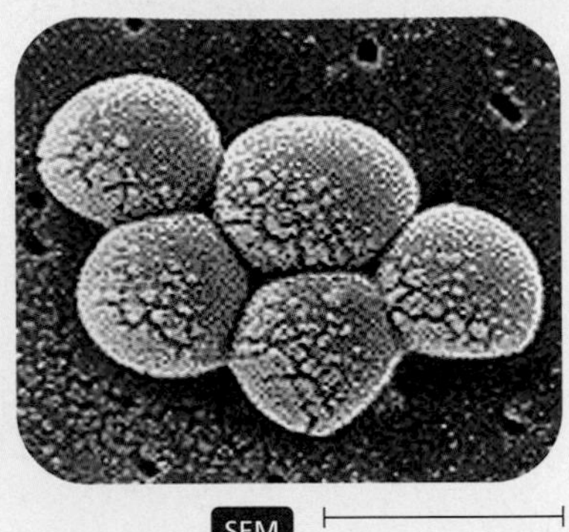

▲ **Methicillin-resistant *S. aureus*.**

A five-year-old girl was admitted to the hospital with a temperature of 103°F and pain in her right hip. After pus was surgically drained from the hip joint, she was treated with a semisynthetic cephalosporin. Her physicians changed the antibiotic regimen after 24-hour cultures of blood and pus revealed the presence of MRSA. On the third day, she suffered respiratory failure and empyema and was placed on mechanical ventilation. She died from pulmonary hemorrhage and pneumonia after five weeks of hospitalization. The girl had been previously healthy with no recent hospitalizations. She had skinned her knee while learning to ride a bicycle two days before admittance to the hospital.

1. How might the girl have been infected?
2. How did her hip joint become infected?
3. Describe the series of diseases she suffered.
4. What was likely the second antibiotic she received?

are able to clot blood, then they are coagulase-positive *S. aureus*. Coagulase-negative staphylococci are usually *S. epidermidis*, which is a normal part of the microbiota of the skin; their presence in a clinical sample is usually not indicative of a staphylococcal infection.

During the latter half of the 20th century, genes for β-lactamase, which convey resistance to natural penicillin, spread among *Staphylococcus* species. Thus, 90% of staphylococci were susceptible to penicillin in 1945, but only 5% are susceptible today. For this reason, the semisynthetic form of penicillin—methicillin, which is not inactivated by β-lactamase—became the drug of choice for staphylococcal infections. Unfortunately, **methicillin-resistant *Staphylococcus aureus* (MRSA)** has emerged as a major problem, initially in health care settings but now increasingly in day care centers, high school locker rooms, and prisons. In fact, more people die from MRSA than from HIV in the United States.

MRSA is also resistant to many other common antimicrobial drugs, including penicillin, macrolides, aminoglycosides, and cephalosporin; as a result, *vancomycin* has been used to treat MRSA infections. Physicians are very concerned about the increasing prevalence of **vancomycin-resistant *staphylococcus aureus* (VRSA).**

Clinical practice has shown how crucial it is that abscesses be drained of pus and cleansed if subsequent antibiotic therapy is to be effective. Systemic infections such as endocarditis and osteomyelitis require long-term therapy with antimicrobial drugs.

Because strains of *Staphylococcus* resistant to antimicrobial agents have become more common, especially in hospitals, it is imperative that health care workers take precautions against introducing the bacterium into patients. Of course, given that *Staphylococcus* is ubiquitous on human skin, staphylococcal infection cannot be eliminated. Fortunately, because a large inoculum is required to establish an infection, proper cleansing of wounds and surgical openings, attention to aseptic use of catheters and indwelling needles, and the appropriate use of antiseptics will prevent infections in most healthy patients. Health care workers with staphylococcal infections may be barred from delivery rooms, nurseries, and operating rooms. The most important measure for protecting against nosocomial infection is frequent handwashing.

Scientists are currently testing the efficacy and safety of a vaccine that has proven effective in protecting dialysis patients from *S. aureus* infections. Such immunization, if it proves safe and effective in preventing other infections, may have a significant effect on nosocomial disease. It would be especially good news for health care providers who are battling resistant strains of *Staphylococcus*.

Streptococcus

Learning Objectives

✓ Describe the classification of streptococcal strains.

✓ Describe seven diseases caused by *S. pyogenes* and the treatments available.

The genus *Streptococcus* (strep-tō-kok'ŭs) is a diverse assemblage of Gram-positive cocci 0.5–1.2 μm in diameter and arranged in pairs or chains. They are catalase negative (unlike *Staphylococcus*), although they do synthesize peroxidase and thus are facultatively anaerobic.

Researchers differentiate species of *Streptococcus* using several different, overlapping schemes, including serological classification based on the reactions of antibodies to specific bacterial antigens, type of hemolysis (alpha, beta, or gamma; see Figure 6.13), cell arrangement, and physiological properties as revealed by biochemical tests. In this chapter we will largely use a serological classification scheme, developed by Rebecca Lancefield (1895–1981), that divides streptococci into serotype groups based on the bacteria's antigens (known, appropriately, as Lancefield antigens). The serotypes in this scheme include Lancefield groups A through H and K through V. Whereas the more significant streptococcal pathogens of humans are in groups A and B, two other significant streptococcal pathogens of humans lack Lancefield antigens. We begin our survey of the streptococci with group A *Streptococcus*.

Group A *Streptococcus*: *Streptococcus pyogenes*

Learning Objectives

- ✓ Describe two structures in *Streptococcus pyogenes* that enable this organism to survive the body's defenses.
- ✓ Identify four enzymes and a type of toxin that facilitate the spread of *S. pyogenes* in the body.
- ✓ Identify the conditions under which group A *Streptococcus* causes disease.

Group A *Streptococcus*, which is synonymously known as *S. pyogenes* (pī-oj′en-ēz), is a coccus that forms white colonies 1–2 mm in diameter surrounded by a large zone of beta-hemolysis after 24 hours on blood agar plates. Pathogenic strains of this species often form capsules. The following sections discuss the pathogenesis and epidemiology of this species, as well as the diagnosis, treatment, and prevention of the diseases it causes.

Pathogenicity

Strains of *Streptococcus pyogenes* have a number of structures, enzymes, and toxins that enable them to survive as pathogens in the body.

Two main structural features enable cells of *S. pyogenes* to evade phagocytosis:

- *M protein.* A membrane protein called *M protein* destabilizes complement, thereby interfering with opsonization and lysis.
- *Hyaluronic acid capsule.* Because hyaluronic acid is normally found in the body, white blood cells may ignore bacteria "camouflaged" by this type of capsule.

Researchers have identified two *streptokinases* that break down blood clots, presumably enabling group A *Streptococcus* to rapidly spread through infected and damaged tissues. Similarly, four distinct *deoxyribonucleases* depolymerize DNA that has been released from dead cells in abscesses, reducing the firmness of the pus surrounding the bacteria and facilitating bacterial spread. *C5a peptidase* breaks down the complement protein C5a, which, as we discussed in Chapter 15, acts as a chemotactic factor. Thus, *S. pyogenes* decreases the movement of white blood cells into a site of infection. Finally, hyaluronidase facilitates the spread of streptococci through tissues by breaking down hyaluronic acid.

Group A *Streptococcus* secretes three distinct *pyrogenic*[7] (pī-rō-jen′ik) *toxins* that stimulate macrophages and helper T lymphocytes to release cytokines that in turn stimulate fever, a widespread rash, and shock. Because these toxins cause blood capillaries near the surface to dilate, producing a red rash, some scientists call them *erythrogenic*[8] (e-rith-rō-jen′ik) *toxins.* The genes for these toxins are carried on temperate bacteriophages, so only lysogenized bacteria—bacteria in which a virus has become part of the bacterial chromosome—secrete the toxins. Fever-stimulating *pyrogenic* toxins should not be confused with pus-producing *pyogenic* toxins.

[7]From Greek *pyr*, meaning fire, and *genein*, meaning to produce.
[8]From Greek *erythros*, meaning red, and *genein*, meaning to produce.

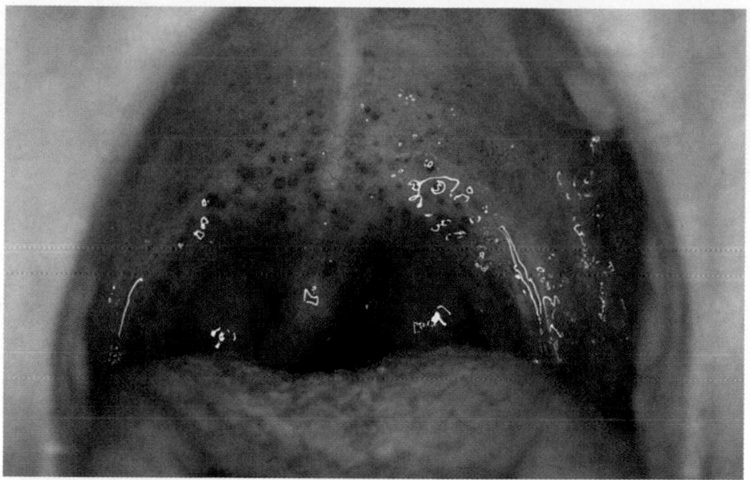

Figure 19.6 Pharyngitis.

S. pyogenes also produces two different, membrane-bound proteins, called *streptolysins,* which lyse red blood cells, white blood cells, and platelets; thus, these proteins interfere with the oxygen-carrying capacity of the blood, immunity, and blood clotting. After they have been phagocytized, group A *Streptococcus* releases streptolysins into the cytoplasm of the phagocyte, causing lysosomes to release their contents, which lyses the phagocyte and releases the bacteria.

Epidemiology

Group A *Streptococcus* frequently infects the pharynx or skin, but the resulting abscesses are usually temporary, lasting only until adaptive immune responses against bacterial antigens (particularly M protein and streptolysins) clear the pathogens. Typically, *Streptococcus pyogenes* causes disease only when normal competing microbiota are depleted; when a large inoculum enables the streptococci to gain a rapid foothold before antibodies are formed against them; or when adaptive immunity is impaired. Following colonization of the skin or a mucous membrane, *S. pyogenes* can invade deeper tissues and organs through a break in such barriers. People spread *S. pyogenes* among themselves via respiratory droplets, especially under crowded conditions such as those in classrooms and day care centers.

Before the discovery of antimicrobials, a major cause of human disease was group A streptococcal infection, which claimed the lives of millions. Because it is sensitive to antimicrobial drugs, its significance as a pathogen has declined, but group A streptococci still sicken thousands of Americans annually.

Group A Streptococcal Diseases

Group A *Streptococcus* causes a number of diseases, depending on the site of infection, the strain of bacteria, and the immune responses of the patient.

Pharyngitis A sore throat caused by streptococci, commonly known as "strep throat," is a kind of **pharyngitis** (far-in-jī′tis)—inflammation of the pharynx (**Figure 19.6**)—that is accompanied

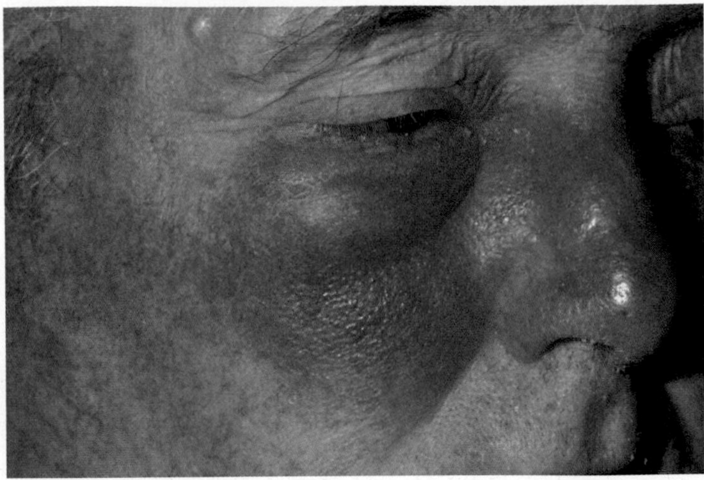

▲ **Figure 19.7 Erysipelas.** The localized, pus-filled lesions are caused by group A streptococci *(Streptococcus pyogenes). What is the meaning of the word* pyogenes?

Figure 19.7 *Pyogenes* means pus-producing.

by fever, malaise,[9] and headache. The back of the pharynx typically appears red, with swollen lymph nodes and *purulent* (pus-containing) abscesses covering the tonsils. Some microbiologists estimate that only 50% of patients diagnosed with strep throat actually have it; the rest have viral pharyngitis. Given that the symptoms and signs for the two diseases are identical, a sure diagnosis requires bacteriological or serological tests. Correct diagnosis is essential because bacterial pharyngitis is treatable with antibacterial drugs, which of course have no effect on viral pharyngitis.

Scarlet Fever The disease known as **scarlet fever** or as *scarlatina* often accompanies streptococcal pharyngitis when the infection involves a lysogenized strain of *S. pyogenes.* After one to two days of pharyngitis, pyrogenic toxins released by the streptococci trigger a diffuse rash that typically begins on the chest and spreads across the body. The tongue usually becomes strawberry red. The rash disappears after about a week and is followed by sloughing of the skin.

Pyoderma and Erysipelas A **pyoderma** (pī-ō-der′mă) is a confined, pus-producing lesion that usually occurs on the exposed skin of the face, arms, or legs. One cause is group A streptococcal infection following direct contact with an infected person or contaminated fomites. This condition is also known as *impetigo* because of its similar appearance to the staphylococcal disease of the same name. After a pus-filled lesion breaks open, it forms a yellowish crust. This stage is highly contagious, and scratching may convey bacteria to the surrounding skin, spreading the lesions.

When a streptococcal infection also involves surrounding lymph nodes and triggers pain and inflammation, the condition is called **erysipelas**[10] (er-i-sip′ĕ-las) **(Figure 19.7)**. Erysipelas occurs most commonly on the faces of children.

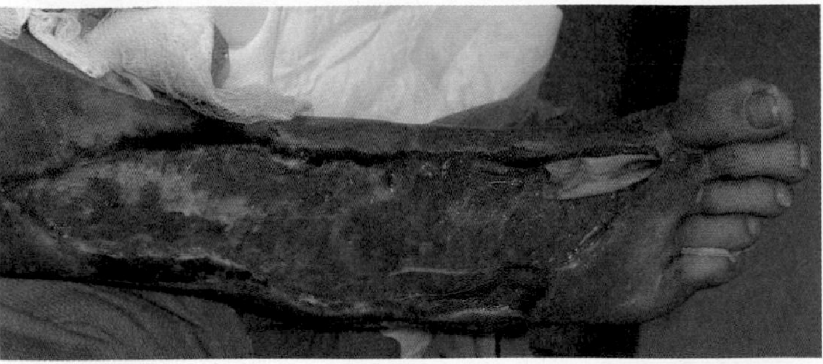

▲ **Figure 19.8 Necrotizing fasciitis.** So-called flesh-eating group A streptococci *(Streptococcus pyogenes)* cause this condition.

Toxic-Shock-Like Syndrome Group A streptococci can spread, albeit rarely, from an initial site of infection, particularly in patients infected with HIV or suffering with cancer, heart disease, pulmonary disease, or diabetes mellitus. Such spread leads to bacteremia and severe multisystem infections producing **toxic-shock-like syndrome (TSLS).** Patients experience inflammation at sites of infection, as well as pain, fever, chills, malaise, nausea, vomiting, and diarrhea. These signs and symptoms are followed by increased pain, organ failure, shock—and over 40% of patients die.

Necrotizing Fasciitis Another serious disease caused by *S. pyogenes* is **necrotizing**[11] **fasciitis** (ne′kro-tī-zing fas-ē-ī′tis), sensationalized by the news media as "flesh-eating bacteria." In this disease, streptococci enter the body through breaks in the skin, secrete enzymes and toxins that destroy tissues, and eventually destroy muscle and fat tissue **(Figure 19.8)**. The bacteria spread deep within the body along the *fascia,* which are fibrous sheets of connective tissue surrounding muscles and binding them to one another—hence the word *fasciitis* in the disease's name. Necrotizing fasciitis also involves *toxemia* (toxins in the blood), failure of many organs, and death of more than 50% of patients.

Rheumatic Fever A complication of untreated *S. pyogenes* pharyngitis is **rheumatic** (rū-mat′ik) **fever,** in which inflammation leads to damage of heart valves and muscle. Though the exact cause of the damage is unknown, it appears that rheumatic fever is not caused directly by *Streptococcus* but instead is an autoimmune response in which antibodies directed against streptococcal antigens cross-react with heart antigens. Damage to the heart valves may be so extensive that they must be replaced when the patient reaches middle age. Rheumatic fever was much more prevalent before the advent of antimicrobial drugs.

Glomerulonephritis For an undetermined reason, antibodies bound to the antigens of some strains of group A *Streptococcus* are not removed from circulation, but instead accumulate in the *glomeruli* (small blood vessels) of the kidneys' *nephrons* (filtering

[9]French, meaning discomfort.
[10]From Greek *erythros,* meaning red, and *pella,* meaning skin.
[11]From Greek *nekros,* meaning corpse.

units). The result is **glomerulonephritis** (glō-mār′ū-lō-nef-rī′tis)—inflammation of the glomeruli and nephrons, which obstructs blood flow through the kidneys and leads to hypertension (high blood pressure) and low urine output. Blood and proteins are often secreted in the urine. Young patients usually recover fully from glomerulonephritis, but progressive and irreversible kidney damage may occur in adults.

Diagnosis, Treatment, and Prevention

Because *Streptococcus* is not a normal member of the microbiota of the skin, the observation of Gram-positive bacteria in short chains or pairs in cutaneous specimens can provide a rapid preliminary diagnosis of pyoderma, erysipelas, and necrotizing fasciitis. In contrast, streptococci are normally in the pharynx, so their presence in a respiratory sample is of little diagnostic value; instead, physicians must use immunological tests that identify the presence of group A streptococcal antigens.

Penicillin is very effective against *S. pyogenes.* Erythromycin or cephalosporin is used to treat penicillin-sensitive patients. *S. pyogenes* is also susceptible to the topical antimicrobial bacitracin—a characteristic that distinguishes it from group B *Streptococcus* (discussed shortly). Necrotizing fasciitis must be treated with aggressive surgical removal of nonviable tissue and infecting bacteria. Because rheumatic fever and glomerulonephritis are the result of an immune response against group A streptococci, they cannot be treated directly; instead, the underlying infection must be arrested.

Antibodies against M protein provide long-term protection against *S. pyogenes.* However, antibodies directed against the M protein of one strain provide no protection against other strains; this explains why a person can have strep throat more than once.

Group B *Streptococcus:* *Streptococcus agalactiae*

Learning Objectives

✓ Contrast group B *Streptococcus* with group A *Streptococcus* in terms of structure.

✓ Discuss the epidemiology, diagnosis, treatment, and prevention of infections with *Streptococcus agalactiae.*

Group B *Streptococcus,* or *S. agalactiae* (a-ga-lak′tē-ī), is a Gram-positive coccus, 0.6–1.2 μm in diameter, that divides to form chains. Like group A *Streptococcus, S. agalactiae* is beta-hemolytic, but it can be distinguished from the former by three qualities: it has group-specific, polysaccharide cell wall antigens; it forms buttery colonies that are 2–3 mm in diameter and have a small zone of beta-hemolysis after 24 hours of growth on blood agar; and it is bacitracin resistant.

Pathogenicity

Even though *S. agalactiae* forms capsules, antibodies target its capsular antigens, so the capsules are not protective. For this reason *S. agalactiae* has a predilection for newborns who have not yet formed type-specific antibodies and whose mothers are uninfected (and so do not provide passive immunity across the placenta or in milk).

Group B streptococci produce enzymes—proteases (that catabolize protein), hemolysins (that lyse red blood cells), deoxyribonuclease, and hyaluronidase—that probably play a role in causing disease, though such a role has not been proven.

Epidemiology

Group B streptococci normally colonize the lower gastrointestinal (GI), genital, and urinary tracts. Diseases in adults primarily follow wound infections and childbirth, though group B *Streptococcus* is emerging as a significant pathogen in the elderly.

Sixty percent of newborns are inoculated with group B streptococcal strains either during passage through the birth canal or by health care personnel. Such infections do not cause disease when maternal antibodies have crossed the placenta; but mortality rates can exceed 50% in children of uninfected mothers. Infections in newborns less than one week old cause *early-onset disease;* infections occurring in infants one week to three months of age cause *late-onset disease.*

Diseases

Even though microbiologists initially described *Streptococcus agalactiae* as the cause of *puerperal*[12] (pyu-er′per-ăl) *(childbirth) fever* in women, today the bacterium is most often associated with neonatal bacteremia, meningitis, and pneumonia, at least one of which occurs in approximately 3 of every 1000 newborns. Mortality has been reduced to about 5% as a result of rapid diagnosis and supportive care, but about 25% of infants surviving group B streptococcal meningitis have permanent neurological damage, including blindness, deafness, or severe mental retardation. Immunocompromised older patients are also at risk from group B streptococcal infections, and about 25% of them die from streptococcal diseases.

Diagnosis, Treatment, and Prevention

Medical laboratory technologists identify group B streptococcal infections by means of ELISA tests utilizing antibodies directed against the bacteria's distinctive cell wall polysaccharides. Samples of clinical specimens can also be incubated in blood media containing the antimicrobial drug bacitracin, which inhibits the growth of other beta hemolytic bacteria.

Penicillin G is the drug of choice for use against group B *Streptococcus,* though some strains tolerate concentrations of the drug more than 10 times greater than that needed to inhibit group A *Streptococcus.* For this reason, physicians may prescribe an aminoglycoside such as streptomycin in addition to penicillin.

The CDC recommends prophylactic administration of penicillin at birth to children whose mothers' urinary tracts are colonized with group B streptococci. Implementation of this guideline in 1996 reduced early-onset disease morbidity and

[12]From Latin *puer,* meaning child, and *pario,* meaning to bring forth.

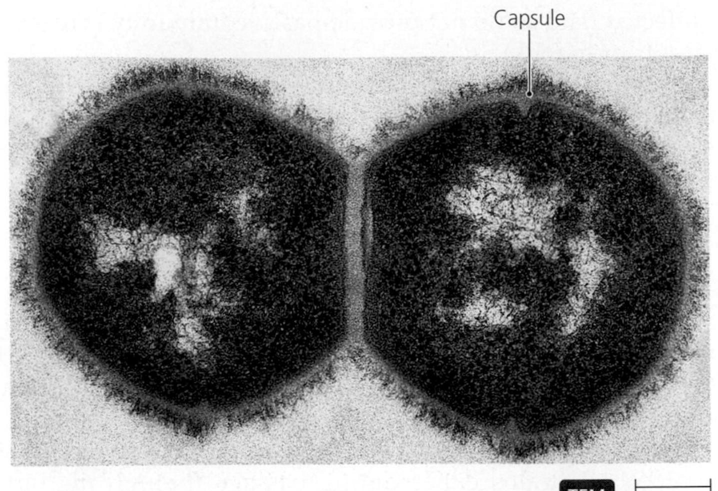

▲ **Figure 19.9** ***Streptococcus pneumoniae.*** Cells of this most common cause of pneumonia are paired and covered with a capsule. *What is pneumonia?*

Figure 19.9 *Pneumonia is inflammation of the lungs, resulting in fluid buildup.*

mortality by 70% by 2001. Additionally, physicians can immunize women against group B streptococci, preventing infection of future children.

Other Beta-Hemolytic Streptococci

Learning Objective

✓ Contrast *Streptococcus equisimilis* and *S. anginosus* in terms of polysaccharide composition, diseases caused, and treatment.

Streptococcus equisimilis (ek-wi-si'mil-is) and *S. anginosus* (an-ji-nō'sŭs) are the only other pathogenic beta-hemolytic streptococci. Although members of both species typically have group C polysaccharides, some strains of *S. anginosus* have group F or group G antigens instead—an example of the confusing status of the classification of the streptococci.

S. equisimilis causes pharyngitis (and occasionally glomerulonephritis) but, unlike group A streptococci, these cases of pharyngitis never lead to rheumatic fever. *S. anginosus* produces pus-containing abscesses. Penicillin is effective against both species.

Alpha-Hemolytic Streptococci: The Viridans Group

Learning Objective

✓ Identify the normal sites of viridans streptococci in the human body, and list three serious diseases they cause.

Many alpha-hemolytic streptococci lack group-specific carbohydrates, and thus they are not part of any Lancefield group. Instead, microbiologists classify them as the **viridans**[13] **streptococci** (vir'i-danz strep'tō-kok'sī) because many of them produce a green pigment when grown on blood media. The taxonomic relationships of these microorganisms are poorly understood; European and American microbiologists do not always agree about the names assigned to species, and some microbiologists place these microbes in a separate genus: *Abiotrophia.* Some of the names given by Americans to viridans streptococci are *S. mitis* and *S. sanguis.* Members of the group are alpha-hemolytic and susceptible to penicillin.

Viridans streptococci normally inhabit the mouth, pharynx, GI tract, genital tract, and urinary tract of humans. They are opportunists that produce purulent abdominal infections, and are one cause of dental **caries**[14] (kār'ēz, cavities). They stick to dental surfaces via an insoluble polysaccharide, *dextran,* which they produce from glucose. Large quantities of dextran allow viridans streptococci and other bacteria to colonize the enamel of teeth, forming biofilms known as *dental plaque* (see Figure 14.6). Viridans streptococci are not highly invasive, entering the blood only through surgical wounds and lacerations of the gums, including undetectable cuts produced by chewing hard candy, brushing the teeth, or dental procedures. Once in the blood they can cause meningitis and endocarditis.

Streptococcus pneumoniae

Learning Objectives

✓ Describe how the structure of *Streptococcus pneumoniae* affects its pathogenicity.

✓ Describe the route of *Streptococcus pneumoniae* through the body, describing the chemical and physical properties that allow it to cause pneumonia.

✓ Discuss the diagnosis, treatment, and prevention of pneumococcal diseases.

Louis Pasteur discovered *Streptococcus pneumoniae* (nū-mō'nē-ī) in pneumonia patients in 1881. The bacterium is a Gram-positive coccus, 0.5–1.2 μm in diameter, that forms short chains or, more commonly, pairs **(Figure 19.9)**. In fact, it was once classified in its own genus, "Diplococcus." Ninety-two different strains of *S. pneumoniae,* collectively called *pneumococci,* are known to infect humans.

Colonies of *S. pneumoniae* grown for 24 hours are 1–3 mm in diameter, round, mucoid, unpigmented, and dimpled in the middle because of the death of older cells. Colonies are alpha-hemolytic on blood agar when grown aerobically, and beta-hemolytic when grown anaerobically. This bacterium lacks Lancefield antigens but does incorporate a species-specific teichoic acid into its cell wall.

Pathogenicity

Streptococcus pneumoniae is a normal member of the pharyngeal microbiota that can colonize the lungs, sinuses, and middle ear. Even though microbiologists have studied the pneumococci extensively—the entire genomes of more than 10 strains have been sequenced—they still do not fully understand their pathogenicity; nevertheless, certain structural and chemical properties are known to be required.

[13]From Latin *viridis,* meaning green.
[14]Latin, meaning dry rot.

The cells of virulent strains of *S. pneumoniae* are surrounded by a polysaccharide capsule, which protects them from digestion after endocytosis. A capsule is required for virulence; unencapsulated strains are avirulent. Microbiologists distinguish 90 unique serotypes based on differences in the antigenic properties of the capsules among various strains.

In addition, cells of *S. pneumoniae* insert into their cell walls a chemical called *phosphorylcholine*. Its binding to receptors on cells in the lungs, in the meninges, and blood vessel walls stimulates the cells to engulf the bacteria. Together, the polysaccharide capsule and phosphorylcholine enable pneumococci to "hide" inside body cells. *S. pneumoniae* can then pass across these cells into the blood and brain.

Pathogenic pneumococci secrete *protein adhesin*, a poorly defined protein that mediates binding of the cells to epithelial cells of the pharynx. From there the bacteria enter the lungs.

The body limits migration of bacteria into the lungs by binding the microbes with the active sites of secretory IgA. The rest of the antibody molecule then binds to mucus, trapping the bacteria where they can be swept from the airways by the action of ciliated epithelium. The bacterium counteracts this defense by secreting *secretory IgA protease*, which destroys IgA, and *pneumolysin*, which binds to cholesterol in the cytoplasmic membranes of ciliated epithelial cells, producing transmembrane pores that result in the lysis of the cells. Pneumolysin also suppresses the digestion of endocytized bacteria by interfering with the action of lysosomes.

MICROBE AT A GLANCE

Streptococcus pneumoniae

Taxonomy: Domain Bacteria, phylum Firmicutes, class "Bacilli," order "Lactobacillales," family Streptococcaceae

Other names: Pneumococcus, formerly "Diplococcus pneumoniae"

Cell morphology and arrangement: Cocci in pairs

Gram reaction: Positive

Virulence factors: Polysaccharide capsule, phosphorylcholine, protein adhesin, secretory IgA protease, pneumolysin

Diseases caused: Pneumococcal pneumonia, sinusitis, otitis media, bacteremia, endocarditis, pneumococcal meningitis

Treatment for diseases: Penicillin, cephalosporin, erythromycin, chloramphenicol

Prevention of disease: Vaccine against 23 common strains is long lasting in adults with unimpaired immunity

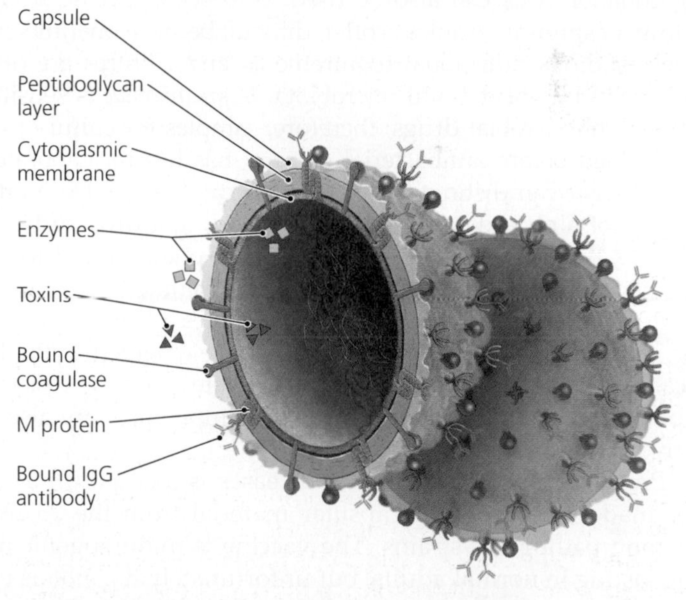

Download the Microbe at a Glance flashcards from the Study Area at www.masteringmicrobiology.com.

Epidemiology

Streptococcus pneumoniae grows in the mouths and pharynges of 75% of humans, without causing harm; however, when pneumococci travel to the lungs, they cause disease. Typically, the incidence of pneumococcal disease is highest in children and the elderly, groups whose immune responses are not fully active.

Pneumococcal Diseases

Streptococcus pneumoniae causes a variety of diseases, which we explore next.

Pneumococcal Pneumonia The most prevalent disease caused by *S. pneumoniae* is **pneumococcal pneumonia** (nū-mō'nē-ă), which constitutes about 85% of all cases of pneumonia. The disease results when pneumococci are inhaled from the pharynx into lungs damaged either by a previous viral disease, such as influenza or measles, or by other conditions, such as alcoholism, congestive heart failure, or diabetes mellitus. As the bacteria multiply in the alveoli (air sacs), they damage the alveolar lining, allowing fluid, red blood cells, and leukocytes to enter the lungs. The leukocytes attack *Streptococcus*, in the process secreting inflammatory and pyrogenic chemicals. The onset of clinical symptoms is abrupt and includes a fever of 39–41°C and severe shaking chills. Most patients have a productive cough, slightly bloody sputum, and chest pain.

Sinusitis and Otitis Media Following viral infections of the upper respiratory tract, *S. pneumoniae* can also invade the sinuses and middle ear, where it causes **sinusitis** (sī-nŭ-sī'tis; inflammation of the nasal sinuses) and **otitis media** (ō-tī'tis mē'dē-ă; inflammation of the middle ear). Pus production and inflammation in these cavities create pressure and pain. Sinusitis occurs in patients of all ages. Otitis media is more prevalent in children because their narrow auditory tubes connecting the pharynx with the middle ears are nearly horizontal, which facilitates the flow of infected fluid from the pharynx into the middle ears. The tubes become wider and more vertical when the shape of the head changes as children grow, making infection less likely in adults.

Bacteremia and Endocarditis *Streptococcus pneumoniae* can enter the blood either through lacerations (as might result from vigorous tooth brushing, chewing hard foods, or dental procedures) or as a result of tissue damage in the lungs during pneumonia. Bacteria generally do not enter the blood during sinusitis or otitis media. As with *Staphylococcus, S. pneumoniae* can colonize the lining of the heart, causing endocarditis. The heart valves, once involved, are typically destroyed.

Pneumococcal Meningitis Pneumococci can spread to the meninges via bacteremia, during sinusitis or otitis media, or following head or neck surgery or trauma that opens a passage between the pharynx and the subarachnoid space of the meninges. The mortality rate of pneumococcal meningitis, which is primarily a disease of children, is up to 20 times that of meningitis caused by other microorganisms.

Diagnosis, Treatment, and Prevention

Medical laboratory technologists can quickly identify pneumococci in Gram stains of sputum smears and confirm their presence with the *Quellung*[15] *reaction,* in which anticapsular antibodies cause the capsule to swell. Antibodies against particular strains trigger Quellung reactions only against those strains. Antibody agglutination tests can also be used to identify specific strains. Culture of sputum samples is often difficult because pneumococci have fastidious nutritional requirements and cultures are often overgrown by normal oral microbiota. *S. pneumoniae* is sensitive to most antimicrobial drugs; therefore, samples for culture must be obtained before antibacterial therapy has begun. Laboratory technologists can distinguish pneumococcal colonies from other colonies of alpha-hemolytic strains by adding a drop of bile to a colony. Bile triggers chemicals present in pneumococci to lyse the cells, dissolving the colony in just a few minutes.

Penicillin has long been the drug of choice against *S. pneumoniae,* though penicillin-resistant strains have emerged in the last decade; about a third of peumococcal isolates are now resistant. Cephalosporin, erythromycin, and chloramphenicol are also effective treatments.

Prevention of pneumococcal diseases is focused on a vaccine made from purified capsular material from the 23 most common pathogenic strains. The vaccine is immunogenic and long lasting in normal adults, but unfortunately it is not as efficacious in patients at greatest risk, such as the elderly, young children, and AIDS patients.

Enterococcus

Learning Objective

✓ Identify two species of *Enterococcus*, and describe their pathogenicity and the diagnosis, treatment, and prevention of their diseases.

We have discussed two Gram-positive cocci that are pathogenic in humans: catalase-positive *Staphylococcus* and catalase-negative *Streptococcus.* Now we turn our attention to another genus of Gram-positive, catalase-negative cocci—*Enterococcus* (en′ter-ō-kok′ŭs), so named because all enterococci are spherical and live in the intestinal tracts of animals. Lancefield classified enterococci with group D streptococci, but they differ significantly from other members of group D in that *Enterococcus* is unencapsulated, produces gas during fermentation of sugars, and is typically nonhemolytic (i.e., gamma-hemolytic). Because of these and other differences, microbiologists now classify *Enterococcus* as a separate genus.

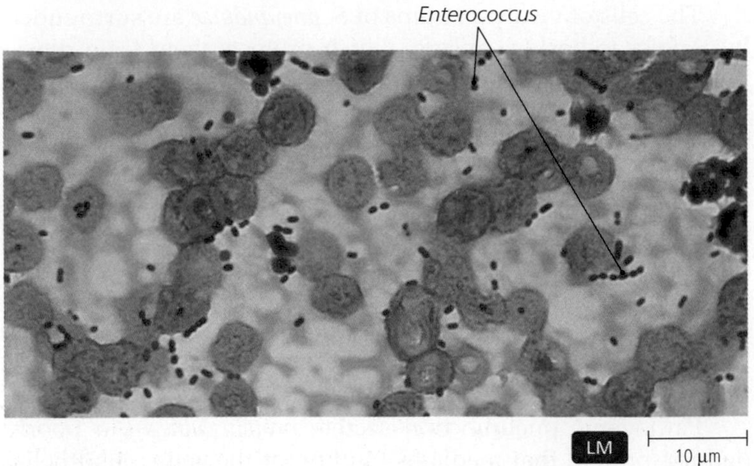

▲ **Figure 19.10** ***Enterococcus* faecalis in lung tissue.** The Gram-positive cells are arranged in pairs and short chains, but (unlike *S. pneumoniae*) lack capsules. *Why are these bacteria called "enterococci"?*

Figure 19.10 They are called "enterococci" because they are spherical bacteria that live in the intestinal tracts of animals.

Structure and Physiology

Enterococci form short chains and pairs; they do not form capsules **(Figure 19.10)**. Enterococci grow at temperatures up to 45°C, at pH as high as 9.6, and in 6.5% NaCl or 40% bile salt broths—all conditions that severely inhibit the growth of *Streptococcus.* The two species that are significant pathogens of humans are *E. faecalis*[16] (fē-kă′lis) and *E. faecium*[17] (fē′sē-ŭm).

Pathogenesis, Epidemiology, and Diseases

E. faecalis is ubiquitous in the human colon. *E. faecium* is found less often. Because both species lack structural and chemical elements that make them virulent, they are rarely pathogenic in the intestinal tract. Nevertheless, they do have the ability to adhere to human epithelial cells, and they secrete *bacteriocins,* which are chemicals that inhibit the growth of other bacteria.

E. faecalis and *E. faecium* can cause serious disease if they are introduced into other parts of the body, such as the lungs, urinary tract, or bloodstream, via poor personal hygiene or intestinal laceration. Enterococci account for about 10% of nosocomial infections and cause bacteremia, endocarditis, and wound infections.

[15]German, meaning swelling.
[16]Latin, meaning pertaining to feces.
[17]Latin, meaning of feces.

Diagnosis, Treatment, and Prevention

Even though a Gram stain of *Enterococcus* looks similar to one of *S. pneumoniae,* the two genera can be distinguished readily in that *Enterococcus* is not sensitive to bile.

Enterococcal infections are very difficult to treat because strains resistant to frequently used antimicrobials—lactams, aminoglycosides, and vancomycin—are relatively common. This is particularly troublesome in that the genes for antimicrobial resistance occur on plasmids that can be transmitted to other bacteria.

It is difficult to prevent enterococcal infections, particularly in health care settings, where patients' immune systems are weakened. Health care workers should always use good hygiene and aseptic techniques to minimize the transmission of these microorganisms.

Table 19.2 summarizes the features of the major pathogenic, Gram-positive streptococci—*Streptococcus* and *Enterococcus.* Now we turn our attention to the Gram-positive pathogenic bacilli, beginning with the low G + C, endospore-forming genera *Bacillus* and *Clostridium.*

Bacillus

Scientists divide Gram-positive bacilli (rod-shaped cells) into endospore-forming and non-endospore-forming genera. The endospore-forming genera are *Bacillus* and *Clostridium.* In this section we examine *Bacillus anthracis*—one of the 51 species of *Bacillus* and a strict pathogen of animals and humans. The following section discusses *Clostridium.*

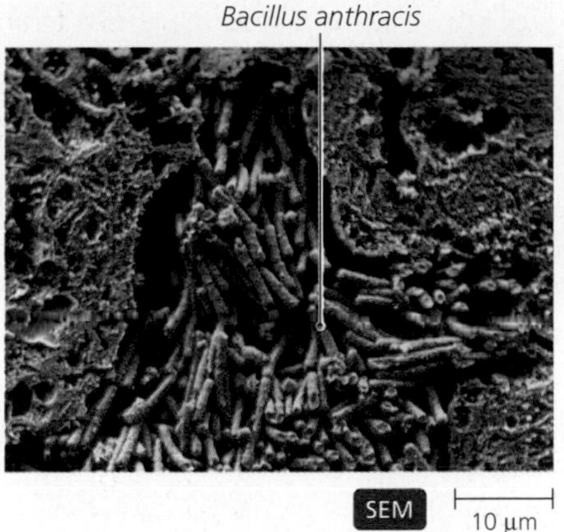

▲ **Figure 19.11** ***Bacillus anthracis* as it appears in tissue.**

Structure, Physiology, and Pathogenicity

Learning Objective

✓ Identify the structural features of *Bacillus* that contribute to its pathogenicity.

Bacillus anthracis[18] (ba-sil'ŭs an-thrā'sis) is a large (1 µm × 3–5 µm) rod-shaped, facultatively anaerobic bacterium that normally dwells in soil. Its cells are arranged singly, in pairs, or in chains **(Figure 19.11)**. In laboratory culture, prominent endospores form in the middle of cells (see Figure 4.19), which lack capsules. In contrast, bacteria in clinical specimens typically lack endospores

[18]Greek, meaning charcoal.

TABLE 19.2 Characteristics of Pathogenic Streptococci

Lancefield Group	Scientific Name	Hemolytic Pattern	Significant Characteristics	Characteristic Diseases
A	*S. pyogenes*	Large zone of beta-hemolysis	1- to 2-mm white colonies on blood agar; bacitracin sensitive	Pharyngitis, scarlet fever, pyoderma, erysipelas, toxic-shock-like syndrome, necrotizing fasciitis, rheumatic fever, glomerulonephritis
B	*S. agalactiae*	Small zone of beta-hemolysis	2- to 3-mm buttery colonies on blood agar; bacitracin resistant	Puerperal fever, neonatal bacteremia, meningitis, pneumonia
C	*S. equisimilis*	Large zone of beta-hemolysis	1- to 2-mm white colonies on blood agar	Pharyngitis, glomerulonephritis
C, F, or G	*S. anginosus*	Small zone of beta-hemolysis	1- to 2-mm white colonies on blood agar	Purulent abscess
—	*S. mutans*	Alpha-hemolysis	Viridans group (produce green pigment when grown on blood agar)	Dental caries; rarely bacteremia, meningitis, endocarditis
—	*S. pneumoniae*	Alpha-hemolysis (aerobic); beta-hemolysis (anaerobic)	Diplococci; capsule required for pathogenicity; bile sensitive	Pneumonia, sinusitis, otitis media, bacteremia, endocarditis, meningitis
D	*Enterococcus faecalis, E. faecium*	None (gamma-hemolysis)	Diplococci; no capsule; bile insensitive	Urinary tract infections, bacteremia, endocarditis, wound infections

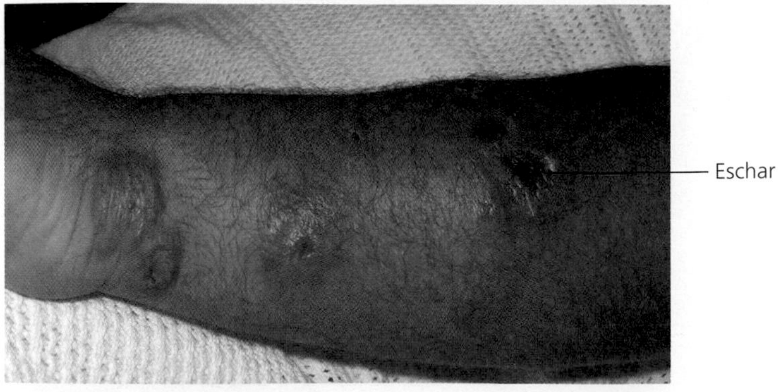

▲ **Figure 19.12 Cutaneous anthrax.** Black eschars are characteristic of cutaneous anthrax.

and possess protective capsules. As we discussed in Chapter 3, the tough external coat and the internal chemicals of endospores make these structures resistant to harsh environmental conditions, enabling *Bacillus* to survive in the environment for centuries or perhaps even longer. A vegetative (non-endospore) cell of *Bacillus* can survive in the body because it has multiple copies of a plasmid coding for a capsule, which is composed solely of glutamic acid. The capsule inhibits effective phagocytosis by white blood cells.

Pathogenic strains of *B. anthracis* cause disease because they contain multiple copies of another plasmid coding for *anthrax toxins*—three distinct polypeptides that work together in a lethal combination. Seven copies of one toxin combine to form a protein pore that breaches the cytoplasmic membrane of a host cell. The other two toxins enter the cell through the pore. Once in the cell, the toxins interfere with intracellular signaling, severely affect cellular metabolism, and ultimately cause the cell to undergo apoptosis (programmed cell suicide). Scientists are investigating the precise mechanisms by which the three components of anthrax toxin work in hopes of developing techniques for neutralizing them. Anthrax can be deadly even after treatment, because antimicrobial drugs do not inactivate accumulated anthrax toxin.

Epidemiology

Learning Objective

✓ List three methods of transmission of anthrax.

Anthrax is primarily a disease of herbivores; humans contract the disease from infected animals. Anthrax is not normally transmitted from individual to individual, but can invade via one of three routes: inhalation of endospores, inoculation of endospores into the body through a break in the skin, or ingestion of endospores. Ingestion anthrax is the normal means of transmission among animals but is rare in humans. In the 25 years between January 1976 and September 2001, only 15 cases of anthrax were reported in the United States; thus epidemiologists suspected bioterrorism when over a dozen cases of anthrax were reported in New York, Florida, and Washington, D.C., in the fall of 2001.

Disease

Learning Objectives

✓ List and describe three clinical manifestations of *Bacillus anthracis* infections.

✓ Identify the diagnosis, treatment, and prevention of anthrax.

Bacillus anthracis causes only one disease—anthrax—but it can have three clinical manifestations. The first, *gastrointestinal anthrax,* is very rare in humans but is common in animals; it results in intestinal hemorrhaging and eventually death.

Cutaneous anthrax begins when a painless, solid, raised nodule forms on the skin at the site of infection. The cells in the affected area die, and the nodule spreads to form a painless, swollen, black, crusty ulcer called an *eschar* (es′kar) **(Figure 19.12)**. It is from the black color of eschars that anthrax, which means "charcoal" in Greek, gets its name. *B. anthracis* growing in the eschar releases anthrax toxin into the blood, producing toxemia. Untreated cutaneous anthrax is fatal for 20% of patients.

Inhalation anthrax is also rare in humans, as it requires inhalation of airborne endospores. After endospores germinate in the lungs, they secrete toxins that are absorbed into the blood-

BENEFICIAL MICROBES

MICROBES TO THE RESCUE?

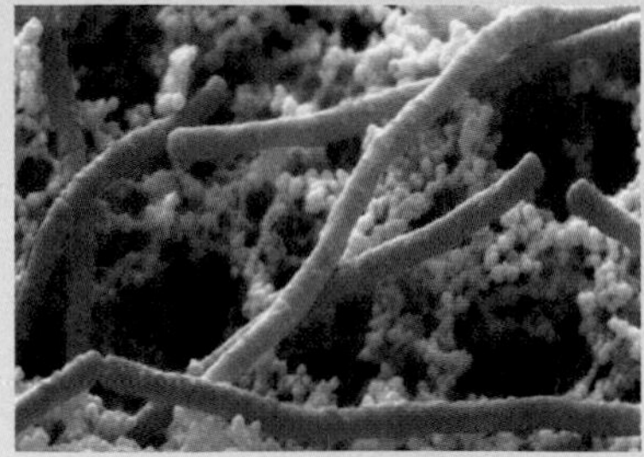

▲ ***Lactobacillus,*** SEM **a potential probiotic.**

The digestive tract is home to viruses, bacteria, protozoa, fungi, and parasitic helminths. The normal microbiota helps protect the body by competing with pathogens for nutrients and space. Some research indicates that consumption of living microbes in food or dietary supplements may change the normal microbiota of the digestive tract. Such microbes may help ward off bowel problems such as irritable bowel syndrome, reduce incidence of yeast infection, alleviate symptoms of gastroenteritis, and shorten the duration of colds by 36 hours. Probiotics, as such microbes are called, are often bacteria used to ferment food, particularly species of *Lactobacillus* or the related genus *Bifidobacterium.* Despite favorable anecdotal evidence and partial support in laboratory studies for the benefits of probiotics, many scientists remain skeptical that consumers can successfully change the makeup of their intestinal microbiota or that probiotics are really helpful.

stream, producing toxemia. Early signs and symptoms include fatigue, malaise, fever, aches, and cough—all of which are common to many pulmonary diseases. In a later phase, victims of inhalation anthrax have a high fever and labored breathing due to localized swelling, and they go into shock. Historically, mortality rates have been high—approaching 100% even when treated—perhaps in part because the disease is often not suspected until it is irreversible, and because antimicrobial drugs do not neutralize toxins released during the course of the disease. During the bioterrorism attack of 2001, physicians learned that early and aggressive treatment of inhalation anthrax with antimicrobial drugs accompanied by persistent drainage of fluid from around the lungs increased the survival rate to greater than 50%.

Diagnosis, Treatment, and Prevention

Large, nonmotile, Gram-positive bacilli in clinical samples from the lungs or skin are diagnostic. As mentioned previously, endospores are not typically seen in clinical samples but are produced after a few days in culture.

Penicillin, erythromycin, chloramphenicol, and many other antimicrobial agents are effective against *B. anthracis.* Ciprofloxacin (Cipro) gained fame in 2001 as physicians were encouraged to use it in the fear that the bioterrorists had genetically altered the microorganism to be penicillin resistant—a fear that proved unfounded.

Prevention of naturally occurring disease in humans requires control of the disease in animals. Farmers in areas where anthrax is endemic must vaccinate their stock and bury or burn the carcasses of infected animals. Anthrax vaccine has proven effective and safe for humans, but requires six doses over 18 months plus annual boosters. Researchers are developing an alternative vaccine.

Clostridium

Learning Objective

✓ Characterize the four major species of *Clostridium*.

Clostridium (klos-trid'ē-ŭm) is an anaerobic, Gram-positive, endospore-forming bacillus that is ubiquitous in soil, water, sewage, and the gastrointestinal tracts of animals and humans. Several species are significant human pathogens. Pathogenicity is due in great part to the ability of endospores to survive harsh conditions, and to the secretion by vegetative cells of potent *histolytic toxins, enterotoxins* (toxins affecting the GI tract), and *neurotoxins*. The most common pathogenic clostridia are *C. perfringens, C. difficile, C. botulinum,* and *C. tetani.*

Clostridium perfringens

Learning Objective

✓ Identify the mechanisms accounting for the pathogenesis of *Clostridium perfringens* infections.

✓ Describe the diagnosis, treatment, and prevention of *Clostridium perfringens* infections.

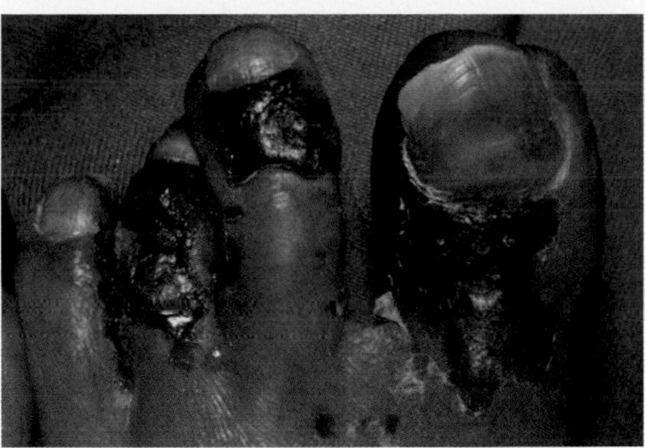

▲ **Figure 19.13 Gas gangrene, a life-threatening disease caused by *Clostridium perfringens*.** The blackening is the result of the death of muscle tissue (myonecrosis), whereas the "bubbling" appearance results from the production of gaseous waste products by the bacteria.

Clostridium perfringens (per-frin'jens), the clostridium most frequently isolated from clinical specimens, is a large, almost rectangular, Gram-positive bacillus. Although it is nonmotile, its rapid growth enables it to proliferate across the surface of laboratory media, resembling the spread of motile bacteria. Endospores are rarely observed either in clinical samples or in culture. *C. perfringens* type A, known by its specific antigens, is the most virulent serotype.

Pathogenesis, Epidemiology, and Disease

C. perfringens produces 11 toxins that lyse erythrocytes and leukocytes, increase vascular permeability, reduce blood pressure, and kill cells, resulting in irreversible damage. Because *C. perfringens* commonly grows in the digestive tracts of animals and humans, it is nearly ubiquitous in fecally contaminated soil and water.

The severity of diseases caused by *C. perfringens* ranges from mild food poisoning to life-threatening illness. Clostridial food poisoning is a relatively benign disease characterized by abdominal cramps and watery diarrhea, but not fever, nausea, or vomiting. It lasts for less than 24 hours. Such food poisoning typically results from the ingestion of large numbers (10^8 or more) of *C. perfringens* type A in contaminated meat.

C. perfringens is not invasive, but when some traumatic event (such as a surgical incision, a puncture, a gunshot wound, crushing trauma, or a compound fracture) introduces endospores into the body, they can germinate in the anaerobic environment of deep tissues. The immediate result is intense pain at the initial site of infection as clostridial toxins induce swelling and tissue death. The rapidly reproducing bacteria can then spread into the surrounding tissue, causing the death of muscle and connective tissue that is typically accompanied by the production of abundant, foul-smelling, gaseous, bacterial waste products—hence the common name for the disease: **gas gangrene**[19] **(Figure 19.13)**. Shock, kidney failure, and death can follow, often within a week of infection.

[19]From Greek *gangraina*, meaning an eating sore.

Diagnosis, Treatment, and Prevention

Medical laboratory technologists show that *Clostridium* is involved in food poisoning by demonstrating more than 10^5 bacteria in a gram of food or 10^6 cells per gram of feces. The appearance of gas gangrene is usually diagnostic by itself, though the detection of large Gram-positive bacilli is confirmatory.

Clostridial food poisoning is typically self-limiting—the pathogens and their toxins are eliminated in the resulting watery stool. In contrast, physicians must quickly and aggressively intervene to stop the spread of necrosis in gas gangrene by surgically removing dead tissue and administering large doses of antitoxin and penicillin. Oxygen applied under pressure may also be effective. Despite all therapeutic care, mortality of gas gangrene exceeds 40%.

It is difficult to prevent infections of *C. perfringens* because the organism is so common; however, refrigeration of food prevents toxin formation and reduces the chance of clostridial food poisoning. Alternatively, reheating contaminated food destroys any toxin that has formed. Given that gas gangrene occurs when endospores are introduced deep in the tissues, proper cleaning of wounds can prevent many cases.

Clostridium difficile

Learning Objectives

- ✓ Discuss the role of antimicrobial drugs in the development of gastrointestinal diseases caused by *Clostridium difficile*.
- ✓ Discuss the diagnosis, treatment, and prevention of *C. difficile* infections.

Clostridium difficile (di-fi'sil-ē) is a motile, anaerobic intestinal bacterium with cells about 1.5 μm in width and 3–6.5 μm in length that form oval, subterminal endospores. The bacterium produces two toxins (called toxins A and B) and the enzyme hyaluronidase.

Pathogenesis, Epidemiology, and Disease

Although so-called *C. diff.* is a common member of the intestinal microbiota, it can be an opportunistic pathogen in patients treated with broad-spectrum antimicrobial drugs, such as penicillin and cephalosporin. In such patients, the normal proportions of different bacteria in the colon can be significantly altered. In many cases the hardy endospores of *C. difficile* germinate, enabling it to become the predominant intestinal bacterium, such that the toxins and enzymes it produces cause hemorrhagic death of the intestinal wall. In minor infections these lesions result in a self-limiting explosive diarrhea; however, in more serious cases, *C. difficile* produces life-threatening **pseudomembranous colitis,** in which large sections of the colon wall slough off, potentially perforating the colon, and leading to massive internal infection by fecal bacteria and eventual death.

Diagnosis, Treatment, and Prevention

Diarrhea in patients undergoing antimicrobial therapy is suggestive of *C. difficile* infection, and laboratory microbiologists confirm the diagnosis either by isolating the organism from feces using selective media or by demonstrating the presence of the toxins via immunoassays.

Discontinuation of the implicated antimicrobial drug, which allows the microbiota to return to normal, usually resolves minor infections with *C. difficile.* More serious cases are treated with vancomycin or metronidazole, though endospores survive such therapy in about a third of patients, causing a relapse. Further treatment of relapses with either vancomycin or metronidazole is often successful.

C. difficile is frequently found in hospitals, and hospital personnel can easily transmit it between patients. Proper hygiene—particularly frequent handwashing—is critical for limiting nosocomial infections.

Clostridium botulinum

Learning Objectives

- ✓ Contrast the three manifestations of botulism.
- ✓ Describe the use of mice in the diagnosis of botulism.
- ✓ Describe three treatments of botulism and explain how to prevent it.

Clostridium botulinum[20] (bo-tū-lī'num) is an anaerobic, endospore-forming, Gram-positive bacillus that is common in soil and water worldwide. Its endospores survive improper canning of food, germinating to produce vegetative cells that grow and release into the jar or can a powerful neurotoxin that causes **botulism** (bot'ū-lizm).

Pathogenesis

Strains of *C. botulinum* produce one of seven antigenically distinct *botulism toxins* (A through G). Some scientists consider botulism toxins among the deadliest toxins known—30 grams of pure toxin would be enough to kill every person in the United States. Botulism toxins are extremely potent; even a small taste of contaminated food can cause full-blown illness or death.

Each of the seven toxins is a quaternary protein composed of a single neurologically active polypeptide associated with one or more nontoxic polypeptides that prevent the inactivation of the toxin by stomach acid. To understand the action of botulism toxins, we must consider the way the nervous system controls muscle contractions.

Each of the many ends of a motor neuron (nerve cell) forms an intimate connection with a muscle cell at a *neuromuscular junction;* however, the two cells do not actually touch—a small gap, called a *synaptic cleft,* remains between them **(Figure 19.14a)**. The neuron stores a chemical, called *acetylcholine,* in vesicles near its terminal cytoplasmic membrane. Acetylcholine is one of a family of chemicals called neurotransmitters that mediate communication among neurons and between neurons and other cells.

When a signal arrives at the terminus of a motor neuron (1), the vesicles containing acetylcholine fuse with the neuron's cytoplasmic membrane, releasing acetylcholine into the synaptic

[20]From Latin, *botulus,* meaning sausage.

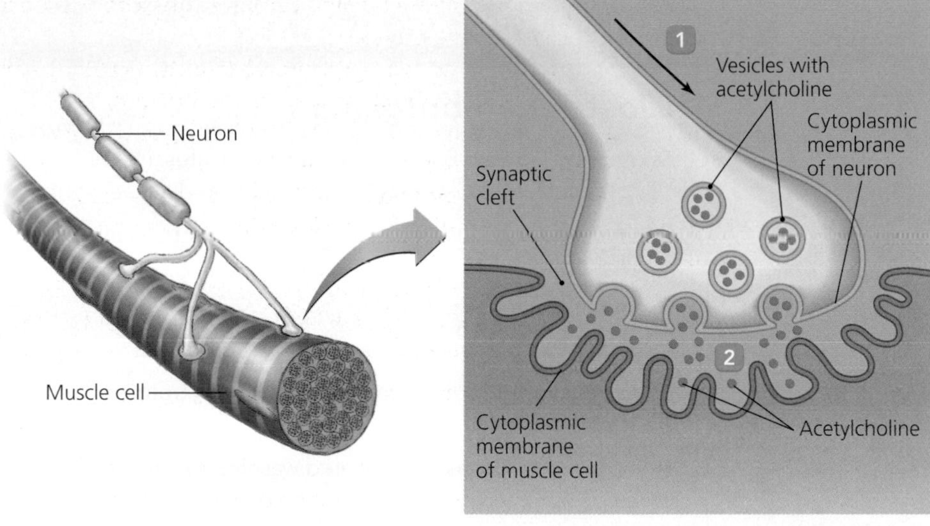

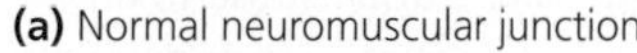

(a) Normal neuromuscular junction

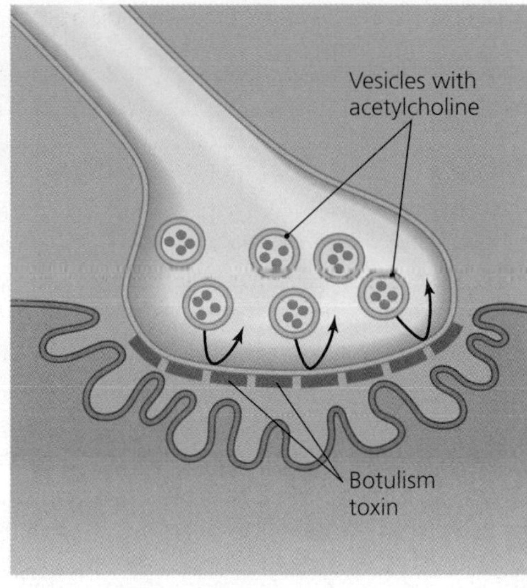

(b) Neuromuscular junction with botulism toxin present

Figure 19.14 How botulism toxin acts at a neuromuscular junction. **(a)** Normal function at a neuromuscular junction. 1 A nerve impulse from the central nervous system causes vesicles filled with acetylcholine to fuse with the neuron's cytoplasmic membrane, 2 releasing acetylcholine into the synaptic cleft. The binding of acetylcholine to receptors on the muscle cell's cytoplasmic membrane stimulates a series of events that result in contraction of the muscle cell (not shown). **(b)** Botulism toxin blocks the fusion of the vesicles with the neuron's cytoplasmic membrane, thereby preventing secretion of the neurotransmitter into the synaptic cleft; as a result, the muscle cell does not contract.

cleft (2). Molecules of acetylcholine then diffuse across the cleft and bind to receptors on the cytoplasmic membrane of the muscle cell. The binding of acetylcholine to muscle cell receptors triggers a series of events inside the muscle cell that results in muscle contraction (not shown).

Botulism toxins act by binding irreversibly to neuronal cytoplasmic membranes, thereby preventing the fusion of vesicles and secretion of acetylcholine into the synaptic cleft **(Figure 19.14b)**. Thus, these neurotoxins prevent muscular contraction, resulting in a *flaccid paralysis.*

Epidemiology and Diseases

Botulism is not an infection, but instead an *intoxication* (poisoning) caused by botulism toxin. Clinicians recognize three manifestations of botulism: foodborne botulism, infant botulism, and wound botulism. Fortunately, all three are rare.

About 25 cases of foodborne botulism occur in the United States each year, usually within one to two days following the consumption of toxin in home-canned foods or preserved fish. Contaminated food may not appear or smell spoiled. Patients are initially weak and dizzy and have blurred vision, dry mouth, dilated pupils, constipation, and abdominal pain, followed by a progressive paralysis that eventually affects the diaphragm. The patient remains mentally alert throughout the ordeal. When fatal, death results from the inability of muscles of respiration to effect inhalation; victims asphyxiate because they cannot inhale. Survivors recover very slowly as their nerve cells grow new endings over the course of months or years, replacing the debilitated termini. (See **Highlight: Botulism in Alaska** on p. 550.)

In contrast to foodborne botulism, infant botulism results from the ingestion of endospores, which then germinate and colonize the infant's gastrointestinal tract. Infants are susceptible to colonization because their GI tracts do not have a sufficient number of benign microbiota to compete with *C. botulinum* for nutrients and space. Botulism toxin is absorbed into the blood from an infected infant's GI tract, causing nonspecific symptoms: crying, constipation, and "failure to thrive." Fortunately, paralysis and death are rare. About 100 cases of infant botulism are reported in the United States each year, though some cases reported as sudden infant death syndrome may be fatal infant botulism instead.

Wound botulism usually begins four or more days following the contamination of a wound by endospores. With the exception that the gastrointestinal system is not typically involved, the symptoms are the same as those of foodborne botulism.

Diagnosis, Treatment, and Prevention

The symptoms of botulism are diagnostic; culturing the organism from contaminated food, from feces, or from the patient's wounds confirms the diagnosis. Further, toxin activity can be detected by using a mouse bioassay. In this laboratory procedure, specimens of food, feces, and/or serum are divided into two portions. Botulism antitoxin is mixed with one of the portions, and the portions are then inoculated into two sets of mice. If the mice receiving the antitoxin survive while the other mice die, botulism is confirmed.

HIGHLIGHT

BOTULISM IN ALASKA

For some indigenous Alaskans living near the Bering Sea, an adult beluga whale beached on the coast in July 2002 seemed a feast fit for a king. The villagers carved up the tail and enjoyed a traditional delicacy known as "muktuk"—skin and pink blubber eaten raw. Within 36 hours, however, 8 of the 14 diners came down with symptoms of dry mouth, blurred vision, swallowing difficulties, and progressive paralysis: telltale signs of botulism.

Sure enough, local authorities discovered that the whale meat had been contaminated with botulism toxin type E, which is found uniquely in marine mammals. Apparently, the whale had been dead for some weeks before it was eaten—enough time for it to become an incubator for *Clostridium botulinum*, an anaerobe that prefers a moist, low-acid environment. That July, the temperature rose high enough to match the lower temperature limit for toxin production, a weather phenomenon that may have been a result of global warming. All patients recovered thanks to the prompt administration of antitoxin and ventilator support.

Treatment of botulism entails three approaches:

- Repeated washing of the intestinal tract to remove *Clostridium.*
- Administration of antibodies against botulism toxin to neutralize toxin in the blood before it can bind to neurons.
- Administration of antimicrobial drugs to kill clostridia in infant and wound botulism cases. (This approach is not effective in treating foodborne botulism, which results from the ingestion of toxin, not of the bacteria themselves.)

Note that damage resulting from any prior binding of toxin to any particular nerve ending is irreversible, so signs and symptoms of botulism may persist despite aggressive medical care.

Foodborne botulism is prevented by destroying all endospores in contaminated food by proper canning techniques, by preventing endospores from germinating by using refrigeration or establishing an acidic environment (pH <4.5), or by destroying the toxin by heating to at least 80°C for 20 minutes or more. Because infant botulism is often associated with the consumption of honey, parents are advised not to feed honey to infants under age one because their intesinal microbiota are not suffciently developed to inhib the germination *C. botulinum* endospores and their growth as toxin-producing vegetative cells.

MICROBE AT A GLANCE

Clostridium botulinum

Taxonomy: Domain Bacteria, phylum Firmicutes, class "Clostridia," order Clostridiales, family Clostridiaceae

Cell morphology and arrangement: Endospore-forming bacillus

Gram reaction: Positive

Virulence factors: Endospore, botulism toxins A–G

Diseases caused: Foodborne botulism, infant botulism, and wound botulism

Treatment for diseases: Repeated washing to intestinal tract to remove *Clostridium*, administration of antitoxin (antibodies against the toxins), antimicrobial drugs used for cases of infant botulism and wound botulism; damage to nerve endings irreversible

Prevention of disease: Proper canning to kill endospores; refrigeration to prevent endospores from germinating; heating to 80°C for 20 minutes to destroy toxin; refrain from feeding honey to infants

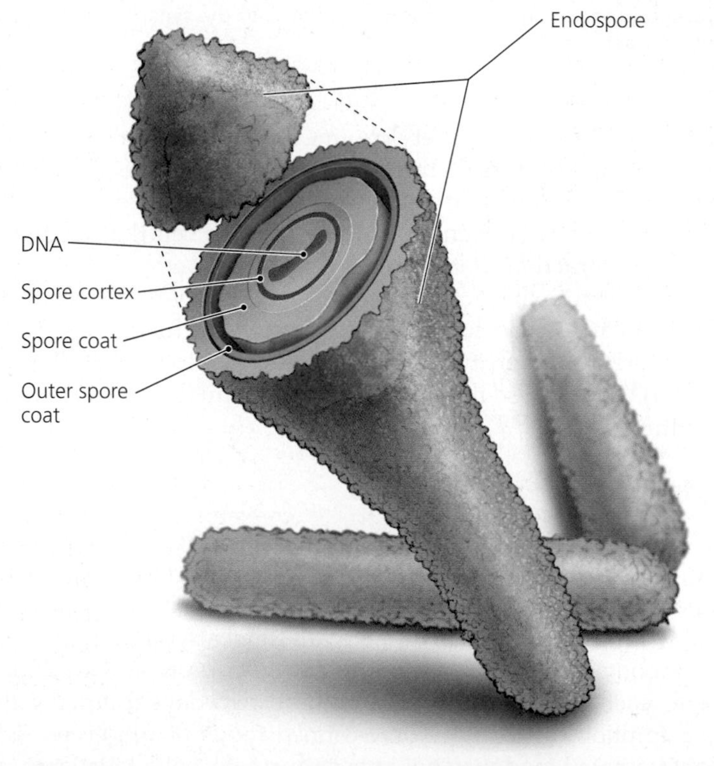

Download the Microbe at a Glance flashcards from the Study Area at www.masteringmicrobiology.com.

Clostridium tetani

Learning Objectives

- ✓ Describe the epidemiology of tetanus.
- ✓ List treatments for and preventative measures against *Clostridium tetani* infections.

Clostridium tetani[21] (te'tan-ē) is a small, motile, obligate anaerobe that produces a terminal endospore, giving the cell a distinctive lollipop appearance **(Figure 19.15)**. *C. tetani* is ubiquitous in soil, dust, and the GI tracts of animals and humans. Its vegetative cells are extremely sensitive to oxygen and live only in anaerobic environments, but its endospores survive for years. Its toxin causes the disease **tetanus.**

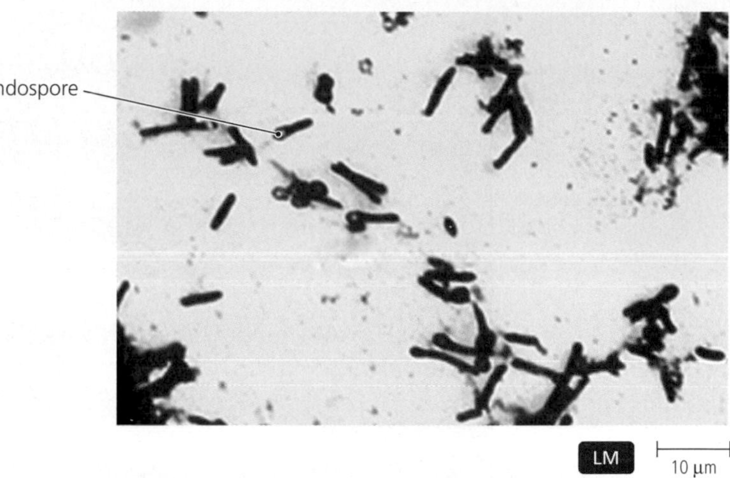

▲ **Figure 19.15 Cells of *Clostridium tetani*, with terminal endospores.** The cells have a distinctive lollipop shape. *How does the location of* C. tetani *endospores distinguish this bacterium from* Bacillus anthracis?

Figure 19.15 *The endospores of* C. tetani *are terminal, whereas those of* Bacillus anthracis *are located centrally.*

Pathogenesis

Contrary to popular belief, deep puncture wounds by rusty nails are not the only (or even primary) source of tetanus. Even a tiny break in the skin or mucous membranes can allow endospores of *C. tetani* access to an anaerobic environment in which they germinate, grow, and produce a fatal disease caused by **tetanospasmin** (tetanus toxin)—a potent neurotoxin released by *C. tetani* cells when they die. To understand the action of tetanospasmin, we must further consider the control of muscles by the central nervous system.

As we have seen, the secretion of acetylcholine at neuromuscular junctions stimulates muscles to contract. Skeletal muscles are typically arranged in competing pairs on opposite sides of a joint: When a motor nerve stimulates one muscle of the pair to contract, the muscle on the other side of the joint relaxes, but there is no "relaxing" neurotransmitter. Relaxation occurs because the muscle is unstimulated by nerve impulses. Inhibitory neurons of the central nervous system release neurotransmitters that inhibit the motor neuron, which therefore cannot stimulate the muscle **(Figure 19.16)**.

Tetanospasmin is composed of two polypeptides held together by a disulfide bond. The heavier of the two polypeptides binds to a receptor on a neuron's cytoplasmic membrane. The neuron then endocytizes the toxin, removes the lighter of the two polypeptides, and transports it to the central nervous system. There the smaller polypeptide enters an inhibitory neuron and blocks the release of inhibitory neurotransmitter. With inhibition blocked, excitatory activity is unregulated, and both muscles in an antagonistic pair are signaled to contract simultaneously **(Figure 19.16b)**. The result is that muscles on both sides of the joint contract and do not relax. Opposing contractions can be so severe they break bones.

Disease and Epidemiology

The incubation period of tetanus ranges from a few days to a week depending on the distance of the site of infection from the central nervous system. Typically the initial and diagnostic sign of tetanus is tightening of the jaw and neck muscles—which is why tetanus is also called *lockjaw*. Other early symptoms include sweating, drooling, grouchiness, and constant back spasms. If the toxin spreads to autonomic neurons, then heartbeat irregularities, fluctuations in blood pressure, and extensive sweating result. Spasms and contractions may spread to other muscles, becoming so severe that the arms and fists curl tightly, the feet curl down, and the body assumes a stiff backward arch as the heels and back of the head bend toward one another **(Figure 19.17)**. Complete, unrelenting contraction of the diaphragm results in a final inhalation; patients die because they cannot exhale.

Over a million cases of tetanus occur annually worldwide, mostly in underdeveloped countries where vaccination is unavailable or medical care is inadequate. The effect of tetanospasmin is irreversible at any particular synapse, so recovery depends on the growth of new neuronal terminals to replace those affected. The mortality rate of tetanus is about 50% among all patients, though the mortality of neonatal tetanus, resulting most commonly from infection of the umbilical stump, exceeds 90%.

Diagnosis, Treatment, and Prevention

The diagnostic feature of tetanus is the characteristic muscular contraction, which is often noted too late to save the patient. The bacterium itself is rarely isolated from clinical samples because it grows slowly in culture and is extremely sensitive to oxygen.

Treatment involves thorough cleaning of wounds to remove all endospores, immediate passive immunization with immunoglobulin directed against the toxin, the administration of antimicrobials such as penicillin, and active immunization with tetanus toxoid. Cleansing and antimicrobials eliminate the bacteria, whereas the immunoglobulin binds to and neutralizes tetanospasmin before it can attach to neurons. Active immunization stimulates the formation of antibodies that neutralize the toxin. Once the toxin binds to a neuron, treatment is limited to supportive care.

[21]From Greek *tetanos*, meaning to stretch.

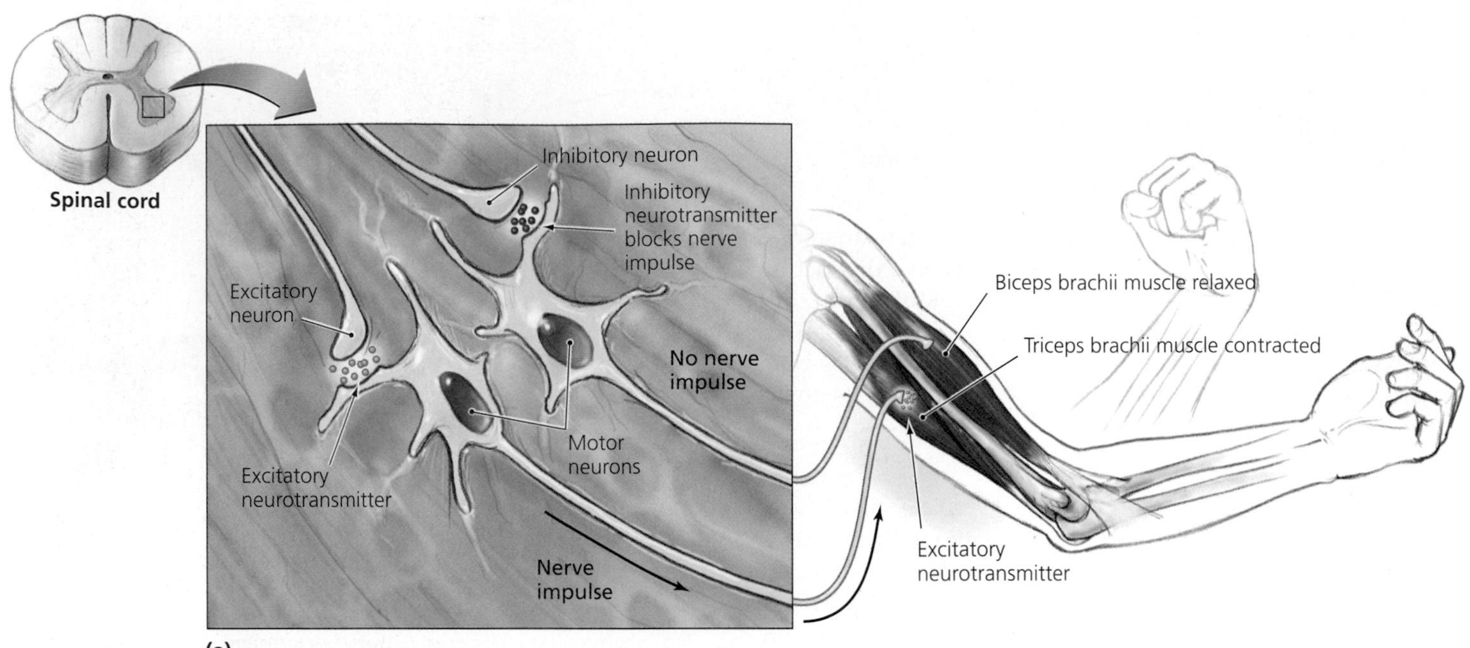

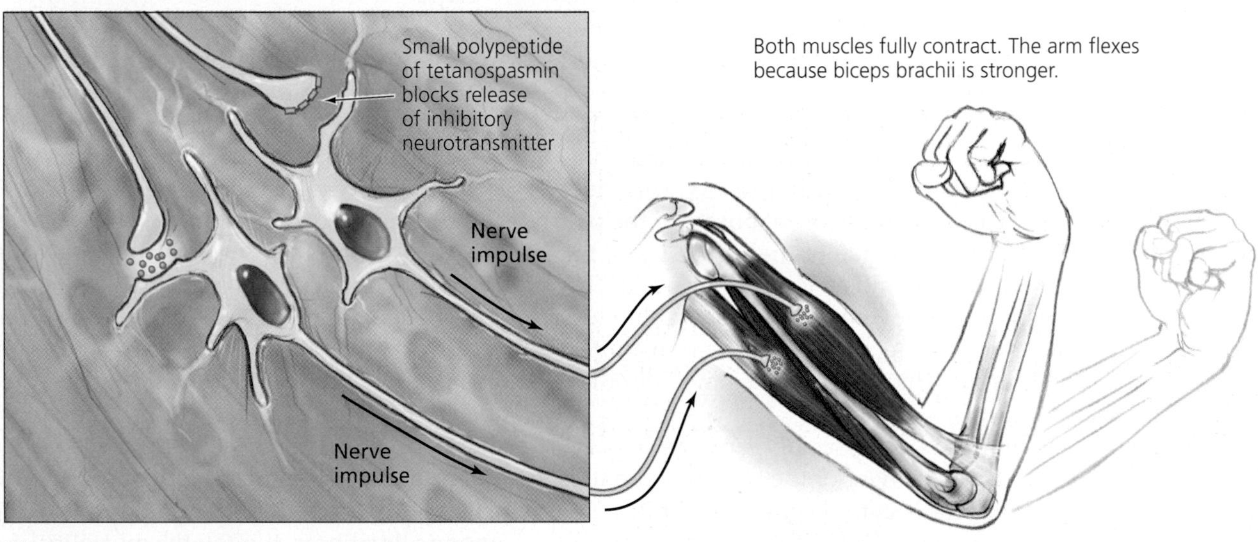

▲ **Figure 19.16 The action of tetanospasmin (tetanus toxin) on a pair of antagonistic muscles. (a)** Normally, when the muscle on one side of a joint is stimulated to contract, neurons controlling the antagonistic muscle are inhibited, so the antagonistic muscle relaxes. **(b)** Tetanospasmin blocks the inhibitory neurotransmitter. As a result, motor neurons controlling the antagonistic muscle are not inhibited, but instead can generate nerve impulses that stimulate muscle contraction.

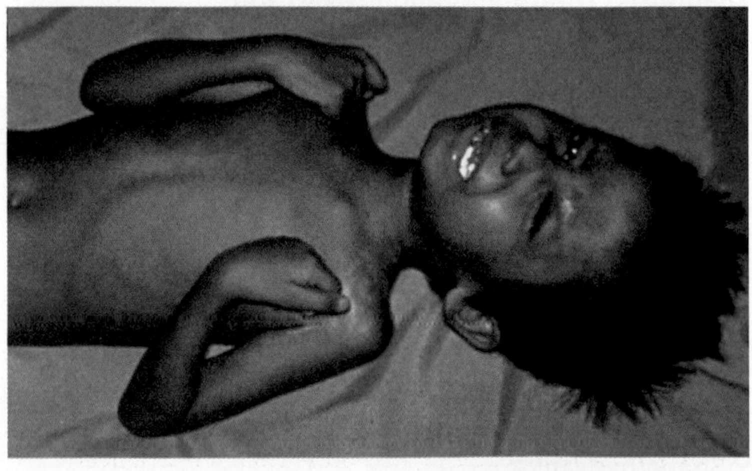

◀ **Figure 19.17 A patient with tetanus.** Prolonged muscular contractions result from the action of tetanus toxin.

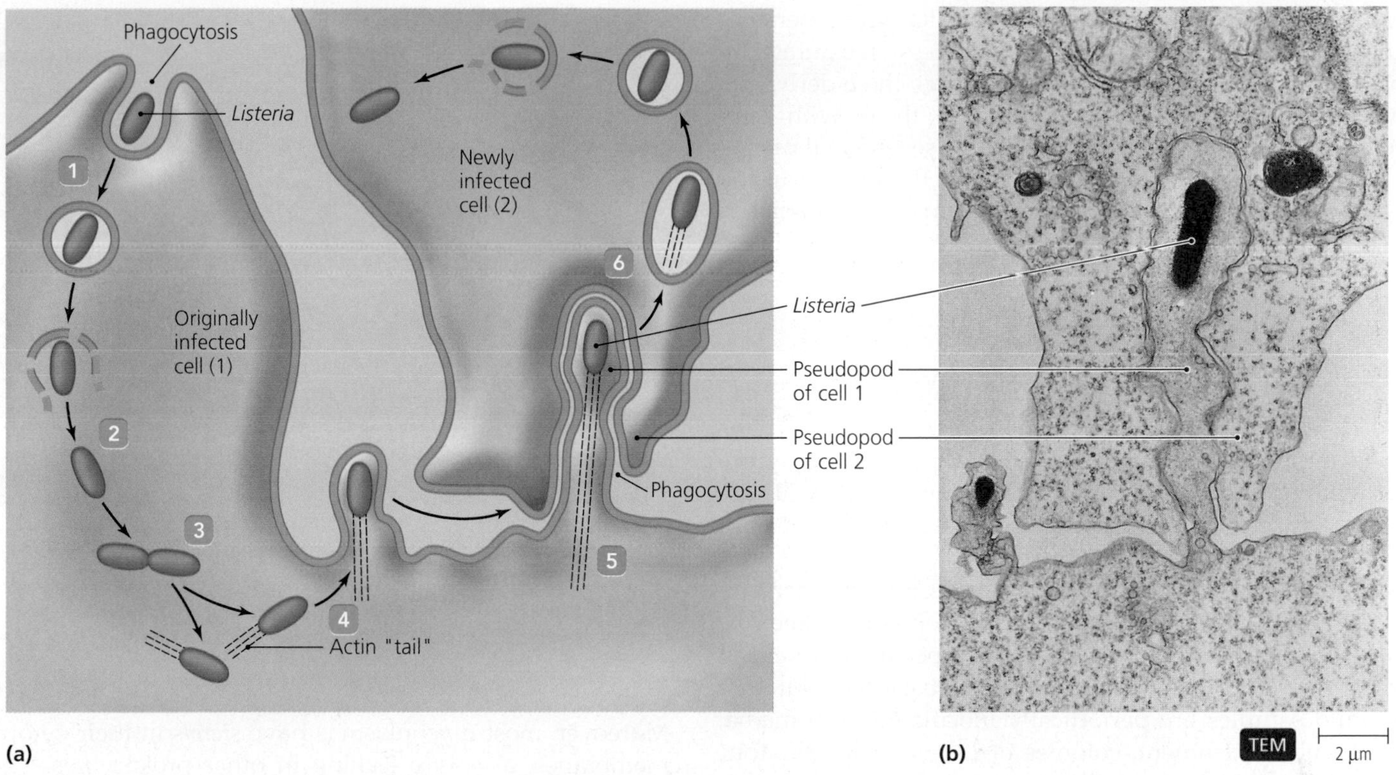

▲ **Figure 19.18 How *Listeria* avoids the host's immune system while infecting new cells.** **(a)** After *Listeria* is initially endocytized (1), the intracellular bacterium escapes the phagosome (2) and reproduces within the phagocyte (3). The bacterium then polymerizes the host cell's actin filaments into a "tail" (4) that pushes the bacterium into a pseudopod (5) that is subsequently endocytized by a new host cell (6). **(b)** A photomicrograph showing the endocytosis of a pseudopod containing a *Listeria* cell.

The number of cases of tetanus in the United States has steadily declined as a result of effective immunization with tetanus toxoid. The CDC currently recommends five doses beginning at two months of age, followed by a booster every 10 years for life.

In the previous two sections we examined two pathogenic, Gram-positive bacilli that form endospores: *Bacillus,* which is facultatively anaerobic, and *Clostridium,* which is anaerobic. Next we consider the low G + C, rod-shaped *Listeria,* and the unusual low G + C mycoplasmas that Gram stain pink because they lack cell walls.

Listeria

Learning Objectives

- ✓ Describe the structures of *Listeria monocytogenes* that account for its pathogenicity.
- ✓ Discuss the signs and symptoms, diagnosis, treatment, and prevention of listeriosis.

Listeria monocytogenes (lis-tēr′ē-ă mo-nō-sī-tah′je-nēz) is a low G + C, Gram-positive, non-endospore-forming bacillus found in soil, water, mammals, birds, fish, and insects. It enters the body in contaminated food and drink.

Pathogenesis, Epidemiology, and Disease

L. monocytogenes binds to the surfaces of a macrophage or liver or gallbladder cell, triggering its own endocytosis to become a facultative intracellular parasite. Once inside a human cell's phagosome (see Figure 15.6), *Listeria* synthesizes a pore-forming protein, called *listeriolysin O,* that punctures the phagosome membrane, releasing the bacterium into the host cell's cytosol before a lysosome can fuse with the phagosome. *Listeria* then grows and reproduces in the cytosol, sheltered from the humoral immune system.

Listeria continues to avoid exposure to the immune system via a unique method of transferring itself to neighboring cells without having to leave host cells **(Figure 19.18)**. The pathogen polymerizes the host cell's actin molecules to form stiff actin filaments that lengthen and push the bacterium through the cytosol to the cell's surface, where it forms a pseudopod. A neighboring macrophage or epithelial cell then endocytizes the pseudopod, and *Listeria* once again "tunnels" its way out of the phagosome to continue its intracellular parasitic existence within a new host cell.

Listeria's virulence is directly related to its ability to live within cells, which is conferred by listeriolysin O and the membrane protein that triggers endocytosis; it produces no toxins or enzymes that make it virulent. Interestingly, some mRNA in *Listeria* is inactive until it is at 37°C in a human body. As a result, the bacterium reproduces rapidly only when it is in a host.

Listeria is rarely pathogenic in healthy adults, who experience either no symptoms or only a mild flulike illness. In contrast, infection in pregnant women, fetuses, newborns, the elderly, and immunocompromised patients (particularly those with suppressed cell-mediated immunity) can be quite severe. In these patients, *Listeria* travels via the bloodstream to the brain, causing meningitis and possibly death. Human-to-human transmission is limited to the transfer of *Listeria* from pregnant women to fetuses, leading to premature delivery, miscarriage, stillbirth, or meningitis in the newborn.

Diagnosis, Treatment, and Prevention

Because *Listeria* most commonly causes meningitis, clinicians look for it in the cerebrospinal fluid (CSF) of people with symptoms of meningitis. Unfortunately, only a few *Listeria* cells are required to produce disease, so the bacterium is rarely seen in Gram-stained preparations of phagocytes.

Listeria can be cultured from blood and CSF specimens on many laboratory media, especially when a technique called *cold enrichment* is used. In this procedure, the specimen is held at 4°C—a temperature that inhibits most bacteria, but not *Listeria*—and samples are periodically inoculated onto media. Even with cold enrichment, colonies of *Listeria* may take four weeks or longer to become visible.

The bacterium exhibits a characteristic end-over-end "tumbling" motility that occurs at room temperature but not at 37°C. Tumbling motility is so distinctive that it can provide the basis for an initial diagnosis, although serological testing is required for positive identification.

Most antimicrobial drugs, including penicillin and erythromycin, inhibit *Listeria,* though the bacterium is resistant to tetracycline and trimethoprim. Prevention of infections is difficult because the organism is ubiquitous. Individuals at greatest risk should avoid undercooked vegetables, unpasteurized milk, undercooked meat, and all soft cheeses such as feta, Brie, and Camembert. Aged cheeses made from raw milk should be particularly avoided because *Listeria* can grow during refrigeration. Further, people at a high risk of infection should thoroughly cook raw meat and heat prepared meats—including cold cuts, luncheon meats, and hot dogs—before consuming them.

Mycoplasmas

Learning Objectives

- ✓ List at least four characteristics of all mycoplasmas.
- ✓ Explain why mycoplasmas have been classified with both Gram-negative and Gram-positive organisms.
- ✓ Compare and contrast mycoplasmas and viruses.

Mycoplasmas (mī′kō-plaz-măz) are unique bacteria named for their most common representative, the genus *Mycoplasma.* Bacteria in the genus *Ureaplasma* (yū-rē′ă-plaz′mă) are also part of this group. Mycoplasmas lack cytochromes (which are present in many other organisms' electron transport chains), enzymes of the Krebs cycle, and cell walls—indeed, they are incapable of synthesizing peptidoglycan and its precursors.

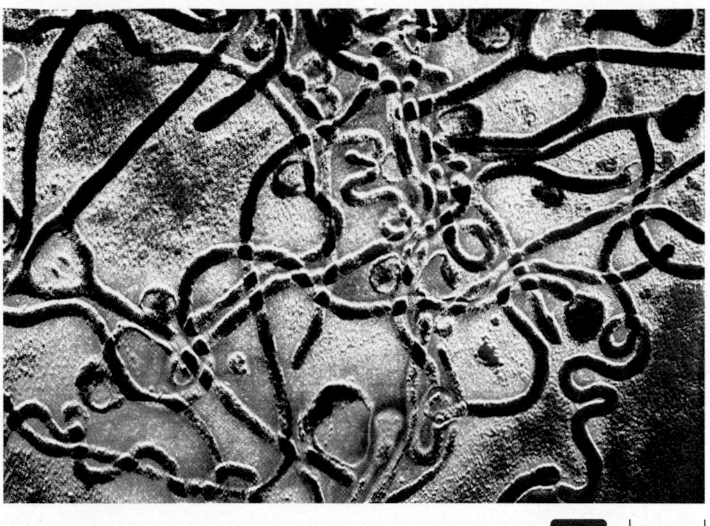

▲ **Figure 19.19 Pleomorphic forms of *Mycoplasma*.** These bacteria can be either coccoid or filamentous because they lack a cell wall.

Moreover, most mycoplasmas have *sterols* in their cytoplasmic membranes, a feature lacking in other prokaryotes. They are distinct from all other organisms in that the codon UGA neither serves as a stop codon nor codes for selenocysteine, but instead codes for the amino acid tryptophan.

Before analysis of the nucleic acid sequences of mycoplasmal rRNA revealed that they are genetically Gram-positive organisms, mycoplasmas were classified in a separate phylum of Gram-negative bacteria—phylum Mollicutes. Modern taxonomists and the second edition of *Bergey's Manual of Systematic Bacteriology* now categorize organisms in Mollicutes as a class of low G + C, Gram-positive bacteria in the phylum Firmicutes. Nevertheless, they appear pink when stained with the Gram stain (because they have no cell walls).

Because they lack cell walls, mycoplasmas[22] are pleomorphic, taking on a variety of shapes, including cocci and thin, unicellular filaments up to 150 μm long that resemble the hyphae of fungi (**Figure 19.19**). Mycoplasmas are able to withstand osmotic stress despite having no walls because they colonize osmotically protected habitats such as animal and human bodies, and the sterols in their membranes convey sufficient strength and rigidity.

Mycoplasmas have diameters ranging from 0.1 μm to 0.8 μm, making them the smallest *free-living* microbes—that is, those that can grow and reproduce independently of other cells. Originally, many mycoplasmas were thought to be viruses because their small, flexible cells enabled them to squeeze through the 0.45-μm pores of filters that were at one time used to remove bacteria from solutions; however, mycoplasmas contain both functional RNA and DNA, and they divide by binary fission—traits that viruses lack.

[22]From Greek *mycos*, meaning fungus, and *plasma*, meaning anything formed or molded.

Mycoplasmas require organic growth factors, including cholesterol, fatty acids, vitamins, amino acids, and nucleotides; these factors must either be acquired from a host or supplied in laboratory media. With exception of the strictly aerobic *Mycoplasma pneumoniae* (nū-mō'nē-ī), mycoplasmas are facultative anaerobes.

Most mycoplasmas form distinctive colonies on solid media. The colonies are so small that often they must be viewed through a microscope, and the colonies of most species resemble fried eggs (because cells at the center of colonies tend to grow into the agar, whereas those at the periphery remain on the surface of the medium; see Figure 11.15). However, two important pathogenic mycoplasmas, *M. pneumoniae* and *Ureaplasma*, are exceptions in that their colonies do not resemble fried eggs.

Mycoplasmas can colonize mucous membranes of the respiratory and urinary tracts and are associated with pneumonia and urinary tract infections. Of the two genera that cause diseases in humans, *Mycoplasma* is unable to utilize urea (that is, it is urease negative), whereas *Ureaplasma* does hydrolyze urea to form ammonia.

Over 100 species of *Mycoplasma* have been identified, but only a few cause significant diseases in humans. We begin our examination of these species with *M. pneumoniae*.

Mycoplasma pneumoniae

Learning Objective

✓ Describe the damage done to respiratory epithelial cells by *Mycoplasma pneumoniae*.

Pathogenesis and Epidemiology

Mycoplasma pneumoniae **(Figure 19.20)** has an adhesive protein that attaches specifically to receptors located at the bases of cilia on epithelial cells lining the respiratory tracts of humans. Attachment causes the cilia to stop beating, and mycoplasmal colonization eventually kills the epithelial cells. This interrupts the normal removal of mucus from the respiratory tract, allowing colonization by other bacteria and causing a buildup of mucus that irritates the upper respiratory tract. Early symptoms of *M. pneumoniae* infections—including fever, malaise, headache, and sore throat—are not typical of other types of pneumonia; thus the disease is called **primary atypical pneumonia.** The body subsequently responds with a persistent, unproductive cough in an attempt to clear the lungs.

Primary atypical pneumonia may last for several weeks, but it is usually not severe enough to require hospitalization or to cause death. Because symptoms can be mild, the disease is also sometimes called *walking pneumonia.*

Nasal secretions spread *M. pneumoniae* among people in close contact, such as classmates and family members. The disease appears to be uncommon in children under age 5 or in adults older than 20, though it is probably the most common form of pneumonia seen in children 5–15 years old. However, because primary atypical pneumonia is not a reportable disease and is difficult to diagnose, the actual incidence of infection is unknown.

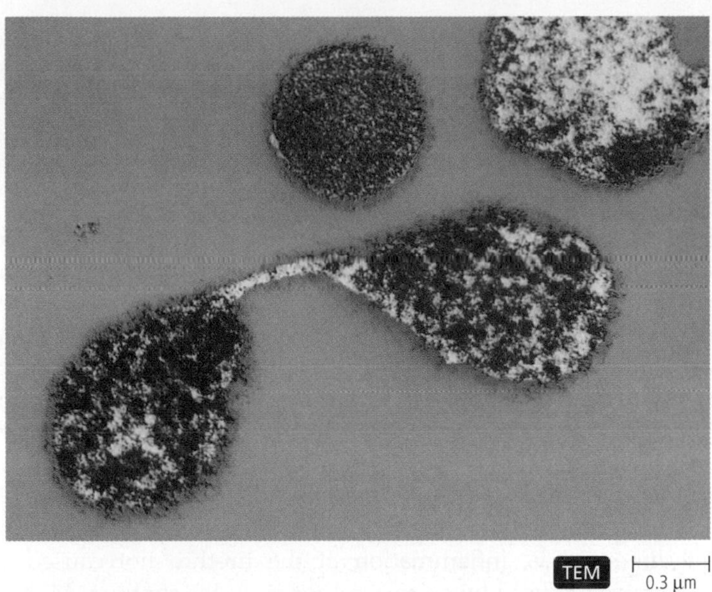

▲ **Figure 19.20** *Mycoplasma pneumoniae.*

Primary atypical pneumonia occurs throughout the year. This lack of seasonality is in contrast to pneumococcal pneumonia (caused by *Streptococcus pneumoniae*), which is more commonly seen in the fall and winter.

Diagnosis, Treatment, and Prevention

Diagnosis is difficult because mycoplasmas are small and difficult to detect in clinical specimens or tissue samples. Further, mycoplasmas grow slowly in culture, requiring two to six weeks before colonies can be seen. As previously noted, the colonies of *M. pneumoniae* differ from the colonies of other mycoplasmas in lacking a fried egg appearance; instead, the colonies have a uniform granular appearance. Complement fixation, hemagglutination, and immunofluorescent tests are sometimes used to confirm a diagnosis, but such tests are nonspecific and are not by themselves positively diagnostic.

Physicians treat primary atypical pneumonia with erythromycin or tetracycline. Prevention is difficult because patients are often infectious for long periods of time without signs or symptoms, and they remain infectious even while undergoing antimicrobial treatment. Nevertheless, frequent hand antisepsis, avoidance of contaminated fomites, and reducing aerosol dispersion can limit the spread of the pathogen and the number of cases of disease. No vaccine against *M. pneumoniae* is available.

CRITICAL **THINKING**

Mycoplasma pneumoniae, like many pathogenic bacteria, is resistant to penicillin; however, unlike most resistant species, *Mycoplasma* does not synthesize β-lactamase. Explain why *Mycoplasma* is resistant to penicillin.

Other Mycoplasmas

Learning Objectives

- ✓ List three mycoplasmas associated with urinary and genital tract infections.
- ✓ Describe pelvic inflammatory disease.

Three other mycoplasmas are associated with diseases of humans: *M. hominis* (ho′mi-nis), *M. genitalium* (jen-ē-tal′ē-ŭm), and *Ureaplasma urealyticum* (ū-rē′ă-li′ti-kŭm) often colonize the urinary and genital tracts of newborn girls, though such infections rarely persist through childhood. However, the incidence of genital reinfections increases after puberty as a result of sexual activity; approximately 15% of sexually active adult Americans (males and females) are currently infected with *M. hominis,* and 70% are infected with *Ureaplasma.*

M. genitalium and *U. urealyticum* cause *nongonococcal urethritis*—that is, inflammation of the urethra not caused by *Neisseria gonorrhoeae* (nī-se′rē-ă go-nor-rē′ī). By contrast, *M. hominis* can cause **pelvic inflammatory disease (PID)** in women. Pelvic inflammatory disease is characterized by inflammation of the organs in the pelvic cavity, fever, and abdominal pain.

Infections of either *M. genitalium* or *U. urealyticum* are treated with erythromycin or tetracycline, whereas clindamycin is used to treat *M. hominis,* which is commonly resistant to tetracycline and erythromycin. Abstinence, mutually faithful monogamy, and proper use of condoms prevent the spread of these sexually transmitted organisms.

CRITICAL **THINKING**

Why do pediatricians refrain from using tetracycline to treat mycoplasmal infections in children?

We have examined low G + C pathogenic, Gram-positive cocci; endospore-forming bacilli; non-endospore-forming, rod-shaped *Listeria;* and wall-less mycoplasmas. Now we turn our attention to the high G + C, Gram-positive pathogens, beginning with *Corynebacterium.*

Corynebacterium

Learning Objectives

- ✓ Characterize the arrangements of *Corynebacterium* cells.
- ✓ Describe the transmission of *Corynebacterium diphtheriae* and the effect of diphtheria toxin.
- ✓ Discuss the diagnosis, treatment, and prevention of diphtheria.

Corynebacterium (kŏ-rī′nē-bak-tēr′ē-ŭm) is a genus of high G + C, pleomorphic, non-endospore-forming bacteria that are ubiquitous on plants and in animals and humans, where they colonize the skin and the respiratory, gastrointestinal, urinary, and genital tracts. The bacteria divide via a type of binary fission called *snapping division,* in which daughter cells remain attached to form characteristic V-shapes and side-by-side *palisade* arrangements **(Figure 19.21)**. Another distinctive feature is that methylene blue or toluidine blue dyes color the cytoplasm of corynebacteria differentially, revealing storage granules called *metachromatic granules.* Although all species of corynebacteria can be pathogenic, the agent of diphtheria is most widely known.

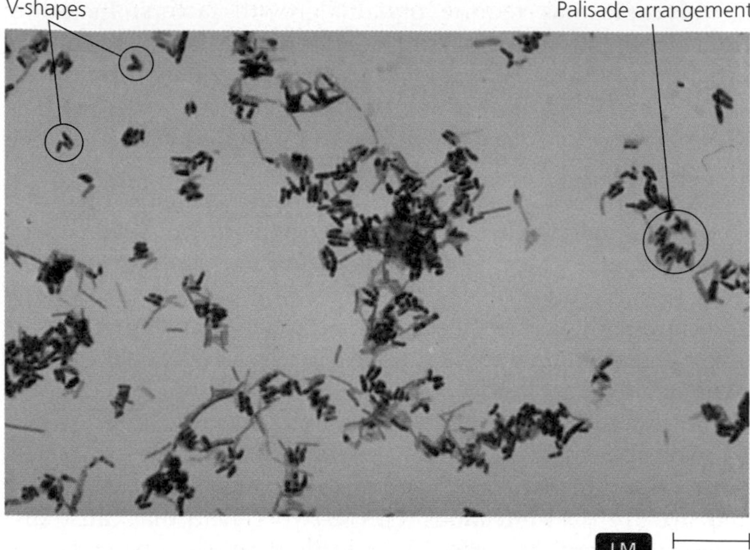

▲ **Figure 19.21 Gram-stained *Corynebacterium diphtheriae.*** The characteristic arrangement of the cells results from the type of binary fission called snapping division.

Pathogenesis, Epidemiology, and Disease

Corynebacterium diphtheriae[23] (dif-thi′rē-ī) is transmitted from person to person via respiratory droplets or skin contact. Diphtheria is endemic in poorer parts of the world that lack adequate immunization. The bacterium normally contains a lysogenic bacteriophage that codes for *diphtheria toxin,* which is directly responsible for the signs and symptoms of diphtheria. Cells lacking the phage are not pathogenic.

Like many bacterial toxins, diphtheria toxin consists of two polypeptides. One polypeptide binds to a growth factor receptor found on many types of human cells, triggering endocytosis of the toxin by these cells. Once inside a cell, proteolytic enzymes cleave the toxin, releasing the second polypeptide into the cytosol. This portion of the toxin enzymatically destroys *elongation factor,* a protein required for synthesis of polypeptides in eukaryotes. Because the action of the toxin is enzymatic, a single molecule of toxin sequentially destroys every molecule of elongation factor in the cell, completely blocking all polypeptide synthesis and resulting in cell death. Diphtheria toxin is thus one of the more potent toxins known.

Infections with *C. diphtheriae* have different effects depending on the host's immune status and the site of infection. Infections in

[23]From Greek *koryne,* meaning club; *bakterion,* meaning small rod; and *diphthera,* meaning leather membrane.

immune individuals are asymptomatic, whereas infections in partially immune individuals result in a mild respiratory disease. Respiratory infections of nonimmune patients are most severe, resulting in the sudden and rapid signs and symptoms of **diphtheria** (dif-thē-rē'ă), including sore throat, localized pain, fever, pharyngitis, and the oozing of a fluid composed of intracellular fluid, blood clotting factors, leukocytes, bacteria, and the remains of dead cells lining the larynx and pharynx. The fluid thickens into a *pseudomembrane* **(Figure 19.22)** that can adhere so tightly to the tonsils, uvula, palate, pharynx, and larynx that it cannot be dislodged without causing bleeding in the underlying tissue. In severe cases the pseudomembrane completely occludes the respiratory passages, resulting in death by suffocation.

In a similar way, cutaneous diphtheria causes cell death and the formation of a pseudomembrane on the skin. With both cutaneous and respiratory infections, toxin absorbed into the blood can kill heart cells and nerve cells throughout the body, resulting in cardiac arrhythmia, coma, and death of 5–10% of patients.

Diagnosis, Treatment, and Prevention

Initial diagnosis is based on the presence of a pseudomembrane. Laboratory examination of the membrane or of tissue collected from the site of infection does not always reveal bacterial cells because the effects are due largely to the action of diphtheria toxin and not the cells directly. Culture of specimens on *Loffler's medium,* which was developed especially for the culture of *C. diphtheriae,* produces three colonial morphologies. Some colonies are large, irregular, and gray; others are small, flat, and gray; and still others are small, round, convex, and black. Even though observations of these distinctive colonies are useful, absolute certainty of diagnosis results from an immunodiffusion assay, called an **Elek test,** in which antibodies against the toxin react with toxin in a sample of fluid from the patient.

The most important aspect of treatment is the administration of antitoxin (immunoglobulins against the toxin) to neutralize toxin before it binds to cells; once the toxin binds to a cell, it enters via endocytosis and kills the cell. Penicillin or erythromycin kills the bacterium, preventing the synthesis of more toxin. In severe cases, a blocked airway must be opened surgically or bypassed with a tracheostomy tube.

Because humans are the only known host for *C. diphtheriae,* the most effective way to prevent diphtheria is immunization. Before immunization, hundreds of thousands of cases occurred in the United States each year; in contrast, only 15 cases total were reported between 1994 and 2009. Toxoid (deactivated toxin) is administered as part of the DTaP vaccine, which combines diphtheria and tetanus toxoids with antigens of the pertussis bacterium, at 2, 4, 6, 18 and 60 months of age, followed by booster immunizations with a slightly different vaccine (Tdap) every 10 years.

CRITICAL **THINKING**

Why must diphtheria and tetanus vaccines be boosted every 10 years?

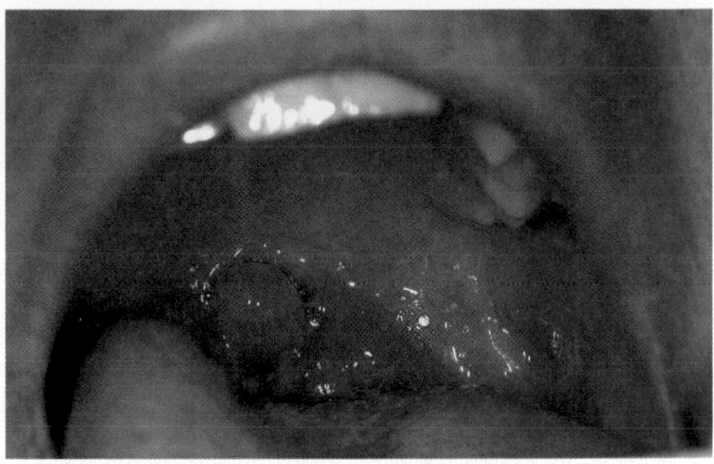

▲ **Figure 19.22 A pseudomembrane.** This feature is characteristic of diphtheria.

Mycobacterium

Learning Objective

✓ Characterize mycobacteria in terms of endospore formation, cell wall composition, growth rate, and resistance to antimicrobial drugs.

Another devastating high G + C, non-endospore-forming pathogen is *Mycobacterium* (mī-kō-bak-tēr'ē-ŭm). Species in this genus have cell walls containing an abundance of a waxy lipid, called **mycolic** (mī-kol'ik) **acid,** that is composed of chains of 60–90 carbon atoms. This unusual cell wall is directly responsible for the unique characteristics of this pathogen. Specifically, mycobacteria:

- Grow slowly (because of the time required to synthesize numerous molecules of mycolic acid). The generation time varies from hours to several days.
- Are protected from lysis once they are phagocytized.
- Are capable of intracellular growth.
- Are resistant to Gram staining, detergents, many common antimicrobial drugs, and desiccation. Because mycobacteria stain only weakly with the Gram procedure (if at all), the acid-fast staining procedure was developed to differentially stain mycobacteria (see Figure 4.18).

Even though almost 75 species of mycobacteria are known, most mycobacterial diseases in humans are caused by two species: *M. tuberculosis,* and *M. leprae,* which cause tuberculosis and leprosy, respectively. *M. avium-intracellulare* and *M. ulcerans* cause emerging mycobacterial diseases.

Tuberculosis

Learning Objectives

✓ Identify two effects of cord factor of *Mycobacterium tuberculosis.*

✓ Describe the transmission of *M. tuberculosis,* and its subsequent action within the human body.

✓ Discuss the diagnosis, treatment, and prevention of tuberculosis.

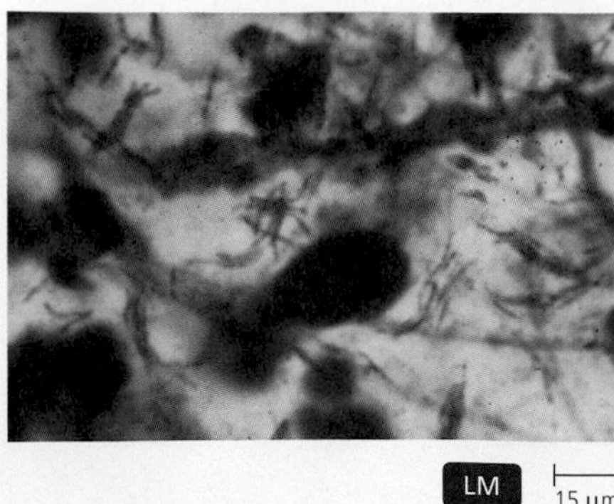

▲ **Figure 19.23** ***Mycobacterium tuberculosis.*** The pink appearance of the bacteria when prepared with an acid-fast stain. Note the corded growth (parallel alignments) of daughter cells, a result of the presence of cord factor in the cell walls of virulent strains.

Tuberculosis (TB), the primary mycobacterial disease, is fundamentally a respiratory disease caused by *Mycobacterium tuberculosis* (too-ber-kyū-lō′sis). This bacterium forms dull-yellow raised colonies after growth for weeks on a special differential medium called Lowenstein-Jensen agar. Virulent strains of *M. tuberculosis* have a cell wall component, called **cord factor,** that produces strands of daughter cells that remain attached to one another in parallel alignments **(Figure 19.23)**. Cord factor also inhibits migration of neutrophils and is toxic to mammalian cells. Mutant mycobacteria that are unable to synthesize cord factor do not cause disease.

Pathogenesis and Disease

The waxy wall protects *Mycobacterium* from desiccation, and it can remain viable in dried aerosol droplets for eight months. The pathogen is not particularly virulent, as only about 5% of people infected with the bacterium develop disease; however, it kills about 50% of untreated diseased patients. Clinicians divide tuberculosis into three types; *primary TB, secondary* (or *reactivated*) *TB,* and *disseminated TB.*

Primary Tuberculosis Although *M. tuberculosis* can infect any organ, 85% of infections remain in the lungs. Primary infection, which typically occurs in children, involves the formation of small, hard nodules in the lungs called **tubercles** (too′ber-klz), which are characteristic of TB and give it its name. The five stages of a primary infection are as follows **(Figure 19.24a)**:

1. *Mycobacterium* typically infects the respiratory tract via inhalation of respiratory droplets formed when infected individuals talk, sing, cough, or sneeze. A respiratory droplet is about 5 mm in diameter and carries 1–3 bacilli. The minimum infectious dose is about 10 cells. *Mycobacterium* has adhesive pili that attach to an extracellular human protein, laminin.
2. Macrophages in the alveoli (air sacs) of the lungs phagocytize the pathogens but are unable to digest them in part because the mycobacteria prevent fusion of lysosomes with phagosomes. *M. tuberculosis* also invades cells lining the alveoli.
3. The bacteria replicate freely within host cells, gradually killing them. Infected cells of the alveolar lining release chemokines that attract more macrophages. Bacteria released from dead macrophages are phagocytized by other macrophages, beginning the cycle anew. This stage of infection, which lasts for a few weeks, is typically asymptomatic or associated with a mild fever.
4. Infected macrophages present antigen to T lymphocytes, which produce lymphokines that attract and activate more macrophages and trigger inflammation. Tightly appressed macrophages surround the site of infection, forming a tubercle.
5. Other cells of the body deposit collagen fibers, enclosing infected macrophages and lung cells within the tubercle. Infected cells in the center of the tubercle die, releasing *M. tuberculosis* and producing *caseous*[24] *necrosis*—the death of tissue that takes on a cheeselike consistency due to the presence of protein and fat released from dying cells. Sometimes, for an unknown reason, the center liquefies and subsequently becomes filled with air. Such a tubercle is called a *tuberculous cavity.*

In most patients the immune system reaches a stalemate with the bacterium at this juncture: The immune system is able to prevent further spread of the pathogen and stop the progression of the disease, but it is not able to rid the body of all mycobacteria. *M. tuberculosis* may remain dormant for decades within macrophages and in the centers of tubercles. If the immune system breaks the stalemate by killing all the mycobacteria, the body deposits calcium around the tubercles, which are then called *Ghon complexes.*

Secondary or Reactivated Tuberculosis Secondary tuberculosis results when *M. tuberculosis* breaks the stalemate, ruptures the tubercule, and reestablishes an active infection in which the bacteria spread through the lungs via the bronchioles **(Figure 19.24b)**. Reactivated TB is a common occurrence in TB-infected individuals with suppressed immune systems.

Disseminated Tuberculosis Disseminated TB results when some macrophages carry the pathogens via the blood and lymph to a variety of sites, including the bone marrow, spleen, kidneys, spinal cord, and brain. The signs and symptoms observed in disseminated TB correspond to complications arising at the various sites involved. The common name for TB in the early 1900s—*consumption*—reflects the wasting away of the body resulting from the involvement of multiple sites in disseminated TB.

Epidemiology

Even though tuberculosis is on the decline in the United States, it is pandemic in other parts of the world, killing almost 2 million people annually. Over one-third of the world's population is infected with *M. tuberculosis,* and 10% of them develop a life-threatening case of tuberculosis. Patients with lowered immunity

[24]From Latin *caseus,* meaning cheese.

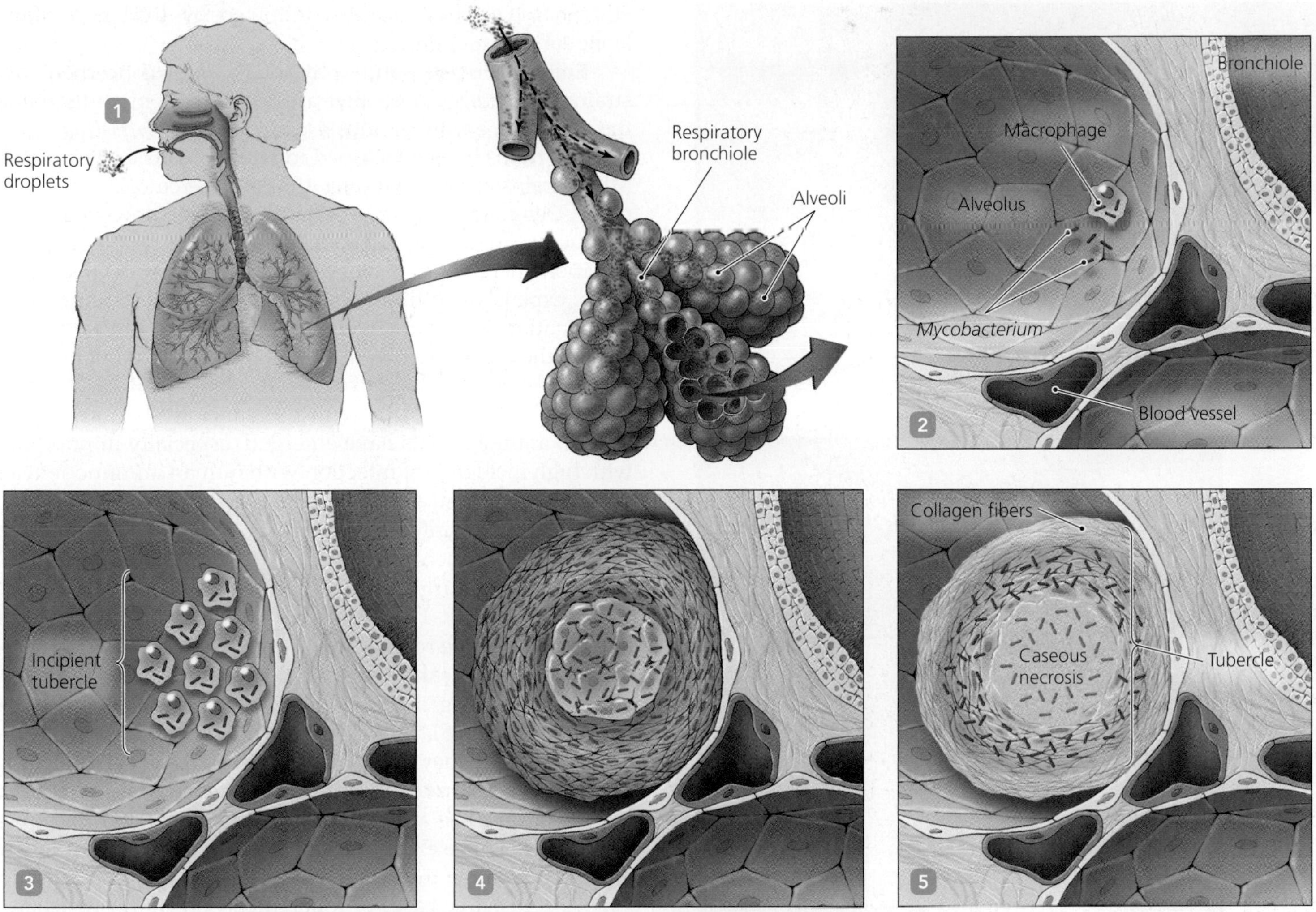

(a) Primary tuberculosis infection

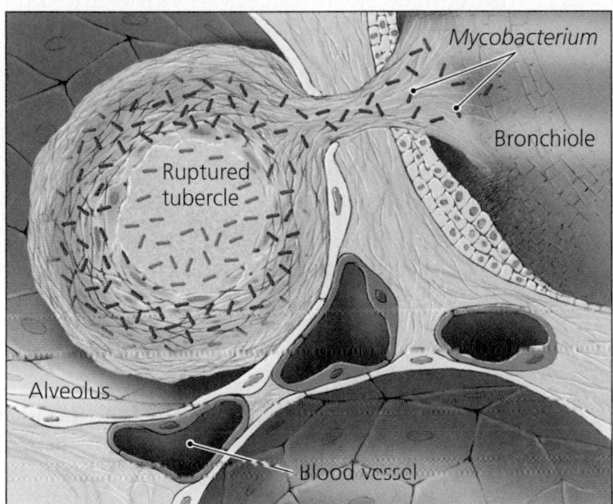

(b) Secondary or reactivated tuberculosis

▲ **Figure 19.24 Development of tuberculosis in the lungs.** **(a)** The formation of tubercules in primary tuberculosis. **(b)** Events in secondary (reactivated) tuberculosis.

are at the greatest risk of infection; other risk factors include diabetes, poor nutrition, stress, crowded living conditions, alcohol and drug abuse, and smoking.

TB is prevalent in the countries of the former Soviet Union, where the breakdown of essential medical services has allowed a resurgence, and in Africa, where TB is the number one killer of AIDS patients. Someone in the world is infected with *M. tuberculosis* every second.

Diagnosis, Treatment, and Prevention

A **tuberculin skin test** is used to screen patients for possible exposure to tuberculosis. In this test, a health care worker injects about 0.1 ml of cell wall antigens from *M. tuberculosis* into a patient's skin. The appearance of a hard, red swelling at the test site within 24 to 72 hours is a positive test **(Figure 19.25a)**. The reaction is a type IV (cell-mediated) hypersensitivity response, indicating past infection or vaccination but not necessarily current disease.

A positive tuberculin skin test is not sufficient to distinguish patients who have been exposed to antigen but are currently

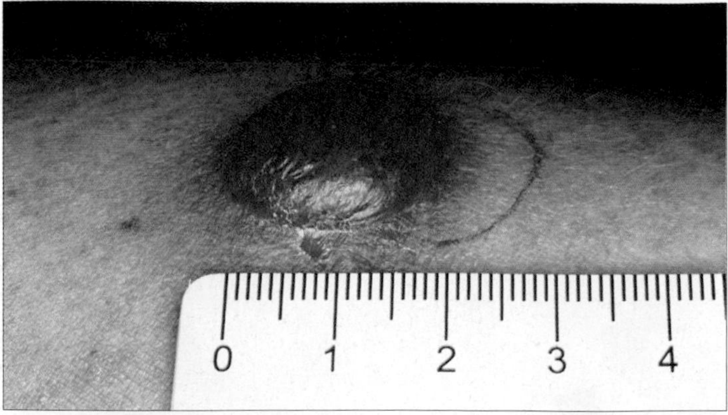

(a)

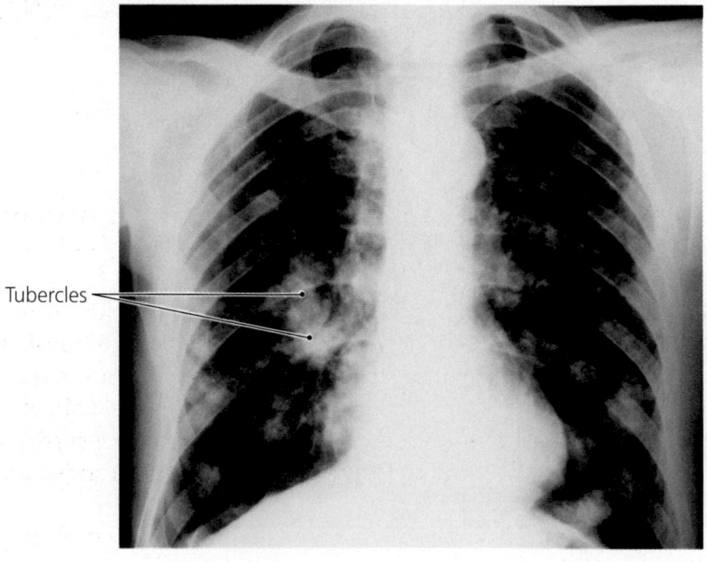

(b)

▲ **Figure 19.25 Diagnosis of tuberculosis.** **(a)** The enlarged, reddened, and raised inoculation site of a positive tuberculin skin test. **(b)** An X-ray of a tuberculosis patient showing white patches indicative of tubercules. *What does a positive tuberculin test indicate?*

Figure 19.25 *A positive tuberculin skin test indicates that a patient has cell-mediated memory against antigens of M. tuberculosis, as a result of either infection or immunization.*

uninfected, chronic carriers from patients with active disease. Chest X-rays can reveal the presence of tubercles in the lungs **(Figure 19.25b)**; primary tuberculosis appears as tubercles in the lower and central areas of the lungs, whereas secondary tuberculosis more commonly appears higher in the lungs. The presence of acid-fast cells and cords in sputum (see Figure 19.23b) confirms an active case of tuberculosis.

Common antimicrobials such as penicillin and erythromycin have little effect on *Mycobacterium tuberculosis,* because it grows so slowly that the drugs are cleared from the body by the liver and kidneys before they have any significant effect. Further, antimicrobial drugs have little impact on *M. tuberculosis* living within macrophages. The currently recommended treatment is a combination of isoniazid (INH), rifampin, pyrazinamide, and ethambutol for two months, followed by INH and rifampin alone for four months.

Epidemiologists and physicians are concerned about strains of *M. tuberculosis* that are resistant to more than one of the standard antituberculosis drugs. Such *multi-drug resistant (MDR)* strains have developed in many countries. For example, MDR strains account for over 10% of tuberculosis cases in Germany, Denmark, and New Zealand. No one knows the extent of MDR cases in developing countries, but it is suspected that the situation there is even worse. MDR-TB must be treated with more expensive antimicrobials, such as fluoroquinolones in combination with kanamycin, for as long as two years. This increases the cost of treatment as much as 100-fold over treatment of drug-sensitive TB.

And now, TB strains with resistance to the most effective anti-TB antimicrobials have emerged, especially in populations with high incidence of infection with human immunodeficiency virus (HIV). These strains, called *extensively drug resistant TB (XDR-TB),* may infect as many as 30,000 people per year in 41 countries, according to the World Health Organization. Researchers are urgently seeking new antimicrobials to treat MDR-TB and XDR-TB lest the progress made toward TB control and eradication be reversed.

The World Health Organization and the CDC recommend a strategy of drug delivery called *Directly Observed Treatment, Shortcourse (DOTS),* in which health care workers observe patients to ensure they take their medications on schedule. Obviously, DOTS treatment is quite labor intensive and expensive.

Physicians use antibacterial drugs prophylactically to treat patients who have either shown a recent conversion from a negative to a positive tuberculin skin test or undergone significant exposure to active cases of tuberculosis. In countries where tuberculosis is common, health care workers immunize patients with *BCG*[25] vaccine, which is composed of attenuated *M. bovis* (bō′vis), a species that causes tuberculosis in cattle and is only rarely transmitted to humans via contaminated milk. The vaccine is not used for immunocompromised patients because it can cause disease.

Studies on the efficacy of TB vaccine vary widely from 80% protected to no protection at all. Nevertheless, the vaccine may reduce the spread of disease, and the World Health Organization (WHO) recommends all children be vaccinated at birth. In the United States, the cost of mass immunization is not warranted because of the relatively low prevalence of tuberculosis. Further, immunized patients may have a positive skin reaction for the rest of their lives, even if they have not been infected with *M. tuberculosis,* and such "false positive" results would hinder the work of epidemiologists trying to track the spread of the disease.

The prediction that tuberculosis would be eliminated in the United States by 2000 was not realized because of a resurgence of tuberculosis among AIDS patients, and its subsequent spread to drug abusers and the homeless. Nevertheless, both renewed efforts to detect and stop infection and the implementation of

[25]For bacillus of Calmette and Guérin, named after the two French developers of the vaccine.

DOTS reduced the number of reported U.S. cases in 2008 to the lowest incidence since reporting began in 1953.

CRITICAL **THINKING**

Why don't physicians try to prevent the spread of TB by simply administering prophylactic antimicrobial drugs to everyone living in endemic areas?

Leprosy

Learning Objectives

- ✓ Compare and contrast tuberculoid leprosy with lepromatous leprosy.
- ✓ Discuss the diagnosis, treatment, and prevention of leprosy.

Mycobacterium leprae (lep′rī) causes **leprosy** (lep′rō-sē)—which is also called by the less dreaded name *Hansen's disease,* after Gerhard Hansen (1841–1912), a Norwegian bacteriologist who discovered its cause in 1873. *M. leprae* is a high G + C, Gram-positive bacillus. Because of the abundance of mycolic acid in the cell wall, bacilli do not Gram stain purple and must instead be stained with an acid-fast stain. *M. leprae* grows best at 30°C, showing a preference for cooler regions of the human body, particularly peripheral nerve endings and skin cells in the fingers, toes, lips, and earlobes. The bacterium does not grow in cell-free laboratory culture, a fact that has hindered research and diagnostic studies. Armadillos, which have a normal body temperature of 30°C, are its only other host and have proven valuable in studies on leprosy and on the efficacy of leprosy treatments.

Pathogenesis, Epidemiology, and Disease

Leprosy has two different manifestations depending on the immune response of the patient: Patients with a strong cell-mediated immune response are able to kill cells infected with the bacterium, resulting in a nonprogressive form of the disease called *tuberculoid leprosy.* Regions of the skin that have lost sensation as a result of nerve damage are characteristic of this form of leprosy.

By contrast, patients with a weak cell-mediated immune response develop *lepromatous* (lep-rō-mă-tŭs) *leprosy* **(Figure 19.26)**, in which bacteria multiply in skin and nerve cells, gradually destroying tissue and leading to the progressive loss of facial features, digits (fingers and toes), and other body structures. Development of signs and symptoms is very slow; incubation may take years before the disease is evident. Death from leprosy is rare and usually results from the infection of leprous lesions by other pathogens.

Lepromatous leprosy is the more virulent form of the disease, but fortunately it is becoming relatively rare: In the 1990s about 12 million cases were diagnosed annually worldwide, but in 2007 this number had decreased to 254,525.

Leprosy is transmitted via person-to-person contact. Given that the nasal secretions of patients with lepromatous leprosy are loaded with mycobacteria, infection presumably occurs via inhalation of respiratory droplets. Alternatively, infection may occur through breaks in the skin. In any case, leprosy is not particularly virulent; individuals are typically infected only after years of intimate social contact with a victim.

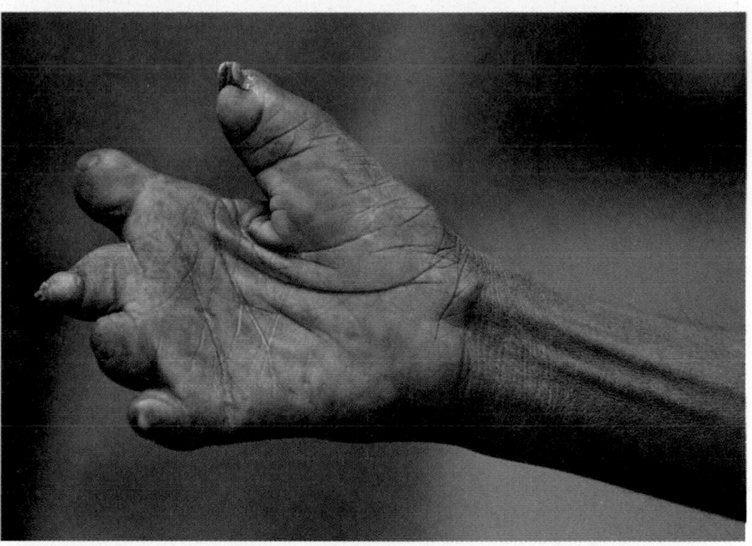

▲ **Figure 19.26** Lepromatous leprosy can result in severe deformities.

Why, then, does a diagnosis of leprosy elicit such dread? Much of the public's aversion to leprosy stems from Biblical references calling for the quarantine of lepers and from the horrible disfigurement lepers can suffer. During the Middle Ages lepers were often required to wear bells to warn passersby of their presence and forced to live in special villages or in isolated areas of the forests.

The 12th-century English king Henry II took lepraphobia (fear of leprosy and lepers) a step further; he ordered that those with the disease be burned at the stake. Not to be outdone, his great-grandson Edward I ordained that lepers be accorded a complete funeral service—before they were led to the cemetery and buried alive. Patients with leprosy are no longer quarantined because the disease is so rarely transmitted and is fully treatable.

Diagnosis, Treatment, and Prevention

Diagnosis of leprosy is based on signs and symptoms of disease—a loss of sensation in skin lesions in the case of tuberculoid leprosy and disfigurement in the case of lepromatous leprosy. Diagnosis is confirmed by a positive skin test with leprosy antigen (similar to the tuberculin skin test) or through direct observation of acid-fast rods (AFRs) in tissue samples or nasal secretions (in the case of lepromatous leprosy).

As with *M. tuberculosis, M. leprae* quickly develops resistance to single antimicrobial agents, so therapy consists of administering multiple drugs, such as clofazimine, rifampin, or dapsone, for 12 months, though treatment can be lifelong for some patients.

BCG vaccine provides some protection against leprosy (as well as against tuberculosis), but prevention is primarily achieved by limiting exposure to the pathogen and by the prophylactic

EMERGING DISEASES

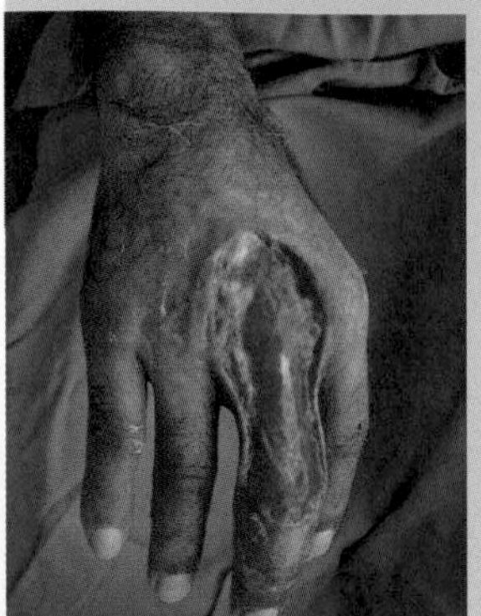

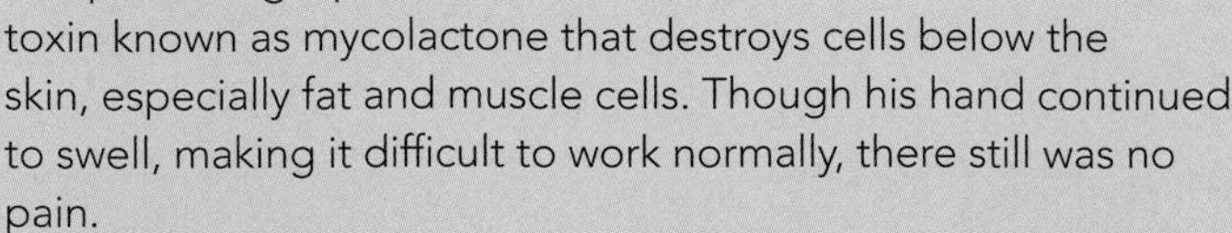

BURULI ULCER

Jacques liked living in the Democratic Republic of the Congo (DRC)—opportunities abounded for exploration, adventure, and wildlife photography, the countryside was beautiful, and he found the people generally friendly. It was on a photographic excursion to the east that Jacques met a not-so-friendly resident of the DRC, an emerging mycobacterial pathogen.

The photographer thought little of the small scratch he received while documenting wildlife in the swamps along the great Congo River, but he should have been concerned; *Mycobacterium ulcerans* had found a new home in his hand. Jacques would pay a grievous price for his lack of circumspection.

He continued to ignore the infection when it produced a small, painless nodule. He even ignored it when his finger swelled to twice its normal size; it was painless and he could still meet his busy schedule. But the bacterium was producing a potent toxin known as mycolactone that destroys cells below the skin, especially fat and muscle cells. Though his hand continued to swell, making it difficult to work normally, there still was no pain.

After six weeks of this condition, pain began suddenly and excruciatingly. The swollen finger ruptured, and a foul-smelling fluid saturated his camera. It was time to see a doctor.

The physician diagnosed Buruli ulcer, an emerging disease that affects more people each year as a result of human encroachment into the swamps where *M. ulcerans* lives. After two surgeries to remove dead tissue and the focus of the infection, several skin grafts, and two months of treatment with the antimicrobial drugs rifampicin and streptomycin, Jacques was released with scars that forever remind him of his adventure with *Mycobacterium ulcerans*.

Track Buruli ulcer online by going to the Study Area at www.masteringmicrobiology.com.

use of antimicrobial agents when exposure occurs. The World Health Organization (WHO) has set a goal to reduce the prevalence of leprosy below one case per 10,000 population. WHO epidemiologists predict that such a low prevalence will permanently interfere with the spread of the bacterium, and leprosy will be eliminated.

Other Mycobacterial Infections

Learning Objective

✓ Explain why the incidence of infections with *Mycobacterium avium-intracellulare*, long thought to be harmless to humans, has been increasing.

Mycobacterium avium-intracellulare (ā'vē-ŭm in'tra-sel-yu-la'rē) which is commonly found in soil, water, and food, was long thought to be a harmless occasional member of the respiratory microbiota. However, the advent of the AIDS epidemic has unmasked it as an important opportunistic pathogen and the cause of an emerging disease, which is the most common mycobacterial infection among AIDS patients in the United States. (*M. tuberculosis* is more common in countries where tuberculosis is epidemic.) Infections are believed to result from the ingestion of contaminated food or water; direct person-to-person transmission does not occur. The bacterium spreads throughout the body via the lymph.

In contrast to infections with other mycobacteria, *M. avium-intracellulare* simultaneously affects almost every organ of the body. In some tissues, every cell is packed with mycobacteria, and during the terminal stages of AIDS the blood is often filled with thousands of bacteria per milliliter. The disease remains asymptomatic until organ failure occurs on a massive scale.

Treatment consists of trial-and-error administration of antimicrobial agents; the disseminated nature of the infection often makes treatment ineffectual.

Emerging Diseases: Buruli Ulcer examines another disease of mycobacteria.

Propionibacterium

Learning Objectives

✓ Identify the species of *Propionibacterium* most commonly involved in infections of humans.

✓ Explain the role of *P. acnes* in the formation of acne.

Propionibacteria are small, Gram-positive rods commonly found growing on the skin. They are so named because they produce propionic acid as a by-product of the fermentation of carbohydrates. The species most commonly involved in infections of humans is *Propionibacterium acnes* (prō-pē-on-i-bak-tēr'-ē-ŭm ak'nēz), which causes **acne** in 85% of adolescents and young adults. The bacterium can also be an opportunistic pathogen in patients with intrusive medical devices such as catheters, artificial heart valves, artificial joints, and cerebrospinal fluid shunts.

In its role in the development of acne **(Figure 19.27)**, *Propionibacterium* typically grows in the sebaceous (oil) glands of

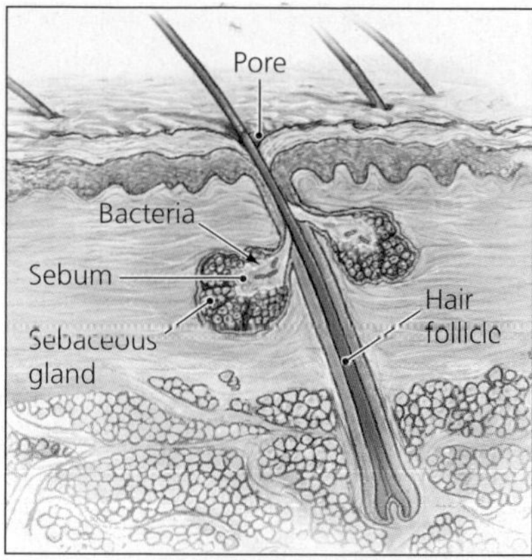

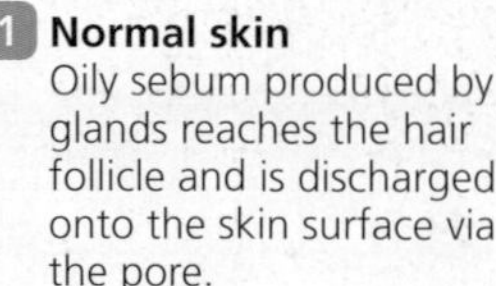

1 Normal skin Oily sebum produced by glands reaches the hair follicle and is discharged onto the skin surface via the pore.

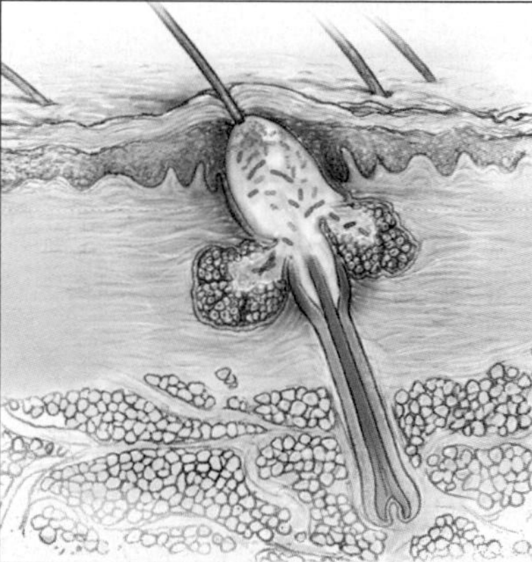

2 Whitehead Inflamed skin swells over the pore when bacteria infect the hair follicle, causing the accumulation of colonizing bacteria and sebum.

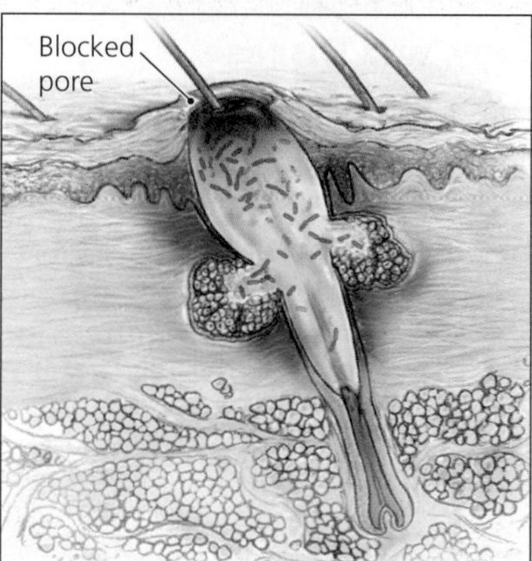

3 Blackhead Dead and dying bacteria and sebum form a blockage of the pore.

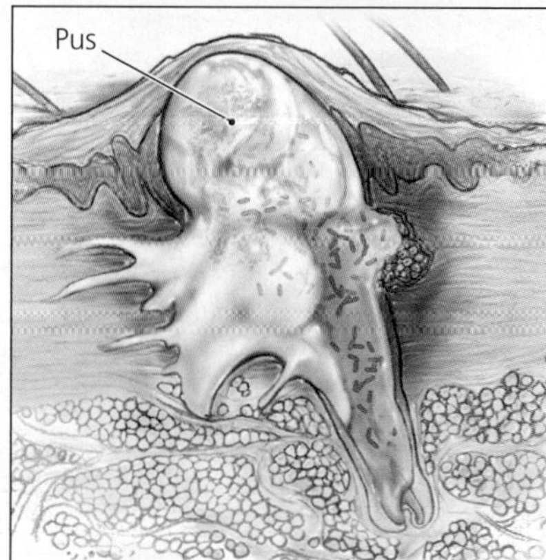

4 Pustule formation Severe inflammation of the hair follicle causes pustule formation and rupture, producing cystic acne, which is often resolved by scar tissue formation.

Figure 19.27 The development of acne. Normally, oily sebum produced by glands reaches the hair follicle and is discharged from the pore and onto the skin surface (1). When inflammation resulting from bacteria infecting the hair follicle causes the skin to swell over the pore, sebum and the colonizing bacteria accumulate to form a whitehead (2). The blockage of the pore by a plug of dead and dying bacteria and sebum produces a blackhead (3). When the inflammation of the follicle becomes severe enough, pustules form (4) and rupture, producing cystic acne, which is often resolved by the formation of scar tissue. *Which bacterial pathogen is a common cause of cystic acne?*

Figure 19.27 Propionibacterium acnes is a common cause of acne.

the skin (1). Excessive oil production triggered by the hormones of adolescence, particularly testosterone in males, stimulates the growth and reproduction of the bacterium, which secretes chemicals that attract leukocytes. The leukocytes phagocytize the bacteria and release chemicals that stimulate local inflammation. The combination of dead bacteria and dead and living leukocytes makes up the white pus associated with the pimples of acne (2). A blackhead is formed when a plug of dead and dying bacteria blocks the gland's pore (3). In *cystic acne,* a particularly severe form of the disease, bacteria form inflamed pustules (cysts) (4) that rupture, triggering the formation of scar tissue.

There are many common misconceptions about acne. Cleaning the surface of the skin is not particularly effective in preventing the development of acne because the bacteria live deep in the sebaceous glands, though frequent cleansing may dry the skin, which helps loosen plugs from hair follicles. Scientists have not shown any connection between acne and diet, including chocolate or oily foods.

In most cases the immune system is able to control *Propionibacterium,* and no treatment is required. Dermatologists may prescribe antimicrobial drugs such as erythromycin and clindamycin, which have proven effective in controlling the bacterium. However, long-term antibiotic use can destroy susceptible normal bacterial microbiota, making the individual more vulnerable to opportunistic fungal and bacterial infections that are resistant to antimicrobial drugs. *Retinoic acid (Accutane),* a derivative of vitamin A, inhibits the formation of oil. Because this drug can cause intestinal bleeding, it is prescribed only for severe cases of acne. Further, it causes birth defects and thus should not be used by pregnant women.

In the previous few sections we have discussed pathogenic, non-endospore-forming, high G + C rods in the genera *Corynebacterium, Mycobacterium,* and *Propionibacterium.* Now we will consider filamentous, high G + C pathogens in the genera *Nocardia* and *Actinomyces.*

Nocardia and *Actinomyces*

Learning Objective

✓ Compare and contrast *Nocardia asteroides* and *Actinomyces* in terms of appearance, cell wall composition, and role in producing disease.

Nocardia and *Actinomyces* have elongated filamentous cells that resemble fungal hyphae. The cell walls of *Nocardia* contain mycolic

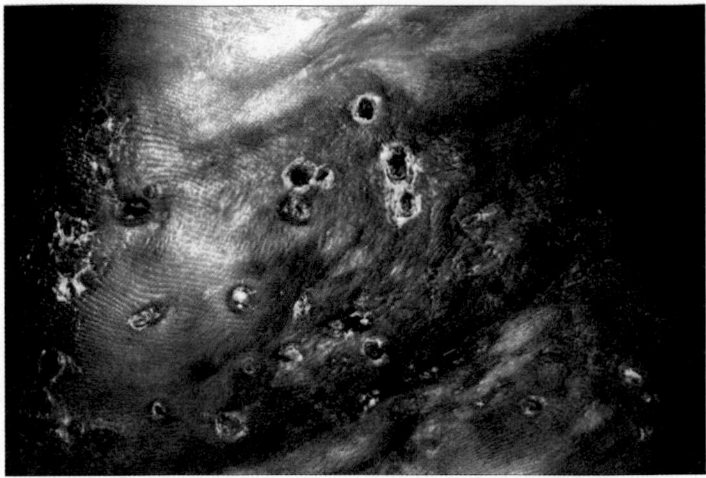

▲ **Figure 19.28** A *Nocardia* infection.

acid, so the cells are difficult to Gram stain, but they are acid-fast. *Actinomyces* stains purple in a Gram stain.

Nocardia asteroides

Nocardia asteroides (nō-kar′dē-ă as-ter-oy′dēz), a common inhabitant of soils rich in organic matter, is the most common pathogen in this genus, accounting for about 90% of infections.

Pathogenesis, Epidemiology, and Disease

N. asteroides is an opportunistic bacterial pathogen that infects numerous sites, including the lungs, skin, and central nervous system. Pulmonary infections develop following inhalation; cutaneous infections result from introduction of the bacteria into wounds; and infections of the central ner vous system and other internal organs follow the spread of the bacterium in the blood.

In the lungs, *Nocardia* causes pneumonia accompanied by signs and symptoms typical of pulmonary infections—cough, shortness of breath, and fever. Cutaneous infections may produce **mycetoma,** a painless, long-lasting infection characterized by swelling, pus production, and draining sores **(Figure 19.28)**.

Diagnosis, Treatment, and Prevention

Microscopic examination of samples of skin, sputum, pus, or cerebrospinal fluid is usually sufficient to suspect infection with *Nocardia;* the presence of long, acid-fast, hypha-like cells is diagnostic. Treatment usually involves six weeks of an appropriate antimicrobial drug; sulfonamides are the drugs of choice. Although such treatment is usually sufficient for localized infections in patients with normal immune systems, the prognosis for immunocompromised patients with widespread infections is poor. Prevention of nocardial diseases involves avoiding exposure to the bacterium in soil, especially by immunocompromised patients.

Actinomyces

Actinomyces is another genus characterized by hypha-like cells **(Figure 19.29a)**, as reflected in the name of the genus, which means "ray fungus" in Greek. Despite the name, *Actinomyces* is not a fungus but a bacterium. Unlike *Nocardia, Actinomyces* is not acid-fast and stains purple when Gram-stained.

Colonies of *Actinomyces* form visible concretions resembling grains of sand. Even though the yellow of these multicellular structures has prompted the name "sulfur granules," they are in fact held together by calcium phosphate.

Pathogenesis, Epidemiology, and Disease

Actinomyces is a normal member of the surface microbiota of human mucous membranes. It can become an opportunistic pathogen of the respiratory, gastrointestinal, urinary, and female

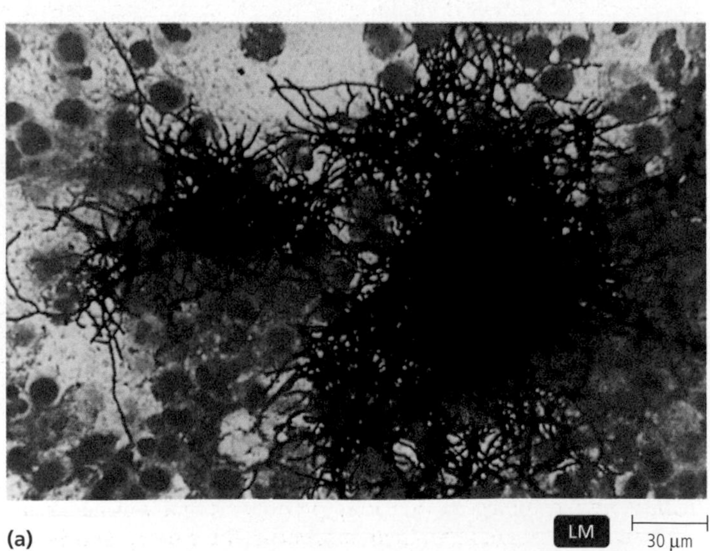

(a)

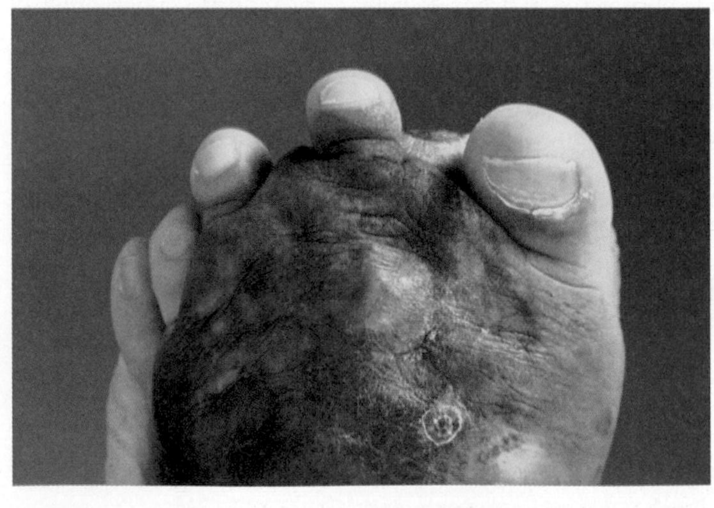

(b)

▲ **Figure 19.29** ***Actinomyces.*** **(a)** A Gram stain of pus containing the bacteria, which resemble fungal hyphae (compare with Figure 12.15a) and may form visible yellow aggregations that look like grains of sand. **(b)** Lesions of actinomycosis.

genital tracts, and it sometimes causes dental caries (cavities). It also causes a disease called *actinomycosis* when it enters breaks in the mucous membranes resulting from trauma, surgery, or infection by other pathogens. Actinomycosis is characterized by the formation of many abscesses connected by channels in the skin or mucous membranes **(Figure 19.29b)**.

Diagnosis, Treatment, and Prevention

Diagnosis of actinomycosis is difficult because other organisms cause similar symptoms, and it is difficult to show that the presence of *Actinomyces* is not merely contamination from its normal site on the mucous membranes. Often a health care worker collects and crushes a "sulfur granule" to reveal the presence of filamentous cells.

Treatment involves surgical removal of the infected tissue and the administration of penicillin, tetracycline, or erythromycin for 4–12 months. Prevention of *Actinomyces* infections involves good oral hygiene, and the prolonged use of prophylactic antimicrobials should trauma or surgery broach the mucous membranes.

Chapter Summary

Staphylococcus (pp. 534–538)

1. *Staphylococcus aureus* and *Staphylococcus epidermidis* are found on the skin and in the upper respiratory, gastrointestinal, and urogenital tracts. *S. aureus* is the more virulent, partly because of its production of various enzymes (including coagulase and staphylokinase) and several toxins.
2. Food poisoning is a noninvasive disease caused by *Staphylococcus.* Cutaneous diseases caused by *Staphylococcus* include **staphylococcal scalded skin syndrome, impetigo, folliculitis, sties, furuncles,** and **carbuncles.**
3. Potentially fatal systemic infections caused by *Staphylococcus* include **staphylococcal toxic shock syndrome (STSS), bacteremia, endocarditis, pneumonia, empyema,** and **osteomyelitis.**
4. **Methicillin-resistant *Staphylococcus aureus* (MRSA),** which is resistant to many antimicrobials including methicillin, is an emerging nosocomial and community-acquired threat. **Vancomycin-resistant *S. aureus* (VRSA)** is also emerging as a threat as vancomycin is used against MRSA.

Streptococcus (pp. 538–544)

1. Cells of **group A *Streptococcus*** *(Streptococcus pyogenes)* produce M protein and a hyaluronic acid capsule, each of which contributes to the virulence of the species. Enzymes and toxins produced by *S. pyogenes* dissolve blood clots, stimulate fever, and lyse blood cells.
2. The diseases caused by group A streptococci include **pharyngitis** ("strep throat"), **scarlet fever** (scarlatina), **pyoderma** (impetigo), **erysipelas, toxic-shock-like syndrome (TSLS), necrotizing fasciitis** ("flesh-eating bacteria"), **rheumatic fever,** and **glomerulonephritis.**
3. Cells of **group B *Streptococcus*** *(S. agalactiae)* occur normally in the lower GI, genital, and urinary tracts. They are associated with neonatal diseases and are treated with penicillin.
4. The **viridans group** of alpha-hemolytic streptococci include members that inhabit the mouth, pharynx, GI tract, and genitourinary tracts, cause dental **caries** (cavities), and can enter the blood to cause bacteremia, meningitis, and endocarditis.
5. Virulent strains of *Streptococcus pneumoniae* are protected by polysaccharide capsules and phosphorylcholine in their cell walls. These bacteria secrete protein adhesin, secretory IgA protease, and pneumolysin, which lyses cells in the lungs.
6. **Pneumococcal pneumonia** is the most prevalent disease caused by *Streptococcus pneumoniae* infection. Other diseases include **sinusitis** (inflammation of the nasal sinuses), **otitis media** (inflammation of the middle ear), bacteremia, endocarditis, and meningitis.

Enterococcus (pp. 544–545)

1. *Enterococci* are normal members of the intestinal microbiota that can cause nosocomial bacteremia, endocarditis, and wound infections.

Bacillus (pp. 545–547)

1. *Bacillus anthracis* secretes anthrax toxin, which can cause three forms of **anthrax:** gastrointestinal anthrax (rare and fatal, with intestinal hemorrhaging), cutaneous anthrax (can be fatal if untreated), and inhalation anthrax (often fatal).

Clostridium (pp. 547–553)

1. *Clostridium perfringens* produces 11 toxins that lyse blood cells and cause diseases such as food poisoning and **gas gangrene.**
2. *Clostridium difficile* is an intestinal bacterium that can cause a self-limiting explosive diarrhea or a life-threatening **pseudomembranous colitis.**
3. *Clostridium botulinum* is an anaerobic, endospore-forming bacterium that can release an extremely poisonous toxin in improperly canned food, causing **botulism.** Infant botulism occurs when the pathogen grows in the gastrointestinal tract of an infant. Wound botulism results when the bacterial endospores germinate in surgical or traumatic wounds.
4. *Clostridium tetani* is a ubiquitous bacterium that enters the body via a break in the skin and causes **tetanus** due to the action of **tetanospasmin.** Immunization is effective in preventing this disease.

Listeria (pp. 553–554)

1. *Listeria monocytogenes* is rarely pathogenic but can cause severe disease in pregnant women, newborns, the elderly, and immunocompromised patients.

Mycoplasmas (pp. 554–556)

1. **Mycoplasmas,** the smallest free-living microbes, lack cell walls. They are mostly facultative anaerobes. The colonies of most mycoplasmas have a "fried egg" appearance.
2. *Mycoplasma pneumoniae* is a human pathogen that affects the epithelial cells of the respiratory tract and cause **primary atypical pneumonia** (walking pneumonia). It does not form "fried-egg" colonies on agar.
3. *Mycoplasma hominis* can cause **pelvic inflammatory disease** in women.
4. *Mycoplasma genitalium* and *Ureaplasma urealyticum* cause nongonococcal urethritis.
5. Venereal mycoplasmal diseases can be prevented by abstinence, mutually faithful monogamy, and the proper use of condoms.

Corynebacterium (pp. 556–557)

1. *Corynebacterium diphtheriae,* transmitted via respiratory droplets, contains a bacteriophage that codes for diphtheria toxin, which causes the symptoms of the potentially fatal disease **diphtheria.** Diagnosis results from an immunodiffusion assay called an **Elek test.**

Mycobacterium (pp. 557–562)

1. Cell walls of *Mycobacterium* contain **mycolic acid.** *Mycobacterium tuberculosis* with a cell wall component called **cord factor** causes **tuberculosis,** a lung disease in which the bacteria replicate within macrophages. The **tuberculin skin test** screens for possible TB exposure.
2. *Mycobacterium leprae* causes **leprosy.** Tuberculoid leprosy is a nonprogressive form of the disease, whereas lepromatous leprosy results in progressive destruction of body structures.
3. *Mycobacterium avium-intracellulare* is an opportunistic pathogen that can become pathogenic in AIDS patients. It adversely affects almost every organ of the body.

Propionibacterium (pp. 562–563)

1. *Propionibacterium acnes* living in the sebaceous glands of the skin causes adolescent **acne.** The bacterium may also infect patients who have intrusive medical devices. Antimicrobial drugs are more effective than topical products in the treatment of acne.

Nocardia and *Actinomyces* (pp. 563–565)

1. The opportunistic pathogen *Nocardia asteroides* has elongated cells resembling fungal hyphae; it is acid-fast because of the presence of mycolic acid in its cell wall. Cutaneous infections may produce **mycetoma.**
2. *Actinomyces* is another opportunistic pathogen with elongated cells, but it is not acid-fast. It causes *actinomycosis.*

Questions for Review

Answers to the Questions for Review (except Short Answer questions) begin on page A-1.

Multiple Choice

1. Which of the following bacteria causes a type of food poisoning?
   a. *Streptococcus sanguis*
   b. *Clostridium perfringens*
   c. *Staphylococcus aureus*
   d. *Streptococcus pyogenes*
2. How does *Staphylococcus aureus* affect the matrix between cells in the human body?
   a. *S. aureus* triggers blood clotting, which coats the matrix and inhibits cellular communication.
   b. *S. aureus* produces an enzyme that dissolves hyaluronic acid and thus enables it to pass between the cells.
   c. *S. aureus* contains a hyaluronic acid capsule that causes leukocytes to ignore the bacterium as if it were camouflaged.
   d. *S. aureus* does not affect the matrix but instead produces a necrotizing agent that dissolves body cells.
3. Which of the following conditions is a systemic disease caused by *Staphylococcus*?
   a. impetigo
   b. folliculitis
   c. carbuncle
   d. toxic shock syndrome
4. A bacterium associated with bacteremia, meningitis, and pneumonia in newborns is
   a. *Staphylococcus aureus.*
   b. *Staphylococcus epidermidis.*
   c. *Streptococcus pyogenes.*
   d. *Streptococcus agalactiae.*
5. Which type of anthrax is much more common in animals than in humans?
   a. cutaneous anthrax
   b. inhalation anthrax
   c. gastrointestinal anthrax
   d. mucoid anthrax
6. Of the following genera, which can survive the harshest conditions?
   a. *Staphylococcus*
   b. *Clostridium*
   c. *Mycobacterium*
   d. *Actinomyces*
7. Pathogenic strains that have become resistant to antimicrobial drugs are found in which of the following genera?
   a. *Staphylococcus*
   b. *Mycobacterium*
   c. *Enterococcus*
   d. all of the above
8. The bacterium causing pseudomembranous colitis is
   a. *Clostridium difficile.*
   b. *Streptococcus pyogenes.*
   c. *Mycobacterium avium-intracellulare.*
   d. *Corynebacterium diphtheriae.*
9. Mycoplasmas
   a. lack cell walls.
   b. are pleomorphic.
   c. have sterol in their membranes.
   d. all of the above

10. In which of the following diseases would a patient experience a pseudomembrane covering the tonsils, pharynx, and larynx?
    a. tuberculoid leprosy
    b. diphtheria
    c. arrhythmia
    d. tetanus
11. Which of the following is *not* characteristic of mycoplasmas?
    a. cytochromes
    b. sterols in cytoplasmic membranes
    c. use of UGA codon for tryptophan
    d. rRNA nucleotide sequences similar to those of Gram-positive bacteria

Matching

For each of the following diseases or conditions, indicate the genus (or genera) of bacterium that causes it.

1. ____	Scalded skin syndrome	A. *Staphylococcus*
2. ____	Osteomyelitis	B. *Streptococcus*
3. ____	Pharyngitis	C. *Mycobacterium*
4. ____	Scarlet fever	D. *Listeria*
5. ____	Pyoderma	E. *Propionibacterium*
6. ____	Rheumatic fever	F. *Corynebacterium*
7. ____	Glomerulonephritis	G. *Bacillus*
8. ____	Sinusitis	H. *Clostridium*
9. ____	Otitis media	I. *Actinomyces*
10. ____	Anthrax	
11. ____	Myonecrosis	
12. ____	Diphtheria	
13. ____	Leprosy	
14. ____	Dental caries	
15. ____	Acne	

Short Answer

1. Why are mycoplasmas able to survive a relatively wide range of osmotic conditions, even though these bacteria lack cell walls?
2. *Mycobacterium avium-intracellulare* was considered relatively harmless until the late 20th century, when it became common in certain infections. Explain how this bacterium's pathogenicity changed.
3. Contrast tuberculoid leprosy with lepromatous leprosy in terms of pathogenesis. How does the cellular immune response of a patient affect the form of the disease?
4. On a trans-Atlantic flight, a Latvian passenger with tuberculosis sits next to a Canadian businessman, who later develops tuberculosis and dies. Trace the path of *Mycobacterium tuberculosis* from a droplet of mucus leaving the European passenger to within the businessman, and explain how the disease became fatal.
5. Explain how mice are used in the diagnosis of botulism poisoning.
6. Why do pediatricians recommend that children under one year never be fed honey?
7. Explain why Gram-positive mycoplasmas appear pink in a Gram stained smear.
8. Explain the different actions of pyogenic and pyrogenic toxins.
9. Explain why *Staphylococcus epidermidis* is rarely pathogenic while the similar *S. aureus* is more commonly virulent.
10. Why did epidemiologists immediately suspect terrorism in the cases of anthrax in the fall of 2001?
11. Explain the action of the toxin of *Clostridium tetani.*
12. Why is mycolic acid a virulence factor for mycobacteria?
13. Compare and contrast mycoplasmas and viruses.

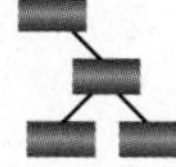

Concept Mapping

Using the following terms, draw a concept map that describes pathogenic clostridia. For a sample concept map, see p. 93. Or, complete this concept map online by going to the Study Area at www.masteringmicrobiology.com.

Antibiotic-associated diarrhea
Botox
Botulism toxin
C. botulinum
C. difficile
C. perfringens
C. tetani
Cosmetic applications
Endospore formers
Food poisoning
Foodborne botulism
Gas gangrene
Genus *Clostridium*
Gram-positive bacilli
Infant botulism
Medical applications
Pseudomembranous colitis
Strict anaerobes
Tetanus
Tetanus toxin
Tetanus toxoid vaccine
Wound botulism

Critical Thinking

1. In Chapter 4 (p. 121) we discussed the identification of microbes using dichotomous taxonomic keys. Design a key for the genera of Gram-positive cocci discussed in this chapter. Design a second key for identifying the Gram-positive bacilli discussed in the chapter.
2. A few days after the death of a child hospitalized with an MRSA infection, another child, who had been admitted to the hospital with viral pneumonia, worsened and died. An autopsy revealed that the second child also died from complications of MRSA. By what route was the second child likely infected? What should hospital personnel do to limit the transfer of MRSA and other bacteria among patients?
3. An elderly man is admitted to the hospital with severe pneumonia, from which he eventually dies. What species is the most likely cause of his infection? What antimicrobial drug is effective against this species? How could the man have been protected from infection? Is the hospital staff at significant risk of infection from the man? Which groups of patients would be at risk if the man had visited their rooms before he died?
4. Botulism toxin can be used as an antidote for tetanus. Can tetanus toxin be used as an antidote for botulism? Why or why not?
5. A blood bank refused to accept blood from a potential donor who had just had his teeth cleaned by a dental hygienist. Why did they refuse the blood?
6. Match the genera of pathogens to their appearance in stained smears: *Actinomyces, Bacillus, Clostridium, Mycobacterium, Staphylococcus, Streptococcus.*

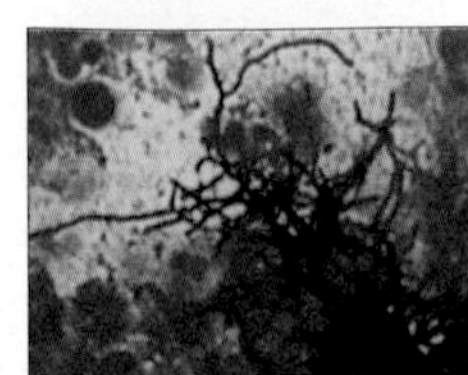
(a) Gram

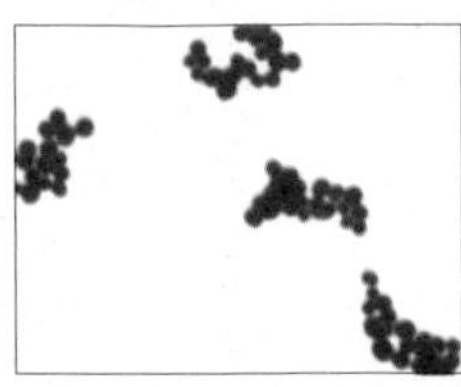
(b) Gram

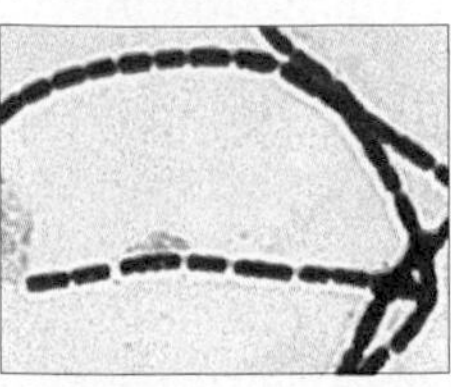
(c) Gram

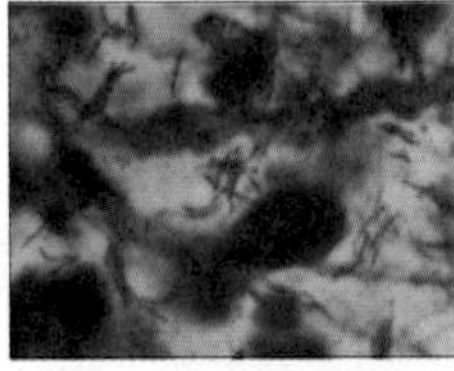
(d) Acid fast

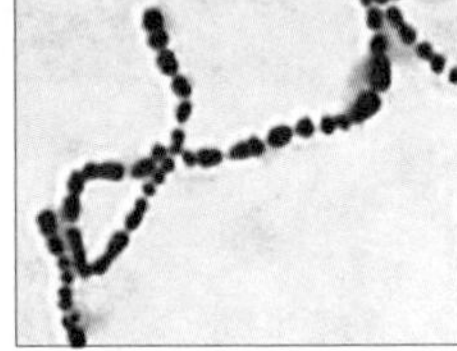
(e) Gram

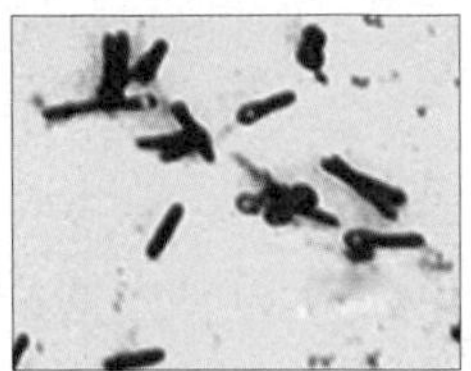
(f) Gram

Access more review material online in the Study Area at **www.masteringmicrobiology.com.** There, you'll find

- **Concept Mapping Activities**
- **Flashcards**
- **Quizzes**

and more to help you succeed.

20 Pathogenic Gram-Negative Cocci and Bacilli

A New Mexico couple vacationing in New York City fell ill with flulike symptoms. A few days later they were admitted to a hospital, suffering from chills and a high fever. The wife had developed a "bubo" (a severely swollen lymph node) on her thigh, leading doctors to suspect bubonic plague—the dreaded "black death" that killed one-third of all Europeans during the 14th century but had not been seen in New York City in over 100 years. Subsequent tests confirmed that they were indeed infected with *Yersinia pestis*, the causative bacterium of plague. The wife was administered antibiotics and recovered quickly, but her husband worsened, requiring amputation of both feet and hospitalization for three months. Authorities later determined that the couple had most likely contracted the disease from infected fleas encountered near their property in rural New Mexico.

Y. pestis is one example of a Gram-negative pathogen—a group that also includes the genera *Escherichia*, *Salmonella*, and *Legionella*. Gram-negative pathogens cause a wide variety of human diseases. In this chapter we study important Gram-negative cocci and bacilli and major human diseases they cause.

Yersina pestis, a Gram-negative, rod-shaped bacterium, causes one of history's more feared diseases—bubonic plague.

Take the pre-test for this chapter online. Visit the Study Area at www.masteringmicrobiology.com.

Numerous Gram-negative, cocci, bacilli, and coccobacilli are bacterial pathogens of humans. In fact, Gram-negative bacteria constitute the largest group of human bacterial pathogens, in part because the outer membrane of a Gram-negative bacterial cell wall contains lipid A (see Figure 3.15). Lipid A triggers fever, vasodilation, inflammation, shock, and **disseminated intravascular coagulation (DIC)**—the formation of blood clots within blood vessels throughout the body. Almost every Gram-negative bacterium that can breach the skin or mucous membranes, grow at 37°C, and evade the immune system can cause disease and death in humans; however, most of the Gram-negative bacteria that cause disease fall into fewer than 30 genera.

We have seen that it is possible to group these organisms according to a variety of criteria. In Chapter 11 we considered the Gram-negative bacteria according to their placement in *Bergey's Manual,* which is based largely on genomic similarities; here we primarily consider them according to a different organizational scheme, one that traditionally has been used for many clinical purposes—according to their shapes, oxygen requirements, and other biochemical properties.

First we survey the primary genus of Gram-negative cocci that are pathogenic to humans; then we consider 15 genera of pathogenic, facultatively anaerobic bacilli in the families Enterobacteriaceae and Pasteurellaceae, which account for almost half of all Gram-negative pathogens. Finally, we examine a heterogeneous collection of 10 genera of pathogenic, Gram-negative aerobic bacilli before concluding by considering two genera of strictly anaerobic pathogenic bacilli.

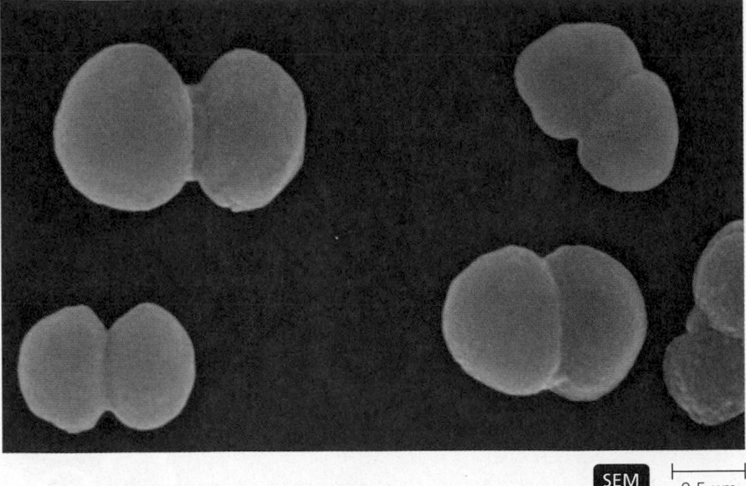

▲ **Figure 20.1 Diplococci of *Neisseria gonorrhoeae*.** *What function of gonococcal fimbriae (not visible here) contributes to the bacterium's pathogenicity?*

Figure 20.1 The fimbriae act as adhesins, enabling gonococci to adhere to cells lining the cervix and urethra and to sperm, which allows Neisseria to travel far into the female reproductive tract.

Pathogenic Gram-Negative Cocci: *Neisseria*

Neisseria (nī-se′rē-ă) is the only genus of Gram-negative cocci that regularly causes disease in humans. *Neisseria* is placed in the class Betaproteobacteria of the phylum Proteobacteria. Morphologically similar pathogenic bacteria in the genera *Moraxella* (mōr′ak-sel′ă) and *Acinetobacter* (as-i-nē′tō-bak′ter), considered coccobacilli, are now known to be more closely related to aerobic bacilli in class Gammaproteobacteria and thus are discussed in a later section.

Structure and Physiology of *Neisseria*

Learning Objective

✓ List three structural features of *Neisseria* that contribute to its pathogenicity.

The cells of all strains of *Neisseria* are nonmotile and typically arranged as diplococci (pairs) with their common sides flattened in a manner reminiscent of coffee beans (**Figure 20.1**). In addition to their shape, these aerobic bacteria are distinguished from many other Gram-negative pathogens by being *oxidase positive* (see Figure 20.4); that is, their electron transport chains contain the enzyme *cytochrome oxidase.* Pathogenic strains of *Neisseria* also have fimbriae and polysaccharide capsules, as well as a major cell wall antigen called *lipooligosaccharide* (lip′ō-ol-i-gō-sak′a-rīd, *LOS*), composed of lipid A (endotoxin) and sugar molecules—all of which enable the bacteria to attach to and invade human cells. Cells of *Neisseria* that lack any one of these three structural features are typically avirulent.

In the laboratory, *Neisseria* is fastidious in its growth requirements. Microbiologists culture this bacterium on either autoclaved blood agar (called *chocolate agar* because of its appearance) or a selective medium called *modified Thayer-Martin medium. Neisseria* is particularly susceptible to drying and extremes of temperature. A moist culture atmosphere containing 5% carbon dioxide enhances their growth. Laboratory personnel may use sugar fermentation tests to distinguish among strains of *Neisseria;* pathogenic strains, for example, are unable to ferment maltose, sucrose, or lactose.

Two species of *Neisseria* are pathogenic to humans: the so-called gonococcus, *N. gonorrhoeae* (go-nor-rē′ī), which causes gonorrhea, and the meningococcus, *N. meningitidis* (me-nin-ji′ti-dis), which causes a type of meningitis.

The Gonococcus: *Neisseria gonorrhoeae*

Learning Objectives

✓ Compare and contrast the symptoms of gonorrhea in men and women.

✓ Discuss the difficulties researchers face in developing an effective vaccine against *Neisseria gonorrhoeae.*

Pathogenesis, Epidemiology, and Disease

Neisseria gonorrhoeae causes **gonorrhea** (gon-ō-rē′ă), which has been known as a sexually transmitted disease for centuries, although the disease was confused with syphilis until the 19th century. In the 2nd century A.D., the Roman physician Claudius Galen named gonorrhea for what he thought was its cause—an excess of semen (gonorrhea means "flow of seed" in Greek).

The disease is sometimes called "clap" from an archaic French word *clapoir,* meaning brothel.

Gonorrhea occurs in humans only. Even though it is one of the more common sexually transmitted diseases in the United States, the number of cases among civilians has been declining over the past few decades **(Figure 20.2a)**. Most U.S. cases of gonorrhea occur in adolescents, particularly among those who engage in a promiscuous sexual lifestyle in several southeastern states **(Figure 20.2b)**. The Centers for Disease Control and Prevention (CDC) has a goal to reduce the incidence of gonorrhea to below 19 cases per 100,000 population. This incidence will theoretically eliminate the disease in the United States.

The disease is 19 times more common among blacks than nonblacks and more common among females than males. Whereas women have a 50% chance of becoming infected during a single sexual encounter with an infected man, men have only a 20% chance of infection during a single encounter with an infected woman. An individual's risk of infection increases with increasing frequency of sexual encounters.

Gonococci adhere, via their fimbriae and capsules, to epithelial cells of the mucous membranes lining the genital, urinary, and digestive tracts of humans. As few as 100 pairs of cells can cause disease. The cocci protect themselves from the immune system by secreting a protease enzyme that cleaves secretory IgA in mucus. Further, endocytized bacteria survive and multiply within neutrophils. As the bacteria multiply, they invade deeper connective tissues, often while being transported intracellularly by phagocytes.

In men, gonorrhea is usually insufferably symptomatic—acute inflammation typically occurs two to five days after infection in the urethra of the penis, causing extremely painful urination and a purulent (pus-filled) discharge. Rarely, gonococci invade the prostate or epididymis, where the formation of scar tissue can render the man infertile. In contrast, recently acquired gonorrhea in women is often asymptomatic; about 50% of infected women have no symptoms or obvious signs of infection. Even when women do have symptoms, they often mistake them for a bladder infection or vaginal yeast infection.

Neisseria cannot attach to cells lining the vagina; instead, the bacteria most commonly infect the cervix of the uterus. Gonococci can infect deep in the uterus and even the uterine (Fallopian) tubes by "hitchhiking" to these locations on sperm to which they attach by means of fimbriae or LOS. In the uterine tubes they can trigger inflammation, fever, and abdominal pain—a condition known as **pelvic inflammatory disease (PID)**. Chronic infections can lead to scarring of the tubes, resulting in *ectopic*[1] *pregnancies* or sterility.

Gonococcal infection of organs outside the reproductive tracts can also occur. Because a woman's urethral opening is close to her vaginal opening, the urethra can become infected with gonococci during sexual intercourse. Anal intercourse can lead to *proctitis* (inflammation of the rectum), and oral sexual intercourse can infect the pharynx or gums, resulting in *pharyngitis* and *gingivitis,* respectively. Oral and anal infections are most commonly seen in men who have sex with men. In very rare cases, gonococci enter the blood and travel to the joints, meninges, or heart, causing arthritis, meningitis, and endocarditis, respectively.

Gonococcal infection of a child during birth can result in inflammation of the cornea, *ophthalmia neonatorum* (inflammation of the conjunctiva in newborns), or blindness. Gonococci can also infect the respiratory tracts of newborns during their passage through the birth canal. Infection of older children with *N. gonorrhoeae* is strong evidence of sexual abuse by an infected adult.

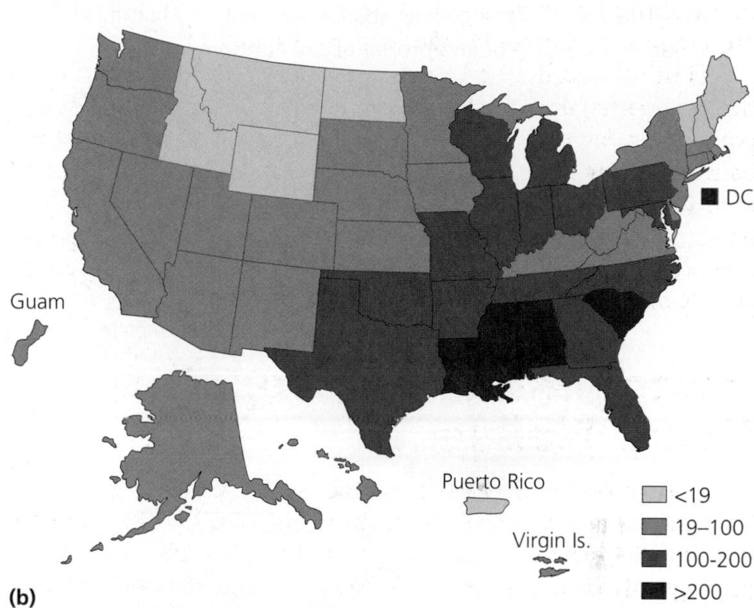

▲ **Figure 20.2 Incidence of gonorrhea in the United States** **(a)** The number of new cases annually, 1977–2008. **(b)** The geographic distribution of cases reported in 2007. The incidences shown are the number of new cases per 100,000 individuals in each state. *What organism causes gonorrhea?*

Figure 20.2 Gonorrhea is caused by *Neisseria gonorrhoeae*

[1]From Greek *ektopos,* meaning out of place.

MICROBE AT A GLANCE

Neisseria gonorrhoeae

Taxonomy: Domain Bacteria, phylum Proteobacteria, class Betaproteobacteria, order Neisseriales, family Neisseriaceae

Other names: Gonococcus

Cell morphology and arrangement: Cocci in pairs

Gram reaction: Negative

Virulence factors: Capsule, fimbriae, and lipooligosaccharide adhere to human cells; IgA protease; can survive inside phagocytic cells; variable surface antigens

Diseases caused: Gonorrhea, pelvic inflammatory disease, ophthalmia neonatorum, proctitis, pharyngitis, gingivitis

Treatment for diseases: Cephalosporins

Prevention of disease: Sexual abstinence, mutually faithful monogamy, consistent and proper use of condoms

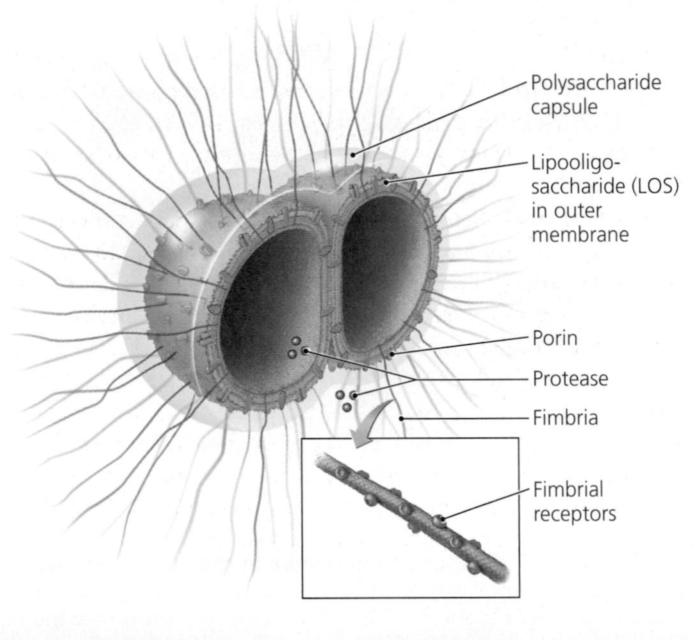

Download the Microbe at a Glance flashcards from the Study Area at www.masteringmicrobiology.com.

Diagnosis, Treatment, and Prevention

The presence of Gram-negative diplococci in pus from an inflamed penis is sufficient for a diagnosis of symptomatic gonorrhea in men. For diagnosis of asymptomatic cases of gonorrhea in men and women, commercially available genetic probes provide direct, accurate, rapid detection of *N. gonorrhoeae* in clinical specimens. Alternatively, the bacterium can be cultured on selective media, such as Thayer-Martin agar.

The treatment of gonorrhea has been complicated in recent years by the worldwide spread of gonococcal strains resistant to penicillin, tetracycline, erythromycin, and aminoglycosides. Currently, the CDC recommends broad-spectrum oral cephalosporins to treat gonorrhea.

Long-term specific immunity against *N. gonorrhoeae* does not develop, and thus promiscuous people can be infected multiple times. This lack of immunity is explained in part by the highly variable surface antigens of this bacterium; immunity against one strain often provides no protection against other strains. Further, the existence of many different strains has prevented the development of an effective vaccine.

Except for the routine administration of antimicrobial agents to newborns' eyes, which successfully prevents ophthalmic disease, chemical prophylaxis is ineffective in preventing disease. In fact, the use of antimicrobials to prevent genital disease may select for hardier resistant strains, worsening the situation. The most effective methods of prevention, in order, are sexual abstinence; monogamy with an uninfected, faithful partner; and consistent and proper use of condoms. Efforts to stem the spread of gonorrhea focus on education, aggressive detection, and the screening of all sexual contacts of carriers.

CRITICAL **THINKING**

Penicillin-resistant strains of *N. gonorrhoeae* produce a plasmid-coded β-lactamase, which degrades penicillin. What changes in structure or metabolism could explain resistance to other antimicrobial drugs in this bacterium?

The Meningococcus: *Neisseria meningitidis*

Learning Objectives

- ✓ Describe how meningococci survive and thrive in humans.
- ✓ Discuss the epidemiology of meningococcal diseases.

Neisseria meningitidis causes life-threatening diseases when it invades the blood or cerebrospinal fluid.

Pathogenicity, Epidemiology, and Disease

Of the 13 known antigenic strains of meningococci, strains known as A, B, C, and W 135 cause most cases of disease in humans. The polysaccharide capsule of *N. meningitidis* resists lytic enzymes of the body's phagocytes, allowing phagocytized meningococci to survive, reproduce, and be carried throughout the body within neutrophils and macrophages. In the United States, the bacterium causes most cases of meningitis in children and adults under 20. Much of the damage caused by *N. meningitidis* results from *blebbing*—a process in which the bacterium sheds extrusions of outer membrane. The lipid A component of LOS thereby released into extracellular spaces triggers fever, vasodilation, inflammation, shock, and DIC.

N. meningitidis is a common member of the normal microbiota in the upper respiratory tracts of up to 40% of healthy people. As a result of crowded living conditions, meningococci are more prevalent in children and young adults from lower economic groups.

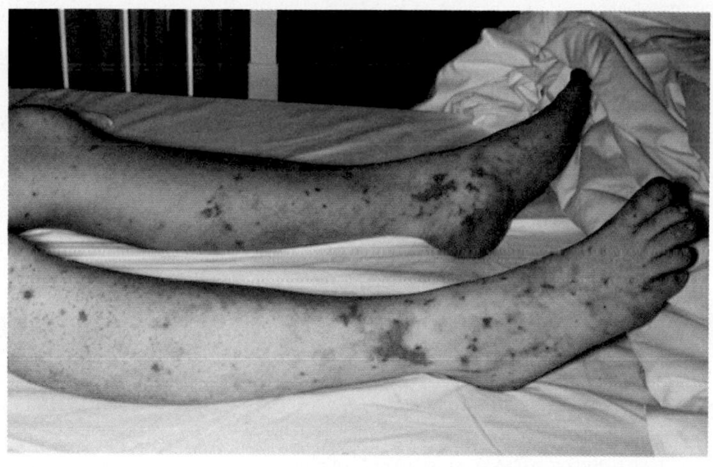

(a)

(b)

▲ **Figure 20.3 Petechiae in meningococcal septicemia.** These lesions can either be small and relatively diffuse **(a)** or may coalesce in areas containing dead cells **(b)**.

Respiratory droplets transmit the bacteria among people living in close contact, especially families, soldiers living in barracks, prisoners, and college students living in dormitories. In fact, meningococcal disease can be 23 times more prevalent in students living in dormitories than in the general population. Individuals whose lungs have been irritated by dust are more susceptible to airborne *Neisseria*. For example, annual outbreaks of meningococcal meningitis can kill tens of thousands in sub-Saharan Africa when dry winter winds create irritating dust storms. Even though meningococcal diseases occur worldwide, cases are relatively rare in countries with advanced health care systems and access to antimicrobial drugs.

The incidence of meningococcal disease in the United States is about 0.3 cases per 100,000 population. Most of these cases are meningitis, usually involving abrupt sore throat, fever, headache, stiff neck, vomiting, and convulsions. (Very young children may have only fever and vomiting.) Arthritis and partial loss of hearing occasionally result. Meningococcal meningitis can progress so rapidly that death results within six hours of the initial symptoms.

Meningococcal septicemia (blood poisoning) can also be life threatening. LOS may trigger shock, devastating blood coagulation in many organs, and *petechiae* (pe-tē′kē-ē)—minute hemorrhagic skin lesions—on the trunk and lower extremities **(Figure 20.3a)**. Petechiae may coalesce and combine with regions of cell death to form large black lesions **(Figure 20.3b)**. In some patients, however, septicemia produces only mild fever, arthritis, and petechiae.

Diagnosis, Treatment, and Prevention

Because meningococcal meningitis constitutes a medical emergency, rapid diagnosis is critical. The demonstration of Gram-negative diplococci within phagocytes of the central nervous system is diagnostic for meningococcal disease. *N. meningitidis* can ferment maltose in laboratory culture, which distinguishes it from *N. gonorrhoeae*. Further, serological tests can demonstrate the presence of antibodies against *N. meningitidis*, though strain B is relatively nonimmunogenic and therefore does not react well in such tests.

Despite the fact that meningococcal meningitis proceeds rapidly and is ordinarily 100% fatal when left untreated, immediate treatment with antimicrobial drugs has reduced mortality to less than 10%. Intravenously administered ceftriaxone, a semisynthetic cephalosporin, is the drug of choice.

Because healthy asymptomatic carriers are common, eradication of meningococcal disease is unlikely. Prevention involves prophylactic treatment with ceftriaxone, ciprofloxacin, or rifampin of anyone exposed to individuals with meningococcal disease, including family, classmates, day care center contacts, roommates, and health care workers. The CDC recommends immunization with a vaccine against strains A, C, and Y, though this vaccine cannot be used with children under two years old. As previously noted, strain B is weakly immunogenic, so immunity to it cannot be induced artificially and must instead develop following a natural infection.

CRITICAL **THINKING**

Investigations of *Mycobacterium tuberculosis* have clearly demonstrated that this pathogen can be transmitted among airline passengers; however, as of 2008, no cases of in-flight transmission of the similarly transmitted *N. meningitidis* had been reported to the CDC. What are some possible explanations for the lack of documented cases of in-flight transmission of meningococci?

Pathogenic, Gram-Negative, Facultatively Anaerobic Bacilli

Learning Objective

✓ Describe how members of the family Enterobacteriaceae are distinguished from members of the family Pasteurellaceae.

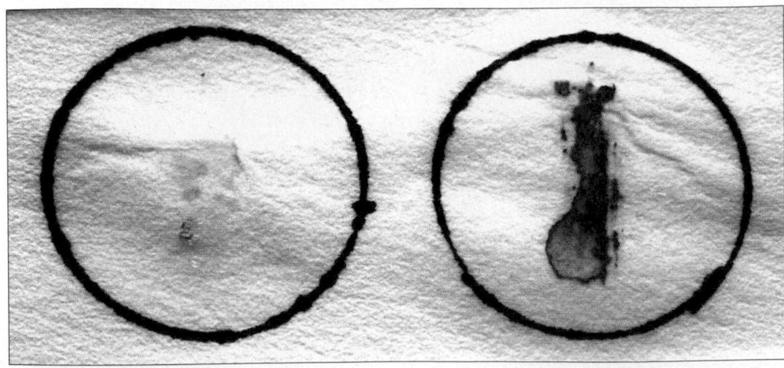

▲ **Figure 20.4 The oxidase test.** The test distinguishes members of the family Enterobacteriaceae (left), which are oxidase negative, from members of the family Pasteurellaceae, which are oxidase positive, as indicated by a purple color.

Two families of facultatively anaerobic bacilli and coccobacilli in the class Gammaproteobacteria—Enterobacteriaceae and Pasteurellaceae—contain most of the Gram-negative pathogens of humans. Scientists distinguish between members of the two families by performing an oxidase test: members of the Enterobacteriaceae are oxidase negative, whereas members of the Pasteurellaceae are oxidase positive **(Figure 20.4)**.

We begin our survey of these microorganisms by examining the family Enterobacteriaceae, which contains about 150 species in 41 genera, including some of the more important nosocomial pathogens of humans **(Figure 20.5)**.

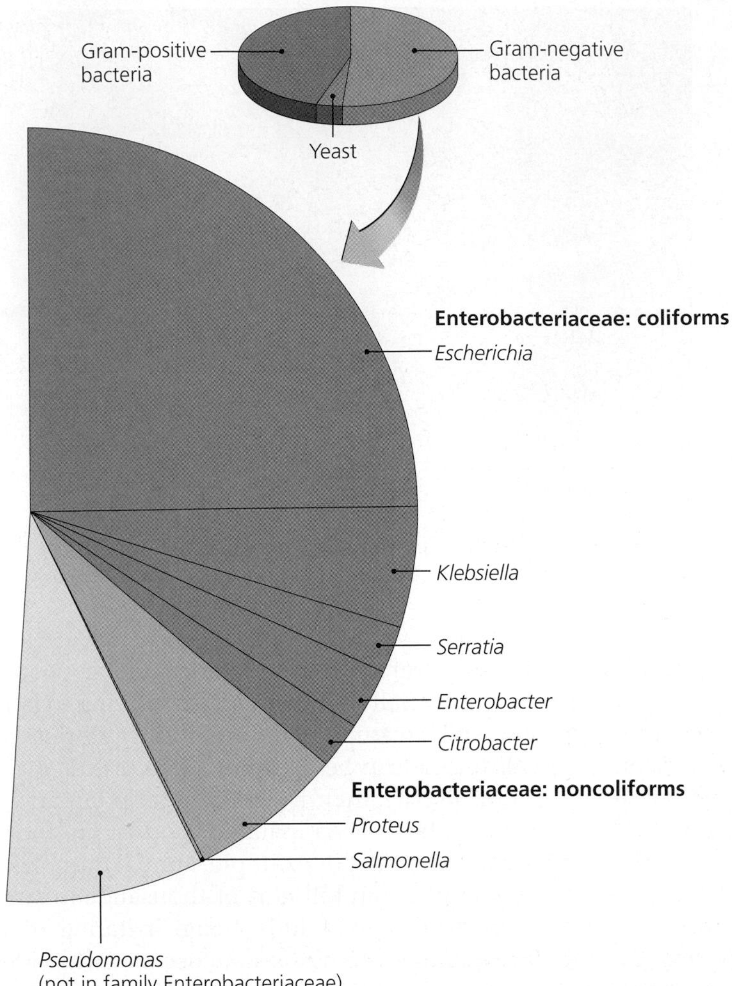

▲ **Figure 20.5 Relative causes of nosocomial infections in the United States.** Gram-negative bacteria in the family Enterobacteriaceae account for most nosocomial infections. *What is a nosocomial infection?*

Figure 20.5 *An infection acquired in a health care setting.*

The Enterobacteriaceae: An Overview

Learning Objectives

✓ Discuss how members of the family Enterobacteriaceae are distinguished in the laboratory.

✓ List six virulence factors found in members of the family Enterobacteriaceae.

✓ Describe the diagnosis, treatment, and prevention of diseases of enteric bacteria.

As their name suggests, prokaryotes in the family **Enterobacteriaceae** (en′ter-ō-bak-tēr-ē-ā′sē-ē)—also called **enteric bacteria**—are members of the intestinal microbiota of most animals and humans. They are also ubiquitous in water, soil, and decaying vegetation. Some species are always pathogenic in humans, whereas others are opportunists normally found in the intestinal microbiota that become pathogenic when introduced into other body sites.

As a group, enteric bacteria are the most common Gram-negative pathogens of humans. Though most enteric bacteria can be pathogenic in humans, over 95% of the medically important species are in the 13 genera we discuss shortly.

Structure and Physiology of the Enterobacteriaceae

Enteric bacteria are coccobacilli or bacilli that measure about 1 μm × 1.2–3 μm **(Figure 20.6)**. Those members that are motile have peritrichous flagella. Some have prominent capsules,

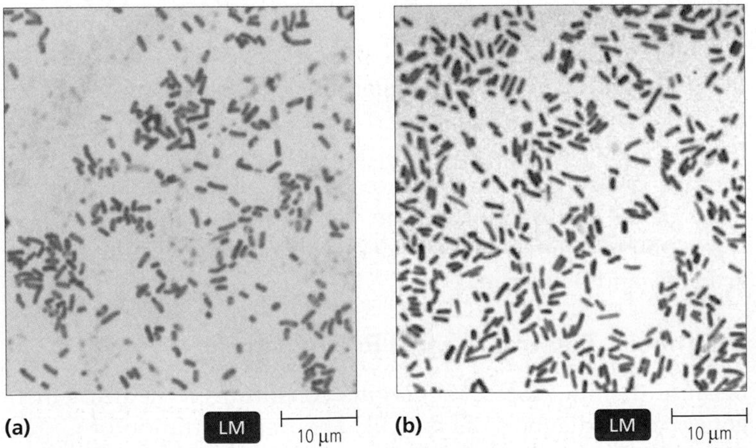

▲ **Figure 20.6 Gram stains of bacteria in the family Enterobacteriaceae. (a)** *Enterobacter aerogenes*, a coccobacillus. **(b)** *Escherichia coli*, a bacillus. Note the similarity in staining properties.

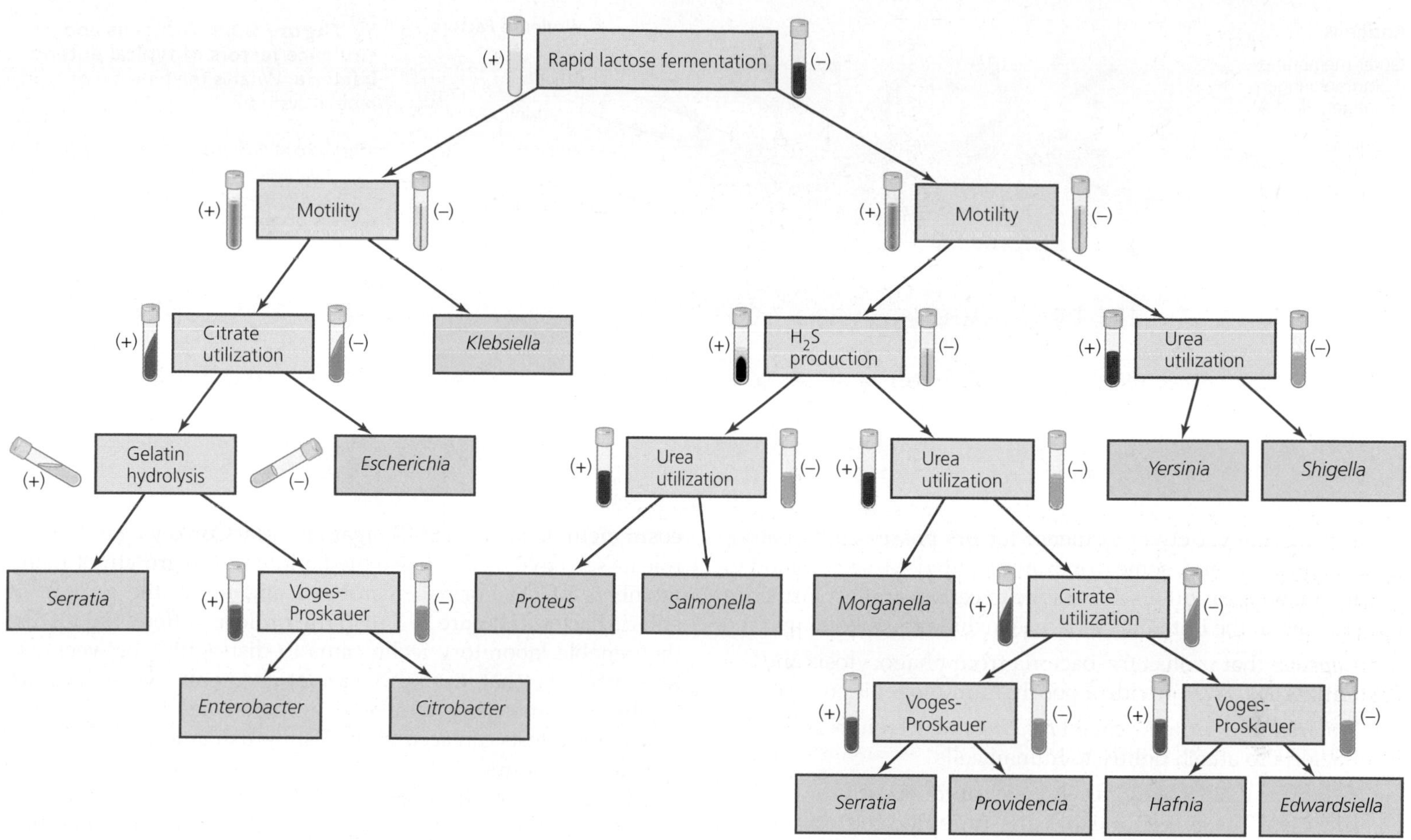

▲ **Figure 20.7 A dichotomous key for distinguishing among enteric bacteria.** This key, based on biochemical tests and motility, is but one of many possible keys. *What characteristics do all enteric bacteria share?*

Figure 20.7 All enteric bacteria are Gram-negative, oxidase negative, and able to metabolize nitrate to nitrite.

whereas others have only a loose slime layer. All members of the family reduce nitrate to nitrite and ferment glucose anaerobically, though most of them grow better in aerobic environments. As previously noted, all of them are also oxidase negative.

Given that all enteric bacteria are similar in microscopic appearance and staining properties, scientists traditionally distinguish among them by using biochemical tests, motility, and colonial characteristics on a variety of selective and nonselective media (for example, MacConkey agar and blood agar). **Figure 20.7** presents one dichotomous key, based on motility and biochemical tests, for distinguishing among the 13 genera of enteric bacteria discussed in this chapter.

Pathogenicity of the Enterobacteriaceae

Members of the Enterobacteriaceae have an outer membrane that contains lipopolysaccharide composed of three antigenic components:

- A core polysaccharide, shared by all of the enteric bacteria and called *common antigen.*
- So-called *O polysaccharide,* which has various antigenic varieties among strains and species.
- Lipid A, which, when released in a patient's blood, can trigger fever, vasodilation, inflammation, shock, and DIC. Because of the presence of lipid A, most of the enteric bacteria can cause serious disease and death.

Other antigens of certain enteric bacteria include protein and polysaccharide capsular antigens called K antigens (called Vi antigens in *Salmonella*) and flagellar proteins called H antigens. By controlling the genetic expression of K and H antigens, alternately producing and not producing the antigens, the bacteria survive by evading their host's immune system. Scientists use serological identification of antigens to distinguish among strains and species of enteric bacteria. For example, *Escherichia coli* O157:H7, a potentially deadly bacterium, is so designated in part because of its antigenic components. The specificity of serological tests can also be used in diagnosis and in epidemiological studies to identify the source of a particular infecting agent.

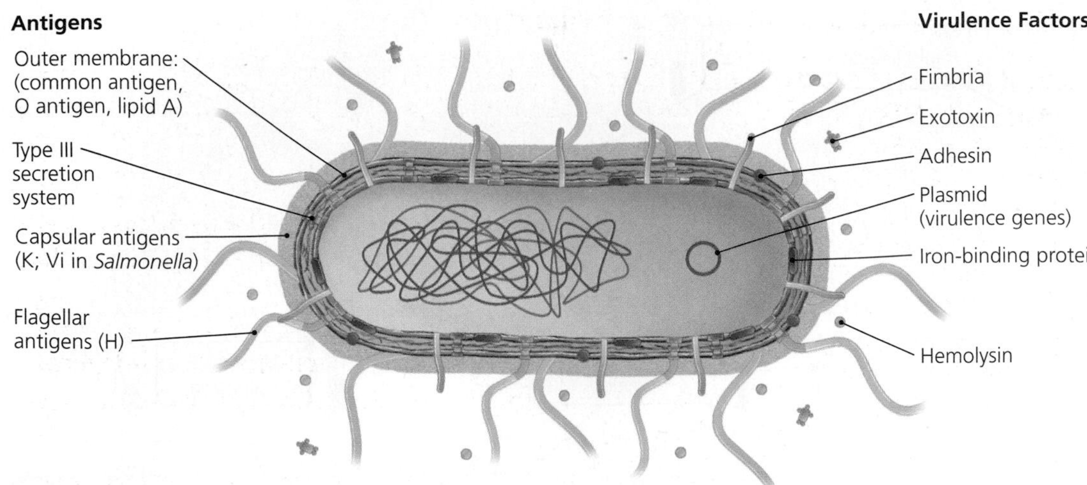

Figure 20.8 Antigens and virulence factors of typical enteric bacteria. *What is the function of a type III system?*

Figure 20.8 A type III secretion system is a virulence factor that enables enteric bacteria to insert chemicals directly into the cytosol of a eukaryotic cell.

Among the variety of virulence factors possessed by pathogenic enteric bacteria, some (for example, lipid A) are common to all genera, whereas others are unique to certain strains. Virulence factors seen in the Enterobacteriaceae include the following:

- *Capsules* that protect the bacteria from phagocytosis and antibodies, and provide a poorly immunogenic surface.
- *Fimbriae* and proteins called *adhesins*, which enable the bacteria to attach tightly to human cells.
- *Exotoxins* that cause a variety of symptoms such as diarrhea. The genes for exotoxins, fimbriae, and adhesins are frequently located on plasmids, which increases the likelihood that they will be transferred among bacteria.
- *Iron-binding compounds* called **siderophores** that capture iron and make it available to the bacteria.
- *Hemolysins,* which release nutrients such as iron by lysing red blood cells.
- A so-called *type III secretion system,* a complex structure composed of 20 different polypeptides that is synthesized by several pathogenic enteric species. Once assembled, the system spans the two membranes and peptidoglycan of the bacterial cell and inserts into a host cell's cytoplasmic membrane like a hypodermic needle. The bacterium then introduces proteins into the host cell. Because bacteria synthesize type III systems only after they have contacted a host cell, the proteins of a type III system are not exposed to immune surveillance.

Figure 20.8 illustrates the locations of the major types of antigens and virulence factors of enteric bacteria. **ANIMATIONS:** ***Virulence Factors: Enteric Pathogens***

Diagnosis, Treatment, and Prevention of Diseases of the Enterobacteriaceae

The presence of enteric bacteria in urine, blood, or cerebrospinal fluid is always diagnostic of infection or disease. To culture members of the Enterobacteriaceae from clinical specimens, clinicians use selective and differential media such as eosin methylene blue (EMB) agar and MacConkey agar. In their role as selective media, these media inhibit the growth of many members of the normal microbiota and allow the growth of enteric bacteria **(Figure 20.9a)**; in their role as differential media, they enable laboratory technicians to distinguish between enteric bacteria that can and cannot ferment lactose **(Figure 20.9b)**. Commercially available systems using sophisticated biochemical tests can accurately distinguish among enteric bacteria in 4–24 hours.

Treatment of diarrhea involves treating the symptoms with fluid and electrolyte replacement. Often, antimicrobial drugs are not used to treat diarrhea because diarrhea is typically self-limiting—expulsion of the organisms from the body is often a more effective therapy than the action of antimicrobials. Also, antimicrobial therapy can worsen the prognosis by

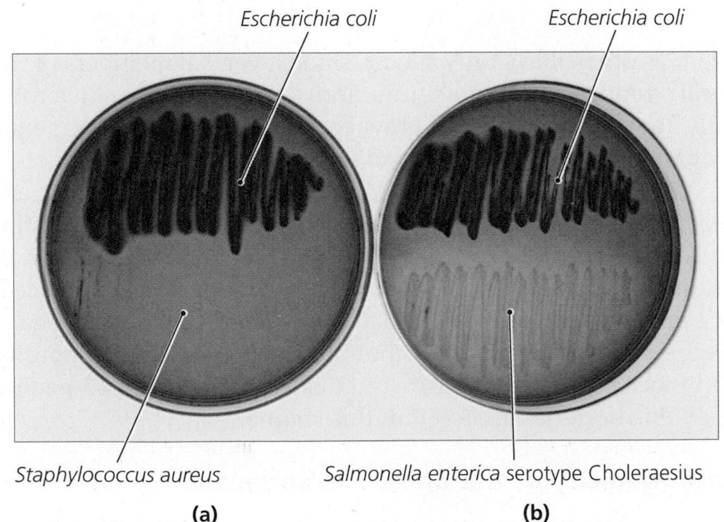

Figure 20.9 The use of MacConkey agar. (a) This selective and differential medium allows growth of Gram-negative bacteria (here, *E. coli*) and inhibits Gram-positive bacteria (here, *Staphylococcus aureus*). **(b)** Additionally, colonies of lactose-fermenters (here, *E. coli*) are pink, differentiating them from other enteric bacteria that do not ferment lactose (here, *Salmonella enterica*).

killing many bacteria at once, which releases a large amount of toxic lipid A.

Treatment of internal infections involves selection of an appropriate antimicrobial drug, which often necessitates a susceptibility test such as a Kirby-Bauer test (see Figure 10.9). Susceptible bacteria may develop resistance to antimicrobial drugs, presumably because enteric bacteria exchange plasmids readily; for example, many enterics can exchange DNA via conjugation. Antibiotics administered to control infections in the blood also kill the normal enteric bacteria, opening a niche for opportunistic resistant bacteria such as the endospore-forming, Gram-positive bacterium *Clostridium difficile* (klos-trid′ē-ŭm di-fi′sil-ē).

Given that enteric bacteria are a major component of the normal microbiota, preventing infections with them is almost impossible. Health care practitioners give prophylactic antimicrobial drugs to patients whenever mucosal barriers are breached by trauma or surgery, while avoiding prolonged unrestricted use of antimicrobials, a practice that selects for resistant strains of bacteria. Good personal hygiene (particularly handwashing) and proper sewage control are important in limiting the risk of infection. Specific methods of control and prevention associated with particular genera of enteric bacteria are discussed in the following sections.

CLINICAL CASE STUDY

A Painful Problem

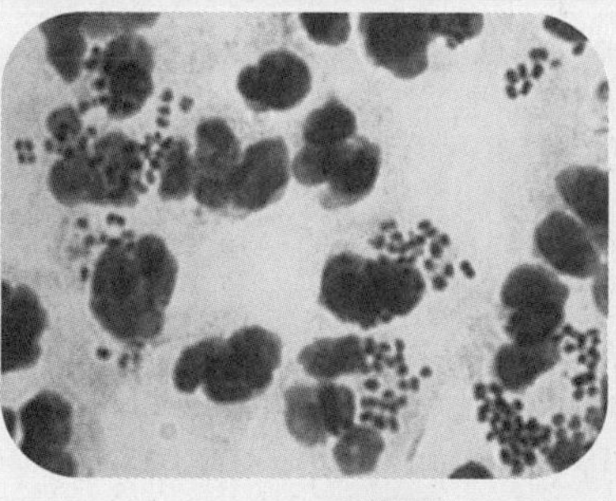

A 20-year-old male reports to his physician that he has experienced painful urination, as if he were urinating molten solder. He has also noticed a puslike discharge from his penis. A stained smear of the discharge is shown in the photo.

The patient reports having been sexually active with two or three women in the previous six months. Because his partners reported being "absolutely sure" that they carried no sexually transmitted diseases, the patient had not used a condom.

1. What disease does this patient most likely have? What are the medical and common names for this disease?
2. How did the patient acquire the disease?
3. What could explain the lack of any history of sexually transmitted disease in his sexual partners?
4. What is the likely treatment?
5. Is the patient immune to future infections with this bacterium?

CRITICAL **THINKING**

Resistance to antimicrobial agents is more commonly seen in hospital-acquired infections with enteric bacteria than in community-based infections with the same species. Explain why this is so.

Scientists, researchers, and health care practitioners find it useful to categorize the pathogenic members of the Enterobacteriaceae into three groups:

- Coliforms, which rapidly ferment lactose, are part of the normal microbiota, and may be opportunistic pathogens.
- Noncoliform opportunists, which do not ferment lactose.
- True pathogens.

In the next three sections we will examine the important members of the Enterobacteriaceae according to this scheme, beginning with opportunistic coliforms.

Coliform Opportunistic Enterobacteriaceae

Learning Objectives

✓ Compare and contrast *Escherichia, Klebsiella, Serratia, Enterobacter, Hafnia,* and *Citrobacter.*

✓ Describe the pathogenesis and diseases of *Escherichia coli* O157:H7.

Coliforms (kol′i-formz) are defined as aerobic or facultatively anaerobic, Gram-negative, rod-shaped bacteria that ferment lactose to form gas within 48 hours of being placed in a lactose broth at 35°C. Coliforms are found in the intestinal tracts of animals and humans, in soil, and on plants and decaying vegetation. The presence of fecal coliforms in water is indicative of impure water and of poor sewage treatment. The most common of the opportunistic coliform pathogens are bacteria in the genera *Escherichia, Klebsiella, Serratia, Enterobacter, Hafnia,* and *Citrobacter.*

Escherichia coli

Escherichia coli (esh-ĕ-rik′ē-ă kō′lē)—the most common and important of the coliforms—is well known as a laboratory research organism.

Scientists have described numerous O, H, and K antigens of *E. coli*, which epidemiologists use to identify particular strains. Some antigens, such as O157, O111, H8, and H7, are associated with virulence. Virulent strains also have genes (located on virulence plasmids) for fimbriae, adhesins, and a variety of exotoxins, enabling these strains to colonize human tissue and cause disease. Such plasmids can be transferred among bacteria.

E. coli is responsible for a number of diseases, including septicemia, urinary tract infections (UTIs), neonatal meningitis, and gastroenteritis. *E. coli* in the blood can colonize the lining of the heart, triggering inflammation and the destruction of heart valves.

Gastroenteritis, the most common disease associated with *E. coli*, is often mediated by exotoxins, called *enterotoxins*, that bind to proteins on cells lining the intestinal tract. A portion of the toxin then enters the cell and triggers a series of chemical reactions that cause the loss of sodium, chloride, potassium, bicarbonate, and

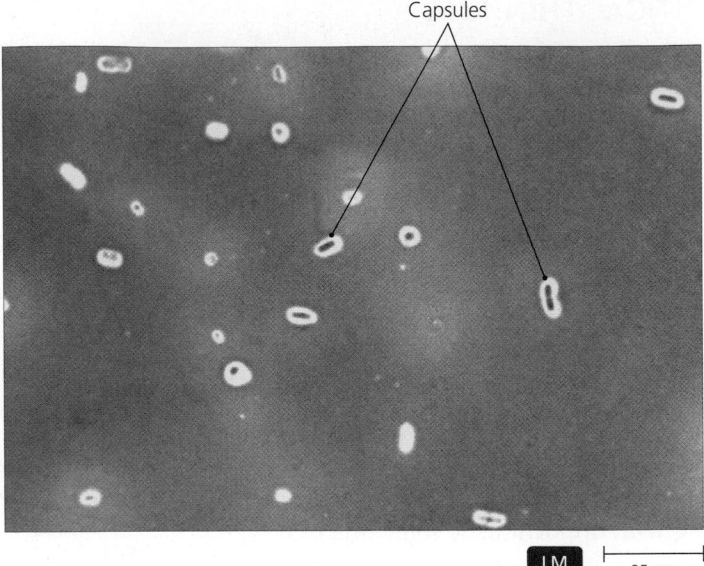

▲ **Figure 20.10 The prominent capsule of *Klebsiella pneumoniae*.** *How does the capsule function as a virulence factor?*

Figure 20.10 Capsules inhibit phagocytosis and intracellular digestion by phagocytic cells.

water from the cells, producing watery diarrhea, cramps, nausea, and vomiting. Enterotoxin-producing strains are common in developing countries and are important causes of pediatric diarrhea, accounting for up to a third of the cases of this life-threatening disease in some countries.

E. coli is the most common cause of non-nosocomial urinary tract infections. Women, with their shorter urethras, are 15 times more likely to acquire UTIs than are men. Surprisingly, girls are less likely to acquire UTIs than are women, perhaps because of the hormonal influences of adulthood or an increased likelihood of UTIs following sexual intercourse. About 32% of adult women will suffer uncomplicated mild bladder infections by *E. coli* sometime during their lifetimes. In some cases, the bacterium ascends the ureters from the bladder, infecting the kidneys and causing a more serious disease called *acute pyelonephritis,* which involves fever, flank pain, bacteriuria (bacteria in the urine), profuse perspiration, nausea, and vomiting.

E. coli O157:H7, first described as a cause of human illness in 1982, is now the most prevalent strain of pathogenic *E. coli* in developed countries. The ingestion of as few as 10 organisms can result in disease, which ranges from bloody diarrhea, to severe or fatal hemorrhagic colitis, to a severe kidney disorder called *hemolytic uremic syndrome* in which renal function ceases.

Urinary tract and blood infections with this organism can be fatal, especially in children and immunocompromised adults. Most epidemics of *E. coli* O157:H7 have been associated with the consumption of undercooked ground beef or unpasteurized milk or fruit juice contaminated with feces. Despite its notoriety, fewer than 1 in 10 million deaths in the United States is caused by a U.S. citizen experiencing death or injury from strain O157:H7 in ground beef.

E. coli O157:H7 produces a type III secretion system through which it introduces two types of proteins into intestinal cells.

CLINICAL CASE STUDY

A Heart-Rending Experience

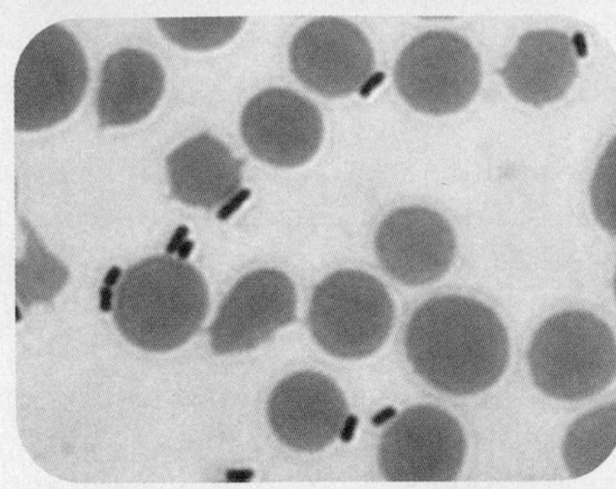

▲ ***E. coli* in a blood smear.**

A construction worker in his mid-30s visited his physician complaining of shortness of breath and a persistent cough that had lasted more than three weeks. Upon examination the doctor noted that the client had a history of asthma and intravenous drug use. Believing him to have an upper respiratory disturbance, the physician ordered a steroid to reduce swelling in the airways.

Two weeks later the construction worker's condition was worse, so he sought help at an emergency room. When he arrived, his heart rate was accelerated, and he appeared to be dehydrated. The ER staff administered a rapid IV infusion of fluids, took blood for analysis, and X-rayed his chest. Shortly after the infusion, the man's respiratory situation deteriorated to the point that he was placed on a ventilator. The X ray revealed an abnormally enlarged heart and fluid throughout both lungs. The blood culture contained *Escherichia coli.*

1. How could the patient have gotten *E. coli* in his blood?
2. What had *E. coli* done to the inner structures of his heart?
3. What factors in the medical treatment may have exacerbated the disease? Why?

Proteins of one type disrupt the cell's metabolism; proteins of the other type become lodged in the cell's cytoplasmic membrane, where they act as receptors for the attachment of additional *E. coli* O157:H7 bacteria. Such attachment apparently enables this strain of *E. coli* to displace normal, harmless strains.

E. coli O157:H7 also produces *Shiga-like toxin,* which inhibits protein synthesis in host cells. Shiga-like toxin, which was first identified in *Shigella* (discussed shortly), attaches to the surfaces of neutrophils and is spread by them throughout the body, causing widespread death of host cells and tissues. Antimicrobial drugs induce *E. coli* O157:H7 to increase its production of Shiga-like toxin, exacerbating the disease.

Klebsiella

Species of the genus *Klebsiella* (kleb-sē-el'ă) grow in the digestive and respiratory systems of humans and animals. They are nonmotile, and their prominent capsules **(Figure 20.10)** give *Klebsiella* colonies a mucoid appearance and protect the bacteria from phagocytosis, enabling them to become opportunistic pathogens.

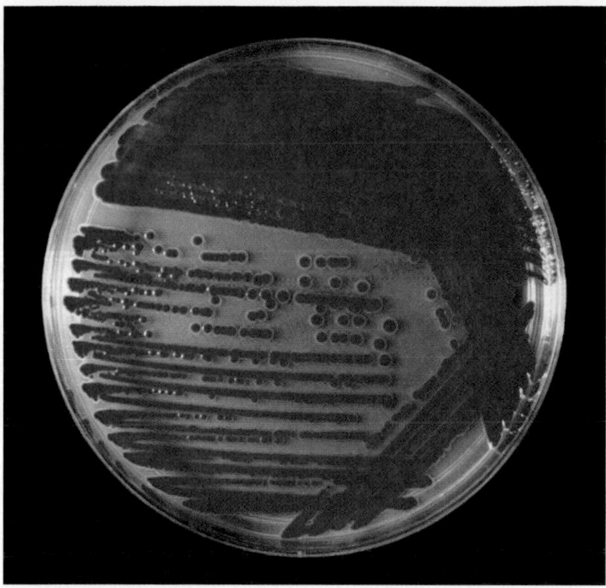

▲ **Figure 20.11 Red colonies of *Serratia marcescens.*** Color develops in colonies grown at room temperature.

▲ **Figure 20.12 The wavelike concentric rings of the swarming *Proteus microbilis.*** This characteristic of the bacterium is beautiful, but it makes isolation of *Proteus* difficult.

The most commonly isolated pathogenic species is *K. pneumoniae* (nū-mō'nē-ī), which causes pneumonia and may be involved in bacteremia, meningitis, wound infections, and UTIs. Pneumonia caused by *K. pneumoniae* often involves destruction of alveoli and the production of bloody sputum. Alcoholics and patients with compromised immunity are at greater risk of pulmonary disease because of their poor ability to clear aspirated oral secretions from their lower respiratory tracts.

Serratia

The motile coliform *Serratia marcescens* (se-rat'ē-a mar-ses'enz) produces a red pigment when grown at room temperature **(Figure 20.11)**. For many years, scientists considered *Serratia* to be totally benign; in fact, researchers intentionally introduced this bacterium into various environments to track the movement of bacteria in general. Using this method, scientists have followed the movement of *Serratia* from the mouth into the blood of dental patients, and throughout hospitals via air ducts. In a series of tests performed between 1949 and 1968, the U.S. government released large quantities of *S. marcescens* into the air over New York City and San Francisco to mimic the movement of a discharged biological weapon.

Now, however, we know that *Serratia* can grow on catheters, in saline solutions, and on other hospital supplies, and that it is a potentially life-threatening opportunistic pathogen in the urinary and respiratory tracts of immunocompromised patients. The bacterium is frequently resistant to antimicrobial drugs.

Enterobacter, Hafnia, and *Citrobacter*

Other motile coliforms that can be opportunistic pathogens are *Enterobacter* (en'ter-ō-bak'ter), *Hafnia* (haf'nē-ă), and *Citrobacter* (sit'rō-bak-ter). These bacteria, like other coliforms, ferment lactose and reside in the digestive tracts of animals and humans as well as in soil, water, decaying vegetation, and sewage. *Enterobacter* can be a contaminant of dairy products. All three genera are involved in nosocomial infections of the blood, wounds, surgical incisions, and urinary tracts of immunocompromised patients. Treatment of such infections can be difficult because these prokaryotes, especially *Enterobacter,* are often resistant to most antibacterial drugs.

Noncoliform Opportunistic Enterobacteriaceae

Learning Objectives

✓ Differentiate between coliform and noncoliform opportunists.

✓ Describe the diseases caused by noncoliform opportunistic enteric bacteria.

The family Enterobacteriaceae also contains genera that cannot ferment lactose; several of these genera are opportunistic pathogens. These noncoliform opportunists include *Proteus, Morganella, Providencia,* and *Edwardsiella.*

Proteus

Like the mythical Greek god Proteus, who could change shape, the Gram-negative, facultative anaerobe *Proteus* (pro'te-ŭs) is a typical rod-shaped bacterium with a few polar flagella when cultured in broth, but differentiates into an elongated cell that swarms by means of numerous peritrichous flagella when cultured on agar. Swarming cells produce concentric wavelike patterns on agar surfaces **(Figure 20.12)**. The roles of the two morphologies in human infections are not clear.

Proteus mirabilis (mi-ra'bi-lis) is the most common species of *Proteus* associated with disease in humans, particularly with urinary tract infections in patients with long-term urinary

MICROBE AT A GLANCE

Escherichia coli

Taxonomy: Domain Bacteria, phylum Proteobacteria, class Gammaproteobacteria, order Enterobacteriales, family Enterobacteriaceae

Cell morphology and arrangement: Bacilli, single

Gram reaction: Negative

Virulence factors: Capsule, fimbriae, endotoxin, adhesins, exotoxins (enterotoxins), type III secretion system; strain O157:H7 produces Shiga-like toxin

Diseases caused: Septicemia; urinary tract infections, including hemolytic uremic syndrome; neonatal meningitis; gastroenteritis, including diarrhea and severe to fatal hemorrhagic colitis

Treatment for diseases: Various antimicrobials

Prevention of disease: Prevent fecal contamination of food and water, good personal hygiene

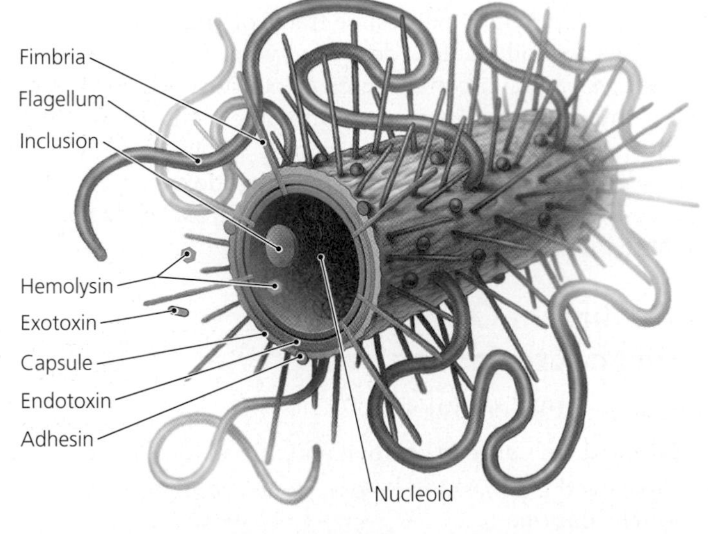

Download the Microbe at a Glance flashcards from the Study Area at www.masteringmicrobiology.com.

catheters. One study showed that 44% of catheterized patients had *P. mirabilis* growing in their urine.

In the presence of urea, as is found in the bladder, this microorganism releases a large amount of the enzyme *urease,* which breaks down urea into carbon dioxide and ammonia. Ammonia raises the pH of urine in the bladder and kidneys such that ions that are normally soluble at the acidic pH of urine (around pH 6.0) precipitate, often around bacterial cells, to form infection-induced kidney stones composed of magnesium ammonium phosphate or calcium phosphate. Kidney stones produced in this manner can contain hundreds of *P. mirabilis* cells.

Because *Proteus* is resistant to many antimicrobial drugs, health care workers must determine proper treatment following a susceptibility test. Researchers have been able to immunize mice against urinary tract infections with *Proteus*. This success is encouraging to those seeking protection for patients who must have long-term urinary catheters.

Morganella, Providencia, and *Edwardsiella*

Nosocomial infections of immunocompromised patients by *Morganella* (mōr'gan-el'-ă), *Providencia* (prov'i-den'sē-ă), and *Edwardsiella* (ed'ward-sē-el'lă) are becoming more frequent. These organisms are primarily involved in urinary tract infections. *Providencia* contains a plasmid that codes for urease and thus triggers the formation of kidney stones.

Truly Pathogenic Enterobacteriaceae

Learning Objectives

- ✓ Describe the diseases caused by truly pathogenic enteric bacteria.
- ✓ Contrast salmonellosis and shigellosis.
- ✓ Describe the life cycle of *Yersinia pestis* and contrast bubonic and pneumonic plague.

Three important non-lactose-fermenting bacteria in the family Enterobacteriaceae are *Salmonella, Shigella,* and *Yersinia.* These genera are not considered members of the normal microbiota of humans, because they are almost always pathogenic due to their numerous virulence factors. All three genera synthesize type III secretion systems through which they introduce proteins that inhibit phagocytosis, rearrange the cytoskeletons of eukaryotic cells, or induce apoptosis. We begin our discussion with *Salmonella,* which is the most common of the three pathogens.

Salmonella

Salmonella (sal'mŏ-nel'ă) is a genus of motile, Gram-negative, peritrichous bacilli that live in the intestines of virtually all birds, reptiles, and mammals and are eliminated in their feces. These bacteria do not ferment lactose, but they do ferment glucose, usually with gas production. They are urease negative and oxidase negative, and most produce hydrogen sulfide (H_2S). Scientists have identified more than 2000 unique serotypes (strains) of *Salmonella.* Analysis of the DNA sequences of their genes indicates that all of them properly belong to a single species—*S. enterica* (en-ter'i-ka). This taxonomically correct approach has not been well received by researchers or medical professionals, so we will consider the various strains of this single species with modern terminology as well as according to their traditional specific epithets, such as *S. typhi* (tī'fē) and *S. paratyphi* (par'a-tī'fē).

Most infections of humans with salmonellae result from the consumption of food contaminated with animal feces, often from pet reptiles. The pathogen is also common in foods containing poultry or eggs—about one-third of chicken eggs are contaminated. *Salmonella* released during the cracking of an egg on a kitchen counter and inoculated into other foods can reproduce into millions of cells in just a few hours. Some strains, such as *S. enterica* serotype Dublin, are acquired through the consumption of inadequately pasteurized contaminated milk. Infections with fewer than 1 million cells of most strains of *Salmonella* are usually asymptomatic in people with healthy immune systems.

Larger infective doses can result in **salmonellosis** (sal'mo-nel-o'sis), which is characterized by nonbloody diarrhea, nausea, and vomiting. Fever, myalgia (muscle pain), headache, and abdominal cramps are also common. The events in the course of salmonellosis are depicted in **Figure 20.13**. After the salmonellae pass through the stomach, they attach to the cells lining the small intestine (1). The bacteria then use type III secretion systems to insert proteins into the host cells, inducing the normally nonphagocytic cells to endocytize the bacteria (2). The salmonellae then reproduce within endocytic vesicles (3), eventually killing the host cells (4) and inducing the signs and symptoms of salmonellosis. Cells of some strains can subsequently enter the blood, causing bacteremia (5) and localized infections throughout the body, including in the lining of the heart, the bones, and the joints.

Humans are the sole hosts of *Salmonella enterica* serotype Typhi *(S. typhi)*, which causes **typhoid fever.** (A mild form of typhoid fever is also produced by serotype Paratyphi.) Infection occurs via the ingestion of food or water contaminated with sewage containing bacteria from carriers, who are often asymptomatic. (For the story of perhaps the most notorious asymptomatic carrier, see **Highlight: The Tale of "Typhoid Mary"** on p. 584.) An infective dose of *S. enterica* Typhi is only about 1000–10,000 cells. The bacteria pass through the intestinal cells into the bloodstream; there they are phagocytized but not killed by phagocytic cells, which carry the bacteria to the liver, spleen, bone marrow, and gallbladder. Patients typically experience gradually increasing fever, headache, muscle pains, malaise, and loss of appetite that may persist for a week or more. Bacteria may reproduce in the gallbladder and be released to reinfect the intestines, producing gastroenteritis and abdominal pain, and then a recurrence of bacteremia. In some patients the bacteria ulcerate and perforate the intestinal wall, allowing bacteria from the intestinal tract to enter the abdominal cavity and cause *peritonitis*. **Figure 20.14** contrasts the incidences of salmonellosis and typhoid fever in the United States since 1935.

Treatment for salmonellosis involves fluid and electrolyte replacement. Typhoid fever is treated with antimicrobial drugs such as ampicillin or ciprofloxacin. Physicians sometimes remove the gallbladder of a carrier to ensure that others are not infected. Researchers have developed vaccines that provide temporary protection for travelers to areas where typhoid fever is still endemic.

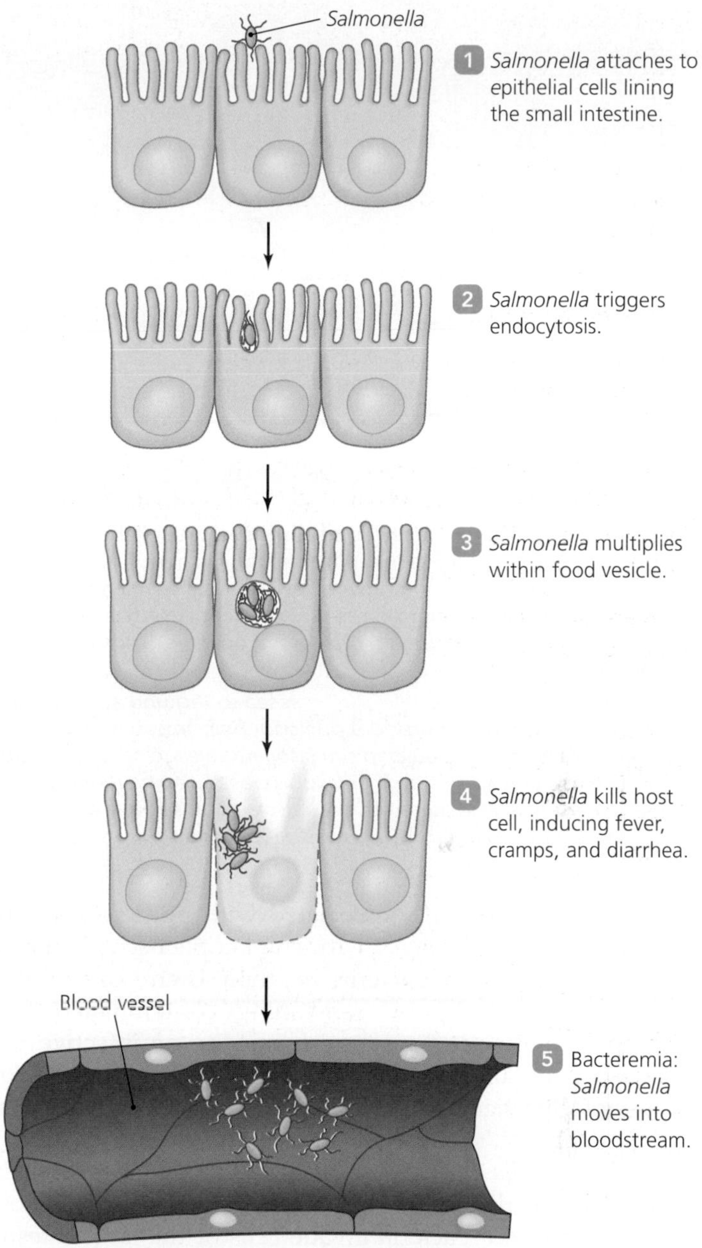

Salmonella
1 Salmonella attaches to epithelial cells lining the small intestine.
2 Salmonella triggers endocytosis.
3 Salmonella multiplies within food vesicle.
4 Salmonella kills host cell, inducing fever, cramps, and diarrhea.
Blood vessel
5 Bacteremia: Salmonella moves into bloodstream.

▲ **Figure 20.13 The events in salmonellosis.**

Shigella

The genus *Shigella* (shē-gel'lă) contains Gram-negative, oxidase-negative, nonmotile pathogens of the family Enterobacteriaceae that are primarily parasites of the digestive tracts of humans. These bacteria produce a diarrhea-inducing enterotoxin, are urease negative, and do not produce hydrogen sulfide gas. Some scientists suggest that *Shigella* may actually be a strain of *E. coli* that has become nonmotile and oxidase negative, while other researchers go so far as to consider *Shigella* an invasive, toxin-producing *E. coli* that is cloaked in *Shigella* antigens. Still other scientists contend the converse—that *E. coli* is really a disguised species of *Shigella* that has acquired genes for flagella!

In any case, taxonomists have historically identified four well-defined species of *Shigella: S. dysenteriae* (dis-en-te'rē-ī), *S. flexneri* (fleks'ner-ē), *S. boydii* (boy'dē-ē), and *S. sonnei* (sōn'ne-ē). *Shigella sonnei* is most commonly isolated in industrialized nations, whereas *S. flexneri* is the predominant species in developing countries. All four species cause a severe form of dysentery called **shigellosis** (shig-ĕ-lō'sis), which is characterized by abdominal cramps, fever, diarrhea, and purulent bloody stools.

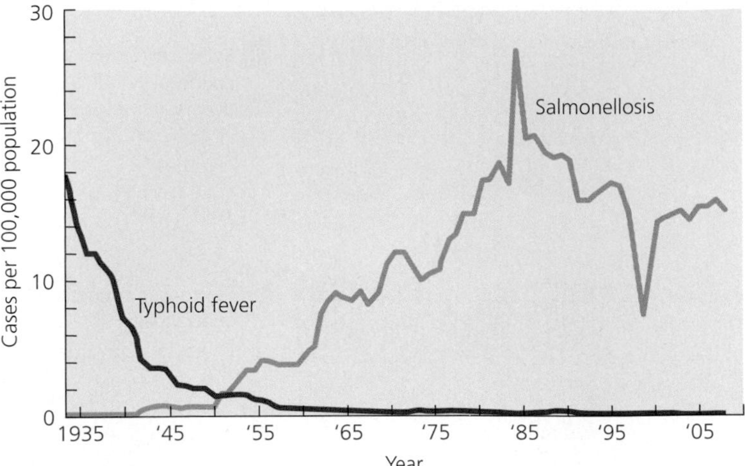
30
20
10
0
Cases per 100,000 population
Salmonellosis
Typhoid fever
1935
'45
'55
'65
'75
'85
'95
'05
Year

▲ **Figure 20.14 The incidences of diseases caused by *Salmonella* in the United States.** Whereas the incidence of typhoid fever has declined because of improved sewage treatment and personal hygiene, the incidence of reported cases of salmonellosis has increased. *What factors might explain the increase in the observed incidence of salmonellosis?*

Figure 20.14 *Since salmonellosis and typhoid fever are caused by the same species and transmitted the same way, and the incidence of typhoid fever has decreased significantly, it is likely that the actual number of cases of salmonellosis has declined. The observed increase may be due to better diagnosis and reporting, rather than to an actual increase in the number of cases.*

Shigellosis is primarily associated with poor personal hygiene and ineffective sewage treatment. People become infected primarily by ingesting bacteria on their own contaminated hands and secondarily by consuming contaminated food. *Shigella* is little affected by stomach acid, so an infective dose may be as few as 200 cells; therefore, person-to-person spread is also possible, particularly among children and among homosexual men.

Initially, *Shigella* colonizes cells of the small intestine, causing an enterotoxin-mediated diarrhea; the main events in shigellosis, however, begin once bacteria invade cells of the large intestine (**Figure 20.15**). When *Shigella* attaches to a large intestine epithelial cell (1), the cell is stimulated to endocytize the bacterium (2), which then multiplies within the cell's cytosol (3). (Note that this differs from salmonellosis, in which bacteria multiply within endocytic vesicles. *Shigella* then polymerizes the host's actin fibers, which push the bacteria out of the host cell and into adjacent cells (4), in the process evading the host's immune system. As the bacteria kill host cells, abscesses form in the mucosa (5); any bacteria that enter the blood from a ruptured abscess are quickly phagocytized and destroyed (6), so bacteremia is rarely a part of shigellosis. About 20,000 cases of shigellosis are reported each year in the United States, but because mild cases are rarely reported, this figure may represent only 1% of the actual total.

Shigella dysenteriae secretes an exotoxin called **Shiga toxin,** which stops protein synthesis in its host's cells. Shiga toxin is similar to the Shiga-like toxin of *Escherichia coli* O157:H7. Shigellosis caused by *S. dysenteriae* is more serious than that caused by other species of *Shigella,* with a mortality rate as high as 20%.

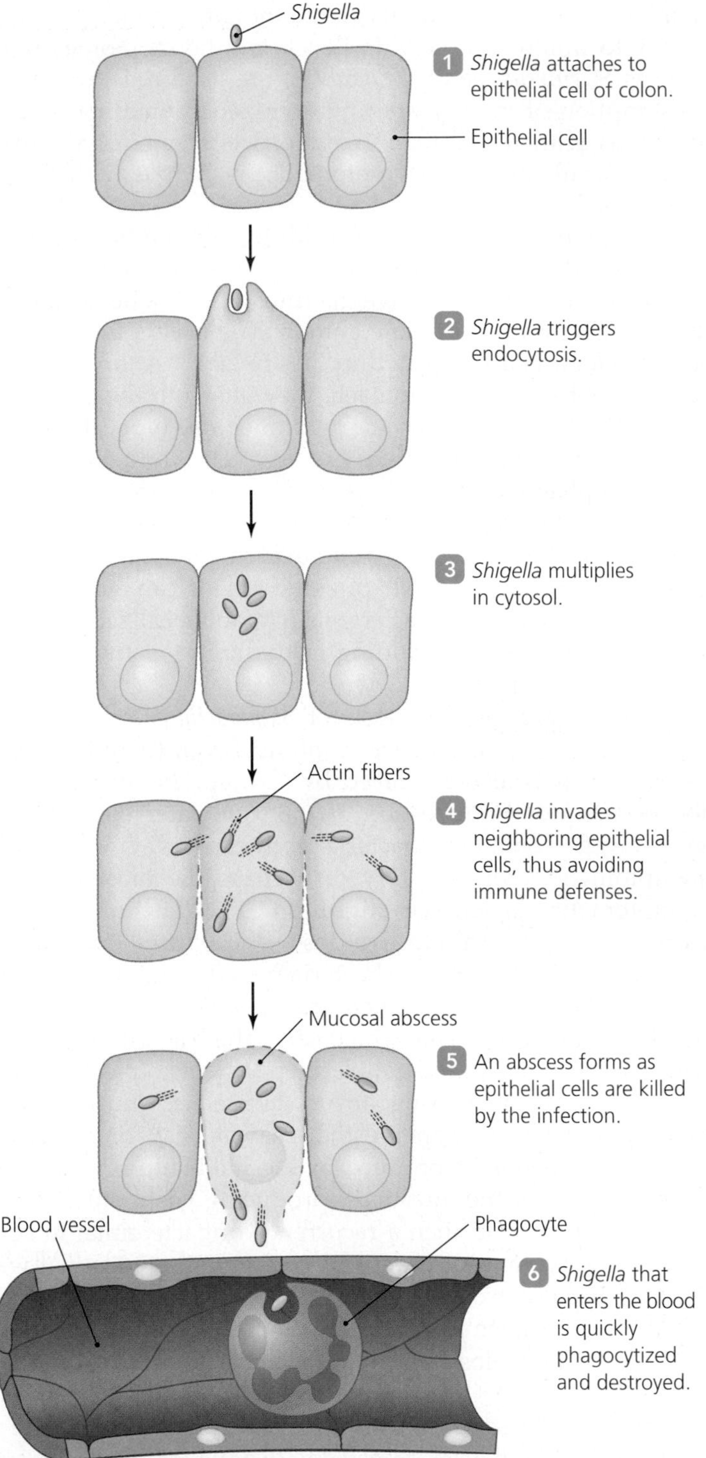
Shigella
1 Shigella attaches to epithelial cell of colon.
Epithelial cell
2 Shigella triggers endocytosis.
3 Shigella multiplies in cytosol.
Actin fibers
4 Shigella invades neighboring epithelial cells, thus avoiding immune defenses.
Mucosal abscess
5 An abscess forms as epithelial cells are killed by the infection.
Blood vessel
Phagocyte
6 Shigella that enters the blood is quickly phagocytized and destroyed.

▲ **Figure 20.15 The events in shigellosis.**

Treatment of shigellosis involves replacement of fluids and electrolytes. The disease is usually self-limiting, but oral antimicrobial drugs such as ciprofloxacin, sulfonamides, penicillin, or cephalosporin can reduce the spread of *Shigella* to close contacts of the patient.

A recent live, attenuated vaccine against *S. flexneri* has been successful in preventing dysentery caused by this species, although the participants in the study experienced mild diarrhea

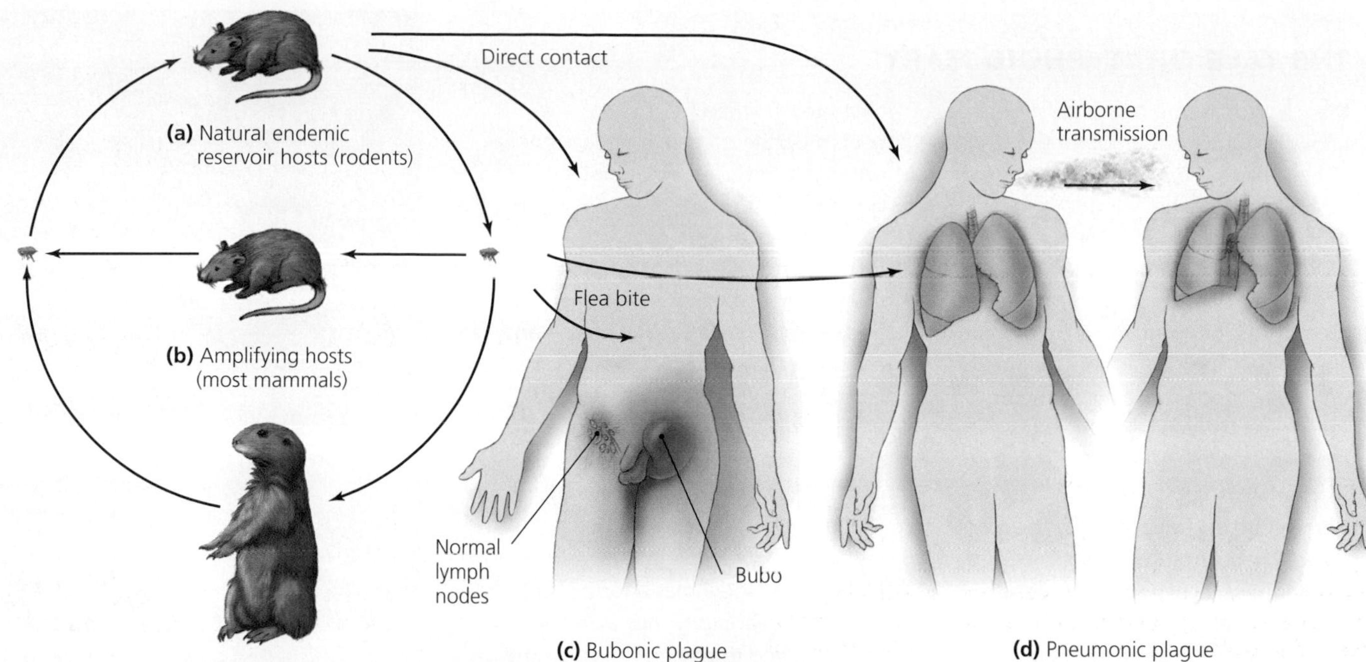

▲ **Figure 20.16 The natural history and transmission of *Yersinia pestis.*** This bacterium causes plague. **(a)** The natural endemic cycle of *Yersinia* among rodents. **(b)** The cycle involving amplifying mammalian hosts. **(c, d)** Humans develop one of two forms of plague as the result of the bite of infected fleas or direct contact with infected hosts. *Yersinia* can move from the bloodstream into the lungs to cause pneumonic plague, which can be spread between humans via the airborne transmission of the bacteria in aerosols. *Why is the plague so much less devastating today than it was in the Middle Ages?*

Figure 20.16 Urban living has reduced contact between people and flea vectors living on wildlife and farm animals; improved hygiene and the use of insecticides have reduced contact with flea vectors living on pets; and antimicrobials are effective against Y. pestis.

and fever as a result of immunization. Researchers are working to perfect a vaccine that will not cause symptoms.

Yersinia

The genus *Yersinia* (yer-sin'ē-ă) contains three notable species—*Y. enterocolitica* (en'ter-ō-ko-lit'ĭ-kă), *Y. pseudotuberculosis* (soo-dō-too-ber-kyū-lō'sis), and *Y. pestis* (pes'tis)—that are normally pathogens of animals. All three species contain virulence plasmids that code for adhesins and type III secretion systems. The adhesins allow *Yersinia* to attach to human cells, after which the type III system is used to inject proteins that trigger apoptosis in macrophages and neutrophils.

Yersinia enterocolitica and *Y. pseudotuberculosis* are enteric pathogens acquired via the consumption of food or water contaminated with animal feces. *Y. enterocolitica* is a common cause of painful inflammation of the intestinal tract; such inflammation is accompanied by diarrhea and fever that can last for weeks or months. Involvement of the terminal portion of the small intestine can result in painful inflammation of the mesenteric lymph nodes, which mimics appendicitis. *Y. pseudotuberculosis* produces a less severe inflammation of the intestines.

Yersinia pestis is an extremely virulent, nonenteric pathogen that has two clinical manifestations: **bubonic** (boo-bon'ik) **plague** and **pneumonic** (noo-mō'nik) **plague.** A major pandemic of plague that lasted from the mid-500s A.D. to the late 700s is estimated to have claimed the lives of more than 40 million people. This devastation was surpassed during a second pandemic that killed about a third of the population of Europe (25 million people) in just five years during the 14th century. A third major pandemic spread from China across Asia, Africa, Europe, and the Americas in the 1860s. So great was the devastation of these three great pandemics that the word "plague" still provokes a sense of dread today.

Rats, mice, and voles are the hosts for the natural endemic cycle of *Yersinia pestis* **(Figure 20.16a)**; they harbor the bacteria but do not develop disease. Among rodents, fleas are the vectors for the spread of the bacteria. As the bacteria multiply within a flea, they block the esophagus such that the flea can no longer ingest blood from a host. The starving flea jumps from host to host seeking a blood meal and infecting a new host with each bite. When other animals—including prairie dogs, rabbits, deer, camels, dogs, and cats—become infected via flea bites, they act as *amplifying hosts* **(Figure 20.16b)**; that is, they support increases in the numbers of bacteria and infected fleas, even though the amplifying hosts die from bubonic plague. Humans become infected either when bitten by infected fleas that have left their normal animal hosts or through direct contact with infected animals **(Figure 20.16c)**. Bubonic plague is not spread from person to person.

HIGHLIGHT

THE TALE OF "TYPHOID MARY"

Mary Mallon is one of the most famous cooks in history—but not for the tastiness of her dishes. An asymptomatic carrier of typhoid, Mary is believed to have caused outbreaks of typhoid fever in at least seven New York families for whom she worked as a private cook during the early 1900s. She first came to attention in 1906, when George Soper, a sanitation engineer hired to investigate an outbreak of typhoid fever in a wealthy Long Island household, became suspicious upon learning that the family's cook—Mary—had mysteriously disappeared three weeks after the illnesses began. Subsequent investigations revealed that Mary had a history of working in households whose members fell ill with typhoid. Public health authorities ultimately tracked Mary down and confirmed the presence of *Salmonella enterica* serotype Typhi in her stool and gallbladder.

Uneducated and headstrong, Mary never quite believed that she could be the cause of so much sickness—perhaps because she never fell ill herself—and she refused to have her gallbladder removed, as authorities suggested. Deemed a public health hazard, Mary was quarantined in 1907 and not released until 1910, with the provision that she was never to work as a cook again. Within a few years, however, she was discovered in Manhattan, working as a food preparer, and still spreading typhoid. She was again detained and remained in quarantine until her death, from a stroke, in 1938.

▲ *A cook from the time of Typhoid Mary.*

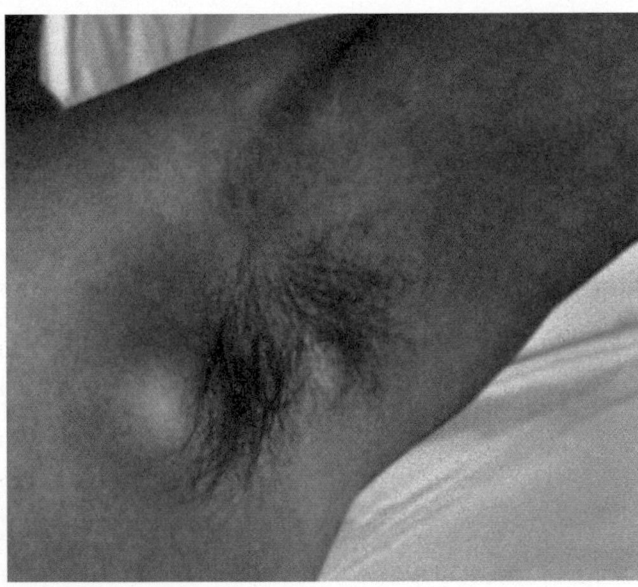

▲ **Figure 20.17 Bubo.** Such lymph nodes swollen by infection are a characteristic feature of bubonic plague.

Bubonic plague is characterized by high fever and swollen, painful lymph nodes called *buboes*[2] **(Figure 20.17)**, which develop within a week of a flea bite. Untreated cases progress to bacteremia, which results in disseminated intravascular coagulation, subcutaneous hemorrhaging, and death of tissues, which may become infected with *Clostridium* (klos-trid′ē-ŭm) and develop gangrene. Because of the extensive darkening of dead skin, plague has been called the "Black Death." Untreated bubonic plague is fatal in 50% of cases. Even with treatment, up to 15% of patients die. In fatal infections, death usually occurs within a week of the onset of symptoms.

Pneumonic plague occurs when *Yersinia* in the bloodstream infects the lungs. Disease develops very rapidly—patients develop fever, malaise, and pulmonary distress within a day of infection. Pneumonic plague can spread from person to person through aerosols and sputum **(Figure 20.16d)**, and if left untreated is fatal in nearly 100% of cases.

Because plague is so deadly and can progress so rapidly, diagnosis and treatment must also be rapid. The characteristic symptoms, especially in patients that have traveled in areas where plague is endemic, are usually sufficient for diagnosis. Rodent control and better personal hygiene have almost eliminated plague in industrialized countries, although wild animals remain as reservoirs. Many antibacterial drugs, including streptomycin, gentamycin, tetracycline, and chloramphenicol, are effective against *Yersinia*.

In summary, **Figure 20.18** illustrates the general sites of infections by the more common members of the Enterobac teriaceae.

The Pasteurellaceae

Learning Objectives

✓ Describe two pathogenic genera in the family Pasteurellaceae.

✓ Identify and describe three diseases caused by species of *Haemophilus*.

In the previous sections we examined opportunistic and pathogenic Gram-negative, facultative anaerobes of the family

[2]From Greek *boubon*, meaning groin, referring to a swelling in the groin.

▲ **Figure 20.18 Sites of infection by some common members of the Enterobacteriaceae.** For each site, genera are listed in their relative order of prevalence.

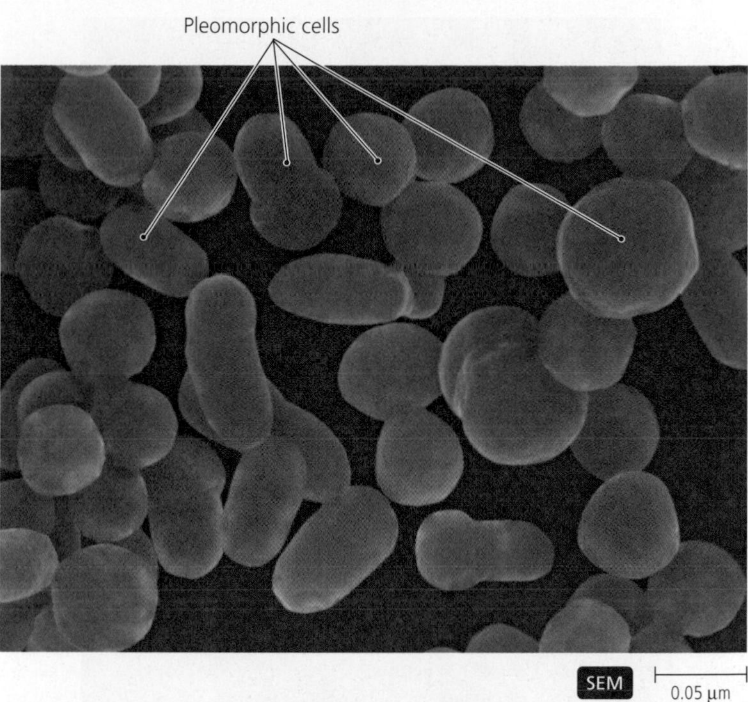

▲ **Figure 20.19** ***Haemophilus influenzae*** **is pleomorphic—it assumes many shapes.** *What simple lab test can be used to distinguish* Haemophilus *from* Escherichia?

Figure 20.19 *An oxidase test can distinguish* Haemophilus, *which is oxidase positive, from* Escherichia, *which is oxidase negative.*

Enterobacteriaceae, all of which are oxidase negative. Now we consider another family of gammaproteobacteria—the **Pasteurellaceae** (pas-ter-el-ă′sē-ē), which contains species that are oxidase positive (see Figure 20.4). These microorganisms are mostly facultative anaerobes that are small, nonmotile, and fastidious in their growth requirements—they require heme or cytochromes. Two genera, *Pasteurella* and *Haemophilus*, contain most of this family's pathogens of humans.

Pasteurella

Pasteurella (pas-ter-el′ă) is part of the normal microbiota in the oral and nasopharyngeal cavities of animals, including domestic cats and dogs. Humans are typically infected via animal bites and scratches or via inhalation of aerosols from animals. Most cases in humans are caused by *P. multocida* (mul-tŏ′si-da), which produces localized inflammation and swelling of lymph nodes at the site of infection. Patients with suppressed immunity are at risk of widespread infection and bacteremia.

Diagnosis of infection with *Pasteurella* depends on identification of the bacterium in cultures of specimens collected from patients. As noted, *Pasteurella* is fastidious in its growth requirements, so it must be cultured on blood or chocolate agar (autoclaved blood agar). A wide variety of antibacterial drugs are effective against *Pasteurella*, including penicillins, sulfonamides, trimethoprim, and fluoroquinolones.

Haemophilus

Haemophilus[3] (hē-mof′i-lŭs) species are generally small, sometimes pleomorphic bacilli **(Figure 20.19)** that require heme and NAD^+ for growth; as a result, they are obligate parasites, colonizing mucous membranes of humans and some animals. *Haemophilus influenzae* is the most notable pathogen in the genus, though *H. ducreyi* is an agent of a sexually transmitted disease. Other species in the genus primarily cause opportunistic infections.

Haemophilus influenzae Most strains of *Haemophilus influenzae* (in-flu-en′zī) have polysaccharide capsules that resist phagocytosis. Researchers distinguish among six strains by differences in the K antigens of the capsules. Type b usually causes 95% of *H. influenzae* diseases.

Belying its name, *H. influenzae* does not cause the flu; in fact, it is rarely found in the upper respiratory tract. Instead, *H. influenzae* b was the most common cause of meningitis in children 3–18 months of age before immunization brought it under control. The pathogen also causes inflammation of subcutaneous tissue, infantile arthritis, and life-threatening epiglottitis in children under four years old. In the latter disease, swelling of the epiglottis and surrounding tissues can completely block the pharynx.

Prompt treatment with antimicrobials is required for meningitis and epiglottitis. Over the past decade, the use of an effective vaccine—Hib, which is composed of capsular antigens conjugated to protein molecules—has virtually eliminated all

[3]From Greek *haima*, meaning blood, and *philos*, meaning love.

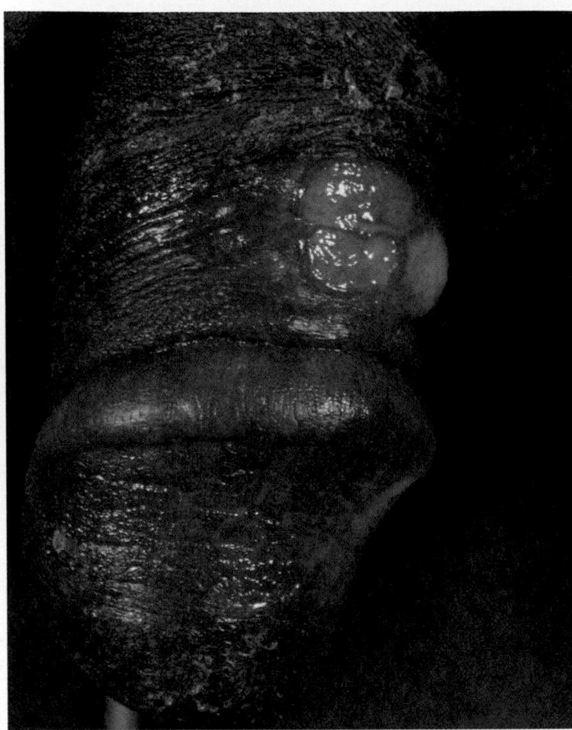

▲ **Figure 20.20 Chancroid (soft chancre).** This painful, soft, genital ulcer is caused by *Haemophilus ducreyi*.

disease caused by *H. influenzae* in the United States. Cuban scientists have produced a synthetic polysaccharide vaccine that is cheaper and purer, which may help poorer countries eliminate the disease as well.

The incidence of infections with strains other than type b has remained fairly constant in the United States. These strains of *H. influenzae* cause a variety of diseases, including conjunctivitis, sinusitis, dental abscesses, middle-ear infections, meningitis, bronchitis, and pneumonia. Strain *aegyptius* causes Brazilian purpuric fever, an extremely rapid pediatric disease characterized by conjunctivitis that within a few days is followed by fever, vomiting, abdominal pain, shock, and death. The pathogenesis of this disease is not understood, but fortunately it is rare in the United States.

CRITICAL **THINKING**

Haemophilus influenzae was so named because researchers isolated the organism from flu patients. Specifically, how could a proper application of Koch's postulates have prevented this misnomer?

Sexually Transmitted *Haemophilus* Sexually transmitted *Haemophilus ducreyi* (doo-krā'ē) causes a genital ulcer, called *chancroid* (shang' kroyd) or *soft chancre* (shang' ker) **(Figure 20.20)**, by producing a toxin that kills human epithelial cells. Chancroid is most often diagnosed in men because the ulceration is more visible and often painful in men, whereas the disease is often asymptomatic in women. Although prevalent in Africa and Asia, chancroid is rare in Europe and the Americas.

Thus far we have examined pathogenic Gram-negative cocci and the more commonly pathogenic, facultatively anaerobic, Gram-negative bacilli. Now we turn our attention to the pathogenic, Gram-negative, aerobic bacilli.

Pathogenic, Gram-Negative, Aerobic Bacilli

Pathogenic, Gram-negative, aerobic bacilli are a diverse group of bacteria in several classes of the phylum Proteobacteria. The following sections examine some of the more common species in the order they are presented in *Bergey's Manual: Bartonella* and *Brucella* in the class Alphaproteobacteria; *Bordetella* and *Burkholderia* in the class Betaproteobacteria; and several members of the class Gammaproteobacteria, including the pseudomonads (*Pseudomonas, Moraxella,* and *Acinetobacter*), *Francisella, Legionella,* and *Coxiella*.

Bartonella

Learning Objective

✓ Distinguish among bartonellosis, trench fever, and cat scratch disease.

Members of *Bartonella* (bar-tō-nel'ă), a genus of aerobic bacilli in the class Alphaproteobacteria, are found in animals, but they are known to cause disease only in humans. They have fastidious growth requirements and must be cultured on special media such as blood agar. Many require prolonged incubation in a humid, 37°C atmosphere supplemented with carbon dioxide before they form visible colonies. Three species (formerly classified in the genus *Rochalimaea*) are pathogenic in humans.

Bartonella bacilliformis (ba-sil'li-for'mis) invades and weakens erythrocytes, causing **bartonellosis**—an often fatal disease characterized by fever, severe anemia, headache, muscle and joint pain, and chronic skin infections. Very small bloodsucking sand flies of the genus *Phlebotomus* (fle-bot'ō-mŭs) transmit the bacterium, so the disease is endemic only in Peru, Ecuador, and Colombia, where such flies live.

Bartonella quintana (kwin'ta-nă) causes **trench fever** *(five-day fever)*, which was prevalent among soldiers during World War I. Human body lice transmit the bacterium from person to person. Many infections are asymptomatic, although in other patients severe headaches, fever, and pain in the long bones characterize the disease. The fever can be recurrent, returning every five days, giving the disease its alternative name. *B. quintana* also causes two newly described diseases in immunocompromised patients: *Bacillary angiomatosis* is characterized by fever, inflamed nodular skin lesions, and proliferation of blood vessels. *Bacillary peliosis hepatitis* is a disease in which patients develop blood-filled cavities in their livers.

Bartonella henselae (hen'sel-ī) causes **cat scratch disease** when the bacterium is introduced into humans through cat scratches and bites. Fleas may also transmit the bacterium from cats to people. Cat scratch disease has emerged as a relatively common and occasionally serious infection of children, affecting an estimated 22,000 children annually in the United States. This disease involves prolonged fever and malaise, plus localized swelling at the site of infection **(Figure 20.21)** and of local lymph nodes for several months. Serological testing of individuals who exhibit these characteristic signs and symptoms following exposure to cats confirms a diagnosis of cat scratch disease.

Gentamicin, erythromycin, azithromycin, or cephalosporin are used to treat most *Bartonella* infections, though there is a high rate of relapse.

Brucella

Learning Objective

✓ Describe brucellosis.

Brucella (broo-sel′lă) is a genus of small, nonmotile, aerobic coccobacilli that lack capsules but survive phagocytosis by preventing lysosomes from fusing with phagosomes containing the bacterium. Analysis of rRNA nucleotide base sequences reveals that *Brucella* is closely related to *Bartonella* in the class Alphaproteobacteria. Although *Brucella*-caused diseases in humans have been ascribed to four species, rRNA analysis has revealed the existence of only a single true species: *Brucella melitensis* (me-li-ten′sis); the other variants are strains of this one species. Because the classical specific epithets are so well known, they are mentioned here.

In animal hosts, the bacterium lives as an intracellular parasite in organs such as the uterus, placenta, and epididymis, but these organs are not infected in humans. Typically, infections in animals are either asymptomatic or cause a mild disease—**brucellosis** (broo-sel-ō′sis)—though they can cause sterility or abortion. Historically named *Brucella melitensis* infects goats and sheep; *B. abortus* (a-bort′us), cattle; *B. suis* (soo′is), swine; and *B. canis* (kā′nis), dogs, foxes, and coyotes.

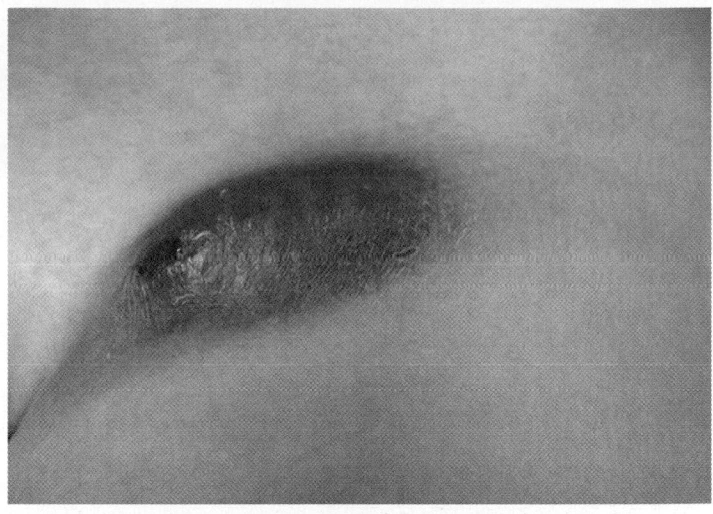

▲ **Figure 20.21 Cat scratch disease.** The localized swelling at the site of cat scratches or bites is characteristic.

Humans become infected either by consuming unpasteurized contaminated dairy products or through contact with animal blood, urine, or placentas in workplaces such as slaughterhouses, veterinary clinics, and feedlots. The bacterium enters the body through breaks in mucous membranes of the digestive and respiratory tracts.

Brucellosis in humans is characterized by a fluctuating fever—which gives the disease one of its common names: *undulant fever*—and chills, sweating, headache, myalgia, and weight loss. The disease in humans has been given a variety of other names, including *Bang's disease,* after microbiologist

BENEFICIAL MICROBES

NEW VESSELS MADE FROM SCRATCH?

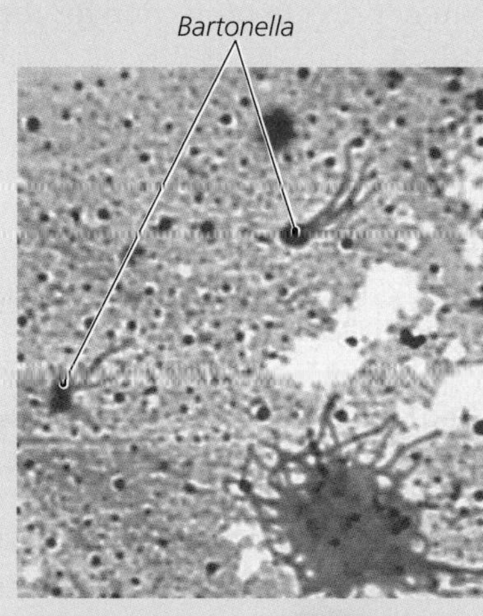

▲ *Bartonella.*

Bartonella henselae causes cat scratch disease in part by being able to live inside human red blood cells as well as in cells lining blood vessels. Scientists at Beth Israel Deaconess Medical Center and the Harvard Medical School in Boston, Massachusetts, discovered recently that the bacterium triggers angiogenesis—the formation of new blood vessels—in infected tissues. Researchers speculate that the pathogen is increasing its food supply and habitat by stimulating the growth of new blood vessels.

How *Bartonella* manages to orchestrate angiogenesis is mysterious. We do know that the bacterium is more efficient in laboratory conditions than is vascular endothelial growth factor—the body's natural angiogenic cytokine. Perhaps it is more efficient in the body as well.

Scientists speculate that a full understanding of *Bartonella*'s method could be harnessed to induce angiogenesis to circumvent blocked arteries in the heart, to prompt tissues to make new blood vessels in damaged limbs, or to speed up wound healing by increasing blood supply to damaged tissues. Investigators continue to probe the genetic basis of the intercellular communication between bacteria and host cells that allow this intriguing phenomenon. Perhaps in the future this pathogenic microbe will benefit patients with blood vessels grown from "scratch."

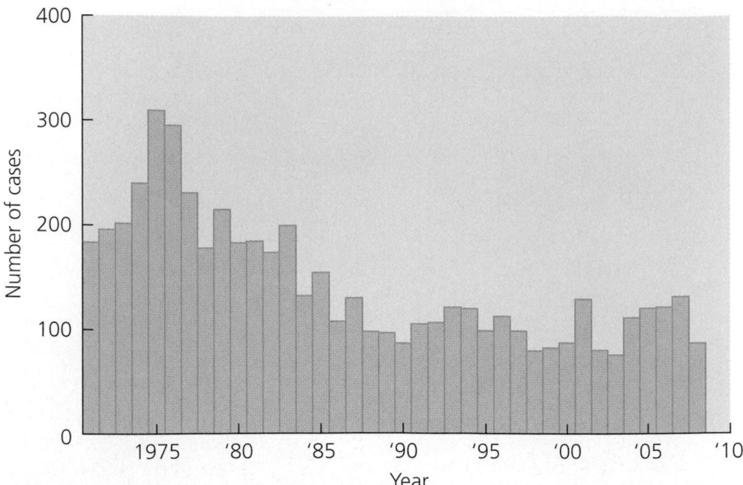

▲ **Figure 20.22 The incidence of brucellosis in humans in the United States, 1971–2008.** The decline in cases is largely the result of improvements in livestock management.

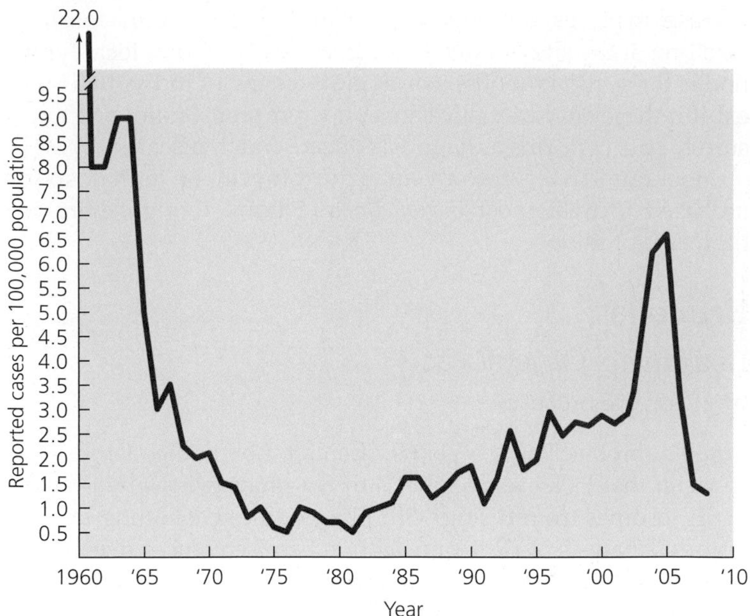

▲ **Figure 20.23 Reported cases of pertussis in the United States, 1960–2008.**

Bernhard Bang (1848–1932), who investigated the disease, and *Malta fever, rock fever of Gibraltar,* and *fever of Crete*, after localized epidemics in those locales.

Physicians treat brucellosis with doxycycline in combination with gentamycin, rifampin, or streptomycin. An attenuated vaccine for animals exists, but is not used in humans because the vaccine can cause disease. Through the vaccination of uninfected domesticated animals and the slaughter of infected ones, the threat of brucellosis has been reduced for U.S. residents **(Figure 20.22)**.

CRITICAL **THINKING**

Why do physicians substitute trimethoprim and sulfanilamide for tetracycline when treating children and pregnant women infected with *Brucella*?

Bordetella

Learning Objectives

✓ Describe five virulence factors of *Bordetella pertussis*.

✓ Identify the four phases of pertussis.

Bordetella pertussis (bōr-dĕ-tel′ă per-tus′is) is a small, aerobic, nonmotile, Gram-negative coccobacillus in the class Betaproteobacteria that is responsible for the disease **pertussis,**[4] commonly called *whooping cough. B. parapertussis* causes a milder form of pertussis.

Pathogenesis, Epidemiology, and Disease

Bordetella pertussis causes disease by interfering with the action of ciliated epithelial cells of the trachea. Various adhesins and toxins mediate the disease.

The bacterium attaches to certain lipids in the cytoplasmic membranes of tracheal cells via two adhesins: *filamentous hemagglutinin* and *pertussis toxin*. Filamentous hemagglutinin also binds to certain glycoproteins on the cytoplasmic membranes of neutrophils, initiating phagocytosis of the bacteria. *B. pertussis* survives within phagocytes, in the process evading the immune system. Pertussis toxin causes infected cells to produce more receptors for filamentous hemagglutinin, leading to further bacterial attachment and phagocytosis.

Four *B. pertussis* toxins are:

- *Pertussis toxin,* a portion of which interferes with the ciliated epithelial cell's metabolism, resulting in increased mucus production. (Note that pertussis toxin is both an adhesin and a toxin.)
- *Adenylate cyclase toxin,* which triggers increased mucus production and inhibits leukocyte movement, endocytosis, and killing. This function may provide protection for the bacterium early in an infection.
- *Dermonecrotic toxin,* which causes localized constriction and hemorrhage of blood vessels, resulting in cell death and tissue destruction.
- *Tracheal cytotoxin,* which at low concentrations inhibits the movement of cilia on ciliated respiratory cells, and at high concentrations causes the expulsion of the cells from the lining of the trachea.

Pertussis is considered a pediatric disease, as most cases are reported in children younger than five years old. More than 60 million children worldwide suffer from pertussis each year. There were over 10,000 reported cases in the United States in 2008 **(Figure 20.23)**. However, these figures may considerably underestimate the actual number of cases, because patients with chronic coughs are not routinely tested for infection with *Bordetella,* and

[4]From Latin *per*, meaning intensive, and *tussis*, meaning cough.

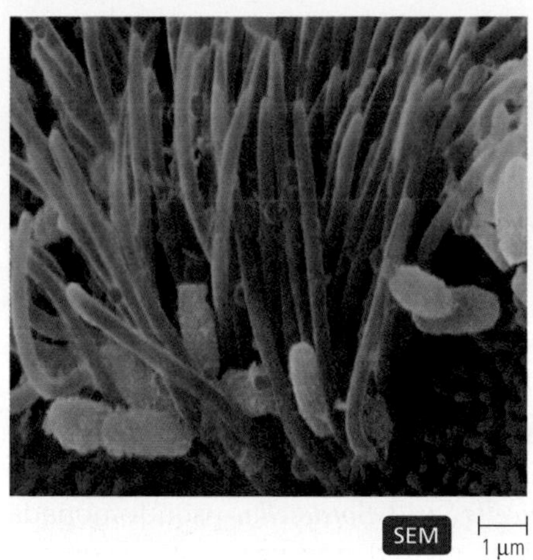

▲ **Figure 20.24** ***Bordetella pertussis.*** The bacteria attach to and infect ciliated epithelial cells, such as these of the trachea, eventually causing the death of the cells.

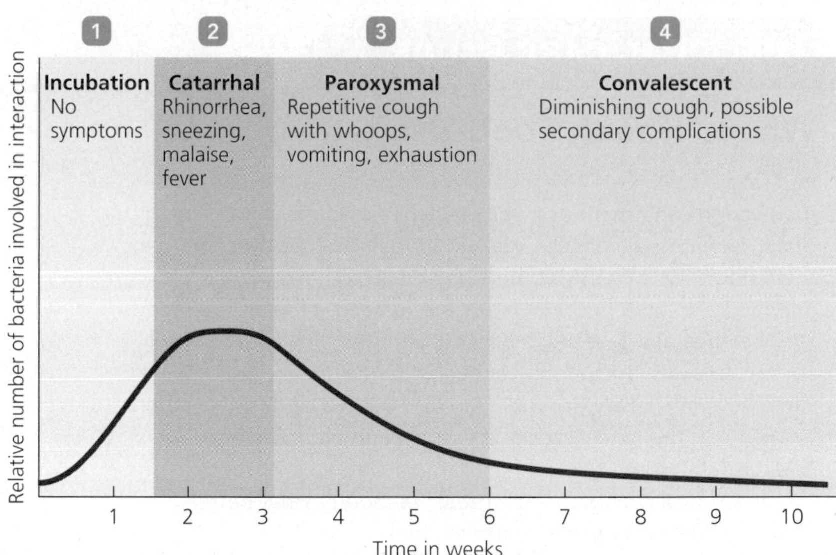

▲ **Figure 20.25 The approximate time course for the progression of pertussis.**

because the disease in older children and adults is typically less severe and is frequently misdiagnosed as a cold or influenza.

Pertussis begins when bacteria, inhaled in aerosols, attach to and multiply in ciliated epithelial cells **(Figure 20.24)**. Pertussis then progresses through four stages **(Figure 20.25)**:

1. During 7–10 days of **incubation** the bacteria multiply, but no symptoms are apparent.
2. The **catarrhal**[5] (kă-tă′răl) **phase** is characterized by signs and symptoms that resemble a common cold. During this phase, which lasts one to two weeks, the bacteria are most abundant and the patient is most infectious.
3. The **paroxysmal**[6] (par-ok-siz′mal) **phase** begins as the ciliary action of the tracheal cells is impaired, even as copious mucus is secreted. The condition worsens as ciliated cells are expelled. To clear the accumulating mucus, the body initiates a series of deep coughs, each of which is followed by a characteristic "whoop" caused by the intake of air through the congested trachea. Each day a patient may experience 40–50 coughing spells that often end with vomiting and exhaustion. During this phase, which can last two to four weeks, coughing may be so severe that oxygen exchange is limited, such that the patient may turn blue or even die. Severe coughing breaks ribs of some patients.
4. During the **convalescent phase,** the number of bacteria present becomes quite small, the ciliated lining of the trachea grows back, and the severity and frequency of coughing diminishes. During this phase, which typically lasts three to four weeks or longer, secondary bacterial infections (such as with *Staphylococcus* or *Streptococcus*) in the damaged epithelium may lead to bacteremia, pneumonia, seizures, and encephalopathy.

[5]From Greek *katarrheo*, meaning to flow down.
[6]From Greek *paroxysmos*, meaning to irritate.

Diagnosis, Treatment, and Prevention

The symptoms of pertussis are usually diagnostic, particularly when a patient is known to have been exposed to *Bordetella*. Even though health care workers can isolate *B. pertussis* from respiratory specimens, the bacterium is extremely sensitive to desiccation, so specimens must be inoculated at the patient's bedside onto *Bordet-Gengou* medium, which is specially designed to support the growth of this bacterium. If this is not practical, clinicians must use special transport media to get a specimen to a laboratory.

Treatment for pertussis is primarily supportive. By the time the disease is recognized, the immune system has often already "won the battle." Recovery depends on regeneration of the tracheal epithelium, not reduction of the number of bacteria; therefore, although antibacterial drugs reduce the number of bacteria and the patient's infectivity to others, they have little effect on the course of the disease.

Given that *B. pertussis* has no animal or environmental reservoir, and that effective vaccines (the P of DTaP and of Tdap) are available, whooping cough could be eradicated. Despite this possibility, over 10,000 cases occur each year in the United States. This is partly due to the refusal of parents to immunize their children and partly due to the fact that immunity, whether acquired artificially or naturally, lasts only about 10 years. The CDC urges parents to immunize their children and now recommends that all adults under age 65 receive one dose of acellular pertussis vaccine.

Burkholderia

Learning Objective

✓ Describe the advantageous metabolic features of *Burkholderia* as well as its pathogenicity.

CLINICAL CASE STUDY

When "Health Food" Isn't

In a single day, two 19-year-old women and one 20-year-old man sought treatment at a university health clinic, complaining of acute diarrhea, nausea, and vomiting. No blood was found in their stools. One of the women was found to have a urinary tract infection. All three had eaten lunch at a nearby health food store the previous day. The man had a turkey sandwich with tomato, sprouts, pickles, and sunflower seeds. One woman had a pocket sandwich with turkey, sprouts, and mandarin oranges; the other woman had the lunch special, described in the menu as a "delightful garden salad of fresh organic lettuces, sprouts, tomatoes, and cucumbers with zesty raspberry vinaigrette dressing." All had bottled water to drink.

1. Which of the foods is the most likely source of the infections?
2. What media would you use to culture and isolate enteric contaminants in the food?
3. Which enteric bacteria could cause these symptoms?
4. How did the woman likely acquire the urinary tract infection?
5. What is the likely treatment of choice?
6. What steps can the food store's manager and the students take to reduce the chance of subsequent infections?

Burkholderia cepacia (burk-hol-der'ē-ă se-pā'se-ă) is soil-dwelling, aerobic, flagellated betaproteobacterium. *Burkholderia* is noteworthy for its ability to decompose a broad range of organic molecules, making it a likely bacterium to assist in the cleanup of contaminated environmental sites. For example, *Burkholderia* is capable of digesting polychlorinated biphenyls (PCBs) and Agent Orange—an herbicide used extensively during the Vietnam War—which persist for long periods in the environment. Further, farmers can use *Burkholderia* to reduce the number of fungal infections of many plant crops, including peas, beans, alfalfa, canola, and cucumbers.

Unfortunately, *Burkholderia* can also grow in health care settings, metabolizing a variety of organic chemicals and resisting many antimicrobial drugs. The bacterium is one of the more important opportunistic pathogens of the lungs of cystic fibrosis patients, in whom it metabolizes the copious mucous secretions in the lungs. Evidence for patient-to-patient spread is clear, making it imperative that infected cystic fibrosis patients avoid contact with uninfected patients. Physicians treat *Burkholderia* infections with beta-lactams (penicillin, ceftriaxone, and aztreonam) and doxycycline.

Another soil-dwelling species, *Burkholderia pseudomallei* (soo-dō-mal'e-ē), causes *melioidosis* (mel'ē-oy-dō'sis), which is an Asian and Australian tropical disease that is emerging as a threat in other locales (see **Emerging Diseases: Melioidosis**). The CDC considers *B. pseudomallei* a potential agent of biological terrorism.

Pseudomonads

Learning Objectives

✓ Reconcile the apparent discrepancy between the ubiquitous distribution of pseudomonads with numerous virulence factors and the fact they cause so few diseases.

✓ Identify three genera of opportunistic pathogenic pseudomonads.

✓ Describe *Pseudomonas aeruginosa* as an opportunistic pathogen of burn victims and cystic fibrosis patients.

Pseudomonads (soo-dō-mō'nadz) are Gram-negative, aerobic bacilli in the class Gammaproteobacteria. Unlike the fastidious members of *Bartonella, Brucella,* and *Bordetella,* pseudomonads are not particular about their growth requirements. They are ubiquitous in soil, decaying organic matter, and almost every moist environment, including swimming pools, hot tubs, washcloths, and contact lens solutions. In hospitals they are frequently found growing in moist food, vases of cut flowers, sinks, sponges, toilets, floor mops, dialysis machines, respirators, humidifiers, and in disinfectant solutions. Some species can even grow using trace nutrients in distilled water. Pseudomonads utilize a wide range of organic carbon and nitrogen sources. The Entner-Doudoroff pathway is their major means of glucose catabolism rather than the more common Embden-Meyerhof pathway of glycolysis.

Here we examine three genera that are commonly isolated opportunistic pathogens—*Pseudomonas, Moraxella,* and *Acinetobacter*—beginning with the species of greatest medical importance: *Pseudomonas aeruginosa.*

Pseudomonas aeruginosa

Pseudomonas aeruginosa (soo-dō-mō' nas ā-roo-ji-nō'să) is somewhat of a medical puzzle in that even though it expresses a wide range of virulence factors, it rarely causes disease. This is fortunate because its ubiquity and inherent resistance to a wide range of antimicrobial agents would render a more virulent *Pseudomonas* a formidable challenge to health care professionals. That *P. aeruginosa* is *only* an opportunistic pathogen is testimony to the vital importance of the body's protective tissues, cells, and chemical defenses.

P. aeruginosa has fimbriae and other adhesins that enable its attachment to host cells. The fimbriae are similar in structure to those of *N. gonorrhoeae.* The bacterial enzyme *neuraminidase* modifies the fimbriae receptors on a host cell in such a way that attachment of fimbriae is enhanced. A mucoid polysaccharide capsule also plays a role in attachment, particularly in the respiratory system of cystic fibrosis patients (as discussed shortly). The capsule also shields the bacterium from phagocytosis.

P. aeruginosa synthesizes other virulence factors, including toxins and enzymes. Lipid A (endotoxin) is prevalent in the cell wall of *Pseudomonas* and mediates fever, vasodilation, inflammation, shock, and other symptoms. Two toxins—*exotoxin A* and *exoenzyme S*—inhibit protein synthesis in eukaryotic cells, contributing to cell death and interfering with phagocytic killing. The enzyme *elastase* breaks down elastic fibers, degrades complement components, and cleaves IgA and IgG. The bacterium also produces a blue-green pigment, called *pyocyanin,* that triggers the formation of superoxide radical (O_2^-) and peroxide anion (O_2^{2-}), two reactive forms of oxygen that contribute to tissue damage in *Pseudomonas* infections.

Although a natural inhabitant of bodies of water and moist soil, *P. aeruginosa* is rarely part of the normal human microbiota. Nevertheless, because of its ubiquity and its virulence factors, this opportunistic pathogen colonizes immunocompromised patients and is involved in about 10% of nosocomial infections. Once it breaches the skin or mucous membranes, *P. aeruginosa* can successfully colonize almost any organ or system. It can be involved in urinary, ear, eye, central nervous system, gastrointestinal, muscle, skeletal, and cardiovascular infections. Infections in burn victims and cystic fibrosis patients are so common they deserve special mention.

Pseudomonas infections of severe burns are pervasive (**Figure 20.26**). The surface of a burned area provides a warm, moist environment, that is quickly colonized by this ubiquitous opportunist; almost two-thirds of burn victims develop environmental or nosocomial *Pseudomonas* infections.

P. aeruginosa also typically infects the lungs of cystic fibrosis patients, forming a biofilm that protects the bacteria from phagocytes. Such infections exacerbate the decline in pulmonary function in these patients by causing certain lung cells to synthesize large amounts of mucus. As the bacteria feed on the mucus, they signal host cells to secrete more of it, creating a positive feedback loop; the result of such *P. aeruginosa* infections is that cystic fibrosis patients are more likely to require hospitalization and more likely to die.

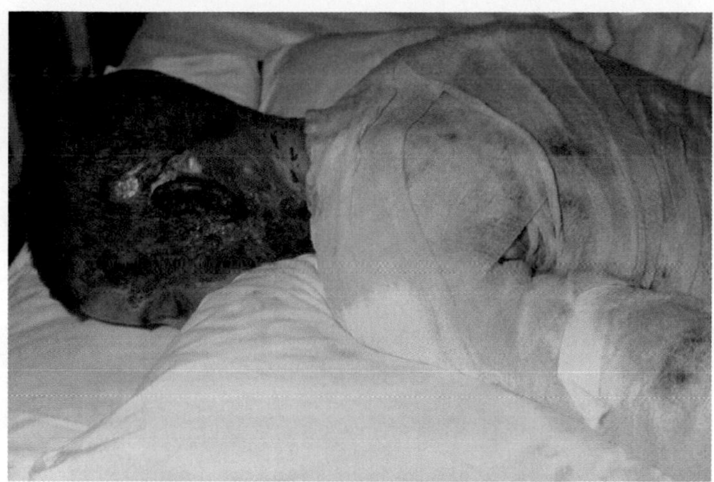

▲ **Figure 20.26 A *Pseudomonas aeruginosa* infection.** Bacteria growing under the bandages of this burn victim produce the green color. *What chemical is responsible for the blue-green appearance of this infection?*

Figure 20.26 Pyocyanin, a blue-green pigment secreted by P. aeruginosa, produces the color.

Diagnosis of *Pseudomonas* infection is not always easy because its presence in a culture may represent contamination acquired during collection, transport, or inoculation. Certainly,

EMERGING DISEASES

MELIOIDOSIS

Isabella felt lucky. How many community college students had the opportunity to help a professor do research in northern Australia's Kakadu National Park for a month, get college credit for the experience, and not have to pay for the trip? Though she was the oldest person on the trip and often felt her 45 years as they hiked the trails, she didn't complain. Of course things would have been better if it didn't seem to rain all the time in Australia and if the thorns didn't tear at her legs quite so regularly; still, the trip was an adventure.

A week after her return to Houston, Texas, Isabella developed a high fever (39.2°C) and general weakness. All other signs, including the results of blood work, appeared normal. The doctors, suspecting flu, told her to get plenty of rest and sent her home. Two days later Isabella was intermittently drowsy and confused, her breathing was labored, and a small cut on her leg was an inflamed, pus-filled lesion. She was admitted to the hospital and died the next day.

She died of melioidiosis, an emerging disease caused by a Gram-negative bacterium, *Burkholderia pseudomallei*, which is endemic to the tropics of Southeast Asia and which appears to be spreading into more moderate climes. Isabella had been infected either via inhalation or through a cut on her leg. Even with treatment, nearly 50% of melioidiosis patients die, including Isabella. For more about melioidiosis, see p. 590.

Track melioidiosis online by going to the Study Area at www.masteringmicrobiology.com.

BENEFICIAL MICROBES

WHEN A BACTERIAL INFECTION IS A GOOD THING

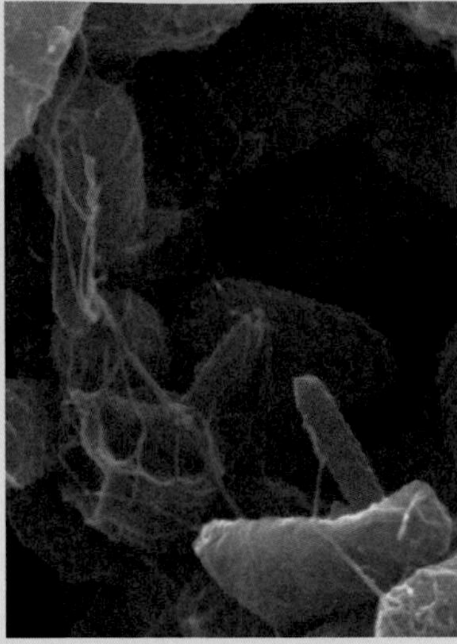

▲ ***Bdellovibrio*** **(pink) attacks** ***E. coli*** **(blue).**

Gram-negative bacteria are common opportunistic and nosocomial pathogens of the cardiovascular system, producing bacteremia, toxemia, endocarditis, and other serious conditions. However, Gram-negative bacteria can themselves be the target of bacterial pathogens, specifically cells of *Bdellovibrio* (del-ō-vib'rē-ō) and *Micavibrio* (mī-kă-vib'rē-ō). These Gram-negative predators are voracious devourers of other Gram-negative bacteria; in fact, their Gram-negative cousins are their only diet!

Bdellovibrio latches onto a Gram-negative bacterium such as *Escherichia coli* or the hard-to-treat *Pseudomonas aeruginosa*, enters its prey's periplasm, digests its host, feasts, replicates, and lyses its victim. *Bdellovibrio* daughters quickly attack other cells, reducing the victim's population a hundredfold in short order.

Micavibrio also attaches to its victim's outer membrane, but it remains outside the cell, replicating by binary fission while literally sucking the life (and cytoplasm) from its target. The predators can attack both free-swimming and biofilm-associated Gram-negative bacteria.

Scientists hope to identify, isolate, and utilize the unusual enzymes that allow *Bdellovibrio* and *Micavibrio* to exclusively attach to and kill Gram-negative bacteria. Alternatively, researchers are considering using the bacterial predators as living antimicrobial poultices on skin or wound infections or as living, intravenous, antimicrobial treatments for cardiovascular infections—a patient would be infected to get rid of an infection.

pyocyanin discoloration of tissues is indicative of massive infection.

Treatment of *P. aeruginosa* is also frustrating. The bacterium is notoriously resistant to a wide range of antibacterial agents, including antimicrobial drugs, soaps, antibacterial dyes, and quaternary ammonium disinfectants. In fact, *Pseudomonas* has been reported to live in solutions of antibacterial drugs and disinfectants.

Resistance of the bacterium is due to the ability of *Pseudomonas* to metabolize many drugs, to the presence of nonspecific proton/drug antiports that pump some types of drugs out of the bacterium, and to the ability of *Pseudomonas* to form biofilms, which resist the penetration of antibacterial drugs and detergents. Physicians treat infections with combinations of aminoglycoside, beta-lactam, and fluoroquinolone antimicrobials that have first proven efficacious against a particular isolate in a susceptibility test. Polymyxin, a drug that is also toxic to humans, is used sometimes as a last resort.

In summary, though *Pseudomonas aeruginosa* is ubiquitous and possesses a number of virulence factors, the bacterium rarely causes disease in healthy individuals because it cannot normally penetrate the skin and mucous membranes, nor ultimately evade the body's other defenses. Only in debilitated patients does this opportunist thrive.

Moraxella and *Acinetobacter*

Moraxella (mōr'ak-sel'ă) and *Acinetobacter* (as-i-nē'tō-bak'ter) are aerobic, short, plump bacilli that formerly were classified in the same family as *Neisseria*. Analysis of rRNA has shown that they are more properly classified as pseudomonads.

Moraxella catarrhalis (kā-tah'răl-is, formerly *Branhamella catarrhalis*) is rarely pathogenic but can cause opportunistic infections of the sinuses, bronchi, ears, and lungs. The bacterium is susceptible to cephalosporins, erythromycin, tetracycline, and most other antibacterial drugs (with the exception of beta-lactams).

Acinetobacter grows in soil, water, and sewage and is only rarely associated with disease in humans, though it is often isolated from clinical specimens. It is an opportunistic pathogen that causes infections of the respiratory, urinary, and central nervous systems. Endocarditis and septicemia have also been reported in infections with *Acinetobacter*. It is often resistant to antimicrobial drugs, so susceptibility tests must guide the choice of an effective treatment.

Francisella

Learning Objectives

✓ Describe the modes of transmission of *Francisella tularensis*.

✓ List practical measures that can be taken to prevent infection by *F. tularensis*.

Francisella tularensis[7] (fran'si-sel'lă too-lă-ren'sis) is a very small (0.2 μm × 0.2–0.7 μm), nonmotile, strictly aerobic, Gram-negative

[7]Named for Tulare County, California.

HIGHLIGHT

THE MYSTERY ON MARTHA'S VINEYARD

Martha's Vineyard, an island near Cape Cod, Massachusetts, is well known as a vacation destination for wealthy New Englanders. But in recent years, it has also caught the attention of the CDC, owing to mysterious outbreaks of tularemia that occur on the island in the summer.

Of special interest to the CDC was the observation that landscape workers seemed to be especially susceptible to contracting the disease, leading to speculation that the use of lawn mowers or other brush-cutting equipment may be releasing contaminated dust, soil, or grass into the air. Some wonder if mowing over grass that contains the contaminated feces of skunks or raccoons, or perhaps mowing over the bodies of dead infected rabbits, may be causing *F. tularensis* to become airborne. Others theorize that mosquitoes or another insect could be implicated in the spread of the disease. To date, no definitive answers have been found.

coccobacillus of the class Gammaproteobacteria that causes a zoonotic disease called **tularemia** (too-lă-rē'mē-ă). *F. tularensis* has a capsule that discourages (by unknown mechanisms) phagocytosis and intracellular digestion.

F. tularensis is found in temperate regions of the Northern Hemisphere living in water and as an intracellular parasite of animals and amoebae. Scientists do not know why it is not found below the equator. *Francisella* has an amazingly diverse assortment of hosts. It lives in mammals, birds, fish, bloodsucking ticks and insects, and amoebae. Indeed, one is hard pressed to find an animal that cannot be its host. The most common reservoirs in the United States are rabbits, muskrats, and ticks, which give the disease two of its common names—*rabbit fever* and *tick fever*.

Francisella is also incredibly varied in the ways it can be spread among hosts. Tularemia in humans is most often acquired either through the bite of an infected tick or via contact with an infected animal. The bacterium, by virtue of its small size, can pass through apparently unbroken skin or mucous membranes. Bloodsucking flies, mosquitoes, mites, and ticks also transmit *Francisella*, and humans can be infected by consuming infected meat, drinking contaminated water, or inhaling bacteria in aerosols produced during slaughter or in a laboratory. Fortunately, human-to-human spread does not occur.

Francisella is one of the more infectious of all bacteria: Infection requires as few as 10 organisms when transmitted by a biting arthropod or through unbroken skin or mucous membranes. For example, only about 10 cells must be inhaled to cause disease, although consumption of 10^8 cells in food or drink is necessary to contract the disease via the digestive tract.

Infections from bites, scratches, or through breaks in the skin cause lesions at the site of infection as well as swollen regional lymph nodes (buboes). Inhalation may produce buboes in the chest that put pressure on lungs and induce dry cough, pain during breathing, and death.

Only 106 cases were reported in 2008, but the actual number of infections was probably much higher considering its virulence, prevalence in animals, and multiple modes of transmission. Tularemia frequently remains unsuspected because its symptoms—fever, chills, headache, sore throat, muscle aches, and nausea—are not notably different from those of many other bacterial and viral diseases, and because it is difficult to confirm tularemia by using laboratory tests. Though it is typically innocuous, respiratory tularemia is fatal to more than 30% of untreated patients. Tularemia was removed from the list of nationally notifiable diseases in 1994, but concern about the possible use of *Francisella* by bioterrorists led officials in 2000 to relist it.

The bacterium produces beta-lactamase, so penicillins and cephalosporins are ineffective, but other antimicrobial drugs have been used successfully. Currently, streptomycin is recommended for use against *Francisella*. A vaccine made from less virulent strains provides protection for people at risk of exposure. The vaccine is not entirely effective but can lessen the severity of the disease.

To prevent infection, people should avoid the major reservoirs of *Francisella* (rabbits, muskrats, and ticks), especially in endemic areas. They can protect themselves from tick bites by wearing long clothing with tight-fitting sleeves, cuffs, and collars, and by using repellent chemicals. Because *Francisella* is not present in tick saliva, but only in its feces, prompt removal of ticks can mitigate infection. Hikers and hunters should never handle ill-appearing wild animals or their carcasses, and should wear gloves and masks when field dressing game. **Highlight: The Mystery on Martha's Vineyard** discusses recent tularemia outbreaks in Massachusetts.

CRITICAL **THINKING**

Most U.S. cases of tularemia occur in the late spring and summer months, and few cases occur in January. Why might this be so?

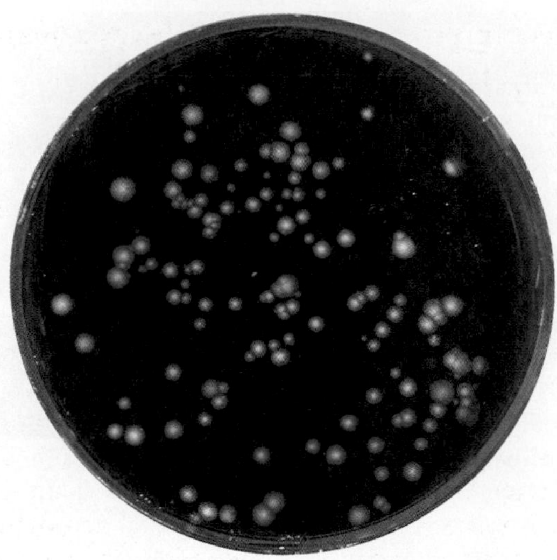

▲ **Figure 20.27** ***Legionella pneumophila*** **growing on buffered charcoal yeast extract agar.** This bacterium cannot be grown in culture without such special media.

Legionella

Learning Objectives

- ✓ Explain why *Legionella pneumophila* was unknown before 1976.
- ✓ Describe the symptoms and treatment of Legionnaires' disease.

In 1976, celebration of the 200th anniversary of the Declaration of Independence was curtailed in Philadelphia when hundreds of American Legion members attending a convention were stricken with severe pneumonia; 29 died. After extensive epidemiological research, this disease—dubbed **Legionnaires' disease** or *legionellosis*—was found to be caused by a previously unknown pathogen, which was subsequently named *Legionella* (lē-jŭ-nel'lă).

Pathogenesis, Epidemiology, and Disease

To date, scientists have identified over 40 species of *Legionella*. These aerobic, slender, pleomorphic bacteria in the class Gammaproteobacteria are extremely fastidious in their nutrient requirements, and laboratory media must be enhanced with iron salts and the amino acid cysteine. **Figure 20.27** shows colonies growing on one commonly used medium—buffered charcoal yeast extract agar. *Legionella* species are almost universal inhabitants of water, but they had not been isolated previously because they stain poorly and cannot grow on common laboratory media. Nineteen species are known to cause disease in humans, but 85% of all infections in humans are caused by *L. pneumophila*[8] (noo-mō'fi-lă).

Legionella pneumophila presented a conundrum for early investigators: How can such a fastidious bacterium be nearly ubiquitous in moist environmental samples? In the original epidemic, for example, *Legionella* was cultured from condensation in hotel air conditioning ducts, an environment that seems unsuitable for a microorganism with such demanding nutritional requirements. Investigations revealed that *Legionella* living in the environment invade freshwater protozoa, typically amoebae, and reproduce inside phagocytic vesicles. Thus the bacteria survive in the environment much as they survive in humans—as intracellular parasites.

Protozoa release bacteria-filled vesicles into the environment; alternatively, *Legionella* forms exit pores through a host cell's vesicular membrane and then through its cytoplasmic membrane. Humans acquire the disease by inhaling *Legionella* in aerosols produced by showers, vaporizers, spa whirlpools, hot tubs, air conditioning systems, cooling towers, and grocery store misters. *Legionella* was not a notable pathogen until such devices provided a suitable means of transmitting the bacterium to humans.

Legionnaires' disease is characterized by fever, chills, a dry nonproductive cough, headache, and pneumonia. Complications involving the gastrointestinal tract, central nervous system, liver, and kidneys are common. If not promptly treated, pulmonary function rapidly decreases, resulting in death in 20% of patients with normal immunity. Mortality is much higher in immunocompromised individuals, particularly kidney and heart transplant recipients.

L. pneumophila also causes a flulike illness called *Pontiac fever* (after the Michigan city where it was first described). This disease has symptoms similar to those of Legionnaires' disease, but it does not involve pneumonia and is not fatal.

Diagnosis, Treatment, and Prevention

Diagnosis is made by fluorescent antibody staining or other serological tests that reveal the presence of *Legionella* in clinical samples. The bacterium can also be cultured on suitable commercially available selective media such as buffered charcoal yeast extract.

Physicians use doxycycline, macrolides, or fluoroquinolones to treat Legionnaires' disease. Pontiac fever is self-limiting and requires no treatment.

Completely eliminating *Legionella* from water supplies is not feasible, as chlorination and heating are only moderately successful. However, the bacterium is not highly virulent, so reducing their number is typically a successful control measure.

Coxiella

Learning Objectives

- ✓ Explain how *Coxiella burnetii* survives desiccation.
- ✓ Describe the mode of transmission of Q fever.

Coxiella burnetii (kok-sē-el'ă ber-ne'tē-ē) is an extremely small, aerobic, obligate intracellular parasite **(Figure 20.28)** that grows and reproduces in the acidic environment within phagolysosomes. Its small size and dependence on the cytoplasm of eukaryotes for growth led early investigators to think it was a virus. However, its bacterial nature is unquestionable: It has a typical Gram-negative cell wall (albeit with minimal peptidoglycan), RNA and DNA (viruses typically have one or the other), functional ribosomes, and Krebs cycle enzymes. *Coxiella*

[8]From Greek *pneuma*, meaning breath, and *philos*, meaning love

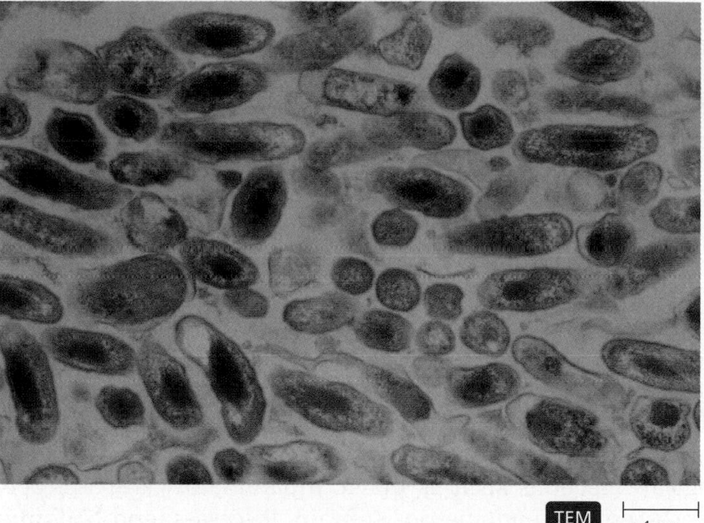

▲ **Figure 20.28** ***Coxiella burnetii.*** This intracellular pathogen grows and reproduces within phagolysosomes, here within one placental cell. An infective body of *Coxiella* can persist in the environment for years.

was originally classified with other obligate intracellular bacteria called *rickettsias* (discussed in Chapter 21), but rRNA analysis reveals that *Coxiella* is more properly classified with *Legionella* in the class Gammaproteobacteria.

Coxiella forms an internal, stable, resistant *infective body* (sometimes called a spore) that is similar in structure and function to the endospores of some Gram-positive species. The infective body enables the bacterium to survive harsh environmental conditions (such as desiccation and heat) for years.

C. burnetii infects a wide range of mammalian and avian hosts and is transmitted among them by feeding ticks. Farm animals and pets are the reservoirs most often associated with disease in humans. Transmission to humans does not occur via feeding ticks; instead, humans are most frequently infected by inhaling infective bodies that become airborne when released from dried tick feces or from the dried urine, feces, or placentas of host animals. Humans can also become infected by consuming contaminated unpasteurized milk.

Most *Coxiella* infections are asymptomatic, though this bacterium can cause **Q fever,** so named because its cause was *questionable* (unknown) for many years. Q fever occurs worldwide, particularly among ranchers, veterinarians, and food handlers, and it may be either an acute or a chronic condition. Acute Q fever follows an incubation period of 20 days or more; besides a high fever, it involves severe headache, chills, muscle pain, and mild pneumonia. In chronic Q fever, months to years may pass from the time of infection until life-threatening endocarditis develops. Inflammation of the lungs and liver may occur simultaneously.

Q fever is diagnosed via serological testing. Oral doxycycline is the preferred antimicrobial. The biggest problem in treating Q fever is getting antimicrobial drugs to work in the acidic environment of phagolysosomes, where the bacteria live. Researchers have developed an effective vaccine for Q fever, but it is not available in the United States. Prevention involves avoiding the inhalation of dust contaminated with barnyard and pet wastes.

Pathogenic, Gram-Negative, Anaerobic Bacilli

Learning Objective

✓ Compare and contrast *Bacteroides* and *Prevotella* with one another, and with other Gram-negative opportunists.

Over 50 species of strictly anaerobic Gram-negative bacteria are known to colonize the human body. Indeed, such anaerobic bacteria are the predominant microbiota of the gastrointestinal, urinary, reproductive, and lower respiratory tracts—they outnumber aerobic and facultatively anaerobic bacteria by a factor of 100 or more. The abundance of anaerobic bacteria in these locations is important for human health: they inhibit the growth of most pathogens, and in the intestinal tract they synthesize vitamins and vitamin precursors and assist in the digestion of food.

Relatively few of these anaerobic normal microbiota cause disease, and then only when they are introduced into other parts of the body by trauma or surgery. Almost all of the opportunistic species are aerotolerant anaerobes; they produce catalase or superoxide dismutase, which are enzymes that inactivate hydrogen peroxide and superoxide radicals. The more important anaerobic opportunists are *Bacteroides* and *Prevotella*.

Bacteroides

Bacteria in the genus *Bacteroides* (bak-ter-oy'dēz) are anaerobes that live as part of the normal microbiota of the intestinal tract and the upper respiratory tract. Being Gram-negative, they have an outer wall membrane containing lipopolysaccharide. However, in contrast to the lipid A of other Gram-negative bacteria discussed in this chapter, the lipid A formed by members of this genus has little endotoxin activity.

The most important pathogen in this genus is *Bacteroides fragilis* (fra'ji-lis). It is associated with about 85% of gastrointestinal diseases, even though it accounts for only about 1% of the bacteria in the colon. *B. fragilis* is a pleomorphic bacillus that produces a number of virulence factors. It attaches to host cells via fimbriae and a polysaccharide capsule, the latter of which also inhibits phagocytosis. Should the bacteria become phagocytized, short-chain fatty acids produced during anaerobic metabolism inhibit the activity of lysosomes, enabling the bacteria to survive within phagocytes.

B. fragilis is frequently involved in a variety of conditions, including abdominal infections (for example, those following a ruptured appendix), genital infections in females (for example, pelvic abscesses), and wound infections of the skin (which can result in life-threatening necrosis of muscle tissue). It is also involved in about 5% of all cases of bacteremia.

If anaerobic bacteria are to be isolated from patients, clinicians must maintain anaerobic conditions while collecting specimens, transporting them to the laboratory, and culturing them. Selective media are often used to reveal the presence of these bacteria. For example, because *B. fragilis* grows well in the presence of bile, it can be cultured on bile-esculin agar in an anaerobic environment—conditions which inhibit aerobic and most

▲ **Figure 20.29** ***Bacteroides fragilis.*** The bacterium produces a black color on bile-esculin agar. Bile suppresses the growth of most aerobes and facultative anaerobes.

anaerobic bacteria (**Figure 20.29**). Metronidazole is the antimicrobial of choice for treating *Bacteroides* infections.

Prevotella

Members of the genus *Prevotella* (prev'ō-tel'ă) are also anaerobic, Gram-negative bacilli. They differ from *Bacteroides* in that they are sensitive to bile; as a result, they do not grow in the intestinal tract, but are instead found in the urinary, genital, and respiratory tracts as part of the normal microbiota. Their virulence factors, including adhesive and antiphagocytic capsules and proteases that destroy antibodies, make them potential pathogens.

Prevotella is involved with other Gram-negative bacteria in about half of all sinus and ear infections and in almost all periodontal infections. Additionally, these bacteria cause gynecological infections such as pelvic inflammatory disease, pelvic abscesses, and endometriosis; brain abscesses; and abdominal infections.

Treatment of *Prevotella* infections involves surgical removal of infected tissue and the use of antimicrobial drugs.

Chapter Summary

1. Lipid A in the outer membranes of Gram-negative bacteria can stimulate symptoms of fever, vasodilation, and shock, as well as blood clots throughout the body, a condition known as **disseminated intravascular coagulation (DIC).**

Pathogenic Gram-Negative Cocci: *Neisseria* (pp. 570–573)

1. *Neisseria* is a pathogenic, Gram-negative, oxidase-positive coccus; its virulence results from the presence of fimbriae, polysaccharide capsules, and lipooligosaccharide (containing lipid A) in its cell walls.
2. *Neisseria gonorrhoeae* causes **gonorrhea,** a sexually transmitted disease of humans. In men it results in acute inflammation of the urethra, whereas in women it is generally asymptomatic. It can infect the uterus and uterine tubes to cause **pelvic inflammatory disease (PID).**
3. *Neisseria meningitidis* causes a type of meningitis; the bacterium is transmitted on respiratory droplets and is life threatening when it enters the bloodstream or central nervous system. Blebbing by the bacterium can release a dangerous level of lipid A.

Pathogenic, Gram-Negative, Facultatively Anaerobic Bacilli (pp. 573–586)

1. Members of the **Enterobacteriaceae,** the **enteric bacteria,** can be pathogenic, are oxidase negative, reduce nitrate to nitrite, and ferment glucose anaerobically. Their outer membranes contain a lipopolysaccharide, contributing to their virulence. Many produce **siderophores,** which capture iron and make it available to the bacteria.
ANIMATIONS: ***Virulence Factors: Enteric Pathogens***
2. Pathogenic enteric bacteria are grouped as coliform opportunists, noncoliform opportunists, and true pathogens. **Coliforms** are found in the intestinal tracts of animals and humans.
3. *Escherichia coli* is the most common and most widely studied coliform. It causes gastroenteritis, non-nosocomial urinary tract infections, fatal hemorrhagic colitis, and hemolytic uremic syndrome.
4. *Klebsiella, Serratia, Enterobacter, Hafnia,* and *Citrobacter* are genera of coliform bacteria. *K. pneumoniae* causes a type of pneumonia. All are involved in nosocomial infections.
5. Noncoliform opportunistic pathogens of the family Enterobacteriaceae include *Proteus* (urinary tract infections) and *Morganella, Providencia,* and *Edwardsiella,* which can cause nosocomial infections in immunocompromised patients.
6. Truly pathogenic enteric bacteria include *Salmonella enterica,* which causes **salmonellosis,** a serious form of diarrhea. *S. enterica* serotypes Typhi and Paratyphi cause **typhoid fever.**
7. Members of the genus *Shigella* cause **shigellosis,** a severe form of diarrhea. **Shiga toxin,** secreted by some *Shigella,* arrests protein synthesis in host cells.
8. Two enteric species, *Yersinia enterocolitica and Y. pseudotuberculosis,* cause different degrees of intestinal distress; the virulent, nonenteric *Y. pestis* causes **bubonic plague** and **pneumonic plague** and has had a major historical impact.
9. Two genera in the family *Pasteurellaceae—Pasteurella* and *Haemophilus*—are significant human pathogens. They differ from enteric bacteria by being oxidase positive.
10. *Haemophilus influenzae* causes infantile meningitis, epiglottitis, arthritis, and inflammation of the skin. Widespread immunization with Hib vaccine has almost eliminated disease caused by this pathogen in the United States. *H. ducreyi* is responsible for a sexually transmitted lesion called chancroid.

Pathogenic, Gram-Negative, Aerobic Bacilli (pp. 586–595)

1. *Bartonella* includes pathogenic aerobic bacilli. *B. bacilliformis,* transmitted by sand flies, causes **bartonellosis.** *B. quintana,* transmitted by lice, causes five-day fever, or **trench fever.** *B. henselae,* transmitted through cat scratches, bites, and fleas, causes **cat scratch disease.**
2. *Brucella,* transmitted in unpasteurized contaminated milk, causes **brucellosis,** also known as Bang's disease, undulant fever, Malta fever, and other names.
3. *Bordetella pertussis* is responsible for **pertussis** (whooping cough), a pediatric disease in which the ciliated epithelial cells of the trachea are damaged. After **incubation,** the disease progresses through three additional phases: the **catarrhal phase** resembles a cold; in the **paroxysmal stage** the patient coughs deeply to expel copious mucus from the trachea; and in the **convalescent phase** the disease subsides, but secondary bacterial infections may ensue.
4. *Burkholderia* can decompose numerous environmental pollutants and can assist farmers by inhibiting fungal pathogens of plants; however, the bacterium also grows in the lungs of cystic fibrosis patients.
5. Pseudomonads are ubiquitous opportunistic pathogens. *Pseudomonas aeruginosa* is involved in many nosocomial infections and is common in burn victims and cystic fibrosis patients. *Moraxella* and *Acinetobacter* are opportunistic pathogenic pseudomonads that rarely cause disease.
6. **Tularemia** (rabbit fever or tick fever) is a zoonotic disease caused by extremely virulent *Francisella tularensis,* which is so small that it can pass through apparently unbroken skin.
7. **Legionnaires' disease** (legionellosis) and Pontiac fever are caused by *Legionella,* a bacterium transmitted in aerosols such as those produced by air conditioning systems.
8. *Coxiella burnetii* live in phagolysosomes of mammal and bird cells; its infective bodies are transmitted via aerosolized dried feces and urine to cause **Q fever.**

Pathogenic, Gram-Negative, Anaerobic Bacilli (pp. 595–596)

1. Gram-negative anaerobes—*Bacteroides* and *Prevotella*—are part of the normal microbiota of the intestinal, urinary, genital, and respiratory tracts.
2. Gram-negative anaerobes may cause disease as a result of virulence factors such as capsules, fimbriae, and proteases that degrade antibodies, particularly when the bacteria are introduced into novel sites in the body.

Questions for Review

Answers to the Questions for Review (except Short Answer questions) begin on page A-1.

Multiple Choice

1. The presence of lipid A in the outer membranes of Gram-negative bacteria
   a. affects the formation of blood clots in the host.
   b. causes these bacteria to be oxidase positive.
   c. triggers the secretion of a protease enzyme to cleave IgA in mucus.
   d. enables enteric bacteria to ferment glucose anaerobically.
2. The only genus of Gram-negative cocci that causes significant disease in humans is
   a. *Pasteurella.*
   b. *Salmonella.*
   c. *Klebsiella.*
   d. *Neisseria.*
3. Which of the following bacterial cells is most likely to be virulent?
   a. a cell with fimbriae and LOS
   b. a cell with a polysaccharide capsule and lipooligosaccharide
   c. a cell with fimbriae, lipooligosaccharide, and a polysaccharide capsule
   d. a cell with fimbriae but no capsule
4. Which of the following statements is true?
   a. PID is a severe type of diarrhea in which infection spreads from the intestines to the bloodstream.
   b. PID can result from *Neisseria* infection.
   c. PID is more common in men than women.
   d. Members of the family Enterobacteriaceae usually cause PID.
5. A coliform bacterium that contaminates dairy products is
   a. *Bartonella.*
   b. *Serratia.*
   c. *Enterobacter.*
   d. *Proteus.*
6. Capsules of pathogenic enteric bacteria are virulence factors because they
   a. capture iron from hemoglobin and store it in the bacteria.
   b. release hemolysins that destroy red blood cells.
   c. produce fimbriae that enable the bacteria to attach to human cells.
   d. protect the bacteria from phagocytosis and from antibodies.
7. Which of the following bacteria might be responsible for the formation of petechiae in a host?
   a. *Neisseria meningitidis*
   b. *Escherichia coli* O157:H7
   c. *Klebsiella*
   d. *Proteus mirabilis*
8. The pathogen *Haemophilus influenzae* b causes
   a. meningitis in children.
   b. upper respiratory flu.
   c. endocarditis.
   d. genital chancroid.
9. Which of the following diseases is typically mild?
   a. Brazilian purpuric fever
   b. bartonellosis
   c. pediatric meningitis
   d. cat scratch disease
10. Which bacterium causes infections in many burn victims?
    a. *Moraxella catarrhalis*
    b. *Pseudomonas aeruginosa*
    c. *Escherichia coli*
    d. *Bartonella bacilliformis*
11. Which of the following statements is true of Q fever?
    a. For many years its cause was questionable.
    b. It was first described in 1976 during an outbreak in Quincy, Massachusetts.
    c. Researchers found it could be effectively treated with quinine.
    d. The sharp spikes of fever on patients' temperature charts resemble porcupine quills.
12. Which of the following is a bile-tolerant anaerobe?
    a. *Bacteroides*
    b. *Escherichia*
    c. *Shigella*
    d. *Prevotella*

Matching

Match the bacteria with the disease (or symptoms) it causes.

1. ____ *Escherichia coli*
2. ____ *Klebsiella pneumoniae*
3. ____ *Proteus mirabilis*
4. ____ *Salmonella enterica* serotype Typhi
5. ____ *Shigella flexneri*
6. ____ *Yersinia pestis*

A. Bubonic plague
B. Typhoid fever
C. Gastroenteritis
D. Kidney stones
E. Purulent bloody stools, cramps, fever, and diarrhea
F. Pneumonia

Labeling

Label the drawing below using these words: adhesin, exotoxin, H antigens, hemolysin, iron-binding protein, K antigens, lipid A, O antigen, fimbria, plasmid virulence genes, type III secretion system.

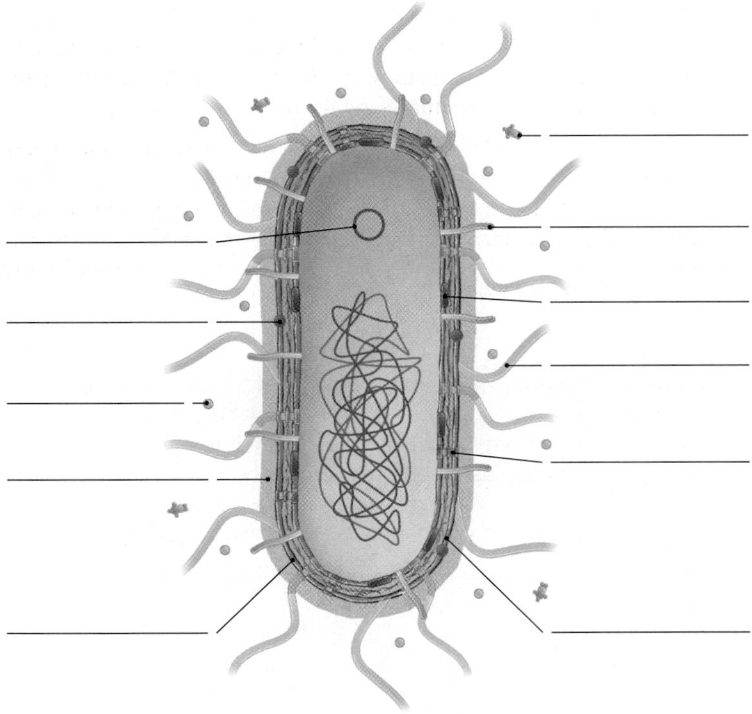

Short Answer

1. A physician prescribes fluid replacement to treat a patient with diarrhea. Although tests showed that a pathogenic enteric bacterium was the cause of the intestinal distress, why was an antimicrobial drug not prescribed?
2. Distinguish among the pathogenicity of coliforms, noncoliforms, and truly pathogenic enteric bacteria.
3. Why do nurses place antimicrobial agents in babies' eyes at birth?
4. Statistics show that meningococcal diseases are more frequent in college dormitories and military barracks than in the population at large. Suggest an explanation of this observation.
5. Why might an alcoholic be susceptible to pulmonary disease?
6. Name six factors that facilitate the production of disease by *Bordetella pertussis*.
7. Given that pseudomonads are present in almost every moist environment, why do they cause less disease than other, less prevalent Gram-negative bacteria?
8. Although an effective vaccine is available to eradicate pertussis in the United States, why has the number of reported cases increased since the 1970s?
9. What two illnesses can be caused by *Legionella pneumophila*?
10. What attribute of *Coxiella burnetii* enables it to survive desiccation and heat for an extended time?
11. What virulence factors allow Gram-negative anaerobes to cause disease?
12. What single biochemical test result distinguishes gammaproteobacteria in the family Enterobacteriaceae from gammaproteobacteria in the family Pasteurellaceae?
13. Describe transovarian transmission of a pathogen.

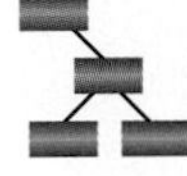

Concept Mapping

Using the following terms, draw a concept map that describes *E. coli*. For a sample concept map, see p. 93. Or, complete this concept map online by going to the Study Area at www.masteringmicrobiology.com.

Blood
Bloody diarrhea
Cystitis
Enterotoxin
Gastroenteritis
Gram negative bacilli
Hemolytic uremic syndrome
Hemorrhagic colitis
Indicators of fecal contamination
Kidneys
Meninges
Meningitis
Normal intestinal microbiota
O157:H7
Pyelonephritis
Septicemia
Shiga-like toxin
Undercooked ground beef
Unpasteurized fruit juices
Unpasteurized milk
Urinary bladder
Water supplies

Critical Thinking

1. A three-year-old boy complains to his day care worker that his head hurts. The worker calls the child's mother, who arrives 30 minutes later to pick up her son. She is concerned that he is now listless and unresponsive, so she drives straight to the hospital emergency room. Though the medical staff immediately treats the boy with penicillin, he dies—only four hours after initially complaining of a headache. What caused the boy's death? Were the day care workers or the hospital staff to blame for his death? What steps should be taken to protect the other children at the day care facility?

2. An epidemiologist notices a statistical difference in the fatality rates between cases of Gram-positive bacteremia treated with antimicrobial drugs and treated cases of Gram-negative bacteremia—that is, patients with Gram-positive bacteremia are much more likely to respond to treatment and survive. Explain one reason why this might be so.

3. In one summer month, local physicians reported 11 cases of severe diarrhea in infants less than three years of age. All the children lived in the same neighborhood and played in the same park, which has a wading pool. Cultures of stool specimens from the children revealed a Gram-negative bacillus that gave negative results for the following biochemical tests: oxidase, lactose, urease, and hydrogen sulfide production. The organism was also nonmotile. What is the organism? How was it transmitted among the children? What could officials do to limit infections of other children?

4. Several years ago, epidemiologists noted that the number of cases of salmonellosis increased dramatically in the summer and was lower the rest of the year. In contrast, the number of cases of typhoid fever, caused by the same bacterium, was constant throughout the year. Explain these observations.

5. Most physicians maintain that the viral "stomach flu" is really a bacterial infection. Which genera of bacteria discussed in this chapter are good candidates for causing this disease?

6. A hunter reports to his physician that he has been suffering with fever, chills, malaise, and fatigue. What Gram-negative bacteria may be causing his symptoms? How can a laboratory scientist distinguish among these species?

7. Ear piercings resulted in a rash of infections in teenagers, all of whom reported having their ears pierced at a kiosk in a local mall. What bacterial species might be responsible? Investigators traced the infections to a bacterium living in a bottle of disinfectant and in a sink used by employees. What bacterium is the likely agent? Why was successful treatment difficult?

8. A 21-month-old child was admitted into the hospital with fever, severe abdominal cramps, and bloody diarrhea. The family revealed they had purchased an iguana one month previously and the child had helped clean the cage. What bacterium was the likely cause of diarrhea? What was the possible treatment? How could the family prevent a recurrence?

Access more review material online in the Study Area at **www.masteringmicrobiology.com.** There, you'll find

- **Animations**
- **Concept Mapping Activities**
- **Flashcards**
- **Quizzes**

and more to help you succeed.

21 Rickettsias, Chlamydias, Spirochetes, and Vibrios

Two days after consuming raw oysters at a cocktail party, a 40-year-old man is admitted to the hospital with a 105°F fever, nausea, myalgia (muscle pain), and circular lesions on his right leg. As part of his medical history he reveals that he consumes more than six bottles of beer a day. The hospital's diagnosis is an infection with *Vibrio vulnificus,* a bacterium that can cause septicemia (blood poisoning), particularly in individuals who are immunocompromised or have liver disease. Within 24 hours of the diagnosis, the man goes into septic shock and dies. Investigations reveal that he had a preexisting alcohol-related liver disease, and that the source of his infection was contaminated oysters.

V. vulnificus is a Gram-negative bacterium, but in some ways an atypical one: It has a slightly curved shape and differs metabolically from other flagellated enteric bacteria. In this chapter we focus on pathogenic Gram-negative bacteria that differ in some way from most other Gram-negative bacteria: the rickettsias, chlamydias, spirochetes, and vibrios.

Take the pre-test for this chapter online. Visit the Study Area at www.masteringmicrobiology.com.

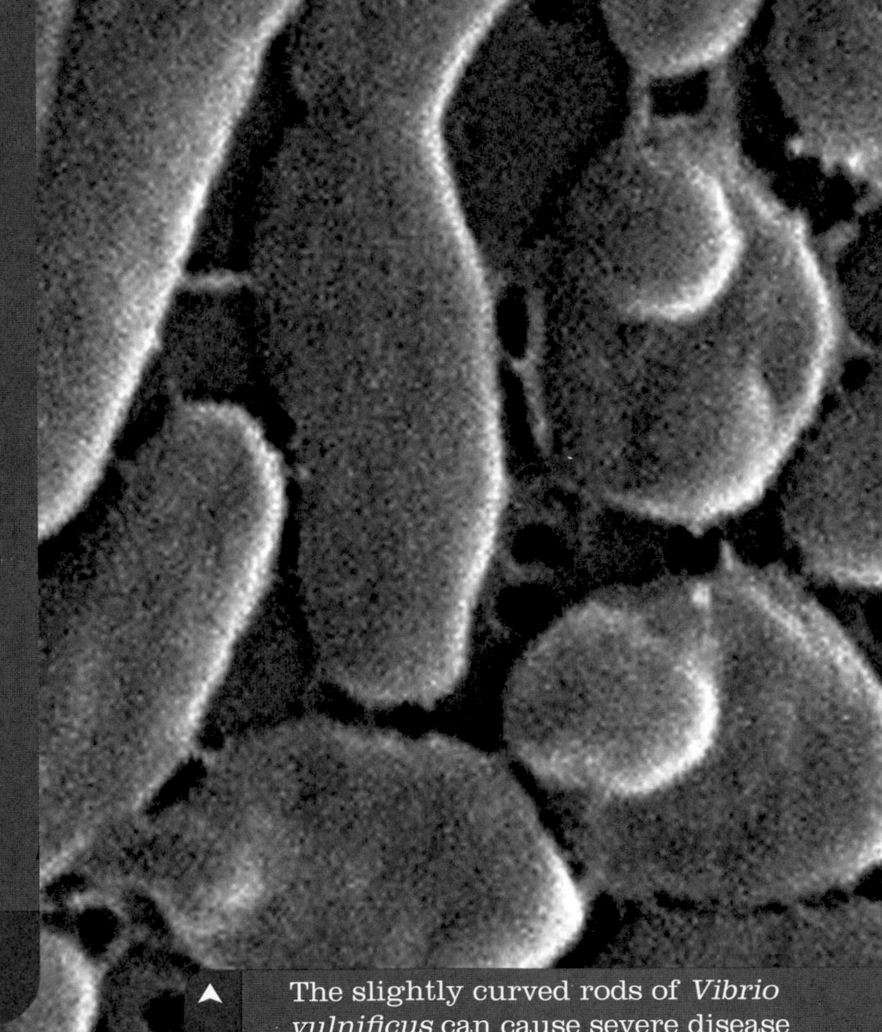

The slightly curved rods of *Vibrio vulnificus* can cause severe disease or death.

In Chapter 20 we explored the more important pathogenic Gram-negative cocci and bacilli. But a number of bacteria that stain pink in a Gram stain differ from typical Gram-negative organisms in morphology, growth habit, or reproductive strategy. In Chapter 19, we examined one such group, the wall-less mycoplasmas, which are low G + C, Gram-positive bacteria that stain pink. Other unusual pathogens include the obligate intracellular rickettsias and chlamydias, the spiral-shaped spirochetes, and the slightly curved vibrios. Because of their unique features, these bacteria have traditionally been discussed separately—a tradition we continue here, beginning with the rickettsias.

Rickettsias

Rickettsias (ri-ket′sē-ăz) are tiny, Gram-negative, obligate intracellular parasites that synthesize only a small amount of peptidoglycan and thus appear almost wall-less. The group as a whole is named after the most common genus of them *(Rickettsia),* named for Howard Ricketts (1871–1910), who first identified rickettsias and described the transmission of one species via its tick vector. Rickettsias are extremely small (0.3 μm × 1.0 μm); in fact, they are not much bigger than a large virus. Because of their small size, rickettsias were originally considered viruses, but closer examination has revealed that they contain both DNA and RNA, functional ribosomes, and Krebs cycle enzymes, and that they reproduce via binary fission—all characteristics of cells, not viruses.

Researchers have proposed several hypotheses to explain why rickettsias are obligate parasites, even though they have functional genes for protein synthesis, ATP production, and reproduction. Primary among these hypotheses is that rickettsias have very "leaky" cytoplasmic membranes and lose small cofactors (such as NAD^+) unless they are in an environment that contains an equivalent amount of these cofactors—such as the cytosol of a host cell.

Scientists classify rickettsias in the class Alphaproteobacteria of the bacterial phylum Proteobacteria based on the sequence of nucleotides in their rRNA molecules. Four genera of rickettsias—*Rickettsia, Orientia, Ehrlichia,* and *Anaplasma*—cause diseases in humans. We begin our discussion with *Rickettsia.*

Rickettsia

Learning Objectives

- ✓ List three species of *Rickettsia* that are responsible for human infections and identify their vectors.
- ✓ Describe the rash and petechiae of Rocky Mountain spotted fever (RMSF).
- ✓ Explain the relationship between epidemic typhus and Brill-Zinsser disease.

Rickettsia is a genus of nonmotile, aerobic, intracellular parasites that live in the cytosol of their host cells. They possess minimal cell walls of peptidoglycan and an outer membrane of lipopolysaccharide, which has little endotoxin activity. A loosely organized slime layer surrounds each cell. Rickettsias do not react well to the Gram stain, coloring only lightly pink,

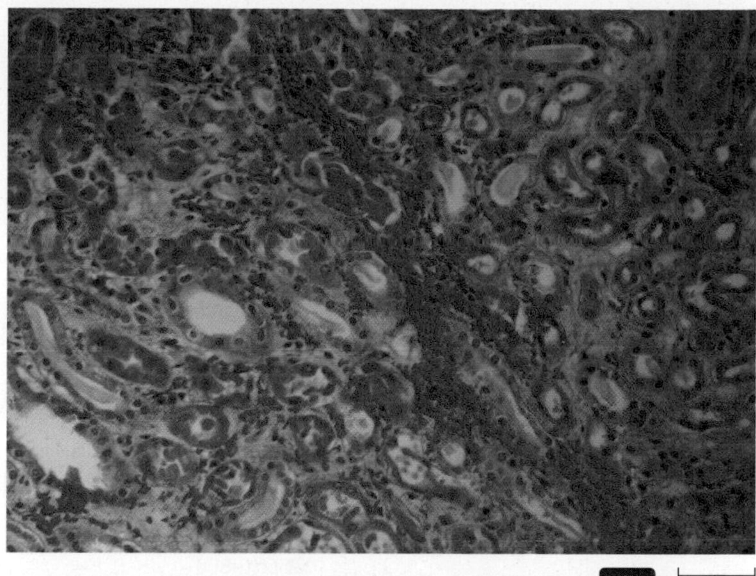

▲ **Figure 21.1** H and E stained ***Rickettsia rickettsii.*** Rickettsias are intracellular parasites, here stained bright pink in cells lining kidney blood vessels.

so scientists use special staining procedures such as *hematoxylin and eosin (H and E)* stain to visualize them **(Figure 21.1)**. Most human infections are by three species: *Rickettsia rickettsii* (ri-ket′-sē-ē), *R. prowazekii* (prō-wă-ze′kē-ē), and *R. typhi* (tī′fē).

Arthropod vectors transmit all three species of *Rickettsia,* which enter host cells by stimulating endocytosis. Once inside a host cell, the microbes secrete an enzyme that digests the endosome membrane, releasing the bacteria into the cytosol. As a result, rickettsias avoid the lysis that would ensue if a lysosome had merged with the endosome.

The pathogens grow and reproduce slowly, producing daughter cells only every 8–12 hours. *R. rickettsii* and *R. typhi* are continually released via exocytosis from long cytoplasmic extensions of host cells. In contrast, *R. prowazekii* gradually fills the host cell until the host cell lyses, releasing the parasites.

Outside of its host's cells, *Rickettsia* is unstable and dies quickly; therefore, rickettsias require vectors for transmission from host to host. Each rickettsial species is transmitted by different vectors, which also act as hosts and reservoirs.

Rickettsia rickettsii

R. rickettsii causes **Rocky Mountain spotted fever (RMSF),** the most severe and most reported rickettsial illness. Even though the earliest reports of RMSF came from the Rocky Mountain states, the disease is actually more prevalent in the Appalachian Mountains, Oklahoma, and the southeastern states **(Figure 21.2)**. Hard ticks in the genus *Dermacentor* (der-mă-sen′ter) transmit *R. rickettsii* among humans and rodents. The latter act as reservoirs. Male ticks infect female ticks during mating. Female ticks transmit the bacteria to eggs forming in their ovaries—a process called *transovarian transmission.*

R. rickettsii is typically dormant in the salivary glands of the ticks, and only when the arachnids feed for several hours are

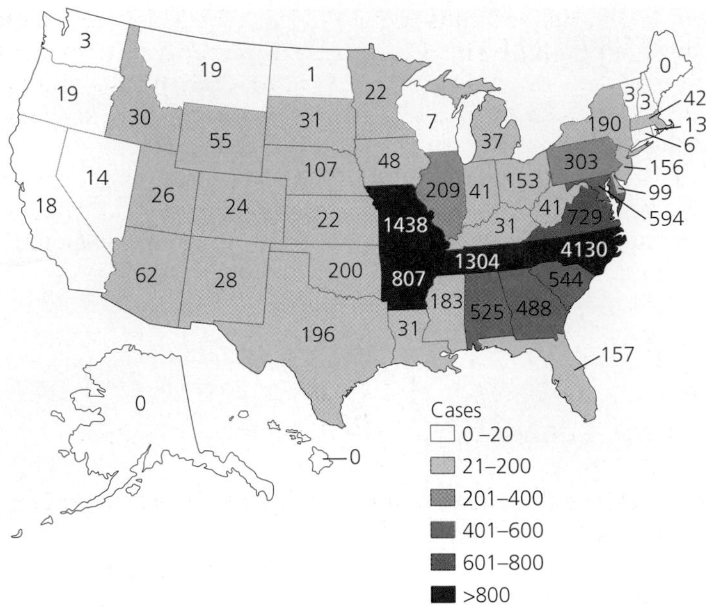

▲ **Figure 21.2** **Incidence of Rocky Mountain spotted fever in the United States, 1999–2008.**

the bacteria activated. Active bacteria are released from the tick's salivary glands into the mammalian host's circulatory system, where they infect endothelial cells lining small blood vessels. In rare cases, humans become infected following exposure to tissues and fluids from crushed ticks or to tick feces.

R. rickettsii secretes no toxins, and disease is not the product of the host's immune response. Apparently, damage to the blood vessels leads to leakage of plasma into the tissues, which may result in low blood pressure and insufficient nutrient and oxygen delivery to the body's organs.

About a week after infection, patients experience fever, headache, chills, muscle pain, nausea, and vomiting. In most cases (90%), a spotted, nonitchy rash develops on the trunk and appendages **(Figure 21.3)**—including the palms and soles, sites not involved in rashes caused by the chickenpox or measles viruses. In about 50% of patients, the rash develops into subcutaneous hemorrhages called *petechiae* (pe-tē'kē-ē). In severe cases, the respiratory, central nervous, gastrointestinal, and renal systems fail. Encephalitis may also occur, producing language disorders, delirium, convulsions, coma, and death. Even with treatment, almost 5% of patients die. Patients recovering from life-threatening acute Rocky Mountain spotted fever may experience paralysis of the legs, hearing loss, and gangrenous secondary infections with *Clostridium* (klos-trid'ē-ŭm) that necessitate the amputation of fingers, toes, arms, or legs.

Serological tests such as latex agglutination and fluorescent antibody stains are used to confirm an initial diagnosis based on sudden fever and headache following exposure to hard ticks, plus a rash on the soles or palms. Nucleic acid probes of specimens from rash lesions provide specific and accurate diagnosis, but such tests are expensive and typically are performed only by trained technicians in special laboratories. Early diagnosis is crucial because prompt treatment often makes the difference between recovery and death.

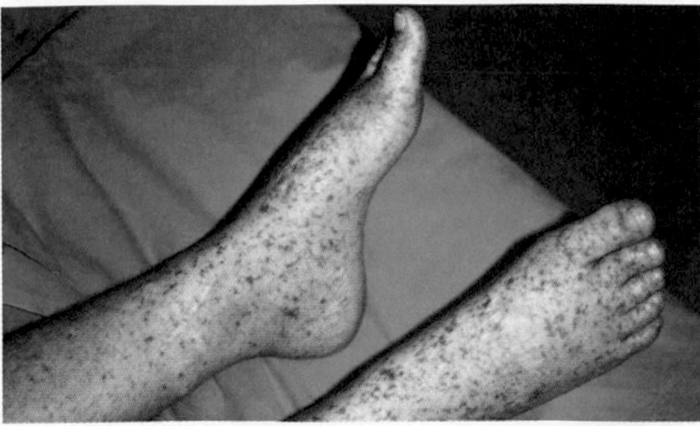

▲ **Figure 21.3**
The rash in a case of Rocky Mountain spotted fever.

Physicians treat Rocky Mountain spotted fever by carefully removing the tick and prescribing doxycycline, tetracycline, or chloramphenicol. An effective vaccine is not available. Prevention of infection involves wearing tight-fitting clothing, using tick repellents, promptly removing attached ticks, and avoiding tick-infested areas, especially in spring and summer, when ticks are most voracious. It is impossible to eliminate the ticks in the wild, in part because they can survive without feeding for more than four years.

CRITICAL **THINKING**

Why do most cases of Rocky Mountain spotted fever occur in May, June, and July?

Rickettsia prowazekii

Rickettsia prowazekii causes **epidemic typhus,**[1] which is also called *louse-borne typhus* because it is vectored by the human body louse, *Pediculis humanus* (pĕ-dik'yu-lŭs hū-man'us) (see Figure 12.33). *R. prowazekii* kills lice within two to three weeks, which prevents transovarian transmission but allows sufficient time for lice to transmit the bacterium between humans. In contrast to other rickettsias, *R. prowazekii* has humans as its primary hosts.

Epidemic typhus occurs in crowded, unsanitary living conditions that favor the spread of body lice; it is endemic in Central and South America and in Africa. It can recur many years (even decades) following an initial episode. The recurrent disease (called *Brill-Zinsser*[2] *disease*) is mild, brief, and resembles murine typhus (discussed shortly).

Diagnosis is based on the observation of signs and symptoms—high fever, mental and physical depression, and a rash that lasts for about two weeks—following exposure to infected lice. Diagnosis must be confirmed by the demonstration of the bacterium in tissue samples using fluorescent antibody tests.

Epidemic typhus is treated with doxycycline, tetracycline, or chloramphenicol. Prevention involves controlling lice populations and maintaining good personal hygiene. An attenuated

[1]From Greek *typhos*, meaning stupor.
[2]After physician Nathan Brill and bacteriologist Hans Zinsser, who studied the condition.

vaccine against epidemic typhus is available for use in high-risk populations.

Rickettsia typhi

Rickettsia typhi causes **murine**[3] (mū′rēn) **typhus,** which is so named because the major reservoir for the bacterium is rodents. The vectors for this disease, which is also known as *endemic typhus,* are fleas. The rat flea *Xenopsylla cheopis* (zen-op-sil′ă chē-op′is) (see Figure 12.33) and the cat flea *Ctenocephalides felis* (tē-nō-se-fal′idez fē′lis) (which also feeds on opossums, raccoons, and skunks) transmit the bacteria among the animal hosts and to humans. About 10 days following the bite of an infected flea, an abrupt fever, severe headache, chills, muscle pain, and nausea occur. A rash typically restricted to the chest and abdomen occurs in less than 50% of cases. The disease usually lasts about three weeks if left untreated and usually is not fatal.

Murine typhus is most often seen in the southern United States from Florida to California, and it is estimated that 50 cases occur annually, though the extent and incidence of the disease is not known because national notification requirements for murine typhus were discontinued in 1994. The disease is still endemic in every continent except Antarctica.

Diagnosis is initially based on signs and symptoms following exposure to fleas. An immunofluorescent antibody stain of a blood smear provides specific confirmation. Treatment is with tetracycline, doxycycline, or chloramphenicol. Prevention, as with other rickettsial diseases, involves avoiding bites by the arthropod vectors, wearing protective clothing, and using chemical repellents. No vaccine is available for murine typhus.

Orientia tsutsugamushi

Learning Objective

✓ Identify the causative agent, vector, and reservoir of scrub typhus.

The rickettsial organism *Orientia tsutsugamushi* (ōr-ē-en′tē-ă tsoo-tsoo-gă-mū′shē) was formerly classified in the genus *Rickettsia,* but taxonomists now assign it to the new genus *Orientia*. It differs from *Rickettsia* by having significantly different rRNA nucleotide sequences, a thicker cell wall, and a minimal slime layer. Mites of the genus *Leptotrombidium* (lep′tō-trom-bid′-ē-ŭm), also known as red mites or chiggers, transmit *Orientia* among rodents and humans. Infected female mites transmit the bacterium to their offspring via transovarian transmission, and the offspring transmit it to a rodent or human while feeding. Because these arachnids feed on a host only once in their lives, an individual mite cannot transmit rickettsia directly from a rodent to a human. In addition to being the vector, mites are the only known reservoir of *Orientia*.

O. tsutsugamushi causes **scrub typhus,** a disease endemic to eastern Asia, Australia, and western Pacific islands, including Japan. It occurs in the United States among immigrants who arrived from endemic areas. Scrub typhus is characterized by fever, headache, and muscle pain, all of which develop about 11 days after a mite bite. Less than half of patients with scrub typhus also develop a spreading rash on their trunks and appendages. In a few cases, death results from failure of the heart and the central nervous system.

Scrub typhus is treated with tetracycline, doxycycline, or chloramphenicol. No vaccine is available. Prevention involves avoiding exposure to mites by wearing appropriate clothing and using repellent chemicals.

[3]From Latin *murinus,* meaning relating to mice.

Ehrlichia and *Anaplasma*

Learning Objectives

✓ Identify the diseases caused by *Ehrlichia* and *Anaplasma,* respectively.

✓ Identify the three developmental stages of *Ehrlichia* and *Anaplasma.*

✓ Discuss the difficulties in diagnosing ehrlichiosis and anaplasmosis.

Ehrlichia chaffeensis (er-lik′ē-ă chaf-ē-en′sis) and *Anaplasma phagocytophilum* (an-ă-plaz′mă fag-ō-sī-to′fil-ŭm, previously *E. equi,* ē′kwē) cause **human monocytic ehrlichiosis (HME)** and **anaplasmosis** (previously human granulocytic ehrlichiosis, HGE), respectively. These two diseases are considered *emerging diseases* in the United States because they were unknown before 1987, and the number of reported cases has increased from a few dozen per year in the 1980s to over 400 cases annually since 2005.

Ticks, including the Lone Star tick *(Amblyomma,* am-blē-ō′ma), the deer tick *(Ixodes,* ik-sō′dēz), and the dog tick *(Dermacentor),* transmit *Ehrlichia* and *Anaplasma* to humans. Once in the blood, each bacterium triggers its own phagocytosis by a white blood cell (either a monocyte or a neutrophil). Unlike *Rickettsia, Ehrlichia* and *Anaplasma* grow and reproduce within the host cell's phagosomes. Because the bacteria are killed if a phagosome fuses with a lysosome, the bacteria must somehow prevent fusion, but the mechanism is unknown. Inside a leukocyte the bacteria grow and reproduce through three developmental stages: an *elementary body,* an *initial body,* and a *morula* **(Figure 21.4)**. The release of the cells from the morula and into the blood makes them available to feeding ticks.

HME and anaplasmosis resemble Rocky Mountain spotted fever but without the rash, which only rarely occurs in ehrlichiosis or anaplasmosis. *Leukopenia* (loo-kō-pē′nē-ă), which is an abnormally low leukocyte count, is also seen. Mortality of untreated HME is about 5%, and that of untreated anaplasmosis approaches 10%, especially in elderly patients.

Diagnosis of ehrlichiosis or anaplasmosis is difficult because the symptoms resemble those of other diseases. Physicians consider ehrlichiosis and anaplasmosis in any case of otherwise unexplained acute fever in patients exposed to ticks in endemic regions **(Figure 21.5)**. Immunofluorescent antibodies against *Ehrlichia* or against *Anaplasma* can demonstrate the bacterium within blood cells, confirming the diagnosis.

Doxycycline and tetracycline are effective against both *Ehrlichia* and *Anaplasma* but chloramphenicol is not. Treatment

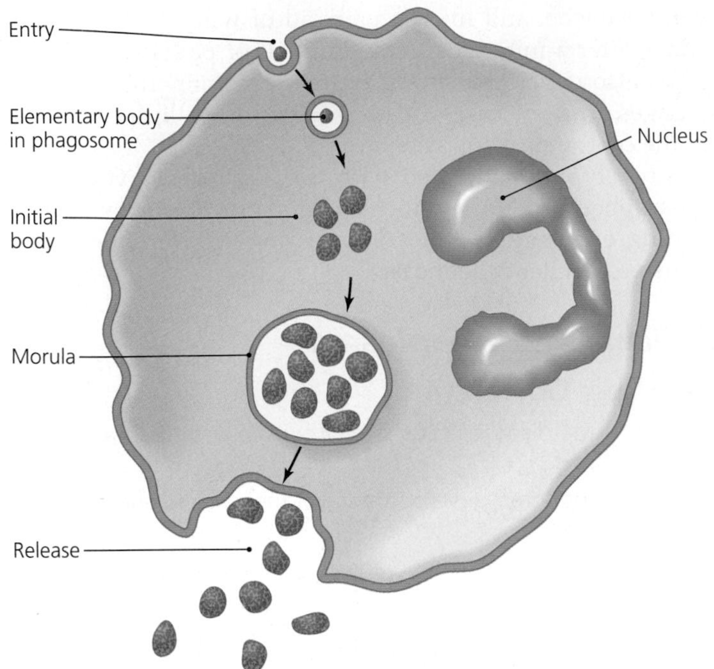

▲ **Figure 21.4 The growth and reproduction cycle of *Ehrlichia* and *Anaplasma* in an infected leukocyte.** *If the rickettsia shown here is* Ehrlichia chaffeensis, *then what kind of leukocyte is it infecting?*

Figure 21.4 *Ehrlichia chaffeensis infects human monocytes.*

should start immediately, even before the diagnosis is confirmed by serological testing, because the complications of infection and mortality rates increase when treatment is delayed. Prevention involves avoiding tick-infested areas, promptly removing ticks, wearing tight-fitting clothing, and using repellent chemicals. Vaccines against the two species are not available.

Clinical Case Study: Blame It on the Chiggers? describes an interesting human rickettsial infection, and Table 21.1 on p. 606 summarizes the rickettsial species, vectors, reservoirs, and diseases.

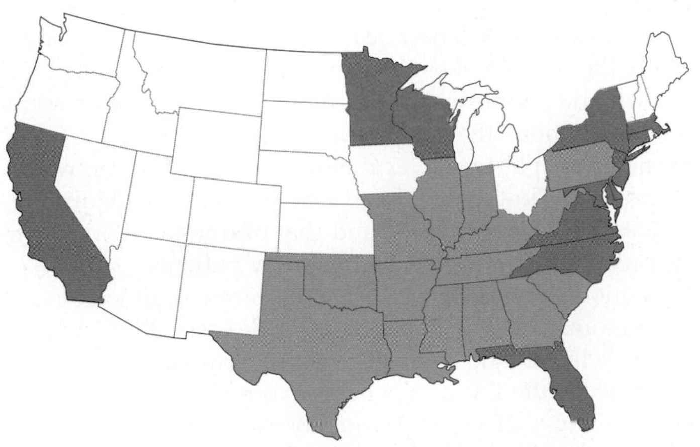

▲ **Figure 21.5 The geographical distribution of ehrlichiosis and anaplasmosis in the contiguous United States.** Blue indicates states where ehrlichiosis is endemic, orange where anaplasmosis is endemic, and purple where both diseases are endemic.

CLINICAL CASE STUDY

Blame It on the Chiggers?

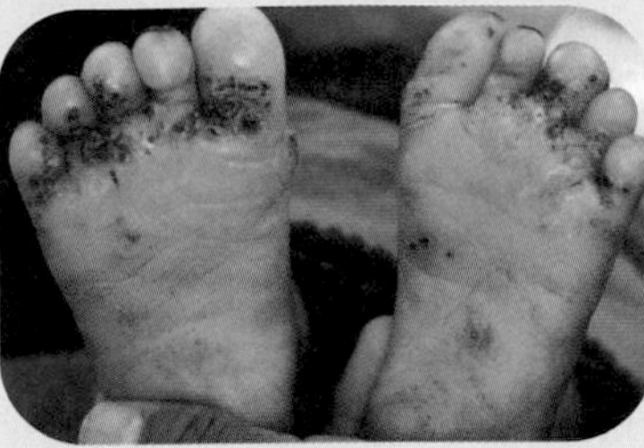

▲ ***Chiggers (mites) embedded in feet.***

A 75-year-old man arrives one summer morning at the hospital with severe headache, chills, muscle pain, and nausea. He reports that the symptoms developed quite suddenly. He has had no rash. During the interview the man states that he is an immigrant from Poland, a survivor of a Nazi concentration camp, and an avid bird watcher who spends many hours in the woods near his home. He had not noticed any fleas, ticks, or lice on his body, but he had been "almost eaten alive by chiggers." He had not been out of the United States in 40 years.

1. What rickettsial diseases might the man have?
2. What are the vectors for these diseases?
3. What diagnostic test(s) might verify the initial diagnosis?
4. Is the man's life in danger?
5. What drug will the physician likely prescribe?

CRITICAL THINKING

Explain the derivation of the names of the diseases caused by *Ehrlichia*.

Chlamydias

Learning Objectives

✓ Describe the life cycle of chlamydias, including the two developmental forms.

✓ Discuss the term *energy parasite* as it relates to chlamydias.

Chlamydias are a group of bacteria that vie with rickettsias for the title "smallest bacterium." Like rickettsias, chlamydias are nonmotile and grow and multiply only within vesicles in host cells. Scientists once considered chlamydias to be viruses because of their small size, obligate intracellular lifestyle, and ability to pass through 0.45-μm pores in filters, which were thought to trap all cells. However, chlamydias are cellular and possess DNA, RNA, and functional 70S ribosomes. Each chlamydial cell is surrounded by two membranes, similar to a typical Gram-negative bacterium, but without peptidoglycan between the membranes—chlamydias lack cell walls. In contrast with rickettsias, chlamydias do not have arthropods as vectors or hosts. Because of their unusual features and unique rRNA nucleotide sequences, taxonomists now classify chlamydias in their own phylum: Chlamydiae.

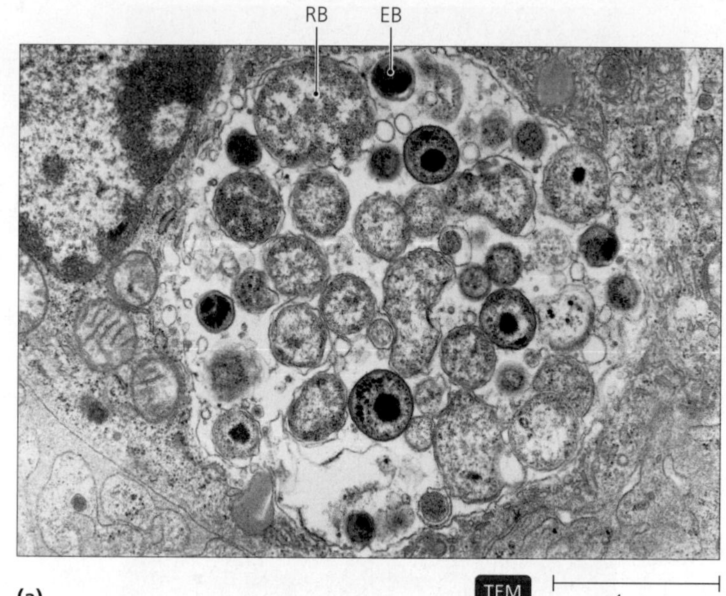
RB
EB
TEM
1 µm
(a)

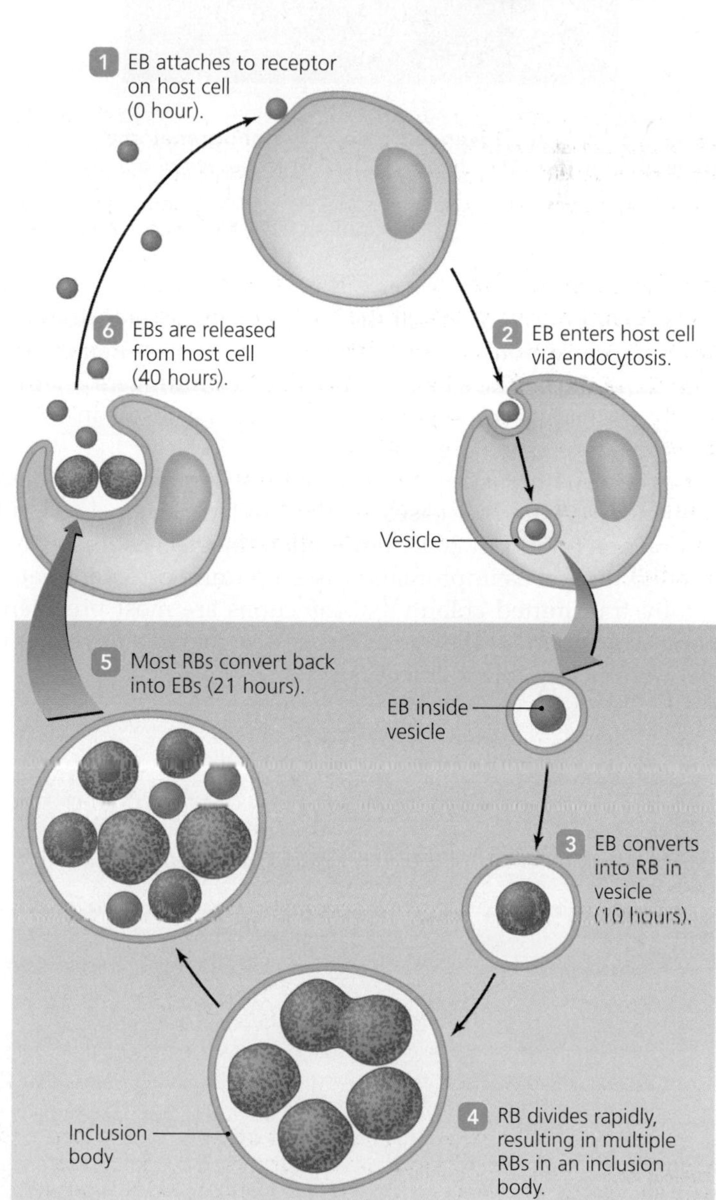
1 EB attaches to receptor on host cell (0 hour).
2 EB enters host cell via endocytosis.
Vesicle
EB inside vesicle
3 EB converts into RB in vesicle (10 hours).
4 RB divides rapidly, resulting in multiple RBs in an inclusion body.
Inclusion body
5 Most RBs convert back into EBs (21 hours).
6 EBs are released from host cell (40 hours).
(b)

▶ **Figure 21.6 Development of chlamydias.** **(a)** Elementary bodies and reticulate bodies, here of *Chlamydia*. **(b)** The life cycle of chlamydias. Times in parentheses refer to hours since infection.

Chlamydias have a unique developmental cycle involving two forms, both of which can occur within the endosome of a host cell **(Figure 21.6a)**: tiny (0.2–0.4 µm) cocci called **elementary bodies (EBs),** and larger (0.6–1.5 µm) pleomorphic **reticulate bodies (RBs).** Elementary bodies are relatively dormant, resistant to environmental extremes, can survive outside cells, and are the infective forms. Reticulate bodies are noninfective, obligate intracellular forms that replicate via binary fission within phagosomes, where they survive by inhibiting the fusion of a lysosome with the phagosome. Chlamydias lack the metabolic enzymes needed to synthesize ATP, so they must depend on their host cells for the high-energy phosphate compounds they require; thus chlamydias have been called *energy parasites.*

In the life cycle of chlamydias **(Figure 21.6b)**, once an EB attaches to a host cell (1), it enters by triggering its own endocytosis (2). Once inside the endosome, the EB converts into an RB (3), which then divides rapidly into multiple RBs (4). Once an infected endosome becomes filled with RBs, it is called an **inclusion body.** About 21 hours after infection, RBs within an inclusion body begin converting back to EBs (5), and about 19 hours after that, the EBs are released from the host cell via exocytosis (6), becoming available to infect new cells and completing the life cycle.

Three chlamydias cause disease in humans. In order of the prevalence with which they infect humans, they are *Chlamydia trachomatis, Chlamydophila pneumoniae,* and *Chlamydophila psittaci.* We begin our discussion with the most common species—*C. trachomatis.*

Chlamydia trachomatis

Learning Objectives

- ✓ List the types of cells in the human body that are most often infected by *Chlamydia.*
- ✓ Explain how sexually inactive children may become infected with *C. trachomatis.*
- ✓ Describe the cause and symptoms of lymphogranuloma venereum.
- ✓ Discuss the prevention of chlamydial infections.

Chlamydia trachomatis (kla-mid′ē-a tra-kō′ma-tis) has a very limited host range. One exceptional strain, which may eventually be classified as a separate species, causes pneumonia in mice, but all other strains are pathogens of humans.

Pathogenesis and Epidemiology

C. trachomatis enters a human body through abrasions and lacerations and infects a limited array of cells—those that have receptors for elementary bodies, including cells of the conjunctiva and cells lining the mucous membranes of the trachea, bronchi, urethra, uterus, uterine (Fallopian) tubes, anus, and rectum. The clinical manifestations of chlamydial infection result from the

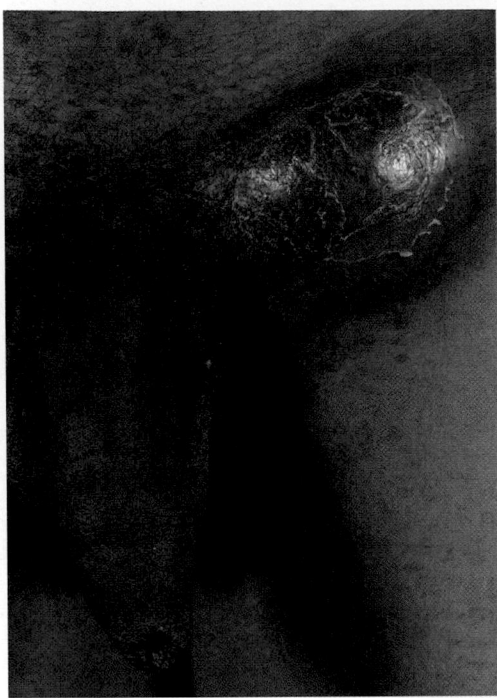

▲ **Figure 21.7 An advanced case of lymphogranuloma venereum in a man.** *Which microorganism causes lymphogranuloma venereum?*

Figure 21.7 *Chlamydia trachomatis.*

destruction of infected cells at the site of infection, and from the inflammatory response this destruction stimulates. Reinfection in the same site by the same or a similar strain triggers a vigorous hypersensitive immune response that can result in blindness, sterility, or sexual dysfunction.

Infection with *C. trachomatis* is the most commonly reported sexually transmitted diseases in the United States; 1,093,514 cases were reported in 2008, but epidemiologists estimate that about 3.5 million asymptomatic cases go unreported annually. Sexually transmitted chlamydial infections are most prevalent among women under the age of 20 because they are physiologically more susceptible to infection.

Researchers further estimate that over 500 million people worldwide, particularly children, contract ocular infections with *C. trachomatis*. Children can be infected as they pass through the birth canal, and the pathogen can be transmitted from eye to eye via droplets, hands, contaminated fomites, or flies. Infected children also harbor the bacterium in their digestive and respiratory tracts, so the bacterium can be transmitted to other children or adults via fecal contamination and respiratory droplets. Chlamydial eye infections are endemic in crowded, poor communities where people have inadequate personal hygiene, inadequate sanitation, or inferior medical care, particularly in the Middle East, North Africa, and India.

Diseases

The diseases caused by the various strains of *Chlamydia trachomatis* are of two main types: sexually transmitted diseases and an ocular disease called trachoma.

Sexually Transmitted Diseases A transient genital lesion and swollen, painfully inflamed, inguinal lymph nodes characterize **lymphogranuloma venereum**[4] (lim′fō-gran-ū-lō′mă ve-ne′rē-ŭm) (**Figure 21.7**), which is caused by the so-called *LGV strain* of *C. trachomatis*. The initial lesion of lymphogranuloma venereum occurs at the site of infection on the penis, urethra, scrotum, vulva, vagina, cervix, or rectum. This lesion is often overlooked because it is small, painless, and heals rapidly. Headache, muscle pain, and fever may also occur at this stage of the disease.

The second stage of the disease involves the development of buboes (swollen lymph nodes) associated with lymphatic vessels draining the site of infection. The buboes, which are accompanied by fever, chills, anorexia, and muscle pain, may enlarge to the point that they rupture, producing draining sores.

In a few cases, lymphogranuloma venereum proceeds to a third stage characterized by genital sores, constriction of the urethra, and genital elephantiasis. Arthritis may also occur during this third stage, particularly in young white males.

[4]From Latin *lympha*, meaning clear water; Latin *granulum*, meaning a small grain; Greek *oma*, meaning tumor (swelling); and Latin *Venus*, the goddess of sexual love.

TABLE 21.1 Characteristics of Rickettsias

Organism	Primary Vectors	Reservoirs	Diseases
R. rickettsii	Hard ticks: wood tick *(Dermacentor andersoni)* and dog tick *(D. variabilis)*	Ticks, rodents	Rocky Mountain spotted fever (RMSF)
R. prowazekii	Human body louse *(Pediculus humanus)*	Humans, squirrels, squirrel fleas	Epidemic (louse-borne) typhus
R. typhi	Rat flea *(Xenopsylla cheopis)* and cat flea *(Ctenocephalides felis)*	Rodents	Murine (endemic) typhus
Orientia tsutsugamushi	Mite (chigger) (*Leptotrombidium* spp.)	Mites	Scrub typhus
Ehrlichia spp.	Hard tick: Lone Star tick *(Amblyomma americanum)*	Ticks	Human monocytic ehrlichiosis (HME)
Anaplasma phagocytophilum	Hard tick: *Ixodes* spp.	Ticks	Anaplasmosis

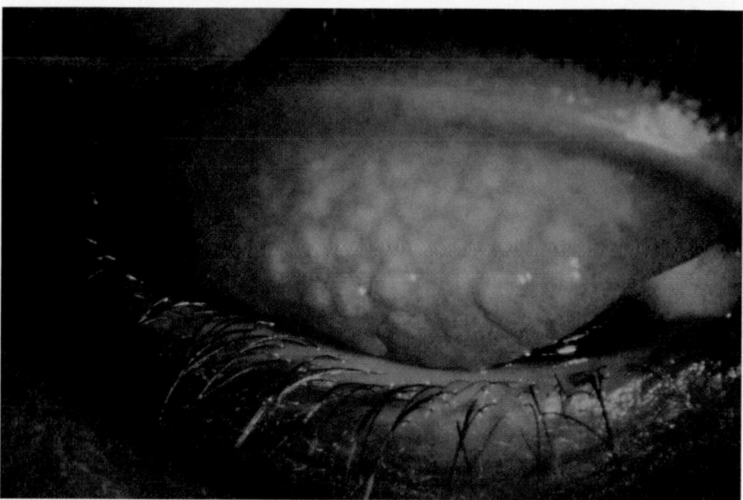

▲ **Figure 21.8** **An eyelid afflicted with trachoma.**

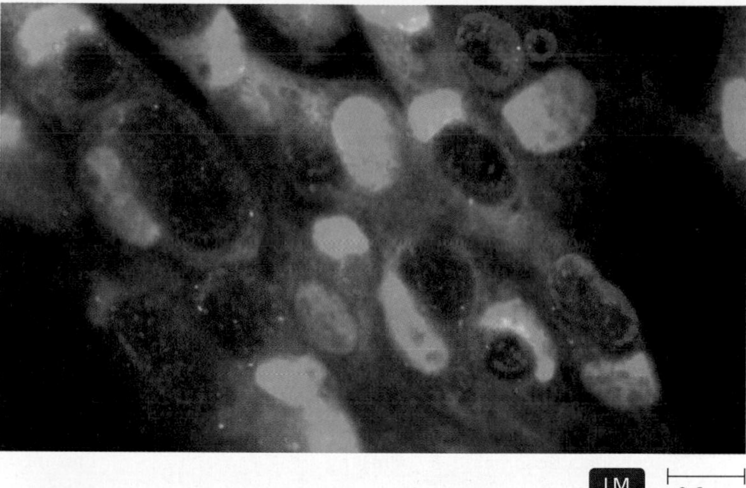

▲ **Figure 21.9** **A direct fluorescent antibody test for *Chlamydia trachomatis*.** The bright red color reveals the presence of *Chlamydia* within cells; this result is diagnostic for a *C. trachomatis* infection. Green indicates human (HeLa) cells.

Although about 85% of genital tract infections in women are asymptomatic, more than 75% of infections in men have symptoms. Infected men also often have urethral inflammation, which cannot be distinguished from gonorrhea based on symptoms alone. Such chlamydial urethritis accounts for about 50% of cases of *nongonococcal urethritis*.

Proctitis[5] may occur in men and women as a result of lymphatic spread of the bacterium from the vagina, vulva, cervix, or urethra to the rectum. About 15% of the cases of proctitis in homosexual men result from the spread of *C. trachomatis* via anal intercourse.

An immune response against reinfections of *C. trachomatis* in women can have serious consequences, causing pelvic inflammatory disease. PID involves chronic pelvic pain; irreversible damage to the uterine tubes, uterus, and ovaries; and sterility.

Trachoma The so-called *trachoma strains* of *C. trachomatis* cause a disease of the eye called **trachoma** (tră-kō′mă; **Figure 21.8**), which is the leading cause of nontraumatic blindness in humans. The pathogen multiplies in cells of the conjunctiva and kills them, triggering a copious, purulent[6] (pus-filled) discharge that causes the conjunctiva to become scarred. Such scarring in turn causes the patient's eyelids to turn inward, such that the eyelashes abrade, irritate, and scar the cornea, triggering an invasion of blood vessels into this normally clear surface of the eye. A scarred cornea filled with blood vessels is no longer transparent, and the eventual result is blindness.

Trachoma is typically a disease of children who have been infected during birth. However, LGV strains of *C. trachomatis* may produce blindness by a similar process in adults when bacteria from the genitalia are introduced into the eyes via fomites or fingers.

[5]From Greek *proktos*, meaning rectum, and *itis*, meaning inflammation.
[6]From Latin *pur*, meaning pus.

Diagnosis, Treatment, and Prevention

Diagnosis of chlamydial infection involves demonstration of bacteria inside cells from the site of infection. Specimens can be obtained from the urethra, vagina, or anus by inserting, rotating, and then removing a sterile swab. Giemsa-stained specimens may reveal bacteria or inclusion bodies within cells, but the most specific method of diagnosis involves amplifying the number of chlamydia by inoculating the specimen into a culture of susceptible cells. Laboratory technicians then demonstrate the presence of *Chlamydia* in the cell culture by means of specific fluorescent antibodies **(Figure 21.9)** or nucleic acid probes.

Physicians prescribe tetracycline or azithromycin for 21 days to eliminate genital infections of LGV strains of *C. trachomatis* in adults. Erythromycin or sulfonamides are recommended for treatment of pregnant women. Trachoma strains of *C. trachomatis* infecting the eyes of newborns are treated with erythromycin cream for 10–14 days, whereas ocular infections with LGV strains in adults are treated with doxycycline cream for 7 days. Surgical correction of eyelid deformities may prevent the abrasion, scarring, and blindness that typically result from ocular infections.

Prevention of sexually transmitted chlamydial infections is best achieved by abstinence or faithful mutual monogamy. Condoms may provide some protection, though some researchers warn that irritation by condoms and their lubricants actually increases the likelihood of infection. Blindness can be prevented only by prompt treatment with antibacterial agents and prevention of reinfection. Unfortunately, genital chlamydial infections are often asymptomatic and frequently occur among populations that have limited access to medical care.

CRITICAL **THINKING**

Why is either erythromycin or sulfonamide substituted for doxycycline in treatment of chlamydial infections in children? Why are penicillins and cephalosporins useless against *Chlamydia*?

TABLE 21.2 Characteristics of the Smallest Microbes

Feature	Rickettsias	Chlamydias	Mycoplasmas	Viruses
Cellular structure	Small cells with little peptidoglycan in cell walls	Small, wall-less cells with two membranes	Small, wall-less, pleomorphic cells with sterol in cytoplasmic membranes	Acellular
Diameter	0.3 μm	EBs: 0.2–0.4 μm RBs: 0.6–1.5 μm	0.1–0.3 μm	0.01–0.3 μm
Lifestyle	Obligate intercellular parasites within cytosol	Obligate intercellular parasites within endosomes	Free-living	Obligate intercellular parasites within cytosol or nuclei
Replication	Binary fission	Binary fission of reticulate bodies	Binary fission	Chemical assembly
Nucleic acid(s)	Both DNA and RNA	Both DNA and RNA	Both DNA and RNA	DNA or RNA
Functional ribosomes	Present	Present	Present	Absent
Metabolism	Present	Present	Limited—lack Krebs cycle, electron transport chains, and enzymes for cell wall synthesis	Completely dependent on metabolic enzymes of host cell
ATP-generating system	Present	Absent	Present	Absent
Phylum	Proteobacteria	Chlamydiae	Firmicutes	None officially recognized

Chlamydophila pneumoniae

Learning Objective

✓ Describe three diseases associated with *Chlamydophila pneumoniae.*

Chlamydophila pneumoniae (kla-mē-dof′ĭl-ă nū-mō′nē-ī) causes about 10% of U.S. cases of community-acquired pneumonia and 5% of the cases of bronchitis and sinusitis. Based primarily on circumstantial and epidemiological data, this chlamydia has also been implicated as a cause of some cases of atherosclerosis—lipid deposits on the walls of arteries, and the first stage of arteriosclerosis, or hardening of the arteries. Most infections with *Chlamydophila pneumoniae* are mild, producing only malaise and a chronic cough, and do not require hospitalization. Some cases, however, are characterized by the development of a severe pneumonia that cannot be distinguished from primary atypical pneumonia, which is caused by *Mycoplasma pneumoniae* (mī′kō-plaz-mă nū-mō′nē-ī).

Fluorescent antibodies demonstrate the intercellular presence of *C. pneumoniae,* which is diagnostic. Tetracycline or erythromycin for 14 days is used to treat infections of *C. pneumoniae,* but the drugs are not always effective, and infections may persist. Prevention of infection is difficult because the bacterium is ubiquitous and spreads via respiratory droplets.

Chlamydophila psittaci

Learning Objective

✓ Describe the diseases caused by *Chlamydophila psittaci.*

Chlamydophila psittaci (sit′ă-sē) causes **ornithosis**[7] (ōr-ni-thō′sis), a disease of birds that can be transmitted to humans, in whom it typically causes flulike symptoms. In some cases severe pneumonia occurs, and rarely nonrespiratory conditions such as endocarditis, hepatitis, arthritis, conjunctivitis, and encephalitis are observed. The disease is sometimes called *psittacosis* or *parrot fever* because it was first identified in parrots (*psitakos* means parrot in Greek). Several dozen cases of ornithosis are reported in the United States annually, often in adults. Most likely this disease is underreported because symptoms are often mild and diagnosis is difficult. Zookeepers, veterinarians, poultry farmers, pet shop workers, and owners of pet birds are at greatest risk of infection.

Elementary bodies of *Chlamydophila psittaci* may be inhaled in aerosolized bird feces or respiratory secretions, or ingested from fingers or fomites that have contacted infected birds. Pet birds may transmit the disease to humans via beak-to-mouth contact. Because very brief exposure to birds can be sufficient for the transmission of elementary bodies, some patients cannot recall having any contact with birds. Symptoms usually occur within 10 days of exposure to the chlamydias. Without treatment, the mortality rate of ornithosis is about 20%, but with treatment death is rare.

Because the symptoms of ornithosis are the same as those of many respiratory infections, diagnosis requires demonstration of *C. psittaci* by serological testing. Doxycycline for two weeks is the preferred treatment. Prevention involves wearing protective clothing when handling infected birds, 30-day quarantine of and tetracycline treatment for all imported birds, and preventive husbandry, in which bird cages are regularly cleaned and are positioned so as to prevent the transfer of feces, feathers, food, and other materials from cage to cage. No vaccine for either birds or humans is available.

Table 21.2 compares and contrasts the features of the smallest microbes—rickettsias, chlamydias, mycoplasmas, and viruses. Next we turn our attention to the spirochetes.

[7]From Greek *ornith,* meaning bird.

Spirochetes

Learning Objective

✓ Describe the morphology and locomotion of spirochetes.

Spirochetes (spī'rō-kētz), which means *coiled hairs* in Greek, are thin (0.1–0.5 μm in diameter), tightly coiled, helically shaped, Gram-negative bacteria that share certain unique features—most notably axial filaments. Axial filaments are composed of endoflagella located in the periplasmic space between the cytoplasmic membrane and the outer (wall) membrane (see Figure 3.8). As its axial filament rotates a spirochete corkscrews through its environment—a type of locomotion thought to enable pathogenic spirochetes to burrow through their hosts' tissues. Mutants lacking endoflagella are rod shaped rather than helical, indicating that the axial filaments play a role in maintaining cell shape.

Taxonomists place spirochetes in their own phylum—Spirochaetes. Three genera, *Treponema, Borrelia,* and *Leptospira,* cause diseases in humans; we begin by discussing *Treponema.*

Treponema

Learning Objectives

✓ Describe the disease caused by *Treponema pallidum.*

✓ Describe the four phases of untreated syphilis and the treatment for each.

✓ Describe the diseases caused by nonvenereal strains of *Treponema.*

Treponema (trep-ō-nē'mă) is a pathogen of humans only. Four types cause diseases, the most widespread of which is *Treponema pallidum pallidum* (i.e., subspecies, strain, or serotype *pallidum*). We will first examine this sexually transmitted pathogen and the venereal disease it causes, and then we will briefly consider the nonvenereal diseases caused by the other three strains—*T. pallidum endemicum, T. pallidum pertenue,* and *T. carateum.*

Treponema pallidum pallidum

Treponema pallidum pallidum (pal'li-dŭm), which is usually simply called *T. pallidum,* causes **syphilis.** Its flattened helical cells are so narrow (0.1 μm) that they are difficult to see by regular light microscopy in Gram-stained specimens. Therefore, scientists use phase-contrast or dark-field microscopy, or they intensify the contrast of the cells with dye linked to antitreponemal antibodies or with special stains using silver atoms **(Figure 21.10)**.

The bacterium lives naturally in humans only. It is destroyed by exposure to heat, disinfectants, soaps, drying, the concentration of oxygen in air, and pH changes, so it cannot survive in the environment. Scientists have not successfully cultured *T. pallidum* in cell-free media, though they have coaxed it to multiply in rabbits, monkeys, and rabbit epithelial cell cultures. Outside of humans it multiplies slowly (binary fission occurs once every 30 hours) and only for a few generations.

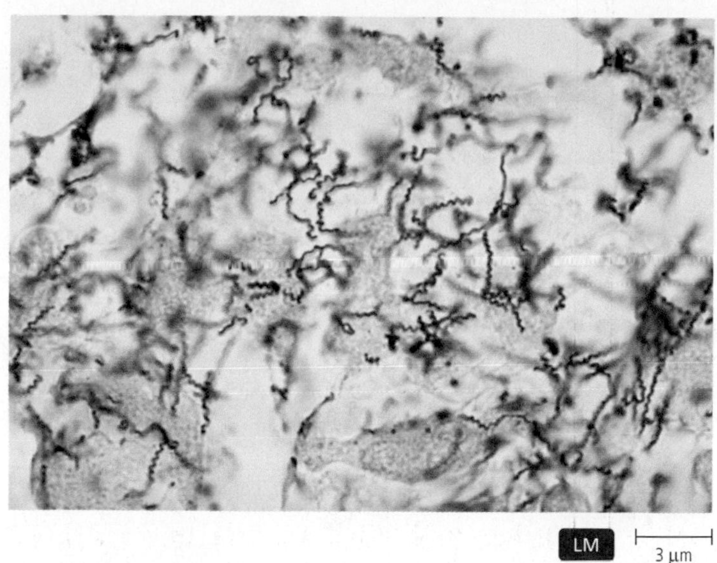

▲ **Figure 21.10 Spirochetes of *Treponema pallidum pallidum.*** Here, stained with a silver stain.

Pathogenicity and Epidemiology Scientists have had difficulty identifying the virulence factors of *T. pallidum* because the pathogen cannot be cultured. Researchers have used recombinant DNA techniques to insert genes from *Treponema* into *Escherichia coli* (esh-ĕ-rik'ē-ă kō'lē) and have then isolated the proteins the genes coded. Apparently, some of these proteins enable *Treponema* to adhere to human cells. Virulent strains also produce hyaluronidase, which may enable *Treponema* to infiltrate intracellular spaces. The bacterium has a glycocalyx, which may protect it from phagocytosis by leukocytes.

Syphilis occurs worldwide. Europeans first recognized the disease in 1495, leading some epidemiologists to hypothesize that syphilis was brought to Europe from the Western Hemisphere by returning Spanish explorers. Another hypothesis is that *T. pallidum* evolved from a less pathogenic strain of *Treponema* endemic to North Africa, and that its spread throughout Europe was coincidental to Europeans' explorations of the New World. In any case, the discovery and development of antimicrobial drugs over four centuries later has greatly reduced the number of syphilis cases **(Figure 21.11a)**. The disease remains prevalent among sex workers, men who have sex with men, and users of illegal drugs. It occurs throughout the United States, particularly in the Southeast **(Figure 21.11b)**.

Because of its fastidiousness and sensitivity, *T. pallidum* is an obligate parasite of humans and is transmitted almost solely via sexual contact, usually during the early stage of infection, when the spirochetes are most numerous. The risk of infection from a single, unprotected sexual contact with an infected partner is 10–30%. *T. pallidum* can rarely spread through blood transfusion; it cannot spread by fomites such as toilet seats, eating utensils, or clothing.

Disease Untreated syphilis has four phases: primary, secondary, latent, and tertiary syphilis. In *primary syphilis,* a small,

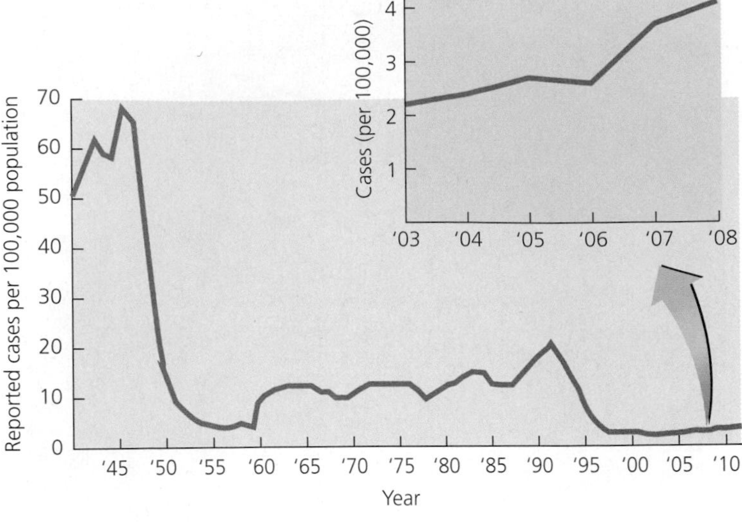

(a)

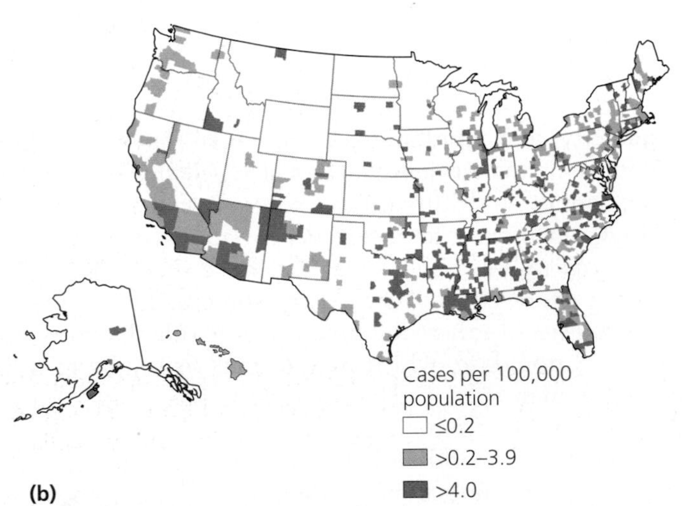

(b)

▲ **Figure 21.11 The incidence of adult syphilis in the United States.** **(a)** Nationwide incidence of syphilis, 1941–2008. **(b)** Reported cases of adult syphilis per 100,000 population, by county, 2006.

painless, reddened lesion called a **chancre** (shan′ker) forms at the site of infection 10–21 days following exposure (**Figure 21.12a**). Although chancres typically form on the external genitalia, about 20% form in the mouth, around the anus, or on the fingers, lips, or nipples. Chancres are often unobserved, especially in women, in whom these lesions frequently form on the cervix. The center of a chancre fills with serum that is extremely infectious because of the presence of millions of spirochetes. Chancres remain for three to six weeks and then disappear without scarring.

In about a third of cases, the disappearance of the chancre is the end of the disease. However, in most infections *Treponema* has invaded the bloodstream and spreads throughout the body to cause the symptoms and signs of *secondary syphilis:* sore throat, headache, mild fever, malaise, myalgia (muscle pain), lymphadenopathy (diseased lymph nodes), and a widespread rash (**Figure 21.12b**) that can include the palms and the soles of the feet. Although this rash does not itch or hurt, it can persist for months, and like the primary chancre, rash lesions are filled with spirochetes and are extremely contagious. People, including health care workers, can become infected when fluid from the lesions enters breaks in the skin, though such nonsexual transmission of syphilis is rare.

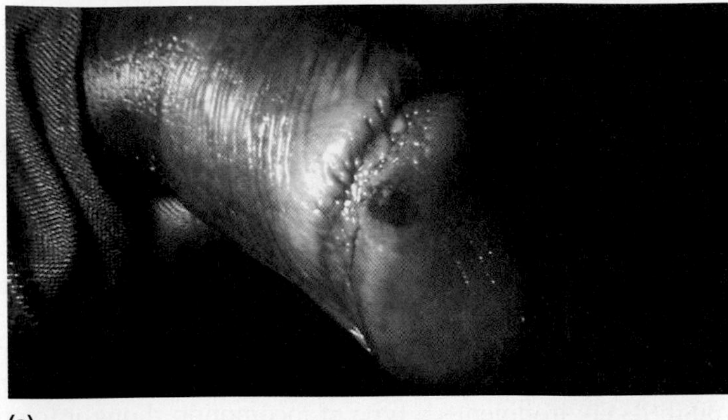

(a)

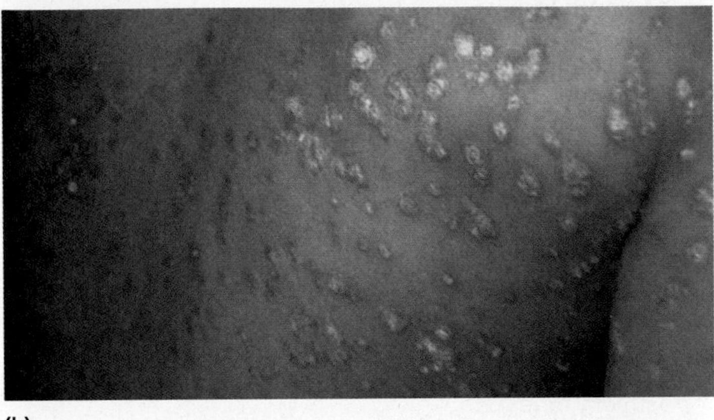

(b)

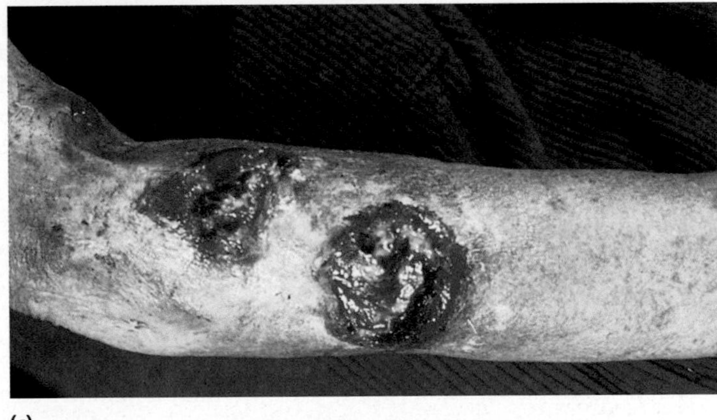

(c)

▲ **Figure 21.12 The lesions of syphilis.** **(a)** A chancre, a hardened and painless lesion of primary syphilis that forms at the site of the infection. **(b)** A widespread rash characteristic of secondary syphilis. **(c)** A gumma, a painful rubbery lesion that often occurs on the skin or bones during tertiary syphilis.

After several weeks or months the rash gradually disappears, and the patient enters a *latent (clinically inactive) phase* of the disease. The majority of cases do not advance beyond this point, especially in developed countries where antimicrobial drugs are in use.

Latency may last 30 or more years, after which perhaps a third of the originally infected patients proceed to *tertiary syphilis*. This phase is not associated with the direct effects of *Treponema*, but rather with severe complications resulting from inflammation and a hyperimmune response against the pathogen. Tertiary syphilis may affect virtually any tissue or organ and can cause dementia, blindness, paralysis, heart failure, and syphilitic lesions called **gummas** (gŭm'ăz), which are rubbery, painfully swollen lesions that can occur in bones, in nervous tissue, or on the skin **(Figure 21.12c)**.

Congenital syphilis results when *Treponema* crosses the placenta from an infected mother to her fetus. Transmission to the fetus from a mother experiencing primary or secondary syphilis often results in the death of the fetus. If transmission occurs while the mother is in the latent phase of the disease, the result can be a latent infection in the fetus that causes mental retardation and malformation of many fetal organs. After birth, newborns with latent infections usually exhibit a widespread rash at some time during their first two years of life.

Diagnosis, Treatment, and Prevention The diagnosis of primary, secondary, and congenital syphilis is relatively easy and rapid using specific antibody tests against antigens of *Treponema pallidum pallidum*. Spirochetes can be observed in fresh discharge from lesions but only when microscopic observations of clinical samples are made immediately—*Treponema* usually does not survive transport to a laboratory. Nonpathogenic spirochetes, which are a normal part of the oral microbiota, can yield false positive results, so clinical specimens from the mouth cannot be tested for syphilis. Tertiary syphilis is extremely difficult to diagnose because it mimics many other diseases, because few (if any) spirochetes are present, and because the signs and symptoms may occur years apart and seem unrelated to one another.

Penicillin is the drug of choice for treating primary, secondary, latent, and congenital syphilis, but it is not efficacious for tertiary syphilis, because this phase is caused by a hyperimmune response, not an active infection. Doxycycline is used for patients with penicillin allergy.

The Centers for Disease Control and Prevention has established a national goal for control of syphilis: maintaining an incidence of fewer than 0.2 cases of syphilis per 100,000 population in every county in the United States. This measure should reduce the incidence of syphilis to the point that the disease will be eliminated in the United States. A vaccine against syphilis is not available, so abstinence, faithful mutual monogamy, and consistent and proper condom usage are the primary ways to avoid contracting syphilis. All sexual partners of syphilis patients must be treated with prophylactic penicillin to prevent spread of the disease.

Nonvenereal Treponemal Diseases

Other members of *Treponema* cause three nonsexually transmitted diseases in humans: bejel, yaws, and pinta. These diseases are primarily seen in impoverished children in Africa, Asia, and South America who live in unsanitary conditions. The spirochetes that cause these diseases look like *Treponema pallidum pallidum*.

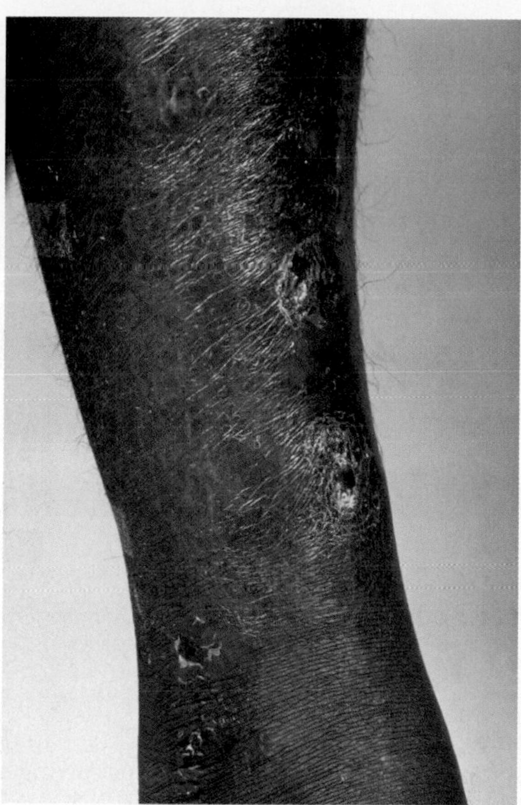

▲ **Figure 21.13 Yaws.** The draining lesions are characteristic of the later stages of the disease.

T. pallidum endemicum (en-de'mi-kŭm) causes **bejel** (be'jel), a disease seen in children in Africa, Asia, and Australia. In bejel, the spirochetes are spread by contaminated eating utensils, so it is not surprising that the initial lesion is an oral lesion, which typically is so small that it is rarely observed. As the disease progresses, larger and more numerous secondary lesions form around the lips and inside the mouth. In the later stages of the disease, gummas form on the skin, bones, or nasopharyngeal mucous membranes.

T. pallidum pertenue[8] (per-ten'ū ē) causes **yaws** (yăz), a disease of tropical South America, central Africa, and Southeast Asia that is characterized initially by granular skin lesions that, although unsightly, are painless. Over time the lesions develop into large, destructive, draining lesions of the skin, bones, and lymph nodes **(Figure 21.13)**. The disease is spread via contact with spirochetes in fluid draining from the lesions.

T. carateum[9] (kar-a'tē-ŭm) causes **pinta** (pēn'tă), a skin disease seen in children in Central and South America. The spirochetes are spread among the children by skin-to-skin contact. After one to three weeks of incubation, hard, pus-filled lesions

[8]Some taxonomists consider this organism to be a separate species: *T. pertenue*.
[9]Some taxonomists consider this organism to be a subspecies: *T. pallidum carateum*.

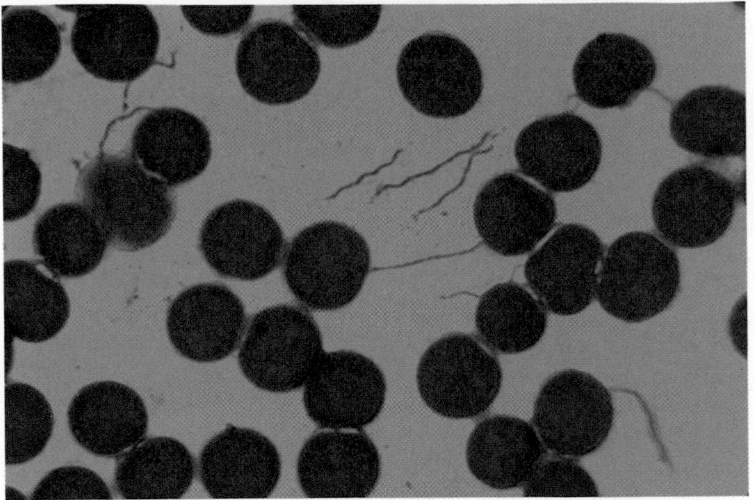

▲ **Figure 21.14** ***Borrelia burgdorferi.*** This Gram-negative spirochete lives in the blood and causes Lyme disease.

called papules form at the site of infection; the papules enlarge and persist for months or years, resulting in scarring and disfigurement.

Physicians diagnose bejel, yaws, and pinta by their distinctive appearance in children from endemic areas. Clinicians can not detect spirochetes in specimens from the lesions of bejel, but they are present in patients with pinta and yaws.

Penicillin, tetracycline, and chloramphenicol are the most favored antimicrobial agents for treating these three tropical diseases. Prevention involves limiting the spread of the bacteria by preventing contact with the lesions.

Borrelia

Learning Objectives

- ✓ Describe Lyme disease, its vector, and its causative agent.
- ✓ Discuss the life cycle of the *Ixodes* tick as it relates to Lyme disease.
- ✓ Compare and contrast the two types of relapsing fever, including their causes and vectors.

Members of the genus *Borrelia* (bō-rē′lē-ă) are lightly staining, Gram-negative spirochetes that are larger than species of *Treponema.* These spirochetes cause two diseases in humans: Lyme disease and relapsing fever.

Lyme Disease

In 1975, epidemiologists noted that the incidence of childhood rheumatoid arthritis in Lyme, Connecticut, was over 100 times higher than expected. Upon investigation, they discovered that ticks transmitted the spirochete *Borrelia burgdorferi*[10] (burg-dōr′-fer-ē; **Figure 21.14**) to human hosts to cause what became known as Lyme disease. **Lyme disease** is characterized by dermatological, cardiac, and neurological abnormalities in addition to the observed arthritis. Infected children may have paralysis of one side of their face (Bell's palsy). Even though its discovery was relatively recent, retrospective epidemiological studies have shown that Lyme disease was present in the United States for decades before its discovery.

B. burgdorferi is an unusual bacterium in that it lacks iron-containing enzymes and iron-containing proteins in its electron

[10]Named for the French bacteriologist Amedee Borrel and a specialist in tick-borne diseases, Willy Burgdorfer.

MICROBE AT A GLANCE

Treponema pallidum

Taxonomy: Domain Bacteria, phylum Spirochaetes, class "Spirochaetes," order Spirochaetales, family Spirochaetaceae

Cell morphology and arrangement: Flattened spirochete

Gram reaction: Negative

Virulence factors: Adhesion proteins, corkscrew motility, glycocalyx, hyaluronidase

Diseases caused: Syphilis: primary syphilis is lesion at site of infection, secondary syphilis is body rash, tertiary syphilis can have widespread and varied signs and symptoms including gummas; strain *endemicum* causes bejel; strain *pertenue* causes yaws; strain *carateum* causes pinta

Treatment for diseases: Penicillin for primary and secondary syphilis; tertiary syphilis not treatable; penicillin, tetracycline, or chloramphenicol to treat bejel, yaws, and pinta

Prevention of disease: Abstinence, mutual monogamy, condom usage for syphilis; avoid contact with patients with other diseases

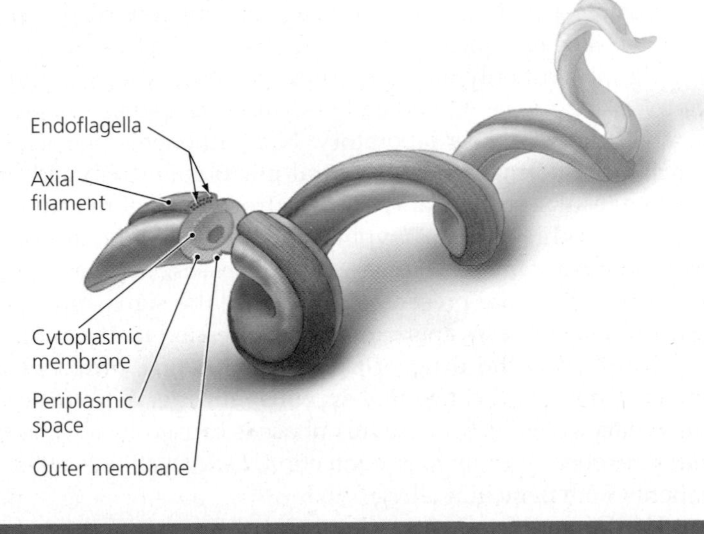

Download the Microbe at a Glance flashcards from the Study Area at www.masteringmicrobiology.com.

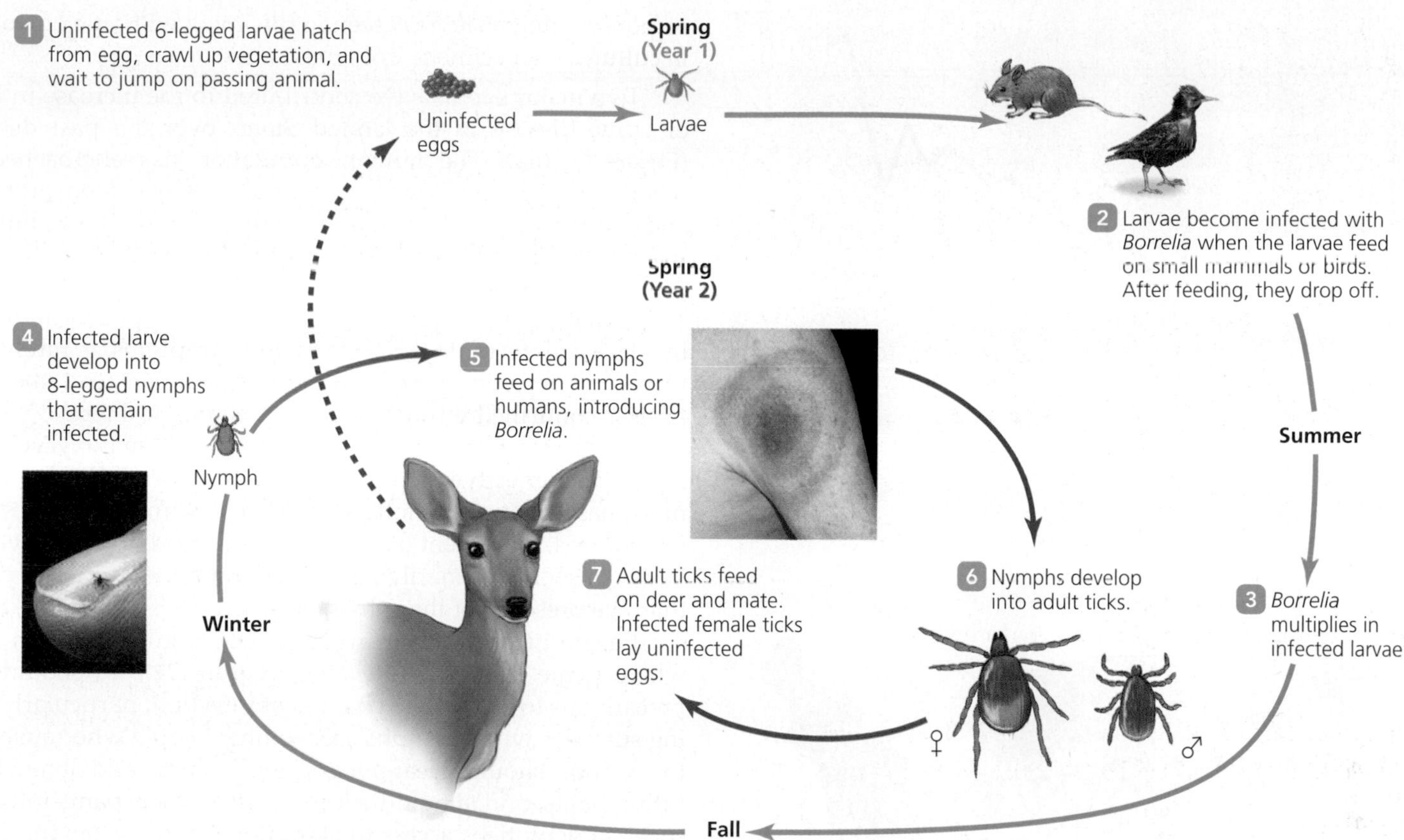
1 Uninfected 6-legged larvae hatch from egg, crawl up vegetation, and wait to jump on passing animal.
Spring (Year 1)
Uninfected eggs
Larvae
2 Larvae become infected with Borrelia when the larvae feed on small mammals or birds. After feeding, they drop off.
Summer
3 Borrelia multiplies in infected larvae.
Fall
Winter
4 Infected larve develop into 8-legged nymphs that remain infected.
Nymph
Spring (Year 2)
5 Infected nymphs feed on animals or humans, introducing Borrelia.
6 Nymphs develop into adult ticks.
7 Adult ticks feed on deer and mate. Infected female ticks lay uninfected eggs.
♀
♂

▲ **Figure 21.15 The life cycle of the deer tick *Ixodes* and its role as the vector of Lyme disease.** *At what other stage, besides the nymph stage, can a tick infect a human with* B. burgdorferi?

Figure 21.15 *Adult ticks infected with* B. burgdorferi *can also infect humans during a blood meal.*

transport chains. By utilizing manganese rather than iron, the spirochete circumvents one of the body's natural defense mechanisms: the lack of free iron in human tissues and fluids.

Hard ticks of the genus *Ixodes*—*I. scapularis* (scap-ū-lar'is) in the northeast and central United States, *I. pacificus* (pas-i'fi-kŭs) on the Pacific coast, *I. ricinus* (ri-ki'nŭs) in Europe, and *I. persulcatus* (per-sool-ka'tŭs) in eastern Europe and Asia—are the vectors of Lyme disease. An understanding of the life cycle of these ticks is essential to an understanding of Lyme disease.

An *Ixodes* tick lives for two years, during which it passes through three stages of development: a six-legged larva, an eight-legged nymph, and an eight-legged adult. During each stage it attaches to an animal host for a single blood meal. After each of its three feedings, the tick drops off its host and lives in leaf litter or on brush.

The different stages of each species of *Ixodes* typically feed on different hosts. For example, the larvae and nymphs of *I. scapularis* tend to feed on deer mice, whereas the adults most frequently feed on deer. In contrast, *I. pacificus* adults feed frequently on lizards, a host that does not support the growth of *Borrelia.* All stages of all species of *Ixodes* may feed on humans.

Transovarian transmission of *Borrelia* is rare, so ticks that hatch in the spring are uninfected **(Figure 21.15)** (1). Larvae become infected during their first blood meal (2). Over the winter, larvae digest their blood meals while *Borrelia* replicates in the ticks' guts (3).

In the spring of their second year, the ticks molt into nymphs (4) and feed a second time, infecting the new hosts with *Borrelia* via saliva (5). Uninfected nymphs can become infected at this time if they feed on an infected host. Laboratory studies have shown that infected ticks must remain on a host for 48 hours in order to transmit enough spirochetes to establish a *Borrelia* infection in that host.

Nymphs drop off and undergo further development into adults (6). In the fall, adult ticks feed a final time, mate, lay eggs, and die (7). Adults infected with *Borrelia* infect their hosts as they feed. Adult ticks are much larger than nymphs, so humans usually see and remove adults before they can transmit *Borrelia;* thus, nymphs most often infect humans.

CRITICAL **THINKING**

For what other pathogen discussed in this chapter is *Ixodes* also the vector?

Lyme disease mimics many other diseases, and its range of signs and symptoms is vast. The disease typically has three phases in untreated patients:

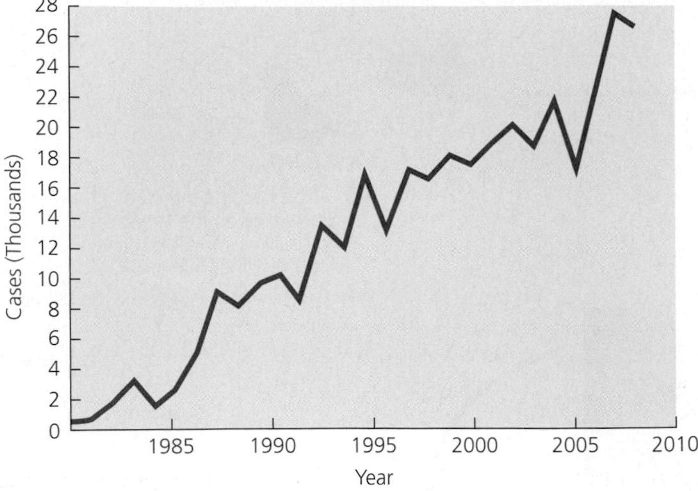

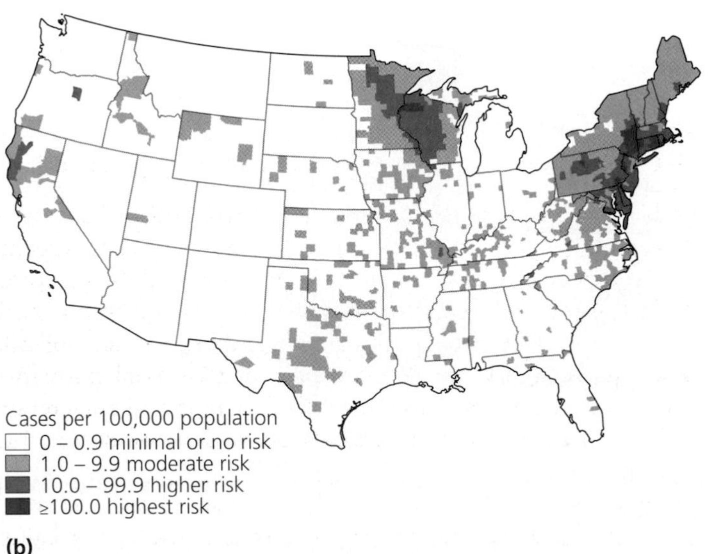

▲ **Figure 21.16 The occurrence of Lyme disease in the United States. (a)** Incidence of cases, 1982–2008. **(b)** The geographic distribution of the disease, 1992–2006.

1. An expanding red rash, which often resembles a bull's-eye, occurs at the site of infection within 30 days. About 75% of patients have such a rash, which lasts for several weeks. Other early signs and symptoms include malaise, headaches, dizziness, stiff neck, severe fatigue, fever, chills, muscle and joint pain, and lymphadenopathy.
2. Neurological symptoms (for example, meningitis, encephalitis, and peripheral nerve neuropathy) and cardiac dysfunction typify the second phase, which is seen in only 10% of patients.
3. The final phase is characterized by severe arthritis that can last for years.

The pathological conditions of the latter phases of Lyme disease are due in large part to the body's immunological response; rarely is *Borrelia* seen in the involved tissue or isolated in cultures of specimens from these sites.

Two major events have contributed to the increase in cases of Lyme disease in the United States over the past decades **(Figure 21.16a)**: The human population has encroached on woodland areas, and the deer population has been protected and even encouraged to feed in suburban yards. Thus, humans have been brought into closer association with deer ticks infected with *Borrelia.*

A diagnosis of Lyme disease, which is typically indicated by observations of its usual signs and symptoms, is rarely confirmed by detecting *Borrelia* in blood smears; instead, the diagnosis is confirmed through the use of serological tests.

Treatment with antimicrobial drugs such as doxycycline or a penicillin effectively cures most cases of Lyme disease in the first phase, though prolonged treatment with large doses may be required. Treatment of later phases is more difficult because later symptoms primarily result from immune responses rather than the presence of the spirochetes.

People hiking, picnicking, and working outdoors in areas where Lyme disease is prevalent **(Figure 21.16b)** should take precautions to reduce the chances of infection, particularly during summer, when nymphs are feeding. People who must be in the woods should wear long-sleeved shirts and long, tight-fitting pants, and should tuck the cuffs of their pants into their socks to deny ticks access to skin. Repellents containing *DEET* (*N,N*-diethyl-*m*-toluamide), which is noxious to ticks, should be used. As soon as possible after leaving a tick-infected area, people should thoroughly examine their bodies for ticks or their bites. Though researchers have developed a vaccine against *B. burgdorferi,* it is not widely used because of the ready availability of other preventive measures, and because the vaccine may produce the symptoms of Lyme disease in some patients.

CRITICAL **THINKING**

Health departments recommend that outdoor enthusiasts wear light-colored pants while hiking in areas where *Ixodes* is endemic. Why?

Relapsing Fever

Other species of spirochetes in the genus *Borrelia* can cause two types of relapsing fever. A disease called **louse-borne relapsing fever** results when *Borrelia recurrentis* (re-kur-ren'tis) is transmitted between humans by the human body louse *Pediculus humanus.* A disease known as **tick-borne relapsing fever** occurs when any of several species of *Borrelia* are transmitted between humans by soft ticks in the genus *Ornithodoros* (or-ni-thod'ō-rŭs).

Lice become infected with *B. recurrentis* when they feed on infected humans, the only reservoir for this spirochete. When infected lice are crushed, spirochetes can infect bite wounds. Because lice live only a few months, maintenance of louse-borne relapsing fever in a population requires crowded, unsanitary conditions such as occur in poor neighborhoods and during wars and natural disasters. This relapsing fever occurs in central and eastern Africa and in South America.

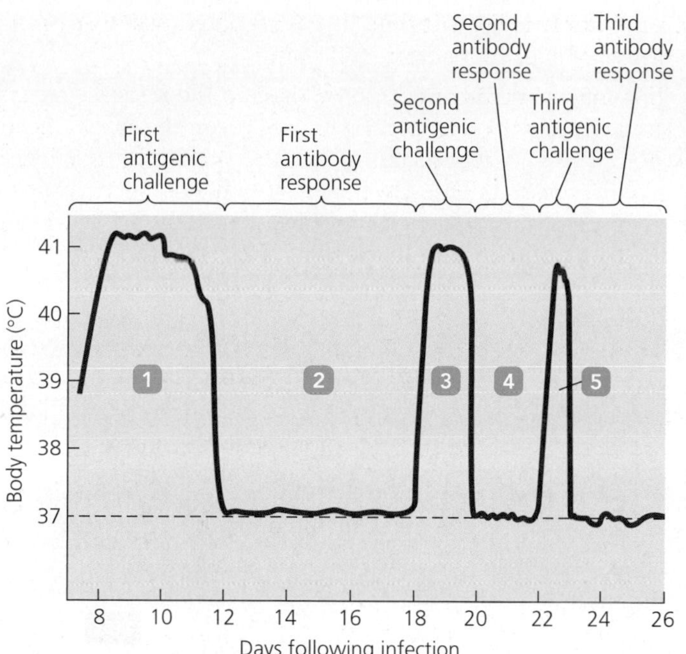

▲ **Figure 21.17 The time course of recurring episodes of fever in relapsing fevers.**

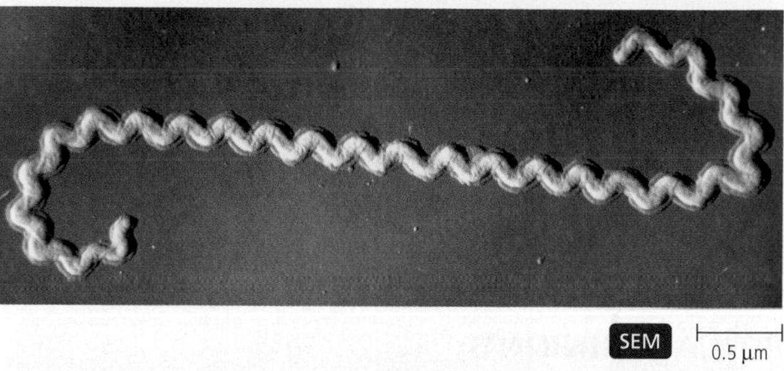

▲ **Figure 21.18** ***Leptospira interrogans.*** The spirochete has one end hooked like a question mark.

In contrast to hard ticks, the soft ticks of the genus *Ornithodoros* that transmit tick-borne relapsing fever can live for years between feedings and pass spirochetes to their offspring via transovarian transmission. Soft ticks can also transmit *Borrelia* spp. (species) to animals, particularly mice and rats. Ticks and rodents are reservoirs of infection. This relapsing fever has a worldwide distribution; in the United States it is found primarily in the West.

Both types of relapsing fever are characterized by recurrent episodes of septicemia and fever separated by symptom-free intervals—a pattern that results from the body's repeated efforts to remove the spirochetes, which continually change their antigenic surface components **(Figure 21.17)**. Within the first week of infection, the spirochetes induce a fever (1), which is associated with chills, myalgia, and headache. In response to infection, the body produces antibodies that are directed against the spirochetes' surface molecules; once the antibodies have targeted the spirochetes, phagocytes clear most pathogens from the blood, and the body's temperature returns to normal (2). However, some *Borrelia* evade the immune response by changing their antigenic surface molecules; when these spirochetes multiply, they induce a relapse of fever (3). The body mounts a second immune response, this time against the new antigens, eventually again clearing the blood of most spirochetes and again returning body temperature to normal (4). But a few spirochetes yet again change their antigenic components, leading to still another recurrence of fever, continuing the pattern (5).

Because of their relatively large size and abundance in the blood, observation of spirochetes in blood smears taken during periods of recurrent fever is the primary method of diagnosis.

Physicians successfully treat the relapsing fevers with doxycycline or, in pregnant women and children, with erythromycin.

Prevention involves avoidance of ticks and lice, good personal hygiene, and the use of repellent chemicals. Effective rodent control is crucial to the control of endemic relapsing fever. Vaccines are not available.

Leptospira

Learning Objective

✓ Describe *Leptospira interrogans* and zoonotic leptospirosis.

The last of the three groups of spirochetes that infect humans is *Leptospira interrogans* (lep′tō-spī′ră in-ter′ră-ganz). The specific epithet *interrogans* alludes to the fact that one end of the spirochete is hooked in a manner reminiscent of a question mark **(Figure 21.18)**. This thin, pathogenic spirochete is an obligate aerobe that is highly motile by means of two axial filaments, each of which is anchored at one end. Clinicians and researchers grow it on special media enriched with bovine serum albumin or rabbit serum.

L. interrogans normally occurs in many wild and domestic animals—in particular, rats, raccoons, foxes, dogs, horses, cattle, and pigs—in which it grows asymptomatically in the kidney tubules. Humans contract the zoonotic disease **leptospirosis** (lep′tō-spī-rō′sis) through direct contact with the urine of infected animals or indirectly via contact with the spirochetes in contaminated streams, lakes, or moist soil, environments in which the organisms can remain viable for six weeks or more. Person-to-person spread has not been observed.

After *Leptospira* gains initial access to the body through invisible cuts and abrasions in the skin or mucous membranes, it corkscrews its way through these tissues. It then travels via the bloodstream throughout the body, including the central nervous system, damaging cells lining the small blood vessels and triggering fever and intense pain. Infection may lead to hemorrhaging and to liver and kidney dysfunction. Eventually, the bacteremia resolves, and the spirochetes are found only in the kidneys. As the disease progresses, spirochetes are excreted in urine. Leptospirosis is usually not fatal.

Leptospirosis occurs throughout the world. National reporting in the United States has ceased in part because the disease is rare; a total of only 89 cases were reported in the last two years (1993–1994) that leptospirosis was listed as a nationally reportable disease.

Because *Leptospira* is very thin and does not stain well with Gram, silver, or Giemsa stain, specific antibody tests revealing the presence of the spirochete in clinical specimens are the preferred method of diagnosis. Intravenous penicillin is used to treat infections. Rodent control is the most effective way to limit the spread of *Leptospira,* but eradication is impractical because of the spirochete's many animal reservoirs. An effective vaccine is available for livestock and pets.

CRITICAL **THINKING**

In Chapter 4 (p. 121) we discussed the identification of microbes using dichotomous taxonomic keys. Design such a key for the spirochetes discussed in this chapter.

Pathogenic Gram-Negative Vibrios

To this point, we have discussed a number of unrelated Gram-negative bacteria: rickettsias, chlamydias, and the spirochetes. We will now turn our attention to a final group—the slightly curved bacteria called *vibrios.* The more important pathogenic vibrios of humans are in the genera *Vibrio, Campylobacter,* and *Helicobacter.*

Vibrio

Learning Objectives

- ✓ Contrast *Vibrio* with enteric bacteria in terms of their flagella and biochemical properties.
- ✓ Describe the action of cholera toxin in causing cholera.
- ✓ Name two species of *Vibrio* and describe the resulting diseases.

Vibrio (vib′rē-ō) is a genus of Gram-negative, slightly curved bacilli in class Gammaproteobacteria, and phylum Proteobacteria **(Figure 21.19)**. *Vibrio* shares many characteristics with enteric bacteria such as *Escherichia* and *Salmonella,* including O polysaccharide antigens that enable epidemiologists to distinguish among strains; however, *Vibrio* is oxidase positive and has a polar flagellum, unlike enteric bacteria, which are oxidase negative and have peritrichous flagella.

Vibrio lives naturally in estuarine and marine environments worldwide. It prefers warm, salty, and alkaline water, and most pathogenic species can multiply inside shellfish. Eleven of the 34 species of *Vibrio* cause diseases in humans; of these, only *V. cholerae* (kol′er-ī) can survive in freshwater, making it the most likely species to infect humans (via contaminated drinking water). In the following sections we examine *V. cholerae* and the serious disease it causes.

Pathogenesis and Epidemiology of *Vibrio cholerae*

Vibrio cholerae, particularly strain O1 El Tor, causes **cholera** (kol′er-ă, sometimes called *epidemic cholera*), one of the world's more pernicious diseases. As discussed in Chapter 14, the

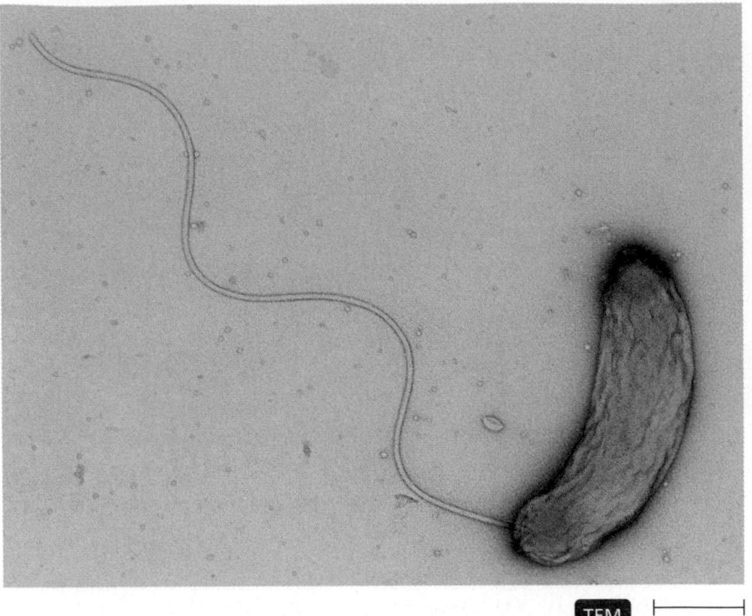

▲ **Figure 21.19** ***Vibrio cholerae.*** This bacterium causes cholera.

science of epidemiology began with Dr. John Snow's efforts to understand and limit the 1854 cholera epidemic in London.

Ships have introduced the pathogen into the harbors of every continent through their practice of taking on ballast water in one port and dumping it in another; two world wars have only exacerbated the situation. In the seven major pandemics that have occurred since 1800, millions of people have been sickened and thousands have died, causing significant socioeconomic upheaval.

The seventh pandemic of O1 El Tor began in 1961 in Asia, spread to Africa and Europe in the 1970s and 1980s, and reached Peru in January 1991. **Figure 21.20** illustrates the progression of this cholera pandemic throughout Latin America in the first half of the 1990s. A new pandemic strain, *V. cholerae* O139 Bengal, arose in India in 1992 and is spreading across Asia; this strain is the first non-O1 strain capable of causing epidemic disease. Other strains of *V. cholerae* do not produce epidemic cholera, only milder gastroenteritis.

Humans become infected with *Vibrio cholerae* by ingesting contaminated food and water. Cholera is most frequent in communities with poor sewage and water treatment. After entering the digestive system, *Vibrio* is confronted with the inhospitable acidic environment of the stomach. Most cells die, which is why a high inoculum—at least 10^8 cells—is typically required for the disease to develop. Fewer cells can achieve infection in people who use antacids.

Recent research indicates that only the environment within a human body activates *Vibrio* virulence genes. Thus, *Vibrio* shed in feces is more virulent than its counterparts in the environment. Scientists hypothesize that such gene activation may explain the rapid, almost explosive, nature of cholera epidemics.

Although cholera infections may be asymptomatic or cause mild diarrhea, some result in rapid, severe, and fatal fluid and

▲ **Figure 21.20 The spread of cholera.** The Latin American epidemics were caused by *Vibrio cholerae* O1 El Tor in the first half of the 1990s.

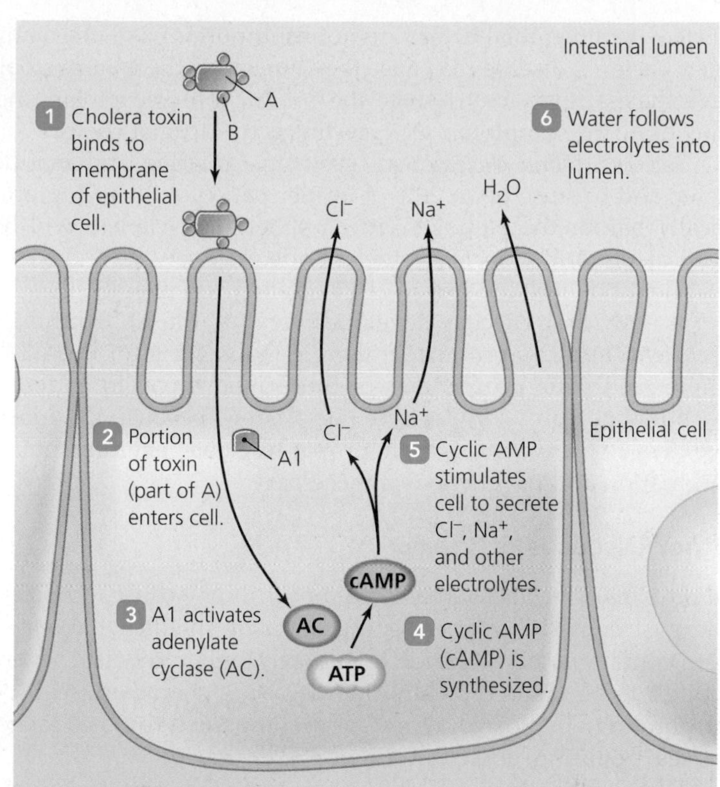

▲ **Figure 21.21 The action of cholera toxin in intestinal epithelial cells.**

electrolyte loss. Symptoms usually begin two to three days following infection, with explosive watery diarrhea and vomiting. As the disease progresses, the colon is emptied and the stool becomes increasingly watery, colorless, and odorless. The stool, which is typically flecked with mucus, is called *rice-water stool* because it resembles water poured off a pan of boiled rice. Some patients lose 1 liter of fluid an hour, resulting in a 50% loss of body weight over the course of infection.

The most important virulence factor of *V. cholerae* is a potent exotoxin called **cholera toxin,** which is composed of five identical B subunits and a single A subunit. **Figure 21.21** illustrates the action of cholera toxin in producing the severe diarrhea that characterizes cholera. The process proceeds as follows:

1. One of the B subunits binds to a glycolipid receptor in the cytoplasmic membrane of an intestinal epithelial cell.
2. The A subunit is cleaved, and a portion (called A1) enters the cell's cytosol.
3. A1 acts as an enzyme that activates adenylate cyclase (AC).
4. Activated AC enzymatically converts ATP into cyclic AMP (cAMP).
5. cAMP stimulates the active secretion of excess amounts of electrolytes (sodium, chlorine, potassium, and bicarbonate ions) from the cell.
6. Water follows the movement of electrolytes from the cell and into the intestinal lumen via osmosis.

Severe fluid and electrolyte losses result in dehydration, metabolic acidosis (decreased pH of body fluids) due to loss of bicarbonate ions, hypokalemia,[11] and hypovolemic shock caused by reduced blood volume in the body. These conditions can produce muscle cramping, irregularities in heartbeat, kidney failure, coma, and death.

Cholera mortality is 60% in untreated patients infected by strain O1; death can occur less than 48 hours after infection. Disease produced by strain O139 is even more severe. In nonfatal cases, cholera ends spontaneously after a few days as the pathogens and toxins are flushed from the system by the severe diarrhea.

Diagnosis, Treatment, and Prevention of Cholera

Rarely, an experienced observer may find the quick darting cells of *Vibrio* in the watery stool of a patient, but diagnosis is usually based on the characteristic diarrhea. *Vibrio* can be cultured on many laboratory media designed for stool cultures, but clinical specimens must be collected early in the disease (before the volume of stool dilutes the number of cells) and inoculated promptly because *Vibrio* is extremely sensitive to drying. Strains are distinguished by means of antibody tests specific for each antigenic variant.

Health care providers must promptly treat cholera patients with fluid and electrolyte replacement before hypovolemic shock

[11]From Greek *hypo*, meaning under; Latin *kalium*, meaning potassium; and Greek *haima*, meaning blood.

ensues. Antimicrobial drugs are not as important as with many other bacterial diseases, because they are lost in the watery stool; nevertheless, they may reduce the production of exotoxin and ameliorate the symptoms. Doxycycline is the drug of choice.

Because *Vibrio cholerae* can grow in estuarine and marine water, and because up to 20% of former patients still asymptomatically harbor the pathogen, it is unlikely that cholera will be eradicated worldwide. Nevertheless, adequate sewage and water treatment can limit the spread of *Vibrio* and prevent epidemics.

A new, oral vaccine developed against the O1 strains of *V. cholerae* appears promising; there is no vaccine for the O139 strain. Antibiotic prophylaxis of those who travel to endemic areas has not proven effective. Fortunately, because the infective dose for *V. cholerae* is high, proper hygiene generally makes vaccination and prophylaxis unnecessary.

Other Diseases of *Vibrio*

Vibrio parahaemolyticus (pa-ră-hē-mō-li'ti-kŭs) causes cholera-like gastroenteritis following ingestion of shellfish harvested from contaminated estuaries; fortunately, only rarely is it severe enough to be fatal. Typically the disease is characterized by a self-limiting, explosive diarrhea accompanied by headache, nausea, vomiting, and cramping for 72 hours.

V. vulnificus (vul-nif'i-kŭs) is responsible for septicemia (blood poisoning) following consumption of contaminated shellfish, and for infections resulting from the washing of wounds with contaminated seawater. Wound infections are characterized by swelling and reddening at the site of infection and are accompanied by fever and chills. Infections with *V. vulnificus* are fatal for 50% of untreated patients; prompt treatment with doxycycline is effective.

CLINICAL CASE STUDY

The Case of the Lactovegetarians

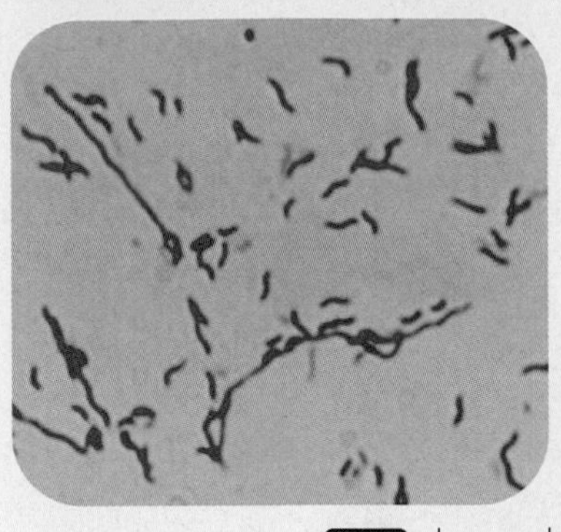

Two patients—a woman and her husband, ages 23 and 22, respectively—arrive at the health clinic one morning. They report having had severe abdominal cramps, grossly bloody diarrhea, nausea, and fever for 48 hours. Cultures of stool samples grown under microaerophilic, capneic conditions contain comma-shaped, Gram-negative bacilli. Both the patients are lactovegetarians and report being part of a "cow leasing" program at a local dairy in which patrons lease part of a cow's milk production so they can drink natural, whole, raw milk. The couple devised the program so that they and several neighbors could circumvent state regulations prohibiting the sale of unpasteurized milk. Investigators obtained and cultured a milk sample from the dairy's bulk milk tank: the cultures contained the bacterium pictured.

1. What is the pathogen?
2. How did the couple become infected?
3. Are the couple's colleagues at work at risk of acquiring an infection from the couple?
4. What other foods that are common sources of this bacterium can be ruled out in this case?

Campylobacter jejuni

Learning Objectives

- ✓ List several possible reservoirs of *Campylobacter jejuni*.
- ✓ Describe gastroenteritis caused by *Campylobacter jejuni*.

Campylobacter jejuni[12] (kam'pi-lō-bak'ter jē-jū'nē) is likely the most common cause of bacterial gastroenteritis in the United States. Like *Vibrio*, it is Gram-negative, slightly curved, oxidase positive, and motile by means of polar flagella; however, genetic analysis has shown that this comma-shaped pathogen is properly classified in the class Deltaproteobacteria (and not in Gammaproteobacteria). *Campylobacter* also differs from *Vibrio* in being microaerophilic and capneic.

Campylobacter infections are zoonotic—many domesticated animals, including poultry, dogs, cats, rabbits, pigs, cattle, and minks, serve as reservoirs. Humans acquire the bacterium by consuming food, milk, or water contaminated with infected animal feces. The most common source of infection is contaminated poultry. One study found that of dozens of Thanksgiving turkeys tested, 100% were contaminated with *Campylobacter*.

The pathogenicity of *C. jejuni* is not well understood, although the bacterium possesses adhesins, cytotoxins, and endotoxins that appear to enable colonization and invasion of the jejunum, ileum, and colon, producing bleeding lesions and triggering inflammation. Interestingly, nonmotile mutants are avirulent. *C. jejuni* infections commonly produce malaise, fever, abdominal pain, and bloody and frequent diarrhea—10 or more bowel movements per day are not uncommon. The disease is self-limiting; as bacteria are expelled from the intestinal tract, the symptoms abate, although a typical infection may last 7–10 days.

Scientists estimate that over 2 million cases of *Campylobacter* gastroenteritis occur each year in the United States—more than those caused by *Salmonella* and *Shigella* combined. However, *Campylobacter* gastroenteritis is not a reportable disease, so its actual incidence is unknown.

The number of infections can be reduced by proper food preparation, including thoroughly washing poultry carcasses, hands, and utensils that have contacted carcasses; cooking food sufficiently to kill bacteria; and pasteurizing cow's milk. Steps must also be taken to prevent contamination of the water supply with feces from stockyards, feedlots, and slaughterhouses.

[12]From Greek *kampylos*, meaning curved, and *jejunum*, the middle portion of the small intestine.

Helicobacter pylori

Learning Objectives

✓ Discuss the major change in medical opinion concerning the cause of peptic ulcers.

✓ Describe the effect of *Helicobacter pylori* on the lining of the human stomach.

Helicobacter pylori (hel′ĭ-kō-bak′ter pī′lō-rē) is a slightly helical, highly motile bacterium that colonizes the stomachs of its hosts **(Figure 21.22)**. Based on rRNA nucleotide sequences, *Helicobacter* is classified in the class Deltaproteobacteria. It was originally placed in the genus *Campylobacter;* however, unlike *Campylobacter, Helicobacter* cannot reduce nitrate, has flagella, and is urease positive, a characteristic that is essential for colonizing the stomach.

Today, an idea that only a few years ago was very controversial—that *H. pylori* causes *gastritis*[13] and most **peptic ulcers,** which are erosions of the mucous membrane of the stomach or of the initial portion of the small intestine—is accepted as fact. What was once the prevailing view—that stress, alcohol consumption, spicy food, or excess stomach acid production caused ulcers—was discredited beginning in 1982, when Australian gastroenterologists Robin Warren

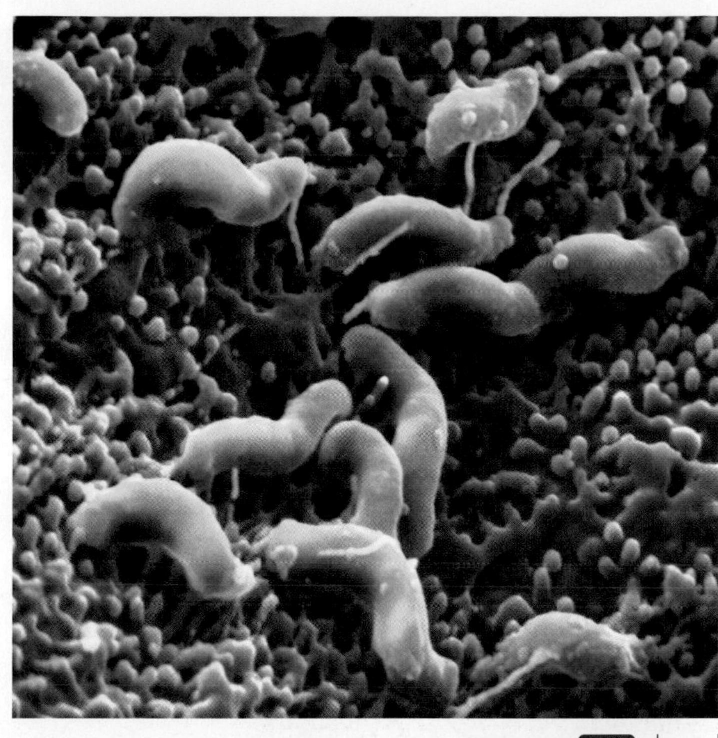

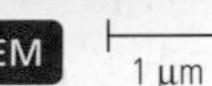

▲ **Figure 21.22** ***Helicobacter pylori.*** The bacterium causes peptic ulcers.

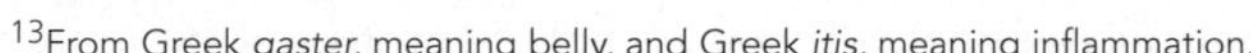

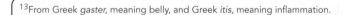

[13]From Greek *gaster,* meaning belly, and Greek *itis,* meaning inflammation.

EMERGING DISEASES

VIBRIO VULNIFICUS INFECTION

Greg enjoyed Florida's beaches on the Gulf side; swimming in the warm water was his favorite pastime. Of course, the saltwater did sting his leg where he had cut himself on some coral, but it didn't sting enough to stop Greg from enjoying the beach. He spent the afternoon jogging in the pure sand throwing a disc, watching the people, drinking a few beers, and of course spending more time in the water.

That evening he felt chilled, a condition he associated with too much sun during the day, but by midnight he thought he must have caught a rare summertime flu. He was definitely feverish, extremely weak and tired. His leg felt strangely tight, as if the underlying muscles were trying to burst through his skin.

The next morning he felt better, except for his leg. It was swollen, dark red, tremendously painful, and covered with fluid-filled blisters. The ugly sight motivated him to head straight for the hospital, a decision that likely saved his life.

Greg was the victim of an emerging pathogen, *Vibrio vulnificus*—a slightly curved, Gram-negative bacterium that is related to *V. cholerae* (cholera bacterium). *V. vulnificus* lives in salty, warm water around the globe and can live in marine organisms such as oysters. Unlike the cholera bacterium, *V. vulnificus* is able to infect a person by penetrating directly into a deep wound, a cut, or even a tiny scratch.

Greg's doctor cut away the dead tissue and prescribed doxycycline and cephalosporin for two weeks, and Greg survived and kept his leg. Half of the victims of *Vibrio vulnificus* are not so lucky; they lose a limb or die. Who knew that a beach could be so dangerous? For more about *Vibrio vulnificus,* see p. 618.

Track *Vibrio vulnificus* online by going to the Study Area at www.masteringmicrobiology.com.

MICROBE AT A GLANCE

Helicobacter pylori

Taxonomy: Domain Bacteria, phylum Proteobacteria, class Epsilonproteobacteria, order Campylobacterales, family Campylobacteraceae

Cell morphology and arrangement: Slightly helical bacillus

Gram reaction: Negative

Virulence factors: Protein that inhibits acid production, urease, flagella, adhesins, antiphagocytic enzymes

Diseases caused: Peptic ulcer

Treatment for diseases: Antibacterial drugs in conjunction with drugs that inhibit acid production in the stomach

Prevention of disease: Good personal hygiene, adequate sewage treatment, water purification, proper food handling

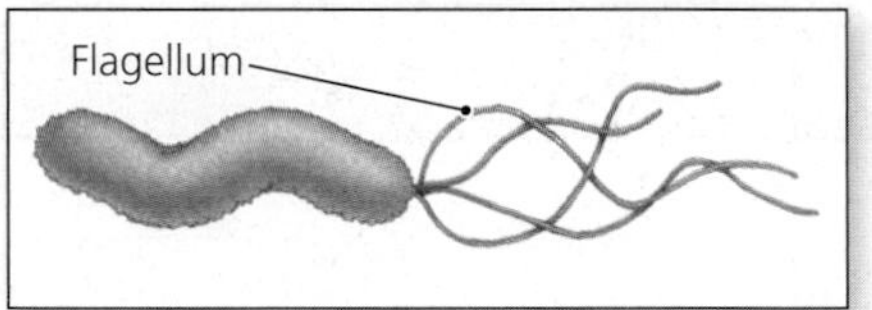

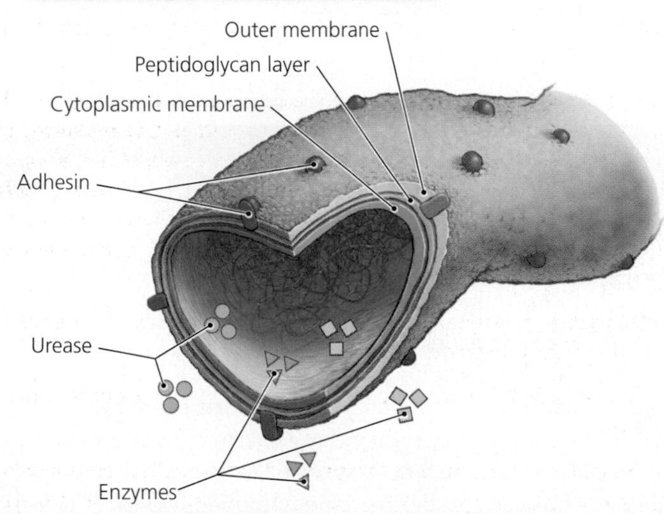

Download the Microbe at a Glance flashcards from the Study Area at www.masteringmicrobiology.com.

(1937–) and Barry Marshall (1951–) detected *Helicobacter* colonizing the majority of their patients' stomachs. Following treatment with antibiotics, the bacteria and the ulcers disappeared. Dr. Marshall provided final proof of *H. pylori's* involvement by fulfilling Koch's postulates himself—he drank one of his cultures of *Helicobacter*! As he expected, he developed painful gastritis and was able to isolate *H. pylori* from his diseased stomach. His sacrifice was worth it: Marshall and Warren received the 2005 Nobel Prize in Physiology or Medicine for their discovery. Since then scientists have shown that long-term infection with *Helicobacter* is a significant risk factor for stomach cancer.

H. pylori possesses numerous virulence factors that enable it to colonize the human stomach: a protein that inhibits acid production by stomach cells; flagella that enable the pathogen to burrow through mucus lining the stomach; adhesins that facilitate binding to gastric cells; enzymes that inhibit phagocytic killing; and urease, an enzyme that degrades urea, which is present in gastric juice, to produce highly alkaline ammonia, which neutralizes stomach acid.

The portal of entry for *H. pylori* is the mouth. Studies have shown that *H. pylori* in feces on the hands, in well water, or on fomites may infect humans. In addition, cat feces may be a source of infection.

The formation of a peptic ulcer occurs as follows **(Figure 21.23)**: The process begins when *H. pylori* (protected by urease) burrows through the stomach's protective layer of mucus to reach the underlying epithelial cells (1), where the bacteria attach to the cells' cytoplasmic membranes and multiply. A variety of factors—the triggering of inflammation by bacterial exotoxin and perhaps the destruction of mucus-producing cells by the bacteria—causes the layer of mucus to become thin (2), allowing acidic gastric juice to digest the stomach lining. Once the epithelial layer has been ulcerated by gastric juice, *H. pylori* gains access to the underlying muscle tissue and blood vessels (3). Those bacteria that are phagocytized survive in part through the actions of catalase and superoxide dismutase, enzymes that neutralize part of the phagocytes' killing mechanism.

The presence of *H. pylori* in specimens from the stomach can be demonstrated by a positive urease test within one to two hours of culturing; the bacterium can also be seen in Gram-stained specimens. Definitive identification is based on a series of biochemical tests.

Physicians treat ulcers with two or three antimicrobial drugs given in combination with drugs that inhibit acid production, allowing the stomach lining to regenerate. Prevention of infection involves good personal hygiene, adequate sewage treatment, water purification, and proper food handling.

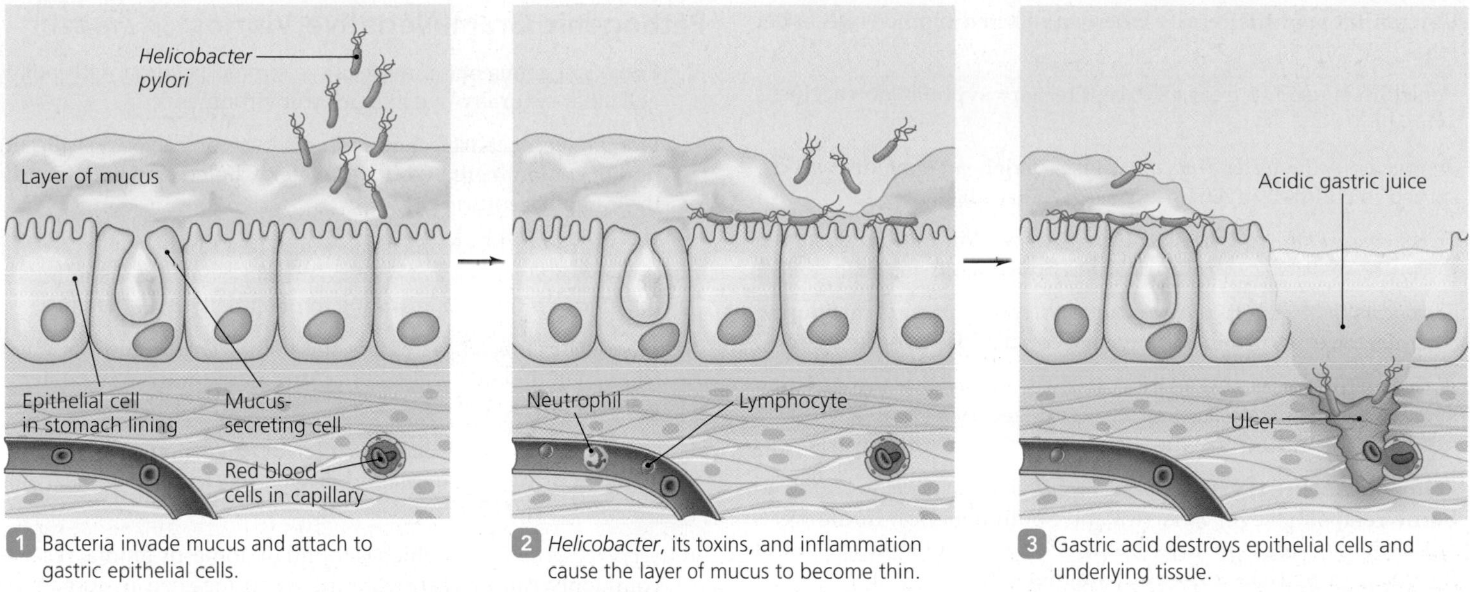

1 Bacteria invade mucus and attach to gastric epithelial cells.

2 *Helicobacter*, its toxins, and inflammation cause the layer of mucus to become thin.

3 Gastric acid destroys epithelial cells and underlying tissue.

▲ **Figure 21.23 The role of *Helicobacter pylori* in the formation of peptic ulcers.**

Chapter Summary

Rickettsias (pp. 601–604)

1. **Rickettsias** are extremely small, Gram-negative, obligate intracellular parasites.
2. *Rickettsia rickettsii* causes **Rocky Mountain spotted fever (RMSF),** a serious illness transmitted by ticks. Rash, malaise, petechiae, encephalitis, and death (in 5% of cases) characterize the disease.
3. *Rickettsia prowazekii* causes louse-borne or **epidemic typhus,** characterized by fever, depression, and rash. The disease may recur as Brill-Zinsser disease.
4. *Rickettsia typhi* causes endemic typhus, also called **murine typhus** because the major reservoir is rodents. Various fleas act as vectors to transmit the bacterium to humans.
5. Mites (chiggers) transmit *Orientia tsutsugamushi* to cause **scrub typhus,** which is characterized by a spreading rash.
6. *Ehrlichia chaffeensis* causes **human monocytic ehrlichiosis (HME),** and *Anaplasma phagocytophilum* causes **anaplasmosis.** These bacteria are transmitted to humans via the Lone Star tick, the deer tick, or the dog tick. The bacteria enter leukocytes and grow through three developmental stages: an elementary body, an initial body, and a morula.

Chlamydias (pp. 604–608)

1. Chlamydias are small, nonmotile obligate intracellular parasites; their developmental cycle includes infectious **elementary bodies (EBs)** and noninfectious **reticulate bodies (RBs).** An endosome full of RBs is called an **inclusion body.**
2. *Chlamydia trachomatis* enters the body through abrasions in mucous membranes of the genitalia, eyes, or respiratory tract. *Chlamydia* causes the most-reported sexually transmitted disease in the United States.
3. *Chlamydia trachomatis* causes **lymphogranuloma venereum.** The majority of infections in women are asymptomatic, whereas the majority of infections in men result in buboes, fever, chills, anorexia, and muscle pain. Other symptoms and signs may include proctitis (in men and women), and pelvic inflammatory disease and sterility (in women). *Chlamydia* infections can be prevented by abstinence or faithful mutual monogamy.
4. **Trachoma** is a serious eye disease that is caused by *C. trachomatis* and may result in blindness.
5. *Chlamydophila pneumoniae* causes bronchitis, pneumonia, and sinusitis.
6. *Chlamydophila psittaci* causes **ornithosis** (also called psittacosis or parrot fever), a respiratory disease of birds that can be transmitted to humans.

Spirochetes (pp. 609–616)

1. **Spirochetes** are helical bacteria with axial filaments that cause the organism to corkscrew, enabling it to burrow into a host's tissues.
2. **Syphilis** is caused by *Treponema pallidum pallidum,* a sexually transmitted obligate parasite of humans.
3. Primary syphilis is characterized by a **chancre,** a red lesion at the infection site. If *Treponema* moves from the chancre to the bloodstream, secondary syphilis results, causing rash, aches, and pains. After a period of latency, progression may occur to tertiary syphilis, characterized by swollen **gummas,** dementia, blindness, paralysis, and heart failure.

4. **Congenital syphilis** results when an infected mother infects her fetus.
5. Penicillin is used to treat all except tertiary syphilis; no vaccine is available.
6. *Treponema pallidum endemicum* causes **bejel,** an oral disease observed in children in Africa, Asia, and Australia.
7. *Treponema pallidum pertenue* causes **yaws,** a skin disease observed in South America, central Africa, and Southeast Asia.
8. *Treponema carateum* (*T. pallidum carateum*) causes **pinta,** a disfiguring skin disease observed in children in Central and South America.
9. *Borrelia burgdorferi* causes **Lyme disease,** a disease transmitted by ticks of the genus *Ixodes* and characterized by a bull's-eye rash, neurological and cardiac dysfunction, and severe arthritis.
10. Human body lice transmit *Borrelia recurrentis,* which causes **louse-borne relapsing fever** in Africa and South America. In the western United States, soft ticks transmit several different species of *Borrelia* causing **tick-borne relapsing fever.**
11. *Leptospira interrogans* causes **leptospirosis,** a zoonotic disease in humans transmitted via animal urine and characterized by pain, headache, and liver and kidney dysfunction.

Pathogenic Gram-Negative Vibrios (pp. 616–621)

1. *Vibrio* is a genus of Gram-negative curved bacteria with polar flagella that naturally live in marine environments.
2. *Vibrio cholerae* causes **cholera,** a disease contracted via the ingestion of contaminated food and water. Cholera has been pandemic through the centuries.
3. Via a series of biochemical steps, **cholera toxin**—an exotoxin produced by *Vibrio cholerae*—causes the movement of water out of the intestinal epithelium, resulting in potentially fatal diarrhea.
4. *Vibrio parahaemolyticus* enters the body via ingestion of shellfish from contaminated waters; it causes a milder form of cholera-like gastroenteritis.
5. *Vibrio vulnificus,* contracted either by ingestion of contaminated shellfish or by contamination of wounds with seawater, causes a potentially fatal blood poisoning.
6. *Campylobacter jejuni,* which is found in domestic animal reservoirs, commonly causes gastroenteritis when ingested in contaminated food, water, or milk.
7. Once *Helicobacter pylori* reduces the amount of mucus produced in the stomach, acidic gastric juice eats away the stomach lining, causing **peptic ulcers.**

Questions for Review

Answers to the Questions for Review (except Short Answer questions) begin on page A-1.

Multiple Choice

1. Most human infections caused by species of *Rickettsia*
   a. are acquired from fomites.
   b. could be prevented by handwashing.
   c. are transmitted via vectors.
   d. are sexually transmitted.
2. The bacterium that causes RMSF is more likely to infect a human
   a. if an infected tick feeds for several hours.
   b. when an infected tick initially penetrates the skin.
   c. when contaminated tick feces dry and become airborne.
   d. if the human is exposed to rodent feces containing the bacterium.
3. The most severe rickettsial illness is caused by
   a. *Rickettsia typhi.*
   b. *Rickettsia rickettsii.*
   c. *Orientia tsutsugamushi.*
   d. *Ehrlichia chaffeensis.*
4. The smallest cellular microbes are
   a. rickettsias.
   b. mycoplasmas.
   c. chlamydias.
   d. both a and c.
5. The most commonly reported sexually transmitted disease in the United States is caused by the bacterium
   a. *Mycoplasma genitalium.*
   b. *Chlamydia trachomatis.*
   c. *Chlamyophila proctitis.*
   d. *Ureaplasma urealyticum.*
6. Which of the following diseases would be *least* likely in rural areas of the United States?
   a. epidemic typhus
   b. Rocky Mountain spotted fever
   c. murine typhus
   d. lymphogranuloma venereum
7. Treatment of chlamydial infections involves
   a. erythromycin cream.
   b. doxycycline creams.
   c. surgical correction of eyelid deformities.
   d. all of the above
8. Which of the following organisms is transmitted via sexual contact?
   a. *Treponema pallidum endemicum*
   b. *Treponema pallidum pertenue*
   c. *Treponema pallidum pallidum*
   d. *Treponema carateum*
9. Which of the following is *not* true of cholera?
   a. The causative agent lives naturally in marine water.
   b. There is an effective vaccine for cholera.
   c. Strain O1 El Tor has been responsible for several pandemics.
   d. Rice-water stool is a symptom.
10. During which stage of syphilis is penicillin ineffective?
    a. primary syphilis
    b. secondary syphilis
    c. tertiary syphilis
    d. all of the above
11. Two weeks after a backpacking trip in Tennessee, a hiker experienced flulike symptoms and noticed a red rash on his thigh. What is the likely cause of his illness?
    a. *Treponema pallidum pertenue*
    b. *Borrelia burgdorferi*
    c. *Borrelia recurrentis*
    d. *Leptospira interrogans*
12. The most common cause of bacterial gastroenteritis in the United States is
    a. *Vibrio parahaemolyticus.*
    b. *Campylobacter jejuni.*
    c. *Helicobacter pylori.*
    d. *Vibrio cholerae.*

13. Historical journals have described gummas on patients. What disease most likely caused these lesions?
    a. ornithosis
    b. syphilis
    c. trachoma
    d. pneumonia

Labeling

Label the following stages and structures of the chlamydia life cycle: *elementary body, endocytosis, vesicle, host cell, inclusion body, reticulate body.* Indicate how many hours typically transpire at each lettered step.

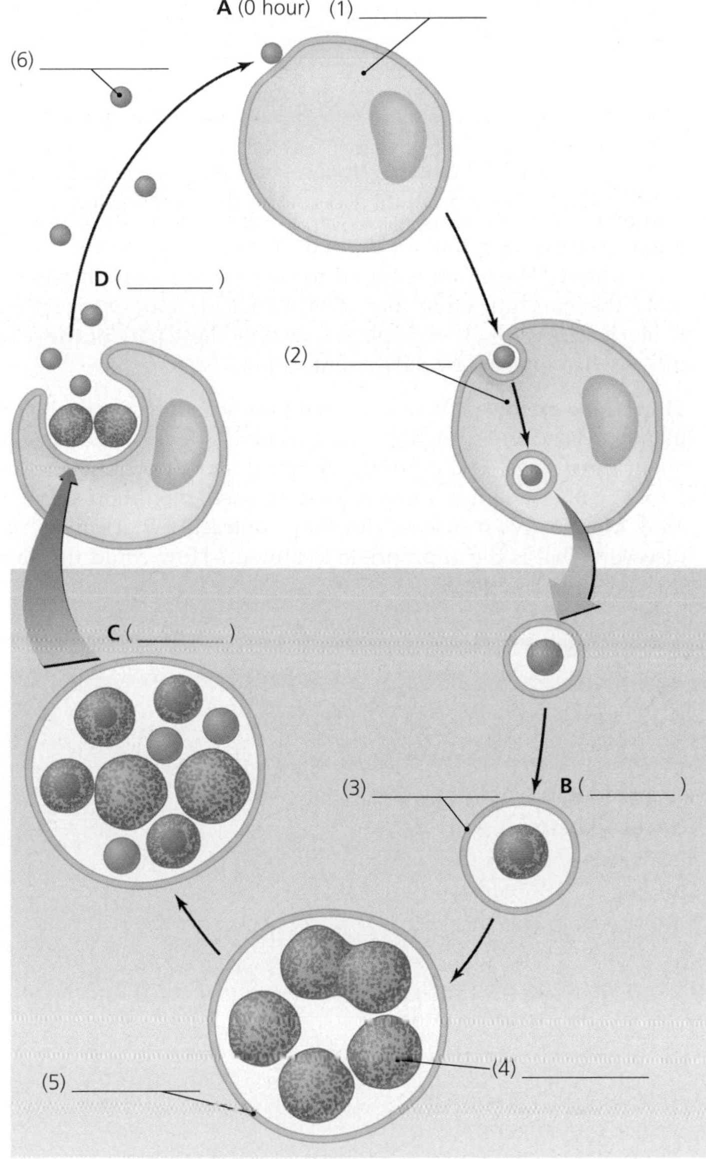

Short Answer

1. Describe a hypothesis to explain why rickettsias are obligate parasites.
2. Describe the three developmental stages of the bacteria *Ehrlichia* and *Anaplasma.*
3. Why have scientists had problems identifying the virulence factors of *Treponema pallidum pallidum*?
4. Describe the phases of untreated syphilis.
5. Discuss the prospects for the eradication of leptospirosis.
6. Beginning with the ingestion of water contaminated with *V. cholerae* O1 El Tor, describe the course of the disease it causes.

Matching

1. Match the disease with the causative pathogen.

____ Rocky Mountain spotted fever — A. *Rickettsia typhi*
____ Endemic typhus — B. *Rickettsia prowazekii*
____ Epidemic typhus — C. *Rickettsia rickettsii*
____ Scrub typhus — D. *Orientia tsutsugamushi*
____ HME — E. *Ehrlichia chaffeensis*

2. Match the pathogen with the vector responsible for transmitting it to humans.

____ *Rickettsia typhi* — A. Rat flea
____ *Rickettsia prowazekii* — B. Body louse
____ *Rickettsia rickettsii* — C. Hard tick
____ *Orientia tsutsugamushi* — D. Mite
____ *Ehrlichia chaffeensis* — E. Soft tick
____ *Borrelia burgdorferi*
____ *Borrelia recurrentis*
____ *Anaplasma phagocytophilum*

3. Match the pathogen with the disease(s) it causes.

____ *Chlamydophila psittaci* — A. Syphilis
____ *Chlamydophila pneumoniae* — B. Trachoma
____ *Chlamydia trachomatis* — C. Sinusitis
____ *Treponema pallidum pallidum* — D. Lymphogranuloma venereum
____ *Treponema pallidum pertenue* — E. Proctitis
____ *Treponema pallidum endemicum* — F. Pelvic inflammatory disease
____ *Treponema carateum* — G. Ornithosis
____ *Borrelia burgdorferi* — H. Yaws

I. Bejel
J. Pinta
K. Lyme disease

4. Match the following diseases with the causative bacterium.

____ Peptic ulcers — A. *Vibrio cholerae*
____ Gastroenteritis (various forms) — B. *Vibrio parahaemolyticus*
____ Blood poisoning — C. *Vibrio vulnificus*
____ Cholera — D. *Campylobacter jejuni*

E. *Helicobacter pylori*

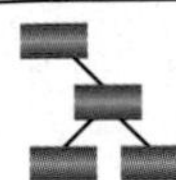

Concept Mapping

Using the following terms, draw a concept map that describes syphilis. For a sample concept map, see p. 93. Or, complete this concept map online by going to the Study Area at www.masteringmicrobiology.com.

Body-wide rash
Cardiovascular syphilis
Cell-free media
Chancre
Congenital syphilis
Darkfield microscopy
Gummas
Infect fetus in pregnant women
Last for years
Latent stage
Neurosyphilis
Never progress
Penicillin
Primary syphilis
Secondary syphilis
Serologic tests
Spirochete
Tertiary syphilis
Treponema pallidum

Critical Thinking

1. Why is it more difficult to rid a community of a disease transmitted by arthropods (for example, Lyme disease or RMSF) than a disease transmitted via contaminated drinking water (for example, cholera)?
2. Some scientists think that syphilis is a New World disease brought to Europe by returning Spanish explorers; others think that syphilis traveled the opposite way. Design an experiment to test the two hypotheses.
3. A patient arrives at a medical emergency room in Austin, Texas, complaining of 48 hours of severe abdominal cramping and persistent diarrhea. While waiting, he falls into a coma. He is admitted into the hospital but never regains consciousness; a week later he dies. No one else in his family is ill. They report that the man has not been out of the country, but that they all dined at a seafood restaurant the night before the man became sick. He alone ate sea urchins. Culture of the man's stool reveals the presence of curved bacilli. What bacterium might account for the man's death?
4. Scientists discover a mutant strain of *Helicobacter pylori* that is urease negative. The strain is found to cause ulcers only in patients that either consume large quantities of antacids and/or take drugs to block acid production. Explain why only these patients develop ulcers when infected by the mutant strain.
5. Thirty-nine members of an extended family sought medical treatment for headaches, myalgia, and recurring fevers, each lasting about three days. The outbreak occurred shortly after a one-day family gathering in a remote, seldom-used mountain cabin in New Mexico. What disease did they contract? What causes this disease? What is the appropriate treatment? How could this family have protected itself?

Access more review material online in the Study Area at **www.masteringmicrobiology.com.** There, you'll find

- **Concept Mapping Activities**
- **Flashcards**
- **Quizzes**

and more to help you succeed.

22 Pathogenic Fungi

The fungus *Microsporum* is beautiful when observed on a microscope slide, but it is much less welcome when found between the toes and on the soles of the feet, where it is one cause of the condition called *tinea pedis*—commonly known as "athlete's foot." Despite its name, this surface fungal disease with symptoms of itching, peeling, and other irritation affects athletes and nonathletes alike; having warm, moist feet is a much more important factor than being physically active. Because fungal spores are constantly being shed from the feet and onto bedding and furniture—and onto moist bathroom floors—athlete's foot is common among college students living in dorms. The spores are hardy; most survive for a long time in the environment, resisting desiccation, heat, cold, salt, acid, and even some antimicrobial cleansers. The best way to deal with athlete's foot is to keep your feet dry, change your socks frequently, use an antifungal cream, and wear a pair of flip-flops in public bathrooms.

In this chapter we examine this and other fungi that cause human diseases.

Take the pre-test for this chapter online. Visit the Study Area at www.masteringmicrobiology.com.

An ascomycete fungus, *Microsporum*, is one cause of athlete's foot among other fungal skin diseases.

Medical mycology is the field of medicine that is concerned with the diagnosis, management, and prevention of fungal diseases, or **mycoses** (mī-kō'sēz). As discussed in Chapter 12, most fungi exist as saprobes (absorbing nutrients from dead organisms) and function as the major decomposers of organic matter in the environment. More than 100,000 species of fungi have been classified in three divisions—Zygomycota, Ascomycota, and Basidiomycota—and of these, fewer than 200 have been demonstrated to cause diseases in humans.

Unlike the other disease chapters in this text, this chapter is not arranged according to any taxonomic scheme, because the classification of fungi is difficult. For example, most fungi have both a sexual stage, known as a *teleomorph,* and an asexual stage, known as an *anamorph.* The taxonomic code allows scientists to give each stage its own unique name. For example, the anamorph *Aspergillus* (as-per-jil'ŭs) is the same fungus as the teleomorph *Neosartorya* (nē-ō-sar-tōr'ya).

To minimize confusion, this chapter considers mycoses using only a single name (usually the anamorph) and according to the site of disease, though this convention is imperfect, as some fungi can invade tissues where they are not typically found. We also consider mycoses based on whether the fungus is a true pathogen or merely an opportunist. First, however, we will briefly consider some of the basic topics in medical mycology.

An Overview of Medical Mycology

Learning Objective

✓ Summarize some of the complexities of identifying and treating fungal infections.

Human mycoses are among the more difficult diseases to diagnose and treat properly. Signs of mycoses are often missed or misinterpreted, and once identified, fungi often prove to be remarkably resistant to antimicrobial agents. Before we begin our discussion of fungal diseases, therefore, it will be helpful to consider some of the characteristics that set fungal infections apart from other microbial infections.

The Epidemiology of Mycoses

Learning Objectives

✓ State the most significant mode of transmission for mycoses.

✓ Explain why the actual prevalence of fungal infections is unknown.

Fungi and the spores they produce are almost everywhere in the environment—in soil, in water, and on or in most multicellular organisms, including plants and animals. They coat the surfaces of almost every object, whether made of wood, glass, metal, or plastic; thus, it is not surprising that most of us will experience a mycosis at some time.

Mycoses are typically acquired via inhalation, trauma, or ingestion; only very infrequently are fungi spread from person to person. Therefore, most mycoses are *not* contagious. Epidemics of mycoses can and do occur, but not as a result of person-to-person contact. Instead, they result from mass exposure to some environmental source of fungi. For example, the cleanup of bird droppings near a building's air conditioning intake vents stirs up fungal spores in the droppings. The airborne spores could then be drawn inside the building and distributed throughout its duct system, potentially infecting numerous people.

One group of fungi that *are* contagious are *dermatophytes* (der'mă-tō-fītz), which live on the dead layers of skin and which may be transmitted between people via fomites. Species of the genera *Candida* (kan'did-a) and *Pneumocystis* (nū-mō-sis'tis) also appear to be transmitted at least some of the time by contact among humans.

Because most mycoses are not contagious, they typically are not reportable; that is, neither local public health agencies nor the Centers for Disease Control and Prevention (CDC) must be notified when they are diagnosed. Again, there are a few exceptions: Pathogenic fungi are usually reported in geographic areas where they are endemic, and certain opportunistic fungal pathogens of AIDS patients are also tracked.

The fact that most mycoses are not reportable creates a problem for epidemiologists. Without reliable data, it is impossible to track the effects of mycoses on the population, to ascertain their risks, and to identify actions to prevent them.

CRITICAL **THINKING**

Discuss the relationship between the following two facts: (1) A huge number of fungal spores are present in the environment, and (2) fungal diseases are not typically acquired via contact with infected individuals. How does not relying on a host for transmission benefit a fungus?

Categories of Fungal Agents: True Fungal Pathogens and Opportunistic Fungi

Learning Objectives

✓ Compare and contrast true fungal pathogens with opportunistic fungi.

✓ Identify factors that predispose people to opportunistic fungal infections.

Of all the fungi known to cause disease in humans, only four—*Blastomyces dermatitidis, Coccidioides immitis, Histoplasma capsulatum,* and *Paracoccidioides brasiliensis*—are considered true pathogens; that is, they can cause disease in otherwise healthy individuals. Other fungi, such as *Candida albicans,* are *opportunistic fungi,* which lack genes for proteins that aid in colonizing body tissues, though they can take advantage of some weakness in a host's defenses to become established and cause disease. Thus, whereas true fungal pathogens can infect anyone, regardless of immune status, opportunists infect only weakened individuals.

Four factors increase an individual's risk of experiencing opportunistic mycoses: invasive medical procedures, medical therapies, certain disease conditions, and specific lifestyle factors (Table 22.1). Surgical insertion of devices such as heart-valve implants can introduce fungal spores and provide a site for fungal colonization, while a variety of medical therapies leave patients with a weakened or dysfunctional immune system. AIDS, diabetes, other serious illnesses, and malnutrition also lessen immunity. Poor hygiene results in reduced skin defenses, and IV drug use can directly introduce fungi into the blood. Note that many of the factors that contribute to opportunistic fungal infections involve the actions of medical personnel in health care settings; thus correct antisepsis and medical procedures by health care providers play a crucial role in reducing the incidence of mycoses of opportunistic fungi.

In addition to differences in pathogenesis and matters related to host susceptibility, fungal pathogens and opportunists differ with respect to geographical distribution. Whereas the four pathogenic fungi are endemic to certain regions, primarily in the Americas, opportunistic fungi are distributed throughout the world.

Dermatophytes—fungi that normally live on the skin, nails, and hair—are the only fungi that do not fall comfortably into either the pathogenic or opportunistic groupings. They are considered by some researchers to be "emerging" pathogens. Dermatophytes can infect all individuals, not just the immunocompromised, which makes them similar to the pathogens. They are not, however, intrinsically invasive, being limited to body surfaces. They also have a tendency to occur in people with the same predisposing factors that allow access by opportunistic fungi. For these reasons, dermatophytes will be discussed as opportunists, rather than as pathogens.

CRITICAL **THINKING**

Suggest a way in which each of the risk factors listed in Table 22.1 can be reduced or eliminated to limit opportunistic infections. What can a patient do? What can health care providers do?

Clinical Manifestations of Fungal Diseases

Learning Objective

✓ Describe three primary clinical manifestations of mycoses.

Fungal diseases are grouped into the following three categories of clinical manifestations:

- *Fungal infections,* which are the most common mycoses, are caused by the presence in the body of either true pathogens or opportunists. As previously noted, this chapter discusses mycoses according to general location within the body. Systemic mycoses are discussed first, followed by superficial, cutaneous, and subcutaneous mycoses.
- *Toxicoses* (poisonings) are acquired through ingestion, as occurs when poisonous mushrooms are eaten. Although relatively rare, toxicoses are discussed briefly near the end of the chapter.

TABLE 22.1 Factors That Predispose Individuals to Opportunistic Mycoses

Factors	Examples
Medical procedures	Surgery; insertion of medical implants (heart valves, artificial joints); catheterization
Medical therapies	Immunosuppressive therapies accompanying transplantation; radiation and other cancer therapies; steroid treatments; long-term use of antibacterial agents
Disease conditions	Inherited immune defects; leukemia and lymphomas; AIDS; diabetes and other metabolic disorders; severe burns; preexisting chronic illnesses
Lifestyle factors	Malnutrition; poor hygiene; IV drug abuse

- *Allergies* (hypersensitivity reactions) most commonly result from the inhalation of fungal spores and are the subject of the chapter's final section.

The Diagnosis of Fungal Infections

Learning Objective

✓ Discuss why the diagnosis of opportunistic fungal infections can be difficult.

Most patients reporting to a hospital emergency room with a severe respiratory illness are routinely tested for influenza, bacterial pneumonia, and perhaps tuberculosis; without some concrete reason to suspect a fungal infection, physicians may not look for them. However, if an emergency-room patient with respiratory distress reveals that she breeds exotic birds, she is much more likely to also be tested for mycoses. Thus a patient's history—including occupation, hobbies, travel history, and the presence of preexisting medical conditions—is critical for diagnosis of most mycoses. Even when fungal infections are relatively obvious, as when distinctive mycelial growth is observed, definitive diagnosis requires isolation, laboratory culture, and morphological analysis of the fungus involved.

Microbiologists culture fungi collected from patients on *Sabouraud dextrose agar,* a medium that favors fungal growth over bacterial growth (see Figure 6.12). The appearance of colonies, and the microscopic appearance of yeast cells, mycelia, or mold spores, are usually diagnostic.

Several techniques are commonly used in identifying fungi. *Potassium hydroxide (KOH) preparations* dissolve keratin in skin scrapings or biopsy specimens, leaving only fungal cells for examination. *Gomori methenamine silver (GMS) stain* is used on tissue sections to stain fungal cells black (other cells remain unstained) **(Figure 22.1)**. *Direct immunofluorescence stain* can also be used to detect fungal cells in tissues; however, immunological tests, though very useful for other microbial infectious

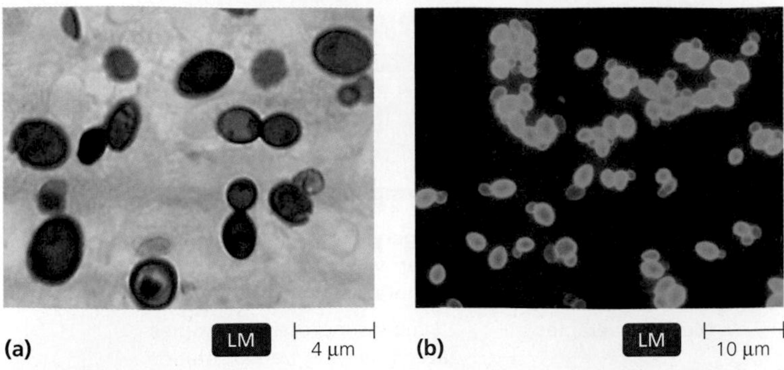

▲ **Figure 22.1 Fungal stains.** **(a)** GMS (Gomori methenamine silver) stain. *Histoplasma capsulatum*, shown here, causes histoplasmosis. **(b)** Direct fluorescent stain of *Candida albicans.*

agents, are not always useful for fungi. Because fungi are so prevalent in the environment and many are part of the normal microbiota, it is often impossible to distinguish between actual infection and simple exposure.

Diagnosis of opportunistic fungal infections is especially challenging. When a fungal opportunist infects tissues in which it is normally not found, it may display abnormal morphology that complicates identification. In addition, the fungi that produce pulmonary infections—the true pathogens and a few opportunists—produce symptoms and imaging profiles (X-ray studies, CT scans) that strongly resemble those of tuberculosis. Fungal masses may also resemble tumors.

Antifungal Therapies

Learning Objective

✓ Discuss the advantages and disadvantages of fungicidal and fungistatic medications.

Mycoses are among the most difficult diseases to heal for two reasons. First, fungi generally possess the biochemical ability to resist T cells during cell-mediated immune responses. Second, fungi are biochemically similar to human cells, which means that most fungicides are toxic to human tissues. The majority of antifungal agents exploit one of the few differences between human and fungal cells—instead of cholesterol, the membranes of fungal cells contain a related molecule, *ergosterol.* Antifungal drugs target either ergosterol synthesis or its insertion into fungal membranes. However, cholesterol and ergosterol are not sufficiently different to prevent some damage to human tissues by such antifungal agents. Serious side effects associated with long-term use of almost all antifungal agents include anemia, headache, rashes, gastrointestinal upset, and serious liver and kidney damage.

The "gold standard" of antifungal agents is the fungicidal drug *amphotericin B,* considered the best drug for treating systemic mycoses and other fungal infections that do not respond to other drugs. Unfortunately, it is also one of the more toxic antifungal agents to humans. The major alternatives to amphotericin B are the azole drugs—*ketoconazole, itraconazole,* and *fluconazole*—which are fungistatic (inhibitory) rather than fungicidal and less toxic to humans.

Two antifungal drugs that do not target ergosterol are *5-fluorocytosine,* which inhibits RNA and DNA synthesis and is given in conjunction with amphotericin B in the treatment of cryptococcal infections, and *griseofulvin,* which interferes with microtubule formation and chromosomal separation in mitosis. Griseofulvin accumulates in the outer epidermal layers of the skin, preventing fungal penetration and growth. Since these skin cells are scheduled to die as they move toward the surface, griseofulvin's toxicity does not permanently damage humans. Patients generally tolerate the drug for the duration of treatment. For example, griseofulvin is usually administered orally for up to a year to clear fungal infections of the nails without harming the patient.

Treatment of opportunistic fungal infections in immunocompromised patients involves two steps: a high-dose treatment to eliminate or reduce the number of fungal pathogens, followed by long-term (usually lifelong) maintenance therapy involving the administration of antifungal agents to control ongoing infections and prevent new infections.

With just a few antifungal drugs being used for long periods and treating more and more patients, scientists predict that the drugs should select drug-resistant strains from the fungal population. Fortunately, this is rarely the case; naturally occurring resistance, especially against amphotericin B, is extremely rare, though researchers cannot explain why resistance does not develop as it does in bacterial populations under similar conditions of long-term use. Nevertheless, strains of *Candida, Cryptococcus,* and *Aspergillus* in AIDS patients, many of whom must remain on antifungal drugs for life, have developed some resistance against 5-fluorocytosine and the azole drugs.

Antifungal Vaccines

Learning Objective

✓ Explain why antifungal vaccines are difficult to develop.

Prevention of fungal infections generally entails avoiding endemic areas and keeping one's immune system healthy. Vaccines against fungi have been difficult to develop because fungal metabolism is similar to our own.

Scientists have developed attenuated vaccines against *Coccidioides,* but there is concern that the weakened microbe could revert to a virulent form and cause disease in immunized individuals. Recently, researchers used recombinant DNA to create a live vaccine that protects mice against a blastomycosis and have conjugated fungal antigens with diphtheria toxoid to create a vaccine against *Candida,* which is the most common fungus infecting AIDS patients. These new vaccines are safer than live, attenuated vaccines and more effective than killed vaccines.

Now that we have considered some of the basics of medical mycology, we turn our attention to the first category of important mycoses: systemic mycoses caused by the four truly pathogenic fungi.

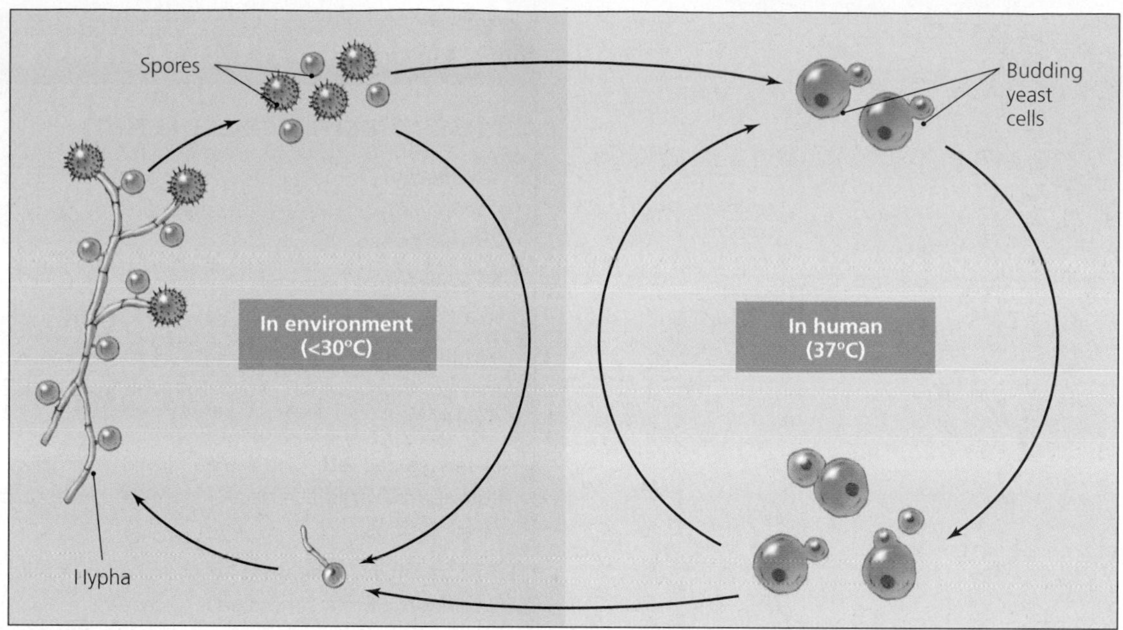

▲ **Figure 22.2 The dimorphic nature of true fungal pathogens.** The mycelial form grows in the environment, and the yeast form grows within a human host. Shown here is *Histoplasma capsulatum. What are the other three genera of pathogenic, dimorphic fungi?*

Figure 22.2 *In addition to* Histoplasma*, members of the genera* Blastomyces*,* Coccidioides*, and* Paracoccidioides *are pathogenic fungi.*

Systemic Mycoses Caused by Pathogenic Fungi

Learning Objectives

- ✓ Compare and contrast the endemic areas for the four genera of pathogenic fungi that cause systemic mycoses.
- ✓ Compare the clinical appearances of the diseases resulting from each of the four pathogenic fungi.
- ✓ Identify the laboratory techniques used to distinguish among the four pathogenic fungi.

Systemic mycoses—those fungal infections that spread throughout the body—result from infections by one of the four pathogenic fungi: *Blastomyces, Coccidioides, Histoplasma,* or *Paracoccidioides.* All are in the fungal division Ascomycota. These pathogenic fungi are uniformly acquired through inhalation, and all begin as a generalized pulmonary infection that then spreads, via the blood to the rest of the body.

All four are also **dimorphic**[1] (dī-mōr′fik); that is, they can have two forms. In the environment, where the temperature is typically below 30°C, they appear as mycelial thalli composed of hyphae, whereas within the body (37°C) they grow as spherical yeasts **(Figure 22.2)**. The two forms differ not only structurally but also physiologically. Yeast forms are invasive because they express a variety of enzymes and other proteins that aid their growth and reproduction in the body. For example, they are tolerant of higher temperatures and are relatively resistant to phagocytic killing.

Dimorphic fungi are extremely hazardous to laboratory personnel, who must take specific precautions to avoid exposure to spores, particularly when culturing the organisms. Biological safety cabinets with HEPA filters (see Figure 9.10), protective clothing, and masks are required when working with these pathogens.

In the following sections we consider each of the systemic conditions caused by true fungal pathogens, beginning with histoplasmosis.

Histoplasmosis

Histoplasma capsulatum (his-tō-plaz′mă kap-soo-lā′tŭm), the causative agent of **histoplasmosis** (his′tō-plaz-mō′sis), is an ascomycete and the most common fungal pathogen affecting humans. *H. capsulatum* is particularly prevalent in the eastern United States along the Ohio River Valley, but endemic areas also exist in Africa and South America **(Figure 22.3)**. *H. capsulatum* is found in moist soils containing high levels of nitrogen such as from the droppings of bats and birds.

Two strains of *H. capsulatum* (a fungal strain referred to as a variety, or var.) are recognized: var. *capsulatum,* which is the most widely distributed, and var. *duboisii* (dū-bwa′sē-ē) which is limited to Africa. Spores of both may become airborne and inhaled when soil containing the fungus is disturbed by wind or by human activities. Cutaneous inoculation can also lead to disease, but such infections are extremely rare.

[1] Form Greek *di,* meaning two, and *morphe,* meaning shape.

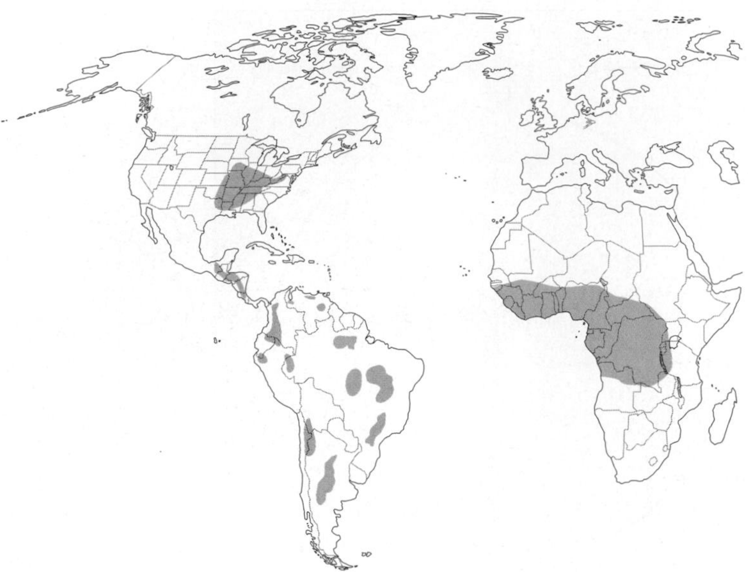

Figure 22.3 Endemic areas for histoplasmosis.

CRITICAL THINKING

Statistically, men are more likely than women to contract histoplasmosis. What might explain this fact?

H. capsulatum is an intracellular parasite that survives inhalation and subsequent phagocytosis by macrophages in air sacs of the lungs. These macrophages then disperse the fungus beyond the lungs via the blood and lymph. Cell-mediated immunity eventually develops, clearing the organism from healthy patients.

About 95% of individuals infected with pulmonary *Histoplasma* are asymptomatic, and their subclinical cases resolve without damage. Some individuals may experience mild or nonspecific respiratory symptoms (for example, pains while breathing or mild coughing) that also disappear on their own. Slightly more pronounced symptoms—fever, night sweats, and weight loss—may occur in a few individuals. About 5% of patients develop clinical histoplasmosis, which manifests as one of four diseases:

- *Chronic pulmonary histoplasmosis* is characterized by severe coughing, blood-tinged sputum, night sweats, loss of appetite, and weight loss. It is often seen in individuals with preexisting lung disease. It can be mistaken for tuberculosis.
- *Chronic cutaneous histoplasmosis,* characterized by ulcerative skin lesions, can follow the spread of infection from the lungs.
- *Systemic histoplasmosis* can also follow if infection spreads from the lungs, but it is usually seen only in AIDS patients. This syndrome, characterized by enlargement of the spleen and liver, can be rapid, severe, and fatal.
- *Ocular histoplasmosis* is a type I hypersensitivity reaction against *Histoplasma* in the eye; it is characterized by inflammation and redness.

MICROBE AT A GLANCE

Histoplasma capsulatum

Taxonomy: Domain Eukarya, Kingdom Fungi, phylum Ascomycota, class Ascomycetes, order Onygenales, family Onygenaceae

Morphology: Thermally dimorphic: <30°C, it forms septate hyphae; 37°C, it forms single-celled yeasts

Virulence factors: Survives phagocytosis to live inside macrophages, which can deliver the fungus throughout the body

Diseases caused: Chronic pulmonary histoplasmosis, chronic cutaneous histoplasmosis, systematic histoplasmosis, ocular histoplasmosis

Treatment for diseases: ketoconazole for mild infections; amphotericin B for more severe cases

Prevention of disease: Ninety percent of people living in endemic areas (Ohio River Valley in North America, equatorial West Africa, and isolated areas of South America) are infected. A healthy immune system is necessary to avoid disease.

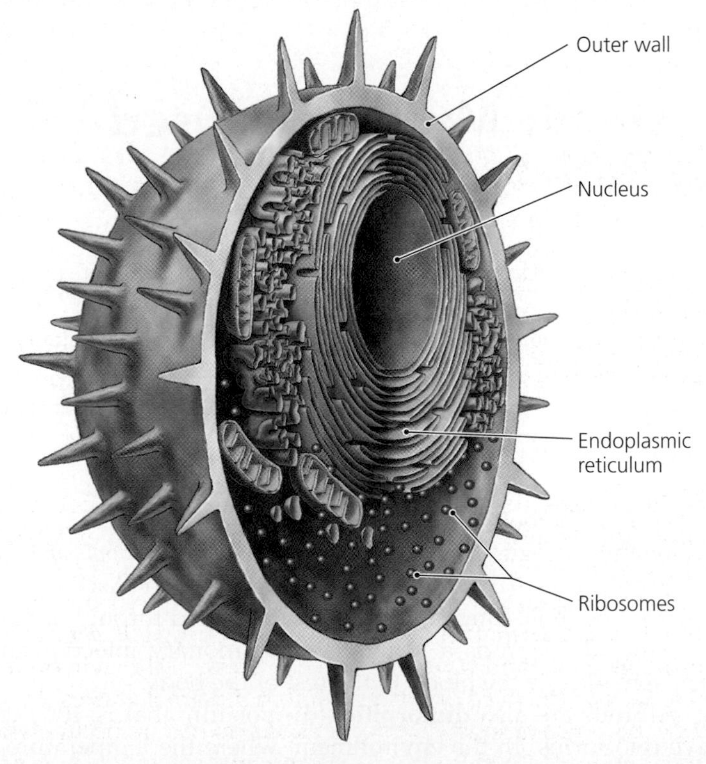

Download the Microbe at a Glance flashcards from the Study Area at www.masteringmicrobiology.com.

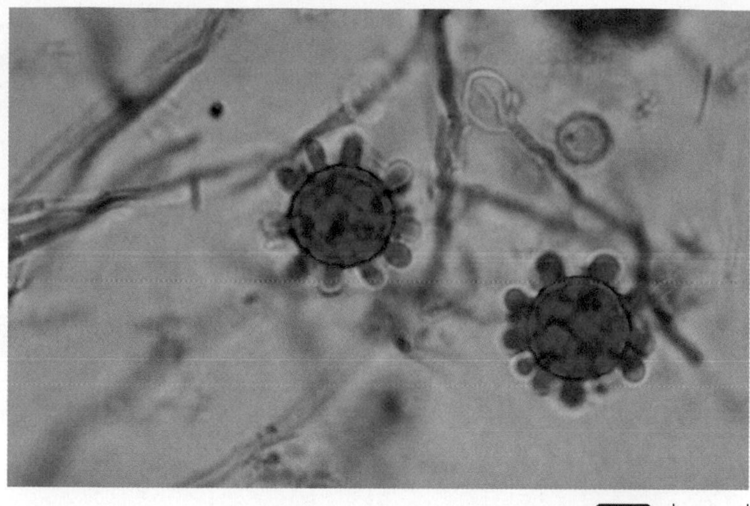

▲ **Figure 22.4** **The characteristic spiny spores of mycelial *Histoplasma capsulatum*.**

Diagnosis of histoplasmosis is based on the identification of the distinctive budding yeast in KOH- or GMS-prepared samples of skin scrapings, sputum, cerebrospinal fluid, or various tissues. The diagnosis is confirmed by the observation of dimorphism in cultures grown from such samples. Cultured *H. capsulatum* produces distinctively spiny spores that are also diagnostic (**Figure 22.4**). Antibody tests are not useful indicators of *Histoplasma* infection because many people have been exposed without contracting disease. In the endemic regions of the United States, close to 90% of the population tests positive for *H. capsulatum* exposure.

Infections in immunocompetent individuals typically resolve without treatment. When symptoms do not resolve, amphotericin B is prescribed. Ketoconazole can be used to treat mild infections. Maintenance therapy for AIDS patients is recommended.

Blastomycosis

Blastomycoses (blas'tō-mī-kō'sēz) are caused by another ascomycete, *Blastomyces dermatitidis* (blas-tō-mī'sēz der-mă-tit'i-dis), which is endemic across the southeastern United States north to Canada (**Figure 22.5**). Outbreaks have also been reported in Latin America, Africa, Asia, and Europe. *B. dermatitidis* normally grows and sporulates in cool, damp soil rich in organic material such as decaying vegetation and animal wastes. In humans, both recreational and occupational exposure occurs when fungal spores in soil become airborne and are inhaled. A relatively small inoculum can produce disease. The incidence of human infection is increasing, in part, because of increases in the number of immunocompromised individuals in the population.

Pulmonary blastomycosis is the most common manifestation of *Blastomyces* infection. After spores enter the lungs, they convert to yeast forms and multiply. Initial pulmonary lesions are asymptomatic in most individuals. If symptoms do develop, they are vague and include muscle aches, cough, fever, chills, malaise, and weight loss. Purulent (pus-filled) lesions develop and expand as the yeasts multiply, resulting in death of tissues and cavity formation. Rarely, a deep productive cough and chest pain accompany the generalized symptoms. In otherwise healthy people, pulmonary blastomycosis typically resolves successfully, although it may become chronic. Respiratory failure and death occur at a high frequency among AIDS patients.

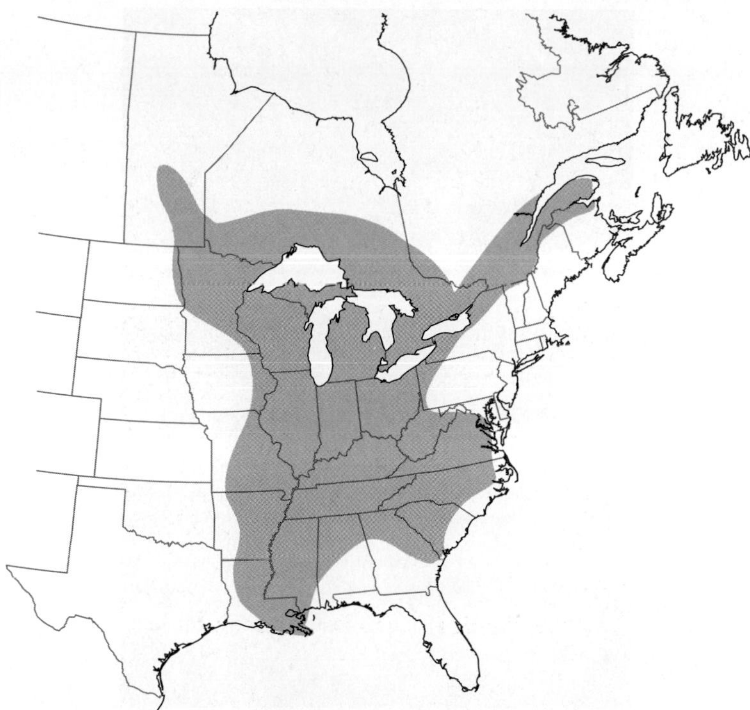

▲ **Figure 22.5** **Geographic distribution of *Blastomyces*.**

The fungus can spread beyond the lungs. *Cutaneous blastomycosis* occurs in 60–70% of cases and consists of generally painless lesions on the face and upper body (**Figure 22.6**). The lesions can be raised and wartlike, or they may be craterlike if tissue death occurs. In roughly 30% of cases, the fungus spreads to the spine, pelvis, cranium, ribs, long bones, or subcutaneous tissues surrounding joints, a condition called *osteoarticular*[2] (os'tē-ō-ar-tik'ū-lar) *blastomycosis*. About 40% of AIDS patients experience meningitis resulting from dissemination of *Blastomyces* to the central nervous system. Abscesses may form in other tissue systems as well.

Diagnosis relies on identification of *B. dermatitidis* following culture, or direct examination of various samples such as sputum, bronchial washings, biopsies, cerebrospinal fluid, or skin scrapings. Observation of dimorphism in laboratory cultures coupled with microscopic examination is diagnostic.

Physicians treat blastomycoses with amphotericin B for 10 weeks or longer. Oral itraconazole may be used as an alternative but must be administered for a minimum of three to six months. Relapse is common in AIDS patients, and suppressive maintenance therapy with itraconazole is recommended.

[2]From Greek *osteon*, meaning bone, and *arthron*, meaning joint.

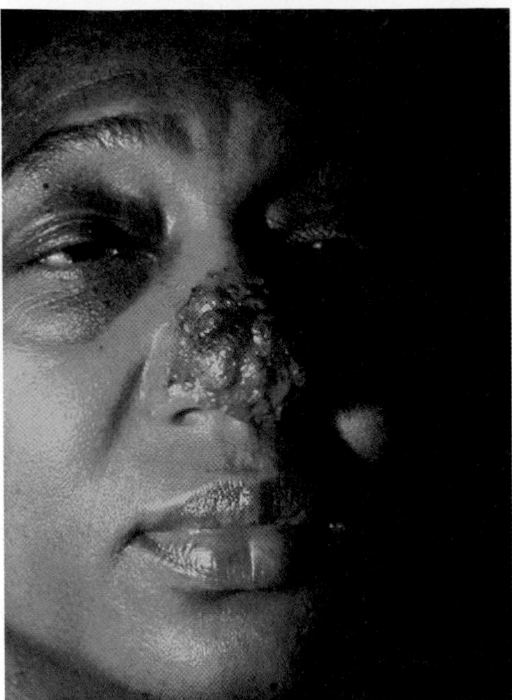

▲ **Figure 22.6 Cutaneous blastomycosis in an American woman.** This condition typically results from the spread of *Blastomyces dermatitidis* from the lungs to the skin.

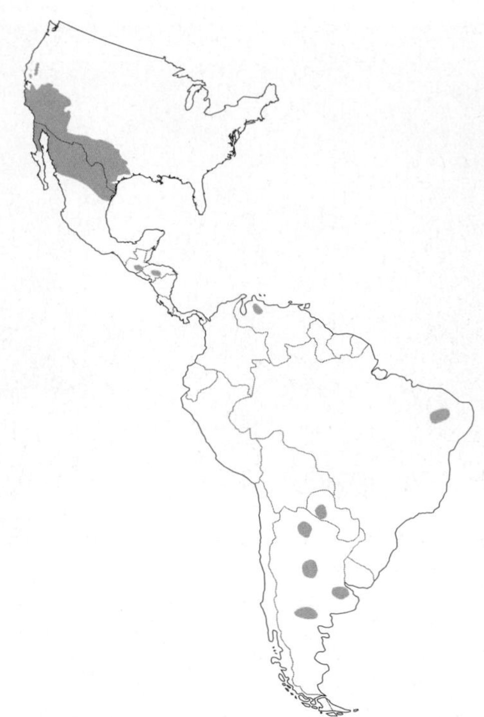

▲ **Figure 22.7 Endemic areas of *Coccidioides*.**

CRITICAL **THINKING**

Outbreaks of blastomycoses have occurred in Latin America even though the pathogen normally is not found there. Based on what you have read, explain why a few cases of blastomycosis might appear outside of Northern Hemisphere endemic areas.

Coccidioidomycosis

Coccidioidomycosis (kok-sid-ē-oy′dō-mī-kō-sis), caused by ascomycete *Coccidioides immitis* (kok-sid-ē-oy′dēz im′i-tis), is found almost exclusively in the southwestern United States (Arizona, California, Nevada, New Mexico, and Texas) and northern Mexico. Small, focally endemic areas also exist in semiarid parts of Central and South America **(Figure 22.7)**. Fungi can be recovered from desert soil, rodent burrows, archeological remains, and mines. Dust that coats materials from endemic areas, including Native American pots and blankets sold to tourists, can serve as a vehicle of infection. Infection rates in endemic areas have risen in recent years as a result of population expansion and increased recreational activities, such as the use of off-road vehicles in the desert. More ominous threats to human health, however, are windstorms and earthquakes, which can disturb large tracts of contaminated soil, spreading the fungal elements for miles downwind and exposing thousands of people.

In the warm and dry summer and fall months, particularly in drought cycles, *C. immitis* grows as a mycelium and produces sturdy chains of asexual spores called *arthroconidia*. When mature, arthroconidia germinate into new mycelia in the environment, but if inhaled, arthroconidia germinate in the lungs to produce a parasitic form called a *spherule* (sfer′ool) **(Figure 22.8)**. As each spherule matures, it enlarges and generates a large number of spores via multiple cleavages, until it ruptures and releases the spores into the surrounding tissue. Each spore then forms a new spherule to continue the cycle of division and release. This type of growth accounts for the seriousness of *Coccidioides* infection.

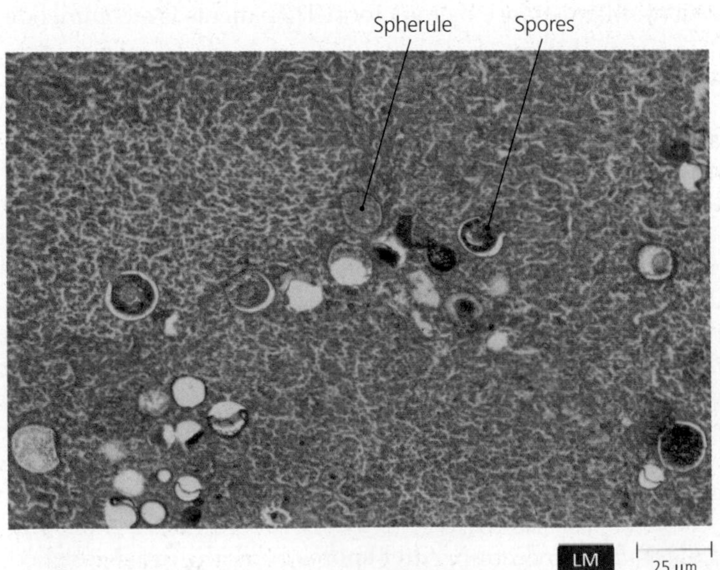

▲ **Figure 22.8 Spherules of *Coccidioides immitis*.** Note the numerous spores within a spherule.

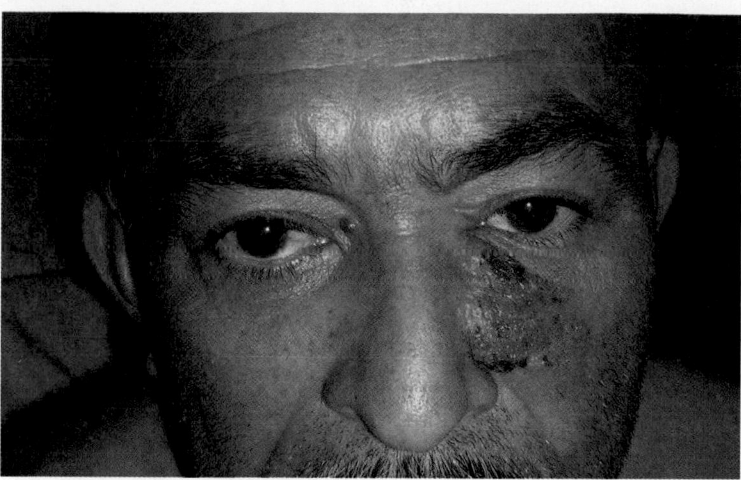

▲ **Figure 22.9 Coccidioidomycosis lesions in subcutaneous tissue.** Painless lesions result from the spread of *Coccidioides immitis* from the lungs.

The major manifestation of coccidioidomycosis is pulmonary. About 60% of patients experience either no symptoms or mild, unremarkable symptoms that go unnoticed and typically resolve on their own. Other patients develop more severe infections characterized by fever, cough, chest pain, difficulty breathing, coughing up or spitting blood, headache, night sweats, weight loss, and pneumonia; in some individuals a diffuse rash may appear on the trunk. Occasionally the mild forms of the pulmonary disease become more chronic, in which cases the continued multiplication of spherules results in large, permanent cavities in the lungs similar to those seen in tuberculosis patients.

In a very small percentage of cases, generally in those who are severely immunocompromised, *C. immitis* spreads from the lungs to various other sites. Invasion of the central nervous system (CNS) may result in meningitis, headache, nausea, and emotional disturbance. Infection can also spread to the bones, joints, and painless subcutaneous tissues; subcutaneous lesions are inflamed masses of granular material **(Figure 22.9)**.

The diagnosis of coccidioidomycosis can be based on the identification of spherules in KOH- or GMS-treated samples collected from patients. Health care workers administer a *coccidioidin skin test* to screen patients for contact with *Coccidioides*. In the test, antigens of *Coccidioides* are injected under the skin. If the body has antibodies against the fungus—that is, the patient has been or is infected—then the site will become inflamed. Diagnosis is confirmed by the isolation of the mycelial form of *C. immitis* in laboratory culture or isolation of encapsulated yeasts in the patients' tissues.

Although infections in otherwise healthy patients generally resolve on their own without damage, amphotericin B is the drug of choice. CNS involvement is fatal without treatment. In AIDS patients, maintenance therapy with itraconazole or fluconazole is recommended to prevent relapse or reinfection. The wearing of protective masks in endemic areas can prevent exposure to spores, but constant, daily wearing of masks may be impractical for all but those whose occupations put them at clear risk of infection.

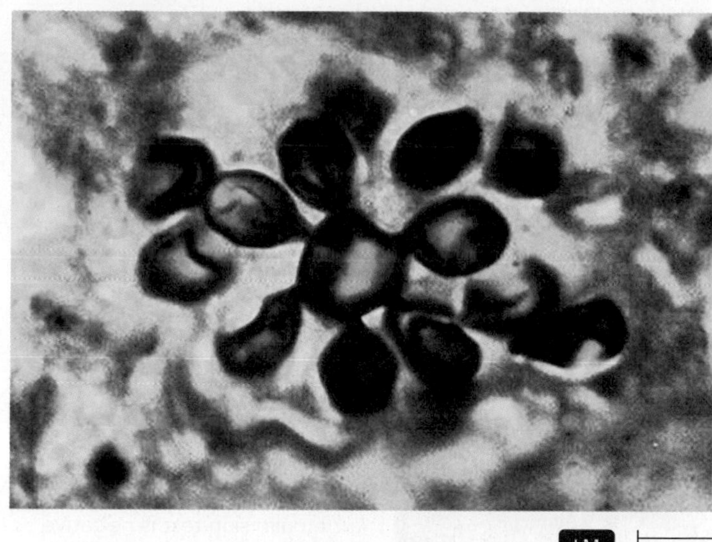

▲ **Figure 22.10 The "steering wheel" formation of buds characteristic of *Paracoccidioides brasiliensis*.**

Paracoccidioidomycosis

Another ascomycete, *Paracoccidioides brasiliensis* (par′ă-kok-sid-ē-oy′dēz bră-sil-ē-en′sis) causes **paracoccidioidomycosis** (par′ă-kok-sid-ē-oy′dō-mī-kō′sis), a chronic fungal disease similar to blastomycosis and coccidioidomycosis. *P. brasiliensis* is found in cool, damp soil from southern Mexico to regions of South America, particularly in Brazil. Because this fungus is far more geographically limited than the other true fungal pathogens, paracoccidioidomycosis is not a common disease. Those most at risk include farm workers in endemic areas.

Infections range from asymptomatic to systemic, and disease first becomes apparent as a pulmonary form that is slow to develop but manifests as chronic cough, fever, night sweats, malaise, and weight loss. The fungus almost always spreads, producing a chronic inflammatory disease of mucous membranes. Painful ulcerated lesions of the gums, tongue, lips, and palate progressively worsen over the course of weeks to months.

KOH or GMS preparations of tissue samples reveal yeast cells with multiple buds in a "steering wheel" formation that is diagnostic for this organism **(Figure 22.10)**. Laboratory culture at 25°C and at 37°C demonstrating dimorphism is confirmatory. Serological identification of antibodies also is useful. Treatment is with amphotericin B or ketoconazole.

CRITICAL **THINKING**

All of the true fungal pathogens manifest initially as a pulmonary disease. Explain how you could ascertain which of the four pathogens a patient has.

CLINICAL CASE STUDY

What's Ailing the Bird Enthusiast?

A 64-year-old man arrives at a hospital emergency room with signs and symptoms of serious pulmonary infection. He complains of a deep cough, blood-tinged sputum, night sweats, and weight loss. His liver and spleen feel enlarged upon physical examination. X-ray studies of his chest suggest tuberculosis, but a tuberculin skin test is negative, as is an HIV test. The man, an Ohio native, has recently traveled to Africa and Asia. He is also an avid bird enthusiast who likes to watch birds from his balcony. He keeps many bird feeders on the balcony, attracts numerous birds, and spends an hour a day cleaning up bird droppings.

Skin tests are positive for histoplasmosis but inconclusive for coccidioidomycosis and blastomycosis.

1. What is the most likely infecting agent? How do you suppose this individual acquired the disease?
2. Is this disease unusual in people who work with birds?
3. Given that some tests were inconclusive, what other tests or lab work would aid in arriving at a specific diagnosis?
4. What treatment would most likely be prescribed?

Systemic Mycoses Caused by Opportunistic Fungi

Learning Objective

✓ Discuss the difficulties in diagnosing opportunistic fungal infections.

Opportunistic mycoses do not typically affect healthy humans, because the fungi involved lack genes for virulence factors that make them actively invasive. Instead, opportunistic mycoses are limited to people with poor immunity. Because of the growing number of AIDS patients, opportunistic fungal infections have become one of the more significant causes of human disease and death. Even though any fungus can become an opportunist, five genera of fungi are routinely encountered: *Pneumocystis, Candida, Aspergillus, Cryptococcus,* and *Mucor.*

Opportunistic infections present a formidable challenge to clinicians. Because they appear only when their hosts are weakened, they often display "odd" clinical signatures; that is, their symptoms are often atypical, or they occur in individuals not residing in endemic areas for a particular fungus. Increased severity of symptoms is often a reflection of the poor immune status of the patient and frequently results in higher fatality rates, even if the fungus is normally nonthreatening.

In the following sections we consider some of the mycoses caused by the "classical" fungal opportunists, beginning with *Pneumocystis pneumonia.*

Pneumocystis Pneumonia

Learning Objective

✓ List the characteristics of *Pneumocystis* that distinguish it from other fungal opportunists.

Pneumocystis jiroveci (nū-mō-sis′tis jē-rō-vĕt′zē), an ascomycete of the normal respiratory microbiota formerly known as *P. carinii,* is one of the more common opportunistic fungal infections seen in AIDS patients. Prior to the AIDS epidemic, disease caused by *Pneumocystis* was extremely rare.

Even though the organism was first identified in 1909, little is known about it. Originally considered a protozoan, it has been reclassified as a fungus based on rRNA nucleotide sequences and biochemistry. Its morphological and developmental characteristics, however, still resemble those of protozoa more than those of fungi. Because *P. jiroveci* is an obligate parasite and cannot survive on its own in the environment, transmission most likely occurs through inhalation.

P. jiroveci is distributed worldwide in humans; based on serological confirmation of antibodies, the majority of healthy children (75%) have been exposed to the fungus by the age of five. In immunocompetent people, infection is asymptomatic, and generally clearance of the fungus from the body is followed by lasting immunity. However, some individuals may remain infected indefinitely; in such carriers the organism remains in the alveoli (air sacs of lungs) and can be passed in respiratory droplets.

Before the AIDS epidemic, the disease ***Pneumocystis* pneumonia** was observed only in malnourished, premature infants and debilitated elderly patients. Now, the disease is almost diagnostic for AIDS. *Pneumocystis* pneumonia is abbreviated ***PCP,*** which originally stood for *Pneumocystis carinii* pneumonia but now can be considered to stand for *Pneumocystis* pneumonia.

Once the fungus enters the lungs of an AIDS patient, it multiplies rapidly, extensively colonizing the lungs, because the patient's defenses are impaired. Widespread inflammation, fever, difficulty in breathing, and a nonproductive cough are characteristic. If left untreated, PCP involves more and more lung tissue until death occurs.

Diagnosis relies on clinical and microscopic findings. Chest X-ray studies usually reveal abnormal lung features. Stained smears of fluid from the lungs or from biopsies can reveal distinctive morphological forms of the fungus **(Figure 22.11)**. The use of fluorescent antibody on samples taken from patients is more sensitive and provides a more specific diagnosis.

Because the fungus has many of the characteristics of protozoa, both primary treatment and maintenance therapy are with

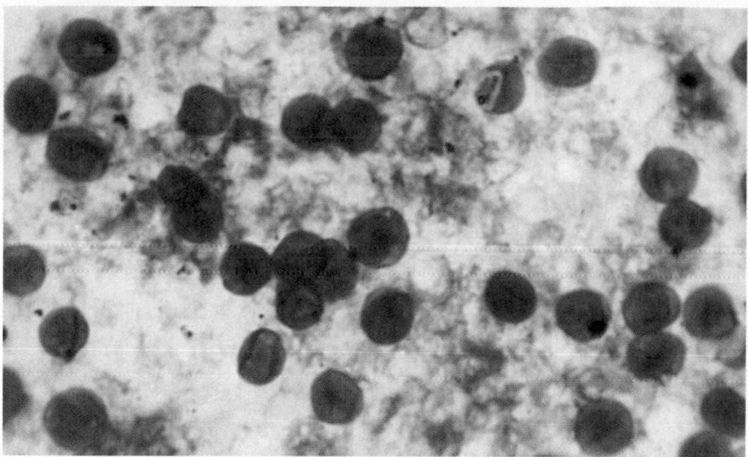

▲ **Figure 22.11** **Cysts of *Pneumocystis jiroveci* in lung tissue.** Such microscopic findings are diagnostic.

the antiprotozoan drugs trimethoprim and sulfanilamide. Pentamidine, another antiprotozoan drug, is an alternative. The drugs may be aerosolized to allow for direct inhalation into the lungs.

Candidiasis

Learning Objective

✓ Explain how candidiasis can develop from a localized infection to a systemic mycosis.

Candidiasis (kan-di-dī'ă-sis) is any opportunistic fungal infection or disease caused by various species of the genus *Candida*—most commonly *Candida albicans* (al'bi-kanz). Fungi in this genus of dimorphic ascomycetes are common members of the microbiota of the skin and mucous membranes; for example, the digestive tracts of 40–80% of all healthy individuals harbor *Candida* species. *Candida* is one of a very few fungi that can be transmitted between individuals. From its site as a normal inhabitant of the female reproductive tract, for example, it can be passed to babies during childbirth and to men during sexual contact.

Although *Candida* species can infect tissues in essentially every body system, producing a wide range of disease manifestations in humans (**Figure 22.12**), in all cases the fungus is an opportunist. It is *Candida* that causes vaginal yeast infections, as the fungus grows prolifically when normal bacterial microbiota are inhibited due to changes in vaginal pH or use of antibacterial drugs. Localized, superficial infections are seen in individuals with impairment of the barrier function of the epithelial layers (generally newborns and the elderly).

Systemic disease is seen almost universally in immunocompromised individuals. Different segments of the population experience a given manifestation of candidiasis at different rates. For example, whereas oral candidiasis (thrush) is rare in

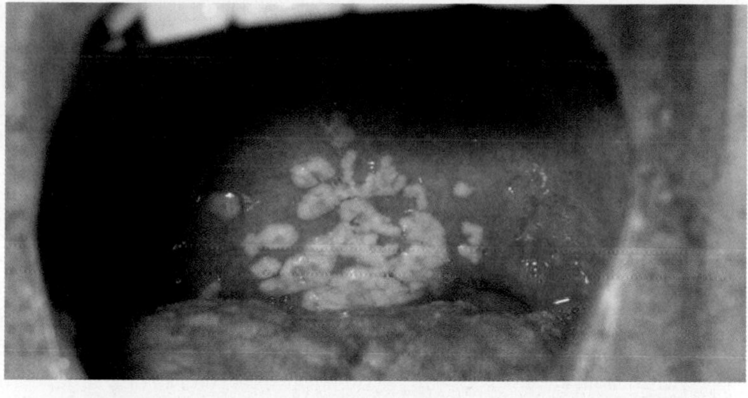
(a)

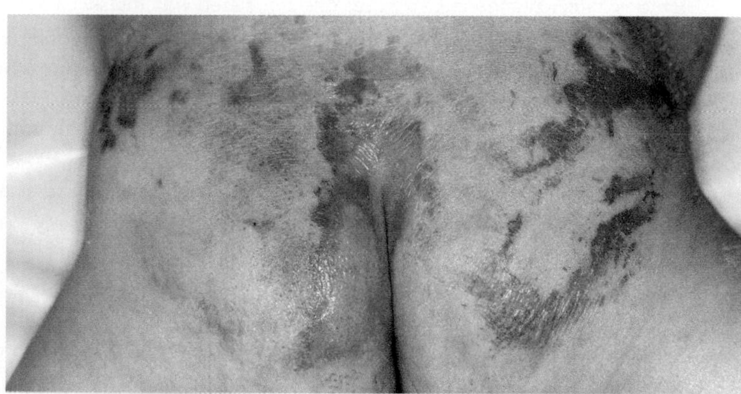
(b)

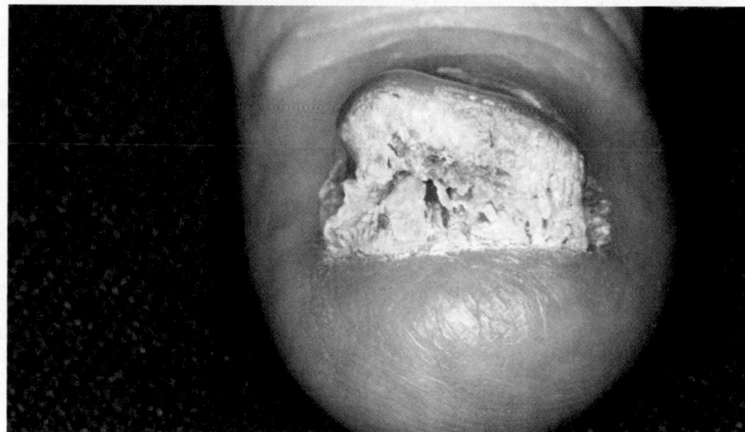
(c)

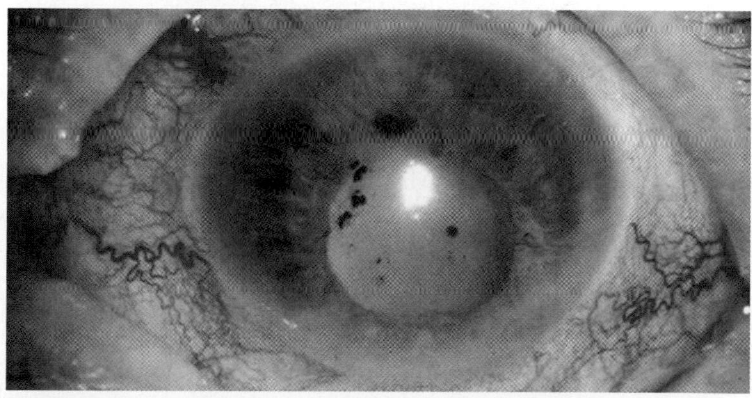
(d)

➤ **Figure 22.12** **Four of the many manifestations of candidiasis.** **(a)** Oropharyngeal candidiasis, or thrush. **(b)** Diaper rash. **(c)** Nail infection (onychomycosis). **(d)** Ocular candidiasis.

TABLE 22.2 The Clinical Manifestations of Candidiasis

Type	Clinical Signs and Symptoms	Predisposing Factors
Oropharyngeal (thrush)	White plaques on the mucosa of the mouth, tongue, gums, palate, and/or pharynx	Diabetes, AIDS, cancers, various chemotherapeutic drug regimens
Cutaneous	Moist, macular red rash between skin folds	Moisture, heat, and friction of skin in skin folds and between digits, particularly in the obese
	Diaper rash: raised red rash, pustules in the gluteal region	Infrequently changed, soiled diapers
	Onychomycosis: painful red swelling around the nails, destruction of nail tissue, and loss of the nail	Immunocompromised individuals
Vulvovaginal	Creamy white, curdlike discharge, burning, redness, painful intercourse	Use of broad-spectrum antibiotics, pregnancy, diabetes, changes to the vaginal microbiota
Chronic mucocutaneous	Lesions associated with the skin, nails, and mucous membranes; severe forms may occur	Various metabolic problems relating to cell-mediated immunity (e.g., diabetes) in children
Neonatal and congenital	Meningitis, renal disorders, or generalized, bodywide, red, vesicular rash	Young age, low birth weight, use of antibiotics by the mother
Esophageal	White plaques along the esophagus, burning pain, nausea, and vomiting	AIDS, immunocompromised status
Gastrointestinal	Ulceration of the stomach and intestinal mucosa	Hematological cancers
Pulmonary	Generalized, nonspecific symptoms; often remains undiagnosed until autopsy	Infection spread from other types of candidiasis
Peritoneal	Fever, pain, tenderness	Indwelling catheters for dialysis or gastrointestinal perforation
Urinary tract	Urinary tract infection: painful urination, possible discharge	Chemotherapeutic drug regimens, catheterization, diabetes, preexisting bladder problems
	Renal: fever, pain, rigors, fungus ball	
Meningeal	Swelling of the meninges, fever, headache, stiffness of the neck	Spread of *Candida* during systemic infection
Hepatic and splenic	Fever, swelling of the liver and spleen, liver dysfunction, lesions and abscesses	Leukemia
Endocardial, myocardial pericardial	Fever, heart murmur, congestive heart failure, anemia	Preexisting heart valve disease, catheterization plus use, of antibiotics, IV drug abuse, valve prosthetics
Ocular	Cloudy vision, lesions within the eye	Spread of candidiasis, indwelling catheters, IV drug abuse, trauma
Osteoarticular	Pain when weight is placed on joint	Spread of candidiasis, prosthetic implants
Candidemia	Antibiotic-resistant fevers, tachycardia, hypotension, skin lesions, small abscesses in various organ systems	IV or urinary catheters, use of antibacterial drugs, surgery, severe burns, antibacterial therapy

healthy individuals and occurs in only about 5% of newborns, it affects 10% of the elderly and nearly all those infected with HIV once they develop clinical AIDS. Table 22.2 summarizes the wide variety of clinical manifestations of candidiasis.

Physicians diagnose candidiasis on the basis of signs and demonstration of clusters of budding yeasts (see Figure 22.1b) and *pseudohyphae,* which are series of buds remaining attached to the parent cell and appearing as a filamentous hypha.

In immunocompetent patients, for whom excessive friction and body moisture in skin folds or preexisting diseases are the reasons for fungal colonization, resolution involves treating these underlying problems in addition to administering topical antifungal agents. Oral candidiasis in infants and children is treated with nystatin. Azole suppositories, creams, and/or oral administration of fluconazole are used for vaginal candidiasis.

The same infections in AIDS patients are much more difficult to treat because most superficial or mucous membrane infections in these patients do not respond well to topical antifungal treatments. Orally administered fluconazole is used for primary and maintenance therapy for candidiasis in AIDS patients. Invasive candidiasis requires treatment with amphotericin B, and often with 5-fluorocytosine; fluconazole may be substituted for either drug.

CRITICAL **THINKING**

Even though *Candida* species are not as virulent as some microbial pathogens, the fungus can still invade every human tissue. Propose a possible explanation for this observation.

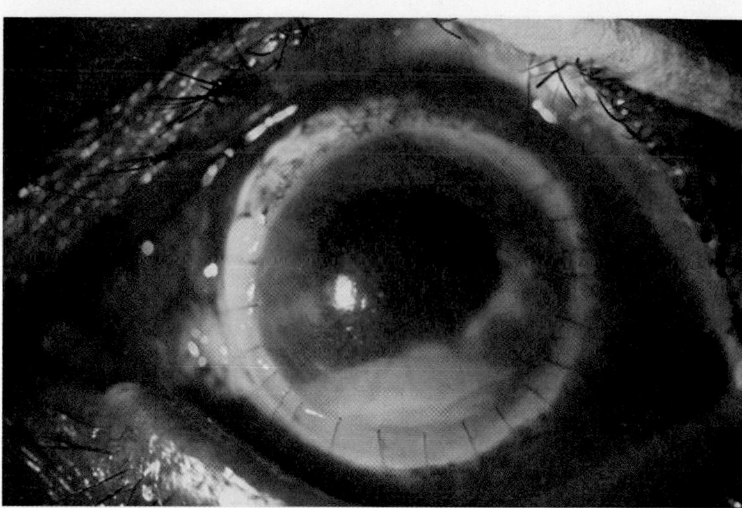

▲ **Figure 22.13 An invasive aspergilloma in the eye.** *In what group of individuals is aspergillosis an emerging disease?*

Figure 22.13 Aspergillosis is an emerging disease of AIDS patients.

MICROBE AT A GLANCE

Aspergillus

Taxonomy: Domain Eukarya, Kingdom Fungi, phylum Ascomycota, class Ascomycetes, order Eurotiales, family Trichocomaceae; principally three species cause disease in humans: *A. fumigatus*, *A. niger*, and *A. flavus*

Morphology: Septate hyphae

Virulence factors: Allergenic surface molecules; opportunistic pathogen with little virulence except in immunocompromised patients

Diseases caused: Hypersensitivity aspergillosis (allergy), noninvasive aspergilloma, acute invasive pulmonary aspergillosis, cutaneous aspergillosis, systemic aspergillosis

Treatment for diseases: Allergy: anti-inflammatory drugs and desensitization to allergens; other disease: amphotericin B and surgical removal of aspergillomas

Prevention of disease: Maintain a healthy immune system.

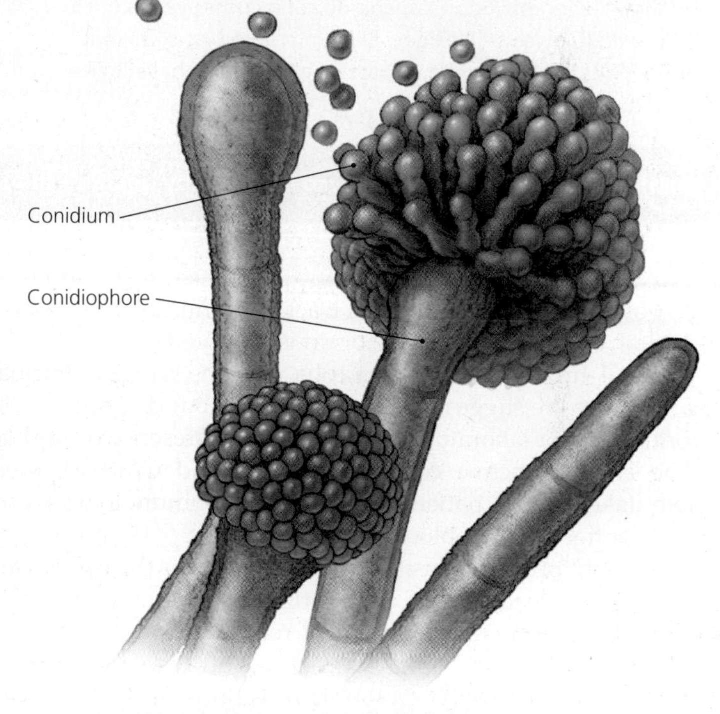

Download the Microbe at a Glance flashcards from the Study Area at www.masteringmicrobiology.com.

Aspergillosis

Learning Objective

✓ Describe the clinical manifestations of aspergillosis.

Aspergillosis (as′per-ji-lō-sis) is not a single disease, but instead a term for several diseases resulting from the inhalation of spores of fungi in the genus *Aspergillus* (as-per-jil′ŭs) in the division Ascomycota. *Aspergillus* is found in soil, food, compost, agricultural buildings, and air vents of homes and offices worldwide. Although exposure to *Aspergillus* most commonly causes only allergies, more serious diseases can occur, and aspergillosis is a growing problem for AIDS patients. Because the fungi are so common in the environment, little can be done to prevent exposure to the spores.

Even though *Aspergillus* species can be opportunistic pathogens of almost all body tissues, these fungi are chiefly responsible for causing three clinical pulmonary diseases:

- *Hypersensitivity aspergillosis* manifests as asthma or other allergic symptoms and results most commonly from inhalation of *Aspergillus* spores. Symptoms may be mild and result in no damage, or they may become chronic, with recurrent episodes leading to permanent damage.
- *Noninvasive aspergillomas*—ball-like masses of fungal hyphae—can form in the cavities resulting from a previous case of pulmonary tuberculosis. Most cases are asymptomatic, though coughing of blood-tinged sputum may occur.
- *Acute invasive pulmonary aspergillosis* is more serious. Signs and symptoms, which include fever, cough, and pain, may present as pneumonia. Death of lung tissue can lead to significant respiratory impairment.

Aspergillus also causes nonpulmonary disease when aspergillomas form in paranasal sinuses, ear canals, eyelids **(Figure 22.13)**, the conjunctivas, eye sockets, or brain. Rarely, *cutaneous aspergillosis* results when the fungus is either introduced into the skin by trauma or spreads from the lungs in AIDS patients. Lesions begin as raised, red papules that progressively die. *Systemic aspergillosis,* which involves invasion of the major organ systems, occurs in AIDS patients and IV drug abusers. The fungus produces abscesses in the brain, kidneys,

EMERGING DISEASES

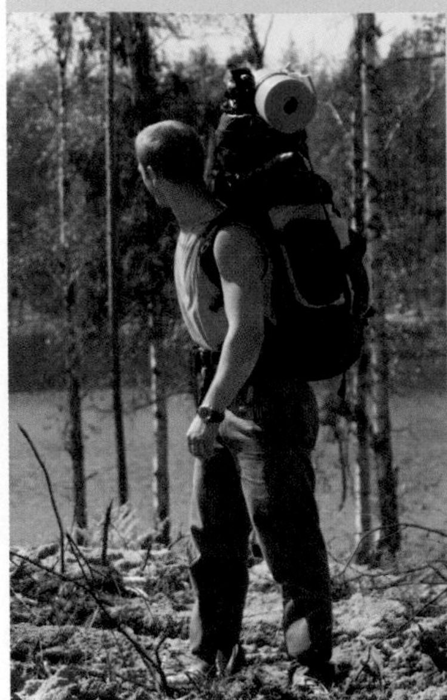

PULMONARY BLASTOMYCOSIS

Phil was glad he had spent the summer fishing, hiking, and camping in western Ontario. Canada is beautiful, and he had needed time off from his two jobs and full schedule of nursing classes. Now he was back in college and remembrances of Canada were helping him cope with the stress.

However, Phil didn't feel well. He was probably coming down with the flu—fever and chills, shivering and coughing, muscle aches, tiredness, a general feeling of yuckiness, and no runny nose; yes, it must be the flu. Or was it? He knew from microbiology class that many diseases of bacteria, protozoa, and viruses have flulike symptoms. Phil had forgotten fungi.

Phil tried every over-the-counter remedy, to no avail. He tried the health clinic on campus, but the antibiotics he received made things worse, not better. He was losing weight, and pus-filled, raised sores had appeared on his face, neck, and legs. More alarming: his testes were swollen and aching. This was getting serious.

Blastomyces, an emergent, dimorphic fungus, was attacking Phil. Inhaled spores from hyphae growing on wet leaves primarily in Wisconsin and Ontario had germinated in Phil's lungs. The yeast phase was multiplying and spreading throughout his body, producing skin lesions that lasted for months, finally resolving into raised, wartlike scars.

Researchers don't know why blastomycosis is becoming more prevalent. Perhaps it has to do with better diagnosis and reporting; perhaps it has to do with a growing number of AIDS patients who are susceptible to infection; or perhaps it has to do with more people like Phil adventuring into the wilderness.

The good news is that with proper diagnosis, Phil got the treatment he needed—itraconazole for six months. He graduated, and now he is a nurse who knows that flulike symptoms can sometimes indicate serious fungal infection. For more about pulmonary blastomycosis, see p. 631.

Track pulmonary blastomycosis online by going to the Study Area at www.masteringmicrobiology.com.

heart, bones, and gastrointestinal tract. Systemic aspergillosis is often fatal, especially when the brain is involved.

Clinical history and radiographs demonstrating abnormal lung structure are suggestive of aspergillosis, but diagnosis must be confirmed via laboratory techniques. The presence of septate hyphae and distinctive conidia in KOH- and GMS-prepared samples taken from a patient are diagnostic. Immunological tests to detect antigens in the blood are confirmatory.

Treatment of hypersensitivity involves either the use of various allergy medications or desensitization to the allergen. Invasive disease is treated by surgical removal of aspergillomas and surrounding tissues, plus high-dose, intravenous administration of amphotericin B. Systemic infections must be treated with amphotericin B in conjunction with other antifungal agents. Maintenance therapy with itraconazole is suggested for preventing relapse in AIDS patients.

Cryptococcosis

Learning Objective

✓ Discuss the characteristics of *Cryptococcus* that contribute to the severity of cryptococcoses in immunocompromised patients.

A basidiomycete, *Cryptococcus neoformans* (krip-tō-kok'ŭs nē-ō-for'manz), is the primary species causing **cryptococcoses** (krip'-tō-kok-ō'sēz). Two strains are known. *C. neoformans gattii* is found in Australia, Papua New Guinea, parts of Africa, the Mediterranean, India, Southeast Asia, parts of Central and South America, and in southern California. It primarily infects immunocompetent individuals. *C. neoformans neoformans* is found worldwide and predominantly infects AIDS patients. Approximately 50% of all cryptococcal infections reported each year are due to strain *neoformans.*

Human infections follow the inhalation of spores or dried yeast in aerosols from the droppings of birds. People who work around buildings where birds roost are at increased risk of infection. Hospitals, convalescent homes, and other such facilities often place devices that deter the roosting of birds near air-intake vents in an effort to prevent *Cryptococcus*-contaminated air from entering a building.

The pathogenesis of *Cryptococcus* is enhanced by several characteristics of this fungus: the presence of a phagocytosis-resistant capsule surrounding the yeast form; the ability of the yeast to produce melanin, which further inhibits phagocytosis; and the organism's predilection for the central nervous system (CNS), which is isolated from the immune system by the so-called blood-brain barrier. Cryptococcal infections also tend to appear in terminal AIDS patients when little immune function remains.

Primary pulmonary cryptococcosis is asymptomatic in most individuals, although some individuals experience a low-grade fever, cough, and mild chest pain. In a few cases, *invasive pulmonary cryptococcosis* occurs, resulting in chronic pneumonia in which cough and pulmonary lesions progressively worsen over a period of years.

Cryptococcal meningitis, the most common clinical form of cryptococcal infection, follows dissemination of the fungus to the CNS. Symptoms develop slowly and include headache, dizziness, drowsiness, irritability, confusion, nausea, vomiting, and neck stiffness. In late stages of the disease, loss of vision and coma occur. Acute onset of rapidly fatal cryptococcal meningitis occurs in individuals with widespread infection.

Cryptococcoma (krip'tō-kok-ō'mă) is a very rare condition in which solid fungal masses form in the cerebral hemispheres, cerebellum, or (rarely) in the spinal cord. The symptoms of this condition, which can be mistaken for cerebral tumors, are similar to those of cryptococcal meningitis but also include motor and neurological impairment.

Cutaneous cryptococcosis may also develop. Primary infections manifest as ulcerated skin lesions or as inflammation of subcutaneous tissues. Infection may resolve on its own, but patients should be monitored for infection spreading to the CNS. Secondary cutaneous lesions occur following spread of *Cryptococcus* to other areas of the body. In AIDS patients, cutaneous cryptococcosis is the second most common manifestation of *Cryptococcus* infection (after meningitis). Lesions are common on the head and neck.

Diagnosis involves collecting specimen samples that correlate with the symptoms. GMS stains revealing the presence of encapsulated yeast in cerebrospinal fluid are highly suggestive of cryptococcal meningitis, even if no overt symptoms are present **(Figure 22.14)**. Recovery of *Cryptococcus* from respiratory secretions is generally not diagnostic because of general environmental exposure, but any patient testing positive for *C. neoformans* should be monitored for systemic disease. The preferred method of confirming cryptococcal meningitis is detection of fungal antigens in cerebrospinal fluid. In AIDS patients, antigens can be detected in serum as well.

Treatment for cryptococcoses is with amphotericin B and 5-fluorocytosine administered together for 6–10 weeks. The synergistic action of the two drugs allows lower doses of amphotericin B to be used, but toxicity is not completely eliminated. Although AIDS patients may appear well following primary treatment, the fungus typically remains and must be actively suppressed by lifelong use of oral fluconazole.

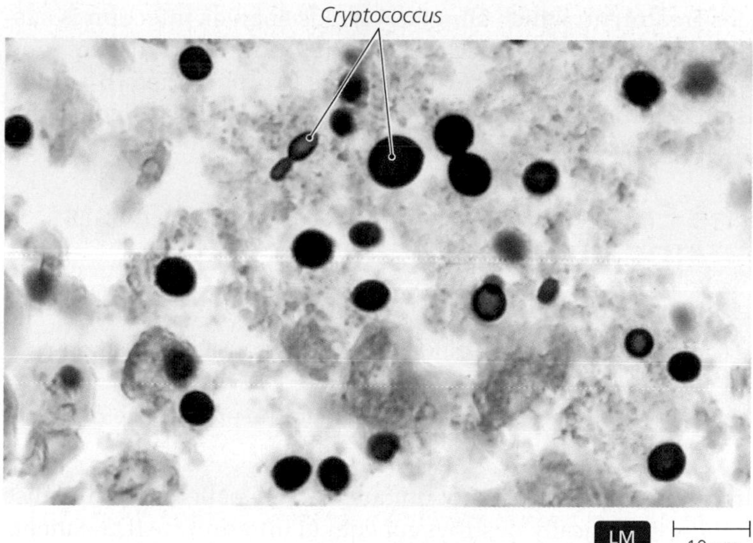

▲ **Figure 22.14 GMS stain of *Cryptococcus*.** Yeasts of this dimorphic fungus multiply in the lungs, as shown, and may spread to other parts of the body.

CRITICAL **THINKING**

Compare and contrast the clinical features of cryptococcosis and PCP. Why are both of these diseases so dangerous to AIDS patients?

Zygomycoses

Learning Objective

✓ Describe the clinical forms of zygomycoses.

Zygomycoses (zī'gō-mī-kō'sez) are opportunistic fungal infections caused by various genera of fungi classified in the division Zygomycota, especially *Mucor* [or sometimes *Rhizopus* (rī-zō'pŭs) or *Absidia* (ab-sid'ē-ă)]. All three have a worldwide distribution and are extremely common in soil, on decaying organic matter, or as contaminants that cause food spoilage (*Rhizopus* is the classic bread mold).

Zygomycoses are commonly seen in patients with uncontrolled diabetes, in people who inject illegal drugs, in some cancer patients, and in some patients receiving antimicrobial agents. Infections generally develop in the face and head area but can develop elsewhere and may spread throughout the bodies of severely immunocompromised individuals, resulting in the following conditions:

- *Rhinocerebral zygomycosis* begins with infection of the nasal sinuses following inhalation of spores. The fungus spreads to the mouth and nose, producing macroscopic cottonlike growths. *Mucor* can subsequently invade blood vessels, where it produces fibrous clots, causes tissue death, and subsequently invades the brain, which is fatal within days, even with treatment.
- *Pulmonary zygomycosis* follows inhalation of spores (as from moldy foods). The fungus kills lung tissue, resulting in the formation of cavities.
- *Gastrointestinal zygomycosis* involves ulcers in the intestinal tract.
- *Cutaneous zygomycosis* results from the introduction of fungi through the skin after trauma (such as burns or needle punctures). Lesions range from pustules and ulcers to abscesses and dead patches of skin.

Diagnosis usually involves microscopic findings of fungus. Samples may be obtained from scrapings of lesions, lung aspirates, or biopsies. KOH- or GMS-stained tissue sections reveal hyphae with irregular branching and few septate divisions.

Treatment, which should begin as soon as infection is suspected, involves physical removal of infected tissues and management of predisposing factors. Amphotericin B administered intravenously for 8–10 weeks is the drug of choice for all zygomycoses.

The Emergence of Fungal Opportunists in AIDS Patients

Learning Objectives

- ✓ Identify several emerging fungal opportunists seen among AIDS patients.
- ✓ Identify the factors that contribute to the emergence of new fungal opportunists.

Immunosuppression is not unique to AIDS patients, but because HIV systematically destroys functional immunity, AIDS patients are extremely vulnerable to opportunistic fungal infections. In otherwise healthy people, immunosuppressive episodes are transitory and result from surgery, chemotherapy, or an acute illness. With AIDS, by contrast, immune dysfunction is permanent. For this reason, once an opportunistic infection is established in an AIDS patient, it is not likely ever to be cured fully. In fact, mycoses account for most deaths associated with AIDS. *Pneumocystis jiroveci, Candida albicans, Aspergillus fumigatus,* and *Cryptococcus neoformans* are so common in HIV-positive individuals that their mycoses are part of what defines end-stage AIDS.

The emergence of new fungal opportunists results from a combination of factors. First, the number of immunocompromised individuals is increasing as a result of the AIDS epidemic. Then, the widespread use of antifungal drugs in this enlarging group of immunocompromised individuals selects fungi resistant to the drugs. Thus as the immunocompromised population grows, fungi can be expected to increasingly cause illness and death. An AIDS patient's lack of immunity allows a normally superficial fungus to gain access to internal systems, with significant, even fatal, results. Over the course of the past decade, several new fungal opportunists have been identified, including *Fusarium* spp. and *Penicillium marneffei,* which are ascomycetes, and *Trichosporon beigelii,* which is a basidiomycete.

Fusarium (fū-zā′rē-ŭm) species cause respiratory distress, disseminated infections, and *fungemia* (fungi in the bloodstream). These fungi also produce toxins that can accumulate to dangerous levels when ingested in food. *Fusarium* spp. are resistant to all antifungal agents, including amphotericin B.

Penicillium marneffei (pen-i-sil′ē-ŭm mar-nef-ē′ī) is a dimorphic, invasive fungus that causes pulmonary disease upon inhalation. It is the third most frequent illness seen in AIDS patients in Southeast Asia (behind tuberculosis and cryptococcosis).

When *Trichosporon beigelii* (trik-ō-spōr′on bā-gĕl-ē′ē) enters an AIDS patient through the lungs, the gastrointestinal tract, or catheters, it causes a drug-resistant systemic disease that is typically fatal.

To combat the threat posed by these opportunists, health care providers maintain rigorous hygiene during medical procedures. Researchers are developing new therapies that limit immunosuppression in all patients, including those with AIDS.

Superficial, Cutaneous, and Subcutaneous Mycoses

Learning Objective

- ✓ Describe the general manifestations of superficial, cutaneous, and subcutaneous mycoses.

All the mycoses we discuss in the following sections are localized at sites at or near the surface of the body. They are the most commonly reported fungal diseases. All are opportunistic infections, but unlike those we have just discussed, they can be acquired both through environmental exposure and more frequently via person-to-person contact. Most of these fungi are not life threatening, but they often cause chronic, recurring infections and diseases.

Superficial Mycoses

Learning Objective

- ✓ Compare and contrast the clinical and diagnostic features of dermatophytosis and *Malassezia* infections.

Superficial mycoses are the most common fungal infections. They are confined to the outer, dead layers of the skin, nails, or hair, all of which are composed of dead cells filled with a protein called *keratin*—the primary food of these fungi. In AIDS patients, superficial mycoses can spread to cover significant areas of skin or become systemic.

Black Piedra and White Piedra

Piedra[3] (pē-ā′drā) is a superficial infection characterized by nodules on hair shafts. Transmission is usually by way of shared hair brushes and combs, and several members of a family are typically infected at the same time.

Piedraia hortae (pī-drā′ă hōr′tī), an ascomycete which causes **black piedra,** is distributed throughout Central and South America and Southeast Asia. It forms hard, black nodules **(Figure 22.15a)**. **White piedra,** which is found worldwide in the tropics and subtropics, is caused by a basidiomycete, *Trichosporon beigelii,* which produces soft, gray to white nodules **(Figure 22.15b)**. In both diseases, the clinical appearance of the hair is diagnostic for the disease.

Treatment for both types of piedra typically involves shaving to remove infected hair; one shave is usually sufficient. To treat difficult-to-cure cases or women in cultures where cutting the hair is socially devastating, physicians prescribe antifungal agents such as terbinafine.

Dermatophytoses

Dermatophytoses (der′mă-tō-fi-tō′sēz) are infections caused by dermatophytes, which are fungi that grow only on skin, nails, and hair. In the past, such infections were called *ringworms,* because dermatophytes produce circular, scaly patches that resemble a worm lying just below the surface of the skin. Even though we

[3]Spanish, meaning a stone.

(a) LM 100 μm

▲ **Figure 22.15 Two forms of piedra.** Piedra is a superficial fungal infection of the hair. **(a)** The hard nodules of black piedra, caused by *Piedraia hortae*. **(b)** The looser, light-colored nodules of white piedra, caused by *Trichosporon beigelii*.

know worms are not involved, the medical names for many dermatophytoses use the term *tinea* (tin′ē-a), which is Latin for worm. Dermatophytes use keratin as a nutrient source and thus colonize only dead tissue. The fungi may provoke cell-mediated immune responses, which can damage living tissues. Dermatophytes are among the few contagious fungi; that is, fungi that spread from person to person. Spores and bits of hyphae are constantly shed from infected individuals, making recurrent infections common.

Three genera of ascomycetes are responsible for most dermatophytoses: (1) *Trichophyton* (trĭk-ō-fī′ton) species, (2) *Microsporum* (mī-kros′po-rŭm) species, and (3) *Epidermophyton floccosum* (ep′i-der-mof′i-ton flŏk′ō-sŭm), which we discussed in the vignette at the beginning of the chapter. All three cause skin and nail infections, but *Trichophyton* species also infect hair. Rare subcutaneous infections occur in AIDS patients.

Dermatophytoses can have a variety of clinical manifestations, some of which are summarized in Table 22.3 on p. 642. Most such diseases are clinically distinctive and so common that they are readily recognized. KOH preparations of skin or nail scrapings or hair samples reveal hyphae and arthroconidia, which confirm a diagnosis. Determination of a specific dermatophyte requires laboratory culture, but is usually unnecessary because treatment is the same for all.

Limited infections are treated with topical antifungal agents, but more widespread infections of the scalp or skin, as well as nail infections, must be treated with oral agents. Terbinafine, administered for 6–12 weeks, is effective in most cases. Chronic or stubborn cases are treated with griseofulvin until cured. **Clinical Case Study: Is It Athlete's Foot?** on p. 642 explores some of the difficulties in dealing with dermatophyte infections.

Malassezia Infections

Malassezia furfur (mal-ă-sē′zē-ă fur′fur) is a basidiomycete that is a normal member of the microbiota of the skin of humans worldwide. It feeds on the skin's oil and causes common, chronic superficial infections.

Infections with *M. furfur* result in **pityriasis** (pit-i-rī′ă-sis), characterized by depigmented or hyperpigmented patches of scaly skin resulting from fungal interference with melanin production **(Figure 22.16)**. This condition typically occurs on the trunk, shoulders, and arms, and rarely on the face and neck. KOH preparations of clinical specimens reveal masses of budding yeast and short hyphal forms that are diagnostic for *M. furfur*.

Superficial *Malassezia* infections are treated topically with solutions of antifungal imidazoles, such as ketoconazole shampoos. Alternatively, topical applications of zinc pyrithione, selenium sulfide lotions, or propylene glycol solutions can be used. Oral therapy with ketoconazole may be required to treat extensive infections or infections that do not respond to topical treatments. Relapses of *Malassezia* infections are common, and prophylactic topical treatment may be necessary. The skin takes months to regain its normal pigmentation following successful treatment of pityriasis.

CRITICAL **THINKING**

Explain why many superficial fungal infections are chronic, recurring problems.

Cutaneous and Subcutaneous Mycoses

Learning Objectives

- ✓ Explain why cutaneous and subcutaneous mycoses are not as common as superficial mycoses.
- ✓ State the specific difference between chromoblastomycosis and phaeohyphomycosis.
- ✓ Describe how the lesions of lymphocutaneous sporotrichosis correlate with its spread through the lymphatic system.

The fungi involved in cutaneous and subcutaneous mycoses are common soil saprobes (organisms that live on dead organisms),

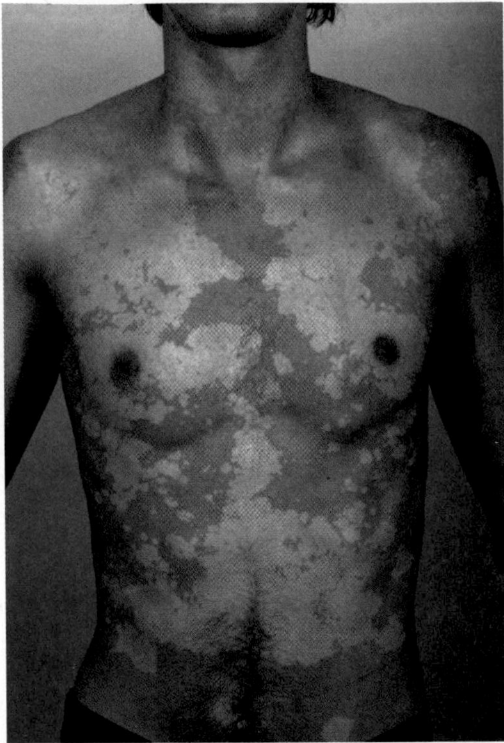

▲ **Figure 22.16 Pityriasis.** The variably pigmented skin patches are caused by *Malassezia furfur.*

CLINICAL CASE STUDY

Is it Athlete's Foot?

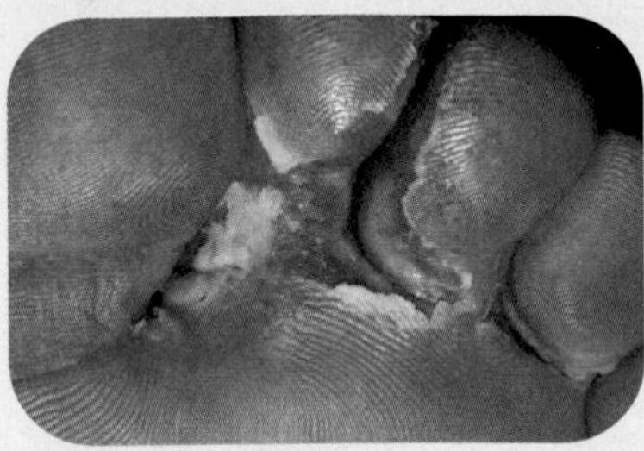

A 30-year-old man comes into a community clinic complaining of persistent redness, itching, and peeling skin on the soles of his feet and between his toes. He states that he works out daily and plays on several sports teams. He concluded he has athlete's foot, but every over-the-counter antifungal agent he has tried has failed to cure his condition. He has finally come to the clinic because his toenails have begun to thicken, yellow, and detach from the nail bed. His condition is pictured.

1. Given the information here, can athlete's foot be diagnosed definitively? If not, what laboratory tests should be performed to identify the infecting organism?
2. What treatment is likely? Would the recommended treatment be different if two fungal species, and not just one fungal species, were present?

but the diseases they produce are not as common as superficial mycoses because infection requires traumatic introduction of fungi through the dead outer layers of skin into the deeper, living tissue. Most lesions remain localized just below the skin, though infections may rarely become systemic.

Chromoblastomycosis and Phaeohyphomycosis

Chromoblastomycosis (krō′mō-blas′tō-mī-kō′sis) and **phaeohyphomycosis** (fē′ō-hī′fō-mī-kō′sis) are similar-appearing cutaneous and subcutaneous mycoses caused by dark-pigmented

TABLE 22.3 Common Dermatophytoses

Disease	Agents	Common Signs	Source
Tinea pedis ("athlete's foot")	*Trichophyton rubrum; T. mentagrophytes* var. *interdigitale; Epidermophyton floccosum*	Red, raised lesions on and around the toes and soles of the feet; webbing between the toes is heavily infected	Human reservoirs in toe webbing; carpeting holding infected skin cells
Tinea cruris ("jock itch")	*T. rubrum; T. mentagrophytes* var. *interdigitale; E. floccosum*	Red, raised lesions on and around the groin and buttocks	Usually spreads from the feet
Tinea unguium (onychomycosis)	*T. rubrum; T. mentagrophytes* var. *interdigitale*	*Superficial white onychomycosis:* patches or pits on the nail surface *Invasive onychomycosis:* yellowing and thickening of the distal nail plate, often leading to loss of the nail	Humans
Tinea corporis	*T. rubrum; Microsporum gypseum; M. canis*	Red, raised, ringlike lesions occurring on various skin surfaces (tinea corporis on the trunk, tinea capitis on the scalp, tinea barbae of the beard)	Can spread from other body sites; can be acquired following contact with contaminated soil or animals
Tinea capitis	*M. canis; M. gypseum; T. equinum; T. verrucosum; T. tonsurans; T. violaceum; T. schoenleinii*	*Ectothrix invasion:* fungus develops arthroconidia on the outside of the hair shafts, destroying the cuticle *Endothrix invasion:* fungus develops arthroconidia inside the hair shaft without destruction *Favus:* crusts form on the scalp, with associated hair loss	Humans; can be acquired following contact with contaminated soil or animals

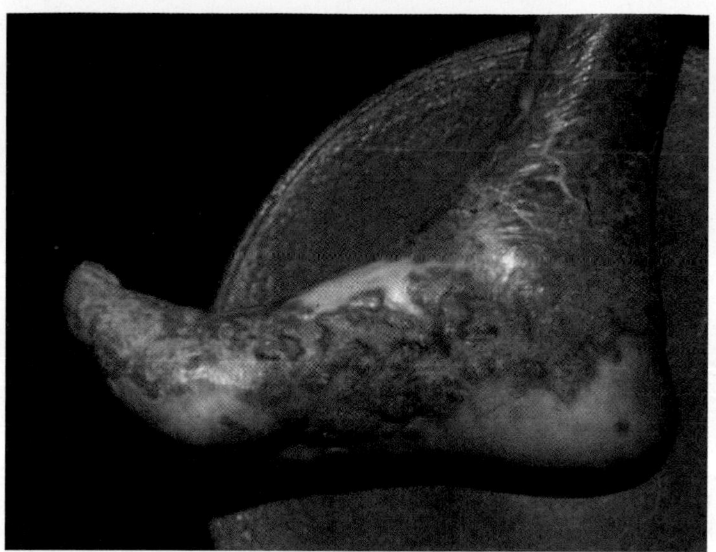

▲ **Figure 22.17 A leg with extensive lesions of chromoblastomycosis.** In this case they are caused by *Fonsecaea pedrosoi.*

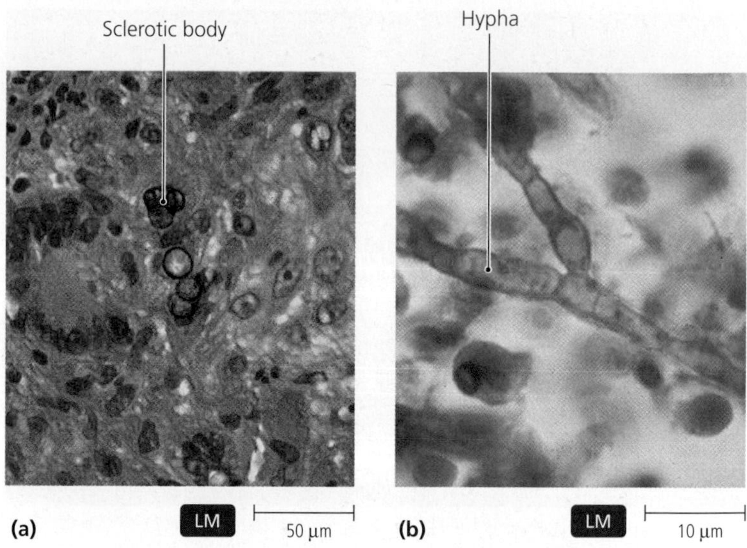

▲ **Figure 22.18 Microscopic differences between chromoblastomycosis and phaeohyphomycosis. (a)** Sclerotic bodies seen in chromoblastomycoses, here caused by *Fonsecaea pedrosoi.* **(b)** Hyphal forms found in tissues during phaeohyphomycosis, here caused by *Cladophialophora bantiana.*

ascomycetes. Table 22.4 on p. 644 lists some ascomycetes that commonly cause these mycoses.

Initially, chromoblastomycosis uniformly presents as small, scaly, itchy but painless lesions on the skin surface resulting from fungal growth in subcutaneous tissues near the site of inoculation. Over the course of several years, the lesions progressively worsen, becoming large, flat to thick, tough, and wartlike. They become tumorlike and extensive if not treated **(Figure 22.17)**. Inflammation, fibrosis, and abscess formation occur in surrounding tissues. The fungus can spread throughout the body.

Phaeohyphomycoses are more variable in presentation, involving colonization of the nasal passages and sinuses in allergy sufferers and AIDS patients, or of the brains of AIDS patients. Fortunately, brain infection is the rarest form of phaeohyphomycosis and occurs only in the severely immunocompromised.

KOH preparations and GMS staining of skin scrapings, biopsy material, or cerebrospinal fluid reveal the key distinguishing feature between the two diseases: the microscopic morphology of the fungal cells within tissues. Tissue sections from chromoblastomycosis cases contain golden brown *sclerotic bodies* that are distinctive and distinguishable from budding yeast forms **(Figure 22.18a)**, whereas tissues from phaeohyphomycosis cases contain brown-pigmented hyphae **(Figure 22.18b)**. Determining the species of fungus causing these two diseases requires laboratory culture, macroscopic examination of colonies on agar, and microscopic examination of spores.

Both diseases are difficult to treat, especially in advanced cases. Some cases of phaeohyphomycosis can be treated with itraconazole, but the disease is permanently destructive to tissues. Chromoblastomycosis requires surgical removal of infected and surrounding tissues followed by antifungal therapy. Extensive lesions may require amputation. Thiabendazole and 5-fluorocytosine, given for 3–12 months, have been effective in some cases. The earlier treatment begins, the more likely it will be successful.

Despite the worldwide occurrence of the relevant fungi, the overall incidence of infection is relatively low. People who work daily in the soil with bare feet are at risk if they incur foot wounds. The simple act of wearing shoes would greatly reduce the number of infections.

Mycetomas

Fungal **mycetomas** (mī-sē-tō′măs) are tumorlike infections of the skin, fascia (lining of muscles), and bones of the hands or feet caused by mycelial fungi of several genera in the division Ascomycota (see Table 22.4 on p. 644). These fungi are distributed worldwide, but infection is most prevalent in countries near the equator. The cases that occur in the United States are almost always caused by *Pseudallescheria* or *Exophiala.*

Mycetoma-producing fungi live in the soil and are introduced into humans via wounds caused by twigs, thorns, or leaves contaminated with fungi. As with chromoblastomycosis and phaeohyphomycosis, those who work barefoot in soil are most at risk, and wearing protective shoes or clothing can greatly reduce incidence.

Infection begins near the site of inoculation with the formation of small, hard, subsurface nodules that slowly worsen and spread as time passes. Local swelling occurs, and ulcerated lesions begin to produce pus. Infected areas release an oily fluid containing fungal "granules" (spores and fungal elements). The fungi spread to more tissues, destroying bone and causing permanent deformity **(Figure 22.19)**.

A combination of the symptoms and microscopic demonstration of fungi in samples from the infected area is diagnostic for mycetomas. Laboratory culture of specimens produces macroscopic colonies that can be identified to species.

Treatment involves surgical removal of the mycetoma that may, in severe cases, involve amputation of an infected limb.

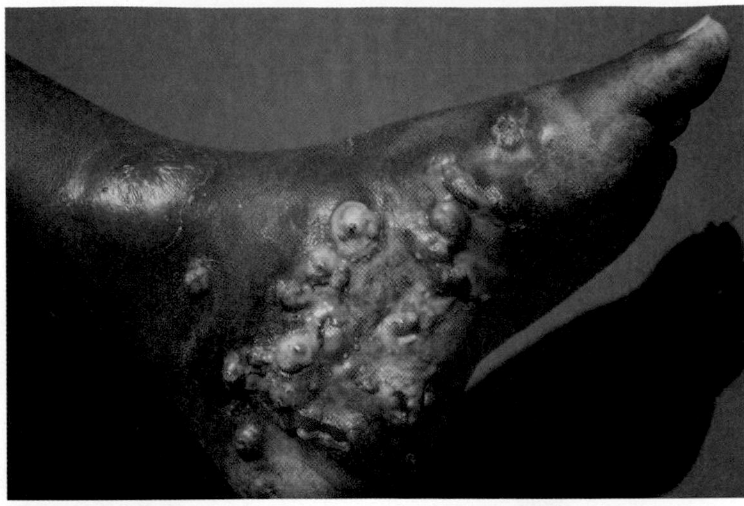

▲ **Figure 22.19 A mycetoma of the ankle.** Here it results from the invasion and destruction of bone and other tissues by the fungus *Madurella mycetomatis*.

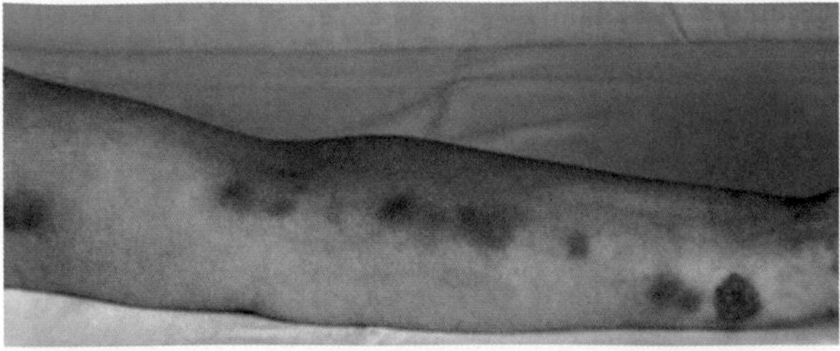

▲ **Figure 22.20 Lymphocutaneous sporotrichosis on the arm.** The locations of these secondary, subcutaneous lesions correspond to the course of lymphatic vessels leading from sites of primary, surface lesions. *How is sporotrichosis contracted?*

Figure 22.20 *Sporotrichosis is contracted when the soilborne fungus is inoculated into the skin by thorn pricks and scratches.*

Surgery is followed by one to three years of antifungal therapy with ketoconazole. Even combinations of surgery and antifungal agents are not always effective.

Sporotrichosis

Sporothrix schenckii (spōr′ō-thriks shen′kē-ī) is a dimorphic ascomycete that causes **sporotrichosis** (spōr′ō-tri-kō′sis), or *rose-gardener's disease,* a subcutaneous infection usually limited to the arms or legs. *S. schenckii* resides in the soil and is most commonly introduced by thorn pricks or wood splinters. Avid gardeners, farmers, and artisans who work with natural plant materials have the highest incidence of sporotrichosis. Though distributed throughout the tropics and subtropics, most cases occur in Latin America, Mexico, and Africa. The disease is also common in warm, moist areas of the United States.

Cutaneous sporotrichosis initially appears as painless, nodular lesions that form around the site of inoculation. With time, these lesions produce a pus-filled discharge, but they remain localized and do not spread. In *lymphocutaneous sporotrichosis,* however, the fungus enters the lymphatic system near the site of a primary lesion, giving rise to secondary lesions on the skin surface along the course of lymphatic vessels **(Figure 22.20)**. The fungus remains restricted to subcutaneous tissues and does not enter the blood.

Microscopic observation of pus or biopsy tissue stained with GMS can reveal budding yeast forms in severe infections, but often the fungus is present at a low density, making direct examination of clinical samples a difficult method of diagnosis. The patient's history and clinical signs, plus the observation of the dimorphic nature of *S. schenckii* in laboratory culture, are considered diagnostic for sporotrichosis.

Cutaneous lesions can be treated successfully with topical applications of saturated potassium iodide for several months. Itraconazole and terbinafine are also useful treatments. Prevention requires the wearing of gloves, long clothing, and shoes to prevent inoculation.

TABLE 22.4 Some Ascomycete Genera of Cutaneous and Subcutaneous Mycoses

Chromoblastomycosis
Fonsecaea *Phialophora* *Cladophialophora*
Phaeohyphomycosis
Alternaria *Exophiala* *Wangiella* *Cladophialophora*
Mycetoma
Madurella *Pseudallescheria* *Exophiala* *Acremonium*
Sporotrichosis
Sporothrix

CRITICAL **THINKING**

Make a dichotomous key to distinguish among all the fungal mycoses covered in the previous sections.

Fungal Intoxications and Allergies

Learning Objectives

✓ Compare and contrast mycotoxicosis, mycetismus, and fungal allergies.

✓ Define *mycotoxin*.

Some fungi produce toxins or cause allergies. Fungal toxins, called, **mycotoxins** (mī′kō-tok-sinz), are low-molecular-weight metabolites that can harm humans and animals that ingest them, causing *toxicosis* (poisoning). **Mycotoxicosis** (mī'kō-tok-si-kō-sis) is caused by eating mycotoxins; the fungus itself is not present. **Mycetismus** (mī'sē-tiz'mŭs) is mushroom poisoning resulting from eating the fungus. Fungal *allergens* are usually proteins or glycoproteins that elicit hypersensitivity reactions in sensitive people who contact them.

Mycotoxicoses

Learning Objective

✓ Identify the most common group of mycotoxins.

Fungi produce mycotoxins during their normal metabolic activities. People most commonly consume mycotoxins in grains or vegetables that have become contaminated with fungi. Some mycotoxins can also be ingested in milk from a cow that has ingested toxin-contaminated feed. Up to 25% of the world's food supply is contaminated with mycotoxins, but only 20 of the 300 or so known toxins are ever present at dangerous levels. Long-term ingestion of mycotoxins can cause liver and kidney damage, gastrointestinal or gynecological disturbances, or cancers; each mycotoxin produces a specific clinical manifestation.

Aflatoxins (af'lă-tok'sinz) produced by the ascomycete *Aspergillus* are the best-known mycotoxins. Aflatoxins are fatal to many vertebrates and are carcinogenic at low levels when consumed continually. Aflatoxins cause liver damage and liver cancer throughout the world, but *aflatoxicosis* is most prevalent in the tropics, where mycotoxins are more common because of subsistence farming, poor food-storage conditions, and warm, moist conditions that foster the growth of *Aspergillus* in harvested foods.

Some mycotoxins are considered useful. Among them are ergot alkaloids, produced by some strains of another ascomycete, *Claviceps purpurea* (klav'i-seps poor-poo'rē-ă). For example, *ergometrine* is used to stimulate labor contractions and is used to constrict the mother's blood vessels after birth (when she is at risk of bleeding excessively). Another mycotoxin, *ergotamine,* is used to treat migraine headaches.

Mushroom Poisoning (Mycetismus)

Learning Objective

✓ Describe at least two types of mushroom poisoning.

Most mushrooms (the spore-bearing structures of certain basidiomycetes) are not toxic, though some produce extremely dangerous poisons capable of causing neurological dysfunction or hallucinations, organ damage, or even death. *Mushroom poisoning (mycetismus)* typically occurs when untrained individuals pick and eat wild mushrooms. Young children are especially attracted to colorful mushrooms, which can be poisonous. Poisonous mushrooms are commonly called *toadstools.*

The deadliest mushroom toxin is produced by the "death cap" mushroom, *Amanita phalloides* (am-ă-nī'tă fal-ōy'dēz) **(Figure 22.21)**. The death cap contains two related polypeptide toxins: *phalloidin,* which irreversibly binds actin within cells, disrupting cell structure; and *alpha-amanitin,* which inhibits mRNA synthesis. Both toxins cause liver damage.

Figure 22.21 *Amanita phalloides*, **the "death cap" mushroom.**

Other deadly mushrooms include the false morel, *Gyromitra esculenta* (gī-rō-mē'tră es-kū-len'tă), which causes bloody diarrhea, convulsions, and death within two days after ingestion, and *Cortinarius gentilis* (kōr'ti-nar-ē-us jen'til-is), which causes excessive thirst, nausea, and kidney failure between three days and three weeks after ingestion.

Psilocybe cubensis (sil-ō-sī'bē kū-ben'sis) produces hallucinogenic *psilocybin,* and *Amanita muscaria* (mus-ka'rē-ă) produces two hallucinogenic toxins—*ibotenic acid* and *muscimol.* These toxins may also cause convulsions in children.

Treatment involves inducing vomiting followed by oral administration of activated charcoal to absorb toxins. Severe mushroom poisoning may necessitate a liver transplant.

Allergies to Fungi

Learning Objective

✓ Identify the types of hypersensitivity reactions that fungal allergens produce.

Fungal allergens are common and can be found both indoors and out. Over 80 genera of fungi have been demonstrated to trigger fungal allergies. Epidemiologists estimate that 3–10% of humans worldwide are affected. However, determining the specific cause of allergies is often difficult—the presence of spores in an environment does not necessarily mean that they are responsible for allergies. Because skin sensitivity testing (discussed in Chapter 18) is not performed on most people, the true impact of fungal allergens remains undetermined, even though they continue to be blamed for a variety of illnesses (see **Highlight: Does "Killer Mold" Exist?**).

Fungal spores dispersed in the air are common allergens. The concentrations of such spores may peak in autumn or spring, although they are present year-round. Fungal allergens typically cause type I hypersensitivities in which immunoglobulin E binds the allergen, triggering responses such as asthma, eczema, hay fever, and watery eyes and nose.

Less frequently, type III hypersensitivities result from chronic inhalation of particular fungal allergens. In these cases, allergens that have penetrated deep into the lungs encounter

HIGHLIGHT

DOES "KILLER MOLD" EXIST?

In 1994, eight infants in Cleveland were hospitalized with life-threatening pulmonary hemorrhage. One child died. Each of the infants had lived in water-damaged buildings, and initial epidemiological investigations pointed at the black mold *Stachybotrys chartarum* as the agent responsible. The cases were widely reported in the media, and *Stachybotrys* became known as "killer black mold." Since then, the mold has been blamed for everything from "sick building syndrome" to asthma to other occurrences of pulmonary hemorrhage in infants.

A subsequent review of the Cleveland cases by the CDC, however, found serious shortcomings in the initial investigations. Significantly, *Stachybotrys* could not be recovered from any of the infants, and in most of the cases the level of mold in the water-damaged homes was below detection limits or statistically no different from levels in homes without water damage. In the numerous studies that have been conducted on *Stachybotrys* since 1994, none has established a direct causal relationship between the mold and infant deaths. Nor have researchers found any compelling relationship between *Stachybotrys* and any increased prevalence of asthma or other respiratory illnesses. Within the scientific community, "killer mold" remains an unproven moniker.

In approaching mold contamination of any kind, individuals should exercise their common sense. Some people are naturally sensitive to mold—*Stachybotrys* and others. Immunocompromised or allergy-prone individuals in particular may be at increased risk of experiencing adverse reactions to mold. Decisions regarding the more drastic methods of mold removal, such as burning down your house, should be cautiously and reflectively considered.

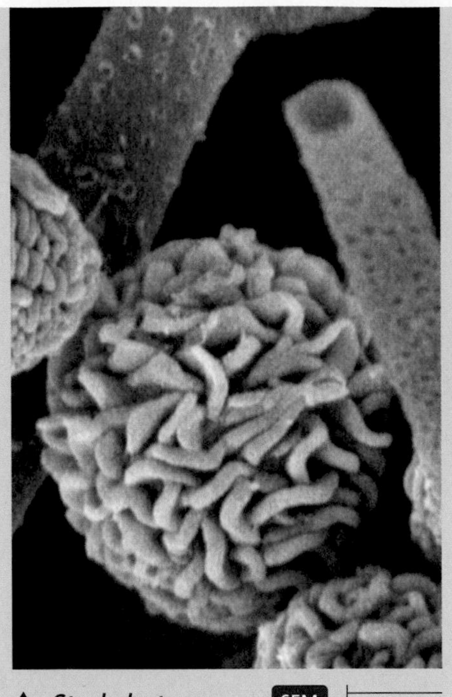

▲ ***Stachybotrys chartarum.*** SEM 30 μm

complementary antibodies and form immune complexes in the alveoli that lead to inflammation, fibrosis, and, in some cases, death. Type III fungal hypersensitivities are associated with certain occupations, such as farming, in which workers are constantly exposed to fungal spores in moldy vegetation.

CRITICAL **THINKING**

Despite the fact that fungi are everywhere, fungal allergies are still not as common as allergies to pollen or dust mites. Propose an explanation to explain this observation.

Chapter Summary

An Overview of Medical Mycology (pp. 626–628)

1. **Medical mycology** is the study of the diagnosis, management, and prevention of **mycoses** (fungal diseases). Only a few fungi cause serious diseases in humans. The actual incidence of most mycoses is unknown because they are not reportable.
2. Most fungi are not considered contagious agents because they do not spread by human-to-human contact. The few exceptions include the **dermatophytes** and species of *Candida* and *Pneumocystis.*
3. True fungal pathogens are capable of actively invading the tissues of normal, healthy people. Opportunists infect only weakened patients.
4. Clinically, mycoses include infections, toxicoses, and hypersensitivity reactions. Infections can be **superficial, cutaneous, subcutaneous,** or **systemic mycoses.**
5. Diagnosis of mycoses usually correlates signs and symptoms with microscopic examination of tissues or laboratory cultures for identifying unique morphological features. Biochemical and immunological tests exist for some fungi.
6. Amphotericin B kills most fungi and can be used to treat most fungal infections, but because it is somewhat toxic to humans, long-term treatment with fungistatic drugs is often used instead.

Systemic Mycoses Caused by Pathogenic Fungi (pp. 629–634)

1. **Systemic mycoses**—fungal infections that spread throughout the body—are caused by one of four pathogenic, dimorphic fungi: *Blastomyces, Coccidioides, Histoplasma,* or *Paracoccidioides.* Whereas each is geographically limited, they all cause similar pulmonary diseases and can spread beyond the lungs.
2. *Histoplasma capsulatum,* which causes **histoplasmosis,** is associated primarily with bat and bird droppings in soil in the Ohio River Valley. Generally, histoplasmosis is an occupationally acquired disease, but recreational exposure does occur. *H. capsulatum* can be carried from the lungs inside macrophages.
3. **Blastomycosis** is found in the eastern United States and is caused by *Blastomyces dermatitidis,* which normally lives in soil rich in

organic material. *B. dermatitidis* almost always spreads beyond the lungs to produce cutaneous lesions.

4. *Coccidioides immitis,* which is limited to deserts, causes **coccidioidomycosis,** which is common in AIDS patients. Contaminated dust is a major source of transmission.
5. *Paracoccidioides brasiliensis* is found in Brazil and some other regions of South and Central America. It causes **paracoccidioidomycosis.** Following a pulmonary phase, the fungus spreads and creates permanently disfiguring lesions on the face and neck.
6. Pathogenic fungi are diagnosed by morphological analysis and by demonstration of dimorphism. Serological tests are also available. If treated promptly, they can be controlled.
7. Opportunistic infections are difficult to diagnose and treat successfully. AIDS patients require both primary treatment to curb infections and maintenance therapy to keep fungal pathogens and opportunists under control.

Systemic Mycoses Caused by Opportunistic Fungi (pp. 634–640)

1. *Pneumocystis jiroveci* causes ***Pneumocystis* pneumonia (PCP),** a leading cause of death in AIDS patients in the United States. The organism shows a blend of characteristics similar to those of both protozoans and fungi. PCP is debilitating because the fungus multiplies rapidly.
2. **Candidiasis,** caused by various species of *Candida,* is one of the most important pathogens of AIDS patients, but it can also cause infections in relatively healthy individuals. Disease ranges from superficial infections to systemic candidiasis.
3. **Aspergillosis** is a group of diseases caused by *Aspergillus* species, including noninvasive fungal balls in the lungs and invasive diseases that can be fatal. *Aspergillus* is very common in the environment, from which it is inhaled.
4. *Cryptococcus neoformans,* common in bird droppings and soil, causes **cryptococcoses,** which most frequently manifest as cryptococcal meningitis in AIDS patients. Other clinical manifestations are possible.
5. **Zygomycoses,** caused by *Mucor* and various other genera of zygomycetes, can involve the brain, resulting in death. Less severe infections also occur.
6. Diseases caused by *Pneumocystis, Candida, Aspergillus* spp., and *Cryptococcus* are defining illnesses of AIDS patients.
7. *Fusarium, Penicillium,* and *Trichosporon* are emerging opportunists. Factors contributing to their emergence include AIDS and the selective pressure of antifungal agents, which select for resistant strains of fungi.

Superficial, Cutaneous, and Subcutaneous Mycoses (pp. 640–644)

1. Superficial, cutaneous, and subcutaneous mycoses are caused by opportunistic fungi.
2. **Black piedra** and **white piedra** are benign, superficial infections of the hair shafts that can be transmitted among family members through shared combs and brushes.
3. **Dermatophytoses** (tineas, ringworms) encompass superficial skin, nail, and hair infections caused by a variety of fungi transmitted from individual to individual. Three genera predominate: *Trichophyton, Microsporum,* and *Epidermophyton.*
4. *Malassezia furfur* is a fungus that infects the skin. Clinical manifestations include **pityriasis,** in which fungal growth disrupts melanin production to produce discolored patches.
5. Most superficial infections are diagnosed by simple clinical observation of symptoms; all can be treated successfully. Rarely are such infections severe or invasive, but they have a tendency to recur.
6. **Chromoblastomycosis** and **phaeohyphomycosis** are similar diseases resulting from infection with dark-pigmented fungi. Both are acquired by traumatic introduction of fungi into the skin. Lesions may be extensive and can spread internally. The two diseases are distinguished by differences in fungal morphology in tissue sections.
7. **Mycetomas** are invasive and destructive infections following introduction of soil fungi through scrapes or pricks from vegetation. Cutaneous lesions can spread to adjacent bone and can be permanently damaging. Surgery or amputation is required to remove infected tissues.
8. **Sporotrichosis** also is acquired from inoculation of soil fungi by thorn pricks. Lymphatic dispersal of the organisms results in multiple lesions along the course of lymphatic vessels.

Fungal Intoxications and Allergies (pp. 644–646)

1. Fungal metabolism may result in **mycotoxins,** which if eaten can cause neurological and physiological damage and lead to death. **Mycetismus** results from eating mycotoxic mushrooms. Other foods contaminated with mycotoxins cause **mycotoxicosis.**
2. **Aflatoxins,** produced by *Aspergillus,* are well-known mycotoxins.
3. Fungal allergens (spores or other fungal elements) cause type I hypersensitivities or, more rarely, type III hypersensitivities.

Questions for Review

Answers to the Questions for Review (except for Short Answer questions) begin on page A-1.

Multiple Choice

1. A fungus that can infect both healthy and immunocompromised patients is called
   a. a true pathogen.
   b. an opportunistic pathogen.
   c. a commensal organism.
   d. a symbiotic organism.
2. Of the following fungi, which is usually transmitted from person to person?
   a. *Blastomyces dermatitidis*
   b. *Coccidioides immitis*
   c. *Tricophyton rubrum*
   d. *Aspergillus fumigatus*

3. Which of the following is not used to identify true fungal pathogens?
   a. growth at 25°C and 37°C to show dimorphism
   b. GMS staining of infected tissues
   c. serological testing
   d. clinical symptoms alone

4. Because amphotericin B is extremely toxic to humans, most clinicians prescribe it only for
   a. dermatophyte infections.
   b. *Malassezia* infections.
   c. systemic infections.
   d. black piedra.

5. Ringworm is caused by a
   a. helminth.
   b. dermatophyte.
   c. dimorphic fungus.
   d. commensal fungus.

6. Which of the following is considered a classical opportunistic fungus?
   a. *Piedraia hortae*
   b. *Histoplasma capsulatum*
   c. *Fonsecaea pedrosoi*
   d. *Aspergillus niger*

7. Subcutaneous infections tend to be acquired through
   a. inhalation and remain localized.
   b. inhalation and become systemic.
   c. trauma and remain localized.
   d. trauma and become systemic.

8. The term *dermatophyte* refers to
   a. pathogenicity.
   b. where a fungus grows.
   c. method of spread.
   d. pigmentation.

9. Which of the following subcutaneous mycoses may exhibit respiratory and cerebral forms?
   a. chromoblastomycosis
   b. mycetoma
   c. phaeohyphomycosis
   d. sporotrichosis

10. Which of the following systemic mycoses is endemic to the deserts of the southwestern United States?
    a. blastomycosis
    b. coccidioidomycosis
    c. histoplasmosis
    d. paracoccidioidomycosis

11. A spherule stage is seen in humans infected with what organism?
    a. *Blastomyces dermatitidis*
    b. *Coccidioides immitis*
    c. *Histoplasma capsulatum*
    d. *Paracoccidioides brasiliensis*

12. Of the following fungal diseases, which is found in almost all terminal AIDS patients?
    a. chromoblastomycosis
    b. blastomycosis
    c. candidiasis
    d. piedra

13. Which of the following predisposing factors would leave a patient with the greatest long-term risk of acquiring a fungal infection?
    a. invasive medical procedures
    b. AIDS
    c. chronic illness such as diabetes
    d. short-term treatment with antibacterial agents

14. Fungal allergens generally stimulate what type of reaction?
    a. type I hypersensitivity
    b. type II hypersensitivity
    c. type III hypersensitivity
    d. type IV hypersensitivity

15. A pathogenic feature of *Cryptococcus neoformans* is
    a. production of destructive enzymes.
    b. production of a capsule.
    c. infection of immune cells.
    d. variation of surface antigens to avoid immune system recognition.

16. The most common manifestation of *Cryptococcus* infection in AIDS patients is
    a. blindness.
    b. cutaneous infection.
    c. meningitis.
    d. pneumonia.

17. Bread mold can cause which disease?
    a. aspergillosis
    b. dermatophytosis
    c. mycetoma
    d. zygomycosis

18. Mycetismus is caused by
    a. inhalation of fungal allergens.
    b. ingestion of mushrooms.
    c. traumatic inoculation of fungi beneath the skin.
    d. close contact with infected individuals.

19. One of the more poisonous mycotoxins is produced by
    a. *Amanita phalloides.*
    b. *Amanita muscaria.*
    c. *Psilocybe cubensis.*
    d. *Claviceps purpurea.*

20. The number of mycoses worldwide is rising, in part, because
    a. the number of fungi in the environment is rising.
    b. the number of immunocompromised individuals in the population is rising.
    c. fungi have become more pathogenic.
    d. fungi are developing a tendency to be spread from person to person.

Modified True/False

Indicate which of the following are true and which are false. Rewrite false statements to make them true by changing the nontaxonomic italicized word(s).

1. ____ Fungi are generally *not transmitted* from person to person.
2. ____ On the whole, fungal infections are relatively *easy* to treat.
3. ____ *Dermatophytes* are always contracted from the environment.
4. ____ Chromoblastomycosis and phaeohyphomycosis are both caused by *dark-pigmented* ascomycetes.
5. ____ *Sporotrichosis* is always caused by traumatic introduction of fungi beneath the skin.
6. ____ Coccidioidomycosis does not occur normally outside the *Western Hemisphere.*
7. ____ Treatment of individuals with broad-spectrum antibacterial agents is a predisposing factor for *opportunistic* fungal infections.
8. ____ *Candida albicans* generally causes localized opportunistic infections but can become systemic, particularly in the *immunocompetent.*
9. ____ Relapse of fungal diseases is common in *AIDS patients.*
10. ____ *Almost everyone has* allergies to fungal elements.

Fill in the Blanks

1. Dimorphic fungi exist as ______________ forms in the environment and as ______________ forms in their hosts.
2. The true fungal pathogens are ______________, ______________, ______________, and ______________.
3. Many antifungal agents target the compound ______________ in fungal cytoplasmic membranes.

4. ________________ piedra is characterized by hard nodules on the hair shaft, whereas ________________ piedra is characterized by soft nodules.
5. Sporotrichosis is caused by the traumatic introduction of ________________ ________________ into the skin (give genus and species).
6. Which systemic mycosis is associated with bird droppings? ________________ ________________ (give genus and species).
7. The five most common agents of opportunistic fungal infections are ________________, ________________, ________________, ________________, and ________________.
8. Thrush is caused by ________________ (genus name).
9. *Pneumocystis* was once classified as a ________________, but now it is classified as a ________________.
10. Ergot alkaloids are produced by some strains of the genus ________________.

Matching

Match each of the following diseases with the manner(s) in which fungi enter the body. Answers can be used more than once.

A. Inhalation
B. Contact
C. Trauma
D. Ingestion

____ Aspergillosis
____ Candidiasis
____ Chromoblastomycosis
____ Coccidioidomycosis
____ Cryptococcosis
____ Dermatophytosis
____ Histoplasmosis
____ Hypersensitivity reactions
____ Mushroom poisoning
____ Mycetoma
____ Sporotrichosis

Short Answer

1. Amphotericin B is considered the "gold standard" of antifungal agents. Technically, its mode of action works against most fungal infections. Why, then, isn't it prescribed for most fungal infections?
2. Discuss why it is difficult in many cases to determine the source of superficial fungal infections (that is, from other humans, animals, or the environment).
3. Given that superficial fungal infections are only on the surface, why is it necessary to even try to identify the source of infection?
4. AIDS patients usually die of bacterial or fungal infections. Why do so many fungal infections appear in these individuals, and why are mycoses severe while fungi, for the most part, are benign residents of the environment?
5. How does mycotoxicosis differ from mycetismus?

6. Color each map below to show the general area where each disease is endemic.

Blastomycosis

Coccidioidomycosis

Histoplasmosis

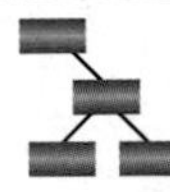

Concept Mapping

Using the following terms, draw a concept map that describes systemic mycoses. For a sample concept map, see p. 93.
Or, complete this concept map online by going to the Study Area at www.masteringmicrobiology.com.

5-fluorocytosine
Amphotericin B (4)
Aspergillus species
Blastomyces dermatitidis
Candida albicans
Coccidioides immitis
Cryptococcal meningitis
Cryptococcus neoformans
Every body system
Fluconazole
Histoplasma capsulatum
Inhalation
Opportunistic fungi
Other organs
Paracoccidioides brasiliensis
Pathogenic fungi
Pneumocystis jiroveci
Pneumocystis pneumonia
Pulmonary disease
Respiratory infection
Trimethoprim/Sulfanilamide

Critical Thinking

1. Correlate the observation that the majority of fungal infections are caused by opportunists with the fact that most infections are acquired from the environment. How does this make the control and diagnosis of fungal infections difficult?
2. The four pathogenic fungi (*Blastomyces, Coccidioides, Histoplasma,* and *Paracoccidioides*) are all dimorphic. Antifungal agents, like all antimicrobials, are generally designed to target something specific about the organism to avoid harming human tissues. Propose how you could use the idea of dimorphism to produce new antifungal agents.
3. What actions could be taken to limit the contamination of foods with fungal toxins? Would these actions differ depending on where in the world the food is grown? Explain.
4. Onychomycoses are nail infections that can be caused by several species of fungi. Explain why these mycoses are so difficult to treat, and why it is generally necessary to treat patients with oral antifungal agents for long periods of time.
5. What factors contribute to the pathogenicity of *Cryptococcus* infections?
6. Clinically, all fungi that cause subcutaneous mycoses produce lesions on the skin around the site of inoculation. What are some of the things you would look for when attempting to distinguish among them?
7. Given the various predisposing factors that make humans susceptible to opportunistic infections, how can health care providers curtail the rising incidence of such infections?

23 Parasitic Protozoa, Helminths, and Arthropod Vectors

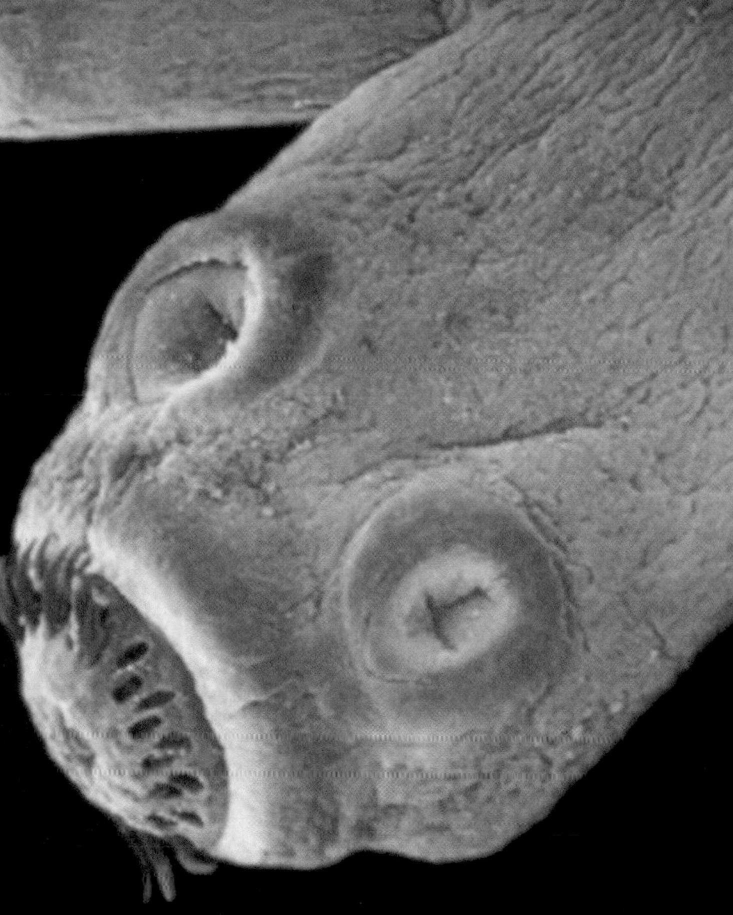

Suckers and hooklets attach *Taenia* to its human host's intestinal wall. The body of the helminth is composed of segments, giving the worm its common name—tapeworm.

The next time you sit down to a nice meal containing pork, it's probably a good idea to make sure the meat is properly cooked—if not well done. One reason is that the muscles of pigs can be infected with cysts that are the immature stage of *Taenia solium,* a parasitic tapeworm commonly called the pork tapeworm. The cysts remain viable in undercooked pork, and in the human small intestine they develop into mature tapeworms, which can grow as long as 800 cm (more than 26 feet). Cysts of a similar tapeworm, *Taenia saginata* (the beef tapeworm) also develop in the intestines of humans who eat insufficiently cooked infected beef.

This chapter provides a brief survey of parasitology, the study of eukaryotic parasites—in particular, various protozoa and helminths (tapeworms, other flatworms called flukes, and roundworms). Because parasitologists also study arthropod vectors (including mites, ticks, flies, mosquitoes, and fleas), which are important in the transmission of a variety of infectious organisms, we consider vectors in the final section of the chapter.

Take the pre-test for this chapter online. Visit the Study Area at www.masteringmicrobiology.com.

As we saw in Chapter 14, parasites live, feed and grow in or on another organism at the expense of their host's metabolism. Sometimes no obvious damage is done to the host, but in other cases parasitism can lead to a host's death. Many protozoan and helminthic parasites exist worldwide, especially in the tropics and subtropics, and particularly among people living in rural, undeveloped, or overcrowded places. Parasitic diseases are also emerging as serious threats among developed nations in nontropical regions. This rise is due to many factors, but can most generally be attributed to human migration, poor regulation and inadequate inspection of food and water supplies, and a changing world climate that alters parasites' transmission patterns, particularly those transmitted by vectors. As a whole, parasites infect billions of humans each year, making parasitic infections medically, socially, and economically important.

Parasitic infections often involve several hosts—a **definitive host** in which mature (often sexual) forms of the parasite are present and usually reproducing, and with many parasites, one or more **intermediate hosts** in which immature parasites undergo various stages of maturation. In general, parasites infect human hosts in one of three major ways: by being ingested, through vector-borne transmission, or via direct contact and penetration of the skin or mucous membranes **(Figure 23.1)**.

We begin our consideration of parasitic infections by discussing the protozoan parasites of humans.

Protozoan Parasites of Humans

Learning Objective

✓ Describe the epidemiology of the more important protozoan parasites.

As discussed in Chapter 12, protozoa are unicellular eukaryotes found in a wide range of habitats. Although the majority of protozoa are free-living and do not adversely affect humans or animals, some are parasites that can cause debilitating and deadly diseases.

Among the many protozoa that enter the body via ingestion, most have two morphological forms: A feeding and reproducing stage called a *trophozoite,* which lives within the host, and a dormant *cyst* stage, which can survive in the environment and is infective to new hosts. Once ingested by a host, cysts

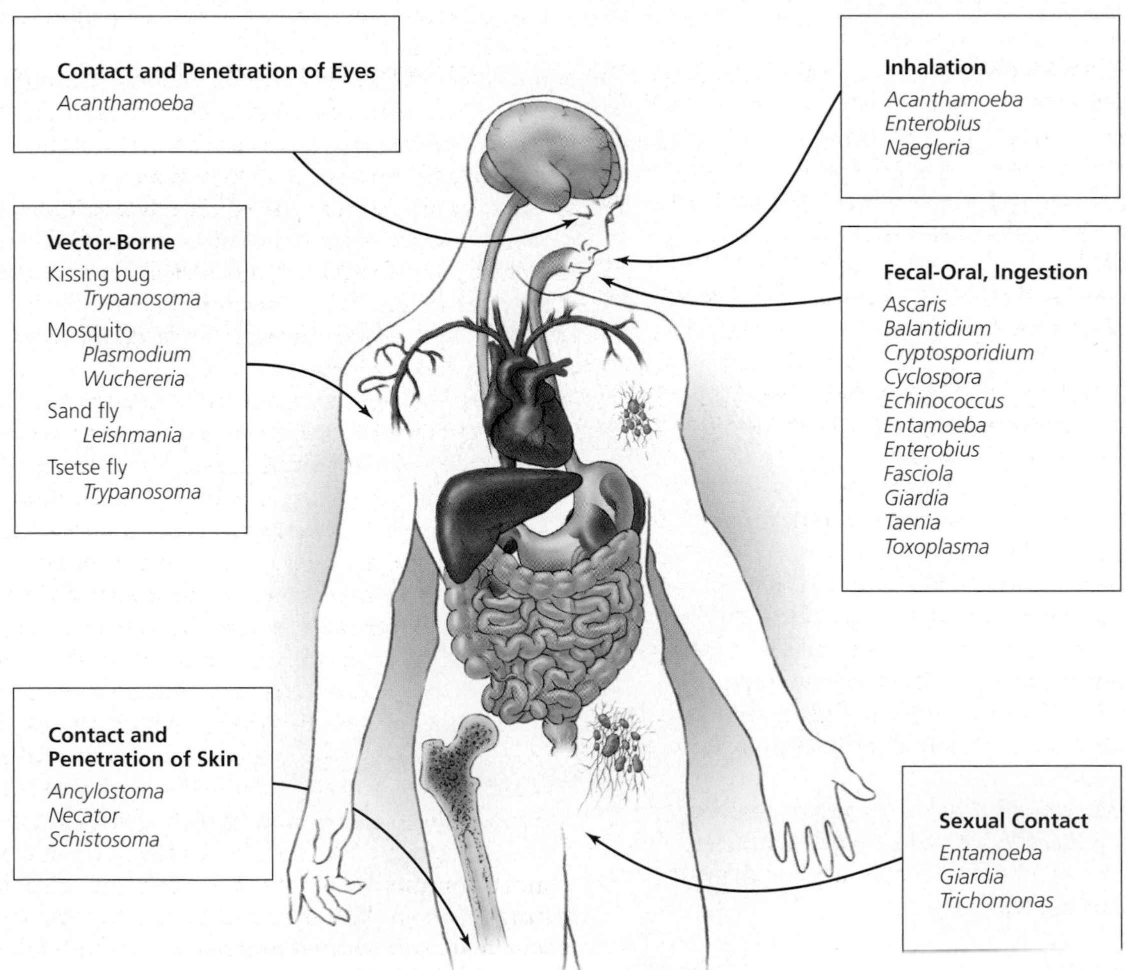

▲ **Figure 23.1 The major routes by which humans acquire parasitic infections.** They are: the fecal-oral route, the vector-borne route, and direct contact. Occasionally parasites are inhaled. Listed genera are discussed in the chapter.

undergo **excystment** and develop into new trophozoites, which resume feeding and reproducing. In most cases, trophozoites undergo **encystment** before leaving the host in the feces, becoming available to infect other hosts.

This chapter presents the protozoa in four groups, an approach that is different from the taxonomic presentation of Chapter 12. The scheme presented in this chapter reflects the largely obsolete, but clinically useful, classical groupings for these parasites, which were based primarily on mode of locomotion: the ciliates, the amoebae, the flagellates, and the typically nonmotile apicomplexans.

Ciliates

Learning Objective

✓ Describe the characteristics of *Balantidium coli.*

Ciliates (sil'ē-āts) are protozoa that in their trophozoite stages use cilia for locomotion, for acquiring food, or both. *Balantidium coli* (bal-an-tid'ē-ŭm kō'lē), a relatively large (50 μm × 100 μm) ciliate commonly found in animal intestinal tracts, is the only ciliate known to cause disease in humans. Pigs are its most common host, but it is also found in rodents and nonhuman primates. Humans become infected by consuming food or water contaminated with feces containing cysts.

Following ingestion, excystment occurs in the small intestine, releasing trophozoites that use their cilia to attach to (and then burrow through) the mucosal epithelium lining the intestine. Eventually, some trophozoites undergo encystment, and both cysts and trophozoites are shed in feces. Trophozoites die outside the body, but cysts are hardy and infective.

In healthy adults, *B. coli* infection is generally asymptomatic. For those in poor health, however, **balantidiasis** (bal'an-ti-dī'ă-sis) occurs. Persistent diarrhea, abdominal pain, and weight loss characterize the disease. Severe infections produce *dysentery* (frequent and painful diarrhea, often containing blood and mucus) and possibly ulceration and bleeding from the intestinal mucosa. *Balantidium* infection is rarely fatal.

Paradoxically, cysts are few and usually cannot be recovered from stool (fecal) samples, although they are the infective stage. Noninfective trophozoites, by contrast, can be detected, and their presence is diagnostic for the disease. Fresh stool samples must be used for diagnostic purposes because the trophozoites do not survive long outside the intestinal tract. The treatment of choice for balantidiasis is the antibacterial drug tetracycline, which does not kill *B. coli* directly but instead modifies the normal microbiota of the digestive tract such that the small intestine is unsuitable for infection by the ciliate.

Prevention of balantidiasis relies on good personal hygiene, especially for those who live around or work with pigs. Additionally, efficient water sanitation is necessary to kill cysts or remove them from drinking water.

CRITICAL **THINKING**

Why are diseases such as balantidiasis generally more severe in those who are already in poor health?

Amoebae

Learning Objective

✓ Compare and contrast three amoebae that cause disease in humans.

Amoebae (ă-mē'bē, singular: *amoeba,* ă-mē'bă) are protozoa that have no truly defined shape and that move and acquire food through the use of pseudopodia. Although amoebae are abundant throughout the world in freshwater, seawater, and moist soil, few cause disease. The most important amoebic pathogen is *Entamoeba*.

Entamoeba

Entamoeba histolytica (ent-ă-mē'bă his-tō-li'ti-kă) is carried asymptomatically in the digestive tracts of roughly 10% of the world's human population. When disease develops, it can be fatal, causing a worldwide annual mortality of over 100,000. Carriers predominate in less-developed countries, especially in rural areas, where human feces are used to fertilize food crops and where water sanitation is deficient. Travelers, immigrants, and institutionalized populations are at greatest risk within industrialized nations. No animal reservoirs exist, but human carriers are sufficiently numerous to ensure continued transmission of the parasite.

Infection occurs most commonly through the drinking of water contaminated with feces that contain cysts. The parasite can also be ingested following fecal contamination of hands or food or during oral-anal intercourse.

Excystment in the small intestine releases trophozoites that migrate to the large intestine and multiply. The organism uses pseudopodia to attach to the intestinal mucosa, where it feeds and reproduces via binary fission. Both trophozoites and cysts are shed into the environment in feces, but the trophozoites die quickly, leaving infective cysts.

Depending on the health of the host and the virulence of the particular infecting strain, three types of **amebiasis** (ă-mē-bī'ă-sis) can result. The least severe form, *luminal amebiasis,* occurs in otherwise healthy individuals. Infections are asymptomatic; trophozoites remain in the lumen of the intestine, where they do little tissue damage. Invasive *amebic dysentery* is a more serious and more common form of infection characterized by severe diarrhea, colitis (inflammation of the colon), appendicitis, and ulceration of the intestinal mucosa. Bloody, mucus-containing stools and pain are characteristic of amebic dysentery, which affects about 500 million people worldwide. In the most serious disease, *invasive extraintestinal amebiasis,* trophozoites invade the peritoneal cavity and the bloodstream, which carries them throughout the body. Lesions of dead cells formed by the trophozoites occur most commonly in the liver but can also be found in the lungs, spleen, kidneys, and brain. Amebic dysentery and invasive extraintestinal amebiasis can be fatal, especially without adequate treatment.

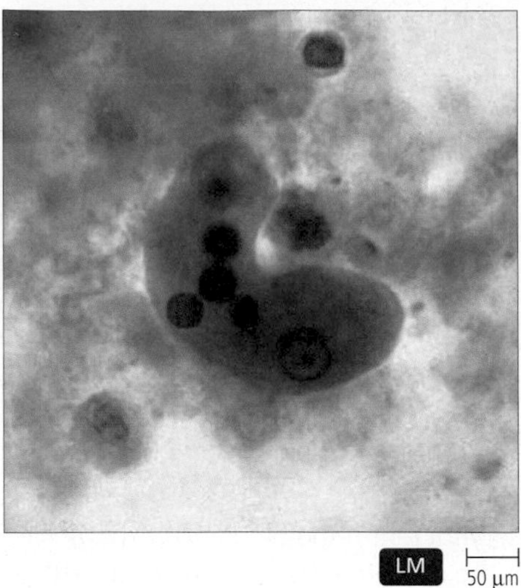

▲ **Figure 23.2** **Trophozoite of *Entamoeba histolytica*.** This specimen has ingested erythrocytes (red blood cells).

Diagnosis is based on the identification of microscopic cysts or trophozoites **(Figure 23.2)** recovered from either fresh stool specimens or intestinal biopsies. Microscopic analysis is necessary to distinguish amebic dysentery from dysentery caused by bacteria. Serological identification using antibodies to detect antigens can aid in distinguishing *E. histolytica* from nonpathogenic amoebae.

For asymptomatic infections, the drug paromomycin is effective. Physicians prescribe iodoquinol for symptomatic amebiasis. They may also prescribe antibacterial agents at the same time to prevent secondary bacterial infections. The coupling of oral rehydration therapy with drug treatment is recommended, and vital in severe cases.

Several preventive measures interrupt the transmission of *Entamoeba*. Discontinuing the use of human wastes as fertilizer reduces the transmission of amebiasis. Normal methods of treating wastewater and drinking water are helpful but not completely effective because the infectious cysts are hardy. Effective processing of water requires extra chemical treatment, filtration, or extensive boiling to eliminate all cysts. Good personal hygiene can eliminate transmission via intimate contact.

Acanthamoeba and *Naegleria*

Two other amoebae—*Acanthamoeba* (ă-kan-thă-mē′bă) and *Naegleria* (nā-glē′rē-ă)—cause rare and usually fatal infections of the brain. These amoebae are common free-living inhabitants of warm lakes, ponds, puddles, ditches, mud, and moist soil. They are also found in artificial water systems such as swimming pools, air conditioning units, humidifiers, and dialysis units. Contact lens wearers who use tap water (as opposed to sterile saline solution) to wash and store their lenses create an additional focal point for infection. *Naegleria* may assume a flagellated form in addition to its amoeboid and cyst stages.

Acanthamoeba usually enters a host through cuts or scrapes on the skin, through the conjunctiva via abrasions from contact lenses or trauma, or through inhalation of contaminated water while swimming. *Acanthamoeba keratitis,* in which trophozoites invade and perforate the eye, results from conjunctival inoculation. Damage can be extensive enough to require corneal replacement. The more common disease caused by infection with *Acanthamoeba* is **amebic encephalitis** (inflammation of the brain), characterized by headache, altered mental state, and neurological deficit. Symptoms progressively worsen over a period of weeks until the patient dies.

Infection with *Naegleria* occurs when swimmers inhale water contaminated with trophozoites, which then invade the nasal mucosa and replicate. The trophozoites migrate to the brain, where they cause an **amebic meningoencephalitis** (inflammation of the brain and its membranes). Severe headache, fever, vomiting, and neurological tissue destruction lead to hemorrhage, coma, and usually death within three to seven days after the onset of symptoms.

Physicians diagnose *Acanthamoeba* and *Naegleria* infections by detecting trophozoites in corneal scrapings, cerebrospinal fluid, or biopsy material. Keratitis can be treated topically with anti-inflammatory drugs. Physicians treat *Acanthamoeba* encephalitis with pentamidine and *Naegleria* infection with amphotericin B, but by the time they have diagnosed the disease, it is almost always too late to begin effective treatment.

As both these amoebae are environmentally hardy, control and prevention of infection can be difficult. Swimmers and bathers should avoid waterways in which the organisms are known to be endemic. Nonsterile solutions should never be used to clean or store contact lenses. Swimming pools should be properly chlorinated and tested periodically to ensure their safety. Air conditioning systems, dialysis units, and other devices that routinely use water should also be cleaned thoroughly and regularly to prevent amoebae from becoming resident.

Flagellates

Learning Objectives

- ✓ Contrast the life cycles of *Trypanosoma cruzi* and *Trypanosoma brucei*.
- ✓ Describe the life cycle of *Leishmania* and the clinical forms of leishmaniasis.
- ✓ Compare *Giardia* infections to those caused by *Entamoeba* and *Balantidium*.
- ✓ Identify the risk factors and preventive measures for *Trichomonas vaginalis* infection.

In 1832 Charles Darwin (1809–1882) explored South America as one of the naturalists sailing aboard HMS *Beagle*. His explorations were to impact him profoundly—intellectually and medically. During the voyage he began to formulate his ideas concerning the evolution of species, and he was bitten by a bloodsucking bug that may have infected him with *Trypanosoma cruzi*, a flagellated protozoan that afflicted him for the rest of his life. The following sections explore this and other parasitic flagellates that infect humans.

Flagellates (flaj'e-lātz) are protozoa that possess at least one long flagellum, which is used for movement. The number and arrangement of the flagella are important features for determining the species of flagellate present within a host.

Here we will examine some parasitic kinetoplastids (*Trypanosoma* and *Leishmania*), *Giardia* (a diplomonad), and *Trichomonas* (a parabasalid). We begin our discussion of the more important flagellate parasites with *Trypanosoma cruzi*.

Trypanosoma cruzi

Trypanosoma cruzi (tri-pan'ō-sō-mă kroo'zē) causes **Chagas' disease.** This disease, named for Brazilian doctor Carlos Chagas (1879–1934), is endemic throughout Central and South America. Localized outbreaks have also occurred in California and Texas. Opossums and armadillos are the primary reservoirs for *T. cruzi*, but most mammals, including humans, can harbor the organism. Transmission occurs through the bite of true bugs—a type of insect—in the genus *Triatoma* (trī-ă-tō'mă). These blood-sucking bugs feed preferentially from blood vessels in the lips, which gives the bugs their common name—kissing bugs.

Figure 23.3 illustrates the basic life cycle of *T. cruzi*. Within the hindgut of the kissing bug, a stage called *epimastigotes*[1] develop into infective trypomastigotes (1), which are shed in the bug's feces while the bug feeds on a mammalian host (2). When the host scratches the itchy wounds created by the bug bites, the *trypomastigotes* in the feces enter the nearby wounds (3). The trypomastigotes are then carried in the bloodstream throughout the body and penetrate certain cells, especially macrophages and heart muscle cells, where they transform into small, nonflagellated forms called *amastigotes* (4). The multiplication of these intracellular amastigotes via binary fission eventually causes the cells to rupture (5), releasing amastigotes that either infect other cells (6) or revert to trypomastigotes that circulate in the bloodstream (7). The circulating trypomastigotes are sub sequently ingested by a kissing bug when it takes a blood meal (8). Within the midgut of the kissing bug, binary fission of the trypomastigotes produces epimastigotes (9), completing the cycle.

Chagas' disease progresses over the course of several months through the following four stages:

1. An acute stage characterized by *chagomas*, which are swellings at the sites of the bites
2. A generalized stage characterized by fever, swollen lymph nodes, myocarditis (inflammation of the heart muscle), and enlargement of the spleen, esophagus, and colon
3. A chronic stage, which is asymptomatic and can last for years
4. A symptomatic stage characterized primarily by congestive heart failure following the formation of *pseudocysts*, which are clusters of amastigotes in heart muscle tissue

Parasite-induced heart disease is one of the leading causes of death in Latin America.

[1]From Greek *epi*, meaning on, and *mastix*, meaning whip. The flagellum of an epimastigote originates on the middle of the cell.

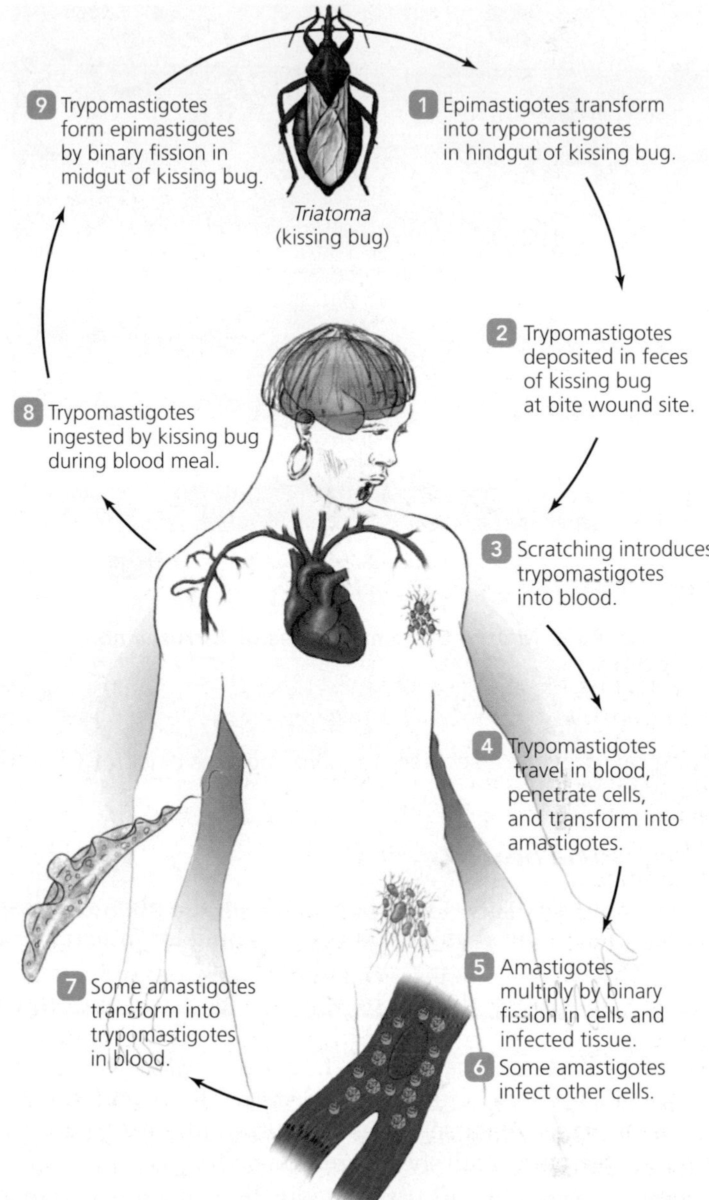
9 Trypomastigotes form epimastigotes by binary fission in midgut of kissing bug.
1 Epimastigotes transform into trypomastigotes in hindgut of kissing bug.
Triatoma (kissing bug)
2 Trypomastigotes deposited in feces of kissing bug at bite wound site.
8 Trypomastigotes ingested by kissing bug during blood meal.
3 Scratching introduces trypomastigotes into blood.
4 Trypomastigotes travel in blood, penetrate cells, and transform into amastigotes.
5 Amastigotes multiply by binary fission in cells and infected tissue.
6 Some amastigotes infect other cells.
7 Some amastigotes transform into trypomastigotes in blood.

▲ **Figure 23.3 The life cycle of *Trypanosoma cruzi*.** *Triatoma*, the vector, is shown life-size.

Microscopic identification of trypomastigotes or their antigens in blood, lymph, spinal fluid, or a tissue biopsy is diagnostic for *T. cruzi* (**Figure 23.4**). Another simple and practical diagnostic method is *xenodiagnosis*. In this procedure, uninfected kissing bugs are allowed to feed on a person suspected of having a *T. cruzi* infection. Four weeks later, the bugs are dissected. The presence of parasites within the hindgut of the bug indicates the person is infected.

In its earliest stages, Chagas' disease can be treated with benznidazole or nifurtimox; late stages cannot be treated. Prevention of Chagas' disease involves replacing thatch and mud building materials, which provide homes for the bugs, with concrete and brick. The use of insecticides, both personally and in the home, and sleeping under insecticide-impregnated

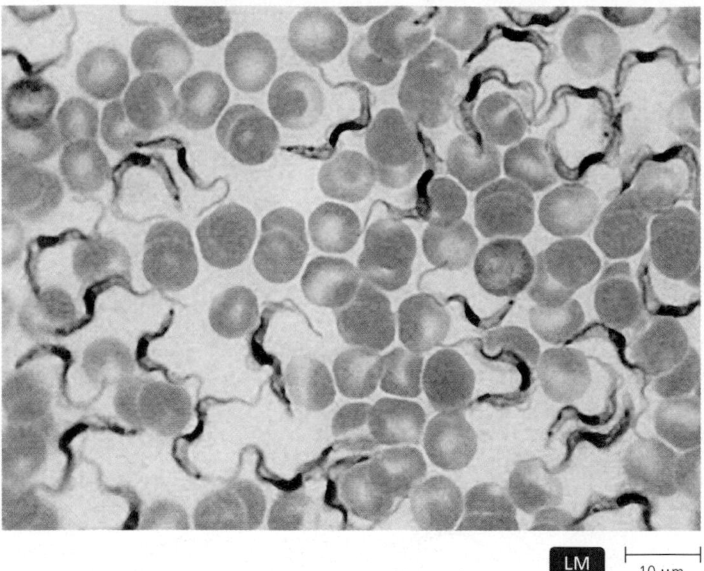
LM
10 μm

▲ **Figure 23.4 Mature trypomastigotes of *T. cruzi* among erthyrocytes.**

netting can prevent insect feeding. No vaccines exist for Chagas' disease.

Trypanosoma brucei

Trypanosoma brucei (brūs'ē) causes **African sleeping sickness,** which afflicts more than 10,000 people annually in equatorial and subequatorial savanna, agricultural, and riverine areas of Africa. Its geographical range depends on the range of its insect vector, the tsetse (tset'sē) fly (*Glossina*, glo-sī'nă). There are two variants of *Trypanosoma brucei*—*T. brucei gambiense* (gam'bē-en'sē), which occurs primarily in western Africa, and *T. brucei rhodesiense* (rō-dē'zē-en'sē) in eastern and southern Africa.

Whereas cattle and sheep are the reservoirs for *T. brucei gambiense,* several species of wild animals apparently serve as reservoirs for *T. brucei rhodesiense*. Although humans usually become infected when bitten by tsetse flies previously infected while feeding on infected animals, flies may transmit both variants of *T. brucei* between humans.

The life cycle of *T. brucei* proceeds as follows **(Figure 23.5)**: Within the salivary gland of a male or female tsetse fly, a form of the parasite called an epimastigote matures into an infective trypomastigote (1). During the course of taking a blood meal from an animal or human, tsetse flies inject trypomastigotes into the wounds (2). Trypomastigotes then travel throughout the lymphatic and circulatory systems to other sites (3), where they reproduce by binary fission (4). Eventually, some trypomastigotes enter the central nervous system (5), while others continue to circulate in the blood, where they can be picked up by feeding tsetse flies (6). In the midgut of the fly, trypomastigotes multiply by binary fission, producing immature epimastigotes that migrate to the salivary glands (7), where they mature and become trypomastigotes that are infective to new hosts when the fly feeds once again.

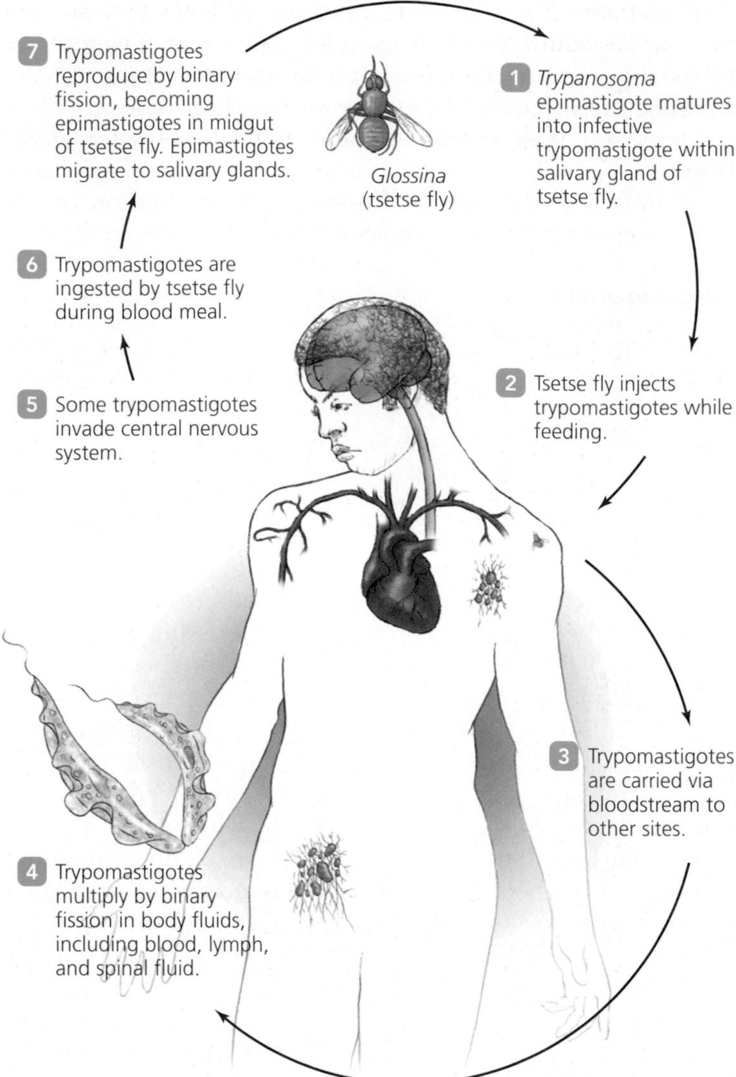
7 Trypomastigotes reproduce by binary fission, becoming epimastigotes in midgut of tsetse fly. Epimastigotes migrate to salivary glands.
Glossina (tsetse fly)
1 *Trypanosoma* epimastigote matures into infective trypomastigote within salivary gland of tsetse fly.
6 Trypomastigotes are ingested by tsetse fly during blood meal.
5 Some trypomastigotes invade central nervous system.
2 Tsetse fly injects trypomastigotes while feeding.
3 Trypomastigotes are carried via bloodstream to other sites.
4 Trypomastigotes multiply by binary fission in body fluids, including blood, lymph, and spinal fluid.

▲ **Figure 23.5 The life cycle of *Trypanosoma brucei*.** *Glossina* is shown life-size. *By which mode of transmission do trypanosomes infect humans?*

Figure 23.5 Trypanosomes are transmitted to humans by arthropod vectors.

The life cycle of *T. brucei* differs from that of *T. cruzi* in several ways:

- *T. brucei* matures in the salivary gland of the tsetse fly, whereas *T. cruzi* matures in the hindgut of the kissing bug.
- Tsetse flies directly inject *T. brucei*. In contrast, the host rubs *T. cruzi* found in a kissing bug's feces into a wound.
- *T. brucei* remains outside its hosts' cells, whereas amastigotes of *T. cruzi* live inside host cells.

Untreated African sleeping sickness progresses through three clinical stages. First, the wound created at the site of each fly bite becomes a lesion containing dead tissue and rapidly dividing parasites. Next, the presence of parasites in the blood triggers fever, swelling of lymph nodes, and headaches. Finally, invasion of the central nervous system results in meningoencephalitis, characterized by headache, extreme drowsiness, abnormal neurological function, and coma. The patient will die perhaps within six months of onset of disease. Symptoms may take years to develop with *T. brucei rhodesiense,* but begin within three to six months with *T. brucei gambiense*.

All *T. brucei* infections are characterized by cyclical waves of *parasitemia* (parasites in the blood) that occur roughly every 7–10 days. Although the presence of parasites in the blood is in itself serious, these cycles are particularly dangerous because with each wave of replication, *T. brucei* changes its surface glycoproteins and thus its surface antigens. The result is that by the time the host's immune system has produced antibodies against a given set of glycoproteins, the parasite has already produced a new set, continually leaving the host's immune system one step behind the parasite. Once infected, a patient is incapable of clearing the infection and never becomes immune.

Microscopic observation of trypomastigotes in blood, lymph, spinal fluid, or a tissue biopsy is diagnostic for both strains of *T. brucei*. Trypomastigotes are long and thin with a single long flagellum running along the cell and extending past the posterior end. Trypomastigotes of *T. brucei* are less curled than those of *T. cruzi*.

African sleeping sickness is one of the few human diseases that is 100% fatal if not treated. Treatment must begin as soon as possible after infection. However, immediate treatment is rare because poor health care infrastructure in endemic areas often prevents identification of the disease at its earliest stages. Pentamidine or suramin is used to treat the early stages. Melarsoprol is used when the disease progresses to the central nervous system because this drug crosses the blood-brain barrier. Eflornithine, a newer and more expensive drug, is remarkably effective against *T. brucei gambiense* (but not against *T. brucei rhodesiense*). In 2001, private pharmaceutical companies guaranteed availability of this drug through the World Health Organization to those who need it.

Clearing of tsetse fly habitats and broad application of insecticides have reduced the occurrence of African sleeping sickness in some localities. However, large-scale spraying of insecticides is impractical, expensive, and may have disastrous long-term environmental consequences. Some countries release up to a million sterile male tsetse flies a week, outnumbering wild flies 10 to 1. Most female flies mate with sterile males and produce no offspring—some scientists call this birth control for tsetse flies. Some regions have eradicated tsetse flies using this method. Personal insecticide use is preferable, as is the use of insecticide-impregnated netting and long, loose-fitting clothing, which can prevent insect feeding. No vaccine currently exists for African sleeping sickness.

Leishmania

Leishmania (lēsh-man'ē-ă) is a genus of kinetoplastid protozoa commonly hosted by wild and domestic dogs and small rodents. *Leishmania* is endemic in parts of the tropics and subtropics, including Central and South America, central and southern Asia, Africa, Europe, and the Middle East. **Leishmaniasis** (lēsh'mă-nī'ă-sis) is a **zoonosis**—a disease of animals transmitted to humans. Twenty-one of the 30 known species of *Leishmania* can infect humans.

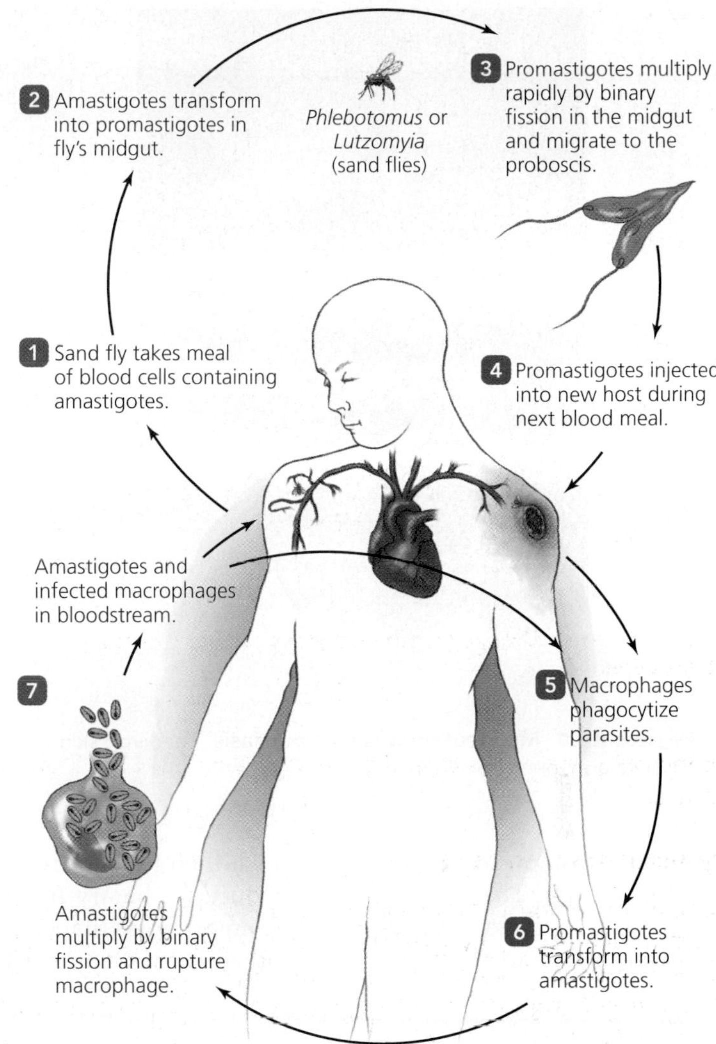

2 Amastigotes transform into promastigotes in fly's midgut.
Phlebotomus or *Lutzomyia* (sand flies)
3 Promastigotes multiply rapidly by binary fission in the midgut and migrate to the proboscis.
1 Sand fly takes meal of blood cells containing amastigotes.
4 Promastigotes injected into new host during next blood meal.
Amastigotes and infected macrophages in bloodstream.
7
5 Macrophages phagocytize parasites.
Amastigotes multiply by binary fission and rupture macrophage.
6 Promastigotes transform into amastigotes.

▲ **Figure 23.6 The life cycle of *Leishmania*.** Sand fly is shown life-size. *What is the effect of amastigote replication on the host's immune system?*

Figure 23.6 *The loss of macrophages severely inhibits the host's immune response.*

Leishmania has two developmental stages: *amastigotes,* which lack flagella and multiply within a mammalian host's macrophages and monocytes (types of white blood cells), and *promastigotes*, each of which has a single anterior flagellum and develops extracellularly within a vector's gut. In the life cycle of *Leishmania* **(Figure 23.6)**, a sand fly of the genera *Phlebotomus* (fle-bot'ō-mŭs) or *Lutzomyia* (lūts-ōm'yē-a) ingests white blood cells containing amastigotes during a blood meal (1). The amastigotes are released from the phagocytes in the fly's midgut, where they transform into promastigotes (2). Rapidly dividing promastigotes fill the fly's digestive tract and migrate to the proboscis (3), so that each time the fly feeds, it injects promastigotes into a new host (4). Macrophages near the bite site

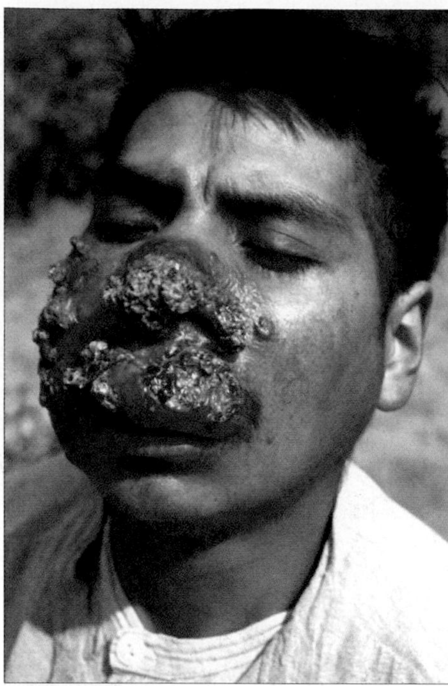

▲ **Figure 23.7 Mucocutaneous leishmaniasis.** The large skin lesions are permanently disfiguring.

phagocytize the promastigotes (5), which then transform into amastigotes (6). The amastigotes reproduce via binary fission until the macrophage ruptures, releasing amastigotes that circulate in the bloodstream and/or infect other macrophages (7), within which they can be ingested by another sand fly.

The clinical manifestations of leishmaniasis depend on the species of *Leishmania* and the immune response of the infected host. Upon initial infection, macrophages become activated, but not sufficiently to kill the intracellular parasites. The macrophages stimulate inflammatory responses that continue to be propagated by the infection of new macrophages. Depletion of macrophage numbers due to the reproduction of amastigotes decreases the efficiency of the immune response. The severity of the resulting immune dysfunction depends on the overall number of macrophages infected. Of the more than 1.5 million cases of leishmaniasis reported each year, over 45,000 are fatal.

Three clinical forms of leishmaniasis are commonly observed. *Cutaneous leishmaniasis* involves large painless skin ulcers that form around the bite wounds. Such lesions often become secondarily infected with bacteria. Scars remain when the lesions heal. *Mucocutaneous leishmaniasis* results when skin lesions enlarge to encompass the mucous membranes of the mouth, nose, or soft palate. Damage is severe and permanently disfiguring **(Figure 23.7)**. Neither of these forms of leishmaniasis is fatal. However, *visceral leishmaniasis* (also known as *kala-azar*) is fatal in 95% of untreated cases. In this disease, macrophages spread the parasite to the liver, spleen, bone marrow, and lymph nodes. Inflammation, fever, weight loss, and anemia increase in severity as the disease progresses. Visceral leishmaniasis is becoming increasingly problematic as an opportunistic infection among AIDS patients.

Microscopic identification of amastigotes in samples from cutaneous lesions, the spleen, or bone marrow is diagnostic of *Leishmania* infection. Immunoassays using antibodies to detect antigen can be used to confirm the diagnosis and to identify the strain. Molecular techniques such as PCR are needed to determine the species.

Most cases of leishmaniasis heal without treatment (though scars may remain) and confer immunity upon the recovered patient. At one time, lesions were purposefully encouraged on the buttocks of small children to induce immunity and prevent lesions from leaving more visible scars. Treatment, which is required for more serious infections and for visceral leishmaniasis, generally involves administering sodium stibogluconate or meglumine antimonate. Pentamidine has been used with some success to treat resistant strains of *Leishmania*.

Prevention is essentially limited to reducing exposure by controlling reservoir host and sand fly populations. For example, rodent nesting sites and burrows can be destroyed around human habitations to reduce contact with potentially infected populations. In some areas, infected dogs are destroyed. Spraying insecticide around homes can reduce the number of sand flies. Personal use of insect repellents, protective clothing, and netting further limits exposure. A vaccine for leishmaniasis does not exist.

CRITICAL **THINKING**

Given the regions of the world where *Leishmania* and HIV are endemic, would you expect the incidence of *Leishmania* to increase or decrease in the next decade? Explain your answer.

Giardia

Giardia intestinalis (jē-ar′dē-ă in-tes′ti-năl′is; previously called *G. lamblia*) is the causative agent of **giardiasis** (jē-ar-dī′ă-sis), one of the more common waterborne gastrointestinal diseases in the United States. *Giardia* lives in the intestinal tracts of animals and humans worldwide. The organism is very hardy and can also be found in water, in soil, on food, and on surfaces that have been contaminated with feces. The organism can survive for months in the environment owing to the protective outer shell of its cyst.

Infection usually results from the ingestion of cysts in contaminated drinking water or accidental ingestion during swimming. Hikers, campers, and their pets are at particular risk because infected wild animals shed *Giardia* into mountain streams. Because beavers are common zoonotic sources of *Giardia*, giardiasis is sometimes referred to as "beaver fever." Even if humans avoid drinking stream water, they usually don't think twice about letting their dogs drink it. Unfortunately, it is often only a matter of time before the dog passes the protozoan to its owner. Alternatively, eating unwashed raw fruits or vegetables that have been contaminated by feces can lead to infection, as can contact with feces during sex. Giardiasis outbreaks in day care facilities are

MICROBE AT A GLANCE

Giardia intestinalis

Taxonomy: Nomenclature is in flux; one possibility, used in this book: domain Eukarya, kingdom Diplomonadida

Other names: Formerly called *Giardia lamblia*

Morphology: Single-celled protozoan with four pairs of flagella and a ventral adhesive disk

Virulence factors: Adhesive disk, resistant cyst

Diseases caused: Giardiasis (form of diarrhea)

Treatment for disease: Metronidazole, oral rehydration, and electrolyte replacement

Prevention of disease: Sterilize water either chemically or via filtration (boiling generally not sufficient because of cysts); good personal hygiene to prevent consumption of cysts

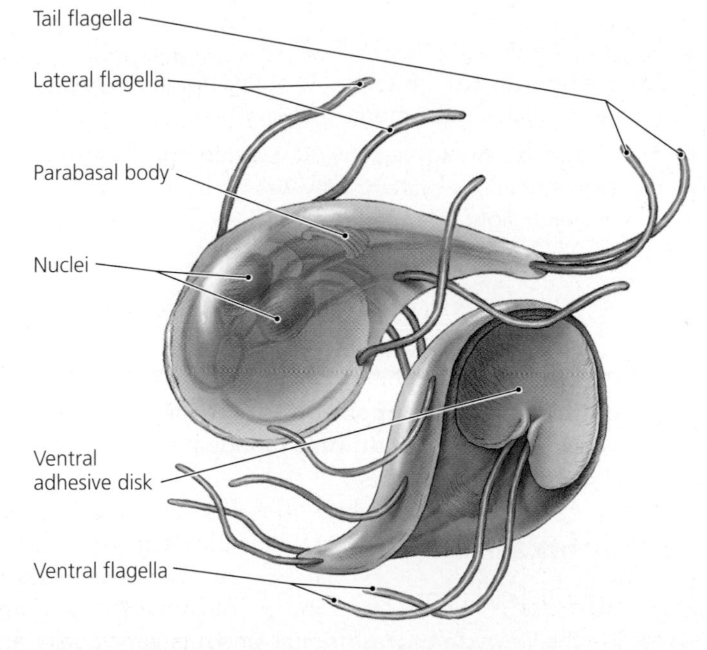

Download the Microbe at a Glance flashcards from the Study Area at www.masteringmicrobiology.com.

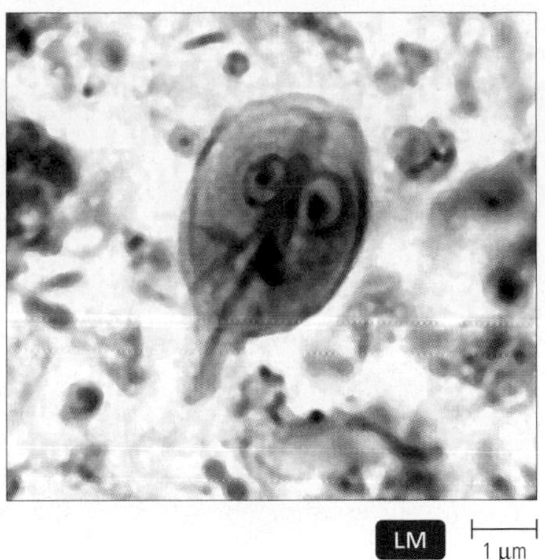

▲ **Figure 23.8 Trophozoite of *Giardia intestinalis*.** Two nuclei make this diplomonad resemble a face.

usually the result of children putting contaminated toys or eating utensils into their mouths.

Giardia has a life cycle similar to that of *Entamoeba*. Following ingestion, each cyst is triggered by acid in the stomach to release a trophozoite into the small intestine that begins multiplying via binary fission. Trophozoites either remain free in the lumen of the small intestine or attach to the intestinal mucosa via a ventral adhesive disk; they do not invade the intestinal wall. As trophozoites pass into the colon, encystment occurs. Cysts are immediately infective upon release, leading to a significant incidence of person-to-person and self-to-self transmission. Cysts survive and remain infective for several months.

Giardiasis can range from an asymptomatic infection to significant gastrointestinal disease. Signs and symptoms, when they occur, include severe watery diarrhea, abdominal pain, bloating, nausea, vomiting, ineffective absorption of nutrients, and low-grade fever. Stools are foul smelling, usually with the "rotten-egg" smell of hydrogen sulfide. Incubation lasts roughly one to two weeks, and symptoms resolve after one to four weeks in normal, healthy adults. In extreme cases, the attachment of parasites to the intestinal mucosa causes superficial tissue damage, and fluid loss becomes life threatening. Chronic giardiasis can occur, often among animals.

Diagnosis of giardiasis, which involves direct microscopic examination of stool specimens, is based on the observation of flat, pear-shaped trophozoites that resemble a face when viewed from below (**Figure 23.8**). In addition to the "face," which results from the pair of nuclei that resemble eyes, four pairs of flagella extending from the ventral surface may also be visible. Clinicians may need to examine several stool samples as the parasites are shed only intermittently.

Metronidazole is the drug of choice for treatment of giardiasis. If diarrhea is not present, treatment is often waived. Oral rehydration therapy may be required for severe cases of giardiasis, or to treat very young children regardless of the severity of infection.

To prevent infection in regions where *Giardia* is endemic, filtering water is necessary. When hiking, neither humans nor their pets should drink unfiltered stream or river water. Most camping and hiking stores sell portable water filtration kits, making it unnecessary to carry bottled water. Filtered water should be used for cooking and cleaning eating utensils. In day care facilities, scrupulous hygiene practices and the separation of feeding and diaper-changing areas are essential to preventing transmission. Recovering patients should be extremely vigilant with their personal hygiene to avoid transmitting *Giardia* to family members and should avoid swimming for several weeks after recovery to ensure they do not shed parasites into the water.

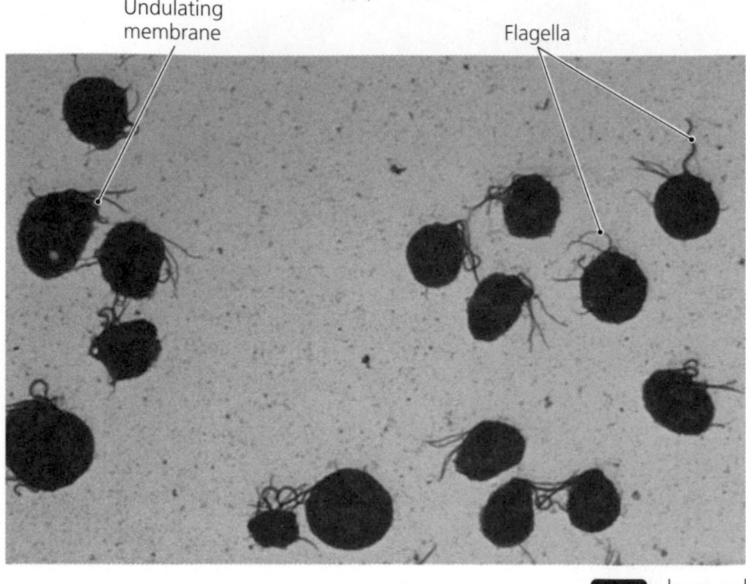

▲ **Figure 23.9** **Trophozoites of *Trichomonas vaginalis*.** Five anterior flagella, one associated with an undulating membrane, characterize this parabasalid.

CRITICAL **THINKING**

How does the visually distinctive appearance of *Giardia* trophozoites improve the success of medical treatment of giardiasis as compared to that for amoebic infections?

CLINICAL CASE STUDY

A Sick Soldier

A 20-year-old soldier serving in Iraq reports to his medic and complains of several large painless sores on his head and face. Some of the ulcers have healed, leaving scars. A sample taken from a lesion and examined microscopically shows the presence of flagellated protozoa.

1. Can a diagnosis be made from this information? If so, what is the diagnosis?
2. What treatment should the medic order?
3. How did the soldier become infected?
4. The soldier asks how he can prevent this disease in the future. How should the medic respond?
5. The soldier is a strong, healthy 20-year-old with a fully functioning immune system. Why was his body not able to sufficiently fight off the parasite?

Trichomonas

Trichomonas vaginalis (trik-ō-mō′nas va-jin-al′is) is globally distributed and is the most common protozoan causing disease in people of industrialized nations. The parasite lives on the vulvas and in the vaginas of women and in the urethras and prostates of men. This obligate parasite, which is incapable of surviving long outside a human host, is transmitted almost exclusively via sex. *T. vaginalis* occurs most frequently in people with a preexisting sexually transmitted disease, such as chlamydial infection, and in people with multiple sex partners.

In women, infection results in **vaginosis** (vaj-i-nō′sis), which is accompanied by a purulent (pus-filled) odorous discharge, vaginal and cervical lesions, abdominal pain, painful urination, and painful intercourse. Since inflammation is not typically involved, it is "vaginosis" rather than "vaginitis." Trophozoites feed on vaginal tissue, leading to erosion of the epithelium. *T. vaginalis* infection may cause inflammation of the urethra or bladder of men, but more typically men are asymptomatic.

Microscopic observation of actively motile trophozoites in vaginal and urethral secretions is diagnostic. The trophozoites are flat and possess five flagella and a sail-like undulating membrane (**Figure 23.9**). An immunofluorescent assay can be performed when infection is suspected and microscopy is too insensitive. Patients and all their sexual partners must be treated to prevent reinfection. Metronidazole is the drug of choice, though some resistant strains of *T. vaginalis* exist. Prevention involves abstinence, mutual monogamy, or consistent and correct condom usage.

Apicomplexans

Learning Objectives

- ✓ Describe the life cycle of *Plasmodium* and relate malarial symptoms to stages in the life cycle.
- ✓ Describe the clinical manifestations of *Toxoplasma gondii* infections.
- ✓ Compare and contrast the intestinal diseases caused by *Cryptosporidium* and *Cyclospora*.

Apicomplexans (ap-i-kom-plek′sănz) are alveolate protozoa whose infective forms are characterized by an ornate complex of organelles at their apical ends, which gives the group its name. They have also been called *sporozoa* because during their life cycles they assume nonmotile, sporelike shapes, but this term is not accurate because the term *spores* correctly refers to reproductive structures of multicellular plants and fungi. All apicomplexans are parasites of animals, and all have complicated life cycles involving at least two types of hosts. A major feature of apicomplexan life cycles is *schizogony*—a form of

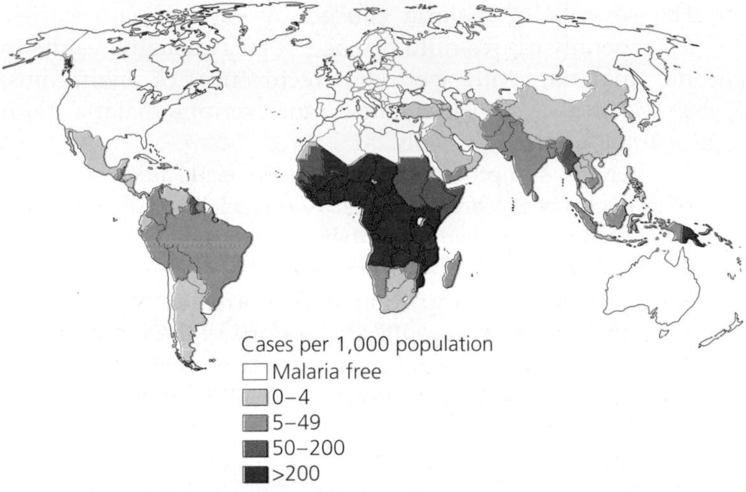
Cases per 1,000 population
Malaria free
0–4
5–49
50–200
>200

▲ **Figure 23.10 The geographical distribution and incidence of malaria.** The disease may be returning to areas from which it was once eradicated, including Europe and the United States.

asexual reproduction in which multinucleate *schizonts* form before the cells divide (see Figure 12.3).

Plasmodium, Toxoplasma, Cryptosporidium, and *Cyclospora* are four important apicomplexan parasites. We begin our discussion with *Plasmodium*, the causative agent of malaria and the most infamous protozoan parasite in the world.

Plasmodium

Four species of *Plasmodium* (plaz-mō′dē-ŭm) cause **malaria** in humans: *P. falciparum* (fal-sip′ar-ŭm), *P. vivax* (vī′vaks), *P. ovale* (ō-vă′lē), and *P. malariae* (mă-lār′ē-ī). Malaria is endemic in over 109 countries and territories throughout the tropics and subtropics, where the parasites' mosquito vector breeds **(Figure 23.10)**. The movement of infected individuals and mosquitoes continues to expand beyond these endemic areas, threatening to reintroduce *Plasmodium* into countries, including the United States, where mosquito eradication programs eliminated the disease decades ago. Over 240 million people are infected with *Plasmodium,* and 880,000 (usually children) die annually, making malaria among the deadlier killers. Females of 60 different species of the mosquito genus *Anopheles* (ă-nof′ĕ-lēz) serve as vectors of *Plasmodium*. (Male *Anopheles* do not feed on blood.)

The life cycle of *Plasmodium* has three prominent stages **(Figure 23.11)**:

1. The **exoerythrocytic phase,** as its name indicates, occurs outside of red blood cells. During a blood meal on a human, an infected female mosquito injects a stage of *Plasmodium* called a *sporozoite* along with saliva containing an anticoagulant into the blood (1). Sporozoites reach the liver via the bloodstream and over the course of one to two weeks undergo schizogony within liver cells (2). Schizogony produces a *Plasmodium* stage called *merozoites* that proceed to rupture the liver cells and enter the blood (3). *P. ovale* and *P. vivax* are further capable of forming a resting stage called a *hypnozoite* (not shown), which can remain dormant in liver cells for years and be reactivated at any time, causing relapses of malaria.
2. The **erythrocytic cycle** (4) begins when free merozoites penetrate erythrocytes and become yet another stage, called *trophozoites,* that endocytize the erythrocytes' hemoglobin protein. Presence of trophozoites inside blood cells is the diagnostic sign of malaria. Trophozoites undergo schizogony to produce more merozoites that lyse erythrocytes, and the cycle continues to repeat. Lysis of erythrocytes occurs nearly simultaneously and cyclically—every 48–72 hours, depending on the species of *Plasmodium* involved. Most merozoites infect new erythrocytes, but some remain in red blood cells, where they develop into male and female *gametocytes* (5).

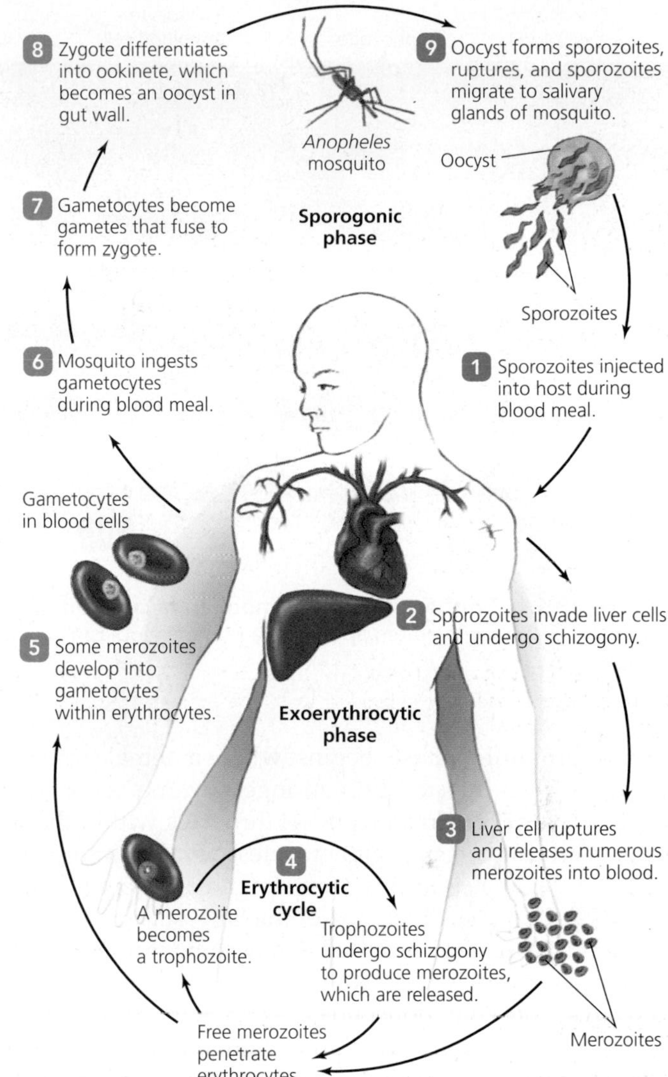
8 Zygote differentiates into ookinete, which becomes an oocyst in gut wall.
9 Oocyst forms sporozoites, ruptures, and sporozoites migrate to salivary glands of mosquito.
Anopheles mosquito
Oocyst
Sporogonic phase
7 Gametocytes become gametes that fuse to form zygote.
Sporozoites
6 Mosquito ingests gametocytes during blood meal.
1 Sporozoites injected into host during blood meal.
Gametocytes in blood cells
2 Sporozoites invade liver cells and undergo schizogony.
5 Some merozoites develop into gametocytes within erythrocytes.
Exoerythrocytic phase
3 Liver cell ruptures and releases numerous merozoites into blood.
4 Erythrocytic cycle
A merozoite becomes a trophozoite.
Trophozoites undergo schizogony to produce merozoites, which are released.
Free merozoites penetrate erythrocytes.
Merozoites

▲ **Figure 23.11 The life cycle of *Plasmodium*.** The exoerythrocytic phase and the erythrocytic cycle occur in humans, and the sporogonic phase in mosquitoes. *During which stage is malaria usually diagnosed in humans?*

Figure 23.11 *Malaria in humans is diagnosed during the parasite's erythrocytic cycle.*

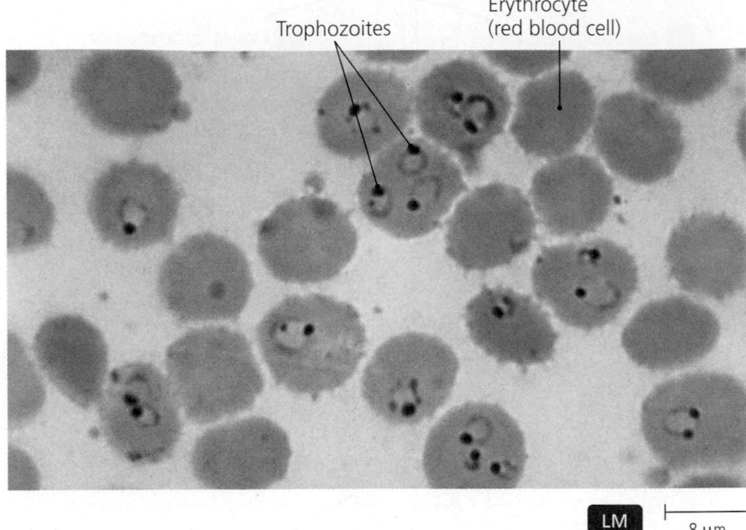

▲ **Figure 23.12 Trophozoites of *Plasmodium falciparum* inside erythrocytes.** Trophozoites often appear as a ring of cytoplasm with a dotlike nucleus.

3. The **sporogonic phase** begins when a female mosquito feeding on an infected human ingests gametocytes within erythrocytes (6). Freed from erythrocytes within the mosquito's digestive tract, gametocytes develop into *gametes*, and male gametes fertilize female gametes to produce *zygotes* (7). A zygote differentiates into an *ookinete*, which penetrates the mosquito's gut wall and becomes an *oocyst*, which undergoes meiosis (8). Some 10–20 days later, the oocyst ruptures, releasing thousands of sporozoites that migrate to the mosquito's salivary glands (9). The mosquito then transmits the sporozoites to a new host during a blood meal, completing the parasite's life cycle.

People living in endemic areas, and their descendants throughout the world, have evolved to have one or more of the following genetic traits that increase their resistance to malaria:

- *Sickle-cell trait*. Individuals with this gene produce an abnormal type of hemoglobin called hemoglobin S (hemoglobin A is normal). Hemoglobin S causes erythrocytes to become sickle-shaped, and somehow it also makes erythrocytes resist penetration by *Plasmodium*.
- *Hemoglobin C*. Humans with two genes for hemoglobin C are invulnerable to malaria. The mechanism by which this mutation provides protection is unknown.
- Genetic deficiency for the enzyme *glucose-6-phosphate dehydrogenase*. Trophozoites must acquire this enzyme from a human host before the trophozoite can synthesize DNA; thus, humans without the enzyme are spared malaria.
- Lack of so-called *Duffy*[2] antigens on erythrocytes. Because *P. vivax* requires Duffy antigens to attach to and infect erythrocytes, Duffy-negative individuals are resistant to this species.

[2]Named for the patient in which the antigen was discovered.

The severity of malaria caused by each species varies. *P. ovale* generally causes mild disease; *P. vivax* usually results in chronic malaria, with periodic recurrence of symptoms; *P. malariae* and *P. falciparum* cause more serious malaria, with *P. falciparum* infection possibly fatal.

The general symptoms of malaria are associated with synchronous cycles of erythrocyte lysis. Two weeks after the erythrocytic cycle begins, sufficient numbers of parasites exist to cause symptoms of fever, chills, diarrhea, headache, and (occasionally) pulmonary or cardiac dysfunction. Fever correlates with erythrocyte lysis and most likely results from efforts of the immune system to remove cellular debris, toxins, and merozoites. Loss of erythrocytes leads to anemia, weakness, and fatigue. The inability of the liver to process the inordinate amount of hemoglobin released from dying erythrocytes results in *jaundice*.

P. falciparum causes a form of malaria called *blackwater fever*, which is characterized by extreme fever, large-scale erythrocyte lysis, renal failure, and dark urine discolored by excreted hemoglobin. Protozoan proteins inserted on the surfaces of infected erythrocytes cause erythrocytes to become rigid and inelastic such that they cannot squeeze through capillaries, blocking blood flow and causing small hemorrhages in various tissues (and ultimately tissue death). *Cerebral malaria* results when tissue death occurs in the brain. Falciparum malaria can be fatal within 24 hours of the onset of symptoms.

If a victim survives the acute stages of malaria, immunity gradually develops. Periodic episodes become less severe over time—unless the victim becomes immunocompromised, in which case episodes can return to original levels of severity.

Microscopy is most commonly used for diagnosing malaria because *Plasmodium* species can be readily identified and distinguished in blood smears. Ringlike trophozoites within erythrocytes are often the diagnostic stage (**Figure 23.12**). Demonstrating antibodies against *Plasmodium* in the serum can also be used for differential diagnosis. However, because the symptoms might be confused with those of less severe infections, diagnosis may be missed in nonendemic areas unless a good case history, including travel history, is obtained.

Treatment varies by species and severity of symptoms. Drug choices are constantly revised based on effectiveness and the existence of resistant strains of *Plasmodium*. Standard antimalarial drugs include chloroquine, mefloquine, or atovaquone in combination with proguanil. These are used for uncomplicated malarial infections in areas where drug resistance does not exist. The World Health Organization recommends new drugs based on a compound called artemisinin, which is derived from a shrub used in traditional Chinese treatment of malaria. Antifever medication and blood transfusions may be required as supportive measures.

Control of malaria involves limiting contact with mosquitoes carrying *Plasmodium*. Widespread use of insecticides, drainage of wetlands, and removal of standing water can reduce mosquito breeding rates. Personal use of insect repellents, netting, and protective clothing reduces mosquito bites. Travelers to endemic areas can take preventive medication to avoid infection. Atovaquone and proguanil are drugs of choice and are usually taken several weeks before and after travel. Chloroquine is

MICROBE AT A GLANCE

Plasmodium falciparum

Taxonomy: Nomenclature is in flux; one possibility, used in this book: domain Eukarya, kingdom Alveolata, phylum Apicomplexa, class Sporozoea, order Eucoccidiida, family Plasmodiidae

Morphology: Sporozoite (slender, elongate single cell); merozoite (oval single cell); trophozoite (ring of cytoplasm with a dotlike nucleus; "ring form"); ookinete (motile wormlike form); oocyte (spherical)

Virulence factors: Lives intracellularly

Diseases caused: Malignant tertian malaria—a form of malaria called blackwater fever

Treatment for disease: Chloroquine, mefloquine, atovaquone, proguanil, artemisinin

Prevention of disease: Mosquito control, avoid mosquito bites

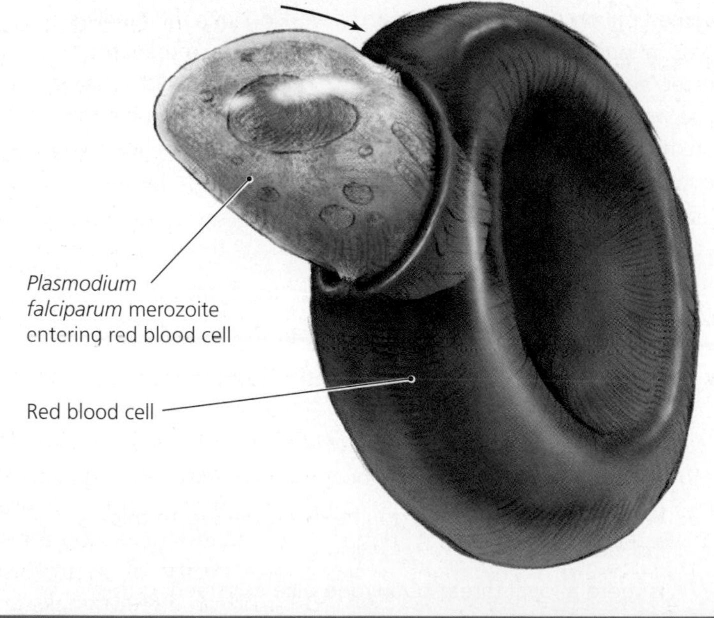

Download the Microbe at a Glance flashcards from the Study Area at www.masteringmicrobiology.com.

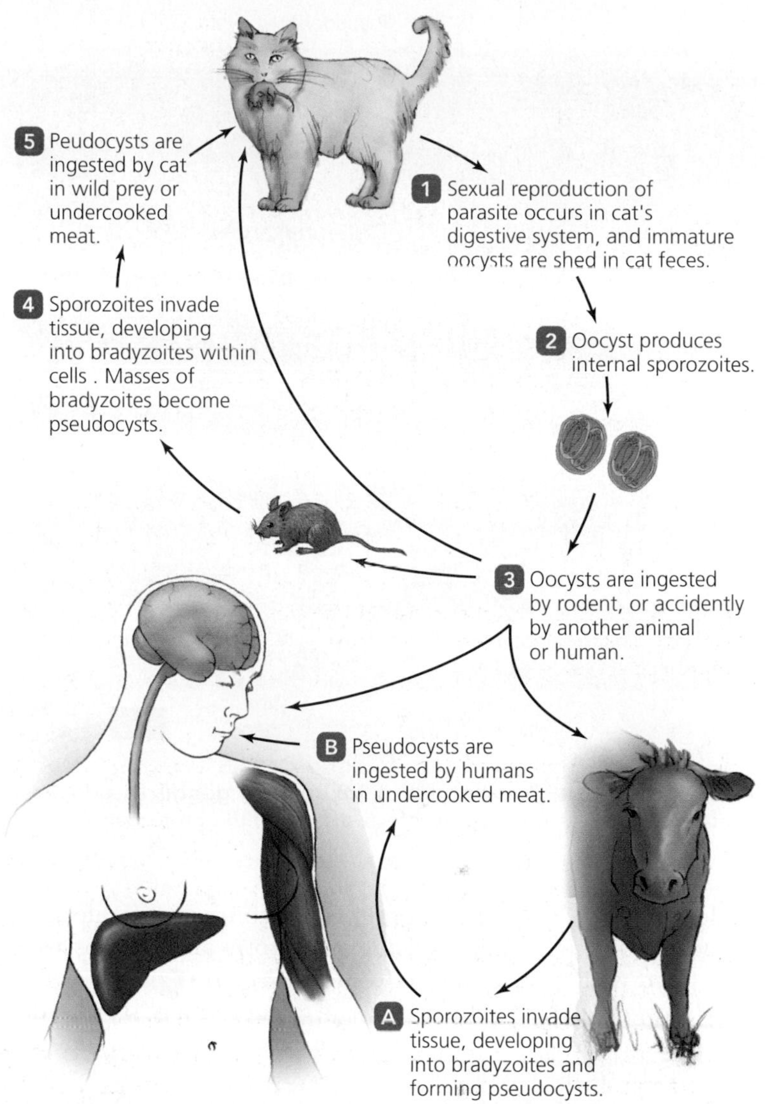

5 Peudocysts are ingested by cat in wild prey or undercooked meat.
1 Sexual reproduction of parasite occurs in cat's digestive system, and immature oocysts are shed in cat feces.
4 Sporozoites invade tissue, developing into bradyzoites within cells . Masses of bradyzoites become pseudocysts.
2 Oocyst produces internal sporozoites.
3 Oocysts are ingested by rodent, or accidently by another animal or human.
B Pseudocysts are ingested by humans in undercooked meat.
A Sporozoites invade tissue, developing into bradyzoites and forming pseudocysts.

▲ **Figure 23.13 The life cycle of *Toxoplasma gondii*.** *Which two groups of humans are most at risk from toxoplasmosis?*

Figure 23.13 AIDS patients and first-trimester fetuses are at greatest risk from toxoplasmosis.

sometimes given prophylactically before travel to areas where *Plasmodium* is still susceptible to this drug. Several malaria vaccines are currently under development.

Toxoplasma

Toxoplasma gondii (tok-sō-plaz′mă gon′dē-ē) is one of the world's most widely distributed protozoan parasites— 25% of the world's human population is infected. Wild and domestic mammals and birds are major reservoirs for *Toxoplasma,* and cats are the definitive host, in which the protozoan reproduces sexually.

Humans typically become infected by ingesting undercooked meat containing the parasite. People at greatest risk include butchers, hunters, and anyone who tastes food while preparing it. Ingestion or inhalation with contaminated soil can also be a source of infection. The protozoan can also cross a placenta to infect the fetus. Historically, contact with infected cats and their feces was proposed as a major risk for infection, but recent studies have shown that cats are not the major source of infection for humans because they shed *Toxoplasma* only briefly.

In the life cycle of *Toxoplasma* (**Figure 23.13**), male and female gametes in a cat's digestive tract fuse to form zygotes, which develop into immature *oocysts* that are shed in the feces (**1**). During the few days a cat sheds oocysts, it can excrete up to 10 million per day. Oocysts survive in moist soil for several months, contaminating vegetation grown in this soil. As each oocyst matures, it produces *sporozoites* internally (**2**). When rodents ingest mature oocysts on vegetation (**3**), digestion releases the sporozoites, which invade the rodent's muscles, lymph

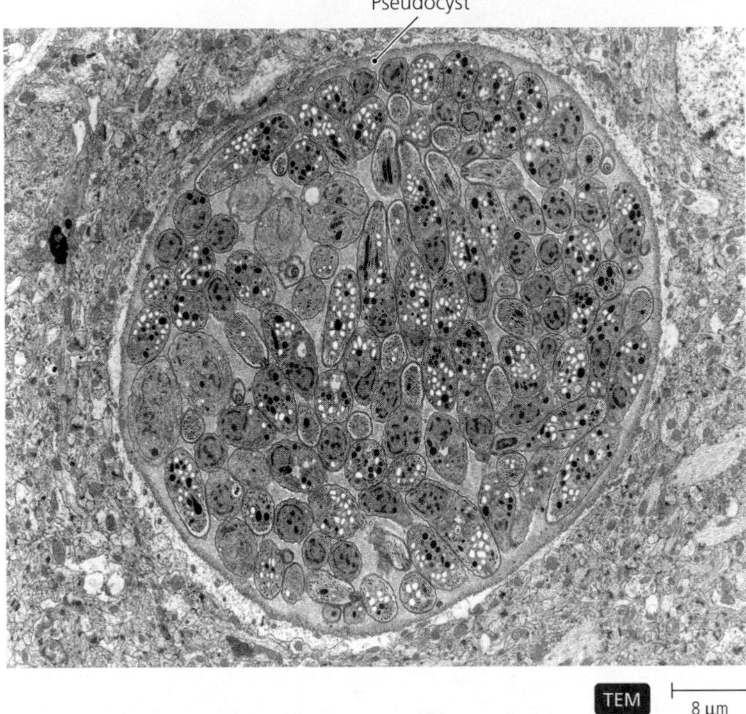

▲ **Figure 23.14 Pseudocysts of *Toxoplasma gondii*.** Pseudocysts contain bradyzoites and are one infective stage of this protozoan.

nodes, digestive organs, and brain. Sporozoites proliferate asexually to produce *bradyzoites*. Thick walls form around masses of cells filled with bradyzoites to form *pseudocysts* (4).

Toxoplasma must return to a cat to reproduce, so pseudocysts preferentially form on and inhibit those parts of a rodent's brain that process cat odors and those that induce fear. With its fear of the smell of cats gone, the rodent is easily caught and eaten (5). Bradyzoites released from digested pseudocysts infect the cat's intestinal cells where the parasites become gametes, completing the life cycle.

Nonrodent animals (and occasionally people) become accidental hosts when they ingest oocysts containing sporozoites (3). These develop into pseudocysts containing bradyzoites (A). Humans most often are infected by eating pseudocysts in undercooked meat (B).

Although the majority of people infected with *T. gondii* have no symptoms, a small percentage develop **toxoplasmosis** (tok'sō-plaz-mō-sis), a fever-producing illness with headache, muscle pain, sore throat, and enlarged lymph nodes in the head and neck. Toxoplasmosis generally results in no permanent damage and is self-limiting, resolving spontaneously within a few months to a year. However, toxoplasmosis is more severe in two populations: AIDS patients and fetuses. In AIDS patients, symptoms are thought to result when tissue pseudocysts reactivate as the immune system fails. Spastic paralysis, blindness, myocarditis, encephalitis, and death result.

Transplacental transfer of *Toxoplasma* from mother to fetus is most dangerous in the first trimester of pregnancy; it can result in spontaneous abortion, stillbirth, or epilepsy, mental retardation, *microcephaly* (abnormally small head), inflammation of the retina, blindness, anemia, jaundice, and other conditions. Ocular infections may remain dormant for years, at which time blindness develops.

Physicians diagnose via microscopic identification of parasites in tissue biopsies **(Figure 23.14)**, or via molecular identification of *T. gondii* genetic material or products in specimens using PCR, Southern blot, or DNA probes. Serology is the most common diagnostic method.

Asymptomatic patients do not need treatment for toxoplasmosis. For those with signs and symptoms, physicians prescribe one of the antimalarial drugs, which all act against

CLINICAL CASE STUDY

A Protozoan Mystery

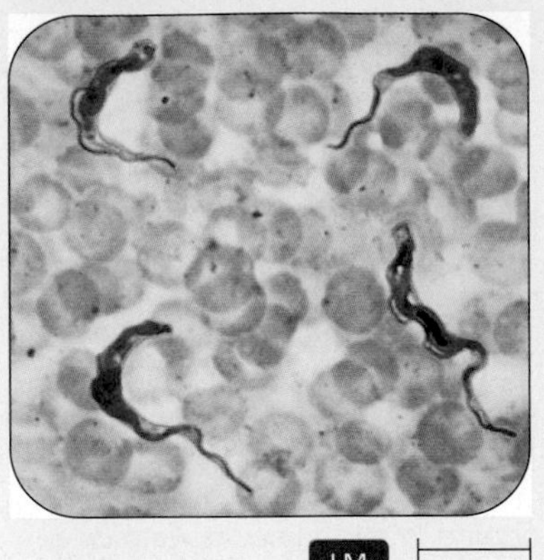

A 20-year-old student was admitted to his college's student health center with headaches and fever shortly after beginning the fall semester. He had spent his summer working with an international aid organization in Nigeria and had returned to the United States only a week earlier. Gross examination revealed a diffuse rash and numerous insect bites. The young man had spent most of his summer outdoors in rural areas, had taken no prophylactic antimalarial medication, and had spent some time on African game reserves working with the families of local guides. The patient could not specifically remember receiving any of the bite wounds on his body, and he did not always use insect repellent in the field. The young man was admitted to the local hospital, where intermittent fever, nausea, and headache continued. Initial blood smears proved negative for malaria.

1. What are some possible protozoan diseases the patient could have contracted in Africa?
2. Can this disease be identified from the symptoms alone?
3. Based on the pictured blood smear, what would you conclude about the cause of the disease?
4. What would the treatment be if the patient had tested positive for malaria?
5. What treatment would you now recommend?
6. What prevention would you have suggested to this individual?
7. Is there a local threat of anyone else contracting this disease from the young man?

Toxoplasma as well. Treatment of infected pregnant women prevents most transplacental infections. More aggressive treatment may be needed for AIDS patients, including the addition of steroids to reduce tissue inflammation.

Controlling the incidence of *T. gondii* infection is difficult because so many hosts harbor the parasites. A vaccine for cats is currently under development to reduce the chance of pet-to-owner transmission. The best prevention is to thoroughly cook or deep-freeze meats and to avoid contact with contaminated soil.

Cryptosporidium

Cryptosporidium enteritis (krip'tō-spō-rid'ē-ŭm en-ter-ī'tis), also known as *cryptosporidiosis* (krip'tō-spō-rid-ē-ō'sis), is a zoonosis; that is, it is a disease of animals that is transmitted to people. *Cryptosporidium parvum* (par-vŭm); an apicomplexan, causes the disease. Once thought to infect only livestock and poultry, *Cryptosporidium* is carried asymptomatically by about 30% of people living in developing nations. It is estimated that most natural waterways in the United States are contaminated with the oocysts of *Cryptosporidium* from livestock wastes.

Infection most commonly results from drinking water contaminated with oocysts, but direct fecal-oral transmission resulting from poor hygienic practices also occurs, particularly in day care facilities. In the intestine, oocysts release sporozoites, which invade the intestinal mucosa and become intracellular parasites. Eventually, new oocysts are produced, released into the lumen of the intestine, and shed in the feces.

Signs and symptoms of *Cryptosporidium* enteritis include severe diarrhea that lasts from one to two weeks accompanied by headache, muscular pain, cramping, and severe fluid and weight loss. In HIV-positive individuals, chronic *Cryptosporidium* enteritis is life threatening and is one of the indicator diseases revealing that a person has AIDS.

Most diagnostic techniques are not very sensitive for *Cryptosporidium,* but microscopic examination of concentrated fecal samples or biopsy material can reveal oocysts **(Figure 23.15)**.

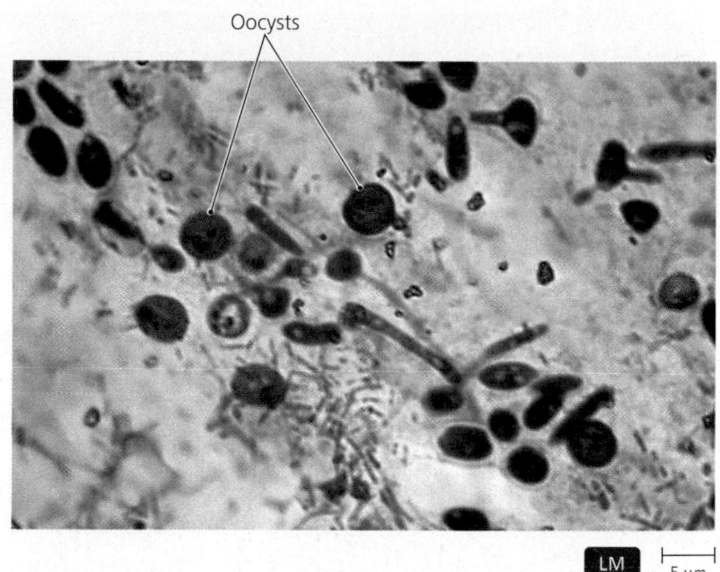

▲ **Figure 23.15 Oocysts of *Cryptosporidium parvum*.** Here they are in a stained fecal smear.

EMERGING DISEASES

BABESIOSIS

Leslie hated being sick, but here she was on a fine summer day with a headache, muscle pains, and sore joints. It was just as well she didn't know that the worst was yet to come.

A small deer tick *(Ixodes scapularis)* had taken a blood meal from Leslie's leg six weeks before. The tick was so small and the bite so painless that Leslie was unaware of the attack. She was equally unaware that the tick had infected her with an apicomplexan parasite normally found in mice. *Babesia microti* was eating her red blood cells from the inside out.

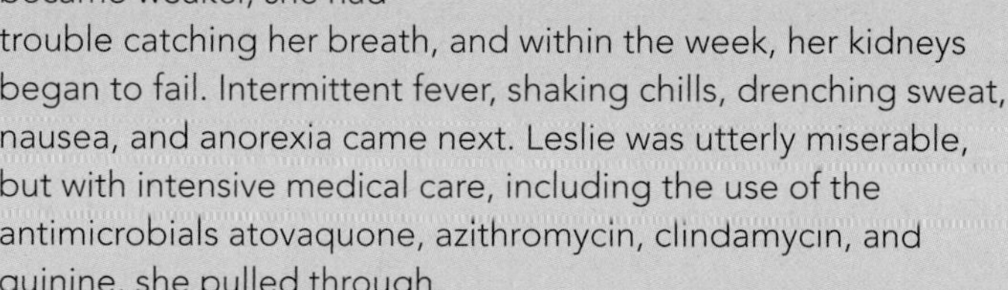

Leslie became anemic, jaundiced, fatigued, and depressed, and she was losing weight. As she became weaker, she had trouble catching her breath, and within the week, her kidneys began to fail. Intermittent fever, shaking chills, drenching sweat, nausea, and anorexia came next. Leslie was utterly miserable, but with intensive medical care, including the use of the antimicrobials atovaquone, azithromycin, clindamycin, and quinine, she pulled through.

In the last decade, the number of reported cases of babesiosis has quadrupled. Perhaps this emerging disease is spreading because its tick vector has an expanded range due to global warming. Perhaps physicians are more aware of a disease whose incidence has been constant. Whatever the reason for our increased awareness of babesiosis, Leslie was grateful that she was diagnosed and treated effectively so her nightmare could finally end.

Track babesiosis online by going to the Study Area at www.masteringmicrobiology.com.

TABLE 23.1 Key Features of Protozoan Parasites of Humans

Organism	Primary Diseases	Geographical Distribution	Mode of Transmission	Host Organisms
Ciliates				
Balantidium coli	Balantidiasis, dysentery	Worldwide	Fecal-oral	Pigs, rodents, primates, humans
Amoebae				
Entamoeba histolytica	Luminal amebiasis, amebic dysentery, invasive extraintestinal amebiasis	Worldwide	Fecal-oral	Humans
Acanthamoeba spp.	Ulcerative keratitis, amebic encephalitis	Worldwide	Contact	Humans
Naegleria	Primary amebic meningoencephalitis	Worldwide	Inhalation	Humans
Flagellates				
Trypanosoma brucei	African sleeping sickness	African subcontinent	Tsetse fly *(Glossina)*	Wild game, pigs, humans
Trypanosoma cruzi	Chagas' disease	Central and South America	Kissing bug *(Triatoma)*	Opossums, armadillos, humans
Leishmania spp.	Cutaneous, mucocutaneous, or visceral leishmaniasis	Tropics, subtropics	Sand flies *(Phlebotomus, Lutzomyia)*	Canines, rodents, humans
Giardia intestinalis (lamblia)	Giardiasis	Developed nations, tropics	Fecal-oral	Humans, wild animals
Trichomonas vaginalis	Vaginitis	Developed nations	Sexual contact	Humans
Apicomplexans				
Plasmodium spp.	Malaria	Tropics, subtropics	Mosquitoes (*Anopheles*)	Humans
Toxoplasma gondii	Toxoplasmosis	Worldwide	Fecal-oral	Cats, livestock, humans
Cryptosporidium parvum	*Cryptosporidium* enteritis (cryptosporidiosis)	Worldwide	Fecal-oral	Livestock, poultry, humans
Cyclospora cayetanensis	Cyclosporiasis, gastrointestinal disorders	North, Central, and South America	Fecal-oral	Humans

Fluorescent-labeled antibodies reveal low concentrations of oocysts in appropriate specimens.

Treatment essentially consists of oral rehydration therapy, because few drugs are effective against *Cryptosporidium.* Drinking from rivers and streams should be avoided in areas where *Cryptosporidium* is found; filtration is required to remove oocysts from drinking water because they cannot be killed by chlorination. Good personal hygiene can eliminate fecal-oral transmission of the parasite.

Cyclospora

Cyclospora cayetanensis (sī-klō-spōr′ă kī′ē-tan-en′sis) is a waterborne apicomplexan that has been responsible for an emerging disease, *cyclosporiasis,* that has affected hundreds in the United States since the early 1980s. The disease is not transmitted between individuals but acquired by eating or drinking oocysts in contaminated food or water. Outbreaks have been linked in particular to raspberries imported from Central and South America. The environmental reservoir for *Cyclospora* remains unknown.

Once *Cyclospora* enters the intestine, it invades the mucosal layer, causing symptoms, which develop after about a week. Manifestations include cramps, watery diarrhea, myalgia (muscle pain), and fever. Symptoms are more severe in AIDS patients, leading to severe dehydration and weight loss. The disease usually resolves in days or weeks in immunocompetent patients.

Microscopic examination of stool samples can sometimes reveal the presence of oocysts in extensive infections, but fluorescent DNA probes more reliably detect the presence of *Cyclospora. Cyclospora* infections are treated with trimethoprim and sulfamethoxazole given in combination for seven days.

Reliable food testing does not currently exist but would be an important tool in preventing contaminated food from reaching U.S. consumers. The only reliable methods for reducing risk of infection in the United States are thoroughly washing fruits and vegetables prior to eating them raw. Cooking or freezing also kills many oocysts.

Table 23.1 lists the general characteristics of the protozoan parasites discussed in this chapter.

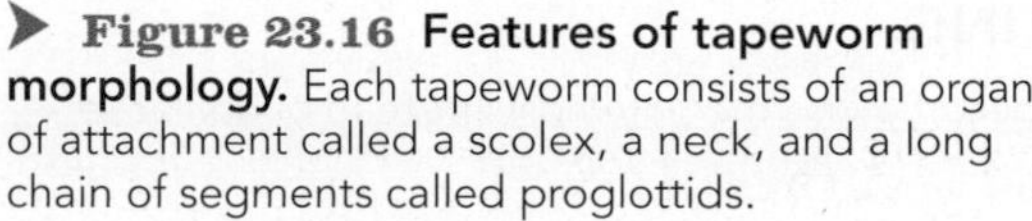
Figure 23.16 Features of tapeworm morphology. Each tapeworm consists of an organ of attachment called a scolex, a neck, and a long chain of segments called proglottids.

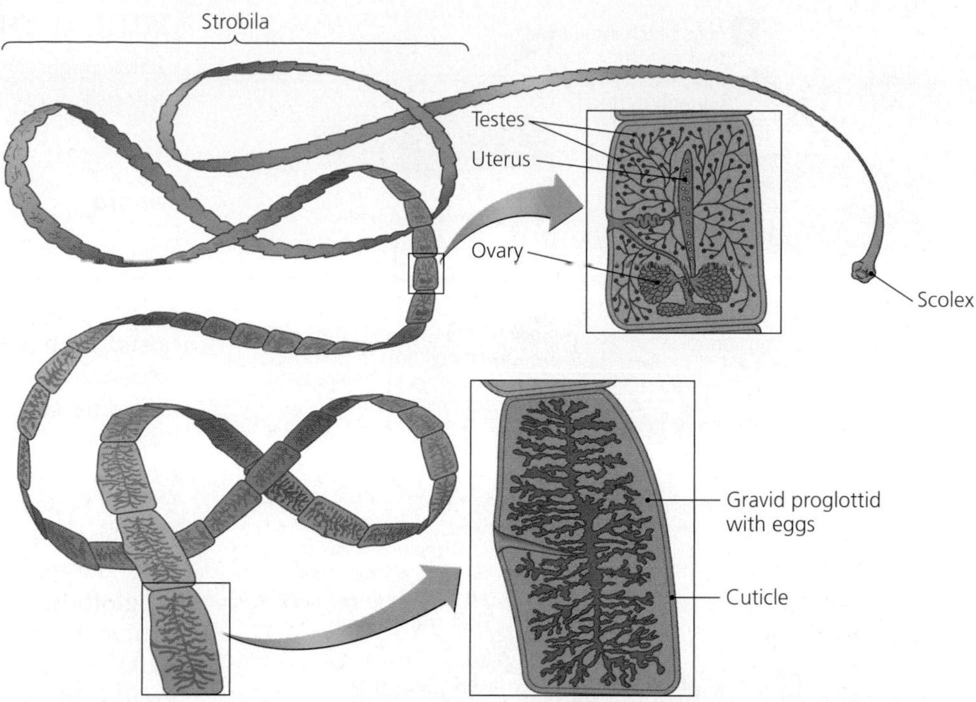

Helminthic Parasites of Humans

Learning Objective

✓ Describe the general morphological and physiological features of parasitic worms.

Helminths are macroscopic, multicellular, eukaryotic worms found throughout the natural world, some in parasitic associations with other animals. They are not microorganisms, though microbiologists generally study them, in part because the diagnostic signs of infestation—eggs or larvae—are microscopic.

Taxonomists divide parasitic helminths into three groups: *cestodes* (ses'tōdz), *trematodes* (trem'a-tōdz), and *nematodes* (nēm'a-tōdz).

The life cycles of parasitic helminths are complex. For many, intermediate hosts are required to support larval (immature) stages needed for the worms to reach maturity. Adult worms are either *dioecious*,[3] meaning that the male and female sex organs are in separate worms, or *monoecious*,[4] meaning that each worm has both sex organs. Monoecious organisms either fertilize each other during copulation or are capable of self-fertilization, depending on the species. Because most parasitic helminths release large numbers of fertilized eggs into the environment, transmission to new hosts is likely to occur.

Cestodes

Learning Objectives

✓ Describe the common features of the life cycles of tapeworms that infect humans.

✓ List the predominant modes of infection for *Taenia* and *Echinococcus*, and suggest measures to prevent infection.

All **cestodes,** commonly called *tapeworms,* are flat, segmented, intestinal parasites that completely lack digestive systems. Though they differ in size when mature, all possess the same general body plan (**Figure 23.16**). The **scolex** (skō'leks) is a small attachment organ that possesses suckers and/or hooks used to attach the worm to host tissues (see the photo at the beginning of the chapter). Anchorage is the only role of a scolex; there is no mouth. Cestodes acquire nutrients by absorption through the worm's *cuticle* (outer "skin").

Behind the scolex is the neck region. Body segments, called **proglottids** (prō-glot'idz), grow from the neck continuously, so long as the worm remains attached to its host. New proglottids displace older ones, moving the older ones farther from the neck. Proglottids mature, producing both male and female reproductive organs. Thus a chain of proglottids, called a *strobila* (strō'bi-lă; plural: *strobilae,* strō'bi-lī), reflects a sequence of development: proglottids near the neck are immature, those in the middle are mature, and those near the end are *gravid*[5]—full of fertilized eggs. Each proglottid is monoecious and may fertilize other proglottids of the same or different tapeworm; a proglottid is not capable of fertilizing itself, although the tapeworm is self-fertile.

After fertilization occurs and the proglottids fill with eggs, gravid proglottids break off the strobila and pass out of the intestine along with feces. In a few cases, the proglottids rupture within the intestine, releasing eggs directly into the feces. Some tapeworms produce proglottids large enough to be obviously visible in stools.

[3]From Greek *di,* meaning two, and *oikos,* meaning house.
[4]From Greek *mono,* meaning one, *and oikos,* meaning house.
[5]From Latin *gravidus,* meaning heavy—that is, pregnant.

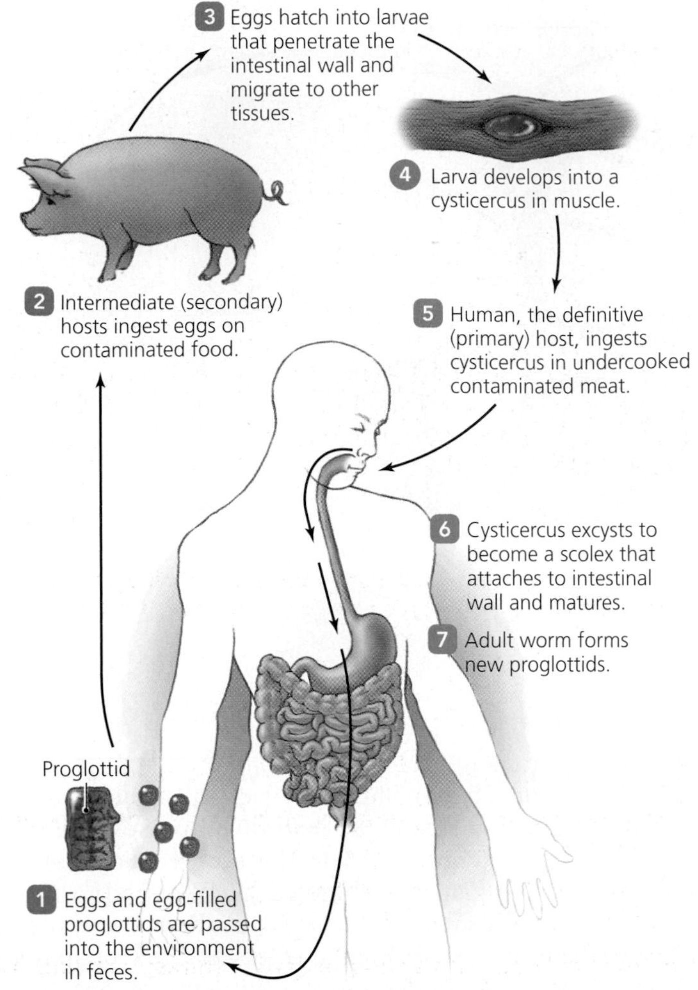

3 Eggs hatch into larvae that penetrate the intestinal wall and migrate to other tissues.
4 Larva develops into a cysticercus in muscle.
2 Intermediate (secondary) hosts ingest eggs on contaminated food.
5 Human, the definitive (primary) host, ingests cysticercus in undercooked contaminated meat.
6 Cysticercus excysts to become a scolex that attaches to intestinal wall and matures.
7 Adult worm forms new proglottids.
Proglottid
1 Eggs and egg-filled proglottids are passed into the environment in feces.

▲ **Figure 23.17 A generalized life cycle of some tapeworms of humans.** *Which species of Taenia is shown in this representation?*

Figure 23.17 Taenia solium is illustrated, as indicated by the *intermediate host being a pig.*

Generalized Tapeworm Life Cycle

The generalized life cycle of most tapeworms of humans is depicted in **Figure 23.17**. Gravid proglottids and/or eggs enter the environment in feces from infested humans, who are the definitive (primary) hosts (1). Intermediate (secondary) hosts, which vary with the species of tapeworm, become infected by ingesting vegetation contaminated with gravid proglottids or eggs (2). Eggs hatch into larvae within the intermediate host's intestine, and the larvae penetrate the intestinal wall and migrate to other tissues, often muscle (3), where the larvae develop into immature forms called **cysticerci** (sis-ti-ser'sī) (4). Cysticerci are generally harmless to intermediate hosts. Humans become infected by consuming undercooked meat containing cysticerci (5). A cysticercus excysts in the human's intestine to become a scolex that attaches to the intestinal wall and matures (6), developing into a new adult tapeworm (7), which eventually sheds gravid proglottids, completing the cycle.

CRITICAL **THINKING**

How might humans become intermediate hosts instead of definitive hosts?

Taenia

Taenia saginata (te'ne-a sa-ji-na'ta), the beef tapeworm, and *Taenia solium* (so'lī-um), the pork tapeworm, are so named because cattle and swine, respectively, serve as intermediate hosts. Both tapeworm species are distributed worldwide in areas where beef and pork are eaten. Poor, rural areas with inadequate sewage treatment and where humans and livestock live in close proximity have the highest incidence of human infection. Overall, *Taenia* infections in humans are rare in the United States.

Tapeworms attached to the intestinal epithelium can grow quite large. At maturity, *T. saginata* worms have 1000–2000 proglottids, each of which can contain 100,000 eggs. *T. solium* adults average 1000 proglottids, each of which contains about 50,000 eggs. An infested human passes strobilae of about six proglottids per day.

Rarely, humans become intermediate hosts of *T. solium* when they ingest eggs or gravid proglottids rather than cysticerci. Larvae released from the eggs become cysticerci in the human, who is thereby an accidental intermediate host. For the parasite this is a dead end. Humans are not intermediate hosts for *T. saginata*.

Most people actively shed strobilae without experiencing symptoms, though in some cases, nausea, abdominal pain, weight loss, and diarrhea accompany infestation. If the tapeworm is particularly large, blockage of the intestine is possible.

Diagnosis is achieved by microscopic identification of proglottids in fecal samples at least three months after infection. Examination of the scolex is required to differentiate between *T. saginata* and *T. solium*.

Normally, treatment is with praziquantel. Thoroughly cooking or freezing meat is the easiest method of prevention. Because cysticerci are readily visible in meat, giving it a "mealy" look, inspecting meat can reduce the chances of eating infected meat.

Echinococcus

Echinococcus granulosus (ĕ-kī'nō-kok'ŭs gra-nū-lō'sŭs) is an unusual tapeworm of canines in that its body consists of only three proglottids—one immature, one sexually mature, and one gravid. A gravid proglottid is released each time the neck forms a new immature proglottid. Canines are infected by eating cysticerci in various herbivorous hosts such as cattle, sheep, and deer. Incidence is highest in areas wherever livestock are kept, including the western United States.

Humans become accidental intermediate hosts by consuming food or water contaminated with *Echinococcus* eggs shed in dogs' feces. The eggs release larvae into the intestine; the larvae invade the circulatory system and are carried throughout the

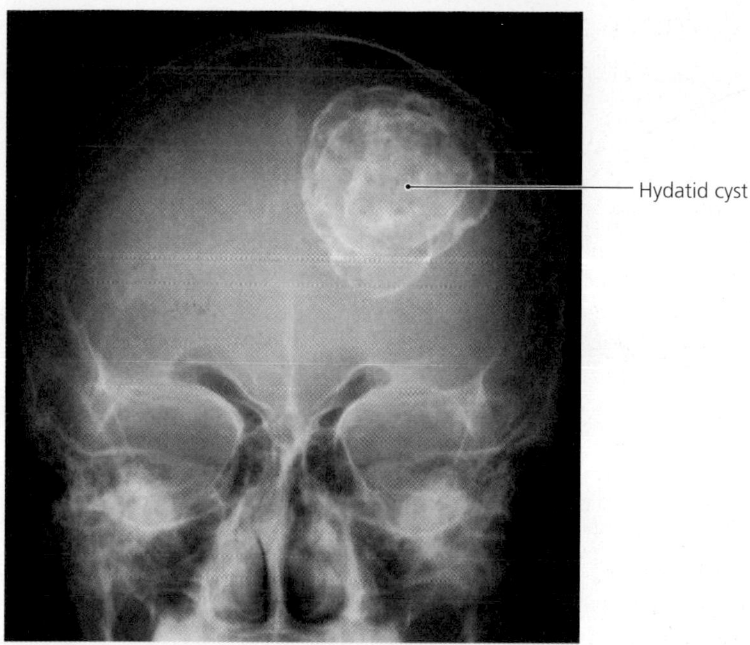

▲ **Figure 23.18 X ray of hydatid cyst.** This large, calcified cyst is in the cerebrum of a 20-year-old man.

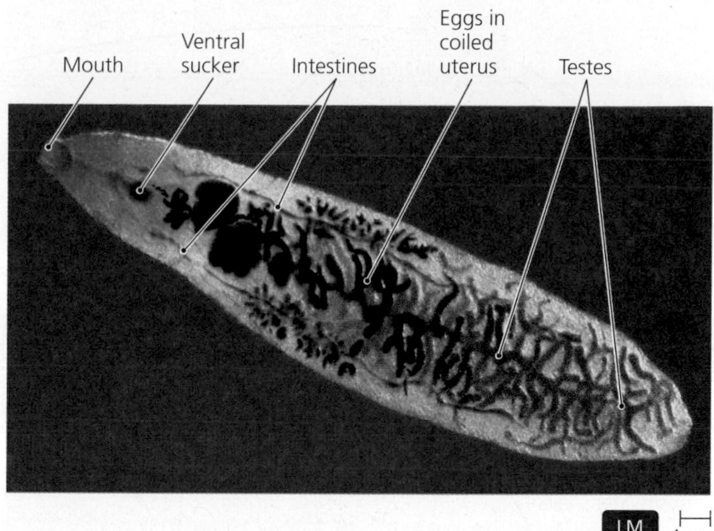

▲ **Figure 23.19 Some features of fluke morphology.** Note the sucker, which keeps the leaf-shaped flatworm attached to its host's tissues.

body, where they form **hydatid**[6] (hī′da-tid) cysts and cause **hydatid disease.** Hydatid cysts form in any organ but occur primarily in the liver.

Over several years, hydatid cysts can calcify and enlarge to the size of grapefruits (20 cm in diameter), filling with fluid and several million granular *protoscoleces*. Each protoscolex is the potential scolex of a new worm, but since dogs do not normally eat humans, the protoscoleces of hydatid cysts in humans are a dead end for *Echinococcus*.

Symptoms of hydatid disease follow the enlargement of cysts in infected tissue and result from tissue dysfunction. For example, hydatid cysts in the liver produce abdominal pain and obstruction of the bile ducts. If a hydatid cyst ruptures, protoscoleces spread throughout the body via the bloodstream, and each protoscolex grows into a new hydatid cyst. Hydatid cysts in the brain, lungs, bones, heart, or kidneys can be fatal.

Diagnosis involves visualization of cyst masses using ultrasound, CT scan, or radiography (**Figure 23.18**), followed by serological confirmation via immunoassays. Biopsies should be avoided as rupture of a cyst can lead to the spread of protoscoleces.

Treatment involves surgery to remove cysts, followed by the anthelmintic drug albendazole. Human infection is prevented by good hygiene practices to avoid fecal-oral transmission of eggs from infected dogs. It is also a good idea to avoid drinking untreated water from streams or other waterways in areas where foxes, coyotes, or wolves roam.

[6]From Greek *hydatis*, meaning a drop of water, presumably in reference to the fluid that accumulates within hydatid cysts.

Trematodes

Learning Objectives

- ✓ Explain how the life cycles of blood flukes differ from those of other flukes.
- ✓ Discuss the life cycles and disease manifestations of *Schistosoma* and *Fasciola*.

Trematodes, commonly known as *flukes* (flūks), are flat, leaf-shaped worms (**Figure 23.19**). A fluke has no anus and so has an incomplete digestive tract. A ventral sucker enables a parasite to attach to host tissues and obtain nutrients. The geographical distribution of flukes is extremely limited by the geographical distribution of the specific species of snails they require as intermediate hosts.

Researchers group flukes together according to the sites in the body they parasitize. The following sections examine some representative flukes, but first we consider common features of fluke life cycles.

Common Features of the Fluke Life Cycle

Flukes that cause disease in humans share similar complex life cycles (**Figure 23.20**). Fluke eggs pass from the body in feces or urine, depending on the site of infestation (**1**). Eggs deposited in freshwater hatch to release free-swimming larvae called *miracidia* (mir a-sid′ē-a) (**2**). Within 24 hours, miracidia actively seek out and burrow into a snail of a specific species (**3**). Two or three cycles of asexual reproduction within the snail produce larvae called *cercariae* (ser-kār′ē-ī) (**4**), which leave the snail as free-swimming forms. The next steps differ depending on whether the parasite is a blood, liver, lung, or intestinal fluke. In blood flukes, the cercariae aggressively seek out and penetrate the skin of humans (**5a**); in some intestinal and liver flukes,

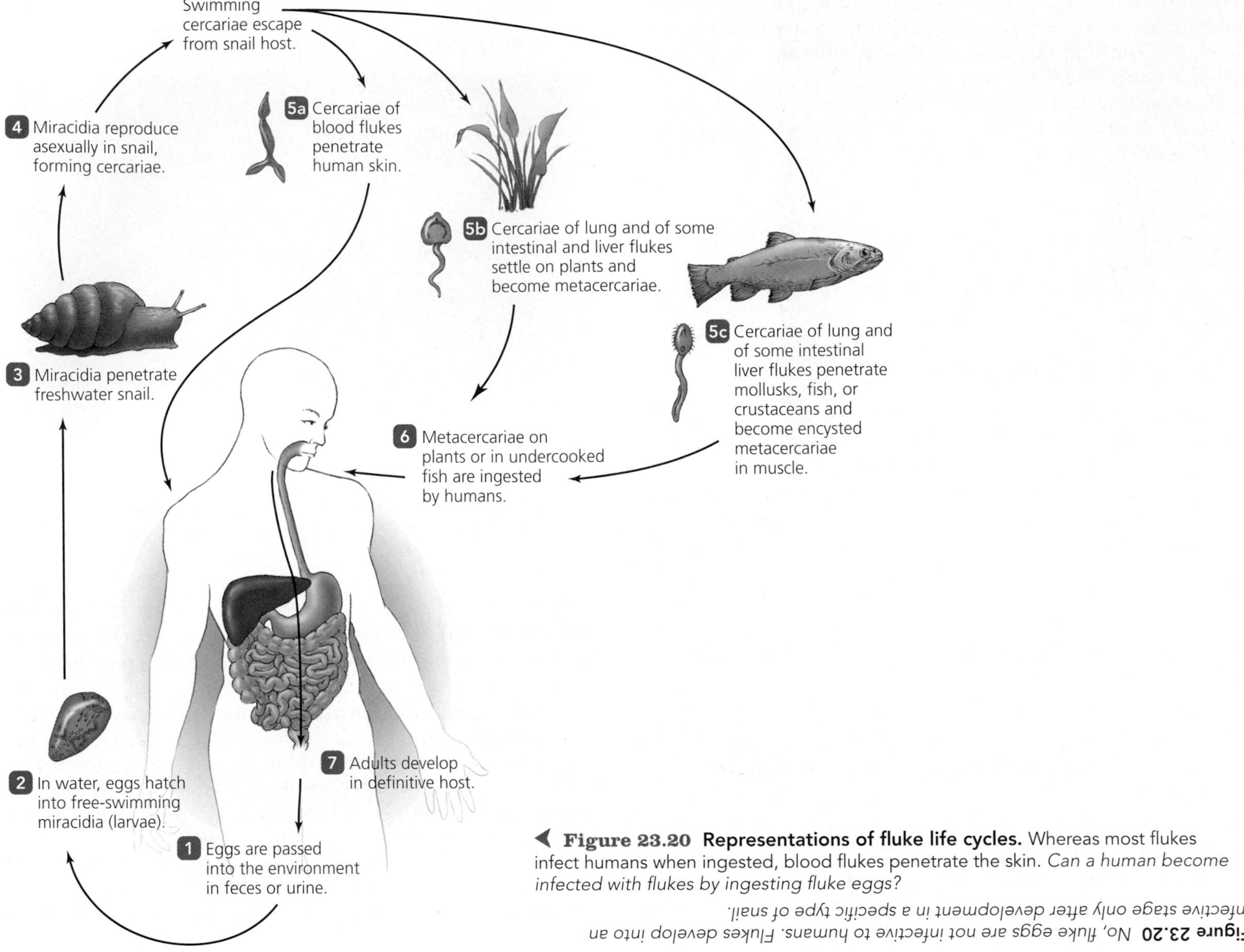
Swimming cercariae escape from snail host.
4 Miracidia reproduce asexually in snail, forming cercariae.
5a Cercariae of blood flukes penetrate human skin.
5b Cercariae of lung and of some intestinal and liver flukes settle on plants and become metacercariae.
5c Cercariae of lung and of some intestinal liver flukes penetrate mollusks, fish, or crustaceans and become encysted metacercariae in muscle.
3 Miracidia penetrate freshwater snail.
6 Metacercariae on plants or in undercooked fish are ingested by humans.
7 Adults develop in definitive host.
2 In water, eggs hatch into free-swimming miracidia (larvae).
1 Eggs are passed into the environment in feces or urine.

Figure 23.20 Representations of fluke life cycles. Whereas most flukes infect humans when ingested, blood flukes penetrate the skin. *Can a human become infected with flukes by ingesting fluke eggs?*

Figure 23.20 *No, fluke eggs are not infective to humans. Flukes develop into an infective stage only after development in a specific type of snail.*

cercariae encyst to become *metacercariae* on vegetation (5b); while in lung flukes and other intestinal and liver flukes, cercariae encyst to become metacercariae within a second intermediate host (5c). For liver, lung, and intestinal flukes, humans become infected by ingesting metacercariae (6). Once inside the definitive host, the parasites migrate to appropriate sites and develop into adults (7), which then produce perhaps 25,000 eggs per day that pass into the environment in feces or urine.

With the exception of one dioecious genus *(Schistosoma)*, all flukes are monoecious. They are not self-fertile.

Now, we consider examples of specific species of flukes, beginning with *Fasciola*, a liver fluke.

CRITICAL **THINKING**

Compare the general tapeworm life cycle to the general fluke life cycle. What is similar? What is different?

Representative Liver Fluke: *Fasciola*

Two species of liver fluke—*Fasciola hepatica* (fa-sē′ō-lă he-pa′-ti-kă) and *F. gigantica* (ji-gan′ti-kă)—infect sheep and cattle worldwide. *Fasciola* also can infect humans, who become accidental definitive hosts when they ingest metacercariae encysted on aquatic vegetation such as watercress. Following excystment in the intestine, the larvae burrow through the intestinal wall, the peritoneal cavity, or the liver to reach the bile ducts, where they mature within three to four months.

Acute disease characterized by tissue death, abdominal pain, fever, nausea, vomiting, and diarrhea accompanies the migration of the parasite from the intestine to the liver; chronic infection begins when flukes take up residence in the bile ducts. Symptoms coincide with episodes of bile duct obstruction and inflammation, and heavy infections cause liver failure. Lesions can occur in other tissues as well.

Diagnosis involves the correlation of symptoms and patient history with microscopic identification of eggs in fecal matter, although *Fasciola* eggs are not always readily distinguishable from those of other intestinal and liver flukes, and symptoms alone mimic other liver disorders. However, liver failure following the consumption of watercress or other raw vegetables grown in endemic areas is suggestive.

Treatment is dependent on accurate diagnosis, as *F. hepatica* does not respond to praziquantel as other flukes do. Either triclabendazole or bithionol is effective.

CRITICAL **THINKING**

Thoroughly cooking raw vegetables prior to consumption would kill the metacercariae of *Fasciola* and prevent infection. Why is this method of prevention not practical?

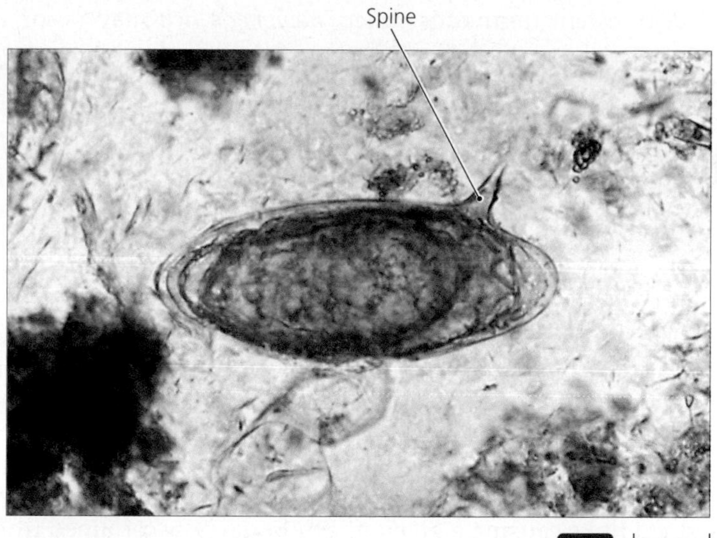

▲ **Figure 23.21**
An egg of *Schistosoma mansoni*, from a stool sample.
Distinctive spines characterize the eggs of blood flukes.

Blood Flukes: *Schistosoma*

Blood flukes in the genus *Schistosoma* (skis-tō-sō′mă) are dioecious and cause **schistosomiasis** (skis′tō-sō-mī′ă-sis)—a potentially fatal disease and one of the major public health problems in the world. The World Health Organization estimates that over 200 million people are infected worldwide, of which 20 million suffer serious disease and about 20,000 die annually. Most cases of schistosomiasis occur in sub-Saharan Africa, but *Schistosoma* is endemic in 76 countries. (See **Highlight: Snail Fever Reemerges in China.**)

Three geographically limited species of *Schistosoma* infect humans:

- *S. mansoni* (man-sō′nē) is common to the Caribbean, Venezuela, Brazil, Arabia, and large areas of Africa.
- *S. haematobium* (hē′mă-tō′bē-ŭm) is found only in Africa and India.
- *S. japonicum* (jă-pon′i-kŭm) occurs in China, Taiwan, the Philippines, and Japan, although infections in Japan are relatively rare.

Humans are the principal definitive host for most species of *Schistosoma*. Cercariae burrow through the skin of humans who contact contaminated water while washing clothes and utensils, bathing, or swimming. The larvae enter the circulatory system, where they mature and mate. Females lay eggs, which have distinctive spines **(Figure 23.21)**, in the walls of the blood vessels. The eggs work their way to the lumen of the intestine (*S. mansoni* and *S. japonicum*) or lumens of the urinary bladder and ureters (*S. haematobium*) to be eliminated into the environment.

HIGHLIGHT

SNAIL FEVER REEMERGES IN CHINA

In China, schistosomiasis is known mainly as "snail fever," since the blood fluke that causes the disease commonly breeds in snails there. The snails live in lakes all over China, and the parasite is transmitted to humans when skin comes into contact with infested water. The flukes travel through the bloodstream and attack the liver, pancreas, and stomach. People are often slow to realize they have the condition; chronic symptoms include lethargy, high fevers, swollen stomachs (as shown), and a lingering and painful death in some cases.

Many of the victims live in villages around lakes and are farmers and fishermen; thus, avoiding contact with infested water is almost impossible. Infection rates in some villages is as high as 80%. The disease was nearly eradicated in the 1950s, when the government regularly swept lakes of snails, but with the waning of such controls today—coupled with frequent floods—the disease is reemerging in rural areas and even extending into urban regions. Nearly 900,000 people now have the disease in China, and an estimated 30 million more are at risk. With proper medication, schistosomiasis can be kept under control; however, many Chinese cannot afford treatment, complicating the challenge China faces in curbing the spread of this disease.

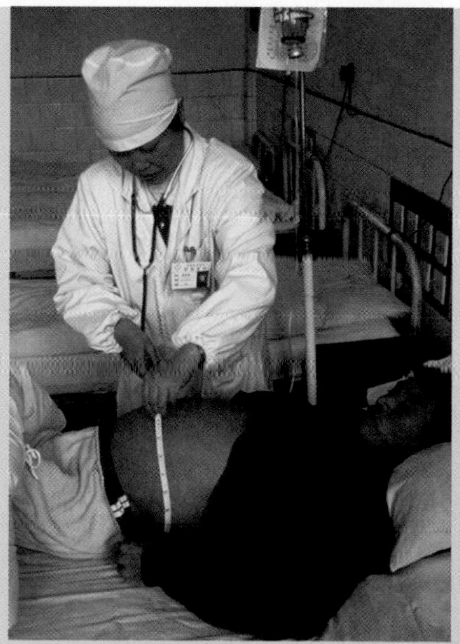

A transient dermatitis called *swimmer's itch* may occur in the area where cercariae burrow into the skin. Migration through the circulatory system is generally asymptomatic.

S. mansoni and *S. japonicum* infections become chronic when eggs lodge in the liver, lungs, brain, or other organs. Trapped eggs die and calcify in the liver, leading to tissue damage that is generally fatal. In *S. haematobium* infections, movement of eggs into the bladder and ureters results in blood in the urine, blockage due to long-term fibrosis and calcification and, in some geographical regions, fatal bladder cancer.

The number of cases of schistosomiasis has increased over the past few decades as economic stability in endemic regions has allowed improved irrigation systems and water reservoirs that provide more habitat for intermediate host snails.

Diagnosis is most effectively made by microscopic identification of spiny eggs in either stool or urine samples. The species of blood fluke causing a given infection can be ascertained from the shape of the egg and the location of its spine. If eggs are not seen but infection is suspected, immunological assays can be used to identify antigen.

The drug of choice for treatment of schistosomiasis is praziquantel. Prevention of infection depends on improved sanitation, particularly sewage treatment, and avoiding contact with contaminated water. A recombinant vaccine for *S. mansoni* is currently in clinical trials.

CRITICAL **THINKING**

Propose some additional methods that could be used to prevent transmission of *Schistosoma* infections to humans.

Nematodes

Learning Objectives

- ✓ Describe the common characteristics of nematodes.
- ✓ Compare and contrast the three most common nematode infections of humans.
- ✓ Discuss the life cycle of filarial nematodes, and contrast it with the life cycles of intestinal nematodes.

Nematodes, or roundworms, are long, cylindrical worms that taper at each end, possess complete digestive tracts, and have a protective outer layer called a *cuticle.*

Features of the Life Cycle of Roundworms

Nematodes are highly successful parasites of almost all vertebrates, including humans. They have a number of reproductive strategies:

- Most intestinal nematodes shed their eggs into the lumen of the intestine, where they are eliminated with the feces. The eggs are then consumed by the definitive host either in contaminated food (particularly raw fruits and vegetables that grow close to the soil) or in contaminated drinking water.
- For a few intestinal nematodes, larvae hatch in the soil and actively penetrate the skin of new hosts. Once in the body, they travel a circuitous route (discussed in a later section) to the intestine.
- Other nematodes encyst in muscle tissue and are consumed in raw or undercooked meat.
- Mosquitoes transmit a few species of nematodes, called *filarial nematodes,* among hosts.

All nematodes are dioecious and develop through four larval stages either within eggs, in intermediate hosts, or in the environment before becoming adults. Adult, sexually mature stages are found only in definitive hosts. Copulation of male and female worms leads to the production of fertilized eggs and perpetuation of the cycle.

We examine representative intestinal and filarial nematodes in the following sections. We begin with the largest nematode parasite of humans, *Ascaris lumbricoides.*

Ascaris

Ascaris lumbricoides (as'kă-ris lŭm'bri-koy'dez), the causative agent of **ascariasis** (as-kă-rī'ă-sis), is the most common nematode infection of humans worldwide, infecting approximately 1 billion people, primarily in tropical and subtropical regions. *Ascaris* is also the largest nematode to infect humans, growing as large as 30 cm in length. The nematode is endemic in rural areas of the southeastern United States.

Adult worms grow and reproduce in the small intestine, where females produce about 200,000 eggs every day. Eggs passed with feces can remain viable for years in moist, warm soil, where an embryo develops within each egg. After eggs are ingested in water or on vegetables, the larvae invade the intestinal wall to enter the circulatory or lymphatic systems. From there the larvae reach the lungs, where they develop through two more larval stages in approximately two weeks. The worms are then coughed up into the pharynx and swallowed. The parasites subsequently develop into adults in the intestine.

Most infections are asymptomatic, though if the worm burden (number of worms) is high, intestinal symptoms and signs can include abdominal pain, nausea, vomiting, and complete intestinal obstruction, which may be fatal. Transitory fever and pulmonary symptoms, including dry cough, difficulty in breathing, and bloody sputum, may occur during larval migration.

Diagnosis is made by the microscopic identification of eggs in the stool, larvae in sputum, or (rarely) by identification of adult worms passed in the stool or exiting via the nose or mouth—a distinctly unpleasant experience with larger worms **(Figure 23.22)**. The treatment of *Ascaris* with albendazole, mebendazole, or pyrantel pamoate over one to three days is 90% effective. Surgery may be required to alleviate intestinal obstruction. Proper sanitation and hygiene, including treatment of sewage and drinking water, are important for prevention. Good personal hygiene and cooking practices are also important methods of preventing infection in endemic areas.

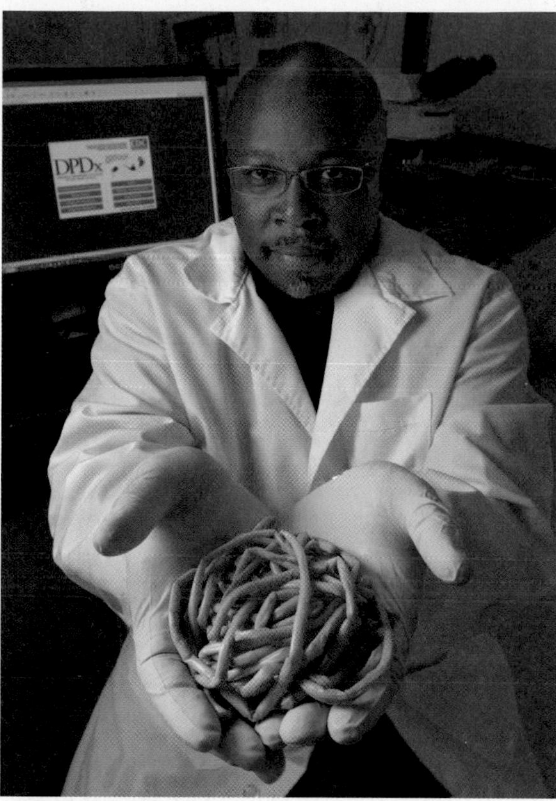

▲ **Figure 23.22** ***Ascaris lumbricoides.*** This mass of worms passed from a child in Kenya, Africa. Infestation also occurs in the southeastern United States.

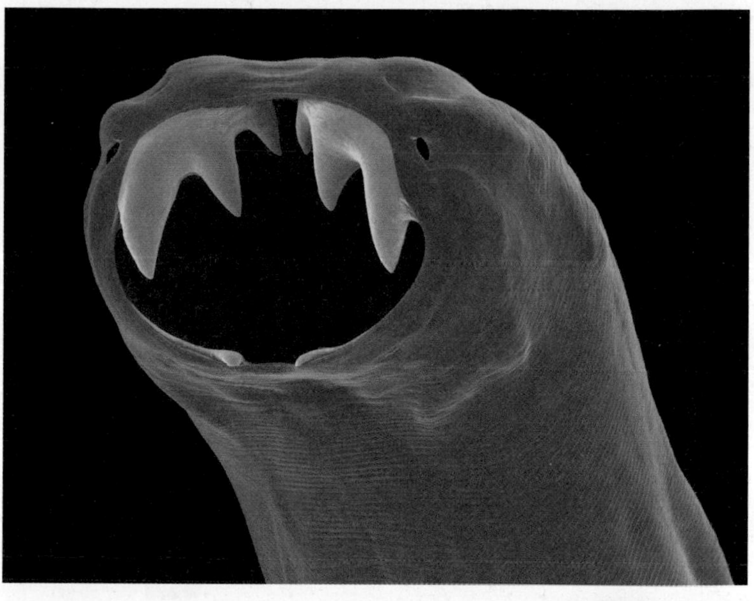

▲ **Figure 23.23 Mouth of a hookworm.** The sharp teeth allow the parasite to attach to the intestinal wall and feed on blood.

Ancylostoma and *Necator*

Hookworms—so called because adults resemble shepherd's hooks—are the second most common nematode infecting humans, responsible for more than 600 million infections worldwide, sucking the equivalent of all the blood from 1.5 million people per day. Two hookworms infect humans. *Ancylostoma duodenale* (an-si-los'tō-mă doo'ō-de-nā'lē) is distributed throughout Africa, Asia, the Americas, the Middle East, North Africa, and southern Europe. *Necator americanus* (nē-kā'tor ă-mer-i-ka'nŭs) predominates in the Americas and Australia but can be found in Asia and Africa as well.

Eggs passed in the stools of infected humans are deposited in soil, where they hatch in a couple of days if the soil is warm and moist. The initial larvae develop over a span of 5–10 days into infective larvae that can survive up to five weeks in soil. The larvae burrow through human skin and are carried by the circulatory system to the heart and the lungs. In the lungs, larvae burrow into the air cavity and migrate up the trachea to the esophagus, where they are swallowed. Larvae attach to the small intestine, mature into adults, and mate. Most adults survive in the intestine for one to two years or longer.

Adult worms have ghastly looking mouths (**Figure 23.23**) and suck the blood of their hosts, consuming as much as 150 μl of blood per day per worm. This results in chronic anemia, iron deficiency, and protein deficiency.

Itching, rash, and inflammation that can persist for several weeks characterize *ground itch,* which occurs at the site of skin penetration by larvae. Ground itch is generally worse with successive infections. Pulmonary symptoms can also occur.

Microscopic identification of eggs in the stool along with anemia and blood in the feces is diagnostic.

Treatment is with albendazole, mebendazole, or pyrantel pamoate for several days. *N. americanus* is more difficult to treat effectively. Proper sanitation (sewage treatment) is essential for prevention. In areas where hookworm is endemic, going barefoot should be avoided to prevent exposure.

Enterobius

Enterobius vermicularis (en-ter-ō'bī-ŭs ver-mi-kū-lar'is) infects about 500 million people worldwide, particularly in temperate climates, in school-age children, and in conditions of overcrowding. *Enterobius* is the most common parasitic worm found in the United States, affecting 40 million Americans. It is commonly known as the pinworm, after the shape of the female worm's tail. Humans are the only host for *Enterobius*.

After mating in the colon, female pinworms migrate at night to the anus, where they deposit eggs perianally. Scratching dislodges eggs onto clothes or bedding, where they dry out, become aerosolized, and settle in water or on food that is then ingested. Alternatively, scratching deposits eggs on the skin and under the fingernails, such that infected individuals can continually reinfect themselves by ingesting the eggs on their hands or in food. Adult worms mature in several weeks and live in the intestinal tract for approximately two months.

One-third of all *Enterobius* infections are asymptomatic. When symptoms do occur, intense perianal itching is the chief complaint. Scratching can lead to secondary bacterial infections.

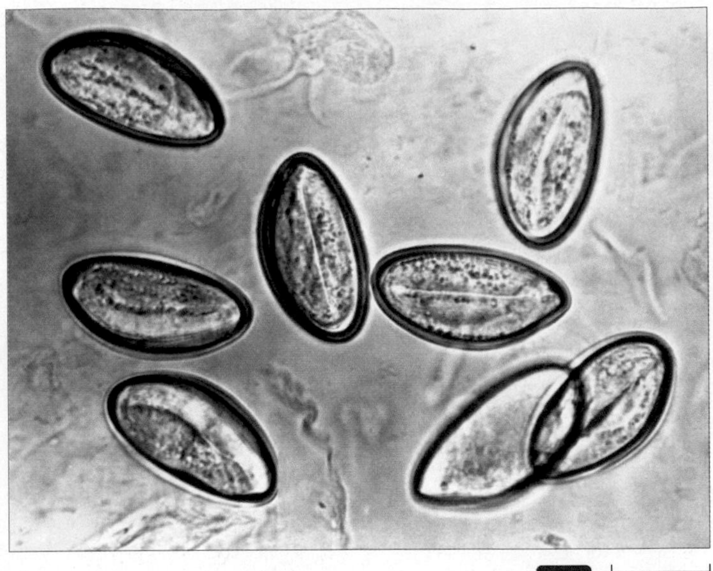

▲ **Figure 23.24 Eggs of *Enterobius vermicularis.*** Note the characteristic flattening of one side of each egg.

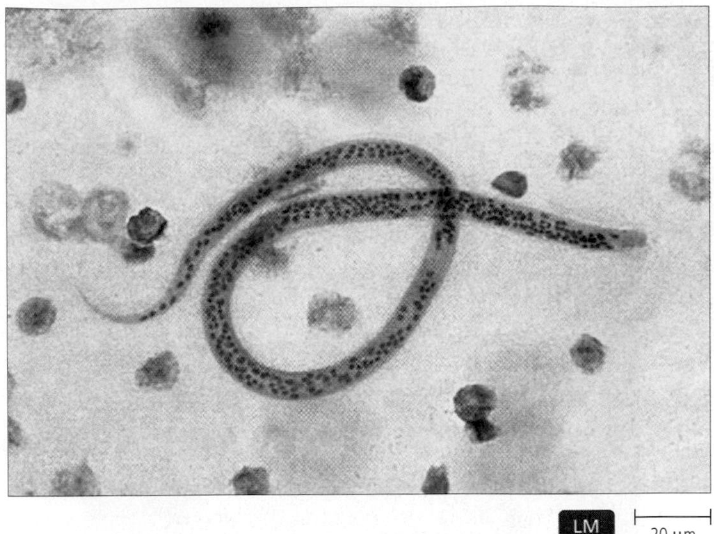

▲ **Figure 23.25 A microfilaria of *Wuchereria bancrofti* in blood.**

CLINICAL CASE STUDY

A Fluke Disease?

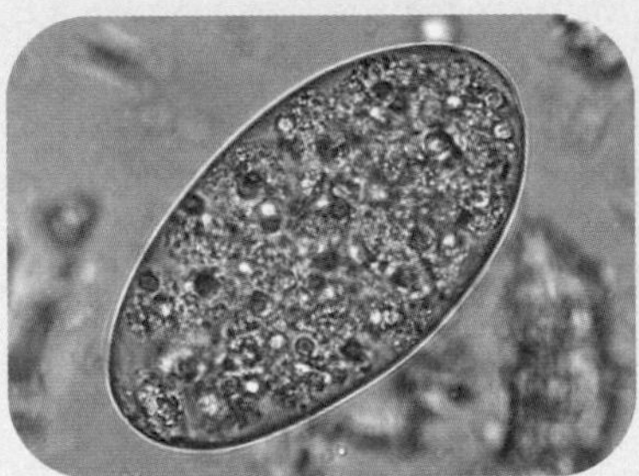

A 40-year-old female of Cambodian descent was admitted to the hospital with fatigue, constipation, weight loss, abdominal pain, and abdominal swelling. Before the onset of symptoms, the woman was healthy, as are the majority of her close relatives. Laboratory testing indicated signs of liver dysfunction, prompting a CT scan. Multiple hepatic lesions and inflammation of the bile ducts were evident. Stool samples examined over several successive days revealed the presence of numerous eggs. The woman had visited Cambodia recently, but her symptoms had begun before she traveled there.

1. What type of patient history would you take? Why is this necessary and important?
2. What would you do to ascertain the parasite causing the woman's illness?
3. What is the treatment for this condition?
4. What type of preventive measures could have kept this woman disease free?
5. Are this woman's relatives at risk of contracting the disease from her?

Rarely, worms enter a female's genital tract, where they can cause inflammation of the vulva.

Microscopy is used for diagnosis. In the morning, before bathing or defecation, transparent sticky tape is applied to the perianal area to collect the readily identifiable microscopic eggs **(Figure 23.24)**. Adult worms, if recovered, are also diagnostic.

Treatment is with an initial dose of pyrantel pamoate or mebendazole followed by a second treatment two weeks later to kill any newly acquired worms. Prevention of reinfection and spread to family members requires thorough laundering of all clothes and bedding of infected individuals. Further, infected individuals should not handle food to be consumed by others.

Wuchereria

Wuchereria bancrofti (voo-ker-e'rē-ă ban-krof'tē) is a *filarial nematode*—a type of nematode that does not infect the intestinal tract of vertebrate hosts but rather the lymphatic system, causing **filariasis** (fil-ă-rī'ă-sis). *W. bancrofti* infects the lymphatic system of about 120 million people throughout the tropics of Africa, India, Southeast Asia, Indonesia, the Pacific Islands, South America, and the Caribbean. Female mosquitoes of the genera *Culex* (kyu'leks), *Aedes* (ā-ē'dēz), and *Anopheles* transmit the worm.

Mosquitoes ingest circulating larvae called *microfilariae* (mī'krō-fi-lar'ē-ī) while feeding on humans **(Figure 23.25)**. The microfilariae develop as larvae in the mosquito and eventually migrate to the salivary glands. When the mosquito next feeds, parasites are injected into a new person.

Larvae migrate via the circulatory system to deeper tissues, where they mature. Adults live and reproduce for up to 17 years in lymphatic vessels. During the day, microfilariae stay in capillaries of internal organs; they swim freely in the bloodstream only at night, coinciding with the feeding habits of most mosquitoes.

Filariasis remains asymptomatic for years. As the disease progresses, lymphatic damage occurs—subcutaneous tissues swell grotesquely because blocked lymphatic vessels cannot drain properly. The end result is **elephantiasis** (el-ĕ-fan-tī'ă-sis), generally in the lower extremities, in which tissues enlarge and harden in areas where lymph has accumulated (**Figure 23.26**). Elephantiasis can be further associated with secondary bacterial infections in affected portions of the body.

Diagnosis is made by the microscopic identification of microfilariae in the blood. Blood samples are collected at night, when microfilariae circulate. In addition to microscopy, immunoassays using antibodies to detect *W. bancrofti* antigens in the blood can be used for diagnosis.

Treatment with diethylcarbamazine effectively kills microfilariae and adult worms.

Prevention relies on using insect repellents or mosquito netting, wearing loose, long, light-colored clothing, or remaining indoors when mosquitoes are most active. In endemic areas, particularly urban areas, eliminating standing water (a requirement of mosquitoes' life cycle) around human dwellings decreases mosquito numbers. Widespread spraying with insecticides to kill mosquitoes has reduced the prevalence of filariasis.

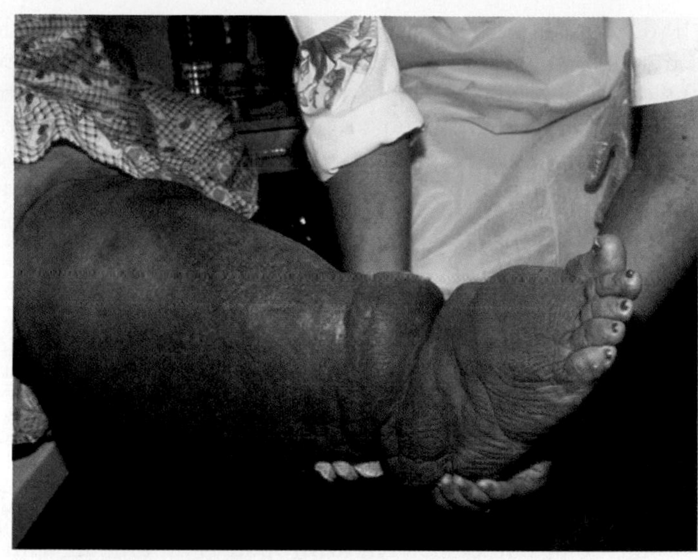

▲ **Figure 23.26 Elephantiasis in a leg.** This condition follows years of infection with *Wuchereria bancrofti. Adult* Wuchereria *are found in what human body system?*

Figure 23.26 Adults of Wuchereria live in the lymphatic system.

Table 23.2 summarizes the key features of the common helminthic parasites of humans.

TABLE 23.2 Key Features of Representative Helminthic Parasites of Humans

Organism	Primary Infection or Disease	Geographical Distribution	Mode of Transmission	Length of Adult Worms
Cestodes (Tapeworms)				
Taenia saginata; Taenia solium	Beef tapeworm infestation; pork tapeworm infestation	Worldwide with local endemic areas	Consumption of undercooked meat	*T. saginata:* 5–25 m; *T. solium:* 2–7 m
Echinococcus granulosus	Hydatid disease	Worldwide with local endemic areas	Consumption of eggs shed in dog feces	3–6 mm
Trematodes (Flukes)				
Fasciola hepatica; F. gigantica	Fascioliasis	*F. hepatica*: Europe, Middle East, Asia; *F. gigantica*: Asia, Africa, Hawaii	Consumption of watercress or lettuce	Up to 30 mm; up to 75 mm
Schistosoma spp.	Schistosomiasis	*S. mansoni*: Caribbean, S. America, Arabia, Africa; *S. haematobium*: Africa, India; *S. japonicum*: east Asia	Direct penetration of the skin	7–20 mm
Nematodes (Roundworms)				
Ascaris lumbricoides	Ascariasis	Tropics and subtropics worldwide	Fecal-oral	Females: 20–35 cm; males: 15–30 cm
Ancylostoma duodenale, Necator americanus	Hookworm disease	*Ancylostoma*: Africa, Asia, the Americas, Middle East, North Africa, southern Europe; *Necator*: the Americas, Australia, Asia, Africa	Direct penetration of the skin	*Ancylostoma* females: 10–13 mm, males: 8–11 mm; *Necator* females: 9–11 mm, males: 7–9 mm
Enterobius vermicularis	Pinworm	Worldwide	Anal-oral and fecal-oral, inhalation	Females: 8–13 mm; males: 2–5 mm
Wuchereria bancrofti	Filariasis, elephantiasis	Worldwide, tropics	Mosquitoes	Females: 80–100 mm; males: 40 mm

Protozoan and helminthic diseases are found worldwide, and much of the world's population is at risk of infection. Many of the drugs used to treat parasitic diseases have been around for decades, and whereas most remain effective, resistant parasites are occurring with greater frequency. No effective vaccines have been produced for any of the diseases discussed in this chapter. The only way to protect oneself, therefore, is to avoid exposure.

Arthropod Vectors

Learning Objective

✓ List the principal arthropod vectors of human pathogens, and give one example of a pathogen transmitted by each.

Vectors are animals—usually *arthropods* (ar'thrō-podz)—that carry microbial pathogens. An **arthropod** is an animal with a segmented body, a hard exoskeleton (external skeleton), and jointed legs. Arthropods are extremely diverse and abundant—insects alone account for more than half of all the known species on Earth.

Some arthropods are *biological vectors,* meaning that they also serve as hosts for the pathogens they transmit. Given that arthropods are usually small organisms (so small that we don't notice them until they bite us), and given that they produce large numbers of offspring, controlling arthropod vectors to eliminate their role in the transmission of important human diseases is an almost insurmountable task.

Disease vectors belong to two classes of arthropods: *Arachnida* (ticks and mites) and the more common *Insecta* (fleas, lice, flies, and true bugs) (see Figure 12.33). Mosquitoes—a type of fly—are the most important vector of human diseases. Most arthropod vectors are found on a host only when they are actively feeding. Lice are the only vectors that may spend their entire lives in association with a single individual.

Chapter 12 discusses vectors in more detail.

Chapter Summary

1. **Parasitology** is the study of protozoan and helminthic parasites of animals and humans. All parasites have a **definitive host** in which adult or sexual stages reproduce, and many parasites also have one or more **intermediate hosts,** in which immature stages live.

Protozoan Parasites of Humans (pp. 652–666)

1. Many protozoa have two life stages: a trophozoite (feeding) stage and a dormant cyst stage. **Encystment** is the process of cyst formation; **excystment** refers to the process by which cysts become trophozoites.
2. *Balantidium coli* is the only **ciliate** to cause disease in humans. It is responsible for **balantidiasis,** usually a mild gastrointestinal illness.
3. **Amoebae** are protozoa that move and feed using pseudopodia. *Entamoeba histolytica* is a relatively common protozoan parasite that causes **amebiasis,** which can be mild luminal amebiasis, more severe amebic dysentery, or very severe invasive extraintestinal amebiasis, in which the parasite causes lesions in the liver, lungs, brain, and other organs.
4. *Acanthamoeba* and *Naegleria* are free-living amoebae that rarely infect humans. Both are acquired from water. *Acanthamoeba* causes keratitis and **amebic encephalitis,** which progresses slowly and is usually fatal. *Naegleria* causes a more rapid **amebic meningoencephalitis,** which is also fatal.
5. Parasitic **flagellates** include *Trypanosoma, Leishmania, Giardia,* and *Trichomonas.*
6. *Trypanosoma cruzi,* transmitted by the kissing bug *(Triatoma),* causes **Chagas' disease** in the Americas, where it is a frequent cause of heart disease. *T. brucei,* transmitted by the tsetse fly *(Glossina),* causes **African sleeping sickness.**
7. Both trypanosomes have similar life cycles. Epimastigotes in the insect gut develop into infective trypomastigotes introduced to the human host by the insect vector. *T. cruzi* trypomastigotes develop intracellularly in macrophages and heart muscle as amastigotes. *T. brucei* trypomastigotes remain in the blood, are antigenically variable, can spread to the CNS, and can be fatal.
8. Many species of *Leishmania* infect humans. Each parasite has two developmental stages: an intracellular amastigote in mammals and a promastigote form in sand flies, which transmit the disease. Three clinical forms of **leishmaniasis** are seen, each with increasing severity: cutaneous, mucocutaneous, and visceral. Leishmaniasis is a **zoonosis**—a disease of animals transmitted to humans.
9. *Giardia intestinalis* is a frequent cause of gastrointestinal disease **(giardiasis)** and is endemic worldwide. Giardiasis is rarely fatal.
10. *Trichomonas vaginalis* is the most common protozoan disease of humans. This sexually transmitted protozoan causes **vaginosis.** Prevention relies on abstinence, monogamy, and condom usage.
11. *Plasmodium* species, *Toxoplasma, Cryptosporidium,* and *Cyclospora* are all **apicomplexans,** in which sexual reproduction produces oocysts that undergo schizogony to produce various forms.
12. Four species of *Plasmodium* cause **malaria** and are transmitted by female *Anopheles* mosquitoes. The complex life cycle of *Plasmodium* has three stages: Sporozoites travel to the liver to initiate the **exoerythrocytic phase.** Rupture of liver cells releases merozoites from intracellular schizonts. Merozoites infect erythrocytes to initiate the **erythrocytic cycle.** The **sporogonic phase** in the mosquito has an oocyst that divides to form sporozoites, which are injected into humans when the mosquito feeds. Dormant hypnozoites are produced by *P. ovale* and *P. vivax. P. falciparum* produces the deadliest form of malaria. Several host traits influence whether or not an individual is susceptible to malaria.
13. *Toxoplasma gondii* causes **toxoplasmosis,** a disease of cats that can be acquired by other animals and humans, primarily via the consumption of meat containing the parasites. **Toxoplasmosis** is generally mild in humans, but *Toxoplasma* can cross the placenta to cause serious harm to fetuses.

14. *Cryptosporidium parvum* and *Cyclospora cayetanensis* both cause waterborne gastrointestinal disease in the United States. *Cryptosporidum,* which causes ***Cryptosporidium enteritis*** (cryptosporidiosis), is resistant to standard water treatment methods. Cyclosporiasis is an emerging disease in the U.S.

Helminthic Parasites of Humans (pp. 667–676)

1. **Helminths** are multicellular eukaryotic worms, some of which are parasitic.
2. **Cestodes** (tapeworms) all share similar morphologies and general life cycles. A **scolex** attaches the tapeworm to the intestine of its definitive host. From this extends a neck region from which **proglottids** grow. When fertilized eggs (sometimes within gravid proglottids) are eaten by an intermediate host, they hatch to release larval stages that invade tissues of the intermediate host and form **cysticerci**. When the cysticerci are eaten by a definitive host, they undergo development into adult tapeworms.
3. Humans are the definitive hosts of *Taenia saginata,* the beef tapeworm, and *Taenia solium,* the pork tapeworm. Humans become infected by eating cysticerci in undercooked beef or pork, respectively.
4. *Echinococcus granulosus* forms **hydatid** cysts to cause **hydatid disease** in humans who are accidental intermediate hosts when they ingest the helminth's eggs. The definitive host is a canine.
5. **Trematodes,** or flukes, are often divided into four groups: blood, intestinal, liver, and lung flukes. Eggs deposited in water hatch to release miracidia, which burrow into freshwater snails. Following asexual reproduction cercariae are released into the water. In intestinal, liver, and lung flukes, cercariae encyst on plants or in a second intermediate host as metacercariae, which are ingested by human hosts. Cercariae of blood flukes directly penetrate the skin of humans. After maturation occurs in the human, flukes release eggs back into the environment via the feces or urine.
6. *Fasciola* is a liver fluke. Humans acquire *Fasciola* by ingesting metacercariae on watercress or other vegetables.
7. Species of *Schistosoma* (blood flukes) affect millions of people worldwide, causing **schistosomiasis.** Three species infect humans: *S. mansoni, S. japonicum,* and *S. haematobium,* each of which is geographically limited. Extensive infections significantly damage the liver.
8. **Nematodes** are round, unsegmented worms with pointed ends.
9. *Ascaris lumbricoides,* the largest nematode to infect humans, causes **ascariasis,** which is the most common nematode disease of humans in the world. Eggs containing infective larvae are consumed with contaminated vegetables. The eggs hatch in the intestine, and larvae enter the lymphatic system. Worm migration through the body may be associated with various symptoms, but infections are usually not fatal.
10. *Ancylostoma duodenale* and *Necator americanus* are both hookworms, the second most common nematodes to infect humans worldwide. Eggs deposited in soil mature to form infective larvae that leave the egg and actively burrow through human skin to initiate infection. The circulatory system carries the worms through the lungs and heart. Eventually worms migrate up the respiratory tract to the pharynx, where they are swallowed. They become adults in the intestine, where they attach to the intestinal mucosa and feed on blood. Hookworm infection can lead to anemia and other blood-related dysfunction.
11. The pinworm *Enterobius vermicularis* is the most common parasitic worm infestation of humans in the United States and the third most common in the world. Eggs deposited in the perianal area are introduced into the mouth following scratching to reinfect the human host. Transmission among family members is also common.
12. *Wuchereria bancrofti* is a filarial (threadlike) nematode of humans, causing **filariasis.** It lives in the lymphatic system, where adult worms block lymph flow, causing **elephantiasis,** in which lymph accumulates, especially in the lower extremities. Bacterial infections may ensue, and subcutaneous tissues harden, producing permanent disfigurement. The parasite is transmitted when microfilariae, which circulate in the blood, are picked up by feeding mosquitoes.

Arthropod Vectors (p. 676)

1. **Arthropods** are organisms with segmented bodies, jointed legs, and a hard exoskeleton. They are found worldwide in nearly all habitats, and some may serve as biological vectors for many organisms. Arthropod vectors are arachnids (adults having eight legs) or insects (adults having six legs).
2. Arachnid vectors include mites and ticks.
3. Insect vectors include fleas, lice, flies, and kissing bugs. Mosquitoes are the most important arthropod vectors of human diseases.

Questions for Review

Answers to the Questions for Review (except for Short Answer questions) begin on page A-1.

Multiple Choice

1. Parasitology is the study of
   a. parasitic viruses.
   b. parasitic prokaryotes.
   c. parasitic fungi.
   d. parasitic eukaryotes.
2. The only ciliate to cause disease in humans is
   a. *Naegleria.*
   b. *Balantidium.*
   c. *Fasciola.*
   d. *Trypanosoma.*
3. Which of the following organisms is regularly transmitted sexually?
   a. *Trichomonas*
   b. *Entamoeba*
   c. *Trypanosoma*
   d. *Enterobius*
4. *Leishmania* species are transmitted by
   a. sand flies.
   b. tsetse flies.
   c. kissing bugs.
   d. mosquitoes.
5. Which of the following is the name of the intracellular infection stage of *Leishmania*?
   a. miracidia
   b. metacercaria
   c. bradyzoite
   d. amastigote

6. In malaria, which stage occurs in red blood cells?
   a. exoerythrocytic phase
   b. erythrocytic cycle
   c. sporogonic phase
   d. amastigote cycle
7. The definitive host for *Toxoplasma gondii* is
   a. humans.
   b. cats.
   c. birds.
   b. mosquitoes.
8. Tapeworms are generally transmitted via
   a. consumption of an intermediate host.
   b. consumption of the definitive host.
   c. vectors such as mosquitoes.
   d. consumption of adult tapeworms.
9. *Cryptosporidium* cannot be killed by routine boiling. Another parasite resistant to such boiling is
   a. *Giardia.*
   b. *Trypanosoma.*
   c. *Toxoplasma.*
   d. *Plasmodium.*
10. The immature fluke stages that infect snail intermediate hosts are called
    a. metacercariae.
    b. cercariae.
    c. cysticerci.
    d. miracidia.
11. The beef tapeworm is known by the scientific name
    a. *Taenia solium.*
    b. *Taenia saginata.*
    c. *Ancylostoma duodenale.*
    d. *Echinococcus granulosus.*
12. *Enterobius vermicularis* is commonly called
    a. hookworm.
    b. pinworm.
    c. whipworm.
    d. tapeworm.
13. The infective larvae of *Necator americanus* must pass through which human organ to mature?
    a. bladder
    b. brain
    c. lung
    d. liver
14. Both *Plasmodium* species and *Wuchereria bancrofti* can be carried by mosquitoes in the genus
    a. *Aedes.*
    b. *Anopheles.*
    c. *Culex.*
    d. *Ctenocephalides.*
15. Which of the following arthropods is responsible for transmitting the most parasitic diseases?
    a. fleas
    b. ticks
    c. mosquitoes
    d. true bugs
16. The majority of cestodes are transmitted via
    a. ingestion.
    b. vectors.
    c. direct contact.
    d. inhalation.
17. Which of the following is most effective in preventing infection by *Giardia?*
    a. sexual abstinence
    b. drinking only bottled water
    c. use of insect repellent
    d. cooking all food
18. The sporogonic phase of *Plasmodium* occurs in
    a. red blood cells.
    b. liver cells.
    c. schizonts.
    d. *Anopheles* mosquitoes.
19. The tapeworm attachment organ is a
    a. scolex.
    b. proglottid.
    c. strobila.
    d. cuticle.
20. Trophozoite-cyst conversion is vital to the life of
    a. *Balantidium.*
    b. *Entamoeba.*
    c. *Giardia.*
    d. all of the above

Modified True/False

Mark each statement as true or false. Rewrite the false statements to make them true by changing the italicized word(s).

____ 1. *Sexual contact* is the most common method of transmitting parasites.

____ 2. Examination of *stool samples* can reveal the presence of *Naegleria* parasites.

____ 3. *Trichomonas vaginalis* is the most common parasitic protozoan of humans in the industrialized world.

____ 4. *Trypanosoma brucei* is transmitted by tsetse flies and *Trypanosoma cruzi* is transmitted by kissing bugs.

____ 5. *Plasmodium falciparum* causes the most serious form of malaria.

____ 6. Toxoplasmosis can be transmitted across a *placenta.*

____ 7. Humans can become intermediate hosts for *Taenia solium.*

____ 8. *Fasciola hepatica* can be acquired by eating infected *sheep.*

____ 9. The number of cases of schistosomiasis has *increased* worldwide owing to improved technology and economic stability in endemic areas.

____ 10. *Wuchereria bancrofti* is a filarial nematode that infects the *lymphatic system.*

Fill in the Blanks

1. *Balantidium coli* can be distinguished from *Entamoeba histolytica* microscopically because *B. coli* has ________________.
2. ________________ may be transmitted to humans from cat litter boxes.
3. African sleeping sickness is caused by *Trypanosoma* ________________ but not by *Trypanosoma* ________________.
4. The parasitic amoeba ________________ can be acquired by ingestion.
5. *Plasmodium vivax* and *P. ovale* can both form dormant ________________.
6. Of the parasitic helminths discussed in this chapter, the only one transmitted by mosquitoes is ________________.
7. The following helminths can directly penetrate the skin of humans to establish infection: ________________, ________________ and ________________ (give genera only).
8. A trematode that can be acquired by eating raw or undercooked vegetables is ________________.
9. Hookworm disease is caused by ________________ in the Middle East.
10. Both of the following parasites demonstrate nocturnal movement, which is important during diagnosis: ________________ and ________________.

Matching

Match the terms with the correct organisms. Answers may be used more than once, and an organism may have more than one answer.

____ 1. *Balantidium coli*	A. Miracidia
____ 2. *Echinococcus granulosus*	B. Bradyzoites
____ 3. *Fasciola* spp.	C. Schizogony
____ 4. *Leishmania* spp.	D. Microfilaria
____ 5. *Plasmodium falciparum*	E. Hydatid cyst
____ 6. *Plasmodium vivax*	F. Cysticerci
____ 7. *Taenia* spp.	G. Trophozoites
____ 8. *Toxoplasma gondii*	H. Amastigotes
____ 9. *Trypanosoma*	I. Hypnozoites
____ 10. *Wuchereria bancrofti*	J. Trypomastigotes

Labeling

Label the three stages of the *Plasmodium* life cycle, and label the forms of the parasite where indicated.

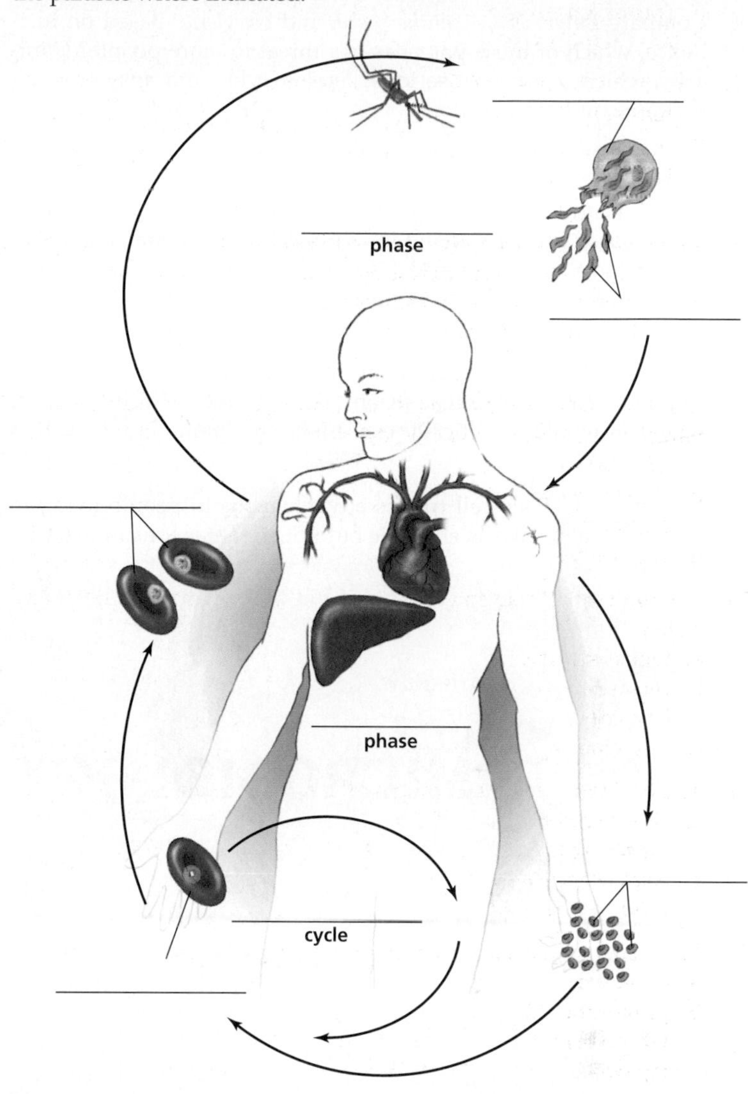

Short Answer

1. Why do insect vectors and animal reservoirs increase the difficulty of preventing and controlling parasitic infections in humans?
2. How is the transmission of the amoeba *Entamoeba* different from the transmission of the amoebae *Acanthamoeba* and *Naegleria*?

Concept Mapping

Using the following terms, draw a concept map that describes intestinal protozoan parasites. For a sample concept map, see p. 93. Or, complete this concept map online by going to the Study Area at www.masteringmicrobiology.com.

Amoebae	Cysts	Flagellates	Reproducing/feeding stage
Apicomplexans	Diarrhea	*Giardia intestinalis*	Trophozoites
Balantidium coli	Dormant stage	Livestock wastes	Two morphological forms
Ciliates	*Entamoeba histolytica*	New host	Unknown
Cryptosporidium parvum	Environment	Oocyst	Water
Cyclospora cayetanensis	Fecal-oral route	People with AIDS	

Critical Thinking

1. Compare *Entamoeba, Acanthamoeba,* and *Naegleria.* Based on incidence, which of these parasites is a threat to more people? Clinically, which is the most serious? Explain why your answer is not the same for both questions.
2. Explain how each of the following could lead to the reemergence of malaria in the United States: (a) global warming, (b) increased travel of individuals from endemic regions to the United States, (c) increased immigration of individuals from endemic regions to the United States, (d) regulations against the use of insecticides, and (e) laws protecting wetlands.
3. People who suffer from AIDS are severely immunocompromised because HIV destroys immune cells; thus, all the diseases in this chapter could be a threat to an AIDS patient. Which diseases, however, would specifically exacerbate the immune dysfunction of AIDS? Why?
4. Discuss why sickle-cell trait is advantageous to people living in areas where malaria is endemic but not advantageous in malaria-free areas.
5. Present a logical argument to explain the differences between the clinical manifestations of *Trypanosoma brucei* and *T. cruzi.* Relate your argument to the respective life cycles.
6. What feature of the life cycle of *T. brucei* makes it difficult to create a successful vaccine?
7. On July 10, a man in Palm Beach, Florida, was admitted to the hospital with a month-long history of fever, chills, headache, and vomiting. A parasite was observed in a blood smear. What parasite causes these signs and symptoms? What is the treatment? Is this a rare disease in Florida?
8. In March 2009, a Peace Corps volunteer was evacuated from Malawi in Africa because of a two-week history of headaches, vision loss, and an episode of unconsciousness. Doctors discovered a granular swelling in his brain and eggs with peripheral spines in his urine. What is the diagnosis? What species is involved? What is the treatment?
9. Humans can accidentally become intermediate hosts of *Taenia solium.* Why is this is a dead end for the parasite?

Access more review material online in the Study Area at **www.masteringmicrobiology.com.** There, you'll find

- **Concept Mapping Activities**
- **Flashcards**
- **Quizzes**

and more to help you succeed.

24 Pathogenic DNA Viruses

Everyone reading this textbook has been infected by **DNA viruses** at one time or another. Fever blisters (cold sores), warts, chickenpox, infectious mononucleosis, and many types of colds are all caused by viruses that have DNA genomes. DNA viruses are also the cause of smallpox, **hepatitis B**, shingles, and genital **herpes**, and they are associated with some cancers. Fortunately, many of the more severe diseases caused by DNA viruses, including hepatitis B and **hepatic cancer**, are relatively rare. However, smallpox—a disease that has been successfully eradicated in nature—is once again a public health concern because of its potential as an agent of **bioterrorism**.

The general characteristics of DNA viruses were covered in Chapter 13. In this chapter we focus on a selection of medically important DNA viruses—the differences among them, the diseases they cause, and the ways they may be transmitted, treated, and prevented.

MM Take the pre-test for this chapter online. Visit the Study Area at www.masteringmicrobiology.com.

Hepatitis B virus is a DNA virus that causes hepatitis and lung cancer. ▲

DNA viruses that cause human diseases are grouped into seven families based on several factors, including the type of DNA they contain, the presence or absence of an envelope, size, and the host cells they attack. Recall from Chapter 13 that DNA viruses have either double-stranded DNA (dsDNA) or single-stranded DNA (ssDNA) for their genomes, and that none of them contains functional RNA. Double-stranded DNA viruses belong to the families *Poxviridae, Herpesviridae, Papillomaviridae, Polyomaviridae,* and *Adenoviridae.* The family *Parvoviridae* contains viruses with ssDNA genomes. Viruses in the family *Hepadnaviridae* are unusual because they have a genome that is partly dsDNA and partly ssDNA. In the following sections we examine the infections and diseases caused by the viruses in each of these families.

Poxviridae

Learning Objectives

- ✓ Name two diseases caused by poxviruses and discuss their signs and symptoms.
- ✓ Describe the progression of disease in poxvirus infections.
- ✓ Discuss the historical importance of poxviruses in immunization and their use in disease eradication.

Poxviruses are double-stranded DNA viruses with complex capsids and envelopes (**Figure 24.1**). They are the second largest of viruses, being up to 300 nm in diameter, which is just visible with high-quality, optical, ultraviolet light microscopes. Their large size allows them to be potential vectors for the introduction of genetic material in vaccinations and gene therapy.

Specific poxviruses infect many mammals, including mice, camels, elephants, cattle, and monkeys. (Note that chickenpox, discussed later in this chapter, is caused neither by a poxvirus nor occurs in chickens; instead, it is caused by a herpesvirus that affects people.) Most animal poxviruses are species specific; they are unable to infect humans because they cannot attach to human cells, though a few poxviruses of monkeys, sheep, goats, and cattle can infect some humans because these people have surface proteins similar to proteins of those animals.

All poxviruses produce lesions that progress through a series of stages (**Figure 24.2**). The lesions begin as flat, reddened **macules**[1] (mak′ūlz), which then become raised sores called **papules**[2] (pap′ūlz); when the lesions fill with clear fluid, they are called **vesicles,** which then progress to pus-filled **pustules**[3] (pŭs′choolz), which are also known as **pocks** or **pox.** These terms for the stages of poxvirus lesions are also used to describe the lesions of other skin infections. Poxvirus pustules dry up to form a crust, and because these lesions penetrate the dermis, they result in characteristic scars.

[1]From Latin *maculatus,* meaning spotted.
[2]From Latin *papula,* meaning pimple.
[3]From Latin *pustula,* meaning blister.

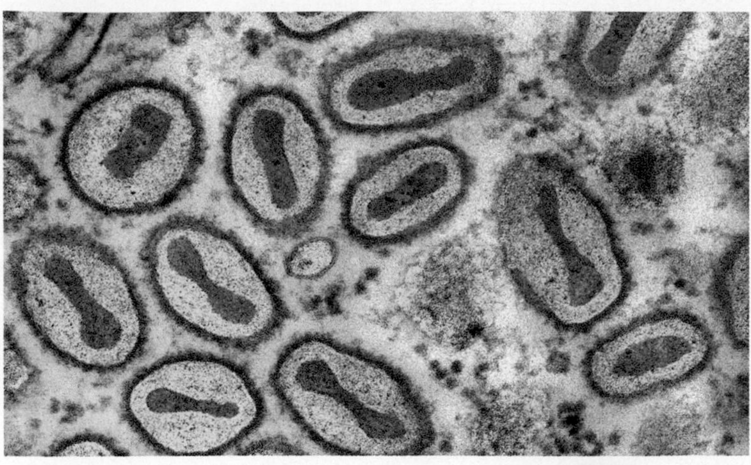

▲ **Figure 24.1 Poxviruses.** Variola, the causative agent of smallpox, and vaccinia, the agent of cowpox, are in the genus *Orthopoxvirus.*

Infection with poxviruses occurs primarily through inhalation of viruses in droplets or dried crusts. Because the envelopes of poxviruses are relatively unstable outside a host's body, close contact is necessary for infection. The two main poxvirus diseases of humans are *smallpox* and *molluscum contagiosum.* Three diseases of animals—*orf* (sheep and goat pox), *cowpox,* and *monkeypox*—may be transmitted to humans as well.

Smallpox

Smallpox has played important roles in the history of microbiology, immunization, epidemiology, and medicine. During the Middle Ages, 80% of the European population contracted smallpox. Later, European colonists introduced smallpox into susceptible Native Americans, resulting in the death of as many as 3.5 million people. Recall from Chapter 1 that Edward Jenner first demonstrated immunization using the relatively mild cowpox virus to protect against smallpox. He was successful because the antigens of cowpox virus are chemically similar to those of smallpox virus, so that exposure to cowpox results in immunological memory and subsequent resistance against both cowpox and smallpox.

Smallpox virus is in the genus *Orthopoxvirus.* Commonly known as **variola** (vă-rī′ō-lă), it exists in two forms, or strains. **Variola major** causes severe disease with a mortality rate of 20% or higher, depending upon the age and general health of the host; **variola minor** causes a less severe disease with a mortality rate of less than 1%. Both types of variola infect internal organs and produce fever—42°C (107°F) has been recorded—malaise, delirium, and prostration before moving via the bloodstream to the skin, where they produce the recognizable pox (**Figure 24.3**). Both strains also leave severe scars on the skin of survivors, especially on the face.

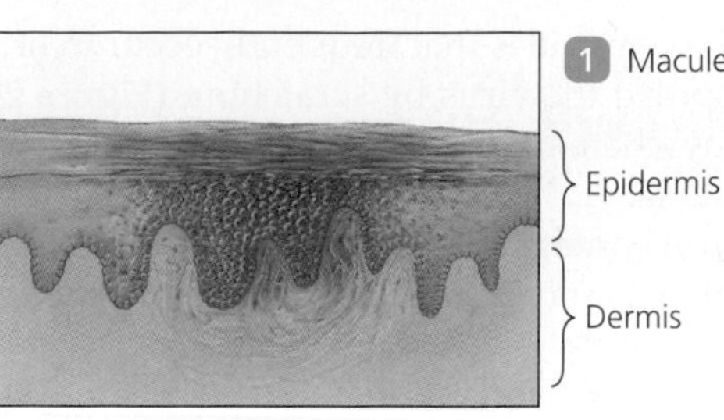

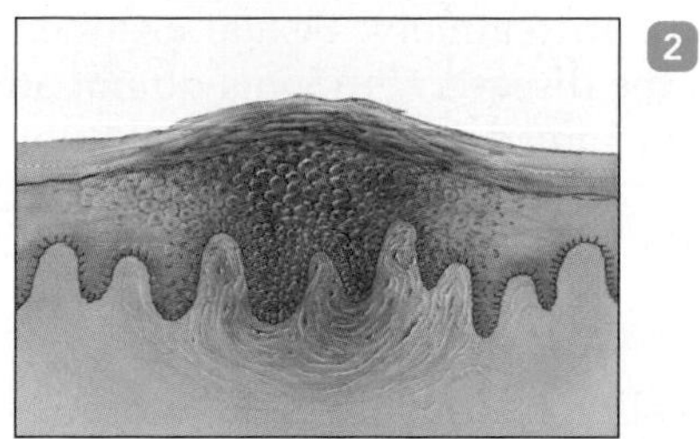

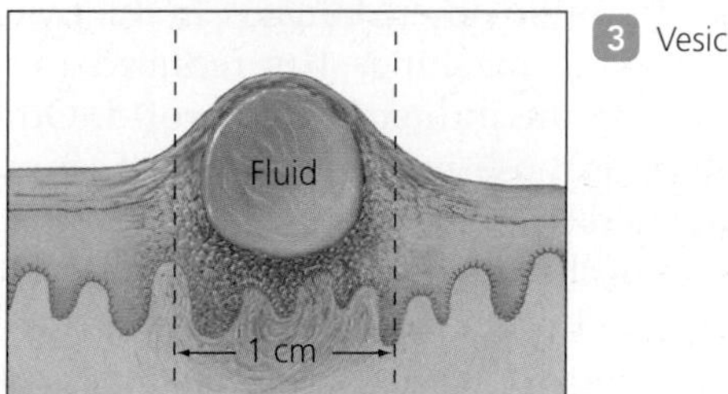

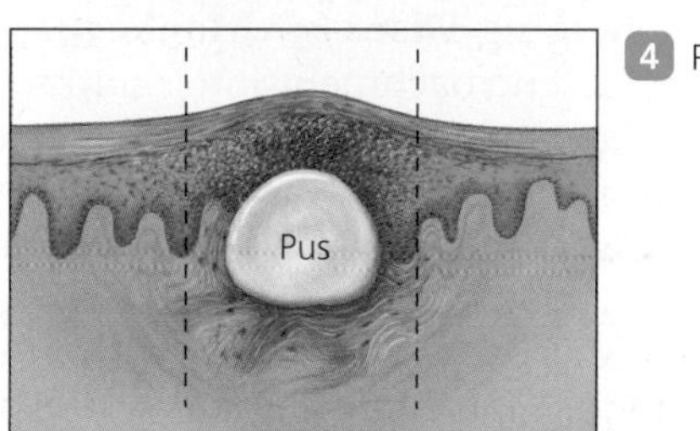

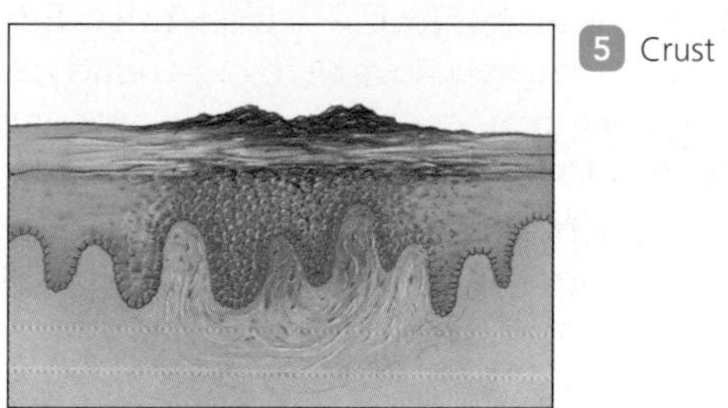

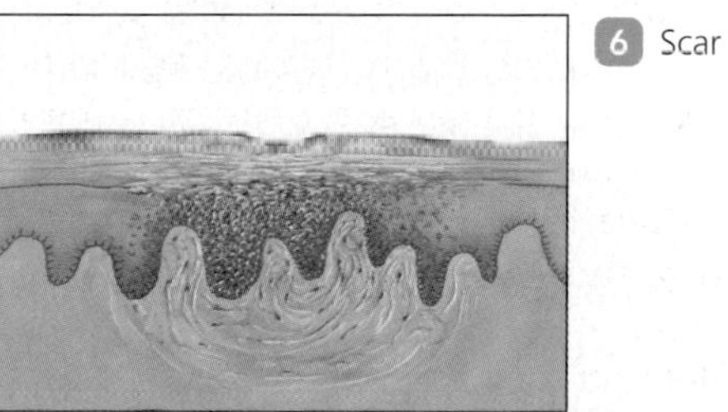

◀ **Figure 24.2 The stages of the lesions in poxvirus infections.** Red macules become hard papules, then fluid-filled vesicles that become pus-filled pustules (pox), which then crust over and leave characteristic scars. *Are the dried viruses in the crusts of smallpox and cowpox lesions infective?*

Figure 24.2 *Yes, the dried crusts of poxvirus lesions may contain active viruses. The Chinese used the viruses in crusts for variolation, an early attempt at vaccination.*

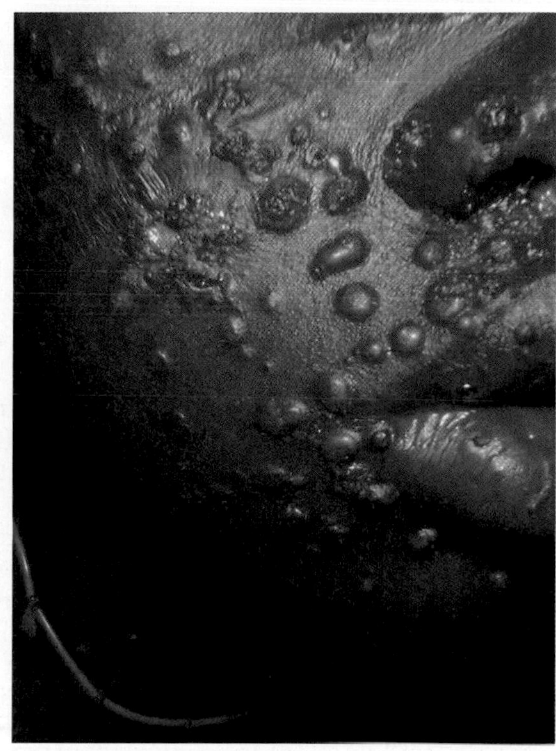

▲ **Figure 24.3 Smallpox lesions.** Facial scarring was often especially severe in smallpox.

Smallpox is the first human disease to be eradicated globally in nature. In 1967, the World Health Organization (WHO) began an extensive campaign of identification and isolation of smallpox victims, as well as vaccination of their contacts, with the goal of completely eliminating the disease. The efforts of WHO were successful, and in 1980 smallpox was declared to be eradicated—one of the great accomplishments of 20th-century medicine. The factors that made it possible for smallpox to be eliminated included the following:

- An inexpensive, stable, and effective vaccine (vaccinia, cowpox virus) is available.
- Smallpox is specific to humans; there are no animal reservoirs.
- The severe, obvious signs and symptoms of smallpox enable quick and accurate diagnosis and quarantine.
- There are no asymptomatic cases.
- The virus is spread only via close contact.

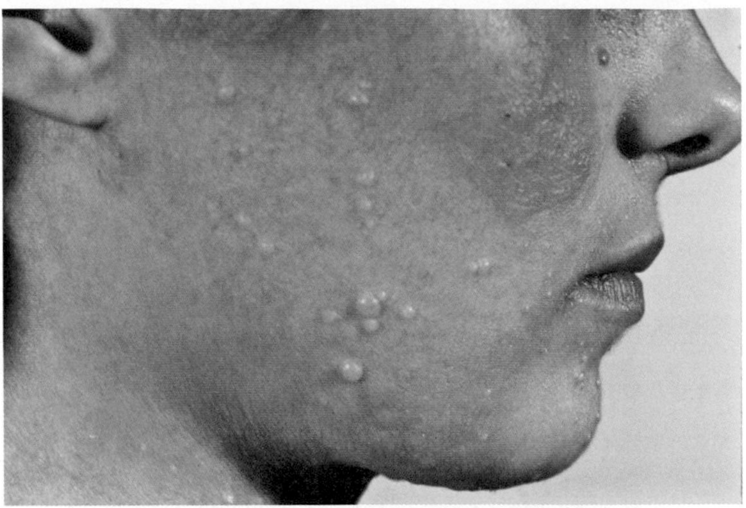

▲ **Figure 24.4 Lesions of molluscum contagiosum.** *Molluscipoxvirus*, a poxvirus, causes the pearly white to light pink, tumorlike lesions.

Stocks of the virus are still maintained in laboratories in the United States and Russia as research tools for investigations concerning virulence, pathogenicity, protection against bioterrorism, vaccination, and recombinant DNA technology.

Routine vaccination against smallpox has ceased in the United States, though some health care workers and military personnel were vaccinated following the 2001 bioterrorism attack. Studies have shown that older citizens, who were vaccinated as children, still retain some immunity. One-tenth the regular vaccine dose is effective for revaccination of these adults.

Now that smallpox vaccination has been discontinued in many countries, experts are concerned that the world's population is once again susceptible to smallpox epidemics if the virus were accidentally released from storage or used as a biological weapon. For this reason, many scientists advocate the destruction of all smallpox virus stocks; other scientists insist that we must maintain the virus in secure laboratories so that it is available for the development of more effective vaccines and treatments if the virus is disseminated in a bioterrorism attack.

CRITICAL **THINKING**

The United States and Russia have repeatedly agreed to destroy their stocks of the smallpox virus, but the deadline for destruction has been postponed numerous times. In the meantime, the entire genome of variola major has been sequenced. What reasons can governments cite for maintaining smallpox viruses? Should all laboratory stores of smallpox viruses be destroyed? Given that the genome of the virus has been sequenced, and that DNA can be synthesized, would elimination of all laboratory stocks really be the extinction of the smallpox virus?

Molluscum Contagiosum

Another poxvirus, *Molluscipoxvirus,* causes **molluscum contagiosum** (mo-lŭs′kŭm kon-taj-ē-ō′sum), a skin disease characterized by smooth, waxy papules that frequently occur in lines where the patient has spread the virus by scratching **(Figure 24.4)**. Lesions may occur anywhere but typically appear on the face, trunk, and external genitalia.

Molluscum contagiosum occurs worldwide and is typically spread by contact among infected children, sexually active individuals (particularly adolescents), and AIDS patients. Treatment involves removing the infected nodules by chemicals or freezing, though people with normal immune systems heal within months without treatment. Sexual abstinence prevents the genital form of the disease. Condoms do not afford protection, as the virus can spread from and to areas not covered by the condom.

Other Poxvirus Infections

Poxvirus infections also occur in animals. Infections of humans with these animal viruses are usually relatively mild, resulting in pox and scars but little other damage. In the case of cowpox, the advent of milking machines has reduced the number of cases to near zero in the industrialized world. Other poxvirus infections, including those in sheep, goats, and monkeys, are less easily contracted by humans.

Transmission of these poxviruses to humans requires close contact with infected animals, similar to the contact Edward Jenner observed for milkmaids and cowpox. Still, an upsurge in the number of monkeypox cases in humans has been seen in the past decade. (See **Emerging Diseases: Monkeypox.**) This may be the result of human encroachment into monkey habitats or changes in viral antigens. Some researchers speculate that the cessation of smallpox vaccination may also be involved, because evidence suggests that smallpox vaccination also protects against monkeypox. Thus, the elimination of smallpox and the cessation of smallpox vaccination may have led to increased human susceptibility to monkeypox. Some scientists have recommended that smallpox vaccination be reinstated in areas where monkeypox is endemic.

Herpesviridae

The family *Herpesviridae* contains a large group of viruses with enveloped polyhedral capsids containing linear dsDNA. A herpesvirus attaches to a host cell's receptor and enters the cell through the fusion of its envelope with the cytoplasmic membrane. After the viral genome is replicated and assembled in the cell's nucleus, the virion acquires its envelope from the nuclear membrane and exits the cell via exocytosis or cell lysis.

Many types of herpesviruses infect humans, and their high infection rates make them the most prevalent of DNA viruses. Indeed, three types of herpesvirus—herpes simplex, varicella-zoster, and Epstein-Barr (discussed shortly)—infect 90% of the adult population in the United States, and several other herpesviruses have infection rates over 50%.

Herpesviruses are often *latent;* that is, they may remain inactive inside infected cells, often for years. Such latent viruses may reactivate as a result of aging, chemotherapy,

EMERGING DISEASES

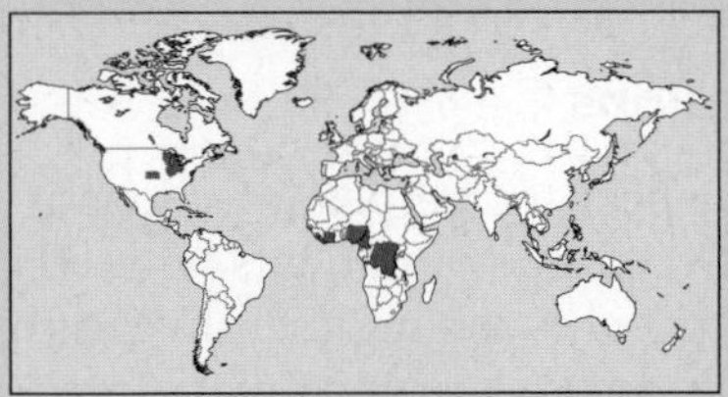

MONKEYPOX

Jamal was excited about the pet he got for his birthday; he was the only boy at school who had a prairie dog! Common dogs were boring; his prairie dog could climb up the drapes, run through plastic tube mazes, and sit on Jamal's shoulder to eat. Doggie was the best. In fact, everything would be just about perfect if only Jamal weren't sick in bed.

Two weeks after his birthday Jamal was exhausted and felt terrible. He had fever accompanied by headache and muscle aches, so his mother had kept him home from school. Then the lymph nodes in his neck and armpits had swollen, and he had gotten frightening, large, raised lesions all over his face and hands. He was really sick.

For the next three weeks, the lesions had become crusty and then fallen off, only to be replaced by new lesions. Jamal felt awful and missed playing with Doggie, who had been confined to his cage, while Jamal stayed in bed. Jamal's doctor had been unable to diagnose the disease, but an infectious disease specialist had finally confirmed monkeypox—a disease named for monkeys who can get the disease. In reality, monkeypox is a disease of rodents, including prairie dogs. Final investigation revealed that the pet wholesaler who had supplied Doggie to Jamal's pet store in Illinois had also supplied African monkeys to stores in Texas. At least one of those monkeys had been infected with monkeypox virus—a virus closely related to smallpox virus. The monkey had infected Doggie who had infected Jamal.

Jamal was sick for four weeks, but monkeypox is not as virulent as smallpox, so Jamal recovered with only some scars as a reminder that not all unusual pets are a blessing. For more about monkeypox, see p. 684.

Track monkeypox online by going to the Study Area at www.masteringmicrobiology.com.

MICROBE AT A GLANCE

Orthopoxvirus variola (Smallpox Virus)

Taxonomy: Family *Poxviridae*

Genome and morphology: Single molecule of double-stranded DNA; enveloped, brick-shaped or pleomorphic, complex capsid, 250 nm $\times$ 250–300 nm $\times$ 150–300 nm

Host: Humans

Virulence factors: Remains very stable in aerosol form; highly contagious between people; can remain viable for as long as 24 hours

Disease caused: Smallpox

Treatment for disease: Administration of vaccine within four days of infection prevents disease in most patients; anticowpox immunoglobulins may be efficacious

Prevention of disease: Smallpox has been eliminated in nature and presumably could reappear only if released from secure storage facilities; if it were released, vaccination of exposed individuals and quarantine of patients would need to be implemented immediately

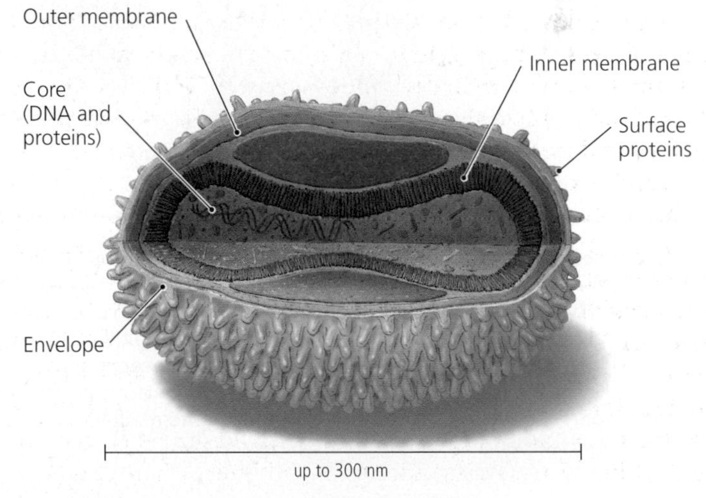

Download the Microbe at a Glance flashcards from the Study Area at www.masteringmicrobiology.com.

immunosuppression, or physical and emotional stress, causing a recurrence of the manifestations of the diseases. Additionally, some latent herpesviruses insert into a host's chromosomes, where they may cause genetic changes and potentially induce cancer.

Besides causing the diseases for which the group is named (oral herpes and genital herpes), herpesviruses cause a variety of other diseases, including chickenpox, shingles, mononucleosis, and some cancers. **Human herpesviruses** have been assigned species names combining "HHV" (for "human herpesvirus")

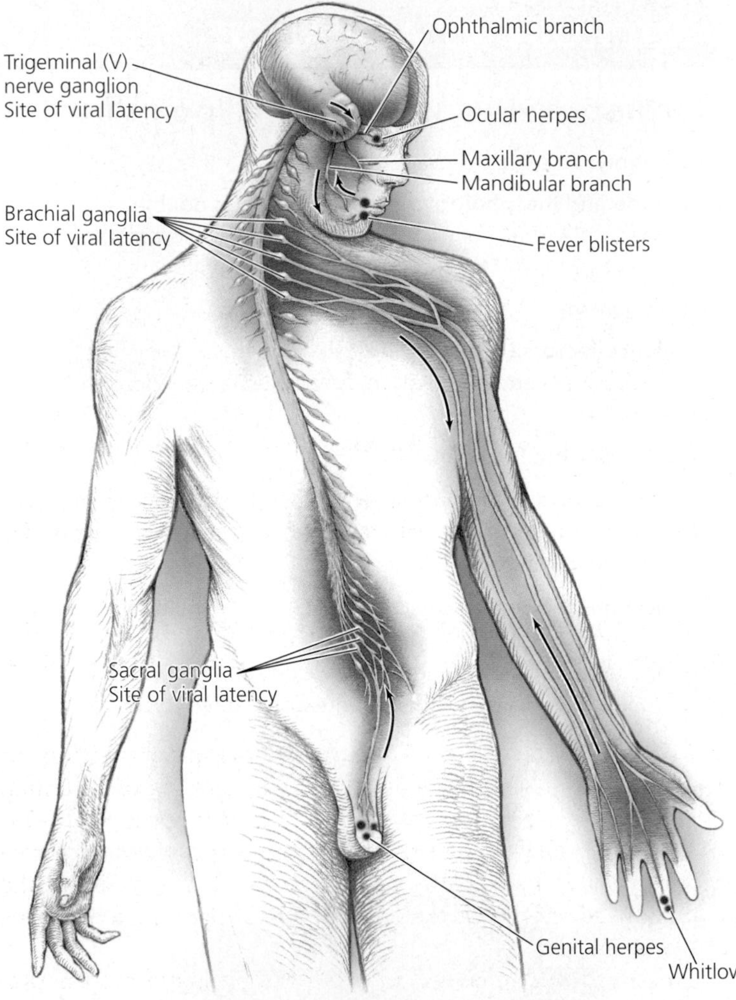

▲ **Figure 24.5 Sites of events in herpesvirus infections.** Infections typically occur when viruses invade the mucous membranes of either the lips (for the fever blisters of oral herpes) or the genitalia (for genital herpes), or through broken skin of a finger (for a whitlow). Viruses may remain latent for years in the trigeminal, brachial, or sacral ganglia before traveling down nerve cells to cause recurrent symptoms in the lips, genitalia, fingers, or eyes. *What factors may trigger the reactivation of latent herpesviruses and the recurrence of symptoms?*

Figure 24.5 *Stress, fever, trauma, sunlight, menstruation, or diseases such as AIDS may cause recurrent symptoms as a result of immune suppression.*

with numbers corresponding to the order in which they were discovered. Thus the genus *Simplexvirus* contains two species (as discussed in the next section) known as HHV-1 and HHV-2. *Varicellovirus* (which causes chickenpox and shingles) is HHV-3; *Lymphocryptovirus,* also known as Epstein-Barr virus, is HHV-4; *Cytomegalovirus* is HHV-5; *Roseolovirus* is HHV-6; and *Rhadinovirus*—HHV-8—is associated with a particular type of cancer in AIDS patients. All these viruses share similar capsids and genes but can be distinguished by genetic variations among them.

What about HHV-7? Scientists isolated this species of *Roseolovirus* in the 1980s from the T cells of an AIDS patient and identified it as HHV-7 on the basis of its genome, envelope, capsid, and antigens. At this time, no disease has been attributed to this herpesvirus. It is an *orphan virus*—a virus without a known disease.

Infections of *Human Herpesvirus 1* and *2*

Learning Objectives

- ✓ Describe the diseases caused by *human herpesvirus 1* and *2*, including their signs, symptoms, treatment, and prevention.
- ✓ List conditions that may reactivate latent herpesviruses.

The name "herpes," which is derived from a Greek word meaning "to creep," is descriptive of the slowly spreading skin lesions that are seen with *human herpesvirus 1* (HHV-1) and *human herpesvirus 2* (HHV-2), which are in the genus *Simplexvirus.* They were known formerly as herpes simplex viruses (HSV-1 and HSV-2).

Latency and Recurrent Lesions

One of the more distressing aspects of herpes infections is the recurrence of lesions. About two-thirds of patients with HHV-1 or HHV-2 will experience recurrences during their lives as a result of the activation of latent viruses. After a primary infection, these herpesviruses often enter sensory nerve cells and are carried by cytoplasmic flow to the base of the nerve, where they remain latent in the trigeminal, sacral, or other ganglia[4] **(Figure 24.5)**. Such latent viruses may reactivate later in life as a result of immune suppression caused by stress, fever, trauma, sunlight, menstruation, or disease and travel down the nerve to produce recurrent lesions as often as every two weeks. Fortunately, the lesions of recurrent infections are rarely as severe as those of initial infections because of immunological memory.

Types of HHV-1 and HHV-2 Infections

Historically, HHV-1 has also been known as "above the waist herpes," whereas HHV-2 has been known as "below the waist herpes." In the following subsections we consider the variety of infections caused by these two species.

Oral Herpes *Human herpesvirus 1* was the first human herpesvirus to be discovered. About two weeks after infection, the virions of HHV-1 produce painful, itchy skin lesions on the lips, called **fever blisters** or **cold sores** **(Figure 24.6)**, which last 7–10 days; initial infections may also be accompanied by flulike signs and symptoms, including malaise, fever, and muscle pain. Severe infections in which the lesions extend into the oral cavity, called *herpetic gingivostomatitis,*[5] are most often seen in young patients and in patients with lowered immune function due to disease, chemotherapy, or radiation treatment. Young adults with sore throats resulting from other viral infections may also develop *herpetic pharyngitis.*

[4] A ganglion is a collection of nerve cell bodies containing the nuclei of the cells.
[5] From Latin *gingival*, meaning gum, Greek *stoma*, meaning mouth, and Greek *itis*, meaning inflammation.

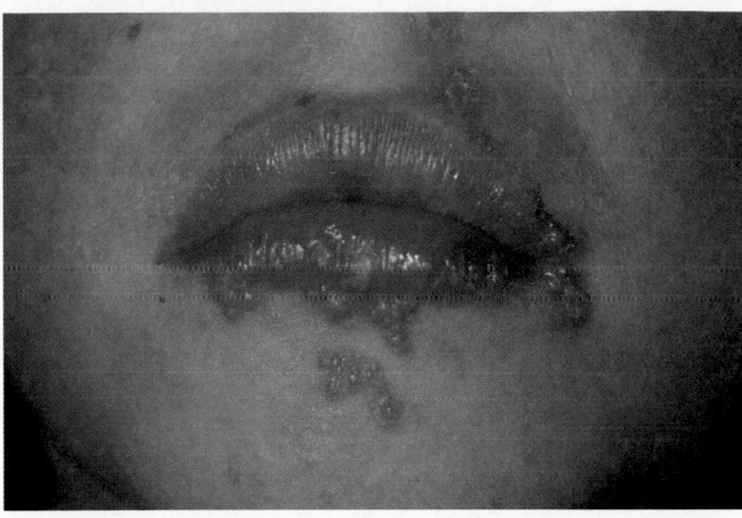

▲ **Figure 24.6 Oral herpes lesions.** *Which is the more likely cause of these lesions, HHV-1 or HHV-2?*

Figure 24.6 Approximately 90% of oral herpes lesions are caused by human herpesvirus 1.

Genital Herpes HHV-1 can infect the genitalia. The virus causes about 15% of genital herpes lesions. The remainder are caused by *human herpesvirus 2.* HHV-2 differs slightly from HHV-1 in its surface antigens, primary location of infection, and mode of transmission. Type 2 herpes is usually associated with painful lesions on the genitalia **(Figure 24.7)** because the virions are typically transmitted sexually. However, HHV-2 can cause oral lesions when it infects the mouth. Thus, about 10% of oral cold sores result from HHV-2. The presence of HHV-2 in the oral region and of HHV-1 in the genital area is presumed to be the result of oral sex.

Ocular Herpes Ocular (ok'yū-lăr) or *ophthalmic* **herpes** is an infection caused by latent herpesviruses **(Figure 24.8a)**. In ocular herpes, latent HHV-1 in the trigeminal ganglion travels down the ophthalmic branch of the trigeminal nerve instead of the mandibular or maxillary branches (see Figure 24.5). Symptoms and signs usually occur in only one eye and include a gritty feeling, conjunctivitis, pain, and sensitivity to light. The viruses may also cause corneal lesions. Recurrent ocular infections can lead to blindness as a result of corneal damage.

Whitlow An inflamed blister called a **whitlow**[6] (hwit'lō) may result if either HHV-1 or HHV-2 enters a cut or break in the skin of a finger **(Figure 24.8b)**. A whitlow usually occurs on only one finger. Whitlow is a hazard for children who suck their thumbs and for health care workers in the fields of obstetrics, respiratory care, gynecology, and dentistry who come into contact with lesions. Such workers should always wear gloves when treating patients with herpes lesions. Since a whitlow is also a potential source of infection, symptomatic personnel should not work with patients until the worker's whitlow has healed. The viruses involved in whitlow may become latent in brachial ganglia and cause recurrent lesions.

[6]From Scandinavian *whick*, meaning nail, and *flaw*, meaning crack.

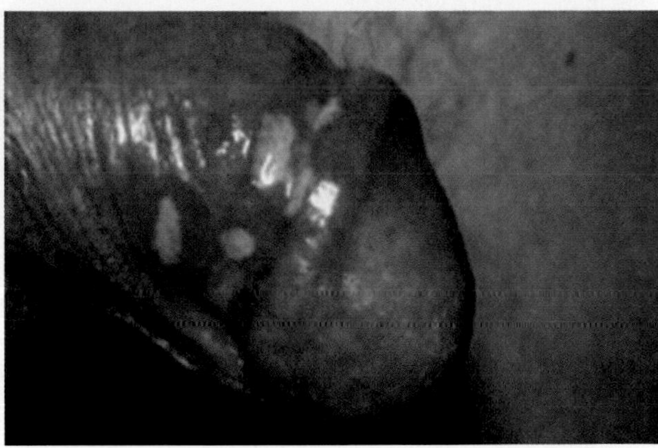

▲ **Figure 24.7 Genital herpes lesions.** *Are these lesions caused by herpes type 1, or type 2?*

Figure 24.7 Either type of herpes can cause genital lesions, but "below the waist" lesions are usually caused by HHV-2.

Neonatal Herpes Although most herpes simplex infections in adults are painful and unpleasant, they are not life threatening. This is not the case for herpes infections in newborns. HHV-2 is usually the species involved in neonatal herpes. Neonatal herpes infections can be severe, with a mortality rate of 30% if the infection is cutaneous or oral, and of 80% if the central nervous system is infected.

A fetus can be infected *in utero* if the virus crosses the placental barrier, but it is more likely for a baby to be infected at birth through contact with lesions in the mother's reproductive tract. Also, mothers with oral lesions can infect their babies if they kiss them on the mouth. Pregnant women with herpes infections, even if asymptomatic, should inform their doctors. To protect the baby, delivery should be by cesarean section if genital lesions are present at the time of birth.

Other Herpes Simplex Infections Athletes may develop herpes lesions almost anywhere on their skin as a result of contact with herpes lesions during wrestling. *Herpes gladiatorum,* as this condition is called, is most often caused by HHV-1.

Human herpesviruses 1 and 2 may also cause other diseases, including encephalitis, meningitis, and pneumonia. Patients with severe immune suppression, such as those with AIDS, are most likely to develop these rare disorders.

Table 24.1 on p. 688 compares infections caused by herpes viruses.

Epidemiology and Pathogenesis of HHV-1 and HHV-2 Infections

HHV-1 and HHV-2 occur worldwide and are transmitted through close body contact. Active lesions are the usual source of infection, though asymptomatic carriers shed HHV-2 genitally. After entering the body through cracks or cuts in mucous membranes, viruses reproduce in epithelial cells near the site of

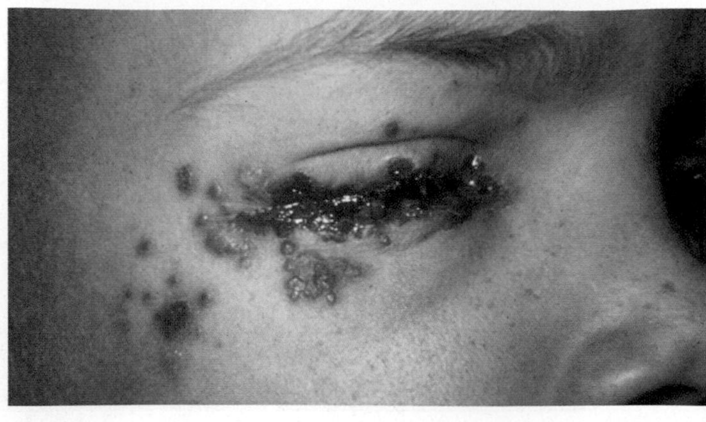

(a)

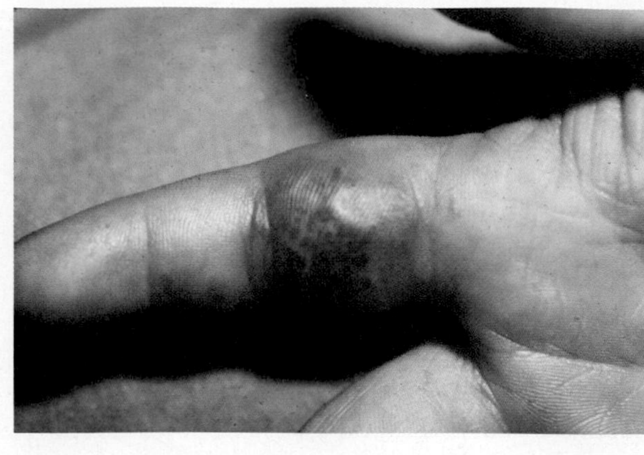

(b)

▲ **Figure 24.8 Two manifestations of herpesvirus infections.** **(a)** Ophthalmic or ocular herpes, the result of the reactivation of latent viruses in the trigeminal ganglion. **(b)** A whitlow on a health provider's finger, resulting from the entry of herpesviruses through a break in the skin of the finger. *How can health care workers protect themselves from a whitlow?*

Figure 24.8 *Health providers should always wear gloves when treating patients who have herpes lesions on any part of their bodies.*

infection and produce inflammation and cell death, resulting in painful, localized lesions on the skin. By causing infected cells to fuse with uninfected neighboring cells to form a structure called a *syncytium,* herpes virions spread from cell to cell, in the process avoiding the host's antibodies.

Initial infections with herpes simplex viruses are usually age specific. Primary HHV-1 infections typically occur via casual contact during childhood and usually produce no signs or symptoms; in fact, by age 2, about 80% of children have been asymptomatically infected with HHV-1. Most HHV-2 infections, by contrast, are acquired between the ages of 15 and 29 as a result of sexual activity, making genital herpes one of the more common sexually transmitted infection in the United States.

Approximately 60% of adult Americans have been exposed to HHV-1. HHV-2 infects about 17% of Americans over age 12.

Diagnosis, Treatment, and Prevention of HHV-1 and HHV-2 Infections

Physicians often diagnose herpes infections by the presence of their characteristic, recurring lesions, especially in the genital region and on the lips. Microscopic examination of infected tissue reveals syncytia. Positive diagnosis is achieved by immunoassay that demonstrates the presence of viral antigen.

Infections with HHV-1 or HHV-2 are among the few viral diseases that can be controlled with chemotherapeutic agents, notably valaciclovir for both cold sores and genital herpes, and other nucleoside analogs (see Figure 10.7) such as iododeoxyuridine or trifluridine for ocular herpes. Topical applications of the drugs limit the duration of the lesions and reduce viral shedding, though none of these drugs cures the diseases or frees nerve cells of latent viral infections. Evidence suggests that a daily oral dose of acyclovir over a period of 6–12 months can be

TABLE 24.1 **Comparative Epidemiology and Pathology of *Human Herpesvirus 1* and *2* Infections**

	HHV-1 (HSV-1)	HHV-2 (HSV-2)
Usual diseases	90% of cold sores/fever blisters; whitlow	85% of genital herpes cases
Mode of transmission	Close contact	Sexual intercourse
Site of latency	Trigeminal and brachial ganglia	Sacral ganglia
Locations of lesions	Face, mouth, and rarely trunk	External genitalia, and less commonly thighs, buttocks, and anus
Other complications	15% of genital herpes cases; pharyngitis; gingivostomatitis; ocular/ophthalmic herpes; herpes gladiatorum; 30% of neonatal herpes cases	10% of oral herpes cases; 70% of neonatal herpes cases

effective in preventing recurrent genital herpes and in reducing the spread of herpes infections to sexual partners. Many obstetricians prescribe acyclovir during the final weeks of pregnancy for women with a history of genital herpes. Chapter 10 examines the actions of these drugs.

Health care workers can reduce their exposure to infection with herpes simplex by wearing latex gloves. Health providers with whitlows should not have contact with patients, especially newborns.

Sexual abstinence or faithful monogamy between uninfected partners is the only sure way of preventing genital infections. People with active lesions should abstain from sexual activity until the crusts have disappeared, and either male or female condoms should be used even in the absence of lesions because asymptomatic individuals still shed viruses. Herpes lesions in women are usually on the external genitalia, so use of a condom during sexual intercourse provides little protection for their partners. Some studies indicate that circumcision reduces by 25% the chance of getting genital herpes.

CRITICAL **THINKING**

After a patient complains that his eyes are extremely sensitive to light and feel gritty, his doctor informs him that he has ocular herpes. What causes ocular herpes? Which human herpesvirus, type I or type 2, is more likely to cause ocular herpes? Why?

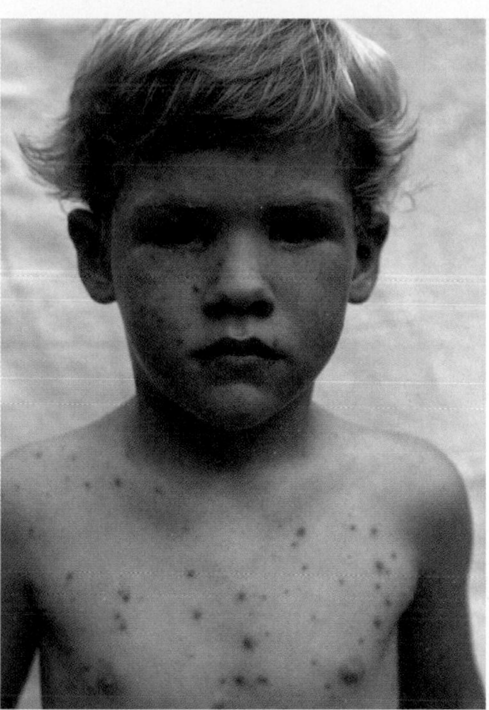

▲ **Figure 24.9 Characteristic chickenpox lesions.** *Varicellovirus* causes these thin-walled, fluid-filled vesicles on red bases.

Human Herpesvirus 3 (Varicella-Zoster Virus) Infections

Learning Objective

✓ Compare and contrast chickenpox in children with shingles in adults.

Human herpesvirus 3, commonly called **varicella-zoster virus (VZV),** takes its name from the two diseases it causes: *varicella,* which is typically a disease of children, and *herpes zoster,* which usually occurs in adults. Taxonomically, VZV is in the genus *Varicellovirus.* Note that the viral name is spelled differently than the disease.

Epidemiology and Pathogenesis of HHV-3 (VZV) Infections

Varicella (var-i-sel'ă), commonly known as **chickenpox,** is a highly infectious disease most often seen in children. Viruses enter the body through the respiratory tract or eyes, replicate in cells at the site of infection, and then travel via the blood throughout the body, triggering fever, malaise, and skin lesions. Viruses are shed before and during the symptoms. Usually the disease is mild, though it is rarely fatal, especially if associated with secondary bacterial infections. Chickenpox is typically more severe in adults than in children, presumably because much of the tissue damage results from the patient's immune response, and adults have a more developed immune system than children do.

About two to three weeks after infection, characteristic skin lesions first appear on the back and trunk and then spread to the face, neck, and limbs (**Figure 24.9**). In severe cases, the rash may spread into the mouth, pharynx, and vagina. Chickenpox lesions begin as macules, progress in one to two days to papules, and finally become thin-walled, fluid-filled vesicles on red bases, which have been called "dewdrops on rose petals." The vesicles turn cloudy, dry up, and crust over in a few days. Successive crops of lesions appear over a period of three to five days, so that at any given time all stages can be seen. Viruses are shed through respiratory droplets and the fluid in lesions; dry crusts are not infective.

Like many herpesviruses, VZV can become latent within sensory nerves and may remain dormant for years. In about 15% of individuals who have had chickenpox, conditions such as stress, aging, or immune suppression cause the viruses to reactivate, travel down the nerve they inhabit, and produce an extremely painful skin rash near the distal end of the nerve (**Figure 24.10**). This rash is known as **shingles** or **herpes zoster** (**Figure 24.11**).

Zoster, a Greek word that means belt, refers to the characteristic localization of shingle lesions along a band of skin, called a *dermatome,* which is innervated by a single sensory nerve (**Figure 24.12**). Shingles lesions form only in the dermatome associated with an infected nerve, and thus the lesions are localized and on the same side of the body as the infected nerve cell. Occasionally, lesions from latent infections of cranial nerves form in the eye, ear, or other part of the head. In some patients, the pain associated with shingles remains for months or years after the lesions have healed.

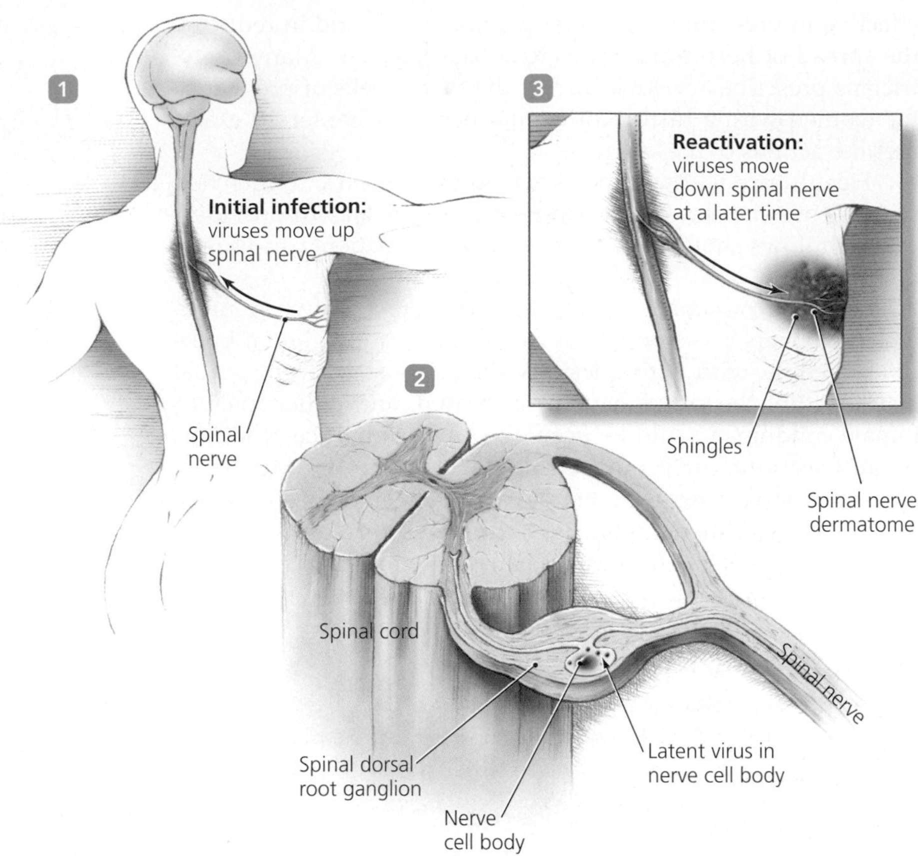

▶ **Figure 24.10 Latency and reactivation of varicella-zoster virus.** After initial infection, VZV can move up the spinal nerve and become latent in nerve cell bodies in spinal root ganglia. After subsequent reactivation (often years later), the viruses move down the spinal nerve to the skin, causing a rash known as shingles.

▼ **Figure 24.11 Shingles, the rash caused by *Varicellovirus*.** These painful lesions occur on the dermatome innervated by an infected nerve.

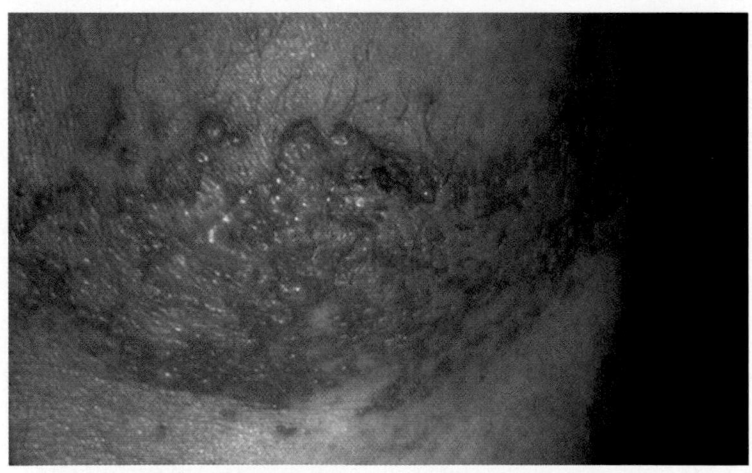

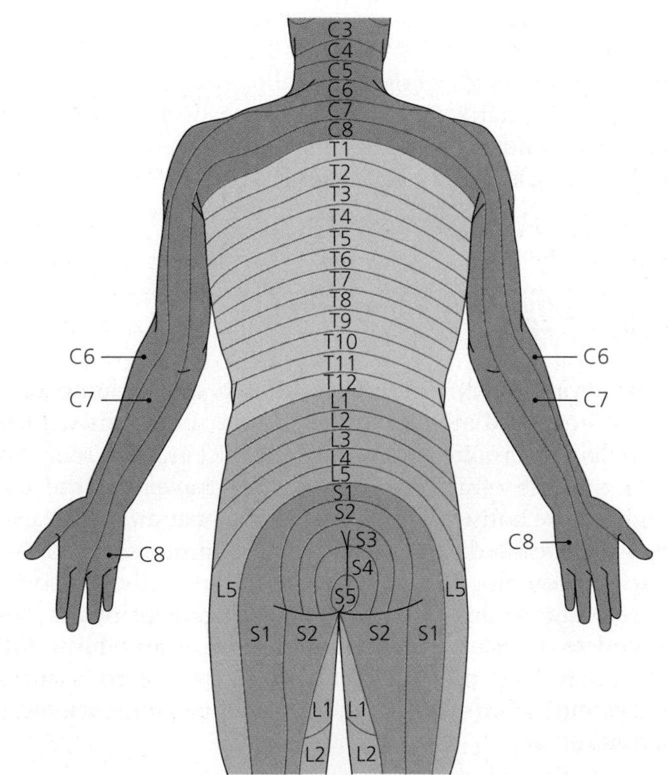

▶ **Figure 24.12 Dermatomes, the bands of skin innervated by a single sensory nerve.** The letters and numbers refer to the spinal nerve that innervates a given dermatome. Dermatomes are generally bilaterally symmetrical.

CLINICAL CASE STUDY

Grandfather's Shingles

The Davises were excited about their newborn twin boys and couldn't wait to take them to see Mr. Davis's father. Grandfather Davis was excited to see his first grandsons as well and thought their visit might help take his mind off the pain of his shingles, which had suddenly appeared only days before.

1. What virus is responsible for Grandfather Davis's shingles?
2. What likely triggered his case of shingles?
3. Are the new parents at risk of catching shingles from Grandfather Davis?
4. Are the newborns at risk of contracting shingles from their grandfather? If Mrs. Davis asked your advice about visiting her father-in-law with the twins, what would you recommend?
5. Could the twins be vaccinated against shingles before the visit?

Unlike the multiple recurrences of herpes simplex lesions, shingles only occurs once or twice in an individual's life, usually after age 45. Before vaccination became routine, the majority of children developed chickenpox, but only a small percentage of adults developed shingles, presumably because most individuals develop complete immunity to VZV following chickenpox and therefore never acquire shingles. The observation that VZV from the lesions of shingles can cause chickenpox in individuals who have not had chickenpox revealed that VZV causes both shingles and chickenpox.

Diagnosis, Treatment, and Prevention of HHV-3 (VZV) Infections

Chickenpox is usually diagnosed from the characteristic appearance of the lesions. In contrast, it is sometimes difficult to distinguish shingles from other herpesvirus lesions, though the localization within a dermatome is characteristic. Antibody tests are available to verify diagnosis of VZV infection.

Uncomplicated chickenpox is typically self limiting and requires no treatment other than relief of the symptoms with acetaminophen and antihistamines. Aspirin should not be given to children or adolescents with symptoms of chickenpox because of the risk of contracting Reye's syndrome, a condition associated with several viral diseases and aspirin usage (see **Highlight: Reye's Syndrome**).

Treatment of shingles involves management of the symptoms and bed rest. Nonadherent dressings and loose-fitting clothing may help prevent irritation of the lesions. Acyclovir provides relief from the painful rash for some patients, but it is not a cure.

HIGHLIGHT

REYE'S SYNDROME: CAN ASPIRIN HARM CHILDREN WITH VIRAL DISEASES?

In the 1900s, Dr. Ralph Reye, an Australian pathologist, observed an unusual syndrome that often followed viral infections of patients under 18 years of age. The syndrome's features include acute encephalopathy (a brain disorder), fatty infiltration of internal organs, rash, and vomiting. Confusion, coma, seizures, respiratory arrest, and death often occur in later stages of the syndrome. Reye's syndrome, as it has come to be known, has followed infections of DNA viruses such as varicella (chickenpox) and Epstein-Barr viruses, as well as those of RNA viruses such as influenzavirus (discussed in Chapter 25.) The cause of the syndrome is unknown, but it is often associated with the use of aspirin to treat symptoms of these viral infections. Consequently, it is recommended that aspirin not be used to treat fever or pain in children or adolescents, particularly when a viral cause is suspected.

It is difficult to prevent exposure to VZV because viruses are shed from patients before obvious signs appear. The U.S. Food and Drug Administration (FDA) has approved an attenuated vaccine, though the vaccine has caused some cases of shingles in Japan, and immunocompromised individuals are at risk of developing disease from the vaccine. Nevertheless, the benefits of vaccination appear to outweigh the risks, and the Centers for Disease Control and Prevention (CDC) recommends two doses of chickenpox vaccine for all children who are least one year old and for adults up to age 50 who do not have evidence of immunity to chickenpox. A single dose of shingles vaccine is recommended for all adults over age 60.

Human Herpesvirus 4 (Epstein-Barr Virus) Infections

Learning Objective

✔ Describe the diseases associated with Epstein-Barr virus.

A fourth human herpesvirus is *Epstein-Barr virus (EBV)*, which is in the genus *Lymphocryptovirus*. EBV is named after its codiscoverers. The official name for the species is *human herpesvirus 4*, or HHV-4.

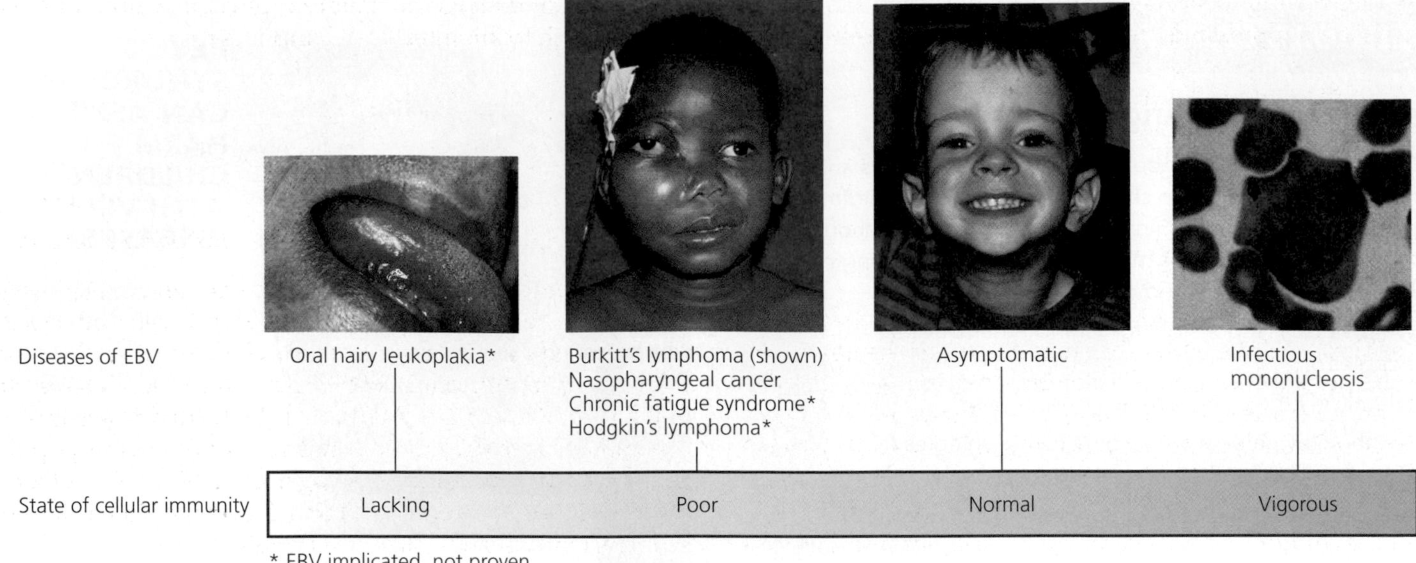

▲ **Figure 24.13 Diseases associated with Epstein-Barr virus.** Which disease results from infection with EBV appears to depend on the relative vigor of the host's cellular immune response, which itself can be related to age. Oral hairy leukoplakia, a precancerous change in the mucous membrane of the tongue, occurs in EBV-infected hosts with severely depressed cellular immunity (for example, AIDS patients). Burkitt's lymphoma, a cancer of the jaw, occurs primarily in African boys whose immune systems have been suppressed by the malaria parasite. Nasopharyngeal cancer is most common in China. EBV is implicated in Hodgkin's lymphoma and chronic fatigue syndrome, which also are associated with impaired immune function. Asymptomatic EBV infections are typical for children exposed at a young age, before the immune response has become vigorous. Infectious mononucleosis, which is characterized by enlarged B lymphocytes with lobed nuclei, results from infection with EBV during adulthood, when the cellular immune response is vigorous.

Diseases Associated with HHV-4 (EBV) Infection

Like other herpesviruses, HHV-4 (EBV) is associated with a variety of diseases. First we consider two diseases EBV is known to cause—Burkitt's lymphoma and infectious mononucleosis—before discussing several other diseases for which EBV is implicated as a cause. As you read about the diseases with which EBV is associated, bear in mind that they are in part the result of the patient's immune status, which itself can be related to the patient's age.

In the 1950s, Denis Burkitt (1911–1993) described a malignant neoplasm of the jaw, primarily in African boys **(Figure 24.13)**, that appeared to be infectious. This cancer is now known as **Burkitt's lymphoma.** Intrigued by the idea of an infectious cancer, Michael Epstein (1921–) and Yvonne Barr (1932–) cultured samples of the tumors and isolated Epstein-Barr virus, the first virus shown to be responsible for a human cancer. The way in which a latent virus such as EBV can trigger cancer was discussed in Chapter 13.

Epstein-Barr virus also causes **infectious mononucleosis** (mon′ō-noo-klē-ō′sis) **(mono).** Discovery of the connection between EBV and mono was accidental. A laboratory technician who worked with EBV and Burkitt's lymphoma contracted mononucleosis while on vacation. Upon her return to work, her serum was found to contain antibodies against EBV antigens. Serological studies of large numbers of college students have provided statistical verification that EBV causes mono.

EBV is an etiological agent of *nasopharyngeal cancer* in patients from southern China and is implicated as a cause of *chronic fatigue syndrome* in the U.S. population, *oral hairy leukoplakia* in AIDS patients, and *Hodgkin's lymphoma.* Chronic fatigue syndrome is associated with debilitating fatigue, memory loss, and enlarged lymph nodes. Hairy leukoplakia is a precancerous change in the mucous membrane of the tongue. Hodgkin's lymphoma is a treatable cancer of B lymphocytes that spreads from one lymph node to another.

CRITICAL THINKING

Whereas many doctors are convinced that Epstein-Barr virus causes chronic fatigue syndrome, others deny the association between EBV and the syndrome. Why is the etiology of chronic fatigue syndrome debated even though Epstein-Barr virus is present?

Epidemiology and Pathogenesis of HHV-4 (EBV) Infections

Transmission of Epstein-Barr viruses usually occurs via saliva, often during the sharing of drinking glasses or while kissing. ("Kissing disease" is a common name for infectious mononucleosis.) After initially infecting the epithelial cells of the pharynx and parotid salivary glands, EBV virions enter the blood. In general, having viruses in the blood is called **viremia** (vī-rē′mē-ă).

EBV invades B lymphocytes,[7] becomes latent, and immortalizes the B cells by suppressing *apoptosis* (programmed cell death). Infected B cells are one source of lymphomas, including Burkitt's and Hodgkin's lymphomas.

Infectious mononucleosis results from a "civil war" between the humoral and cellular branches of specific immunity: cytotoxic T lymphocytes of the cell-mediated branch kill infected B lymphocytes. This "civil war" is responsible for the symptoms and signs of mono, including sore throat, fever, enlargement of the spleen and liver, and fatigue.

Seventy percent of Americans over the age of 30 have antibodies against EBV. As with other herpesviruses, the age of the host at the time of infection is a determining factor in the seriousness of the disease, in this case because the competence of cytotoxic T cells is in part related to age (see Figure 24.13). Infection during childhood, which is more likely to occur in countries with poor sanitation and inadequate standards of hygiene, is usually asymptomatic because a child's cellular immune system is immature and cannot cause severe tissue damage. Where living standards are higher, childhood infection is less likely, and the postponement of infection until adolescence or later results in a more vigorous cellular immune response that produces the signs and symptoms of mononucleosis.

Cofactors appear to play a role in the development of cancers by EBV. For example, Burkitt's lymphoma is almost exclusively limited to young males exposed to malaria parasites, and the restriction of EBV-induced nasopharyngeal cancer to certain geographical regions is suggestive of cofactors in food, environment, or the genes of regional populations.

Extreme disease, such as oral hairy leukoplakia, arises in individuals with a T cell deficiency, as occurs in malnourished children, the elderly, AIDS patients, and transplant recipients. These individuals are more susceptible to EBV because infected cells are not removed by cytotoxic T lymphocytes and therefore remain a site of virus proliferation.

Diagnosis, Treatment, and Prevention of HHV-4 (EBV) Infections

Large, lobed, B lymphocytes with atypical nuclei and neutropenia are characteristic of EBV infection. Some diseases associated with EBV, such as hairy leukoplakia and Burkitt's lymphoma, are easily diagnosed by their characteristic signs. Other EBV infections, such as infectious mononucleosis, have symptoms common to many pathogens. Fluorescent antibodies directed against anti-EBV immunoglobulins or ELISA tests provide specific diagnosis.

Burkitt's lymphoma responds well to chemotherapy, and tumors can be removed from affected jaws of patients in geographic areas where surgery is available. Care of mono patients involves relief of the symptoms. Radiation or chemotherapy often successfully treats Hodgkin's lymphoma. There is no effective treatment for other EBV-induced conditions.

Prevention of EBV infection is almost impossible because the virus is widespread and transmitted readily in saliva. However, only a small proportion of EBV infections result in disease because the immune system is either too immature to cause cellular damage (in children) or cellular immunity is efficient enough to kill infected B lymphocytes.

Human Herpesvirus 5 (Cytomegalovirus) Infections

Learning Objective

✓ Describe the epidemiology of cytomegalovirus infection.

A fifth herpesvirus that affects humans is *Cytomegalovirus* (sī-tō-meg'ă-lō-vī'rŭs) (CMV), so called because cells infected with this virus become enlarged. *Cytomegalovirus* infection is one of the more common infections of humans. Studies have shown that 50% of the adult population of the United States is infected with CMV; in some other countries, 100% of the population tests positive for antibodies against CMV. Like other herpesviruses we have studied, HHV-5 becomes latent in various cells, and infection by CMV lasts for life.

Epidemiology and Pathogenesis of HHV-5 (CMV) Infections

CMV is transmitted in bodily secretions, including saliva, mucus, milk, urine, feces, semen, and cervical secretions. Individual viruses are not highly contagious, so transmission requires intimate contact involving a large exchange of secretion. Transmission usually occurs via sexual intercourse, but can result from *in utero* exposure, vaginal birth, blood transfusions, and organ transplants. CMV infects 7.5% of all neonates, making it the most prevalent viral infection in this age group.

Most people infected with CMV are asymptomatic, but fetuses, newborns, and immunodeficient patients are susceptible to severe complications of CMV infection. About 10% of congenitally infected newborns develop signs of infection, including enlarged liver and spleen, jaundice, microcephaly, and anemia. CMV may also be **teratogenic**[8] (ter'ă-tō-jen'ik)—that is, cause birth defects—when the virus infects *stem cells*[9] in an embryo or fetus. In the worst cases, mental retardation, hearing and visual damage, or death may result.

AIDS patients and other immunosuppressed adults may develop pneumonia, blindness (if the virus targets the retina), or *cytomegalovirus mononucleosis,* which is similar to infectious mononucleosis caused by Epstein-Barr virus. CMV diseases result from initial infections or from latent viruses.

Diagnosis, Treatment, and Prevention of HHV-5 (CMV) Infections

Diagnosis of CMV-induced diseases is dependent on laboratory procedures that reveal the presence of abnormally enlarged cells and inclusions within the nuclei of infected cells **(Figure 24.14)**. Viruses and antibodies against them can be detected by ELISA

[7]These lymphocytes are also known as mononuclear leukocytes, from which the name mononucleosis is derived.

[8]From Greek *teratos,* meaning monster.

[9]Formative cells, the daughter cells of which develop into several different types of mature cells.

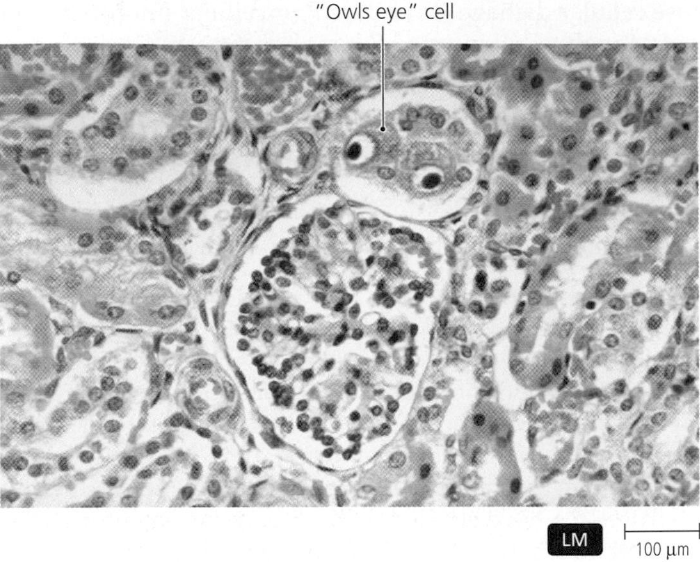

▲ **Figure 24.14 An owl's eye cell.** The cell is diagnostic for cytomegalovirus infection. Such abnormally enlarged cells have enlarged nuclei and contain inclusion bodies that are sites of viral assembly.

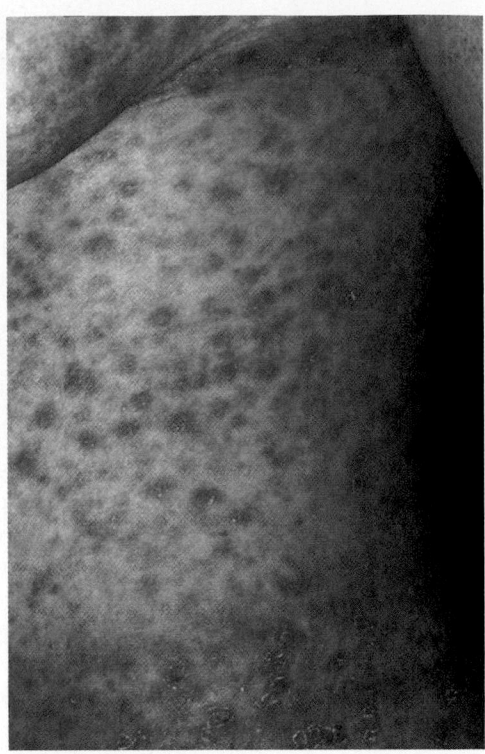

▲ **Figure 24.15 Roseola.** This rose-colored rash on the body results from infection with *human herpesvirus 6*.

tests and DNA probes. Treatment for fetuses and newborns with complications of CMV infection is difficult because in most cases damage is done before an infection is discovered.

Treatment of adults can also be frustrating. Interferon and gammaglobulin treatment can slow the release of CMV from adults but does not affect the course of disease. Physicians inject fomivirsen into an infected eye to inhibit CMV replication in retinal cells. Fomivirsen is an RNA molecule complementary to CMV mRNA. It works by binding to the mRNA, preventing translation of polypeptides critical for CMV replication. This stops progression of disease but does not eliminate the infection. There is no vaccine for preventing infection.

Abstinence, mutual monogamy, and use of condoms can reduce the chances of infection. Organs harvested for transplantation can be made safe by treatment with monoclonal antibodies against CMV, which passively reduces the CMV load before the organs are transplanted.

Other Herpesvirus Infections

Learning Objective

✓ Describe the diseases that are associated with HHV-6 and HHV-8.

Human herpesvirus 6 (HHV-6), in the genus *Roseolovirus,* causes **roseola** (rō-zē′ō′lă). Roseola is an endemic illness of children characterized by an abrupt fever, sore throat, enlarged lymph nodes, and a faint pink rash on the face, neck, trunk, and thighs **(Figure 24.15)**. The name of the disease is derived from the rose-colored rash.

Some researchers have linked HHV-6 to multiple sclerosis (MS). These investigators point to the fact that over 75% of MS patients have antibodies against the virus, that patients with higher antibody levels are more likely to develop MS later in life, and that the brain lesions of MS patients typically contain HHV-6. Further, MS patients respond well to treatment with interferon. This evidence is not proof of causation, but it provides tantalizing ideas for further research.

HHV-6 may also cause mononucleosis-like symptoms and signs, including enlargement of lymph nodes. There is some evidence that *human herpesvirus 6* may also be a cofactor in the pathogenesis of AIDS; that is, infection with HHV-6 may make individuals more susceptible to AIDS.

Human herpesvirus 8 (HHV-8, Rhadinovirus) causes Kaposi's sarcoma, a cancer often seen in AIDS patients **(Figure 24.16)**. The virus is not found in cancer-free patients or in normal tissues of victims and has been shown to make cells lining blood vessels cancerous.

Papillomaviridae and *Polyomaviridae*

Papillomaviruses and polyomaviruses each have a single molecule of double-stranded DNA contained in a small, naked, icosahedral capsid. They were previously classified in a single family, *Papovaviridae,* but have recently been separated into two families, *Papillomaviridae* and *Polyomaviridae,* based in part on the different diseases they cause.

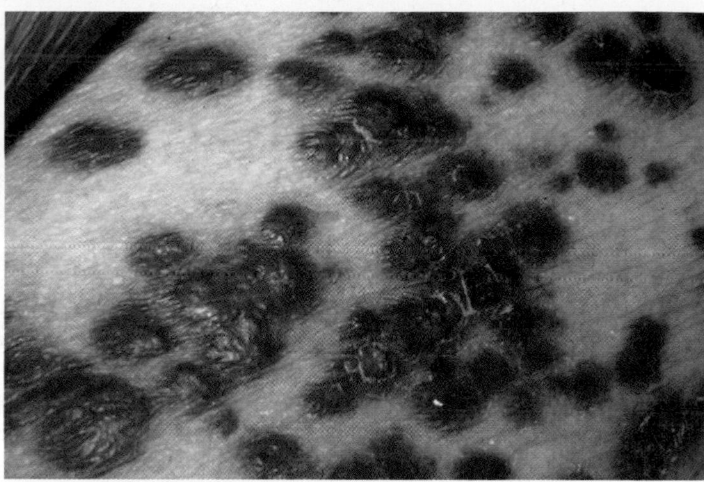

▲ **Figure 24.16 Kaposi's sarcoma.** A rare and malignant neoplasia of blood and blood vessels, this cancer is most often seen in AIDS patients. *What observations support the hypothesis that Kaposi's sarcoma is caused by HHV-8?*

Figure 24.16 *HHV-8 is not found in healthy patients, but it is found in Kaposi tumor cells. The virus also causes some cells to become cancerous.*

Papillomavirus Infections

Learning Objectives

- ✓ Describe four kinds of warts associated with papillomavirus infections.
- ✓ Describe the pathogenesis, treatment, and prevention of genital warts.

Papillomaviruses cause **papillomas** (pap-i-lō'măz), which are benign growths of the epithelium of the skin or mucous membranes commonly known as **warts.** Papillomas form on many body surfaces, but are most often found on the fingers or toes *(seed warts);* deep in the soles of the feet *(plantar warts);* on the trunk, face, elbow, or knees *(flat warts);* or on the external genitalia *(genital warts)* **(Figure 24.17)**. Genital warts range in size from almost undetectable small bumps to giant, cauliflower-like growths called **condylomata acuminata**[10] (kon-di-lō'mah-tă ă-kyū-mi-nah'tă). Over 40 varieties of papillomaviruses cause warts in humans.

Although warts are often painful and unsightly, genital warts are even more distressing because of their association with an increased risk of cancer. Papillomaviruses of genital warts precipitate anal, vaginal, penile, and oral cancers, as well as nearly all cervical cancer. Papillomavirus infections appear to

[10]From Greek *kondyloma*, meaning knob, and Latin *acuminatus*, meaning pointed.

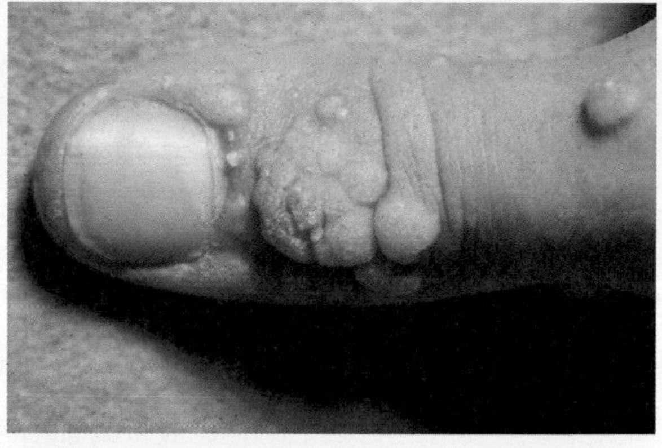

(a)

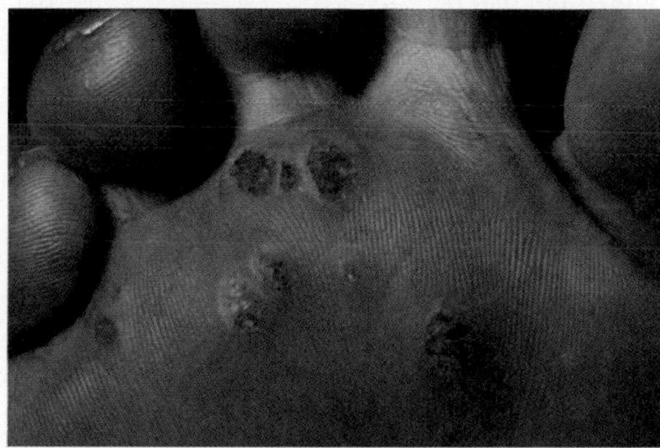

(b)

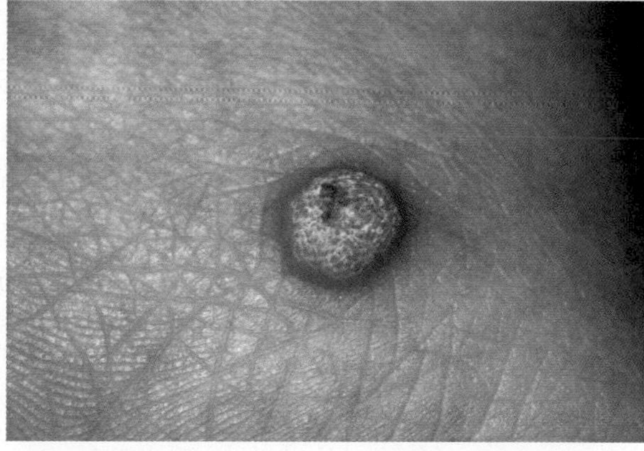

(c)

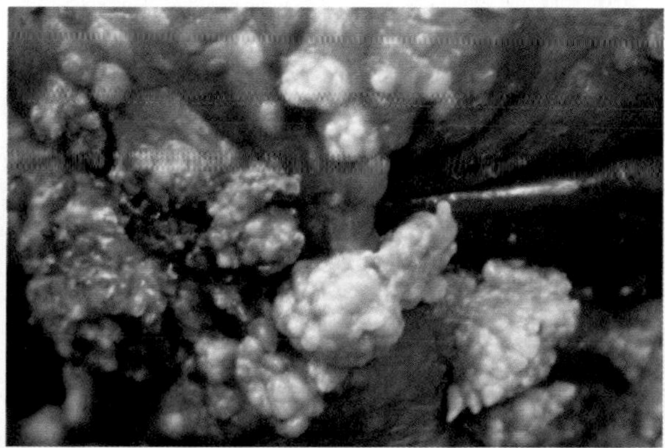

(d)

➤ **Figure 24.17 The various kinds of warts, the lesions caused by infections with papillomaviruses. (a)** Seed warts of the fingers. **(b)** Plantar warts on the sole of a foot. **(c)** Flat warts, which occur on the trunk, face, elbows, or knees. **(d)** Condylomata acuminata on a vulva. Such genital warts are considered precancerous.

CLINICAL CASE STUDY

A Child with Warts

Ten-year-old Rudy has several large warts on the fingers of his right hand. They do not hurt, but their unsightly appearance causes him to shy away from people. He is afraid to shake hands, or to play with other children, out of fear that he may transfer the warts to them. Initially, his mother tells him not to worry about them, but Rudy cannot help feeling self-conscious. Furthermore, Rudy fears that the warts may somehow spread on his own body. After consulting with a physician, Rudy and his mother decide to have the warts surgically removed.

1. Is it possible that Rudy's warts will spread to other parts of his body?
2. Is it possible for someone else to "catch" Rudy's warts by shaking his hand?
3. If Rudy's warts disappear without treatment, are they likely to return to the same sites on his hand?
4. Are Rudy's warts likely to become cancerous?
5. Following surgical removal of the warts, is there a chance that Rudy will develop warts again?

double the risk of developing head and neck cancer and increase the risk of cancer in the tonsils fourteenfold.

Epidemiology and Pathogenesis of Papillomavirus Infections

Papillomaviruses are transmitted via direct contact and, because they are stable outside the body, via fomites. They can also be spread from one location to another on a given person by a process called *autoinoculation.*

Viruses that cause genital warts invade the skin and mucous membranes of the penis (particularly when it is uncircumcised), vagina, and anus during sexual intercourse. The incubation time from infection to the development of the wart usually is three to four months. Genital warts are among the more common sexually transmitted diseases. Over 500,000 new cases are reported each year in the United States, most among young adults.

Diagnosis, Treatment, and Prevention of Papillomavirus Infections

Diagnosis of warts is usually a simple matter of observation, though only DNA probes can elucidate the exact strain of papillomavirus involved. Warts usually regress over time as the cellular immune system recognizes and attacks virally infected cells. Cosmetic concerns and the pain associated with some warts may necessitate removing infected tissue via surgery, freezing, cauterization (burning), laser, or the use of caustic chemicals. These techniques are not entirely satisfactory because viruses may remain latent in neighboring tissue and produce new warts at a later time. Laser surgery has the added risk of causing viruses to become airborne; some physicians have developed warts in their noses after inhaling airborne viruses during laser treatment. Over-the-counter remedies containing salicylic acid often reduce and remove warts. One of the stranger treatments is also one of the more effective—cover a wart with duct tape for several weeks, and it will likely disappear. Scientists do not know why this method works.

Effective treatment of cancers caused by papillomaviruses often depends on early diagnosis resulting from thorough inspection of the genitalia in both sexes, and in women by a Papanicolaou (Pap) smear to screen for cervical cancer. Treatment involves radiation or chemical therapy directed against reproducing tumor cells. Advanced cases of genital cancer necessitate removal of the entire diseased organ.

Prevention of most types of warts is difficult, but prevention of genital warts is possible by abstinence or mutual monogamy. Using recombinant DNA techniques, scientists have created a vaccine (Gardasil) that successfully prevents infection by the most common strains of sexually transmitted papillomaviruses, including the strains that cause most penile and cervical cancers. The CDC recommends three doses of HPV vaccine for all females 11–26 years of age.

The evidence for the effectiveness of condoms in preventing the spread of genital wart viruses is equivocal—one study suggests that condoms reduce the risk of acquiring genital warts by 25–50%, whereas other studies have revealed no decrease in the acquisition of warts by condom users or their partners. Researchers have found that circumcised men and their female sexual partners have about 35% less chance of getting genital cancer than do uncircumcised men or their partners. This research suggests that if all men were circumcised, the worldwide incidence of cervical cancer would be reduced at least 43%.

Clinical Case Study: A Child with Warts concerns a young patient with papillomas.

Polyomavirus Infections

Learning Objective

✓ Describe two diseases caused by polyomavirus infection in humans.

The search for cancer-causing viruses led scientists to the discovery of a group of viruses capable of causing several different tumors in animals and humans. Researchers named these viruses **polyomaviruses** by combining *poly* (meaning "many") and *oma* (meaning "tumor").

Polyomaviruses cause other diseases as well. Two human polyomaviruses, the *BK* and *JC* viruses, are endemic worldwide. The initials are derived from the names of the patients from which the viruses were first isolated. Most everyone appears to be infected with BK and JC by age 15, probably via the respiratory route. The viruses initially infect lymphocytes, and what happens subsequently depends at least in part on the state of an individual's immune system. If the immune response is normal, latent infections are typically prevented; if the immune

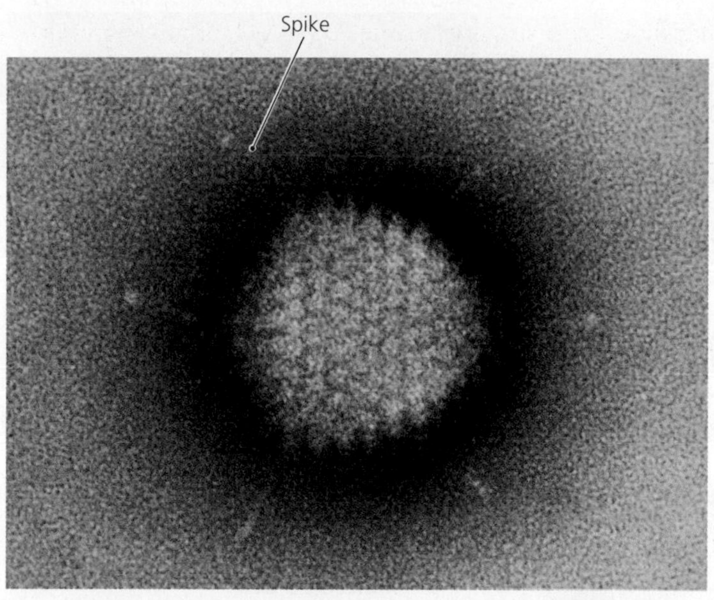

▲ **Figure 24.18 Adenovirus.** Notice the distinctive spikes.

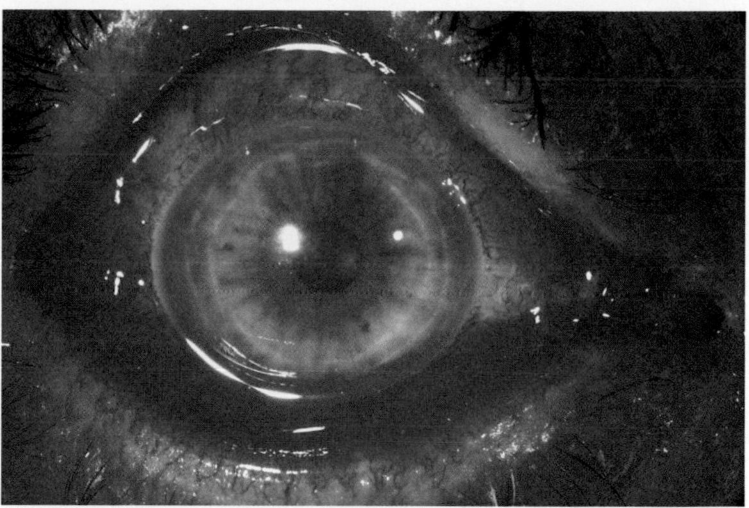

▲ **Figure 24.19 Adenoviral conjunctivitis (pinkeye).** *What signs and symptoms can result from infections with various adenoviruses?*

Figure 24.19 *Respiratory adenoviruses cause sore throat, cough, fever, headache, and malaise; intestinal adenoviruses cause mild diarrhea in children; still other adenoviruses cause conjunctivitis (pinkeye).*

system is compromised, however, then latent infections can become established in the kidneys, where reactivation of the viruses leads to the release of new virions. In the case of BK virus, the new virions can cause potentially severe urinary tract infections.

In individuals with latent infections of JC virus, reactivated virions are shed from kidney cells into the blood, spreading the virions throughout the body. Virions that reach the brain of an immunosuppressed individual cause a rare and fatal disease called **progressive multifocal leukoencephalopathy (PML),** in which the viruses infect and kill cells called oligodendrocytes in the white matter of the central nervous system. Oligodendrocytes produce myelin, a white lipid that surrounds and insulates some nerve cells. As myelin in multiple sites is lost, PML patients progressively lose various brain functions, as the name of the disease implies, because the conduction of nerve impulses becomes increasingly impaired. Vision, speech, coordination, and cognitive skills are affected. Eventually paralysis and death result.

Beta interferon treatments can prevent kidney damage by BK. The receptor for JC virus is a serotonin receptor, so researchers hope that serotonin receptor inhibitors will block JC virus. Unfortunately, by the time a diagnosis can be made, damage to the brain is often severe and irreversible.

Adenoviridae

Learning Objective

✓ Discuss the epidemiology and pathogenicity of diseases caused by adenoviruses.

The last family of double-stranded DNA viruses we will discuss is *Adenoviridae.* Adenoviruses have a single, linear dsDNA genome contained in a naked polyhedral capsid with spikes (**Figure 24.18**). Adenoviruses are so named because they were first discovered infecting cultures of human adenoid[11] cells.

Adenoviruses are but one of the many causative agents of "common colds." Even though colds typically have a common set of signs and symptoms, their causes are quite varied. At least 30 different respiratory adenoviruses (DNA viruses) and over 100 RNA viruses cause colds.

Adenoviruses, which are spread via respiratory droplets, are very stable outside the body and can survive on fomites and in improperly chlorinated drinking water. Respiratory adenovirus infections become established when adenoviruses are taken into cells lining the respiratory tract via endocytosis. The resulting infections have a variety of manifestations, including sneezing, sore throat, cough, headache, and malaise. (Table 25.1 on p. 708 compares the signs and symptoms of colds with those of other respiratory infections.) For some reason, epidemics of respiratory adenoviruses occur on military bases but rarely under the similar conditions found in college dormitories.

Adenoviruses can also cause infections in other systems of the body. After entering cells lining the intestinal tract, intestinal adenoviruses can cause mild diarrhea in children. Still other adenoviruses infect the conjunctiva, resulting in conjunctivitis, which is also called *pinkeye* (**Figure 24.19**). Human adenovirus 36 is suspected of being a cause of obesity in humans. **Highlight: Catch a Cold and Catch Obesity** on p. 698 examines this interesting hypothesis.

Adenoviruses often produce semi-crystallized viral masses within the nuclei of infected cells (see Highlight box), a feature that is diagnostic for adenoviral infections. Although this feature

[11]Adenoids are swollen pharyngeal tonsils, which are masses of lymphatic tissue located in the pharynx, posterior to the nose.

HIGHLIGHT

CATCH A COLD AND CATCH OBESITY?

Your roommate sneezes, and you get infected with obesity. Sound far-fetched? If researchers at Louisiana State University in Baton Rouge are correct, many cases of obesity may be blamed on adenoviruses—a family of DNA viruses more usually associated with common colds.

Researchers have shown that human adenovirus 36 (Ad-36) (and to a lesser extent its relatives Ad-37 and Ad-5) directly affects adipocytes (fat cells) in animals and humans. Ad-36 is present in about 30% of obese people but in only about 11% of the non-obese, and infected laboratory animals gain more weight than their identical siblings eating the same amount of food.

In identical human twins where one sibling is infected with Ad-36 and the other is not, the infected twins are heavier and fatter than their uninfected cotwins. At the cellular level, stem cells infected with Ad-36 more frequently become adipocytes than do uninfected stem cells, and infected adipocytes make and store more fat than uninfected fat cells.

Even more exciting than this circumstantial evidence, the researchers have shown that Ad-36 regulates a single gene in humans, a gene that is necessary and solely sufficient for fat production. Moreover, Ad-36 enhances a cell's ability to take up glucose, which can then be anabolized into fat and stored.

Perhaps someday soon, more effective anti-cold remedies may also help you lose weight.

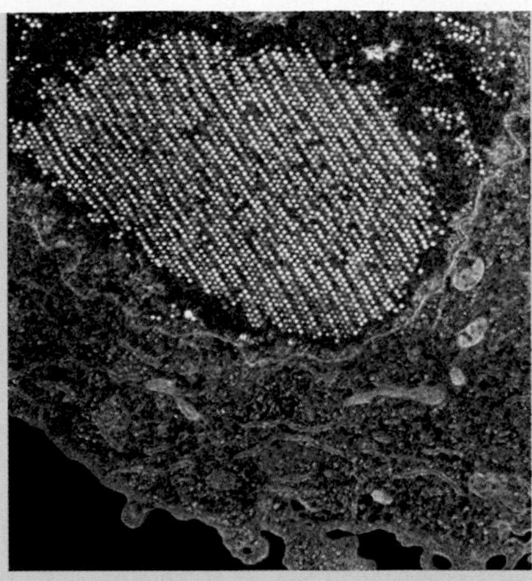

▲ ***Semi-crystalline masses of adenoviruses replicating in the nucleus of an infected cell.***

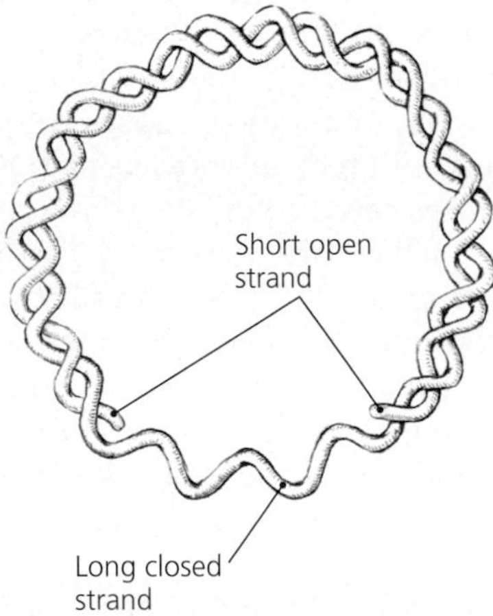

▲ **Figure 24.20 The genome of a hepadnavirus.** This unusual genome is composed of incompletely double-stranded DNA resulting from incomplete DNA replication.

is similar in appearance to the nuclear inclusions seen in cytomegalovirus infections, adenoviruses do not cause infected cells to enlarge.

Adenovirus infection can be treated during the early stages with gamma interferon, and a live, attenuated vaccine is available. The vaccine is used only for military recruits, because some adenoviruses of animals are known to be oncogenic, and there is concern about widespread use of a potentially oncogenic vaccine.

Hepadnaviridae

Viruses of the family *Hepadnaviridae*[12] are enveloped DNA viruses with icosahedral capsids that invade and replicate in liver cells. From these characteristics, human **hepatitis B virus (HBV)** acquires the name of its genus, *Orthohepadnavirus.* Its scientific species name, *hepatitis B virus,* is also its common name.

The genome of a hepadnavirus is unique in that it is partly double-stranded DNA and partly single-stranded DNA (**Figure 24.20**). The amount of ssDNA in a hepadnaviral genome varies. Some virions are almost entirely dsDNA, whereas others have a significant amount of ssDNA. How does this variation in genomic structure occur?

HBV replicates through an RNA intermediary, a phenomenon that is unique among DNA viruses. During replication, one strand of the DNA genome of hepatitis B virus is transcribed into RNA by an enzyme called *reverse transcriptase.* This process is the opposite of normal transcription, thus the name of the enzyme. The RNA molecule then serves as a template for a new DNA strand. The result is an RNA-DNA hybrid molecule. Next, the RNA is removed and a DNA complement of the DNA strand is synthesized, forming dsDNA. Incomplete dsDNA results when virions are assembled before the complementary DNA strand is completely copied. Thus the earlier that assembly occurs, the more ssDNA will remain in the genome; the later that assembly occurs, the more complete replication will be, and the more dsDNA will be in the genome.

[12]From Greek *hepar,* meaning liver, and DNA.

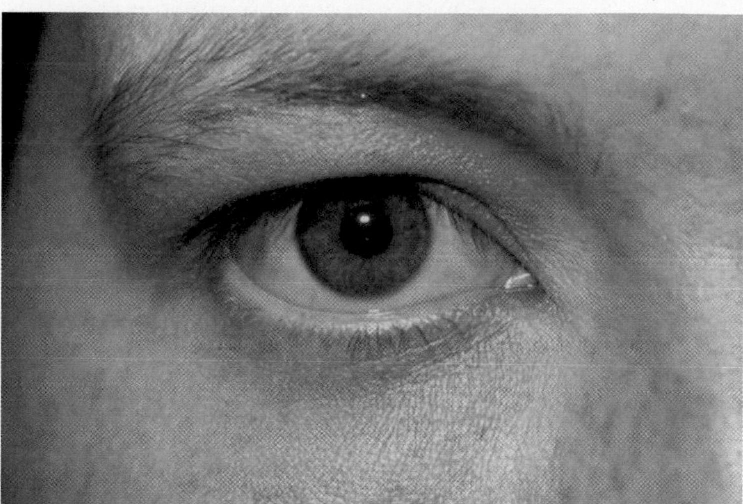

▲ **Figure 24.21 Jaundice.** This condition is characterized by yellow skin and eyes. One cause of jaundice is liver damage by hepatitis B virus.

Hepatitis B Infections

Learning Objective

✓ Describe the epidemiology, treatment, and prevention of hepatitis B infections.

Hepatitis (hep-ă-tī′tis) is an inflammatory condition of the liver. The liver has many functions, including making blood-clotting factors, storing sugar and other nutrients, assisting in the digestion of lipids, and removing wastes from the blood. When a patient's liver is severely damaged by viral infection, all these functions are disturbed. **Jaundice** (jawn′dis), a yellowing of the skin and eyes, occurs when a greenish-yellow waste product called *bilirubin* accumulates in the blood **(Figure 24.21)**. Hepatitis is also characterized by enlargement of the liver, abdominal distress, and bleeding into the skin and internal organs. The patient becomes stuporous and eventually goes into a coma as a result of the accumulation of nitrogenous wastes in the blood.

Hepatitis B virus is the only DNA virus that causes hepatitis; all other cases of viral hepatitis result from RNA viruses (see Table 25.3 on p. 711). Hepatitis B is also called *serum hepatitis.* HBV infection results in serious liver damage in less than 10% of cases, but co-infection with hepatitis D virus (an RNA virus also called *delta agent*) increases the risk of permanent liver damage.

Epidemiology and Pathogenesis of HBV Infections

Hepatitis B viruses replicate in liver cells and are released by exocytosis rather than cell lysis. Infected liver cells thus serve as a source for the continual release of virions into the blood, resulting in billions of virions per milliliter of blood. Virions in the blood are shed into saliva, semen, and vaginal secretions.

HBV is transmitted when infected bodily fluids, particularly blood, come into contact with breaks in the skin or mucous membranes. The viral load is typically so high and the infective dose so low that infection can result from sharing a razor or toothbrush, making it difficult to determine the source of infection. Improperly sterilized needles used in acupuncture, ear piercing, tattooing, intravenous drug usage, and blood transfusion have spread the virions, which can remain infective for at least a week outside the body. They may also be spread by sexual intercourse, especially anal intercourse, which can cause tearing of the rectal lining. Babies can become infected during childbirth.

Many infected individuals are asymptomatic, and most manifest only mild jaundice, low-grade fever, and malaise. Severe liver damage is most likely to result from such asymptomatic, chronic infections after 20–40 years. The CDC estimates that approximately 350 million people worldwide have chronic

MICROBES AT A GLANCE

Adenovirus

Taxonomy: Family *Adenoviridae*

Genome and morphology: Single molecule of double-stranded DNA; naked icosahedral capsid, 60–90 nm with spikes

Host: Various species infect mammals, birds, frogs, or fungi

Virulence factors: Infection through aerosol, ingestion, hand-to-eye transfer, or sexual contact

Diseases caused: Respiratory illness, such as common cold, mild diarrhea, conjunctivitis

Treatment for disease: Gamma interferon for early stages

Prevention of disease: Get attenuated vaccine; avoid contact with infected people or contaminated food or beverage

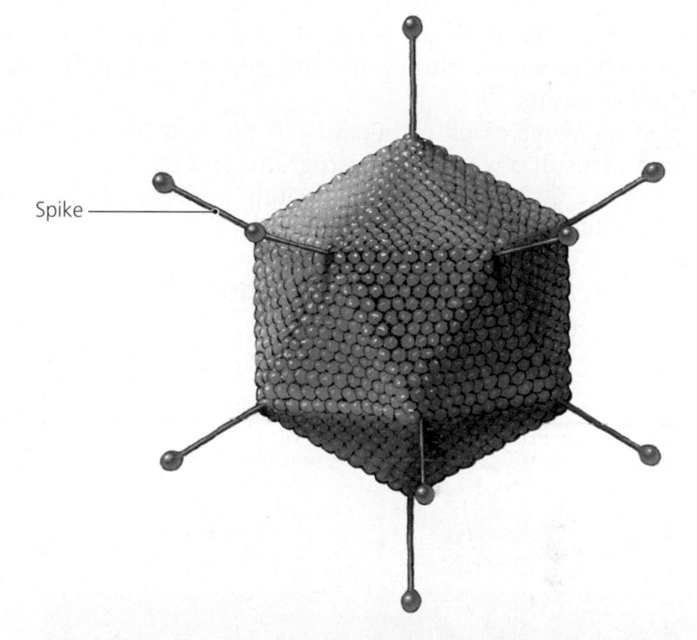

Download the Microbe at a Glance flashcards from the Study Area at www.masteringmicrobiology.com.

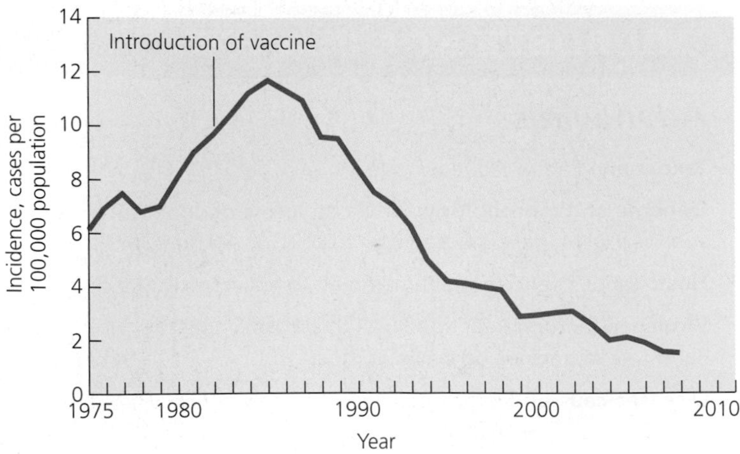

▲ **Figure 24.22 Estimated incidence of acute hepatitis B in the United States.**

infections. The carrier state is age related: newborns are much more likely to remain chronically infected than are individuals infected as adults.

Today, fewer cases of hepatitis B are seen in the United States as a result of vaccination programs and safer sexual practices (**Figure 24.22**). Still, approximately 3000 people die each year from hepatitis B infection.

Diagnosis, Treatment, and Prevention of HBV Infections

Laboratory diagnosis of hepatitis B infection involves the use of labeled antibodies to detect the presence of viral antigens released from HBV-infected cells. The bodily fluids of infected individuals contain copious amounts of viral protein of three forms: so-called *Dane particles, spherical particles,* and *filamentous particles* (**Figure 24.23**). Dane particles are complete and infective virions, whereas spherical and filamentous particles are merely capsids composed of surface antigens without genomes. Within infected individuals, the release of so much excess viral surface antigen in the form of spherical and filamentous particles in effect serves as a decoy: the binding of antibody to empty capsids reduces the effective antibody response against infective Dane particles. The large amount of viral antigen released provides one benefit to a patient: it ensures a plentiful substrate for binding labeled antibodies in diagnostic tests.

There is no universally effective treatment for hepatitis B, though alpha interferon, lamivudine, or entecavir help in 40% of cases. Liver transplant is the only treatment for end-stage chronic hepatitis B.

Prevention is possible because an effective vaccine is available. Three doses of HBV vaccine result in protection against the virus in 95% of individuals. Several studies indicate that immunity against HBV lasts for at least 23 years, and probably for life. Vaccination is recommended for high-risk groups, including health care workers, illegal drug users, people receiving blood or blood products, and the sexually promiscuous, especially those who engage in anal intercourse. The CDC recommends HBV immunization for all children because children are at a greater risk of chronic infection, life-threatening liver damage, and (as we will see shortly) liver cancer.

To prevent the spread of HBV, special care should be taken with needles and other sharp instruments, especially in health care settings. Unlike most enveloped viruses, HBV is resistant to detergents, ether, freezing, acid, and moderate heat, but contaminated materials can be disinfected with 10% bleach solutions.

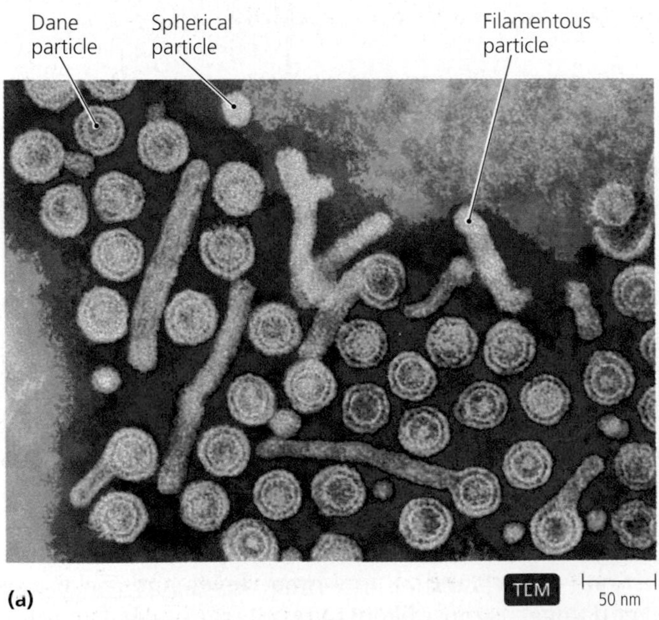

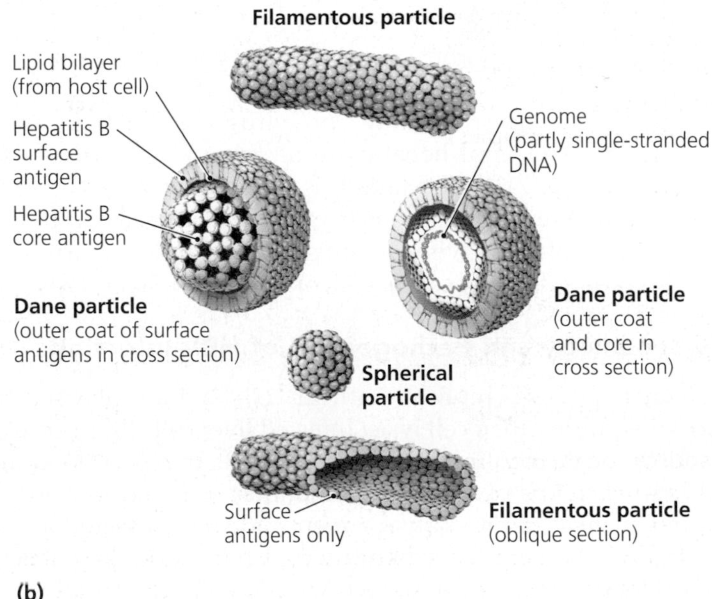

▲ **Figure 24.23 The three types of viral particles produced by hepatitis B viruses.** Dane bodies are complete virions, whereas filamentous particles and spherical particles are capsomeres that have assembled without genomes. **(a)** Micrograph. **(b)** Artist's rendition.

Abstinence or mutually faithful monogamy is the only sure way to prevent sexually transmitted infection.

The Role of Hepatitis B Virus in Hepatic Cancer

Learning Objective

✓ Describe evidence that hepatitis B virus causes hepatic cancer.

HBV has been shown to be associated with hepatic cancer based on the following evidence:

- Epidemiological studies reveal large numbers of hepatic cancer cases in geographic areas that also have a high prevalence of HBV.
- The HBV genome has been found integrated into hepatic cancer cells.
- Hepatic cancer cells typically express HBV antigen.
- Chronic carriers of HBV are 200 times more likely to develop hepatic cancer than are noncarriers.

How does HBV produce cancer? Perhaps the gap in the double strand of DNA allows HBV to easily integrate into the DNA of a liver cell, in the process possibly activating oncogenes or suppressing oncogene repressor genes. Another theory is that repair and cell growth in response to liver damage proceeds out of control, resulting in cancer. In any case, because of its association with HBV, hepatic cancer may become the first cancer eliminated as a result of vaccination.

Parvoviridae

Learning Objective

✓ Describe the pathogenesis of erythema infectiosum.

Parvoviruses are the only pathogens of humans with a single-stranded DNA (ssDNA) genome. They are also the smallest of the DNA viruses and have an icosahedral capsid. They cause a number of endemic diseases of animals, including a potentially fatal disease of puppies.

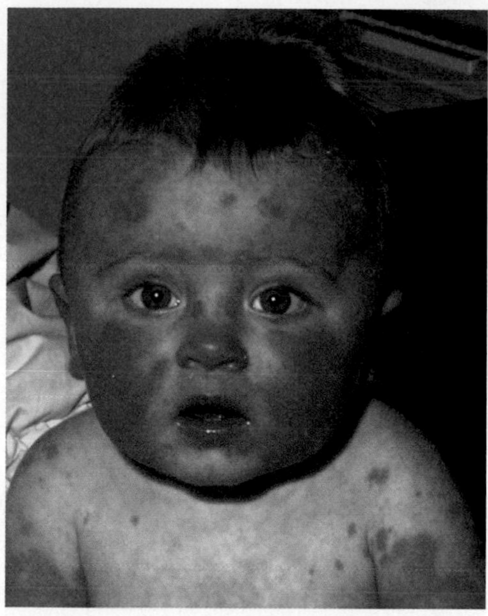

▲ **Figure 24.24 A case of erythema infectiosum (fifth disease).** Distinct reddening of facial skin that resembles the result of a slap is characteristic of this parvovirus-caused disease.

The primary parvovirus of humans is *B19 virus* in the genus *Erythrovirus,* which is the cause of **erythema infectiosum** (er-ĭ-thē′mă in-fek-shē-ō′sŭm), also known as **fifth disease** because it was the fifth childhood rash[13] on a 1905 list. Erythema infectiosum, as the name implies, is an infectious reddening of the skin, beginning on the cheeks **(Figure 24.24)** and appearing later on the arms, thighs, buttocks, and trunk. The rash is distinctive in appearance and usually sufficient for diagnosis. As the rash progresses, earlier lesions fade, though they can reappear if the patient is exposed to sunlight. Sunlight aggravates the eruptions at all stages. No treatment is available.

Table 24.2 on p. 702 summarizes features of DNA viruses of humans.

[13]The other four rashes are those of scarlet fever, rubella (German measles), roseola, and measles.

BENEFICIAL MICROBES

COCAINE NO BRAINER

Cocaine addiction is a major health and societal problem in America. The drug allows the accumulation of a neurotransmitter that stimulates an intense sense of joy and feelings of increased confidence. Simply understood, addiction results when people sense a physical or mental need for the effect. Chemical agents have proved disappointing for treating cocaine addiction, but a microbe may change this state of affairs.

Scientists at the Scripps Research Institute in La Jolla, California, have engineered a filamentous bacteriophage that displays cocaine-binding antibodies on its surface. In the central nervous system, the phage absorbs cocaine, preventing the drug from interacting with neurons and mitigating the cocaine's effect. The scientists chose a bacteriophage to deliver the antibodies because bacteriophages can enter directly into the brain through the nose. Perhaps soon addicts can snort a treatment.

TABLE 24.2

Taxonomy and Characteristics of DNA Viruses of Humans

Family	Strand Type	Enveloped or Naked	Capsid Symmetry	Size (diameter, nm)	Representative Genera (disease)
Poxviridae	Double	Enveloped	Complex	200–300	*Orthopoxvirus* (smallpox, cowpox), *Molluscipoxvirus* (molluscum contagiosum)
Herpesviridae	Double	Enveloped	Icosahedral	150–200	*Simplexvirus*—herpes, type 1 (fever blisters, respiratory infections, encephalitis), type 2 (genital infections), *Varicellovirus* (chickenpox), *Lymphocryptovirus* Epstein-Barr virus (infectious mononucleosis, Burkitt's lymphoma), *Cytomegalovirus* (birth defect), *Roseolovirus* (roseola)
Papillomaviridae	Double	Naked	Icosahedral	45–55	*Papillomavirus* (benign tumors, warts, cervical and penile cancers)
Polyomaviridae	Double	Naked	Icosahedral	45–55	*Polyomavirus* (progressive multifocal leukoencephalopathy)
Adenoviridae	Double	Naked	Icosahedral	60–90	*Mastadenovirus* (conjunctivitis, respiratory infections)
Hepadnaviridae	Partial single and partial double	Enveloped	Icosahedral	42	*Orthohepadnavirus* (hepatitis B)
Parvoviridae	Single	Naked	Icosahedral	18–26	*Erythrovirus* (fifth disease)

Chapter Summary

Poxviridae (pp. 682–684)

1. Poxviruses, among the largest of viruses, cause skin lesions that begin as flat, red **macules,** enlarge into **papules,** fill with clear fluid to become **vesicles,** and then become pus-filled **pustules,** also called **pocks** or **pox.** These crust over and may scar.
2. **Smallpox,** eradicated in nature by 1980, was caused by a poxvirus **(variola).** The smallpox virus strain commonly known as **variola minor** caused less-severe cases of smallpox than **variola major.** Governments maintain smallpox viruses in secure laboratories as research tools for studies in pathogenicity, vaccination and recombinant DNA technology, and as material for developing protection against bioterrorism attacks.
3. The poxvirus *Molluscipoxvirus* causes **molluscum contagiosum,** which results in tumorlike growths on the skin.
4. Although monkeypox and cowpox can infect humans, such infections are rare.

Herpesviridae (pp. 684–694)

1. Herpesviruses include viruses that cause fever blisters, genital herpes, chickenpox, shingles, mononucleosis, and cancer. Individual species are named **human herpesviruses 1** through **8** (HHV-1 through HHV-8).
2. *Human herpesvirus 1* (HHV-1), which was known previously as herpes simplex type 1 (HSV-1), and *human herpesvirus 2* (HHV-2) (previously HSV-2) typically produce painful skin lesions on lips and genitalia, respectively. Oral herpes lesions are called **fever blisters** or **cold sores.**
3. HHV-1 and HHV-2 enter sensory nerve cells, where they may remain latent for years. Lesions recur in about two-thirds of patients in response to emotional stress, sunlight, immune suppression, or other factors that stimulate the latent viruses to reactivate.
4. **Ocular herpes** usually affects one eye to various degrees, from recurring discomfort and light sensitivity to blindness.
5. A **whitlow** is an inflamed blister resulting from infection of HHV-1 and HHV-2 via a cut or break in the skin.
6. Infections of HHV-1 and HHV-2 are usually treatable and are often preventable.
7. **Varicella (chickenpox)** in children and **herpes zoster (shingles)** in adults are both caused by *human herpesvirus 3,* which is also known as **varicella-zoster virus (VZV),** in the genus *Varicellovirus.* Skin lesions progress from macules to papules to vesicles to crusts. Latent viruses in the sensory nerves can be activated in adults by stress, age, or immune suppression.
8. **Burkitt's lymphoma** is an infectious cancer caused by *human herpesvirus 4,* which is also known as Epstein-Barr virus (EBV). This herpesvirus also causes **infectious mononucleosis (mono).** EBV is also implicated in chronic fatigue syndrome, Hodgkin's lymphoma, nasopharyngeal cancer, and in hairy leukoplakia seen in AIDS patients.
9. The *human herpesvirus 5,* cytomegalovirus (CMV), causes infected cells to enlarge. This virus is widespread, transmitted by bodily secretions, and often asymptomatic, though infections in immunosuppressed adults and congenitally infected newborns can be severe and may be fatal.
10. *Human herpesvirus 6 (HHV-6)* causes **roseola,** a rose-colored rash on the face of children.
11. HHV-7 is an orphan virus; that is, it causes no known disease.
12. *Human herpesvirus 8 (HHV-8)* causes Kaposi's sarcoma, a cancer often seen in AIDS patients.

Papillomaviridae and *Polyomaviridae* (pp. 694–697)

1. A **wart,** also known as a **papilloma,** is a benign growth of epithelium or mucous membrane caused by a papillomavirus. Seed warts are most often found on fingers or toes; plantar warts on the soles of the feet; flat warts on the trunk, face, elbow, or knees; and genital warts on the external genitalia. **Condylomata acuminata** are large, cauliflower-like genital warts.
2. **Polyomaviruses,** such as BK and JC viruses, infect the kidneys of most people, but cause disease only in immunosuppressed patients. In these cases, BK viruses compromise kidney function, and JC viruses cause **progressive multifocal leukoencephalopathy (PML),** which results in severe, irreversible brain damage.

Adenoviridae (pp. 697–698)

1. Adenoviruses are one cause of the "common cold" and a cause of one form of conjunctivitis (pinkeye).

Hepadnaviridae (pp. 698–701)

1. Hepadnaviruses are unique in being partially dsDNAand partially ssDNA—a condition resulting from incomplete DNA replication involving an RNA intermediary.
2. **Hepatitis B virus (HBV)** infects the liver, is transmitted in blood and other bodily fluids, and is the only DNA virus that causes **hepatitis.** A vaccine against HBV is available, but no treatment is universally acceptable.
3. HBV is implicated as a cause of liver cancer.

Parvoviridae (pp. 701–702)

1. **Parvoviruses,** the only genus of human pathogens with ssDNA genomes, include a virus that causes a fatal disease in puppies, and *B19 virus,* which causes **erythema infectiosum (fifth disease),** a harmless red rash occurring in children.

Questions for Review

Answers to the Questions for Review (except Short Answer questions) begin on p. A-1.

Multiple Choice

1. DNA viruses in which of the following families are relatively large and thus potentially well suited for vaccinations and the introduction of genetic material in gene therapy?
   a. *Herpesviridae*
   b. *Poxviridae*
   c. *Papillomaviridae*
   d. *Hepadnaviridae*
2. For which of the following reasons are most animal poxviruses unable to infect humans?
   a. Affected animals are not in frequent contact with humans.
   b. The human immune system makes it impossible for the foreign viral particles to reproduce effectively.
   c. Attachment to human cells is unlikely.
   d. Human cells lack the necessary enzymes for infection.
3. The initial flat, red skin lesions of poxviruses are called
   a. macules.
   b. papules.
   c. pustules.
   d. pocks.
4. Which of the following statements is true concerning variola major?
   a. It carries a mortality rate of less than 1%.
   b. It affects internal organs before appearing on the skin.
   c. It has been totally eradicated from the Earth.
   d. The skin lesions it causes are smooth, waxy, tumorlike nodules on the face.
5. Which of the following herpesvirus infections would be potentially most serious?
   a. a whitlow
   b. ocular herpes
   c. shingles
   d. cytomegalovirus in fetuses
6. A man experienced a laboratory accident in which he was infected with adenovirus. What signs or symptoms might he exhibit?
   a. fever
   b. headache
   c. cough
   d. pinkeye
   e. all of the above
7. Which of the following is an inflammatory condition of the liver?
   a. fifth disease
   b. PML
   c. seed warts
   d. hepatitis
8. Which of the following statements is *false*?
   a. B19 virus is the primary parvovirus of humans.
   b. Erythema infectiosum is caused by a parvovirus.
   c. In children, parvovirus infections are accompanied by a high mortality rate.
   d. Parvovirus infection in humans results in infectious reddening of the skin.
9. A distinguishing feature of poxvirus is
   a. its large size.
   b. a polyhedral capsid.
   c. the type of RNA it contains.
   d. the production of several types of warts.
10. Monkeypox has been diagnosed in several humans in Zaire. What might be recommended to prevent further risk of infection?
    a. Catch the monkeys for inoculation with monkeypox vaccine and then release them.
    b. Remove infected tissues from humans with chemicals, by surgery, or by freezing.
    c. Reinstate smallpox vaccinations for the population of Zaire.
    d. The diagnosis must have been incorrect, because humans are unaffected by monkeypox.

11. Which of the following viral families is most likely to contain viruses that exist in a latent state in humans?
    a. *Herpesviridae*
    b. *Poxviridae*
    c. *Adenoviridae*
    d. *Parvoviridae*

12. If the envelope of a particular virus were unstable outside the host's body, which of the following statements would you expect to be true concerning this virus?
    a. It would be a dsRNA virus.
    b. It would be transmitted by intimate contact.
    c. Touching a doorknob would easily transmit it.
    d. That virus would eventually cease to be a threat to the population.

13. Being habitually careful not to touch or rub your eyes with unwashed hands would reduce your risk of contracting
    a. chickenpox.
    b. infectious mononucleosis.
    c. seed warts.
    d. a cold.

14. *Human herpesvirus 2*
    a. can cause genital herpes.
    b. may infect a baby at birth.
    c. causes about 10% of cold sores.
    d. all of the above

15. Which of the following viruses is latent and can reactivate to cause further disease?
    a. herpesvirus
    b. papillomavirus
    c. adenovirus
    d. parvovirus

Labeling

Label the successive stages of skin lesions as exemplified by smallpox.

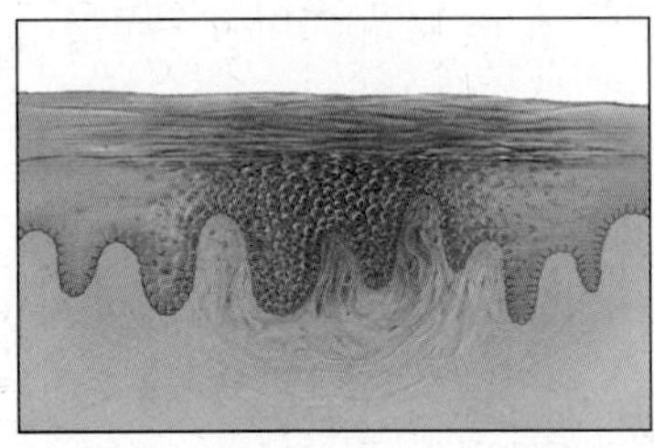

1 ____________

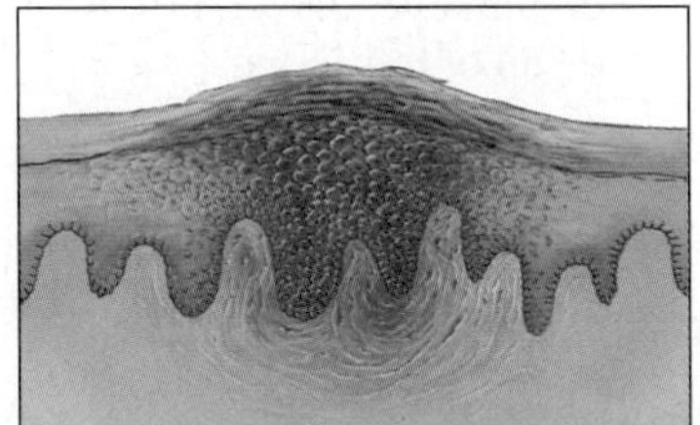

2 ____________

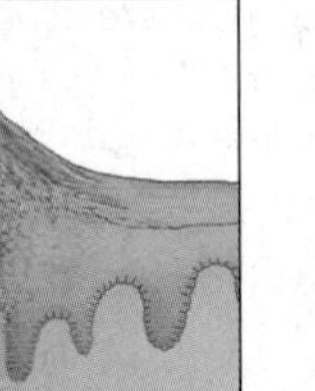

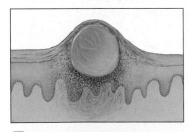

3 ____________

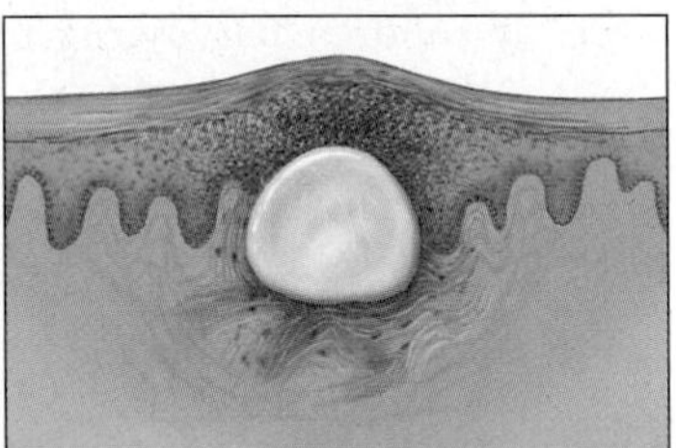

4 ____________

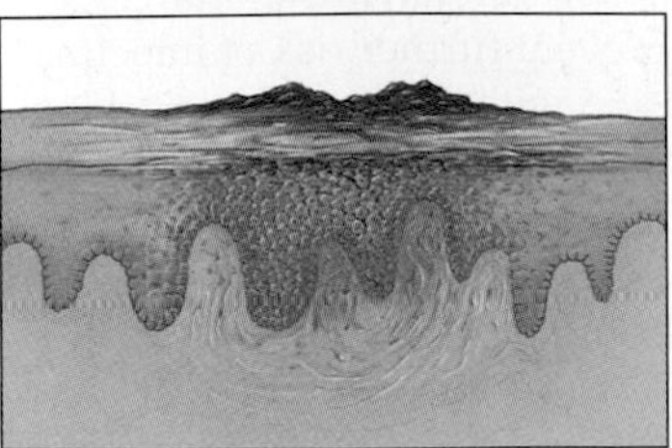

5 ____________

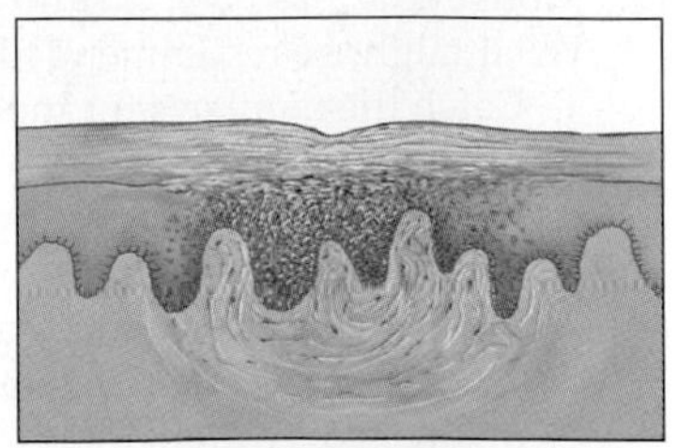

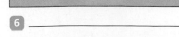

6 ____________

Short Answer

1. The Smith family seems to get fever blisters regularly. Suggest an explanation for this observation.
2. You have been given a large grant to do postgraduate research on live smallpox viruses. Where in the world would you find samples?
3. What was the difference in the effects of variola major and variola minor?
4. What observation led scientists to understand the relationship between shingles and chickenpox?
5. Most of the world's population has been infected with EBV by age 1 and shows no ill effects, even where medical care is poor. In contrast, individuals in industrialized countries are ordinarily infected after puberty, and these older patients tend to have more severe reactions to infection despite better overall health and access to medical care. Explain this apparent paradox.

Matching

In each of the blanks beside the diseases in the left column, write the letter of the viral family to which that disease's causative agent belongs. Answers may be used more than once.

____ Chickenpox
____ Smallpox
____ Cowpox
____ Molluscum contagiosum
____ HHV-1 infection
____ Whitlow
____ Shingles
____ Burkitt's lymphoma
____ Infectious mononucleosis
____ Chronic fatigue syndrome
____ Cytomegalovirus infection
____ Genital warts
____ Roseola
____ Plantar warts
____ Progressive multifocal leukoencephalopathy
____ Common cold
____ Hepatitis B
____ Fifth disease

A. *Poxviridae*
B. *Herpesviridae*
C. *Papillomaviridae*
D. *Adenoviridae*
E. *Hepadnaviridae*
F. *Parvoviridae*
G. *Polyomaviridae*

Concept Mapping

Using the following terms, draw a concept map that describes herpesvirus. For a sample concept map, see p. 93.
Or, complete this concept map online by going to the Study Area at www.masteringmicrobiology.com.

Acyclovir
Cold sores
dsDNA
Envelope
Fever blisters
Finger infection
Genital herpes
Gingivostomatitis
Herpetic whitlow
Human herpesvirus-1
Human herpesvirus-2
Latency
Life-threatening
Neonatal herpes
Ocular herpes
Oral herpes
Polyhedral capsid
Primary infection
Recurrences
Simplexvirus

Critical Thinking

1. Most DNA viruses replicate within nuclei of host cells, using host enzymes to replicate their DNA. In contrast, poxviruses replicate in the cytoplasm. What problem does this create for poxvirus replication? How could the virus overcome this problem?
2. Mrs. Rathbone called the pediatrician concerning her young daughter Rene, who had a rosy facial rash and coldlike sniffles for two weeks. What is the most likely cause of Rene's problem?
3. Certain features of smallpox viruses allowed them to be eradicated in nature. Which other DNA viruses are suitable candidates for eradication, and what features of their biology make them suitable candidates?
4. A 10-year-old girl at summer camp complains of fever, runny nose, cough, sore throat, and tiredness. Within several hours, 36 other girls report to the infirmary with the same signs and symptoms. Although the girls were assigned to several different cabins and ate in two different dining halls, all of them had participated in outdoor archery, had gone horseback riding, and had been swimming in the camp pool. Infection with what DNA virus could account for their symptoms? The facts point to a common source of infection; what is it? What could the camp management do to limit such an outbreak in the future?
5. A week after spending their vacation rafting down the Colorado River, all five members of the Chen family developed cold sores on their lips. Their doctor told them that the lesions were caused by a herpesvirus. Mr. and Mrs. Chen were stunned: Isn't herpes a sexually transmitted disease? How could it have affected their young children?

Access more review material online in the Study Area at **www.masteringmicrobiology.com**. There, you'll find

- **Concept Mapping Activities**
- **Flashcards**
- **Quizzes**

and more to help you succeed.

25 Pathogenic RNA Viruses

A patient lies delirious and near death in an African hospital, **bleeding** from his eyes and gums and into his digestive tract as a result of Marburg hemorrhagic fever. An AIDS patient in Europe continues his expensive daily **"cocktail"** of antiviral drugs. Passengers on a cruise ship suffering from severe diarrhea wish they had never eaten food infected with **noroviruses**. Thousands of people around the world go to bed with fever, sore throat, cough, and malaise, hoping to survive the **flu** season.

What do these diseases have in common? They are all caused by viruses that use RNA as their genetic material. The manifestations of **RNA** viral infection are widely variable and pose great challenges in diagnosis, treatment, and prevention. Although RNA viruses can also cause diseases in animals and plants, in this chapter we will focus on RNA viruses that cause human **diseases**.

Take the pre-test for this chapter online. Visit the Study Area at www.masteringmicrobiology.com.

▲ The filamentous Marburg virus, which is native to Africa, causes usually fatal hemorrhaging (bleeding).

Recall from previous chapters that all cells and all DNA viruses use DNA as their genetic material. In contrast, RNA viruses and viroids are the only agents that use RNA molecules to store their genetic information.

As we saw in Chapter 13, there are four basic types of RNA viruses:

- Positive single-stranded RNA (+ssRNA) viruses
- Retroviruses (which are +ssRNA viruses that convert their genome to DNA once they are inside a cell)
- Negative single-stranded RNA (−ssRNA) viruses
- Double-stranded RNA (dsRNA) viruses

Recall that positive RNA is RNA that can be used by a ribosome to translate protein; it is essentially mRNA. Negative RNA cannot be processed by a ribosome; it must first be transcribed as mRNA. (Chapter 13 examined the replication strategies of these four types of RNA viruses, and Table 13.3 on p. 389 summarizes the information.) In addition to grouping RNA viruses by the kind of RNA they contain, scientists also note that many RNA viruses have more than one molecule of RNA—in other words, their genomes are *segmented.*

RNA viruses of humans are categorized into 15 families on the basis of:

- Their genomic structure
- The presence of an envelope
- The size and shape of their capsid

Positive single-stranded RNA viruses of humans are classified into eight families: *Picornaviridae, Caliciviridae, Astroviridae, Hepeviridae, Togaviridae, Flaviviridae, Coronaviridae,* and *Retroviridae.* Negative single-stranded RNA viruses are in six families: *Paramyxoviridae, Rhabdoviridae, Filoviridae, Orthomyxoviridae, Bunyaviridae,* and *Arenaviridae.* The only family containing double-stranded RNA viruses is *Reoviridae.* We begin our discussion of pathogenic RNA viruses by considering four families of positive ssRNA viruses that lack envelopes.

Naked, Positive ssRNA Viruses: *Picornaviridae, Caliciviridae, Astroviridae,* and *Hepeviridae*

The families *Picornaviridae, Caliciviridae, Astroviridae,* and *Hepeviridae* contain positive single-stranded RNA viruses with naked polyhedral capsids. The family *Picornaviridae* is a large viral family containing many human pathogens, including viruses that cause common colds, poliomyelitis, and hepatitis A. Picornaviruses range in size from 22–30 nm in diameter, making them the smallest of animal viruses. Five hundred million picornaviruses could sit side by side on the head of a pin. Their small size and RNA genome is reflected in the name, **picornavirus**[1] (pi-kōr-nă-vī'rus). Representative picornaviruses that cause human diseases are in the genera *Rhinovirus, Enterovirus,* and *Hepatovirus.*

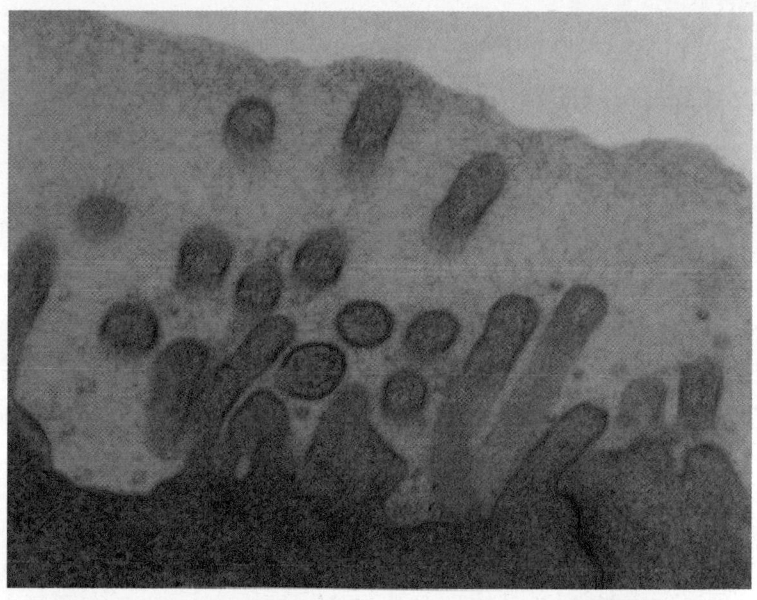

TEM 240 nm

▲ **Figure 25.1 Rhinoviruses, the most common cause of colds.** The viruses (stained blue) are able to infect microvilli of nasal cells (pink) despite the layer of mucus (beige). *What other types of viruses cause common colds?*

Figure 25.1 *Besides rhinoviruses (in the family Picornaviridae), adenoviruses, coronaviruses, reoviruses, and paramyxoviruses cause colds.*

Caliciviruses, astroviruses, and hepeviruses are generally larger than picornaviruses (27–40 nm in diameter) and have a six-pointed star shape; they cause gastrointestinal diseases.

In the following sections we will consider in turn three diseases caused by picornaviruses—common cold, polio, and hepatitis A—before examining diseases caused by caliciviruses, astroviruses, and hepeviruses.

Common Colds Caused by Rhinoviruses

Learning Objective

✓ Discuss treatment of the common cold.

Many Americans sniffle and sneeze their way through at least two colds each year. Though the symptoms of all colds are similar—sneezing, *rhinorrhea* (runny nose), congestion, mild sore throat, headache, malaise, and cough for 7–10 days—there is no single common cause of the "common cold." Rather, many different viruses, including picornaviruses, adenoviruses, coronaviruses, reoviruses, and paramyxoviruses, cause colds. However, the over 100 *serotypes* (varieties) of picornaviruses in the genus *Rhinovirus,* commonly known as **rhinoviruses**[2] (rī'nō-vī'rŭs-ĕz), cause most colds **(Figure 25.1)**. Table 25.1 on p. 708 compares the symptoms of colds and other respiratory infections.

[1]From Latin *pico,* meaning small, and RNA virus.
[2]From Greek *rhinos,* meaning nose.

TABLE 25.1 Manifestations of Respiratory Infections

Ailment	Manifestations
Common cold (viral)	Sneezing, rhinorrhea, congestion, sore throat, headache, malaise, cough
Influenza (viral)	Fever, rhinorrhea, headache, body aches, fatigue, dry cough, pharyngitis, congestion
"Strep" throat (bacterial)	Fever, red and sore throat, swollen lymph nodes in neck
Viral pneumonia	Fever, chills, mucus-producing cough, headache, body aches, fatigue
Bacterial pneumonia	Fever, chills, congestion, cough, chest pain, rapid breathing, and possible nausea and vomiting
Bronchitis (viral or bacterial)	Mucus-producing cough, wheezing
Inhalation anthrax (bacterial)	Fever, malaise, cough, chest discomfort, vomiting
Severe acute respiratory syndrome (SARS)	High fever (>38°C), chills, shaking, headache, malaise, myalgia

Epidemiology of *Rhinovirus* Infections

Rhinoviruses are limited to infecting the upper respiratory tract. They replicate best at a temperature of 33°C, which is the temperature of the nasal cavity.

Rhinoviruses are extremely infective—entry of a single virus into a person is sufficient to cause a cold in 50% of individuals. Symptomatic or not, an infected person can spread viruses by releasing them into the surrounding environment, where they are transmitted in aerosols produced by coughing or sneezing, via fomites (nonliving carriers of pathogens), or via hand-to-hand contact. Direct person-to-person contact is the most common means of transmitting these infections.

When cold symptoms are most severe, over 100,000 virions may be present in each milliliter of nasal mucus. They remain viable for hours outside the body. Infection often results from inoculation by hand into the mucous membranes of the eyes, where the viruses are washed by tears into the nasal cavity and readily infect nasal cells.

Although people of all ages are susceptible to rhinoviruses, people acquire some immunity against serotypes that have infected them in the past. For this reason, children typically have six to eight colds per year, younger adults have two to four, and adults over age 60 have one or fewer. An isolated population may acquire a certain amount of *herd immunity* (see Chapter 17) to specific varieties of rhinoviruses; however, new serotypes introduced into the population by outsiders or by mutations in the RNA of the viruses ensure that no group is free of all colds.

Diagnosis, Treatment, and Prevention of *Rhinovirus* Infections

The manifestations of rhinoviruses are usually diagnostic (see Table 25.1), and laboratory tests are required only if the serotype is to be identified.

While many home remedies and over-the-counter medicines exist to treat the common cold, none prevents colds or provides a cure. Studies of one "cure"—large amounts of vitamin C—have yielded conflicting results: some studies indicate no benefit, whereas others suggest a slight benefit for patients who are treated with vitamin C. A prescription remedy, pleconaril, taken at the onset of symptoms, reduces the seriousness and duration of colds caused by rhinoviruses. Antihistamines, decongestants, and pain relievers relieve the symptoms but do not reduce the duration of the disease. Rest and fluids allow the body to mount an effective immune response.

Because rhinoviruses do not share any common antigens that are accessible to the immune system, an effective vaccine for the common cold would have to immunize against hundreds of serotypes, which is impractical. Antisepsis is probably the most important preventive measure, especially if you have touched the hands of an infected person. Disinfection of fomites is also effective in limiting the spread of colds.

CRITICAL **THINKING**

As you have probably noticed, colds occur more frequently in the fall and winter. One explanation for this observation is that more people are crowded together in buildings when school starts and the weather cools. Design an experiment or epidemiological survey to test the hypothesis that crowded conditions explain the prevalence of colds in the fall and winter.

Diseases of Enteroviruses

Learning Objectives

- ✓ Describe the effects of polioviruses on humans.
- ✓ Describe the contributions of Jonas Salk and Albert Sabin toward eliminating polio.
- ✓ Compare and contrast the diseases caused by coxsackieviruses and echoviruses.

A second genus of picornaviruses is *Enterovirus*. Contrary to their name, *enteroviruses* do not usually cause diseases of the digestive system. They infect the pharynx and intestine, where they multiply in the mucosa and lymphatic tissue without causing symptoms. Instead, enteroviruses are so named because they are transmitted via the fecal-oral route. Enteroviruses spread from their initial sites of infection through the blood (viremia) and infect different target cells, depending on the particular virus. Enteroviruses are cytolytic,[3] killing their host cells. The clinical manifestations of enteroviruses depend on the viral serotype, the size of the infecting dose, the target organ, and the

[3]From Greek *cytos*, meaning cell, and *lysis*, meaning dissolution.

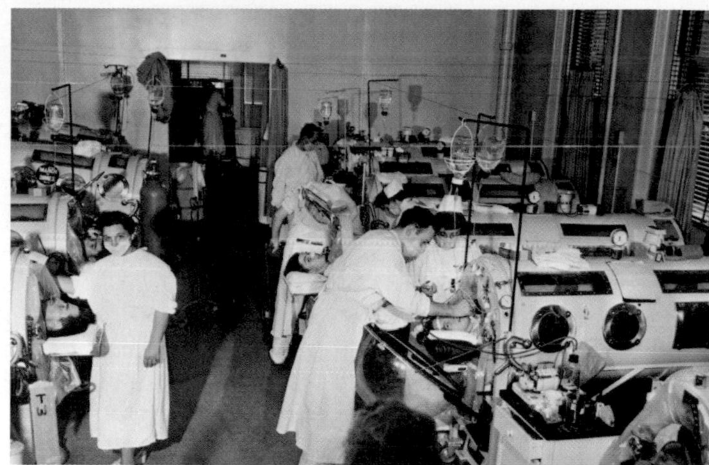

▲ **Figure 25.2 A hospital ward full of mechanical respirators.** These "iron lungs" assisted the paralyzed respiratory muscles of polio patients and were common before 1955.

patient's health, gender, and age. There are three main types of enteroviruses: polioviruses, coxsackieviruses, and echoviruses.

Poliomyelitis

Many Americans still remember the dreaded **poliomyelitis** (pō'lē-ō-mī'ĕ-lī'tis), or **polio,** epidemics of the past, when the floors of hospitals were filled with iron lungs (Figure 25.2) and schoolchildren donated their change to the March of Dimes to develop a polio vaccine. Because polioviruses are stable for prolonged periods in swimming pools and lakes and can be acquired by swallowing contaminated water, parents in the 1930s and 1940s often feared to let their children swim. Thankfully, those days appear to be over. The last case of wild-type poliomyelitis in the Americas occurred in 1979 (though vaccine-induced polio occurred as recently as 2001). The World Health Organization (WHO) has worked for years to eradicate polio worldwide (Figure 25.3). Then, like smallpox virus, poliovirus will exist only in laboratories.

There are three serotypes of poliovirus. After being ingested and infecting pharyngeal and intestinal cells, any of these serotypes could cause one of the following four conditions:

- *Asymptomatic* infections. Most infections (almost 90%) are asymptomatic.
- *Minor polio,* which includes nonspecific symptoms such as temporary fever, headache, malaise, and sore throat. Approximately 5% of cases are minor polio.
- *Nonparalytic polio,* resulting from polioviruses invading the meninges and central nervous system, producing muscle spasms and back pain in addition to the general symptoms of minor polio. Nonparalytic polio occurs in about 2% of cases.
- *Paralytic polio,* in which the viruses invade cells of the spinal cord and motor cortex of the brain, producing paralysis by limiting nerve impulse conduction. The degree of paralysis varies with the type of poliovirus involved, the infective dose, and the health and age of the patient. In **bulbar poliomyelitis,** the brain stem and medulla are infected, resulting in paralysis of muscles in the limbs or of respiratory muscles, in which case iron lungs were used to assist in respiration. In most paralytic cases, complete recovery resulted after 6–24 months, but in some cases paralysis was lifelong. Paralytic polio occurs in less than 2% of infections. Famous victims of paralytic polio include novelist Sir Walter Scott, President Franklin Delano Roosevelt, and violinist Itzhak Perlman (Figure 25.4).

Postpolio syndrome is a crippling deterioration in the function of polio-affected muscles that occurs in up to 80% of recovered polio patients some 30–40 years after their original bout with poliomyelitis. This condition is not caused by a reemergence of polioviruses, as viruses are not present. Instead, the effects appear to stem from an aging-related aggravation of nerve damage that occurred during the original infection.

The near elimination of polio stands as one of the great achievements of 20th-century medicine. It was made possible by the development of two effective vaccines. Jonas Salk (1914–1995) developed an **inactivated polio vaccine (IPV)** in 1955. Six years later, it was replaced in the United States by a live, attenuated, **oral polio vaccine (OPV)** developed by Albert Sabin (1906–1993). Both vaccines are effective in providing immunity against all three strains of poliovirus, though OPV occasionally

▶ **Figure 25.3 Reports of naturally occurring polio in the first quarter of 2009.** The goal of global eradication of polio is within reach.

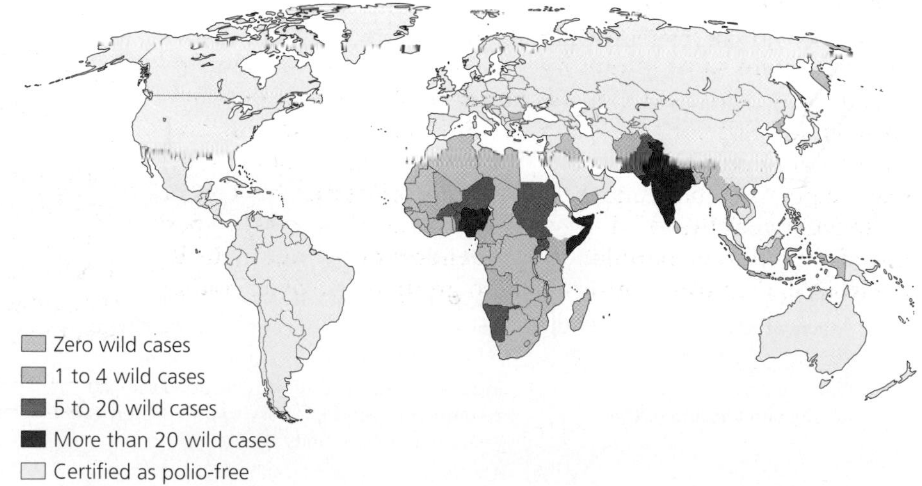

Figure 25.4 Violinist Itzhak Perlman. This world-famous violinist did not let paralytic polio stop him from achieving in his field of endeavor.

mutates into a virulent form and causes polio. Table 25.2 compares the advantages and disadvantages of the two vaccines.

CRITICAL **THINKING**

Smallpox is the only disease that has been eradicated worldwide, though scientists are close to eradicating polio. What features do the smallpox and polio viruses share that has allowed medical science to rid the world of these diseases? Compare these features to those of rhinoviruses in discussing the likelihood of eliminating colds caused by rhinoviruses.

Other Diseases of Enteroviruses

The other enteroviruses that cause human disease are **coxsackieviruses**[4] and **echoviruses.** As with all enteroviruses, infection is by the fecal-oral route. Most infections are subclinical or result in mild fever, muscle aches, and malaise. There are two types of coxsackieviruses (type A and type B), each with several serotypes, as well as numerous serotypes of echoviruses, which cause a variety of human diseases.

Coxsackieviruses Coxsackie A viruses are associated with lesions and fever that last for a few days to weeks and are self-limiting. Several serotypes cause lesions of the mouth and pharynx called *herpangina* because of their resemblance to herpes lesions. Sore throat, pain in swallowing, and vomiting accompany herpangina.

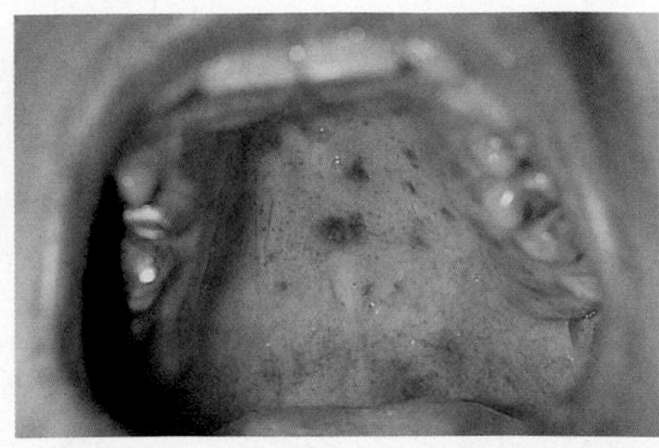

(a)

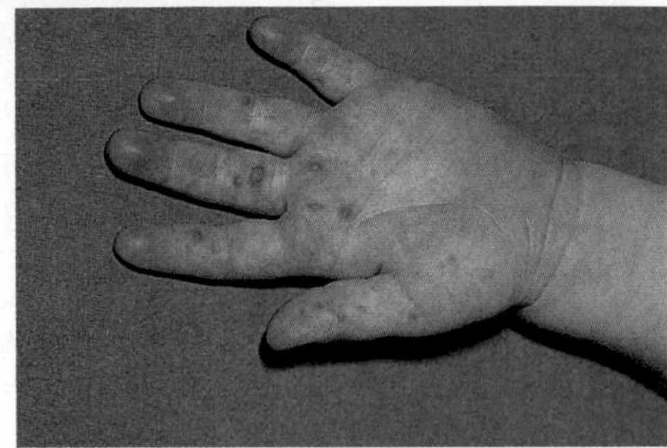

(b)

Figure 25.5 Lesions characteristic of hand-foot-and-mouth disease. The malady is caused by a coxsackie A virus. **(a)** Lesions on the hard palate (roof of the mouth). **(b)** Lesions on a hand. *Why are coxsackieviruses considered enteroviruses when they do not cause infections of the gastrointestinal tract?*

Figure 25.5 *All enteroviruses, including coxsackievirus, are infective through the gastrointestinal tract, though they may affect other organ systems.*

Another serotype of coxsackie A virus causes *hand-foot-and-mouth disease*, an apt name because it involves lesions on the extremities and in the mouth (Figure 25.5). Yet another coxsackie A virus causes an extremely contagious *acute hemorrhagic conjunctivitis.* Coxsackie A viruses also cause some colds.

Coxsackie B viruses are associated with *myocarditis* (inflammation of the heart muscle) and *pericardial*[5] *infections.* The symptoms in young adults may resemble myocardial infarction (heart

[4]From Coxsackie, New York, where the virus was first isolated.

[5]The pericardium is the membrane that surrounds and protects the heart.

TABLE 25.2 Comparison of Polio Vaccines

	Advantages	Disadvantages
Salk vaccine: Inactivated polio vaccine (IPV)	Is effective and inexpensive; is stable during transport and storage; poses no risk of vaccine-related disease	Requires booster to achieve lifelong immunity; is injected and can be painful; requires higher community vaccination rate than does OPV
Sabin vaccine: Oral polio vaccine (OPV)	Induces secretory antibody response similar to natural infection; is easy to administer; can result in herd immunity	Requires boosters to achieve immunity; is more expensive than IPV; is less stable than IPV; can mutate to disease-causing form; poses risk of polio developing in immunocompromised patients

TABLE 25.3 Comparison of Hepatitis Viruses

Feature	Hepatitis A	Hepatitis B	Hepatitis C	Hepatitis D	Hepatitis E
Common names of disease	Infectious hepatitis	Serum hepatitis	Non-A, non-B hepatitis; chronic hepatitis	Delta agent hepatitis	Hepatitis E, enteric hepatitis
Virus family	*Picornaviridae*	*Hepadnaviridae*	*Flaviviridae*	*Arenaviridae*	*Hepeviridae*
Genome	+ssRNA	dsDNA	+ssRNA	−ssRNA	+ssRNA
Envelope?	Naked	Enveloped	Enveloped	Enveloped	Naked
Transmission	Fecal-oral	Needles; sex	Needles; sex	Needles; sex	Fecal-oral
Severity (mortality rate)	Mild (<0.5%)	Occasionally severe (1–2%)	Usually subclinical (0.5–4%)	Requires coinfection with hepatitis B virus; may be severe (high)	Mild (1–2%) except in pregnant women (20%)
Chronic carrier state?	No	Yes	Yes	No	No
Other disease associations	—	Hepatic cancer	Hepatic cancer	Cirrhosis	—

attack), though fever is associated with a coxsackie infection. Newborns are particularly susceptible to coxsackie B myocarditis resulting in a rapid onset of heart failure with a high mortality rate.

One type of coxsackie B virus causes *pleurodynia* (ploor-ō-din′ē-ă), also known as *devil's grip*. This disease involves the sudden onset of fever and unilateral, severe, low thoracic pain, which may be excruciating. Coxsackie B virus can be transmitted across a placenta to produce severe and sometimes fatal disseminated disease of the brain, pancreas, and liver in the fetus. Infection of the pancreas by coxsackie B is suspected to be a cause of *diabetes mellitus* because the viruses destroy cells in the islets of Langerhans that produce insulin.

Both coxsackie A and B viruses can cause *viral meningitis*, an acute disease accompanied by headache and, with type A infections, a skin rash. Coxsackievirus meningitis is usually self-limiting and uneventful unless the patient is less than one year old.

Echoviruses The name *echovirus* is derived from *enteric cytopathic human orphan* virus, because these viruses are acquired intestinally and were not initially associated with any disease—they were *orphan viruses*. It is now known that echoviruses cause viral meningitis and some colds.

Epidemiology of *Enterovirus* Infections

Enteroviruses have a worldwide distribution and occur particularly in areas with inadequate sewage treatment. Transmission is by the fecal-oral route involving ingestion of contaminated food or water or oral contact with infected hands or fomites. For some reason, enterovirus diseases are more common in summer.

As we have seen, enteroviruses pose the greatest risk to fetuses and newborns, with the exception of poliovirus, which results in more serious symptoms in older children and young adults.

Diagnosis, Treatment, and Prevention of *Enterovirus* Infections

Enterovirus infections are usually benign and have mild symptoms, so they are not often diagnosed except in severe cases such as paralytic polio and viral meningitis. The cerebrospinal fluid (CSF) of viral meningitis patients has a normal glucose level, in contrast to that seen with bacterial meningitis. Viruses are rarely found in the CSF, so their presence there is diagnostic for viral meningitis; serological testing can confirm enterovirus infection.

No antiviral therapy is effective against enterovirus infection. Treatment involves support and limitation of pain and fever. Good hygiene and adequate sewage treatment can prevent infection with enteroviruses. No vaccines exist for coxsackievirus and echovirus infections, but effective vaccines do exist for polio. Figure 17.3 summarizes current vaccination recommendations.

Hepatitis A

Learning Objectives

- ✓ Describe the signs and symptoms of hepatitis.
- ✓ Compare and contrast the five viruses that cause hepatitis.

Like enteroviruses, **hepatitis A virus** (genus *Hepatovirus*), is transmitted through the fecal-oral route, but unlike enteroviruses it is not cytolytic. Hepatitis A virus can survive on surfaces such as countertops and cutting boards for days and resists common household disinfectants such as chlorine bleach.

Hepatitis A has an incubation period of about one month before fever, fatigue, nausea, anorexia, and jaundice abruptly start. As with hepatitis B (caused by a DNA virus), the patient's own cellular immune system kills infected liver cells, resulting in the signs and symptoms. Children are less likely than adults to develop symptoms, because their cellular immune responses are still immature. Hepatitis A virus does not cause chronic liver disease, and complete recovery occurs 99% of the time. Patients release virions in their feces and are infective even without developing symptoms. To prevent hepatitis A, two doses of hepatitis A vaccine are recommended for all children and adults (see Figure 17.3).

Table 25.3 compares five viruses known to cause viral hepatitis. Four of them (hepatitis A, C, D, and E) are RNA viruses discussed in this chapter. The fifth is hepatitis B virus, which was discussed in Chapter 24.

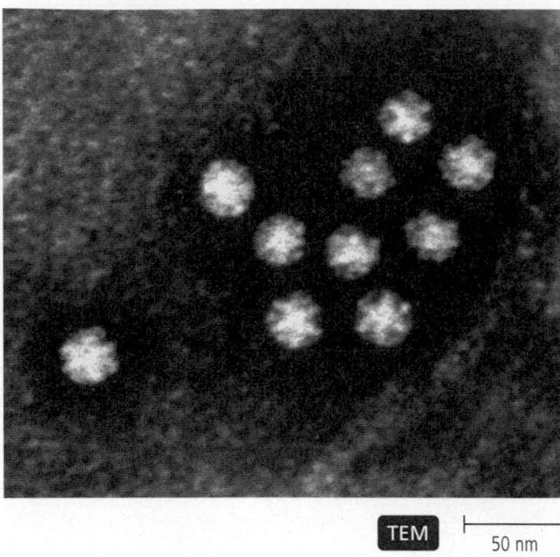

▲ **Figure 25.6 Viruses of the families *Caliciviridae* and *Astroviridae* have naked, star-shaped capsids.**

Acute Gastroenteritis

Learning Objective

✓ Compare and contrast caliciviruses and astroviruses.

Two groups of small (about 30–40 nm in diameter), round viruses that cause acute gastroenteritis are **caliciviruses** (kă-lis′i-vī′rŭs-ĕz) and **astroviruses** (as′trō-vī′rŭs-ĕz), which are slightly larger than picornaviruses and have naked, star-shaped, polyhedral capsids (**Figure 25.6**). Like enteroviruses, caliciviruses and astroviruses enter the body through the digestive system, but in contrast to enteroviruses they cause gastrointestinal disease. The capsids of caliciviruses have indentations, whereas those of astroviruses do not. Antigens of viruses in the two groups can be distinguished through serological tests. The incubation period for these two types of viruses is about 24 hours. Both caliciviruses and astroviruses have caused outbreaks of gastroenteritis in day care centers, schools, hospitals, nursing homes, restaurants, and on cruise ships. The symptoms resolve within 12–60 hours.

Caliciviruses cause diarrhea, nausea, and vomiting, though the symptoms of any given serotype vary from patient to patient. The best studied of the caliciviruses are **noroviruses,** first discovered in the stools of victims during an epidemic of diarrhea in Norwalk, Ohio. Noroviruses cause 90% of viral cases of gastroenteritis.

Astroviruses also cause diarrhea, but no vomiting, and they are less likely to infect adults. In the United States, about three-fourths of children have antibodies against astroviruses by age seven.

There is no specific treatment for caliciviral or astroviral gastroenteritis except support and replacement of lost fluid and electrolytes. Prevention of infection involves adequate sewage treatment, purification of water supplies, frequent handwashing, and disinfection of contaminated surfaces and fomites.

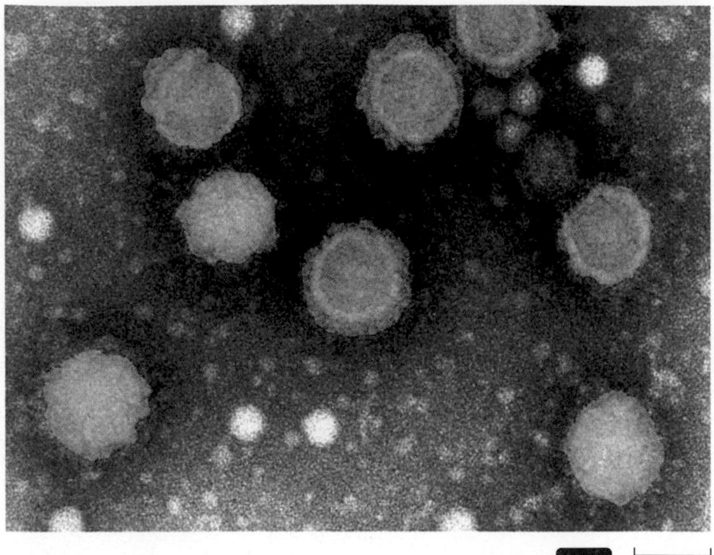

▲ **Figure 25.7 Togaviruses.** Each virus has a closely appressed envelope around its capsid.

Hepatitis E

Learning Objective

✓ Describe the prevention of hepatitis E viral infection.

Hepatitis E virus (genus *Hepevirus*) was formerly considered a calicivirus based on structure and size (27–34 nm), but now it is placed in its own family, *Hepeviridae*. Hepatitis E virus infects the liver and causes **hepatitis E,** also known as *enteric hepatitis*. Hepatitis E is fatal to 20% of infected pregnant women. There is no treatment for hepatitis E. Prevention involves interrupting the fecal-oral route of infection with good personal hygiene, water purification, and sewage treatment.

So far we have considered naked, positive ssRNA viruses in the families *Picornaviridae, Caliciviridae, Astroviridae,* and *Hepeviridae.* Now we turn our attention to enveloped, positive ssRNA viruses.

Enveloped, Positive ssRNA Viruses: *Togaviridae, Flaviviridae,* and *Coronaviridae*

Members of the *Togaviridae* and *Flaviviridae* are enveloped, icosahedral, positive, single-stranded RNA viruses. Flaviviruses are smaller and have a protein matrix between the capsid and envelope. Both flaviviruses and togaviruses differ from the viruses we examined previously in being enveloped (*toga* is Latin for "cloak"); however, the envelopes of togaviruses and flaviviruses are often tightly appressed to the capsids (like shrink wrapping), which allows individual capsomeres to be visualized (**Figure 25.7**). Because most togaviruses and flaviviruses are transmitted by arthropods (including mosquitoes, ticks, flies, mites, and lice), these viruses are designated **arboviruses**[6] (ar′bō-vī′rŭs-ĕz), and this shared characteristic is the reason taxonomists originally included the flaviviruses in family *Togaviridae*. However, differences in antigens, replication strategy, and RNA sequence indicate that flaviviruses belong in their own family. Thus, the term arbovirus has no taxonomic significance; some arboviruses are +ssRNA, some are −ssRNA, and some are dsRNA.

Viruses of the family *Coronaviridae* are also enveloped, positive, single-stranded RNA viruses. However, in contrast to togaviruses and flaviviruses, they have helical capsids (**Figure 25.8**), and none are arthropod borne.

In this section we consider diseases of the arboviruses from the families *Togaviridae* and *Flaviviridae,* discuss the togaviruses and flaviviruses that are not transmitted via arthropods, and finally discuss the coronaviruses.

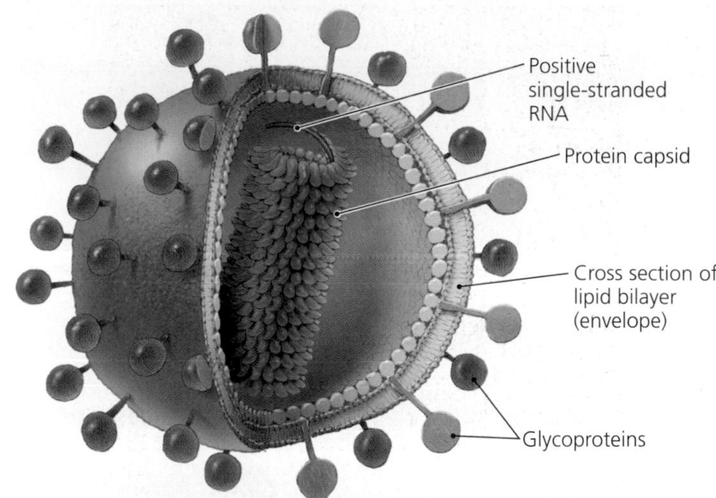

▲ **Figure 25.8 Enveloped +ssRNA coronavirus.** Viruses in the family *Coronaviridae* have helical capsids; whereas, viruses in the families *Togaviridae* and *Flaviviridae* have icosahedral capsids.

Diseases of +RNA Arboviruses

Learning Objectives

- ✓ Define *zoonosis* and *arbovirus.*
- ✓ Compare and contrast EEE, WEE, and VEE.
- ✓ Describe the transmission and spread of West Nile virus encephalitis.
- ✓ Contrast the two types of dengue fever.
- ✓ Discuss the signs, symptoms, and prevention of yellow fever.

Specific genera of mosquitoes and ticks transmit arboviruses among animal hosts, usually small mammals or birds, and into humans. Animal diseases that spread to humans are called **zoonoses,**[7] so these diseases are both arboviral and zoonotic. Arthropod vectors remain infected with arboviruses and are a continual source of new infections when they bite vertebrate hosts. In animals, virions are released from infected cells into the blood (viremia), but persistent viremia does not occur as readily in humans. Thus, humans are dead-end hosts for these zoonotic viruses.

Arboviruses enter target cells through endocytosis and replicate within them. Most cause mild, flulike symptoms in humans within three to seven days of infection. Arboviral disease does not usually proceed beyond this initial manifestation. Occasionally, however, arboviruses in the blood infect the brain, liver, skin, or blood vessels. Diseases associated with such second-stage infections include several kinds of encephalitis, dengue fever, and yellow fever.

Encephalitis

Different togaviruses cause **Eastern equine encephalitis (EEE), Western equine encephalitis (WEE),** and **Venezuelan equine encephalitis (VEE).** As the names indicate, viral replication occurs in the brains of horses as well as in humans. The normal host for these viruses is either a bird (EEE, WEE, VEE) or a rodent (VEE) (**Figure 25.9**). Of the three, EEE causes the most severe disease in humans, though any of them may produce fatal brain infections.

In 1999, hundreds of birds—mostly crows—suddenly began dying in the New York City area. At the same time, there were reports of people feeling ill with flulike symptoms after being bitten by mosquitoes. Seven of these patients developed viral encephalitis and died. As it turned out, all these events were related.

The culprit was West Nile virus. An arbovirus in the family *Flaviviridae* and endemic to Africa and Israel, West Nile virus had never before been seen in the Americas. After the initial outbreak, infected migratory birds carried the virus across the lower 48 United States. Mosquitoes further spread West Nile virus among bird populations, between birds and horses, and between birds and humans. Animals feeding on dead birds also become infected. Significant viremia with West Nile virus does not develop in humans, so humans are dead-end hosts except in cases of blood transfusion or organ transplant.

Eighty percent of infected humans have no symptoms; 20% have fever, headache, fatigue, and body aches. In the most severe cases—about 1 of 750 human infections—West Nile virus invades the nervous system to cause encephalitis, which may be fatal. **Figure 25.10** illustrates the number of human cases and deaths due to West Nile virus.

Other flaviviruses also cause usually benign viral encephalitis. These viruses include *St. Louis, Japanese,* and *Russian spring-summer* viruses.

Dengue Fever

A flavivirus transmitted by *Aedes* mosquitoes causes **dengue** (den′gā) **fever.** It afflicts approximately 100 million people in tropical and subtropical areas of the world (**Figure 25.11**). The disease usually occurs in two phases separated by 24 hours of remission. First, the patient suffers from fever, weakness, edema (swelling) of the extremities, and severe pain in the

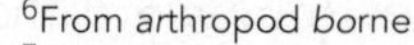

[6]From arthropod *borne*.
[7]Plural of Greek *zoos*, meaning animal, and *nosis*, meaning disease.

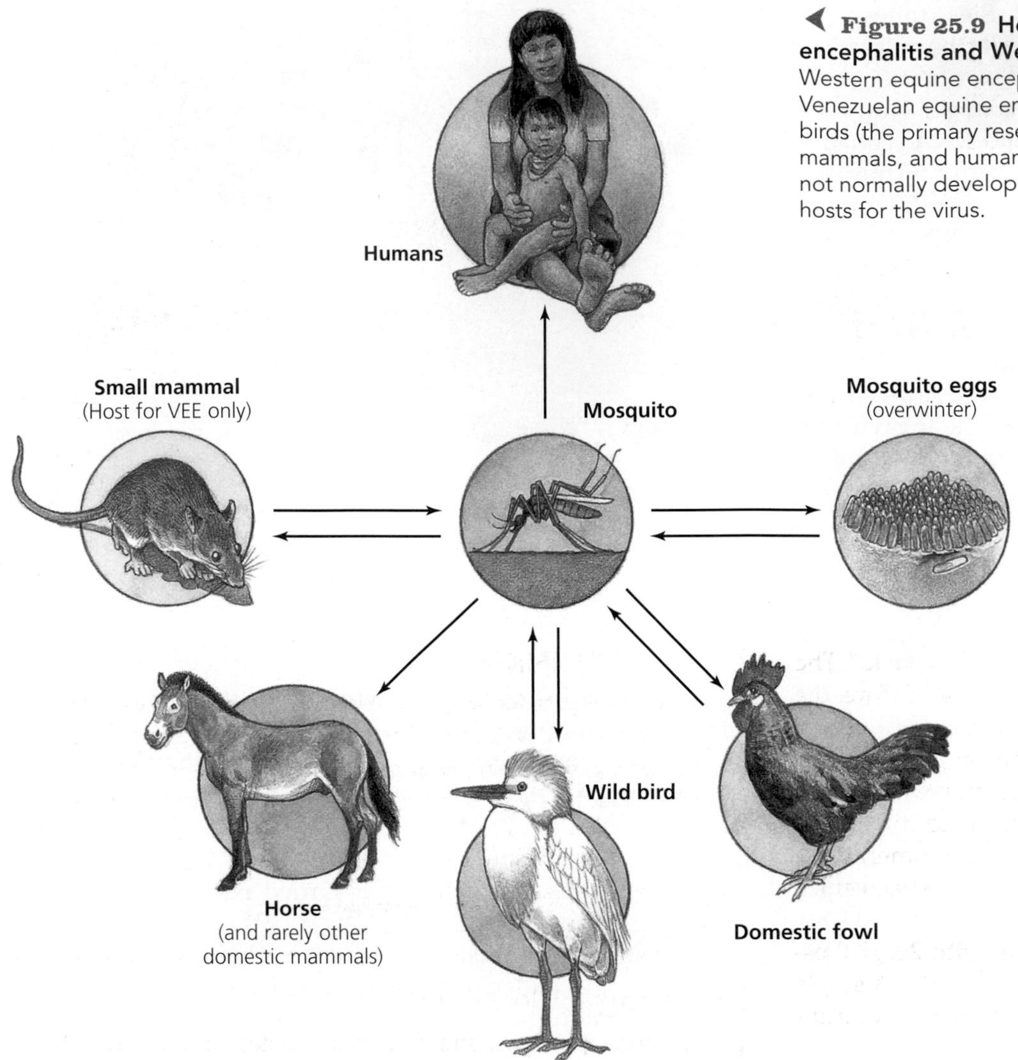

◀ **Figure 25.9 Hosts and transmission of viruses of equine encephalitis and West Nile encephalitis.** Mosquitoes carry Western equine encephalitis (WEE), Eastern equine encephalitis (EEE), Venezuelan equine encephalitis (VEE) and West Nile viruses from wild birds (the primary reservoir) to domesticated birds, horses, other mammals, and humans. Except for rodents carrying VEE, mammals do not normally develop viremia with these viruses, making them dead-end hosts for the virus.

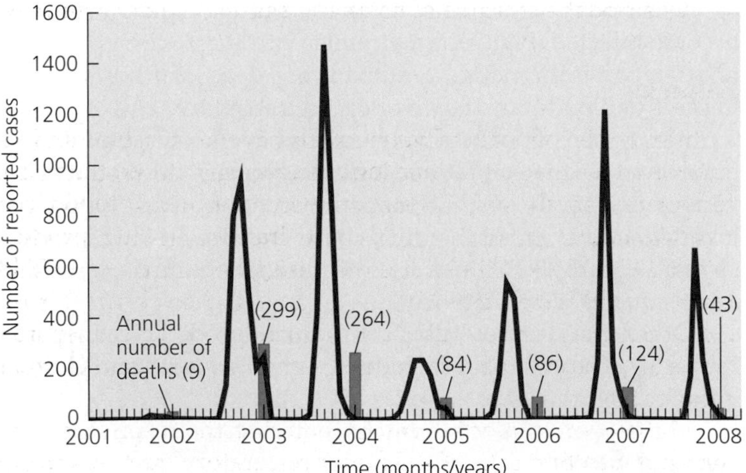

▲ **Figure 25.10 Human West Nile virus infections in the United States.** Note the seasonal nature of the disease.

head, back, and muscles. The severity of the pain is indicated by the common name for the disease, *breakbone fever.* The second phase involves a return of the fever and a bright red rash. Dengue fever is self-limiting and lasts six or seven days.

Dengue hemorrhagic fever is a more serious disease caused by reinfection with the dengue virus, and it involves a hyperimmune response **(Figure 25.12)**. Inflammatory cytokines released by activated memory T cells cause rupture of blood vessels, internal bleeding, shock, and possibly death.

Dengue and dengue hemorrhagic fever are epidemic in South America. Given that *Aedes* mosquitoes already live in the southern United States, dengue could potentially become established there if the virus were introduced into the United States by immigrants or returning travelers.

Yellow Fever

Another flavivirus causes **yellow fever,** a disease involving degeneration of the liver, kidneys, and heart, as well as massive hemorrhaging. Hemorrhaging in the intestines may result in "black vomit." Liver damage causes jaundice, from which the

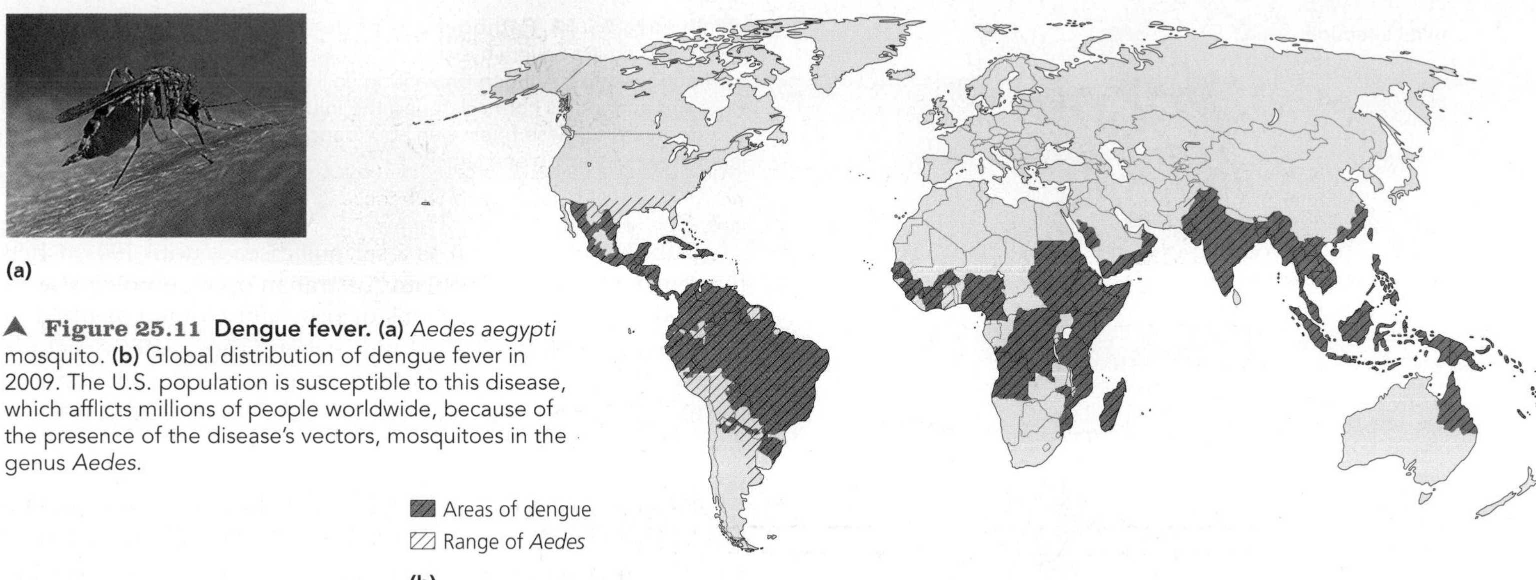

Figure 25.11 Dengue fever. **(a)** *Aedes aegypti* mosquito. **(b)** Global distribution of dengue fever in 2009. The U.S. population is susceptible to this disease, which afflicts millions of people worldwide, because of the presence of the disease's vectors, mosquitoes in the genus *Aedes*.

disease acquires its name and its nickname, "Yellow Jack." The mortality rate of yellow fever can approach 20%. Before the 1900s, yellow fever epidemics ravaged the Americas, killing thousands. In 1793, it killed over 4000 people in Philadelphia (then the capital of the United States), and during the Spanish American War of 1898, it killed more American soldiers than bullets did. With mosquito control and the development of a vaccine in the 1900s, however, yellow fever has since been eliminated from the United States.

Yellow fever remains a significant disease, with 200,000 estimated cases and 30,000 deaths annually worldwide. A single dose of yellow fever vaccine provides protection to 95% of individuals, probably for their lifetimes.

Table 25.4 on p. 718 summarizes the arboviruses and their diseases, including additional viruses in families *Bunyaviridae* and *Reoviridae*.

Diagnosis, Treatment, and Prevention of Arbovirus Infections

Serological tests, such as ELISA and agglutination of latex beads to which arbovirus antigens are affixed, are used for the diagnosis of arboviral infections. The only treatment for arboviral diseases is to provide supportive care. Only acetaminophen is recommended for managing pain and fever associated with dengue infections, as the anticoagulant properties of aspirin could aggravate the hemorrhagic properties of dengue virus infections.

Prevention of infection involves control of the vectors, which has resulted in elimination of many arboviral diseases in certain geographic areas. Insect repellents and netting reduce infections with arboviruses.

Vaccination is recommended for people traveling to areas where arboviral diseases are prevalent. Vaccines for humans are available against yellow fever, Japanese encephalitis, and Russian spring-summer encephalitis viruses. Animal vaccines against VEE, EEE, and West Nile encephalitis are also available. No vaccine is available for dengue.

CRITICAL **THINKING**

Suppose a vaccine for dengue that induced the production of memory T cells could be developed. After reviewing the characteristics of infection and reinfection, would you argue for or against the use of such a vaccine? Why or why not?

Other Diseases of Enveloped +ssRNA Viruses

Learning Objectives

- ✓ Describe the signs, effects, and prevention of rubella.
- ✓ Compare and contrast hepatitis C with the other types of viral hepatitis.
- ✓ Describe the structure of coronaviruses and the diseases they cause.

Rubella

Rubella (rū-bel'ă) virus *(Rubivirus)* shares the structure and reproductive strategy of other togaviruses, but unlike them it is transmitted by the respiratory route and not by arthropod vectors. **Rubella** is one of five childhood viral diseases that produce skin lesions. (The other four are measles, caused by a −ssRNA virus and discussed later in this chapter; and chickenpox, roseola, and fifth disease.) Rubella, first distinguished as a separate disease by German physicians, is commonly known as "German measles" or "three-day measles."

Rubella only infects humans, entering the respiratory system and infecting cells of the upper respiratory tract. It spreads from there to lymph nodes and into the blood, and then throughout the body. Afterward, the characteristic rash of flat, pink to red spots (macules) develops **(Figure 25.13)** and lasts about three days. Rubella in children is usually not serious, but adults may develop arthritis or encephalitis. Patients shed virions in respiratory droplets for approximately two weeks before and two weeks after the rash.

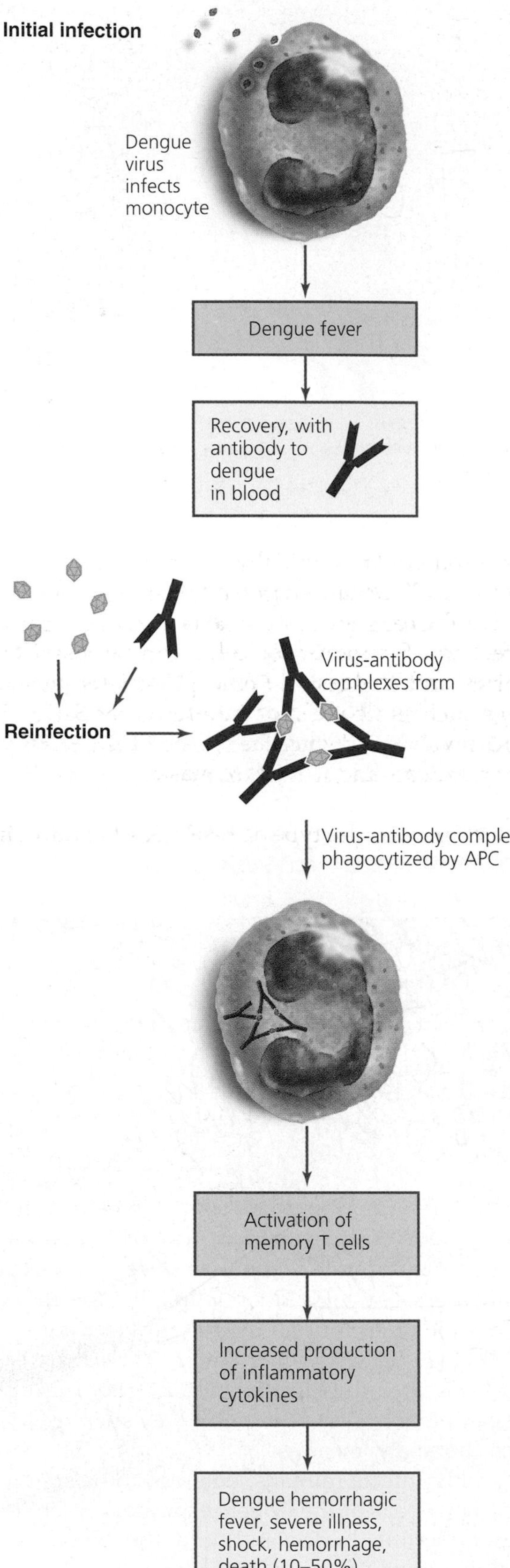

Figure 25.12 Pathogenesis of dengue hemorrhagic fever. The condition is a severe hyperimmune response to a second infection with dengue virus. Antigen-presenting cells phagocytize complexes of virus and antibodies (formed during the initial infection), which activates T memory cells. These release an abundance of lymphokines that induce hemorrhaging and shock.

Rubella was not seen as a serious disease until 1941, when Norman Gregg (1892–1966) an Australian ophthalmologist, recognized that rubella infections of pregnant women resulted in severe congenital defects in their babies. These effects include cardiac abnormalities, deafness, blindness, mental retardation, microcephaly,[8] and growth retardation. Death of the fetus is also common. It is now known that a mother is able to transmit the virus across the placenta even if she is asymptomatic.

Diagnosis of rubella is usually made by observation and by serological testing for IgM against rubella. No treatment is available, but immunization has proven effective at reducing the incidence of rubella in industrialized countries (Figure 25.14). The rubella vaccine is made from a live, attenuated virus, and therefore should never be given to pregnant women or immunocompromised patients. Immunization is aimed at reducing the number of rubella cases that might serve to introduce rubella to pregnant women.

Hepatitis C

A flavivirus called **hepatitis C virus (HCV)** accounts for 90% of the cases of so-called non-A, non-B hepatitis in the United States (see Table 25.3 on p. 711). About 3.2 million Americans are chronically infected by HCV, which is spread via needles, organ transplants, and sexual activity, but not by arthropod vectors. In the past, blood transfusions accounted for many cases of hepatitis C, but testing blood for HCV has considerably reduced the risk of infection by this means.

Hepatitis C is a chronic disease, often lasting for 20 years or longer, with few if any symptoms. Severe liver damage (20% of

[8]From Greek *mikro*, meaning small, and *kephalos*, meaning head.

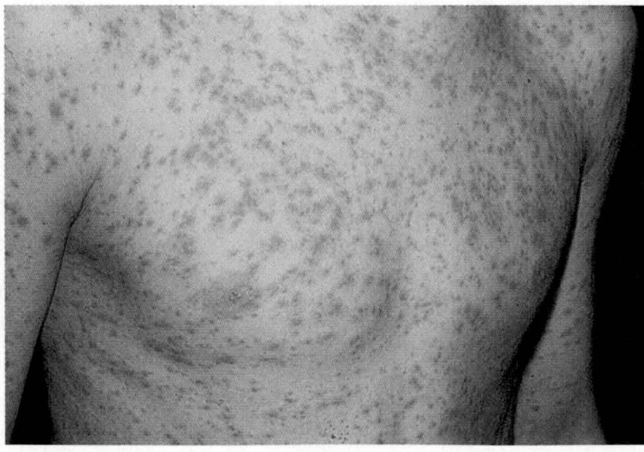

Figure 25.13 The rash of rubella. The disease, also known as German measles or three-day measles, is characterized by flattened red spots (macules).

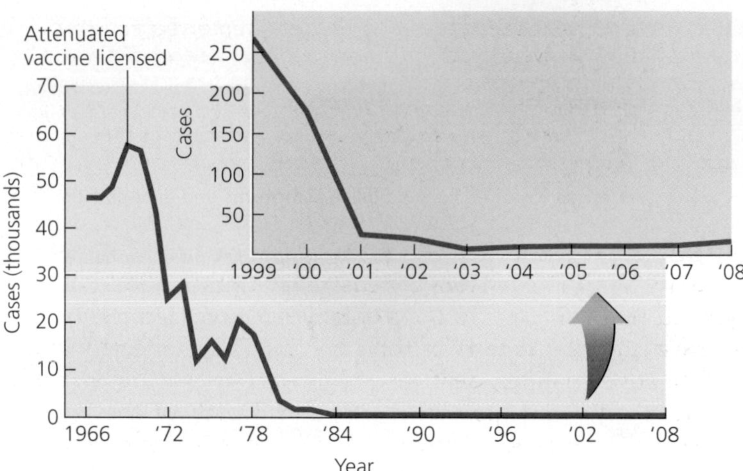

▲ **Figure 25.14 The efficacy of vaccination against rubella.** The use of a live, attenuated virus vaccine has practically eliminated rubella in the United States.

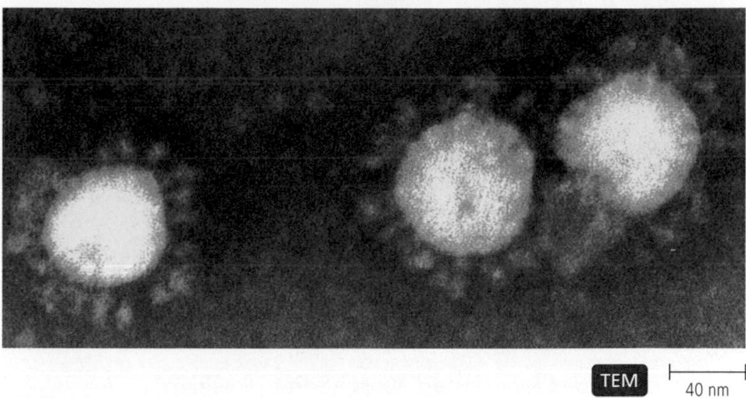

▲ **Figure 25.15 Coronaviruses.** These +ssRNA viruses have distinctive envelopes with glycoprotein spikes that resemble flares from the corona of the sun.

cases) and liver failure (5% of cases) occur over time. Hepatic cancer can also result from HCV infection.

Alpha interferon and ribavirin provide some relief for hepatitis C but are not cures. There is no other treatment or vaccine for hepatitis C.

Diseases of Coronaviruses

Coronaviruses (kō-rō′nă-vī′rŭs-ĕz) are enveloped, positive, single-stranded RNA viruses with helical capsids. Their envelopes form corona-like halos around the capsids, giving these viruses their name **(Figure 25.15)**. Two coronaviruses cause colds and are the second most common cause of colds after picornavirus rhinoviruses. Coronaviruses are transmitted from epithelial cells of the upper respiratory tract in large droplets (sneezes and coughs) and, like rhinoviruses, replicate best at 33°C, which is the temperature of the nasal cavity.

In the winter of 2002–2003, a newly emerging disease of a previously unknown coronavirus made headlines worldwide (**Highlight: Sequencing the SARS Virus** on p. 718). **Severe acute respiratory syndrome (SARS)** was first identified in China's Guangdong province. SARS initially taxed health care systems in several Asian countries, but unprecedented worldwide cooperation allowed epidemiologists to identify the virus, track its spread, sequence its genome, and investigate effective quarantine procedures faster than with any previous disease. SARS is characterized by symptoms of fever above 100.4°F (38°C), headache, general discomfort, and respiratory distress. By spring of 2003, the disease had spread to Singapore, Vietnam, Taiwan, Canada, and other parts of the world. Amazingly, epidemiologists were able to trace the disease's global spread back to a single individual: an infected physician who had traveled from China to Hong Kong in order to attend his nephew's wedding and unwittingly spread the disease to other travelers.

SARS created widespread alarm because of its sudden emergence, highly infectious nature (it is spread by close person-to-person contact), initially unknown origin, and 10% mortality rate. SARS remains a potential problem in China.

No antiviral treatment or vaccine is available against coronaviral infections. Symptoms of colds can be alleviated with antihistamines and analgesics. Some physicians prescribe antiviral drugs such as ribavirin or interferon for SARS. Prevention of SARS involves reducing the spread of the SARS virus by quarantining patients and using face masks **(Figure 25.16)**.

Now we examine a unique type of +ssRNA—the retroviruses.

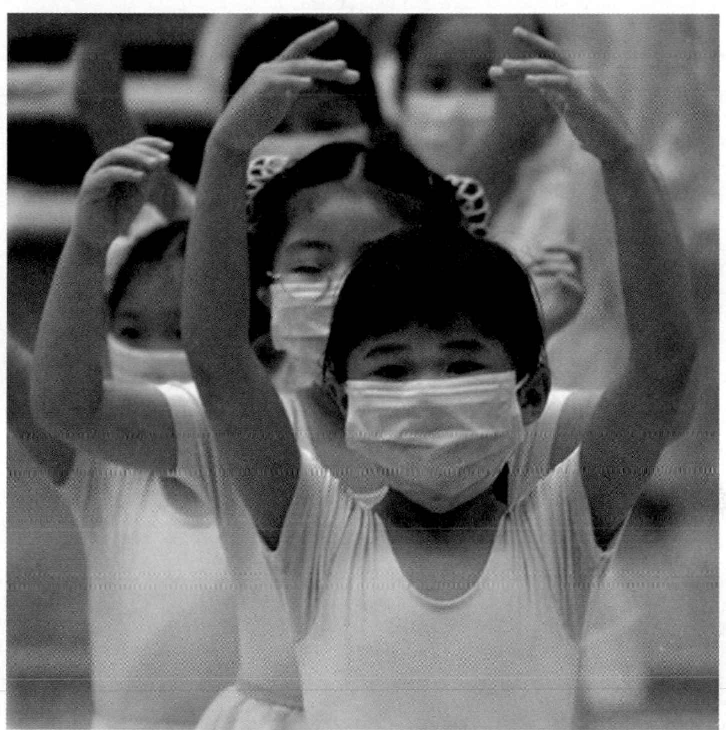

▲ **Figure 25.16 Prevention of SARS.** Ballet students in Hong Kong in 2003 donned face masks to slow the spread of SARS virus.

TABLE 25.4 Diseases Caused by Arboviruses, by Viral Family

Disease	Vector	Natural Host(s)	Distribution	Symptoms
***Togaviridae* (enveloped, icosahedral, +ssRNA)**				
Eastern equine encephalitis (EEE)	*Aedes, Culex,* and *Culiseta* mosquitoes	Birds	Americas	Flulike symptoms and encephalitis
Western equine encephalitis (WEE)	*Aedes* and *Culex* mosquitoes	Birds	Americas	Flulike symptoms and encephalitis
Venezuelan equine encephalitis (VEE)	*Aedes* and *Culex* mosquitoes	Rodents, horses	Americas	Flulike symptoms and encephalitis
***Flaviviridae* (enveloped, icosahedral, +ssRNA)**				
Japanese encephalitis	*Culex* mosquitoes	Birds, pigs	Asia	Flulike symptoms and encephalitis
West Nile encephalitis	*Aedes, Anopheles,* and *Culex* mosquitoes	Birds	Africa, Europe, Asia, North America	Flulike symptoms and potentially fatal encephalitis
St. Louis encephalitis	*Culex* mosquitoes	Birds	North America	Flulike symptoms and encephalitis
Russian spring-summer encephalitis	*Ixodes* and *Dermacentor* ticks	Birds	Russia	Flulike symptoms and encephalitis
Dengue and dengue hemorrhagic fever	*Aedes* mosquitoes	Monkeys, humans	Worldwide, especially tropics	Severe pain, hemorrhaging, hepatitis, shock
Yellow fever	*Aedes* mosquitoes	Monkeys, humans	Africa, South America	Hepatitis, hemorrhagic fever, shock
Chikungunya	*Aedes* mosquitoes	Monkeys	Africa, Asia	Rash, nausea, joint pain
***Bunyaviridae* (enveloped, filamentous segmented, −ssRNA)**				
LaCrosse encephalitis	*Aedes* mosquitoes	Rodents, small mammals, birds	North America	Fever, rash, encephalitis
Rift Valley fever	*Aedes* mosquitoes	Sheep, goats, cattle	Africa, Asia	Hemorrhagic fever, encephalitis
Sand fly fever	*Phlebotomus* flies	Sheep, cattle	Africa	Hemorrhagic fever, encephalitis
Crimean-Congo hemorrhagic fever	*Hyalomma* ticks	Horses, cattle, goats, seabirds	Africa, Crimea	Hemorrhagic fever
***Reoviridae* (naked, dsRNA)**				
Colorado tick fever	*Dermacentor* ticks	Small mammals, deer	Western North America	Fever, chills, headache, photophobia, rash, and, in children, hemorrhage

HIGHLIGHT

SEQUENCING THE SARS VIRUS

When SARS came to the world's attention in late 2002, the virus that causes it was unknown. To combat the growing epidemic, scientists needed to rapidly isolate the virus.

Scientists generated small slides spotted with numerous tiny DNA pieces from various viruses. This DNA microarray acted as a library that the scientists could use to compare with gene sequences from known and unknown viruses. The key to this comparison was that the DNA pieces on the array were all highly conserved—that is, the same sequence existed among many members of a taxon of viruses. So, if an unknown virus contained one of these gene pieces, it could be placed in the same group as the virus with which it shared a common DNA segment.

Once the array was generated, researchers purified a viral sample from a SARS patient. The resulting RNA was coated onto the array. The idea: any SARS RNA pieces that matched the DNA pieces on the array would stick to the slide; all other nonspecific segments would wash off. These common pieces were easily identified because they lit up via fluorescence imaging. Using this technique, the team quickly found out that SARS virus is a member of the coronavirus family. There are three types of coronaviruses, but SARS didn't fit into any of these three: it was unique. Researchers have since learned much more about the SARS virus and are more prepared to face new outbreaks.

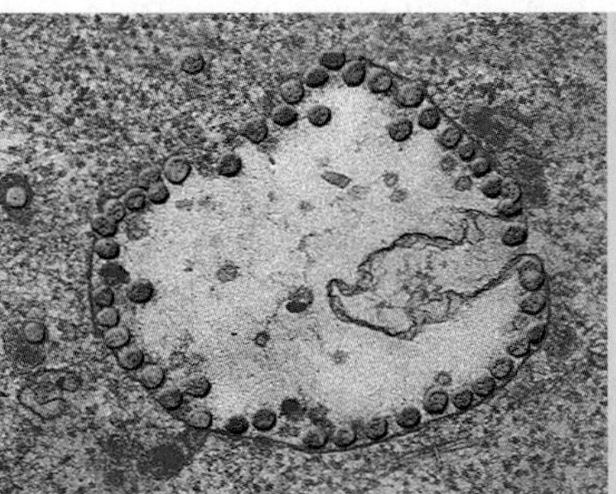

▲ **SARS virus, a coronavirus.**

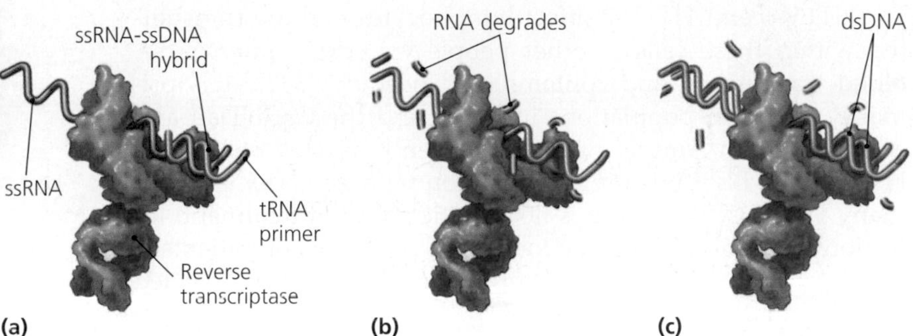

Figure 25.17 The process by which reverse transcriptase transcribes dsDNA from ssRNA. Three distinct steps occur rapidly and nearly simultaneously: 1 **(a)** A complementary −ssDNA molecule is transcribed to form a DNA-RNA hybrid. Virion tRNA is used as a primer for this synthesis. 2 **(b)** The RNA portion of the hybrid is degraded, leaving −ssDNA. 3 **(c)** A complementary +DNA strand is synthesized to form dsDNA.

Enveloped, Positive ssRNA Viruses with Reverse Transcriptase: *Retroviridae*

Learning Objectives

- ✓ Explain how retroviruses do not conform to the central dogma of molecular biology.
- ✓ Describe the steps of reverse transcription.
- ✓ List two types of retroviruses.

Scientists have studied **retroviruses** (re′trō-vī′rŭs-ĕz) more than any other group of viruses because of their unique features and the diseases they cause. These viruses have polyhedral capsids with spiked envelopes 80–146 nm in diameter and genomes composed of two molecules of positive, single-stranded RNA. Each virion also contains two tRNA molecules and 10–50 copies of the enzymes *reverse transcriptase, protease,* and *integrase,* whose function will be described shortly. The box on p. 726 illustrates these features as exemplified by HIV, the most studied of the retroviruses.

The unique characteristic of retroviruses is that they do not conform to what is considered the *central dogma* of molecular biology; that is, that genetic information flows from DNA to RNA to translated proteins. Retroviruses[9] reverse this flow of information; that is, they transcribe DNA from RNA.

Retroviruses accomplish this amazing feat by means of a complex enzyme, **reverse transcriptase,** which transcribes double-stranded DNA from single-stranded RNA. Transcription occurs as the enzyme moves along an RNA molecule and can best be understood as occurring in three steps, though in reality the events occur simultaneously **(Figure 25.17)**:

1. An RNA-DNA hybrid is made from the +RNA genome using tRNA carried by the virion as a primer for DNA synthesis. The DNA portion of the hybrid molecule is negative DNA.
2. The RNA portion of the hybrid molecule is degraded by reverse transcriptase, leaving −ssDNA.
3. The reverse transcriptase transcribes a complementary +DNA strand to form dsDNA, and the tRNA primer is removed.

The discovery of reverse transcriptase was a momentous event in biology, and it made possible an ongoing revolution in recombinant DNA technology and biotechnology. For example, it is much easier to isolate mRNA for a particular protein (e.g., human insulin) and then use reverse transcriptase to make a gene than it is to find the gene itself amid the DNA of 23 pairs of human chromosomes.

In the following sections we consider two types of retroviruses, both of which infect humans: those that are primarily oncogenic (genus *Deltaretrovirus*), and those that are primarily immunosuppressive (genus *Lentivirus*).

Oncogenic Retroviruses

Learning Objective

- ✓ Describe how oncogenic viruses may induce cancer.

Retroviruses have been linked with cancer ever since they were first isolated from chicken neoplasias. In 1981, Robert Gallo (1937–) and his associates isolated the first human retrovirus, **human T-lymphotropic virus 1 (HTLV-1),** from a patient with *adult acute T-cell lymphocytic leukemia,* a cancer of humans. Since then, two other human retroviruses of this type have been discovered: HTLV-2, which causes a rare cancer called *hairy cell leukemia* because of distinctive extensions of the cytoplasmic membrane of infected cells **(Figure 25.18)**, and HTLV-5, which has not been linked to cancer or any other disease.

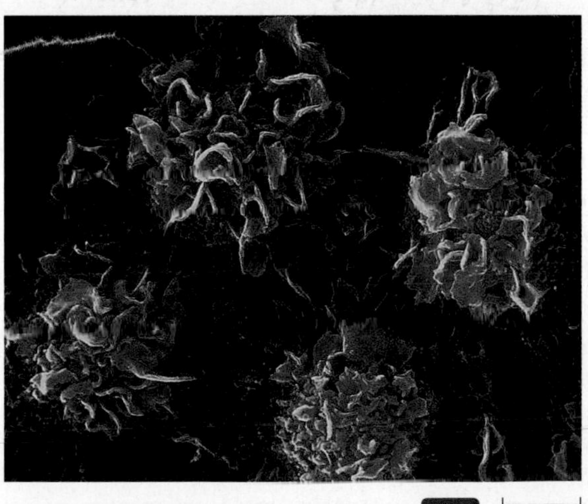

Figure 25.18 The characteristic extensions of the cytoplasmic membrane in hairy cell leukemia.

[9] From Latin *retro,* meaning reverse.

HTLV-1 and HTLV-2 infect lymphocytes and are transmitted within these cells to other people via sexual intercourse, blood transfusion, and contaminated needles. HTLV-1 is primarily found in populations in Japan and the Caribbean and among African Americans. HTLV-2 is not as well studied as HTLV-1. An ELISA can reveal infection with an HTLV. As with many viral diseases, there is no specific antiviral treatment. Infections are chronic, and the long-term prognosis of patients is poor. Prevention involves the same changes in behavior needed to prevent HIV infection (see p. 725).

Oncogenic retroviruses cause cancer in various ways. For example, HTLV-1 codes for a protein, called *Tax,* that activates cell growth and division genes in helper T lymphocytes.

Acquired Immunodeficiency Syndrome

Learning Objectives

- ✓ Differentiate between a disease and a syndrome, using AIDS as an example.
- ✓ Describe the virions that cause AIDS.
- ✓ Explain the roles of gp41 and gp120 in HIV infection.
- ✓ Describe the relationship of helper T cells to the course of AIDS.
- ✓ List four measures that can be effective in preventing AIDS.

From the time of its discovery in 1981 among homosexual males in the United States to its emergence as a worldwide pandemic, no affliction has affected modern life as much as **acquired immunodeficiency syndrome (AIDS).** AIDS is not a disease but a **syndrome;** that is, a complex of signs, symptoms, and diseases associated with a common cause. AID syndrome is several opportunistic or rare infections, listed in Table 25.5, in the presence of antibodies against the **human immunodeficiency virus (HIV)** and a CD4 white blood cell count below 200 cells per microliter of blood.

HIV is restricted to replicating in humans and, as its name indicates, this pathogen destroys the human immune system. Its many characteristics that make it particularly difficult to combat are summarized in Table 25.6 on p. 722.

There are two major types of HIV. **HIV-1,** which is more prevalent in the United States and Europe, was discovered by Luc Montagnier (1932–) and his colleagues at the Pasteur Institute in 1983. **HIV-2,** the prevalent strain in West Africa, reproduces more slowly than HIV-1; therefore, HIV-1 will be the major focus of our discussion.

Structure of HIV

HIV is a typical retrovirus in size and shape. Its envelope is characterized by two antigenic glycoproteins (see p. 726). The larger of these, named **gp120,**[10] is the primary attachment molecule of HIV and when shed into the blood triggers inflammation. Its antigenicity can change during the course of prolonged infection, making an effective antibody response against it very difficult. The smaller glycoprotein, **gp41,** promotes fusion of the viral envelope to a target cell. The effects of these structural characteristics—antigenic variability and the ability to fuse with host cells—impede immune clearance of HIV from a patient.

[10]gp120 is short for glycoprotein with a molecular weight of 120,000 daltons.

TABLE 25.5 Opportunistic Infections and Tumors of AIDS Patients

Type	Manifestations
Bacterial infections (19–21)[a]	Tuberculosis, especially extrapulmonary *(Mycobacterium)*
	Rectal gonorrhea *(Neisseria)*
	Recurrent fever and septicemia due to *Salmonella, Haemophilus,* or *Streptococcus*
Fungal infections (22)	*Pneumocystis* pneumonia
	Thrush, disseminated in trachea, lungs, esophagus *(Candida)*
	Histoplasmosis *(Histoplasma)*
Protozoal infections (23)	Toxoplasmosis *(Toxoplasma)*
	Chronic diarrhea *(Cryptosporidium, Isospora)*
Viral-induced tumors (24)	Kaposi's sarcoma (especially when associated with HHV-8)
	Lymphoma (induced by Epstein-Barr virus)
Viral infections (24)	Cytomegalovirus disseminated in brain, lungs, retina, etc.
	Human herpesvirus 1 and *2* disseminated in lungs, GI tract, etc.
	Progressive multifocal leukoencephalopathy (PML) (JC virus)
	Oral hairy leukoplakia (Epstein-Barr virus)
Others	Wasting disease; called *slim* in Africa (cause unknown)
	Dementia

[a]Numbers in parentheses refer to chapters where relevant material is discussed.

Origin of HIV

Evidence suggests that human immunodeficiency viruses arose from mutations of a similar virus, *simian immunodeficiency virus (SIV),* found in African primates. The nucleotide sequence of SIV is similar to that of AIDS viruses. Antibodies against HIV have been found in human blood stored since 1959, though the first cases of AIDS were not documented until 1981. Based on mutation rates and the rate of antigenic change in HIV, it is estimated that HIV emerged in the human population in Africa about 1930.

The relationship between the two types of HIV is not clear. HIV-1 and HIV-2 are likely derived from different strains of SIV. We may never know the exact temporal relationships among the immunodeficiency viruses.

Replication of HIV

The replication cycle of HIV is shown in **Figure 25.19.**

Figure 25.19 The replication cycle of HIV. The artist's rendition depicts the steps involved in the replication of the virus within a helper T cell: 1 attachment and penetration, 2 uncoating, 3 synthesis of dsDNA and entry into nucleus, 4 latency, 5 synthesis, 6 assembly and budding, and 7 maturation of the virion. Photos **(a), (b),** and **(c)** show attachment and penetration; photos **(d), (e),** and **(f)** show the budding and release of a virion.

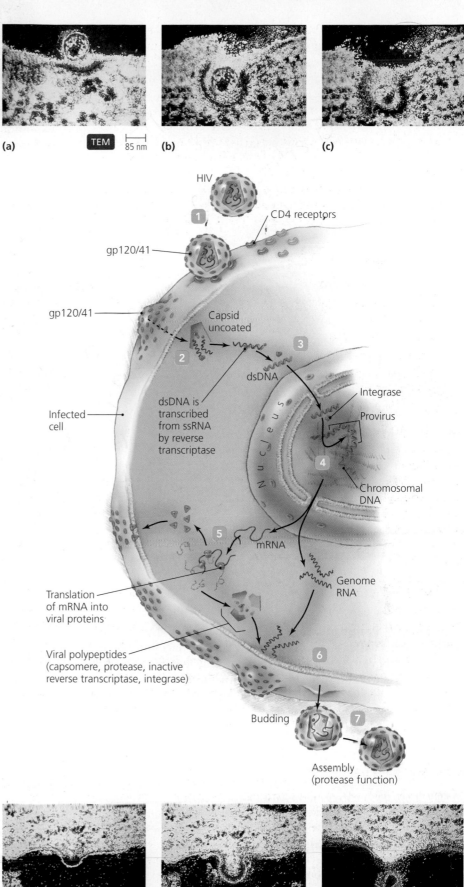

(a) TEM 85 nm (b) (c)

HIV
1
CD4 receptors
gp120/41
gp120/41
Capsid uncoated
2
3
dsDNA
dsDNA is transcribed from ssRNA by reverse transcriptase
Infected cell
Nucleus
Integrase
Provirus
4
Chromosomal DNA
5
mRNA
Genome RNA
Translation of mRNA into viral proteins
Viral polypeptides (capsomere, protease, inactive reverse transcriptase, integrase)
6
Budding
7
Assembly (protease function)

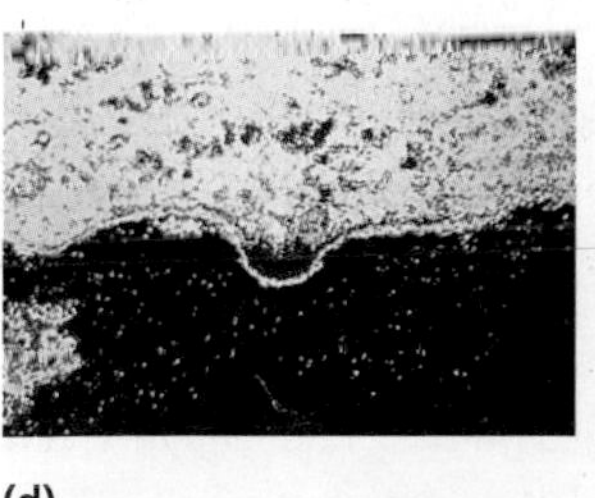

(d)

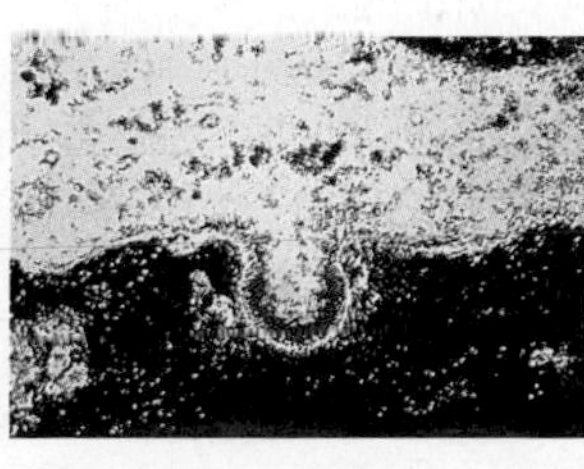

(e)

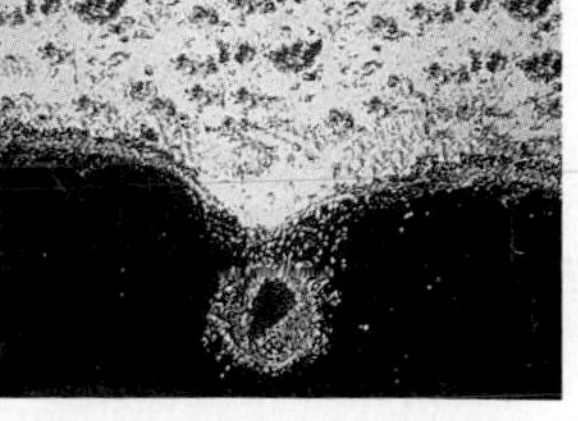

(f)

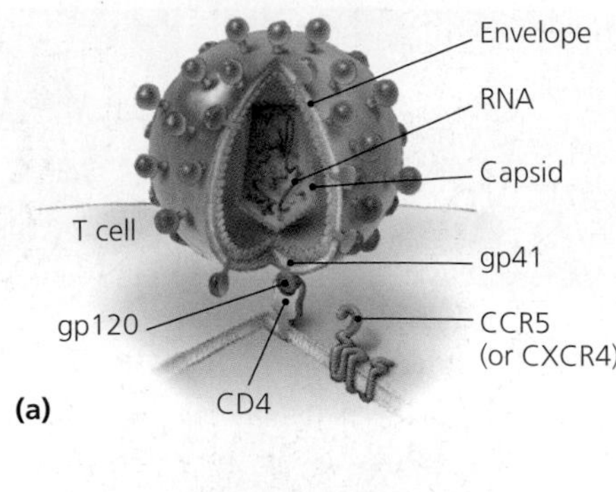

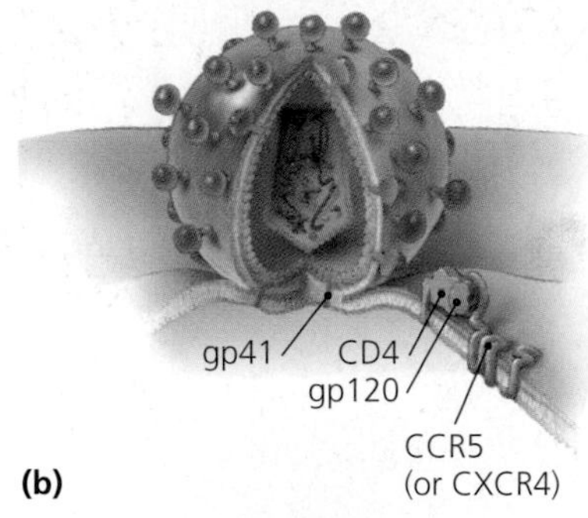

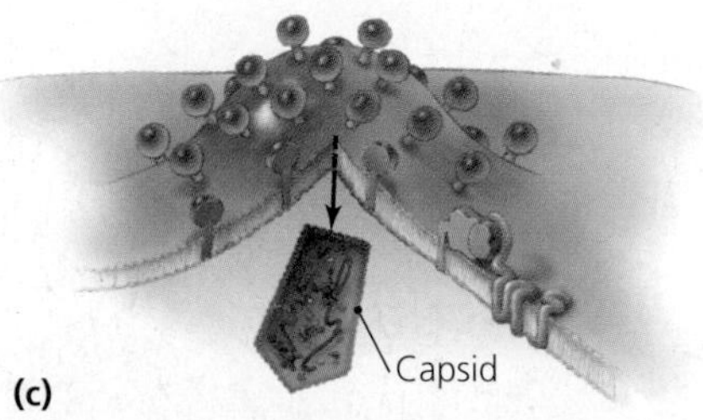

▲ **Figure 25.20 The process by which HIV attaches to and enters a CD4 cell.** **(a)** Binding of gp120 to a CD4 receptor on the cell membrane. **(b)** Removal of the CD4-gp120 complex, allowing gp41 to attach to the cytoplasmic membrane. **(c)** Fusion of the lipid bilayers, introducing the capsid into the cytoplasm.

Attachment, Penetration, and Uncoating HIV infects four kinds of cells: helper T cells; cells of the macrophage lineage, including monocytes, macrophages, and *microglia* (special phagocytic cells of the central nervous system); smooth muscle cells, such as those in arterial walls; and *dendritic cells,* which are antigen-presenting cells of the skin, mucous membranes, and lymph nodes. B cells do not themselves become infected with HIV, but they adhere to HIV that has been covered with complement proteins (see Chapter 15). They then deliver HIV to the lymphoid tissues, where T cells mature.

Attachment of HIV and entry into T cells and macrophages involves three steps **(Figure 25.20)**. First, gp120 binds to CD4 receptor molecules on a host cell. Second, the gp120-CD4 complex binds to another receptor, called a *chemokine receptor,* which removes gp120 from the virion. Different chemokine receptors exist on T cells and on macrophage derivatives—*CCR5*[11] on most cells and *CXCR4* on T cells. Finally, once a virion binds to a cell, its envelope contacts the cytoplasmic membrane via gp41 and fuses with it. The HIV capsid is introduced into the cytoplasm while its envelope remains behind as part of the cell's cytoplasmic membrane. Glycoprotein 41 remaining on the cytoplasmic membrane induces the cell to fuse to as many as 500 neighboring cells to form a **syncytium** (sin-sish'ē-ŭm), a giant multinucleate cell. Cellular fusion allows HIV to move from cell to cell without being exposed to antibodies in the blood.

In contrast, dendritic cells are infected in a single step, when the gp120 of HIV attaches to a receptor (DC-SIGN) found on their cytoplasmic membranes. HIV fuses with the target cell and introduces its capsid into the cytoplasm. Infected dendritic cells deliver HIV into T cells during antigen presentation.

Latency and Synthesis In the cytoplasm, reverse transcriptase uses cellular tRNA (packaged in the virion when it was assembled) as a primer to transcribe dsDNA. Reverse transcriptase is very error prone, making about five errors per genome, and this generates many new antigenic variations of HIV during the course of infection. As many as 100 variants may develop in a single patient over the course of the syndrome.

HIV is a latent virus. The dsDNA made by reverse transcriptase, which is known as a **provirus,** enters the nucleus and is inserted into a human chromosome by the virion-carried enzyme **integrase.** Once integrated, the provirus remains a part of the cellular DNA for life. Integrated HIV is transcribed and replicated in the same way as any cellular gene. It may remain dormant for years or be activated immediately, depending on its location in the human genome and the availability of enhancer and promoter sequences (see Figure 7.8a). Macrophages and monocytes are major reservoirs as well as means of distribution of HIV.

[11]Membrane proteins are often designated by letter-number combinations.

TABLE 25.6 Characteristics of HIV That Challenge the Immune System

Characteristic	Effect(s)
Retroviral genome consists of two copies of +ssRNA	Reassortment of viral genes possible; virus is highly mutable, reverse transcription produces much genetic variation; genome can integrate into host's chromosomes
Targets helper T lymphocytes especially, but also macrophages, dendritic cells, smooth muscle cells	Infects key cells of host's immune system
Antigenic variability	Large amount of antigenic drift due to mutations helps virus evade host's immune response
Induces formation of syncytia	Increases routes of infection; intracellular site helps virus evade immune detection

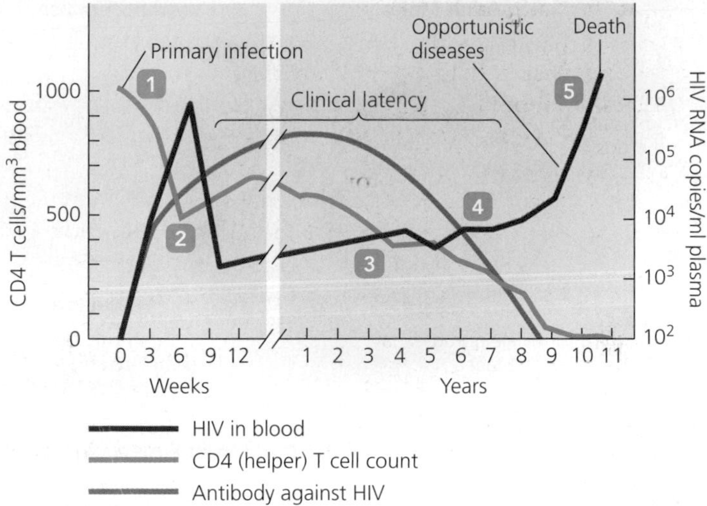

▲ **Figure 25.21 The course of AIDS.** The course follows helper T cell destruction. The circled numbers correspond to the steps described in the text.

Assembly and Release RNA transcribed from a provirus may be used in either of two ways:

- A molecule of RNA may be used as mRNA by ribosomes for the translation of viral proteins, including capsomeres, gp120, and reverse transcriptase.
- Two molecules of RNA are incorporated into a new virion along with copies of reverse transcriptase, integrase, and two molecules of tRNA (the latter are coded by the host's genome).

Virions subsequently bud from the cytoplasmic membrane (see Figure 25.19, step 6). At this point in its replication cycle, a virion is nonvirulent because its capsid is not assembled and reverse transcriptase is inactive. **Protease** (prō′tē-ās), a viral protein packaged in the virion, releases reverse transcriptase and capsomeres after the virion buds from the cell. This action of protease allows final assembly, rendering HIV virulent.

Pathogenesis of AIDS

The course of AIDS is directly related to the destruction of helper T cells in a patient **(Figure 25.21)**.

1. Initially, there is a burst of virion production and release from infected cells. This primary infection is accompanied by fever, fatigue, weight loss, diarrhea, and body aches.
2. The immune system responds by producing antibodies, and the number of free virions plummets. The body and HIV are waging an invisible war with few signs or symptoms. During this period, almost a billion virions are destroyed each day, while about 100 million CD4 helper T cells are killed by viruses and by cytotoxic T cells. No specific symptoms accompany this stage, and the patient is often unaware of the infection.
3. As latent viruses continue to replicate and virions are released into the blood, the body cannot adequately replenish helper T cells. Over the course of 5–10 years, the number of helper T cells declines to a level that severely impairs the immune response.
4. The rate of antibody formation falls precipitously as helper T function is lost.
5. HIV production climbs, and the patient dies from infections the body can no longer resist.

Many of the diseases associated with the loss of immune function in AIDS (see Table 25.5 on p. 720) are nonlethal infections in other patients, but AIDS patients cannot effectively resist them. Diseases such as disseminated herpes and Kaposi's sarcoma (a cancer of blood vessels) **(Figure 25.22)**, as well as toxoplasmosis and *Pneumocystis* pneumonia, are typical in AIDS patients. AIDS often results in dementia during the final stages.

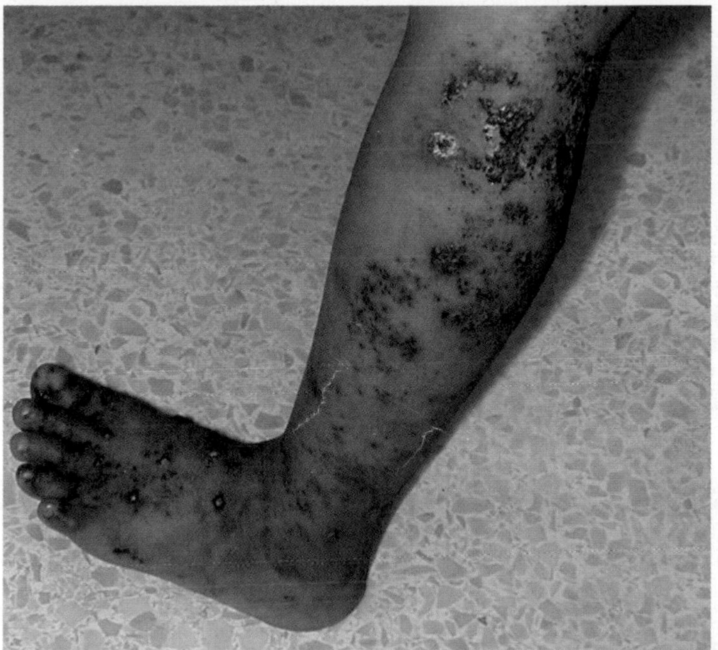
(a)

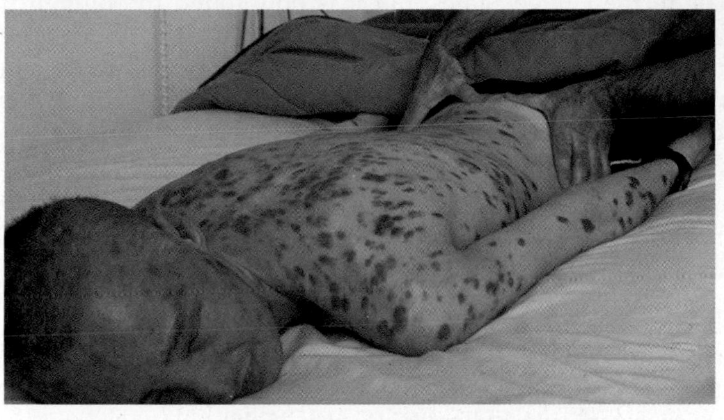
(b)

▲ **Figure 25.22 Diseases associated with AIDS.** (a) Disseminated herpes. (b) Kaposi's sarcoma. *Which viruses cause herpes?*

Figure 25.22 *Herpes is caused by human herpesvirus 1 and 2.*

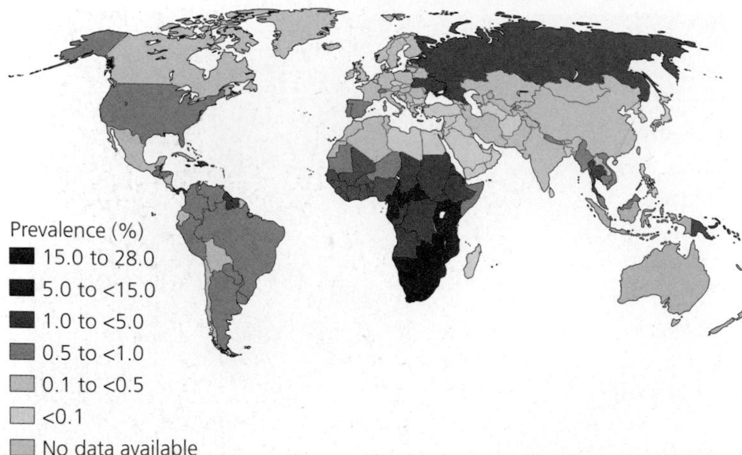

▲ **Figure 25.23 The global prevalence of HIV (2007).** Over 33 million were HIV positive. Note the alarming epidemic in sub-Saharan Africa.

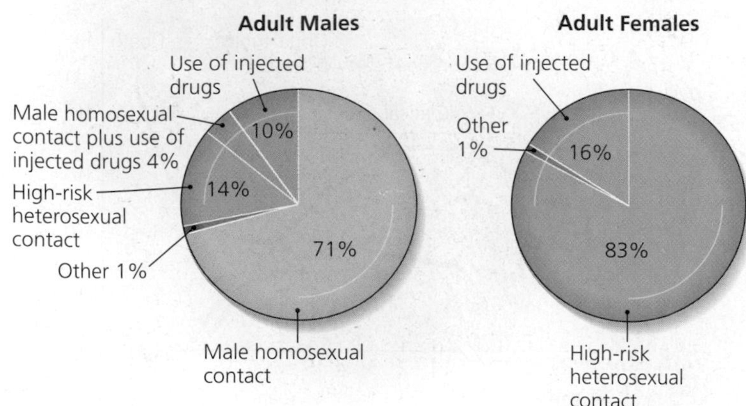

▲ **Figure 25.24 Modes of HIV transmission in Americans over 12 years of age during 2007.** Percentages represent the estimated proportions of HIV infections that result from each type of transmission. High-risk heterosexual behavior is sexual contact with a person known to have, or to be at high risk for, HIV infection.

Epidemiology of AIDS

Epidemiologists first recognized AIDS in young male homosexuals in the United States, but it is now found throughout the world's population. The United Nations Program on AIDS estimates that there are about 33 million HIV-infected individuals worldwide. Approximately 7400 new infections occur every day. It is disturbing that although the rate of new infections declined throughout most of the 1990s, it is now rising again, presumably because of a false sense of security in richer nations with better access to availability of drugs that slow the progress of the disease, and because of a relaxation of precautions to prevent infection.

About a third of those infected have already developed AIDS; it is presumed that most of the rest will eventually succumb to the syndrome. The spread of AIDS in sub-Saharan Africa is particularly horrific **(Figure 25.23)**.

HIV is found in all secretions of infected individuals; however, it typically exists in sufficient concentration to cause infection only in blood, semen, saliva, vaginal secretions, and breast milk. Virions are found both free and inside infected white blood cells. Sufficient numbers of virions must be transmitted to target cells to establish an infection, though the exact number of virions needed to establish an infection is unknown and probably varies with the strain of HIV and the overall health of the person's immune system. Infected blood contains 1000–100,000 virions per milliliter, and semen has 10–50 virions per milliliter. Other secretions have lower concentrations and are less infective than blood or semen.

Infected fluid must come in contact with a tear or lesion in the skin or mucous membranes, or it must be injected into the body. Because HIV is small—about 90 nm in diameter—a break in the protective membranes of the body large enough to allow entry of HIV may be too small to see or feel.

HIV is transmitted primarily via sexual contact (including vaginal, anal, and oral sex, either homosexual or heterosexual) and intravenous drug abuse **(Figure 25.24)**. The risk of infection from a needlestick injury involving an HIV-infected patient is less than 1%. HIV can also be transmitted from mother to baby across the placenta and in breast milk. Approximately one-third of babies born to HIV-positive women are infected with HIV at birth. HIV is rarely acquired from blood transfusions, organ transplants, hemophiliac therapy, and tattooing. Behaviors that increase the risk of infection include the following:

- Sexual promiscuity; that is, sex with more than one partner
- Anal intercourse, especially receptive anal intercourse
- Intravenous drug use
- Sexual intercourse with anyone who engages in the previous three behaviors

The mode of infection is unknown for about 1% of AIDS patients. These patients are often unable or unwilling to answer questions about their sexual and drug abuse history. The Centers for Disease Control and Prevention (CDC) has documented a few cases of casual spread of HIV, including infections from sharing razors and toothbrushes and from mouth-to-mouth kissing. In all cases of casual spread, it is suspected that small amounts of the donor's blood may have entered abrasions in the recipient's mouth.

Diagnosis, Treatment, and Prevention of AIDS

The signs and symptoms of AIDS vary according to the diseases present. The syndrome indicates that patients have one or more rare diseases, antibodies against HIV, and fewer than 200 CD4 lymphocytes per microliter of blood.[12] Patients also typically have unexplained weight loss, fatigue, and fever.

HIV itself is difficult to locate in patient's secretions and blood because it becomes a provirus inserted into the patient's chromosomes, and because for years HIV is kept in check by the immune system. Polymerase chain reaction (PCR) can be used to reveal the presence of HIV, and serological diagnosis can reveal antibodies against HIV. The latter involves detecting antibodies against HIV using ELISA, agglutination, or western blot

[12]The normal value for CD4 cells is 800–1200 cells/μl.

HIGHLIGHT

AIDS AND TUBERCULOSIS

In the early 1900s, millions of people suffered from tuberculosis (TB). Patients were treated with months of bed rest in a sanatorium. In the 1940s and 1950s, the introduction of effective antibiotic treatments brought about a dramatic decline in the disease. In recent years, however, tuberculosis has been on the rise once again. The World Health Organization (WHO) estimates that TB killed 1.8 million people worldwide in 2007. This resurgence of tuberculosis is linked to the spread of AIDS.

WHO estimates that one-third of the world's population is infected with *Mycobacterium tuberculosis*, the bacterium that causes tuberculosis. In most infected individuals, the immune system fights the infection and stops the bacterium from reproducing. The person doesn't feel sick, has no symptoms, and can't spread the disease. Among people with healthy immune systems, only 5–10% of those infected develop TB symptoms. But because HIV weakens the immune system, HIV-positive individuals are much more susceptible to developing the disease.

An HIV patient who contracts tuberculosis and is not treated has a 90–95% chance of dying in a few months. The leading cause of death among the HIV-infected population, tuberculosis accounts for 13% of AIDS deaths worldwide.

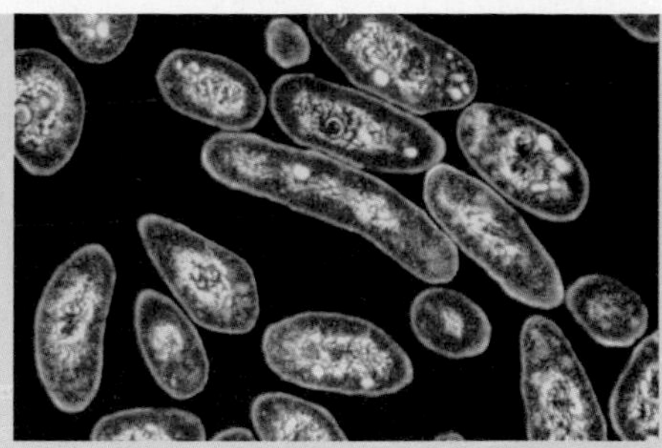

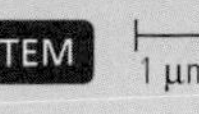

▲ *Mycobacterium tuberculosis.*

The good news is that tuberculosis can be treated successfully and inexpensively with a course of antibiotics taken faithfully for two to six months. Proper treatment and cure of tuberculosis can give an HIV patient years of additional life.

testing. Definitive diagnosis of infection is by polymerase chain reaction assay. Most individuals develop antibodies within six months of infection, though some remain without detectable antibodies for three years or more. A positive test for antibodies does not mean that the patient is contagious, that HIV is currently present, or that the patient has or will develop AIDS; it merely indicates that the patient has been exposed to HIV.

A small percentage of infected individuals, called *long-term nonprogressors,* have not yet developed AIDS. It appears that these individuals are either infected with defective virions, lack effective coreceptors (CXCR4 and CCR5) needed to maintain an infection, or have unusually well-developed immune systems.

Discovering more effective treatment for AIDS is an area of intense research and development. Currently physicians prescribe **highly active antiretroviral therapy (HAART)**—a cocktail of antiviral drugs, including nucleoside analogs, protease inhibitors, and reverse transcriptase inhibitors—to reduce viral replication. Chapter 10 examined the actions of antiviral drugs. HAART is expensive and must be taken on a fairly strict schedule. Studies indicate that HAART can stop the replication of HIV because strains of HIV are not likely to develop resistance to all the drugs simultaneously. As long as treatment continues, the patient can live a relatively normal life; however, treatment is not a cure, because the infection remains. Individual diseases associated with AIDS must be managed and treated on a case-by-case basis.

Progress in developing a vaccine against HIV has been disappointing. Among the problems that must be overcome in developing an effective vaccine are the following:

- A vaccine must generate both secretory IgA to prevent sexual transmission and infection, and cytotoxic T lymphocytes to eliminate infected cells.
- IgG induction, while a necessary part of a vaccine, can actually be detrimental to a patient. Because IgG-viral complexes remain infective inside phagocytic cells, a vaccine must stimulate cellular immunity more than humoral immunity so that infected cells can be killed.
- HIV is highly mutable, generating antigenic variants that enable it to evade the immune response. Indeed, every nucleotide site of AIDS RNA is mutated with every incorrect nucleotide every day in an AIDS patient.
- HIV can spread through syncytia (formed by gp41) and thus can evade immune surveillance.
- HIV infects and inactivates macrophages, dendritic cells, and helper T cells—cells that combat infections and the very cells a vaccine would stimulate.
- Testing a vaccine presents ethical and medical problems because HIV is a pathogen of humans only. Scientists cannot ethically challenge vaccinated patients with active HIV.

Notwithstanding the challenges, scientists have developed a vaccine that protects monkeys from SIV disease and are hopeful that an effective vaccine for humans will become available. In the meantime, the principal way the AIDS epidemic can be slowed and eventually stopped is by effecting changes in behavior:

- Abstinence or faithful monogamy between uninfected individuals are the only truly safe behaviors. Studies have shown that although condoms reduce a heterosexual individual's risk of acquiring HIV by about 69%, the benefit of condom usage to a population can be offset by an overall increase in sexual activity. This finding presumably is due to a false sense of security that condoms provide.
- Use of new, clean needles and syringes for all injections, as well as caution in dealing with sharp, potentially contaminated objects, can reduce HIV infection rates. If clean supplies are not available, undiluted bleach or other antiviral agents can deactivate HIV.

MICROBE AT A GLANCE

Lentivirus human immunodeficiency virus (HIV)

Taxonomy: Family *Retroviridae*

Genome and morphology: Two molecules of positive single-stranded RNA; polyhedral capsid with spiked envelope, about 110 to 145 nm in diameter

Host: Human

Virulence factors: Attaches to CD4 of helper T lymphocytes, vigorous mutation rate due to transcription errors of reverse transcriptase, latency

Syndrome caused: Acquired immunodeficiency syndrome

Treatment for disease: Highly active antiretroviral therapy (HAART), the so-called AIDS cocktail, composed of a variety of antiviral drugs such as nucleoside analogs, protease inhibitors, and inhibitors of reverse transcriptase

Prevention of disease: Sexual abstinence; monogamy. Refrain from sharing intravenous needles. Correct, consistent condom usage can reduce but not eliminate risk of infection. Male circumcision reduces risk of infection.

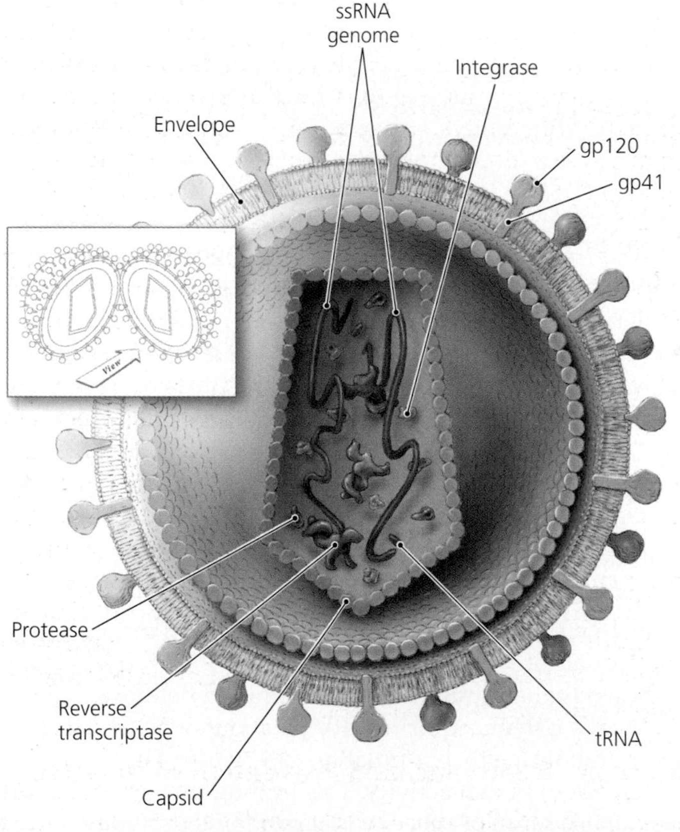

Download the Microbe at a Glance flashcards from the Study Area at www.masteringmicrobiology.com.

- AZT given to pregnant women has reduced transplacental transfer of HIV, but infants can still be infected via breast milk. HIV-infected mothers should not breast-feed their infants.
- Screening of blood, blood products, and organ transplants for HIV and anti-HIV antibodies has virtually eliminated the risk of HIV infection from these sources.
- The CDC is considering recommending voluntary circumcision as a way to reduce the HIV transmission rate.

We have considered positive single-stranded RNA viruses, including retroviruses. Now we turn our attention to negative single-stranded RNA viruses.

Enveloped, Unsegmented, Negative ssRNA Viruses: *Paramyxoviridae, Rhabdoviridae,* and *Filoviridae*

Viruses of the families *Paramyxoviridae, Rhabdoviridae,* and *Filoviridae* are enveloped, helical, negative, single-stranded RNA viruses. Recall that a negative RNA genome cannot be directly translated because its information is nonsensical to a ribosome; however, it is the complement of a readable sequence of nucleotides. Therefore, a −ssRNA virus must make a positive copy of its genome. The +ssRNA copy serves as mRNA for protein translation. The positive copy also serves as a template for transcription of more −ssRNA to be incorporated into new virions.

Viruses in the family *Paramyxoviridae* have similar morphologies and antigens. They also share the ability to cause infected cells to fuse with their neighbors, forming giant, multinucleate syncytia (**Figure 25.25**), which enable virions to pass from an infected cell into neighboring cells and to evade immune surveillance and antibodies. The family *Paramyxoviridae* contains four genera that infect humans: *Morbillivirus* (measles virus), *Respirovirus* (two species of parainfluenza viruses), *Rubulavirus* (mumps virus and two other species of parainfluenza viruses), and *Pneumovirus* (respiratory syncytial virus). **Highlight: Nipah Virus From Pigs to Humans** on p. 728 describes an emerging fatal disease caused by a newly identified genus in the *Paramyxoviridae.*

Rhabdoviruses *(Rhabdoviridae)* have bullet-shaped envelopes and include a variety of plant and animal pathogens. Rabies is the most significant pathogen among the rhabdoviruses.

Filoviruses *(Filoviridae)* are particularly significant because they are pathogens that cause a number of emerging and frightening diseases, including Ebola and Marburg hemorrhagic fevers.

Measles

Learning Objective

✓ Describe the signs and symptoms of rubeola and SSPE, and some ways of preventing them

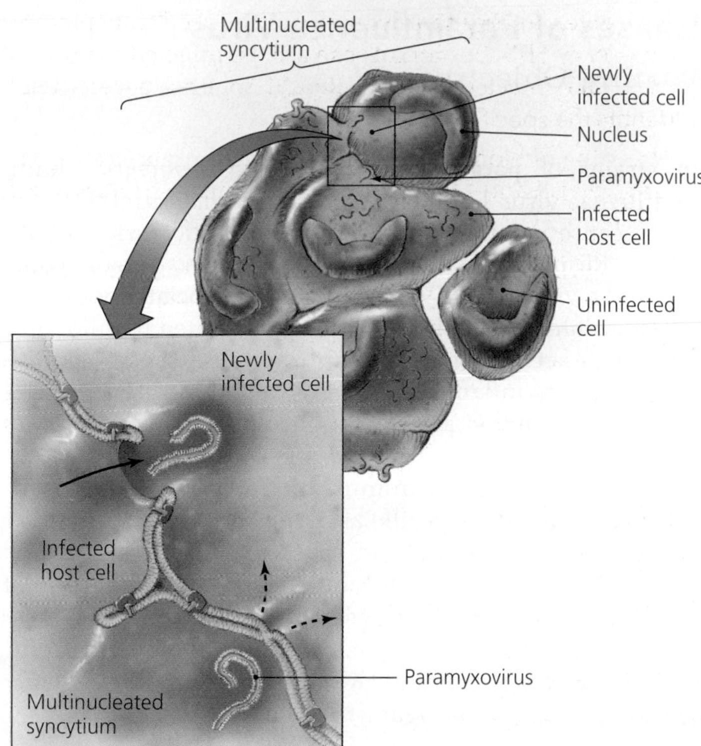

▲ **Figure 25.25 Syncytium formation.** Paramyxoviruses trigger infected cells to fuse with uninfected neighboring cells. Virions moving through these large multinucleate cells infect new cells, all the while evading the host's immune system.

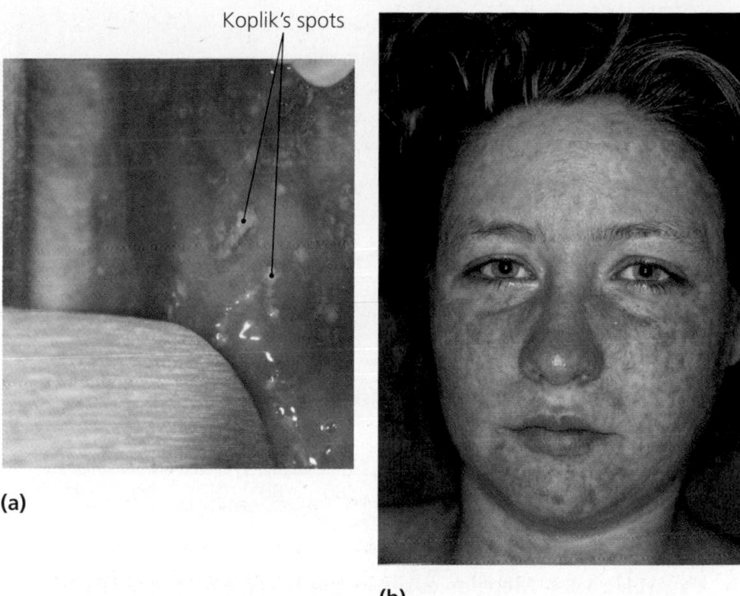

▲ **Figure 25.26 Signs of measles.** **(a)** White Koplik's spots on the oral mucous membrane. **(b)** Raised lesions begin on the face and spread across the body. They may grow and fuse into large reddish patches.

Measles virus causes one of five classical childhood diseases. **Measles,** also known in the United States as *rubeola*[13] or *red measles,* is one of the more contagious and serious childhood diseases, and it should not be confused with the generally milder rubella (German measles). Table 25.7 on p. 729 compares and contrasts measles and rubella.

Epidemiology of Measles Infections

Coughing, sneezing, and talking spread measles virus in the air via respiratory droplets. The virus infects cells of the respiratory tract before spreading via lymph and blood throughout the body. In addition to the respiratory tract, the conjunctiva, urinary tract, small blood vessels, lymphatics, and central nervous system become infected.

Humans are the only host for measles virus, and a large, dense population of susceptible individuals must be present for the virus to spread. In the United States before vaccination was instituted, measles outbreaks occurred in one- to three-year epidemic cycles as the critical number of susceptible individuals increased in the population. More than 80% of susceptible patients exposed to the virus developed symptoms 7–13 days after exposure.

Signs and symptoms include fever, sore throat, headache, dry cough, and conjunctivitis. After two days of illness, lesions called **Koplik's spots** appear on the mucous membrane of the mouth (Figure 25.26a). These lesions, which have been described as grains of salt surrounded by a red halo, last one to two days and provide a definitive diagnosis of measles. Red, raised (maculopapular) lesions then appear on the head and spread over the body (Figure 25.26b). These lesions are extensive and often fuse to form red patches, which gradually turn brown as the disease progresses. Death may result.

Rare complications of measles include pneumonia, encephalitis, and the extremely serious **subacute sclerosing panencephalitis (SSPE).** SSPE is a slow, progressive disease of the central nervous system that involves personality changes, loss of memory, muscle spasms, and blindness. The disease begins 1–10 years after infection with measles virus and lasts a few years before resulting in death. A defective measles virus, which cannot make a capsid, causes SSPE. The virus replicates and moves from brain cell to brain cell via syncytia formation, limiting the functioning of infected cells and resulting in the symptoms. SSPE afflicts fewer than 7 measles patients in 1 million and is becoming much rarer as a result of vaccination of children.

An effective, live, attenuated vaccine for measles introduced in 1963 has eliminated measles as an endemic disease in the United States (Figure 25.27); however, measles remains a cause of death in other countries.

Diagnosis, Treatment, and Prevention of Measles

The signs of measles, particularly Koplik's spots, are sufficient for diagnosis, but serological testing can confirm the presence of measles antigen in respiratory and blood specimens. No antiviral treatment is available.

Measles vaccine, in combination with vaccines against mumps and rubella (MMR vaccine), is given to children in the

[13]Unfortunately, rubeola is the name given to rubella (German measles) in some other countries.

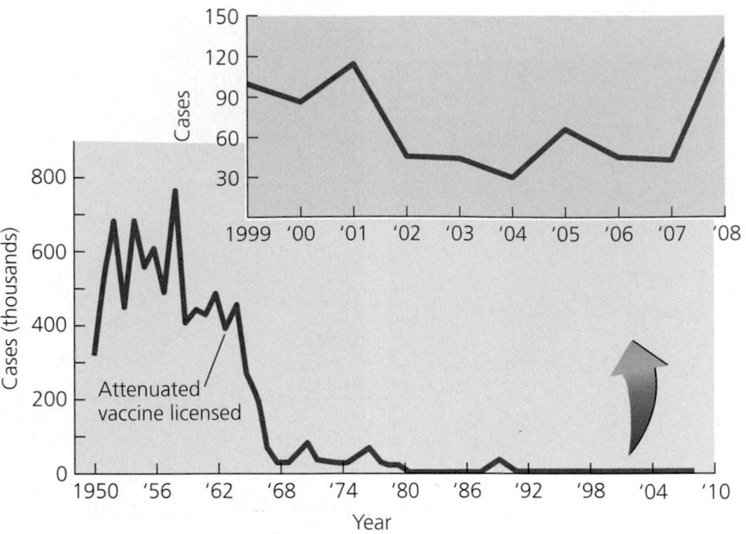

▲ **Figure 25.27 Measles cases in the United States since 1950.** The number of cases has declined dramatically since vaccination began in 1963. *Why does the measles vaccine pose risks for immunocompromised contacts of vaccinated children?*

Figure 25.27 Measles vaccine is a live attenuated vaccine and can cause disease in the immunocompromised.

United States who are less than 15 months old. The measles portion of the vaccine is effective 95% of the time but is boosted before grade school to ensure protection against measles introduced from outside the country. Eradication of measles worldwide is possible, but there is no consensus among politicians and scientists that eradication would be worth the cost.

Diseases of Parainfluenza Virus

Learning Objective

✓ Identify the specific cause of croup.

Four strains of **parainfluenza viruses** (*Respirovirus*: human parainfluenza virus 1 and HPIV-3; and *Rubulavirus*: HPIV-2 and HPIV-4) cause respiratory tract disease, particularly in young children. Respiratory droplets and person-to-person contact transmit the virions. HPIV-1, 2, and 3 are associated with lower respiratory infections, whereas HPIV-4 is limited to mild, upper respiratory tract infections. **Croup** (kroop), a severe condition characterized by inflammation and swelling of the larynx, trachea, and bronchi and a "seal bark" cough, is a childhood disease of HPIV-1 and 2 and rarely of other respiratory viruses.

Most patients recover from parainfluenza infections within two days. There is no specific antiviral treatment beyond support and careful monitoring to check that airways do not become completely occluded. If they do, *intubation,* insertion of a tube into the airways, is necessary. Vaccination with deactivated virus is not effective, possibly because it fails to produce significant secretory antibody. Researchers have not developed a live attenuated vaccine against parainfluenza.

Mumps

The **mumps** virus *(Rubulavirus)* is another −ssRNA virus. Mumps virions are spread in respiratory secretions and infect cells of the upper respiratory system. As virions are released from these areas, viremia develops, and a number of other organs are infected, resulting in fever and pain in swallowing. The parotid salivary glands, located just in front of the ears under

HIGHLIGHT

NIPAH VIRUS: FROM PIGS TO HUMANS

Many diseases become known to science either when people encroach on a pathogen's home territory and come into contact with natural hosts, or when pathogens are introduced into geographical areas outside their historical range. Diseases resulting under such conditions are called *emerging diseases.* One example of an emerging disease is a new form of encephalitis that has appeared in Malaysia. The culprit: a never-before-identified virus now known as Nipah virus (genus *Henipavirus*) after the region in Malaysia in which it was discovered.

The victims of Nipah encephalitis experience high fever, severe headache, muscle pain, drowsiness, disorientation, convulsions, and coma. Forty percent of them die. Serological investigations and RNA sequencing of the genome revealed that *Henipavirus* belongs to the family *Paramyxoviridae.* Victims contract Nipah virus from infected animals: 93% of Malaysian victims report occupational exposure to pigs.

How, then, do pigs become infected with Nipah virus? The prevailing theory involves human encroachment on bat habitats. Certain species of fruit bats are the natural hosts of Nipah virus. Highway workers pushing into these bats' rainforest habitats have disturbed their roosts, driving the bats into proximity with pig farms. The virus then "jumped" species and infected the pigs. Individuals working on pig farms or in slaughterhouses were subsequently infected either via the respiratory route or through breaks in their skin and mucous membranes.

▲ ***Removing pigs to prevent the spread of Nipah virus.***

There is no cure for Nipah encephalitis, but scientists have inserted genes for *Henipavirus* proteins into cowpox virus, which thereby becomes an effective recombinant vaccine that protects against infection.

▲ **Figure 25.28 Parotitis.** This inflammation of the parotid salivary glands is a common sign of mumps.

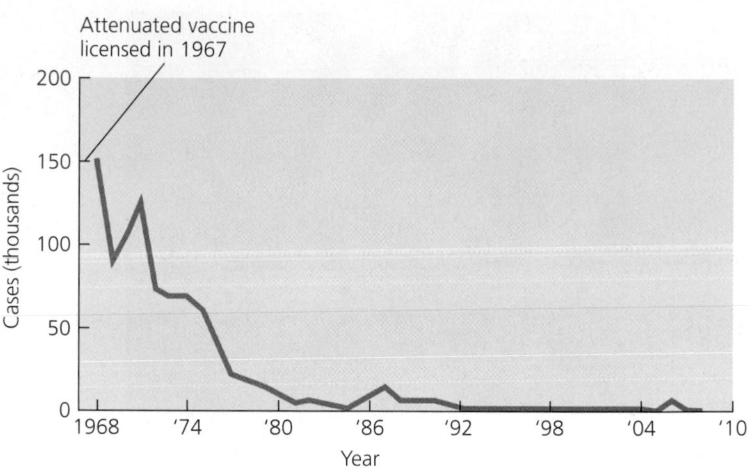

▲ **Figure 25.29 Incidence of mumps in the United States.** Mumps cases have declined since vaccination began in 1967. *With which other vaccines is mumps vaccine administered?*

Figure 25.29 *The mumps vaccine is administered with the measles and rubella vaccines.*

the skin covering the lower jaw, are particularly susceptible to the virus and become painfully enlarged, a condition known as *parotitis* **(Figure 25.28)**. In some patients, the mumps virus causes inflammation of the testes (orchitis), meningitis, pancreatitis, or permanent deafness in one ear. In many other patients, infection is asymptomatic. Recovery from mumps is typically complete, and the patient has effective lifelong immunity.

There is no specific treatment for mumps. However, because an effective attenuated vaccine (MMR) is available and humans are the only natural host for the virus, mumps has almost been eradicated in the industrialized world **(Figure 25.29)**. Epidemics of mumps in the late winter and early spring still occur in countries without vaccination programs, especially in densely populated areas.

Disease of Respiratory Syncytial Virus

Learning Objective

✓ Compare the effects of RSV infection in infants and adults.

Respiratory syncytial virus (RSV) in the genus *Pneumovirus* of the family *Paramyxoviridae* causes a disease of the lower respiratory tract, particularly in young children. As its name indicates, the virus causes formation of syncytia in the lungs. Additionally, plugs of mucus, fibrin, and dead cells fill the smaller air passages, resulting in *dyspnea* (disp-nē′ă)—difficulty in breathing—and sometimes croup. RSV is the leading cause of fatal respiratory disease in infants and young children worldwide, especially those who are premature, immune impaired, or otherwise weakened. The disease in older children and adults is manifested as a cold or is asymptomatic. Most otherwise healthy people fully recover in one to two weeks.

Damage to the lungs is increased by the action of cytotoxic T cells and other specific immune responses to infection. Attempts at developing a vaccine with deactivated RSV have proven difficult because the vaccine enhances the severity of the cellular immune response and lung damage.

Epidemiology of RSV Infection

RSV is prevalent in the United States: epidemiological studies reveal that 65–98% of children in day care centers are infected by age three. Approximately 125,000 infants require hospitalization each year, and about 2000 die. The virus is transmitted on fomites, hands, and less frequently via respiratory droplets. Because the virus is also extremely contagious, its introduction

TABLE 25.7 A Comparison of Measles and Rubella

Disease	Causative Agent	Primary Patient(s)	Complications	Skin Rash	Koplik's Spots
Measles (also knownas rubeola and red measles)	*Paramyxoviridae: Morbillivirus measles virus*	Child	Pneumonia, encephalitis, subacute sclerosing panencephalitis	Extensive	Present
Rubella (also known as German measles, rubeola, and three-day measles)	*Togaviridae: Rubivirus rubella virus*	Child, fetus	Birth defects	Mild	Absent

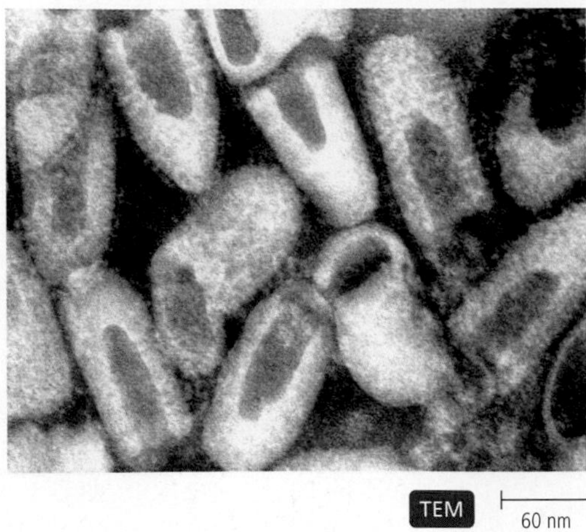

▲ **Figure 25.30 A rabies virus.** Note the bullet-shaped envelope.

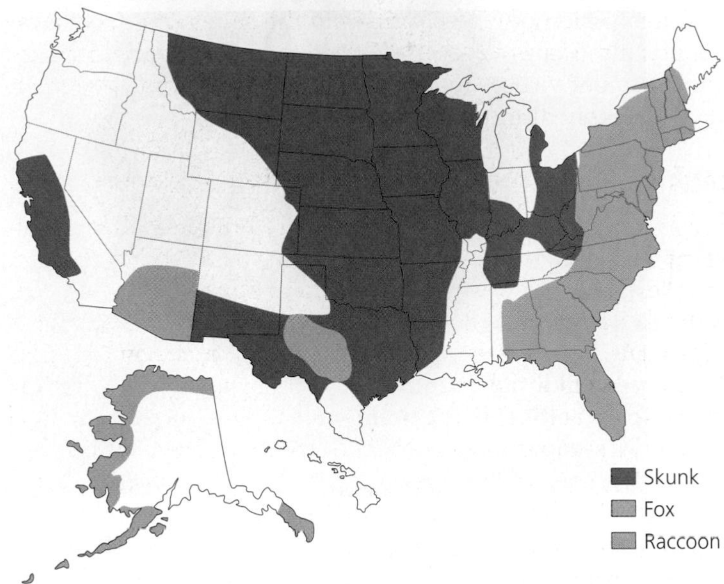

▲ **Figure 25.31 Predominant wildlife reservoirs for rabies by area.** Absence of a color does not indicate the absence of rabies in that region, only that no wildlife reservoir is predominant. Bats, for instance, harbor rabies throughout the country but are rarely the predominant reservoir in an area. Nevertheless, bats are the major source of human infection.

into a nursery can be devastating; therefore, good aseptic technique by health care workers, and good hygiene by day care center employees, is essential. Handwashing and the use of gowns, goggles, masks, and gloves are important measures to reduce nosocomial infections.

Diagnosis, Treatment, and Prevention of RSV Infection

Prompt diagnosis is essential if infected infants are to get the care they need. The signs of respiratory distress provide some diagnostic clues, but verification of RSV infection is made by immunoassay. Specimens of respiratory fluid may be tested by immunofluorescence, ELISA, or complementary DNA probes.

Supportive treatment involves administration of oxygen. Antisense RNA has shown promise in reducing the spread of RSV infection.

Control of RSV is limited to attempts to prevent the spread of RSV by frequent handwashing.

Rabies

Learning Objective

✓ Describe the effects, treatment, and prevention of rabies.

Although the cry "Mad dog! Mad dog!" may not make you apprehensive, in the 19th century—before Pasteur developed a rabies vaccine and treatment—a bite from a rabid dog was a death sentence, and the cry "Mad dog!" often spawned panic. Few people survived rabies before Pasteur's pivotal work.

Rabies[14] virus is a negative, single-stranded RNA (−ssRNA) virus in the genus *Lyssavirus*, family *Rhabdoviridae*. Like viruses in the family *Paramyxoviridae*, all of the rhabdoviruses have helical capsids, but those of rhabdoviruses are supercoiled into cylinders, which gives them a striated appearance, and are surrounded by bullet-shaped envelopes (**Figure 25.30**).

Glycoprotein spikes on the surface of the envelope serve as attachment proteins. Attachment to nerve cells triggers endocytosis, and rabies viruses then replicate in the cytoplasm of these cells. The virus is carried to the central nervous system via cytoplasmic flow within the neurons. The spinal cord and brain degenerate as a result of infection, though infected cells show little damage when examined microscopically. Viruses travel back to the periphery, including the salivary glands, through nerve cells. Viruses secreted in the saliva of infected mammals are typically the infective agents.

Initial signs and symptoms of rabies include pain or itching at the site of infection, fever, headache, malaise, and anorexia. Once viruses reach the central nervous system, neurological manifestations characteristic of rabies develop: **hydrophobia**[15] (triggered by the pain involved in attempts to swallow water), seizures, disorientation, hallucinations, and paralysis. Death results from respiratory paralysis and other neurological complications.

Epidemiology of Rabies

Rabies is a classical zoonosis of mammals, though not all mammals are reservoirs. Rodents, for instance, rarely get rabies. The distribution of rabies in humans follows the distribution in animals. The primary animals involved differ from locale to locale and change over time as a result of dynamics of animal populations and interactions among animals and humans. The main reservoir of rabies in urban areas is the dog. In the wild, rabies can be found in foxes, badgers, raccoons, skunks, cats, bats, coyotes, and feral dogs (**Figure 25.31**). Bats are the source of most human cases of rabies in the United States, causing about 75% of cases between 1990 and 2008.

[14]From Latin *rabere*, meaning rage or madness.
[15]From Greek *hydro*, meaning water, and *phobos*, meaning fear.

Transmission of rabies viruses in the saliva of infected animals usually occurs via a bite. Infection is also possible via the introduction of viruses into breaks in the skin or mucous membranes or rarely through inhalation.

Diagnosis, Treatment, and Prevention of Rabies

The neurological symptoms of rabies are unique and generally sufficient for diagnosis. Serological tests for antibodies can confirm a diagnosis. Postmortem laboratory tests are often conducted to determine whether a suspected animal in fact carries rabies virus. These tests include antigen detection by immunofluorescence and the identification of aggregates of virions (called **Negri bodies**) in the brain **(Figure 25.32)**. Unfortunately, by the time symptoms and antibody production occur, it is too late to intervene, and the disease will follow its natural course **(Highlight: Rabies: A Threat from the Wild)**.

The rabies vaccine—*human diploid cell vaccine (HDCV)*—is prepared from deactivated rabies viruses cultured in human diploid cells. It is administered intramuscularly on days 0, 3, 7, and 14 after exposure. The vaccine can also be administered prophylactically to workers who regularly come into contact with animals (veterinarians, zookeepers, and animal control workers) and to people traveling to areas of the world where rabies is prevalent.

Treatment of rabies begins with treatment of the site of infection. The wound should be thoroughly cleansed with water and soap or another substance that deactivates viruses. The World Health Organization recommends anointing the wound with antirabies serum. Initial treatment also involves injection of *human rabies immune globulin (HRIG)*. Subsequent treatment involves five vaccine injections. Rabies is one of the few infections that can be treated with active immunization, because the progress of viral replication and movement to the brain is slow enough to allow effective immunity to develop before disease develops.

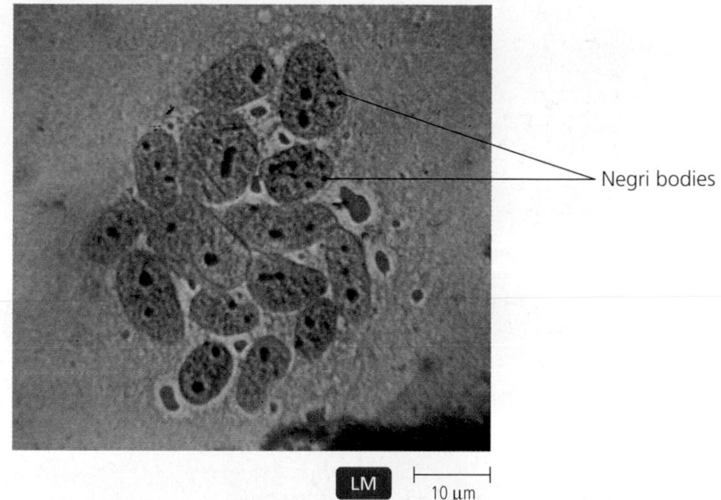

▲ **Figure 25.32 Negri bodies, a characteristic of rabies infection.** Here the viruses are in human cerebellum brain cells. *What are Negri bodies?*

Figure 25.32 Negri bodies are stained aggregates of rabies virions.

HIGHLIGHT

RABIES: A THREAT FROM THE WILD

Vaccination of cats and dogs has sharply reduced the number of cases of human rabies in the United States; however, infections from wild animals still result in a few cases of rabies each year.

In one case, a man entered the hospital with neurological disorders that produced uncontrolled facial twitching, anxiety, and feelings of fear. He also had a sore throat, difficulty in swallowing, and complained of itching all over his body. Though the symptoms were typical of rabies, it was not diagnosed because there was no history of an animal bite.

The patient remained alive for several days, but became increasingly agitated and refused to drink because of the pain involved in swallowing. He vomited repeatedly and his temperature rose to 106°F. He died one week after being admitted to the hospital.

An autopsy revealed circulating antibodies against rabies virus as well as Negri bodies in the victim's brain. Epidemiologists determined from interviews with the man's neighbors that a bat had landed on his face about a month before the symptoms began, though the bat had not bitten him. Possibly the bat scratched the man during the encounter and transmitted the virus. This story reinforces the recommendation of the Centers for Disease Control and Prevention (CDC) that all contacts with wild animals (particularly bats, skunks, foxes, and feral dogs) be taken seriously and rabies prophylaxis administered.

Not all cases of rabies exposure are so tragic. In 1885, a rabid dog savagely mauled a young boy, Joseph Meister. His distraught mother brought the boy to Louis Pasteur, who was in the process of perfecting the rabies vaccine. The vaccine had never been tested on humans, but with assurance from doctors that the boy would surely die without treatment, Pasteur injected the 14 doses of attenuated virus that made up the vaccine at that time. Joseph survived and Pasteur's fame and fortune were further secured.

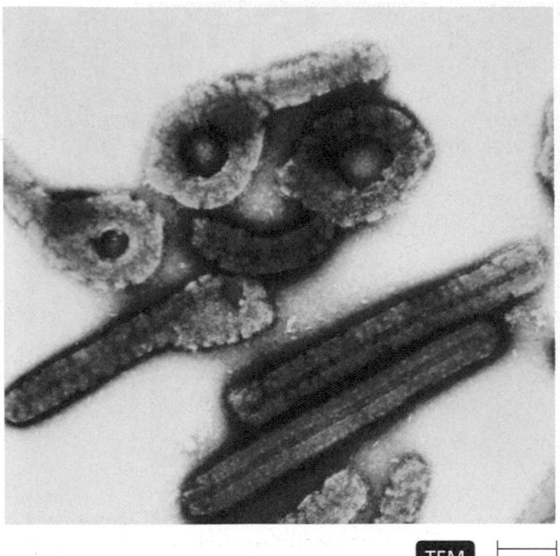

▲ **Figure 25.33** **Filamentous Ebola viruses.** Filoviruses can be up to 14,000 nm long.

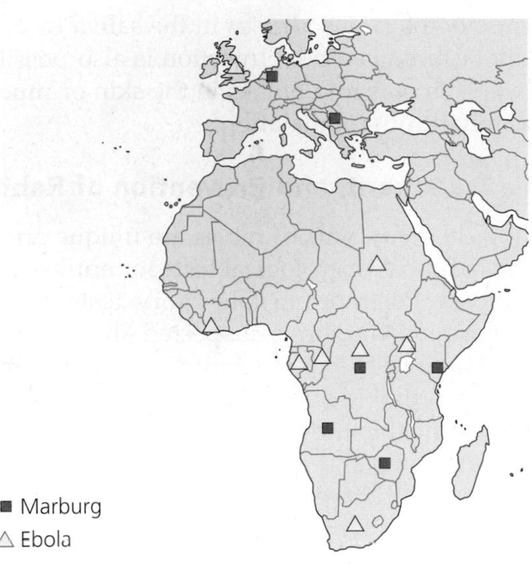

▲ **Figure 25.34** **Sites in which localized human disease cases of Marburg and Ebola viruses are known to have occurred.**

Ultimately, control of rabies involves immunization of domestic dogs and cats and the removal of unwanted strays from urban areas. Little can be done to completely eliminate rabies in wild animals because rabies virus can infect so many species, though in the late 1990s an epidemic of rabies was stopped in southern Texas by the successful vaccination of the wild coyote population. This was accomplished by lacing meat with an oral vaccine and dropping it from airplanes into the coyotes' range.

Hemorrhagic Fevers

Learning Objective

✓ Describe the effects of filoviruses on the human body.

Filamentous viruses that cause **hemorrhagic fever** were originally placed in the *Rhabdoviridae* but have now been assigned to their own family, *Filoviridae,* primarily on the basis of the symptoms they cause. These enveloped viruses form long filaments, sometimes up to 14,000 nm long, that sometimes curve back upon themselves (**Figure 25.33**).

There are two known filoviruses: Marburg virus and Ebola virus. **Marburg virus** was first isolated from laboratory workers in Marburg, Germany, and it has also been found in Zimbabwe, Kenya, Uganda, Angola, and in the Democratic Republic of the Congo (DRC). Outbreaks of **Ebola virus** hemorrhagic fever were first recorded in the DRC in 1976 near the Ebola River. Ebola cases have also occurred in humans in Sudan, Ivory Coast, Gabon, Uganda, South Africa, Congo, and in a single laboratory worker in England via an accidental needlestick (**Figure 25.34**).

The natural reservoirs of filoviruses appear to be fruit bats, but the initial method of transmission to humans is unknown. The virions attack many cells of the body, especially macrophages and liver cells. One predominant manifestation is uncontrolled bleeding under the skin and from every body opening as the internal organs are reduced to a jelly-like consistency. Hemorrhaging appears to be due to the synthesis of a virally coded glycoprotein that is expressed on the surface of infected cells lining blood vessels. The glycoprotein prevents neighboring cells from adhering to one another, allowing blood to leak out of the vessels. Headache and severe fever accompany this "meltdown." Up to 90% of human victims of Ebola die, and

CLINICAL CASE STUDY

The Sick Addict

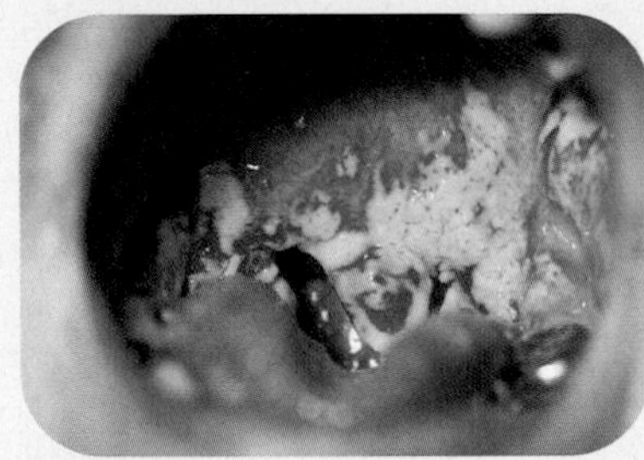

▲ *Thrush.*

A 25-year-old male was admitted to the hospital with thrush, diarrhea, unexplained weight loss, and difficulty in breathing. Cultures of pulmonary fluid revealed the presence of *Pneumocystis jiroveci.* The man admitted to being a heroin addict and to sharing needles in a "shooting gallery."

1. What laboratory tests could confirm a diagnosis in this case?
2. How did the man most likely acquire the infection?
3. What changes to the man's immune system allowed the opportunistic infections of *Candida* (thrush) and *Pneumocystis* to arise?
4. What precautions should be taken with the patient's blood?

▲ **Figure 25.35 Working in a Biosafety Level 4 lab.** The "space suit," which is designed to completely prevent contact with pathogens, constitutes more than a physical barrier; should the material of the suit be breached, positive air pressure within the suit ensures that pathogens present will be blown back from the opening.

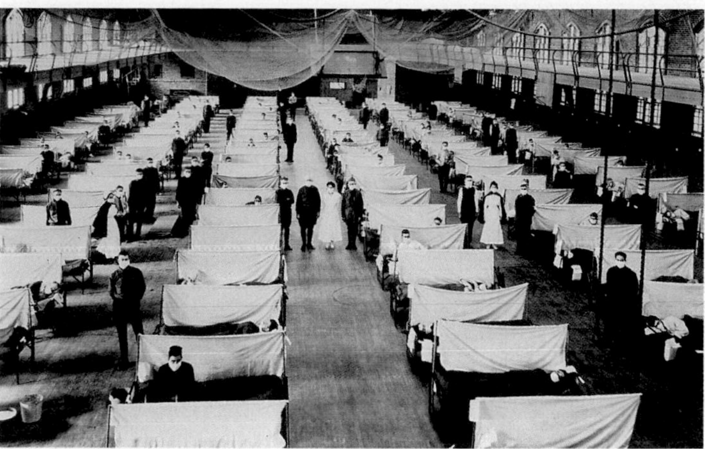

▲ **Figure 25.36 A scene from the influenza pandemic of 1918–19.** Flu afflicted so many people in the United States that gymnasiums were used as hospital wards.

about 25% of Marburg victims die. Epidemics sometimes cease only when there are not enough living humans to sustain viral transmission.

Marburg and Ebola viruses spread from person to person via contaminated bodily fluids, primarily blood, and through the use of contaminated syringes. Even a small amount of infected blood contacting a mucous membrane or conjunctiva can induce disease. Ebola virus can also spread among monkeys through the air, but aerosol transmission from monkeys to humans or among humans has not been observed.

The symptoms of filovirus infection are diagnostic. Additionally, the typical curved virus filaments are found in the victim's blood. Investigators detect antigens and antibodies using immunofluorescence, ELISA, or radioactive immunoassay. Extreme caution must be exercised in dealing with Ebola and Marburg patients and their bodily fluids. In fact, Biosafety Level 4—which involve the use of complete "space suits" to protect laboratory workers from contact with blood, vomit, or any other bodily fluids that might be contaminated with virus (**Figure 25.35**)—is mandatory. The only treatment for Ebola and Marburg fevers involves fluid replacement.

In 2003, scientists developed a vaccine that protects monkeys from Ebola in as little as 28 days. Immunization involves injecting pieces of viral nucleic acid (both by themselves and inserted into adenovirus capsids) into the monkeys. It is not known whether the vaccine will protect humans.

CRITICAL **THINKING**

Ebola hemorrhagic fever belongs to a group of diseases called "emerging diseases"—diseases that were previously unidentified or had never been identified in human populations. Emerging diseases are often first seen in less-developed countries such as the DRC. What factors may explain the emergence of new diseases in these countries?

Enveloped, Segmented, Negative ssRNA Viruses: *Orthomyxoviridae*, *Bunyaviridae*, and *Arenaviridae*

A second group of negative, enveloped, single-stranded RNA viruses differ from those in the previous section in having a **segmented genome;** that is, their capsids contain more than one molecule of nucleic acid. These viruses include the flu viruses of the family *Orthomyxoviridae* and hundreds of viruses in the families *Bunyaviridae* and *Arenaviridae* that normally infect animals but can be transmitted to humans.

Influenza

Learning Objective

✓ Describe the roles of hemagglutinin and neuraminidase in the replication cycle of influenzaviruses.

Imagine being the only elementary schoolchild of your sex in your mid-sized town because all of your peers died six winters ago. Or imagine returning to college after a break, only to learn that half of your fraternity brothers had died during the previous two months. These vignettes are not fictional; they happened to relatives of this author during the great flu pandemic in the winter of 1918–19 (**Figure 25.36**). During that winter, half the world's 1.9 billion people were infected with a new, extremely virulent strain of influenza virus, and an estimated 50 million died during that one flu season. In some U.S. cities, 10,000 people died each week for several months. Could it happen again? In this section we will learn about the characteristics of influenzaviruses that enable flu—a common infection, second only to common colds—to produce such devastating epidemics, and some ways to protect ourselves from the flu.

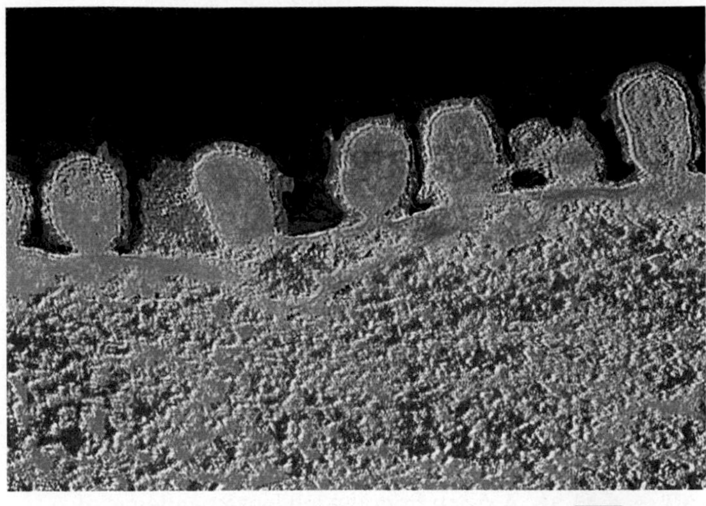

▲ **Figure 25.37 Influenzaviruses budding from an infected cell.** *What are the types of glycoprotein spikes on influenzaviruses, and what is their relationship to viral strain?*

Figure 25.37 *Glycoprotein spikes on influenzaviruses are neuraminidase (NA) and hemagglutinin (HA); variations in these glycoproteins determine the strain of the virus.*

The causes of **influenza**[16] (in-flū-en′ză), or **flu,** are two species of orthomyxoviruses, designated types A and B. An influenza virion contains a genome consisting of eight different −ssRNA molecules surrounded by a pleomorphic (many-shaped) lipid envelope studded with prominent glycoprotein spikes composed of either **hemagglutinin**[17] (hē′mă-glū′ti-nin) **(HA)** or **neuraminidase** (nūr-ă-min′i-dāz) **(NA)** (**Figure 25.37**). Both HA and NA play roles in the attachment (and thus the virulence) of the viruses: NA spikes provide the virus access to cell surfaces by hydrolyzing mucus in the lungs, whereas HA spikes attach to pulmonary epithelial cells and trigger endocytosis. Because influenzaviruses rarely attack cells outside the lungs, so-called stomach flu is probably caused by enteroviruses and not by orthomyxoviruses.

The genomes of flu viruses are extremely variable, especially with respect to the genes that code for HA and NA. Mutations are responsible for the production of new strains of influenzavirus, via processes known as antigenic drift and antigenic shift. **Antigenic** (an-ti-gen′ik) **drift** (**Figure 25.38a**) refers to the accumulation of HA and NA mutations within a single strain of virus in a given geographic area. Such mutations produce relatively minor variations in spike glycoproteins. As a result of these minor changes in the virus's antigenicity, localized increases in the number of flu infections occur about every two years. **Antigenic shift** (**Figure 25.38b**) is a major antigenic change in influenza A resulting from the reassortment of genomes from different influenzavirus strains infecting the same host cells (either human hosts or animal hosts, including birds and pigs). On average, antigenic shift occurs in influenza A every 10 years. Asia is a major site of antigenic shift, and the source of most pandemic strains, because the very high population densities of humans, domesticated birds and pigs there; however, the 2009 H1N1 pandemic strain resulted from genetic reassortment of four influenza A viruses in animals and humans, most likely in Mexico.

In 2005, researchers determined the genetic sequence of the 1918–19 pandemic strain by analyzing preserved tissue from flu victims. They discovered that a combination of genes from influenza A viruses from birds, pigs, and humans caused the pandemic. The 2009 H1N1 strain contains many of the same genetic sequences as the 1918–19 strain, leading epidemiologists to estimate that one third of the world's population could be infected in this new pandemic. The outcome of the pandemic could be as severe as the 1918–19 outbreak, despite vaccines, antiviral drugs, and modern health care facilities, because there are so many more people today, we live closer together in urban areas, and international air travel make it easier for influenzaviruses to spread. WHO officials consider 2 million deaths worldwide a "best-case" scenario, though some epidemiologists estimated that over 360 million people could die from H1N1. **Emerging Diseases: H1N1 Influenza** on p. 737 illustrates one patient's story.

Once virions are taken into epithelial cells lining the lung, the viral envelopes fuse with the membranes of cellular vesicles. Viruses multiply using positive-sense RNA molecules both for translation of viral proteins and as templates for transcription of new −ssRNA genomes.

During the process by which virions bud from the cytoplasmic membrane of infected cells (see Figure 25.37), NA keeps viral proteins from clumping. During budding, genomic segments are enveloped in a random manner, such that each virion ends up with about 11 RNA molecules. However, to be functional, a virion must have at least one copy of each of the eight genomic segments. Numerous defective particles are released for every functional virion formed.

Deaths of epithelial cells infected with influenzaviruses eliminate the lungs' first line of defense against infection, its epithelial lining. As a result, flu patients are more susceptible to secondary bacterial infection. One common infecting bacterium is *Haemophilus influenzae* (hē-mof′i-lŭs in-flu-en′zī), which was mistakenly named because it was found in so many flu victims.

The signs and symptoms of flu—fever, malaise, headache, and myalgia (body aches)—are induced by cytokines released as part of the immune response. Rarely, flu viruses cause croup. The symptoms of influenza are compared to those of other respiratory infections in Table 25.1 on p. 708.

Epidemiology of Influenza

The changing antigens of orthomyxoviruses guarantee that there will be susceptible people, especially children, each flu season. Infection occurs primarily through inhalation of airborne viruses but can also occur via fomites.

Strains of influenza are named by type (A or B), location and date of original identification, and antigen (HA and NA). For example, A/Singapore/1/09 (H1N2) is influenza type A,

[16]*Influenza,* which is Italian for "influence," derives from the mistaken idea that the alignment of celestial objects caused or "influenced" the disease.

[17]The word *hemagglutinin* refers to these spikes' ability to attach to and clump (agglutinate) red blood cells.

(a) Antigenic Drift
3 Influenzavirus 1' differs slightly from virus 1.
1 Influenza-virus 1 enters host cell.
2 Mutations in antigen genes occur during replication within host cell.
(b) Antigenic Shift
3 Influenza-virus 3 very different from viruses 1 and 2.
1 Influenza-viruses 1 and 2 enter host cell.
Virus 2
Virus 1
2 Genes and antigens from both viral types are incorporated into new virions.
Number of cases
2 4 6 8 10 12 14 16 18 20
Time (years)
Biennial outbreaks of mild influenza
Number of cases
2 4 6 8 10 12 14 16 18 20
Time (years)
Occasional outbreaks of very severe influenza

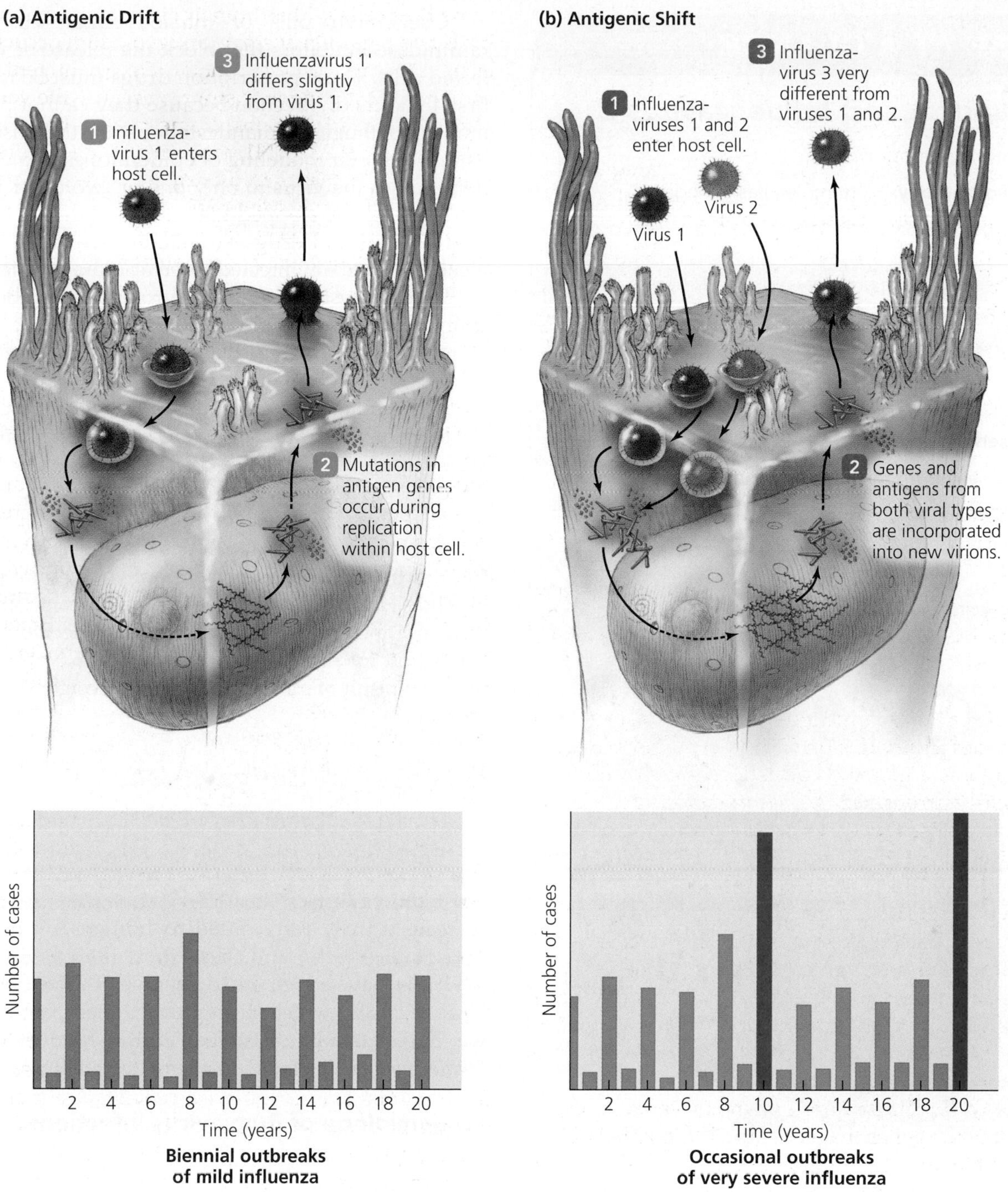

▲ **Figure 25.08 The development of new strains of flu viruses. (a)** Antigenic drift, which results from variation in the NA and HA spikes of a single strain of influenzavirus. **(b)** Antigenic shift, which occurs when RNA molecules from two or more influenzaviruses infecting a single cell are incorporated into a single virion. Because antigenic shift produces significantly greater antigenic variability than occurs in antigenic drift, antigenic shifts can result in major epidemics or pandemics.

isolated in Singapore in January 2009, that contains HA and NA antigens of type 1 and type 2, respectively. If the virus is isolated from an animal, the animal name is appended to the location. From these names, common names such as "Hong Kong flu" or "swine flu" arise. The number of different flu strains is almost infinite.

Every year millions of Americans are infected, and an estimated 64,000 Americans die annually of flu-related illness. Patients with weak immune systems, including the very young, the elderly, and those immunocompromised as a result of cancer therapy or disease, are particularly vulnerable.

Diagnosis, Treatment, and Prevention of Influenza

The manifestation of flu signs and symptoms during a community-wide outbreak of the flu is often sufficient for

MICROBE AT A GLANCE

Influenzavirus A and *Influenzavirus B*

Taxonomy: Family *Orthomyxoviridae*

Genome and morphology: Eight molecules of negative single-stranded RNA in helical capsids, surrounded by an envelope

Hosts: Vertebrates, including humans, pigs, birds, horses, sea mammals

Virulence factors: Hemagglutinin and neuraminidase glycoproteins allow attachment to pulmonary cells where viruses induce their own endocytosis

Syndrome caused: Influenza (flu)

Treatment for disease: Amantadine, oseltamivir, zanamivir

Prevention of disease: Annual vaccine provides protection against some strains

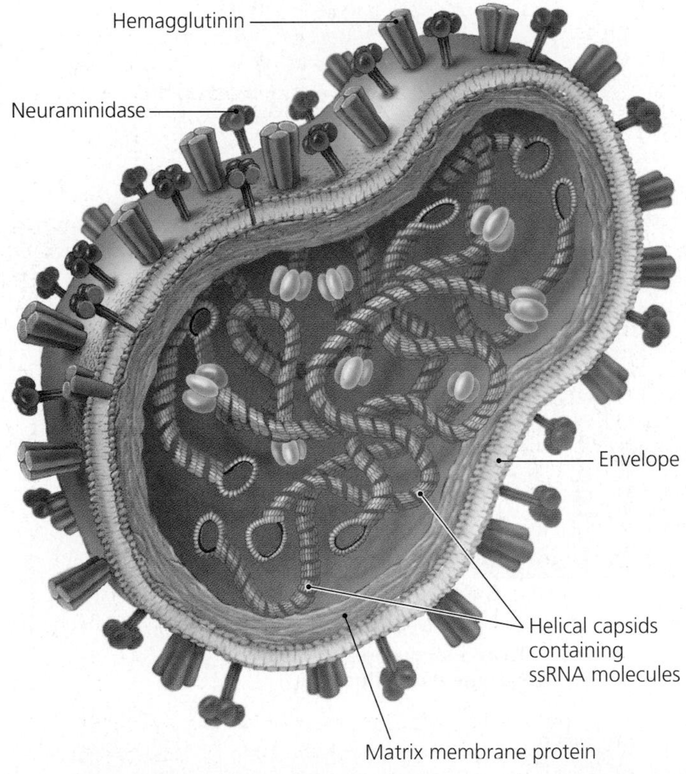

Download the Microbe at a Glance flashcards from the Study Area at www.masteringmicrobiology.com.

diagnosis of influenza. Laboratory tests such as immunofluorescence or ELISA can distinguish strains of flu virus.

Hundreds of millions of dollars are spent each year in the United States on antihistamines and pain relievers to alleviate flu symptoms. As discussed in Chapter 24, aspirin and aspirin-like products should not be used to treat the symptoms of flu in children and teenagers because of increased risk of *Reye's syndrome*.

Oseltamivir pills or inhaled zanamivir mist are neuraminidase inhibitors that block the release of virions from infected cells. These prescription drugs must be taken during the first 48 hours of infection because they cannot prevent the later immunopathogenic manifestations of the disease. Scientists using computer modeling of epidemics estimate that if a flu patient passes the virus to only one or two other people, then an epidemic can be stopped with antiviral drugs. However, health care workers and drug manufacturers would be overwhelmed if each sick person infected three or more contacts.

The greatest success in controlling flu epidemics has come from culling infected birds and from immunization by injection or inhalation with multivalent vaccines, that is, vaccines that contain several antigens at once. The WHO has personnel in Asia who track changes in the HA and NA antigens of emerging flu viruses. Newly discovered antigens are then used to create flu vaccine in advance of the next flu season in the United States. The vaccines are at least 70% effective but only against the viral antigens they contain.

Flu vaccines are made of deactivated viruses; therefore, contrary to many people's perception, you cannot get flu from the vaccine. However, you may get flu from strains of influenzavirus not included in the vaccine. Flu vaccine usually provides protection for no more than three years. Natural active immunity as a result of infection lasts much longer but is probably not lifelong.

Diseases of Bunyaviruses

Learning Objective

✓ Contrast hantaviruses with the other bunyaviruses.

Most **bunyaviruses** (bun′ya-vī′rŭs-ĕz) are zoonotic pathogens that are usually transmitted to humans by biting arthropods (mosquitoes, ticks, and flies); thus, they are arboviruses. Bunyaviruses have a segmented genome of three –ssRNA molecules (**Figure 25.39**), which distinguishes them from the arboviruses we have already discussed in the families *Togaviridae* and *Flaviviridae*, which have unsegmented +ssRNA.

Epidemiology of Bunyavirus Infections

Four genera of *Bunyaviridae* contain human pathogens. The viruses of *Rift Valley fever, California encephalitis,* and *hemorrhagic fevers* are arboviruses and thus enter the blood through the bite of an arthropod, which establishes an initial viremia. The blood carries viruses to target organs, such as the brain, kidney, liver, or blood vessels. Each bunyavirus has a specific target, and different viruses cause encephalitis, fever, or rash. Most patients have mild symptoms, but symptomatic cases can be severe and fatal.

Hantaviruses are an exceptional type of bunyaviruses: they are not arboviruses, but are transmitted to humans via inhalation of virions in dried mouse urine or feces. Two American strains[18] of hantaviruses infect the lungs only, causing a rapid, severe, and often fatal pneumonia called ***Hantavirus*** (han′tă-vī-rŭs) **pulmonary syndrome (HPS).** This syndrome was first

[18] Sin Nombre and Convict Creek strains.

recognized during an epidemic in the Four Corners[19] area of the United States in May 1993.

Bunyaviruses cause human diseases around the world. Each virus is typically found only in certain geographic areas. However, with modern travel patterns, viruses can be introduced into new areas relatively easily.

Diagnosis, Treatment, and Prevention of Bunyavirus Infections

Diseases caused by bunyavirus infections are indistinguishable from illnesses caused by several other viruses, so laboratory tests must be employed for diagnosis. ELISAs are used to detect viral antigens associated with Rift Valley fever, the LaCrosse strain of California encephalitis, and hemorrhagic fevers. Hantaviruses in the United States were first identified in patients' tissue using the *reverse transcriptase polymerase chain reaction*: viral RNA was transcribed to DNA using retroviral reverse transcriptase, multiple copies of the DNA were generated through polymerase chain reaction, and the resulting strands were compared to known hantavirus sequences.

No specific treatment exists for most bunyavirus infections. Human disease caused by arboviruses is prevented by limiting

[19]Four Corners is the geographic area where Arizona, Utah, New Mexico, and Colorado meet.

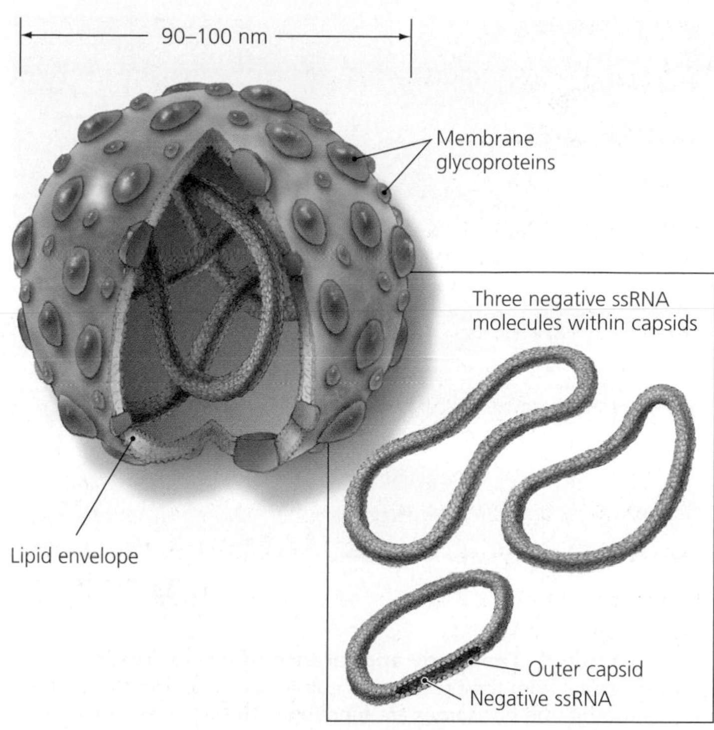

▲ **Figure 25.39 A bunyavirus is an enveloped helical virion.** The genome consists of three −ssRNA molecules.

EMERGING DISEASES

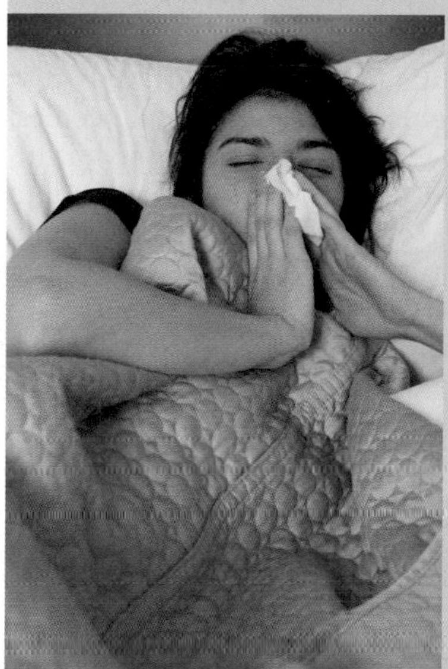

H1N1 INFLUENZA

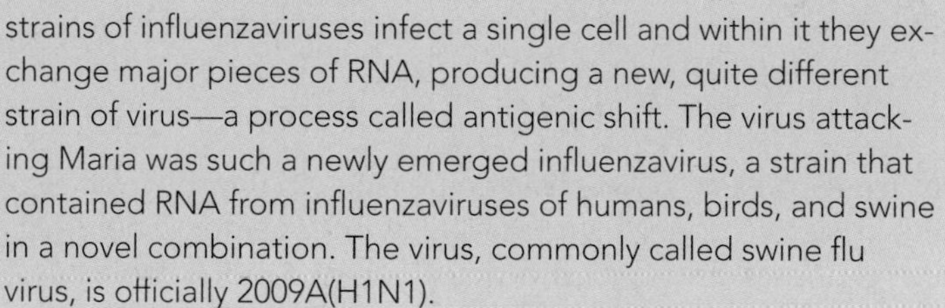

Middle school is supposed to be a time of exploration, learning, and fun, but Maria was too sick to enjoy it right now. Her 104°F fever had lasted two very unpleasant days, but at last it had gone. For another two days she ached all over and was dizzy; her head felt as if it could explode at any moment; and the nausea, frequent vomiting, and extreme tiredness were unrelenting. For over a week her tired red eyes stared listlessly as she struggled to cope with a constantly running nose, severe sore throat, and a dry hacking cough. Would this onslaught never end? A newly emerging strain of influenza type A had a victim in its grasp.

Flu viruses infect birds, pigs, horses, and humans. These viruses normally mutate, producing slightly different strains every year—a process called antigenic drift. About once a decade, however, different strains of influenzaviruses infect a single cell and within it they exchange major pieces of RNA, producing a new, quite different strain of virus—a process called antigenic shift. The virus attacking Maria was such a newly emerged influenzavirus, a strain that contained RNA from influenzaviruses of humans, birds, and swine in a novel combination. The virus, commonly called swine flu virus, is officially 2009A(H1N1).

This new strain can have devastating and sometimes fatal effects on hosts because it has antigenic combinations that the adaptive immune system has never seen before, necessitating a prolonged defensive response before the body can conquer the infection. When H1N1 flu cases reached epidemic levels on multiple continents by June 2009, the World Health Organization declared a pandemic—the first flu pandemic since 1968. During this period, the well-known signs and symptoms of flu play out.

Maria was exhausted and weak for another two weeks, but she survived the flu, though many other patients were not so fortunate.

Track H1N1 influenza online by going to the Study Area at www.masteringmicrobiology.com.

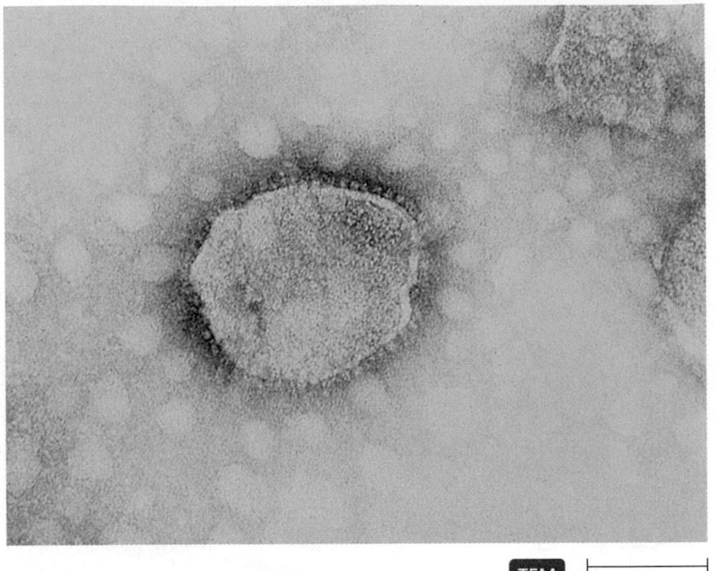

▲ **Figure 25.40 The sandy appearance of arenaviruses.** The graininess results from the presence of numerous ribosomes within the capsid. Although the ribosomes are functional, the virions apparently do not use them.

contact with the vectors (the use of netting and insect repellents) and by reducing arthropod abundance (elimination of breeding sites and the use of insecticides). Rodent control is necessary to prevent *Hantavirus* pulmonary syndrome. A vaccine has been developed to protect humans and animals against Rift Valley fever, and ribavirin has proved effective in treating Crimean-Congo hemorrhagic fever.

Diseases of Arenaviruses

Learning Objectives

- ✓ Identify the distinguishing characteristics of arenaviruses.
- ✓ Describe the unique feature of satellite viruses such as hepatitis D virus.

Arenaviruses[20] (ă-rē′nă-vī′rŭs-ĕz) are the final group of segmented, negative ssRNA viruses we will consider. Their segmented genomes are each composed of two RNA molecules. Most arenaviruses contain ribosomes, which gives them a sandy appearance in electron micrographs (**Figure 25.40**). The ribosomes are functional *in vitro*,[21] but there is no evidence that the viruses utilize them in nature.

Zoonoses

Arenaviruses cause zoonoses that include hemorrhagic fevers named for locales where they occur (**Lassa** in Africa, **Junin** in Argentina, **Sabiá** in Brazil, and **Machupo** in Bolivia) and **lymphocytic choriomeningitis** (lim-fō-sit′ik kō-rē-ō-men-in-jī′tis) **(LCM)**. Hemorrhagic fever viruses in the family *Arenaviridae* are endemic in rodent populations in Africa and South America. They cause severe bleeding under the skin and into internal organs. LCM virus has been found infecting pet hamsters in the United States. Contrary to its name, LCM is rarely associated with meningitis but more typically manifests with flulike symptoms.

Humans become infected through the inhalation of aerosols or consumption of contaminated food or from fomites. Lassa fever virus can be spread from human to human through contact with bodily fluids.

Diagnosis of arenaviral diseases is based on a combination of symptoms and immunoassay. Fluid samples from hemorrhagic patients are too dangerous for normal laboratories and instead must be processed in maximum biocontainment facilities. LCM samples can be handled with Biosafety Level 3 procedures (gown, mask, gloves, and labs with laminar flow hoods).

Supportive care is usually all that can be done for patients with arenavirus zoonoses, though ribavirin has shown some benefit in treating Lassa fever. Prevention of arenavirus disease involves rodent control and limiting contact with rodents, their droppings, and their dried urine.

Hepatitis D

Hepatitis D virus *(Deltavirus)*, which is also known as *delta agent*, is an arenavirus that causes hepatitis D and plays a role along with hepatitis B virus in triggering liver cancer. Delta agent is unusual in that it does not possess genes for the glycoproteins it requires to attach to liver cells. To become infective, it "steals" the necessary glycoprotein molecules from hepatitis B viruses that are simultaneously infecting the same liver cell. For this reason, hepatitis D virus is called a **satellite virus** because its replication cycle "revolves around" a helper virus.

Like hepatitis B virus, hepatitis D virus is spread in bodily fluids via sexual activity and the use of contaminated needles. Prevention involves the same precautions used to prevent infection with hepatitis B virus—abstinence, monogamy, condom usage, the use of sterile needles, and avoidance of accidental needlesticks. Since hepatitis D virus requires hepatitis B virus to become virulent, hepatitis B vaccine limits the spread of hepatitis D viruses.

Naked, Segmented dsRNA Viruses: *Reoviridae*

Learning Objective

- ✓ Discuss the meaning of the name "reovirus."

Reoviruses (rē′ō-vī′rŭs-ĕz) are respiratory and enteric viruses with naked icosahedral capsids. They are unique in being the only microbes, viral or cellular, with genomes composed of double-stranded RNA (dsRNA). Each virus contains 10–12 segments of dsRNA.

Reoviruses were originally orphan viruses—not initially associated with any diseases—though now many have been linked to specific conditions. The acronym *reo* is derived from *respiratory, enteric, orphan*. Reoviruses include *Rotavirus* and *Coltivirus*.

[20]From Greek *arenosa*, meaning sandy.
[21]From Latin *in*, meaning within, and *vitreus*, meaning glassware; in other words "in the laboratory."

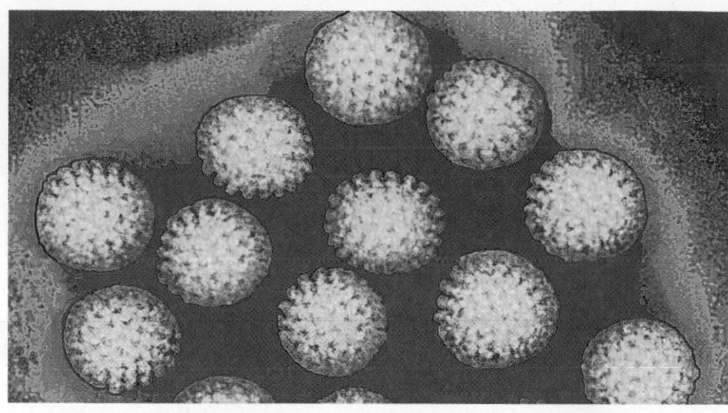

▲ **Figure 25.41 Rotaviruses.** The wheel-like appearance of rotaviruses, from which they get their name.

Rotaviruses

Learning Objective

✓ Describe the effects of rotavirus infections.

Rotavirus[22] (rō'tă-vī'rŭs) is an almost spherical reovirus with prominent glycoprotein spikes **(Figure 25.41)**. During replication, rotaviruses acquire and then lose an envelope. Even without envelopes, they function like enveloped viruses because glycoprotein spikes on their capsids act as attachment molecules that trigger endocytosis, a process common to enveloped viruses. Rotaviruses infect almost all children everywhere in the first few years of life via the fecal-oral route.

Rotaviruses are the most common cause of infantile gastroenteritis and account annually in the U.S. for about 55,000 cases of diarrhea in children requiring hospitalization as a result of fluid and electrolyte loss. In developing countries, rotaviruses account for an estimated 600,000 deaths annually from uncontrolled diarrhea **(Figure 25.42)**. Infected children may pass as many as 100 trillion virions per gram of stool.

[22]From Latin *rota*, meaning wheel.

Rotavirus infection is self-limiting, and treatment involves supportive care such as replacement of lost water and electrolytes. Prevention of infection involves good personal hygiene, proper sewage treatment, and vaccination. The vaccine approved for use in the U.S. prevents 75% of rotavirus cases and reduces hospitalizations and deaths by more than 96%.

Coltiviruses

Learning Objective

✓ Discuss the most common disease caused by a coltivirus.

Coltivirus (kol'tē-vī'rŭs) is an arbovirus that causes a zoonosis, **Colorado tick fever** (see Table 25.4 on p. 718), for which these viruses are named. Colorado tick fever is usually a mild disease involving fever and chills, though severe cases can manifest as headache, photophobia, myalgia, conjunctivitis, rash, and, in children, hemorrhaging. Colorado tick fever should not be confused with Rocky Mountain spotted fever, a tick-borne bacterial disease discussed in Chapter 21. Colorado tick fever is self-limiting, but red blood cells in all stages of development retain the virus for some time. Patients should not donate blood for several months after recovery.

Health care providers diagnose Colorado tick fever by detecting antigens in the blood using ELISA or other immunoassays. As with many viral diseases, there is no specific treatment. Prevention involves limiting contact with infected ticks.

RNA viruses are a heterogeneous group of virions capable of causing diseases of varying severity, including several life-threatening ones. Table 25.8 on p. 740 summarizes the physical characteristics of most RNA viruses of humans, as well as the diseases they cause. As we have seen, RNA viruses pose significant problems for health care personnel with respect to diagnosis, treatment, and prevention, and some of these viruses also pose threats to health care workers who must handle them. Emerging RNA viruses are of particular concern as people move deeper into rainforests and come into contact with new viral agents, but such viruses have also provided the impetus for expanded research into the fields of immunization and treatment of viral diseases.

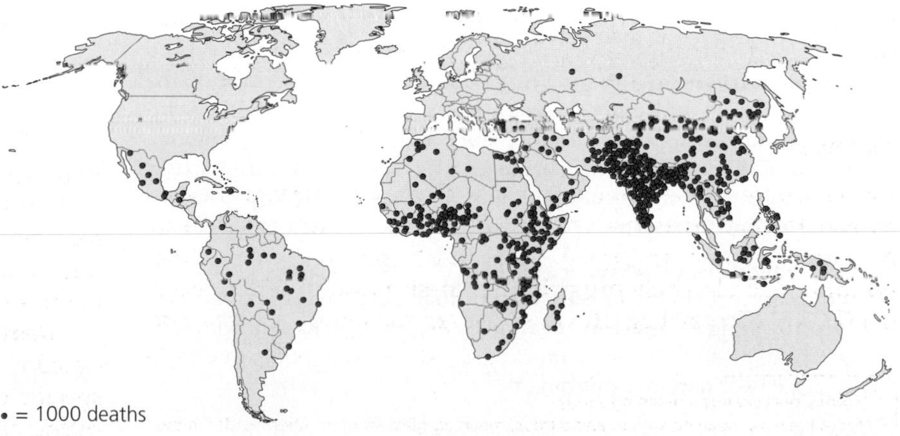

▸ **Figure 25.42 Deaths from rotaviral diarrhea (2008).** Mortality is most common in developing countries.

TABLE 25.8 Taxonomy and Characteristics of Human RNA Viruses

Family	Strand Type	Enveloped or Naked	Capsid Symmetry	Size (nm)	Representative Genera (Diseases)
Picornaviridae	Single, positive	Naked	Icosahedral	22–30	*Enterovirus* (polio) *Rhinovirus* (common cold) *Hepatovirus* (hepatitis A)
Caliciviridae	Single, positive	Naked	Icosahedral	35–40	*Norovirus* (acute gastroenteritis)
Hepeviridae	Single, positive	Naked	Icosahedral	27–34	*Hepevirus* (hepatitis E)
Astroviridae	Single, positive	Naked	Icosahedral	30	*Astrovirus* (gastroenteritis)
Togaviridae	Single, positive	Enveloped	Icosahedral	40–75	*Alphavirus* (encephalitis) *Rubivirus* (rubella)
Flaviviridae	Single, positive	Enveloped	Icosahedral	37–50	*Flavivirus* (yellow fever) *Hepacivirus* (hepatitis C)
Coronaviridae	Single, positive	Enveloped	Helical	80–160	*Coronavirus* (common cold, severe acute respiratory syndrome)
Retroviridae	Single, positive, segmented	Enveloped	Icosahedral	80–146	*Deltaretrovirus* (leukemia) *Lentivirus* (AIDS)
Paramyxoviridae	Single, negative	Enveloped	Helical	125–250	*Paramyxovirus* (colds, respiratory infections) *Pneumovirus* (respiratory syncytial disease, rarely croup or common cold) *Morbillivirus* (measles) *Rubulavirus* (mumps)
Rhabdoviridae	Single, negative	Enveloped	Helical	75 × 130–240	*Lyssavirus* (rabies)
Filoviridae	Single, negative	Enveloped	Helical	790–970	*Ebolavirus* (Ebola hemorrhagic fever) *Marburgvirus* (Marburg hemorrhagic fever)
Orthomyxoviridae	Single, negative, segmented	Enveloped	Helical	80–120	*Influenzavirus* (flu, rarely croup)
Bunyaviridae	Single, negative, segmented	Enveloped	Helical	90–100	*Bunyavirus* (encephalitis) *Hantavirus* (pneumonia)
Arenaviridae	Single, negative, segmented	Enveloped	Helical	50–300	*Lassavirus* (hemorrhagic fever) *Deltavirus* (hepatitis D)
Reoviridae	Double, segmented	Naked	Icosahedral	78–80	*Rotavirus* (diarrhea) *Coltivirus* (Colorado tick fever)

Chapter Summary

Naked, Positive ssRNA Viruses: *Picornaviridae, Caliciviridae, Astroviridae,* and *Hepeviridae*

(pp. 707–712)

1. RNA viruses and viroids are the only infective agents that use RNA molecules as their genetic material.
2. **Picornaviruses** are the smallest of animal viruses and include the **rhinoviruses** and **enteroviruses.** Rhinoviruses are extremely infective in the upper respiratory tract and cause most "common" colds. Enteroviruses, which are mostly cytolytic picornaviruses that infect via the fecal-oral route, include poliovirus, coxsackievirus, and echovirus.
3. **Poliomyelitis,** or **polio,** is a disease of varied degrees of severity from asymptomatic infections to **bulbar poliomyelitis** (requiring an iron lung to assist in breathing), and postpolio syndrome occurring 30–40 years after infection. Eradication of polio, due to effective vaccination, is imminent.
4. Jonas Salk developed **inactivated polio vaccine (IPV)** in 1955. It was replaced by live **oral polio vaccine (OPV),** which was developed by Albert Sabin in 1961. Now, health professionals in many countries have switched back to IPV because OPV, which is a live, attenuated vaccine, can revert to a disease-causing form.
5. **Coxsackie A virus** causes lesions in the mouth and throat, respiratory infections, hand-foot-and-mouth disease, viral meningitis, and acute hemorrhagic conjunctivitis. **Coxsackie B virus** causes myocarditis and pericardial infections, which may be fatal in newborns. In addition, coxsackie B viruses cause pleurodynia and viral meningitis, and may be a cause of diabetes mellitus.
6. So-called orphan viruses are initially unassociated with a specific disease.
7. **Echoviruses** (enteric cytopathic human orphan viruses) cause viral meningitis and colds.
8. **Hepatitis A virus** is a noncytolytic picornavirus that infects liver cells but is rarely fatal.

9. **Caliciviruses** and **astroviruses** are small viruses that enter through the digestive system, multiply in the cells of the intestinal tract, and cause gastroenteritis and diarrhea. **Noroviruses** are representative caliciviruses.
10. **Hepatitis E virus** causes **hepatitis E,** which is fatal to about 20% of infected, pregnant women.

Enveloped, Positive ssRNA Viruses: *Togaviridae, Flaviviridae,* and *Coronaviridae* (pp. 712–718)

1. Flaviviruses and most togaviruses are **arboviruses** (viruses transmitted by arthropods) and are enveloped, icosahedral, positive, ssRNA viruses that cause flulike symptoms, encephalitis, dengue, and yellow fever.
2. Various viruses, most transmitted via mosquito vectors, cause zoonoses—diseases of animals that are spread to humans. Those caused by togaviruses are **Eastern equine encephalitis (EEE), Western equine encephalitis (WEE),** and **Venezuelan equine encephalitis (VEE).** A flavivirus causes West Nile encephalitis. All these may be fatal in humans.
3. **Dengue** and **dengue hemorrhagic fever** are endemic diseases in South America caused by flaviviruses and carried by the *Aedes* mosquito.
4. **Yellow fever,** which is contracted through the bite of a mosquito carrying a flavivirus, results in internal hemorrhaging and damage to liver, kidneys, and heart.
5. *Rubivirus,* a togavirus that is not transmitted by an arthropod vector, causes **rubella,** commonly known as "German measles" or "three-day measles." It can cause serious birth defects if the virus crosses the placenta into a fetus.
6. **Hepatitis C virus (HCV),** caused by a flavivirus and transmitted via the fecal-oral route, is a disease that often results in serious liver damage.
7. **Coronaviruses** have a corona-like envelope and cause colds as well as gastroenteritis in children. A newly discovered coronavirus causes **severe acute respiratory syndrome (SARS).**

Enveloped, Positive ssRNA Viruses with Reverse Transcriptase: *Retroviridae* (pp. 719–726)

1. **Retroviruses** have **reverse transcriptase** and **integrase** enzymes. The first allows retroviruses to make dsDNA from RNA templates; the latter allows them to integrate into a host cell's chromosomes.
2. **Human T-lymphotropic viruses** are oncogenic retroviruses associated with cancer of lymphocytes. HTLV-1 causes adult acute T-cell lymphocytic leukemia; HTLV-2 causes hairy cell leukemia. Sexual intercourse, blood transfusion, and contaminated needles transmit them.
3. **Acquired immunodeficiency syndrome (AIDS)** is a condition marked by the presence of antibodies against **HIV** in conjunction with certain opportunistic infections. A **syndrome** is a complex of signs and symptoms with a common cause.
4. **Human immunodeficiency viruses (HIV-1** or **HIV-2)** destroy the immune system. HIV is characterized by glycoproteins such as **gp120** and **gp41,** which enable attachment, antigenic variability, and formation of **syncytia** (fused cells). The destruction of the immune system results in vulnerability to any infection.
5. HIV affects helper T cells, macrophages, and dendritic cells. Using gp120, HIV fuses to a cell, uncoats, enters the cell, and transcribes dsDNA to become a **provirus,** which inserts into a cellular chromosome. After HIV is replicated and released, an internal viral enzyme, **protease,** makes HIV virulent. HIV is spread primarily through contact with an infected individual's secretions or blood.
6. **Highly active antiretroviral therapy (HAART)** is a combination of antiviral drugs for the treatment of AIDS.

Enveloped, Unsegmented, Negative ssRNA Viruses: *Paramyxoviridae, Rhabdoviridae,* and *Filoviridae* (pp. 726–733)

1. Paramyxoviruses can cause infected cells to fuse together into giant, multinucleate syncytia. One virus causes **measles** ("rubeola" or "red measles"), recognized by **Koplik's spots** in the mouth followed by lesions on the head and trunk. Complications of measles include **subacute sclerosing panencephalitis (SSPE),** a progressive fatal disease of the central nervous system. Vaccination is effective and eradication is possible.
2. **Parainfluenza** viruses usually cause mild coldlike diseases, including **croup.** There is no effective vaccine.
3. **Mumps** virus is a paramyxovirus that causes painfully enlarged parotid glands. Mumps vaccine is very effective, and eradication is possible.
4. **Respiratory syncytial virus (RSV)** is a paramyxovirus that affects the lungs and can be fatal in infants and young children, while in older children and adults the symptoms are similar to those of a cold.
5. **Rabies** is caused by a rhabdovirus—a bullet-shaped virus with glycoprotein spikes—that attaches to muscle cells, replicates, moves into nerve cells and eventually causes brain degeneration. Infection usually results from bites by infected mammals. Symptoms include pain in swallowing (which results in **hydrophobia**), seizures, paralysis, and death. Rabies may be identified by aggregates of virions called **Negri bodies** in the brain of an affected mammal. Human and animal vaccines are effective.
6. **Marburg virus** and **Ebola virus** are filamentous viruses that cause hemorrhagic fever. There is no known treatment or prevention for hemorrhagic fever, though a vaccine has been effective in preventing disease in monkeys.

Enveloped, Segmented, Negative ssRNA Viruses: *Orthomyxoviridae, Bunyaviridae,* and *Arenaviridae* (pp. 733–738)

1. A **segmented genome** consists of more than one molecule of nucleic acid. The term usually refers to viruses, as the genomes of most of the known viral families are not segmented.
2. **Influenza (flu)** is caused by type A or B orthomyxoviruses, which enter lung cells by means of **hemagglutinin (HA)** and **neuraminidase (NA)** glycoprotein spikes. Gene mutations for HA and NA account for continual formation of new variants of flu viruses. **Antigenic drift** occurs every 2 or 3 years when a single

strain mutates within a local population. **Antigenic shift** of influenza A virus occurs about every 10 years and results in reassortment of genomes from different strains of animal and human viruses infecting a common cell. Vaccines for the flu in any given flu season are prepared against newly identified variants.

3. **Bunyaviruses,** with three −ssRNA molecules as a genome, are zoonotic pathogens. They are usually arboviruses. Each bunyavirus has a specific target and may cause encephalitis, fever, rash, or death.
4. **Hantaviruses** are bunyaviruses that are transmitted to humans via inhalation of virions in dried mouse urine. They cause ***Hantavirus* pulmonary syndrome.**
5. **Arenaviruses** are segmented −ssRNA viruses with apparently nonfunctional ribosomes. They cause **Lassa, Junin,** and **Machupo** hemorrhagic fevers; **lymphocytic choriomeningitis (LCM);** and hepatitis D.
6. **Hepatitis D virus,** also known as delta agent, is a **satellite virus;** that is, it requires proteins coded by another virus (in this case hepatitis B virus) to complete its replication cycle.

Naked, Segmented, dsRNA Viruses: *Reoviridae* (pp. 738–740)

1. The **reoviruses** are unique in having double-stranded RNA as a genome. They include ***Rotavirus,*** which has glycoprotein spikes for attachment and causes potentially fatal infantile gastroenteritis, and *Coltivirus,* which causes **Colorado tick fever.**

Questions for Review

Answers to the Questions for Review (except for Short Answer questions) begin on page A-1.

Multiple Choice

1. A segmented genome is one that
   a. has more than one strand of nucleic acid.
   b. has double-stranded RNA.
   c. has both RNA and DNA strands.
   d. has both +ssRNA and −ssRNA molecules.
2. What do viruses in the families *Picornaviridae, Caliciviridae, Astroviridae, Coronaviridae, Togaviridae, Flaviviridae,* and *Retroviridae* have in common?
   a. They are arboviruses.
   b. They are nonpathogenic.
   c. They have positive single-stranded RNA genomes.
   d. They have negative single-stranded RNA genomes.
3. The smallest animal viruses are in the family
   a. *Caliciviridae.* c. *Togaviridae.*
   b. *Astroviridae.* d. *Picornaviridae.*
4. Which of the following viruses cause most colds?
   a. rhinoviruses c. pneumoviruses
   b. parainfluenza viruses d. bunyaviruses
5. Arboviruses are
   a. zoonotic pathogens.
   b. deactivated viruses used in vaccines.
   c. viruses that are transmitted to humans via the bite of an arthropod.
   d. found in arbors.
6. Negri bodies are associated with which of the following?
   a. Marburg virus c. coltivirus
   b. hantavirus d. rabies virus
7. If mosquitoes were eradicated from an area, which of the following diseases would be most affected?
   a. mumps
   b. hantavirus pulmonary syndrome
   c. hepatitis E
   d. breakbone fever
8. Koplik's spots are oral lesions associated with
   a. mumps. c. flu.
   b. measles. d. colds.
9. Which of the following is an accurate statement concerning zoonoses?
   a. They are animal diseases that spread to humans.
   b. They are diseases specifically transmitted by mosquitoes and ticks.
   c. They are mucus-borne viruses, which are transmitted in the droplets of moisture in a sneeze or cough.
   d. They are diseases that can be transmitted from humans to an animal population.
10. A horror movie portrays victims of biological warfare with uncontrolled bleeding from the eyes, mouth, nose, ears, and anus. What actual virus causes these symptoms?
   a. Ebola virus c. hantavirus
   b. bunyavirus d. human immunodeficiency virus
11. Reoviruses, such as rotaviruses and coltiviruses, are unique in
   a. being naked.
   b. having double-stranded RNA.
   c. being both arboviruses and zoonotic.
   d. having protein spikes.

Matching

Match the diseases on the left with the infecting virus on the right.

____ Myocarditis	A. Rhabdoviridae
____ Colorado tick fever	B. Paramyxoviridae
____ Rabies	C. Reoviridae
____ Influenza	D. Coronaviridae
____ Dengue fever	E. Togaviridae
____ German measles	F. Flaviviridae
____ Acute gastroenteritis	G. Orthomyxoviridae
____ Ebola virus	H. Orphan virus
____ RSV	I. Caliciviridae
____ Western equine encephalitis	J. Filoviridae
____ No known disease	K. Picornaviridae

True/False

1. ____ A single virion is sufficient to cause a cold.
2. ____ All infections of polio are crippling.
3. ____ Postpolio syndrome is due to latent polioviruses that become active 30–40 years after the initial infection.
4. ____ Because the oral polio vaccine contains live attenuated viruses, mutations of these viruses can cause polio.
5. ____ A typical host for a togavirus is a horse.

Labeling

Label the steps in retroviral replication shown for HIV.

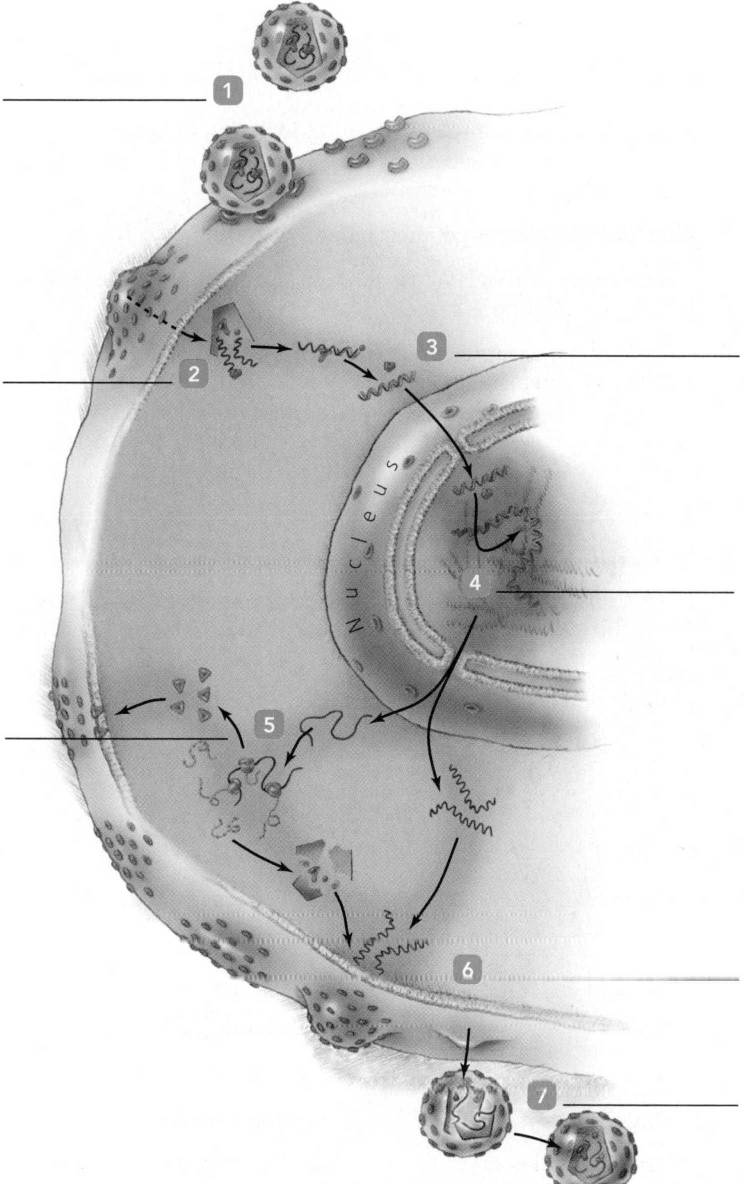

Short Answer

1. Why are humans considered "dead-end" hosts for many arboviruses?
2. Young Luis has skin lesions. His mother knows from microbiology class that five childhood diseases can produce spots. Name those five diseases and the viruses that cause each. List some questions to ask to determine which of these viruses Luis has.
3. The patient in room 519 exhibits yellowing skin and eyes, and it is suspected among the nursing staff that the diagnosis will be some kind of viral hepatitis. Make a chart of five kinds of hepatitis mentioned in this chapter, the infecting pathogen, how the patient might have become infected, and the relative degree of seriousness.
4. Why is AIDS more accurately termed a "syndrome" instead of a "disease"?
5. Consider the viruses you studied in this chapter. Which three would you rank as the deadliest?
6. Support or refute the following statement: "Rubeola is common and of little concern as a childhood disease."
7. Translate the following identification label on a vial of influenza virus: B/Kuwait/6/10 (H1N3).
8. Polio and smallpox have been eliminated as natural threats to human health in the United States. (Some risk from bioterrorism remains.) You have considered the features of these diseases that allowed them to be eliminated. From your studies of other viruses, what other viral diseases are candidates for elimination? Why hasn't AIDS been eliminated?
9. Several laboratory tests are used to identify viruses. From your study of this chapter alone, which tests would you surmise are the most common?
10. Why are there more cases of West Nile virus encephalitis in summer than in winter of every year?
11. Compare influenzavirus A 2009 (H1N1) to the 1918–1919 pandemic influenzavirus.

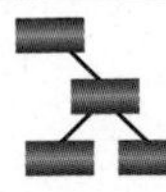

Concept Mapping

Using the following terms, draw a concept map that describes polioviruses. For a sample concept map, see p. 93. Or, complete this concept map online by going to the Study Area at www.masteringmicrobiology.com.

<2% of cases
2% of cases
5% of cases
90% of cases
Asymptomatic infection
Bulbar poliomyelitis
Central nervous system
Enteroviruses
Fecal-oral route
Inactivated polio vaccine (IPV)
Intestine
Limb paralysis
Minor polio
Nonparalytic polio
Oral polio vaccine (OPV)
Paralytic polio
Pharynx
Respiratory paralysis
Sabin vaccine
Salk vaccine
Three serotypes

Critical Thinking

1. Chapter 4 presented the use of dichotomous keys to identify microorganisms (see p. 121). Design a key for the RNA viruses discussed in this chapter.
2. A 20-year-old man is brought to a South Carolina hospital's emergency room suffering from seizures, disorientation, hallucinations, and an inordinate fear of water. His family members report that he had suffered with fever and headache for several days. The patient had been bitten by a dog approximately 12 days before admission. What disease does the patient have? Will a vaccine be an effective treatment? If the dog is found to be disease free, what wild animals are the likely source of infection in South Carolina? What treatment should the patient's family, coworkers, friends, and caregivers receive?
3. How did coltiviruses get their name?
4. Retroviruses such as HIV use RNA as a primer for DNA synthesis. Why is a primer necessary?
5. In 2003, alligator farms reported significant loss of animals to West Nile virus. Since alligator hide is generally too thick for mosquitoes to feed, how can you explain the alligators' infections?

Access more review material online in the Study Area at **www.masteringmicrobiology.com.** There, you'll find

- **Concept Mapping Activities**
- **Flashcards**
- **Quizzes**

and more to help you succeed.

26 Applied and Environmental Microbiology

To most of us, *Saccharomyces cerevisiae* is familiar as the yeast that makes bread rise and that produces alcoholic beverages, and these are two important roles of this microorganism in applied microbiology.

However, for millions of cancer sufferers, *Saccharomyces* may become a source of a life-saving drug—taxol. Taxol is made by yew trees (genus *Taxus*), but there are not enough trees producing this anticancer drug to provide it to all patients who need it. Scientists using recombinant DNA technology are attempting to insert genes from *Taxus* into *Saccharomyces* so that the yeast is able to make the drug.

This chapter examines the astounding variety of ways we use microbes today, from food production to medical treatments to waste management to defense against bioterrorism—all topics in the fields of applied and environmental microbiology.

Take the pre-test for this chapter online. Visit the Study Area at www.masteringmicrobiology.com.

The yeast *Saccharomyces cerevisiae* not only ferments sugar into alcohol and makes bread rise but also can be genetically modified to produce other useful products.

The metabolic activities of microorganisms shape much of our environment, and microbial reactions are essential to life on Earth. Bacteria and fungi in particular are capable of many metabolic processes that are useful to humans. Chapter 8 introduced techniques of recombinant DNA technology that enhance the natural processes of microorganisms and, in some cases, give them new functions that make them valuable in industry, agriculture, or medicine. In this chapter we examine other areas of applied microbiology, as well as topics in the field of environmental microbiology.

Applied microbiology, the commercial use of microorganisms, encompasses two distinct fields: **food microbiology,** which includes the use of microorganisms in food production and the prevention of food spoilage and food-related illnesses, and **industrial microbiology,** which involves both the application of microbes to industrial manufacturing processes and solutions of environmental, health, and agricultural problems. **Environmental microbiology** explores where microorganisms are found in nature, and how their activities affect other organisms (including humans) and the environment itself. We begin by considering food microbiology.

Food Microbiology

Microorganisms are involved in producing many of our favorite foods and beverages, from bread to wine to yogurt. Indeed, fermented foods are among some of the oldest foods known and are culturally very diverse. The characteristic flavors, aromas, and consistencies of such foods result from the presence of acids or sugars made by microbes during fermentation. Besides conferring taste and aroma, microbial metabolism also acts as a preservative, destroys many pathogenic microbes and toxins, and, in some cases, adds nutritional value in the form of vitamins or other nutrients. In the following sections, we explore how our knowledge of microbial growth and metabolism enables us to use microbes in food production and to control microbial activity that results in food spoilage.

The Roles of Microorganisms in Food Production

Learning Objective

✓ Describe how microbial metabolism can be manipulated for food production.

Bread

Leavened[1] breads result from metabolism by the yeast *Saccharomyces cerevisiae* (sak-ă-rō-mī'sēz se-ri-vis'ē-ī). Warm milk or water is added to dried yeast powder to revive it; flour, salt, and other ingredients are added to make dough, which is kneaded to introduce oxygen. The dough rises when metabolic reactions release CO_2, producing expanding pockets within the dough. Ethanol produced by fermentation evaporates during baking. Sourdough bread is made using starter cultures consisting of yeast and lactic acid bacteria. Lactic acid produced by the bacteria gives sourdough its characteristic taste.

As we saw in Chapter 5, biochemists use the word *fermentation* to refer to the partial oxidation of sugars to release energy using organic molecules as electron acceptors. In food microbiology, however, **fermentation** may refer to any desirable changes that occur to a food or beverage as a result of microbial growth. In contrast, **spoilage** denotes unwanted change to a food that occurs from undesirable metabolic reactions, the growth of pathogens, or the presence of unwanted microorganisms.

Fermentative microbes naturally occur on grains, fruits, and vegetables. In antiquity, people relied on these naturally occurring microbes to produce fermented foods and drinks. However, the same microbes are not always present on a food from harvest to harvest, yielding varying results. Most modern commercial food and beverage production relies on **starter cultures** composed of known microorganisms that perform specific fermentations consistently. In addition to an initial starter culture, *secondary cultures* may be added to further modify the flavor or aroma of foods. For example, the same starter culture is used to initiate formation of both Swiss cheese and blue cheese; the two cheeses differ because different secondary cultures are used in their production.

Fermented Vegetables

People around the world ferment many types of vegetables. Most of these vegetable products are the result of the actions of lactic acid bacteria such as *Streptococcus* (strep-tō-kok'ŭs), *Leuconostoc* (loo'kō-nos-tŏk), *Lactobacillus* (lak'tō-bă-sil'ŭs), or *Lactococcus* (lak-tō-kok'us), which specifically produce lactic acid during fermentation. Lactic acid acidifies the food and produces a "sour" flavor.

Food products derived from the fermentation of cabbage include kimchi (kim-chē') from Korea and sauerkraut. Soy sauce is made by the fermentation of soybeans and wheat by lactobacilli, yeast *(S. cerevisiae),* and the fungus *Aspergillus oryzae* (as-per-jil'ŭs o'ri-zī). As mentioned in Chapter 5, chocolate is derived from the fermentation of cacao seeds. Coffee production relies on natural fermentation to release the coffee bean from the outer layers of the coffee berry, so that the bean can then be dried and roasted.

Pickles are another common fermented food. While people often equate "pickles" with cucumbers, other foods such as beets and eggs can also be pickled. *Pickling* refers to the process of preserving or flavoring foods with brine (saturated salt water) or acid. Microbial fermentation can be the source of the acid—as it is with dill pickles. Various spices can be added to the pickling solution to enhance taste. Because pickled foods are acidic, few pathogenic microorganisms survive, making pickling an excellent preservation method.

Fermented foods are not made for human consumption alone. *Silage,* a product used as animal feed on many farms, is made by the natural fermentation of potatoes, corn, grass, grain stalks, or other types of green foliage. The vegetation is cut up and stored in silos that are closed to air. Under such moist,

[1]From Latin *levare,* meaning to raise.

anaerobic conditions, the vegetation in the silos ferments, producing many organic compounds that make the silage both aromatic and tasty to livestock. Some farmers produce silage from baled hay and grasses wrapped in plastic.

Fermented Meat Products

Unless dried, cured, smoked, or fermented in some way, meat has a tendency to spoil rapidly. The combination of fermentation and drying or smoking has been used for centuries to preserve meats, particularly pork and beef. Dry sausages such as salami and pepperoni are made by grinding meat, mixing in various spices and starter cultures, allowing the mixture to ferment in the cold, and then stuffing it into casings. Fermented fish—common in Asian countries—are prepared by grinding fish in brine. Naturally occurring microbes ferment the mixture. The solids are removed and pressed to form a fish paste, while the liquid portion is drained off and mixed with flavoring agents to form fish sauce.

Fermented Dairy Products

Milk fermentation relies, for the most part, on the activities of lactic acid bacteria. Milk in an udder is sterile, but because milking introduces microorganisms, we pasteurize "raw" milk. The metabolic products of starter cultures give fermented milk products their characteristic textures and aromas.

Buttermilk is made from fat-free milk (skim milk) by adding a starter culture of *Lactococcus lactis* (lak′tis) subspecies *cremoris* (kre-mōr′is) and *Leuconostoc citrovorum* (sit-rō-vō′rum). Yogurt production utilizes starter cultures of *Streptococcus thermophilus* (ther-mo′fil-us) and *Lactobacillus bulgaricus* (bul-gă′ri-kŭs). Yogurt manufacturers mix pasteurized milk, milk solids, sweeteners, and other ingredients to a uniform consistency, and then add the starter culture. They ferment the mix at 43°C and then cool it to stop microbial activity. They may add flavors prior to packaging.

Cheeses are classified as unripened or ripened. Ripened cheeses can be hard or soft and mild or sharp. Regardless of the end product, the cheese-making process typically begins with pasteurized milk (**Figure 26.1**). A pound of cheese requires about 5 gallons of milk. Cultures of *Lactococcus lactis* coagulate protein in milk to form *curds* (solids) and *whey* (liquids). For unripened cheeses such as cottage cheese, the curd is removed, cut into small pieces, and packaged. Alternatively, the curdling process can be accelerated by the addition of the enzyme *rennin*, a type of protease (protein-digesting enzyme).

Ripened cheeses are made from milk curds that have been pressed to remove water. Hard cheeses, such as Parmesan and cheddar, are pressed such that little water remains, while soft cheeses retain enough moisture to make them spreadable. Once the curds are pressed, the cheese is ripened (aged) for months to years while continued microbial activity imparts characteristic smells and tastes. A secondary fungal culture of *Penicillium roqueforti* (pen-i-sil′ē-ŭm rōk′fōr-tē) is added to make blue cheese.

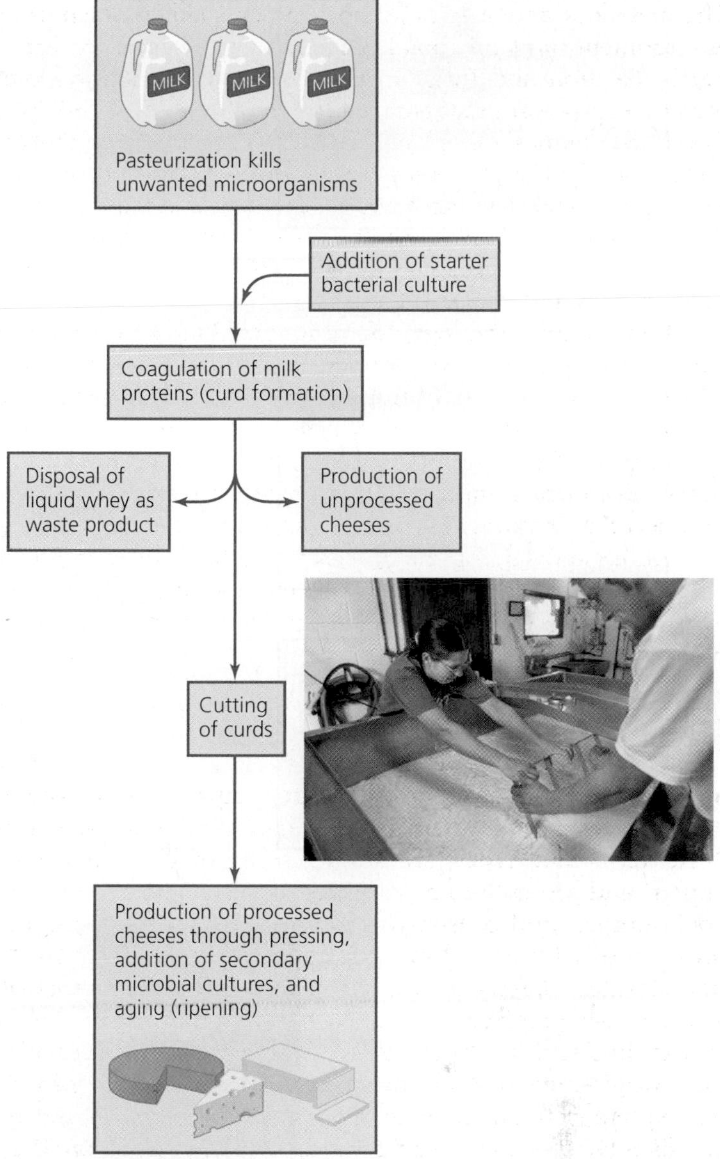

▲ **Figure 26.1 The cheese-making process.** The commercial production of all cheeses begins with the pasteurization of milk; the wide variety of cheeses available on the market results from the nature of subsequent processing steps. *Why do cheeses that are aged longer have more acidic, "sharper" tastes?*

Figure 26.1 *Fermentation during aging results in continued production of acid; the longer the cheese is aged, the more acid produced and the "sharper" the taste.*

CRITICAL **THINKING**

Even though microbes are naturally present in raw milk, raw milk is not generally used for the production of cheeses. Why not?

Products of Alcoholic Fermentation

Alcoholic fermentation is the process by which various microorganisms convert simple sugars such as glucose into ethanol (drinking alcohol) and carbon dioxide (CO_2). These two products

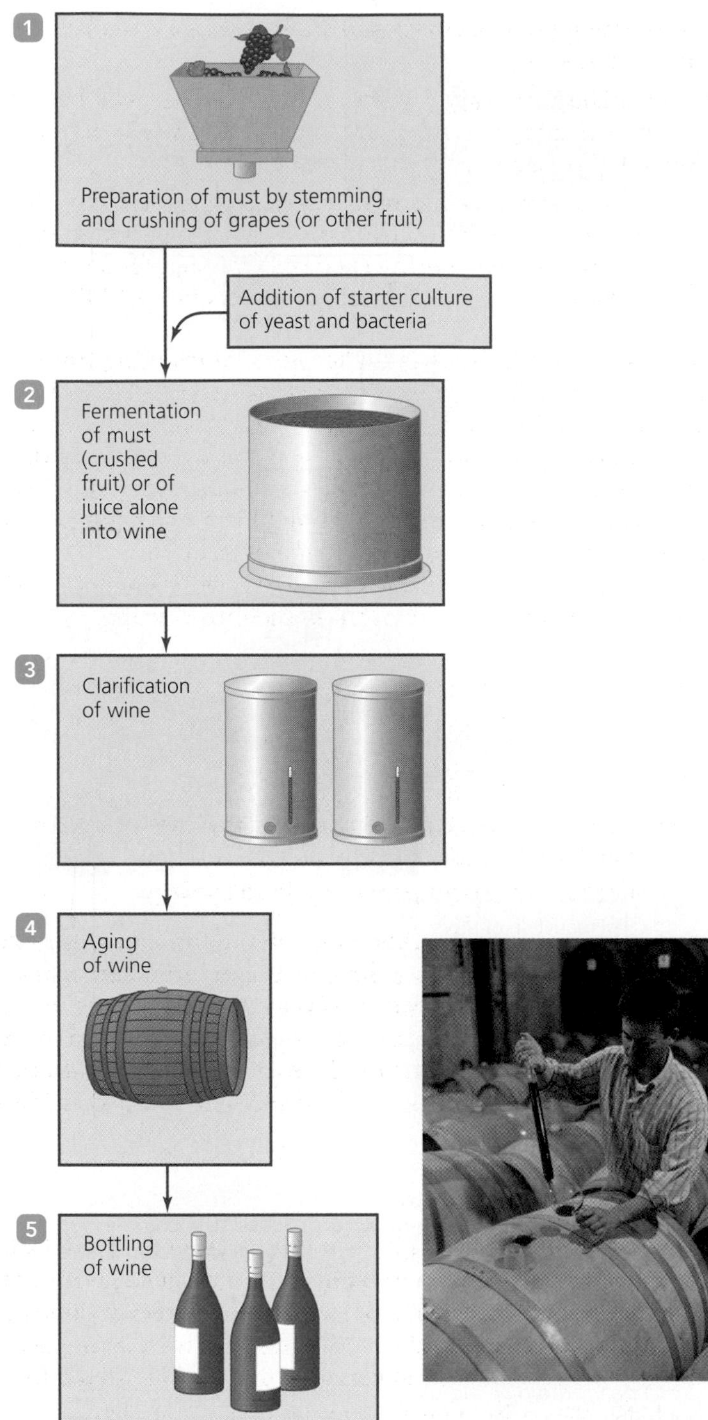

▲ **Figure 26.2 The wine-making process.** Red wine is produced by fermenting must (fruit solids and juice) from dark grape varieties, whereas white wine results from the fermentation of juice alone from either dark or light grape varieties.

of fermentation take on different importance for different foods and beverages. As is the case for the production of fermented dairy products, manufacturers use specific starter cultures in the large-scale commercial applications of alcohol fermentation. In the next sections we consider the role of fermentation in the production of wine, spirits, beer, sake, and vinegar.

Wine and Spirits Generally, wine production proceeds as follows **(Figure 26.2)**:

1. *Preparation of must.* Wine makers crush the fruit and remove the stems to form *must* (fruit solids and juices). They make red wines from the entire must of dark grapes, whereas they make white wines using only the juice of either dark or light grapes. Weather conditions and soil composition affect the sugar content of fruit, which in turn greatly affects the quality of the wine.
2. *Fermentation.* A thin film of bacteria and yeast naturally covers grapes and other fruit, and in some wineries these natural microbes produce the wine. However, in most commercial operations, vintners add sulfur dioxide (SO_2) to retard growth of naturally occurring bacteria and then add starter cultures of yeast and bacteria to ensure that wine of a consistent quality is generated from year to year. *Saccharomyces* ferments sugar in fruit juice to alcohol to make wine. *Leuconostoc* removes acids naturally present in grapes.
3. *Clarification.* Filtration or settling removes solids—a process called clarification.
4. *Aging.* Wine is aged in wooden barrels. Continued yeast metabolism and the leaching of compounds from the wood add taste and aroma.
5. *Bottling.* The wine is then bottled and distributed.

Dry wines are those in which all the sugar has fermented, whereas sweet or dessert wines retain some sugar. Most table wines have a final alcohol content of 10–12%, but fortified wines have an alcohol content of about 20% because distilled spirits (discussed next) are added. Sparkling wine is produced by the addition of sugar and a second fermentation that produces CO_2 inside the bottle.

Distilled spirits are made in a process similar to wine making in that *Saccharomyces* ferments fruits, grains, or vegetables. The difference is that during the aging process, the alcohol is concentrated by distillation. In this process, the liquid is heated to drive off the alcohol, which is then condensed and added back. The net result is an increase in the alcohol content (100-proof spirits are 50% alcohol). Brandies are made from fruit juices, whiskeys are made from cereal grains, and vodka is often made from potatoes.

Beer and Sake Beer is made from barley in the following steps **(Figure 26.3)**:

1. *Malting.* Barley is moistened and germinated, a process that produces catabolic enzymes, which convert starch into sugars, primarily maltose. Drying halts germination, and the barley is crushed to produce *malt.*
2. *Mashing.* Malt and additional sources of carbohydrate called adjuncts (such as starch, sugar, rice, sorghum, or corn) are combined with warm water, allowing enzymes released during malting to generate more sugars. The sugary liquid is called *wort.*
3. *Preparation for fermentation.* The brewer removes the spent grain from the wort, and adds *hops*—the dried flower-bearing

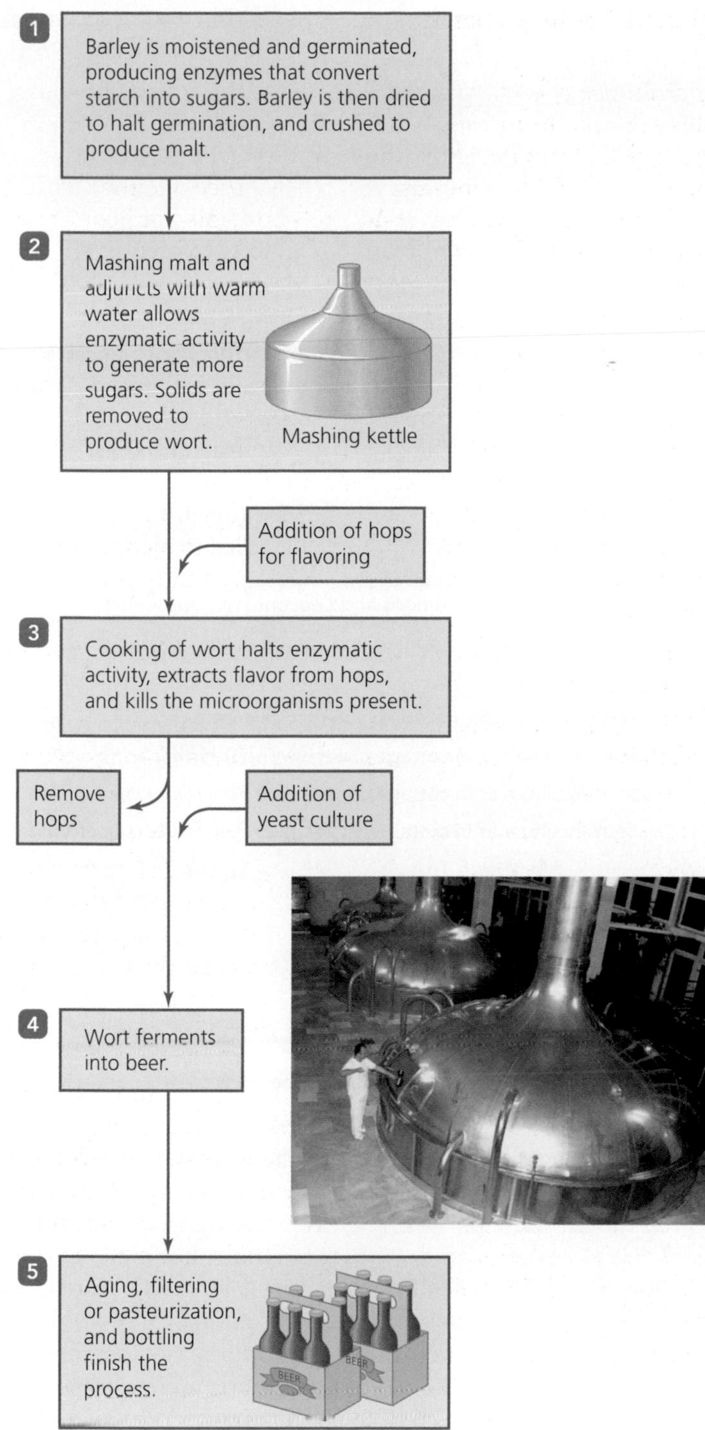

1
Barley is moistened and germinated, producing enzymes that convert starch into sugars. Barley is then dried to halt germination, and crushed to produce malt.
2
Mashing malt and adjuncts with warm water allows enzymatic activity to generate more sugars. Solids are removed to produce wort.
Mashing kettle
Addition of hops for flavoring
3
Cooking of wort halts enzymatic activity, extracts flavor from hops, and kills the microorganisms present.
Remove hops
Addition of yeast culture
4
Wort ferments into beer.
5
Aging, filtering or pasteurization, and bottling finish the process.

▲ **Figure 26.3 The beer-brewing process.** The type of beer produced (ale, lager) is largely determined by the type of yeast used (top-fermenting or bottom-fermenting, respectively).

parts of the vinelike hops plant. The wort mixture is boiled to halt enzymatic activity, extract flavors from the hops, kill most microorganisms, and concentrate the mixture.

4 *Fermentation.* A starter culture of *Saccharomyces* is added to the wort. Most beers—called lagers—are made with the bottom-fermenting yeast *S. carlsbergensis* (karlz-bur-jen′sis). Other beers—called ales—are produced with the top-fermenting yeast *S. cerevisiae* (so called because the yeast floats in the vat).

5 *Aging.* The beer is aged, pasteurized or filtered, and bottled or canned. Beer has a considerably lower alcohol content (about 4%) than either wine or distilled spirits.

Sake, sometimes called rice wine, is more correctly "rice beer" because the starch in cooked rice is first converted to sugar (by the fungus *Aspergillus oryzae*) and then the sugar is fermented by *Saccharomyces.* Sake has an alcohol content of about 14%.

Vinegar Vinegar[2] is produced when ethanol resulting from the fermentation of a fruit, grain, or vegetable is oxidized to acetic acid. *Acetic acid bacteria* such as *Acetobacter* (a-sē′tō-bak-ter) or *Gluconobacter* (gloo-kon′ō-bak-ter) perform this secondary metabolic conversion, generating about 4% acetic acid. Different flavors of vinegar are derived from different starting materials. Cider vinegar, for example, is made from apples.

Table 26.1 on p. 750 summarizes types of fermented foods along with the organisms involved in their production.

The Causes and Prevention of Food Spoilage

Learning Objectives

- ✓ Describe how food characteristics and the presence of microorganisms in food can lead to food spoilage.
- ✓ List several methods for preventing food spoilage.

Whereas many chemical reactions involving food are desirable because they enhance taste, aroma, or preservation, other reactions are not. Spoilage of food involves adverse changes in nutritive value, taste, or appearance. Spoilage leads not only to economic loss but potentially to illness and even death. In this section we examine the causes of food spoilage before considering ways to prevent it.

Causes of Food Spoilage

Foods may spoil because of inherent properties of the food itself. Such *intrinsic factors* include nutrient content, water activity, pH, and physical structure of the food, as well as the competitive activities of microbial populations. Alternatively, food spoilage results from *extrinsic factors* that have nothing to do with the food itself but instead with the way it is processed or handled.

Intrinsic Factors in Food Spoilage The nutritional composition of food determines both the types of microbes present and whether they will grow; for example, high-starch foods tend to support a different group of microbes than do high-protein foods. Some foods contain natural antimicrobial agents, such as the benzoic acid in cranberries, and thus are not prone to spoilage. In contrast, *fortified foods*—those enriched in vitamins and minerals to improve the health of humans—may inadvertently facilitate growth of microorganisms by providing more nutrients.

[2]From French *vinaigre,* meaning sour wine.

TABLE 26.1 Some Fermented Foods and the Microorganisms Used in Their Production

Food	Starting Material	Representative Culture Microorganisms
Fermented Vegetables		
Sauerkraut/kimchi	Cabbage	Various lactic acid bacteria
Pickle	Cucumbers, peppers, beets	Various lactic acid bacteria
Soy sauce	Soybeans and wheat	*Aspergillus oryzae* and *Lactobacillus* spp.
Miso	Rice and soybeans or rice and other grains	*Aspergillus oryzae, Lactobacillus* spp., and *Torulopsis etchellsii*
Fermented Meat Products		
Dry salami	Pork, beef, chicken	Various lactic acid bacteria
Fish sauce/paste	Ground fish	Various naturally occurring bacteria
Fermented Dairy Products		
Milks		
Buttermilk	Pasteurized skim milk	*Lactococcus lactis cremoris* and *Leuconostoc citrovorum*
Yogurt	Pasteurized skim milk	*Streptococcus thermophilus* and *Lactobacillus bulgaricus*
Cheeses		
Cottage cheese	Pasteurized milk	*Lactococcus lactis*, including subspecies *cremoris*
Hard cheese (e.g., cheddar)	Milk curd	Starter culture as in cottage cheese without further additions
Soft cheese (e.g., Camembert)[a]	Milk curd	Starter culture as in cottage cheese plus *Penicillium camemberti*
Mold-ripened (e.g., Roquefort)	Milk curd	Starter culture as in cottage cheese plus *Penicillium roqueforti*
Animal Feed		
Silage	Corn, grains, vegetation	Various naturally occurring bacteria
Alcoholic Fermentations		
Wine	Grapes	*Saccharomyces cerevisiae*
Distilled spirits	Fruits, vegetables, grains	*Saccharomyces cerevisiae*
Beer	Barley	*Saccharomyces cerevisiae* or *S. carlsbergensis*
Sake	Cooked rice	*Aspergillus oryzae* and *Saccharomyces* spp.
Vinegar	Fruits, vegetables, grains	*Saccharomyces cerevisiae* and *Acetobacter* or *Gluconobacter*
Bread	Flour, salt, etc.	*Saccharomyces cerevisiae*

[a]In addition to being a soft cheese, Camembert is also a mold-ripened cheese.

Water activity refers to water that is not bound physically by solutes or to surfaces and is thus available to microbes. The water activity of pure water is set at 1.0, and most microbes require environments with a water activity of at least 0.90. Moist foods, such as fresh meat, with water activities near 0.90 support microbial growth, whereas dry foods (for example, uncooked pasta) with water activities near zero do not support microbial growth. Food processors reduce water activity by drying foods or by adding salts or sugars. Thus even though jam is moist, its sugar content is very high, and its water activity is too low to support much microbial growth.

Acidity of food can either be an intrinsic chemical property of the food, as in the case of citrus fruit, or result from fermentation, as with dill pickles. In either case, a pH below 5.0 typically reduces microbial growth except by a few molds or lactic acid bacteria. A pH closer to neutral (7.0) supports a wider range of microbial growth. Since the pH of most foods is close to neutral, pH is generally not a deterrent to spoilage microorganisms.

Physical structure is a visible intrinsic factor. Rinds or thick skins naturally protect many fruits and vegetables, and eggs have shells. These coverings are dry and nutritionally poor, and thus they support little microbial growth. If the outer covering is broken or cut, however, microbes can reach the moist interior of the food and cause it to rot. Additionally, moisture left on the rind or skin can create conditions that support microbial growth.

Ground hamburger supports microbial growth, and thus spoils faster, than whole steaks or roasts, because grinding creates air pockets and more surfaces that can harbor microbes. In uncut meats, the largest volume of the meat is anaerobic and not exposed to microorganisms.

To some extent, microbial competition can also be an intrinsic factor in food spoilage, especially in fermented foods. Fermented foods are populated with large numbers of fermentative bacteria but few pathogens, because the latter do not readily grow in the environment produced by fermentation. Furthermore, pathogens that require the rich nutrient levels found in the human body are not typically capable of growing on "lower nutrient" foods.

Table 26.2 on p. 752 summarizes the effects of intrinsic factors on food spoilage.

Extrinsic Factors in Food Spoilage Extrinsic factors, including how food is processed, handled, and stored, also govern food spoilage. Microorganisms can enter food in a variety of ways; some are introduced accidentally during harvesting from soil or contaminated water, and commercial food processing introduces others. Mechanical vectors such as flies also deposit microbes onto food. However, most microbial contaminants leading to spoilage can be traced to improper handling by consumers. Methods to prevent the introduction of microbes into food or to limit their growth will be presented shortly.

CRITICAL **THINKING**

Describe the intrinsic and extrinsic factors that would contribute to the spoilage of butter, an apple, and a steak.

Classifying Foods in Terms of Potential for Spoilage

Based on the previous considerations, foods can be grouped into three broad categories based on their likelihood of spoilage. Highly *perishable* foods, such as milk and meat, tend to be nutrient rich, moist, and unprotected by rinds or coverings. They need to be kept cold or they spoil relatively quickly (within days). *Semiperishable* foods, such as tomato sauce, can be stored in sealed containers for months without spoiling so long as they are not opened. Once opened, however, they may spoil within several weeks. Many fermented foods are semiperishable. *Nonperishable* foods are usually dry foods, such as pasta or uncooked rice, or canned goods that can be stored almost indefinitely without spoiling. Nonperishable foods are typically either nutritionally poor, dried, fermented, or preserved (discussed shortly).

Prevention of Food Spoilage

Food spoilage can begin during production, processing, packaging, or handling. Spoilage results in significant economic losses, especially to producers—in the form of lost productivity, food recalls, and even the loss of jobs resulting from the shutdown of unsafe food processing facilities—but also to consumers in the form of medical expenses and time lost from work. In the next sections we examine some of the many techniques used to prevent food spoilage: food processing methods, the use of preservatives, and attention to temperature and other storage conditions.

Food Processing Methods *Industrial canning* is a major food packaging methodology for preserving foods (**Figure 26.4**)

▲ **Figure 26.4 Industrial canning.** Food is processed, put into cans or jars, and heated under pressure for a length of time and a temperature sufficient to kill most microbes, including endospore-forming bacteria.

After food is prepared by washing, sorting, and processing, it is packaged into cans or jars and subjected to high heat (115°C) under pressure for a given amount of time, depending on the food. This is followed by rapid cooling. The heat kills vegetative mesophilic bacteria and destroys endospores formed by *Bacillus* (ba-sil'lŭs) and *Clostridium* (klos-trid'ē-ŭm). Although canning results in the elimination of most contaminating microorganisms, it does not sterilize food; hyperthermophilic microbes remain, but because they cannot grow at room temperature, they do not pose a threat.

Spoilage can occur if the food is underprocessed or if mesophilic microbes contaminate the can during cooling or thereafter. The two most frequent contaminants found in canned goods are *Clostridium* spp. and coliforms, organisms that grow well anaerobically in low-acid foods. Contamination with the resistant endospores of *Clostridium* is especially problematic because if the cans or jars are not heated sufficiently, the endospores germinate, grow, and release toxins inside the sealed container.

Pasteurization is less rigorous than canning and is used primarily with beer, wine, and dairy products because it does not degrade the flavor of these more delicate foods. By heating foods only enough to kill mesophilic, non-endospore-forming microbes (including most pathogens), pasteurization lowers the overall number of microbes, but because some microbes survive, pasteurized foods will spoil without refrigeration.

Moisture is essential for microbial growth; therefore, desiccation (drying) is an excellent food-preservation technique. Drying greatly reduces or eliminates microbial growth, but it does not kill bacterial endospores. Although drying by leaving foods in the sun is still used for some fruits, today most commercially dried foods are processed in ovens or heated drums that evaporate the water from the foods.

TABLE 26.2 Factors Affecting Food Spoilage

	Foods at Greatest Risk of Spoilage	Food at Least Risk of Spoilage
Intrinsic Factors		
Nutritional composition	Chemically rich or fortified foods (steak, bread, yogurt)	Chemically limited foods (flour, cereals, grains)
Water activity	Moist foods (meat, milk)	Dry foods or those with low water activity (pasta, jam)
pH	Foods with neutral pH (bread)	Foods with low pH (orange juice, pickles)
Physical structure	Foods without rinds, skins, or shells; ground meat	Foods with rinds, skins, or shells; intact foods
Microbial competition	Foods that lack natural microbe populations (ground beef)	Foods with resident microbial populations (pickles)
Extrinsic Factors		
Degree of processing	Unprocessed foods (raw milk, fruit)	Processed foods (pasteurized foods)
Amount of preservatives	Foods without preservatives (meats, natural foods)	Foods with either naturally occurring or added preservatives (garlic, spices, sulfur dioxide)
Storage temperature	Foods left in warm conditions	Foods kept cold
Storage packaging	Foods stored exposed	Foods wrapped or sealed

Lyophilization (lī-of′i-li-zā′shŭn), or *freeze-drying,* involves freezing foods and then using a vacuum to draw off the ice crystals. Freeze-dried foods such as soups and sauces can be reconstituted by mixing with water.

Gamma radiation, produced primarily by the isotope cobalt-60, penetrates fruits, vegetables, and meats, including fish, poultry, and beef, to cause irreparable and fatal damage to the DNA of microbes. Such ionizing *irradiation* is controversial because some consumers believe erroneously that irradiated foods become radioactive, are less nutritious, or contain toxins. Nevertheless, because irradiation can achieve complete sterilization, it is used routinely to preserve some foods such as spices.

UV light is not ionizing and does not penetrate very far. It is used to treat packing and cooling water used in industrial canning and to treat work surfaces and utensils in industrial meat processing plants.

In *aseptic packaging,* paper or plastic containers that cannot withstand the rigors of canning or pasteurization are sterilized with hot peroxide solutions, UV light, or superheated steam. Then a conventionally processed food is added to the aseptic package, which is sealed without the need for additional processing.

The Use of Preservatives Humans have preserved foods with salt or sugar throughout history. Both chemicals draw water by osmosis out of foods and microbes alike, killing microbes present initially and retarding the growth of any subsequent microbial contaminants. Bacon is an example of a high-salt food; jams and jellies are examples of high-sugar foods.

Whereas salt and sugar act by removing water, some natural preservatives actively inhibit microbial enzymes or disrupt cytoplasmic membranes. For example, garlic contains *allicin,* which inhibits enzyme function. Benzoic acid, produced naturally by cranberries, also interferes with enzymatic function. Cloves, cinnamon, oregano, and thyme (and to some degree, sage and rosemary) produce oils that interfere with the functions of membranes of microorganisms.

As we have seen, fermentation preserves some foods by producing an acidic environment that is inhospitable to most microbes. For other foods such as meats, the use of wood smoke during the drying process introduces growth inhibitors that help preserve the food. Other naturally occurring and synthetic chemicals can be purposely added to foods as preservatives. Acceptable preservatives are harmless and do not alter the taste or appearance of the food to any great extent. Organic acids such as benzoic acid, sorbic acid, and propionic acid are commonly used in beverages, dressings, baked goods, and a variety of other foods. Gases such as sulfur dioxide and ethylene oxide are used to preserve dried fruit, spices, and nuts. All such chemical preservatives inhibit some aspect of microbial metabolism, but many do not actually kill microbes; in other words, they are *germistatic* rather than *germicidal.* Some chemicals work better against bacteria than against fungi, and vice versa. Benzoic acid, for example, has a largely antifungal function and does not affect the growth of many bacteria.

Attention to Temperature During Processing and Storage In general, higher temperatures are desirable during food processing and preparation to prevent food spoilage, whereas lower temperatures are desirable for food storage. High temperatures such as those of pasteurization, canning, or cooking kill potential pathogens because proteins and enzymes become irreversibly denatured. However, heat does not inactivate many toxins; for example, botulism toxin may remain in cooked foods even when the bacteria that produced it are dead.

Unlike heat, cold rarely kills microbes, but instead merely retards their growth by slowing metabolism. Even freezing fails to kill all microorganisms and may only lower the level of microbial contamination enough to reduce the likelihood of food poisoning after frozen foods have been thawed.

TABLE 26.3 Most Common Bacterial and Protozoan Agents of Foodborne Illnesses

Organism	Affected Food Products	Comments
Campylobacter jejuni	Raw and undercooked meats; raw milk; untreated water	Most common cause of diarrhea of all foodborne agents
Clostridium botulinum	Home-prepared foods	Produces a neurotoxin
Escherichia coli O157:H7	Meat; raw milk	Produces an enterotoxin
Listeria monocytogenes	Dairy products; raw and undercooked meats; seafood; produce	Common in soils and water, contamination from these sources occurs easily; grows at refrigerator temperature
Salmonella spp.	Raw and undercooked eggs; meat; dairy products; fruits and vegetables	Second most common cause of foodborne illness in the United States
Shigella spp.	Salads; milk and other dairy products; water	Third most common cause of foodborne illness in the United States
Staphylococcus aureus	Cooked high-protein foods	Produces a potent toxin that is not destroyed by cooking
Toxoplasma gondii	Meat (pork in particular)	Parasitic protozoan
Vibrio vulnificus	Raw and undercooked seafood	Causes primary septicemia (bacteria in the blood)
Yersinia enterocolitica	Pork; dairy products; produce	Causes generalized diarrhea and severe cramping that mimics appendicitis; grows at refrigerator temperature

To prevent food spoilage, foods should be prepared at sufficiently high temperatures and then stored under conditions that do not facilitate microbial growth. Wrapping leftovers or putting them in containers enables foods to be stored away from exposure to air or contact that could result in contamination. Storing foods in the cold of a refrigerator or freezer combines the protection of packaging with growth-inhibiting temperatures.

One microbe for which cold storage does not suppress growth is *Listeria monocytogenes* (lis-tēr′ē-ă mo-nō-sī-tah′je-nēz), the causative agent of listeriosis and a common environmental bacterium, which is prevalent in certain dairy products, such as soft cheeses. This bacterium grows quite well under refrigeration; therefore, it is best to prevent its entry into foods.

Foodborne Illnesses

Learning Objective

✓ Discuss the basic types of illnesses caused by food spoilage or food contamination, and describe how they can be avoided.

Spoiled food, if consumed, can result in illness, but not all food-borne illnesses result from actual food spoilage. Food-borne illnesses may also result from the consumption of harmful microbes or their products in food.

Foodborne illnesses (*food poisoning*) can be divided into two types: **food infections,** caused by the consumption of living microorganisms, and **food intoxications,** caused by the consumption of microbial toxins instead of the microbes themselves. Typical signs and symptoms of food poisoning are generally the same regardless of the cause and include nausea, vomiting, diarrhea, fever, fatigue, and muscle cramps. Symptoms occur within 2 to 48 hours after ingestion, and the effects of the illness can linger for days. Most outbreaks of food poisoning are *common-source epidemics,* meaning that a single food source is responsible for many individual cases of illness.

According to the Centers for Disease Control and Prevention (CDC), roughly 87 million cases of food poisoning occur each year. Of these, about 371,000 people require hospitalization and 5700 die. Researchers identify the microbes involved in only about 14 million of the total cases. The United States Department of Agriculture estimates the economic cost of food poisoning—due to loss of productivity, medical expenses, and death—at roughly $5–10 billion per year.

More than 250 different foodborne diseases have been described. Table 26.3 lists 10 common causes of foodborne illness and their sources. Except for the protozoan *Toxoplasma gondii* (tok-sō-plaz′mă gon′dē-ē), all are bacterial agents. A few fungi (*Aspergillus* and *Penicillium*) and viruses (e.g., hepatitis A, noroviruses) also cause food poisoning.

CRITICAL **THINKING**

Suggest why neither the original food involved nor the organism responsible can ever be identified in more than 80% of food poisoning cases.

Industrial Microbiology

The potential uses of microorganisms for producing valuable compounds, as environmental sensors, and in the genetic modification of plants and animals makes industrial microbiology one of the more important fields of study within the microbiological sciences. In the sections that follow we will examine the use of microbes in industrial fermentation, in the production of several industrial products, in the treatment of water and wastewater, and in the disposal and cleanup of biological wastes.

▲ **Figure 26.5 Fermentation vats.** Such large containers are used for growing microorganisms in the vast quantities needed for the large-scale production of many industrial, agricultural, and medical products. *Why are industrial fermentation vats made of stainless steel instead of wood?*

Figure 26.5 *Fermentation vats are made of stainless steel so that they can be cleaned more easily to avoid contamination of the product being produced.*

The Roles of Microbes in Industrial Fermentations

Learning Objective

✓ Describe the role of genetically manipulated microorganisms in industrial and agricultural processes and the basics of industrial-scale fermentation.

In industry, the word *fermentation* is used a bit differently than it is used in food microbiology. *Industrial fermentations* involve the large-scale growth of particular microbes for producing beneficial compounds, such as amino acids and vitamins. Temperature, aeration, and pH are all regulated to maintain optimal microbial growth conditions. Generally, industrial fermentations start with the cheapest growth medium available, often the waste product from another process (such as whey from cheese production). Because of its scale, industrial fermentation is performed in huge vats that can be readily filled, emptied, and sterilized **(Figure 26.5)**. Vats are typically made from stainless steel so that they can be cleaned and sterilized more easily.

There are two types of industrial fermentation. In *batch production,* organisms ferment their substrate until it is exhausted and then the end product is harvested all at once. In *continuous flow production,* the vat is continuously fed new medium while wastes and product are continuously removed. For this setup to work, the organism must secrete its product into the surrounding medium.

Industrial products are produced as either primary or secondary metabolites of the microorganisms. *Primary metabolites* such as ethanol are produced during active growth and metabolism because they are either required for reproduction or are by-products of active metabolism. *Secondary metabolites* such as penicillin are produced after the culture has moved from log phase of growth and entered stationary phase, during which time the substances produced are not immediately needed for growth.

Chapter 8 examined recombinant DNA techniques used for the production of *recombinant microorganisms* and some of their uses. Most genetically modified organisms used in industry have been specifically designed to produce a stable, high-yield output of desirable chemicals or to perform certain novel functions.

CRITICAL **THINKING**

Which process would you expect to yield more product from the same sized vats over the same amount of time, batch production or continuous flow production? Why?

Industrial Products of Microorganisms

Learning Objective

✓ List some of the various commercial products produced by microorganisms.

Microorganisms, particularly bacteria, metabolically produce an incredible array of industrially useful chemicals, including enzymes, food additives and supplements, dyes, plastics, and fuels. Recombinant organisms add to this diversity by producing pharmaceuticals, such as human insulin, which are not normally manufactured by microbial cells. In the sections that follow we consider a variety of industrial products produced by microbes, including enzymes, alternative fuels, pharmaceuticals, pesticides, agricultural products, biosensors, and bioreporters.

Enzymes and Other Industrial Products

Enzymes are among the more important products made by microbes. Most are naturally occurring enzymes for which industrial uses have been devised. For example, amylase, produced by *Aspergillus oryzae,* is used as a spot remover. Pectinase, obtained from species of *Clostridium,* enzymatically releases cellulose fibers from flax, which are then made into linen. Proteases from a variety of microbes are used in meat tenderizers, spot removers, and cheese production. Streptokinases and hyaluronidase are used in medicine to dissolve blood clots and enhance the absorption of injected fluids, respectively. Microbes also make the enzymatic tools of recombinant DNA technology, including restriction enzymes, ligases, and polymerases.

Some products made naturally by microorganisms are useful to humans as food additives and food supplements. *Food additives* generally enhance a food in some way, such as by improving color or taste, whereas *food supplements* make up for nutritional deficiencies. Amino acids and vitamins are two important microbial products used as supplements. Vitamins are added directly to foods or sold as vitamin tablets. Amino acids are either sold in tablet form or are combined to make new compounds, such as the sweetener aspartame (made from the amino acids phenylalanine and aspartic acid). Other organic acids, such as citric acid, gluconic acid, and acetic acid (vinegar), are also microbially produced to be used in food manufacturing. Citric acid is used as an antioxidant in foods, while gluconic acid is used medically to facilitate calcium uptake.

Other industrial products made by microbes include dyes, such as indigo, which makes denim jeans blue (see **Highlight: Making Blue Jeans "Green"** on p. 756), and cellulose fibers used in woven fabrics. Microbially produced biodegradable plastics can replace nonbiodegradable, petroleum-based plastics. This is possible because many bacteria produce a carbon-based storage molecule called polyhydroxyalkanoate (PHA), a polymer with a structure similar to petroleum-based plastics. Even though such biodegradable plastics are expensive to manufacture, they have been commercially available since the early 1990s.

Alternative Fuels

Photosynthetic microorganisms use the energy in sunlight to convert CO_2 into carbohydrates that can be used as fuels—so-called biofuels. Other microorganisms convert biomass (organic materials such as plants or animal wastes) into renewable biofuels, including ethanol, methane, and hydrogen.

Ethanol, which is made during alcoholic fermentation and is the simplest alternative biofuel to synthesize, can be mixed with gasoline to make *gasohol*, which is used in existing cars. Currently, in the U.S., food crops such as corn are fermented to ethanol, but a more efficient production of alternative fuel utilizes crop wastes—for example, corn stalks.

All microorganisms synthesize hydrocarbons as part of their normal metabolism, but only some microbes produce hydrocarbons that could be useful fuels. The colonial alga *Botryococcus braunii* (bot′rē-ō-kok′ŭs brow′nē-ē), for example, produces hydrocarbons that account for 30% of its dry weight. The technology to harvest such hydrocarbons does not yet exist, but one day this alga may be an important source of fuel. One exception is the harvesting of methane gas, which can be collected and piped through natural gas lines to be used for cooking and heating. The largest potential source of methane is from landfills where methanogens anaerobically convert wastes into methane via anaerobic respiration **(Figure 26.6).** Some communities already use methane from landfills to produce energy. Methane can also be used as a fuel in properly equipped cars.

Although a variety of microbes release hydrogen as part of their normal metabolism, the hydrogen-producing metabolic pathway does not cost-effectively produce fuel. An economically efficient method for the production of hydrogen fuel using sunlight and photosynthetic microbes would have widespread applications and enormous appeal, but much research is needed to develop technologies to produce hydrogen as a viable energy source.

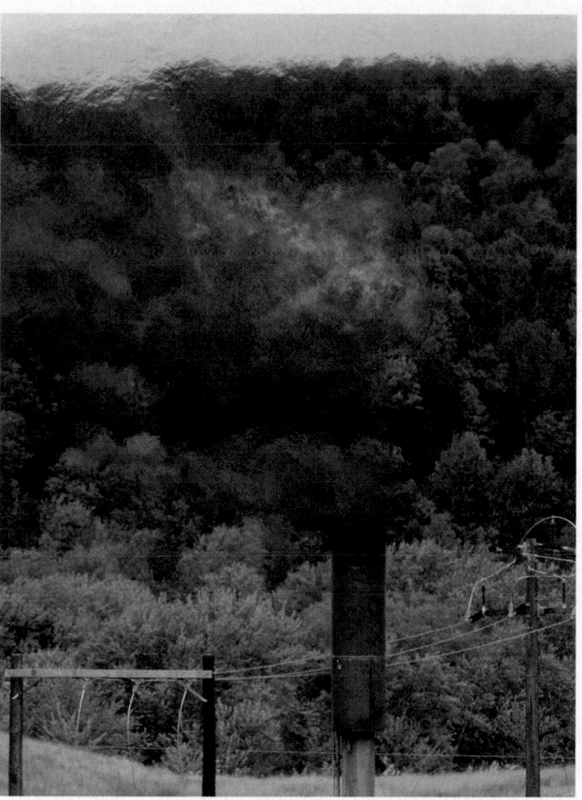

▲ **Figure 26.6 Burning methane gas released from a landfill.** Such microbially produced methane could be an important alternative fuel source. *What are some potential benefits of renewable energy produced by microbes?*

Figure 26.6 *Microbial fuels may be easier to produce and less harmful to the environment.*

Highlight: Bacterial Batteries on p. 757 explores the potential use of bacteria in directly generating electricity without having to burn fuel.

Pharmaceuticals

Foremost among the pharmaceutical substances microorganisms produce are antimicrobial drugs. About 6000 antimicrobial substances have been described since penicillin was first produced during World War II. Of these, 100 or so have current medical applications. The bacteria *Streptomyces* (strep-tō-mī′sēz) and *Bacillus* and the fungus *Penicillium* synthesize the majority of useful antimicrobials, though recombinant DNA techniques enable the modification of other microbes to produce new versions of old drugs.

In addition to antimicrobials, genetically modified microbes are producing hormones and other cell regulators. Table 26.4 on p. 756 lists some products of recombinant DNA technology used in medicine.

Pesticides and Agricultural Products

Farmers use microbes and their products in a variety of agricultural applications, particularly with regard to crop management.

TABLE 26.4

Some Products of Recombinant DNA Technology Used in Medicine

Product	Modified Cell	Uses of Product
Interferons	*Escherichia coli, Saccharomyces cerevisiae*	To treat cancer, multiple sclerosis, chronic granulomatous disease, hepatitis, and warts
Interleukins	*E. coli*	To enhance immunity
Tumor necrosis factor	*E. coli*	In cancer therapy
Erythropoietin	Mammalian cell culture	To stimulate red blood cell formation; to treat anemia
Tissue plasminogen activating factor	Mammalian cell culture	To dissolve blood clots
Human insulin	*E. coli*	For diabetes therapy
Taxol	*E. coli*	In ovarian cancer therapy
Factor VIII	Mammalian cell culture	In hemophilia therapy
Macrophage colony stimulating factor	*E. coli, S. cerevisiae*	To stimulate bone marrow to produce more white blood cells; to counteract side effects of cancer treatment
Relaxin	*E. coli*	To ease childbirth
Human growth hormone	*E. coli*	To correct childhood deficiency of growth hormone
Hepatitis B vaccine	Carried on a plasmid of *S. cerevisiae*	To stimulate immunity against hepatitis B virus

Bacillus thuringiensis (ba-sil′ŭs thur-in-jē-en′sis) is one of the more widely used organisms because during sporulation it produces *Bt toxin* that, when digested by the caterpillars of such pests as gypsy moths, diamondback moths, and tomato hornworms, destroys the lining of the insect's gut wall, eventually killing it. Bt toxin isolated from the bacterium can be spread as a dust on plants, but with recombinant DNA technology the Bt gene can be added to a plant's genome. In this way plants such as corn, soy beans, and cotton protect themselves by manufacturing their own Bt toxin. There is no evidence of insect resistance to this Bt toxin despite the fact that large tracts of U.S. farmland are devoted to raising Bt toxin–expressing plants. Some people are leery of the prospect of transgenic foods derived from these plants, whereas others contend that the genetic manipulation of crops is crucial if we are to continue to feed the world's population. In any case, one thing is clear: the genetic manipulation of plants is big business.

Pseudomonas syringae (soo-dō-mō′nas sēr′in-jī) is another example of a bacterium with agricultural applications. This bacterium produces a protein that serves in the formation of ice crystals, but the protein is not essential to the survival of the organism, and scientists can remove its gene. When sprayed on crops such as strawberries, the strain lacking the gene, known as *P. syringae* ice^{-} ("ice-minus"), inhibits the formation of ice, protecting the plants from freeze damage.

Table 26.5 lists selected industrial products produced by microorganisms and some of their uses.

HIGHLIGHT

MAKING BLUE JEANS "GREEN"

Most blue denim today is colored with an indigo dye that has been synthetically created via a petroleum-based process, because extracting the dye from indigo plants is labor intensive and expensive. Environmentalists are concerned that the potentially toxic by-products of the coal- and oil-dependent process pose risks to the environment.

To address such concerns, scientists have been investigating an alternative method of producing indigo using *E. coli,* which naturally produces the amino acid tryptophan, with a chemical structure very similar to that of indigo. Since 1983, scientists have known how to take advantage of this similarity and genetically alter *E. coli* so that the bacterium produces indigo instead. The problem? Bacterial synthesis of indigo was both slow and resulted in an unwanted red pigment that remained visible when the dye was used on denim.

These problems may soon be resolved. Scientists have succeeded in modifying the *E. coli* genome so that the red pigment is no longer produced.

They have also boosted bacterial production of indigo by 60%. Though more work is needed to make the process economical, these results provide hope for environmentally sensitive blue jeans in the future.

TABLE 26.5 Selected Industrial Products Produced by Microorganisms

Products Made	Use of Product
Enzymes	
Amylase, proteases	Spot removers
Streptokinase	Breakdown of blood clots
Restriction enzymes, ligases, polymerases	Molecular biology, recombinant DNA technology
Food Additives/Supplements	
Amino acids, vitamins	Health supplements
Citric acid	Antioxidant
Sorbic acid	Food preservative
Other Industrial Products	
Indigo	Dye used in manufacturing clothes
Plastics	Biodegradable substitutes for petroleum-based plastics
Alternative Fuels	
Ethanol	Used in gasohol
Methane	Burned to generate heat and electricity
Hydrogen, hydrocarbons	Potential fuels
Pharmaceuticals	
Antimicrobial drugs	Treatment of bacterial infections
Insulin, human growth hormone	Replacement hormones
Taxol	Cancer treatment
Pesticides and Agricultural Products	
Bt toxin	Insecticide

Biosensors and Bioreporters

Among the relatively new applications of microorganisms to solve environmental problems are biosensors and bioreporters. **Biosensors** are devices that combine bacteria or microbial products (such as enzymes) with electronic measuring devices to detect other bacteria, bacterial products, or chemical compounds in the environment. **Bioreporters** are somewhat simpler sensors that are composed of microbes (again usually bacteria) with innate signaling capabilities, such as the ability to glow in the presence of biological or chemical compounds.

Currently, biosensors and bioreporters are used to detect the presence of environmental pollutants (for example, petroleum) and to monitor efforts to remove harmful substances; they may also be useful in detecting bioterrorist attacks. Because bacteria are very sensitive to their environments, they can detect compounds in very small amounts. Biosensors and bioreporters could serve as early warning systems to give officials more time to respond by quickly detecting the metabolic waste products of weaponized biological agents.

Water Treatment

Learning Objectives

- ✓ Contrast water pollution and water contamination.
- ✓ Describe two waterborne illnesses.
- ✓ Explain how water for drinking and wastewater are treated to make them safe and usable.

Water Pollution

Water becomes polluted in three basic ways: *physically,* through the presence of particulate matter; *chemically,* from the presence of inorganic and organic compounds, usually derived from industrial activities or agricultural runoff; or *biologically,* through an overabundance of organisms or the presence of non-native

HIGHLIGHT

BACTERIAL BATTERIES

Some bacteria and archaea have an astonishing ability: they can generate electricity from underwater mud. Researchers have observed that several genera, such as *Geobacter* and *Desulfuromonas,* are able to oxidize organic compounds in underwater sediments or sewage and transfer a current of electrons to graphite electrodes placed in the water by scientists. Essentially, these bacteria act as parts of batteries that use mud or sewage as a fuel to produce enough electricity to power a small calculator.

The current methods of using bacteria to generate electricity are neither practical nor efficient enough for large-scale energy production. Intriguing as the idea is, we will likely never have bacteria-powered cars. However, scientists believe that bacteria could be used as inexpensive batteries. In addition, because these bacteria use organic sediments as fuel, they may be useful in cleaning up organic pollution on the seafloor by converting pollutants to electricity. Sound too good to be true? Microbiologists are working to turn this possibility into reality.

▲ ***Powered by bacteria?***

TABLE 26.6 **Selected Waterborne Agents and the Diseases They Cause**

Organism	Disease
Bacteria	
Vibrio cholerae	Cholera
Campylobacter jejuni	Acute gastroenteritis
Salmonella spp.	Salmonellosis, typhoid fever
Shigella spp.	Shigellosis (bacterial dysentery)
Escherichia coli	Acute gastroenteritis
Viruses	
Hepatitis A virus	Infectious hepatitis
Norovirus	Acute gastroenteritis
Poliovirus	Poliomyelitis
Eukaryotic Parasites	
Giardia intestinalis	Giardiasis
Schistosoma spp.	Schistosomiasis
Entamoeba histolytica	Amebic dysentery
Toxin Producers	
Gonyaulax (saxitoxin)	Paralytic shellfish poisoning
Gambierdiscus (ciguatoxin)	Fish poisoning

microorganisms. Generally, polluted water is obvious in that one can either see, taste, or smell the pollutant. *Contaminated water,* by contrast, contains physical, chemical, or biological pollutants that are not readily visible. Most people wouldn't use polluted water, but contaminated water is more difficult to avoid.

"Clean" water is generally low in nutrients (that is, it contains few organic molecules) and thus does not readily support microbial growth. The microbes that are present come from soil runoff or are naturally present primary producers. Polluted waters, in contrast, are high in organic compounds (particularly nitrogen and phosphorus) and support a greater than normal microbial load. Pollution is generally more problematic in closed lake systems (those that lack a constant influx and efflux of water), but slow-moving rivers and coastal waters can also experience problems with microbial overgrowth due to the dumping of industrial wastes or sewage.

CRITICAL **THINKING**

Following eutrophication of a lake, microbes reproduce prolifically, die, and then sink to the bottom, "feeding" anaerobes in the sediment. Even though the surface water may then look clear, why is it still unsafe to drink?

Although physical and chemical pollutants are important, of greater concern is biological contamination of water with human pathogens. They can cause significant human diseases, which can be prevented only through water treatment.

Waterborne Illnesses

Consumption of contaminated water, either as drinking water or in water added to foods, can result in a variety of bacterial, viral, or protozoan diseases (see **Emerging Diseases: *Norovirus* Gastroenteritis** on p. 761). Fungal water contaminants do not cause diseases.

Each year contaminated drinking water results in roughly 3–5 billion episodes of diarrheal disease worldwide, including over 3 million deaths among children ages five years or younger. Intoxication can also occur from the presence of microbial toxins in the water.

Water treatment removes most waterborne pathogens, so waterborne illnesses are generally rare in the United States as compared to countries with inadequate water-treatment facilities. Outbreaks that do occur in the United States are *point-source infections,* in which a single source of contaminated water leads to illness in individuals that consume the water.

In some marine environments, eutrophic blooms of marine dinoflagellates such as *Gonyaulax* (gon-ē-aw'laks) and *Gambierdiscus* (gam'bē-er-dis-kŭs) chemically pollute water with their toxins. These blooms are one cause of *red tides* because of the color often imparted to the water by the huge number of dinoflagellates. The toxins are absorbed and concentrated by shellfish, and human intoxication results from food consumption rather than water consumption.

Table 26.6 lists some common human waterborne pathogens and toxins.

Clean water is vital for people and their activities. In the following sections we consider the treatment of drinking water and the treatment of wastewater (sewage). In both cases, treatment is designed to remove microorganisms, chemicals, and other pollutants to prevent human illness.

Treatment of Drinking Water

Potable water is water that is considered safe to drink, but the term *potable* (pō'tăbl) does not imply that the water is devoid of all microorganisms and chemicals. Rather, it implies that the levels of microorganisms or chemicals in the water are low enough that they are not a health concern. Water that is not potable is **polluted;** that is, it contains organisms or chemicals in excess of acceptable values.

The permissible levels of microbes and chemicals in potable water varies from state to state. Nationally, the U.S. Environmental Protection Agency (EPA) requires that drinking water have a count of 0 (zero) coliform bacteria per 100 ml of water and that recreational waters have no more than 200 coliforms per 100 ml of water. Recall that coliforms are intestinal bacteria such as *E. coli*. The presence of coliforms in water indicates fecal contamination and thus an increased likelihood that disease-causing microbes are present.

(a)

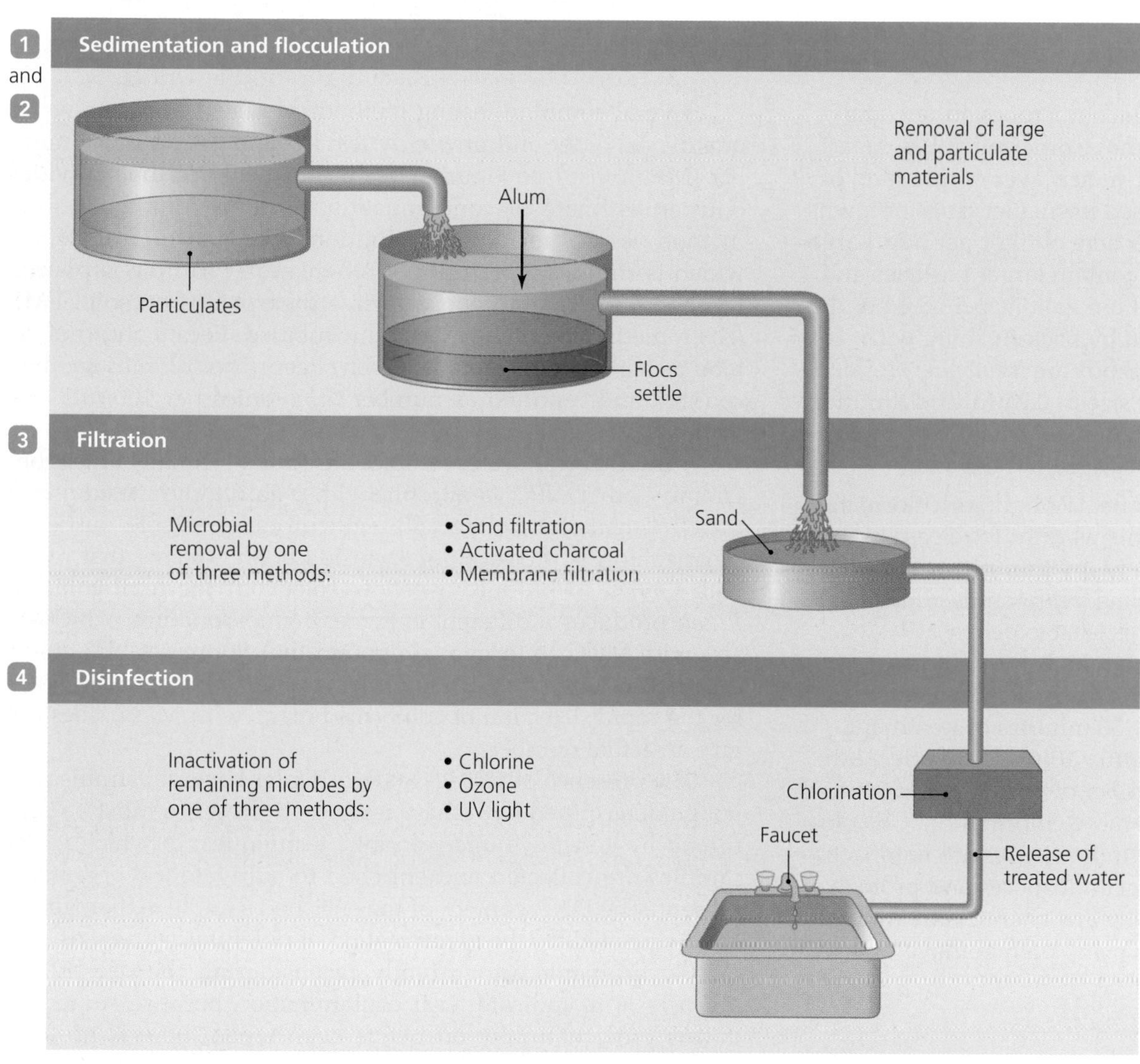

(b)

Figure 26.7 The treatment of drinking water. **(a)** A water treatment facility. **(b)** The stages of water treatment: sedimentation, flocculation, filtration, and disinfection. *Why is it that chemical treatment cannot destroy most viruses?*

Figure 26.7 *Most chemicals are designed either to inhibit some aspect of active metabolism or damage cellular structures; therefore, they do not damage acellular and nonmetabolizing viruses.*

The treatment of drinking water can be divided into four stages **(Figure 26.7)**:

1. **Sedimentation.** During sedimentation, water is pumped into holding tanks where particulate materials (sand, silt, and organic material) settle.
2. **Flocculation.** The partially clarified water is then pumped into a secondary tank for *flocculation,* in which alum (aluminum potassium sulfate) added to the water joins with suspended particles and microorganisms to form large aggregates, called *flocs,* which also settle to the bottom of the tank. The water above the sediment is then pumped into a different tank for filtration.
3. **Filtration.** In this stage, the number of microbes is reduced by about 90% in one of several ways. One method uses

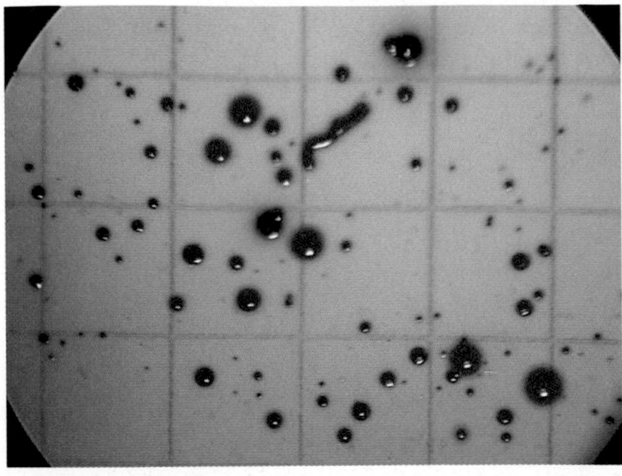

(a)

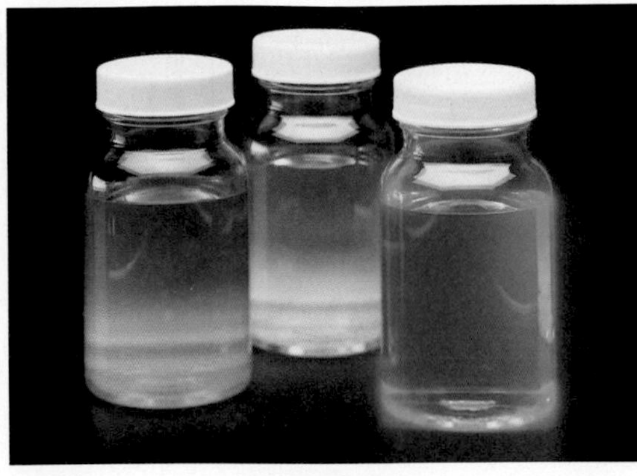

(b)

Figure 26.8 Two water quality tests. **(a)** Membrane filtration. The grid on the membrane makes it easier to count the colonies of fecal coliforms. **(b)** The ONPG and MUG tests. The yellow color of the ONPG bottle indicates the presence of coliforms, whereas the blue fluorescence in the MUG bottle indicates the presence of the coliform *E. coli;* the clear bottle is the negative control.

sand and other materials to which microbes adhere and form biofilms that trap and remove other microbes. *Slow sand filters* are composed of a 1-meter layer of fine sand or diatomaceous earth and are used in smaller cities or towns to process 3 million gallons per acre of filter per day. Large cities use *rapid sand filters* that contain larger particles and gravel and can process 200 million gallons per acre per day. Both types of filters are cleaned by back-flushing with water. Two other filtration methods are *membrane filtration,* which uses a filter with a pore size of 0.2 μm, and filtration with *activated charcoal,* which provides the added benefit of removing some organic chemicals from the water.

4 **Disinfection.** In this stage, ozone, UV light, or chlorination is used to kill most microorganisms prior to release of the water for public consumption. Many European communities use ultraviolet light. Chlorine treatment is widely used in the United States because it is least expensive.

Chlorine gas, an oxidizing agent, is thought to kill bacteria, algae, fungi, and protozoa by denaturing their proteins within approximately 30 minutes of treatment. Chlorine levels must be constantly adjusted to reflect estimates of *microbial load,* the number of microbes in a unit of water—a higher load requires more chlorination. Chlorination does not kill all microbes: most viruses are not inactivated by chlorine, and bacterial endospores and protozoan cysts are generally unaffected by any chemical treatment. Only mechanical filtration can completely remove viruses, endospores, and cysts.

Water Quality Testing

Water quality testing is a technique that uses the presence of certain **indicator organisms** to indicate the possible presence of pathogens in water. Because the majority of waterborne illnesses are caused by fecal contamination, the presence of *E. coli* (the most commonly used indicator organism) or other coliforms in water indicates a high probability that pathogens are present as well. *E. coli* clearly meets the criteria for a good indicator: it is consistently prevalent in human wastes, survives in water as long as (if not longer than) most pathogens, and is easily detected by simple tests.

Several common testing methods are used to assess water quality. An older and unwieldy test is the most probable number (MPN) test (see Figure 6.24), a statistical test that provides only an estimate of contamination. A more common testing method is the membrane filtration method **(Figure 26.8a)**, which is simple to perform: A 100-ml water sample is poured through a fine membrane, which is then placed on solid EMB (eosin methylene blue) agar and incubated. Fecal coliforms exhibit a characteristic metallic green sheen. The colonies are then counted and reported as number of colonies per 100 ml. This method gives an actual coliform count.

In another test, water samples are added to small bottles containing both *ONPG* (*o*-nitrophenyl-β-D-galactopyranoside) and *MUG* (4-methylumbelliferyl-β-D-glucuronide) as sole nutrients. Most coliforms produce β-galactosidase, an enzyme that reacts with ONPG to produce a yellow color, but the fecal coliform *E. coli* produces a different enzyme, β-glucuronidase, which reacts with MUG to form a compound that fluoresces blue when exposed to long-wave UV light **(Figure 26.8b)**. This test allows for the rapid detection of coliforms but, as with MPN, does not give an actual number.

The presence of viruses and particular bacterial strains cannot be determined with these tests; their presence must be confirmed by genetic "fingerprinting" techniques, in which water samples are collected and enriched to cultivate the organisms present. The DNA content of the enriched sample is then genetically screened for the identification of potential pathogens.

Governments are currently reconsidering the use of coliform tests to indicate fecal contamination because some coliforms grow naturally on plants even when there is no fecal contamination, giving a false positive result for fecal pollution. Regulators are considering replacing coliform tests with genetic assays that would specifically indicate the presence of *E. coli.*

CRITICAL **THINKING**

Explain how biosensors might be used for water quality testing. Would biosensors give more accurate identifications of the microorganisms present in water than an ONPG or MUG test? Is such specificity truly necessary?

EMERGING DISEASES

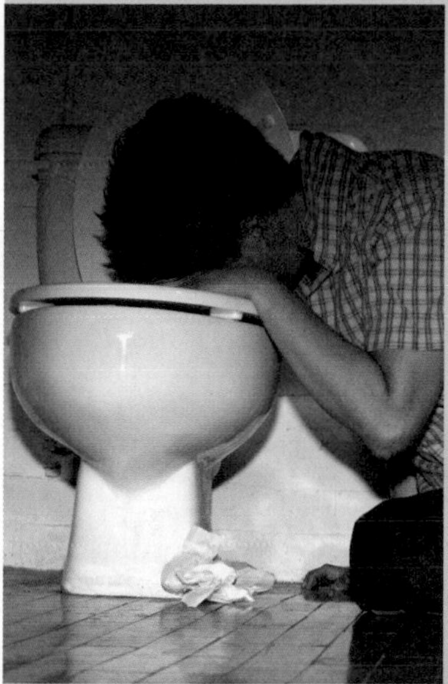

NOROVIRUS GASTROENTERITIS

Zack, Tran, Roy, and Justin were not happy roommates; in fact, things could not be much worse for the college friends. Each of the men had stomach cramps, fever, chills, muscle aches, extreme tiredness, nausea, and, most distressing, horrible diarrhea and persistent vomiting. They had been fighting over the toilet, and whoever wasn't in the bathroom often had his head in a trashcan. None of the four had left their suite for two days, being unable to venture more than a few feet from the bathroom. The friends had never experienced such an attack of gastroenteritis. *Norovirus* had arrived in the dorm.

Norovirus gastroenteritis afflicts people living in close quarters: prisoners, nursing home residents, vacationers on cruise ships, and students in college dormitories. People spread the virus on contaminated hands and fomites due to poor personal hygiene. Students often blame food services for the gastrointestinal distress, but noroviruses rarely travel in food, though they are often found in contaminated water.

The four roommates recovered as their bodies eliminated viruses from their digestive tracts. They also learned the value of handwashing, disinfecting bathrooms, and keeping the dorm room sanitized. For more about *Norovirus*, see p. 712.

Track *Norovirus* online by going to the Study Area at www.masteringmicrobiology.com.

Treatment of Wastewater

Sewage, or **wastewater,** is typically defined as any water that leaves homes or businesses after being used for washing or flushed from toilets. (Some municipalities also include industrial water and rainwater as wastewater.) Wastewater contains a variety of contaminants, including suspended solids, biodegradable and nonbiodegradable organic and inorganic compounds, toxic metals, and pathogens. The objective of wastewater treatment is to remove or reduce these contaminants to acceptable levels.

At one time, "raw" (unprocessed) sewage was simply dumped into the nearest river or ocean; the idea was that wastes would be diluted to a point at which they would be harmless. Burgeoning populations and the realization that waterways were becoming increasingly polluted led to greater use of effective wastewater treatment processes.

Because sewage is mostly water (less than 1% solids), most sewage treatment involves the removal of microorganisms. A key concept in the processing of wastewater is reducing **biochemical oxygen demand (BOD),** which is a measure of the amount of oxygen required by aerobic bacteria to fully metabolize organic wastes in water. This amount is proportional to the amount of waste in the water; the higher the concentration of degradable chemicals, the more oxygen is required to catabolize them and the higher the BOD. Effective wastewater treatment reduces the BOD to levels too low to support microbial growth, thus reducing the likelihood that pathogens will survive.

The following sections consider wastewater treatments of various types: the traditional sewage treatment used in municipal systems, treatments used in rural areas, a treatment used for agricultural wastes, and the use of artificial wetlands.

Municipal Wastewater Treatment Today, people living in larger U.S. towns and cities are usually connected to municipal sewer systems—pipes that collect wastewater and deliver it to sewage treatment plants for processing. Traditional sewage treatment consists of four phases **(Figure 26.9)**:

1. **Primary treatment.** Wastewater is pumped into settling tanks, where lightweight solids, grease, and floating particles are skimmed off and heavier materials settle onto the bottom as **sludge.** After alum is added as a flocculating agent, the sludge is removed, and the partially clarified water is further treated. Primary treatment removes 25–35% of the BOD in the water.
2. **Secondary treatment.** The biological activity in this phase reduces the BOD to 5–25% of the original. Most pathogenic microorganisms are also removed. The water is aerated to facilitate the growth of aerobic microbes that oxidize dissolved organic chemicals to CO_2 and H_2O. In an *activated sludge system,* aerated water is seeded with primary sludge containing a high concentration of metabolizing bacteria; flocculation also occurs during this step. Any remaining solid material settles and is added to the sludge from primary treatment. The combined sludge is pumped into anaerobic holding tanks. Some smaller communities accomplish secondary treatment using a *trickle filter system,* which is similar to the slow sand filters used in treating drinking water but less effective in removing BOD than activated sludge systems.

(a)

Figure 26.9 Traditional sewage treatment. **(a)** A municipal wastewater treatment facility. **(b)** The traditional sewage treatment process. Microbial digestion during secondary treatment removes most of the biochemical oxygen demand (BOD) before the water is chemically treated and released; dried sludge is recycled as landfill.

1 Primary Treatment: Sedimentation

Sewer line

Water effluent

Removal of 25–35% BOD

Primary sludge (particulates + flocs)

Primary sludge

2 Secondary Treatment

Activated sludge or trickle filter system promotes microbial degradation of organic material

Aerated sludge

Secondary sludge

Removal of 75–95% BOD

3 Chemical Treatment

Cl

Chlorination and/or filtration prior to release

4 Sludge Treatment

Anaerobic digestion of sludge

Methane burned off or trapped for fuel

Dried sludge for landfill or agriculture

(b)

3 **Chemical treatment.** Water from secondary treatment is disinfected, usually by chlorination, after which it is either released into rivers or the ocean or, in some states, used for spray irrigation of crops and highway vegetation. Some communities remove nitrates, phosphates, and any remaining BOD or microorganisms from the water by passing it over fine sand filters and/or activated charcoal filters. Nitrate is converted to ammonia and discharged into the air (removes roughly 50% of the nitrogen content), whereas phosphorus is precipitated using lime or alum (removes 70% of the phosphorus content). Such tertiary treatment is generally used in environmentally sensitive areas or in areas where the only outlet for the water is a closed-lake system.

4 **Sludge treatment.** Sludge is digested anaerobically in three steps: First, anaerobic microbes ferment organic materials to produce CO_2 and organic acids. Second, microbes metabolize these organic acids to H_2, more CO_2, and

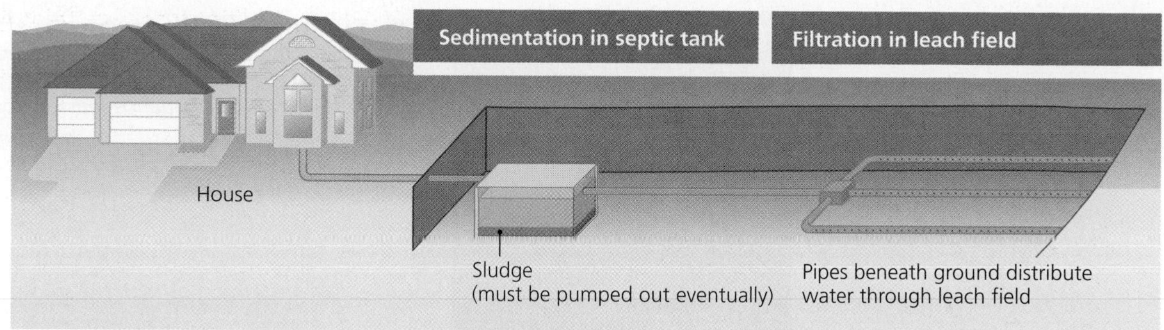

Figure 26.10 A home septic system. After wastewater from the house enters the septic tank, solids settle out as sludge, and the effluent liquid is filtered by the soil in the leach fields.

simpler organic acids such as acetic acid. Finally, the simpler organic acids, H_2, and CO_2 are converted to methane gas. Any leftover sludge is then dried for use as landfill or fertilizer.

Nonmunicipal Wastewater Treatment Houses in rural areas, which typically are not connected to sewer lines, often use **septic tanks,** essentially the home equivalent of primary treatment **(Figure 26.10)**. Sewage from the house enters a sealed concrete holding tank. Solids settle to the bottom and the liquid flows from the top of the tank into an underground *leach field* that acts as a filter. Sludge in the tank and organic chemicals in the water are digested by microorganisms. However, because the tank is sealed, it must occasionally be pumped out to remove sludge buildup. **Cesspools** are similar to septic tanks except that they are not sealed. As wastes enter a system of porous concrete rings buried underground, the water is released into the surrounding soil; solid wastes accumulate at the bottom and are digested by anaerobic microbes.

Treatment of Agricultural Wastes Farmers and ranchers often use **oxidation lagoons** to treat animal waste from livestock raised in feedlots—a penned area where animals are fattened for market. Oxidation lagoons accomplish the equivalent of primary and secondary sewage treatment. Wastes are pumped into deep lagoons and left to sit for up to three months; sludge settles to the bottom of the lagoon; and anaerobic microorganisms break down the sediment. The remaining liquid is pumped into shallow, secondary lagoons, where wave action aerates the water. Aerobic microorganisms, particularly algae, break down organic chemicals suspended in the water. Eventually, the microbes die, and the clarified water is released into rivers or streams. One problem with oxidation lagoons is that they are open, which can be dangerous if floodwaters inundate the lagoons and spread largely untreated animal wastes over a wide area.

Artificial Wetlands Since the 1970s, small planned communities and some factories have constructed **artificial wetlands** to treat wastewater. Wetlands use natural processes to break down wastes and to remove microorganisms and chemicals from water before its final release. Individual septic tanks are not needed; instead, wastewater flows into successive ponds where microbial digestion occurs **(Figure 26.11)**. The first pond in the series is aerated to allow aerobic digestion of wastes; anaerobic digestion occurs in the sludge at the bottom. The water then flows through marshland where soil microbes further digest organic chemicals. A second pond, which is still and contains algae, removes additional organic material, and the water then passes through open meadowland, where grasses and plants trap pollutants. By the time the water reaches a final pond, most of the BOD and microorganisms have been removed, and the water can be released for recreational purposes or irrigation. One drawback of an artificial wetland is that it requires considerable space—an artificial wetland to serve a small community can cover 50 acres of land or more.

Environmental Microbiology

Environmental microbiology is the study of microorganisms as they occur in their natural **habitats**—the physical localities in which organisms are found. Because of their vast metabolic capacities and adaptability, microbes flourish in every habitat on Earth, from Antarctic ice to boiling hot springs to bedrock.

In the following sections, we will explore the roles of microbes in the cycling of chemical elements in soil and aquatic habitats. First, however, we turn our attention to the relationships among microbes and between microbes and their habitats.

Microbial Ecology

Learning Objectives

- ✓ Define the terms used to describe microbial relationships within the environment.
- ✓ Explain the influences of competition, antagonism, and cooperation on microbial survival.

Microorganisms use a variety of energy sources and grow under a variety of conditions. They adapt to changing conditions, compete with other organisms for scarce resources, and change their habitat in many ways. In some cases, their effects on the environment are undesirable from a human perspective, but in most cases they are beneficial and essential. The study of the interrelationships among microorganisms and the environment is called **microbial ecology.** The first aspects of microbial ecology we will examine are the levels of microbial associations in the environment.

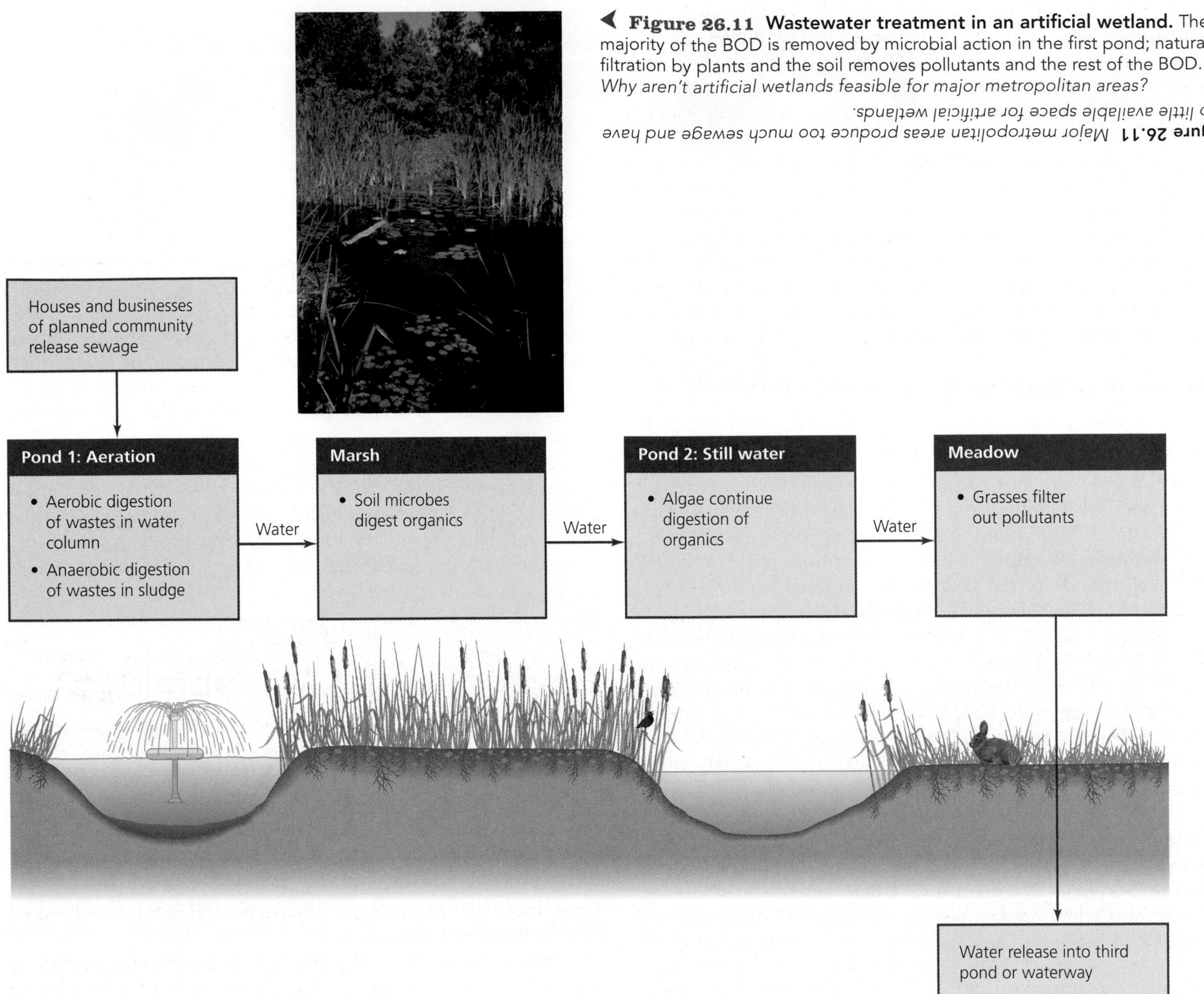

◀ **Figure 26.11 Wastewater treatment in an artificial wetland.** The majority of the BOD is removed by microbial action in the first pond; natural filtration by plants and the soil removes pollutants and the rest of the BOD. *Why aren't artificial wetlands feasible for major metropolitan areas?*

Figure 26.11 *Major metropolitan areas produce too much sewage and have too little available space for artificial wetlands.*

Levels of Microbial Associations in the Environment

A variety of terms are used to describe levels of microbial associations in the environment:

- An individual organism produces a *population* (all the members of a single species) through reproduction.
- Populations of microorganisms performing metabolically related processes make up groups called *guilds*. For example, anaerobic fermenters in a habitat make up a guild.
- Sets of guilds constitute *communities* that are typically quite heterogeneous; a variety of relationships exist among the various populations and guilds. Only rarely—and usually only in extreme environments—does a community consist of a single population.
- Populations and guilds within a community typically reside in their own distinct *microhabitats*—specific small spaces where conditions are optimal for survival.
- Groups of microhabitats form habitats in which microorganisms interact with larger organisms and the environment. Together, the organisms, the environment, and the relationships between the two constitute an **ecosystem.**
- All of the ecosystems taken together constitute the *biosphere,* that region of Earth inhabited by living organisms.

To illustrate these concepts, consider a plot of garden soil **(Figure 26.12)**. Soil is composed of tiny particles of rock and organic material, each of which forms a microhabitat for populations of microorganisms. Populations are distributed among the soil particles according to the available light, moisture, and nutrients. Populations of photosynthetic microbes, for example, live near the top and form a photosynthetic guild, which provides nutrients to other guilds living deeper within the soil. A soil community is composed of all of these populations and guilds, whose activities are a major factor affecting soil quality. Garden soil is just one type of soil occupying the soil ecosystem, which in turn is just one ecosystem within the biosphere.

Ecologists use the term **biodiversity** to refer to the number of species living within a given ecosystem, whereas the term **biomass** refers to the mass of all organisms in an ecosystem. By either measure, microorganisms are the most abundant of all living things: the number of species of microbes exceeds the number of larger organisms, and the biomass of microorganisms living throughout the biosphere—including places that are uninhabitable by plants, animals, or humans—is enormous.

The Role of Adaptation in Microbial Survival

Lush ecosystems support great biodiversity, but the vast majority of microorganisms live in harsh environments, where nutrient levels are low enough to limit growth. The harsher the environment, the more specially adapted a microbe must be to survive. Some environments cycle between periods of excess and depletion, and the microbes that live in them must be capable of adapting to such constantly varying conditions. *Extremophiles*—microbes adapted to extremely harsh temperatures, pH levels, and salt concentrations—are so adapted to their habitats that they cannot survive anywhere else.

Biodiversity is held in check by *competition.* The best-adapted microorganisms have traits that provide them advantages in nutrient uptake, reproduction, response to environmental changes, or some other factor. *Antagonism,* in which a microbe makes some product that actively inhibits the growth of another, may also occur.

Although competition is a major factor in limiting biodiversity, rarely does a microorganism outcompete all other rivals; heterogeneity is the norm, not the exception. Even though cooperation (at least that characterized by a specific and close-knit relationship) is rare among organisms, microbes readily use the waste products of other microbes for their own metabolism. Moreover, one microbe's metabolic activities sometimes make the environment more favorable for other microbes. Biofilms are examples of microbial cooperation.

The relationships among microorganisms and their environment constantly change; each microbe adapts to subtle changes in its environment. Most habitats support successions of microbial populations within a community. By studying how these successions occur, we learn much about the microbes' effects on the environment.

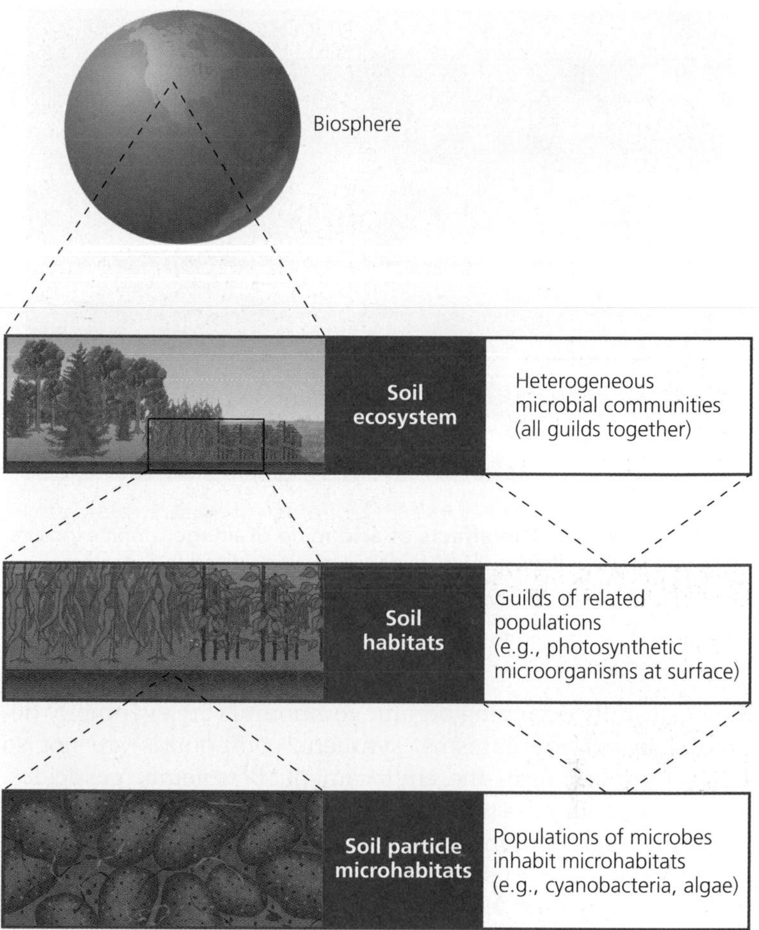

▲ **Figure 26.12 The basic relationships among microorganisms and between microorganisms and the environment.**

Bioremediation

Learning Objective

✓ Describe the process of bioremediation.

Each year Americans produce more than 150 million tons of solid wastes, accumulated from household, industrial, medical, and agricultural sources; most of it ends up in landfills. A landfill is essentially a large, open pit into which wastes are dumped, compacted, and buried. Soil microbes anaerobically break down biodegradable wastes; methanogens degrade organic molecules to methane. When a landfill is full, it is covered with soil and revegetated.

To prevent leaching of potentially hazardous materials into the soil and groundwater, a landfill pit is lined with clay or plastic, and sand and drainage pipes lining the bottom filter out small particulates and some microorganisms. Unfortunately, chemicals and hazardous compounds might still leak from landfills. Some of these substances are potentially carcinogenic (benzene, phenol, petroleum products); others are toxic (gasoline, lead); and still others are *recalcitrant* (resistant to decay, degradation, or reclamation by natural means).

The reason that one synthetic molecule is biodegradable while another is recalcitrant relates to chemical structure; sometimes a subtle variation—perhaps in a single atom or bond—can be enough to make a difference. Most recalcitrant molecules resist microbial degradation because the microbes lack enzymes capable of degrading them; after all, until very recently synthetic compounds didn't even exist.

Bioremediation is the use of organisms, particularly microorganisms, to clean up toxic, hazardous, or recalcitrant compounds by degrading them to harmless compounds. Although

▲ **Figure 26.13 The effects of acid mine drainage.** Upon exposure to air, iron in water leached from mine tailings is oxidized to Fe^{3+}; the activity of iron-sulfur bacteria reduces the pH of the water to a level that is destructive to plants and animals.

most naturally occurring organic compounds are eventually degraded by microorganisms, synthetic compounds are not so easily removed from the environment. Petroleum, pesticides, herbicides, and industrial chemicals that accumulate in soil and water are degraded only slowly by naturally existing microbes.

Bioremediation uses either natural or artificially enhanced microbes. In *natural bioremediation,* environmentally extant microbes are "encouraged" to degrade toxic substances in soil or water. The addition of nutrients (carbon, nitrogen, or phosphorus) stimulates growth. In *artificial bioremediation,* microbes are genetically modified by recombinant DNA technology to specifically degrade certain pollutants. Recombinant microbes also benefit from nutrient enrichment.

The most widely known application of bioremediation is the use of bacteria to clean up oil spills. Species of *Pseudomonas* have proven particularly useful in the degradation of crude oil, and these microbes played a role in the cleanup of the 1989 *Exxon Valdez* spill in Alaska (see **Beneficial Microbes: Oil-Eating Bacteria to the Rescue** on p. 768).

The Problem of Acid Mine Drainage

Learning Objective

✓ Discuss the problem of acid mine drainage.

Acid mine drainage is a serious environmental problem resulting from the exposure of certain metal ores to oxygen and microbial action **(Figure 26.13)**. Coal deposits are often found associated with reduced metal compounds such as pyrite (FeS_2). Strip-mining for coal exposes pyrite to oxygen in the air, which oxidizes the iron, and bacteria such as *Thiobacillus* (thī-ō-bă-sil′ŭs) oxidize the sulfur. Rainwater then leaches the oxidized compounds from the soil to form sulfuric acid (H_2SO_4) and iron hydroxide [$Fe(OH)_3$], which are carried into streams and rivers, reducing the pH enough (pH 2.5–4.5) to kill fish, plants, and other organisms. Such acidic water is also unfit for human consumption or recreational use. The EPA requires that strip mines be reburied as soon as possible to halt the processes.

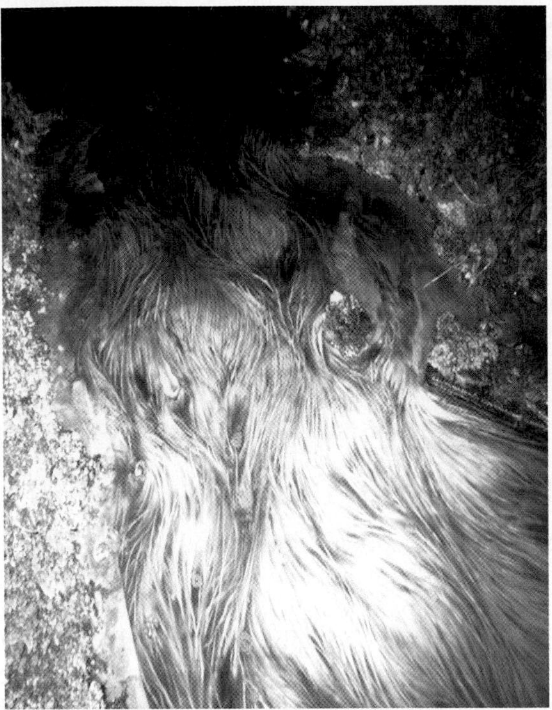

▲ **Figure 26.14 An acid-loving microbe.** The filamentous archaeon *Ferroplasma acidarmanus* growing as long filaments in acid mine runoff (pH 0) in California. The ore from this mine is rich in iron.

Underground mining operations pose similar (but somewhat less severe) problems resulting from runoff from mine tailings, which are the low-grade ores remaining after the extraction of higher-grade ores. Typically, subsurface mines are backfilled as richer veins of minerals are depleted, thus limiting the exposure of iron sulfides to oxygen.

While acid mine drainage is generally devastating to the environment, some microbes—mostly archaea—actually flourish in acidic conditions. One unique archaeal species, found in mine drainage in California, is *Ferroplasma acidarmanus* (fe′rō-plaz-ma a-sid′ar-mă-nŭs) **(Figure 26.14)**, which lives in a pH near zero and obtains its energy from oxidizing pyrite in mine sediments. Such an organism is just one example of the incredible diversity of microorganisms that colonize every habitat on Earth.

The Roles of Microorganisms in Biogeochemical Cycles

Learning Objective

✓ Compare and contrast the processes by which microorganisms cycle carbon, nitrogen, sulfur, phosphorus, and trace metals.

Most chemical elements are tied up in forms unavailable to organisms. The release of elements from rock, for example, requires thousands of years of degradation by rain, wind, and mi-

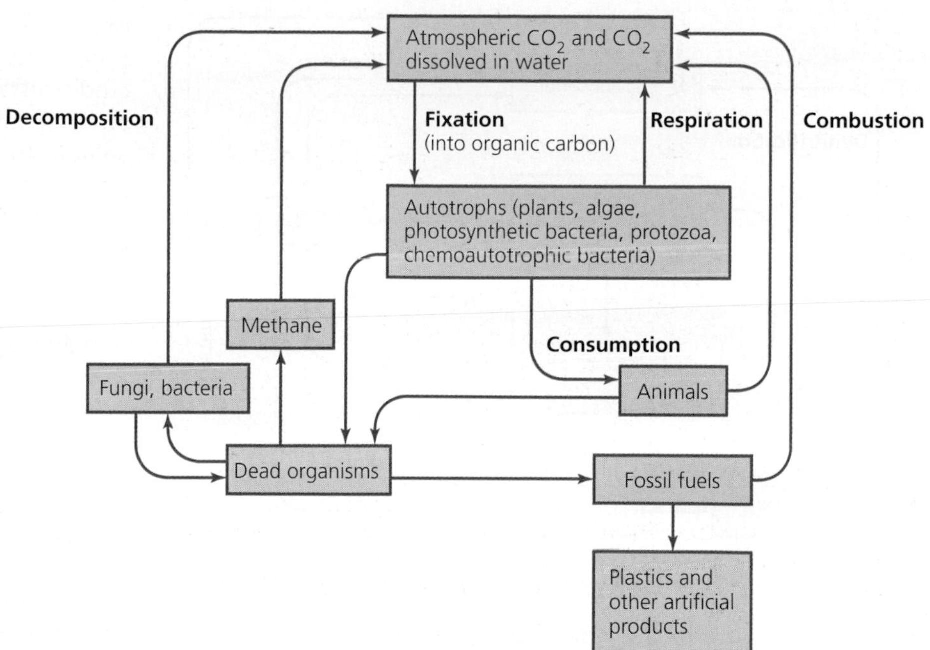

▶ **Figure 26.15 Simplified carbon cycle.** The most mobile form of carbon in the cycle is CO_2; this inorganic molecule is fixed by autotrophs, which incorporate its carbon into organic molecules.

crobes, so these processes contribute little to the day-to-day availability of nutrients for living things. As a consequence, the actions of organisms in recycling elements are the major components of **biogeochemical cycles**—the processes by which organisms convert elements from one form to another, typically between oxidized and reduced forms.

Biogeochemical cycling essentially entails three processes: (1) *production,* in which organisms called producers convert inorganic compounds into the organic compounds of biomass; (2) *consumption,* in which organisms called consumers feed on producers and other consumers, converting organic molecules to other organic molecules, and (3) *decomposition,* in which organisms called decomposers convert organic molecules in dead organisms back into inorganic compounds. As with all cycles, balance is critical. Elements must be cycled continuously among organisms. When one part of a cycle becomes skewed relative to other parts, the cycle becomes inefficient, often with detrimental effects.

The following sections examine the biogeochemical cycles for carbon, nitrogen, sulfur, phosphorus, and some trace metals.

The Carbon Cycle

Carbon is the fundamental element of all organic chemicals. The continual cycling of carbon in the form of organic molecules constitutes the majority of the **carbon cycle (Figure 26.15)**. Carbon in rocks and sediments has a very low *turnover rate* (rate of conversion to other forms)—this form of carbon is incorporated into organic chemicals only slowly over long periods of time.

The start of the carbon cycle is autotrophy. Photoautotrophic *primary producers*—cyanobacteria, green and purple sulfur bacteria, green and purple nonsulfur bacteria, algae, photosynthetic protozoa, and plants—convert CO_2 to organic molecules via carbon fixation (see Figure 5.28). Photoautotrophs are restricted to the surfaces of soil and water systems because phototrophs require light. Chemoautotrophs can also fix carbon but acquire energy from H_2S or other inorganic molecules; therefore, chemoautotrophs are found in a greater variety of habitats. However, they do not fix as much carbon as photoautotrophs and are not as important in the carbon cycle.

Heterotrophs catabolize some organic molecules for energy, resulting in the release of CO_2. Other organic molecules made by autotrophs are subsequently incorporated into the tissues of heterotrophs. There they remain until the organism dies, and decomposers catabolize the organic materials, releasing CO_2—the reverse of autotrophy. Waste products are also broken down by decomposers.

The release of CO_2 starts the cycle over again as primary producers fix CO_2 once more into organic material. A rough balance exists between CO_2 fixation and CO_2 release.

Many scientists are concerned by a growing imbalance in the carbon cycle due to the overabundance of CO_2 in the atmosphere. The burning of fossil fuels and wood sends tons of CO_2 into the atmosphere each year. Furthermore, in waterlogged soils, sewage treatment plants, landfills, and the digestive systems of ruminants, methanogens actively release methane gas (CH_4), which can be photochemically transformed into CO (carbon monoxide) and CO_2. Methane-oxidizing bacteria, living in conjunction with methanogens, can also directly convert methane to CO_2. Worldwide, the rate of CO_2 production exceeds the rate at which it is being incorporated into organic material.

Carbon dioxide is called a "greenhouse gas" because its presence in the atmosphere prevents the escape of some infrared radiation into space, redirecting heat back to Earth, much as the glass panes of a greenhouse trap heat. Such global warming causes climate change.

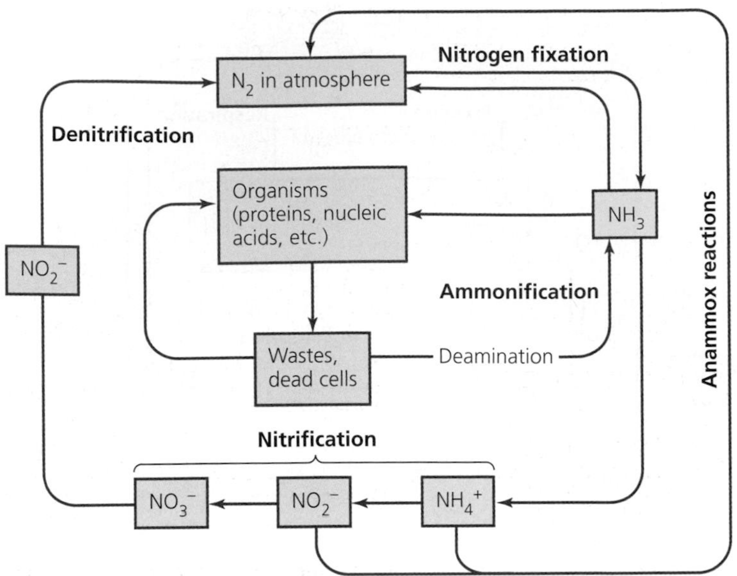

▲ **Figure 26.16 Simplified nitrogen cycle.** Even though nitrogen gas (N_2) is the most common form of nitrogen in the environment, most organisms cannot use it; to become available, nitrogen from the atmosphere must be incorporated into organic compounds by the very few species of nitrogen-fixing organisms.

The Nitrogen Cycle

Nitrogen is an important nutritional element required by organisms as a component of proteins, nucleic acids, and other compounds. Most nitrogen in the environment is in the atmosphere as dinitrogen gas (N_2), which is unusable by most organisms. The majority of organisms acquire nitrogen as part of organic molecules or from soluble inorganic nitrogen compounds found in limited quantities in soil and water. These include nitrate, nitrite, and ammonia. Microbes cycle nitrogen atoms from dead organic materials and animal wastes to these soluble forms of nitrogen. They also cycle nitrogen between the biosphere and the atmosphere. **Figure 26.16** summarizes the basic processes involved in the **nitrogen cycle**—nitrogen fixation, ammonification, nitrification, denitrification, and anammox reactions.

Nitrogen fixation, a process whereby gaseous nitrogen (N_2) is reduced to ammonia (NH_3), is an energy-expensive process in which the extremely stable nitrogen-nitrogen triple bond is broken in a reaction catalyzed by the enzyme *nitrogenase.* A limited number of prokaryotes, but no eukaryotes, fix nitrogen.

Nitrogenase functions only in the complete absence of oxygen, a condition that presents no problem for anaerobes. Aerobic nitrogen fixers, on the other hand, must protect nitrogenase from oxygen. They can do this in a number of ways. For example, some aerobes use oxygen at such a high rate that it does not diffuse into the interior of the cell where nitrogenase is sequestered. Some cyanobacteria form thick-walled, nonphotosynthetic cells called *heterocysts* to protect nitrogenase from oxygen in the environment as well as from the oxygen generated during photo synthesis (see Figure 11.13a). Other cyanobacteria fix nitrogen only at night, when oxygen-producing photosynthesis does not occur, thereby separating nitrogen fixation from photosynthesis in time rather than in space.

Nitrogen fixers may be free-living or symbiotic. Among the free-living nitrogen fixers are aerobic species of *Azotobacter* (ā-zō-tō-bak′ter) and anaerobic species of *Bacillus* and *Clostridium.* When they die, these soil-dwelling microbes release fixed nitrogen into the soil, where it becomes available to

BENEFICIAL MICROBES

OIL-EATING BACTERIA TO THE RESCUE

▲ ***Cleaning up after the* Exxon Valdez *oil spill.***

In March 1989, the *Exxon Valdez* ran aground on Alaska's Bligh Reef, spilling 11 million gallons of crude oil into the waters of Prince William Sound. It was the worst oil spill in U.S. history, affecting hundreds of miles of shoreline and killing untold numbers of birds, sea otters, and other wildlife. The subsequent cleanup involved the efforts of over 10,000 workers from Exxon, the Environmental Protection Agency, and the state of Alaska—and trillions of oil-eating bacteria.

Consisting largely of species native to the contaminated areas (including *Pseudomonas*), these bacteria metabolized the oil, converting it into harmless end products such as carbon dioxide and water. To boost bacterial activity, scientists added nitrogen, phosphorus, and various trace elements to polluted shoreline. The addition of nutrients sped up the rate at which the bacteria degraded the oil and cleaned up the shore.

The long-term effects of the oil spill on Prince William Sound continue to be debated, though the sound's ecosystem has proved to be surprisingly resilient. The area now appears much as it did before the spill, thanks in part to the bioremediation efforts of oil-eating bacteria.

plants and to animals that eat the plants. None of the free-living genera are directly associated with the plants they fertilize.

Symbiotic nitrogen fixers, in contrast, live in direct association with plants, forming *root nodules* on legumes (for example, peas and peanuts) (see Figure 11.19). The predominant genus of nitrogen-fixing, symbiotic bacteria is *Rhizobium* (rī-zō'bē-ŭm). In the symbiosis, the plants provide nutrients and an anaerobic environment in the nodules, while the bacteria provide usable nitrogen to the plant. Farmers do not have to apply nitrogen fertilizers to legume crops because nitrogen fixers provide enough.

Bacteria and fungi in the soil decompose wastes and dead organisms, disassembling proteins into their constituent amino acids, which then undergo *deamination* (removal of their amino groups). Amino groups are converted to ammonia (NH_3)—a process called **ammonification.** In dry or alkaline soils, NH_3 escapes as a gas into the atmosphere, but in moist soils, NH_3 is converted to ammonium ion (NH_4^+), which organisms absorb, or ammonium is oxidized.

In **nitrification,** ammonium is oxidized to nitrate (NO_3^-) via a two-step process requiring autotrophic archaea and bacteria. In the most well-studied nitrification pathway, species of the bacteria *Nitrosomonas* (nī-trō-sō-mō'nas) convert NH_4^+ to nitrite (NO_2^-), which is toxic to plants. Fortunately, *Nitrosomonas* spp. are usually found in association with species of the bacterium *Nitrobacter* (nī-trō-bak'ter) that rapidly convert nitrite to nitrate (NO_3^-), which is soluble and can be used by plants. The soluble nature of nitrate, however, means that it is leached from the soil by water and accumulates in groundwater, lakes, and rivers. Certain microorganisms in waterlogged soils perform **denitrification,** in which NO_3^- is oxidized to N_2 by anaerobic respiration. N_2 gas escapes into the atmosphere.

Scientists have recently discovered an important new aspect of the nitrogen cycle—*anaerobic ammonium oxidation,* or **anammox.** Anammox prokaryotes oxidize 30–50% of the world's ammonium into nitrogen gas using nitrite as an electron acceptor.

The Sulfur Cycle

The **sulfur cycle** involves moving sulfur between several oxidation states (**Figure 26.17**). Bacteria decompose dead organisms, which releases sulfur-containing amino acids into the environment. Sulfur released from amino acids is converted to its most reduced form, hydrogen sulfide (H_2S), by microorganisms via a process called sulfur *dissimilation.* H_2S is oxidized to elemental sulfur (S^0) and then to sulfate (SO_4^{2-}) under various conditions and by various organisms, including nonphotosynthetic autotrophs such as *Thiobacillus* and *Beggiatoa* (bej'jē-a-tō'a) and photoautotrophic green and purple sulfur bacteria. Sulfate is the most readily usable form of sulfur for plants and algae, which animals then eat. Anaerobic respiration by the bacterium *Desulfovibrio* (dē'sul-fō-vib'rē-ō) reduces SO_4^{2-} back to H_2S. Thus, the two major inorganic constituents of the sulfur cycle are H_2S and SO_4^{2-}.

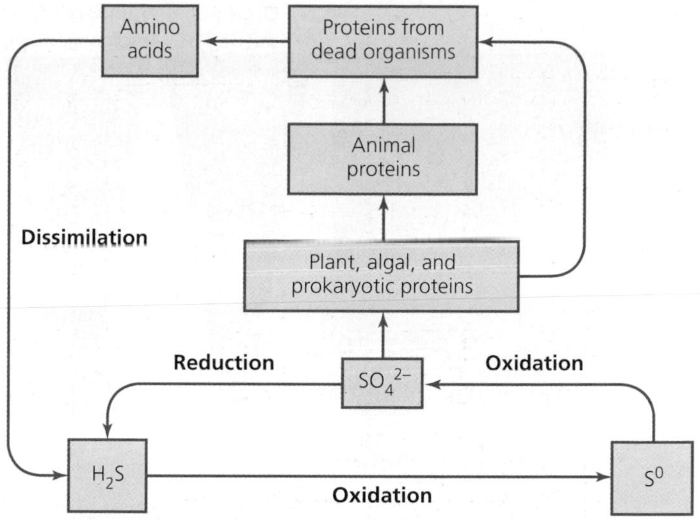

▲ **Figure 26.17 Simplified sulfur cycle.** The two main constituents of this cycle are hydrogen sulfide (H_2S) and sulfate (SO_4^{2-}), the fully reduced and oxidized forms of sulfur, respectively.

The Phosphorus Cycle

Unlike nitrogen and sulfur, phosphorus undergoes little change in oxidation state in the environment; it usually exists in the environment and is utilized by organisms as phosphate ion (PO_4^{3-}). The **phosphorus cycle** involves the movement of phosphorus from insoluble to soluble forms available for uptake by organisms and the conversion of phosphorus from organic to inorganic forms by pH-dependent processes. No gaseous form exists to be lost to the atmosphere, but dissolved phosphates do accumulate in water, particularly the oceans, and organic forms of phosphorus are deposited in surface soils following the decomposition of dead animals and plants.

Too much phosphorus can be a problem in a habitat; for example, agricultural fertilizers rich in phosphate are easily leached from fields by rain. The resulting runoff into rivers and lakes can result in **eutrophication** (yū-trō'fi-kā'shŭn)—the overgrowth of microorganisms (particularly algae and cyanobacteria) in nutrient-rich waters. Such overgrowth, called a *bloom,* depletes oxygen from the water, killing aerobic organisms such as fish. Anaerobic organisms then take over the water system, leading to an increased production of H_2S and the release of foul odors. When excess phosphate (and nitrogen) are removed, such a water system recovers over time.

The Cycling of Metals

Metal ions, including Fe^{2+}, Zn^{2+}, Cu^{2+}, Cd^{2+}, and Mg^{2+}, are important microbial nutrients. Though they are needed only in trace amounts, they can nonetheless be limiting factors in the growth of organisms. Many metals—including the most important trace metal, iron—are present in the environment in insoluble forms in rocks, soils, and sediments, and are generally unavailable for uptake by organisms. The cycling of metal ions primarily involves a transition from an insoluble to a soluble

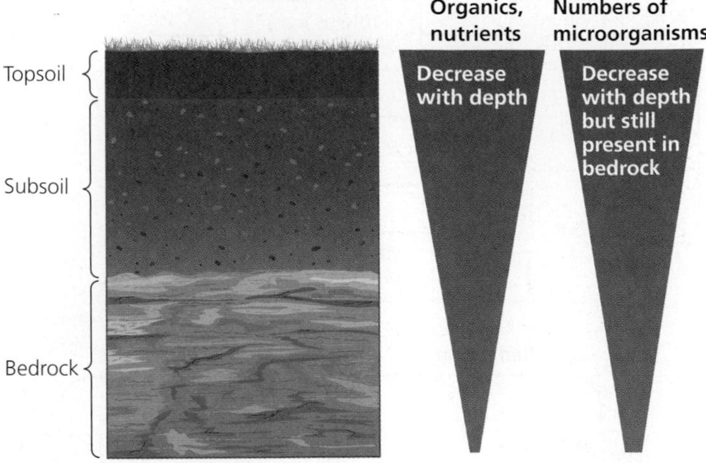

▲ **Figure 26.18 The soil layers, and the distributions of nutrients and microorganisms within them.** Although topsoils in general are richer in nutrients and microbes than are subsoils, the nutrient and microbial content of topsoils is highly variable.

form, allowing them to be used by organisms and to move through the environment.

Biogeochemical cycles are sustained by microorganisms, most of which live in soil. Next we examine aspects of soil microbes and their habitats.

Soil Microbiology

Learning Objectives

- ✓ Identify five factors affecting microbial abundance in soils.
- ✓ Describe several human and plant diseases caused by soil microbes.

Soil microbiology examines the roles played by organisms living in soil. They rarely cause human disease, though plant pathogens are prevalent in soils and are agriculturally and economically important.

The Nature of Soils

Soil arises both from the weathering of rocks and through the actions of microorganisms, which produce wastes and organic materials needed to support more complex life forms such as plants. Soil is composed of two major layers (**Figure 26.18**): *topsoil,* which is rich in *humus* (organic chemicals), and *subsoil,* which is composed primarily of inorganic materials. Soil overlies bedrock, which is solid rock and contains little organic material. Most microorganisms are found in topsoil, where the richness of the organic deposits sustain a large biomass. Topsoil itself, however, is highly heterogeneous, and therefore different kinds and amounts of microbes are found in various soils around the world.

Factors Affecting Microbial Abundance in Soils

Several environmental factors influence the density and the composition of the microbial population within a soil, including the amount of water, oxygen content, acidity, temperature, and the availability of nutrients.

Moisture and oxygen content are closely linked in soils. Moisture is essential for microbial survival; microbes exhibit lower metabolic activity, are present in lower numbers, and are less diverse in dry soils than in moist soils. Because oxygen dissolves poorly in water, moist soils have a lower oxygen content than drier soils. When soil is waterlogged, microbial diversity declines and anaerobes predominate, even at the surface. Weather patterns also affect oxygen content, as the presence or absence of rain water determines moisture, and thus dissolved oxygen.

The pH of a soil determines in part whether it is rich in bacteria or rich in fungi. Highly acidic and highly basic soils favor fungi over bacteria, though fungi typically prefer acidic conditions. Bacteria dominate when soil pH is closer to 7.

Most soil organisms are mesophiles and prefer temperatures between 20°C and 50°C. Thus, most soil microbes live quite well in areas where winters and summers are not too extreme. Psychrophiles grow only in consistently cold environments and cannot survive in soils that experience spring thawing, the opposite is true for thermophiles, which cannot survive where winters are harsh.

Nutrient availability also affects microbial diversity in soil habitats. Most soil microbes are heterotrophic, utilizing organic matter in the soil. The size of a microbial community is determined more by the amount of organic material than by the kind of organic material: any soil that has a relatively constant input of organic material, such as agricultural land, supports a wider array of microorganisms than soil that is more barren.

Microbial Populations in Soils

Because of the variety of soils, microbial populations differ tremendously from soil to soil, and even within the same soil over the course of a season. Bacteria are numerous and diverse inhabitants of soil and are found in all soil layers, where they often form biofilms. Archaea are present in soils, but the inability to culture many of them has limited our ability to study them. The fungi are also a populous group of soil microorganisms. Free-living and symbiotic fungi are found only in topsoil, where they can form gigantic mycelia that cover acres. Viruses are active within soil microorganisms; they are rarely found free.

Some algae and protozoa also live in soils. Soil algae live on or near the surface because as photoautotrophs they require light. Protozoa are mobile and move through the soil, grazing on other microbes. For the most part, protozoa require oxygen and remain in the topsoil. Neither algae nor protozoa can withstand dramatic environmental changes or the introduction of pollutants.

Wherever present, microbes perform a variety of necessary functions. They cycle nitrogen, sulfur, phosphorus, and other elements, converting them into usable forms. Microbes degrade dead organisms and their wastes, and some can clean up industrial pollutants. Further, microbes produce an incredible variety of compounds that have potential human uses. **Biomining** involves identifying and isolating natural populations of microbes that produce valuable chemicals or have useful biological functions.

CRITICAL **THINKING**

Microbes are found mostly in topsoil, but some are found miles deep in bedrock. Nutritionally, how do deeply buried microbes survive?

Soilborne Diseases of Humans and Plants

Although the majority of soil microorganisms are harmless, there are exceptions. Soilborne infections of humans generally result from either direct contact with, ingestion of, or inhalation of microorganisms deposited in soil in animal or human feces or urine. In some cases the microbes live and replicate in the soil, but in most cases soil is simply a vehicle for moving the pathogen from one host to another. The majority of soilborne disease agents are fungal or bacterial. Few soil protozoa or viruses cause disease.

Soil pathogens include the bacterium *Bacillus anthracis* (an-thrā′sis), the causative agent of anthrax, which produces endospores shed from the skins of infected livestock. Endospores may remain dormant in soil for decades or centuries. Disturbing the soil can lead to infection if endospores enter cuts or abrasions on the skin (cutaneous anthrax) or are inhaled into the lungs (inhalation anthrax).

Histoplasma capsulatum (his-tō-plaz′mă kap-soo-lā′tŭm) is a fungus that causes histoplasmosis—a serious respiratory tract infection. *Histoplasma* grows in soil and is also deposited there as spores in the droppings of infected birds and bats. The spores can be inhaled by humans when contaminated soil is disturbed.

Hantavirus (han′tă-vī-rŭs) pulmonary syndrome is a life-threatening, viral respiratory disease acquired via the inhalation of soil contaminated by mouse droppings and urine containing *Hantavirus* (see **Emerging Diseases** on p. 429). *Hantavirus* has been found throughout North America.

Soil contains many more plant pathogens than human pathogens. Microbial plant infections are generally characterized by one or more of the following signs: necrosis (rot), cankers/lesions, wilt (droopiness), blight (loss of foliage), galls (tumors), growth aberrations (too much or too little), or bleaching (loss of chlorophyll). Bacteria, fungi, and viruses all cause diseases in plants and spread either as airborne spores, through roots or wounds, or by insects.

Table 26.7 lists selected bacterial, fungal, and viral soilborne diseases of humans and plants.

TABLE 26.7 Selected Soilborne Diseases of Humans and Plants

Microorganism	Host	Disease
Bacteria		
Bacillus anthracis	Humans	Anthrax
Clostridium tetani	Humans	Tetanus
Agrobacterium tumefaciens	Plants	Crown gall disease
Ralstonia solanacearum	Plants	Potato wilt
Streptomyces scabies	Plants	Potato scab
Fungi		
Histoplasma capsulatum	Humans	Histoplasmosis
Blastomyces dermatitidis	Humans	Blastomycosis
Coccidioides immitis	Humans	Coccidioidomycosis
Polymyxa spp.	Plants	Root rot in cereals
Fusarium oxysporum	Plants	Root rot in many plants
Phytophthora cinnamomi	Plants	Potato blight; root rot in many plants
Viruses		
Hantavirus	Humans	*Hantavirus* pulmonary syndrome
Tobacco mosaic virus	Plants	Necrotic spots in various plants
Soilborne wheat mosaic virus	Plants	Mosaic disease in winter wheat and barley

Aquatic Microbiology

Learning Objective

✓ Compare the characteristics and microbial populations of freshwater and marine ecosystems.

Aquatic microbiologists study microorganisms living in freshwater and marine environments. Compared to soil habitats, water ecosystems support fewer microbes overall because nutrients are diluted. Many organisms that live in aquatic systems exist in biofilms attached to surfaces. Biofilms allow aquatic organisms to concentrate enough nutrients to sustain growth; without forming biofilms, they likely would starve.

Types of Aquatic Habitats

Aquatic habitats are divided primarily into freshwater and marine systems. *Freshwater* systems, which are characterized by low salt content (about 0.05%), include ground water, water from deep wells and springs, and surface water in the form of lakes, streams, rivers, shallow wells, and springs. *Marine* environments, characterized by a salt content of about 3.5%, encompass the open ocean and coastal waters such as bays, estuaries, and lagoons.

Natural aquatic systems can be greatly affected by the release of so-called *domestic water,* which is water resulting from the treatment of sewage and industrial waste. Domestic water released into the environment affects water chemistry and the microorganisms living in the water. Changing levels of chemicals cause increases or decreases in microbial numbers. Furthermore, faulty treatment of sewage leads to contamination of natural water systems with pathogenic microorganisms.

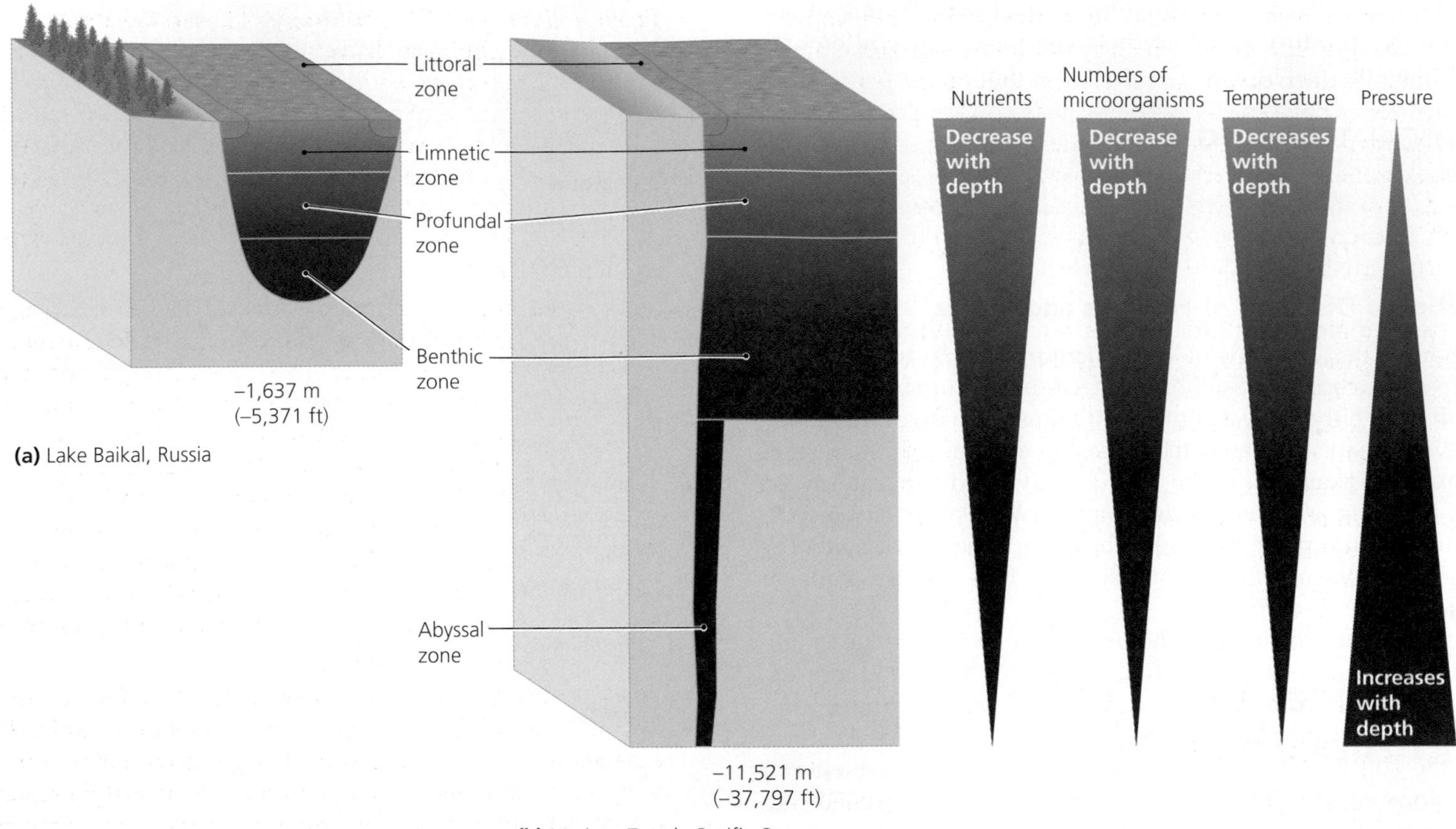

▲ **Figure 26.19 Vertical zonation in deep bodies of water: (a)** Lake Baikal, Russia, the world's deepest freshwater lake, and **(b)** the Mariana Trench in the Pacific Ocean, the deepest place in the ocean. Both deep lakes and oceans can be divided into zones that vary with respect to light penetration, concentration of nutrients, temperature, and pressure—and thus the types and abundance of microorganisms. *Why would a bacterium from the bottom of Lake Baikal die in the Mariana Trench?*

Figure 25.19 *A bacterium from the bottom of Lake Baikal could not survive the even greater pressure and the salinity of the seawater deep in the Mariana Trench.*

Freshwater Ecosystems Microorganisms become distributed vertically within lake systems according to oxygen availability, light intensity, and temperature. Surface waters are high in oxygen, well lit, and warmer than deeper waters. In large lakes, wave action continually mixes nutrients, oxygen, and organisms, which allows efficient utilization of resources. In stagnant waters, oxygen is readily depleted, resulting in more anaerobic metabolism and poorer water quality.

Scientists observe four zones in deep lakes (**Figure 26.19a**): The **littoral** (li'ter-al) **zone** is the area along the shoreline, where nutrients enter the lake. The littoral zone is shallow, and light penetrates it; most microbes live here. The **limnetic** (lim-net'ik) **zone** is the upper layer of water away from the shore. Photoautotrophs reside here and in the littoral zone. The **profundal** (prō-fun'dal) **zone** is the deeper water beneath the limnetic zone. It has a lower oxygen content and more diffuse light than the previous two zones. Some photosynthetic organisms, such as purple and green sulfur bacteria, perform anaerobic photosynthesis here. Beneath this is the **benthic** (bĕn'thik) **zone,** which encompasses deeper lake water and sediments. Anaerobic bacteria in the sediments produce H_2S, which is used by organisms nearer the surface.

In contrast to lakes, streams and rivers usually lack stratification because organisms and nutrients are swept along and mixed. Biofilms are particularly important in moving waterways, and the majority of organisms live toward the edges, where currents are less severe and organic materials enter the water.

Marine Ecosystems Marine ecosystems are typically nutrient poor, dark, cold, and subject to great pressure. Photoautotrophic prokaryotes, diatoms, dinoflagellates, and algae are found near the surface. Most marine waters are extreme environments inhabited only by highly specialized microorganisms. All microbes in marine systems must be salt tolerant and possess highly efficient nutrient-uptake mechanisms to compensate for the scarcity of nutrients.

As with freshwater lake systems, scientists delineate zones in the oceans (**Figure 26.19b**). The majority of marine microorganisms are found in the littoral zone, where nutrient levels are high and light is available for photosynthesis. The benthic zone makes up the majority of the marine environment. Oceans have a fifth zone—the **abyssal** (a-bis'sal) **zone,** which encompasses deep ocean trenches. Even though the benthic and abyssal zones have sparse nutrients, they still support microbial growth, particularly around **hydrothermal vents** located in the abyssal zone. Such vents spew superheated, nutrient-rich water, providing nutrients and an energy source for thermophilic chemoautotrophic anaerobes, which in turn support a variety of invertebrate and vertebrate animals.

Specialized Novel Aquatic Ecosystems In addition to the two broad categories of water systems just described, many distinctive aquatic ecosystems also exist, including salt lakes, iron springs, and sulfur springs. Each of these systems is inhabited by specialized microorganisms that are highly adapted to the conditions. The Great Salt Lake in Utah, for example, has a salt concentration of 5–7%, depending on its level, and contains the extreme halophile *Halobacterium salinarium* (hā-lō-bak-tēr'ē-ŭm sa-lē-nar'ē-ŭm), an archaeal prokaryote that thrives in highly saline water.

Biological Warfare and Bioterrorism

The properties of microorganisms that allow scientists to manipulate organisms to make advancements in medicine, food production, and industry also enable humans to fashion microbes into *biological weapons,* which can be directed at people, livestock, or crops. **Bioterrorism,** the use of microbes or their toxins to terrorize human populations, is a topic of major concern in today's world. A topic of growing concern is **agroterrorism**—the use of microbes to terrorize humans by destroying the food supply. International treaties and laws of the United States, Great Britain, and other countries prohibit the use of biological weapons.

Assessing Microorganisms as Potential Agents of Warfare or Terror

Learning Objectives

- ✓ Identify the criteria used to assess microorganisms for potential use as biological weapons or agents of bioterrorism.
- ✓ List the characteristics of microbes that make them threats as agents of biological warfare and bioterrorism.

Many microorganisms cause disease, but not all disease-causing organisms have potential as biological weapons. Governments establish criteria for evaluating the potential of microorganisms to be "weaponized." Establishing such criteria helps focus research and defense efforts where they are needed most and facilitates efforts to develop better response and deterrence capabilities.

Criteria for Assessing Biological Threats to Humans

In the United States, the assessment of a potential biological threat to humans is based on the following four criteria:

- *Public health impact:* This criterion relates to the ability of hospitals and clinics to deal effectively with numerous casualties. The more casualties, the more difficult it is for hospitals and clinics to effectively respond to the needs of all patients. If an agent causes numerous serious cases, emergency response systems could be overwhelmed and even cease to function. If an agent is highly lethal, proper disposal of bodies might become difficult, which would contribute to the spread of disease.
- *Delivery potential:* This criterion evaluates how easily an agent can be introduced into a population. The more people that can be infected at one time, the more devastating a primary attack. If an introduced agent spreads on its own through a population, secondary and tertiary waves of illness will augment the number of casualties. Also assessed as part of delivery potential are ease of mass production (the easier an agent is to produce in quantity, the greater its potential threat), availability (the more prevalent it is in the environment, the easier it is to obtain and weaponize), and environmental stability (the longer it remains infective once released, the greater the threat).
- *Public perception:* This criterion evaluates the effect of public fear on the ability of response personnel to control a disease outbreak following an attack. Agents with high mortality, few treatment options, and no vaccine instill greater fear into a populace, making quarantine (isolating infected and sick patients from the rest of the population) difficult to enforce. The resulting chaos could dramatically decrease the ability of response personnel to control disease transmission and treat patients.
- *Public health preparedness:* This criterion assesses existing response measures and attempts to identify improvements needed in the health care infrastructure to prepare for a biological attack. Diagnosis and recognition involve proper surveillance and training medical personnel to ascertain whether an attack has occurred. Once an attack has been confirmed, predetermined responses are required to reduce confusion and allow rapid control of the situation. Public health preparedness also involves funding for research and development of new vaccines, treatments, and diagnostic capabilities.

When assessing a threat, each potential bioterrorist agent is given a score for each of the criteria. Agents with the highest total scores are considered the most serious threats.

Criteria for Assessing Biological Threats to Livestock and Poultry

The criteria used to evaluate biological threats to livestock and poultry are very similar to those used to evaluate potential threats to humans and include agricultural impact, delivery potential, and plausible deniability.

Infectious agents prove the most devastating for agricultural livestock kept in large herds or flocks. Many animal pathogens exist naturally in soil or are already endemic in livestock and poultry, so they can be easily obtained and readily

TABLE 26.8 Bioterrorist Threats to Humans in Order of Concern

Disease	Agent	Natural Source
Category A Threats: Highest Priority		
Smallpox	Variola major *(Orthopoxvirus)*	None
Anthrax	*Bacillus anthracis* (bacterium)	Soil
Plague	*Yersinia pestis* (bacterium)	Small rodents
Botulism	*Clostridium botulinum* toxin (bacterial)	Soil
Tularemia	*Francisella tularensis* (bacterium)	Wild animals
Viral hemorrhagic fevers	Filoviruses and arenaviruses	Unknown in most cases
Category B Threats: Moderate		
Q fever (fever + flulike syndrome)	*Coxiella burnetii* (bacterium)	Sheep, goats, cattle
Brucellosis (severe flulike syndrome)	*Brucella* spp. (bacteria)	Livestock
Glanders (pulmonary syndrome)	*Burkholderia mallei* (bacterium)	Horses
Melioidosis (severe pulmonary syndrome)	*Burkholderia pseudomallei* (bacterium)	Horses, livestock, rodents, soil
Viral encephalitis	Alphaviruses	Rodents, birds
Typhus fever	*Rickettsia prowazekii* (bacterium)	Humans
Toxins	Various bacteria	Various
Psittacosis (pneumonia-like syndrome)	*Chlamydophila psittaci* (bacterium)	Birds
Foodborne diseases	Various bacteria and viruses	Soil or animals
Waterborne diseases	Various bacteria and viruses	Water
Category C Threats: Low Risk		
Nipah virus (encephalitis)	Paramyxovirus	Pigs
Hantavirus pulmonary syndrome	*Hantavirus*	Rodents, soil

grown in large quantities. The highly infectious nature of some agents means that by the time a disease is recognized, much of a herd is infected already, so all must be destroyed in an attempt to control the outbreak. Some pathogens persist even after the animals are destroyed. Many animal diseases are spread either by contact or inhalation, thus making attack easier for a terrorist. Though highly contagious among animals, most are not infectious to humans, making them "safe" for terrorists to handle.

Criteria for Assessing Biological Threats to Agricultural Crops

Plant diseases are generally not as contagious as animal or human diseases. Threats to crops are evaluated on predicted extent of crop loss, delivery and dissemination potential, and containment potential.

Plant pathogens that either cause severe crop loss or produce toxins are considered the greatest threats. Such agents already exist in the environment and can be readily obtained; however, plant pathogens are not as easily mass-produced as animal agents. Plant pathogens that can be spread systematically through fields by natural means, such as through contaminated soil or by insects, could remain in the environment even after destruction of the target crop. Successive plantings in contaminated fields could result in continued crop loss for as long as the agent persists. Because the causes of many plant diseases are not easily diagnosed, widespread dissemination of such a pathogen could occur before response measures were instituted. Economic losses would be staggering, particularly given that embargos on affected crops would likely remain in place for years after an attack.

Known Microbial Threats

In the following sections we briefly discuss some of the microorganisms currently considered threats as agents of bioterrorism. Note that as threat assessments become more refined and technology advances, the list of known bioterrorist agents is likely to change.

Human Pathogens

The U.S. government categorizes biological agents that could be used against humans into three categories (Table 26.8). *Category A agents* are those with the greatest potentials as weapons. *Category B agents* have some potential as weapons, but for various reasons are not as dangerous as category A agents (most lack the potential to cause mass casualties). *Category C agents* are potential threats; not enough is currently known about them to determine their true potential as weapons.

Smallpox currently tops the list of bioterrorist threats. Fortunately, it is difficult for would-be terrorists to acquire viral samples for propagation; it also takes a high degree of skill, in addition to specific containment facilities, to work with smallpox virus. An effective vaccine is available, and the vaccine is effective when administered soon after infection.

Animal Pathogens

Biological agents against animals are also divided into categories, with category A agents being the most dangerous. Whereas many potential agents are spread via inhalation, others are spread by insect vectors, making them less likely to be used as weapons. Some agents infect wild animal populations in addition to livestock, potentially amplifying any outbreak that might occur.

Foot-and-mouth disease virus, the most dangerous of potential agroterrorism agents, affects all wild and domestic cloven-hoofed animals. The virus is spread by aerosols and by direct or indirect contact. Humans can transport it from herd to herd on their person, on farm equipment, or through the movement of animals between auctions and farms. Any appearance of foot-and-mouth disease on a farm requires destruction of entire herds, complete disinfection of all areas occupied by the herd, and disposal of all animals by burning or burial. A vaccine exists, but not in sufficient quantities to protect all animals.

Plant Pathogens

Many plant pathogens exist, but the categorization of plant pathogens as terrorist agents lags behind similar efforts for humans and animals. Most potential agents are fungi whose dissemination could easily result in contamination of soils, which would be difficult to neutralize. Attacks against grains, corn, rice, and potatoes are considered the most dangerous, as they would have significant negative impacts on national economies and food supplies. All of the agents are naturally present, so detecting the difference between a natural outbreak and an intentional attack would be difficult.

Defense Against Bioterrorism

No defense can completely prevent a carefully planned biological attack against any group or nation. However, much can be done to limit the impact of an attack. The key is coupling *surveillance*—the active diagnosis and tracking of human, animal, and plant diseases—with *effective response protocols*.

Because category A human biological agents are not common, the appearance of more than a few scattered cases is highly suggestive that an attack of some kind has occurred. This is one reason the diseases of category A are *reportable*, that is, they must be reported to state departments of health whenever they occur. Active monitoring of reportable diseases allows epidemiologists to quickly determine when unusual outbreaks are occurring. Once an unexpected pattern is seen, diagnostic confirmation can be obtained and, if an attack is deemed to have occurred, appropriate responses implemented **(Figure 26.20)**. Such responses may entail forced quarantine, distribution of antimicrobial drugs, or mass vaccination.

▲ **Figure 26.20 One aspect of the response to a bioterrorist attack.** Shown here are biohazard-suited personnel investigating at the time of anthrax-containing mail delivered in Washington, D.C., in October 2001.

Agroterrorism has become more of a concern with the realization that very little security protects the nation's agricultural enterprises. Livestock and poultry are routinely moved around the country without being tested for disease and without being quarantined prior to introduction into new herds or flocks. Infected animals could therefore spread disease as they pass from facility to facility. Compounding the problem is the fact that farms, ranches, auction houses, livestock shows, and irrigation facilities are all open to the public and impose few security measures to prevent purposeful infection of animals. It has been suggested that a step in defending against agroterrorism would be to restrict public access to such facilities. Additionally, effective screening of imported animals and plants would help prevent the introduction of foreign pathogens into the United States. Further, better diagnostic techniques, vaccines, and treatments need to be developed for animal pathogens.

The Roles of Recombinant Genetic Technology in Bioterrorism

Recombinant genetic technology could be used to create new biological threats or modify existing ones, so that, for example, vaccines against them no longer work. Traits of various agents could be combined to create novel agents for which no immunity exists in the population.

In addition to the manipulation of existing threat agents, the techniques of recombinant DNA technology enable the synthesis of agents from scratch. Terrorists could, in theory, make their own microbes. To learn about the production of a completely manufactured poliovirus, see **Highlight: Could Bioterrorists Manufacture Viruses from Scratch?** on p. 776. Poliovirus is a relatively simple virus. There is no guarantee that such a process would work for more complex agents such as smallpox.

The techniques of recombinant genetic technology may also be used to thwart bioterrorism. Scientists can identify unique

HIGHLIGHT

COULD BIOTERRORISTS MANUFACTURE VIRUSES FROM SCRATCH?

Researchers at the State University of New York at Stony Brook have managed an alarming achievement: They synthesized a fully functional poliovirus from materials that can be readily obtained from any of a number of molecular biology supply companies. After they pieced together sequences of RNA to form a full-length poliovirus genome, they successfully replicated and translated this material in cell-free extracts in test tubes. The resulting nucleic acids and proteins were then able to assemble spontaneously into fully infectious viral agents. The scientists began their work from genetic blueprints that exist in the public domain—in published journal articles and on Internet databases.

The ability to manufacture an infectious agent from scratch using preexisting, published knowledge is an unsettling development. Terrorists may be able to manufacture their own agents in similar fashion—rather than needing to steal agents from research facilities or isolate them from natural sources.

As a result of studies like that at Stony Brook, an ethical debate has arisen over whether such research should be pursued—and if so, whether the details of such research should be published. Some argue that the pursuit and publication of such research unwittingly aids would-be terrorists; others argue that the dissemination of information is necessary for the effective sharing of research within the scientific community and for science to progress. What do you think?

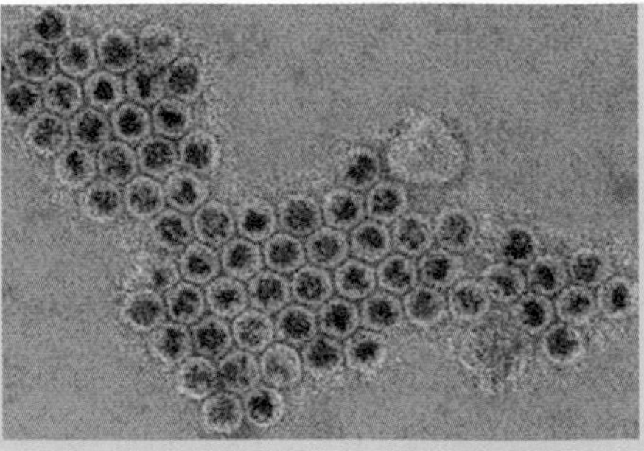

▲ *Polioviruses.* TEM

genetic sequences—"fingerprints" or signatures—of recombinants, which may aid in tracking biological agents and determining their source. Genetic techniques may also help in developing vaccines and treatments, and recombinant DNA technology could be used to create pathogen-resistant crops.

CRITICAL THINKING

Compare human, animal, and plant pathogens that could be used as biological agents in terms of environmental survivability. Why are animal and plant pathogens more common in environmental reservoirs than are human pathogens?

Chapter Summary

Food Microbiology (pp. 746–753)

1. The commercial use of microorganisms is referred to as **applied microbiology** and includes two distinct fields: food microbiology and industrial microbiology.
2. **Food microbiology** involves the use of microorganisms in food production and the prevention of foodborne illnesses. In this context, **fermentations** involve desirable changes to food; **spoilage** involves undesirable changes to food.
3. Food fermentations involve the use of **starter cultures**—known organisms that carry out specific and reproducible fermentation reactions. For most of the great variety of fermented vegetables, meats, and dairy products, starter cultures of lactic acid bacteria are used. The acid produced results in "sour" flavors.
4. Alcoholic fermentations, usually performed by yeasts, convert sugars to ethanol and carbon dioxide. Alcoholic fermentation is used in the production of wine, distilled spirits, beer, vinegar, and bread.
5. Intrinsic factors of food spoilage are properties of the food itself, such as moisture content and physical structure, that determine how susceptible a food is to spoilage. Extrinsic factors of food spoilage include ways in which the food is handled.
6. Industrial processes preserve food via canning, pasteurization, drying, freeze-drying **(lyophilization),** irradiation, and aseptic packaging techniques. Natural and artificial preservatives are added to some foods to inhibit microbial growth. In stores and at home, foods should be properly stored in appropriate containers, cold foods should be kept cold, foods should be cooked thoroughly, and leftovers should be refrigerated to reduce spoilage.
7. Food poisoning is a general term that describes instances of **food infections** (illnesses due to the consumption of living microbes) or **food intoxications** (illnesses due to the consumption of microbial toxins). Food poisoning frequently follows poor food handling.

Industrial Microbiology (pp. 753–763)

1. **Industrial microbiology** is concerned with the use of microorganisms for the production of commercially valuable materials. Such industrial fermentations synthesize desired products and can use genetically modified microbes.
2. Batch production is the growth of organisms followed by harvesting of the entire culture and its products. Continuous flow production involves the constant addition of nutrients to a culture and the removal of the products formed. Products may either be

primary metabolites (produced during active growth) or secondary metabolites (produced during the stationary phase).

3. Microorganisms produce a variety of useful products, including enzymes, dyes, alternative fuels, plastics, pharmaceuticals, pesticides, biosensors, and bioreporters. Alternative fuels (biofuels) can be produced from products of photosynthesis or by fermentation of biomass into fuel. **Biosensors** combine microbes and electronics to detect microbial activity in the environment. **Bioreporters** use microbes alone as sensors.
4. **Potable** drinking water is derived via water treatment, which involves the removal of microbes and of organic and inorganic contaminants from water. **Polluted** water contains organisms or chemicals at unsafe levels. Some diseases result from consumption of polluted water or of food harvested from polluted water.
5. Water treatment involves four steps: **sedimentation, flocculation, filtration,** and **disinfection.** Sedimentation removes large materials. In flocculation, alum combines with suspended materials to make them precipitate. Filtration, using either slow or rapid sand filters, removes microorganisms and chemicals. Disinfection, usually chlorination, kills most microbes that remain after filtration. Water is potable following treatment if it has zero coliforms per 100 ml of water, as determined by one of several testing methods (MPN, membrane filtration, ONPG/MUG test).
6. **Wastewater** (sewage) refers to water used for washing or flushed from toilets. Wastewater treatment involves the removal of solids, organic chemicals, and microorganisms. **BOD (biochemical oxygen demand)** is a measure of the amount of oxygen required to fully metabolize organic wastes.
7. Municipal wastewater treatment involves four phases. Primary treatment entails sedimentation of large materials (primary **sludge**) and flocculation. Secondary treatment involves sedimentation of secondary sludge as well as the removal of microorganisms and organic material using activated sludge systems or trickle filters. In the third phase, effluent water is chemically treated (chlorinated) and released. In the fourth phase, primary and secondary sludge is digested and dried to produce landfill. Methane gas can also be recovered during the processing of sludge.
8. **Septic tanks** and **cesspools** are home equivalents of municipal wastewater treatment. After wastewater leaves the home, it is deposited in underground tanks. Sludge settles, and the water is released into the soil, where natural processes remove organic chemicals and microorganisms.
9. **Oxidation lagoons** are used by farmers and ranchers to process animal wastes. Waste is pumped into successive lagoons, where wastes are digested by microorganisms prior to the release of the water into natural water systems.
10. In **artificial wetlands**—found in some planned communities and industrial sites—ponds, marshes, and meadowland remove organic compounds, chemicals, and microorganisms from sewage as the water moves through them.

Environmental Microbiology (pp. 763–773)

1. Microorganisms live in microhabitats within larger **habitats,** or physical localities, in the environment. Organisms and habitats together form **ecosystems.** Single cells give rise to populations; populations performing similar functions form guilds; and many guilds together form a community of organisms living in a habitat. The study of the interactions of microorganisms among themselves and with their environment is **microbial ecology,** which is a part of **environmental microbiology** along with studies of microbial habitats.
2. All the ecosystems on Earth form the biosphere. **Biodiversity** describes the number of species living in a given ecosystem, whereas **biomass** refers to the quantity of all these species.
3. Microbes compete for the scarce resources that characterize the majority of habitats. Some microbes actively oppose the growth of other microbes (antagonism), but many microbes cooperate, forming complex biofilms.
4. **Bioremediation** is the use of microorganisms to metabolize toxins in the environment to reclaim soils and waterways. Industrial products are either biodegradable or recalcitrant (resistant to degradation by natural means). Recombinant DNA technology enables scientists to create microbes to degrade some recalcitrant chemicals.
5. Acid mine drainage is an environmental problem in places where ores contain iron. Microbial action on leached iron in water from mines results in the production of acid and ferric iron deposits that acidify water, which is destructive to most plant and animal life.
6. **Biogeochemical cycling** involves the movement of elements and nutrients from unusable forms to usable forms by the activities of microorganisms. These processes involve production of new biomass, consumption of existing biomass, and decomposition of dead biomass for reuse in the cycle. The four major biogeochemical cycles are the carbon, nitrogen, sulfur, and phosphorus cycles. The cycling of trace metals is also important.
7. In the **carbon cycle,** CO_2 is fixed by photoautotrophs and chemoautotrophs into organic molecules, which are used by other organisms. Organic carbon is converted back to CO_2 in aerobic respiration, by decomposition, and by combustion.
8. In the **nitrogen cycle,** nitrogen gas in the atmosphere is converted to ammonia via a process called **nitrogen fixation.** Ammonia may be converted to nitrate via a two-step process called **nitrification.** Organisms also use ammonia and nitrate to make nitrogenous compounds. Such compounds in wastes and dead cells are converted back to ammonia via **ammonification.** Nitrate can be converted to nitrogen gas by **denitrification. Anammox** prokaryotes oxidize ammonium anaerobically into nitrogen.
9. In the **sulfur cycle,** sulfur moves between several inorganic oxidation states (primarily H_2S, SO_4^{2-}, and S^0) and proteins.
10. The **phosphorus cycle** involves the conversion of PO_4^{2-} among organic and inorganic forms.
11. **Eutrophication**—the overgrowth of microorganisms in aquatic systems—can result from the presence of excess nitrogen and phosphorus, which act as fertilizers. The overgrowth of microbes depletes the oxygen in the water, resulting in the death of fish and other animals.
12. Soil microbiology is the study of the roles of microbes in the ground. Soils are fairly diverse and differ greatly in nutrients, water content, pH, oxygen content, and temperature. Microbes inhabit topsoil in high numbers and are less abundant in deeper rock and sediments. **Biomining** is the science involved in searching natural populations of soil microbes for useful organisms.

Pathogenic microorganisms found in soil can be acquired through contact, but often disease follows the consumption of contaminated soil.

13. Aquatic habitats include freshwater and marine water systems. Scientists recognize four zones in freshwater based on temperature, light, and nutrient levels: the nutrient-rich **littoral zone** along the shore, the sunlit **limnetic zone** at and near the surface, the **profundal zone** just below the limnetic zone, and the **benthic zone** on the bottom, which is devoid of light and nutrients. Marine environments have the same four zones plus an **abyssal zone** (below the benthic zone), which is virtually devoid of life except around **hydrothermal vents.**
14. Microorganisms living in aquatic environments typically form biofilms to better accumulate nutrients that are limiting in most nonpolluted water systems.

Biological Warfare and Bioterrorism (pp. 773–776)

1. Of the relatively few microorganisms that can cause disease in humans, animals, and plants, some might be used to purposely infect individuals and are thus potential agents of biological warfare and **bioterrorism.** It is illegal to deploy such weapons. **Agroterrorism** is the deliberate infection of livestock or crops.
2. The degree to which an organism is considered a biological threat depends on several criteria concerning public health impact, dissemination or delivery potential, public perception, and public health preparedness.
3. Human, animal, and plant pathogens are categorized by threat level, with category A agents having the greatest potential to be used for bioterrorism, and category C referring to agents whose threat potential needs further study.
4. Defense against bioterrorism begins with surveillance—the reporting and monitoring required for effective response to biological attacks. Diagnoses must be reliable, and efficient control measures must exist to limit the impact of any attack that occurs.
5. Recombinant genetic technology could potentially lead to the development of novel agents or the modification of existing agents to make them more difficult to control in the event of an attack. Such technology could also lead to potential vaccines, cures, or pathogen-resistant crops.

Questions for Review

Answers to the Questions for Review (except Short Answer questions) begin on page A-1.

Multiple Choice

1. Food fermentations do all of the following except
   a. give foods a characteristic taste.
   b. lower the risk of food spoilage.
   c. sterilize foods.
   d. increase the shelf life of the food.
2. Commercially produced beers and wines are usually fermented with the aid of
   a. naturally occurring bacteria.
   b. naturally occurring yeast.
   c. specific cultured bacteria.
   d. specific cultured yeast.
3. Which of the following lists foods in order, from perishable to nonperishable?
   a. pasta, cheese, fruit, uncooked ground beef
   b. pasta, fruit, uncooked ground beef, cheese
   c. uncooked ground beef, fruit, cheese, pasta
   d. uncooked ground beef, fruit, pasta, cheese
4. Which of the following would be the best growth medium to use for industrial fermentations?
   a. corn
   b. synthetic medium made by hand
   c. whey from cheese production
   d. brewing mash
5. Biodegradable plastics are made from which of the following microbial metabolites?
   a. sludge
   b. PHA
   c. BOD
   d. alum
6. Strains of the bacterium *Pseudomonas syringae* have been identified as being capable of
   a. producing plastics.
   b. producing alternative fuels.
   c. fermenting foods.
   d. preventing ice formation.
7. Which of the following is added during water or sewage treatment to promote flocculation?
   a. sludge
   b. PHA
   c. BOD
   d. alum
8. During chemical treatment of drinking water and wastewater, which of the following microbes is least likely to be inactivated or killed?
   a. algae
   b. viruses
   c. fungal spores
   d. bacteria
9. In which step is most of the organic content of sewage removed?
   a. primary treatment
   b. secondary treatment
   c. tertiary treatment
   d. sludge treatment
10. Microbial communities are composed of
    a. single, pure populations.
    b. all organisms in a locale.
    c. mixed populations of organisms.
    d. a biosphere.

11. In the environment, nutrients are generally
    a. limiting.
    b. present in excess.
    c. stable.
    d. artificially induced.
12. Most chemical elements exist in the environment as
    a. usable forms in soil and rock.
    b. usable forms in water.
    c. unusable forms in soil and rock.
    d. unusable forms in water.
13. In the carbon cycle, microbes
    a. convert CO_2 into organic material for consumption.
    b. convert CO_2 into inorganic material for storage.
    c. convert fossil fuels into usable organic compounds.
    d. convert oxygen into water as a by-product of photosynthesis.
14. Nitrification
    a. converts organic nitrogen to NH_3.
    b. converts NH_3 to NH_4^+.
    c. converts NH_4^+ to NO_3^-.
    d. converts NO_3^- to N_2.
15. In aquatic environments, most microbial life is found in the
    a. littoral zone.
    b. limnetic zone.
    c. profundal zone.
    d. benthic zone.
    e. abyssal zone.
16. Which of the following diseases is *not* caused by category A biological weapons agents?
    a. smallpox
    b. plague
    c. Q fever
    d. tularemia
17. Of the following characteristics, which would contribute most to making a microorganism an effective biological warfare agent?
    a. is readily available in the environment
    b. can be spread by contact after original dissemination
    c. cannot be treated well outside of a hospital
    d. is easily identified by symptoms
18. Anammox reactions are
    a. anaerobic and part of nitrogen cycling.
    b. anaerobic and part of carbon cycling.
    c. aerobic and part of sulfur cycling.
    d. aerobic and part of metal ion oxidation.
19. Industrial fermentation
    a. always involves alcohol production.
    b. involves the large-scale production of any beneficial compound.
    c. refers to the oxidation of sugars using organic electron acceptors.
    d. is any desirable change to food by microbial metabolism.
20. Lyophilization in food preservation is by
    a. cell lysis.
    b. gamma radiation.
    c. rapid heating.
    d. freeze-drying.

Matching

Match each term with its correct definition.

1. ____ Organisms whose presence in water indicates contamination from feces
2. ____ Compound produced by a bacterium that kills insects
3. ____ Refers to water that is fit to drink
4. ____ Community of organisms surrounded by polysaccharides and attached to surfaces
5. ____ Used in the processing of animal wastes; mimics primary and secondary wastewater treatment
6. ____ Refers to compounds that are resistant to microbial degradation
7. ____ Water that is not bound by solutes
8. ____ Refers to quantity of all organisms present in an environment
9. ____ Process whereby pollutants accumulate to high levels in waterways, causing overgrowth and anaerobic conditions
10. ____ Process of reducing nitrogen from the atmosphere
11. ____ The process whereby organisms actively inhibit the growth of other organisms
12. ____ Undesirable fermentation reactions in food leading to poor taste, smell, or appearance
13. ____ Brief heating of foods during processing
14. ____ Descriptor of the level of organic material present in wastewater
15. ____ Fermentative products produced by microorganisms during stationary phase

A. Spoilage
B. Water activity
C. Coliforms
D. Pasteurization
E. Secondary metabolites
F. Bt toxin
G. Potable
H. BOD
I. Oxidation lagoon
J. Recalcitrant
K. Biomass
L. Antagonism
M. Nitrogen fixation
N. Eutrophication
O. Biofilm

Modified True/False

Indicate whether each of the following statements is true or false. Rewrite the phrase in italics to make a false statement true.

1. ____ The fermentation of dairy products relies on *mixed acid* fermentation.
2. ____ Sauerkraut production involves the *alcoholic* fermentation of cabbage.
3. ____ Pasteurization kills *mesophilic* microorganisms except endospore formers.
4. ____ Methane is a gas produced by microbial metabolism that can be used directly as a *fuel source*.
5. ____ The treatment of drinking water and sewage involves *similar* processes.
6. ____ *Recalcitrant* molecules can be degraded by naturally occurring microorganisms.
7. ____ Biofilms of microorganisms form in *aquatic* environments only.
8. ____ *Cooperation* is common among microorganisms living in microhabitats.
9. ____ Aquatic microorganisms are *more* prevalent near the surface than at the bottom of waterways.
10. ____ *Abyssal* organisms are found near shores of oceans.

Fill in the Blanks

1. Intrinsic factors affecting food spoilage are properties of ________________ rather than ________________.
2. Leaving foods out at room temperature ________________ the likelihood of food spoilage.
3. The two types of industrial fermentation equipment are designed for ________________ production or ________________ production.
4. Potable water is allowed to have ________________ coliforms per 100 ml of water tested.
5. Leaching of compounds from mine tailings often results in the oxidation of two elements: ________________ and ________________.
6. Biogeochemical cycling involves three primary steps: ________________, ________________, and ________________.
7. Nitrogen exists primarily as ________________ in the environment.
8. Phosphorus exists primarily as ________________ in the environment.
9. A ________________ is a device composed of microbes and electronics used to detect other microbes or their products.
10. ________________ is the amount of oxygen required by aerobic organisms to fully metabolize organic waste in water.

Labeling

Label the general phases in the carbon cycle.

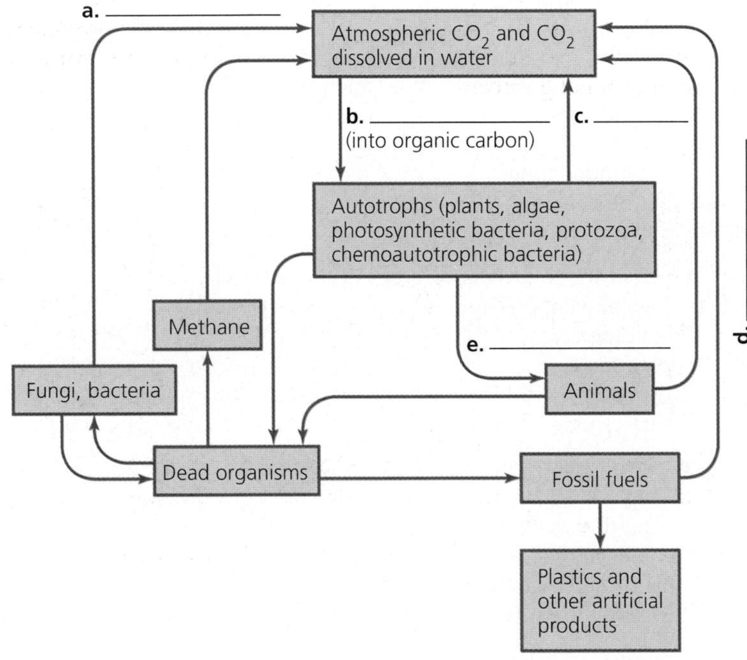

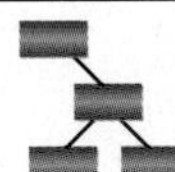

Concept Mapping

Using the following terms, draw a concept map that describes microbial roles in food production. For a sample concept map, see p. 93. Or, complete this concept map online by going to the Study Area at www.masteringmicrobiology.com.

Acetic acid in vinegar
Acetobacter
Alcohol in beer
Alochol in wine
Aspergillus spp. and *Lactobacillus* spp.
Bread to rise
Fermentation
Flavor
Gluconobacter
Holes
Lactobacillus delbrueckii bulgaricus
Malt from grains
Milk
Propionibacterium spp.
Saccharomyces cerevisiae (yeast)
Soy sauce
Soybeans and wheat
Streptococcus thermophilus
Sugars in bread dough
Sugars in fruit juice
Swiss cheese
Yogurt

Critical Thinking

1. Why does the application of recombinant DNA technology to food production have the potential to enhance not only food quality but also food output? Given that it has the potential to feed more people, why are some people opposed to genetic modification of foods?
2. Given what you know about microbial nutrition and metabolism, explain why it is technically more difficult to achieve high yields of a secondary metabolite than of a primary metabolite.
3. Compare the types of alternative fuels that could be produced by microbes. Based on starting materials, which would provide the most renewable energy?
4. One way that farmers are attempting to prevent the development of widespread insect resistance to Bt toxin is by planting non-Bt-producing crops in fields adjacent to Bt-containing crops. Insects can infest both fields, but the insects eating the non-Bt plants should survive at a higher rate than those in the Bt-producing field. Why would this prevent the dissemination of resistance to Bt toxin among insects?
5. Even though water and wastewater undergo essentially the same forms of treatment, treated wastewater usually is not put into the water system from which drinking water is derived. Why not?
6. Take a critical look at the garbage in all of your wastebaskets. How much of the material present could possibly be degraded by microbes? How much could be recycled either at a recycling center or in compost? What is left if the degradable or recyclable materials were removed?
7. Given the amount of pollutants and disease-causing microbes that are in soil and water, why don't we see higher incidences of soil-borne and waterborne illnesses?
8. Explain why influenzaviruses could be potentially devastating biological weapons.
9. Explain why sake—sometimes called rice wine—would be more accurately described as "rice beer."
10. Inexpensive bulk wines and wines that have been left exposed to air for too long often acquire a "vinegary" smell or taste. Explain why this is so.

Access more review material online in the Study Area at **www.masteringmicrobiology.com.** There, you'll find

- **Concept Mapping Activities**
- **Flashcards**
- **Quizzes**

and more to help you succeed.

Answers
TO END-OF-CHAPTER QUESTIONS FOR REVIEW

Answers to multiple-choice, fill-in-blank, labeling, matching, and true/false questions are listed here. Answers for Short Answer and Critical Thinking questions are available for instructors only in the Instructor's Manual that accompanies this text.

CHAPTER 1

Multiple Choice
1. a; 2. c; 3. d; 4. a; 5. c; 6. d; 7. a; 8. b; 9. d; 10. d

Fill in the Blanks
1. Martinus Beijerinck and Sergei Winogradsky; 2. Louis Pasteur and Eduard Buchner; 3. Paul Ehrlich; 4. Edward Jenner; 5. John Snow; 6. Robert Koch; 7. John Snow; 8. Louis Pasteur; 9. Louis Pasteur

Labeling
1. cilium; 2. flagellum; 3. pseudopod; 4. nucleus

Matching
1. J; 2. H; 3. C; 4. C, H, K; 5. B; 6. A; 7. C; 8. E; 9. D; 10. D; 11. I; 12. L

CHAPTER 2

Multiple Choice
1. b; 2. d; 3. c; 4. c; 5. b; 6. c; 7. a; 8. a; 9. a; 10. c

Fill in the Blanks
1. valence; 2. nonpolar covalent; 3. ATP; 4. fat; 5. functional groups; 6. hydrolysis; 7. exothermic; 8. products; 9. pH; 10. ribose

Labeling
Primary structure: light blue strands. Secondary structure: green α-helices, gray β-pleated sheets. The entire molecule represents tertiary structure.

CHAPTER 3

Multiple Choice
1. b; 2. b; 3. c; 4. c; 5. c; 6. d; 7. a; 8. a; 9. a; 10. b; 11. c; 12. d; 13. d; 14. b; 15. c

Matching
1. D Glycocalyx; B,H,I Flagella; F Axial filaments; H Cilia; A,E Fimbriae; C,G Pili; A,E,G Hami; 2. A Ribosome; D Cytoskeleton; F Centriole; E Nucleus; I Mitochondrion; G Chloroplast; C Endoplasmic reticulum; H Golgi body; B Peroxisome

CHAPTER 4

Multiple Choice
1. c; 2. d; 3. c; 4. d; 5. d; 6. d; 7. b; 8. a; 9. a; 10. d

Fill in the Blanks
1. 600×; 2. heat fixation; 3. increases, increases, more; 4. Contrast; 5. negatively

Labeling
1. scanning electron; 2. bright-field light; 3. phase-contrast light; 4. fluorescent light; 5. transmission electron; 6. differential interference contrast (Nomarski)

CHAPTER 5

Multiple Choice
1. c; 2. a; 3. c; 4. a; 5. a; 6. c; 7. b; 8. d; 9. a; 10. c; 11. d; 12. d; 13. a; 14. a; 15. a; 16. c; 17. a; 18. c; 19. a; 20. c

Matching
1. C; 2. B; 3. E; 4. A

Fill in the Blanks
1. the original reaction center, chlorophyll; 2. 2; 3. pentose phosphate, Entner-Doudoroff; 4. The Krebs cycle; 5. O_2; 6. NO_3^-, SO_4^{2-}, CO_3^{2-}; 7. inorganic
8.

Category of Enzyme	Description
Hydrolase	Catabolizes substrate by adding water
Isomerase	Rearranges atoms
Ligase/polymerase	Joins 2 molecules together
Transferase	Moves functional groups
Oxidoreductase	Adds or removes electrons
Lyase	Splits large molecules

9. chemiosmosis; 10. NAD^+, FAD

CHAPTER 6

Multiple Choice
1. b; 2. c; 3. b; 4. a; 5. b; 6. b; 7. d; 8. a; 9. c; 10. a; 11. b; 12. d; 13. c; 14. a; 15. a

Fill in the Blanks
1. carbon, energy, electrons; 2. singlet; 3. nitrogen; 4. Growth factors; 5. minimum growth temperature; 6. osmotic; 7. halophiles; 8. Carotenoid; 9. fixation; 10. streak plate

Labeling
See Figure 6.3.

CHAPTER 7

Multiple Choice
1. a; 2. c; 3. d; 4. c; 5. d; 6. a; 7. a; 8. d; 9. a; 10. c; 11. c; 12. d; 13. a; 14. d; 15. c; 16. b; 17. d; 18. c; 19. b; 20. c; 21. b; 22. b; 23. a; 24. d; 25. b

Fill in the Blanks
1. initiation of transcription, elongation of the RNA transcript, termination of transcription; 2. codon; 3. silence, missense, nonsense; 4. frameshift; 5. promoter, operator, a series of genes; 6. inducible; 7. semiconservative; 8. transformation, transduction, bacterial conjugation; 9. Transposons; 10. Crossing over; 11. Transfer; 12. Short interference, micro

Labeling
1. replication fork; 2. stabilizing proteins; 3. nucleotide (triphosphate); 4. leading strand; 5. helicase, 6. primase, 7. DNA polymerase III; 8. RNA primer; 9. Okazaki fragment; 10. DNA polymerase I; 11. lagging strand; 12. ligase

CHAPTER 8

Multiple Choice
1. d; 2. b; 3. c; 4. d; 5. a; 6. c; 7. c; 8. b; 9. a; 10. d

Modified True/False
1. *cut DNA at specific sites*; 2. True; 3. *Electrophoresis*; 4. True; 5. *Southern blotting*

Labeling
Step 1, denaturation: 94°C; step 2, priming: DNA primer, deoxyribonucleotide triphosphates, DNA polymerase, cool to 65°C; step 3, extension: same reagents as step 2, 72°C; step 4, repeat

CHAPTER 9

Multiple Choice
1. a; 2. b; 3. d; 4. d; 5. d; 6. a; 7. d; 8. d; 9. c; 10. a; 11. d; 12. d; 13. a; 14. c; 15. c; 16. b; 17. a; 18. d; 19. d; 20. a

CHAPTER 10

Multiple Choice
1. d; 2. a; 3. a; 4. c; 5. d; 6. c; 7. a; 8. d; 9. a; 10. d

Labeling
See Figure 10.4.

CHAPTER 11

Modified True/False
1. *asexually*; 2. *vibrio*; 3. True; 4. *Elementary*; 5. *rRNA sequences*; 6. True; 7. True; 8. True; 9. *Epulopiscium*; 10. True

Matching
1. E; 2. B; 3. C; 4. D; 5. L; 6. K; 7. H; 8. Q; 9. I; 10. N; 11. P; 12. O; 13. M; 14. F; 15. A

Multiple Choice
1. c; 2. a; 3. c; 4. d; 5. a; 6. d; 7. a; 8. b; 9. c; 10. d

Labeling
See Figure 11.1.

CHAPTER 12

Multiple Choice
1. a; 2. d; 3. c; 4. b; 5. a; 6. d; 7. b; 8. c; 9. a; 10. c; 11. c; 12. a; 13. d; 14. a; 15. d

Matching
First section: 1. E; 2. B; 3. D; 4. C; 5. A.
Second section: 1. A; 2. F; 3. E; 4. C; 5. D; 6. B.
Third section: 1. C; 2. E; 3. B; 4. D; 5. A

Labeling
1. ascospore, sexual; 2. basidiospore, sexual; 3. conidia, asexual; 4. chlamydospore, asexual

Fill in the Blanks
1. protozoology; 2. mycology; 3. phycology; 4. mycoses; 5. radiolarians

CHAPTER 13

Multiple Choice
1. c; 2. c; 3. a; 4. a; 5. b; 6. c; 7. a; 8. d; 9. d; 10. b

Matching
1. H; 2. G; 3. C; 4. B; 5. D; 6. E; 7. F; 8. A; 9. J; 10. I

Labeling
See Figure 13.8.

CHAPTER 14

Multiple Choice
1. a; 2. b; 3. b; 4. a; 5. d; 6. a; 7. d; 8. c; 9. a; 10. d; 11. d; 12. b; 13. c; 14. d; 15. b

Fill in the Blanks
1. pathogen; 2. asymptomatic or subclinical; 3. etiology; 4. epidemiology; 5. zoonoses; 6. fomites;

7. Nosocomial; 8. prevalence; 9. biological; 10. lipid A

Labeling
1. Endemic (or sporadic); 2. Epidemic; 3. Pandemic

CHAPTER 15
Multiple Choice
1. d; 2. d; 3. b; 4. a; 5. d; 6. c; 7. d; 8. b; 9. a; 10. d

Modified True/False
1. *dead*; 2. True; 3. True; 4. True; 5. *monocytes*; 6. True; 7. *pathogen*; 8. *ingestion*; 9. *phagolysosomes*; 10. *pathogen's cytoplasmic membrane*; 11. *inflammation*; 12. True; 13. True; 14. *antimicrobial peptides*; 15. True

Matching
First section: 1. B; 2. B; 3. B; 4. B; 5. A; 6. B; 7. A; 8. B; 9. A; 10. A; 11. A; 12. B; 13. A; 14. C; 15. A
Second section: 1. J; 2. E; 3. G; 4. D; 5. C; 6. H; 7. A; 8. B; 9. F; 10. I

Labeling
See Figure 15.6.

CHAPTER 16
Multiple Choice
1. b; 2. e; 3. e; 4. b; 5. a; 6. a; 7. c; 8. d; 9. e; 10. a

Modified True/False
1. *antigen-presenting cells*; 2. True; 3. *cytotoxic*; 4. *Plasma cells*; 5. *humoral*

Matching
1. D Plasma cell; C Cytotoxic cell; B Th2 cell; A Dendritic cell;
2. D Artificially acquired passive immunity; A Naturally acquired active immunity; B Naturally acquired passive immunity; C Artificially acquired active immunity

Labeling
See Figures 16.4 and 16.5.

CHAPTER 17
Multiple Choice
1. d; 2. c; 3. b; 4. c; 5. e; 6. e; 7. e; 8. a; 9. b; 10. e; 11. a; 12. d; 13. d; 14. d; 15. c

True/False
1. False; 2. False; 3. True; 4. True; 5. False

Matching
1. D; 2. C; 3. A; 4. B; 5. C; 6. A

Labeling
See Figure 17.14.

CHAPTER 18
Multiple Choice
1. e; 2. c; 3. d; 4. c; 5. c; 6. e; 7. b; 8. d; 9. c; 10. e

Modified True/False
1. *Histamine, kinins, and/or proteases are*; 2. True; 3. *red blood*; 4. *type IV*; 5. *allograft*

Matching
1. A; 2. D; 3. E; 4. D; 5. E; 6. D; 7. A; 8. C; 9. A; 10. A

Labeling
See Figure 18.13.

CHAPTER 19
Multiple Choice
1. c; 2. b; 3. d; 4. d; 5. c; 6. b; 7. d; 8. a; 9. d; 10. b; 11. a

Matching
1. A; 2. A; 3. B, F; 4. B; 5. B; 6. B; 7. B; 8. B; 9. B; 10. G; 11. B, H; 12. F; 13. C; 14. B, I; 15. E

CHAPTER 20
Multiple Choice
1. a; 2. d; 3. c; 4. b; 5. c; 6. d; 7. a; 8. a; 9. d; 10. b; 11. a; 12. a

Matching
1. C; 2. F; 3. D; 4. B; 5. E; 6. A

Labeling
See Figure 20.8.

CHAPTER 21
Multiple Choice
1. c; 2. a; 3. b; 4. d; 5. b; 6. a; 7. d; 8. c; 9. b; 10. c; 11. b; 12. b; 13. b

Labeling
See Figure 21.6.

Matching
1. C Rocky Mountain spotted fever; A Endemic typhus; B Epidemic typhus; D Scrub typhus; E HME; 2. A *Rickettsia typhi*; B *Rickettsia prowazekii*; C *Rickettsia rickettsii*; D *Orientia tsutsugamushi*; C *Ehrlichia chaffeensis*; C *Borrelia burgdorferi*; B *Borrelia recurrentis*; C *Anaplasma phagocytophilum*; 3. G *Chlamydophila psittaci*; C *Chlamydophila pneumoniae*; B,D,E,F *Chlamydia trachomatis*; A *Treponema pallidum pallidum*; H *Treponema pallidum pertenue*; I *Treponema pallidum endemicum*; J *Treponema carateum*; K *Borrelia burgdorferi*; 4. E Peptic ulcers; A,B,D Gastroenteritis; C Blood poisoning; A Cholera

CHAPTER 22
Multiple Choice
1. a; 2. c; 3. d; 4. c; 5. b; 6. d; 7. c; 8. b; 9. c; 10. b; 11. b; 12. c; 13. b; 14. a; 15. b; 16. c; 17. d; 18. b; 19. a; 20. b

Modified True/False
1. True; 2. *difficult*; 3. *Systemic mycoses*; 4. True; 5. True; 6. True; 7. True; 8. *immunocompromised*; 9. True; 10. *3–10% of individuals have*

Fill in the Blanks
1. mycelial forms, yeast forms; 2. *Blastomyces dermatitidis, Coccidioides immitis, Histoplasma capsulatum, Paracoccidioides brasiliensis*; 3. ergosterol; 4. Black, white; 5. *Sporothrix schenckii*; 6. *Cryptococcus neoformans*; 7. *Aspergillus, Candida, Cryptococcus, Pneumocystis, Mucor*; 8. *Candida*; 9. protozoan, fungus; 10. *Claviceps*

Matching
A,C Aspergillosis; B Candidiasis; C Chromoblastomycosis; A Coccidioidomycosis; A Cryptococcosis; B Dermaphytosis; A,C Histoplasmosis; A Hypersensitivity reactions; D Mushroom poisoning; C Mycetoma; C Sporotrichosis

CHAPTER 23
Multiple Choice
1. d; 2. b; 3. a; 4. a; 5. d; 6. b; 7. b; 8. a; 9. a; 10. d; 11. b; 12. b; 13. c; 14. b; 15. c; 16. a; 17. b; 18. d; 19. a; 20. d

Modified True/False
1. *Ingestion*; 2. *corneal scrapings, cerebrospinal fluid, or biopsy material*; 3. True; 4. True; 5. True; 6. True; 7. True; 8. water plants; 9. True; 10. True

Fill in the Blanks
1. cilia; 2. *Toxoplasma gondii*; 3. *brucei, cruzi*; 4. *Entamoeba*; 5. hypnozoites; 6. *Wuchereria*; 7. *Ancylostoma, Necator, Schistosoma*; 8. *Fasciola*; 9. *Ancylostoma*; 10. *Enterobius vermicularis* and *Wuchereria bancrofti*

Matching
1. G; 2. E; 3. A; 4. H; 5. C, G; 6. C, G, I; 7. F; 8. B; 9. J; 10. D

Labeling
See Figure 23.11.

CHAPTER 24
Multiple Choice
1. b; 2. c; 3. a; 4. b; 5. d; 6. e; 7. d; 8. c; 9. a; 10. c; 11. a; 12. b; 13. d; 14. d; 15. a

Labeling
See Figure 24.2.

Matching
B Chickenpox; A Smallpox; A Cowpox; A Molluscum contagiosum; B HHV-1; B Whitlow; B Shingles; B Burkitt's lymphoma; B Infectious mononucleosis; B Chronic fatigue syndrome; B Cytomegalovirus; C Genital warts; B Roseola; C Plantar warts; G Progressive multifocal leukoencephalopathy; D Common cold; E Hepatitis B; F Fifth disease

CHAPTER 25
Multiple Choice
1. a; 2. c; 3. d; 4. a; 5. c; 6. d; 7. d; 8. b; 9. a; 10. a; 11.b

Matching
K Myocarditis; C Colorado tick fever; A Rabies; G Influenza; F Dengue fever; D German measles; I Acute gastroenteritis; J Ebola virus; B RSV; E Western equine encephalitis; H No known disease

True/False
1. True; 2. False; 3. False; 4. True; 5. True

Labeling
See Figure 25.19.

CHAPTER 26
Multiple Choice
1. c; 2. d; 3. c; 4. c; 5. b; 6. d; 7. d; 8. b; 9. b; 10. c; 11. a; 12. c; 13. a; 14. c; 15. a; 16. c; 17. b; 18. a; 19. b; 20. d

Matching
1. C; 2. F; 3. G; 4. O; 5. I; 6. J; 7. B; 8. K; 9. N; 10. M; 11. L; 12. A; 13. D; 14. H; 15. E

Modified True/False
1. *lactic acid*; 2. *lactic acid*; 3. True; 4. True; 5. True; 6. *Biodegradable*; 7. True; 8. *Competition*; 9. True; 10. *Littoral*

Fill in the Blanks
1. the food, processing or handling; 2. increases; 3. batch production, continuous flow production; 4. zero; 5. iron, sulfur; 6. production, consumption, decomposition; 7. dinitrogen gas (N_2); 8. phosphate ion (PO_4^{3-}); 9. biosensor; 10. BOD (biochemical oxygen demand)

Labeling
See Figure 26.15.

APPENDIX A
Metabolic Pathways

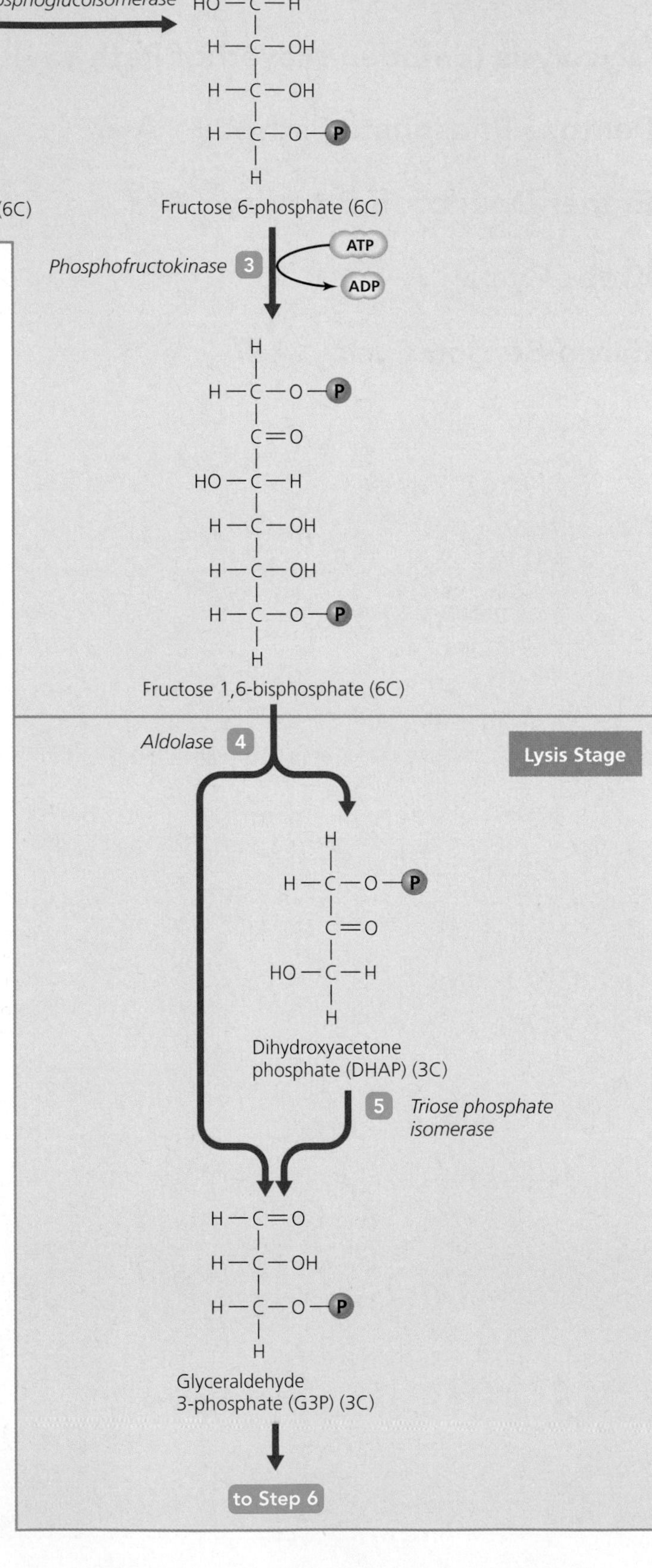

GLYCOLYSIS (Embden-Meyerhof Pathway):

Energy Investment Stage

Step 1. *Hexokinase* transfers phosphate from ATP to carbon 6 of glucose, forming glucose 6-phosphate—a charged molecule. Since the cytoplasmic membrane is impermeable to ions, phosphorylated glucose cannot diffuse out of a cell.

Step 2. *Phosphoglucoisomerase* rearranges the atoms of glucose to form an isomer—fructose 6-phosphate.

Step 3. *Phosphofructokinase* invests more energy by adding another phosphate group from ATP to form fructose 1,6-bisphosphate.

Lysis Stage

Step 4. *Aldolase* (fructose 1,6-bisphosphate aldolase) splits six-carbon fructose 1,6-bisphosphate into three-carbon glyceraldehyde 3-phosphate (G3P) and three-carbon dihydroxyacetone phosphate (DHAP). It is this cleavage that gives glycolysis its name.

Step 5. *Triose phosphate isomerase* rearranges the atoms of DHAP to form another molecule of glyceraldehyde 3-phosphate. From this point, every step of glycolysis occurs twice—once for each of the three-carbon molecules.

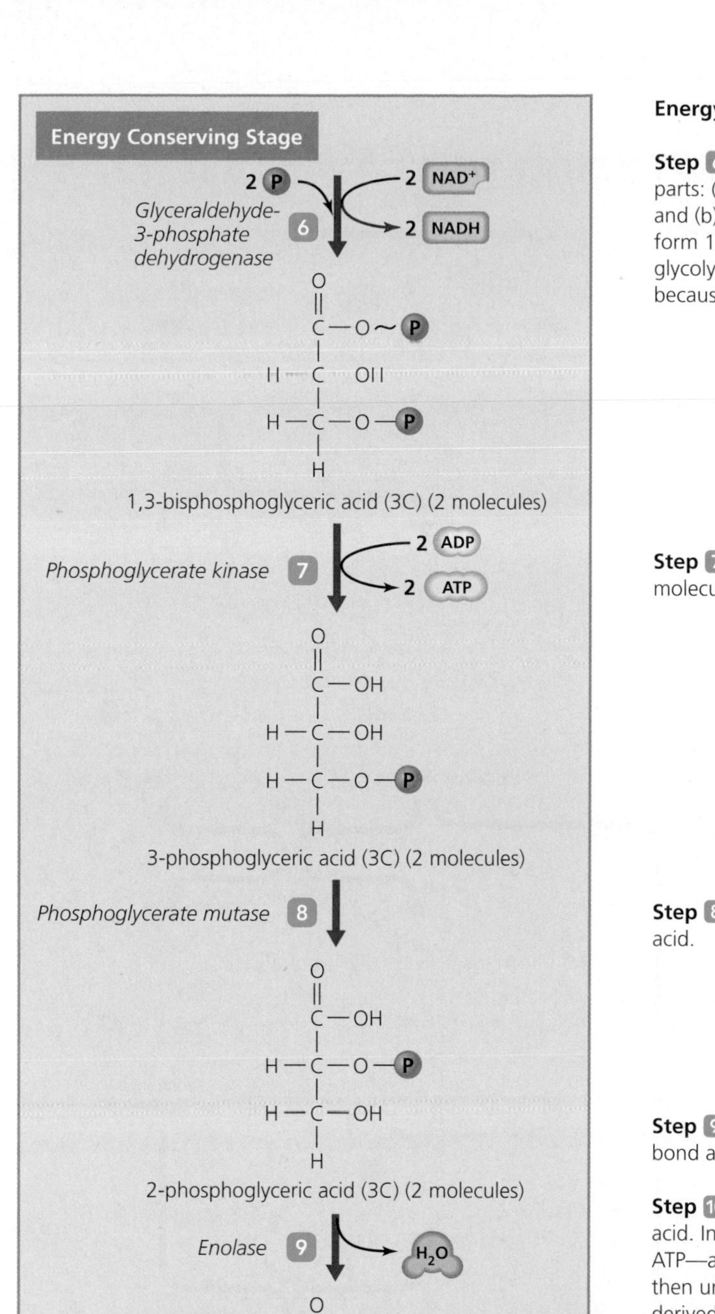

Energy Conserving Stage

Step 6. *Glyceraldehyde-3-phosphate dehydrogenase* catalyzes a reaction with two parts: (a) It oxidizes G3P, transferring electrons (and hydrogen) to NAD^+ to form NADH and (b) it adds inorganic phosphate from the cytosol to G3P with a high-energy bond to form 1,3-bisphosphoglyceric acid. This two-part reaction is among the more important in glycolysis because it generates a molecule of NADH for each molecule of G3P and because it creates the first high-energy intermediate.

Step 7. *Phosphoglycerate kinase* conserves the energy in the high-energy bonds in two molecules of ATP, yielding 3-phosphoglyceric acid molecules.

Step 8. *Phosphoglycerate mutase* rearranges the atoms to form 2-phosphoglyceric acid.

Step 9. *Enolase* removes a molecule of water from each substrate, forming a double bond and a high-energy bond with phosphate.

Step 10. *Pyruvate kinase* ends glycolysis by transferring energy to ATP, forming pyruvic acid. In the final analysis, two molecules of ATP are invested to yield four molecules of ATP—a net gain of two molecules of ATP—and two molecules of NADH. Pyruvic acid then undergoes respiration (when there is an inorganic, or rarely an extracellularly derived organic, final electron acceptor) or fermentation (when the final electron acceptor is an organic molecule from the cell).

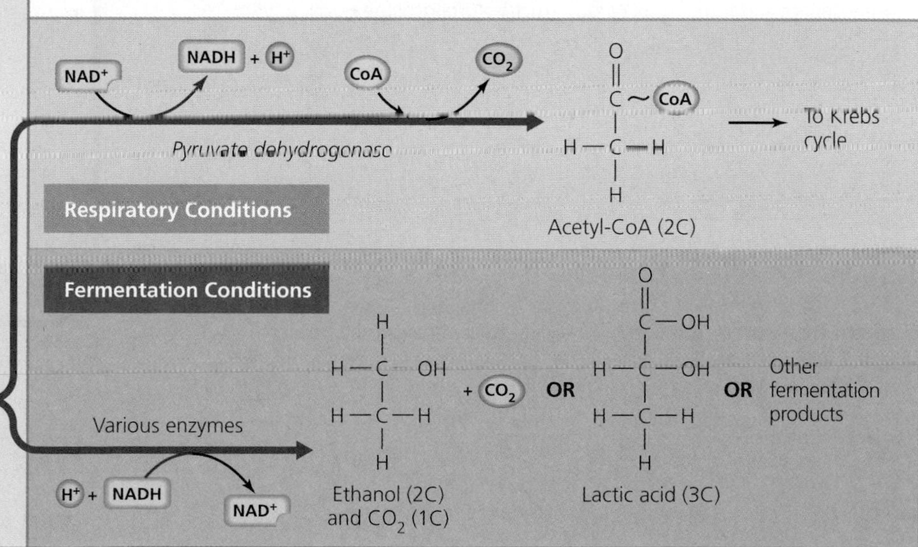

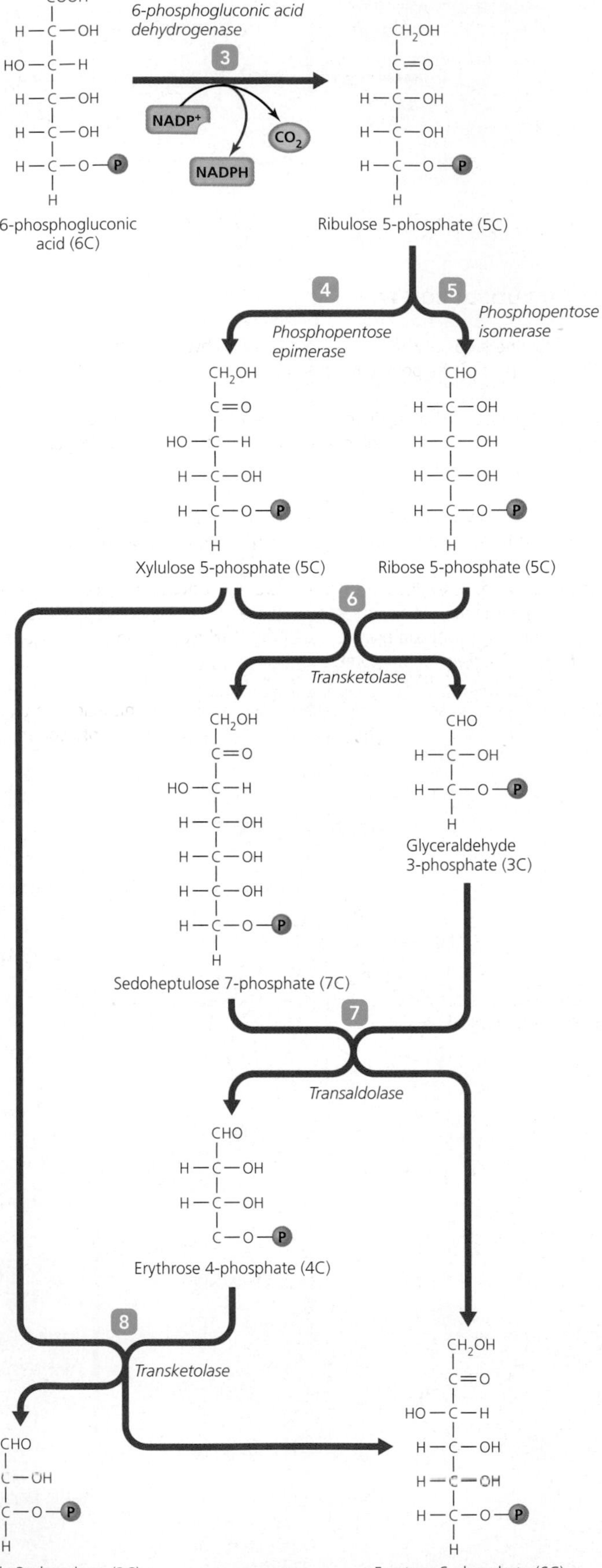

PENTOSE PHOSPHATE PATHWAY

Step 1. *Glucose-6-phosphate dehydrogenase* oxidizes glucose 6-phosphate to 6-phosphogluconolactone by transferring hydrogen to $NADP^+$, forming NADPH.

Step 2. *Lactonase* adds hydroxide from a water molecule, forming 6-phosphogluconic acid.

Step 3. *6-phosphogluconic acid dehydrogenase* oxidizes this intermediate (transferring hydrogen to another molecule of NADPH) and removes a molecule of carbon dioxide to form ribulose 5-phosphate. This five-carbon phosphorylated sugar is used in the synthesis of nucleotides, certain amino acids, and glucose (via photosynthesis).

Step 4. *Phosphopentose epimerase* converts some molecules of ribulose 5-phosphate to xylulose 5-phosphate.

Step 5. Simultaneously, *phosphopentose isomerase* converts other molecules of ribulose 5-phosphate to ribose 5-phosphate.

Step 6. *Transketolase* catalyzes a reaction in which a two-carbon fragment from xylulose 5-phosphate is transferred to ribose 5-phosphate, forming seven-carbon sedoheptulose 7-phosphate and three-carbon glyceraldehyde 3-phosphate (G3P).

Step 7. *Transaldolase* then transfers a three-carbon fragment from sedoheptulose 7-phosphate to G3P, yielding four-carbon erythrose 4-phosphate and forming six-carbon fructose 6-phosphate.

Step 8. *Transketolase* transfers another two-carbon fragment from another molecule of xylulose 5-phosphate to erythrose 4-phosphate, forming another molecule of fructose 6-phosphate and another molecule of G3P. G3P enters glycolysis at step 6; fructose 6-phosphate can enter glycolysis at step 1 or may be converted into glucose 6-phosphate, which reenters the pentose phosphate pathway.

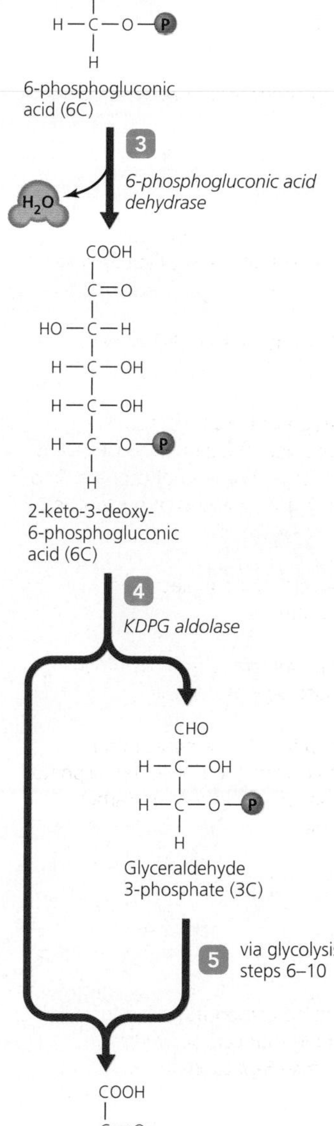

ENTNER-DOUDOROFF PATHWAY

The first two steps of the Entner-Doudoroff pathway are the same as the first two steps of the pentose phosphate pathway:

Step 1. *Glucose-6-phosphate dehydrogenase* oxidizes glucose 6-phosphate to 6-phosphogluconolactone by transferring hydrogen to $NADP^+$, forming NADPH.

Step 2. As in step 2 of the pentose phosphate pathway, *lactonase* adds hydroxide from a water molecule, forming 6-phosphogluconic acid.

Step 3. *6-phosphogluconic acid dehydrase* removes a molecule of water to form a six-carbon molecule. (The enzyme in this step should not be confused with the similarly named 6-phosphogluconic acid dehydrogenase of the pentose phosphate pathway.)

Step 4. *Aldolase* splits 2-keto-3-deoxy-6-phosphogluconic acid into two three-carbon compounds: pyruvic acid and glyceraldehyde 3-phosphate (G3P).

Step 5. G3P is converted to pyruvic acid by steps 6 through 10 of Embden-Meyerhof glycolysis.

Answers and Appendices

KREBS CYCLE

In a complex reaction with several parts, pyruvate dehydrogenase decarboxylates and oxidizes pyruvic acid (from the Embden-Meyerhof or Entner-Doudoroff pathways) and adds the remaining two-carbon acetic acid to coenzyme A, forming acetyl-CoA. It is this molecule that enters the Krebs cycle. Recall that for every molecule of glucose that enters glycolysis, two molecules of pyruvic acid are produced, necessitating two sets of Krebs cycle reactions.

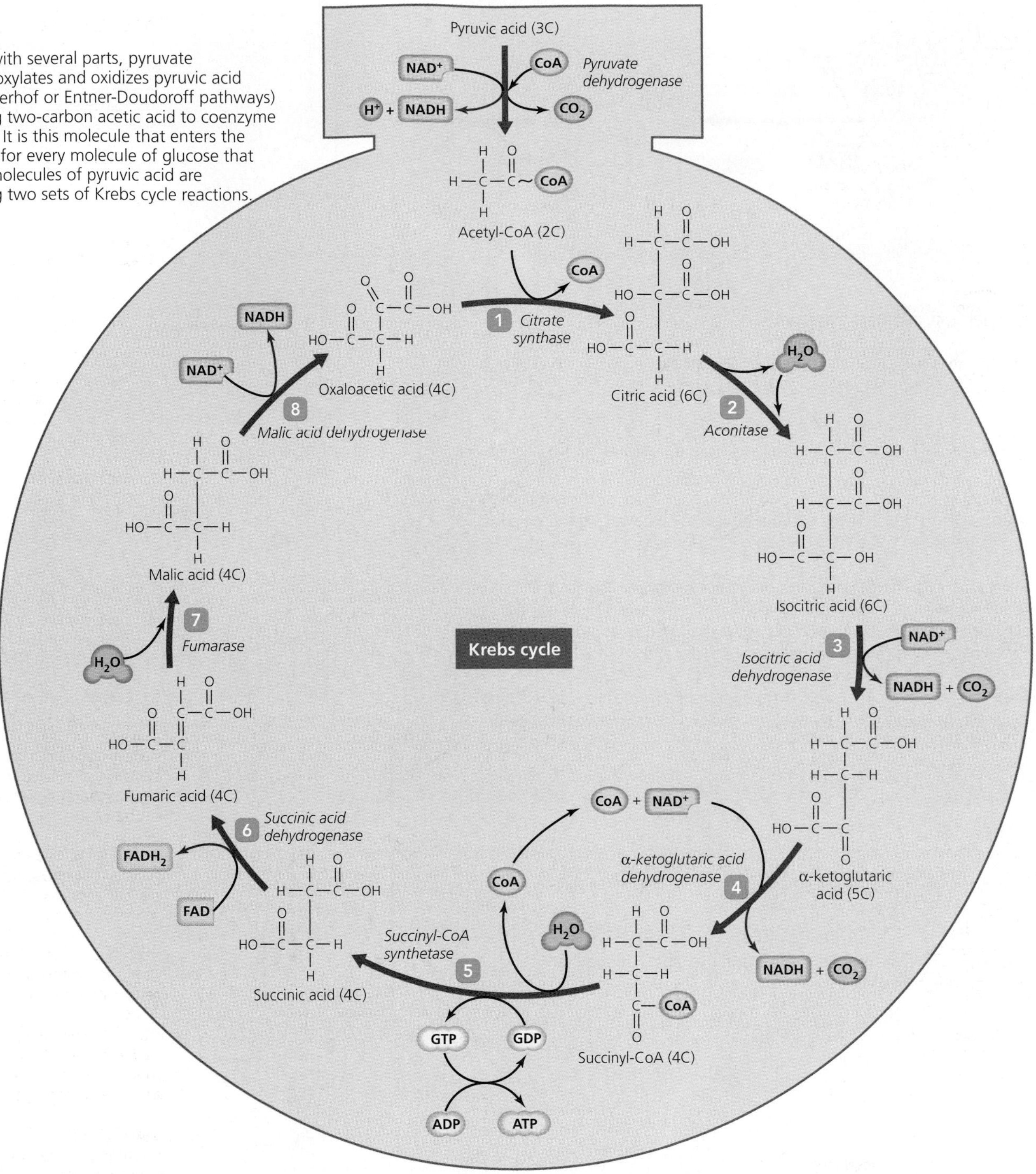

Pyruvic acid (3C)
NAD^+ CoA *Pyruvate dehydrogenase*
H^+ + NADH CO_2
Acetyl-CoA (2C)
Krebs cycle
1 *Citrate synthase* CoA
Citric acid (6C)
2 *Aconitase* H_2O
Isocitric acid (6C)
3 *Isocitric acid dehydrogenase* NAD^+ NADH + CO_2
α-ketoglutaric acid (5C)
4 *α-ketoglutaric acid dehydrogenase* CoA + NAD^+ NADH + CO_2
Succinyl-CoA (4C)
5 *Succinyl-CoA synthetase* H_2O CoA GTP GDP ADP ATP
Succinic acid (4C)
6 *Succinic acid dehydrogenase* FAD $FADH_2$
Fumaric acid (4C)
7 *Fumarase* H_2O
Malic acid (4C)
8 *Malic acid dehydrogenase* NAD^+ NADH
Oxaloacetic acid (4C)

Step 1. *Citrate synthase* adds the two carbons of acetic acid from acetyl-CoA to oxaloacetic acid, forming citric acid. This step gives the Krebs cycle its alternate name—the citric acid cycle.

Step 2. *Aconitase* forms an isomer, isocitric acid, by removing a molecule of water and then adding another molecule of water back.

Step 3. *Isocitric acid dehydrogenase* oxidizes and decarboxylates six-carbon isocitric acid to five-carbon α-ketoglutaric acid; a molecule of NADH is formed in the process.

Step 4. *α-ketoglutaric acid dehydrogenase* oxidizes and decarboxylates α-ketoglutaric acid to form a four-carbon fragment that is attached to coenzyme A, forming succinyl-CoA.

Step 5. *Succinyl-CoA synthetase* forms four-carbon succinic acid and simultaneously phosphorylates GDP to form GTP. The latter molecule subsequently phosphorylates ADP to ATP.

Step 6. *Succinic acid dehydrogenase* oxidizes succinic acid to four-carbon fumaric acid, forming a molecule of $FADH_2$.

Step 7. *Fumarase* rearranges the four carbons to form malic acid.

Step 8. *Malic acid dehydrogenase* oxidizes malic acid to reform oxaloacetic acid, completing the cycle and forming another molecule of NADH.

NADH and $FADH_2$ carry electrons to an electron transport chain.

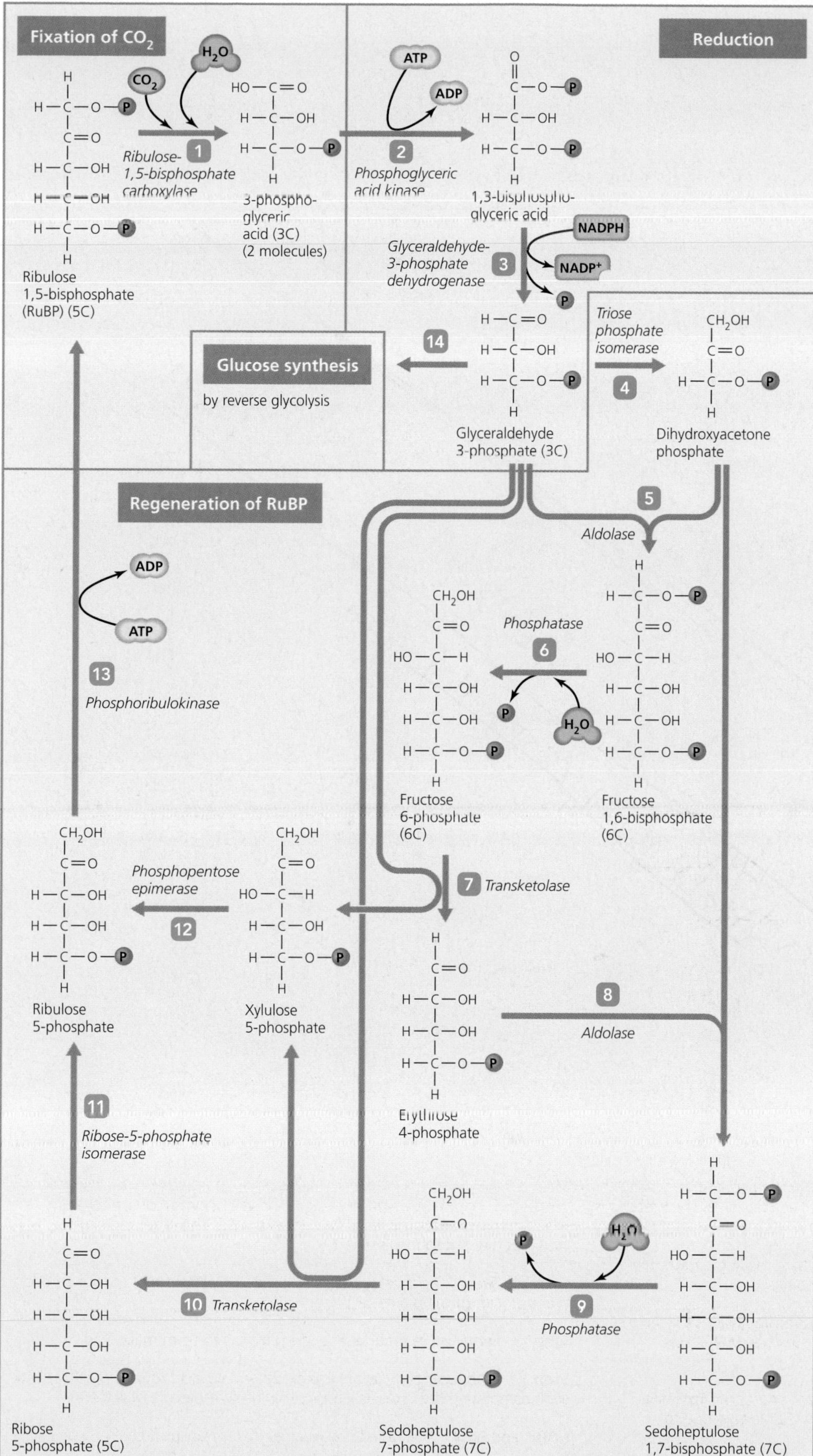

CALVIN-BENSON CYCLE

Fixation of CO_2

Step 1. *Ribulose-1,5-bisphosphate carboxylase* adds CO_2 and H_2O to five-carbon ribulose 1,5-bisphosphate (RuBP), forming a six-carbon intermediate (not shown) that is immediately split into two, three-carbon molecules of 3-phosphoglyceric acid.

Reduction

Step 2. *Phosphoglyceric acid kinase* phosphorylates phosphoglyceric acid at the expense of ATP to form 1,3-bisphosphoglyceric acid.

Step 3. *Glyceraldehyde-3-phosphate dehydrogenase* reduces (using NADPH) and removes one phosphate from 1,3-bisphosphoglyceric acid, forming glyceraldehyde 3-phosphate (G3P).

Regeneration of RuBP

Step 4. *Triose phosphate isomerase* changes some molecules of G3P to dihydroxyacetone phosphate (DHAP).

Step 5. *Aldolase* combines a molecule of G3P and a molecule of DHAP to form fructose 1,6-bisphosphate.

Step 6. *Phosphatase* adds water and removes a phosphate group, forming fructose 6-phosphate.

Step 7. *Transketolase* combines fructose 6-phosphate with a molecule of G3P (from step 3) to form five-carbon xylulose 5-phosphate and four-carbon erythrose 4-phosphate.

Step 8. *Aldolase* rearranges the carbons of erythrose 4-phosphate and of a molecule of DHAP (from step 4), forming sedoheptulose 1,7-bisphosphate.

Step 9. *Phosphatase* adds water and removes a phosphate group to form sedoheptulose 7-phosphate.

Step 10. *Transketolase* combines this molecule with a molecule of G3P (from step 3) to form five-carbon ribose 5-phosphate and five-carbon xylulose 5-phosphate.

Step 11. *Ribose-5-phosphate isomerase* rearranges the atoms of ribose 5-phosphate to form its isomer—ribulose 5-phosphate.

Step 12. *Phosphatase epimerase* similarly rearranges the atoms of xylulose 5-phosphate, forming another molecule of ribulose 5-phosphate.

Step 13. *Phosphoribulokinase* phosphorylates ribulose 5-phosphate to regenerate ribulose bisphosphate (RuBP).

Step 14. For every three molecules of CO_2 that enter the Calvin-Benson cycle (at step 1), one molecule of G3P leaves the cycle to be used for the synthesis of glucose via the reversal of the early reactions of glycolysis.

APPENDIX B
Some Mathematical Considerations in Microbiology

Scientific Notation

Scientific notation (also called exponential notation) is a mathematical device developed to express numbers, particularly very small and very large numbers, in a manner that is convenient and readable. For example, 1.234×10^{-8} is much less cumbersome than 0.00000001234. Similarly, 5.67×10^{17} is more easily read than 567,000,000,000,000,000.

A number expressed in scientific notation is composed of a *coefficient,* which is always a number with only one whole number digit to the left of the decimal place, times 10 to some *exponential power.* In the examples above, the coefficients are 1.234 and 5.67 respectively. The exponents are −8 and 17 respectively.

To write a number in scientific notation, move the decimal point either right or left so that there is only one nonzero number to the left of it. For example,

454 becomes 4.54

In this case we moved the decimal point (which is understood to be at the end of a whole number) two places to the left. Since we moved two places left, the exponent will be positive 2, and the scientific notation is

$$4.54 \times 10^2$$

Similarly, 4,500,264.75 becomes 4.50026475×10^6 since we moved the decimal six places to the left.

As you would expect, when we have to move the decimal point to the right, as in 0.00562 (which becomes 5.62), then the exponent will be a negative number, in this case −3; therefore, $0.00562 = 5.62 \times 10^{-3}$.

Logarithms

A **logarithm** is the number of times a number, called the *base number,* must be multiplied by itself to get a certain number. For example, the number 10 must be multiplied by itself three times to get 1000 ($10 \times 10 \times 10$); therefore, the logarithm of 1000 in base 10 is 3. Similarly, the logarithm of 100,000 is 5. Though any number can be used as the base number, in microbiology, base 10 is commonly used. Microbiologists use base 10 logarithms to express pH values and the size of bacterial populations in culture.

When a number is expressed in scientific notation, the logarithm is the exponent when the coefficient is exactly 1. When the coefficient is a number other than 1, the logarithm must be calculated using the logarithm function on a calculator.

Generation Time

When bacteria and other microbes reproduce by binary fission, the number in the population doubles with each division cycle (generation). Such growth is called *logarithmic* (or *exponential*) *growth.*

The number in a population can be calculated as 2 (because each cell produces two offspring) multiplied by itself as many times as there are generations. In other words, the number of cells in a population arising from a single individual is expressed mathematically as

$$2^{\text{number of generations}}$$

A cell dividing for five generations would produce a population of 32 ($2^5 = 2 \times 2 \times 2 \times 2 \times 2 = 32$). When the population begins with more than one individual, then the final population equals

$$\text{original number of cells} \times 2^{\text{number of generations}}$$

For example, seven cells reproducing for five generations would produce a population of 224 cells ($7 \times 2^5 = 7 \times 2 \times 2 \times 2 \times 2 \times 2 = 224$).

In most cases, microbiologists are not concerned with the number of generations required to produce a given population, but they do want to know the **generation time;** that is, the time required for a bacterial cell to grow and divide. This is also the time required for a population of cells to double in number. To calculate generation time, scientists must first calculate how many generations have been produced, knowing the beginning and ending population sizes. When population sizes are converted to logarithms, the number of generations is calculated as

$$\text{number of generations} = \frac{\text{logarithm (log) number of cells at the end of reproduction} - \text{log number of cells initially}}{\log 2}$$

Log 2 is used in the formula because each cell produces two offspring each time it divides. Log 2 = 0.301.

The number of generations is used to calculate generation time:

$$\text{generation time (min / generation)} = \frac{60 \text{ minutes} \times \text{number of hours}}{\text{number of generations}}$$

For example, if 100 bacteria multiply to produce a population of 3.28×10^6 cells in seven hours, then the generation time for this bacterium is calculated as follows:

$$\begin{aligned}\text{number of generations} &= \frac{\text{logarithm (log) number of cells at the end of reproduction} - \text{log number of cells initially}}{\log 2}\\ &= \frac{\log (3.28 \times 10^6) - \log (100)}{\log 2}\\ &= 15 \text{ generations}\end{aligned}$$

$$\begin{aligned}\text{generation time (min/generation)} &= \frac{60 \text{ minutes} \times \text{hours}}{\text{number of generations}}\\ &= \frac{60 \text{ min} \times 7 \text{ h}}{15}\\ &= 28 \text{ minutes/generation}\end{aligned}$$

Glossary

A site In a ribosome, a binding site that accommodates tRNA delivering an amino acid.

Abscess An isolated site of infection such as a pimple, boil, or pustule.

Abyssal zone In marine habitats, the zone of water beneath the benthic zone, virtually devoid of life except around hydrothermal vents.

Acellular Noncellular.

Acetyl-CoA Combination of two-carbon acetate and coenzyme A.

Acid Compound that dissociates into one or more hydrogen ions and one or more anions.

Acid-fast rod (AFR) Bacilli that retain stain during decolorization by acid-alcohol, particularly species of *Mycobacterium*.

Acid-fast stain In microscopy, a differential stain used to penetrate waxy cell walls.

Acidic dye In microscopy, an anionic chromophore used to stain alkaline structures. Works most effectively in acidic environments.

Acidophile Microorganism requiring acidic pH.

Acne Skin disorder characterized by presence of whiteheads, blackheads, and in severe cases, cysts; typically caused by infection with *Propionibacterium acnes*.

Acquired (secondary) immunodeficiency diseases Any of a group of immunodeficiency diseases that develop in older children, adults, and the elderly as a direct consequence of some other recognized cause, such as infectious disease.

Acquired immunodeficiency syndrome (AIDS) Cluster of characteristic signs and symptoms that develop several years after infection with the human immunodeficiency virus (HIV), which destroys helper T cells.

Actinomycetes High G + C Gram-positive bacteria that form branching filaments and produce spores, thus resembling fungi.

Activation energy The amount of energy needed to trigger a chemical reaction.

Active site Functional site of an enzyme, the shape of which is complementary to the shape of the substrate.

Active transport The movement of a substance against its electrochemical gradient via carrier proteins and requiring cell energy from ATP.

Acute anaphylaxis Condition in which the release of inflammatory mediators overwhelms the body's coping mechanisms.

Acute disease Any disease that develops rapidly but lasts only a short time, whether it resolves in convalescence or death.

Acute inflammation Type of inflammation that develops quickly, is short lived, and is usually beneficial.

Adaptive immunity Resistance against pathogens that acts more effectively upon subsequent infections with the same pathogen.

Adenine Ring-shaped nitrogenous base found in nucleotides of DNA and RNA.

Adenosine triphosphate (ATP) The primary short-term, recyclable energy molecule fueling cellular reactions.

Adherence Process by which phagocytes attach to microorganisms through the binding of complementary chemicals on the cytoplasmic membranes.

Adhesins Molecules that attach pathogens to their target cells.

Adhesion The attachment of microorganisms to host cells.

Adhesion factors A variety of structures or attachment proteins by which microorganisms attach to host cells.

Adjuvant Chemical added to a vaccine to increase its ability to stimulate active immunity.

Aerobe An organism that uses oxygen as a final electron acceptor.

Aerobic respiration Type of cellular respiration requiring oxygen atoms as final electron acceptors.

Aerosol A cloud of water droplets, which travels more than 1 meter in airborne transmission and less than 1 meter in droplet transmission.

Aerotolerant anaerobe Microorganism which prefers anaerobic conditions but can tolerate exposure to low levels of oxygen.

Aflatoxin Carcinogenic mycotoxin produced by *Aspergillus*.

African sleeping sickness Potentially fatal disease caused by a bite from a tsetse fly carrying *Trypanosoma brucei* and characterized by formation of a lesion at the site of the bite, followed by parasitemia and central nervous system invasion.

Agar Gel-like polysaccharide isolated from red algae and used as thickening agent.

Agglutination Aggregation (clumping) caused when antibodies bind to two antigens, perhaps hindering the activity of pathogenic micro organisms and increasing the chance that they will be phagocytized.

Agglutination test In serology, a procedure in which antiserum is mixed with a sample that potentially contains its target antigen.

Agranulocyte Type of leukocyte having a uniform cytoplasm lacking large granules.

Agroterrorism The use of microbes to terrorize humans by destroying the livestock and crops that constitute their food supply.

Airborne transmission Spread of pathogens to the respiratory mucous membranes of a new host via the air or in droplets carried more than 1 meter.

Alcohol Intermediate-level disinfectant that denatures proteins and disrupts cell membranes.

Aldehyde Compound containing terminal CHO groups; used as a high-level disinfectant because it cross-links organic functional groups in proteins and nucleic acids.

Algae Eukaryotic unicellular or multicellular photosynthetic organisms with simple reproductive structures.

Alginic acid Cell wall polysaccharide of brown algae.

Alkalinophile Microorganism requiring alkaline pH environments.

Allergen An antigen that stimulates an allergic response.

Allergic contact dermatitis Type of delayed hypersensitivity reaction in which chemically modified skin proteins trigger a cell-mediated immune response.

Allergy An immediate hypersensitivity response against an antigen.

Allograft Type of graft in which tissues are transplanted from a donor to a genetically dissimilar recipient of the some species.

Allylamines Class of antifungal drugs that disrupt cytoplasmic membranes.

Alpha interferons (IFN-α) Interferons secreted by virally infected monocytes, macrophages, and some lymphocytes within hours after infection.

Alphaproteobacteria Class of aerobic Gram-negative bacteria in the phylum Proteobacteria capable of growing at very low nutrient levels.

Alternation of generations In algae, method of sexual reproduction in which diploid thalli alternate with haploid thalli.

Alveolar macrophage Fixed macrophage of the lungs.

Alveolates Protozoa with small membrane-bound cavities called alveoli beneath their cell surfaces.

Amebiasis A mild to severe dysentery that, if invasive, can cause the formation of lesions in the liver, lungs, brain, and other organs; caused by infection with *Entamoeba histolytica*.

Ames test Method for screening mutagens that is commonly used to identify potential carcinogens.

Amination Reaction involving the addition of an amine group to a metabolite to make an amino acid.

Amino acid A monomer of polypeptides.

Aminoglycoside Antimicrobial agent that inhibits protein synthesis by changing the shape of the 30S ribosomal subunit.

Ammonification Process by which microorganisms disassemble proteins in soil wastes into amino acids, which are then converted to ammonia.

Amoebae Protozoa that move and feed by pseudopodia.

Amphibolic reaction A reversible metabolic reaction; that is, a reaction which can be catabolic or anabolic.

Anabolism All of the synthesis reactions in an organism taken together.

Anaerobe An organism that cannot tolerate oxygen.

Anaerobic respiration Type of cellular respiration not requiring oxygen atoms as final electron acceptors.

Analytical epidemiology Detailed investigation of a disease, including analysis of data to determine the probable cause, mode of transmission, and possible means of prevention.

Anammox Anaerobic ammonium oxidation, which is one aspect of the nitrogen cycle.

Anaphase Third stage of mitosis, during which sister chromatids separate and move to opposite poles of the spindle to form chromosomes. Also used for the comparable stage of meiosis.
Anaphylactic shock Condition in which the release of inflammatory mediators overwhelms the body's coping mechanisms, causing suffocation, edema, smooth muscle contraction, and often death.
Anaplasmosis *(human granulocytic anaplasmosis, HGA)* Tick-borne disease caused by a rickettsia, *Anaplasma phagocytophilum*, manifesting with flu-like signs and symptoms. Formerly called human granulocytic ehrlichiosis.
Anion A negatively charged ion.
Anthrax Gastrointestinal, cutaneous, or pulmonary disease that is usually fatal without aggressive treatment; caused by ingestion, inoculation, or inhalation of spores of *Bacillus anthracis*.
Antibiotic Antimicrobial agent that is produced naturally by an organism.
Antibody *(immunoglobulin)* Proteinaceous antigen-binding molecule secreted by plasma cells.
Antibody-dependent cellular cytotoxicity (ADCC) Process whereby natural killer lymphocytes (NK cells) lyse cells covered with antibodies.
Anticodon Portion of tRNA molecule that is complementary to a codon on mRNA.
Antigen Molecule that triggers a specific immune response.
Antigen-binding site Site formed by the variable regions of a heavy and light chain of an antibody.
Antigen-presenting cell (APC) Dendritic cells, macrophages, and B cells, which process antigens and activate cells of the immune system.
Antigenic determinant *(epitope)* The three-dimensional shape of a region of an antigen that is recognized by the immune system.
Antigenic drift Phenomenon that occurs every 2–3 years when a single strain of influenzavirus mutates within a local population.
Antigenic shift Major antigenic change that occurs on average every 10 years and results from the reassortment of genomes from different influenzavirus strains within host cells.
Antihistamines Drugs that specifically neutralize histamine.
Antimicrobial Any compound used to treat infectious disease; may also function as intermediate-level disinfectant.
Antimicrobial agent Chemotherapeutic agent used to treat microbial infection.
Antimicrobial enzyme Enzyme that acts against microbes.
Antimicrobial peptide *(defensin)* Chain of about 20–50 amino acids that acts against microorganisms.
Antisense nucleic acid RNA or single-stranded DNA with a nucleotide sequence complementary to a molecule of mRNA; used to control translation of polypeptide.
Antisense RNA RNA with a nucleotide sequence complementary to a molecule of mRNA; used to control translation of polypeptide.
Antisepsis The inhibition or killing of microorganisms on skin or tissue by the use of a chemical antiseptic.
Antiseptic Chemical used to inhibit or kill microorganisms on skin or tissue.
Antiserum In serology, blood fluid containing antibodies which bind to the antigens that triggered their production.
Antitoxin Antibodies formed by the host that bind to and protect against toxins.
Antiviral proteins Proteins triggered by alpha and beta interferons that prevent viral replication.
Apicomplexans In protozoan taxonomy, group of pathogenic alveolate protozoa characterized by the complex of special intracellular organelles located at the apices of the infective stages of these microbes.
Apoenzyme The protein portion of protein enzymes that is inactive unless bound to one or more cofactors.
Apoptosis Programmed cell suicide.
Applied microbiology Branch of microbiology studying the commercial use of microorganisms in industry and foods.
Arachnid Group of arthropods distinguished by the presence of eight legs, such as spiders, ticks, and mites.
Arboviral encephalitis Inflammation of the brain and/or meninges caused by viruses transmitted by bloodsucking arthropods.
Arboviruses Viruses that are transmitted by arthropods; they include members of several viral families.
Archaea (*Archaeon*, sing.) In Woese's taxonomy, domain which includes all prokaryotic cells having archaeal rRNA sequences.
Archaezoa Protozoa that lack mitochondria, Golgi bodies, chloroplasts, and peroxisomes.
Arenaviruses Group of segmented, negative ssRNA viruses that cause zoonotic diseases.
Arthropod Animal with a segmented body, hard exoskeleton, and jointed legs, including arachnids and insects.
Artificial wetlands Use of successive ponds, marshes, and meadowlands to remove wastes from sewage as the water moves through the wetlands.
Artificially acquired active immunity Type of immunity which occurs when the body receives antigens by injection, as with vaccinations, and mounts a specific immune response.
Artificially acquired passive immunotherapy Treatment in which patient receives via injection preformed antibodies in antitoxins or antisera, which can destroy fast-acting and potentially fatal antigens such as rattlesnake venom.
Ascariasis Symptomatic but not typically fatal disease caused by infection with the nematode *Ascaris lumbricoides*.
Ascomycota Division of fungi characterized by the formation of haploid ascospores within sacs called asci.
Ascospore Haploid germinating structure of ascomycetes.
Ascus Sac in which haploid ascospores are formed and from which they are released in fungi of the division Ascomycota.
Aseptate Lacking cross walls.
Aseptic Characteristic of an environment or procedure that is free of contamination by pathogens.
Aspergillosis Term for several localized and invasive diseases caused by infection with *Aspergillus* species.
Assembly In virology, fourth stage of the lytic replication cycle, in which new virions are assembled in the host cell.
Asthma Hypersensitivity reaction affecting the lungs and characterized by bronchial constriction and excessive mucus production.
Astroviruses Group of small, round enteric viruses, which cause diarrhea, typically in children.
Asymptomatic *(subclinical)* Characteristic of disease that may go unnoticed because of absence of symptoms, even though clinical tests may reveal signs of disease.
Atom The smallest chemical unit of matter.
Atomic force microscope (AFM) Type of probe microscope that uses a pointed probe to traverse the surface of a specimen. A laser beam detects vertical movements of the probe, which a computer translates to reveal the atomic topography of the specimen.
Atomic mass *(atomic weight)* The sum of the masses of the protons, neutrons, and electrons in an atom.
Atomic number The number of protons in the nucleus of an atom.
ATP synthase (ATPase) Enzyme that phosphorylates ATP in oxidative phosphorylation and photophosphorylation.
Attachment In virology, first stage of the lytic replication cycle, in which the virion attaches to the host cell.
Attenuated vaccine Inoculum in which pathogens are weakened so that, theoretically, they no longer cause disease; residual virulence can be a problem.
Attenuation The process of reducing vaccine virulence.
Autoantigens Antigens on the surface of normal body cells.
Autoclave Device that uses steam heat under pressure to sterilize chemicals and objects that can tolerate moist heat.
Autograft Type of graft in which tissues are moved to a different location within the same patient.
Autoimmune disease Any of a group of diseases which result when an individual begins to make autoantibodies or cytotoxic T cells against normal body components.
Autoimmune hemolytic anemia Disease resulting when an individual produces antibodies against his or her own red blood cells.
Avirulent Harmless.
Axenic Having only one organism present.
Axial filament In cell morphology, structure composed of rotating endoflagella that allows a spirochete to "corkscrew" through its medium.
Azoles Class of antifungal drugs that disrupt cytoplasmic membranes.

B cell B lymphocyte.
B cell receptor (BCR) Antibody integral to the cytoplasmic membrane and expressed by B lymphocytes.
B lymphocyte *(B cell)* Lymphocyte that arises and matures in the red bone marrow in adults and

is found primarily in the spleen, lymph nodes, red bone marrow, and Peyer's patches of the intestines and which secretes antibodies.

Bacillus Rod-shaped prokaryotic cell.

Bacitracin Antimicrobial which blocks NAG and NAM secretion from the cytoplasm, prompting cell lysis.

Bacteremia The presence of bacteria in the blood; often caused by infection with *Staphylococcus aureus* or *Streptococcus pneumoniae.*

Bacteria Prokaryotic microorganisms typically having cell walls composed of peptidoglycan. In Woese's taxonomy, domain which includes all prokaryotic cells having bacterial rRNA sequences.

Bacterial intoxication *(toxification)* Food poisoning caused by bacterial toxin.

Bacteriophage *(phage)* Virus that infects and usually destroys bacterial cells.

Bacteriorhodopsin Purple protein synthesized by *Halobacterium* that absorbs light energy to synthesize ATP.

Bacteroid Diverse group of Gram-negative microbes that have similar rRNA nucleotide sequences, e.g., *Bacteroides.*

Balantidiasis A mild gastrointestinal illness caused by *Balantidium coli.*

Barophile Microorganism requiring the extreme hydrostatic pressure found at great depth below the surface of water.

Base Molecule that binds with hydrogen ions when dissolved in water.

Base pair (bp) A complementary arrangement of nucleotides in a strand of DNA or RNA. For example, in both DNA and RNA, guanine and cytosine pair.

Base-excision repair Mechanism by which enzymes excise a section of a DNA strand containing an error, and then DNA polymerase fills in the gap.

Basic dye In microscopy, a cationic chromophore used to stain acidic structures. Works most effectively in alkaline environments.

Basidiocarp Fruiting body of basidiomycetes; includes mushrooms, puffballs, stinkhorns, jelly fungi, bird's nest fungi, and bracket fungi.

Basidiomycota Division of fungi characterized by production of basidiospores and basidiocarps.

Basophil Type of granulocyte that stains blue with the basic dye methylene blue.

Bejel Childhood disease caused by the spirochete *Treponema pallidum endemicum;* characterized by rubbery oral lesions.

Benign tumor Mass of neoplastic cells that remains in one place and is not generally harmful.

Benthic zone Bottom zone of freshwater or marine water, devoid of light and with scarce nutrients.

Beta interferons (IFN-β) Interferons secreted by virally infected fibroblasts within hours after infection.

Beta-lactam Antimicrobial whose functional portion is composed of beta-lactam rings, which inhibit peptidoglycan formation by irreversibly binding to the enzymes that cross-link NAM subunits.

Beta-lactamase Bacterial enzyme that breaks the beta-lactam rings of penicillin and similar molecules, rendering them inactive.

Beta-oxidation A catabolic process in which enzymes split pairs of hydrogenated carbon atoms from a fatty acid and join them to coenzyme A to form acetyl-CoA.

Betaproteobacteria Class of diverse Gram-negative bacteria in the phylum Proteobacteria capable of growing at very low nutrient levels.

Binary fission The most common method of asexual reproduction of prokaryotes, in which the parental cell disappears with the formation of progeny.

Binomial nomenclature The classification method used in the Linnaean system of taxonomy, which assigns each species both a genus name and a specific epithet.

Biochemical oxygen demand (BOD) A measure of the amount of oxygen aerobic bacteria require to metabolize organic wastes in water.

Biochemistry Branch of chemistry which studies the chemical reactions of living things.

Biodiversity The number of species living in a given ecosystem.

Biofilm A slimy community of microbes growing on a surface.

Biogeochemical cycling The movement of elements and nutrients from unusable forms to usable forms due to the activities of microorganisms.

Biological vector Biting arthropod or other animal that transmits pathogens and serves as host for the multiplication of the pathogen during some stage of the pathogen's life cycle.

Biomass The quantity of all organisms in a given ecosystem.

Biomining The science involved in searching natural populations of microbes for useful organisms.

Bioremediation The use of microorganisms to metabolize toxins in the environment to reclaim soils and waterways.

Bioreporter Type of biosensor composed of microbes with innate signaling capabilities.

Biosensor Device that combines bacteria or microbial products such as enzymes with electronic measuring devices to detect other bacteria, bacterial products, or chemical compounds in the environment.

Biosphere The region of Earth inhabited by living organisms.

Biotechnology Branch of microbiology in which microbes are manipulated to manufacture useful products.

Bioterrorism The use of microbes or their toxins to terrorize human populations.

Black piedra Benign scalp disease caused by infection with *Piedraia hortae.*

Blastomycosis Pulmonary disease found in the southeastern United States, caused by infection with *Blastomyces dermatitidis.*

Blood group antigens The surface molecules of red blood cells.

Bodily fluid transmission Spread of pathogenic microorganisms via blood, urine, saliva, or other bodily fluids.

Botulism Potentially fatal intoxication with botulism toxin; three types include foodborne botulism, infant botulism, and wound botulism.

Bradykinin Peptide chain of nine amino acids that is a potent mediator of inflammation.

Broad-spectrum drug Antimicrobial that works against many different kinds of pathogens.

Bronchitis Inflammation of the bronchi.

Broth A liquid, nutrient-rich medium used for cultivating microorganisms.

Broth dilution test Test for determining the minimum inhibitory concentration in which a standardized amount of bacteria is added to serial dilutions of antimicrobial agents in tubes or wells containing broth.

Brucellosis Disease caused by *Brucella;* usually asymptomatic or mild, though it can result in sterility or abortion in animals.

Bruton-type agammaglobulinemia An inherited disease in which affected babies cannot make immunoglobulins and experience recurrent bacterial infections.

Bt toxin (Bt) Insecticidal poison produced by *Bacillus thuringiensis* bacteria.

Bubo Swollen inflamed lymph node.

Bubonic plague Severe systemic disease, fatal if untreated in 50% of patients, and characterized by fever, tissue necrosis, and the presence of buboes; caused by infection with *Yersinia pestis.*

Budding In prokaryotes and yeasts, reproductive process in which an outgrowth of the parent cell receives a copy of the genetic material, enlarges, and detaches. In virology, extrusion of enveloped virions through the host's cell membrane.

Buffer A substance, such as a protein, that prevents drastic changes in pH.

Bulbar poliomyelitis Infection of the brain stem and medulla resulting in paralysis of muscles in the limbs or respiratory system; caused by infection with poliovirus.

Bunyaviruses Group of zoonotic pathogens that have a segmented genome of three –ssRNA molecules and are transmitted to humans via arthropods.

Burkitt's lymphoma Infectious cancer of the jaw caused by infection with Epstein-Barr virus.

Caliciviruses Group of small, round enteric viruses, which cause diarrhea, nausea, and vomiting.

Calor Heat.

Calvin-Benson cycle Stage of photosynthesis in which atmospheric carbon dioxide is fixed and reduced to produce glucose.

Cancer Disease characterized by the presence of one or more malignant tumors.

Candidiasis Term for several opportunistic diseases caused by infection with *Candida* species.

Candin Antifungal drug that inhibits cell wall synthesis.

Capnophile Microorganism that grows best with high levels of carbon dioxide in addition to low levels of oxygen.

Capsid A protein coat surrounding the nucleic acid core of a virion.

Capsomere A proteinaceous subunit of a capsid.

Capsule Glycocalyx composed of repeating units of organic chemicals firmly attached to the cell surface.

Capsule stain *(negative stain)* In microscopy, a staining technique used primarily to reveal bacterial capsules and involving application of an acidic dye that leaves the specimen colorless and the background stained.

Carbohydrate Organic macromolecule consisting of atoms of carbon, hydrogen, and oxygen.

Carbon cycle Biogeochemical cycle in which carbon is cycled in the form of organic molecules.
Carbon fixation The attachment of atmospheric carbon dioxide to ribulose 1,5-bisphosphate (RuBP).
Carbuncle The coalescence of several *furuncles* extending deep into underlying tissues; caused by infection with *Staphylococcus aureus*.
Carcinogen Chemical capable of causing cancer.
Caries *(cavities)* Tooth decay; caused by viridans streptococci and other bacteria.
Carotenoid Plant pigment that acts as an antioxidant.
Carrageenan Gel-like polysaccharide isolated from red algae and used as thickening agent.
Carrier In human pathology, continuous asymptomatic human source of infection.
Catabolism All of the decomposition reactions in an organism taken together.
Catarrhal phase In pertussis, initial phase lasting 1–2 weeks and characterized by signs and symptoms resembling those of a common cold.
Cation A positively charged ion.
Cat scratch disease Common and occasionally serious infection in children; characterized by fever and malaise plus localized swelling; caused by infection with *Bartonella henselae*.
CD4 Distinguishing cytoplasmic membrane protein of helper T cells, which is the initial binding site of HIV.
CD8 Distinguishing cytoplasmic membrane protein of cytotoxic T cells.
CD95 pathway In cell-mediated cytotoxicity, pathway involving CD95 protein which triggers apoptosis of infected cells.
Cell culture Cells isolated from an organism and grown on the surface of a medium or in broth. Viruses can be grown in a cell culture.
Cell wall In most cells, structural boundary composed of polysaccharide or protein chains that provides shape and support against osmotic pressure.
Cell-mediated immune response Immune response used by T cells to fight intracellular pathogens and abnormal body cells.
Cellular respiration Metabolic process that involves the complete oxidation of substrate molecules and production of ATP via a series of redox reactions.
Cellular slime mold Individual haploid myxamoeba that phagocytizes bacteria, yeasts, dung, and decaying vegetation.
Central dogma In genetics, fundamental description of protein synthesis which states that genetic information is transferred from DNA to RNA to polypeptides, which function alone or in conjunction as proteins.
Centrioles Nonmembranous organelles in animal cells which appear to function in the formation of flagella and cilia and in cell division.
Centrosome Region of a cell containing centrioles.
Cesspool Home equivalent of primary wastewater treatment in which wastes enter a series of porous concrete rings buried underground and are digested by microbes.
Cestodes (tapeworms) Group of helminths that are long, flat, and segmented and lack digestive systems.
Chagas' disease Potentially fatal disease caused by a bite from a kissing bug carrying *Trypanosoma cruzi* and characterized by the formation of swellings at the site of the bite, followed by fever, swollen lymph nodes, myocarditis, organ enlargement, and eventually congestive heart failure.
Chancre Painless red lesion that appears at the site of infection with *Treponema pallidum*, the agent of syphilis.
Chancroid Soft, painful, venereal ulcer at site of infection by *Haemophilus ducreyi*.
Chemical bond An interaction between atoms in which electrons are either shared or transferred in such a way as to fill their valence shells.
Chemical fixation In microscopy, a technique that uses methyl alcohol or formalin to attach a smear to a slide.
Chemical reaction The making or breaking of a chemical bond.
Chemiosmosis Use of ion gradients to generate ATP.
Chemoautotroph Microorganism that uses carbon dioxide as a carbon source and catabolizes organic molecules for energy.
Chemoheterotroph Microorganism that uses organic compounds for both energy and carbon.
Chemokine An immune system cytokine that signals leukocytes to rush to the site of inflammation or infection and activate other leukocytes.
Chemotactic factors Chemicals, such as peptides derived from complement and cytokines, that attract cells.
Chemotaxis Cell movement that occurs in response to chemical stimulus.
Chemotherapeutic agent Chemical used to treat disease.
Chemotherapy A branch of medical microbiology in which chemicals are studied for their potential to destroy pathogenic microorganisms.
Chickenpox *(varicella)* Highly infectious disease characterized by fever, malaise, and skin lesions, caused by infection with varicella-zoster virus.
Chitin Strong, flexible nitrogenous polysaccharide found in fungal cell walls and in the exoskeletons of insects and other arthropods.
Chlamydia Any of the small Gram-negative pathogenic cocci that grow and reproduce within the cells of mammals, birds, and a few invertebrates and that spread as elementary bodies.
Chloramphenicol Antimicrobial drug that blocks the enzymatic site of the 50S ribosomal subunit, inhibiting polypeptide synthesis.
Chlorophyll Pigment molecule that captures light energy for use in photosynthesis.
Chlorophyta Green-pigmented division of algae that have chlorophylls *a* and *b*, store sugar and starch as food reserves, and have rRNA sequences similar to plants. Considered the progenitors of plants.
Chloroplast Light-harvesting organelle found in photosynthetic eukaryotes.
Cholera Disease contracted through the ingestion of food and water contaminated with *Vibrio cholerae* and characterized by vomiting and watery diarrhea.
Cholera toxin Exotoxin produced by *Vibrio cholerae* that causes the movement of water out of the intestinal epithelium.
Chromatin Threadlike mass of DNA and associated histone proteins that becomes visible during mitosis as chromosomes.
Chromatin fiber An association of nucleosomes and proteins found within the chromosomes of eukaryotic cells.
Chromoblastomycosis Cutaneous and subcutaneous disease characterized by lesions that can spread internally; caused by traumatic introduction of ascomycete fungi into the skin.
Chromosome A molecule of DNA associated with protein. In prokaryotes, typically circular and localized in a region of the cytosol called the nucleoid. In eukaryotes, chromosomes are threadlike and are most visible during mitosis and meiosis.
Chronic disease Any disease that develops slowly, usually with less severe symptoms, and is continual or recurrent.
Chronic granulomatous disease Primary immunodeficiency disease in which children have recurrent infections characterized by the development of large masses of inflammatory cells in lymph nodes, lungs, bones, and skin.
Chronic inflammation Type of inflammation that develops slowly, lasts a long time, and can cause damage (even death) to tissues, resulting in disease.
Chrysophyta Division of algae including the golden algae, yellow-green algae, and diatoms.
Cilia Short, hairlike, rhythmically motile projections of some eukaryotic cells.
Ciliate In protozoan taxonomy, group of alveolate protozoa characterized by the presence of cilia in their trophozoite stages.
Class Taxonomical grouping of similar orders of organisms.
Class switching The process in which a plasma cell changes the type of antibody F_c region (stem) that it synthesizes and secretes.
Clinical specimen Sample of human material, such as feces or blood, that is examined or tested for the presence of microorganisms.
Clonal deletion Process by which cells with receptors that respond to autoantigens are selectively killed via apoptosis.
Clonal expansion In immunology, the reproduction of activated lymphocytes.
Clonal selection In humoral immunity, recognition and activation only of B lymphocytes with BCRs complementary to a specific antigenic determinant.
Coccidioidomycosis Pulmonary disease found in the southwestern United States, caused by infection with *Coccidioides immitis*.
Coccobacillus A prokaryotic cell intermediate in shape between a sphere and a rod, e.g., an elongated coccus.
Coccus Spherical prokaryotic cell.
Codon Triplet of mRNA nucleotides that codes for specific amino acids. For example, AAA is a codon for lysine.
Coenocyte Multinucleate cell resulting from repeated mitosis but postponed or absent cytokinesis.
Coenzyme Organic cofactor.
Cofactor Inorganic ions or organic molecules that are essential for enzyme action.

Glossary

Coinfection Condition in which a patient is infected simultaneously with hepatitis B and D viruses.
Cold enrichment Incubation of a specimen in a refrigerator to enhance the growth of cold-tolerant species.
Coliforms Enteric Gram-negative bacteria that ferment lactose to gas and are found in the intestinal tracts of animals and humans.
Colony Visible population of microorganisms living in one place; an aggregation of cells arising from a single parent cell.
Colony-forming unit (CFU) A single cell or group of related cells that produce a colony.
Colorado tick fever Zoonosis caused by *Coltivirus* that is typically characterized by mild fever and chills.
Combination vaccine Inoculum composed of antigens from several pathogens that are administered simultaneously.
Commensalism Symbiotic relationship in which one member benefits without significantly affecting the other.
Communicable disease Any infectious disease that comes either directly or indirectly from another host.
Competence Ability of a cell to take up DNA from the environment.
Competitive inhibitor Inhibitory substance that blocks enzyme activity by blocking active sites.
Complement fixation test A complex assay used to determine the presence of specific antibodies in serum.
Complement system Set of blood plasma proteins that act as chemotactic attractants, trigger inflammation and fever, and ultimately effect the destruction of foreign cells.
Complementary DNA (cDNA) DNA synthesized from an mRNA template using reverse transcriptase.
Complex medium Culturing medium that contains nutrients released by the partial digestion of yeast, beef, soy, or other proteins; thus, the exact chemical composition is unknown.
Complex transposon Transposon containing genes not connected with transposition.
Compound A molecule containing atoms of more than one element.
Compound microscope Microscope using a series of lenses for magnification.
Concentration gradient The difference in concentration of a chemical on the two sides of a membrane. Also called a *chemical gradient*.
Condenser lens In a compound microscope, a lens that directs light through the specimen as well as one or more mirrors that deflect the light's path.
Condyloma acuminata Large, cauliflower-like genital warts caused by infection with a papillomavirus.
Confocal microscope Type of light microscope that uses ultraviolet lasers to illuminate fluorescent chemicals in a single plane of the specimen.
Congenital syphilis Disease characterized by mental retardation, organ malformation, and in some cases, death of the fetus of a woman infected with *Treponema pallidum*, the agent of syphilis.
Conjugation In genetics: method of horizontal gene transfer in which a bacterium containing a fertility plasmid forms a conjugation pilus that attaches and transfers plasmid genes to a recipient; in reproduction of ciliates: coupling of mating cells.
Conjugation pilus Proteinaceous, rodlike structure extending from the surface of a cell; mediates conjugation.
Conjunctivitis *(pinkeye)* Inflammation of the lining of an eyelid.
Consumption Tuberculosis; refers to wasting away of a body affected with TB at several sites.
Contact immunity Immunity conferred to an unvaccinated individual following contact with an individual vaccinated with an attenuated vaccine.
Contagious disease A communicable disease that is easily transmitted from a reservoir or patient.
Contamination The presence of microorganisms in or on the body or other site.
Continuous cell culture Type of cell culture created from tumor cells.
Contrast The difference in visual intensity between two objects, or between an object and its background.
Convalescence In the infectious disease process, final stage during which the patient recovers from the illness, and tissues and systems are repaired and return to normal.
Convalescent phase In pertussis, final phase lasting 3–4 weeks during which the ciliated lining of the trachea grows back and frequency of coughing spells diminishes, but secondary bacterial infections may ensue.
Cord factor A cell-wall component of pathogenic *Mycobacterium tuberculosis* that produces strands of daughter cells, inhibits migration of neutrophils, and is toxic to body cells.
Coronaviruses Group of enveloped ssRNA viruses which cause colds as well as severe acute respiratory syndrome.
Corticosteroids Another name for immunosuppressive agents that suppress the action of T cells.
Counterstain In a Gram stain, red stain that provides contrasting color to the primary stain, causing Gram-negative cells to appear pink.
Covalent bond The sharing of a pair of electrons by two atoms.
Coxsackieviruses Group of enteroviruses which cause a variety of diseases in humans, ranging from mild fever and colds to myocar ditis and heart failure.
Crenation Shriveling of a cell caused by osmosis in a hypertonic environment.
Cristae Folds within the inner membrane of a mitochondrion, which increase its surface area.
Cross resistance Phenomenon in which resistance to one antimicrobial drug confers resistance to similar drugs.
Crossing over Process in which portions of homologous chromosomes are recombined during the formation of gametes.
Croup Inflammation and swelling of the larynx, trachea, and bronchi and a "seal bark" cough, often caused by infection with a parainfluenza or rarely by other respiratory viruses.
Cryptococcosis Disease caused by the dimorphic fungus *Cryptococcus;* typically manifests as meningitis.
***Cryptosporidium* enteritis** *(cryptosporidiosis)* A gastrointestinal disease caused by infection with *Cryptosporidium parvum*; in humans, characterized by diarrhea and fluid and weight loss; may be fatal in HIV-positive patients.
Culture Act of cultivating microorganisms or the microorganisms that are cultivated.
Cuticle Outer protective "skin" of a nematode.
Cyanobacteria Gram-negative photosystem bacteria that vary greatly in shape, size, and method of reproduction.
Cyclic photophosphorylation Return of electrons to the original reaction center of a photosystem after passing down an electron transport chain.
Cycloserine Semisynthetic antibiotic used to treat infections with Gram-positive bacteria.
Cyclosporine Immunosuppressive drug that inhibits action of activated T cells.
Cyst In protozoan morphology, the hardy resting stage characterized by a thick capsule and a low metabolic rate.
Cysticercus Immature tapeworm, usually in muscle of intermediate host.
Cytokines Proteins secreted by many types of cells that regulate adaptive immune responses.
Cytokinesis Division of a cell's cytoplasm.
Cytoplasm General term used to describe the semiliquid, gelatinous material inside a cell.
Cytoplasmic membrane Membrane surrounding all cells, and composed of a fluid mosaic of phospholipids and proteins.
Cytosine Ring-shaped nitrogenous base found in nucleotides of DNA and RNA.
Cytoskeleton Internal network of fibers contributing to the basic shape of eukaryotic and rod-shaped prokaryotic cells.
Cytosol The liquid portion of the cytoplasm.
Cytotoxic drugs Group of drugs that inhibit cells.
Cytotoxic T cell *(Tc cell, CD8 cell)* In cell-mediated immune response, type of cell characterized by CD8 cell-surface glycoprotein; secretes perforins and granzymes which destroy infected or abnormal body cells.

Dark-field microscope Microscope used for studying pale or small specimens; deflects light rays so that they miss the objective lens.
Dark repair Mechanism by which enzymes cut damaged DNA sections from a molecule, creating a gap that is repaired by DNA polymerase and DNA ligase.
Deamination Process in which amine groups are split from amino acids.
Death phase Phase in a growth curve in which the organisms are dying more quickly than they are being replaced by new organisms.
Decimal reduction time (D) The time required to destroy 90% of the microbes in a sample.
Decline In the infectious disease process, period in which the body gradually returns to normal as the patient's immune response and any medical treatments vanquish the pathogens.
Decolorizing agent In a stain, a solution that washes the primary stain away.
Decomposition reaction A chemical reaction in which the bonds of larger molecules are broken to form smaller atoms, ions, and molecules.

Glossary

Deep-freezing Long-term storage of cultures at temperatures ranging from −50°C to −95°C.
Deeply branching bacteria Prokaryotic autotrophs with rRNA sequences and growth characteristics thought to be similar to those of earliest bacteria.
Defensins *(antimicrobial peptides)* Small peptide chains that act against a broad range of pathogens.
Defined medium (synthetic medium) Culturing medium of which the exact chemical composition is known.
Definitive host In the life cycle of parasites, host in which mature and sometimes sexual forms of the parasite are present and usually reproducing.
Degerming The removal of microbes from a surface by scrubbing.
Dehydration synthesis Type of synthesis reaction in which two smaller molecules are joined together by a covalent bond, and a water molecule is formed.
Delayed hypersensitivity reaction *(type IV hypersensitivity)* T cell–mediated inflammatory reaction that takes 24–72 hours to reach maximal intensity.
Deletion Type of mutation in which a nucleotide base pair is deleted.
Deltaproteobacteria Group of Proteobacteria that includes *Desulfovibrio*, *Bdellovibrio*, and myxobacteria.
Denaturation Process by which a protein's three-dimensional structure is altered, eliminating function.
Dendritic cells Cells of the epidermis and mucous membranes that devour pathogens.
Dengue fever Self-limiting but extremely painful disease caused by a flavivirus transmitted by *Aedes* mosquitoes.
Dengue hemorrhagic fever Potentially fatal disease involving a hyperimmune response to reinfection with dengue virus and causing ruptured blood vessels, internal bleeding, and shock.
Denitrification The conversion of nitrate into nitrogen gas by anaerobic respiration.
Deoxyribonucleic acid (DNA) Nucleic acid consisting of nucleotides made up of phosphate, a deoxyribose pentose sugar, and an arrangement of the bases adenine, guanine, cytosine, and thymine.
Dermatophyte Fungus that normally lives on skin, nails, or hair.
Dermatophytoses Any of a variety of superficial skin, nail, and hair infections caused by dermatophytes.
Dermis The layer of the skin deep to the epidermis and containing hair follicles, glands, and nerve endings.
Descriptive epidemiology The careful recording of data concerning a disease.
Desiccation Inhibition of microbial growth by drying.
Detergent Positively charged organic surfactant.
Deuteromycetes Informal grouping of fungi having no known sexual stage.
Diapedesis *(emigration)* Process whereby leukocytes leave intact blood vessels by squeezing between lining cells.
Diatom Type of alga in the division Chrysophyta; has cell walls made of silica arranged in nesting halves called frustules.
Dichotomous key Method of identifying organisms in which information is arranged in paired statements, only one of which applies to any particular organism.
Differential interference contrast microscope Type of phase microscope that uses prisms to split light beams, giving images a three-dimensional appearance.
Differential medium Culturing medium formulated such that either the presence of visible changes in the medium or differences in the appearances of colonies help microbiologists differentiate among kinds of bacteria growing on the medium.
Differential stain In microscopy, a stain using more than one dye so that different structures can be distinguished. The Gram stain is the most commonly used.
Differential white blood cell count Lab technique that indicates the relative numbers of leukocytes.
Diffusion The net movement of a chemical down its concentration gradient.
Diffusion susceptibility test *(Kirby-Bauer test)* Simple, inexpensive test widely used to reveal which drug is most effective against a particular pathogen. Procedure involves inoculating a Petri plate uniformly with a standardized amount of the pathogen in question and arranging on the plate disks soaked in the drugs to be tested.
DiGeorge syndrome Failure of the thymus to develop, and thus, absence of T cells.
Dimorphic Having two forms, e.g., dimorphic fungi have both yeastlike and moldlike thalli.
Dinoflagellate In protozoan taxonomy, group of unicellular, flagellated, alveolate protozoa characterized by photosynthetic pigments.
Dioecious Male and female sex organs are in separate individuals.
Diphtheria Mild to potentially fatal respiratory disease caused by diphtheria toxin following infection with *Corynebacterium diphtheriae*.
Diphtheroids Generally nonpathogenic pleomorphic bacilli named for the similarity of their appearance to *Corynebacterium diphtheriae*.
Diplococcus A pair of cocci.
Diploid A nucleus with two copies of each chromosome.
Diploid cell culture Type of cell culture created from embryonic animal, plant, or human cells that have been isolated and provided appropriate growth conditions.
Dipstick immunochromatographic assay Rapid modification of ELISA test in which an antigen solution flows through a porous strip, encountering labeled antibody; used for pregnancy testing and for rapid identification of infectious agents.
Direct antibody test Immune test allowing direct observation of the presence of antigen.
Direct contact transmission Spread of pathogens from one host to another involving body contact between the hosts.
Disaccharide Carbohydrate consisting of two monosaccharide molecules joined together.
Disease Any adverse internal condition severe enough to interfere with normal body functioning.
Disease process Definite sequence of events following contamination and infection.
Disinfectant Physical or chemical agent used to inhibit or destroy microorganisms on inanimate objects.
Disinfection The use of physical or chemical agents to inhibit or destroy microorganisms on inanimate objects. In water treatment, ozone, UV light, or chlorination kill most microorganisms.
Disseminated intravascular coagulation (DIC) The formation of blood clots within blood vessels throughout the body; triggered by lipid A.
DNA microarray Numerous distinct ssDNA molecules bound to a substrate and used to probe for complementary sequences.
Dolor Pain.
Domain Any of three basic types of cell groupings distinguished by Carl Woese, containing the Linnaean taxon of *kingdoms*.
Donor cell In horizontal gene transfer, a cell that contributes part of its genome to a recipient.
Droplet transmission Spread of pathogens from one host to another via aerosols, which exit the body during exhaling, coughing, and sneezing and travel less than 1 meter.
Dysentery Disease characterized by severe diarrhea often with stools containing blood and mucus.
Dyspnea Difficulty in breathing.

E site In translation, site at which tRNA exits from the ribosome.
Eastern equine encephalitis (EEE) Potentially fatal infection of the brain caused by a togavirus.
Ebola virus Virus of Africa causing a type of hemorrhagic fever fatal in 90% of cases.
Echoviruses *(enteric cytopathic human orphan viruses)* Group of enteroviruses that cause viral meningitis and colds.
Ecosystem All of the organisms living in a particular habitat, and the relationships between the two.
Ehrlichiosis *(human monocytic ehrlichiosis, HME)* Tick-borne disease caused by a rickettsia, *Ehrlichia chaffeensis*, manifesting with flulike signs and symptoms. Also previously used to refer to human granulocytic ehrlichiosis, now called *anaplasmosis*.
Electrical gradient Voltage across a membrane created by the electrical charges of the chemicals on either side.
Electrochemical gradient The chemical and electrical gradients across a cell membrane.
Electrolyte Any hydrated cation or anion; can conduct electricity through a solution.
Electron A negatively charged subatomic particle.
Electron transport chain Series of redox reactions that pass electrons from one membrane-bound carrier to another, and then to a final electron acceptor.
Electronegativity The attraction of an atom for electrons.
Elek test Immunodiffusion assay used to detect the presence of diphtheria toxin in a fluid sample.
Element Matter that is composed of a single type of atom.
Elementary bodies Infectious stage in the life cycle of chlamydias.
Elephantiasis Enlargement and hardening of tissues, especially in the lower extremities, where lymph has accumulated following infection with *Wuchereria bancrofti*.

Empyema In patients with staphylococcal pneumonia, the presence of pus in the alveoli of the lungs.
Encephalitis Inflammation of the brain.
Encystment In the life cycle of protozoa, stage in which cysts form in host tissues.
Endemic In epidemiology, a disease that occurs at a relatively stable frequency within a given area or population.
Endemic typhus *(murine typhus)* Disease transmitted by fleas and characterized by high fever, headache, chills, muscle pain, and nausea; caused by infection with *Rickettsia typhi.*
Endocarditis Potentially fatal inflammation of the endocardium; typically caused by infection with *Staphylococcus aureus* or *Streptococcus pneumoniae.*
Endocytosis Active transport process, used by some eukaryotic cells, in which pseudopodia surround a substance and move it into the cell.
Endoflagellum A special flagellum of spirochetes that spirals tightly around a cell, rather than protruding from it.
Endogenous antigen Antigen produced by microbes that multiply inside the cells of the body.
Endogenous infection An infection arising within the patient from opportunistic pathogens.
Endoplasmic reticulum (ER) Netlike arrangement of hollow tubules continuous with the outer membrane of the nuclear envelope and functioning as a transport system.
Endosome A sac formed during endocytosis containing the endocytized substance.
Endospore Environmentally resistant structure produced by the transformation of a vegetative cell of the Gram-positive genera *Bacillus* or *Clostridium.*
Endosymbiotic theory Proposal that eukaryotes were formed from the phagocytosis of small prokaryotes by larger prokaryotes, forming organelles.
Endothermic reaction Any chemical reaction that requires energy.
Endotoxin *(lipid A)* Potentially fatal toxin released from the cell wall of dead and dying Gram-negative bacteria.
Enrichment culture Technique used to enhance the growth of less abundant microorganisms by using a selective medium.
Enterobacteriaceae *(enteric bacteria)* Family of oxidase-negative Gram-negative bacteria, which can be pathogenic.
Enteroviruses Group of picornaviruses that are transmitted via the fecal-oral route but cause disease in any of a variety of target organs.
Entner-Doudoroff pathway Series of reactions that catabolize glucose to pyruvic acid using different enzymes from those used in either glycolysis or the pentose phosphate pathway.
Entry In virology, second stage of the lytic replication cycle, in which the virion or its genome enters the host cell.
Envelope In virology, membrane surrounding the viral capsid.
Environmental microbiology Branch of microbiology studying the role of microorganisms in soils, water, and other habitats.
Environmental specimen Sample of material taken from such sources as ponds, soil, or air and tested for the presence of microorganisms.
Enzyme An organic catalyst.
Enzyme-linked immunosorbent assays (ELISAs) A family of simple immune tests that use enzymatic products as a label and that can be readily automated and read by machine.
Eosinophil Type of granulocyte that stains red to orange with the acidic dye eosin.
Eosinophilia An abnormal blood condition in which the number of eosinophils is greater than normal.
Epidemic In epidemiology, a disease that occurs at a greater than normal frequency for a given area or population.
Epidemiology Study of the occurrence, distribution, and spread of disease in humans.
Epidermis The outermost layer of the skin.
Epididymitis Inflammation of the epididymis.
Epitope *(antigenic determinant)* The three-dimensional shape of a region of an antigen that is recognized by the immune system.
Epsilonproteobacteria Group of Gram-negative rods, vibrios, and spiraled bacteria in the phylum Proteobacteria.
Erysipelas Impetigo spreading to lymph nodes, accompanied by pain and inflammation and caused by infection with group A *Streptococcus.*
Erythema infectiosum *(fifth disease)* Harmless red rash occurring in children and caused by infection with B19 virus.
Erythrocyte Red blood cell.
Erythrocytic cycle In the life cycle of *Plasmodium,* stage during which merozoites infect and cause lysis of erythrocytes.
Eschar Black, swollen, crusty, painless skin ulcer of anthrax.
Etest Test for determining minimum inhibitory concentration; a plastic strip containing a gradient of the antimicrobial agent being tested is placed on a plate inoculated with the pathogen of interest.
Ethambutol Antimicrobial drug that disrupts formation of arabinogalactan-mycolic acid by mycobacteria.
Etiology The study of the causation of disease.
Euglenids Protozoa that store food as paramylon, lack cell walls, and have eyespots used in positive phototaxis.
Eukarya In Woese's taxonomy, domain which includes all eukaryotic cells.
Eukaryote Any organism made up of cells containing a nucleus composed of genetic material surrounded by a distinct membrane. Classification includes animals, plants, algae, fungi, and protozoa.
Eutrophication The overgrowth of microorganisms in aquatic systems.
Evolution Changes in the genetic makeup of a population leading to the production of new varieties.
Exchange reaction Type of chemical reaction in which atoms are moved from one molecule to another by means of the breaking and forming of covalent bonds.
Excystment In the life cycle of protozoa, stage following ingestion by the host, in which cysts become trophozoites.
Exfoliative toxins Toxins of certain strains of *Staphylococcus aureus* that break down desmosomes in the skin, causing the outer layers of skin to slough off.
Exocytosis Active transport process, used by some eukaryotic cells, in which vesicles fuse with the cytoplasmic membrane and export their substances from the cell.
Exoerythrocytic phase In the life cycle of *Plasmodium,* stage during which infected mosquito injects sporozoites into the blood.
Exogenous antigen Antigen produced by microorganisms that multiply outside the cells of the body.
Exogenous nosocomial infection An infection caused by pathogens acquired from the health care environment.
Exon Coding sequence of mRNA. Exons are connected to produce a functional mRNA molecule.
Exothermic reaction Any chemical reaction that releases energy.
Exotoxin Toxin secreted by a pathogenic microorganism into its environment.
Experimental epidemiology The testing of hypotheses resulting from analytical epidemiology concerning the cause of a disease.
Exponential (logarithmic) growth Increase in size of a microbial population in which the number of cells doubles in a fixed interval of time.
Extremophile Microbe that requires extreme conditions of temperature, pH, and/or salinity to survive.

F (fertility) plasmid (F factor) Small, circular, extrachromosomal molecule of DNA coding for conjugation pili. Bacterial cells that contain an F plasmid are called F^+ cells and serve as donors during conjugation.
F_C region The stem region of an antibody.
Facilitated diffusion Movement of substances across a cell membrane via protein channels.
Facultative anaerobe Microorganism which can live with or without oxygen.
Family Taxonomical grouping of similar genera of organisms.
Fats Compounds composed of three fatty acid molecules linked to a molecule of glycerol.
Fecal-oral infection Spread of pathogenic microorganisms in feces to the mouth, such as results from drinking sewage-contaminated water.
Feedback inhibition *(negative feedback)* Method of controlling the action of enzymes in which the end-product of a series of reactions inhibits an enzyme in an earlier part of the pathway.
Fermentation In metabolism, the partial oxidation of sugar to release energy using an endogenous organic molecule rather than an electron transport chain as the final electron acceptor. In food microbiology, any desirable change to food or beverage induced by microbes.
Fever Body temperature above 37°C.
Fever blisters (cold sores) Painful, itchy lesions on the lips; characteristic of infection with *human herpesvirus 1.*
Filariasis Infection of the lymphatic system caused by a filarial nematode.
Filtration The passage of air or liquid through a material that traps and removes microbes. In water treatment, a process in which microbial biofilms on sand particles trap and remove other microbes.
Fimbriae Sticky, proteinaceous extensions of some bacterial cells that function to adhere cells to one another and to environmental surfaces.

Firmicutes Phylum of bacteria that includes clostridia, mycoplasmas, and low G + C Gram-positive bacilli and cocci.

Flagellates Group of protozoa that possess at least one long flagellum, generally used for movement.

Flagellum A long, whiplike structure protruding from a cell.

Flavin adenine dinucleotide (FAD) Important vitamin-derived electron carrier molecule.

Flea Vertically flattened, bloodsucking, wingless insect, vector of some pathogens.

Flocculation Process in water treatment in which alum (aluminum ammonium sulfate) added to the water forms sediments with particles and microorganisms.

Fluid mosaic model Model describing the arrangement and motion of the proteins within the cytoplasmic membrane.

Fluorescent microscope Type of light microscope that uses an ultraviolet light source to fluoresce objects.

Fly Insect with transparent wings that are not hidden or covered, including mosquitoes; vectors for many pathogens.

Folliculitis Infection of a hair follicle by *Staphylococcus aureus*.

Fomes (pl. fomites) Objects inadvertently used to transfer pathogens to new hosts, e.g., glass, towel.

Food infection Type of food poisoning in whch living organisms are consumed.

Food intoxication Type of food poisoning resulting from consumption of microbial toxin.

Food microbiology The use of microorganisms in food production and the prevention of foodborne illnesses.

Food vesicle Sac formed during endocytosis of a solid, also called an endosome or phagosome.

Foodborne transmission Spread of pathogenic microorganisms in or on foods that are poorly processed, undercooked, or improperly refrigerated.

Foraminifera Type of armored marine amoeba.

Formed elements Cells and cell fragments suspended in blood plasma.

Frameshift mutation Type of mutation in which nucleotide triplets subsequent to an insertion or deletion are displaced, creating new sequences of codons that result in vastly altered polypeptide sequences.

Functional group An arrangement of atoms common to all members of a class of organic molecules, such as the amine group found in all amino acids.

Fungi Eukaryotic organisms that have cell walls and obtain food from other organisms.

Furuncle A large, painful, nodular extension of folliculitis into surrounding tissue; may be caused by infection with *Staphylococcus aureus*.

Gametocyte In sexual reproduction of protozoa, cell that can fuse with another gametocyte to form a diploid zygote.

Gamma interferons (IFN-γ) Interferon produced by T lymphocytes and NK lymphocytes; activates macrophages and neutrophils days after an infection.

Gammaproteobacteria Largest and most diverse class of Proteobacteria, including purple sulfur bacteria, methane oxidizers, pseudomonads, and others.

Gas gangrene Death of muscle and connective tissues accompanied by gaseous waste, caused by *Clostridium perfringens*.

Gaseous agent High-level disinfecting gas used to sterilize heat-sensitive equipment and large objects.

Gastroenteritis Inflammation of the mucous membrane of the stomach and intestines.

Gel electrophoresis Technique used in recombinant DNA technology to separate molecules by size, shape, and electrical charge.

Gene A specific sequence of nucleotides that codes for a polypeptide or an RNA molecule.

Gene library Collection of bacterial or phage clones, each of which carries a fragment of an organism's genome.

Gene therapy The use of recombinant DNA technology to insert a missing gene or repair a defective gene in human cells.

Genera Plural of genus.

Generation time Time required for a cell to grow and divide.

Genetic engineering The manipulation of genes via recombinant DNA technology for practical applications.

Genetic fingerprinting (DNA fingerprinting) Technique which identifies unique sequences of DNA to: determine paternity; connect blood, semen, or skin cells to suspects in criminal investigations; or identify pathogens.

Genetic mapping Application of recombinant DNA technology in which genes are located on a nucleic acid molecule.

Genetic recombination The exchange of segments, typically genes, between two DNA molecules.

Genetic screening Procedure by which laboratory tests are used to screen patient and fetal DNA for mutant genes.

Genetics The study of inheritance and heritable traits as expressed in an organism's genetic material.

Genome The sum of all the genetic material in a cell or virus.

Genomics The sequencing, analysis, and comparison of genomes.

Genotype Actual set of genes in an organism's genome.

Genus Taxonomical grouping of similar species of organisms.

Germ theory of disease Hypothesis formulated by Pasteur in 1857 that microorganisms are responsible for disease.

Giardiasis A mild to severe gastrointestinal illness caused by ingestion of cysts of *Giardia intestinalis*.

Gingivitis Inflammation of the gums.

Glomerulonephritis Deposition of immune complexes in the walls of the glomeruli—networks of minute blood vessels in the kidneys—which may result in kidney failure; typically caused by infection with group A *Streptococcus*.

Glucocorticoids (corticosteroids) Immunosuppressive agents, including prednisone and methylprednisolone, that suppress the response of T cells.

Glycocalyx Sticky external sheath of prokaryotic and eukaryotic cells.

Glycolysis (Embden-Meyerhof pathway) First step in the catabolism of glucose via respiration and fermentation.

Goblet cells Mucus-secreting cells in the epithelium of mucous membranes.

Golgi body In eukaryotic cells, a series of flattened, hollow sacs surrounded by phospholipid bilayers and functioning to package large molecules for export in secretory vesicles.

Gonorrhea A sexually transmitted disease caused by infection with *Neisseria gonorrhoeae*.

gp41 Antigenic HIV glycoprotein that promotes fusion of the viral envelope with a target cell.

gp120 Antigenic glycoprotein that is the primary attachment molecule of HIV.

Graft Tissue or organ transplanted to a new site.

Graft rejection Rejection of donated tissue or organs by a transplant recipient.

Graft-versus-host disease Disease resulting when donated bone marrow cells mount an immune response against the recipient's cells.

Gram-negative cell Generally, a prokaryotic cell having a wall composed of a thin layer of wall material, an external membrane, and a periplasmic space between; appears pink after the Gram staining procedures.

Gram-positive cell Prokaryotic cell having a thick wall; in bacteria, composed of a thick layer of peptidoglycan containing teichoic acids; Gram-positive cells retain the crystal violet dye used in the Gram staining procedure, appearing purple.

Gram stain Technique for staining microbial samples by applying a series of dyes that leave some microbes purple and others pink. Developed by Christian Gram in 1884.

Granulocyte Type of leukocyte having large granules in the cytoplasm.

Granzyme Protein molecule in the cytoplasm of cytotoxic T cells that causes an infected cell to undergo apoptosis.

Graves' disease Production of autoantibodies that stimulate excessive production of thyroid hormone and growth of the thyroid gland.

Group A *Streptococcus* (*S. pyogenes*) A coccus that produces protein M and a hyaluronic acid capsule, both of which contribute to the pathogenicity of the species.

Group B *Streptococcus* (*S. agalactiae*) A Gram-positive coccus that normally resides in the lower GI, genital, and urinary tracts but can cause disease in newborns.

Group translocation Active process, occurring in some prokaryotes, by which a substance being actively transported across a cell membrane is chemically changed during transport.

Growth An increase in size; in bacteriology, an increase in population.

Growth curve Graph that plots the number in a population over time.

Growth factor Organic chemical such as a vitamin required in very small amounts for metabolism. In immunology, an immune system cytokine that stimulates stem cells to divide, ensuring that the body is supplied with sufficient leukocytes of all types.

Guanine Ring-shaped nitrogenous base found in nucleotides of DNA and RNA.

Glossary

Gumma Lesion that occurs in bones, nervous tissue, or on skin in patients with tertiary syphilis.

HAART (highly active antiretroviral therapy) A cocktail of antiviral drugs including nucleoside analogs, protease inhibitors, and reverse transcriptase inhibitors.
Habitat The physical localities in which organisms are found.
Halogen One of the four very reactive, nonmetallic chemical elements: iodine, chlorine, bromine, and fluorine. Used in disinfectants and antiseptics.
Halophile Microorganism requiring a saline environment (greater than 9% NaCl).
Hamus Proteinaceous, filamentous, helical extension of some archaeal cells that functions to attach the cells to one another and environmental surfaces.
Hansen's disease (leprosy) Disease caused by infection with *Mycobacterium leprae* that produces either a nonprogressive tuberculoid form or a progressive lepromatous form that destroys tissues, including facial features, digits, and other structures.
***Hantavirus* pulmonary syndrome (HPS)** Rapid, severe, and often fatal pneumonia caused by infection with a *Hantavirus*.
Hantaviruses Group of bunyaviruses that are transmitted to humans via inhalation of virions in dried deer-mouse excreta and that cause *Hantavirus* pulmonary syndrome.
Haploid A nucleus with a single copy of each chromosome.
Haustoria Modified hyphae that penetrate the tissue of the host to withdraw nutrients.
Hay fever Allergic reaction localized to the upper respiratory tract and characterized by nasal discharge, sneezing, itchy throat and eyes, and excessive tear production.
Heat fixation In microscopy, a technique that uses the heat from a flame to attach a smear to a slide.
Heavy-metal ions The ions of high-molecular-weight metals such as arsenic that are used as antimicrobial agents because they denature proteins. They have largely been replaced because they are also toxic to human cells.
Helminths Multicellular eukaryotic worms, some of which are parasitic.
Helper T cell *(Th or CD4 cell)* In cell-mediated immune response, a type of cell characterized by CD4 cell-surface glycoprotein. It regulates the activity of B cells and cytotoxic T cells during an immune response.
Hemagglutinin (HA) Component of glycoprotein spikes in the lipid envelope of influenza viruses that help them attach to pulmonary epithelial cells.
Hemolytic disease of the newborn Disease that results when antibodies made by an Rh-negative woman cross the placenta and destroy the red blood cells of an Rh-positive fetus.
Hemorrhagic fever Viral syndrome characterized by fever, bleeding in the skin and mucous membranes, low blood pressure, and shock.
HEPA (high-efficiency particulate air) filter Filters built into biological safety cabinets; they prevent exposure to microbes by maintaining a barrier of moving filtered air across the cabinet's openings.
Hepatitis Inflammation of the liver.
Hepatitis A Inflammation of the liver, resulting from infection with hepatitis A virus.
Hepatitis B Inflammatory condition of the liver caused by infection with hepatitis B virus and characterized by jaundice.
Hepatitis C Chronic inflammation of the liver that can cause permanent liver damage or hepatic cancer; caused by infection with hepatitis C virus.
Hepatitis D Inflammation of liver that can cause severe liver damage or hepatic cancer; caused by infection with hepatitis D virus.
Hepatitis E *(enteric hepatitis)* Inflammation of the liver, which is fatal in 20% of infected pregnant women; caused by hepatitis E virus.
Herd immunity Protection against illness provided to a population when a pathogen cannot spread because the majority of the group are resistant to the pathogen.
Herpangina Lesions of the mouth and pharynx caused by coxsackie A virus; resemble those of herpesvirus.
Herpes Painful, itchy skin lesions caused by herpesviruses.
Herpes zoster *(shingles)* Extremely painful skin rash caused by reactivation of latent varicella-zoster virus.
Heterocyst Thick-walled nonphotosynthetic cell of cyanobacteria; reduces nitrogen.
Hfr (high frequency of recombination) cell Cell containing an F plasmid that is integrated into the prokaryotic chromosome. Hfr cells form conjugation pili and transfer cellular genes more frequently than normal F^+ cells.
Highly active antiretroviral therapy (HAART) A cocktail of antiviral drugs including nucleotide analogs, protease inhibitors, and reverse transcriptase inhibitors.
Histamine Inflammatory chemical released from damaged cells that causes vasodilation of capillaries.
Histone Globular protein found in eukaryotic and archaeal chromosomes.
Histoplasmosis Pulmonary, cutaneous, ocular, or systemic disease found in the Ohio River Valley and caused by infection with *Histoplasma capsulatum*.
Holoenzyme The combination of an apoenzyme and its cofactors.
Horizontal (lateral) gene transfer Process in which a donor cell contributes part of its genome to a recipient cell, which may be a different species or genus from the donor.
Host In symbiosis, member of a parasitic relationship that supports the parasite.
Human herpesviruses Group of viruses of humans that cause skin lesions, which are often creeping; diseases include herpes, chickenpox, mononucleosis, and roseola.
Human immunodeficiency viruses Retro viruses that destroy the immune system.
Human T-lymphotropic viruses Group of oncogenic retroviruses associated with cancer of lymphocytes.
Humoral immune response The antibody immune response used by B cells to fight exogenous antigens.
Hybridomas Tumor cells created by fusing antibody-secreting plasma cells with cancerous plasma cells called *myelomas*.
Hydatid Fluid-filled.
Hydatid disease Potentially fatal disease caused by infection with the canine tapeworm *Echinococcus granulosus* and characterized by the presence of fluid-filled cysts in the liver or other tissues.
Hydrogen bond The electrical attraction between a partially charged hydrogen atom and a full or partial negative charge on a different region of the same molecule or another molecule. Hydrogen bonds confer unique properties to water molecules.
Hydrolysis A decomposition reaction in which a covalent bond is broken, and the ionic components of water are added to the products.
Hydrophilic Attracted to water.
Hydrophobia Literally, a fear of water; symptom caused by painful swallowing characteristic of rabies infection.
Hydrophobic Insoluble in water.
Hydrothermal vent Vent in marine abyssal zone that spews superheated, nutrient-rich water.
Hydroxyl radical Most reactive of the toxic forms of oxygen.
Hypersensitivity Any immune response against a foreign antigen that is exaggerated beyond the norm.
Hypersensitivity pneumonitis A form of pneumonia.
Hyperthermophile Microorganism requiring temperatures above 80°C.
Hypertonic Characteristic of a solution having a higher concentration of solutes than another.
Hyphae Long, branched, tubular filaments in the thalli of molds.
Hypotonic Characteristic of a solution having a lower concentration of solutes than another.

Iatrogenic infections A subset of nosocomial infections that are the direct result of a medical procedure or treatment, such as surgery.
Illness In the infectious disease process, the most severe stage, in which signs and symptoms are most evident.
Immune complexes Antigen-antibody complexes.
Immune thrombocytopenic purpura Disease resulting when drugs bound to platelets bind antibodies and complement, causing the platelets to lyse.
Immunization Administration of an antigenic inoculum to stimulate an adaptive immune response and immunological memory.
Immunochromatographic assay Immune test in which antigen molecules form visible immune complexes with antibodies labeled with a colored substance.
Immunodiffusion An immune test in which antibodies and antigens diffuse from separate wells in agar to form a line of precipitate.
Immunofiltration assay Rapid modification of ELISA test using membrane filters rather than plates.

Immunoglobulin (Ig) *(antibody)* Proteinaceous antigen-binding molecule secreted by plasma cells.
Immunoglobulin A (IgA) The antibody class most commonly associated with various body secretions, including tears and milk. IgA pairs with a secretory component to form *secretory IgA*.
Immunoglobulin D (IgD) A membrane-bound antibody molecule found in some animals as a B cell receptor.
Immunoglobulin E (IgE) Signal antibody molecule that triggers the inflammatory response, particularly in allergic reactions and infections by parasitic worms.
Immunoglobulin G (IgG) The predominant antibody class found in the bloodstream and the primary defender against invading bacteria.
Immunoglobulin M (IgM) The second most common antibody class and the predominant antibody produced first during a primary humoral immune response.
Immunological synapse Interface between cells of the immune system that involves cell-to-cell signaling.
Immunology Study of the body's specific defenses against pathogens.
Immunophilins Immunosuppressive drugs, such as cyclosporine, that inhibit T cell function.
Immunotherapy Administration of antibodies (passive immunization) or dilute antigen so as to provide immunological protection against antigens.
Impetigo Presence of red, pus-filled vesicles on the face and limbs of children; caused by infection with *Staphylococcus aureus* or *Streptococcus pyogenes*.
Inactivated polio vaccine (IPV) Inoculum developed by Jonas Salk in 1955 for vaccination against poliovirus.
Inactivated vaccine Inoculum containing either whole agents or subunits and often adjuvants.
Incidence In epidemiology, the number of new cases of a disease in a given area or population during a given period of time.
Inclusion Deposited substance such as a lipid, gas vesicle, or magnetite, stored within the cytosol of a cell.
Inclusion bodies In the life cycle of chlamydias, eukaryotic phagosomes full of chlamydial reticulate bodies.
Incubation period Stage in infectious disease process between infection and occurrence of the first symptoms or signs of disease. In a laboratory culture, the period between adding a sample to a plate and the development of colonies.
Index case In epidemiology, the first instance of the disease in a given area or population.
Indicator organisms Fecal microbe found in the environment that reveals potential contamination by feces.
Indirect contact transmission Spread of pathogens from one host to another via inanimate objects called *fomites*.
Indirect fluorescent antibody test Immune test allowing observation through a fluorescent microscope of the presence of antigen in a tissue sample flooded with labeled antibody.
Indirect selection *(negative selection)* Process by which auxotrophic mutants are isolated and cultured.
Induced-fit model Description of way in which an enzyme changes its shape slightly after binding to its substrate so as to bind it more tightly.
Inducible operon Type of operon that is not normally transcribed and must be activated by inducers.
Induction In virology, excision of a prophage from the host chromosome, at which point the prophage reenters the lytic phase.
Industrial microbiology Branch of microbiology in which microbes are manipulated to manufacture useful products.
Infection Successful invasion of the body by a pathogenic microorganism.
Infection control Branch of microbiology studying the prevention and control of infectious disease.
Infectious mononucleosis *(mono)* Disease characterized by sore throat, fever, fatigue, and enlargement of the spleen and liver; caused by infection with Epstein-Barr virus.
Influenza (flu) Infectious disease caused by two species of orthomyxoviruses and characterized by fever, malaise, headache, and myalgia; certain strains can be fatal.
Initial body *(reticulate body)* Reproductive structure of chlamydias that undergoes repeated binary fissions until the host cell is filled.
Innate immunity Resistance to pathogens conferred by barriers, chemicals, cells, and processes that remain unchanged upon subsequent infections with the same pathogens.
Inoculum Sample of microorganisms.
Inorganic chemical Molecule lacking carbon.
Insect Arthropod with three distinct body divisions: head, thorax, abdomen; vectors for some helminthic, protozoan, bacterial, and viral pathogens.
Insertion Type of mutation in which a base pair is inserted into a genetic sequence.
Insertion sequence A simple transposon consisting of no more than two inverted repeats and a gene that encodes the enzyme transposase.
Integrase Enzyme carried by the virions of HIV which allows integration into a human chromosome.
Interferons (IFNs) Protein molecules that inhibit the spread of viral infections.
Interleukins (ILs) Immune system cytokines that signal among leukocytes.
Intermediate host In the life cycle of parasites, host in which immature forms of the parasite are present and undergoing various stages of maturation.
Intoxication (bacterial) Food poisoning caused by bacterial toxin.
Intron Noncoding sequence of mRNA, which, is removed to make functional mRNA.
In-use test Method of evaluating the effectiveness of a disinfectant or antiseptic which tests efficacy under specific, real-life conditions.
Inverted repeat Palindromic sequence found at each end of a transposon.
Ion An atom or group of atoms that has either a full negative charge or a full positive charge.
Ionic bond A type of bond formed from the attraction of opposite electrical charges. Electrons are not shared.
Ionizing radiation Form of radiation with wavelengths shorter than 1 nm that are energetic enough to create ions by ejecting electrons from atoms.
Ischemia Local anemia due to interruption of blood supply by mechanical blockage.
Isograft Type of graft in which tissues are moved between genetically identical individuals (identical twins).
Isoniazid Antimicrobial drug that disrupts formation of arabinogalactan-mycolic acid by mycobacteria.
Isotonic Characteristic of a solution having the same concentration of solutes and water as another.
Isotopes Atoms of a given element that differ only in the number of neutrons they contain.

Jaundice Yellowing of skin and eyes due to accumulation of bilirubin in the blood.

Kelsey-Sykes capacity test Standard assessment approved by the European Union to determine the ability of a given chemical to inhibit microbial growth.
Keratitis Inflammation of the cornea.
Kinetoplastid Euglenozoan protozoan with a single large mitochondrion that contains an apical region of mitochondrial DNA called a *kinetoplast*.
Kingdom Taxonomical grouping of similar phyla of organisms.
Kinins Powerful inflammatory chemicals released by mast cells.
Kissing bug Blood-eating insect of family Reduviidae that seemingly prefers oral blood vessels.
Koch's postulates A series of steps, elucidated by Robert Koch, that must be taken to prove the cause of any infectious disease.
Koplik's spots Mouth lesions characteristic of measles.
Korarchaeota Phylum of archaea; known only from environmental RNA samples.
Krebs cycle Series of eight enzymatically catalyzed reactions that transfer stored energy from acetyl-CoA to coenzymes NAD^+ and FAD.

Lag phase Phase in a growth curve in which the organisms are adjusting to their environment.
Lagging strand Daughter strand of DNA synthesized in short segments that are later joined. Synthesis of the lagging strand always moves away from the replication fork, and lags behind synthesis of the leading strand.
Laryngitis Inflammation of the larynx.
Latency In virology, process by which an animal virus, sometimes not incorporated into the chromosomes of the cell, remains inactive in the cell, possibly for years.
Latent disease Any disease in which a pathogen remains inactive for a long period of time before becoming active.
Latent virus *(provirus)* An animal virus that remains inactive in a host cell.
Lateral (horizontal) gene transfer Process in which a donor cell contributes part of its genome to a recipient cell, which may be a different species or genus from the donor.

Leading strand Daughter strand of DNA synthesized continuously toward the replication fork as a single long chain of nucleotides.

Legionnaires' disease *(legionellosis)* Severe pneumonia caused by infection with a *Legionella species,* usually *L. pneumophila.*

Leishmaniasis Any of three clinical syndromes caused by a bite from a sand fly carrying *Leishmania* and ranging from painless skin ulcers to disfiguring lesions to visceral leishmaniasis, which is systemic and fatal in 95% of untreated cases.

Lepromin test Assay utilizing antigens of *Mycobacterium leprae* used in diagnosis of leprosy.

Leprosy (Hansen's disease) Disease caused by infection with *Mycobacterium leprae* that produces either a nonprogressive tuberculoid form or a progressive lepromatous form that destroys tissues, including facial features, digits, and other structures.

Leptospirosis Zoonotic disease contracted by humans upon exposure to infected animals; characterized by pain, headache, and liver and kidney disease; caused by infection with *Leptospira interrogans.*

Leukocyte White blood cell.

Leukopenia Decrease in the number of white blood cells in the blood.

Leukotrienes Inflammatory chemicals released from damaged cells that increase vascular permeability.

Lichen Organism composed of a fungus living in partnership with photosynthetic microbes, either green algae or cyanobacteria.

Light-dependent reaction Reaction of photosynthesis requiring light.

Light-independent reaction Reaction of photosynthesis not requiring light and synthesizing glucose from carbon dioxide and water.

Light repair Mechanism by which prokaryotic DNA photolyase breaks the bonds between adjoining pyrimidine nucleotides, restoring the original DNA sequence.

Limnetic zone Sunlit, upper layer of freshwater or marine water away from the shore.

Lincosamides Antimicrobial drugs that bind to the 50S subunit of bacterial ribosomes, preventing ribosomal movement.

Lipid Any of a diverse group of organic macromolecules not composed of monomers and insoluble in water.

Lipid A The lipid component of lipopolysaccharide, which is released from dead Gram-negative bacterial cells and can trigger shock and other symptoms in human hosts.

Lipopolysaccharide (LPS) Molecule composed of lipid A and polysaccharide found in the external membrane of Gram-negative cell walls.

Listeriosis Disease caused by *Listeria monocytogenes* and usually manifesting as meningitis and bacteremia.

Lithotroph Microorganism that acquires electrons from inorganic sources.

Littoral zone Shoreline zone of freshwater or marine water.

Log phase Phase in a growth curve in which the population is most actively growing.

Logarithmic (exponential) growth Increase in size of a microbial population in which the number of cells doubles in a fixed interval of time.

Louse (pl. lice) Sucking or biting parasitic insects, which vector some bacterial pathogens.

Louse-borne relapsing fever Disease caused by *Borrelia recurrentis* transmitted between humans by the body louse *Pediculus humanus.*

Lyme disease Disease carried by ticks infected with *Borrelia burgdorferi* and characterized by a "bull's-eye" rash, neurologic and cardiac dysfunction, and severe arthritis.

Lymph Fluid found in lymphatic vessels, which is similar in composition to blood serum and intercellular fluid.

Lymph nodes Organs that monitor the composition of lymph.

Lymphangitis Condition in which inflamed lymphatic vessels become visible as red streaks under the skin.

Lymphatic system Body system composed of lymphatic vessels and lymphoid tissues and organs.

Lymphatic vessels Tubes that conduct lymph.

Lymphocyte Type of small agranulocyte which originates in the red bone marrow and has nuclei that nearly fill the cell.

Lymphocytic choriomeningitis (LCM) Zoonosis caused by an arenavirus; characterized by flulike symptoms and rarely by meningitis.

Lymphogranuloma venereum Sexually transmitted disease caused by infection with *Chlamydia trachomatis* and leading in some cases to proctitis or, in women, pelvic inflammatory disease.

Lyophilization Removal of water from a frozen culture or other substance by means of vacuum pressure. Used for the long-term preservation of cells and foods.

Lysogenic conversion Change in phenotype due to insertion of a lysogenic bacteriophage into a bacterial chromosome.

Lysogenic phage Bacteriophage that does not immediately kill its host cell.

Lysogenic replication cycle (lysogeny) Process of viral replication in which a bacteriophage enters a bacterial cell, inserts into the DNA of the host, and remains inactive. The phage is then replicated every time the host cell replicates its chromosome. Later, the phage may leave the chromosome.

Lysosome Vesicle in animal cells that contains digestive enzymes.

Lysozyme Antibacterial protein secreted in sweat.

Lytic replication cycle Process of viral replication consisting of five stages ending with lysis of and release of new virions from the host cell.

Macrolide Antimicrobial agent that inhibits protein synthesis by inhibiting the ribosomal 50S subunits.

Macrophage Mature form of monocyte, which is a phagocyte of bacteria, fungi, spores, and dust, as well as dead cells.

Macule Any flat, reddened skin lesion; characteristic of early infection with a poxvirus.

Magnification The apparent increase in size of an object viewed via microscopy.

Major histocompatibility complex (MHC) A cluster of genes, located on each copy of chromosome 6 in humans, that codes for membrane-bound glycoproteins called major histocompatibility antigens.

Malaise Feeling of general discomfort.

Malaria A mild to potentially fatal disease caused by a bite from an *Anopheles* mosquito carrying any of four species of *Plasmodium;* characterized by fever, chills, hemorrhage, and potential destruction of brain tissue.

Malignant tumor Mass of neoplastic cells that can invade neighboring tissues and may metastasize to cause tumors in distant organs or tissues.

Marburg virus Filamentous virus causing a type of hemorrhagic fever and fatal in 25% of cases.

Margination Process by which leukocytes stick to the walls of blood vessels at the site of infection.

Mast cells Specialized cells located in connective tissue that release histamine when they are exposed to complement.

Matter Anything that takes up space and has mass.

MDR TB (multi-drug-resistant TB) Tuberculosis caused by *Mycobacterium* resistant to at least isoniazid and rifampin.

Measles *(rubeola)* Contagious disease characterized by fever, sore throat, headache, dry cough, conjunctivitis, and lesions called Koplik's spots; caused by infection with *Morbillivirus.*

Mechanical vector Housefly, cockroach, or other animal that passively carries pathogens to new hosts on its feet or other body parts and is not infected by the pathogens it carries.

Medium A collection of nutrients used for cultivating microorganisms.

Meiosis Nuclear division of diploid eukaryotic cells resulting in four haploid nuclei.

Membrane attack complexes (MACs) The end products of the complement cascade, which form circular holes in a pathogen's membrane.

Membrane filters Thin circles of nitrocellulose or plastic containing specific pore sizes, some small enough to trap viruses.

Membrane filtration Direct method of estimating population size in which a large sample is poured through a filter small enough to trap cells.

Membrane raft In a eukaryotic membrane, a distinct assemblage of lipids and proteins that remains together as a functional group.

Memory B cell B lymphocyte that migrates to lymphoid tissues to await a subsequent encounter with antigen previously encountered.

Memory T cell Type of T cell that persists in lymphoid tissues for months or years awaiting subsequent contact with an antigenic determinant matching its TCR, at which point it produces cytotoxic T cells.

Memory response The rapid and enhanced immune response to a subsequent encounter with a familiar antigen.

Meningitis Inflammation of the meninges, which can be caused by bacteria, viruses, fungi, or protozoa.

Meningoencephalitis Inflammation of the brain and of its meninges.

Mesophile Microorganism requiring temperatures ranging from 20°C to about 40°C.
Messenger RNA (mRNA) Form of ribonucleic acid that carries genetic information from DNA to a ribosome.
Metabolism The sum of all chemical reactions, both anabolic and catabolic, within an organism.
Metachromatic granules Inclusions of *Corynebacteria* that store phosphate and stain differently from the rest of the cytoplasm.
Metaphase Second stage of mitosis, during which chromosomes line up and attach to microtubules of the spindle. Also used for the comparable stage of meiosis.
Metastasis The spreading of malignant cancer cells to nonadjacent organs and tissues, where they produce new tumors.
Methane oxidizer Any Gram-negative bacterium that utilizes methane both as a carbon and as an energy source.
Methanogen Obligate anaerobe that produces methane gas.
Methicillin-resistant *Staphylococcus aureus* (MRSA) Strain of *S. aureus* that is resistant to many common antimicrobial drugs, and has emerged as a major nosocomial problem.
Methylation Process in which a cell adds a methyl group to one or two bases that are part of specific nucleotide sequences.
Microaerophile Microorganism that requires low levels of oxygen.
Microbe An organism or virus too small to be seen without a microscope.
Microbial antagonism (microbial competition) Normal condition in which established microbiota use up available nutrients and space, reducing the ability of arriving pathogens to colonize.
Microbial death Permanent loss of reproductive capacity of a microorganism.
Microbial death rate A measurement of the efficacy of an antimicrobial agent.
Microbial ecology The study of the interactions of microorganisms among themselves and their environment.
Microbiota The group of microbes that normally inhabit the surfaces of the body without causing disease.
Microglia Fixed macrophages of the nervous system.
Micrograph A photograph of a microscopic image.
Microorganism An organism too small to be seen without a microscope.
MicroRNA (miRNA) Short (about 21-nucleotide) RNA molecule that binds to complementary segment of messenger RNA (mRNA), preventing translation.
Microscopy The use of light or electrons to magnify objects.
Microsporidia Unicellular, intracellular, parasitic fungi previously classified as protozoa.
Minimum bactericidal concentration (MBC) test An extension of the MIC test in which samples taken from clear MIC tubes are transferred to plates containing a drug-free growth medium and monitored for bacterial replication.
Minimum inhibitory concentration (MIC) The smallest amount of a drug that will inhibit a pathogen.
Mismatch repair Mechanism by which enzymes scan newly synthesized, nonmethylated DNA for mismatched bases, remove them, and replace them.
Missense mutation A substitution in a nucleotide sequence resulting in a codon that specifies a different amino acid: what is transcribed makes sense, but not the right sense.
Mite Minute arachnid, which vectors *Orientia*, the agent of scrub typhus.
Mitochondria Spherical to elongated structures found in most eukaryotic cells that produce most of the ATP in the cell.
Mitosis Nuclear division of a eukaryotic cell resulting in two nuclei with the same ploidy as the original.
Mold A typically multicellular fungus that grows as long filaments called *hyphae* and reproduces by means of spores.
Molecular biology Branch of biology combining aspects of biochemistry, cell biology, and genetics to explain cell function at the molecular level, particularly via the use of genome sequencing.
Molecular mimicry Process in which microorganisms with epitopes similar to self-antigens trigger autoimmune tissue damage.
Molecule Two or more atoms held together by chemical bonds.
Molluscum contagiosum Skin disease caused by *Molluscipoxvirus;* characterized by smooth, waxy papules.
Monoclonal antibodies Identical antibodies secreted by a cell line originating from a single plasma cell.
Monocyte Type of agranulocyte which has slightly lobed nuclei.
Monoecious One individual contains both male and female organs.
Monomer A subunit of a macromolecule such as a protein.
Monosaccharide *(simple sugar)* A monomer of carbohydrate, such as a molecule of glucose.
Morbidity Any change from a state of health.
Mordant In microscopy, a substance that binds to a dye and makes it less soluble.
Mosquito Type of fly with bloodsucking females; vector for many pathogens.
Most probable number (MPN) method Statistical estimation of the size of a microbial population based upon the dilution of a sample required to eliminate microbial growth.
Multiple drug resistance Lack of sensitivity to three or more antimicrobials by so-called "superbugs."
Multiple sclerosis (MS) Autoimmune disease in which cytotoxic T cells attack and destroy the myelin sheath that insulates neurons.
Mumps Disease caused by infection with the mumps virus and characterized by fever, parotitis, pain in swallowing, and in some cases, meningitis or deafness.
Mutagen Physical or chemical agent that introduces a mutation.
Mutant A cell with an unrepaired genetic mutation, or any of its descendants.
Mutation In genetics, a permanent change in the nucleotide base sequence of a genome.
Mutualism Symbiotic relationship in which both members benefit from their interaction.
Myalgia Muscle pain.
Mycelium Tangled mass of hyphae.
Mycetismus Mushroom poisoning.
Mycetoma Destructive, tumorlike infection of the skin, fascia, and/or bones of the hands or feet caused by mycelial fungi of several genera in the division Ascomycota.
Mycolic acid Long carbon-chain waxy lipid found in the walls of cells in the genus *Mycobacterium* that makes them resistant to desiccation and staining with water-based dyes.
Mycology The scientific study of fungi.
Mycoplasmas Class of low G + C bacteria which lack cytochromes, enzymes of the Krebs cycle, and cell walls and are pleomorphic.
Mycosis Fungal disease.
Mycotoxicosis Poisoning caused by eating food contaminated with fungal toxins.
Mycotoxins Secondary metabolites produced by fungi and toxic to humans.
Myxobacteria Gram-negative, aerobic, soil-dwelling bacteria with a unique life cycle including a stage of differentiation into fruiting bodies containing resistant myxospores.

Nanoarchaeum Small archaeon genus that possibly represents a fourth phylum of archaea; known only from environmental RNA samples.
Narrow-spectrum drug Antimicrobial that works against only a few kinds of pathogens.
Natural killer (NK) lymphocyte Type of defensive leukocyte of innate immunity that secretes toxins onto the surfaces of virally infected cells and neoplasms.
Naturally acquired active immunity Type of immunity that occurs when the body responds to exposure to antigens by mounting specific immune responses.
Naturally acquired passive immunity Type of immunity that occurs when a fetus, newborn, or child receives antibodies across the placenta or within breast milk.
Necrosis Death of a tissue of organ.
Necrotizing fasciitis Potentially fatal condition marked by toxemia, organ failure, and destruction of muscle and fat tissue following infection with group A *Streptococcus.*
Negative feedback *(feedback inhibition)* Method of controlling the action of enzymes in which the end-product of a series of reactions inhibits an enzyme in an earlier part of the pathway.
Negative selection *(indirect selection)* Process by which auxotrophic mutants are isolated and cultured.
Negative stain *(capsule stain)* In microscopy, a staining technique used primarily to reveal bacterial capsules and involving application of an acidic dye that leaves the specimen colorless and the background stained.
Negative-strand RNA (–RNA) Viral single-stranded RNA transcribed from the +ssRNA genome by viral RNA polymerase.
Negri bodies Aggregates of virions in the brains of rabies patients.

Nematodes Group of round, unsegmented helminths with pointed ends that have a complete digestive tract.

Neoplasia Uncontrolled cell division in a multicellular animal.

Neuraminidase (NA) Component of glycoprotein spikes in the lipid envelope of influenzaviruses that provides access to cell surfaces by hydrolyzing mucus in the lungs.

Neutralization Antibody function in which the action of a toxin or attachment of a pathogen is blocked.

Neutralization test Immune test that measures the ability of antibodies to neutralize the biological activity of pathogens and toxins.

Neutron An uncharged subatomic particle.

Neutrophil Type of granulocyte that stains lilac with a mixture of acidic and basic dyes.

Neutrophile Microorganism requiring neutral pH.

Nicotinamide adenine dinucleotide (NAD^+) Important vitamin-derived electron carrier molecule.

Nicotinamide adenine dinucleotide phosphate ($NADP^+$) Important vitamin-derived electron carrier molecule.

Nitrification The process by which bacteria convert reduced nitrogen compounds such as ammonia into nitrate, which is more available to plants.

Nitrifying bacteria Chemoautotrophic bacteria that derive electrons from the oxidation of nitrogenous compounds.

Nitrogen cycle Biogeochemical cycle involving nitrogen fixation, ammonification, nitrification, denitrification, and anammox reactions.

Nitrogen fixation The conversion of atmospheric nitrogen to ammonia.

NOD protein In innate immunity, intracellular receptor for microbial component.

Noncommunicable disease An infectious disease that arises from outside of hosts or from normal microbiota.

Noncompetitive inhibitor Inhibitory substance that blocks enzyme activity by binding to an allosteric site on the enzyme other than the active site.

Noncyclic photophosphorylation The production of ATP by noncyclic electron flow.

Nonionizing radiation Electromagnetic radiation with a wavelength greater than 1 nm.

Nonliving reservoir of infection Soil, water, food, or inanimate object that is a continuous source of infection.

Nonpolar covalent bond Type of chemical bond in which there is equal sharing of electrons between atoms with similar electronegativities.

Nonsense mutation A substitution in a nucleotide sequence that causes an amino acid codon to be replaced by a stop codon.

Normal microbiota Microorganisms that colonize the surfaces of the human body without normally causing disease. They may be resident or transient.

Noroviruses Group of caliciviruses that cause diarrhea.

Nosocomial disease A disease acquired in a health care facility.

Nosocomial infection An infection acquired in a health care facility.

Nuclear envelope Double membrane composed of phospholipid bilayers surrounding a cell nucleus.

Nuclear pores Spaces in the nuclear envelope that function to control the transport of substances through it.

Nucleoid Region of the prokaryotic cytosol containing the cell's chromosome(s).

Nucleolus Specialized region in a cell nucleus where RNA is synthesized.

Nucleoplasm The semiliquid matrix of a cell nucleus.

Nucleosome Bead of DNA bound to histone in a eukaryotic chromosome.

Nucleotide Monomer of a nucleic acid.

Nucleotide analog Compound structurally similar to a normal nucleotide that can be incorporated into DNA; may result in mismatched base pairing.

Nucleus Spherical to ovoid membranous organelle containing a eukaryotic cell's primary genetic material.

Numerical aperture Measure of the ability of a lens to gather light.

Nutrient Any chemical such as carbon, hydrogen, etc., required for growth of microbial populations.

Objective lens In microscopy, the lens immediately above the object being magnified.

Obligate aerobe Microorganism that requires oxygen as the final electron acceptor of the electron transport chain.

Obligate anaerobe Microorganism that cannot tolerate oxygen and uses a final electron acceptor other than oxygen.

Obligate halophile Microorganism requiring high osmotic pressure.

Occult septicemia The condition of an unidentified bacterial pathogen being present in the blood and causing signs of illness.

Ocular herpes *(ophthalmic herpes)* Disorder characterized by conjunctivitis, a gritty feeling in the eye, and pain, and characteristic of latent *human herpesvirus 1*.

Ocular lens In microscopy, the lens closest to the eyes. May be single *(monocular)* or paired *(binocular)*.

Operator Regulatory element in an operon where repressor protein binds to stop transcription.

Operon A series of genes, a promoter, and often an operator sequence controlled by one regulatory gene. The operon model explains gene regulation in prokaryotes.

Opportunistic pathogens Microorganisms that cause disease when the immune system is suppressed, when microbial antagonism is reduced, or when introduced into an abnormal area of the body.

Opsonin Antimicrobial protein that enhances phagocytosis.

Opsonization The coating of pathogens by proteins called *opsonins*, making them more vulnerable to phagocytes.

Optimum growth temperature Temperature at which a microorganism's metabolic activities produce the highest growth rate.

Oral polio vaccine (OPV) Inoculum developed by Albert Sabin in 1961 for vaccination against poliovirus.

Orchitis Inflammation of a testis.

Order Taxonomical grouping of similar families of organisms.

Organelle Cellular structure that acts as a tiny organ to carry out one or more cell functions.

Organic compounds Molecules that contain both carbon and hydrogen atoms.

Organotroph Microorganism that acquires electrons from organic sources.

Ornithosis *(parrot fever)* A respiratory disease of birds that can be transmitted to humans and is caused by infection with *Chlamydophila psittaci*.

Orphan virus A virus that has not been specifically linked to any particular disease.

Osmosis The diffusion of water molecules across a selectively permeable membrane.

Osmotic pressure The pressure exerted across a selectively permeable membrane by the solutes in a solution on one side of the membrane. The osmotic pressure exerted by high-salt or high-sugar solutions can be used to inhibit microbial growth in certain foods.

Osteomyelitis Inflammation of the bone marrow and surrounding bone; often caused by infection with *Staphylococcus*.

Otitis media Inflammation of the middle ear, often caused by *Streptococcus pneumoniae*.

Oxazolidinone Antibacterial drug that inhibits initiation of polypeptide synthesis in Gram-positive bacteria.

Oxidase test Chemical test for presence of cytochrome oxidase in a cell.

Oxidation lagoons Successive wastewater treatment areas (lagoons) used by farmers and ranchers to treat animal wastes and from which water is released into natural water systems.

Oxidation-reduction reaction *(redox reaction)* Any metabolic reaction involving the transfer of electrons from an electron donor to an electron acceptor. Reactions in which electrons are accepted are called *reduction* reactions, whereas reactions in which electrons are donated are *oxidation* reactions.

Oxidative phosphorylation The use of energy from redox reactions to attach inorganic phosphate to ADP.

Oxidizing agent Antimicrobial agent that releases oxygen radicals.

P site In a ribosome, a binding site that holds a tRNA and the growing polypeptide.

Palisade In cell morphology, a folded arrangement of bacilli.

Pandemic In epidemiology, the occurrence of an epidemic on more than one continent simultaneously.

Papilloma *(wart)* Benign growth of the epithelium of the skin or mucous membranes.

Papule Any raised, reddened skin lesion that progresses from a macule; characteristic of infection with a poxvirus.

Parabasalid Group of single-celled, animal-like microorganisms that contain a Golgi-like parabasal body.

Paracoccidioidomycosis Pulmonary disease found from southern Mexico to South America caused by infection with *Paracoccidioides brasiliensis*.

Parainfluenzaviruses Group of enveloped, negative ssRNA viruses that cause respiratory disease, particularly in children.

Parasite A microbe that derives benefit from its host while harming it or even killing it.

Parasitism Symbiotic relationship in which one organism derives benefit while harming, or even killing, its host.

Parasitology The study of parasites.

Parenteral route A means by which pathogenic microorganisms can be deposited directly into deep tissues of the body, as in puncture wounds and hypodermic injections.

Paroxysmal phase In pertussis, second phase lasting 2–4 weeks and characterized by exhausting coughing spells.

Parvoviruses Group of extremely small, pathogenic ssDNA viruses.

Passive immunotherapy *(passive immunization)* Delivery of preformed antibodies against pathogens to patients.

Pasteurellaceae Family of gammaproteobacteria, two genera of which—*Pasteurella* and *Haemophilus*—are pathogenic.

Pasteurization The use of heat to kill pathogens and reduce the number of spoilage microorganisms in food and beverages.

Pathogen A microorganism capable of causing disease.

Pathogen-associated molecular patterns (PAMPs) Molecules that are shared by a variety of microbes, are absent in humans, and trigger immune responses.

Pathogenicity A microorganism's ability to cause disease.

Pelvic inflammatory disease (PID) Infection of the uterus and uterine tubes; may be caused by infection with any of several bacteria.

Pentose phosphate pathway Enzymatic formation of phosphorylated pentose sugars from glucose 6-phosphate.

Peptic ulcer Erosion of the mucous membrane of the stomach or duodenum, usually caused by infection with *Helicobacter pylori*.

Peptide bond A covalent bond between amino acids in proteins.

Peptidoglycan Large, interconnected polysaccharide composed of chains of two alternating sugars and crossbridges of amino acids. Main component of bacterial cell walls.

Perforin Protein molecule in the cytoplasm of cytotoxic T cells which forms channels (perforations) in an infected cell's membrane.

Periodontal disease Inflammation and infection of the tissues surrounding and supporting the teeth.

Periplasmic space In Gram-negative cells, the space between the cell membrane and the outer membrane containing peptidoglycan and periplasm.

Peritrichous Term used to describe a cell having flagella covering the cell surface.

Peroxide anion Toxic form of oxygen which is detoxified by catalase or peroxidase.

Peroxisome Vesicle found in all eukaryotic cells that degrades poisonous metabolic wastes.

Pertussis (whooping cough) Pediatric disease characterized by development of copious mucus, loss of tracheal cilia, and deep "whooping" cough; caused by infection with *Bordetella pertussis*.

Petechiae Subcutaneous hemorrhages.

Petri plate Dish filled with solid medium used in culturing microorganisms.

pH scale A logarithmic scale used for measuring the concentration of hydrogen ions in a solution.

Phaeohyphomycosis Cutaneous and subcutaneous disease characterized by lesions that can spread internally; caused by traumatic introduction of ascomycetes into the skin.

Phaeophyta Brown-pigmented division of algae having cell walls composed of cellulose and alginic acid, a thickening agent.

Phage *(bacteriophage)* Virus that infects and usually destroys bacterial cells.

Phage typing Method of classifying microorganisms in which unknown bacteria are identified by observing plaques.

Phagocytes Cells, often leukocytes, that are capable of phagocytosis.

Phagocytosis Type of endocytosis in which solids are moved into the cell.

Phagolysosome Digestive vesicle formed by the fusing of a lysosome with a phagosome.

Phagosome A sac formed by a phagocyte's pseudopodia; an intracellular food vesicle.

Pharyngitis *(strep throat)* Inflammation of the throat, often caused by infection with group A *Streptococcus*.

Phase microscope Type of microscope used to examine living microorganisms or fragile specimens.

Phase-contrast microscope Type of phase microscope that produces sharply defined images in which fine structures can be seen in living cells.

Phenol coefficient Method of evaluating the effectiveness of a disinfectant or antiseptic that compares the agent's efficacy to that of phenol.

Phenolic Compound derived from phenol molecules that have been chemically modified to denature proteins and disrupt cell membranes in a wide variety of pathogens.

Phenotype The physical features and functional traits of an organism expressed by genes in the genotype.

Phospholipid Phosphate-containing lipid made up of molecules with two fatty acid chains.

Phospholipid bilayer Two layered structure of a cell's membranes.

Phosphorus cycle Biogeochemical cycle in which phosphorus is cycled between oxidation states.

Photoautotroph Microorganism which requires light energy and uses carbon dioxide as a carbon source.

Photoheterotroph Microorganism that requires light energy and gains nutrients via catabolism of organic compounds.

Photophosphorylation The use of energy from light to attach inorganic phosphate to ADP.

Photosynthesis Process in which light energy is captured by chlorophylls and transferred to ATP and metabolites.

Photosystem Network of light-absorbing chlorophyll molecules and other pigments held within a protein matrix on thylakoids.

Phototaxis Cell movement that occurs in response to light stimulus.

Phycoerythrin Red accessory pigment of photosynthesis in red algae.

Phycology Study of algae.

Phylum Taxonomical grouping of similar classes of organisms.

Picornaviruses Family of viruses that contain positive single-stranded RNA with naked polyhedral capsids; many are human pathogens.

Piedra Firm, irregular nodules on hair shafts caused by aggregates of fungal hyphae and spores.

Pilus *(conjugation pilus)* A tubule involved in bacterial conjugation.

Pinocytosis Type of endocytosis in which liquids are moved into the cell.

Pinta Childhood disease caused by the spirochete *Treponema carateum;* characterized by hard, pus-filled lesions.

Pinworm Common name of *Enterobius vermicularis*, whose adult female has a tail like a straight pin.

Pityriasis Condition characterized by depigmented or hyperpigmented patches of scaly skin resulting from infection with *Malassezia furfur*.

Plague Disease caused by *Yersinia pestis*, often manifesting with enlarged lymph nodes (bubonic plague) or with severe pulmonary distress (pneumonic plague).

Plaque In phage typing, the clear region within the bacterial lawn where growth is inhibited by bacteriophages.

Plaque assay Technique for estimating phage numbers in which each plaque corresponds to a single phage in the original bacterium/virus mixture.

Plasma The liquid portion of blood.

Plasma cells B cells that are actively fighting against exogenous antigens and secreting antibodies.

Plasmid A small, circular molecule of DNA that replicates independently of the chromosome. Each carries genes for its own replication and often for one or more nonessential functions such as resistance to antibiotics.

Plasmodial slime mold *(acellular slime mold)* Streaming, coenocytic, colorful filaments of cytoplasm that phagocytize organic debris and bacteria.

Platelet Cell fragments involved in blood clotting.

Platelet activating factor (PAF) Cytokine that is a potent trigger for blood coagulation.

Pleomorphic In cell morphology, term used to describe a variably shaped prokaryotic cell.

Pneumococcal pneumonia Inflammation of the lungs caused by *Streptococcus pneumoniae*—the pneumococcus.

Pneumococcus Common name of *Streptococcus pneumoniae*.

***Pneumocystis* pneumonia (PCP)** Debilitating fungal pneumonia that is a leading cause of death in AIDS patients and is caused by opportunistic infection with *Pneumocystis jiroveci*.

Pneumonia Inflammation of the lungs; typically caused by infection with *Streptococcus pneumoniae*.

Pneumonic plague Fever and severe respiratory distress caused by infection of the lungs with *Yersinia pestis*; fatal if untreated in nearly 100% of cases.

Glossary

Point mutation A genetic mutation affecting only one or a few base pairs in a genome. Point mutations include substitutions, insertions, and deletions.
Polar In cell morphology, pertaining to either end of a cell, e.g., polar flagella.
Polar covalent bond Type of bond in which there is unequal sharing of electrons between atoms with opposite electrical charges.
Poliomyelitis (polio) Infection of varied degrees of severity from asymptomatic to crippling and caused by infection with poliovirus.
Polluted Containing microorganisms or chemicals in excess of acceptable values.
Polyenes Group of antimicrobial drugs such as amphotericin B that disrupt the cytoplasmic membrane of targeted cells by becoming incorporated into the membrane and damaging its integrity.
Polymer Repeating chains of covalently linked monomers found in macromolecules.
Polymerase chain reaction (PCR) Technique of recombinant DNA technology that allows researchers to produce a large number of identical DNA molecules *in vitro*.
Polyomavirus A cancer-causing virus.
Polysaccharide Carbohydrate polymer composed of several to thousands of covalently linked monosaccharides.
Polyunsaturated fat Triglyceride with several double bonds between adjacent carbon atoms in its fatty acids.
Portal of entry Entrance site of pathogenic microorganisms, including the skin, mucous membranes, and placenta.
Portal of exit Exit site of pathogenic microorganisms, including the nose, mouth, and urethra.
Positive selection Process by which mutants are selected by eliminating wild-type phenotypes.
Positive-strand RNA (+RNA) Viral single-stranded RNA that can act directly as mRNA.
Postpolio syndrome Crippling deterioration of muscle function, likely due to aging-related aggravation of nerve damage by poliovirus.
Potable Fit to drink.
Pour-plate Method of culturing microorganisms in which colony-forming units are separated from one another using a series of dilutions.
Pox *(pocks; pustule)* Any raised, pus-filled skin lesion; characteristic of infection with a poxvirus.
Precursor metabolite Any of 12 molecules typically generated by a catabolic pathway and essential to the synthesis of organic macromolecules in cells.
Prevalence In epidemiology, the total number of cases of a disease in a given area or population during a given period of time.
Primary amebic meningoencephalopathy Often fatal inflammation of the brain characterized by headache, vomiting, fever, and destruction of neurological tissue; caused by infection with *Naegleria* or *Acanthamoeba*.
Primary atypical pneumonia So-called *walking pneumonia* characterized by mild respiratory symptoms that last for several weeks; caused by infection with *Mycoplasma pneumoniae*.
Primary immunodeficiency diseases Any of a group of diseases detectable near birth and resulting from a genetic or developmental defect.
Primary response The slow and limited immune response to a first encounter with an unfamiliar antigen.
Primary stain In staining, the initial dye, which colors all cells.
Prion Proteinaceous infectious particle that lacks nucleic acids and replicates by converting similar normal proteins into new prions.
Probe Nucleic acid molecule with a specific nucleotide sequence that has been labeled with a radioactive or fluorescent chemical so that its location can be detected.
Prodromal period In the infectious disease process, the short stage of generalized, mild symptoms that precedes illness.
Products The atoms, ions, or molecules that remain after a chemical reaction is complete.
Profundal zone Zone of freshwater or marine water beneath the limnetic zone and above the benthic zone.
Proglottids Body segments of a tapeworm, produced continuously as long as the worm remains attached to its host.
Progressive multifocal leukoencephalopathy (PML) Progressive, fatal disease in which JC virus (a polyomavirus) kills cells of the central nervous system.
Prokaryote Any unicellular microorganism that lacks a nucleus. Classification includes bacteria and archaea.
Promoter Region of DNA where transcription begins.
Prophage An inactive bacteriophage, which is inserted into a host's chromosome.
Prophase First stage of mitosis, during which DNA condenses into chromatids and the spindle apparatus forms. Also used for the comparable stage of meiosis.
Prostaglandins Inflammatory chemicals released from damaged cells that increase vascular permeability.
Protease Enzyme secreted by microorganisms that digests proteins into amino acids outside a microbe's cell wall; in inflammatory reactions, chemicals released by mast cells that activate the complement system; in virology, an internal viral enzyme that makes HIV virulent.
Protein A complex macromolecule consisting of carbon, hydrogen, oxygen, nitrogen, and sulfur and important to many cell functions.
Proteobacteria Phylum of prokaryotes that includes five classes (designated alpha, beta, gamma, delta, and epsilon) of Gram-negative bacteria sharing common 16S rRNA nucleotide sequences.
Proton A positively charged subatomic particle, which is also the nucleus of a hydrogen atom.
Proton gradient Electrochemical gradient of hydrogen ions across a membrane.
Protozoa Single-celled eukaryotes that lack a cell wall and are similar to animals in their nutritional needs and structure.
Provirus *(latent virus)* Inactive virus in an animal cell.
Pseudohyphae Long cellular extension of the yeast *Candida* that look like the filamentous hyphae of molds.
Pseudomembranous colitis Disease of *Clostridium difficile* in which sections of the colon wall slough off.
Pseudomonad Any Gram-negative, aerobic, rod-shaped bacterium in the class Gamma proteobacteria that catabolizes carbohydrates by the Entner-Doudoroff and pentose phosphate pathways.
Pseudopodia Movable extensions of the cytoplasm and membrane of some eukaryotic cells.
Psychrophile Microorganism requiring cold temperatures (below 20°C).
Pure culture (axenic culture) Culture containing cells of only one species.
Purple sulfur bacteria Group of gammaproteobacteria, which are obligate anaerobes and oxidize hydrogen sulfide to sulfur.
Pustule *(pox)* Any raised, pus-filled skin lesion; characteristic of infection with a poxvirus.
Pyoderma Any confined, pus-producing lesion on the exposed skin of the face, arms, or legs; often caused by infection with group A *Streptococcus*.
Pyrimidine dimer Mutation in which adjacent pyrimidine bases covalently bond to one another; caused by nonionizing radiation in the form of ultraviolet light.
Pyrogen Chemical that triggers the hypothalamic "thermostat" to reset at a higher temperature, inducing fever.

Q fever Fever caused by the bacterium *Coxiella burnetii;* cause was questionable (thus "Q") for many years.
Quaternary ammonium compound (quat) Detergent antimicrobial that is harmless to humans.
Quorum sensing Process by which bacteria respond to changes in microbial density by utilizing signal and receptor molecules.

Rabies Neuromuscular disease characterized by hydrophobia, seizures, hallucinations, and paralysis; fatal if untreated. Caused by infection with the rabies virus.
Radial immunodiffusion Type of immunodiffusion test in which an antigen solution is allowed to diffuse into agar containing specific concentrations of antibodies, causing a ring of precipitate to form.
Radiation The release of high-speed subatomic particles or waves of electromagnetic energy from atoms.
Reactants The atoms, ions, or molecules that exist at the beginning of a chemical reaction.
Reaction center chlorophyll In a photosystem, a chlorophyll molecule in which electrons excited by light energy are passed to an acceptor molecule which is the initial carrier of an electron transport chain.
Recipient cell In horizontal gene transfer, a cell that receives part of the genome of a donor cell.
Recombinant Cell or DNA molecule resulting from genetic recombination between donated and recipient nucleotide sequences.
Recombinant DNA technology Type of biotechnology in which scientists change the genotypes and phenotypes of organisms.
Recombinant vaccine Vaccine produced using recombinant genetic technology.
Red tide Abundance of red-pigmented dinoflagellates in marine water.

Redox reaction *(oxidation-reduction reaction)* Any metabolic reaction involving the transfer of electrons from an electron donor to an electron acceptor. Reactions in which electrons are accepted are called *reduction* reactions, whereas reactions in which electrons are donated are *oxidation* reactions.
Reducing medium Special culturing medium containing compounds that combine with free oxygen and remove it from the medium.
Regulatory T cell *(Tr cell, suppressor T cell)* Thymus-derived lymphocyte that serves to repress adaptive immune responses and prevent autoimmune diseases.
Release In virology, final stage of the lytic replication cycle, in which the new virions are released from the host cell, which lyses.
Reoviruses Group of naked, segmented, dsRNA viruses that cause respiratory and gastrointestinal disease.
Repressible operon Type of operon that is continually transcribed until deactivated by repressors.
Reproduction An increase in number.
Reservoir of infection Living or nonliving continuous source of infectious disease.
Resolution The ability to distinguish between objects that are close together.
Respiratory syncytial virus (RSV) infection Disease caused by a virus in genus *Pneumovirus* that causes fusion of cells in the lungs and difficulty in breathing.
Responsiveness An ability to respond to environmental stimuli.
Restriction enzyme Enzyme that cuts DNA at specific nucleotide sequences and is used to produce recombinant DNA molecules.
Reticulate bodies Noninfectious stage in the life cycle of chlamydias.
Retrovirus Any +ssRNA virus that uses the enzyme reverse transcriptase carried within its capsid to transcribe DNA from its RNA.
Reverse transcriptase Complex enzyme that allows retroviruses to make dsDNA from RNA templates; used in recombinant DNA technology to make cDNA.
Revolving nosepiece Portion of a compound microscope on which several objective lenses are mounted.
Rh antigens Cytoplasmic membrane proteins common to the red blood cells of 85% of humans as well as rhesus monkeys.
Rheumatic fever A complication of untreated group A streptococcal pharyngitis in which inflammation leads to damage of the heart valves and muscle.
Rheumatoid arthritis A crippling, systemic autoimmune disease, resulting from a type III hypersensitivity reaction in which antibody complexes are deposited in the joints, causing inflammation.
Rhinoviruses Group of picornaviruses that cause upper respiratory tract infection and the "common cold."
Rhodophyta Red algae, generally containing the pigment phycoerythrin, the storage molecule floridean starch, and cell walls of agar or carrageenan.
Ribonucleic acid (RNA) Nucleic acid consisting of nucleotides made up of phosphate, a ribose pentose sugar, and an arrangement of the bases adenine, guanine, cytosine, and uracil.
Ribosomal RNA (rRNA) Form of ribonucleic acid which, together with polypeptides, makes up the structure of ribosomes.
Ribosome Nonmembranous organelle found in prokaryotes and eukaryotes that is composed of protein and ribosomal RNA and functions to make polypeptides.
Riboswitch RNA molecule that changes shape in response to shifts in environmental conditions, which results in genetic regulation.
Ribozyme RNA molecule functioning as an enzyme.
Rickettsias Group of extremely small, Gram-negative, obligate intracellular parasites that appear almost wall-less.
RNA polymerase Enzyme that synthesizes RNA by linking RNA nucleotides that are complementary to genetic sequences in DNA.
RNA primer RNA molecule used by DNA polymerase or reverse transcriptase as a starting point for DNA synthesis.
Rocky Mountain spotted fever (RMSF) Serious illness caused by infection with *Rickettsia rickettsii* transmitted by ticks and characterized by rash, malaise, petechiae, encephalitis, and death in 5% of cases.
Roseola Endemic illness of children characterized by an abrupt fever, sore throat, enlarged lymph nodes, and faint pink rash; caused by infection with human herpesvirus 6.
Rotaviruses Group of reoviruses that cause a potentially fatal infantile gastroenteritis.
Rough endoplasmic reticulum (RER) Type of endoplasmic reticulum that has ribosomes adhering to its outer surface; these produce proteins for transport throughout the cell.
R-plasmid Extrachromosomal piece of DNA containing genes for resistance to antimicrobial drugs.
Rubella *(German measles)* Disease caused by infection with *Rubivirus* resulting in characteristic rash lasting about three days; mild in children but potentially teratogenic to fetuses of infected women.
Rubeola *(measles, red measles)* Contagious disease characterized by fever, sore throat, headache, dry cough, conjunctivitis, and lesions called Koplik's spots, caused by infection with *Morbillivirus*.
Rubor Redness.

Salmonellosis A serious diarrheal disease resulting from consumption of food contaminated with the enteric bacterium *Salmonella*.
Salt A crystalline compound formed by ionic bonding of metallic with nonmetallic elements.
Sanitization The process of disinfecting surfaces and utensils used by the public.
Saprobe Fungus that absorbs nutrients from dead organisms.
Sarcina A cuboidal packet of cocci.
Satellite virus A virus, such as hepatitis D virus, that requires glycoproteins coded by another virus to complete its replication cycle.
Saturated fat A triglyceride in which all but the terminal carbon atoms are covalently linked to two hydrogen atoms.
Scabies Skin disease cause by a burrowing mite.
Scalded skin syndrome Reddening and blistering of the skin caused by infection with *Staphylococcus aureus*.
Scanning electron microscope (SEM) Type of electron microscope that uses magnetic fields within a vacuum tube to scan a beam of electrons across a specimen's metal-coated surface.
Scanning tunneling microscope (STM) Type of probe microscope in which a metallic probe passes slightly above the surface of the specimen, revealing surface details at the atomic level.
Scarlet fever Diffuse rash and sloughing of skin caused by infection with group A *Streptococcus*.
Schaeffer-Fulton endospore stain In microscopy, staining technique that uses heat to drive a malachite green primary stain into an endospore.
Schistosomiasis A potentially fatal disease caused by infection with a blood fluke in the genus *Schistosoma*; may cause tissue damage in the liver, lungs, brain, or other organs.
Schizogony Special type of asexual reproduction in which the protozoan *Plasmodium* undergoes multiple mitoses to form a multinucleate schizont.
Schizont Multinucleate body that undergoes cytokinesis to release several cells.
Scientific method Process by which scientists attempt to prove or disprove hypotheses through observations of the outcomes of carefully controlled experiments.
Scolex Small attachment organ that possesses suckers and/or hooks used to attach a tapeworm to host tissues.
Sebum Oily substance secreted by the sebaceous glands of the skin that lowers pH.
Secondary immune response Enhanced immune response following a second contact with an antigen.
Secretory IgA The combination of IgA and a secretory component, found in tears, mucous membrane secretions, and breast milk, where it agglutinates and neutralizes antigens.
Secretory vesicle In eukaryotic cells, vesicles containing secretions packaged by the Golgi body that fuse to the cytoplasmic membrane and then release their contents outside the cell via exocytosis.
Sedimentation Settling of particulate matter; the first step in treating water for drinking.
Segmented genome Genetic material consisting of more than one molecule of nucleic acid, used particularly for viruses.
Selective medium Culturing medium containing substances that either favor the growth of particular microorganisms or inhibit the growth of unwanted ones.
Selective toxicity Principle by which an effective antimicrobial agent must be more toxic to a pathogen than to the pathogen's host.
Selectively permeable In cell physiology, characteristic of a membrane that allows some substances to cross while preventing the crossing of others.
Semisynthetic antimicrobial Antimicrobial that has been chemically altered.
Septate Characterized by the presence of cross walls.
Septic shock Extremely low blood pressure resulting from dilation of blood vessels triggered by bacteria or bacterial toxins.

Septic tank The home equivalent of primary wastewater treatment, consisting of a sealed concrete holding tank in which solids settle to the bottom and the effluent flows into a leach field that acts as a filter.

Septicemia (sepsis) The condition of pathogens being present in the blood and causing signs of illness.

Serology The study and use of immunological tests to diagnose and treat disease or identify antibodies or antigens.

Serum Blood plasma with clotting factors removed.

Serum sickness Type III hypersensitivity resulting from antibodies directed against antisera.

Severe acute respiratory syndrome (SARS) Manifestation of infection by a coronavirus called SARS virus.

Severe combined immunodeficiency disease (SCID) Primary immunodeficiency disease in children that affects both T cells and B cells and causes recurrent infections.

Shiga toxin Exotoxin secreted by *Shigella dysenteriae* that stops protein synthesis in host cells.

Shigellosis A severe form of dysentery caused by any of four species of *Shigella*.

Shingles *(herpes zoster)* Extremely painful skin rash caused by reactivation of latent varicella-zoster virus.

Shock Severe disturbance of blood circulation resulting in insufficient delivery of oxygen to vital organs.

Short interference RNA (siRNA) RNA molecule complementary to a portion of a molecule of mRNA, tRNA, or a gene, rendering the target ineffective.

Siderophore An iron-binding molecule released by some bacteria and fungi.

Signs In pathology, objective manifestations of a disease that can be observed or measured by others.

Silent mutation Mutation produced by base-pair substitution that does not change the amino acid sequence, because of the redundancy of the genetic code.

Simple microscope Microscope containing a single magnifying lens.

Simple stain In microscopy, a stain composed of a single dye such as crystal violet.

Singlet oxygen Toxic form of oxygen, neutralized by pigments called carotenoids.

Sinusitis Inflammation of the nasal sinuses; typically caused by *Streptococcus pneumoniae*.

Slant tube (slant) Test tube containing agar media that solidified while the tube was resting at an angle.

Slime layer Loose, water-soluble glycocalyx.

Slime mold Eukaryotic microbe resembling a filamentous fungus but lacking a cell wall and phagocytizing rather than absorbing nutrients.

Sludge After primary treatment of wastewater, the heavy material remaining at the bottom of settling tanks.

Smallpox Infectious disease eradicated in nature by 1980 and characterized by high fever, malaise, delirium, pustules, and death in about 20% of untreated cases.

Smear In microscopy, the thin film of organisms on the slide.

Smooth endoplasmic reticulum (SER) Type of endoplasmic reticulum that lacks ribosomes and plays a role in lipid synthesis and transport.

Snapping division A variation of binary fission in Gram-positive prokaryotes in which the parent cell's outer wall tears apart with a snapping movement to create the daughter cells.

SOS response Mechanism by which prokaryotic cells with extensive DNA damage use a variety of processes to induce DNA polymerase to copy the damaged DNA.

Southern blot Technique used in recombinant DNA technology that allows researchers to stabilize specific DNA sequences from an electrophoresis gel and then localize them using DNA dyes or probes.

Species Taxonomic category of organisms that can successfully interbreed.

Species resistance Property that protects a type of organism from infection by pathogens of other, very different organisms.

Specific epithet In taxonomy, latter portion of the descriptive name of a species.

Specific immunity The ability of a vertebrate to recognize and defend against distinct species or strains of invaders.

Spectrum of action The number of different kinds of pathogens a drug acts against.

Spinal tap Collection of cerebrospinal fluid from the lumbar regions for diagnostic purposes.

Spiral In cell morphology, a spiral-shaped prokaryotic cell.

Spirillus A stiff spiral-shaped prokaryotic cell.

Spirochetes Group of helical, Gram-negative bacteria with axial filaments that cause the organism to corkscrew, enabling it to burrow into a host's tissues.

Spliceosome Protein-RNA complex that removes introns from eukaryotic RNA.

Spoilage Any unwanted change to a food.

Spontaneous generation The theory that living organisms can arise from nonliving matter.

Sporadic In epidemiology, a disease that occurs in only a few scattered cases within a given area or population during a given period of time.

Spore Reproductive cell of actinomycetes and fungi.

Sporogonic phase In the life cycle of *Plasmodium*, stage during which sporozoites are produced in the mosquito's digestive tract, and migrate into the mosquito's salivary glands.

Sporotrichosis Subcutaneous infection usually limited to the arms and legs; lesions form around the site of infection with *Sporothrix schenckii*.

Spp Abbreviation used to indicate several species of a genus.

Staining Coloring microscopy specimens with stains called dyes.

Staphylococcal scalded skin syndrome (SSSS) Disease caused by exfoliative toxin of *Staphylococcus aureus* in which epidermis peels off.

Staphylococcal toxic shock syndrome Potentially fatal syndrome characterized by fever, vomiting, red rash, low blood pressure, and loss of sheets of skin, usually caused by systemic infection with strains of *Staphylococcus* that produce toxic shock syndrome toxins.

Staphylococcus A cluster of cocci.

Starter culture Group of known microorganisms that carry out specific and reproducible fermentation reactions.

Stationary phase Phase in a growth curve in which new organisms are being produced at the same rate at which older organisms are dying.

Stem cells Generative cells capable of dividing to form daughter cells of a variety of types.

Sterile Free of microbial contamination.

Sterilization The eradication of all organisms, including bacterial endospores and viruses, although not prions, in or on an object.

Steroid Lipids consisting of four fused carbon rings attached to various side chains and functional groups.

Streak-plate Method of culturing microorganisms in which a sterile inoculating loop is used to spread an inoculum across the surface of a solid medium in Petri dishes.

Streptococcal toxic-shock-like syndrome (TSLS) Potentially fatal condition resembling *staphylococcal* toxic shock syndrome.

Streptococcus A chain of cocci.

Streptogramins Antimicrobial drugs that bind to the 50S ribosomal subunit and prevent ribosome movement along messenger RNA.

Structural analog Chemical that competes with a structurally similar molecule.

Sty Inflamed bacterial infection of the base of an eyelid.

Subacute disease Any disease that has a duration and severity that lies somewhere between acute and chronic.

Subacute sclerosing panencephalitis (SSPE) Slow, progressive disease of the central nervous system that results in memory loss, muscle spasms, and death several years after infection with a defective measles virus.

Subclinical *(asymptomatic)* Characteristic of disease that may go unnoticed because of absence of symptoms, even though clinical tests may reveal signs of disease.

Substitution Type of mutation in which a nucleotide base pair is replaced.

Substrate The molecule upon which an enzyme acts.

Substrate-level phosphorylation The transfer of phosphate to ADP from another phosphorylated organic compound.

Subunit vaccine Type of vaccine developed using recombinant DNA technology, which exposes the recipient's immune system to a pathogen's antigens but not the pathogen itself.

Sulfonamide Antimetabolic drug that is a structural analog of para-aminobenzoic acid (PABA).

Sulfur cycle Biogeochemical cycle in which sulfur is cycled between oxidation states.

Superficial mycoses Fungal infections of the surface of the skin.

Superinfection Condition in which a patient infected with hepatitis B virus is subsequently infected with hepatitis D virus.

Superoxide radical Toxic form of oxygen that is detoxified by superoxide dismutase.

Surfactant Chemical that acts to reduce the surface tension of solvents such as water by decreasing the attraction among solvent molecules.

Symbiosis A continuum of close associations between two or more organisms that ranges from mutually beneficial to associations in which one member damages the other member.

Symptoms Subjective characteristics of a disease that can be felt by the patient alone.

Synapse In immunology, the interface between cells of the immune system that involves cell-to-cell signaling.

Syncytium Giant, multinucleated cell formed by fusion of virally infected cell to neighboring cells.

Syndrome A group of symptoms, signs, and diseases that collectively characterizes a particular abnormal condition.

Synergism Interplay between drugs that results in efficacy that exceeds the efficacy of either drug alone.

Synthesis In virology, the production of new viral proteins and nucleic acids using the metabolic machinery of the host cell; third stage of lytic replication cycle.

Synthesis reaction A chemical reaction involving the formation of larger, more complex molecules.

Synthetic drug Antimicrobial that has been completely synthesized in a laboratory.

Synthetic medium (defined medium) Culturing medium of which the exact chemical composition is known.

Syphilis Sexually transmitted disease caused by infection with *Treponema pallidum*.

Systemic diseases Diseases caused by microbes spread via the blood and lymph that affect other body systems.

Systemic lupus erythematosus (SLE) *(lupus)* A systemic autoimmune disease in which the individual produces autoantibodies against numerous antigens, including nucleic acids.

T cell T lymphocyte.

T cell receptor (TCR) Antigen receptor generated in the cytoplasmic membrane of T lymphocytes.

T lymphocyte *(T cell)* Lymphocyte that matures in the thymus and acts directly against endogenous antigens in cell-mediated immune responses.

T-dependent antigens Molecules that stimulate an immune response only with the involvement of a helper T cell.

T-dependent humoral immunity Adaptive immune response resulting in antibody production that requires the action of a specific helper T cell (Th2).

T-independent antigens Large molecules with repeating subunits that trigger a humoral immune response without the activation of T cells.

T-independent humoral immunity Adaptive immune response resulting in antibody production following cross-linking of BCRs on numerous B cells and lacking involvement of helper T cells.

Taxa Nonoverlapping groups of organisms sorted on the basis of mutual similarities.

Taxis Cell movement that occurs as a positive or negative response to light or chemicals.

Taxonomic system A system for naming and grouping similar organisms together.

Taxonomy The science of classifying and naming organisms.

Telophase Final stage of mitosis, during which nuclear envelopes form around the daughter nuclei. Also used for the comparable stage of meiosis.

Temperate phage *(lysogenic phage)* Bacteriophage that does not immediately kill its host cell.

Teratogenic Characterized by an ability to cause birth defects.

Terminator Region of DNA where transcription ends.

Tetanospasmin Neurotoxin of *Clostridium tetani* that blocks the release of inhibitory neurotransmitters in the central nervous system.

Tetanus Potentially fatal infection with *Clostridium tetani*, which produces tetanospasmin, a potent neurotoxin.

Tetracycline Antimicrobial agent that inhibits protein synthesis by blocking the tRNA docking site.

Tetrad In genetics: two chromosomes, which are each made up of two DNA molecules, physically associated together during prophase I and metaphase I of meiosis; in cellular arrangements four cocci remaining attached following cell division.

Thallus Body of a fungus or alga.

Thermal death point The lowest temperature that kills all cells in a broth in 10 minutes.

Thermal death time The time it takes to completely sterilize a particular volume of liquid at a set temperature.

Thermophile Microorganism requiring temperatures above 45°C.

Thrombocytopenia Decrease in the number of platelets in the blood.

Thylakoid In photosynthetic cells, portion of cellular membrane containing light-absorbing photosystems.

Thymine Ring-shaped nitrogenous base found in nucleotides of DNA.

Tick Bloodsucking arachnid, which vectors a number of bacterial and viral pathogens.

Tick-borne relapsing fever Disease caused by *Borrelia* spp. transmitted between humans by soft ticks.

Tincture Solution of antimicrobial chemical in alcohol.

Titer In serology, a measure of the level of antibody in blood serum, determined by titration.

Toll-like receptors (TLRs) Integral membrane proteins that bind to specific microbial chemicals.

Total magnification A multiple of the magnification achieved by the objective and ocular lenses of a compound microscope.

Toxemia Presence in the blood of poisons called *toxins*.

Toxic shock syndrome (TSS) Potentially fatal condition characterized by fever, vomiting, red rash, low blood pressure, and loss of sheets of skin, usually caused by systemic infection with strains of *Staphylococcus* or *Streptococcus*.

Toxic-shock-like syndrome (TSLS) Shock produced by toxins of *Streptococcus* with manifestations similar to toxic shock syndrome of *Staphylococcus*.

Toxin Chemical that either harms tissues or triggers host immune responses that cause damage.

Toxoid vaccine Inoculum using modified toxins to stimulate antibody-mediated immunity.

Toxoplasmosis A disease affecting animals and caused by infection with *Toxoplasma gondii*. In humans, characterized by mild, febrile symptoms, but may be fatal in AIDS patients, and transplacental transmission may result in miscarriage, stillbirth, or severe birth defects.

Trace element Element required in very small amounts for microbial metabolism.

Trachoma Serious eye disease caused by *Chlamydia trachomatis*.

Transamination Reaction involving transfer of an amine group from one amino acid to another.

Transcription Process in which the genetic code from DNA is copied as RNA nucleotide sequences.

Transducing phage Virus that transfers bacterial DNA from one bacterium to another.

Transduction Method of horizontal gene transfer in which DNA is transferred from one cell to another via a replicating virus.

Transfer RNA (tRNA) Form of ribonucleic acid that carries amino acids to the ribosome.

Transformation Method of horizontal gene transfer in which a recipient cell takes up DNA from the environment.

Transgenic organism Plant or animal that has been genetically altered by the inclusion of genes from other organisms.

Translation Process in which the sequence of genetic information carried by mRNA is used by ribosomes to construct polypeptides with specific amino acid sequences.

Transmission electron microscope (TEM) Type of electron microscope which generates a beam of electrons that passes through the specimen and produces an image on a fluorescent screen.

Transport medium A special type of medium used to move clinical specimens from one location to another while preserving the relative abundance of organisms and preventing contamination of the specimen or environment.

Transposition Mutation in which a genetic segment is transferred to a new position through the action of a DNA segment called a transposon.

Transposon Segment of DNA found in most prokaryotes, eukaryotes, and viruses that codes for the enzyme transposase and can move from one location in a DNA molecule to another location in the same or a different molecule.

Trematodes *(flukes)* Group of helminths that are flat, leaf-shaped, have incomplete digestive systems, and have oral and ventral suckers.

Trench fever A disease common among World War I soldiers; caused by the bacterium *Bartonella quintana*.

Trichomoniasis Inflammation of the genitalia caused by *Trichomonas vaginalis*.

Trophozoite The motile feeding stage of a protozoa.

Tubercle Hard pulmonary nodule resulting from infection with mycobacteria.

Tuberculin response Type of delayed hypersensitivity reaction in which the skin of an individual exposed to tuberculosis or tuberculosis vaccine reacts to a subcutaneous injection of tuberculin.

Tuberculin skin test Test for a delayed hypersensitivity reaction to a subcutaneous injection of tuberculin.

Tuberculosis (TB) A respiratory disease caused by infection with *Mycobacterium tuberculosis*; its disseminated form can result in wasting away of the body and death.

Tularemia Zoonotic disease causing fever, chills, malaise, and fatigue, and caused by infection with *Francisella tularensis*.

Tumor In the pathology of cancer, a mass of neoplastic cells. In inflammation, a symptom of swelling *(edema)*.
Tumor necrosis factor (TNF) An immune system cytokine secreted by macrophages and T cells to kill tumor cells and to regulate immune responses and inflammation.
Type I diabetes mellitus Immunological attack on the islets of Langerhans cells in the pancreas, resulting in the inability to produce the hormone insulin.
Type III secretion systems Complex proteinaceous structure that inserts into target cells, forming a channel for the secretion of bacterial toxin or enzymes.
Typhoid fever Fever, headache, and malaise produced by infection with *Salmonella enterica* serotypes Typhi and Paratyphi; severe infections may cause peritonitis.
Typhus (epidemic typhus, murine typhus, scrub typhus) A group of diseases caused by rickettsias transmitted by arthropod vectors.

Uncoating In animal viruses, the removal of a viral capsid within a host cell.
Unsaturated fat A triglyceride with at least one double bond between adjacent carbon atoms, and thus at least one carbon atom bound to only a single hydrogen atom.
Uracil Ring-shaped nitrogenous base found in nucleotides of RNA.
Urticaria Hives.
Use-dilution test Method of evaluating the effectiveness of a disinfectant or antiseptic against specific microbes in which the most effective agent is the one that entirely prevents microbial growth at the highest dilution.

Vaccination Active immunization; specifically against smallpox.
Vaccine The inoculum used in active immunization.
Vacuole General term for membranous sac that stores or carries a substance in a cell.
Vaginosis Noninflammatory infection of the vagina.
Valence The combining capacity of an atom.
Vancomycin Antimicrobial drug that disrupts formation of Gram-positive bacterial cell walls by interfering with alanine-alanine crossbridges linking *N*-acetylglucosamine subunits.
Vancomycin-resistant *Staphylococcus aureus* (VRSA) Strain of *S. aureus* that is resistant to vancomycin and usually resistant to many common antimicrobial drugs as well.
Variant Creutzfeldt-Jakob disease (vCJD) Dementia caused by a prion that destroys brain tissue such that the brain appears sponge-like—full of holes.
Varicella *(chickenpox)* Highly infectious disease characterized by fever, malaise, and skin lesions, and caused by infection with varicella-zoster virus.
Varicella-zoster virus (VZV) Virus that causes chickenpox (varicella) and shingles (herpes zoster).
Variola Common name for the smallpox virus.
Variola major Variant of smallpox virus, which causes severe disease with a mortality rate of 20% or higher.
Variola minor Variant of smallpox virus, which causes less severe disease and mortality rate of less than 1%.
Vector In genetics and recombinant DNA technology, nucleic acid molecule such as a viral genome, transposon, or plasmid that is used to deliver a gene into a cell. In epidemiology, an animal (typically an arthropod) that transmits disease from one host to another.
Vegetations Bulky masses of platelets and clotting proteins that surround and bury the bacteria involved in endocarditis.
Vehicle transmission Spread of pathogens via air, drinking water, and food, as well as bodily fluids being handled outside the body.
Venezuelan equine encephalitis (VEE) Potentially fatal infection of the brain caused by a togavirus.
Vesicle General term for membranous sac that stores or carries a substance in a cell; in human pathology, any raised skin lesion filled with clear fluid.
Viable plate count Estimation of the size of a microbial population based upon the number of colonies formed when diluted samples are plated onto agar media.
Vibrio A slightly curved rod-shaped prokaryotic cell.
Viral hemagglutination inhibition test Immune test commonly used to detect antibodies against influenza, measles, and other viruses that naturally agglutinate red blood cells.
Viral neutralization Test of serum for presence of antibodies against a particular virus in which test serum is mixed with the virus, and then the mixture is added to a cell culture. Survival of the cells indicates antibodies in the serum neutralized the viruses.
Viremia Viral infection of the blood.
Viridans streptococci Group of alpha-hemolytic streptococci, which produce a green pigment when grown on blood media and normally inhabit the mouth and throat, and the GI, genital, and urinary tracts.
Virion A virus outside of a cell, consisting of a proteinaceous capsid surrounding a nucleic acid core.
Viroid Extremely small, circular piece of RNA that is infectious and pathogenic in plants.
Virulence A measure of pathogenicity.
Virulence factors Enzymes, toxins, and other factors that affect the relative ability of a pathogen to infect and cause disease.
Virus Tiny infectious acellular agent with nucleic acid surrounded by proteinaceous capsomeres that form a covering called a capsid.
Viviparity Process by which live offspring are produced in the body of a mother.

Wandering macrophage Type of macrophage that leaves the blood via diapedesis to travel to distant sites of infection.
Warts *(papillomas)* Benign epithelial growths caused by papillomaviruses.
Wastewater *(sewage)* Any water that leaves homes or businesses after being used for washing or flushed from toilets.
Water mold Eukaryotic microbe resembling a filamentous fungus but having tubular cristae in their mitochondria, cell walls of cellulose, two flagella, and true diploid thalli.
Waterborne transmission Spread of pathogenic microorganisms via water.
Wavelength The distance between two corresponding points of a wave.
Wax Alcohol-containing lipid made up of molecules with one fatty acid chain.
Western blot test *(immunoblot)* Variation of an ELISA test that can detect the presence of antibodies against multiple antigens; used to verify the presence of antibodies against HIV in the serum of individuals who have tested positive by ELISA.
Western equine encephalitis (WEE) Potentially fatal infection of the brain caused by a togavirus.
White piedra Benign scalp disease caused by infection with *Trichosporon beigelii*.
Whitlow Inflamed blister that may result from infection with *human herpesvirus 1* or HHV-2 via a cut or break in the skin.
Whooping cough (pertussis) Pediatric disease characterized by development of copious mucus, loss of tracheal cilia, and deep "whooping" cough; caused by infection with *Bordetella pertussis*.
Wild-type cell A cell normally found in nature (in the wild); a nonmutant.
Wound Trauma to body's tissue.

XDR-TB Tuberculosis caused by extensively drug-resistant *Mycobacterium*.
Xenodiagnosis Method of diagnosing Chagas' disease in which an uninfected *Triatoma* vector is allowed to feed on a patient. Subsequent presence of trypanosomes in the bug's gut indicates the patient is infected.
Xenograft Type of graft in which tissues are transplanted between individuals of different species.
Xenotransplant Technique involving recombinant DNA technology in which human genes are inserted into animals to produce cells, tissues, or organs that are then introduced into the human body.

Yaws Large, destructive, pain-free lesions of the skin, bones, and lymph nodes caused by *Treponema pallidum pertenue*.
Yeast A unicellular, typically oval or round fungus that usually reproduces asexually by budding.
Yellow fever Often fatal hemorrhagic disease contracted through a mosquito bite carrying a flavivirus.

Zone of inhibition In a diffusion susceptibility test, a clear area surrounding the drug-soaked disk where the microbe does not grow.
Zoonoses Diseases that are naturally spread from usual animal host to humans.
Zygomycoses Opportunistic fungal infections caused by various genera of fungi classified in the division Zygomycota.
Zygomycota Division of fungi including coenocytic molds called zygomycetes. Most are saprobes.
Zygosporangium Thick, black, rough-walled sexual structure of zygomycetes that can withstand desiccation and other harsh environmental conditions.
Zygospores Haploid spores formed from the surviving nuclei within zygosporangia.
Zygote In sexual reproduction, diploid cell formed by the union of gametes.

Credits

Illustration Credits

All illustrations have been rendered by Precision Graphics unless noted otherwise.

CHAPTER 1 **1.10, 1.14:** J.B. Woolsey Associates, LLC; **1.12:** J.B. Woolsey Associates, LLC/Precision Graphics.

CHAPTER 2 **2.13, 2.14:** J.B. Woolsey Associates, LLC.

CHAPTER 3 **3.2, 3.3, 3.6, 3.8, 3.14–3.16, 3.31, 3.34b–3.37, 3.40, 3.41,** end-of-chapter **01** and **02:** Kenneth Probst/Precision Graphics; **3.4, 3.32:** Darwen Hennings/Kenneth Probst/Precision Graphics; **3.19:** J.B. Woolsey Associates, LLC; **3.26:** Kenneth Probst; **3.39:** Darwen Hennings.

CHAPTER 4 **4.2, 4.11, 4.12:** J.B. Woolsey Associates, LLC/Precision Graphics; **4.4, 4.6, 4.7:** J.B. Woolsey Associates, LLC; **4.22, 4.28ab:** Darwen Hennings/Precision Graphics.

CHAPTER 5 **5.18, 5.25–5.27:** Kenneth Probst/Precision Graphics.

CHAPTER 6 **6.3, 6.9, 6.10, 6.22, 6.24–6.26:** J.B. Woolsey Associates, LLC; **6.17, 6.23,** End-of-chapter: J.B. Woolsey Associates, LLC/Precision Graphics.

CHAPTER 7 **7.29–7.30, 7.32:** J.B. Woolsey Associates, LLC; **7.28:** J.B. Woolsey Associates, LLC/Precision Graphics.

CHAPTER 9 **9.7, 9.10:** J.B. Woolsey Associates, LLC/Precision Graphics.

CHAPTER 10 **10.2:** Kenneth Probst/Precision Graphics; **10.12:** J.B. Woolsey Associates, LLC.

CHAPTER 11 **11.1, 11.5, 11.20, 11.25,** End-of-chapter: Kenneth Probst/Precision Graphics.

CHAPTER 12 **12.6, 12.10, 12.13–12.15, 12.19, 12.22, 12.24, 12.25, 12.27, 12.29:** Kenneth Probst/Precision Graphics.

CHAPTER 13 **13.4, 13.6–13.8, 13.11, 13.12, 13.14, 13.18,** End-of-chapter: Kenneth Probst/Precision Graphics.

CHAPTER 14 **14.3, 14.4, 14.11,** Table **2:** Kenneth Probst/Precision Graphics; **14.7:** J.B. Woolsey Associates, LLC.

CHAPTER 15 **15.2–15.4, 15.13–15.15:** Kenneth Probst/Precision Graphics.

CHAPTER 16 **16.2:** Kenneth Probst/Precision Graphics.

CHAPTER 17 **17.4, 17.7, 17.8:** J.B. Woolsey Associates, LLC /Precision Graphics; **17.9–17.10:** J.B. Woolsey Associates, LLC.

CHAPTER 18 **18.5, 18.8:** Cassio Lynm/Precision Graphics; **18.13:** J.B. Woolsey Associates, LLC.; End-of-chapter: J.B. Woolsey Associates, LLC/Precision Graphics.

CHAPTER 19 **19.16:** Cassio Lynm/Precision Graphics; **19.27:** Cassio Lynm; Microbes at a Glance **19.1** and **19.2:** Kenneth Probst/Precision Graphics.

CHAPTER 20 **20.8, 20.16, 20.18,** Microbes at a Glance **20.1** and **20.2,** End-of-chapter: Kenneth Probst/Precision Graphics; Highlight **20.1:** Minnesota Historical Society/Corbis; Highlight **20.2:** Russell Enscore, CDC.

CHAPTER 21 **21.15,** Microbes at a Glance **21.1** and **21.2:** Kenneth Probst/Precision Graphics.

CHAPTER 22 **22.2,** Microbes at a Glance **22.1** and **22.2:** Kenneth Probst/Precision Graphics.

CHAPTER 23 **23.1, 23.3, 23.5, 23.6, 23.13, 23.17, 23.20,** Microbes at a Glance **23.1** and **23.2:** Kenneth Probst/Precision Graphics.

CHAPTER 24 **24.2, 24.5, 24.10, 24.20, 24.23,** Microbes at a Glance **24.1** and **24.2,** End-of-chapter: Kenneth Probst/Precision Graphics.

CHAPTER 25 **25.8, 25.12, 25.17, 25.19, 25.20, 25.25, 25.38, 25.39,** Microbes at a Glance **25.1** and **25.2,** End-of-chapter: Kenneth Probst/Precision Graphics; **25.9:** Darwen Hennings/Precision Graphics.

Photo Credits

Author Photo: Elizabeth McBride. Preface Photos: (1) GoGo Images/Almay. (2) Jack Hollingsworth/Photodisc/Getty Images.

CHAPTER 1 **Opener:** Peter Parks/imagequestmarine.com. **1.1:** Pfizer. **1.2:** Al Shinn. **1.3:** Rich Robison, Pearson Science. **1.4a:** Manfred Kage/Peter Arnold. **1.4b:** L. Brent Selinger, Pearson Science. **1.5a:** M. I. Walker/NHPA. **1.5b:** M. I. Walker/Photo Researchers. **1.5c:** Steve J. Upton, Parasitology Research, Division of Biology, Kansas State University. **1.6a:** M. I. Walker/Photo Researchers. **1.6b:** Manfred Kage/Peter Arnold. **1.7:** L. Brent Selinger, Pearson Science. **1.8:** Sinclair Stammers/Photo Researchers. **1.9:** Lee D. Simon/Photo Researchers. **1.11:** Albert Edelfelt, Louis Pasteur in his library. Oil on canvas, 1889/The Granger Collection. **1.15:** Culver Pictures. **1.16:** Kirk Hartwein. **1.17:** L. Brent Selinger, Pearson Science. **1.18:** French Colored Engraving, 1883/The Granger Collection. **1.20:** C. James Webb/Phototake. **Beneficial Microbe 1.1:** Marc Vermeirsch/iStockPhoto. **Emerging Diseases 1.1:** Keith Weller, Agricultural Research Service, USDA. **Highlight 1.1:** Chung Sung-Jun/Getty Images. **End-of-chapter 01:** M. I. Walker/Photo Researchers; **02:** Steve J. Upton, Parasitology Research, Division of Biology, Kansas State University; **03:** M. I. Walker/NHPA.

CHAPTER 2 **Opener:** NASA/JPL/Space Science Institute. **2.12b:** H. Eisenbeiss/FLPA/Minden Pictures. **Beneficial Microbe 2.1:** Adam Woolfitt/Corbis. **Clinical Case Study 2.1:** Jonathan Littlejohn/Alamy.

CHAPTER 3 **Opener:** Dennis Kunkel Microscopy/Phototake. **3.1a:** Gopal Murti/Photo Researchers. **3.1b:** Michael Ross/Photo Researchers. **3.1c:** Dennis Kunkel/Phototake. **3.1d:** Stephen Durr. **3.5a:** Dennis Kunkel Microscopy. **3.5b:** Bergey's Manual Trust, Michigan State University. **3.7a:** A. Barry Dowsett/Photo Researchers. **3.7b:** Dennis Kunkel Microscopy. **3.7c:** The American Phytopathological Society. **3.8a:** Bill Schwartz, CDC. **3.10:** Dennis Kunkel Microscopy. **3.11:** Linda Stannard, UCT/Photo Researchers. **3.23:** From: Influence of phenylacetic acid on poly-β-hydroxybutyrate (PHB) polymerization and cell elongation in Azotobacter chroococcum Beij. M. P. Nuti, M. De Bertoldi, A. A. Lepidi. *Canadian Journal of Microbiology*. 1972 Aug;18(8):1257–61. **3.25:** Rut Carballido-López. **3.26:** From: The unique structure of archaeal 'hami', highly complex cell appendages with nano-grappling hooks. C. Moissl, R. Rachel, A. Briegel, H. Engelhardt, R. Huber, *Mol. Microbiol.* 2005 Apr; 56(2):361–70; Fig. 1 and 2. **3.27a:** From: An archaeal bi-species biofilm formed by Pyrococcus furiosus and Methanopyrus kandleri. S. Schopf, G. Wanner, R. Rachel, R. Wirth. *Arch. Microbiol.* 2008 Sep; 190(3):371–7; Epub 2008 Apr 26. **3.27b:** William Hixon and Dennis G. Searcy. **3.27c:** Mike Dyall-Smith (www.haloarchaea.com). **3.28:** Robert Bauman. **3.29:** Don W. Fawcett/Photo Researchers. **3.30abcd:** Mike Abbey/Visuals Unlimited. **3.31a:** SPL/Photo Researchers. **3.31b:** Omnikron/Science Library/Photo Researchers. **3.31c:** Steve Gschmeissner/Photo Researchers. **3.33b:** Albert Tousson/Phototake. **3.34a:** David M. Phillips/Visuals Unlimited. **3.35:** D. W. Fawcett/Photo Researchers. **3.36:** R. Bolender and D. Fawcett/Visuals Unlimited. **3.37:** Biophoto Associates/Photo Researchers. **3.38:** From: *Plant Cell Vacuoles: An Introduction*. D. N. De. (2000) CSIRO Publishing, Melbourne, Australia. Reproduced with permission from CSIRO. Photo by Celia Miller, CSIRO Plant Industry, Canberra, Australia. **3.40:** D. W. Fawcett/Photo Researchers. **3.41:** G. Chapman/Visuals Unlimited. **Beneficial Microbe 3.1:** Josh Reynolds/AP Images. **Highlight 3.1:** Gopal Murti/SPL/Photo Researchers. **Highlight 3.2:** Esther R. Angert/Phototake. **Highlight 3.3:** From: Critical Role of Bcr1-Dependent Adhesins in C. albicans Biofilm Formation In Vitro and In Vivo. C. J. Nobile, D. R. Andes, J. E. Nett, F. J. Smith Jr., F. Yue, et al. *PLoS Pathog*. 2006; 2(7):e63, Fig.8c.

CHAPTER 4 **Opener:** Robert Bauman. **4.4a:** Charles D. Winter/Photo Researchers. **4.8abcd:** Rich Robison, Pearson Science. **4.9ab:** Rich Robison, Pearson Science. **4.10b:** Larry Stauffer, Oregon State Public Health Laboratory/CDC. **4.11b:** Seelevel.com. **4.11c:** Stanley C. Holt, University of Texas Health Science Center. **4.13a:** Steve Gschmeissner/Photo Researchers. **4.13b** and **d:** Eye of Science/Photo Researchers. **4.13c:** Andrew Syred/Photo Researchers. **4.14ab:** Digital Instruments. **4.16ab:** L. Brent Selinger, Pearson Science. **4.17:** Rich Robison, Pearson Science. **4.18, 4.19:** Rich Robison, Pearson Science. **4.20:** P. Birn/Custom Medical Stock Photo. **4.21** and **4.24ab:** L. Brent Selinger, Pearson Science. **4.25ab:** Dade Behring. **4.26a:** L. Brent Selinger, Pearson Science. **4.27:** Microbial Diseases Laboratory, Berkeley, CA. **Beneficial Microbe 4.1:** Yasunori Tanji, Tokyo, Institute of Technology, Dept. of Biotechnology. **Emerging Diseases 4.1:** Lawrence B. Stack, Emergency Medicine, Vanderbilt University, Nashville. **Highlight 4.1:** From: In Situ Microspatial Imaging Using Two-Photon and Confocal Laser Scanning Microscopy of Bacteria and Extracellular Polymeric Secretions (EPS) Within Marine Stromatolite. T. Kawaguchi, A. W. Decho. *Marine Biotechnology*. 2002 March; 4(2):127–131, Fig.1A. © Springer-Verlag New York, LLC. **Highlight 4.2:** Phillipe Fauchet, ASM News, Volume 68 No.2, 2002. **End-of-chapter 01:** Meckes and Ottawa/Eye on Science/Photo Researchers; **02:** John Durhan/Photo Researchers; **03:** Rich Robison, Pearson Science; **04:** Larry Stauffer, Oregon State Public Health Laboratory/CDC; **05:** M. Wurtz, Biozentrum, University of Basil/SPL/Photo Researchers; **06:** M. I. Walker/Photo Researchers.

CHAPTER 5 **Opener:** bobo/Alamy. **5.5b:** Ken Eward, Biografx.com. **5.25b:** Eye of Science/Photo Researchers. **Beneficial Microbe 5.1:** Morley Read/SPL/Photo Researchers. **Highlight 5.1:** Frederick R. McConnaughey/Photo Researchers. **Highlight 5.2:** ImageSource/age fotostock.

CHAPTER 6 **Opener:** Image from the IMAX film "Volcanoes of the Deep Sea". Rutgers University/The Stephen Low Company. **6.2:** Brenda Wellmeyer, North Harris College Biology Dept., Houston, TX. **6.4b:** L. Brent Selinger, Pearson Science. **6.6a:** Doug Allan/NPL/Minden Pictures. **6.6b:** Ronald W. Hoham. **6.7:** SciMAT/Photo Researchers. **6.8b** and **6.9b:** L. Brent Selinger, Pearson Science. **6.10b:** Kirk Hartwein. **6.11** and **6.12ab:** L. Brent Selinger, Pearson Science. **6.13:** Rich Robison, Pearson Science. **6.14** and **6.15abc:** L. Brent Selinger, Pearson Science. **6.18b:** Lee D. Simon/Photo Researchers. **6.23b:** Pr. Courtieu/BSIP/Phototake. **6.23c:** L. Brent Selinger, Pearson Science. **6.26a:** Fundamental Photographs. **6.26b:** TOPAC. **Beneficial Microbe 6.1:** U.S. Department of Energy/Savannah River National Laboratory. **Clinical Case Study 6.1:** Martin S. Spiller. **Highlight 6.1:** Douglas Faulkner/Photo Researchers.

CHAPTER 7 **Opener:** Alfred Pasieka/SPL/Photo Researchers. **7.2a:** Ralph Slepecky/Visuals Unlimited/Getty Images. **7.2b:** Huntington Potter and David Dressler. **7.3a:** Barbara Hamkalo. **7.3b:** J. R. Paulsen and U. K. Laemmli, *Cell*, © Cell Press. **7.3cd:** G. F. Bahr/Armed Forces Institute of Pathology. **7.18b:** Visuals Unlimited. **7.34a:** Charles C. Brinton, Jr., University of Pittsburgh. **Beneficial Microbe 7.1:** Ed Austin and Herb Jones/National Park Service. **Emerging Diseases 7.1:** Richard Bartz, Munich/Wikimedia Commons. **Highlight 7.1:** Will and Deni McIntyre/Photo Researchers.

CHAPTER 8 **Opener:** David Sanger Photography/Alamy. **8.6b:** Science Source/Photo Researchers. **8.8b:** CSC, IC Microarray Centre, London. **8.9d:** Sovereign/Phototake. **8.10:** AdvanDx. **8.12:** D. Parker/SPL/Photo Researchers. **Highlight 8.1:** Michael Gadomski/Animals Animals. **Highlight 8.2:** Marc Moritsch/National Geographic/Getty Images.

CHAPTER 9 **Opener:** Dynamic Graphics/age fotostock. **9.6a:** Photofusion Picture Library/Alamy. **9.8:** Jonathon Blair/Corbis. **9.9b:** Tina Carvalho/Visuals Unlimited. **9.11:** Richard Megna/Fundamental Photographs. **9.12:** Ramon Flick. **9.14:** Richard Megna/Fundamental Photographs. **9.16:** Talaro/Visuals Unlimited. **Beneficial Microbe 9.1:** From: Cryo-electron microscopy of vitrified specimens. J. Dubochet, M. Adrian, J-J Chang, J-C Homo, J. Lepault, A. W. McDowall, and P. Schultz. *Quarterly Review of Biophysics*. 1988; 21:129–228, fig. 49b. **Emerging Diseases 9.1:** P. Garg & G. N. Rao/International Centre for Eye Health (ICEH). **Highlight 9.1:** Michael Rosenfeld/Photographer's Choice/Getty Images. **Highlight 9.2:** Rob Reed. **Highlight 9.3:** Stockbyte Platinum/Getty Images.

CHAPTER 10 Opener: CDC/Science Source/Photo Researchers. **10.1:** Rich Robison, Pearson Science. **10.3be:** From: Surface manifestations of antibiotic-induced alterations in protein synthesis in bacterial cells. A. S. Klainer, R. L. Perkins. *Antimicrob. Agents Chemother*. 1972 Feb; 1(2):164–70; Fig. 1. **10.9:** L. Brent Selinger, Pearson Science. **10.10:** Craig Warner/Albert Einstein College of Medicine. **10.11:** Tim Pietzcker, Universitätsklinikum Ulm, Germany. **10.14a:** Barts Medical Library/Phototake. **10.14b:** Beth Yarbrough. **10.17:** E. Vercauteren, Dept. of Microbiology, University Hospital, Antwerp, Belgium. **Beneficial Microbe 10.1:** SciMAT/Photo Researchers. **Emerging Diseases 10.1:** Toledo-Lucas County Health Department. **Highlight 10.1:** Corel.

CHAPTER 11 Opener: Iconotec/Alamy. **11.3ab:** T. A. Krulwich. **11.4:** Kim Findlay, John Innes Centre, Norfolk, UK. **11.6a:** Mike Dyall-Smith, Dept. of Microbiology and Immunology, University of Melbourne, Australia. **11.6b:** Eastman Kodak. **11.6cde:** L. Brent Selinger, Pearson Science. **11.7ab:** William A. Clark, CDC. **11.7c:** CDC. **11.7d:** William A. Clark, CDC. **11.8ab:** Rich Robison, Pearson Science. **11.10a:** From: Extending the upper temperature limit for life. K. Kashefi, D. R. Lovley. *Science*. 2003 August 15; 301(5635): 934, fig1A. Reprinted with permission. © 2003 AAAS. **11.10b:** Reinhard Rachel. **11.11:** Sascha Burkard/Shutterstock. **11.12:** *San Francisco Chronicle*. **11.13a:** Microbial Culture Collection, MCC-NIES, Japan. **11.13b:** Yuuji Tsukii, Hosei University, Tokyo, Japan. **11.13c:** Biodisc/Visuals Unlimited/Alamy. **11.14:** Paul Johnson/Biological Photo Service. **11.15:** F. Thiaucourt, Cirad, Montpellier France. **11.16:** SciMAT/Photo Researchers. **11.17:** David Berd, CDC. **11.18:** Yves Brun, University of Indiana. **11.19:** Visuals Unlimited/Getty Images. **11.21:** Leonard Lessin/Peter Arnold. **11.22:** David J. Patterson, http://micro*scope.mbl.edu. **11.23:** K. S. Kim/Peter Arnold. **11.24b:** From: Bdellovibrio bacteriovorus—a predatory bacterial parasite. *Naturwissenschaften*. H. Stolp. 1968 February; 55(2):57–63. With permission of Springer Science and Business Media. **11.25b:** Martin Dworkin/Phototake. **Beneficial Microbe 11.1:** Rick Gomez/Corbis. **Beneficial Microbe 11.2:** Yves Brun/Indiana University. **Emerging Diseases 11.1:** MIXA Co., Ltd./Alamy. **Highlight 11.1:** Gudrun Herzner.

CHAPTER 12 Opener: Arco Images GmbH/Alamy. **12.2a:** Dartmouth Electron Microscope Facility. **12.2b:** Michael Danilchik, Program in Molecular and Cellular Biosciences, Oregon Health and Science University. **12.2c:** SciMAT/Photo Researchers. **12.5a:** M. Abbey/Visuals Unlimited. **12.5b:** Cabisco/Visuals Unlimited. **12.7:** Kevin Carpenter and Patrick Keeling. **12.8:** Guy Brugerolle, Universite Clermont-Ferrand. **12.9:** Meckes and Ottawa/Eye on Science/Photo Researchers. **12.11:** R. Kessell and G. Shih/Visuals Unlimited. **12.12:** Manfred Kage/Peter Arnold. **12.13a:** Patrick W. Grace/Photo Researchers. **12.13b:** Eye of Science/Photo Researchers. **12.14b:** Myron G. Schultz, CDC. **12.15b:** David Scharf/Photo Researchers. **12.15c:** Christine L. Case, Skyline College. **12.16:** Jeremy Burgess/Photo Researchers. **12.17:** N. Allin and G. L. Barron, University of Guelph/Biological Photo Service. **12.18a:** K. Talaro/Visuals Unlimited. **12.18b:** Buckman Laboratories International. **12.18c:** David Scharf/Peter Arnold. **12.21:** Matt Meadows/Peter Arnold. **12.22:** Nino Santamaria. **12.23:** Phyzome/Wikimedia Commons. **12.24:** Merton F. Brown and Harold G. Brotzman from the APS Slide collection. **12.25:** V. Ahmakojian/Visuals Unlimited. **12.26:** Fred Rhoades/Mycena Consulting. **12.28:** Wim Van Egmond. **12.30:** Bob Evans/Peter Arnold. **12.31:** Eric Condliffe/Visuals Unlimited. **12.32:** Fred Rhoades/Mycena Consulting. **12.33a:** Kent Wood/Photo Researchers. **12.33b:** Parasitology Laboratory, Dept. of Medical Technology, Faculty of Health Sciences, Kobe University School of Medicine. **12.33c:** World Health Organization (WHO). **12.33d:** From: Genetic Analysis of Lice Supports Direct Contact between Modern and Archaic Humans. D. L. Reed, V. S. Smith, S. L. Hammond, A. R. Rogers, D. H. Clayton. *PLoS Biology*. 2(11); e340. Photo by Vincent S. Smith. **12.33e:** Steve Mihok. **12.33f:** From the website of the Food and Environmental Hygiene Dept., Hong Kong, reproduced under a license from the Government of Hong Kong Special Administrative Region. All rights reserved. **12.33g:** Marcelo de Campos Pereira, The Veterinary Parasitology Images Gallery. **Beneficial Microbe 12.1:** Agence France Press/Getty Images. **Emerging Diseases 12.1:** From: Pulmonary Aspergilloma: A Significant Cause of Life-Threatening Massive Hemoptysis. Z. A. Ali, A. Mahmoud, F. R. Krueger, S. Kapre, C. Mead. *Resident & Staff Physician*. 2008; 54(1):27–32; Fig. 2. **Highlight 12.1:** Weizmann Institute of Science/*Nature* Magazine. **End-of-chapter 01:** Kenneth Probst/Precision Graphics; **02:** Merton F. Brown and Harold G. Brotzman. from the APS Slide collection; **03:** David Scharf/Peter Arnold; **04:** Buckman Laboratories International.

CHAPTER 13 Opener: CAMR/A. Barry Dowsett/Photo Researchers. **13.1b:** Yu. G. Kuznetsov and A. McPherson, University of California, Irvine. **13.2:** Huntington Potter, Byrd Alzheimers Institute and University of South Florida and David Dressler, Oxford University and Balliol College. **13.3a:** From: Programmed cell death, mitochondria and the plant hypersensitive response. E. Lam, N. Kato, M. Lawton. *Nature*. 2001 June; 411:848–53, Fig.1. Reprinted with permission from Nature Publishing Group (www.nature.com). **13.3b:** Dennis Kunkel Microscopy/Phototake. **13.3c:** Eye of Science/Photo Researchers. **13.5a:** Robley C. Williams. **13.5b:** Hans Gelderblom/Visuals Unlimited. **13.5c:** CDC. **13.5d:** Frederick A. Murphy, University of Texas Medical Branch, Galveston. **13.6a:** Harold Fisher/Visuals Unlimited. **13.7:** Klaus Boller/Photo Researchers. **13.10:** M. Wurtz, Biozentrum, University of Basil/SPL/Photo Researchers. **13.17:** Wikimedia Commons. **13.19:** OriGen Biomedical. **13.20:** Theodor O. Diener, Agricultural Research Service, USDA. **13.21:** Barry Fitzgerald and Peggy Greb, Agricultural Research Service, USDA. **13.23:** James King-Holmes/SPL/Photo Researchers. **Beneficial Microbe 13.1:** NASA. **13.BM2:** Juergen Berger/SPL/Photo Researchers. **Emerging Diseases 13.1:** James Gathany, CDC. **Highlight 13.1:** AP Images.

CHAPTER 14 Opener: Gopal Murti/Visuals Unlimited/Getty Images. **14.1:** Morley Read/SPL/Photo Researchers. **14.2:** Tony Brain/Photo Researchers. **14.5b:** J. Willam Costerton, Montana State University. **14.6:** David Scharf/Peter Arnold. **14.12:** Andrew Davidhazy, Photo Arts and Sciences at Rochester Institute of Technology. **14.13:** Namit Arora/Shunya.net. **Beneficial Microbe 14.1:** Ralph E. Berry. **Clinical Case Study 14.3:** Ben Edwards/Stone/Getty Images. **Clinical Case Study 14.4:** Vincent P. Walter/Prentice Hall. **Emerging Diseases 14.1:** Rob and Ann Simpson/Visuals Unlimited.

CHAPTER 15 Opener: SPL/Photo Researchers. **15.1:** David Scharf/Peter Arnold. **15.5a.1:** Phototake. **15.5a.2:** Ed Reschke/Peter Arnold. **15.5a.3:** Carolina Biological/Phototake. **15.5b:** J.O. Ballard. **15.6:** Alison K. Criss/University of Virginia Health Sciences Center. **15.10:** Sucharit Bhakdi. **Beneficial Microbe 15.1:** Steve Gschmeissner/SPL/Photo Researchers. **Highlight 15.1:** CDC. **End-of-chapter:** Peter Parks/imagequestmarine.com.

CHAPTER 16 Opener: CNRI/Photo Researchers. **16.1:** Michael Ross/Photo Researchers. **16.5b:** Tim Evans/Photo Researchers. **16.11:** Cécile Chalouni, Genentech. **16.17:** Dorothea Zucker-Franklin/Phototake. **Emerging Diseases 16.1:** iStockPhoto. **Highlight 16.2:** NIBSC/Photo Researchers. **Highlight 16.3:** Andrejs Liepins/SPL/Photo Researchers. **Table 16.4(1):** PBWPIX/Alamy; **(2):** Diane Macdonald/Stockbyte/Getty Images; **(3):** SUPRI/Reuters/Landov; **(4):** Barbara Rice, CDC.

CHAPTER 17 Opener: Andrew Brookes/Corbis. **17.9b:** Karen E. Petersen, Dept. of Biology, University of Washington. **17.11:** Maxine Jalbert and Leo Kaufman, CDC. **17.12b:** Maxine Jalbert, CDC. **17.14b:** L. Brent Selinger, Pearson Science. **17.15b:** Museum of Science and Industry, Chicago. **Beneficial Microbe 17.1:** CDC. **Highlight 17.1:** Barry Dowsett and David A. J. Tyrrell.

CHAPTER 18 Opener: Peter Arnold/Alamy. **18.2a:** Microfield Scientific Ltd./Photo Researchers. **18.2b:** Carolina Biological Supply/Phototake. **18.2c:** Andrew Syred/Photo Researchers. **18.3:** ISM/Phototake. **18.4:** SUI/Visuals Unlimited. **18.9:** P. Marazzi/Photo Researchers. **18.10:** A. Doria and R. Rondinone, Dept. of Rheumatology, University of Padova, Italy. **18.11:** Princess Margaret Hospital/Dept. of Pediatrics. **18.12:** Seelevel.com. **Highlight 18.1:** Elizabeth Bauman. **Highlight 18.2:** Tom and Dee Ann McCarthy/Corbis. **Highlight 18.3:** NASA.

CHAPTER 19 Opener: Dennis Kunkel/Phototake. **19.1:** David M. Phillips/Photo Researchers. **19.2:** From: Images in clinical medicine. Staphylococcal scalded skin syndrome. L. A. Schenfeld. *New England Journal of Medicine*. 2000 Apr 20; 342(16):1178, fig. 1. Used by permission of the Massachusetts Medical Society. **19.3:** ISM/Phototake. **19.4:** From: Evaluating the Febrile Patient with a Rash. H. McKinnon, T. Howard. *American Family Physician*. 2000; 62:804–16, Fig. 6. **19.6:** CDC. **19.7:** P. Marazzi/Photo Researchers. **19.8:** Ben Barankin, Dermatlas. **19.9:** Electron micrograph by Vincent A. Fischetti, Rockefeller University, NY. **19.10:** Mike Miller, CDC. **19.11:** A. Dowsett/Health Protection Agency/SPL/Photo Researchers. **19.12:** Greater Southern Area Health Service, NSW Australia. **19.13:** SPL/Photo Researchers. **19.15:** Rich Robison, Pearson Science. **19.17:** CDC. **19.18b:** L. T. Tilney, P. S. Connelly and D. A. Portnoy. **19.19:** Institut Pasteur/Phototake. **19.20:** Kari Lounatmaa/Photo Researchers. **19.21:** G. W. Willis. **19.22:** Medical-on-Line/Alamy. **19.23:** Biophoto Associates/Photo Researchers. **19.25a:** B. Lopez Oblare' (www.fotogeriatria.net). **19.25b:** Biophoto Associates/Photo Researchers. **19.26:** Reuters/Rupak De Chowdhuri/Landov. **19.28:** DERMNET. **19.29a:** From: *Infections in Medicine*, 1992; 9(9). **19.29b:** Lester V. Bergman/Corbis. **Beneficial Microbe 19.1:** Andrew Syred/Photo Researchers. **Clinical Case Study 19.1:** CDC/Photo Researchers. **Emerging Diseases 19.1:** Fernando Caro, Dermatologist, Lima, Perú. **Highlight 19.1:** Flip Nicklin/Minden Pictures. **End-of-chapter a:** From: *Infections in Medicine*, 1992; 9(9); **b:** L. Brent Selinger, Pearson Science; **c:** CDC; **d:** Biophoto Associates/Photo Researchers; **e:** Eastman Kodak; **f:** Rich Robison, Pearson Science.

CHAPTER 20 Opener: A. Barry Dowsett/Photo Researchers. **20.1:** Dennis Kunkel/Phototake. **20.3a:** John Radcliffe Hospital/SPL/Photo Researchers. **20.3b:** Ken Greer/Visuals Unlimited. **20.4:** L. Brent Selinger, Pearson Science. **20.6a:** Mike Peres, RBP SPAS/Custom Medical Stock Photo. **20.6b:** Department of Pathology/University of Cambridge. **20.9a:** David M. Phillips/Visuals Unlimited. **20.9b:** Gladden Willis/Visuals Unlimited. **20.10:** Gladden Willis/Visuals Unlimited. **20.11:** James A. Shapiro, University of Chicago. **20.12:** John J. Farmer, CDC. **20.17:** Margaret Parsons and Karl F. Meyer, CDC. **20.19:** Dennis Kunkel Microscopy/Phototake. **20.20:** M. A. Ansary/Photo Researchers. **20.21:** Division of Pediatric Surgery, Brown Medical School. **20.24:** NIBSC/SPL/Photo Researchers. **20.26:** Leonard J. Morse. **20.27:** Jim Feeley, CDC. **20.28:** Moredun Animal Health Ltd./SPL/Photo Researchers. **20.29:** www.troybio.com, BD Microbiology. **Beneficial Microbe 20.1:** Jenny Merello/Instituto de Medicina Tropical Alexander von Humboldt, Universidad Peruana Cayetano Heredia. **Beneficial Microbe 20.2:** SEM by Daniel Kadouri and George O'Toole, colorized by Russell Monds, Dartmouth Medical School. **Clinical Case Study 20.1:** Norman Jacobs, CDC. **Clinical Case Study 20.2:** L. Brent Selinger, Pearson Science. **Emerging Diseases 20.1:** Tetra Images/SuperStock.

CHAPTER 21 Opener: Janice Carr, colorized by James Gathany, CDC. **21.1:** J. L. Carson/Custom Medical Stock Photo. **21.3:** Kenneth E. Greer/Visuals Unlimited. **21.6a:** SPL/Photo Researchers. **21.7:** M. A. Ansary/SPL/Photo Researchers. **21.8:** Western Ophthalmic Hospital/SPL/Photo Researchers. **21.9:** From: Infectivity Assays for Chlamydia Trachomatis. M. Singla and B. Bal. *The Internet Journal of Microbiology*. 2006; 2(2); Fig. 2. **21.10:** Edwin P. Ewing, Jr., CDC. **21.12a and c:** CDC. **21.12b:** Lester V. Bergman/Corbis. **21.13:** St. Bartholomew's Hospital/SPL/Photo Researchers. **21.14:** Carolina Biological Supply Company/Phototake. **21.15.2:** American Lyme Disease Foundation (aldf.com). **21.15.3:** James Marshall/The Image Works. **21.18:** Ben Adler, Australian Research Council Centre of Excellence in Structural and Functional Microbial Genomics, Monash University VIC, Australia. **21.19:** Gopal Murti/Photo Researchers. **Clinical Case Study 21.1:** Vaughan Fleming/SPL/Photo Researchers. **Clinical Case Study 21.2:** Robert Weaver, CDC. **Emerging Disease 21.1:** Shutterstock.

CHAPTER 22 Opener: E. Gueho/Photo Researchers. **22.1a:** Libero Ajello, CDC. **22.1b:** Maxine Jalbert and Leo Kaufman, CDC. **22.4:** Libero Ajello, CDC. **22.6:** John W. Rippon, University of Chicago. **22.8:** E. C. Chan/Visuals Unlimited. **22.9:** From: Dermatology Photo Quiz. M. Miller. *Resident & Staff Physician*. May 2006; (52)5; Fig. 1. **22.10:** Visuals Unlimited. **22.11:** Lois Norman, CDC. **22.12a:** P. Marazzi/Photo Researchers. **22.12b:** Phototake. **22.12c:** Edward H. Gill/Custom Medical Stock Photo. **22.12d:** Atlas of Ophthalmology (www.atlasophthalmology.com). **22.13:** From: Aspergillus fumigatus keratitis after laser in situ keratomileusis: a case report and review of post-LASIK fungal keratitis. F. Rahimi, M. N. Hashemian and M. T. Rajabi. *Eye*. 2007 Jun; 21(6):843–5; Fig. 1c. **22.14:** Edwin P. Ewing, Jr., CDC. **22.15a:** M. McGinnis/Doctor Fungus/Visuals Unlimited. **22.15b:** Samuel Freire da Silva, www.dermatlas.org. **22.16:** Barts Medical Library/Phototake. **22.17:** Everett Benke/Visuals Unlimited. **22.18a:** M. McGinnis/Doctor Fungus/Visuals Unlimited. **22.18b:** Everett Benke/Visuals Unlimited. **22.19:** From: Prise en charge des mycétomes en Afrique de l'Ouest. M. Develoux, M. T. Dieng, A. Kane and B. Ndiaye. *Bull. Soc. Pathol. Exot*. 2003; 96(5):376–82. **22.20:** Custom Medical Stock Photo. **22.21:** Peter Arnold. **Clinical Case Study 22.1:** Alamy. **Clinical Case Study 22.2:** SPL/Photo Researchers. **Emerging Disease 22.1:** iStockPhoto. **Highlight 22.1:** Electron Microscopy Center/North Dakota State University.

CHAPTER 23 Opener: Andrew Syred/Photo Researchers. **23.2:** N. J. Wheeler, Jr., CDC. **23.4:** Biology and Life Sciences Dept., Thiel College, Greenville, PA. **23.7:** Zuño Burstein, Dermatologia Sanitaria, Instituto de Medicina Tropical Daniel A. Carrion, Universidad Nacional Mayor de San Marcos, Lima, Peru. **23.8:** Steve J. Upton, Parasitology Research, Division of Biology, Kansas State

University. **23.9:** J.L. Carson/Custom Medical Stock Photo. **23.12:** Steven Glenn, Laboratory and Consultation Division/CDC. **23.14:** ISM/Phototake. **23.15:** From: Three-step stool examination for cryptosporidiosis in 10 homosexual men with protracted watery diarrhea. P. Ma, R. Soave. *Journal of Infectious Diseases*. 1983 May; 137(5):824–28. Provided by the CDC. **23.18:** From: The Imaging of Tropical Diseases, by P. E. S. Palmer and M. M. Reeder. Springer. **23.19:** Eye of Science/Photo Researchers. **23.21:** CDC. **23.22:** Henry Bishop, CDC. **23.23:** Dennis Kunkel/Visuals Unlimited/Getty Images. **23.24:** CDC. **23.25:** Steve J. Upton, Parasitology Research, Division of Biology, Kansas State University. **23.26:** Andy Crump, W.H.O./SPL/Photo Researchers. Clinical **Case Study 23.1:** Leslie E. Kossoff/Associated Press. **Clinical Case Study 23.2** and **23.3:** Mae Melvin, CDC. **Emerging Disease 23.1:** Rebecca Ash. **Highlight 23.1:** A. Crump, W.H.O./Photo Researchers.

CHAPTER 24 **Opener:** Linda M. Stannard, University of Cape Town/Photo Researchers. **24.1:** Klaus Boller/Photo Researchers. **24.3:** CDC. **24.4:** Reprinted with permission from the American Academy of Dermatology. All rights reserved. **24.6:** P. Marazzi/Photo Researchers. **24.7:** Biophoto Associates/Photo Researchers. **24.8a:** Science VU/Visuals Unlimited. **24.8b:** Ken Greer/Visuals Unlimited. **24.9:** Custom Medical Stock Photo. **24.11:** Reprinted with permission from the American Academy of Dermatology. All rights reserved. **24.13(1):** SPL/Photo Researchers; **(2):** Custom Medical Stock Photo; **(3):** Robert Bauman; **(4):** National Institute of Health (nlm.nih.gov). **24.14:** CDC. **24.15:** CNRI/Phototake. **24.16** and **24.17a:** Science VU/Visuals Unlimited. **24.17b:** Barts Medical Library/Phototake. **24.17c:** Photo Researchers. **24.17d:** Biophoto Associates/Photo Researchers. **24.18:** M. Wurtz, Biozentrum, University of Basel/SPL/Photo Researchers. **24.19:** ISM/Phototake. **24.21:** Garry Watsib/Photo Researchers. **24.23a:** Linda Stanard/UCT/Photo Researchers. **24.24:** H. C. Robinson/Photo Researchers. **Emerging Disease 24.1:** The Image Bank/Getty Images. **Highlight 24.1:** Kenneth Probst. **Highlight 24.2:** Frederick A. Murphy, University of Texas Medical Branch, Galveston.

CHAPTER 25 **Opener:** Scott Camazine/Alamy. **25.1:** Dennis Kunkel Microscopy/Phototake. **25.2:** Corbis. **25.4:** Kathy Willnes/AP Images. **25.5ab:** P. Marazzi/SPL/Photo Researchers. **25.6:** F. P. Williams/US EPA. **25.7:** Marion Sourisseau, Marie-Christine PrÇvost, Olivier Schwartz, Institut Pasteur Paris. **25.11a:** USDA/Science Source/Photo Researchers. **25.13:** CNRI/Phototake. **25.15:** Kjell-Olof Hedlund/Smittskyddsinstitutet (SMI). **25.16:** AP Images. **25.18:** Microworks/Phototake. **25.19abcdef:** C. Dauguet/Institute Pasteur/Photo Researchers. **25.22a:** A. Ramey/PhotoEdit. **25.22b:** Virat Sirisanthana, Dept. of Pediatrics, Chiang Mai University, Thailand. **25.26a:** P. Marazzi/Photo Researchers. **25.26b:** ISM/Phototake. **25.28:** Barbara Rice, NIP/CDC. **25.30:** Frederick A. Murphy, University of Texas Medical Branch, Galveston. **25.32:** Biophoto Associates/Photo Researchers. **25.33:** Science Source/Photo Researchers. **25.35:** James Gathany, CDC. **25.36:** Iowa State University Library, University Archives. **25.37:** NIBSC/Photo Researchers. **25.40:** Phototake. **25.41:** Linda Stanard/Photo Researchers. **Clinical Case Study 25.1:** CDC. **Emerging Disease 25.1:** iStockPhoto. **Highlight 25.1:** John Nicholls/Visuals Unlimited. **Highlight 25.2:** From: Critical Role of Bcr1-Dependent Adhesins in C. albicans Biofilm Formation In Vitro and In Vivo. C. J. Nobile, D. R. Andes, J. E. Nett, F. J. Smith Jr., F. Yue, et al. *PLoS Pathog*. 2006; 2(7):e63. **Highlight 25.3:** Andy Wong/AP Images. **Highlight 25.4:** USGS Biological Resources Division, Western Ecological Research Center.

CHAPTER 26 **Opener:** London Scientific Films/Oxford Scientific/Photolibrary. **26.1:** Joe Herritt/AP Images. **26.2:** David Pollack/Corbis. **26.3:** Jeffrey Ullbrich/AP Images. **26.4:** EPA Photo/Attila Kisbenedek. **26.5:** Novo Nordisk. **26.6:** Toby Talbot/AP Images. **26.7a:** Paul Glendell/Peter Arnold. **26.8a:** Central Virginia Governor's School for Science and Technology. 26.8b: IDEXX Laboratories. **26.9a:** JRC/Alamy. **26.11:** Ross Coleman/*Albuquerque Tribune*/AP Images. **26.13:** Inga Spence/Visuals Unlimited. **26.14:** K. J. Edwards, Woods Hole Oceanographic Institution. **26.20:** Kenneth Lambert/AP Images. **Beneficial Microbe 26.1:** AP Images. **Emerging Disease 26.1:** Roderick Chen/First Light/Getty Images. **Highlight 26.1:** Seelevel.com. **Highlight 26.2:** Photodisc Green/Getty Images. **Highlight 26.4:** Klaus Boller/Photo Researchers.

Subject Index

NOTE: A *t* following a page number indicates tabular material, and an *f* following a page number indicates a figure.

Subject Index

Subject Index

Subject Index

Subject Index

Subject Index

Subject Index

Subject Index

Subject Index

Subject Index

Subject Index

Subject Index

Subject Index

Subject index

Subject Index

Subject Index

PRONUNCIATIONS OF SELECTED ORGANISMS AND VIRUSES

Absidia (ab-sid′ē-ă)
Acanthamoeba (ă-kan-thă-mē′bă)
Acetobacter (a-sē′tō-bak-ter)
Acinetobacter (as-i-nē′tō-bak′ter)
Acremonium (ak′re-mō′nē-ŭm)
Actinomyces israelii (ak′ti-nō-mī′sez is-rā′el-ē-ē)
Agrobacterium tumefaciens (ag′rō-bak-tēr′ē-um tū′me-fāsh-enz)
Alternaria (al-ter-nā′rē-ă)
Amanita muscaria (am-ă-nī′tă mus-ka′rē-ă)
Amanita phalloides (am-ă-nī′tă fal-ōy′dēz)
Amoeba (am-ē′bă)
Amycolatopsis orientalis (am-ē-kō′la-top-sis o-rē-en-tal′is)
Anaplasma phagocytophilum (an-ă-plaz′mă fag-ō-sī-to′fil-ŭm)
Ancylostoma duodenale (an-si-los′tō-mă doo′ō-de-nā-lē)
Aquaspirillum magnetotacticum (ă-kwă-spī′ril-ŭm mag-ne-tō-tak′ti-kŭm)
Aquifex (ăk′wē-feks)
Ascaris lumbricoides (as′kă-ris lŭm′bri-koy′dēz)
Aspergillus oryzae (as-per-jil′ŭs o′ri-zī)
Azomonas (ā-zō-mō′nas)
Azospirillum (ā-zō-spī′ril-ŭm)
Azotobacter (ā-zō-tō-bak′ter)

Bacillus anthracis (ba-sil′ŭs an-thrā′sis)
Bacillus cereus (ba-sil′ŭs se′rē-ŭs)
Bacillus licheniformis (ba-sil′ŭs lī-ken-i-for′mis)
Bacillus polymyxa (ba-sil′ŭs po-lē-miks′ă)
Bacillus popilliae (ba-sil′ŭs pop-pil′ē-ī)
Bacillus sphaericus (ba-sil′ŭs sfe′ri-kŭs)
Bacillus stearothermophilus (ba-sil′ŭs ste-rō-ther-ma′fil-ŭs)
Bacillus subtilis (ba-sil′ŭs sŭt′i-lis)
Bacillus thuringiensis (ba-sil′ŭs thur-in-jē-en′sis)
Bacteroides fragilis (bak-ter-oy′dēz fra′ji-lis)
Balantidium coli (bal-an-tid′ē-ŭm kō′lē)
Bartonella bacilliformis (bar-tō-nel′ă ba-sil′li-for′mis)
Bartonella henselae (bar-tō-nel′ă hen′sel-ī)
Bartonella quintana (bar-tō-nel′ă kwin′ta-nă)
Bdellovibrio (del-lō-vib′rē-ō)
Beggiatoa (bej′jē-a-tō′ă)
Blastomyces dermatitidis (blas-tō-mī′sēz der-mă-tit′i-dis)
Bordetella pertussis (bōr-dĕ-tel′ă per-tus′is)
Borrelia burgdorferi (bō-rē′lē-ă burg-dōr′fer-ē)
Borrelia recurrentis (bō-rē′lē-ă re-kur-ren′tis)
Botryococcus braunii (bot′rē-ō-kok′ŭs brow′nē-ē)
Brucella abortus (broo-sel′lă a-bort′us)
Brucella canis (broo-sel′lă kā′nis)
Brucella melitensis (broo-sel′lă me-li-ten′sis)
Brucella suis (broo-sel′lă soo′is)
Burkholderia cepacia (burk-hol-der′ē-ă se-pā′se-ă)
Burkholderia pseudomallei (burk-hol-der′ē-ă soo-dō-mal′-e-ē)

Campylobacter jejuni (kam′pi-lō-bak′ter jē-jū′nē)
Candida albicans (kan′did-ă al′bi-kanz)
Carsonella ruddii (kar-son-el′ă rŭd′-ē-ē)
Caulobacter (kaw′lō-bak-ter)
Cephalosporium (sef′ă-lō-spor′ē-ŭm)
Chlamydia trachomatis (kla-mid′ē-ă tra-kō′ma-tis)
Chlamydophila pneumoniae (kla-mē-dof′ĭ-lă noo-mō′nē-ī)
Chlamydia psittaci (kla-mē-dof′ĭ-lă sit′ă-sē)
Chondrus crispus (kon′drŭs krisp′ŭs)
Chromatium buderi (krō-ma′tē-ŭm bū′de-rē)
Citrobacter (sit′rō-bak-ter)
Cladophialophora carrionii (klă-dŏf′ē-ă-lof′ŏ-rā kar-rē-on′ē-ē)
Claviceps purpurea (klav′i-seps poor-poo′-rē′ă)
Clostridium botulinum (klos-trid′ē-ŭm bo-tū-li′num)
Clostridium difficile (klos-trid′ē-ŭm di-fi′sil-ē)
Clostridium perfringens (klos-trid′ē-ŭm per-frin′jens)
Clostridium tetani (klos-trid′ē-ŭm te′tan-ē)
Coccidioides immitis (kok-sid-ē-oy′dēz im′mi-tis)
Codium (kō′dē-ŭm)
Coltivirus (kol′tē-vī′rŭs)
Cortinarius gentilis (kōr′ti-nar-ē-us jen′til-is)
Corynebacterium diphtheriae (kŏ-rī′nē-bak-tēr′ē-ŭm dif-thi′rē-ī)
Coxiella burnetii (kok-sē-el′ă ber-ne′tē-ē)
Cryptococcus neoformans (krip-tō-kok′ŭs nē-ō-for′manz)
Cryptosporidium parvum (krip-tō-spō-rid′ē-ŭm par′vŭm)
Cyclospora cayetanensis (sī-klō-spōr′ă kī-ē-tan-en′sis)
Cytomegalovirus (sī-tō-meg′ă-lō-vī′rŭs)
Cytophaga (sī-tof′ă-gă)

Deinococcus radiodurans (dī-nō-kok′ŭs rā-dē-ō-dur′anz)
Desulfovibrio (dē′sul-fō-vib′rē-ō)
Dictyostelium (dik-tē-ō-stē′lē-um)
Didinium (dī-di′nē-ŭm)
Diplococcus pneumoniae (dip′lō-kok′ŭs nū-mō′nē-ī)

Echinococcus granulosus (ĕ-kī′nō-kok′ŭs gra-nū-lō′sŭs)
Edwardsiella (ed′ward-sē-el′ă)
Ehrlichia chaffeensis (er-lik′ē-ă chaf-ē-en′sis)
Entamoeba histolytica (ent-ă-mē′bă his-tō-li′ti-kă)
Enterobacter (en′ter-ō-bak′ter)
Enterobius vermicularis (en-ter-ō′bī-ŭs ver-mi-ku-lar′is)
Enterococcus faecalis (en′ter-ō-kok′ŭs fē-kă′lis)
Enterococcus faecium (en′ter-ō-kok′ŭs fē-sē′ŭm)
Epidermophyton floccosum (ep′i-der-mof′i-ton flŏk′ō-sŭm)
Epulopiscium fishelsoni (ep′yoo-lō-pis′sē-ŭm fish-el-sō′nē)
Escherichia coli (esh-ĕ-rik′ē-ă kō′lē)
Euglena granulata (yū-glēn′ă gran-yū-lă′tă)
Eupenicillium (yū-pen-i-sil′ē-ŭm)
Exophiala (ek-sō-fī′ă-lă)

Fasciola gigantica (fa-sē′ō-lă ji-gan′ti-kă)
Fasciola hepatica (fa-sē′ō-lă he-pa′ti-kă)
Ferroplasma acidarmanus (fe′rō-plaz′ma a-sid′ar-ma-nŭs)
Fonsecaea compacta (fon-sē-sē′ă kom-pak′ta)
Fonsecaea pedrosoi (fon-sē-sē′ă pe-drō′sō-ē)
Francisella tularensis (fran′si-sel′ă too-lă-ren′sis)
Fusarium (fū-zā′rē-ŭm)

Gambierdiscus (gam′bē-er-dis-kŭs)
Gardnerella vaginalis (gărd′ner-el′ă va-ji-nă′lis)
Gelidium (jel-li′dē-ŭm)
Geogemma barossii (jē′ō-jem-a ba-rōs′ē-ē)
Giardia intestinalis (jē-ar′dē-ă in-tes′ti-năl′is)
Gluconobacter (gloo-kon′ō-bak-ter)
Gonyaulax (gon-ē-aw′laks)
Gymnodinium (jīm-nō-din′ē-ŭm)
Gyromitra esculenta (gī-rō-mē′tră es-kū-len′tă)

Haemophilus ducreyi (hē-mof′i-lŭs doo-krā′ē)
Haemophilus influenzae (hē-mof′i-lŭs in-flū-en′zī)
Hafnia (haf′nē-ă)
Halobacterium salinarium (hā′lō-bak-tēr′ē-ŭm sal-ē-nar′ē-ŭm)
Hantavirus (han′tā-vī-rŭs)
Helicobacter pylori (hel′ĭ-kō-bak′ter pī′lō-rē)
Histoplasma capsulatum (his-tō-plaz′mă kap-soo-lā′tŭm)

Isabella abbottae (iz-ă-bel′ă ab′ot-tī)

Klebsiella pneumoniae (kleb-sē-el′ă nū-mō′nē-ī)

Lactobacillus bulgaricus (lak′-tō-bă-sil′ŭs bul-gā′ri-kŭs)
Lactococcus lactis (lak-tō-kok′ŭs lak′tis)
Legionella pneumophila (lē-jŭ-nel′lă noo-mō′fi-lă)
Leishmania (lēsh-man′ē-ă)
Leptospira interrogans (lep′tō-spī′ră in-ter′ră-ganz)
Leuconostoc citrovorum (loo′kō-nos-tŏk sit-rō-vō′rum)
Listeria monocytogenes (lis-tēr′ē-ă mo-nō-sī-tah′je-nēz)
Lyssavirus (lis′ă-vī-rŭs)

Madurella (mad′ū-rel′ă)
Malassezia furfur (mal-ă-sē′zē-ă fur′fur)
Methanobacterium (meth′a-nō-bak-tēr′ē-ŭm)
Methanopyrus (meth′a-nō-pī′rŭs)
Micavibrio (mī-kă-vib′rē-ō)
Microsporidium (mī-krō-spor-i′dē-ŭm)
Microsporum (mī-kros′po-rŭm)
Moraxella catarrhalis (mōr′ak-sel′ă kă-tah′răl-is)
Morganella (mōr′gan-el′ă)
Mucor (mū′kōr)
Mycobacterium avium-intracellulare (mī′kō-bak-tēr′ē-ŭm ā′vē-ŭm in′tra-sel-yu-la′rē)
Mycobacterium bovis (mī′kō-bak-tēr′ē-ŭm bō′vis)
Mycobacterium leprae (mī-kō-bak-tēr′ē-ŭm lep′rī)
Mycobacterium tuberculosis (mī-kō-bak-tēr′ē-ŭm too-ber-kyū-lō′sis)
Mycoplasma genitalium (mī′kō-plaz-mă jen-ē-tal′ē-ŭm)
Mycoplasma hominis (mī′kō-plaz-mă ho′mi-nis)
Mycoplasma pneumoniae (mī′kō-plaz-mă nū-mō′nē-ī)

Naegleria (nā-glē′rē-ă)
Necator americanus (nē-kā′tor ă-mer-i-ka′nus)
Neisseria gonorrhoeae (nī-se′rē-ă go-nor-rē′ī)
Neisseria meningitidis (nī-se′rē-ă me-nin-ji′ti-dis)
Neurospora crassa (noo-ros′pōr-ă kras′ă)
Nitrobacter (nī-trō-bak′ter)
Nitrosomonas (nī-trō-sō-mō′nas)
Nocardia asteroides (nō-kar′dē-ă as-ter-oy′dēz)
Nosema (nō-sē′mă)

Orientia tsutsugamushi (ōr-ē-en′tē-ă tsoo-tsoo-gă-mū′shē)
Orthopoxvirus variola (ōr-thō-poks′vī-rŭs vă-rī′-ō-lă)

Paracoccidioides brasiliensis (par′ă-kok-sid-ē-oy′dēz bră-sil-ē-en′sis)
Paramecium (par-ă-mē′sē-ŭm)
Pasteurella haemolytica (pas-ter-el′ă hē-mō-lit′i-kă)
Pasteurella multocida (pas-ter-el′ă mul-tŏ′si-da)
Penicillium chrysogenum (pen-i-sil′ē-ŭm krī-so′jĕn-ŭm)
Penicillium marneffei (pen-i-sil′ē-ŭm mar-nef-ē′ī)
Penicillium roqueforti (pen-i-sil′ē-ŭm rok′for-tē)
Pfiesteria (fes-tēr′ē-ă)
Phialophora verrucosa (fī-ă-lof′ŏ-ră ver-ū-kō′să)
Physarum (fī-sar′-um)
Phytophthora infestans (fī-tof′tho-ră in-fes′tanz)
Piedraia hortae (pī-drā′ă hōr′tī)
Plasmodium falciparum (plaz-mō′dē-ŭm fal-sip′ar-ŭm)
Plasmodium malariae (plaz-mō′dē-ŭm mă-lār′ē-ī)
Plasmodium ovale (plaz-mō′dē-ŭm ō-vă′lē)
Plasmodium vivax (plaz-mō′dē-ŭm vī′vaks)
Pneumocystis jiroveci (nū-mō-sis′tis jē-rō-vĕt′zē)
Prevotella (prev′ō-tel′ă)
Propionibacterium acnes (prō-pē-on-i-bak-tēr′ē-ŭm ak′nēz)
Proteus mirabilis (prō′tē-ŭs mi-ra′bi-lis)
Prototheca (prō-tō-thē′kă)
Providencia (prov′i-den′sē-ă)
Pseudallescheria (sood′al-es-kē-rē-ă)
Pseudomonas aeruginosa (soo-dō-mō′nas ā-roo-ji-nō′să)
Pseudomonas putida (soo-dō-mō′nas pyoo′ti-dă)
Pseudomonas syringae (soo-dō-mō′nas sēr′in-jī)
Psilocybe cubensis (sī-lō-sī′bē kū-benz′is)
Pyrodictium (pī rō dik′tē-um)

Rhizobium (rī-zō′bē-ŭm)
Rhizopus nigricans (rī-zō′pŭs ni′gri-kanz)